D0078987

The Science of GENETICS

ALAN G. ATHERLY
Iowa State University

JACK R. GIRTON
Iowa State University

JOHN F. MCDONALD
University of Georgia

SAUNDERS COLLEGE PUBLISHING
HARCOURT BRACE COLLEGE PUBLISHERS
Fort Worth Philadelphia San Diego New York Orlando Austin San Antonio Montreal London Sydney Tokyo

Publisher: Emily Barrosse
Executive Editor: Edith Beard Brady
Product Manager: Erik Fahlgren
Developmental Editor: Gabrielle Goodman
Project Editor: Robin C. Bonner
Production Manager: Charlene Catlett Squibb
Art Director: Ruth A. Hoover
Text Designer: Ruth A. Hoover

Cover Credit: The cover illustrates the importance of experimental organisms in the study of genetics. On the left is a diagram of the *Escherichia coli* chromosome, where each arrow represents the area of the chromosome transferred by various *Hfr* strains. On the lower right is a photograph of a wild-type *Arabidopsis thaliana* plant, shortly after the plant has produced the flowering stem, or inflorescence. One can see flowers of many developmental ages, from the small undeveloped flowers in the center of the photograph to the open flowers, which are undergoing self-pollination, and outward to the maturing fruits, called "siliques," which contain the developing seeds. Shown here is a plant approximately 26 days after seed germination. At the top is a photograph of *Drosophila melanogaster,* the premiere model organism for the study of genetics. *Drosophila melanogaster* has a short life span, high fertility, and grows well in small bottles in laboratories. Many techniques have been developed for identifying and analyzing mutations in *Drosophila* and for the molecular analysis of gene products and gene structure. Many genes initially discovered in studies of *Drosophila* have later been found to have counterparts in higher organisms, including humans. (Drosophila melanogaster, © *Darwin Dale/ScienceSource/Photo Researchers, Inc.;* Arabidopsis thaliana, © *1998 by Steve Jacobsen)*

Printed in the United States of America

The Science of Genetics

ISBN 0-03-033222-2

Library of Congress Catalog Card Number: 98-85629

9012345678 048 10 987654321

We dedicate this book to all our genetics students, past and future.
We thank our former students for providing us with insight
into how to be better teachers,
and we hope that this book will inspire future students
with an appreciation of this exciting science.

Preface

When our descendants debate the 20th century's greatest contributions to scientific advancement, they will likely decide that unraveling the mysteries of genetics is high on the list. Genetics is certainly one of the most exciting and vital fields of study in the biological sciences. In fact, about 80% of all research in biology today is done in the area of genetics or molecular genetics. An integral part of current events, news of genetic breakthroughs in medicine, agriculture, and basic research fills today's headlines. It is truly an exciting time for the study of genetics, and this book is designed to aid the student in its understanding. Our primary goal in writing this book is to present clearly the fundamental principles of genetics that every student of biology as well as every well-educated citizen of our society should possess. In addition, we hope that this textbook will stimulate students to pursue genetics and other basic sciences as a career goal.

Genetics is an analytical science, and many of our students have difficulty with this aspect of the course. For this reason, we have developed a number of learning tools that we hope will help students understand the complexity of genetics, make the material more accessible, and convey the excitement of the field. Using an inquiry-based approach, each chapter opens with a series of questions that represent key concepts to be explored in the chapter. To help students develop analytical and problem-solving skills, we have created a series of in-text problem sets that are unique to this book. Within each chapter are several example problems with detailed solutions. Each of these "Solved Problems" is followed by a few unsolved "Section Review Problems." These problems encourage students to test their understanding of concepts as they progress through the chapter. At the end of each chapter, two additional levels of questions are also presented: some that ask students to integrate ideas from the whole chapter ("Chapter Review Problems"), and others that require integration of material from previous chapters ("Challenge Problems").

We have also developed a series of boxed essays, some of which provide a more in-depth focus on particularly complex concepts, others correlate current research to our everyday lives, while others profile the pioneers, both past and present, of genetic research. Finally, we have organized this book in a logical sequence by grouping related subjects. We start by explaining basic inheritance patterns. This is followed by an explanation of our current understanding of modern molecular genetics as well as how some specialized techniques are used to gain an understanding of genes and how they function in complex organisms. The book ends by considering how genes and genomes evolve through time.

This textbook is intended for undergraduate students in the biological sciences. It is designed to ground students in the fundamentals of genetics, yet also enable them to explore more advanced and specialized subjects. Although the text does not presume an advanced knowledge of biology and chemistry, it does contain numerous examples of how the study of modern genetics rests upon these basic life sciences.

The Author Team

Genetics can be separated into three distinct yet overlapping areas of study: transmission genetics, molecular genetics, and population/evolutionary genetics. Each of the three authors of this textbook has extensive expertise in one or more of the above areas, and they have among them an accumulated 70 years of experience in teaching genetics. Alan Atherly obtained his Ph.D. at the University of North Carolina and did postdoctoral work in molecular genetics at the University of Oregon before going to Iowa State University. He has been on the faculty at Iowa State since 1969 and served as Chair of the Genetics Department for 10 years. Jack Girton obtained his Ph.D. in genetics at the University of Alberta and has studied *Drosophila* developmental genetics since 1980. Dr. Girton came from the University of Nebraska, where he was an Assistant Professor, to Iowa State University in 1985. John McDonald obtained his Ph.D. at the University of California, Davis, in 1977. He is currently the Chair of the Genetics Department at the University of Georgia, where he studies transposable elements and molecular evolution in *Drosophila*.

Learning Aids

Learning genetics for the first time can be challenging. To understand genetics, students must understand facts, learn the meaning of many terms, comprehend complex principles, visualize molecular structures in space, and apply mathematics. A variety of aids that promote these forms of learning are incorporated into each chapter to help students master the concepts presented.

1. Each chapter begins with a **set of clearly stated questions** designed to outline the important points that will be covered in the chapter.

2. The **statement heads** are presented in complete sentences so that the objective of every text section can be easily understood.

3. Unique to this book is the inclusion of **Solved Problems** and **Section Review Problems** integrated throughout the text of each chapter. Solved Problems and Section Review Problems address only the material immediately preceding the questions, allowing students to quiz themselves as they progress through the material. The within-chapter positioning of these problems makes it easier to link the answers with the text. The Solved Problems delineate each step in problem solving and help students understand both the material and how to solve the Section Review and Chapter Review Problems. In addition, there are two sets of chapter-ending questions: approximately 30 **Chapter Review Problems** that integrate concepts presented throughout the chapter and range in difficulty from relatively easy to more challenging, and approximately three **Challenge Problems** for more advanced practice. Knowledge of previous chapters and how earlier information integrates with the present chapter is frequently required to answer the Challenge Problems.

4. The text material includes three types of **Boxed Essays.** The information in these boxes enlarges upon key topics in the text to help broaden the students' base of knowledge. Nearly all chapters include at least one **Historical Profile** of a prominent geneticist whose work has affected genetics significantly. These profiles introduce students to the people who have done the actual work behind important scientific findings. Nearly all chapters include one or more **Concept Close-Up** boxes that explain in plain language especially difficult concepts introduced in the chapter. Finally, nearly all chapters contain **Current Investigation** boxes, which highlight important, timely, and/or newsworthy areas of research.

5. Carefully rendered **illustrations** support concepts in the text. Color pedagogy has been painstakingly developed so that visual elements, such as specific genes, are represented with the same colors from figure to figure, chapter to chapter. In many figures, we have included **boxed explanatory labels** on the artwork itself to help clarify step-by-step processes without the necessity of glancing back and forth between the figures and the legends.

6. The **tables** in the book summarize and organize material related to the textual material, making it easier for students to comprehend the information.

7. Each chapter has a **Summary** that provides a brief review of the material covered.

8. Mastering genetics also involves learning a new vocabulary. To facilitate this process, alphabetized **Selected Key Terms** are included at the end of each chapter. The definitions of these words can be found in the text (they are highlighted in bold) or in the Glossary at the end of the book.

9. A list of **Suggested Readings** is included so that students may easily find related material to extend their knowledge in specific subjects. These listings include both current and classic papers related to the chapter material. Each Suggested Readings list ends with a reminder to visit our **Web site,** where students will find links to chapter-related material on the World Wide Web.

10. An extensive **Glossary** is provided at the end of the book to help students rapidly locate terms and definitions used throughout the book.

11. Short answers to all Section Review Problems are found at the back of the book in the **Appendix: Answers to Section Review Problems.** This gives students an opportunity to check their solutions immediately or to work through a problem with knowledge of the answer. (Short answers to all problems in the book are on our Web site, and complete solutions are given in the Student Solutions Manual; see "Supplements.")

The Organization of the Book

One of the strongest features of this book is its organization. There are two general schools of thought on the teaching of genetics: the Mendel-first approach and the DNA-first approach. The organization of this text accommodates either teaching preference. This book is composed of an introductory chapter, followed by three easily recognizable parts on transmission genetics; molecular genetics; and genetic change, population genetics, and evolution. Chapter 1 introduces the discipline of genetics, explains how it is organized, and gives a brief history of the science of genetics. This chapter also introduces some of the organisms commonly used in genetic studies. Instructors may begin their course with either Part 1 or 2 depending upon their teaching preference. The material in Part 3 presumes a working knowledge of the information in Parts 1 and 2. In the typical introductory class, not all chapters in Part 3 will be covered. The material in Part 3 is arranged so that the chapters can be presented independently, allowing instructors to select those advanced topics deemed most relevant to their courses and/or to assign topics for individual research.

PART 1: How Genes Are Organized and Transmitted Through Generations

Part 1 begins with Chapter 2, which focuses on the principles of Mendelian inheritance and its role in current genetics. Chapters 3 and 4 present a comprehensive overview of the fundamentals of transmission genetics, how genes interact in their expression, and how a gene can take different forms. Chapter 5 introduces the subject of how multiple genes affect the inheritance of a single trait in an individual, and Chapter 6 describes how it is possible to locate genes on a chromosome and precisely map their positions. Chapter 7 demonstrates that not all genes are located in the nucleus of higher organisms; some genes reside on extrachromosomal elements and have a very distinctive inheritance pattern. Finally, Chapter 8 introduces the subject of the genetics of prokaryotes and viruses and how mutants are isolated and mapped to their positions on the chromosome.

PART 2: How Genes Function at the Molecular Level

Part 2 focuses on the molecular aspects of genetics as it is studied today. Chapter 9 covers our present understanding of DNA and RNA structure and the process of DNA replication. This leads naturally into Chapter 10, which presents our current understanding of how DNA is folded to form chromosomes in both eukaryotes and prokaryotes. This complex structure has the coded information that determines the phenotype of an organism and must be precisely unfolded to express genes. The mechanisms underlying this phenomenon are discussed in Chapter 11. Chapters 12 and 13 are devoted to an explanation of the technology of molecular biology and how it can be used to clone and characterize a specific gene even from complex organisms such as humans. Chapter 14 presents a discussion of how genes are expressed and how this expression is regulated. Chapters 15 and 16 present the basic principles of development and how genes are involved in the control of this complex pattern of interactions. By the end of Part 2, students should have a broad grasp of the physical nature of genes and genomes and an appreciation of the complex nature of their expression.

PART 3: How Genes and Genomes Change and Evolve

Part 3 of the book investigates how gene mutations occur, how they are repaired, and how genes and genomes change and evolve through time. Chapter 17 explores the molecular explanation of how mutations in DNA occur and how they are detected and repaired. Because mutations can also occur as major chromosomal alterations, Chapter 18 examines how inversions, duplications, translocations, and deletions of chromosomes occur and discusses the genetic consequences of these changes and of changes in chromosome number. Chapter 19 follows with a discussion of two additional methods by which genomes change, i.e., through movement of sequences within the genome and by recombination. Chapter 20 examines one major genetic consequence of mutations, the development of cancers. Finally, Chapters 21 and 22 deal with the mechanisms that underlie the evolutionary process. By the end of Part 3, students should have a sound understanding of how mutations affect individuals and populations and an appreciation for some of the unanswered questions in the field of genetics.

Supplements

To facilitate learning and teaching, a series of supplemental materials are available. These include:

An *Instructor's Manual/Test Bank* with suggested formats for teaching, outlines of the chapters, key concepts, and approximately 800 test questions.

A set of full-color *Overhead Transparencies* for 150 of the figures, including the more complex figures that are difficult for the instructor to draw easily.

A *Student Solutions Manual* by Lois Girton of Grand View College, which gives complete solutions to all of the Section Review Problems, Chapter Review Problems, and Challenge Problems in the main text.

Instructor's Resource CD-ROM (IRC) for Biology 1999, which contains all of the illustrations from *The Science of Genetics,* as well as those from Solomon/Berg/Martin: *Biology 5/e.* This CD-ROM is available for Windows™ and Macintosh platforms as a presentation tool to be used with commercial presentation packages, such as PowerPoint™ and Persuasion™. In addition, the IRC imagery can also be used as masters to custom-make additional overhead transparencies.

Saunders College Publishing's new **General Life Sciences Web site,** an online magazine, is updated monthly and features articles, reviews of recent research findings, stories focusing on applications and public policy, interactive online experiments, a calendar of events, a searchable glossary, and an extensive quizzing and testing program. *The Science of Genetics* also has its own book-specific section of the magazine, which includes links to the chapter-related Web resources described at the end of each chapter; short answers to all Section Review Problems, Chapter Review Problems, and Challenge Problems; as well as more instructor's presentation tools and students' study tools.

Saunders College Publishing may provide complementary instructional aids and supplements or supplement packages to those adopters qualified under our adoption policy. Please contact your sales representative for more information. If as an adopter or potential user you receive supplements you do not need, please return them to your sales representative or send them to

Attn: Returns Department
Troy Warehouse
465 South Lincoln Drive
Troy, MO 63379

Acknowledgments

The development and production of a book such as this requires the interaction and cooperation of many individuals. We greatly appreciate the valuable input of the editors, colleagues, and students.

We are particularly grateful to the editorial and production staff at Saunders College Publishing for their help throughout this project. We thank our publisher, Emily Barrosse, and our acquisitions editor, Julie Levin Alexander, for their early support, ideas, and enthusiasm. Developmental Editor James Funston was also very helpful during the early stages of this project. Julie Alexander was succeeded by Executive Editor Edith Beard Brady, who has been instrumental in seeing the project through to its final stages. Special thanks are due to Senior Developmental Editor Gabrielle Goodman for her patience and guidance through many difficult periods. Gay Gragson provided outstanding editorial and technical assistance in preparing the final manuscript. We would also like to express our thanks to Matt Lux (Iowa State University) for verifying the accuracy of all the problems in the book. Special thanks also go to Robin Bonner, our project editor, for shepherding this manuscript through all the complexities of production, and to Art Director/Designer Ruth Hoover for the beautiful internal design and also the cover. We also would like to thank illustration and design consultants John and Bette Woolsey and their associates for developing and rendering an excellent illustration program.

We thank our families for their patience during what at times seemed to all of us a project without end. We hope that the final product will serve as an engaging and thorough beginning for those about to enter the fascinating and ever-changing science of genetics.

We express our appreciation to the many geneticists who read the manuscript during various stages of writing and provided helpful suggestions on improvement. Their input has greatly improved the final product. These include:

Karen Anderson, Hofstra University
L. Rao Ayyargi, Lindenwood University
Norbert Belzer, La Salle University
Dorthy Bennett, University of Texas–Austin
Edward M. Berger, Dartmouth College
Roberta Berlani, Santa Clara University
R. L. Bernstein, San Francisco State University
James Berry, State University of New York at Buffalo
Paul E. Bibbins, Jr., Kentucky State University
Margaret Bird, Marshall University
Paul J. Bottino, University of Maryland
Nicole Bournias, California State University–San Bernardino

James Boyd, University of California–Davis
Carl Candiloro, Pace University
Laurie F. Caslake, State University of New York at Fredonia
Ruth Chestnut, Eastern Illinois University
John A. Chisler, Glenville State College
Bruce Cochrane, University of South Florida
Bart Cook, Texas A&M University–Corpus Christi
Donald Cronkite, Hope College
John David, University of Missouri–Columbia
F. Paul Doerder, Cleveland State University
Veronica Dougherty, Cecil Community College
Frank Einhelling, Southwest Missouri State University
DuWayne C. Englert, Southern Illinois University at Carbondale
Nancy H. Ferguson, Clemson University
Edward Fliss, Missouri Baptist College
Robert G. Fowler, San Jose State University
Teryl K Frey, Georgia State University
David Futch, San Diego State University
Michael Goldman, San Francisco State University
Elliott S. Goldstein, Arizona State University
Miriam Golomb, University of Missouri–Columbia
Nels H. Granholm, South Dakota State University
Richard Halliburton, Western Connecticut State University
Jenna J. Hellack, University of Central Oklahoma
Deborah K. Hoshizaki, University of Nevada–Las Vegas
Austin Hughes, Pennsylvania State University
Lynne Hunter, University of Pittsburgh
David R. Hyde, University of Notre Dame
Santha Jeyabalan, University of Michigan
Margaret Jefferson, California State University–Los Angeles
William W. Johnson, University of New Mexico
Spencer Johnston, Texas A&M University
Duvall Jones, St. Joseph's College
Phil Keeting, University of West Virginia
Robert Kitchin, University of Wyoming
Keith Klein, Mankato State University
Barbara Liedl, Central College
David Lofsvold, Franklin & Marshall College
Clint Magill, Texas A&M University
Pat McCreary, USAF Academy
Lauren McHenry, Loyola College–Maryland
Philip Meneely, Haverford College
Robert Morell, Michigan State University
John C. Osterman, University of Nebraska
E. David Peebles, Mississippi State University
Kent Reed, University of Wisconsin–Milwaukee
Richard Richardson, University of Texas–Austin
Steven Runge, University of Central Arkansas
Henry E. Schaffer, North Carolina State University
Brian Schwartz, Colombus State University
Ann Spencer, Charles County Community College
Beat Suter, Magill University
Chuck Staben, University of Kentucky
Edwin C. Stephenson, University of Alabama
Martin Tracey, Florida International University
Craig Tuerk, Moorehead State University
Judith Van Houten, University of Vermont
Daniel E. Wivagg, Baylor University
Patrick Woolley, East Central College

Despite the best efforts of these reviewers in correcting our errors of fact or omission, we are, of course, responsible for any remaining errors in this book. We encourage readers to send their comments or corrections directly to Alan Atherly via e-mail at agatherl@iastate.edu.

Alan Atherly
Jack Girton
John McDonald

November 1998

Contents in Brief

1 An Introduction to the Science of Genetics *1*

PART 1
How Genes Are Organized and Transmitted Through Generations *15*

2 Transmission Genetics: Mendelian Analysis of Inheritance *16*

3 Genes, Chromosomes, and the Mechanism of Mendelian Inheritance *46*

4 Multiple Alleles and Gene Interaction *85*

5 Quantitative Traits and Polygenic Inheritance *114*

6 Linkage and Gene Mapping In Eukaryotes *142*

7 Extranuclear Inheritance *187*

8 The Genetics of Bacteria and Their Viruses *210*

PART 2
How Genes Function at the Molecular Level *249*

9 The Physical Nature of DNA: Structure and Replication *250*

10 The Molecular Structure of Prokaryotic and Eukaryotic Chromosomes *282*

11 Transcription, Translation, and The Genetic Code *309*

12 Gene Cloning and Analysis *348*

13 Genomic Analysis and Modification *382*

14 Regulation of Gene Expression *410*

15 Developmental Genetics: Genetic Regulation of Cell Fate *451*

16 Developmental Genetics: Hierarchies of Genetic Regulation *485*

PART 3
How Genes and Genomes Change and Evolve *507*

17 Gene Mutation and Repair *508*

18 Changes in Chromosome Number and Structure *540*

19 Recombination and Transposable Elements *578*

20 The Genetics of Cancer *612*

21 Population and Evolutionary Genetics *643*

22 Molecular Evolution *678*

Appendix: Answers to Section Review Problems *A.1*

Glossary *G.1*

Index *I.1*

Contents

1 An Introduction to the Science of Genetics *1*

Genetics Is the Study of Heredity and Variation 2
Genetics Is Central to All Biology 2
Genetics Research Is Focused on Three Areas 3
Genetics Has a Rich History 5
Genetics Is an Experimental Science 7
The Scientific Method Is Used in Genetics 8
Choosing an Appropriate Experimental Organism Is Important 9
The Desirable Characteristics for a Genetic Research Organism Are Varied 9
Commonly Used Genetic Research Organisms Have Many of These Features 10

Current Investigations: Adenosine Deaminase Deficiency and Gene Therapy 4

Current Investigations: Evolutionary Relationships Through DNA Analysis 6

Current Investigations: Genetics and the Internet 12

PART 1

How Genes Are Organized and Transmitted Through Generations 15

2 Transmission Genetics: Mendelian Analysis of Inheritance *16*

Mendel Discovered the Rules of Inheritance 17
Monohybrid Crosses Demonstrate that Alleles Segregate 19
Dihybrid and Trihybrid Crosses Demonstrate that Alleles of Different Genes Segregate Independently 22
Mendel's Rules Are Used To Predict Genetic Combinations 26
Probability and Chi-Square Are Used To Evaluate Data 27
The Multiplication and Addition Rules of Probability Are Widely Used in Genetics 28
The Chi-Square Test Is Used To Evaluate How Closely Observed Results Fit Expectations 32
Genetic Variation Is Necessary for the Study of Genetics 35
Drosophila *Wild-Type Strains Are Used To Measure the Effects of Mutations on the Phenotype 35*
Variant Alleles Are Used as Genetic Markers and as Tools To Study Gene Function 35
Mendelian Principles Have Many Applications 36
Inherited Human Diseases Are Analyzed Using Pedigree Analysis 36
Single Genes Are Responsible for a Variety of Human Disorders 39
Mendel's Rules Are Widely Used in Agriculture 41

Historical Profile: Johann Gregor Mendel (1822–1884) 18

Concept Close-Up: Gene Symbols 23

Concept Close-Up: The Binomial Theorem 30

Concept Close-Up: Are Mendel's Results Too Good To Be True? 34

3 Genes, Chromosomes, and the Mechanism of Mendelian Inheritance *46*

The Cell Cycle and Chromosome Behavior Are Fundamentally Important 47
The Somatic Cell Cycle Is Divided into Three Stages 47
Mitosis Produces Cells that Are Genetically Identical to the Parental Cell 49
The Morphology of Eukaryotic Chromosomes Can Be Studied During Mitosis 52
Chemical Staining Reveals Differences in the Structure of Chromosomes 55

Chromosomal Abnormalities Can Be Detected by Karyotype Analysis 57
Meiosis Is the Basis for Mendel's Rules 60
Homologous Chromosomes Segregate from Each Other During Anaphase of Meiosis I 60
Sister Chromatids Separate During Meiosis II 61
Chromosome Segregation in Meiosis I Is the Mechanism for Mendel's Rules 65
Meiosis Occurs at Different Stages in Different Species' Life Cycles 66
Sex-Determining Genes Are Located on Specialized Chromosomes 72
Specific Genes Segregate with Specific Chromosomes 72
The Final Proof of the Chromosome Theory Came from Sex-Chromosome Nondisjunction 77
Two Different Methods of Sex Determination Are Common in Animals 78
Dosage Compensation Occurs in Mammals Through X Inactivation 79
Concept Close-Up: Using Chromosome Banding To Establish Species Relationships 57
Historical Profile: Theodore Boveri (1862–1915) 65

4 Multiple Alleles and Gene Interaction 85

Alleles Have Many Different Variations 86
Alleles of the Same Gene Can Show Complete Dominance, Incomplete Dominance, Overdominance, or Codominance 86
One Allele May Show Different Types of Dominance 89
Codominant Alleles Are Often Detected Using Antigen–Antibody Reactions 91
Segregation and Complementation Tests Are Used To Determine Whether Different Mutations Are Alleles of the Same Gene 94
H. J. Muller Devised a Functional Classification System for Mutant Alleles 97
Gene Interactions Change the Ratios of Dihybrid Crosses 99
Alleles of Different Genes that Produce the Same Phenotype Give a Modified Dihybrid Ratio 99
A Mutant Allele of One Gene May Mask an Allele of Another Gene 100
Modifier Genes Enhance or Suppress the Phenotype Produced by Other Mutations 102
Mutant Alleles Can Show Pleiotropic Effects 102
Mutant Alleles Can Show Partial Penetrance and Have Variable Expressivity 103
Some Alleles Cause the Death of an Organism 104
Lethal Alleles Cause Death at Particular Stages of the Life Cycle 105
The Phenotypes Produced by Conditional Alleles Are Affected by the Environment 106
Current Investigations: How Can Genetics Be Used To Solve a Human Rights Crime? 93
Historical Profile: Hermann Joseph Muller (1890–1967) 98

5 Quantitative Traits and Polygenic Inheritance 114

The Inheritance of Quantitative Traits Is Mendelian 115
Quantitative Trait Inheritance Has a Predictable Pattern 115
Quantitative Traits Are Controlled by Multiple Genes 117
Genetic Analysis of Quantitative Traits Uses Statistics 119
Means and Variances of Frequency Distributions Define the Phenotypes of Populations 119

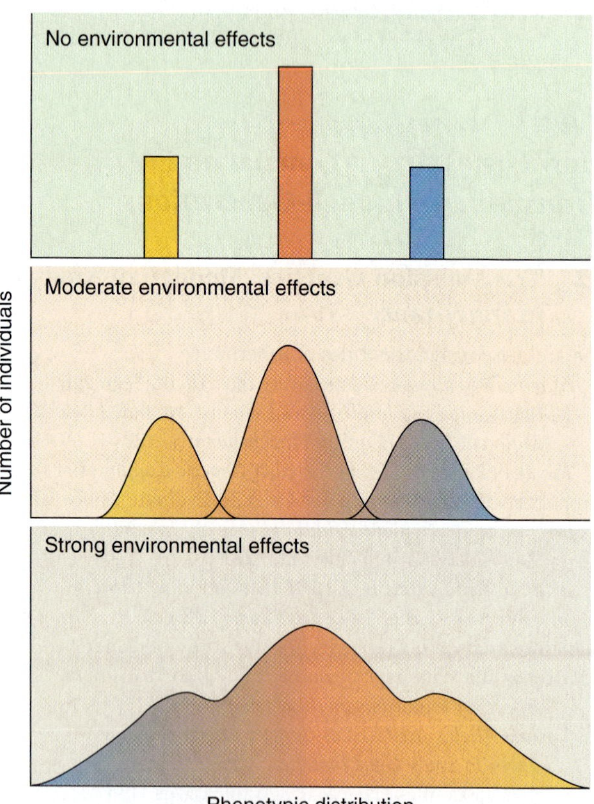

The Means of Two Distributions Can Be Compared Using the Student's t Statistic 122
The Strength of the Association Between Two Traits Is Determined Using Correlation 124
The Values of Correlated Traits Can Be Predicted Using Regression Analysis 125
The Phenotypic Variance Can Be Divided into Genetic and Environmental Components 126
Heritability Is Useful in Predicting the Phenotypes of Offspring 129
Heritability Is a Measure of the Genetic Component of the Variance 129
Heritability in the Broad Sense Measures the Effects of All Genetic Variation on the Phenotype 134

Historical Profile: Sir Ronald Aylmer Fisher (1890–1962) 120

Current Investigations: Is It Important To Conserve Genetic Diversity? 135

Current Investigations: How Are Identical Twins Used To Study the Inheritance of Human Behavior? 137

6 Linkage and Gene Mapping in Eukaryotes 142

Recombination Is the Basis of Gene Mapping 143
The Alleles of Two Linked Genes Usually Segregate Together in Meiosis 143
Linkage Is Easily Detected in an F_1 Testcross 147
Linkage Can Be Detected in an $F_1 \times F_1$ Cross 148
Meiotic Crossing Over Involves a Physical Exchange Between Homologous Chromosomes 150
Gene Mapping Creates a Road Map of the Chromosome 153
Linkage Mapping Depends Upon the Frequency of Crossing Over Between Two Linked Loci 154
The Three-Point Testcross Is the Preferred Mapping Cross 154
Interference Increases or Decreases the Chance of a Second Crossover 156
Linkage Maps of Chromosomes Can Be Constructed by Combining the Results of Many Mapping Experiments 157
Tetrad Analysis Is Used for Chromosome Mapping 159
Linkage Maps Are Determined by Analyzing Crossover Tetrads 160
Ordered Tetrad Species Aid in Determining the Map Distance Between a Locus and Its Centromere 163
Gene Mapping Can Be Done in the Absence of Meiosis 166
The Genes Can Be Located on a Chromosome Using Mitotic Crossing Over 166
Human Genes Can Be Mapped by Somatic Cell Fusion 170
Deficiency Complementation Can Map a Gene to a Region of a Chromosome 174
Genetic Fine-Structure Mapping Defines a Gene Genetically 176
Genetic Fine-Structure Maps Are Made by Measuring Crossing Over Within a Locus 176
Complementation Mapping Defines the Cistron, a Unit of Genetic Function 177

Historical Profile: Thomas Hunt Morgan (1866–1945) 144

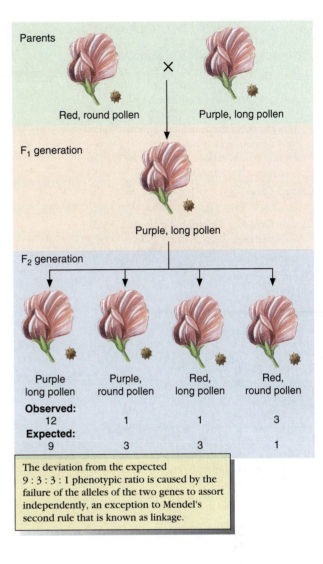

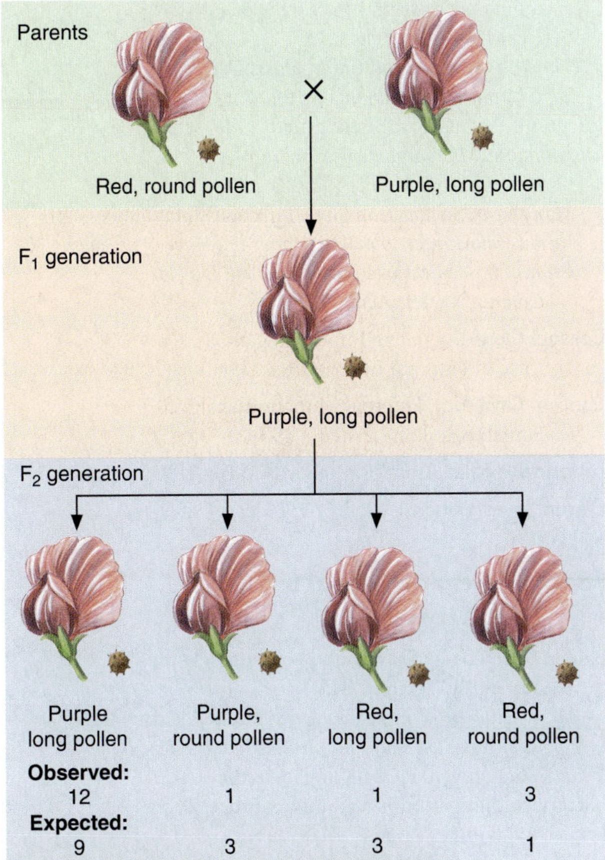

Concept Close-Up: Detecting Linkage in an $F_1 \times F_1$ Cross 151

Concept Close-Up: How Can Linkage Maps Be Used to Predict the Results of a Cross? 160

7 Extranuclear Inheritance 187

Maternal Genes Can Affect the Phenotype of the Offspring 188
Maternal Effects Are Produced by Gene Products Packaged in the Egg 188
Maternal-Effect Genes Control Early Stages of Embryonic Development 189
Gene Imprinting and Maternal Effects Are Not the Same 191
Some Cytoplasmic Organelles Have Chromosomes 192
Mitochondria and Chloroplasts Provide Energy and Carbohydrates to Cells 192
The Genomes of Mitochondria and Chloroplasts Are Organized as Single, Circular Chromosomes 193
Mutations in Mitochondria and Chloroplast Genes Show Maternal Inheritance 194

Genes in the Organelles Interact with Genes in the Nucleus 197
The Mitochondria and Chloroplasts Originated as Cytoplasmic Symbionts 199
Infectious Agents Can Be Inherited 199
Symbiotic Microorganisms Can Show Infectious Inheritance 199
Many Viral Strains Can Show Infectious Inheritance 201
Viroids Show Infectious Inheritance but Have No Genes 202
Prions Show Infectious Inheritance but Do Not Contain Nucleic Acid 203
Concept Close-Up: Gene Imprinting and a Case of Paternal Inheritance 191
Concept Close-Up: Paternal Mitochondria Can Sometimes Be Inherited 195
Historical Profile: Tracy Sonneborn (1904–1981) 202
Current Investigations: Mad Cow Disease 205

8 The Genetics of Bacteria and Their Viruses 210

Bacteria Are Easily Grown, and Bacterial Mutants Are Highly Varied 211
Many Different Kinds of Bacterial Mutants Exist 212
Spontaneous Bacterial Mutants Are Not the Result of Adaptation to a Changing Environment 213
Bacterial Conjugation Is a Complex Process Involving Plasmids 214
Gene Transfer in Bacteria Was Discovered in 1946 214
Conjugation Requires the Presence of an F-factor 216
Conjugation Enables Construction of Genetic Maps of Bacterial Chromosomes 219

Use of Different Hfr Strains Reveals a Circular Genetic Map of the E. coli Chromosome 221
F'-Plasmids Are Formed Upon Imprecise Excision of an F-Plasmid 223
F'-Plasmids Are Used in Complementation Analysis 223
Many Types of Plasmids Are Present in Bacteria 226
Transformation Involves Cellular Uptake of DNA 227
Transformation Involves the Uptake of DNA by a Bacterial Cell 228
Transformation Is Useful in Fine-Scale Genetic Mapping 228
Phage Genetics Has Revealed Information About the Gene 230
The Phage Life Cycle May Include Lytic and Lysogenic Cycles 230
Mapping Phage Genes Has Proved Enormously Revealing 231
The rII Locus of Phage T4 Has Been Mapped in Great Detail 235
Deletion Mapping Allows Rapid Location of Mutational Events 237
Transduction Is Useful for Mapping Closely Linked Genes 239
Transduction Can Be Used for Genetic Mapping 239
Specialized Transduction Transmits Specific Genes from Cell to Cell 242
Genetic Mapping Has Its Limitations 244
Historical Profile: Joshua Lederberg (1925–) 215
Current Investigations: Is It Safe to Feed Antibiotics to Cattle, Swine, and Poultry? 227

PART 2
How Genes Function at the Molecular Level 249

9 The Physical Nature of DNA: Structure and Replication 250

Scientists Needed Proof that DNA Is the Genetic Material 250
DNA Is the Primary Origin of Transformation 251
The Amount of DNA Per Cell Is Constant, and DNA Is Chemically Stable 252
Phage Were Used To Confirm that DNA Is the Genetic Material 252
RNA Can Also Be the Genetic Material 254
The Chemical Nature of DNA and RNA Gives Clues to Their Function 256
The Monomeric Units of DNA and RNA Are Nucleotides 256
DNA Is Composed of Only Four Different Nucleotides 257
DNA Is a Double-Helical Molecule 258
DNA Is a Long, Thin Molecule 261
The Structure of RNA Is Very Similar to That of DNA 262
DNA Replicates Using Itself as a Template 264
Meselson and Stahl Proved that DNA Replicates Semiconservatively 264
DNA Replication Is Bidirectional 265

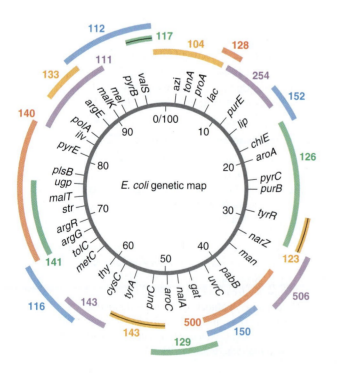

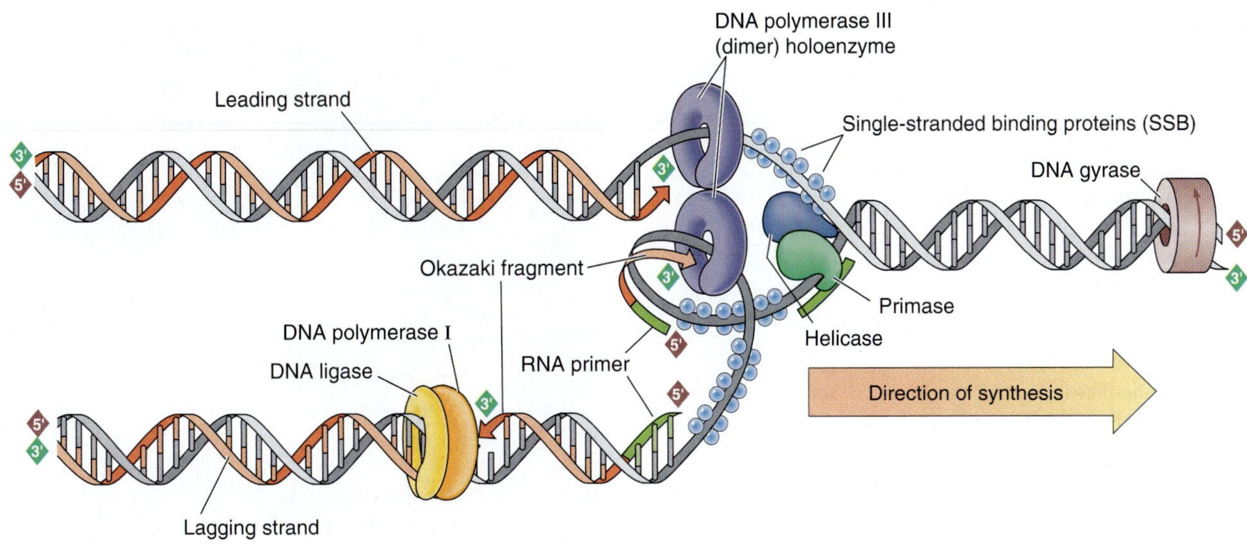

DNA Polymerase Is Responsible for DNA Replication 269
DNA Synthesis Occurs Using Both Strands as a Template 271
DNA Replication in E. coli *Is a Good General Model* 272
DNA Replication Must Be Extremely Accurate 274
DNA Replication in Eukaryotes Must Deal with Many Chromosomes 276
Replication of Virus and Plasmid DNA Uses a Rolling Circle 277
Circular DNA Molecules Are Sometimes Replicated as a Rolling Circle 277
Between Organisms, DNA Replication Has Similarities and Differences 278

Current Investigations: How Did Life on Earth Begin? 263

Historical Profile: Arthur Kornberg (1918–) 270

10 The Molecular Structure of Prokaryotic and Eukaryotic Chromosomes 282

A Prokaryotic Chromosome Is Packed into a Small Space 283
The Chromosome of Prokaryotes Is Highly Coiled 283
Most Double-Stranded DNA Molecules Contain Supercoils 285
Supercoils Are Introduced and Removed from DNA by Topoisomerases 286
The Sequence of the Eukaryotic Chromosome Is Complex 287
A Chromosome Has Only One Molecule of DNA 287
Eukaryotic DNA Has Repetitive Sequences 288
The Eukaryotic Chromosome Is Tightly Packed 293
Chromosomes Contain Two Major Classes of Proteins 293
First-Order DNA Coiling Creates an 11-nm Filament of Nucleosomes 295
The Second-Order DNA Coiling Creates a 34-nm Solenoid 295
The Chromatin Is Composed of 300-nm Coiled Domains 297
The Structure of the Chromosome Changes During the Cell Cycle 298
The Eukaryotic Chromosome Has Specialized Sequences 299
Euchromatin and Heterochromatin Have Different Compositions 299
Telomeres and Centromeres Possess Unique Sequences 301
Unique Chromosomes Are Useful in Understanding Genetic Mechanisms and Structure 304

Concept Close-Up: Reassociation Kinetics and Repetitive DNA 290

Current Investigations: Telomeres Grow and Shrink Each Generation 302

11 Transcription, Translation, and the Genetic Code 309

Proteins Determine the Phenotype of an Organism 310
Proteins Are Linear Polymers of Amino Acids 310
One Gene–One Polypeptide is a Fundamental Principle 311
Transcription Is the Passing of Information from DNA to RNA 315
Transcription in Prokaryotes Is Catalyzed by One Polymerase 315
Transcription in Eukaryotes Uses Three Different Polymerases 320
Eukaryotes Modify mRNA After Transcription 321
The 5′ Cap and a 3′ Tail Are Added After Transcription 322
A Complex Splicing Reaction Removes Sequences from Pre-mRNA 323
Ribosomal RNAs Are Also Processed After Synthesis 327
mRNA Sequences Are Translated into Sequences of Amino Acids 328
Ribosomes and tRNAs Play an Essential Role in Protein Synthesis 331
Proteins Are Assembled on the Ribosome 334
Mistakes Can Sometimes Occur During Protein Synthesis 335

The Genetic Code Is Simple 338
Genetics and Biochemistry Were Used To Decipher the Genetic Code 338
More than One Code Word Determines a Single Amino Acid 339
Not All Codons Are Used to the Same Extent 340
There Are a Few Exceptions to the Nearly Universal Genetic Code 342
One DNA Sequence Can Determine the Sequence of More than One Protein 343

Current Investigations: What Is the Function of an Intron? 326

Current Investigations: Not All Enzymatic Reactions Are Catalyzed by Proteins 329

Historical Perspective: Francis Crick (1916–) 340

12 Gene Cloning and Analysis 348

Standardized Techniques Are Used for Cloning Genes 348
DNA Can Be Cut into Sized Fragments Using Restriction Endonucleases 349
Plasmids Make Excellent Cloning Vectors 353
Some Vectors Can Replicate in Multiple Hosts 357
A Library Is a Collection of Cloned DNA Fragments from an Organism 358
Libraries Can Be Prepared from mRNA (cDNA Library) 359

Different Approaches Are Used To Find a Gene in a Library 360
A Probe Is Used To Identify a Clone in a Library 360
Genes Can Be Cloned in the Absence of a Probe 361
A DNA Sequence Can Be Amplified Using the Polymerase Chain Reaction 362

A Cloned Gene Can Be Characterized To Reveal Basic Information 365
Cloned DNA Sequences Have Characteristic Restriction Maps 365
Cloned DNA Fragments Can Be Sequenced 368
DNA Fragments Can Be Identified After Transfer to a Membrane 370

Specialized Cloning Vectors Are Widespread 373
Prokaryotic Vectors Permit the Expression of Eukaryotic Genes 373
Cloning Vectors Permit the Study of Gene Regulatory Regions 374

Historical Profile: The Discovery of Restriction Endonucleases 354

Current Investigations: The Human Genome Will Be Sequenced Soon 372

Current Investigations: Many Different Proteins Are Currently Manufactured Using Cloned Genes 377

13 Genomic Analysis and Modification 382

DNA Sequence Polymorphisms Make Useful Genetic Markers 383
Methods Are Available To Detect Genomic DNA Differences 383

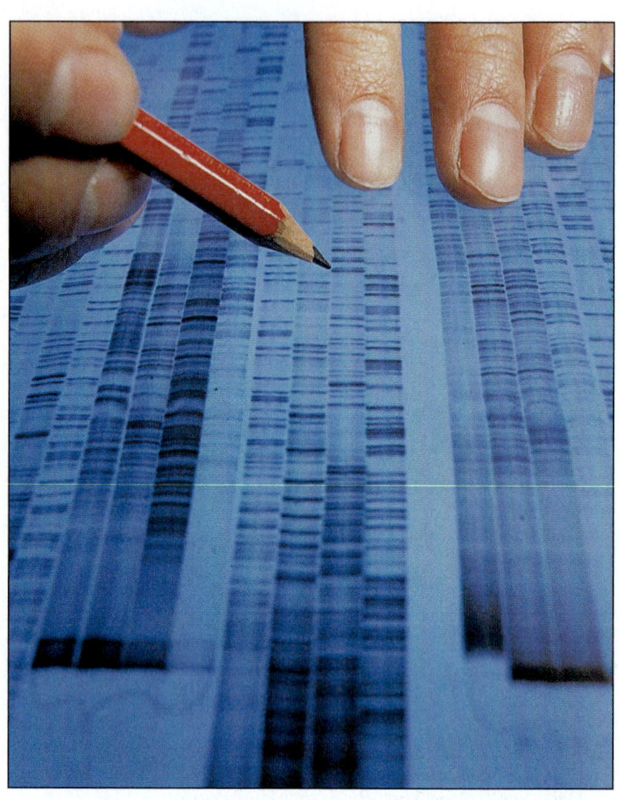

RFLP and Microsatellite Maps of Chromosomes Can Be Generated for Any Organism 386
RFLP and Microsatellite Maps Are Useful in Studying Genome Organization and in Breeding Programs 388
RFLP and Microsatellite Markers Have Many Advantages over Morphological Markers 389

Large DNA Fragments Can Be Isolated and Cloned 390
Phage Lambda and Cosmids Are Useful for Cloning Medium-Sized DNA Fragments 391
Artificial Chromosomes Are Useful for Cloning Large DNA Fragments 393
PFGE Permits Large DNA Fragments To Be Separated 396

Genome Sequences Have Revealed Some Interesting Findings 398
Bacterial Gene Numbers Vary from 500 to 8000 398
Eukaryotes Have Very Similar Gene Numbers 398

Cloned Genes Have Many Applications 399
Foreign Genes Can Be Introduced into Plants 400
Transformed Drosophila *Is Useful for Studying Genes and Genomes 400*
Genes Can Be Introduced into Human Cells To Reverse the Effect of Inherited Diseases 401
Mutations in Mice that Mimic Human Diseases Can Be Made 402

Concept Close-Up: Taking the Witness Stand with DNA Fingerprinting 385

Concept Close-Up: Eugenics: Are We Seeing a Rebirth? 388

Historical Profile: Herbert Boyer (1936–) and Stanley Cohen (1935–) 395

14 Regulation of Gene Expression *410*

The Regulation of Gene Expression Occurs at Many Levels *411*

Transcriptional Control I: The *lac* Operon Is an Excellent Model of Transcriptional Control *412*

Enzyme Synthesis Is Induced by Environmental Signals 414

Mutations Were the Key to Understanding the lac *Operon 414*

Catabolite Repression Affects lac *Operon Expression 421*

The Three-Dimensional Structure of Some Proteins Is Changed by Binding Small Molecules 422

Transcriptional Control II: Attenuation, Antitermination, Methylation, and DNA-Binding Proteins Control Gene Expression *423*

Attenuation and Positive Control Regulate Tryptophan Synthesis 423

Antiterminators and Repressors Control Gene Expression in Lambda Phage 426

The Yeast GAL Regulatory Pathway Is a Good Model of Eukaryotic Gene Regulation 429

Steroid Hormones Can Activate Gene Expression in Mammals 431

DNA Methylation May Control Some Aspects of Cell Development 432

DNA-Binding Proteins Have Unique Structural Regions that Enable Transcription 435

Gene Regulation Can Occur at the Posttranscriptional Level *436*

One Gene May Code for More than One Protein 436

The Coding Sequences of mRNAs Can Be Changed After Synthesis 438

The Stability of mRNA Can Control Gene Expression 439

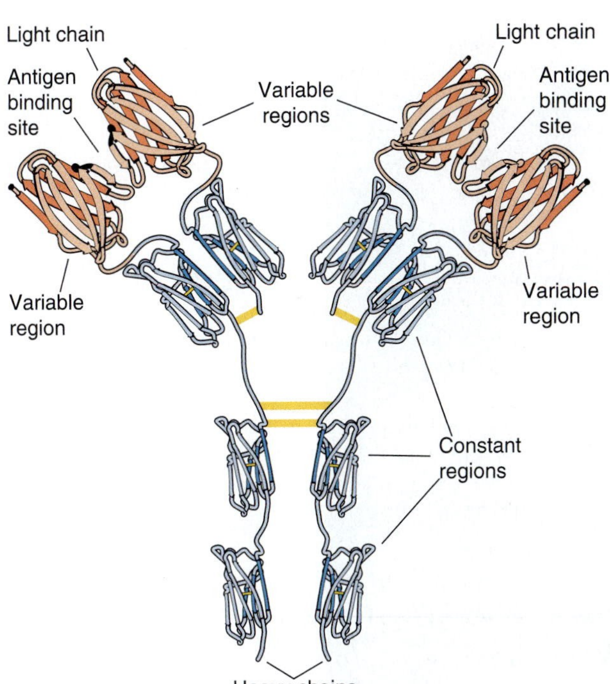

DNA Sequence Rearrangements Can Permanently Alter Gene Expression *439*

Salmonella *Strains Undergo DNA Rearrangements To Evade the Immune Response 440*

Trypanosomes Use DNA Rearrangement To Alter Their Surface Coats To Evade Immune Responses 441

Antibody Diversity Is Required for Hosts To Survive 443

Concept Close-Up: Why Can't Some People Digest Milk? *413*

Historical Profile: Jacques Monod (1910–1976) *416*

Current Investigations: How Do Bacteria, Animals, and Plants Adapt to Stress? *433*

15 Developmental Genetics: Genetic Regulation of Cell Fate *451*

Cell Determination Is an Essential Developmental Process *452*

Cell-Fate Control Is Essential for Multicellular Development 452

Mosaic and Regulative Development Are Two General Mechanisms for Cell Determination 454

The Developmental Genetics of *Drosophila* *456*

The Drosophila *Life Cycle Has Five Stages 457*

In Drosophila *Embryos, Cell Determination and Pattern Formation Occur at the Cellular Blastoderm Stage 460*

Fate Mapping Reveals that Cell Determination Is Spatially Organized in the Blastoderm 461

A Cell's Lineage Controls Its Fate in *C. elegans* *467*

Cell Determination Is Controlled by Cell Lineage in C. elegans *467*

Adult Development in C. elegans *Builds on the Pattern Established During Embryonic Development 468*

Genetic Mechanisms Control Cell-Fate Decisions *469*

Maternal Gene Products Concentrated in the Egg Control Developmental Regulatory Gene Action 469

Germ-Line Cell Determination Is Controlled by Cytoplasmic Determinants Packaged in the Polar Granules 471

Homeotic Genes Control Segmental Identity in Drosophila *472*

Cell Determination in C. elegans *Is Controlled by Maternal Genes and Homeotic Genes 478*

Cell–Cell Communication Regulates Individual Cell-Fate Decisions 478

Current Investigations: A Sheep Called Dolly *455*

Current Investigations: Using Somatic Mosaics in Developmental Genetic Analysis *462*

Historical Profile: Edward B. Lewis (1918–) *474*

Current Investigations: The Homeobox *477*

16 Developmental Genetics: Hierarchies of Genetic Regulation *485*

Gene Hierarchies Control Embryonic Pattern Formation *485*

The Drosophila *Body Axis Is Established by Gradients of Maternal-Gene Products 486*

Zygotic Segmentation Genes Organize the Embryo into Segments and Compartments 488

Homeotic Genes Control Cell Identity in Animals *493*

The Homeotic Gene Complex in Drosophila *Contains ANT-C and BX-C Genes* 493
The Hox Genes Control Determination in Mammals 495
Sex Determination Is Controlled by a Gene Hierarchy 497
Sex-Specific RNA Processing Controls Sex Determination in Drosophila 497
Sex-Specific Transcriptional Regulation Controls Sex Determination in C. elegans 499
Historical Profile: Christiane Nüsslein-Volhard and Eric Wieschaus 492
Concept Close-Up: The *X*: A Primary Signal in *Drosophila* 498
Current Investigations: Human Sex Determination 502

PART 3
How Genes and Genomes Change and Evolve 507

17 Gene Mutation and Repair 508

There Are Many Different Categories of Mutations 509
Spontaneous Mutations Are Caused by Natural Phenomena 513
DNA Replication Errors May Cause Mutations 513
Spontaneous Chemical Changes Can Cause Mutations 517
Transposons and Insertion Sequences Act as Mutagens 517
Unequal Crossing-Over Produces Mutations 518
Spontaneous Mutations Occur at a Low Rate 519
Induced Mutations Are the Result of Our Environment 521
Radiation Can Cause Mutations 521
Chemicals Can Induce Mutations 522
Mutagenicity Can Be Measured 526
Biological Repair Reverses Many Mutations 530
Mutations Can Be Directly Repaired 530
The Complementary Strand of DNA Is Used in Repair 531
DNA Recombination Can Repair Mutations 534
Some DNA Repair Mechanisms Produce Errors 535
Concept Close-Up: Antisense Genes Can Cause a Mutant Phenotype 511
Historical Profile: Lewis J. Stadler (1896–1954) 520
Current Investigations: Many Natural Products Are Mutagenic 528
Current Investigations: Accumulated Mutations May Cause Death, Aging, and Cancer 532

18 Changes in Chromosome Number and Structure 540

Polyploidy Is an Increase in the Number of Chromosome Sets 541
Plants and Animals Differ in Their Ability To Tolerate Ploidy Changes 542
Autopolyploids Contain Extra Copies of the Same Set of Chromosomes 543
Allopolyploids Contain Chromosome Sets from More than One Species 544
Segregation of Chromosomes in Autopolyploids Gives Different Mendelian Ratios 546
Polyploid Animal Species Have Special Problems with Sex Determination 550
Aneuploidy Is a Gain or Loss of Individual Chromosomes 550
Meiotic Chromosome Segregation in Aneuploids Produces Unbalanced Gametes 552

Aneuploidy for Sex Chromosomes in Animals Produces Abnormal Sexual Development 554
Aneuploidy for Autosomes Is Usually Lethal in Animal Species 555
Chromosome Rearrangements Are Changes in Chromosome Structure 558
Chromosomal Breakage Can Produce Deficiencies and Duplications 558
Inversions Change the Order of Genes Within Chromosomes 560
Translocations Are Produced by Exchanges Between Nonhomologous Chromosomes 568
Robertsonian Translocations Can Produce Compound Chromosomes and Ring Chromosomes 571

Concept Close-Up: The Rabbage 548

Current Investigations: Generation of Polyploids by Cell Fusion 551

Historical Profile: Calvin B. Bridges (1889–1938) 556

Concept Close-Up: *Bar* Eye, a Stable Position Effect 562

Concept Close-Up: Double Crossovers in Inversion Loops 566

19 Recombination and Transposable Elements 578

Homologous Recombination Utilizes Sequence Complementarity 579
Generalized Recombination Involves an Exchange Between Homologous DNA Molecules 579
Site-Specific Recombination Is Exemplified by Lambda Phage Integration and Excision 583
Transposable Elements Are Ubiquitous 584
Early Genetic Evidence Indicated that Genomic Instability May Be Caused By Transposable Elements 585
Transposable Elements Were First Isolated from Bacteria 587
Bacterial Transposable Elements Share Common Features 589
Insertion Sequences Are the Simplest Class of Bacterial Transposable Elements 589
Some Transposons Are Made Up of Two IS Elements 590
Bacterial Transposable Element Transposition Occurs by a Conservative, Replicative Mechanism 592
Eukaryotic Transposable Elements Are Highly Varied 592
Some Eukaryotic Transposable Elements Require a Transposase for Transposition 593
Some Eukaryotic Transposable Elements Require Reverse Transcriptase for Transposition 597
The Retroviral and Retrotransposon Life Cycles Share Many Common Features 597
Some Eukaryotic Transposable Elements Are Repetitive Sequences 606
Biological Significance of Recombination and Transposable Elements 608

Historical Profile: Barbara McClintock: A Scientific Hero (1902–1992) 586

Current Investigations: AIDS, a Disease Associated with the Human Immunodeficiency Virus 601

Current Investigations: Retrotransposon Insertions Can Result in Novel Regulatory Mutations 607

Current Investigations: The Use of Transposable Elements in Genetic Engineering 608

20 The Genetics of Cancer 612

Cancer Is a Genetic Disease 612
Cancers Can Be Classified According to Tissue Type 613
The Major Distinguishing Characteristic of Cancer Cells Is Uncontrolled Growth 614
The Evidence that Cancer Has a Genetic Basis Is Multifaceted 616
Some Cancers Are Inherited 616
Many Carcinogens Are also Mutagens 617

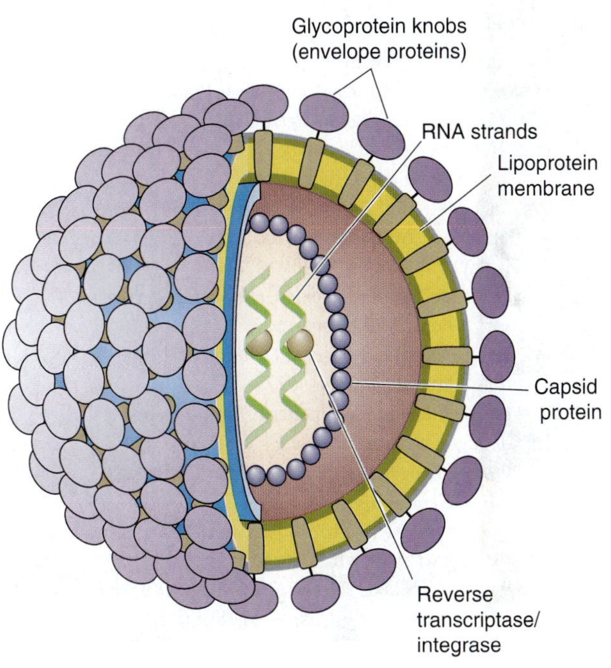

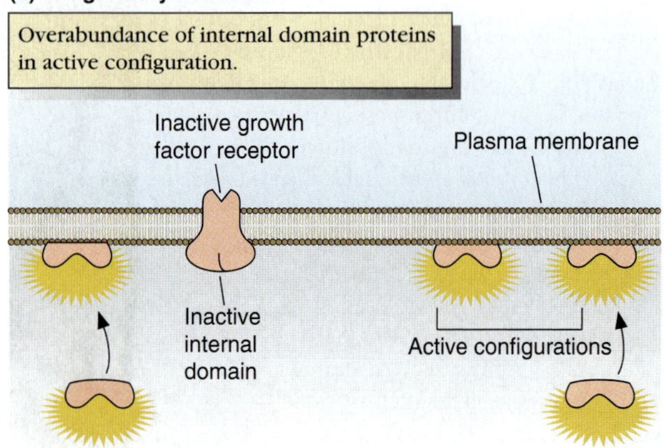

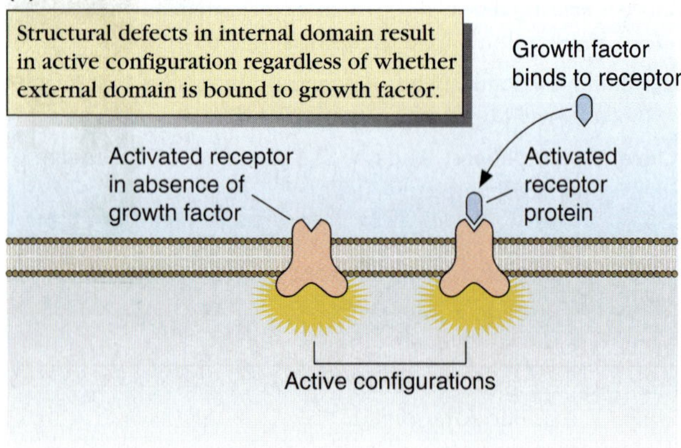

Inherited Defects in DNA Repair Processes Cause Cancer 621
Viruses Can Carry Viral Oncogenes that Are Responsible for Cancer 621
Viral Oncogenes Are Homologous to Genes Contained in Normal Cells 623
Nonacutely Transforming Retroviruses Produce Acutely Transforming Oncogenic Retroviruses 623
Genes Homologous to Viral Oncogenes Are Present in the Genomes of Normal Cells 624
Viral Oncogenes and Proto-Oncogenes Are Different 625
Proto-Oncogenes Can Be Converted into Cellular Oncogenes 626
Some Proto-Oncogenes Become Cellular Oncogenes Through Point Mutations 627
Some Proto-Oncogenes Become Cellular Oncogenes Through Translocations 627
A Retroviral Element Can Induce a Proto-Oncogene To Become a Cellular Oncogene 629
Some Proto-Oncogenes Become Cellular Oncogenes Through Gene Amplifications 629
The Proteins Encoded by Proto-Oncogenes Function as Cell-Growth Regulators 631
Some Proto-Oncogenes Encode Growth Factors 631
Some Proto-Oncogenes Encode Growth-Factor Receptor Proteins 631
Some Proto-Oncogenes Encode Intracellular Signaling Proteins 633
Some Proto-Oncogenes Encode Transcription Factors 634
Tumor-Suppressor Genes Prevent Uncontrolled Cell Growth 636
Some Tumor-Suppressor Genes Encode Regulatory Proteins 637
Some Tumor-Suppressor Genes Encode Intracellular Signaling Proteins 637
Some Tumor-Suppressor Genes Encode Cell Adhesion Proteins 637
Cancer Development Is a Multistep Process 638
Prospects for the Future Are Promising 638
Historical Profile: Harold Varmus and J. Michael Bishop 626
Current Investigations: Suppression of a Malignant Tumor by Gene Transfer 639

21 Population and Evolutionary Genetics 643

Genetic Variation Is the Raw Material for Evolutionary Change 644
Different Techniques Are Used To Detect Genetic Variation in Populations 644
Geneticists Are Interested in Understanding Genetic Variation in Natural Populations 647
The Amount of Genetic Variation in a Natural Population Can Be Quantified 649
The Level of Genetic Variation in Natural Populations Can Be Estimated 651
The Levels of Genetic Variation in Natural Populations May Be Expressed in Two Ways 651
DNA Sequence Variation Provides Information on Natural Populations 652
Different Factors Shape the Genetic Variation in Natural Populations 655
The Chi-Square Test Will Determine Whether a Population Is in Hardy-Weinberg Equilibrium 656

Equilibrium Values of X-Linked Genes Are Different in Males and Females 657
Nonrandom Mating May Alter the Genotypic Frequencies in a Population 657
Migration Affects Gene Frequencies in Populations 661
Mutation Is the Ultimate Source of All Genetic Variation 663
The Frequency of Genes in Populations May Fluctuate by Chance 663
Natural Selection Directs Changes that Increase a Population's Adaptiveness to Its Environment 666
Gene Frequency Changes in Natural Populations Are the Result of a Combination of Processes 669
Speciation Has a Genetic Basis 671
Speciation Is the Acquisition of Reproductive Isolating Mechanisms 672
Speciation May Occur When Populations Are Geographically Isolated 672
Speciation May Occur by Sudden Genetic Changes 673
Many Questions Remain Concerning the Genetic Basis of Speciation 674

Historical Profile: J. B. S. Haldane (1892–1967) 648

Concept Close-Up: The Neutralist–Selectionist Debate 670

Current Investigations: *Clarkia biloba* and *Clarkia lingulata:* Two Plant Species that Arose by Sympatric Speciation 674

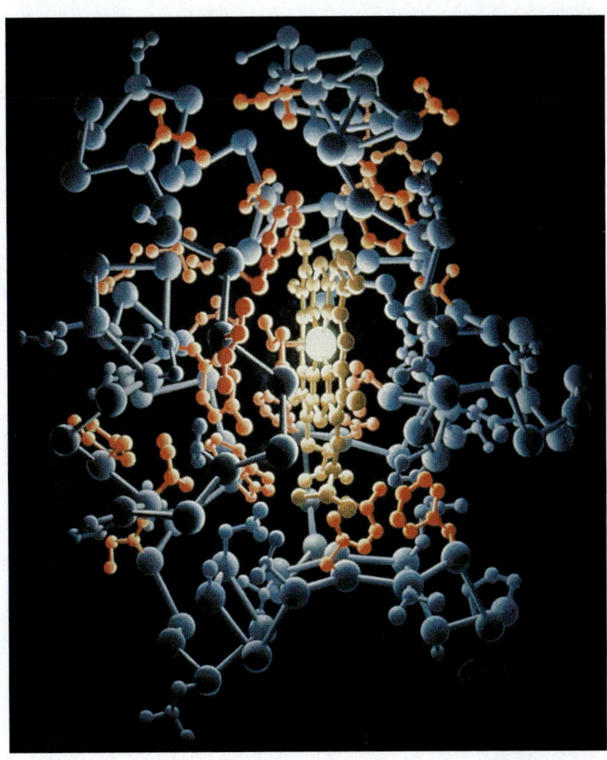

22 Molecular Evolution 678

Multigene Families Have Arisen Over Evolutionary Time 679
Gene Families Arise by Gene Duplication 679
The Globin Gene Family Evolved by Gene Duplications 680
Some Duplicated Genes Are the Product of Reverse Transcription 681
Some Gene Duplications May Be of Adaptive Evolutionary Significance 682
The Evolutionary Origin of Introns Is Controversial 683
Some Introns May Be Older than the Genes in Which They Are Contained 683
Some Introns May Be Younger than the Genes in Which They Are Contained 684
Molecular Mechanisms May Explain How Introns Were Acquired Over Evolutionary Time 684
Sequence Changes Are Used To Predict Evolutionary Time 686
Nucleotide Substitutions Change at Different Rates in Different Regions of the Gene 686
High Substitution Rates Characterize Genes that Are Evolving a New Function 687
Amino Acid Substitution Rates Are Constant Among Functionally Homologous Groups of Proteins 689
Nucleotide and Amino Acid Differences Are Used To Establish Molecular Phylogenies 690
Sequence Differences Can Be Used to Estimate Times of Evolutionary Divergence 695
Sequence Evolution Rates Are Not Always Correlated with Rates of Morphological Evolution 700

Current Investigations: The Utility of mtDNA Surveys in Conservation Biology 696

Current Investigations: Are Neanderthals Our Ancestors? 699

Historical Profile: Allan Wilson—A Pioneer in the Application of Molecular Data to Evolutionary Questions (1937–1991) 701

Appendix: Answers to Section Review Problems *A.1*

Glossary *G.1*

Index *I.1*

CHAPTER

1

Modern techniques allow geneticists to manipulate cells in ways considered science fiction only a few years ago. For example, the cloning of farm animals promises to revolutionize the agriculture industry. Shown here is Dolly, the first mammal to be successfully cloned from an adult cell, with her first lamb, Bonnie, who was naturally conceived. *(Roslin Institute/PA News Photo Library)*

An Introduction to the Science of Genetics

You are about to embark on an exciting journey into the world of genetics. Genetics is the study of how traits are inherited from generation to generation, how genes are organized and expressed, and how genes behave in populations and evolve over time. As you progress through this book, you will discover that genetics plays a central role in all modern biology and is essential for an understanding of the living world. A true understanding of biology cannot be obtained or appreciated without a knowledge of genetics. In addition, genetics has many applications to our everyday lives and to improvement of our ecosystem. For example, ancient and modern strategies to improve the yield and quality of plant and animal food products have relied heavily upon genetic knowledge and technology. Many of the miracles of modern medicine are derived from discoveries made in the field of genetics. Moreover, the very understanding of what we are as a species and how we are related to other organisms on this planet is based upon our knowledge of the genetic differences that exist among species.

This chapter will introduce you to the science of genetics and provide basic information that will help you to appreciate the genetic and molecular biological principles presented in the following chapters. By the end of this chapter, we will have addressed five questions:

1. What is genetics and why do we study genetics?
2. What are the major areas of research in genetics?

3. What are the historical origins of modern genetics?
4. What are the features geneticists look for in an experimental organism?
5. What are some useful experimental organisms used in genetic research?

GENETICS IS THE STUDY OF HEREDITY AND VARIATION.

Genetics is a science devoted to the study of the underlying basis of heredity and variation (Figure 1.1). The term **heredity** indicates "like begets like," that is, offspring generally tend to resemble their parents (for example, dogs have puppies and not kittens). Today we realize that the basis of this observation lies in the fact that how we look and function biologically, or our **phenotype,** is determined in large measure by our genetic makeup, or **genotype,** which results from equal contributions of genetic material from both of our parents at fertilization. A major focus of the science of genetics is the study of how information is stored in the basic unit of heredity, the **gene,** and how this information changes and becomes expressed during our development from an embryo to a sexually mature adult. The term **variation** refers to the fact that despite the similarity between parents and offspring, heritable differences do exist (for example, brown dogs may produce black or white puppies). These differences are ultimately based on **mutations,** or informational changes that occur in genes. Geneticists want to learn the molecular basis of these changes and how these changes become an established feature among groups of related organisms over time.

FIGURE 1.1 Heredity and variation. Although every individual is genetically distinct (except identical twins!), similarity in genetically determined traits between parents and offspring are often obvious. These puppies have many similarities, but they are also distinctively individual. *(Ron Spomer/Visuals Unlimited)*

GENETICS IS CENTRAL TO ALL BIOLOGY.

Genetics is central to nearly every area of modern biology. New techniques and discoveries are emerging almost daily from genetic research laboratories, and they directly or indirectly affect nearly every aspect of contemporary society. It is virtually impossible to pick up a newspaper today and not find a story related in some way to genetics, whether it be the introduction of genetic evidence in a murder trial or the application of a recent genetic discovery or technique to the treatment of a previously incurable disease (Figure 1.2). Indeed, modern genetics even plays a key role in scripts of many of our most popular books and movies.

If you plan to pursue a career in any area of modern biology, you will certainly need a firm grounding in the principles of modern genetics. In addition, every responsible citizen, whether a professional biologist or not, needs to have a grasp of at least the rudiments of modern genetics. We all need to be able to deal intelligently with the many political, social, and moral issues that will inevitably arise as the consequences of contemporary genetic research continue to penetrate the fabric of our society.

As you progress through this book, you will begin to realize the extent to which genetics has an impact on society. Throughout history, the primary impact of genetics has been on agriculture. Even prehistoric farmers realized that if they saved seed from their best plants for the next year's crop, they could improve their yields. Today, genetics plays an important role in medicine, crime prevention, and law. A knowledge of genetics may allow us to make wiser decisions in our everyday lives (for example, knowing that we have a specific genetic disease trait in our family may influence whether we choose to have children).

In recent years scientists have made incredible advances in the science of genetics, and more can be anticipated in the future. For example, a number of laboratories are currently engaged in **genome projects** to determine the genetic composition of many important organisms, including humans. The genetic composition of many pathogenic bacteria, yeast, and viruses has already been completely determined, and it is estimated that by the year 2005 we will have completely determined the genetic composition of humans as well as that of many other plants and animals. These genome projects will give us unprecedented knowledge about how organisms work and develop and about the causes of many diseases. They will also enable us to compare the complete genetic makeups, or **genomes,** of different organisms, which will in turn provide a better understanding of how evolution works at the molecular level. The future of genetics is indeed exciting.

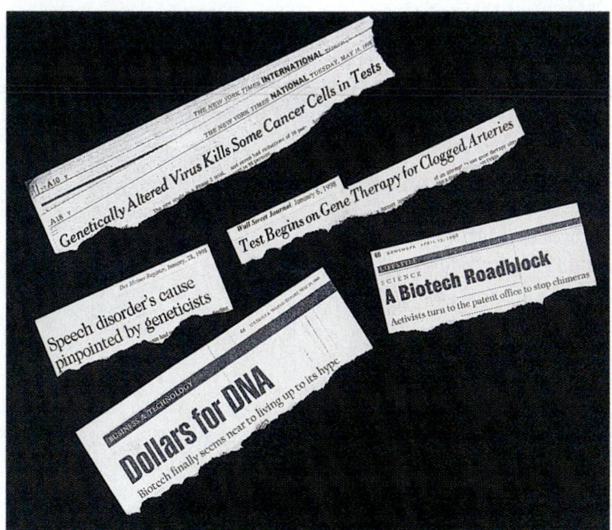

FIGURE 1.2 Genetics findings influence modern society. Shown are a collection of headlines taken from popular magazines and newspapers. *(George Semple)*

GENETIC RESEARCH IS FOCUSED ON THREE AREAS.

There are three primary areas of research focus in genetics: transmission genetics, molecular genetics, and population and evolutionary genetics (Table 1.1). **Transmission genetics** (sometimes referred to as **Mendelian genetics**) is the oldest area of genetic investigation. The principles of transmission genetics are considered in Part 1. This area of genetic research has historically focused on

1. Inferring the function of genes by observing the effect of heritable changes or mutations on the phenotype
2. Establishing the precise rules by which heritable traits are passed on from parents to their offspring
3. Mapping the location of genes on chromosomes and determining whether physical associations or linkage relationships exist between genes located on the same chromosome

Molecular genetics is devoted to studying the biochemical and molecular mechanisms by which hereditary information is stored in DNA and subsequently transmitted to the proteins. As you know from introductory biology, **DNA** (deoxyribonucleic acid) is the molecule that stores genetic information within the cell. Much of contemporary genetic research involves biochemical analysis of DNA. The principles of molecular genetics are presented in Part 2 of this book.

Population and evolutionary genetics elucidate the factors that determine the genetic composition of groups of interbreeding or related organisms, called **populations,** and how genetic changes in populations can lead to the formation of new, reproductively isolated groups of populations, or **species.** The underlying mechanisms and processes of how genes change and evolve are discussed in Part 3 of this book.

Traditionally, each of these three areas of focus has used its own distinct methodologies and techniques. However, the demarcation between these areas has been eroding in recent years because of advances in molecular biology. In addition to their own distinct methods, all areas now use molecular biological techniques such as **DNA sequencing** (the ability to determine the sequence of different nucleotides in a linear DNA molecule). For example, changes in the sequence in DNA can provide us with a rapid method of diagnosing many human genetic diseases such as cystic fibrosis, Huntington disease, and muscular dystrophy. The isolation of genes associated with hereditary defects may eventually lead to the elimination of at least some of these diseases through gene therapy (see *CURRENT INVESTIGATIONS: Adenosine Deaminase Deficiency, an Inherited Disease Treated by Gene Therapy*).

Many contemporary evolutionary geneticists also use molecular techniques such as DNA sequencing to investigate genetic changes that accompany the speciation process. Application of another recently developed molecular technique called **polymerase chain reaction (PCR)** has allowed evolutionary geneticists to recover and amplify (expand) DNA from fossil specimens and thus to establish likely genetic relationships even between species that have been extinct for a long time (see *CURRENT INVESTIGATIONS: Evolutionary Relationships Through DNA Analysis*). This same procedure also allows forensic analysis of vanishingly small samples of tissue to help determine the genetic fingerprints of violent criminals.

TABLE 1.1 *Areas of Research Focus in Genetics*

Area	Questions of Interest
Transmission genetics	Establish phenotypic effect of genes and rules of inheritance
	Localize genes on chromosomes
Molecular genetics	Establish the nature of the hereditary material
	Determine the structure of DNA
	Establish how hereditary information is stored in DNA and is manifested in the organism
Population and evolutionary genetics	Establish the factors that determine the genetic composition of populations and species and how they change over time

CURRENT INVESTIGATIONS

Adenosine Deaminase Deficiency and Gene Therapy

The technology necessary for the transfer of "foreign" genes into experimental organisms is well developed today. Such gene transfer experiments have been of tremendous value in helping geneticists understand the molecular basis of a number of basic biological processes such as cancer and cell development. A logical extension of gene transfer technology is to apply it to humans for the treatment and possible cure of inherited genetic diseases. Although the reality of being able to cure all genetic diseases with gene therapy is still years away, in recent years scientists have made significant strides that demonstrate that such goals are not merely science fiction but are very much within the realm of possibility in the not-too-distant future.

An example of human gene therapy accomplished recently is the work spearheaded by Dr. French Anderson at the University of Southern California. Over the past several years, Dr. Anderson's group has focused on developing a strategy to treat the human inherited disorder adenosine deaminase deficiency (ADD) by genetic intervention. Adenosine deaminase is an enzyme that plays a key role in the human immune system. Patients suffering from ADD carry defective genes for adenosine deaminase, and as a consequence their immune systems are defective, rendering them unable to fight off opportunistic infections. ADD is primarily a childhood disease because patients usually die within the first two years of life.

In the protocol initiated by Dr. Anderson's group, blood is removed from the ADD patient and the defective immune cells are isolated from the white cell fraction of the blood. A normal, functional ADD gene is introduced into the chromosomes of these defective cells by procedures discussed in Chapter 12. The final step is the introduction of the engineered immune cells back into the patient by transfusion.

This protocol was initiated several years ago on a young female ADD patient who had been permanently confined to quarantine to reduce chances of being exposed to the opportunistic infections common to childhood. The results of the procedure were dramatic. After the initial transfusion with the engineered cells, there was a gradual increase observed in adenosine deaminase levels with a concomitant loss of ADD symptoms. Although the patient continues to be closely monitored and periodically receives additional transfusions, she is now able to participate in many normal childhood activities.

Other human diseases that researchers hope to be able to treat with gene therapy include diabetes, various inherited blood disorders, cystic fibrosis, and certain cancers of the brain, breast, lungs, kidneys, and liver. The pioneering efforts by Dr. Anderson and his colleagues for the treatment of patients with ADD suggest that the future is promising for the application of modern genetic techniques to the treatment of human disease.

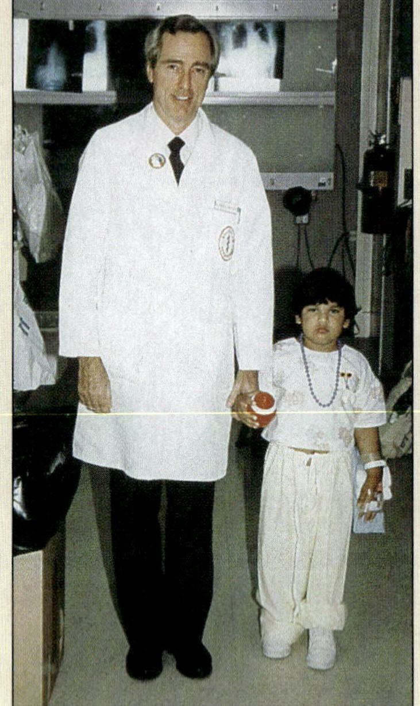

FIGURE 1.A Dr. French Anderson with Ashanti DiSilva, the first gene therapy patient, in 1995. *(Courtesy of Dr. W. French Anderson, Director, Gene Therapy Laboratories, University of Southern California)*

SOLVED PROBLEM

Problem

PCR is a technique that allows geneticists to *amplify* minute quantities of DNA through a series of replications to levels that can be analyzed by diagnostic procedures such as DNA sequencing. How do you think this procedure might be applied in:

a. forensic science?
b. the early diagnosis of a viral-based disease such as AIDS?

Solution

a. PCR can be used to amplify DNA contained in minute quantities of blood or other body fluids collected at a crime scene. Analysis of this DNA can determine whether it matches the victim and/or the alleged perpetrator of the crime. Such evidence has proven to be conclusive in some trials.
b. Tests to determine whether an individual has been exposed to a virus are usually dependent upon antibodies to the virus being present in the individual's blood. A certain period of time is required for antibodies to be produced and to reach high enough titer (concentration) to be detected in immunological tests. Thus, some recently infected individuals may test negatively on immunologically based tests. In addition, it's possible that an individual may have been exposed to a virus and thus carry antibodies to the virus and yet not actually be infected with the

virus. PCR can be used to test directly whether even minute quantities of viral DNA (or RNA—ribonucleic acid) are present in the blood or other body fluids of individuals suspected of being infected by viruses such as HIV.

SECTION REVIEW PROBLEMS

1. What is meant by the terms "heredity" and "variation," and how do these terms relate to the modern science of genetics?
2. Compare and contrast the three areas of research focus in modern genetics.

GENETICS HAS A RICH HISTORY.

As you will learn in Chapter 2, the beginning of modern genetics can be traced to a rather obscure 1865 publication by an Augustinian monk named **Gregor Mendel.** However, interest in heredity and variation goes back much further than the mid-19th century and can even be traced to prehistoric times.

Artifacts such as dried seeds and primitive wall paintings found within caves inhabited by early humans indicate that efforts to domesticate plants and animals were among the earliest of human activities. Plant and animal husbandry endeavors are recorded in the earliest cultures of Egypt and the Middle East (Figure 1.3). For example, there are records of gazelles, and, later, sheep and goats being maintained and bred in ancient Egypt for purposes of domestic consumption. Strains of wild bees are known to have been domesticated in the ancient Nile valley to facilitate the pollination of crops. Selection programs were instituted in ancient China to develop strains of moths capable of producing high yields of silk fiber for use in the manufacture of clothing. Many crops, such as wheat, corn, rice, and the date palm, were developed by farmers as early as 5000 BC.

FIGURE 1.3 Ancient cultures used selective breeding. Preserved wall paintings show that genetic breeding played a critical role in ancient cultures. Shown is a portion of a wall painting in the tomb of Nakhat (18th Dynasty), Luxor-Thebes, Egypt. *(Erich Lessing/Art Resource)*

FIGURE 1.4 Long-necked giraffes. Jean Baptist Lamarck proposed an evolutionary theory whereby traits acquired over the lifetime of the parents could be passed on to their progeny. He hypothesized that giraffes acquired their long necks by stretching to eat leaves in the high trees. *(Walt Anderson/Visuals Unlimited)*

The first records we have of formal explanations of heredity and variation come from the ancient Greek philosophers around 500 BC. These early theories were based on what today seem like rather vague concepts. For example, hereditary information was believed to be contained in ether-like "humors," which were believed to flow from the various parts of the body, carrying hereditary information to the sexual organs and ultimately to the developing fetus. Inherent in this theory was the idea that structural and functional characteristics, such as big muscles or the ability to play a musical instrument, which may have been acquired over the lifetime of one parent or the other, could be transmitted to the offspring. The notion that acquired characteristics can be inherited is one that recurs in evolutionary theories of the 19th century, most notably in the writings of the French evolutionist **Jean Baptiste Lamarck** (1744–1829) (Figure 1.4).

Current Investigations

Evolutionary Relationships Through DNA Analysis

Polymerase chain reaction (PCR) is a molecular technique that allows scientists to amplify minute quantities of DNA to levels that can be subjected to qualitative analyses such as DNA sequencing (Chapter 12). The PCR technique has recently been used to amplify trace quantities of DNA isolated from museum specimens of extinct species. Perhaps information on the DNA sequences carried by extinct species will allow scientists to establish precisely the evolutionary relationships between extinct and contemporary species. Such data might also help determine whether species that are genetically less variable are more prone to extinction. Even though students of ancient DNA thus far have been most successful in amplifying DNA from the pelts of extinct mammalian species, much attention has recently been turned toward isolating DNA from long-extinct species of dinosaurs.

Unlike the scenario portrayed in the film adaptation of Michael Crichton's best-seller *Jurassic Park,* there is slim chance of regenerating any of these prehistoric hulks for display in mega-theme-parks. Any molecules of dinosaur DNA that might have been preserved in dinosaur fossils (or amber-preserved mosquitos) will almost certainly be too fragmented to allow reconstruction of anything approaching a complete genome. Nevertheless, even fragments of dinosaur DNA, if amplified and sequenced, might allow scientists to reconstruct evolutionary relationships between these extinct animals and contemporary species. For example, some experts propose that dinosaurs are most closely related to crocodiles, whereas others believe that birds are the closest living relatives to dinosaurs. Comparison of DNA sequences of these species with dinosaur DNA could help resolve this question.

The limiting step in being able to amplify DNA from any extinct species, including dinosaurs, is finding preserved cells in fossils that contain DNA. Some researchers have been looking carefully at bones of *Tyrannosaurus rex* and other dinosaur species for preserved blood cells and have identified some promising material (Figure 1.B). Other researchers have taken the approach portrayed in *Jurassic Park,* attempting to isolate blood from the guts of amber-encased insect fossils, hoping that the last blood meal enjoyed by these insects may have been from a dinosaur (Figure 1.C).

After putative dinosaur cellular material is isolated from fossils, it is treated with enzymes and detergents to extract DNA. The DNA is further purified and then amplified by the PCR technique. When workable quantities of DNA are available, the DNA is sequenced, and the data are entered into a computer, where they are systematically compared with all other existing DNA sequences in the data banks.

One of the problems with the PCR technique is that it is so sensitive that it can easily amplify trace quantities of "contaminating" bacterial or human DNA that may become inadvertently introduced during the extraction process. Thus, if the sequence of the amplified DNA shows similarity to anything other than crocodile or bird DNA, it is considered to be an artifact, and the whole experiment is discounted. If the amplified sequence does seem bird- or crocodile-like, it is considered to be putative dinosaur DNA, and efforts are initiated to repeat the entire experiment to verify the outcome.

FIGURE 1.C Fossilized biting insects. Biting insects that became "fossilized" by being encased in amber during the age of the dinosaurs may contain blood from which DNA can be isolated. Purified DNA from such fossil material can be amplified by PCR techniques and sequenced. *(Jan Hinsch/Science Photo Library/Photo Researchers, Inc.)*

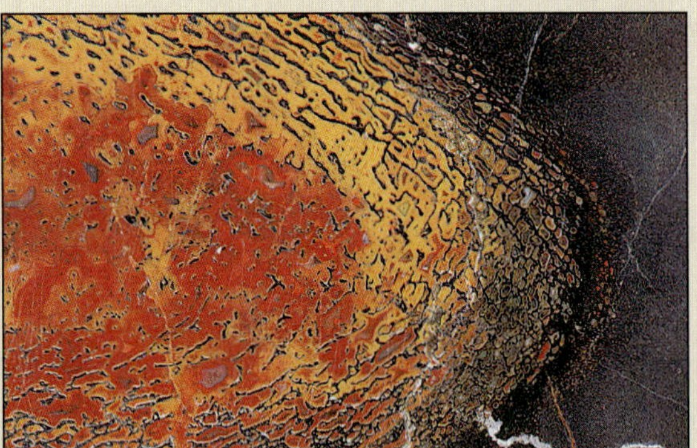

FIGURE 1.B A bone section from a dinosaur. Blood cells within dinosaur bones such as these may contain DNA. *(François Gohier/Photo Researchers, Inc.)*

Aristotle, a Greek philosopher, believed that the female provided the substance from which the fetus developed, whereas the male provided the "form," or information dictating how this matter was to be shaped into the developing child. The Aristotelian view was widely accepted during the Middle Ages, although it was modified somewhat by the 16th-century biologist **William Harvey,** who introduced his theory of **epigenesis.** According to this theory, organisms develop from the material contained in eggs, which is stimulated to develop by the contribution of the male. A contrasting theory, called **preformationism,** arose in the 17th century.

According to the preformationists, sex cells contained a fully developed (albeit miniaturized) adult called the **homunculus,** which had only to increase in size within the mother's womb until birth (Figure 1.5). There were two versions of this theory, depending upon one's perspective. The "spermists" believed that the homunculus resided in sperm, and the "ovists" believed that it was to be found within the egg. Each homunculus was believed to contain another homunculus within its germ cells, and so on. According to preformation theory, Adam or Eve (depending upon whether you were a spermist or an ovist) contained within his or her loins all future human generations.

By the 18th century, microscopes and their use (called microscopy) had developed to the point at which some of the more fanciful theories of heredity could be tested. For example, embryologists such as **Casper Wolff** (1733–1794), having access to good-quality microscopes, were able to disprove the doctrine of preformationism. Wolff showed that not all adult structures were present in very early embryos but rather appeared gradually as the embryo developed. This and similar findings led embryologists to re-embrace Harvey's epigenetic view of development, but it still left open the question of the original contributions of the parents to the process.

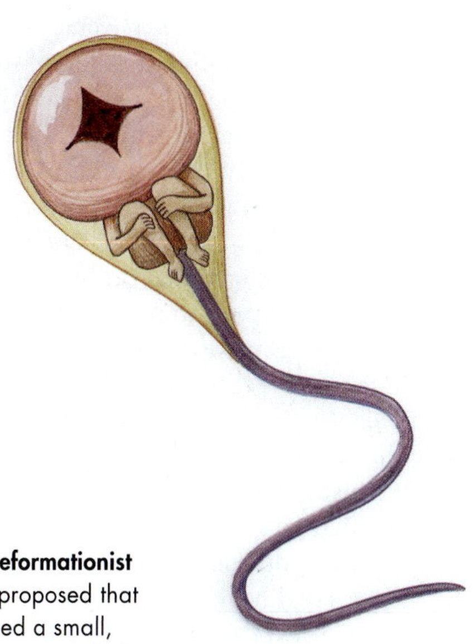

FIGURE 1.5 Preformationist theory. This theory proposed that each sperm contained a small, fully developed organism called a *homunculus*.

Blending inheritance was the popular hereditary theory during the late 18th and early 19th centuries. This theory postulated that hereditary material is supplied more or less equally by both parents and becomes mixed or blended in the offspring, like different colored paints. By the mid-19th century, a strong experimental approach was being taken to nearly all scientific questions, including those of heredity. The blending theory initially withstood empirical scrutiny because it was consistent with the observations of many plant and animal breeders that the progeny of different strains, called **hybrids,** were often intermediate in appearance between their parents. Thus the blending theory was well entrenched in the minds of most students of heredity by the mid-1800s, when Gregor Mendel performed his now-classic experiments, which are today viewed as the beginning of modern genetics.

In the next chapter, you will examine Mendel's experiments in some detail and learn how they disproved the blending theory and led to our present understanding of how hereditary information is passed from parent to offspring. One of the primary reasons Mendel was successful in establishing his set of genetic principles is that he followed a logical procedure in the testing of his ideas, which today we call the scientific method.

SECTION REVIEW PROBLEMS

3. What are the essential differences between the epigenetic and preformationist theories of heredity?
4. Based upon your own personal observations of the way parents and children look and/or act, can you cite any examples that you think falsify the blending theory of inheritance?

GENETICS IS AN EXPERIMENTAL SCIENCE.

One of the most distinguishing characteristics of modern genetics is that it is an experimental science. Biological disciplines such as anatomy are largely descriptive sciences that involve little hypothesis testing. The tremendous progress that has been made in the field of genetics over the last century can be attributed in large measure to the careful application of the scientific method to the issues of heredity and variation. In this section we will discuss the process of scientific inquiry using the scientific method, with special emphasis on genetic research.

The Scientific Method Is Used in Genetics.

The mark of an experimental science is its devotion to the **scientific method.** The scientific method consists of three logical steps:

1. Observation
2. Formation of alternative explanations, called **hypotheses,** to account for the observations
3. Testing of these alternative hypotheses

The careful and objective collection of data is the foundation upon which experimental science is based. After data are recorded, scientists look for trends in the data that need to be explained. The next step is to formulate hypotheses to explain the trend. For example, suppose that you note through careful observation that the average number of blueberries per bush is greater for plants located at the top of a nearby hill than for plants located at the bottom of the same hill. Your next step as an experimental scientist would be to propose alternative hypotheses to explain the observation. The number of plausible hypotheses to explain your observation may be very large and is limited only by your imagination. Indeed, the ability to form novel alternative hypotheses is one of the most creative aspects of experimental science and is one of the characteristics that separates good scientists from great scientists.

In the example of the high- and low-yielding blueberry bushes, we might hypothesize that (1) the bushes on the top of the hill are more productive because they receive more light, (2) the bushes at the bottom of the hill are growing in poor soil, or (3) genetic differences between the bushes on the top and bottom of the hill explain the differences in yield. Of course, these three possibilities by no means exhaust the number of possible hypotheses that could account for the differences in yield between the two populations of blueberry bushes. When all plausible hypotheses have been formed to account for an observation, it is time to subject the hypotheses to experimental test (Table 1.2).

The scientific testing of an hypothesis is an effort to find or generate data that are inconsistent with the hypothesis. Hypothesis testing is based upon the simple logical fact that no amount of data consistent with a hypothesis can prove that the hypothesis is universally correct. This is true because it is always possible that if we did just one more experiment, we might uncover a result or observation that is inconsistent with our hypothesis. Therefore, it takes only one experimental result or observation inconsistent with our hypothesis to unequivocally prove that the hypothesis is not universally correct. Thus, although hypotheses can be logically falsified, they can never be proven correct. An hypothesis that fails to be falsified after repeated attempts is held as tentatively true. Indeed, the more times an hypothesis withstands experimental tests, the more confidence scientists have in the hypothesis, which may result in its being elevated to the status of a **scientific theory.** The meaning of the word "theory" in common use is not as precisely defined as it is in science. In everyday use, a theory is a possible explanation, something more akin to the scientific concept of hypothesis. In science, a theory is an explanatory principle that has withstood extensive tests and is currently held by nearly all scientists as almost certainly true. Scientists always leave room for doubt because absolute truths lie outside the realm of science.

To test our hypotheses about why the hilltop blueberry plants yield more fruit, we might do some transplant experiments in which hilltop seedlings are moved to an experimental plot on the bottom of the hill and vice versa. If the yield of the transplanted plants in these experimental plots remained the same as their parents', we could consider hypotheses *a* and *b* falsified because changing the growing conditions did not influence yield.

As you progress through this book, you will see many elaborate descriptions of how genes function and how they have evolved over time. Unfortunately, because of the constraints of space and time, it will not always be possible to describe in detail the careful observations and extensive hypothesis testing that underlie the various discoveries that have been made in genetics over the past century. However, you can be certain that careful application of the scientific method underlies each of these discoveries.

TABLE 1.2 *The Three Stages of the Scientific Method*

1. Carefully observe and recognize trends or generalities.

The average number of blueberries per bush is greater for plants growing on the hilltop.

2. Form alternative hypotheses to explain the observed trends.

There are a number of possible hypotheses to explain differences in fruit yield:

a. Plants on hilltop receive more light.

b. Plants at bottom of hill are growing in poor soil.

c. Plants on hilltop are genetically determined to produce more fruit.

3. Test the alternative hypotheses by setting up experiments designed to falsify each hypothesis.

Transfer hilltop plants to bottom of hill and observe yield of progeny plants. If progeny plants produce a low yield of fruit, then hypothesis *c* is falsified, and either hypothesis *a* or *b* may be correct. If progeny plants produce the same high yield of fruit, then hypotheses *a* and *b* are falsified, and hypothesis *c* may be correct.

SECTION REVIEW PROBLEMS

5. Devise some additional experiments to test the three hypotheses proposed in this section to account for hilltop blueberry bushes having a higher yield than those growing at the bottom of the hill.
6. Propose a fourth hypothesis to account for the difference in blueberry yield and devise an experiment to test your hypothesis.
7. List three examples from the history of science in which preconceived notions served as an intellectual barrier to scientific progress.

Choosing an Appropriate Experimental Organism Is Important.

In the case of the blueberries, discussed in the preceding section, the investigator observed a natural phenomenon and then attempted to support or disprove an hypothesis. Another approach is to first pose a question and then select an experimental organism that is well suited to answer the question. The importance of an experimental organism to the success of genetic research is an issue often overlooked by students. Many students seriously question why scientists spend so much time studying "useless" organisms such as fruit flies and worms. As you proceed through this book, you will consider the details and significance of experiments carried out in a variety of experimental organisms, some that seem to have no practical applications. You need to understand that experimental organisms are not chosen haphazardly but rather because they possess specific, biological characteristics judged to be desirable for obtaining answers to the particular question being investigated.

The Desirable Characteristics for a Genetic Research Organism Are Varied.

Although there is no such thing as "the ideal experimental organism," there are certain biological characteristics that most researchers consider to be desirable in an experimental genetic system. Any one characteristic may be judged more or less important for the success of a particular experiment, but the following characteristics are considered generally desirable for any experimental organism used in genetic research (Table 1.3).

The Organism Should Be Small. It is almost always important that a genetic experimental organism be small because large organisms consume large amounts of food, require large amounts of space for containment and growth, and are frequently difficult to handle on a day-to-day basis. For these reasons, dogs, cats, or rabbits, for example, are not considered as desirable as fruit flies or bacteria for the study of fundamental genetic principles.

The Organism Should Produce a Large Number of Progeny. For a geneticist to study inheritance patterns and variation, it is usually necessary to have a large number of progeny (offspring) per generation. One reason humans make poor experimental organisms for the study of genetics is the very limited number of progeny they have per generation per individual. Plants, on the other hand, frequently produce millions of progeny per generation. For reliable statistical analysis of patterns of inheritance, an organism that produces large numbers of progeny is usually desirable.

The Organism's Life Cycle Should Be Short. To study inheritance over consecutive generations, a short life cycle is extremely important. One popular organism, *Caenorhabditis elegans,* a small nematode, has a life cycle of only 3 days, which allows the study of many generations in a relatively short time. Bacteria, yeasts, and some plants also have short life cycles, making them useful experimental organisms. However, as important as a short life cycle is, it does not always dictate the organism for study. Maize (corn), for example, has about a 5-month life cycle and is a popular study organism because of its other attributes, not the least of which is its use as an economically important crop.

The Organism's Genome Should Be Small, with Relatively Large Chromosomes. In the early days of genetics, a small genome was not considered as important as it is today. Smaller genomes are more easily studied on the

TABLE 1.3 *Desirable Characteristics of Genetic Research Organisms*

Characteristic	Advantage
Small size	Easy to handle, feed, breed
Large number of progeny	Allows reliable statistical analysis of patterns of inheritance
Short life cycle	Allows patterns of inheritance over consecutive generations
Small genome/large chromosomes	Less DNA to analyze; easier to inspect chromosomes under light microscope
Developmental complexity	Allows genetic analysis of complex developmental processes
Previously established information/techniques	Many genetic mutants available for analysis; experimental protocols already established

molecular level simply because there is less DNA to analyze. Large chromosomes are easier for a researcher to see under a light microscope in order to detect aberrations visually in chromosome structure. Large chromosomes also allow the investigator to localize genes more easily by molecular techniques that you will learn about in Chapter 12.

The Organism Should Be Developmentally Complex. For multicellular organisms, many important questions deal with the developmental processes that control organ formation and function. To study these types of processes, the model organism must have a certain degree of developmental complexity. During the initial years in the study of genetics, developmental complexity was not important, but as many of the basic foundations of genetics have been defined, scientists have turned their attention to how development occurs. This is now possible because of recent advances in molecular genetic techniques. *Caenorhabditis elegans* has a relatively simple muscular and nervous system and a low level of complexity and is an excellent model to study basic concepts of development. In contrast, *Drosophila* is considerably more complex, and the mouse is even more complex. As a consequence, the geneticist interested in development will choose an organism displaying the level of complexity suitable to the question being investigated.

Accumulated Knowledge About the Organism Is Essential. A very important consideration in choosing an experimental organism is the amount of information already known about its genetics, the number of genetic mutants available to the investigator, the number of techniques that have already been developed to facilitate the use of the organism in genetic research, and the number of other researchers working in the field. For example, there may be a number of insect species with shorter generation times and smaller genomes than *Drosophila*. However, these advantages will not override the fact that more than 3000 genetic mutants and a detailed map of the chromosomal position of hundreds of genes are available for *Drosophila*. Such abundant information would not be available to someone choosing to work with a new experimental organism such as a mosquito. A similar argument can be made for the choice of maize as an experimental organism despite the fact that its generation time is rather long.

Another advantage of working with an established experimental organism is the fact that there is likely to be a large community of researchers available for consultation and collaboration. For example, it has been estimated that currently over 10,000 scientists are actively involved in *Drosophila* research on a worldwide basis. Thousands of these researchers meet yearly at international meetings to exchange recent findings and discuss new ideas. Such an established support system is an invaluable asset in facilitating scientific progress.

Working with an established experimental organism has distinct advantages, but new experimental organisms have been established in recent years and will continue to be developed in the future. For example, the flowering plant species *Arabidopsis thaliana* has recently been established as an extremely attractive research organism for plant geneticists. This has occurred because this organism possesses many of the characteristics we have just discussed. For example, it has a relatively small genome, its generation time is only about 6 weeks, its progeny are abundant, and considerable knowledge has been accumulated about this weedy plant in a relatively short time.

Commonly Used Genetic Research Organisms Have Many of These Features.

Some organisms have been chosen for genetic study because of their economic importance, but many have been selected for the reasons cited previously. When studying an organism at the molecular level, simple organisms were studied first (for example, viruses and bacteria). As many genetic and molecular questions were answered with these organisms, more complex organisms came under scrutiny. A discussion of the attributes that make certain organisms the experimental systems of choice for most genetic studies follows.

Bacteria Are Among the Best-Understood Organisms. As experimental organisms, bacteria have proved to be extremely well suited for the study of many basic genetic and molecular biological questions. In the 1940s and 1950s, the common nonpathogenic bacterium *Escherichia coli* became one of the most important organisms for the study of genetic mechanisms including biochemical pathways and processes and for the regulation of gene expression. Indeed, much of what we currently know about the detailed molecular biology of DNA replication and other basic genetic processes was initially discovered using *E. coli* as an experimental system.

The general success of bacteria, and *E. coli* in particular, as an experimental genetic organism is because they possess many of the characteristics listed in the previous section. Large numbers of *E. coli* (and other bacterial species) can be grown in a small area (a billion bacteria in a small flask), allowing for the easy identification and study of rare mutational events at a low cost. In addition, all bacteria have small genomes, and, as a consequence, the sequence of bases in the DNA of a number of bacteria have been determined. Because of this, most of the genes of *E. coli* have been located on its chromosome. This informational base is of tremendous advantage to a researcher wishing to establish the function and position of a newly discovered gene of *E. coli*.

Some bacterial species are studied because they are commercially or medically important. For example, *Rhizobium* is a bacterium studied intensively by geneticists

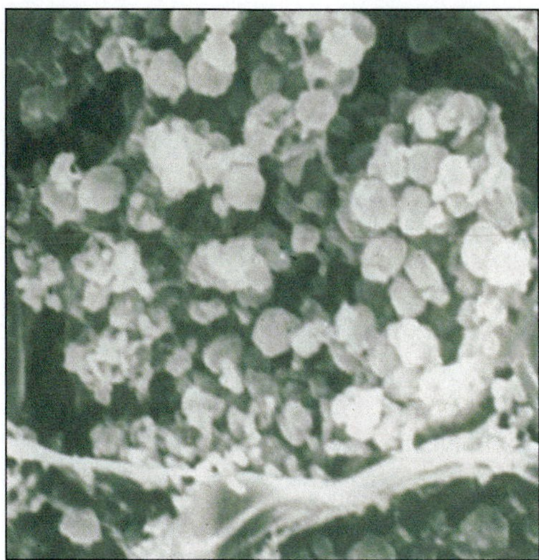

FIGURE 1.6 Nitrogen-fixing bacteria. These *Rhizobium* bacteria infect the roots of specific legumes (soybeans, peas, alfalfa) and form nodules that convert atmospheric nitrogen into fertilizer for the plant. *(Alan Atherly and James Berry)*

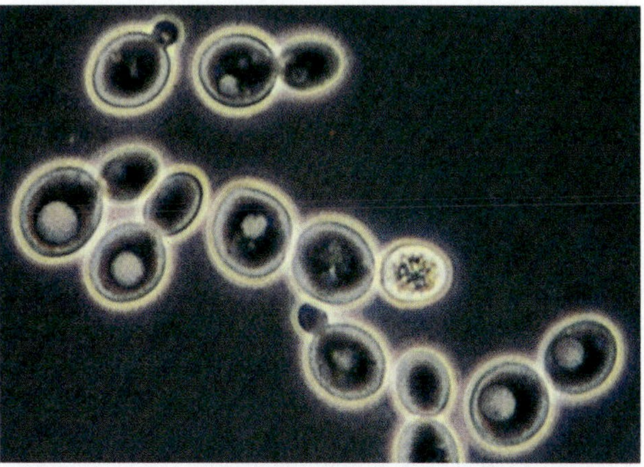

FIGURE 1.7 The yeast *Saccharomyces cerevisiae*. Yeast is a popular genetic research organism because it combines the advantages of the small size typical of bacteria with the genetic complexity characteristic of eukaryotes. The entire genome has also been sequenced, making it extremely useful for molecular genetic analysis. *(Manfred Kage/Peter Arnold, Inc.)*

because it is able to convert atmospheric nitrogen (N_2) into usable nitrogen (ammonia) for use by legume plants (Figure 1.6). Many geneticists are working to isolate the genes controlling this process in *Rhizobium* with the expectation that it may be possible to insert them into agriculturally important plants so that they may be grown without chemical fertilizers. You will study the methods by which bacterial and other foreign genes can be introduced into plant and animal genomes in Chapter 13.

Yeast Has Proven To Be an Excellent Eukaryote for Study. *Saccharomyces cerevisiae,* a species of yeast, is a single-celled eukaryotic organism that has been grown for centuries to make bread, beer, and wine (Figure 1.7). As a consequence, yeast is of significant economic importance, and a considerable amount of knowledge has been accumulated about the biology of yeast cells. Like *E. coli,* yeast is a single-celled organism with a very rapid growth rate. As in *E. coli,* yeast mutants can be easily obtained. The genetic map of *S. cerevisiae* is very well developed, which facilitates the chromosomal localization of newly discovered genes. Finally, as of 1997, yeast is the first eukaryote to have its genome completely sequenced, and this information is easily available to any investigator with a computer and access to the World Wide Web. (see *Current Investigations: Genetics and the Internet*).

***Arabidopsis thaliana* Is Useful for the Study of Plant Genetics.** Genetic research on plants has traditionally focused almost exclusively on agronomically important crop species such as maize, cotton, tomatoes, and wheat. This situation has recently changed because of the small plant *Arabidopsis thaliana,* which has become a popular experimental system over the past decade, mostly because it has many of the characteristics of an ideal experimental organism. *Arabidopsis thaliana* is a flowering, dicotyledonous plant sufficiently small that several plants can be grown in a few centimeters of soil (Figure 1.8). In addition, seeds are small enough to allow the germination of 5000 seeds in one small laboratory dish. The plant completes its life cycle in about 6 weeks under laboratory conditions, and each plant produces over 10,000 seeds, which greatly facilitates the screening of rare mutational events. Another attractive feature of *Arabidopsis* is that its seeds can be stored under cool, dry conditions for many years.

Arabidopsis is believed to possess the smallest genome among higher plants, less than 1/10 the size of the maize genome. Mutants are easily induced in this species by soaking seeds in mutagenic chemicals or exposing seeds to X-rays. Using these approaches, scientists have produced and isolated many mutants that are

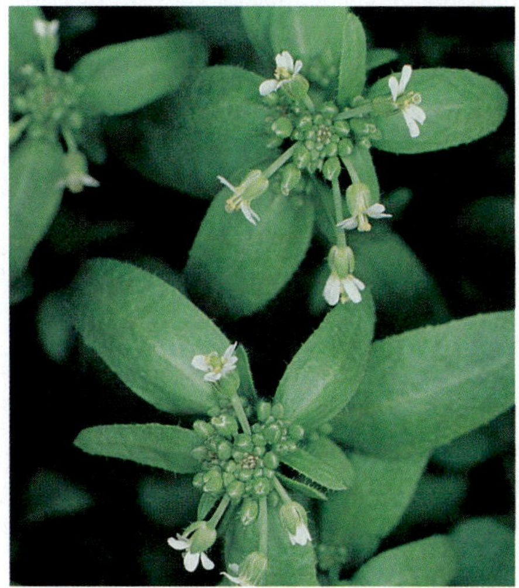

FIGURE 1.8 *Arabidopsis thaliana*. This small plant is related to cabbage (*Brassica*) and has developed into a popular experimental organism for the study of plant genetics and molecular biology. *(Dr. Jeremy Burgess/Science Photo Library/Photo Researchers, Inc.)*

Current Investigations

Genetics and the Internet

Genetics, like all sciences, depends on a free exchange of information. To plan experiments efficiently, geneticists must be aware of the work of others. If they are not, either they will needlessly duplicate experiments already done or their experiments will be poorly designed. In the past, geneticists gained information mostly from reading publications and from conversations with colleagues at meetings. These still remain the primary means of scientific exchange, but a new dimension has been added. With today's powerful personal computers, geneticists can access a vast communications web that links them to laboratories around the world. This nearly instantaneous exchange of information is made possible by the Internet and interface programs such as Netscape.

A key use of the Internet is to find information to answer specific questions. For example, suppose that you have a question about a specific topic and wish to discover recent articles written about it in the scientific literature. You could go to the library and hunt through back issues of journals until you eventually find a list of articles (this may take several days), or you could type a few key words into a Medline search program (current address is: http://www.ncbi.nlm.nih.gov/medline/query_form.html). This will give you a list of articles, as well as abstracts of their contents, within a few minutes. You could also search computer bulletin boards devoted to exchanging scientific information on your topic, or you could post a question on a bulletin board and receive an answer from a researcher anywhere in the world.

Geneticists can also use the Internet to gain access to immense amounts of genetic data stored in global databases. This information covers all aspects of genetics, including descriptions of mutations, DNA sequences, genetic maps, and promising new experimental techniques. This information can become available months or even years before it appears in the printed literature, if it ever does.

The key is to gain access to the right database. Many genetics databases are organized by organism. For example, researchers working with *Drosophila* pool their information in a database called Flybase (http://morgan.harvard.edu/). In Flybase, you can find very recent information from more than 4000 researchers who have listed descriptions of particular mutant strains that they are willing to share and the results of their most recent experiments on more than 15,000 *Drosophila* genes. Over 70,000 literature citations on *Drosophila* are listed, as are detailed descriptions of more than 11,000 abnormal chromosomes or mutations that have been characterized. For many of these, actual pictures or diagrams describing the abnormality in detail are available. Detailed genetic maps are listed, and these are updated daily. A researcher seeking a particular mutation, a cloned DNA sequence or a strain with a particular combination of mutant genes, can search the data bank or post a request on a computer bulletin board and receive an answer from any of the thousands of researchers who scan the bulletin board every day.

Similar databases have been set up for yeast (http://genome-www.stanford.edu/sgd.html), maize, *E. coli* (http://www.mbl.edu/html/ecoli/html), *Arabidopsis, C. elegans,* mice, cats, dogs, and many important agronomic crop plants and animals such as wheat, rye, soybeans, barley, sorghum, chickens, pigs, and cattle. Information on humans consumes a large part of the Internet (http://gdb.www.gdb.org/maps, for example). These databases all have only one purpose: to increase the communication of information between researchers.

In addition to information about their own research organism, geneticists can also gain information from databases that contain DNA and protein sequence information from all organisms. For example, if you have determined the DNA sequence of a gene and you want to find out if this sequence is similar to any known sequence, you can access a supercomputer at the National Center for Biotechnology Information (http://www.ncbi.nlm.nih.gov/Recipon/bs_seq.html). You can then enter your sequence and discover in minutes whether similar sequences have been discovered, and, if so, what these sequences are and where they were found.

All these uses of computer networks have been developed only in the past few years. Other uses that will have far-ranging effects on genetics and all other fields of science will certainly be found in the future.

defective in flowering, leaf shape, and flower color as well as other morphological and biochemical characteristics. Another important feature is that foreign DNA can be readily introduced into *Arabidopsis,* which makes the plant an attractive experimental organism from the perspective of molecular biology.

***Caenorhabditis elegans* Is Frequently Used To Study Development.** The roundworm *C. elegans* has been the subject of biological experiments since before 1900 (Figure 1.9). It was first studied by experimental embryologists who were studying cell determination in early embryonic development. The modern use of *C. elegans* began with the British geneticist Sydney Brenner in the mid-1960s. Brenner was seeking a multicellular animal for developmental, neurobiological, and genetic studies that could be analyzed with the ease and resolution characteristic of microorganisms.

Caenorhabditis elegans is a small (1.0-mm-long), free-living nematode with an extremely short life cycle,

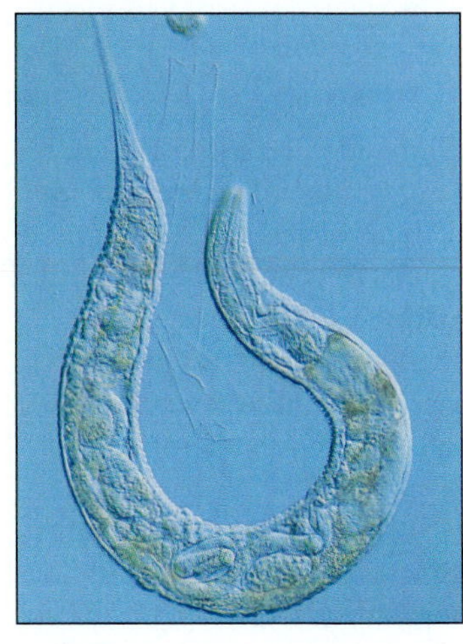

◀FIGURE 1.9 The roundworm *Caenorhabditis elegans.* This organism is a popular experimental organism, especially among geneticists interested in the developmental process. *(Sinclair Stammers/Science Photo Library/Photo Researchers, Inc.)*

▶FIGURE 1.10 The "fruit fly" *Drosophila melanogaster.* *Drosophila* has been one of the most universally popular experimental organisms used in modern genetics, and you will read about it repeatedly throughout this text. *(Sinclair Stammers/Science Photo Library/Photo Researchers, Inc.)*

about 52 hours at room temperature. Large numbers can be easily reared in a laboratory dish using *E. coli* as a food source. The body is transparent, and all developmental events can be observed in living animals with a light microscope. The genome of *C. elegans* is the smallest known for any multicellular animal. Living individuals may be frozen for storage and later revived without being killed. This allows large numbers of strains to be maintained in the laboratory without continual subculturing.

The origin and developmental fate of each cell in *C. elegans* does not vary from individual to individual. Detailed maps (called fate maps) have been made to trace the precise lineage of each of the 959 cells that constitute the adult organism back to the embryonic stem cells from which they descended. The transparent body of *C. elegans* allows lasers to be used to destroy particular cells at any stage of development and to assess the consequence on the development of adult structures and functions. This provides the researcher with the unique ability to study the role of specific cells in the developmental process. Another attractive feature of *C. elegans* is that foreign DNA can be introduced into its genome by the technique of DNA microinjection (the direct injection of foreign DNA by a microneedle). You will learn more about this technique in Chapter 13.

***Drosophila melanogaster* Has a Rich History as a Research Organism.** *Drosophila* is often referred to as a fruit fly because it can be commonly found buzzing around fruit stands at the markets or around fruit bowls at home. Taxonomically, however, *Drosophila* is not classified as a true fruit fly because it does not breed in living fruit (fruit still growing on a tree) but only in fallen or rotting fruit. For this reason, *Drosophila* is not considered an agricultural pest and is suitable as an experimental animal (Figure 1.10).

Drosophila is a dipteran, one of the most highly evolved of the insects. It has well-developed nervous, muscular, digestive, and visual systems, as well as complex epidermal sense organs. The developmental complexity of *Drosophila* is between that of nematodes and mammals. Its small size (adults are about 2 mm in length) allows large numbers to be reared in the laboratory in small bottles, with easily prepared and inexpensive media. *Drosophila* has a relatively short life cycle (14 days) and a relatively small genome, about 1/20 the size of the mammalian genome.

Genetic studies using *Drosophila* began with T. H. Morgan in the early 1900s. In subsequent years, this genus has provided us with much of our current knowledge of fundamental aspects of transmission genetics. During the course of these studies, many special *Drosophila* strains have been derived, and many useful experimental techniques have been developed. This wealth of genetic material is of tremendous value to contemporary investigators.

Mutations can be easily isolated and mapped in *Drosophila*. The range of mutant phenotypes extends from behavior and learning to a variety of biochemical and molecular processes. Of the estimated 12,000 total genes in the *Drosophila* genome, over 4000 have been identified and at least partially characterized.

Summary

Genetics is a science devoted to the study of heredity and variation. The term "heredity" refers to the fact that offspring generally resemble their parents. The term "variation" refers to the fact that despite this resemblance, heritable differences do continue to arise because of informational changes in DNA called mutations, and these changes can increase in frequency in populations and species. Genetics is central to nearly every field of modern biology.

Interest in heredity and variation can be traced back to prehistoric times and was motivated by the desire to improve methods of producing plants and animals useful to humans. The first records we have of formal proposals to explain heredity and variation were those proposed by the Greek philosophers. These primitive theories were modified during the Middle Ages. The common observations of plant and animal breeders that offspring are often intermediate in appearance between their parents led to the general acceptance of the blending theory of inheritance among most 19th-century students of heredity.

Genetics can be subdivided into three distinct areas of research focus. Transmission genetics focuses on establishing the precise rules by which heritable traits are passed down from parents to offspring, as well as on defining the location of genes on chromosomes. Molecular genetics is devoted to the study of the biochemical mechanisms by which hereditary information is stored in nucleic acids and subsequently used to direct the synthesis of proteins that make up living organisms. Population and evolutionary genetics are devoted to the study of the factors that determine the genetic composition of populations and how genetic changes in populations may lead to the formation of new species. The methodological differences that have traditionally characterized these subdivisions of genetics have eroded in recent years as a result of the near-universal application of techniques used in molecular biology to all areas of genetic research.

Genetics is an experimental science. The scientific method consists of (1) making careful observations; (2) forming alternative hypotheses derived from these observations; and (3) testing these hypotheses by conducting experiments designed to falsify one or more of the alternative hypotheses.

The appropriate choice of an experimental organism is essential to successful genetic research. Several biological characteristics are commonly associated with the experimental organisms used in genetic research, including small size, rapid generation time, and large number of offspring. Some popular genetic research organisms are the bacterium *Escherichia coli,* the yeast *Saccharomyces cerevisiae,* the nematode *Caenorhabditis elegans,* the "fruit fly" *Drosophila melanogaster,* and the small flowering plant *Arabidopsis thaliana.*

Selected Key Terms

Aristotle
blending inheritance
DNA
DNA sequencing
epigenesis
gene
genetics
genome
genome project
genotype
William Harvey
heredity
homunculus
hybrid
hypothesis
Jean Baptiste Lamarck
Gregor Mendel
Mendelian genetics
molecular (biology) genetics
phenotype
polymerase chain reaction (PCR)
population
population and evolutionary genetics
preformationism
scientific method
scientific theory
species
transmission genetics
variation
Casper Wolff

Suggested Readings

Anderson, W. F. 1995. "Human gene therapy." *Scientific American,* 273 (September): 124–129.

Dunn, L. 1965. *A Short History of Genetics.* McGraw-Hill, New York.

Kuhn, T. 1962. *The Structure of Scientific Revolutions.* University of Chicago Press, Chicago.

Olby, R. 1966. *Origins of Mendelism.* Schocken Books, New York.

Popper, K. 1959. *The Logic of Scientific Discovery.* Basic Books, New York.

Provine, W. 1971. *The Origins of Theoretical Population Genetics.* University of Chicago Press, Chicago.

Stubbe, H. 1972. *History of Genetics: From Prehistoric Times to the Rediscovery of Mendel.* MIT Press, Cambridge, MA.

On the Web

Visit our Web site at **http://www.saunderscollege.com/lifesci/titles.html** and click on **A/G/M Genetics** for links to the following chapter-related resources on the World Wide Web:

1. **Medline search program.** This site allows you to search the current literature in genetics, biology, and medicine.
2. **Flybase.** This is the database for all *Drosophila* information, including maps, researchers, DNA sequence data, and much more.
3. **Yeast.** This is the database for all information relating to *Saccharmyces* (yeast), including the complete DNA sequence of its genome.
4. **E. coli.** This is the database for *Escherichia coli* and its complete DNA sequence.
5. **National Center for Biotechnology Information.** This site will link you to other biotechnology databases, as well as allow you to search for any DNA sequence that has been deposited.

Open pods filled with ripe garden peas. The pea was used by Gregor Mendel in his pioneering studies of genetics. *(Adam Hart-Davis/Science Photo Library/Photo Researchers, Inc.)*

PART

1

How Genes Are Organized and Transmitted Through Generations

CHAPTER 2
Transmission Genetics: Mendelian Analysis of Inheritance

CHAPTER 3
Genes, Chromosomes, and the Mechanism of Mendelian Inheritance

CHAPTER 4
Multiple Alleles and Gene Interaction

CHAPTER 5
Quantitative Traits and Polygenic Inheritance

CHAPTER 6
Linkage and Gene Mapping In Eukaryotes

CHAPTER 7
Extranuclear Inheritance

CHAPTER 8
The Genetics of Bacteria and Their Viruses

CHAPTER

2

Wrinkled and round seeds of the garden pea, *Pisum sativum*. The characteristics of this plant were critical to the success of Gregor Mendel's investigations of the nature of heredity. *(R. W. Van Norman/Visuals Unlimited)*

Transmission Genetics: Mendelian Analysis of Inheritance

It has been known for thousands of years that biological traits are inherited from generation to generation. This fact was probably first recognized by farmers, and, as a consequence, they saved seeds from the best plants for the next year's crop and used the best animals for breeding. During all this time, there were no significant insights into how individual traits are actually inherited. This understanding had to await the discoveries of Gregor Mendel, an Augustinian priest who first published his treatise on the rules of inheritance in 1866. In this publication, Mendel proposed the first theory about the units of inheritance (what we now call genes) and described two fundamental rules governing how traits are inherited, even though at the time chromosomes had not been discovered. Mendel's work is recognized today as one of the greatest breakthroughs in the history of science, and his paper is regarded as the true beginning of the modern science of genetics. The science of genetics has served as the foundation for the incredible advances we have seen in crop yields and animal production and has provided us with an understanding of how human diseases are inherited. Since the early part of the 20th century, genetics has been at the forefront of biological research and is today the basis of a majority of the research done in the biological sciences.

In this chapter, we will examine the details of Mendel's experiments and how the rules of inheritance are used today to investigate the nature of genes. The development of Mendel's rules of inheritance is the basis for what is now known as **transmission genetics** or **Mendelian inheritance.** By the end of this chapter, we will have addressed four specific questions:

1. Through what experiments did Gregor Mendel uncover the rules governing how traits are passed from one generation to the next?
2. What are the rules of inheritance that Mendel deduced from his results?
3. How are the Mendelian rules, the laws of probability, and statistical analysis used to solve genetics problems?

4. How can we apply the Mendelian rules to understand the genetics of inherited traits and diseases in humans?

Mendel Discovered the Rules of Inheritance.

In 1866, Gregor Mendel (see *Historical Profile: Johann Gregor Mendel*) published a paper reporting the results of breeding experiments done using the garden pea, *Pisum sativum*. From the results of these experiments, he described two basic rules governing the inheritance of traits. Since the early part of the 1900s, researchers have wondered why Mendel was successful in discovering how traits are inherited when earlier scientists were unable to make this basic conclusion. The complete answer will never be known, but there are several important factors to consider. First, Mendel had many years of experience breeding plants. As a result, he selected the garden pea as an experimental organism. This was an excellent choice because peas grow well in small garden spaces, produce large numbers of seeds, and are easily pollinated. Pea plants are usually self-pollinating because the anthers (which produce pollen) and the stigma (which receives the pollen) are enclosed within the same flower (Figure 2.1). Self-pollination is called **selfing,** whereas

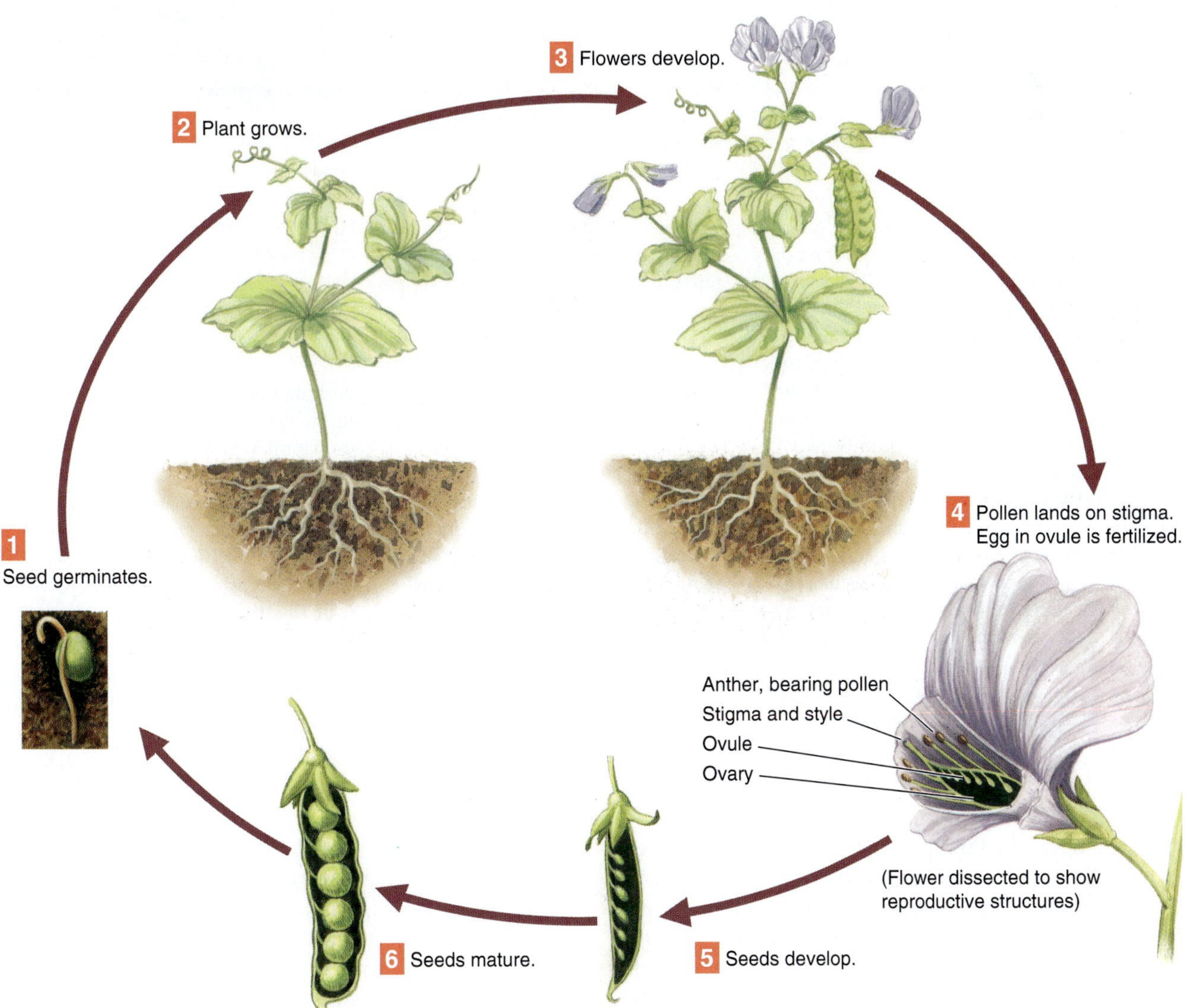
3 Flowers develop.
2 Plant grows.
4 Pollen lands on stigma. Egg in ovule is fertilized.
1 Seed germinates.
Anther, bearing pollen
Stigma and style
Ovule
Ovary
(Flower dissected to show reproductive structures)
6 Seeds mature.
5 Seeds develop.

FIGURE 2.1 Life cycle of the pea, *Pisum sativum*. A feature of peas that was very important to Mendel was self-pollination. Over many generations, true-breeding strains result.

HISTORICAL PROFILE

Johann Gregor Mendel (1822–1884)

Johann Gregor Mendel was born on July 22, 1822, in the village of Heinzendorf in the northwest corner of Moravia, in what is now the Czech Republic. The Mendels were German peasant farmers, gardeners, and local magistrates who had lived in this region for many generations. Johann was the second child and only son of Anton Mendel, a farmer with a large orchard. Johann worked on the farm and in the orchard as a boy, a life of hard physical labor. He attended the village school, where he learned reading, writing, mathematics, and practical subjects such as beekeeping and grafting fruit trees. Because he was an exceptional student, his father was persuaded to allow him to attend middle school and high school, where he was consistently at the top of his classes. The large school fees were almost beyond the means of his family, so high school was a period of desperate poverty for young Johann. After Johann completed high school, he planned to enter the two-year course at the Olmutz Philosophical Institute, a preparatory school for university study. However, ill health forced his father to retire, and without his father's financial support, Mendel could not continue his education. At this critical point in his life, Johann's younger sister gave him her dowry money, an act of kindness he never forgot. With these funds, he completed the course at the Olmutz Institute, where he was again an excellent student. Johann realized that the next step, attending a university, was completely beyond his means and that he must instead find some way of earning a living. His physics instructor at the Olmutz Institute recommended him to the Augustinian Order in Brünn as a candidate for the priesthood, and on September 7, 1843, he joined the order as a Novice and adopted the name Gregor.

Brünn (today called Brno) was the capital of the province of Moravia, an important part of the Austo-Hungarian empire that was noted for industry, agriculture, and commerce. The Augustinian Order in Brünn was wealthy, and the monastery had a library of over 20,000 volumes, as well as a school. The monastery was the spiritual and intellectual center of the city. The order encouraged intellectual pursuits, and several of the priests had national or international reputations as scholars, plant breeders, musicians, or philosophers. Mendel was not the only intellectually gifted youth from a poor family who preferred joining a religious order to living a life of hard physical toil on a peasant farm. Mendel entered the program of religious training for the priesthood and was again an exceptional student, completing the 4-year program in 3 years.

After being ordained a priest, he began serving as a parish priest (which he did not like) and also

FIGURE 2.A Johann Gregor Mendel. This photograph was taken about 1860 and shows Mendel holding a flower from a hybrid fuchsia, called the Mendel fuchsia. *(© V. Orel, Mendelian Museum of Moravian Museum/Biological Photo Service)*

working in botany and plant breeding (which he enjoyed). A well-known photograph of Mendel, taken during the time that he was performing his experiments with peas (Figure 2.A), shows him holding a hybrid fuchsia, known as the "Mendel fuchsia," that he had bred. Realizing that he was not suited to work as a parish priest, Mendel's Abbot arranged for him to try teaching by serving as a temporary or "supply" teacher at a local high

the pollination of a flower from a different plant is called cross-pollination, or **crossing.** Over time, many generations of selfing produces pea lines in which all the plants are identical. These lines are called **strains.** When a strain produces identical progeny generation after generation, it is called a **true-breeding strain.** Controlled crossing is easily done in peas by opening the flower bud, removing the anthers before they are mature, and depositing pollen from the flower of a different plant on the stigma (Figure 2.2). As an experienced plant breeder, Mendel was well aware of the techniques of controlled crossing in plants and of how the characteristics of peas could be used in experiments.

For his experiments, Mendel selected only true-breeding strains of peas. In addition, he wisely selected only a few traits for study. When Mendel started his experiments, he tested each pea strain by selfing plants and growing the progeny for 2 years to ensure their true-breeding nature. When he actually crossed true-breeding strains, he made certain that the two parents differed by

school. Mendel was immediately popular with the students, so popular that he was encouraged to take the examination for a permanent teaching license, even though a university education was normally required. Mendel failed the teaching license examination, but the board of examiners, noting his lack of formal education, did not dismiss him outright. Instead, they suggested that he complete his education by studying at the university at Vienna and then try the examination again. The Abbot of Mendel's monastery agreed and sent him to Vienna University to study for 2 years.

Between 1851 and 1853, Mendel studied physics, chemistry, zoology, botany, and mathematics at Vienna. His instructors included first-rate scientists, including the famous physicist Doppler. He returned to Brünn in the summer of 1853 and again became a temporary teacher of high school physics and natural history in the Brünn Modern School. Although he did not have an official teaching license, Mendel did not need to attempt the examination again. The local school authorities often made exceptions for priests, believing that their religious education gave them unique qualifications for instructing children, and many priests taught regular class loads for years as temporary teachers. Mendel was determined to get his license, however, and in 1856 he took the qualifying examination a second time. He passed the written portion, but, under circumstances that have never been made clear, he withdrew during the oral examination. Years later, members of his order claimed that Mendel argued with the examiners about the nature of heredity and then withdrew rather than being failed for disagreeing with current dogma. His fellow priests believed that the need to prove his ideas correct was the reason why he began his pea experiments. The truth will never be known, although it is true that Mendel had been actively breeding plants, mice, birds, and bees for some years before he took this exam and that he did begin his experiments with peas almost immediately after the exam. Mendel continued as a temporary teacher and carried out genetics experiments in the monastery garden. In 1865, he presented the results of his pea crosses to the Brünn Society of Natural History. The members did not understand his findings but agreed to publish them in their journal. His paper was published in the proceedings of the society in 1866, and although the journal was circulated to over 120 universities and other societies, the paper was almost completely ignored. No one in the scientific establishment understood the significance of the work.

Mendel continued doing breeding experiments on 26 different species of plants, some of which confirmed the pea results. His experimental work ended in 1868, when, at the age of 46, he was elected abbot of his monastery. He was now the administrator of the order's extensive property and an important member of the community. A few years later, he became involved in a protracted legal dispute with the national government over taxes on the monastery, which required even more of his time. The lack of recognition of his discoveries in genetics bothered Mendel throughout his life. He believed that sooner or later his work would be appreciated, and he reportedly told his nephew late in his life that "my time will come." His later years were busy and fulfilling. He was a popular, well-respected figure in the city. He made extensive improvements in the monastery gardens, made tours of inspection of the monastery's properties, was a noted chess player, served on the boards of several societies and a bank, and kept extensive weather records. On January 6, 1884, Mendel died of heart failure, complicated by kidney disease. After his death, his personal records (including his experimental notebooks) were destroyed. He may have requested this himself, as it was a common practice in religious orders at that time.

a small number of carefully chosen traits. Scientists working before Mendel made the mistake of not limiting the number of traits between the parents and not using true-breeding strains. A final factor in Mendel's success was his diligence in subjecting his results to careful mathematical analysis. When Mendel made his crosses, he noted the exact number of different progeny types produced by each cross. This careful numerical analysis had never been done in plant-breeding experiments before and was an essential factor in Mendel's success.

Monohybrid Crosses Demonstrate That Alleles Segregate.

In his first series of experiments, Mendel made seven crosses between true-breeding pea strains that had alternative forms of one particular plant or seed characteristic (Figure 2.3). For example, in one experiment he made a cross between two true-breeding strains, one with wrinkled seeds and another with round seeds (Figure 2.3). This is called a **monohybrid cross** because the two

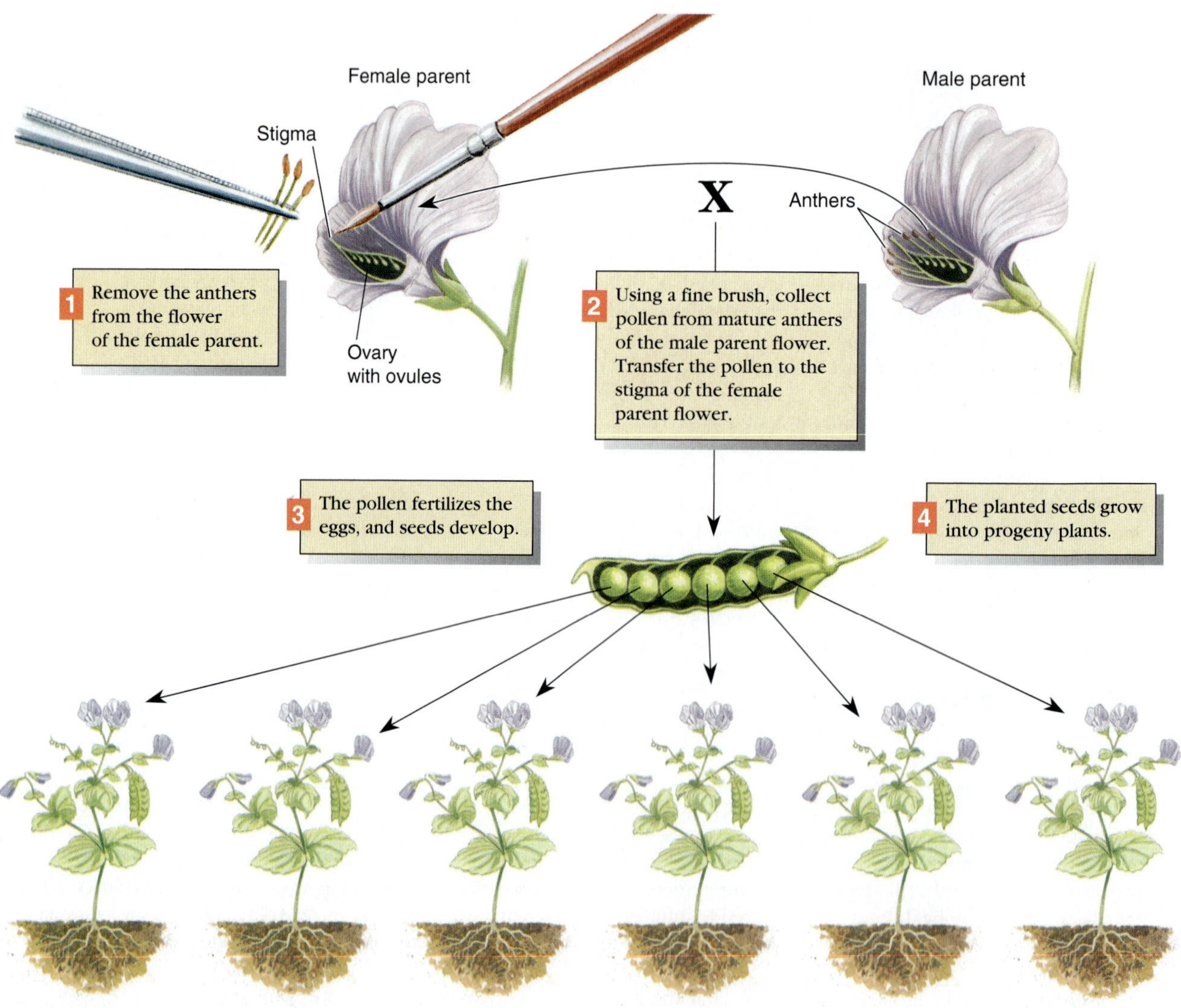

Female parent
Male parent
Stigma
X
Anthers
1 Remove the anthers from the flower of the female parent.
Ovary with ovules
2 Using a fine brush, collect pollen from mature anthers of the male parent flower. Transfer the pollen to the stigma of the female parent flower.
3 The pollen fertilizes the eggs, and seeds develop.
4 The planted seeds grow into progeny plants.

FIGURE 2.2 Cross-pollination of peas. Mendel was able to make genetic crosses by taking the pollen from one plant's anther and placing it on the stigma of another plant.

plants involved in the cross differ only in one character (in this case, seed shape). The plants involved in the original cross are called the **parental** or **P generation.** In this experiment, Mendel found that all the progeny plants, the **first filial** or **F_1 generation,** produced only round seeds. The wrinkled trait seemed to have been masked, or "dominated," by the round trait in the F_1 seeds. Mendel called the round trait **dominant** and the wrinkled trait **recessive.** All seven monohybrid crosses behaved in a similar manner; they produced F_1 progeny possessing the trait of one of the parents. Mendel planted the F_1 seeds for each cross, raised the plants, and allowed them to self-pollinate to produce a **second filial** or **F_2 generation** of seeds. He found that both dominant and recessive types appeared in the F_2 generation in a numerical ratio of approximately three dominant to one recessive (Table 2.1). For example, in the cross between the pea strains with round seeds and wrinkled seeds, the F_2 contained 5474 round and 1850 wrinkled seeds.

Mendel continued each cross for another generation by collecting F_2 seeds, planting them, rearing the plants, and allowing them to self-pollinate to produce a **third filial** or **F_3 generation.** In each cross, the results were the same. Every F_2 individual with a recessive trait bred true, whereas the F_2 with dominant traits were of two types: one third bred true and two thirds produced an F_3 with a ratio of 3 : 1 (Table 2.1). For example, in the cross between round and wrinkled seed strains, all F_2 wrinkled seeds bred true, producing F_3 plants that produced only wrinkled seeds. In contrast, the F_2 round seeds were of two types. About one third of the F_2 round seeds grew into plants that produced only round seeds, and about two thirds of the F_2 round seeds grew into plants that produced a ratio of three round seeds to one wrinkled seed.

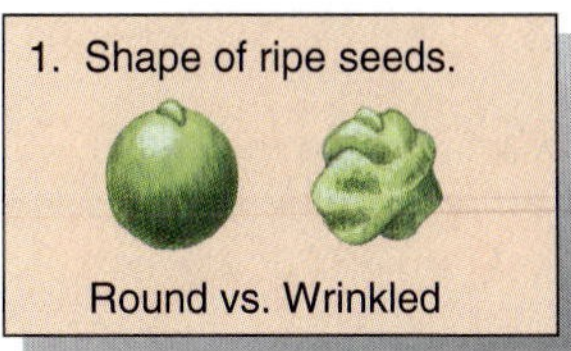

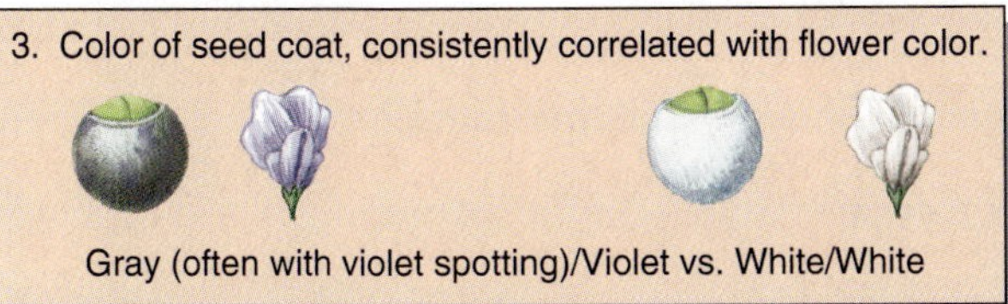

3. Color of seed coat, consistently correlated with flower color.
Gray (often with violet spotting)/Violet vs. White/White

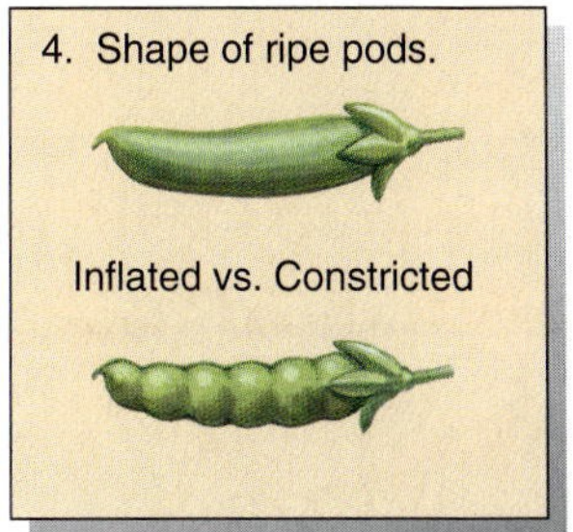

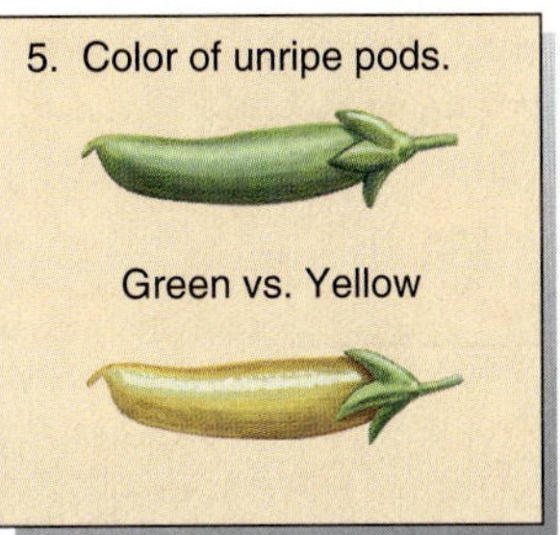

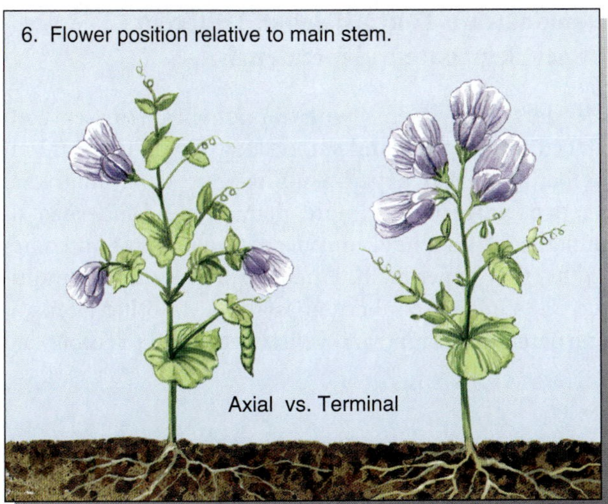

After carefully collecting and quantifying his data, Mendel turned to the task of interpreting the results of his monohybrid crosses. He hypothesized that alternative traits, such as wrinkled or round seeds, are determined by hereditary "factors." Today we refer to these hereditary factors as **genes.** Mendel proposed that genes exist in different forms and can consequently give different traits. Today we call different forms of one gene **alleles.** For example, the seed shape characteristic is determined by one gene but exists in two allelic forms. Mendel's original true-breeding wrinkled seed strain had one allele of this gene (represented by *r*) and his true-breeding round seed strain had a different allele (represented by *R*).

Before we discuss Mendel's interpretation of his monohybrid results, we need to introduce some basic terms. After doing his experiments, Mendel postulated that each individual pea plant carried two copies (two alleles) of each gene. A **homozygous** individual, or **homozygote,** has two identical alleles. Thus, a plant that is true-breeding for wrinkled seeds is homozygous for the wrinkled-seed determining allele (*rr*), and a plant that is true-breeding for round seeds is homozygous for the round-seed determining allele (*RR*). A **heterozygous** individual, or **heterozygote,** possesses two different alleles of one gene. Mendel also postulated that each plant passed one allele to each of its offspring through the gametes. As a consequence, each individual inherits one allele from each of its parents. Thus, the F_1 plants from the cross between homozygous round plants (*RR*) and homozygous wrinkled plants (*rr*) are heterozygous for the round and wrinkled alleles (*Rr*). The **phenotype** of an individual is the observable form of the trait (e.g., round seeds or wrinkled seeds). The **genotype** of an individual is its genetic makeup (e.g., *RR*, *Rr*, *rr*). Each allele is given a short (one to three letters) italicized symbol. By convention, the symbol of a dominant allele begins with an uppercase letter, and the symbol of a recessive allele with a lowercase letter (see *Concept Close-Up: Gene Symbols*).

To explain the results of his monohybrid crosses, Mendel devised an hypothesis that contained several key assumptions. First, contrary to the blending theory of inheritance (see Chapter 1), Mendel proposed that alleles do not mix or "blend" in heterozygotes. He proposed this based on the fact that recessive traits reappear in the progeny of heterozygotes. This suggested that the dominant and recessive alleles are passed intact from heterozygotes to their offspring. Second, Mendel proposed that the two alleles segregate from one another during

◀**FIGURE 2.3** **Pea traits studied by Mendel.** Mendel chose seven easily distinguishable and visible traits to study. Mendel selected the traits from true-breeding strains that had very different forms of the trait.

TABLE 2.1 *The Results of Mendel's Monohybrid Crosses*

Parents	F_1	F_2	F_2 Ratio
tall × short plants	tall	787 tall 277 short	2.84 : 1
axial × terminal flowers	axial	651 axial 207 terminal	3.15 : 1
green × yellow pods	green	428 green 152 yellow	2.82 : 1
inflated × constricted pods	inflated	882 inflated 299 constricted	2.95 : 1
green × yellow seeds	yellow	6022 yellow 2001 green	3.01 : 1
round × wrinkled seeds	round	5474 round 1850 wrinkled	2.96 : 1
gray × white seed coat	gray	705 gray 224 white	3.15 : 1

the formation of gametes. This means that a heterozygous plant (*Rr*) produces gametes containing either the *round* allele (*R*) or the *wrinkled* allele (*r*), but never both. On average, 1/2 of the gametes contain the *wrinkled* allele and 1/2 contain the *round* allele (Figure 2.4). Of course, plants homozygous for the *wrinkled* allele (*rr*) would produce only gametes containing the *wrinkled* allele (*r*), and plants homozygous for the *round* allele (*RR*) would produce only gametes carrying the *round* allele (*R*). Mendel's proposition that alleles segregate from one another in the formation of gametes is called **Mendel's Rule of Segregation,** or Mendel's first rule.

Dihybrid and Trihybrid Crosses Demonstrate That Alleles of Different Genes Segregate Independently.

Mendel's experiments with monohybrid crosses convinced him that he understood how single traits are inherited. His next experiments were to determine what happens when two or more characters are inherited simultaneously. Mendel considered two possible outcomes to this new experiment. One possibility was that multiple traits might be inherited as a unit. In other words, if one parent had round and yellow seeds and the other had

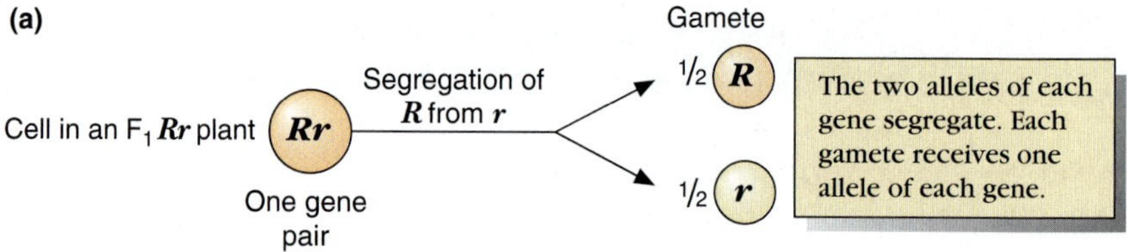

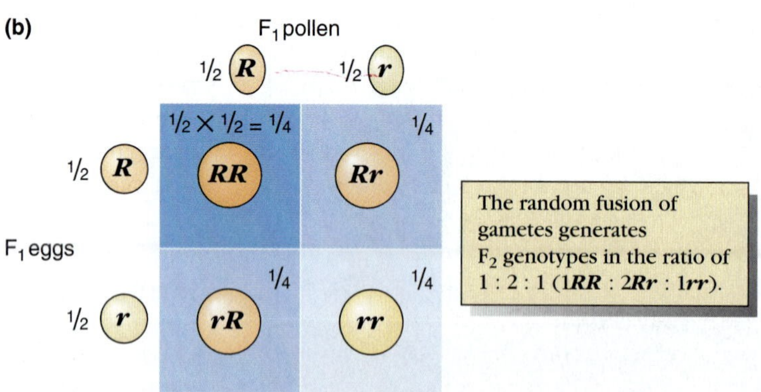

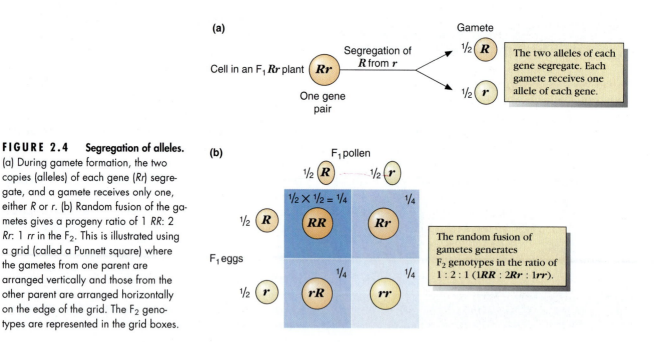

FIGURE 2.4 Segregation of alleles. (a) During gamete formation, the two copies (alleles) of each gene (*Rr*) segregate, and a gamete receives only one, either *R* or *r*. (b) Random fusion of the gametes gives a progeny ratio of 1 *RR*: 2 *Rr*: 1 *rr* in the F_2. This is illustrated using a grid (called a Punnett square) where the gametes from one parent are arranged vertically and those from the other parent are arranged horizontally on the edge of the grid. The F_2 genotypes are represented in the grid boxes.

Concept Close-Up

Gene Symbols

Mendel used a series of arbitrary letter symbols to represent the different genes he was studying, but as more mutations were recovered, it became necessary to devise systems of rules for naming genes and assigning symbols to alleles. An advantage of having one set of rules is that investigators can provide important information about the nature of a particular allele by the form of the symbol. For most eukaryotic organisms, mutant alleles are given a name (written in italics) that describes in some way the mutant phenotype, for example, *rosy, white,* or *defective kernel.* Each gene is given a name that often also reflects the phenotype produced by mutant alleles (which is italicized) and a short (1, 2, or 3 letters) italicized designation (*ry, w, dek*). Each allele of a gene is designated by a short series of letters and/or numbers that follow the gene designation, either on the same line or as a superscript (ry^1, w^{67g}, *dek*12). Gene symbols never contain Greek letters, subscripts, or spaces. Geneticists working with different organisms, unfortunately, have derived slightly different gene-naming systems, which we will explain at different points throughout this book.

The *Drosophila* genetic nomenclature system is based on the concept of the wild-type allele of the gene. The first strains of *Drosophila* studied in laboratories were produced by breeding individuals caught in the wild (hence the name "wild-type strain"), and mutations were detected in these populations as visible deviations from some aspect of the wild-type phenotype. It was assumed that the individuals in these wild-type populations represented the normal *Drosophila* and that their alleles were thus normal alleles. If the first mutant allele of a gene to be discovered is dominant to the wild-type allele, then that gene is given a name that begins with a capital letter (e.g., *Cy* for the *Curly* gene, whose mutant alleles give a dominant curly winged phenotype). If the first mutant allele of a gene to be discovered is recessive to the wild-type allele, then the gene is given a name that begins with a lowercase letter (e.g., *y* for the *yellow* gene, whose mutant alleles give a yellow body color). Recessive alleles of a gene originally identified as a dominant mutation have a superscript that begins with r (for example Dfd^r for the recessive allele of the *Deformed* gene), and dominant alleles of a gene originally identified by a recessive allele have a superscript that begins with D (for example, ey^D and ci^D for dominant alleles of the *eyeless* and *cubitus interruptus* genes). The wild-type allele is always represented by a + symbol, either as a superscript (w^+) or alone (+), and the complete absence of a gene is indicated by a − superscript (y^-).

wrinkled and green seeds, these pairs of traits might be passed together to the F_1 and F_2 generations. A second possibility was that traits are inherited independently. If this is correct, then the two traits would be passed in all possible combinations to future generations.

Mendel investigated this question by crossing plants that differed in two separate traits; this is called a **dihybrid cross.** He crossed plants from a true-breeding strain with round, yellow seeds (*RR YY*) with plants from a true-breeding strain with wrinkled, green seeds (*rr yy*). The F_1 seeds (*Rr Yy*) were all round and yellow. This finding agreed with the results of the monohybrid crosses because round was dominant to wrinkled, and yellow was dominant to green. The F_1 plants were grown and selfed, and the phenotypes of 556 F_2 seeds were recorded. These seeds had four different phenotypes: 315 round, yellow seeds; 108 round, green seeds; 101 wrinkled, yellow seeds; and 32 wrinkled, green seeds. These numbers are very close to a 9/16 : 3/16 : 3/16 : 1/16 ratio and represent all possible combinations of the two traits. From this and other experiments, Mendel concluded that seed shape and seed color are not inherited as a unit.

To explain the ratio of F_2 phenotypes, Mendel proposed that the alleles of these two genes segregated independently. This means that in the heterozygous F_1 individuals (*Rr Yy*), each gamete would have a 1/2 chance of containing an *R* allele and a 1/2 chance of containing an *r* allele. Each gamete would also have a 1/2 chance of containing a *Y* allele and a 1/2 chance of containing a *y* allele. Thus, a gamete would have a $1/2 \times 1/2 = 1/4$ chance of having any one of the four genotypes, and the four gamete genotypes (*RY, Ry, rY, ry*) would be produced in equal frequency (Figure 2.5). Mendel also proposed that when the F_1 plants were selfed, each gamete combined randomly with another gamete. This means there would be $4 \times 4 = 16$ possible gamete combinations. It should be noted that some of these gamete combinations will give the same genotype. For example, combining an *RY* gamete and an *ry* gamete gives an individual with an *Rr Yy* genotype, and combining an *Ry* gamete with an *rY* gamete also gives an *Rr Yy* genotype. Of the 16 gamete combinations, nine give different F_2 genotypes. Because of dominance, these nine genotypes produce four different F_2 phenotypes: 9/16 round yellow; 3/16 round

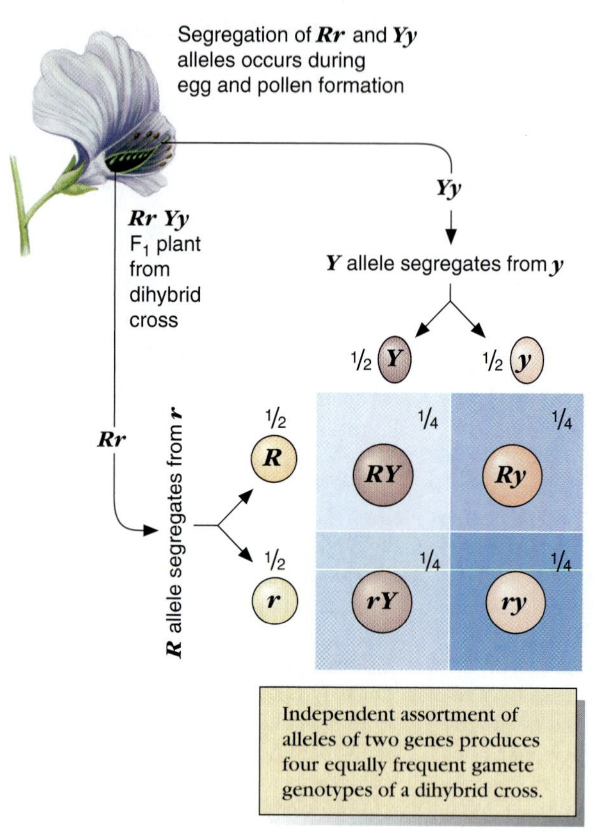

green; 3/16 wrinkled yellow; 1/16 wrinkled green (Figure 2.6). The exact genotypes and phenotypes of all the progeny of a dihybrid cross are easily analyzed by constructing a **Punnett Square,** named after the British geneticist R. C. Punnett, who employed it to predict the outcome of genetic crosses (Figure 2.6). The close agreement between what Mendel observed in the F_2 of his dihybrid cross and what he expected convinced him that the alleles of different genes did segregate independently. The independent segregation of alleles of different genes is today called **Mendel's Rule of Independent Assortment,** or Mendel's second rule.

To confirm the Rule of Independent Assortment, Mendel examined the inheritance pattern of traits in a cross between plants that differed for three characters. This is called a **trihybrid cross.** He crossed plants from a true-breeding strain with round, yellow seeds and a gray

◀**FIGURE 2.5 Independent assortment of alleles.** In a dihybrid cross for seed shape and color, independent segregation of the alleles of the two genes produces four gamete genotypes (*RY, Ry, rY, ry*) in equal frequencies.

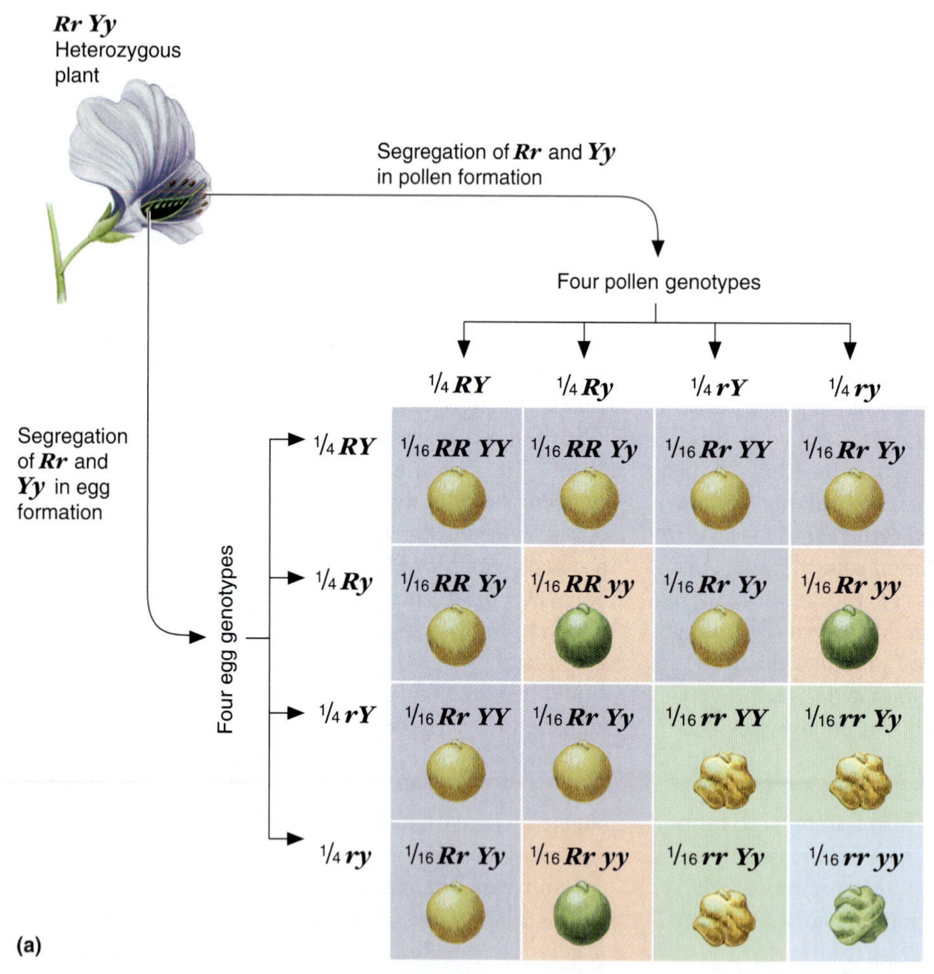

FIGURE 2.6 Determination of dihybrid F_2 genotypic ratios from a Punnett square. (a) Each parent (*Rr Yy*) generates gametes as described in Figure 2.5. The genotypes of the F_2 progeny from this cross can now be determined by randomly combining the four equally frequent gametes using a Punnett square. (b) Given the genotypes, and knowing which allele is dominant, the phenotypic ratios can be determined. The offspring of this dihybrid cross will occur in a 9 : 3 : 3 : 1 phenotypic ratio.

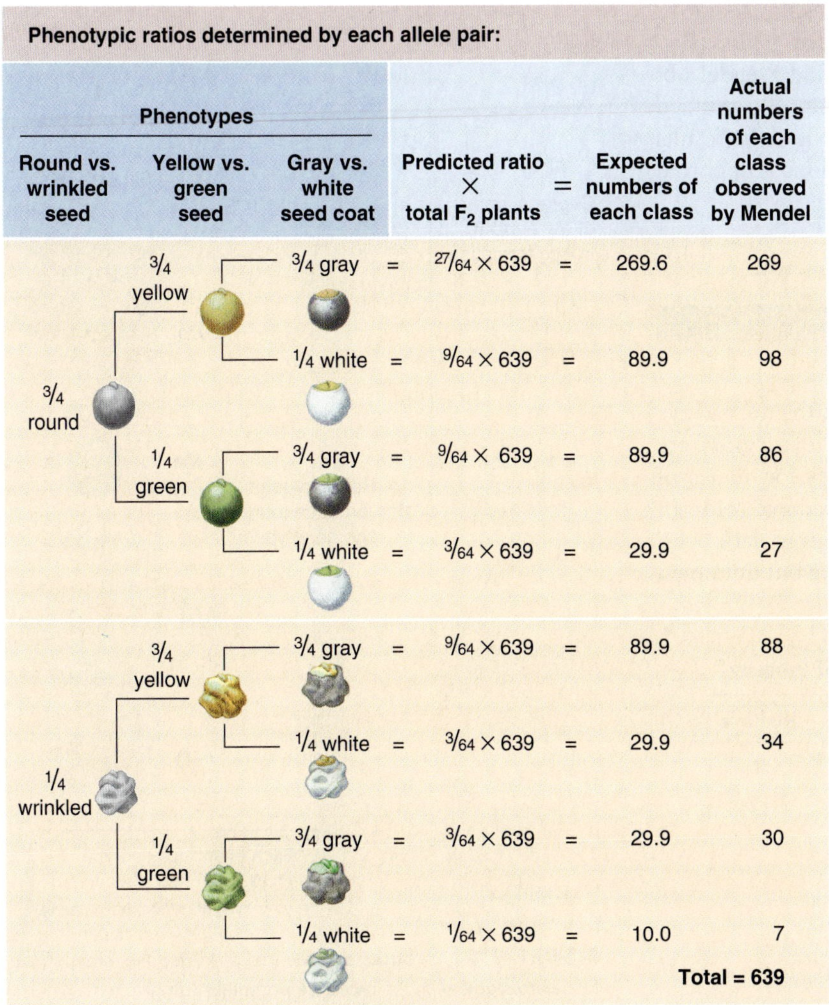

Phenotypic ratios determined by each allele pair:

Round vs. wrinkled seed	Yellow vs. green seed	Gray vs. white seed coat		Predicted ratio × total F_2 plants	=	Expected numbers of each class	Actual numbers of each class observed by Mendel
3/4 round	3/4 yellow	3/4 gray	=	27/64 × 639	=	269.6	269
		1/4 white	=	9/64 × 639	=	89.9	98
	1/4 green	3/4 gray	=	9/64 × 639	=	89.9	86
		1/4 white	=	3/64 × 639	=	29.9	27
1/4 wrinkled	3/4 yellow	3/4 gray	=	9/64 × 639	=	89.9	88
		1/4 white	=	3/64 × 639	=	29.9	34
	1/4 green	3/4 gray	=	3/64 × 639	=	29.9	30
		1/4 white	=	1/64 × 639	=	10.0	7
							Total = 639

FIGURE 2.7 Predicting progeny ratios using the "branching method." Using the Punnett square to analyze this trihybrid cross (seed shape, seed color, seed-coat color) would be very cumbersome. Alternatively, using the branching method allows easy prediction of expected ratios of genotypes or phenotypes. The expected outcomes for the first gene are listed first, then the second, and so on. The outcome of each combination is one branch of the diagram. Mendel's results in this trihybrid cross were very close to his expectations.

seed coat (*RR YY GG*) with plants from a strain with wrinkled, green seeds and a white seed coat (*rr yy gg*). The F_1 seeds (*Rr Yy Gg*) were all round, yellow, and gray, in agreement with the dominance relationships established for these traits in the monohybrid crosses (Table 2.1). According to Mendel's rules of Segregation and Independent Assortment, each F_1 plant should produce gametes with eight different genotypes. Selfing the F_1 plants should give $8 \times 8 = 64$ possible combinations of gametes producing 27 different F_2 genotypes. Because of dominance, these 27 genotypes should give eight different F_2 phenotypes in a 27 : 9 : 9 : 9 : 3 : 3 : 3 : 1 ratio (Figure 2.7). When Mendel performed the trihybrid cross, he observed eight unique phenotypes in the F_2 progeny, and the frequencies of these phenotypes were very close to the expected 27 : 9 : 9 : 9 : 3 : 3 : 3 : 1 ratio. These results supported his conclusion that dominance, allelic segregation, and independent assortment were fundamental rules of inheritance of all traits. Figure 2.7 shows another graphical approach to determining the outcome of genetic crosses, known as the branching method. This approach allows for easy analysis of complex crosses and is sometimes a better tool than a Punnett square.

Mendel derived his rules by observing the phenotypes produced in the F_2 and F_3 generations of monohybrid, dihybrid, and trihybrid crosses. To test his rules further, Mendel carried out additional crosses whose results he could predict but that were not monohybrid, dihybrid, or trihybrid crosses. For example, in one experiment he crossed heterozygous F_1 plants from his dihybrid cross (*Rr Yy*) to plants from the parental wrinkled, green true-breeding strain (*rr yy*) as well as to plants from the parental round, yellow (*RR YY*) strain. A cross between an F_1 individual and an individual from one of the parental strains is called a **backcross** (Figure 2.8). If a homozygous recessive parent is used in this cross, it is called a **testcross** because it allows a determination or "test" of the genotype of the heterozygous individual. Mendel's rules predicted that the heterozygous F_1 plants should produce four types of gametes in equal frequency (1/4 *RY*; 1/4 *Ry*; 1/4 *rY*; 1/4 *ry*). The homozygous plants will produce gametes with only one genotype, *ry* or *RY*. Thus, the progeny of each cross should consist of four genotypic classes in equal frequency. For example, in the cross of the F_1 progeny (the heterozygote) to the wrinkled, green parent (*rr yy*), the progeny should be 1/4 *Rr*

Yy (round yellow), 1/4 *Rr yy* (round green), 1/4 *rr Yy* (wrinkled yellow), and 1/4 *rr yy* (wrinkled green) (Figure 2.8). This ratio is almost exactly what Mendel obtained: 55 round yellow, 51 round green, 49 wrinkled yellow, and 52 wrinkled green. The close fit of the results of the backcrosses with his predictions supported Mendel's belief that his rules were correct.

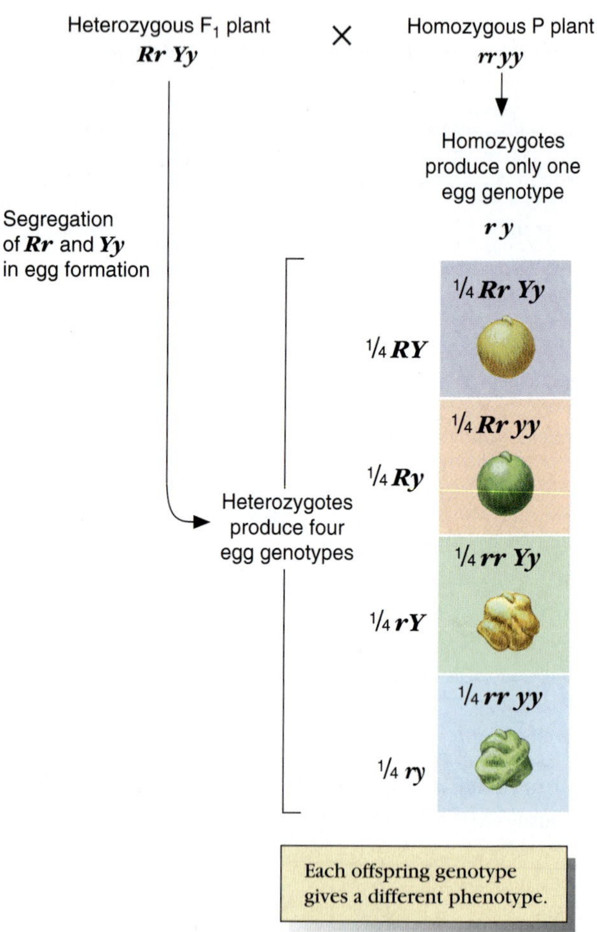

FIGURE 2.8 **A backcross.** Mendel predicted that crossing the F_1 heterozygote with the recessive parent (P) will give a 1 : 1 : 1 : 1 ratio of genotypes and phenotypes in the offspring. This figure illustrates the gametes and progeny from such a cross, using alleles for wrinkled seeds and yellow color. Because the parent is homozygous for recessive alleles, this backcross is also a testcross. In a testcross, the homozygote produces gametes containing only recessive alleles, so the phenotype of the offspring is determined by the genotype of the other parent.

Problem

In the fruit fly, *Drosophila melanogaster*, the normal eye color is red, but different strains of eye colors are known. These strains are homozygous for alleles of several different genes that cause altered eye color. In an analysis of one such stain, which gives dark eyes, several individuals of known phenotype but unknown genotype were crossed.

Parental Phenotypes	Number of Progeny	
	Dark eye	Red eye
1. dark × red	10	0
2. red × red	0	11
3. red × dark	6	5
4. dark × dark	10	0
5. dark × dark	8	3

Is the dark-eye allele dominant or recessive?
What are the most likely genotypes for the parents in each cross?

Solution

The results of cross 1 suggest that progeny with one allele for red and one allele for dark have a dark phenotype. This means that the dark phenotype must be a dominant phenotype given by a dominant allele. This is confirmed by cross 5, a monohybrid cross. Using the symbol *D* for the *dark* allele and *d* for the *red* allele, the genotypes are

1. *DD* × *dd*
2. *dd* × *dd*
3. *dd* × *Dd*
4. *DD* × *D*–
5. *Dd* × *Dd*

In cross 4, it is not possible to determine whether both parents are homozygous for dark or whether one is heterozygous. In cases in which an individual could be either homozygous or heterozygous, it is the convention to use a dash (–) to indicate that this could be either allele. In this cross, either homozygous or heterozygous alleles will give the observed results.

Mendel's Rules Are Used To Predict Genetic Combinations.

In his 1866 paper, Mendel discussed what he believed were the reasons why no previous plant breeder had discovered the rules of inheritance. He suggested one major reason was that previous breeders had not used pure breeding, or homozygous, lines in their crosses. Using randomly selected strains as parents meant that the plants being crossed would have many genetic differences. He performed a calculation based on his rules that showed that individuals differing for only a few genes would give an F_2 generation with a very large number of different genotypes and phenotypes. If two plants that differ by *n* genes are crossed, their progeny, the F_1, will

TABLE 2.2 *Numbers of Genotypes and Phenotypes in Mendelian Crosses*

Heterozygous Gene Pairs in the F_1	F_1 Gamete Genotypes	F_2 Genotype Combinations	F_2 Unique Genotypes	F_2 Phenotypes
n	2^n	4^n	3^n	2^n
1	$2^1 = 2$	$4^1 = 4$	$3^1 = 3$	2
2	$2^2 = 4$	$4^2 = 16$	$3^2 = 9$	4
3	$2^3 = 8$	$4^3 = 64$	$3^3 = 27$	8
10	$2^{10} = 1024$	$4^{10} = 1{,}048{,}576$	$3^{10} = 59{,}049$	1024

produce 2^n different types of gametes. When the F_1 are inbred or selfed, the number of genotypic combinations in the F_2 generation will be equal to $2^n \times 2^n = 4^n$. Because some of these F_2 will have identical genotypes, the number of unique F_2 genotypes will be equal to 3^n, and the number of unique phenotypes (assuming one allele is dominant) will be equal to 2^n (Table 2.2). Based on these calculations, Mendel noted that if two plants that differ by only ten genes are crossed, then the F_2 will contain $3^{10} = 59{,}049$ unique genotypes and 1024 different phenotypes. As a consequence, many of these F_2 phenotypes will never be seen in small-scale experiments. Mendel believed that most plant populations contain more than ten genetic differences and would produce so many different F_2 genotypes and phenotypes that the constant phenotypic and genotypic ratios could not be seen.

SECTION REVIEW PROBLEMS

1. Suppose that Mendel had selected *Cc* plants (the F_1 from his monohybrid cross of colored seed coat [*CC*] with white seed coat [*cc*]) and had crossed them to plants from the parental strain with white seed coat, *cc*.
   a. What genotypes and phenotypes would he have found in the progeny?
   b. How many of each genotype should he have found in 1000 backcross progeny?
2. A red-orange Mexican swordtail fish with a crescent spot at the base of the caudal fin was crossed to a fish from a true-breeding strain that is olive green with no spot. The phenotypes of the progeny are shown in the following table.

Number	Color	Spot
15	red-orange	crescent
12	red-orange	none
18	olive green	crescent
14	olive green	none

   a. What can you conclude about the inheritance of these traits and about the genotype of the original parents?
   b. Produce symbols for the genotypes of the red, spotted parent and the olive, nonspotted parent.
   c. Using the symbols from Problem 2b, indicate the genotype of each category of progeny.
3. A *Drosophila* with a curved wing phenotype was crossed with one from a strain with normal, straight wings. The F_1 all had curved wings and were inbred (F_1 mated to F_1) to produce an F_2 generation. The F_2 included 20 flies with straight wings and 75 with curved wings. What can you conclude about the inheritance of the wing shape phenotype in this cross?
4. An *Arabidopsis* plant with stunted flowers was crossed with a plant from a true-breeding strain of long flowers.
   a. If stunted flowers are a recessive trait, what phenotype should the F_1 have?
   b. If you generated 100 F_2 plants, how many would you expect to be stunted?
5. A tomato with oblong, red fruit was crossed with a plant with round, yellow fruit. The F_1 were inbred to generate the following F_2:

oblong red	60
oblong yellow	180
round red	20
round yellow	60

Explain the inheritance of these traits.

Probability and Chi-Square Are Used To Evaluate Data.

The Mendelian genotypic and phenotypic F_2 ratios from crosses we discussed previously are averages expected from large progeny populations. These ratios do not tell us what any one F_2 individual will be, but rather they give us the **probability** that the individual will have a particular genotype or phenotype. For example, if two individuals with *Aa* genotypes are crossed and have four offspring, the Mendelian rules do not insist that the offspring will be exactly one *aa*, two *Aa*, and one *AA* (1 : 2 : 1). Instead, they predict only the probability that the offspring will have an *aa* genotype of 1/4, an *Aa* genotype of 1/2, and an *AA* genotype of 1/4. In this section, we will discuss some of the rules of probability and statistical tests that geneticists use to evaluate the results of genetic experiments.

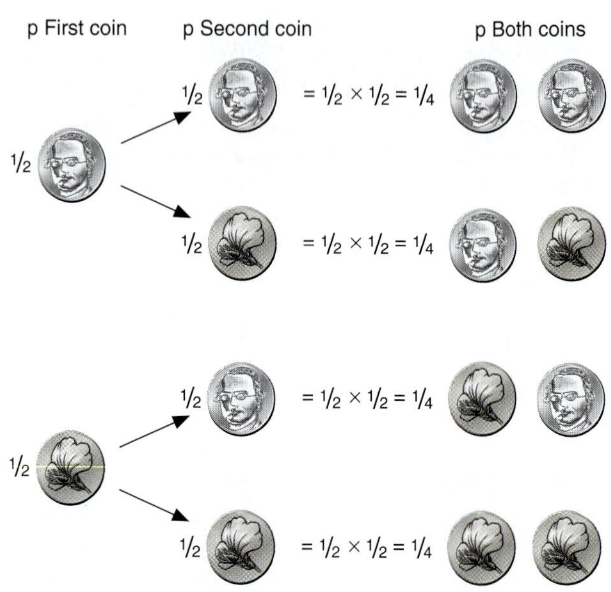

FIGURE 2.9 Multiplication rule. The first coin can be either heads or tails with equal probability, as can the second coin. To find the probability that both coins will come up heads (or any combination of the two coins) when you flip both coins, you multiply their individual probabilities.

The Multiplication and Addition Rules of Probability Are Widely Used in Genetics.

Two basic rules of probability widely used in making genetic predictions are the multiplication rule and the addition rule. The **multiplication rule** states that if two events are independent, meaning that one does not affect the other, then the probability that both will occur is the product of each single probability. That is, the probability that both A and B will occur is equal to the probability that A will occur times the probability that B will occur. This can be written as an equation: $p(\text{A and B}) = p(\text{A}) \times p(\text{B})$. The **addition rule** states that if two outcomes are mutually exclusive, meaning that they cannot both occur at the same time, then the probability that either will occur is the sum of their probabilities. This can also be written as an equation: $p(\text{A or B}) = p(\text{A}) + p(\text{B})$.

A simple example that illustrates these rules of probability is the flipping of a coin. When a coin is flipped, it may land either head side up or tail side up because heads and tails are mutually exclusive outcomes. The coin has a probability of 1/2 of landing head side up and a probability of 1/2 of landing tail side up. Thus, $p(\text{heads}) = 1/2$, and $p(\text{tails}) = 1/2$. The probability that a coin will land as either a head or a tail is given by the addition rule: $p(\text{head or tail}) = (1/2) + (1/2) = 1$. If two coins are flipped, the two have independent outcomes because how the first coin lands has no effect on how the second coin lands. Thus the probability that both coins will land head side up is given by the multiplication rule: $p(\text{head and head}) = (1/2) \times (1/2) = 1/4$ (Figure 2.9). The multiplication rule can be applied to any number of multiple outcomes. If we flipped 10 coins, the probability all will land head side up is: $p(\text{head})^{10} = (1/2)(1/2)(1/2)(1/2)(1/2)(1/2)(1/2)(1/2)(1/2)(1/2) = (1/2)^{10} = 1/1024$.

The addition rule is used when more than one possible outcome of a situation is possible or acceptable. For example, if two coins are flipped, what is the probability that they both will land with the same face up? There are two ways that the faces could be the same: either they are both heads or they are both tails. Using the multiplication rule, we can calculate the probability that both are heads, $p(\text{hh}) = 1/4$, and the probability that both are tails, $p(\text{tt}) = 1/4$. The probability that both coins land with the same face up is the sum of these two probabilities: $p(\text{hh or tt}) = p(\text{hh}) + p(\text{tt}) = (1/4) + (1/4) = 1/2$.

To illustrate how the multiplication and addition rules are used in genetics, consider the example of crossing two F_1 individuals in a dihybrid cross (*Aa Bb* × *Aa Bb*). According to Mendel's rules of Segregation and Independent Assortment, all four types of gametes produced by these two parents will be equally frequent (1/4*AB* : 1/4*Ab* : 1/4*aB* : 1/4*ab*). The probability that any two gametes will combine (*AB* and *ab*) is given by the multiplication rule: $p(AB \text{ and } ab) = p(AB) \times p(ab) = (1/4)(1/4) = 1/16$. Because more than one combination of gametes can give the same offspring genotype, the overall offspring genotypic ratios are given by the addition rule. For example, four combinations give the *Aa Bb* genotype, so the frequency of this genotype is $(1/16) + (1/16) + (1/16) + (1/16) = 4/16$ (Figure 2.10). The multiplication and addition rules are often used in genetics to predict gamete genotypes, offspring genotypes, or combinations of genotypes or phenotypes. We will use these rules repeatedly in this book.

In some genetic analyses, it is important to know the probability that particular combinations of genotypes will occur. For example, consider the cross between the

F_1 heterozygotes (*Rr*) in Mendel's monohybrid cross of round and wrinkled strains. If a pod on one plant contained five seeds, what would be the probability that these five seeds are in the ratio three round to two wrinkled? To answer this question, we must determine the probabilities of all possible orders in which three round and two wrinkled can occur in five seeds and sum these probabilities. The answer can be calculated much more easily using the **binomial theorem,** a mathematical formula known to Mendel. The binomial theorem and how it is used to determine the probability of multiple events is discussed in the CONCEPT CLOSE-UP: *The Binomial Theorem.*

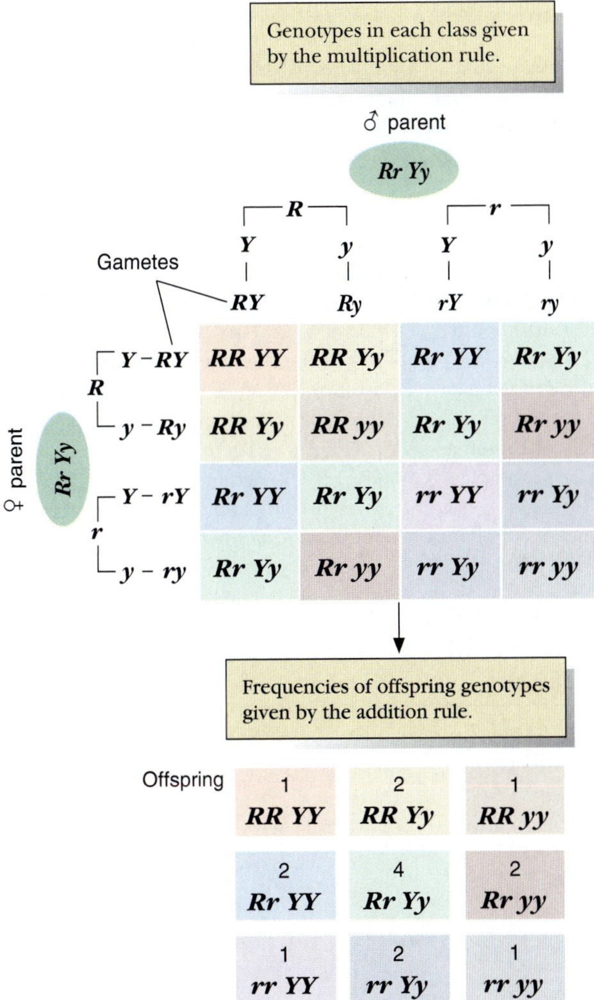

FIGURE 2.10 The addition rule and multiplication rule are used to determine progeny in a cross. The probability of obtaining a particular genotype in the F_2 of a dihybrid cross is demonstrated using both the multiplication and addition rules of probability.

Problem

What is the probability that selfing an F_1 plant from Mendel's dihybrid cross (*Rr Yy* × *Rr Yy*) will produce an offspring with an *rr YY* genotype?

Solution

An individual with an *rr YY* genotype is produced by the fusion of an *rY* gamete with another *rY* gamete. The probability of a gamete receiving the *r* allele from an *Rr Yy* individual is 1/2, and the probability that it also receives the *Y* allele is 1/2. These are independent events, so the probability of one gamete receiving both the *r* allele and the *Y* allele is: $p(rY) = (1/2)(1/2) = 1/4$. The probability that two *rY* gametes fuse to produce an *rr YY* individual is given by the multiplication rule: $p(rr\ YY) = (1/4)(1/4) = 1/16$.

Problem

If a corn plant homozygous for a dominant allele (*R*) of the *ragged leaf* gene and a recessive allele (*p*) of the *angular pollen* gene is crossed with a corn plant homozygous for a recessive allele (*r*) of the *ragged leaf* gene and a dominant allele (*P*) of the *angular pollen* gene, the F_1 will all have ragged leaves and angular pollen (*Rr Pp*).

a. If the F_1 is selfed, what is the probability an F_2 individual will have a ragged leaf and an angular pollen phenotype?

b. What is the probability, if we collected four F_2 individuals at random, that the first would have an *Rr PP* genotype, the second an *RR Pp* genotype, the third an *Rr Pp* genotype, and the fourth an *Rr pp* genotype?

Solution

a. To answer this question, we must determine the probability of each of the different F_2 genotypes that give this phenotype and then sum these probabilities. Four genotypes give this phenotype (*RR AA*, *Rr AA*, *RR Aa*, and *Rr Aa*), and each has a particular probability: $p(RR\ AA) = 1/16$, $p(Rr\ AA) = 2/16$, $p(RR\ Aa) = 2/16$, and $p(Rr\ Aa) = 4/16$. The probability that an F_2 individual has any one of these genotypes is given by the addition rule: $p(\text{ragged angular}) = (1/16) + (2/16) + (2/16) + (4/16) = 9/16$.

b. This probability is given by determining the probability of each individual and then multiplying these probabilities. The probabilities of the four genotypes are $p(Rr\ PP) = 2/16$, $p(RR\ Pp) = 2/16$, $p(Rr\ Pp) = 4/16$, and $p(Rr\ pp) = 2/16$. The probability (p) that in a group of four F_2 the first would have an *Rr PP* genotype, the second an *RR Pp* genotype, the third an *Rr Pp* genotype, and the fourth an *Rr pp* genotype is given by the multiplication rule: $p = (2/16)(2/16)(4/16)(2/16) = 1/2048$.

CONCEPT CLOSE-UP

The Binomial Theorem

Using the multiplication and addition rules to answer some genetics questions requires a large number of calculations, especially when calculating the probability that particular combinations of genotypes or phenotypes will occur. For example, in Mendel's first monohybrid cross (round with wrinkled), a pod on one F_1 plant contained five seeds. You could ask, What would be the probability that these five seeds would include two round seeds and three wrinkled seeds? To answer this question, we must calculate the probability of all possible arrangements of two round and three wrinkled seeds in a group of five and then sum these probabilities. This is not a simple calculation, and in fact, a much easier way to obtain the answer is to use a mathematical formula well known to Mendel, the **binomial theorem.** The binomial theorem deals with the probabilities of two mutually exclusive events, called p and q, whose combined probabilities add up to one (i.e., $p + q = 1$). In Mendel's first monohybrid cross, the round and wrinkled phenotypes are mutually exclusive events; that is, a seed may be round or it may be wrinkled, but it cannot be both. If we let p represent the probability of an F_2 being round and q represent the probability of being wrinkled, then any F_2 seed in Mendel's cross has a probability of 3/4 of being round ($p = 3/4$) and 1/4 of being wrinkled ($q = 1/4$).

The binomial theorem can be used to determine the probability that any size group of F_2 individuals will have a particular combination of phenotypes by calculating the probabilities of all possible combinations of individuals that could make up this group and then summing these probabilities. All the possible combinations of individuals can be determined by calculating the terms of the binomial expression $(p + q)^n$, where n is the total number of individuals in the group. Calculating all the terms you get when this expression is multiplied out is called expanding the binomial, and it produces a series of terms, each of which gives the probability of a different combination of individuals. For example, for a group of two F_2 seeds ($n = 2$), all possible combinations of phenotypes are given by expanding the binomial raised to the power 2 = $(p + q)^2 = (p + q) \times (p + q) = p^2 + 2pq + q^2 = 1$. Each of the three terms of this expanded binomial gives the probability of a different combination of two seeds, with p^2 being the probability that both seeds are round [(3/4)(3/4) = 9/16], with $2pq$ being the probability that one seed is smooth and the other is wrinkled [(2(3/4)(1/4) = 6/16], and with q^2 being the probability that both seeds are wrinkled [(1/4)(1/4) = 1/16]. The coefficients for each term in a binomial expansion can be obtained from standard tables for any expansion, as shown in the table.

To solve our problem of the group of five seeds, we need to determine the number of possible combinations in a group of five seeds ($n = 5$), which is done by expanding the binomial raised to the power 5, $(p + q)^5$. We note that for $n = 5$, the coefficients of the terms are 1, 5, 10, 10, 5, and 1, and using these we calculate that $(p + q)^5 = p^5 + 5p^4q + 10p^3q^2 + 10p^2q^3 + 5pq^4 + q^5$. Each of these terms gives the probability of a different combination of five seeds. The probability that all five seed are

n	Binomial Coefficients
0	1
1	1 1
2	1 2 1
3	1 3 3 1
4	1 4 6 4 1
5	1 5 10 10 5 1
6	1 6 15 20 15 6 1
7	1 7 21 35 35 21 7 1
8	1 8 28 56 70 56 28 8 1
9	1 9 36 84 126 126 84 36 9 1
10	1 10 45 120 210 252 210 120 45 10 1

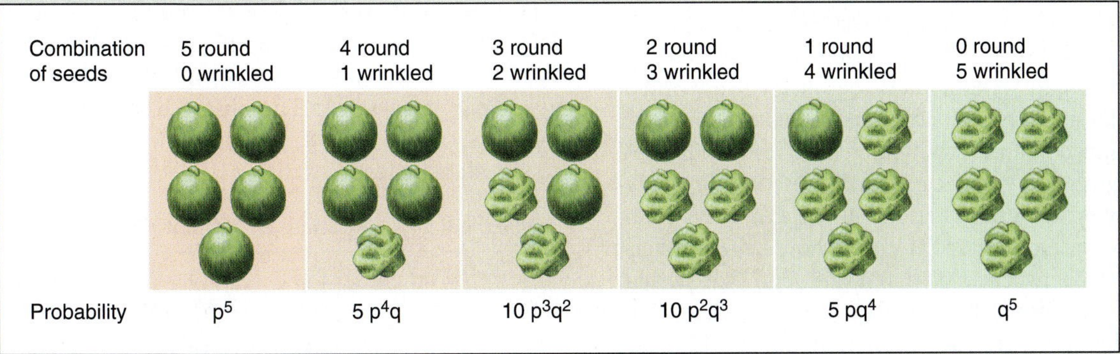

FIGURE 2.B Expanding the binomial. Expanding the binomial raised to the fifth power, $(p + q)^5$, gives the numbers of each possible combination of round and wrinkled seeds in a pod with five seeds.

smooth is p^5, and the probability that four are smooth and one is wrinkled is $5p^4q$, and the probability of two smooth and three wrinkled is $10p^2q^3 = 10(3/4)^2(1/4)^3 = 0.088$ (Figure 2.B). This means that if we were to examine a large number of pods containing five seeds, then 0.088, or 8.8%, would contain exactly two smooth seeds and three wrinkled seeds.

Using the binomial expansion can be cumbersome if the entire expansion must be calculated to determine a single term. An easier method is to use the following formula to calculate the probability of a single term directly:

$$\text{probability} = \frac{n!}{w!\,x!}p^w q^x$$

In this formula, w is the number of individuals of one type, x is the number of individuals of the other type, n is the total number in the group (i.e., $n = w + x$), p is the probability of the first type, and q is the probability of the other. The ! symbol stands for factorial, meaning the multiplication of a number by all of the integers between it and one. For example, $4! = 4 \times 3 \times 2 \times 1 = 24$. In our example of a group of five seeds containing two smooth and three wrinkled, $n = 5$, $w = 2$, $x = 3$, $p = 3/4$, and $q = 1/4$, and the probability of such a group is

$$p = \frac{5!}{3!\,2!}(3/4)^2(1/4)^3$$

$$= \frac{5 \times 4 \times 3 \times 2 \times 1}{3 \times 2 \times 1 \times 2 \times 1}(9/16)(1/64)$$

$$= 10(9/1024) = 0.088$$

This formula can be used to calculate the probabilities of all types of combinations, provided that the probabilities of the two mutually exclusive events are known. For events such as dihybrid crosses that have more than two possible outcomes, the binomial can be used by expressing the several probabilities as two mutually exclusive events. The probability (p) that an F_2 seed from Mendel's dihybrid cross has a wrinkled green phenotype is 1/16, so the probability (q) that it has any other phenotype is $q = 1 - (1/16) = 15/16$, and we could use these probabilities in the binomial formula to calculate the probability that any group of F_2 seeds would contain a certain number that had a wrinkled green phenotype. For example, if we collected a group of 25 F_2 seeds from Mendel's dihybrid cross what would be the probability that only 2 seeds would have a wrinkled green phenotype? This can be calculated using the binomial formula with p = wrinkled green = 1/16, q = all other phenotypes = 15/16, $n = 25$, $w = 2$, $x = 23$, $p = 1/16$, and $q = 15/16$, giving a value of

probability of 2 wrinkled green seeds in 25 =

$$\frac{25!}{2!\,23!}(1/16)^2(15/16)^{23} = 0.266$$

If we selfed a large number of F_1 plants from Mendel's dihybrid cross and collected 25 F_2 seeds from each of them, then 26.6% of the groups of 25 seeds would have only 2 wrinkled green seeds.

The Chi-Square Test Is Used To Evaluate How Closely Observed Results Fit Expectations.

The actual ratios of phenotypes and genotypes that Mendel observed in his experiments appears to be in good agreement with the predicted ratios. In each cross, the observed results and the expected results were nearly the same. Although his results were not absolutely identical to those expected, Mendel knew that in the real world a certain amount of error is expected in all experiments, and he assumed that the small differences between his observed results and his expected results were simply due to chance errors. Very important to all experimental measurements is a method to determine how large a difference can exist between an observed value and an expected value. Mendel did not have a method at his disposal for estimating whether his deviations from expected results were too large to be the result of chance errors.

Today, statistical tests are done to evaluate how well experimental results agree with the expected results. This is an essential part of the scientific method in which all hypotheses must be compared with reality. When an experiment is done, the observed results are compared with the expected results using a statistical test. This test determines the probability that the difference between the observed and the expected results are the result of chance errors alone. If the probability is high enough, we can conclude that the differences were the result of chance errors, and we accept the hypothesis as correct. If the probability is low, we must conclude that the differences are too great for chance errors, and we reject the hypothesis.

These types of statistical tests use a **test statistic,** which is a numerical measure whose value depends on the amount of difference between the observed and the expected results. One widely used test statistic is **chi square** (χ^2). The value of chi square is calculated by squaring the difference between each observed and expected value, dividing this result by the expected value, and summing all values:

$$\chi^2 = \Sigma \frac{(\text{observed value} - \text{expected value})^2}{\text{expected value}}$$

If the differences between the observed and expected results are small, then the value of the chi-square statistic will be small. If the differences are large, the value of the chi-square statistic will be large. Values of the chi-square statistic have been calculated for differences of various sizes and have been published as tables. Thus, the probability that the difference between the observed and expected results in an experiment is the result of chance errors alone can be determined. First, the chi-square value is calculated, and then this value is compared with values in the table.

To illustrate how the chi-square test works, let us compare the phenotypic ratio in the 556 F_2 seeds from Mendel's dihybrid cross to what would be expected. Mendel observed 315 round, yellow seeds; 101 round, green seeds; 108 wrinkled, yellow seeds; and 32 wrinkled, green seeds. The predicted phenotypic ratio is 9 : 3 : 3 : 1 in the F_2 so the expected numbers of each phenotype are 556(9/16) = 312.75 round, yellow seeds; 556(3/16) = 104.25 round, green seeds; 556(3/16) = 104.25 wrinkled, yellow seeds; and 556(1/16) = 34.75 wrinkled, green seeds. The observed and the expected numbers can be used to compute the value of the chi-square statistic for this cross:

Observed, O	Expected, E	$(O - E)^2$	$(O - E)^2/E$
315	312.75	5.06	0.016
101	104.25	10.56	0.101
108	104.25	14.06	0.135
32	34.75	7.56	0.218
Total 556	556		$0.470 = \chi^2$

To use the chi-square value, we need one additional factor, called the **degrees of freedom** (df). The degrees of freedom are a measure of the number of independent variables present in a given experiment. This must be taken into account because chance errors affect only an independent variable. An experiment with a large number of independent variables will thus have more random error than an experiment with only one or two variables. In Mendel's cross, the variables are the four phenotypic classes. The degrees of freedom are equal to the number of phenotypic classes minus one: df = 4 – 1 = 3. One is subtracted because the total number of F_2 individuals is known. When the number in any three phenotypic classes has been determined, the number in the fourth class is fixed.

To determine the probability associated with this chi-square value, consult a chi-square table (Table 2.3), where the degrees of freedom are listed along one axis, the values of the chi-square statistic are listed in columns, and the probability associated with each chi-square value is listed at the top of each column. The probability associated with our chi-square value can be found by finding the proper number of degrees of freedom on the left (3), reading across to the chi-square value (0.47), and then reading up the column to the probability. For an experiment with three degrees of freedom, a chi-square value of 0.47 has a probability of greater than 90%. If we were to repeat this experiment many times, more than 90% of the experiments would have this much (or more) difference between the observed and the expected values because of chance error alone. A generally accepted standard for acceptance or rejection of an hypothesis using the chi-square test is 5%. If the probability associated with the chi-square statistic is less than 5%, we reject the hypothesis. If the probabil-

TABLE 2.3 *Probabilities Associated with Values of the Chi-Square Statistic*

	Probabilities				
df	.90	.50	.10	.05	.01
1	0.016	0.46	2.71	3.84	6.64
2	0.21	1.39	4.61	5.99	9.25
3	0.58	2.37	6.25	7.82	11.35
4	1.06	3.36	7.78	9.49	13.28
5	1.61	4.35	9.24	11.07	15.09
6	2.2	5.35	10.65	12.59	16.81
7	2.83	6.35	12.02	14.07	18.48
8	3.49	7.34	13.36	15.51	20.09
9	4.17	8.34	14.68	16.92	21.67
10	4.87	9.34	15.99	18.31	23.21
11	5.58	10.34	17.28	19.68	24.73
12	6.3	11.34	18.55	21.03	26.22
13	7.04	12.34	19.81	22.36	27.69
14	7.79	13.34	21.06	23.69	29.14
15	8.55	14.34	22.31	25.00	30.58
20	12.44	19.34	28.41	31.41	37.57
25	16.47	24.34	34.38	37.65	44.31
50	37.69	49.34	63.17	67.51	76.15

Adapted from Table IV of Fisher and Yates, 1963. *Statistical Tables for Biological, Agricultural and Medical Research,* 13th ed., Oliver and Boyd Ltd., Edinburgh.

ity is greater than 5%, we accept the hypothesis as correct. For Mendel's dihybrid cross, the probability was greater than 90%, far above the acceptance level of 5%, so we can conclude that Mendel was correct. With an acceptance threshold of 5%, we may reject a true hypothesis 5% of the time. In these cases, we will be in error to reject the hypothesis. This is one reason why multiple tests of an hypothesis are necessary.

The English mathematician Ronald A. Fisher performed a series of statistical tests on all Mendel's data. He found that Mendel's results have an extraordinarily good fit to the expected ratios, far above the 5% threshold value. He was struck by how good a fit Mendel obtained, and he published his analyses in a 1936 article entitled "Has Mendel Been Rediscovered?" Fisher suggested the overall difference between the observed and expected results for Mendel's experiments was so extraordinarily small that the data must have been adjusted to give a closer fit to the expected results. This suggestion was highly controversial, even though Fisher's findings did not necessarily mean that Mendel had been dishonest (see CONCEPT CLOSE-UP: *Are Mendel's Results Too Good To Be True?*). The real reason why Mendel's data show such a good fit will probably never be known.

Problem

In Mendel's trihybrid cross, plants from a strain with round, yellow seeds and gray seed coats (*RR YY GG*) were crossed with plants with wrinkled, green seeds and white seed coats (*rr yy gg*). The F_1 were selfed, and the F_2 were examined. The results Mendel obtained are shown here. Mendel concluded that these results have an acceptable fit to the expected 27 : 9 : 9 : 9 : 3 : 3 : 3 : 1 ratio. Use a chi-square test to determine whether he was correct.

F2 phenotypes	Number
round yellow gray	269
round yellow white	98
round green gray	86
wrinkled yellow gray	88
round green white	27
wrinkled yellow white	34
wrinkled green gray	30
wrinkled green white	7
Total	639

Solution

To determine the value of the chi-square statistic, we first calculate the expected number of progeny types in each phenotypic class. This is done by multiplying the total by the fraction expected in each class. We then use the chi-square formula to determine the value of the chi-square statistic.

	Observed, O	Expected, E	$(O - E)^2$	$(O - E)^2/E$
	269	269.58	0.002	0.002
	98	89.86	65.61	0.730
	86	89.86	15.21	0.169
	88	89.86	3.61	0.040
	27	29.95	8.41	0.281
	34	29.95	16.81	0.562
	30	29.95	0.01	0.0003
	7	9.98	9.00	0.900
Total	639	639		$2.684 = \chi^2$

There are eight classes in the offspring, giving seven degrees of freedom, df = 8 − 1 = 7. From the chi-square table we find a value of $p > 95\%$. This means that the probability that this much difference between the observed and the expected results could be the result of chance is 95%. This value is well above our threshold value (5%). We conclude that Mendel was most likely correct.

SECTION REVIEW PROBLEMS

6. If we examined the F_2 progeny of Mendel's dihybrid cross (see the preceding solved problem), what is the probability that the first seed we examined would have
   a. an *Rr Yy* genotype?
   b. an *RR Yy* genotype?
   c. an *rr Yy* genotype?
7. If we were to select five F_2 seeds from Mendel's dihybrid cross, what is the probability
   a. they would all be *Rr Yy*?
   b. they would all be *RR yy*?
   c. the first would be *Rr Yy*, the second *RR YY*, the third *rr YY*, the fourth *rr yy*, and the fifth *rr Yy*?
   d. the first three would all be *Rr Yy*, the fourth would be *RR YY*, and the fifth would be *rr YY*?
8. If we were to repeat Mendel's backcross to the recessive parent (*Rr Yy* × *rr yy*), what is the probability
   a. the first progeny seed would be either yellow and round or round and green?
   b. the first two seeds would be either *rr yy* or *rr Yy*?
   c. the first three seeds would either be all *Rr Yy* or all *rr yy*?
   d. the first five seeds would all be round and yellow or the first five seeds would have a wrinkled, yellow phenotype?
9. The genotypes of the F_2 individuals from a monohybrid cross follow.

   Homozygous dominant 420, Heterozygous 820, Homozygous recessive 390

   Test these data with the chi-square test to see if they fit a 1: 2: 1 ratio.
10. Corn plants with a genotype of (*Ww Cc*) were testcrossed to a plant with a genotype of *ww cc*. The numbers and genotypes of the progeny follow.

Progeny Genotype	Number
Ww Cc	125
ww Cc	118
Ww cc	129
ww cc	106
Total Progeny	478

Test these data with a chi-square test to determine whether the two genes show independent assortment.

CONCEPT CLOSE-UP

Are Mendel's Results Too Good To Be True?

All Mendel's observed results show remarkably close fits to the expected values. In 1936, the noted English statistician Ronald A. Fisher examined Mendel's data and published an article entitled "Has Mendel's Work Been Rediscovered?" In this article he did chi-square tests on all Mendel's reported results and calculated that the probability of getting as much or more deviation as Mendel reported for his entire set of experiments was 0.99993. This means that, considering the work as a whole, observed results with this close a fit to the expectations would occur fewer than once in 14,000 times. Fisher suggested this was too close a fit to be the result of chance and that Mendel's data must have been adjusted to fit the ratios he was expecting. There is, of course, no possible way of determining exactly what Mendel did, and there need not have been any deliberate attempt at deception by Mendel in his experiments. It is possible that he simply collected data until the numbers were in good agreement with the ratio he was expecting and then stopped the experiment. His data would thus be the result of an accurate reporting of what he found but would be biased toward his expectations. A second possibility is that Mendel, who was a popular figure in the monastery, had young assistants who knew the results he wanted and who attempted to please him by classifying doubtful specimens in the way that gave the best ratio. A third possibility is that the results reported in Mendel's paper represent only a portion of the experiments he performed. Mendel stated that he started with 34 varieties of peas and that he did experiments with 22 varieties, but in his paper he reported the results from only 7 monohybrid crosses, 1 dihybrid cross, and 1 trihybrid cross. Reporting the results of only the crosses that gave the best fit with his expectations would also cause the results to appear to be biased toward his expectations.

GENETIC VARIATION IS NECESSARY FOR THE STUDY OF GENETICS.

A large part of modern genetics is concerned with the generation and study of new alleles of genes. Heritable changes in genes that produce new alleles are called **mutations,** and the new alleles are called **mutant alleles** or **variant alleles.** In this section, we will introduce some basic principles of mutations and will discuss the terminology used to describe different types of mutations.

Drosophila Wild-Type Strains Are Used To Measure the Effects of Mutations on the Phenotype.

In some species, such as the fruit fly *Drosophila melanogaster,* thousands of different mutations have been collected and analyzed. Early in the study of mutations in *Drosophila,* it was decided that a reference standard was needed, and so a few strains were selected and designated "normal" strains. These strains were started by capturing individuals from wild populations, and so they were called **wild-type** strains. It was assumed that the phenotypes of these strains were normal for *Drosophila.* The alleles these strains contain are called **wild-type alleles.** The wild-type strains are used extensively in genetics research today to determine the effects of mutant alleles. A phenotype produced by a new mutation is defined in terms of how the individual's phenotype differs from the wild-type phenotype. For example, the wild-type eye color of *Drosophila* is bright red. Thus, the phenotypes of mutant eye color alleles are defined in terms of how these alleles change the normal red eye color. Flies homozygous for the *white* (*w*) mutation have white eyes, and flies homozygous for the *brown* (*bw*) mutation have dark brown eyes. By using the same wild-type strain, geneticists all around the world can define the effects of their mutations against the same standard.

As more and more mutations were discovered, geneticists recognized that it was necessary to adopt rules governing the symbols used to describe genes and different types of alleles (see CONCEPT CLOSE-UP: *Gene Symbols*). *Drosophila* geneticists adopted a system of rules designed around the concept of the wild-type. In *Drosophila,* the wild-type allele of any gene is always given the symbol +, either by itself (+) or as a superscript (e.g., the wild-type allele of the *white* gene is represented as w^+). The term "mutant allele" or "mutation" is used to describe any allele that gives a phenotype recognizably different from the wild-type phenotype for that gene.

In many species, there is no accepted wild-type, and the rules governing what is called a "mutant" allele are not as clear. Alleles that produce a crippling genetic disease or a phenotype that is strikingly abnormal or unhealthy are generally recognized as mutant alleles. On the other hand, if two alleles of one gene give only slightly different phenotypes and neither is clearly abnormal or unhealthy, these alleles are called **variant alleles.** Good examples of variant alleles are the pea alleles studied by Mendel that produced yellow or green seeds and human alleles that produce brown or blue eye color or pale or dark skin. In each case, neither phenotype is an abnormal phenotype.

Variant Alleles Are Used as Genetic Markers and as Tools To Study Gene Function.

Two important uses for mutant or variant alleles are as genetic markers and as tools for the study of gene function. When alleles are used as genetic markers, the goal of the experiment is to reveal the details of biological processes. For example, the goal of Mendel's crosses was not to determine how each allele produced its particular phenotype but to illustrate how inheritance works. This concept of using variant alleles as markers whose inheritance can be followed through several generations is very important, and we will return to it repeatedly in this book.

Variant alleles can also provide information about the function of a gene in an organism. The Mendelian concept of a gene is that each gene has a particular function, and that a mutation produces a different phenotype because it changes this function. By studying the details of a mutant phenotype, the normal function of that gene can sometimes be deduced. For example, consider the *white* mutation in *Drosophila.* This mutation gives a white-eyed mutant phenotype (*ww* = white), whereas the wild-type has bright red eyes, suggesting that the normal function of the *white* gene is required for the production of color in the eye. Likewise, in Mendel's seed color variants, one allele produced yellow seeds and the other produced green seeds. The normal function of this gene must be involved in the generation of color in the seed. The use of mutations to study the function of a gene is called **mutational analysis,** and it plays a very important role in modern genetics.

SECTION REVIEW PROBLEMS

11. For each of the following mutant phenotypes, suggest a possible normal function of the gene.

Mutant Phenotype	Wild-Type Phenotype
a. No wings (a recessive mutation)	Two wings
b. Loss of bristles (a recessive mutation)	Body covered with bristles
c. No storage protein in kernels (a recessive mutation)	Kernel filled with storage protein

12. For each of the mutations in Problem 11, suggest a possible name and a one-, two-, or three-letter gene designation.

MENDELIAN PRINCIPLES HAVE MANY APPLICATIONS.

Mendel's rules have had an enormous impact on all areas of the biological sciences, but especially on medicine and agriculture. Long before Mendel did his experiments, physicians as well as plant and animal breeders recognized that many traits and disorders were inherited in families. For example, in Jewish legal writings thousands of years old, references are found to the patterns of inheritance of hemophilia, which causes improper blood clotting in humans. We now know that we can apply Mendel's rules to observed patterns of inheritance in families and show that many disorders are caused by variant alleles of a particular gene. In this section, we will consider some of the ways in which Mendel's rules can be used to determine the genetic basis of traits in human populations.

Inherited Human Diseases Are Analyzed Using Pedigree Analysis.

As humans, some of our most important genetic questions are about the genes that control human traits. However, Mendelian genetic analysis of humans presents special problems because the most powerful genetic investigative technique, controlled matings, cannot be used in human genetic analysis. Even if we could make controlled matings between humans, it would take too long for the offspring to mature, and the number of offspring per cross is very limited. However, Mendel's rules can be used to determine the genetic basis for inherited human traits through **pedigree analysis.** A pedigree symbolically represents the history of matings and the offspring they produce in a family, with reference to specific genetic traits. Analysis of these pedigrees can help reveal the genetic basis of a particular trait.

In a pedigree analysis, the information about family relationships and the phenotypes of family members is summarized in the pedigree in a graphical format. Each individual is represented by a symbol (Figure 2.11); a male is a square and a female is a circle. Individuals that have the trait of interest are represented by filled squares or circles. The symbols for individuals are arranged in rows by generation, with each generation indicated by a roman numeral (I, II, III, etc.). Each individual within a generation is numbered, left to right. In this manner, it is easy to refer to an individual as "individual II-3," which means the third person from the left in the second row of the pedigree. For example, in the partial pedigree in Figure 2.11, II-3 is a male who expresses the trait being studied. When two individuals mate, they are connected by a horizontal line. Their offspring are arranged together under the connecting parental line. Siblings are usually arranged in birth order with the oldest on the left. With this information, the Mendelian rules, and the rules of probability, it is possible to generate an hypothesis about the alleles that are causing the trait. Conclusions from pedigree data usually include whether the allele is dominant or recessive as well as predictions as to the genotype of the individuals in the pedigree. This hypothesis can then be used to make predictions about the genotypes and phenotypes of future offspring.

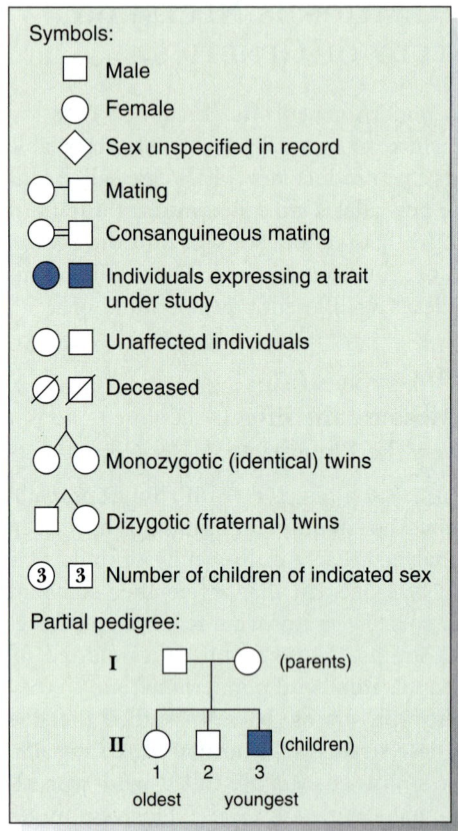

FIGURE 2.11 Commonly used symbols in human pedigrees. Each symbol contains phenotypic information about the individuals to enable conclusions about their genotype.

Traits produced by recessive or dominant alleles show a defined pattern of inheritance in pedigrees. For example, individuals heterozygous for recessive alleles do not show the phenotype. As a consequence, recessive alleles may be present in a family for many generations before a homozygous individual appears. Homozygous individuals result from **inbreeding,** or the mating of two related individuals, also known as a **consanguineous** marriage. In Figure 2.12a, individual IV-3 is homozygous for the recessive allele causing phenylketonuria. Individual IV-3 is the offspring of two related individuals (III-2 and III-3) who do not show the trait. Therefore, these individuals must be heterozygous for the same recessive allele. Both III-2 and III-3 probably inherited this allele from one of their common grandparents (I-1 or I-2). In

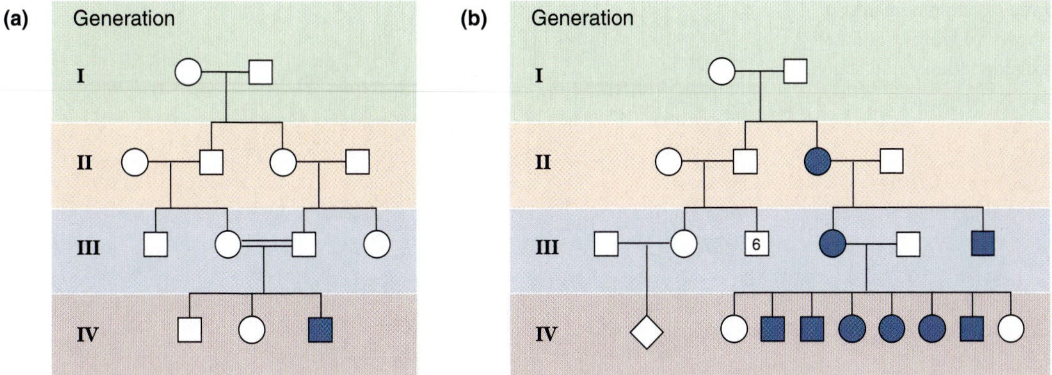

FIGURE 2.12 Pedigrees of recessive and dominant traits. (a) This pedigree shows the pattern of inheritance of the recessive disorder PKU. (b) This pedigree shows the inheritance of a dominant form of eye cataracts. Individuals in each pedigree are arranged in rows by generation (I, II, III, etc.) and within rows by birth order (1, 2, 3, etc.) with the eldest on the left. *(Adapted from C. Stern. 1973.* Principles of Human Genetics, *3rd ed., W.H. Freeman, San Francisco.)*

making this conclusion, we have assumed that individuals marrying into the family (II-1 and II-4) are not heterozygous for the allele causing phenylketonuria. This assumption is usually valid for rare traits caused by alleles present only in a few families in a population. However, if the phenylketonuria allele were common in the population, then this assumption could not be made.

Traits produced by a dominant allele usually appear in all individuals that possess one or more copies of the allele. For example, the pedigree in Figure 2.12b shows the pattern of inheritance of a form of human eye cataracts produced by a dominant allele. The trait appears in one female (II-3), in both of her children (III-9 and III-11), and in six out of eight of her grandchildren (row IV). Because the trait appears in every generation and every individual with the trait is the offspring of an affected individual, the allele is dominant. This allele probably arose as a mutation in one of the original parents (row I) and was inherited by II-3.

After an hypothesis has been made about the nature of the allele producing a trait, the genotypes of the individuals in the pedigree can be predicted. This is especially easy for a dominant allele because everyone who is homozygous or heterozygous for the allele has the trait. For recessive alleles, it is not always possible to predict exactly the genotype of an individual. If they do not manifest the trait, they can be either heterozygous for the allele or homozygous for the normal allele. On the other hand, it is usually possible to calculate the probability that an individual is heterozygous. To do this we use Mendel's first rule, which predicts that the offspring of a known heterozygote has a 1/2 chance of inheriting the mutant allele. The probability that this individual would pass the allele to his or her offspring is given by the multiplication rule: (probability the individual has the allele) × (probability of passing the allele) = (1/2)(1/2) = 1/4. The key to pedigree analysis is the information in the pedigree, Mendel's rules, and the rules of probability.

To illustrate how pedigree analysis works, let us analyze the pedigree showing the inheritance of a rare form of human panhypopituitary dwarfism (shown in Figure 2.13). Initially, we form an hypothesis as to whether the mutant allele is dominant or recessive. Examining the pedigree, we see that the offspring of an individual with the dwarfism trait (I-2) do not show the trait, and the trait does appear in the offspring of two related individuals (V-3 and V-4) who have a normal phenotype. These characteristics suggest that the allele is recessive. Thus, our hypothesis would state that individuals who have the dwarfism trait are homozygous for the recessive dwarfism allele. Using this hypothesis, we can now calculate the probability that any individual in this pedigree is heterozygous for the dwarfism allele. For simplicity, we will assume that all individuals marrying into the family are homozygous for a normal (nondwarfism) allele. If this assumption is correct, then both individuals V-3 and V-4 must be heterozygous because they produced children (VI-6, VI-7, VI-8) who were homozygous. Because V-3 and V-4 have inherited the dwarfism allele from their parents, we can trace the inheritance of this allele back up the pedigree. Individual V-4 inherited the allele from IV-6, who inherited it from III-6, who inherited it from II-6, who inherited it from I-2. Therefore, individuals II-6, III-6, IV-6, V-4, IV-5, III-4, and II-3 are all heterozygous for the dwarfism allele.

We can also trace the inheritance down the pedigree from known homozygotes or heterozygotes to their offspring. In each generation, Mendel's first rule states that the probability is 1/2 that a heterozygous individual will pass a mutant allele to one of their offspring. Thus, the

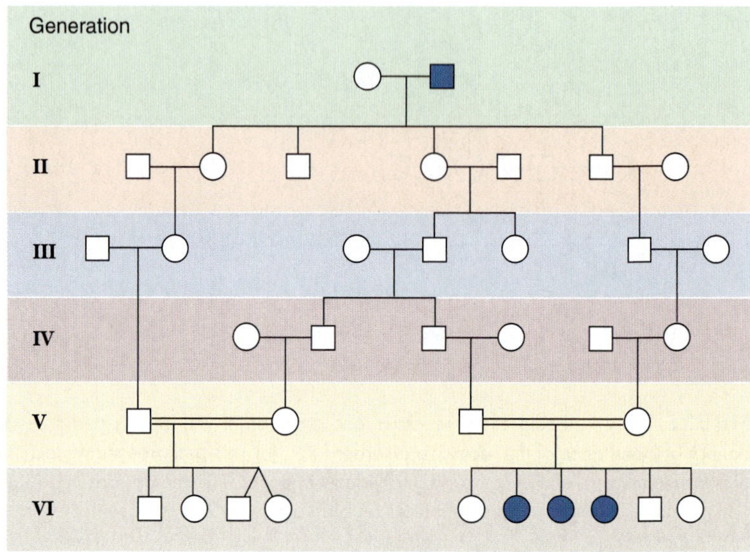

FIGURE 2.13 Pedigree of panhypopituitary dwarfism. This pedigree shows the pattern of inheritance of dwarfism in an extended family and suggests that it is caused by a recessive allele.

probability that an individual will inherit the allele is $1/2^n$, where n is the number of generations separating that individual from a known heterozygote. For example, individual II-2 must be heterozygous because her father is homozygous. The probability her daughter (III-2) is heterozygous is 1/2, and the probability that her grandson (V-1) is heterozygous is (1/2)(1/2) = 1/4.

Determining the probability that an individual is heterozygous is complicated by inbreeding because offspring of related parents have more than one parent from whom they can inherit a particular allele. To estimate the total probability that an individual is heterozygous for an allele, we must use the addition rule to sum all the possible ways the offspring may have inherited the allele. For example, in our sample pedigree (Figure 2.13), we might ask, What is the probability that individual VI-1 is heterozygous? Individual VI-1 could inherit the dwarfism allele from his father (V-1), or he could inherit the allele from his mother (V-2), who may have inherited the allele from IV-2, who may have inherited it from III-4, a known heterozygote. The probability that VI-1 is a heterozygote is the sum of all possible ways he could inherit the dwarfism allele from one parent and the normal allele from the other parent. There are three possible ways this might occur. First, V-1 might be a heterozygote (p = 1/4), V-2 might be homozygous normal (p = 3/4), and VI-1 might inherit the dwarfism allele from his father (p = 1/2). The probability of this first possibility is $p(1)$ = (1/4)(3/4)(1/2) = 3/32. Second, V-1 might be homozygous normal (p = 3/4), V-2 might be heterozygous (p = 1/4), and VI-1 might inherit the dwarfism allele from his mother (p = 1/2). The probability of this second possibility is $p(2)$ = (3/4)(1/4)(1/2) = 3/32.

The third possibility is that V-1 might be heterozygous (p = 1/4), V-2 might also be heterozygous (p = 1/4), and VI-1 might inherit a dwarfism allele from only one of them (p = 2/3). You might think that the probability the offspring of two heterozygotes is heterozygous should be 1/2. Mendel's first rule indicates that the offspring of two heterozygotes have a 1/4 chance of being homozygous normal, a 2/4 chance of being heterozygotes, and a 1/4 chance of being homozygous for the mutant allele. However, in this case, you can eliminate the possibility that VI-1 is homozygous for the dwarfism allele because he does not show the trait. If both his parents are heterozygotes, he must be one of the three remaining combinations. He thus has a 2/3 chance of being heterozygous. The probability of this third possibility is $p(3)$= (1/4)(1/4)(2/3) = 1/24. The total probability that VI-1 is heterozygous is the sum of these three probabilities: $p(\text{h}) = p(1) + p(2) + p(3)$ = (3/32) + (3/32) + (1/24) = 11/48 = 0.2292. This example illustrates how inbreeding can increase the complexity of probability calculations in pedigree analysis. It also illustrates how even complex pedigree calculations can be solved using Mendel's rules and the rules of probability.

Another question that can often be answered using pedigree analyses is the probability that future offspring may be heterozygous (carriers) or may be homozygous and have the trait. In fact, pedigree analysis is often done because two individuals discover that a particular trait is inherited in their families and they wish to know the probability they will have a child with the trait. For example, suppose individuals VI-1 and VI-5 from our example pedigree in Figure 2.13 marry and wish to know the probability they will have a child with the dwarfism trait. To have a child with the trait, they must both be heterozygous and must both pass the allele to their child. The probability that their first child will be homozygous is calculated using the multiplication rule: p(homozygous) = (probability VI-1 is heterozygous) × (probability VI-5 is heterozygous) × (probability they both will pass the dwarfism allele to the offspring). The probability that VI-1 is heterozygous is 11/48, and because VI-5 does not have the trait and is the daughter of two known heterozygotes (V-3 and V-4), the probability that she is heterozygous is 2/3. Mendel's rules predict that the probability of two heterozygotes having a homozygous child is 1/4. Therefore, the probability that VI-1 and VI-5 will have a dwarf child is p(homozygous) = (11/48)(2/3)(1/4) = 11/288 = 0.0382. If this couple's first child did have the dwarfism trait, then the probability that their next child will also have the trait is 1/4, not 0.0382, because we would then know that they are both heterozygotes.

SECTION REVIEW PROBLEMS

13. The following pedigree shows the inheritance of a rare trait in a human family. From the information in this pedigree, what can you say about the inheritance of this trait? If individual IV-4 married IV-7, what is the probability their first child will show the trait?

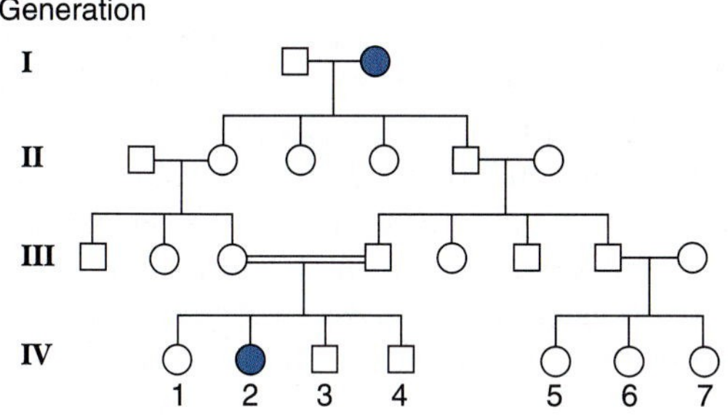

14. In a large family, one individual has a trait given by a recessive allele. What is the probability each of the other individuals in this family (listed below) has inherited this allele? For each individual, their relationship with the affected individual is given.
    a. great grandfather
    b. uncle
    c. sister
    d. brother-in-law
    e. first cousin

Single Genes Are Responsible for a Variety of Human Disorders.

In human medicine, a collection of deleterious phenotypes that usually appear together is called a **syndrome.** For example, an individual may be mentally retarded, have stunted growth, and numerous other little changes in phenotype. Mutant alleles that cause syndromes are known as **pleiotropic** alleles; that is, a single mutation can cause multiple phenotypes. An important application of Mendel's rules is to discover the genetic basis of inherited human syndromes. Before Mendel's rules, the wide range of phenotypes in most syndromes suggested that they must have complex causes. There appeared to be little hope of ever understanding these syndromes or predicting the probability of their appearance in future generations. The application of Mendel's rules has identified the mutant alleles responsible for hundreds of syndromes.

The discovery that a human syndrome is caused by a single mutant allele is an important step toward understanding the nature of the disorder and its eventual cure. After the mutant allele has been identified, biochemical and molecular genetic investigations can be done to determine how the mutation has altered the normal structure of the gene. This often leads to an understanding of how this alteration has changed the function of the gene product, and this information sometimes leads to treatments that reduce or eliminate the effects of the mutant allele.

Hundreds of human syndromes are caused by recessive mutations. These mutations alter the ability of genes to carry out important metabolic processes and lead to the accumulation of metabolic products. This accumulation can cause the individuals with these syndromes to develop a wide variety of physical ailments. In 1902, it was suggested that the human disorder **alkaptonuria** was produced by a recessive mutant allele. This suggestion was based on information from pedigrees showing that alkaptonuria is inherited as a recessive trait. Alkaptonuria was one of the first human disorders whose cause was determined by the application of Mendel's rules. Alkaptonuria was originally called an "inborn error of metabolism." Homozygous mutant individuals accumulate homogentisic acid, which is a byproduct of the metabolism of the amino acid tyrosine (see Chapter 11 for a description of the amino acids). Homogentisic acid turns black when exposed to oxygen and is relatively harmless unless it accumulates at high levels in the body. If it does, it can lead to severe arthritis and damage to joint tissues. Before the development of modern blood tests, individuals with alkaptonuria were diagnosed by the black color of their urine, cartilage, joints, and tendons.

Another recessive syndrome is **phenylketonuria,** or PKU, which was discovered in 1934. In this syndrome, large amounts of the amino acid phenylalanine accumulate in the body. Humans cannot synthesize phenylalanine, but a small amount of phenylalanine is vital for constructing proteins. However, most diets have an excess of phenylalanine, and it is normally metabolized into tyrosine by the enzyme phenylalanine hydroxylase. In individuals homozygous for the PKU allele, phenylalanine hydroxylase is faulty, producing high levels of phenylalanine in body tissues. This, in turn, produces defects in the development of the central nervous system, mental retardation, and epileptic seizures. Today, most newborn children are given blood tests to determine the level of phenylalanine in their blood. An increased level, which is easily detected, is a reliable sign of PKU. A reasonably effective treatment for PKU consists of a carefully controlled diet with low levels of phenylalanine. For individuals with PKU, the critical period of central nervous system development is during the first 6 years. After this time, dietary restrictions are relaxed somewhat, and the individual may lead a relatively normal life. However, nervous system damage sustained during the early years cannot be repaired, so early diagnosis and treatment of PKU are essential.

The **Lesch-Nyhan** syndrome, which was formally described in 1964, is also produced by a recessive allele. Children with Lesch-Nyhan syndrome appear normal at birth but soon begin to overproduce uric acid, which is followed by frequent vomiting and compulsive-aggressive behavior. This often turns into self-mutilation, and they must be restrained to prevent harm to themselves. Late in the disease, the excess production of uric acid also leads to symptoms of gout and mental retardation. A wide variation in the severity of this disease exists, and many individuals are intelligent enough to be fully aware of their condition but are unable to control their self-destructive behavior. Because individuals with Lesch-Nyhan syndrome seldom live beyond childhood, they rarely reproduce.

The syndromes just described are produced by mutant alleles that prevent normal metabolism of dietary chemicals, leading to the destructive accumulation of these products. Other syndromes are produced by mutant alleles that block the production of important metabolic products. For example, individuals with **albinism** suffer from a lack of production of a metabolic product, the pigment melanin, which is normally derived from the amino acid tyrosine by a series of biochemical reactions. Mutant alleles of several different genes can block the production of melanin, so all albinos do not have the same genotype.

Some human syndromes are caused by dominant alleles. One is **Huntington disease** (previously called Huntington's chorea), a syndrome whose principal phenotype is a progressive degeneration of the nervous system. This leads to a loss of memory, higher-order thought, and eventually the ability to control bodily functions. Most individuals with Huntington disease are heterozygotes who eventually lose all mental capacity and may require complete and constant care. One characteristic of Huntington disease is that the symptoms usually do not appear until the individual is entering middle age (40 to 50), so children of an individual with Huntington disease must often make vital decisions about marriage and having a family, knowing that they have a 50% chance of having the mutant allele. The gene whose mutation causes Huntington's disease has been identified and cloned. Several disease-causing alleles have been isolated, and their molecular structure has been analyzed. These disease-causing alleles all have similar molecular defects (repeated copies of a particular portion of the gene). Diagnostic genetic tests that can detect the presence of a mutant allele of the gene for Huntington disease are currently available. However, there is considerable debate over how such tests should be made available to the public. Because all tests have a certain rate of error, serious emotional damage could be caused by a false-positive test. In addition, there is the potential for the misuse of the results of these and other genetic tests by employers and insurance companies. The questions of who should do genetic screening, what types of screens should be done, and who should have access to the results are serious questions that have not yet been answered.

Not all inherited human diseases produce crippling abnormalities. For example, **Marfan syndrome** is produced by a dominant allele that alters skeletal structure and connective tissues. About 30,000 Americans today have Marfan syndrome. Individuals with Marfan syndrome tend to have long, thin limbs; rubbery joints; narrow chests; displacement of the eye lens; abnormal heart valves; and often develop an aneurysm of the aorta. These are not immediately lethal symptoms, and many individuals with Marfan syndrome live into their fifties and sixties, often dying of heart failure or rupture of the aorta. The syndrome is thought to be related to a reduction in production of fibrillin, a protein that supports the elastic fibers in normal tissues. It has been suggested that Abraham Lincoln, the 16th president of the United States, had Marfan syndrome. This cannot be confirmed by pedigree analysis because Lincoln has no known living descendants. The suggestion was originally made because of Lincoln's physique (he was very tall, gangly, and loose limbed and described as walking with a shuffling gait), which is typical of individuals with Marfan syndrome. However, Lincoln was a noted athlete in his youth and had extraordinary physical strength even in his fifties, traits rarely seen in individuals with Marfan syndrome.

The question about Lincoln and Marfan syndrome was of only academic interest until the rapid advance of molecular genetic techniques. These procedures now allow the allele responsible for Marfan syndrome to be detected in samples of blood or other tissues. Recently the National Museum of Health and Medicine in Washington, DC, gave approval for researchers to use samples of bone, blood, and hair taken from Lincoln during his autopsy. The results suggest that Lincoln did not have Marfan syndrome. That such a test could even be considered on tissue samples over 100 years old—samples that had not been treated or specially preserved—as a tribute to the increasing sophistication of molecular genetic techniques. The techniques used in such tests will be discussed in Chapter 12.

Mendel's Rules Are Widely Used in Agriculture.

For thousands of years, farmers have sought to enhance desirable traits and eliminate undesirable traits in their herds and crop strains by selecting individuals with the most desirable phenotypes to serve as parents for the next generation. After Mendel's rules were understood, scientific principles could be applied to the breeding and selection of improved animals and plants. Many alleles responsible for desirable traits have been identified and are selected for producing favorable genotypes.

The application of Mendelian principles to plant and animal agricultural species has contributed to the enormous success of modern agriculture. Current plant seed catalogs list numerous varieties with specific traits, such as resistance to specific pathogens, increased yield, well-defined time to reach maturity, and resistance to frost or drought. Animal breeding has produced more efficient, healthier, faster-growing animals with greater disease resistance. Modern cattle, sheep, chickens, turkeys, and pigs are all the result of extensive controlled breeding programs. More recent technology, including artificial insemination, cloning and freezing embryos, and use of frozen semen, provide modern agricultural breeders with powerful tools for agricultural improvement by expanding the range of breeding possibilities. The key ingredient is still the discovery of particular alleles that give favorable phenotypes, and the discovery of these alleles depends on an understanding and application of Mendel's rules.

Summary

Mendel's discoveries were one of the most important advances in the history of biological science. His rules stimulated many biologists to undertake investigations of the mechanism of inheritance in a variety of different organisms. This has led to the development of the modern science of genetics. Mendel's work was based on the concept of particulate inheritance, with the fundamental unit of inheritance being the gene. Mendel's concept was that a gene is a heritable factor that controls a particular trait, and different forms of a gene (called alleles) may produce differences in this trait. The sum total of an individual's alleles is his or her genotype, and the physical appearance produced by the genotype is his or her phenotype. In his famous experiments, Mendel crossed pea plants from two strains that differed for one trait (monohybrid cross), two traits (dihybrid cross), or three traits (trihybrid cross). He examined the phenotypes of the progeny of these crosses over several generations and discovered that they had constant numerical ratios of phenotypes. To explain these ratios Mendel proposed two rules of inheritance that govern the transmission of genes from parent to offspring. The Segregation Rule states that individuals have two copies of each gene (a gene pair) and that the two copies (alleles) segregate during gamete formation. The Rule of Independent Assortment states that the segregation of two alleles is independent of the segregation of any other pair of alleles.

Mendel's rules are widely used in conjunction with the addition and multiplication rules of probability and with statistical tests to deduce the nature of the alleles controlling particular traits. One important statistical test is the chi-square test, which is used to determine whether the difference between an observed set of results and the expected set is reasonably explained by chance alone.

The Mendelian concept of the gene and Mendel's two rules have been applied to a large number of species of plants and animals, and the results have been of enormous importance in many practical areas of medicine and agriculture. In humans, in whom controlled matings are not possible, Mendel's rules are applied to the study of pedigrees. Pedigree analyses show how particular traits are inherited in families, enabling scientists to determine the genetic basis for many human disorders. In agriculture, Mendelian genetics provides the theoretical framework for understanding the genetic basis for numerous traits in agriculturally important species.

Selected Key Terms

allele
addition rule
backcross
chi-square test
consanguinity
degrees of freedom
dihybrid cross
dominant
heterozygote
homozygote
monohybrid cross
multiplication rule
mutant
pedigree
phenotype
recessive
testcross
variant
wild-type

Chapter Review Problems

1. The inheritance of a very rare trait in humans is shown in the pedigree that follows.

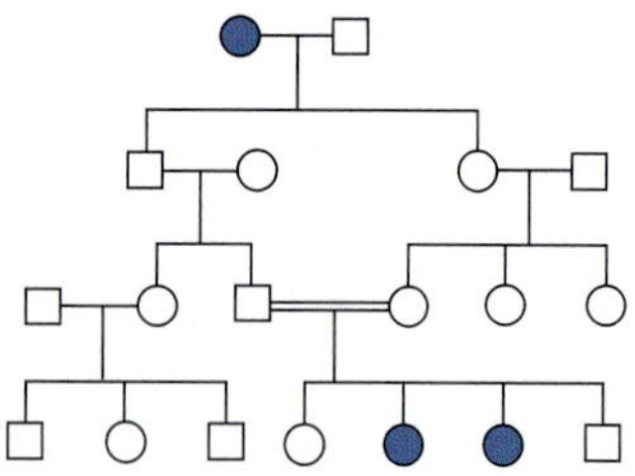

Generation

I

II

III

IV

From the information in this pedigree:

   a. Is this trait caused by a dominant or a recessive allele?
   b. If individual IV-3 married IV-4, what would be the probability that their first child would show the trait?

2. Two individuals who experience a very bitter taste when sampling the chemical phenylthiourea discover that two of their six children are unable to taste this chemical. Assuming that the ability to taste this chemical is controlled by one gene pair, how would you explain the inheritance of this trait?

3. The coat color of cattle can be red and white or black and white. A farmer, wishing to improve his herd, made a series of matings using artificial insemination. Bull 1 was mated with 45 black-and-white cows, and 40 black-and-white F_1 offspring were produced. One of the F_1 bulls was mated with 20 red-and-white cows, and 10 red-and-white and 10 black-and-white offspring were produced.
   a. Which colors are given by a dominant allele and which by a recessive allele?
   b. What was the genotype of Bull 1?
   c. What was the color of Bull 1?

4. Twenty female rabbits from a true-breeding strain with the Himalayan coat color were mated to one gray rabbit of unknown origin. The progeny contained 61 Himalayan and 56 gray rabbits.
   a. Is Himalayan a dominant or a recessive phenotype?
   b. What is the genotype of the unknown rabbit?
   c. Test the fit of your hypothesis using the chi-square test.

5. A male *Drosophila* homozygous for a mutant allele of the *rosy* gene had dark brown eyes. This male was mated to 10 females with normal red eye color. Several F_1 female flies were collected and backcrossed to the original parental male. The offspring were 630 with red eyes and 581 with dark brown eyes.
   a. Is the mutant allele of the *rosy* dominant or recessive?
   b. Design symbols for these alleles and give the genotypes of the parents and the F_1 from this cross.
   c. Test the fit of your hypothesis using the chi-square test.

6. Humans with the inherited Huntington disease usually begin to show symptoms in middle age. The disease is caused by a rare dominant allele. A woman has just learned that her mother has developed symptoms of Huntington disease.
   a. What is the probability that she will eventually have this disease?
   b. If she marries, what is the probability that she will pass the disease to her offspring?
   c. If she has three offspring, what is the probability that all three will have the disease?

7. Molecular tests have recently been developed to detect the presence of alleles responsible for Huntington disease, although the tests are not 100% accurate. If the woman in Problem 6 were tested and the test was negative (the allele is not present) what are the chances of her son developing the disease if the test is
   a. 75% accurate?
   b. 99% accurate?

8. Two parents with the following genotypes were crossed: *Aa YY* and *aa Yy*.
   a. What is the probability of the first offspring having an *aa Yy* genotype?
   b. What is the probability that the first two offspring will have an *Aa Yy* genotype?

9. A fly from a true-breeding strain of *Drosophila* with short wings and black body was crossed with a fly with long wings and a gray body. The progeny all had long wings and black bodies.
   a. If these F_1 are inbred, what will be the probability of obtaining an F_2 fly with short wings and a gray body?
   b. What is the probability that the first three F_2 flies will all have long wings and black bodies?

* *To answer Problems 10–13, you must read the concept box on the binomial theorem.*

10. If 10 F_2 flies were collected from the cross in Problem 9, what is the probability that only five of them will have long wings and black bodies?

11. Several black-bodied *Drosophila* were mated to an unknown, gray male. The F_1 (all gray) were inbred and produced an F_2 containing 245 black flies and 710 gray flies.
    a. Is black body a dominant or a recessive trait?
    b. What is the probability that the next F_2 fly will have a black body?
    c. What is the probability of obtaining exactly 5 black flies and 15 gray flies in the next 20 F_2 flies?

12. Some of the F_1 female flies from Problem 11 were crossed to black-bodied males from a true-breeding strain.
    a. What is the probability that the first offspring will be black?
    b. What is the probability that the first three offspring will all be black?
    c. What is the probability that the first offspring will be black, the second gray, the third gray, and the fourth black?
    d. What is the probability that in a group of 20 offspring, 10 will be black?

13. A tomato plant with compound leaves was crossed with a plant with simple leaves. The F_1 were inbred and produced 90 F_2 with compound leaves and 310 F_2 with simple leaves. If these F_1 had been crossed with the compound parent, what is the probability that 9 of the first 20 offspring would have had compound leaves?

14. A series of crosses were made between bean plants that either have smooth-edged leaves or rough-edged leaves. Using your own symbols for the alleles involved, indicate the genotypes of the parents in each of the following crosses.

Parents	**Progeny**	
	smooth	rough
a. smooth × smooth	100	45
b. smooth × rough	55	60
c. smooth × smooth	75	0
d. smooth × rough	40	0
e. rough × rough	0	85

15. In the leopard frog (*Rana pipiens*), the color pattern on the back can be either mottled or clear. A mottled male was mated with a clear female, and the F_1 were inbred. There were 128 mottled frogs and 49 clear frogs in the F_2.
    a. Design symbols for the alleles involved and indicate the genotypes of the two parents and of the F_1 in this cross.
    b. Do a chi-square test to determine whether the data fit your expectations.
    c. What phenotype would you expect the F_1 to have?

16. An F_1 male frog from Problem 15 was crossed to a clear female.
    a. What is the probability that the first three offspring will be clear?
    b. What is the probability that the first offspring will be clear, the second mottled, the third clear, and the fourth mottled?

17. A male *Drosophila* with a dark eye color and a black body was mated with several females that had normal red eyes and normal gray bodies. The F_1 all had red eyes and gray bodies. The F_1 were inbred, and the F_2 contained 33 flies with dark eyes and black bodies, 95 with dark eyes and gray bodies, 106 with red eyes and black bodies, and 288 with red eyes and gray bodies.
    a. How are these two traits inherited?
    b. Using symbols of your own design, indicate the genotypes of each type of individual in this cross.
    c. Do your predicted numbers of F_2 agree with the observed F_2?
    d. If you could do one mating to test your hypothesis further, what would it be, and what results would you expect?

18. A tomato with the genotype *Rr Cc Hh* is selfed, and a large number of progeny are raised. What proportion of the progeny would you expect to be of the *RR cc Hh* genotype?

19. The following pedigree shows the inheritance of a rare trait in humans, malabsorption of vitamin B_{12}. The filled symbols represent individuals showing the trait. Is this trait produced by a dominant or a recessive allele? Explain the evidence that lead you to your conclusions.

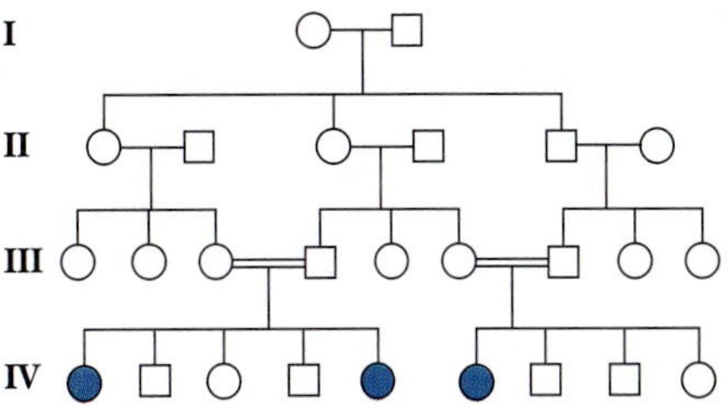

20. A number of crosses were made between *Drosophila* differing in body color (black or gray) and wing shape (straight or curled). For each cross listed here, what is the most likely genotype of the parents?

Parents		**Progeny**			
Male	Female	black, straight	black, curled	gray, straight	gray, curled
black, straight	gray, curled	100	0	0	0
black, straight	black, curled	75	0	25	0
gray, curled	gray, curled	0	0	0	100
black, curled	gray, straight	50	0	50	0
black, straight	gray, straight	75	25	75	25

21. Two guinea pigs of *Aa Bb Cc Dd* genotypes were crossed. Assuming that these are all simple Mendelian genes,
    a. what is the probability the first offspring will be *DD?*
    b. what is the probability the first offspring will be *AA Bb Cc dd?*
    c. what is the probability the first offspring will be *Aa Cc Dd?*
    d. what is the probability the first offspring will be *AA Cc Dd* and the second *Aa CC dd?*

22. The following pedigree shows the inheritance of a very rare human trait in one family.

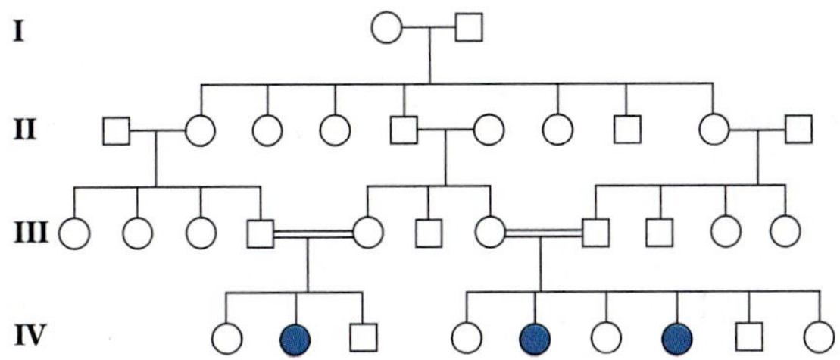

    a. Is this trait recessive or dominant?
    b. What are the genotypes of individuals II-2, II-5, III-4, and III-7?

c. If individual IV-3 marries IV-4, what is the probability that their first child will show the trait?

23. The following pedigree shows the inheritance of the rare human trait hypercystinuria in one family.

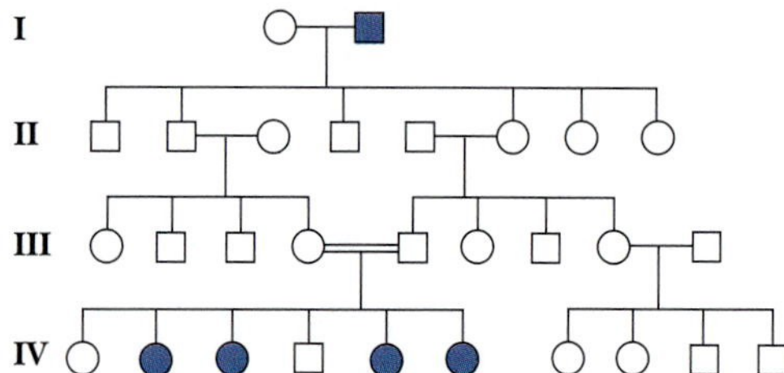

a. What is the nature of inheritance of this trait?
b. What are the genotypes of individuals II-2, II-6, III-4, IV-3, and III-2?
c. If individual IV-4 marries IV-8, what is the probability their first child will have the trait?

24. The following pedigree shows the inheritance of a common form of myopia in a human family.

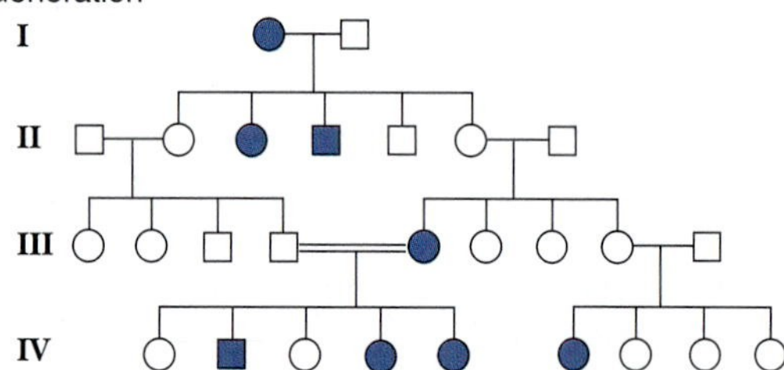

a. What is the nature of the allele causing this trait?
b. What is the genotype of individual II-7?
c. Would your answer to the first question (a) change if this were a very rare trait?

25. In *Drosophila,* a recessive allele of the *black* gene (b) gives a dark body color, and the wild-type gray color is given by the dominant wild-type allele, b^+. A recessive allele of the *sepia* gene (s) gives a sepia (dark brown) eye color, and the wild-type allele of this gene (s^+) gives the normal red eye color. In a cross between a homozygous *black, sepia-eyed* female and a homozygous wild-type male, what would be the genotype(s) and phenotype(s) of the offspring in the F_1 generation and the F_2 generation, and the proportion of each type?

26. In humans, assume that brown eye color is controlled by a dominant allele, *B*, and blue eye color by its recessive allele, *b*. If a heterozygous (*Bb*) man marries a heterozygous (*Bb*) woman and they have five children,
a. what is the probability they will have five boys?
b. what is the probability they will have five blue-eyed children?
c. what is the probability all five children will be blue-eyed daughters?

27. What was Mendel's most important contribution to the modern understanding of biology?
a. The concept of meiosis
b. The concept of the chromosome
c. The concept that hereditary information comes in discrete units
d. The recognition that populations of organisms are more important than individuals in biology
e. The concept that genes are ordered along chromosomes

28. Assume that you have two recessive traits in dogs: *Gg*, in which *G* fetches and *g* does not, and *Dd*, in which *D* sits on command and *d* does not. You cross a male and female that both fetch and sit and have the genotypes: *Gg Dd* and *Gg Dd*.
a. What fraction of the offspring will neither fetch nor sit?
b. What fraction of the offspring will be *GG dd?*
c. What fraction of the offspring will fetch but not sit on command?

29. In pumpkins, *yellow* (*Y*) is dominant over *white* (*y*); *big* (*B*) is dominant over *small* (*b*). What are the phenotypes and their ratios of the offspring of the mating: *YY bb* × *yy BB*?

30. In pea plants, genes *A, B, C, D,* and *E* segregate independently. A plant with genotype *Aa Bb Cc DD ee* is crossed with a plant with genotype *AA bb Cc dd EE.* The hybrid F_1 plant is grown. What fraction of the F_1 plants will have the genotype *Aa bb CC Dd Ee?*

31. The dominant autosomal trait polycystic kidney disease causes death due to kidney failure at about age 50. A man marries a woman whose father has the disease. Only one of her grandparents had the disease.
a. What is the probability that the woman has the gene that causes polycystic kidney disease (she is age 25 and it cannot be diagnosed until age 40)?
b. What is the probability that their first child will develop the disease?

Challenge Problems

1. In *Drosophila,* a recessive allele of the *ebony* gene (e) gives a dark body color, and the wild-type gray color is given by the dominant wild-type allele, e^+. A dominant allele of the *Plum* gene (*Pm*) gives a brown eye color, and the wild-type allele of this gene (Pm^+) gives the normal red eye color. A dominant allele of the *cubitus interuptus* (ci^D) gene gives abnormal wing veins. The wild-type allele of this gene (ci^+) gives normal wing veins.
a. In a cross between a homozygous dark-bodied, dark-eyed, abnormal-winged female and a homozygous wild-type male, what would be the genotype and phenotype of the F_1 generation?
b. What would be the phenotypic ratio of the F_2 generation?

2. In a trihybrid cross, plants from a strain of beans with oblong, red seeds and white seed coats (*OO RR WW*) were

crossed with plants with round, gray seeds and clear seed coats (*oo rr ww*). The F_1 were selfed, and the F_2 were examined. The results follow. Use a chi-square test to determine whether these results fit the expected Mendelian ratio.

F_2 Phenotypes	Number
oblong red white	1130
oblong red clear	378
oblong gray white	349
round red white	351
round gray white	102
round red clear	111
oblong gray clear	109
round gray clear	30
Total	2560

3. If the F_1 plant from Problem 2 is crossed with an F_2 plant with an oblong, gray, white phenotype, what is the probability that the first offspring will have an oblong, gray, white phenotype?

Suggested Readings

Bateson, W. 1909. *Mendel's Principles of Heredity.* Cambridge University Press, Cambridge, England.

Hernon, P. 1994. *Statistics: A Component of the Research Process.* Ablex Publishing Corporation, Norwood, NJ.

Iltis, H. 1932. *Life of Mendel* (E. Paul and C. Paul, trans.) W. W. Norton, New York.

King, R. C., and W. D. Standsfield. 1990. *Dictionary of Genetics,* 5th ed. Oxford University Press, New York.

McKusick, V. A. 1990. *Mendelian Inheritance in Man: Catalogs of Autosomal Dominant, Autosomal Recessive, and X-Linked Phenotypes.* Johns Hopkins University Press, Baltimore.

McKusick, V. A. 1994. *Mendelian inheritance in Man: A Catalog of Human Genes and Genetic Disorders.* Johns Hopkins University Press, Baltimore.

Mendel, G. 1866. *Experiments in Plant Hybridization* (W. Bateson, trans.). Harvard University Press, Cambridge, reprinted in Peters, J. A. (ed.). 1959. *Classic Papers in Genetics.* Prentice-Hall, Englewood Cliffs, NJ.

Orel, V. 1996. *Gregor Mendel: The First Geneticist.* Oxford University Press, New York.

Stern, C., and Sherwood, E. R. (eds.). 1966. *The Origin of Genetics, A Mendel Source Book.* W. H. Freeman, San Francisco.

Stewart, A. (ed.). 1995. *Genetic Nomenclature Guide,* Supplement to Trends in Genetics. Elsevier Science, New York.

Sturtevant, A. H. 1965. *A History of Genetics.* Harper & Row, New York.

On the Web

Visit our Web site at **http://www.saunderscollege.com/lifesci/titles.html** and click on A/G/M Genetics for links to the following chapter-related resources on the World Wide Web:

1. **Online Mendelian Inheritance in Man.** A site maintained by the National Institutes of Health that contains descriptions of Mendelian alleles that cause human diseases.
2. **Mendelweb Homepage.** This site contains a variety of information about Gregor Mendel and his work. It is edited by Roger B. Blumberg and maintained by The University of Washington, in Seattle, WA, and also by Brown University in Providence, RI.
3. **Life of Mendel.** A biography of Gregor Mendel maintained by Villanova University, Villanova, PA.

CHAPTER

3

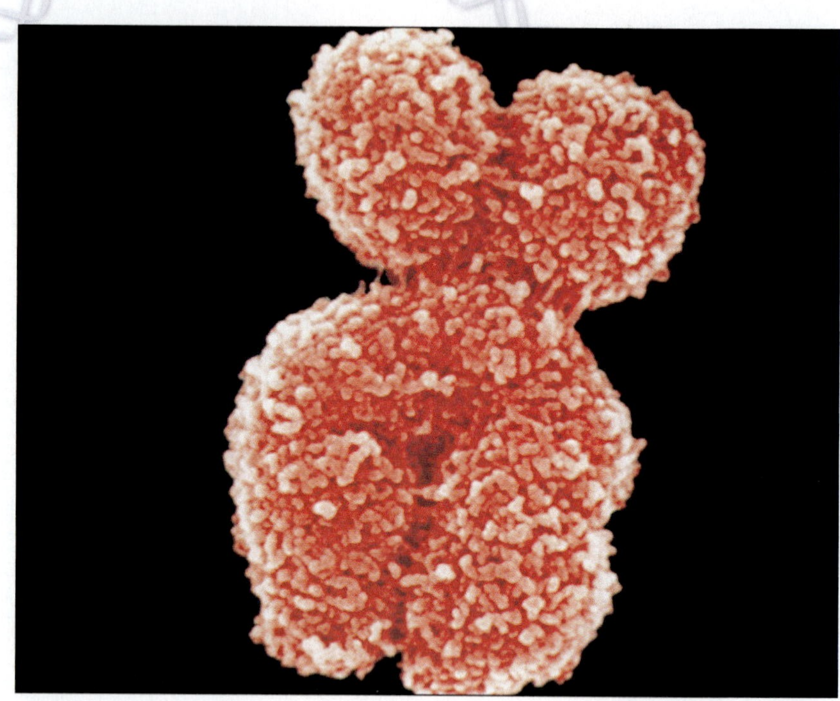

A scanning electron microscope view of a human chromosome in metaphase of the cell cycle. The behavior of chromosomes is the physical basis for Mendel's rules of inheritance *(Biophoto Associates/Photo Researchers, Inc.)*.

Genes, Chromosomes, and the Mechanism of Mendelian Inheritance

Gregor Mendel's work is today recognized as one of the greatest breakthroughs in the history of the biological sciences. Yet when Mendel published his paper in 1866, it was ignored, and it was not until 1900 that its importance was recognized. Considering some of the possible reasons why other scientists failed to recognize the importance of Mendel's paper tells us something about how science progresses. One possible reason is that Mendel's use of mathematical data analysis was difficult for his contemporaries to understand. Mendel reported the discovery of constant numerical ratios in the progeny of his monohybrid and dihybrid crosses as his key finding. His hypotheses depended on his mathematical interpretation of these ratios and the ratios from other crosses. His peers, who were trained in qualitative, observational science, may have found his ideas too abstract. A second reason was that neither Mendel nor his contemporaries knew about the physical nature of genes, about where they might be located in cells, or about how they might be passed from parent to offspring. Chromosomes, mitosis, and meiosis had not been described in 1866. Without a physical basis for genes and their transmission, other scientists considered Mendel's ideas only an abstract theory.

During the latter part of the 19th century, the study of cells became one of the most productive areas of biological research. Investigators in **cytology** (cyto- = cell; -ology = the study of) used improved light microscopes and staining techniques to examine cells taken from a variety of plant and animal tissues. They described the visible appearance and structure of many cell types and the visually elegant events of cell division by which one cell divides to produce two progeny cells. They discovered that **somatic cells** (those not forming gametes) of all types divided by a consistent series of events. The most striking events involved the cell **nucleus** (nucleus = kernel). During cell division, densely staining structures called **chromosomes** (chromo- = colored; -somes = bodies) became visible in the nucleus, divided, and eventually were distributed to the nuclei of progeny cells.

Three investigators, Carl Correns, Hugo De Vries, and Erich von Tschermak had become convinced by the constancy of the events of cell division that inheritance was controlled by a physical substance transmitted from parent cell to progeny cells by the events of cell division. Each began to study the inheritance of single characters in controlled crosses, and they obtained the same constant ratios of phenotypes in the offspring that Mendel observed. Searching the literature for other similar studies, they discovered Mendel's paper. Amazed that Mendel had made his discoveries so long before, they widely publicized his work in a series of papers published in 1900.

In 1902 Walter Sutton, a graduate student at Columbia University, and Theodore Boveri, a distinguished cytologist, independently noted that chromosome segregation during the specialized cell divisions that occur during gamete formation paralleled the behavior of genes. They proposed that genes must reside on chromosomes, and their hypothesis became known as the **chromosome theory of heredity.** The proposal of a physical basis for Mendel's rules greatly aided their acceptance.

The final step that incorporated Mendel's work into 20th-century science was made by Thomas Hunt Morgan. Morgan was skeptical of the chromosome theory and began an extensive study of the inheritance of mutations in the fruit fly *Drosophila melanogaster.* He initially intended to disprove the theory. Aided by an outstanding group of students, Morgan discovered numerous mutations in *Drosophila* and studied their inheritance in large numbers of controlled crosses. Morgan provided a large amount of data that indicated that not only do genes reside on chromosomes but each gene is also normally located at one particular place on its chromosome. He published his ideas in *The Mechanism of Mendelian Heredity* (1915). As a testament to the significance of his work, Morgan was awarded the Nobel prize for physiology and medicine in 1933. Genetics had reached the juncture where the physical basis for Mendel's rules was understood, and this made it easier for others to understand exactly how inheritance occurred. As a result, genetics matured into the separate science that we know today.

In this chapter, we will focus on one key aspect of the acceptance of Mendel's rules, the chromosome theory of heredity. We will discuss the physical nature of eukaryotic chromosomes, the processes of mitotic and meiotic cell division, and how chromosome behavior during meiosis is the physical basis for Mendel's rules. We will also discuss the evidence produced by Morgan and his colleagues that is considered the proof of the chromosome theory of heredity. Finally, we will discuss some special properties of sex chromosomes and their roles in sex determination. In these discussions we will seek to answer five questions:

1. What are the physical events that occur during the growth and division of a eukaryotic cell?
2. What are the physical events that occur during mitosis and meiosis?
3. How does chromosome division and segregation during meiosis provide the physical basis for Mendel's Rules of Gene Inheritance?
4. What is the evidence that genes reside on chromosomes, and how does this explain the production of Mendel's phenotypic ratios?
5. What are the roles of the sex chromosomes in determining the sex of animals?

THE CELL CYCLE AND CHROMOSOME BEHAVIOR ARE FUNDAMENTALLY IMPORTANT.

The Somatic Cell Cycle Is Divided into Three Stages.

The 19th-century cytologists noted that somatic cells move through cycles of cell growth and division. One cell cycle was defined as the period between one stage of cell division to the same stage of the next division. The early cytologists divided the cell cycle into three parts or stages that could be distinguished using a light microscope: mitosis (nuclear division), which has four subphases; cytokinesis (cytoplasmic division); and interphase (the period between divisions), which has three subphases (Figure 3.1). These early investigators also recognized that a mature multicellular organism consists of billions of cells that are all derived from the original single-celled zygote.

Under a light microscope, cells in interphase appear to be quiet, and the nuclei are featureless, which is why early cytologists considered this a "resting" stage between the major events of cell division. Today we understand that many vital cellular functions take place during this period. Modern scientists divide interphase into three parts, termed G_1, S, and G_2, based on the events that occur in the cells. Cells enter the first part of interphase, called the **G_1 or first gap phase,** immediately after the preceding cell division is completed. During G_1, many genes become active, RNA and proteins are synthesized, and the cell expands. This is the major synthetic activity period of cells when most of the materials

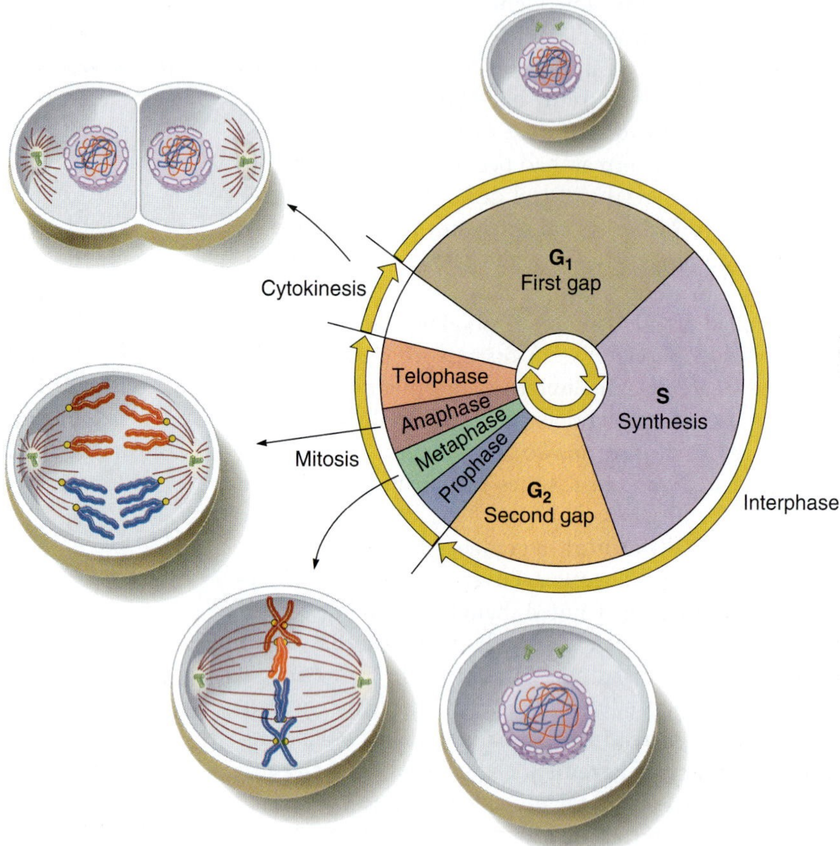

FIGURE 3.1 The mitotic cell cycle. The mitotic cycle is divided into three phases: mitosis, cytokinesis, and interphase. Interphase has three parts: G_1 (first gap), S (synthesis), and G_2 (second gap). The relative time spent in each stage by an average growing and dividing cell is roughly indicated by the size of the sectors. Most cells actually spend less time in mitosis (5–10% of the cell cycle) than indicated.

necessary for the next cell division are produced or accumulated. The length of G_1 varies greatly in different cell types. Rapidly proliferating embryonic cells, especially those drawing nutrients from yolk or from the mother, spend little time in G_1. Slowly dividing cells or cells requiring a specific stimulation to divide may spend most or all their time in G_1. This condition is known as the **G-zero (G_0) state,** or **G_0 arrest.** For most cells, a point comes when the cell becomes committed to division and begins to advance through the rest of the cell cycle.

The commitment of a cell to division is controlled by an interacting set of genes that are collectively known as the **mitotic trigger.** Individual genes are termed **cell division cycle (CDC)** genes. These genes are identified by isolating mutations that block or otherwise alter cell division. For example, yeast cells that have a mutation inactivating the *CDC2* gene remain permanently in G_1 phase. These cells enter G_1 normally but cannot advance through the rest of the cycle. The structure of the *CDC2* gene is very similar or identical in all eukaryotic species. In fact, the normal *CDC2* genes of yeast, *Drosophila,* and humans are so similar in structure that if a mutant yeast gene is replaced by a functional copy from *Drosophila* or humans, the yeast cells will divide. These replacement experiments were done using recombinant DNA technology, which is discussed in Chapters 12 and 13. The results of these experiments suggest that the *CDC2* gene must be very old in evolutionary terms, or it would not be found in all eukaryotes. The *CDC2* gene structure must be very important for its function, or it would not have remained so unchanged over time. The study of how mitotic trigger genes function is of great importance to medicine because some mutations of mitotic trigger genes cause cells to divide continuously. This continuous division is a common feature of cancerous cells, suggesting that some forms of cancer may be caused by faulty mitotic trigger genes.

When a cell leaves G_1, it enters into the **S phase** or **synthesis phase,** which in rapidly growing cells can occupy 35–45% of the entire cell cycle. The S phase received its name from the fact that DNA synthesis occurs during this phase. DNA synthesis during chromosome replication produces two identical copies of each chromosome, called **sister chromatids,** which remain attached at the centromere. The **centromere** is a specialized region of the chromosome that is required for proper segregation of the chromosome during division.

After the S phase, eukaryotic cells enter into the **G_2 phase** or **second gap phase,** which is usually shorter than G_1. Dividing cells finish many metabolic processes in G_2 and will remain in G_2 until ready for mitosis. For instance, if yeast cells that have completed the S phase have their DNA damaged by ionizing radiation, they will remain in G_2 until the damage is repaired. This type of cell is considered to be in G_2 arrest.

Mitosis Produces Cells that Are Genetically Identical to the Parental Cell.

Physical division of the nucleus occurs during the phase of the cell cycle called **mitosis,** or the **M phase.** The length of time spent in mitosis is usually short (see Figure 3.1), occupying only 5–10% of the cell cycle. Mitosis has been divided into four stages based on observed chromosomal changes: prophase, metaphase, anaphase, and telophase. To understand mitosis, we must examine not only these chromosomal changes but also several important nonchromosomal events.

During the initial stage of mitosis, called **prophase** (Figure 3.2), the chromosomes condense from the diffuse, elongated state they are in during interphase to a more compact, condensed form. They now become visible under the microscope. Prophase begins when the chromosomes first become visible within the nucleus as long, thin, intertwined strands that become progressively shorter and more condensed. By late prophase, the two sister chromatids of each chromosome with their connecting centromere are clearly visible.

In animal cells, there are two important cytoplasmic organelles called **centrioles,** which are composed of complexes of microtubules containing the protein called tubulin. During interphase, the centrioles are visible adjacent to the nucleus, replicating during S phase. At the beginning of prophase, the two centrioles separate, and each migrates to one of the cell's poles (Figure 3.2). Cells have a complex internal array of microtubules collectively called the cellular **cytoskeleton.** During interphase, cytoskeletal elements are distributed throughout the cytoplasm. During prophase, the cytoskeleton reorganizes in preparation for mitotic activity. Some microtubules reorganize around the migrating centrioles to form two new structures called **asters** (Figure 3.2). Each aster consists of a pair of centrioles surrounded by a halo of microtubules.

As prophase proceeds, the asters migrate to opposite ends of the cell. There they become sites for the organization of the **mitotic spindle** from long bundles of microtubules, associated proteins, and other cytoskeletal elements. Many of the spindle microtubules stretch from one aster to the center of the cell, and some reach the other aster. By the end of prophase, the mitotic spindle resembles two inverted cones that are joined at their bases in the center of the cell and have their points at the asters (Figure 3.2). Spindle fibers constantly increase and decrease in length by adding and substracting tubulin molecules. In contrast to animal cells, higher plant cells do not contain centrioles and do not form asters during prophase. This is presumably because plant cells have rigid cell walls and do not need asters to anchor the spindle apparatus. In addition to their connections to the asters, spindle fibers attach to each condensing chromosome at a specialized region called the **kinetochore** (Figure 3.3). In most animal and plant chromosomes, the kinetochore is within the centromere. At the end of prophase, the nuclear membrane disappears and the spindle fibers begin to move the chromosomes toward a central region midway between the two asters.

The second stage of mitosis is **metaphase,** which is usually short in duration. During metaphase, the highly condensed chromosomes are pulled to a central region in the spindle, called the **metaphase plate** (Figure 3.2). Chromosome movement is controlled by the lengthening and shortening of the spindle fibers, which are under considerable tension. If the fibers connecting a chromosome to one aster are cut with a fine glass needle or a laser microbeam, the chromosome is pulled toward the opposite aster.

The third stage of mitosis, **anaphase,** is short but active. Anaphase begins with the division of the centromeres. The centromeres of all the chromosomes divide at about the same time, changing each chromosome from a single chromosome with two chromatids and one centromere into two separate progeny chromosomes (Figure 3.2). Each progeny chromosome remains anchored to the mitotic spindle via its kinetochore. Even though the centromere and kinetochore are usually located in the same region of the chromosome, they are physically distinct and serve different purposes (Figure 3.3). The centromere couples the two sister chromatids during prophase and metaphase, and the kinetochore connects each chromatid with the mitotic spindle from prophase through anaphase.

Immediately after centromere division, the two progeny chromosomes migrate toward opposite asters, drawn by the spindle fibers. The division and migration of the chromosomes during anaphase produces two sets of progeny chromosomes, each containing one copy of all chromosomes present in the parental cell (Figure 3.2). This means that each progeny cell will have a genetic makeup identical to that of the parent cell.

The fourth stage of mitosis, **telophase,** begins as the migrating chromosomes approach the asters. During telophase, new nuclear envelopes form around the two sets of chromosomes, the chromosomes return to the diffuse form characteristic of interphase, and the microtubules of the mitotic spindle dissociate to reorganize as interphase cytoskeletons (Figure 3.2).

Most mitotic nuclear divisions are followed by **cytokinesis.** During cytokinesis the cytoplasm divides, and the cell physically separates into two progeny cells. In

(a) Mitosis in an animal cell

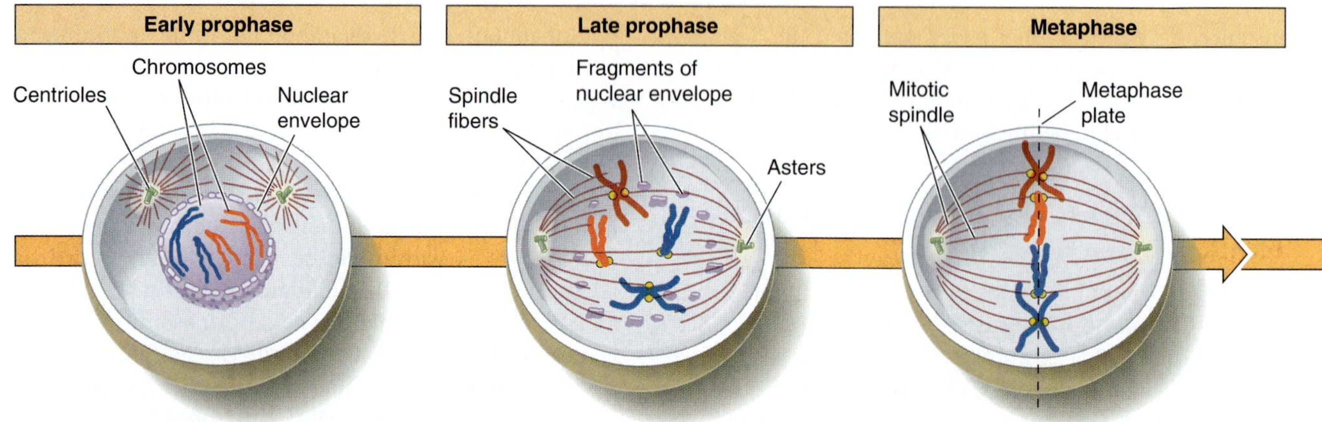

(b) Mitosis in a plant cell

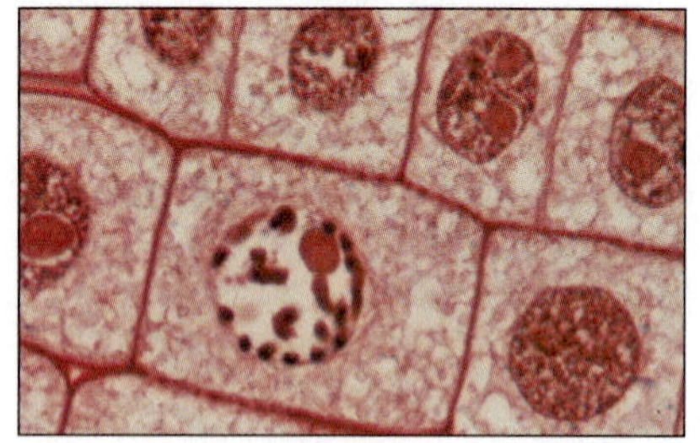

During early prophase the chromosomes first become visible as long thin strands that are rapidly condensing and contracting. Each chromosome has replicated during the S stage of interphase and has two sister chromatids held together by one centromere.

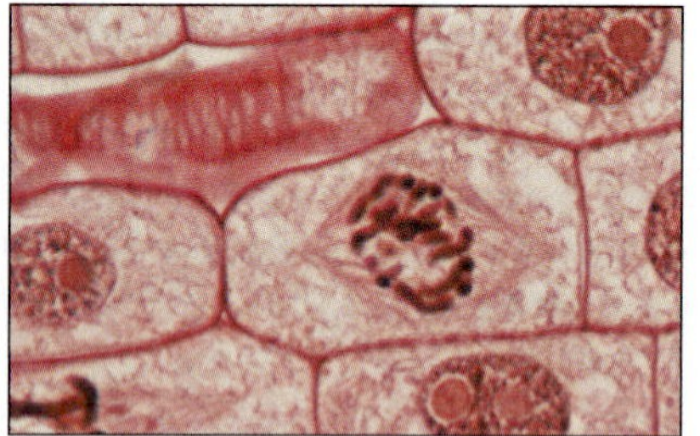

Late in prophase, condensation of the chromosomes is complete, and each chromosome has a characteristic size and shape. The nuclear envelope disappears and the spindle fibers from both ends of the cell attach to the kinetochores of each chromosome.

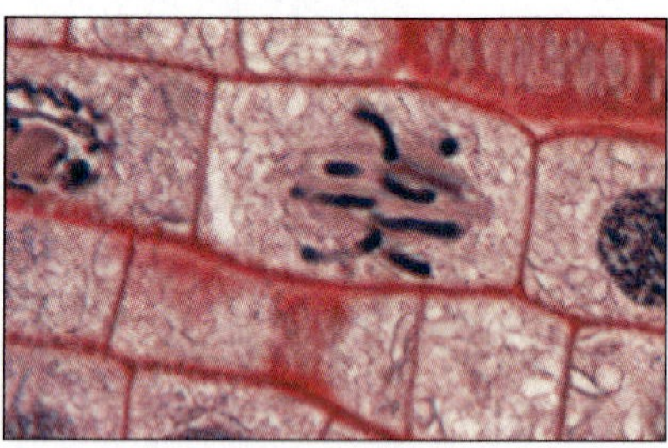

During metaphase, the spindle fibers pull the chromosomes to the center of the cell, where they align along a region called the metaphase plate.

FIGURE 3.2 **Mitosis.** Each of the phases of mitosis is defined by characteristic chromosomal activity. (a) Diagrams interpreting chromosomal activity in an animal cell undergoing mitosis as seen under a light microscope. Vital nonchromosomal activity involving the centrioles and mitotic spindle is also shown. (b) Light micrographs of onion root tip cells undergoing mitosis. The cells have been stained to show the chromosomes in red. *(Early prophase–telophase, Ed Reschke; cytokinesis, Carolina Biological Supply Co./Phototake.)*

animal cells (Figure 3.4a), cytokinesis begins with the formation of a **cell furrow,** which forms at the location of the metaphase plate through the contraction of a ring of microtubules called the **contractile ring** (Figure 3.4a). In plants, cells are surrounded by both a plasma membrane and a rigid cell wall. Plant-cell cytokinesis begins with the formation in the middle of the cell of a **cell plate** that forms by the aggregation and fusion of membrane vesicles. The vesicles form new plasma membranes that fuse with the existing plasma membrane, dividing the parental cell into separate progeny cells (Figure 3.4b). The formation of new cell walls completes cytokinesis.

From a genetic point of view, the key aspect of mitotic cell division is that each progeny cell receives one copy of all the chromosomes present in the parental cell. This means that each progeny cell will receive an exact copy of the set of alleles that were in the parental chromosomes (Figure 3.5). If the parental cell is heterozygous for one or more genes, then all the mitotic progeny of that cell throughout the body will also be heterozygous. This is accomplished by the accurate replication of the chromosomes during S phase and the segregation of the progeny chromosomes during anaphase.

SECTION REVIEW PROBLEMS

1. List one similarity and one difference between the G_1 and G_2 phases of the mitotic cell cycle.
2. You have been given a sample of mouse tissue culture cells containing a new mutation that blocks cell division in either G_1 or G_2. Describe how you might determine at which stage these cells are blocked.

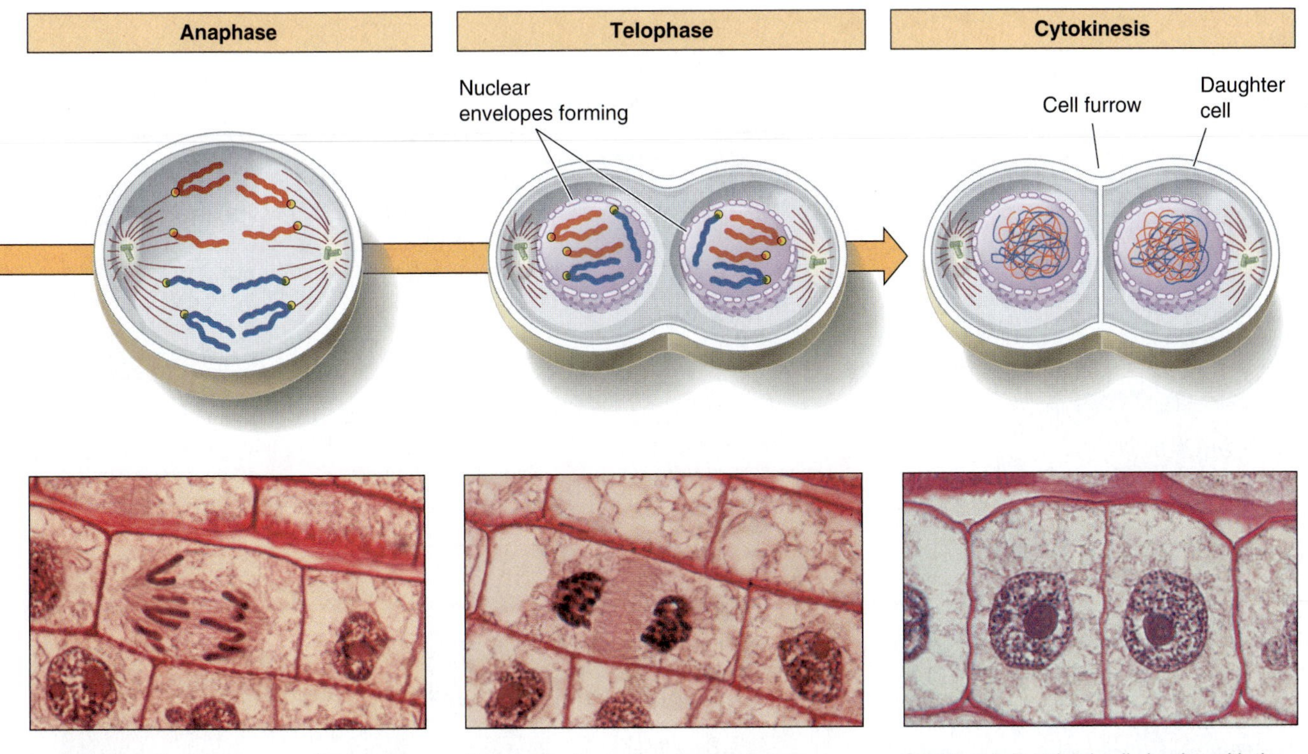

In anaphase, the centromeres divide and each sister chromatid becomes a separate daughter chromosome. The two daughter chromosomes migrate to opposite poles, ensuring that each set of chromosomes contains one copy of each chromosome.

During telophase, the chromosomes elongate, returning to their interphase condition. Nuclear envelopes form around each set of chromosomes, producing two separate daughter nuclei.

Cytokinesis in animal cells begins with the formation of a cell furrow that gradually constricts the cell, dividing the cytoplasm and separating the two daughter cells. In plant cells, cytokinesis occurs by the growth of a cell plate between the cells.

▼FIGURE 3.3 Chromosome structure during mitotic metaphase. (a) An electron microscope photograph showing the compact folding of a metaphase chromosome. (b) A diagram showing how the spindle fibers attach to each sister chromatid at the kinetochore. In this chromosome, the kinetochore is located within the centromere. *(E. J. DuPraw)*

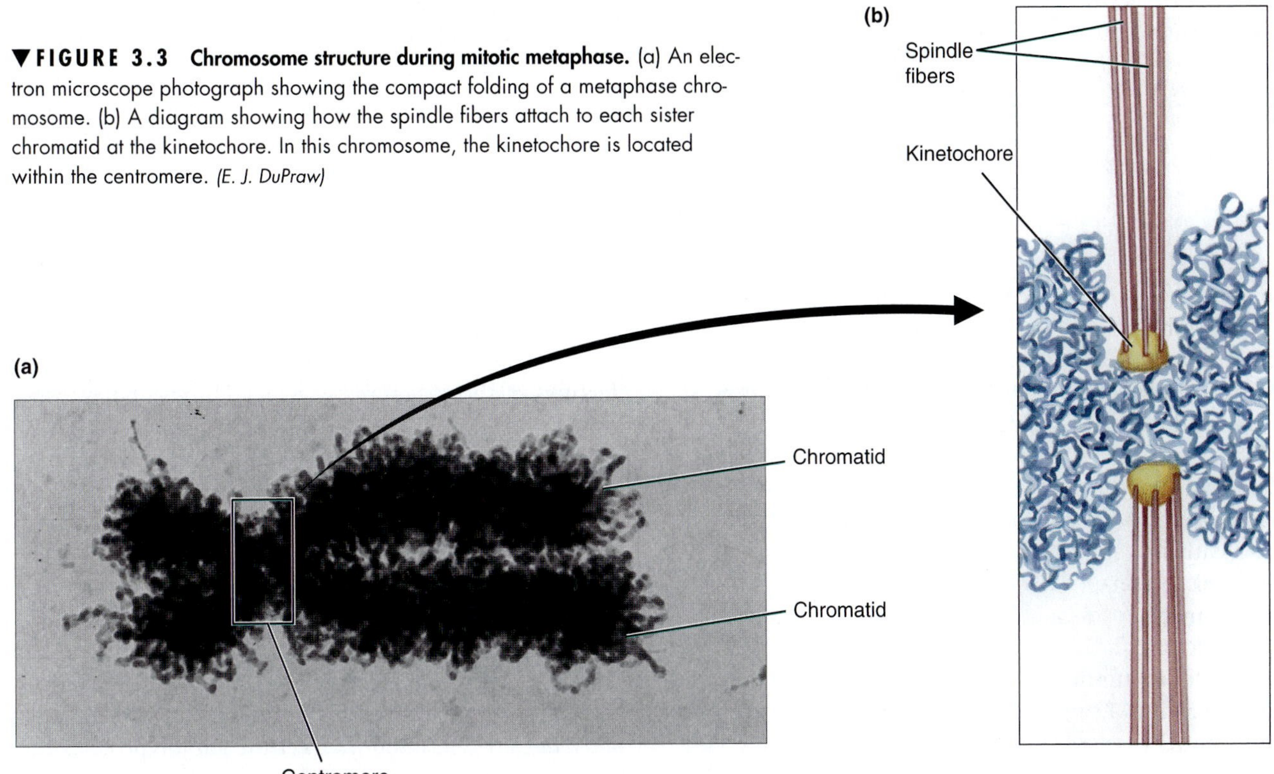

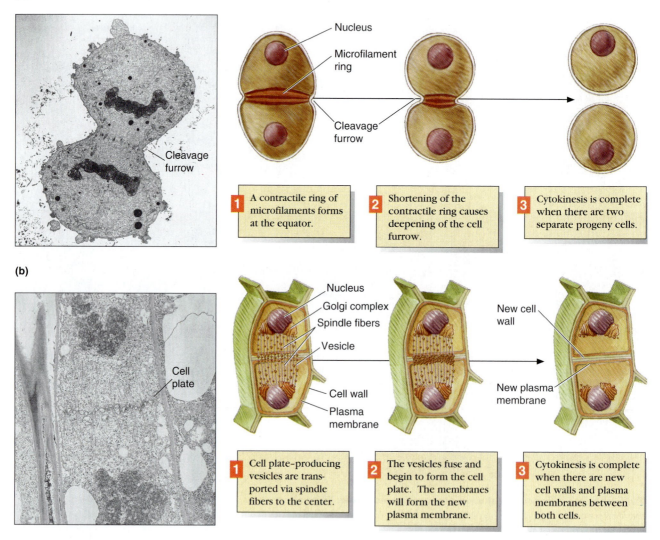

FIGURE 3.4 Cytokinesis in animal cells and plant cells.
(a) Cytokinesis during the first mitotic division of an animal cell shows how the cleavage furrow divides the cytoplasm. (b) Cytokinesis in a plant cell involves the formation of a cell plate that eventually gives rise to new plasma membranes and cell walls. *(a, T. E. Schroeder, University of Wisconsin/Biological Photo Service; b, E. H. Newcomb and B. A. Palevitz, University of Wisconsin/Biological Photo Service)*

The Morphology of Eukaryotic Chromosomes Can Be Studied During Mitosis.

The study of the physical size and structure of chromosomes is part of the field of **cytogenetics.** Most cytogenetic studies use cells that are near the end of prophase or in early metaphase of mitosis because the chromosomes are compact and densely staining and have a characteristic size and shape. Understanding the physical appearance of normal chromosomes can be vital for understanding a variety of genetic abnormalities that are produced by mutations that alter chromosomes (see Chapter 18). In this section, we will discuss some key features of the structure of normal chromosomes, especially those used to distinguish one chromosome from another. Such features include chromosome size, the location of the centromere, and patterns of light and dark staining that occur when chromosomes are treated with different chemical dyes.

These features are collectively referred to as **chromosome morphology.** The number and morphology of an individual's chromosomes is called that individual's **karyotype.** Each eukaryotic species has a characteristic normal karyotype. Examination of an individual's chromosomes to determine whether the karyotype conforms to the species norm is called **karyotype analysis.**

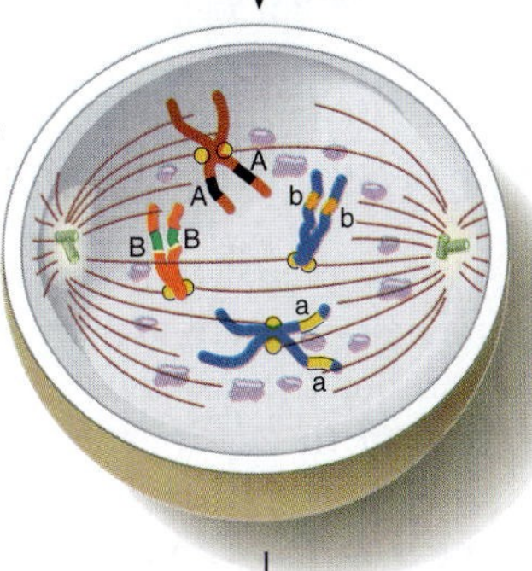

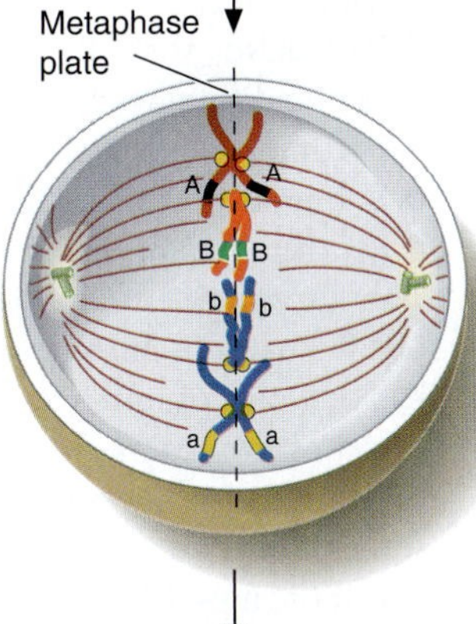

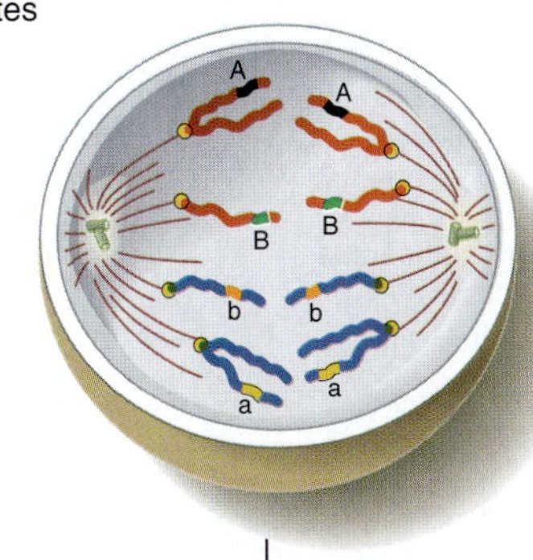

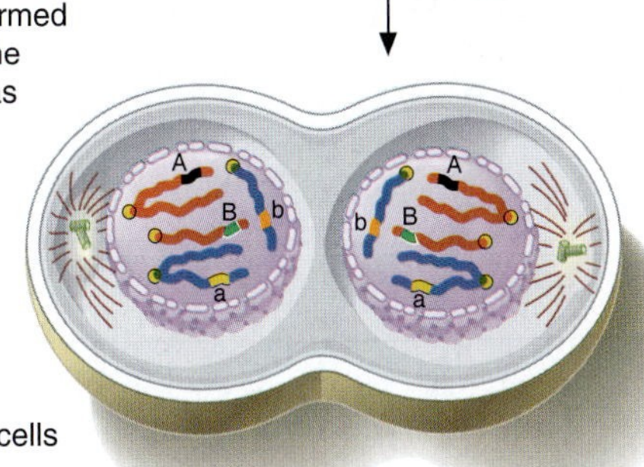

◀**FIGURE 3.5 Chromosome behavior during mitosis and the genetic makeup of the progeny cells.** In the somatic cell cycle, the replication of each chromosome occurs during the S phase, producing two sister chromatids with identical copies of each allele. The separation of sister chromatids followed by segregation of the progeny chromosomes ensures that each progeny cell receives the same genetic makeup as the parent cell.

One key feature of chromosome morphology is the location of the centromere. During metaphase, the centromere is visible as the region where the two sister chromatids are attached to each other. The portions of a chromatid on either side of the centromere are called the **chromosome arms** of that chromatid. A chromosome arm is not the same as a chromatid. A **chromatid** is one complete copy of the entire length of the chromosome (Figures 3.3 and 3.6), and a chromosome arm is only that portion of the chromosome on one side of the centromere. Each chromosome arm terminates in a special structure called a **telomere,** which we will discuss in the next section and also in Chapter 10. Particular locations on the chromosome arms can be identified as **proximal**

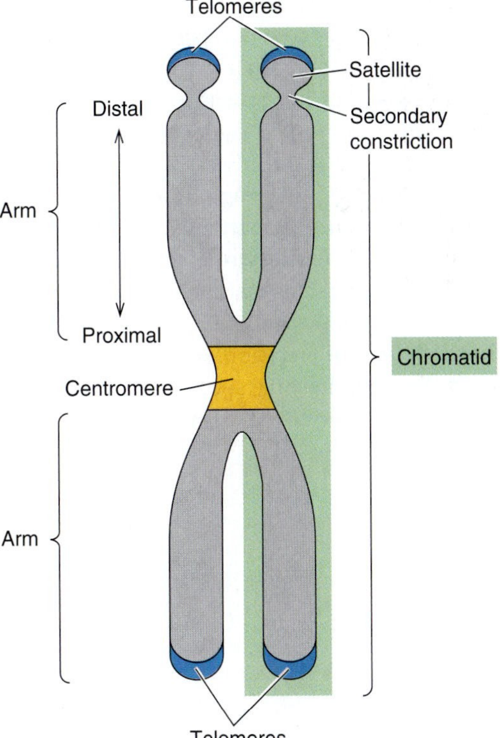

▲**FIGURE 3.6 The structure of replicated chromosomes.** This diagram of a replicated chromosome shows the sister chromatids (single copies of an entire chromosome), the centromere (site of attachment of the sister chromatids), the telomeres (specialized ends of chromosomes), the chromosome arms (the portion of the chromatid on one side of the centromere), and satellites (small end regions connected by a stalk).

(closer to) or **distal** (farther from) the centromere. When geneticists state that "gene A is proximal to gene B," they mean that the normal position of gene A in the chromosome arm is closer to the centromere than the normal position of gene B.

Most eukaryotic chromosomes can be placed into one of four morphological classes based on centromere location: metacentric, submetacentric, acrocentric, or telocentric (Figure 3.7). **Metacentric** chromosomes have the centromere in the middle, and the two arms are of equal length. **Submetacentric** chromosomes have arms that are of similar, but not equal, length. **Acrocentric** chromosomes have the centromere very near one end, and so have one short and one long arm. The distinction between submetacentric and acrocentric is subjective and can vary from investigator to investigator. **Telocentric** chromosomes have the centromere adjacent to the telomere, and these chromosomes have only one visible arm.

Another key feature of metaphase chromosome morphology is chromosome size, usually measured as the length of the chromosome at metaphase. Chromosome size can vary widely, with some chromosomes being several times as long as other chromosomes in the same nucleus. To examine an individual's chromosomes, cells in late prophase or in metaphase are placed on a microscope slide, the cell membranes are ruptured, and the condensed chromosomes are spread out on the slide (Figure 3.8a). The chromosomes are then treated with one of a variety of different dyes, many of which specifically stain DNA. Photographs are taken of the stained chromosomes, and the chromosomes are sorted according to size (called **karyotyping**). The sorting was once done by hand with cutouts made from the photographs, but today computer programs sort the chromosomes with a high level of precision. A human karyotype is shown in Figure 3.8b. In some species, centromere locations and sizes are sufficient to identify reliably all the chromosomes. However, in other species, including humans, several chromosomes are nearly equal in size and shape, and these features are not sufficient to distinguish all chromosomes.

A third distinguishing feature of metaphase chromosomes is the presence or absence of **satellites,** loosely associated, distal regions that are attached to the proximal portion of the chromosome arm by thin stalks called secondary constrictions (Figure 3.6). Not all chromosomes have satellites, but for those that do, satellites can be reliable features for identification.

SECTION REVIEW PROBLEMS

3. A new animal species you are studying has one pair of metacentric and one pair of submetacentric chromosomes in its somatic cells. Draw a diagram of a cell of this species (a) during metaphase and (b) during anaphase of mitosis. In addition to the chromosomes, include the mitotic spindle and the asters.
4. Briefly define the following terms, and for each pair give one difference between the two:
   a. chromosome arm and chromatid
   b. submetacentric and telocentric chromosomes
   c. centromere and kinetochore
5. The karyotype of *Drosophila melanogaster* consists of four pairs of chromosomes. A female *Drosophila* has one pair of mid-sized telocentric chromosomes (the X chromosomes), two pairs of large metacentric chromosomes (the second and third chromosomes), and one pair of very small submetacentric chromosomes (the fourth chromosomes). Draw a diagram of a somatic cell from a female *Drosophila* in anaphase of mitosis.
6. What stages of the mitotic cell cycle are best for karyotype analysis? Briefly explain your answer.

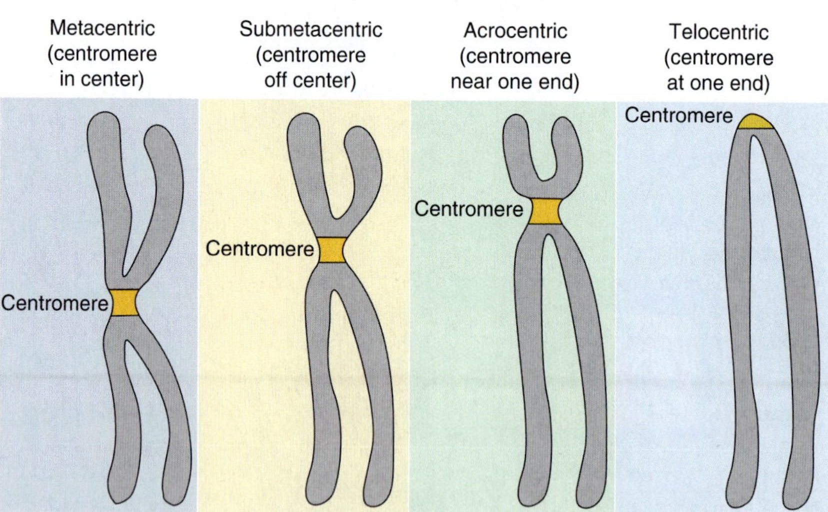

FIGURE 3.7 Classes of metaphase chromosomes. The metaphase chromosome types are shown distinguished by the position of the centromere.

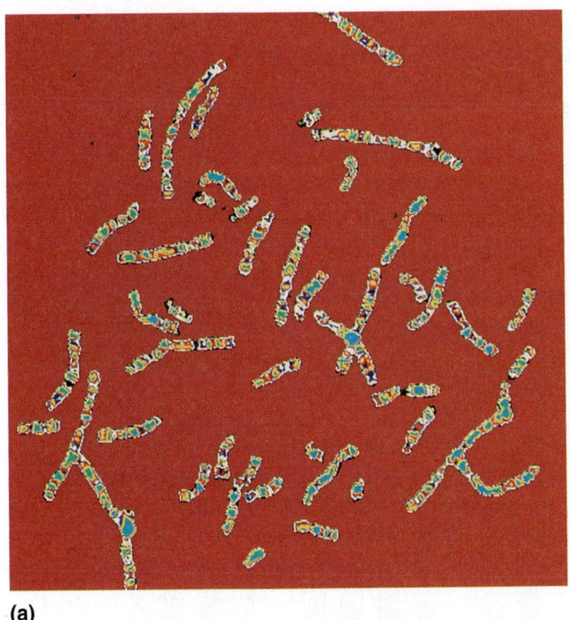

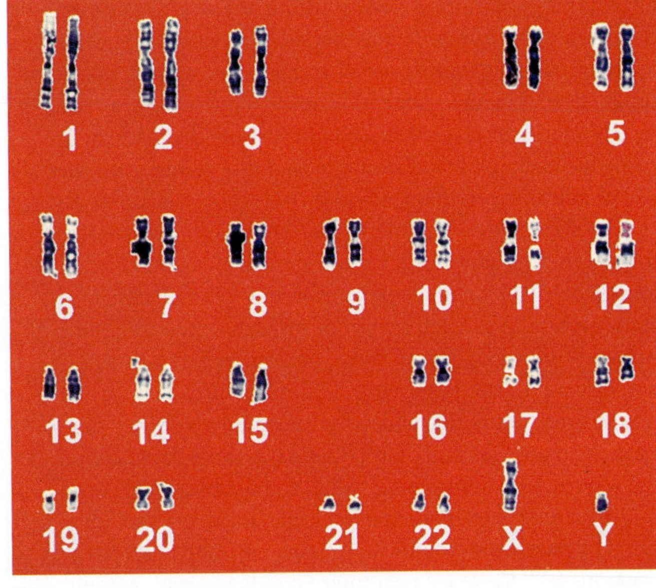

(a) (b)

FIGURE 3.8 Normal human chromosomes. (a) Human chromosomes are prepared for viewing by spreading the chromosomes from a cell in mitotic metaphase onto a microscope slide. The chromosomes are then stained and photographed. (b) The chromosomes are arranged according to size. *(a, b, Custom Medical Stock Photo)*

Chemical Staining Reveals Differences in the Structure of Chromosomes.

A very important feature of chromosomes is that they show characteristic patterns of staining when treated with certain chemical reagents or dyes. The analysis of chromosome morphology has been greatly aided by improved techniques for chromosome staining. Chromosomes can be stained at any time during mitosis, but most examinations are done on cells isolated in late prophase or very early metaphase, before the chromosomes reach maximum condensation.

One of the most important staining techniques is the **Feulgen reaction,** discovered in 1912. In this procedure, cells are treated with a warm acid and a chemical known as Schiff's reagent, which causes a reddish purple staining that is specific to DNA. When the Feulgen reaction is used on cells in metaphase, the chromosomes stain uniformly. However, if cells are stained in mid- to late prophase, each chromosome has characteristic regions of heavy and light staining. The regions that condense latest in prophase stain lightly and are called **euchromatin.** The regions that condense early in prophase stain heavily and are called **heterochromatin.** Because individual chromosomes have different amounts of heterochromatin and euchromatin, the light and heavy pattern produced by the Feulgen reaction is different for each chromosome and can be used to identify particular chromosomes. When chromosomes from several different types of cells are compared, some regions of some chromosomes are always heterochromatic, and these regions are called **constitutive heterochromatin.** Other regions are heterochromatic in some cells but euchromatic in others and are called **facultative heterochromatin.** The functional differences between heterochromatin and euchromatin are discussed in detail in Chapter 10. We will note here, however, that actively expressed genes are located in euchromatic regions, and inactivate genes are located in heterochromatic regions.

Techniques other than the Feulgen reaction are also used to stain chromosomes. Some dyes produce characteristic patterns of alternating dark and light bands of staining in chromosomes. Chromosomes stained with fluorescent quinacrine dyes, for example, show bands of intense fluorescence, called **Q bands,** when exposed to ultraviolet light. Chromosomes treated with trypsin and then stained with Giemsa stain show alternating regions of heavy staining (called **G bands**) and light staining (called **interbands**). These characteristic banding patterns can be used to identify each chromosome in cell preparations. The banding pattern is particularly useful when chromosomes have the same approximate size, centromere location, and presence or absence of satellites.

These dye techniques do not work for every species, but for many plant and animal species, accurate karyotyping cannot be done without the use of staining techniques. In human cytogenetics, for example, the exact number and identity of chromosomes was in doubt until 1956 because several human chromosomes are nearly identical in size and shape. Precise identification of all the human chromosomes was not possible until the use of staining techniques was introduced.

In 1971, the Fourth International Congress of Human Genetics adopted a standard system, based on the locations of G bands, for numbering human chromosomes that allows each region of each chromosome to be uniquely defined. Human chromosome pairs are numbered from 1 (the largest) to 22 (the smallest), with the sex chromosomes (X and Y) considered the 23rd pair. Each human chromosome contains a short arm, called the **p-arm,** and a long arm, called the **q-arm.** The p- and q-arms are subdivided into regions based on the locations of particular G bands. Starting on each side of the centromere, a series of prominent bands in each chromosome arm were chosen and numbered (1, 2, 3, etc.), with band number 1 being the band in each arm closest to the centromere. Within each region between numbered bands, individual small bands were chosen to subdivide the region; these were also numbered.

This system divided each human chromosome into a series of uniquely numbered regions, allowing investigators to indicate the locations of specific genes using the bands as landmarks. For example, the human gene whose mutation causes the disease cystic fibrosis is located in the region 7q31, meaning that this gene is on the long arm of chromosome 7, in region 3, at band 1. The precise, detailed pattern of bands given by G-banding and other staining techniques allows highly accurate gene localizations. In addition to their usefulness in gene localization, chromosome banding patterns from closely related species can be studied to help evaluate the extent to which evolutionary changes have occurred between the two species. For example, studies reveal that human and other primate chromosomes have many similarities is size, shape, and banding patterns. This cytogenetic evidence suggests how closely humans are related to other primate species (see CONCEPT CLOSE-UP: *Using Chromosome Banding to Establish Species Relationships* and Figure 3.9).

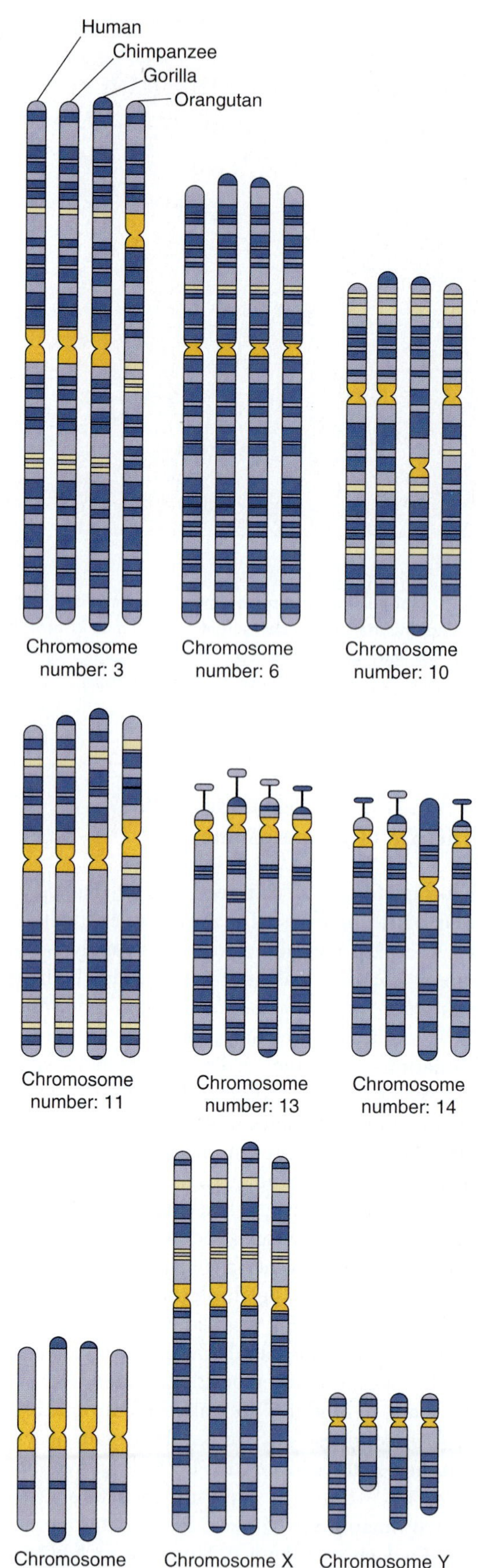

FIGURE 3.9 G-banding patterns of human and primate chromosomes. The G-banding patterns of chromosomes from humans and three other primate species (chimpanzee, gorilla, and orangutan). Note the many regions of identical banding patterns in some chromosomes, suggesting that these species are closely related.

CONCEPT CLOSE-UP

Using Chromosome Banding To Establish Species Relationships

Chromosome banding patterns often reveal important features about related species of plants or animals. This is the case for a comparison of the chromosome banding patterns of humans, chimpanzees, gorillas, and the orangutan (Figure 3.9). These four species were compared using G-banding of late prophase chromosomes (when the chromosomes are not as highly condensed as in metaphase), and a remarkable number of similarities were found. Using these findings, J. J. Yunis and O. Prakash were able to extrapolate backward to hypothesize the existence of a common 24 chromosome-pair ancestor of the four species. In the 24-pair ancestor, 18 pairs were similar to those found in humans, and 15 pairs were similar to those found in gorillas, chimpanzees, and orangutans. Based on these findings, they concluded that the gorilla separated from the common ancestor first, and further chromosomal rearrangements led to the divergence of chimpanzees from humans. These conclusions, discussed in later chapters, are consistent with other methods of uncovering evolutionary relationships, including DNA reassociation kinetics (Chapter 9), comparison of amino acid sequences for specific proteins (Chapter 21), and blood group antigens (Chapter 4). All the results suggest that these four species (humans, chimpanzees, gorillas, and orangutans) have substantial common ancestry. In fact, the similarities between two of these species (humans and chimpanzees) are so remarkable that in 1975, Mary-Claire King and A. C. Wilson proposed that the obvious phenotypic differences between humans and chimps must be a result of the differences in the action of the developmental regulatory genes rather than the structural genes. In humans and chimps, 13 of the 24 chromosomes are presumably identical (3, 6, 8, 10, 11, 13, 14, 19, and XY) if constitutive heterochromatin is excluded. These studies illustrate the usefulness of chromosome banding.

SOLVED PROBLEM

Problem

Draw a diagram of a cell from a female mammal that has three pairs of chromosomes in prophase of mitosis. Include in your diagram

a. one pair of metacentric chromosomes with satellites
b. one pair of submetacentric chromosomes
c. one pair of acrocentric chromosomes

Solution

a. metacentric chromosomes with satellites
b. submetacentric chromosomes
c. acrocentric chromosomes

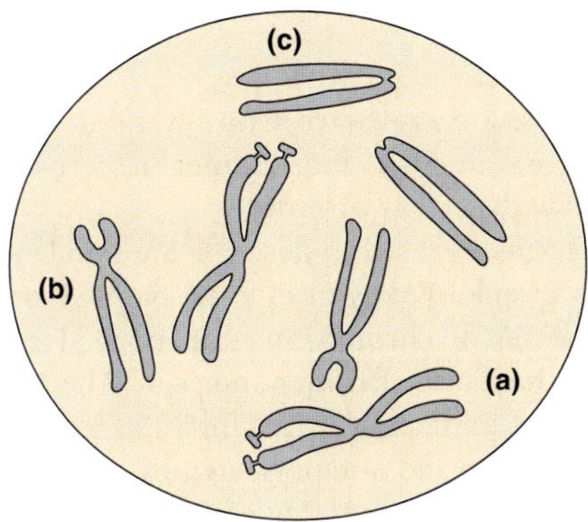

SECTION REVIEW PROBLEMS

7. Describe one difference between euchromatin and heterochromatin.
8. What is the difference between facultative heterochromatin and constitutive heterochromatin?

Chromosomal Abnormalities Can Be Detected by Karyotype Analysis.

Standard karyotypes have been described for many different species, including humans. Techniques to obtain cells for karyotypic analysis of human fetuses have been developed to allow physicians to diagnose chromosomal abnormalities before birth. These techniques involve isolating a small sample of fetal or fetal-derived cells via the techniques of **amniocentesis** or **chorionic villi sampling.** Amniocentesis is generally done at 14–16 weeks of pregnancy because the volume of amniotic fluid available for sampling is usually adequate by then. A small sample of amniotic fluid containing fetal cells is obtained with a syringe (Figure 3.10), the cells are isolated and placed in culture to undergo mitotic division. Samples of cells in metaphase are collected, the chromosomes are stained and spread on microscope slides for examination, and the number and morphology of the fetal chromo-

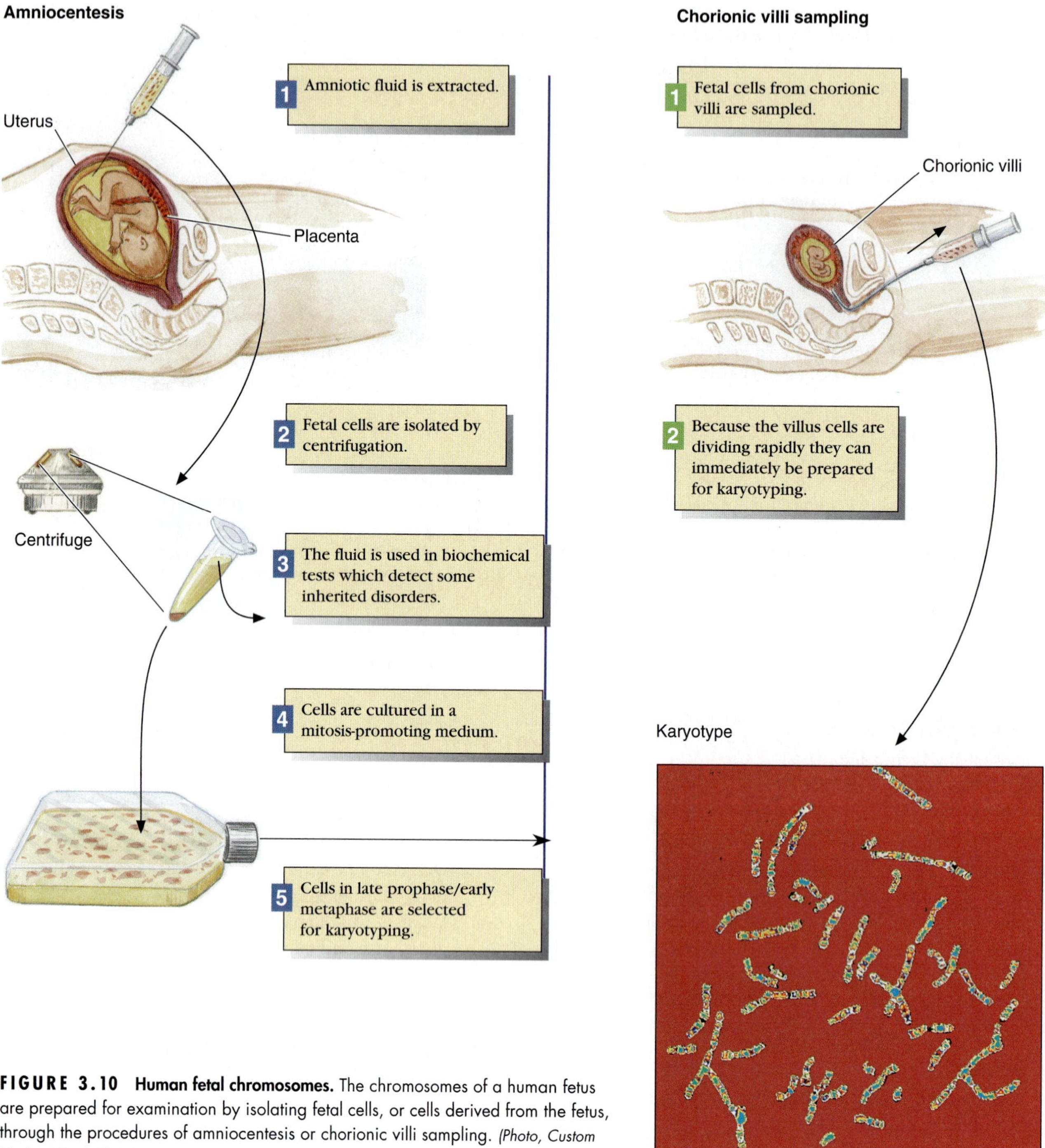
Amniocentesis
Chorionic villi sampling
1 Amniotic fluid is extracted.
Uterus
Placenta
2 Fetal cells are isolated by centrifugation.
Centrifuge
3 The fluid is used in biochemical tests which detect some inherited disorders.
4 Cells are cultured in a mitosis-promoting medium.
5 Cells in late prophase/early metaphase are selected for karyotyping.
1 Fetal cells from chorionic villi are sampled.
Chorionic villi
2 Because the villus cells are dividing rapidly they can immediately be prepared for karyotyping.
Karyotype

FIGURE 3.10 Human fetal chromosomes. The chromosomes of a human fetus are prepared for examination by isolating fetal cells, or cells derived from the fetus, through the procedures of amniocentesis or chorionic villi sampling. *(Photo, Custom Medical Stock Photo)*

somes are determined. Chorionic villi sampling (CVS) can be performed earlier in pregnancy (8–10 weeks). In CVS, a small sample of cells are obtained from the chorionic villi (Figure 3.10), the fetal-derived tissues that will form the placenta. Because these cells are actively dividing, chromosome preparations can be made almost immediately after collecting them. Karyotype analysis can determine whether the fetus has an unusual number of chromosomes, chromosomes with structural abnormalities, or both. Abnormal numbers or morphologies of chromosomes are associated with several developmental abnormalities that produce nonviable fetuses or live newborns with serious disorders.

Two terms are used to describe the number of chromosomes in eukaryotic karyotypes: haploid and monoploid. The set of chromosomes in normal gametes is called the **haploid chromosome set.** The number of chromosomes in the haploid set for a species is called the **haploid number** and is represented by the symbol N. A single set of chromosomes that contains one copy of each

TABLE 3.1 *Numbers of Chromosomes in Somatic Cells of Different Species*

Animal Species	Number of Chromosomes	Plant Species	Number of Chromosomes
alligator (*Alligator mississippiensis*)	32	barley (*Hordeum vulgare*)	14
cat (*Felis domesticus*)	38	bean (*Phaseolus vulgaris*)	22
cattle (*Bos taurus*)	60	bread wheat (*Triticum aestivum*)	42
chicken (*Gallus domesticus*)	78	cabbage (*Brassica oleracea*)	18
chimpanzee (*Pan troglodytes*)	48	clover (*Trifolium repens*)	32
cockroach (*Blatta germanica*)	23 (male) 24 (female)	corn (*Zea maize*)	20
dog (*Canis familiaris*)	78	cotton (*Gossypium hirstum*)	52
donkey (*Equus asinus*)	62	emmer wheat (*Triticum turgidum*)	28
flatworm (*Planaria torva*)	16	evening primrose (*Oenothera biennis*)	14
frog (*Rana pipiens*)	26	green algae (*Acetabularia mediterranes*)	20
fruit fly (*Drosophila melanogaster*)	8	jimson weed (*Datura stramonium*)	24
gorilla (*Gorilla gorilla*)	48	oak (*Ouercus alba*)	24
grasshopper (*Melanoplus differentialis*)	24	oat (*Avena sativa*)	42
guinea pig (*Cavia cobaya*)	64	onion (*Allium cepa*)	16
hamster (*Mesocricetus auratus*)	44	pea (*Pisum sativum*)	14
honey bee (*Apis mellifera*)	32	pine (*Pinus* species)	24
horse (*Equus calibus*)	64	potato (*Solanum tuberosum*)	48
house fly (*Musca domestica*)	12	radish (*Raphanus sativus*)	18
human (*Homo sapiens*)	46	rice (*Oryza sativa*)	24
hydra (*Hydra vulgaria attenuata*)	32	snapdragon (*Antirrhinum majus*)	16
mosquito (*Culex pipiens*)	6	squash (*Cucurbita pepo*)	40
mouse (*Mus musculus*)	40	tobacco (*Nicotiana tabacum*)	48
nematode (*Caenorhabditis elegans*)	11 (male) 12 (female)	tomato (*Lycopersicon esculentum*)	24
pigeon (*Columbia livia*)	80	weed (*Arabidopsis thaliana*)	10
rabbit (*Oryctolagus cuniculus*)	44	**Fungi**	
rat (*Rattus norvegicus*)	44	black bread mold (*Aspergillus nidulans*)	8
Rhesus monkey (*Macaca mulatta*)	42	brewer's yeast (*Saccharomyces cerevisiae*)	17
silkworm (*Bombyx mori*)	56	penicillin mold (*Penicillium* species)	4
starfish (*Asterias forbesi*)	36	pink bread mold (*Neurospora crassa*)	7
toad (*Bufo americanus*)	22	slime mold (*Dictostelium discoideum*)	7

type of chromosome from that species is called the **monoploid chromosome set,** and it is represented by the symbol X. The number of monoploid sets of chromosomes present in the somatic cells of an individual is called the **ploidy** of that individual. Most animal species and many plant species are **diploid,** meaning that they have two copies of each chromosome in each somatic cell (2N or 2X). In diploid species, the haploid set and the monoploid set are the same.

Species that contain more than two monoploid sets of chromosomes are called **polyploid** species. In these species, the haploid and monoploid sets of chromosomes are not the same (polyploidy is discussed in greater detail in Chapter 18). Polyploid species are usually identified by a term that reflects the number of monoploid chromosome sets in the somatic cells: Triploids are 3X, tetraploids 4X, pentaploids 5X, and so forth. Humans are diploids with N = X = 23, which means there are 23 pairs of chromosomes in normal human somatic cells (2N = 2X = 46 chromosomes). The number of chromosomes present in several common species of plants and animals is given in Table 3.1.

SECTION REVIEW PROBLEMS

9. How many chromosomes would be present in the following types of cells?
   a. diploid human cells
   b. triploid human cells
   c. triploid *Drosophila melanogaster* cells
   d. haploid *Drosophila melanogaster* cells
   e. triploid maize cells
10. The following diagram represents the karyotype of a female *Drosophila melanogaster.* Is there anything unusual about this individual?

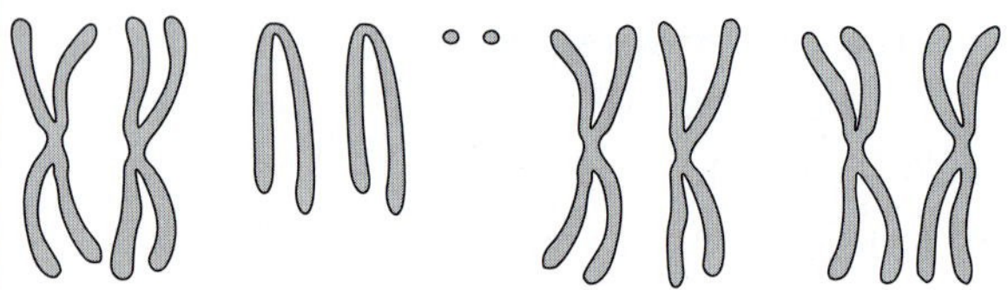

MEIOSIS IS THE BASIS FOR MENDEL'S RULES.

The physical basis of Mendel's rules lies in the two specialized meiotic divisions germ cells undergo. In all species, the key event in meiosis is that the number of sets of chromosomes is reduced by half. A diploid nucleus (2N) that goes through the two meiotic divisions gives rise to four haploid (N) nuclei. This reduction in the number of sets of chromosomes is essential for genetic stability in species that reproduce sexually. If chromosome reduction did not occur, each gamete produced by a diploid individual would also have a diploid (2N) chromosome number, and the fusion of two such gametes would produce a tetraploid (2N + 2N = 4N) zygote. If a viable individual arose from that zygote, it would produce tetraploid (4N) gametes. If two such gametes fused and produced a viable individual, that individual would have 8N chromosomes. This process of doubling the chromosome number would continue until eventually there would be so many chromosomes in each cell that no viable progeny would be produced.

Reduction in chromosome number occurs during the first of the two meiotic nuclear divisions, which is termed **meiosis I,** or the **reduction division.** The second meiotic division is termed **meiosis II,** or the **equational division,** because the chromosome number does not change during this nuclear division. The reduction in chromosome number during meiosis I is *the physical basis* for the segregation of alleles (Mendel's first rule). Other than the special behavior of the chromosomes in terms of number reduction, the cellular events (such as centriole division and spindle formation) are very similar or identical to those of mitosis. Because of the similarity of nonchromosomal events, our discussion of meiosis will focus primarily on chromosome behavior.

Homologous Chromosomes Segregate from Each Other During Anaphase of Meiosis I.

The first meiotic division is preceded by an interphase period containing G_1, S, and G_2 phases. A germ cell becomes committed to meiotic division during the G_1 phase of this interphase period. After that decision point, the cell enters the S phase, where the chromosomes replicate, and then it enters the G_2 phase. Following G_2, the cell enters meiosis I. Both meiotic nuclear divisions are divided into prophase, metaphase, anaphase, and telophase, just as in mitosis. But to distinguish the two meiotic divisions, the phases of meiosis I are termed prophase I, metaphase I, anaphase I, and telophase I, while the phases of the second meiotic division are called prophase II, metaphase II, anaphase II, and telophase II (Figure 3.11).

The prophase stage of meiosis I is longer and more complex than it is in meiosis II. During prophase I, the replicated chromosomes condense and undergo a series of complex actions that are essential for the upcoming reduction in chromosome number. For example, a delay in centromere replication is the key to the eventual segregation of **homologous chromosomes** (the two copies of a chromosome in diploids). The complexity of these events has led to a further subdivision of prophase I into five distinct substages: leptotene, zygotene, pachytene, diplotene, and diakinesis. Each of these substages was first defined by the behavior of the chromosomes.

The first phase of prophase I, **leptotene,** begins when the chromosomes become visible as long, thin threads that become progressively shorter and thicker (Figure 3.11a). Each replicated chromosome consists of two sister chromatids that are tightly associated and are not separately visible. The second phase of prophase I, **zygotene** (Figure 3.11a), begins when the condensing homologous chromosomes (or **homologues**) make contact and start to pair. Chromosome pairing starts at or near the telomeres, which are anchored to the nuclear envelope, and spreads along the length of the chromosomes, zipper-like, until the homologous chromosomes are precisely aligned. This physical pairing of homologous chromosomes during meiosis I is called **synapsis,** and it is unique to the reductional division. Synapsis is accompanied by the formation of a **synaptonemal complex** between the paired chromosomes. The complex consists of two dense lateral elements and a thin central element (Figure 3.12). The synaptonemal complex brings the alleles of the genes on each homologous chromosome into precise alignment. This precise pairing is essential for the exchange that occurs between the paired chromosomes later in prophase I.

During the next phase of prophase I, called **pachytene,** the synaptonemal complex formation is completed so that it extends from one end of the homologous chromosomes to the other. During pachytene the two sister chromatids of each homologous chromosome become visible (Figure 3.11b). Each pair of homologous chromosomes is termed a **bivalent.** The four chromatids in each bivalent are collectively termed a **tetrad.** At the end of pachytene, the synaptonemal complex dissociates, and the homologous chromosomes are held together by special points of contact called **chiasmata** (singular, chiasma). Chiasmata are the points where recombination has occurred, that is, where a physical exchange has occurred between one chromatid of one chromosome and another chromatid of its homologue. Recombination between homologous chromosomes in meiosis I has profound genetic and evolutionary implications. The use of recombination in genetic studies and the molecular mechanisms of recombination will be discussed later in this book (Chapters 6 and 19).

The next phase of prophase I, **diplotene** (Figure 3.11b), begins when the synaptonemal complex can no longer be seen under the microscope. During diplotene, the chromosomes continue to condense, and all four chromatids of the tetrad can be clearly seen. During the next and last phase of prophase I, **diakinesis,** the chromosomes reach their most condensed state, and the chiasmata appear to migrate to the ends of the chromosomes. During diakinesis, the nuclear envelope disappears, and the spindle fibers reach the chromosomes and connect to the kinetochore.

Metaphase I begins when the bivalents begin to migrate to the metaphase plate (Figure 3.11c). Each bivalent is attached by spindle fibers to both poles of the spindle. As the bivalents line up along the metaphase plate, one kinetochore in each bivalent faces one of the spindle poles. The most important event in meiosis occurs during the following phase, **anaphase I** (Figure 3.11d). During anaphase I the chiasmata dissociate, releasing the two homologous chromosomes of each bivalent. Unlike mitotic anaphase, the centromeres of each chromosome do not divide during anaphase I, and therefore the two sister chromatids do not separate to form new progeny chromosomes. Instead, the two chromosomes of each bivalent segregate from each other by moving toward opposite spindle poles.

As the migrating chromosomes approach the spindle poles, the cell enters **telophase I** (Figure 3.11e). Nuclear envelopes form around each set of chromosomes to produce two haploid nuclei. The events that occur between meiosis I and meiosis II vary widely among different species, and often between males and females in the same species. After meiosis I, cytokinesis may or may not occur. If two distinct cells are formed, they may remain paired as they undergo meiosis II.

The progeny nuclei produced at the end of meiosis I have half the number of chromosomes the parent cell had, and each chromosome has two chromatids. The precise pairing of homologous chromosomes during prophase I ensures that the progeny cells receive one complete haploid set of chromosomes and the alleles that are on them. This segregation of homologous chromosomes in anaphase I is the basis for Mendel's rule of segregation.

SECTION REVIEW PROBLEMS

11. Give a brief definition of each of the following pairs of terms. For each pair, give one similarity between them.
    a. leptotene and pachytene
    b. synaptonemal complex and chiasma
    c. tetrad and bivalent
12. Give one similarity and one difference between anaphase I of meiosis and anaphase of mitosis.

Sister Chromatids Separate During Meiosis II.

The equational meiotic division, meiosis II, begins at the end of the interphase that follows meiosis I. This interphase period can vary significantly in length in different species. There is no S phase because the chromosomes do not replicate, and there is only a single gap period (the equivalent of G_1) immediately followed by meiosis II.

Meiosis II is divided into prophase II, metaphase II, anaphase II, and telophase II; it more closely resembles a mitotic division than does meiosis I. In prophase II, the chromosomes condense and shorten, and the two sister chromatids of each chromosome are visible (Figure 3.11f). In many species the chromosomes remain in a relatively condensed state after meiosis I, and prophase II is short because the chromosomes quickly reach metaphase size and attach to the spindle fibers. During prophase II, the two chromatids of each chromosome are held together by a single centromere, and there is no chromosome pairing because no homologous chromosomes are present. During metaphase II, the chromosomes migrate to the metaphase plate and become aligned. Each chromosome has the same morphological characteristics in metaphase II as it had in mitotic metaphase (Figure 3.11g). A key difference between mitotic metaphase and metaphase II is that in metaphase II the cells have half as many chromosomes because of the chromosome reduction that occurred in meiosis I. In diploid species, there is only one copy of each chromosome in the cell during metaphase II.

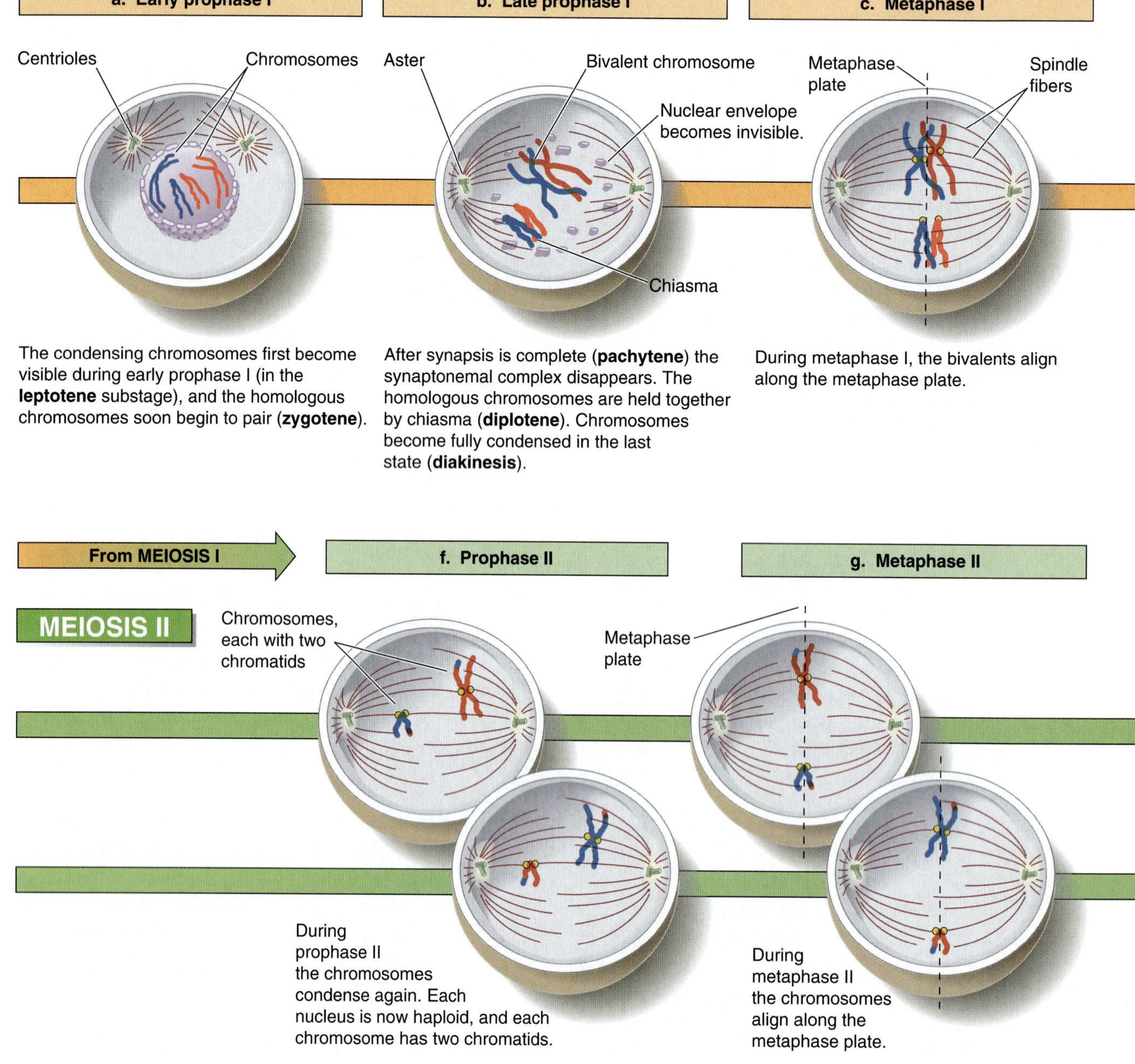

FIGURE 3.11 The nuclear divisions of meiosis I and meiosis II. Each meiotic division has four stages: prophase, metaphase, anaphase, and telophase. The complex prophase of meiosis I is subdivided into five phases based on visible chromosome behavior. Meiosis II has a number of similarities with mitotic division (compare this figure with Figure 3.2).

The cells enter anaphase II when the centromeres divide, and the two sister chromatids of each chromosome separate (Figure 3.11h). Each chromatid becomes one separate, progeny chromosome. One haploid set of chromosomes then migrates to each pole of the spindle. Telophase II begins as the chromosomes approach the spindle poles. During telophase II (Figure 3.11i), the chromosomes elongate and become diffuse, the nuclear envelopes form, and the chromosomes in the four haploid nuclei enter interphase.

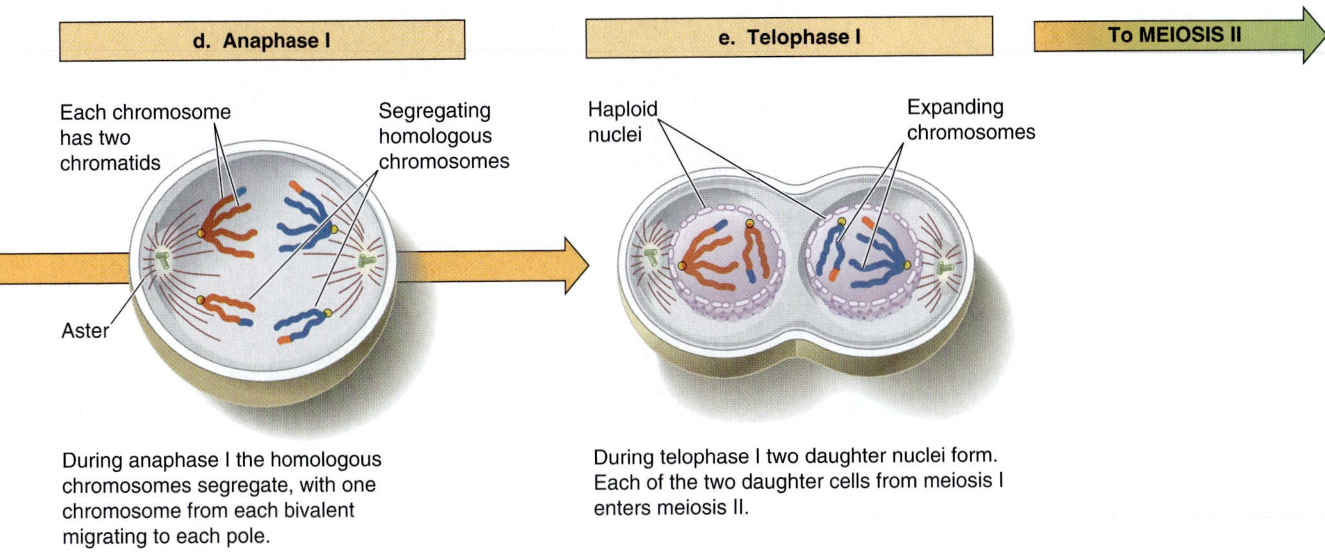
d. Anaphase I
Each chromosome has two chromatids
Segregating homologous chromosomes
Aster
During anaphase I the homologous chromosomes segregate, with one chromosome from each bivalent migrating to each pole.
e. Telophase I
Haploid nuclei
Expanding chromosomes
During telophase I two daughter nuclei form. Each of the two daughter cells from meiosis I enters meiosis II.
To MEIOSIS II

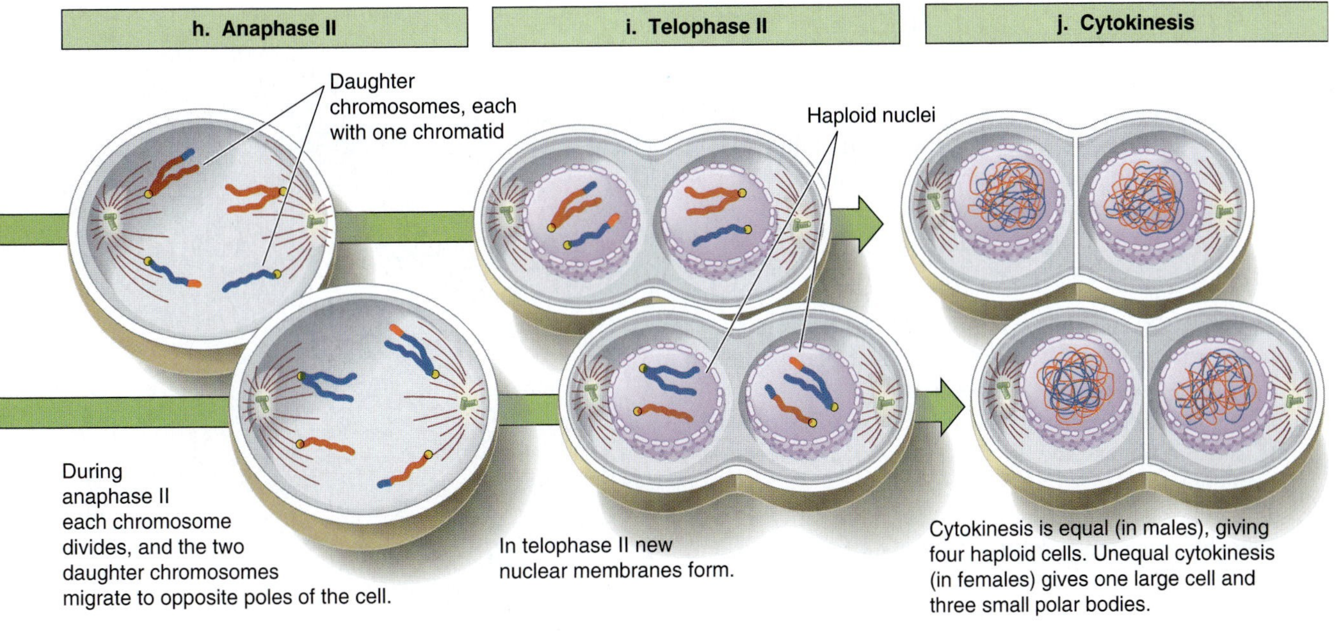
h. Anaphase II
Daughter chromosomes, each with one chromatid
During anaphase II each chromosome divides, and the two daughter chromosomes migrate to opposite poles of the cell.
i. Telophase II
Haploid nuclei
In telophase II new nuclear membranes form.
j. Cytokinesis
Cytokinesis is equal (in males), giving four haploid cells. Unequal cytokinesis (in females) gives one large cell and three small polar bodies.

The four nuclei produced from a single premeiotic nucleus are individually called **meiotic products** and are collectively termed a **tetrad.** Each of the four nuclei contains one of the four chromatids that were present in each bivalent before anaphase I. The fates of the four meiotic products vary in different species and in different sexes, but eventually at least one of the four products gives rise to the nucleus of a gamete. The behavior of the chromosomes during these divisions and what it meant was worked out in careful cytogenetic analyses by dedicated cytogeneticists such as Boveri [see *Historical Perspective: HP Boveri* (1862–1915)].

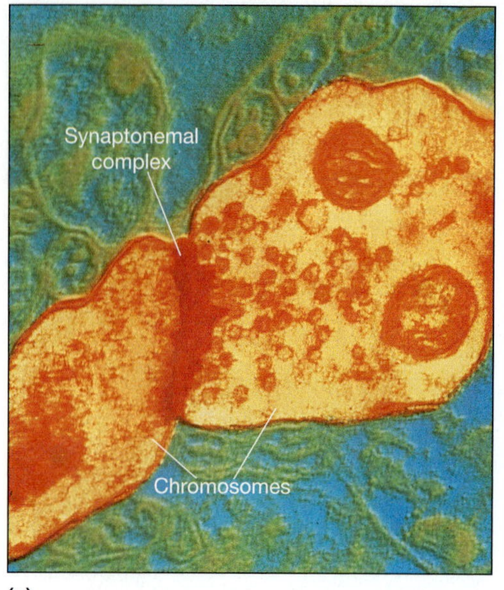

(a)

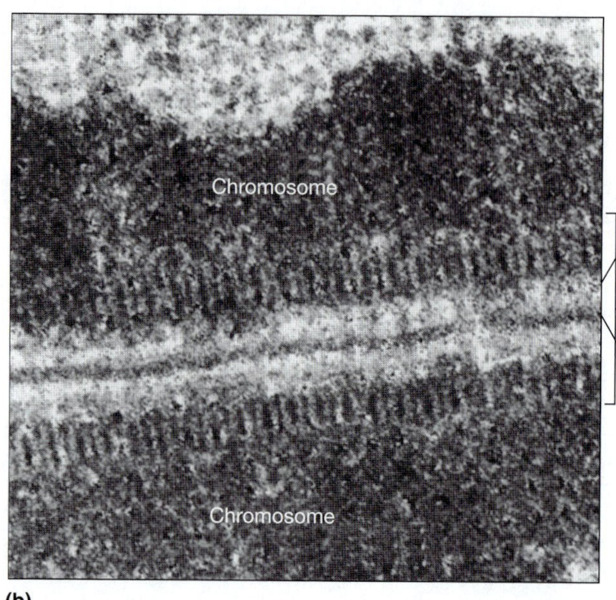

(b)

FIGURE 3.12 The synaptonemal complex. Within the synaptonemal complex, the DNA in the two homologous chromosomes is precisely aligned. (a) A false-color photograph of a synaptonemal complex between two homologous chromosomes, magnified 17,600 ×. (b) A higher magnification (36,000 ×) of the synaptonemal complex in the lily, showing the central and lateral elements. *(a, Custom Medical stock Photo; b, D. von Wettslein)*

SOLVED PROBLEM

Problem

The diagrams at right for animal species (2N = 4) show chromosomes from cells isolated in different stages of division. For each diagram, indicate what stage in mitosis or meiosis the cell is in.

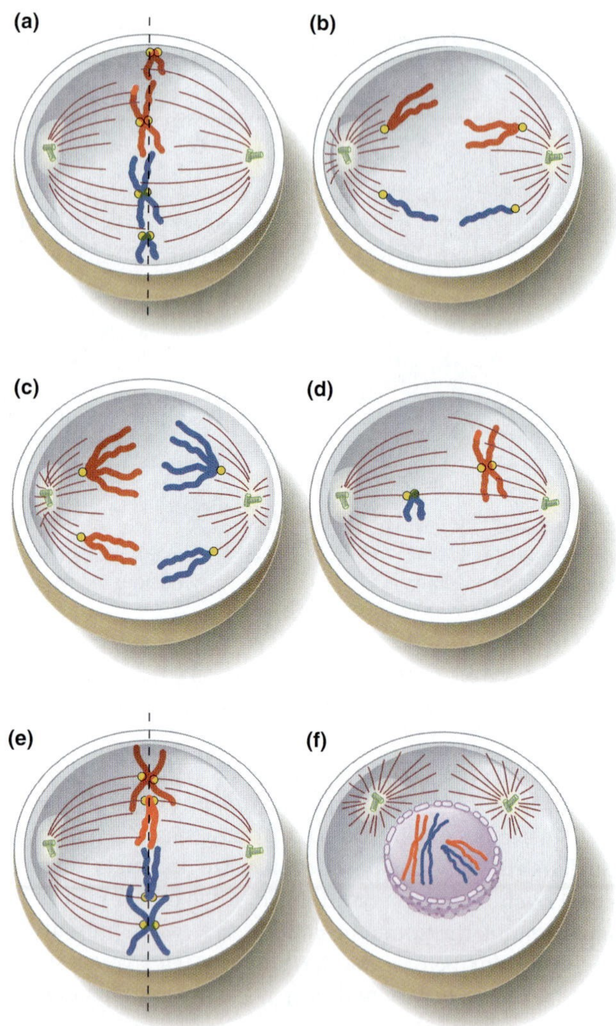

Solution

a. This cell is in metaphase I of mitosis. All four chromosomes are aligned on the metaphase plate, and no chromosome pairing has occurred.
b. This cell is in anaphase II of meiosis. Only one type of each chromosome is present, and each centromere has divided.
c. This cell is in anaphase I of meiosis. Each chromosome has two chromatids and is segregating from its homologue.
d. This cell is in prophase II of meiosis. Only two chromosomes are present, and the chromosomes are not yet aligned on the metaphase plate.
e. This cell is in metaphase II of meiosis. The chromosomes are paired, and the bivalents are aligned on the metaphase plate.
f. This cell is in the pachytene stage of prophase I of meiosis. The chromosomes are paired; the four chromatids in each bivalent are visible.

Historical Profile

Theodore Boveri (1862–1915)

Theodore Boveri is widely recognized as an outstanding cytologist and developmental biologist who, along with Walter Sutton, was one of two biologists to propose the Chromosome Theory of Heredity. Boveri was born on October 12, 1862, in Bamberg, Germany. The son of a physician, he studied anatomy and biology at Munich, Germany. In 1887, he was appointed a lecturer in zoology and comparative anatomy at Munich. His abilities were recognized, and in 1893 he was appointed to the Chair of Zoology and Comparative Anatomy at the University of Wurtzburg, a position he held until his death in 1915. Boveri made a number of important scientific contributions to the cytological study of inheritance. In addition, as an important biologist, his espousal of the value of experiments as opposed to simple observations influenced many other investigators, especially those with whom he worked during the summers at the Naples research station. These included his great friend and admirer, Thomas Hunt Morgan, who patterned his own summer program at the marine biological station at Wood's Hole, Massachusetts, on Boveri's organization at Naples.

Boveri's greatest scientific contributions came in two areas. First, working with the roundworm *Ascaris,* he observed that each chromosome in *Ascaris* had a unique morphology and that chromosomes with these same features reappeared during each cell division. He proposed that each chromosome was in some way unique, as opposed to the prevailing view that all chromosomes were identical. He also proposed that chromosomes did not physically dissolve during mitotic interphase but rather that they remained as discrete entities throughout the cell cycle. In 1888, working with *Ascaris,* Boveri discovered the existence of the centriole and defined its function in the generation of the aster and the spindle.

Boveri also worked with sea urchin embryos. Rather than being simply observations on biological processes, this work involved experimental intervention in fertilization and early development. His aim was to deduce rules of normal growth from the response of the system. Along with other contributions to genetics and development, he also demonstrated that artificially triggered development in sea urchins yielded relatively normal haploid (N) larvae. This was proof that both sets of chromosomes (maternal and paternal) contained all the hereditary information necessary for development. Boveri's work is considered to have established a vital connection between cytology and genetics. He is considered a founder of the modern school of experimental embryology. His example of rigorous experimental investigations had a great impact at a time when much of cytology was purely observational.

FIGURE 3.A Theodore Boveri. Recognized as an outstanding cytogeneticist, Boveri is a codiscoverer of the Chromosome Theory of Heredity. *(© Science Photo Library/Photo Researchers, Inc.)*

Chromosome Segregation in Meiosis I Is the Mechanism for Mendel's Rules.

As we discussed in the chapter introduction, Sutton and Boveri proposed the chromosome theory because they observed that the pattern of segregation of homologous chromosomes during anaphase I of meiosis resembled the segregation of Mendelian alleles. Mendel had proposed that the two copies of each gene found in each cell segregate during gamete formation to produce gametes with only one copy each. Sutton and Boveri observed that during anaphase I, the two homologous chromosomes found in each diploid cell separate, with one chromosome moving toward each pole of the cell. The alleles on each of the two homologues go into different progeny nuclei and ultimately into different gametes. Hence, chromosome segregation is the mechanism for physically segregating the alleles during gamete formation.

Consider a diploid individual that is heterozygous for one gene (*Aa*) (Figure 3.13). In prophase of meiosis I, the chromosome containing the *A* allele will pair with its homologue containing the *a* allele. During anaphase I, the homologues segregate, and during telophase I, two haploid progeny nuclei are formed, one with the chromosome containing the *A* allele and the other with the

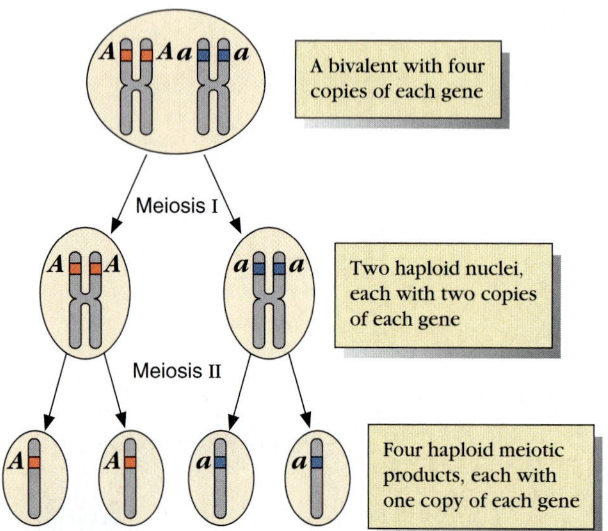

FIGURE 3.13 Chromosome behavior and Mendel's first rule. The segregation of homologous chromosomes during meiosis I is the mechanism underlying the segregation of alleles.

chromosome containing the *a* allele. After meiosis II, there will be four haploid meiotic products: two will contain only the *A* allele, and two will contain only the *a* allele.

Mendel's second rule states that alleles of two different genes assort independently. This is paralleled by the segregation behavior of different pairs of chromosomes in metaphase I and anaphase I. Each bivalent orients itself at the metaphase plate, unaffected by any other bivalent. Thus, any two *different* chromosomes will move to the same pole half the time and to the opposite pole half the time. The alleles of the genes on these chromosomes will also move to the same pole half the time and to the opposite pole half the time.

For example, consider a diploid individual that is heterozygous for two genes (*Aa* and *Bb*) that are located on different chromosomes (Figure 3.14). During prophase I, the homologous chromosome pair, forms two bivalents that align independently at the metaphase plate. During anaphase I, the chromosomes of each bivalent segregate, and each progeny nucleus will eventually receive one copy of each chromosome. There are two possible ways in which the bivalents may align at the metaphase plate; hence, there are two possible patterns of segregation. One progeny nucleus may receive the chromosomes containing dominant alleles (*AB*), whereas the other progeny nucleus receives the chromosomes that contain recessive alleles (*ab*). Alternatively, both progeny nuclei may receive one chromosome containing one dominant allele (*A* or *B*) and one chromosome containing one recessive allele (*a* or *b*). Each progeny nucleus would then contain either the genotype *Ab* or *aB* (Figure 3.14). Because these two patterns of chromosomal alignment occur equally often, all four nuclear genotypes (*AB*, *Ab*, *aB*, *ab*) will be present in the same frequency in the progeny nuclei. When these nuclei form gametes, the gametes will contain the genotypes *AB*, *Ab*, *aB*, and *ab* in equal proportions. Independent alignment and assortment of chromosomes in meiosis I provide the mechanism for Mendel's second rule, the independent assortment of alleles.

The independent assortment of chromosomes plays an important role in the generation of genetic diversity. The probability that any gamete will receive one particular chromosome from each parental homologous pair is $(1/2)^N$ in diploid organisms, where N is the number of chromosomes in the haploid set. For example, in a diploid species with two pairs of chromosomes ($N = 2$), the probability of one gamete receiving any two particular chromosomes is $(1/2)^2 = 1/4$. As the number of chromosomes increases, the number of different possible combinations also increases. Humans have 23 pairs of chromosomes, so there are 2^{23}, or 8,388,608, possible gamete chromosome combinations. Each of us receives 23 chromosomes from each parent, and the probability is very low [$(1/2)^{23}$, or 0.000000119] that we have received exactly the same set of chromosomes as a sibling of the same sex. Chromosome segregation during meiosis generates immense genetic variation in a population. More genetic variation makes it more likely that at least some individuals will survive under changing or variable environmental conditions.

Meiosis Occurs at Different Stages in Different Species' Life Cycles.

The life of every individual in a sexually reproducing diploid species includes a haploid phase and a diploid phase. At some point in the life cycle of a diploid species, germ cells undergo meiosis to generate haploid cells. Eventually, one or more of these haploid cells form gametes that fuse with another haploid gamete to generate a diploid zygote. The fusion of gametes is called **syngamy.** Note that the amount of the life span that individuals of different species spend in the diploid phase and in the haploid phase varies significantly (Figure 3.15). Complex higher organisms, such as *Drosophila*, humans, or flowering plants, spend almost all their life cycles in the diploid phase. Often the haploid meiotic products undergo no or only a very few mitotic divisions. At the other extreme, molds such as *Neurospora* and algae spend the vast majority of their life cycles in the haploid phase. Ferns and yeasts fall into an intermediate category, with about half the life cycle spent in the diploid and half in the haploid phase. The sexual reproduction process is a vital part of a species' life cycle, and complex genetic systems have evolved to regulate it.

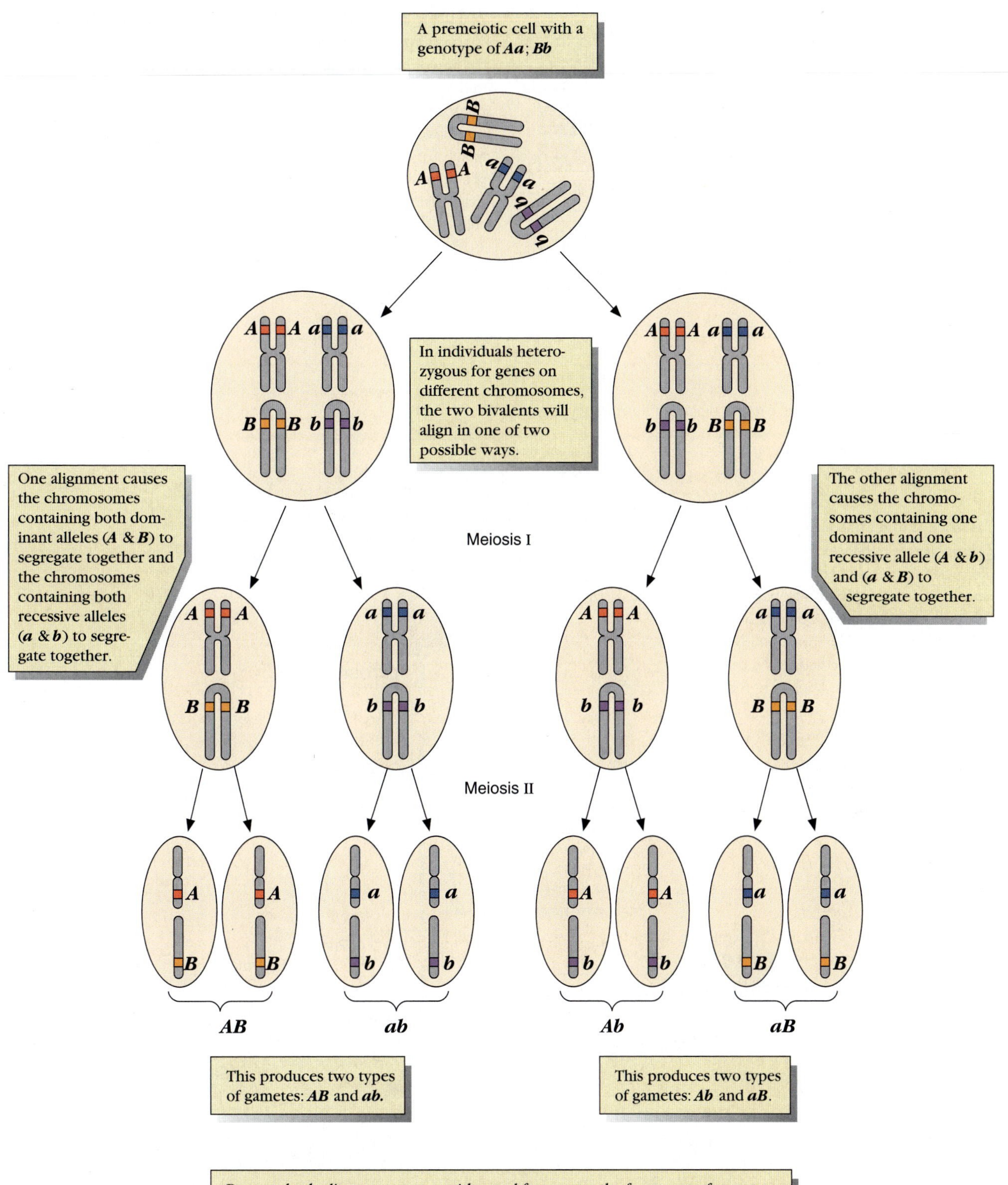

FIGURE 3.14 Chromosome behavior and Mendel's second rule. The independent alignment of each bivalent at the metaphase plate is the mechanism underlying the independent assortment of alleles of different genes.

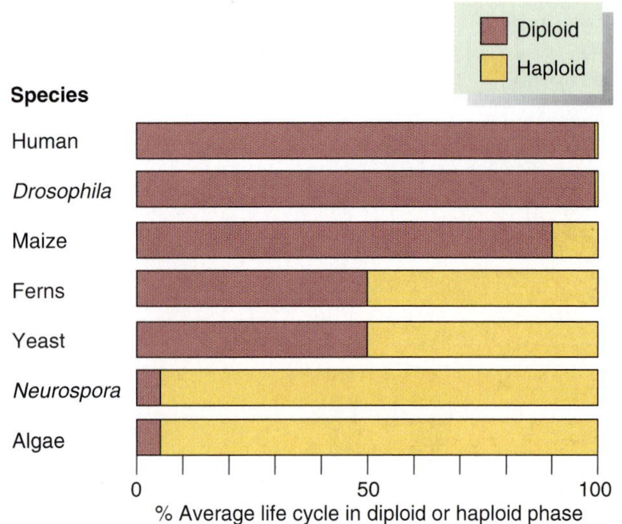

FIGURE 3.15 The haploid and diploid phases of different species life cycles. The average percentage of the life cycle spent in the haploid and diploid phases varies from species to species. Plant and fungal species spend significantly more time in the haploid phase. The percentages shown represent averages and can vary from species to species. For example, some algal species spend months as haploids and only a few days as diploids, whereas others spend most of their lives as diploids. Yeast can spend years as mitotically dividing haploid cells or years as mitotically dividing diploid cells.

FIGURE 3.16 The life cycle of *Drosophila melanogaster*. The *Drosophila* life cycle includes diploid and haploid stages. The embryonic, larval, pupal, and adult stages are diploid. Gametes are the only haploid stage. The haploid and diploid stages are indicated by the colored arrows (diploid = blue, haploid = red).

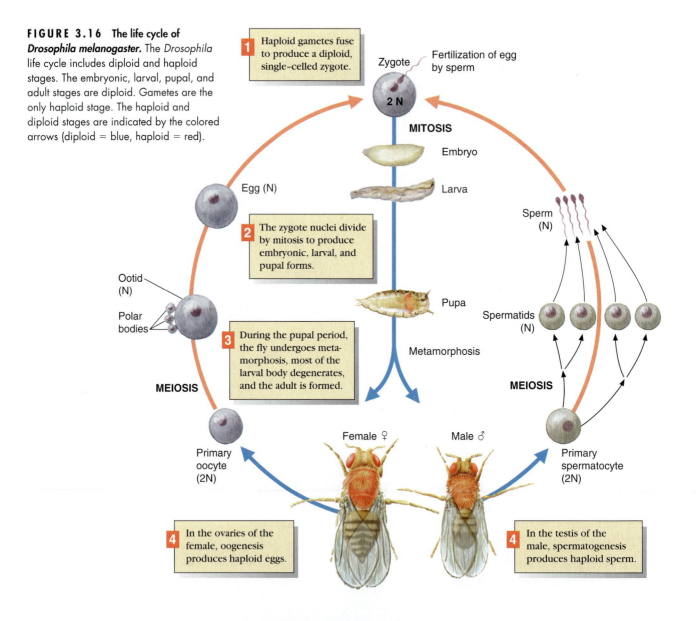

1 Haploid gametes fuse to produce a diploid, single-celled zygote.

Zygote

Fertilization of egg by sperm

2 N

MITOSIS

Embryo

Larva

Egg (N)

Sperm (N)

2 The zygote nuclei divide by mitosis to produce embryonic, larval, and pupal forms.

Ootid (N)

Polar bodies

Pupa

Spermatids (N)

3 During the pupal period, the fly undergoes metamorphosis, most of the larval body degenerates, and the adult is formed.

Metamorphosis

MEIOSIS

MEIOSIS

Primary oocyte (2N)

Female ♀

Male ♂

Primary spermatocyte (2N)

4 In the ovaries of the female, oogenesis produces haploid eggs.

4 In the testis of the male, spermatogenesis produces haploid sperm.

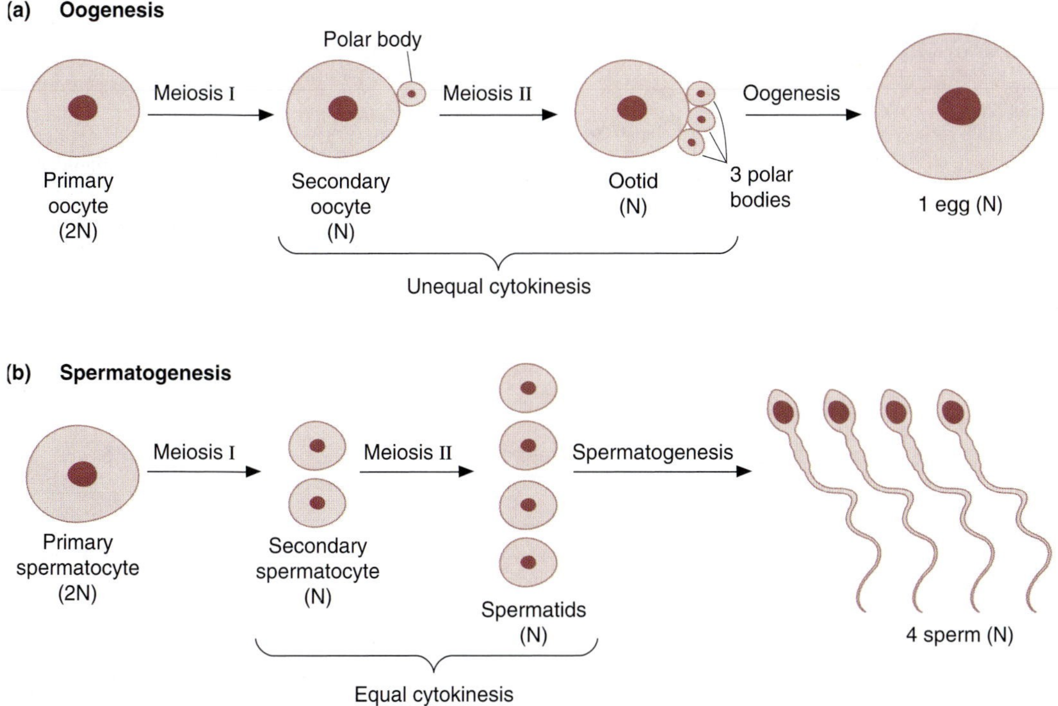

FIGURE 3.17 Gametogenesis in animals. (a) In female animals, the primary oocyte undergoes meiosis to produce one haploid ootid and three haploid polar bodies. Cytokinesis is unequal, and the ootid receives most or all the cytoplasm. (b) In male animals, the primary spermatocyte undergoes meiosis to produce four haploid spermatids. During spermatogenesis each spermatid matures into a single sperm.

Animal species spend most of their life cycle in the diploid stage, with the haploid stage represented by only the meiotic products that form the gametes. For example, the complete life cycle of *Drosophila* showing the haploid and diploid phases is shown in Figure 3.16. In animal species, eggs form from diploid germ cells in the ovaries of females. The germ cells divide mitotically to form **primary oocytes** that undergo meiosis (Figure 3.17). In meiosis I and meiosis II, cytoplasmic division is unequal, and one of the meiotic products receives almost all of the cytoplasm. The two meiotic divisions produce three small haploid **polar bodies** and one large **ootid.** The ootid contains the haploid nucleus of the female gamete and eventually develops into a mature egg. In males, sperm are formed in testes from diploid germ cells called **primary spermatocytes.** The diploid primary spermatocytes undergo meiosis to produce four haploid spermatids that mature into sperm (Figure 3.17). The fusion of one sperm and one egg generates a new diploid zygote.

Higher plants spend most of their life cycle in the diploid phase (called the **sporophyte**) and only a small part in the haploid phase (called the **gametophyte**). For example, in corn the gametophyte phase is limited to a few mitotic divisions in the reproductive structures of the flowers (Figure 3.18). Pollen develops in the anthers from premeiotic male germ cells, which undergo meiosis to produce four haploid **microspores,** each of which forms one **pollen grain** (the male gametophyte). The haploid nucleus of each pollen grain divides by mitosis to produce a **generative nucleus** and a **pollen tube nucleus.** The generative nucleus divides once again to produce two gametic nuclei. In the embryo sac, female germ cells undergo meiosis to produce four haploid **megaspore nuclei.** One of these becomes the female gametophyte, and the other three degenerate. The gametophyte nucleus undergoes three mitotic divisions to produce eight haploid nuclei. Two of these nuclei fuse to form the diploid nucleus of the central cell, one becomes the gametic nucleus, two form synergids, and three form antipodal cells. A germinating pollen grain sends a pollen tube into the embryo sac where a **double fertilization** occurs (Figure 3.18). One pollen gametic nucleus fuses with the female gametic nucleus to form a diploid zygotic nucleus. The other pollen gamete nucleus fuses with the diploid nucleus of the central cell to form a **triploid nucleus.** The central cell containing this triploid nucleus develops into a triploid tissue in the seed called the **endosperm.** The endosperm accumulates nutrients during the maturation of the seed that will support the young seedling after germination.

Yeasts, such as *Saccharomyces cerevisiae,* are eukaryotic fungi that, in marked contrast to higher plants and

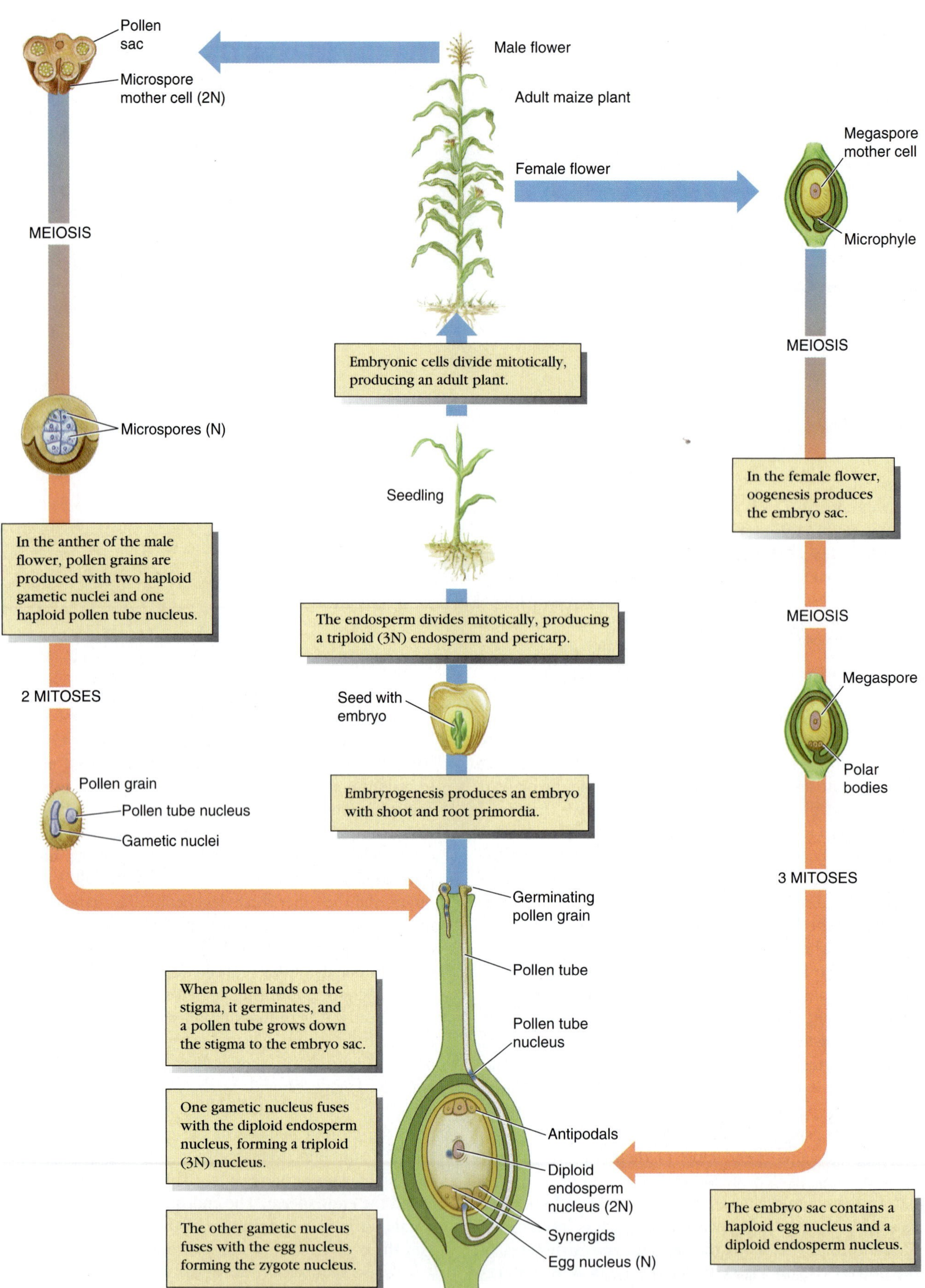
Pollen sac
Microspore mother cell (2N)
Male flower
Adult maize plant
Female flower
Megaspore mother cell
Microphyle
MEIOSIS
Embryonic cells divide mitotically, producing an adult plant.
Microspores (N)
MEIOSIS
Seedling
In the female flower, oogenesis produces the embryo sac.
In the anther of the male flower, pollen grains are produced with two haploid gametic nuclei and one haploid pollen tube nucleus.
The endosperm divides mitotically, producing a triploid (3N) endosperm and pericarp.
MEIOSIS
2 MITOSES
Seed with embryo
Megaspore
Polar bodies
Pollen grain
Pollen tube nucleus
Gametic nuclei
Embryrogenesis produces an embryo with shoot and root primordia.
3 MITOSES
Germinating pollen grain
Pollen tube
When pollen lands on the stigma, it germinates, and a pollen tube grows down the stigma to the embryo sac.
Pollen tube nucleus
One gametic nucleus fuses with the diploid endosperm nucleus, forming a triploid (3N) nucleus.
Antipodals
Diploid endosperm nucleus (2N)
Synergids
Egg nucleus (N)
The other gametic nucleus fuses with the egg nucleus, forming the zygote nucleus.
The embryo sac contains a haploid egg nucleus and a diploid endosperm nucleus.

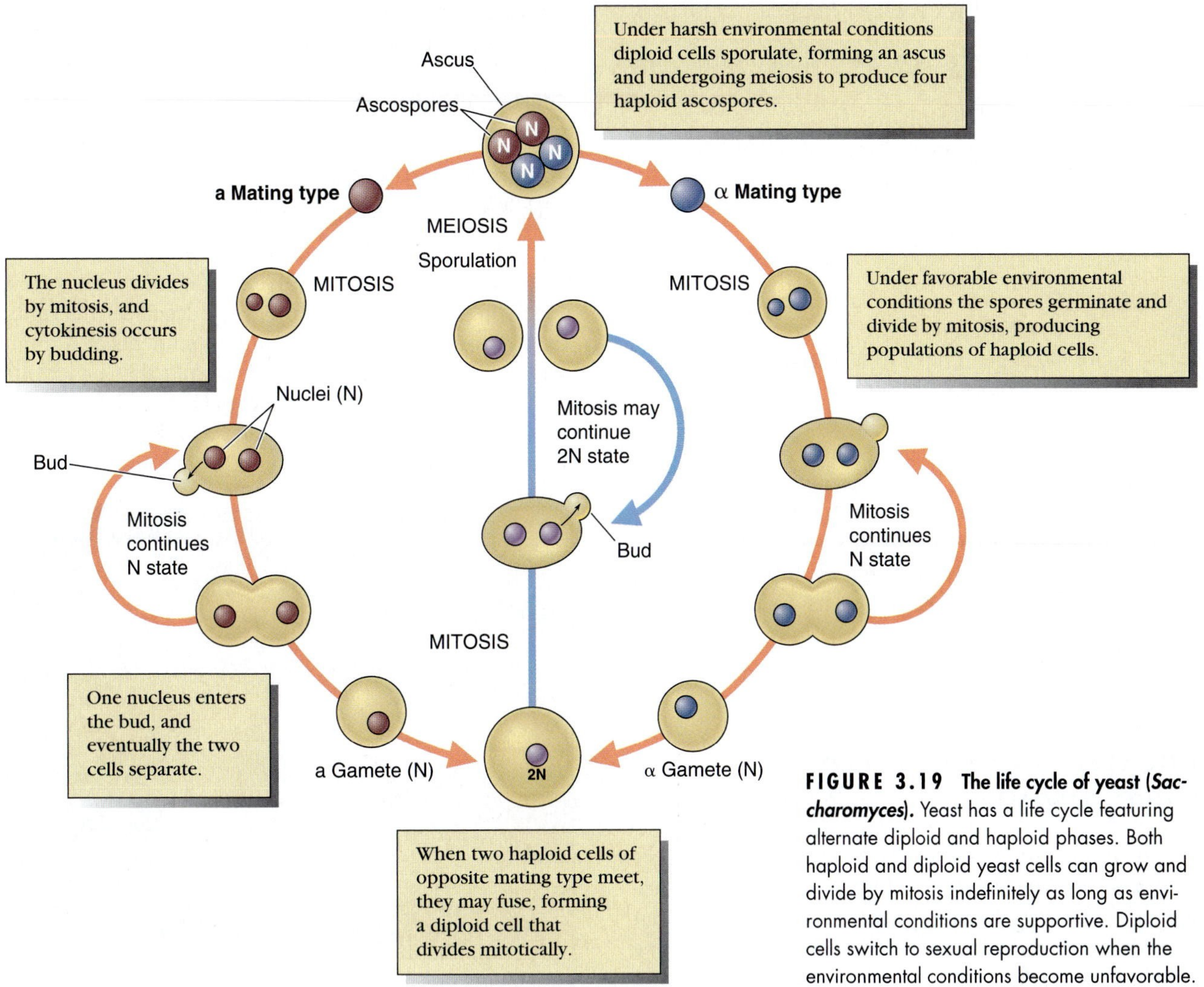

Under harsh environmental conditions diploid cells sporulate, forming an ascus and undergoing meiosis to produce four haploid ascospores.
Ascus
Ascospores
N
N
N
N
a Mating type
α Mating type
MEIOSIS
Sporulation
The nucleus divides by mitosis, and cytokinesis occurs by budding.
MITOSIS
MITOSIS
Under favorable environmental conditions the spores germinate and divide by mitosis, producing populations of haploid cells.
Nuclei (N)
Mitosis may continue 2N state
Bud
Bud
Mitosis continues N state
Mitosis continues N state
MITOSIS
One nucleus enters the bud, and eventually the two cells separate.
a Gamete (N)
2N
α Gamete (N)
When two haploid cells of opposite mating type meet, they may fuse, forming a diploid cell that divides mitotically.

FIGURE 3.19 The life cycle of yeast (*Saccharomyces*). Yeast has a life cycle featuring alternate diploid and haploid phases. Both haploid and diploid yeast cells can grow and divide by mitosis indefinitely as long as environmental conditions are supportive. Diploid cells switch to sexual reproduction when the environmental conditions become unfavorable.

animals, can exist as either haploid or diploid organisms. Yeasts are usually unicellular, but many species can also exist in multicellular, filamentous forms. Unicellular yeast divide mitotically in a process of vegetative reproduction called **budding.** As a cell enters mitosis, a small cytoplasmic extension, called a bud, forms and increases in size. After mitosis, one of the progeny nuclei enters the bud. The bud separates shortly after it becomes equal in size to the original cell and becomes an independent cell. Diploid and haploid yeast cells can continue vegetative reproduction as long as environmental conditions are supportive.

Saccharomyces has two mating types, a and α (Figure 3.19). When haploid a and α cells come into contact, they may fuse to form a single diploid cell. This diploid cell can reproduce vegetatively under favorable conditions, but when conditions become unfavorable, diploid yeast cells switch to sexual reproduction. The diploid cell undergoes meiosis and produces four meiotic products that mature into four haploid **ascospores.** These are packaged in a specialized structure called an **ascus** (plural, asci). When culture conditions improve, the ascospores germinate, and the four haploid cells begin vegetative reproduction. This ability to shift between asexual (vegetative) and sexual reproduction offers yeast the ability to adapt quickly to changing conditions. The study of yeast reproduction and the complex genetic systems that control it is an important part of yeast genetics.

◀**FIGURE 3.18 The life cycle of maize, a flowering plant.** In maize, the embryonic and adult plants are diploid. The haploid stage includes the two mitotic divisions in the pollen grains and the three mitotic divisions in the ovary. These divisions produce the two haploid pollen gametic nuclei that fuse with the haploid egg nucleus and diploid central cell nucleus in a double fertilization.

SECTION REVIEW PROBLEMS

13. Give one difference between each of the following pairs of terms.
    a. oogonia and polar body
    b. primary spermatocyte and spermatid
    c. gametophyte and sporophyte
    d. synergids and antipodals
    e. endosperm and embryo sac
    f. bud and ascospore
14. a. Briefly explain one similarity between meiosis II and mitosis.
    b. Briefly explain three differences between meiosis and mitosis.
15. a. Briefly explain two differences between male and female gametogenesis in animals.
    b. Briefly explain two differences between male and female gametogenesis in flowering plants.
16. Briefly explain the events of sexual and vegetative reproduction in yeast. What evolutionary advantages might exist because of its ability to reproduce in alternative ways?

SEX-DETERMINING GENES ARE LOCATED ON SPECIALIZED CHROMOSOMES.

When Sutton and Boveri first proposed the chromosome theory of inheritance, not all biologists accepted it immediately. Although many investigators believed that chromosomes did contain hereditary information, the physical assignment of Mendelian genes to chromosomes needed rigorous proof (e.g., documented cases where specific Mendelian alleles could be unmistakably associated with a particular chromosome). This proof was found in a series of studies of a special class of chromosomes, the sex chromosomes.

Sex chromosomes were first discovered in 1902 when karyotype analyses revealed that in several species of grasshoppers, females and males have different numbers of copies of one particular chromosome. This chromosome was termed the **sex,** or **X, chromosome,** and the other chromosomes were termed **autosomes.** In mammals, including humans, karyotypically normal females contain two copies of the X chromosome, whereas males contain one X chromosome and one structurally distinct Y chromosome. Because females have two X chromosomes, they produce gametes with the same sex chromosome composition (one X chromosome). Therefore, female mammals are considered the **homogametic sex.** Because males have one X and one Y chromosome, they produce gametes with two different sex chromosome compositions (half contain an X chromosome and half contain a Y chromosome). Male mammals are termed the **heterogametic sex.** Two types of studies were extremely influential in the acceptance of the chromosome theory: studies of sex-linked genes and studies of sex chromosome nondisjunction.

Specific Genes Segregate with Specific Chromosomes.

Key evidence supporting the chromosome theory came from studies of mutant alleles of **sex-linked genes,** which are inherited differently in males and females. In 1910, Thomas Hunt Morgan and his students at Columbia University found a male *Drosophila* that had white eyes instead of the normal red eyes (Figure 3.20). Today this mutation is called *white* and has the symbol *w;* the

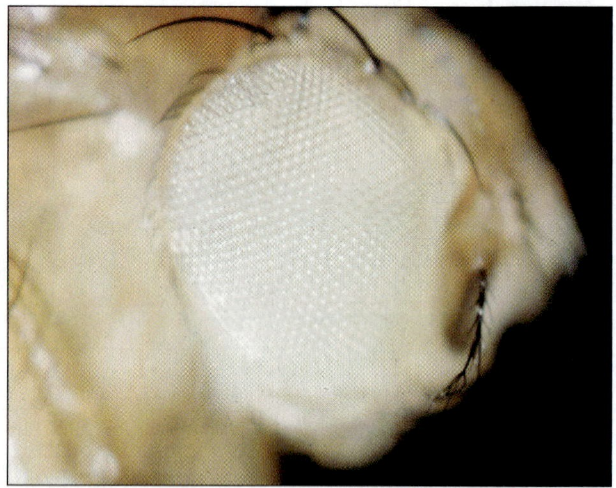

(a)

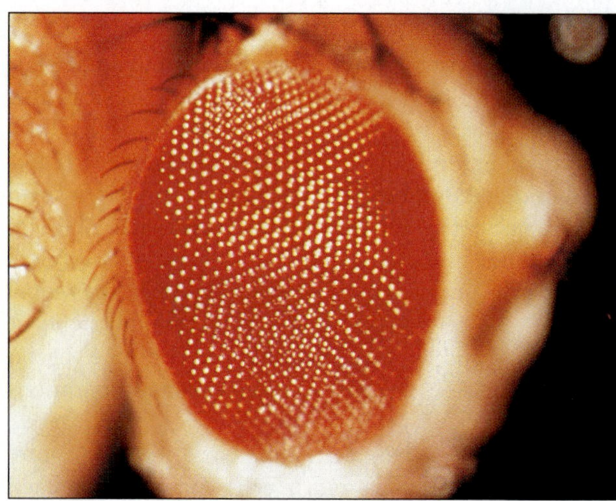

(b)

FIGURE 3.20 The *white* gene in *Drosophila.* (a) Flies that are homozygous for the *white* mutation have no pigment in their eyes. (b) Flies homozygous or heterozygous for the wild-type allele of this gene have red eyes. *(a and b, © Cabisco/Visuals Unlimited)*

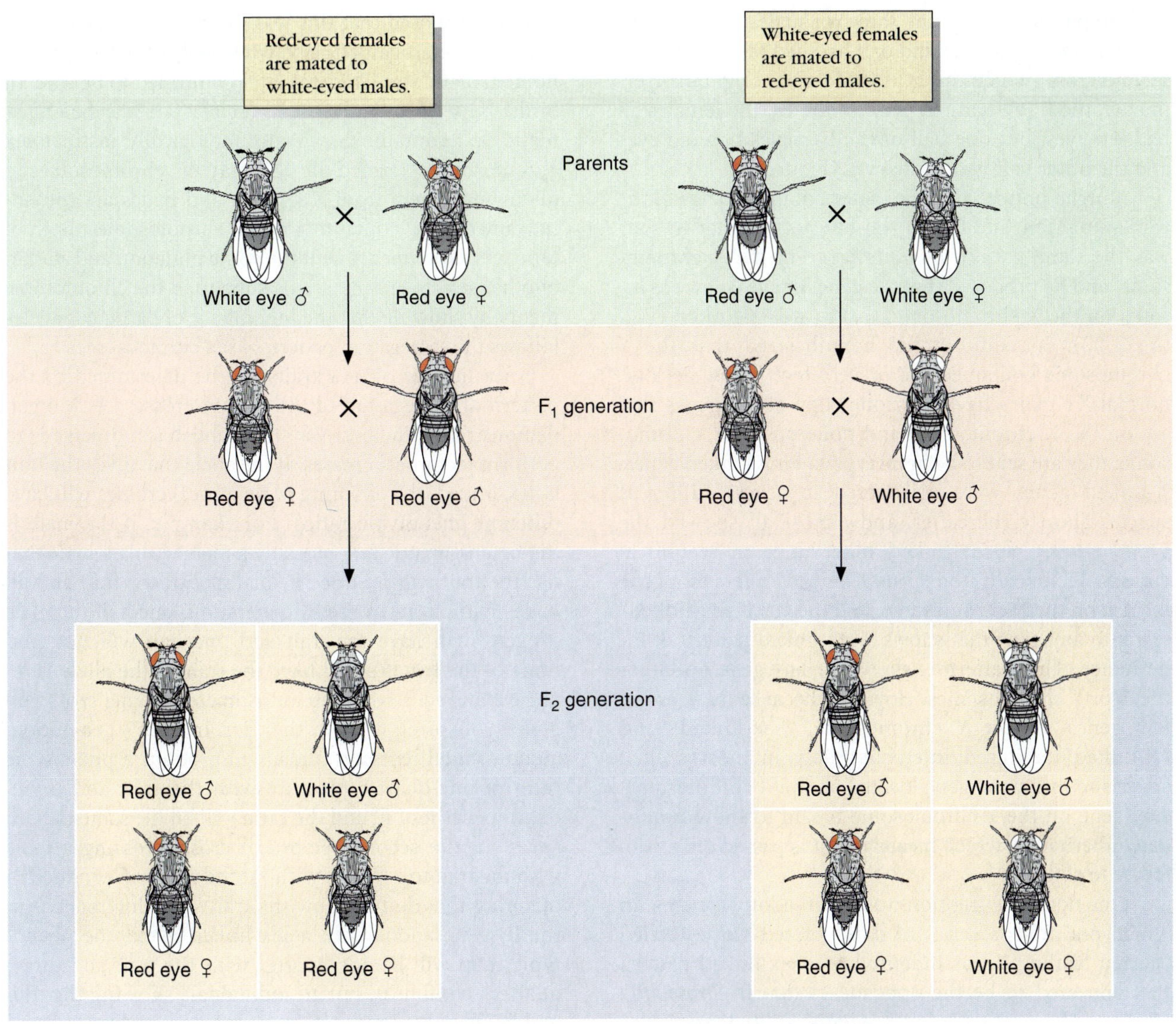

FIGURE 3.21 The *white* mutation demonstrates sex-linked inheritance. The pattern of inheritance of the sex-linked *white* mutation can be seen by tracing the white-eyed phenotype in reciprocal crosses.

normal allele of this gene has the symbol w^+, or simply + (the terminology used in *Drosophila* to name alleles and allele symbols is discussed in Chapter 2). When Morgan crossed a white-eyed male to a normal red-eyed female, all the progeny had red eyes. They concluded that w was recessive to w^+. He then inbred the F_1 flies to produce an F_2 generation. The F_2 flies had an approximately 3 : 1 ratio of red eyes to white eyes, as he had expected for a recessive allele. However, the ratios in the F_2 flies were different for each sex. One half of the F_2 males had white eyes and one half had red eyes, but all the F_2 females had red eyes.

To investigate this further, Morgan inbred the F_2 flies and established a strain of true-breeding flies with white eyes. He then performed two crosses (Figure 3.21). In one cross, males from the white-eyed strain were crossed with females from the red-eyed strain. In the other cross, females from the white-eyed strain were mated with males from the red-eyed (wild-type) strain. Such pairs of crosses, in which only the sex of the parents is different, are called **reciprocal crosses.** For each of Morgan's crosses, the F_1 flies were collected and inbred to produce an F_2 generation. In the first cross, all F_1 males had white eyes, and all F_1 females had red eyes. In the F_2, the phenotypic ratios were the same for males and females, one half had white eyes and one half had red eyes. In the second cross (which duplicated Morgan's original experiment), the F_1 all had red eyes. In the F_2, the females all had red eyes, but one half of the F_2 males had red eyes and the other half had white eyes (Figure 3.21).

Morgan noticed that the pattern of inheritance of the *white* mutation (mother to son but never father to son) was the same pattern of inheritance of the X chromosome, and he proposed that the gene for eye color was located on the X chromosome, that is, it is **X-linked** (Figure 3.22). In addition, he hypothesized that the Y chromosome contained no, or very few, genes. Because normal XY males have only one copy of the genes that are on the X chromosome and none on the Y chromosome, they are said to be **hemizygous** for X-linked genes. X-linked genes were discovered in several different species shortly thereafter, and almost all showed the same pattern of sex-linked inheritance in reciprocal crosses. Technically, the term "X-linked" refers to a gene located on the X chromosome, and the term "sex-linked" refers to any gene that shows a sex-linked pattern of inheritance. This pattern is shown by any gene on either the X or Y chromosomes. However, because there are so few genes on the Y chromosome, "sex-linked" and "X-linked" are used interchangeably in most animal species to mean any gene located on the X chromosome. Any gene on the Y chromosome is said to show **holandric** inheritance, which means that it is passed only from father to son.

One notable exception was observed in chickens, in which one trait, a series of dark-colored stripes called "barred feathers," was inherited in a sex-linked pattern that appeared to be the opposite of that in *Drosophila* (Figure 3.23). The German geneticist Richard Goldschmitt proposed that this was the result of female birds being the heterogametic sex, whereas male birds were the homogametic sex. This was later confirmed to be true. In birds, butterflies, and moths, females are the heterogametic sex, containing two morphologically distinct sex chromosomes (termed the "Z" and "W" chromosomes to distinguish them from X and Y), and males are the homogametic sex, containing two Z chromosomes. Both of Morgan's experiments with the white mutation and studies of other sex-linked mutations supported the chromosome theory by showing that the inheritance of particular alleles follows the segregation pattern of sex chromosomes.

Sex linkage of new traits can be determined by the pattern of segregation of the trait in crosses. As Morgan demonstrated, the best way to establish sex linkage is to perform reciprocal crosses. If the allele that gives the trait is sex-linked, the offspring of reciprocal crosses will have different phenotypic ratios. For example, if the male is the heterogametic sex, crossing a male with a female that has the trait will produce F_1 males with the trait and females without it. In the F_2 generation, one half the male progeny will have the trait and one half will not, and none of the females will have the trait. If the allele is recessive and located on an autosome, the same cross will give F_1 offspring without the trait. In the F_2 generation, an autosomal recessive allele will produce a phenotypic ratio of one-quarter progeny with the trait and three-quarters without it, and the ratio will be the same in both sexes. In the second reciprocal cross, crossing a male with the trait to a female without the trait will produce F_1 offspring that do not show the trait. The difference is in the F_2 generation. With a sex-linked allele, the phenotypic ratio will be one quarter with the trait and three-quarters without it, but all individuals showing the trait will be male.

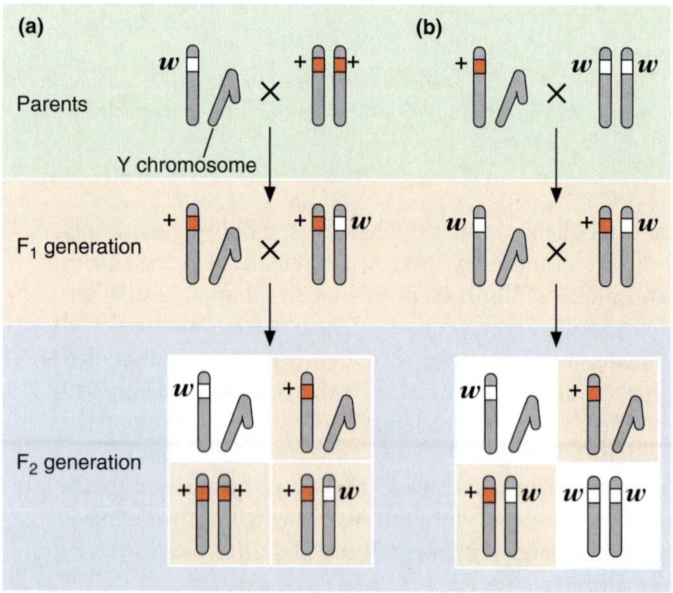

FIGURE 3.22 The chromosomal basis for sex-linked inheritance patterns. Sex-linked inheritance results from the inheritance pattern of the sex chromosomes. (a) In this cross, the sons of *ww* mothers (the F_1 males) have a single X chromosome with a *w* allele, whereas their progenies (the F_1 females) are heterozygous (+*w*). The F_1 females pass an X with a *w* allele to half of their sons (the F_2 males) and an X with a + allele to the other half. The F_1 males pass an X with a *w* allele to all their progeny (the F_2 females). (b) In this cross, the sons of ++ mothers (the F_1 males) have a single X chromosome with a + allele, whereas their progeny (the F_1 females) are heterozygous (+*w*). The F_1 females pass an X chromosome with a + allele to half their sons (F_2 males) and an X chromosome with a *w* allele to the other half. The F_1 males pass an X chromosome with a + allele to all their progeny (the F_2 females). Note that the Y chromosome is represented by a symbol that resembles the cytological appearance of the Y chromosome in *Drosophila* metaphase cells.

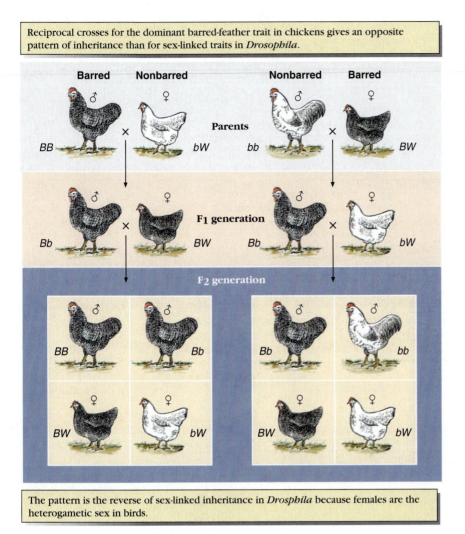

Reciprocal crosses for the dominant barred-feather trait in chickens gives an opposite pattern of inheritance than for sex-linked traits in *Drosophila*.

The pattern is the reverse of sex-linked inheritance in *Drosphila* because females are the heterogametic sex in birds.

FIGURE 3.23 Inheritance of feather pattern in chickens. Reciprocal crosses show the sex-linked inheritance for the barred-feather trait in chickens. Female chickens are heterogametic (WZ) and male chickens are homogametic (ZZ). Compare this figure with Figure 3.21 to see that the inheritance pattern of the recessive trait white feathers is the reverse of that of the recessive trait white eye color in *Drosophila*.

When geneticists suspect that a human trait or disease may be sex-linked, they use the tools of pedigree analysis that we discussed in Chapter 2 to determine the trait's inheritance pattern. In pedigrees, alleles of sex-linked genes show identifiable patterns of inheritance. Women heterozygous for a recessive, sex-linked mutant allele have a 50% chance of passing the mutant allele to their offspring. If their sons inherit the allele, they will show the mutant phenotype. Such men cannot pass the allele to their sons but always pass it to their daughters. An historically significant case of inheritance of a sex-linked disease is that of hemophilia A in the descendants of British Queen Victoria. Men who inherit the hemophilia A allele produce a nonfunctional form of a protein vital for proper blood clotting (called factor VIII). Hemophiliacs who lack a functional factor VIII bleed excessively from even minor wounds. Victoria had 9 children and 40 grandchildren (Figure 3.24). One of her sons had hemophilia and three did not, which suggests that she was heterozygous for the recessive, sex-linked allele that gives hemophilia A. Two of her daughters were healthy but passed the allele to their sons and daughters. Heterozygotes for deleterious alleles (also called **carriers,** see Chapter 2) appear healthy but may pass a recessive genetic disease to their children. The pattern of inheritance of hemophilia suggests that it is sex-linked. The disease appeared in the sons of carriers, but not in the daughters. In addition, we can see that individual II-8 (Leopold, Duke of Albany) was a hemophiliac, his daughter Alice (III-10) was healthy, but her son Rupert (IV-12) had hemophilia. This three-generation pattern is classic for a recessive, sex-linked allele.

A number of other human diseases are caused by sex-linked alleles, including the metabolic disorder Lesch-Nyhan syndrome, Duchenne muscular dystrophy, testicular feminization syndrome, and fragile X syndrome (the last of which will be discussed in Chapter 17). A form of color blindness is also caused by a sex-linked recessive allele, which shows that not all mutations producing obvious phenotypes cause medically significant disorders.

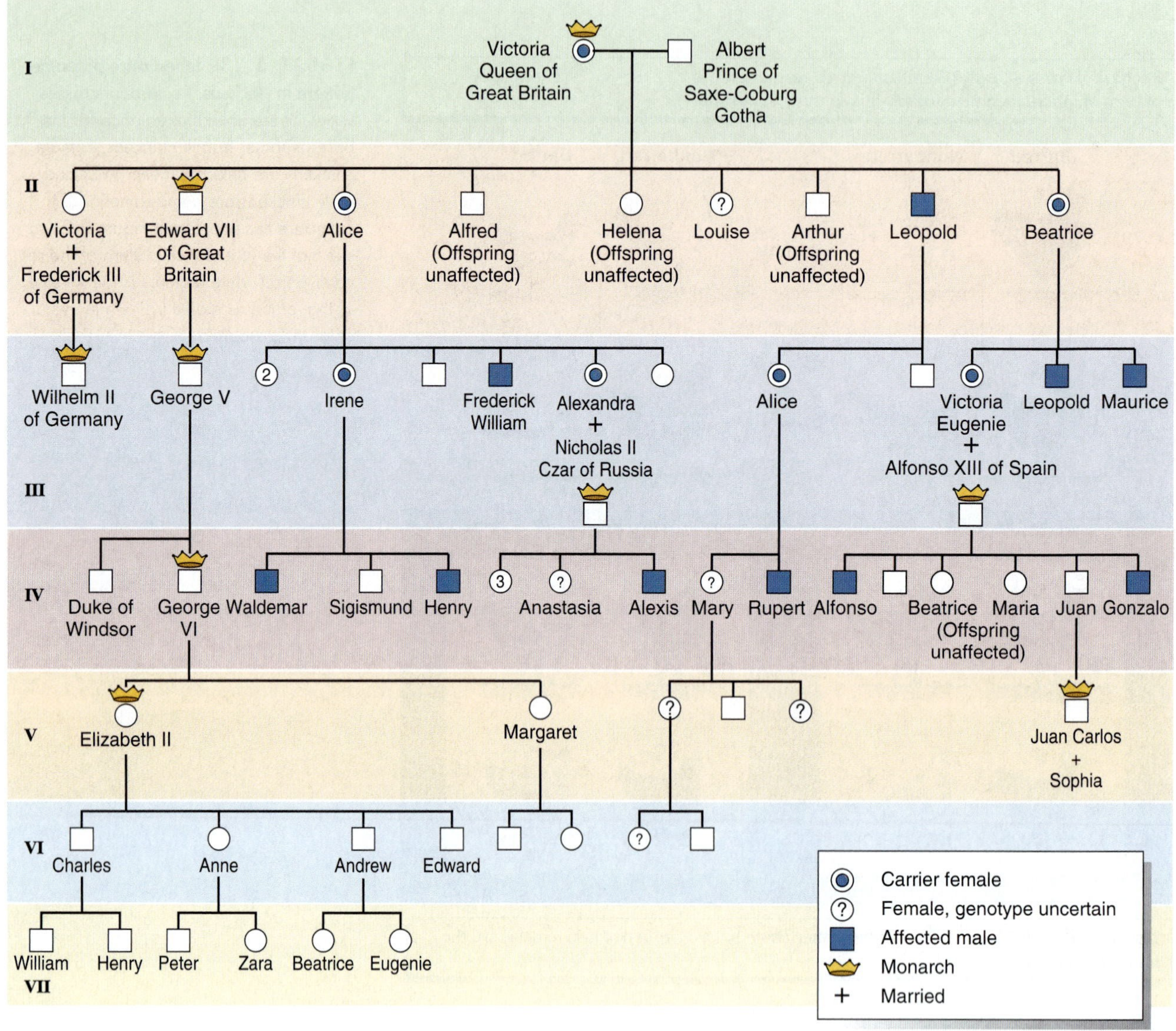

FIGURE 3.24 Inheritance of hemophilia in humans. This pedigree shows the inheritance of hemophilia in the descendants of Queen Victoria of Great Britain. The inheritance pattern suggests that she was heterozygous for the recessive, sex-linked allele that causes the disease.

SECTION REVIEW PROBLEMS

17. A rare, dark-bodied male moth was crossed with a common light-bodied female moth. The F_1 progeny were all dark. The F_1 were inbred, and half the F_2 females were dark and half were light, and all the F_1 males were dark. Give one hypothesis that would account for these results, and describe one test you could do to confirm your hypothesis.
18. Suppose you accidentally mixed several parental dark male moths from Problem 17 with dark F_1 male moths. How would you determine which is which?
19. Female *Drosophila* with yellow bodies and rosy eyes were crossed with wild-type males with gray-brown bodies and red eyes. The F_1 flies were inbred to produce an F_2 generation. The phenotypes of the F_2 were as follows:

	yellow rosy	yellow red	gray-brown rosy	gray-brown red
F_2 males	1/8	3/8	1/8	3/8
F_2 females	1/8	3/8	1/8	3/8

a. How are these traits inherited?
b. Using symbols of your own choice, draw a diagram showing the inheritance of the alleles producing the yellow and rosy traits in this cross.
c. What were the phenotypes of the F_1 males and females from this cross?

The Final Proof of the Chromosome Theory Came from Sex-Chromosome Nondisjunction.

The work considered the final proof of the chromosome theory came from a continuation of Morgan's study of the *white* allele in *Drosophila*. It was carried out by one of Morgan's students, Calvin Bridges, who had noticed that when white-eyed females were mated with red-eyed males, most of the offspring were red-eyed females and white-eyed males, as expected, but a few (roughly 1/2000) were white-eyed females and red-eyed males. The rare red-eyed males were sterile. However, when Bridges mated the rare white-eyed females to normal red-eyed males, progeny were produced. He found that about 4% of these progeny were also the unusual white-eyed females and red-eyed males.

To determine the cause of these unusual results, Bridges analyzed the karyotypes of the flies. He discovered that the unusual white-eyed females contained normal autosomes but had three sex chromosomes, two X chromosomes and a Y chromosome (XXY), and that the unusual red-eyed males had normal autosomes but no Y chromosome (X). He hypothesized that in the original cross, the two X chromosomes in the white-eyed females (XX) were carrying *white* alleles and that the unusual offspring were the result of rare cases where the two X chromosomes failed to segregate from each other during anaphase I of meiosis.

The segregation of homologous chromosomes (disjunction) during anaphase I of meiosis is highly accurate, but it is not perfect. In a low but detectable percentage of germ cells, both X chromosomes segregate together. The failure of homologous chromosomes to segregate properly during anaphase I is called **nondisjunction** (Figure 3.25). In Bridges' experiment, fertilization of the XX eggs by a normal sperm with a Y chromosome (Y) produced an XXY female with white eyes. Fertilization of an egg with no X chromosome ("O" denotes the absence of a chromosome) by a normal sperm with an X chromosome (X) produced a male with no Y chromosome (XO).

When nondisjunction occurs in individuals with a normal chromosome composition (such as the white-eyed XX females), it is called **primary nondisjunction** (Figure 3.25a), but if improper separation occurs in the

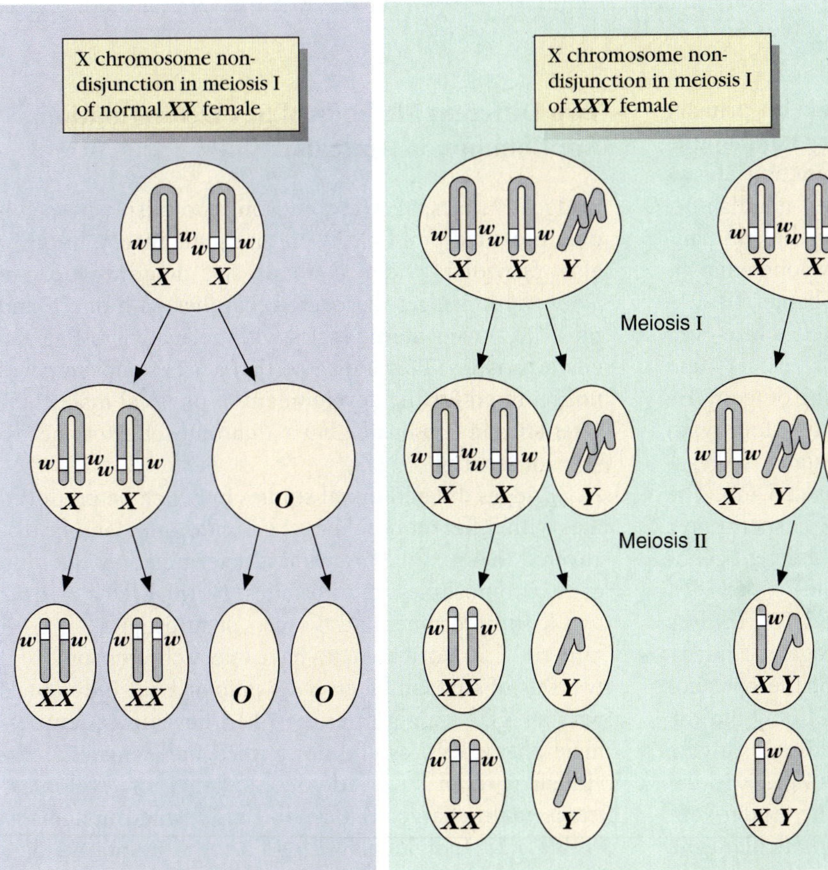

FIGURE 3.25 X chromosome nondisjunction in *Drosophila melanogaster*. (a) Primary X chromosome nondisjunction in a normal, diploid XX female produces gametes with either two X chromosomes (XX) or no X chromosome (O). (b) Secondary X chromosome nondisjunction in a female whose germ cells have an XXY genotype produces gametes with four different chromosome compositions: XX, Y, X, or XY.

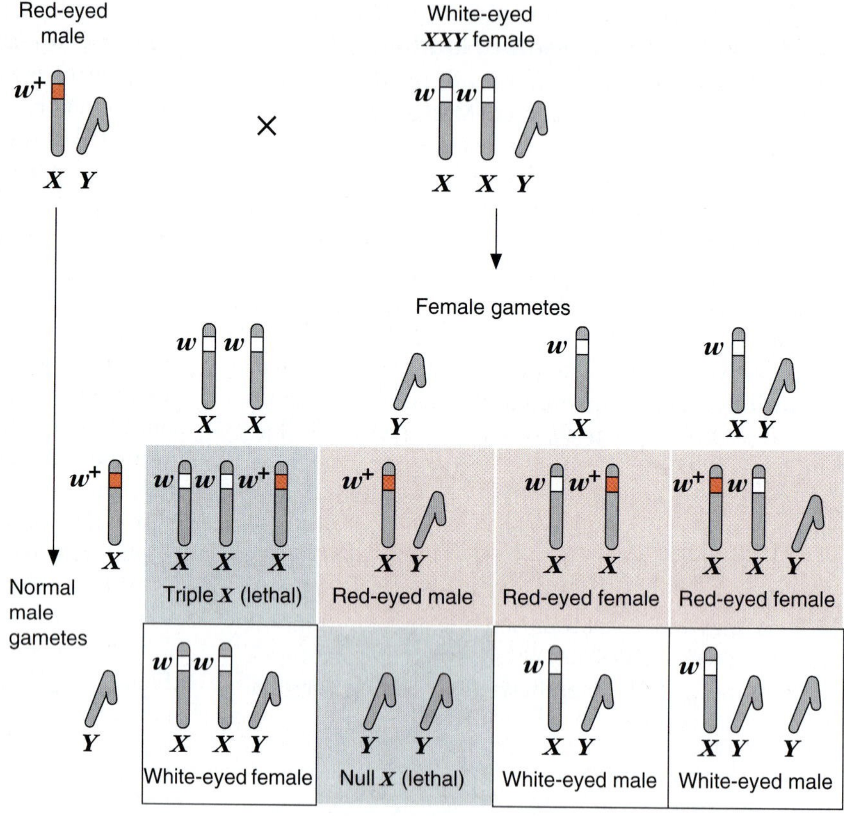

FIGURE 3.26 Bridges' test of the nondisjunction hypothesis. Bridges predicted that white-eyed XXY females would produce progeny with eight different combinations of X and/or Y chromosomes because of secondary X chromosome nondisjunction.

progeny of females that were produced by primary nondisjunction, then it is called **secondary nondisjunction** (Figure 3.25b). Secondary nondisjunction always occurs at a higher frequency than primary nondisjunction because the extra chromosomes have problems pairing during prophase of meiosis I. The developmental effects of nondisjunction are discussed in Chapter 18.

To test his hypothesis, Bridges mated white-eyed XXY females (*ww* Y) to red-eyed XY males (w^+ Y) and examined the karyotypes of all the surviving progeny. He predicted that the XXY females would produce four types of eggs based on all the possible chromosomal segregation patterns: XX (*ww*), XY (*w*Y), X (*w*), and Y (Y). The random fertilization of these eggs by X (w^+) sperm or Y sperm would produce progeny with eight different genotypes, six of which would be viable: (w^+Y), (ww^+), (ww^+Y), (*ww*Y), (*w*Y), and (*w*YY) (Figure 3.26). Bridges found each of these karyotypes in the progeny. Bridges' experiments are considered final proof of the chromosome theory because they showed that not only did the *white* allele segregate with the X chromosome during normal meiosis, but it also segregated with the X in the rare instances when the X chromosome underwent nondisjunction. This was a clear case of a particular gene always segregating with a particular chromosome. Nondisjunction and its consequences are discussed further in Chapter 18.

Two Different Methods of Sex Determination Are Common in Animals.

Bridges' studies of X chromosome nondisjunction not only provided proof of the chromosome theory, but they also provided evidence about sex determination in *Drosophila.* Bridges discovered that flies with one X and no Y (XO) were sterile males with normal visible sexual characteristics. This indicated that a Y chromosome was not required for the development of physical male characteristics in *Drosophila,* but a Y chromosome was necessary for fertility.

Bridges did additional studies in which he produced flies with different numbers of chromosomes, and he discovered that sex in *Drosophila* is determined by the ratio of X chromosomes to autosomes. Normal females have two X chromosomes and two sets of autosomes, an X : A ratio of 1 : 1. Normal males have one X chromosome and two sets of autosomes, an X : A ratio of 1 : 2. Individuals with an X : A ratio greater than 1 : 1 develop extreme female characteristics and are termed metafemales. Individuals with an X : A ratio of less than 1 : 2 develop extreme male characteristics, and are termed metamales (Table 3.2). Individuals with an X : A ratio intermediate between 1 : 1 and 1 : 2, such as individuals with two X chromosomes and three sets of autosomes (an X : A ratio of 2 : 3) have an intermediate "intersex" phenotype.

TABLE 3.2 *Chromosomal Determination of Sex in Drosophila*

Karyotype to Autosomes (X : A)	Ratio of Sex Chromosomes	Sex
X : AAA	1/3	Metamale
X : AA	1/2	Normal male
XX : AAA	2/3	Intersex
XX : AA	1	Normal female
XXX : AAA	1	Triploid female
XXX : AA	3/2	Metafemale

These studies demonstrated that the X : A ratio, and not the presence of a Y chromosome, is the primary signal for sexual development in *Drosophila*. Bridges' observations about the role of the sex chromosomes in *Drosophila* suggested that there must be a number of genes on the X and on the autosomes that control sexual development. Many of the genes that control sex determination in *Drosophila* have been identified, and the expression of these genes is discussed in detail in Chapter 16.

In some other animals, including mammals, chromosome composition is also the first or primary signal that initiates sexual development, and the Y chromosome has a critical role in sexual development. Evidence for the role of the Y chromosome in sexual development first came from studies of individuals who had abnormal karyotypes that contained normal numbers of autosomes but unusual numbers of sex chromosomes. Around 1940, the Klinefelter and Turner syndromes were described in humans. Each is caused by abnormal sex chromosome numbers. Individuals with Klinefelter syndrome have normal autosomes, two X chromosomes, and a Y chromosome for a total of 47 chromosomes (usually indicated 47,XXY). In designating human karyotypes, the number before the comma indicates the total number of chromosomes that are present in somatic cells and the abnormality is indicated after the comma. Klinefelter individuals usually develop male sexual characteristics at puberty, although they are often sterile because they fail to produce active sperm. Some female sexual characteristics, such as enlarged breasts, are often present. Individuals with Turner syndrome have normal autosomes, one X chromosome, and no Y chromosome (45,X). Individuals with Turner syndrome develop physically as females. They may have a range of distinctive physical characteristics, including rudimentary ovaries, a short stature, a webbed neck, and a broad ("shield") chest. Individuals with Klinefelter or Turner syndrome often suffer from psychological problems that are associated with appearing "different" from other children or adults, but the frequency of mental retardation or other developmental problems is low.

The Klinefelter and Turner syndromes indicate that the Y chromosome has an active role in sex determination in humans and that the presence of a Y promotes male sexual development. If a Y chromosome is not present, even if the individual has only one X, male sexual development does not occur. If a Y is present, even if the individual also has two X chromosomes, male development occurs. This development is not completely normal, as one Y chromosome cannot completely overcome the effects of two X chromosomes. The role of the Y chromosome in the Klinefelter and Turner syndromes was discovered in 1959 when karyotypes of these individuals were made. These discoveries indicated that the human Y chromosome must contain genes whose function is to promote male sexual development. The human Y chromosome is largely heterochromatic and was originally thought to be genetically inert, so this discovery stimulated many geneticists to begin investigations to find the sex-determining genes on the Y. The role of the Y chromosome in human sex determination is discussed in more detail in Chapter 16.

Many searches have been made for genes on the Y chromosome by searching for traits with a holandric pattern of inheritance. One example of such a trait is the H–Y antigen, which was discovered in 1955 during studies of transplantation of organs in mice. Transplantation of organs between individual mammals that are not identical twins is often unsuccessful because the donor and recipient cells have different cell surface antigens, and the recipient's immune system attacks the transplant as a foreign body. In highly inbred strains of mice that have similar alleles for most cell surface antigen genes, transplants can be made from one male to another male or from one female to another female. However, if male tissue is transplanted into a female, the female will reject the tissue. This indicated that males have an antigen that females do not, and the gene for this was found to be on the Y chromosome. Additional evidence for a Y antigen comes from studies of cells from XYY and XXYY individuals, whose cells bind twice as much antibody as XY or XXY individuals, suggesting that they have twice as many copies of the gene producing the antigen.

Dosage Compensation Occurs in Mammals Through X Inactivation.

The sex chromosomes are unique in that they are normally present in different numbers in males and females. This is very unlike the autosomal chromosomes, which normally have two copies in each somatic cell (for diploid species) of both sexes. Individuals with extra

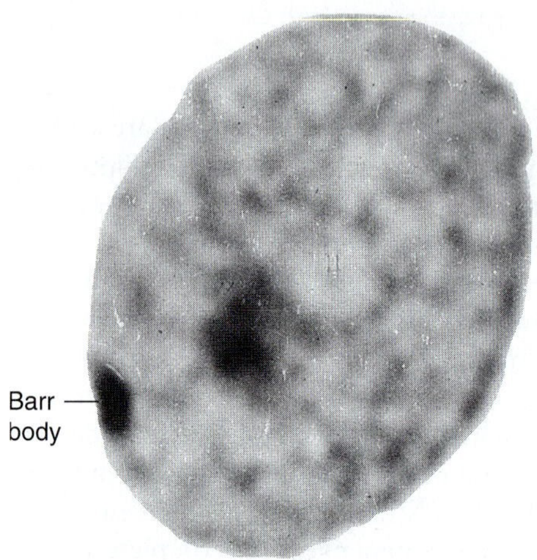
Barr body

FIGURE 3.27 A Barr body. A Barr body is a darkly staining, condensed, heterochromatic X chromosome found in the cells of mammals that contain two or more X chromosomes. The number of Barr bodies in each cell is one less than the number of X chromosomes. *(Dr. Dorothy Warburton/Peter Arnold, Inc.)*

copies of an autosome almost never survive (Down syndrome is an exception), yet humans with one, two, three, or four copies of the X chromosome survive. This is the result of a system of **dosage compensation** that regulates the activities of genes on the X chromosome to compensate for the difference in X chromosome number in males and females. Species that have X and Y sex chromosomes have some form of dosage compensation that ensures that most X-linked genes have about the same level of activity in males and females. Most, but not all, genes on the X chromosome are subject to dosage compensation.

In mammals, dosage compensation is accomplished by inactivating most of the genes on all but one of the X chromosomes. X chromosome inactivation was discovered by Mary Lyon and Lianne Russell. Physical evidence supporting this finding was discovered by Murray Barr and her coworkers, who observed a darkly staining body attached to the nuclear envelope in the somatic cells of female humans, but not to the somatic cells of male humans. This structure is called a **Barr body**, and it consists of an X chromosome that has become completely heterochromatic and nearly genetically inert (Figure 3.27). The discovery of Barr bodies suggested that dosage compensation is accomplished in female mammals by reducing the number of active copies of X-linked genes in females.

The dosage compensation hypothesis is supported by the results of studies of individuals with abnormal numbers of sex chromosomes that demonstrated that individuals with more X chromosomes have more Barr bodies. Individuals with a 47,XXX karyotype have two Barr bodies in each somatic cell, and individuals with a 48,XXXX karyotype have three Barr bodies. Individuals with Klinefelter syndrome (47,XXY) have one Barr body in their somatic cells, and individuals with Turner syndrome (45,X) have no Barr bodies. In all cases, the number of Barr bodies in the somatic cells is one fewer than the number of copies of the X chromosome present. This indicates that dosage compensation can occur in either sex if two or more X chromosomes are present, and that X inactivation ensures that only one X is functional in each somatic cell.

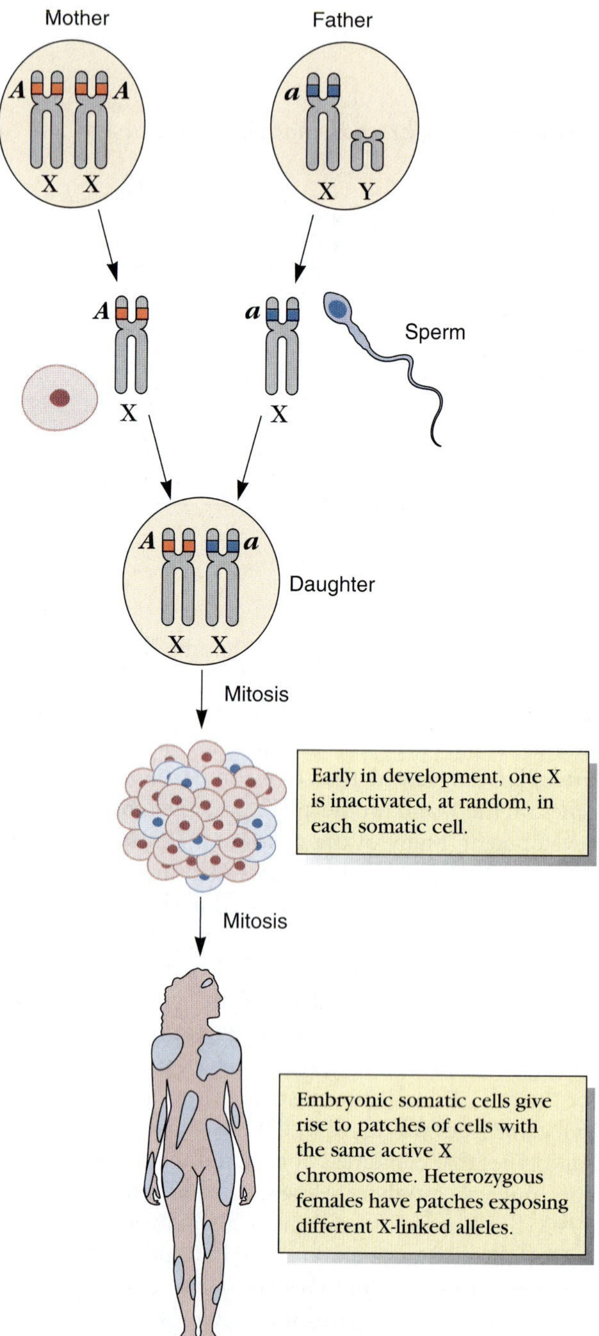
Mother
Father
A
A
X X
a
X Y
A
X
a
X
Sperm
A
a
X X
Daughter
Mitosis
Early in development, one X is inactivated, at random, in each somatic cell.
Mitosis
Embryonic somatic cells give rise to patches of cells with the same active X chromosome. Heterozygous females have patches exposing different X-linked alleles.

FIGURE 3.28 X inactivation in female mammals. As predicted by the Lyon hypothesis, cells in female mammals inactivate different X chromosomes. This is demonstrated by testing skin cells from individuals heterozygous for *G6Pd* alleles. Some patches of skin have one X chromosome (and its *G6Pd* allele) active and other patches have the other X chromosome (and its *G6Pd* allele) active.

Lyon proposed what has become known as the "Lyon hypothesis about X inactivation." She proposed that at a particular stage in embryonic development, one of the two X chromosomes is randomly inactivated in each somatic cell. This hypothesis is supported by the observation that in normal females some of the somatic cells have one X chromosome inactivated and the other cells have the other X chromosome inactivated. This had been documented in women who are heterozygous for alleles of an X-linked gene whose product can be detected in somatic cells. For example, the *G6Pd* gene in humans produces a protein that can be detected by gel electrophoresis of skin cells. Individuals known to be heterozygous for this gene produce two detectably different gene products. When small samples of skin cells from heterozygous women are tested for *G6Pd,* patches of skin are found to have only one type of gene product (Figure 3.28). Some skin cells have the product produced by one allele, and other cells have the product produced by the other allele. Clearly these women began development with both X chromosomes functional, and at some point each cell inactivated one X chromosome, which formed a Barr body. After X inactivation, the cells continued to divide by mitosis, producing clones of cells that all had the same active X.

Our discussion of the developmental abnormalities in persons with abnormal numbers of X chromosomes indicates that dosage compensation in humans is not perfect. If it were, 47,XXY individuals would be normal males, and 45,X individuals would be normal females. These individuals show defined syndromes because X inactivation is not complete. Either some genes on the X chromosome in a Barr body are active, or else X inactivation is complete but occurs after the abnormal number of X chromosomes has already disturbed the process of sexual development. If the latter is the case, then the abnormalities seen in individuals with Klinefelter or Turner syndrome are simply the results of a long chain of events that were set in motion early in embryonic development. The mechanism of X inactivation and the molecular nature of dosage compensation are areas of active research today. We will discuss the genetic and molecular basis of these phenomena in more detail in Chapter 16.

SECTION REVIEW PROBLEMS

20. If you crossed a large number of the parental dark-bodied male moths from Problem 17 with light-bodied females, and you observed that a very few of the progeny were light-bodied females, how would you explain the appearance of these females?
21. A new mutation that causes a recognizable change in the shape of earlobes has been discovered in humans. What characteristics would you expect this trait to show in pedigree analysis if the gene is located on the Y chromosome?

Summary

A mature, multicellular organism consists of billions of somatic cells and many germ cells. All these cells are derived from the original zygote through many rounds of mitotic cell division. During mitotic nuclear division, each replicated chromosome divides into two, producing two genetically identical progeny chromosomes. One chromosome is incorporated into each of the two progeny nuclei and ultimately, after cytokinesis, into two progeny cells. The equal division of chromosomes means that the genetic composition of each progeny cell is identical to that of the parent cell. The precision of the mitotic process allows each cell to retain all the genetic information in the original zygote.

Cytologists originally defined the stages of mitosis—prophase, metaphase, anaphase, and telophase—through chromosomal activity observed under the light microscope. When cytologists examined cells from various species of plants and animals, they realized that the events of mitosis were consistent among all species. Furthermore, they could identify different species by the number and structure of their chromosomes. Cytogeneticists have determined the composition of the normal set of chromosomes, the karyotype, for many species. By observing large numbers of cells from normal and abnormal individuals under the microscope, geneticists have detected many genetic abnormalities in which the size, shape, or number of chromosomes is changed.

The nuclei of germ cells undergo two special meiotic divisions. Chromosomal replication occurs before the first meiotic division, meiosis I, but not the second, meiosis II. During prophase of meiosis I (termed the "reduction division"), two copies of each chromosome pair to form a bivalent. Later, during anaphase I, the two homologous chromosomes segregate so that each of the progeny nuclei receives one copy of each chromosome. This chromosomal segregation reduces the number of chromosomes by one half and ensures that each progeny nucleus receives one complete set of chromosomes (termed the haploid set). During meiosis II, the centromeres of each chromosome divide, separating the two sister chromatids and changing them into separate progeny chromosomes that segregate to opposite poles of the cell. Each progeny nucleus receives one copy of each chromosome. The number of chromosomes does not change in meiosis II, so it is commonly termed the "equational division." The net result of the two meiotic divisions is that a single diploid nucleus divides twice to produce four haploid nuclei (termed "meiotic products"). Depending on the sex and species of the individual, one or more of the meiotic products will eventually give rise to gametes.

Sutton and Boveri first proposed the chromosome theory of heredity because the behavior of chromosomes in meiosis parallels the behavior of Mendel's alleles. Two alleles of the same gene located on homologous chromosomes segregate with their chromosomes during meiosis I, which is the physical basis of Mendel's first rule. Because the chromosomes in each bivalent assort independently, the alleles on different chromosomes assort independently (Mendel's second rule). Experimental proof of the chromosome theory came from studies of alleles of genes located on the sex chromosomes (sex-linked genes). Because many species of animals have different sex chromosome compositions in males and females, sex-linked genes show a special pattern of inheritance. The final proof of the chromosome theory came from studies of rare cases in which sex chromosomes underwent nondisjunction. Studies of nondisjunction of X chromosomes in *Drosophila melanogaster* carrying mutant alleles showed that the mutant allele always segregated with the X chromosome. This was a clear example of a Mendelian allele that always segregated with a particular chromosome.

In mammals, all but one X chromosome in each somatic cell forms a heterochromatic Barr body to become completely, or nearly completely, genetically inactive. This ensures that males and females have the same dose of active X-chromosomal genes, no matter how many X chromosomes they actually have in their somatic cells. Dosage compensation is imperfect, however, as abnormalities are seen in persons with abnormal numbers of X chromosomes, such as 47,XXY (Klinefelter syndrome) or 45,X (Turner syndrome).

Selected Key Terms

autosome
Barr body
bivalent
centromere
chiasma
chromosome
chromatid
diploid
euchromatin
haploid chromosome set
heterochromatin
homologous chromosome
karyotype
meiosis
mitosis
nondisjunction
sex chromosome
sex-linked genes
somatic cell
synapsis
telomere

Chapter Review Problems

1. State how many chromosomes you would expect to find in a nucleus of a cell of maize (2N = 20) in each of the following stages:
   a. prophase of mitosis
   b. zygotene of meiosis I
   c. prophase of meiosis II
   d. mature pollen grain
   e. endosperm of a mature seed
2. The following diagrams represent cells from a diploid species (2N = 4). In what stage are each of the cells?

(a)
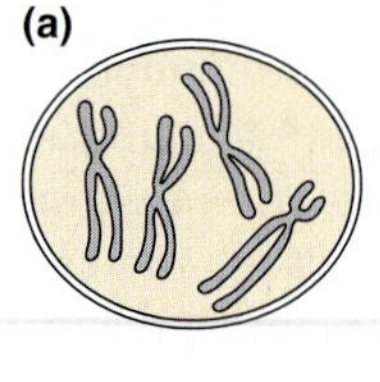

(b)
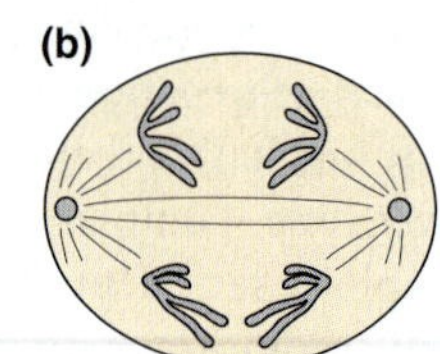

(c)
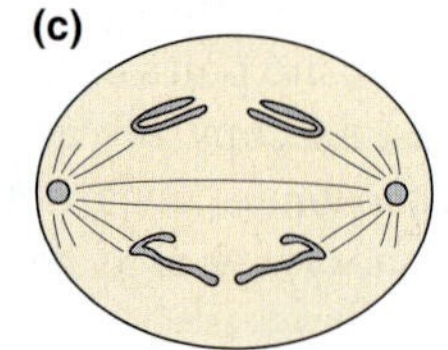

(d)
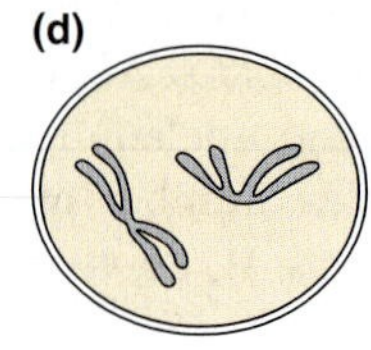

3. A mouse with a new dominant trait (small tail) was examined and found to also have a small knob on chromosome number 3. Describe how you would determine whether or not the cytological abnormality may be responsible for the observed trait?
4. Meiosis in a species of predominantly haploid fungi takes place in the diploid cells formed by the fusion of haploid gametes. In a species with $n = 16$, what is the probability that a haploid spore will receive all maternal or all paternal chromosomes?
5. Diagram the following stages of mitosis and meiosis for a cell from a diploid organism with two pairs of chromosomes, one pair telocentric and the other submetacentric.
   a. meiosis: zygotene, metaphase I, anaphase I, anaphase II
   b. mitosis: late prophase, metaphase, anaphase

6. The following pedigree shows the inheritance of a very rare trait. Solid symbols indicate individuals with the trait.

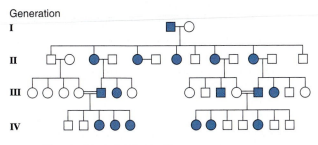

Generation
I
II
III
IV

   a. How is this trait inherited?
   b. What were the genotypes of the original parents?
   c. If IV-2 married IV-6, what would be the probability of their first offspring showing the trait?

7. A male butterfly with brown wings was crossed with a female butterfly with dark green wings. The F_1 were all brown. The F_2 males were 3/4 brown and 1/4 green, and the F_2 females were 3/4 brown and 1/4 green. Is this a sex-linked trait? Explain why or why not.

8. The following pedigree shows the inheritance of retinitis pigmentosum, a human condition that causes the progressive loss of night vision.

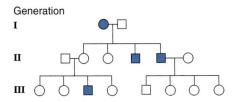

Generation
I
II
III

   a. How is this trait inherited?
   b. If individual III-4 marries III-5, what is the probability that they will have a son with retinitis pigmentosum?
   c. What is the probability that they will have a child with retinitis pigmentosum?

9. A brown female *Tribolium* (flour beetle) was crossed with a gray male *Tribolium*. The F_1 were collected and crossed to produce an F_2 generation. From the following results, explain how this color trait is inherited.

	Female		Male
Parents	brown	×	gray
F_1	all gray	×	all gray
F_2	35 brown		0 brown
	39 gray		78 gray

10. Unusual XXXX:AA humans are observed infrequently. How might such an individual be produced?

11. You have discovered a new plant species that grows well in the laboratory. Explain how you might use the location of meiosis in this plant's life cycle to determine whether it is more closely related to ferns or to maize.

12. Describe one major similarity and one major difference in the results of meiosis and mitosis.

13. In *Drosophila* the recessive *y* mutation is sex-linked and the dominant *Pm* mutation is autosomal. In a cross between a heterozygous female (*y*+; *Pm*+) and a male hemizygous for *y* and heterozygous for *Pm* (*yY*; *Pm*+), what are the different types of gametes that will be formed by
   a. the female?
   b. the male?

14. Considering only the autosomes, what is the probability that you and your brother would both have inherited
   a. one particular copy of chromosome 1 from your mother?
   b. the same complete set of autosomes from your mother?

15. A somatic cell in a garden pea (2N = 14) undergoes mitosis.
   a. How many chromosomes does each progeny cell have?
   b. If a cell in the same plant undergoes meiosis, how many chromosomes does each cell have after the first division?
   c. How many after the second division?

16. Explain the difference between primary nondisjunction and secondary nondisjunction.

17. Analyze the following pedigree to determine whether the mutant trait (shaded symbols) is recessive or dominant and whether it is caused by a sex-linked or an autosomal mutation.

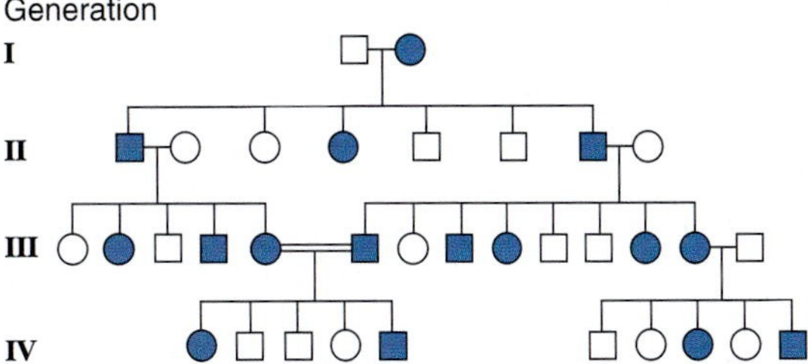

18. You have discovered that a new strain of banana plant that produces excellent fruit. How might you determine whether this strain is diploid?

19. The following diagrams represent somatic cells from a diploid organism (2N = 6) in different stages of mitosis or meiosis. Indicate in what stage each cell is.

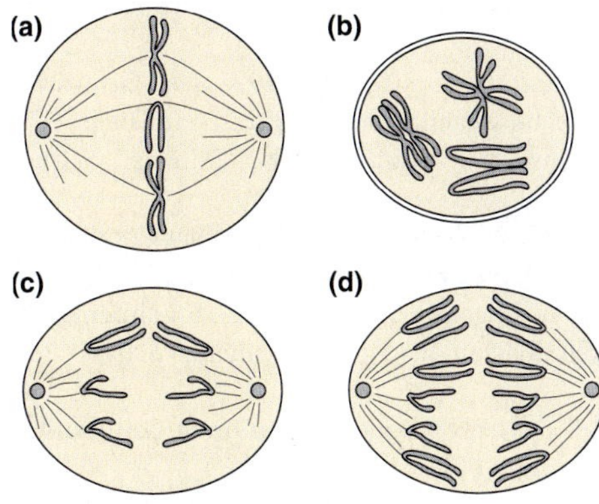

20. a. How many sperm would be produced by five primary spermatocytes?
   b. How many ootids would be produced by five primary oocytes?

21. How many chromosomes would you expect to find in each of the following types of cells from tomato plants (2N = 12)?

a. nondividing leaf cell
b. leaf cell in anaphase of mitosis
c. one meiotic product

22. For each of the cells in Problem 21, how many centromeres would be present?
23. For each of the cells in Problem 21, how many chromatids would be present?
24. An unreplicated haploid set of human chromosomes has approximately 3.2×10^{-12} DNA. How much DNA would you expect to find in the nuclei of the following human cells?
    a. nondividing skin cell
    b. primary spermatocyte in anaphase of meiosis I
    c. primary oocyte in pachytene of meiosis I
    d. skin cell in G_2
    e. secondary oocyte in prophase of meiosis II
25. The karyotype of a mitotic nucleus from a female cat shows 76 sister chromatids. What are the diploid and haploid chromosome numbers of the cat? How many homologous chromosome pairs are present?
26. How many telomeres are present in the cat of Problem 25?
27. Which of the following would you consider *not* to be typical of sex-linked dominant inheritance?
    a. affected males → 100% affected offspring
    b. affected females about twice as common as affected males
    c. affected males → 50% affected sons
28. Red-green color blindness is sex-linked and recessive. A woman with normal color vision marries a man with normal color vision. They have one son who is color-blind. What are the genotypes of the two parents?
29. During which phases of the mitotic cycle would each chromosome have only a single chromatid?

Challenge Problems

1. In *Drosophila* and humans, normal females and males are determined by the XX and XY genotypes. The abnormal genotype XXY is a female in *Drosophila,* but a male in humans.
    a. What does this tell you about the sex-determining genes in the two species?
    b. What phenotypes would you predict for an XO individual in the two species?
2. In a *Drosophila* experiment, a large number of females with white eye color and wild-type body color (*ww*; ++) were crossed to males with vermillion eyes and black body color (*vY*; *bb*). The F_1 contained a very small number of females with white eyes and wild-type body color. These females were backcrossed to parental males (*vY*; *bb*).
    a. What is the genotype and chromosome composition of these F_1 females?
    b. What genotypes and phenotypes of offspring will result from the backcross?
3. What frequencies would you expect for the progeny genotypes listed in the answer to Problem 2b?

Suggested Readings

Hartwell, L. H., and T. A. Weinert. 1989. "Checkpoints: Controls that ensure the order of cell cycle events." *Science,* 246: 629–634.

Lindsley, D. L., and G. G. Zimm. 1992. *The Genome of* Drosophila melanogaster. Academic Press, San Diego.

McIntosh, J. R., and M. P. Koonce. 1989. "Mitosis." *Science,* 246: 622–634.

Murray, A. W,. and M. W. Kerschmer. 1991. "What Controls the Cell Cycle." *Scientific American,* 265 (March): 56–63.

O'Farrell, P. H., B. A. Edger, D. Lakich, and C. F. Lehner. 1989. "Directing cell division during development." *Science,* 246: 635–640

Pardee, A. B. 1989. "G_1 Events and Regulation of Cell Proliferation." *Science,* 246: 603–614.

Stubbe, H. 1972. *History of Genetics.* MIT Press, Cambridge, MA.

On the Web

Visit our Web site at **http://www.saunderscollege.com/lifesci/titles.html** and click on A/G/M Genetics for links to the following chapter-related resources on the World Wide Web:

1. **Genetics search site.** This is a search site for genetics links all over the world. You can find almost anything that is available in any area of genetics.
2. **Chromosome link site.** This site links to other sites that have information on chromosomes from various organisms, including *Homo sapiens*.

CHAPTER

4

A *Blackpatch* eye spot phenotype in *Drosophila melanogaster.* The large, dark patch contains degenerating cells, dying as the result of an interaction between the *Blackpatch* mutation and the *facet* mutation. *(Jack R. Girton, Iowa State University)*

Multiple Alleles and Gene Interaction

When doing experiments in science, it is fundamental that you repeat your experiments to obtain reproducibility. There is an old saying in science, however: "Cherish your exceptions." This means that when an experimental result does not agree with the predictions of the standard theory, there is probably a good reason. By investigating this deviation from the predicted further, you may learn more about the phenomenon itself. Studying an exception may reveal that the standard theory is wrong, and your "exceptional" data may allow you to modify or replace the old theory. This is how science advances.

After the rediscovery of Mendel's paper, many scientists performed similar genetic experiments and confirmed many times over the correctness of Mendel's rules. In the process, they also discovered that mutant alleles did not always behave in the predicted Mendelian manner. Most of these deviations were detected when the results of a cross deviated from what Mendel's rules predicted. Upon further study, scientists discovered that these exceptions revealed many interesting properties associated with some mutant alleles. The unusual patterns of inheritance of the newly discovered mutant alleles were found to have their own consistent characteristics, and so new patterns of inheritance were uncovered.

In this chapter we will discuss the patterns of inheritance that some of these exceptional alleles revealed. In Chapter 2, we discussed genes having only two alleles: one dominant and the other recessive. Today we know that a gene may have many alleles, and that alleles are not restricted to a dominant–recessive relationship. In fact, alleles can have many different allelic relationships. All Mendel's pea alleles gave consistent, single phenotypes, and all the different genotypes he studied were equally healthy. Today we understand that mutant alleles can give a range of different phenotypes. We also understand that some mutant alleles produce deleterious or lethal phenotypes. Finally, all the alleles Mendel studied produced their phenotypic effect independently of mutant alleles of other genes. Today we know that the products of different genes interact, and that alleles of one gene generate different phenotypes depending upon whether the individual also has particular alleles of another gene.

In this chapter, we will also discuss some of the characteristics of mutant alleles and the experiments geneticists do to discover them. By the end of this chapter, we will have addressed five specific questions:

1. What types of dominance relationships occur between different alleles of the same gene?
2. What are the different phenotypes that mutant alleles produce?
3. What are the different types of interactions that can occur between alleles of different genes?
4. How do interactions between alleles of different genes affect the phenotype of an organism?
5. How can we discover the nature of an interaction between two genes?

ALLELES HAVE MANY DIFFERENT VARIATIONS.

Genes are lengths of DNA that contain thousands or hundreds of thousands of nucleotide pairs, and the sequence of nucleotide pairs in this DNA contains the information that makes it a gene (see Chapter 11). Any change in the sequence of nucleotide pairs in this length of DNA generates a new allele, so a gene could potentially have millions of different alleles. For example, more than 350 alleles of the human cystic fibrosis gene have been discovered in the last few years.

In practice, new alleles of a gene are usually detected only when they produce a new phenotype, a phenotype that is different from that produced by other alleles of the gene. In this section, we will discuss how different alleles of one gene can produce different phenotypes, and how different alleles of one gene can have different dominance relationships. We will also discuss how these different types of dominance relationships can be detected in monohybrid and dihybrid crosses. Our understanding of alleles has come a long way from Mendel's experiments with his seven different crosses, for which each gene had only one dominant allele and one recessive allele.

Alleles of the Same Gene Can Show Complete Dominance, Incomplete Dominance, Overdominance, or Codominance.

When an individual is heterozygous for two different alleles of one gene, the individual's phenotype is determined by the dominance relationship between the alleles. If heterozygotes are indistinguishable from homozygous individuals, the allele controlling the phenotype is **completely dominant** and the other allele is **completely recessive.** As noted earlier, one allele in each of the Mendel's seven crosses was completely dominant and the other allele was completely recessive. Other examples of completely dominant or completely recessive alleles are the alleles of the *white* gene in *Drosophila* studied by Morgan and discussed in Chapter 2. Since the time of Morgan's experiments, dozens of mutant alleles of the *white* gene have been found, including *tinged, apricot, blood, eosin, cherry,* and *wine.* Each of these gives a different eye color phenotype because each gives different amounts of pigment in the eye. These alleles collectively constitute an **allelic series.** If we did not know that these were all alleles of the *white* gene, their different phenotypes would suggest that they were alleles of several different genes. They illustrate an important point that we have learned from the study of different alleles of many genes: It is essential to study more than one mutant allele of a gene because all mutant alleles of one gene do not give the same phenotype. Each of the alleles of the *white* gene is completely recessive to the wild-type allele (w^+). Homozygotes of the first mutant gene found (w^1w^1) have pure white eyes. Individuals that are heterozygous for w^1 and the wild-type allele of the white gene (w^1w^+) have eyes with a normal red color. If some of these heterozygotes are placed in a container with flies that are homozygous for the wild-type allele, it is impossible to distinguish the heterozygotes from the homozygotes. All the members of the *white* gene allelic series are also completely recessive to the wild-type allele.

Dominance relationships in which heterozygotes do not have the same phenotype as homozygotes include incomplete dominance (sometimes called partial dominance), overdominance, and codominance (Table 4.1). Two alleles are **incompletely dominant** if the heterozygotes have a phenotype intermediate between the two

TABLE 4.1 *The Different Types of Dominance Relationships Between Alleles*

Dominance Relationship	Phenotype of Heterozygotes
Complete dominance	The same as individuals homozygous for one allele
Incomplete dominance	Intermediate between individuals homozygous for the alleles
Overdominance	More extreme than individuals homozygous for either allele
Codominance	Both phenotypes of the two homozygotes are expressed

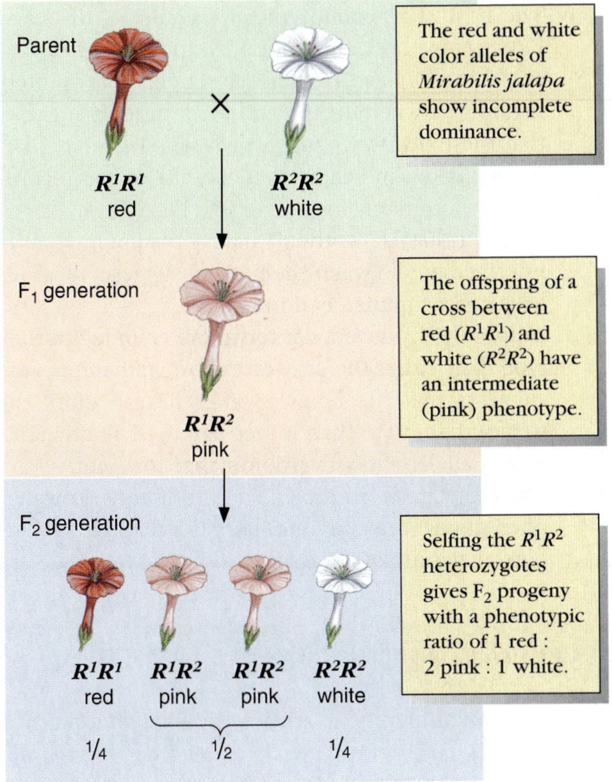

FIGURE 4.1 Incomplete dominance in *Mirabilis jalapa* (four-o'clocks). Heterozygous individuals have a phenotype intermediate between that of the homozygotes.

homozygotes. This relationship between alleles is very common. Carl Correns discovered a case of incomplete dominance in his study of flower color in *Mirabilis jalapa* (four-o'clocks). He crossed a red-flowered strain with a white-flowered strain and observed that the F_1 had pink flowers, an intermediate color. When these F_1 individuals were inbred, the phenotypes in the F_2 were 1/4 red, 2/4 pink, and 1/4 white (Figure 4.1). Heterozygous individuals showed a phenotype intermediate between that of the two homozygotes. However, alleles showing incomplete dominance do not always give a phenotype exactly halfway between the parents. The heterozygote phenotype may range from nearly equal to one homozygote to nearly equal to the other homozygote (Figure 4.2).

Overdominance is another dominance relationship in which the phenotype of the heterozygote is not equal to that of either homozygote. Heterozygotes with overdominant alleles have a phenotype more extreme than either homozygote (Table 4.1 and Figure 4.2). An example of overdominant alleles was discovered by G. P. Redi in *Arabidopsis thaliana.* He discovered a mutant allele of the *An* gene (called *an*) that caused altered leaf morphology. Individuals heterozygous for this allele (*An an*) were larger and had a greater total plant weight than individuals homozygous for either allele.

If the phenotypes of both alleles in a pair are present in a heterozygous individual at the same time, the alleles are said to be **codominant.** Codominance is often exhibited by alleles of genes that produce identifiable protein products. For example, several mutant alleles that alter the structure of seed proteins have been discovered in *Phlox pilosa.* These differences are detectable by separating the proteins by electrophoresis. If two plants homozygous for different alleles of one gene are crossed, the seed proteins found in their heterozygous offspring are a mixture of both parental types. If we had the task of characterizing an unknown individual, we could use electrophoresis to determine what forms of seed proteins were present. If two different types were present, we would conclude that the plant was heterozygous for the alleles that produce these forms. As we will discuss later, the alleles responsible for different human blood groups are excellent examples of codominance.

The dominance relationship between two alleles is readily determined in a monohybrid cross by examining the phenotype of the F_1 individuals or by determining the phenotypic ratios in the F_2 generation. When alleles show complete dominance, the F_1 individuals have the same phenotype as one of the homozygous parents, and

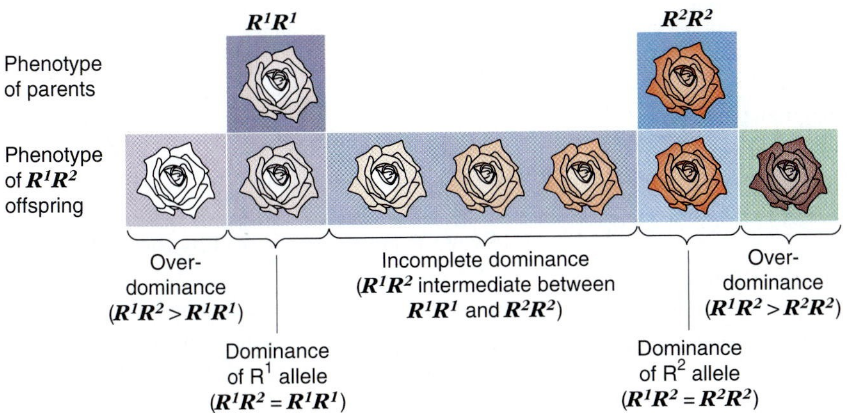

FIGURE 4.2 Degrees of dominance. Dominance is a relationship between two alleles that is defined by the phenotype of individuals heterozygous for the two alleles. If the phenotype of the heterozygote is the same as that of the dominant homozygote, the dominant allele shows complete dominance. If the phenotype is intermediate between the two homozygotes, both alleles show incomplete dominance. If the heterozygote's phenotype is more extreme than that of either homozygote, the alleles show overdominance.

the F_2 individuals have a 3 : 1 phenotypic ratio. When alleles show incomplete dominance, overdominance, or codominance, the F_1 individuals are not the same as either parent. Because each F_2 genotype produces a distinct phenotype, the F_2 phenotypic ratio is the same as the F_2 genotypic ratio (1 : 2 : 1).

Determining incomplete dominance, codominance, or overdominance relationships is more difficult in dihybrid or trihybrid crosses. In a dihybrid cross involving two genes that assort independently, the final F_2 ratio is a multiple of the ratios of the two genes. If one allele of each gene is completely dominant, then the overall F_2 phenotypic ratio is $(3:1) \times (3:1) = (9:3:3:1)$, the ratio observed by Mendel. If one allele of one gene is completely dominant (3 : 1) and the alleles of the second gene are incompletely dominant, overdominant, or codominant (1 : 2 : 1), then the F_2 phenotypic ratio is $(3:1) \times (1:2:1) = (3:6:3:1:2:1)$. If the alleles of both genes are incompletely dominant, overdominant, or codominant, then the F_2 phenotypic ratio is the same as the genotypic ratio $(1:2:1) \times (1:2:1) = (1:2:1:2:4:2:1:2:1)$. Phenotypic ratios that deviate from the 9 : 3 : 3 : 1 ratio observed by Mendel are called **modified dihybrid ratios.** Because these types of dominance relationships are very common, we will discuss a number of examples of modified dihybrid ratios later in the chapter.

SOLVED PROBLEM

Problem

The F_2 phenotypes of several different crosses are shown. For each cross, indicate what the dominance relationship is between the alleles.

Parents	F_2 Phenotypic Ratios
a. white × yellow flower	1 white 3 yellow
b. red × white fruit	1 red 2 pink 1 white
c. tall × short	1 tall 2 medium 1 short
d. gray × brown fur	1 gray 2 black 1 brown
e. red × blue flowers	1 red 2 red and blue 1 blue

Solution

In all five cases, we assume that the original parents were homozygous for different alleles of a single gene (a^1 and a^2). The F_2 will have a genotypic ratio of 1/4 homozygous (a^1a^1): 1/4 heterozygous (a^1a^2): 1/4 homozygous (a^2a^2).

a. The F_2 individuals show a phenotypic distribution of 3/4 yellow and 1/4 white. Both homozygous and heterozygous individuals are yellow, so the *yellow* allele must be completely dominant.

b. The 1 : 2 : 1 F_2 phenotypic ratio indicates that each F_2 genotype has a different phenotype. The fact that 1/2 of the F_2 have a phenotype (pink) that is intermediate between the red and white parents suggests that these heterozygotes have an intermediate phenotype. This indicates that the alleles are incompletely dominant.

c. The fact that 1/2 of the F_2 have a phenotype that is intermediate between the parents suggests these alleles are incompletely dominant.

d. Half of the F_2 have a darker (more colored) phenotype than either the gray or the brown parent. This suggests that the heterozygotes have a more extreme phenotype than either parent. The gray and brown alleles show overdominance for color.

e. The fact that 1/2 of the F_2 have both of the parental phenotypes (red and blue) suggests that the red and blue alleles are codominant.

SECTION REVIEW PROBLEMS

1. Give one difference between the following pairs of terms:
   a. complete dominance and overdominance
   b. incomplete dominance and codominance
2. A female *Drosophila* homozygous for a new recessive mutation that produced eyes with irregular facets was mated to a male homozygous for a recessive allele of the *sparkling* gene (*spa*). This allele also gives an irregular eye-facet phenotype. The facets in the eyes of the F_1 were much more irregular than the facets in the eyes of either parent. What would you conclude about the new mutation?
3. A morning glory with purple flowers and dark stems was crossed with a plant with white flowers and light stems. The F_1 were inbred, and the following F_2 were produced:

F_2 Phenotype	Fraction of F_2
purple flowers, dark stems	3/16
purple flowers, light stems	1/16
blue flowers, dark stems	6/16
blue flowers, light stems	2/16
white flowers, dark stems	3/16
white flowers, light stems	1/16

   a. How many genes are controlling these characters in this cross?
   b. How many alleles does each gene have, and what are the phenotypes produced by each allele?
   c. What was the phenotype of the F_1 individuals?

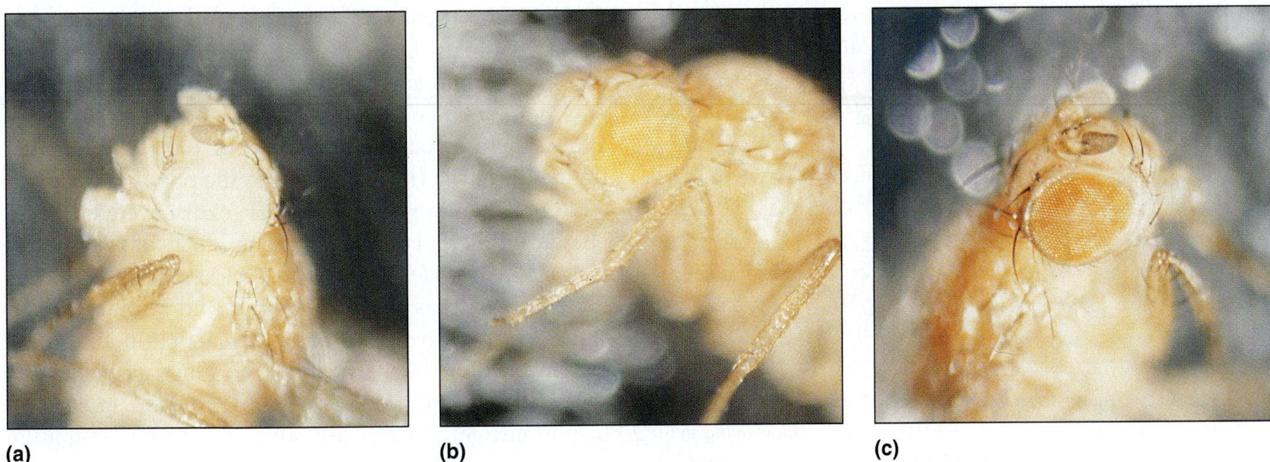

FIGURE 4.3 Dominance relationships and multiple alleles. The dominance relationships between members of the allelic series of the *white* gene is determined by the phenotype of the heterozygous individuals, and it can vary. The *white-apricot* allele is recessive to the wild-type allele, but it is incompletely dominant to the original *white* allele. (a) An individual homozygous for the *white* allele. (b) An individual with a heterozygous (*white/white-apricot*) genotype. (c) An individual homozygous for *white apricot*. *(Jack R. Girton/Iowa State University)*

One Allele May Show Different Types of Dominance.

Because dominance is a relationship between alleles, one allele can be recessive to a second allele but be dominant or incompletely dominant to a third allele. For example, individuals homozygous for the *white-apricot* (w^a) allele of the sex-linked *white* gene in *Drosophila* have dark orange-colored eyes, and females heterozygous for *white-apricot* and the wild-type allele ($w^a w^+$) have red eyes. From this observation, we would conclude that the *white-apricot* allele is completely recessive to the wild-type allele. However, females heterozygous for the original *white* (w^1) allele and the *white-apricot* allele ($w^a w^1$) have light orange-colored eyes, a phenotype intermediate between w^a and w^1. From this information, we would conclude that *white-apricot* is incompletely dominant to *white* (Figure 4.3). This illustrates how the *white-apricot* allele could be considered a recessive or an incompletely dominant allele, depending on the other alleles with which it is being compared. We can conclude that the designation of an allele as recessive or dominant is always relative to another allele.

How we classify the dominance relationship between two alleles is greatly influenced by the manner in which the phenotype of the individual is determined. An example that illustrates the relative nature of dominance relationships in humans is the allele responsible for sickle-cell anemia. Sickle-cell anemia is an inherited human disease characterized by severe anemia that is often fatal at an early age. It is caused by a mutation (Hb^s) that alters the molecular structure of the hemoglobin molecule. This allele can be classified as recessive, incompletely dominant, overdominant, or codominant, depending on how we measure the phenotype. Individuals homozygous for the sickle-cell hemoglobin allele (Hb^sHb^s) have blood cells that tend to change from the normal round blood cell shape into an unusual "sickle" shape (Figure 4.4). This sickle shape does not function well in oxygen transport, thus depriving tissue of oxygen. Also, sickled cells do not move easily through capillaries, and they degenerate more quickly than normal cells.

Although normal red blood cells will sickle when severely deprived of oxygen, the red blood cells of

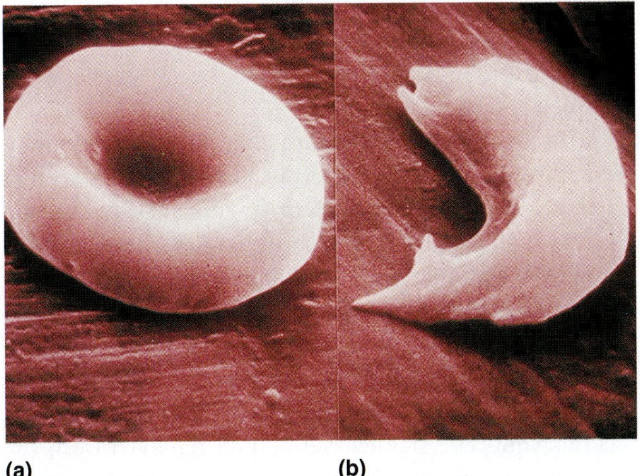

FIGURE 4.4 Sickle-cell anemia. The phenotype of normal red blood cells (a) compared with sickled cells (b). *(Stanley Flegler/Visuals Unlimited)*

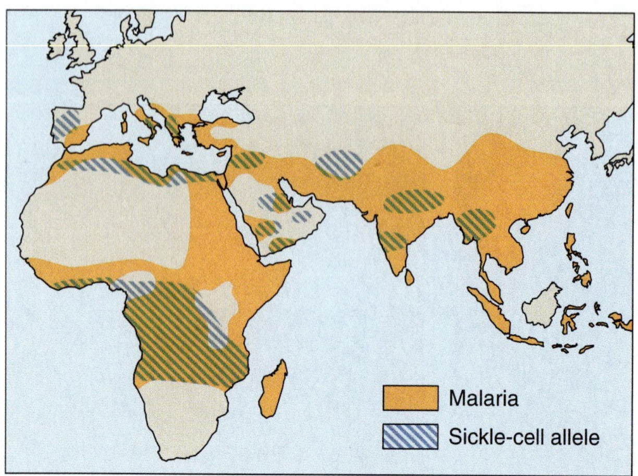

Malaria
Sickle-cell allele

FIGURE 4.5 The distribution of the Hb^s allele and malaria. The Hb^s allele occurs much more frequently among individuals living in areas of the world that have had malarial infestations over a long time period.

individuals homozygous for Hb^s are especially prone to sickling in a reduced oxygen environment. For example, vigorous exercise depletes the oxygen in the muscles and can cause a sudden increase, or "attack," of sickling. The blood cells of individuals heterozygous for the Hb^s allele and for the normal allele Hb^A (Hb^AHb^s) do not sickle, even after vigorous exercise. Individuals with this genotype are nearly as healthy as individuals homozygous for Hb^A. If we based our determination of dominance on the visible phenotype of heterozygous individuals, we would say that the Hb^s allele was completely recessive to the Hb^A allele. However, if blood cells taken from Hb^AHb^s heterozygous individuals and from Hb^A homozygous individuals are placed in a low-oxygen environment in a test tube, the cells from the heterozygous individual will sickle but the homozygous cells will not. Only when the oxygen concentration is lowered even further will the Hb^AHb^A cells sickle. In this test, the heterozygotes show an intermediate phenotype; that is, the blood cells assume the sickle shape at an intermediate oxygen concentration. This finding suggests that the Hb^s allele is incompletely dominant to Hb^A.

The frequency of the Hb^s allele in human populations is high in some areas of Asia and Africa, as well as in Americans of African descent (Figure 4.5). The reason for this high frequency is that individuals heterozygous for Hb^s have an increased resistance to malaria, a disease caused by a blood parasite that infests red blood cells, and malaria is indigenous to these areas of the world. In individuals heterozygous for Hb^s, the parasite's invasion of a red blood cell often causes the cell to assume the sickle shape. The sickled cell is then removed from the blood by the spleen and degraded, frequently destroying the parasite at the same time. In areas where the incidence of malaria is high, heterozygotes (Hb^sHb^A) are healthier than individuals homozygous for either allele. They have neither the anemia of the homozygous Hb^s individuals nor the malarial symptoms of the homozygous Hb^A individuals. Under these conditions, the Hb^s allele shows overdominance for the survival of the host in this environment. The increased health of the heterozygotes allows them to have more children, and this keeps the frequency of the Hb^s allele high in the population.

Linus Pauling and his coworkers discovered the biochemical difference between the products of the Hb^s and Hb^A alleles in 1949. The Hb^s allele produces a globin molecule that has a neutral amino acid (valine) substituted for a negatively charged amino acid (glutamic acid) at one position in the hemoglobin molecule, changing the rate of migration of the globin molecule in an electrophoretic gel. Pauling used this difference in migration rate to identify the many different types of globin molecules among individuals in the population. He isolated globin from the blood of heterozygotes (Hb^AHb^s), placed purified globin in an electrophoretic gel, and showed that in their globin both the Hb^s and Hb^A types were present (Figure 4.6). If we use the presence of both types of

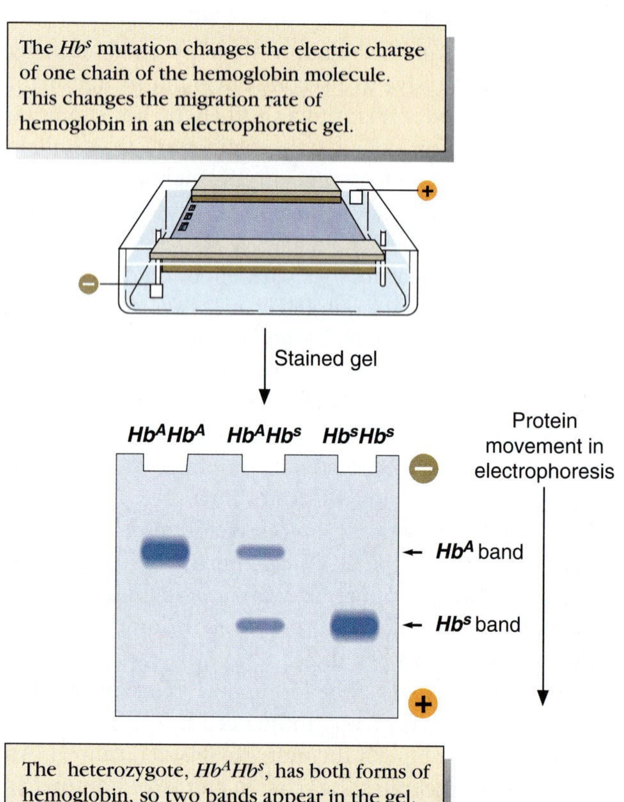

The Hb^s mutation changes the electric charge of one chain of the hemoglobin molecule. This changes the migration rate of hemoglobin in an electrophoretic gel.
Stained gel
Hb^AHb^A Hb^AHb^s Hb^sHb^s
Protein movement in electrophoresis
Hb^A band
Hb^s band
The heterozygote, Hb^AHb^s, has both forms of hemoglobin, so two bands appear in the gel.

FIGURE 4.6 Codominance of Hb^s based on physical detection. The differences in the globin molecule produced by the Hb^A and Hb^s alleles can be detected using gel electrophoresis. Blood isolated from Hb^sHb^A heterozygotes contains both types of globin. If we use the presence of the protein as a measure of phenotype, Hb^A and Hb^s are codominant alleles.

hemoglobin molecule as a measure of phenotype, Hb^s is a codominant allele because the heterozygote contains both Hb^s and Hb^A hemoglobin molecules.

Codominant Alleles Are Often Detected Using Antigen–Antibody Reactions.

A widely used technique for detecting codominant alleles is the antigen–antibody response. Organisms with an immune system (all mammals) respond to the introduction of a foreign substance into their bloodstream by generating **antibodies,** proteins that recognize and attach to specific structures called **antigens.** Each antigen is a particular part of a protein or other molecule, and each antibody recognizes and attaches to only one specific antigen. Most people have thousands of different antibodies in their bloodstream at all times and can produce many more in response to infections. Because of their high specificity of antigen binding, antibodies are used to detect the presence or absence of allele-specific proteins. For example, if human red blood cells are injected into a mammal (for example, a mouse), the animal will produce antibodies that bind specifically to human blood cell antigens. If the serum (the fluid remaining after the blood cells have been removed) from the mouse is then extracted, it will contain antibodies that specifically bind to human blood cell antigens. Thus, if the mouse serum is mixed with a sample of human blood cells, the antibodies will bind to the cells. Because each antibody has two binding sites, the blood cells clump and form a precipitate, which is referred to as **agglutination** (Figure 4.7). An antibody whose antigen is produced by one

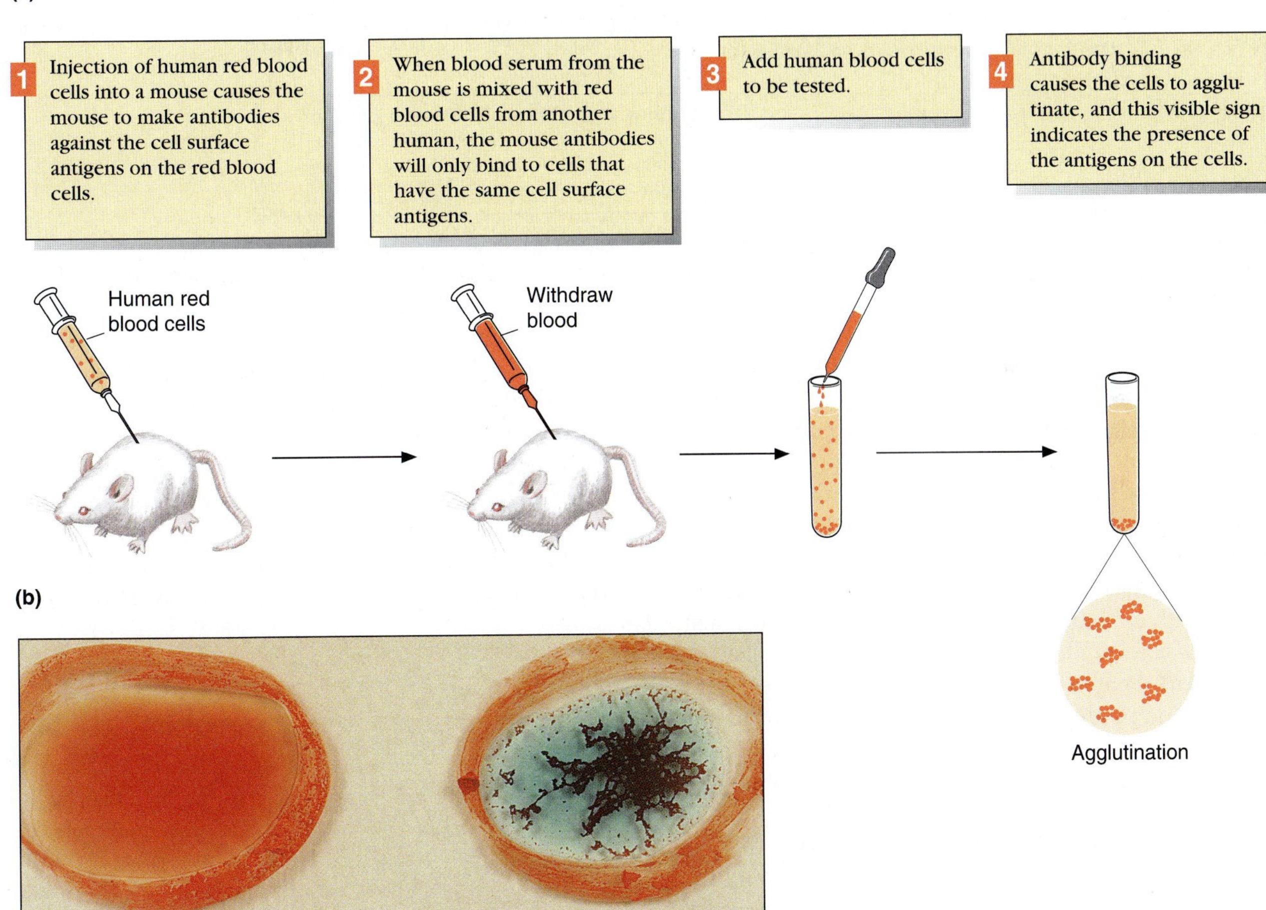

(a)
1 Injection of human red blood cells into a mouse causes the mouse to make antibodies against the cell surface antigens on the red blood cells.
2 When blood serum from the mouse is mixed with red blood cells from another human, the mouse antibodies will only bind to cells that have the same cell surface antigens.
3 Add human blood cells to be tested.
4 Antibody binding causes the cells to agglutinate, and this visible sign indicates the presence of the antigens on the cells.
Human red blood cells
Withdraw blood
Agglutination
(b)

FIGURE 4.7 Blood cell agglutination. Blood typing depends on the presence on red blood cells of antigens that are bound by specific antibody molecules. (a) Agglutination is the clumping of cells that occurs when antibodies bind to red blood cell surface antigens. Because each antibody has two binding sites that can bind to the surface antigens of two different cells, antibody binding pulls cells together into clumps. (b) A photograph showing agglutination. A blood sample placed on a slide *(left)* reacts to the addition of the antibody *(right)*. *(Christine L. Case/Visuals Unlimited)*

FIGURE 4.8 The MN blood-typing system. Individuals with different alleles of the MN blood type genes produce antigens on the surface of their red blood cells as well as antibodies in their serum. Agglutination occurs when the antibody reacts with a specific antigen on the surface of red blood cells.

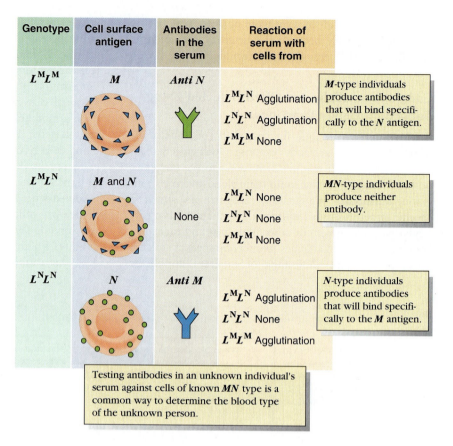

Genotype	Cell surface antigen	Antibodies in the serum	Reaction of serum with cells from
$L^M L^M$	*M*	*Anti N*	$L^M L^N$ Agglutination $L^N L^N$ Agglutination $L^M L^M$ None
$L^M L^N$	*M* and *N*	None	$L^M L^N$ None $L^N L^N$ None $L^M L^M$ None
$L^N L^N$	*N*	*Anti M*	$L^M L^N$ Agglutination $L^N L^N$ None $L^M L^M$ Agglutination

M-type individuals produce antibodies that will bind specifically to the *N* antigen.

MN-type individuals produce neither antibody.

N-type individuals produce antibodies that will bind specifically to the *M* antigen.

Testing antibodies in an unknown individual's serum against cells of known *MN* type is a common way to determine the blood type of the unknown person.

allele of a blood cell surface protein gene will bind to blood cells from individuals that have that allele, but not to blood cells from individuals that do not have that allele.

In 1927, the agglutination reaction was first used to demonstrate the genetic basis of blood types by Karl Landsteiner. He tested human blood samples using the agglutination reaction and discovered that each individual could be classified into one of three **blood types,** called M, MN, and N. People with M blood type have the M antigen on the surface of their blood cells, and they make antibodies that bind to cells with N-type antigens. People with N blood type have N-type antigens on their blood cells and produce antibodies that bind to cells with M antigens. People with the MN blood type have both antigens and produce neither antibody. Serum from MN individuals will not agglutinate cells from either N or M individuals, but serum from either M- or N-type individuals will agglutinate cells from MN individuals (Figure 4.8). The MN blood antigen type is controlled by a single gene that has two codominant alleles, called L^M and L^N. The gene symbol *L* was used in honor of the discoverer of blood types, Karl Landsteiner. M-type individuals are homozygous for the L^M allele, N-type individuals are homozygous for the L^N allele, and MN-type individuals are heterozygotes ($L^M L^N$). The blood type of an individual can be determined by extracting a small sample of blood cells, exposing the cells to serum from individuals of known MN type, and observing whether the cells agglutinate. Because L^M and L^N are codominant, an individual who has one of these alleles will always produce the appropriate antibodies and will show the appropriate phenotype in a blood test. This makes it possible to determine directly the genotype of any individual, whether they are homozygous or heterozygous. For this reason, codominant blood-type alleles are widely used in human genetics. The phenotype of each individual indicates what its genotype is (Figure 4.9).

Analyzing blood samples for codominant alleles is widely used in cases of disputed parentage, particularly

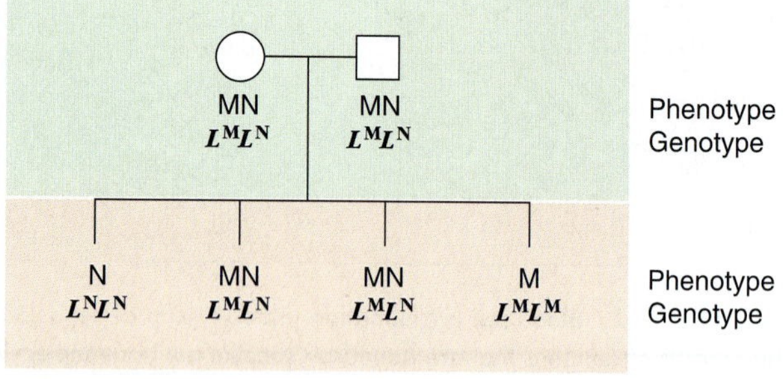

FIGURE 4.9 Inheritance of the MN alleles. The presence of codominant alleles can be determined by reacting both antibodies (M and N) with the blood cells. Codominant blood-type allele antigens are widely used in pedigree analyses and paternity tests.

CURRENT INVESTIGATIONS

How Can Genetics Be Used To Solve a Human Rights Crime?

Most people associate parentage testing with determining whether a particular man is the father of a child. However, the genetic techniques used in parentage testing can also help resolve questions about the genetic identity of children, including kidnapped or adopted children who desire to determine whether they are members of a particular family. Several particularly serious instances of this occurred in Argentina beginning in 1977. In 1976, a military government seized power in Argentina and soon developed a history of human rights abuses. Political opponents of the government were kidnapped, tortured, or sometimes just "disappeared." Estimates of the numbers of individuals killed range as high as 12,000 for the 6 years of military rule. In 1977, a human rights group called the Abuelas de Plaza de Mayo (the grandmothers of the Plaza of May) was formed to protest these abuses. This group of mostly older women demonstrated daily in the plaza opposite the government headquarters. One of their main concerns was to trace children who were kidnapped with their parents or who were born in prison and to return them to their families. Many of these children had been secretly given up for adoption or had been kept by families of military officers.

The women collected a great deal of evidence and traced the fate of many of the children. In 1983, when a civilian government was reestablished, the grandmothers began to ask the courts to return the children. Some of these cases were highly emotional, involving grandmothers claiming children who had been kept by the military officers charged with murdering the child's parents. To help establish the identity of the children, an American geneticist, Mary-Claire King (Figure 4.A), employed a blood test using an antibody reaction to determine the HLA (human leukocyte antigen) blood cell type. The HLA antigens are part of the self-recognition immune response system. The surviving grandparents and other relatives who claimed relationship to a child could be typed, and the results then could be compared with the type of the child. The results often showed that the child had a very high probability of being a member of the family. The courts accepted such arguments and returned many children to their families. For cases where HLA antigens did not give a clear indication, a second test involving mitochondrial DNA sequences was used. Because mitochondria are maternally inherited, children have mitochondria identical to their mothers and to their maternal grandmothers. A 600-base pair region of mitochondrial DNA was chosen as the target. The nucleotide sequence from each child was compared with the same sequence from the proposed maternal grandmother. Most families had

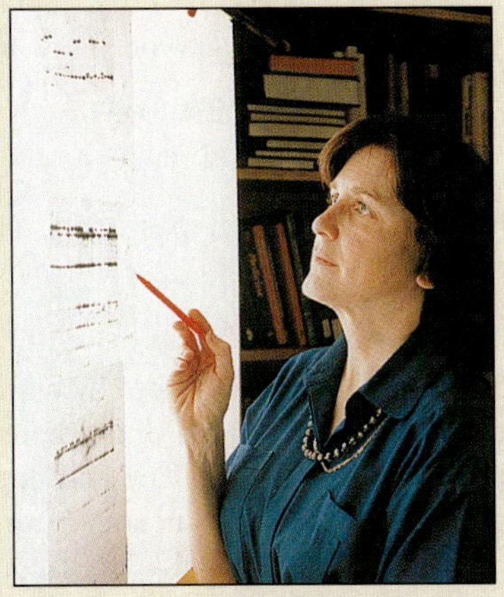

FIGURE 4.A Mary-Claire King. A leading American geneticist. *(Cindy Charles/Science Photo Library/Photo Researchers, Inc.)*

a unique sequence. The absence of a paternal mitochondrial contribution made the typing much easier. Any uncle or aunt on the maternal side could provide the mitochondria sequence.

Genetic testing provided the grandmothers with a vital tool needed in their fight for justice by scientifically establishing relationships with the children. For those children who could not be traced, the organization has established a bank of HLA and mitochondrial sequence information. They believe that many missing children, as they mature and become adults, may eventually seek to determine their own heritage. If so, the information will be available to help them determine their true identities.

of disputed paternity (see *CURRENT INVESTIGATIONS: How Can Genetics Be Used To Solve a Human Rights Crime?*). For example, if a child of MN blood type is born to a woman of N blood type, then the child must have inherited its L^M allele from its father, so the father must have either an M or an MN blood type. If one of the possible fathers of the child has N blood type, we can conclude that he cannot be the father because he is homozygous for L^N, and he cannot have contributed an L^M allele to the child. It is important to note that this test does not prove parentage; rather it identifies individuals who are not the parents. The MN, ABO, and Rh systems are widely known examples of some of the more than 100 different human blood antigen–antibody systems currently known. Each represents different cell surface antigens produced by one gene.

SOLVED PROBLEM

Problem

The following three children are the offspring of a single set of parents:

child 1 = M blood type

child 2 = MN blood type

child 3 = N blood type

What are the genotypes of the parents of these children?

Solution

The problem can be solved by working backwards, predicting the phenotype of the parents from the phenotypes of the children. The first child must be homozygous for L^M, which means both parents must have contributed an L^M allele. The second child must be heterozygous ($L^M L^N$). This means that one parent must have contributed an L^M allele and the other an L^N allele. The third child must be homozygous for L^N. This means that both parents must have contributed an L^N allele. To be the parents of all these children, both parents must be heterozygotes.

SECTION REVIEW PROBLEMS

4. Explain the difference between incomplete dominance and codominance.
5. The MN blood types of several sets of men and women were determined. For each of the following couples, indicate the phenotypes of the possible children they might have.

Woman	Man
a. M	N
b. MN	M
c. MN	MN
d. N	MN

6. In a paternity test, a mother and child were found to differ for a number of codominant alleles that produce blood antigens.

Mother: $A^1A^1\ B^1B^1\ C^2C^2\ D^2D^2\ E^2E^2\ F^1F^1\ G^1G^1\ H^1H^1$

Child: $A^1A^2\ B^1B^2\ C^1C^2\ D^1D^2\ E^1E^2\ F^1F^2\ G^1G^2\ H^1H^2$

a. What alleles must the father have contributed?

b. A possible father was told that because he has all the correct blood antigen alleles, the genetic match proves that he is the father. Do you agree with this statement?

Segregation and Complementation Tests Are Used to Determine Whether Different Mutations Are Alleles of the Same Gene.

The members of an allelic series do not always give the same phenotype, and mutant alleles of other genes may appear to be members of an allelic series when they are not. This sometimes makes it difficult to determine from phenotypes whether two mutations are alleles of the same gene. Two genetic tests for allelism that are widely used are the segregation and complementation tests. The **segregation test** is based on Mendel's first rule and is relatively easy to use. If two mutations are alleles, they must segregate from each other during gamete formation. This means that an individual heterozygous for these mutations will always produce gametes that contain either one allele or the other, but never both or neither. To do a segregation test, individuals heterozygous for the mutations being tested are crossed, and the genotypes of their gametes are deduced from the genotypes or phenotypes of their progeny. If the two mutations are alleles, then the gametes produced by the heterozygous F_1 individuals will always contain either one allele or the other, but never both. If the two mutations are not alleles, then some of the gametes will contain both alleles, and some will contain neither allele (Figure 4.10).

The segregation test was used to demonstrate that the ABO blood type in humans is controlled by one gene with multiple alleles. Like the MN blood type, the ABO blood type is another antigen–antibody system, also discovered by Karl Landsteiner in 1927. At that time, when human blood transfusions were done, recipients sometimes had a serious or even fatal reaction to the transfusion. These reactions were caused by an antibody–antigen reaction between antigens on the red blood cells of the donor and antibodies in the serum of the recipient, resulting in an agglutination reaction. There are many antigenic differences among individual humans, but only a few cause a strong blood agglutination reaction. The ABO antigen system is one of these. Individuals have one of four possible ABO blood types:

Blood Type	Antigen on Cell	Antibody in Blood
A	A	anti B
B	B	anti A
AB	A and B	none
O	none	anti A, anti B

In most transfusions, the antibodies in the donor serum do not generate a serious reaction because they are quickly absorbed or diluted. The major concern is whether the recipient's antibodies will bind to the donor blood cells and cause an agglutination reaction. If they

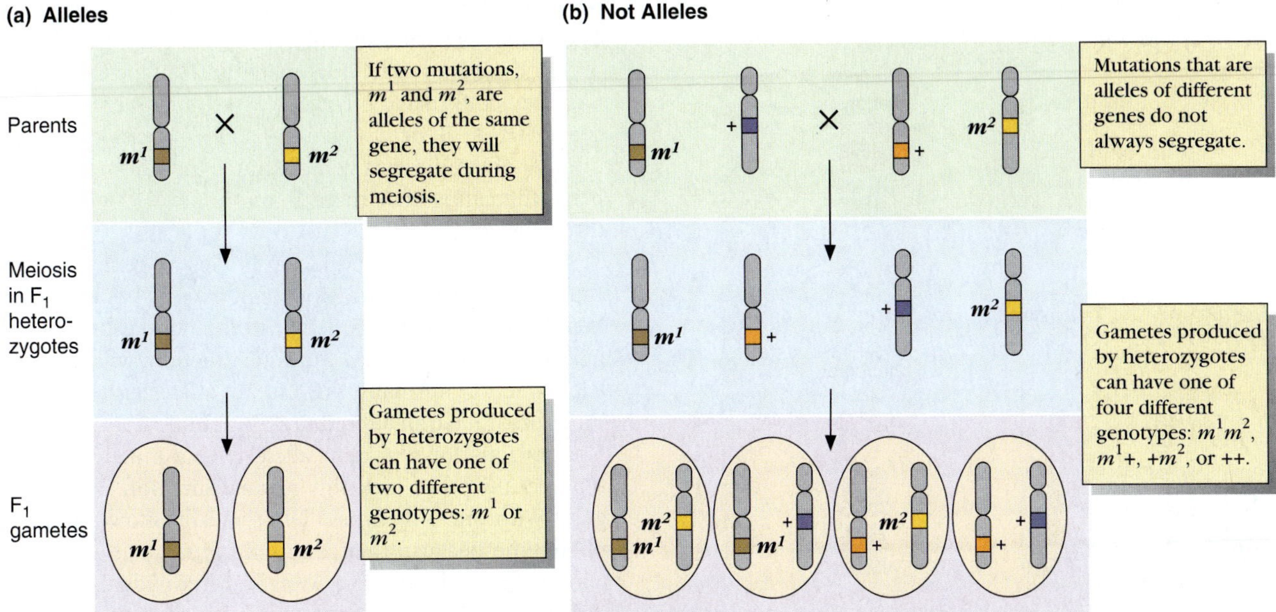

FIGURE 4.10 Segregation test for two alleles. A technique used to determine whether two mutations are allelic to each other is whether they segregate in meiosis. (a) Alleles that segregate will not appear in the same gamete. (b) Alleles that fail to segregate will sometimes be found in the same gamete. Failure to segregate in meiosis is evidence that two mutations are not alleles of the same gene.

do, the clumping of blood cells in the recipient's body can cause severe problems. Safe transfusion can be made between two individuals with the same blood cell antigens (A to A, B to B, AB to AB, and O to O). Also, type AB individuals can receive transfusions from type A, type B, type O, or type AB individuals because type AB individuals do not have antibodies to either the A or the B antigen. Finally, type O individuals can donate blood to A, B, or AB individuals because type O blood cells do not have either the A or the B antigen.

The ABO antigens are produced by multiple alleles of one gene (I), and two of these alleles are codominant (I^A *and* I^B). One of these alleles produces the A-specific antigen, and the other produces the B-specific antigen. The third allele is a recessive allele that does not produce any antigen, and it is called i (or I^O). The products of these three alleles have different properties, so there are several possible ABO phenotypes. Because of the complexity of the multiple phenotypes (Table 4.2), initially there was confusion as to whether the ABO system was controlled by one gene with multiple alleles or by two genes. The answer was discovered by observing the segregation of the alleles responsible for the ABO antigens in the offspring of matings between individuals with different genotypes. According to the segregation test, if the alleles responsible for different ABO antigens always segregate from each other, they must be alleles of the same gene. If there are two or more genes, the alleles need not segregate.

To illustrate how the segregation test works, consider a mating between two individuals, one with type AB blood and one with type O. If there is only one gene, the AB individual must be heterozygous (I^AI^B) and will produce gametes with either an I^A allele or an I^B allele but never with both. The type O individual must be homozygous for i and will produce gametes with i only. The offspring of this mating will either have type B blood (I^Bi) or have type A blood (I^Ai). These progeny will never have either type AB or type O blood (Figure 4.11).

If the ABO system were controlled by two genes with two alleles each (*Aa* and *Bb*), the type O individuals would be homozygous for the recessive alleles of both genes (*aa bb*). Therefore, AB individuals would contain at least one dominant allele of both genes. AB individuals

TABLE 4.2 *The ABO Phenotypes Are Produced by Different Genotypes*

Phenotypes	Genotypes
A	I^AI^A or I^Ai
B	I^BI^B or I^Bi
AB	I^AI^B
O	ii

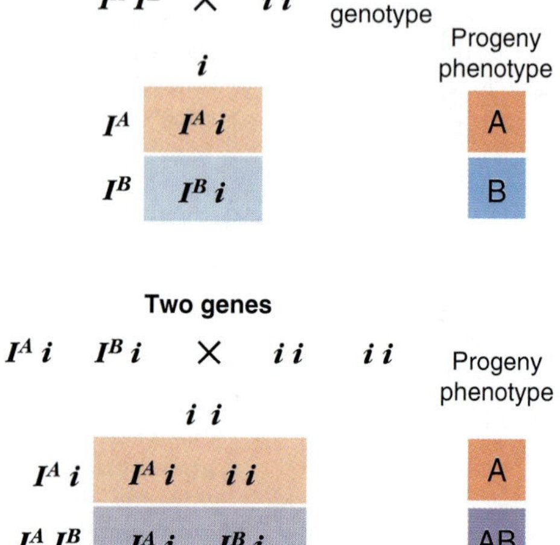

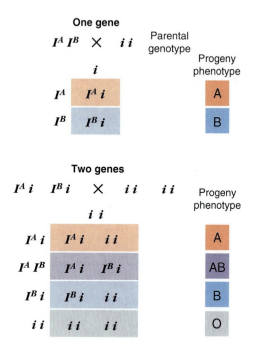

FIGURE 4.11 Segregation test for the ABO blood-type system. The one-gene hypothesis for the ABO blood antigen system predicts that a mating between a type AB individual and a type O individual will give only type A and type B offspring but will never give type AB or type O offspring. The two-gene hypothesis predicts that a mating between a type AB and a type O individual may give type AB or type O offspring. The results of actual family studies indicate that the one-gene hypothesis is correct.

who are heterozygous for both genes (*Aa Bb*) would produce gametes with four different genotypes (*AB*, *Ab*, *aB*, or *ab*). If such an individual mates with a type O individual, they could have offspring with four different blood types: AB, A, B, or O (Figure 4.11). Analyses of a large number of families has demonstrated that this never happens, indicating that there is only one ABO gene with three alleles.

The second test for allelism is a test of gene function called the **complementation test.** This is based on the Mendelian concept of a gene as a functional unit of inheritance. Each gene has a unique function: wild-type alleles function normally, but mutant alleles either do not function or have an altered function. Individuals who are homozygous for a mutant allele have a mutant phenotype because they lack this normal function. Heterozygotes with one wild-type allele typically show a normal phenotype because their one wild-type allele produces enough normal function. When an individual has two *different* mutant alleles of the same gene, neither mutant allele has a normal function, and the lack of normal function produces a mutant phenotype.

To determine whether two recessive mutations are alleles, we make crosses that produce individuals with both alleles and examine their phenotypes. If the mutations are alleles of the same gene, heterozygous individuals will have a mutant phenotype. If so, we can say that the two mutations **fail to complement.** But, if the heterozygous individuals have a normal phenotype, the two mutations **complement** each other and are not alleles (Figure 4.12). The complementation test works well for most recessive mutations.

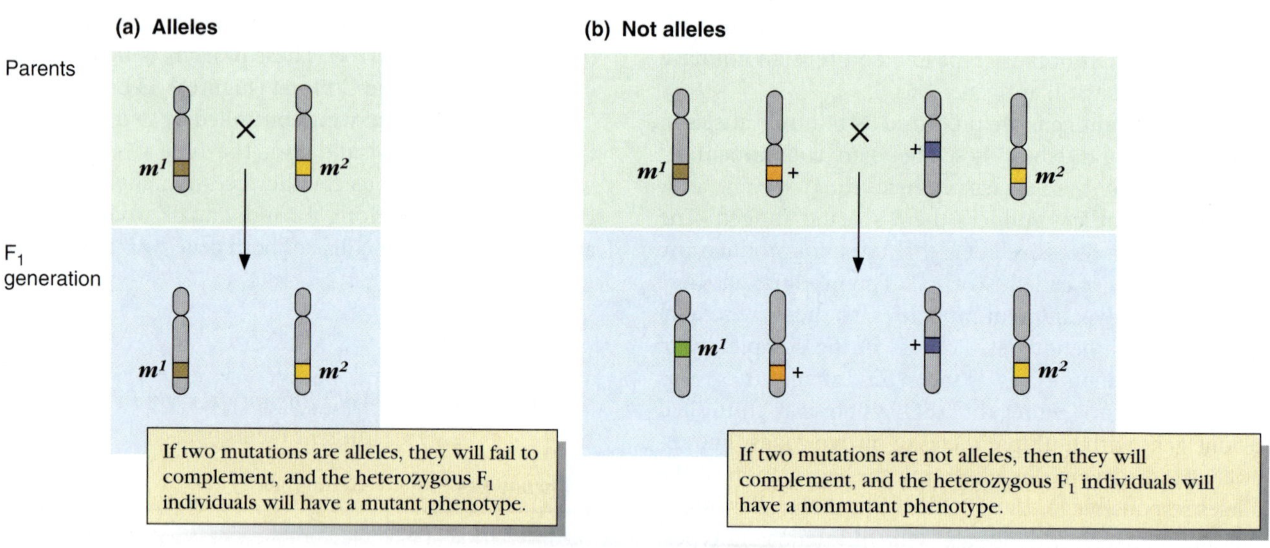

FIGURE 4.12 The complementation test. If two mutations are alleles of the same gene, heterozygotes will have a mutant phenotype and thus not complement each other. In this instance, (a) no wild-type alleles are present. (b) If two mutations are alleles of different genes, heterozygotes will have a wild-type phenotype, which is the result of the presence of one wild-type allele for each gene.

As an example, let's examine the sex-linked *yellow* gene in *Drosophila.* Mutant alleles of the *yellow* gene (y) are recessive and, when homozygous, change the wild-type black bristle color into a light brown or yellow color and lighten the normally gray/brown body cuticle. If a fly with a sex-linked mutation that produces a light bristle color (m) is crossed with a fly from a strain homozygous for a known *yellow* allele (y), the F_1 females will have the y allele on one X chromosome and m on the other. If the new mutation is also an allele of the *yellow* gene, these females will have a yellow phenotype because neither chromosome has a normal copy of the *yellow* gene. If the m light bristle color mutation is an allele of a different gene, then these females will have a normal phenotype because they are actually heterozygous for both genes ($y\ m^+\ y^+m$). Thus, using a complementation test, these alleles are established as being in different genes and not alleles of the same gene.

SOLVED PROBLEM

Problem

Two mutations in *Drosophila, black body* and *ebony body,* are recessive mutations that produce a similar phenotype, a black body color instead of the normal gray-brown body. The results of a cross between a black strain and an ebony strain are shown. Are these two mutations alleles of the same gene?

Parents: black × ebony

F_1: gray-brown

Solution

The F_1 have a gray-brown body color, which is the normal or wild-type color. This means that the two mutations complement, and because the mutations complement, they must not be alleles of the same gene.

SECTION REVIEW PROBLEMS

7. Results of a series of crosses between different mutant alleles follow. All alleles are recessive, and for each cross the normal or wild-type phenotype is listed. For each cross, determine which mutations are alleles of the same gene.

Wild-type	Parents	Offspring
a. gray-brown body	black × black	black body
b. green fruit	white × white	green fruit
c. blue flower	white × white	white flower
d. red eye	orange eye × white eye	orange eye
e. round seed	wrinkled seed × shriveled seed	shriveled seed

H. J. Muller Devised a Functional Classification System for Mutant Alleles.

The geneticist Hermann J. Muller [see HISTORICAL PROFILE: *Hermann Joseph Muller (1890–1967)*] received the Nobel Prize for his pioneering work on the nature of mutation and for his discovery that X-rays can cause mutations. Muller proposed a classification system for mutant alleles that is still widely used today. His system was based on how the mutation altered the gene's function. He found that the great majority of mutant alleles are **loss-of-function alleles,** that have either a reduced level of function or no normal function at all. Many mutations fall somewhere between no function and complete function. Alleles that make a product with a reduced level of function are called **hypomorphic** (Table 4.3). Individuals homozygous for a hypomorphic allele have a mutant phenotype because they have too little gene function to produce the normal phenotype. For most genes, there is a minimum level of gene function, called the **threshold level.** This minimal level of gene function is required to produce the normal phenotype, and if a mutation reduces the gene's function below the threshold level, the individual will have a mutant phenotype. Mutations that completely eliminate the wild-type function of a gene are called **amorphic** alleles or amorphs. Individuals homozygous for an amorphic allele have no gene function and have the same phenotype as individuals homozygous for a physical deletion or physical loss of the gene. This phenotype is called the complete loss of function phenotype, or amorphic phenotype. Alleles that produce no gene product are sometimes called **null** alleles. An allele that produces a nonfunctional gene product would be an amorph, but not a null. Whether an amorphic allele is also a null can be determined only by using a test

TABLE 4.3 *Muller's Classification System for Mutant Alleles*

Loss-of-Function Alleles	Usually Recessive
Hypomorphic	Less than the normal amount of gene function
Amorphic	No gene function
Null	No gene product
Gain-of-Function Alleles	**Usually Dominant**
Hypermorphic	More than normal amount of gene function
Neomorphic	New gene function
Antimorphic	New function that is antagonistic to normal gene function

HISTORICAL PROFILE

Hermann Joseph Muller (1890–1967)

Hermann Joseph Muller was one of the pioneers of modern genetics, especially of modern mutational analyses. Muller (Figure 4.B) was born on December 21, 1890, in New York City in a working class family. He grew up knowing the trials and financial difficulties of ordinary working men and women. As a youth, he was attracted to socialism and became convinced that this political philosophy offered hope for the improvement of the lot of the working classes. This belief remained with Muller for most of his life. An excellent student, Muller joined the research group of T. H. Morgan as a graduate student at Columbia University in 1910, and he received his Ph.D. degree in 1915.

Muller's great interest was in determining what genes were and how they functioned. At that time, the study of genes depended on the isolation of new mutations. Muller excelled in this, deriving a number of techniques for the recovery of mutations that are standard practice today. From 1915 to 1918, Muller worked at the Rice Institute in Houston and then returned to Columbia briefly (1918–1920) before accepting a faculty position at the University of Texas in Austin, where he remained for 12 years (1920–1932). While at Texas, Muller discovered that ionizing radiation (X-rays and gamma rays) could produce mutations. The discovery that mutations could be induced was revolutionary. Using radiation, Muller found large numbers of new mutations in *Drosophila* and launched a new era in the study of genetics. In recognition of this accomplishment, he was elected to the National Academy of Sciences in 1931 and was awarded a Nobel Prize in 1946.

Unhappy with the large amount of teaching at his position at Texas, Muller spent a year in Berlin (1932–1933) on a Guggenheim Fellowship and considered offers to leave Texas permanently. In Germany, he was persuaded by the Soviet geneticist N. I. Vavilov to accept a position as senior geneticist at the Institute of Genetics of the USSR Academy of Sciences. Muller believed, as did many with faith in socialism, that in the Soviet Union he would see the power of the state used to improve in the lives of the working classes. Muller spent 4 years in the Soviet Union (1933–1937). Conditions were not what he had expected. The rise of T. D. Lysenko to political power led to a suppression of genetics and the persecution of geneticists. Muller was persuaded to attend the International Congress of Genetics in Edinburgh, Scotland, in 1937, knowing that he might never be allowed to return to the USSR. He remained in Edinburgh for 3 years (1937–1940). Back in the USSR, many of his colleagues were arrested and perished in labor camps. In 1940, Muller returned to America and took a temporary position at Amherst College. He had difficulty finding a permanent position and believed that his time in the USSR was a factor. In 1945, he accepted a position at Indiana University and remained there until his retirement in 1964. Muller died April 5, 1967, in Indianapolis, Indiana.

FIGURE 4.B Hermann Joseph Muller. One of America's greatest geneticists. *(Science Photo Library/Photo Researchers, Inc.)*

Muller made several major contributions to genetics. He was an important part of the Morgan Group, which established the basic facts for the chromosomal location of genes. His own work pioneered the mutational analysis of genes, an essential part of genetics today. Muller published over 350 articles and authored or coauthored several books. He is one of a handful of scientists who shaped modern genetics.

that can detect the physical presence of the gene product. Functionally, null alleles and amorphic alleles both have no gene function, and so both give a complete loss of function phenotype. Some of the alleles of the *white* gene in *Drosophila* are hypomorphic, and others are amorphic. The w^1 allele is amorphic, and flies homozygous for this allele have no eye pigment. Other alleles like w^a or w^{co} are hypomorphic; flies homozygous for these alleles have a reduced level of eye pigment, giving them a light-colored eye.

Muller also found that some mutations produce **gain-of-function alleles.** He divided these into three classes: hypermorphic, neomorphic, and antimorphic. Muller called gain-of-function alleles that produce an abnormal increase in the amount of the gene's normal function **hypermorphic.** Individuals with hypermorphic mutations have a mutant phenotype because they have an excess of the gene's function. Muller called those mutations that cause a gene to have a new function **neomorphic.** Individuals with a neomorphic mutation have a mutant phenotype because of the presence of the new or altered genetic function. For example, some neomorphic alleles of the *Antennapedia* gene in *Drosophila* cause this gene to function in regions of the body where it is not normally active. The presence of the gene product in this new location causes a new mutant phenotype (*Antennapedia* is discussed in more detail in Chapter 15). The final category of gain-of-function mutations included mutations that cause the allele's function to become antagonistic to the wild-type gene function. He called these mutations **antimorphic.** Individuals heterozygous for an antimorphic mutation have a mutant phenotype because the product of the mutant allele interacts with and reduces the effectiveness of the product of the normal alleles, leading to a loss of normal function.

Muller's classification system is widely used in many genetic investigations of mutant alleles because it is very important to understand how an allele is functioning. For example, let us consider a mutation in humans that causes an hereditary disease, such as lactose intolerance. If the nature of the normal gene product is known, then treatment procedures can be developed to counteract the effects of the mutational change. If the mutant allele is hypomorphic, amorphic, or antimorphic, the disease symptoms are caused by a lack of gene product. By providing additional amounts of the normal gene product as a dietary supplement, it might be possible to alleviate the symptoms of the disease. However, for a hypermorphic allele, the symptoms are the result of an excess of gene product, and the correct treatment would be to reduce the amount of gene product in the individual.

SECTION REVIEW PROBLEMS

8. Compare the following:
    a. amorph and antimorph
    b. null allele and amorph
    c. hypermorph and hypomorph
9. You have been given several strains of mice that have different forms of the blood group antigen T7. You have antibodies that bind specifically to the antigens in each strain. You have been asked to determine whether these antigenic differences result from a series of codominant alleles of a single gene or from mutations of several different genes. Explain how you could do this.

GENE INTERACTIONS CHANGE THE RATIOS OF DIHYBRID CROSSES.

In the previous section, we discussed interactions between alleles of the same gene. Alleles of different genes may also interact, and this interaction may have an effect on the phenotype of the individual. Many such interactions have been detected in dihybrid crosses because they alter the phenotypic ratios in the F_2. The F_2 progeny of a dihybrid cross have a $(1:2:1) \times (1:2:1) = (1:2:1:2:4:2:1:2:1)$ genotypic ratio. If both genes have completely dominant alleles and there are no interactions between the alleles, the F_2 will have the $9:3:3:1$ phenotypic ratio that Mendel observed. In honor of Mendel's achievement, any F_2 phenotypic ratio that is different from $9:3:3:1$ is called a modified dihybrid ratio. The ratios of a dihybrid cross vary depending upon the interactions between the gene products. These unusual ratios allow geneticists to deduce the nature of the interaction between newly discovered alleles. Analyzing the ratios of phenotypes in crosses is one of the most important techniques of transmission genetics. In this section, we will discuss some examples of different types of allelic interactions and the modified phenotypic ratios that result.

Alleles of Different Genes That Produce the Same Phenotype Give a Modified Dihybrid Ratio.

One type of modified dihybrid ratio results from a cross between strains that have mutant alleles of two different genes that produce the same phenotype. When the F_2 of such a cross are counted, different genotypes can have the same phenotype. All individuals with this phenotype will be placed in the same class, and this inability to distinguish between individuals with different genotypes gives the modified dihybrid ratio. When the indistinguishable phenotype is given by recessive alleles of both genes, the F_2 will have two phenotypic classes: 9/16 and 7/16. The 7/16 class contains individuals homozygous for the recessive allele of the first gene (3/16), individuals who are homozygous for the recessive allele of the second gene (3/16), and individuals who are homozygous for the recessive alleles of both genes (1/16).

An example of two genes with recessive alleles that produce similar phenotypes was discovered in *Lathyrus odoratus* (sweet pea) by Bateson, Saunders, and Punnett in 1906. They crossed plants from two strains with white flowers and obtained an F_1 with normal, blue-colored flowers. When the F_1 were crossed, they produced an F_2 with 2132 blue-colored flowers and 1593 white flowers. These results are very close to 9/16 and 7/16. Similar examples have been discovered in many species, including several *Drosophila* mutations that affect eye pigment (Figure 4.13).

SECTION REVIEW PROBLEMS

10. A plant with blue, round flowers was crossed with a plant with white, long flowers. The F_1 progeny all had pale blue, round flowers. The F_2 progeny were

    1/16 white, long flowers
    2/16 pale blue, long flowers
    1/16 blue, long flowers
    3/16 white, round flowers
    6/16 pale blue, round flowers
    3/16 blue, round flowers

    a. Using symbols of your own choosing, indicate the genotypes of the parents, the F_1, and each of the phenotypic classes of the F_2 progeny.
    b. Describe any gene interactions in this cross.

11. A summer squash plant with long fruit was crossed with one with disc-shaped fruit. The F_1 all gave disc-shaped fruit. The F_2 were

disc	round	long
93	60	11

    a. How many genes are segregating in this cross?
    b. What types of interactions are occurring?
    c. What were the genotypes for the parental plants?

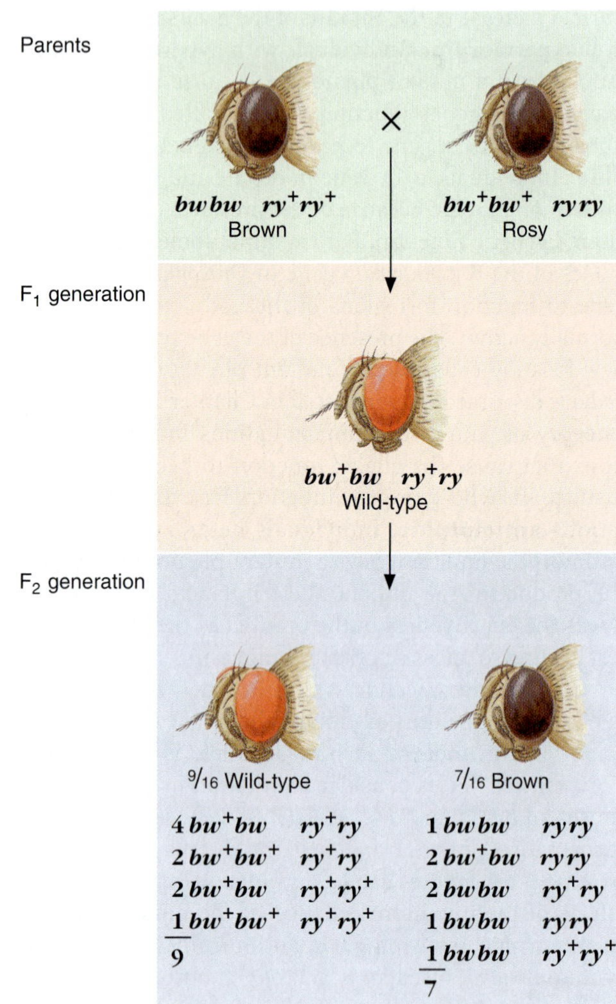

FIGURE 4.13 Distinguishing two mutations with similar phenotypes. Recessive mutant alleles of the *brown* gene (*bw*) and the *rosy* gene (*ry*) in *Drosophila* give similar brown eye phenotypes. Double mutant individuals (*bw bw ry ry*) have brown eyes. In a dihybrid cross, the F_2 progeny have a modified dihybrid phenotypic ratio (9/16 red and 7/16 brown) because individuals homozygous for *bw*, for *r*, or for both have similar phenotypes. This segregation pattern suggests that the two mutations are not allelic, even though they have a nearly identical phenotype.

A Mutant Allele of One Gene Can Mask an Allele of Another Gene.

One well-studied type of gene interaction occurs when the phenotype produced by a mutant allele of one gene blocks or masks the phenotype produced by an allele of another gene. This type of interaction is called **epistasis,** and the allele that blocks or masks is called an epistatic allele. Epistatic alleles can be either dominant or recessive, and each type gives a different modified dihybrid ratio. A recessive epistatic allele hides the phenotypic effects of alleles of the other gene when the recessive epistatic allele is homozygous. The coat color gene of mice is a good example of this type of inheritance. The *albino* allele (*a*) eliminates all color in homozygotes, and it is recessive to the normal pigment allele (*A*). The *black* allele (*b*) causes the normal gray-brown coat color (called agouti) to be much darker, and it is recessive to the normal allele (*B*). If we make a dihybrid cross between a black strain and an albino strain, the F_2 will be 9/16 agouti, 3/16 black, and 4/16 white. All individuals that have one dominant allele of both genes (9/16 = $A{-}B{-}$) will have a normal agouti coat. Individuals that are homozygous for the black allele and have at least one dominant allele of the albino gene (3/16 = $A{-}\,bb$) will be black. Individuals that are homozygous for the *albino* allele will be white, no matter what alleles of the *black* gene are present (1/16 *aa BB* + 2/16 *aa Bb* + 1/16 *aa bb* = 4/16 white) (Figure 4.14).

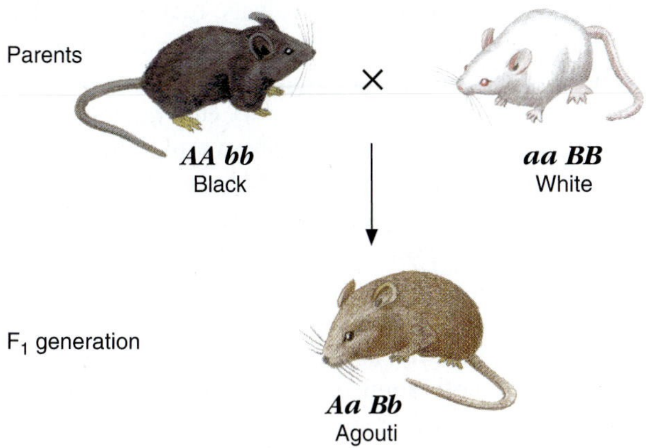

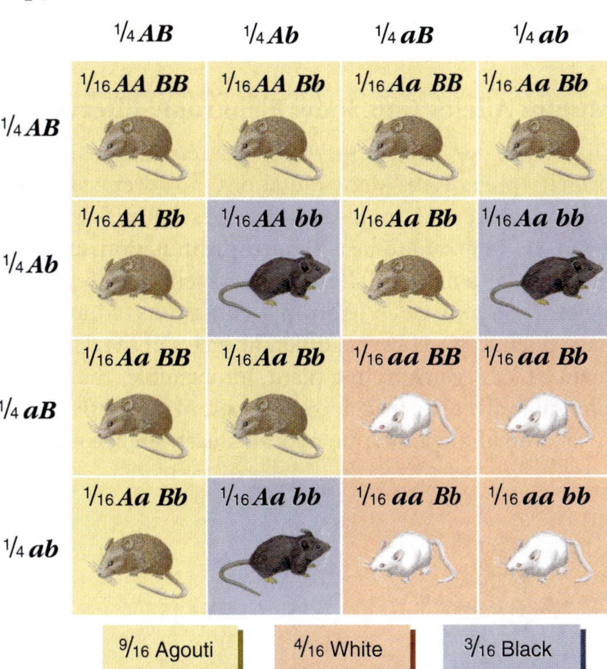

FIGURE 4.14 Recessive epistasis in mice. The recessive *albino* mutation in mice is epistatic to the recessive *black* mutation. Individuals homozygous for both mutations have the same fur color (white) as individuals homozygous for only the *albino* mutation. In a dihybrid cross, the F_2 have a modified phenotypic ratio (9/16 agouti, 4/16 white, and 3/16 black).

Mutations can also show dominant epistasis. A dominant epistatic mutation will produce a phenotype in heterozygotes or in homozygotes and masks the phenotype produced by alleles of another gene. In a dihybrid cross, this will give an F_2 phenotypic ratio of 12 : 3 : 1. Consider the example of dominant epistasis in fruit color of summer squash. A dominant allele (*W*) of one gene gives a white fruit color, and the recessive allele of that gene (*w*) gives the normal yellow color. A recessive allele of a second gene (*g*) gives a green fruit color when homozygous, and the dominant allele of this gene (*G*) gives a normal yellow color. If we cross a white strain with a green strain, the F_1 will all be white, and the F_2 will be 12/16 white, 3/16 yellow, and 1/16 green, a 12 : 3 : 1 ratio.

	white	×	**green**
Parents:	*WW GG*		*ww gg*
F_1		white	
		Ww Gg	
F_2	white (12/16)	yellow (3/16)	green (1/16)
	WW GG 1/16	*ww GG* 1/16	*ww gg* 1/16
	Ww GG 2/16	*ww Gg* 2/16	
	WW Gg 2/16		
	Ww Gg 4/16		
	WW gg 1/16		
	Ww gg 2/16		

SOLVED PROBLEM

Problem

Chickens with white feathers were crossed with birds from a black-feathered strain. The F_1 were all brown, a normal feather color. The F_2 were 9/16 brown, 3/16 black, and 4/16 white.

a. Using symbols of your own choosing, indicate how many pairs of alleles are involved in this cross.
b. What are the genotypes of the parents, the F_1, and each phenotypic class in the F_2?

Solution

a. The F_2 phenotypic ratio (9 : 3 : 4) suggests that there are two genes, each with two alleles, and that the allele giving the white color is showing recessive epistasis. The black strain appears to be homozygous for a recessive allele that gives a black feather color. We can use the symbol *a* for the recessive white allele and *A* for the dominant allele of the white gene. We can use *b* for the recessive black color allele and *B* for the dominant allele of this same gene.
b. Using the – symbol to stand for either allele of the gene, the genotypes of each class would be

	white	×	**black**
Parents	*aa BB*	×	*AA bb*
F_1		brown	
		Aa Bb	
F_2	brown	black	white
	A– B–	*A– bb*	*aa ––*

SECTION REVIEW PROBLEMS

12. Compare the following:
    a. epistasis and complete dominance
    b. epistasis and codominance
13. A black mouse was bred with a white mouse. The F_1 were all agouti colored. The F_2 were

agouti	black	white
9/16	3/16	4/16

    a. Using symbols of your own choice, indicate the genotypes of the parents, the F_1, and the F_2.
    b. Describe the dominance relationships between alleles and any interactions that are occurring.
14. A strain of shepherd's purse with oval seed pods is crossed with a strain with triangular seed pods. The F_1 all had triangular seed pods. The F_2 had phenotypes of 120 triangular, 30 round, and 10 oval seed pods.
    a. How many genes are controlling seed pod shape in this cross?
    b. Describe the characteristics and interactions that are occurring between the alleles of these genes.

Modifier Genes Enhance or Suppress the Phenotype Produced by Other Mutations.

Unlike epistatic mutations, whose mutant phenotype masks the phenotype of alleles of other genes, **modifier mutations** increase or decrease the phenotypic effect of other mutations. Modifiers do not affect other genes at random. Instead, each modifier has a specific set of target genes that it affects. Modifiers can be suppressors or enhancers. **Suppressors** decrease the mutant phenotype of the target mutation, making it less severe, and **enhancers** increase the severity of the phenotypic effect of the target mutation. Individuals with a double mutant combination of modifier and target mutation have either a reduced (suppressor) or more severe (enhancer) mutant phenotype. Many modifier genes produce a different phenotype from their target mutation, and some produce no mutant phenotype at all when the mutant allele of the target gene is not present. In a dihybrid cross between a strain with a recessive allele of an autosomal target gene and a recessive, autosomal modifier, the F_2 will be 9/16 normal, 3/16 target mutation, 3/16 modifier, and 1/16 target-modifier double mutant phenotype.

An example of a suppressor mutation in *Drosophila* is the sex-linked *suppressor of forked* gene. Mutant alleles of this gene, such as the $su(f)^1$ allele, suppress the phenotypic effects of mutant alleles of the *forked* gene, such as f^1. Individuals homozygous for f^1 have bent, forked bristles. Individuals homozygous for $su(f)^1$ are indistinguishable from wild-type individuals; that is, $su(f)^1$ has no visible mutant phenotype. Individuals homozygous for both $su(f)^1$ and f^1 have straight, nearly normal bristles (Figure 4.15). An example of an enhancer mutation is the autosomal dominant *Enhancer of white-apricot* ($En(w^a)$) mutation in *Drosophila*. The sex-linked *white-apricot* mutation produces an orange-eyed phenotype, and the *Enhancer of white-apricot* gene, when mutant, has no visible phenotype. Individuals that have *Enhancer of white-apricot* and are homozygous for *white-apricot* have nearly pure white eyes, a more extreme phenotype. Some alleles can be both suppressors and enhancers. For example, the $su(f)^1$ allele suppresses the *forked* gene, and this same allele also enhances *white-apricot*. The mutation is called "suppressor of forked" because that phenotype was discovered first.

Mutant Alleles Can Show Pleiotropic Effects.

Up to now, we have dealt with mutations that have one primary phenotype. Many mutations, however, can have multiple effects. A mutation that gives multiple phenotypic effects is said to be a **pleiotropic** mutation, and the multiple, phenotypic effects collectively represent the mutant phenotype. Pleiotropic mutations commonly have phenotypic effects at different stages of the development of an organism. In a plant, for example, the mutation may affect the seed and also a part of the adult plant. In an animal, the mutation may alter early embryonic de-

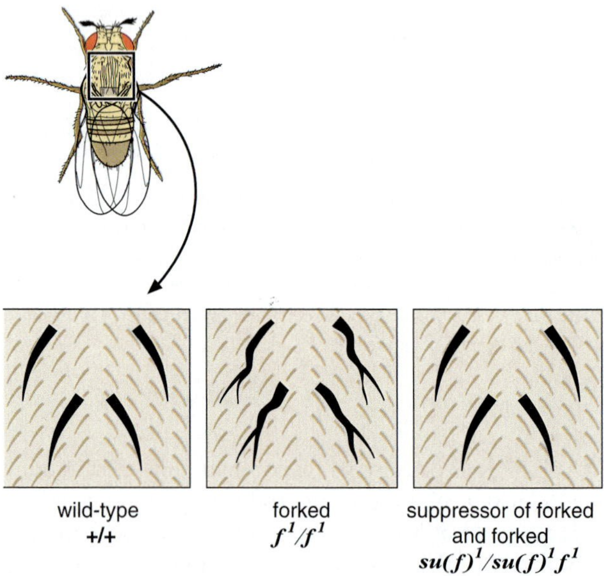

FIGURE 4.15 Modifiers: *suppressor of forked*. The *suppressor of forked* mutation is a modifier that reduces the severity of the mutant phenotype of certain mutant alleles of the *forked* gene. The phenotype of mutant alleles of the *forked* gene is to have bent, forked bristles. The suppressed phenotype seen in individuals with both mutations is to have straight bristles that look normal.

velopment and also adult development. Other cases may involve a mutation that affects different parts of the individual.

An example of a pleiotropic allele can be found in Mendel's own paper. In his paper, Mendel presented a list of the pairs of traits that he was using in his crosses. One of these was seed-coat color. One of his original strains of peas had seeds with white seed coats and another strain had colored seed coats. Mendel noted that these strains also differed for color of flower, with the white seed-coat strain having white flowers and the colored seed-coat strain having purple flowers. Mendel noted that the flower color and seed-coat color traits were always found together. He stated:

> . . . *difference in the color of the seed-coat.* This is either white, with which character white flowers are constantly correlated; or it is gray, gray-brown, leather-brown, with or without violet spotting, in which case the color of the standards is violet, that of the wings purple, and the stem in the axils of the leaves is of a reddish tint.

In all his crosses, Mendel always found seed-coat color and flower color to be associated. Seeds with a colored coat always gave rise to plants with colored flowers, and seeds with white-colored coats always gave rise to a plant with white flowers. This is the characteristic effect of a pleiotropic gene that functions in two stages of the life of the organism, in this case controlling seed-coat color and also flower color.

Mendel also noted that there were differences in the actual color of the seed coat in individuals with a colored coat. There were different shades of color as well as cases of spotting or plain color. To truly understand the phenotype of the colored allele would require a careful study of the range of these color variations, and a determination of whether they did in fact represent pleiotropic effects of one single allele. Some variations in phenotype between individuals with the same alleles can be produced by other factors. Examples of these other factors are discussed in this chapter, including interaction with modifier alleles of other genes (discussed earlier) or variations in the environment (discussed later in the chapter).

Mutant Alleles Can Show Partial Penetrance and Have Variable Expressivity.

When Mendel examined the progeny of his crosses, all the individuals with the same genotype had the same phenotype. This is not the case for all mutations. There are two types of expression of mutant phenotypes. First, some individuals in a population may have a mutant genotype and not show any mutant phenotype, whereas others will show the mutant phenotype. The percentage of genetically mutant individuals showing a mutant phenotype is the **penetrance** of that mutation. A good example is the *Blackpatch* (*Bpt*) mutation in *Drosophila.* This is a dominant mutation that produces black spots in the adult eye. Because it is dominant, we would expect that all individuals that are either homozygous or heterozygous for *Bpt* would have black eye spots. However, when a sample of heterozygous individuals were examined, only 45/100 males and 85/100 females had eye spots, and the remaining individuals had a normal phenotype (Figure 4.16). The penetrance of this dominant

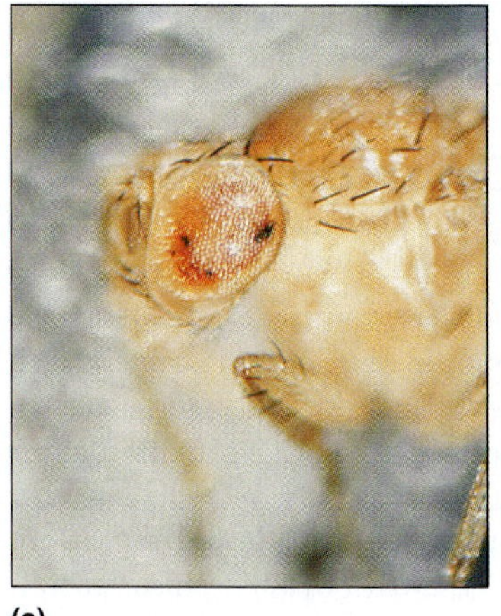
(a)

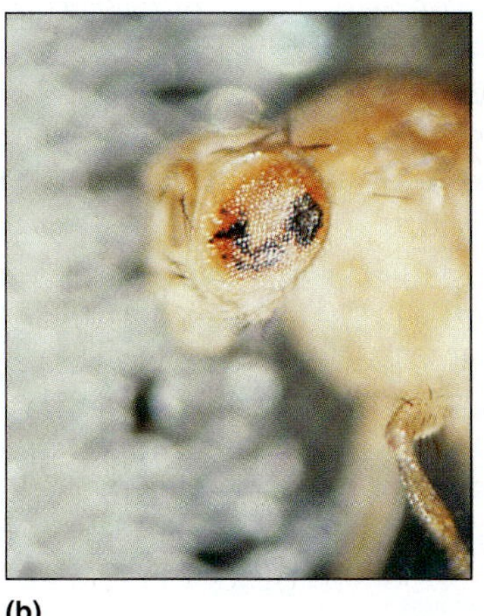
(b)

(c)

FIGURE 4.16 Partial penetrance and variable expressivity. The *Blackpatch* mutation produces black spots in the eyes of adult *Drosophila.* This mutation has both partial penetrance and variable expressivity. Individuals with the same genotype may have (a) small eye spots, (b) medium-sized eye spots, or (c) large eye spots. The individuals shown here are also homozygous for the *facet-glossy* mutation, giving the surface of their eyes a smooth, glossy texture. *(Jack R. Girton/Iowa State University)*

mutation is thus 85% in females and 45% in males. The *Bpt* mutation is an example of a mutation that has **partial penetrance.** An example of an allele with **complete penetrance** is the original mutant allele of the *ebony* gene (e^1) in *Drosophila.* All flies homozygous for e^1 have a dark body. In crosses, alleles with partial penetrance produce modified phenotypic ratios because some of the individuals with a mutant genotype do not produce a mutant phenotype, making the normal phenotypic class larger and the mutant class smaller. Unlike the earlier examples of modified phenotypic ratios, these types of alleles do not produce characteristic modified ratios because the amount of the modification depends on the penetrance of the allele, and this varies from allele to allele.

A second common phenotypic difference between individuals with the same mutant genotype is in the degree, or severity, of the mutant phenotype. In a collection of mutant individuals, all showing a mutant phenotype, some individuals may have a stronger mutant phenotype and others a weaker mutant phenotype. The range in severity of the mutant phenotype is called the **expressivity** of the mutation. Mutations with a wide range of expression are said to have **variable expressivity.** *Bpt* mutations have variable expressivity as well as reduced penetrance. In a collection of *Bpt* individuals, a range of sizes and shapes of eye spots is always observed. Some eyes are nearly completely covered by one spot, others have only a very small spot, and still others have eyes with several small spots (Figure 4.16). On the other hand, the *yellow* mutation has **constant expressivity,** and all bristles on all mutant individuals are light-colored.

A mutant allele may show partial penetrance and/or variable expressivity for a variety of reasons. One common reason, which is discussed later in this chapter, is that some mutant changes cause gene products to become sensitive to changes in the environment and in different environments may give different phenotypes. In a population raised in a variable environment, individuals in a favorable environment will have a normal or less mutant phenotype. A second common reason for partial penetrance and variable expressivity is gene interaction. The phenotypes given by many mutant alleles are affected by mutations of other genes. In a population that contains many different alleles of many different genes, individuals may have combinations of modifier alleles that alter or completely eliminate the effect of a mutant allele (as discussed earlier).

SOLVED PROBLEM

Problem

For each of the following mutations, a homozygous strain was isolated and a sample of individuals was taken. What is the penetrance of each mutation?

Mutation	Phenotype	
a. Short stems in *Arabidopsis*	short 62	long 38
b. Curved wings in *Drosophila*	curved 81	straight 19
c. Brown coat in mice	brown 73	agouti 27
d. Shrunken kernels in corn	shrunken 67	full 33

Solution

a. 62%
b. 81%
c. 73%
d. 67%

SECTION REVIEW PROBLEMS

15. Compare the following:
    a. partial penetrance and incomplete dominance
    b. variable expression and codominance
    c. suppression and epistasis
    d. enhancers and overdominance
16. Down syndrome in humans results from a dominant genetic defect (the presence of an extra copy of part of chromosome 21). In a study of families with Down syndrome, several sets of monozygotic (identical) twins were examined for several of the phenotypes associated with Down syndrome. Both twins did not always have identical phenotypes. For all phenotypes, the following frequencies were found:

    Twins with identical phenotypes: 89%

    Twins with different phenotypes: 11%

    Given that these monozygotic twins have identical genotypes, explain these results.

SOME ALLELES CAUSE THE DEATH OF AN ORGANISM.

All the alleles that Mendel used in his experiments, no matter what phenotypes they gave, had the same viability. This is not the case for all alleles. Some genes perform vital functions in the individual, and if a mutation alters or eliminates this function, the individual dies prematurely. This type of mutation is called a **lethal mutation.** Genes whose mutation is lethal obviously have an important function in the organism, and this makes lethal mutations of great interest to geneticists. The use of lethal mutations in analyzing developmental processes is discussed further in Chapter 15. In this section, we will discuss what effects lethal mutations may have on phenotypic ratios in the F_2 of crosses.

Lethal Alleles Cause Death at Particular Stages of the Life Cycle.

Lethal mutations have a unique effect on Mendelian ratios in genetic crosses. Gene interactions produce modified phenotypic ratios by causing more than one genetic class to have the same phenotype, but lethal mutations cause modified phenotypic ratios by eliminating (killing) the individuals that have a particular genotype. Vital genes do not all function at the same stage of the life cycle of an individual, so different lethal mutations cause death at different stages. The stage at which a particular lethal mutation causes death is called the **lethal period** of that mutation. The lethal period of a particular mutation can, in fact, serve as a phenotype and can be used to distinguish among different lethal mutations. A lethal mutation that causes death during embryonic development has a different phenotype than a mutation that causes premature death of mature adults.

Understanding the lethal period is important when examining the effects of a lethal mutation on the phenotypic ratios of the offspring of crosses. At the moment of conception, the offspring have a proper Mendelian genotypic ratio. However, in most instances where a lethal gene is acting, the phenotypic ratio of the F_2 is usually not determined at the very beginning of life but rather later, when the individuals have completed growth and development. If a mutation has a lethal period early in embryonic development, then these individuals will die before birth and not be counted. The absence of the lethal genotypic class in the progeny gives the modified phenotypic ratio. However, if the lethal period is late in life, after the stage when the offspring are usually counted, the lethal individuals will be counted, and their lethality will be recognized as a phenotype of a particular genotype. In this case, the offspring will not have a modified phenotypic ratio.

An example of a recessive lethal mutation with an early lethal period is the *yellow* mutation in mice. The *yellow* mutation (*Y*) is a dominant mutation that produces a yellow fur color, instead of the agouti fur color given by the normal allele (+) (Figure 4.17). Crossing a heterozygous yellow-colored mouse (*Y*+) with a normal mouse gives progeny that are half yellow-colored and half agouti, as expected. However, crossing two yellow-colored mice gives two unusual results. First, such crosses would be expected to give 1/4 homozygous mice (*YY*), 1/2 heterozygous mice (*Y*+), and 1/4 agouti mice (++). In contrast to these expected results, the offspring consist of yellow-colored (heterozygotes) and agouti mice in a 2 to 1 ratio. And, when the offspring from such crosses are test-crossed, no mice homozygous for the *yellow* mutation are found. In addition, when the number of offspring is counted, the average number of offspring per litter is 1/4 smaller than expected. The explanation for these observations is that the *yellow* mutation gives both a dominant and a recessive phenotype. The dominant phenotype is a visible change in the fur

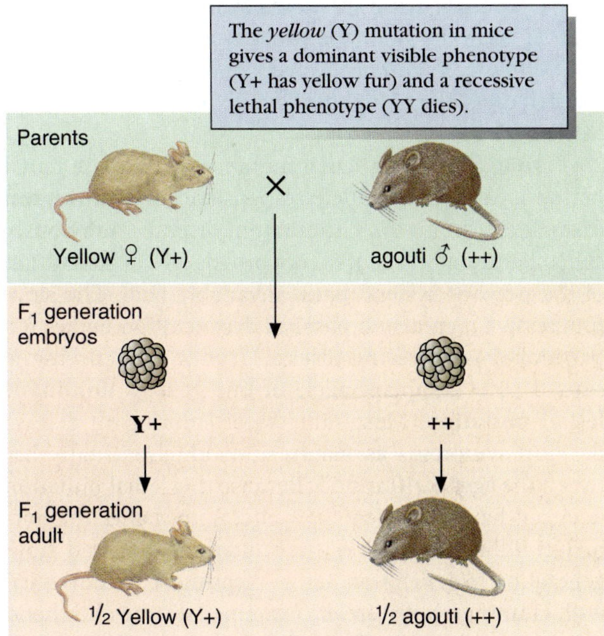
The *yellow* (Y) mutation in mice gives a dominant visible phenotype (Y+ has yellow fur) and a recessive lethal phenotype (YY dies).
Parents
×
Yellow ♀ (Y+)
agouti ♂ (++)
F1 generation embryos
Y+
++
F1 generation adult
1/2 Yellow (Y+)
1/2 agouti (++)

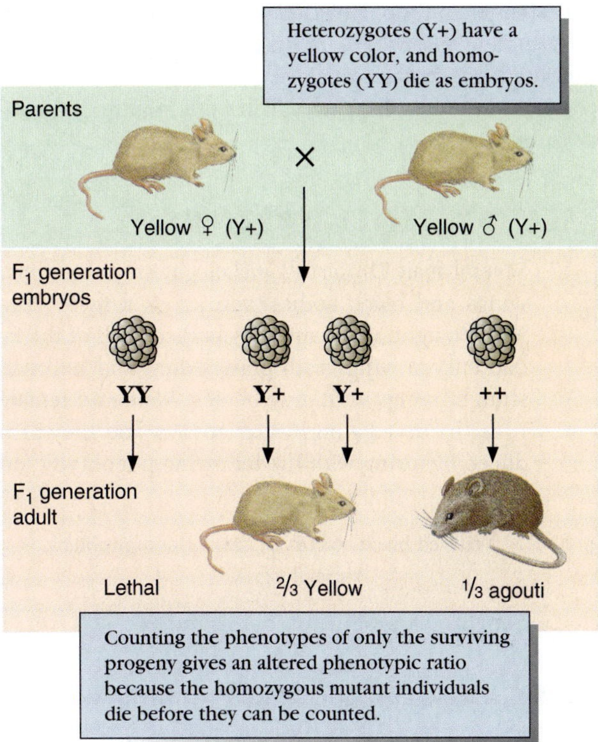
Heterozygotes (Y+) have a yellow color, and homozygotes (YY) die as embryos.
Parents
×
Yellow ♀ (Y+)
Yellow ♂ (Y+)
F1 generation embryos
YY
Y+
Y+
++
F1 generation adult
Lethal
2/3 Yellow
1/3 agouti
Counting the phenotypes of only the surviving progeny gives an altered phenotypic ratio because the homozygous mutant individuals die before they can be counted.

FIGURE 4.17 Recessive lethal alleles. Modified phenotypic ratios of the recessive, lethal *yellow* mutation are seen in mice that have died at an early stage of development.

color ($Y+$ have yellow fur) and the recessive phenotype is lethality (*YY* die as embryos). Because the individuals homozygous for the *yellow* mutation die during embryonic development, they are not present when the offspring phenotypes are counted after birth. The absence of the homozygous class yields a modified phenotypic ratio (Figure 4.17). If the progeny of a cross between two yellow-colored mice were examined during the first few cell divisions of life, they would be present in a ratio of 1 *YY* : 2 *Y*+ : 1 ++.

An example of a lethal mutation with a late lethal period is the mutant allele responsible for **Huntington disease** in humans. Huntington disease (previously called Huntington's chorea) is a progressive degeneration of the nervous system that is invariably fatal. The slow, progressive mental and physical degeneration makes it a particularly distressing disease. Huntington disease is caused by a dominant allele of one gene, a **dominant lethal mutation.** Most sufferers of Huntington disease develop symptoms as mature individuals, usually between the ages of 40 and 55. Because this lethal mutation does not have any phenotypic effects until after the normal childbearing period, the mutation is passed from generation to generation. If the offspring of an individual with Huntington disease are examined during childhood or early in adult life, they will have no apparent abnormal phenotype; all phenotypes will be present in a normal Mendelian ratio. However, if the offspring are examined at age 60 or 70, most of the individuals who inherited the disease-causing allele will have died, and consequently the offspring will have a modified phenotypic ratio.

SECTION REVIEW PROBLEMS

17. Several male *Drosophila melanogaster* with curved wings and black bodies were crossed to females with straight wings and gray bodies. Half of the F_1 had curved wings with gray bodies, and half had straight wings with gray bodies. Male and female F_1 with curved wings were crossed, and they produced offspring with the following phenotypic ratio:

    2 curved black : 6 curved gray : 1 straight black : 3 straight gray

    Explain the production of this ratio.

The Phenotypes Produced by Conditional Alleles Are Affected by the Environment.

In Mendel's studies, the phenotype of the organism was determined entirely by the genotype. In fact, the phenotype given by a mutant allele is often strongly affected by the environment in which the individual lives. Mutations that give different phenotypes in different environments are called **conditional mutations.** In this section we will discuss some of the different types of conditional mutations and some of the properties of each type. Conditional mutations may be sensitive to a variety of environmental effects; the most common are temperature and diet. The analysis of conditional mutations presents special difficulties because they will give different phenotypic ratios in different environments. Conditional mutations also present special opportunities. They allow the investigator to alter the effect of a mutation by altering the environmental conditions. Conditional mutations often are used to study the role of vital genes in developmental processes.

A common type of conditional mutation is the **temperature-sensitive mutation.** These mutations give a normal phenotype at one temperature (the **permissive temperature**) and a mutant phenotype at another (the **restrictive temperature**). Mutations with a high restrictive temperature are called **heat-sensitive mutations,** and mutations with low restrictive temperatures are called **cold-sensitive mutations.** Mutations that show some form of mutant phenotype at all temperatures but have different mutant phenotypes at different temperatures are sometimes called mutations with a **temperature effect.** Temperature-sensitive mutations are known in a wide range of species. They are especially common in species that do not regulate their body temperature, such as bacteria, yeast, or *Drosophila*. *Drosophila* homozygous for a temperature-sensitive allele of the *Curly* gene, for example, have tightly curled wings at the restrictive temperature (29°C, which is a high temperature for *Drosophila melanogaster*) and straight wings at permissive temperatures (18°C, which is a low temperature for *Drosophila melanogaster*). Crosses involving temperature-sensitive mutations that are carried out at a restrictive temperature will give a different phenotypic ratio from the same cross carried out at a permissive temperature (Figure 4.18).

Conditional mutations that prevent individuals from producing biochemical compounds normally produced during growth are called **auxotrophic mutations.** Auxotrophic mutations are especially important in bacterial and fungal genetics, as described in Chapter 8. Normal strains of the bacteria *E. coli,* for example, grow well on a simple chemical medium containing salts, a carbon source, and minerals. From this "minimal medium," they are able to synthesize all the compounds needed for growth. A strain with an auxotrophic mutation unable to synthesize a particular vital compound, such as the amino acid arginine, will not grow because it cannot synthesize proteins (containing arginine). This inability to synthesize proteins is a lethal condition, so the mutant bacterial cells will not grow on a minimal medium. However, if the mutant bacteria are able to get arginine from the medium,

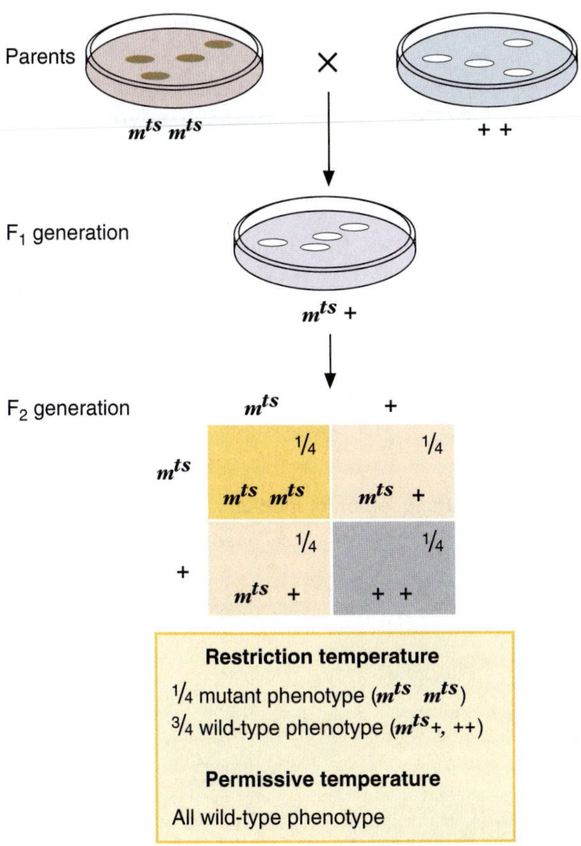

FIGURE 4.18 Phenotypic ratios for temperature-sensitive alleles. The F_2 ratios of a monohybrid cross involving a temperature-sensitive mutation are different at different temperatures. For a recessive allele at restrictive temperatures, the offspring are 1/4 mutant and 3/4 normal. At permissive temperatures, all offspring are normal.

FIGURE 4.19 Auxotrophic mutations. Auxotrophic mutations in *E. coli* affect the ability to grow on media that do not contain specific nutritional supplements. An auxotrophic mutation in *E. coli* that cannot synthesize the amino acid arginine does not grow on media lacking arginine, but it can grow if supplied with arginine. *(Jack R. Girton/Iowa State University)*

they will be able to grow (Figure 4.19). For many auxotrophic mutations in microorganisms, the ability to grow on a medium with a particular supplement but not on a minimal medium defines the phenotype of the mutation.

Auxotrophic mutations can give either a normal Mendelian ratio in crosses or a modified phenotypic ratio, depending on the environmental conditions. Consider a dihybrid cross between a diploid yeast that is homozygous for a recessive arginine auxotrophic mutation (arg^-) and a strain that is homozygous for a recessive methionine auxotrophic mutation (met^-). The heterozygous F_1 ($arg^-arg^+\ met^-met^+$) will grow on minimal medium, but the phenotypic ratio in the F_2 generation will depend on the composition of the growth medium. If the F_2 are grown on a supplemented medium that contains both arginine and methionine, then all the F_2 will grow normally, and all will appear to be wild-type. If the F_2 are grown on a medium supplemented with arginine but not methionine, then the F_2 homozygous for the methionine mutant would not grow normally, but all others would. This would give a phenotypic ratio of 12/16 growing to 4/16 not growing. If the F_2 were grown on a minimal medium that has neither supplement, then none of the homozygotes would grow, giving an F_2 phenotypic ratio of 9/16 growing to 7/16 not growing (Figure 4.20). The genotype of auxotrophic mutations in microorganisms is often determined by testing the ability of the strain to grow on media with different supplements. These techniques are discussed further in Chapter 8.

It is important to distinguish between mutations whose expression is affected by environmental conditions and phenotypic effects caused by the environment alone. Unusual environmental conditions can produce a phenotypic effect that resembles a mutant phenotype. This is known as a **phenocopy** of the mutant phenotype. In humans, one well-known case of phenocopy is one effect of the drug thalidomide. In the 1950s and early 1960s, thalidomide was prescribed to pregnant women for treatment of morning sickness. Before authorities fully realized its adverse effects, thousands of babies in Europe and other parts of the world were born with severe birth defects. The drug produced a phenocopy of the inherited human disease phocomelia. This disease causes a shortening or truncation of the limbs. Individuals suffering this disease, or the children of women who take thalidomide, are often born without arms or legs, or with only vestigial limbs. Thalidomide was later found to be an effective treatment for some of the symptoms of leprosy, and it has recently been shown to relieve some of the symptoms of individuals with AIDS. This has led to a resurgence in production and use of the drug, which could have serious consequences. Recent cases have occurred in which pregnant women have taken thalidomide either because they did not know what they were taking or because they were not adequately warned of the dangerous side effects, and they have had children with severe birth defects.

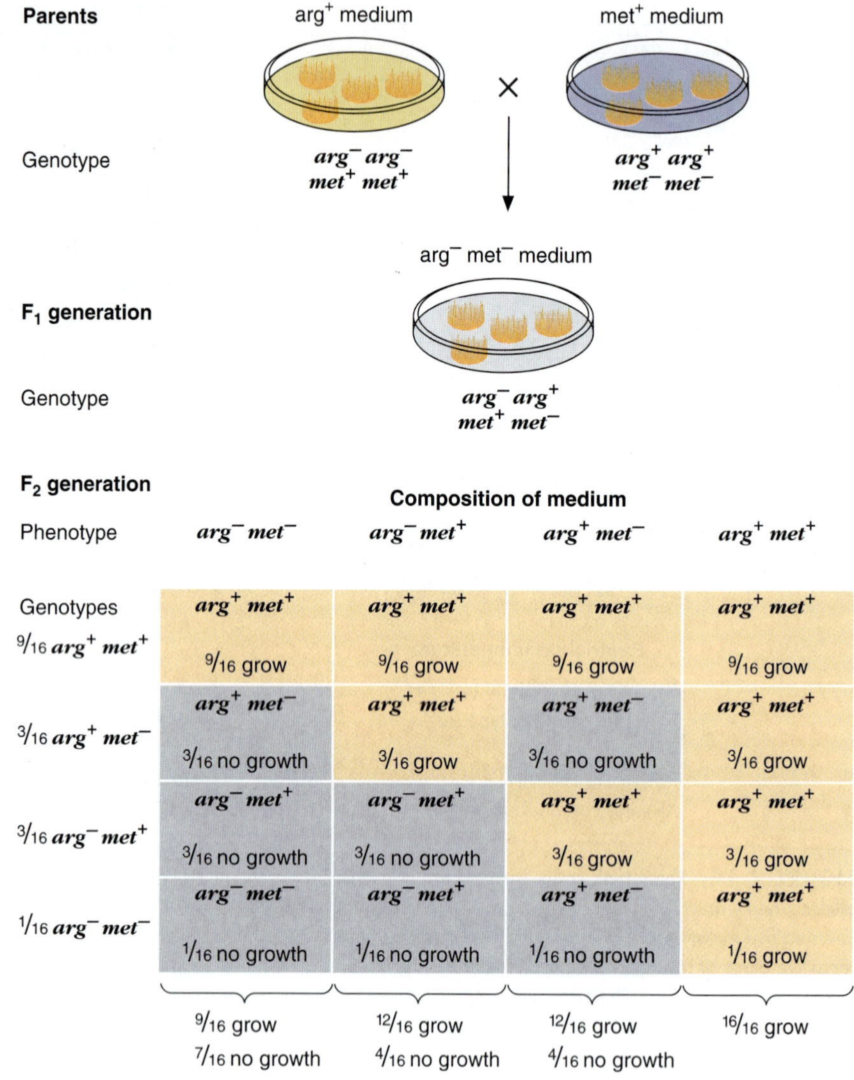

FIGURE 4.20 Dihybrid ratios for auxotrophic mutants. The F_2 phenotypic ratios in a dihybrid cross with auxotrophic mutations depends on the composition of the medium. If two strains with different auxotrophic mutants are crossed and the F_1 inbred, the F_2 will give a ratio of 9/16 growers, 7/16 nongrowers, or 12/16 growers, 4/16 nongrowers, or all will grow, depending on the medium.

SOLVED PROBLEM

Problem

Several collections of 100 eggs were made from a new mutant strain of *Drosophila* at 22°C. The collections were shifted to 29°C at different stages of development. The table shows the time of shift and the number of the eggs in each collection that formed viable adults. What can you conclude about this mutation?

			Age at Shift to 29°C (in hours)						
	22°C Control	29°C Control	24	48	72	96	120	144	148
Viable adults	100	0	10	8	11	95	97	95	99

Solution

Because few viable adults are produced when the individuals are shifted at an early age (24, 48, 72 hours), the mutation appears to be a temperature-sensitive lethal with a restrictive temperature of 29°C whose active period is early in development, before 96 hours.

SECTION REVIEW PROBLEM

18. Two varieties of chickens were crossed, and the F_1 were inbred to produce an F_2. Two crosses were made. For each cross, 1000 eggs were collected and placed in incubators. In one cross, 63 eggs failed to hatch. Each unhatched egg contained a poorly formed embryo. In the second cross, all eggs hatched. Consulting their records, the workers discovered that the second incubator had been set 5°C cooler than the first. Explain these results.

Summary

In this chapter, we have discussed a variety of allele characteristics. The Mendelian concept of a gene is that it is fundamentally a unit of genetic function that gives a defined phenotype. However, alleles of the same gene may interact to give a variety of different dominance relationships. These include complete dominance, which Mendel first described; incomplete dominance, where the heterozygote is phenotypically halfway between the parents; codominance, where each allele is expressed in the heterozygote and can be detected; and overdominance, where the heterozygote has a more extreme phenotype than either parent. Alleles can also come in many different variations, depending upon the mutation in the gene. Dominance is a relationship between alleles, not a fixed characteristic of an allele. This means that an allele's dominance is always relative to another allele. An allele that is dominant to a second allele may be recessive to a third allele. The determination of dominance relationships can be done only by examining the phenotypes of heterozygous individuals. Mutational alteration of normal gene function may take several forms, including partial or complete loss of function, increase in function, generation of a new function, or establishment of the gene's function dependent upon its environment.

A gene's function can interact with the function of other genes. Mendel did not report any interactions between alleles of the different genes in his studies. All the alleles Mendel studied produced independent phenotypic effects. Since then, many exceptions have been discovered. Mutant alleles of one gene may mask the effects of alleles of another gene, or they may increase, decrease, or modify another gene's phenotype. These modifications are often detected because they give modified phenotypic ratios in the progeny of a dihybrid cross. Different types of interaction each produce a particular modified ratio, so a knowledge of the modified ratios produced by different types of interaction can be used to deduce the different types of interactions. Analysis of a modified ratio may be complicated if the mutations produce conditional or lethal phenotypes. Another complicating effect is that modified ratios are also produced when individuals with a mutant genotype do not always show a mutant phenotype (partial penetrance) or when individuals with the same mutant genotype have different degrees of mutant phenotypes (variable expressivity).

Selected Key Terms

allelic series
amorphic
antimorphic
auxotrophic mutation
codominance
complementation test
conditional mutation
enhancers
epistasis
expressivity
hypermorphic
hypomorphic
incomplete dominance
lethal mutation
modified dihybrid ratios
neomorphic
overdominance
penetrance
phenocopy
suppressor

Chapter Review Problems

1. A white-flowered, long-leafed tulip was crossed with a red-flowered, short-leafed plant. The F_1 were all pink-flowered and long-leafed. Two F_1 plants were crossed to produce an F_2, whose phenotypes were:

Phenotype	Number
red-flowered, long-leafed	49
white-flowered, long-leafed	41
pink-flowered, long-leafed	91
red-flowered, short-leafed	16
pink-flowered, short-leafed	28
Total	225

Explain these results, given that the average number of offspring in a tulip cross is 240.

2. A female rabbit of the genotype *Aa Bb Cc Dd* produced a litter whose genotypes follow:

Progeny Genotypes
AA Bb cc Dd
AA Bb cc dd
aa Bb cc dd
aa Bb CC DD

a. What are the possible genotypes of the father of this litter?

b. Given the following frequency data on individuals with different genotypes, what percentage of males in this population are potential fathers?

Frequencies of Genotypes

AA = 1/4	*Aa* = 1/4	*aa* = 1/4
BB = 1/4	*Bb* = 1/2	*bb* = 1/4
CC = 1/4	*Cc* = 1/2	*cc* = 1/4
DD = 1/2	*Dd* = 1/4	*dd* = 1/4

3. Red-flowered plants from a newly developed strain of *Phaseolus* were crossed with a standard purple-flowered strain. The F_1 were all purple and were inbred to produce an F_2. Of the 34 F_2 plants recovered, 32 were purple flowered and 2 were red flowered. Explain how this might occur.
4. A strain of cotton with dark red leaves was crossed with a virescent yellow strain, and the F_1 were inbred. The F_2 follow:

Phenotype	Number
dark red	36
light red	95
dark bronze	18
light bronze	34
green	46
virescent yellow	13

   a. Using symbols of your choosing, indicate what the genotypes of the parental strains were, and describe any interactions that are occurring.
   b. What would you predict was the F_1 phenotype?
5. Describe one experiment that would test your hypothesis in Problem 4. Explain how your experiment would test the hypothesis, and indicate the result you would expect if the hypothesis is correct.
6. A summer squash plant with white flowers and white fruit was crossed with a plant with yellow flowers and brown fruit. The F_1 all had yellow flowers and purple fruit. The F_1 were inbred, and the following F_2 were produced:

yellow flowers, brown fruit	56
yellow flowers, purple fruit	98
yellow flowers, red fruit	49
white flowers, white fruit	69

   Using your own choice of gene symbols, propose a model for the inheritance of flower and fruit color in this cross. Your model should state how many genes are segregating and what the genotypes are of each class in the F_2 and describe any interactions that are occurring.
7. Two strains of summer squash (knobby-skinned fruit and smooth-skinned fruit) were crossed. The F_1 gave knobby-skinned fruit. The F_2 follow:

knobby	rough	smooth
93	59	11

   a. How many genes are segregating in this cross?
   b. What types of interactions are occurring?
   c. Using symbols of your own devising, indicate the genotypes of the parents, the F_1, and the F_2.
8. A cross between a smooth and a rough-haired guinea pig gave offspring with partly rough hair. Inbreeding these F_1 produced an F_2 with the following phenotypes:

smooth	partly rough	rough
7/16	4/16	5/16

   a. Assuming that two genes are involved, what types of interactions would produce this ratio?
   b. How would this ratio change if the allele for rough were a recessive lethal with 50% penetrance?
9. In a dihybrid cross between two strains of chickens (rose comb with pea comb), the F_1 were all walnut combed. The F_2 follow:

walnut	rose	pea	single
9/16	3/16	3/16	1/16

   a. If the F_2 single-combed birds were backcrossed with the F_1, what would be the probability of obtaining 3 single-combed birds in a brood of 10?
   b. How would your answer change if *single comb* was known to show 50% penetrance?
10. The following human pedigree shows the inheritance of the ability to taste a specific compound.

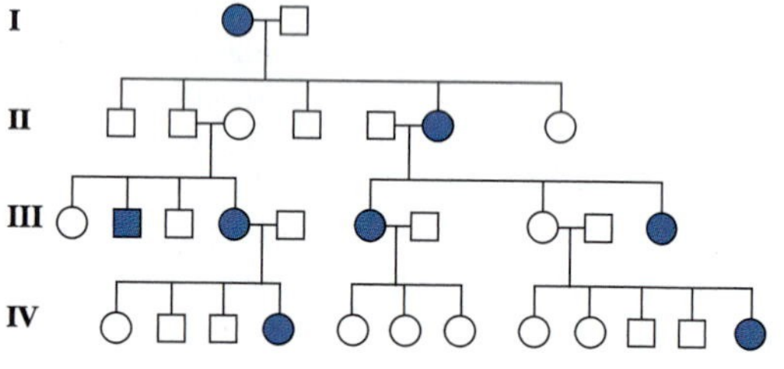

   a. Assuming this is a very rare trait in this population, explain the inheritance of this trait.
   b. Would your answer change if this were a very common trait?
11. Humans with Huntington disease usually begin to show symptoms in middle age. The disease is caused by a rare autosomal dominant allele. A woman's mother has developed Huntington disease, but her father has not.
   a. What is the probability that she will develop this disease if the allele has 50% penetrance?
   b. If this allele has 100% penetrance, and the woman has inherited the allele, is there any way that she might not develop Huntington disease?
12. Assume that two modifier genes that have alleles that suppress the Huntington disease allele have been discovered. Suppressor alleles of both genes are dominant, and both show partial penetrance. Fifty percent of individuals with the Huntington disease allele and one suppressor allele de-

velop the disease. Individuals with the Huntington disease allele and suppressor alleles of both genes never develop the disease. What is the probability that the woman in Problem 11 will develop the disease if she has inherited the Huntington disease allele and

a. her father is heterozygous for suppressor alleles of both genes, and her mother has neither suppressor allele.
b. her mother is heterozygous for a suppressor allele of one gene, and her father is heterozygous for a suppressor allele of the other gene.

13. Consider the following hypothetical scheme of inheritance of coat color in a mammal. Gene *A* controls the conversion of a colorless compound J to a gray pigment J1; the dominant allele *A* produces the enzyme necessary for this conversion, but the recessive allele *a* produces an enzyme with no activity. Gene *B* controls the conversion of the gray pigment J1 to a black pigment J2; the dominant allele *B* produces the active enzyme that catalyzes the conversion of J1 to J2. The recessive allele *b* produces a defective enzyme with no activity. The dominant allele *C* of a third gene produces a product that completely inhibits the activity of the enzyme produced by gene *A*, that is, it prevents the reaction of J to J1. Allele *c* produces a defective product that does not inhibit this same reaction. The three genes (*A*, *B*, *C*) assort independently, and no other genes are involved.

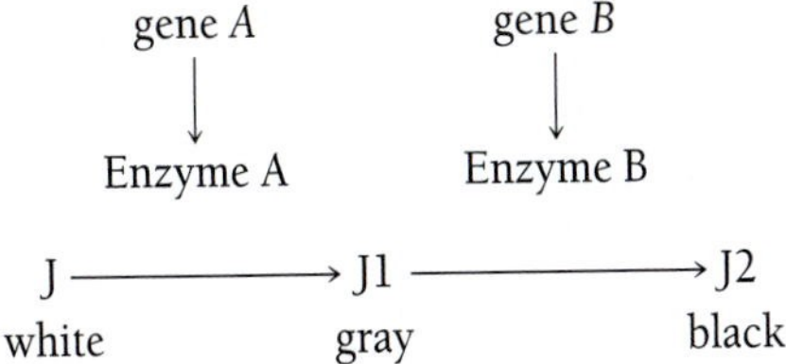

In the F_2 of the cross *AA bb CC* × *aa BB cc*, what is the expected phenotypic segregation ratio?

14. In the F_2 of the cross *AA bb cc* × *aa BB cc*, what is the expected phenotypic segregation ratio and the genotypes of each phenotype?

15. In cats, a cross between a cream-colored cat and a red cat always yields a coat color that is pink. However, crosses between pink cats yield cream, pink, and red in an approximate 1 : 2 : 1 ratio. This type of inheritance is referred to as
a. overdominance.
b. codominance.
c. incomplete dominance.
d. complete dominance.
e. multiple alleles.

16. In the ABO blood typing system, which of the following progeny are *not* possible?
a. an O child from a mating of an A and O individual
b. an O child from a mating of two A individuals
c. an AB child from a mating of an A and an O individual
d. an O child from a mating of a B individual to an A individual
e. an A child from a mating of an AB individual to a B individual

17. A woman whose blood type is A has a baby whose blood type is B. The woman has three suitors: Type O, Type B, and Type AB. The woman marries Type B and has another child whose blood type is O. Type B states that he could not possibly be the father of the type-O child and accuses Type O of being the father. Is Type B justified in his assumption and accusation?

18. The following pathway to produce red fruit is given:

$$\text{white pigment} \xrightarrow{\text{gene } A} \text{blue pigment} \xrightarrow{\text{gene } B} \text{red pigment}$$

A and *B* are dominant alleles of the two unlinked genes that produce an enzyme that catalyzes the conversion of one pigment to another, and *a* and *b* are recessive alleles that are nonfunctional. A fully homozygous blue plant is crossed with a white plant of genotype *aa BB*. The F_1 are allowed to self mate. What proportions would you expect in the F_2 progeny?
a. 9/16 red, 4/16 white, 3/16 blue
b. 9/16 red, 4/16 blue, 3/16 white
c. 9/16 red, 7/16 blue
d. 9/16 red, 7/16 white
e. 3/16 red, 13/16 white

19 A homozygous blue-eyed mouse ($b_1 b_1$) is crossed to a homozygous blue-eyed mouse ($b_2 b_2$) from another stock. Wild-type mice have gray eyes. Because this cross produced all gray-eyed progeny, then
a. b_1 and b_2 do not complement each other and are not alleles.
b. b_1 and b_2 complement each other and are allelic.
c. b_1 and b_2 complement each other and are not alleles.
d. b_1 and b_2 do not complement each other and are allelic.

20. If a dominant trait shows a penetrance of 50%, then
a. half the individuals who have the gene show the trait.
b. half the individuals in the population have the gene.
c. individuals who have the gene show half the trait.
d. individuals with one allele show only half the trait.
e. the gene is on X chromosomes and half the males will get the gene.

21. Assume that eye color in humans is controlled by two genes, *G* and its recessive allele g and *B* and its recessive allele b. Individuals with the *B* allele always have brown eyes, irrespective of the *G* allele. However, *G– bb* individuals have green eyes, and *gg bb* individuals have blue eyes. Two brown-eyed individuals marry and have a brown-eyed son named David, who marries a green-eyed girl named Jill, and David and Jill have a blue-eyed daughter named Nicole.
a. What are the probable genotypes of the two grandparents, the parents, and Nicole?
b. What type of inheritance is present in this situation?

22. For each of the crosses, indicate what the dominance relationship is between the alleles.

Parents	F₂ Phenotypic Ratios		
a. white eye × red eye	17 white	32 pink	14 red
b. red flower × white flower	36 red	11 white	
c. long stem × short stem	8 long	26 short	
d. black fur × white fur	12 white	24 gray	10 black
e. white flowers × blue flowers	22 white	68 blue	

23. Explain the difference between the following:
    a. incomplete dominance and complete dominance
    b. overdominance and codominance
    c. epistasis and complete dominance
    d. variable expression and codominance
    e. suppression and epistasis

24. A female *Drosophila* homozygous for the sex-linked recessive mutation *facet,* which gives eyes with irregular facets, was mated to a male hemizygous for a new recessive mutation that gives irregular facets. The facets in the eyes of the F_1 females were irregular—not as irregular as females homozygous for facet, but more irregular than females homozygous for the new mutation. What would you conclude about the new mutation?

25. A Siberian iris with purple flowers and long stems was crossed with an iris with white flowers and short stems. The F_1 were inbred, and the following F_2 were produced:

F_2 Phenotype	Fraction of F_2
purple flowers, long stems	3/16
purple flowers, short stems	1/16
blue flowers, long stems	6/16
blue flowers, short stems	2/16
white flowers, long stems	3/16
white flowers, short stems	1/16

    a. How many genes are controlling these characteristics in this cross?
    b. How many alleles does each gene have, and what are the phenotypes produced by each allele?
    c. What was the phenotype of the F_1 individuals?

26. A couple had three children with the following blood types:

    child 1 = MN blood type, child 2 = M blood type, child 3 = N blood type

    What are the genotypes of the parents of these children?

27. The ABO blood types of several sets of men and women were determined. Indicate the phenotypes of the possible children each couple might have.

Woman	Man
a. AB	B
b. O	A
c. AB	AB
d. B	A

28. In a paternity test, a mother and child were found to differ for a number of codominant alleles that produce blood antigens.

    Mother: $A^1A^1\ B^1B^1\ C^2C^2\ D^2D^2\ E^2E^2\ F^1F^1\ G^1G^1\ H^1H^1$
    Child: $A^1A^4\ B^1B^2\ C^3C^2\ D^6D^2\ E^8E^2\ F^1F^4\ G^1G^3\ H^4H^1$

    a. What alleles must the father have contributed?
    b. A man was told that because he has all these alleles, he must be the father. Is this statement correct?

29. A series of crosses were made between mice with different coat colors. Assuming that all alleles are recessive, determine which mutations are alleles of the same gene. The wild-type color in mice is agouti.

Parents	Offspring
a. black × black	black
b. white × white	agouti
c. black × white	agouti
d. gray × white	gray
e. tan × brown	agouti

30. Compare the following:
    a. amorph and null allele
    b. amorph and hypermorph
    c. hypermorph and antimorph

31. You have been given strains of guinea pig with different antigens on the surface of their red blood cells. You have been asked to determine whether these antigens are produced by (1) a series of codominant alleles of a single gene or (2) dominant mutations of several different genes. Explain how you could do this.

32. A mouse with white fur was bred with a mouse with black fur. The F_1 were all agouti-colored (the normal color for mice). The F_2 follow:

agouti	black	white
94	28	39

    a. Using symbols of your own choice for the black and white mutations, indicate the genotypes of the parents, the F_1, and the F_2.
    b. Describe the dominance relationships between these mutations, and describe any interactions between them.

Challenge Problems

1. An *Arabidopsis* with white stems and double flowers appeared in a strain with green stems and double flowers. This plant was crossed with plants from a strain with green stems and single flowers. A few F_1 plants were recovered, and these had yellow stems and double flowers. The F_1 were selfed, and the following F_2 were produced:

white stems, double flowers	120
yellow stems, single flowers	40
yellow stems, double flowers	240
green stems, single flowers	40
green stems, double flowers	120

a. Propose a model for the inheritance of stem color and flower shape in these crosses. In your answer, explain how many genes are involved, devise symbols for the alleles of each gene, and explain what the genotypes are of the parents, the F_1, and each F_2 class.
b. Describe a cross or series of crosses you could do to test your model.

2. A chicken with white, split feathers was found on a farm near Chernobyl and was crossed to highly inbred, wild-type jungle fowl with red, flat feathers. Half the eggs failed to hatch. The F_1 that hatched had white, flat feathers. The F_1 were inbred, and 1000 eggs were collected. They produced the following:

dead	375
white flat	281
white split	94
red flat	188
red split	62

a. Propose a model for the pattern of inheritance in these chickens. In your answer, explain how many genes are involved, devise symbols for the alleles of each gene, and explain what the genotypes were for the parents, the F_1, and each F_2 class.
b. Describe a cross or series of crosses that could test your model.

3. Plants from the vitellina strain of *Antirrhinum majus* that are short and have medium green leaves were crossed with plants from the spectabilis strain that were short and had light green leaves. The F_1 were tall and had dark green leaves. The F_1 were inbred and produced the following F_2:

Height	Color	Number
short	green	40
short	dark green	80
short	light green	40
tall	green	40
tall	dark green	80
short	light green	40

Propose a model for the pattern of inheritance of these traits in these plants. In your answer explain how many genes are involved, devise symbols for the alleles of each gene and explain what the genotypes were of the parents, the F_1, and each F_2 class.

4. Two strains of *Gossypium* (new world cotton) were crossed. A barbadense strain with spotted flowers and red stems was crossed with an hirsutum strain with no spots and green stems. The F_1 had lightly spotted flowers and light red stems. The F_2 follow:

Color	Flower	Number
green	no spot	40
light red	light spot	60
green	light spot	20
red	spotted	30
green	spotted	10

Design a genetic model for the inheritance of these two traits in these crosses. In your explanation, indicate the number of genes and alleles involved, as well as the genotypes of the parents, the F_1, and each class of F_2.

Suggested Readings

Castle, W. E., and C. C. Little. 1910. "On a modified Mendelian ratio among yellow mice." *Science,* 32: 868–870.

Hadorn, E. 1961. *Developmental Genetics and Lethal Factors.* John Wiley & Sons, New York.

Laughnan, J. R., and S. Gabay-Laughnan. 1983. "Cytoplasmic male sterility in maize." *Ann. Rev. Genet.,* 17: 27–48.

Lewis, R. 1997. *Human Genetics: Concepts and Applications,* 2nd ed. W. M. Brown, Dubuque, IA.

Strickberger, M. W. 1985. *Genetics,* 3rd ed. Macmillian, New York.

Suzuki, D. T. 1970. "Temperature-sensitive mutations in *Drosophila melanogaster.*" *Science,* 170: 695–706.

On the Web

Visit our Web site at **http://www.saunderscollege.com/lifesci/titles.html** and click on A/G/M Genetics for links to the following chapter-related resources on the World Wide Web:

1. **Cardiff Gene Database.** This human gene mutation database is maintained by the Institute of Medical Genetics at the University of Wales College of Medicine.
2. **A genetic glossary.** This site contains definitions and references useful for analysis of complex genetic interactions. It is maintained by Cambridge University.

CHAPTER

5

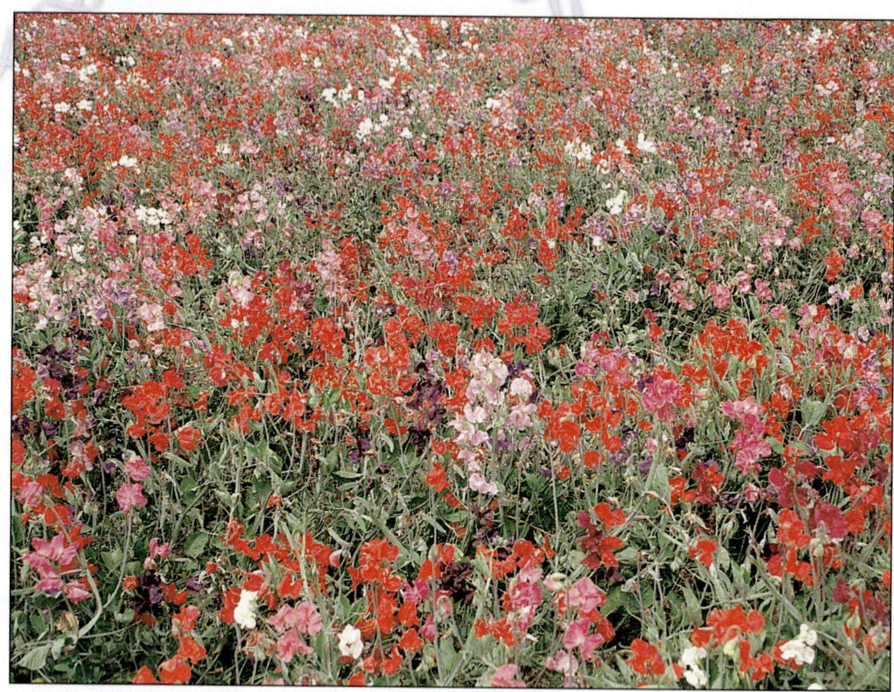

A field of sweet peas showing diversity of flower color, a polygenic trait. *(Tim Hauf/Visuals Unlimited)*

Quantitative Traits and Polygenic Inheritance

The different alleles of peas that Mendel used in his experiments (discussed in Chapter 2) gave one of two distinct phenotypes. In Chapter 3, we discussed some variations of this theme, but the end result was always one of two distinct phenotypes. For example, the white eyes produced by mutations of the *white* gene in *Drosophila* are easily distinguished from the red eyes of a wild-type. Such phenotypes are called discontinuous or **qualitative traits** because the mutant phenotype can be described in qualitative terms and because all individuals fit into one of a few distinct phenotypic categories. If we cross a white-eyed *Drosophila* with a red-eyed *Drosophila,* the progeny will either have white eyes or red eyes. Each phenotype is caused by one genotype or, in cases of dominance or epistasis, by a small number of different genotypes. Because the relationship is simple, it is usually possible to deduce the genotype of each individual from their phenotype or from the phenotypes of offspring.

After Mendel's publication was rediscovered in the early 1900s and investigators repeated Mendel's experiments, they also experimented with other traits and with species other than peas. They found that some traits did not behave in a qualitative manner. These traits showed a different pattern of inheritance. The offspring of crosses did not fall into distinct phenotypic classes or phenotypic ratios. Height, weight, and IQ are examples of this type of trait. These three traits are examples of **quantitative traits.** An individual's phenotype for these traits is measured in quantitative terms: inches, pounds, or IQ points. Quantitative traits have three general characteristics:

1. They are continuously variable. This means that there are no discrete phenotypic classes in the progeny of crosses and the phenotypes of individuals in a population from one continuous distribution.
2. Quantitative traits are polygenic ("poly" means many, "genic" means genes), in other words, the phenotype of a single trait (for example, weight) is controlled by the alleles of many different genes. A population of individuals may contain a very large number of different genotypes, and individuals with different genotypes may have the same phenotype.

3. An individual's phenotype for a quantitative trait may be influenced by the environment. Two individuals with identical genotypes raised in different environments can have very different phenotypes for quantitative traits.

Quantitative traits present a unique challenge for geneticists because Mendelian genetic analysis gives little information about their genetic basis. A quantitative phenotype is produced by the simultaneous action of many alleles; consequently, it is usually not possible to deduce the genotypes of individuals from their phenotype. However, since the early part of the 1900s, many geneticists have studied this seeming noncompliance with Mendel's rules, and now the genetics of quantitative trait inheritance is a key part of modern biology. In contrast to Mendelian genetic analysis, which relies on determining the discrete phenotype of each individual and on numerical analysis of offspring phenotypic classes, quantitative trait analysis relies on statistical techniques and analysis of populations of individuals. Statistical techniques can tell us what phenotypes are present in a given population, whether two populations have similar or different phenotypes, and whether two traits are related, and they can make predictions about what the phenotypes of future generations will be. By the end of this chapter, we will have addressed six specific questions:

1. How do we know that Mendelian inheritance and quantitative inheritance are both controlled by Mendelian genes?
2. How are quantitative traits measured?
3. How can we determine whether two populations differ for a quantitative trait?
4. How can we determine whether two quantitative traits are related?
5. What can we learn by studying the phenotypic variation of a trait in a population?
6. How can we use the quantitative trait phenotypes of parents to predict offspring phenotypes?

The Inheritance of Quantitative Traits Is Mendelian.

Quantitative Trait Inheritance Has a Predictable Pattern.

The analysis of quantitative traits began long before Mendel performed his experiments. Between 1761 and 1766, Josef Kölreuter crossed tall tobacco plants with short plants, inbred the offspring (F_1), and examined the phenotypes of their progeny (F_2). Kölreuter observed that the F_1 plants were all intermediate in height and that most F_2 plants were also intermediate, although some were as short as the shortest parents and others were as tall as the tallest parents. If the heights of the parents, the F_1, and the F_2 individuals were plotted on a graph, they formed a frequency distribution that was a single continuous distribution. Interestingly, both the F_1 and F_2 populations had the same average value, but the F_1 and F_2 frequency distributions had very different shapes. The F_2 distribution was much broader than that of the F_1 (Figure 5.1). This same pattern of frequency distributions in the F_1 and F_2 generations was later observed for the inheritance of many other quantitative traits in plants and animals. The observation that many inherited traits seemed to combine to give a single, intermediate value in the F_1 and F_2 was accepted as evidence in favor of the blending theory of inheritance (Chapter 1). When Mendel did his work, it was generally assumed that all traits showed this pattern of inheritance (Table 5.1).

In the late 1800s, biologists performed many studies with quantitative traits, largely in attempts to understand how evolution occurred. Prominent among these investigators was Sir Francis Galton, an English scientist who attempted to deduce the rules of genetic inheritance by statistically analyzing the phenotypic distributions of

TABLE 5.1 *Examples of Continuous, Quantitative Traits in a Variety of Species*

Species	Traits
Cattle	Body weight
	Milk yield
	Muscle composition
Humans	Height
	Weight
	Life span
	IQ
	Skin color
Pigs	Back-fat thickness
	Weight gain
Mice	Tail length (6 weeks)
	Average litter size
Fruit flies	Body size
	Egg production
Horses	Racing speed
	Trotting speed
Poultry	Body weight (8 weeks)
	Egg size

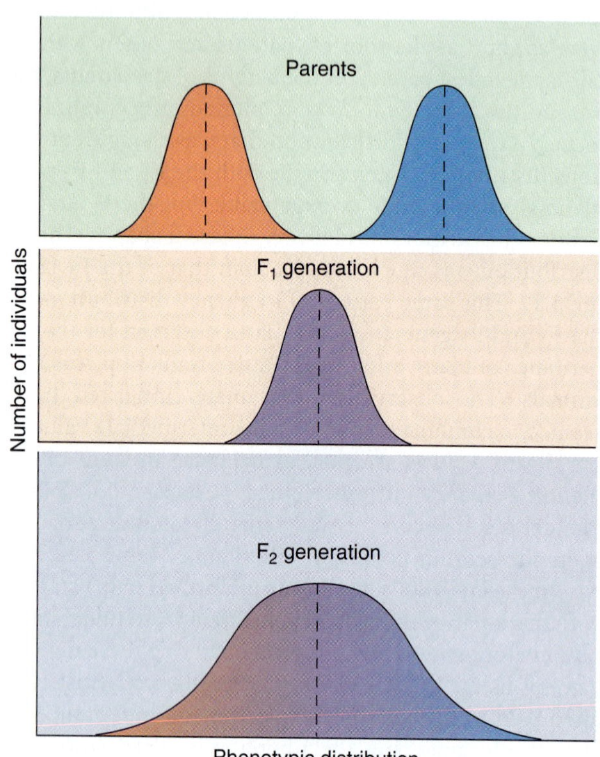

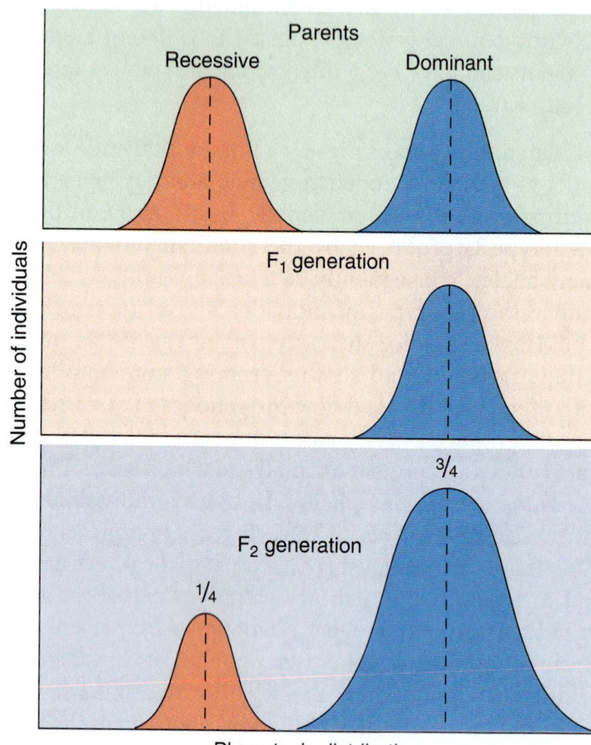

FIGURE 5.1 Inheritance patterns of quantitative and qualitative traits. (a) The F_1 distribution for quantitative traits has an intermediate phenotype, but the F_2 distribution has continuous distribution of phenotypes, with no discrete phenotypic classes. (b) The F_1 distribution for qualitative traits, however, resembles one parent, and the F_2 have discrete phenotypic classes. For a monohybrid cross, the F_2 have a 3/4 : 1/4 distribution.

quantitative traits in several species. Galton's work demonstrated that many human traits were inherited, and that they showed the same quantitative inheritance pattern observed by Kölreuter. Galton and his followers (called biometricians) invented a number of statistical techniques for use in their analysis of quantitative traits, and these methods are still used today. We will discuss these techniques later in this chapter.

After the rediscovery of Mendel's paper, the biometricians argued that genes inherited in a Mendelian pattern, being discrete units of inheritance, could not control continuously varying quantitative traits. An important fact supporting their argument was that quantitative traits showed continuous distributions in the F_2 of crosses, unlike the discrete phenotypic classes characteristic of Mendelian traits. Another group of biologists (called the Mendelians), led by the English geneticist William Bateson, argued that all inherited phenotypic traits were caused by Mendelian alleles. The Mendelians argued that the continuous phenotypic distribution of quantitative traits in the F_2 of crosses was the result of environmental influences that blurred the distinctions between the phenotypic classes. Whether quantitative traits were controlled by Mendelian genes was a hotly debated question between 1901 and 1918. Eventually the work of a number of investigators, especially Ronald A. Fisher, demonstrated that quantitative traits are controlled by Mendelian genes.

SECTION REVIEW PROBLEM

1. For each of the traits listed here, state whether you believe the trait would be inherited as a qualitative or quantitative trait. For each trait, explain why you reached that conclusion.
   a. body weight in cattle
   b. skin color in humans
   c. red vs. white flower color in roses
   d. red vs. white coat color in cattle
   e. rate of growth of mice

Quantitative Traits Are Controlled by Multiple Genes.

The proof that quantitative traits are controlled by Mendelian genes came from a number of genetic studies done between 1903 and 1918. These studies demonstrated that quantitative traits are **polygenic,** that is, the phenotype of the trait is determined by the alleles of many genes acting together. A landmark study to demonstrate this principle was done in 1909 by the Swedish geneticist Herman Nilsson-Ehle. Nilsson-Ehle crossed a strain of wheat that had white kernels with a strain that had dark red kernels. The F_1 kernels were intermediate in color, as expected for a quantitative trait. Nilsson-Ehle then planted the kernels, inbred the resulting F_1 plants, and examined the F_2 kernels. The F_2 contained both white and red kernels, in a ratio of 1/64 white to 63/64 red (Figure 5.2). Nilsson-Ehle recognized that this 1/64 : 63/64 ratio was the expected ratio for the F_2 of a trihybrid cross where homozygous recessive alleles for all three loci yield white kernels. The red kernels were hypothesized to possess at least one color-producing allele of any of the three genes. From this analysis, he concluded that kernel color was a polygenic trait, and that his original white-kernel strain was homozygous for recessive alleles at three loci ($r_1\,r_1$, $r_2\,r_2$, $r_3\,r_3$) and his original dark-red strain was homozygous for alleles that produced red color ($R_1\,R_1$, $R_2\,R_2$, $R_3\,R_3$). These data also demonstrated that quantitative traits were governed by Mendelian rules of inheritance.

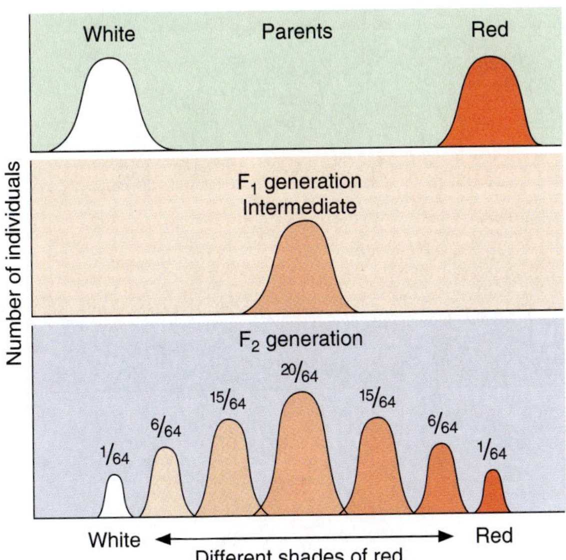

FIGURE 5.2 Inheritance of kernel color in wheat. Nilsson-Ehle crossed a white-kerneled wheat strain with a red-kerneled strain and produced F_1 plants with seeds that were an intermediate-red color. Inbreeding the F_1 produced an F_2 with 63/64 having red kernels and 1/64 having white kernels. Careful inspection of the F_2 revealed seven different color classes in a 1/64 : 6/64 : 15/64 : 20/64 : 15/64 : 6/64 : 1/64 ratio.

The number of genotypic classes in the F_2 of a cross involving polygenic traits depends on the number of genes that have different alleles segregating in the cross. If there are n genes with segregating alleles, then there will be 3^n genotypes in the F_2. [Recall that when one pair of alleles ($A\,a$) controls a trait, there are four allelic combinations in the F_2, $1\,A\,A : 2\,A\,a : 1\,a\,a$. Each allelic combination is 1/4 of the total, but because two combinations are identical ($A\,a$), there are three unique genotypes]. In a monohybrid cross ($n = 1$), alleles of only one gene are segregating and there are $3^1 = 3$ genotypes. In a dihybrid cross ($n = 2$), there are $3^2 = 9$ genotypes, and in a trihybrid cross there are $3^3 = 27$ genotypes. Only one of these F_2 genotypes will be the same as one of the original parents. This fraction corresponds to $(1/4)^n$, or in the case of trihybrid cross, 1/64 of the total. We can determine the number of genes in a polygenic cross by observing the fraction of the F_2 that have the same phenotype as one of the original parents. In Nilsson-Ehle's wheat cross, 1/64 of the F_2 were white. Because $(1/4)^3 = 1/64$, the parental strains had different alleles in three color-controlling genes. If the original parental strains had different alleles in two genes, there would have been 1/16 white F_2 kernels. If the parents had different alleles in four genes, then there would have been $1/4^4 = 1/256$ white kernels. As you can imagine, trying to identify the extreme classes when they represent only 1/256, or even 1/64, of the total population is a difficult task. Practically, when the number of genes is greater than 3, it is usually not possible to identify the extreme phenotypic classes because of the inherent rareness of the extreme genotypic classes.

After identifying the 1/64 extreme class (white), Nilsson-Ehle closely inspected the remaining F_2 red kernels and observed that they did not all have the same shade of red. In fact, he was able to identify a total of seven separate color classes in the F_2, including the previously identified white. The F_2 kernels were distributed in the seven shades of white to red in a ratio of 1 : 6 : 15 : 20 : 15 : 6 : 1 (Figure 5.2). Nilsson-Ehle realized that in a trihybrid cross this ratio represents the number of individuals in the F_2 that have 0, 1, 2, 3, 4, 5, or 6 color-producing alleles, as illustrated in the Punnett square in Figure 5.3. He performed several testcrosses, which demonstrated that each phenotypic class of the F_2 kernels did indeed have a different number of color-producing alleles in its genotype. The amount of color contributed by the color-producing alleles of each locus was the same, so individuals that had one color-producing allele of any of the three genes (R_1, or R_2, or R_3) had the same lightest red color. Kernels with any combination of two color alleles had the next darkest color. Individuals

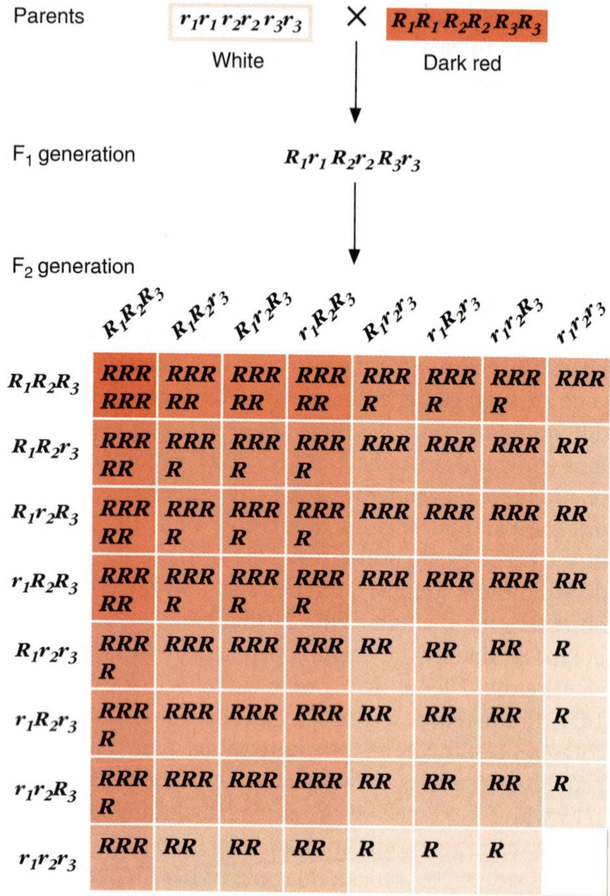

Parents
$r_1r_1\,r_2r_2\,r_3r_3$ × $R_1R_1\,R_2R_2\,R_3R_3$
White
Dark red
F_1 generation
$R_1r_1\,R_2r_2\,R_3r_3$
F_2 generation
$R_1R_2R_3$ $R_1R_2r_3$ $R_1r_2R_3$ $r_1R_2R_3$ $R_1r_2r_3$ $r_1R_2r_3$ $r_1r_2R_3$ $r_1r_2r_3$
$R_1R_2R_3$: RRR RRR | RRR RR | RRR RR | RRR RR | RRR R | RRR R | RRR R | RRR
$R_1R_2r_3$: RRR RR | RRR R | RRR R | RRR R | RRR | RRR | RRR | RR
$R_1r_2R_3$: RRR RR | RRR R | RRR R | RRR R | RRR | RRR | RRR | RR
$r_1R_2R_3$: RRR RR | RRR R | RRR R | RRR R | RRR | RRR | RRR | RR
$R_1r_2r_3$: RRR R | RRR | RRR | RRR | RR | RR | RR | R
$r_1R_2r_3$: RRR R | RRR | RRR | RRR | RR | RR | RR | R
$r_1r_2R_3$: RRR R | RRR | RRR | RRR | RR | RR | RR | R
$r_1r_2r_3$: RRR | RR | RR | RR | R | R | R

FIGURE 5.3 Phenotype and genotype of kernel color in wheat. The kernel color classes in Nilsson-Ehle's wheat cross are produced by the segregation of additive, color-producing alleles of three different genes. Independent assortment in the F_1 produces eight different gamete genotypes. Random fusion of gametes produces 64 possible combinations. The color intensity of F_2 kernels is determined by the number of color-producing, contributing alleles present.

homozygous for color alleles of one gene had the same phenotype as individuals heterozygous for color alleles of two genes ($R_1\,R_1$, $r_2\,r_2$, $r_3\,r_3 = R_1\,r_1$, $R_2\,r_2$, $r_3\,r_3$). Individuals with three color alleles were still darker, and so forth. The darkest class, with six color alleles, had the same color as the original parental dark red. This pattern of phenotypes is characteristic of a class of alleles known as **additive alleles.** Additive alleles each make an independent contribution to the phenotype, and the final phenotype is the sum of the contributions of all the alleles. This is a very different relationship among alleles than the dominant–recessive relationship Mendel observed in his crosses.

If two individuals that are homozygous for different, additive alleles at n different genes are crossed, the number of different phenotypic classes in the F_2 is equal to $2n + 1$. For one gene there are three classes, for two there are five classes, for three there are seven classes, and so forth. The difference between each phenotypic class becomes smaller as the number of classes gets larger. In Nilsson-Ehle's cross, alleles of only three genes were segregating, so the F_2 had only seven phenotypic classes. Because there were relatively few classes in this cross, the phenotypic differences among the classes were sufficiently large to be perceived. However, if color-producing alleles of four, five, or more genes had been segregating in the cross, the number of phenotypic classes in the F_2 would have been larger (9, 11, or more). The larger the number of phenotypic classes, the smaller the color differences among classes and thus distinguishing among the classes becomes harder (Figure 5.4). Eventually it becomes impossible to distinguish among classes, and the F_2 have a continuous distribution of colors. For example, if we made a cross between two wheat strains, one with a red kernel and another with a white kernel, and observed that $1/1{,}048{,}576 = 1/4^{10}$ of the F_2 were white, we would conclude that the two parental strains had different alleles of 10 color genes. In theory, these F_2 individuals would be distributed into $2n + 1 =$

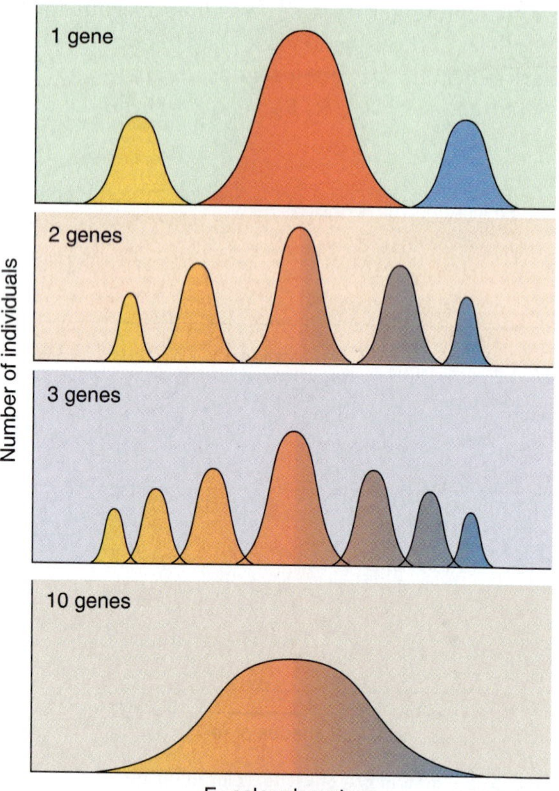

FIGURE 5.4 Phenotypic distributions of F2 populations. The phenotypic distributions in the F_2 of crosses between strains with different alleles of 1, 2, 3, or 10 genes have different patterns and shapes. When the number of F_2 phenotypic classes becomes large enough, the classes overlap and the population appears to have a continuous distribution.

21 phenotypic classes. In reality, the differences among classes would be so small that the F_2 would appear to have a smooth, continuous color distribution from white to red. The number of classes necessary for a distribution to appear continuous depends on the particular trait and also on how much the phenotypes are affected by environmental differences. For many quantitative traits, environmental effects cause individuals with the same genotype to have slightly different phenotypes. If environmental effects strongly influence a trait, then a continuous distribution is produced when the alleles of only a few genes are segregating (Figure 5.5).

In 1918, the English geneticist Sir Ronald Fisher [see *HISTORICAL PROFILE: Sir Ronald Aylmer Fisher (1890–1962)*] published a landmark paper that largely settled the controversy over how quantitative traits are inherited. Fisher examined the results of many different studies with quantitative traits and concluded that quantitative trait inheritance is controlled by the alleles of many genes. He performed mathematical analyses, which clearly demonstrated that the segregation of multiple Mendelian alleles in a population could explain the observed inheritance of quantitative traits. He concluded that the Mendelian alleles affecting quantitative traits generally show additive effects, although examples of all types of allelic interactions (dominance, epistasis, suppression, enhancement, etc.) could be found. He also concluded that most quantitative traits are affected by the environment, and strong environmental effects that blur the distinction between phenotypic classes produce the observed continuous distribution of phenotypes.

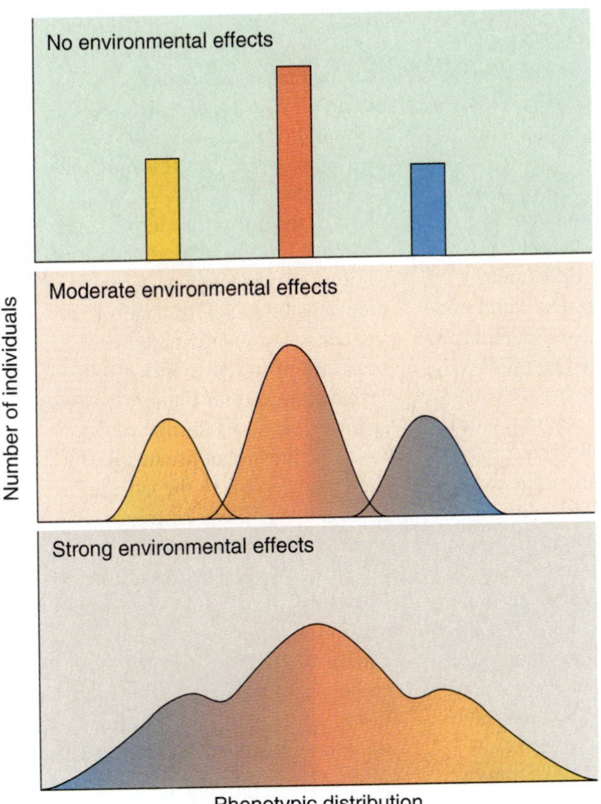

FIGURE 5.5 Environmental effects and F2 phenotypic distributions. If the environmental effects are small, the distribution of the F_2 phenotypic classes is visible. If the environmental effects are more pronounced, the distribution of the F_2 phenotypic classes is broad and they overlap.

SECTION REVIEW PROBLEMS

2. If you were given two samples of beans that differed for a genetically determined trait, how would you ascertain whether this was a quantitative or qualitative trait?
3. In a population with four genes controlling a quantitative trait, how many different phenotypic classes would be present if there were two alleles of each gene, each with additive effects?
4. How would you answer Problem 3 if there were 20 genes controlling the trait?

GENETIC ANALYSIS OF QUANTITATIVE TRAITS USES STATISTICS.

Using statistical techniques to analyze quantitative genetic data is now a standard technique, largely as a result of Fisher's influence. In Chapters 2 and 3, when we did a Mendelian analysis of a cross, we counted the individuals in each phenotypic class and deduced the genotypes. In quantitative analysis, there are no discrete phenotypic classes; consequently, the phenotype of the trait is defined in terms of a numerical measurement. Thus, statistical terms are used to describe the phenotype of the population, to compare two different populations, and to predict the expected phenotypes of offspring populations. In this section, we will discuss some of the statistical techniques that are commonly used in quantitative genetic analysis and review how each is used.

Means and Variances of Frequency Distributions Define the Phenotypes of Populations.

The analysis of a quantitative trait begins by first describing the phenotype of the target population. The phenotype of a population is described by organizing the phenotypic values of the individuals in the population into a **frequency distribution.** To describe a population completely, the values of every member of the population must be included in the distribution. In reality, it is not practical to measure all the individuals in large

HISTORICAL PROFILE

Sir Ronald Aylmer Fisher (1890–1962)

Ronald A. Fisher was born on February 17, 1890, to a family of wealthy English merchants, and he grew up in a well-to-do Victorian setting. Early in life he demonstrated a remarkable intelligence and a talent for mathematics that caused him to be advanced several grades in school. Because he had extremely poor eyesight, he was tutored verbally and developed an ability to solve mathematical problems in his head. In later life, this allowed him to confound others by producing solutions without any indication of the intermediate steps in his analysis. He had a strong personality with a fiery temper that could erupt in violent outbursts, but these passed as quickly as they came. He had a passionate loyalty for his friends and delighted in being the center of a social group.

Fisher attended Cambridge University on a scholarship from 1909 to 1913, his family having lost their fortune in business reverses. In his undergraduate years, he published his first paper in mathematics and developed a profound interest in genetics. Because of this interest, he became a founding member of the Cambridge Eugenics Society. Throughout his life, he believed in the use of genetic knowledge to improve the human species. However, his first love was mathematics, and he obtained a first-class degree in mathematics. He then attempted to enlist in the army to fight in World War I, but he was rejected because of his poor eyesight. In 1919, he accepted a position as a statistician at the Rothamsted Agricultural Experiment Station. In this position, he carried out numerous statistical analyses of experimental and agricultural research data and assisted in the design of experiments. He quickly realized that appropriate statistical tests did not exist for many of the needs of this work, and so he produced his own. His work on tests of significance, correlation, regression, and analysis of variance in small samples had a revolutionary impact on agricultural experimentation. His techniques were embodied in his classic text, *Statistical Methods for Research Workers,* published in 1925. Fisher had a great talent for quickly grasping problems of experimental design in any area of research and for rapidly producing statistical tests that could improve the experiment.

FIGURE 5.A Sir Ronald Aylmer Fisher. An outstanding statistician and geneticist. *(A. Barrington Brown/Science Photo Library/Photo Researchers, Inc.)*

Fisher's fame grew rapidly. He received a doctorate in mathematics from Cambridge in 1926 and the Weldon medal for genetics and evolutionary research in 1928, and in 1929 he was elected a member of the Royal Society as a mathematician. In 1933, he accepted the Galton Professorship in Eugenics at Cambridge and began a wide-ranging program of research into human heredity. There was a growing interest in and appreciation of his techniques for statistical analysis among plant and animal breeders. This led to Fisher being in great demand as a lecturer, and, because he loved to travel, he gave many lectures each year, often traveling to America. Additionally, a growing stream of international visiting scholars and research students found their way to Fisher.

Fisher found his position as Galton Professor a mixed blessing. The professorship was accompanied by funds for research, which he used to expand his experimental investigations, especially into human genetics. He also developed friendly relations with leading research workers such as J. B. S. Haldane. However, Fisher had numerous disagreements with members of the statistics department and conflicts with the administration. This was a period of lively and sometimes heated debate on the subjects of proper experimental design and statistical analysis in the biological and agricultural sciences. Fisher's own work expanded into several areas, ranging from genetic studies of Mendelian characters in mice and chickens to the evolutionary analysis of primroses to human genetics.

The advent of World War II nearly caused the disbanding of Fisher's research group. He continued to work with reduced resources until 1943, when he was offered the Balfour Professorship of Genetics and made head of the Genetics Department. This brought him increased resources for research, and he enlarged his activities, expanding into the area of bacterial genetics. With his administrative duties as head of the Genetics Department, his biological and statistical interests, and his traveling, he was busy and productive. In 1957, Fisher retired from Cambridge and spent a year as a visiting professor at the University of Michigan. He then made a visit to the University of Adelaide and was so impressed with Australia that he decided to remain. As a Senior Fellow at the University of Adelaide, he continued to teach, research, and travel widely. In 1962, Fisher died of an embolism after surgery for cancer. Fisher made many contributions to statistics and genetics in his 50-year career. Perhaps primary among these were his efforts to incorporate statistical principles in the design of biological and agricultural experiments and in the statistical analysis of experimental results.

populations, so a **random sample** of individuals is selected to represent the population. The phenotypes of the individuals in the sample are then measured and considered to represent the phenotypes of the entire population. For example, it is not practical to measure the height of all men everywhere. But, if the heights of samples of men are measured, and the sample is taken randomly, then the sample will represent the total population. For example, the heights of 1000 French soldiers are given in Table 5.2. These same heights are organized into a frequency distribution in Figure 5.6. This sample is relatively large; the frequency distribution appears as a smooth curve.

A great deal can be learned about a quantitative trait in a population by analyzing the frequency distribution of the phenotypes of the individuals in the population. The first step is to determine the values of two descriptive statistics of the frequency distribution: the arithmetic average or **mean** ($\bar{x}$) and the **phenotypic variance** (V_p). The mean represents the center of the distribution of the phenotypes and gives a simple statistical description of the phenotype of this sample for the trait. The mean is calculated by adding all the individual phenotypic values in the sample and dividing the sum by the number of individuals. A simple formula for calculating the mean is

$$\text{Mean} = \bar{x} = \Sigma x_i / n$$

TABLE 5.2 *The Heights of a Sample of French Soldiers*

Number of Soldiers	Height (inches)	(cm)
2	60	152.4
2	61	155.0
20	62	157.5
48	63	160.0
75	64	162.5
117	65	165.1
134	66	167.6
157	67	170.2
140	68	172.7
121	69	175.3
80	70	177.8
57	71	180.3
26	72	182.9
13	73	185.4
5	74	188.0
2	75	190.5
1	76	193.0

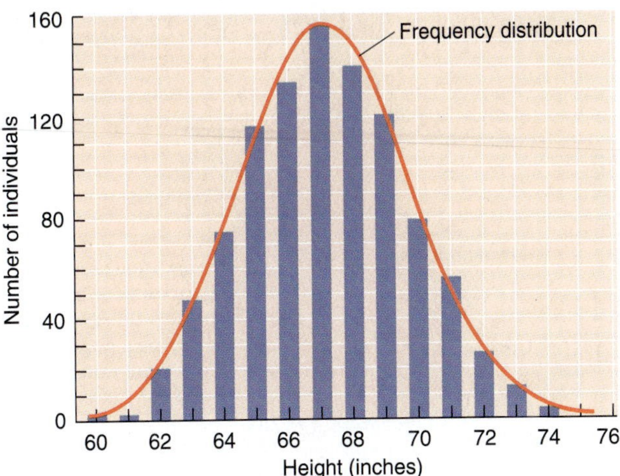

FIGURE 5.6 Height distribution of a sample of men. The heights of a sample of French soldiers form a frequency distribution when graphed. The heights can be presented as a histogram, with the number of individuals in each height class indicated by a bar. A frequency distribution can also be represented as a smooth curve.

The symbol Σx_i represents the sum (Σ) of all individual values in the sample (x_i), and n is the number of individuals in the sample. For example, from the data in Table 5.2, the mean height of the sample of French soldiers is 67.2 in (170.7 cm).

The phenotypic variance represents a measure of how much the phenotypes of the individuals in the sample differ; that is, how much variation exists between individuals in the sample population. This is represented graphically by the width of the frequency distribution (Figure 5.6). A sample where all individuals have values close to the mean will have a small phenotypic variance, and the distribution around the mean will be narrow. In contrast, a sample where individuals differ widely from the mean will have a large phenotypic variance, and the sample distribution will be broad (Figure 5.7). The value of the phenotypic variance is calculated by taking the value of each individual and subtracting it from the mean, squaring this difference, and summing all the squared differences. This sum is then divided by $n - 1$.

$$\text{Phenotypic variance} = V_p = \Sigma(x_i - \bar{x})^2 / (n - 1)$$

In calculating the phenotypic variance, we divide the summed deviation by $n - 1$, the degrees of freedom (degrees of freedom are defined in Chapter 2). This is done because the phenotypic variance of a sample of individuals (V_p) is only an estimate of the phenotypic variance of the population from which the sample was drawn. In making this estimation, we have lost one degree of freedom. We square the differences between the individual values and the mean to prevent positive and negative differences from canceling each other out. As a result, the

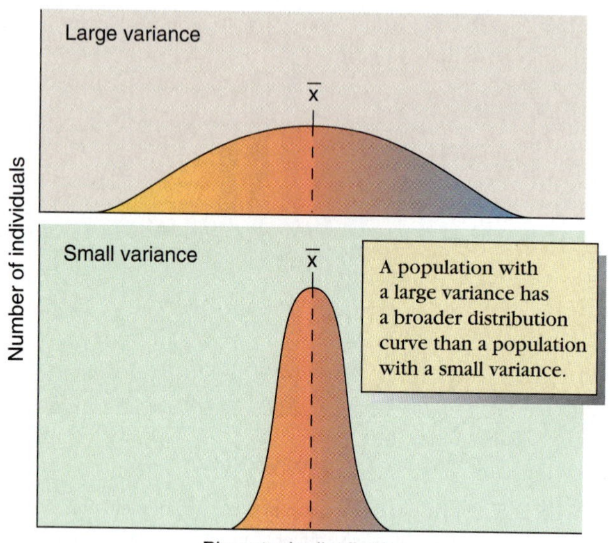

FIGURE 5.7 Phenotypic variance of a population. The variance is a measure of the differences between the phenotypes of the individuals in the population. In these examples, both populations have the same mean ($\bar{x}$) but very different variances.

phenotypic variance is measured in squared units. Because means are not measured in squared units, it is often convenient to calculate the square root of the phenotypic variance, which has the same units as the mean, and to use that in descriptions of the distribution. The square root of the phenotypic variance is called the **standard deviation** (s):

$$\text{Standard deviation} = s = \sqrt{V_P}$$

In our sample of French soldiers, the phenotypic variance $V_P = 6.50 \text{ in}^2$; therefore, the standard deviation $s = 2.55$ in.

SECTION REVIEW PROBLEM

5. The following numbers represent the weight (in pounds) of a random sample of male college students:

   155, 126, 188, 166, 155, 160, 145, 170, 166, 140

   Calculate the mean weight, the phenotypic variance, and the standard deviation for this sample.

The Means of Two Distributions Can Be Compared Using the Student's t Statistic.

One important use of a population's mean and variance is to determine whether two populations have the same phenotype. For example, a geneticist may wish to compare an existing strain with a new mutant strain or to compare the offspring of a cross with the parents. A standard technique for comparing populations is to determine whether the two populations have the same mean value for the trait being studied. For example, we might ask whether French soldiers have the same height as American male college students. To answer this question, we compare the mean height of the sample of French soldiers (Table 5.2) with the mean height of a sample of American students (Table 5.3). If these two samples have the same mean height, then we would conclude that the populations of men from which they were drawn also have the same height. However, whenever samples are taken from a population, there is the possibility of a sampling error; that is, the sample is not an exact representation of the population. Because of sampling error, we should not expect that even the means of two samples drawn from the same population would be identical. When comparing the means of two samples, our question is: Is the difference between the two means sufficiently large to be statistically significant? In other words, we want to know whether the difference between the two means is large enough that the probability of obtaining this much difference by chance sampling error alone is unacceptably low. We can answer this question by using a test statistic called **Student's t** or simply t.

TABLE 5.3 *The Heights of 175 Male Students from Connecticut Agricultural College*

Number of Students	Height (in.)	Height (cm)
1	58	147.5
0	59	149.9
0	60	152.4
1	61	154.9
5	62	157.5
7	63	160.0
7	64	162.6
22	65	165.1
25	66	167.6
26	67	170.2
27	68	172.7
17	69	175.3
11	70	177.8
17	71	180.3
4	72	182.9
4	73	185.4
1	74	188.0
Total 175		

Data taken from the Library of Congress.

The t test is based on the probability that samples within each population will have a certain amount of random error, and that the two derived means will thus not necessarily be identical, even when the samples come from the same population. To use it, we begin by formulating a **null hypothesis** (H_0). In this example, our hypothesis will be that the two means are actually the same. If this hypothesis is correct, then the difference between the two means will be entirely the result of chance sampling error, and there will be a high probability that the difference between the means will be small. If our null hypothesis is not correct, then there will be a high probability that the difference between the means will be large. The value of the t statistic is calculated as

$$t = (\bar{x} - \bar{y}) / (V_x / n_x + V_y / n_y)$$

In this formula, $\bar{x}$ and $\bar{y}$ are the means of the two samples, V_x and V_y are the phenotypic variances of the two samples, and n_x and n_y are the numbers in the samples. The probability that a given value of t occurs depends on the number of degrees of freedom, which is determined by the size of the samples, df = $(n_x - 1) + (n_y - 1)$. We determine the probability associated with different values of t by using a t table (Table 5.4) in the same way that we used the chi-square table in Chapter 2. We determine the row corresponding to the degrees of freedom of this example, read across until we find the value of t, and then read up the column to find the probability that this value of t has occurred entirely by chance error. If the measured value of t is so large that it has less than a 5% probability of occurring by chance error alone, then we reject the null hypothesis and conclude that the two means actually are different. Note that if we reject at the 5% probability level, then 5% of the time the deviation will actually be the result of chance error, and we will be rejecting a true hypothesis. This is one reason why tests like this are often done several times.

To illustrate the use of the t test, let us determine whether our two samples of men differ in height. Our null hypothesis (H_0) is that the mean heights of French soldiers and American college students are the same, and that the means of the two samples differ only by chance sampling error. To test this hypothesis, we calculate the value of t and then determine from the t table that the probability that this value of t arose by chance. For these two samples,

$$t = (67.1 - 67.2)/(6.50/1000 + 7.21/175) = 0.45$$

The sample of French soldiers contained 1000 men, so n_x = 1000. The sample of American college students contained 175 men, so n_y = 175. The degrees of freedom (df) = 999 + 174 = 1173. Consulting the table of values of t, we see that for samples with this many degrees of freedom our value of t, 0.45, has a probability of a little less than 70%, meaning that nearly 70% of the time we would expect this much deviation even if the two samples were drawn from the same population. This high degree of probability indicates that the difference between the two means could easily be the result of chance sampling error, so we accept the null hypothesis and conclude that there is no significant difference in height between French soldiers and American college students.

TABLE 5.4 *Probabilities Associated with Values of t for Different Degrees of Freedom*

	Probabilities							
df	.90	.70	.50	.30	.10	.05	.01	.001
1	.16	.51	1.00	2.00	6.31	12.70	63.66	636.60
2	.14	.44	.82	1.39	2.92	4.30	9.92	31.60
3	.14	.42	.76	1.25	2.35	3.18	5.84	12.92
4	.13	.41	.72	1.19	2.13	2.78	4.60	8.60
5	.13	.41	.73	1.16	2.02	2.57	4.03	6.87
10	.13	.40	.70	1.09	1.81	2.23	3.17	4.59
15	.13	.39	.69	1.07	1.75	2.13	2.95	4.07
20	.13	.39	.69	1.06	1.72	2.09	2.84	3.85
25	.13	.39	.68	1.06	1.71	2.06	2.79	3.71
40	.13	.39	.68	1.05	1.68	2.02	2.70	3.55
120	.13	.39	.68	1.04	1.66	1.98	2.62	3.37

Abridged from Table III of Fisher and Yates. 1963. *Statistical Tables for Biological, Agricultural and Medical Research*. Oliver and Boyd Ltd., Edinburgh.

SECTION REVIEW PROBLEMS

6. Data sets A and B represent mean heights (in inches) of two strains of tomato at 2 weeks of age. Do the two strains have the same phenotype for this trait?

   A: 15, 15, 14, 13, 12, 12, 12, 12, 12, 11, 10, 10, 10, 10, 10

   B: 15, 15, 15, 15, 13, 13, 13, 13, 10, 10, 10, 10, 10, 9, 9, 9

7. Data sets A2 and B2 represent the heights (in inches at 2 weeks) of these same two strains in Problem 6 after five generations of selective breeding. Are the two strains the same now?

   A2: 17, 16, 16, 15, 15, 14, 14, 14, 13, 13, 12, 12, 11, 10

   B2: 20, 20, 19, 18, 17, 16, 16, 16, 15, 15, 14, 14, 13, 13, 12, 12

TABLE 5.5 *The Weights and Heights of a Sample of College Students*

	Weight (lb)	Height (in.)	Height (cm)
1	130	66	167.6
2	125	65	165.1
3	150	70	177.8
4	140	68	172.7
5	160	72	182.9
6	165	71	180.3
7	155	69	175.3
8	150	69	175.3
9	160	71	180.3
10	170	73	185.4
$\bar{x}$	150.50	69.40	176.3
s^2	219.17	6.49	16.5
s	14.80	2.55	6.5

The Strength of the Association Between Two Traits Is Determined Using Correlation.

When a population of individuals is examined for two different quantitative traits, the traits sometimes show a tendency to vary together. That is, as one increases the other increases, or as one increases the other decreases. Traits that show such a tendency are said to be correlated. Two traits are said to have a positive, or direct, **correlation** if they increase or decrease together. Alternatively, if one trait increases as the other decreases, the traits are said to have a negative or inverse correlation. For example, in Table 5.5, the values of two traits, height and weight, of a hypothetical sample of college students are given. If we plot these data on a graph (called a **scatter plot**) with height on the x axis and body weight on the y axis, we see that the traits appear to be correlated (Figure 5.8a). As height increases, weight also increases

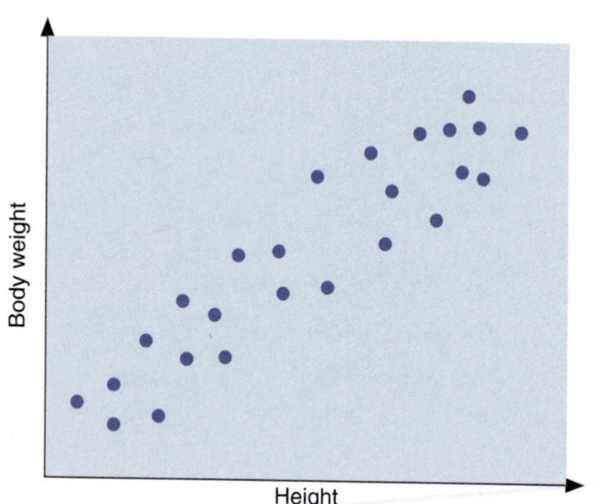

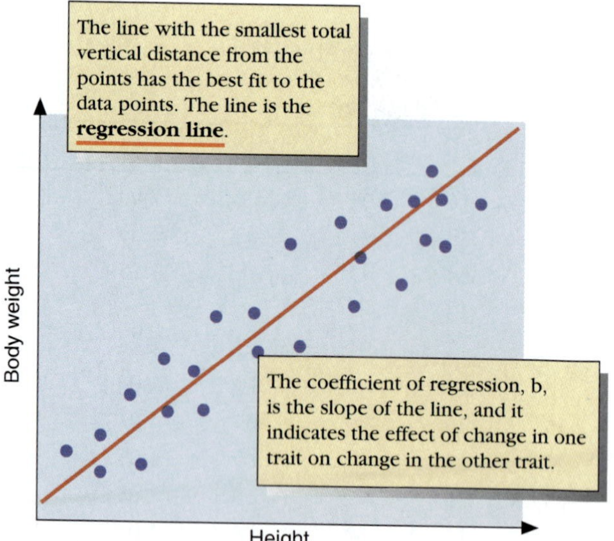

The line with the smallest total vertical distance from the points has the best fit to the data points. The line is the regression line.
The coefficient of regression, b, is the slope of the line, and it indicates the effect of change in one trait on change in the other trait.
Body weight
Height

FIGURE 5.8 Body weight and height in college students. (a) A scatter plot is produced when weight is plotted against height for a sample of American college students. These points indicate a positive correlation between these two factors. (b) A regression line that shows the correlation more clearly can be drawn through the scatter points.

(a positive correlation). We can determine a numerical value for the strength of the correlation between these two traits by calculating the **correlation coefficient** (r_{xy}). The value of the correlation coefficient depends on the product of the deviation of each value of one trait (x) from its mean value ($\bar{x}$) and the deviation of each value of the other trait (y) from its mean value ($\bar{y}$). The average product of these deviations is called the **covariance of x and y,** [cov(xy)]:

$$\text{Covariance} = \text{cov}(xy) = \Sigma(x_i - \bar{x})(y_i - \bar{y})/(n - 1)$$

The covariance is used to calculate the value of the correlation coefficient. The value of the correlation coefficient is the covariance divided by the product of the standard deviations of x and y:

$$\text{Correlation coefficient} = r_{xy} = \text{cov}(xy)/(s_x s_y)$$

In our example, we can calculate the value of the covariance of these two traits and use this result to calculate the value of the correlation coefficient:

$$\text{cov}(xy) = 328/9 = 36.4$$

$$r_{xy} = 36.4/[(14.8)(2.55)] = 0.98$$

The value of r can range from -1, (for a perfect negative correlation) to 0 (for no correlation) to 1 (for a perfect positive correlation). The large positive value of r_{xy} in our example indicates that these two traits show a strong positive correlation. In almost every case, whenever one increases, the other also increases.

It is important to understand just what correlation means. The observation of a positive or negative correlation tells us that in the study population the two traits tend to vary together, but this does not give us any information about the reason why the two traits vary together. A correlation does *not* mean that there is a causal relationship between two traits, that is, that a change in one trait causes the other to change. The correlation might be the result of a third factor that is acting on both traits. Assuming a causal relationship on the basis of a correlation can lead to serious false assumptions. For example, a significant percentage (up to 21%) of individuals in the United States who have acquired immune deficiency syndrome (AIDS) resulting from infection by the human immunodeficiency virus (HIV) develop Kaposi's sarcoma (spreading, purple-brown sores caused by tumors growing in the walls of blood vessels). Individuals in the United States who do not have AIDS develop Kaposi's at such a low frequency that it is barely measurable. There is thus a very strong positive correlation between having AIDS and having Kaposi's. This strong correlation led to the hypothesis that there was a causal relationship between AIDS and Kaposi's, namely, that Kaposi's was a symptom of AIDS. This implied causal relationship sometimes convinced individuals that they had AIDS because they had Kaposi's. Recently, however, increasing numbers of individuals have been found with Kaposi's who do not have AIDS. These cases suggest that Kaposi's is not always caused by infections of HIV.

How can we explain this correlation? There are several possible explanations. The high frequency of Kaposi's in AIDS patients may mean that individuals whose immune systems have already been weakened by AIDS are more susceptible to the agent that causes Kaposi's. It is also possible that individuals whose behavior places them at risk of contracting AIDS also places them at higher risk of developing Kaposi's whether or not they have AIDS. Whatever the reason for the correlation, the misdiagnosis of AIDS based on the appearance of Kaposi's can be a serious problem for patients who falsely believe they have AIDS because they have Kaposi's.

SECTION REVIEW PROBLEM

8. The following set of parents and children were measured for height. What is the correlation between parents and children for this trait?

	Offspring	
Parent (in.)	**(in.)**	**(cm)**
59	62	157.5
61	62	157.5
61	60	152.4
63	63	160.0
62	60	152.4
64	64	162.6
64	65	165.1
65	63	160.0
65	66	167.6
67	65	165.1
67	66	167.6
68	66	167.6
69	67	170.2
70	67	170.2
70	69	175.3

The Values of Correlated Traits Can Be Predicted Using Regression Analysis.

The correlation coefficient indicates the strength of the correlation between two traits and tells us the direction of the change, but it does not indicate how much the second trait will change when the first trait changes by a certain amount. This type of prediction can be made using **regression analysis.** In regression analysis, the relationship between two traits is expressed in the form of a regression line. Consider our example of the heights and weights of college students. The regression line for these two traits represents the best-fit line to the data points on the scatter diagram. By best-fit line, we mean that the regression line is the line with the smallest possible vertical distance to all the data points (Figure 5.8b).

The mathematical formula we will use for the regression line is the formula for a straight line:

$$y_i = \bar{y} + b(x_i - \bar{x})$$

In this formula, b is called the **coefficient of regression,** and it represents the slope of the line. The value of b indicates how much the value of one trait changes for a given change in the other trait. The value of b for any set of data can be calculated using the covariance:

$$b = \text{cov}(xy)/V_x$$

Using this formula, we can calculate the value of b for the sample in Table 5.5:

$$b = 36.4/219.17 = 0.17$$

Substituting this value into the equation for the regression line, we get

$$\text{height} = 69.4 + 0.17(\text{weight} - 150.5)$$

We can now estimate the height of college students from their weight. For example, an individual with a weight of 200 lb would be expected to have a height of 77.8 in.:

$$77.8 = 69.4 + 0.17(200 - 150.5)$$

There are two important points to remember about this estimate. First, the estimated value is an average value. If we were to examine all the students in this population with a weight of 200 lb, we would not expect them all to have a height of exactly 77.8 in. However, we would expect them to have a distribution of heights with a mean of 77.8 in. Second, in our regression analysis, we have used the formula for a straight line. Regression done using a straight line is called **linear regression.** It is possible that the two traits being analyzed are related in a nonlinear fashion, and the formula for a curved line is a better representation of their relationship. If this is the case, then using a linear regression line will not give the most accurate prediction (Figure 5.9).

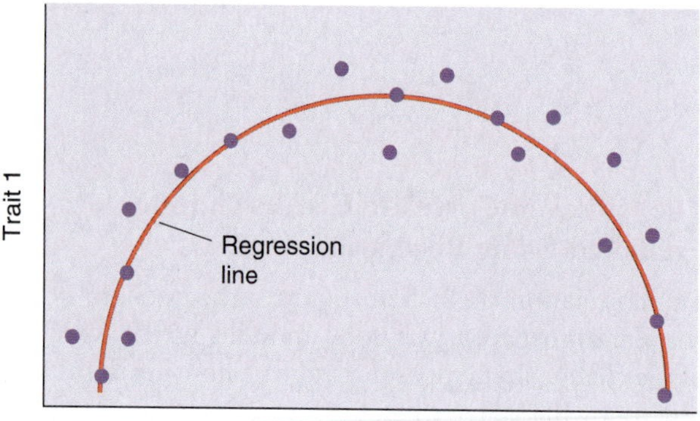

FIGURE 5.9 Traits with nonlinear relationships. Some traits have a nonlinear relationship, and the line with the best fit to the data is not a straight line.

SECTION REVIEW PROBLEMS

9. What is the coefficient of regression for the parents and offspring in Problem 8?
10. From Problem 8, what would be the expected height of the offspring of the two parents with heights of 75 in.? What would be the expected height of offspring from the parents with heights of 55 in.?

The Phenotypic Variance Can Be Divided into Genetic and Environmental Components.

One important statistical technique widely used in quantitative genetics is the analysis of the components of the phenotypic variance. The goal of an analysis of variance is to determine the relative contribution that different factors make to the total phenotypic variance. Variances are additive, meaning that the different factors affecting the phenotype in the population each produce a variance that is a component, or portion, of the phenotypic variance. The total phenotypic variance is equal to the sum of all these component variances. For example, the phenotypic variance (V_p) is often divided into three components: the **genetic variance** (V_g), the **environmental variance** (V_e), and the **interaction variance** (V_i) (Figure 5.10). The relationship among these components of phenotypic variance is a simple one:

$$V_p = V_g + V_e + V_i$$

The genetic variance (V_g) is a measure of the differences in the phenotype caused by differences in genotype. In a population that has been inbred until it is homozygous for all genes that affect this trait, such as Mendel's highly inbred peas, there is no genetic variation because all individuals have the same genotype. In such a population, the genetic variance is zero. Although experimental populations that have little or no genetic variation can sometimes be produced, most natural populations have considerable genetic variation. In this case, each individual in the population possesses different alleles of many genes, and this has an effect on the phenotype of the individuals. The amount of phenotypic differences caused by these genetic differences is the genetic variance.

The environmental variance (V_e) is a measure of the phenotypic differences that are produced by the different environments in which the individuals in the population live. Environmental differences do not need to be extreme (such as a tropical rain forest vs. a desert) to have

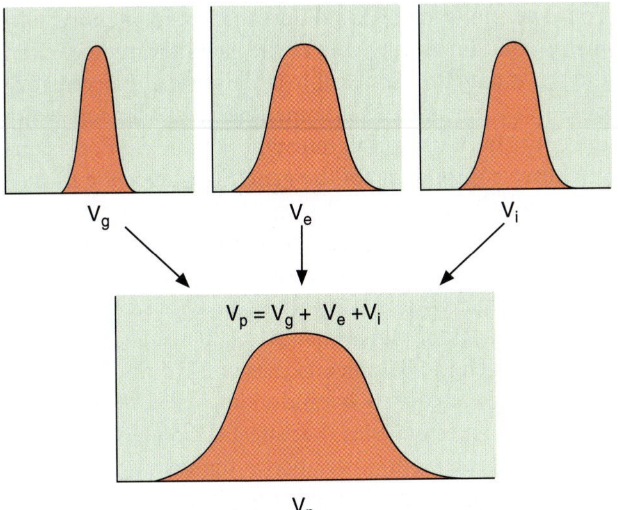

FIGURE 5.10 Variances are additive. The total phenotypic variance of a population is the sum of all the different variances. Important components of the V_p variance are the genotypic (V_g), the environmental (V_e), and the interaction (V_i) variances. These three variances constitute the total phenotypic variance (V_p).

an effect on the phenotype. Individual plants in one field may be in different environments if one part of the field has a different soil type, is wetter, or receives more sunlight. If all individuals were reared in an identical environment, the environmental variance would be zero. This can sometimes be arranged by careful experimental design for plants and animals in laboratories and selected field plots, but natural populations living in the wild are always exposed to a range of environments.

The third component, the interaction variance (V_i) measures the nonrandom interactions of the different genotypes in the population with the different environments. If all the different genotypes present in the population were distributed evenly throughout all the environments, so that each genotype was equally exposed to each environment, then the interaction variance would be 0, and the phenotypic variance would consist of only genetic and environmental variance. In natural populations, however, the interaction variance is rarely zero because many factors prevent different genotypes from dispersing freely throughout all environments. In human populations, for example, many cultural differences (such as wealth or social status) keep groups separate and prevent each genotype from being equally exposed to each environment. In experimental populations in which the environment is controlled ($V_e = 0$), the genotype is uniform ($V_g = 0$), or the different genotypes are randomly distributed, the interaction component is zero. Because we have no way of measuring the interaction variance, it is usually assumed that the interaction variance is zero.

In properly designed experiments, it is possible to measure the value of the components of phenotypic variance. For example, the environmental variance may be measured using a genetically uniform population that is produced by inbreeding (such as Mendel's peas). All individuals in this population have the same genotype, so the population has no genetic variance ($V_g = 0$), and all phenotypic variation must be caused by the environment ($V_p = V_e$). Consider the cross between two lines of inbred beans shown in Figure 5.11. The P_1, P_2, and F_1 populations are all genetically uniform, and the total phenotypic variance is an estimate of the environmental variance. Assuming that all three populations are raised in the same environment, a good estimate of the environmental variance is the average of the variances of the parents and the F_1:

$$V_e = (V_{p1} + V_{p2} + V_{F1})/3$$

If the F_1 individuals from this cross are inbred, the phenotypic variance of the F_2 population will be the sum of the genetic and environmental variances ($V_p = V_g + V_e$). If the F_2 individuals are reared in the same environment as the P_1, P_2, and F_1 individuals, they will have the same environmental variance. Subtracting the previously calculated value for the environmental variance from the

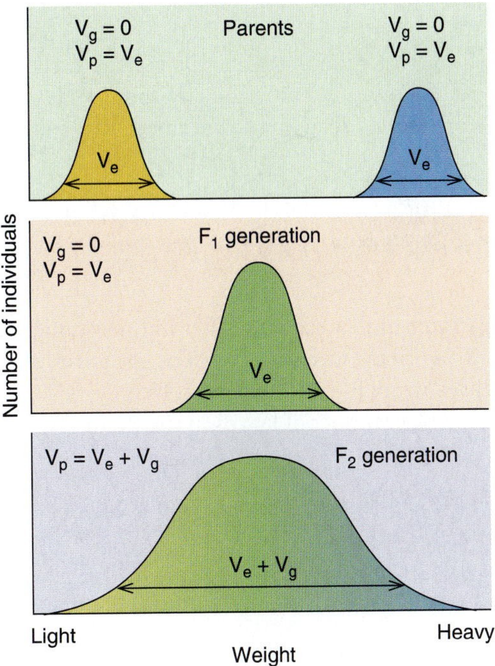

FIGURE 5.11 Homozygous strains and environmental variance. A cross is shown between two pure (homozygous) lines of beans that differ in weight of seed. The parental populations are genetically uniform ($V_g = 0$), as is the F_1 population. Thus, all variation from the mean is caused by the environment. The F_2 population is not genetically uniform; thus variation from the mean is caused by both genetics and the environment.

measured value of the phenotypic variance gives the value of the genetic variance:

$$V_g = V_p - V_e$$

SOLVED PROBLEM

Problem

Two pure lines of beans that produce seeds with different weights were crossed. The weights and variances of seeds from the parental, F_1, and F_2 plants are listed here.

$\bar{x}$ Weight	(cg)	V_p
P_1	45.9	1.7
P_2	50.2	2.1
F_1	48.1	2.0
F_2	48.0	4.1

a. What is the value of the environmental variance (V_e) for this trait?

b. What is the value of the genetic variance (V_g) for the F_2 population?

Solution

a. The value of the environmental variance can be determined by averaging the variances of the parents and the F_1: $V_e = (1.7 + 2.1 + 2.0)/3 = 1.9$.

b. If we ignore interaction effects, the genetic variance is calculated by subtracting the environmental variance from the F_2 phenotypic variance: $V_g = 4.1 - 1.9 = 2.2$.

SECTION REVIEW PROBLEMS

11. Two strains of tobacco that differ in the number of leaves per plant were crossed. The mean number of leaves were: $P_1 = 15.0$, $V = 1.40$; $P_2 = 17.5$, $V = 1.25$. The number of leaves in the F_1 and F_2 follow:

    F_1 = 18, 15, 16, 18, 15, 16, 14, 16, 18, 17, 16, 13, 16, 14, 16, 15, 16, 15, 15, 16, 15, 16, 16, 15, 16

    F_2 = 16, 20, 19, 17, 14, 16, 14, 14, 15, 17, 20, 13, 12, 15, 16, 21, 18, 15, 14, 18, 14, 17, 13, 15, 13

    Given that these plants were raised in the same environment and that the interaction variance (V_i) is 0, calculate the environmental variance (V_e) and the genotypic variance (V_g).

The genetic and environmental variances are themselves composed of a number of different components. For example, each allelic difference in each genotype in the population contributes to the genetic variance. Even though it may not be possible to determine the contribution of any one allele, it is sometimes possible to measure the contributions of different types of alleles. For example, three components of the genetic variance are the additive genetic variance (V_a), the dominant genetic variance (V_d), and the genetic interaction variance (V_i). The additive genetic variance consists of all the phenotypic variation that is caused by individuals having different additive alleles of genes that affect the phenotype. The dominant genetic variance is that portion of the phenotypic variance caused by individuals having different dominant alleles of genes that affect the phenotype. And, the genetic interaction variance is the result of all other types of genetic interactions (epistasis, enhancement, suppression, etc.) that cause individuals to have different phenotypes. Because all variances are additive, the genetic variance is equal to the sum of these three components: $V_g = V_a + V_d + V_i$

The values of the additive genetic variance (V_a) and the dominant genetic variance (V_d) can sometimes be estimated by measuring the variances of groups of individuals that have known genetic relationships. The values of these variances can then be used to make deductions about the alleles present in the population. The two types of F_1 backcrosses (F_1 individuals crossed with both sets of parents) are especially valuable for this type of analysis. For example, one quantitative trait is the length of time it takes *Drosophila* to complete development. The values of the variance for this trait for several different crosses are given in Table 5.6. The environmental variance (V_e) can be calculated as the average of the variances of the P_1, P_2, and F_1, assuming that the parental

TABLE 5.6 *The Means and Variances for the Offspring from a Series of Crosses Between Two* Drosophila *Strains That Vary with Respect to the Length of Time To Complete Development*

	Number of Individuals	Mean Hours	Phenotypic Variance
P_1	62	218	23
P_2	49	306	28
F_1	127	281	21
B_1	202	235	56
B_2	188	299	50
F_2	336	268	76

strains were homozygous. The genetic variance (V_g) can be calculated by subtracting the environmental variance from the variance of the F_2 individuals ($V_g = V_{F2} - V_e$). The additive variance can be calculated by subtracting the backcross variances (V_{B1} and V_{B2}) from the F_2 variance using the following formula:

$$V_a = 2\,[V_{F2} - (V_{B1} + V_{B2})/2]$$

The dominant variance (V_d) can be calculated by subtracting the additive variance from the genetic variance. In these calculations, the interactive component is ignored. If we do these calculations using the data in Table 5.6, we see that the additive genetic variance (46) is considerably larger than the dominant genetic variance (6). This result indicates that the individuals in this population have different additive alleles and dominant alleles that affect this trait, but that the differences between additive alleles make a larger contribution to the overall phenotypic variance than the differences between dominant alleles. In natural populations, the additive variance is the largest component of the genetic variance for most quantitative traits.

SECTION REVIEW PROBLEM

12. The F_1 plants from Problem 11 were backcrossed and the progeny were:

    B_1 = 18, 17, 20, 15, 17, 17, 11, 13, 16, 19, 17, 12, 16, 16, 16, 16, 17, 17, 14, 16, 17

    B_2 = 17, 16, 15, 13, 14, 19, 17, 21, 17, 18, 18, 20, 20, 17, 18, 19, 18, 16, 18, 15, 18

    Calculate the additive genetic variance and the dominant genetic variance.

Heritability Is Useful in Predicting the Phenotypes of Offspring.

An important goal of genetic analysis is to gain the information to predict the phenotypes of offspring from a set of parents. For qualitative traits, this is relatively easy. A series of controlled crosses, or a pedigree analysis, provides information about whether the mutant alleles are dominant or recessive and how they interact. This information is used to define the parental genotypes and, using the Mendelian rules, to predict the ratios of genotypes and phenotypes in the progeny. For quantitative traits, however, genetic analyses usually do not define the genotypes of the parental individuals, and it is not possible to predict the phenotype of offspring, at least not in terms of discrete genotypic and phenotypic classes. However, information about the mean and variance of a quantitative trait obtained from the parental population can be used to make predictions about the mean of the phenotypic distribution in the offspring generation. The key to this prediction is determining the **heritability** of the trait. Heritability is a statistical measure of how strongly the phenotype of the offspring will resemble the phenotype of the parents.

Why is heritability important? For thousands of years, plant and animal breeders have relied on **phenotypic selection** to improve their strains of crop plants and animals. The breeders selected from their existing strains individuals that had a particular desired phenotype and used these individuals as parents of the next generation. The assumption was that the offspring would inherit some of the desirable phenotype of the parents. However, phenotypic selection works only when the phenotypic differences between individuals in a population are the result of genetic differences. If the parents have a desirable phenotype because they have a particular collection of alleles, then they may pass these favorable alleles on to their offspring, and the offspring will have the desired phenotype. However, if the phenotypes of the parents have been strongly influenced by the environment, then the favorable phenotype of the parents may not be passed on to the offspring. Because selection often requires a great deal of time and money, it is a tremendous advantage to a breeder to be able to estimate in advance how a particular population might respond to selection. Determining the heritability of a trait allows the breeder to make this type of estimate.

Heritability Is a Measure of the Genetic Component of the Variance.

There are two widely used numerical measures of heritability. One is the **coefficient of heritability** (h^2), also called **heritability in the narrow sense.** The other is the **degree of genetic determination** (H^2), also called **heritability in the broad sense.** Both values depend on the ratio of genetic variance to phenotypic variance. h^2 is the ratio of the additive variance to the total phenotypic variance, and H^2 is the ratio of total genetic variance to total phenotypic variance:

$$h^2 = V_a/V_p$$

$$H^2 = V_g/V_p$$

Both of these values indicate what portion of the variation in a population is the result of genetic variation. Values of h^2 and H^2 range from 1 to 0, with a high value indicating that a large portion of the phenotypic variation is caused by genetic variation.

Both h^2 and H^2 have some important limitations. Their values are calculated for one population in one environment, so they cannot be used for other generations of this same population raised in different environments nor for other populations. Each population has a different set of genotypes and so will have a different proportion of the phenotypic variance caused by genetic variance and different values of h^2 and H^2. If this same population were raised in a different environment, it would have a different value of V_p and different values of h^2 and H^2. If a different population of this same species were examined, it may contain different alleles of the genes controlling the phenotype, and thus different values of h^2 and H^2. In the remainder of this section, we will discuss some examples for the calculation and uses of h^2 and H^2.

The values of h^2 for a number of traits are given in Table 5.7 and demonstrate differences in h^2 for different traits. A high value of h^2, approaching 1, means that the phenotypic variance is largely the result of additive alleles; consequently, traits with high values of h^2 respond well to selection. This is because the variation between individuals is caused by having different additive alleles, and these alleles can be passed on to the next generation. In future generations, each allele will have its same additive effect on the phenotype, and segregation will generate many new combinations of alleles that will give new phenotypes. Conversely, a low value of h^2 means that the differences in phenotype between individuals are not caused by different additive alleles. Instead, it means that the offspring will have fewer new combinations of additive alleles that will generate change in the phenotype. If the other factors, such as environmental effects, are not present, the offspring will not have the same desirable phenotype as their parents.

TABLE 5.7 *Values of Heritability (h^2) for Selected Traits of Different Species*

Species	Selected Traits	Heritability
Cattle	Body weight	0.65
	Milk yield	0.35
	Frequency of twin birth	0.02
Humans	Height	0.65
	Serum immunoglobulin level	0.45
	IQ	0.75–0.80
Pigs	Back-fat thickness	0.70
	Weight gain	0.40
	Litter size	0.05
Mice	Tail length at 6 weeks of age	0.40
	Litter size	0.20
Fruit Flies	Body size	0.40
	Egg production	0.20
Horses	Racing speed	0.60
	Trotting speed	0.40
Poultry	Body weight at 8 weeks of age	0.31
	Egg production	0.10
Maize	Plant height	0.70
	Ear length	0.55
	Ear diameter	0.14

As an example of how h^2 might be used to predict the phenotype of offspring, consider a characteristic important in modern aquaculture, the rate of growth of fish. A new strain of fish called the saugeye (which is a cross between a sauger and a walleye) is being studied by Dr. Robert Summerfelt for possible commercial use on fish farms. Suppose that we had a population of these fish with a mean length of 16 cm after 41 days of growth, and the variances for this population were $V_p = 8.6\ cm^2$, $V_g = 8.0\ cm^2$, $V_a = 6.5\ cm^2$. Using these values, we can calculate the value of h^2 for the quantitative trait rate of growth:

$$h^2 = V_a/V_p = 6.5\ cm^2/8.6\ cm^2 = 0.76$$

This means that 76% of the observed phenotypic variation for this trait in this population is caused by additive alleles. Using selected parents from this population, we can use the value of h^2 to make predictions about the phenotypes of the offspring. For example, if we were to breed two individuals, a male with a length of 21 cm and a female with length of 22 cm, what would be the average length of the offspring after 41 days of growth? The average length of the parents is (21 cm + 22 cm)/2 = 21.5 cm. The difference between this parental value and the population mean (21.5 cm − 16 cm = 5.5 cm) is called the **selection differential** (S). The value of h^2 is used to estimate how much of this difference is passed to the offspring using the formula

$$R = Sh^2$$

In this formula, R is the **response to selection** and represents the difference of the offspring of the cross from the mean of the population. For our fish, R = 5.5 cm × 0.76 = 4.2 cm. We would expect the offspring of these parents to have a phenotype of 16 cm + 4.2 cm = 20.2 cm after 41 days of growth. Note that this estimate is valid only if the offspring are raised in the same environment as the parents. A change in food, temperature, water purity, or other conditions might affect the growth rate.

In this example, we first determined the components of genetic variance and then used them to calculate the value of h^2. This value was then used to estimate R. But, we can also do the reverse. We can determine the value of R and S experimentally and then use these values to de-

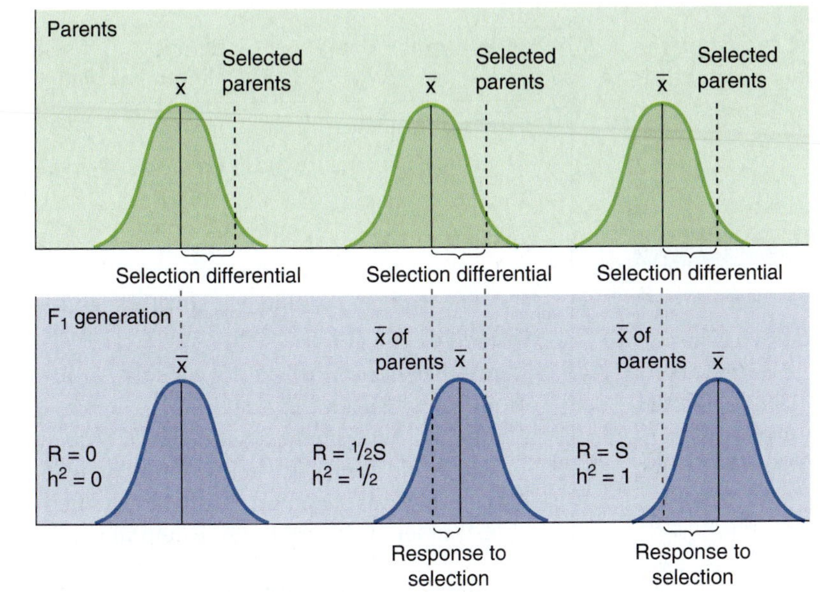

FIGURE 5.12 Heritability and selection. Three populations with identical phenotypic distributions but different heritabilities (h^2) give different responses to selection. When heritability is zero, no response is seen, but as the heritability increases, so does the response to selection.

termine the value of h^2. This is done with a rearranged form of this same formula:

$$h^2 = R/S$$

This estimate of h^2 is sometimes referred to as the **realized heritability.** If the heritability is large, then the mean of the offspring's phenotype will be closer to that of the selected parents than to the original population. If, however, the value of heritability is low, then the mean of the offspring's phenotype will be closer to the mean of the original population, not the selected parents (Figure 5.12). Consider the saugeye population again. Suppose that we selected a sample of fish with a mean length at 41 days of 25 cm from an original population with a mean length of 16 cm at 41 days. We allow the selected fish to breed randomly among themselves and observe that their progeny have a mean length of 22.8 cm at 41 days of growth (Figure 5.13). S for this sample of parents is 25

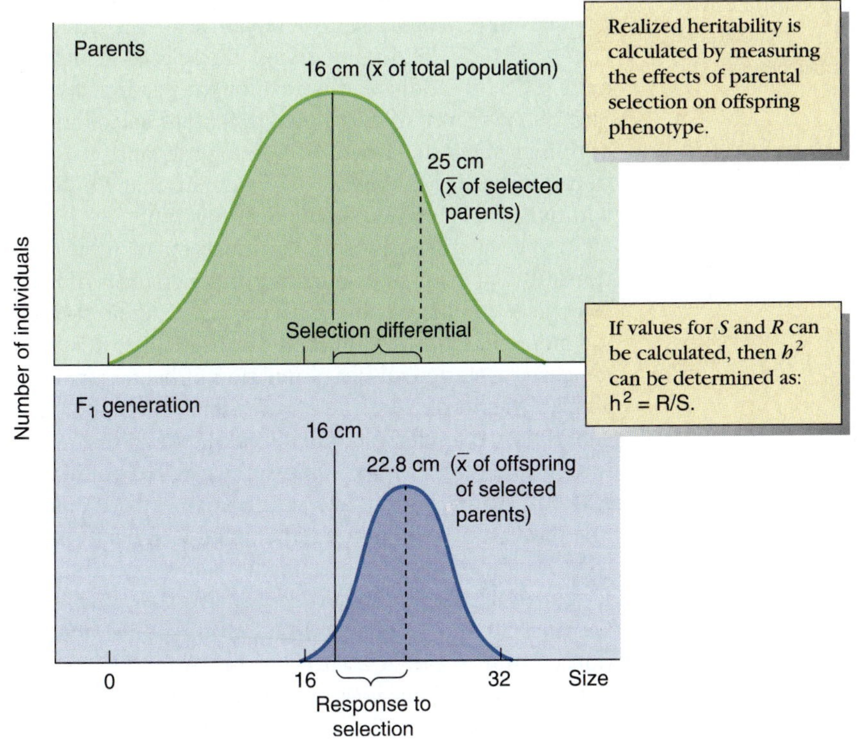

FIGURE 5.13 Realized heritability. Realized heritability (h^2), which is an estimate of the true coefficient of heritability, is calculated for a sample of fish by dividing the response to selection (R) by the selection differential (S).

cm − 16 cm = 9 cm. R is 22.8 cm − 16 cm = 6.8 cm. Using these values we can determine that the value of h^2 is 6.8 cm/9 cm = 0.76.

SOLVED PROBLEM

Problem

A geneticist studying the number of bristles on the second leg of *Drosophila melanogaster* determined that a wild-type strain has a mean number of 486.3 bristles per leg. A sample of males and females from this population with 420.0 bristles were bred and the offspring had a mean bristle number of 432.0.

a. What is h^2 for this population?

b. If we were to select a group of flies with 500.0 bristles and breed them, what would be the mean bristle number in the progeny?

c. If we wished to generate a population of flies that had a mean leg bristle number of 450.0, what should we do?

Solution

a. The value of heritability can be calculated as h^2 = R/S, where S = 486.3 bristles − 420.0 bristles = 66.3 bristles and R = 486.3 bristles − 432.0 bristles = 54.3 bristles. Thus, h^2 = 54.3/66.3 = 0.82.

b. The expected number of bristles in the progeny can be calculated using the value of h^2. The selection differential is 500.0 − 486.3 = 13.7 bristles. The expected response to selection is 13.7 × 0.82 = 11.2. The expected mean bristle number in the progeny is 486.3 + 11.2 = 497.5.

c. To produce progeny with a phenotype of 450 bristles, we must do a selection experiment in which the response to selection (R) is equal to 486.3 − 450.0 = 36.3. Using the value of h^2 calculated previously (h^2 = 0.82) to obtain this large R, we must have a selection differential of S = R/h^2 = 36.3/0.82 = 44.3. This means we must select a parental population with a mean bristle number of 486.3 − 44.3 = 442.0.

SECTION REVIEW PROBLEMS

13. Calculate h^2 for the plants in Problems 11 and 12.

14. The lengths of portions of the leg (tibia and tarsus) were measured in a sample of amphibians, and the following variances were observed:

	Tibia Length (mm)	Tarsus Length (mm)
Phenotypic variance	36.4	34.5
Environmental variance	11.1	8.7
Dominant variance	20.1	5.9
Additive variance	5.3	19.9

a. What is the heritability (h^2) for these traits?

b. Which trait would respond quickest to selection?

15. Another portion of the leg (femur) in these amphibians had a mean length of 102.3 mm. Two individuals with femur length of 175.0 mm were selected and bred. The mean femur length in the offspring was 136.0 mm. What is the heritability (h^2) of femur length in these amphibians?

The short-term response to selection of most quantitative traits can be accurately predicted using h^2. This indicates that in most populations the phenotype of many quantitative traits is largely determined by additive alleles. Sometimes effective selection may proceed beyond expectations. A good example of this is found during selection of ear length in corn. This trait is of considerable importance in commercial corn strains because plants that have longer ears will have larger yields of kernels. Arnel Hallauer in 1986 reported the results of a selection experiment started in 1963 for length of ear. This was a **divergent selection** experiment, meaning that two lines were started from one original population. One line was selected for increased ear length, and the other was selected for decreased ear length. For 15 generations, the populations showed a linear response to the selection (Figure 5.14). The original population clearly contained a great deal of additive variance for the ear length trait. Surprisingly, even after 15 generations of selection, the two lines show no sign of decreasing response to selection. This indicates that they still contain many different additive alleles for genes that affect this trait.

However, if selection is continued for many generations, it may cease to be effective. In an analysis of long-term selection in an animal species, F. D. Enfield selected for increased body size in the flour beetle *Tribolium castaneum*. He started with a population of beetles that had a mean weight of 2.4 g and a variance of 4.0 g^2. For each generation, the selection differential (S) was 0.022 g. The initial value of h^2 for body size in the original population was 0.3, so the response to selection (R) was 0.0066 g. For the first 50 generations, the mean weight of the selected population increased steadily, showing a response to selection close to the predicted 0.0066 g (Figure 5.15). Between 50 and 125 generations, the response to selection continued but at a reduced rate. After 125 generations of selection, the mean weight had increased to 5.8 g,

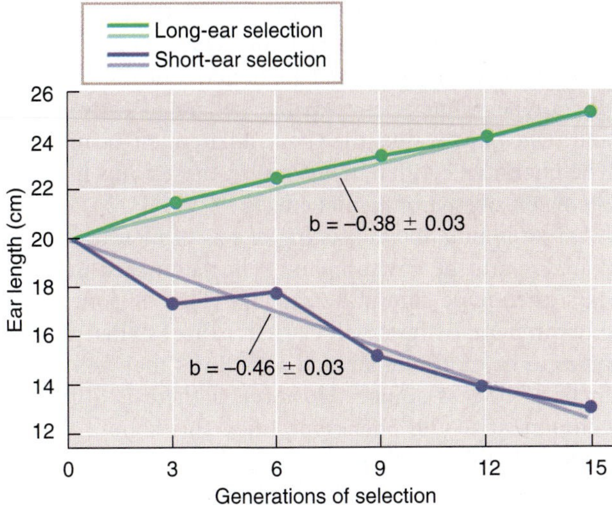

FIGURE 5.14 Selection for ear length in corn. Two divergent selections were started from one population of corn. One population was selected for increased ear length, and the other was selected for decreased ear length. The response to selection is a result of the segregation of additive alleles that increase and decrease ear length. b = slope of the regression line. *(From A. Hallauer, Iowa State University)*

more than twice the original mean, and additional selection did not result in a further increase in size. The lightest individuals in the selected *Tribolium* population were heavier than the heaviest individuals in the original population.

How could the mean of the population be shifted to a value outside the original range of the population? Additive alleles have both positive and negative effects on phenotype, and the mean weight of the original population represented a balance between alleles that increased body size and alleles that decreased it. During the selection experiment, these additive alleles segregated, producing some progeny with fewer negative alleles and more positive alleles than any individuals in the original population. Consequently, these individuals were larger than any individuals in the original population.

Enfield determined h^2 for the selected population after 125 generations and discovered that it was only slightly less than for the original population. Even after 125 generations, the population still contained a great deal of additive genetic variance for this trait. The failure of the population to respond to further selection was not because the population was now homozygous for the genes controlling body size. The reason selection was no longer effective was that the larger beetles produced few offspring, indicating that the alleles that increased body weight also decreased fertility. Eventually the point was reached where any further increase in weight of an individual resulted in sterility. When selection was stopped, the body size dropped, and fertility increased. The body size in this species thus represents a balance between the opposing actions of different selection pressures. When we impose a new, artificial selection for increased size on the population, the equilibrium shifts, but it will shift only so far, until countervailing natural selection balances the artificial selection. This balance may be reached when the population still has a great deal of genetic variation. For many species, such natural selection pressures are why it is difficult to generate stable strains of highly selected plants and animals. After the artificial selection is removed, the natural selective forces shift the phenotype of the population back toward the original value.

Each natural population contains unique alleles, and selection in one population is limited by the range of alleles in that population. In the beetles, for example, the selected population reached the point where no combinations of the alleles in that population could give larger beetles that were also fertile. However, it is possible that alleles exist in other populations that could give larger, fertile beetles. If so, crossing individuals from the two populations, followed by selection of the offspring, would give a new population with different combinations of alleles for size and fertility, and these new alleles would produce still larger individuals.

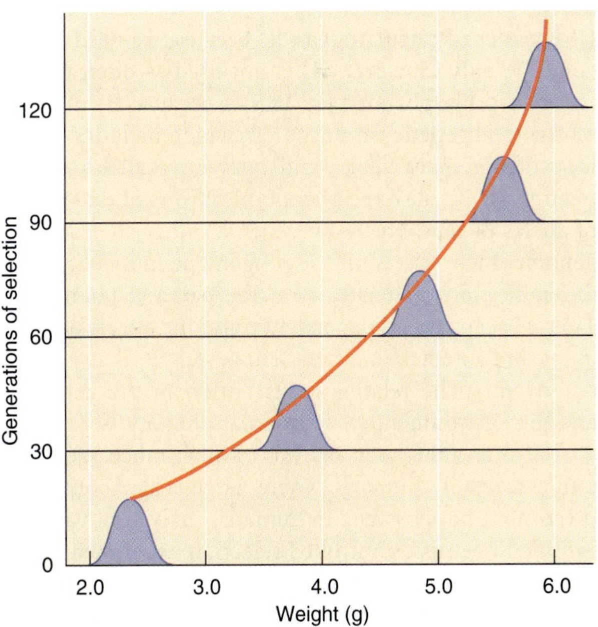

FIGURE 5.15 Long-term selection. The results of a long-term selection experiment on body size in *Tribolium*. Individuals were selected from each population that had increased body size. The final population has a mean body weight twice as great as the mean of the original population.

A key point about selection for a quantitative trait is that selection of a population is limited by the alleles present in that population. A constant worry of quantitative geneticists working with highly selected strains of commercial plants and animals is that other populations may have unique alleles that affect the trait being selected. These populations may be smaller plants or animals or unexplored wild populations of the species that are invaluable for future selection experiments but that face the possibility of extinction. If farmers and breeders abandon all but the currently most popular strains, these rare alleles will be lost. An international effort has begun to preserve samples of many populations of important species for precisely this reason. Some minor strain may contain unique alleles that prove to be of great importance in the future (see CURRENT INVESTIGATIONS: *Is It Important To Conserve Genetic Diversity?*).

Heritability in the Broad Sense Measures the Effects of All Genetic Variation on the Phenotype.

There are many different ways to calculate the value of heritability in the broad sense (H^2). The most common is the ratio of the total genetic variance to the phenotypic variance:

$$H^2 = V_g/V_p$$

H^2 is easier to calculate than h^2 because we need to calculate V_g only. However, H^2 is not used as often to estimate offspring phenotypes. This is because H^2 depends on the total genetic variance (V_g), which includes the effects of dominant alleles and interactive alleles on the phenotype. Because the phenotypic effects of these types of alleles depend on the presence of other alleles in the genome, their effects on the progeny phenotype may be different from their effects on the parental phenotype. This is because H^2 is dependent upon the effects of dominant and interactive alleles. However, H^2 is useful for determining the relative contribution of the genotype and the environment to a particular phenotype.

H^2 is useful for the analysis of inheritance of quantitative traits in humans where large-scale, controlled breeding is not possible. In humans, analysis of variance is difficult because controlled selection experiments cannot be done to determine the value of h^2. Thus, calculations of H^2 may be used to estimate the genetic contribution to a phenotype. In humans, this is done by studying the phenotypes of sets of identical twins. Identical (monozygotic) twins have the same genotype, so phenotypic differences between two twins must be the result of environmental influences. However, most twins are reared in the same family, and so they have very similar environments. In such cases, it may be impossible to separate the environmental and genotypic contributions to a phenotype. Sometimes twins are separated early in life and reared apart; thus they have different environments. The phenotypic differences between these twins must be the result of environmental effects.

In a population consisting of sets of identical twins reared apart, the environmental effects on each individual's phenotype should be random. That is, sometimes environmental effects will cause a pair of twins' phenotypes to be more similar, and sometimes they will cause them to be less similar. However, if there is a genetic component to the phenotype, then this should be the same in both individuals. In a population of twins, each set of twins will be genetically different from the other sets of twins, and this genetic difference will be reflected in the overall population genetic variance. Each individual's genotype will cause him or her to have a certain amount of deviation from the population mean. This genetic deviation will be the same for that individual's twin. Two twins should thus always show the same genetic deviation from the population mean. This can be measured by calculating the **covariance** of the sets of twins. For a set of identical twins reared apart, the value of heritability in the broad sense is

$$H^2 = \text{cov}(\text{twin}_1\ \text{twin}_2) / V_p$$

If twins represent a genetically random sample of individuals in a population, then for this trait H^2 is an estimate of the percentage of the phenotypic variation caused by the genotypic variation in the population.

The most controversial use of H^2 is in the determination of the genetic component of psychological and behavioral traits in humans. There is little argument that physical characteristics like height, weight, or rate of growth are influenced by both environment and inheritance. However, there is less agreement on traits such as schizophrenia or intelligence. One of the most controversial issues is intelligence in humans. Determining the relative contribution of genotype to intelligence has been a favorite subject of twin studies for many years. In theory, this can be done by calculating the value of H^2 by measuring the intelligence of a number of sets of twins that have been reared apart. However, such studies have a number of practical problems that must be overcome. For example, it is often not possible to know for certain that the twins have truly been reared independently. A second problem is with the measurement of intelligence. Most studies use standardized IQ tests, but there is considerable controversy about such tests. It is not certain just what such tests actually measure. Given this reservation, we can examine studies of IQ in sets of human twins as examples of the calculation of heritability.

Current Investigations

Is It Important To Conserve Genetic Diversity?

An important concern in modern agriculture is whether the selection and inbreeding techniques used to produce modern crop strains is causing an excessive loss of genetic diversity. In the past, humans relied on between 3000 and 5000 species of plants for nutrition. Today's agriculture relies heavily on only about 15 species in three main groups: cereals, grain legumes, and potatoes. Modern plant breeding has been enormously successful in increasing yield and modifying the characteristics of plant species to meet consumer demands. However, these increases have been accompanied by reductions in genetic diversity. Consider maize, for example. There are over 300 different races of maize in the western hemisphere. Each race is a well-defined population with inherited, unique characteristics. Of these 300, only 6 are used in commercial agriculture, and all the modern field corn lines have been derived from a single race. Within this race, there were over 10,000 different varieties in the early 1900s. These varieties were mostly developed by independent farmers who each year kept a part of their crop to serve as seed for the next year. Many farmers developed unique varieties by continued inbreeding and selection. However, today's high-yielding hybrids have been developed by large seed companies largely from two varieties known as Reid's and Lancaster. More than 70% of the US field corn is derived from just six inbred lines, three from each variety. Thus a very large industry depends on a few, highly inbred, elite lines. These lines contain only a small fraction of the alleles present in the total corn population, and further inbreeding and selection will result in more homozygosity and the loss of more alleles. Corn is not unique in this respect. For all other commercial plant species, the initial diversity of varieties has been greatly reduced. As modern breeders have developed strains with higher yields, farmers have discarded their old strains, and a tremendous amount of genetic diversity has been lost.

Is this reduction in diversity necessarily a bad thing? Potentially it might have a very serious effect. Pests (insects and diseases) will continue to evolve mechanisms to overcome plant defenses. With large fields containing genetically uniform populations of plants, a pest that develops an effective means of attacking those plants could spread rapidly through the population, causing major losses. This is, in fact, what happened in the mid-1800s to the Irish potato crop, causing a major famine, and in the 1970s to the US corn crop. Natural populations contain many alleles that have evolved to provide plants with resistance to pests. If these alleles have negative effects on a trait being selected by breeders, then these alleles will be eliminated, producing a line with superior performance—as long as that pest is not present. However, if the pest attacks, that line will have no resistance. In addition, as we have seen with laboratory populations, eventually agricultural populations will reach the limits of selection. When this happens, the only way to continue to improve the strains will be to cross them to new populations that have different alleles.

Recognizing that potentially valuable alleles may be present in strains that otherwise would be discarded, several countries have started to maintain diverse populations of many valuable plants in seed banks. The US Department of Agriculture established a national seed storage laboratory in 1956, and in 1986 it began a plant germplasm operation to oversee its national plant germplasm system. By 1988, over 370,000 strains were being preserved. Over 80 countries participate in these efforts. In India, for example, a national bureau of plant genetic resources was established in 1976 to oversee long-term and medium-term seed storage of over 45,000 strains. The goal of these efforts is to preserve as wide a variety of strains as possible and to provide catalogs that describe these strains and many of the alleles they contain. These resources are used by breeders seeking to add new alleles to their strains, by geneticists seeking alleles that affect different aspects of plant growth and development, and by molecular biologists seeking alleles of particular genes. Maintaining germplasm collections is expensive and requires skilled geneticists. Each strain must be planted at regular intervals to produce fresh seed, and care must be taken to prevent inbreeding that would lead to the loss of alleles. As the world becomes more and more industrialized, great care must be taken to maintain and expand these germplasm collections, or we may find that our agricultural industry is resting on an extremely small and vulnerable genetic base in the future.

Three well-known twin studies are those of Horatio H. Newman and coworkers (1937), James Shields (1962), and Niels Juel-Nielsen (1965).* These studies used sets of twins who were separated early in life, reared separately, and later reunited. All used different types of IQ tests. Some of the data from these twin studies are shown in Table 5.8. For example, consider the 37 sets of twins that were given the Dominoes IQ test. The value of the covariance for these twins is 214.20. The value of V_p, the total phenotypic variance, is 264.69. Substituting these values into our formula for H^2 gives

$$H^2 = V_g/V_p = 214.20/264.69 = 0.81$$

The result suggests that 81% of the phenotypic variance in IQ between sets of twins in the population is the result of genetic variance. This implies that genotype differences are the largest factor in determining the difference in IQ in this population of humans. We must remember, however, that H^2 is concerned only with how much variance is the result of genetic differences and not whether any particular genotype gives a higher or lower IQ value. We must also remember that this value of H^2 is valid only for this population. If we perform the same calculations on the sets of twins given the Stanford-Binet IQ test, we see that for them H^2 is 0.68, a significant but lower value. What overall conclusion can we draw about heritability and IQ? The best conclusion is that genotype has a large role in the variation in IQ, although precisely what percentage is not certain. It is certain, however, that the environment also has a large role. Susan Farber, in a 1981 review of all major twin studies done up to 1980, suggests that for human IQ measured by standard tests, H^2 is between 0.75 and 0.80.† More recent twin studies tend to agree with this (see *CURRENT INVESTIGATIONS: How Are Identical Twins Used To Study the Inheritance of Human Behavior?*).

SECTION REVIEW PROBLEM

16. Calculate H^2 for the IQ of the twins in Table 5.8 who were given the Wais IQ test.

TABLE 5.8 *The IQ Scores for Three Samples of Identical Twins Reared Apart*

Dominoes IQ Test		Standford-Binet Test		Wais Test	
Twin 1	**Twin 2**	**Twin 1**	**Twin 2**	**Twin 1**	**Twin 2**
97	104	124	107	99	103
98	93	99	101	119	121
112	103	85	84	100	94
103	91	92	116	91	98
107	83	107	110	105	97
72	67	94	95	111	117
100	79	105	106	104	103
108	97	92	77	125	111
109	107	122	127	91	100
109	93	96	77	108	109
69	74	89	93	111	117
58	55	116	109		
123	121	115	105		
107	106	90	88		
97	81	91	90		
91	88	66	78		
97	86	85	97		
117	119	100	104		
90	72	103	102		
91	88	102	94		
97	103	89	106		
102	109	102	96		
114	110	88	79		
103	110				
98	95				
112	102				
109	97				
84	67				
95	103				
107	100				
58	74				
90	100				
98	88				
121	117				
67	77				
74	90				
69	62				

* Shields, J. 1962. *Monozygotic Twins Brought Up Apart and Brought Up Together: An Investigation into the Genetic and Environmental Causes of Variation in Personality.* Oxford University Press, New York.

Newman, H. H., Freeman, F. N., Holzinger, K. J. 1937. *Twins: A Study of Heredity and Environment.* The University of Chicago Press, Chicago, IL.

Juel-Nielsen, N. 1965. "Individual and Environment: A Psychiatric-Psychological Investigation of MZ Twins Reared Apart." *Acta Psychiat. Scand. Suppl.* 183 Copenhagen: Munksgaard.

† Farber, S. 1981. *Identical Twins Reared Apart.* Basic Books, New York,.

CURRENT INVESTIGATIONS

How Are Identical Twins Used To Study the Inheritance of Human Behavior?

Most individuals readily accept that physical human characteristics such as hair color, skin color, and height are controlled by genes. However, the question of the importance of genetic and environmental factors in the control of human psychological characteristics, such as behavior and intelligence, is more controversial. This question has a long history and is often referred to as the Nature vs. Nurture controversy. The question is not whether the genotype has an effect; clearly genotype and environment both influence personality and behavior. The questions are how large a contribution each makes and which is the dominant influence. In Western society, the prevailing opinion has long been that behavior is most strongly influenced by environment. This implies that personality disorders have their cause in childhood experiences and societal influence, not in the genome. This controversy is hard to test scientifically because doing quantitative genetic studies of human behavior is difficult. Large-scale experiments involving controlled human matings cannot be made, nor is it possible to control the environment of human populations. It is possible, however, to study cases in which humans with identical genotypes are exposed to different environments by studying cases of monozygotic twins who have been reared separately. Identical twins have the same genotype so, in principle, all differences in behavioral characteristics will result from environmental influences, whereas similarities may result from genetic influences.

T. J. Bouchard at the University of Minnesota performed an important twin study. Since 1979, Bouchard has studied over 100 pairs of monozygotic twins raised together (MZT) or raised apart (MZA). Twins participating in the study undergo over 50 hours of intelligence and psychological testing. The results suggest that genetics has a large influence on intelligence and behavior. Calculations of H^2 for IQ, for example, suggest that more than 70% of the variance in IQ can be traced to genetics. Comparing MZA with MZT suggests that being reared together actually has only a minor effect on IQ. The togetherness factor was more pronounced in childhood and early adolescence but declined greatly during adulthood.

Some of the most interesting results came from tests measuring personality and behavior characteristics. For characteristics such as aptitude, leisure time activities, and vocational interests, the twins showed a strong genetic influence, as large as for IQ. No aspect of behavior was unaffected. Overall, the genetic influence on all behavioral characteristics was between 50 and 70%. Major personality disorders (manic depression, schizophrenia) also have significant genetic basis.

Individual study cases cannot be the basis for a strong conclusion. But taken together, the many cases suggest that genotypic differences may account for many, or even a majority of, the differences in behavior between individuals in human populations, including behaviors normally assumed to be under conscious control. It is becoming clear that the brain, like other organs of the body, is strongly influenced by the genotype, and these influences may shape our behavior to a much greater extent than was previously believed.

Summary

Quantitative genetics is the study of the genetic regulation of continuously variable, polygenic traits such as weight, height, or IQ. An individual's phenotype for a quantitative trait is influenced by the actions of the alleles of many different genes, and each gene makes a small contribution to the phenotype. On the other hand, the phenotype is also affected by the environment. The observed continuous distributions of phenotypes for quantitative traits reflects the complexity of the control of these traits in natural populations. Individuals in a population may have different alleles for hundreds of genes. The segregation of individual alleles that affect quantitative traits usually cannot be followed from generation to generation by Mendelian analysis because each individual allele makes such a small contribution to the overall phenotype. As a result, quantitative traits are studied using statistical analysis of numerical measurements of the phenotype of populations of individuals or samples taken from these populations.

Two important descriptive statistics of populations and samples are the mean and the phenotypic variance. Mean values can be compared using t tests to determine whether two populations are similar or different for the trait. In addition, comparisons of two traits, or of one trait in two populations, using correlation and regression analysis indicate whether two traits are related, and if so, they allow the value of one trait to be predicted from the phenotype of the other trait.

Analysis of the phenotypic variance gives information about the different influences that are contributing to the overall differences in phenotype in the population. Two main components of phenotypic variance are the genetic variance and the environmental variance. The genetic variance can be subdivided into different components that reflect the different types of alleles in the population. For example, the additive and dominant components of the genetic variance reflect the contributions of alleles with additive or dominant effects on the trait. The genetic variance can be used to calculate the heritability of the trait. Heritability measures the ratio of genetic variance to total variance and is used to predict the response of the trait to selection based on the phenotype. Heritability is widely used in analyses of agricultural plant and animal populations to predict the phenotypes of the offspring of selected parents. Heritability is also used in human genetics to estimate what percentage of the variation in complex traits, such as IQ or psychological disorders, might be genetically controlled.

Selected Key Terms

additive alleles
correlation coefficient
coefficient of regression
correlation
covariance of x and y
degree of genetic determination
environmental variance
frequency distribution
genetic variance
heritability
interaction variance
mean
phenotypic selection
phenotypic variance
polygenic inheritance
qualitative trait
quantitative trait inheritance
random sample
scatter plot
standard deviation

Chapter Review Problems

1. Samples of 20 individuals were taken from two populations of *Drosophila*. The number of bristles on the first leg tarsus was determined for each individual. Given the following results, is there a significant difference between these two populations?

	Mean	s_x	*n*
Pop. 1	213.1	4.4	20
Pop. 2	198.0	3.2	20

2. In a study of the inheritance of weight in swine, a cross was made between two breeds, Ph and Pl. The F_1 were inbred to produce an F_2, which were also backcrossed to the Ph strain (B_1) and to the Pl strain (B_2). The means and variances of samples of each genotype follow:

	Number	Mean	Variance
Ph	621	218	23
Pl	499	306	28
F_1	802	281	29
B_1	708	256	56
B_2	656	295	48
F_2	922	275	76

From these data, calculate the values of V_e, V_a, V_d, H^2, and h^2.

3. Two inbred lines of corn that differ in the number of leaves per plant were crossed. The mean number of leaves and the variances for different generations follow:

	Mean	Variance
P_1	11.0	2.6
P_2	14.5	2.8
F_1	12.6	2.7
F_2	12.8	7.9
B_1	11.9	5.5
B_2	13.1	5.8

Given that these plants were raised in the same environment, calculate the values of V_e, V_a, V_d, V_g, H^2, and h^2.

4. In a population of *Drosophila* isolated from the wild, the mean number of sternoplural bristles was 20.0. Two flies with 25 sternoplural bristles were selected from this population, mated, and the number of sternoplural bristles on the progeny were

19, 20, 20, 20, 20, 21, 21, 21, 21, 22, 22, 22, 22, 22, 22, 22, 22, 23, 23, 23, 23, 24, 24

What is the heritability (h^2) for bristle number?

5. In an effort to produce smaller horses, 100 individuals were selected with a mean height of 48 in. from a population with a mean height of 60 in. The progeny of the selected individuals had a mean height of 55 in.
   a. What is the selection differential?
   b. What is the selection response?
   c. What is the realized heritability?
6. Explain why an estimate of h^2 is valid only for the particular population and environment in which it is measured.
7. Three genes that increase height in snouters above the base height of 24 in. have been identified. Each has two alleles; one increases height by 3 in., and the other increases height by 1 in. If these alleles are additive, and all alleles are equally frequent in the population, what heights will be present in this population, and what will be the frequency of each height?
8. The following two samples were drawn from two populations.

   Sample A = 12, 12, 8, 14, 12, 10, 9, 7, 13, 9, 10, 12, 9, 10, 11

   Sample B = 13, 9, 14, 12, 10, 13, 11, 11, 14, 12, 9, 13, 8, 10, 11

   Do these two samples differ statistically for this trait?
9. Three pairs of genes with three alleles each (*X x*, *Y y*, and *Z z*) determine corn ear length additively in a population. The homozygote *X X Y Y Z Z* has ears 24 cm long, and the homozygote *x x y y z z* has ears 12 cm long.
   a. What is the F_1 ear length in a cross between the two homozygous stocks?
   b. Assuming all alleles contribute equally to ear length, how much does each allele contribute in centimeters?
10. You would like to improve the length of maize ears, and you know that length is a polygenic trait with an heritability of 0.70. Your maize population has a mean ear length of 20 cm. You select two individuals from this population that have ear lengths of 22 and 24 cm and breed them. What ear length would you expect in the population resulting from the mating of these two maize plants? (Hint: $R = S \times h^2$.)
    a. 22.1 cm
    b. 21.1 cm
    c. 22.5 cm
    d. 23 cm
    e. 24 cm
11. Which of the following is *not* a quantitative trait?
    a. ear length in corn
    b. skin color in humans
    c. leaf size in tobacco
    d. height in humans
    e. eye color in humans
12. You are a soybean breeder and want to increase the yield of soybeans from 40 to 50 bu/acre. You know from previous experiments that yield has an heritability of 0.50. Assuming that you can find a population of plants that has a yield 20% greater than the mean of the previous population, how many generations would it take you to obtain a yield that exceeds 50 bu/acre?
    a. one generation
    b. two generations
    c. three generations
    d. four generations
    e. greater than five generations
13. If a polygenic trait is controlled by four genes, each with one additive allele and one recessive allele, and individuals heterozygous at all loci are self-mated, what proportion of the total progeny is expected to be homozygous for all additive alleles?
14. You would like to improve the weight of hamsters, and you know that weight is a polygenic trait with an heritability of 0.85. Your hamster population has a mean weight of 65 g. You select two individuals from this population that have weights of 86 and 84 g and breed them. What weight would you expect in the population resulting from the mating of these two hamsters?
15. Some farmers living in northern Iowa bought from their local supplier seed corn that was guaranteed to have a high mean yield, and each of them planted the corn. At the end of the season, they noted that the yields varied widely, from very high to relatively low. They calculated the variance from these data, submitted it to the seed corn dealer, and demanded their money back. Do they have a case?
16. What is the broad-sense heritability of a population of individuals homozygous for all genes determining a trait?
17. What is the difference between the variance and the covariance?
18. In a cross between two inbred lines of tomatoes, the variance of leaf number per plant in the F_1 generation (V_{F1}) is 1.46, and in the F_2 generation (V_{F2}) it is 5.97. What are the genotypic and environmental variances? What is the broad-sense heritability in leaf number?
19. A flock of chickens has a mean weight gain of 700 g at age 8 weeks. The narrow-sense heritability is 0.80. Assuming a weight gain of 50 g at 8 weeks for each selection cycle, what will be the weight at 8 weeks after five selection cycles.
20. Which of the following traits would you expect to be inherited as quantitative traits? Briefly explain why.
    a. body weight in humans
    b. eye color in humans
    c. red vs. white flower color in roses
    d. milk yield in cattle
    e. rate of growth of mice
21. If two strains of sunflowers differ for a genetically determined trait that gives an altered visible phenotype, how would you determine whether this was a quantitative or a qualitative trait?
22. Two populations of mice (A and B) are homozygous for different, additive alleles of four genes that control body weight. The alleles from the A strain each cause an addition of 1 g to the final weight, and the alleles from the B strain cause no addition. If these strains are crossed, and the F_1 are inbred, how many phenotypic classes would be present in the F_2, and what would they be?

23. The weights (in pounds) of a random sample of men follow.

    145, 128, 178, 162, 155, 161, 175, 191, 182, 148

    Use these weights to calculate the mean weight, the phenotypic variance, and the standard deviation for these men.

24. The mean heights (in inches) of two strains (A and B) of Siberian Iris measured on May 12 follow.

    A: 20, 18, 19, 15, 20, 21, 18, 19, 19, 21, 20, 20, 19, 18, 22

    B: 17, 17, 16, 18, 19, 16, 15, 18, 15, 17, 18, 16, 18, 16, 17

    Are these two strains significantly different for height? Explain your answer.

25. The following data table shows the final mean heights of a series of controlled crosses of inbred lines of sunflowers. What is the correlation between the parents and offspring for this trait?

Parental Mean (in.)	Offspring Mean (in.)	Parental Mean (in.)	Offspring Mean (in.)
69	66	69	68
68	67	67	65
71	68	68	67
66	64	72	69
66	65	69	67
69	67	70	69
71	68	70	66
69	65		

26. a. What is the coefficient of regression for the parents and offspring in Problem 25?
    b. What would be the expected height of the offspring of parent plants with height of 75 in.?

27. The variances in size of feet and middle fingers were measured in a sample of college students. What is the heritability (h^2) for these traits? Which trait would respond more quickly to selection?

	Foot Length (in.)	Finger Length (in.)
Phenotypic variance	3.46	0.67
Environmental variance	0.44	0.11
Dominant variance	0.57	0.28
Additive variance	2.45	0.28

28. A population of experimental fish (saugeye) were determined to have a mean length of 14 cm at 41 days of growth. A small group that had a mean length of 18 cm was selected from this population. These fish were bred and their offspring had a length of 17 cm after 41 days of growth.
    a. What is h^2 for growth for this population?
    b. If a set of fish were selected from the original population with a length of 21 cm after 41 days, what would be the length of their offspring?
    c. A small group of fish escaped from the original population and spawned in a different pond. When these offspring were discovered, they had a length of 19.5 cm at 41 days. What was the length of the group of fish that escaped?

29. Human twin studies have suggested that human IQ is partly determined by genotype. If you were performing a large-scale analysis of human IQ, describe how you might determine an estimate for h^2 for this trait.

Challenge Problems

1. A desirable trait in pigs is a high percentage of lean muscle (%L). A farmer wishes to improve his herd by selecting for %L of 65. An undesirable trait in pigs is susceptibility to porcine stress syndrome (PSS). Pigs show symptoms of this disease at different ages, but an animal cannot be sold after it shows symptoms. Given the following characteristics of the herd, would you recommend the farmer do this selection? If not, how would you recommend that he get his herd to %L of 65? The market age for these pigs is 12 months.

L (%)	Age of Onset (months)	L (%)	Age of Onset (months)
48	20	63	12
52	18	65	10
55	20	67	11
57	15	72	9
60	14	75	8

2. Two lines of rye were compared with respect to their maturity times. One line matured in 30 days, whereas another matured in 24 days. The breeder conducting this experiment crossed these two lines, and then selfed the F_1 generation. She examined the F_2 plants for maturity rates and found that 50 of the 3 million plants matured at 30 days, whereas the remainder all matured at an earlier time. From these data, estimate the number of additive alleles that contribute to maturation in rye.

3. Two inbred lines of corn both had a height of 55 in., under the same environmental conditions as in Challenge Problem 2. These two lines were crossed and found to yield plants that were nearly uniform in height at 80 in. This F_1 line was inbred and yielded a few plants that were 100 in. in height. Propose an hypothesis to explain these data using the additive gene concept as well as the dominant gene concept.

Suggested Readings

Cavalli-Sforza, L. L., and W. F. Bodmer. 1971. *The Genetics of Human Populations*. W. H. Freeman, San Francisco.

Falconer, D. S. and T. F. C. Mackey. 1996. *Introduction to Quantitative Genetics*, 4th ed. Longman Scientific and Technical, London.

Farber, S. 1981. *Identical Twins Reared Apart: A Reanalysis*. Basic Books, New York.

Fisher, R. A. 1925, 1970. *Statistical Methods for Research Workers*. Oliver and Boyd, Edinburgh.

Goodman, M. M. 1990. "Genetic and Germ Plasm Stocks Worth Conserving." *Journal of Heredity*, 81: 11–19.

Herrnstein, R. J., and C. Murray. 1994. *The Bell Curve: Intelligence and Class Structure in American Life*. The Free Press, New York.

Sokal, R. R., and F. J. Rohlf. 1981. *Biometry: The Principles and Practice of Statistics in Biological Research*, 2nd ed. W. H. Freeman, San Francisco.

Shands, H. L., 1990. "Plant Genetic Resources Conservation: The Role of the Germ Bank in Delivering Useful Genetic Materials to the Research Scientist." *Journal of Heredity*, 81: 7–10.

Stern, C. 1973. *Principles of Human Genetics*, 3rd ed. W. H. Freeman, San Francisco.

On the Web

Visit our Web site at **http://www.saunderscollege.com/lifesci/titles.html** and click on A/G/M Genetics for links to the following chapter-related resources on the World Wide Web:

1. **Resource.** This is a quantitative genetics resource.
2. **Center for Quantitative Genetics.** This is the center for quantitative genetics at North Carolina University.
3. **Research Center.** This is the home page for the Roslin Institute, a major center for research on molecular and quantitative genetics of farm animals.

CHAPTER

6

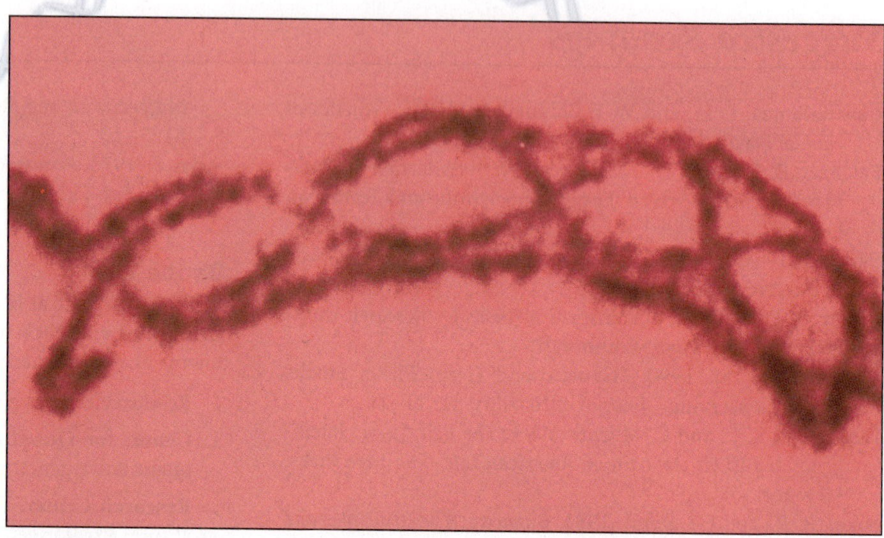

A photograph showing chiasma, the physical evidence of recombination between chromosomes. Recombination is the basis for gene mapping. *(B. John/Cabisco/Visuals Unlimited)*

Linkage and Gene Mapping in Eukaryotes

The first years of the 20th century were a time of extensive genetics investigation. Partly sparked by rediscovery of Mendel's paper and partly by Sutton and Boveri's proposal of the chromosome theory, many biologists began to do genetic crosses using many different alleles, sometimes for one, two, or more genes. In most of these experiments, the alleles of different genes showed independent assortment in crosses, as predicted by Mendel's second rule. However, in some cases they did not, and again, as mentioned in Chapter 4, these exceptions to the rule resulted in some revolutionary findings as to how genes are arranged on chromosomes. Sometimes the alleles of two different genes tended to pass together from parent to offspring with no segregation, as predicted by Mendel's second rule. When this happened, the genes were said to be linked in some way. Thomas Hunt Morgan and his students suggested that the alleles of linked genes were inherited together because they were on the same chromosome and consequently could not segregate. This and other important observations about the relationship between genes and chromosomes were made by Morgan and his students. The phenomenal success of Morgan's group revolutionized genetics and established *Drosophila melanogaster* as one of the premiere organisms for genetic research, and it remains as one of the most thoroughly studied and understood higher eukaryotes to this day.

Morgan proposed that each gene resides at a particular location on a chromosome, and that during meiosis physical recombination (called meiotic crossing over) occurs between paired homologous chromosomes. This crossing over sometimes leads to an exchange of alleles between the chromosomes. Beginning in 1912, Morgan's studies of meiotic crossing over led to the development of techniques for defining the distance between genes on chromosomes, which in turn led to making genetic maps of the location of genes on chromosomes. The mapping of genes on chromosomes of eukaryotes has had an enormous impact on genetics and biology. Biologists stopped envisioning genes as abstract, invisible entities associated with chromosomes in some mysterious way and began to understand them as entities that had precise, physical locations on chromosomes. This, in turn, led to the revolutionary conclusion that genes had a physical structure. This recognition initiated an enormous effort to determine what the physical structures of genes and chromosomes are and how they function, an effort that continues today. Today there are genetic maps showing the location and function of known genes for many organisms. Any analysis of the genes of any fungi, plant, or animal today invariably includes determining the location of that gene on the genetic map of its chromosome. Accurate genetic maps allow geneticists to plan sophisticated analyses of genes and chromosomes. Like any road map, genetic maps provide a means of locating the gene in future experiments. Genetic maps also link the physical structure of DNA in a chromosome with the genes.

In this chapter, we will discuss gene linkage and techniques for using linkage to map the location of genes on chromosomes. This was the original genetic mapping technique. We will also discuss a number of other techniques that are used to make genetic maps today, including techniques for mapping human genes to chromosomal locations and for mapping the distances between alleles within genes. By the end of this chapter, we will have addressed six specific questions:

1. What is meant by genetic linkage?
2. How are linkage and crossing over between chromosomes related to meiotic events?
3. How can meiotic crossing over be used to make maps showing the location of a gene?
4. How do geneticists use genetic maps?
5. What other techniques are available for making genetic maps?
6. How can we map the internal structure of a gene?

Recombination Is the Basis of Gene Mapping.

Genetic linkage of two genes was first discovered by W. Bateson and R. C. Punnett in 1905 using the sweet pea. They crossed a plant with purple flowers and long pollen to a plant with red flowers and round pollen. According to Mendel's rules, the F_2 should have contained 9/16 purple and long, 3/16 purple and round, 3/16 red and long, and 1/16 red and round. Instead, they observed the F_2 contained 12/16 purple and long, 1/16 purple and round, 1/16 red and long, and 3/16 red and round. This ratio indicated that the F_1 individuals had produced about seven times as many gametes with red-round and purple-long genotypes as red-long and purple-round genotypes. This was a clear violation of Mendel's rule of independent assortment (Figure 6.1). The tendency of alleles of two different genes to segregate together during meiosis is called **linkage,** and genes whose alleles show linkage are said to be linked genes. Geneticists working with other organisms, such as mice, corn, and humans, soon discovered other cases of linked genes, and linkage was recognized as a widespread and important phenomenon.

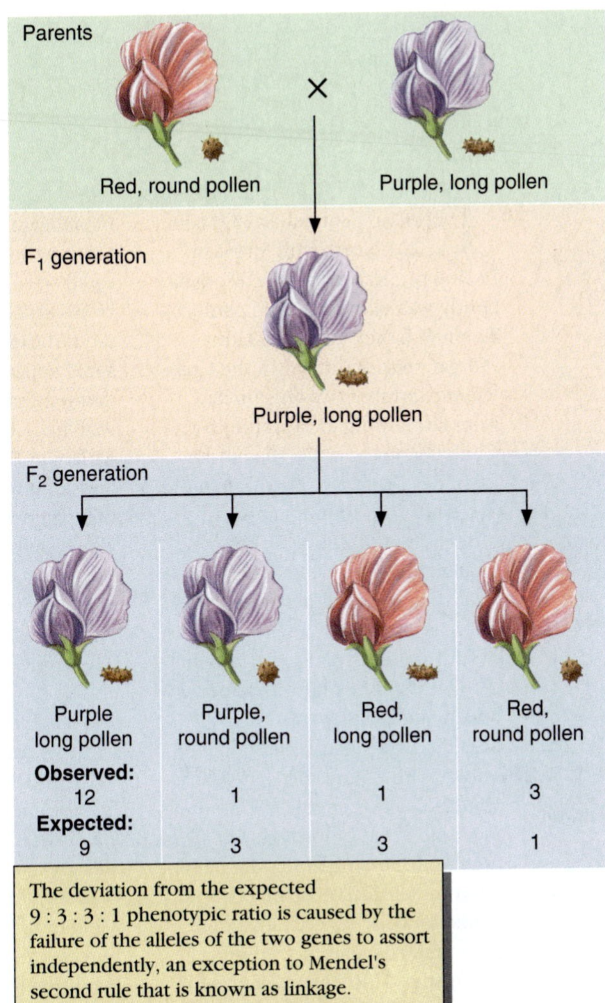

FIGURE 6.1 **Linkage in a dihybrid cross.** Bateson and Punnett discovered linkage in a cross between sweet pea plants. The phenotypic ratio of the F_2 generation was very different from the expected values according to Mendel's rules.

Beginning in 1910, a large number of genetic studies done by T. H. Morgan and his group [see HISTORICAL PROFILE: *Thomas Hunt Morgan (1866–1945)*] demonstrated that linkage was the result of two genes being on the same chromosome, and that the tendency of pairs of alleles to segregate together was the natural consequence of chromosomal segregation in meiosis.

The Alleles of Two Linked Genes Usually Segregate Together in Meiosis.

To illustrate the basic principle of genetic linkage, let us consider a dihybrid cross involving two genes, *A* and *B*, that are located on the same chromosome. The heterozygous alleles of these genes can be arranged in two different ways on the chromosome pairs. For example, if an individual homozygous for the dominant alleles of both genes (*AA BB*) is crossed with an individual homozygous

HISTORICAL PROFILE

Thomas Hunt Morgan (1866–1945)

Thomas Hunt Morgan was born on September 25, 1866, at 201 North Mill Street in Lexington, Kentucky. The Morgan family was of high social position. Thomas' father, Charlton Hunt Morgan, was an officer in the Confederate army, as was his uncle, John Hunt Morgan (a well-known Confederate general). His mother, Ellen Key Howard, was from an old aristocratic Baltimore family. Although they had suffered serious financial losses during the Civil War, they were not destitute, and Thomas grew up in an atmosphere of southern aristocracy. During his early years in Lexington, he developed a life-long love for the outdoors. He attended Lexington public schools and the Kentucky State College at Lexington, graduating at the head of his class with highest honors in 1886. Morgan entered Johns Hopkins University for graduate work, where he studied embryonic development of marine organisms. This was an exciting period in biology because a transition was underway from the earlier purely descriptive type of biological studies to an active, experimental approach. Morgan became a dedicated experimentalist.

In 1891, after completing his Ph.D. thesis on the development of sea spiders, Morgan accepted a position on the faculty of Bryn Mawr College, where he remained for 13 years. During this period, he published 103 articles and 3 books, most dealing with embryology and regeneration. Morgan was a dedicated scientist who was well liked by his students and colleagues. He did not care greatly about his personal appearance, and he often dressed shabbily, even though he was financially well off. He traveled widely in Europe during the summers and was especially fond of doing experimental work at the Naples marine research station. This led to his being an early and active supporter of the marine biological station at Wood's Hole, where for many years he and his entire lab spent every summer. He was a member of the Wood's Hole board of trustees from 1897 until 1937. In 1900, Morgan traveled to Europe and visited De Vries and the Naples station. In both places there was great excitement over the rediscovery of Mendel's paper, which led Morgan to begin thinking about heredity and the genetic basis for evolution and developmental phenomena such as sex determination. He was impressed by the suggestion that mutations could lead to evolutionary changes, and he decided to begin to study mutations.

In 1904, Morgan made two major changes in his life. First, he married Lilian Sampson, who had been a student and graduate student at Bryn Mawr. Although she was an intelligent and talented research scientist, Lilian devoted herself to her husband and to her children. Later, after the children were older and in school, she returned to lab work and made a number of important discoveries. Second, he accepted a position as professor of Biology at Columbia University, where he became interested in questions of Mendelian genetics, chromosomes, sex determination, and evolution. It was characteristic of Morgan that his interests ranged over many subjects, and he was always ready to take up a new problem. He began to search for ways to induce mutations, and for a suitable organism for such studies. In 1908, he began to experiment with *Drosophila*, being attracted by the ability to rear many generations quickly and cheaply in a small laboratory space. In 1910 he recovered his first mutation, *white*. Other mutations quickly followed, and Morgan was on his way to a revolution in biology.

At Columbia, Morgan and his talented students A. H. Sturtevant, C. B. Bridges, and H. J. Muller, worked in a small (16 × 23 feet) room in Schermerhorn Hall known as "the fly room." Sturtevant and Bridges remained collaborators with Morgan for many years. Together with a steady stream of graduate students and postdoctoral associates, they isolated and analyzed hundreds of *Drosophila* mutations. Much has been said about the spirit of cooperation and easy informality

for the recessive alleles of both genes (*aa bb*), the F_1 individuals will have two dominant alleles on one chromosome and two recessive alleles on the other. This arrangement is called the ***cis* configuration,** or coupling. The alternative arrangement, in which each chromosome has one recessive allele and one dominant allele, is called the ***trans* configuration,** or repulsion.

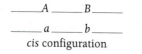

___*A*___*B*___ ___*a*___*b*___	___*A*___*b*___ ___*a*___*B*___
cis configuration	*trans* configuration

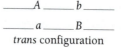

When two alleles are in the *cis* configuration, during anaphase I of meiosis, the homologous chromosomes will segregate, and their genes will segregate with them. The *A* and *B* alleles, being on the same chromosome, will segregate together, and the *a* and *b* alleles will also segregate together (Figure 6.2). When these cells complete meiosis and form gametes, the gametes will have either an *AB* or *ab* genotype. Because these are the same combinations of alleles present in the chromosomes of the parental individuals, they are called **parental type** gametes. If these alleles do not segregate together, then ga-

FIGURE 6.A T. H. Morgan. This photograph was taken on the morning that Morgan learned he had won the Nobel Prize. He insisted on having his picture taken with the group of neighborhood children who had come to congratulate him. *(UPI/Corbis Bettmann)*

that existed in this group, and their scientific output was phenomenal (by the end of 1912 they had characterized over 40 mutations that produced visible phenotypes). This period culminated with the publication of a landmark book, *The Mechanism of Mendelian Heredity,* that provided definitive proof of the chromosome theory and presented for the first time the essentials of modern transmission genetics. From 1915 to 1930, Morgan's group extended their studies, establishing *Drosophila* as the premier organism for genetic research. As his many students left and established *Drosophila* labs throughout the world, Morgan received much recognition for his work. In 1927 he was elected president of the National Academy of Sciences, and in 1933 he was awarded the Nobel Prize.

In 1928, at the age of 62, Morgan left Columbia University to establish the Biology Division at the California Institute of Technology. Cal Tech was an outstanding research university in physics, chemistry, and engineering, and the opportunity to build a strong biology division from the bottom up appealed to Morgan. He had also been impressed with California and had often considered retiring to a place with such a mild climate. Morgan insisted that the division be founded on the principle of basic research and worked hard to attract the finest scientists to work there. He naturally brought the members of his fly room with him, although he himself had begun to leave the increasingly complex details of *Drosophila* genetic experiments to others and to return to his earlier experiments in embryology and regeneration in marine animals. Between 1928 and 1942, Morgan built the Biology Division at Cal Tech into a world-renowned research department. He greatly enjoyed California life, especially the opportunity for year-round outdoor activity. He formally retired in 1942 at the age of 76, but was active in research through the fall of 1945. On December 4, 1945, Morgan died of a ruptured artery in an old stomach ulcer.

In the course of his long career, Morgan made numerous contributions to science. The three greatest are his contributions to the establishment of the experimental school of biological inquiry, his founding and leading the *Drosophila* group that revolutionized genetics, and his development of a strong department dedicated to basic research at Cal Tech. Much of the experimental work associated with Morgan's lab was done by his students, whom he never formally tried to direct. He led by example, providing inspiration and enthusiasm, framing important questions, and then allowing the students to proceed without interference. He was not afraid to have his students be ahead of him, and on more than one occasion it was the students who explained important advances to him. His students remarked that he always knew where he wished to go but was prepared to leave the details of technique to others. He also provided an example of unceasing devotion to his work. One associate of his said, "I have never seen a man who wasted so little time."

metes will be formed that have either an *aB* or an *Ab* genotype. This is not the same combination of alleles present on the chromosomes of the parental individuals, and so these are called **recombinant type** gametes. It is important to understand that the terms "parental" and "recombinant" refer only to the arrangement of alleles in the chromosomes in one particular cross. If individuals that had alleles of these genes in *trans* (*Ab, aB*) were used as the parents in a cross, then gametes with *Ab* and *aB* genotypes would be the parental type, and gametes with *AB* and *ab* genotypes would be the recombinant type. Thus, the parental type of one cross may be the recombinant type of another cross. As a result of linkage of genes on chromosomes, there will be a higher frequency of parental type gametes than recombinant type gametes. The *cis* or *trans* arrangement of alleles in chromosomes does not affect our ability to detect linkage in a cross.

Morgan and others noticed that linked genes usually did not show complete linkage; that is, heterozygous individuals formed mostly gametes with parental genotypes but also formed a few with recombinant genotypes. Morgan proposed that chromosomes containing

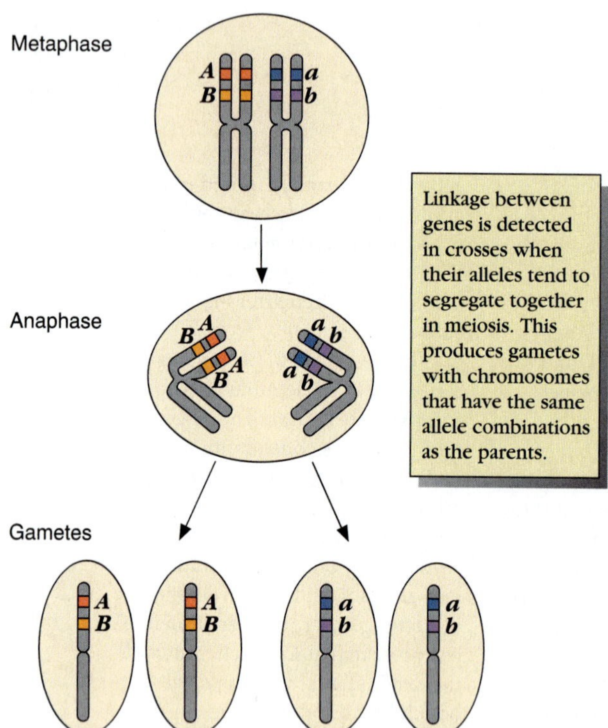

Metaphase
A
B
a
b
Linkage between genes is detected in crosses when their alleles tend to segregate together in meiosis. This produces gametes with chromosomes that have the same allele combinations as the parents.
Anaphase
Gametes

FIGURE 6.2 Segregation of linked genes in meiosis. The alleles of two linked genes (*AB* and *ab*) that are on the same chromosome segregate together in meiosis.

recombinant type allele arrangements were produced during meiosis by a recombination event that involved a physical exchange between the two homologous chromosomes. This recombination event is called **meiotic crossing over.** A meiotic crossover generates chromosomes with new combinations of alleles, but only if the recombination event occurs **between** two genes on a chromosome (Figure 6.3). If the *A* and *B* alleles from our previous example are in *cis* (*AB* and *ab*) configuration, then meiotic crossing over on the chromosome between the *A* and *B* alleles will generate recombinant type chromosomes with *aB* and *Ab* allele arrangements. If complete linkage occurred, then only *AB* and *ab* allele arrangements would be found on the chromosomes.

If crossing over occurs between two linked heterozygous genes, then gametes with four different genotypes will be produced: *AB, ab, Ab,* and *aB*. If the frequency of crossing over between *A* and *B* is low, then most gametes will be parental types and only a few gametes will be recombinant types. This is called **partial linkage,** or incomplete linkage. By definition, genes show partial linkage when more parental type gametes are produced than recombinant type gametes. Most linked genes show partial linkage, and in fact, the ratio of parental to recombinant type gametes is used as the definitive test for linkage. The actual ratio of parental to recombinant type gametes defines the strength of the linkage between the two genes.

Crossing over occurs in most organisms, often with high frequencies. If the frequency of crossing over is high enough, heterozygous individuals will produce an equal number of parental and recombinant type gametes; that is, the four different genotypes of gametes (*AB, ab, Ab,* and *aB*) will be present in equal numbers. In this example, crossing over has occurred at high frequency, and the parental types equal the recombinant types. This is the same ratio of gamete types that occurs when two genes are on different chromosomes and are assorting independently. Thus, these two genes will appear to be unlinked. This explains how two genes could be on the same chromosome and yet still show independent assortment in crosses. There are many examples of genes that are on the same chromosome but appear to be unlinked because the frequency of crossing over between them is very high.

The discovery of linkage and crossing over required the invention of new symbolism for describing geno-

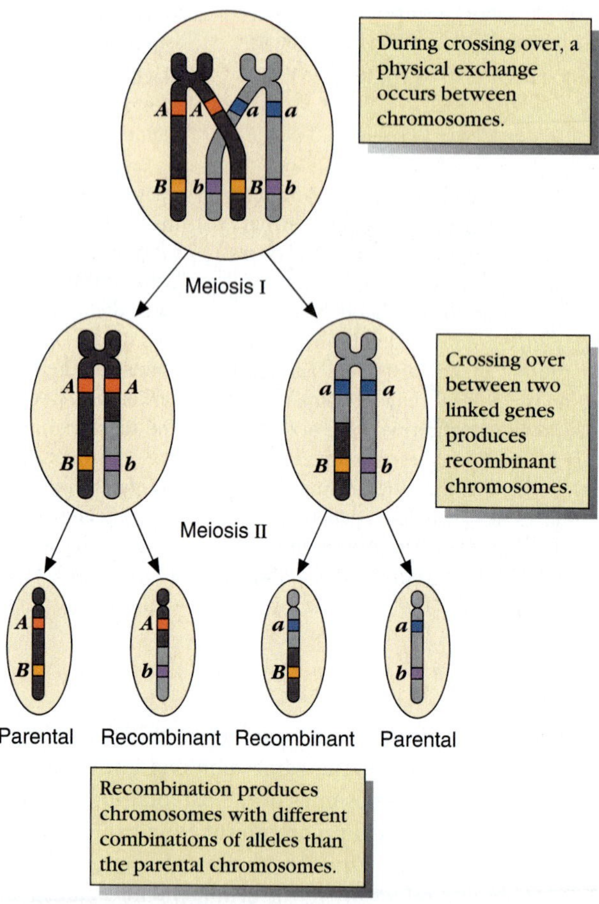

FIGURE 6.3 Chromosome crossing over produces new allele combinations. Recombinant chromosomes are produced by crossing over between homologous chromosomes in prophase of meiosis.

types. This new symbolism includes information about the arrangement of alleles on the chromosomes. For example, suppose that we have a female *Drosophila* that has recessive alleles of the *yellow* and *forked* genes on one X chromosome and recessive mutations of the *white* and *singed* genes on the other. We could describe this genotype graphically by drawing lines that indicate the chromosomes and indicating the position of each mutant allele and wild-type allele:

___ *y* ___ + ___ + ___ *f* ___

___ + ___ *w* ___ *sn* ___ + ___

When describing genotypes in a written text it is inconvenient to draw graphic symbols, so mutant alleles of genes on the same chromosome are simply written next to each other, separated by a space. The symbols of the mutant alleles on one chromosome are separated from the symbols of the alleles on the other homologous chromosome by a slash (/). For example, the heterozygote shown earlier would have a wild-type phenotype, and its genotype would be shown as: *y f / w sn*. The symbols for the wild-type alleles (+) are sometimes not written because it is assumed that all alleles not mentioned are wild-type. Thus we need not write the symbols for the w^+ and sn^+ alleles on the *y f* chromosome, nor for the y^+ and f^+ alleles on the *w sn* chromosome. For species such as corn (*Zea mays*), where the + symbol is not used, or for cases where the individual has two different mutant alleles of a gene, the allele designations of both alleles are listed. When describing genotypes that contain mutations on more than one pair of chromosomes, the symbols for the mutant alleles on one pair of homologous chromosomes are separated from the alleles on different chromosomes by a semicolon (;). For example, if the female from the preceding example was also heterozygous for mutant alleles of the autosomal *black* (*b*) and *vestigial* (*vg*) genes and these mutant alleles were in *trans*, the genotype could be written as: *y f / w sn; b / vg*.

SOLVED PROBLEM

Problem

The following chromosome diagrams represent the genotypes of several different strains of different species. Indicate what the genotype would be for each using proper allelic notation.

a. *Drosophila*

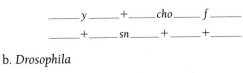

___ *y* ___ + ___ *cho* ___ *f* ___

___ + ___ *sn* ___ + ___ + ___

b. *Drosophila*

___ f^{36} ___ ___ bw^1 ___ + ___ ___ e^s ___

___ f^1 ___ ___ + ___ *Cy* ___ ___ e^2 ___

c. *Zea mays*

___ bz^1 ___ ___ def^2 ___

___ bz^3 ___ ___ def^{27} ___

Solution

For each genotype the alleles on each homologous chromosome are listed and separated by a slash. For the *Zea mays* strains the + symbol is not used, so all alleles are listed.

a. *y cho f / sn*

b. f^{36} / f^1; bw^1 / *Cy* ; e^s / e^2

c. bz^1 / bz^3 ; def^a / def^{27}

SECTION REVIEW PROBLEM

1. For each of the individuals whose genotype is listed here, draw a diagram showing the chromosomes and arrangement of the alleles on the chromosomes. Assume that the genes are in the order listed.

a. *Drosophila*	*y w / os* ; p^P / *bx pbx*
b. *Zea mays*	*A D rg / a d Rg ; y sh C / Y Sh c*
c. *Danio rerio*	*wnt1 / cyc ; br3 / +*
d. *Drosophila*	*Cy tri / wtl ; ttr / wz ; pho / ey*

Linkage Is Easily Detected in an F_1 Testcross.

To determine whether any two genes are linked, a heterozygote must be generated, and the pattern of allelic segregation in meiosis must be determined. If the genes are linked, more parental type gametes will be produced than recombinant type gametes. However, it is usually not possible to determine directly which alleles are present in gametes. For most plant and animal species, the heterozygous F_1 individuals must be crossed, the phenotypes of the resulting progeny determined, and the genotypes of the F_1 gametes deduced from the progeny phenotypes. The cross that is greatly preferred for determining whether two genes are linked is the F_1 testcross. In an F_1 testcross, one parent is heterozygous for alleles of the genes being tested, and the other parent is homozygous for the recessive alleles. The homozygous parent produces gametes containing only recessive alleles, so the phenotypes of the progeny will be determined by the genotypes of the gametes contributed by the heterozygous parent. If the gamete contributed by the heterozygous parent contains a dominant allele, then the offspring will have a dominant phenotype, and if the gamete contains a recessive allele, the offspring will have a recessive phenotype. This direct relationship between the genotype of the gametes produced by the heterozygous parent and the phenotype of the offspring makes it much easier to deduce the progeny genotype from their phenotype. This, in turn, can be used to deduce the genotypes of the gametes produced by the heterozygous parent.

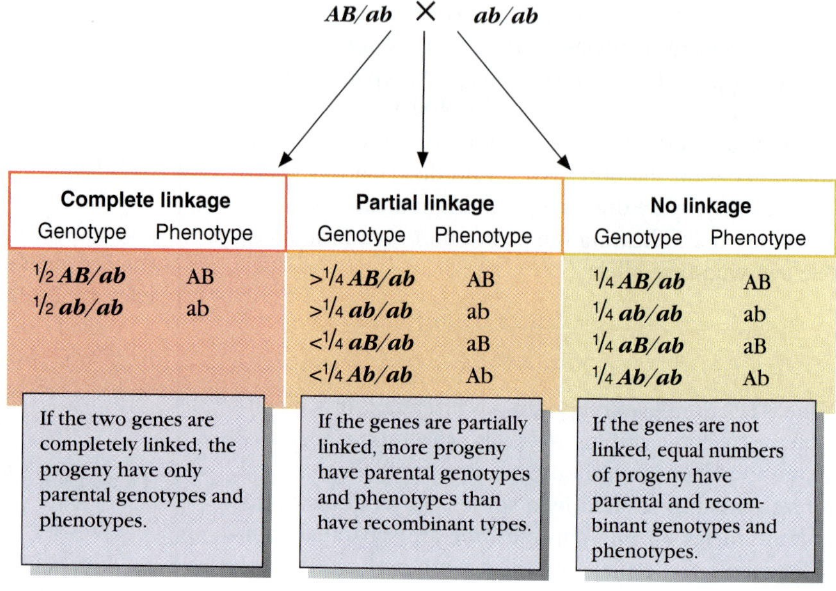
AB/ab × ab/ab
Complete linkage
Genotype Phenotype
1/2 AB/ab AB
1/2 ab/ab ab
If the two genes are completely linked, the progeny have only parental genotypes and phenotypes.
Partial linkage
Genotype Phenotype
>1/4 AB/ab AB
>1/4 ab/ab ab
<1/4 aB/ab aB
<1/4 Ab/ab Ab
If the genes are partially linked, more progeny have parental genotypes and phenotypes than have recombinant types.
No linkage
Genotype Phenotype
1/4 AB/ab AB
1/4 ab/ab ab
1/4 aB/ab aB
1/4 Ab/ab Ab
If the genes are not linked, equal numbers of progeny have parental and recombinant genotypes and phenotypes.

FIGURE 6.4 Linkage detection using a testcross. Linkage can be detected by observing the numbers of each of the different types of progeny of a testcross. Note the difference between complete, partial, and no linkage with respect to the ratio of the genotypes.

To illustrate how a testcross works, consider our example of the *AB* / *ab* heterozygotes. These heterozygous individuals are testcrossed by mating them to *ab* / *ab* individuals (Figure 6.4). If the two genes are completely linked, half of the progeny will have an *AB* phenotype, and half will have an *ab* phenotype. In this example, the parental types are *AB* and *ab*, so if the *A* and *B* genes are partially linked, the progeny will show more of the parental type phenotypes than the recombinant types: $(AB + ab) > (aB + Ab)$. Alternatively, the testcross could have been done equally well if the parental types had been *aB* and *Ab* and the recombinant types, *AB* and *ab*. In this cross, the genotypes of the F_1 individuals are *aB* / *Ab*, and if the genes are linked, these individuals will produce an excess of *aB* and *Ab* gametes, and there will be more progeny with *aB* and *Ab* phenotypes than with *AB* and *ab* phenotypes. If these two genes are not linked, then there will be four equally frequent phenotypes in the progeny from both types of testcross: (1/4 *AB*, 1/4 *Ab*, 1/4 *aB*, 1/4 *ab*). This direct relationship between the genotypes of the gametes in the heterozygous individual and the phenotypes of the progeny is what makes the testcross such a useful tool for determining whether two genes are linked.

To examine a real example of linkage, let us look at one of Morgan's experiments in which he studied the linkage between alleles of the *black body* (*b*) gene and the *vestigial wing* (*vg*) gene in *Drosophila*. Two crosses were done in this experiment, one with heterozygous females with the mutant alleles in the *cis* arrangement, *vg b* and + +, and the other with heterozygous females with the alleles in the *trans* arrangement, *b* + and + *vg*. In both experiments, the heterozygous F_1 females were testcrossed by mating them to males homozygous for both recessive mutant alleles (*vg b* / *vg b*). Morgan was careful to always use heterozygous females because he had discovered that if males are used as the heterozygous parent, the genes always show complete linkage. This is a curious fact about the biology of *Drosophila*. Meiotic recombination simply does not occur in male *Drosophila*. The results of Morgan's two crosses were nearly identical (Figure 6.5). About 83% of the progeny of the testcross had parental type phenotypes (*vg b* and + + in the first cross and *vg* + and + *b* in the second cross), and 17% had recombinant phenotypes (*vg* + and + *b* in the first cross and *vg b* and + + in the second cross). These values represented a significant deviation from the phenotypic ratios expected for independent assortment, so Morgan concluded that the genes are partially linked. These two crosses also illustrate the point made earlier that either arrangement of alleles (*cis* or *trans*) can serve as the parental type and not affect the outcome of the results.

Linkage Can Be Detected in an $F_1 \times F_1$ Cross.

Linkage is easily detected in F_1 testcrosses because a testcross gives simple expected phenotypic ratios in the progeny (1 : 1 : 1 : 1), and deviations are readily apparent. Linkage can, however, sometimes be detected in $F_1 \times F_1$ crosses. If we were to cross two heterozygous individuals who had the mutant alleles in *cis* (*ab* / *AB* with *ab* / *AB*), the F_2 will be 9/16 *AB*, 3/16 *Ab*, 3/16 *aB*, and 1/16 *ab*, assuming *A* and *B* show complete dominance. If *A* and *B* are linked, the heterozygotes will produce an excess of gametes with parental genotypes (*ab* and *AB*) and a decreased number of gametes with recombinant genotypes (*aB* and *ab*). This will change the frequencies of each of the four phenotypic classes in the progeny. A significant deviation of the progeny phenotypes from the expected ratio caused by an excess of the parental types is evidence for linkage. A chi-square (χ^2) test can be used to determine whether the observed phenotypic ratio differs significantly from the expected ratio (see CONCEPT CLOSE-UP: *Detecting Linkage in an* $F_1 \times F_1$ *Cross*).

Experiments in different species have demonstrated that very few, if any, pairs of linked genes show complete linkage. Nearly every pair of linked genes shows partial linkage. Each pair of partially linked genes has a characteristic frequency of recombinant progeny that occurs again and again in crosses. Pairs of genes that give a low frequency of recombinant progeny are said to be tightly linked, and pairs of genes that give a higher frequency of recombinant progeny are said to be loosely linked. Each pair of linked genes falls into a group of genes that shows linkage to each other. These sets of genes are called **linkage groups,** and each linkage group corresponds to the genes located on one chromosome. *Drosophila* has four chromosomes and thus four linkage groups.

SOLVED PROBLEM

Problem

The *Drosophila* mutants *black body* (*b*) and *pupal wing* (*pu*) are both recessive alleles. Flies with the black body and the pupal wing phenotypes were crossed with wild-type flies (gray bodies and long wings). The F_1 females were crossed to males with black bodies and pupal wings. The phenotypes of the progeny of this cross follow. Are these genes linked?

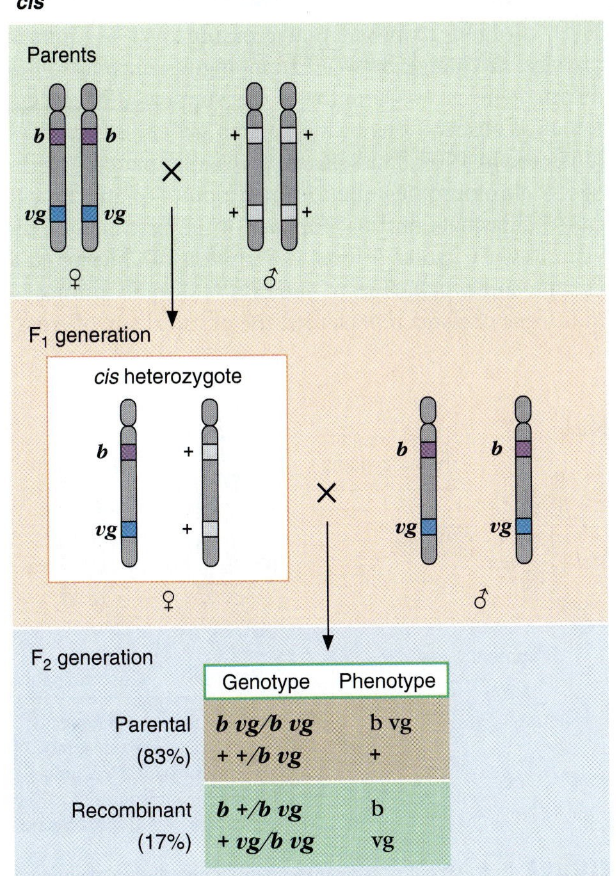

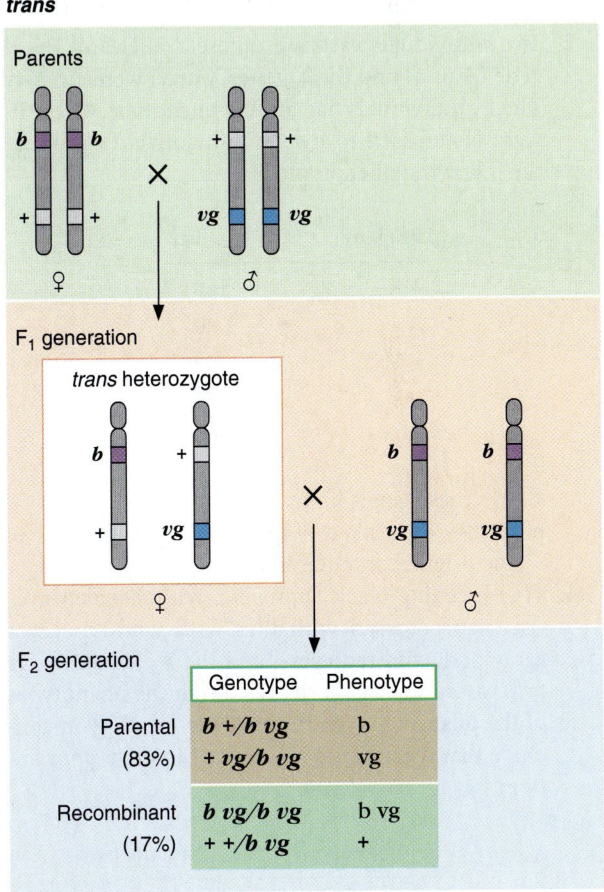

FIGURE 6.5 The *cis* and *trans* arrangements of alleles. The *cis* or *trans* arrangement of alleles does not affect linkage. In testcrosses of *Drosophila* heterozygous for *black body* (*b*) and *vestigial wings* (*vg*) done by T. H. Morgan, both arrangements gave 83% parental type progeny and 17% recombinant type progeny.

Progeny Phenotype	Number
Black pupal	485
Black long	13
Gray pupal	12
Gray long	490
Total	1000

Solution

If these genes assorted independently, then the four phenotypic classes of the progeny should be present in equal numbers (1/4 black pupal : 1/4 black long : 1/4 gray pupal : 1/4 gray long). We see that there are more parental types (black pupal and gray long) than recombinant types (gray pupal and black long). A chi-square (χ^2) test indicates these four progeny classes show a significant deviation from the 1 : 1 : 1 : 1 ratio expected from independent assortment ($\chi^2 = 169.47$, degrees of freedom = 3, probability < 0.001), so we conclude that these genes are linked.

SECTION REVIEW PROBLEMS

2. Two individuals carrying different alleles of the *A* gene (*A* or *a*) and the *X* gene (*X* or *x*) were crossed. The F_1 individuals had an A X phenotype. These F_1 were testcrossed to *a a x x* individuals to produce the following phenotypic classes.

Progeny	Number
A X	160
A x	40
a X	50
a x	150
Total	400

   a. Are these genes linked?
   b. If they are linked, what were the genotypes of the original parental individuals?

3. True-breeding black *Drosophila* with *cinnabar* eyes (*cn*) were crossed with flies with wild-type eyes (gray body and red eyes), and the F_1 females were testcrossed to *b cn* / *b cn* flies. From the phenotypes of the offspring given in this table, would you conclude that these genes are linked? Explain your answer.

F_2 Phenotype	Number
Black cinnabar	455
Black red	45
Gray cinnabar	45
Gray red	455
Total	1000

4. F_1 male *Drosophila* from a trihybrid cross between an *aa bb cc* strain and a wild-type were testcrossed. From the following results, which genes (if any) would you say are linked? (Hint: In male *Drosophila*, linkage is always complete.)

Phenotype	Number
+ + +	200
+ + c	0
+ b +	200
+ b c	0
a + +	0
a + c	200
a b +	0
a b c	200
Total	800

Meiotic Crossing Over Involves a Physical Exchange Between Homologous Chromosomes.

T. H. Morgan proposed that crossing over involves a physical exchange between homologous chromosomes during meiosis. His hypothesis was supported by the cytological observations of the Belgian geneticist Franz A. Janssens. In 1909, Janssens observed that paired homologous chromosomes often formed points of attachment called chiasmata during prophase of the first meiotic division (see Chapter 3 for a description of chiasma and chromosome behavior in meiosis). Morgan proposed that these chiasma represented the physical site of cross-

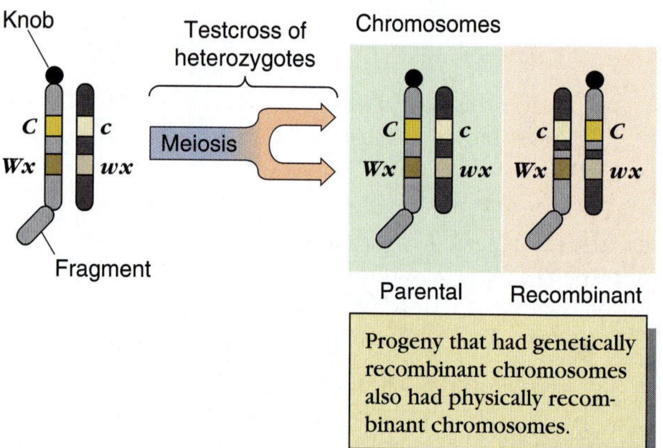

FIGURE 6.6 Proof of chromosome exchange during meiosis. McClintock and Creighton proved that physical exchange occurs between homologous chromosomes during meiotic crossing over. They testcrossed heterozygotes that had one chromosome with a cytologically distinct knob at one end and a chromosome fragment attached to the other end and observed that the physical attachment was inherited with the nearby gene.

Concept Close-Up

Detecting Linkage in an $F_1 \times F_1$ Cross

If two genes are showing independent assortment, then in an $F_1 \times F_1$ dihybrid cross each parent will produce four different types of gametes, and each will be equally frequent. The random combination of these gametes produces the 16 equally frequent genotypes in the F_2. If the genes are linked, then the gametes with parental genotypes will be present in greater numbers than gametes with recombinant type genotypes. This will change the numbers of progeny in each of the different genotypic classes. This can be illustrated using a Punnett square. Consider the results for two linked genes that have 10% crossing over between them. The two recombinant-type gametes will each be 5% of the total gametes. When the $F_1 \times F_1$ cross is made, the genotypes and frequencies of the progeny will be as shown in Figure A.

If we add up the genotypic classes that give the same phenotype, we see that this amount of linkage has had a significant effect on the frequencies of phenotypes in the progeny (Figure B).

FIGURE A

	a b	*a* +	+ *b*	+ +	Gamete genotypes
	0.45	0.05	0.05	0.45	Frequency
a b	*ab* /*ab*	*a*+ /*ab*	+*b* /*ab*	++ /*ab*	Offspring genotype
	(ab)	(a)	(b)	(+)	Offspring phenotype
0.45	0.2025	0.0225	0.0225	0.2025	Frequency
a +	*ab* /*a*+	*a*+ /*a*+	+*b* /*a*+	++ /*a*+	
	(a)	(a)	(+)	(+)	
0.05	0.0225	0.0025	0.0025	0.0225	
+ *b*	*ab* /+*b*	*a*+ /+*b*	+*b* /+*b*	++ /+*b*	
	(b)	(+)	(b)	(+)	
0.05	0.0225	0.0025	0.0025	0.0225	
+ +	*ab* /++	*a*+ /++	+*b* /++	++ /++	
	(+)	(+)	(+)	(+)	
0.45	0.2025	0.0225	0.0225	0.2025	

FIGURE B

	a b	*a B*	*A b*	*A B*
No linkage	0.0625	0.1875	0.1875	0.5625
Linkage	0.2025	0.0475	0.0475	0.7025

ing over between chromosomes. Evidence supporting this hypothesis was obtained in both corn and *Drosophila* in 1935.

Barbara McClintock and Harriet Creighton reported the first evidence for a physical exchange between chromosomes in a study of two linked genes in corn using the genetic markers *colored endosperm* (*C*) and its recessive allele *colorless* (*c*) and *starchy endosperm* (*Wx*) and its recessive allele *waxy* (*wx*). Both of these genes are on chromosome 9. In their study, McClintock and Creighton used a mutant chromosome 9 that contained a large, densely staining knob at one end and a fragment of chromosome 8 attached to the other end. Both the knob and the attachment could be observed in cytological preparations of maize cells undergoing meiosis or mitosis, and the attachments thus served as cytological markers for identifying the ends of chromosome nine. Plants containing the *Wx* and *C* alleles on the chromosome with the knob and the attachment were crossed to plants carrying the *wx* and *c* alleles on a chromosome without a knob or attachment. McClintock and Creighton reasoned that if there were a physical exchange between these two chromosomes, the exchange would generate recombinant chromosomes that were physically different. That is, chromosomes either with the knob but not the attachment or with the attachment but not the knob (Figure 6.6) would be formed during crossing over. Any chromosome that was genetically recombinant would also have to be physically recombinant. The heterozygous plants were testcrossed, and progeny with genetically recombinant chromosomes (*Wx c* or *wx C*) were selected. Cytological preparations were made in which the structure of the chromosomes in these genetically recombinant individuals could be determined. They observed that these individuals did indeed have physically recombinant chromosomes (Figure 6.6). Other progeny from this

Two-strand stage recombination

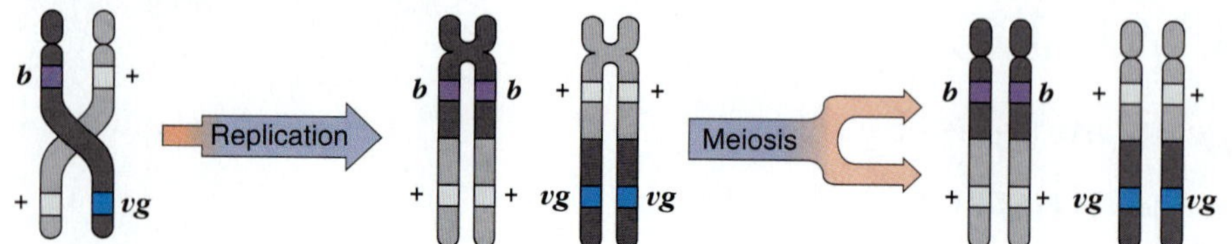

Four-strand stage recombination

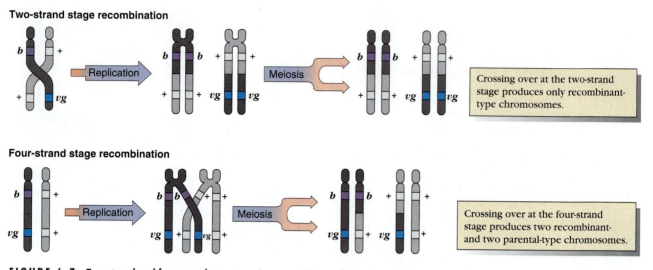

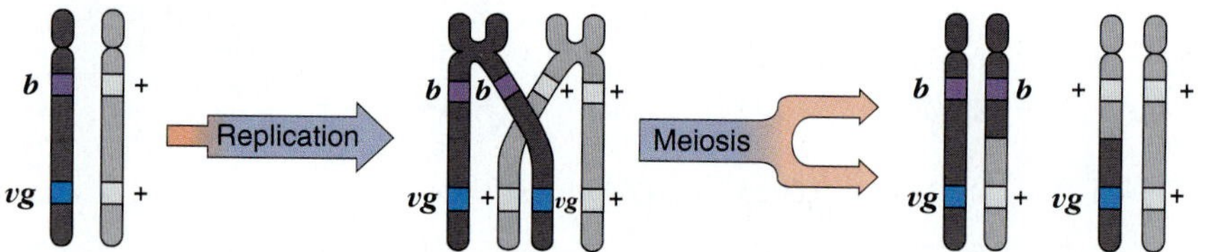

FIGURE 6.7 Two-strand and four-strand stage crossing over. We now know that crossing over occurs in cells at the four-strand stage of meiosis I.

cross that had parental genotypes (*Wx C* or *wx c*) had chromosomes with the original structure. These results indicated that genetic crossing over is accompanied by a physical exchange between homologous chromosomes.

These experiments did not, however, indicate when in meiosis crossing over occurs. Crossing over might occur before chromosome replication. At this stage, each chromosome contains only a single chromatid (containing one DNA strand), and each pair of homologous chromosomes contains two chromatids (two DNA strands). This is called the **two-strand stage.** Alternatively, crossing over might occur after chromosome replication. At this stage, each chromosome has two chromatids, and each pair of homologous chromosomes has four chromatids (four DNA strands). This stage is called the **four-strand stage.** If crossing over occurs at the two-strand stage in a cell, replication will occur after the crossover event, and the cell will produce four recombinant chromosomes that eventually form four recombinant gametes (Figure 6.7). If crossing over occurs at the four-strand stage, only two of the four chromatids will be involved in the crossover. This means that the cell will produce two

Two-strand stage recombination

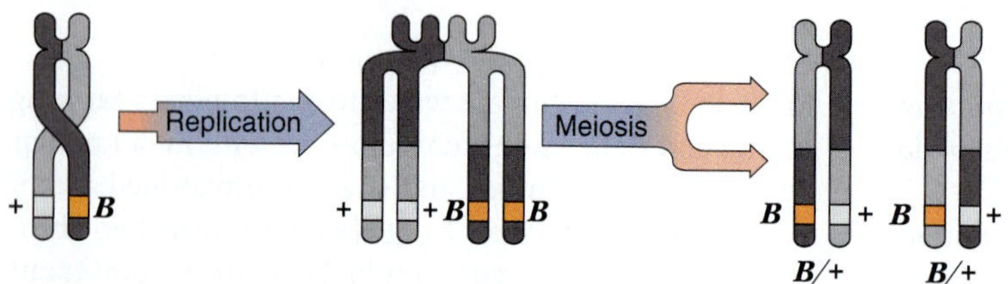

Four-strand stage recombination

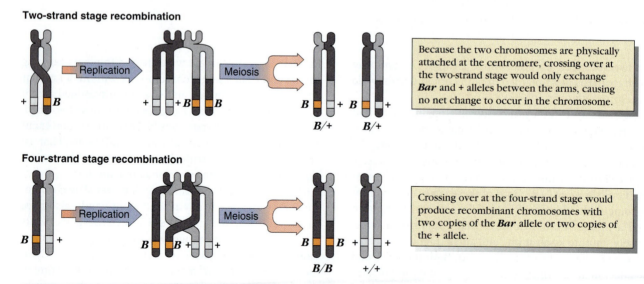

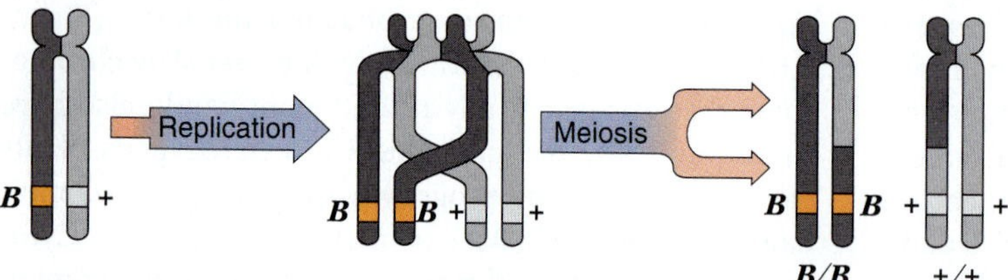

FIGURE 6.8 Proof that crossing over occurs at the four-strand stage. Anderson demonstrated that crossing over occurs at the four-strand stage using attached-X chromosomes in *Drosophila* that were heterozygous for the *Bar* (*B*) mutation. If crossing over had occurred during the two-strand stage, all the gametes would be *B* / + instead of the + / + observed.

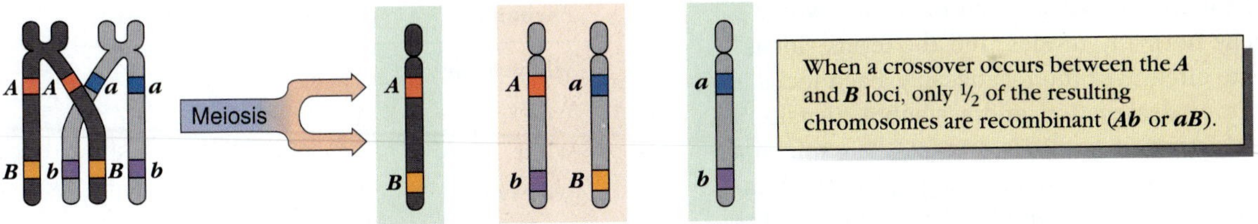

FIGURE 6.9 Crossing over and the frequency of recombination. The frequency of recombination is half the frequency of crossing over because each crossover involves only two of the four chromatids. Thus, 100% crossing over yields 50% recombinant gametes.

parental and two recombinant chromosomes. Both models predict that during gametogenesis, parental and recombinant type gametes will be formed in equal frequencies. Because progeny are formed by a random fusion of gametes, it is impossible to distinguish between these two models by examining progeny.

An elegant experiment done by Russell Anderson provided proof that crossing over occurs at the four-strand stage. Anderson used a strain of *Drosophila* that had an **attached-X** chromosome (see Chapter 3). Normal X chromosomes in *Drosophila* are acrocentric with one large arm. Attached-X chromosomes are two nearly complete X chromosomes attached to a single centromere. Because they share a single centromere, the two X chromosomes always segregate together. Anderson produced an attached-X chromosome that contained a dominant mutation, *Bar* (*B*), on one of the X chromosomes and the wild-type allele (+) on the other. He reasoned that if crossing over occurs at the two-strand stage, the *B* and the + alleles would be exchanged, but there would be no net change in the chromosome. The chromosome would still have + and *B* alleles. However, if crossing over occurs at the four-strand stage, then crossing over would produce an attached-X chromosome having two *B* alleles or two + alleles (Figure 6.8). This was, in fact, what he observed. Later experiments in several species of fungi in which all four of the chromosomes produced by a single meiotic cell can be recovered confirmed that crossing over occurs at the four-strand stage.

Because crossing over occurs at the four-strand stage and only two chromatids are involved in a crossing over event, each crossover produces two recombinant chromosomes and two parental chromosomes. This means that 1/2 of the chromosomes, and ultimately 1/2 the gametes, are recombinant. It also means that the frequency of recombinant gametes will be equal to 1/2 of the frequency of crossing over (Figure 6.9). To illustrate this point, consider a hypothetical example of two genes that are on the same chromosome. If the frequency of crossing over between these genes is 80%, then a heterozygous individual will produce gametes that are 40% recombinant types (1/2 of 80%) and 60% parental types (100% − 40%). If this heterozygote is testcrossed, 40% of the progeny will have a recombinant phenotype, and 60% will have a parental phenotype.

SECTION REVIEW PROBLEM

5. The overall frequency of chiasmata in the X chromosome of *Drosophila* is 100%. What recombination frequency would you expect between two genes located near opposite ends of the chromosome?

GENE MAPPING CREATES A ROAD MAP OF THE CHROMOSOME.

T. H. Morgan believed that genes are arranged in a linear order along the chromosomes and that each gene has a fixed location, called the **locus** (plural, loci), on the chromosome. He also believed that meiotic crossover events occur randomly along the chromosome, so that the frequency of crossovers that occur between two loci is proportional to the physical distance between them. If two loci are far apart, crossovers will often occur between them, giving a high frequency of recombinant progeny in crosses. If two loci are close together, crossovers will occur between them only rarely, giving a low frequency of recombinant progeny in crosses. In 1911, Alfred Sturtevant, an undergraduate student working in T. H. Morgan's lab, recognized that this meant that the frequency of crossing over between two loci could be used as a numerical measure of the distance between loci. If enough of these distances could be calculated, he could use the distances between genes to make a map of the chromosome. The end result would be a **linkage map** showing the order of the loci on the chromosome and the distances between them. Sturtevant's idea was widely accepted, and the generation of linkage maps has become a key part of genetics. The discovery that genes have fixed loci resulted in a change in terminology. The term "locus" is used for the portion of a chromosome where a particular gene normally resides, and the terms "locus" and "gene" are often used interchangeably. Each locus is given a name or designation and a symbol, usually a short series of letters or numbers. In text, the name of a locus is italicized, just as the name of a gene is. The symbol is also italicized. For example, the symbol for the *white* locus is *w* and for the *yellow* locus is *y*. The names of particular alleles reflect their locus, and the symbols for mutant alleles are the locus symbol with a superscript to designate the particular allele. For example, two mutant alleles at the *white* locus are $white^1$ and $white^2$, whose symbols are w^1 and w^2. The terminology for naming loci varies from species to species, as does the terminology for naming genes. Additional terminology will be introduced in later chapters, especially in Chapter 8.

Linkage Mapping Depends Upon the Frequency of Crossing Over Between Two Linked Loci.

The key to the construction of a linkage map is determining the frequency of recombinant gametes produced by crossing over between linked genes. This is usually done by testcrossing heterozygous individuals and counting the number of progeny that have recombinant phenotypes. For example, if female *Drosophila* heterozygous for alleles of the *white eye* and *miniature wing* genes are testcrossed, about 35% of the progeny have recombinant phenotypes. If individuals heterozygous for alleles of the *yellow* and *white* genes are testcrossed, about 1.5% of the progeny have recombinant phenotypes. The differences in strength of linkage indicate that the *white eye* and *miniature wing* loci are farther apart than the *yellow* and *white* loci. Sturtevant used the frequency of recombinant progeny in a testcross as the unit of distance in his calculations. He devised a linkage **map unit**, with one map unit being equal to 1% recombinant progeny in a testcross. The map unit is called the **centimorgan** (cM) in honor of T. H. Morgan, with 1 cM being equal to 1% recombinant progeny in a testcross. Sturtevant realized that this was also equal to the frequency of recombinant gametes produced by the heterozygote, so he calculated the value of the map distance (called R) between two linked genes using a simple formula:

$$R = (\text{\# recombinant progeny}/\text{\# total progeny}) \times 100 = \text{cM}$$

To illustrate how Sturtevant's formula is used, let us consider an example of a two-point testcross. In a two-point testcross, individuals heterozygous for mutant alleles are testcrossed, and the map distance between the loci is calculated. If we testcross female *Drosophila* heterozygous for recessive mutations at the *yellow body* (*y*) and the *white eye* (*w*) loci, we obtain progeny with four different genotypes, each with a different phenotype:

Phenotype	Genotype	Number
Yellow white	*y w / y w*	490
Yellow	*y + / y w*	7
White	*+ w / y w*	8
Wild-type	*+ + / y w*	495

We can draw three conclusions from these results. First, these data tell us whether or not the genes are linked. If the genes are not linked, Mendel's rule of independent assortment predicts that four types of progeny will be present in a 1 : 1 : 1 : 1 ratio. Because these progeny clearly do not fit this ratio, we can conclude that these two genes are linked. Second, we can determine the arrangement of the alleles (*cis* or *trans*) in the parental type chromosomes. If the genes are linked, the types that are present in excess (*y w* and + +) in the progeny are the parental types. Thus, our second conclusion is that the mutant alleles are in the *cis* configuration in the heterozygous females (*y w* / + +). Finally, we can calculate the map distance between the *yellow* locus and the *white* locus using Sturtevant's formula. This map distance is: $(7 + 8)/1000 \times 100 = 1.5$ cM and we can draw a linkage map as

y ——— *w*
1.5

The Three-Point Testcross Is the Preferred Mapping Cross.

The most useful linkage mapping cross is the trihybrid testcross, or **three-point testcross.** The three-point testcross allows the order of the three loci on the chromosome and the map distances between them to be determined in a single cross. In a three-point testcross, individuals heterozygous for alleles of three genes are crossed, the progeny are examined to determine the different phenotypes present, and the number in each phenotypic class is determined. For example, in one of Morgan's crosses of loci on the X chromosome of *Drosophila* (*scute* [*sc*] that causes a loss of scutelar bristles on the thorax, *crossveinless* [*cv*] that reduces the crossveins in the wings, and *echinus* [*ec*] that affects the shape of the eye), he testcrossed females heterozygous for all three mutant alleles and determined the phenotypes of the progeny. There are eight possible combinations of phenotypes, or phenotypic classes of progeny, in a three-point testcross. These classes are two parental types, four recombinant types produced by single crossovers between the two chromosomes, and two recombinant types produced by double crossovers (Figure 6.10). The numbers of progeny of each type observed by Morgan in his crosses follow.

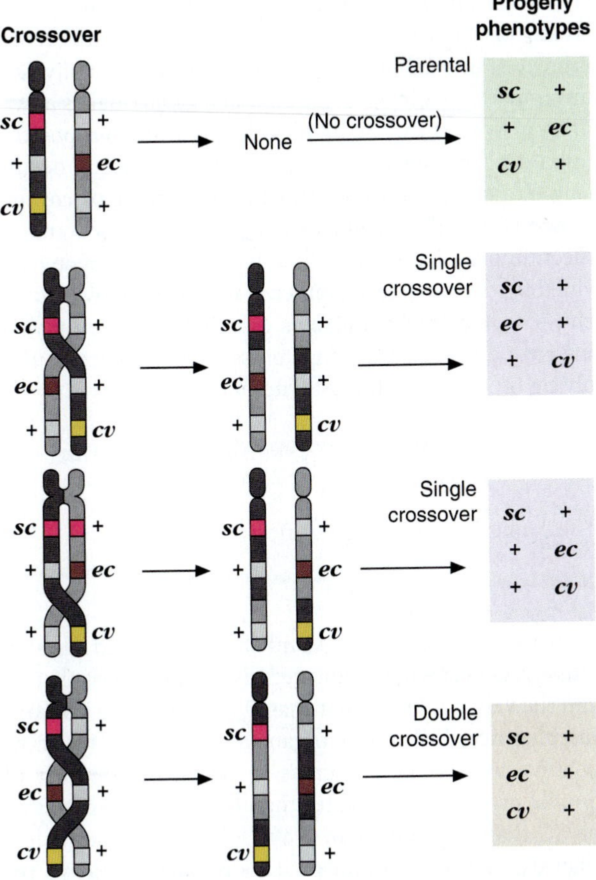

FIGURE 6.10 The three-point testcross. In a three-point testcross, zero, one, or two crossing over events in different chromosomal locations produce the eight classes of progeny.

Progeny Phenotype	Progeny Genotype	Number
Echinus crossveinless	*sc ec cv* / *sc ec cv*	4
Wild type	+ + + / *sc ec cv*	1
Scute	*sc* + + / *sc ec cv*	997
Echinus crossveinless	+ *ec cv* / *sc ec cv*	1002
Scute echinus	*sc ec* + / *sc ec cv*	681
Crossveinless	+ + *cv* / *sc ec cv*	716
Scute crossveinless	*sc* + *cv* / *sc ec cv*	8808
Echinus	+ *ec* + / *sc ec cv*	8576
Total		20,785

We can use these results to answer a number of questions. First, we can determine whether the three loci are linked. Mendel's rule of independent assortment predicts that if the loci are not linked, the eight types of progeny will have a 1 : 1 : 1 : 1 : 1 : 1 : 1 : 1 ratio. If they are linked, the parental types will be more frequent, and the recombinant types will be less frequent, in violation of Mendel's rule of independent assortment. The data show a significant deviation from the independent assortment ratio, and the *sc* + *cv* and + *ec* + types are far more frequent than any of the other types. This pattern is typical of linkage in a three-point testcross. After we have determined that the loci are linked, we can determine other facts about these loci. If we did not know what the original parental types were, we could determine them by examining the progeny. In a testcross involving linked loci, there will always be more progeny with parental types than recombinant types, so the most frequent progeny classes must be the parental types. In this cross, the parental genotypes must be *scute crossveinless* (*sc* + *cv*) and *echinus* (+ *ec* +). The heterozygous F_1 females must have had *sc* and *cv* on one chromosome and *ec* on the other (*sc* + *cv* / + *ec* +).

We can now determine the correct order of these loci in the chromosome by comparing the parental types with the double crossover types. Double crossover events are much rarer than single crossover events because in a double crossover both types of single crossovers must occur between two chromosomes. The probability that two crossovers will both occur is equal to the multiple of the frequencies of the two single crossovers. The least-frequent genotypes in the progeny of this cross are *sc ec cv* / *sc ec cv* and + + + / *sc ec cv*; therefore, the double crossover chromosomes must be *sc ec cv* and + + +. A double crossover exchanges the allele of the middle locus on one chromosome with the allele on the other homologue. This means that the double crossover chromosomes have the same alleles at the two outer loci as the parental type chromosomes and different alleles at the middle locus. Comparing these double crossover chromosomes with the parental types (*sc ec cv* with *sc* + *cv* and + + + with + *ec* +), we see that both double crossover chromosomes have the same *sc* and *cv* alleles as the parental chromosomes, and different alleles of *ec*. This indicates that the *ec* locus is in the middle. The true order is *sc ec cv*.

We can produce a linkage map showing the distances between these three loci using Sturtevant's formula. Map distances are calculated between pairs of loci, so we first determine the distance between *sc* and *ec* and then determine the distance between *ec* and *cv*. For each calculation, we examine each of the eight progeny classes to determine whether they are parental or recombinant for that pair of loci. The map distance is then calculated by dividing the total number of recombinant types by the total number of progeny and multiplying the result by 100. The map distance between *sc* and *ec* in Morgan's cross is $(681 + 716 + 4 + 1)/20{,}785 \times 100 = 6.74$ cM. The map distance between *cv* and *ec* is $(1003 + 997 + 4 + 1)/20{,}785 \times 100 = 9.65$ cM. In each of these calculations, we include the 4 and 1 because they represent a single crossover between each of the two loci, but are represented in the double crossover classes. The map distance between *sc* and *cv* is the sum of the two internal distances (6.74 cM + 9.65 cM = 16.39 cM). The final map therefore is

*sc*______*ec*______*cv*
6.74 9.65

SECTION REVIEW PROBLEM

6. Three linked loci in tomato are *mottled* (*m*) or normal (*M*) leaf, *smooth* (*P*) or *pubescent* (*p*) epidermis, and *purple* (*Aw*) or *green* (*aw*) stem. Individuals heterozygous for all three genes were testcrossed, and the results follow.

Progeny Phenotype	Number
Normal smooth purple	18
Mottled pubescent green	15
Normal smooth green	180
Mottled pubescent purple	187
Normal pubescent purple	1880
Mottled smooth green	1903
Mottled smooth purple	400
Normal pubescent green	417
Total	5000

a. Determine the genotypes of the original parental types.
b. Determine the order of the genes on the chromosome.
c. Make a linkage map showing the distances between the three loci.

Interference Increases or Decreases the Chance of a Second Crossover.

In our previous three-point testcross mapping calculations, we assumed that all crossovers were occurring independently from each other, that is, that one crossover does not increase or decrease the probability that a second crossover will occur nearby. If this assumption is correct, then the frequency of double crossovers will be equal to the product of the frequencies of the two single crossovers. However, in most diploid organisms, the frequency of double crossovers is not equal to the multiple of the single crossovers. This was originally noted by Herman J. Muller, who observed that there were consistently fewer double crossover types than expected in his mapping crosses. Muller proposed the term **interference** **(I)** for this phenomenon because it appeared that the first crossover event interfered with the second. Interference is greatest in small intervals, as though there is a physical hindrance or stiffness in chromatids, which makes a second crossover near the first less likely. Muller proposed a numerical measure of how much interference is occurring in a cross. He called this the **coefficient of coincidence** (CC). The coefficient of coincidence is equal to the number of observed double crossover progeny divided by the expected number of double crossover progeny. The expected number is calculated by multiplying the frequencies of the single crossovers and then multiplying the result by the total number of progeny:

$$\text{CC} = \text{observed double crossovers}/(\text{frequency of single 1})(\text{frequency of single 2})(\text{total progeny})$$

The numerical value for interference is

$$\text{I} = 1 - \text{CC}$$

A cross in which the number of observed double crossovers is less than expected gives a CC of less than 1, and the value of I is greater than 0. This is called **positive interference.** Eukaryotic organisms almost always show positive interference. A cross in which the number of double crossovers is greater than expected gives a value of CC that is greater than 1 and a value of I that is less than 0 (a negative number). This is called **negative interference.** Genetic crosses in bacteria and viruses often show negative interference.

SOLVED PROBLEM

Problem

What is the coefficient of coincidence for T. H. Morgan's three-point testcross? Is there interference in this cross?

Solution

In this cross, the expected number of double crossover progeny is 0.0965 × 0.067 × 20,785 = 133.7 The number of observed double crossover progeny is 5.0. Thus, CC = 5.0/133.7 = 0.037. The amount of interference may be expressed as = 1 − CC = 1 − 0.037 = 0.963. This means that in this cross 96% of the expected double crossovers were not observed. Yes, there is interference in this cross.

SECTION REVIEW PROBLEM

7. Calculate the coefficient of coincidence for the cross in Problem 6.

Linkage Maps of Chromosomes Can Be Constructed by Combining the Results of Many Mapping Experiments.

Linkage map distances are additive, meaning that additional loci can be added to a map by calculating the distance of the new loci from loci that have been mapped. For example, in our two-point testcross, we determined the map distance between the *yellow* and *white* loci on the X chromosome of *Drosophila*. The *miniature wing* locus (*m*) is also on the X chromosome. The map distance from the *yellow* locus to the *miniature wing* locus is 36.1 cM, and the map distance from the *white* locus to the *miniature wing* locus is 34.6 cM. Combining these values with our existing map gives

*y*_____*w*_____*m*
1.5 34.6

Because linkage maps are additive, the distance between two genes is equal to the sum of the distances between the intervening loci. This means that large maps showing the locations of many genes can be constructed by combining the results of different mapping experiments. For example, the map distances of several additional X-linked loci, such as *singed* (*sn*), *forked* (*f*), and *carnation* (*car*) from the *yellow*, *white*, and *miniature wing* loci have been determined. Adding these to the map gives the following:

*y*_____*w*_____*sn*_____*m*_____*f*_____*car*
1.5 19.5 16.1 20.6 6.0

In this map we have indicated the map distance between each locus and its nearest neighbor. When maps show many different loci, it is more convenient to orient the map so that one locus at one end of the chromosome has map position 0.0, and all other distances are given in cM from this gene. In *Drosophila*, the *y* locus is near the end of the chromosome and is assigned map position 0.0. Adjusting the map numbers gives a map in which each locus has a unique map position:

*y*_____*w*_____*sn*_____*m*_____*f*_____*car*
0.0 1.5 21.0 37.1 57.7 63.7

By convention, the end of a chromosome arm that is attached to the centromere is called the **proximal** end, and the end that is attached to the telomere is called the **distal** end. Proximal and distal often indicate direction or orientation of loci that are on the same chromosome arm. If two loci are on the same chromosome arm, then the one that is closest to the centromere is said to be located proximal to the other. The X chromosome of *Drosophila* is an acrocentric chromosome; the *carnation* locus is near the proximal end, and the *yellow* locus is located near the distal end.

A newly discovered gene may be added to this map by determining the map distances between its locus and any two loci on the map. For example, the *almondex* eye (*amx*) locus is between *sn* and *m*. Testcrosses indicate *amx* is 6.7 cM from *sn* and 9.4 cM from *m*. The *amx* locus can be placed on the map at position 27.7 ($21.0 + 6.7 = 27.7$, or $37.1 - 9.4 = 27.7$). In this way, complex genetic maps are constructed by different investigators, each determining the location of new genes. In a few well-studied species like *Drosophila*, linkage maps have been constructed showing thousands of loci on all the chromosomes; some of these are shown in Figure 6.11. Linkage maps can be an extremely valuable tool for a geneticist, not only for locating the position of a newly discovered gene but also for planning experiments and for using crossing over to generate chromosomes with particular combinations of alleles (see CONCEPT CLOSE-UP: *How Can Linkage Maps Be Used to Predict the Results of a Cross?*).

Map distances are additive only over short distances. The map distance between two loci calculated from a testcross is usually smaller than the sum of the intervening distances because of the occurrence of double crossovers. If two crossovers occur within a region between two loci, then the second crossover cancels the effect of the first, giving a chromosome with a parental genotype (Figure 6.12). Progeny that develop from gametes containing such double crossover chromosomes will have parental phenotypes and will thus not be counted as recombinant types in the map calculation. This means that some of the crossover events that occur between two loci are not counted, making the calculated map distance smaller than it really is. This effect will be larger for loci that are far apart and will be insignificant for loci that are so close together that double crossovers occur with very low frequencies. Linkage maps today are generally produced by calculating the map distances between loci that are relatively close together (5–10 cM) and then combining the distances.

Sometimes two loci are far apart on a chromosome, and no loci between them can be used to subdivide the interval. In such cases, the true map distance can be estimated from the observed recombination frequency using a mathematical formula that corrects for the effects of the uncounted double crossovers. Such a formula is called a **mapping function.** Mapping functions are derived by comparing experimental map distances calculated by summing several short distances with observed recombination frequencies. To use a map function, the function is plotted as a curve, such as that shown in Figure 6.13. The corrected map distance between two loci is given on the horizontal axis, and the observed map distance is given on the vertical axis. To correct a map distance, the observed distance is located on the vertical axis. We then

Chromosome X

0 yellow body, achaete bristles, Hairy wing, scute bristles, silver body
0.8 suppressor of sable
1.0 minute bristle (1)
3.0 facet eyes
5.5 echinus eyes
6.9 bifid wing veins
7.5 ruby eyes
13.7 crossveinless wings
17.9 shifted wing veins
18.9 carmine eyes
20.0 cut wings
23.0 ocelliless
27.7 lozenge eyes
29.2 Hexokinase enzyme
32.8 raspberry eyes
36.0 nod meiosis
36.9 furrowed eyes
41.9 wavy wings
43.0 sable body
44.4 garnet eyes
44.5 tiny bristles
51.5 scalloped wings
53.5 small-wing
56.7 forked bristles
59.2 outstretched wings
59.4 Beadex wings
62.5 carnation eyes
64.8 maroonlike eyes
67.7 bobbed bristles
Centromere

Chromosome 2

0 net wing veins
0.3 dachsous wings
1.3 star eyes
4.0 heldout wings
5.0 female-sterile
6.1 Curly wings
7.6 Phosphoglycerokinase
10.0 Detached vein
12.0 fat body
13.0 dumpy wings
16.5 clot eyes
22.0 Sternopleural bristles
22.9 lysine accumulation
31.0 dachs legs
38.0 abnormal oocyte
41.0 jammed wings
41.5 daughterless
44.0 abrupt wing veins
50.5 Suppressor of hairless
53.9 hook bristles
54.8 Bristle short
light eyes
55.0 **Centromere**
Segregation distorter
55.3 thick legs
56.1 lightoid eyes
62.0 engrailed scutellum
67.0 vestigial wings
69.7 waxy wings
71.1 comb-gap bristles
75.5 curved wings
77.3 Amylase enzyme
81.1 root wings
83.0 abero
91.5 smooth abdomen
99.2 arc wings
100.5 plexus wing veins
104.5 brown eyes
107.4 balloon wings

Chromosome 3

0 roughoid eyes
0.2 veinlet wings
1.4 Roughened eyes
5.4 meiotic
19.2 javelin bristle
20.0 Moiré eyes
23.0 Henna eyes
26.0 sepia eyes
26.5 hairy body
35.5 eye-gone
37.0 Dichaete wings
40.4 Lyra wings
41.4 Glued eyes
44.0 scarlet eyes
46.0 **Centromere**
47.9 Antennapedia
48.7 blistery wings
51.3 tetraltera
52.0 rosy eyes
54.8 Suppressor of Hairy-Wing
58.8 Stubble bristle
59.0 spineless bristle
63.1 bithorax
64.0 kidney eyes
66.2 Delta wing veins
70.7 ebony body
72.5 detached veins
75.7 cardinal eyes
76.2 white ocelli
88.0 mahogany eyes
90.0 Prickly bristle
91.1 rough eyes
93.8 Beaded wings
95.5 suppressor of purple
100.7 claret eyes
101.0 Minute bristle (3) 1
102.9 Killer of prune
104.3 brevis bristle
106.2 Minute bristle (3) g

Chromosome 4

Centromere
cubitus-interruptus vein
0 Minute bristle (4)
0–0.2 abdomen rotatum
0.2 grooveless scutellum
1.4 bent wings
2.0 eyeless
3.0 shaven bristles
4.0 sparkling eyes

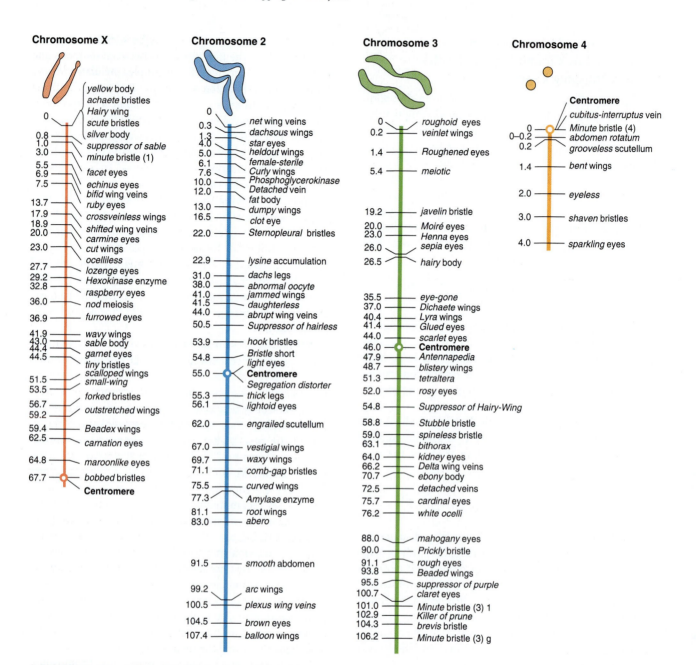

▲**FIGURE 6.11 A *Drosophila* linkage map.** A linkage map showing the arrangements of some of the known loci in *Drosophila*. The numbers refer to linkage map positions. *Drosophila melanogaster* has four chromosomes, one midsized acrocentric (X), two large metacentrics (2 and 3), and one very small chromosome (4). The sizes of the four linkage groups reflect the differences in the size of the chromosomes. *(These maps were produced from information in The genome of Drosophila melanogaster by Lindsley and Zimm, 1992, Academic Press, San Diego.)*

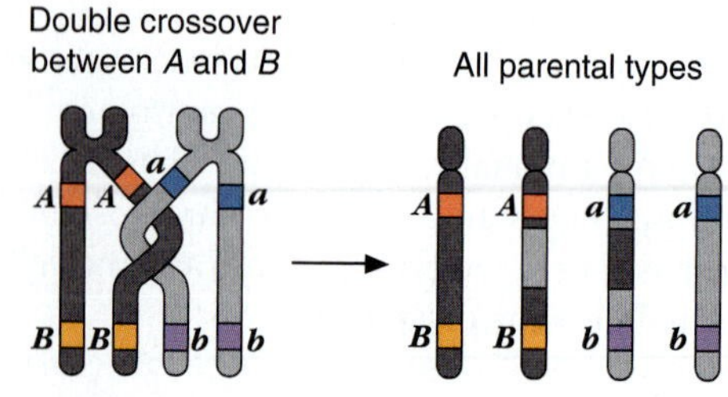

►**FIGURE 6.12 Double crossovers affect estimations of map distance.** If two crossovers occur between two loci, the second crossover may cancel the effect of the first, giving parental type arrangements of the alleles. This results in parental type progeny, and these crossovers are not counted in a mapping experiment. Uncounted double crossovers make calculated map distances smaller than actual map distances.

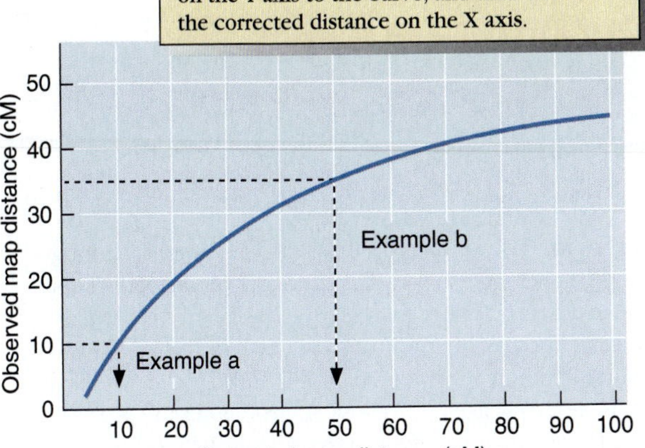

FIGURE 6.13 Mapping functions. Mapping functions are used to correct for undetected double crossovers in large map distances.

read across to the curve and down from the point of intersection to the horizontal axis. The corrected map distance is only an estimate of the true map distance, but it is often the best that can be obtained.

SECTION REVIEW PROBLEMS

8. The results of a series of dihybrid testcrosses follow (R = % recombinant progeny). From these data, make a map showing the order of the loci and the distances between them in map units. On the final map, show only the distance between adjacent genes.

Genes	R	Genes	R
a–b	50	*b–d*	5
a–c	40	*b–e*	48
a–d	50	*c–d*	23
a–e	10	*c–e*	30
b–c	18	*d–e*	50

9. In corn, the *shrunken endosperm* locus (*sh* = shrunken or *Sh* = full) and *aleurone color* locus (*c* = colorless or *C* = colored) are both on chromosome 9. F_1 plants heterozygous for different alleles at both loci were testcrossed, and the phenotype of the progeny kernels recorded. From the following data, calculate the map distance between these two loci.

Phenotype	Number
Full colored	117
Full colorless	3334
Shrunken colored	3139
Shrunken colorless	118
Total	6708

10. The following map shows the distances between three loci in tomatoes, *shiny leaf* (*s*), *purple stem* (*aw*), and *pubescent leaf* (*p*). Plants differing in all three genes are crossed (*s aw p* × *S Aw P*), and the F_1 were testcrossed.

s —— *aw* —— *p*
21 17

   a. What genotypes would you expect to find in the progeny of the testcross?
   b. Assuming that there is no interference, how many of each progeny genotype would you expect if a total of 15,000 progeny were collected?
   c. How would your answer to (b) change if the coefficient of coincidence for this cross was 0.25?

TETRAD ANALYSIS IS USED FOR CHROMOSOME MAPPING.

In previous discussions of genetic mapping, we have considered only eukaryotic species (such as *Drosophila*, corn, mice, chickens, or humans) that spend the majority of their life cycle as diploids. A number of other eukaryotic species, including yeasts, fungi, and algae, spend a large part of their life cycle as haploids, and many of these have been used in genetic mapping studies. The advantages of these species are that they are easily cultured on defined media in the laboratory and they grow rapidly. Also, because haploid species have only a single set of genes, the phenotype of an individual always reflects the individual's genotype. This eliminates the need to do testcrosses to determine the genotypes of the progeny of a cross. Crosses are usually easily done in these species; two haploid cells from different strains are fused to form a heterozygous diploid cell, and this cell undergoes meiosis in a specialized fruiting body. The fruiting body can be isolated and each of the haploid cells can be cultured to determine their phenotypes and genotypes (Figure 6.14).

In yeast and other fungi, meiosis takes place in a sack-like structure called an **ascus** (plural, asci). Within each ascus, all the haploid products of a single meiotic cell develop into spores. The four haploid cells that are formed from one meiotic cell are called a **tetrad,** and the genetic analysis of tetrads is called **tetrad analysis.** Unlike testcrosses in higher organisms, which involve the random recovery of individual gametes, tetrad analysis allows recovery and examination of all the gametes produced by one meiotic cell. Several different species of fungi are used for tetrad analysis, such as *Neurospora*, where the four haploid meiotic products also undergo a mitotic division before maturing into spores. As a result,

CONCEPT CLOSE-UP

How Can Linkage Maps Be Used to Predict the Results of a Cross?

An important use of linkage maps is to predict the genotypes of the progeny of a cross and to predict the expected numbers of each genotype. A geneticist who wishes to generate a new strain by isolating a particular recombinant type from a cross can use a linkage map to calculate in advance how frequent that recombinant type will be and can plan the size of the cross accordingly. Consider an example of how this can be done. Here is a linkage map of a portion of the X chromosome in *Drosophila*.

y	*sn*	*m*
0.0	21.0	37.1

If females with the genotype *y sn* / *m* are crossed with males with the genotype *y sn m* / *Y*, and 10,000 female progeny were collected, what would be the genotypes of the progeny, and how many of each type would you expect? If you wished to isolate 100 females with a *y m* / *y sn m* genotype, how many progeny should you collect?

To answer the first question, we determine the frequency with which the different crossovers will occur in these females, and the genotypes of the chromosomes these females will pass to their progeny. We can then use these map distances to predict how many double crossover, single crossover, and parental type X chromosomes the females will produce and how many of each there will be. According to the map, the distance between the *y* and *sn* loci is 21.0 − 0.0 = 21.0 cM, and the distance between the *sn* and *m* loci is 37.1 − 21.0 = 16.1 cM. This means that 21.0% of the progeny should receive a chromosome that is recombinant for *y* and *sn*, and 16.1% of the progeny should receive a chromosome that is recombinant for *sn* and *m*. Some of these recombinant chromosomes will have only a single crossover, and some will have double crossovers. If there is no interference, then the expected number of progeny with a double crossover chromosome would be 0.210 × 0.161 × 10,000 = 338.1. One half of these individuals (169.05) will have a yellow phenotype (*y* + + / *y sn m*), and the other half (169.05) will have a singed miniature phenotype (+ *sn m* / *y sn m*). A single crossover between *y* and *sn* is 0.21 × 10,000 − 338.1 = 1761.9. One half of these individuals (880.95) will have a yellow miniature phenotype, and one half (880.95) will have a singed phenotype. The expected number of progeny in the classes produced by single crossovers between *sn* and *m* is 0.161 × 10,000 − 338.1 = 1271.9. One half of these (635.95) will have a yellow, singed, miniature phenotype, and one half (635.95) will have a wild-type phenotype. The expected number of parental type progeny (the remainder) is 10,000 − 338.1 − 1761.9 − 1271.9 = 6628.1. One half of these (3314.05) will have a yellow singed phenotype, and one half (3314.05) will have a miniature phenotype.

How would interference alter these predicted numbers? If the coefficient of coincidence for this cross were 0.65, the number of double crossover progeny would be reduced to 0.65 × 338.1 = 219.8. The number of progeny in each of the single crossover classes would be increased to 0.161 × 10,000 − 219.8 = 1390.2 and 0.21 × 10,000 − 219.8 = 1880.2. The parental classes would be decreased to 3254.9 each. It is important to note that interference does not change the map distance or the recombination frequency. The only effect of increasing interference is a change in the fraction of the total recombinants that are double crossover and single crossover types.

---

eight spores are produced in each ascus (Figure 6.14), but these meiotic products are still called a tetrad.

Linkage Maps Are Determined by Analyzing Crossover Tetrads.

In a tetrad analysis, we determine whether two genes were linked, and if so, what the map distance is between them. This is done by measuring the frequency with which crossovers occur between the two genes' loci. What is unique about tetrad analysis is that the number of meiotic cells that contain recombinant type cells is counted by simply counting the number of tetrads. Map distances are not measured in terms of the frequency of recombinant gametes (or progeny of testcrosses) but in terms of the frequency of recombinant tetrads.

To determine whether two genes are linked by tetrad analysis, a dihybrid cross is made, and the resulting tetrads are collected. The genotypes of the spores in each tetrad are then determined. If the alleles of the two genes being tested segregate together in one meiotic cell, then the tetrad produced by that cell will contain only parental type spores. This type of tetrad is called a **parental ditype** (PD). If the alleles do not segregate together in a cell, then the tetrad produced by that cell will contain recombinant type spores. This type of tetrad is

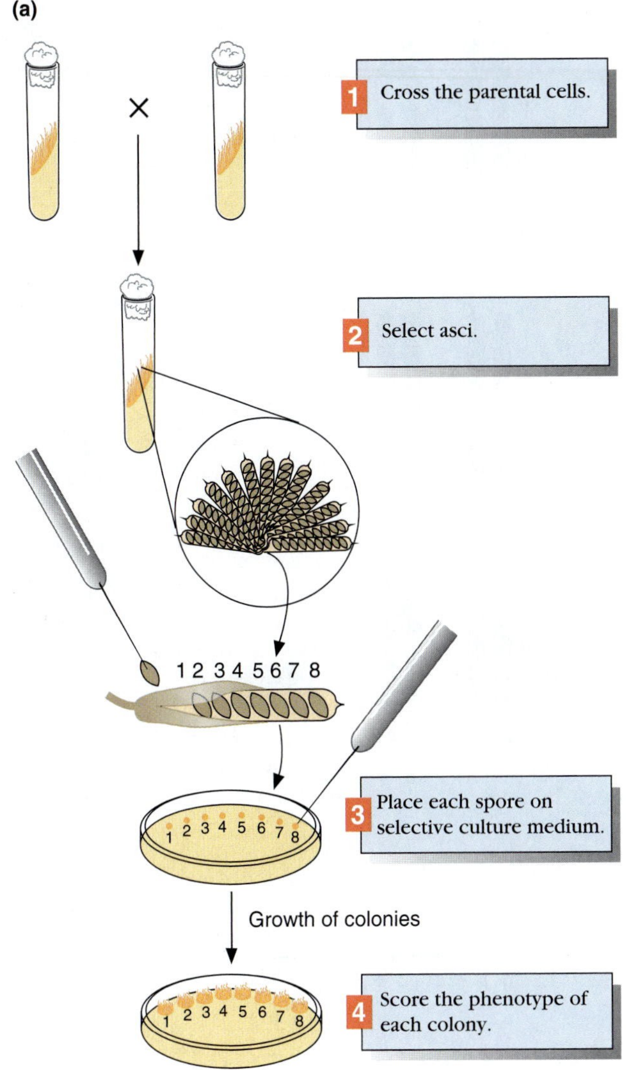
(a)
1 Cross the parental cells.
2 Select asci.
1 2 3 4 5 6 7 8
3 Place each spore on selective culture medium.
Growth of colonies
4 Score the phenotype of each colony.

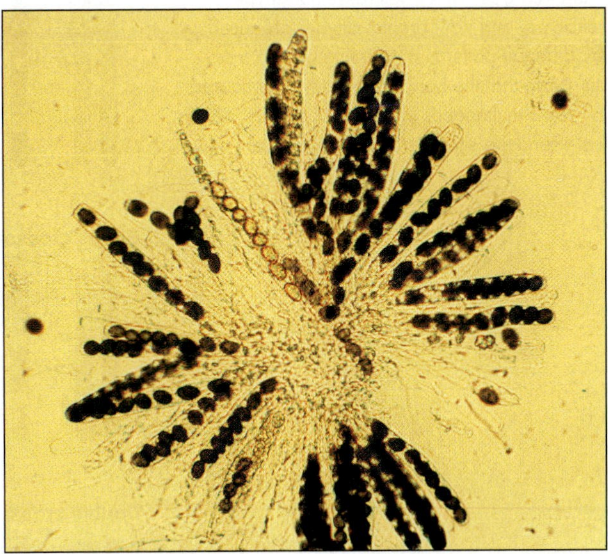
(b)

FIGURE 6.14 Tetrad analysis in *Neurospora*. Genetic crosses are easily done in species with an alternate haploid and diploid life cycle. (a) The procedure for isolating and analyzing the genotypes of individual spores. (b) A photograph of asci from *Neurospora*. The individual that produced these asci was heterozygous for an allele affecting spore color. *(James W. Richardson/Visuals Unlimited)*

called a **nonparental ditype** (NPD). If two genes are not linked, then Mendel's rule of independent assortment predicts that a dihybrid cross will result in an equal number of parental and nonparental type tetrads. If the two genes are linked, then the number of parental type tetrads will be greater than the number of nonparental type tetrads (PD > NPD). To determine linkage, we simply compare the numbers of the two different types of tetrads.

If two genes are linked, then recombinant spores are formed by crossing over between the loci. If a single crossover occurs between the loci in one cell, then the tetrad formed by that cell will have four different types of spores: two parental type spores and two recombinant type spores. This type of tetrad is called a **tetratype** (TT) (Figure 6.15). The map distance between linked loci is determined by counting the numbers of PD, NPD, and TT tetrads. PD tetrads are produced when no crossovers occur between the two linked loci. A single crossover between the loci gives rise to a TT tetrad, and two crossovers between the loci gives rise to an NPD tetrad. The number of TT and NPD tetrads represents the number of crossovers that have taken place, and by counting them we can determine the frequency of recombination between the two loci. The formula used to calculate map distance (R) from the numbers of different tetrads is

$$R = [\text{NPD} + (1/2)\ \text{TT}]/(\text{total tetrads} \times 100) = \text{map units}$$

In this calculation, each tetratype ascus is multiplied by 1/2 because only 1/2 (two of four) of the chromatids in the tetrad have crossed over. In NPD tetrads, all chromatids have crossed over.

As an example of how map distances are calculated in tetrad analysis, consider the cross between an *arg* + strain of *Chlamydomonas* and a + *pab* strain. Individuals from each strain are crossed, and the following tetrads are collected:

PD	NPD	TT
+ *pab*	+ +	+ *pab*
+ *pab*	+ +	+ +
arg +	*arg pab*	*arg pab*
arg +	*arg pab*	*arg* +
110	2	68

FIGURE 6.15 PD, TT, and NPD tetrads. The three types of tetrads, parental ditype, tetratype, and nonparental ditype, produced by different patterns of crossing over in meiosis. Note that the nonparental ditype is caused by two simultaneous crossovers between different strands, a situation not discussed until now.

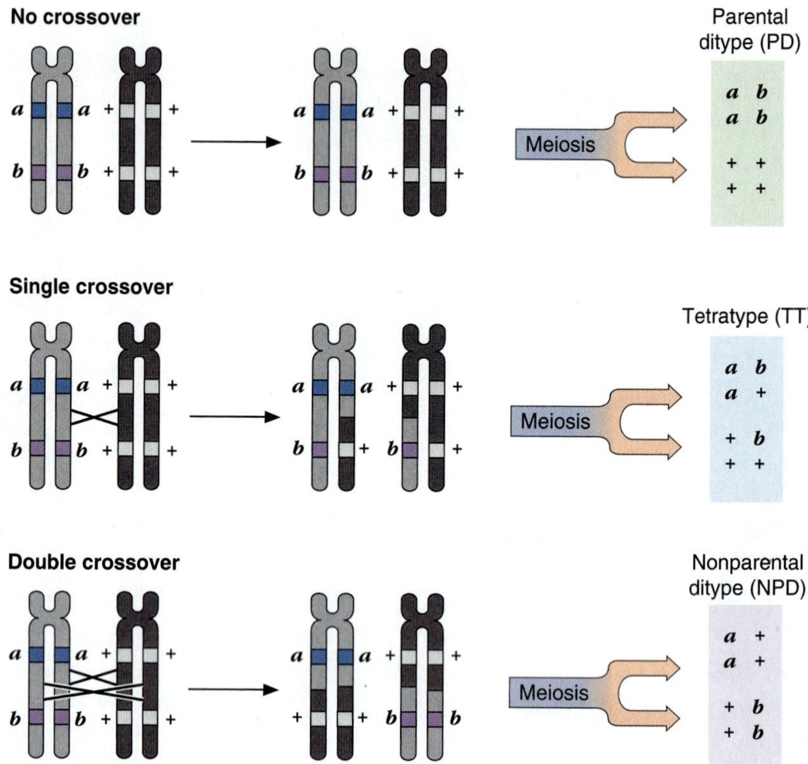

To determine whether these two loci are linked, we compare the numbers of PD and NPD tetrads. The PD tetrads are clearly much more frequent than the NPD, so the loci are linked. The map distance between the loci is calculated using the formula described earlier:

$$R_{arg-pab} = [\text{NPD} + (1/2)\ \text{TT}]/(\text{Total} \times 100) = [2 + (1/2)(68)]/(180 \times 100) = 20 \text{ cM}$$

In haploid organisms, linkage maps are made using dihybrid crosses, but the preferred mapping cross is the three-point testcross. For example, let us consider an hypothetical three-point testcross in which *a b c* and + + + strains of yeast are crossed. The heterozygous cells undergo meiosis, and the resulting tetrads are collected. A sample of these tetrads are collected, and the spores in each are germinated and tested to determine their genotypes. In a sample of 1000 asci, the following genotypes are observed:

A	B	C	D
a b c	*a b* +	*a* + +	*a b c*
a b c	*a b c*	*a b c*	*a* + *c*
+ + +	+ + *c*	+ *b c*	+ *b* +
+ + +	+ + +	+ + +	+ + +
837	80	75	2

E	F	G
a b +	*a* + +	*a b* +
a + +	*a* + +	*a b* +
+ *b c*	+ *b c*	+ + *c*
+ + *c*	+ *b c*	+ + *c*
2	2	2

Preparing a linkage map from these data is similar to preparing linkage maps in other types of organisms. First, we can determine if the genes are linked. To do this, we determine whether each ascus is a PD or NPD for each pair of genes. You need to remember that whether doing tetrad analyses or testcrosses in *Drosophila*, genes are always scored two at a time in linkage mapping (as they really represent a single crossover event). If any pair of genes has significantly more PD than NPD asci, then the two genes are linked. In this cross, the parental strains were *a b c* and + + +, so the parental ditype tetrads contain two spores of the recessive alleles and two spores of the + alleles. For example, for the *a* and *c* loci, parental ditypes contain two *a c* spores and two + + spores. For these loci, A and D appear to be parental ditypes and E, F, and G are nonparental ditypes. Classifying and counting all the tetrad types gives the following:

Loci	Parental Ditypes	Nonparental Ditypes
a–b	A + B + G 837 + 80 + 2 = 919	F 2
a–c	A + D 837 + 2 = 839	E + F + G 2 + 2 + 2 = 6
b–c	A + C + F 837 + 75 + 2 = 914	G 2

For all three pairs, there are more PD tetrads, so we conclude that all three loci are linked. The next step is to determine the linkage distance between each pair of loci using the formula described earlier {R = [NPD + (1/2) TT]/Total × 100}.

$$R_{a-b} = [2 + (1/2)(75 + 2 + 2)]/1000 \times 100 = 4.15 \text{ cM}$$

$$R_{a-c} = [2 + 2 + 2 + (1/2)(80 + 75)]/1000 \times 100 = 8.35 \text{ cM}$$

$$R_{b-c} = [2 + (1/2)(80 + 2 + 2)]/1000 \times 100 = 4.40 \text{ cM}$$

The correct order of the three loci in the chromosome can be deduced directly from these map distances. The *a* and *c* loci have the largest map distance, so these must be the furthest apart. The order of the loci is thus *a–b–c*, and the linkage map is

*a*______*b*______*c*
4.15 4.4

For each of the tetrads, we can determine the number of crossovers and the locations of the crossovers in the meiotic cell that produced that tetrad. For example, the D class of tetrads must have been produced by a double crossover involving the second and third chromatids, with one crossover between the *a* and *b* loci and the second between the *b* and *c* loci (Figure 6.16).

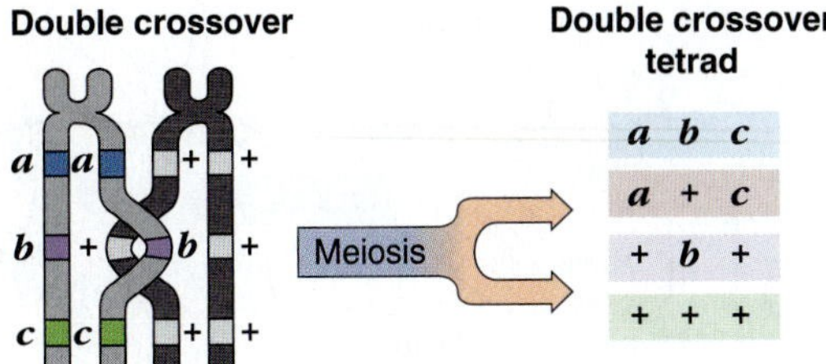

FIGURE 6.16 Mapping crossover location in a three-point testcross. The location of the crossovers that produce a particular tetrad can be determined in a three-point testcross. Examine the data from left to right to see how the double crossover produces the four different genotypes.

SOLVED PROBLEM

Problem

A cross was made between two strains of yeast (*Saccharomyces cerevisiae*) that had different alleles at three loci: *arg, trp,* and *lys*. A sample of asci was collected from the heterozygous F_1, and the genotypes of the spores in each ascus were determined. The results are given here. From these data determine whether the three genes are linked, and if they are, make a linkage map showing the gene order and distances between them.

1	2	3	4	5	6	7	
arg trp lys	*arg trp lys*	*arg trp lys*	*arg trp lys*	+ *trp lys*	*arg trp* +	*arg* + *lys*	
arg trp lys	+ *trp lys*	+ + *lys*	+ *trp* +	+ *trp lys*	*arg trp* +	*arg* + *lys*	
+ + +	*arg* + +	*arg trp* +	*arg* + *lys*	*arg* + +	+ + *lys*	+ *trp* +	
+ + +	+ + +	+ + +	+ + +	*arg* + +	+ + *lys*	+ *trp* +	Total =
7989	1365	559	20	56	10	1	10,000

Solution

To determine whether the genes are linked, we determine the numbers of the three types of ascus (PD, NPD, and TT) for each pair of loci (*arg–trp, trp–lys,* and *arg–lys*). If two loci are linked, then one ditype (PD) will be present in excess. Examining the data, we see that for all three pairs the ditype with the recessive alleles in *cis* is present in great excess. This indicates that all three loci are linked and that the two original parental strains must have been *arg trp lys* and + + +.

To generate a linkage map we must determine the map distances between the loci:

$$R_{arg-trp} = [(56 + 1) + (1/2)(1365 + 20)]/10{,}000 \times 100 = 7.5 \text{ cM}$$

$$R_{arg-lys} = [(56 + 10) + (1/2)(1365 + 559)]/10{,}000 \times 100 = 10.3 \text{ cM}$$

$$R_{trp-lys} = [(10 + 1) + (1/2)(559 + 20)]/10{,}000 \times 100 = 3.0 \text{ cM}$$

These results indicate that *arg* and *lys* are the farthest apart and that the map is

*arg*______*trp*______*lys*
7.5 3.0

Ordered Tetrad Species Aid in Determining the Map Distance Between a Locus and its Centromere.

In this section, we will discuss how *Neurospora* and certain other fungal species are especially useful in mapping studies because they allow us to calculate the map distance between a locus and its centromere. During the

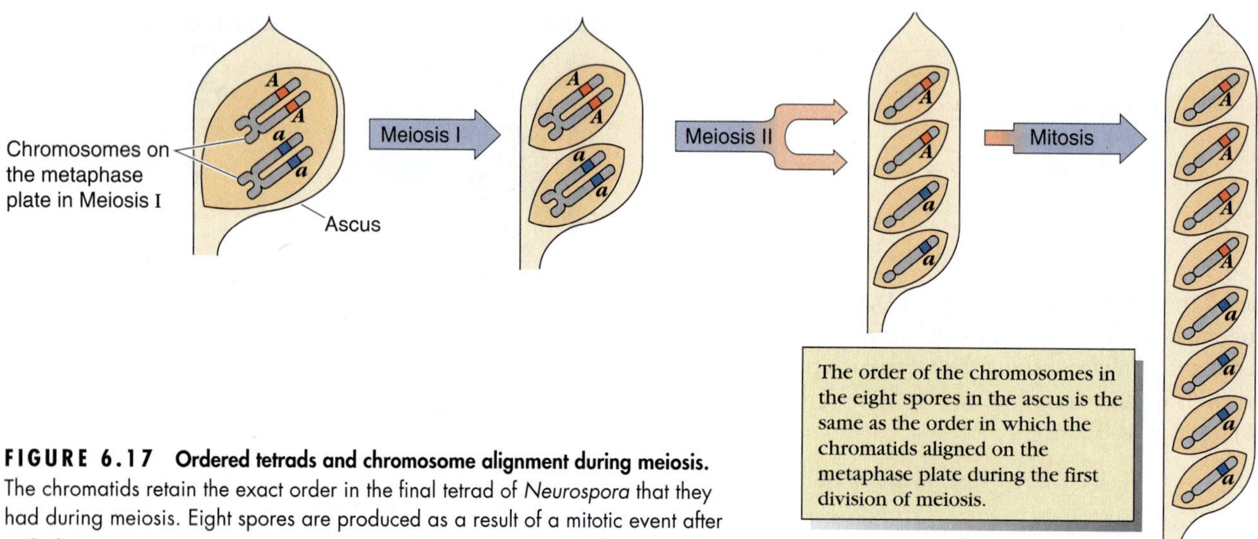
Chromosomes on the metaphase plate in Meiosis I
Ascus
Meiosis I
Meiosis II
Mitosis
The order of the chromosomes in the eight spores in the ascus is the same as the order in which the chromatids aligned on the metaphase plate during the first division of meiosis.

FIGURE 6.17 Ordered tetrads and chromosome alignment during meiosis. The chromatids retain the exact order in the final tetrad of *Neurospora* that they had during meiosis. Eight spores are produced as a result of a mitotic event after meiosis.

meiotic and mitotic divisions that take place in an ascus of these fungal species, the nuclei maintain a strict linear order. Accordingly, these types of tetrads are called **ordered tetrads.** In an ordered tetrad, the alignment of the chromosomes in the nuclei of the spores is the same as the alignment of the chromatids in bivalents at the metaphase plate during the first meiotic division (Figure 6.17). Centromere mapping is possible in ordered tetrads because a crossover between a locus and the centromere changes the order of the spores in the tetrad. Consider the hypothetical example of tetrads formed by *Neurospora* that are heterozygous for the *A* allele (*aA*). If there are no crossovers between the *A* locus and the centromere, the segregation of the chromosomes during the first meiotic division results in two nuclei, each containing one of the two alleles (*a* or *A*), but not both. The alleles of the locus have segregated from each other during the first meiotic division. This pattern of segregation is called **first division segregation.** The second meiotic division produces in four nuclei, arranged in a characteristic order. The two cells with the *a* allele are at one end of the ascus, and the two cells with the *A* allele are at the other end. In *Neurospora,* a mitotic division follows meiosis, and this produces eight nuclei arranged in a 4 *a* : 4 *A* pattern. This pattern is characteristic of a first division segregation tetrad. If a crossover occurs between the *A* locus and the centromere during the first meiotic division, two nuclei are produced, each with one *a* allele and one *A* allele. Because the nuclei contain one of each type of allele, the alleles have not yet segregated from each other. During the second meiotic division, each chromosome divides, and the resulting four nuclei each have only one allele, either *a* or *A*. Thus, the two alleles finally segregate from each other during the second, rather than the first, meiotic division. The mitotic division that follows meiosis in *Neurospora* produces an ascus that does not have a 4 : 4 arrangement of nuclei. Instead, it has an ascus with eight nuclei arranged in one of several different patterns, such as a 2 : 2 : 2 : 2 pattern or a 2 : 4 : 2 pattern. The precise pattern depends on which of the four chromatids were involved in the crossover. These types of tetrad are called **second-division segregation** tetrads (Figure 6.18).

Because second-division segregation tetrads are produced by crossovers between a locus and the centromere, the frequency of crossing over can be determined by counting the number of second-division segregation tetrads. These values are then used to calculate the linkage map distance between the locus and the centromere:

$$R = (1/2)(\text{\# second-division segregation tetrads})/(\text{\# total tetrads} \times 100) = \text{cM}$$

The number of second-division segregation tetrads is multiplied by 1/2 because they are produced by a single crossover and only two of the four chromatids are involved. The calculation of the locus-to-centromere distance is not as accurate as the locus-to-locus calculation because four-strand double crossovers cannot be counted. A four-strand double crossover between two loci produces an NPD tetrad, but a four-strand double crossover between a locus and the centromere can produce a first-division segregation tetrad, with the second crossover reversing the effect of the first.

To illustrate how centromere mapping is done, consider the following group of tetrads, all produced by *Neurospora* heterozygous for the *b* allele:

A	B	C	D	E
b	+	+	*b*	*b*
+	*b*	*b*	+	*b*
b	+	*b*	+	+
+	*b*	+	*b*	+
10	10	10	10	60

Crossover between chromatids 2 and 3

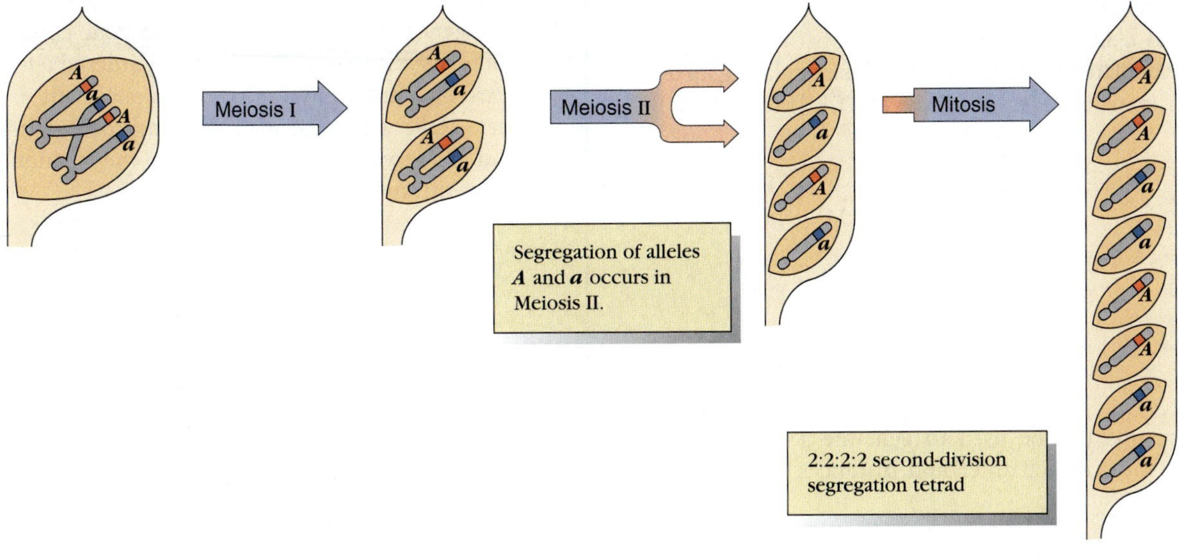

A crossover between chromatids 2 and 4

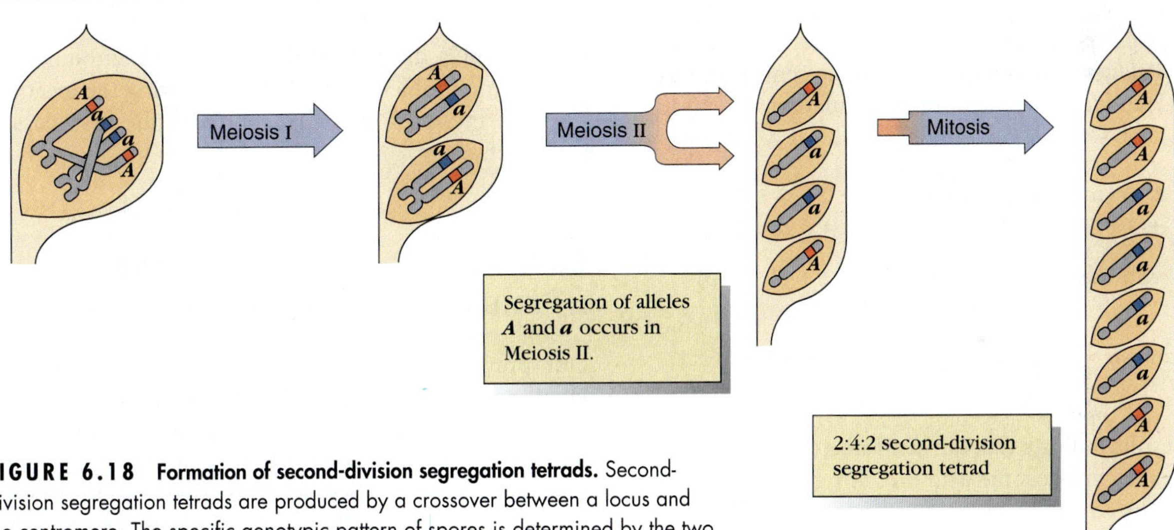

FIGURE 6.18 Formation of second-division segregation tetrads. Second-division segregation tetrads are produced by a crossover between a locus and the centromere. The specific genotypic pattern of spores is determined by the two chromatids involved.

The first four classes of tetrads (A, B, C, D) are second-division segregation tetrads, and the last class (E) is made up of first-division segregation tetrads. The map distance between the *b* locus and the centromere can be calculated using the formula:

$$R_{b-\text{centromere}} = (1/2)\ (40)/(100 \times 100) = 20 \text{ cM}$$

In species with ordered tetrads, a three-point mapping cross combines the calculation of distances between loci and between the loci and the centromere. Consider the following example. Two strains of *Neurospora* that differ in mating type (*A* or *a*), adenine auxotrophy (ad^+ or ad^-), and slow or fast growth (*s* or *f*) were crossed, and 1161 tetrads were recovered from the F_1:

A	B	C	D	E	F
A ad^- *f*	*A* ad^- *f*	*A* ad^- *f*	*A* ad^- *f*	*A* ad^- *s*	*A* ad^- *f*
A ad^- *f*	*a* ad^- *f*	*a* ad^+ *f*	*A* ad^- *s*	*A* ad^+ *s*	*a* ad^+ *s*
a ad^+ *s*	*A* ad^+ *s*	*A* ad^- *s*	*a* ad^+ *f*	*a* ad^- *f*	*A* ad^- *f*
a ad^+ *s*	*a* ad^+ *s*	*a* ad^+ *s*	*a* ad^+ *s*	*a* ad^+ *f*	*a* ad^+ *s*
888	85	43	126	1	2

G	H	I	J	K	L
A ad^- *f*	*A* ad^- *s*	*A* ad^- *f*	*A* ad^- *f*	*A* ad^- *s*	*A* ad^- *s*
a ad^+ *s*	*a* ad^+ *f*	*a* ad^- *s*	*a* ad^- *s*	*a* ad^- *f*	*a* ad^- *f*
A ad^- *s*	*A* ad^- *s*	*A* ad^+ *f*	*A* ad^+ *s*	*A* ad^+ *f*	*A* ad^+ *s*
a ad^+ *f*	*a* ad^+ *f*	*a* ad^+ *s*	*a* ad^+ *f*	*a* ad^+ *s*	*a* ad^+ *f*
2	2	3	5	3	1

The great excess of the A-class tetrads shows that these three loci are linked. The parental types must be *A* ad^- *f* and *a* ad^+ *s*. The map distances between the loci can be calculated using the formula for calculating locus-to-locus map distances: [NPD + (1/2)TT)]/(total × 100) = cM.

$$A\text{–}ad = (1/2)(85 + 1 + 3 + 5 + 3 + 1)/(1161 \times 100) = 4.22 \text{ cM}$$

$$ad\text{–}f = 2 + (1/2)(43 + 126 + 1 + 2 + 3 + 5 + 3 + 1)/(1161 \times 100) = 8.10 \text{ cM}$$

$$A\text{–}f = 1 + 1 + 2 + (1/2)(85 + 43 + 126 + 2 + 5 + 3)/(1161 \times 100) = 11.71 \text{ cM}$$

These results can be used to generate a linkage map showing the correct order and distances between the loci:

A_____ *ad* _____ *f*
4.22 8.10

To this map we can now add the centromere by calculating the distance between the loci and the centromere using the locus-to-centromere map calculation: (1/2)(second-division segregation tetrads)/(total × 100) = cM.

$$A\text{–centromere} = (1/2)(85 + 43 + 2 + 2 + 2 + 3 + 5 + 3 + 1)/(1161 \times 100) = 6.28 \text{ cM}$$

$$ad\text{–centromere} = (1/2)(43 + 1 + 2 + 2 + 2)/(1161 \times 100) = 2.15 \text{ cM}$$

$$f\text{–centromere} = (1/2)(126 + 2 + 2 + 2 + 3 + 5 + 3 + 1)/(1161 \times 100) = 6.20 \text{ cM}$$

These results indicate that the centromere (–*o*–; the centromere of a chromosome with one chromatid) is located between the *ad* and *f* loci:

A_____ *ad* _____ *o* _____ *f*
4.22 2.15 6.28

After the arrangement of the loci and the centromere are known, the location of the crossover that gave rise to each class of tetrad is determined. For example, the 43 tetrads in the C class were produced by single crossovers. These crossovers involved chromatids 2 and 3 and occurred between the *ad* locus and the centromere (–*o*–; the centromere of a chromosome with two chromatids):

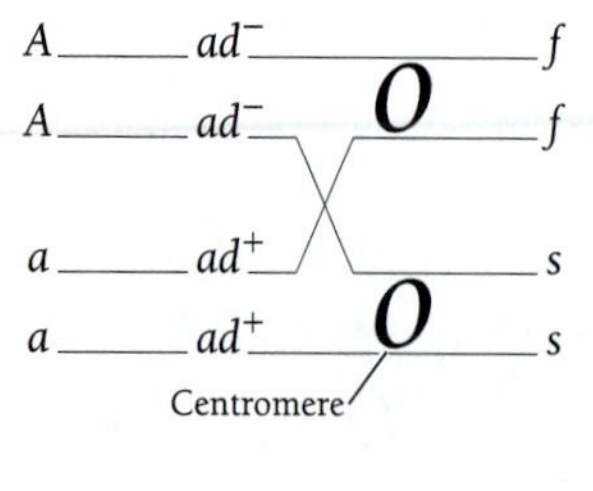

A _____ ad^- __________ *f*
A _____ ad^- ___ O ___ *f*
a _____ ad^+ __________ *s*
a _____ ad^+ ___ O ___ *s*
Centromere

SECTION REVIEW PROBLEMS

11. The genes *ala*, *mt*, and *p* are on the same chromosome in *Neurospora*. The following ordered tetrads were obtained from a cross of *ala p* + with + + *met*. Make a linkage map showing the locations and distances between the loci and the location of the centromere.

ala p +	*ala p* +	*ala p* +	*ala p* +
+ *p mt*	*ala* + +	*ala p* +	+ + +
ala + +	+ *p mt*	+ + *mt*	*ala p mt*
+ + *mt*	+ + *mt*	+ + *mt*	+ + *mt*
24	16	203	2
ala + +	*ala p* +	*ala p* +	*ala p* +
+ *p* +	+ + *mt*	*ala mt p*	+ *p* +
ala + *mt*	*ala p* +	+ + +	*ala* + *mt*
+ *p mt*	+ + *mt*	+ + *mt*	+ + *mt*
1	1	1	58

12. Draw a diagram of the *Neurospora* chromosomes from Problem 11 in meiosis showing the locations of the crossovers that produced the first class of tetrads.

GENE MAPPING CAN BE DONE IN THE ABSENCE OF MEIOSIS.

Our previous mapping discussions have all involved mapping by measuring the frequency of meiotic recombination. Other mapping techniques do not use the frequency of meiotic recombination as a measure of chromosomal distance. These techniques are used in species that reproduce asexually, or where meiotic crossing over is difficult or inconvenient to measure. In this section, we will discuss three such mapping techniques: (1) using the frequency of mitotic crossing, (2) using cell fusion and chromosomal elimination, and (3) using deficiency complementation.

The Genes Can Be Located on a Chromosome Using Mitotic Crossing Over.

Curt Stern discovered in 1936 that homologous chromosomes recombine during mitotic as well as meiotic cell division in *Drosophila*. This mitotic crossing over has since been shown to occur in other species, and it is an important mapping tool in some fungal species. Stern observed that female *Drosophila* heterozygous for the recessive, X-linked mutations *yellow* (*y*), *forked* (*f*), or *singed* (*sn*) would appear mostly wild-type but would occasionally have small patches of tissue showing mutant pheno-

FIGURE 6.19 Mitotic crossing over cell clone. This scanning electron micrograph shows a somatic cell clone in the abdomen of *Drosophila*. The cells in the clone (within the dotted line) are homozygous for mutant alleles that cause the large bristles to bend and twist and cause the small hairs to have multiple shafts. The surrounding cells have a wild-type phenotype. *(From J. H. Postlethwait. 1976. The Genetics and Biology of* Drosophila *Vol. 2C Ch. 24, p. 359.)*

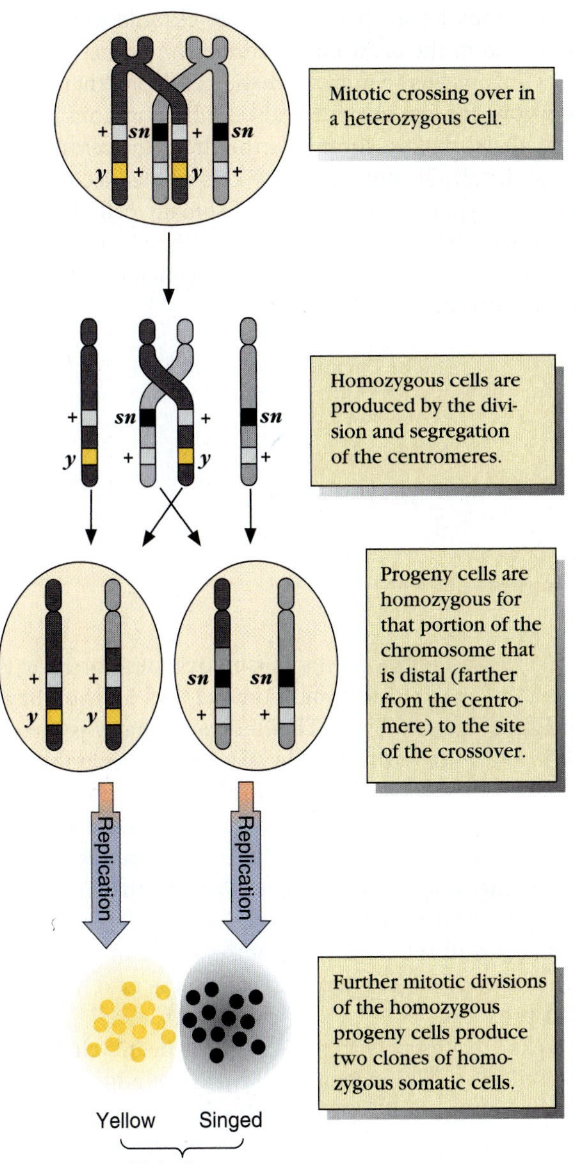

FIGURE 6.20 Mitotic crossing over. Mitotic crossing over in heterozygous individuals can produce somatic clones of homozygous cells. In this example, a *y* + / +*sn Drosophila* produces two somatic clones, one *yellow* (*y* + / *y* +) and one *singed* (+*sn* / +*sn*).

types (Figure 6.19). He also noticed that when a fly had two mutations in *cis* (*y sn* / +), patches of mutant tissue often showed both phenotypes (yellow and singed), but that when the mutant alleles were in *trans* (*y* / *sn*), two separate patches of mutant tissue usually appeared adjacent to each other in the same individual, one patch showing yellow and the other patch singed. Stern called the double patches twin spots. He concluded that these somatic patches resulted from a reciprocal exchange between the homologous chromosomes carrying the mutant alleles. Mitotic crossing over occurs far less frequently than meiotic crossing over (less than 1/1000 as often), although the frequency can be increased by irradiation.

The patches that Stern was observing were, in fact, somatic clones produced by mitotic crossing over. If a mitotic crossover occurs between a heterozygous locus and the centromere while the cell is undergoing mitosis, each of the two progeny cells may end up homozygous for the alleles (Figure 6.20). If mitotic crossing over occurs early in development, the two homozygous cells will undergo further mitotic divisions, and each will give rise to a population of homozygous cells, that is, the patch that Stern observed. Because they are all descended from one original progeny cell, these patches are called **somatic cell clones.** The study of mutations in somatic clones is called somatic clonal analysis. Somatic clonal analyses are often done in developmental genetic studies (Chapter 15). The frequency and phenotype of somatic cell clones reflects the frequency and location of mitotic crossing over events along the chromosome. By counting the phenotype and frequency of somatic clones, we can determine the location and frequency of mitotic crossing over events, and we can use these frequencies to generate chromosome maps. The goals of mitotic mapping are to determine the order of loci on the chromosome, the relative distances between the loci, and the location of the centromere.

The order of loci on a chromosome can be determined by examining the phenotypes of somatic cell clones. Mitotic crossing over generates cells that are homozygous for all portions of the chromosome distal (farther from the centromere) to the site of the crossing over. This means that any somatic cell clone that is homozygous for one allele will also be homozygous for all other alleles that are located farther from the centromere (more distal). To illustrate how this works, let us consider the hypothetical example of somatic cell clones in individuals heterozygous for alleles at five loci, *a b c d e*. In a sample of such individuals, 300 somatic cell clones were observed.

Clone Phenotype	Number
a b c d e	100
b c d e	60
c d e	75
d e	40
e	25
Total	300

Whenever a clone of cells is homozygous for *a*, it is also homozygous for *b*, but some clones are homozygous for *b* but not for *a*. This indicates that *a* is closer to the centromere than *b*. The clones that are homozygous for *a* and *b* are the result of crossover events that occur between *a* and the centromere. Those clones that are homozygous for *b* but not for *a* result from crossing over events that occur between *a* and *b* (that is, crossing over distal to *a* but proximal to *b*) (Figure 6.21). Examining all the clone phenotypes, we can see that e is the most distal, d is next, c is next, b is next, and a is most proximal.

The frequency with which each somatic clone occurs reflects the frequency with which each different mitotic crossing over occurs, and this frequency is used to calculate mitotic map distances between loci. Mitotic map distance is calculated as the percentage of clones resulting from crossovers between two genes. For example, we can use the collection of 300 clones listed previously to calculate the mitotic map distance between these loci. For example, crossovers between *b* and *c* produce clones whose phenotype is c d e. The distance between *b* and *c* is equal to the percentage of such clones, or (75 / 300) × 100 = 25. Calculating the distances between all the genes gives a mitotic chromosome map:

*o*______*a*______*b*______*c*______*d*______*e*______
33.3 20 25 13.3 8.3

The distance between a gene and the centromere in a somatic clonal analysis is equal to the percentage of the somatic clones in which that gene is homozygous. The distance between *a* and the centromere is equal to (100/300) × 100 = 33.3. The distance between *b* and the centromere is (160/300) × 100 = 53.3. Because mitotic map distances are calculated as a percentage of a particular population of somatic clones, the distances always sum to 100. Also, unlike meiotic map distances, mitotic map distances calculated from different experiments are not additive. However, by mapping several mutations against a standard mutation that is far from the centromere, mitotic maps that show the locations and distances between many genes on the chromosome can be made.

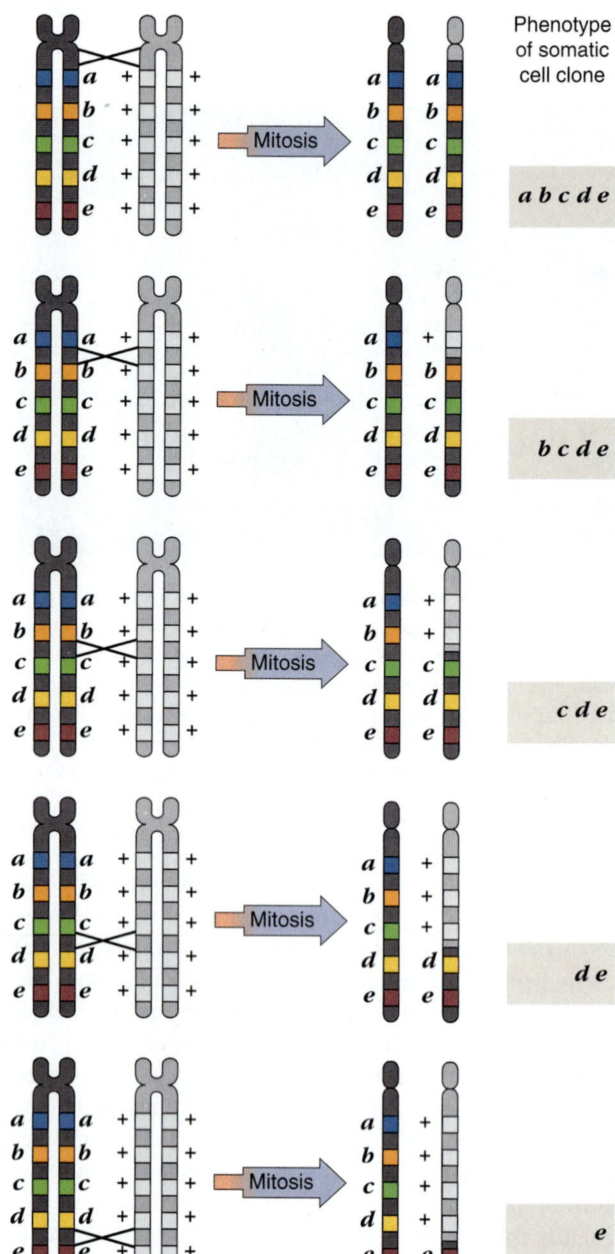

FIGURE 6.21 Mapping based on mitotic crossing over. Mitotic crossing over at different locations between chromosomes that have different alleles for several loci (a–e) produces clones with different phenotypes. These can be used to generate mitotic maps. Note that for simplicity only the progeny cell from the mutant clone is shown.

In this particular example, all the mutant alleles were on the same homologue (in *cis*). If some of the mutant alleles were in *trans*, then a single crossover could produce two clones, or a twin spot. Twin-spot clones are

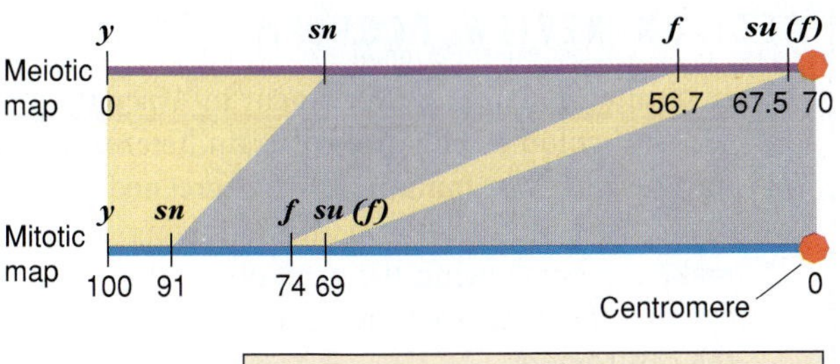
y
sn
f
su (f)
Meiotic map
0
56.7
67.5
70
y
sn
f
su (f)
Mitotic map
100
91
74
69
Centromere
0
The mitotic and meiotic linkage maps of the X chromosome in *Drosophila* give the same gene order but different relative spacing.

FIGURE 6.22 Comparison of mitotic and meiotic maps. The X chromosome meiotic and mitotic maps of *Drosophila* demonstrate that meiotic and mitotic mapping give the same linear arrangement of loci in the chromosome, but they give different distances between loci. *(Data from Becker, H. J. 1976.* The Genetics and Biology of Drosophila *Vol. 1C, Ch. 25, p. 1069.)*

produced when both progeny cells of the original mitotic cell are homozygous for mutant alleles and, as a result, give them a visible phenotype. Consider the following arrangement of the alleles in our preceding example:

	a	+	*c*	+	*e*
O	*a*	+	*c*	+	*e*
	1	2	3	4	5
	+	*b*	+	*d*	+
O	+	*b*	+	*d*	+

A mitotic crossover in interval 1 will produce a twin-spot clone, with one spot having a phenotype of a c e and the other spot having a phenotype of b d.

Comparing a mitotic and a meiotic map of the X chromosome in *Drosophila* (Figure 6.22) shows that the maps have some important similarities and some important differences. The order of the loci is the same in both maps, but relative spacing between the loci is different. Some loci that are relatively close together on one map are farther apart on the other. The most striking difference between mitotic and meiotic maps is in the heterochromatic region of the chromosome near the centromere. The majority of mitotic crossing over events (70% in the X chromosome) occur in the heterochromatic region, whereas very few, if any, meiotic crossing over events occur in the heterochromatin. These differences may represent a very real difference in the molecular mechanism of meiotic and mitotic crossing over. The molecular mechanism of crossing over is discussed further in Chapter 19.

Mitotic crossing over has been an important tool for mapping genes in a number of fungal species that lack a sexual cycle. Some species of *Aspergillus* are normally haploid, but hyphae of two strains occasionally fuse to form cells with nuclei from both strains, called **heterokaryons.** Two nuclei in a heterokaryon will occasionally fuse to form a diploid nucleus. Mitotic crossing over occurs in the mitotic divisions that follow nuclear fusion, resulting in homozygous sectors of diploid hyphae. Asexual sporulation of these cells produces homozygous diploid individuals. Geneticists interested in crossing two strains of *Aspergillus* can select for the formation of heterokaryons by crossing two strains containing different auxotrophic mutations. For example, if a strain requiring histidine (his^-) is crossed with a strain requiring tryptophan (trp^-), then the heterozygous heterokaryons and the resulting diploids will grow on media containing neither histidine nor tryptophan (Figure 6.23).

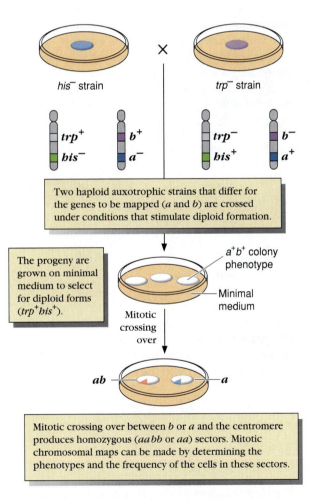

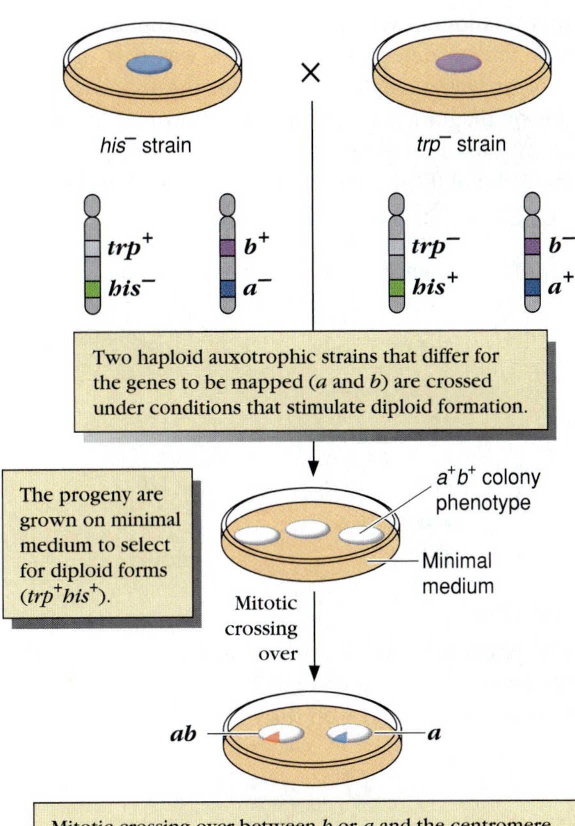
×
his⁻ strain
trp⁻ strain
trp⁺
his⁻
b⁺
a⁻
trp⁻
his⁺
b⁻
a⁺
Two haploid auxotrophic strains that differ for the genes to be mapped (a and b) are crossed under conditions that stimulate diploid formation.
The progeny are grown on minimal medium to select for diploid forms (trp⁺his⁺).
a⁺b⁺ colony phenotype
Minimal medium
Mitotic crossing over
ab
a
Mitotic crossing over between b or a and the centromere produces homozygous (aabb or aa) sectors. Mitotic chromosomal maps can be made by determining the phenotypes and the frequency of the cells in these sectors.

FIGURE 6.23 Mitotic mapping in *Aspergillus*. Mitotic mapping is done in *Aspergillus* by crossing two strains with different mutations, selecting diploid heterozygotes, and then determining the frequency of mitotic crossing over by detecting the frequency of recombinant, homozygous cells. These appear as sectors in the colonies of *Aspergillus*. Each sector contains homozygous cells produced by one mitotic crossing over. Detecting recombinant sectors can be made easier by using selective media on which the nonrecombinant heterozygous cells cannot grow but the homozygous recombinant cells can.

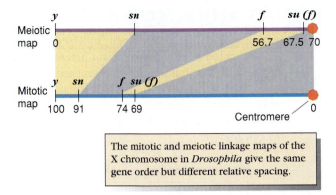

Mitotic crossing over in fungi, as in *Drosophila*, generates clones of cells that are homozygous for all genes on the chromosome that are distal to the crossover. Following the procedures already discussed, mitotic maps showing the order and relative distances between genes in fungi can be generated by analyzing the phenotype and frequency of somatic clones. These clones will appear as regions or sectors of homozygous cells in a population of heterozygous cells.

SOLVED PROBLEM

Problem

A number of *Drosophila* that were heterozygous for four recessive mutations, *sn*, *y*, *v*, and *cv*, were examined for the presence of somatic cell clones. Each clone that was found was examined to determine its phenotype and to determine whether it was part of a twin spot or was a single clone. From the following data, draw a diagram showing the chromosomes in the heterozygous individuals, and generate a map showing the order of these genes, the relative distances between them, and the location of the centromere.

Clone	Phenotype	Number
Twin	v cv sn y	1005
Twin	cv sn y	180
Twin	cv y	120
Single	y	195
Total		1500

Solution

The phenotypes of the twin-spot clones suggest that the heterozygous individuals had *v* and *cv* on one homologue and *sn* and *y* on the other:

a + *c*

0 *a* + *c*

1 2 3

\+ *b* +

0 + *b* +

The map distances are calculated by determining the percentages of somatic clones that are produced by crossovers between loci. For example, the distance between the *v* and *sn* loci is the percentage of somatic clones that are *sn* but not *v* [180/(1500 × 100) = 12.0]. Calculating the map distances between each locus gives us a mitotic map:

o ___ *v* ___ *sn* ___ *cv* ___ *y*

67.0 12.0 8.0 13.0

SECTION REVIEW PROBLEM

13. In a mitotic mapping experiment in *Aspergillus*, a series of colonies of a diploid strain heterozygous for five recessive mutations were screened for recombinant sectors. From the following results, make a map showing the gene order, the relative distance between each gene, and the location of the centromere:

Phenotype of Sector	Number
a b c d e	450
a b c d	100
a c d	150
a d	200
a	100
Total	1000

Human Genes Can Be Mapped by Somatic Cell Fusion.

Mapping human genes on chromosomes presents special problems. Obviously, the controlled matings and testcrosses normally used for linkage mapping cannot be done with humans. However, some linkage data can be obtained from pedigree analyses of families where different alleles of known genes are segregating. But this type of analysis is difficult, reflected in the fact that prior to 1965 fewer than 25 human genes had been mapped. This difficulty led to a new technique for determining the location of genes on human chromosomes, known as **somatic cell hybridization.** This, combined with improved cytological techniques for distinguishing chromosomes (discussed in Chapter 3), has enabled researchers to determine the chromosomal location of thousands of human genes.

Somatic cell hybridization, as its name suggests, involves fusing (also called hybridizing) two tissue-culture-grown somatic cells. Many different types of tissue-culture cells can be fused. A fusion of cells from the same species is called **monospecific hybridization,** and fusion of cells from two different species is called **interspecific hybridization.** If two cells fuse but the nuclei remain separate, the binucleate cell is called a **heterokaryon.** If the nuclei fuse, the cell becomes uninucleate and is called a **synkaryon.** Cells in tissue culture fuse spontaneously only very rarely, less than 1 in 10^6, but several experimental treatments that greatly increase the rate of cell fusion are known. Agents commonly used to induce cell fusion are the inactivated *Sendai* virus and polyethylene glycol. The *Sendai* virus stimulates fusion by binding to two cells and holding them close together, which promotes fusion. Polyethylene glycol alters cell membranes so that two cells that come in contact are more likely to fuse.

Human gene mapping is usually done using mouse–human or hamster–human cell hybrids. First, cell lines are established containing populations of the mitotic descendants of a single fused cell. A key feature of a hybrid cell line is the behavior of the chromosomes when the cells undergo mitotic division. Although all the mouse or hamster chromosomes are faithfully passed to progeny cells during mitosis, the human chromosomes are not. Individual human chromosomes are lost from the cells, seemingly as random events. By carefully screening cells, it is possible to derive cell lines that contain different human chromosomes. Because many human genes are expressed in fused cells, the chromosomal location of a human gene may be determined by correlating the presence of the human gene product with the presence of a particular human chromosome. However, this technique is limited to the mapping of human genes whose product is known and can be identified. If the product of a human gene is present in all cell lines that contain one particular human chromosome, and if it is always absent from lines that do not have that particular human chromosome, we can conclude that the gene resides on that chromosome.

For example, human cells that produce the enzyme thymidine kinase (TK^+) can be fused with mouse cells that cannot produce this enzyme (TK^-). After fusion, several fused cells are selected and used to start separate populations, called cell culture lines, or simply cell lines. Each cell line is tested cytologically to determine what human chromosomes are present (mouse and human chromosomes can be readily distinguished in cytological preparations), and it is also tested to determine whether the human TK enzyme is present. Consider the results of the following series of hypothetical cell fusion lines:

Human Chromosome	Line A	B	C	D	E	F	G	H	I	J
TK	+	+	+	+	+	−	−	−	−	−
1	+	+	+	+	−	+	−	−	−	+
2	+	−	+	−	−	+	−	−	−	+
3	+	+	+	−	−	+	−	−	−	−
4	+	+	+	−	−	−	+	−	−	+
5	+	−	−	+	−	−	+	−	−	+
6	−	+	−	+	+	+	+	+	−	−
7	−	−	−	+	+	−	−	+	−	+
8	−	−	−	+	+	+	−	+	+	−
9	−	−	+	−	+	+	−	+	+	−
10	+	−	+	−	−	+	−	−	+	+
11	+	−	+	−	+	+	+	−	+	−
12	+	+	+	+	−	−	−	+	+	+
13	+	+	−	+	+	+	+	−	+	+
14	−	+	+	+	−	+	+	+	+	−
15	+	−	−	−	−	+	−	+	+	+
16	−	+	+	+	−	−	+	+	−	+
17	+	+	+	+	+	−	−	−	−	−
18	−	−	−	−	−	−	+	−	−	+
19	−	−	−	−	−	+	−	+	−	−
20	−	+	−	+	+	−	−	−	+	−
21	−	−	+	−	+	−	+	−	+	−
22	−	−	−	+	+	−	−	−	+	+
X	−	−	−	−	+	+	+	−	+	−

(continued)

In this sample of cell lines, human chromosome 17 is always present in TK^+ lines and always absent in TK^- lines, indicating that the TK locus is on chromosome 17.

Even using the *Sendai* virus or polyethylene glycol as fusing agents, the majority of mouse and human cells in a culture do not fuse, so a variety of selection techniques have been developed to eliminate the unfused cells. These techniques usually involve the use of selective media in which the fused cells can grow, but the unfused cells cannot grow. A human cell line that cannot produce a vital enzyme, for example, may be fused with cells from a mouse cell line that cannot produce a different vital enzyme. Because the two lines complement each other, fused cells will have both enzymes and be able to grow in selective media. One widely used selection technique involves fusing mouse cells that cannot produce thymidine kinase (TK^-) with human cells that cannot produce the enzyme hypoxanthine–guanidine phosphoribosyltransferase ($HGPRT^-$). If these cells are placed in a medium containing hypoxanthine, aminopterin, and thymidine (called HAT medium) only fused cells will grow (Figure 6.24). The HAT selection technique is based on the two biochemical pathways that both mouse and human cells use to produce nucleotides required for DNA synthesis. Aminopterin is an antimetabolite that blocks the pathway used to synthesize nucleotides from simple sugars and amino acids (this is called the *de novo* synthesis pathway). Cells can use a second pathway to produce nucleotides from existing pyrimidine and purine bases, such as hypoxanthine and thymidine in the media (this is called the salvage pathway). The TK and HGPRT enzymes catalyze two key steps in the salvage pathway, and the loss of either enzyme can block the production of nucleotides. Thus, in HAT media neither the original TK^-

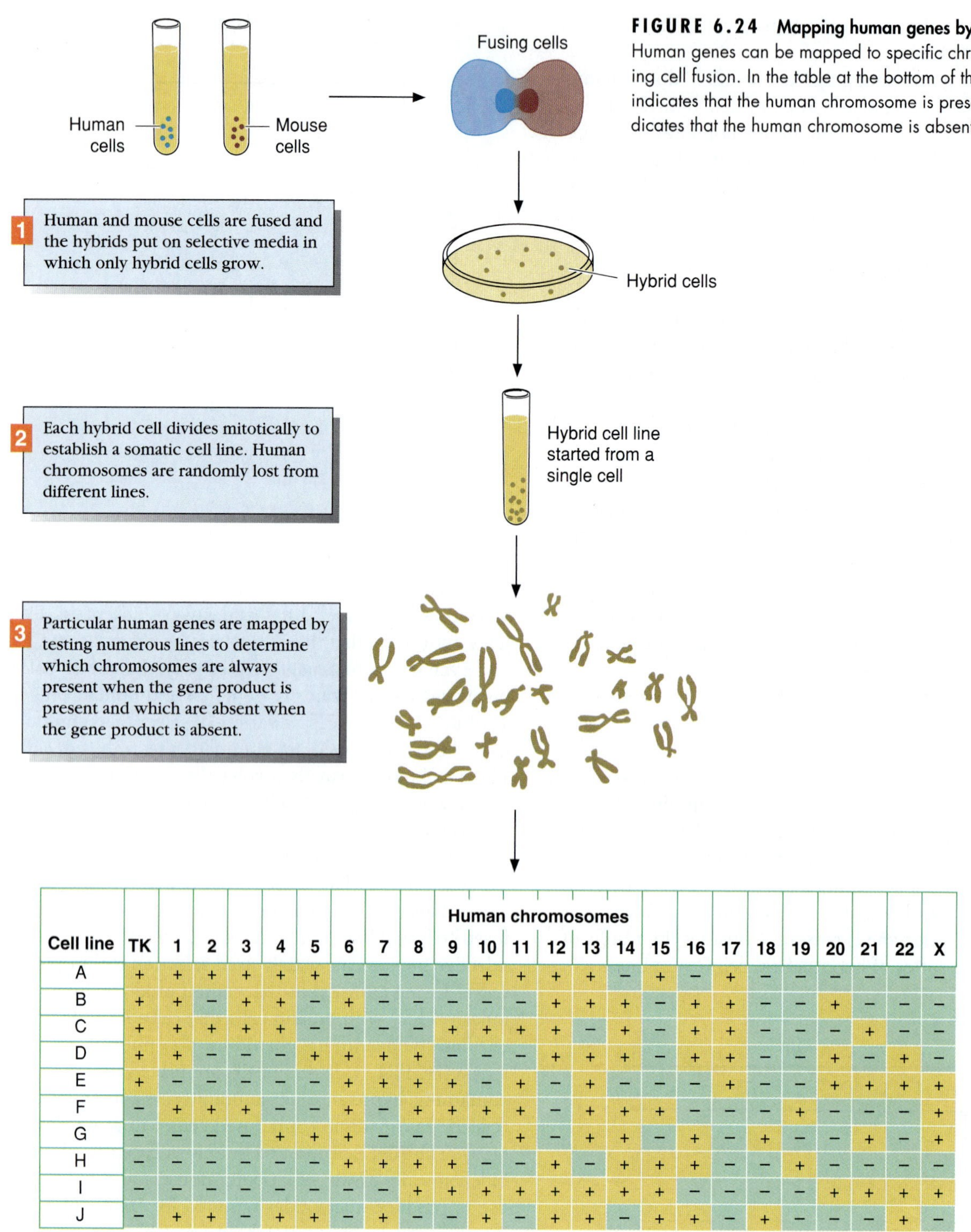
Human cells
Mouse cells
Fusing cells
1 Human and mouse cells are fused and the hybrids put on selective media in which only hybrid cells grow.
Hybrid cells
2 Each hybrid cell divides mitotically to establish a somatic cell line. Human chromosomes are randomly lost from different lines.
Hybrid cell line started from a single cell
3 Particular human genes are mapped by testing numerous lines to determine which chromosomes are always present when the gene product is present and which are absent when the gene product is absent.

Cell line	TK	Human chromosomes 1	2	3	4	5	6	7	8	9	10	11	12	13	14	15	16	17	18	19	20	21	22	X
A	+	+	+	+	+	+	−	−	−	−	+	+	+	+	−	+	−	+	−	−	−	−	−	−
B	+	+	−	+	+	−	+	−	−	−	−	−	+	+	+	−	+	+	−	−	+	−	−	−
C	+	+	+	+	+	−	−	−	−	+	+	+	+	−	+	−	+	+	−	−	−	+	−	−
D	+	+	−	−	−	+	+	+	+	−	−	−	+	+	+	−	+	+	−	−	+	−	+	−
E	+	−	−	−	−	−	+	+	+	+	−	+	−	+	−	−	−	+	−	−	+	+	+	+
F	−	+	+	+	−	−	+	−	+	+	+	+	−	+	+	+	−	−	−	+	−	−	−	+
G	−	−	−	−	+	+	+	−	−	−	−	+	−	+	+	−	+	−	+	−	−	+	−	+
H	−	−	−	−	−	−	+	+	+	+	−	−	+	−	+	+	+	−	−	+	−	−	−	−
I	−	−	−	−	−	−	−	−	+	+	+	+	+	+	+	+	−	−	−	−	+	+	+	+
J	−	+	+	−	+	+	−	+	−	−	+	−	+	+	−	+	+	−	+	−	−	−	+	−

FIGURE 6.24 Mapping human genes by cell fusion. Human genes can be mapped to specific chromosomes using cell fusion. In the table at the bottom of the figure, + indicates that the human chromosome is present, and − indicates that the human chromosome is absent.

mouse nor HGPRT$^-$ human cells will be able to produce DNA nucleotides, and neither will be able to grow. Only the fused cells that contain both enzymes will grow.

Using HAT and other similar selective media, large numbers of fusion cell lines have been established, and over 4000 human genes have been mapped in the last 30 years. A limitation of cell fusion mapping is that it can establish only whether a gene is on a chromosome, not its exact location on the chromosome. However, by combining this procedure with cytology of chromosomes

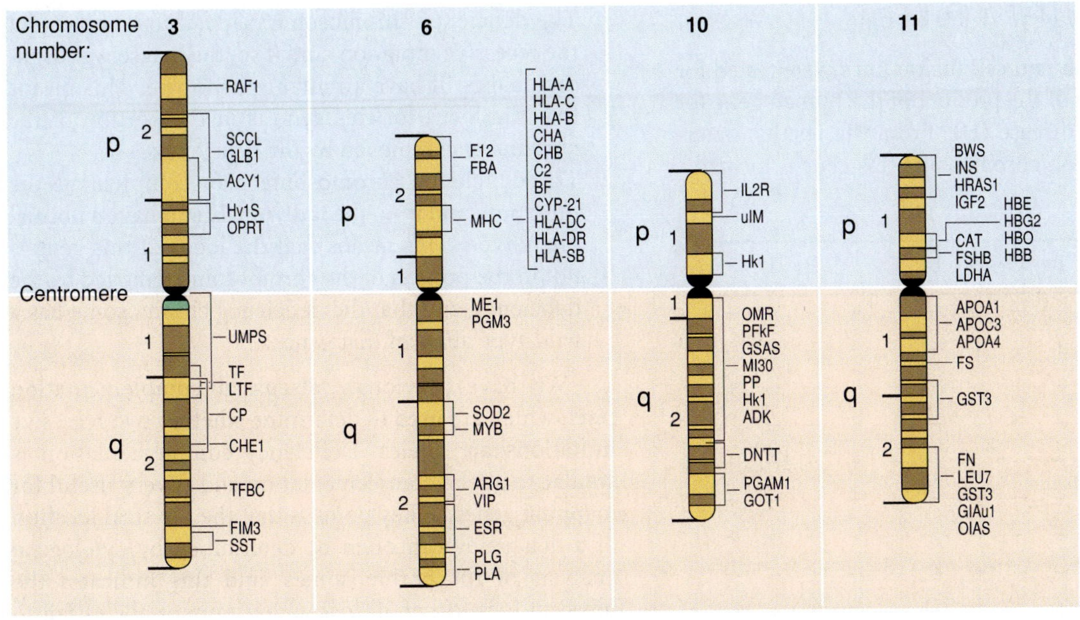

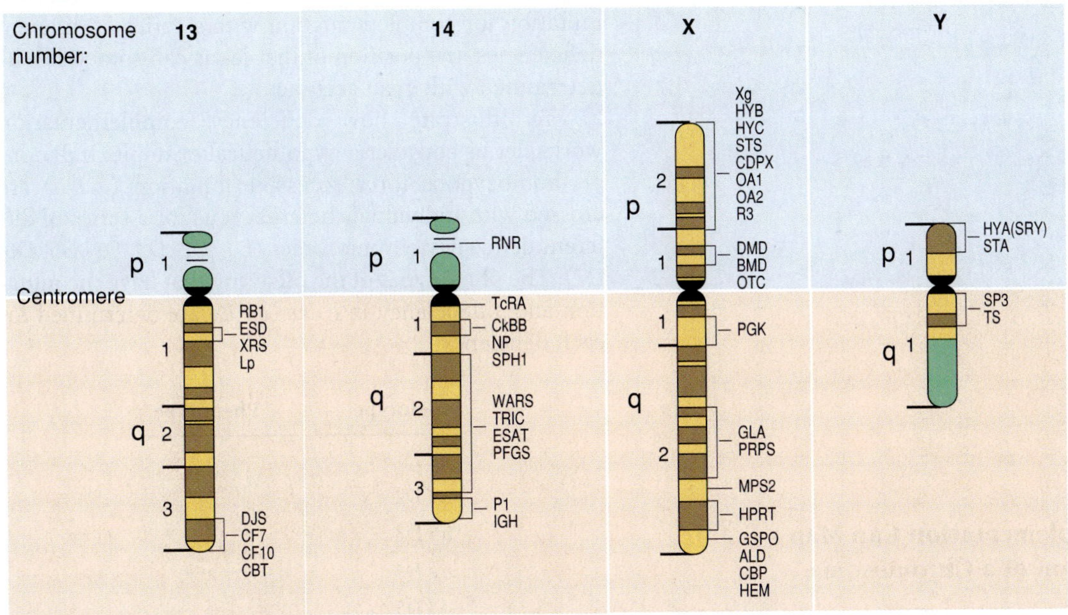

FIGURE 6.25 A map of some human chromosomes. This map shows the location of some of the more than 7000 human genes that have been mapped. Human chromosomes have a short (*p*) and a long (*q*) arm. Each arm is divided into numbered regions based on cytological markers (discussed in Chapter 3). Because gene localization depends on chromosome morphology, good cytological preparations are essential for this gene mapping technique.

with missing parts, the location of genes can also be established. This type of deficiency mapping is discussed in greater detail in the next section. For human gene mapping, after a gene has been mapped to a particular chromosome, cell fusion lines in which portions of that chromosome are missing can be screened. If the product of that gene is present when a particular region is present, but missing when that region is missing, the gene must be located in that region. The locations of a number of human genes are shown in Figure 6.25.

SECTION REVIEW PROBLEM

14. A series of human cell fusion lines were tested for the presence of the product of the human gene for Huntington disease (H). From the results, determine on which chromosome this gene is located.

Human Chromosome	Line A	B	C	D	E	F	G	H	I	J
H	+	+	−	+	+	−	+	−	−	+
1	+	+	−	+	−	+	−	+	+	−
2	−	−	+	−	−	+	−	+	−	+
3	+	−	+	−	−	+	−	−	−	−
4	+	+	−	+	+	−	+	−	−	+
5	+	+	−	+	−	−	+	−	−	+
6	+	+	−	+	+	−	+	−	+	+
7	−	−	−	+	+	−	−	+	−	+
8	+	+	−	+	+	−	−	+	+	−
9	+	−	+	+	−	−	−	+	+	−
10	+	+	−	−	−	+	+	−	−	−
11	−	+	+	−	−	−	+	−	+	+
12	−	+	+	−	−	+	−	+	−	−
13	+	+	−	+	−	−	−	−	+	−
14	−	+	+	+	−	−	−	−	+	−
15	+	+	+	+	−	+	−	−	+	+
16	−	+	−	−	−	+	−	−	−	+
17	+	−	−	+	+	−	+	+	−	−
18	−	−	+	−	+	−	−	+	−	+
19	−	+	−	−	−	+	−	+	−	−
20	+	+	−	+	+	−	+	−	+	−
21	+	−	−	−	−	−	−	−	+	−
22	+	−	−	+	−	−	−	−	−	+
X	+	−	−	+	+	+	+	−	−	+

Deficiency Complementation Can Map a Gene to a Region of a Chromosome.

A third technique for mapping genes involves the use of chromosomes that contain deficiencies. Deficiencies and other chromosomal abnormalities are discussed in greater detail in Chapter 18. Here we are simply concerned with the fact that a deficiency, a physical removal of a portion of a chromosome, removes all the loci that are normally in that portion of the chromosome. A chromosome with a deficiency will, in fact, behave as if it had null alleles for these genes. This allows geneticists to determine whether a particular locus is in the deficiency region by **deficiency complementation.** When an individual is heterozygous for a chromosome containing a recessive allele and a deletion, there are two possible situations:

1. The deficiency chromosome may fail to complement the recessive mutation, and if so, the heterozygous individuals will have a mutant phenotype. This means that the locus for this gene is in the portion of the chromosome removed by the deficiency.
2. The deficiency chromosome may complement the mutation, and if so, the individuals will have a normal phenotype. This means that the locus of this gene is not in the portion of the chromosome removed by the deficiency, and that the deficiency chromosome has a wild-type allele of that gene.

We have previously discussed complementation tests, which are used to determine whether two recessive mutations are alleles. Deficiency complementation is similar to allelic complementation and is very useful for mapping genes. A key point is that the physical location of a deficiency can often be determined by cytological analysis of the chromosomes, and this indicates the physical location of all loci whose mutant alleles fail to complement that deficiency. Thus, deficiency complementation allows the correlation of genetic maps and physical maps of chromosomes. By testing a recessive mutation for complementation with a series of different deficiencies, the position of that locus can sometimes be determined with great accuracy.

To illustrate how deficiency complementation works, let us consider a hypothetical example. Individuals homozygous for a recessive mutation (*a* / *a*) are crossed with individuals heterozygous for a series of different deficiency chromosomes (*D1, D2, D3, D4, D5, D6, D7*). The phenotypes of the offspring that have the mutation and a deficiency in *trans* (*a* / *D*) are determined for each deficiency.

Genotype	Phenotype
a / *D1*	a
a / *D2*	+
a / *D3*	a
a / *D4*	a
a / *D5*	+
a / *D6*	+
a / *D7*	a

The results of this experiment indicate that the *a* locus is in the chromosome portion that is removed by the *D1, D3, D4,* and *D7* deficiencies and is not removed by the *D2, D5,* or *D6* deficiencies. A cytological examination indicates the physical location of the region removed by each deficiency, and this allows us to determine the chromosomal position of the *a* locus.

The accuracy of deficiency complementation varies from species to species and depends on how accurately the location of deficiencies can be determined in cytolog-

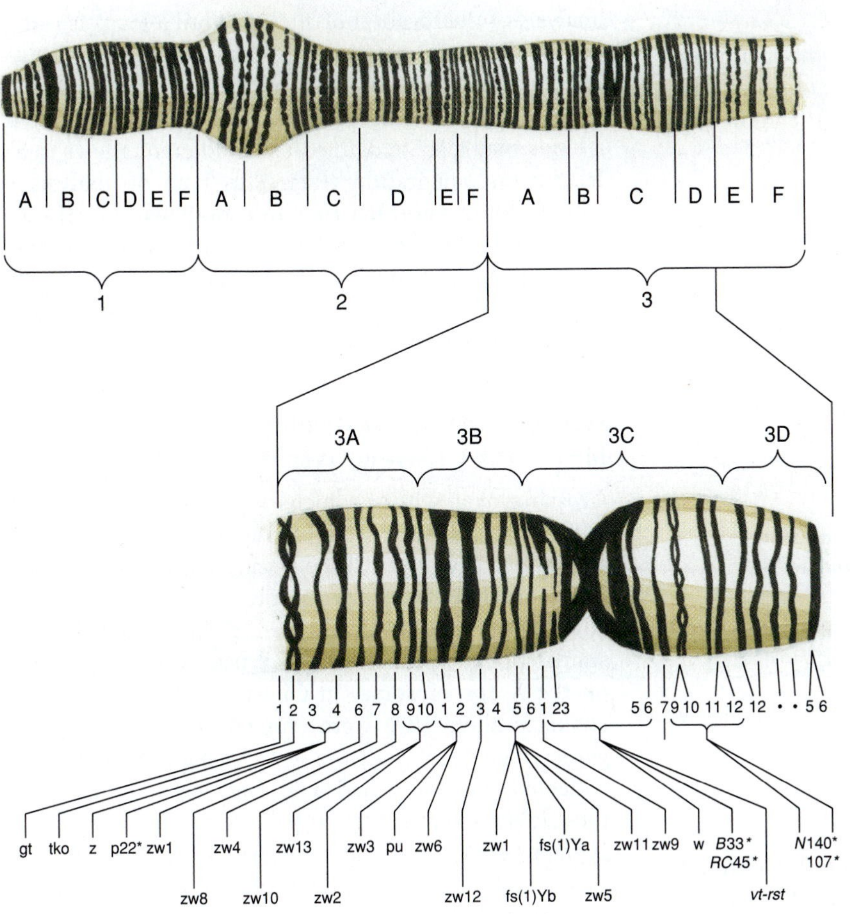

FIGURE 6.26 A polytene chromosome map of *Drosophila*. A portion of C. B. Bridges' salivary gland chromosome diagram of the X chromosome showing the bands in the first three numbered regions (1–3). The loci in a small portion of the 3A–D region of the X chromosome have been mapped by deficiency analysis and are shown in the lower part of the figure. The X chromosome linkage map has been correlated with the salivary gland chromosome band map by determining the salivary gland chromosomal location of a number of loci in deletion mapping experiments. *(Figure modified from Judd, B. H., M. W. Shen, and T. C. Kaufman. 1972. "The anatomy of a segment of the X chromosome of* Drosophila melanogaster." Genetics, *71: 139–156.)*

ical analysis of chromosomes. Techniques for staining and cytological analyses of chromosomes are discussed in detail in Chapter 18. One of the best species for deficiency complementation is *Drosophila melanogaster* because the presence of polytene chromosomes in the cells of the salivary glands. Each polytene chromosome contains up to 2000 tightly paired copies of the chromosomal DNA, and when stained with a DNA-specific stain, the chromosomes show a series of dark bands. The pattern of these bands is unique to each chromosome and allows the location of deficiencies to be determined with great accuracy. There are over 5000 bands in the total genome of *Drosophila*, and C. B. Bridges prepared detailed maps showing the pattern of bands in each chromosome. Bridges prepared diagrams in which each band was assigned a unique number-letter-number designation (Figure 6.26). The portion of the genome removed by a deletion in *Drosophila* can be identified with great precision by examining the salivary gland chromosomes from individuals that are heterozygous or homozygous for that deletion. In such individuals, the region that is deleted is missing in the region where the bands are normally found in the wild-type chromosome. Some genes have been mapped to very small regions, for example within a band or between two bands (Figure 6.26).

SOLVED PROBLEM

Problem

The following data table represents the results of a series of complementation tests in *Drosophila*. Five different recessive mutations (*a, b, c, d, e*) were tested for complementation with a series of deficiencies (*1, 2, 3, 4, 5, 6, 7*) that remove portions of the X chromosome. The results of these complementation tests are given in the complementation matrix, where + represents complementation and m represents no complementation. Determine the chromosomal location of each mutation.

Deficiency	Complementation	Bands Removed by Each Deficiency
	a b c d e	
1	m + + + +	3C3–3C4
2	+ + m + m	3C8–3C11
3	m m + m +	3B3–3C6
4	+ m + + +	3B6–3C3
5	m m + + +	3C1–3C5
6	+ + + m m	3C6–3C9
7	+ + + + +	3B2–3B5

Solution

Each mutation is located in the region that is missing from all deletions that fail to complement the mutation and is not missing from all deletions that complement the mutation. *a* = 3C3, *b* = 3C1–3C2, *c* = 3C10–3C11, *d* = 3C6, *e* = 3C8–3C9.

SECTION REVIEW PROBLEM

15. *Drosophila* with a chromosome containing multiple recessive alleles in the following order: *th st cp in ri* p^P *ss bxd k* e^s, were crossed to a series of strains containing deficiencies. The progeny were examined to determine which mutations were complemented by which deficiencies.

Strains	Progeny Mutations That Fail to Complement
Df1	*in ri* p^P *ss*
Df2	*th st cp in*
Df3	*cp in ri* p^P *ss*
Df4	*st cp in ri* p^P *ss bxd*
Df5	*cp in ri* p^P *ss bxd k*
Df6	p^P *ss bxd k* e^s

Draw a complementation map for this chromosome showing the location of these deficiencies.

GENETIC FINE-STRUCTURE MAPPING DEFINES A GENE GENETICALLY.

The success of mapping studies in determining the position of loci on chromosomes caused a major change in how geneticists viewed chromosomes. By the early 1930s, chromosomes were seen as linear collections of genes, with the genes arranged like beads on a string. Crossing over was believed to occur only between loci, and each locus was believed to have only one genetic function that could be altered by mutation. In this section, we will discuss how additional information gained from linkage mapping and complementation tests changed this concept of genes and their arrangement on the chromosome. We now know that genes are sequences of the chromosomal DNA and crossing over can occur within as well as between genes. Furthermore, some loci contain several units of genetic function that can be mutated independently of each other. The discovery that crossing over could occur within a locus led to the use of linkage mapping techniques to make genetic maps of the internal genetic structure of loci. These maps are called **fine-structure maps** and show the order and distances between the different alleles in a locus. The analysis of the location of alleles within a locus is called **genetic fine-structure analysis.** In addition to using recombination frequency to map the internal structure of loci, the number and locations of functional subunits of a locus may also be defined using different alleles in allelic complementation tests. This type of analysis is called **allelic complementation analysis,** and the results can be used to generate a **complementation map** showing the distribution of the genetic functional sites within a locus. In this section, we will discuss how these two techniques are used.

Genetic Fine-Structure Maps Are Made by Measuring Crossing Over Within a Locus.

Crossing over within a locus was discovered by C. P. Oliver in 1940 in a study of the *lozenge* (*lz*) locus of *Drosophila*. Mutations at the *lozenge* locus cause the normally ordered array of facets in the compound eyes of adults to be rough and irregular. All *lozenge* alleles have similar phenotypes, and they all map to the same locus on the X chromosome. In Oliver's study, flies with the *spectacle* allele (lz^s) were crossed with flies with the *glassy* allele (lz^g), and the lz^s/lz^g female progeny were collected. All these had a lozenge phenotype. When these lz^s/lz^g females were crossed to *lz* males, a small percentage (0.2%) of the offspring had wild-type eyes. Further testing indicated that these individuals had one wild-type allele of *lozenge*. Oliver proposed that these wild-type alleles had arisen by crossing over within the *lozenge* locus. This conclusion was supported by further studies done by Melvin Green and others (Figure 6.27).

Oliver and Green used other mutations that were closely linked to *lozenge* to demonstrate that the rare wild-types they recovered were produced by crossing over. Their procedure was to have one of these mutations (called outside markers, or flanking markers) on each side of the *lozenge* locus. Crossing over within the locus would produce a recombinant chromosome for the flanking markers, whereas other types of genetic change, such as a new mutation, would not be associated with crossing over for the flanking markers (Figure 6.27). This technique of using flanking markers has become a standard practice for detecting crossing over within a locus.

The frequency of recombination between alleles can be used to make a fine-structure map showing the locations and distances between alleles within a locus. The same procedure is used for making a fine-structure map as for making a map showing the locations and distances between loci on a chromosome. Fine-structure maps have been made for numerous loci in many different eukaryotic organisms including *Drosophila*, maize, yeast, and other fungal species, as well as for a number of bacteria and viruses (discussed in Chapter 8). Recombination distances between alleles within a locus are far

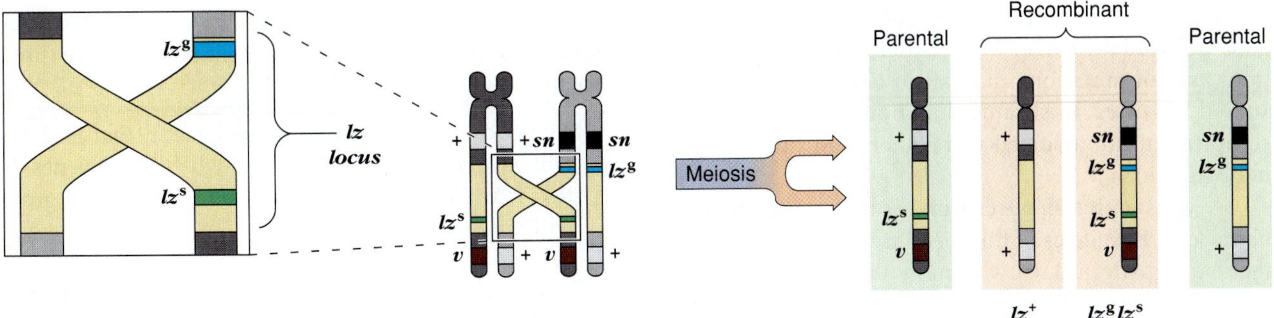

FIGURE 6.27 Meiotic crossing over within the *lozenge* locus. Crossing over between the lz^g and the lz^s alleles produces two recombinant chromosomes, one containing a *lozenge* locus with two mutant lesions and the other containing a *lozenge* locus with no lesions. Recombination within the *lozenge* locus also gives a recombinant arrangement of the flanking markers, *singed* (*sn*) and *vermilion* (*v*).

smaller than recombination distances between different loci. For example, the frequency of recombination between the two *lozenge* alleles that are farthest apart is only 0.14 cM. The small frequency explains why crossing over within loci was not detected in early genetic studies. It was such a rare an event that it was overlooked.

Fine-structure mapping revealed important new aspects of genes. It revealed that loci had length and complex internal structures. Each mutant allele was different from the wild-type allele at one particular site in the locus. This difference was called the **mutant lesion** associated with that mutant allele. Mutant lesions are any type of change in the DNA sequence of the locus that alters the function of the gene and generates the mutant allele. Consider the example of the *Notch* locus in *Drosophila.* Null alleles that eliminate the entire *Notch* locus have a dominant visible phenotype (notches in the wings) and a recessive lethal phenotype (homozygotes die during embryonic development). Recessive mutations at the *Notch* locus have a variety of phenotypes, including abnormal structures of eyes, bristles, wings, or legs. Fine-structure mapping of a number of recessive *Notch* alleles indicated that the mutational lesion associated with each allele maps to one particular region of the locus. This means that changes in the DNA sequence of the *Notch* locus in different regions give different mutant phenotypes. For example, several *facet* alleles, which all give a similar rough eye phenotype, map to the same region within the *Notch* locus (Figure 6.28). This suggests that this region of the locus contains a particular genetic function whose alteration by mutation produces the facet phenotype. Subsequent molecular analysis of the locus and its normal and mutant products may indicate the nature of these functions in the organism.

Complementation Mapping Defines the Cistron, a Unit of Genetic Function.

Prior to the 1940s, all recessive mutations that mapped to the same locus failed to complement, that is, gave a mutant phenotype in *trans* heterozygotes. This convinced many geneticists that each locus contained only one genetic function. However, as increasing numbers of alleles were isolated, it was discovered that this was not true. Recessive mutations were found that mapped to the same locus but that complemented each other, that is, individuals with these two mutations in *trans* had a wild-type phenotype. This discovery meant that a locus could have multiple genetic functions. Mutations that fail to complement are said to be in the same **complementation group**, or **cistron.** The term "cistron" was coined by Seymour Benzer from the names of the two possible

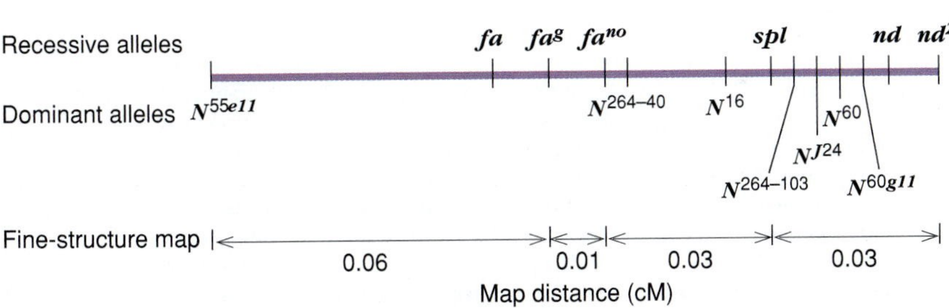

FIGURE 6.28 A fine-structure map of the *Notch* locus. This map shows the locations of several different alleles within the locus. Recessive visible mutant sites are listed above the horizontal line, and recessive lethal *Notch* mutant sites are shown below. *(Figure produced using data provided by W. J. Welshons.)*

types of allelic arrangement (*cis* and *trans*) in heterozygous individuals. Each cistron defines one genetic function, and all mutations that fail to complement are assumed to alter this same function. Mutations that complement are assumed to be altered in different genetic functions.

A locus with a single cistron is called a simple locus because all the alleles of this locus give a simple complementation pattern. Loci with more than one cistron are called complex loci because alleles of the locus give a complex complementation pattern. By analyzing the complementation of a number of alleles at one locus, it is possible to determine how many cistrons are in that locus. It is also possible to make a complementation map showing the arrangement of the cistrons in the locus as well as the location of each allele in each cistron. To illustrate how such complementation mapping experiments are done, suppose we have isolated ten arginine-requiring auxotrophic mutations in yeast that all map to the same locus. To test these mutations for complementation, all *trans*-heterozygous combinations of the mutations are generated, and the heterozygotes are plated on a medium that does not contain arginine. There are three possible outcomes of each test:

1. The two mutations might complement, that is, each will have one wild-type allele of two different genes. If this is the case, then all the heterozygotes will grow without arginine, producing colonies of heterozygous cells.
2. The two mutations might fail to complement (the mutations are in the same gene), but meiotic crossing over between them could generate a few wild-type recombinants. If this is the case, then the heterozygotes will not grow, and the plates will contain only a few wild-type recombinant colonies.
3. The two mutations might fail to complement, and meiotic crossing over will not generate any wild-type recombinants. If this is the case, there will be no colonies on the plates (Figure 6.29).

The results of the series of complementation tests between a set of alleles are presented in a **complementa-**

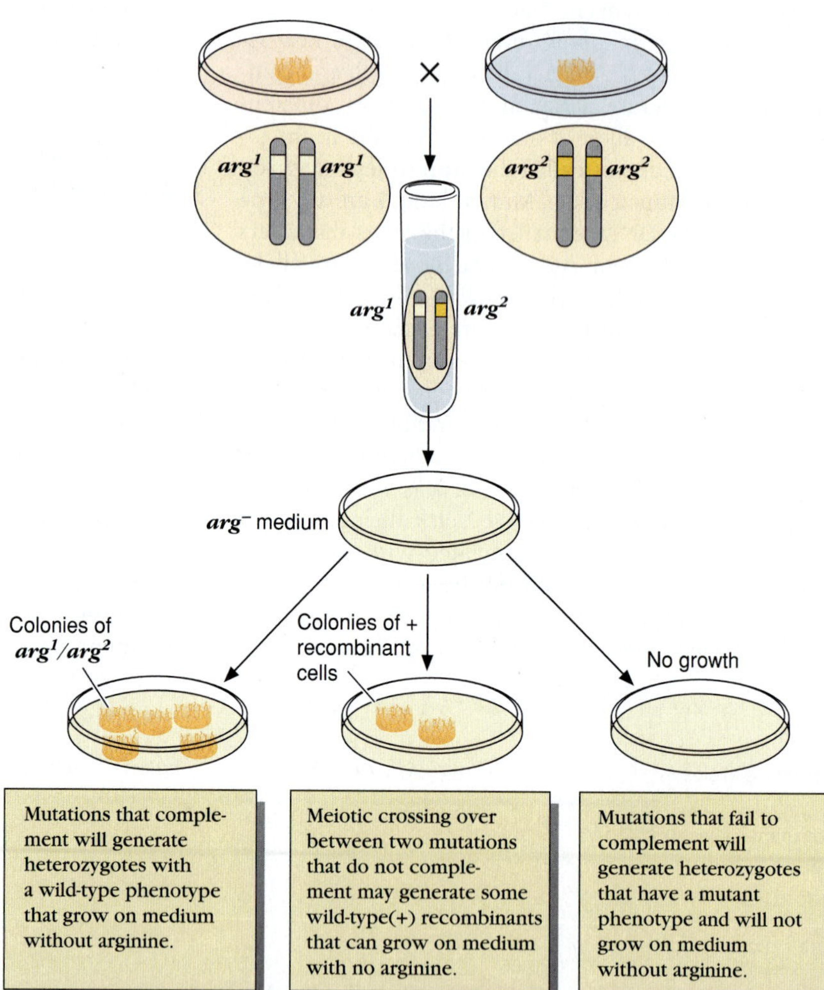

×
arg^1
arg^1
arg^2
arg^2
arg^1
arg^2
arg^- medium
Colonies of arg^1/arg^2
Colonies of + recombinant cells
No growth
Mutations that complement will generate heterozygotes with a wild-type phenotype that grow on medium without arginine.
Meiotic crossing over between two mutations that do not complement may generate some wild-type(+) recombinants that can grow on medium with no arginine.
Mutations that fail to complement will generate heterozygotes that have a mutant phenotype and will not grow on medium without arginine.

FIGURE 6.29 Complementation tests with auxotrophic mutations. Auxotrophic mutant alleles are tested for complementation using selective media, in this a case medium that does not contain arginine, an essential amino acid.

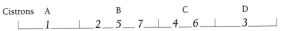

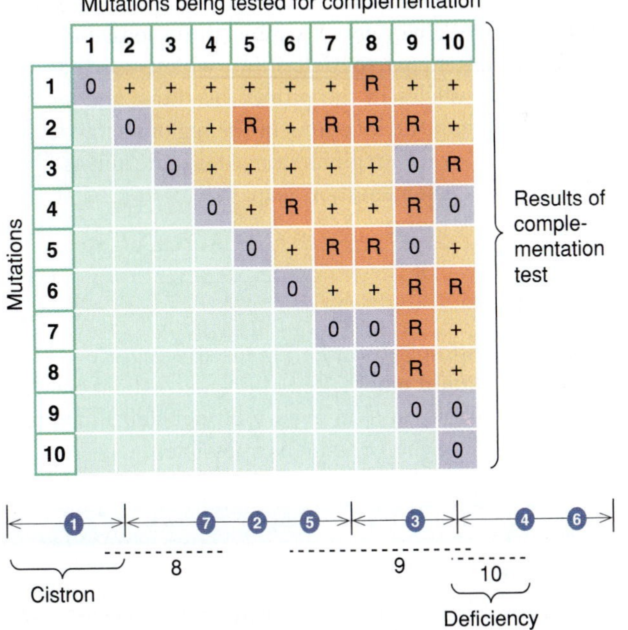

FIGURE 6.30 A complementation matrix. The results of complementation tests among a series of recessive mutations are presented in a complementation matrix. A + indicates that the two mutations complement each other, and an R indicates that the two mutations fail to complement but that a few wild-type recombinants are recovered. An 0 indicates that the two mutations fail to complement and no wild-type recombinants are recovered.

tion matrix. In the complementation matrix, the three possible results of a complementation test are written as: + = complementation; R = no complementation, wild-type recombinants are recovered; 0 = no complementation, no recombination. The complementation matrix showing the results of the complementation tests between our 10 arginine auxotrophic mutations is shown in Figure 6.30. Let us examine the results of the complementation tests between mutations *1* and *7* first, and then return to mutations *8, 9,* and *10.* Mutation *1* and mutation *3* complement. This means that these two mutations must be in separate cistrons. Mutations *2, 5,* and *7* fail to complement each other, so these three must be in one cistron. Mutations *4* and *6* fail to complement each other but do complement mutations *1, 2, 3, 5,* and *7,* so they must be in another cistron. The complementation tests indicate that this locus must have at least four cistrons, and a complementation map showing these cistrons (which we have labeled A, B, C, and D) can be drawn. In this map the location of each mutation is indicated by its number, and the boundaries between cistrons are indicated by vertical lines:

Cistrons A B C D

|___*1*___|___*2*__*5*__*7*___|___*4*__*6*___|___*3*___|

It is important to understand that a complementation map is simply a model of what we believe to be the structure of the locus. Unlike linkage maps, there may be more than one complementation map that will explain a given set of data. For example, the complementation tests between mutations *1* and *7* give no information about what the correct order of the cistrons is in this locus, so the preceding map is drawn with the cistrons arranged in an arbitrary order. Any other order would be equally valid, for example,

|___*3*___|___*2*__*5*__*7*___|___*4*__*6*___|___*1*___|

Also, complementation tests give no information about the order of mutations within the cistrons. Obviously, the cistrons are arranged in one order, and the mutant lesions for each of the alleles are in particular locations within the cistrons, so only one of the many possible complementation maps is correct. Complementation tests simply do not give the information necessary to determine which map is correct. Genetic fine-structure mapping would give us the information needed to determine the correct map with the order of the alleles within each cistron as well as the order of the cistrons. By measuring the frequency of recombination between alleles, we could determine the order of the alleles. Combining fine-structure mapping and complementation tests is a widely used procedure in microorganisms such as bacteria and virus, where it is possible to screen large numbers of progeny from crosses (Chapter 8). Another approach is complementation mapping with deletion mutations.

In complementation tests and in fine-structure mapping, some mutations behave as **point mutations,** that is, they occupy a single point in the locus. Mutations that alter a single nucleotide pair of DNA are point mutations. Point mutations fail to complement other mutations in the same cistron, and crossing over can occur between any two different point mutations. For example, the recovery of wild-type recombinant progeny from *trans* heterozygotes indicates that two separate lesions of the two alleles are within the locus. However, not all mutations are point mutations. In Chapters 17 and 18, we will discuss several different types of mutations that are not point mutations. Here we are concerned with only one type: deletions. Deletions that remove more than one DNA base pair are not point mutations. In our previous discussion of deficiency complementation, we noted how a deficiency that removes several loci will fail to complement with alleles of all the loci. In fine structure analysis, deficiencies have similar properties. If the deletion removes two cistrons, then the deletion will fail to complement with all the mutations in both cistrons (Figure 6.31). A point mutation in the cistron will not complement with the deletion, and the heterozygote will not produce any wild-type recombinant progeny. This is

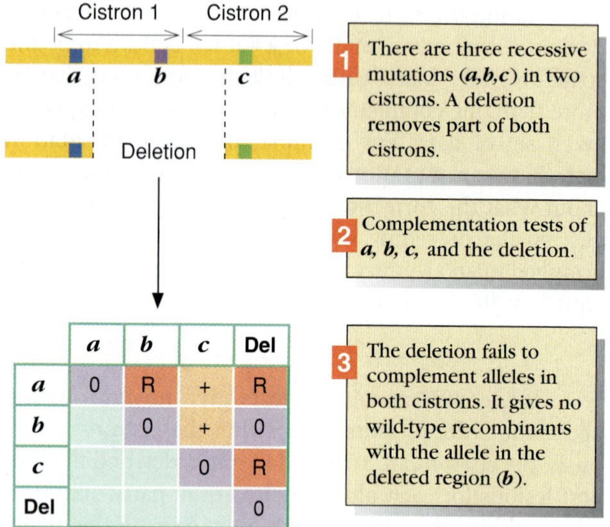

FIGURE 6.31 Complementation mapping with deletions. This figure shows complementation, or the lack of it, between mutant alleles and deletion mutants. In the matrix, + is complementation, R is no complementation but some recombinants, and 0 is no complementation.

because it is impossible for a crossover to generate an intact, wild-type copy of the locus.

To illustrate the special characteristics of deletions in complementation mapping, let us return to our example of the hypothetical arginine auxotrophic mutations mapped in Figure 6.30 and consider mutations *8*, *9*, and *10*. These three mutations do not behave like point mutations. They fail to complement with mutations that are in different cistrons, and they fail to give wild-type recombinants with several different point mutations. For example, *8* fails to complement both *1* and *2*, and our previous analysis indicates that *1* and *2* are in different cistrons. Mutations *9* and *10* give similar results, so we can conclude that all three are deletions that remove different portions of this locus. The complementation patterns of the three deletions may be used to define the order of these cistrons within the locus and the order of the point mutations within the cistrons. First, if a deletion removes parts of two or more cistrons, these cistrons must be adjacent to each other. Because mutation *8* fails to complement mutations in the two cistrons containing point mutations *1* and *2*, these cistrons must be adjacent. Mutation *9* fails to complement point mutations *2*, *3*, and *4*, so the cistrons containing these mutations must also be adjacent. Mutation *10* fails to complement mutations *3* and *4* so these two cistrons must be together. Second, if a mutation gives no wild-type recombinant progeny with a deletion, the site of that mutation is located within the region removed by the deletion. The deficiency mutation *8* must encompass the sites of mutation *7*; deficiency mutation *9* must encompass the site of mutations *3* and *5*; and deficiency mutation *10* must encompass the site of mutation *4*. We can now examine all the possible complementation maps to determine which one has the cistrons and alleles in the proper locations, and we can add to this map the locations of the regions removed by each deficiency. The region removed is indicated as a line under the complementation map:

|___1___|___7__2__5___|___3___|___4__6___|
__8__ ______9______
______10______

There is no simple formula for discovering the correct complementation map. Often all the needed deficiencies are not available, and so even with deficiency complementation more than one map is possible.

SOLVED PROBLEM

Problem

A series of pairwise crosses were made between eight new histidine auxotrophic mutations in yeast (*a*, *b*, *c*, *d*, *e*, *f*, *g*, *h*). The results are presented here in a complementation matrix. In this matrix, 0 indicates that there was no complementation and no wild-type recombinant progeny were recovered, R indicates that there was no complementation but a few wild-type recombinant progeny were obtained, and + indicates that the mutations complemented. From this data, make a complementation map showing the numbers of cistrons, the locations of the mutations, and the order of the cistrons within the locus.

	a	*b*	*c*	*d*	*e*	*f*	*g*	*h*
a	0	+	0	R	R	R	+	0
b		0	+	0	+	+	R	+
c			0	+	R	+	+	+
d				0	+	0	R	R
e					0	+	+	+
f						0	+	R
g							0	+
h								0

Solution

It is sometimes best to begin an analysis of a complementation matrix by making a provisional map using only the first few complementation tests, and then use the remaining data to modify the map. A brief examination of the results suggests that the *a* and *d* mutations are deficiencies. Both fail to complement several other mutations and also give no recombinants with two mutations. Using only the data from the *a* and *d* complementation tests, we can draw a preliminary map:

└─g─b─┴─f─c─h─e─┘
 ─d─ ─a─

The results of the other tests indicate that there must be three cistrons:

└─g─b─┴─f─h─┴─c─e─┘
 ─d─ ─a─

SECTION REVIEW PROBLEMS

16. The complementation matrix shows the pattern of complementation among a series of auxotrophic mutations at the *histidine-3* locus. From these results, make a complementation map showing the number of cistrons in this locus and the position of each mutation.

	a	*b*	*c*	*d*	*e*
a	0	R	0	0	R
b		0	+	0	+
c			0	R	+
d				0	+
e					0

17. This complementation map shows the location of seven mutations in a gene containing three cistrons. Draw the complementation matrix you would expect if all seven alleles were tested against each other in complementation tests.

└─*a1*─*a3*─┴─*a2*─*a7*─*a5*─┴─*a4*─┘
 ─*a6*─

Summary

The discovery that genes reside on chromosomes led to the development of genetic mapping techniques for determining where a particular gene resides on the chromosome. The most widely used mapping technique uses the frequency with which meiotic crossing over occurs between two loci as a measure of how far apart the loci are on the chromosome. This frequency can be measured by examining either the progeny of testcrosses or $F_1 \times F_1$ crosses in diploid species or the spores in haploid species. In many fungal species, the frequency of recombination is determined in tetrad analyses, in which all the cells produced by a single meiotic cell can be examined together. In fungal species with ordered tetrads, the location of the centromere can be determined as can the precise location of each crossover and the identity of the particular chromatids involved in that crossover. Linkage maps showing the locations of many genes on a chromosome may be made by combining the results of different linkage mapping experiments.

Other techniques have been devised for chromosomal mapping that do not use the frequency of meiotic recombination. Mitotic maps can be made using the frequency of mitotic recombination, determined by measuring the frequency and phenotype of somatic cell clones. The locations of many human genes have been deduced in human /mouse cell fusion experiments. Cell lines derived from fused cells spontaneously lose human chromosomes, and genes can be mapped by correlating the loss of a particular gene with the loss of a particular chromosome, or part of a chromosome. The physical location of a gene can also be determined by deficiency complementation.

Mapping techniques devised for estimating the distances between loci are also applied to mapping the distances within loci. Mapping the distance between alleles within a gene is done by measuring the frequency of recombination between them, called fine-structure mapping. Fine-structure maps show the locations and distances between the mutant lesions that are associated with each different mutant allele. Complementation tests of different alleles in a locus indicate that some loci contain multiple cistrons. Combining complementation tests with fine-structure mapping, or with deficiency complementation, allows the production of complementation maps showing the location of each cistron and the location of each allele within each cistron. Deficiency complementation also defines the physical location in the DNA of the genetic functions in a locus.

Selected Key Terms

centimorgan
cistron
complementation group
crossing over
fine-structure map
genetic fine-structure analysis
interference
linkage
linkage groups
linkage map
locus
tetrad analysis

Chapter Review Problems

1. A tomato plant heterozygous for alleles at three loci (*Aa Bb Cc*) was testcrossed to an *aa bb cc* plant. From the progeny listed here, which loci would you say are linked?

Phenotype	Number
A B C	2250
A B c	2225
A b C	240
A b c	255
a B C	260
a B c	250
a b C	2250
a b c	2275

2. A wild-type *Drosophila* with gray body and straight wings was crossed with a black-bodied, curled-winged fly. The F_1 were all gray bodied with straight wings. The F_1 were backcrossed to black-bodied, curled-winged flies. The progeny were as shown here.

Cross	Progeny	Number
F_1 female ×		
Black curled male	Black curled	762
	Black straight	210
	Gray curled	234
	Gray straight	784
F_1 male ×		
Black curled female	Black curled	1013
	Black straight	0
	Gray curled	0
	Gray straight	987

a. What is the map distance between these loci?
b. What do you conclude about crossing over in male *Drosophila*?

3. This map shows the distances between three loci (*v*, *b*, and *lg*) in corn. If individuals with a *v b lg* / + + + genotype were testcrossed, what classes of phenotypes would you expect to find in the progeny, and how many of each would you expect if you examined 1000 progeny?

*v*______ *b*______ *lg*
18 28

4. The following complementation matrix contains the results of a series of complementation tests between a series of hypothetical auxotrophic mutations.

	2	3	4	5	6	7	12
2	0	+	+	R	0	R	+
3		0	+	+	0	+	R
4			0	+	+	+	+
5				0	0	R	+
6					0	R	R
7						0	+
12							0

a. From these results, make a complementation map showing the numbers of cistrons and the locations of each mutation.
b. Is there more than one possible map that could be drawn from these data?

5. Several female *Drosophila* heterozygous for four mutations were testcrossed. The mutations and their map positions were *engrailed* (*en*) map position 62.0, *scabrous* (*sca*) 66.7, *droopy* (*dr*) 71.2, and *curved* (*cu*) 75.5.
   a. If the genotype of the original heterozygotes was *en sca dr c* / + + + +, how many different phenotypic classes would you expect to find in the progeny?
   b. If 25,000 progeny were examined, how many progeny with an engrailed, droopy phenotype would you expect to find, if there is no interference?

6. A series of cell lines were selected from a human/mouse cell fusion. Each line had lost a number of human chromosomes. Each line was tested for the presence of a series of human enzymes. From the results given in this table, determine the chromosomal location of each human gene.

Line	Human Chromosomes						LDH	6GPGD	PGK
X	3	6	9	12	17	22				
A	+	+	+	−	+	−	−	+	+	+
B	−	−	−	−	+	+	+	+	−	−
C	+	+	−	−	−	−	+	−	+	+
D	+	−	−	+	−	−	+	−	+	+
E	−	+	+	+	−	−	−	−	−	−
F	+	−	−	−	+	+	+	+	+	+

7. In cell fusion experiments, human chromosomes that have large deficiencies or other abnormalities can be used to map genes. In a such a cell fusion experiment, a series of cell lines with human X chromosomes that contained deletions were tested for the presence of the human enzyme phosphoglycerate kinase (PGK). In the table here, the region of the X chromosome that is deficient is listed. From the results in this table, where is this gene located?

Line	Deficiency	PGK
a	*Xq11*	+
b	*Xq12*	+
c	*Xq13*	−
d	*Xq11–12*	+
e	*Xq11–13*	−
f	*Xq12–13*	−

8. From a three-point testcross mapping experiment, the following gamete genotype frequencies were obtained:

X Y Z	365
x y z	367
x Y z	110
X y Z	105
x Y Z	3
X y z	4
X Y z	25
x y Z	21
	1000

From these data, what can be said of the genes?
a. All are unlinked.
b. *XYZ* is the gene order.
c. *YZX* is the gene order.
d. *YXZ* is the gene order.
e. Two of the genes are linked; one is independently assorting.

9. From the data in Problem 8, the distance between gene *X* and gene *Y* is
a. 21.5 cM
b. 4.6 cM
c. 22.2 cM
d. 5.3 cM
e. 27.5 cM

10. From the data in Problem 8, the distance between *Y* and *Z* is
a. 21.5 cM
b. 4.6 cM
c. 22.2 cM
d. 5.3 cM
e. 27.5 cM

11. From the data in Problem 8, the coefficient of coincidence is
a. 0.7
b. 1
c. 2.3
d. 0.59

12. If genes *A* and *B* are linked and separated by 14 map units, what percentage of the progeny produced (in a testcross) will be parental?
a. 7%
b. 14%
c. 43%
d. 86%
e. 68%

13. Three mutations of yeast, *m1*, *m2*, and *m3*, are known to cause a requirement for the amino acid tryptophan in the growth medium. Heterozygotes containing both *m1* and *m2* do not require tryptophan; heterozygotes containing both *m2* and *m3* do require tryptophan, and heterozygotes containing both *m1* and *m3* do not require tryptophan. From these data, how many genes have been identified that are required for the synthesis of tryptophan in yeast cells?
a. 0
b. 1
c. 2
d. 3
e. 4

14. A cross between females heterozygous for three loci (*x*, *y*, *z*) and wild-type males yielded the following progeny:

Females:		Males:	
+ + +	800	+ + +	48
		+ + *z*	6
		+ *y z*	290
		+ *y* +	68
		x y z	40
		x y +	2
		x + +	266
		x + *z*	80

a. On which chromosome do these three loci reside?
b. Which gene is in the middle, and what are the distances between *x* and *y*, and *y* and *z*?
c. What is the coefficient of coincidence for this cross?

15. Use the following linkage data to construct a chromosome map of these loci.

Loci	Map Units
a–b	10
b–c	6
a–c	16
b–d	13
a–d	3
b–e	2
a–e	8

16. The genes *dumpy wing* (*dp*), *clot eye* (*cl*), and *apterous wing* (*ap*) are linked on chromosome 2 of *Drosophila*. In a series of two-point mapping crosses, the following genetic distances were determined. What is the sequence of the three genes?

dp–ap	42
dp–cl	3
ap–cl	39

17. In *Drosophila*, the mutants *black body* (*b*) and *vestigial* (*vg*) are both recessive to the wild-type alleles (+). Flies with a black body color and vestigial wings were crossed with wild-type flies with gray bodies and normal wings. The F_1 females were crossed with black vestigial males. The phenotypes and genotypes of the progeny of this cross follow.

Progeny Phenotype	Genotype	Number
Black vestigial	*b vg/b vg*	410
Black long	*b +/b +*	90
Gray vestigial	*+ vg/+ vg*	95
Gray long	+ +/+ +	410

a. What are the genotypes of the parents?
b. What is the map distance between *b* and *vg*?

18. Consider *a*, *b*, and *c* to be three recessive mutations in cockroaches. The following data are the results of a testcross in which F_1 females heterozygous at all three loci were crossed to males homozygous for all three recessive mutants. Construct a linkage map showing the correct order of the three genes and the map distances between adjacent genes. Which genotypic class is missing?

+ + +	75
+ + *c*	348
+ *b c*	96
a + +	110
a b +	306
a b c	65
Total	1000

19. The genotypes of different individuals are diagrammed here. Write the genotype for each strain using proper allelic notation.

a. *Drosophila*

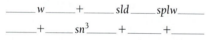

——*w*——+——*sld*——*splw*——
——+——sn^3——+——+——

b. *Drosophila*

——el^3——*pu*——+——*Sd*——
——el^1——+——*Cy*——Sd^2——

c. *Zea mays*

——bz^2——ex^{3c}——dek^2——
——bz^7——ez^2——dek^7——

20. Draw a diagram showing the arrangement of the alleles on the following chromosomes:

a. *Drosophila* — *y w / os ; p^P / bx Ubx*
b. *Zea mays* — *a d Rg / a D rg ; Y C / y c*
c. *Danio rerio* — *unc1 / cyc ; dis3 / +*
d. *Drosophila* — *Pm sp / cui ; db / cp ; ey^D / ci*

21. A cross between individuals with an Lg n phenotype and an lg n phenotype produced F_1 individuals with an Lg N phenotype. The F_1 were crossed with individuals with lg n phenotypes. The F_2 follow.

Progeny	Number
Lg N	180
Lg n	50
lg N	60
lg n	190
Total	480

a. Are these genes linked?
b. If they are linked, what were the genotypes of the original parental individuals?

22. Flies from a true-breeding strain of *Drosophila* with yellow bodies and notched wings were crossed with flies from a wild-type strain (gray body and full wings). The F_1 females were testcrossed.

a. From the phenotypes of the offspring given here, would you conclude that these genes are linked?
b. If these genes are linked, what is the linkage map distance between these genes?

F_2 Phenotype	Number
Yellow notched	455
Gray notched	45
Yellow full	45
Gray full	455
Total	1000

23. Sunflowers with genotypes of *aa bb cc* and *AA BB CC* were crossed. The F_1 were testcrossed to give the progeny listed here. Which (if any) of these genes are linked? Briefly explain your answer.

Phenotype	Number
+ + +	303
+ + c	8
+ b +	309
+ b c	15
a + +	13
a + c	304
a b +	10
a b c	306
Total	1268

24. Make a map of the chromosome showing the order of the loci and the distances between them from the results of the series of dihybrid mapping crosses shown here (R = % recombinant progeny).

Genes	R	Genes	R
a–b	33	*b–d*	18
a–c	21	*b–e*	43
a–d	15	*c–d*	36
a–e	10	*c–e*	11
b–c	54	*d–e*	25

25. A series of *Aspergillus* colonies from a diploid strain heterozygous for five recessive mutations was examined for recombinant sectors. The results are shown here. Use these results to make a map showing the order of the genes, the distances between the genes, and the location of the centromere.

Phenotype of Sector	Number
l n p o m	200
l n p o	300
l n p	300
l n	50
l	150
Total	1000

26. A series of human / mouse hybrid cell lines were screened for the presence of the human *p53* gene (+ = gene product present). From these results determine the location of this gene.

	Line									
	A	**B**	**C**	**D**	**E**	**F**	**G**	**H**	**I**	**J**
p53	+	+	+	+	+	−	−	−	−	−
1	+	+	−	+	+	+	−	+	−	+
2	+	+	+	−	−	+	−	−	−	+
3	−	−	−	−	+	+	−	+	+	−
4	+	−	+	−	+	+	+	−	−	−
5	+	−	−	+	−	−	+	−	−	+
6	−	+	+	+	+	+	+	+	−	−

(continued)

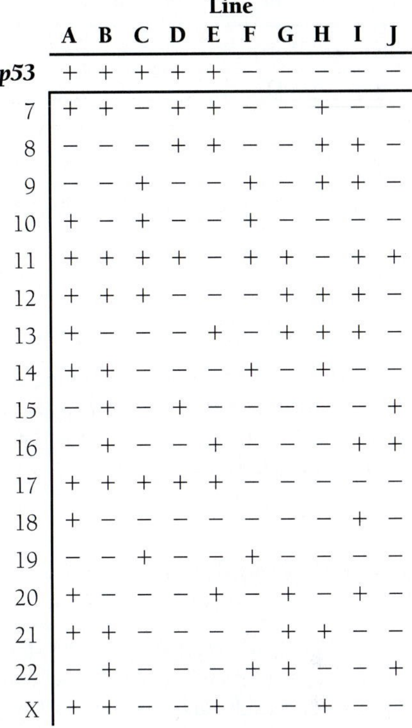

	Line									
	A	**B**	**C**	**D**	**E**	**F**	**G**	**H**	**I**	**J**
p53	+	+	+	+	+	−	−	−	−	−
7	+	+	−	+	+	−	−	+	−	−
8	−	−	−	+	+	−	−	+	+	−
9	−	−	+	−	−	+	−	+	+	−
10	+	−	+	−	−	+	−	−	−	−
11	+	+	+	+	−	+	+	−	+	+
12	+	+	+	−	−	−	+	+	+	−
13	+	−	−	−	+	−	+	+	+	−
14	+	+	−	−	−	+	−	+	−	−
15	−	+	−	+	−	−	−	−	−	+
16	−	+	−	−	+	−	−	−	+	+
17	+	+	+	+	+	−	−	−	−	−
18	+	−	−	−	−	−	−	−	+	−
19	−	−	+	−	−	+	−	−	−	−
20	+	−	−	−	+	−	+	−	+	−
21	+	+	−	−	−	−	+	+	−	−
22	−	+	−	−	−	+	+	−	−	+
X	+	+	−	−	+	−	−	+	−	−

27. The complementation map shows the location of eight mutations in a gene containing three cistrons. Draw the complementation matrix you would expect if all eight alleles were tested against each other in complementation tests.

m^4 m^7 | m^3 m^6 m^5 | m^1

m^2 m^8

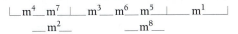

Challenge Problems

1. In the nematode *Caenorhabditis elegans, d* (*dumpy*), *u* (*uncoordinated*), and *k* (*knobby*) are all recessive genes located on the same chromosome. Females heterozygous normal for all three traits were mated with dumpy, uncoordinated, knobby males and the following progeny were observed.

Phenotypes	Number
dumpy uncoordinated knobby	3
uncoordinated knobby	392
knobby	34
uncoordinated	61
dumpy uncoordinated	32
dumpy knobby	65
dumpy	410
wild-type	3
Total	1000

a. In the heterozygous normal females, what is the *cis/trans* arrangement of the following pairs of genes:
d and *u*
d and *k*
u and *k*

b. What is the order of the three genes on the chromosome?

c. What is the map distance between the three genetic markers and the coefficient of coincidence?

2. A series of pairwise crosses were made between eight new methionine auxotrophic mutations in yeast (*l, m, r, p, q, s, w, z*). The results are presented in the following complementation matrix. In this matrix, 0 indicates that there was no complementation and no wild-type recombinant progeny were recovered, R indicates there was no complementation but a few wild-type recombinant progeny were obtained, and + indicates the mutations complemented. From these data make a complementation map showing the numbers of cistrons, the locations of the mutations, and the order of the cistrons within the locus.

	l	*m*	*r*	*p*	*q*	*s*	*w*	*z*
l	0	+	R	+	+	+	R	+
m		0	R	+	R	+	+	+
r			0	+	0	+	0	+
p				0	+	R	+	0
q					0	+	+	+
s						0	+	0
w							0	+
z								0

3. *Drosophila* having a chromosome containing multiple recessive alleles in the following order, *roughoid (ru), hairy (h), thread (th), scarlet (st), curled (cu), stripe (sr), ebony (es), Prickly (Pr),* and *claret (ca),* were crossed to a series of strains containing deficiencies. The progeny were examined to determine which mutations were complemented by which deficiencies.

Strains	Progeny Phenotype
Df1	hairy thread scarlet curled stripe
Df2	scarlet curled stripe ebony Prickly
Df3	thread scarlet curled
Df4	curled stripe ebony Prickly claret
Df5	stripe ebony
Df6	curled stripe ebony Prickly

Draw a diagram showing the location of these deficiencies.

Suggested Readings

Allen, G. E. 1978. *Thomas Hunt Morgan: The Man and His Science.* Princeton University Press, Princeton, NJ.

Carlson, E. A. 1981. *Genes, Radiation, and Society: The Life and Work of H. J. Muller.* Cornell University Press, Ithaca and London.

Donis-Keller, H., P. Green, C. Helms, S. Cartinhour, B. Weiffenbach, K. Stephens, T. P. Keith, D. W. Bowden, D. R. Smith, E. S. Lander, D. Botstein, G. Akots, K. S. Rediker, T. Gravius, V. A. Brown, M. B. Rising, C. Parker, J. A. Powers, D. E. Watt, E. R. Kauffman, A. Bricker, P. Phipps, H. Muller-Kahle, T. R. Fulton, S. Ng, J. W. Schumm, J. C. Braman, R. G. Knowlton, D. F. Barker, S. M. Crooks, S. E. Lincoln, M. J. Daly, and J. Abrahamson. 1987. "A Genetic Linkage Map of the Human Genome." *Cell,* 51, October 23: 319–337.

Helentjaris, T., G. King, M. Slocum, M., C. Siedenstrang, and S. Wegman. 1985. "Restriction Fragment Polymorphisms as Probes for Plant Diversity and Their Development as Tools for Applied Plant Breeding." *Plant Molecular Biology,* 5: 109–118.

Helentjaris, T., and B. Burr. 1989. "Development and Application of Molecular Markers to Problems in Plant Genetics." *Plant Cell,* 2: 163–165.

Lindsley, D. L., and G. G. Zimm. 1992. *The Genome of* Drosophila melanogaster. Academic Press, San Diego.

Ott, J., and H. Donis-Keller. 1994. "Statistical Methods in Genetic Mapping." *Genomics,* 22: 496–497.

Shine, I., and Wrobel, S. 1976. *Thomas Hunt Morgan, Pioneer of Genetics.* The University Press of Kentucky, Lexington, Kentucky.

Stewart, A. 1995. "Genetic Nomenclature Guide." *Trends in Genetics* 11 (3): Supplement 1–43.

Sturtevant, A. H. 1965. *A History of Genetics.* Harper & Row, New York.

On the Web

Visit our Web site at **http://www.saunderscollege.com/lifesci/titles.html** and click on A/G/M Genetics for links to the following chapter-related resources on the World Wide Web:

1. **The CMS Molecular Biology Resource.** This is a source page with links to a large number of organism-specific linkage and gene map databases.

CHAPTER

7

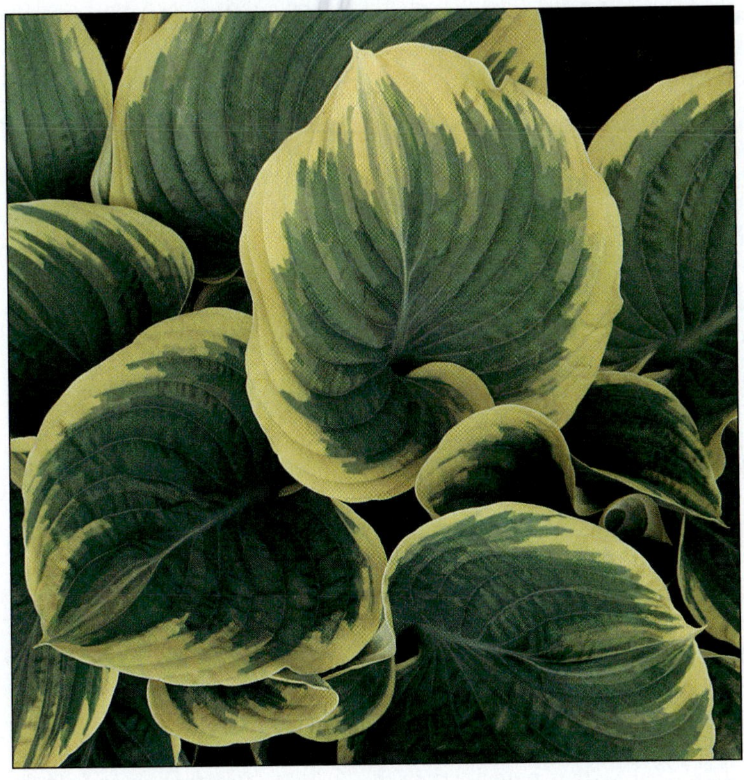

Leaves of a *Hosta* plant of the *Sagae* strain. The variegated leaves result from the maternal inheritance of mutant chloroplasts. *(Janet Mills)*

Extranuclear Inheritance

In the previous chapter, we discussed examples of genes that violate Mendel's second rule, that is, alleles fail to segregate independently because they are linked on a single chromosome. Investigation of these exceptional cases of linkage led to some major advances in genetics, illustrating the importance of investigating observations that appear to be exceptions to current scientific theories.

As early as 1909, geneticists discovered examples of mutations that violated Mendel's first rule. These mutations showed non-Mendelian inheritance patterns and could not be mapped to any chromosome in the nucleus. Additional examples of this type of inheritance continued to appear in the literature for most of the early years of the 20th century, and for the most part, they were met with great skepticism. We now know that these reports were examples of what is known as **extranuclear inheritance.** As its name suggests, extranuclear inheritance refers to cases where a specific phenotype is not controlled by genes on the chromosomes in the nucleus. In this chapter, we will discuss three types of extranuclear inheritance: maternal effects, genes residing in cytoplasmic organelles (maternal inheritance), and infectious inheritance. Each of these has its own characteristics, but they all are examples of non-Mendelian inheritance.

Maternal effects are instances in which an individual's phenotype is controlled by genes in the genome of his or her mother. In many animal and plant species, the mother's reproductive system prepares a large, complex egg containing gene products essential for the normal development of the offspring. Mutations of the maternal genes controlling these gene products can give abnormal phenotypes to the offspring by altering the composition of the egg cytoplasm. The study of maternal-effect genes is currently a very active area of genetics because these genes can tell us a great deal about the events of early embryonic development (see Chapter 16).

Mitochondria and chloroplasts are essential components of the cytoplasm of eukaryotic cells (the DNA of these organelles is discussed in detail in Chapter 9). These cytoplasmic organelles are usually inherited with the egg cytoplasm from the maternal parent. This is termed **maternal inheritance,** or sometimes cytoplasmic inheritance. Mitochondria and chloroplasts have their own genomes, and mutant alleles in these genomes can have striking effects on the organism's phenotype. Such phenotypes will show maternal inheritance of cytoplasmic organelles in crosses. Genes in cytoplasmic organelles are widely studied today to determine how they function and how they interact with nuclear genes, as well as because they are useful tools in the study of evolution.

The cytoplasm of animal and plant cells is complex. In addition to the normal cellular components, numerous parasitic entities can invade the cytoplasm and replicate there. These parasites include bacteria, viruses, and other entities that can be inherited along with the cytoplasm but that also infect cells. This is termed **infectious inheritance.** The presence of such entities can alter the function or survival of the cells they inhabit, and this in turn can change the phenotype of the organism. Agents that show infectious inheritance pose immense problems for human medicine and modern agriculture.

The study of extranuclear inheritance is an important and growing part of genetics today. Extranuclear inheritance can tell us about important aspects of inheritance, and it has had a very significant impact on human society. By the end of this chapter, we will have addressed six questions about extranuclear inheritance:

1. How are maternal-effect inheritance patterns different from Mendelian inheritance patterns?
2. What are some of the functions of mitochondria and chloroplasts in eukaryotic cells?
3. How are mitochondria and chloroplast genes inherited?
4. How can bacteria and viruses show infectious inheritance?
5. What are viroids and prions, and how are they inherited?
6. Why are viroids and prions a problem for human medicine and agriculture?

MATERNAL GENES CAN AFFECT THE PHENOTYPE OF THE OFFSPRING.

Maternal Effects Are Produced by Gene Products Packaged in the Egg.

As we discussed in Chapter 2, higher eukaryotic animal and plant species produce large, complex eggs or seeds. These eggs and seeds contain a variety of gene products from the mother that are necessary for the growth and development of the zygote. These cytoplasmic components are made and distributed to the egg by the female reproductive system under the control of genes in the female's genome. If one of the mother's genes is mutated, the contents of the egg cytoplasm will be altered, and this alteration may affect the phenotype of the developing zygote. When this happens, and we see a change in the phenotype of the offspring, it is called a maternal effect. A mutation that causes a maternal effect is called a **maternal-effect mutation.** Maternal effects have a unique pattern of inheritance in crosses because the phenotype of the individual is not determined by its own genotype but by the genotype of its mother.

A well-studied example that illustrates the inheritance of a maternal-effect mutation is the pattern of shell coiling in the fresh water snail *Limnaea peregra*. The shells of these snails grow in a coiled spiral, which can have either a right-handed spiral (dextral) or a left-handed spiral (sinistral). The direction of coiling is controlled by a single gene that has two alleles. A dominant allele (*D*) gives dextral coiling, and a recessive allele (*d*) gives sinistral coiling. The direction an individual's shell coils is controlled by the genotype of the mother, and neither the genotype of the father nor of the individual itself has any effect.

The pattern of maternal-effect inheritance of shell coiling can be seen in reciprocal crosses (Figure 7.1). When homozygous *DD* females are crossed with *dd* males, all the offspring are heterozygous (*Dd*), and all have dextral coiling. When the heterozygous (*Dd*) F_1 individuals are inbred, the F_2 all have dextral coiling, even though these F_2 individuals have *DD, Dd,* or *dd* genotypes. If the F_2 are inbred to produce an F_3, the F_2 females with *DD* and *Dd* genotypes produce offspring with dextral coiling, and the F_2 females with *dd* genotypes produce offspring with sinistral coiling. The genotype of the male, the phenotype of the F_2 females, and the genotype of the F_3 individuals have no effect on the F_3 phenotypes. The direction of shell coiling is entirely controlled by the genotype of the female parent.

The pattern of inheritance of coiling can clearly be seen in a testcross involving snails. If the homozygous (*DD*) females with dextral coiling are crossed to males from a true-breeding strain of sinistral snails (*dd*) (Figure

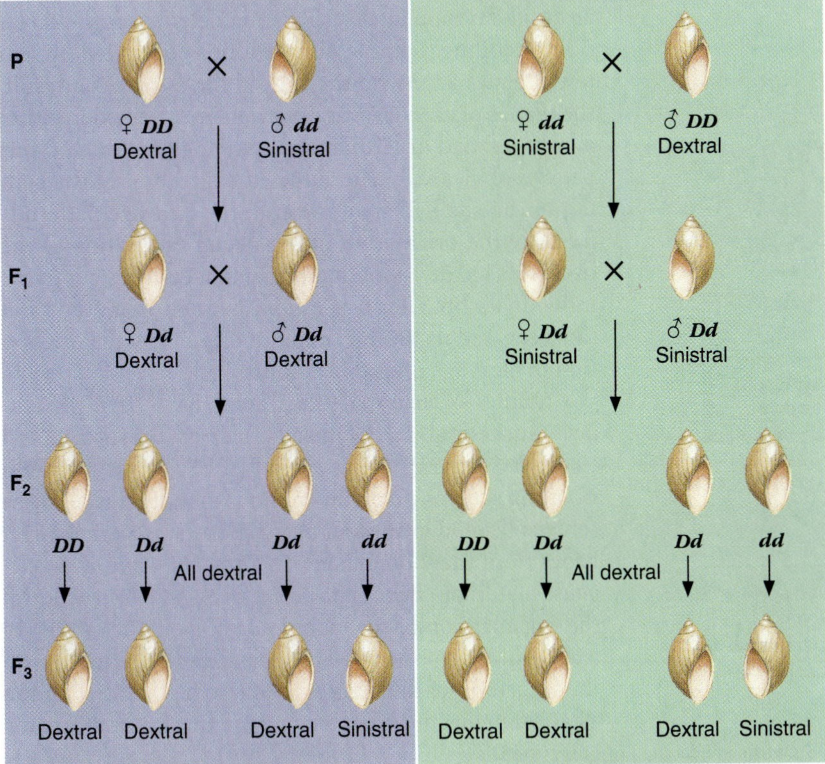

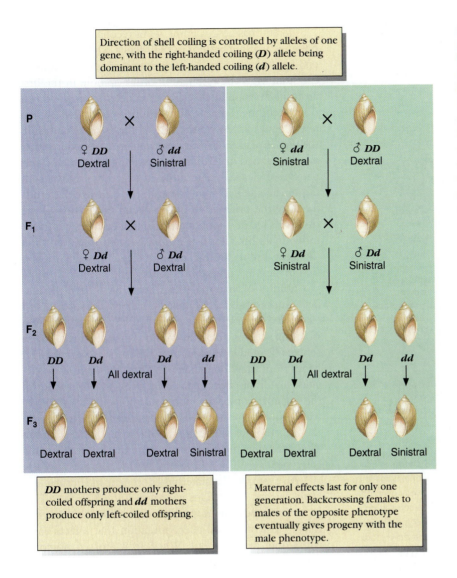

FIGURE 7.1 Maternal-effect inheritance of shell coiling. Reciprocal crosses illustrate how the direction of shell coiling in the snail *Limnaea peregra* is controlled by an allele with a dominant maternal effect. Females with a *DD* or *Dd* genotype produce only offspring that have a right-handed spiral (dextral) and females with a *dd* genotype produce only offspring with a left-handed spiral (sinistral). In reciprocal crosses, the offspring always have the genotype and phenotype dictated by their mother's genotype.

7.2), the offspring (generation G_1) all have a *Dd* genotype and dextral coiling. If the G_1 females are testcrossed to sinistral males (*dd*), the next generation (generation G_2) will have a genotypic ratio of 1 *Dd* : 1 *dd*, the expected Mendelian ratio. However, they will all have a dextral phenotype. If the G_2 females are testcrossed to *dd* males, their offspring (generation G_3) will have a phenotypic ratio of 1 dextral : 1 sinistral, reflecting the genotypic ratio of the G_2 females. This is the characteristic inheritance pattern for maternal effects in testcrosses. This pattern reflects the Mendelian segregation of the alleles of the maternal-effect gene, which are located on chromosomes in the mother's nucleus. The one-generation delay in expressing the phenotype is characteristic of maternal-effect genes. The ability to generate offspring with the male's genotype and phenotype by repeated backcrossing is characteristic of maternal effect genes and distinguishes maternal effect inheritance from other types of maternal inheritance in crosses. In this cross, the generations are labeled G_1, G_2, G_3, and not F_1, F_2, F_3 (Figure 7.2). This is because, by definition, only the offspring of an $F_1 \times F_1$ cross can be called an F_2, and only the offspring of an $F_2 \times F_2$ cross can be called an F_3. In many multigenerational crossing experiments that involve backcrosses or outcrosses, the generations are simply numbered G, with the first generation called P or G_0, the first offspring generation G_1, the second G_2, and so forth.

Maternal-Effect Genes Control Early Stages of Embryonic Development.

The study of maternal-effect genes is one of the most active and dynamic areas of genetics today. Geneticists who are interested in how genes control development have long sought to identify maternal-effect genes because of their function during the earliest stages of development. This has been successfully done by mutational analyses that identify maternal-effect genes by their characteristic

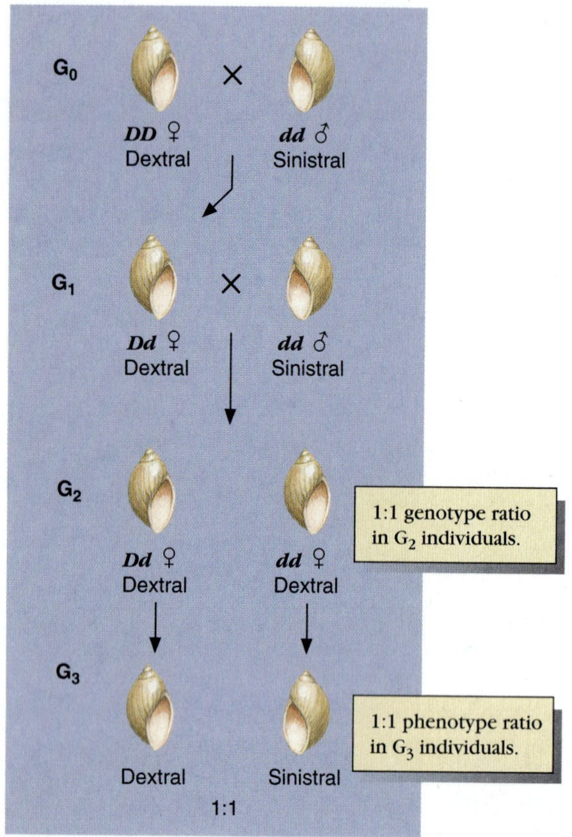

FIGURE 7.2 Maternal effects in backcrosses. Maternal effects last only one generation, so repeated backcrossing of females to males homozygous for the recessive allele will eventually give individuals with the recessive phenotype. Crossing *DD* females to *dd* males gives *Dd* G_1 individuals with the dextral phenotype. Backcrossing the *Dd* G_1 females to *dd* males gives *Dd* and *dd* G_2 individuals that also have the dextral phenotype. Crossing the *dd* G_2 females will finally produce individuals with a sinistral phenotype.

pattern of inheritance. Careful analyses of the phenotype produced by these mutations can indicate what the normal role of these genes is in development, and eventually, the genes' products can be identified and their molecular and biochemical functions understood.

To emphasize this point, let us reconsider the maternal-effect coiling pattern in snail shells. The basis for this coiling effect lies in the earliest embryonic divisions in the snail egg. The orientation of the spindle in the first cleavage division after fertilization determines the direction of coiling. Spindle orientation is controlled by maternal genes acting on the eggs in the ovary. As a result, the planes of each cleavage division are oriented in either a right-handed or left-handed spiraling pattern, and this pattern of cleavage determines the direction of the coiling of the shell. The direction of the cleavage pattern is fixed by the orientation of the planes of division of the first cleavage divisions and is maintained throughout the individual's life after it is established. Further studies of this gene and its product could indicate how the orientation of cleavage divisions is controlled.

Studies of maternal-effect genes have been particularly successful in organisms such as *Drosophila* and *Caenorhabditis elegans*, as a result of large-scale mutation isolation screens involving millions of individuals. These screens have identified all or most of the maternal-effect genes from these organisms. In *Drosophila* and in *C. elegans*, maternal-effect genes have been shown to control the organization of the embryonic body along two body axes (anterior/posterior and dorsal/ventral), to control the formation of the germ cells and to be essential for the establishment and maintenance of the first embryonic cell types.

An example of a maternal-effect gene in *Drosophila* is *pleiohomeotic* (*pho*). This gene was studied by Sang-Hak Jeon. Jeon demonstrated that one allele of this gene (pho^{cv}) is a maternal-effect mutation that affects early embryonic development. When Jeon crossed homozygous females (pho^{cv}/pho^{cv}) with normal (+ / +) males, the females produced only abnormal offspring, even though these offspring had a pho^{cv}/+ genotype. A few of these offspring survived because pho^{cv} is a hypomorphic or "leaky" allele (these types of alleles are discussed in Chapter 4). The lethal offspring usually died early in development because of a number of severe developmental defects. When Jeon crossed homozygous (pho^{cv}/pho^{cv}) males with normal (+ / +) females, most of the progeny

TABLE 7.1 *Progeny Ratios of* Drosophila *Offspring from Various* pho *Mutation Crosses*

Genotypes of Cross		Fertilized Eggs	Offspring Lethal	Offspring Surviving
female	male	number	number	number
pho^{cv} / pho^{cv}	pho^{cv} / pho^{cv}	460	460	0
pho^{cv} / pho^{cv}	+ / +	332	308	24
+ / +	pho^{cv} / pho^{cv}	469	67	402

CONCEPT CLOSE-UP

Gene Imprinting and a Case of Paternal Inheritance

Rare instances have been found in which phenotypes can show paternal inheritance instead of the better-known maternal inheritance. This type of inheritance is known as gene imprinting and indicates that nuclear maternal and paternal gene contributions are functionally unequal. The standard Mendelian assumption is that for autosomal genes both parents contribute one allele of each gene, and unless one allele is a nonfunctional mutant allele, both will contribute equally to the offspring's phenotype. Recent discoveries suggest that for some genes this is not always the case. For these genes, only one of the offspring's alleles is expressed, and the other is inactivated. Unlike the random inactivation of X chromosomes (and most of their genes) in females (discussed in Chapter 4), the inactivated allele is always the allele inherited from one parent. For some genes, the allele inherited from the mother is always inactivated, and for other genes the allele inherited from the father is always inactivated. This parent-specific inactivation is believed to be the result of methylation of components of the DNA that changes the ability of the gene to function. Methylation and how it can affect gene function is discussed in Chapter 14.

A systematic survey of the entire genome was done in the mouse for genes that show single-allele expression, and several were found. In this experiment, genetic rearrangements were used to generate mice that had the normal number of two copies of each gene but that had inherited both copies of some genes from one parent and none from the other. For several genes, these mice showed abnormal phenotypes similar to those shown by mice with no copies or extra copies of the gene. For each gene, the copies inherited from one parent were always both inactive, and the copies from the other parent were always both active. This represents the characteristic pattern of inheritance of an imprinted gene.

In humans, imprinting is believed to be responsible for unusual patterns of inheritance of several autosomal genetic diseases. For example, a tendency to develop glomus tumors (benign tumors that develop in the inner ear) is caused by a mutant allele. Although the gene is located on an autosome, pedigree analyses indicate that the phenotype can show paternal inheritance; it is passed from father to offspring. Molecular analyses indicate that this pattern of inheritance is the result of imprinting. In the affected individuals, the paternally inherited allele is always expressed, and the maternally inherited allele is always inactivated. If a man who is homozygous or heterozygous for the tumor-causing allele mates with a woman with normal alleles, they will have heterozygous offspring. In these offspring, the paternal, disease-causing allele is expressed, and the offspring develops tumors. If a woman who is homozygous for the tumor-causing allele mates with a man with normal alleles, they will also have heterozygous offspring. However, these offspring express only the normal allele inherited from their father and will not develop tumors.

In this example, the paternal pattern of inheritance of tumors is not the result of a paternal-specific inheritance of the tumor-causing allele but rather an inherited pattern of gene expression. Imprinting, like X chromosome inactivation, is not permanent, and an allele that is inactivated in one generation may be the active allele in the next.

survived (Table 7.1). These studies indicate that the normal function of the *pho* gene in the mother is essential for early embryonic development.

Gene Imprinting and Maternal Effects Are Not the Same.

Maternal effects are an example of extranuclear inheritance because a maternal effect causes an individual's phenotype to be determined by gene products inherited in the egg cytoplasm. This is quite unlike maternal inheritance, another type of extranuclear inheritance discussed in this chapter. Maternal effect is also unlike gene imprinting, yet another type of inheritance. **Gene imprinting** is a recently discovered phenomenon in which alleles of certain genes function differently depending on whether they are inherited from the paternal or maternal parent.

Imprinting of genes can alter the phenotype of an individual because either the paternally inherited or the maternally inherited allele is repressed. This means that different individuals heterozygous for alleles of an imprinted gene will have different phenotypes depending on which allele they inherit from their mother and which allele they inherit from their father (see CONCEPT CLOSE-UP: *Gene Imprinting and a Case of Paternal Inheritance*). Imprinting is thought to involve the methylation of DNA and is reversible. As a consequence, the paternal and maternal chromosomes are differentially marked with methylation patterns and allow gene expression on chromosomes to be regulated differently. Gene imprinting is thus a mechanism of regulation of nuclear gene expression rather than a case of extranuclear inheritance. DNA methylation and its effects on gene expression will be discussed in Chapter 14.

SECTION REVIEW PROBLEMS

1. A strain of plants with twisted leaves was crossed with a strain with straight leaves. The F_1 all had twisted leaves. Explain how you would determine whether this was caused by a maternal effect or a dominant mutation.
2. Female plants homozygous for a dominant maternal-effect mutation short leaf (*Sl / Sl*) were crossed to male plants with long leaves that were homozygous for the recessive allele (*sl / sl*). The progeny were collected, and the females were backcrossed for two generations to *sl / sl* males. What would be the phenotypic ratio of long to short leaves in the progeny of the second backcross?
3. Several normal-looking females from a strain of *Drosophila* treated with a mutagenic chemical have produced lethal offspring with severe developmental abnormalities. How could you determine if these abnormalities were the result of a maternal-effect mutation?

SOME CYTOPLASMIC ORGANELLES HAVE CHROMOSOMES.

Eukaryotic cells contain a number of cytoplasmic organelles, each with a special function. Two of the most important organelles are the **mitochondria** and **chloroplasts.** All eukaryotic cells contain mitochondria, and all green plant cells contain both mitochondria and chloroplasts. Mitochondria and chloroplasts provide cells with energy, and chloroplasts are the site of photosynthesis. One unique feature that distinguishes mitochondria and chloroplasts from other cytoplasmic organelles is that they contain their own genomes. As a result, mutations of mitochondrial and chloroplast genes show a very characteristic pattern of inheritance that is easily distinguished from both maternal effects and nuclear inheritance. In this section, we will discuss the structure and function of mitochondria and chloroplasts, the structure and organization of their genomes, and the special patterns of inheritance of genes on organellar DNA.

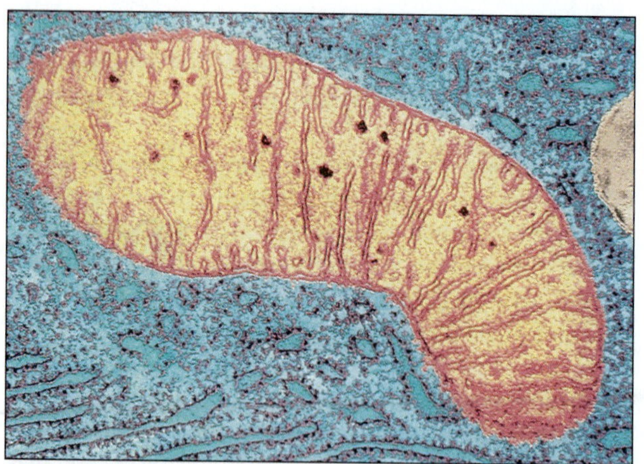

FIGURE 7.3 The structure of a mitochondria. This electron micrograph of a mitochondria shows the organization of the internal membrane and the cristae. *(Bill Longacre/Photo Researchers, Inc.)*

Mitochondria and Chloroplasts Provide Energy and Carbohydrates to Cells.

Mitochondria are bounded by two membranes, an outer and an inner. The outer membrane is smooth and highly permeable, and the inner membrane is less permeable and is folded into complex internal structures called **cristae** (Figure 7.3). In the cristae, sugars and fatty acids are broken down into CO_2 and H_2O in a series of chemical reactions collectively called oxidative phosphorylation. These reactions yield energy that is captured in the form of high-energy electrons. The energy of these electrons is converted into high-energy chemical bonds, such as the high-energy bond joining a third phosphate group to ADP: ADP + P + energy ⇒ ATP. The energy stored in the high-energy phosphate bond of ATP is the main "energy currency" of the cell, and the energy stored in this bond is used throughout the cell to power a wide variety of cellular processes. The production of ATP is carried out by large enzyme complexes imbedded in the mitochondrial inner membrane. Some of the components of these complexes are determined by genes in the mitochondrial genome, and others, by genes in the nucleus. Assembling these complexes requires a close interaction between these two genomes.

Chloroplasts have a number of similarities to mitochondria. Chloroplasts also have a permeable outer membrane and a convoluted, less permeable inner membrane. The space within the chloroplast inner membrane is termed the **stroma** (Figure 7.4). Within the stroma is a third membrane, termed the **thylakoid membrane,** which is folded into flattened disks called **thylakoids.** Thylakoids are assembled into stacked complexes called **grana,** where the enzyme complexes that carry out photosynthetic reactions are located. Chloroplasts exist in an immature form in young plant embryos called **proplastids.** Proplastids contain no chlorophyll and do not carry out photosynthesis. After the embryonic plant begins to grow and is exposed to light, some of the proplastids develop the thylakoid membrane structure and produce the enzyme complexes necessary for photosynthesis. If the seedling is kept in complete darkness, however, the proplastids never mature, and the plant remains white.

The photosynthetic reactions carried out within chloroplasts are divided into two general categories. One category, termed the **light-dependent reactions,** are the reactions by which energy from sunlight is captured and stored as high-energy phosphate bonds in molecules of ATP. Electrons in chlorophyll absorb energy from sunlight, which shifts them into a high-energy state. These high-energy electrons are used by chloroplasts to generate ATP by a series of chemical reactions similar to those used by mitochondria to generate ATP. The ATP produced by chlorophyll is used to power general cellular reactions and also to power the second set of photosynthetic reactions.

The second set of photosynthetic reactions is termed the **dark reactions.** The dark reactions use ATP to convert carbon dioxide into carbohydrate, a process termed **carbon fixation.** The central reaction of carbon fixation is the conversion of a molecule of carbon dioxide to organic carbon. This reaction is catalyzed by the enzyme ribulose bisphosphate carboxylase. This enzyme makes up more than 50% of the total chloroplast protein and is believed to be the most abundant protein on earth. The overall result of the two sets of photosynthetic reactions is to convert water and carbon dioxide into carbohydrate and oxygen:

$$H_2O + CO_2 \xrightarrow[\text{energy}]{\text{light}} \text{carbohydrates} + O_2$$

All animals and plants, and most microorganisms, require a steady supply of organic compounds derived from carbohydrate to survive. Thus, nearly all life on earth depends on photosynthesis. The genetic investigation of the genes that regulate photosynthesis is an exciting and very important part of plant genetics today.

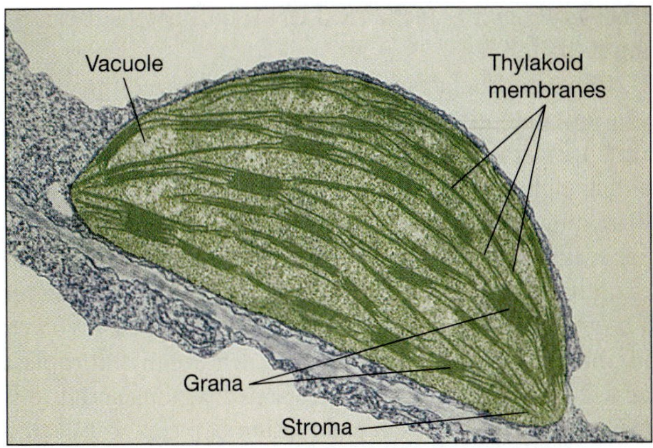

FIGURE 7.4 The structure of a chloroplast. This electron micrograph shows the organization of the stroma and the grana. *(Dr. Jeremy Burgess/Science Photo Library/Photo Researchers, Inc.)*

The Genomes of Mitochondria and Chloroplasts Are Organized as Single, Circular Chromosomes.

The mitochondrial genome is almost always organized into a single, circular chromosome, very similar to the chromosome found in bacteria (Chapter 8). The size of this chromosome varies from species to species but is constant within a species. Animal mitochondrial chromosomes, for example, range in size from 13,000 to 18,000 base pairs of DNA, or 13–18 kilobases (kb). Fungal mitochondrial chromosomes are about 75 kb, and higher plants have mitochondrial chromosomes of 300–500 kb. The reason plants have such large mitochondrial chromosomes is not understood. Mitochondrial genomes contain only a small number of genes. The human mitochondrial chromosome, for example, contains 37 genes that produce only 13 of the more than 80 subunits of the ATP production complexes (Figure 7.5). The remaining genes are required for mitochondrial gene expression or replication (discussed further in Chapters 9 and 11).

Each mitochondrion contains between 5 and 20 copies of the mitochondrial chromosome, and most cells contain many mitochondria. Human liver cells contain about 1000 mitochondria, skin cells contain about 100 copies, and a mature egg may contain 10 million mitochondria. The total number of copies of the mitochondrial genome in human cells is thus variable, ranging from 1000 to 50 million copies. Thus, although an individual mitochondrial chromosome may be small, mitochondrial DNA may comprise a significant fraction of the total DNA in a cell. In animal liver cells, mitochondrial

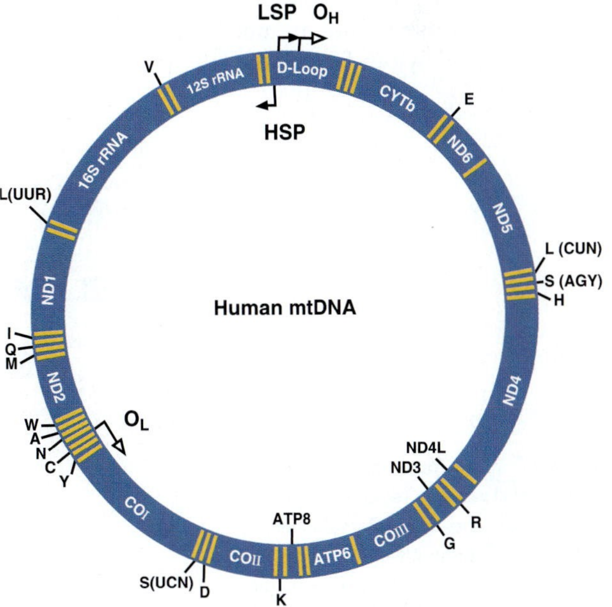

FIGURE 7.5 Map of the human mitochondrial genome. This map shows the arrangement of the genes in the chromosome of the human mitochondria. The human mitochondrial chromosome contains 16.6 kb of DNA and has 37 genes.

DNA is about 1% of the total DNA, but in eggs it can be more than 99%.

The chloroplast genome is also organized into a single, circular chromosome. Chloroplast chromosomes vary in size from species to species but are generally about eight to ten times larger than animal mitochondria, ranging from about 130 to 150 kb. Chloroplasts in general have more genes in their genome than do mitochondria. A typical chloroplast genome contains about 110 genes, of which about 50 produce products involved in photosynthesis. A characteristic feature of chloroplast genomes is that they contain repeated regions and duplicate copies of genes. Recombination between duplicated regions leads to a great deal of variability of chloroplast chromosome structure (the effects of recombination between duplicated regions of a circular chromosome are discussed in Chapter 18).

Chloroplasts also contain multiple copies of the chloroplast chromosome, and each plant cell contains multiple chloroplasts. The numbers of chloroplasts per cell varies from species to species and in different tissues within a plant. For example, there are about 80 chromosomes per chloroplast and 2 chloroplasts per cell in the single-celled alga *Chlamydomonas,* whereas in maize each chloroplast contains between 20 and 40 chromosomes, and each leaf cell contains between 20 and 40 chloroplasts. The chloroplast DNA is about 15% of the total DNA in a typical maize leaf cell.

Mutations in Mitochondria and Chloroplast Genes Show Maternal Inheritance.

The inheritance patterns of genes on mitochondria and chloroplast chromosomes are not Mendelian, and they are determined by two factors: the inheritance pattern of the organelle and the number of copies of organelle chromosomes in the cell. Mitochondria and chloroplasts are permanent residents of the cytoplasm, and in most species, individuals inherit their cytoplasmic organelles from only one parent (the mother). In these species, the mitochondria and/or chloroplasts are inherited along with the egg cytoplasm. This type of inheritance is termed maternal inheritance, or uniparental inheritance, and it has several distinguishing features. First, the genotype of the offspring is determined entirely by the parent that contributes the cytoplasmic organelle. If a female with a mutant organelle genome is crossed to a male with a normal organelle genome, the progeny will all have mutant organelle genomes. Recent investigations suggest that male mitochondria are sometimes passed to the progeny, but are excluded or degraded (see CONCEPT CLOSE-UP: *Paternal Mitochondria Can Sometimes Be Inherited*).

Let us consider the hypothetical example in Figure 7.6 of reciprocal crosses between two strains of animals, one with a mutant allele (*M*) of a mitochondrial gene and a second strain with a normal (+) allele of this gene. All the progeny from the cross of an *M* female with a + male have the same genotype and phenotype (*M*) as the mother. If the female offspring from this cross are backcrossed to + males for many generations, the result is always the same: all the progeny have an *M* genotype and phenotype. In the reciprocal cross between a + female and an *M* male, all the progeny have a + genotype and phenotype. If the female offspring from this cross are backcrossed to *M* males for many generations, the result is always the same: the progeny are all +. This is characteristic of maternal inheritance—backcrossing females for many generations will not produce progeny with the male genotype or phenotype.

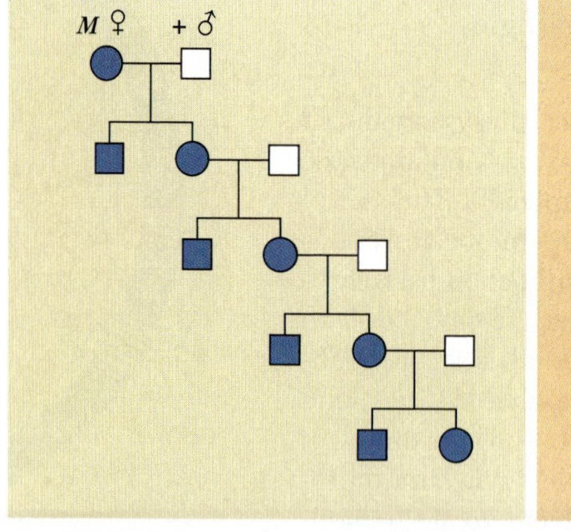

The genotype and phenotype of the offpring are the same as that of the female parent.

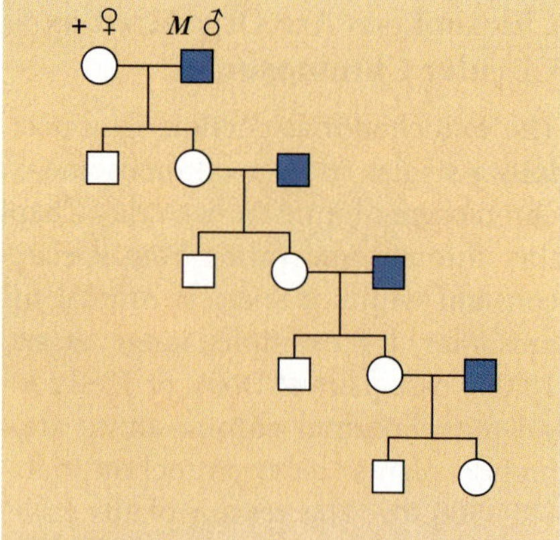

Backcrossing normal females to mutant males for many generations will not produce mutant offspring.

FIGURE 7.6 Maternal inheritance in reciprocal crosses. In reciprocal crosses involving traits showing maternal inheritance, the offspring always have the same phenotype as their mothers. Thus, in the first cross, the offspring always have the mutant (*M*) genotype and phenotype and in the second cross the offspring always have the + genotype and phenotype.

Concept Close-Up

Paternal Mitochondria Can Sometimes Be Inherited

In most plant and animal species, an offspring's cytoplasmic organelles are maternally inherited. There are exceptions to this rule, however, and investigating these exceptions has given new insight into the nature of organelle inheritance. In animals, sperm contains functional mitochondria, and these mitochondria enter the egg during fertilization. However, the paternal mitochondria are normally never found in the cells of the offspring. The key question is why these mitochondria are never included in the zygote's cells. One possibility is that sperm mitochondria are somehow modified during sperm maturation so that they die, or otherwise cannot function, after fertilization. The alternative is that sperm mitochondria are functional but that they are somehow excluded or degraded.

In the past, the only way to investigate this question was to cross individuals with different mitochondrial genotypes and to try to recover offspring with the paternal genotype and phenotype. The large numbers of maternal mitochondria packaged in the egg cytoplasm has made detecting any paternal mitochondria difficult. The heteroplasmic zygotes would, at best, have only a few paternal mitochondria and many maternal mitochondria. Generating a homoplasmic offspring would require the random segregation of mitochondria to somehow generate a germ cell that was homoplasmic for the rare paternal mitochondria. The rare paternal mitochondria could just as easily be lost, and at best could not become homoplasmic for several generations. More recently, molecular genetic techniques that can detect rare mitochondrial DNA markers have been developed, and this has led to numerous recent experiments to test this old question.

In *Drosophila* and mice, offspring from crosses between individuals from the same species do not have paternal mitochondria. However, in crosses between individuals from separate but closely related species, paternal mitochondria are found in the zygote. In experiments involving interspecies *Drosophila* crosses and interspecies mouse crosses, heteroplasmic offspring were detected at a low (10^{-4}) frequency. The low frequency suggests that these heteroplasmic offspring are indeed exceptions. However, in crosses involving other species, the paternal contribution is much greater. For example, in crosses between *Mytilus* (marine mussels) species, more than 50% of the offspring were heteroplasmic. These results suggest that male mitochondria are capable of functioning in the zygote but are normally excluded. In interspecies crosses, the male mitochondria have species-specific differences that can confuse the system and allow them to survive more often. The overall conclusion is that maternal inheritance of mitochondria is caused by a system of deliberate selection.

The second major factor affecting the pattern of inheritance of an organelle gene is the numbers of copies of the chromosome in the organelle. Consider the events that must occur to generate an individual with a mutant mitochondrial or chloroplast genome. When a new mutation of a mitochondrial gene first arises, it is present on a single chromosome in an organelle that contains several other chromosomes containing a different allele. This organelle is present in a cell with perhaps thousands of other organelles, all possessing the normal (or different) allele. Several rounds of chromosome replication and mitochondrial division will produce a population of organelles that contain the mutant allele, but each organelle contains a mixed population of chromosomes. Eventually, random segregation of the chromosomes during organelle division may produce an organelle containing only chromosomes with the new mutation. Replication of this organelle, and random segregation of organelles during oogenesis, may eventually produce a gamete containing only mitochondria with the new mutation.

An individual whose organelles contain only one allele of a gene is termed **homoplasmic.** An individual whose organelles have more than one allele is termed **heteroplasmic.** A heteroplasmic individual may have

1. individual organelles that contain multiple alleles
2. different alleles in different organelles within one cell
3. cells that contain different alleles in different tissues

Heteroplasmic females may produce progeny with different genotypes and phenotypes, depending on how the organelles are inherited. During oogenesis in many animal species, the number of organelles is reduced and then undergoes expansion. This means that only a sample of the mitochondria in the female are passed to each egg. Sampling error in a heteroplasmic female may result in offspring with significantly different mitochondrial genotypes.

An example of a human mitochondrial mutation that is responsible for the disease known as myoclonic epilepsy and ragged-red fibers (MERRF) illustrates the

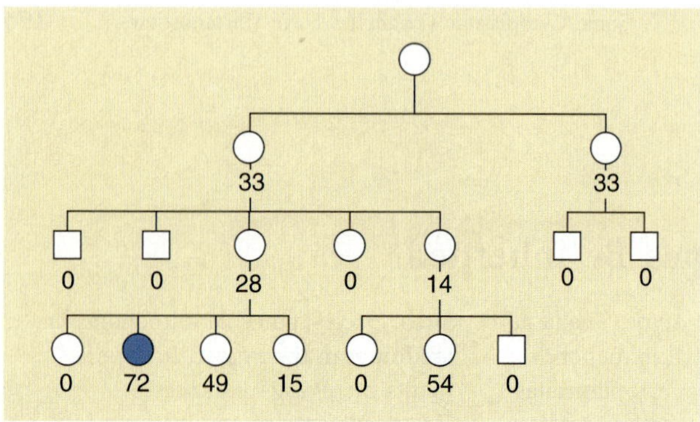

FIGURE 7.7 Pedigree of the human MERRF disorder. The MERRF disorder shows a complex maternal inheritance. Females who are heteroplasmic for this disease-causing mitochondrial gene have progeny with different percentages of normal and disease-causing mitochondria. The numbers indicate the percentage of disease-causing mitochondria.

inheritance of mitochondrial mutations in heteroplasmic individuals. This disease has a collection of symptoms all relating to the failure of the mitochondria to produce sufficient amounts of ATP. Lack of energy causes cells to die and tissues to degenerate, especially neural cells and muscle cells. Heteroplasmic individuals have a threshold of 65% mutant mitochondria before they show symptoms, so most heteroplasmic individuals appear normal. Molecular genetic analysis can determine the percentage of mutant mitochondria in an individual, and this information can be used in pedigree analysis. The pedigree in Figure 7.7 shows how offspring of a heteroplasmic individual may have different mitochondrial genotypes. The disease-causing mutation always shows maternal inheritance, but the percentage of mitochondria with the disease-causing allele varies widely. Offspring may always inherit their mitochondria from one parent, but that does not mean that they have exactly the same mitochondrial genotype as that parent or as their siblings.

Many plant species have visible mutant phenotypes that are caused by mutations in chloroplast genes. Mutations that eliminate the ability of the chloroplast to function often cause it to contain no chlorophyll or reduced amounts of chlorophyll. A cell that is homoplasmic for such mutant plastids will be white, and a plant that is homoplasmic for nonfunctional plastids will die. However, many heteroplasmic plants survive as mosaics, having some white cells and some green cells. In some individuals, the leaves and/or stems may be **variegated,** having patches or stripes of white and green cells.

Numerous strains of variegated ornamental plants, such as the *Hosta montana Aureomarginata* (Figure 7.8) that have striking patterns of white and green regions in the leaf, have been produced. Such strains are highly prized by breeders and collectors. The green/white variegation is maternally inherited, but the offspring of a variegated plant are variable and can be entirely green, entirely white, or variegated with a different pattern. Plants with an interesting or valuable pattern must be propagated asexually to ensure faithful replication of the color pattern. There is no good method for intentionally breeding to produce particular patterns of variegation. Developing commercially valuable new varieties is a matter of making many crosses with variegated plants and searching the offspring for desirable patterns.

A variegated leaf phenotype was the first trait to be shown to have maternal inheritance. In 1909, Carl Correns studied the inheritance of leaf color in some variegated strains of *Mirabilis* (four o'clocks). These plants had an irregular pattern of white and green variegation in the leaves and stems that was caused by the plants being heteroplasmic for mutant chloroplasts that did not contain chlorophyll. These plants sometimes produced entire branches that were entirely green or entirely white. Correns used the flowers from white branches, green branches, and variegated branches to make a series of crosses to test the inheritance of the color variegation (Figure 7.9). Pollinated flowers on green branches produced offspring that were entirely green, and pollinated flowers on white branches produced offspring that were entirely white, no matter what the genotype of the pollen. Pollination of flowers of variegated branches gave offspring with a mixture of phenotypes, including green, variegated, or white. Again, the genotype of the pollen had no influence on the offspring color. Pollen from variegated flowers gave offspring whose color was deter-

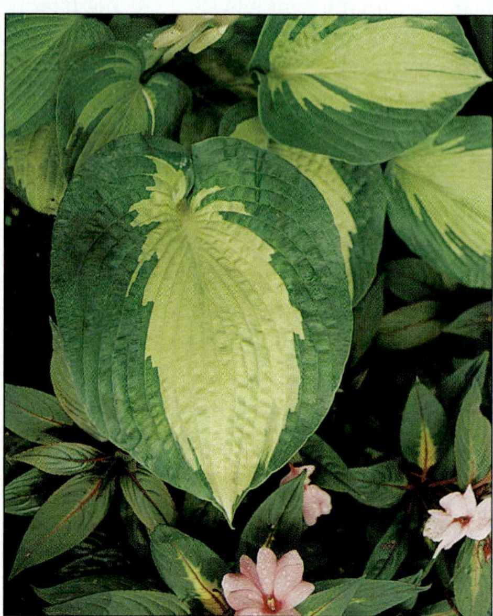

FIGURE 7.8 Variegated leaves in *Hosta*. The light-colored regions of the variegated leaves of *Hosta montana Aureomarginata* have defective chloroplasts that do not contain chlorophyll. The arrangement of the light-colored regions differs from strain to strain and can only be reproduced asexually. *(Jack R. Girton, Iowa State University)*

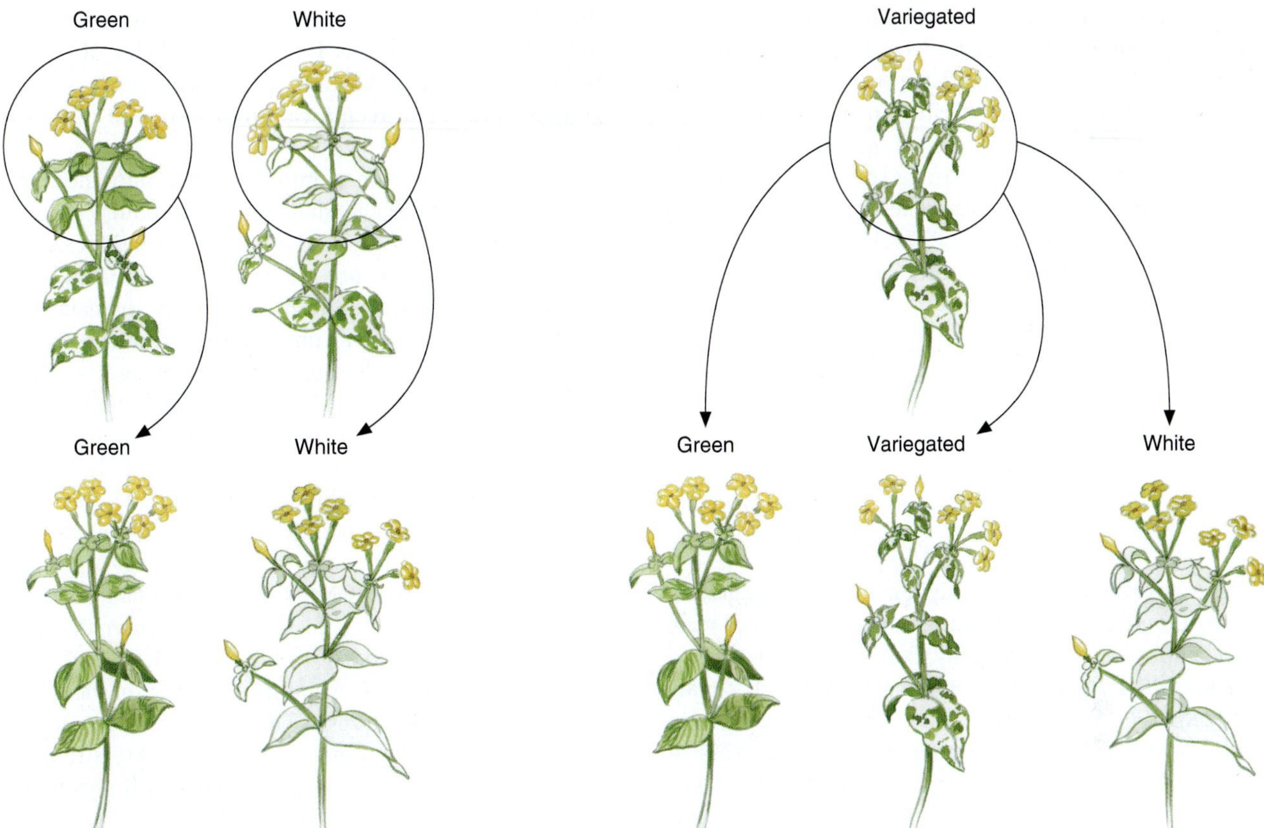

FIGURE 7.9 Inheritance of variegation in *Mirabilis.* Carl Correns made a series of crosses with flowers found on different colored sectors of variegated *Mirabilis* plants. The flowers in a green sector on a variegated plant give rise to only green plants, flowers from white sectors on a variegated plant give rise to white plants, but variegated sectors on a variegated plant give rise to all three phenotypes. He concluded that the phenotype of the offspring is controlled by the genotype of the female gamete.

mined by the maternal plant. This evidence strongly suggested that the mutant chloroplasts responsible for the different color of plants were inherited maternally.

SECTION REVIEW PROBLEMS

4. Compare the F_2 phenotypic ratios produced by the following:
   a. cytoplasmic inheritance and a dominant mutation
   b. cytoplasmic inheritance and dominant epistasis
5. A strain of the plant *Mirabilis jalapa* that is called variegated can have variegated, white, or green leaves. This is a cytoplasmically inherited phenotype. What offspring do you expect from the following crosses?
   a. (male) white leaf × (female) green leaf plants
   b. (female) variegated leaf × (male) white leaf plants

Genes in the Organelles Interact with Genes in the Nucleus.

There is a great deal of genetic interaction between the genes of the organelles and the nuclear genes. Because of this interaction, mutations in nuclear genes may interact with mutant alleles of mitochondria or chloroplasts in all the same ways that mutant alleles of two nuclear genes interact (Chapter 4). One well-known example of this occurs in certain types of cytoplasmic male sterility (CMS) in maize. CMS was first observed in maize in 1933 by Marcus Rhoades, and since that time, three types of maize CMS have been identified: CMS-S, CMS-T, and CMS-C. A CMS plant fails to produce viable pollen in its tassel (the maize male flowers) and is thus male sterile. However, these plants are fertile as females (maize has both male and female parts on each plant).

The CMS phenotype shows maternal inheritance. All progeny of CMS plants have CMS, and crossing CMS plants to male fertile plants for many generations does not produce fertile male progeny. The CMS trait is not unique to maize. CMS strains have been discovered in virtually all plant species that have been investigated. Although cytoplasmic male sterility in maize has all the characteristics of a mutation in an organelle gene, in certain crosses male fertility can be restored to CMS-T plants by alleles of two nuclear genes known as *restorer of fertility* (*Rf*) genes, *Rf1* and *Rf2*. Individuals with dominant alleles of both *Rf* genes produce viable pollen.

Cytoplasm	Nuclear Genes		Phenotype
CMS-T	*Rf1* / +	*Rf2* / +	Viable pollen (male fertile)
CMS-T	+ / +	*Rf2* / +	Inviable pollen (male sterile)
CMS-T	*Rf1* / +	+ / +	Inviable pollen (male sterile)

Crosses involving CMS-T individuals that are segregating for the *Rf* genes can produce complex ratios of male fertile and male sterile progeny. For example, crossing a female CMS-T *Rf1* / + *Rf2* / + female plant with a male fertile *Rf1* / + *Rf2* / + results in a 9 fertile : 7 sterile ratio in the progeny. This ratio is that of a dihybrid cross in which all individuals with dominant alleles of both unlinked genes have one phenotype (male fertile), and all individuals homozygous for recessive alleles of one or both genes have a different phenotype (male sterile) (Figure 7.10). The reciprocal cross results in all fertile progeny.

CMS results from a mutation of a mitochondrial gene whose function is vital for proper pollen maturation. The precise molecular basis for CMS and for the action of the *Rf* genes is not known, but it is an area of active research. Cytoplasmic male sterility in maize has considerable commercial importance. Producers of hybrid seed corn, for example, some years ago began to make extensive use of CMS-T in their breeding programs. If, in a controlled cross, the breeder wants to prevent unwanted pollination of certain plants, he or she must manually remove the tassel (male part) on one of the individuals in the cross. Male sterility eliminates the need for manual removal of tassels during large-scale crosses. Unfortunately, CMS-T plants were later found to be highly susceptible to infestation by Southern corn leaf blight (a fungal infection). In the early 1970s, outbreaks of the blight destroyed a significant portion of the American hybrid corn harvest. This additional phenotype caused a great reduction in the use of CMS strains in commercial corn breeding. Today there is still a great deal

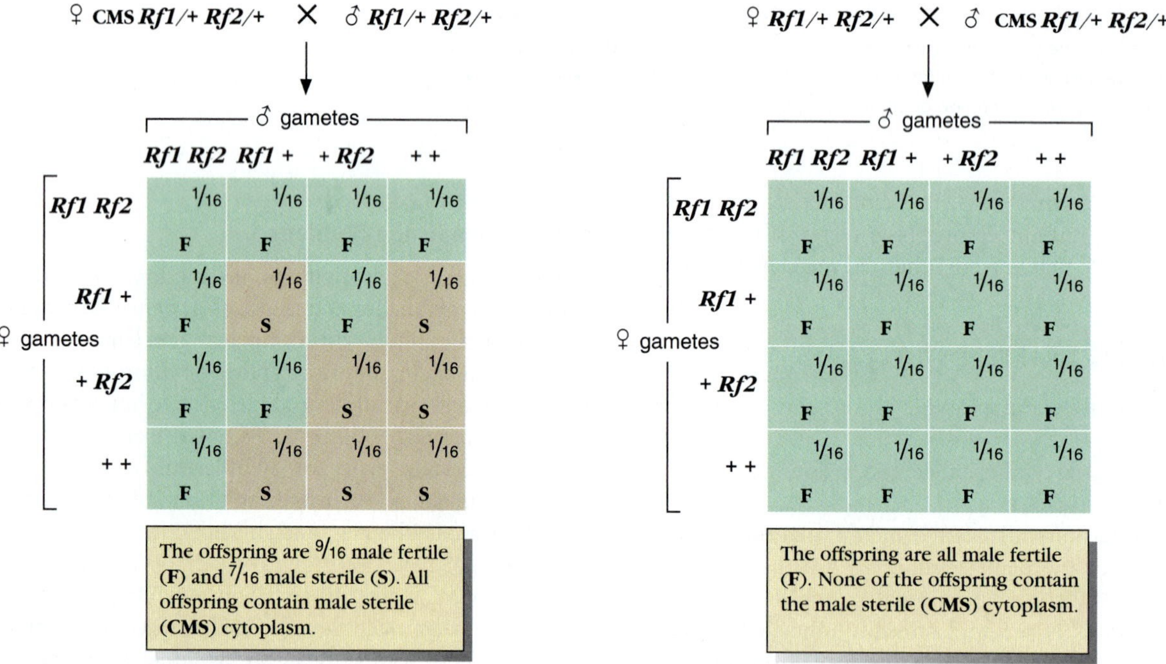

FIGURE 7.10 CMS and the *Rf* genes in maize. Crosses between individuals heterozygous for *Rf1* and *Rf2* that have CMS cytoplasm give different ratios in reciprocal crosses. In the cross with a CMS female, the Mendelian segregation of the two dominant restorer alleles gives a 9 male fertile : 7 male sterile ratio. In the cross with a female that does not have CMS, all the progeny are male fertile.

of research being done on how CMS results in blight sensitivity.

The Mitochondria and Chloroplasts Originated as Cytoplasmic Symbionts.

The endosymbiont theory on the origin of mitochondria and chloroplasts is that they were originally free-living bacteria that developed symbiotic relationships with primitive eukaryotic cells. This theory is supported by several important facts about mitochondria and chloroplasts. For example, these organelles are not assembled by cells but arise by the division of existing mitochondria and chloroplasts. Division of both types of organelles occurs by a simple fission process, similar to the division of bacteria (discussed in Chapter 8).

Organelle division occurs throughout the cell cycle with organelle DNA synthesis occurring in all stages of the interphase period (G_1, S, and G_2). This continued division indicates that organelle division is not regulated by the same genes that regulate the nuclear and cell division cycles. On the other hand, mitochondria and chloroplast division is not completely independent of the cell cycle. Both types of organelles are present in characteristic numbers in different cell types, and both reproduce enough to exactly double their numbers in one cell cycle, maintaining the characteristic number. This suggests that the reproductive cycles of organelles are at least coordinated with, if not completely controlled by, the cell.

Several other facts support the endosymbiont theory. Mitochondrial membrane structure resembles bacterial membrane structure. Mitochondria and chloroplasts both have single, circular chromosomes, like bacteria. The molecular structure of the organelle enzymes that control DNA synthesis, RNA synthesis, and gene function resemble bacterial enzymes more closely than they do eukaryotic enzymes. One additional fact is the discovery of bacteria and single-celled algae that can establish symbiotic relationships with eukaryotic cells.

SECTION REVIEW PROBLEMS

6. If you had a maize plant that possessed the male sterility mutation in its mitochondria, and if this plant had the nuclear alleles *Rf1* / *Rf1*, what would be the fertility phenotype of this plant?
7. What would be the phenotype with respect to fertility if the nuclear genes were *Rf1* / *rf1*; *Rf2* / *rf2*?
8. What would be the phenotypic ratio of the F_1 and F_2 from a cross between these two plants: one (the female) is CMS-T in the mitochondria and *rf1* / *rf1*; *rf2* / *rf2* and the other (the male) has male fertile mitochondria and is *Rf1* / *Rf1*; *Rf2* / *Rf2*?

Infectious Agents Can Be Inherited.

A variety of parasitic agents can reproduce in the cytoplasm of plant or animal cells, and their presence gives their host an altered phenotype. Such agents generally are inherited along with the cytoplasm, yet they can also infect new cells and/or new organisms. This pattern of inheritance is termed infectious inheritance. Phenotypes showing infectious inheritance do not depend on the inheritance of an intact cytoplasmic organelle, and they can be transferred from cell to cell in a small amount of cytoplasm, or by the infectious agent alone. In this section, we will discuss four different types of agents that show infectious inheritance: bacteria, viruses, viroids, and prions.

Symbiotic Microorganisms Can Show Infectious Inheritance.

An example of infectious inheritance was discovered in *Paramecium* by Tracy Sonneborn [see Historical Profile: *Tracy Sonneborn (1904–1981)*]. *Paramecium* are single-celled, diploid eukaryotes that undergo a complex process of sexual reproduction. *Paramecium* have three nuclei, two diploid micronuclei, and one large vegetative macronucleus. The macronucleus is a partial polyploid, containing many copies of some genes but no copies of others. *Paramecium* reproduce by sexual reproduction and also by fission, with the micronuclei dividing mitotically. Sexual reproduction begins with the pairing of cells of different mating types. After the cells pair, the two micronuclei in each cell divide meiotically, producing eight haploid nuclei. Seven of these haploid nuclei and the macronucleus then degenerate and the one surviving nucleus divides mitotically, producing two haploid nuclei. The two cells then exchange one haploid micronucleus during conjugation. Following conjugation, the donor and host micronuclei fuse to produce a diploid nucleus. This diploid nucleus divides mitotically, and one of the nuclei forms a new macronucleus. The other nucleus divides mitotically again to produce two diploid micronuclei (Figure 7.11).

Sonneborn discovered that the cells of one strain of *Paramecium*, called killer, can destroy cells of other strains, called sensitive, by secreting paramecin, a toxic chemical substance. The cells of the killer strain contain small, bacteria-like entities that Sonneborn called kappa particles; they produce paramecin and also render their hosts resistant to paramecin. Studies of the kappa bacteria-like membrane, its chromosome, the metabolism of isolated kappa, and the process of kappa replication indicate that kappa is a symbiotic bacterium. A single *Paramecium* cell may accumulate large numbers (> 1000) of kappa particles.

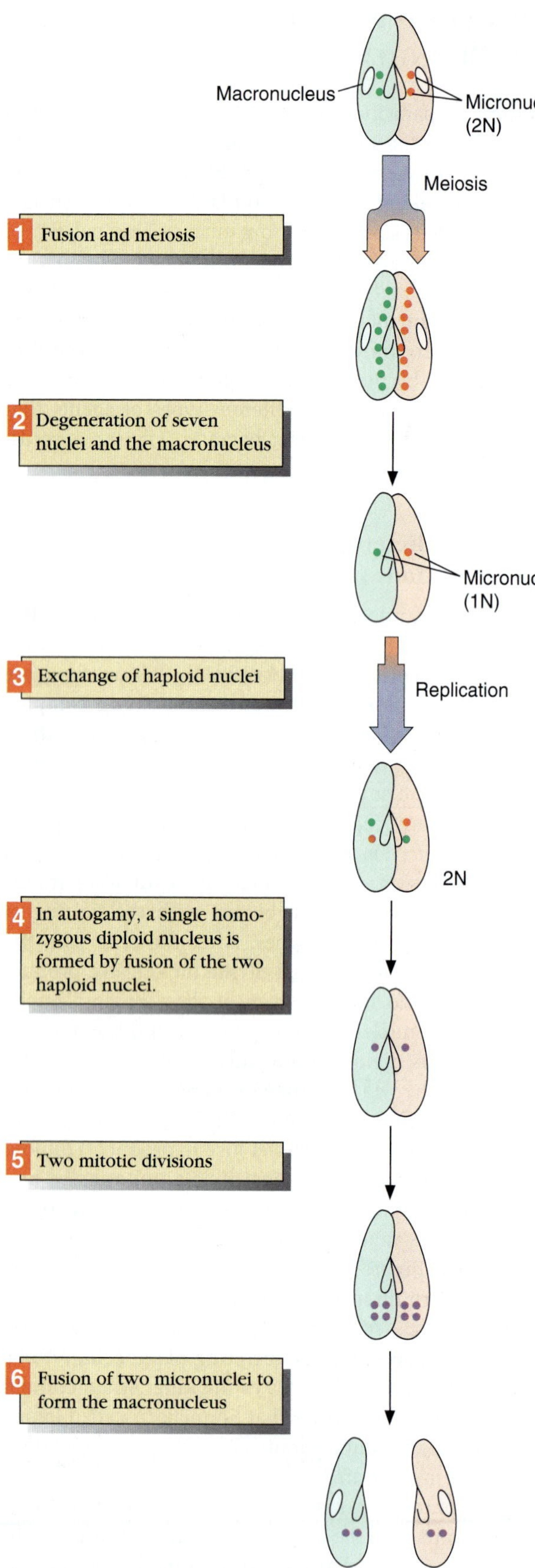
Macronucleus
Micronuclei (2N)
Meiosis
1 Fusion and meiosis
2 Degeneration of seven nuclei and the macronucleus
Micronuclei (1N)
3 Exchange of haploid nuclei
Replication
2N
4 In autogamy, a single homozygous diploid nucleus is formed by fusion of the two haploid nuclei.
5 Two mitotic divisions
6 Fusion of two micronuclei to form the macronucleus

◀ **FIGURE 7.11 Sexual reproduction in *Paramecia.*** Sexual reproduction in *Paramecia* is a complex process that involves meiosis, the exchange of haploid nuclei, fusion of haploid nuclei to generate diploid offspring, and mitotic division to generate new macronuclei.

Sonneborn was able to induce conjugation between killer strains and sensitive stains. During normal conjugation in *Paramecium,* only a very small amount of cytoplasm is transferred with the nuclei, and killer-type cells often do not pass kappa particles. In such crosses, the progeny had a phenotypic ratio of 1/2 sensitive and 1/2 killer. However, if conjugation is experimentally prolonged so that kappa are able to be passed, all the progeny cells have kappa particles along with a killer phenotype (Figure 7.12).

Paramecia with a kappa-sensitive phenotype can be transformed into killer types by ingesting kappa released from fragmented killer cells. Sensitive cells can also be transformed into killer cells by being injected with purified kappa particles. This transformation occurs without the transfer of any organelles or nuclear genes, just the kappa particles. Kappa particles are a not a normal cytoplasmic organelle, and *Paramecium* do not need them to function. They are an infectious bacterial parasite that can reproduce in the cellular cytoplasm, and they can infect new cells. They have somehow evolved a mechanism to resist digestion by *Paramecium's* cytoplasmic enzymes, and they have found a comfortable niche in the cell's environment.

Examples of infectious cytoplasmic chloroplast-like entities have also been found. Several species of hydra, such as *Hydra viridis,* contain cytoplasmic bodies that contain chlorophyll. These bodies are transmitted through the egg cytoplasm and were originally thought to be chloroplasts. However, when the bodies were isolated, they were found to be single-celled green algae, *Chlorella vulgaris.* Removal of the algae does not always kill the hydra, and alga-less hydra can reproduce. Likewise, isolated algae can be cultured and will reinfect alga-less hydra. Similar cases of symbiotic algae living in species of *Paramecium* have also been discovered.

These examples are reminiscent of the proposed origin of mitochondria and chloroplasts. Initially, bacteria and algae that develop a symbiotic relationship with another organism retain the ability to reproduce independently, as do the host individuals. The free-living algae and bacteria also have the ability to infect new host organisms. As time passes, the host and symbiont develop a closer relationship through the accumulation of mutations that give the combined organism an advantage, but that reduce the ability of host and symbiont to survive alone. Eventually, neither can survive alone, and the symbiont becomes a cytoplasmic organelle.

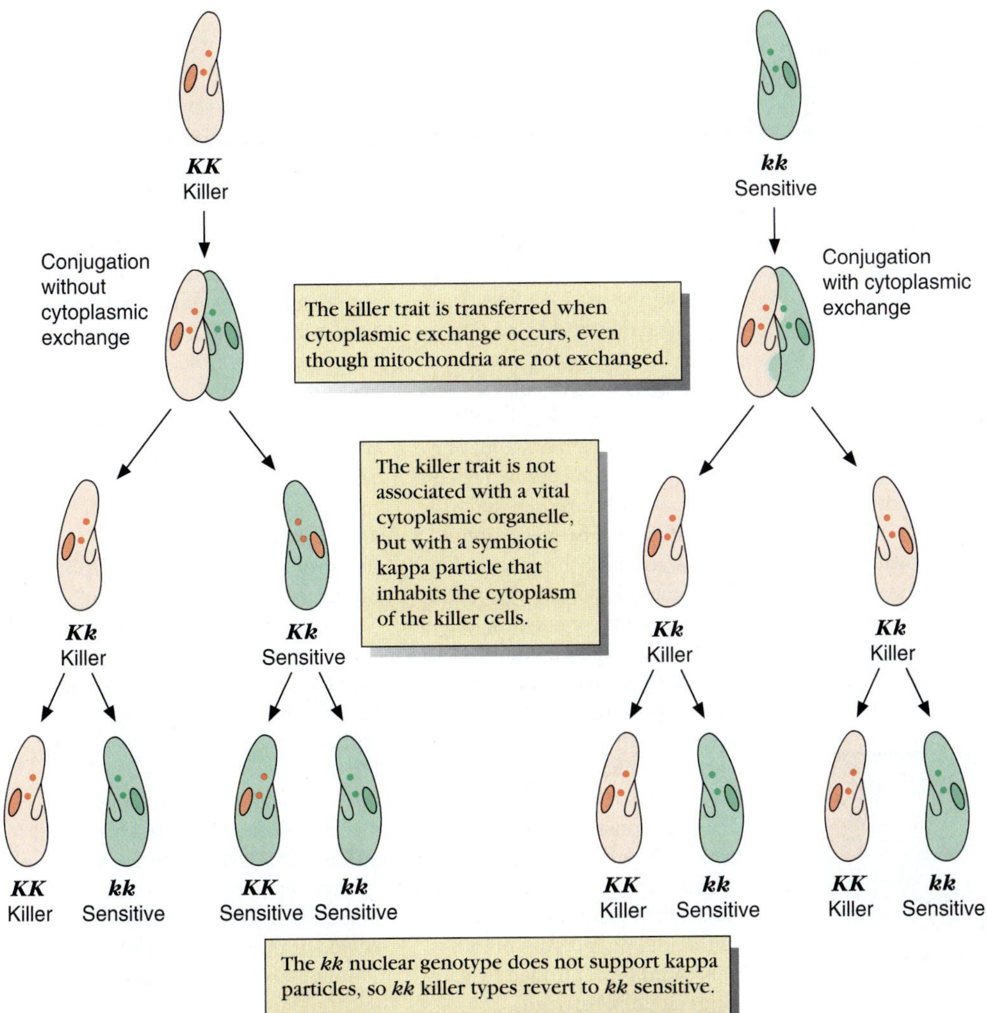

FIGURE 7.12 Cytoplasmic inheritance of kappa. Conjugation of killer and sensitive strains without cytoplasmic exchange does not transfer kappa or the killer phenotype. Conjugation with cytoplasmic exchange does transfer kappa and the killer phenotype. Transfer of the killer phenotype does not require transfer of normal cytoplasmic organelles.

Many Viral Strains Can Show Infectious Inheritance.

The life cycle of many strains of viruses lends itself to infectious inheritance. A common viral life cycle is to infect a cell, reproduce in the cell's cytoplasm using the cell's synthetic machinery, and then cause the death of the host cell, releasing new viral particles (viral reproduction is discussed in detail in Chapter 8). However, some viruses reproduce for many generations in the cytoplasm of a cell without killing it, and these can show cytoplasmic inheritance. For example, an RNA tumor virus (described in Chapter 20) can infect a cell and cause it to show a cancerous phenotype. The virus reproduces in the cell for many cell generations (as part of the nuclear genome), and progeny viral particles are continually extruded from the cell. When an infected cell divides, progeny cells inherit viral chromosomes with their cytoplasm. The cancerous phenotype conveyed to the cells is cytoplasmically inherited and can also be generated by infection of new cells by viruses.

Viruses reproducing in female mammals often can infect their offspring before birth, giving a pattern of maternal inheritance in pedigree analyses. For example, consider the human immunodeficiency virus (HIV), the virus that causes AIDS. A high percentage of the offspring born to females infected with HIV are also infected and develop AIDS. This causes the symptoms of AIDS to show maternal inheritance in pedigrees. This occurs not because HIV is inherited along with the egg cytoplasm, but because the virus infects the unborn offspring before birth. The pedigree, however, gives a maternal pattern of inheritance. To discover the true nature of the inheritance pattern requires additional

HISTORICAL PROFILE

Tracy Sonneborn (1904–1981)

Tracy Sonneborn is recognized as one of the outstanding geneticists of his generation. Sonneborn was born on October 19, 1904, in Baltimore, Maryland. He attended Johns Hopkins University, receiving his undergraduate degree in 1925 and his Ph.D. in 1928. He remained at Johns Hopkins for 11 years until 1939, when he left to take a faculty position at Indiana University, where he remained until his retirement in 1975. After his formal retirement, Sonneborn continued an active research program until shortly before his death in 1981.

Sonneborn was a dedicated and productive researcher who produced 230 publications between 1928 and 1980. Most of these dealt with the genetics of ciliated protozoa, especially *Paramecium*. He was a pioneer in the use of ciliates as research organisms, and he was responsible for developing many of the techniques used to culture *Paramecium* in the laboratory. For example, he discovered the mating types of *Paramecium* and perfected the culture conditions necessary for controlled mating between strains. His major accomplishments came in his studies of the complex genetics of cytoplasmic organelles, and in his analysis of kappa. He performed sophisticated analyses of the inheritance of kappa and the killer phenotype. In addition, he discovered nuclear genes that are essential for the maintenance of kappa in the cytoplasm, demonstrating that the nucleus is not a passive spectator in the symbiosis but rather is an active partner. These analyses revealed that inherited phenotypes of individuals could be controlled by complex interactions between cytoplasmic and nuclear genes.

FIGURE 7.A **Tracy M. Sonneborn.** An outstanding American geneticist who made many discoveries in the field of extranuclear inheritance. *(D. M. Nanney. 1981. "T. M. Sonneborn: An Interpretation,"* in Annual Review of Genetics, *Vol. 15, pp 1–9. Annual Reviews, Inc., Palo Alto, CA)*

The outstanding quality of Sonneborn and his work was widely recognized, and he received many honors during his long career. He was elected to numerous prestigious scientific societies, including the National Academy of Science (1946), the American Academy of Arts and Sciences (1948), and the American Philosophical Society (1952), and he was made a Foreign member of the British Royal Academy of Sciences (1964). He served as president of the Genetics Society of America and the American Society of Naturalists and as an officer of the American Society of Zoologists and the American Institute of Biological Sciences. He received five honorary degrees and numerous other recognitions. Sonneborn's honors were given as much for his ability as a geneticist and his widely recognized scientific integrity as for his actual scientific discoveries.

Sonneborn had a dynamic personality that made a strong impression on his colleagues and students. He was highly regarded as a lecturer with a riveting lecture presence that often led to spontaneous applause in classroom lectures and public audiences. His intensity and his unquestioning devotion to scientific research were communicated to his numerous graduate students and postdoctoral research associates. Through his own work and that of the students he trained, he is credited with having a major role in establishing *Paramecium* as a research tool in genetics and cell biology. In many of his investigations, he was far ahead of his time, and the problems he attempted to solve were beyond the techniques of classic genetics and cell biology available to him. Some of these problems are only now yielding to the combined procedures of genetics and molecular biology.

evidence that HIV is infective. For example, treatment of females with combinations of antiviral drugs that slow HIV replication greatly decreases the incidence of prenatal infection, and HIV can infect individuals who are not related. Geneticists studying cases of apparent maternal inheritance in a small family for which only a small pedigree is available may find it impossible to determine whether the phenotype is infective or maternally inherited.

Viroids Show Infectious Inheritance but Have No Genes.

Anther example of infectious inheritance is the small virus-like RNA particles known as **viroids.** Viroids are small, circular nucleic acid molecules that replicate in the cytoplasm of plant and animal cells. Viroids do not contain any genes but are simply naked RNA molecules. They are replicated by the cell's machinery and can accumulate to enormously high numbers in the cytoplasm of an infected cell. Viroids normally show maternal inheritance but can also infect damaged cells. Viroids are a serious problem in plant agriculture; they can be transmitted on the hands or tools of workers who are pruning or handling plants, and they can quickly infect entire fields of plants. Unfortunately, there is no effective treatment for plants infected with viroids.

One example of a viroid that shows infectious inheritance is the potato spindle tuber viroid (PSTV). This viroid is a small circular RNA molecule of only 359 nucleotides. In the cytoplasm of a potato cell, PSTV accumulates to very high concentrations, and eventually it may kill the cell. Infected plants produce small and spindly potatoes and will pass this trait (along with the viroid) maternally to their progeny. PSTV is highly infectious (Figure 7.13). It is not entirely clear how this viroid with no genes generates the visible phenotype in the host plant. Most likely the unchecked replication of the viroid interferes with the production of vital cellular RNAs.

SECTION REVIEW PROBLEMS

9. A worker in a greenhouse notices that an entire row of plants in one bed is wilting and showing signs of stress. Her supervisor suggests that this symptom is caused by an infection, either by a viroid or a virus. How can she determine whether these plants have been infected by a virus or by a viroid?
10. How would you distinguish between a maternally inherited trait and a viroid-induced trait?

Prions Show Infectious Inheritance but Do Not Contain Nucleic Acid.

Other infectious agents that show infectious inheritance are the **prions.** Prions are small proteinaceous infectious particles that do not contain nucleic acids. Prions cause neural degeneration diseases in a variety of animal species, including humans. These diseases are all classed

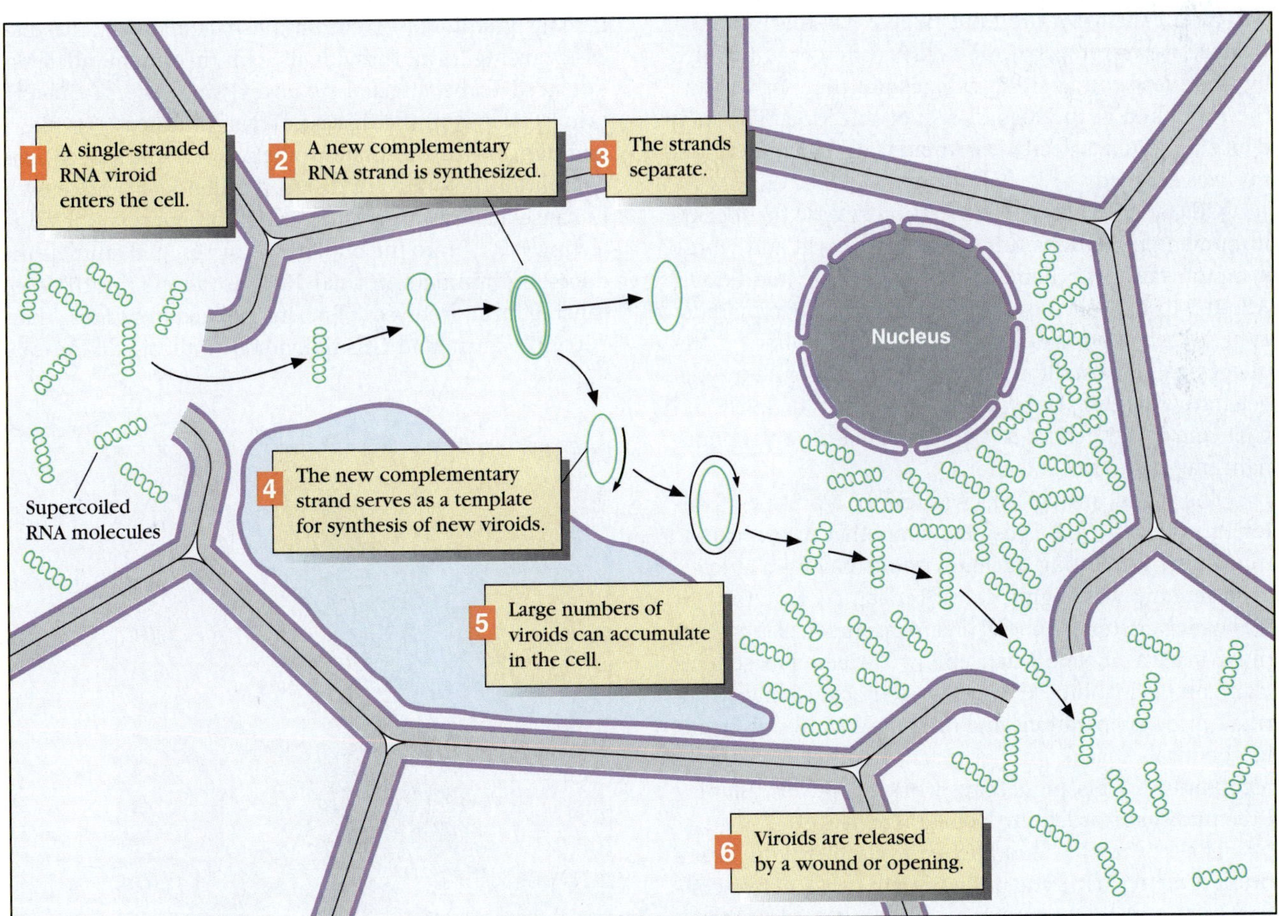

FIGURE 7.13 The viroid life cycle. The potato spindle tuber viroid is a circular, single-stranded RNA molecule that can enter a cell through a wound. It reproduces in the cytoplasm, producing very large numbers of progeny viroids. These can leave the cell through a wound or a cytoplasmic channel into another cell.

as **transmissible spongiform encephalopathies** (TSE). All cause degeneration of the brain leading to increasing loss of mental capacity, dementia, paralysis, and death. They are invariably lethal, and there is no treatment and no cure. Prions were originally called slow viruses because of the long incubation times of these diseases.

Three prion diseases in humans have been well characterized and illustrate the inheritance of prions: kuru, Creutzfeld–Jakob disease (CJD), and Gerstmann–Staussle Syndrome (GSS). Kuru is a fatal neural degeneration disease of the Fore highlanders of Papua, New Guinea. Initial studies of kuru indicated that there was a strong correlation between maternal infection and offspring infection, suggesting that kuru showed maternal inheritance. However, careful pedigree analyses revealed a number of exceptional cases, such as infection of adopted children, which suggested that kuru could be caused by an infectious agent.

The agent that causes kuru is a prion that is present in high concentrations in the brains and central nervous systems of infected individuals. The maternal transmission pattern of kuru was the result of the practice of ritual cannibalism by the Fore people. In this practice, women always prepared the bodies, often accompanied by their very young children. This put the mothers and their children at the highest risk of infection. The incidence of kuru has decreased dramatically since this practice was stopped.

CJD is a rare neural degeneration disease that occurs in human populations throughout the world and whose symptoms do not become apparent until late middle age. About 5 to 10% of cases are inherited, and the remainder appear to be spontaneous. CJD occurs at about a frequency of 1 in 1 million individuals. GSS is a neural degenerative disease with symptoms similar to kuru and CJD, and it is inherited as a Mendelian autosomal dominant trait and is not infectious.

The agent causing TSEs was termed a prion by Stanley Prusiner in 1982. Prusiner's claim that prions can be infectious, replicating agents without having a nucleic acid genome was highly controversial. At that time, it was widely assumed that all organisms that cause disease must have a nucleic acid genome. Prusiner based his claim on his inability to detect nucleic acid in highly purified prion preparations and on the resistance of prions to chemicals that destroy or inactivate nucleic acids. In recognition of his pioneering work on prions, Stanley Prusiner was awarded the Nobel Prize in 1997.

The key to the reproductive cycle of prions is the prion protein (PrP). The prion protein exists in two configurations. One is a normal cellular form (PrP^C) that is found in the brains of all adult mammals and does not cause disease. The second form is a structural variation of the normal form that causes disease and is found only in infected cells. The infected form is given a symbol consisting of PrP with a superscript that indicates the source of the prion. For example, prions from sheep infected with the prion disease scrapie are PrP^{sc}, and prions from humans with CJD are PrP^{CJD}. The structure of the normal form, PrP^C, varies only slightly in different animal species.

The current hypothesis for prion diseases is that the disease-causing form of the prion acts as a template, or seed crystal, that induces the normal form to change into the disease-causing form. The newly altered prion proteins join together to form a large, rod-like crystalline aggregate in the cell (Figure 7.14). Eventually these aggregates become so large that they disrupt cell function and kill the cell. The dead cell degenerates, releasing the disease-causing prions, which are taken up by neighboring cells, and the cycle repeats.

Molecular genetic analyses of prion diseases support the prion hypothesis. Individuals with GSS have a high concentration of PrP^{GSS} in their brains. This prion has an altered structure that is caused by a mutation in the gene that produces the normal form of the prion. This mutation changes the normal prion structure to make it spontaneously shift to, or continuously remain in, the disease-causing form. Individuals with the mutant allele do not need to be infected by an existing disease-causing prion to develop the disease. Thus the disease is inherited along with the mutant allele. PrP^{GSS} has been isolated from the brains of diseased individuals and shown to cause TSE in animals of several different species. Injecting PrP^{GSS} into the brains of experimental animals induces the animal's normal PrP^C to adopt the disease-causing form. Analyses of prions isolated from more than 100 CJD, kuru, and GSS individuals indicate that prions

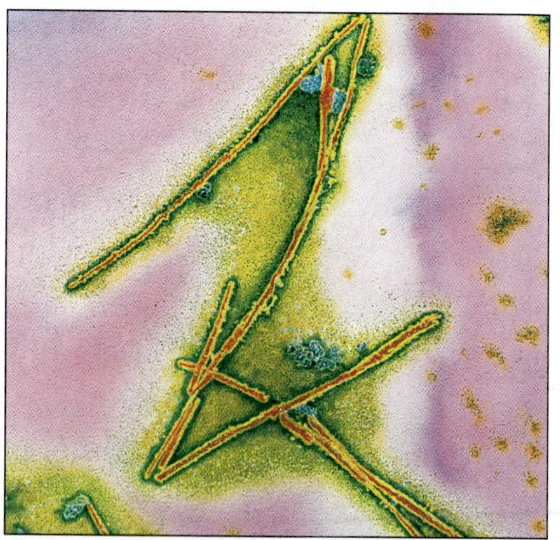

FIGURE 7.14 Prions. This photomicrograph shows crystalline aggregates of disease-causing prions. Prions are infectious protein molecules that cause fatal neural degeneration diseases in animals. *(E. M. Unit, VLA/Science Photo Library/Photo Researchers, Inc.)*

CURRENT INVESTIGATIONS

Mad Cow Disease

The prion disease of cattle, bovine spongiform encephalopathy, or BSE, is also known as mad cow disease because of the irritable, irrational actions of infected animals. This disease is currently a major problem for animal agriculture in Great Britain, and there is concern that it may cause a major human health problem in the future. BSE began when cattle were fed protein supplements produced by processing the carcasses of sheep infected with the sheep prion disease, scrapie. The scrapie prion (PrP^{SC}) was not inactivated by the processing, and it was able to infect some cattle. This cross-species infection was originally thought to be impossible, and the mechanism by which it occurs is still being investigated. After the cattle were infected, the infection generated large numbers of disease-causing cattle prions (PrP^{BSE}) in the brains of the diseased animals. Because they were cattle prions, they were more efficient at causing BSE than sheep PrP^{SC}. Unfortunately, the widespread practice in British agriculture of slaughtering diseased cattle and processing their carcasses into protein supplements to be added to the diets of healthy cattle lead to a rapid increase in the disease. First reported in 1985, BSE spread rapidly through British beef herds. By 1990, approximately 300 cattle a week were dying throughout the country. By 1995, virtually all herds were infected, over 1000 cattle were dying every week, and it was concluded that widespread destruction of beef herds was the only remedy.

BSE would have simply been an expensive agricultural problem were it not for the fact that at this same time a number of domestic pets (cats) and zoo animals that had been fed processed meat from BSE-infected cows developed BSE-like diseases. This led to widespread alarm that prions from BSE-infected cattle could infect other animals and that humans might develop BSE-like symptoms by eating meat from infected cattle. This concern was so great that nations around the world banned the import of British livestock, beef, and beef by-products, and the British population drastically reduced its consumption of beef. These steps resulted in an economic loss of millions of dollars a year to the British beef industry and drove many farms out of the beef business.

The concern that prions from beef might infect humans is not unreasonable. Prions in brain extracts from BSE-infected cattle have been shown to infect mice and hamsters, as well as cats and zoo animals. Prions have a well-documented ability to infect different species, and they are resistant to cooking or other food preparation procedures. The US beef herds do not have BSE, and a strict ban on importing British livestock is in place. However, in the US herds, there is currently a disease called downers that resembles BSE. The early symptoms are different from BSE (the infected animals are not irrational but simply fall down and cannot rise), but the end result is the same: death from brain degeneration. The practice of processing slaughtered cattle carcasses, including those of diseased cattle, into protein supplements for feeding to healthy livestock is widespread in the United States. If downers is a prion disease, or if a prion disease should develop in the United States, it could quickly become widespread. The normal incubation time for BSE-like diseases in humans is 10–30 years, so that if a prion disease were to infect humans, the transmission might not be noticed until large numbers of people had been infected.

have a wide host range. All human prions can infect monkeys and chimpanzees. Some are able to infect mice, hamsters, mink, and cats. In each case, the human prion recruits animal prions in the PrP^{C} form to shift to the disease-causing form. These altered prions then recruit others, causing the disease.

Current concerns about prions stem from an epidemic the TSE referred to as bovine spongiform encephalopathy (BSE) in British cattle. This disease is also called mad cow disease because of the irritable, aggressive behavior of infected animals. BSE was caused by feeding cattle dietary protein supplements produced by processing the carcasses of sheep that were infected with the TSE scrapie. The agricultural practice of processing the carcasses of diseased cattle into protein supplements generated a recurring cycle of prion infection that led to a rapid buildup of the disease. Concern in the general human population arose when several other animal species contracted BSE by eating protein supplements prepared from infected cows. This raised the possibility that humans consuming beef from infected animals might also develop TSE (see *Current Investigations: Mad Cow Disease*). This concern led to a drastic reduction in the consumption of British beef in the 1990s and a ban on British beef imports by nations around the world.

SECTION REVIEW PROBLEM

11. Flies from a strain of *Drosophila melanogaster* sensitive to CO_2 die when placed in a high CO_2 atmosphere. When sensitive flies are mated with nonsensitive flies, the F_1 are all sensitive. For each of the following possible modes of inheritance, propose a test that would determine whether this might be the mode of inheritance of the CO_2-sensitive phenotype:
    a. maternal effect
    b. cytoplasmic inheritance
    c. dominant mutation
    d. infectious inheritance

Summary

In this chapter we have discussed a number of examples of genes and other agents that produce phenotypes showing nonnuclear inheritance. The investigation of these exceptions to Mendel's rules has provided a great deal of information about the role of cytoplasmic agents in cells and has led to several advances in genetics. Extranuclear inheritance can be divided into three categories: maternal effects, genes in cytoplasmic organelles, and agents showing infectious inheritance.

Maternal effects are phenotypes controlled by the genes of the mother. These maternal-effect genes produce gene products that are packaged in the egg cytoplasm and control early stages of the organism's development. Mutant alleles in the mother's genome can produce defective or nonfunctional maternal products in the egg, and this will cause an abnormal phenotype in the offspring by altering early steps in development. Maternal-effect genes are nuclear genes that show Mendelian patterns of inheritance but that express their phenotypic effect in the next generation.

Mitochondria and chloroplasts are cytoplasmic organelles that possess their own genomes. These genomes contain genes that are essential for energy production and photosynthesis. Mutations of mitochondrial or chloroplast genes can have serious effects on the organisms. Cytoplasmic organelles usually show maternal inheritance because they are inherited from the mother, who has the organelles packaged in the cytoplasm of the egg or seed. Maternal inheritance can be distinguished from nuclear inheritance by reciprocal crosses; the offspring always show the phenotype and genotype of the maternal parent.

A variety of parasitic agents that can infect cells and reproduce in the cytoplasm also show an infectious inheritance pattern. These include bacteria and algae that can establish symbiotic relationships with hydra and paramecia. To be classified as an infectious symbiont, the bacteria and algae must be able to reproduce outside the host, and the host must be able to reproduce without the symbiont. In some cases, organisms that have been shown to have this ability also have the ability to infect new hosts and reestablish the symbiotic relationship.

Several viral strains can show infectious inheritance by reproducing in the cytoplasm of cells without killing the cells. In the case of RNA tumor viruses, the infection gives the cell a new phenotype, for example, uncontrolled cell growth.

Agents that show infectious inheritance but do not easily fit into the classification of living organisms are the viroids and the prions. Viroids are small nucleic acid molecules that do not contain any genes but that can be reproduced in the cytoplasm of cells. Several examples of plant viroids that cause infected plants to have altered phenotypes are known. These phenotypes show maternal inheritance but are also highly infectious. Another infectious agent that does not possess genetic material is the prion. Prions are small, disease-causing agents that consist entirely of protein and have no nucleic acid genome. They are thought to be normal proteins that can adopt different configurations, and they are found in the brains of all animal species. In the disease-causing configuration, they can form long, crystalline aggregates that can disrupt and kill cells. This causes fatal brain degeneration diseases known as transmissible spongiform encephalopathy (TSE). Prions are inherited maternally in the cytoplasm and can also be infectious in cases where animals consume meat from infected animals.

Selected Key Terms

chloroplast
cristae
extranuclear inheritance
gene imprinting
grana
heteroplasmic
homoplasmic
infectious inheritance
maternal effect
maternal-effect mutation
maternal inheritance
mitochondria
prion
protoplastids
thylakoid
transmissible spongiform encephalopathy
variegation
viroid

Chapter Review Problems

1. Compare the following, giving one similarity and one difference:
   a. maternal inheritance and maternal effect
   b. maternal effect and epistasis
2. An unusual pattern of inheritance was noted in *Archirrhinos haeckelii* (snouters):

   yellow (female) × brown (male)
   ⇓
   all yellow F_1

   F_1 yellow (female) × brown (male)
   ⇓
   all yellow

   Describe the experiments you would do to determine whether this pattern of inheritance was caused by
   a. cytoplasmic inheritance of a yellow mutation
   b. a dominant yellow mutation with a maternal effect
3. In the plant *Antirrhinum*, a yellow leaf phenotype was found, and it was inherited as follows:

   normal female ⇒ yellow leaf male ⇒ 10,000 normal and 12 variegated leaf plants

   yellow leaf females ⇒ normal males ⇒ 10,000 yellow and 12 variegated leaf plants

   This is a cytoplasmic inheritance pattern. Explain these results and the unusual ratios.
4. A human male with the disease myoclonic epilepsy and ragged-red fibers (MERRF) had three children before he was diagnosed with the affliction. His wife was normal. Predict whether the children will have this disease and explain your prediction.
5. In another case, the female had MERRF and had two children before diagnosis of the affliction. The husband was normal. Predict the phenotype of the children, if possible, and explain your answer.
6. What is the function of the mitochondria and chloroplasts?
7. It is proposed that mitochondria and chloroplasts were derived from an ancestral symbiosis with bacteria, and this has evolved into an obligate symbiosis. What is some evidence that would support this hypothesis?
8. Females from a strain of *Drosophila* with short wing veins were crossed with a strain with normal wing veins. The F_1 all had short wing veins. Explain how you would determine whether this was the result of a dominant maternal-effect mutation.
9. Explain how you would determine if the wing vein phenotype in Problem 8 was caused by imprinting.
10. Sunflower plants homozygous for a recessive maternal-effect mutation causing twisted stems (*tsm / tsm*) were fertilized using pollen from a plant with normal straight stems (*Tsm / Tsm*). The F_1 were backcrossed by fertilizing them with pollen from *tsm / tsm* plants. What would be the phenotype of the F_1? What would be the phenotypic ratios in the progeny of the backcross?
11. Several normal-looking male and female mice were collected from the neighborhood of the Chernobyl nuclear reactor. These mice were bred in the lab, and one of the females produced offspring with developmental abnormalities. How would you determine whether these abnormalities were the result of a maternal-effect mutation?
12. For each of the following, describe the F_2 phenotypic or genotypic ratios that you would expect to be produced in a cross if the phenotype were controlled by
    a. cytoplasmic inheritance
    b. a dominant maternal-effect mutation
    c. a recessive maternal-effect mutation
13. A corn plant that was CMS-T *Rf1* / +; *Rf2* / + was crossed as a female with a male fertile plant that is *Rf1* / +; *Rf2* / +.
    a. What ratio of phenotypes (sterile/fertile) would you expect in the progeny?
    b. What ratio would you expect in the progeny if you had done the reciprocal cross?
14. In humans, a tendency to develop glomus tumors (benign tumors that develop in the inner ear) is caused by a mutant allele. Although the gene is located on an autosome, pedigree analyses indicate that the phenotype shows paternal inheritance, with affected men having affected offspring. Explain how this might occur.
15. Give a brief definition of the following terms:
    a. cristae
    b. stroma
    c. light reactions
    d. dark reactions
    e. homoplastic
16. Give one difference and one similarity between the genomes of mitochondria and chloroplasts.
17. A phenotypically normal couple had a child with MERRF. There is no history in either parent's family of this disease. Explain how this might occur.
18. Two green hostas were crossed, and although most of the offspring were green, one variegated hosta was produced. This variegated hosta was crossed as a female and gave green and variegated offspring. When this same variegated plant was crossed as a male, all the progeny were green.
    a. Explain how this might occur.
    b. Design an experiment or series of experiments that would test your hypothesis.
19. In the text of this chapter, it was stated that it is impossible to design a breeding plan to develop a strain of hostas that breeds true for the variegated leaf trait. Explain why this statement is true.
20. A hydra from a green strain was crossed as a female with a hydra from a normal clear strain. All the F_1 were green. The F_1 were backcrossed to the clear strain, and all the F_2 were initially green, but 1/4 lost the green color during development. The green color in hydra is caused by the

presence of the symbiont *Chlorella vulgaris*. Explain the results of these crosses.

21. Women infected with HIV have children who are often, but not always, born infected. The current hypothesis is that the virus is transmitted prenatally. Explain why, if this hypothesis is correct, some children might not become infected.
22. A worker in a greenhouse noticed that one row of plants in a large bed developed a new wasting disease. The plants do not die, but they have wilted leaves and do not grow properly. How would you determine whether this is caused by a viroid infection or by a new dominant mutation?
23. Reciprocal crosses were done between two strains of potato. The offspring of one cross were stunted and produced stunted tubers. The offspring of the other cross were vigorous plants that produced large tubers. Give an hypothesis as to how this could occur and design an experiment, or experiments, that would test your hypothesis.
24. Mad cow disease has had a serious impact on agriculture in Britain. Explain how you might determine whether a herd was infected.
25. One of the most serious questions about mad cow disease is whether humans can become infected by eating meat from infected cattle. This question became even more important when a number of individuals who were in contact with farm animals developed Crutzfield-Jacob disease. Based on what you know about the cause of mad cow disease, could you determine whether these individuals became infected from contact with infected cattle?
26. A variegated hosta was crossed as a female with a normal green plant. The offspring were all green. An F_1 plant was backcrossed as a male with a parental green plant. The offspring were all green. An F_1 plant was crossed as a female with a plant from the parental green strain. The offspring of this cross were 1/2 variegated and 1/2 green. Explain the inheritance of variegation in these crosses, and devise one test of your hypothesis.

Challenge Problems

1. A strain of *Drosophila* is extraordinarily sensitive to exposure to CO_2. This trait is inherited but is not sex-linked. Females from a sensitive strain were crossed with males from a normal strain, and the F_1 were all sensitive. A reciprocal cross was done in which normal females were crossed to males from the sensitive strain. All the offspring of this cross were normal. Explain how you would determine whether this trait shows cytoplasmic inheritance or whether it is caused by a gene that shows imprinting.
2. A corn plant that produced no viable pollen was crossed as a female with a normal corn plant. The F_1 produced no pollen. Backcrossing these F_1 plants to normal corn for many generations never produced plants with viable pollen. One of the F_1 plants was crossed with a plant from a South American strain. All the offspring of this cross produced pollen. Selfing these offspring produced some plants that produced pollen and some that did not.
   a. How would you determine if the South American strain contains alleles that restore male fertility to the original sterile line?
   b. How would you determine whether the restoration of fertility by the South American line requires dominant alleles of one, two, or three genes?
3. The pedigree in this figure is for a family that has a rare disease that causes blindness, but only in adulthood.

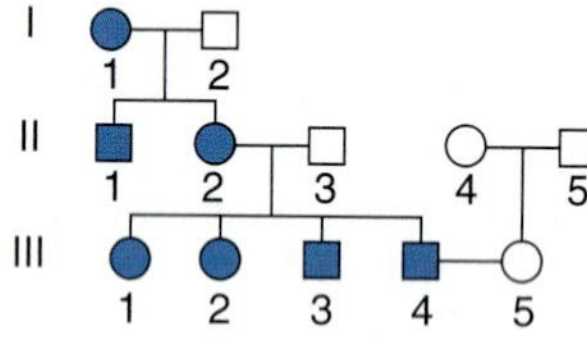

   a. From this pedigree information, can you determine what mode of inheritance is acting?
   b. Assume that this disease is cytoplasmically inherited from defective mitochondria. What proportion of the progeny will be affected in the marriage of III-4 and III-5 if you assume that III-5 comes from a normal family?
   c. Making the same assumption as in b, what proportion of the progeny from a marriage of III-1 into a normal family will have the disease? Will the affliction be different for the sons (compared to the daughters) of this marriage?

Suggested Readings

Aldhous, P. 1990. "BSE causing public alarm." *Nature*, 343: 196.

Barlow, D. P. 1995. "Genotypic imprinting in mammals." *Science*, 270: 1610.

Girton, J. R., and S-H. Jeon. 1994. "Novel embryonic and adult phenotypes are produced by Pleiohomeotic Mutations in *Drosophila*." *Developmental Biology*, 161: 393–407.

Kingsbury, D. T. 1990. "Genetics of response to slow virus (prion) infection." *Annual Review of Genetics*, 24: 115–132.

Larsson, N-G., and D. A. Clayton. 1995. "Molecular genetic aspects of human mitochondrial Disorders." *Annual Review of Genetics*, 29: 151–178.

Mestel, R. 1996. "Putting prions to the test." *Science,* 273: 184–189.

Preer, J. R., Jr. 1971. "Extrachromosomal inheritance: Hereditary symbiont, mitochondria, chloroplast." *Annual Review of Genetics,* 5: 361–406.

Prusiner, S. B. 1995. "The prion diseases." *Scientific American,* 272: 48–51.

Sonneborn, T. 1959. "Kappa and related particles in *Paramecium.*" *Advances in Virus Research,* 6: 229–356.

Wallace, D. C. 1997. "Mitochondrial DNA in aging and disease." *Scientific American,* 277: 40–59.

Zouros, E., K. Freeman, A. Ball, and G. Pogson. 1992. "Direct evidence for extensive paternal mitochondrial DNA inheritance in the marine mussel *Mytilus.*" *Nature,* 359: 412–414.

On the Web

Visit our Web site at **http://www.saunderscollege.com/lifesci/titles.html** and click on A/G/M Genetics for links to the following chapter-related resources on the World Wide Web:

1. **Fungal Mitochondria.** This site hass links to other sites for comparative mapping and sequence data for a number of fungal mitochondria.
2. **Mitochondrial Sequences.** This gopher site provides links to the sequences of a large number of plant, fungal, and animal organelle genomes.
3. **Mapping and Sequence Data.** This site links to sequence and map data for many of the genome projects, including the sequencing and mapping of mitochondria and chloroplasts.

CHAPTER

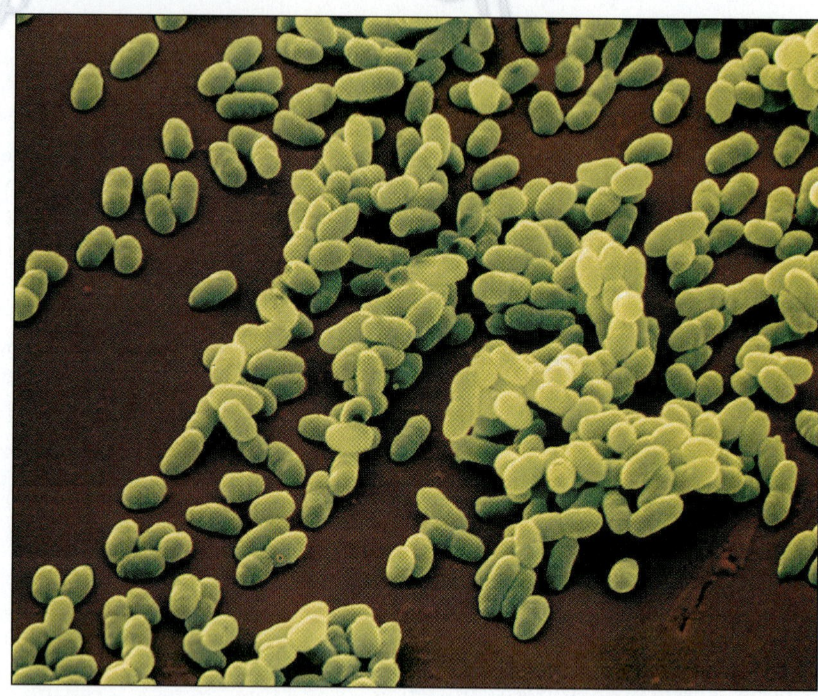

Haemophilus influenzae, a bacteria found in the mouth flora of about half the human population. It can cause pneumonia under the right conditions. *(Oliver Meckes/Photo Researchers, Inc.)*

The Genetics of Bacteria and Their Viruses

In the early years of genetics, from 1900 to the 1930s, the favorite organisms of study were all eukaryotes: maize, mice, *Drosophila,* and *Neurospora crassa.* Notably absent amongst the organisms for genetic study were bacteria. As we have learned in preceding chapters, most early genetic studies concentrated on how a trait was passed from one generation to another, the role of meiosis and crossing over in the inheritance of traits, and the inheritance of polygenic traits. To be a subject for study, the organism must be able to undergo sexual reproduction. In this respect, bacteria were largely ignored because it was believed that bacteria did not possess any means of gene transfer between individuals. They reproduced by cell division and gave rise to two daughter cells with traits identical to the parent cell.

This all changed in 1946 when Joshua Lederberg and Edward Tatum discovered that transfer of genetic information could occur between two bacterial cells. When they mixed two bacterial strains differing in genotype, where each strain had different nutrient requirements for growth, they obtained some cells that did not possess *any* growth requirement. Even though these new cells occurred at a very low frequency (1 in every 10 million cells, or 1×10^{-7}), Lederberg and Tatum correctly interpreted these results to mean that some type of genetic transfer had occurred between the two bacterial strains.

We now know that bacteria can transfer genetic information between strains by three clearly different means:

1. **Conjugation** requires actual physical contact between the bacterial donor cell and the recipient cell. A physical tube between the two cells allows unidirectional transfer of segments of the chromosome. After

the genetic material has been transferred, it can be incorporated into the chromosome of the recipient cell by recombination.

2. **Transformation** is a process of genetic transfer that does not require physical contact between the donor and recipient cells. In transformation, the donor cell lyses (breaks open), releasing its DNA. The free DNA can then be taken up by the recipient cell and integrated into its chromosome.
3. **Transduction** requires the participation of a third party, a **bacteriophage** (a virus that attacks only bacteria), to transfer DNA from a donor cell to a recipient cell. Again, the donor cell does not actually need to come in physical contact with the recipient cell. In transduction, pieces of DNA from the donor cell are mistakenly packaged into a phage head during infection. In a subsequent infection with the mispackaged phage, the DNA is injected into the recipient cell and may be integrated into its genome.

Viruses play a very important role in genetics and molecular genetics. Viruses are very small forms that lie between life and nonlife and can be as small as one thousandth the size of a bacterial cell. They possess either RNA or DNA as their genetic material and are enveloped in a protein coat. They do not possess enzymes for cellular metabolism and are incapable of reproduction in the absence of a host cell. To reproduce, they must parasitize living cells and use their metabolic machinery. In spite of these properties, viruses can be subjected to very refined genetic analysis and have revealed many key insights into the nature of the genetic material, the genetic code, and mutations. There are other reproducing elements that are even smaller than viruses; however, viroids and prions (see *Current Investigations: Mad Cow Disease,* Chapter 7) have not been extensively exploited in genetic studies.

Bacteria and their viruses have a long history of yielding useful genetic information and have contributed huge amounts of information on how the cell operates. And, there is no doubt that bacteria will continue to yield useful information for years to come. More recently, with the advent of genetic engineering, bacteria have found a new use: they are the workhorses of the genetic engineer. For these and other reasons, it is important to understand the genetics of bacteria and their viruses. By the end of this chapter, we will have addressed five specific questions:

1. What are the details of the three methods whereby bacteria transfer genetic information from one bacterium to another?
2. How do geneticists use the methods of bacterial gene transfer to construct genetic maps of the bacterial chromosome?
3. What are the life cycles and characteristics of bacterial viruses?
4. How are some bacterial viruses, or bacteriophage, useful in mapping the location of genes in bacteria?
5. How does the geneticist construct a genetic map of the bacteriophage chromosome?

Bacteria Are Easily Grown, and Mutants Are Highly Varied.

Bacterial cells are relatively small when compared to eukaryotic cells (about 1/1000 the volume; Figure 8.1). *Escherichia coli* cells are about 1 μm long and 0.5 μm in diameter and, given the appropriate nutrients and temperature, are capable of doubling every 20–60 minutes. Most bacteria, including *E. coli,* can grow either in liquid medium or on a semisolid surface such as agar supplemented with nutrients. A single *E. coli* cell with a doubling time of about 1 hour will yield over 10 million progeny within 1 day. This is a sufficient number of bacteria to form an easily seen 3–4 mm diameter colony on the surface of nutrient agar in a Petri dish. All members of this colony have an identical genotype because they originated from a single cell dividing by binary fission. For this reason, the colony is called a **clone.** Two methods are available for isolating genetically pure clones of

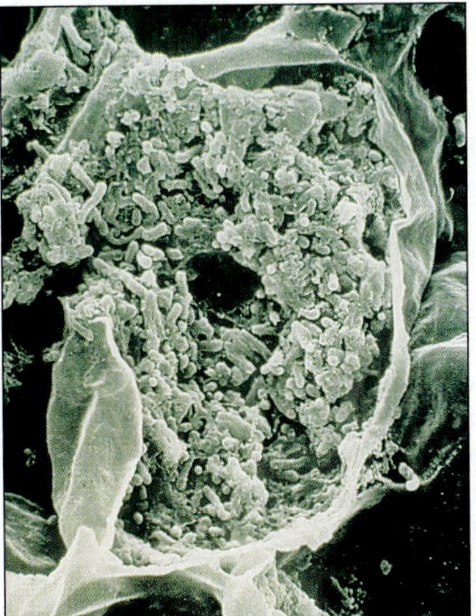

FIGURE 8.1 Bacterial cells in a plant cell. This scanning electron micrograph of *Bradyrhizobium japonicum* shows cells released from a soybean cell after gentle lysis. These bacteria invade the root cells of several different plants, multiply, and take this form. *Bradyrhizobium* is capable of converting atmospheric nitrogen to plant-usable nitrogen. Note the large difference in size of the bacterial cells and the plant cell. Each bacterial cell is about 1–2 μm in length. *(Courtesy of James Berry and Alan Atherly)*

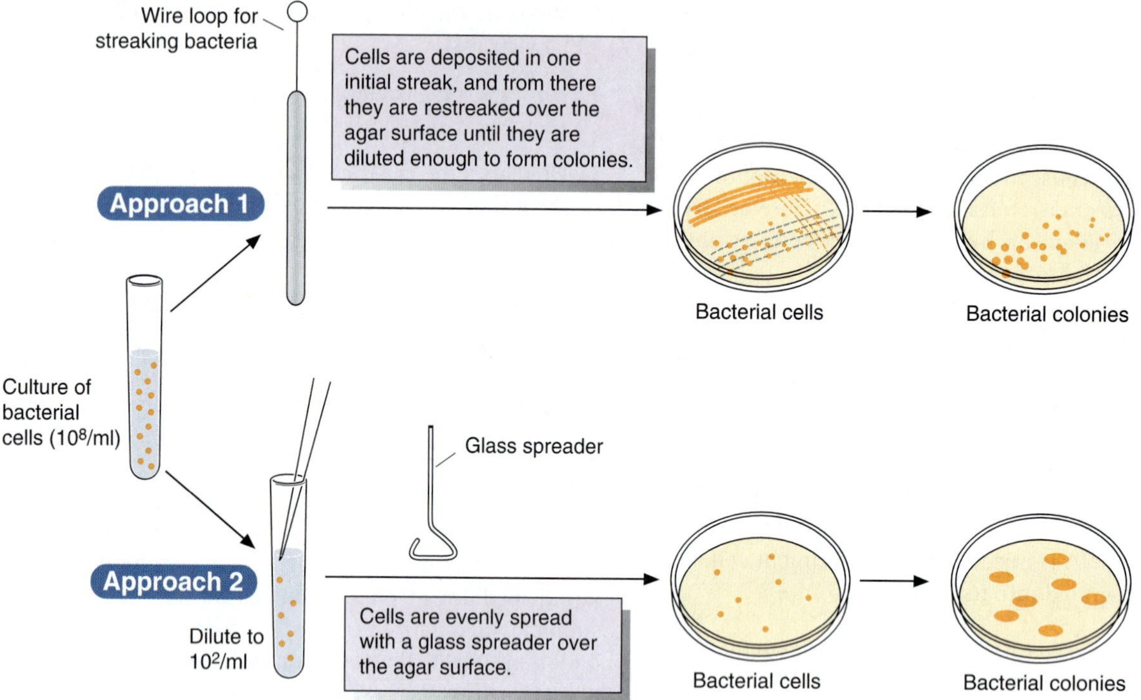

FIGURE 8.2 Separating individual clones of bacteria. Individual colonies of bacterial cells can be isolated using two different methods. Both approaches 1 and 2 are commonly used and yield pure clones of individual cells.

bacterial cells (Figure 8.2). In the first procedure, a wire loop is dipped into the mixed culture of bacteria. A mixture of the bacteria from the culture sticks to this wire loop. These bacteria are streaked out on the surface of the agar medium, thinning them out, and they will eventually grow into individual colonies. These colonies represent pure clones of a bacterial genotype.

A second method of isolating a clone from a population of bacterial cells is by serial dilution and **plating.** For example, a culture of bacterial cells that may possess 10^8 cells per milliliter is diluted (0.1 ml into 9.9 ml sterile water, then 0.1 ml of the first dilution is added to 9.9 ml sterile water, and so on to the desired final concentration) so that 1 ml possesses several hundred cells. One tenth of a milliliter of cells is spread on a nutrient agar surface with a sterile bent glass rod (Figure 8.2) so that individual cells are evenly distributed. The bent glass rod acts as a convenient instrument to distribute the liquid over the Petri dish. After incubation at the appropriate temperature, individual colonies form from the cells.

Many Different Kinds of Bacterial Mutants Exist.

Bacteria do not have many visible phenotypes to distinguish one strain from another. As a result, a number of defined growth conditions have been devised to distinguish genotypes. Some of the distinguishing conditions as well as the genetic notations that denote the change from the wild-type strain follow.

1. *Bacteria unable to utilize a nutrient for growth. E. coli* can synthesize nearly everything required for growth from a simple carbon source such as glucose or galactose, a nitrogen source such as nitrate, and a few inorganic constituents: K^+, Na^+, and PO_4. This type of medium is referred to as a **minimal medium.** However, some strains cannot use certain carbon sources (for example, galactose), whereas others can. These mutant strains are easily identified by their inability to use this carbon source; they cannot grow at all on galactose and are thus referred to as *gal* or *gal*$^-$. *Genetic notation for bacterial genes always uses three lowercase letters.* A plus superscript indicates a wild-type (normal) genotype (*gal*$^+$) associated with the ability to use a specific ingredient that the mutant strain cannot use. Negative superscripts are not usually used to indicate a mutant genotype but are sometimes included for clarity. The lack of a superscript indicates that it is a negative or mutant genotype. Thus, *gal* is the mutant counterpart of *gal*$^+$.
2. *Bacteria dependent on a nutrient for growth.* If a mutant

strain cannot grow on the minimal medium, it may have acquired a dependence on the addition of a nutrient for growth. A good example is a requirement for one of the amino acids (for example, alanine). The mutant strain would then be *ala*, distinguishing it from the wild-type, which would be ala^+. In this case, the mutant has lost the ability to synthesize a required nutrient and is now known as an **auxotroph.**

3. *Bacteria sensitive or resistant to the presence of a drug or phage.* Bacteria are sensitive to a large number of chemical compounds, especially antibiotics such as penicillin, tetracycline, streptomycin, and chloramphenicol. When a bacterium shows resistance to a drug (for example, tetracycline), it is designated tet^r. The parent strain would be tet^s, where the superscript r refers to resistance and the s refers to sensitivity. The same notation is used if a strain develops a resistance to specific phage particle (for example, resistance to infection by phage T4). A bacterial strain resistant to T4 infection is mutant and is designated *tfr* (*tfr* refers to T-four resistance), whereas tfr^+ is the wild-type.

4. *Bacteria possess conditional mutations.* A conditional mutation reveals a mutant phenotype under some growth conditions and a wild-type or nearly wild-type phenotype under other growth conditions. The most frequently encountered conditional mutation is for temperature sensitivity. The normal growth temperature for *E. coli* is 37°C, but it can grow at temperatures up to about 42°C and as low as 10°C. A conditional mutant strain might have a wild-type phenotype at 30°C but express its mutant phenotype at 40°C. Conditional mutations are useful when present in genes that code for required functions, such as RNA polymerase (an enzyme responsible for the synthesis of RNA) because a mutation in a required function will normally kill the cell. Thus, conditional mutations permit proliferation under altered environmental conditions. The genetic notation for conditional mutations uses the three-letter notation for the gene with a "ts" superscript (for *t*emperature *s*ensitive), for example $dnaA^{ts}$, which is a gene involved in initiation of DNA replication.

The three-letter gene notation in bacteria is frequently followed by a capital letter (as with *dnaA*). This is done because many metabolic pathways that seem to have one phenotype may have a defect in any of a series of steps involved in the synthesis of the end product and thus can be caused by a series of different mutations. For example, a number of different enzymes are involved in DNA synthesis. Thus, a gene for one step in DNA synthesis is *dnaA*, another is *dnaB*, and so forth. But what happens when two mutant strains are isolated with mutations in the same gene? First of all, it is very likely that the mutation will be at a different location within the gene and therefore be a different allele. These two mutations are distinguished from each other by using a number after the capital letter, for example *dnaA1* and *dnaA2*.

When the gene symbol is written, it is always italicized; however, when the *product* of the gene is referred to, such as the DnaA protein, the first letter is capitalized and it is not italicized. This allows the reader to distinguish between the gene and the gene product.

Finally, as we learned in Chapter 6, a bacterial strain that requires a substance for growth on minimal media is called an auxotroph. The wild-type counterpart of that gene is the **prototroph.** Therefore, if a strain is defective in the synthesis of arginine, it is said to be auxotrophic for arginine. Bacteria contain only one copy of the their genetic material, thus a mutation in a gene always manifests itself (it is not recessive or dominant). The only exception to this rule is when a second copy of the gene is present on the chromosome or a plasmid.

Spontaneous Bacterial Mutants Are Not the Result of Adaptation to a Changing Environment.

During the early years of research on bacteria, the majority of bacteriologists believed that environmental factors induced some unknown change in bacteria that led to their survival or adaptation to new growth conditions. They did not suspect that these changes were caused by mutations in the bacterial DNA. In 1943, Salvadore Luria and Max Delbrück presented the first direct evidence that bacteria, like eukaryotic organisms, are capable of spontaneous mutation. In these experiments they used *E. coli* and the bacteriophage T1, which will grow on and lyse *E. coli* cells. If a Petri dish containing many *E. coli* cells growing on a nutrient agar is sprayed with phage T1, almost all the *E. coli* cells are killed. Rare *E. coli* cells survive, and these cells can be cultured and are resistant to lysis by phage T1 infection. Early investigators attributed this observation to **adaptation,** assuming that the bacteria somehow had adapted to the presence of phage T1. It seemed that this adaptation was the direct result of the interaction of the phage with the bacteria. In other words, the bacteria had acquired immunity to infection by the presence of the phage by some physiological adaptation, and this would not occur spontaneously in the absence of the phage.

It can be experimentally determined whether or not T1-resistant bacteria are present within a population of cells, irrespective of the presence of T1 phage. If a bacterial culture is initiated from a small number of cells (100 or so) and spontaneous mutations occur during growth, then as the population increases, T1-resistant mutations could appear in any cell at any time, and all the daughter cells would be resistant to T1 infection. However, the number of T1-resistant cells will vary in the final population, depending upon *when* during growth the change occurred. If the mutation occurred early in the growth period, many of its progeny will be present in the final population; however, if the mutation occurred late in the growth period, then only a few T1-resistant mutant cells will be present in the final population. However, if each bacterial cell has the same probability of becoming T1-resistant through physiological adaptation, then separate populations would all have the same frequency of T1-resistant cells. Using this logic, Luria and Delbrück grew a series of bacterial cultures, starting from a few hundred cells, to a concentration of 10 million cells per milliliter. As a control, they grew a separate culture from which they took ten separate samples to measure T1 phage resistance. If the adaptation hypothesis were true, then the frequency of mutants found in ten separate cultures would correspond to the frequency of mutants found in the ten separate samples taken from one culture. What they found was that the frequency of T1-resistant mutants varied considerably among the ten separately grown samples, whereas the frequency of T1-resistant cells from ten samples from a single culture were nearly identical (Table 8.1).

The simplistic explanation for the tremendous variation among the separately grown populations is random genetic change, which either occurred late in the growth of the cells (a low number of mutants) or early in the growth of the cells (a large number of mutants). The lack of variation in frequency of mutants in the control culture is simply the result of the original mutation(s) giving rise to all the separate samples. Thus, using this simple experiment, Luria and Delbrück showed that random genetic change occurs in a population of bacteria and the presence of spontaneous mutations are not the result of adaptation to the environment, which selected for the presence of the mutation.

TABLE 8.1 *Results from the Experiment of Luria and Delbrück*

Culture Number or Sample Number	Number of T1-Resistant Colonies in Individual Cultures	Number of T1-Resistant Colonies in Separate Samples of the Control
1	1	14
2	0	15
3	0	15
4	3	16
5	107	21
6	1	13
7	64	20
8	0	14
9	5	13
10	35	26
Mean:	21.6	16.7

SECTION REVIEW PROBLEMS

1. If you had a single bacterial cell on an agar surface that doubled every hour, how many bacterial progeny would be present after 24 hours?
2. Write the exact genetic notations for the following bacterial strains, all of which are mutant:
   a. sensitive to tetracycline (*tet*)
   b. requires phenylalanine for growth (*phe*)
   c. will not metabolize lactose (*lac*)
   d. resistant to kanamycin (*kan*)
   e. temperature-sensitive for rRNA synthesis (*rrn*)
   Which of these strains are auxotrophs and which are prototrophs?
3. Clearly distinguish between conjugation, transduction, and transformation.

BACTERIAL CONJUGATION IS A COMPLEX PROCESS INVOLVING PLASMIDS.

The genetics of bacteria was largely unexplored before World War II. Because of several key discoveries just after the war, bacterial genetics blossomed and has paved the way for a more complete understanding of eukaryotic genetics. In the late 1940s, it was established that bacteria had a chromosome, and techniques were discovered that allowed construction of genetic maps of the chromosome. One extremely important finding was the discovery of plasmids in bacteria and the fact that plasmids are capable of transferring themselves as well as promoting the transfer of the bacterial chromosome from strain to strain. In this section, we will discuss the discovery of plasmids and their role in DNA transfer between bacteria.

Gene Transfer in Bacteria Was Discovered in 1946.

Very little was known about the genetics of bacteria before Joshua Lederberg and Edward Tatum made their monumental discovery in 1946 [see HISTORICAL PROFILE: *Joshua Lederberg (1925–)*]. It was known that certain strains of bacteria remained true to their phenotype over many generations (with occasional mutations), but it was not known whether bacteria had one or more chromosomes. To answer the fundamental question of whether bacteria could transfer genetic information, Lederberg and Tatum designed an elegant experiment

HISTORICAL PROFILE

Joshua Lederberg (1925–)

Joshua Lederberg is noted for two landmark discoveries in bacterial genetics: bacterial conjugation and transduction. Lederberg, along with Beadle and Tatum, was awarded the Nobel Prize for these findings in 1958 for their studies with *Neurospora* that led to the "one-gene, one-enzyme" hypothesis.

Lederberg was born in Montclair, New Jersey, in 1925 to Zwi, a rabbi, and Esther Lederberg; both had emigrated from Palestine 2 years earlier. Joshua was educated in the public school system of New York and entered Columbia University at the precocious age of 16. Much to his father's chagrin, he was interested in science, at this time the physical sciences, but his interests soon turned toward biological and biomedical sciences. Biology seemed to be an area where science could help humankind with its many problems. At Columbia, he came under the tutelage of a zoology instructor, Francis Ryan. Ryan had just returned to Columbia from Stanford, where he had learned of the pivotal studies of Beadle and Tatum, who were then just starting their work with *Neurospora*. Lederberg learned that Beadle and Tatum were using mutational markers in *Neurospora* that conferred nutritional requirements on the organism. When Lederberg first learned that bacteria were thought not to be able to transfer genetic information, the work of Beadle and Tatum suggested to him that the same techniques could be applied to bacteria. At the age of 20, Lederberg entered Columbia's medical school, but continued to moonlight in Ryan's laboratory. It was at this time that he learned that Tatum was starting a new program at Yale University, and as a consequence, Lederberg left medical school and entered graduate school at Yale in microbiology to study the problem of transfer in bacteria. He was only 21 years old when he made the momentous discovery of conjugation in *E. coli*. After receiving his Ph.D. from Yale, Lederberg migrated west to the University of Wisconsin. At Wisconsin, Lederberg, along with his wife Esther, developed a unique method of studying bacterial mutants, now known as "replica plating." With this technique, which is still a mainstay of bacterial genetics, it is possible to transfer bacterial colonies from one agar growth plate to others so that each new plate is an exact replica of the original. In this way, it is possible to detect and isolate rare mutants with differing nutritional requirements. Using this technique, Lederberg showed that mutations in bacteria occur randomly and spontaneously, thus confirming a long-held hypothesis in the field of evolutionary genetics.

In collaboration with Norton Zinder, a student in his laboratory at Wisconsin, Lederberg made the second of his famous discoveries in bacterial genetics: that genetic information could be transferred between bacteria by bacteriophage. He dubbed this new process transduction. Transduction and conjugation have "made" the science of bacterial genetics and have subsequently spawned many advances, including aspects of modern molecular genetics of gene cloning. Lederberg continued on at Wisconsin, where he was asked in 1957 to organize and head the Department of Human Genetics. In 1959, he went to Stanford to start a department of genetics, and during the early years of the space program, he was a consultant to the American Space Program and speculated on the scientific and medical consequences of space exploration. In 1978, Lederberg left Stanford to become the president of Rockefeller University in New York, his hometown, where he remained president until 1990. In addition to receiving the Nobel Prize in physiology and medicine, Lederberg has received many other awards and honors. These include memberships in the prestigious National Academy of Sciences and the Royal Society of London. It can truly be said that Joshua Lederberg singlehandedly changed the nature of bacterial genetics and changed the course of both genetics and biochemistry.

FIGURE 8.A **Joshua Lederberg.**
(Ingbert Gruttner)

that strongly suggested that bacteria had a sex life of their own. Lederberg and Tatum started with two different strains of *Escherichia coli*, each with different nutritional requirements. Neither strain could grow on minimal medium. However, one strain (A) could grow if the minimal medium was supplemented with threonine, leucine, and thiamine (amino acids and a vitamin), and strain B grew if supplemented with methionine and biotin (an

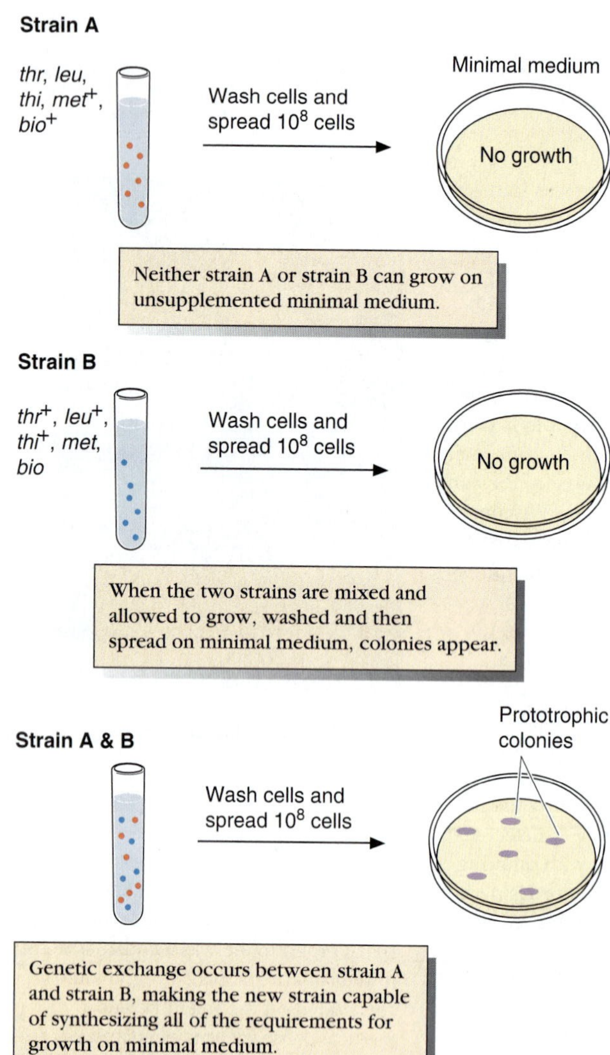

FIGURE 8.3 Bacterial conjugation. Lederberg and Tatum showed that bacteria are capable of exchanging genetic information in the experiment outlined here. This was a landmark experiment in bacterial genetics.

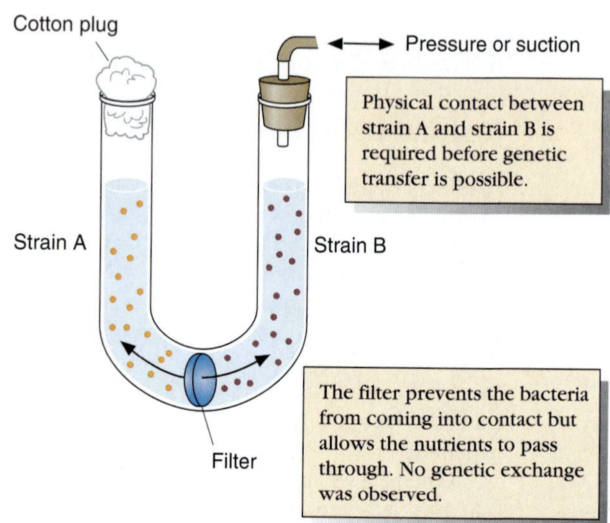

Cotton plug
Pressure or suction
Physical contact between strain A and strain B is required before genetic transfer is possible.
Strain A
Strain B
Filter
The filter prevents the bacteria from coming into contact but allows the nutrients to pass through. No genetic exchange was observed.

FIGURE 8.4 Physical contact is needed for conjugation. This experiment was performed by Bernard Davis and showed that physical contact between strain A and strain B is required before conjugation is possible.

amino acid and a vitamin). In this experiment, strain A is designated as *thr, leu, thi* (but prototrophic for methionine and biotin, or met^+ and bio^+) and strain B is designated as *met, bio* (prototrophic for threonine, leucine, and thiamine, or thr^+, leu^+, and thi^+). At that time (1946), Lederberg and Tatum did not know that each of these traits was governed by a single gene, so the notation is for convenience.

When strain A or strain B was spread on unsupplemented minimal medium, no colonies appeared after the Petri dishes were incubated for several days. In contrast, if strain A and strain B were mixed and allowed to grow on a medium containing all the supplements (Figure 8.3) and then spread on unsupplemented minimal medium, one colony appeared for every 1×10^7 bacteria spread. Because only prototrophic bacteria (met^+, bio^+, thr^+, leu^+, thi^+) grow on this medium, this finding strongly suggested that wild-type colonies were produced by transfer of genetic material between the two strains. This experiment was, in fact, the first observation of transfer of genetic information between bacteria.

For the transfer of genes between bacterial strains, physical contact between the two strains is necessary. This was shown by an elegant experiment performed by Bernard Davis. He constructed a U-tube containing a filter separating the two strains (Figure 8.4) where nutrients and other small objects (bacteriophage) could pass through, but bacteria could not. When samples from each arm of the tube were plated on a minimal medium, there was no growth, indicating that no genetic transfer had occurred. Davis thus concluded that physical contact between two bacterial strains was necessary for genetic transfer.

Conjugation Requires the Presence of an F-Factor.

When Lederberg and Tatum first observed genetic transfer between two different bacterial strains, the frequency of the transfer was extremely low, only one in several million bacterial cells. The reason for the genetic transfer was later found to be caused by the presence in the bacterial population of a specialized type of bacterial cell now known as an *Hfr*, which refers to **High frequency of recombination.** In their experiment, Lederberg and Tatum were actually observing gene transfer by *Hfr* cells that were present as a very small fraction of the bacterial population (about 1 in 1 million cells were *Hfr* cells). It is now possible to obtain a population of bacteria composed of 100% *Hfr* cells. All cells in the *Hfr* population simultaneously transfer their genetic information to a recipient cell in a coordinated manner.

The generation of an *Hfr* is dependent upon a certain type of plasmid. A **plasmid** is a small, circular, self-replicating, dispensable segment of DNA that is separate from the bacterial chromosome. Plasmids are generally about 1/50 to 1/500 the size of the bacterial chromosome. By definition, a plasmid can be removed or lost from a bacterial cell without harming the host; they usually add extra genetic information to help the bacterial cell grow under specialized circumstances. However, some plasmids have specialized functions that assist in chromosome transfer and are known as **conjugative plasmids.** The first of these to be described was the **F-plasmid,** which has 94.5 kilobase pairs (vs. 4800 kilobase pairs for the *E. coli* chromosome) in its DNA. The F-plasmid is just one example of a great variety of bacterial plasmids that can promote the transfer of bacterial chromosomes. A bacterial cell possessing an F-plasmid (sometime known as the F-factor) in its cytoplasm is referred to as an F^+ strain; the F^- strains lack the F-plasmid. Note the use of a minus superscript in contrast to bacterial gene symbolism. F^+ bacterial cells possess hair-like projections from their surfaces known as **F-pili** or sex-pili (Figure 8.5), which are involved in chromosome and plasmid transfer. When F^+ and F^- cells are mixed together, pairs of F^+ and F^- cells become attached via a **conjugation tube** derived from the F-pili. After a conjugation tube is established, the F-factor replicates its DNA, and a single-stranded DNA molecule is transferred from the F^+ cell to the F^- cell (Figure 8.6). The newly arrived, single-stranded DNA is quickly converted to a circular double-stranded DNA molecule, changing the F^- cell into an F^+ cell. This process repeats itself many times over until a large percentage of the F^- cells is converted to F^+ cells. In this transfer, only the F-plasmid, and not the bacterial chromosome, is transferred. A special event must take place for the F-plasmid to be converted to *Hfr* cells and promote chromosome transfer from the *Hfr* cell to F^- cells.

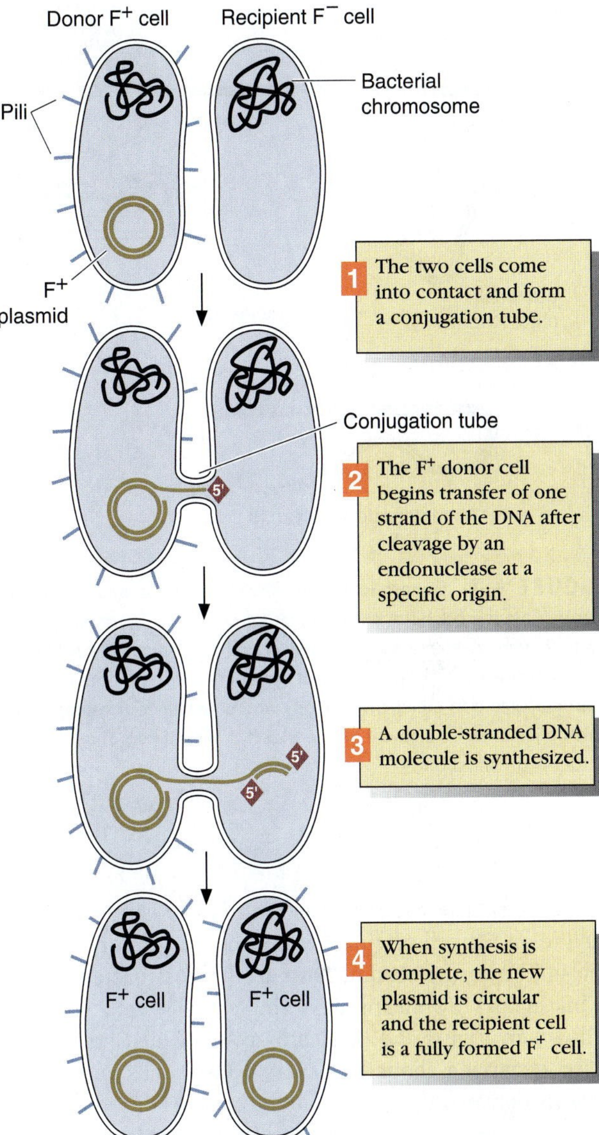

FIGURE 8.6 Transfer of an F-plasmid between cells. The two cells come in contact and form a conjugation tube, whereupon the F^+ cell transfers its F-plasmid to the F^- cell. As a consequence of this transfer, the F^- cell becomes an F^+ cell.

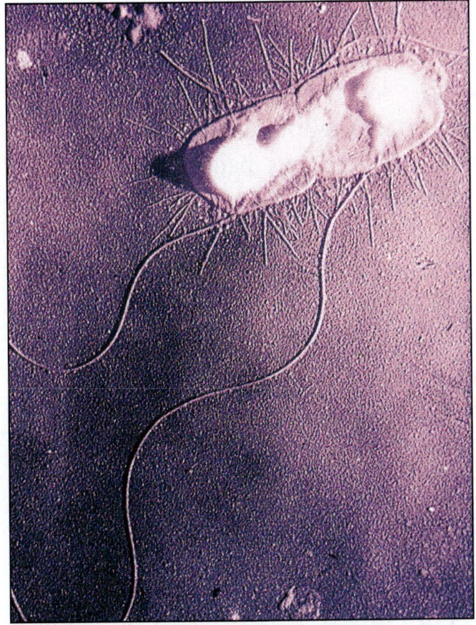

FIGURE 8.5 Electron micrograph of *E. coli*. This electron micrograph of an F^+ *E. coli* cell shows pili and flagella (the long tail-like structures). *(Fred Hossler/Visuals Unlimited)*

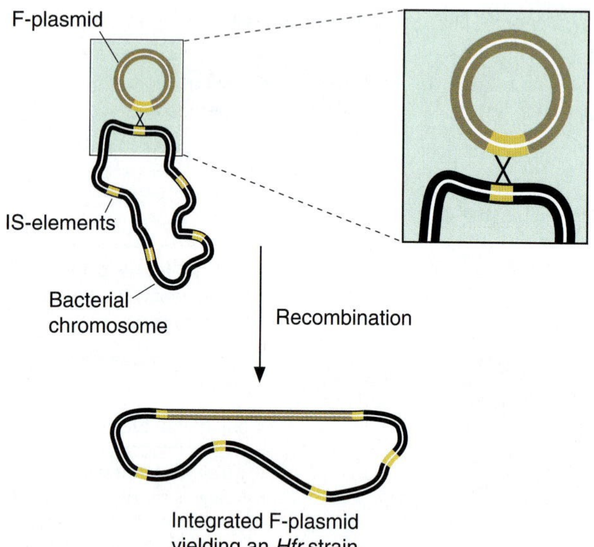

FIGURE 8.7 Generation of an *Hfr* cell. The F-plasmid is integrated into the bacterial chromosome of an F^+ cell by recombination between IS-element sequences on both the F-plasmid and the chromosome. Integration occurs at any location of the bacterial chromosome and either orientation, where sequence homology exists between the IS-element on the plasmid. After integration the strain is an *Hfr* strain.

The special event that is required to convert an F^+ cell to an *Hfr* cell originates from the integration of an F-plasmid into the chromosome of the host bacterial cell. This occurs by means of a single crossover event, as shown in Figure 8.7. The crossover event is relatively rare, occurring about 1 in every 50,000 cells. Crossover events between the F-plasmid and the host chromosome occur at sequences known as **IS elements,** where IS stands for insertion sequence. IS elements are discussed in greater detail in Chapter 19. Because IS elements occur at many locations in the bacterial genome, the integration of F-plasmids occurs at different locations. For example, the F-plasmid has been found to integrate in at least 20 different sites in the *E. coli* K-12 chromosome (strain K-12 is a well-established laboratory strain). When the F-factor is integrated into the chromosome, it is replicated as part of the chromosome every time the cell divides (Figure 8.7). A cell population with an integrated F-plasmid is known as an *Hfr* strain, and every individual in this population has the F-plasmid integrated in exactly the same location. A specific *Hfr* strain has a name associated with it depending upon the location of integration of the F-plasmid, for example *Hfr*H was named after its discoverer, William Hayes, and the F-plasmid is always located adjacent to the *uxu*AB genes (genes involved in mannonate metabolism).

Because they possess conjugative pili and genes necessary for chromosome transfer, *Hfr* cells can conjugate with F^- cells. After contact between an *Hfr* and an F^- cell is established, a conjugation tube is formed and a nick in one strand of the *Hfr* chromosome opens the circular DNA to allow transfer of a single strand of the *Hfr* DNA to the F^- cell. After the single-stranded DNA is transferred to the recipient cell, the DNA may undergo recombination to integrate a fragment of donor DNA into the chromosome of the F^- cell DNA (Figure 8.8). Recombination does not always happen but depends upon a number of factors.

The bacterial cell receiving the chromosomal segment is called the recipient, and, after the cells separate, it is known as an **exconjugant.** The transfer of the *Hfr* bacterial chromosome begins at the site of integration of the F-plasmid and goes in a clockwise or counterclockwise direction, depending upon the orientation of the F-plasmid. Different *Hfr* strains have different F-plasmid integration sites, thus each *Hfr* transfers its chromosome starting at a different location (Figure 8.9). For example, *Hfr*H transfer always starts near the *uxu*AB locus, whereas *Hfr* KL99 begins transfer near *pyr*D, a gene involved in pyrimidine synthesis.

The transfer of the DNA from the donor cell to the recipient cell is time-dependent; the longer the conjugation takes, the more DNA will be transferred from one strain to another. But, the conjugation is usually interrupted before transfer of the entire donor chromosome resulting from Brownian movement of the two cells, which breaks the conjugation tube. As the bulk of the integrated F-plasmid is transferred last during conjugation, it rarely ends up in the F^- cell because the conjugation tube is broken before the F-plasmid can enter the recipient cell. Thus, F^- cells are rarely converted to F^+ cells by *Hfr* conjugation.

Of what benefit is conjugation to bacteria? If bacteria were not able to undergo conjugation, an individual cell would not be able to benefit from the genetic variability that exists in a bacterial population. When means of DNA transfer are available, one bacterial cell can obtain DNA from another bacterial cell, increasing its chances of survival. In this way, the population is not dependent upon random mutations to individual cells but can share genes with other bacteria. Also, conjugation is not restricted to bacterial cells of the same species. Widely different types of microorganisms, including yeast (which is a eukaryote) can undergo conjugation. Yeast can even conjugate with bacteria.

SECTION REVIEW PROBLEMS

4. What is the definition of *conjugative plasmid?*
5. What event converts an F^- cell into an F^+? What event converts an F^+ cell into an *Hfr*?
6. Why is it that only rarely does an *Hfr* transfer the entire chromosome during conjugation?

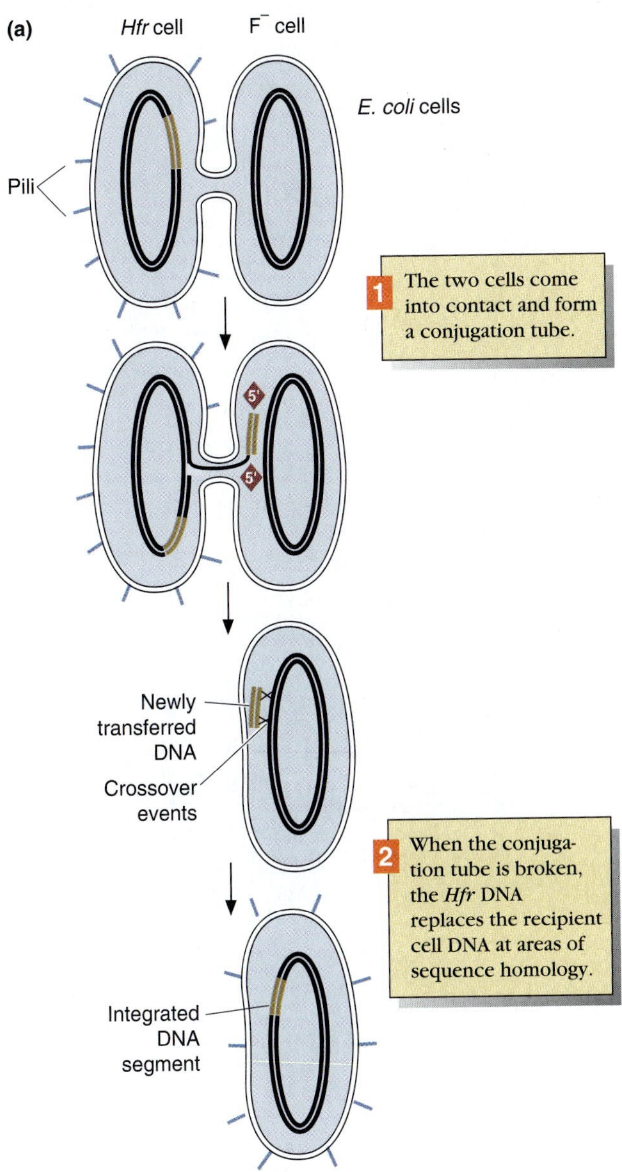

◀FIGURE 8.8 Chromosome transfer from an *Hfr* strain. (a) The *Hfr* cell physically comes in contact with an F⁻ cell, forms a conjugation tube, and starts transfer of the chromosome at the site where the F-plasmid is integrated. After transfer, the new strand is integrated into the genome of the F⁻ recipient cell by recombination. (b) This electron micrograph shows an F⁻ cell and an *Hfr* cell forming a conjugation tube for transfer of the DNA. *(Dr. L. Card/Science Photo Library/Custom Medical Stock Photo)*

(b)

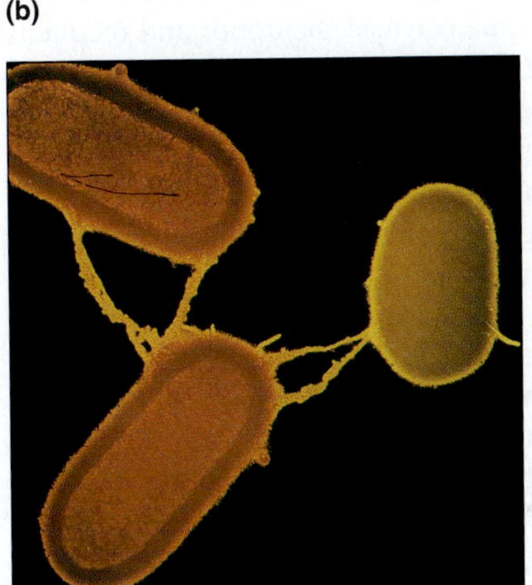

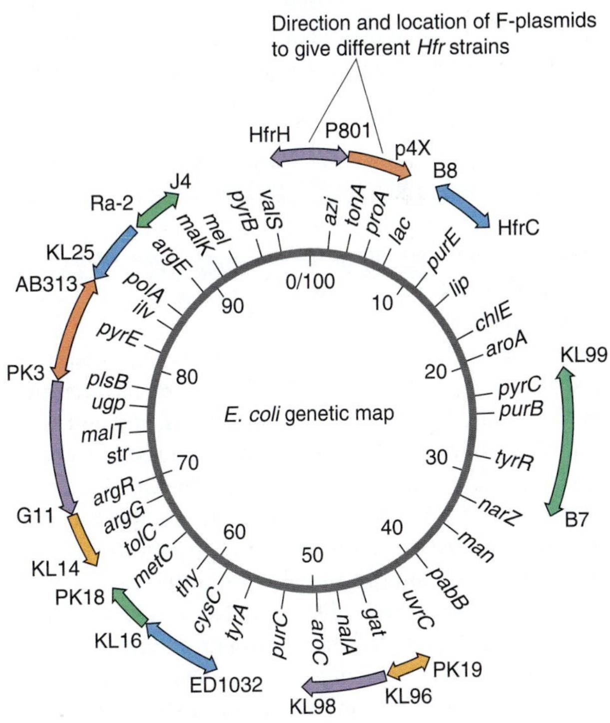

FIGURE 8.9 The map positions of some integrated F-plasmids. In the outer ring, each arrowhead indicates the position and orientation of integration of an F-plasmid. The inner ring shows the location of some genes on the *E. coli* genome and the map distance between them in minutes. The sequence of genes transferred during conjugation from a given *Hfr* strain begins after the arrowhead (adjacent to the name of the *Hfr* strain).

Conjugation Enables Construction of Genetic Maps of Bacterial Chromosomes.

In the following discussion, we will consider construction of a genetic map only of the bacterium *E. coli*. Keep in mind that the same principles apply to other bacteria, although not all bacteria can undergo conjugation.

A genetic map of the entire chromosome can be constructed by comparing the order of entry of genetic markers (genes) into an F⁻ cell using different *Hfr* strains. Different insertion points of the F-plasmids results in different *Hfr* strains and gives the bacterial geneticist great opportunity for controlling what genes will enter an F⁻ cell during conjugation. This capacity is essential to performing the many different crosses that permit the construction of a map of the *E. coli* chromosome. Such a map reveals that the *E. coli* chromosome is continuous and is referred to as circular.

In establishing a gene map of the *E. coli* chromosome, or in any mapping experiment, several factors must be established. First, a method must be available for differentiating between the donor and recipient cells during a conjugation experiment. Second, the *rate* of transfer of the chromosome between the donor and recipient cells will vary depending upon environmental conditions so experiments are usually conducted in a rich growth medium at 37°C without shaking of the culture (shaking disrupts the conjugation tube). When these factors are established, conjugation can be used for gene mapping in *E. coli* with reproducible results.

In performing a conjugation experiment, it is important to have conditions that will selectively recover only the F$^-$ exconjugants of a mating. This is usually accomplished by selecting for recipient strain cells containing an antibiotic resistance gene. Thus, after conjugation has taken place, the donor cells (*Hfr*) are selected against (killed) by the presence of an antibiotic in the growth medium. For example, if the recipient is *str*^{r}, and the donor is *str*^{s}, only the recipient cells that are resistant to the presence of streptomycin will grow. In addition, easily identifiable genetic markers (mutations in specific genes) must be present on both the donor and recipient strains. To illustrate, let us consider an idealized pair of strains for genetic mapping of a series of markers:

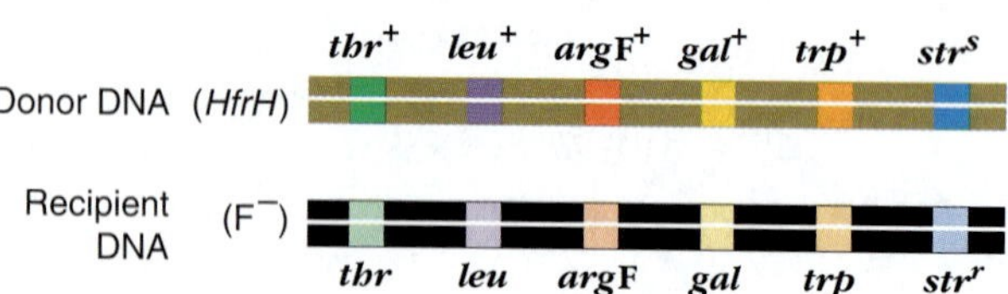

In this example, each gene has an alternate genotype and phenotype in the two strains (*thr* refers to threonine synthesis, *leu* refers to leucine synthesis, *arg* refers to arginine synthesis, *gal* refers to galactose catabolism, and *trp* refers to tryptophan synthesis). The streptomycin gene is distant from the other five genes and is used only as a selective marker to differentiate between the *Hfr* and the F$^-$ cells as conjugation proceeds. After conjugation, samples of exconjugate cells are spread on a series of different media in Petri dishes that contain all but one of the required growth factors, plus streptomycin. The exconjugate cells of the auxotrophic strain cannot grow on the selected medium unless they acquire the wild-type gene from the *Hfr* cells. Thus, the *str*^{r} exconjugants will grow on the selected medium only if they acquire the wild-type genes from the donor strain. As time progresses and more of the DNA is transferred from the donor to the recipient, the F$^-$ strain acquires each of the genes from the donor strain, permitting growth on the selected medium (Fig-

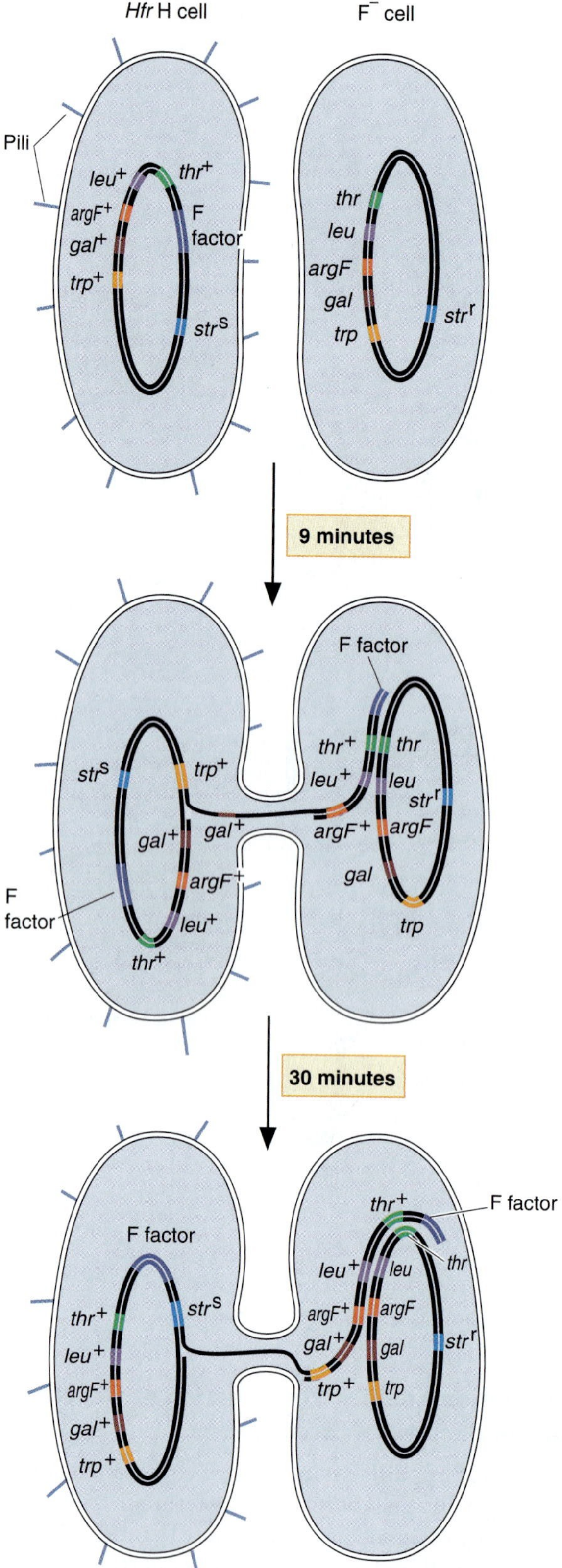

FIGURE 8.10 Conjugation with *Hfr*H. *Hfr*H is adjacent to the *thr* (threonine) marker on the *E. coli* chromosome and begins transfer of the chromosome to the F$^-$ strain at that point. After entry of the chromosomal segment, recombination occurs, replacing the genes that possess sequence homology. If conjugation is interrupted at any point, only those markers that have transferred up to that time point will be integrated.

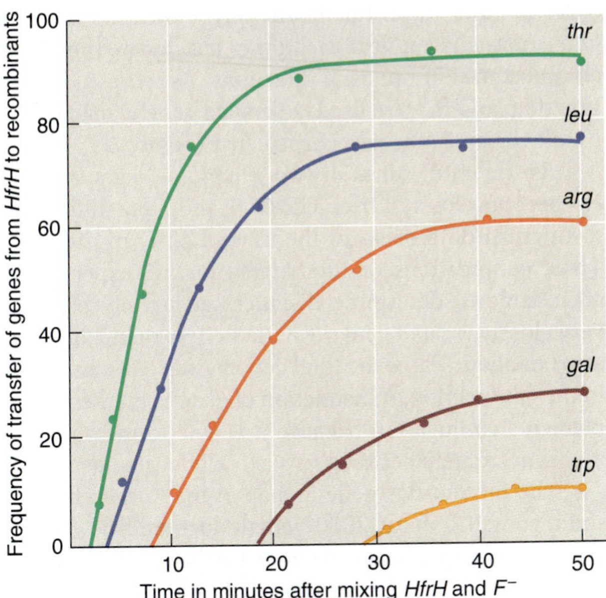

FIGURE 8.11 The time of entry of genes from *Hfr*H. The genes nearest to the F-plasmid are transferred to the F^- cell first and, consequently, are present with the highest frequency after entry into the recipient. More distant markers arrive later and at a lower frequency because the conjugation tube has been broken during mating.

ure 8.10). The number of F^- cells acquiring a specific wild-type gene from the *Hfr* will increase with time as more and more cells complete transfer of that specific gene.

Starting with the site of integration of the F-plasmid in the *Hfr,* the chromosome is progressively transferred into the F^- cell (Figure 8.10). Complete transfer of the chromosome requires about 100 minutes at 37°C, and the rate of transfer is constant. In this experiment, after conjugation is initiated, the *thr*$^+$ gene will begin to appear at 2 minutes, the *leu*$^+$ gene at 4 minutes, the *arg*F^+ gene at 9 minutes, the *gal*$^+$ gene at 19 minutes, and the *trp*$^+$ gene at 30 minutes. The number of individuals that acquire a specific gene increases with time (Figure 8.11). The time of gene transfer is consistent with a specific *Hfr* and remains the same each time the experiment is done. If this same experiment were done with a different *Hfr* strain [for example, KL25 instead of *Hfr*H (see Figure 8.9 for the location of KL25)], which is 13 minutes more distant from the *thr* gene, all these times would be shifted, delaying each entry time by 13 minutes.

Use of Different *Hfr* Strains Reveals a Circular Genetic Map of the *E. coli* Chromosome.

Because the rate of entry of genetic markers is constant, a genetic map can be constructed using the entry time in minutes as the map location. Thus, in the preceding experiment, if *thr* is designated as minute 0, then *leu* is located at 2 minutes (the actual time of entry of *thr,* 2 minutes, is subtracted from the actual time of entry of *leu,* 4 minutes, yielding 2 minutes on the map). It follows that *arg* is at 9 − 2, or 7, minutes, *gal* is at 19 − 2, or 17, minutes, and *trp* is at 30 − 2, or 28, minutes. Taking the entry times as the physical distance of the genes from each other it is possible to construct a genetic map as follows:

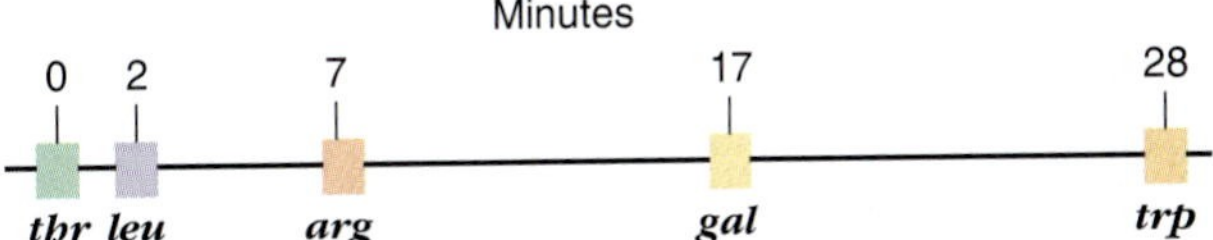

In the preceding experiment, the distance between genes, in minutes, was established. This distance will always remain the same no matter which *Hfr* is used. These data can then be used to construct a genetic map of the genes on the chromosome. To determine the relationship of map positions of these genes to other genetic markers, it is necessary to map additional marker genes using several different *Hfr* strains. Eventually, by using overlapping maps, a complete map of the entire chromosome can be constructed. This is, in fact, what François Jacob and Elie Wollman did in 1958 to show that the genetic map of *E. coli* was not a linear map, as shown earlier, but rather a circular map. They used six different *Hfr* strains and followed the order of entry of genetic markers in each of these six strains. An important feature of their experiment was that some of the *Hfr* strains had opposing orders of entry of the genetic markers into the F^- cells. The opposing order of entry of different *Hfr* strains was important to establish overlap of genetic markers and led to the construction of a complete map of the *E. coli* chromosome. Another important feature of their experiment was the breakage of the conjugation tube to stop mating. They did this by placing the bacteria in a blender for a few seconds before spreading on the selective media. This mechanically disrupted the mating process so that mating could not proceed after spreading the bacteria on selective media. As a consequence of this aspect of the experiment, conjugation mapping is also referred to as **interrupted mating.** They found the following entry order for a series of genes:

*Hfr*H. *thr leu azi ton lac tsx gal lam*
*Hfr*1. *leu thr thi met mtl xyl mal str*
*Hfr*2. *ton azi leu thr thi met mtl xyl mal str*
*Hfr*3. *tsx lac ton azi leu thr thi met mtl xyl mal str*
*Hfr*4. *thi met mtl xyl mal str lam gal tsx*
*Hfr*5. *met thi thr leu azi ton lac tsx*

Figure 8.12 shows the location of each of the genetic markers used in this experiment and the direction of entry for each of the *Hfr* strains. Some of the *Hfr* strains enter in one direction, whereas others enter in the opposite direction. For example, *Hfr* strains 4 and 5 have inverted

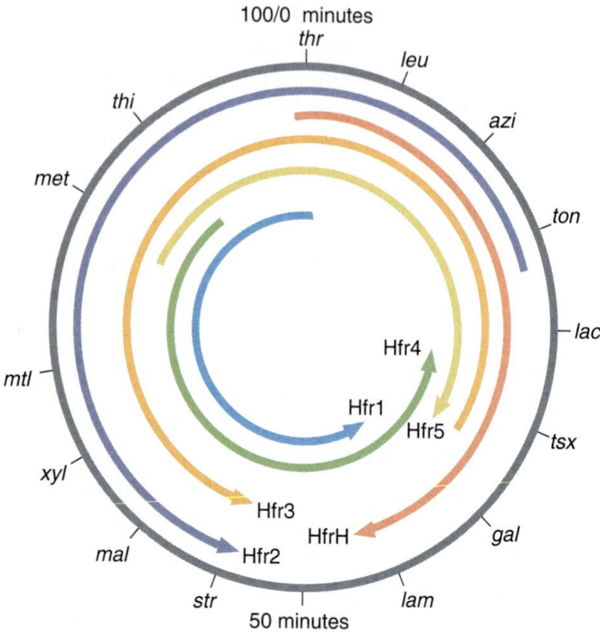

FIGURE 8.12 Different *Hfr* strains are used to map different areas of the chromosome. A diagram of the areas of chromosome transferred by various *Hfr* strains. The head of the arrow indicates the direction of entry of the chromosome from the donor *Hfr* strain, thus the first marker to enter the recipient is at the tail of the arrow. Use of these strains demonstrated the circular nature of the *E. coli* chromosome.

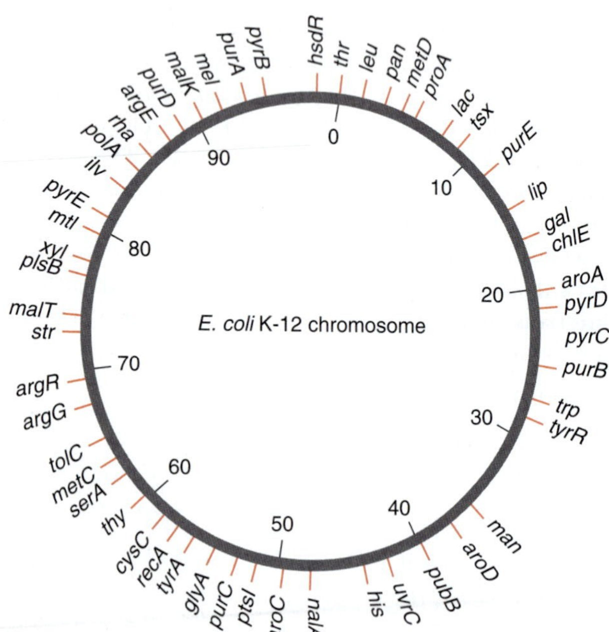

FIGURE 8.13 The *E. coli* map. A partial linkage map of the *E. coli* K-12 chromosome. The total time scale is 100 minutes, beginning arbitrarily with zero at the *thr* locus. The total number of known genetic markers is over 2000.

order of entry times for *thi* and *met*, indicating that the two *Hfr* strains are near each other but donate their host chromosomes in opposite directions. As more and more genetic markers were used, a detailed genetic map of the *E. coli* chromosome was constructed (Figure 8.13).

As the entry times of two genetic markers from an *Hfr* become closer, it becomes more and more difficult to distinguish differences in the time of entry of the gene. Thus, conjugations or interrupted mating experiments are unable to determine distances accurately between very closely located genetic markers (less than 2 minutes apart or about 2% of the total genome size). As discussed in the next section, transduction of genetic markers complements conjugation because it is able to measure distances accurately between very closely located genes.

The present-day genetic map of the *E. coli* chromosome contains about 2000 genetic loci, and new genes are being added at a rate of about 100 per year. The physical size of the *E. coli* genome is close to 4800 kilobase pairs, and a complete sequence of bases in the DNA has been determined. If an average gene occupies about 1500 base pairs, then the *E. coli* chromosome could have as many as 3000 genes. An extremely abbreviated version of the *E. coli* circular genetic map is shown in Figure 8.13, only because it is not possible to show a complete circular map on one page and indicate the positions of all the genes.

SOLVED PROBLEM

Problem

You have independently isolated three separate *Hfr* strains from a single F^+ strain of *E. coli*. Subsequently, you have used these three *Hfr* strains in matings with F^- strains carrying mutations in many different genes to map them in *E. coli*. The following results were obtained, showing the genetic markers and their entry times (in parentheses) in minutes. The donor strain was always wild-type for the mutations.

Mutant Type and Minute of Transfer

Hfr Strain		
A	**B**	**C**
mal^+ (2)	ade^+ (12)	pro^+ (3)
str^s (12)	his^+ (27)	met^+ (29)
ser^+ (17)	gal^+ (37)	xyl^+ (32)
ade^+ (37)	pro^+ (43)	mal^+ (37)
his^+ (51)	met^+ (69)	str^s (47)

Draw a map of the chromosome showing the order of the genes and the distances between each gene in minutes. Also, show the approximate insertion point of the F-plasmid in the *E. coli* chromosome that formed *Hfr*A, *Hfr*B, and *Hfr*C.

Solution

Each of these matings can be dealt with independently, and then the collective data can be used to construct a single map of the genome. With *Hfr*A, *Hfr*B, and *Hfr*C, the following map can be constructed:

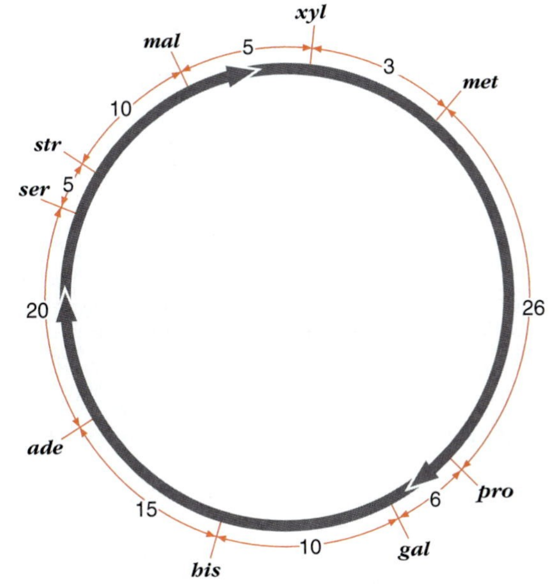

The insertion point of the F-plasmid into the genome is before the entry time of the nearest genetic marker or gene. In the case of *Hfr*A, the F-plasmid has inserted just before the *mal* gene, whereas in *Hfr*B the insertion point is 12 minutes before *ade*. Taking all the data together, the total map of the genome can be constructed showing the sites and orientation of the F-plasmid insertion, as well as the distances between each marker.

SECTION REVIEW PROBLEMS

7. How does *Hfr* KL16 differ from *Hfr*H? (Hint: See Figure 8.9.)
8. How were *Hfr* strains used to establish that the chromosome of *E. coli* is circular?
9. In conjugation experiments, the recipient strain (F^-) usually has an str^r or some other antibiotic resistance marker. Why?
10. Why does conjugation have to be interrupted before plating the cells on selective media?

F′-Plasmids Are Formed Upon Imprecise Excision of an F-Plasmid.

Just as F-plasmids can integrate into the chromosome, they can also excise and become independent components in the cytoplasm. After excision of an F-plasmid from the chromosome, the cell will be an F^+ cell rather than an *Hfr*. However, F-plasmid excision is not always precise; sometimes part of the bacterial chromosome is excised together with the F-plasmid. When this happens, the F-plasmid is called an **F′-plasmid** (F-prime plasmid). An imprecise excision is illustrated in Figure 8.14 where the F-plasmid was originally located between the *thi* and *thr* genes. During the excision event, the *thr* gene is removed together with the F-plasmid so that the bacterial chromosome no longer contains the *thr* gene. The genotype of the bacterial strain remains the same because the *thr* gene is still present on the F-plasmid in the cell. However, upon conjugation of an F′-plasmid into a wild-type F^- cell, a **merozygote** or **partial diploid** is obtained. This new strain has the *thr* gene in two copies, one on the F′-plasmid and one on the chromosome.

A considerable number of different F′-plasmids have been generated by bacterial geneticists because they are extremely useful for mapping genes using complementation analysis. Individual F′-plasmids have been found that collectively cover the entire *E. coli* genome (Figure 8.15). The size of F′-plasmids can be as large as one quarter of the bacterial chromosome.

F′-Plasmids Are Used in Complementation Analysis.

In bacterial genetics, when searching for a particular mutant, a frequently encountered problem is finding a series of different mutants that give the same phenotype. For example, 12 different genes are involved in the synthesis of arginine (*arg*A, *arg*B, *arg*C, etc.), and they all give the same phenotype—a requirement for arginine for growth. Furthermore, four of these genes map at minute 90 on the *E. coli* chromosome, making it difficult to separate the mutations by recombination or by genetic mapping. Thus, if bacterial geneticists found two different arginine-requiring mutants in *E. coli,* they would not know whether the two mutations were in the same gene or different genes. F′-plasmids can be used to distinguish between these confusing alternatives and to determine whether two different mutations that give rise to the same phenotype are in the same gene or in separate genes.

To illustrate, let us consider an example where the two mutations (in separate strains) are in separate genes, *arg*C and *arg*B. To determine whether complementation occurs for either of these genes, an F′-plasmid is prepared (or obtained from another source) that possesses an *arg*C^+ wild-type gene and a mutant *arg*B gene (*arg*C^+ / *arg*B). The F′-plasmid is conjugated into the two mutant strains defective in arginine synthesis (*arg*). If the mutation is in the *arg*C gene, the product of the *arg*C^+ gene on the plasmid (*arg*C^+ / *arg*B) will complement the defective *arg*C gene on the chromosome, giving a wild-type phenotype. However, if the mutation on the chromosome is in the *arg*B gene, no complementation will occur (because the plasmid also has a defective *arg*B gene), and the strain will still require arginine for growth. Thus, one mutant strain will grow in the absence of arginine, and the other will not (the one with the *arg*B mutation)

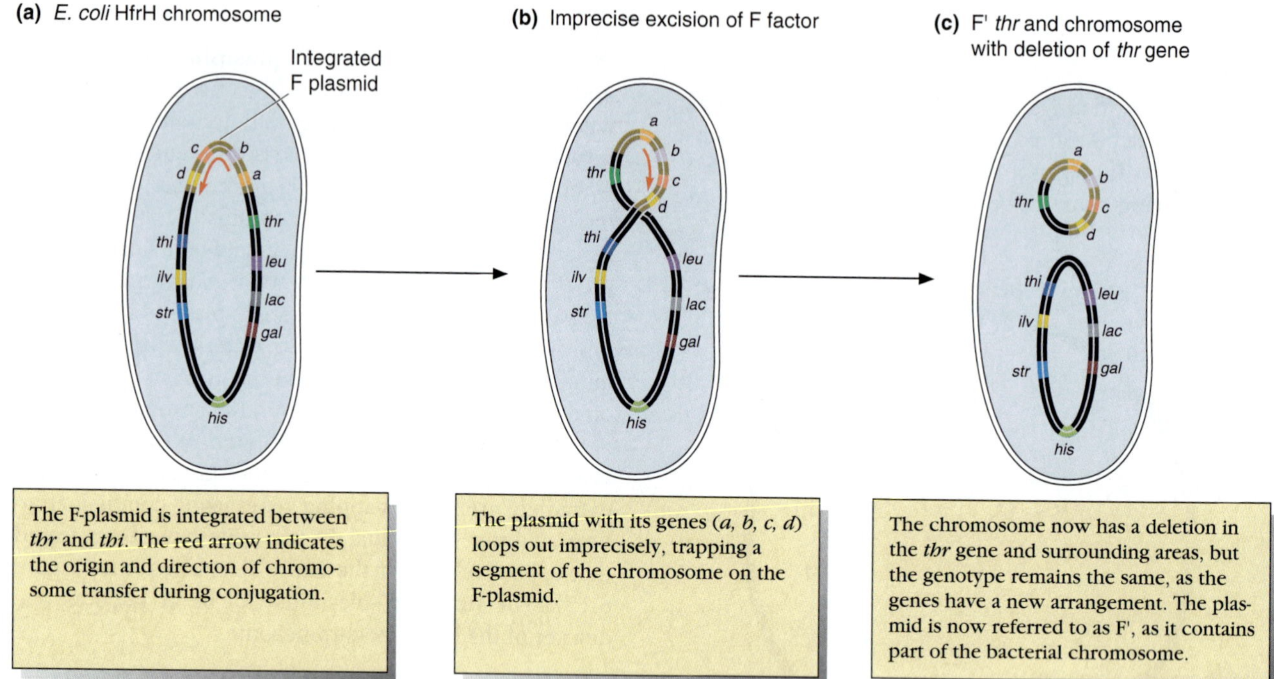

FIGURE 8.14 Formation of F′-plasmids. F′-plasmids are formed by imprecise excision of an F-plasmid from the *E. coli* chromosome, carrying along a segment of the chromosome as it leaves. This new F′-plasmid can be transferred to an F⁻ *E. coli* cell and forms a partial diploid for those genes on the plasmid.

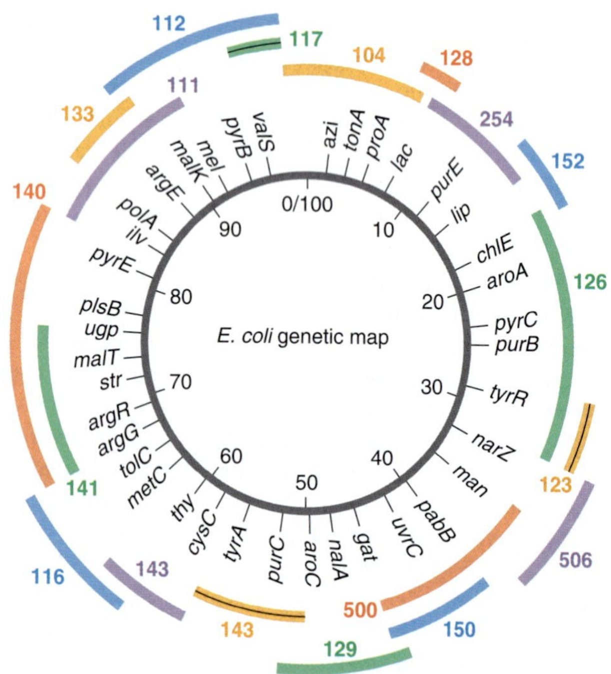

FIGURE 8.15 Commonly used F′ plasmids. The genetic area spanned by some F′-plasmids is shown. The inner circle shows an abbreviated version of the *E. coli* genetic map. The numbered colored lines around the outer edge are the DNA segments associated with individual F′-plasmids; each has a specific name to identify it. Different strains, each containing a different F′-plasmid, can be obtained from the *E. coli* genetic stock center at Yale University.

when the F′-plasmid is present in each strain with *arg*C⁺, *arg*B genes. In contrast, if the two mutations in the two different strains were in the same gene (for example, *arg*B), neither would grow. Complementation can be visualized by the following diagram where part of the pathway for arginine synthesis is depicted:

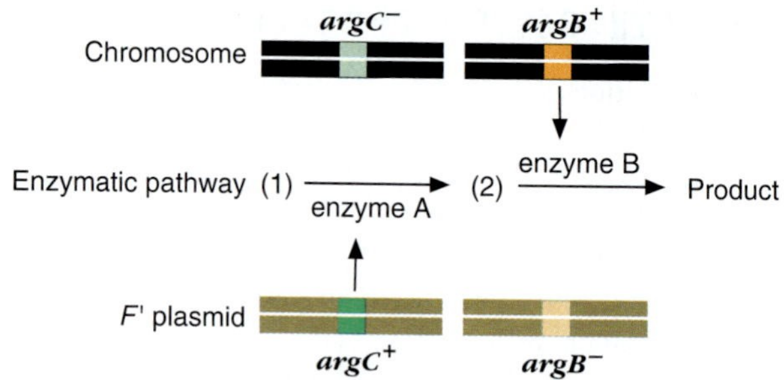

This same example is also illustrated in Figure 8.16a. The *arg*B gene on the F′-plasmid produces a defective gene product, but the *arg*B⁺ gene on the chromosome makes a diffusible normal gene product, thus giving a wild-type phenotype. Likewise, the *arg*C⁺ gene on the F′-plasmid makes a diffusible product that complements the defective product made by the chromosomal *arg*C gene. These genes are said to **complement** each other, and this genetic configuration is part of a complementation test or *cis–trans* test discussed in Chapter 6. In this example, we can conclude that the two *arg* mutations are in different genes. If complementation did not occur, a mutant phenotype would be observed, and we could conclude that the two mutations were in the same gene (Figure 8.16b). In one configuration a wild-type phenotype is observed, indicating the two mutations are in separate genes, whereas in the other case a mutant phenotype is observed, indicating that the two mutations are in the same gene.

To further clarify this example, consider the following analogy. Assume that you have two production groups making computers in a factory. For some reason, one production group has a breakdown in the making of monitors (gene A), and the other group has a simultaneous breakdown in the production of microprocessors (gene B). Production can continue when the two groups exchange monitors and microprocessors (the diffusible products), giving a wild-type phenotype—or normal computers. They thus have complementation. But, if both groups are unable to make the same component, such as microprocessors, they will not be able to produce computers (and consequently have a defective phenotype). They have no product to exchange and thus do not complement each other.

As might be expected, some complications arise in interpreting a complementation test. Earlier we stated that two different mutations in the same gene, present in the same cell, would not give a wild-type phenotype (because no complementation occurs). This is not completely true, however, because a crossover event may occur between the two mutant genes, generating a wild-type gene and double mutation gene. However, if the two mutations are very close to each other, the likelihood of a crossover event is extremely low. With widely separated mutations, the probability of a crossover event rises significantly. Usually, however, crossover events can be distinguished from complementation because of large differences in frequency. Another approach to ensuring that no recombination has occurred is to perform the experiment in a *recA*$^-$ strain where generalized recombination is not possible because the product of the *recA* gene is required for recombination.

SECTION REVIEW PROBLEMS

11. Two independently isolated mutants in *E. coli* have been found that cannot grow on medium lacking the amino acid leucine. These two mutants have been designated *leu*1 and *leu*2. We find that an F′-plasmid containing an *leu* gene complements the *leu*1 mutation (assume that we know that only one *leu* gene is present on the plasmid) as a partial diploid (merozygote) but does not complement the strain carrying the *leu*2 mutation (it will not grow on a medium lacking leucine). Are these two mutations in the same gene or in separate genes? Explain.
12. In an experiment you have an F′-plasmid with *leuA6*$^-$ in a *leuA2*$^-$strain, and you find *leu*$^+$ colonies at a rate of one in several thousand cells when you plate the cells on a medium lacking leucine. What can you conclude from this observation, assuming a reversion rate for each mutation of 1×10^{-6}?

(a) Complementation:
wild-type phenotype; ***argB***$^+$***/argC***$^+$ genotype

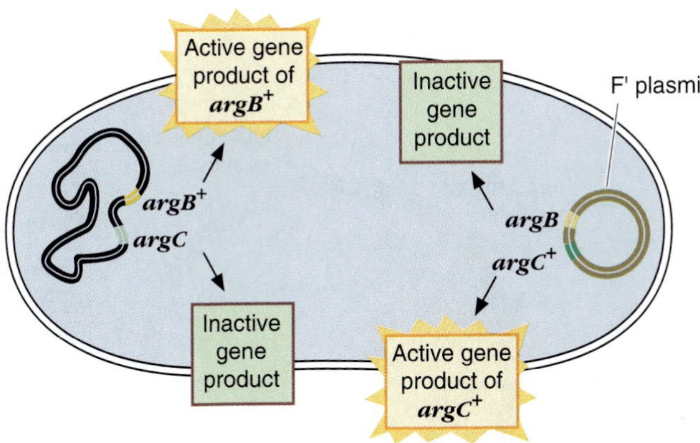

(b) No complementation:
mutant phenotype; ***argB***$^+$***/argB***$^+$ genotype

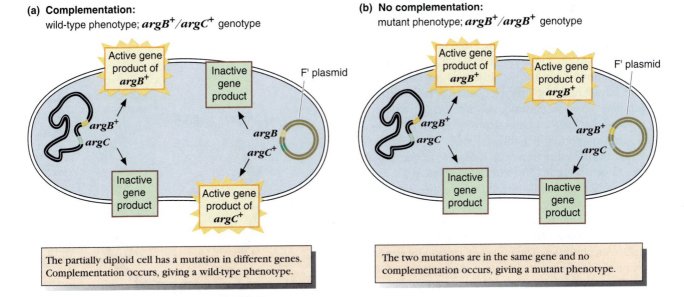

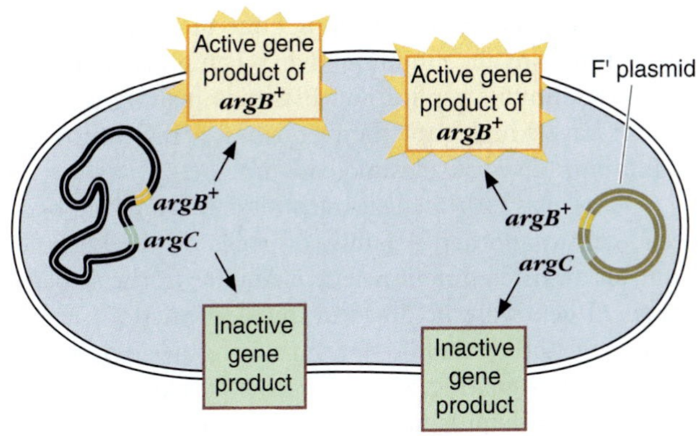

FIGURE 8.16 The *cis-trans* complementation test. Depending upon the presence of an active allele of the two genes on the F′-plasmid, the strain can give either a wild-type phenotype or a mutant phenotype. (a) If an active allele is present on the plasmid, this allele complements the inactive allele on the chromosome and gives a wild-type phenotype. (b) If the active allele is absent from the plasmid, no complementation occurs.

Many Types of Plasmids Are Present in Bacteria.

Many types of plasmids exist, some of which are listed in Table 8.2. In general, plasmids are placed into several categories, some of which are as follows:

1. *F-plasmids, sometimes called sex-plasmids.* As discussed in the preceding paragraphs, F-plasmids have the ability to transfer segments of chromosomal material from cell to cell.
2. *Col-plasmids.* These plasmids have genes that code for toxic proteins, called colicins, which are capable of killing other closely related bacteria that do not possess the *col*-plasmids.
3. *R-plasmids.* These plasmids carry genes that code for resistance to antibiotics and are frequently able to transfer between bacterial strains.
4. *Specialized function plasmids.* Many bacteria contain plasmids that have genes coding for specialized functions. These functions include the ability to fix nitrogen in symbiotic relationships with legume plants, the ability to degrade aromatic hydrocarbons, and the ability to infect plants and form galls. Many plasmids have genes that allow self-transmission to other closely related bacteria; however, others do not.

TABLE 8.2 *Some Natural and Artificially Made Plasmids in Bacteria*

Plasmid Name	Bacterial Host	Size (base pairs × 1000)	Genetic Functions
Col E1	*E. coli*	6.3	Colicin E1 synthesis
F	*E. coli*	95	F-pili
pBR322	*E. coli*	4.3	Recombinant plasmid; tet^r, amp^r
R100	*E. coli*	106	R-plasmid; cam^r, str^r, tet^r, sul^r
Sym-a	*Rhizobium meliloti*	1600	Nitrogen fixation and nodulation of legumes
Ti	*Agrobacterium tumefaciens*	180	Causes crown gall on dicotyledonous plants
TOL	*Pseudomonas putida*	116	Degrades toluene and xylenes

Plasmids rely heavily on the host replication apparatus for their own replication. All plasmids have a defined origin of replication on their DNA (the region governing replication is called *rep* and contains several genes and an *ori* site, Figure 8.17) associated with genes that control the number of copies of the plasmid in an individual cell. Some plasmids are present in multiple copies (10–30) per cell, whereas others are present at one copy per cell. The mechanism of control of copy number varies from plasmid to plasmid. For example, one method involves the synthesis of an "inhibitor" that, when present in high amounts, inhibits replication. Plasmids making less inhibitor have a high copy number, whereas high amounts of inhibitor give low plasmid copy numbers.

Many plasmids are self-transmissible from one bacterial cell to another. A family of genes on the plasmid controls transfer function. For example, in the F-plasmids, 23 genes are involved in transfer function. These genes have the symbol *tra,* for transfer, followed by a capital letter designating a specific gene (*tra*A, *tra*B, *tra*C, etc.). Some plasmids have been modified, using genetic engineering, to accommodate specific needs in recombinant DNA techniques. The *tra* genes are usually removed from plasmids when they are used in genetic manipulations. Removal of the *tra* genes does not affect the replication of the plasmid, but it does ensure that inserted foreign genes present in the plasmid are not transferred to other bacteria should an engineered strain escape into the environment. The self-transmissibility of plasmids has caused many medical problems because R-plasmids sometimes accumulate antibiotic resistant genes, which then can be transmitted to other strains. As a consequence, multiple drug resistant strains are more and more frequently found in bacteria and pose medical problems (see CURRENT INVESTIGATIONS: *Is It Safe To Feed Antibiotics to Cattle, Swine, and Poultry?*).

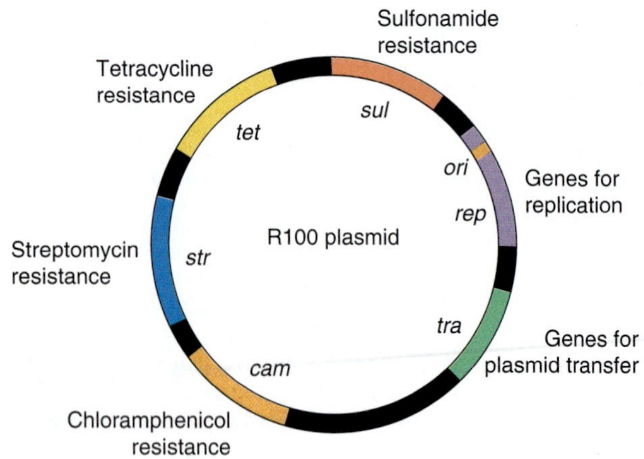

FIGURE 8.17 A map of a plasmid. The genetic map of the R100 plasmid showing the general location of some genes. This is a large plasmid (58,000 base pairs) and is capable of conjugation.

Current Investigations

Is It Safe To Feed Antibiotics To Cattle, Swine, and Poultry?

In the 1950s, it was discovered that addition of antibiotics, especially tetracycline and penicillin, to animal feeds was very effective in increasing the rate of growth of farm animals. This was primarily the result of more efficient conversion of food into meat and the prevention of many diseases that occur in livestock. It is now common practice to add antibiotics to livestock feed, and, in fact, over 50% of all antibiotics produced in the United States are used for addition to animal feeds. Adding antibiotics to animal feeds has two conflicting aspects: the danger of increased spread of bacterial antibiotic resistance, which may endanger treatment of human diseases, and an estimated cost to the consumer of over $4 billion per year in increased meat prices if this practice is stopped.

Bacterial resistance to an antibiotic can occur through a chromosomal mutation or the gain of a plasmid that contains an antibiotic resistance gene. In the latter case, it is known that some plasmids carry the genes necessary for plasmid transfer to other related bacteria. It follows then that a self-transmissible plasmid with an antibiotic-resistant gene will spread throughout a bacterial population rather rapidly, especially if the bacteria's environment is consistently exposed to low levels of antibiotics. In a test case with two cattle herds over a 13-year period, one herd received antibiotics with their feed and the other herd did not. Antibiotic-resistant bacteria were found in over 85% of the antibiotic-treated cattle but in only 50% of the nontreated cattle. The real question is, Does the prevalence of antibiotic-resistant bacteria in the animal population affect human health? This question was the subject of a National Academy of Sciences study in the early 1980s. The study concluded that transfer of antibiotic-resistant bacteria from animals to humans is difficult to determine accurately, and under natural conditions it is probably infrequent. Because of the complex sequence of events from selection for resistant bacteria in the farm environment to contamination of meats in the marketplace, an accurate assessment of transfer of resistant bacteria is difficult. More recently, a study reported in the *Journal of the American Medical Association* traced a strain of *Salmonella* from the farm to an illness associated with an individual whose only association with the farm was via the meat consumed, thus giving credence to the idea that feeding antibiotics to animals is not a good idea.

All in all, it is not likely that antibiotics will be banned from use in US livestock feeds in the near future, as was done in Great Britain in 1971. Only circumstantial evidence is available to indicate that the practice is a health hazard, and as one animal scientist said, "We have been doing it for 40 years, and no documented deaths or injuries to humans have been substantiated yet." However, this leaves unaddressed the extensive proliferation of resistant bacteria, which pose a genuine threat to effective antibiotic treatment in humans. Much of this increase is the result of widespread overprescription of antibiotics to humans by physicians. In the 1990s we saw an increase in the number of bacterial infections that could not be treated with *any* or just one single antibiotic. In Spain, where penicillin or its analogues are prescribed very frequently, nearly 80% of the bacteria are resistant to penicillin. We are reaching a point in our history when antibiotics are no longer going to be the miracle drug that they were 40 years ago.

Section Review Problem

13. What features distinguish a plasmid from the chromosome? Describe some characteristics that you would expect to find associated with a plasmid.

Transformation Involves Cellular Uptake of DNA.

Transformation is a general term that refers to the uptake of free DNA into an organism and is another process by which bacteria can transfer genetic information that might increase their chances for survival in an uncertain environment. In this section, we will discuss transformation of bacteria, but remember that fungus, plant, and animal cells can also be transformed. One of the most useful aspects of transformation is the introduction of plasmid DNAs into *E. coli* or foreign DNA into plant and animal cells. In genetic engineering, it is frequently necessary to introduce a genetically engineered plasmid into an *E. coli* host for subsequent propagation and growth. This DNA can then be studied and transferred by transformation into other organisms. Chapters 12 and 13 on genetic engineering discuss introduction of foreign DNA into bacteria by transformation of plasmids possessing cloned sequences.

Transformation Involves the Uptake of DNA by a Bacterial Cell.

In bacteria, transformation occurs naturally in a variety of species including *Bacillus subtilis, Hemophilus influenzae, Pseudomonas aeruginosa,* and *Streptococcus pneumoniae.* In contrast, many bacteria do not undergo natural transformation but can be artificially induced to take up DNA. Natural transformation does not occur in *E. coli,* but it can be artificially induced by treating cells with calcium ion and temperature changes or with electric shock (known as electroporation, see Chapter 12). Within a population of bacterial cells, usually only a very small fraction (1 in 1000) is able to take up DNA. These transformed cells are referred to as **competent,** and they are altered in some poorly understood fashion that allows transfer of DNA through the cell wall and membrane.

In the first step of natural transformation (excluding *E. coli* and other artificially induced forms of transformation), the double-stranded DNA is transferred through the bacterial membrane, probably assisted by a protein specifically for this purpose. Immediately after uptake, one DNA strand is digested by a nuclease. The remaining single strand aligns itself with a similar sequence of DNA on the chromosome of the recipient cell, displacing one strand of the recipient's DNA. The displaced strand is excised, leaving the donor strand in its place. The incoming strand from the donor is sealed into the recipient's DNA by covalently bonding the two ends. This process is outlined in graphic form in Figure 8.18.

During transformation of DNA into a bacterial cell, the entering DNA may be identical to the host DNA, or they may differ in a few bases. If the entering DNA sequence differs from the recipient and is incorporated into the recipient's DNA, the two strands will not pair properly (one from the donor and the other from the recipient) and are known as a **heteroduplex.** Upon duplication of the heteroduplex DNA, the two daughter molecules will also differ. Thus, when the donor and recipient cells differ genetically, the daughter cells from the transformed cell will inherit information from the donor-transformed strand and the recipient strand. The end result is a transformed bacterial cell that has DNA from another bacteria.

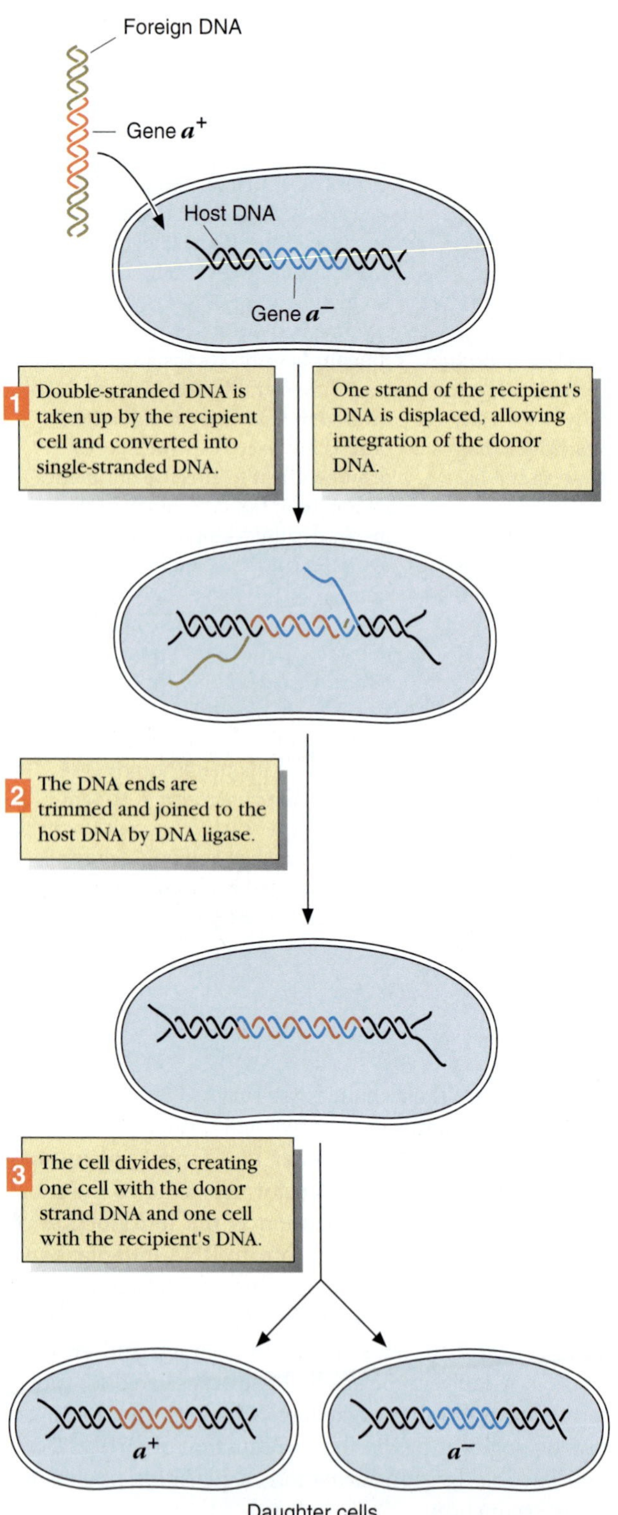
Foreign DNA
Gene a^+
Host DNA
Gene a^-
1 Double-stranded DNA is taken up by the recipient cell and converted into single-stranded DNA.
One strand of the recipient's DNA is displaced, allowing integration of the donor DNA.
2 The DNA ends are trimmed and joined to the host DNA by DNA ligase.
3 The cell divides, creating one cell with the donor strand DNA and one cell with the recipient's DNA.
a^+
a^-
Daughter cells

◄ **FIGURE 8.18 Natural DNA uptake.** During transformation, DNA binds to the cell surface and is taken up by the recipient cell. One of the two strands is integrated into the genome at a site of sequence homology.

Transformation Is Useful in Fine-Scale Genetic Mapping.

Using natural transformation, it is possible to determine the map location of closely linked genes. Transformation is not useful in determining the location of genes that are spaced more than several minutes apart on the genetic map because a transforming fragment of DNA is rarely larger than 100,000 base pairs (about 2 minutes or about 50 genes on a typical bacterial genetic map). As we learned in the previous section, conjugation is superbly useful in mapping distantly spaced markers but poor in mapping closely linked marker genes. Transformation is useful for mapping closely linked genes. In addition, transduction is extremely useful in mapping closely linked genes and is discussed in detail in the next section.

To describe how transformation mapping is actually done, let us assume that we have prepared DNA from a wild-type strain of bacteria and sheared it into pieces of about equal size, say 50,000 base pairs long. Shearing is commonly done by vigorously shaking the DNA sample or forcing the DNA through a hypodermic needle. The recipient bacterial strain is *arg*H (let us say that this gene is located at minute 50 on the map of this genus) and *his*E (at minute 40). Transformation is an infrequent event, and each event covers a small area on the chromosome. Distantly linked genetic markers have been chosen in this example to demonstrate the result of independent events because there is little chance that these distantly located genetic markers will be co-transformed. **Co-transformation** occurs when crossover events incorporate two or more genetic markers simultaneously. For this to happen in a single event, they have to be close enough so that both are present on the same fragment of DNA. If the two markers are distantly located—for example, if the *arg*H gene is transformed at a frequency of 1 in 500 (or 0.002) and *his*E is transformed at an equal frequency—*the co-transformation frequency would be the product of each independent event,* or $0.002 \times 0.002 = 0.000004$ (4×10^{-6}). On the other hand, a higher co-transformation frequency, say 0.008, would suggest that the two markers are close enough to be carried on one piece of DNA. Thus, if two genes are transformed simultaneously on one fragment of DNA (co-transformed), this frequency must be significantly greater than the product of the frequencies of the independent events.

However, if instead of having a recipient strain that is of this genotype, suppose the recipient strain were *trp*E and *tyr*R, which are only 1 minute of map distance apart. In this case, the two genes are very close to each other and will be contained on the same fragment of DNA a certain percentage of the time. This will increase the frequency of simultaneous integration (into the recipient bacterial genome) of the two genetic markers. But not all integrated DNA fragments will carry both genetic markers at the same time. The transforming DNA fragment of about 50,000 base pairs may contain one or both genetic markers (Figure 8.19). However, the closer the two genetic markers, the greater the probability they will both be on the same transformed DNA fragment. The distance between two genetic markers can thus be calculated on the basis of simultaneous incorporation of two markers versus the incorporation of individual markers.

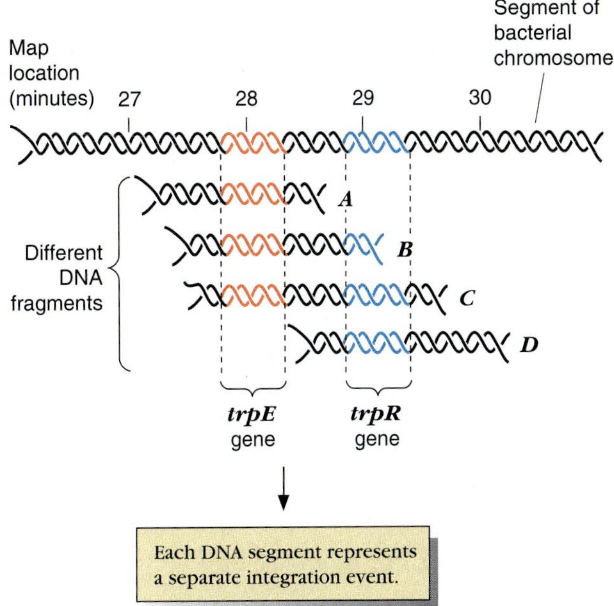

FIGURE 8.19 Co-transformation of two markers. During transformation, one fragment may span two genetic markers if they are sufficiently close. In this example, only fragment C contains both the *trp*E and *tyr*R genetic markers and thus will be the only single fragment integrated that contains both genetic markers. In contrast, integration of fragments A and D in the same cell would be independent events and would occur at a much lower frequency.

Transformation is most useful in establishing that two individual markers are closely linked, but it can also be used to establish the gene order when three genes are closely linked. For example, if three genes *m*, *n*, and *o* are closely linked, and in that order, then the co-transformation frequency will be greater for adjacent genes (*m* with *n* and *n* with *o*, but not for *m* and *o*). More distantly located genes will not be on the same DNA fragment, thus making a co-transformation event less likely. If *m* and *n* had a co-transformation frequency of 0.003%, *n* and *o* had a co-transformation frequency of 0.005%, and *m* and *o* had a co-transformation frequency of 0.0008%, these data would suggest that *n* is in the middle, between *m* and *o*. Why? Because the co-transformation frequency between *m* and *o* is much lower than for the other two markers, indicating a greater distance between them.

SOLVED PROBLEM

Problem

You have discovered three genes that are located near minute 98 on the *Bacillus cereus* genetic map, and you have named them *pil*A, *pil*B, and *pil*C. However, you do not know their order. To find out which gene is in the middle, you determine the co-transformation frequency for the three possible combinations of the two genes, with the following results.

Experiment	Co-Transformed Genes	Frequency
1	*pil*A & *pil*B	0.12%
2	*pil*A & *pil*C	0.9%
3	*pil*B & *pil*C	0.78%

What is the gene order?

Solution

*pil*C is in the middle, so the gene order is *pil*A, *pil*C, *pil*B. This conclusion comes from the observation that the co-transformation frequency for *pil*A and *pil*B is lower than for the other two markers; therefore, *pil*A and *pil*B must be farther apart than either of the other two markers.

SECTION REVIEW PROBLEM

14. In a transformation experiment, we have a strain of *B. subtilis* that is auxotrophic for *phe*A and *leu*C. You want to determine the relationship of these two markers on the chromosome using transformation. Using the DNA of a wild-type strain in an experiment, we find that the frequency of transformation of *phe*A is 0.016% and *leu*C is 0.009%. There are no co-transformants (to phe^+leu^+) found among the 10^8 cells screened. What do these data tell you about the linkage distances between *phe*A and *leu*C? What would be the predicted frequency of co-transformation if these transformation events were completely independent?

PHAGE GENETICS HAS REVEALED INFORMATION ABOUT THE GENE.

Transduction is the transfer and integration of genetic information from one bacterial cell to a second bacterial cell using a bacteriophage as a vector. To accomplish this, a bacterial virus or bacteriophage must accidentally package some bacterial DNA into its head, in place of or in addition to phage DNA, and transmit the DNA to another bacterial cell where integration into its genome follows. To clearly understand how transduction works, it is first necessary to examine the morphology, life cycle, and some genetics of bacteriophage because they are responsible for the transduction process.

The Phage Life Cycle May Include Lytic and Lysogenic Cycles.

In this section, we will discuss only the bacteriophage particles ("phage," for short) that invade *E. coli,* but nearly all bacteria are susceptible to attack by phage. In fact, the word *bacteriophage* literally means "eaters of bacteria." They are sometimes referred to as bacterial viruses. Bacteriophage are ubiquitous in the soil, sewage, intestines, and anywhere else bacteria may roam. A phage particle is very small in relation to a bacterial cell (1/100 to 1/500 the size; Figure 8.20) and consequently contains very little genetic information. Phage cannot replicate on their own but depend upon the bacterial host for the machinery to reproduce. A phage particle is composed of a chromosome (from about 5000 to 100,000 base pairs) that can be either DNA or RNA and is surrounded by a protein coat. Many phage have extensions from the protein coat to specifically attach the virus to the surface of a bacterial cell. Usually phage are very specific about the bacterial type to which they attach. A phage that attacks *E. coli* cannot attack *Pseudomonas* spp. and vice versa. Some of the most thoroughly studied phage are the "T-even" phage of *E. coli* (T2, T4, T6, and so on); two other well-studied *E. coli* phage are P1 and lambda (Figure 8.21). In fact, lambda phage is probably the most thoroughly understood biological entity on the face of the earth, and we shall discuss it in some detail in this book.

During the infection of a bacterial cell by T4, the phage first attaches to the outer surface of the cell, releasing an enzyme that digests a hole in the bacterial cell wall to allow injection of the genetic material. The phage genetic information then takes over, turning off the synthesis of bacterial DNA and proteins. All the cell's energies are directed to reproducing the phage components. After infection, the phage also confers "immunity" to further infection by other phage particles. Ultimately, many phage descendants are constructed and stored temporarily until the bacterial cell wall is destroyed (lysed) by the enzyme lysozyme. This releases several hundred new phage into the environment to attack other bacteria and repeat the cycle. This whole process is called the

FIGURE 8.20 Electron micrograph of phage-infected cell. This electron micrograph shows T4 phage attached to an *Escherichia coli* cell. *(Lee D. Simon/Science Source/Photo Researchers, Inc.)*

T4 phage

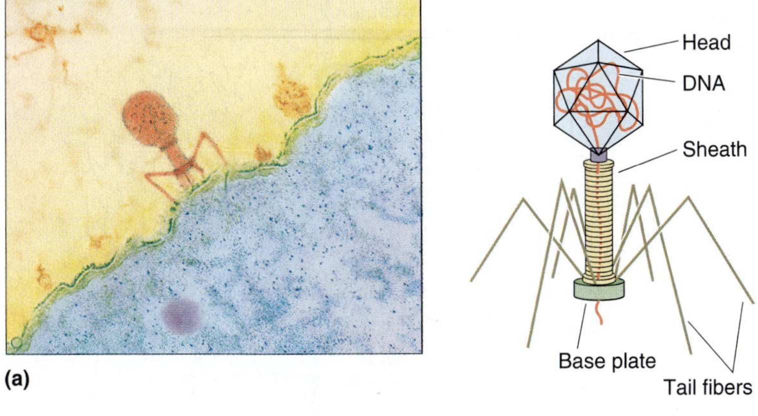

(a)

λ phage

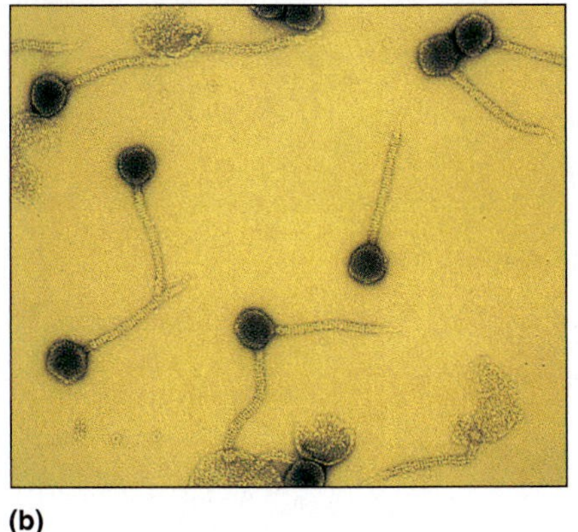

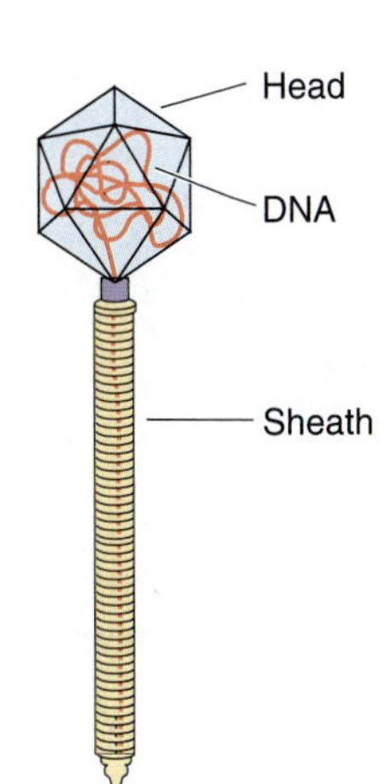

(b)

FIGURE 8.21 Phage structure. These electron micrographs and drawings show T4 phage (a) and lambda phage (b), both of which attack *E. coli*. Note the presence of a head and a tail, both composed of proteins. The head encloses the genetic material, and the tail acts as a mechanism for attachment to the host bacterial cell and provides a tube for injection of the DNA (or RNA) into the bacterial cell. *(a, Biozentrum/University of Basel/Science Photo Library/Photo Researchers, Inc.; b, M. Wurtz/Biozentrum/University of Basel/Science Photo Library/Photo Researchers, Inc.)*

lytic cycle and is characteristic of phage T4. This type of phage is called a **virulent phage** because it always kills the host.

But phage do not always kill their bacterial host cells. Instead, the phage genome may become integrated into the bacterial genome and be replicated as if they were part of the genetic information of the bacterial cell. This type of infection is called a **lysogenic infection.** When the phage is integrated into the bacterial genome, it is called a **prophage.** Lambda phage is an example of a phage that integrates into the *E. coli* genome and becomes a prophage. Both lytic and lysogenic cycles are diagrammed in Figure 8.22.

Some phage can alternate between a pathway of destroying the host cell or integrating into the bacterial genome. These are referred to as **temperate phage.** In temperate phage, the lysogenic state is not permanent. After integration of the phage into the genome, the prophage (integrated phage) can be disrupted and the phage DNA released. Release from the prophage state is induced by a variety of environmental challenges, including ultraviolet light. This results in a lytic response that generates new phage particles and destroys the host cell. Induction by ultraviolet light is likely an escape mechanism by the prophage because the host cell (the bacterium) may be killed by the UV irradiation. Thus, the phage starts replication and lysis to ensure its survival.

Mapping Phage Genes Has Proved Enormously Revealing.

It would seem that the genomes of bacteriophage might be difficult to map because phage are so small and seemingly do not have any recognizable phenotypes. In fact, the opposite is true. Even though phage can be seen only with the aid of an electron microscope, the number of phage particles can be easily counted by observing the effects on their hosts. Also, phage mutants affect their hosts in different ways, giving the phage many easily distinguishable phenotypes. Phage are counted and phenotypically assayed by use of a **plaque assay.** In this procedure, a small number of phage (100–500) are mixed with millions of bacterial cells, and the mixture is poured onto an agar surface in a Petri dish for growth of the bacteria and phage. The bacteria grow rapidly and form a "lawn" on the agar surface. Individual phage particles infect individual bacterial cells, multiply, and eventually

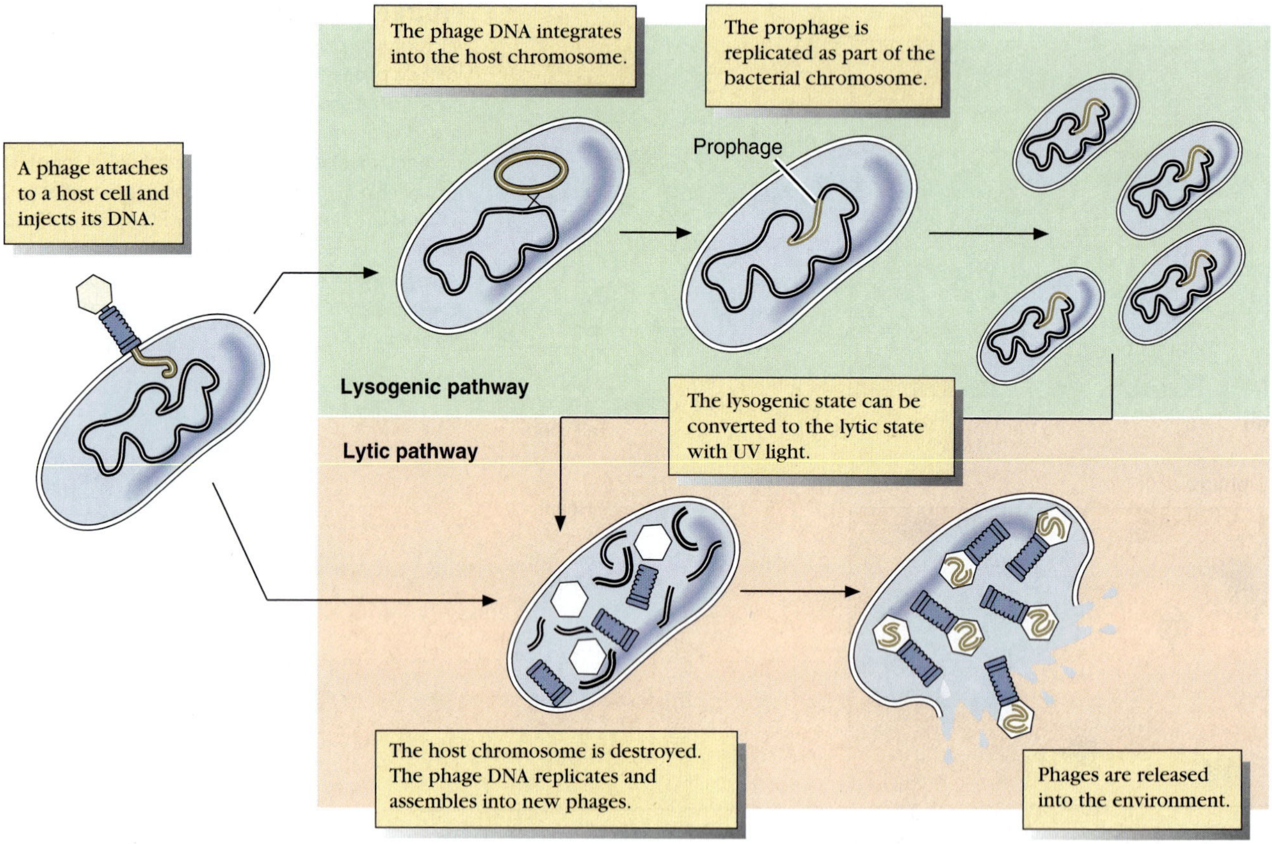

A phage attaches to a host cell and injects its DNA.

The phage DNA integrates into the host chromosome.

The prophage is replicated as part of the bacterial chromosome.

Prophage

Lysogenic pathway

Lytic pathway

The lysogenic state can be converted to the lytic state with UV light.

The host chromosome is destroyed. The phage DNA replicates and assembles into new phages.

Phages are released into the environment.

FIGURE 8.22 Phage infection of *E. coli*. This diagram shows the two alternate pathways for phage infection. If the phage integrates into the bacterial genome, a lysogenic cycle occurs, and the phage is replicated with the bacterial genome during growth. Alternatively, a lytic cycle can occur when the phage replicates and lyses the cell.

lyse the bacterial cells to release several hundred new phage particles. These, in turn, infect neighboring bacteria on the lawn, and the cycle repeats itself over and over again. Eventually, a hole forms in the bacterial lawn where all (or most) of the bacteria have lysed. This hole can be observed by the naked eye and is called a **plaque** (Figure 8.23). If the phage in the original mixture are dilute enough, each individual phage produces a plaque that can be counted and observed for phenotypic differences. Thus, one means of discriminating between different phage particles is plaque morphology. The number of **plaque-forming units** (pfu) can then be related to the number of phage in the original stock, by knowing the exact dilution. For example, assume that you diluted a phage stock by 10^6 and mixed 0.10 ml with bacteria and poured the mixture on the agar surface of a Petri dish. If 254 plaques are counted, then the phage stock has 2.54×10^9 pfu/ml [multiply the dilution of $10^6 \times 10$ (because you used 0.10 ml) $\times$ 254 to get the answer].

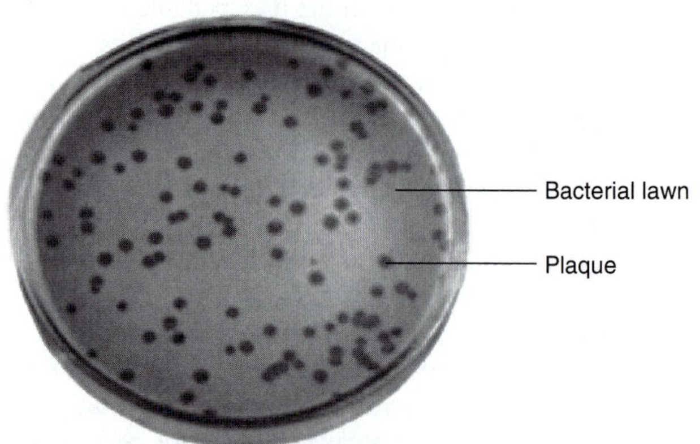

FIGURE 8.23 Phage plaques. Phage T2, when grown on a lawn of *E. coli,* infects a single bacterial cell, forming new phage particles that infect neighboring bacteria. Eventually a clear circular area, or plaque, forms representing lysed bacteria. One phage creates one plaque; thus the number of phage can be determined from the number of plaques. *(Stent, G. S. 1963.* Molecular Biology of Bacterial Viruses, *W. H. Freeman and Company, San Francisco.)*

Nearly all bacteria are subject to phage infection, but a specific phage usually attacks one species of bacteria. In fact, frequently a specific phage type will infect only one strain of a bacterial species. As a result, our environment is strewn with a bewildering array of different phage types that attack every imaginable type of bacteria. But, in scientific study, it is very common for scientists to focus all their attention on the study of only one organism. This is what happened when Max Delbrück started the science of phage genetics by focusing his studies on a small group of phage known as the **T-phage** (T1–T7, where T stands for type), which infect *E. coli*. Most of his studies were directed at a closely related group of the T-phage known as the T-even (T2, T4, and T6) phage. Subsequent studies by many other investigators have focused on these phage, allowing findings to be compared. This work has produced a detailed picture of how one phage type is genetically organized and how it grows. A modern genetic map of phage T4 is shown in Figure 8.24 and demonstrates the sophistication of one small reproducing entity. More recent analysis reveals the exact sequence of bases in the DNA of this phage and other phage.

Three types of phage mutants have proved extremely useful in genetic mapping. These are the host-range, plaque-morphology, and temperature-sensitive mutants.

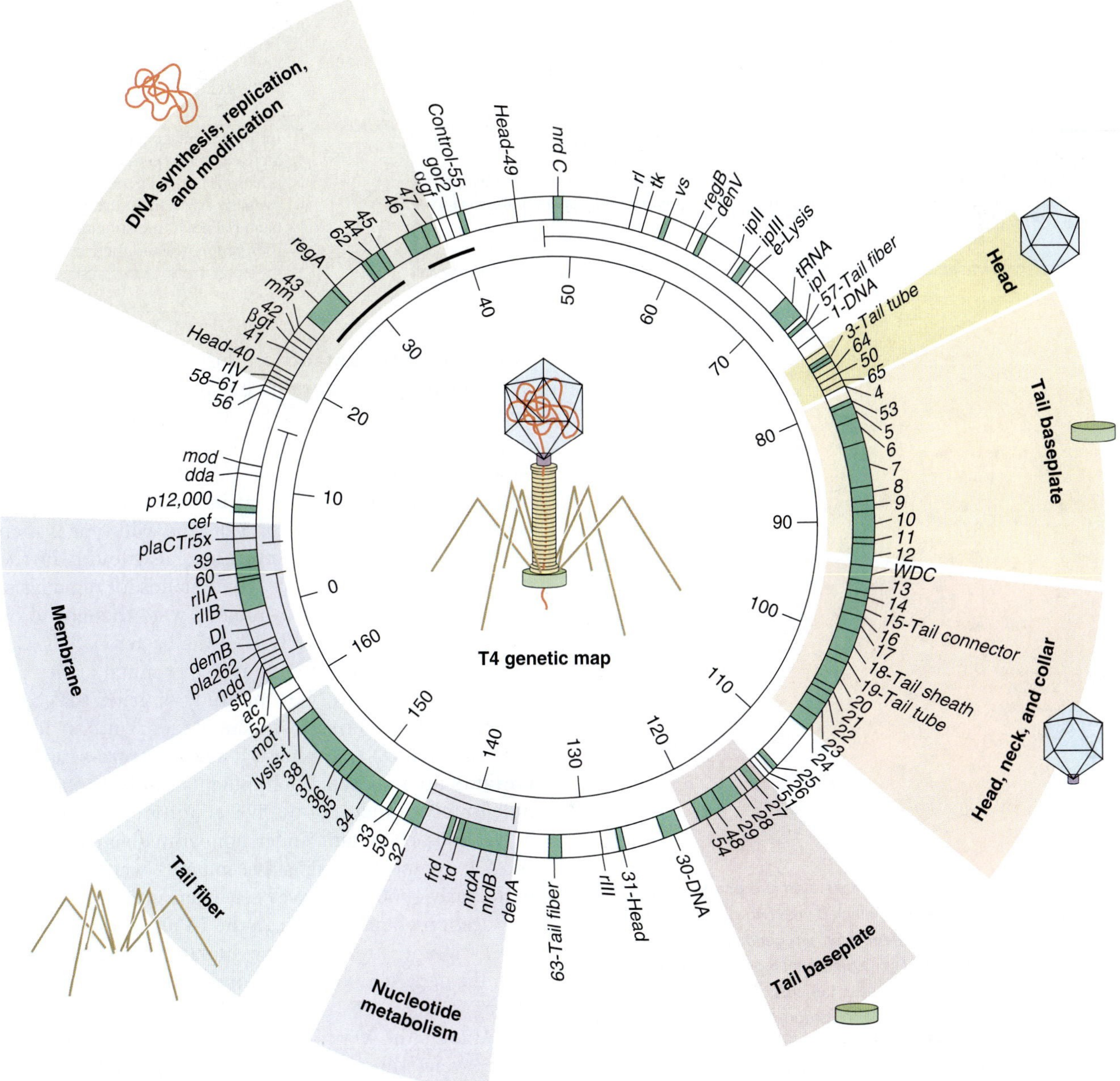

FIGURE 8.24 Phage T4 genetic map. Genes with related functions are clustered together on the phage map. The inside circle shows the map distance between the genes. Gene symbols are shown in the outer circle. Sometimes a name has not been assigned to a gene, so a number is assigned until a function has been determined for the gene product; then a name is assigned to the gene.

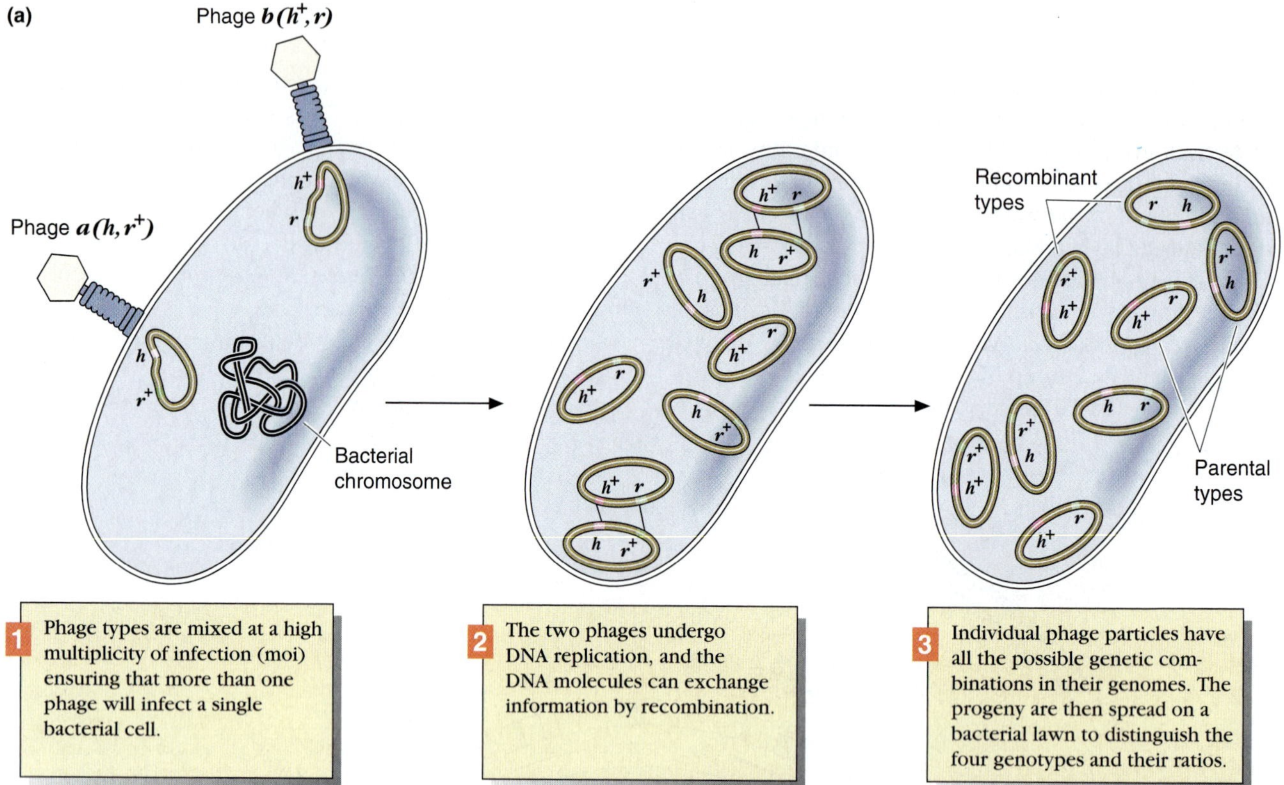

(a)
Phage $b(h^+, r)$
Phage $a(h, r^+)$
Bacterial chromosome
Recombinant types
Parental types
1 Phage types are mixed at a high multiplicity of infection (moi) ensuring that more than one phage will infect a single bacterial cell.
2 The two phages undergo DNA replication, and the DNA molecules can exchange information by recombination.
3 Individual phage particles have all the possible genetic combinations in their genomes. The progeny are then spread on a bacterial lawn to distinguish the four genotypes and their ratios.

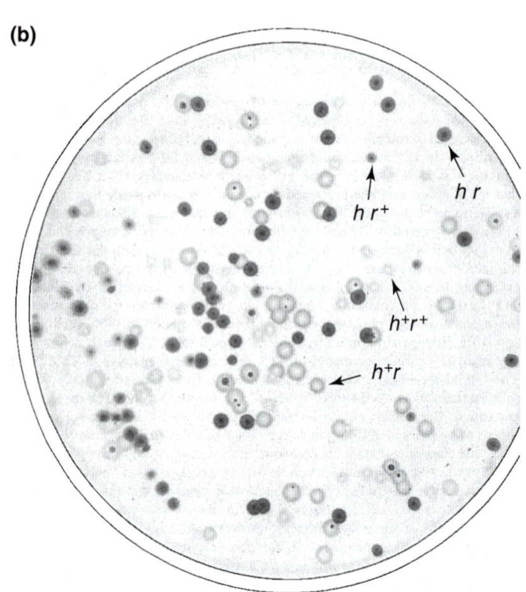

FIGURE 8.25 Recombination between two phage DNAs. (a) A bacterial cell is simultaneously infected with two different phages, allowing recombination between their two genomes. (b) This photograph shows the four different recombinant types that arose from the genetic cross shown in (a). After the cross was performed, the newly formed phage particles were spread on a mixture of the two different bacterial host strains (the permissive and nonpermissive strains). The four different kinds of plaques are visible. *(Stent, G. S. 1963.* Molecular Biology of Bacterial Viruses, *W. H. Freeman and Company, San Francisco.)*

The **host-range mutants** are just what the name implies; the phage can grow on one bacterial host but cannot grow on another, and thus the mutant is easily distinguished from the wild-type by the bacterial host that supports phage infection and growth. When host-range mutants were first discovered in the mid-1940s, they were designated with an *h* or T2*h*. The wild-type is then h^+. **Plaque-morphology mutants**, also found in the mid-1940s, are designated *r*, which stands for rapid lysis. The large plaques of the *r*-mutants were thought to be caused by the very rapid lysis of the bacterial cells. Because there is more than one type of *r*-mutant, they are designated *r*1, *r*2, etc. For the study of genes that code for essential functions (for example, enzymes in RNA, protein, and DNA synthesis), **temperature-sensitive mutants** are required. Such mutants can be propagated indefinitely under permissive conditions (growth at 30°C) and are mutant under nonpermissive conditions (40°C). Temperature-sensitive mutants were first used in the early 1960s and have been extensively used to map the location of many genes on the T4 phage map (Figure 8.24). Temperature-sensitive mutants are designated with the gene symbol with a superscript of ts, for example $regA^{ts}$.

Among the very first phage genetic crosses ever made were those between host-range and plaque-morphology mutants. To do a genetic cross between two genetically different phage types, single bacterial cells are infected with both phage particles simultaneously (this is done by increasing the ratio of phage to bacteria). This allows the DNA from both phage particles to occupy the same cytoplasm and permits a recombination event to occur. In a sample experiment, the following two T4 phage genetic types (the 7 in *r7* refers to a particular isolate of a rapid lysis mutant) were used to simultaneously infect a single cell:

$$h^+, r7 \times h, r7^+$$

If recombination occurred, the progeny from such a cross would give

$h^+, r7$	Parental type	Cloudy and large
$h, r7^+$	Parental type	Clear and small
$h, r7$	Recombinant type	Clear and large
$h^+, r7^+$	Recombinant type	Cloudy and small

This phage cross is schematically shown in Figure 8.25a. Because one of these mutations affects host-range, two different hosts are used (a mixture of the two make up the lawn of bacteria) for assaying the progeny of such a cross. The different phage types clearly distinguish the four different progeny arising from this genetic cross. These different plaque types are easily distinguished by examining the plates with the naked eye. The four different genotypes, with their four different phenotypes, are shown in Figure 8.25b.

Genetic crosses or recombination between different mutant phage types, as with the *h* and *r* mutants, have allowed construction of genetic maps of the T4 genome (Figure 8.24). These maps were constructed in much the same way that the maps in *Drosophila* and other organisms were constructed. The parental types are the most frequent events, and the recombinant types are the least frequent. The farther apart the two genes are on the phage chromosome, the more frequent the crossover types will be. For example, to calculate the recombination frequency in the preceding cross, the recombinant types are added and then divided by the total number of all plaques

$$\frac{\text{(Recombinant type 1)} + \text{(Recombinant type 2)}}{\text{Total plaques}} \times 100$$

more specifically,

$$\frac{(h^+\ r^+) + (h\ r)}{\text{Total plaques}} \times 100$$

This equation gives the percentage of recombinant types, and this is translated into map distance, where 1% = 1 m. u. Many crosses with different *r*-mutants revealed the surprising fact that the genetic map was continuous, that is, circular. Subsequently, it has been shown that nearly all prokaryotic organisms and viruses have circular genomes.

SECTION REVIEW PROBLEMS

15. What features distinguish a virulent phage from a temperate phage?
16. How do a phage and prophage differ?
17. You have received a stock of phage T4 from a friend and want to know the exact concentration of phage particles per milliliter (the titer). You dilute your phage stock 1×10^6 and mix 0.010 ml with *E. coli* in soft agar and pour the mixture onto the agar surface in Petri dishes. In a triplicate assay you get 83, 78, and 92 plaques per plate. What is the concentration of the original phage stock?

The *rII* Locus of Phage T4 Has Been Mapped in Great Detail.

Phage T4 also gives rise to plaque mutants that are large, small, fuzzy, clear, rapid lysis, and so forth, which have been extremely useful in the analysis of specific genes. In the early 1960s, Seymour Benzer reported on the detailed genetic analysis of one of the rapid lysis or *r*-mutants of phage T4, the ***rII* locus** that was first described by A. Hershey in the 1940s. A feature that made the study of the *rII* locus very attractive is that phage mutants in the *rII* locus cannot grow on *E. coli* K-12 when the bacterial chromosome contains an integrated phage lambda [designated *E. coli* K-12(λ)] but *can* grow on *E. coli* strain B, or on K-12 strains lacking the integrated lambda prophage. Thus, *rII* mutants can be propagated on one strain and tested for the presence of an *rII* mutation in another strain. In the following paragraphs, we shall describe the details of Benzer's experiments and how he applied them to understand new and interesting features of the *rII* locus and more generally for genes and gene mutations.

For a detailed analysis of the *rII* locus, Benzer needed a very large number of *rII* mutants. These were obtained by serially transferring plaques of mutagenized and nonmutagenized phage (with sterile toothpicks) simultaneously to specific locations on lawns of *E. coli* B and *E. coli* K-12(λ). As previously mentioned, *rII* locus mutants grow on *E. coli* B but not on *E. coli* K-12(λ) and thus can be easily identified as *rII* mutations. Using this approach, Benzer isolated over 3000 independent *rII* mutants of phage T4.

Two questions that Benzer wanted to address were, How many genes are in the *rII* locus, and where are his mutations located within the gene(s) with respect to each other? The first question was answered using a *cis–trans* or complementation test, similar to that discussed in Chapter 6. Complementation tests with many different mutants will eventually establish whether the locus produces one, two, or more gene products. When doing the complementation test, a bacterial cell must have two different phage genomes present at the same

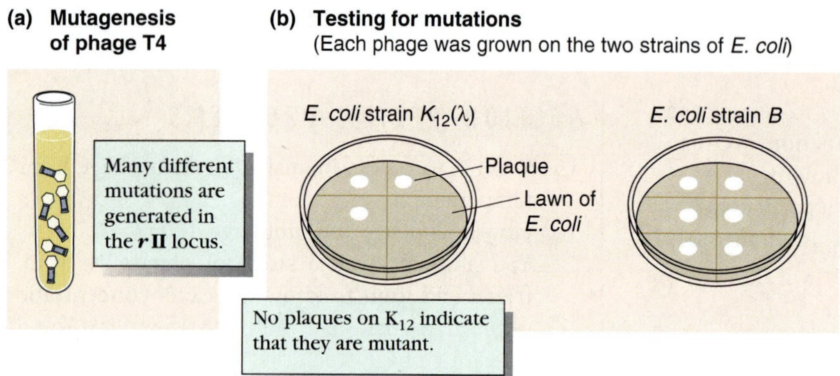

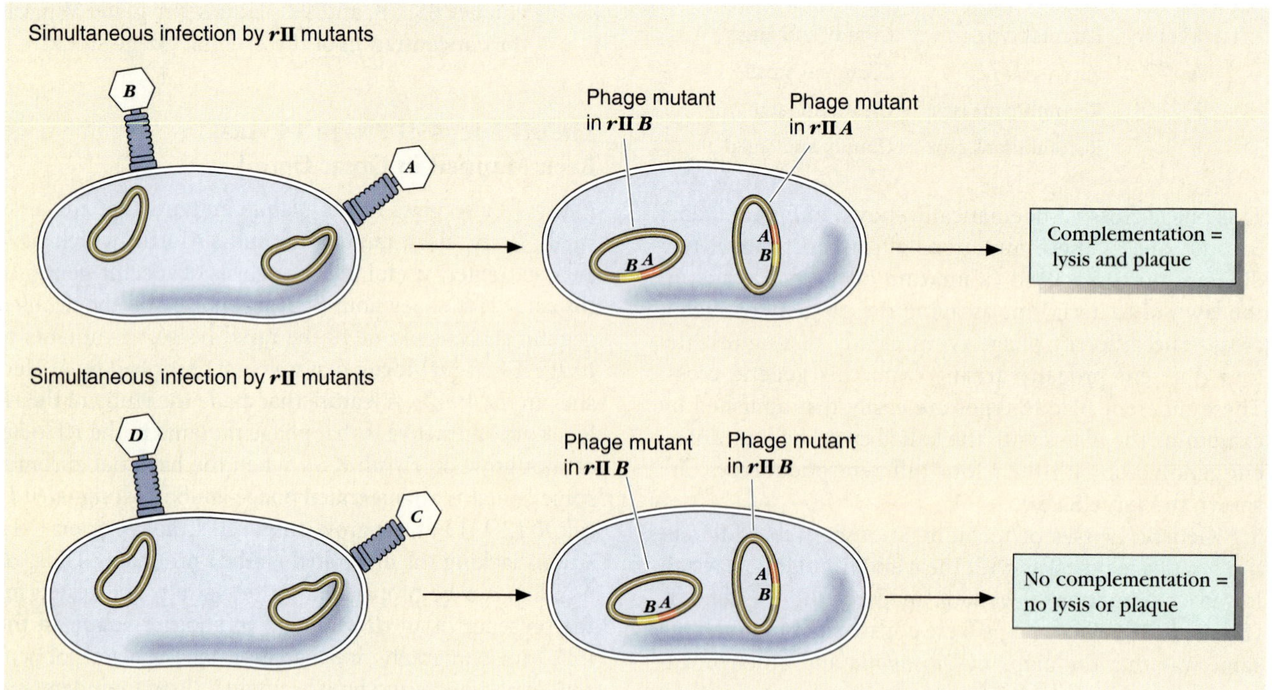

FIGURE 8.26 Genetic analysis of the *rII* locus. This schematic drawing shows how Benzer demonstrated that the *rII* locus was composed of two cistrons (genes). Using the complementation test, he was able to define whether each mutation was in the A or B cistron.

time, each with a different mutation in the *rII* locus. This is accomplished by simultaneous infection with two different phage particles. If complementation takes place, the different mutations are in separate genes. Otherwise, if two different mutations are in the same gene (on the two phage chromosomes), no functional gene product will be made in the *trans* test, and a mutant phenotype is observed. For many of Benzer's *rII* mutants, plaques were observed at a high frequency upon simultaneous infection with some phage mutant pairs of *E. coli* K-12(λ), but no plaques were observed upon simultaneous infection with other pairs. This is illustrated in Figure 8.26. From the many different complementation tests, Benzer concluded that the *rII* locus was composed of two different cistrons or genes (in fact, he coined the term cistron from the use of the *cis–trans* test).

Genetic mapping studies were necessary to localize many of the independently isolated mutations within these two cistrons. Benzer used an elegant method for mapping that required relatively little work, considering the task. Benzer first co-infected *E. coli* B with two different *rII* mutant phage and measured the frequency of recombination from the progeny. Only recombinants will grow on *E. coli* strain K-12(λ) because *rII* mutants will not grow on this strain. If recombination occurred between the two mutations, it would generate a wild-type phage genome, and a plaque would appear on *E. coli* K-12(λ). (Note that the frequency of plaque formation, or recombination, was much lower than complementation and is related to the distance between the two mutations.) Knowing the frequency of recombination between two markers enabled him to determine the genetic distance between the two mutations. Extremely rare crossover events (between very closely linked markers) were detected—it was easier to see a wild-type plaque on a clear lawn of *E. coli* K-12(λ) cells than to pick one out of a sea of mutant plaques. In fact, this technique is so powerful, recombination could be detected between adjacent bases in the DNA. In reality, the lowest recombination event detected was 0.01% or 1 in 10,000 (the limit of the technique was 1 in 1 million). In some of the crosses, Benzer could not detect *any* recombination and concluded that the independently derived mutations were within 3 base pairs of each other (as we will learn in Chapter 11, the code word for one amino acid in a protein is 3 base pairs). We now know that the genetic map of phage T4 is 1500 m. u., where 1 m. u. equals 1% recombination. Recombination of 0.01% really represents 0.02% because the reciprocal events were undetected. The T4 phage genome contains approximately 165,000 base pairs, so 0.02% recombination represents 0.02/1500 m. u. or about 1.3×10^{-5} of the total phage genome. Thus, if you assume that the frequency of recombination is constant throughout the genome (which it probably is not), then the lowest percent recombination observed was about 1.2 base pairs. We know that this is not possible, so Benzer probably observed some recombination events separated by two pairs, although some events were also between adjacent base pairs.

One other important concept came from these studies, aside from the location of the individual mutations. If you remember from Chapter 6, using *Drosophila* it was established that recombination could be obtained between alleles within a gene. Likewise, Benzer established that this also occurred in phage T4.

Deletion Mapping Allows Rapid Location of Mutational Events.

Using the technique of genetic crosses, Benzer localized the relative positions of many mutations. But, to locate all the mutations with respect to each other, Benzer would have had to cross every mutant with every other mutant. With 3000 different mutants, this would be 4,498,500 crosses or $(N - 1)(N/2)$ where N = the number of crosses. Even the most dedicated geneticist might be daunted by this task. Fortunately for genetics and for Benzer, all the mutations that he found were not caused by just single-base changes. Some were the result of deletions of large segments of the *rII* gene, and this finding facilitated the generation of a precise genetic map of all of the 3000 mutants that he had accumulated. In total, Benzer found 47 different deletions. To accurately use the deletions, Benzer first ordered the deletions on the map, locating the exact ends. He then used these deletions to do an initial screening of each mutant. This general technique is referred to as **deletion mapping,** which we discussed in Chapter 6 using eukaryotes. The deletions that Benzer located to the *rII* locus are shown in Figure 8.27.

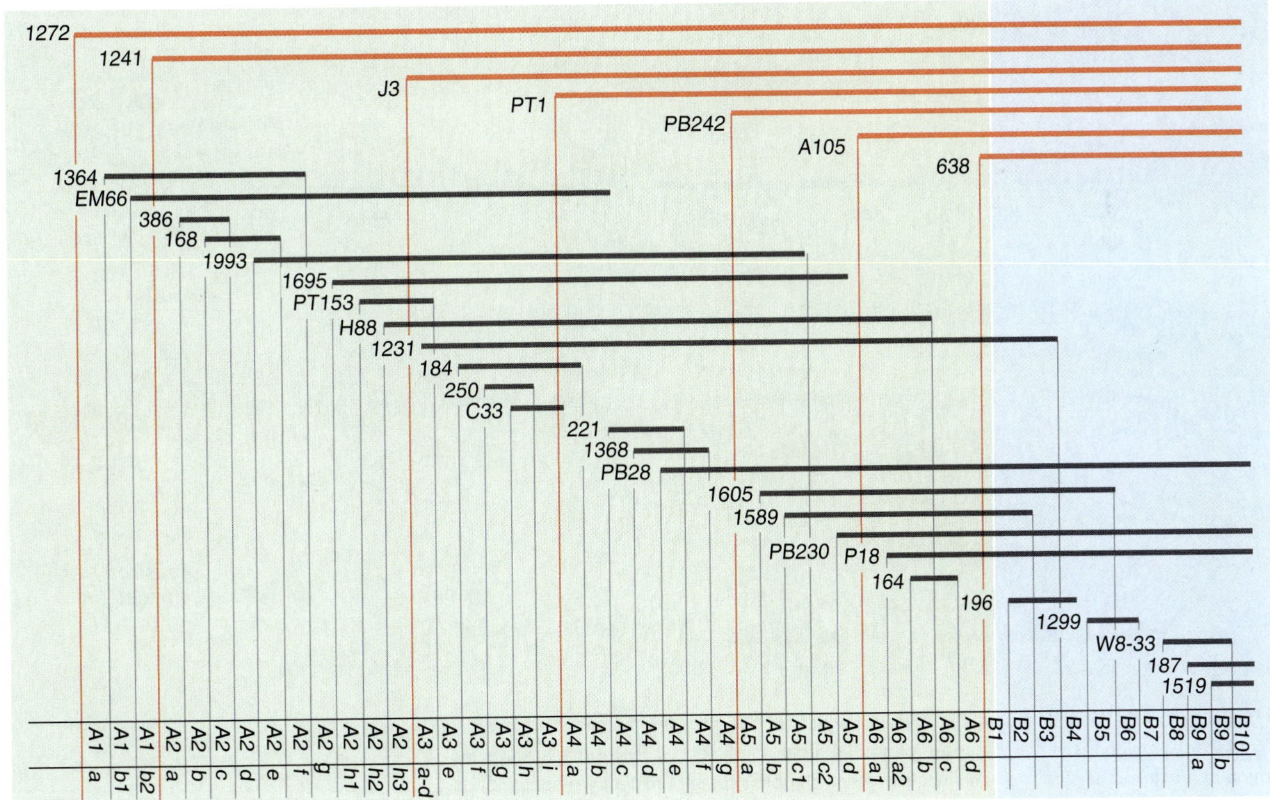

FIGURE 8.27 Deletions of the T4 *rII* locus. This drawing shows 32 overlapping deletions identified by Benzer and coworkers. The deletions divide the *rII* locus into 47 segments. The designation of the segments are given in the boxes at the bottom. The limits of the A and B cistrons are designated by the two colors. *(Adapted from Benzer, S. 1961.* Proceedings of the National Academy of Sciences USA, *47: 410.)*

Recognizing that a deletion represents a loss of DNA, Benzer crossed each of his point mutations with the 47 deletion mutants to localize the point mutation to a specific segment of the map. During the cross between a point mutation and any of the deletions, if the deletion encompassed the same area of the genome as the point mutation, no recombination is observed (no wild-type progeny grow). But, if the point mutation and the deletion *did not* encompass the same area, wild-type progeny are recovered.

Actually, Benzer first crossed each new point mutation with each of seven large deletions (shown in red in Figure 8.27), allowing a rough location of the mutation. He then did crosses with deletions covering subregions of the larger deletion to more precisely locate the mutation. Finally, a third cross was done with a closely linked point mutation, allowing a precise location of the mutation. With this approach, he quickly mapped all 3000 mutations, creating a detailed picture of the *rII* locus.

Presented in Figure 8.28 are the locations of 1612 of Benzer's spontaneous mutations, precisely located within the *rII* cistrons. We can now answer the second question concerning the *rII* locus, where are each of the 3000 mutations, and are some of them same-site events? From the map, it is apparent that the mutations are not equally distributed along the locus. Some sites are referred to as hot spots, where mutations occur over and over again, whereas other areas do not have any mutations. In one spot in the *rIIB* cistron, over 500 mutations occurred. We now know that hot spots are commonly found in many

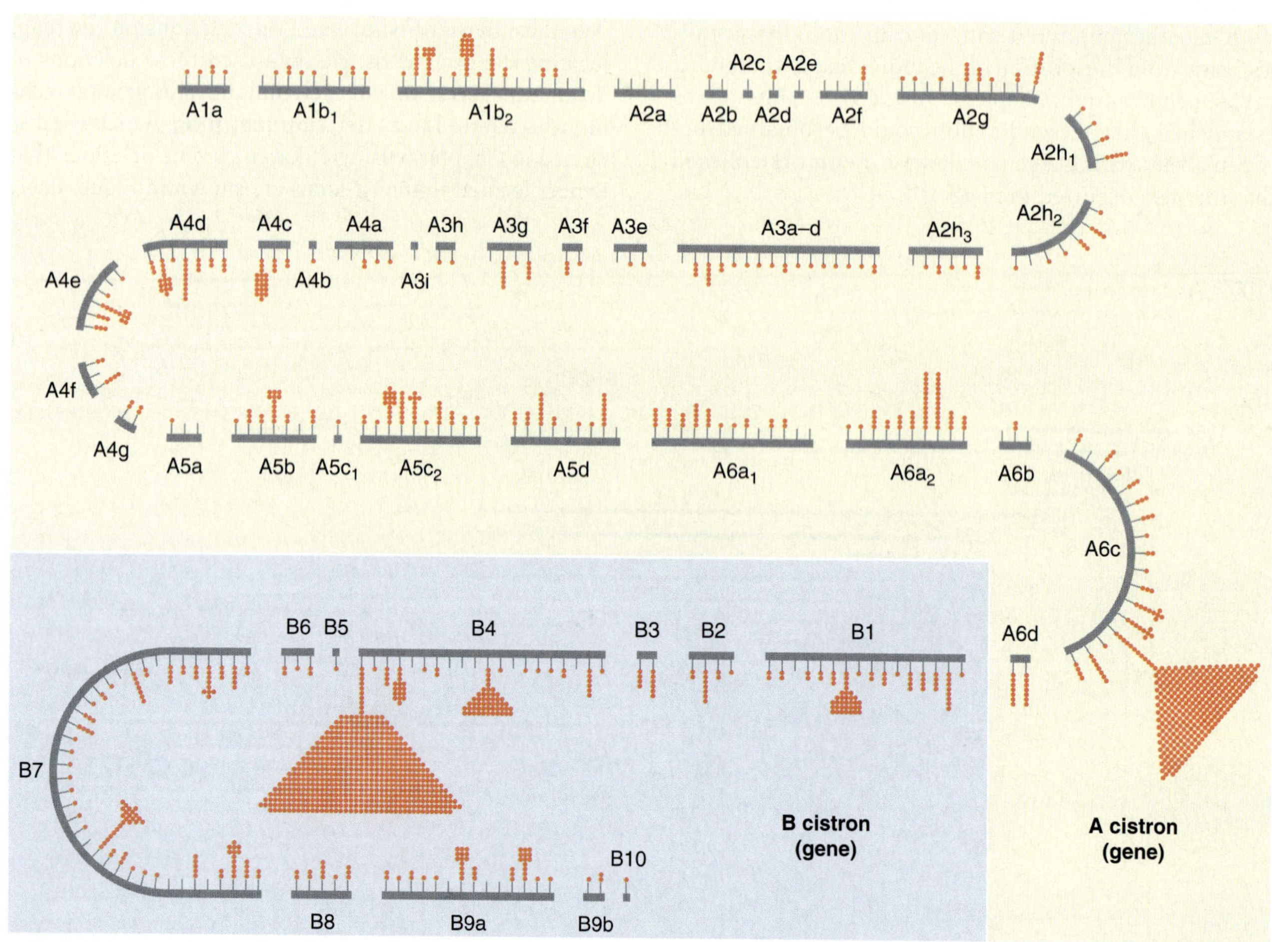

FIGURE 8.28 Genetic map in the *rII* locus. S. Benzer and his coworkers located these spontaneous mutations in the *rII* locus of phage T4. Each small square represents a single mutation at that site. The small vertical lines represent mutable sites separable by recombination. The discontinuity in the map represents the segments identified by each of the 32 deletion mutants (Figure 8.27). It should be noted that several areas of the locus have what are known as "hot spots." At these locations mutations occur over and over again. In contrast, some areas of the genes had no identifiable mutations, either spontaneous or induced. *(Adapted from Benzer, S. 1961.* Proceedings of National Academy of Sciences USA, *47: 410.)*

other genes. Hot spots are not always at the same location but depend upon whether they are induced by different mutagens or are spontaneous mutations. Why one spot is more susceptible to mutation than another is only partially understood. The sequence of bases in the DNA of the entire *rIIA* and *rIIB* cistrons is now known, and the base pairs at the hot spots can be examined. What is found is that many hot spots (for example, at site 131 and 117 in the *rIIB*) consist of stretches of five or six consecutive adenine : thymine base pairs (see Chapter 9 for more information on bases and base pairing). The frequent mutations that arise from these areas are thought to be caused by slippage along the base pairs when they are traversed by the replication apparatus. Another reason for the existence of a hot spot is the loss of an amino group from 5-methylcytosine creating thymine in its place (see Chapter 17 for details as to how this change can cause a mutation).

SECTION REVIEW PROBLEMS

18. You have identified 567 independent mutants of phage T4 and wish to map all the mutant loci. How many crosses would you have to do to cross every mutant with every other mutant?
19. Explain how the use of a series of deletions covering various areas of the loci to be mapped would speed the process of mapping the large number of mutations described in Problem 18.

Transduction Is Useful for Mapping Closely Linked Genes.

In the preceding section the life cycle and growth of bacteriophage were discussed. This section will discuss transduction, which uses the phage particle as a carrier of genetic information from one bacterial cell to another. As with conjugation and transformation, transduction confers upon bacteria a means of genetic transfer and facilitates their survival in an uncertain environment. The following discussion on transduction is limited to *E. coli*, but it occurs in a wide range of bacteria.

Transduction Can Be Used for Genetic Mapping.

Transduction requires the participation of a phage to transfer bacterial DNA from one bacterial cell to another. During the packaging of DNA in the phage growth cycle inside the bacterial cell, bacterial host DNA is sometimes packaged inside the coat protein instead of the phage DNA. This produces a phage particle that is unable to reproduce in the next infection cycle (Figure 8.29), but it allows the transfer of host DNA to the next infected bacterial cell. This general phenomenon is called transduction. Accidental packaging of bacterial DNA is a relatively rare occurrence, but because of the small size of phage and the large numbers of phage particles generated, it happens frequently enough to be useful in the transfer of genetic information from one bacterial cell to another. The phage particles that contain bacterial DNA still retain their ability to infect other bacterial cells. But, because the injected DNA does not contain any phage genetic information, the infected cells do not undergo lysis. Instead, the bacterial DNA takes one of two routes after it is injected into a new cell: abortive transduction or complete transduction. In an **abortive transduction,** which occurs 90–95% of the time, the foreign bacterial DNA is neither replicated nor integrated into the host cell DNA and is eventually lost. In **complete transduction,** the foreign bacterial DNA molecule is integrated into the host by two crossover events replacing the original copy and giving rise to a new genotype.

In a typical transduction experiment, a phage infects the donor bacterial cell (for example, an *argA*$^+$ host) and random fragments of the donor DNA are accidentally packaged into the phage heads, forming new phage particles. By chance, some of these fragments will contain the gene for *argA*$^+$. For example, the lytic *E. coli* phage P1 has a genome of 94,000 base pairs, which corresponds to about 2 minutes of the *E. coli* genetic map. The size of the P1 genome represents 2% of the bacterial genome, so 1 in 50 of the accidentally packaged phage particles may contain the *argA*$^+$ gene. These phage particles are then used to infect a recipient strain, say an *argA* strain. The phage particle containing *bacterial* DNA will not destroy the new bacterial cells because no phage DNA is present. However, most of the recipient bacterial cells are destroyed through the normal phage infection cycle. The surviving bacterial cells are then spread on agar lacking arginine, and the only cells that grow are transductants that have received the *argA*$^+$ gene from the donor strain via the phage particle. The frequency of transduction for a specific gene is 1 in 10^5 to 10^6 bacterial cells infected. The mutation rate of *argA* to *argA*$^+$ is less than 1 in 10^8; thus the two events can usually be easily distinguished from each other.

As mentioned before, conjugation is very useful in mapping distantly located genes but not helpful in precisely locating closely placed genes. Thus, transduction is most useful for mapping two or more genes that are closely linked. Transduction is especially useful in determining gene order of three linked genes. The correct sequence of three genes can be established by performing three-factor crosses between a donor strain and a recipient strain.

In a three-factor analysis, a cross is set up to determine accurately the map distance between each of three genes. Relative map distance can be determined from the

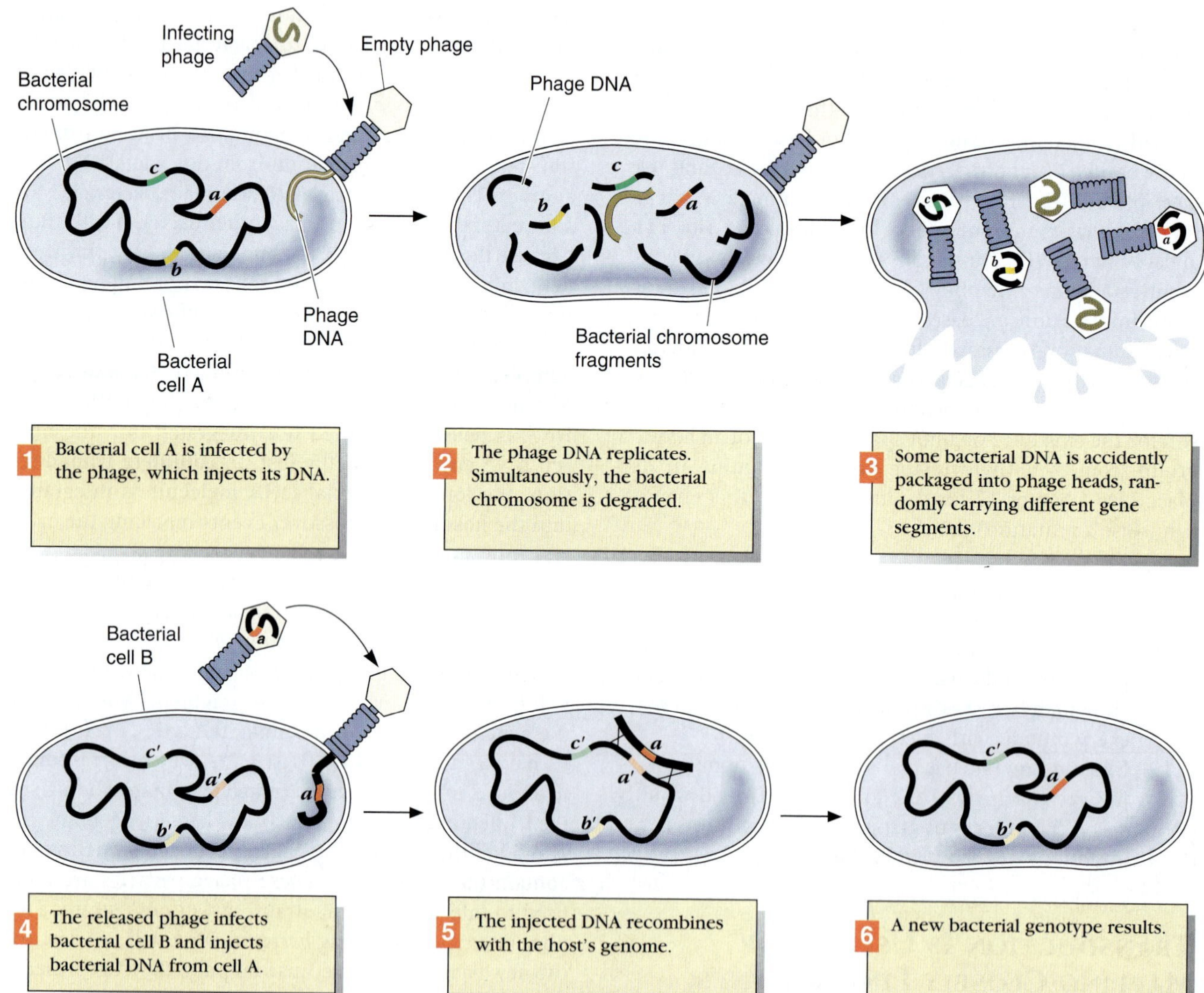
Infecting phage
Empty phage
Bacterial chromosome
Phage DNA
Phage DNA
Bacterial cell A
Bacterial chromosome fragments
1 Bacterial cell A is infected by the phage, which injects its DNA.
2 The phage DNA replicates. Simultaneously, the bacterial chromosome is degraded.
3 Some bacterial DNA is accidently packaged into phage heads, randomly carrying different gene segments.
Bacterial cell B
4 The released phage infects bacterial cell B and injects bacterial DNA from cell A.
5 The injected DNA recombines with the host's genome.
6 A new bacterial genotype results.

FIGURE 8.29 The formation of a transducing phage. After infection of the bacterial cell by the phage, a fragment of the bacterial chromosome is accidentally encapsulated in a phage head. This phage can now infect another bacterial cell and transfers bacterial DNA instead of phage DNA. This fragment can integrate into the new host's DNA by recombination.

co-transduction frequency of two markers. For example, *ara*C, *leu*B, and *ilv*H are located at minutes 1.0, 1.8, and 1.9 on the *E. coli* genetic map. If the gene order and map distance between these three genes are to be established by transduction, a recipient bacterial strain that possesses mutations in all three genes is needed. The transducing phage stock is prepared on a bacterial strain wild-type for the genes, and the wild-type markers are then transduced into the triple-mutant strain. If the recipient strain were *ara*C, *leu*B, and *ilv*H, the transduction would be accomplished by selecting for one gene marker at a time (for example, *ilv*H$^+$). The recipient strain is spread on a medium lacking isoleucine / valine (the gene governs one step in the pathway of the synthesis of these two related amino acids), and only *ilv*H$^+$ transductants grow. Isolated individual bacterial cells grow, forming colonies on the surface of the medium, and are then tested for co-transduction of the *leu*B$^+$ gene and the *ara*C$^+$ gene simultaneously with the selected *ilv*H$^+$ gene. This is done by individually transferring colonies to test media to determine the presence or absence of the active genes [for example, on arabinose-supplemented media (for *ara*C$^+$)] to determine whether growth occurs. The map distance in *E. coli* is calculated from the formula

$$\text{Frequency of co-transduction} = (1 - d/L)^3$$

where d is the distance between markers in minutes and L is the length of the transducing fragment in minutes. The frequency of co-transduction is given as a fraction of 1.0, where 1 equals 100% co-transduction. L is usually given a value of 2.0 minutes for the generalized transducing phage P1 (its genome is nearly exactly 2% of the size of the *E. coli* genome). In this experiment, the co-transduction frequency between *ilv*H and *leu*B will be very high (about 85%) because they are 0.1 minute apart, whereas the co-transduction frequency between *ilv*H and *ara*C will be low (about 10%) because they are relatively far apart. After doing this experiment, we find the results presented in Figure 8.30.

From these data, the gene order and the co-transduction frequency (and thus the map distance) between these three gene markers can be established. The gene order is found by determining which of these genotypes can arise by just two crossover events because these will be the most frequent events. Any genotype arising from four crossovers will be a rare event. By inspection, it would appear that the *leu* gene is in the middle because a recombination event yielding a ilv^+, *leu*, ara^+, which is the rarest genotype, requires four crossover events (see Figure 8.30). The map distance can be calculated using the co-transduction frequencies. For example, the co-transduction frequency between *ilv* and *ara* is about 16%; therefore, from the formula,

$$0.16 = (1 - d/2)^3$$

or, rearranging the equation,

$$d = 2(1 - \sqrt[3]{0.16})$$

Solving for *d* gives a map distance of 0.92 m. u. between *ilv*H and *ara*C. Using mapping data from many transduction experiments, geneticists have constructed a very detailed map of the *E. coli* genome.

SOLVED PROBLEM

Problem

Three genetic markers in *E. coli* are to be mapped using phage P1, a generalized transducing phage. The donor strain is $araE^+$, $galR^+$, *gly*U, and the recipient strain is *ara*E, *gal*R, $glyU^+$. After the transduction, $araE^+$ was selected for (on a medium lacking arabinose but possessing galactose and glycine) and 100

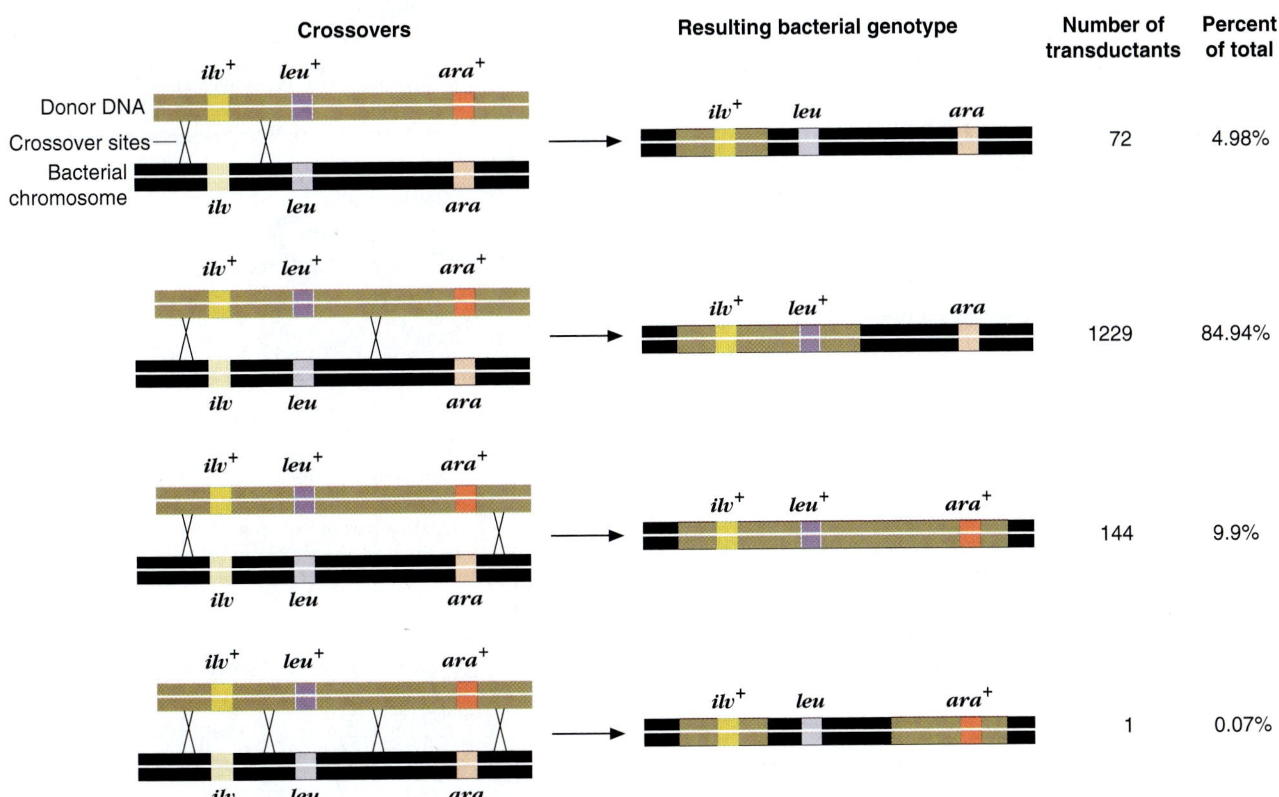

FIGURE 8.30 **Mapping by transduction.** Gene order and map distances can be determined using transduction. Recombination occurs more frequently between closely linked markers, and double crossovers are the least frequent. The actual number of individuals in each recombinant class is shown with these values converted to percentages. Map distances can be converted into base pairs because we know that one map unit on the *E. coli* genome represents approximately 48,000 base pairs.

transductants were tested for the other genes. The results follow (all are *ara*E$^+$).

Genotype	Number of Colonies
*gal*R$^+$, *gly*U$^+$	6
*gal*R$^+$, *gly*U	20
*gal*R, *gly*U$^+$	48
*gal*R, *gly*U	26
Total colonies	100

What is the co-transduction frequency between *ara*E and *gal*R and between *ara*E and *gly*U? Which is closer to *ara*E, *gly*U, or *gal*R?

Solution

The co-transduction frequency can be found by adding the number of colonies that co-transduce with the selected marker: for example, *gal*R$^+$, *gly*U$^+$, and *gal*R$^+$, *gly*U represent all the *gal*R$^+$ co-transductants with the selected marker *ara*E$^+$. Thus, the co-transduction of *ara*E and *gal*R$^+$ is (20 + 6)/100 = 26%. Likewise, the co-transduction frequency of *gly*U with araE$^+$ is (20 + 26)/100 = 46%. Because the co-transduction frequency between *gly*U and *ara*E$^+$ is greatest (46% vs. 26% for *ara*E$^+$ and *gal*R), *ara*E is closer to *gly*U than to *gal*R.

SECTION REVIEW PROBLEMS

20. Outline the steps involved in a phage infection, and indicate at what step the "accident" occurs that permits transduction.
21. Why isn't the recipient bacterial cell in the transduction process lysed upon infection?

Specialized Transduction Transmits Specific Genes from Cell to Cell.

In addition to generalized transducing phage such as P1 for *E. coli* and P22 for *Salmonella typhimurium*, which randomly carry small sections of the chromosome, there are also specialized transducing phage that carry only specific bacterial genes. The *E. coli* bacteriophage lambda (λ) is an example of a specialized transducing phage, and it will occasionally carry sections of the bacterial chromosome with it during excision from the prophage state. To understand exactly how lambda is capable of specialized transduction, it is first necessary to examine how phage lambda integrates into the bacterial chromosome and occasionally excises imprecisely, producing a defective phage particle containing a DNA fragment from the bacterial genome (Figure 8.31).

Lambda is a temperate phage that can be maintained in the prophage state in the bacterial chromosome. For

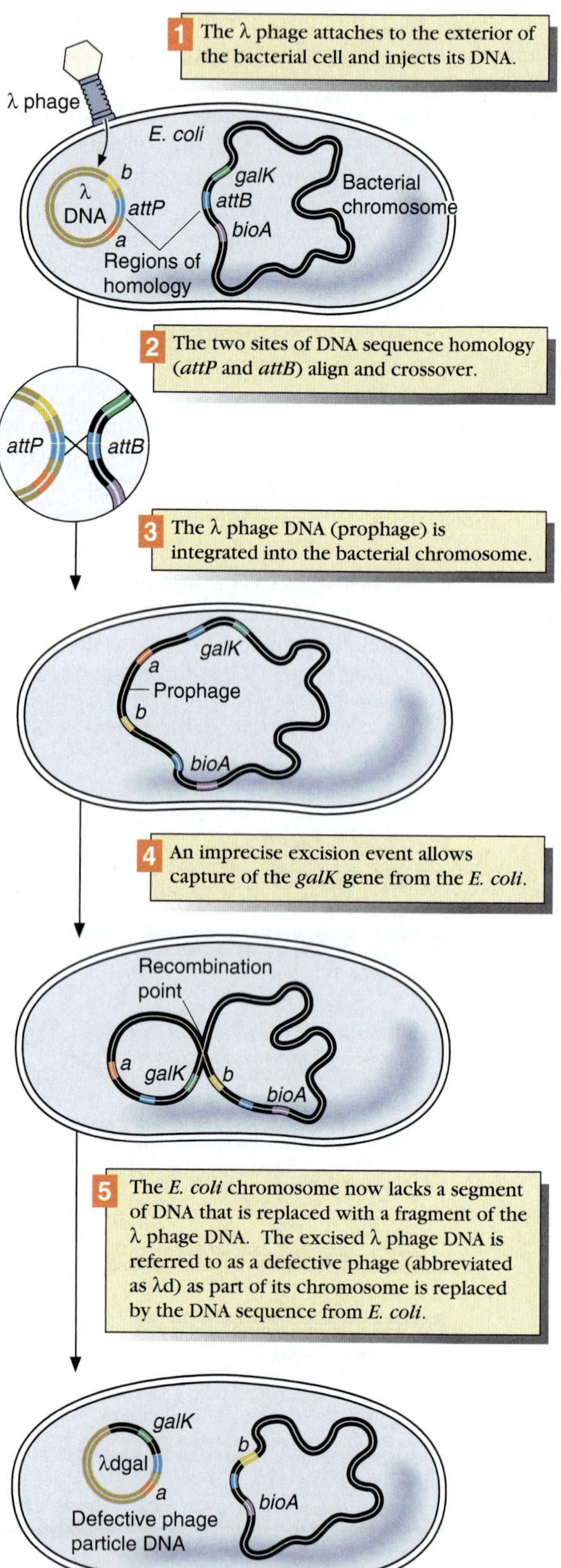

FIGURE 8.31 Genomic integration and excision of lambda phage. This shows how phage lambda can integrate into the *E. coli* chromosome and occasionally excise imprecisely. Note that the imprecise excision results in loss of some of the phage genome.

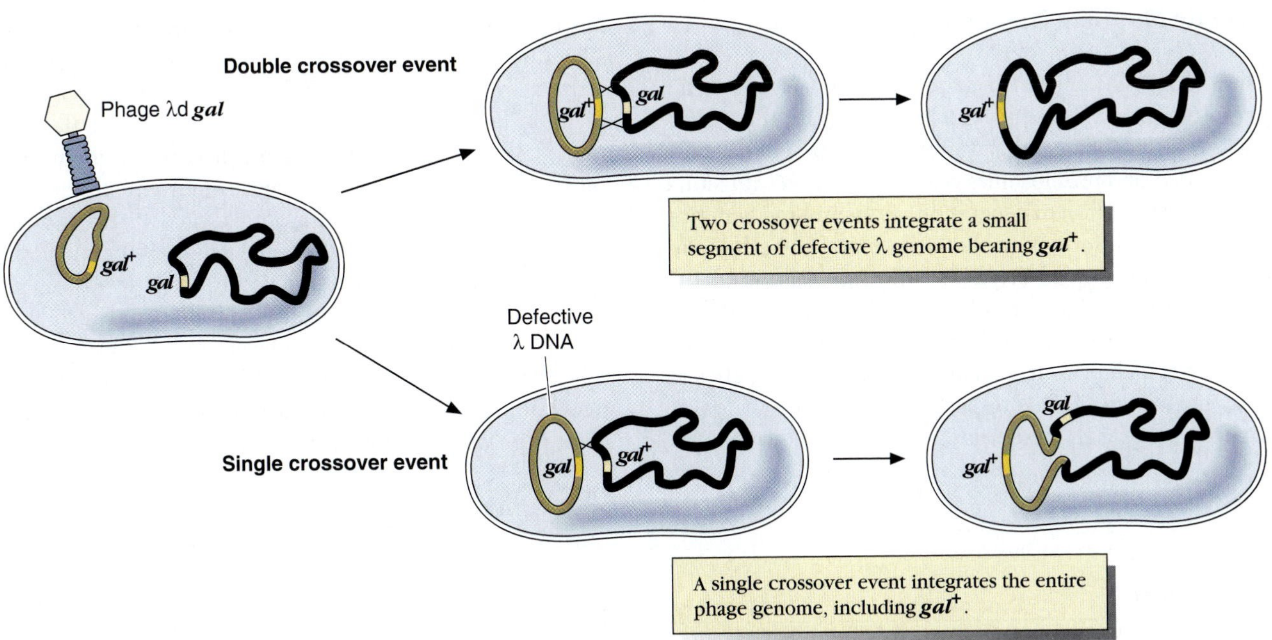

FIGURE 8.32 Integration of lambda into the bacterial chromosome. These steps show how lambda (d*gal*) can integrate into the bacterial genome when it carries a gene from the bacteria. The phage DNA carrying the bacterial DNA sequence can integrate two different ways, depending upon the number of recombination sites. Progeny from both pathways yield a *gal*$^+$ phenotype.

integration to occur, regions of the phage and the *E. coli* DNA align (the regions are referred to as *att*B in *E. coli* and *att*P in lambda), followed by a single crossover event to integrate the lambda chromosome into the bacterial chromosome (for details of the molecular mechanism, see Chapter 19). The integrated phage is now in the prophage state. The region on the *E. coli* genetic map where lambda phage normally integrates is at minute 17.2. As a consequence, lambda only transduces genes adjacent to its integration site, explaining why lambda is a **specialized transducing phage.** Only certain strains of *E. coli* allow integration of the lambda prophage, specifically a strain commonly used in research laboratories called K-12. When the K-12 strain is in the lysogenic state, it is designated *E. coli* K-12(λ).

When the lytic cycle, or excision cycle, is initiated, the sequence of events is a reversal of the integration events, and the phage chromosome reforms the exact lambda circular chromosome by a single crossover event at the *att*B / *att*P site. In rare cases, the excision event is not precise, and crossing over occurs between the lambda DNA and the bacterial DNA. This leaves a segment of the lambda DNA in the bacterial chromosome and removes a fragment of the bacterial DNA, placing it into the lambda DNA. In the example presented in Figure 8.31, the lambda *b* gene region is left in the bacterial chromosome, whereas the bacterial *gal* gene region is transferred to the lambda chromosome. This rare phage particle is present in a mixture of normal phage particles at a frequency of about 1×10^5. However, it can still be used to infect bacteria to insert the *gal*$^+$ gene into a recipient *gal* strain.

The integration of a defective phage particle can take several pathways, as outlined in Figure 8.32. After the phage particle injects its genome into the bacterial cell, the phage DNA can be integrated into the recipient bacterial chromosome by one crossover event, or the homologous bacterial sequence can be replaced by the phage DNA sequence with two simultaneous crossover events. Both pathways yield a *gal*$^+$ (in a *gal* cell) phenotype.

SECTION REVIEW PROBLEMS

22. In *E. coli*, lambda phage is an excellent transducing phage but is usually restricted to transducing those bacterial genes surrounding minute 17 on the *E. coli* genetic map. Why is lambda transduction restricted to the area surrounding minute 17?
23. After filtering some sewage through a very fine membrane that excludes all organisms larger than 1 μm in diameter, you find a new phage that lyses *E. coli*. In testing this phage for transducing ability, you find that it transduces only one genetic marker from your test strain (*art*) and not any of 45 other scattered markers tested. What tentative conclusion might you make? How would you prove or disprove this hypothesis?

Genetic Mapping Has Its Limitations.

The study of inheritance patterns along with genetic mapping with many different kinds of gene mutations has revealed considerable information about where genes are located on chromosomes and how they are transmitted from generation to generation. Transmission genetics can also yield information about the genetic length of the gene, but this approach yields little information about the structure and nature of genes. To understand just how a gene functions at the molecular level requires another approach of genetic analysis. Molecular genetics is extremely useful in this respect and came into wide use for gene analysis in the 1970s. Molecular genetic analysis has revealed considerable information about the detailed structure of genes, repeated sequences in the DNA, DNA sequences between genes, and expression of genes. Some of these findings were unsuspected until molecular genetics exposed them; for example, it is now known that most genes in higher organisms are not constructed in the same manner as genes in prokaryotes. Before molecular genetics, there was no reason to suspect that this was the case. Thus, we are at a junction in the study of genetics in this textbook. The following chapters make the transition to molecular genetic techniques to understand the exact nature of the gene and indeed extend our understanding of gene structure and expression far beyond what is possible with transmission genetics alone.

Summary

In the mid-1940s, it was discovered that bacteria could transfer genetic information from one cell to another by several different mechanisms. One method, conjugation, requires physical contact between bacterial cells. Cells possessing an F-plasmid are known as F^+ cells and can transfer the plasmid to F^- cells that do not possess an F-plasmid. If the F-plasmid is integrated into the genome by a single recombination event, it becomes an *Hfr* strain and can transfer the bacterial chromosome into an F^- cell. During conjugation, the chromosome is transferred from the donor cell to the recipient cell and incorporated into the recipient's genome by recombination. Using conjugation, it is possible to determine the order and spatial relationship of genes that are distantly removed from each other. One significant finding that arose from conjugation experiments is that the *E. coli* genome is circular.

A second mechanism for transfer of genetic information from one bacterial cell to another is transformation. In transformation, single-stranded DNA enters the cell through the cell membrane from the environment and is then incorporated into the recipient cell's DNA by recombination. Transformation is especially useful for incorporating DNA into a bacterial cell but has limited usefulness in genetic mapping.

For closely linked genes, map positions can be precisely determined using transduction. In transduction, a segment of the bacterial host DNA is packaged into a phage head instead of the phage's DNA. Upon release of this phage from the first host and infection of a second host, this bacterial DNA is injected into the second bacterial cell whereupon it can undergo recombination to incorporate new genetic information into the second host genome. However, only a limited amount of bacterial DNA, representing about 2% of the bacterial genome, can be packaged into a phage head. Thus, genetic markers that are closely linked can be precisely located in relation to each other by determining the co-transduction frequency of the two markers.

Bacteriophage reproduce by attaching to a bacterial cell, injecting their DNA (or RNA), synthesizing new DNA and phage proteins, encapsulating the DNA into a complete phage particle, and releasing new particles from the lysed bacterial cell to repeat the cycle. The genomes of bacteriophage can also be mapped by genetic crosses. In mapping, two phage of different genotypes simultaneously infect a single bacterial cell. The two different genomes can then undergo recombination, generating new phage types. Using this approach, very detailed genetic maps of the many different phage have been obtained.

Selected Key Terms

auxotroph
conjugation
co-transformation
F′-plasmid
High frequency of recombination
lysogenic infection
lytic cycle
merozygote
plaque
plasmid
prophage
prototroph
temperate phage
transduction
transformation
virulent phage

Chapter Review Problems

1. In an *E. coli* mutant strain just isolated, you discover a new mutation that inhibits the synthesis of a type of RNA (ribosomal), but only when grown at 41°C. Growth is normal at 34°C. Because this is such an important mutant, you want to map it to determine if the mutation maps close to any other known genes for RNA synthesis. How would you go about mapping this new mutant to locate it precisely on the map?
2. In a series of experiments, you find that the following sets of genes co-transform in a *B. subtilis* strain: *X* with *B*, *X* with *C* and *Y*, and *C* with *E* and *D*. What is the order of these genes?
3. Using three different *Hfr* strains, the order of transfer of seven genes was as follows: *X Z J K R M L*; *R M L X Z J K*; *R K J Z X L M*. Draw a map of these genes and explain the results.
4. You have infected *E. coli* cells with two strains of T2 phage. One phage strain is mutant for three genetic markers, *cat*, *dog*, and *fly*; the other is wild-type for all three markers. After lysis of the cells, you examine the progeny and find the following:

Genotype	Frequency
cat dog fly	3520
+ + +	3896
cat dog +	945
+ + *fly*	854
cat + *fly*	155
+ *dog* +	167
+ *dog fly*	567
cat + +	602
Total	10,706

   a. Determine the distance between each of the genetic markers.
   b. What is the order of the genes?
5. You have isolated three mutants (*joe*, *sam*, and *bob*) in *E. coli* and have tentatively mapped them using conjugation. Much to your surprise, these three mutants map very close to each other. You then attempt to determine the map distances between each of these markers using transduction and get the following results when selecting for sam^+, using a triple-mutant recipient.

Genotype	Quantity
sam^+ *bob joe*	1602
sam^+ bob^+ *joe*	786
sam^+ *bob* joe^+	10
sam^+ bob^+ joe^+	250

   a. What is the gene order?
   b. What are the map distances between *sam* and *bob* and between *sam* and *joe*?
6. Wild-type phage T4 grows on both strains K-12(λ) and B of *E. coli*, but *rII* mutants of phage T4 cannot grow on strain K-12(λ) while retaining its ability to grow on strain B. You have isolated two different mutations in the *rII* locus (*rIIa* and *rIIb*) and wish to determine whether they are in the same cistron and find the percent recombination between the two mutations. You doubly infect with these phage mutants in both a B strain and a K-12(λ) strain. You then titer the progeny virus and get 9.4×10^7 phage particles from the B strain lysate (a lysate is the phage that results from an infection of a bacterial host) and 4.5×10^5 phage particles from the K-12(λ) strain lysate.
   a. Are they in the same or different cistrons?
   b. Calculate the map distance between the two markers.
7. You have simultaneously infected an *E. coli* strain with two different strains of phage T6, sue^+ pat^+ and *sue pat* and observe the following frequencies of progeny:

Genotype	Frequency	Type
sue^+ pat^+	465	Parental type
sue pat	382	Parental type
sue^+ *pat*	28	Recombinant type
sue pat^+	34	Recombinant type

   Calculate the map distance between these two markers.
8. In *E. coli*, the following *Hfr* strains donated the markers shown and in the order given:

Hfr Strain	Order of Gene Transfer
1	A D F C J E
2	F D A M L K
3	D F C J E K
4	K L M A D F

   Construct a genetic map using these data.
9. If most single-gene mutations revert to wild-type (go back to wild-type) at a frequency of 1×10^{-7}, or 1 in 10 million cells, and Lederberg and Tatum observed a recombination rate of 1×10^{-7} between different mutant genes in their experiment showing transmission of genetic material, how were they justified in concluding that they were obtaining recombination rather than reversion?
10. Lederberg and Tatum observed a recombination frequency of only 1 in 10 million cells in their initial experiments, but we know now that *Hfr* strains transfer their DNA at frequencies of nearly 100%, although only about 50% of the DNA is integrated. Why was their observed frequency so low?
11. Using the data for entry times from Figure 8.11, how would these entry times change if *Hfr*Ra-2 (Figure 8.9) were used as a donor instead of *Hfr*H?
12. Gene R^+ is a regulatory gene that somehow is involved in the control of the expression of genes A^+, B^+, and C^+ in *E. coli*. The level of expression of genes A^+, B^+, and C^+ can be easily monitored by assay of the function of gene C^+, which converts a colorless chemical to a blue compound. We want to establish whether or not gene R^+ controls

genes A^+, B^+, and C^+ by a cytoplasmically produced diffusible product. What genetic combinations do we need to answer this question and how do we do the experiment?

13. You have an *E. coli* strain (strain A) that possesses a plasmid (plasmid *a*). But, when you attempt to transfer plasmid *a* to strain B (also *E. coli*), you cannot find any evidence for the presence of plasmid *a* in strain B. In contrast, when you mix strain A with strain C (which contains a self-transmissible plasmid *b*) and strain B, strain A now transfers its plasmid to strain B. Propose a reasonable explanation for why strain A could transfer its plasmid *a* only in the presence of strain C containing plasmid *b*.

14. In an experiment, you have three genetic markers that you know are very closely linked to each other on the chromosome. The genes are *a, b,* and *c*. Upon co-transformation of the markers, we obtain the following data.

Experiment	Co-Transformed Markers	Frequency of Co-Transformation
1	*a* & *b*	0.001%
2	*a* & *c*	0.008%
3	*c* & *b*	0.012%

What is the gene order?

15. Why do you need a high ratio of phage to bacteria during a phage genetic cross, but a very low ratio during a phage assay?

16. Approximately how many DNA base pairs does a 0.5% recombination rate represent if the map of the phage T4 genome is 1500 m. u. long and the genome is 165,000 base pairs?

17. Using the formula for calculating map distance given in the text, estimate the map distance between the genes *dum* and *dmr* in *E. coli* that have a co-transduction frequency of 2% with the generalized transducing phage P1.

18. If a mutation in *E. coli* had a frequency of reversion from *phe* to a wild-type of 1×10^{-5} and another mutation in *E. coli* had a frequency of reversion from *thi* to a wild-type of 2×10^{-6}, what would be the reversion frequency to the wild-type of a *phe, thi* double mutant?

19. In the following timed mating experiment in *E. coli*, four different *Hfr* strains were tested for the order of entry in the following genes. The order of transmission is shown for each strain.

Hfr Strain	Order of Entry
1	B Y F I G
2	B C O M L
3	O C B Y F
4	M L N G I

What is the correct order of these genes on the chromosome?

20. Transduction in bacteria
    a. is transfer of the plasmid only
    b. is time-dependent
    c. involves a virus intermediate
    d. is the uptake of DNA from the surrounding media
    e. is none of the above

21. Complementation tests are used to establish
    a. the distance between two genes
    b. whether different mutations in one gene each can cause a distinct phenotype
    c. whether different mutations that cause an identical phenotype are located in a single gene
    d. whether genes code for enzymes

22. Transformation of gene *a* to a^+ (in bacteria) occurs at a frequency of 1 in 10^2, and transformation of *b* to b^+ occurs at a frequency of 1 in 10^4. If the genes are distant from each other on the bacterial genome, what will be the frequency of a^+b^+ transformants in the *ab* recipient strain?

23. During a conjugation experiment, it is found that thr^+ enters the recipient strain at minute 22, leu^+ at minute 24, $argF^+$ at minute 29, gal^+ at minute 39, and trp^+ at minute 40. Draw a linkage map of these genes using *thr* as 0 on your map.

24. You have isolated two mutants from *E. coli,* both requiring cysteine for growth. You are in a quandary as to whether these two mutants are in the same gene governing cysteine biosynthesis or in more than one gene. To attempt to answer this question, you decide to map the location of these genes using interrupted mating, and you find that both mutations map to minute 59 on the *E. coli* genetic map. You examine the literature and find that three genes for cysteine biosynthesis map at minute 59: *cys*G, *cys*H, and *cys*I. How would you distinguish whether your *E. coli* mutations are in one or more of these genes, and which ones?

25. Three mutants have been isolated from *E. coli: trk*E, *trp*D, and *tyr*R. From an interrupted mating experiment, you know that they all map near minute 29 on the *E. coli* map. All three mutations have been placed in one strain. You do not know the order of the three genes but wish to establish their order using transformation. You prepare your DNA from a wild-type strain and transform the triple mutant strain, selecting for *trp*D. You find that *tyr*R is co-transformed at a frequency of 3.2% and *trk*E at a frequency of 0.11%. In a second experiment, you select for *tyr*R and find that *trp*D co-transforms at a frequency of 3.2%, but *trk*E co-transforms at a frequency of 0.02%. Which gene is in the middle, and what are the relative positions of the three genes?

26. You have two mutant strains of phage T4; one is mutant in host range (*h2*) and the other in plaque morphology (*r5*). In an attempt to map the distance between these two genes, you do a phage cross by co-infecting a host bacterial strain with the two mutant phage lines. The progeny follow:

Number of Plaques	Genotype	
867	$h2^+\ r5^-$	Parental type
927	$h2^-\ r5^+$	Parental type
38	$h2^+\ r5^+$	Recombinant type
45	$h2^-\ r5^-$	Recombinant type
Total 1877		

What is the map distance between *h2* and *r5*?

27. You have just prepared a stock of *E. coli* phage P1 and wish to determine the exact concentration of phage in your stock solution. You successively dilute your phage 100-fold three times and then plate 0.10 ml on a lawn of *E. coli*. You find 87, 95, and 102 plaques on three different plates. What is the concentration of the phage in the original stock in plaque-forming units per milliliter?

28. Two mutants of *E. coli* are found (*upp* and *ung*), and it is determined that they map close to each other using conjugation mapping. Using phage P1, a stock of phage is prepared on a wild-type strain, and a double mutant strain is transduced with this stock. The co-transduction frequency of the two markers (*upp* and *ung*) is done by selecting for *upp* and measuring the frequency that *ung* co-transduces and is found to be 32%. What is the map distance between these two genes?

29. The following genes were found to co-transform in a series of transformation experiments:
    a. *uup* with *ugp*
    b. *tyr* and *trp*
    c. *tyr* and *ugp*
    d. *trp* and *rna*

    What is the order of these genes?

30. Is it possible to have a plasmid with no genes?

Challenge Problems

1. *Hfr*J4 strain of *E. coli* is used to map three genes in an interrupted mating experiment. The cross is *Hfr*J4 / *argH*$^+$, *lysX*$^+$, *leuR*$^+$, *str*^{s} × F$^-$ / *argH*$^-$, *lysX*$^-$, *leuR*$^-$, *str*^{r} (no map order is implied). The *argH* allele is required for the synthesis of arginine, the *lysX* allele is required for synthesis of lysine, and the *leuR* allele is required for synthesis of leucine. All are amino acids. The minus alleles are auxotrophs for these amino acids.

    A cross is initiated at time = 0, and at various times samples are removed from the mating mixture, shaken vigorously, and plated on a minimal medium containing streptomycin plus the amino acids indicated in the table. After incubation of the plate containing the spread cells, colonies appeared and the number is also given in the table.

Number of Colonies on Selective Media

Amino Acids Added to Minimal Media	Time After Interruption of Mating (min)							
	5	10	15	20	25	30	35	40
Arginine and leucine	0	0	0	0	0	0	28	89
Lysine and arginine	0	0	12	45	90	150	200	220
Leucine and lysine	8	45	90	180	280	300	305	310

   a. What is the purpose of the streptomycin in the media?
   b. Using these data, determine the approximate distance between each of the genetic markers.
   c. What is the order of the genes on the map?
   d. What is the relative position of *Hfr*J4 entry with respect to the genetic markers?

2. *E. coli* was simultaneously infected by two genetically different strains of phage T2. Strain A had the genotype *x y z*, and strain B had the genotype x^+ y^+ z^+. The progeny of this multiple infection were analyzed for the two possible alleles of each of these genes and the following data were collected.

Genotype	Frequency	Genotype	Frequency
x y z	0.400	x y^+ z	0.010
x^+ y^+ z^+	0.385	x^+ y z^+	0.009
x^+ y z	0.100	x y z^+	0.002
x y^+ z^+	0.090	x^+ y^+ z	0.004

   a. What is the sequence of these three genes on the T2 genome?
   b. Calculate the distances separating these three genes.
   c. What is the coefficient of coincidence?
   d. What is the meaning of this coefficient of coincidence?

3. You have found three genes in a strain of bacteria that are too close together to map accurately using interrupted mating, so you have decided that you will try and determine the map distance and gene order of these three genes using transformation. Your strain does not transform naturally, but you can use electroporation to incorporate the DNA into the cells. These three genes are *mar*, *bar*, and *car*, and the donor strain is *mar*$^+$, *bar*$^+$, and *car*$^+$. The recipient possesses mutant alleles of all three genes. Each gene responds to specific selective media so that you can distinguish all allele combinations of these three genes. After doing the transformation, you collect the following data.

	Allele			Frequency
Class	*mar*	*bar*	*car*	
1	+	+	+	5000
2	+	–	–	300
3	–	+	+	600
4	+	–	+	1300
5	–	+	–	1800
6	–	–	+	200
7	+	+	–	50
			Total	9250

a. What is the gene order?
b. What is the linkage distance between each pair of genes?
c. Does the distance between each of the pairs of markers add to the total distance of the most distant markers? If not, why not?

Suggested Readings

Benzer, S. 1961. "On the topography of the genetic fine structure." *Proceedings of National Academy of Science USA,* 47: 403–415.

Benzer, S. 1962. "The fine structure of the gene." *Scientific American,* 206 (January): 70–87.

Drlica, K., and M. Riley. 1990. *The Bacterial Chromosome.* American Association for Microbiology, Washington, DC.

Fox, M. S., and M. K. Allen. 1964. "On the mechanism of DNA integration in pneumococcal transformation." *Proceedings of the National Academy of Science USA,* 52: 412.

Goodgal, S. H. 1982. "DNA uptake in haemophilus transformation." *Annual Review of Genetics,* 16: 169.

Hayes, W. 1968. *The Genetics of Bacteria and Their Viruses.* John Wiley and Sons, New York.

Lederberg, J. 1987. "Genetic recombination in bacteria: A discovery account." *Annual Review of Genetics,* 21: 23–46.

Miller, R. 1998. "Bacterial gene swapping in nature." *Scientific American,* 278 (January): 66–71.

Neidhardt, F. C. 1996. Escherichia coli *and* Salmonella typhimurium, *Cellular and Molecular Biology,* 2nd Ed. American Association for Microbiology, Washington, DC.

Notani, N. K., and J. K. Setlow. 1974. "Mechanism of bacterial transformation and transfection." *Progress in Nucleic Acid Research and Molecular Biology,* 14: 39.

Novick, R. P. 1980. "Plasmids." *Scientific American,* 243 (December): 102–127.

On the Web

Visit our Web site at **http://www.saunderscollege.com/lifesci/titles.html** and click on A/G/M Genetics for links to the following chapter-related resources on the World Wide Web:

1. **Culture Collections.** This site includes links to access culture collection sites, stock centers, and germplasm repositories.

2. **Bacterial Genomes.** This site includes links to access sites to bacteria, bacterial genomes, and other organisms.

PART 2

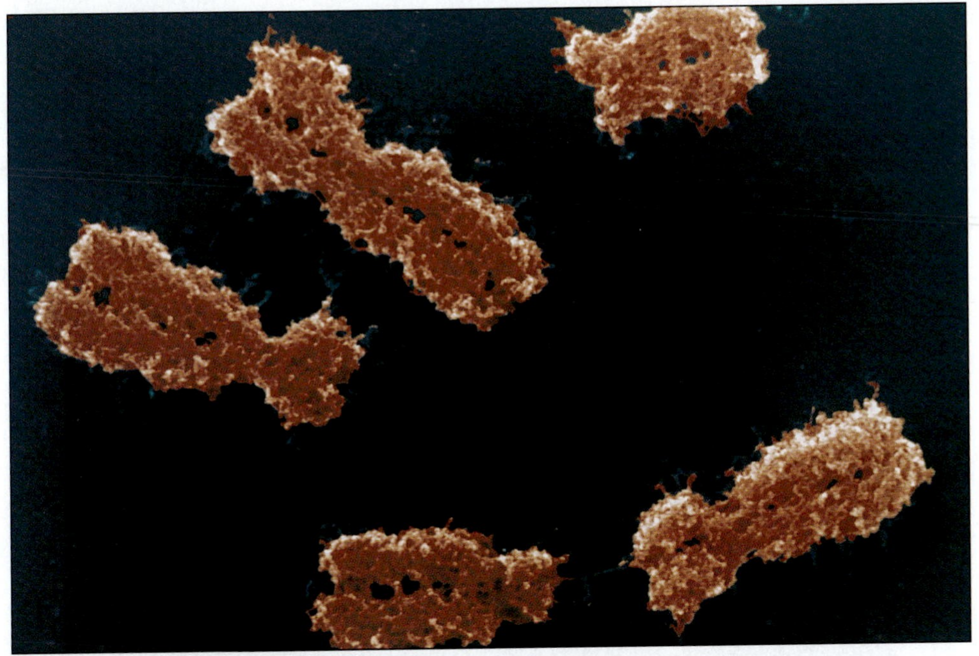

A colorized scanning electron micrograph (4500×) of human metaphase chromosomes. *(Oliver Meckes/Photo Researchers, Inc.)*

How Genes Function at the Molecular Level

CHAPTER 9
The Physical Nature of DNA: Structure and Replication

CHAPTER 10
The Molecular Structure of Prokaryotic and Eukaryotic Chromosomes

CHAPTER 11
Transcription, Translation, and the Genetic Code

CHAPTER 12
Gene Cloning and Analysis

CHAPTER 13
Genomic Analysis and Modification

CHAPTER 14
Regulation of Gene Expression

CHAPTER 15
Developmental Genetics: Cell Fate and Pattern Formation

CHAPTER 16
Developmental Genetics: Gene Regulation and Differentiation

CHAPTER

9

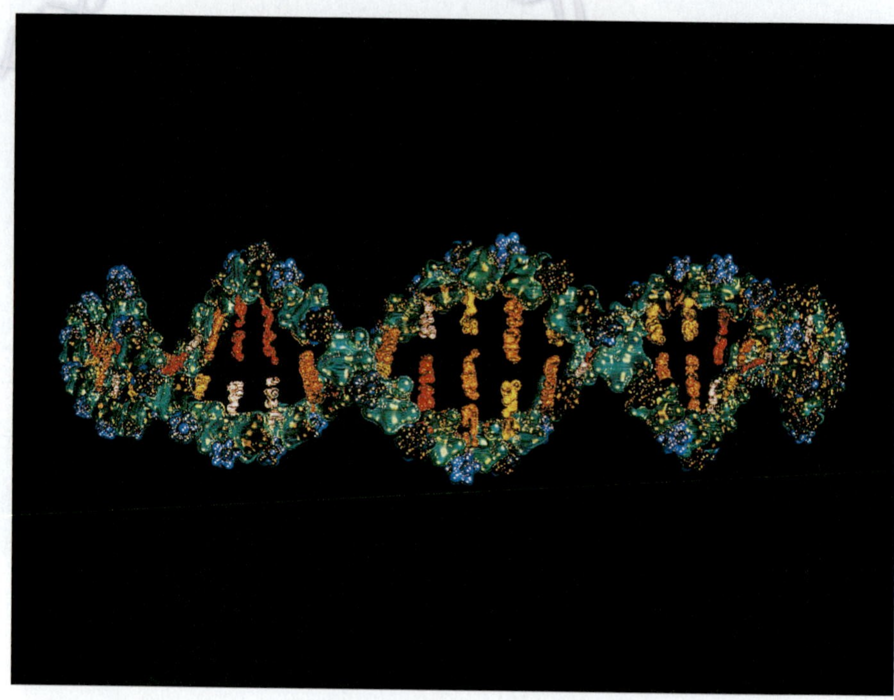

A computerized rendering of a double-helical DNA molecule. *(Douglas Struthers/Tony Stone Images)*

The Physical Nature of DNA: Structure and Replication

The big question in genetics during the early part of the 20th century was whether nucleic acids or proteins contained the genetic code because both are present in chromosomes in abundant quantities. In the 1800s, biologists observed a darkly staining material, termed chromatin, in the nuclei of all cells and believed this to be the genetic material. Prior to cell division, the chromatin was seen to condense into discrete, darkly staining bodies, the chromosomes, which were distributed to the progeny cells. Scientists knew that chromosomes contained DNA as well as proteins, but they did not know that deoxyribonucleic acid (DNA) is the genetic material. Discovering this fact occupied the time and efforts of many people for the first 50 years of this century.

To understand how genes act in an organism, we need to know the nature and composition of the chemical components of the gene. By the end of this chapter, we will have addressed the following three questions to define more clearly the chemical composition of a gene and the chromosome:

1. What evidence do biologists have that DNA is the genetic material?
2. What is the chemical composition and three-dimensional structure of DNA and RNA?
3. How does DNA replicate while keeping the original genetic information intact?

Scientists Needed Proof that DNA Is the Genetic Material.

By the early 1930s, scientists knew that DNA is a long polymer composed of four different repeating units, called nucleotides, whereas proteins are shorter polymers composed of 20 different repeating units called amino acids. DNA did not seem to be a likely candidate for the genetic material because it was thought that DNA was not sufficiently complex. DNA contained only four repeating units, with each repeating unit erroneously thought to be present in about equal amounts. In fact, in the first decade of this century, it was widely believed that DNA (and RNA) were rather simple molecules, with the four nucleotides repeating in a monotonous pattern over and over again. Four repeating units seemingly could not code all the complex information needed in a chromosome. This left protein as the likely candidate for the genetic material because its 20 different amino acids can occur in a virtually limitless variety of sequences.

The erroneous but widely held belief that proteins contained the genetic information was based primarily on the seemingly too-simple nature of the chemical constituents of DNA. Just what are the tasks the genetic material must be capable of performing? Let us consider, from a purely theoretical point of view, what properties a molecule might need to act as the genetic material.

1. The genetic material must have a chemical composition that allows the molecule to code information for many different characteristics.
2. Sperm and eggs have one half the genetic material (and one half the number of chromosomes), so the chemical in the gametes responsible for heredity should be one half the amount of somatic cells.
3. The hereditary material should be stable during the life of a cell; otherwise, information might be gained, lost, or altered.
4. The molecule responsible for genetic information should have the capability to act as a template for an accurate copy of itself during duplication.
5. The hereditary material should also be subject to small changes, or mutations, because genetic information changes slightly over time.

Next, we will examine the experimental information that supported the hypothesis that DNA is the genetic material and learn how proteins were eliminated from consideration. After this discussion, we will examine the structure of DNA molecules. DNA's structure supports the conclusion that it, and not protein, is the genetic material.

DNA Is the Primary Origin of Transformation.

Evidence to prove that DNA is the genetic material emerged somewhat unexpectedly from studies on a bacterium now known as *Streptococcus pneumoniae*. When grown in the laboratory on a nutrient (agar) medium in a Petri dish, the wild-type bacteria form a colony that is large and smooth. If these smooth-colony bacteria are injected into a mouse, they are capable of killing the mouse. However, a mutant of *S. pneumoniae* produces a small, rough colony (as a result of the loss of an extracellular polysaccharide coat), and this mutant is not lethal to mice. In 1928, Frederick Griffith laid the foundation for demonstrating that DNA was the genetic material by making the startling discovery that rough-nonvirulent *Streptococcus pneumoniae* became virulent (and smooth) when mixed with heat-killed virulent bacteria (Figure 9.1). This change was passed on to subsequent generations of bacteria and, thus, was heritable. This process came to be known as **transformation.**

These experiments raised the possibility that heat-killed virulent bacteria were liberating some genetic material into the medium and that this material was passing through the cell walls of the nonvirulent bacteria. Within the nonvirulent cell, this substance permanently changed the bacteria by becoming incorporated into the host's hereditary material (Figure 9.1). However, this interpretation did not constitute evidence that the genetic material is DNA, nor did it explain how transformation occurs.

In 1944, after 10 years of work, the American microbiologists Oswald T. Avery, Colin M. MacLeod, and Maclyn McCarty (working at the Rockefeller Institute in New York) put forth evidence that DNA from the dead *S. pneumoniae* cells was responsible for the change from a nonvirulent to a virulent state (Figure 9.2). The basis of this startling conclusion was the result of the following series of experiments they conducted. First, they purified DNA from the *S. pneumoniae* virulent strain that transformed the nonvirulent strain to a virulent strain. Second, they destroyed the protein in this extract by digesting it with enzymes that degrade proteins (proteases). After this treatment, the extract from the virulent bacteria was still capable of transforming nonvirulent bacteria, strongly suggesting that proteins could not be the genetic material. Third, they digested the virulent extract with ribonucleases that destroy RNA. Again, they found that the extract could still transform nonvirulent bacteria to the virulent form, strongly suggesting that RNA was not the genetic material either. Then they found that deoxyribonuclease (DNase), an enzyme that digests DNA, destroyed the transforming material. Furthermore, they found that DNA from nonvirulent bacteria and DNA from other organisms were ineffective in transformation. Finally, they purified the DNA and established that it is a polymer consisting of at least 1000–2000 bases, which they believed might be a sufficient length to encode the information required for inherited traits (although nothing was known at the time about DNA information coding). From these experiments, they concluded that DNA was the transforming material.

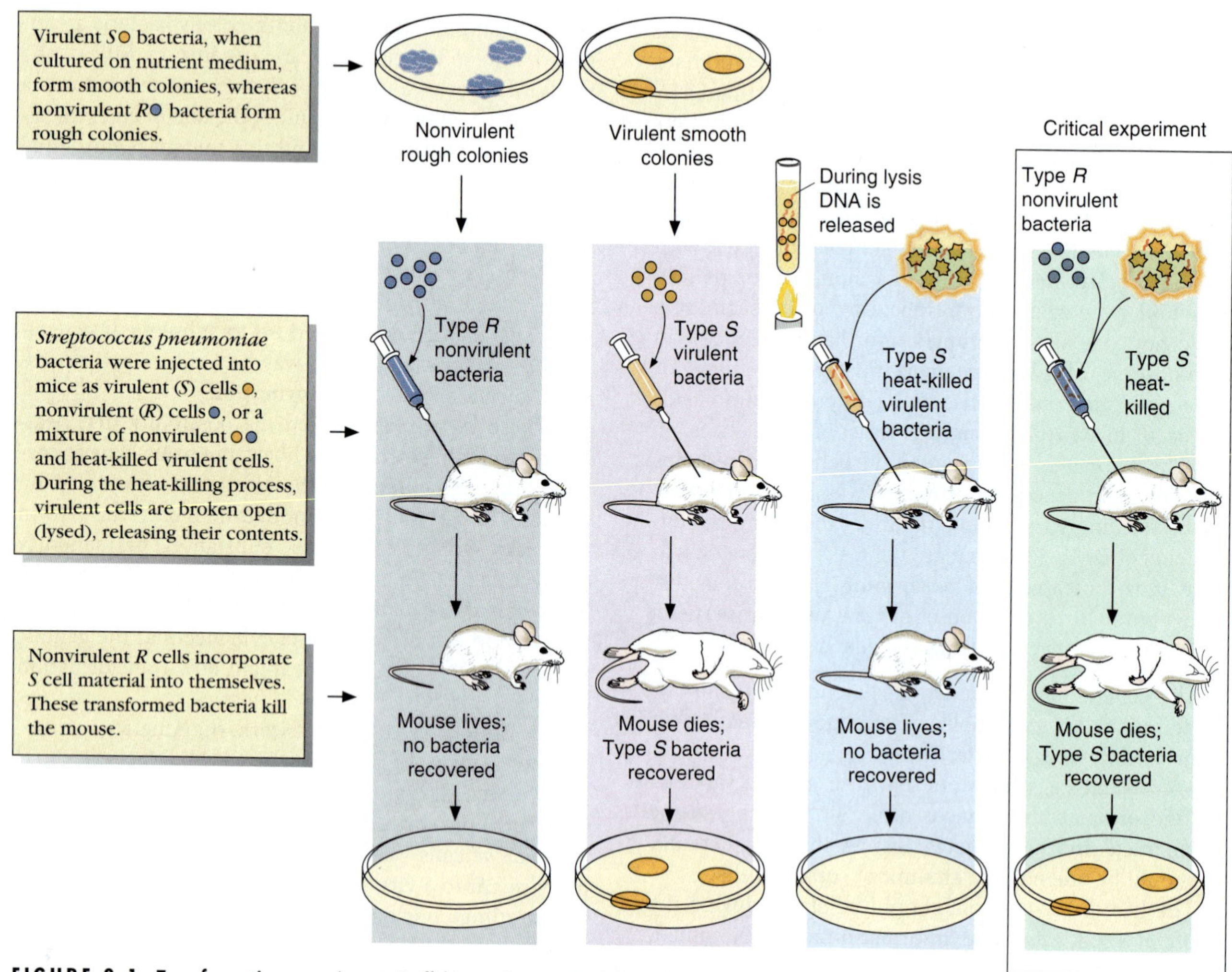
Virulent S bacteria, when cultured on nutrient medium, form smooth colonies, whereas nonvirulent R bacteria form rough colonies.
Nonvirulent rough colonies
Virulent smooth colonies
During lysis DNA is released
Critical experiment
Type R nonvirulent bacteria
Streptococcus pneumoniae bacteria were injected into mice as virulent (S) cells, nonvirulent (R) cells, or a mixture of nonvirulent and heat-killed virulent cells. During the heat-killing process, virulent cells are broken open (lysed), releasing their contents.
Type R nonvirulent bacteria
Type S virulent bacteria
Type S heat-killed virulent bacteria
Type S heat-killed
Nonvirulent R cells incorporate S cell material into themselves. These transformed bacteria kill the mouse.
Mouse lives; no bacteria recovered
Mouse dies; Type S bacteria recovered
Mouse lives; no bacteria recovered
Mouse dies; Type S bacteria recovered

FIGURE 9.1 Transformation experiment. Griffith's work suggested that a factor was capable of transforming one bacterial cell type to another. This experiment suggested other experiments that eventually led to the understanding that DNA is the genetic material.

The Amount of DNA Per Cell Is Constant, and DNA Is Chemically Stable.

Even though the evidence supporting DNA as the genetic material was strong, there were still critics. Chemical analysis of the purified transforming DNA showed the presence of a tiny amount of protein, which prolonged the argument about whether DNA was actually the genetic material. A small minority of scientists still insisted that contaminating protein might be the genetic material. For example, it could not be ruled out that the DNA was merely an agent causing a change (mutation) in the "real" genetic material, creating new virulent bacteria.

However, other supporting evidence pointed to DNA, rather than protein, as the genetic material. For example, the amount of DNA in the diploid cells of an organism was shown to be constant and equal to twice the amount present in the haploid cells (Table 9.1), whereas the amount of protein per cell varied widely. Also, DNA was shown to be metabolically stable. For example, after a radioactive atom (such as ^{14}C, an isotope of the normal ^{12}C) is incorporated into DNA, it remains there as long as the cell is healthy. The same is not true for proteins. Recall that these are two of the characteristics we speculated would need to be present in the genetic material.

Phage Were Used To Confirm that DNA Is the Genetic Material.

The controversy over whether or not DNA was the genetic material was laid to rest by experiments published

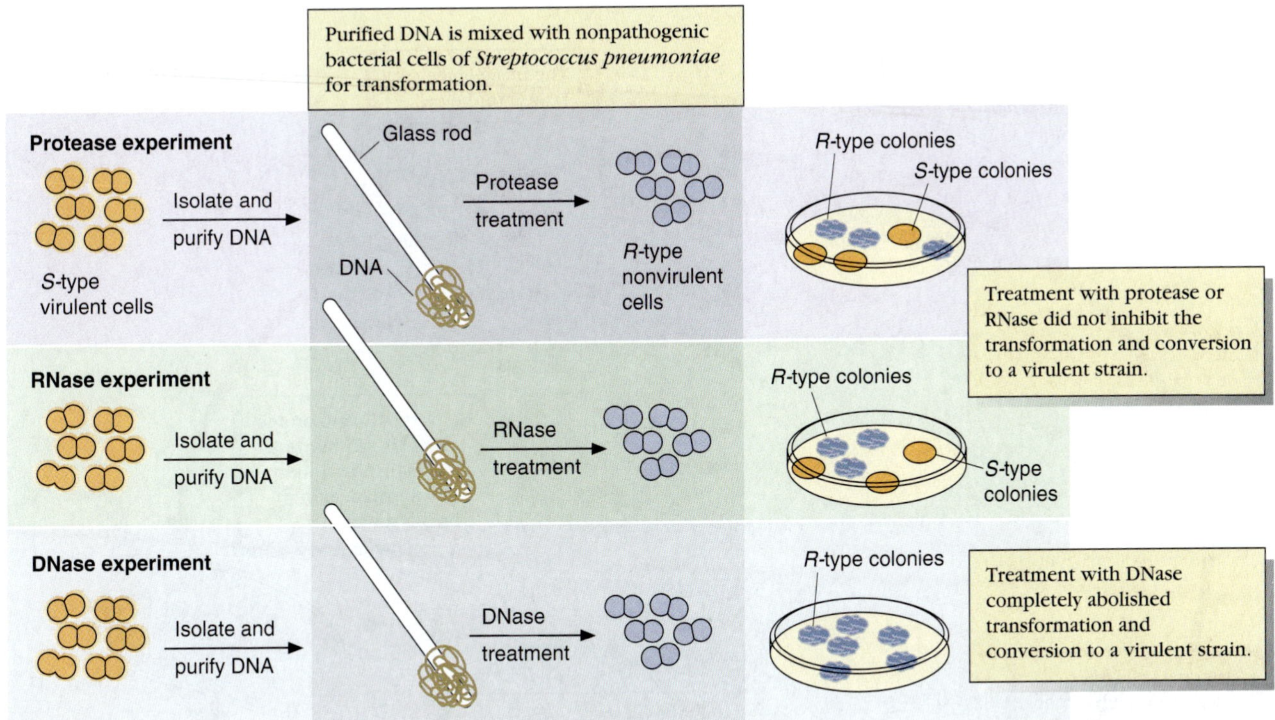

FIGURE 9.2 Strong proof that DNA is the genetic material. Avery, MacLeod, and McCarty showed that DNA is the genetic material by isolating and purifying DNA from pathogenic bacteria, combining this DNA with nonpathogenic bacterial cells of *Streptococcus pneumoniae,* and observing cells transform to pathogenic cells. This experiment convinced most scientists that DNA was the genetic material.

by Alfred D. Hershey and Martha Chase in 1952. These experiments made elegant use of the bacterial virus T2 (Chapter 8). In the early 1950s, it was established that the viruses that infect bacteria, called **bacteriophage** or simply phage, are composed solely of protein and nucleic acids, usually DNA. Phage T2 is a bacterial virus composed of a DNA molecule surrounded by a protein coat, referred to as the head. Attached to the head is a tail with tail fibers that allow the virus to attach to a bacterial cell's outer coat (Figure 9.3). A great deal was also known about phage growth patterns, and it was possible to obtain pure virus particles.

In the experiments performed by Hershey and Chase, the protein coat of the infecting phage was labeled with ^{35}S, a radioactive isotope of sulfur. Proteins contain sulfur in the amino acids methionine and cysteine, but DNA does not contain sulfur. The DNA of the phage was labeled with the radioactive phosphorus isotope ^{32}P because phosphorus is abundant in DNA but not found in the protein coat of the phage. These labeled phage were then used to study the fates of the phage protein and DNA during infection and multiplication. The researchers specifically asked, Is ^{35}S (protein) or ^{32}P (DNA) incorporated into the progeny phage after infection by a radioactively labeled phage? By measuring the radioactivity in progeny cells, they could conclude whether the DNA or the protein was passed on to

TABLE 9.1 *The Amount of DNA Per Nucleus of Bovine Tissue*

Organ	DNA (pg)*	Probable Chromosome Number
Thymus	6.6	Diploid
Liver	6.4	Diploid
Pancreas	6.9	Diploid
Kidney	5.9	Diploid
Sperm	3.3	Haploid

*The variance of the first four values reflects errors in measurement. Only the last value is significantly different and represents approximately one half of the other values. Picograms (pg) are equal to 1×10^{-12} g.

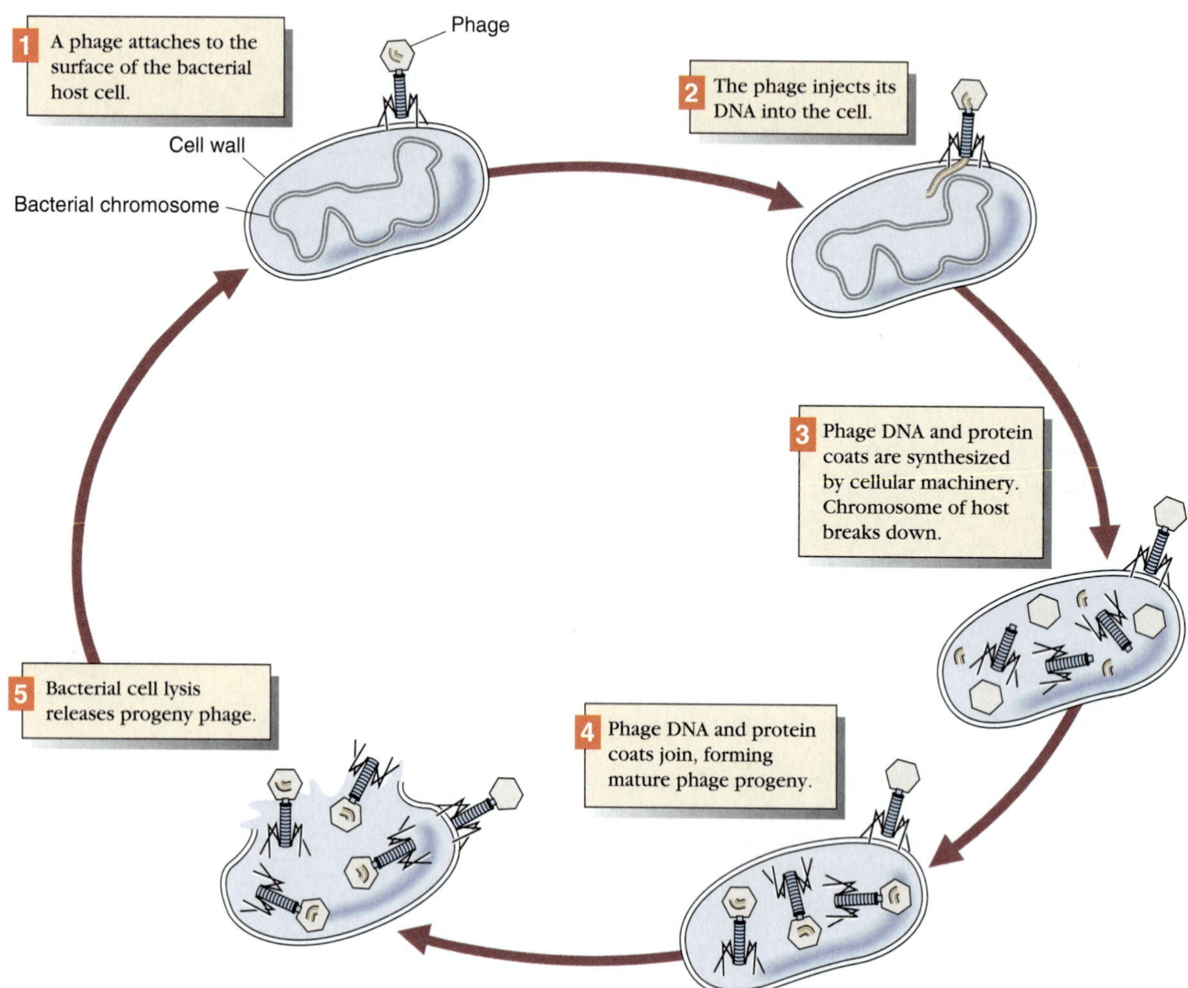

Phage
1 A phage attaches to the surface of the bacterial host cell.
Cell wall
Bacterial chromosome
2 The phage injects its DNA into the cell.
3 Phage DNA and protein coats are synthesized by cellular machinery. Chromosome of host breaks down.
4 Phage DNA and protein coats join, forming mature phage progeny.
5 Bacterial cell lysis releases progeny phage.

FIGURE 9.3 The life cycle of a bacteriophage. Phage reproduce by injecting their DNA into a bacterial cell to form new DNA, heads, and tails. The old phage particle remains on the exterior of the cell.

succeeding generations of phage particles. The details of the experiment are described in Figure 9.4.

The results of these experiments very convincingly demonstrated that DNA is the genetic material. Most of the labeled parental DNA appeared in the offspring, but none of the labeled protein was found in the offspring. The results clearly showed that very little or no phage parental protein even entered the bacterial cell. After infection, violent agitation of the infected bacteria removed the empty phage heads and tails (ghosts) from the bacteria. The loss of virus protein from the surface of the bacterial cells did not prevent the bacteria from synthesizing new phage particles. On the other hand, ^{32}P-DNA from the infecting phage particle was found in the progeny. This elegant, but simple, experiment finally put to rest the notion that protein could possibly be the genetic material and convinced all scientists that DNA is the genetic material.

RNA Can Also Be the Genetic Material.

All organisms (for example, animals, plants, and fungi) and most viruses have DNA as their genetic material. However, some viruses that infect bacteria, plants, and animals have RNA as their genetic material. One such virus, the tobacco mosaic virus (TMV), is composed of an RNA core surrounded by a protein coat and has an entirely different organization than phage T2. The protein coat is composed of a monomeric unit repeated many times into a large multimeric coat arranged in a helical manner over the RNA core (Figure 9.5). The protein monomers and the RNA of the virus are easily separated from each other in the laboratory and can be reassem-

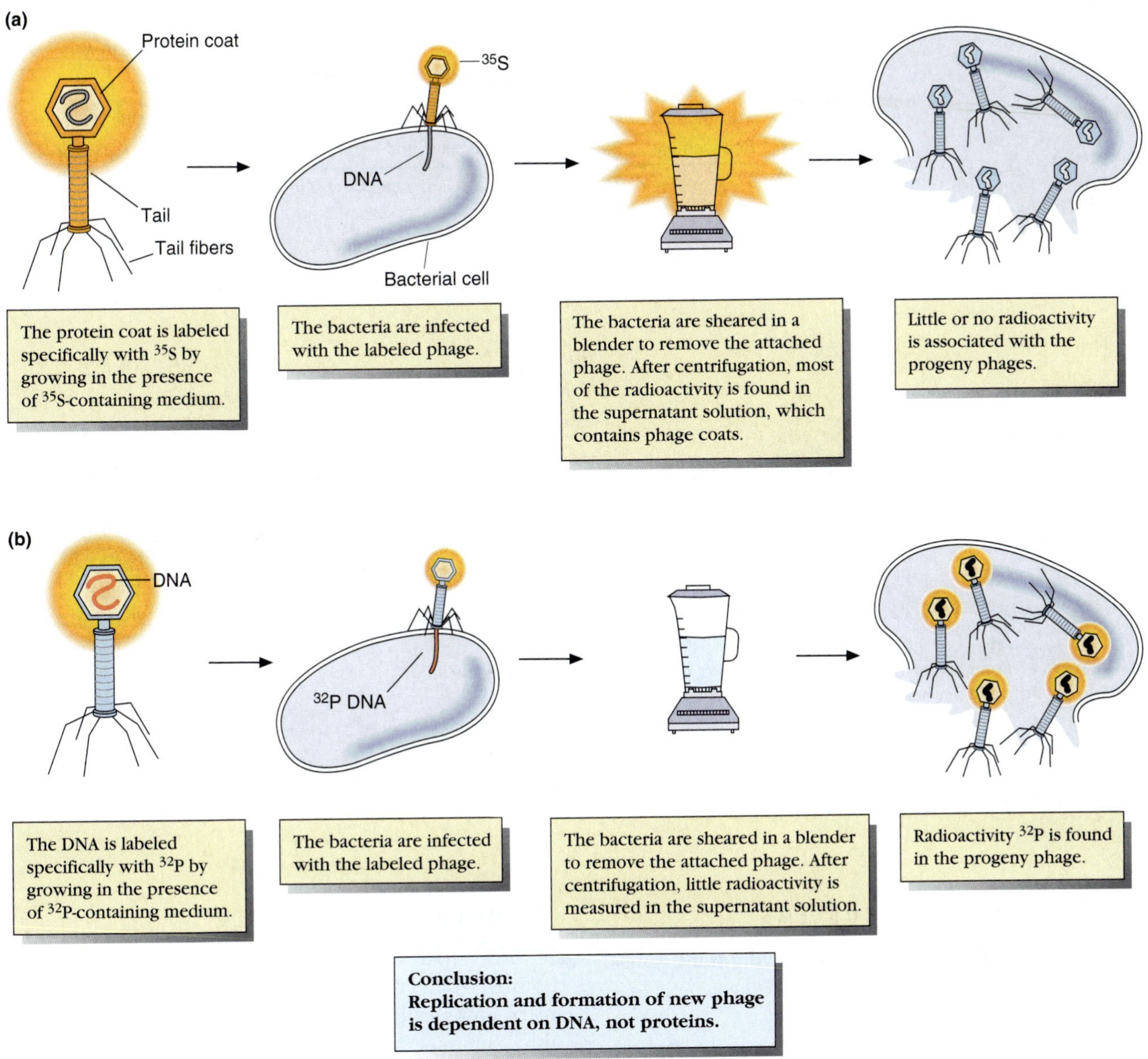

(a)
Protein coat
Tail
Tail fibers
35S
DNA
Bacterial cell
The protein coat is labeled specifically with 35S by growing in the presence of 35S-containing medium.
The bacteria are infected with the labeled phage.
The bacteria are sheared in a blender to remove the attached phage. After centrifugation, most of the radioactivity is found in the supernatant solution, which contains phage coats.
Little or no radioactivity is associated with the progeny phages.
(b)
DNA
32P DNA
The DNA is labeled specifically with 32P by growing in the presence of 32P-containing medium.
The bacteria are infected with the labeled phage.
The bacteria are sheared in a blender to remove the attached phage. After centrifugation, little radioactivity is measured in the supernatant solution.
Radioactivity 32P is found in the progeny phage.
Conclusion:
Replication and formation of new phage is dependent on DNA, not proteins.

FIGURE 9.4 Final proof that DNA is the genetic material. The Hershey–Chase experiment convincingly demonstrated that DNA is the genetic material. The ^{35}S protein-labeled phage or the ^{32}P DNA-labeled phage were allowed to attach to the bacterial hosts and inject their DNA. They were then vigorously shaken in the blender (shearing) to remove the phage tails from the exterior of the hosts. The mixture was centrifuged to separate the small phage (in the supernatant) from the large bacterial cells (pellet), and radioactivity was measured in each.

bled into an infectious TMV particle. In 1956, Heinz Fraenkel-Conrat and B. Singer separated the RNA and protein components of two very distinct TMV strains and reconstituted the RNA of one virus type with the protein of the another virus type, giving rise to two hybrid virus particles. They then infected tobacco with each hybrid virus and found that the progeny were always of the type specified by the RNA and had nothing to do with the protein present on the infecting particle (Figure 9.5). These results were again consistent with the hypothesis that nucleic acids and not proteins are the hereditary material. Many other viruses also contain RNA as their genetic material, most notably the human immunodeficiency virus (HIV).

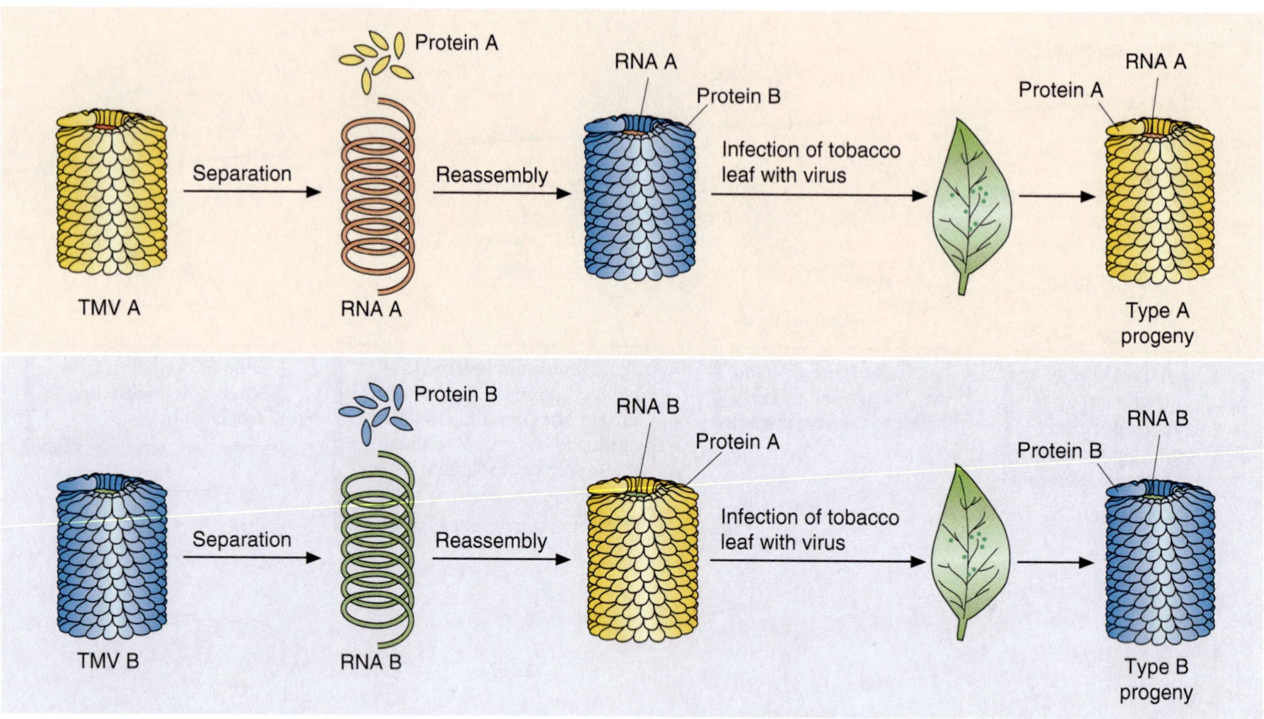

FIGURE 9.5 Proof that RNA can be genetic material. Fraenkel-Conrat and Singer used two very distinctly different strains of tobacco mosaic virus (TMVA and TMVB) to prepare hybrid virus particles. They prepared the hybrid particles by reconstituting the protein coat of strain A with the RNA from strain B, and vice versa. When they infected tobacco with their hybrid virus particles, they found that the progeny generated were always like the RNA of the virus, not the protein.

SECTION REVIEW PROBLEMS

1. What is transformation? What role did transformation play in understanding whether the genetic material is DNA or protein?
2. Why did scientists in the early part of the century think protein rather than DNA was the genetic material?
3. Why didn't all scientists accept the results of Avery, MacLeod, and McCarty as conclusive proof that DNA is the genetic material?

THE CHEMICAL NATURE OF DNA AND RNA GIVES CLUES TO THEIR FUNCTION.

To understand the structural organization of the chromosome and the cell machinery, we need to examine the molecular components in more detail. First, we will examine the chemical and physical nature of DNA and RNA and then how DNA replicates.

The Monomeric Units of DNA and RNA Are Nucleotides.

Both DNA and RNA are polymers composed of monomeric units called **nucleotides.** Nucleotides consist of three subunits:

1. a nitrogen-containing ring structure referred to as a **base**
2. a pentose sugar
3. a phosphate group (PO_4) positioned on the pentose sugar

The nitrogenous base may be either a **purine** or a **pyrimidine;** purines are two-ring bases, and pyrimidines are single-ring bases. The chemical structure of the bases and pentose sugar are shown in Figure 9.6. Figure 9.7 shows the combination of the base, sugar, and phosphate that forms a nucleotide. If the phosphate group is missing, the sugar–nitrogen base is called a **nucleoside.** The numbering system of the atoms in the nucleoside is of special interest. Note in Figure 9.6 that the purine and pyrimidine rings are numbered, as is the deoxyribose molecule. When the two molecules are attached (Figure

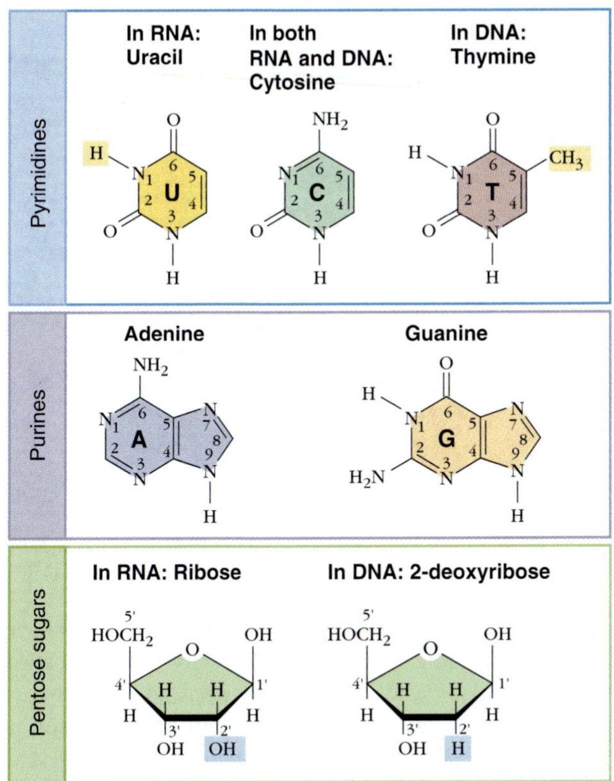

FIGURE 9.6 Structural formulas for the constituents of RNA and DNA. The carbon atoms in pentose sugars are numbered 1′, 2′, 3′, 4′, and 5′ when bonded to a purine or pyrimidine to distinguish them from the carbon numbering system in the purines and pyrimidines.

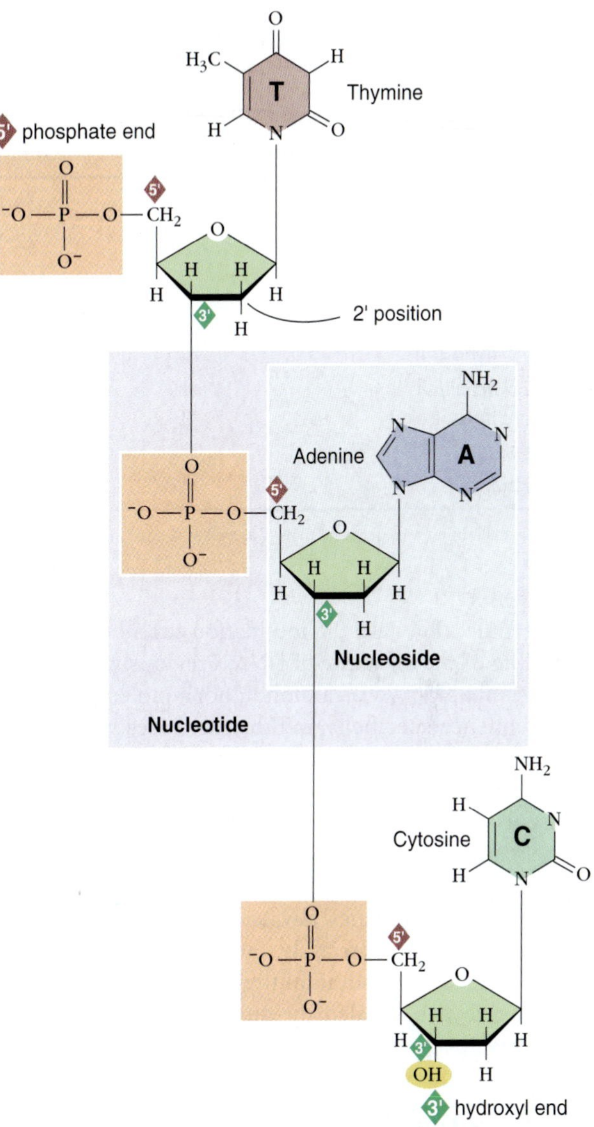

FIGURE 9.7 Chemical structure of a DNA strand. The pentose sugar plus the base is called a nucleoside; when the phosphate is added, it is a nucleotide. Note that the 5′ phosphate of a deoxyribose is attached to the 3′-OH of the adjacent deoxyribose to form a phosphodiester linkage. Also note that the molecule has a 5′ and 3′ end, which indicates polarity.

9.7), the same numbering system cannot be used on both molecules, so the deoxyribose atoms are given a prime indication. Thus, for example, the 3 position of the deoxyribose is referred to as 3 prime, or 3′. The phosphate group is attached at the 5′ position and the base is attached at the 1′ position of the deoxyribose.

The purines **adenine (A)** and **guanine (G)** and the pyrimidines **cytosine (C)** and **thymine (T)** are found in DNA (Figure 9.6). In RNA, the pyrimidine **uracil (U)** replaces thymine, but guanine, cytosine, and adenine are also present. RNA also differs from DNA in that the pentose sugar in RNA nucleotides is **ribose** bonded to the pyrimidine or purine, not **deoxyribose** as in DNA (Figure 9.6).

The seemingly small differences between nucleotides of RNA and DNA actually confer significantly different properties to the polymers. For example, RNA is sensitive to degradation by alkali because of the 2′-OH on the pentose ring, whereas DNA is more resistant to alkali degradation. In addition, specific enzymes that degrade, synthesize, or repair RNA and DNA can readily distinguish between the two types of nucleic acids. These and other properties enable researchers to separate RNA from DNA for purification and study.

DNA Is Composed of Only Four Different Nucleotides.

By 1949, Erwin Chargaff, then working at Columbia University, had published data on the nucleotide composition of DNA from a variety of organisms (Table 9.2). Although the implications of the data were not appreciated

TABLE 9.2 *Base Composition of DNA from Various Organisms (given as a percentage of total DNA)**

Organism	Adenine	Thymine	Guanine	Cytosine	$\frac{A+T}{G+C}$
Human	31.0	31.5	19.1	18.4	1.67
Drosophila	27.3	27.6	22.5	22.5	1.22
Yeast	31.3	32.9	18.7	17.1	1.79
Rat	28.6	28.4	21.4	21.5	1.32
Escherichia coli	26.0	24.0	25.0	25.0	1.00
Mycobacterium tuberculosis	15.1	14.6	34.9	35.4	0.42
Rhizobium japonicum	17.9	18.0	32.1	32.0	0.56

*The small differences between A and T as well as G and C percentages represent errors in measurement.

at the time, they laid the foundation for a better understanding of the structure of DNA. Previously held thinking was that DNA was a monotonous molecule without any sequence specificity and that the bases were present in equal amounts. Chargaff's data clearly put this thinking to rest. He showed that the four nucleotides were not present in equal amounts and that the amounts varied widely among organisms. Another important point from Chargaff's data was that, in every organism examined, the amount of purines always equaled the amount of pyrimidines. More precisely, the amount of adenine always equaled the amount of thymine, and the amount of guanine always equaled the amount of cytosine. These relationships are referred to as **Chargaff's rule**.

DNA Is a Double-Helical Molecule.

In addition to nucleotide composition, the chemical structure of the DNA polymer is essential to understanding its properties and function. In 1952, a group of organic chemists showed that the backbone of each chain of the DNA molecule was composed of deoxyriboses linked end to end to each other with a single phosphate between them. The purines and pyrimidines stick out to one side, attached to the deoxyriboses. The nucleotides are covalently linked by an intervening phosphate from the 5′-hydroxyl group (on the deoxyribose) of one nucleotide to the 3′-hydroxyl group (also on the deoxyribose) of the adjacent nucleotide (Figure 9.7). This linkage is called a **phosphodiester linkage** because the phosphate group is linked to the two sugars by means of two ester (–O–) bonds. Two nucleotides are joined when a water molecule is removed between the 3′-hydroxyl of one nucleotide and the 5′-phosphate of another. Hydrolysis (the addition of a water molecule) is a reversal of this reaction. Also, because of the chemical nature of the phosphate group, the nucleotides and the polymers have a distinctly acidic nature, which is why they are called nucleic acids. Finally, it is important to note that a polynucleotide has a polarity. One end has a nucleotide that usually possesses a 5′-phosphate group on the exposed deoxyribose, and the other end has a free 3′-OH

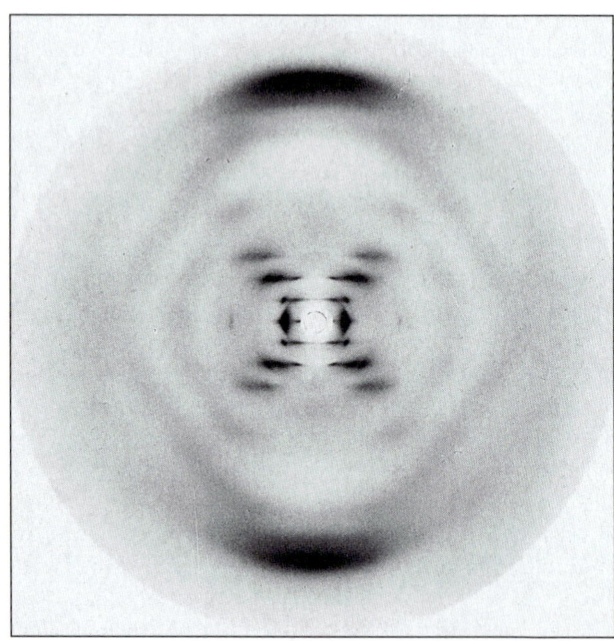

FIGURE 9.8 R. Franklin's X-ray picture of purified DNA. The simple and regular pattern of the spots indicates that the molecule is a repeating structure, now known to be a helix. The spacing between the bands at the top and the bottom indicates that the spacing between the elements of the helical molecule or the base pairs (0.34 nm). The repeat of the helix (one turn) is 3.4 nm and is indicated by the spacing between neighboring lines in the pattern. *(Dr. S. D. Dover, Division of Biomolecular Sciences, Kings College, London)*

group on the deoxyribose of the nucleotide. By convention, the sequence is written in the 5′ to 3′ direction (the phosphate and hydroxyl groups are included here for illustrative purposes only):

$$\overrightarrow{5'\text{-}PO_4\text{–TGCACGA–OH-}3'}$$

A final clue to unraveling the structure of DNA came from X-ray diffraction analysis of fibers of purified DNA. The first high-quality X-ray diffraction patterns were obtained by Rosalind Franklin with the DNA supplied by M. H. F. Wilkins, working in London at King's College, between 1950 and 1952. When an X-ray beam encounters groups of atoms in regular arrays, the path of the X-ray is bent (or diffracted), creating a pattern that can be captured on photographic film (Figure 9.8). An analysis of this pattern suggested to James Watson and Francis Crick (at Cambridge University in England), as well as to Rosalind Franklin, that the DNA molecule must be a relatively simple molecule with a repeating pattern because the spacing of the spots on the photograph is mathematically related to the spacing of the atoms. The most straightforward interpretation was that DNA has a repeating pattern of helical turns. However, these data did not tell them how the molecule was put together (that is, guanine pairs with cytosine, and adenine pairs with thymine). These conclusions were derived from Chargaff's data. The quantitative data was derived from the X-ray defraction data, including size of the helix, spacing between base pairs, and repeats in the helix.

In 1953, with all this information about DNA structure, Watson and Crick constructed a double-stranded model of DNA with the two polynucleotide chains running in antiparallel directions (Figure 9.9) (Wilkins, Watson, and Crick received the Nobel Prize in 1962 for solving the three-dimensional structure of DNA). In other words, the two chains of the helical molecule run in opposite directions (or antiparallel directions), where one strand goes up (in the 5′ end to the 3′ direction) and

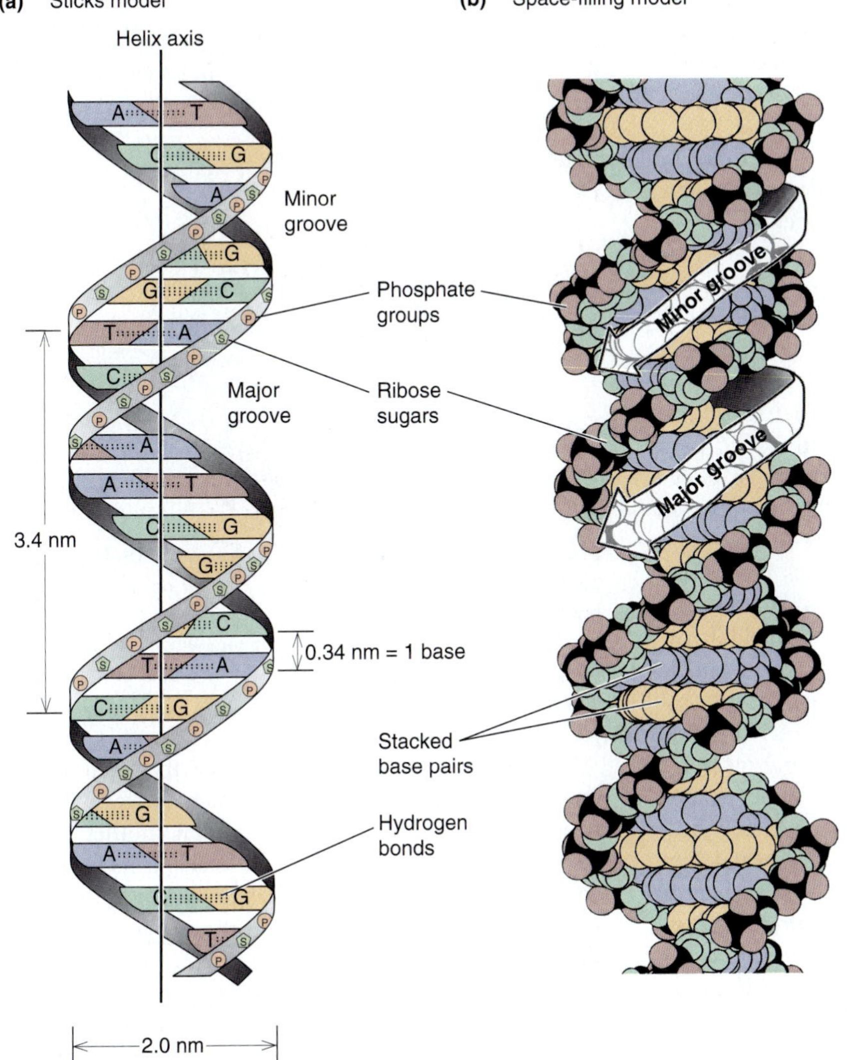

FIGURE 9.9 Structure of the DNA double helix. (a) The general form of the molecule, called the B form, in which the sugar-phosphate backbone is represented by two ribbons. Hydrogen bonds are shown as dots and hold the two strands together. (b) A space-filling model shows how stacking forces contribute to maintaining the double-helical structure.

the other goes down (in the 3′ end to the 5′ direction). Thus, each end of a double-helical molecule has a 3′-hydroxyl protruding from one strand and a 5′-phosphate protruding from the other strand; the two strands are said to be **antiparallel** to each other.

$$\begin{matrix} 3'3' \\ \uparrow\uparrow \text{ parallel} \\ 5'5' \end{matrix} \qquad \begin{matrix} 3'5' \\ \uparrow\downarrow \text{ antiparallel} \\ 5'3' \end{matrix}$$

The base-pairing is very specific. The purine guanine pairs with the pyrimidine cytosine only, and the purine adenine always pairs with the pyrimidine thymine. Consequently, Chargaff's rule was obeyed; the sequence of bases on one strand can be predicted from the sequence of bases on the other strand, and they are said to be **complementary** to each other. In the cell, DNA takes a very specific configuration, known as the B form (Figure 9.9). In this form, each chain makes a complete turn every 3.4 nm, the helix is right-handed, and each base is spaced 0.34 nm apart, which means that each turn of the molecule has ten nucleotide pairs. The two grooves of the helix are not symmetrical. One groove is larger and is known as the **major groove,** and the other is called the **minor groove** (Figure 9.10). DNA can exist in two other configurations, the A form and the Z form, but these exist only under laboratory conditions.

An important feature of the DNA model is that the two strands are held together by noncovalent **hydrogen bonds** (see Figures 9.9 and 9.10). A hydrogen bond is a weak electrostatic bond between two atoms that can be broken and reformed with relative ease. The complementary nucleotide pairs on opposing DNA strands are held together by the hydrogen bonds, a key feature of DNA structure. The A : T pair and the G : C pair differ, however; two hydrogen bonds occur in the former and three in the latter. The energy required to break the G : C pair is greater; thus DNA molecules with stretches of G : C pairs are less susceptible to separation of the two strands.

Hydrogen bonds can separate and re-form under certain conditions, sometimes referred to as breathing. This can be demonstrated by reacting formaldehyde with intact, double-stranded DNA. Normally formaldehyde does not react with double-stranded DNA, but if internally facing amino groups of the bases are exposed (by breathing), a reaction can occur. When DNA is reacted with formaldehyde, it slowly and irreversibly loses its hydrogen-bonding ability. The hydrogen bonding between the two strands of DNA can also be reversibly disrupted by heat, low pH, and a variety of other means. For example, if a sample of DNA is heated in boiling water, single strands of DNA appear. The separation of DNA into single-stranded molecules is referred to as **denaturation.** Slow cooling of denatured DNA can result in reformation of double-stranded DNA as the complementary strands eventually find each other, and this is referred to as **renaturation.**

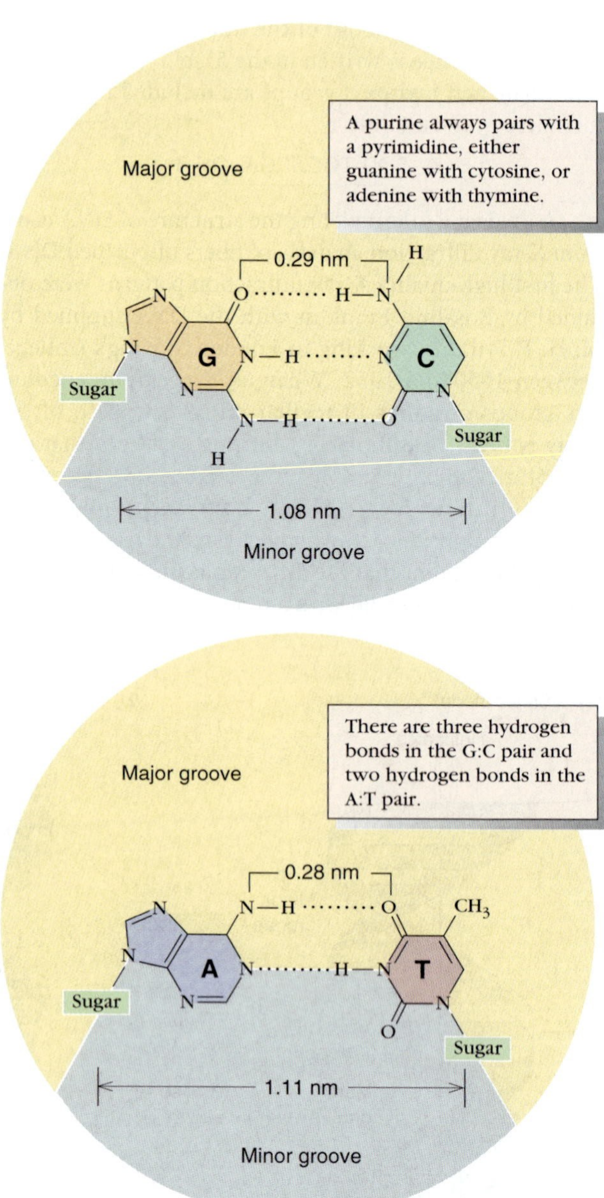

FIGURE 9.10 Complementary base-pairing in DNA. A purine always pairs with a pyrimidine. The G : C pair has three hydrogen bonds and is more stable than an A : T pair. Note the configuration of the two bases when they are paired, which is also reflected in double-stranded DNA. They are set on an angle, giving DNA a minor and a major groove.

In addition to the hydrogen bonds that hold the two strands together, another noncovalent force exists between adjacent bases stacked one upon another (Figure 9.9). The base pairs appear to be stacked like a pile of coins in the space-filling model, which depicts the actual size of each molecule. The atoms in one pair of bases are in close association with those of the adjacent base pairs. They are held together by a force known as van der Waals's forces, or **stacking forces.** Cumulatively, these forces are very important for maintaining the double-stranded helical structure of DNA.

SOLVED PROBLEMS

Problem

From the following sequence, predict the sequence and polarity of the complementary strand, and calculate the G + C content of the entire molecule and the percent of each base in the molecule.

5′-ATTTAACGGAAAATTTGGGATATAT
GCATATACATTAAATTGCAAA-3′

Solution

The complementary strand can be predicted because A always pairs with T, and G always pairs with C. Thus, the sequence would be

3′-TAAATTGCCTTTTAAA....etc.

The G + C content can be determined by adding the total number of G + Cs and dividing by the total number of base pairs. Thus, G + C = 11 / 46 × 100 = 23.9%. Therefore, the G content of the double-stranded molecule is about 12%, C = 12%, and A = 100% − 24% / 2 or 38% and C = 38%. The polarity of the complementary strand is antiparallel, that is, it is 3′ to 5′ in relation to the given strand.

SECTION REVIEW PROBLEMS

4. How does deoxyribose differ from ribose?
5. When writing the sequence of nucleotides in DNA, what is the convention, 5′ to 3′ or 3′ to 5′?
6. What is the difference between a nucleotide and a nucleoside? To what atom is the phosphate bonded?
7. You have found that a DNA preparation from a new bacteria has 22% guanine. Predict the percentage of cytosine, adenine, and thymine.

Naturally occurring, single-stranded forms of DNA also exist. But, because the stability of double-stranded DNA is much greater, single-stranded forms of DNA tend to fold back upon themselves whenever possible to form double-stranded helices with loops (Figure 9.11). These antiparallel duplex structures are called **hairpins** with a **stem** and a **loop.** The stem has the paired bases and the loop has unpaired bases. A number of small DNA bacterial viruses were found to possess single-stranded DNA, much to the surprise of the early investigators. These include phage ϕX174 and S13 viruses. Some parvoviruses, minute viruses that infect many vertebrate organisms, are also single-stranded. For single-stranded DNA to replicate, it goes through an intermediate stage when the molecule is temporarily double-stranded. One advantage of having a single-stranded DNA molecule is that it is more flexible than double-stranded DNA and thus can be more compacted and occupy less space in the virus particle. Double-stranded DNA is more rigid, although it is still a very flexible molecule. As you might expect, Chargaff's rules are not followed in single-stranded DNA.

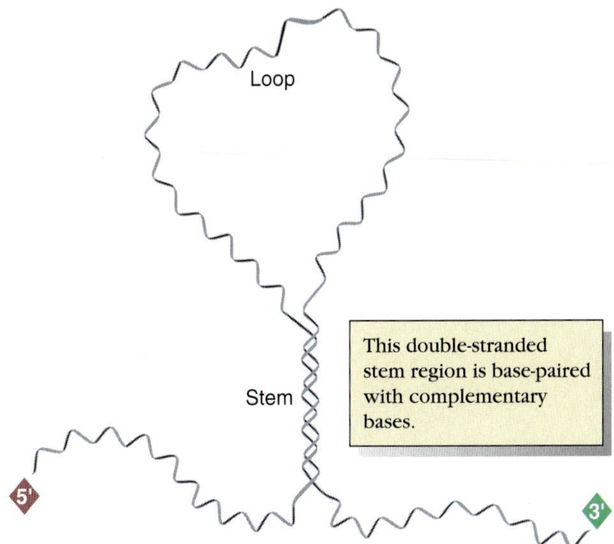

Loop
Stem
This double-stranded stem region is base-paired with complementary bases.
5'
3'

FIGURE 9.11 The hairpin (stem–loop) structure of single-stranded DNA. Single-stranded DNA frequently has complementary sequences within it that allow it to form these structures.

DNA Is a Long, Thin Molecule.

In eukaryotic cells, which have a nucleus, DNA is partitioned into chromosomes. Each chromosome contains one large molecule of DNA (Figure 9.12). Thus, the mouse (*Mus musculus*), which has 20 pairs of chromosomes (40 total), has 40 large DNA molecules that comprise its genetic material. In prokaryotes (bacteria) and viruses, the genetic material is usually composed of one copy of double-stranded DNA.

The length of a DNA molecule can be estimated and is enormous in relation to its diameter. One micrometer (μm; 1 m = 1×10^6 μm) of DNA corresponds to 2×10^6 Da (dalton or unit of molecular weight; one dalton equals the molecular weight of the hydrogen atom, or 1). The molecular weight of the largest human chromosome is 8×10^{10} Da; thus, it is $(8 \times 10^{10}$ Da$)/(2 \times 10^6$ Da$)$ = 40,000 μm (0.04 m or about 4 cm) in length. If each human chromosome has an average length of 0.03 m and humans have 46 chromosomes, then the total length of the human genome is about 1.3 m (about 4.5 ft). All this DNA is packed into the nucleus of each of our cells, which have a diameter of about 100 μm. In contrast, the *E. coli* genome is about 0.001 m in length. Compare these relatively long distances with the 2×10^{-9} m diameter of DNA. Because of this extreme length : diameter ratio, DNA is very subject to breakage during pouring, mixing, or pipetting. DNA molecules greater than 3×10^7 Da break very easily. However, the smaller DNA molecules from bacteria and viruses can be spread out for electron microscopic examination and measurement

(a)

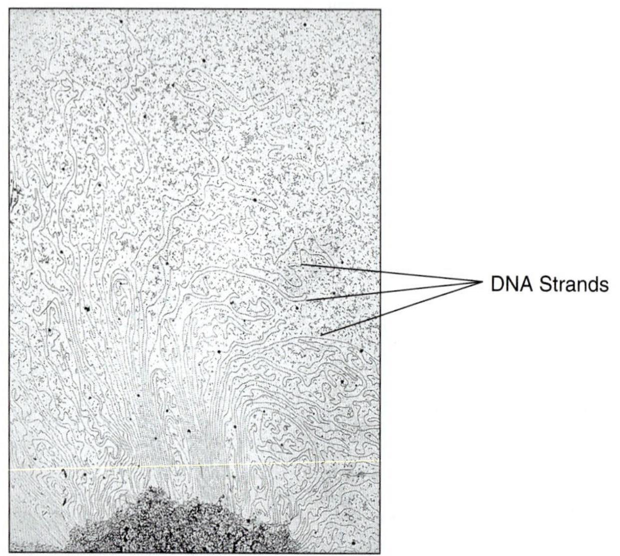

(b)

FIGURE 9.12 Eukaryotic and prokaryotic electron micrographs of DNA. (a) Mouse chromosomal and (b) *Escherichia coli* DNA gently spread from the cell without breaking the thin long molecule. The mouse DNA is very tightly coiled and compact, making it very difficult to measure the total length of the molecule. However, bacterial DNA can be measured. *(a, courtesy of U. K. Laemmli, University of Geneva, Switzerland, 1988. From Cell, 12: 817; b, Dr. Gopal Murti/Science Photo Library/Photo Researchers, Inc.)*

(Figure 9.12). Other means of measuring the length of DNA molecules are available, and the total DNA content of a number of different organisms has been determined (Table 9.3).

The total molecular weight of a DNA molecule can also be converted into number of nucleotide base pairs as well. This is done by dividing the molecular weight of DNA by 660, the molecular weight of a nucleotide pair (Table 9.3).

TABLE 9.3 *The Sizes of DNA Molecules from Various Organisms*

Organisms	Molecular Weight (Da)* ($\times 10^6$)	Base Pairs ($\times 10^4$)
Polyoma virus	3.2	0.48
Phage lambda	32	4.8
Phage T4	106	16
Mycoplasma homina	530	80
Rhizobium meliloti plasmid	990	150
Escherichia coli	2,900	439
Yeast (largest chromosome)	1,500	227
Human (largest chromosome)	80,000	12,120
Human (total)	2,000,000	300,000
Drosophila melanogaster (total)	100,000	15,500
Maize (average chromosome)	165,000	25,000
Onion (total haploid genome)	10,000,000	1,500,000

*One base pair weighs approximately 660 Da; thus, molecular weight in daltons is converted into the total number of base pairs by dividing by 660.

The Structure of RNA Is Very Similar to That of DNA.

We have already discussed two major differences between RNA and DNA; that is, RNA contains uracil instead of thymine and ribose instead of deoxyribose in each of the nucleotides. In addition, unlike most DNAs, RNA is single-stranded and can form the hairpin loops found in single-stranded DNA. Because RNA lacks the regular double-helical structure, it does not follow Chargaff's rules.

RNA has a number of specialized functions in the cell, which we will discuss later in the text. For example, RNA is able to catalyze some chemical reactions, but DNA cannot (see CURRENT INVESTIGATIONS: *How Did Life on Earth Begin?*). RNA comes in a large number of molecular sizes. The size of RNA molecules is usually much smaller than DNA, and as a consequence, it is possible to isolate intact RNA without shearing damage. However, all cells have enzymes known as ribonucleases that degrade RNA, and they make it difficult to isolate intact RNA.

CURRENT INVESTIGATIONS

How Did Life on Earth Begin?

There are a variety of scientific opinions about how life on earth began. Some believe that life on earth arose from outer space, either as a primitive cell or as DNA deposited from another planet. The hypothesis that life began on some far-off planet, at best, simply begs the question. How did life begin on *that* planet? A more scientific explanation is that life slowly formed under conditions present on our planet 4 to 5 billion years ago. Evidence supporting this idea has come from experiments in the laboratory that reconstruct the conditions of primitive earth. Many experiments along this line have been done within the last 20 years. When plausible conditions on the primitive earth are simulated in reaction vessels, many simple organic compounds of biological significance form spontaneously and in relative abundance. These include amino acids, aliphatic hydrocarbons, sugars, purines, and pyrimidines, which are the precursors of proteins and nucleic acids. To carry this to the next logical step, investigators have produced peptides, but in relative small amounts. If dry amino acids are heated for a week or so at 120°C, large numbers of protein-like molecules with molecular weights of about 10,000 can be produced. Similarly, polymers of nucleotides can be produced.

This brings up an important question, Which came first, the protein (the chicken) or the nucleic acid (the egg)? The proteins catalyze biological reactions including the synthesis of nucleic acids, but nucleic acids provide the information required to make proteins. It would seem that you cannot have one without the other. This was indeed a big quandary until the early 1980s, when Thomas Cech (at the University of Colorado) discovered for the first time that nucleic acids, specifically RNA, could also catalyze reactions. This was truly an astounding discovery, and in 1989 he was awarded the Nobel Prize.

It is now postulated that evolution began with RNA molecules performing the catalytic activities necessary to assemble themselves from nucleotide precursors. The RNA molecules evolved in self-replicating patterns, using recombination and mutation (RNA mutates much more rapidly than DNA) to explore new functions and adapt to new niches. RNA molecules then began to synthesize proteins, eventually arranging them according to an RNA template. Eventually proteins were able to carry out catalytic functions better than RNA and replaced this function of RNA. At some point, single-stranded DNA appeared on the scene, coded from RNA, and then double-stranded DNA appeared with its unique properties of increased stability and faithful self-replication.

How did cells arise? Most scientist believe that protocells, the precursors of modern-day cells, developed first. These were clusters of proteins and nucleic acids. Billions of years ago, nucleic acids evolved more capacity to encode new proteins that stabilized the protocell, and primitive bacteria arose. Further evolution over about 3 billion years gave rise to what we see around us.

SOLVED PROBLEMS

Problem

DNA is very long in relation to its diameter. If a strand of DNA were 0.1 mm in length, what is the ratio of length to diameter, or the axial ratio?

Solution

0.1 mm = 10^5 nm

$$\text{diameter of DNA} = 2 \text{ nm}$$
$$\text{axial ratio} = 1 \times 10^5 / 2 = 50{,}000$$

Problem

Assume that you are studying a new bacterial virus you found in the sewer. The first thing you analyze is the DNA base composition. You find it to be A = 22%, T = 28%, G = 20%, and C = 30%. How would you explain these data?

Solution

Double-helical DNA would have a base composition in which A = T and G = C. This DNA does not; thus the virus must not have double-stranded DNA as its genetic material. It must have single-stranded DNA.

Problem

You have isolated a new virus and found that it is 6.6×10^7 Da. You want to convert this number into nucleotide pairs because this virus has double-stranded DNA. How do you convert daltons to kilobase pairs?

Solution

Each nucleotide pair has a molecular weight of about 660 Da. Thus, if you divide the total molecular weight by 660, you will get the number of base pairs, or 100,000. This number can be converted into kilobase pairs by dividing by 1000, which is equal to 100 kilobase pairs (kb).

SECTION REVIEW PROBLEMS

8. If you had two preparations of DNA from two different organisms in which one had a G + C content of 52% and the other had a G + C content of 38%, what prediction could you make about the stability of the DNA to denaturation? Which is more stable and why?
9. What are some of the characteristics of DNA?
10. Why is it difficult to isolate and purify DNA with molecular weights above 3×10^7 Da?

DNA Replicates Using Itself as a Template.

The discovery that DNA is a double helix was profound and suggested how genetic information could be transferred from generation to generation. In this section, we will first discuss how the complementary nature of DNA enables it to duplicate and the proof of the duplication mechanism. We will then delve into the details of how DNA replicates itself, which enzymes are required, and how they work to ensure accurate replication of DNA.

Meselson and Stahl Proved That DNA Replicates Semiconservatively.

In 1953, it did not escape the notice of Watson and Crick that their proposed structure for DNA immediately suggested a mechanism for its replication. The two strands are complementary; thus one strand could serve as the template and determine the sequence of the other strand. For example, if the sequence of one strand is 5′-AGCACCT-3′, then the other strand will be 3′-TCGTGGA-5′. The chemical nature of the hydrogen bonds that hold the two strands together allows the helix to be "unzipped" and zipped up again, or to form a new complementary structure along each naked single strand. This form of replication is referred to as **semiconservative replication.** In this model, one of the parental strands is retained and gives rise to two daughter molecules consisting of one parental strand and one newly synthesized strand (a hybrid molecule of one new and one old strand; Figure 9.13). Another model of replication, called **conservative replication,** was also possible and was seriously considered. In the conservative model, the two parental strands of DNA remain together and serve as a template for the synthesis of a new daughter double helix.

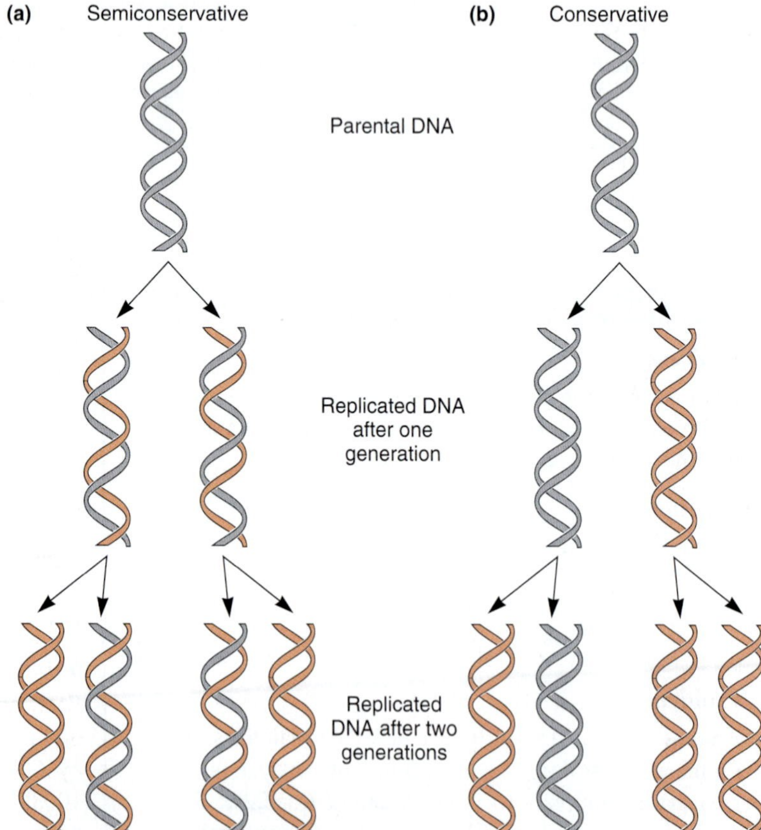

FIGURE 9.13 Two models of DNA replication. (a) Semiconservative replication passes one parent strand to each daughter molecule. (b) In conservative replication, both parent strands are passed to only one daughter molecule. The parental strands are shown in gray, and the newly synthesized stands are shown in red.

In 1958, Matthew Meselson and Franklin Stahl, then at the California Institute of Technology, published the results of an experiment that differentiated between the conservative and semiconservative replication models. They allowed bacteria (*E. coli*) to grow for several generations in a medium containing the heavy isotope of nitrogen, ^{15}N. These bacteria, and their DNA, were denser than normal because heavy nitrogen had been incorporated into the bacteria. They then transferred these bacteria to normal lighter isotope ^{14}N medium and harvested the DNA from the bacteria at various intervals.

To differentiate heavy DNA from light DNA, they used the technique of **equilibrium density gradient centrifugation.** This technique uses a cesium chloride solution because cesium is a very heavy atom (molecular mass of 132 and a specific gravity of 1.873 g/ml) and its chloride salt is very soluble in water. Thus, CsCl concentrations greater than 6 M can be prepared and have a very high density. DNA has a density of about 1.70 g/cm^3 and thus the CsCl solution is prepared at this density. However, upon centrifugation at very high speeds, the atoms of cesium are sufficiently heavy to move in response to the high *g*-forces and form a density gradient in the centrifuge tube (from 1.6 g/cm^3 at the top of the centrifuge tube to 1.8 g/cm^3 at the bottom). The DNA molecules equilibrate at a density that equals the density of the surrounding CsCl to form a sharp band within the tube. If a DNA molecule has a high percentage of ^{15}N, it will form a distinctly different band with a higher density than normal ^{14}N-containing DNA. Intermediate amounts of ^{15}N to ^{14}N in DNA will form intermediate bands of DNA. In the actual experiment (Figure 9.14), bacteria were grown in a medium containing the heavy isotope of nitrogen (^{15}N) for a sufficient time to allow all the DNA molecules to become completely labeled (refer to this DNA as heavy–heavy DNA because both strands contain the isotope). Excess unincorporated ^{15}N is washed from the bacteria, the growth medium is replaced with ^{14}N-containing medium, and the cells are allowed to grow. Cells are removed immediately (these contain the heavy–heavy DNA) after one and two generations. To analyze the DNA for density differences, the cells were broken open, and the DNA was extracted for density gradient centrifugation. The control, which has only ^{14}N in its DNA, equilibrates toward the top of the centrifuge tube, whereas heavy–heavy DNA equilibrates toward the bottom of the centrifuge tube. Meselson and Stahl found that after one generation of bacterial growth in ^{14}N (light) medium of the heavy–heavy labeled DNA, only intermediate density DNA were present. After the second DNA doubling, DNA equilibrated at both intermediate and light density DNA.

The results of this experiment were inconsistent with a conservative model of DNA replication. Meselson and Stahl found intermediate densities of DNA, which is not predicted from the conservative model of replication. Under the conservative model of DNA replication, you would still expect to find a band containing heavy–heavy DNA after one and two generations of growth in a ^{14}N-containing medium. However, this was not observed in Meselson's and Stahl's experiment. In contrast, the semiconservative model of replication, in which each strand pairs with a newly synthesized strand, predicts *only* that an intermediate density molecule would be found after one generation. This molecule would contain one strand with the heavy isotope of nitrogen (^{15}N) and one strand with the lighter isotope of nitrogen (^{14}N) and can be explained only by the semiconservative model of replication. Thus, Meselson's and Stahl's results were consistent with the prediction of the Watson–Crick model that DNA replicates in a semiconservative manner.

DNA Replication Is Bidirectional.

Many of the early investigations on DNA were done with simple organisms such as bacteria and their viruses. Individual bacteria are single cells that lack a nucleus, are easily grown in enormous numbers, and are ideal for analyzing molecular mechanisms such as DNA replication. It was expected that findings with bacteria could be extrapolated to more complex organisms. As a result, in the years after the proof of semiconservative replication, bacteria were widely used to study DNA replication.

In this respect, *E. coli* was the first bacterium used to determine how DNA initiated its replication. The question was, if DNA replication always begins at a specific point, does replication occur in only one direction, or simultaneously in both directions? To answer this question, in 1963, John Cairns used replicating DNA from *E. coli* and a technique known as **autoradiography.** He first incorporated radioactive hydrogen (tritium or ^3H) into replicating DNA and then placed the ^3H–containing DNA against a photographic emulsion. The radiation emitted by the isotope converts the silver in the emulsion into a visible form, after developing, in the same way that visible light activates an image on film. At this time, it was known that the chromosomes of bacteria and some viruses take the shape of a continuous circle. This information came from both genetic experiments (Chapter 8) and autoradiographic pictures of the *E. coli* chromosome (Figure 9.15). Cairns did his experiment on replicating DNA hoping to "catch" DNA during the replication cycle. He obtained the now-famous image shown in Figure 9.16 and used it to interpret the mechanism by which the *E. coli* chromosome undergoes replication. Careful examination of the photograph in Figure 9.16 shows two different levels of labeling in different loops of the DNA suggesting that the DNA has undergone one complete round of replication and part of a second. In the first round of replication, one strand of the DNA is labeled,

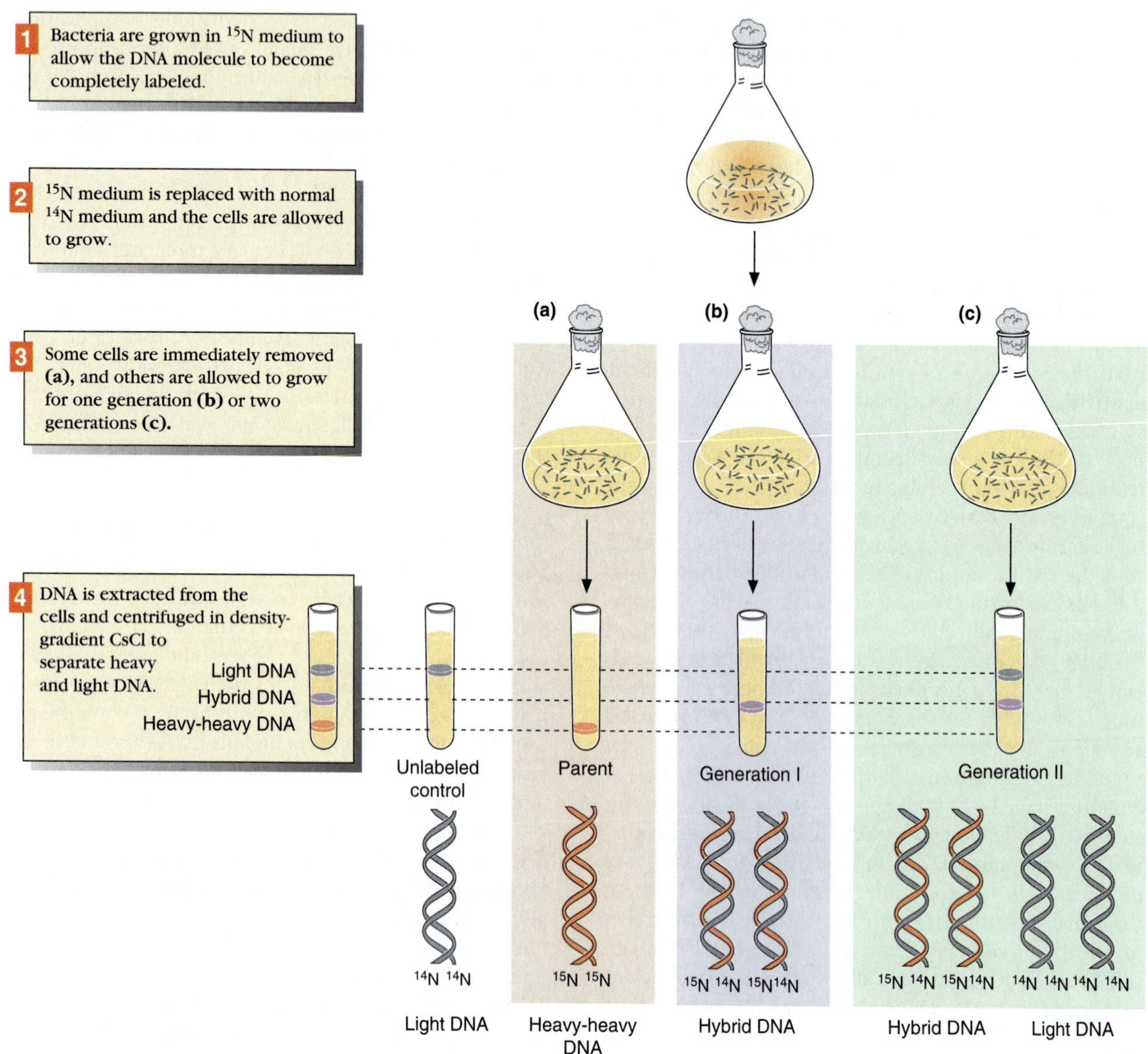
1 Bacteria are grown in ¹⁵N medium to allow the DNA molecule to become completely labeled.
2 ¹⁵N medium is replaced with normal ¹⁴N medium and the cells are allowed to grow.
3 Some cells are immediately removed (a), and others are allowed to grow for one generation (b) or two generations (c).
4 DNA is extracted from the cells and centrifuged in density-gradient CsCl to separate heavy and light DNA.
Light DNA
Hybrid DNA
Heavy-heavy DNA
(a)
(b)
(c)
Unlabeled control
Parent
Generation I
Generation II
14N 14N
15N 15N
15N 14N 15N14N
15N 14N 15N14N 14N 14N 14N 14N
Light DNA
Heavy-heavy DNA
Hybrid DNA
Hybrid DNA
Light DNA

FIGURE 9.14 Proof that DNA replicates semiconservatively. Meselson and Stahl, using normal heavy radioisotopes of nitrogen (nitrogen is abundant in DNA) to label DNA strands, showed that the DNA strands followed the semiconservative model (Figure 9.13) rather than the conservative model of replication.

but in the second round, two strands are labeled. The double-labeled part of the DNA exposes the film to a greater extent, resulting in more silver particles. These experiments were interpreted to mean that the *E. coli* chromosome has one point of origin for replication. At the beginning of the experiment, the entire molecule was labeled with ^{3}H, and only after completion of one replication cycle did another round of replication begin.

Cairns' experiment found how many origins of replication there are in the *E. coli* genome (and very likely all circular DNAs from bacteria), but it did not determine whether replication is **unidirectional** or **bidirectional.** In unidirectional replication, the replication site expands in one direction, whereas in bidirectional replication the replication site expands in both directions at the same time. In an attempt to answer this question, the ^{3}H label was incorporated into bacterial DNA for short intervals, followed by removal of the unincorporated radioactivity. This is known as **pulse labeling.** In 1972, David Prescott and Peter Kuempel at the University of Colorado at Boulder grew *E. coli* on a medium containing ^{3}H-labeled thymine that incorporated into the DNA. The ^{3}H was present at a low level in the DNA; there was just enough ^{3}H to visualize the track of the DNA from the origin

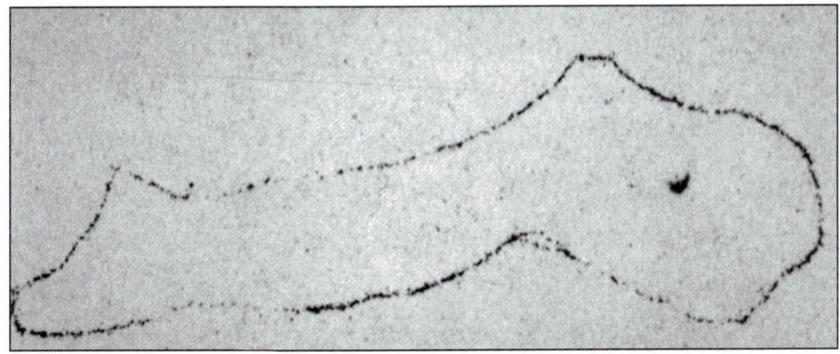

FIGURE 9.15 Replicating *E. coli* DNA. This autoradiograph shows the *E. coli* chromosome labeled with ^{3}H-medium during replication. The lines are actually spots of silver grains. It is apparent that the chromosome is a continuous circle. *(From Cairns, J. 1963.* Endeavour, *22: 144. Copyright 1963. Pergamon Press, Ltd.)*

when later exposed to X-ray film (light labeling). They then gave the cells a short exposure to a very high concentration of ^{3}H-thymine (the pulse-label, which is the dark labeling in Figure 9.17). After the lysed cells were exposed to photographic emulsion, Prescott and Kuempel observed two short tracks of dense silver grains at each end of the growing chromosome (Figure 9.17). The only interpretation of these results is that DNA is replicating in both directions simultaneously from one origin, proving that DNA replication in *E. coli* is bidirectional.

Today we know that the origin of replication in *E. coli* is governed by a sequence of bases in DNA referred to as *oriC*. This sequence is composed of a region of DNA about 250 base pairs long that contains a common 9-bp [base pairs; kilobase pairs (kb) is used when referring to thousands of base pairs] sequence, 5′-TTAT(C/A)CA(C/A)A, repeated four times (the bases in parenthesis can be either C or A). This sequence is also found in many other gram-negative bacteria and is the binding site of a protein responsible for initiating DNA replication (the DnaA protein).

After the Prescott and Kuempel experiments with *E. coli*, other investigators asked the same question of eukaryotes. Investigation of DNA replication in *Xenopus*

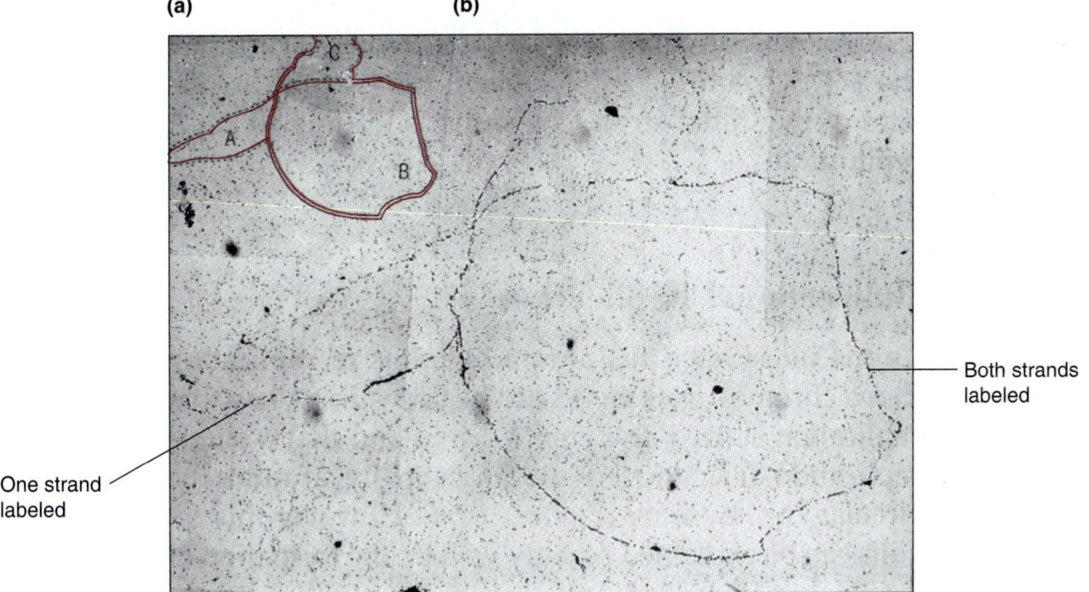

FIGURE 9.16 DNA replication begins at one point. Cairns exposed *E. coli* cells to an ^{3}H medium during DNA replication, lysed the cells, and fixed the expanded chromosome in photographic emulsion for autoradiographic analysis. The DNA molecule has been completely labeled once, and the darker, second round of replication shows both strands labeled, giving more silver particles in a given area. This is shown in part (b). An interpretative drawing (micrograph) is shown (a), where the solid line depicts ^{3}H-thymidine-labeled DNA and the dotted line is unlabeled DNA. A and C represent the incompletely labeled segments of the replicated chromosome, whereas B is completely labeled because of growth in ^{3}H medium for two generations. *(From Cairns, J. 1963.* Cold Spring Harbor Symposium on Quantitative Biology, *28: 44.)*

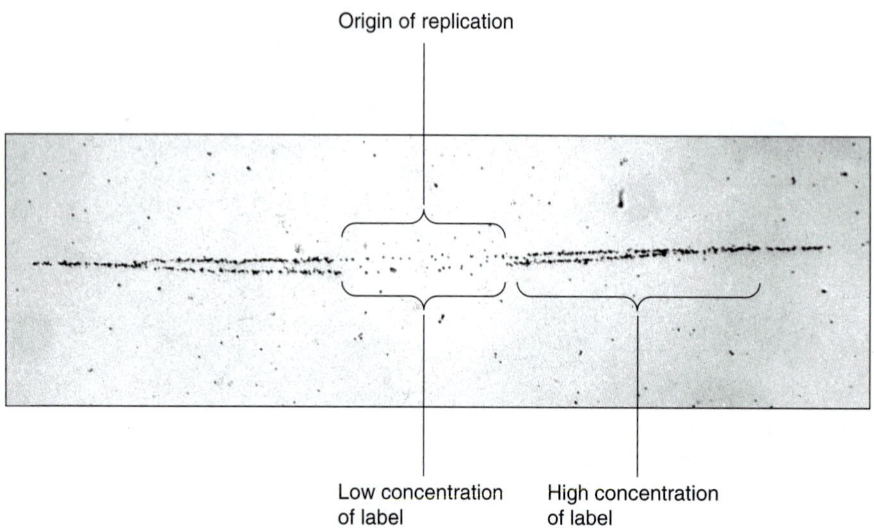

FIGURE 9.17 Proof of bidirectional replication. In the Prescott and Kuempel experiment, *E. coli* chromosomes are first labeled for 13 minutes with a low concentration of ^{3}H-thymine, followed by labeling for 6 minutes with a high concentration of ^{3}H-thymine and ^{3}H-thymidine. The cells are beginning a new round of DNA replication when the experiment is begun. The DNA is extracted and examined by autoradiography and produces a grain track on the developed film. Note that both ends of the molecule are heavily labeled and that the separated DNA strands (after replication) are visible, which can be explained only by bidirectional replication. *[Taken from Kuempel, P., D. M. Prescott, and P. Maglothin. 1973. "Autoradiographic demonstration of bidirectional replication in* Escherichia coli." DNA Synthesis in Vitro *(R. D. Wells and R. B. Inman, Eds.), pp 463–471. University Park Press, Baltimore, MD.]*

laevis (an African aquatic toad) showed that, as in prokaryotes, replication is bidirectional. However, unlike prokaryotes, eukaryotic replication has multiple origins on each chromosome. In a typical experiment (Figure 9.18), cells were first briefly labeled with ^{3}H-thymidine; then the radioactive material was removed and replaced with a more dilute ^{3}H-thymidine medium, and DNA replication was allowed to continue. The cells were broken open and exposed to film for autoradiographic analysis. A denser track of labeling was observed during the initial phase of replication, and a less-dense track was seen in the second phase of labeling. This pattern radiated from one fixed point, suggesting that the replication was bidirectional. The observation of multiple tracks on the same molecule suggested that **multiple origins of replication** were present on each chromosome.

Multiple origins of replication are clearly required if the large amount of eukaryotic DNA is to be replicated during the time of one cell division. A large chromosome of *Drosophila melanogaster* contains as much as 7×10^7 nucleotide pairs. The rate of replication in *Drosophila* is about 2600 bp/minute, compared to 50,000 bp/minute for *E. coli*. At a rate of 2600 bp/minute, a single origin of replication would require a little over 2 weeks to replicate one chromosome. In reality, it takes only 4 minutes for a *Drosophila* cell to replicate its DNA, indicating the presence of 6000–7000 origins of replication. Different cell types have different cell division times and, consequently, different DNA replication times. The replication time is governed by the number of *active* origins of replication. In more slowly growing cells, the number of origins of replication per chromosome is greatly decreased.

SECTION REVIEW PROBLEMS

11. What is semiconservative replication? What is the proof of semiconservative replication?
12. What is bidirectional replication? Why is it important that organisms with large genomes (eukaryotes) have multiple origins of replication?
13. If the human genome has 3 billion base pairs, and polymerization occurs at 3000 bp/minute (in both directions) per origin, and replication requires 6 hours, how many origins of replication, at a minimum, are required to complete synthesis of the entire genome for each cell division? How far apart in base pairs will origins of replication be from each other (assuming equal spacing)?

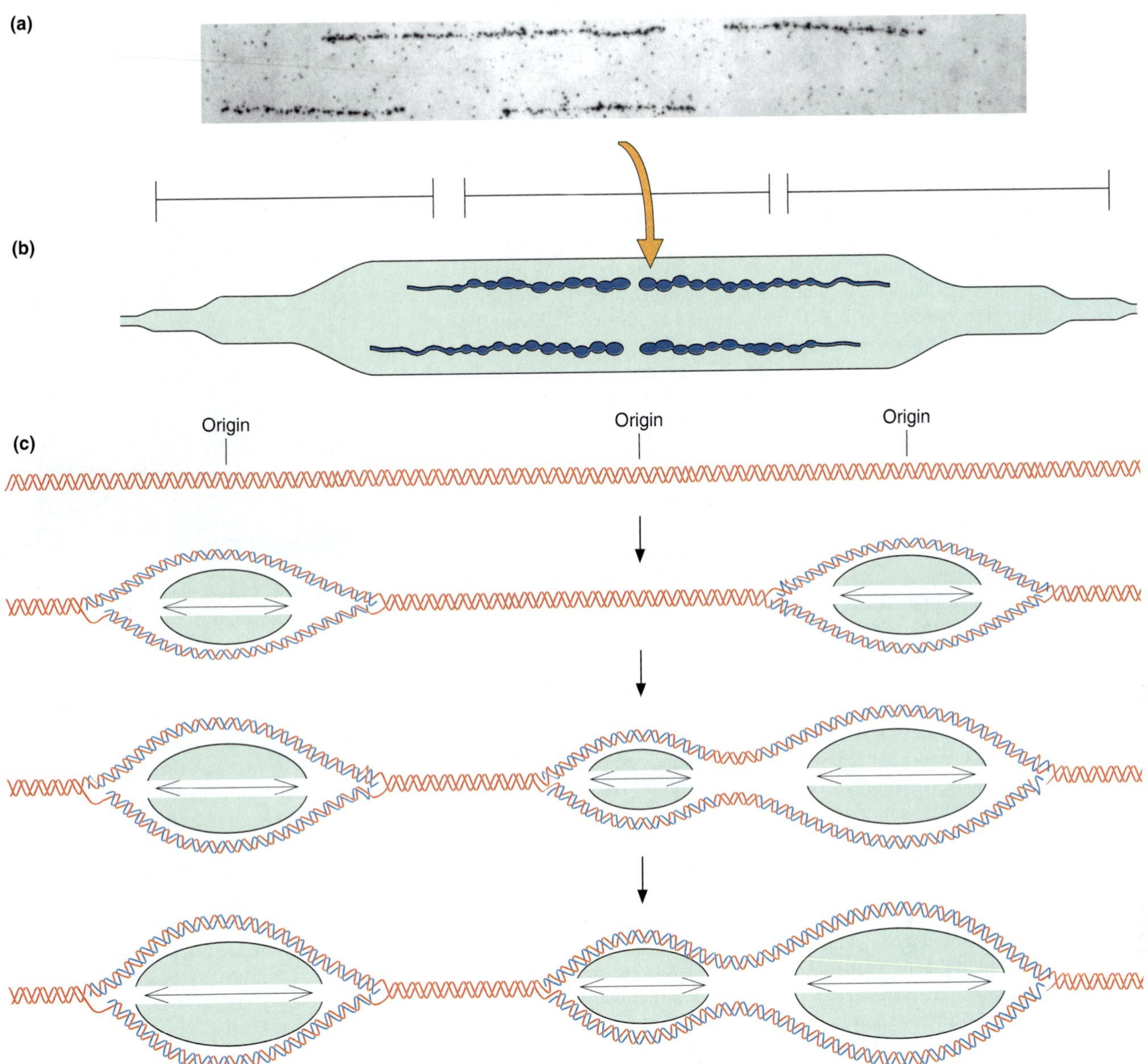

FIGURE 9.18 Eukaryotic chromosomes have multiple origins of replication. (a) The grain tracks produced by a replicating chromosome of *Xenopus laevis* cells. The cultured cells were pulse-labeled with very high levels of radioactive ^{3}H-thymidine. The gap between the start points represents unlabeled DNA. An interpretation of one replication is presented in (b), and a schematic representation of the progress of multiple events is shown in (c). *(Taken from Callan, H. G. 1973.* Cold Spring Harbor Symposium on Quantitative Biology, *38: 196.)*

DNA Polymerase Is Responsible for DNA Replication.

In 1956, shortly after Watson and Crick proposed their model for the structure of DNA, Arthur Kornberg at Stanford University discovered and began to characterize an enzyme from *E. coli* that could synthesize DNA. Although Kornberg's enzyme was capable of synthesizing DNA (DNA polymerase I, see *Historical Profile: Arthur Kornberg (1918–)*, DNA synthesis is actually catalyzed by **DNA polymerase III,** which is responsible for the covalent addition of nucleotides to a growing chain of DNA. The enzyme requires deoxynucleoside triphosphates (dATP, dCTP, dGTP, and TTP), Mg^{++} ions, and preexisting DNA molecules known as a **template** and a **primer** (Figure 9.19). A template is the DNA strand supplying the information necessary to synthesize the complementary strand, whereas a primer provides a free 3′-OH group and acts as a starter for new strand synthesis. *E. coli* has two other DNA polymerases—I and II, which

HISTORICAL PROFILE

Arthur Kornberg (1918–)

Arthur Kornberg (Figure 9.A) has been instrumental in unraveling the mechanism of DNA replication and received the Nobel Prize for his efforts. He was born (1918) and grew up in Brooklyn and graduated at an early age because he was promoted through 3 years of school. He went on to the City College of New York, continuing an interest in chemistry that developed in high school. After receiving his M.D. during the first part of World War II at the University of Rochester, he did research at The Nutrition Laboratory at the National Institutes of Health. This was his first real introduction to research. He spent the next ten years pursuing the study of enzymes at the National Institutes of Health.

One of the crucial steps in unraveling how DNA is replicated is to understand what enzymes are critical for its synthesis. The first paper on this subject was published by Kornberg in 1956 (*Biochim. Biophys. Acta,* 21: 197). This was only 3 years after Watson and Crick had unraveled the structure of DNA and had proposed a model for its replication. In this first paper, A. Kornberg, I. R. Lehman, M. J. Bessman, and E. S. Simms showed that ^{14}C-thymidine could be incorporated into DNA. In these studies, Kornberg and his colleagues had added DNA to prevent nuclease action, which they knew was rampant in the reaction mixtures. As it turned out, the DNA fulfilled two essential roles: it served as a template, and it supplied nucleotides that were converted to triphosphates from the degraded DNA.

Kornberg and his colleagues were responsible for the discovery and understanding of many of the features of DNA polymerase I, including the requirement for a primer (in 1970) and the presence of exonuclease activity. In later reflections, Kornberg admitted that it made no sense to him that DNA polymerase should be able to degrade the very product that it synthesized. Kornberg and his colleagues discovered that the polymerase's 3′-exonuclease action removed frayed or mismatched ends of the primer, permitting correct nucleotides to be added. After defining 3′-exonuclease action, Kornberg's group was flabbergasted to find additional 5′-exonuclease action. Through careful investigation, they defined the 5′-exonuclease action as necessary for removing damaged DNA segments and for removing RNA that initiates DNA chains.

After Kornberg had received the Nobel Prize for his landmark discovery of what was supposed to be *the* enzyme that synthesized DNA, the scientific community was astonished when John Cairns found a mutant of *E. coli* that seemingly lacked this polymerase function, yet was still able to reproduce. How could a cell reproduce without a DNA polymerase? The conclusion was rapidly drawn that the polymerase described by Kornberg and his colleagues was not, in fact, the critical enzyme. His DNA polymerase, which had been so carefully characterized, was called a "red herring" by the editors of *Nature New Biology,* a widely read, new journal in molecular biology. Shortly afterward, polymerase II and III were found, and the exact role of polymerase I was established in the replication of DNA.

FIGURE 9.A Arthur Kornberg, 1994. *(Courtesy of A. Kornberg)*

Did Kornberg and company ever prove "genetic activity" for the *in vitro* synthesized DNA? Kornberg was uniformly unsuccessful in obtaining transforming ability with *in vitro* synthesized DNA until 1967, when DNA ligase was discovered. This discovery finally allowed the complete synthesis of biologically active DNA. With this accomplishment, Kornberg felt confident that all impediments had been removed for the synthesis of genes and the manipulation of the genetic material. Since then, Kornberg has defined other aspects of DNA replication and repair.

have similar requirements. The main function of bacterial polymerase I *in vivo* is DNA repair (Chapter 17), but polymerase I also has a dispensable function in DNA replication. Little is known about polymerase II. In eukaryotes, four DNA polymerases have been characterized; the eukaryotic enzymes are referred to as α (alpha), β (beta), γ (gamma), and δ (delta).

Before we go into the details of DNA replication, you should be aware that DNA synthesis by DNA polymerases has the following characteristics:

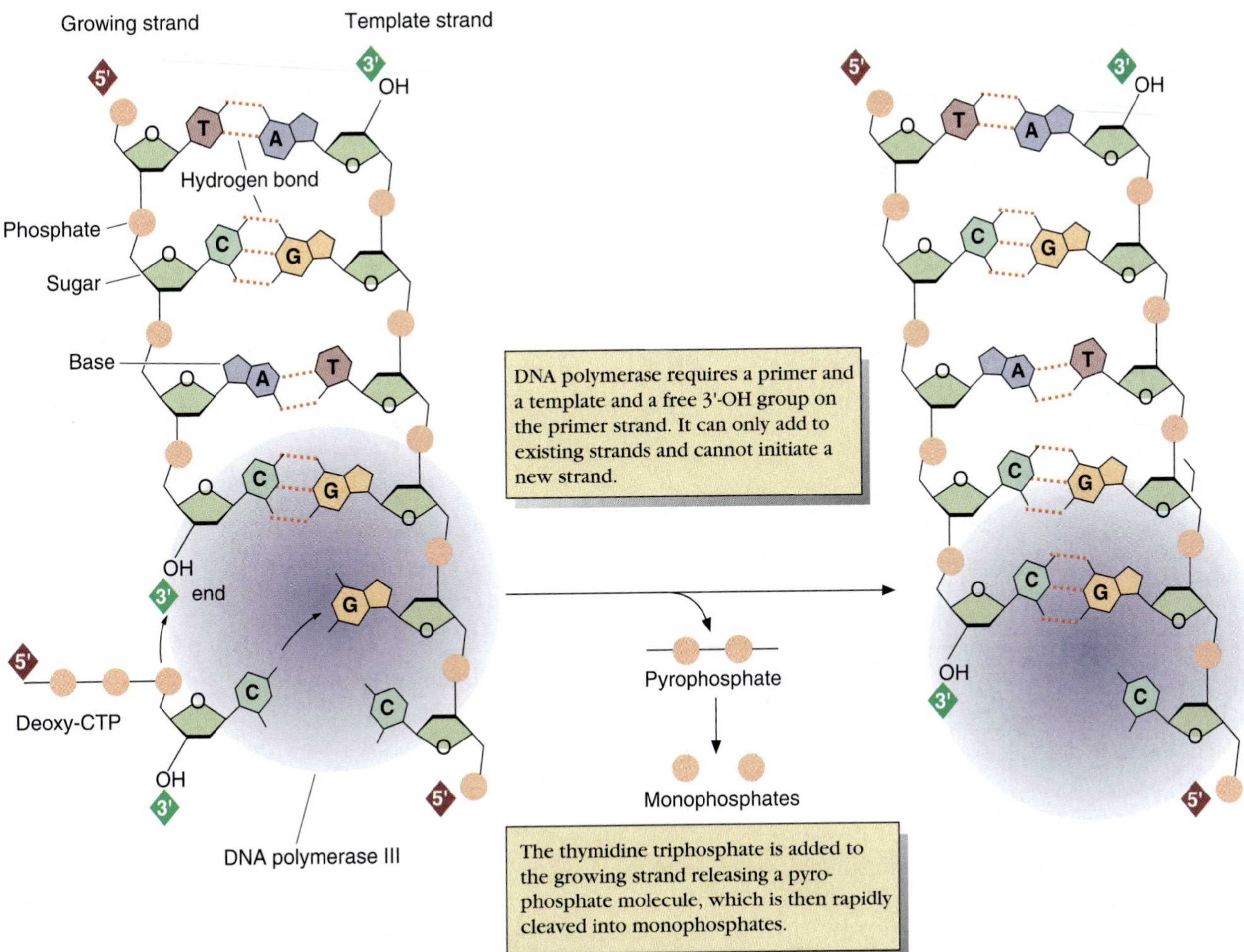

FIGURE 9.19 The reaction catalyzed by DNA polymerase III. This figure shows how a new nucleotide is added to the growing 3′-end of a double-stranded DNA molecule. Note that the new nucleotide is added to the 3′-OH end of the growing chain and follows Chargaff's rules.

1. There is an absolute requirement for a template strand, and new base selection is template driven.
2. Polymerization is always in the 5′→3′ direction on the new strand (the new nucleotide is always added to the 3′-OH group) and in the opposite direction (antiparallel) to the template strand.
3. Initiation of new chain growth is not possible without a primer with a 3′-OH group.

DNA Synthesis Occurs Using Both Strands as a Template.

The two template strands of a replicating DNA molecule are antiparallel (5′→3′ and 3′→5′) at the unwinding replication fork. If DNA polymerase can synthesize in the 5′→3′ direction only, how can both template strands be used for DNA replication simultaneously? This problem of DNA replication is resolved by **discontinuous** DNA synthesis on the one strand and **continuous** synthesis on the other strand (Figure 9.20). Discontinuous DNA synthesis results in small (1000–2000 bp) segments of DNA using the 5′→3′ template strand, where DNA polymerase reinitiates new short segments. These segments are subsequently linked. Because of the "hopping" nature of the initiation of DNA synthesis on the discontinuous strand, it is also referred to as the **lagging strand**, and the other continuously synthesized strand is the **leading strand.**

Discontinuous synthesis of DNA was discovered by Reiji Okazaki in the 1960s, and the short fragments generated during discontinuous DNA synthesis are sometimes referred to as **Okazaki fragments.** Okazaki discovered these precursor fragments by adding radioactive ^{3}H-thymidine to cultures of *E. coli* for a brief time and

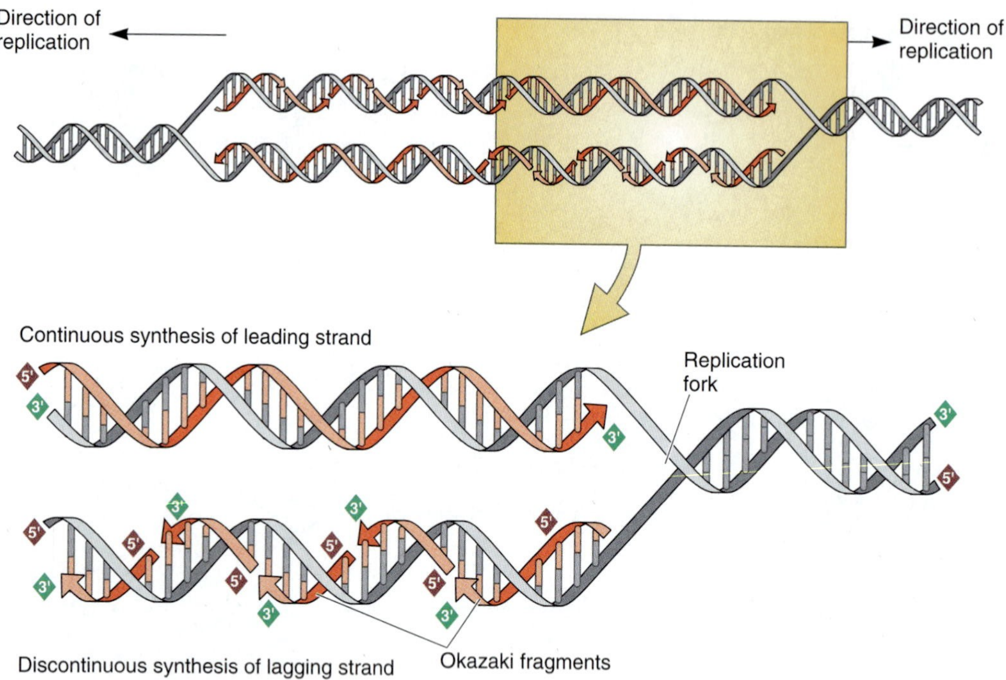

FIGURE 9.20 Bidirectional DNA replication. This model shows how DNA replication proceeds in both directions at once, whereas DNA polymerase synthesizes new strands from the antiparallel strands of the parent molecule at each replication fork.

then adding nonradioactive thymidine in great excess to prevent further incorporation of radioactivity. Samples were removed at timed intervals, and the DNA was extracted, denatured into single-strand segments, and separated according to molecular size in an ultracentrifuge. Immediately after adding the radioactive label, small single-stranded fragments of labeled DNA appeared. Larger DNA molecules appeared after longer time intervals because the short fragments were linked together. Thus, this finding explained the paradox of how DNA synthesis occurs simultaneously on both strands at a replication site, while still retaining the unidirectional nature of DNA polymerase synthesis, which is always in the 5′→3′ direction (Figure 9.20).

A problem arises for discontinuous synthesis because DNA polymerization requires a primer to start each small precursor fragment. Recall that a primer is a segment of DNA (or RNA) possessing a free 3′-OH group on the deoxyribose to initiate synthesis. The continuously synthesized strand needs a primer only once when its synthesis begins because nucleotides are added continuously thereafter. However, for discontinuous synthesis to occur, a new primer must be present every 1000–2000 bp to start the new strand. This problem was solved when it was discovered that the newly synthesized Okazaki fragments transiently possess a small RNA fragment attached to their 5′-end. The primer for discontinuous synthesis turns out to be RNA, not DNA. We now know that a special enzyme called **primase** synthesizes an RNA primer, about 5–10 bp long, which permits initiation for each new DNA segment. In *E. coli,* primase is part of a complex of about 15 subunits referred to as a **primosome.** The primer RNA fragment is removed from the discontinuously synthesized DNA segment by **5′→3′exonuclease** activity, which DNA polymerase I possesses, but another exonuclease, known as RNase H, can function in a similar manner. The exonuclease removes nucleotides from the end of DNA one at a time and plays a very important role in the metabolism of DNA. After removing the RNA primer, DNA polymerase I (which can be replaced by DNA polymerase III if polymerase I is mutant) resynthesizes DNA in its place so that the Okazaki fragments can be joined to each other. DNA polymerase I has both exonuclease ability and synthetic ability. The final step in elongation of DNA on the lagging strand is the joining of adjacent Okazaki fragments. This is done by the enzyme **DNA ligase,** which catalyzes the formation of a covalent bond between a 3′-OH of the growing chain to the 5′-PO_4 of the previous Okazaki fragment. It requires an energy source (ATP or NADH, depending on the organism) and Mg^{++} to complete the reaction.

DNA Replication in *E. coli* Is a Good General Model.

The exact nature of DNA replication in eukaryotes is not understood as completely as in *E. coli,* but the general features are likely very similar. A model for *E. coli* DNA replication is presented here; additional features of DNA replication in eukaryotes will be discussed in the next section. In both cases, DNA replication is accomplished by many different enzymes and proteins acting in concert and represents a very complex series of reactions (Figure 9.21). To understand how these enzymes and proteins function, we will dissect the different reactions one by one and analyze their role in DNA replication.

DNA Polymerase III Is Responsible for DNA Replication. Polymerase III is really an aggregate of ten different polypeptides and is referred to as a **holoenzyme.** The assembled aggregate has a total molecular weight of 760 kDa. The subunits are given Greek letters (α, ϵ, θ,γ, δ, δ', χ, ψ, τ, β; alpha, epsilon, theta, gamma, delta, delta prime, chi, psi, tau, and beta, respectively) and are present in the holoenzyme in equal ratios. There are only 10–15 holoenzyme molecules per cell, and when the holoenzyme binds to the DNA, it remains firmly attached and proceeds along the growing chain until the strand is complete. The holoenzyme is a dimer with two active sites, one for synthesizing the leading strand and one for extending the lagging strand. A model for this process is outlined in Figure 9.22. How the holoenzyme bypasses the previously synthesized strand to get to the next primer on the lagging strand is not exactly understood, but it likely involves folding the DNA at the site of synthesis.

DNA Helicases Are Responsible for Unwinding the DNA. The strands of DNA must be unwound for DNA

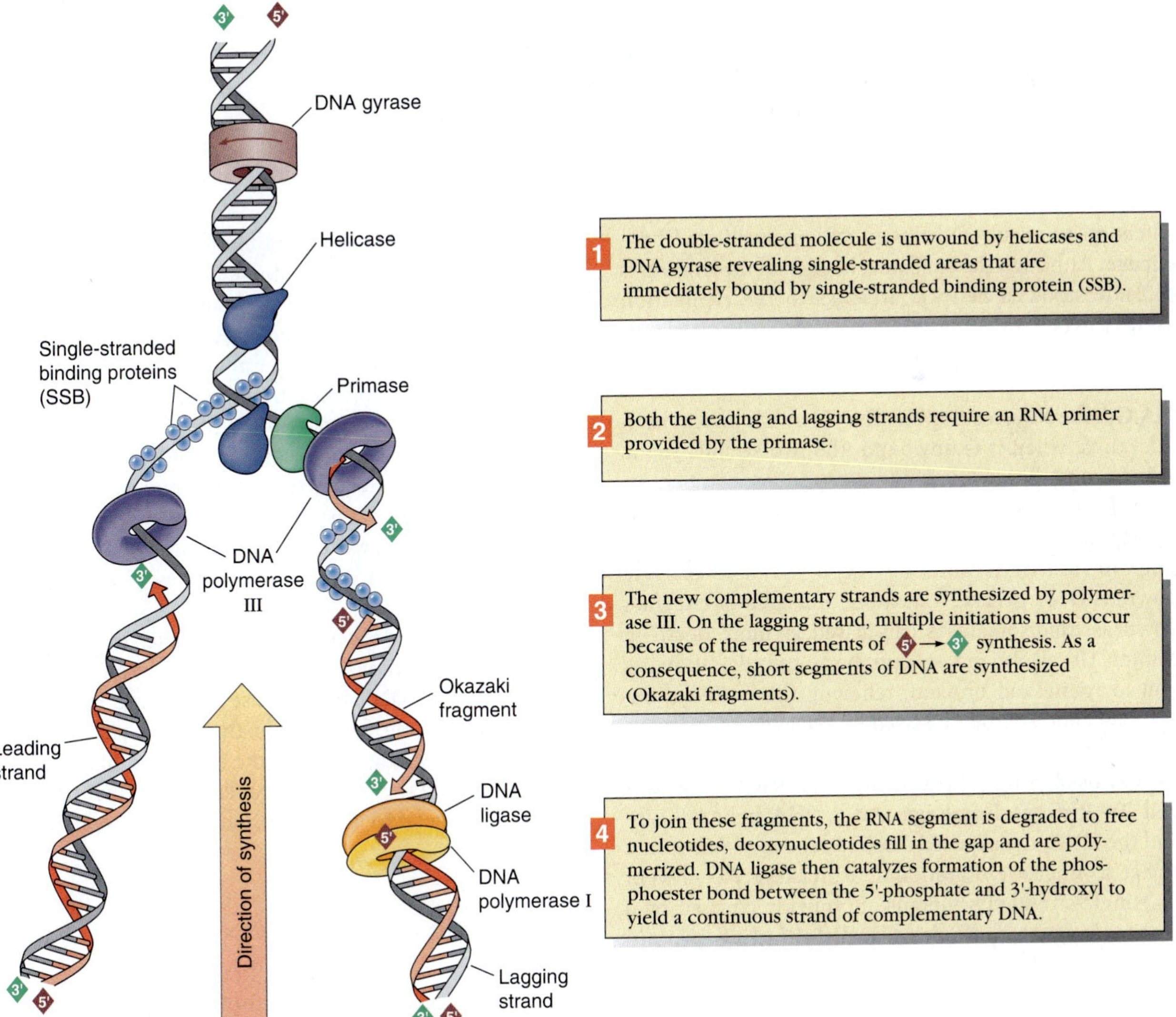

3'
5'
DNA gyrase
Helicase
Single-stranded binding proteins (SSB)
Primase
DNA polymerase III
Okazaki fragment
Leading strand
Direction of synthesis
DNA ligase
DNA polymerase I
Lagging strand
1 The double-stranded molecule is unwound by helicases and DNA gyrase revealing single-stranded areas that are immediately bound by single-stranded binding protein (SSB).
2 Both the leading and lagging strands require an RNA primer provided by the primase.
3 The new complementary strands are synthesized by polymerase III. On the lagging strand, multiple initiations must occur because of the requirements of 5' → 3' synthesis. As a consequence, short segments of DNA are synthesized (Okazaki fragments).
4 To join these fragments, the RNA segment is degraded to free nucleotides, deoxynucleotides fill in the gap and are polymerized. DNA ligase then catalyzes formation of the phosphoester bond between the 5'-phosphate and 3'-hydroxyl to yield a continuous strand of complementary DNA.

FIGURE 9.21 A general model for the replication of DNA in *E. coli.* This drawing presents an unfolded model of DNA replication to show the participants in the reaction clearly. The true model is presented in Figure 9.22.

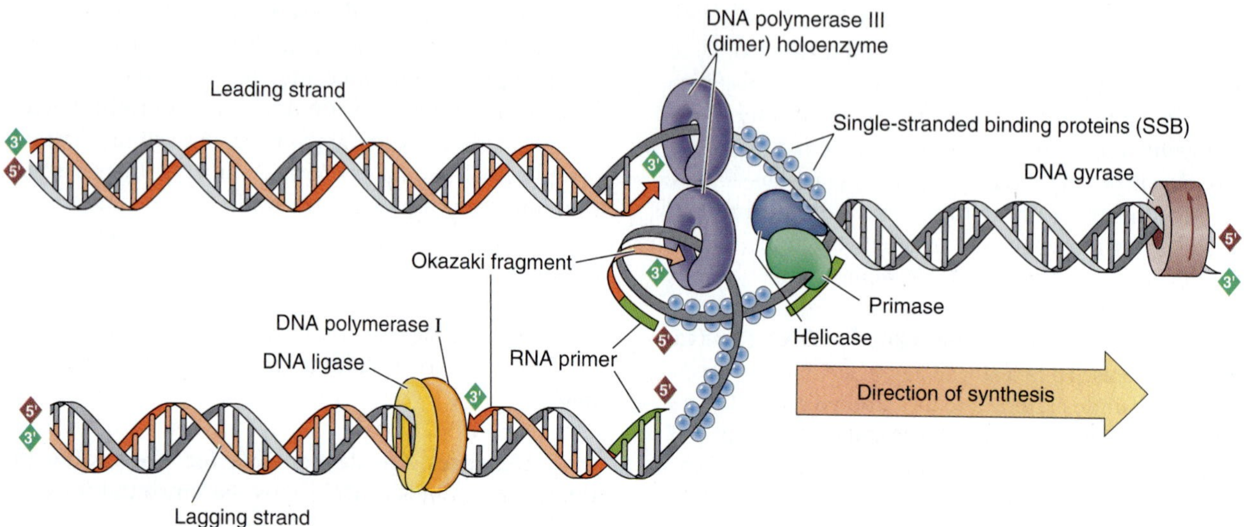

FIGURE 9.22 DNA replication model showing leading and lagging synthesis. The polymerase III holoenzyme is a dimer with two active sites, one for the synthesis of the leading strand and another for extending the lagging strand from the primer. The primer sequence becomes accessible to the polymerase by looping a small segment of DNA. The loop moves down the strand as the enzyme synthesizes a new Okazaki fragment, which is subsequently joined to the previously formed fragment (by DNA ligase) after RNase H removes the RNA primer.

replication to occur, and this process is facilitated by a **helicase.** Although the exact mechanism of unwinding is not understood, a helicase attaches to the replication fork and uses energy derived from hydrolysis of ATP to drive the unwinding of the DNA molecule.

DNA Gyrase Relaxes Supercoils. DNA is a helical molecule; thus, when it is unwound, the two strands rotate around each other. As unwinding continues, compensating DNA turns accumulate downstream of the replication fork. This intolerable tension must be relieved somehow, and it is done by the remarkable enzyme known as **DNA gyrase,** a member of a class of enzymes known as topoisomerases (discussed in more detail in Chapter 10). DNA gyrase cuts both strands, allowing them to rotate and unwind, relieving the built-up tension. The ends are then resealed by the gyrase.

The Exposed Single Strand of DNA Must Be Stabilized To Prevent Breakage. After DNA helicases unwind the DNA, the exposed single-stranded segments are subject to nuclease digestion and breakage. To prevent this, single-stranded segments are quickly covered by a protein designated **single-stranded binding protein (SSB),** which also stabilizes the short flexible molecule. This protein is small (177 amino acids) and does not have any enzymatic function. It occurs as a tetramer and binds to groups of 32 nucleotides. DNA bound to SSB protein is semirigid with no bends, which facilitates synthesis of a new complementary strand of DNA.

The Primosome Is Responsible for DNA Strand Initiation. The primosome is a huge protein composed of 15 subunits responsible for the initiation of each new Okazaki fragment. As mentioned earlier, DNA polymerase cannot initiate new strand synthesis, but RNA polymerases can initiate strand synthesis *de novo.* The primosome accomplishes initiation of strand synthesis by synthesizing a short segment of RNA for the DNA polymerase to use as a primer. The primase component of the primosome is responsible for synthesis of the short RNA fragment. However, it cannot act alone; it must be present in the primosome complex.

DNA Ends Are Joined by DNA Ligase. In *E. coli,* the RNA primer is removed by the 5′→3′ exonuclease activity of either DNA polymerase I or RNase H. As mentioned earlier, the last step in the polymerization reaction on the lagging strand is the formation of a covalent bond between the preexisting Okazaki fragment and the newly synthesized fragment. This is accomplished by DNA ligase.

DNA Replication Must Be Extremely Accurate.

The maintenance of life as we know it depends upon the accurate transmission of information with each cell growth and division cycle. Even small mistakes can be life threatening to a cell. In dividing cells that give rise to subsequent generations, a single base-pair change can result in a genetic disease that is inherited. For example, if one mistake in DNA replication occurred in humans just once every million synthesized base pairs (humans have 3×10^9 or 3 billion bp per genome), approximately 9000 mistakes would be made during each cell division cycle. This is clearly an unacceptable number, and in fact the actual rate of mistakes is about 1 in 10 billion, or about 1 mistake every 3 cell divisions. How is this tremendous accuracy attained in complex organisms?

Although base-pairing between A : T and G : C is normally quite accurate, during DNA synthesis mismatch errors of the bases do occur. For example, during synthesis an added guanine may accidentally pair with a thymine, forming mispaired bases. Proper base-pairing is primarily responsible for error-avoidance during synthesis. It is possible to insert any base opposite any other base, but the correct pairing is the most energetically favorable. Incorrect pairings occur about 1 in 100,000 added base pairs. If a mispairing occurs, a correction mechanism comes into play involving the $3' \rightarrow 5'$ exonuclease activity of DNA polymerase III. Remember that polymerases synthesize DNA only in the $5' \rightarrow 3'$ direction. Thus, if a mistake is made during synthesis, the enzyme recognizes the mistake and can "backup" and clip out the mistake using its $3' \rightarrow 5'$ exonuclease activity. This process is referred to as **proofreading.** This characteristic of DNA polymerase guarantees a very high degree of fidelity during DNA synthesis. If this proofreading capability did not exist, a high number of mistakes (mutations) would be created with every cycle of DNA replication and cell division. The organism could not function with such a high level of mistakes (Figure 9.23). A discussion of the detailed mechanisms as to how mismatching

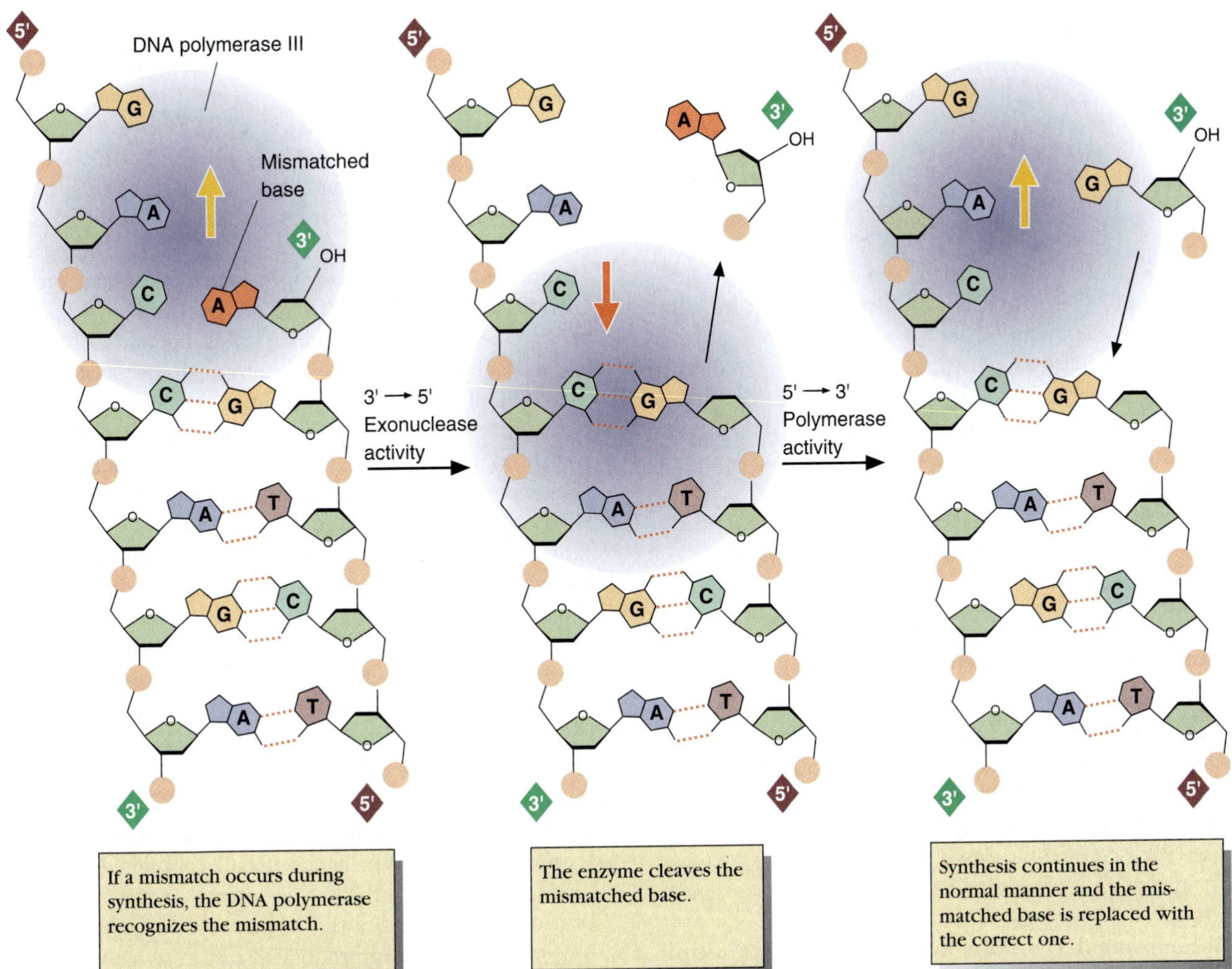

FIGURE 9.23 Proofreading during DNA replication. DNA is proofread by the polymerase's $3' \rightarrow 5'$ exonuclease activity, which can remove an incorrectly paired base. This feature of DNA polymerase ensures that the complementary base will be added to the growing chain.

of bases occurs during DNA replication and their repair can be found in Chapter 17.

Error correction also involves the 5′→3′ exonuclease activity of DNA polymerase I. In addition to removing the RNA primer during synthesis, the 5′→3′ exonuclease activity of polymerase I has two main functions: removing mispaired bases and removing damaged DNA (for example, DNA damaged by exposure to ultraviolet light). This error correction mechanism is discussed in detail in Chapter 17 along with other types of mutations and their repair. The consequence of these several mechanisms is that very few errors in DNA replication go uncorrected.

SOLVED PROBLEM

Problem

If the DNA of humans contains 3×10^9 bp, and if an Okazaki fragment, on average, is 2000 bp long, how many initiation points for RNA primase are present in the genome?

Solution

Divide the total number of base pairs by the size of the Okazaki fragment, or $(3 \times 10^9) / (2 \times 10^3) = 1.5 \times 10^6$ initiation sites.

SECTION REVIEW PROBLEMS

14. What is the difference between a primer and a template during DNA replication? What function does each play?
15. During DNA replication, both growing chains are synthesized simultaneously, but DNA polymerase can synthesize in the 5′→3′ direction only. How is this possible when the two chains of DNA are antiparallel?
16. What are the two mechanisms that maintain the high degree of fidelity in DNA replication?

DNA Replication in Eukaryotes Must Deal with Many Chromosomes.

Eukaryotic organisms differ from prokaryotes in that they have multiple copies of DNA—one per chromosome. Furthermore, the eukaryotic organelles—mitochondria and chloroplasts—have DNA that is quite different and separate from the DNA of the chromosomes. We will discuss replication of both the chromosomal and organellar DNA in eukaryotes.

Chromosomal DNA Replication Requires Multiple Replication Origins. For DNA replication to take place in a short period of time (the cell cycle usually takes less than 24 hours), each eukaryotic chromosome must have multiple replication origins. However, the exact nature of replication origin in most eukaryotes is not completely understood. Evidence suggests that some organisms possess a specific nucleotide sequence that initiates DNA replication. This sequence was first isolated from *Saccharomyces cerevisiae* and is 11 bp long. It represents the origin of replication and is referred to as an **autonomous replication sequence,** or **ARS element.** Although found in *Drosophila* and seemingly functional, this sequence is not functional as an origin of replication in most eukaryotes, suggesting that ARS elements are not the only mechanism for initiating DNA replication in eukaryotes. In fact, initiation of DNA replication in higher eukaryotes is quite different, but is poorly understood.

At least four DNA polymerases are present in eukaryotes, designated α, β, γ, and δ. DNA polymerase α is located in the nucleus and is responsible for DNA synthesis on the discontinuous (lagging) strand during replication. DNA polymerase δ is responsible for continuous (leading) strand synthesis. They are comparable to polymerase III from *E. coli* and have associated 3′→5′ exonuclease activity. Polymerase β functions in DNA repair and does not possess any exonuclease activity but can associate with another enzyme, DNase V, a 12,000-Da enzyme that possesses both 3′→5′ and 5′→3′ exonuclease activity. DNA polymerase γ is responsible for replication of DNA in the mitochondria and chloroplasts.

Mitochondrial and Chloroplast DNA Replication Is Unique. As described in Chapter 7, all animal and plant cells possess organelles known as mitochondria, which are the sites of synthesis of many essential molecules, most notably ATP and its derivatives. ATP acts as an energy source for many enzymatic reactions in the cell. Nearly all plant cells possess chloroplasts, in which light energy is converted to chemical energy for the cell to use. Both mitochondria and chloroplasts possess single DNA molecules, which are usually circular, although linear molecules occur in some plants, fungi, and protozoa. This DNA is not organized like a nuclear chromosome, but more like the DNA in a bacterial cell with sizes ranging between 15 and 2000 kb. DNA replication of mitochondria and chloroplasts is different from both chromosomal DNA replication and bacterial DNA replication. DNA replication in chloroplasts and mitochondria is initiated by an RNA primer, and as one strand is replicated, it displaces the complementary strand (displacement loop) until replication is about half completed. At this time, replication of the second strand is initiated, and synthesis proceeds in the opposite direction (Figure 9.24). The replication product results in two circles that are intertwined. This whole process is referred to as D-loop expansion. The two circles are separated by an enzyme known as **topoisomerase II.** Topoisomerase II is

1 Replication of one strand is initiated at a specific site, displacing the second strand.

Initiation site

New strand

Mitochondrial or chloroplast DNA

Displacement loop

2 After replication is about half completed, a second strand is initiated to complete the circle without generation of Okazaki fragments

Second initiation site

Interlinked circles of double-stranded DNA

3 Topoisomerase II separates the two completed circles into individual molecules.

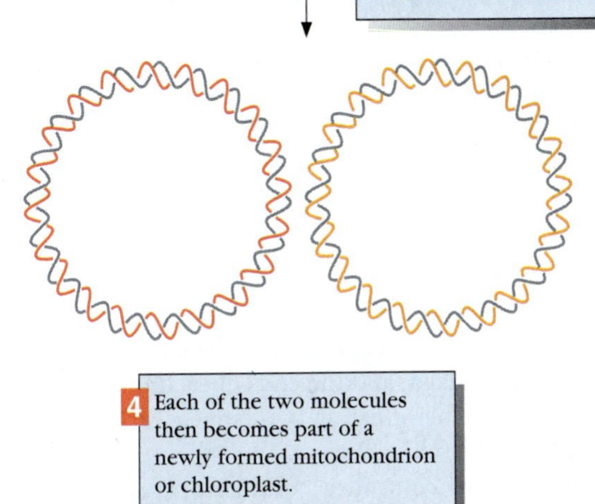

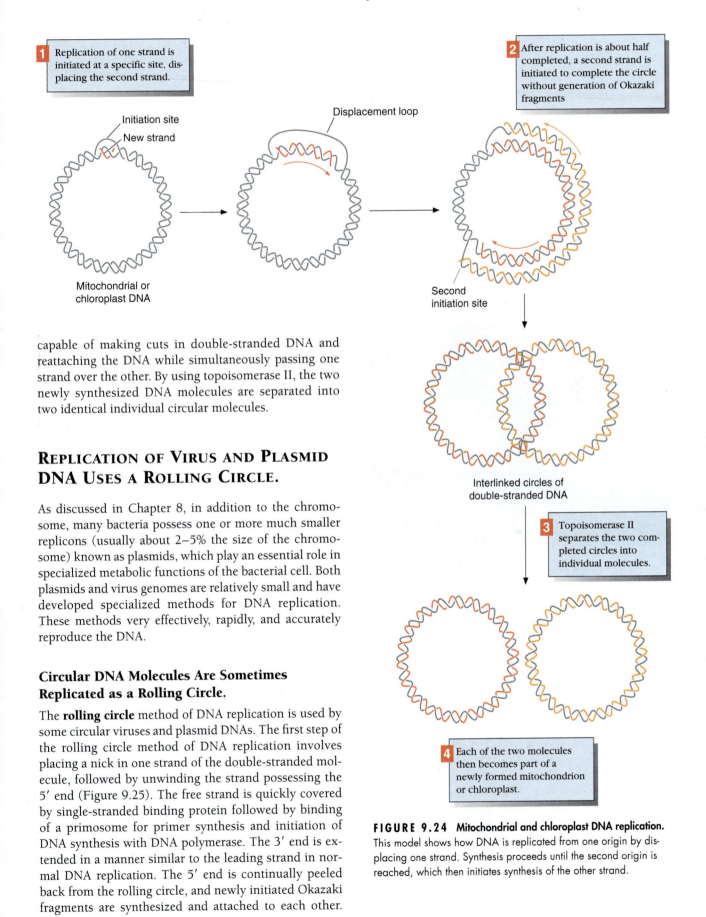

FIGURE 9.24 Mitochondrial and chloroplast DNA replication. This model shows how DNA is replicated from one origin by displacing one strand. Synthesis proceeds until the second origin is reached, which then initiates synthesis of the other strand.

capable of making cuts in double-stranded DNA and reattaching the DNA while simultaneously passing one strand over the other. By using topoisomerase II, the two newly synthesized DNA molecules are separated into two identical individual circular molecules.

Replication of Virus and Plasmid DNA Uses a Rolling Circle.

As discussed in Chapter 8, in addition to the chromosome, many bacteria possess one or more much smaller replicons (usually about 2–5% the size of the chromosome) known as plasmids, which play an essential role in specialized metabolic functions of the bacterial cell. Both plasmids and virus genomes are relatively small and have developed specialized methods for DNA replication. These methods very effectively, rapidly, and accurately reproduce the DNA.

Circular DNA Molecules Are Sometimes Replicated as a Rolling Circle.

The **rolling circle** method of DNA replication is used by some circular viruses and plasmid DNAs. The first step of the rolling circle method of DNA replication involves placing a nick in one strand of the double-stranded molecule, followed by unwinding the strand possessing the 5′ end (Figure 9.25). The free strand is quickly covered by single-stranded binding protein followed by binding of a primosome for primer synthesis and initiation of DNA synthesis with DNA polymerase. The 3′ end is extended in a manner similar to the leading strand in normal DNA replication. The 5′ end is continually peeled back from the rolling circle, and newly initiated Okazaki fragments are synthesized and attached to each other.

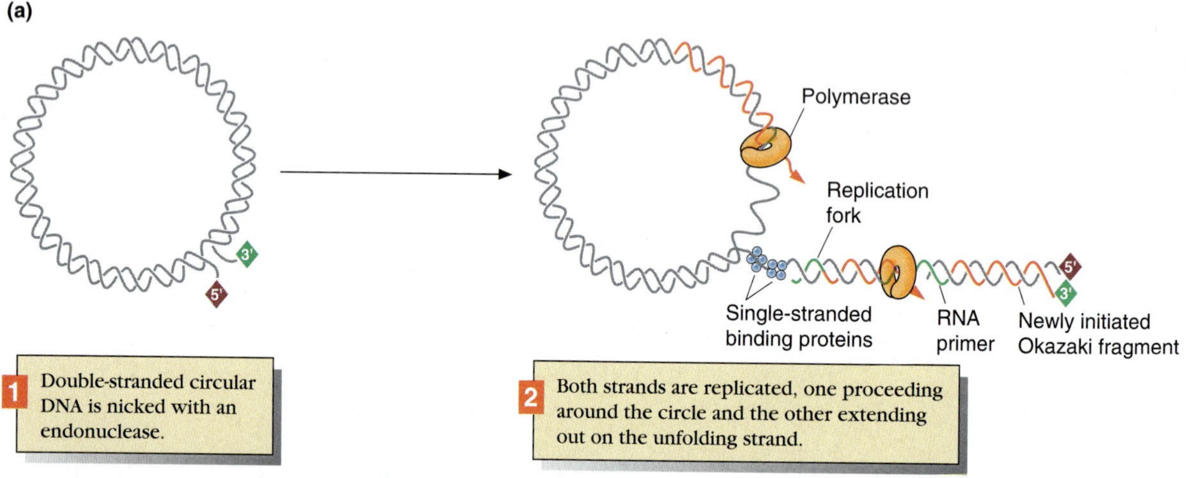
(a)
Polymerase
Replication
fork
Single-stranded
binding proteins
RNA
primer
Newly initiated
Okazaki fragment
3'
5'
1 Double-stranded circular DNA is nicked with an endonuclease.
2 Both strands are replicated, one proceeding around the circle and the other extending out on the unfolding strand.

(b)

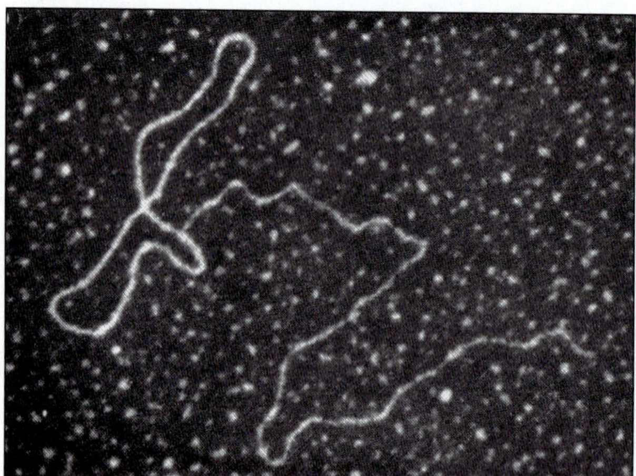

FIGURE 9.25 The rolling circle model of DNA replication. (a) Double-stranded, circular DNA shows the mechanism of DNA replication using the rolling circle method. (b) This electron micrograph shows the circular form of the molecule attached to an extended and replicating tail. This model produces numerous copies of DNA using only one template. *(Photo from Koths, K., and D. Dressler. 1978.* Proceedings of the National Academy of Sciences USA, *75: 605.)*

After completion of a full turn of the circle, a new circle will be synthesized, and the ends of each linear copy will be joined together creating two circles, each having one strand from the old circle.

The circle continues rolling to create a double-stranded molecule that is many times longer than the contour length of the initial circular DNA. An excellent example of rolling circle replication is the much-studied lambda phage particle that replicates in *E. coli* (Chapter 8). A long DNA molecule is first synthesized; then it is cut at specific sequences and packaged into phage heads, creating new lambda phage particles.

SECTION REVIEW PROBLEMS

17. How do the DNA polymerase(s) of prokaryotes differ from the polymerase(s) of eukaryotes?
18. What are the differences between replication of mitochondrial DNA and the DNA from the phage lambda?

Between Organisms, DNA Replication Has Similarities and Differences.

In this chapter, we have examined the details of the structure of DNA and how it is replicated in a variety of organisms. We can conclude from this survey that the structure of DNA is nearly identical in all organisms with the exception that it is occasionally single-stranded. However, the method of replication of double-stranded DNA varies somewhat between organisms. All DNA polymerases require a primer and a template, and synthesis is in only one direction (adding nucleotides to the exposed 3′-OH end), but they differ in the general method of how they replicate the double-stranded molecule. In addition, very large DNA molecules from eukaryotes need multiple replication origins to complete replication in a timely manner and to replicate bidirectionally. Smaller DNA molecules need only one replication origin and do not always replicate bidirectionally. The end result is always the same, however, because all organisms and organelles faithfully produce daughter DNA molecules semiconservatively.

In the next chapter, we will examine how these huge DNA molecules are packed into the limited space of the cell, cell nucleus, or virus particle. Also, we will investigate what interesting sequences are present in DNA and how and what information present within the DNA that the cell uses to govern its structure and metabolism.

Summary

During the early part of this century, it was known that two polymers were present in chromosomes—proteins and nucleic acids. Avery, MacLeod, and McCarty provided the first proof that DNA is the genetic material by transforming a nonvirulent strain of bacteria with DNA from a virulent strain, converting the nonvirulent strain to a virulent strain. Later, other investigators showed that RNA could also serve as the genetic material in some viruses.

DNA is composed of four deoxynucleotides, each containing a purine or pyrimidine, a deoxyribose sugar (missing an oxygen atom at carbon 2), and a phosphate group. Two different pyrimidines (thymine and cytosine) and two different purines (adenine and guanine) make up the four nucleotides. In RNA, uracil replaces thymine, and the sugar ribose replaces deoxyribose.

DNA is a double-helical polymer characterized by phosphodiester bonds between the 5′-phosphate of one nucleotide and the 3′-OH of the adjacent nucleotide. The two complementary strands are antiparallel to each other with adenines (A) in one strand always pairing with thymines (T) in the other strand and guanines (G) always pairing with cytosines (C). The two complementary strands are held together by hydrogen bonds.

The complementary structure of DNA provides a mechanism for semiconservative replication yielding two identical progeny molecules where one of the parent strands ends up in each. When circular *E. coli* DNA is replicated, new molecules are formed in both directions of the enclosed circle, allowing two growing points. In eukaryotes, multiple sites of replication are present, allowing DNA to be duplicated in a shorter period of time. The enzyme largely responsible for DNA replication in *E. coli* is DNA polymerase III, which requires deoxynucleotide triphosphates, magnesium ions, a preexisting DNA or RNA for a primer, and a template strand. Eukaryotes have a similar enzyme called DNA polymerase α. The DNA molecule to be synthesized is first unwound by a helicase, which separates the two strands into single-stranded molecules. Single-stranded binding proteins quickly bind to the single-stranded region, and a primosome synthesizes a short RNA sequence complimentary to the template. Attaching deoxyribonucleotides to the RNA primer, the DNA molecule is then extended with polymerase III in prokaryotes. Short DNA segments are initiated from the primer RNA sequence on the lagging strand. The RNA sequences on the Okazaki fragments are removed by DNA polymerase I (in prokaryotes) which has $5' \rightarrow 3'$ exonuclease activity. Polymerase I then replaces the removed bases with its polymerization ability. Finally, adjacent fragments are connected by DNA ligase, which joins a 3′-OH to a 5′-PO_4.

Three factors make DNA synthesis remarkably accurate. First, proper base-pairing selects the most energetically stable base-pair complex. Second, proofreading by DNA polymerase III using a $3' \rightarrow 5'$ exonuclease activity allows it to remove mismatched base pairs during synthesis. And finally, DNA polymerase I possesses a $5' \rightarrow 3'$ exonuclease activity that can remove mismatched bases after the molecule has been synthesized.

Small circular DNA molecules are present in viruses and in bacteria as plasmids and have been established to replicate differently. These small replicons duplicate by a rolling circle form of replication, making a nick in one strand and allowing the circular molecule to roll while synthesizing new strands.

Selected Key Terms

antiparallel
autoradiography
bacteriophage
bidirectional replication
complementary base-pairing
conservative replication
denaturation
DNA gyrase
DNA ligase
exonuclease
hairpin structure
helicase
hydrogen bond
lagging strand
leading strand
nucleosides
nucleotides
Okazaki fragments
phosphodiester linkage
primase
primer
proofreading
purine
pyrimidine
semiconservative replication
template

Chapter Review Problems

1. What are three main chemical differences between DNA and RNA?
2. An animal virus was found to contain a circular, double-stranded DNA with 145,000 bp. How many helical turns does this DNA contain? How many phosphorous atoms does the DNA contain?
3. What four criteria would you expect to find in a molecule to make it an ideal candidate for the genetic material?
4. If an organism had a genome size of 1×10^9 bp and the number of mistakes made by DNA polymerase during replication were 1 in every 10^6, how many mistakes would be expected in each cell division?
5. In Problem 4, if an error correction mechanism that made corrections 1000 times above the background level (the natural rate of errors without correction) were operative ($3' \rightarrow 5'$ exonuclease activity), how many errors would be expected during each cell division?
6. You have a rapidly growing culture of bacteria to which you add high specific activity radioactive thymine triphosphate (TTP). After 2 minutes, you abruptly stop the incorporation of TTP into the DNA by adding trichloroacetic acid (which stops all reactions in the cell and inactivates all proteins). You isolate the DNA from the cells and examine their molecular weights by sedimentation in a sucrose density gradient. You find that some relatively low molecular-weight fragments are labeled with radioactive material, but some high molecular-weight labeled DNA is also present. Knowing what you do about DNA replication, what do these size fragments represent and why did you find them?
7. How were ^{32}P and ^{35}S useful in proving that DNA is the genetic material?
8. What is an autonomously replicating sequence (ARS)?
9. Match the following enzymes with their functions by placing the correct capital letter in front of the enzyme.

a. ____ exonuclease	A. introduces negative supercoils
b. ____ primase	B. completes polymerization by catalyzing final phosphodiester bond
c. ____ DNA polymerase III	C. degrades DNA from the ends
d. ____ gyrase	D. degrades DNA by internal cuts
e. ____ ligase	E. repeatedly initiates lagging-strand synthesis
f. ____ helicase	F. major enzyme for DNA synthesis
g. ____ endonuclease	G. removes RNA primers and fills in with DNA
h. ____ DNA polymerase I	H. unwinds DNA

10. You isolate double-stranded DNA from two viruses, A and B. DNA from virus A is 150,000 bp in length; DNA from virus B is 5×10^7 Da. Which DNA is larger? What is the size of each DNA molecule in micrometers?
11. Suppose that DNA was single-stranded but that adenine and thymine always occurred in pairs 5′-AT-3′ and cytosine and guanine always occurred in pairs 5′-CG-3′ (for example, 5′-CGCGATATCGATCG-3′). Would this model explain Chargaff's rules?
12. A single-stranded DNA molecule has the following sequence:

    5′-AAGCTCTTACCTATCGGTCATCTAGGGCCCAATTT-3′

    What is the percentage of G + C of this molecule? What is the percentage of A + T?
13. Indicate the order of the following events that lead to DNA replication, starting with the event in c.
    a. mitosis
    b. binding of polymerase to the origin of replication
    c. G_1 phase of the cell cycle
    d. leading and lagging strand synthesis
    e. termination of synthesis by separation of the two newly formed molecules
    f. G_2 phase of the cell cycle
14. Why is radioactivity a crucial aspect of the Hershey–Chase experiment?
15. What is the evidence that the genetic material in TMV virus is RNA?
16. The melting temperatures for several different DNAs isolated from different organisms are 36°C, 78°C, 55°C, 83°C, and 44°C. Arrange the DNA with respect to G + C content, with the greatest G + C last.
17. Which carbon atoms in the deoxyribose and ribose of DNA and RNA are involved in formation of the phosphodiester bond?
18. Predict the complementary base sequence to the following sequence:

    5′-GCATCATCATTTAAACCCGGG-3′

    Which chemical groups protrude from each end of this DNA chain or the complimentary strand?
19. What is a hydrogen bond (H-bond) and how does it differ from a covalent bond and a stacking force?
20. The molecular weight of the DNA from *E. coli* is 2.7×10^9 Da. Calculate the length of the *E. coli* genome from this value.
21. The DNA from the bacteriophage φX174 is single-stranded. Would you expect the DNA base composition to follow Chargaff's rules? Why?
22. What is the significance of the $3' \rightarrow 5'$ exonuclease activity associated with DNA polymerase III?
23. Explain the functions of helicase, SSB (single-stranded binding protein), the primosome, and $5' \rightarrow 3'$ exonuclease activity during DNA replication.

Challenge Problems

1. You have just isolated a new bacterial strain and determined the molecular weight of its DNA. Much to your surprise, the DNA is relatively small for a bacterium, only 5×10^7 bp in length. Further analysis shows that the base ratio is 40% A + T and 60% G + C. You would like to cut this DNA into small fragments, and you have an enzyme that cuts double-stranded DNA only when it recognizes the sequence 5′-AAGG-3′. How many fragments would this DNA be cut into by the enzyme if you assume that this sequence occurs randomly throughout one of the strands?
2. Suppose that you isolated a phage particle from an *E. coli* strain. Upon determination of the G, C, T, and A concentrations in this new phage, you find that G = 20%, C = 15%, A = 25%, and T = 40%. What can you conclude about the structure of the phage's DNA from these data?
3. Suppose that DNA replication were "dispersive" such that each progeny DNA molecule consisted of two strands of DNA each with partly old and partly newly synthesized DNA. What would the cesium chloride gradients of a Meselson and Stahl experiment look like with this model of replication?

Suggested Readings

Alberts, B., D. Bray, J. Lewis, M. Raff, K. Roberts, and J. D. Watson. 1994. *Molecular Biology of the Cell,* 3d ed. Garland Publishing, New York.

Benbow, R. M. 1992. "On the nature of origins of DNA replication in eukaryotes." *BioEssays,* 14: 661–670.

Cairns, J. 1966. "The bacterial chromosome." *Scientific American,* 214: 36–51.

Lewin, B. 1994. *Genes V.* Cell Press, Cambridge, MA; Oxford University Press, Oxford.

Radman, M., and R. Wagner. 1988. "The high fidelity of DNA duplication." *Scientific American,* 258 (February): 40–46.

Singer, M., and P. Berg. 1991. *Genes and Genomes.* University Science Books, Mill Hill, CA.

Stillman, B. 1994. "Smart machines at the DNA replication fork." *Cell,* 78: 725–728.

Watson, J. D., and F. H. C. Crick. 1953. "A Structure of deoxyribose nucleic acid." *Nature,* 171: 737–738.

Zimmer, C. 1995. "First cell." *Discovery,* 16: 71–78.

Zimmerman, S. B. 1982. "The three-dimensional structure of DNA." *Annual Review of Biochemistry,* 51: 395–427.

On the Web

Visit our Web site at **http://www.saunderscollege.com/lifesci/titles.html** and click on A/G/M Genetics for links to the following chapter-related resources on the World Wide Web:

1. **DNA structure.** This site is tended by faculty at Massachusetts Institute of Technology and discusses much that is in this chapter using nice graphics.
2. **DNA replication.** This site has some nice pictures of the enzymes and how DNA replicates.
3. **Book on DNA replication.** This site shows the table of contents of a comprehensive book on DNA replication written by Arthur Kornberg (1991).

CHAPTER

10

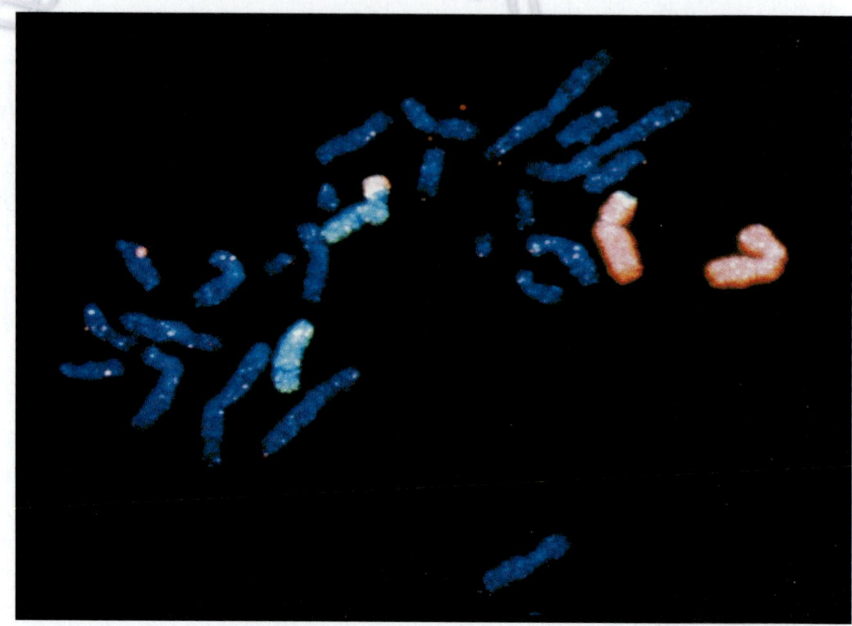

Whole chromosome paints. This new technology can detect chromosomal abnormalities such as an exchange of DNA between chromosomes. A normal chromosome 7 is shown with a red stain (*far right*). To its left, the abnormal 7 has a green tip, indicating the presence of DNA exchanged from chromosome 12. The lower green-stained chromosome is the normal 12, while the upper green chromosome has a red tip exchanged from chromosome 7. *(Department of Energy/Photo Researchers, Inc.)*

The Molecular Structure of Prokaryotic and Eukaryotic Chromosomes

DNA is a long, linear molecule that contains genetic information and is replicated in a semiconservative manner. Each DNA molecule is packaged into a chromosome. These properties of DNA raise a number of new questions, such as, how can a molecule as large as DNA be packaged into such a small space, and how is chromatin unfolded to express individual genes? To answer the first question, we will first consider the packaging of the DNA in prokaryotes (cells that have no true nucleus—bacteria) because they are simpler to understand. Second, we will consider the packaging of DNA into chromosomes in eukaryotes (cells with a true nucleus). Scientists do not have a complete answer to the second question, so it is only briefly discussed in this chapter. We will also examine a few research techniques for studying chromosomes and DNA, some of which may be new to you.

One of the first things we learned about eukaryotic DNA organization was that the genes were not tightly arranged on the chromosome and that the chromosomes possessed a considerable amount of seemingly useless DNA. This was not the case in the simpler organisms, such as bacteria and viruses, where the genes are closely spaced and no "extra" DNA is present. Why this is the case is still not clearly understood, but rapid strides are being made to understand the detailed organization of eukaryotic chromosomal DNA. This understanding has tremendous implications in both medicine and agriculture (for example, in engineering new organisms and understanding how genes are expressed and how their expression is regulated).

Specifically, in this chapter we will address the following three questions:

1. How is the DNA molecule condensed to fit into a small space such as an organelle, nucleus, or a bacterial cell while still allowing the information in the molecule to be expressed and replicated?

2. Chromosomes have large amounts of DNA that do not seem to have any specific genetic function. How can nongenetic DNA be distinguished from genetic DNA, and what is the nature of nongenetic DNA sequences?

3. The DNA of both prokaryotes and eukaryotes is associated with proteins in the chromosome. How are these proteins associated with DNA, and what roles do they play in packaging DNA?

A Prokaryotic Chromosome Is Packed into a Small Space.

The prokaryotic genome is an excellent example of how DNA can be packaged into a highly organized and condensed structure. Most prokaryotes possess only one double-stranded DNA molecule. In this section, we will learn how this molecule is packed to reduce its volume while still retaining the ability to replicate and express the information present in the nucleotide sequence.

The Chromosome of Prokaryotes Is Highly Coiled.

The prokaryotic bacteria cell is much smaller than a eukaryotic cell, ranging from 1 to 5 μm in diameter. Most eukaryotic cells range in size from 10 to 100 μm. In fact, the size of a bacteria cell is even smaller than the nucleus of most eukaryotic cells. (Figure 10.1). The chromosome of a bacterial cell can be clearly seen in an electron micrograph (Figure 10.2) and is referred to as a **nucleoid.** A nucleoid is composed of about 80% DNA, 10% protein, and 10% RNA and occupies about 25% of the volume of the cell. The total length of the chromosome in *E. coli* is 1400 μm. Clearly, the nucleoid must be a highly coiled and compacted structure to fit within a cell less than 1/1000 its length. Most bacterial cells have only one circular chromosome with no free ends. As mentioned in Chapters 8 and 9, a great many bacteria have one or more smaller circular DNA components called plasmids, which confer specialized functions to the cell (for example, resistance to an antibiotic such as penicillin). Plasmids are between 1 and 10% of the size of the chromosome and are not required for the survival of the bacterial cell under normal growth conditions. On the other hand, plasmids may occupy a fair amount of space in a cell because they can be present in as many as 500 copies per cell.

The *E. coli* chromosome can be isolated by gently removing the cell wall and disrupting the plasma membrane in the presence of cations that neutralize the negatively charged phosphate groups of DNA. Under these conditions, the nucleoid can be isolated in its native state as a highly compacted structure. Using an electron microscope, the nucleoid structure can be seen to consist of tightly coiled DNA strands of about 40 kb each that are looped out from a central point (Figure 10.2 for the electron micrograph and Figure 10.3 for an interpretive drawing of how the chromosome is folded). This compact coiled structure can be disrupted by RNase or DNase action, enzymes that cut RNA and DNA, respectively. Thus, the nucleoid structure is somehow dependent upon the presence of intact RNA molecules, and the DNA cannot have cuts in it and maintain its coiling. How RNA maintains coiling is not understood.

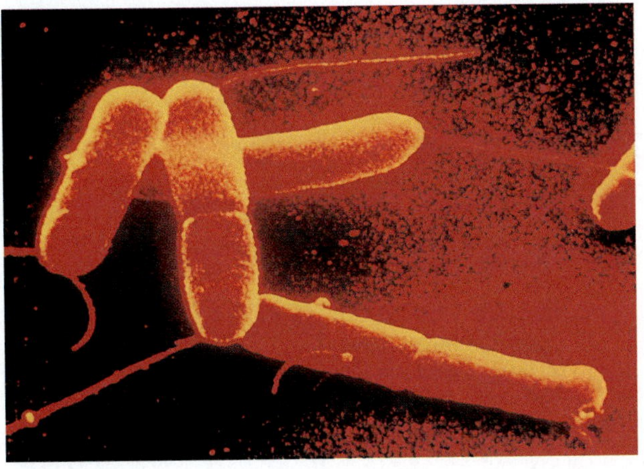

(a)

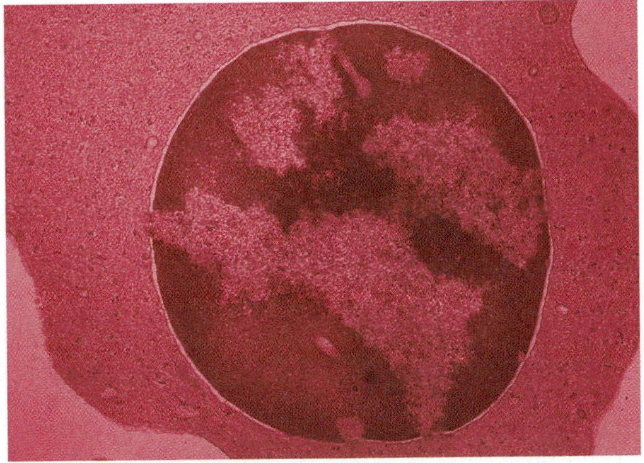

(b)

FIGURE 10.1 Comparison of a bacterial cell and a mammalian nucleus. (a) An electron micrograph of an *Escherichia coli* cell (1 μm × 3 μm). (b) The nucleus of a mammalian cell (10 μm × 10 μm). *(a, David M. Phillips/Visuals Unlimited; b, G. B. Chapmans-Williams-Blake/Visuals Unlimited)*

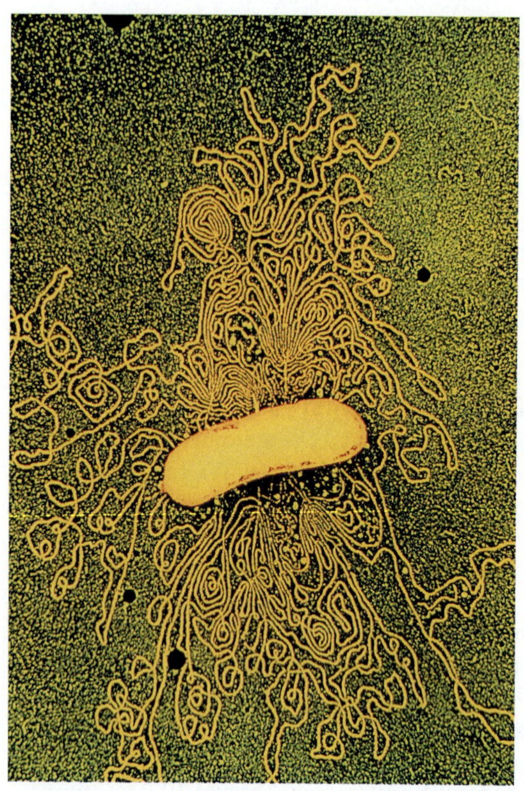

FIGURE 10.2 The *E. coli* nucleoid structure. An electron micrograph of the nucleoid that has been spilled out of cell by lysing the cell with enzymes that degrade the cell wall. The DNA would not be intact unless the cell was very gently lysed to release the DNA. *(K. G. Murti/Visuals Unlimited)*

A schematic illustration of nucleoid structure in various stages of coiling is shown in Figure 10.3. What holds the nucleoid structure together besides RNA? Interestingly, proteases (enzymes that hydrolyze proteins) do not disrupt the nucleoid structure. This is because the proteins involved in maintaining nucleoid structure are buried deeply within the complex and are inaccessible to protease action. A protein, called the HU protein, plays a central role in holding DNA in a tightly compact complex. The HU protein is a dimer composed of 9535 and 9225 Da subunits, each cell possessing about 120,000 copies. HU has a positive charge that binds to the negative charges present in DNA. Researchers have shown that the HU protein binds to DNA and causes the DNA to form tightly packed structures.

SECTION REVIEW PROBLEMS

1. How many nucleoid structures are present in a growing bacterial cell just before division if you assume that the bacterial cell normally has one chromosome?
2. What does digestion of the purified nucleoid structure with DNase, RNase, and protease do to the coiling of the nucleoid? What are the implications of this experiment?
3. What is a plasmid and how does it differ from the chromosome?

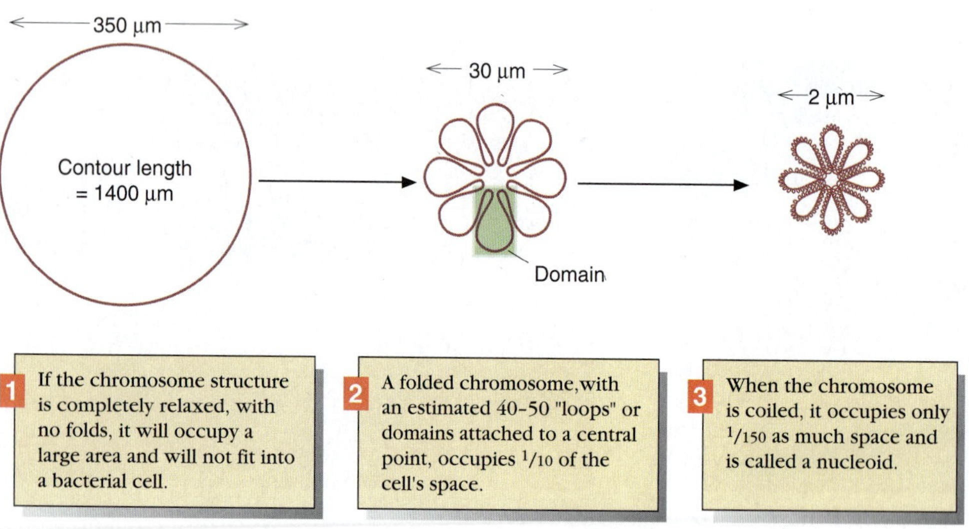

350 μm
Contour length = 1400 μm
30 μm
Domain
2 μm
1 If the chromosome structure is completely relaxed, with no folds, it will occupy a large area and will not fit into a bacterial cell.
2 A folded chromosome, with an estimated 40–50 "loops" or domains attached to a central point, occupies 1/10 of the cell's space.
3 When the chromosome is coiled, it occupies only 1/150 as much space and is called a nucleoid.

FIGURE 10.3 *E. coli* chromosome packing into a nucleoid. The bacterial chromosome is packed into the nucleoid. *(Adapted from R. R. Sinder and D. E. Pettijohn. 1981.* Proceedings of the National Academy of Sciences USA, *78: 224.)*

Most Double-Stranded DNA Molecules Contain Supercoils.

A key feature of packaging large, circular prokaryotic chromosomes into small volumes is the twisting of the DNA molecule. This twisting is referred to as **supercoiling** and involves rotating a free end of one of the two strands around the other and reattaching the ends. This can be simulated by twisting two ropes about each other until they are tightly coiled and holding the ends so that they do not rotate. If you then cut or release one of the ropes, it will spin until it is uncoiled. Supercoiling is an ideal configuration for tightly packing a long slender molecule into a small space because supercoiled DNA molecules fold up into much less physical space than uncoiled, or **relaxed,** molecules. Only covalently closed, double-stranded, circular DNA molecules or linear DNA that is tightly anchored can supercoil because the ends are not free to rotate (Figure 10.4). A cut in one strand of a double-stranded DNA molecule allows the strain of supercoiling to be released by rotation. This fact is the basis for an important deduction about the native structure of the DNA molecule in nucleoids. That is, it takes about 40–50 single-strand cuts to relax the DNA molecule in the *E. coli* nucleoid completely. This finding indicates that about 40–50 independent regions of supercoiling (referred to as domains) are present in the nucleoid (Figure 10.3), with each domain attached at a central point.

Before we discuss how DNA is packed into chromosomes, we need to first consider supercoiling of DNA because it is such an important feature of DNA structure. Supercoiled DNA can be characterized in two ways. First, it can be characterized by the number of twists, referred to as the linking number. The second characteristic is the direction, or handedness, of the twisting (Figure 10.4). DNA can be overtwisted (referred to as positive or right-handed twisting) or undertwisted (negative or left-handed twisting). The direction of twisting has an important impact on the structure of the DNA because the stress of negative supercoiling tends to make it easier to open the DNA into single-stranded regions, whereas positive supercoiling makes such unwinding more difficult. The supercoiling in *E. coli* nucleoids is negative because it is in all biologically active DNA and is likely an important factor for genetic processes such as DNA replication, gene expression, and breakage and rejoining of DNA molecules. All these processes require the opening of the double helix.

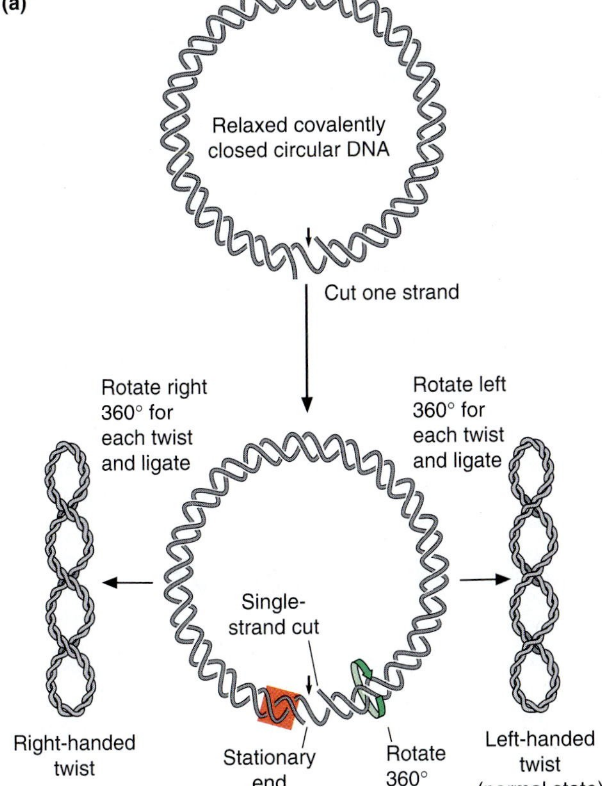

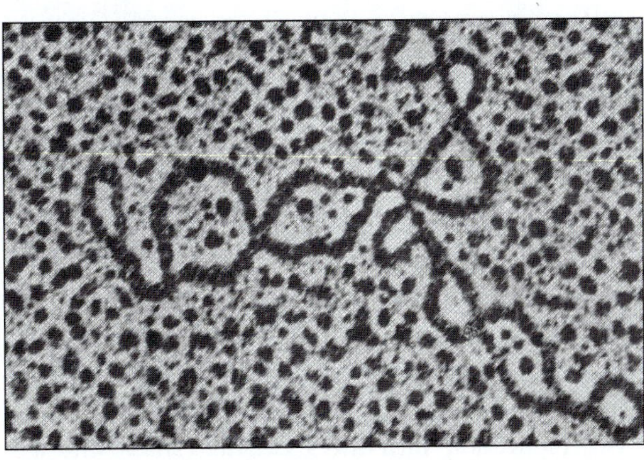

FIGURE 10.4 Supercoiling of DNA. (a) The model shows how supercoiling can be accomplished in a circular, double-helical DNA molecule and results in a very compact structure. (b) This electron micrograph of bacteriophage PM2 DNA shows the supertwisted form. *(Courtesy of Wang, J. C. 1982.* Scientific American, *247: 97.)*

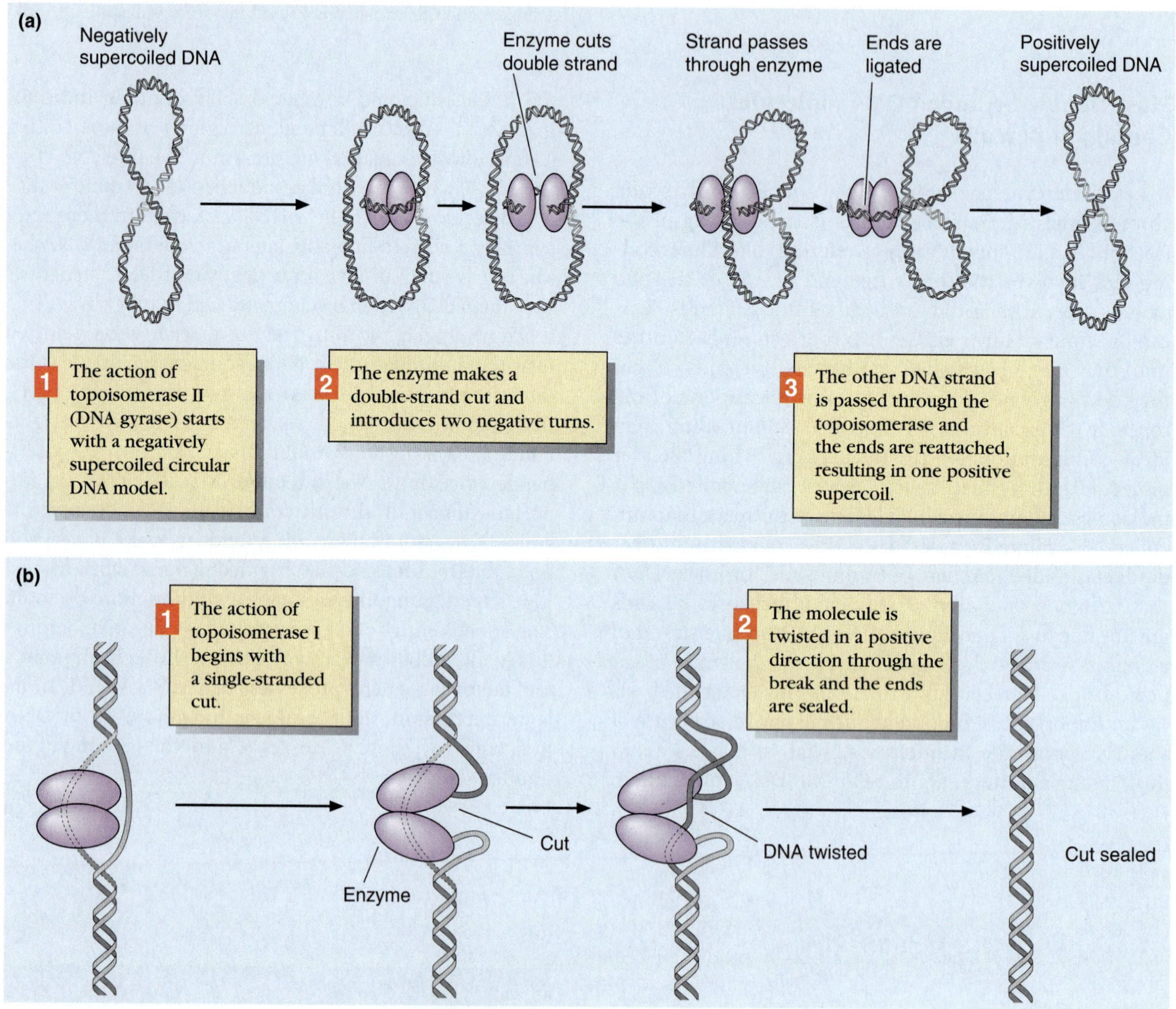
(a)
Negatively supercoiled DNA
Enzyme cuts double strand
Strand passes through enzyme
Ends are ligated
Positively supercoiled DNA
1 The action of topoisomerase II (DNA gyrase) starts with a negatively supercoiled circular DNA model.
2 The enzyme makes a double-strand cut and introduces two negative turns.
3 The other DNA strand is passed through the topoisomerase and the ends are reattached, resulting in one positive supercoil.
(b)
1 The action of topoisomerase I begins with a single-stranded cut.
2 The molecule is twisted in a positive direction through the break and the ends are sealed.
Cut
Enzyme
DNA twisted
Cut sealed

FIGURE 10.5 The action of topoisomerase I and II. Topoisomerase I (b) only cuts one strand of the DNA, whereas topoisomerase II (a) cuts both strands. Both enzymes relax or induce coiling in the DNA.

Supercoils Are Introduced and Removed from DNA by Topoisomerases.

Supercoiling of DNA is controlled by a class of enzymes called **topoisomerases,** which we referred to briefly in Chapter 9 when discussing DNA replication. Recall that DNA replication introduces twisting stress into the coiled molecule as it unwinds, and this tension must be relieved somehow. As mentioned in Chapter 9, DNA gyrase is responsible for removing the twisting. Two types of topoisomerases have been identified in both prokaryotes and eukaryotes: topoisomerase I, which introduces temporary, single-stranded cuts in DNA, and topoisomerase II (also referred to as DNA gyrase), which introduces temporary, double-stranded cuts in DNA. The mechanisms of action of these two enzymes are illustrated in Figure 10.5.

Topoisomerases are required in all reactions that deal with double-stranded DNA. For example, when strands of double-helical DNA are pulled apart (for replication or gene expression), the result is an increase in supercoiling farther down the molecule. This twisting problem is overcome by introducing a transient cut in one strand (as in Figure 10.5b), allowing the free end to rotate about the remaining intact strand. The cut is then resealed. These steps are repeated over and over again as DNA unwinds (for example, in DNA replication).

SECTION REVIEW PROBLEMS

4. What is the definition of *linking number*?
5. What are the two classes of topoisomerases, and how do they differ?
6. What happens to supercoiled DNA when it is cut by a nuclease? Why?
7. What function does supercoiling play with respect to packaging DNA?

THE SEQUENCE OF THE EUKARYOTIC CHROMOSOME IS COMPLEX.

As might be expected, the amount of DNA in eukaryotes is much greater than in prokaryotes, but both must be packed in a systematic manner to gain access to the information in DNA. Packing of DNA in prokaryotes and eukaryotes has many of the same features: supercoiling, the use of proteins to aid in coiling, and only one DNA molecule per chromosome. Additional features of eukaryotic chromosomes include the fact that they have added proteins with specialized functions (eukaryotic chromosomes are about 50% DNA) and have repeated sequences within their DNA. In this section, we will examine the nature of the "repeated DNA sequences" and the many proteins of the chromosome and how they interact with the DNA to accommodate coiling.

A Chromosome Has Only One Molecule of DNA.

Most bacteria and viruses have one circular chromosome with a single, double-helical molecule of DNA. If we extend this knowledge to eukaryotic genomes, we would expect that each chromosome would have only one double-helical molecule of DNA. Several lines of circumstantial evidence support this assumption, although rigorous proof is not available. First, electron micrographs of chromosomes, where the DNA has been carefully spread out over a large area, reveal no free ends (see Figures 10.11 and 10.12, later in this chapter). Second, attempts to measure the molecular size of an intact chromosome give values that are within the predicted molecular weight range for the DNA of that chromosome. For example, in 1974, Ruth Kavenoff and Bruno Zimm measured the molecular weight of the longest *Drosophila melanogaster* chromosome and found a value of 0.41×10^{11} Da (remember that the *E. coli* chromosome is 0.025×10^{11} Da). The haploid content of *D. melanogaster* is 1.2×10^{11} Da and it has four chromosomes. Dividing this haploid content by 4 yields 0.3×10^{11}, so these two values are sufficiently close (0.41×10^{11} and 0.30×10^{11}) to conclude that the largest chromosome from *Drosophila* is composed of one double-helical molecule of DNA.

How many base pairs are present in an average DNA molecule of a eukaryotic chromosome, and how big is the chromosome? The number of base pairs present in the haploid genomes of a representative group of organisms is shown in Table 10.1. [To convert molecular weight (in daltons) to DNA length, use the conversion factor 6.6×10^5 Da = 340 nm of double-stranded DNA and 1000 bp = 6.6×10^5 Da.] The length of the DNA per chromosome for most of these organisms ranges from 10 to 4000 mm. In contrast, the average chromosome varies from 0.002 to 0.020 mm in length when in its highly condensed state for mitotic or meiotic division. From these two facts, we can deduce that DNA must be condensed several thousand times to form the compact structure of a single visible chromosome.

Another interesting observation from the values presented in Table 10.1 is that the DNA content of higher organisms varies considerably. Some species, like the lily plant, have 100 m of DNA in each genome, whereas a more complex organism, for example *Homo sapiens*, has only about 1 m of DNA per genome. A single lily chromosome has more than four times the amount of DNA that the entire human genome has! It would seem from this observation that no simple correlation exists between the complexity of an organism and the amount of DNA in its genome.

The total amount of DNA per haploid genome is constant within each living organism and is called the

TABLE 10.1 *DNA Content of Different Organisms*

Organisms	Number of Chromosomes	Genome Size (bp)	Genome Molecular Weight (Da)	Genome Length (m)	Average Chromosome Length (mm)
Escherichia coli	1	0.0048×10^9	0.0025×10^{12}	0.0014	1.4
Drosophila melanogaster	8	0.175×10^9	0.125×10^{12}	0.06	7.5
Mouse	40	2.2×10^9	1.45×10^{12}	0.74	18.5
Humans	46	3×10^9	2×10^{12}	0.95	20.6
Maize	20	6.6×10^9	4.4×10^{12}	2.2	110
Lily	24	300×10^9	200×10^{12}	100	4160

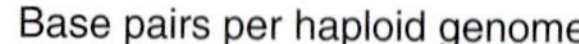

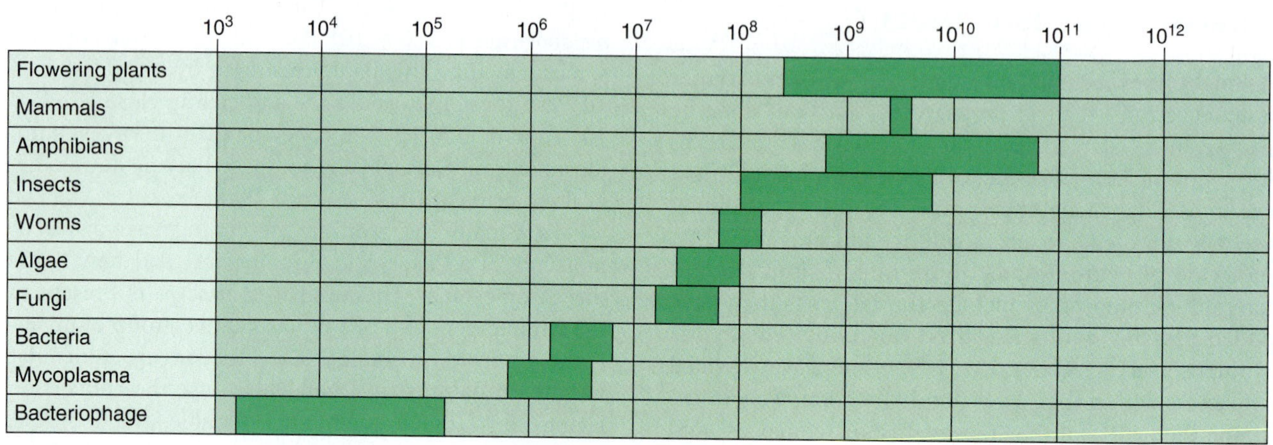

FIGURE 10.6 The relationship between DNA content and organism complexity. This graph shows the relationship between groups of related organisms and the C-value of the haploid genomes. The C-value is not precisely correlated with the morphological complexity of the species, although genome size does consistently increase from bacteria to higher eukaryotes. However, beyond this point, a general increase is not seen; only variation within groups is seen. The shaded area indicates the range of DNA values.

C-value. The amount of DNA per haploid genome varies enormously between organisms (Figure 10.6) and does not always correlate with the complexity of the different organisms. This lack of correlation between the complexity of the organism and the amount of DNA per haploid genome is known as the C-value paradox. We now know that the complexity of the DNA, as described later in the chapter, explains why there is no logical correlation between the DNA content and the complexity of an organism.

Eukaryotic DNA Has Repetitive Sequences.

A eukaryotic gene can vary in size from a few thousand to more than 1 million bp in length. However, if we assume an average size of 8000 bp, then by dividing the total number of base pairs (3×10^9 bp) in the human genome by 8000, we find that humans could potentially possess about 375,000 genes. How close is this estimate to the real number of genes? One way to determine the true number of genes is to count the number of different gene products produced by the different types of cells in an organism (the direct product of a gene is a special type of RNA, but usually the number of proteins are counted). Using this approach, about 10,000 different gene products can be detected in most cells. If we assume that this is an underestimate by a factor of 10 because it is very likely that not all genes are expressed in all cells at any specific time, it still allows for 100,000 genes in a human (or mammalian) genome, far fewer than the calculated 375,000.

What is the explanation for this large discrepancy between the calculated and experimentally estimated number of genes in a genome? A significant fraction of this "excess or nongene" DNA is composed of multiple copies of short DNA sequences. The existence of these **repetitive DNA** sequences was discovered in studies of the sequence complexity of genomic DNA using reassociation kinetics. For an explanation of this method for estimating how much repetitive DNA is present in a genome, see *CONCEPT CLOSE-UP: Reassociation Kinetics and Repetitive DNA*.

The sequence complexity of DNA in organisms is spoken of as unique, moderately repetitive, and highly repetitive. Unique sequences are present in the genome as a single copy, moderately repetitive sequences are present at 1000 to 100,000 copies, and highly repetitive sequences are present at values above 100,000 and up to 10 million copies. The data in Table 10.2 list the percentage of repetitive DNA sequences in some organisms. These values include both moderately and highly repetitive sequences.

The **unique** or **single-copy DNA sequences** represent most of the genes of an organism. It is, however, very likely that not all unique sequences are active genes or genes at all. In prokaryotes, nearly all the genes are single copy, with the exception of ribosomal RNA and transfer RNA genes.

Moderately repetitive sequences can also be genes. For example, cells contain many structures (ribosomes, tubulin) that are present in many copies. For the cell to produce these structures in large quantities, multiple copies of the gene are present. For example, humans have 260 copies of the ribosomal RNA gene, and the gar-

TABLE 10.2 *The Percentage of Repetitive DNA Sequences in Some Eukaryotic Organisms*

Organism	Repetitive DNA (%)
Escherichia coli	0.3
Saccharomyces cerevisiae (yeast)	5.0
Arabidopsis thaliana (plant)	10
Human	21
Tobacco	33
Mouse	35
Soybean	40
Wheat	42
Rye	42
Calf	42
Cotton	44
Snail	45
Pea	52
Maize	60
Drosophila melanogaster (fruit fly)	70
Lily	95

den pea has over 3900 copies of this gene. Usually multiple-copy genes are clustered in one location, although some multiple-copy genes are dispersed throughout the genome.

One group of genetically important, moderately repetitive sequences that are dispersed within the genome are the transposable elements. Transposable elements are relatively short DNA sequences capable of moving about in the genome (Chapter 19) and range in size from 500–6500 bp. They are present in all eukaryotes and prokaryotes. Transposable elements are usually present in families, with each family consisting of a number of elements with closely related but not necessarily identical sequences. Such families are believed to have arisen by duplication of a single transposable element, with the differences arising later as random base changes occurred in individual elements. Another moderately repetitive DNA sequence is a type of short sequence of 100–2000 bp that does not contain any information useful to the organism. An example of such a sequence is the *Alu* family of DNA sequences in humans. *Alu* sequences are 150–300 bp in length and have enough sequence similarity to be recognized as a family. *Alu* sequences represent about 5% of the human genome and are always found interspersed in the genome, adjacent to unique sequences. The function, if any, of this type of sequence is not known, but moderately repetitive sequences make up a significant fraction of the genome; for example, the human genome contains 5–10% moderately repetitive sequences, and the *Drosophila* genome contains 10–12%.

Highly repetitive sequences represent a large percentage of the genome of many organisms. For example, highly repetitive sequences comprise 15% of the human genome and about 40% of the bovine genome. Many of these highly repetitive sequences are very short, usually about 6–12 bp, although some are several hundred base pairs long. The shorter sequences can be repeated millions of times. Often, highly repetitive sequences have a very different percentage of guanine and cytosine, changing the density of these sequences. This difference in density facilitates their purification from moderately repetitive and unique sequences using density gradient centrifugation (discussed in the Meselson–Stahl experiment in Chapter 9). Sequences that have distinctly different densities are referred to as **satellite DNA.** This term arose when sheared DNA from an organism was centrifuged in a cesium chloride gradient and DNA subbands appeared (Figure 10.7) at a lower density, representing a significant percentage of the total DNA. Because these bands were usually subordinate in amount to most of the DNA, this DNA was called satellite DNA.

We now know the sequence of many of these highly repetitive DNAs. For example, the major satellite species in the mouse is composed of about 10^6 copies of a

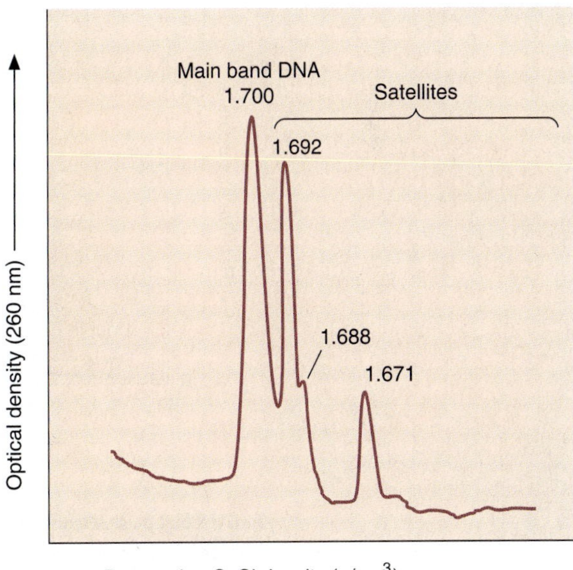

FIGURE 10.7 DNA separated by centrifugation in a cesium gradient. *Drosophila virilis* DNA is first sheared into fragments of about 25 kb and then centrifuged to separate the fragments according to density. The majority of the DNA appears at 1.700 g/cm^3 and represents the bulk of the single-copy sequences. Several satellite bands appear at a lower buoyant density and represent the moderately repetitive and highly repetitive sequences. *(Adapted from Gall, J. G., and D. D. Atherton. 1974.* Journal of Molecular Biology, *85: 633.)*

CONCEPT CLOSE-UP

Reassociation Kinetics and Repetitive DNA

The principle of reassociation kinetics is simple and can be used to estimate the percentage of repeated DNA sequences within a genome. If we start with a solution of double-stranded DNA molecules and heat it to boiling, each molecule will separate into two complementary single-stranded molecules when the hydrogen bonds break. This process is known as denaturation. If the solution is cooled, the complementary single-stranded molecules will rejoin to form double-stranded molecules. This process is called reassociation or renaturation (Figure 10.A). To reassociate, the complementary single-stranded molecules must collide with each other. If the solution contains many molecules, all single-stranded molecules with complementary sequences will eventually collide to reform double-stranded molecules. If a specific sequence is present in only one or a few copies, this sequence will collide less frequently with its complementary sequence than if many copies of the same sequence are present. When one sequence is repeated many times, any single-stranded molecule with this sequence will have many complementary partners and will reassociate more quickly than the single-copy sequences. Thus, the presence of a significant fraction of repetitive sequences in a genome will be indicated by a fraction of the genomic DNA that reassociates more quickly. This is exactly what happens with the DNA from most eukaryotic organisms, indicating that much of the DNA is repetitive.

Renaturation of DNA can be examined quantitatively, giving a more precise understanding of the number of repetitive DNA sequences within a mixture of DNA molecules. In addition, DNA reassociation kinetics can be used to estimate the total molecular weight of a genome. To understand how this works, we first need to look at the renaturation process as a bimolecular reaction, which depends upon random collisions between two molecules (the complementary strands). Thus, the rate of the reac-

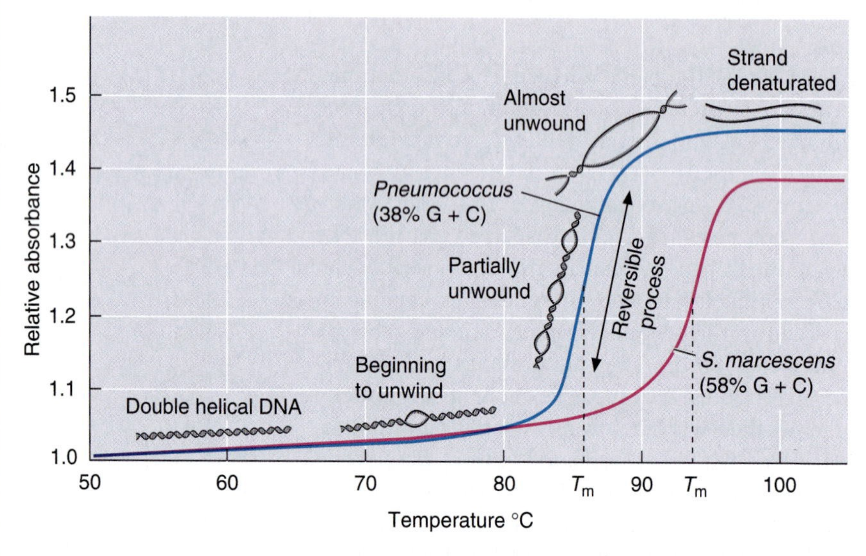

FIGURE 10.A Denaturation and renaturation of DNA. The conversion into double-stranded DNA can be experimentally followed by measuring absorbency at 260 nm because double-stranded DNA and single-stranded DNA at the same concentration absorb differently. The DNA from two stains of bacteria, *Pneumococcus* and *S. marcesens,* is first sheared into short fragments and then the solution is slowly heated to separate the double-helical molecules, eventually yielding single-stranded DNA. The melting temperature (T_m) is distinct for each DNA and depends upon the G + C content. Denaturation can be reversed by slowly cooling the solution and allowing time for the complementary sequences to locate each other randomly.

tion is governed by second-order kinetics, described in the following equation:

$$\frac{C}{C_0} = \frac{1}{1 + kC_0t}$$

where C is the concentration of single-stranded DNA at time t, C_0 is the concentration of single-stranded DNA at $t = 0$, and k is a second-order rate constant in liters per mole seconds.

The reassociation reaction is the product of the DNA concentration (C_0) at time zero and the time of incubation (t, or the time required to reform the double helical molecule), and is described by C_0t. Thus, the fraction of DNA remaining in the reaction mixture (C/C_0) is a function of C_0t, the product of the time elapsed to reform double-stranded DNA and the initial concentration. The kinetics of this reaction are usually graphed as shown in Figure 10.B, where C/C_0 is plotted against C_0t. The greater the $C_0t_{1/2}$ (where the reaction is half completed), the slower the renaturation reaction because it represents a lower concentration of a specific DNA sequence in the reaction mixture. Thus, the $C_0t_{1/2}$ will decrease if the concentration of a specific DNA sequence is increased.

To generate a C_0t curve like that shown in Figure 10.B, genomic DNA is first sheared into fragments of about 400–500 bp in length. The short fragments are necessary to allow repetitive copies of DNA to be on separate fragments, thus effectively increasing the concentration of any particular sequence if it is represented more than once. The short DNA fragments are heated to separate them into single strands and are then cooled very slowly. At several times during the renaturation reaction, the fraction of single-stranded molecules is determined. The percentage of repetitive DNA sequences within a genome can be determined by measuring the fraction of the total DNA that renatures fastest. The single-copy species (the sequences that likely represent actual genes) renature at the slowest rate because they are present at the lowest concentration.

The renaturation rates of DNAs from a variety of species reveal that a great amount of diversity exits between genomes. The results from a few organisms are shown in Table 10.2. The reassociation kinetics for the human genome, for example, is shown in Figure 10.C. The curve indicates that our genome contains several categories of repetitive sequences: highly repetitive (more than 50,000 copies), moderately repetitive (fewer than 50,000 copies), and unique sequences (single copies). This is a general characteristic of eukaryotic genomes, although the amount of DNA of each type varies widely. Plant species also contain large amounts of highly repetitive and moderately repetitive DNA in their genomes (remember that the lily has more DNA in a single chromosome than the entire human genome, Table 10.1).

(continued)

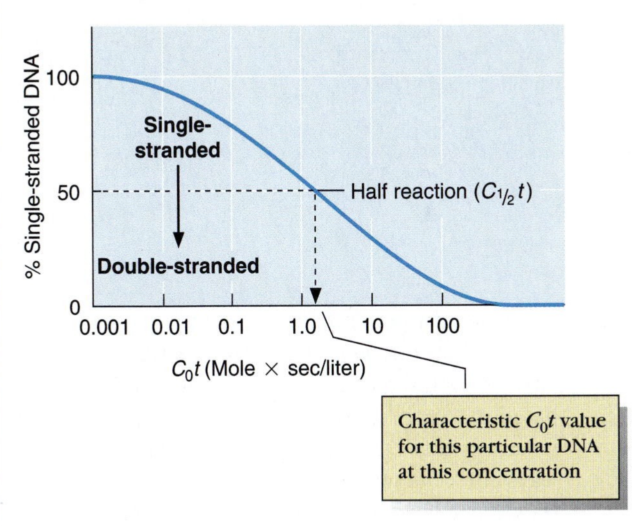

FIGURE 10.B The reassociation of single-stranded DNA to double-stranded DNA. The rate of this reaction is dependent upon the concentration of the complimentary strands in the medium. The conversion into double-stranded DNA can be experimentally followed by measuring absorbency at 260 nm because double-stranded DNA and single-stranded DNA at the same concentration absorb differently.

(continued)

The overall molecular weight of the genome can be calculated by measuring the reassociation rate of the unique sequences. The concentration of a specific single-copy sequence is dependent upon how many other different single-copy sequences are present in the total genome. Thus, a unique (single-copy) sequence will reassociate faster if present at a concentration of 1/100 of the total, for example, than if present at a concentration of 1/10,000. Thus at a specific concentration of DNA, say 10 mg/ml, a larger genome (10 million daltons) will have a lower concentration of a specific unique sequence than a smaller genome (1 million Da), and a slower renaturation rate of its unique DNA fragments.

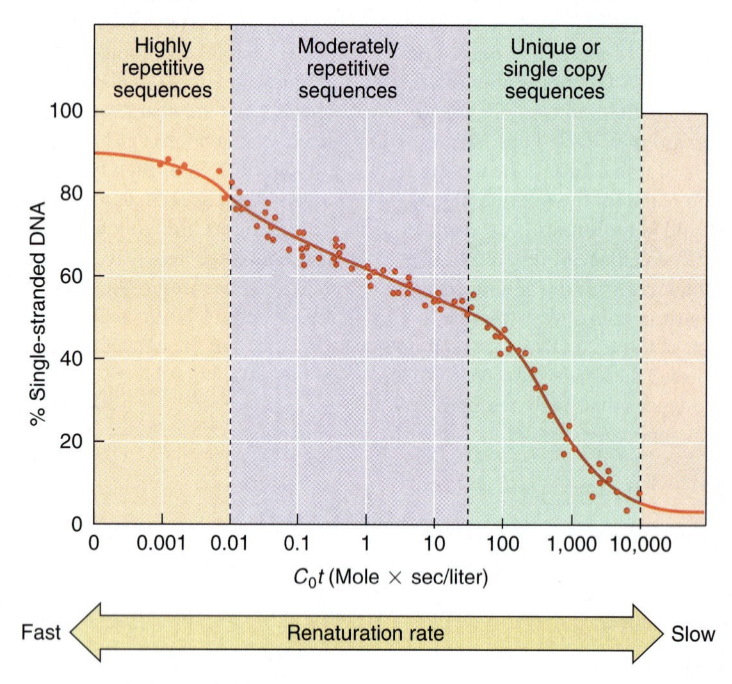

FIGURE 10.C Renaturation kinetics of human DNA. These data reveal the presence of highly repetitive sequences, moderately repetitive sequences, and unique or single-copy sequences. *(From Schmid, C. W., and P. O. Deininger. 1975.* Cell, *6: 345–358.)*

234-bp repetitive sequence with a very high AT content, and it comprises about 10% of the total DNA. Unfortunately, this information does not reveal much about the function of these DNAs. One proposed function of the satellite DNA is in spindle fiber attachment during meiosis and mitosis. This conclusion derives from the fact that most satellite DNA is located within and near the centromere region of the chromosome. Absolute proof of this role of repetitive DNA sequences in metaphase chromosome movement is not available at present.

What can we conclude about the sequence complexity of eukaryotic genomes? Are repetitive sequences the answer to the excess DNA question or to the C-value paradox? Recall that the C-value paradox is that the C-value or total DNA content of the genome does not always correlate with the complexity of the organism. If all the excess DNA in the genome were repetitive sequences, species with exceptionally large C-values would have a correspondingly large amount of repetitive sequences. This is usually, but not always, the case. We must conclude that although much of the excess DNA in eukaryotic genomes is repetitive DNA, there is still an excess of unique sequences over what appears to be necessary for the genes. This strongly suggests that not all the unique sequence DNA is necessary for the functioning of the organism.

The conclusion that not all unique sequences are required is supported by genetic evidence. On average, only one essential gene occurs in every 30 kb of *Drosophila* DNA (we expect one gene to occupy, on average, about 5–10 kb of DNA). In yeast, a eukaryotic organism with a small (12,800 kb) genome that is 90% unique, genetic studies suggest that only 12% of the genome is essential for growth.

It would seem that, in many organisms, significant portions of the unique sequences in the genome are not essential for proper growth, development, or function of the organism. It also appears that only a small fraction of the repetitive sequences are required.

SOLVED PROBLEM

Problem

Estimate the number of single-copy genes in an organism that has eight chromosomes, each containing 1×10^7 bp of DNA. This organism's genome also contains 22% repetitive DNA. Assume that an average gene has 8×10^3 bp.

Solution

This organism has eight chromosomes and each has 10 million bp; thus the total genome has 8×10 million, or 80 million, base pairs. Of the genome, 22% is repetitive DNA and can be excluded from our calculation; thus 0.22×80 million yields 17.6 million bp, which can be subtracted from the total (80 million less 17.6 million yields 64.2 million bp that are not repetitive). We can assume that all these are single-copy genes. By dividing the average size of a gene into this number $[(64.2 \times 10^6)/(8 \times 10^3) = 8 \times 10^3]$, we get the possible number of single-copy genes (8025) present in this organism's genome.

SECTION REVIEW PROBLEMS

8. What is the evidence that eukaryotic chromosomes have a single continuous, double-stranded DNA molecule?
9. What is the ratio of the length of the average DNA molecule in a human chromosome to the actual length of the condensed chromosome, assuming a length of 0.01 mm for an average condensed chromosome?
10. What is the function of highly repetitive DNA?

THE EUKARYOTIC CHROMOSOME IS TIGHTLY PACKED.

The chromosome is not always tightly compacted like the metaphase chromosome (Chapter 3). Chromosomes exist in a relatively dispersed pattern during most stages of the cell cycle, making the details of individual chromosomes invisible under a light microscope. However, during mitosis and meiosis, the chromatin takes on a highly ordered structure (the condensed chromosome) that is easily seen under a light microscope. The increasing levels of coiling that make up the compacted structure are necessary to pack the huge length of the DNA into the small volume of the visible chromosome. Tight packing is also vital for the proper functioning of genes at various stages of the cell cycle. A diagrammatic representation of the different levels of coiling seen in condensed chromosomes is shown in Figure 10.8. DNA is first coiled into an 11-nm-diameter fiber; then this fiber is coiled into a 34-nm solenoid (discussed in the next section), which, in turn, is looped to form a 300-nm fiber, finally forming a coil representing an arm of a chromosome. We will examine each level of coiling of the chromosome in the following sections.

Chromosomes Contain Two Major Classes of Proteins.

Eukaryotic chromosomes contain more protein, on a weight per weight basis, than DNA. The protein component of chromatin aids in packing the huge length of the DNA into the compact structure known as the chromosome. Proteins in chromatin are divided into two major classes: **histone proteins** and **nonhistone proteins.** There are five types of histone proteins: **H1, H2A, H2B, H3,** and **H4.** All are relatively small proteins that contain a high proportion of positively charged amino acids (lysine and arginine), which help them bind to negatively charged DNA, similar to the HU protein in bacteria. Table 10.3 gives the relative amounts of arginine and lysine in the five different types of histones found in all eukaryotic cells (except yeast, which has no H1 histone). All histones occur in equal molar ratios, except for H1, which is present in half the amounts of the others. Histone proteins have very nearly the same amino acid sequences in all the species from which they have been isolated and characterized. When amino acid sequences are nearly identical in different species, they are said to have **conserved sequences.** The similarity of amino acid sequences among histones (and thus the three-dimensional structure of the protein) implies that the amino acid sequence must be extremely important for histone function. The structure of the H1 protein varies somewhat between species, but it contains regions that

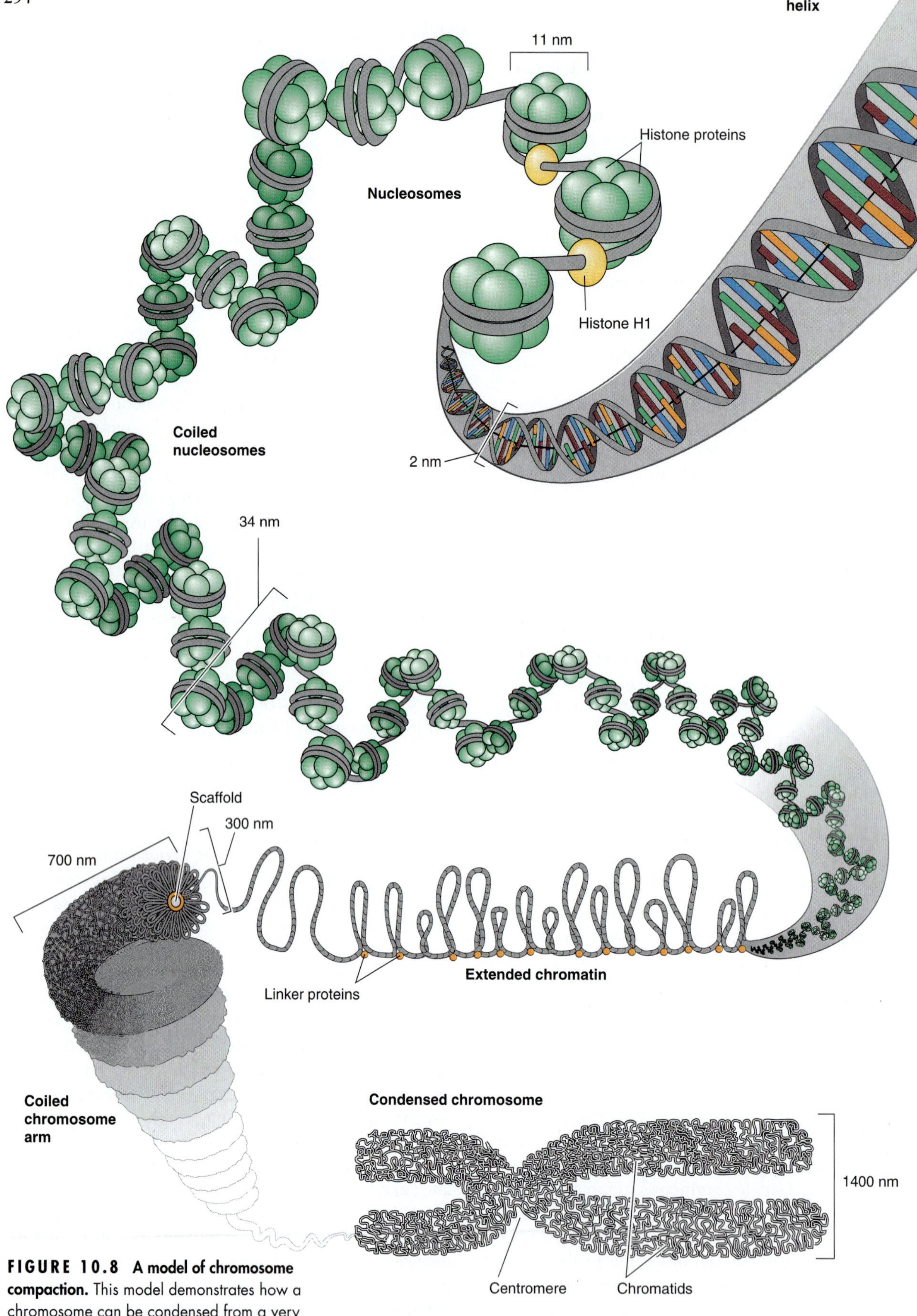

FIGURE 10.8 A model of chromosome compaction. This model demonstrates how a chromosome can be condensed from a very long molecule into a compact space.

TABLE 10.3 *Properties of Histones*

Histone	Lysine (%)	Arginine (%)	Molecular Weight* (Da)	Relative Molar Abundance
H1	29	1	22,000	1
H2A	11	9	14,000	2
H2B	16	6	13,700	2
H3	10	13	15,300	2
H4	11	14	11,300	2

*The actual molecular weight and percent of basic amino acids varies slightly among species.

are relatively invariant. Protein H1 varies sufficiently that in the cells of some species it is even given a different name: H1 is called H5 in avian red blood cells.

An equally abundant group of proteins in chromatin are nonhistone proteins. The nonhistone proteins are a heterogeneous group of proteins and are not usually positively charged, unlike histones. Many nonhistone proteins participate in maintaining the coiled structure of the chromosome, but many are associated with regulation of gene expression or replication of the DNA; for example, DNA polymerase, DNA ligase, single-stranded binding protein, and RNA polymerase are just a few examples. The nonhistone protein content of a cell varies depending upon the cell type and the organism, suggesting that these proteins also play an important role in expression of specific genes during the development of the organism. As a class of proteins, however, the nonhistone proteins are not well characterized, with the exception of the specific enzymes necessary for replication and expression of DNA.

First-Order DNA Coiling Creates an 11-nm Filament of Nucleosomes.

The structure of chromatin is complex, but highly regular. It consists of repetitive units of protein complexed with DNA that is packaged into several levels or orders of coiling. The first order of packaging results in a protein-DNA filament 11 nm in diameter. This filament is produced when about 200 bp of DNA coil around a small protein body containing two each of histones H2A, H2B, H3, and H4. This DNA-protein unit is called a **nucleosome core particle.** A single molecule of histone H1 is bound to the DNA between the core particles (Figure 10.9). The H1 linking protein and the core particle together are called a **nucleosome.** When chromatin with this first-order coiling is viewed under an electron microscope, it resembles a thin string of beads, about 11 nm in diameter (Figure 10.9). If the nucleosome is briefly treated with a specialized DNase (micrococcal nuclease) derived from the bacteria *Staphylococcus aureus,* the DNA connecting the nucleosomes is cleaved. This generates segments of DNA about 200 bp long, most of which are coiled around the nucleosome. Further nuclease digestion removes an additional 50–60 bp (depending on the species), leaving 146 bp, which represent the coiled DNA around the nucleosome core. This DNA is protected from nuclease digestion because of its association with the histones. We can conclude from this experiment that the DNA between nucleosomes is 50–60 bp long and that the DNA surrounding the nucleosome is 146 bp in length. Throughout the chromosome, however, areas (genes) that are being expressed will not have histones associated with them. The first-order coiling around the histone proteins reduces the length of the chromosomal DNA about fivefold, from about 50 nm of linear DNA to about 10 nm per nucleosome.

The Second-Order DNA Coiling Creates a 34-nm Solenoid.

If you were to break open an interphase nucleus gently, you would find that most of the chromatin was not the beads-on-a-string seen in the electron micrograph of Figure 10.9, but rather the filament about 34 nm in diameter represented in Figure 10.10. This filament represents a second-order coiling of nucleosomes, known as a **solenoid,** and further condenses the chromosome. Each turn of the solenoid contains six to eight nucleosomes. Histone H1 is important in the coiling of the 11-nm filament into the 34-nm solenoid. Histone H1 is an elongated molecule that acts to seal the two turns of the DNA that are coiled around the core histone particle; however, the exact nature of the second-order structure is not fully understood. We do know that the two ends of the H1 protein have extended amino-terminal and carboxyl-terminal arms that attach adjacent nucleosome core particles, allowing the nucleosomes to arrange in a regular repeating array that facilitates second-order coiling of DNA.

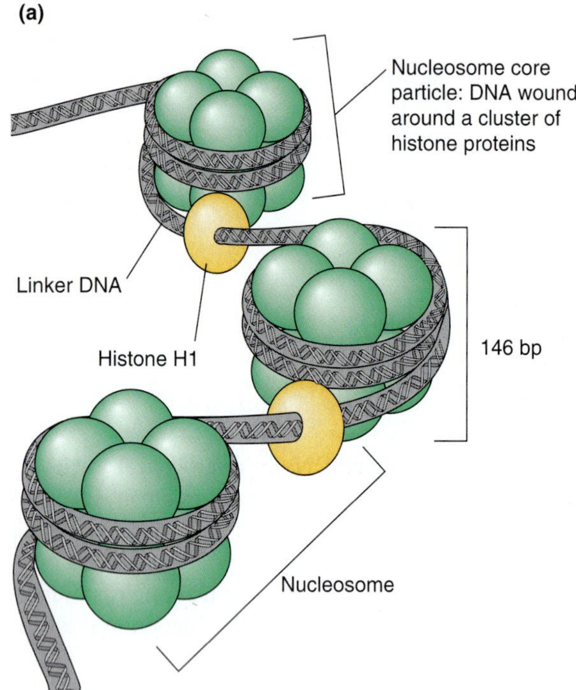

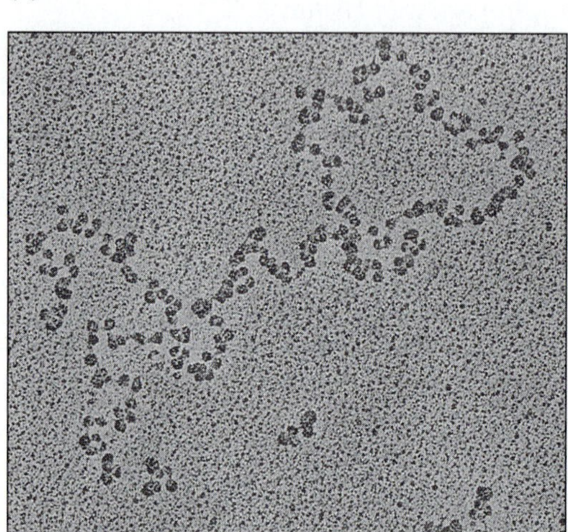

FIGURE 10.9 Nucleosome structure. (a) Each nucleosome core particle contains eight histone proteins, two each of H2A, H2B, H3, and H4, with 146 bp of double-stranded DNA coiled around the core. A DNA segment of 50–60 bp associated with histone H1 links adjacent core nucleosome particles, creating the nucleosome. (b) These nucleosomes are from the nucleus of a chicken red blood cell. The spherical structures are the nucleosome core particles and the filament linking them together is the 50-to-60-bp DNA sequence. *(From Koller, T., Swiss Federal Institute of Technology, Zurich, 1977.* Cell, *12: 101.)*

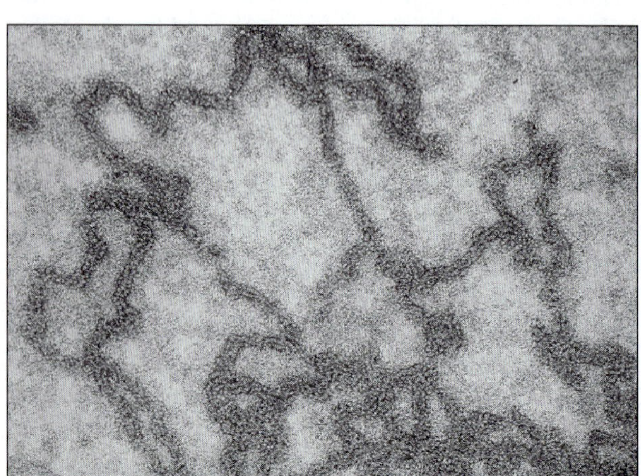

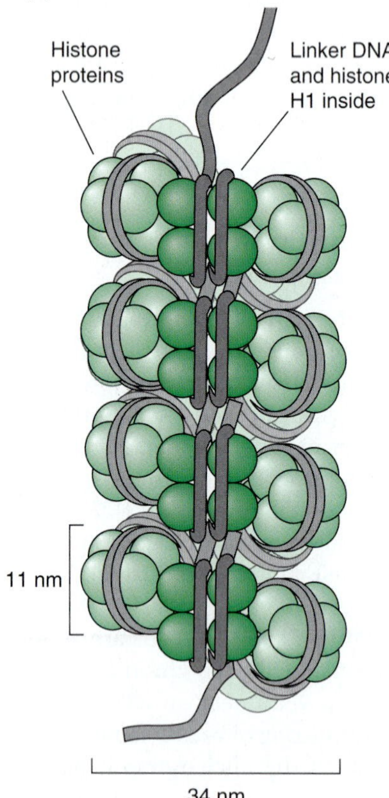

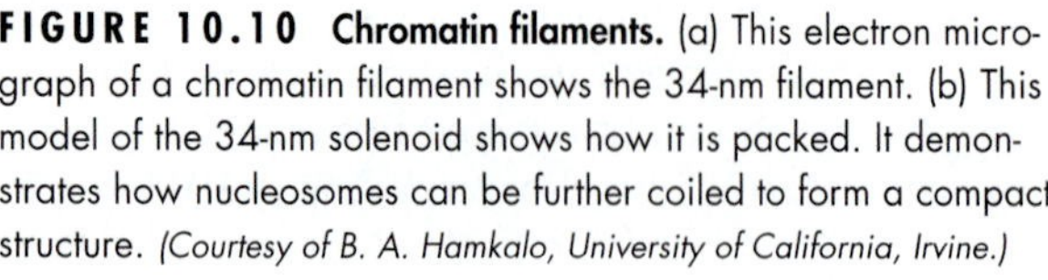

FIGURE 10.10 Chromatin filaments. (a) This electron micrograph of a chromatin filament shows the 34-nm filament. (b) This model of the 34-nm solenoid shows how it is packed. It demonstrates how nucleosomes can be further coiled to form a compact structure. *(Courtesy of B. A. Hamkalo, University of California, Irvine.)*

The Chromatin Is Composed of 300-nm Coiled Domains.

As you may recall, the *E. coli* chromosome is arranged in a series of DNA loops (about 50) that allow for a compact, coiled structure. Similarly, the eukaryotic chromosome is also composed of domains. Figures 10.8 through 10.10 illustrate the manner in which coiling may occur. The DNA is separated into loops that are attached to a protein linker every 20,000 to 100,000 bp of double-stranded DNA. The protein linkers and the base to which they are attached are called a **scaffold.** The scaffold represents about 8% of the total protein of the chromosome and can be seen in Figure 10.11 as a darkly staining material. In this example, the histones have been removed during isolation so that the coiling of the chromosome into 11-nm filaments is not visible. Also note that the length of the scaffold is approximately the same as a metaphase chromosome, strongly suggesting that the scaffold is in fact the primary structure holding the metaphase chromosome into a compact unit.

A more gentle isolation of the metaphase chromosome reveals both the 34- and 300-nm filaments (Figure 10.12). These structures are believed to be accompanied by nonhistone adhesive proteins whose properties have not yet been thoroughly investigated. The end result is a greatly condensed structure, reducing the length of the DNA molecule several thousand-fold.

FIGURE 10.11 Scaffolding in chromosomes. An electron micrograph of a mouse chromosome depleted of histones showing the looped domains present in the 300-nm coils and the scaffold structure (the darkly staining structure at the bottom). *(From Paulson, R. F., and H. R. Laemmli. 1977.* Cell, *12: 820.)*

When metaphase chromosomes are carefully isolated from cells and visualized under the scanning electron microscope, additional levels of coiling can be seen. The 300-nm fibers are coiled within the chromosome and represent the final level of coiling (Figures 10.13 and 10.14). The total diameter of a chromatid is about 700 nm.

(a)

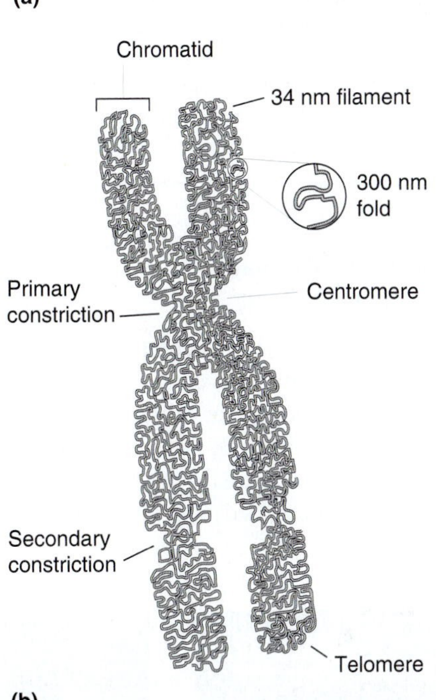

(b)

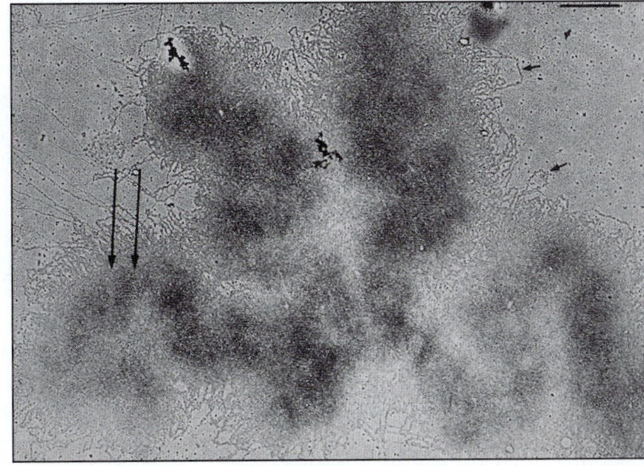

FIGURE 10.12 The compacted chromosome. (a) This drawing of a chromosome shows different levels of coiling and some of the parts. (b) This electron micrograph of a human chromosome shows the chromatids uncoiled. The path of the 300-nm fiber is somewhat obscure, but the general outline can be seen (arrows). The 34-nm fibers are looped out from the 300-nm fiber. *(From Rattner, J. B., and C. C. Lin. 1985.* Cell, *42: 291–296.)*

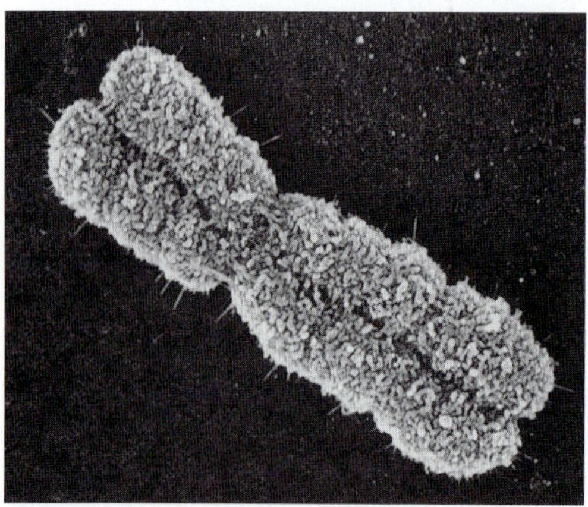

FIGURE 10.13 A scanning electron micrograph of the end of a mitotic chromosome. Each of the knobbed projections are believed to represent the tip of the 300-nm coiled filaments. *(Biophoto Associates)*

The Structure of the Chromosome Changes During the Cell Cycle.

It is important to recognize that the chromosome is not a static body. Coiling and uncoiling occurs continuously throughout the cell division cycle, and different portions of a chromosome can be in quite different states of packaging at the same time. It can be easily understood that for genes to be expressed (activated) the DNA must be uncoiled. Also, the chromosome must unwind for DNA replication. Nucleosomes are transiently displaced to allow DNA copying with minimal structural disturbances. In addition, precise regulation of the packaging of DNA in eukaryotic chromosomes is a vital part of the regulation of gene expression and cell division. As the cell progresses from DNA replication through the subsequent stages of mitotic or meiotic division, the chromosomes undergo profound structural changes in preparation for their interaction with the spindle apparatus. During this time, recombination (the exchange of genetic material between two homologous chromosomes) occurs. Unfortunately, the coiling and uncoiling of DNA is not understood in detail.

SOLVED PROBLEM

Problem

If the largest *Drosophila* chromosome is 6.2×10^7 Da, and you assume that all nucleotide pairs are packaged in nucleosomes, how many nucleosomes are in this chromosome?

Solution

The number of nucleosomes can be calculated by dividing by 200 (which is the number of base pairs per nucleosome), which equals 3.1×10^5 nucleosomes for this chromosome.

(a)

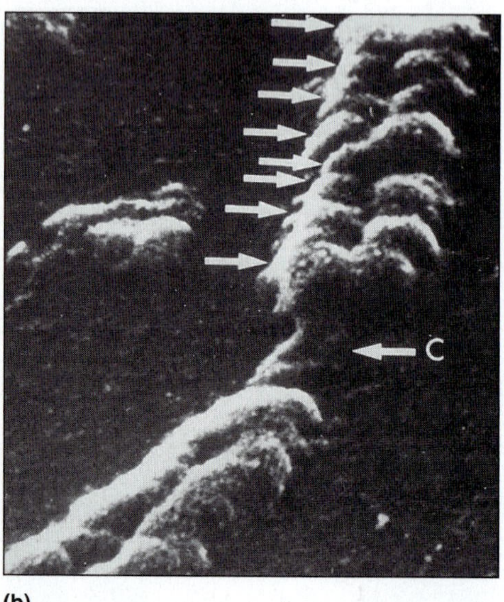

(b)

FIGURE 10.14 Scanning electron micrographs of human chromosomes. (a) The areas between the coiled 300-nm filaments can be seen (small arrows). The 300-nm filament is extended from the telomere or tip of the upper chromatid (large arrow). (b) The centromere is the lower region in the middle of the chromosome and is indicated by the arrow. The ridges within the two chromatids (arrows) are the areas between the coiled 300-nm chromatin material. In this photograph, the already-coiled DNA is in the final stage of being coiled again. *(From Rattner, J. B., and C. C. Lin. 1985.* Cell, *42: 291–296.)*

SECTION REVIEW PROBLEMS

11. Calculate the number of nucleosomes that you would expect to find in the genome of *Drosophila* using the information in Table 10.1.
12. What is the function of histone proteins?
13. What is the scaffold, and what is its function?

THE EUKARYOTIC CHROMOSOME HAS SPECIALIZED SEQUENCES.

Some chromosomes or specific areas of chromosomes have specialized functions that are essential for the viability and continued propagation of the cell. For example, each chromosome has a special area where the spindle fibers attach (centromere), as well as specialized areas at the ends of the chromosomes (the telomeres). In Chapter 3, we discussed how the sex chromosomes inactivate in females mammals to balance the amount of DNA between the two sexes. In this section, we will discuss these structure in more depth.

Euchromatin and Heterochromatin Have Different Compositions.

As mentioned in Chapter 3, light microscope examination of stained metaphase chromosomes of higher eukaryotes reveals the presence of highly condensed chromatin that appears darker than surrounding chromatin. This condensed chromatin is referred to as **heterochromatin,** and it stains darkly with Fuelgen stain. The less-condensed chromatin is called **euchromatin,** and it stains lightly with Fuelgen stain. Euchromatin is active genetic material (the genes within euchromatin are expressed), whereas heterochromatin is mostly inactive genetic material. Heterochromatin can be further subdivided into **constitutive heterochromatin** and **facultative heterochromatin.** Constitutive heterochromatin is always inactive and highly condensed, whereas facultative heterochromatin can shift back and forth between heterochromatin and euchromatin, depending upon developmental stage and cell type. An example of facultative heterochromatin is the Barr body, or inactivated X chromosome, in females. The biochemical difference between euchromatin and heterochromatin is not fully understood, but we do know that some euchromatin is converted into heterochromatin during some stages of mitosis and meiosis. We also know that much of the constitutive heterochromatin is composed of repetitive DNA sequences.

The location of constitutive heterochromatin can be determined within individual chromosomes by a method known as ***in situ* hybridization.** In this technique, cells are spread on a glass microscope slide, where they are broken open and fixed so that they cannot be easily washed off. They are treated with alkali to denature the DNA. Either a radioactive label (with ^{3}H-labeled sequence) or chemical treatment (so that a fluorescent dye can bind to it) is placed on the disrupted cells and allowed to hybridize with the chromosomal DNA. Nonhybridized material is removed by washing. The slide is then dipped in a photographic emulsion (or the fluorescent dye is bound). When the slide is developed, it produces an image that reflects the location of the bound DNA on the chromosome. When radioactive DNA is used and the location of the bound DNA visualized on film, the process is called **autoradiography,** and the image is an autoradiograph. When dye is used, this procedure is known as **fluorescent *in situ* hybridization** (FISH). Using these techniques, repetitive DNA sequences have been localized to constitutive heterochromatic regions of the chromosome near the centromere (Figure 10.15). Why this is true in not clear, but we do know that artificial chromosomes will not replicate in some cells without added heterochromatin near the centromere. Genes can also be localized to specific area of a chromosome as well.

Euchromatin can be converted to heterochromatin when a euchromatic segment of DNA is physically placed within the heterochromatin region of a chromosome. This suggests that a physical change occurs within the active genetic material to inactivate the genes. This physical change is believed to be the product of a process where methyl groups are added to certain nucleotides, producing sequences of **methylated DNA.** In methylated DNA, most of the methyl groups are found in CG doublets where methylation occurs on the 5-position of the cytidine ring (Figure 10.16). The shorthand representation of methylation is often written

$$5'\ {}^{m}CG\ 3'$$
$$3'\ G^{m}C\ 5'$$

Satellite DNA is frequently extensively methylated, but some unique sequences are also methylated. Methylation is associated with inactive genetic material, and methylation-free sequences are associated with active genetic material, although exceptions to this rule are known.

In facultative heterochromatin, the controlling region adjacent to an active gene in one tissue will be undermethylated, whereas the same controlling region in another tissue where the gene is inactive will be highly methylated. Removal or addition of methyl groups from the CG dinucleotides is a method of conversion between genetically inactive chromatin to genetically active chromatin.

A striking example of facultative heterochromatin is the sex chromosomes in female mammals that determine the sex of the individual. Recall from Chapter 3 that all

FIGURE 10.15 ***In situ* hybridization.** (a) This autoradiogram shows a metaphase chromosome of the kangaroo rat *Diplodomys ordii*. ^{3}H-labeled RNA prepared from purified satellite DNA was hybridized to the chromosomes using *in situ* hybridization. (b) One of the chromosomes shown in (a) has been enlarged. Note that the radioactivity occurs near the regions adjacent to the centromeres. This experiment shows that satellite DNA is present near the centromeres. *(Courtesy of David Prescott, University of Colorado, Boulder.)*

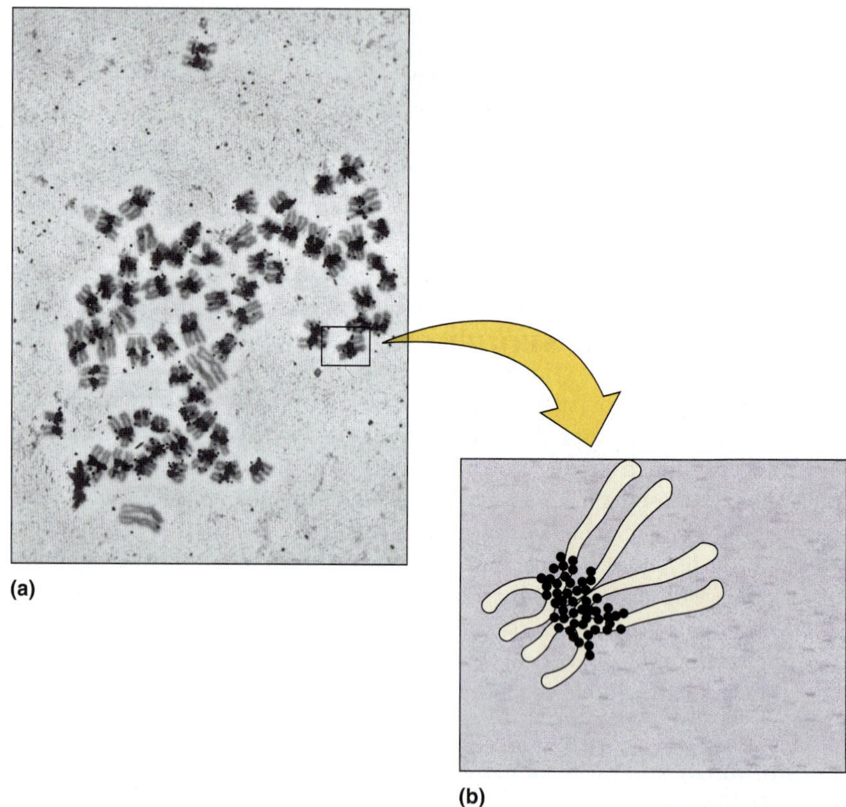

female mammals have two large sex chromosomes, identical in size but very different in appearance. Male mammals have one large sex chromosome (the X chromosome) and one small sex chromosome (the Y chromosome). The male Y chromosome is almost completely composed of heterochromatin and is, for the most part, genetically inactive. In female mammals, one chromosome appears highly condensed and has all the features of heterochromatin (the Barr body), whereas the other chromosome is clearly euchromatin. In females, one or the other of the sex chromosomes is active by chance, so female tissue is actually a mosaic of two cell types, each with different active X chromosomes. We know from biochemical tests on individual cells that only one of the sex chromosomes is active and the other is almost completely inactive. Methylation of the GC sequences occurs extensively in the inactive X chromosome and may account for the inactivation. An example of sex chromosome inactivation producing somatic mosaics is shown in Figure 3.27.

Analyses of different tissues in the body further support the fact that the extent of CG-sequence methylation affects gene expression. Human placental DNA is 47% CG methylated (a very active tissue), and thymus DNA is 85% methylated (a less active tissue). A direct correlation exists between the level of methylation of CG sequences and the general genetic activity of the tissue.

Recall that plants have extremely large genomes and a very high percentage of repetitive sequences. Plant DNA also has an extremely high level of methylation, which is consistent with the explanation that methylated

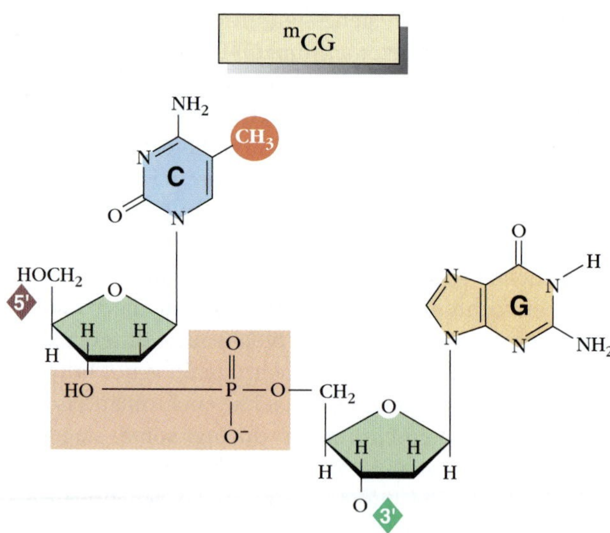

FIGURE 10.16 A methylated CG pair. The dinucleotide 5′-^{m}CG-3′ structure shows where the methyl group is present on the cytidine ring. Methylated DNA is thought to be mostly genetically inactive.

DNA is heterochromatin. In some species, the level of 5-methylcytosine amounts to about 30% of all cytosines and is found in sequences ^{m}CG and ^{m}CXG (animal tissue does not have ^{m}CXG; where X = any base). Vertebrate DNA is also relatively highly methylated and contains clusters of nonmethylated CG associated with genes that are active. However, *Drosophila* does not contain *any* methylated DNA, indicating that methylation is not a universal mechanism of gene inactivation. In general, we can conclude that

1. DNA methylation in eukaryotic cells may serve as a gene inactivation mechanism.
2. This function is accomplished by a tissue-specific and chromosome-specific pattern.

SECTION REVIEW PROBLEMS

14. What are the structural and functional difference between euchromatin and heterochromatin?
15. What is the difference between facultative heterochromatin and constitutive heterochromatin?
16. What is the presumed function of 5-methylcytosine in determining whether genes are expressed or not?

Telomeres and Centromeres Possess Unique Sequences.

As we have just seen, most repetitive DNA sequences do not have an identifiable function. However, two specialized structures of the chromosome that do have repeated sequences and essential functions are the telomeres and centromeres. Telomeres are absolutely essential for chromosome function and are composed of very specific repeated elements. Centromeres also contain characteristic sequences and are usually located within a chromosome (the exception being telocentric chromosomes). These sequences are essential for attaching the spindle fibers to the chromosome for movement during mitosis and meiosis.

Telomeres. It has been known for a very long time that the ends of eukaryotic chromosomes are specialized structures. This conclusion is inferred from a long-standing observation of what happens when a cell is exposed to X-rays, yielding broken chromosome fragments. Usually these fragments are unstable; however, occasionally they fuse end to end and form abnormal chromosomes or parts of chromosomes. Why don't normal, unbroken chromosomes fuse end to end? Earlier, it was hypothesized that the ends of each chromosome had a specialized structure called a **telomere** that prevented the ends from fusing with the ends of other chromosomes. The most prominent feature of chromosome ends that have telomeres is that they are not "sticky." The ends of chromosomes must be blocked by some mechanism, although the mechanism is not certain at present, but it probably involves two proteins that tightly bind to telomere sequences.

TABLE 10.4 *The Sequence of Representative Telomeric Repeats*

Sequence*	Organism	Common Name
T_4–G_2	*Tetrahymena*	Ciliate
T_4–G_4	*Oxytricha*	Ciliate
T_2–AG_3	*Trypanosoma*	Ciliate
T_2–AG_3	*Homo sapiens*	Humans
$(TG)_{1\text{-}3}A(G)_{1\text{-}3}$	*Saccharomyces*	Yeast
$AG_{1\text{-}8}$	*Dictyostelium*	Slime mold
T_3AG_3	*Arabidopsis*	Plant

*Sequences all read 5′ to 3′ toward the chromosomal terminus.

Examination of the sequences of telomeres shows that they consist of short, tandemly repetitive DNA sequences (Table 10.4). These sequences are repeated between 5 and 350 times in a telomere. It is clear from the sequences (shown in Table 10.4) that one strand is guanosine-rich (the strand shown) and its complement is therefore cytosine-rich. The G-rich strand extends beyond the C-rich strand, constituting a G-strand 3′ overhang at the chromosomal terminus. This orientation of strands is absolutely required because an inverted telomere will not function. This structure is highly conserved between species, suggesting that the molecular mechanisms responsible for the replication and function of telomeres are similar in most eukaryotic organisms (see CURRENT INVESTIGATIONS: *Telomeres Grow and Shrink Each Generation*). Molecular hybridization studies show that telomeric sequences from most species are similar to the sequence found in *Arabidopsis* (a small plant) and humans, the only higher eukaryotes listed in Table 10.4.

Centromeres. All stable eukaryotic chromosomes have centromeres, and they are responsible for accurate segregation of the replicated chromosome into the progeny cells during mitosis and meiosis. If a chromosome lacks a centromere, the DNA will be replicated but the chromosome will not segregate properly to the progeny cells. Functional centromere DNA segments have been sequenced. Because the complete genome of *Saccharomyces cerevisiae* (yeast) has been sequenced, all the centromere sequences are known. The centromere sequences are referred to as *CEN1. . . CEN16,* where the number refers to the chromosome from which the centromere was sequenced. Centromere DNA sequences share a number of common properties. First, they all are relatively small segments of DNA, less than about 250 bp

CURRENT INVESTIGATIONS

Telomeres Grow and Shrink Each Generation

In addition to preventing the ends of eukaryotic chromosomes from indiscriminately binding to each other, telomeres have a second very important function: They provide the ends of the chromosomes with a nonessential sequence of DNA that can be shortened during DNA replication. Recall from Chapter 9 that all DNA-dependent DNA polymerases require a primer to initiate 5′⟶3′ strand synthesis. After chromosome replication, an unreplicated single-stranded sequence is left at the ends of the lagging strand as a result of the removal of the short-terminal RNA primer (Figure 10.D). As a result, the lagging strand shortens with each DNA replication cycle, shortening the telomere by 10–30 bp. The amount of shortening is equivalent to the length of the short RNA primer because there is no mechanism to replace the RNA primer with DNA.

If chromosomes are shortened with each DNA replication cycle, how is the full length of the telomeric sequence maintained? In most muticellular eukaryotes studied thus far, the telomeric sequences maintain their length through **telomerase,** a ribonucleoprotein polymerase that contains its own RNA component. An interesting

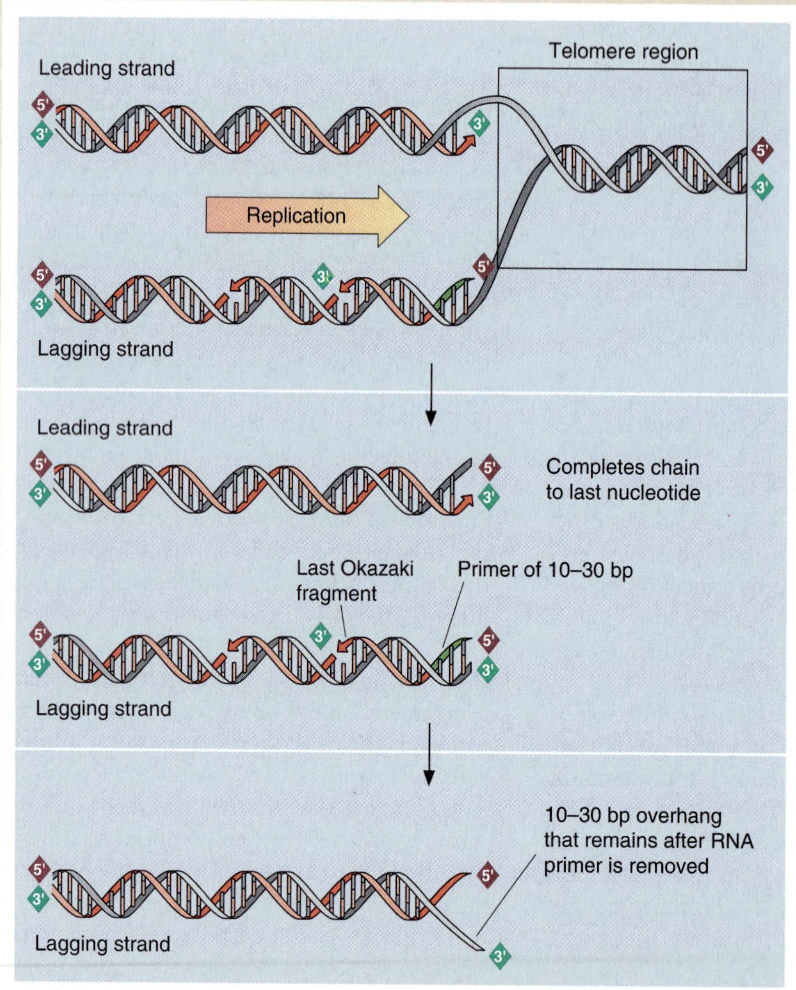

FIGURE 10.D Telomere shortening by DNA replication. The model shows how DNA synthesis at the end of a chromosome results in shortening (10–30 bp) of the lagging strand every round of replication. The DNA sequence complementary to the RNA primer is lost during every cell division, shortening the chromosome by 10–30 bp.

feature of telomeric synthesis is that it is characteristic of a particular organism, irrespective of the source of the telomere. For example, when *Tetrahymena* telomeres are put into yeast cells and grown, sequences are added to the *Tetrahymena* telomeres that are representative of yeast sequences, not *Tetrahymena.* This is caused by the RNA component of the polymerase, which has complementary sequences that direct the synthesis of the extended 3′ strand of the telomere (Figure 10.E). The mechanism that governs the number of repeated units of the telomere is not known, but it is characteristic of higher eukaryotic organisms and must be genetically controlled. A model has been proposed for telomerasic action and is known as the inchworm model (Figure 10.E). This model shows how a 3′ terminus is extended by about 1.5 telomere repeats per synthesis cycle. The enzyme and its RNA component provide a means of extending the 3′-end strand. The other strand can be extended using DNA polymerase.

An interesting feature of telomeres is that they decrease in size as an organism ages, presumably because of the loss of the short sequences at each replication cycle. This is consistent with the observation that essentially no telomerase activity is found in normal somatic cells. On the other hand, a great deal of telomerase activity is found in germ cells. Thus, it would appear that telomeres are extended to their full length during germ cell formation, and they decrease in size incrementally as replication cycles ensue throughout the life of the eukaryotic organism. An interesting application of this latter observation is that cancer cells seemingly *have* telomerase activity, which may explain why they possess immortality. A cancer cell's telomere does not shorten as cell division continues. The search is now on for inhibitors of telomerase action because such inhibitors would not affect normal cell division but would limit the number of divisions of cancer cells.

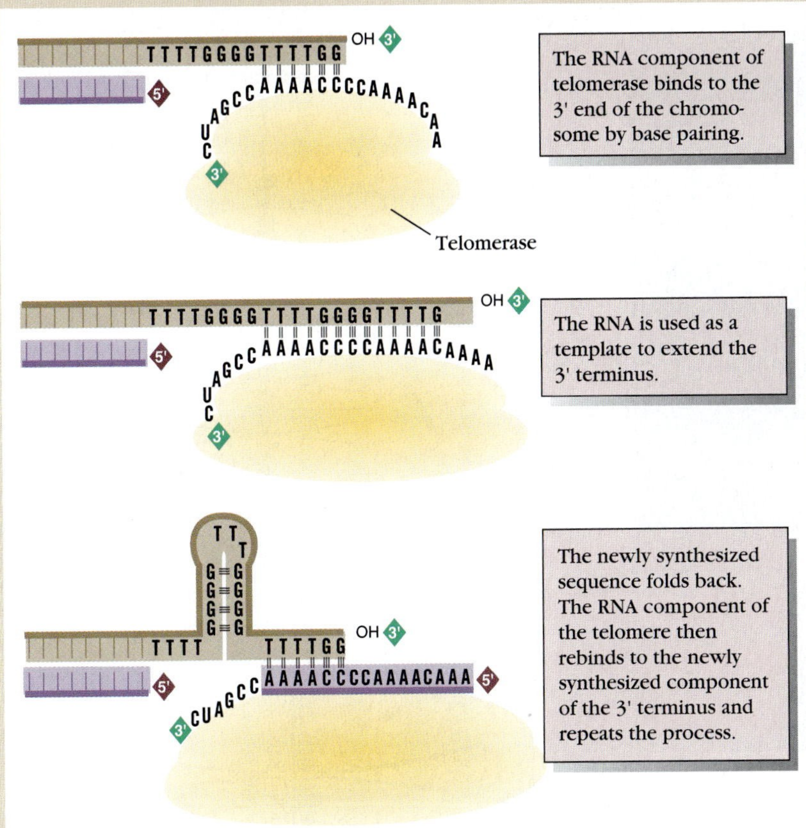

TTTTGGGGTTTTGG
OH 3'
5'
CUAGCCAAAACCCCAAAACAA
3'
The RNA component of telomerase binds to the 3' end of the chromosome by base pairing.
Telomerase
TTTTGGGGTTTTGGGGTTTTG
OH 3'
5'
The RNA is used as a template to extend the 3' terminus.
TTTT
TTTTGG
OH 3'
AAAACCCCAAAACAAA
5'
The newly synthesized sequence folds back. The RNA component of the telomere then rebinds to the newly synthesized component of the 3' terminus and repeats the process.

FIGURE 10.E Telomere synthesis. The model shows the extension of the 3′ terminus of a telomere by telomerase. The RNA component of the enzyme acts like an inchworm to extend a short sequence repeatedly. The 5′ strand is extended by normal replication mechanisms. This occurs only in germ cells of higher eukaryotes.

FIGURE 10.17 Centromeres. The sequence of the centromere region of chromosomes 3 (*CEN3*) and 4 (*CEN4*) of yeast are shown. All known centromeres are composed of three central DNA element regions (I, II, and III) which are highly conserved between organisms and chromosomes. These regions are essential for chromosome movement during mitosis and meiosis. The telomere is shown at the end of the chromosome.

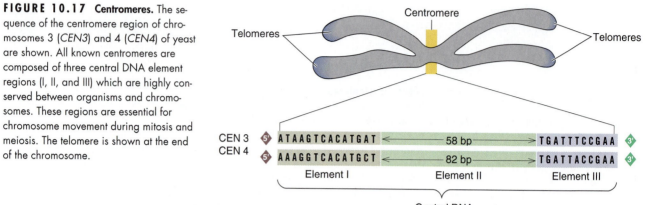

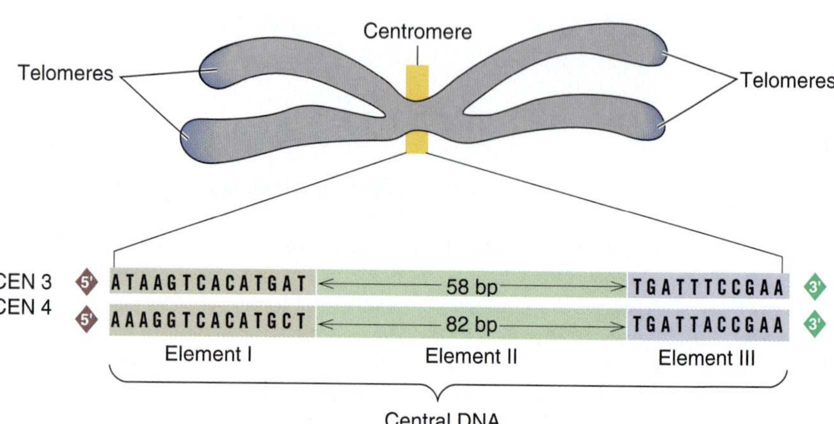

in length. Secondly, centromere sequences possess a region of 78–89 bp that is very A + T rich (93–94%). This region is referred to as CDEII (for central DNA element). CDEII is bordered on one side by a highly conserved 11-bp sequence (element III in Figure 10.17) and a less highly conserved 14-bp sequence (element I) on the other side. These regions are essential for *CEN* function in mitotic stabilization.

Unique Chromosomes Are Useful in Understanding Genetic Mechanisms and Structure.

So far we have discussed mitotic chromosomes that are typically found in every eukaryotic species. There are exceptions, however. Some chromosomes have special structures or unusual features. Often these chromosomes

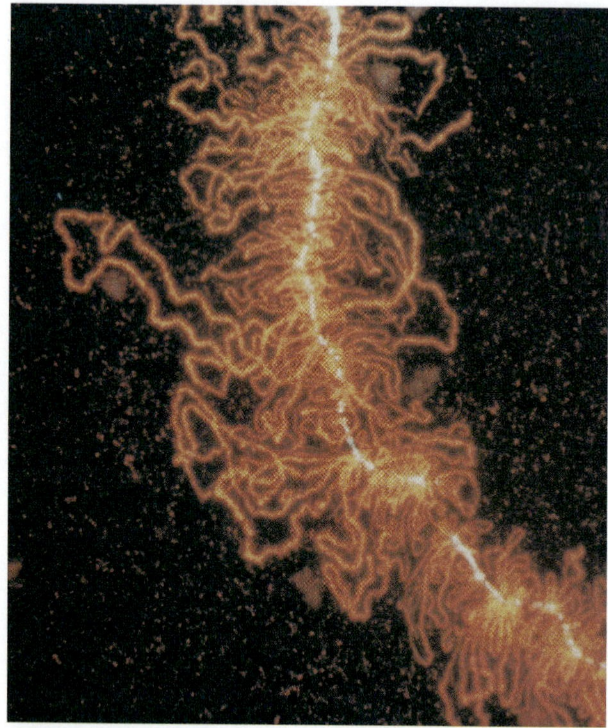

(a)

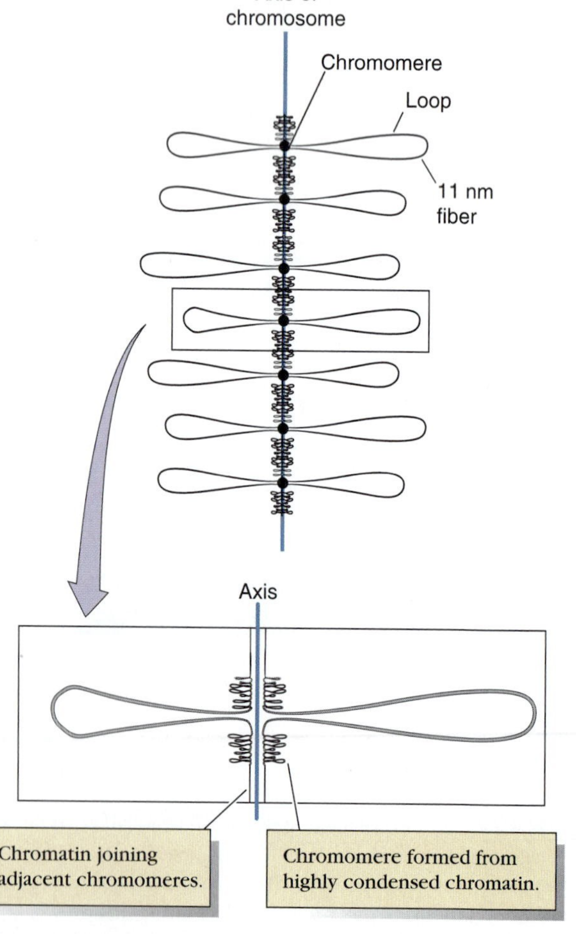

(b)

FIGURE 10.18 Lampbrush chromosomes. (a) This light micrograph shows a lampbrush chromosome from an amphibian oocyte *Notophthalmus viridescens.* (b) The diagrammatic representation of the lampbrush region of the chromosome shows looped domains that are the expressed genes. Each loop has a condensed region surrounding it known as a chromomere. *(Courtesy of J. G. Gall, Carnegie Institution of Washington.)*

occur in only certain cell types or in particular developmental stages. Examining these chromosomes can sometimes yield valuable information about the structure of chromosomes in general.

Further evidence that chromosomes are composed of looped structures attached to a central scaffold is found in the so-called **lampbrush chromosomes.** They are given this name because their structure under the light microscope resembles the brush once used to clean the chimneys of oil-burning lamps. Normally it is not possible to see chromosomes during interphase because they are too fine and tangled. However, in lampbrush chromosomes, looped domains can be seen directly with a light microscope and were first observed by early cytologists (Figure 10.18) during early oogenesis (development of the egg) in amphibians. A detailed analysis of lampbrush chromosomes shows that the loops consist of 50,000–200,000 bp that are attached to a protein core and are similar in structure to the looped domains pictured in Figure 10.18. Adjacent to the attachment point of the domain are **chromomeres,** or tightly coiled or condensed regions of DNA (Figure 10.18). Note that the nucleosomes are still intact in the coiled loops; in fact, the lampbrush chromosomes are strings of nucleosomes. This highly uncoiled structure is believed to represent active DNA synthesizing RNA for the expression of genes required for early development. This conclusion is based on the finding that RNA is synthesized from lampbrush loops at a high rate, and as we will see in subsequent chapters, RNA polymers formed this way have base sequences complementary to specific segments of the chromosomal DNA and are the intermediate products in expression of genes into proteins.

Additional evidence that uncoiled DNA is, in fact, active genetic material comes from radioactive labeling of RNA in **chromosome puffs** of polytene chromosomes. **Polytene chromosomes** are specialized chromosomes found in the salivary glands of the family Diptera which includes *Drosophila melanogaster* (Figure 10.19). Polytene chromosomes are extremely large chromosomes, easily seen under the light microscope, and are the result of ten replications of DNA without cell division. The many-replicated DNA strands (2^{10} = 1024 strands) are aligned side by side forming an immense chromosome. When these chromosomes are viewed with a light microscope, distinct alternating dark and light bands are visible (Figure 10.19). No staining is required to see the banding pattern, but staining emphasizes the details of the banding patterns. The darker bands actually represent highly condensed DNA similar to lampbrush chromosome chromomeres. Approximately 85% of the DNA is in the banded regions and the remaining 15% is in the interband regions. Individual bands range from 3000 to 300,000 bp. Figure 10.20 shows a highly magnified region of a band and interband region clearly displaying extended DNA loops in the interband region. The

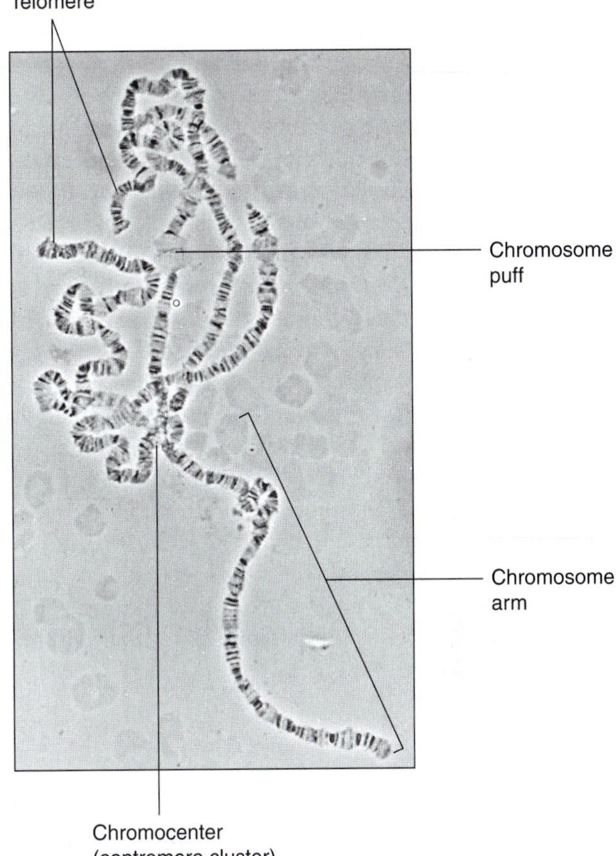

FIGURE 10.19 Polytene chromosome. This photograph shows a *Drosophila melanogaster* polytene chromosome viewed under phase optics. The banding patterns that serve as markers for cytogeneticists to locate specific areas of the chromosome are clearly shown. *(Courtesy of L. Ambrosio, Iowa State University.)*

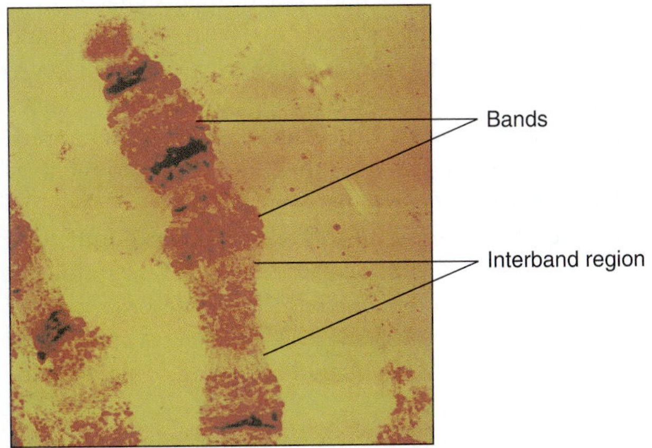

FIGURE 10.20 A band in a polytene chromosome. A false color section of the *Drosophila* salivary gland chromosome revealed by an atomic force microscope. The darker areas indicate more highly condensed DNA from the banded region. *(Courtesy of J. Vesenka, C. Mosher, J. Schaus, L. Ambrosio, and E. Henderson, Iowa State University.)*

banded region has elevated concentrations of DNA (the chromomeric region).

Expanded regions of the polytene chromosomes (called puffs) can be seen under the microscope. The puffs seen on polytene chromosomes are also active in the production of RNA polymers. This has been established by labeling the RNA with ^{3}H-uridine (recall that uracil replaces thymine as a pyrimidine in RNA). The cells containing the puffed chromosomes are incubated in the presence of ^{3}H-uridine briefly, the excess uridine is removed, and the chromosome is exposed to a photographic emulsion to form an autoradiograph of the region surrounding the puff (Figure 10.21). The formation of puffs suggests that active DNA undergoes major changes in packing to allow gene expression. There are approximately 5000 bands (and interbands) in the *Drosophila* polytene chromosome. Each band contains about 30,000 bp and has enough genetic information to code for up to seven genes. We now know that each band represents about one or more genes, but the genes are both in the bands and the interband space (the light areas shown in Figure 10.21).

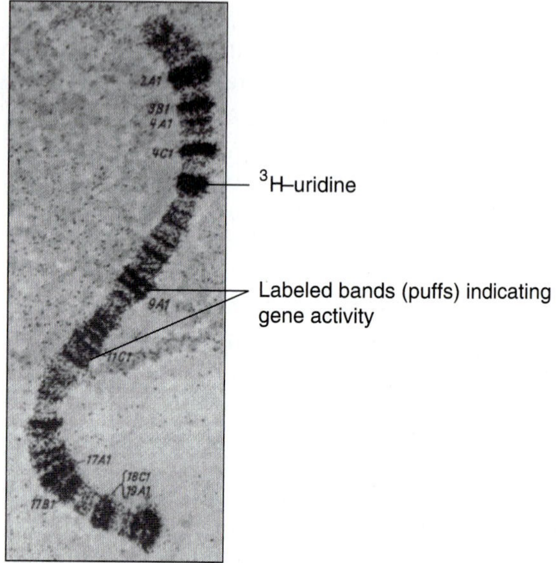

FIGURE 10.21 Expression of a gene in a polytene chromosome. This autoradiograph shows a *Drosophila* polytene chromosome that has the RNA labeled with ^{3}H-uridine. The puffed regions of the chromosome are associated with active genetic material. *(From Pelling, C. 1964.* Chromosoma, *15: 71–122.)*

Summary

The cell must pack an immensely long molecule (DNA) into a small space and allow for its replication and expression. Prokaryotes solve this problem by packaging DNA in the nucleoid, a small compact structure of protein and nucleic acids. In the nucleoid, the long DNA molecule is looped into domains (about 50) attached to the central core, with each domain separately supercoiled. Supercoiling introduces twists in the DNA helix, tightening the structure and decreasing its volume. Supercoiling is introduced into DNA by topoisomerases I or II, which make single-strand cuts (I) or double-strand cuts (II) followed by resealing. To further compact DNA, eukaryotic chromosomes are coiled around an octamer of histone proteins to form a nucleosome. Histone proteins are very basic proteins that bind tightly to the negatively charged phosphodiester backbone of DNA. Eukaryotic chromosomes also contain many nonhistone proteins, some of which form the protein scaffold, the central core structure that holds the second-order coiling of the chromosome together. Chromosomes are further packed by a series of superfolds to give a highly condensed structure.

Bacteria possess one chromosome, composed of a single DNA molecule, as does each eukaryotic chromosome. The amount of DNA per haploid genome in eukaryotes widely varies and does not always correlate with the complexity of the organism. This inconsistency, called the C-value paradox, is partly explained by the fact that most eukaryotes have a large percentage of highly repetitive sequences within their DNA. Some of these sequences (5–300 bp) are repeated millions (highly repetitive sequences) of times, whereas others (300–500 bp in length) are repeated 50,000 times (moderately repetitive sequences). Heterochromatin, an inactive, darkly staining area of a chromosome, is largely composed of these repeated sequences. In contrast, euchromatin is primarily single-copy sequences and represents the active genes of an organism.

Specialized DNA sequences within each eukaryotic chromosome include the telomere and centromere. The telomere is composed of repeated sequences and protects the ends of the chromosomes, and the centromere is composed of conserved sequences to facilitate DNA attachment to the spindle fibers. Telomere sequences are synthesized by a special enzyme, called telomerase.

Some organisms have highly specialized chromosomes that aid in identifying the location of genes and also in understanding the structure of the chromosomes. Polytene chromosomes and lampbrush chromosomes are examples of highly specialized chromosomes. Polytene chromosomes are large densely packed chromosomes that are the result of DNA duplication without cell division, and they permit easy visualization of the chromosome. Lampbrush chromosomes have uncoiled DNA that reveals a looping structure. The uncoiled DNA is correlated with the synthesis of RNA polymers, using DNA as the complementary strand, and represents sites of gene expression.

Selected Key Terms

C-value
chromosome puffs
euchromatin
histone protein
in situ hybridization
nonhistone proteins
nucleosome
polytene chromosome
renaturation
repetitive DNA
satellite DNA
scaffold
supercoiling
telomere
topoisomerase

Chapter Review Problems

1. Most human genes have been assigned to chromosomes by linkage analysis, which was covered in Chapter 6. However, it is possible to assign a gene to a chromosome using molecular techniques. Assume that you have isolated a pure form of a unique DNA sequence that you know is associated with a specific dysfunction (for example, cystic fibrosis) and you want to locate it on a chromosome. Explain how you would accomplish this goal. What specific problems would you expect to encounter? How can you enhance this finding by assigning the gene to a specific location on the chromosome?
2. Explain the difference in banding patterns seen in polytene chromosomes compared to patterns seen in stained metaphase chromosomes such a those used in karyotyping human cells.
3. What do a lampbrush chromosome and a chromosome puff have in common?
4. Histone H3 has essentially an identical amino acid sequence in mammals, insects, and worms. What can you conclude from this finding?
5. During routine isolation of DNA from *E. coli*, the size of the DNA obtained is usually much less than the estimated size of the chromosome of 5×10^6 bp. Why?
6. If a gene is defined as a portion of DNA equal to 2000 bp, how many genes would you expect to find in an *E. coli* cell (see Problem 5)?
7. What is the relationship between the density of DNA and its GC content?
8. How is the density of DNA related to its melting temperature (T_m)?
9. Which of the following DNA molecules will have the lower T_m?

   a. 5′-AATTGCTTTA-3′
   TTAACGAAAT

   b. 5′-AGGTCCTAGA-3′
   TCCAGGATCT

   Explain.
10. Calculate the number of base pairs present in the genome DNA of the SV40 virus and T4 phage if their $C_0t_{1/2}$ values are 0.013 and 0.20, respectively (you must understand reassociation kinetics to answer this problem).
11. Would you expect to find the protein-coding genes in highly repetitive, moderately repetitive, or unique DNA sequences? Explain.
12. Describe the essential differences or similarities between a centromere and a telomere. What happens *in vivo* when you remove a telomere from a chromosome? What happens when you delete a centromere?
13. Propose a mechanism by which telomeres can be synthesized by telomerase to create tandemly arranged repeating elements.
14. In each of the following sentences, insert a word or term that most appropriately completes the sentence.
    a. The ends of chromosomes in eukaryotes contain a repetitive sequence of DNA known as a ________, which protects the ends from sticking to other chromosomes.
    b. A structural component of the eukaryotic chromosome that is composed of nonhistone proteins and is where superhelical loops of DNA are attached is called the ________.
    c. When sheared DNA is denatured and then renatured, the rate of formation of double-stranded, complementary sequences depends upon the concentration of individual subpopulations. The subpopulation that reassociates most slowly with this mixture is the ________ sequences.
    d. When eukaryotic DNA is wound around the histones, the whole structure is known as a ________.
    e. Examination of tracts or spots of decomposed ^3H, after radioactive labeling of some biological molecule, in a photographic emulsion is referred to as ________.
15. What is the ratio of the fully extended length of the DNA molecule in an *E. coli* cell to the diameter of the cell?
16. Calculate the length (in meters) of the genome of an organism that has a C-value of 1×10^{10} bp of DNA.
17. Determine the theoretical number of genes in a genome of 5×10^9 bp with 65% repetitive DNA sequences. (Assume that an average gene is composed of 5000 bp.)
18. What is the correlation between the $C_0t_{1/2}$ value (the rate of renaturation) of a mixture of DNA molecules and the concentration of individual sequences within a reaction mixture? (See CONCEPT CLOSE-UP: *Reassociation Kinetics and Repetitive DNA* for an explanation.)
19. Explain the levels of coiling of a eukaryotic chromosome, and compare this to how the prokaryotic chromosome is packed.
20. Give one example each of facultative heterochromatin and constitutive heterochromatin.

Challenge Problems

1. You have isolated a plasmid from a bacterial strain using the usual procedures for DNA isolation. You would like to determine the molecular weight of this plasmid and decide to use electrophoresis to accomplish this goal. You have an agarose gel with several wells in it, and in lane one you put a standard DNA of known molecular weight. This standard is linear dsDNA varying from 1000 to 10,000 bp in 1000-bp increments. You put your plasmid DNA in the second well. After applying a voltage gradient for 4 hours, you stain the separated DNA in the gel with ethidium bromide and examine it under a UV light. In lane one, each of the ten bands for the standard are nicely separated, but in your sample lane, you find three bands, one at 8000 bp, one at 6000 bp, and one at 4000 bp. Knowing what you do about the structure of supercoiled, linear, and double-stranded circular DNA, how would you interpret these results?
2. Estimate the number of single-copy genes (assume an average size of 10,000 bp) in an organism that has five chromosome pairs, each containing 1×10^8 bp of DNA. This organism's genome also contains 78% repetitive DNA sequences interspersed throughout.
3. The following C_0t curve was found for a newly discovered eukaryotic organism. (You need to read CONCEPT CLOSE-UP: *Reassociation Kinetics and Repetitive DNA* to do this problem.)

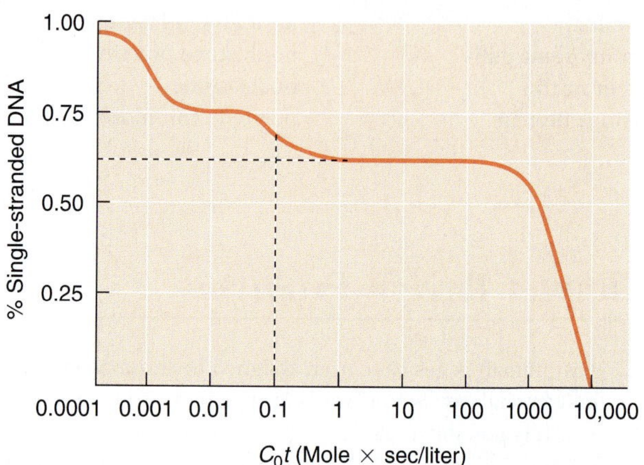

a. What percentage of the genome is highly repetitive, moderately repetitive, and unique sequences?
b. If the length constant equals 5×10^5, what is the sequence size of the different components?
c. How many copies are there for the highly repetitive and middle repetitive components?
d. What is the total number of base pairs in this new organism?

Suggested Readings

Blackburn, E. H. 1991. "Structure and Function of Telomeres." *Nature,* 350: 569–572.

Britten, R. J., and D. E. Kohne. 1970. "Repeated Segments of DNA." *Scientific American,* 222: 24–31.

Chuang, P.-T., J. D. Lieb, and B. J. Meyer. 1996. "Sex-Specific Assembly of a Dosage Compensation Complex on the Nematode X Chromosome." *Science,* 274: 1736.

Igo-Kaemenes, T., W. Horz, and H. G. Zachau. 1982. "Chromatin." *Annual Review of Biochemistry,* 51: 89–121.

Kornberg, R. D., and A. Klug. 1981. "The Nucleosome." *Scientific American,* 244: 52–64.

Lewin, B. 1997. *Genes VI.* Cell Press, Cambridge, MA/Oxford University Press, Oxford.

Lodish, H., D. Baltimore, A. Berk, S. L. Zipursky, P. Matsupaira, and J. Darnell. 1995. *Molecular Cell Biology,* 3rd ed, Scientific American Books, New York.

Ronemus, M., J. Galbiati, M. Ticknor, J. C. Chen, and S. L. Dellaporta. 1996. "Demethylation-Induced Developmental Pleiotropy in *Arabidopsis.*" *Science,* 273: 654.

Singer, M., and P. Berg. 1991. *Genes and Genomes.* University Science Books, Mill Valley, CA.

Zakian, V. A. 1995. "Telomeres: Beginning to Understand the End." *Science,* 270: 1601.

On the Web

Visit our Web site at **http://www.saunderscollege.com/lifesci/titles.html** and click on A/G/M Genetics for links to the following chapter-related resources on the World Wide Web:

1. **Human telomeres.** This site provides links to other sites on human telomeres.
2. **Genomes.** This site provides links to many other sites on the taxonomy, structure, sequence, and mapping of human and other genomes.

CHAPTER

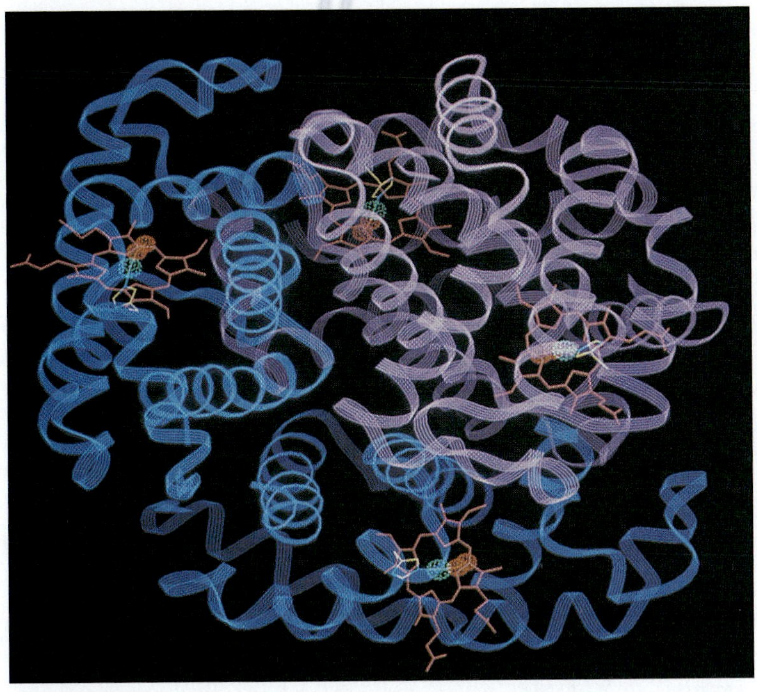

A computer graphics representation of a hemoglobin molecule. The molecule is composed of two different subunits (purple and blue) as well as the heme molecule (red). Hemoglobin is the oxygen- and carbon dioxide–carrying molecule in red blood cells. *(David Parker/Science Photo Library/Photo Researchers, Inc.)*

Transcription, Translation, and the Genetic Code

Knowing the chemical structure of the genetic material does not tell us how the information in DNA is used to determine the nature of the organism (that is, its shape, species identity, life span, and a myriad of other characteristics). To find the answers to these questions, we must learn how information is stored in DNA and expressed in the cell and how the expressed information ultimately determines the phenotype of the organism.

Extraordinary progress has been made in recent years in understanding how the information from DNA is expressed. In the early years of molecular genetics, DNA was found to be the template for the production of RNA (transcription), and RNA dictated the sequence of amino acids in proteins (translation). Until recently, the "central dogma" of molecular biology stated that "information flows from DNA to RNA to proteins"; in other words, the information flow was considered to be unidirectional. We now know that exceptions exist to the central dogma; for example, in some situations the information can flow from RNA to DNA.

We need to know how the information in DNA—in the form of genes—is converted into proteins and how proteins determine the phenotype of the organism. By the end of this chapter, we will have addressed the following four sets of questions:

1. How is the information in DNA passed to RNA? What steps are involved in synthesizing an RNA molecule, and what enzymes participate in these steps?
2. How is the sequence of nucleotides in mRNA translated into the sequence of amino acids in proteins?
3. How is the level and timing of gene expression controlled? What DNA sequences control gene expression?
4. What are the characteristics of the genetic code?

PROTEINS DETERMINE THE PHENOTYPE OF AN ORGANISM.

We know that the genetic information in a cell is stored in DNA within the chromosome. In a series of steps, this information is converted into amino acid sequences in proteins. These proteins determine the phenotype of the cell and, consequently, the whole organism. Proteins are remarkable molecules that are largely responsible for the structure and function of individual cells. Each cell type has a characteristic set of proteins that determines the phenotype of that cell type. Proteins also have myriad other functions, some of which we will touch upon in this chapter.

Proteins Are Linear Polymers of Amino Acids.

The chemical nature of the protein molecule was understood by the 1930s. Proteins are linear polymers composed of 20 different subunits called amino acids (Figure 11.1). These 20 different amino acids have two common features: one end of every amino acid has an amino group ($-NH_2$) and the other end has a carboxyl group ($-COOH$). The amino group and carboxyl group are attached to a common carbon atom that is bonded to a side group (commonly called the R group; R refers to radical). The common amino acids have 20 different R groups, which vary from merely a hydrogen atom (the amino acid glycine) to more complex R groups that contain sulfur or ring structures. R groups vary in their chemical properties; some are acidic, some are basic, and some are neutral. Each type confers specific properties to a protein.

The amino group of one amino acid can form a covalent bond with the carboxyl group of another amino acid, releasing water and forming a **peptide bond** (Figure 11.2). Formation of many such peptide bonds creates a linear polymer known as a protein or polypeptide. A specific protein is characterized by its amino acid sequence, which can vary in size from as few as 30–40 amino acids to enormous proteins with several thousand amino acids. The sequence and number of amino acids within a protein, therefore, define the specific characteristics of that protein. The amino acid sequence of thousands of different proteins are now known. Proteins can be structural components of the chromosome or cell but are more often enzymes that catalyze a specific chemical reaction within the cell. The roles of thousands of different enzymes are known, and we will discuss many of them throughout this text. Other functions of proteins include transport (hemoglobin), defense (antibodies), and motion (plant tubulin, actin, myosin).

The sequence of amino acids in a protein also determines the three-dimensional structure of the entire molecule (Figure 11.3). The sequence of amino acids in a protein is referred to as the **primary structure.** A protein chain can form twists and turns upon parts of itself to form an α-helix, which represents an aspect of the **secondary structure.** An α-helix is a very stable, rod-like structure composed of a spiral chain of amino acids stabilized by hydrogen bonds between adjacent –NH and –CO groups. A series of α-helical secondary structures in the protein can fold to produce the characteristic shape known as the **tertiary structure** (the tertiary structure is the endpoint of protein formation). If two or more proteins in tertiary form associate with each other, they then possess what is called a **quaternary structure.** A protein's tertiary structure is its conformation after the α-helix and other secondary structures have taken shape. An example of each of these structures, including the quaternary structure, is shown in Figure 11.3. For example, the enzyme glyceraldehyde 3-phosphate dehydrogenase

L-Amino acids	Abbreviation	Properties of side chains
Alanine	Ala (A)	Nonpolar
Valine	Val (V)	
Isoleucine	Ile (I)	
Leucine	Leu (L)	
Phenylalanine	Phe (F)	
Proline	Pro (P)	
Methionine	Met (M)	Contains sulfur
Glycine	Gly (G)	Polar
Serine	Ser (S)	
Threonine	Thr (T)	
Tyrosine	Tyr (Y)	
Tryptophan	Trp (W)	
Asparagine	Asn (N)	
Glutamine	Gln (Q)	
Cysteine	Cys (C)	Contains sulfur
Aspartic acid	Asp (D)	Acidic
Glutamic acid	Glu (E)	
Lysine	Lys (K)	Basic
Arginine	Arg (R)	
Histidine	His (H)	

FIGURE 11.1 The amino acids found in proteins. The amino acids in proteins can be grouped into four categories depending upon the chemical nature of the R group. The nonpolar amino acids are relatively insoluble in water, whereas the polar amino acids are more soluble. The basic and acidic amino acids are soluble in water and may confer a charge to the protein (depending on the pH of the medium) when present in high percentages, and they therefore can be considered polar.

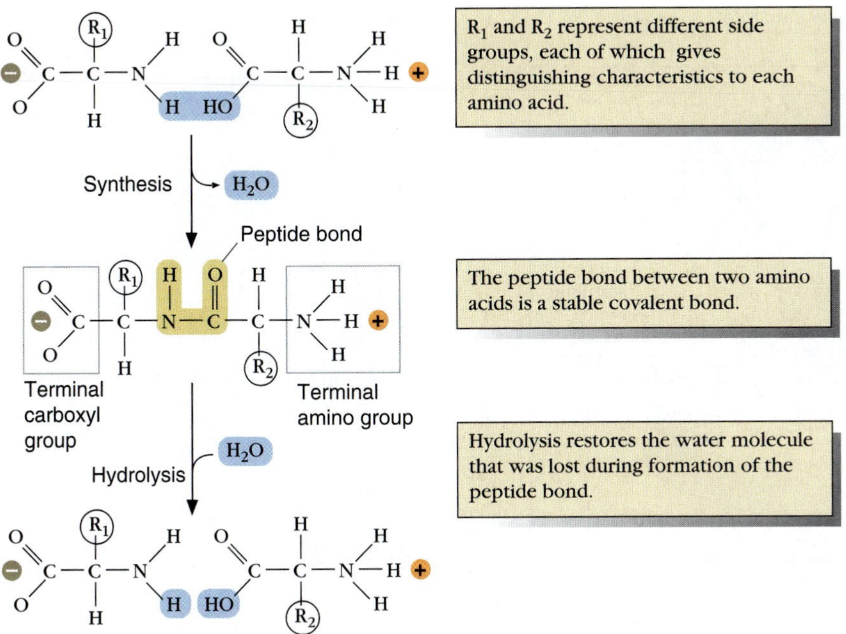

FIGURE 11.2 Formation of a peptide bond. The formation of a peptide bond (a covalent bond) between two amino acids results in the loss of a water molecule. Hydrolysis of a peptide bond reinserts a water molecule. Amino acids are joined together in proteins by means of a peptide bond.

possesses all four levels of structure. This enzyme is composed of two nonidentical polypeptides, and the molecule is active only when it is in this structure. The primary structure of a protein dictates the outcome of the secondary, tertiary, and quaternary structures and therefore is extremely important for function (enzymes and structural components of the cell). As we will see, the primary structure of a protein is determined by a sequence of nucleotides in the coding region of the DNA of a gene.

One Gene–One Polypeptide Is a Fundamental Principle.

The role of proteins in determining phenotype was not established until the 1940s. But the fundamental idea that they do, indeed, play a role began in the first decade of the 20th century with the physician–biochemist Sir Archibald Garrod during his study of a genetically inherited metabolic disorder of humans known as **alkaptonuria.** Alkaptonuria is easily detected because urine from affected individuals turns black upon exposure to air. The chemical compound responsible for this blackening is **homogentisic acid,** an intermediate in the degradation of the aromatic amino acids tyrosine and phenylalanine. Garrod's experiments provided the first clue that enzymes or proteins determine phenotype. The pathway for phenylalanine–tyrosine degradation is illustrated in Figure 11.4 and shows the block in the degradation pathway that results in alkaptonuria. From family pedigrees, Garrod reasoned that alkaptonuria was caused by a recessive mutation. The recessive mutation blocked the normal pathway of homogentisic acid metabolism. Although Garrod did not know the exact step in the blockage at that time, he correctly surmised that an enzyme was involved. In 1909 Garrod published a book entitled *Inborn Errors of Metabolism,* in which he described alkaptonuria and other genetically inherited traits.

We now know the detailed steps in the metabolism of phenylalanine–tyrosine degradation (Figure 11.4) and that alkaptonuria is caused by a recessive mutation. The mutation is in the gene for homogentisic acid oxidase, and it prevents homogentisic acid from being converted to acetoacetic acid and fumaric acid. As a consequence, homogentisic acid is excreted and is rapidly oxidized by oxygen in the air to give a black color to the urine of an afflicted individual.

Another recessive defect in phenylalanine–tyrosine metabolism is albinism (Figure 11.4). In albinism, steps in the pathway leading to melanin are affected. A defect in any one of these steps blocks the formation of melanin, a dark-colored pigment that darkens the hair, eyes, and skin. Albinism is clearly another example of how a defect in a gene causes a genetically inherited trait. Albinism is also an example of how mutations in several

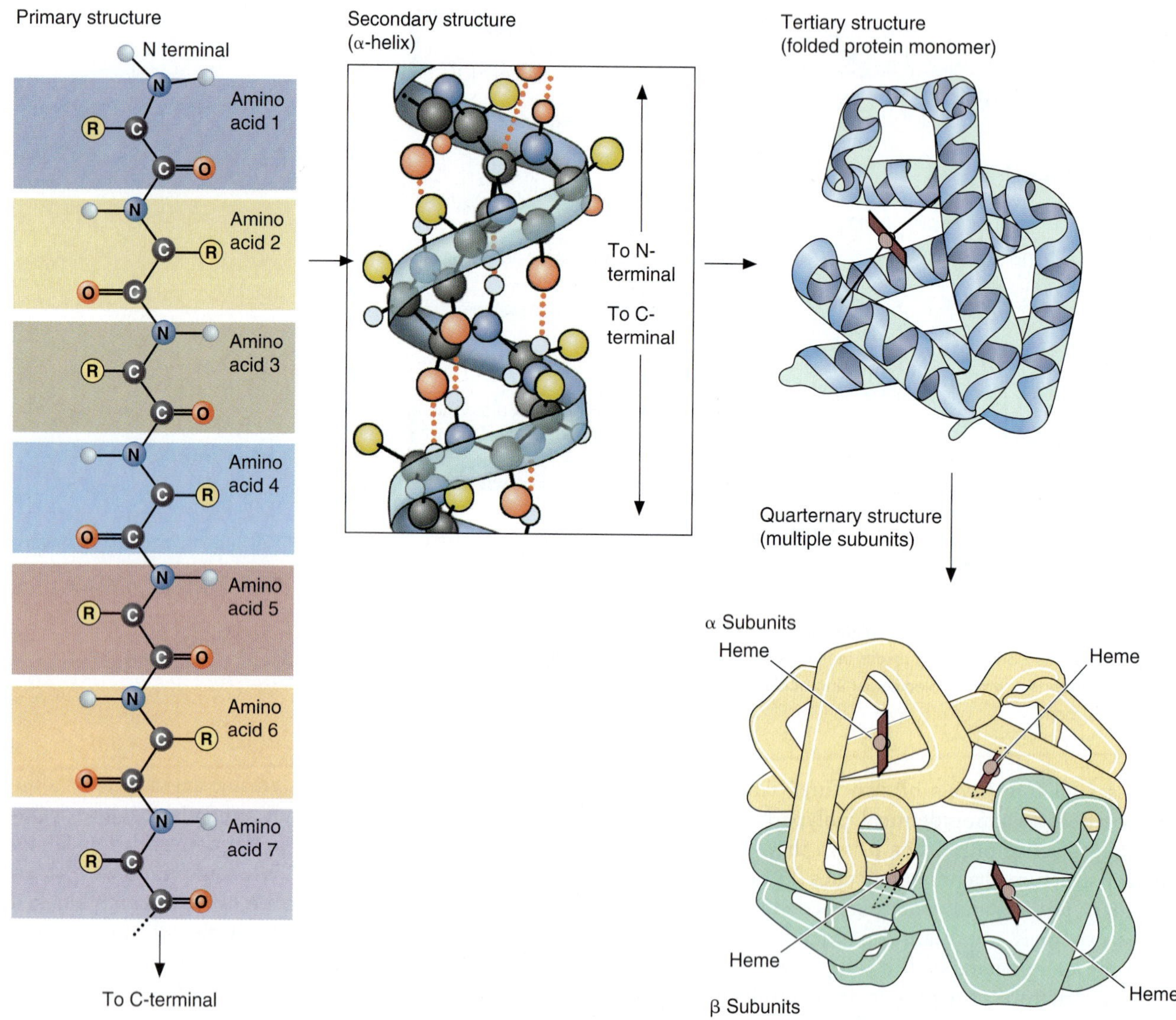

FIGURE 11.3 The folding of proteins. The different levels of protein folding are shown in this sequence of drawings. The primary structure of a protein (the amino acid sequence) determines its final three-dimensional structure. The first level of folding of a protein is referred to as the secondary structure; and in this example an α-helix is shown (the secondary structure can take on other forms as well). An α-helix is a component of the tertiary structure (the third diagram from the left), which is the final form of the protein when it is a monomer. However, many proteins (hemoglobin in this example) are dimers, trimers, tetramers, and so on. When two or more single proteins come together to form an active protein, it is referred to as the quaternary structure. The subunits of the quaternary structure can be identical or nonidentical, which is the case in this example.

different genes involved in a biochemical pathway can result in the same general phenotype.

Although Garrod did not understand the relationship between a gene and the protein product made by the gene (in these cases, enzymes), his work was consistent with the idea that a single gene is responsible for a genetic defect. Unfortunately, Garrod's work was largely ignored until the 1940s, when the concept of one gene–one enzyme was elucidated by George W. Beadle and Edward L. Tatum, who were working with the fungus *Neurospora crassa*. Using mutants of *Neurospora*, Beadle and Tatum demonstrated that there was a direct relationship between a single gene and a single enzyme. *Neurospora crassa* lent itself to understanding the one

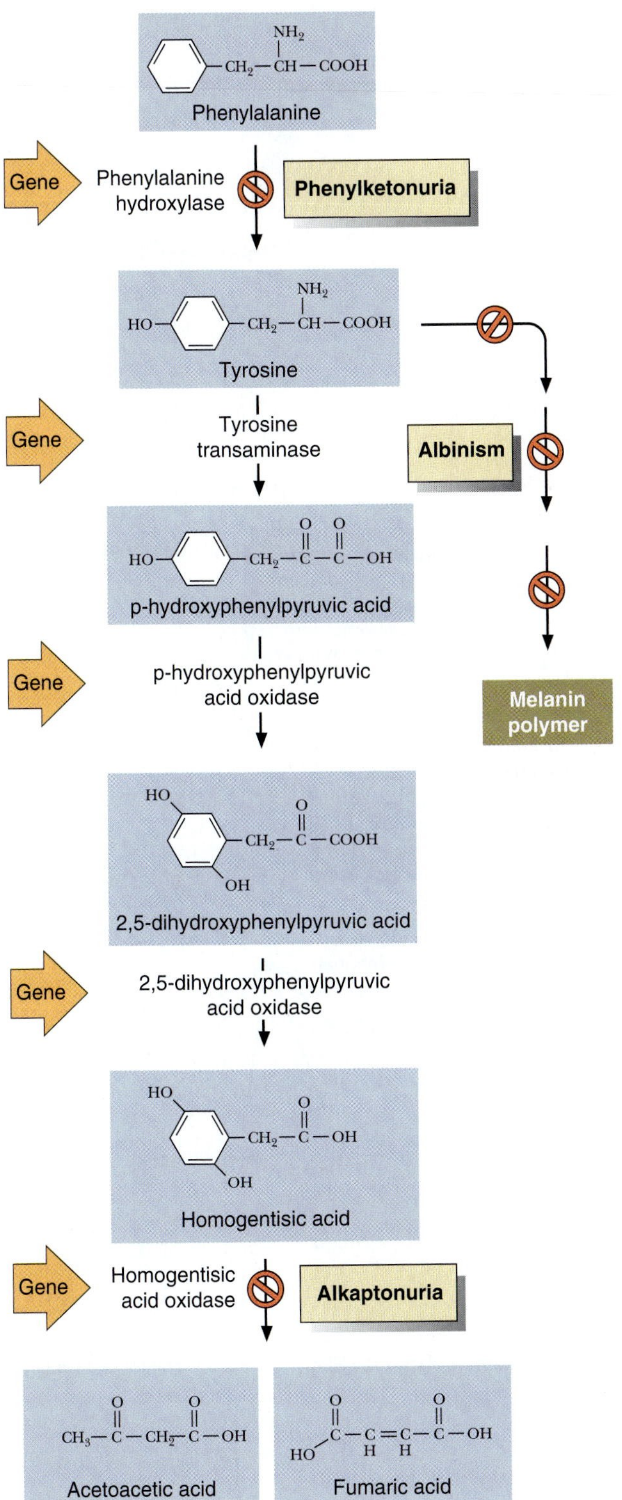

◀ **FIGURE 11.4 Phenylalanine and tyrosine metabolic pathway.** Disruption of the metabolic pathway for the degradation of phenylalanine and tyrosine results in a number of metabolic disorders in humans. Loss of phenylalanine hydroxylase causes phenylketonuria, and loss of homogentisic acid oxidase causes alkaptonuria. Defects in any one of the steps from tyrosine to the melanin polymer cause albinism. The chemical compound that immediately precedes the affected step accumulates as a result of the block in its normal pathway. Sometimes by-products that can cause serious side effects are formed (as in the case of phenylketonuria and alkaptonuria).

gene–one enzyme concept because it is able to grow on media containing only very minimal ingredients: inorganic salts, sugars, and the vitamin biotin. Wild-type *Neurospora* can synthesize all metabolites (such as amino acids, purines, pyrimidines, and vitamins) from these basic ingredients. Beadle and Tatum reasoned that the synthesis of the compounds essential for cell growth must be under genetic control. They tested this prediction by exposing asexual spores (called conidia) of wild-type *Neurospora* to X-rays or ultraviolet light, which will induce mutations affecting specific steps in metabolism (Figure 11.5). Beadle and Tatum took irradiated spores of *Neurospora* and crossed them to wild-type spores of the opposite mating type. Some yielded progeny that could not grow on a "minimal medium" but could grow on an enriched medium. However, if the mutants were grown on a minimal medium supplemented with one or more amino acids, purines, pyrimidines, or vitamins, some mutants grew. These growth requirements were inherited, suggesting that mutagenized *Neurospora* cells had a defective gene for a metabolic step in the synthesis of the factor. Beadle and Tatum were able to show that each mutation resulted in a requirement for one growth component. Further work on the biochemical steps of the pathways for the synthesis of individual growth components identified which enzyme was defective in each mutant. They concluded that a defect in one gene produced a single defect in an essential enzyme, resulting in the growth factor requirement (that is, one gene–one enzyme).

In some cases, different mutational events cause the same nutrient requirement (for example, the synthesis of the essential amino acid valine). The pathway for the synthesis of valine is shown in Figure 11.6; valine synthesis is a multistep process requiring the correct action of a series of enzymes encoded by different genes. A defect in any of these genes results in the requirement for valine in the growth medium.

Since the work of Beadle and Tatum, defective enzymes from many different organisms have been associated with mutations in single genes. We now know that the one gene–one enzyme hypothesis is not precisely correct. Many enzymes are dimers or tetramers in which each subunit (an individual polypeptide) is determined by a single gene. Many other types of multisubunit proteins are structural components of the cell or have other functions. For example, hemoglobin transports oxygen

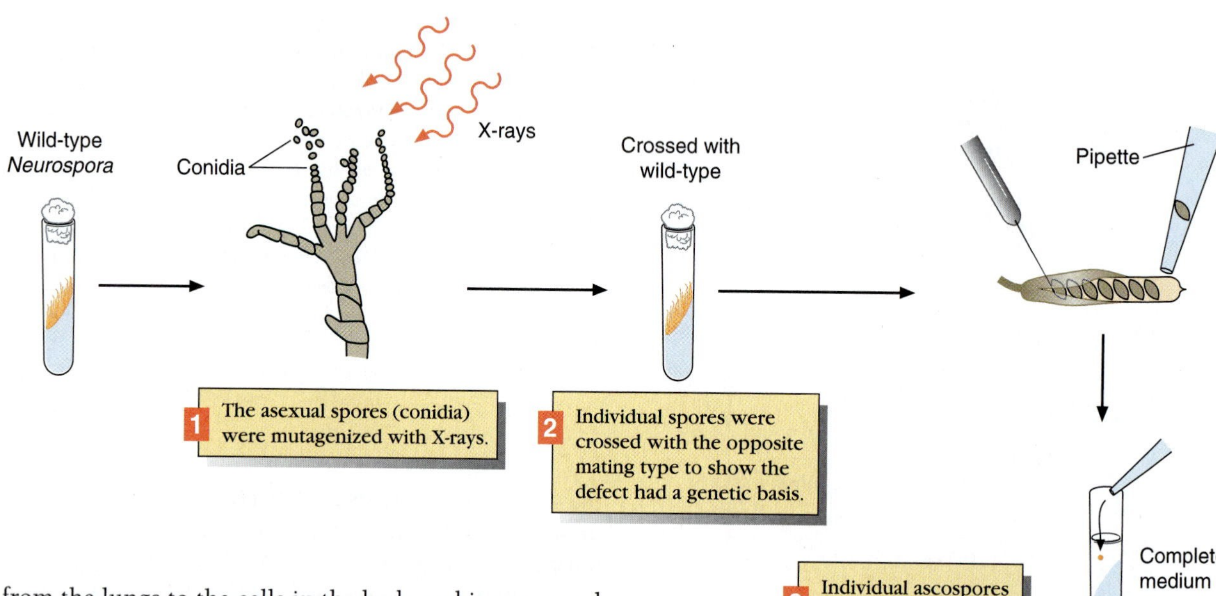

from the lungs to the cells in the body and is composed of four subunits: two identical α-polypeptides and two identical β-polypeptides; thus, two genes are required to produce functional hemoglobin. This changes the concept of one gene–one enzyme to one gene–one polypeptide.

Literally hundreds of inherited metabolic defects are known in humans. One of the first to be understood (1954) was **phenylketonuria** or PKU (Figure 11.4). PKU is a defect in the metabolism of phenylalanine resulting in phenylalanine and its metabolic by-products accumulating in the body. The accumulated phenylalanine is converted to phenylpyruvic acid in a secondary pathway that drastically affects the central nervous system, causing mental retardation, slow growth, and early death. This defect results from the loss of the enzyme phenylalanine hydroxylase (Figure 11.4). PKU occurs in about 1 in every 11,000 live births in the United States and is easy and inexpensively screened for in newborn infants. Early detection within the first day or two after birth allows physicians to place these infants on a low phenylalanine diet. A low phenylalanine diet prevents phenylalanine and its by-products of the metabolism from accumulating in the body tissues. For example, diet drinks containing the artificial sweetener Nutrasweet have caution labels for phenylketonurics because Nutrasweet is composed of the dimer phenylalanine-

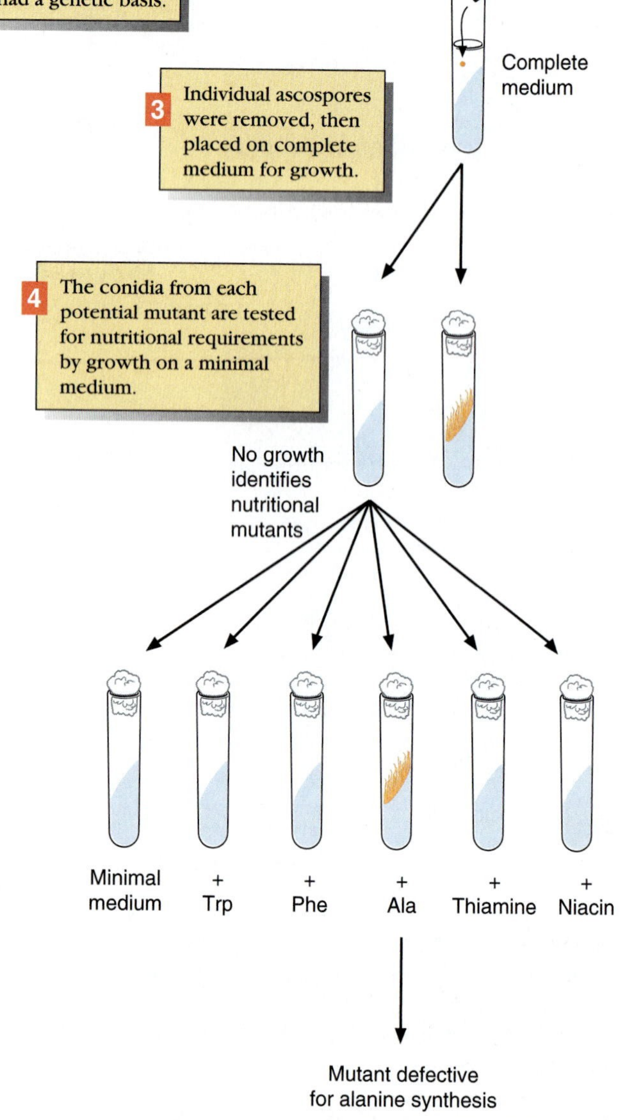

▶ **FIGURE 11.5 Identifying mutants in metabolic pathways.** The general approach to identifying mutants in metabolic pathways involves mutagenesis of the spores and then searching for mutants having nutritional requirements that differ from the wild-type parent. Beadle and Tatum used this approach to isolate *Neurospora* mutants that possessed nutritional requirements, but a similar approach can also be used with bacteria. The mutants obtained were fundamental in the development of the one gene–one enzyme hypothesis.

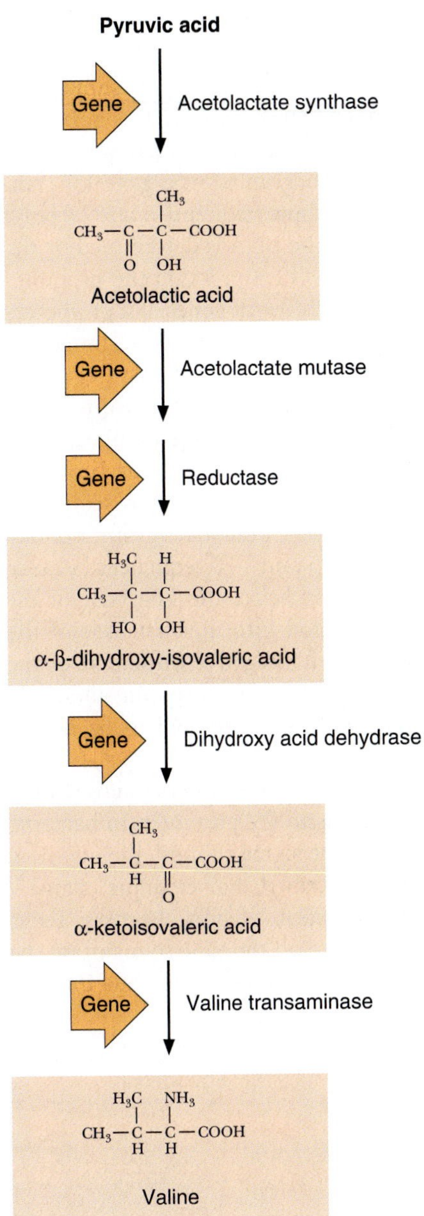

FIGURE 11.6 Valine synthesis. The five metabolic steps are catalyzed by five different enzymes, which are determined by five different genes. A defect in any one of the genes will result in a valine requirement for growth.

glutamic acid. PKU individuals placed on a low phenylalanine diet are never completely cured because they always show some signs of the disease, but they can lead near-normal lives.

SECTION REVIEW PROBLEMS

1. What is the basic relationship between a gene and an enzyme?
2. Assume that you found a mutant bacterial strain that requires the amino acid cysteine to be supplied in the medium for it to survive. What conclusion can be made from the finding that not only does adding cysteine to the growth medium restore growth to the mutant strain, but adding L-cystathionine (the immediate precursor of cysteine) to the growth medium also restores growth?

Transcription Is the Passing of Information from DNA to RNA.

The first step in making a gene into a protein is synthesizing RNA from DNA. All RNA is synthesized using a DNA template, and this copying process is called **transcription.** In this section, we will show that the information in DNA molecules is transcribed into RNA from the coding DNA strand. This process requires enzymes, known as RNA polymerases, that are responsible for the synthesis of all RNA species in the cell. Some of the RNAs made in the cell are **messenger RNA** (mRNA), which specify the information for making proteins. Others are **transfer RNA** (tRNA), **ribosomal RNA** (rRNA), and **small nuclear RNA** (snRNA). Transfer and ribosomal RNA are involved in the synthesis of proteins, and small nuclear RNA is found only in eukaryotes. RNA molecules come in different sizes, depending upon the gene. The amount of RNA produced from different genes also varies. In this section, we will also explore some of the signals present near genes that determine how much RNA is made at different times.

RNA transcription in prokaryotes and eukaryotes is similar, but several fundamental differences exist. As a consequence, we will discuss prokaryotic and eukaryotic transcription separately. The most striking difference between prokaryotes and eukaryotes is that prokaryotic cells lack a nucleus, whereas eukaryotic cells contain a true nucleus. Thus, eukaryotic RNA is transported from the nucleus to the cytoplasm before the information within it can be converted into a protein product.

Transcription in Prokaryotes Is Catalyzed by One Polymerase.

The enzyme **RNA polymerase** is responsible for the synthesis of RNA. It acts much like DNA polymerase, requiring a template, nucleotide triphosphates (ATP, CTP, GTP, and UTP), and magnesium ions. A distinguishing feature of RNA polymerase is its ability to initiate chain growth without the need of a primer, unlike DNA polymerase.

RNA polymerase from *Escherichia coli* is composed of six polypeptide subunits: two alpha (α) subunits, one

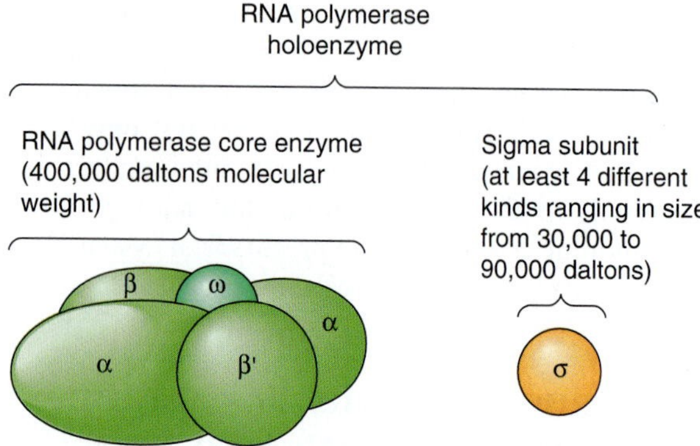

FIGURE 11.7 Structure of the RNA polymerase. The core enzyme and the sigma subunit (shown) form a holoenzyme when they associate. Each bacterial cell has between 4000 and 7000 RNA polymerase molecules and 80–90% are engaged in active RNA synthesis in a rapidly growing cell. *E. coli* has at least four different sigma factors, each specifying a different group of genes for expression.

beta (β) subunit, one beta prime (β′) subunit, one omega (ω) subunit, and a sigma (σ) subunit (Figure 11.7). Five of these subunits (α, α, β, β′, and ω) are tightly bound together and constitute the core enzyme. The core enzyme in itself is capable of synthesizing RNA from DNA, but it is not capable of identifying the initiation site of a gene. The sigma subunit is essential for initiating transcription at the beginning of a gene and binds to the core enzyme to form the **holoenzyme.** Bacterial cells have several different sigma subunits (at least four), and each is responsible for initiating RNA synthesis for specific groups of genes (Table 11.1). Consequently, to a geneticist, the most interesting aspect of RNA polymerase is the sigma subunit. A specific sigma subunit (in complex with the core enzyme) is able to recognize and bind to a specific DNA sequence called a **promoter.** The promoter site defines the beginning of a gene for transcription. How tightly and how frequently an RNA polymerase molecule binds to the DNA is determined by the promoter sequence. Figure 11.8 presents the promoter sequence for a number of different genes. All promoters recognized by a single sigma subunit have similar sequences, and the nucleotides that occur most frequently in the promoter are known as a **consensus sequence.** A consensus sequence is thus the average base sequence found after examining many sequences. The promoter consensus sequence is composed of two short sequences separated by about 20 bp. These two groups of sequences are labeled the −35 and −10 sequences because they occur 35 and 10 bases *before* the first base of the DNA that will be transcribed into the first base of the RNA. DNA bases to the left (negative numbers) of the transcription start site are said to be upstream, and bases to the right (positive numbers) are said to be downstream (Figure 11.8).

RNA polymerase binds to DNA in two steps. In the first step, the enzyme binds to the promoter. In bacteria, the promoter, along with the sigma subunit, dictates how frequently a gene is transcribed. For example, sigma^{70} serves to initiate transcription of housekeeping genes (those genes that are needed all the time to maintain the metabolism and structure of the cell) and thus initiates

TABLE 11.1 *Bacterial Sigma Subunits*

Subunit (superscript gives MW* × 10^3)	Function	Consensus Sequence for Positions −35	−10	Gene Symbol
σ^{70}	Housekeeping genes	TTGACA	TATAAT	*rpo*D
σ^{54}	Nitrogen metabolism	CTGGPyAPyPu	TTGCA	*ntr*A
σ^{32}	Heat-shock genes	CTTGAA	CCCCATTA	*rpo*H
$\sigma^{?}$	Flagellar synthesis and chemotaxis	TAAA	GCCGATAA	*flb*B

The consensus sequence for the nitrogen metabolism genes is actually at −25 and −10 from the point of initiation of RNA synthesis. Py and Pu refer to pyrimidine and purine, respectively.
*MW refers to molecular weight.

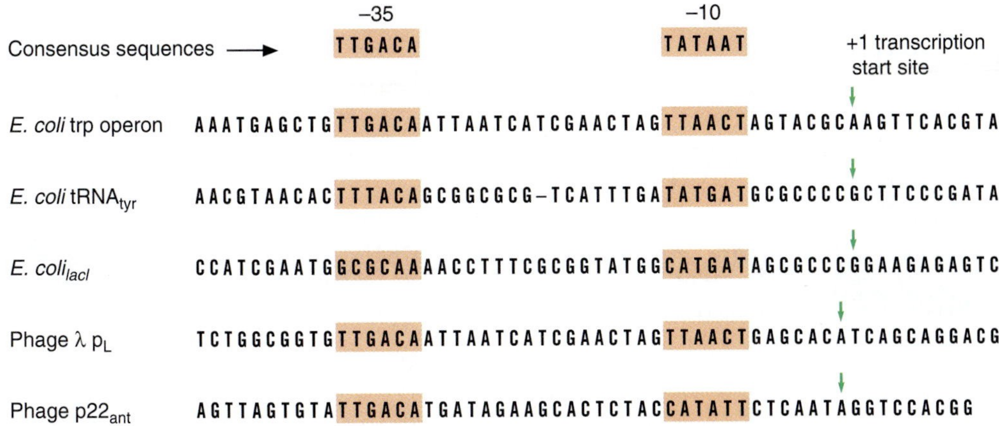

FIGURE 11.8 Promoter sequences. Shown here are promoter sequences of the coding strand of genes from different bacteria and bacterial viruses. The consensus sequence is given at the top for the –35 and –10 regions of the promoter, and it represents the average sequences found in promoters.

the synthesis of mRNAs from the largest class of genes. However, there is variation in the promoter sequence among genes recognized by the sigma^{70} subunit, and this variation affects the frequency of transcription initiation. As a result, one gene may be transcribed with one initiation event every second, whereas another may be transcribed every 5 minutes. This is a fundamental means of regulating the level of gene expression in prokaryotes.

In the second step of binding, the helix unwinds to allow base-pair recognition to the template—DNA sequence for the RNA polymerization reaction. DNA unwinding starts at the –10 sequence and proceeds to the right of the start point, or downstream. The sigma subunit is required only for recognition of the consensus sequence and thus detaches from the core enzyme. The steps involving RNA polymerase binding and opening of the double helix are illustrated in Figure 11.9.

The product of transcription is always RNA (tRNA, rRNA, snRNA, or mRNA). The precursors for RNA synthesis are ATP, GTP, CTP, and UTP, collectively referred to as NTPs, where the N stands for any base. The greatest variety of RNAs are found in the messenger RNAs. Messenger RNA gets its name from the fact that it contains the information for producing the protein product. To start RNA synthesis, a ribonucleotide triphosphate is properly base-paired at the RNA start site on the DNA. This represents the 5′ end of the new molecule, and it grows when ribonucleotides are added to the 3′ end. The 5′ end retains the triphosphate precursor. As the chain elongates, the enzyme proceeds down the DNA helix, unwinding the DNA as it goes along and adding ribonucleoside monphosphates to the growing chain and releasing the diphosphate. About 15 bp are unwound at one time because the generated upstream twist would otherwise become too intense for the DNA. Only one strand of the DNA, called the **template strand,** acts as a template. The other strand, with the same 5′⟶3′ orientation as the RNA transcript, is called the **coding strand.** In *E. coli,* the polymerization reaction occurs at a rate of about 40 nucleotides per second at 37°C; thus a gene of 1000 bp is transcribed in about 30 seconds. Eventually the RNA polymerase will reach a **terminator sequence** that tells the enzyme to stop polymerization of the RNA and to dissociate from the DNA.

One type of termination sequence is composed of a stretch of AT nucleotides preceded by two symmetrical, 8-bp GC sequences (a palindrome) followed by a run of AT base pairs. Other termination sequences do not possess the run of AT base pairs (ρ-dependent termination sequences). A representative terminator sequence is shown in Figure 11.10. A hairpin-like secondary structure in the RNA forms when the inverted repeats pair. The string of U residues combined with the hairpin structure probably signals the RNA molecule to dissociate from the DNA, resulting in termination. RNA termination sometimes requires the participation of a protein factor known as rho (ρ), depending upon the gene. The ρ-factor functions by combining with the RNA transcript and using the energy provided by the breakdown of ATP to dissociate the RNA and the polymerase from the DNA template.

Do the start and stop signals for RNA synthesis define a gene? Geneticists have a tradition of defining a gene by genetic tests. On the other hand, a gene can also

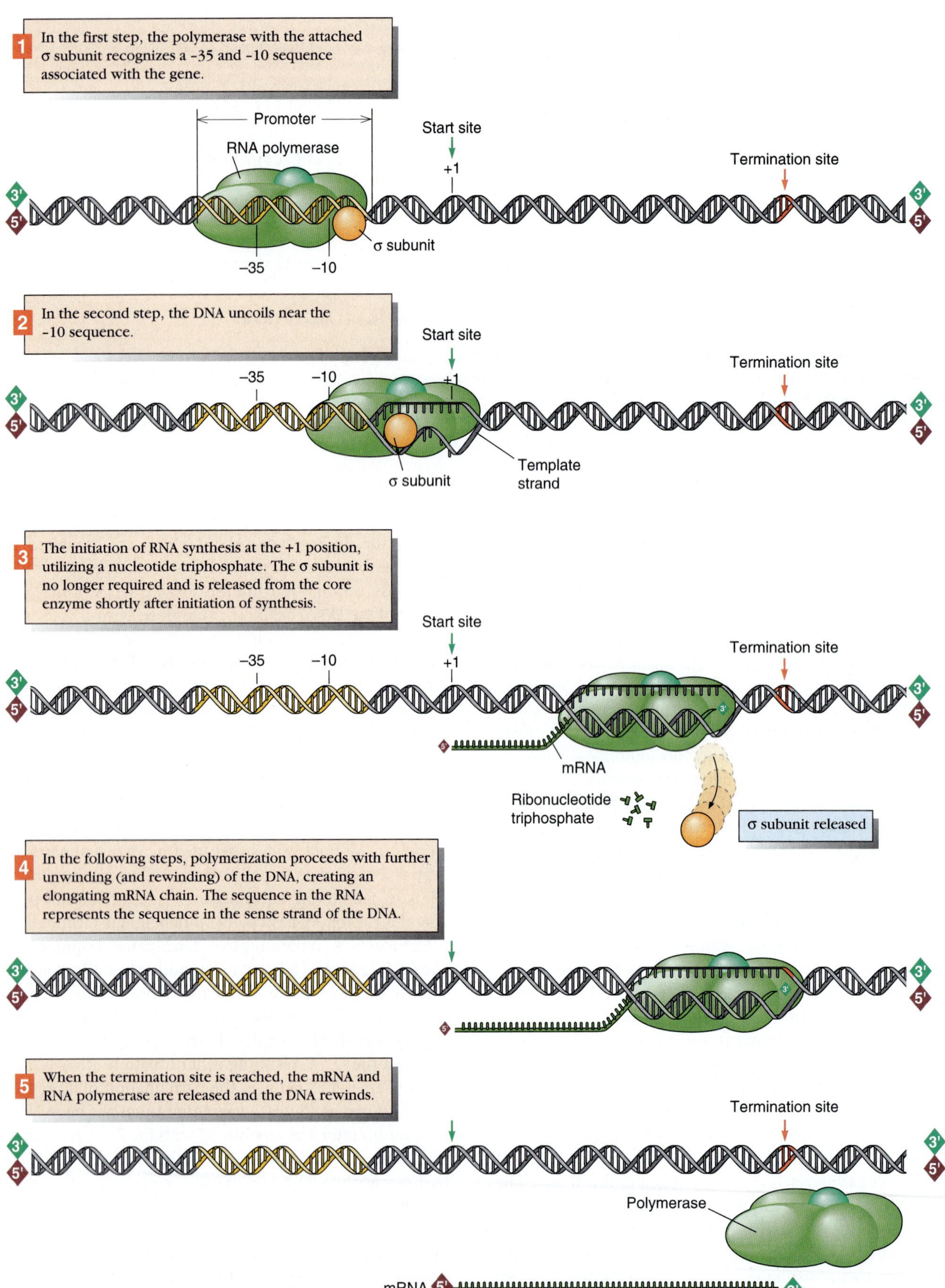
1
In the first step, the polymerase with the attached σ subunit recognizes a -35 and -10 sequence associated with the gene.
Promoter
RNA polymerase
Start site
+1
Termination site
3'
5'
σ subunit
–35
–10
2
In the second step, the DNA uncoils near the -10 sequence.
Start site
Termination site
–35
–10
+1
σ subunit
Template strand
3
The initiation of RNA synthesis at the +1 position, utilizing a nucleotide triphosphate. The σ subunit is no longer required and is released from the core enzyme shortly after initiation of synthesis.
Start site
Termination site
–35
–10
+1
mRNA
Ribonucleotide triphosphate
σ subunit released
4
In the following steps, polymerization proceeds with further unwinding (and rewinding) of the DNA, creating an elongating mRNA chain. The sequence in the RNA represents the sequence in the sense strand of the DNA.
5
When the termination site is reached, the mRNA and RNA polymerase are released and the DNA rewinds.
Termination site
Polymerase
mRNA
5'
3'

◀ **FIGURE 11.9 Transcription of genes by RNA polymerase.** Shown here is the general mechanism of gene transcription in prokaryotes. The RNA polymerase holoenzyme recognizes the promoter and begins transcription of the template strand (running 5′⟶3′), synthesizing an RNA molecule in the 5′ to 3′ direction. Upon encountering the termination sequence (see Figure 11.10), the RNA is released, and the polymerase is reused to synthesize another RNA molecule.

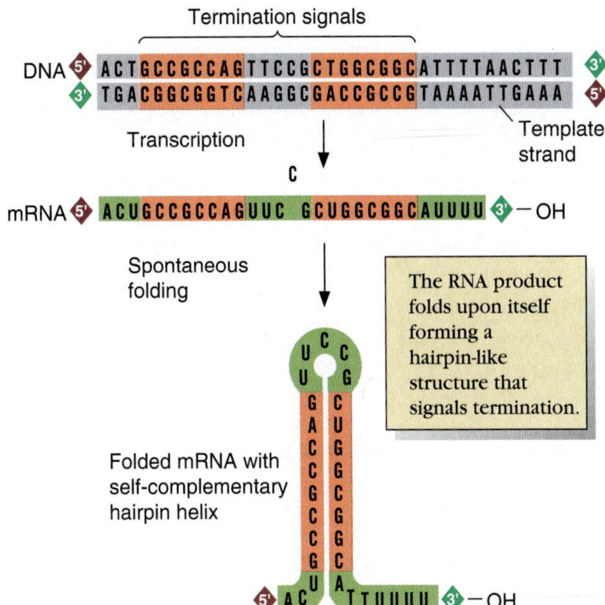

FIGURE 11.10 Transcription termination. One mechanism of transcription termination uses a palindromic sequence high in G and C, which allows the RNA product to form a hairpin-like loop that can disassociate from the DNA template strand. This sequence is surrounded by sequences high in A and T, which probably aid in the dissociation from the template strand. Another mechanism of termination uses a protein (ρ) to assist in the disassociation of the RNA from the template.

be defined as a physical entity like the DNA sequence that includes a start site (promoter), the coding information, and a termination sequence. Another term for a gene is a **cistron,** which is the information required to make one polypeptide. Unfortunately, in bacteria and organelles of eukaryotic cells, this neat description is complicated in many instances because more than one polypeptide is often encoded in a single message. These are called **polycistronic mRNAs** because more than one cistron is present in an mRNA. An example is illustrated in Figure 11.11. The function of polycistronic mRNAs will become apparent during the discussion of protein synthesis.

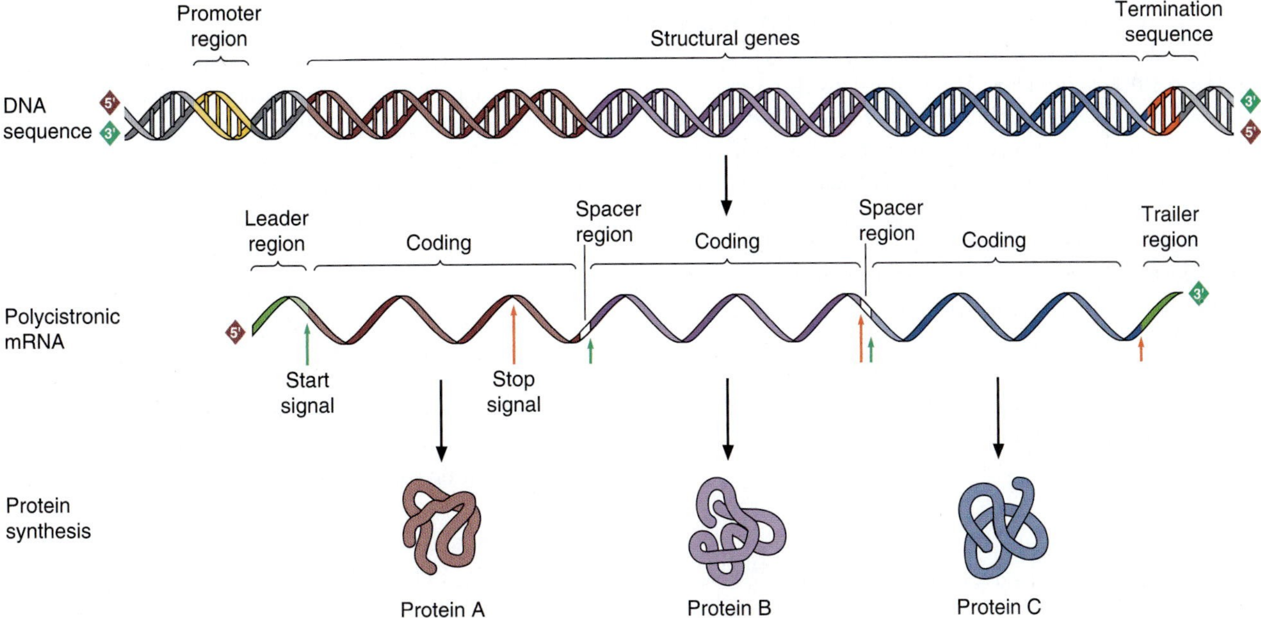

FIGURE 11.11 A polycistronic mRNA. This polycistronic messenger RNA codes for three polypeptides, although more are possible. Polycistronic mRNAs are produced only in prokaryotes and usually involve genes that have related functions (for example, the genes involved in lactose metabolism). The long RNA molecule has sequences within it for initiation and termination of the synthesis (translation) of the individual gene proteins. The short sequence at the beginning of the RNA (before the first start signal) is called the leader region, and similarly the end has a trailer region. Sequences between cistrons are called spacer, or noncoding, regions.

SECTION REVIEW PROBLEMS

3. What is the function of the sigma subunit of RNA polymerase holoenzyme?
4. What is the basic structure of a bacterial promoter, and what is its function? Where is the promoter located with respect to the coding sequences in the prokaryotic gene?
5. What is a polycistronic mRNA? What is a cistron? Are the terms "gene" and "cistron" interchangeable?

Transcription in Eukaryotes Uses Three Different Polymerases.

Although many aspects of transcription in eukaryotes are identical to prokaryotes, several important features are different:

1. Eukaryotes have three different RNA polymerases, whereas prokaryotes have only one RNA polymerase.
2. Eukaryotes do not synthesize polycistronic mRNAs.
3. mRNA from eukaryotes undergoes chemical modification before it is used to make proteins.

Three different RNA polymerases are found in all eukaryotic cells, and each is responsible for the synthesis of a different class of RNA molecules. The characteristics of the different RNA polymerases in eukaryotic cells are summarized in Table 11.2. RNA polymerase I is responsible for the synthesis of small and large subunit ribosomal RNA. RNA polymerase III synthesizes tRNA as well as 5S RNA ribosomal molecules (S is a function of the size and shape of the molecule) and some of the snRNAs. RNA polymerase II synthesizes mRNA and the remainder of the snRNAs. All the RNA polymerases are constructed of multiple subunits, and the holoenzymes have molecular weights up to half a million daltons.

The most active RNA-synthesizing ability is associated with RNA polymerase I. Its product, ribosomal RNA, is a major component of the ribosome. The ribosome is a large ribonucleoprotein particle necessary for protein synthesis. Many copies of rRNA must be made for protein synthesis to occur; thus polymerase I accounts for most of the RNA synthesized in the cell. The second largest amount of RNA product is produced by RNA polymerase III, which synthesizes tRNA and 5S RNA, another participant in protein synthesis.

TABLE 11.2 *Characteristics of Eukaryotic RNA Polymerases*

Enzyme	Cellular Location	Product
RNA polymerase I	Nucleolus	rRNA
RNA polymerase II	Nucleus	mRNA and snRNA
RNA polymerase III	Nucleus	tRNA, 5S RNA, and snRNA

RNA polymerase II is responsible for the synthesis of the precursor of mRNA, or more precisely, **heterogeneous nuclear RNA** (or **pre-mRNA),** which is usually much larger than mRNA. Pre-mRNA is the primary transcript, but before pre-mRNA can be used by the cell, it must first be enzymatically processed into mRNA.

Like prokaryotic genes, eukaryotic genes have promoter and terminator sequences that direct the transcriptional machinery to start and stop. However, the signal sequences in the DNA for starting and stopping RNA synthesis are variable in eukaryotes. It is thus impossible to specify a single sequence responsible for stops and starts for all genes. Some commonality exists, however. In eukaryotes, most promoters have a sequence called the **TATA box,** which actually has the consensus sequence 5′-TATAAA-3′. The TATA box is located about 25 bp upstream (−25) from the starting point and determines the start site for transcription. It is nearly always located at this spot and tends to be surrounded by GC sequences. Two other sequences upstream of the start point also affect gene expression and are present in many eukaryotic promoters. These are the **CAAT box** and the **GC box.** The CAAT box is named for its consensus sequence (which is GGCCAATCT) and is often located at about 80 bp upstream (−80) of the transcription starting point. As with the TATA box, mutational changes in the CAAT box affect downstream gene activity or RNA transcription. The GC box has the consensus sequence GGGCGG, and multiple copies are often present. Figure 11.12 presents a representation of the controlling sequence of a eukaryotic gene. Just upstream of the transcriptional start site is a TATA box at its proper location, a GC box, a CAAT box, and another GC box. Although it is not known for certain, the CAAT and GC boxes probably direct the initial binding of the RNA polymerase to the TATA box region. However, RNA polymerase cannot bind by itself; it requires the participation of a group of proteins known as **transcription factors.** Transcription factors facilitate the binding of RNA polymerase, and in some cases these ancillary proteins direct RNA polymerase to transcribe very specific genes. Figure 11.12 shows one example of how sequences are arranged in the regulatory region of a eukaryotic gene, but many variations of this general theme exist among different genes. Some upstream areas do not even have CAAT boxes or GC boxes, whereas others have multiple copies of each.

The hallmark characteristic of eukaryotic transcription is that it requires multiple protein binding sites for initiation of transcription and a number of different transcription factors. In prokaryotes, RNA polymerase re-

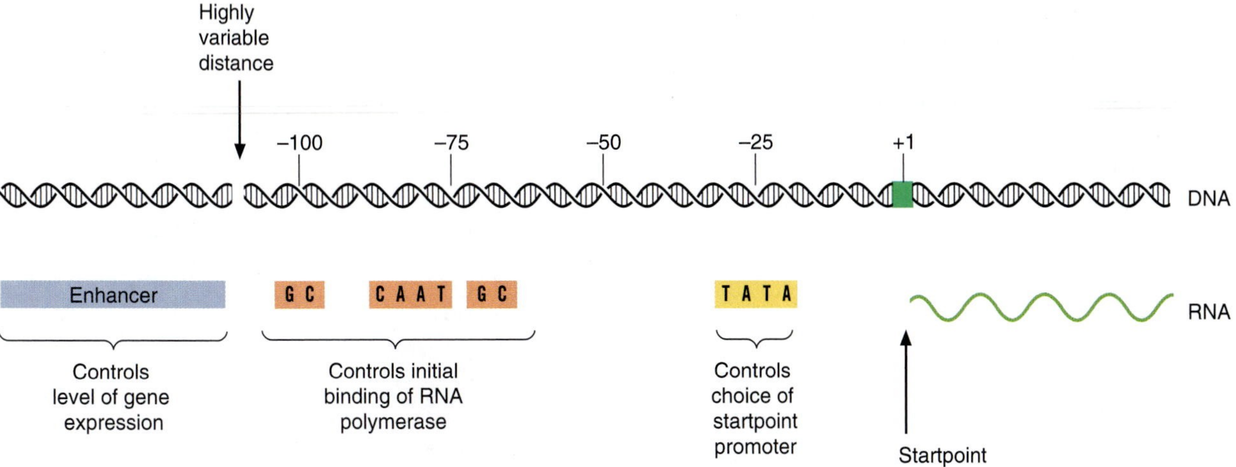

FIGURE 11.12 The controlling region of an average eukaryotic gene. The location of important regulatory sequences is shown, but the position of these controlling elements relative to the mRNA starting point is highly variable from gene to gene. For example, the enhancer region can also follow the gene, and not all genes have TATA boxes.

quires only the sigma factor to bind to the holoenzyme, which can then bind to the promoter region. In eukaryotes, promoters do not act alone in regulating the amount of RNA transcript synthesized. Another very important class of DNA sequences that control RNA transcription are the **enhancer sequences** or **enhancer elements** (Figure 11.12). Enhancer sequences are binding sites for a wide variety of transcription factors. The enhancer sequences, in conjunction with the proper transcription factors, can have an enormous effect on the level of synthesis of an RNA transcript, increasing it by as much as a thousand-fold. A distinctive characteristic of **enhancers** is that they can be at variable distances from the genes that they control, sometimes even several thousand base pairs from the promoter. Another interesting feature of enhancers is that they function just as well whether placed upstream or downstream of the gene. Also, some of the sequences frequently found in promoters are also found in enhancers. Like eukaryotic promoters, enhancer sequences vary considerably, and enhancers are sometimes interchangeable between genes. For example, when an enhancer taken from a gene that is very frequently transcribed is placed in front of a cloned β-globin gene (which is not as frequently transcribed), β-globin expression is increased more than 200 times. Likewise, this same sequence can be placed thousands of base pairs upstream or downstream of the gene and still have the same dramatic effect on β-globin gene transcription. Just exactly how an enhancer works is not known, but it is speculated that the chromatin containing the DNA to be expressed is unfolded, transcription factors are bound at the enhancer site, and the enhancer–transcription factor–DNA complex bends toward the promoter to enhance expression. This may occur by loading additional transcription factors as well as RNA polymerase to the promoter site.

Information about termination of transcription with RNA polymerase II in eukaryotes is less well understood than termination in prokaryotes. It is possible that the signal for termination of RNA polymerase II transcription is loosely specified and may occur 1000 or so nucleotides downstream of the actual site corresponding to the end of the mature pre-mRNA.

SECTION REVIEW PROBLEMS

6. What are the signal sequences in eukaryotes that direct binding of RNA polymerase II for the initiation of pre-mRNA synthesis?
7. What is the difference between an enhancer and a promoter?

EUKARYOTES MODIFY mRNA AFTER TRANSCRIPTION.

There are several basic differences between transcriptional products from eukaryotes and prokaryotes, one of which has to do with stability. First, nearly all prokaryotic mRNAs are extraordinarily unstable, with a half-life of just a few minutes. In contrast, eukaryotic mRNA can be very stable, lasting hours, days, or weeks, or it can be unstable and last only minutes. Unstable RNAs permit more responsiveness in the expression of gene products (when mRNA is rapidly degraded, it allows for rapid changes in gene expression), which is the hallmark of prokaryotes who must survive in a rapidly changing environment. Second, eukaryotes synthesize their RNA in the nucleus and transport it into the cytoplasm. Three major alterations are introduced into pre-mRNA after its synthesis: first, the 5′ end is capped by adding a modified guanosine, usually a 7-methyl group. Second, the 3′ end is modified by adding multiple adenosines. And third, pre-mRNA is usually processed to remove specific internal nucleotide sequences by a splicing mechanism, giving rise to mRNA.

The 5′ Cap and a 3′ Tail Are Added After Transcription.

The first nucleotide of all pre-mRNA transcripts is a nucleotide triphosphate with the three phosphates projecting from the 5′ end. All subsequent nucleotides are added to the 3′ end. Immediately after synthesis of pre-mRNA, a guanosine in reverse orientation is added to the 5′ end of the transcript joined by a 5′–5′ bond (Figure 11.13). The 5′ terminal guanosine is added by the enzyme guanylyl transferase, and is referred to as a cap. Subsequent to capping, adjacent nucleotides are methylated by another enzyme, guanine-7-methyl transferase, which usually adds a methyl group to the 7 position of guanosines. This capped end is abbreviated ^{7}mG.

In eukaryotes, the cap structure is required to translate the RNA into proteins. Bacteria do not add a cap to the mRNA, and bacterial mRNA will function only very weakly in reconstructed *in vitro* eukaryotic protein synthetic systems unless a cap is added.

In addition to the ^{7}mG cap added after transcription of the pre-mRNA in eukaryotes, a poly(A) tail is added to the 3′ end after an enzyme clips some of the 3′ end from the primary transcript. The enzyme poly(A) polymerase adds about 200 adenosines (called polyadenylation) immediately after pre-mRNA synthesis in eukaryotes. Some mRNAs are polyadenylated in prokaryotes as well, but it is not as common. In eukaryotes, most pre-mRNAs contain a poly(A) tail, with the notable exception of the pre-mRNA for the histone genes. The presence of the poly(A) tail increases the stability of the mRNA.

The presence of a poly(A) tail on pre-mRNA (and mRNA) has a practical significance to the investigator who is interested in purifying mRNA from a cell. The poly(A) tail of mRNA will bind to poly(T) bound to an inert solid support, such as sepharose, as illustrated in Figure 11.14. In practical terms, the RNA is first extracted from the cells and separated from other components of the cell. The mRNA can then be separated from rRNA and tRNA by passing it through a poly(T)-sepharose column, where only RNAs containing a poly(A) tail will be retained. All other RNAs pass through the column. The retained RNAs can then be released by elution with a buffer that breaks the hydrogen bonds between the poly(A) tail and the immobilized

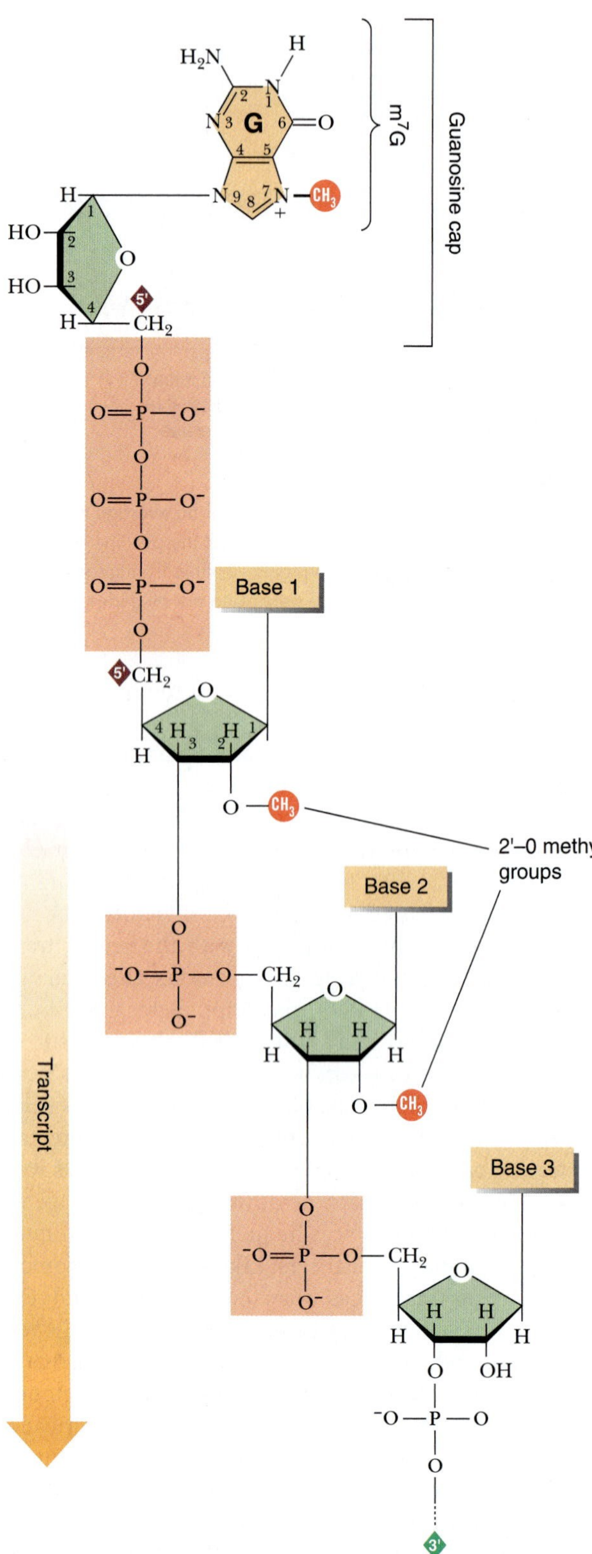

FIGURE 11.13 The 5′ end of eukaryotic mRNA. Eukaryotic mRNAs possess on the 5′ end a cap that is added immediately after synthesis. This cap is composed of an inverted guanosine connected by triphosphates. In addition, the riboses are methylated after synthesis. These additions to pre-mRNA add to its stability and are required for translation.

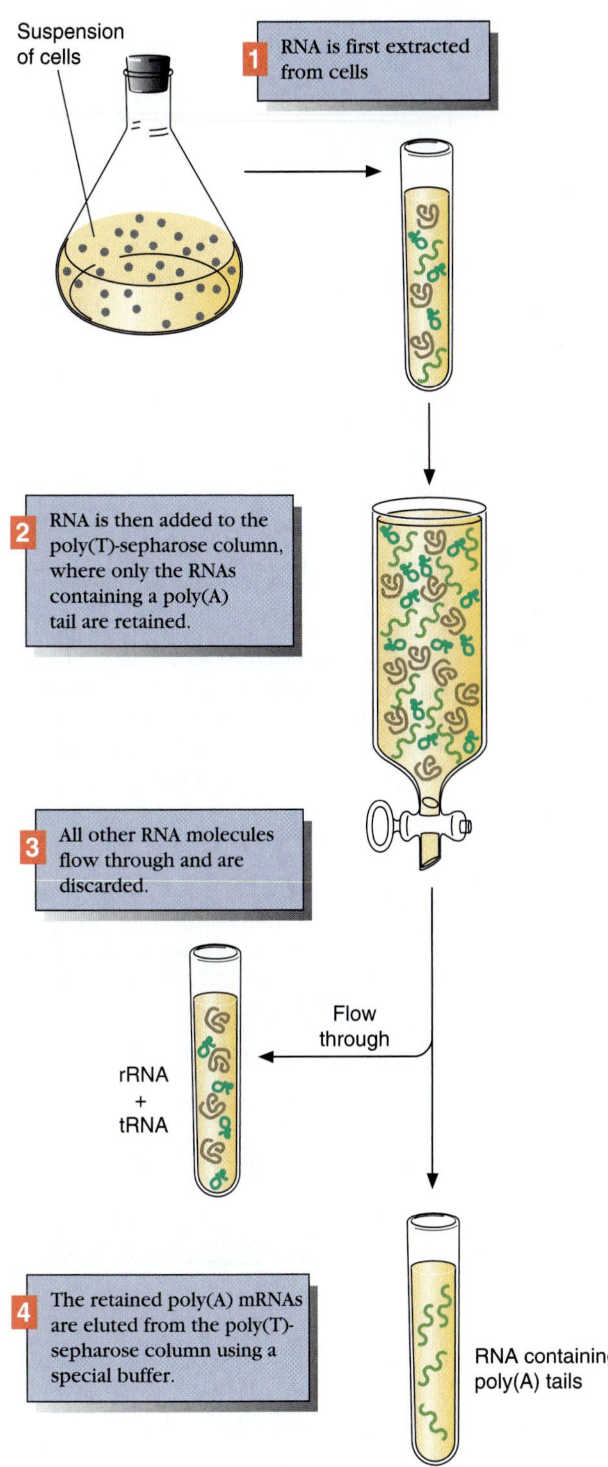
Suspension of cells
1 RNA is first extracted from cells
2 RNA is then added to the poly(T)-sepharose column, where only the RNAs containing a poly(A) tail are retained.
3 All other RNA molecules flow through and are discarded.
Flow through
rRNA + tRNA
4 The retained poly(A) mRNAs are eluted from the poly(T)-sepharose column using a special buffer.
RNA containing poly(A) tails

FIGURE 11.14 **Eukaryotic mRNA purification.** Because most mRNAs possess a poly(A) tail, they can be immobilized on a poly(T)-containing compound in a column. This is possible because the poly(A) tails of the mRNA will hydrogen-bond to the poly(T) immobilized on beads in the column, allowing rapid and easy separation of mRNA from rRNA and tRNA.

poly(U). Thus, when an investigator wants to isolate a specific gene, the mRNA product from transcription of the gene should be present in the mRNA population removed from the column.

A Complex Splicing Reaction Removes Sequences from Pre-mRNA.

During the 1960s, it was known that RNA destined to code for proteins in eukaryotes was much larger than predicted (based on the size of the protein products). In fact, some RNAs contained ten times as much genetic information as expected. In addition, these large RNA molecules were always found in the nucleus and never in the cytoplasm. An explanation as to why the pre-mRNA was much larger than expected had to await 1977, when it was discovered that genes contain internal, noncoding sequences. This discovery was made possible by comparing the sequences of the mRNA and the actual product of the gene (pre-mRNA). For example, the chicken ovalbumin polypeptide is 386 amino acids long, but the actual gene is 7700 bp in length (Figure 11.15). Because three nucleotides code for one amino acid, a 386 amino acid polypeptide should be 1158 nucleotides long. Thus, the gene for ovalbumin appears to have an excess of 6542 nucleotides, or more simply stated, 85% of the gene is noncoding information. These noncoding sequences are called **introns.** The reason for the existence of introns within eukaryotic genes is a hotly debated subject (see *CURRENT INVESTIGATIONS: What Is the Function of an Intron?*).

We now know that the DNA coding sequences in a pre-mRNA are interrupted by these noncoding introns. Coding DNA sequences are called **exons.** Transcribed intron sequences in the pre-mRNA are removed, enabling exon sequences to join. This completed molecule is the mRNA found in the cytoplasm. The process of removing introns from pre-mRNA is termed **RNA splicing** or **RNA processing.** The removed intron sequences are rapidly degraded into individual nucleotides and reused to synthesize more RNA.

It is possible to compare visually the organization of DNA sequences within a gene and the actual sequences in the spliced RNA through a technique called **heteroduplex mapping.** In fact, this procedure was used to establish the fact that introns are actually present in pre-mRNA. Heteroduplex mapping involves hybridization between two molecules (DNA : DNA or DNA : RNA) that are not entirely complementary. The noncomplementary regions of the sequences do not hybridize to each other and instead form single-stranded loops. These loops can be seen with an electron microscope. An electron micrograph of a heteroduplex between the ovalbumin mRNA and its gene shows loops (representing the

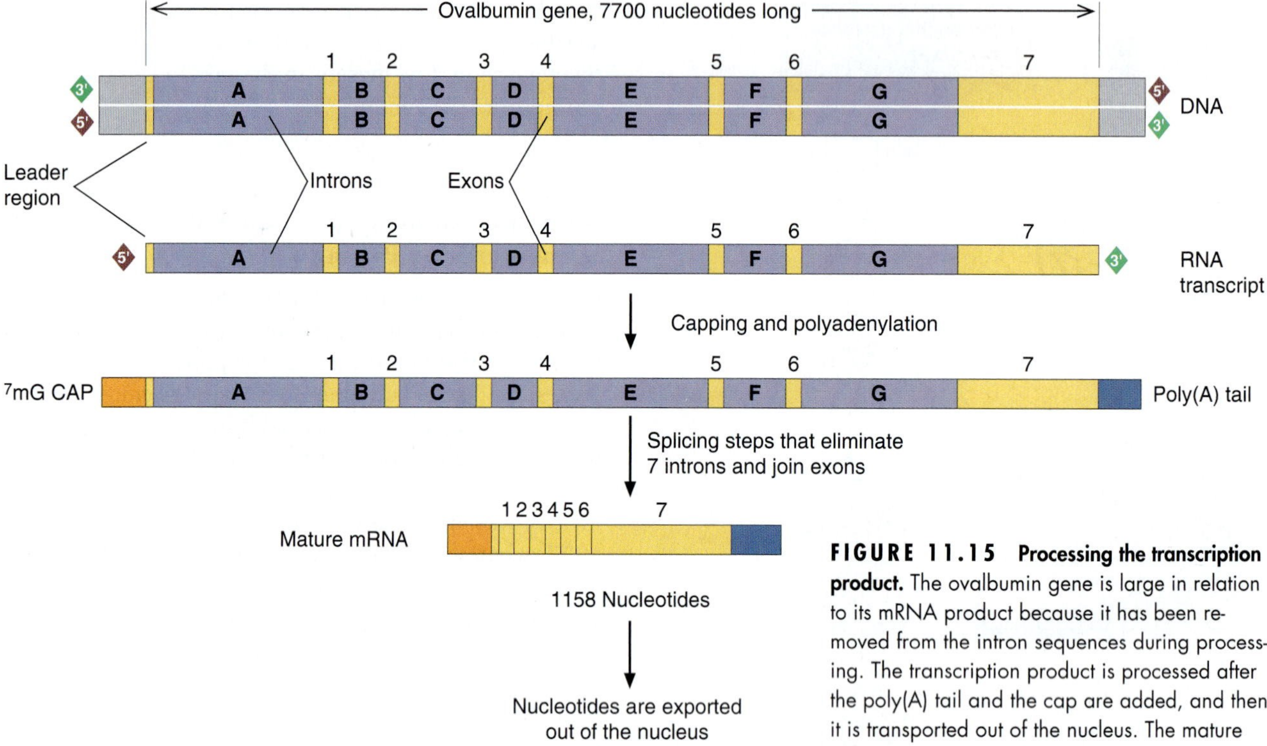

FIGURE 11.15 Processing the transcription product. The ovalbumin gene is large in relation to its mRNA product because it has been removed from the intron sequences during processing. The transcription product is processed after the poly(A) tail and the cap are added, and then it is transported out of the nucleus. The mature mRNA is now ready for translation.

introns) that have large noncomplementary sequences (Figure 11.16).

After the discovery that introns were present in pre-mRNA, the next logical question concerned the molecular mechanism(s) of splicing out the introns. A number of splicing pathways for pre-mRNA exist, but only two will be considered here: splicing *without* the participation of enzymes, also known as **autocatalytic splicing**, and splicing *with* the participation of an RNA-enzyme complex known as the **spliceosome.** Spliceosomes mediate removal of introns in the nucleus of most eukaryotes, whereas autocatalytic splicing has so far been found to occur only in nuclear genes of some lower eukaryotics, fungal mitochondrial genes, and some viruses.

Spliceosomes Mediate Processing of Pre-mRNA. In removing introns from pre-mRNA, the enzymes responsible for the breakage and reformation of nucleotide–

(a)

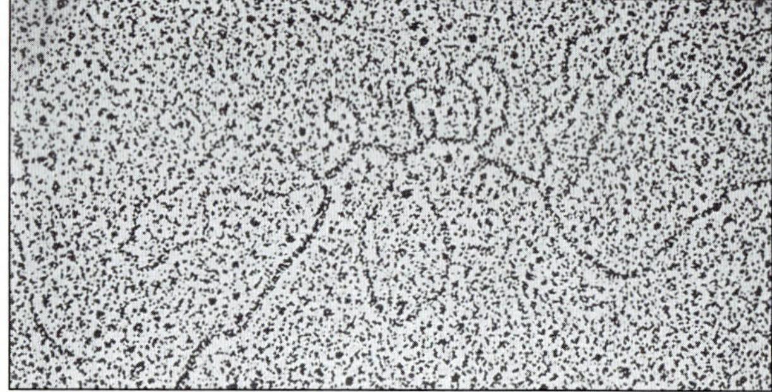

(b)

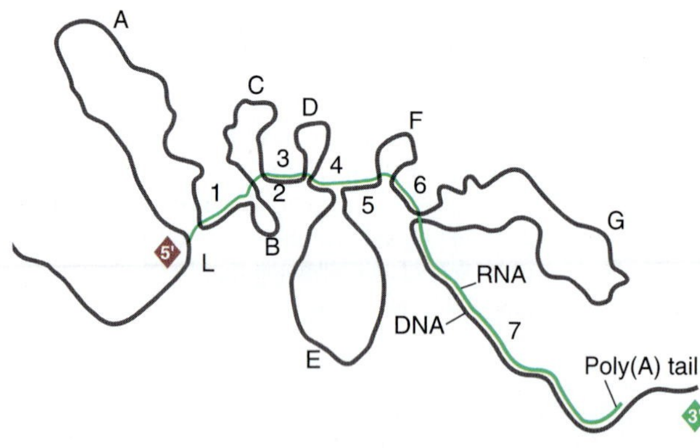

▶ **FIGURE 11.16 Micrograph of unspliced RNA complexed to its coding sequence.** (a) The electron micrograph shows a heteroduplex between ovalbumin mRNA and a single strand of DNA coding for ovalbumin. (b) This interpretative drawing mirrors the information shown in (a). The looped-out, single-stranded segments represent the introns, and the regions that hybridize represent the exons. Seven introns of varying sizes are visible (see Figure 11.15). *(From Chambon, P., "Split Genes." Copyright, 1981 by Scientific American, Inc. New York, NY)*

nucleotide bonds recognize a conserved sequence present at intron–exon junctions. This sequence is GU, present on the 5′ side of the intron, and AG, present on the 3′ side of the intron. The **GU–AG rule** is nearly always obeyed. The vast majority of exon–intron–exon junctions contain the sequence: 5′-AG**GU**AAGU. . . (intron nucleotides). . . PyPyPyPyPyPyC**AG**NN-3′, where the intron–exon junctions are adjacent to the bolded GU–AG base pairs. In this sequence, the GU. . . AG base pairs occur 100% of the time, whereas the other bases vary somewhat in frequency. The presence of a conserved sequence implies a common mechanism for splicing in eukaryotes. *In vitro* experiments established that the splicing reaction occurs in the nucleus and is independent of capping and addition of a poly(A) tail. The first step of the splicing reaction involves the enzymatic cleavage of the left junction site (just before the GU in the RNA sequence) followed by formation of a lariat structure between the released 5′ end of the intron and the 2′-carbon in the sugar ring of an adenosine within the intron (Figure 11.17). The right junction is then cleaved, releasing the branched lariat and simultaneously joining the two exons. The released intron RNA fragment is degraded, and the nucleotides are recycled into new RNA.

The splicing reaction is catalyzed by a large complex called a spliceosome, which consists of nucleic acids and eight to ten enzymes (Figure 11.18). Present within the spliceosome are snRNAs. Eukaryotic cells have many different types of snRNAs that range between 100 and 300 nucleotides in length (excluding yeast). A prominent snRNA is U1 (they are called U1, U2, and so on, because they have an abundance of uridines), which has

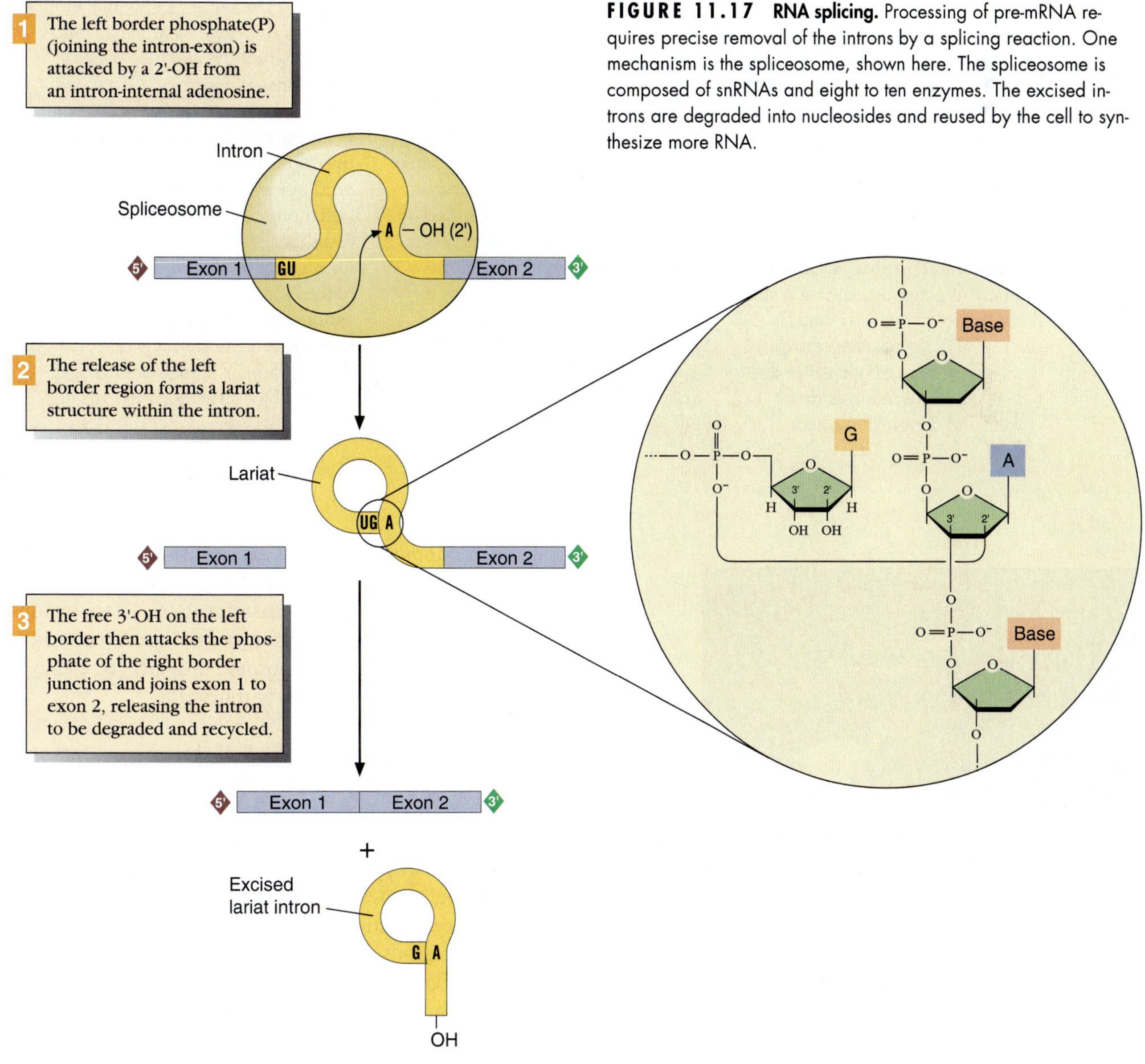
1 The left border phosphate(P) (joining the intron-exon) is attacked by a 2'-OH from an intron-internal adenosine.
Intron
Spliceosome
A – OH (2')
5'
Exon 1
GU
Exon 2
3'
2 The release of the left border region forms a lariat structure within the intron.
Lariat
UG
A
Exon 1
Exon 2
3 The free 3'-OH on the left border then attacks the phosphate of the right border junction and joins exon 1 to exon 2, releasing the intron to be degraded and recycled.
Exon 1
Exon 2
+
Excised lariat intron
G
A
OH
Base
G
A
Base
OH OH

FIGURE 11.17 RNA splicing. Processing of pre-mRNA requires precise removal of the introns by a splicing reaction. One mechanism is the spliceosome, shown here. The spliceosome is composed of snRNAs and eight to ten enzymes. The excised introns are degraded into nucleosides and reused by the cell to synthesize more RNA.

CURRENT INVESTIGATIONS

What Is the Function of an Intron?

At first glance, it would seem that introns are just nuisances to remove and discard, and in fact this may be the case. Because bacteria do not have introns (with some exceptions) and survive quite well, introns are clearly not essential for gene function. Do introns have any function in eukaryotes, and if they do, why don't prokaryotes have introns?

An approach to the question of why prokaryotes do not have introns comes from the study of the structure of genes from mitochondria and chloroplast genomes. Mitochondria and chloroplasts probably originated when primitive bacteria were inserted into the cytoplasm of early eukaryotic cells. This theory is called endosymbiosis and is supported by the many striking similarities between bacteria and these organelles. If we compare genes in mitochondria and chloroplasts with genes that function similarly in prokaryotes, we find that the organellar genes have introns, whereas the prokaryotic genes do not. If the endosymbiosis theory is correct and the organelle gene structure represents primitive prokaryotic gene structure, then the progenitor cells of bacteria and organelles may have had introns, and the ability to splice out introns has been lost during the evolution of modern bacteria. An alternate hypothesis states that mitochondria and chloroplasts acquired introns from their eukaryotic host late in evolutionary time and did not possess them in the first place.

Why do eukaryotes and organelles retain introns? One suggestion is that they have been conserved during evolution (that is, they have not changed much because they are so important). Some evidence supports this suggestion. If radioactively labeled DNA is prepared from an exon and is hybridized to genomic DNA cut with a restriction endonuclease, the labeled exon frequently hybridizes to other exons that are in genes with related functions. This result implies that two (or more) genes were generated by the duplication of a progenitor gene and then diverged (accumulated mutational base sequence changes), creating a new but related function. Usually the size of the exons between duplicated genes does not change significantly, and the location of the introns and exons within the gene are somewhat conserved during evolution.

What could the function of introns be? The answer may lie in the study of exons. Exons are often small, coding for an average of 20–80 amino acids. This is a little bigger than the size of a stable folded amino acid domain within a protein, but it suggests that exons represent functional domains in proteins. Domains are subunits of proteins that have a particular function or structure, and modern proteins consist of a series of domains. This explains why, in many cases, the location of introns remains relatively constant. Individual domains need not encode a complete functional protein, just a sequence that could be combined with others to form a functional protein. This hypothesis is supported by the observation that some exons encode sequences of amino acids that either are similar in many proteins or that produce structures that are similar (β-sheets, α-helices, DNA

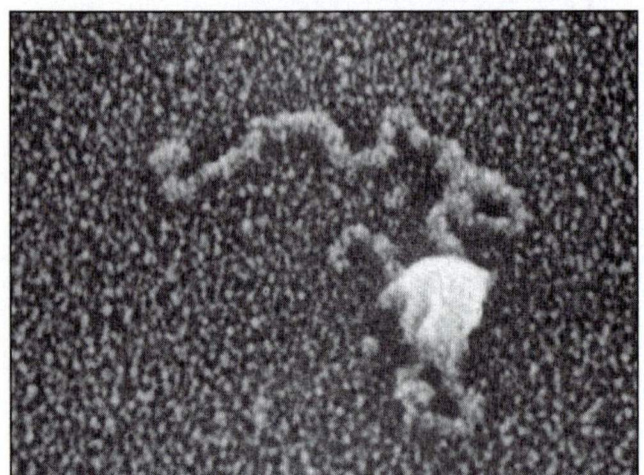

FIGURE 11.18 Electron micrographs of purified spliceosomes. This spliceosome is associated with pre-mRNA and shows the large size of the spliceosome. *(From Reed, R., J. Griffith, and T. Maniatis. 1988.* Cell, *53: 949–961.)*

sequence homology to the exon–intron left junction. The role of the snRNAs is to recognize certain regions by hydrogen-bonding in the transcript RNA. These regions include the consensus regions at the intron–exon borders. This orients the spliceosome properly for cleavage of the intron from the primary transcript.

Autocatalytic Splicing Is the Self-Removal of Introns. In autocatalytic splicing, no enzymes are required in *in vitro* reactions; however, recent evidence suggests that enzymes are required *in vivo.* Short, conserved sequences are required within the intron for autocatalytic splicing but can be located a considerable distance from the intron–exon junctions. In autocatalytic splicing, double-stranded regions form between intron consensus sequences (Figure 11.19). The folded secondary structure allows an attacking guanosine residue with a free hy-

binding domains). According to this theory, evolution of genes can occur by adding or subtracting domains, and introns represent the spaces between domains. Introns would thus function as regions where recombination can occur without danger of disrupting the sequence of the domain. At first, evidence supporting this hypothesis accumulated. However, more recent evidence indicates that, in fact, introns may be randomly inserted and do not separate protein domains. These findings exacerbate the debate about the function of introns and suggest that they may just be "selfish DNA" that has found a spot in the genome that is safe from loss.

An example of intron placement conservation comes from the globin family of genes. Globins bind iron-containing heme groups and transport oxygen. Two members of the globin protein family are found in hemoglobin, which is composed of four globin proteins, two α globins and two β globins. These two different globin genes are similar enough in sequence to indicate that they arose by duplication and diverged during evolution. Two introns are present in the α- and β-globin genes, located at positions that divide the gene into a central, heme-binding domain and two flanking domains. The myoglobin gene is another member of the globin gene family, and the protein product is a monomer that binds oxygen in muscles. The myoglobin gene has many sequence differences when compared to α or β globin, indicating that it diverged very early in evolution. But, the myoglobin gene has two introns, and they are in exactly the same place as the introns of the α- and β-globin genes. Another member of the globin gene family is present in plants (mostly legumes) and also binds oxygen to leghemoglobin. Amazingly, the leghemoglobin gene has introns in the same place as the globin genes in animals, but it also has an additional intron that splits the heme-binding domain into two exons. The similarity of these genes suggests an ancient origin for the oxygen-binding globin genes. It also suggests that the heme-binding domain may have originated as two domains that were combined into one domain early in animal evolution.

This evidence does not provide any clear answer to the question, Why don't bacteria have introns? We can only speculate. One hypothesis suggests that there is great evolutionary pressure on bacteria to replicate quickly, in order to outgrow competing species of bacteria. In such a race for survival, species with the smallest possible genome would have an advantage. This type of selection pressure against any excess DNA may have led to the elimination of introns. Eukaryotes, with their much longer life span and more complex development, may have a greater need for flexibility and the ability to generate new genes than bacteria. It is also true that prokaryotes, with their short division time, have undergone many more generations than eukaryotes, and thus may have evolved faster. Alternatively, it is also possible that neither bacteria nor eukaryotes possessed introns in the beginning, and only eukaryotes gained them in recent evolutionary time. Further investigation may clarify this very controversial issue.

droxyl group (as a cofactor) to release the 5′ side of the intron, joining and guiding the two intron–exon junctions into close proximity. The guanosine becomes linked to the intron, and the free 3′-hydroxyl on the left exon then attaches to the right 3′ exon-intron junction, releasing the intron.

Autocatalytic intron splicing is found in the nuclear genes coding for rRNA in *Tetrahymena* and *Physarum* (a ciliate and a slime mold), in the mitochondria of fungi, and in the mitochondria and chloroplasts of many plants. Self-splicing RNA sequences were first viewed with great scepticism, but we now know that other RNAs are capable of self-splicing, not just intron excision. Catalysis by RNA may be quite widespread and may have significant evolutionary implications (see Current Investigations: *Not All Enzymatic Reactions Are Catalyzed by Proteins*).

Ribosomal RNAs Are Also Processed After Synthesis.

Eukaryotes possess many copies of rRNA genes because they need considerable amounts of rRNA to assemble ribosomes for the synthesis of proteins. The genes for rRNA are usually clustered in the nucleolus of the cell and arranged end to end with short stretches that are not transcribed. These regions are referred to as **nontranscribed spacers.** Each gene cluster for ribosomal RNA is composed of sequences that eventually result in 18S, 5.8S, and 28S rRNA molecules (S is explained in the next section), but they are transcribed as a unit. Each of the rRNA sequences is separated by a **transcribed spacer,** which must be processed out before the separate rRNA molecules are functional. Ribosomal RNA gene clusters do not contain any introns.

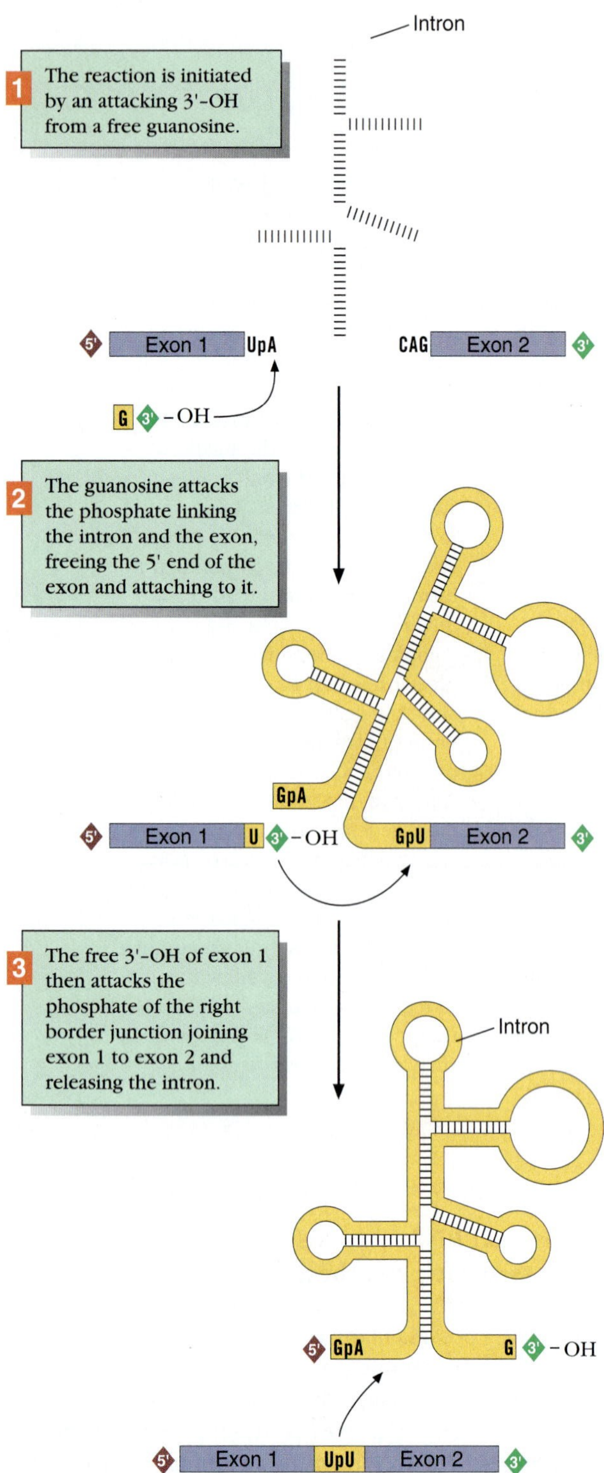
Intron
1
The reaction is initiated by an attacking 3'-OH from a free guanosine.
5'
Exon 1
UpA
CAG
Exon 2
3'
G
3'
-OH
2
The guanosine attacks the phosphate linking the intron and the exon, freeing the 5' end of the exon and attaching to it.
GpA
5'
Exon 1
U
3'
-OH
GpU
Exon 2
3'
3
The free 3'-OH of exon 1 then attacks the phosphate of the right border junction joining exon 1 to exon 2 and releasing the intron.
Intron
5'
GpA
G
3'
-OH
5'
Exon 1
UpU
Exon 2
3'

FIGURE 11.19 **Autocatalytic splicing.** A second mechanism for removing introns involves a self-splicing reaction. This splicing mechanism is less common than the spliceosome reaction, but it was revolutionary when it was first discovered because it does not require the participation of an enzyme.

As an example, the human rRNA precursor, or immediate transcript, of the rRNA gene cluster is a large 45S molecule. This cluster contains the 18S, 5.8S, and 28S rRNA molecules and three transcribed spacers—one between the 18S and 5.8S sequence, one between the 5.8S and 28S, and one large sequence at the 5′ end of the molecule. rRNA transcript processing occurs in the nucleolus. In the first step, the transcribed 5′ end spacer segment is removed. The second step makes a cut between the 18S sequence and the 5.8S sequence within the spacer region, and in the third step the remaining spacer regions are trimmed out, leaving each component of the ribosomal RNA. This process is not universal in all eukaryotes; it varies a little from species to species, but the basic mechanism of rRNA processing is preserved throughout the eukaryotic kingdom.

SECTION REVIEW PROBLEMS

8. How is the product of RNA polymerase II action altered after synthesis?
9. Explain the difference between an exon and an intron.
10. How does the enzyme-catalyzed splicing reaction differ from autocatalytic splicing?

mRNA Sequences Are Translated into Sequences of Amino Acids.

Before we can discuss how the information in mRNA is used for the synthesis of proteins, it is first necessary to understand that the information stored in the mRNA is in the form of a genetic code composed of three bases. Each code word, called a **codon,** specifies a single amino acid in the protein. The sequence of code words thus controls the sequence of the 20 amino acids found in proteins. At this point, we will not discuss the details of the code, codon usage, nor how the code was deciphered. We will delay a discussion of this subject until we understand exactly how the information in mRNA is converted into proteins. But, we do need to know a little about the nature of the genetic code.

Proteins are composed of 20 different amino acids, and nucleic acids are composed of four different nucleotides; thus, single nucleotides alone are insufficient to specify all the amino acids. Would two nucleotides per codon be enough to specify all amino acids? Two nucleotides per codon would yield only 4^2 or 16 possible codons—still not enough to accommodate all 20 amino acids. On the other hand, three nucleotides per codon would yield 4^3 or 64 different codons, an excess of codons. Three nucleotides are used in the codons, and we now know that most amino acids employ more than one codon in protein synthesis (Figure 11.20). It is

CURRENT INVESTIGATIONS

Not All Enzymatic Reactions Are Catalyzed by Proteins.

It had been a firmly established belief in biology that catalysis is reserved for proteins. However, in 1989 the Nobel Prize was presented to Sidney Altman and Thomas Cech for discovering that RNA can catalyze a reaction. In retrospect, catalytic RNA makes a lot of sense. This is based on the old question regarding the origin of life, Which comes first, enzymes that do the work of the cell or nucleic acids that carry the information required to produce the enzymes? Nucleic acids as catalysts circumvent this problem.

Research leading to the discovery that RNA can act as a catalyst started in the 1970s. Thomas Cech, at the University of Colorado at Boulder, was studying the excision of introns in a ribosomal RNA gene in *Tetrahymena thermophila*. In attempting to purify the enzyme responsible for the splicing reaction, he instead found, much to his amazement, that the intron could be spliced-out in the absence of any added cell extract. Much as they tried, Cech and his colleagues could not identify any protein associated with the splicing reaction. After much work, Cech proposed that the intron sequence portion of the RNA had properties of an enzyme enabling it to break and reform phosphodiester bonds. At about the same time, Sidney Altman, who is a Professor at Yale University, was studying the way tRNA molecules are processed in the cell when he and his colleagues isolated an enzyme called RNase-P, which is responsible for the conversion of a precursor tRNA into the active tRNA. Much to their surprise, they found that RNase-P contained RNA in addition to protein and that RNA was an essential component of the active enzyme. This was such a foreign idea that they had difficulty publishing their findings. The following year, Altman demonstrated the final bit of evidence establishing that RNA can act as a catalyst by showing that the RNase-P RNA subunit could catalyze the cleavage of precursor tRNA into active tRNA in the absence of the protein component.

Since Cech's and Altman's discovery, other investigators have discovered other examples of self-cleaving RNAs or catalytic RNAs. Most notable among these later discoveries are a group of RNA molecules dubbed ribozymes. These were first identified in plant viral pathogens. All ribozymes have either a hairpin- or hammerhead-shaped active center and a unique secondary structure allowing them to cleave other RNA molecules at specific sequences. It is possible to produce in the laboratory ribozymes that will specifically cleave any RNA molecule. These RNA catalysts may have pharmaceutical applications. For example, a ribozyme has been designed to cleave the RNA of HIV (the virus responsible for AIDS). By placing a ribozyme in the cell, all incoming virus particles that express this particular gene will have the RNA product cleaved by the ribozyme, which, in the end, would kill all invading virus particles.

1st letter (5' end)	2nd letter U	2nd letter C	2nd letter A	2nd letter G	3rd letter (3' end)
U	Phe UUU	Ser UCU	Tyr UAU	Cys UGU	U
	Phe UUC	Ser UCC	Tyr UAC	Cys UGC	C
	Leu UUA	Ser UCA	STOP UAA	STOP UGA	A
	Leu UUG	Ser UCG	STOP UAG	Trp UGG	G
C	Leu CUU	Pro CCU	His CAU	Arg CGU	U
	Leu CUC	Pro CCC	His CAC	Arg CGC	C
	Leu CUA	Pro CCA	Gln CAA	Arg CGA	A
	Leu CUG	Pro CCG	Gln CAG	Arg CGG	G
A	Ile AUU	Thr ACU	Asn AAU	Ser AGU	U
	Ile AUC	Thr ACC	Asn AAC	Ser AGC	C
	Ile AUA	Thr ACA	Lys AAA	Arg AGA	A
	Met AUG	Thr ACG	Lys AAG	Arg AGG	G
G	Val GUU	Ala GCU	Asp GAU	Gly GGU	U
	Val GUC	Ala GCC	Asp GAC	Gly GGC	C
	Val GUA	Ala GCA	Glu GAA	Gly GGA	A
	Val GUG	Ala GCG	Glu GAG	Gly GGG	G

FIGURE 11.20 The genetic code. This codon table is the same (that is, universal) for nearly all organisms. A few rare exceptions to these assigned codons do exist. The nucleotide sequence of each codon, including the start and stop codons, are shown. The table is read from left to right, drawing one letter from each column to obtain a codon. For example, phenylalanine has two codons, UUU and UUC, whereas methionine has only one, AUG.

important to emphasize that the codons are read without any punctuation. Extra bases are not present between the codons, nor are the codons overlapping. Also, four specific codons in the RNA serve as start and stop signals for the reading of genes (Figure 11.20), thus explaining the usage of all 64 codons.

After the information from DNA is transcribed into an mRNA sequence, it is then converted into the amino acid sequence of a protein by a series of chemical reactions referred to as cellular protein synthesis or **translation.** In eukaryotes, transcription and translation are very separate events. Translation occurs in the cytoplasm, and mRNA must be transported there from the nucleus (Figure 11.21). In contrast, in prokaryotes, which do not have a nucleus, transcription and translation are coupled, occurring nearly simultaneously (Figure 11.21). For protein synthesis to occur in either prokaryotes or eukaryotes, both ribosomes and tRNA are

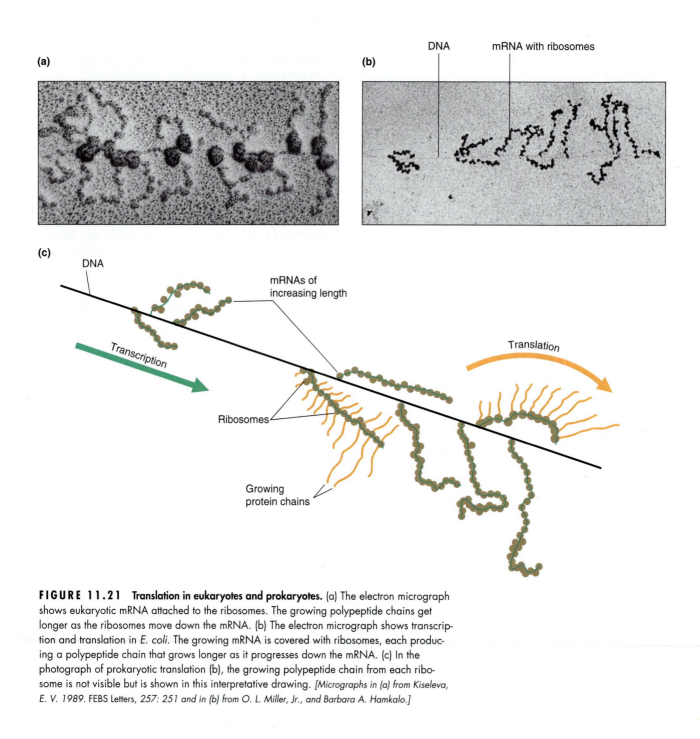

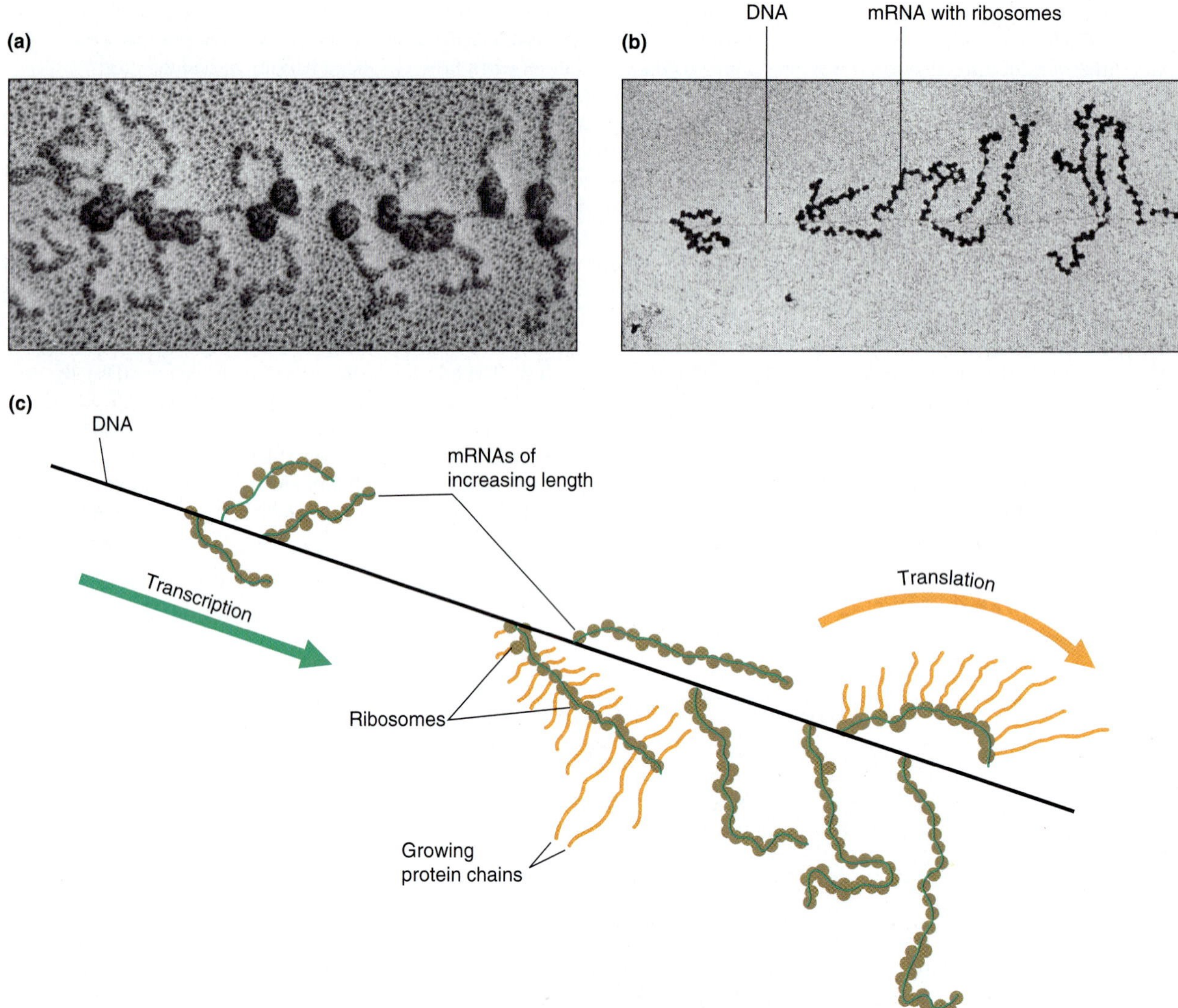

FIGURE 11.21 Translation in eukaryotes and prokaryotes. (a) The electron micrograph shows eukaryotic mRNA attached to the ribosomes. The growing polypeptide chains get longer as the ribosomes move down the mRNA. (b) The electron micrograph shows transcription and translation in *E. coli.* The growing mRNA is covered with ribosomes, each producing a polypeptide chain that grows longer as it progresses down the mRNA. (c) In the photograph of prokaryotic translation (b), the growing polypeptide chain from each ribosome is not visible but is shown in this interpretative drawing. *[Micrographs in (a) from Kiseleva, E. V. 1989.* FEBS Letters, *257: 251 and in (b) from O. L. Miller, Jr., and Barbara A. Hamkalo.]*

required. The ribosomes are clearly visible in Figure 11.21, but the tRNAs are too small to be seen with an electron microscope.

Next we will examine the structure of the ribosome and its component parts as well as the structure of tRNA. The tRNA and ribosome complexes are very important because they are key ingredients in converting the genetic message (mRNA, which specify codons) into protein or enzymes, the workhorses of the cell. Proteins determine phenotype; thus, they are the link between genotype and phenotype in all organisms.

Ribosomes and tRNAs Play an Essential Role in Protein Synthesis.

The **ribosome** is an extremely large aggregation of ribosomal proteins and ribosomal RNAs and is the site of protein synthesis. The total molecular weight of a ribosome is about 3 million Da, about 50 to 60 times larger than an average protein (about 50,000 Da). In prokaryotes, and specifically in *E. coli,* three different ribosomal RNAs are present in each ribosome: one each of species called 5S RNA, 16S RNA, and 23S RNA. The S characterization for these RNA molecules derives from the rate of sedimentation of the RNA molecule in a centrifugal field. This sedimentation coefficient, measured in Svedberg units (S), reflects the three-dimensional structure and molecular weight of the molecule. As a rule, the larger the S value, the larger the mass of the molecule; however, a linear relationship does not exist between molecular mass and S values.

Separating molecules with different S values is a valuable experimental tool and is known as **sucrose gradient centrifugation.** In this procedure, a sucrose gradient is made in a centrifuge tube by layering successively lower concentrations of sucrose solutions, one on top of the other, from the bottom of the tube to the top (Figure 11.22). The material to be studied (for example, ribosomes dissociated into their subcomponents) is gently placed in a uniform layer on top of the uppermost sucrose layer in the centrifuge tube. When the tube is centrifuged at a high speed and allowed to swing free, the sedimenting material travels through the gradient at different speeds depending upon the size and shape of the aggregates or structures; larger structures will be found near the bottom of the tube. The centrifugation run is stopped before the structures or molecules reach the bottom of the tube so that they are "caught" in flight. Individual fractions can be separated from each other by punching a pinhole in the bottom of the tube and collecting drops in separate test tubes. These fractions can then be analyzed for their component molecules. This procedure is used for a variety of separation applications, including separation of intact macromolecules or

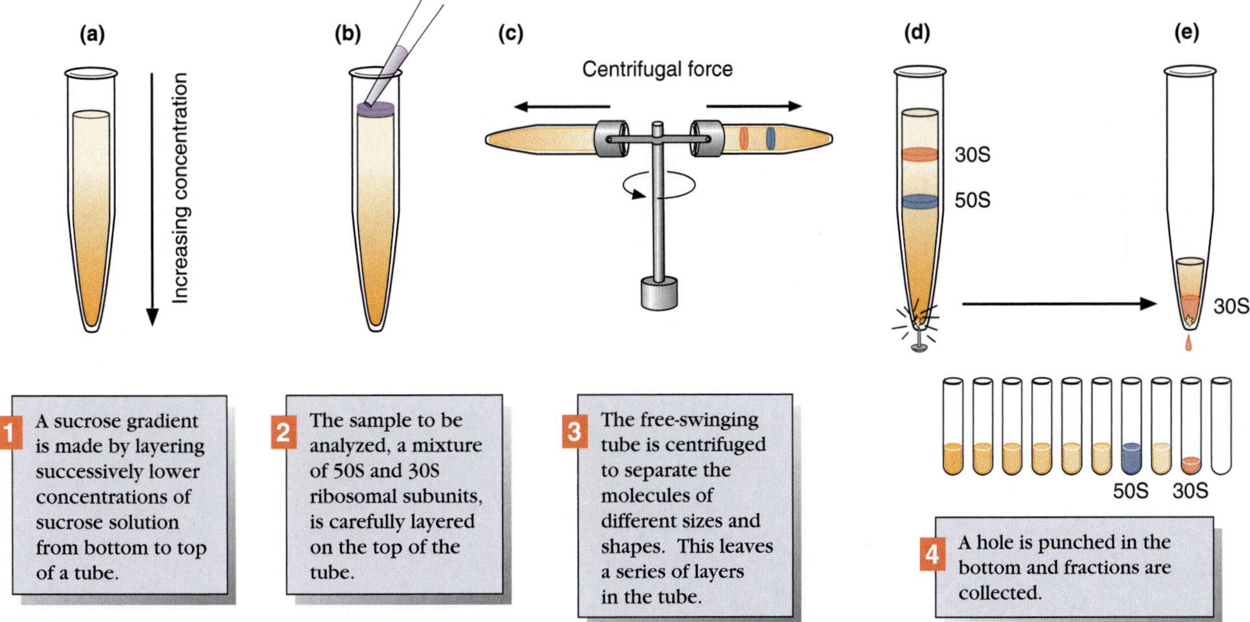

FIGURE 11.22 Sucrose gradient centrifugation. Sucrose gradient centrifugation is a general method used to separate large molecules or components of the cell. The larger molecules or aggregates of cell components sediment more rapidly, and they will eventually accumulate on the bottom of the tube if centrifuged long enough. Usually the separated components are "caught" in flight and collected by puncturing the bottom of the tube and collecting fractions.

organelles from large aggregate structures following disruption of a cell.

The ribosomes from both prokaryotes and eukaryotes share the same general structure and are composed of two major subunits of different sizes; the 50S and 30S subunits in prokaryotes, which combine to form a 70S ribosome, and the 60S and 40S subunits in eukaryotes, which combine to form an 80S ribosome. The ribosome can be dissociated into these component parts in solutions of urea or in the presence of mild detergents that disrupt weak chemical interactions. Each ribosome subunit contains a large, single-stranded RNA molecule complexed with a large number of proteins. In *E. coli,* the 50S subunit contains a 23S rRNA molecule 2904 bp in length, and the small 30S subunit contains a 16S rRNA molecule 1541 bp in length. The large subunit also contains a 120-bp RNA molecule known as 5S rRNA. The role of each of the RNA molecules is an area of active research; it is known that they play a pervasive role in the structure and function of the ribosome. Each component of the ribosome includes a large number of basic proteins; the ribosome is about 40% protein (Figure 11.23). An important feature of ribosomal structure is that the proteins and the rRNAs can self-assemble *in vitro* into a functionally complete ribosome. Thus, the architecture of the ribosome is an inherent result of the physical properties of its component rRNAs and proteins.

Ribosomal RNAs in prokaryotes and eukaryotes are transcribed from DNA in a manner similar to the transcription of mRNA (Figure 11.24). You will recall from the last section that in eukaryotes the rRNA transcript is composed of several rRNA components, and the polymerase that transcribes rRNA is RNA polymerase I. In prokaryotes, all RNA is transcribed by one RNA polymerase. A bacterial cell contains many more rRNA (and ribosomal proteins) than mRNA molecules (80% of the RNA in a bacterial cell is rRNA) because rRNA is much more stable than mRNA. In order for a bacterial cell to maintain a large number of ribosomes, it possesses multiple copies of the genes responsible for their synthesis. In *E. coli,* seven copies of rRNA genes are present. In eukaryotes, hundreds of copies of the rRNA genes are present. In some eukaryotic species, the number of copies of the rRNA genes may be in the thousands. In contrast, only one copy of each of the ribosomal protein genes is present in *E. coli,* but each of these genes is expressed at

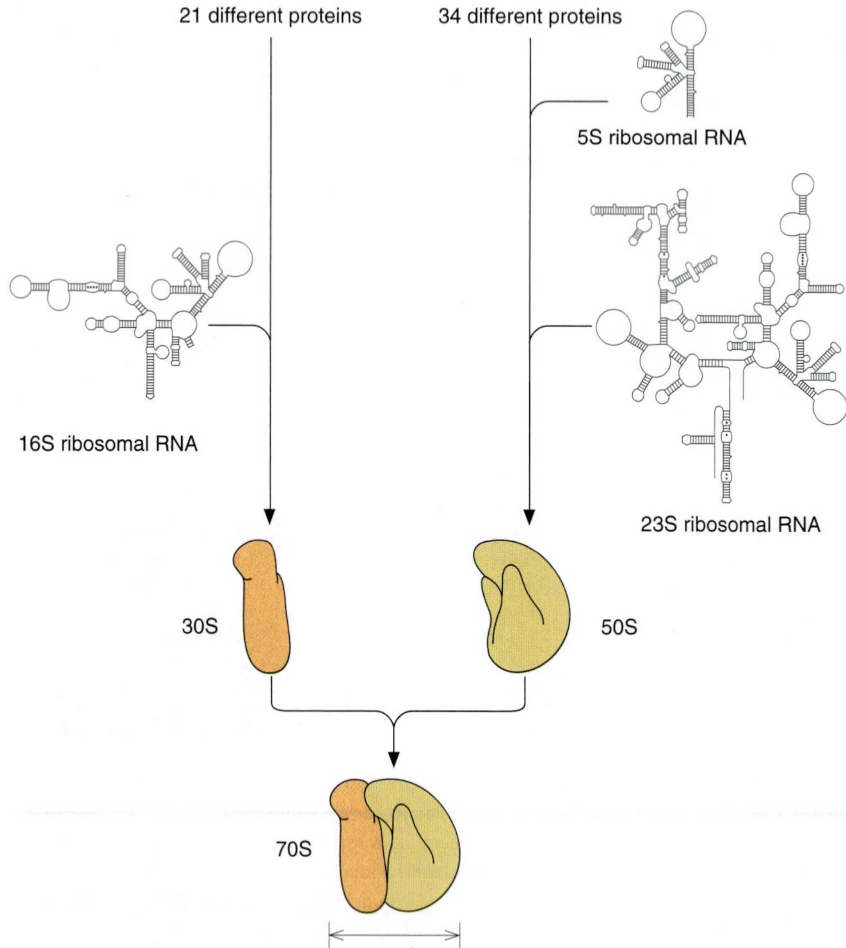

FIGURE 11.23 The *E. coli* ribosome and its parts. All bacteria have similarly sized ribosomes and subunits, but in higher organisms the ribosome is somewhat larger (80S), or 4.2 instead of 2.7 million molecular weight. This demonstrates how complex a ribosome is, but surprisingly, if the component parts are mixed together, they will self-assemble into a complete ribosome.

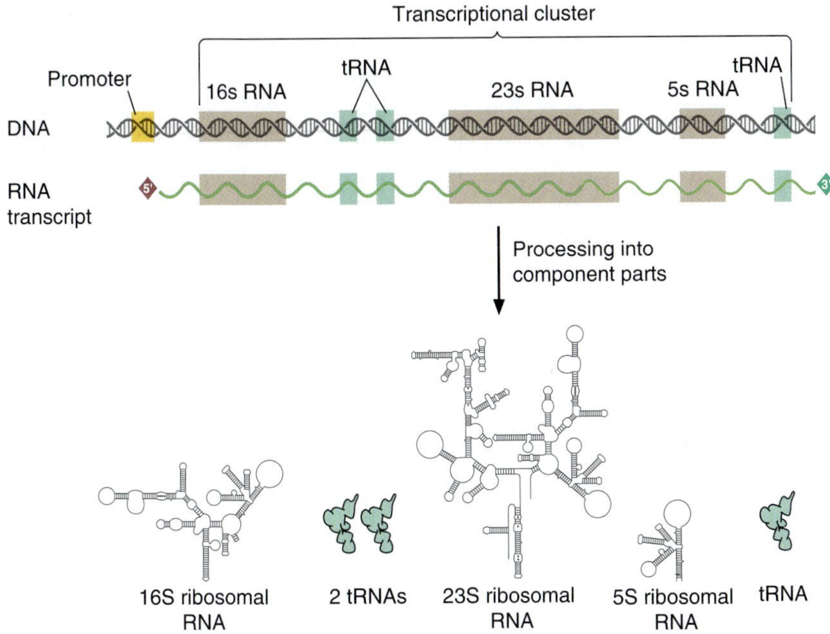

FIGURE 11.24 Ribosomal RNA genes. The general organization of a typical rRNA transcriptional cluster for prokaryotes. Three rRNA genes are arranged from 16S, 23S, and 5S starting at the 5′ end and going to the 3′ end. tRNA genes are present in the spacer regions and sometimes at the 3′ end. The whole sequence is transcribed into one molecule and then processed into individual components using the enzyme RNase III.

a very high level to supply a sufficient quantity of the proteins for assembly of the ribosome.

An example of a transcriptional unit for an *E. coli* rRNA gene is shown in Figure 11.24. Usually two different tRNA genes are present between the 16S rRNA gene and the 23S rRNA gene. In three of the rRNA gene clusters in *E. coli*, tRNA genes are present at the 3′ end of the cluster. The enzyme RNase III cleaves the rRNA cluster into its component parts.

An amino acid cannot directly recognize a codon on the mRNA. Thus, a mechanism must exist that allows the amino acids to recognize their respective codons in the mRNA. An amino acid recognizes its proper codon in the mRNA by first being attached to tRNA, which acts as an adapter molecule to recognize the proper codon on the mRNA. Transfer RNA is a small molecule containing about 75 nucleotides, and it is sometimes referred to a 4S RNA (because of its sedimentation characteristics). Most importantly, tRNAs possess the important **anticodon loop** (Figure 11.25), which recognizes the codon on the mRNA. Each codon in the mRNA has a complementary anticodon on a tRNA. The anticodon sequence recognizes the codon sequence and positions the amino acid to be added to the growing polypeptide chain. Because

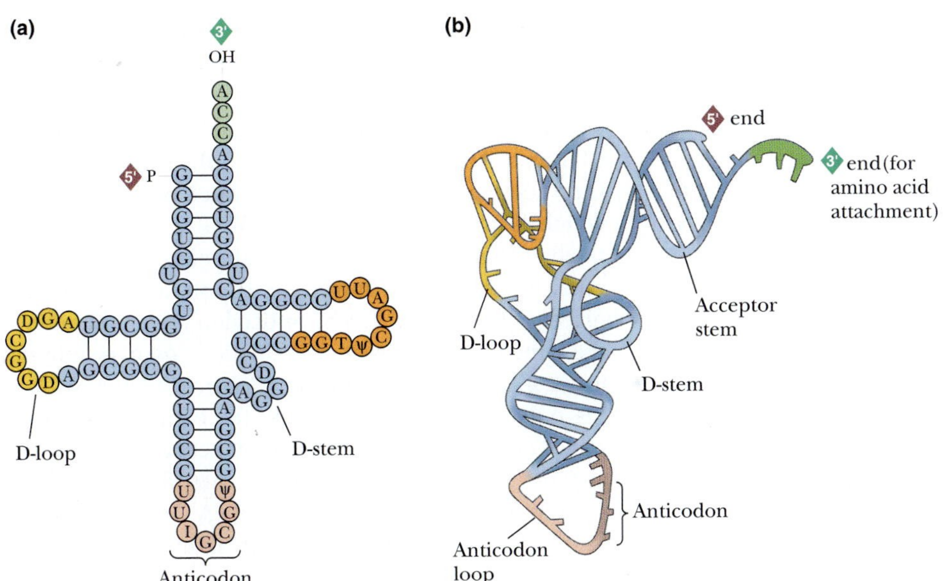

FIGURE 11.25 tRNA structure. (a) The two-dimensional drawing of alanine-tRNA shows the three loops present in all tRNA molecules: the anticodon triplet, several paired regions, the 3′ terminal region where the amino acid is covalently attached, and some of the modified bases (I for inosine, D for dihydrouridine, T for ribothymidine, and ψ for pseudouridine). (b) This three-dimensional drawing indicates the positions of the loops and how the molecule actually looks.

64 possible codons exist (that is, DNA has four different bases, which when taken three at a time equals 64 different possibilities), it should follow that 64 different kinds of tRNAs are potentially present in each cell, each having a different complementary sequence in the anticodon loop. This is partly true, but because three of the codons specify termination signals for translation, no tRNAs exist for the stop signals. Thus, 61 anticodons remain. Each of the different tRNAs is coded by a different gene, and thus each tRNA has a different nucleotide sequence. As indicated in Figure 11.20, some amino acids have more than one codon; these amino acids will have several different tRNAs with different anticodons.

In *E. coli,* the majority of the 61 different possible tRNAs are represented, some in multiple copies, but in eukaryotes, multiple copies of tRNA genes are present for each of the tRNAs. *Drosophila melanogaster* has 10 copies of each tRNA gene, whereas the African clawed toad *Xenopus levis* has about 180 copies of each tRNA gene. These genes are sometimes clustered in one region of the chromosome, but in other cases they are scattered around the genome.

Transfer RNA is modified in various ways after it is transcribed (Figure 11.25a). All tRNAs have a 3′-CCA-terminus that is added after tRNA transcription. In addition, some of the bases within the tRNA are modified to give unusual bases; for example, uridine is sometimes post-transcriptionally converted to pseudouridine by attaching a ribose at the 5 position on the ring instead of on the 1 position. These modified bases are thought to stabilize the tRNA from nuclease attack to give them a much longer life span in the cell than mRNA.

Proteins Are Assembled on the Ribosome.

Now that we have examined the structure and character of the three major players in translation (mRNA, ribosomes, and tRNA), we are ready to examine the details of the actual assembly of a polypeptide. The process of protein synthesis or translation is primarily a series of biochemical reactions, but this process must be understood to some extent to appreciate its genetic implications. First of all, we know that proteins are the primary determinants of phenotype. However, if mistakes occur in protein synthesis, what are the genetic and phenotypic implications? If mistakes occur in protein synthesis, what are the safeguards? What happens when a mutation occurs in a gene for tRNA, rRNA, or rRNA, and how does this affect protein synthesis? These are among the questions answered by understanding protein synthesis and how it determines phenotype.

The first step in the process of protein synthesis occurs when the amino acid attaches to its specific tRNA. Attachment is achieved by a family of at least 20 different enzymes called **aminoacyl tRNA synthetases.** Each of the 20 different enzymes is responsible for attachment of only 1 of the 20 different amino acids to the tRNA that has the correct anticodon specified in the genetic code. For example, phenylalanine-tRNA synthetase is responsible for attaching phenylalanine to $tRNA^{phe}$, which carries the anticodon –AAA– triplet for phenylalanine. The phenylalanyl-tRNA synthetase will not attach any other amino acid to the $tRNA^{phe}$. The reaction requires ATP as an energy source plus magnesium ion as a cofactor. The overall reaction follows:

$$\text{Amino acid}_1 + \text{tRNA}_1 + \text{ATP} \underset{\text{Synthetase}_1}{\leftrightharpoons} \text{Amino acid}_1\text{-tRNA}_1 + \text{AMP} + \text{PP}$$

Synthetase refers to the 20 different aminoacyl tRNA synthetases, each catalyzing the reaction for one specific amino acid; thus, the enzyme for proline, for example, would be prolyl-tRNA synthetase. When the amino acid is attached to its correct tRNA, the tRNA recognizes the correct codon on the mRNA via its anticodon loop (through hydrogen bonding). This process occurs on the ribosome, which coordinates the events of protein synthesis.

Protein synthesis requires the coordinated operation of many enzymes together on the ribosome in combination with aminoacyl-tRNA and mRNA. The ribosome provides the site, the enzymes and the ribosome catalyze the reactions, and the tRNA brings the proper amino acid to the codon specified by the mRNA. GTP provides the energy for the reactions, and mRNA provides the sequential information for translation (Figure 11.26).

Before translation can begin, the mRNA must first bind to the ribosome. In actuality, the mRNA binds directly to a free 30S subunit (in bacteria), which then binds to the initiator tRNA; then the 50S subunit binds to give a 70S ribosome. The assembly of a protein is a continuous process on the ribosome/mRNA complex, but for convenience it is easier to divide translation into discrete steps. We will consider each step of protein synthesis: initiation, elongation, and termination.

In the initiation step of translation, the mRNA provides the coding information that allows the first tRNA with an attached amino acid to bind to the P site on the ribosome. The first tRNA to bind to the ribosome recognizes the first complementary codon on the mRNA, which is nearly always AUG (GUG is also used, but rarely). This first codon specifies the amino acid methionine. It is called the **initiation codon,** and it is recognized by met-$tRNA^{init}$. The anticodon for methionine is the same whether it is at the initiation codon or internal to the mRNA coding sequence. However, met-$tRNA^{init}$ is the only tRNA that can occupy the P site at the initial stages of protein synthesis. The sequence of nucleotides and the odd bases present in the met-$tRNA^{init}$ make it structurally different from the tRNA that adds amino acids after the initiation codon. In addition, but only in prokaryotes, the amino group of the methionine has a formyl group attached to it (which simulates a peptide bond) that is added after the methionine is attached to the tRNA.

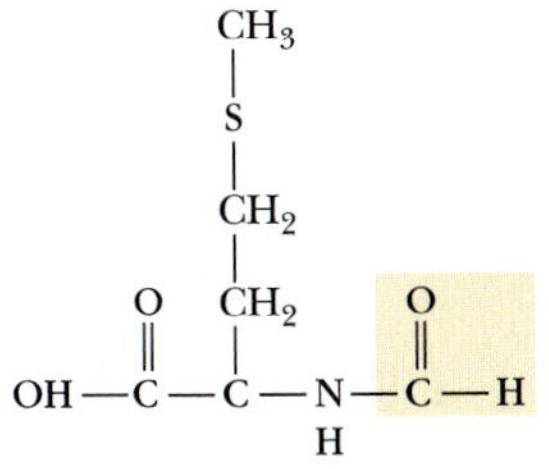

$$
\begin{array}{ccccccc}
 & & CH_3 & & & & \\
 & & | & & & & \\
 & & S & & & & \\
 & & | & & & & \\
 & & CH_2 & & & & \\
 & & | & & & & \\
 & O & CH_2 & & O & & \\
 & \| & | & & \| & & \\
OH- & C- & C- & N- & C- & H \\
 & & & H & & &
\end{array}
$$

The formyl group is removed from the completed protein after synthesis, but sometimes the methionine and the formyl group are removed, making the second amino acid the first amino acid in the completed protein product.

Initiation of translation requires enzymes or **initiation factors** (IF1, IF2, and IF3), which are released from the ribosome immediately after the initiation reaction. Note that the initiation codon is not the first codon on the mRNA molecule, but is internal from the 5′ end by 20–30 or 1000 bases, depending upon the gene. The untranslated region of the mRNA before the start codon is called the **leader region** and, in prokaryotes, contains a ribosome-binding sequence. This region in the mRNA contains a sequence known as the **Shine–Dalgarno sequence** (after its discoverers) and forms hydrogen bonds by base pairing to the 16S rRNA of the 30S ribosomal subunit in combination with the initiator protein IF3 (Figure 11.26a). Eukaryotic mRNAs do not contain a Shine–Dalgarno sequence; instead, the cap at the 5′ end of the mRNA seems to provide a docking site for an initiation factor and cap-binding protein, leading to formation of the initiation complex.

In the elongation phase of protein synthesis, the growing chains always add the incoming amino acid to the C-terminal end of the polypeptide. In the first step of the elongation process, an amino acyl-tRNA enters the A site (or aminoacyl-tRNA site) on the ribosome, and its anticodon hydrogen bonds with the complementary codon (remember that the P site is occupied by a methionyl-$tRNA^{init}$) on the mRNA. In the second step, a covalent peptide bond is formed between the carboxyl group of the amino acid attached to the tRNA in the P site and the amino group of the amino acid attached to the tRNA in the A site (see Figure 11.26b). This reaction is catalyzed by peptidyl transferase, which is not a protein but is part of the 23S ribosomal subunit. The growing peptide is now attached to the tRNA in the A site, and the tRNA in the P site is ejected. As the ribosome shifts downstream on the mRNA, the growing polypeptide attached to the tRNA at the A site is immediately transferred to the P site, whereas an incoming aminoacyl-tRNA is added to the unoccupied A site. This transfer process is catalyzed by a specific elongation enzyme called EF–G. The cycle repeats itself over and over until the information in the message is completely translated into a protein product. Two other enzymes involved in elongation are EF–Tu and EF–Ts, which are involved in bringing the aminoacyl-tRNA into the A site.

As mentioned earlier, three codons of the 64 possible codon combinations do not specify an amino acid and consequently do not have a complementary tRNA. These three codons (UAG, UAA, and UGA) are known as **termination codons** or **stop codons.** For historical reasons, these three triplets have names associated with them. The UAG triplet is known as the amber codon, UAA is the ochre codon, and UGA is the opal codon. When protein synthesis comes across one or more of these three codons, termination of translation occurs immediately. In *E. coli*, two enzymes called release factors RF1 and RF2 (Figure 11.26c) are involved in termination, whereas eukaryotes have only one release factor, eRF.

The information between a start codon and a stop codon is referred to as an **open reading frame** (ORF). As we will see in Chapters 12 and 13, DNA can be sequenced and open reading frames can be identified in the genome by the presence of a start codon and a stop codon. Not all AUG codons and stop codons define the ends of ORFs; only those that are in the proper reading frame, are adjacent to promoters, and have sufficient information for a putative protein can do this job.

Mistakes Can Sometimes Occur During Protein Synthesis.

Clearly, it is important that very few mistakes occur during protein synthesis. Otherwise, the cell would accumulate incorrectly assembled proteins that would presumably be inactive and serve no useful function. There are two opportunities in protein synthesis for the wrong amino acid to be inserted. The first point where mistakes can occur is the attachment of the amino acid to the proper tRNA by the aminoacyl-tRNA synthetase. The aminoacyl-tRNA synthetase must recognize its specific tRNA and attach the correct amino acid to the tRNA. The synthetase recognizes the correct tRNA through a combination of the sequence of nucleotides, the location of unique bases, and the anticodon triplet. Specifically, the sequence of nucleotides in the D stem and the acceptor stem of tRNA are important (Figure 11.25). A single nucleotide change may alter the specificity of the tRNA so that it can be recognized by a different aminoacyl-tRNA

FIGURE 11.26 Translation in prokaryotes. (a) The protein assembly is initiated with the combination of an initiator tRNA attached to an N-formylated methionine; initiation factors 1, 2, and 3; the mRNA; and the small ribosomal subunit. (b) During elongation, the assembly of a protein occurs at two sites on the ribosome, one for the binding of the complementary tRNA (the A site or acceptor site) and the other for formation of the polypeptide bond (the P site or peptidyl site). The mRNA moves from the 5′ end to the 3′ end and provides the sequential information for each incoming tRNA, thus specifying the sequence of the protein. (c) Termination occurs when an mRNA termination triplet is in the A site and is recognized by termination factors. This series of reactions consumes considerable energy and requires the participation of multiple components, but it is essential for the life of the cell.

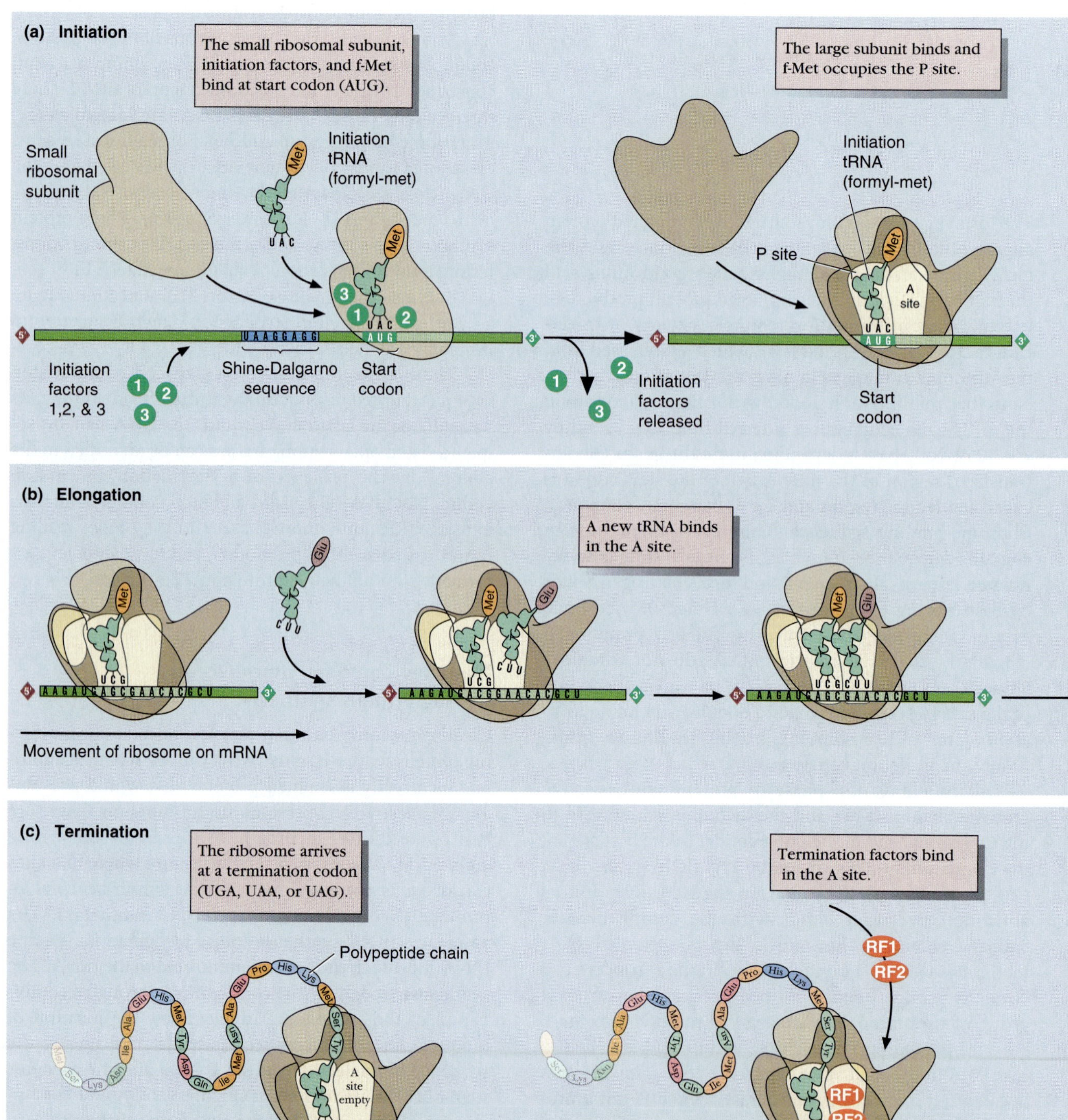

synthetase. Even with the high degree of specificity between an aminoacyl-tRNA synthetase and its tRNA, mistakes do occasionally occur. Fortunately, many of the substitutions are of like-structured amino acids that do not affect the activity of the protein significantly; for example, an isoleucine substituted for a valine will have no significant effect on protein structure. The overall mistake rate in prokaryotes caused by aminoacyl-tRNA synthetases is less than 1 in 10^5 amino-acylations.

The second important point at which errors can occur in protein synthesis is the complementary hydrogen bonding between the codon and the anticodon. Anticodon recognition of a codon is, in fact, the weak point in protein synthesis and is probably responsible for a significant number of mistakes. It is estimated that mistakes occur as frequently as 1 in 100 amino acids because incorrect triplets are recognized in many instances by only two out of the three codon bases. As a consequence, it is hypothesized that ribosomes have a proofreading mechanism that ejects incorrectly formed proteins when the codon–anticodon recognition is incorrect. Evidence for this hypothesis comes from a mutant in *E. coli* defective in the enzyme peptidyl-tRNA hydrolase that hydrolyzes only partially formed, and presumably incorrectly formed, proteins from their attached tRNA. Hydrolysis of the peptidyl-tRNA bond by this enzyme allows the tRNA to be recycled and the partially completed protein to be degraded into amino acids. If a bacteria lacks this enzyme (by a mutation in the gene), it rapidly accumulates partially completed proteins attached to tRNAs and dies. At present, no one knows how the ribosome recognizes the incorrectly paired tRNA for ejection from the synthetic apparatus.

SOLVED PROBLEM

Problem

What is the minimum number of nucleotides in the mRNA coding for the β-chain of hemoglobin, which has 146 amino acids? What other sequences would also be present in this mRNA?

Solution

The code is 3, so $3 \times 146 = 438$. With the termination triplet, the total comes to 441. Although you cannot tell from the way the problem is stated, the initial methionine is cleaved from the globin, and thus the precise answer is 444. An mRNA also has a leader sequence and a trailer sequence that can add several hundred nucleotides.

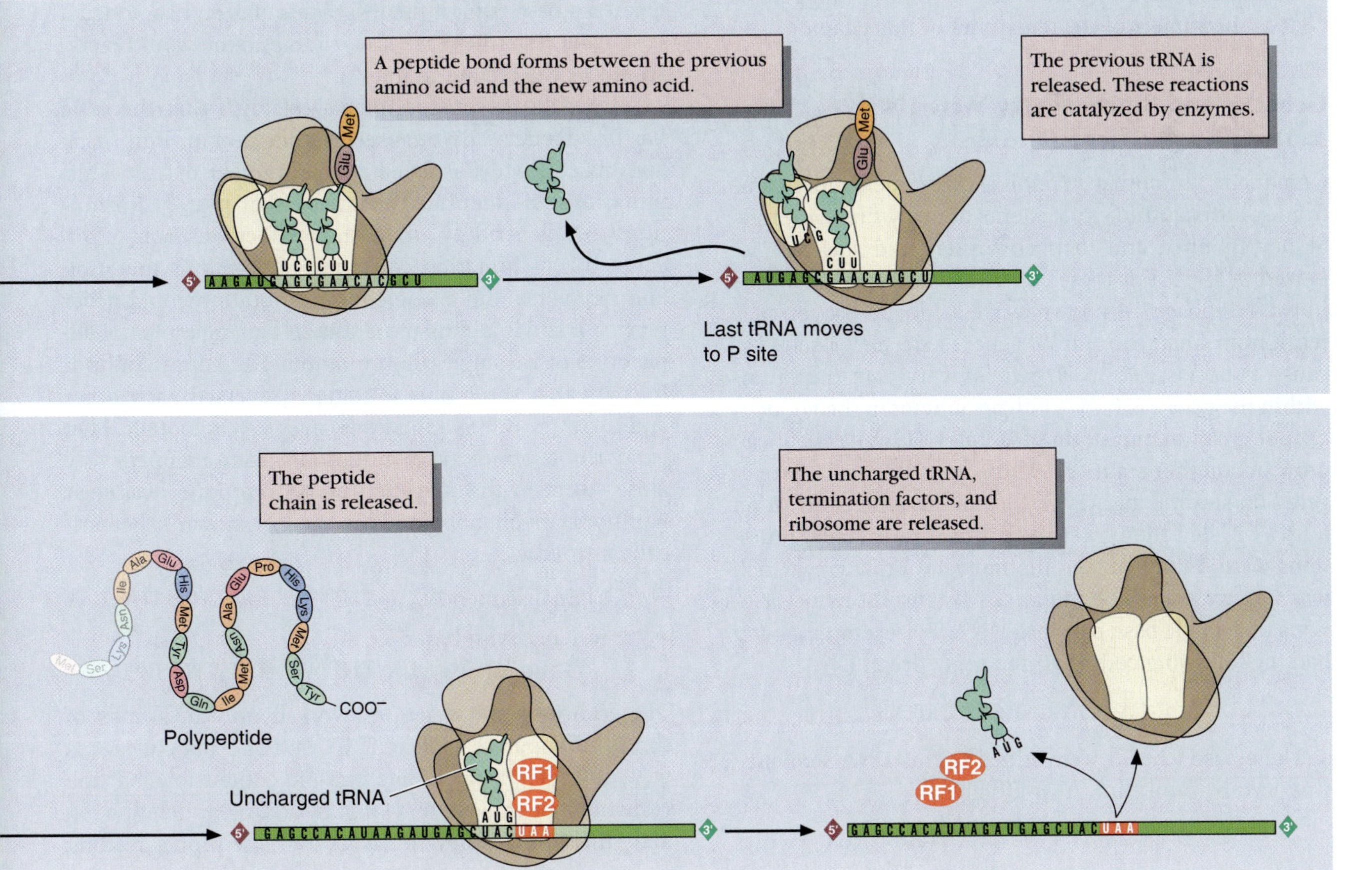

A peptide bond forms between the previous amino acid and the new amino acid.
The previous tRNA is released. These reactions are catalyzed by enzymes.
Last tRNA moves to P site
The peptide chain is released.
The uncharged tRNA, termination factors, and ribosome are released.
Polypeptide
Uncharged tRNA
COO⁻
RF1
RF2

SECTION REVIEW PROBLEMS

11. What is the difference between transcription and translation?
12. How do mRNA, tRNA, and rRNA differ and what is the function of each?
13. What is the function of methionyl-tRNA^{init} in translation?
14. Why is it important that translation have a high level of fidelity?

THE GENETIC CODE IS SIMPLE.

Early in the 1950s, scientists reasoned that each codon is composed of a minimum of three nucleotides to ensure that each amino acid would have at least one codon. However, at that time we did not know whether the code had overlapping adjacent triplets or had punctuation between triplets. Another important question was whether the genetic code was the same in all organisms; that is, was the genetic code universal? The unraveling of the genetic code was one of the more exciting scientific events of the 1960s and revealed fascinating information (for example, how protein synthesis starts and stops). At this time, investigators answered all the questions that had been originally posed about the code in the 1950s. The story of the discovery and understanding of the genetic code is the subject of the remainder of this chapter.

Genetics and Biochemistry Were Used to Decipher the Genetic Code.

Convincing proof that a codon is three nucleotides came from some beautiful experiments done by Francis Crick, Sidney Brenner, and their coworkers and was first reported in 1961. For these experiments, they used a gene in bacteriophage T4 known as the *rII* locus. As you may recall from Chapter 8, the *rII* gene is extremely useful because it is possible to screen for very rare mutations within the gene easily. A T4 phage mutant in the *rII* locus cannot grow on one strain of *E. coli* [K12(λ)], but it can grow on another stain (B). Thus, mutants can be propagated on strain B and tested for rare mutations on strain K12(λ). Using the powerful tool of the *rII* locus, mutations were induced using the mutagen proflavin, which was known to induce mutations by insertions or deletions of a single base pair into DNA. For example, if you had the following coding strand sequence of DNA:

5′-CAT CAT CAT CAT CAT-3′

and one base (the T) were added to this DNA sequence (induced by proflavin), it would give

5′-CAT CAT T*CA TCA TCA-3′

or if one base were deleted (the C) from the original DNA sequence, it would give

5′-CAT CAT ATC ATC ATC-3′

By adding a base pair or deleting a base pair from the DNA, the information downstream (to the 3′ end) is changed to give entirely different codons. This type of mutation is called a **frameshift** mutation and affects all amino acids downstream of the changed codon. Thus a frameshift mutation usually completely inactivates the protein product of the gene. Crick and his colleagues found many proflavin-induced mutants in the *rII* locus and took one of these mutants, called FCO, and looked for a "revertant" or back mutation to wild-type gene function, again induced by proflavin. What they found was very surprising. This second mutation restored the function of the *rII* gene product, but not quite to wild-type levels. In fact, after determining the location of the reversion by genetic mapping, the reversion was found to be caused by the presence of a second mutation within the *rII* gene very close to the first mutation, but not in the same nucleotide that gave the original mutation. This second mutation was separated from the original mutation by genetic means (recombination), and the separated mutation was found to confer an *rII* mutant phenotype. This was interpreted to mean that the second mutation had either added a base pair to the reading frame (if the original mutation had deleted a base pair) or deleted a base pair (if the original mutation had added a base pair). In either case, the reading frame would be restored by the second mutation.

How do these experiments establish that the code has three letters? By themselves, they do not. But, Crick and his colleagues isolated a large number of frameshift mutations and identified them as either plus (+) or minus (−), depending upon their effect on each other. Thus, a plus mutation could reverse a minus mutation and vice versa. But a single minus mutation could never reverse a single minus mutation and a single plus could never reverse a single plus mutation. The important finding was that three plus mutations reversed each other and the three minus mutations reversed each other. That is, if three minus-type mutations (that mapped very close) were in the *rII* gene, the gene product was near wild-type in phenotype. This can be shown schematically as follows:

Original sequence: THE FAT RAT ATE THE BAT
Sequence with bases 3, 7, and 12 deleted: THF ATA TAT THE BAT

The additions and deletions worked only in groups of threes to suppress each other. Crick and his colleagues were then left with the inescapable conclusion that the code must be three letters long. By adding or deleting exactly three bases, they could get back the proper reading frame so that all downstream information was then restored to the proper information. Later, George Streisenger at the University of Oregon did similar experiments with the gene for the enzyme lysozyme and found that the amino acid sequence in the protein changed as Crick's experiments predicted. This finding by Crick and colleagues was a landmark, just one of many in his incredible science career [see HISTORICAL PROFILE: *Francis Crick (1906 –)*].

This information did not tell us what the codons were for each amino acid, but it did tell us that the code is composed of three letters. These experiments also tell us that the code is read from a fixed starting point and continues to the end of the coding sequence. We know this because a single frameshift mutation anywhere in the coding sequence alters the codon alignment for the rest of the sequence. It also tells us that the code uses all of the 61 possible codon triplets. If only 20 triplets are used and the rest are nonsense, then most frameshift mutations would produce nonsense triplets for which there would be no tRNA anticodon, and protein synthesis would stop in the middle of the protein. But, if all triplets specify an amino acid, then a frameshift mutation would cause incorrect amino acids to be inserted downstream.

The actual "cracking" of the genetic code occurred shortly after Crick's experiments with the *rII* locus. The big discovery was made in 1961 by Marshal Nirenberg and Heinrich Mathaei. They mixed in a test tube some cellular extracts from *E. coli*, ATP, ribosomes, aminoacyl-tRNA synthetases, and amino acids (this is known as a cell-free translation system). These are the necessary ingredients for the synthesis of proteins, but lacking the mRNA. To mimic the mRNA, they added poly-uridylic acid. What they found was very exciting: the reaction mixture produced poly-phenylalanine. This information, when combined with Crick's finding that the codon is three letters, could only mean that the codon for phenylalanine is 5′-UUU-3′. These initial experiments led to a series of other experiments where the *in vitro* system was given different synthetic messages. After the obvious use of the synthetic polymers of the four nucleotides was done, the investigators used synthetic messenger RNAs prepared from fixed proportions of two, three, or four nucleotides. For example, if a polymer was prepared from a mixture of 1/8 A and 7/8 U, the proportions of triplets in the polymer can be calculated as follows: the probability of the codon UUU in a synthetic messenger would be $7/8 \times 7/8 \times 7/8 = 348/512$ (or approximately two thirds of the codons would be UUU). Similarly, the codon UAA would be present at the rate of $7/8 \times 1/8 \times 1/8 = 7/512$ of the total codons. Using the predicted ratios of codons in synthetic messenger RNAs and determining the actual frequency of amino acids within the synthesized protein, it was possible to establish the ratio of bases in a triplet that coded for groups of amino acids (for example, two Us and one A code for leucine, tyrosine, and isoleucine). But, the specific sequence of bases for each of these amino acids could not be established with certainty.

The final codon assignment was determined by chemically synthesizing short trinucleotides (for example, UUA) and binding each to the ribosome (in the presence of high magnesium to stabilize the binding). The short sequences acted like a small mRNA, but only the aminoacyl-tRNA specific for that trinucleotide bound to the ribosome-codon complex. Thus, leucyl-tRNA would bind to ribosomes with bound UAA, but no other tRNA would bind. These experiments quickly led to a complete understanding of the genetic code. Marshall Nirenberg, Severo Ochoa, and Gobind Khorana received the Nobel prize for their efforts in establishing the genetic code.

It should be noted that codon assignments are generally written in mRNA language, not DNA language, and are always written from left to right, from the 5′ end to the 3′ end. This is the same direction in which the mRNA is read.

More than One Code Word Determines a Single Amino Acid.

Close examination of the codon assignments reveals that more than one codon is assigned to a single amino acid (Figure 11.20). For example, isoleucine is assigned three codons, and leucine is assigned six codons. A codon that has more than one codon for an amino acid is called a **degenerate code.** If more than one codon exists for a single amino acid, then more than one tRNA exists, one for each codon. Transfer RNAs that code for the same amino acid but that have different anticodons are called **isoaccepting tRNAs.** The exceptions are methionine and tryptophan, which each have a single codon assignment.

Further inspection of the codon assignments shows that in many cases the first two bases are the same. For example, the codon assignments for isoleucine are AUU, AUC, and AUA. Therefore, most of the degeneration occurs in the third base of the codon. Francis Crick proposed an explanation for this observation in 1966 that he referred to as the **wobble hypothesis.** The hypothesis states that the pairing between the third base of the codon and the anticodon of the tRNA can withstand some unusual configurations. In other words, base-pairing is not always true at the third base. For example, according to the wobble hypothesis, a U in the anticodon may pair with an A or G in the third position in the mRNA. Inosine, a modified base that occurs frequently in tRNA, may pair with U, C, or A in the third position of an mRNA codon. Also, the fact that amino acids with similar chemical characteristics (for example, all have basic side groups) are represented by related codons

HISTORICAL PROFILE

Francis Crick (1916–)

Francis Crick is largely known (along with James Watson) for the discovery of the structure of DNA. But, in addition to this, Crick has made several other important contributions during a productive career in science. Among these are his proposal that an "adapter molecule" must exist for each amino acid (these were found by Mahlon Hoagland and shown to be tRNA), his series of experiments demonstrating that the fundamental unit in the genetic code is composed of three nucleotides, his proposal (with Sydney Brenner) that a messenger RNA must exist, and his hypothesis (with Leslie Orgel) for the existence of "selfish" DNA (DNA that replicates just to maintain itself and does not serve any other useful purpose). More recently, he has been in the news for suggesting that the soul is just a wiring pattern in the nervous system.

Francis Crick was born into a middle class family near Northampton, England, in 1916. During his early years, Crick loved science and scientific experiments that ranged from performing home chemistry experiments to collecting wildflowers. He attended University College, London, and obtained a Bachelor of Science degree with second class honors in physics just before World War II. During the war, Crick held a civilian job with the British Admiralty where he worked designing circuits for magnetic and acoustic mines. After the war, he did not know which field to choose for graduate work. Uncertain of his true interests, he decided to follow what he called the "rule of gossip," that is, he kept track of the subjects he felt most inclined to talk about with his colleagues, reasoning these would be where his interest lay. This quickly narrowed the possibilities to "the border between life and nonlife," what we now call molecular biology. After a year or so of working on tissue culture, Crick obtained a position working toward a PhD in the Medical Research Council unit of the Cavendish Laboratories studying the molecular structure of proteins using X-ray diffraction. The lab was headed by Max Perutz (who later also received the Nobel Prize) under the general direction of Sir Lawrence Bragg (the formulator of Bragg's law for X-ray diffraction). At this time, Crick met the brash young American (James Watson) who was determined to discover the structure of DNA.

Both Watson and Crick worked on the structure of DNA in their spare time. Crick's job and thesis project involved studying the structure of proteins, and Watson was tentatively working on myoglobin and the structure of RNA in TMV. Neither Watson nor Crick performed experiments on DNA, concentrating instead on building structural models that theoretically should fit the known data about DNA. As Crick stated, "chance favors the prepared mind," and when they hit upon the correct structure, they immediately recognized it for what it was. Crick has explained why so much attention has been given to the discovery of the structure of DNA when collagen, for example, makes up more of the body: The answer lies in the function of DNA, the carrier of the hereditary information that determines the uniqueness of each species and each individual.

After discovering the structure of DNA, Crick, and others, soon became interested in the nature of the genetic code. Two major questions to be answered concerned the nature of the code itself and the molecular machinery that used the code to generate proteins. Many believed

minimizes the effects of mutations. It increases the probability that a single random nucleotide change will result in no amino acid substitution or in one involving amino acids of similar character. This observation may have evolutionary implications to minimize deleterious mutations.

Amino acids with multiple codon assignments are found more frequently in proteins, and amino acids with only one or two codon assignments are found less frequently in proteins. For example, arginine, leucine, and serine, which each have six codon assignments, occur much more frequently in proteins than methionine and tryptophan, which each have only one codon assignment.

Not All Codons Are Used to the Same Extent.

The fact that many of the amino acids have more than one codon assignment presents the following question: Are each of the alternative codons within a set used with equal frequency, or are some of the codons in a set used more frequently than others? Now that the sequences of many genes are known, it is possible to determine which codons are used more frequently than others in a wide variety of organisms. From these studies, it is apparent that some codons are used repeatedly, whereas others are hardly ever used. Surprisingly, in different species, the preferred codon usage is not the same for the same amino acids (Table 11.3). For example, even though leucine has

FIGURE 11.A Francis Crick (1956), displaying his double-helix tie. *(Courtesy of Francis DiGennaro & Sons, Baltimore, MD; from Francis Crick,* What Mad Pursuit, *Basic Books, Inc., New York, NY, 1988).*

that a combination of three nucleotides must code for each amino acid, but a key question was whether the code was **nonoverlapping,** with each code word adjacent to the next, or **overlapping,** with each nucleotide being part of more than one code word. Crick performed a series of experiments involving the generation of sets of one, two, or three mutations in a single gene, which demonstrated that the code was a three-nucleotide code. In 1956, Crick wrote a paper suggesting the existence of an "adapter molecule" that could physically interact with DNA to read the code, as amino acids are incapable of directly reading DNA base sequences. This paper was never published but was widely circulated. Shortly thereafter, the existence and function of tRNA was discovered. The existence had been proposed as a "messenger RNA," which would serve as an intermediate between DNA and the cytoplasmic site of protein synthesis. Crick, and others, originally believed that rRNA must be the messenger because it was so abundant in cells. In 1960 Crick, Sydney Brenner, and others realized that this was not correct after hearing of the experiments of François Jacob, Arthur Pardee, and Jacques Monod on the synthesis of β-galactosidase (an enzyme that degrades lactose) in *E. coli.* This group had discovered that β-galactosidase is produced within minutes after the bacterial cell encounters lactose. They rightly concluded from this finding that the messenger must be unstable, rapidly being made and degraded. Because rRNA is very stable, it was probably not the intermediate. A class of unstable RNA molecules discovered by Lazarus Astrachan and Elliot Volkin fit this description, and it was not long before these fleeting molecules were shown to be mRNA.

In 1976, Crick moved to the Salk Institute in La Jolla, California. After moving to California, Crick collaborated with Leslie Orgel to write a stimulating paper on selfish DNA. In this paper, they suggested that much of the DNA that does not code for proteins (which can be as much as 90% of the total DNA in some organisms) originates as DNA parasites. These parasites move from place to place in the chromosomes, leaving replicas of themselves embedded in the host DNA. This concept has been widely discussed, but the issue remains unresolved. Crick's current interests are in neurobiology, in particular with neural networks and how such networks in the brain can generate conscious thought.

six anticodon assignments, *E. coli* uses CUG over 85% of the time, whereas yeast uses UUG over 85% of the time. As expected, preferential codon usage is correlated with the abundance of the respective tRNA species. The more frequently a codon is used, the more that tRNA species is present in the cell.

Additionally, within an organism, some codons are used more frequently for certain types of genes. For example, genes that are expressed at high levels (more abundant gene products) preferentially use one codon in a set (for example, only one of the six codon assignments for leucine). Genes for less abundant proteins use a larger set of codons and show less preference toward the set encoding the more abundantly synthesized proteins. The codon usage shown in Table 11.3 is representative of the more abundant proteins. The reason for this difference in codon usage is not clearly understood.

Another observation concerning codon usage is that codons with the forms NCG and NUA (where N is any nucleotide) are avoided. Although the reason these codons are avoided is not known, it is speculated that it may be related to the frequent methylation of cytosine in CG dinucleotides and the tendency for methyl-CG to mutate by deamination to TG. However, this hypothesis would apply only to eukaryotes because DNA methylation is minimal in prokaryotes.

Although codon usage may seem like an esoteric subject, it may have some practical applications when it

TABLE 11.3 *Comparison of Preferential Codon Usage for Some Amino Acids*

Codon for	*E. coli*	Yeast	*Euglena chloroplast*
Leucine (6)*	CUG	UUG	UUA/UUG/CUU
Arginine (6)	CGC	AGA	CGU/CGC/AGA
Serine (6)	UCU/UCC/AGC	UCU/UCC	UCU/UCA/AGU
Proline (4)	CCG	CCA	CCU/CCA
Tyrosine (4)	UAC	UAC	UAU/UAC
Lysine (2)	AAA	AAG	AAA

*The number in parentheses represents the number of codons for that amino acid. Preferred codons are those used more than 85% of the time. For example, UCU, UCC, and AGC are used more than 85% of the time for serine in *E. coli*.

comes to transferring genes between different organisms and species. It is now a common procedure to insert genes from eukaryotes into *E. coli* to facilitate the production of a specific gene product. For example, the insulin gene from humans has been put into fast-growing *E. coli* to produce human insulin cheaply and in large quantities. If certain tRNA species are in limited supply in the *E. coli* cell, then a highly used codon is inadequately read, consequently limiting production of the cloned gene's product.

There Are a Few Exceptions to the Nearly Universal Genetic Code.

The genetic code shown in Figure 11.20 is nearly universal among all living organisms, but there are a few exceptions. This fact was discovered when the entire nucleotide sequence of the human mitochondrial genome of 16,569 bp was determined. Mitochondria are the small organelles present in all eukaryotes and are the sites for energy transformation and ATP production. A mitochondrion is about the same size as an *E. coli* cell, and usually a eukaryotic cell has several hundred mitochondria per cell. Each mitochondrion has its own DNA separate from the nuclear DNA, and replication occurs independently of nuclear DNA. Also, the mitochondrion has its own ribosomes and tRNAs, which are very similar to those found in bacteria. In the human mitochondrial genome, there is enough information to code for 15 different genes (including two rRNAs) and 22 tRNA genes. However, most of the mitochondrial enzymes are in fact coded for by nuclear DNA, not mitochondrial DNA, and they are transported into the mitochondria from the cytoplasm.

After careful analysis of mitochondrial protein products and the DNA sequences of some mitochondrial genes, it was found that in mitochondria:

1. UGA is not a stop signal but codes for tryptophan.
2. Internal methionine is coded by both AUG and AUA, and initiation codons for methionine are specified by AUG, AUA, AUU, and AUC.
3. AGA and AGG are not arginine codons but specify chain termination. Thus, there are four stop codons: UAA, UAG, AGA, and AGG.

We now know that other exceptions to the universal genetic code exist. For example, the prokaryote *Mycoplasma capricolum* also substitutes the termination signal UGA for tryptophan. In fact, it is the predominant tryptophan codon. Also, ciliates (unicellular eukaryotes) read UAA and UAG as glutamine instead of termination signals. Exceptions to the universal genetic code also occur in the mitochondria from several species (yeast, *Drosophila*, and mammals). Why is this so? It could be that the mitochondrial codon assignments are simplified and represent a more primitive pattern or that they may have evolved as a result of some special selective pressure. Another reason is RNA editing, which is covered in Chapter 14. In this process, the codon present in the initial transcript is modified or changed, and consequently the codons are changed to the universally used codons.

SOLVED PROBLEM

Problem

If you had a synthetic message composed of 1/4 A and 3/4 G present in a cell-free *in vitro* system with all the necessary ingredients to synthesize proteins, what amino acids would you expect to find in this protein and in what proportions?

Solution

The proportion of a single trinucleotide sequence (or codon) can be determined by multiplying by the individual frequencies of the nucleotide present in the mixture, thus GGG is present $(3/4)^3$ or 27/64 of the total. GGA, GAG, and AGG are all present at $3/4 \times 3/4 \times 1/4 = 9/64$ each, and AAG, AGA, and GAA are present at $1/4 \times 1/4 \times 3/4 = 3/64$ each. AAA is present $(1/4)^3$ of the total or 1/64. Because GGG and GGA both code for glycine, you must add 27/64 + 9/64 (their respective codon frequencies) to get the total of 36/64, which is then the proportion of glycines found in the final protein. This same logic is used with the remaining amino acids for arginine (AGG and AGA), glutamic acid (GAA and GAG), and lysine (AAA and AAG).

SECTION REVIEW PROBLEMS

15. Using the genetic code, predict the amino acid sequence of the protein synthesized from the following nucleotide sequence:

    AUG CCG GGA AAC ACA GCA GGG AAU AGC UAG

    What would the amino acid sequence be if a G were placed between nucleotides 8 and 9?
16. How many codons would exist in the genetic code if a codon was four nucleotides long?

One DNA Sequence Can Determine the Sequence of More than One Protein.

Any set of bases can simultaneously code for six different proteins. For example, in the following sequence, you could start coding from the first G of the top strand, the second G of the top strand, or the first C of the top strand. Similarly, the bottom strand could also be started in any of three frames:

5′-GGCATTTGCAACT-3′
3′-CCGTAAACGTTGA-5′

Both strands of the DNA can code for a mRNA with the reading frame offset by one base on each strand, yielding six different protein sequences. All the protein products of this DNA sequence would have different sequences. On the other hand, a one-base-change mutation would result in all six proteins being changed. Evolutionary selection against such a situation should be high because it is more favorable for a single base change to result in a change in a single protein.

In 1977 Fred Sanger and his colleagues determined the sequence of the small single-stranded DNA bacteriophage ϕX174, which has 5386 bases. Much to their surprise, they found that a number of genes within the phage overlapped each other (Figure 11.27). The ϕX174 genome codes for 11 proteins, and it is possible to calculate the number of bases required to code for the proteins. They found that the coding information required for the 11 proteins was more than the phage DNA possessed. Thus, the overlapping genes explained how the

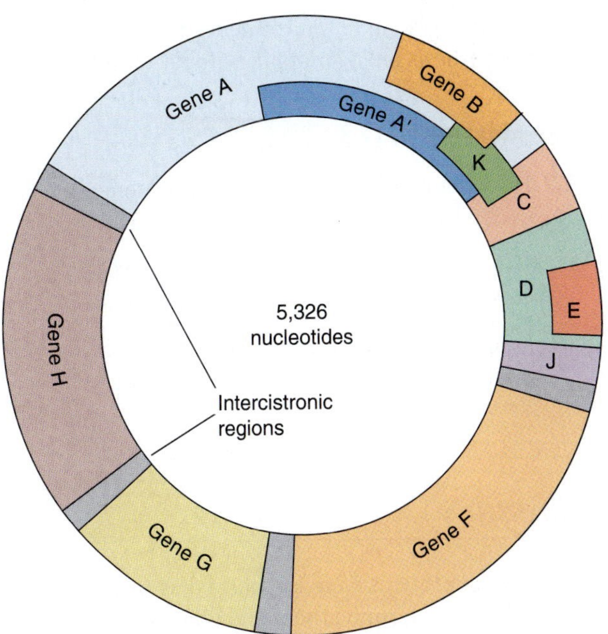

FIGURE 11.27 The genetic map of phage ϕX174. The bacteriophage has 11 different genes. Considerable overlap occurs in genes A, A′, B, K, C, D, and E. Overlapping of genes allows more genetic information to be packed in a small sequence. The distances are only approximate. *(Based on data from Sanger, F., G. M. Air, G. G. Garrell, N. L. Brown, A. R. Coulson, J. C. Fiddes, C. A. Hutchison III, P. M. Slocombe, and M. Smith. 1977.* Nature, *265: 687, and Tessman, E. S., I. Tessman, and T. J. Pollack. 1980.* Journal of Virology, *33: 557.)*

"excess" amino acids could be coded. The phage accomplished this feat by using overlapping sequences to code for different gene products in different reading frames, and in the same frame. For example, in genes D and E, the reading frame is shifted one base. In addition to extensive overlapping protein-coding regions, other types of signals were found within the genes. For example, promoters are present within the A and C protein-coding regions, as well as an origin of DNA replication within the A gene.

Overlapping genes are relatively rare, so why would an organism have overlapping genes? It is reasoned that organisms that have a very limited amount of genetic material have a distinct advantage if they are able to produce more gene products than just one sequence–one product. They expend less energy and time in replication, which gives them this advantage. Using this logic, it would follow that higher organisms that have large amounts of DNA would not have overlapping genes. In fact, this is nearly always true. Only a few examples of overlapping genes have been discovered in higher organisms having large amounts of DNA and in human mitochondria. Most overlapping genes occur in bacteria and involve very minor changes, usually only one or two base pairs. For example, in *E. coli* the genes coding for tryptophan biosynthesis are overlapping (Table 11.4).

SECTION REVIEW PROBLEMS

17. How many different amino acid sequences can be coded from the following DNA sequence?

    5′-TAGGGATTTCGCAATCGCAAATTTTTTCG-3′
    3′-ATCCCTAAAGCGTTAGCGTTTAAAAAAGC-5′

18. What are the disadvantages of having overlapping coding sequences? What are the advantages?

TABLE 11.4 *Two Overlapping Gene Sequences in* E. coli

Adjacent Genes	Overlapping Sequences
*trp*E–*trp*D	CAG GAG ACT TTC **TG A*** TG GCT stop start
*trp*B–*trp*A	CGA GGG GAA ATC **TG A*** TG GAA stop start

*Indicates overlap. One adenine is serving in two codons (TGA and ATG).

Summary

In the 1940s, proteins were established as the molecules in the cell that determine the phenotype of an organism, and the experiments of Beadle and Tatum suggested that one gene determines each polypeptide. For a gene to make a protein, the information present in the gene must be converted into mRNA. Usually one kind of messenger is made for each gene. The synthesis of RNA from DNA in prokaryotes is accomplished by the enzyme RNA polymerase, which is a large multimeric protein. When combined with the sigma factor, RNA polymerase can initiate RNA synthesis at specific sequences, called promoters, adjacent to each gene. Promoter sequences vary, and some sequences are recognized more frequently by RNA polymerase than others, resulting in some genes being transcribed more frequently than other genes. At the end of each gene is a terminator sequence the RNA polymerase recognizes, stopping RNA synthesis.

In contrast to prokaryotes, several different RNA polymerases are present in eukaryotes. In eukaryotes, RNA is synthesized as pre-mRNA, which contains noncoding sequences known as introns interspersed among coding sequences called exons. The intron sequences are removed by a splicing reaction giving rise to mRNA. The 5′ end of the mRNA is capped with the addition of a 7-methylguanosine, and the 3′ end is modified by adding multiple adenosines. The level of transcription in eukaryotes is controlled by promoters, but enhancer elements located upstream or downstream of the regulated gene also have a regulatory role.

The genetic code is comprised of codons, each three nucleotides long, giving 64 different combinations of the four nucleotides. One of the codons (AUG) is a start signal that ensures that the gene is read in the correct frame. Three other codons are stop signals (UAA, UGA, and UAG) and do not have amino acid assignments. Because 61 codons have amino acid assignments, some of the 20 amino acids have more than one codon (for example, leucine has six codons). The genetic code is almost universal, with a few exceptions found in ciliates, prokaryotes, and mitochondria. For example, UGA is not a stop signal but codes for tryptophan in human mitochondrial DNA. Where multiple codons exist for a single amino acid, not all codons are used to the same extent. For example, leucine, which has six codons, uses CUG over 85% of the time in *E. coli*.

The coded genetic information in DNA is transcribed into mRNA, which is then converted into a protein by a process known as translation, using ribosomes and tRNAs. Translation requires the assistance of tRNAs as adapter molecules to recognize the codons on the mRNAs. Transfer RNAs are small RNA molecules made by tRNA genes. One tRNA may exist for each codon except for the termination triplets. An amino acid specific for a certain tRNA is attached to it by aminoacyl-tRNA synthetases. Twenty different enzymes exist, one for each amino acid. After the amino acid is attached to a specific tRNA possessing the correct anticodon, the AA-tRNA anticodon binds to the corresponding codon of the mRNA, which is itself bound to the ribosome. The ribosome acts as a coordinating site for protein synthesis and has two sites of tRNA attachment, the A site and the P site. The growing polypeptide chain shuttles back and forth between these two sites as new, incoming aminoacyl-tRNA and new codons are brought into juxtaposition. After a protein is complete, the growing chain is terminated by the involvement of several enzymes and the termination triplets, which do not have a cognate tRNA.

Selected Key Terms

anticodon loop
codon
enhancer sequence
exon
heterogeneous nuclear RNA
holoenzyme
intron
leader region
messenger RNA (mRNA)
promoter
ribosome
ribosomal RNA (rRNA)
RNA splicing
sigma factor
transcription
transfer RNA (tRNA)
translation
transcription factors
wobble hypothesis

Chapter Review Problems

1. What would be the effect on reading frame and gene function under the following conditions?
   a. Three bases were inserted together in the middle of the gene.
   b. Three bases were deleted from the gene.
   c. One base was inserted and another one deleted five bases downstream of the insertion.
2. Explain the main differences between the organization and expression of prokaryotic genes and eukaryotic genes.
3. Why is it not correct to say "one gene makes one enzyme"?
4. How many isoaccepting tRNAs (tRNAs that accept the same amino acid) would you predict exist for the amino acid valine?
5. You have prepared an artificial messenger RNA to be used in determining the meaning of the individual codons. This mRNA will be used in a cell-free translation system. The synthetic mixture used to prepare the mRNA had 1/3 A and 2/3 G and was polymerized by an enzyme. What would the proportion of the triplets AGG and AGA be in the artificial mRNA?
6. Correspondingly, what proportion of the protein synthesized in a cell-free system would be arginine, using the artificial mRNA prepared for Problem 5?
7. What is the anticodon for tryptophan?
8. You know the amino acid sequence of a protein coded for by a gene in *E. coli* and the very end of the sequence is

   Pro–Try–Ser–Glu

   You find a mutant of this gene and the preceding sequence of the protein has changed to

   Pro–Gly–Val–Lys–Met–Arg–Val

   Explain what has happened. What is your prediction as to the effect on protein function?
9. What is the function of the terminator sequence with respect to a gene and RNA polymerase? Where is the terminator sequence located with respect to the coding information in the gene?
10. What is the definition of a gene in physical terms?
11. How can you easily purify mRNA from a eukaryotic cell and avoid contamination by other RNAs such as rRNA and tRNA?
12. You have cloned the DNA of the same gene from two different yeast cell lines. In one of the cell lines, the gene has a deleted segment, and the other is wild-type. How can you physically define exactly where the deleted segment is within the mutant gene?
13. Two aggregate complex molecular structures have S values (Svedbergs) of 1150S and 780S. During centrifugation in a sucrose density gradient, which will reach the bottom of the tube first if both are layered together on the top of the tube before centrifugation? Explain.
14. In *E. coli*, 80% of the RNA present in the cell at any one instant is rRNA, not mRNA, even though only 7 genes produce rRNA and about 2500 genes synthesize mRNA. Why?
15. How does aminoacyl-tRNA synthetase^{leu} recognize only the tRNA^{leu} family (of which there are six) to specifically attach leucine to the proper tRNA?
16. If the methionine attached to tRNA^{met} were chemically modified to give alanyl-tRNA^{met} and was used in a cell-free system to synthesize proteins, what result would you predict? Will the alanine be placed in the protein where a methionine should be or where an alanine should be?
17. Crick and coworkers isolated many proflavin-induced mutants of the *rII* locus in phage T4. Using genetic means, they classified the mutants as either plus (+) or minus (−). When three plus mutations or three minus mutations were together in the same gene, they were found to yield a wild-type phenotype, but not when one or two plus or minus mutations were combined. What can you conclude from these experiments since you know that proflavin induces only base additions or deletions?
18. Frameshift mutations frequently cause the translated protein to terminate downstream of the mutation. How do you explain this phenomenon?
19. Using a cell-free system to synthesize proteins, what product would you get if you added poly(C) as an mRNA?
20. In translation of mRNAs, the start codon
    a. indicates only an amino acid.
    b. indicates an amino acid and the start of translation.
    c. indicates an amino acid and the start of translation and sets the reading frame.
    d. indicates the start of translation and sets the reading frame.
    e. indicates only the start of translation.

21. If an mRNA were composed of poly U and C mixed randomly, how many possible codons (of three letters) could these two bases specify?
a. 8 b. 9 c. 10 d. 6 e. 12

22. Using the table of code words, what is the amino acid sequence of the polypeptide produced from the following mRNA strand:

5′-UAGAUGUCCAUUGUGGGGUAACAU-3′

a. Arg–Cys–Pro–Leu–Trp–Gly
b. Met–Ser–Ile–Val–Gly
c. Ser–Ile–Val–Gly
d. Met–Gly–Val–Ile–Ser
e. Gly–Trp–Leu–Pro–Cys–Arg

23. An α-helix is an example of what type of protein structure?
a. primary b. secondary c. tertiary d. quaternary

24. In prokaryotes, translation is stopped when
a. the ρ protein reaches the ribosome.
b. a stop codon pairs with a stop tRNA.
c. termination or release factors recognize a stop codon.
d. fMet binds to the AUG codon.
e. the sigma protein reaches the stop codon.

25. In a dipeptide, which bond is the peptide bond?
a. 1 (between the left end amino group and the first carbon)
b. 2 (between the second carbon atom and the carbonyl group)
c. 3 (between the carbonyl group of one and the amino group of a second)
d. 4 (between the second amino group and the third carbon)
e. 5 (between the third and fourth carbon atoms)

26. An mRNA of the sequence AUGUUUUUUUUUUUU produces the following polypeptide in a cell-free, protein-synthesizing reaction: fMet–Phe–Phe–Phe–Phe–Phe. If an antibiotic (minonycin) is added to the reaction, however, the protein Met–Phe–Phe–Phe–Phe attached to tRNA is found. Which step in translation does minonycin block?
a. the formation of the initiation complex
b. peptide bond formation
c. translocation
d. amino-acetylation
e. termination

27. Which of the following is involved in the joining of a tRNA to its correct amino acid?
a. the ribosome
b. the anticodon loop
c. the three-dimensional structure of the specific tRNA
d. aminoacyl-tRNA synthetases
e. b, c, and d

28. In the wobble hypothesis put forth by F. Crick,
a. the first base of a codon is inconsequential.
b. the second base of a codon can be any of the four nucleotides.
c. the first two bases are usually the same for a specific amino acid and the third base is varied.
d. the last two bases are usually the same for a specific amino acid and the first base is varied.
e. the tRNA wobbles.

29. During the elongation of a polypeptide,
a. a free amino acid-tRNA enters the P site on the ribosome.
b. the aminoacyl-tRNA synthetase attaches the amino acid to the tRNA.
c. the EF–Tu catalyzes the movement of the ribosome along the mRNA.
d. the termination tRNA ejects the finished polypeptide and disassembles the ribosome, tRNA, and protein.
e. a free amino acid-tRNA enters the A site on the ribosome.

30. The following DNA sequence contains a gene. The mRNA is made from this sequence of nucleotides and then translated into a small protein of five amino acids. Which strand is the template strand?

strand X: 5′-CCGACTATGCCCTTTAAACGATAACCTATG-3′
strand Y: 3′-GGCTGATACGGGAAATTTGCTATTGGATAC-5′

a. strand X
b. both strand X and strand Y
c. strand Y
d. neither strand

Challenge Problems

1. During the replication of DNA in one cell of a horned toad, the DNA within an rRNA gene is not correctly replicated. Because of the presence of a long stretch of 5′-AAAAAAA-3′ within the gene, a portion of this sequence is looped out during the replication cycle, and 14 bp are deleted from a critical portion of the gene, rendering its product nonfunctional. What effect will this have on the cell? Explain.

2. If you had a synthetic message composed of 1/3 A and 2/3 C present in a cell-free *in vitro* system with all the necessary ingredients to synthesize proteins, what amino acids would you expect to find in this protein and in what proportions?

3. All cells possess enzymes known as DNases, which degrade DNA. DNases will not degrade the DNA into mono- or dinucleotides when protein is bound to the DNA, however. Suppose that you have a cloned copy of a gene that has an upstream sequence that binds to a transcription factor. Using this cloned sequence, agarose gel electrophoresis, and this transcription factor, speculate how you would identify the specific sequence that binds to the protein.

Suggested Readings

Chambron, P. 1981. "Split genes." *Scientific American,* 244 (May): 60–71.

Crick, F. H. C. 1966. "Codon–Anticodon pairing: The wobble hypothesis." *Journal of Molecular Biology,* 19: 548–555.

Lake, J. 1981. "The ribosomes." *Scientific American,* 245 (August): 84–97.

Lewin, B. 1997. *Genes VI,* 6th ed. Oxford University Press, New York.

Miller, O. L., Jr. 1973. "The visualization of genes in action." *Scientific American,* 228 (March): 34–42.

Nirenberg, M. 1963. "The genetic code." *Scientific American,* 208 (March): 80–95.

Schimmel, P. 1987. "Aminoacyl tRNA synthetases: General scheme of structure–function relationships in polypeptides and recognition of tRNAs." *Annual Review of Biochemistry,* 56: 125–158.

Singer, M., and P. Berg. 1991. *Genes and Genomes.* University Science Books, Mill Valley, CA.

On the Web

Visit our Web site at **http://www.saunderscollege.com/lifesci/titles.html** and click on A/G/M Genetics for links to the following chapter-related resources on the World Wide Web:

1. **Protein sequences.** This site is an extremely large database of known protein sequences.
2. **Molecular biology.** This is the home page to launch molecular biology-related search and analysis services by function, and a single point of entry for related searches.

CHAPTER

12

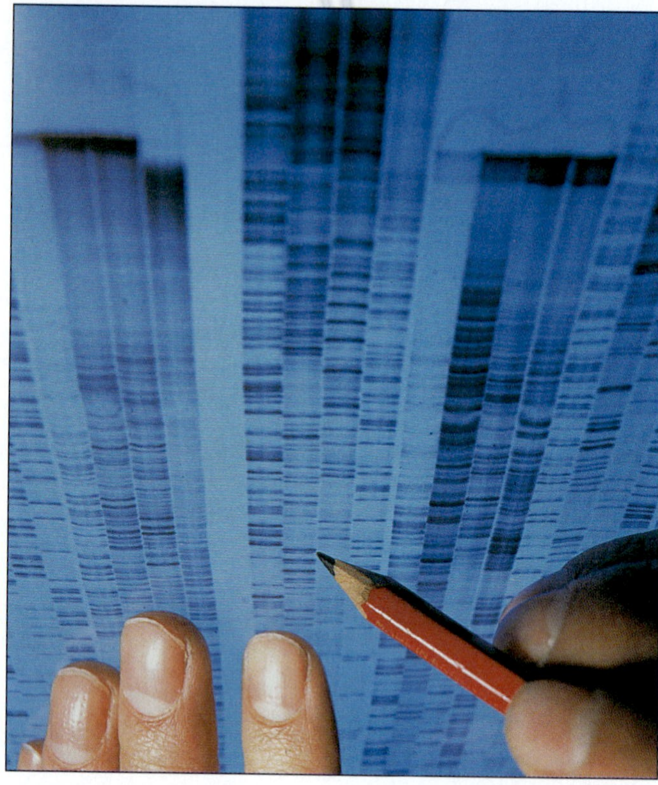

DNA Sequencing. DNA fragments are separated using electrophoresis. The fragments are labeled with radioactivity and then exposed to X-ray film, making an autoradiograph (shown here). Each lane of strips shows the nucleotide sequence of the original DNA. *(Sinclair Stammers/Science Photo Library/Photo Researchers, Inc.)*

Gene Cloning and Analysis

Recombination analysis has the tremendous ability to define distances between genes. This ability has allowed us to understand gene structure as well as the many characteristics of genomic DNA. To a large extent, these conclusions were drawn from just a few organisms (for example, *Drosophila, Escherichia coli,* and bacterial viruses). Because the genomes of these organisms are so small, very rare events could be identified. In the past, this has not been possible with humans, large animals, and most plants. Now, with the advent of recombinant DNA technology, this situation has changed, and we can analyze the genes of very complex organisms. Before we reach the end of this chapter, we will have discussed four of the techniques and applications of gene cloning that have enabled investigators to analyze the detailed structure of genes and genomes in complex organisms. Specifically, we will have addressed the following four questions:

1. What are the steps involved in cloning DNA and identifying a specific gene?
2. How do you replicate and maintain a cloned DNA sequence?
3. How is a cloned DNA fragment characterized?
4. How do you study gene expression with a cloned gene?

Standardized Techniques Are Used for Cloning Genes.

Before the 1970s, investigators working with DNA had to be content with heterogeneous populations of DNA fragments. This, of course, prevented the study of individual DNA sequences unless they had been chemically synthesized. Remember that most genes exist in only two copies in diploid organisms and one copy in bacteria. Thus, trying to isolate a copy of a gene was a nearly insurmountable task, and even if a copy was isolated, such small quantities were available that little could be learned about it. The ability to clone DNA has changed the face of genetics and molecular biology tremendously. A **clone** is a collection of identical DNA fragments. The cloning of DNA, thus, results in a population of identical DNA sequences that allow it to be studied in detail. In the late 1950s and the early 1960s, several very prominent molecular biologist/geneticists declared molecular biology "dead," or nearly dead; they left the field for other promising frontiers—especially neurobiology. What they could not have foreseen was the astounding series of events that unfolded following the discovery of the techniques for gene cloning.

Recombinant DNA technology or the manipulation of cloned DNA fragments, as we know it today, is dependent upon several significant findings that occurred between 1950 and 1975. They included

- an understanding of the structure of DNA;
- the finding that the genetic code is universal, or nearly so, in all organisms;
- the discovery of plasmids, how they replicate, and how to put plasmid DNA back into bacteria using transformation, especially with *E. coli;*
- the discovery and understanding of a group of enzymes called restriction endonucleases; and
- the ability to join together two DNA molecules from completely different sources.

Other later discoveries that enhanced and broadened our ability to clone DNA were the ability to

- purify large and small quantities of DNA easily;
- separate pieces of DNA from each other based on size; and
- detect very small quantities of DNA.

The technology of gene cloning has expanded our understanding of how the cell and the organism work to a depth that was unimaginable only a few years ago. For example, biology has several "holy grails" such as understanding cancer, how the brain works, and how organisms express genes during development. All these processes are composed of complex events that produce specialized products. All the events in these processes are governed by genes, whose precise patterns of expression are almost impossible to understand unless they can be examined individually. Gene cloning enables precisely this type of examination (for example, the time and cellular location of expression of a specific gene during development). We are a long way from a complete understanding of these processes, but at least gene cloning provides a clear path and a mechanism to achieve the stated goal.

Collectively, technology and new knowledge in molecular genetics has enabled scientists to uncover considerable information about how the cell functions in both simple and complex organisms. In the following sections, we will discuss this technology and some of the findings that are a direct result of these new discoveries of recombinant DNA analysis.

DNA Can Be Cut into Sized Fragments Using Restriction Endonucleases.

Endonucleases are a large family of enzymes that hydrolyze the phosphodiester bonds of DNA. **Restriction endonucleases** are a subgroup of the endonuclease family that recognizes specific nucleotide sequences within DNA and cleaves both strands. As a result, DNA can be cut into differently sized fragments, and any one group of fragments is a collection of identical sequences. This is the starting point for most recombinant DNA experiments (see Figure 12.1); thus the discovery of restriction

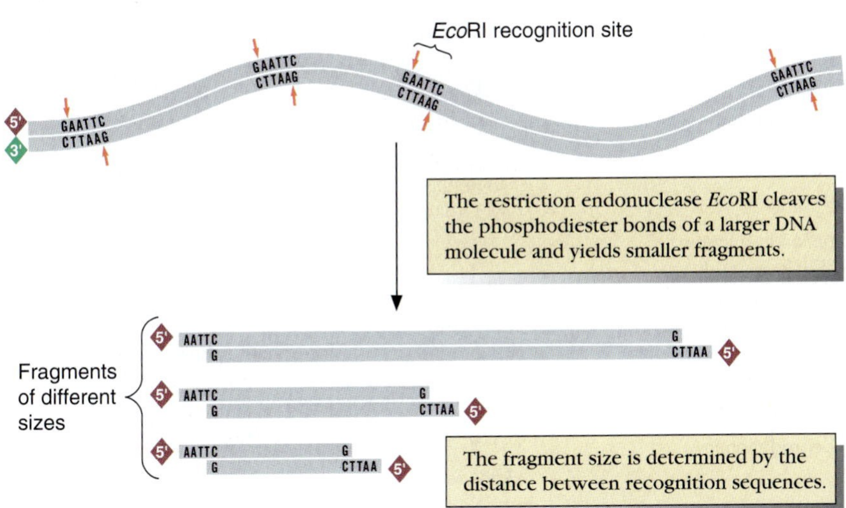
*Eco*RI recognition site
5'
3'
GAATTC
CTTAAG
The restriction endonuclease *Eco*RI cleaves the phosphodiester bonds of a larger DNA molecule and yields smaller fragments.
Fragments of different sizes
AATTC
G
CTTAA
The fragment size is determined by the distance between recognition sequences.

FIGURE 12.1 Digestion of DNA with a restriction endonuclease. The fragment size is determined by the how far apart the recognition sequences are; for example, the *Eco*RI recognition sites are on the average of 4096 nucleotides apart, but in reality they can be spaced anywhere from just a few nucleotides to thousands of nucleotides.

endonucleases was a turning point in the ability to clone DNA.

Restriction endonucleases were discovered during studies of an old but puzzling phenomenon: bacteriophage host restriction (see *HISTORICAL PROFILE: The Discovery of Restriction Endonucleases*). Host restriction refers to the inability of bacteriophages grown on one strain of bacteria to infect a different strain, even if that strain is the normal host for that species of bacteriophage. To illustrate, consider two populations of the phage P grown on two different *E. coli* host strains, strain B and strain K12 (Figure 12.2). The P phage replicates without any problems in both B strain and K12 cells. But, when P phage grown in strain B cells tries to infect strain K12 cells, the DNA of most P phage infections is rapidly degraded (after injection into the strain K12 cells). Only a few of the P phage produce progeny. However, if these few progeny phage reinfect more strain K12 cells, they can now replicate without any difficulty. Similarly, if the phage grown in strain K12 tries to infect strain B cells, they again replicate poorly, and in most cases the DNA is immediately degraded. Replication in the strain K12 cells has somehow "conditioned" or modified the phage so that they replicate well only in strain K12 cells. The fact that phage transferred between two host strains repeatedly show host restriction suggests that host restriction does not involve changes in the DNA sequence of the phage but rather that the DNA is modified in some manner.

In the late 1960s, Stewart Linn and Werner Arber, working in Geneva, found that host restriction in *E. coli* involved two enzymes: one that methylated DNA (called DNA methylases) and one that cleaved unmethylated DNA, now known as a restriction endonuclease. The methylated form of DNA is not digested by the cell's restriction endonuclease; methyl groups are added to specific bases and protect the DNA. The bacterial cell's own DNA and any foreign DNA in the cell become methylated, protecting the DNA from restriction endonuclease digestion. Newly injected foreign DNA, such as the phage DNA, is digested unless the methylase adds methyl groups to particular bases. These data explain why P phage could grow on strain B or strain K12, but not on both without losing viability. The P phage was methylated at specific sequences in *E. coli* strain B, and at other sequences in strain K12, protecting its DNA from restriction endonuclease digestion for that particular strain. When the P phage was subsequently grown in another strain, its DNA was subject to digestion by another restriction endonuclease at unprotected sites on the DNA (Figure 12.3).

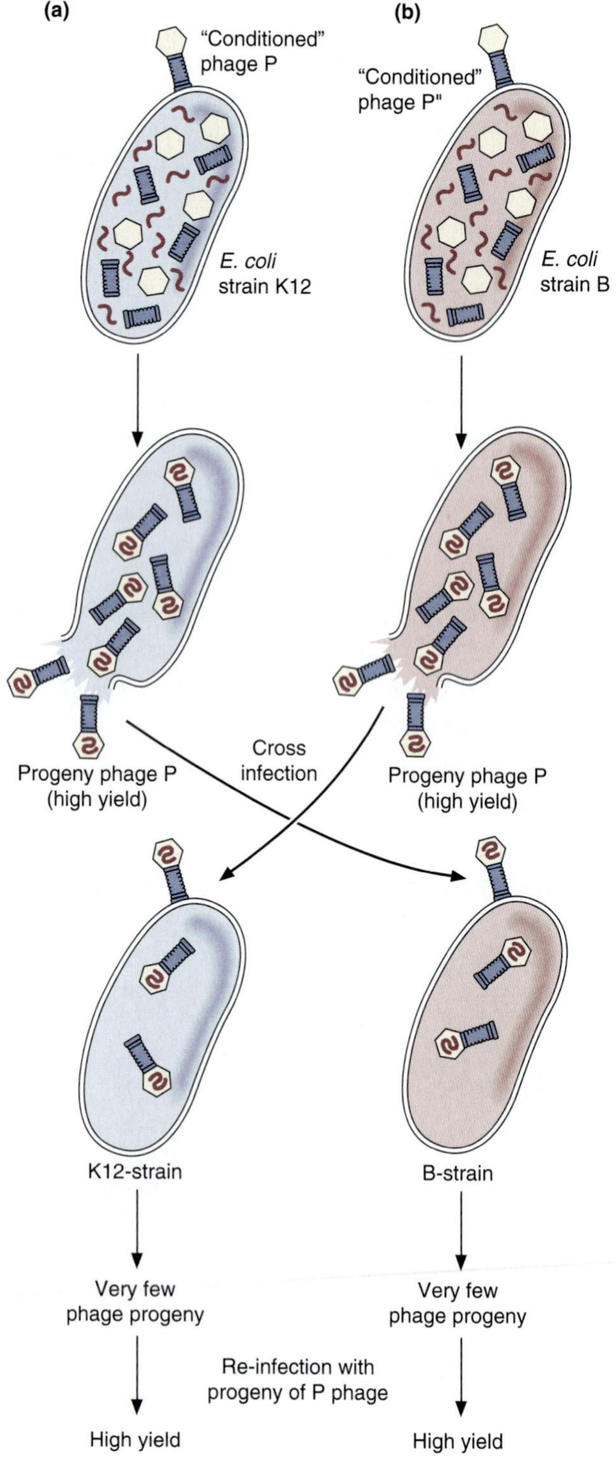

FIGURE 12.2 Restriction in bacteria. Restriction in bacteria results from "conditioning" of the phage DNA so that it can replicate in one bacterial strain but not another. In this example (a), phage P can grow on *E. coli* strain K12 and strain B, after it has been conditioned. (b) But, if the phage progeny of the K12 strain is grown on the B strain (and vice versa), very few phage are produced until the second generation. In the first generation, the few phage DNA that survive are modified to prevent degradation in subsequent infections.

Hundreds of different restriction endonucleases have been found in hundreds of different strains of bacteria (Table 12.1). The first restriction endonuclease that possessed true sequence recognition specificity was found by Hamilton Smith in 1970. This particular enzyme, called *Hind*II, cleaved DNA at this sequence:

$$\begin{matrix} \downarrow & & & \\ \text{5}'\text{-G-T-Py-Pu-A-C-3}' & \longrightarrow & \text{5}'\text{-G-T-Py} & \text{Pu-A-C-3}' \\ \text{3}'\text{-C-A-Pu-Py-T-G-5}' & & \text{3}'\text{-C-A-Pu} & \text{Py-T-G-5}' \\ \uparrow & & & \end{matrix}$$

Py and Pu represent pyrimidine and purine residues, respectively, and the arrows indicate where the cut occurs.

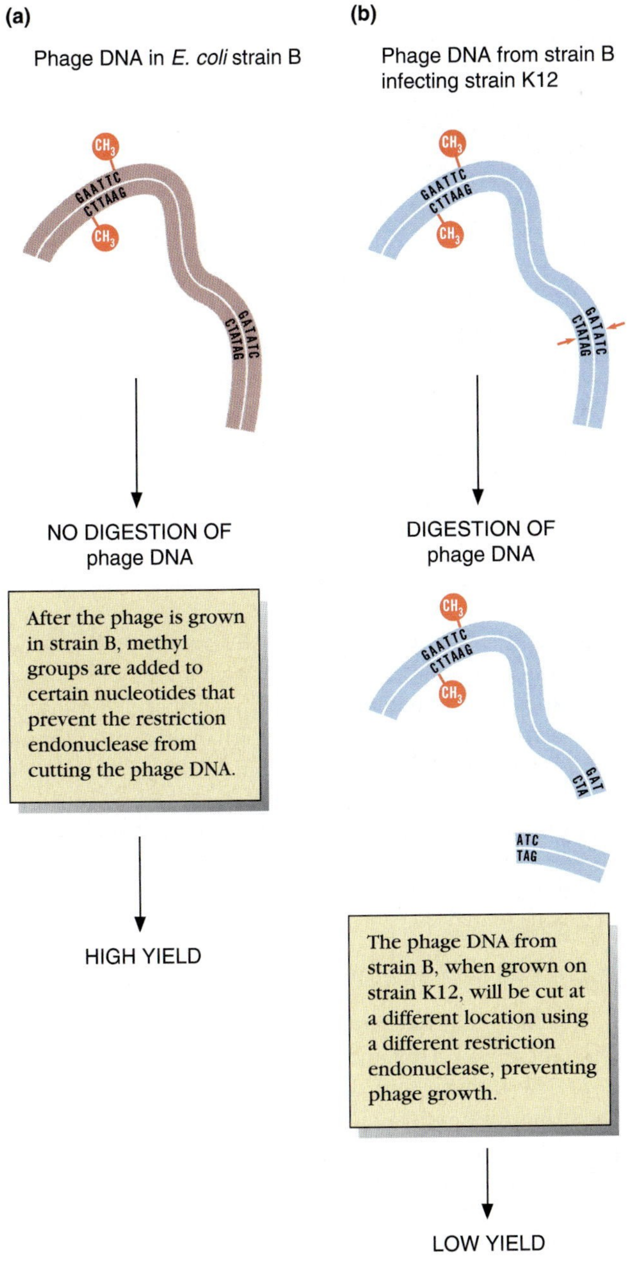

(a) Phage DNA in *E. coli* strain B

NO DIGESTION OF phage DNA

After the phage is grown in strain B, methyl groups are added to certain nucleotides that prevent the restriction endonuclease from cutting the phage DNA.

HIGH YIELD

(b) Phage DNA from strain B infecting strain K12

DIGESTION OF phage DNA

The phage DNA from strain B, when grown on strain K12, will be cut at a different location using a different restriction endonuclease, preventing phage growth.

LOW YIELD

FIGURE 12.3 Methylation prevents digestion. Conditioning or restriction of phage when grown in bacteria can be explained at the molecular level. The conditioned DNA has methyl groups added to protect it from digestion by strain B restriction endonucleases (a), which target the 5′-GAATTC-3′ site, but not by strain K12 endonucleases, which recognize the 5′-GATATC-3′ site (b).

Restriction enzymes commonly recognize short DNA sequences that are **palindromes** and cleave the DNA somewhere within the palindromic sequence. A palindrome reads the same backward or forward, from a point of symmetry, such as the following words and phrases:

A man, a plan, a cat, a canal; Panama
madam, I'm Adam
mom
bob

In DNA, which has two strands, the enzyme recognizes a twofold rotational symmetry surrounding a centerpoint. For example, the sequence

5′-GGCC-3′
3′-CCGG-5′

reads the same in both directions and has a defined centerpoint between the G and C.

Some restriction enzymes recognize a four-base sequence, some recognize a six-base sequence, and a few recognize an eight-base sequence. Some restriction enzymes cut DNA to produce blunt ends (both strands are cut at the same place), and others make a staggered cut (Table 12.1 and Figure 12.4). Some of the enzymes that make a staggered cut leave 3′ overhangs, whereas others leave 5′ overhangs.

The frequency with which a given sequence occurs in DNA, such as the recognition site for *Hind*II, can be easily calculated. If you assume that each of the four nucleotides occurs randomly in DNA, then there is an equal chance of finding one of the four nucleotide pairs at any one position (they are: G : C, C : G, A : T, or T : A). Thus, a tetranucleotide sequence will occur at a frequency of $1/4 \times 1/4 \times 1/4 \times 1/4$ or 1 in 256 nucleotides, a hexanucleotide sequence will occur at a frequency of 1 every $(1/4)^6$ or 1 in 4096 nucleotides, and an octanucleotide sequence will occur on the average of 1 in every $(1/4)^8$ or 1 in every 65,556 nucleotides. However, as you know, the distribution of nucleotides in DNA is not random. Some sequences are GC-rich and others are AT-rich; thus, an AT-rich sequence will not have many recognition sequences for the restriction endonuclease *Not*I because it recognizes GC-rich sequences. This can sometimes be useful to the investigator if large DNA fragments are desired. For example, if a specific bacterial strain had a genome rich in As and Ts and it was digested with *Not*I, average sized fragments much longer than the

TABLE 12.1 *Representative Restriction Endonucleases Showing the Sequence of DNA Where Cleavage Occurs*

Enzyme Name	Pronounciation	Bacterial Source	Recognition Sequence	Recognition Sequence Length (nucleotides)
Enzymes That Leave a 5′ Overhang				
*Sau*3A	"sow-three-A"	*Staphylococcus aureus* 3A	5′-↓GATC-3′ 3′-CTAG↑-5′	4
*Eco*RI	"echo-R-one"	*Escherichia coli*	5′-G↓AATT C-3′ 3′-C TTAA↑G-5′	6
*Not*I	"not-one"	*Nocardia otitidis-caviarum*	5′-G↓CGGCCG C-3′ 3′-C GCCGGC↑G-5′	8
*Bam*HI	"bam-H-one"	*Bacillus amyloliquefaciens* H	5′-G↓GATC C-3′ 3′-C CTAG↑G-5′	6
Enzymes That Leave a Blunt End				
*Hpa*I	"hepa-one"	*Haemophilus parainfluenzae*	5′-GTT↓AAC-3′ 3′-CAA↑TTG-5′	6
*Sca*I	"scaa-one"	*Streptomyces caespitosus*	5′-AGT↓ACT-3′ 3′-TCA↑TGA-5′	6
Enzymes That Leave a 3′ Overhang				
*Sac*I	"sack-one"	*Streptomyces achromogenes*	5′-G AGCT↓C-3′ 3′-C↑TCGA G-5′	6
*Hae*III	"hay-three"	*Haemophilus aegyptius*	5′-G CG↓C-3′ 3′-C↑GC G-5′	4

The cleavage sites are indicated by the arrows. Enzyme names come from the first letter of the genus and the second two letters of the species, followed by the strain designation, if one is available, as well as the number of the enzyme (for example, Roman numeral II) identified in that particular strain. Some strains have several different restriction endonucleases.

predicted $(1/4)^8$ or 65,556 nucleotides would be generated.

To form recombinant DNA molecules, DNA from two different sources is subjected to digestion with the same restriction endonuclease, producing DNA fragments with the complementary overlapping ends or blunt ends. Subsequently, the DNA fragments from the two different sources are mixed, and under the proper conditions, fragments will hydrogen-bond by base-pairing of their single-stranded overhanging ends (sometimes called **sticky ends**). The joined recombinant molecules are covalently bonded by the action of the enzyme DNA ligase, which catalyzes the formation of a phosphodiester bond between adjacent 3′-OH and 5′-P termini in DNA (Figure 12.5).

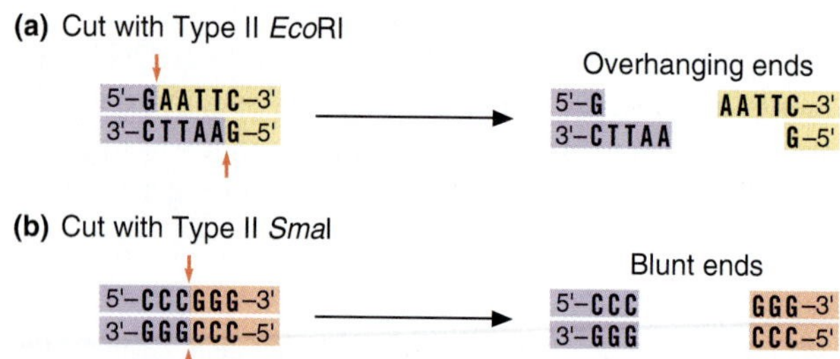

FIGURE 12.4 How restriction endonucleases cut DNA. (a) *Eco*RI cuts DNA in a staggered manner, releasing overhanging termini, whereas (b) *Sma*I cuts at the same place on both strands, giving blunt ends.

SECTION REVIEW PROBLEMS

1. What properties of restriction endonucleases make them especially attractive for cloning DNA?
2. The transfer and expression of cloned genes from one species to another is made possible by a very essential feature of the genetic code. What is it?
3. If you found that a DNA sample could not be cut with *Eco*RI and *Hin*dIII but could be cut with *Sal*I, how would you explain this observation?

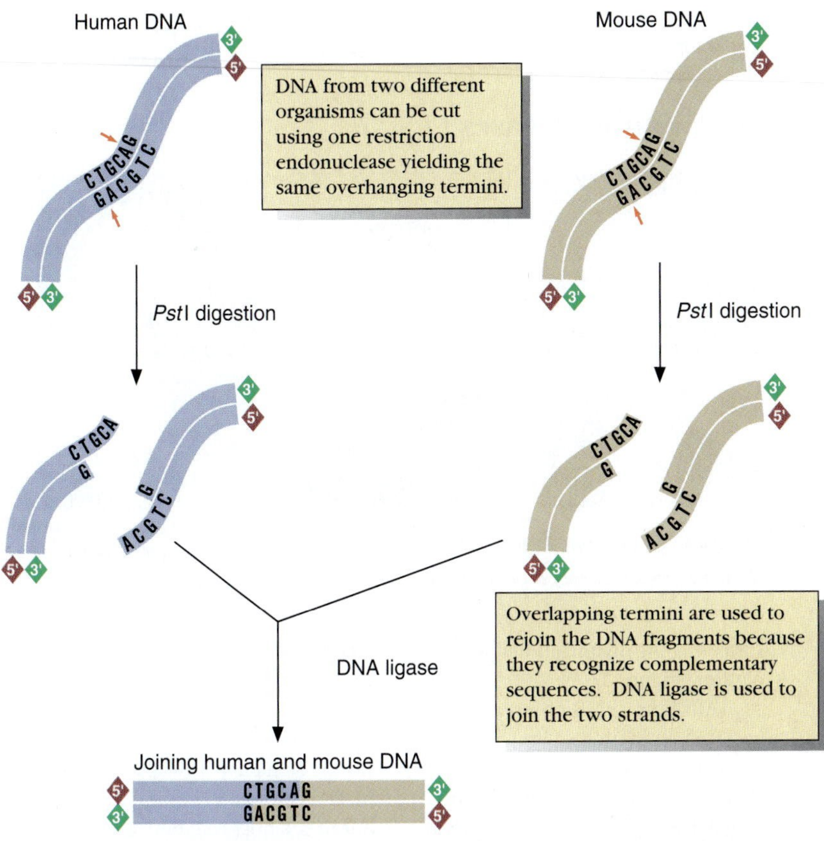

FIGURE 12.5 Recombinant DNA. This example shows how DNA from two different organisms is cut and rejoined using identical overhanging termini generated by a single restriction endonuclease. The ends are the same irrespective of the source of the DNA.

Plasmids Make Excellent Cloning Vectors.

The foundation of recombinant DNA technology is the generation of large numbers of identical copies, or clones, of specific DNA fragments. Using the overhanging complementary ends, two DNA fragments, such as one containing a human gene specifying the β-globin protein of hemoglobin, can be joined to another fragment of DNA. This recombinant molecule must then be replicated to obtain sufficient copies to make it useful. One approach to obtaining many copies of a cloned fragment of DNA is to place it into a plasmid and allow it to replicate in a host organism. Recall from Chapter 8 that plasmids are small (usually less than 15 kb), circular DNA molecules that replicate independently from the host chromosome. Plasmids can replicate in a host cell and produce many copies of the cloned DNA fragment. Some plasmids have been specifically designed for cloning and are known as **cloning vectors.** A DNA fragment cloned into a plasmid can be replicated for further study, for example, for expression analysis of the gene or for determining the nucleotide sequence of the gene.

Many restriction enzymes generate DNA fragments that can be joined together because of their overhanging sticky ends (Figure 12.4), irrespective of the source of the original DNA. However, some restriction enzymes produce blunt-ended fragments after digestion. Blunt-ended fragments can be joined through several approaches. The most common approach uses higher concentrations of DNA ligase from phage T4 in the reaction mixture with the blunt-ended DNAs to be joined. The disadvantage of blunt-ended cloning is that the ligase is nonspecific and cannot discriminate between which two blunt-ended DNA fragments are joined. Unwanted products are a consequence. However, procedures are available to select against unwanted clones or products resulting from blunt-ended ligation.

For a cloning vector or plasmid to be useful, it should possess (1) several unique restriction endonuclease cleavage sites that allow for site-specific insertion of a DNA fragment, and (2) a selectable marker, usually a gene conferring antibiotic resistance, which is used to select and maintain the plasmid in the host cell. One of the first commonly used cloning vectors for *E. coli* was pBR322 (Figure 12.6a), although it is no longer used to any great extent. It is of historical interest because many of the vectors commonly used today were derived from pBR322. Plasmid pBR322 has two antibiotic resistance

HISTORICAL PROFILE

The Discovery of Restriction Endonucleases

It all started back in the early 1950s with the description of a strange phenomenon in bacteria. In these experiments, when bacterial viruses (bacteriophage) were grown on a particular strain of bacteria and then isolated and regrown on the same bacterial strain, the number of infected bacteria (plaques) increased 10,000-fold with the second infection. For some reason the original infection did not work well, but the subsequent infection did. This experiment can be repeated with different bacterial strains with the same results; the first infection is inefficient, but the second infection is very efficient. This is the original observation that led to present-day genetic engineering technology. In 1978, three scientists received the Nobel Prize for deciphering this puzzle. Werner Arber (Biozentrum in Basel, Switzerland) is credited with having first predicted the existence of restriction endonucleases, Hamilton Smith (Johns Hopkins) isolated the restriction endonuclease, and Daniel Nathans (Johns Hopkins) first applied these enzymes to the study of gene organization and regulation (Figure 12.A).

FIGURE 12.A From left to right: Hamilton Smith, Werner Arber, and Daniel Nathans. *(Photos courtesy of Smith and Nathans)*

After the 1950s observation of bacterial restriction of bacteriophage infection, W. Arber and Daisy Dussoix, a graduate student, published a paper (1962) showing that the host degraded the DNA of the bacteriophage, thus resulting in the low efficiency of infection. They also showed that the host could modify the DNA somehow, allowing the survivors of the original infection to infect the same strain efficiently in the second round of infections. They predicted a model (Figure 12.B) suggesting that the "modification" was produced by methylation of the DNA at a specific sequence, thus protecting the DNA from an endonuclease present in the host cell, which also cut DNA only at the same specific sequence. In 1968, Hamilton Smith and his coworkers reported the purification and characterization of the first restriction endonuclease, *Hin*dII, which recognized a six-nucleotide sequence. Having direct access to Smith's newly discovered *Hin*dII restriction endonuclease, Daniel Nathans and a coworker, Kathleen Danna, found that double-stranded, circular DNA from Simian virus 40 could be cleaved into 11 separable specific fragments. They then constructed a restriction map of the virus. With the map, they proved that DNA replication begins at a unique origin and proceeds bidirectionally around the circle, terminating roughly 180° from the initiation point.

Few could have predicted that the initial observation on "restriction" and "modification" in bacterial viruses, a seemingly esoteric subject, would be the basis for revolutionizing genetics to the point where today it has had substantial commercial application with a direct impact on economics and human health and well-being.

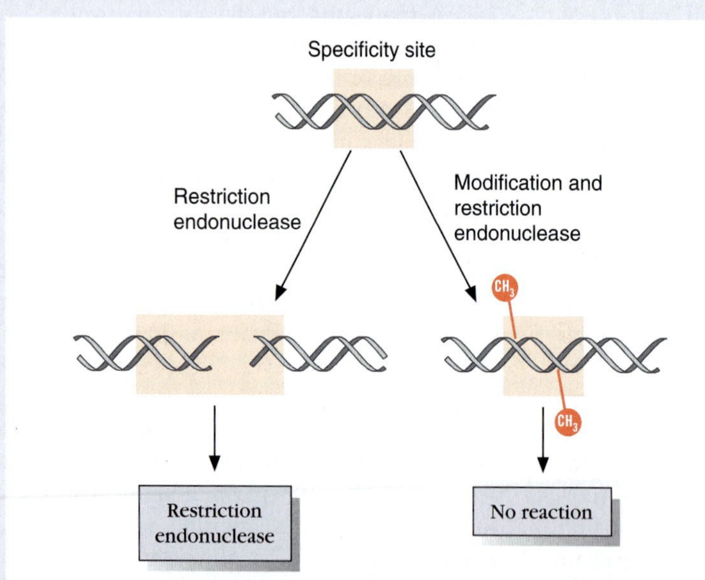

FIGURE 12.B A model proposed for the restriction and modification of DNA during phage infection.

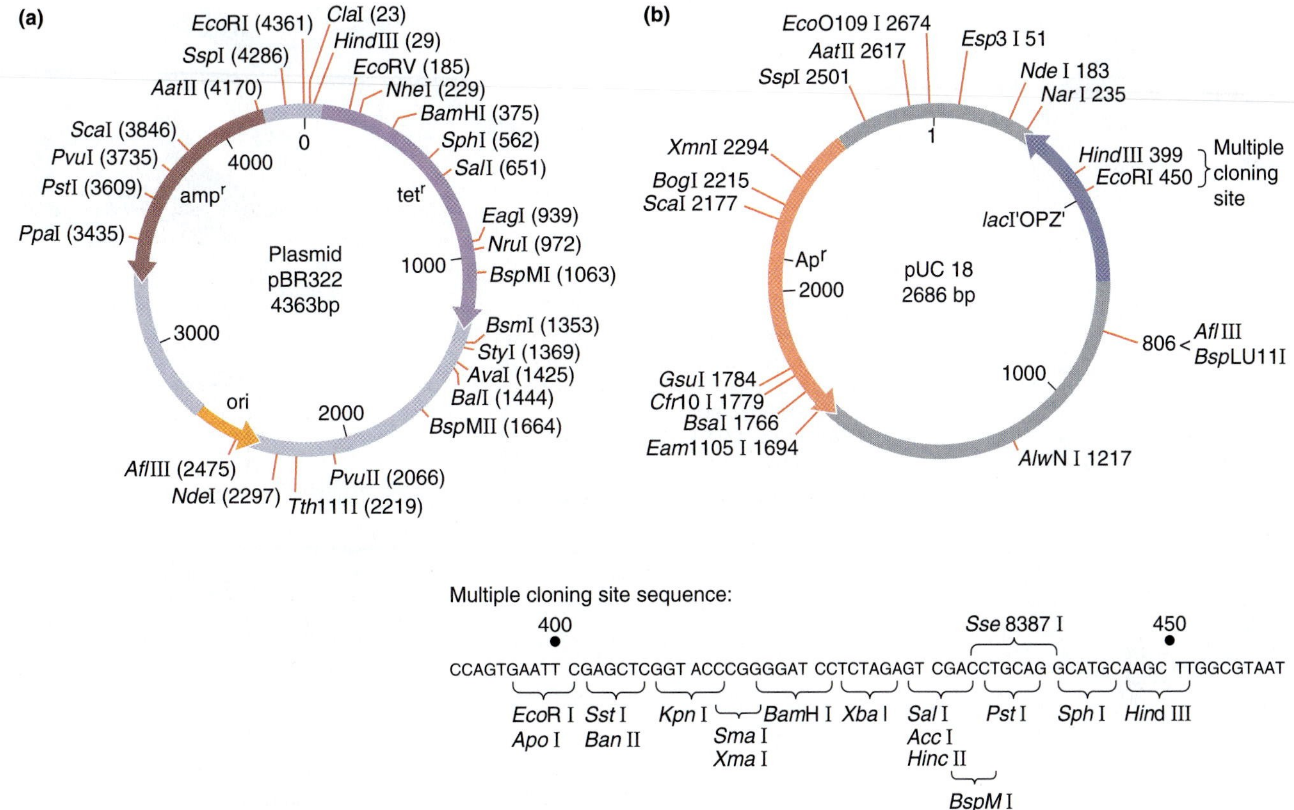

FIGURE 12.6 Plasmid cloning vectors. The commonly used restriction enzyme cut sites are listed in their abbreviated form. Their location on the plasmid (in nucleotide pairs) from an arbitrary reference point is given adjacent to each restriction site. Arrows indicate the direction of the transcription of the genes: (a) the original pBR322 cloning vector; (b) the more recently constructed and versatile cloning vector pUC18. Derivatives of these vectors are in common use in laboratories around the world.

genes (tetracycline and ampicillin) and 21 unique sites for restriction endonuclease cleavage but has largely been abandoned as newer cloning vectors have incorporated more useful features. For example, the plasmid cloning vector pUC18 (Figure 12.6b) was derived from pBR322 but has incorporated two important features that make gene cloning relatively straightforward: (1) a **polylinker, multiple cloning,** or **polycloning site** containing multiple, unique restriction endonuclease cleavage sites that enable restriction fragments generated with different enzymes to be inserted at nearly the same site (a selectable marker, *amp*^{R}, is also included), and (2) a *lacZ′* gene with its upstream regulatory regions (Chapters 8 and 14). When a foreign DNA fragment is present in the polycloning site, β-galactosidase is not expressed because the foreign DNA interrupts the reading frame of the β-galactosidase gene. β-galactosidase cleaves the colorless substrate 5-bromo-4-chloro-3-indolyl-β-D-galactoside (X-gal for short) to galactose and 5-bromo-4 chloroindigo, which is blue. Consequently, cells containing β-galactosidase produce blue colonies on an X-gal-containing agar medium, and non-β-galactosidase-producing colonies are white. In this way, bacterial cells containing foreign DNA inserts are readily distinguished from non-insert-containing colonies by their color.

An example of DNA cloning using plasmid pBR322 is given in Figure 12.7, and this general approach is still used in many modern cloning vectors. In this example, the restriction endonuclease *Sal*I cleaves pBR322 within the tetracycline resistance gene, yielding a linear molecule with 5′ overhang tails. Similarly, purified human DNA cleaved with the same enzyme generates the same tails. When the two are mixed in the presence of DNA ligase, new plasmids that now contain human DNA fragments are generated. However, not all plasmids form recombinant DNA molecules; some merely rejoin their own sticky ends (they self-close). These competing processes generate two populations of plasmids: pBR322 plasmids and plasmids bearing the human DNA fragments. Both populations of plasmids are introduced into host cells (such as *E. coli*) by transformation (Chapter 8), and the transformed bacteria are grown on media

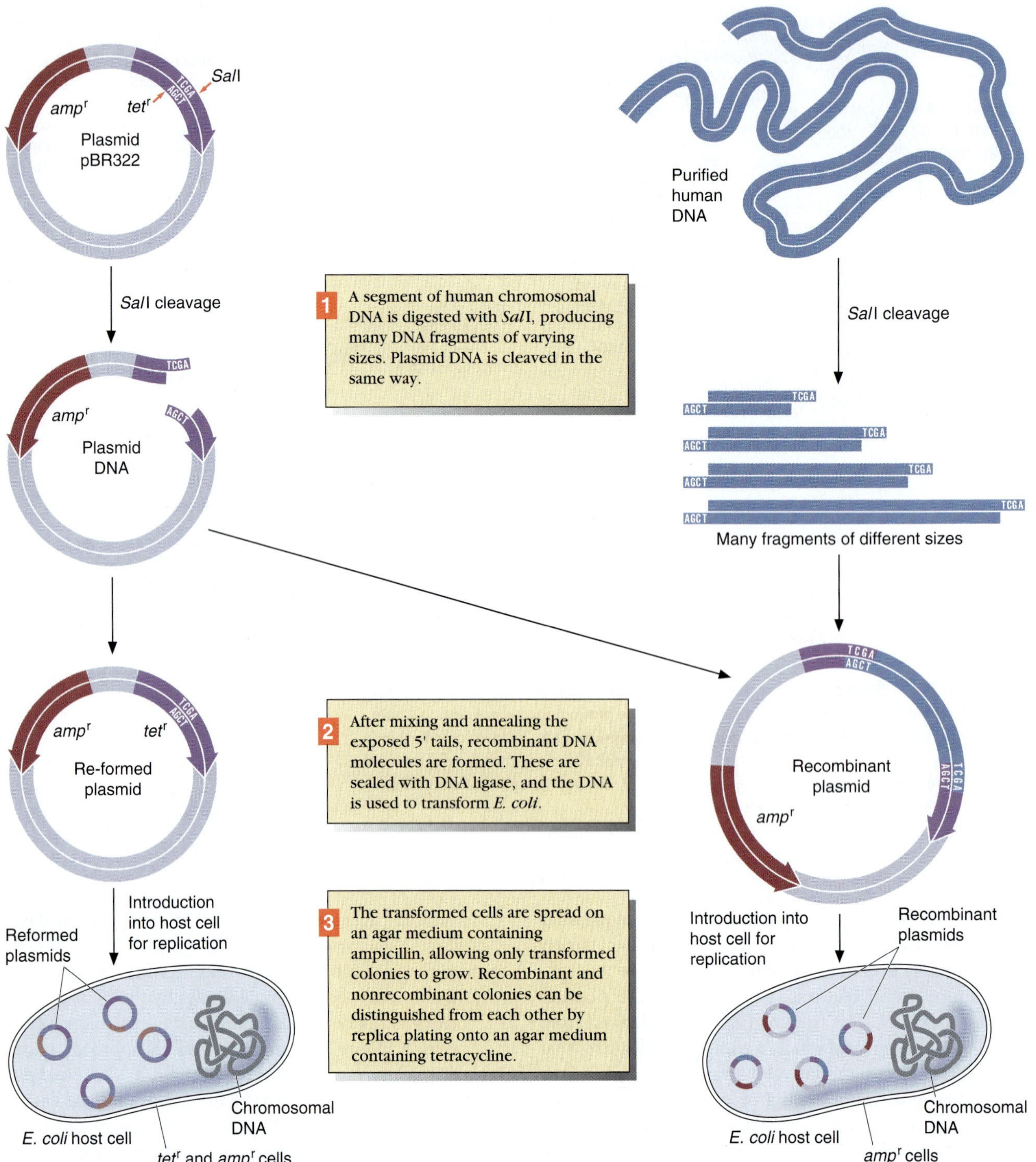

FIGURE 12.7 Cloning a DNA sequence. This illustration shows how human DNA fragments can be cloned into a plasmid. The *E. coli* plasmid pBR322 is used as a cloning vector. pBR322 contains two important genes that are expressed in the host cell: tetracycline resistance and ampicillin resistance. Additionally, pBR322 has a single *Sal*I cleavage site in the tetracycline-resistance gene, facilitating the selection of a foreign DNA fragment into this site, because the *tet*^{R} gene is inactivated upon insertion of a DNA fragment.

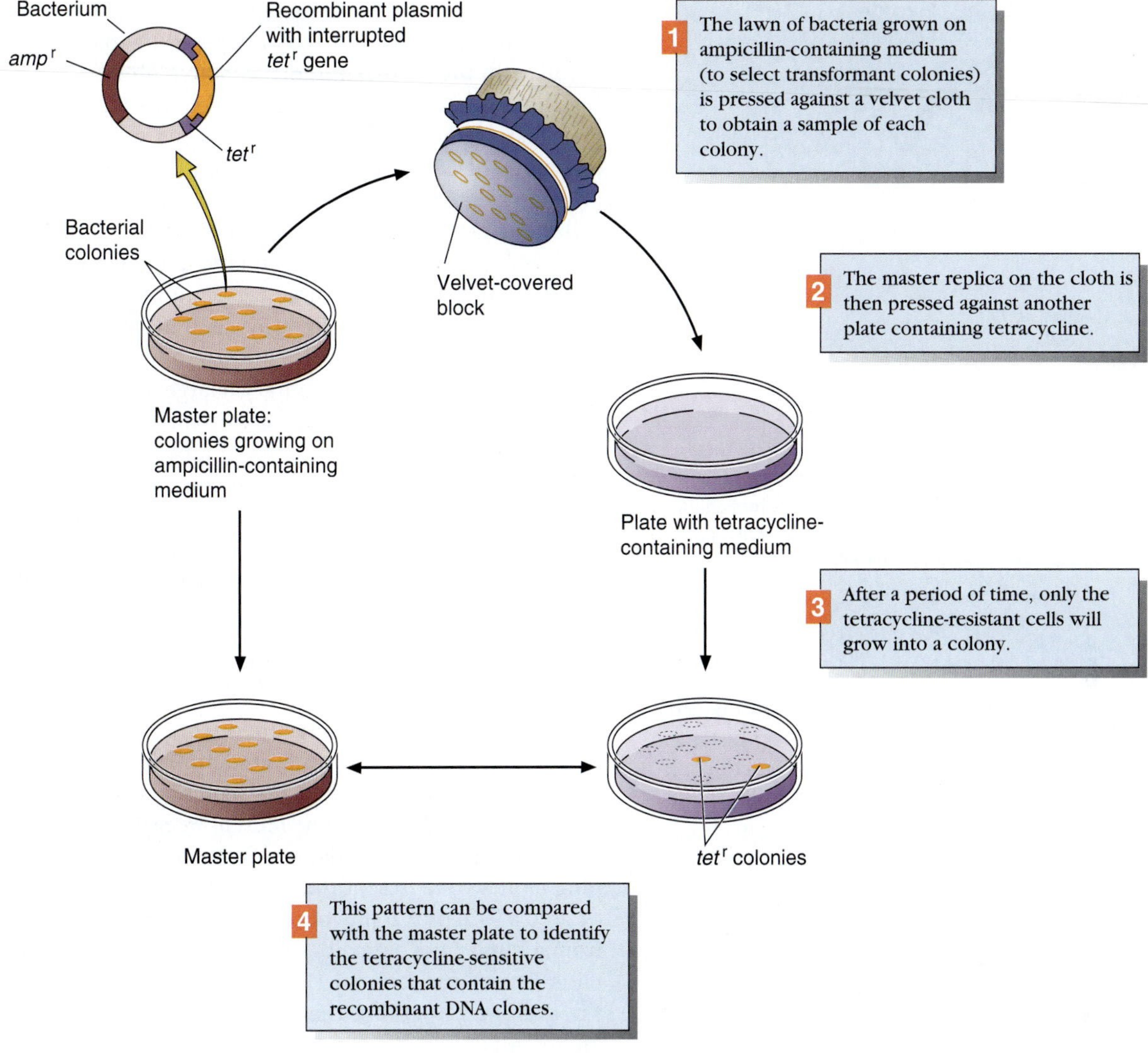
Bacterium
amp^r
Recombinant plasmid with interrupted tet^r gene
tet^r
Bacterial colonies
Velvet-covered block
1 The lawn of bacteria grown on ampicillin-containing medium (to select transformant colonies) is pressed against a velvet cloth to obtain a sample of each colony.
2 The master replica on the cloth is then pressed against another plate containing tetracycline.
Master plate: colonies growing on ampicillin-containing medium
Plate with tetracycline-containing medium
3 After a period of time, only the tetracycline-resistant cells will grow into a colony.
Master plate
tet^r colonies
4 This pattern can be compared with the master plate to identify the tetracycline-sensitive colonies that contain the recombinant DNA clones.

FIGURE 12.8 Replica plating. This technique is used to distinguish tetracycline-resistant bacterial colonies from tetracycline-sensitive colonies (which contain the inserted DNA fragment; see Figure 12.7).

containing ampicillin. Colonies containing these two different populations can be distinguished from each other because the recombinant plasmid will contain an inactive tetracycline-resistance gene where the DNA fragment is inserted. The self-closed pBR322 will be resistant to tetracycline. To distinguish between these two kinds of clones, the fully grown colonies are replica-plated (Figure 12.8) onto media containing tetracycline. The cells possessing the recombinant plasmid will not grow on tetracycline-containing media, thus these recombinant DNA-containing bacterial colonies are identified on the original agar dish for further characterization.

Plasmids are not the only cloning vectors. Many other vectors are available, including λ phage (Chapter 8), BACs, YACs, and derivatives of plasmids and λ phage (cosmids). These vectors are described in Chapter 13, but it should be noted that they all have one clear advantage over plasmids; they can contain and replicate larger cloned DNA fragments (20 kb and up). Most DNA fragments cloned into plasmids are less than 20 kb because very large inserts are frequently unstable during bacterial growth.

Some Vectors Can Replicate in Multiple Hosts.

The vectors discussed thus far can replicate in only one host (bacteria, specifically *E. coli*). Other vectors are designed to replicate in more than one host [for example, *E. coli* and eukaryotic cells (animal and plant)]. These vectors have replication origins that specifically allow replication in both cell types and are referred to as **shuttle vectors.** A shuttle vector is a cloning vector that can

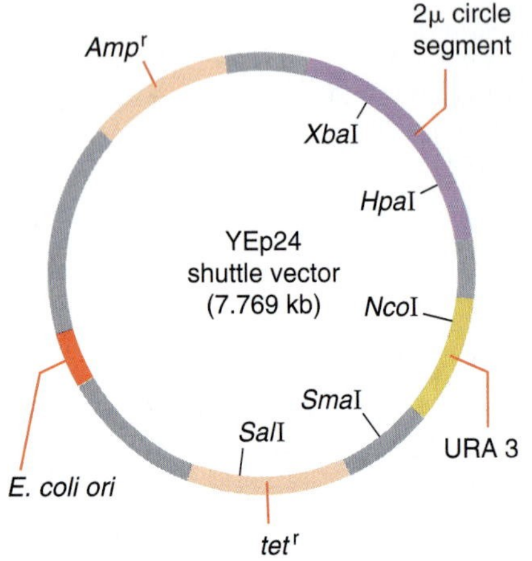

FIGURE 12.9 Yeast–*E. coli* shuttle vector. This vector will replicate in both *E. coli* and yeast cells because it possesses replication origins for both cell types. Some of the 16 unique restriction endonuclease cut sites are shown scattered around the vector and allow introduction of recombinant DNA segments.

replicate in two or more host organisms. Among the most developed shuttle vectors are those that replicate in yeast cells and *E. coli*. Figure 12.9 shows the yeast/*E. coli* shuttle vector YEp24. This vector has both *E. coli ori* and a segment of the 2μ circle found in yeast. The tet^R and amp^R markers are used for positive selection in *E. coli*, and the *ura3* gene (a requirement for uracil) is a positive selection marker in yeast. Thus, when the host yeast cell is mutant for *ura3*, the YEp24 shuttle vector will supply the necessary active *ura3* gene product for growth, and only cells that contain this vector will grow on medium lacking uracil.

What are some uses of shuttle vectors? *E. coli* is an easy organism to grow and maintain, so scientists prefer to use *E. coli* as a host cell whenever possible to propagate DNA clones. Thus, a gene can be cloned into a shuttle vector using *E. coli* as the host cell and then transformed into another cell type to determine aspects of gene function. For example, a gene involved in leucine metabolism in yeast could be cloned into YEp24 using *E. coli* as the host. When the desired gene is found and characterized, it is directly transferred to yeast cells to determine whether it is active and functions as predicted. This same rationale is used for shuttle vectors between *E. coli* and animal or plant cells. Some shuttle vectors integrate into the genome of the host for replication.

A Library Is a Collection of Cloned DNA Fragments from an Organism.

Gene cloning usually requires the initial preparation of multiple clones, each composed of a different DNA fragment, from the entire genome of an organism. A random collection of clones from a specific genome is called a **genomic DNA library** and may represent the total DNA of the organism. We have already discussed the basic steps involved in the preparation of a genomic library: digesting the organism's DNA with a restriction endonuclease and ligating the fragments into a cloning vector. These recombinant DNA molecules are then transformed into an appropriate host cell. A genomic library may contain a desired gene, but identifying this clone among the thousands of clones from the library can sometimes be difficult. In recent years, some very innovative methods have been developed to simplify gene identification in genomic libraries.

In constructing a gene library, a large number of different clones must be generated to ensure that all DNA fragments are represented. This number is dependent upon the size of the cloned fragments, the size of the genome, and the probability that an individual clone will be in the library. The number of clones necessary to represent an entire genome can be calculated from the equation

$$N = \frac{\ln(1 - p)}{\ln(1 - I/G)}$$

where N is the number of required recombinant clones, ln is the natural log, p is the desired probability of recovering a given fragment, and I/G represents the fraction (in base pairs) of the genome present in each clone. This equation takes into consideration the need for overlapping clones to represent the library. For example, if you desired a 99% probability of obtaining an individual gene from a human genomic library (the human genome contains 3×10^9 bp), and the DNA fragment insert size is on the average 20,000 bp, then 660,000 individual clones are required:

$$N = \frac{\ln(1 - 0.99)}{\ln[1 - (2 \times 10^4/3 \times 10^9)]} = 6.6 \times 10^5$$

Just dividing 20,000 bp (the size of the insert) into 3 billion bp (the size of the genome) would yield a value of 150,000 clones, a serious underestimate. This underestimate occurs because it represents only end-to-end clones, whereas in a library overlapping clones are needed to obtain complete representation of the genome.

From this calculation, we can see that extremely large numbers of clones are required to represent the entire human genome with an average insert size of 20 kb. In reality, a plasmid library of 660,000 clones is nearly unmanageable because of the tremendously large numbers. If you were to search a library of this size, you would need to examine each colony on a standard Petri dish. At a maximum, about 500 colonies can be grown individually; thus you would have to search through 660,000/500 or 1350 Petri dishes to find your desired gene. The number of clones can be greatly reduced if the average insert size in the cloning vector is increased. For example, if the average insert size is increased to 200 kb, the number of clones required to get a 99% probability of finding any gene sequence in the library falls to 6900 clones. Plasmid cloning vectors are impractical when preparing libraries from large genomes because the average size of a cloned fragment is usually less than 20 kb. As a consequence, new cloning vectors have been developed to increase the size of the insert and reduce the number of clones to a manageable number. These are discussed in Chapter 13.

Libraries Can Be Prepared from mRNA (cDNA Library).

In many instances, an investigator may be interested in examining only the genes that are expressed in a certain organism or tissue. In this case, it is possible to prepare a gene library from mRNA instead of from the DNA of the whole organism. A gene library prepared from mRNA is referred to as a **complementary DNA** or **cDNA library** because the mRNA must be converted into DNA to make the library.

Methods are available for purifying mRNA (see Figure 11.14) from eukaryotic cells using the poly(A) tails on the 3′ end of most mRNAs. To prepare a cDNA library, the information in purified mRNA is used to make a copy of DNA. This is done with **reverse transcriptase,** an enzyme capable of synthesizing DNA from an RNA template *in vitro*. Reverse transcriptase is more accurately called RNA-dependent DNA polymerase because it catalyzes the conversion of RNA into DNA. The reverse transcript (called the first strand) is used as a template to synthesize a complementary strand to form double-stranded DNA. Because all DNA polymerases require a primer for synthesis, oligo(T) sequences act as a primer by hydrogen bonding to the poly(A) tails. After synthesis of the first strand, the RNA strand is removed with the enzyme RNase H, which degrades complexes of RNAs and DNA but leaves short pieces of RNA behind. These short sequences conveniently act as a primer for second-strand synthesis with DNA polymerase I from *E. coli* (Figure 12.10). DNA ligase is also added to seal the remaining single-strand breaks in the double-stranded DNA molecule. These double-stranded DNAs can be inserted into plasmid, cosmid, or λ phage cloning vectors by blunt-ended ligation.

A cDNA library is significantly different from a genomic library. A cDNA library contains only transcribed sequences and does not contain promoters, termination signals, or enhancer elements. Furthermore, the members of a cDNA library vary depending upon the tissue from which the mRNA is prepared because not all genes are transcribed in all tissues. In addition, not all mRNAs are transcribed to the same level (some are rare and others very abundant); thus the frequency of a cloned gene in a cDNA library will reflect the amount of mRNA that is synthesized by that gene. Another difference between a cDNA library and a genomic library is that the cloned genes do not contain introns because the introns have been spliced out during processing of the pre-RNA. The absence of introns permits expression of the gene in a bacterial cell (which cannot process introns from the pre-RNA), and it also allows scientists to locate the introns exactly when they compare the genomic clone to the cDNA clone.

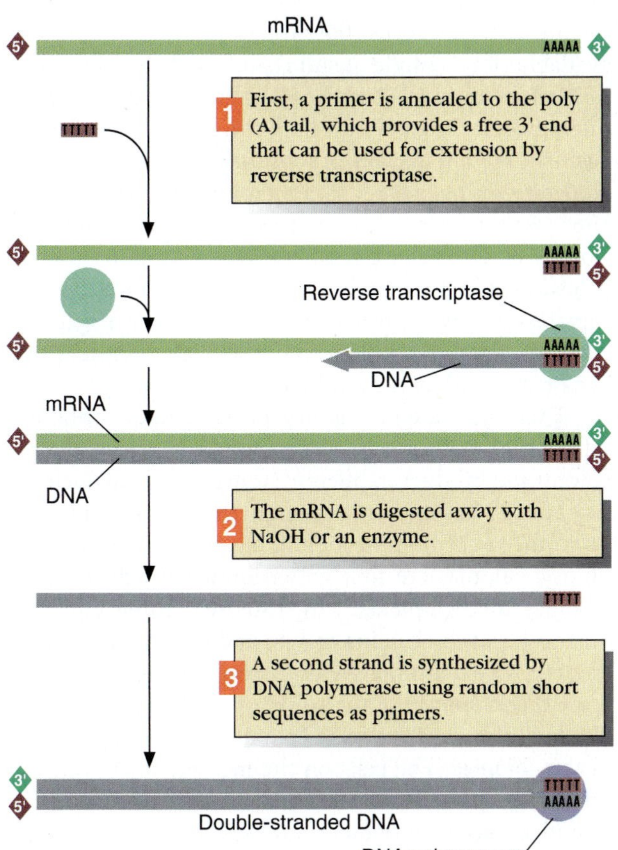
mRNA
5'
AAAAA
3'
TTTTT
1 First, a primer is annealed to the poly (A) tail, which provides a free 3' end that can be used for extension by reverse transcriptase.
Reverse transcriptase
DNA
mRNA
DNA
2 The mRNA is digested away with NaOH or an enzyme.
3 A second strand is synthesized by DNA polymerase using random short sequences as primers.
Double-stranded DNA
DNA polymerase

FIGURE 12.10 Preparation of complementary DNA (cDNA) from mRNA. Messenger RNA can be converted into DNA by using reverse transcriptase, an enzyme that uses RNA as a template to make DNA. The DNA is then cloned into a vector for propagation.

SECTION REVIEW PROBLEMS

4. What are some features of plasmids that make them excellent cloning vectors?
5. Blunt-ended DNA fragments do not possess overlapping sticky ends. How are they joined?
6. What are the advantages and disadvantages of cDNA clones over genomic clones?

7. The equation $N = \ln(1 - p)/\ln(1 - I/G)$ does not apply to cDNA cloning. Why?
8. If the average insert size of a clone bank were 12 kb, and the size of the genome were 2.1×10^6 bp, how many clones would be needed to give a 99% probability that any given gene will be present in the library?

DIFFERENT APPROACHES ARE USED TO FIND A GENE IN A LIBRARY.

Because a specific DNA sequence may occur in only 1 of 100,000 clones, a means must be available to easily identify the bacterial colony or phage plaque that carries a specific clone. Several techniques that depend upon the circumstances and the available information are currently in use. In the following section, we discuss several of these methods and their limitations.

A Probe Is Used to Identify a Clone in a Library.

One method for identifying a specific gene sequence in a library uses a probe. A **probe** is a DNA or RNA sequence that has base sequence complementarity so that it will hybridize to the desired gene. Probes may be cloned genes from other organisms with similar sequences, and these are known as **heterologous probes.** For example, the insulin gene from pigs is very similar to the insulin gene in humans. Thus, if the pig insulin gene were cloned first, this gene could then be used as a heterologous probe to search a human genomic library for the human insulin gene. A probe does not need to consist of the entire gene but can contain just enough genetic material to hybridize to a section of the desired gene in the library. For example, if the sequence of the protein product of the gene is known, or if the sequence of a protein from a gene with a similar function in a related organism is known, a short oligonucleotide probe can be synthesized from knowledge of the amino acid sequence. When a probe is synthesized from knowledge of the amino acid sequence of a protein, it must be remembered that the code is degenerate. To account for this, a mixed population of probes is synthesized, usually about 20 nucleotides long, and this population contains all possible sequences from the codons of the amino acid sequence. This mixed population has one sequence that will hybridize to the desired clone in the library. RNA can also act as a probe; for example, ribosomal RNA or its DNA clone from a rat can act as a probe to find a maize rRNA gene from the maize genomic library.

A probe uses DNA : DNA or RNA : DNA hybridization to find a sequence of interest among a population of other DNA sequences. To find a DNA sequence from a library prepared in a bacterial host, the bacterial cells containing the cloned DNA sequences are grown into colonies on a master plate (Figure 12.11). The colonies are then replica-plated onto a nitrocellulose or nylon membrane laid upon nutrient agar in a Petri dish and allowed to grow. The membrane with its newly formed colonies is removed from the Petri dish, and the DNA from each colony is released by lysing the cells. Under denaturing conditions, the single-stranded DNA is then bound to the thin nylon or nitrocellulose membrane. Radioactively labeled probe DNA is then combined with the membrane in a sealed plastic bag, and the labeled, single-stranded probe DNA is given sufficient time to hybridize to complementary DNA strands bound to the membrane. Excess radioactive probe is then washed from the membrane, and the bound radioactive DNA on the filter is exposed to a photographic film to determine the location of the colony containing the desired DNA sequence(s) (Figure 12.11). This information is used to identify the original colony on the master plate that contains the desired DNA sequence. The original colony is then saved and used to produce large quantities of the gene for study or production of a gene product. Several variations of this procedure have been developed, depending upon the vector used to clone the DNA, such as λ phage, but the principles are identical in every case.

In some instances, the protein product of a gene is available (for example, insulin) and can be used to identify a clone in a cDNA library. In this procedure, enough pure protein must be available to prepare antibodies against the protein. This procedure also requires that the cloned DNA be present in an **expression vector** (see a later section for a more complete discussion of expression vectors). In an expression vector, the cDNA clone is placed next to a promoter sequence and translation start signal so that in the insert mRNA will be transcribed. The transcribed mRNA is then translated into a protein that can be detected by the antibody.

Each individual clone in a cDNA library contains a different insert, and when cells are plated, each bacterial colony will produce a different protein, each characteristic of the DNA sequence present in the expression vector. As in the procedure for identifying clones in a genomic library, colonies are first grown on a Petri dish. These colonies are then picked with a sterile toothpick and put on another Petri dish for growth and in a microtiter dish (a dish with multiple wells to contain the media) for growth and storage of the colonies. A membrane is then overlaid onto the Petri dish colonies to make an imprint of the colonies. The cells on the membrane are lysed to release the proteins within the cells, and the protein is bound to the membrane. Radioactive antibody is then mixed with the membrane to react with the specific protein, and excess radioactivity is washed off. An autoradiograph is then prepared and the colony that synthesized the protein is identified. The clone can then be identified in the microtiter dish and used for study.

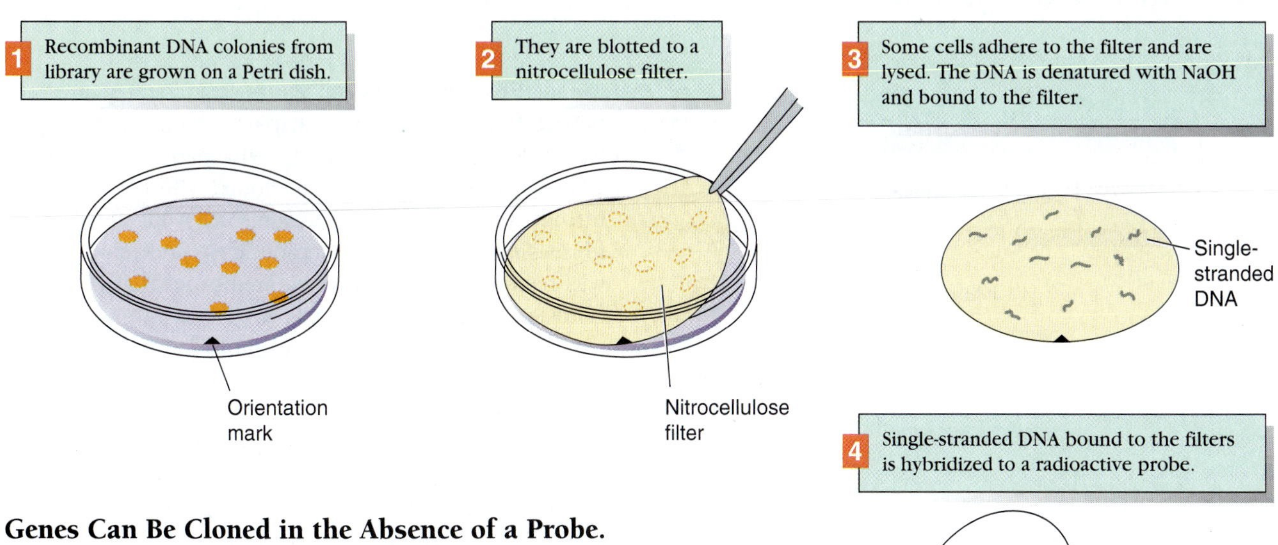

Genes Can Be Cloned in the Absence of a Probe.

In many instances, a probe is not available for a gene, for example, a gene where only the phenotype of the mutant is known, such as the mutant gene responsible for the disease cystic fibrosis. In this example, the product of the cystic fibrosis gene is not identified, and consequently no probe is available to help find the gene from a human genomic library. Cystic fibrosis is one of the more common autosomal recessive genetic diseases affecting about 1 in 2500 people; 1 in 25 people are heterozygotes for this mutant allele. The major symptoms of the disease are serious respiratory and digestive problems and very salty sweat.

To find the cystic fibrosis gene, investigators used a technique known as **chromosome walking** and **chromosome jumping,** which are forms of **positional cloning.** Chromosome walking depends upon generating a series of relatively large clones that possess overlapping sequences (Figure 12.12a) that eventually contain the gene of interest. Chromosome jumping provides a means of jumping across potentially unclonable areas of DNA (like repeated sequences) and also generates widely spaced landmarks along the sequence (Figure 12.12b).

Linkage analysis using molecular markers (see the discussion of RFLPs in Chapter 13) has located the cystic fibrosis gene to the long arm of chromosome 7, between bands 7q22 and 7q31.1. This gene is flanked by the *Met* gene on one side and the D788 DNA marker on the other side, but these two markers are nearly 1.5 million bp apart. Additional efforts found closer genetic markers, but there were still significant distances between the closest marker and the cystic fibrosis gene. Using chromosome jumping and walking, a series of overlapping clones were obtained between the two flanking markers, and their DNA sequence was determined. The sequence of nucleotides in the DNA within the area of the gene gave a clue as to the location of the cystic fibrosis gene because of the presence of **GC islands.** It is known that genes from humans generally are preceded at the 5′ end by clusters of GC sequences, and several of these were found. Near the GC islands, DNA sequences were found that possessed genes that were similar to those of related

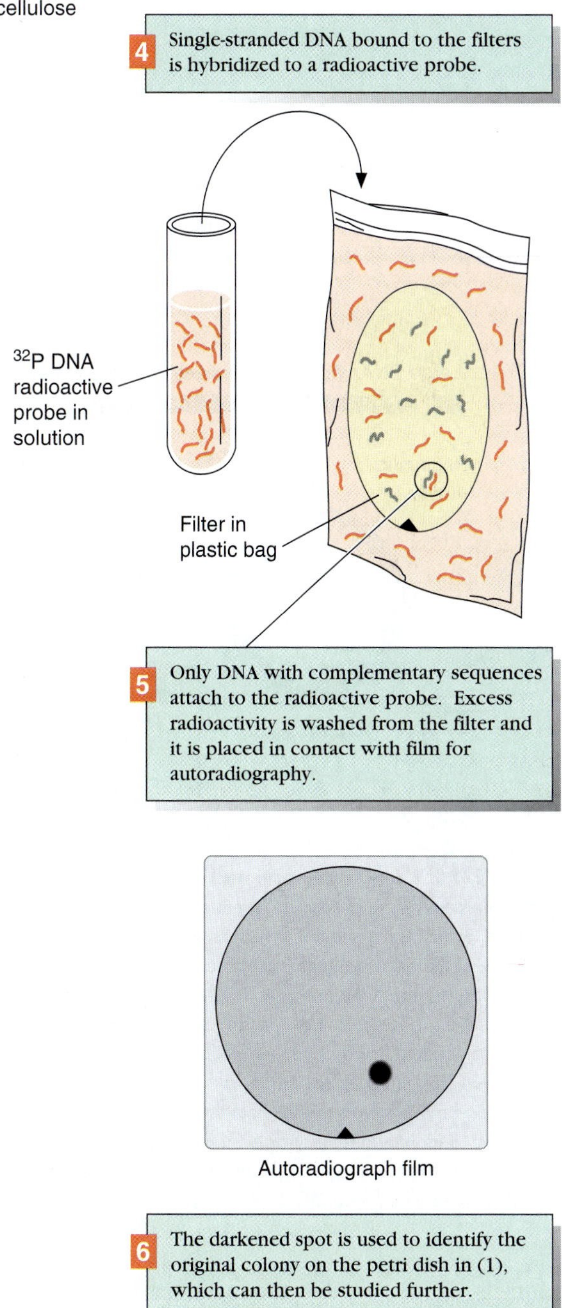

4
Single-stranded DNA bound to the filters is hybridized to a radioactive probe.
32P DNA radioactive probe in solution
Filter in plastic bag
5
Only DNA with complementary sequences attach to the radioactive probe. Excess radioactivity is washed from the filter and it is placed in contact with film for autoradiography.
Autoradiograph film
6
The darkened spot is used to identify the original colony on the petri dish in (1), which can then be studied further.

FIGURE 12.11 Identifying a cloned gene. An *E. coli* colony containing a DNA clone can be identified when it possesses sequences similar to a DNA probe. The colony is then used to propagate the clone and study the sequence.

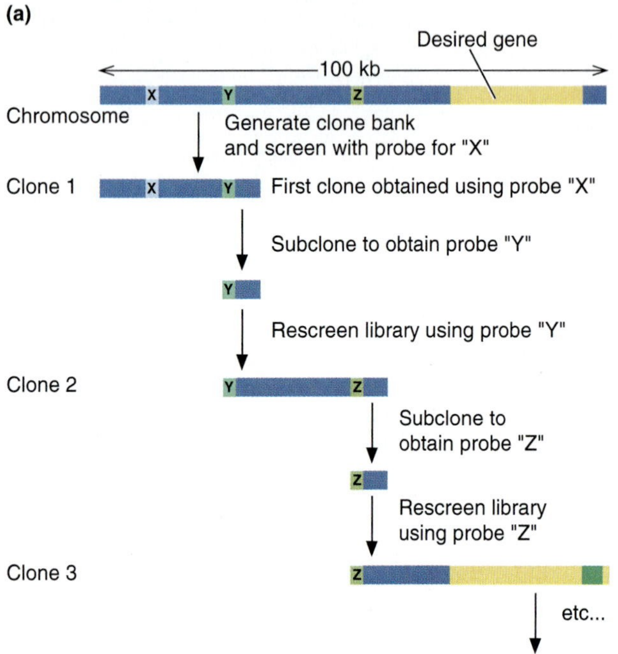

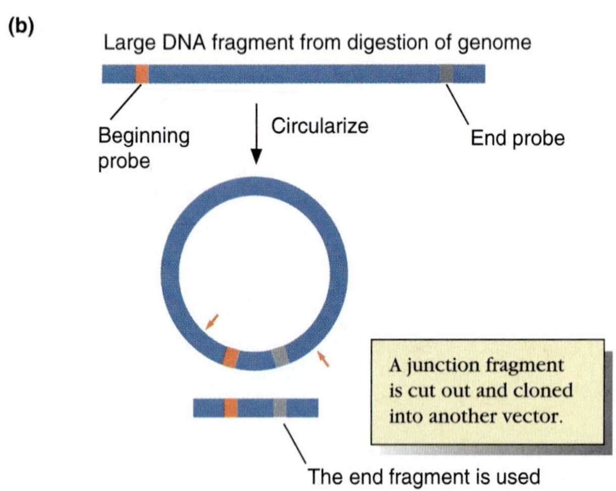

FIGURE 12.12 Chromosome walking and jumping. (a) In the chromosome walking procedure, the first identified DNA fragment from the library is fragment X, which is near the desired gene. A subclone of the first fragment is prepared, and the subclone is used as a probe to search the library to find a neighboring clone. This procedure is repeated until the desired sequence is located. (b) In chromosome jumping, a beginning probe is used to identify a large DNA fragment. By circularizing the fragment, a second probe can be generated distant from the first. This second probe is then used to identify another fragment in the genomic library, and so forth. This procedure permits spanning large regions of DNA that otherwise would be difficult to survey.

animals, suggesting that the investigators had found an evolutionarily conserved important gene. The appropriate translational start and stop signals were then found, and when they examined the same coding frame from an individual with cystic fibrosis, a 3-bp deletion in the 6500 nucleotide-long normal coding sequence was found. From the deduced amino acid sequence of the cystric fibrosis gene, similarities to ion-transport proteins were found, explaining why patients have such salty sweat. Finally, when the cloned wild-type gene was introduced into cultured mutant cells line from cystic fibrosis patients, researchers found that normal function was restored. This was considered proof that the cystic fibrosis gene had been found.

Chromosome walking and jumping can be done in both directions from the start point, depending upon which end fragment is used to start the walk. One drawback to chromosome walking and jumping is the requirement that the end fragment, used to search the genome library for a nearby overhanging clone, must be present in the genome only one time. If the end fragment is a repeated sequence, it will be present in many clones of the library, but only one of these clones will represent the adjacent sequence. Thus, this procedure is most useful in organisms that have fewer repeated sequences (e.g., the small plant *Arabidopsis* or the fruit fly *Drosophila*). When a large number of overlapping clones are obtained, it is called a **contig set** (for contiguous clones) and represents an area of the genome. Contig sets can be joined when one set possesses overlap with an adjacent set, yielding a complete contig genomic library. A contig library is extremely useful for finding a gene and orienting it in relation to other markers and genes. Creating a contig library is one of the primary goals of the human genome project as well as other genome projects.

A DNA Sequence Can Be Amplified Using the Polymerase Chain Reaction.

The **polymerase chain reaction** (PCR) is a method of greatly increasing the quantity of a specific DNA sequence from as little as only one DNA or RNA molecule. This method for amplifying DNA sequences was developed in 1986 by K. B. Mullis (who later received the Nobel Prize for his discovery) and has found application in a wide range of disciplines, including DNA sequencing, evolution, and forensics. PCR has extended the range of recombinant DNA techniques and replaced some earlier methods.

The goal of gene cloning is to increase the quantity of a desired segment of DNA. As we have learned thus far, this often tedious and time-consuming process usually involves cloning a DNA fragment into a vector. The PCR permits a more direct amplification of a specific DNA or RNA sequence. The PCR can be used on DNA or RNA fragments that are present at extremely low concentrations, which is very difficult with traditional gene cloning. The PCR is based upon the use of DNA polymerase and two primers to delineate the ends of a sequence to be amplified (Figure 12.13). The primers are made synthetically and are usually about 20 nucleotides long, so information must be available concerning the

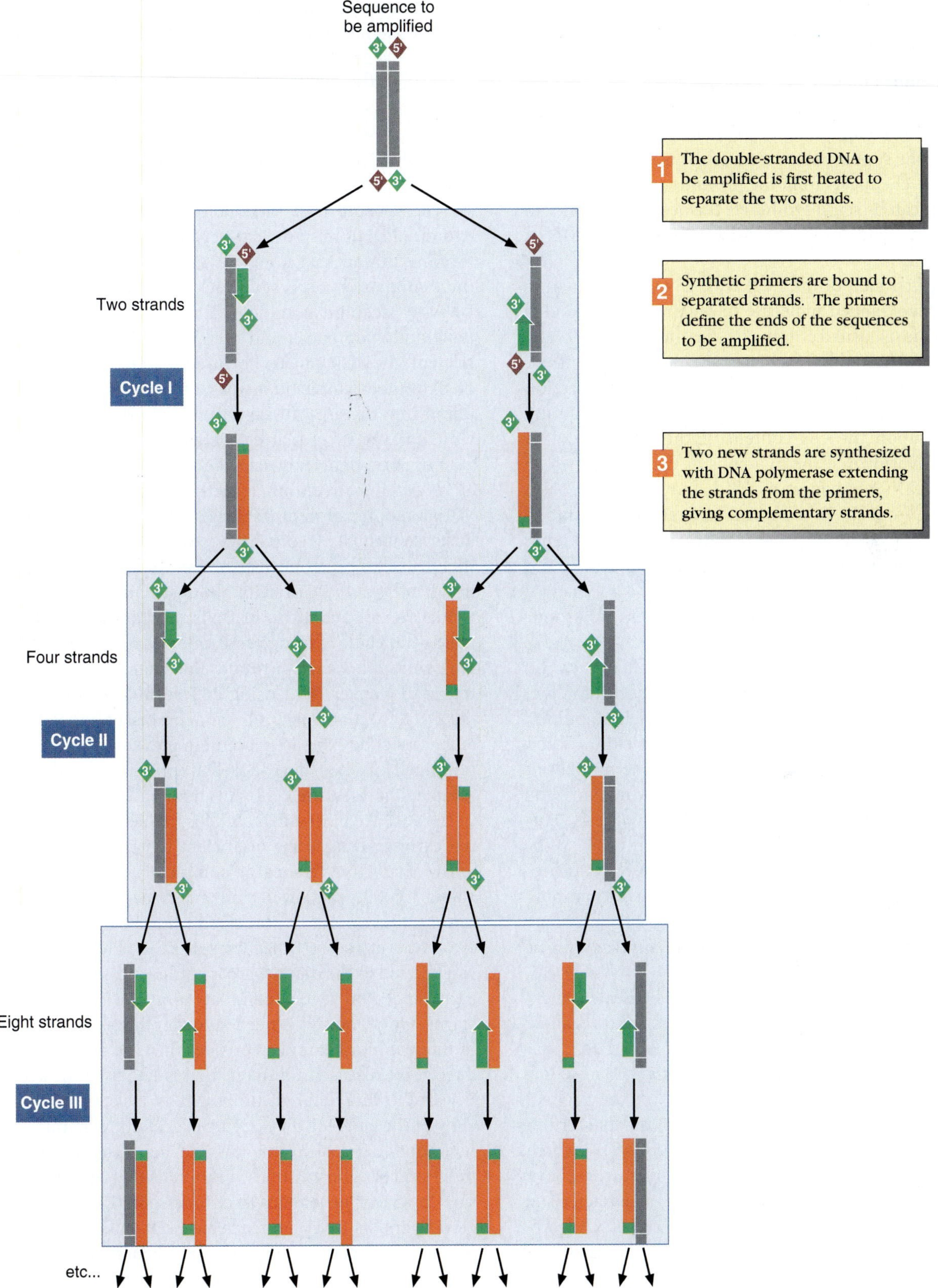

Sequence to be amplified
3'
5'
Two strands
Cycle I
Four strands
Cycle II
Eight strands
Cycle III
etc...
1 The double-stranded DNA to be amplified is first heated to separate the two strands.
2 Synthetic primers are bound to separated strands. The primers define the ends of the sequences to be amplified.
3 Two new strands are synthesized with DNA polymerase extending the strands from the primers, giving complementary strands.

FIGURE 12.13 The polymerase chain reaction. The PCR can be used to amplify a specific DNA sequence if the flanking sequences are known. The flanking DNA sequence gives information for the chemical synthesis of the primers. Each cycle is repeated 20–30 times, doubling the amount of DNA each time.

DNA sequence of the amplified region. The area between the two primers is then amplified in a geometric manner.

It is possible to synthesize primers chemically up to 80 nucleotides long with commercially available instruments. These instruments attach the first nucleotide to a solid matrix or support and add nucleotides in a stepwise fashion. The support matrix is washed between each step to remove chemicals from the previous reaction. The instrument can be set to synthesize the desired sequence in the evening, and the next morning the product will be ready for use.

The first step of the PCR with DNA is to denature the double-stranded molecule to obtain single-stranded DNA. This is usually done by heating to 94°C. The primers are then added (one for the $5' \longrightarrow 3'$ strand and one for the $3' \longrightarrow 5'$ strand), and when the reaction mixture is cooled to 40–50°C (the cooling temperature depends upon the GC content of the sequence), the primers hybridize to the two ends of the segment to be amplified. DNA polymerase is added to synthesize DNA, extending from the primers and generating two double-stranded copies of the gene or DNA sequence (Figure 12.13). In the second step, DNA is again denatured and cooled (the primers are still present), and the polymerase synthesizes new strands; this time the process yields four double-stranded sequences. These sequences, in turn, are used as templates to synthesize 16 strands, then 32, and 64, and so on, during each cycle of heating, cooling, and polymerization. Twenty cycles will yield 1 million new DNA sequences, and 30 cycles will yield 1 billion DNA sequences from the original DNA. The number of final copies can be calculated using 2^n, where n is the number of cycles of heating, cooling, and synthesis. This same procedure is used to amplify RNA with the exception that RNA must first be converted into DNA using reverse transcriptase. The PCR is usually used to amplify sequences of less than 1500 bases long, but with some modifications of the procedure, segments as long as 40,000 bases can be amplified.

During the PCR, the mixture must be heated every cycle to separate the DNA strands and cooled to allow the primers to anneal. Heating to 94°C inactivates *E. coli* DNA polymerase, which makes it necessary to add new polymerase at each step. Improvements in the original technique use DNA polymerase from the bacterium *Thermus aquaticus* (abbreviated *Taq* polymerase), which normally lives in hot springs at temperatures above 80°C. The *Taq* polymerase is heat-stable, a characteristic that gives it an advantage over the *E. coli* DNA polymerase. One disadvantage of *Taq* polymerase is that it does not possess proofreading properties, thus mistakes in synthesis occur at a rate of about 1 in 1000 bases incorporated. These mistakes are amplified in subsequent cycles, thus producing copies that are not true to the original. Other heat-stable polymerases have subsequently been discovered that possess proofreading properties, but they are usually more expensive, and investigators use them when they need an accurate PCR-generated copy of a gene.

Using the PCR, researchers were able to obtain enough DNA to sequence a portion of the DNA from a woolly mammoth, freed from its icy grave of 40,000 years. The DNA had been greatly degraded, but sufficient template DNA remained to support the PCR. Researchers, using PCR, were able to obtain enough mitochondrial DNA from the extinct species to eventually determine the entire sequence. This DNA sequence was then compared to present-day elephant mitochondrial DNA to ascertain homology. The findings showed that the woolly mammoth and the elephant are very closely related. The PCR can also be used to detect the presence of extremely small quantities of DNA, such as DNA found in a disease-causing virus. For example, an individual may suspect that he/she has the HIV-1 virus (human immunodeficiency or AIDS virus) but may test negative using conventional methods (a false-negative). With a modification of the PCR method (HIV is an RNA virus, so it must be converted to DNA first), one copy of the virus can be replicated for easy detection. In fact, a disadvantage of PCR is its ability to amplify such extremely small quantities of DNA that just one contaminating molecule (during preparation of the sample in the lab) will give a false-positive. For example, a scientist trying to recover DNA using PCR from 10-million-year-old DNA from an ancient bone or an insect imbedded in amber must be *extremely* careful to prevent even one bacterial cell or shed human cell from contaminating the sample. If it does, the PCR will amplify the bacterial DNA in the absence of fossil DNA. This, in fact, occurred during early investigations of ancient bone DNA. We now know that DNA is nearly completely degraded after about 30,000 years, even under the most desirable conditions, and cannot be recovered using PCR.

What are some other uses of PCR? This procedure can be used in the diagnosis of hereditary diseases in very early stages of development (even at the 8- or 16-cell stage). Only one cell is needed to amplify a specific gene of interest. For example, consider the case of Tay-Sachs, a recessive disease in humans that is lethal by the age of 3 years. Individuals with Tay-Sachs lack a functional copy of the gene for the enzyme *N*-acetylhexosaminidase A. Individuals with this disease accumulate an unprocessed ganglioside in the brain cells that causes death. The Tay-Sachs gene maps to chromosome 15, it has been cloned, and its DNA sequence has been determined. Tay-Sachs disease is relatively rare in the general population, but it occurs at an incidence of 1 in 3600 in Ashkenazi Jews of central European origin. A relatively straightforward assay for heterozygosity of this gene can be done on a blood sample of an individual without using the PCR. If two high-risk individuals were to marry and find that they were both heterozygotes, they could then have fertilization done *in vitro* and allow development of the embryo to the 16-cell stage. They could then test one of the 16 cells for homozygosity or heterozygosity using the PCR. The test is done by designing two primers that flank the ends of the gene. The gene is then amplified, and the DNA sequence is determined. If no defective gene is found, the embryo can be implanted in the mother. Some other uses of the PCR include amplifying DNA to determine sex, in forensics, diagnosis of infections, and determination of DNA sequences, which is discussed in the next section.

A Cloned Gene Can Be Characterized To Reveal Basic Information.

After a segment of DNA is cloned, it can be characterized in a variety of ways. For example, a cloned DNA fragment will have a distinctive combination of restriction endonuclease recognition sequences, and these sites can be located to create a map. The ultimate characterization of a cloned DNA fragment is, of course, the sequence of bases of the clone, which may indicate the presence of genes and where the gene(s) start and stop.

Cloned DNA Sequences Have Characteristic Restriction Maps.

After digesting a DNA fragment with restriction endonucleases, DNA fragments of different sizes are generated, and these differently sized fragments can be used to measure the distance between the locations of restriction sites. **Gel electrophoresis** (Figure 12.14) is a commonly used procedure for estimating the size of DNA fragments. This is done by placing the sample in a precast well at one end of the gel and applying an electric current to the medium surrounding the gel. Because DNA has a net negative charge, it will migrate to a positive pole. The gel is composed of either agarose or polyacrylamide and acts as an inert matrix that retards the migration of DNA molecules in the electric field, with larger fragments of DNA traveling slowly and smaller fragments traveling more rapidly. DNA fragments of known size are included to accurately estimate the length of DNA fragments during electrophoresis. By varying the concentration of

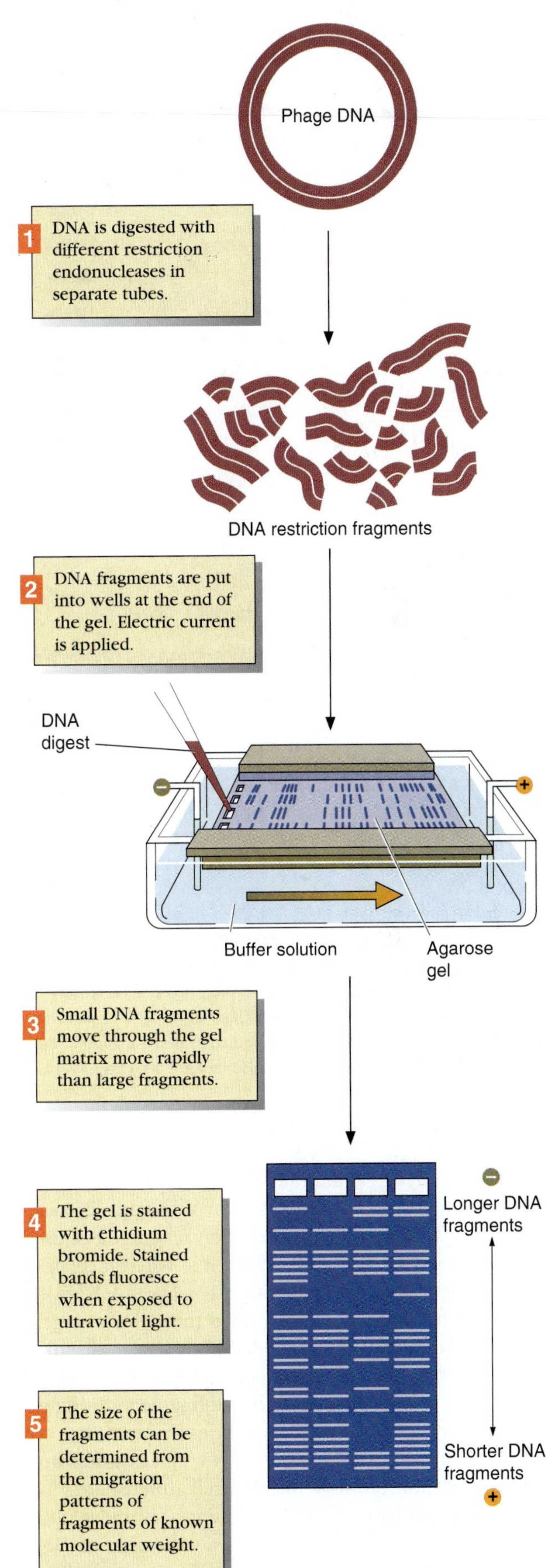

Phage DNA
1 DNA is digested with different restriction endonucleases in separate tubes.
DNA restriction fragments
2 DNA fragments are put into wells at the end of the gel. Electric current is applied.
DNA digest
Buffer solution
Agarose gel
3 Small DNA fragments move through the gel matrix more rapidly than large fragments.
4 The gel is stained with ethidium bromide. Stained bands fluoresce when exposed to ultraviolet light.
5 The size of the fragments can be determined from the migration patterns of fragments of known molecular weight.
Longer DNA fragments
Shorter DNA fragments

▶**FIGURE 12.14 Agarose gel electrophoresis.** The DNA sample is digested with different restriction endonucleases and separated according to size by agarose gel electrophoresis. Larger DNA molecules travel through the gel more slowly. Different banding patterns are seen for each enzyme as the cut sites, and thus fragment sizes vary. Each band represents a collection of fragments of a particular size.

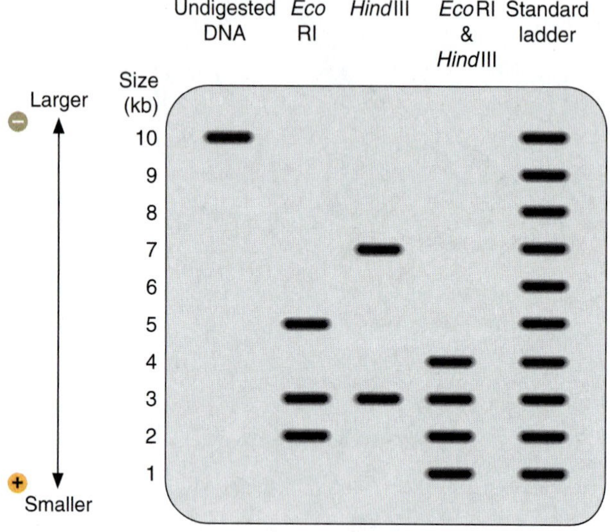

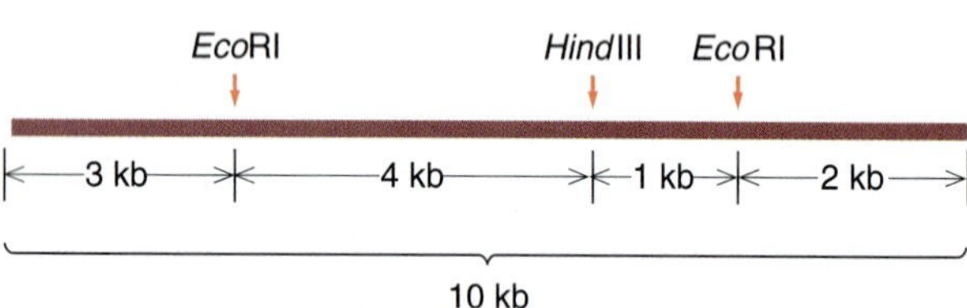

FIGURE 12.15 Construction of a restriction map. A restriction map of a DNA fragment can be prepared using information from agarose gel electrophoresis. The size of the DNA fragments generated from restriction endonuclease digestion is indicated in kilobases of DNA and is determined from simultaneously run standards. The linear segment of DNA was digested with *Eco*RI, *Hin*dIII, or *Eco*RI + *Hin*dIII. *Eco*RI digestion generates three fragments of 5, 3, and 2 kb, whereas *Hin*dIII generates only two fragments of 7 and 3 kb. Digestion by both enzymes simultaneously generates a new band at 1 kb and another at 4 kb, which equals 5 kb (the total is always 10 kb). This indicates that the *Hin*dIII site is within the 5-kb *Eco*RI fragment, and also that the 5-kb fragment is internal to the other two fragments.

agarose or polyacrylamide and the electrophoresis conditions, fragments that differ in size by thousands of bases or just one base can be separated. Polyacrylamide is used for separating smaller DNA fragments, whereas agarose gels are used for larger fragments. Also, because DNA restriction fragments move unharmed through these gels, they can be eluted as intact, double-helical molecules. After digestion of a DNA fragment with a restriction endonuclease and separation of the subfragments using gel electrophoresis, they appear as a series of bands that can be visualized after staining the DNA with **ethidium bromide.** Ethidium bromide inserts itself (intercalates) between adjacent bases of DNA and fluoresces under ultraviolet light. Each band represents a collection of DNA molecules of the same size class.

The pattern of DNA fragments, after separation by agarose gel electrophoresis, is dependent upon the frequency and placement of restriction endonuclease recognition sites within the original sequence. For example, a hypothetical linear DNA fragment of 10 kb that has been digested with *Eco*RI and separated on an agarose gel reveals three bands of 2, 3, and 5 kb (Figure 12.15). With just this information, it is not possible to determine the order of the fragments. To determine the order of the fragments, additional digestions with other enzymes and simultaneous digestion with two enzymes (double digestions) are necessary. In this example, *Hin*dIII digestion of the same DNA yields two fragments of 7 and 3 kb. However, a double digestion with both enzymes gives four fragments of 1, 2, 3, and 4 kb. From this information, we can deduce that the 7-kb *Hin*dIII fragment was cleaved into a 4- and 3-kb fragment by *Eco*RI (4 plus 3 add to 7 kb), and the 3-kb *Hin*dIII fragment is cleaved into a 1- and 2-kb fragment by *Eco*RI (1 plus 2 add to 3 kb, and the total adds to 10 kb). Thus, the 10-kb fragment contains one *Hin*dIII cleavage site, and each of the *Hin*dIII fragments contains one *Eco*RI cleavage site. This accounts for all the fragments generated by the double digestion.

Another, simpler means of ordering fragments in a DNA molecule is partial digestion. The order of the fragments can be determined by slowing down the rate of restriction endonuclease digestion, either by digesting at a lower temperature or decreasing the incubation time or enzyme concentration. This yields both partially digested and completely digested DNA fragments (Figure 12.16). Using this approach, it is possible to construct a restriction map of a DNA fragment or plasmid with the

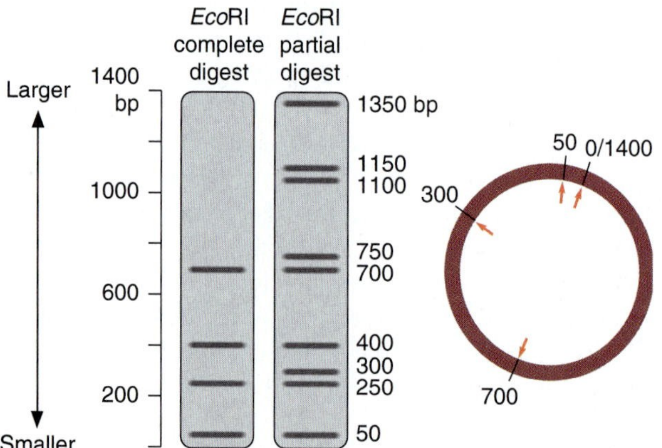

FIGURE 12.16 Mapping by partial digestion. A restriction map of a small, 1400-bp circular plasmid prepared by partial digestion of the DNA. The DNA was digested with *Eco*RI, either completely or partially. By adding the fragment sizes and comparing them to the map, fragment sizes adjacent to each other on the plasmid map appear on the gel (by adding the size of contiguous fragments, nonadjacent fragments are not present; i.e., no 950-bp molecule cleaved into 250- and 700-bp fragments).

relative order of the fragments unambiguously established. During partial digestion of the DNA, fragments of all sizes are generated (including undigested and completely digested DNA). If two fragments are attached to form one fragment, they must be adjacent. Thus, fragments of increasing size represent the distances from the ends of cut sites. If the molecule is circular, as in Figure 12.16, complete digestion produces fragments that add to 1400 bp (adding the size of all the fragments in the first lane) and has four restriction sites. Partially digested DNA must yield fragments that equal in size two or more of the individual fragments. Thus, if the enzyme cuts between the 700- and 400-bp fragments, a 1100-bp fragment will be found. However, if the 700-bp fragment were adjacent to the 250-bp fragment, a 950-bp fragment would be found. Because they are not adjacent, no 950-bp fragment is observed. This same logic is used to place each fragment adjacent to their respective adjacent fragments.

By combining these different approaches, detailed restriction maps have been constructed for many plasmids, genes, chromosomes fragments, and whole chromosomes. For example, the restriction map of the *E. coli* chromosome was completed in 1987 using six different enzymes that recognize six-base sequences. This represents a monumental achievement because the *E. coli* chromosome is a 4.6-million-bp circle.

SOLVED PROBLEM

Problem

A linear DNA molecule is digested with two restriction endonucleases, separately and simultaneously, with the following results.

Enzyme	Fragment Sizes (kb)
*Bam*HI	8.0, 7.5, 4.5, 2.9
*Bgl*II	13, 6, 3.9
*Bam*HI & *Bgl*II	7.5, 6, 3.5, 2.9, 2, 1

With this information, draw a restriction map of the linear fragment.

Solution

In the double digestion with *Bam*HI and *Bgl*II, three fragments are not cleaved into smaller fragments (7.5-, 2.9-, and 6-kb fragments), meaning that they do not possess internal cleavage sites for the other enzyme (for example, the 7.5-kb *Bam*HI fragment does not possess an internal *Bgl*II site). Digestion with *Bgl*II results in three fragments, meaning that two *Bgl*II cleavage sites exist in the linear fragment, and *Bam*HI has three cleavage sites. The 6-kb *Bgl*II fragment remains after the double digestion, suggesting that it does not have an internal *Bam*HI site, but the 13- and 3.9-kb *Bgl*II fragments disappear after the double digestion. The 2.9- and 1-kb fragments add to 3.9 kb, suggesting that the 3.9-kb *Bgl*II fragment has an internal *Bam*HI site. The 2-, 7.5-, and 3.5-kb fragments add to 13 kb, suggesting that the 13-kb *Bgl*II fragment has two *Bam*HI sites within it. The 3.5- and 1-kb fragments add to 4.5 kb, suggesting that the 4.5-kb *Bam*HI fragment has a *Bgl*II site. Taking these observations together, the following map is consistent with the digestion data.

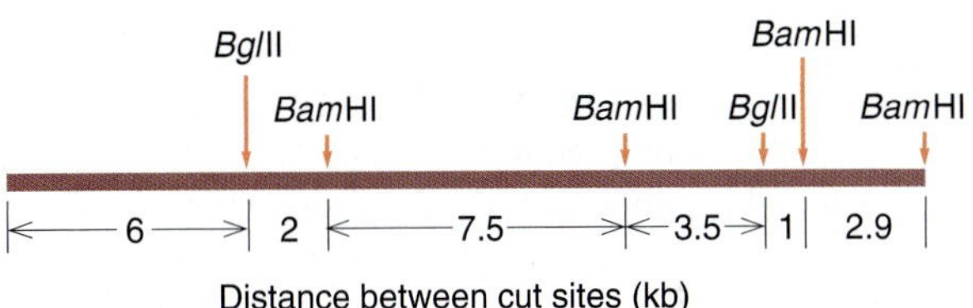

SECTION REVIEW PROBLEMS

9. A digest of a linear fragment of DNA separated on an agarose gel follows. The DNA was first digested with *Hin*dIII alone, then *Eco*RI alone, and both *Eco*RI and *Hin*dIII. Construct a restriction map of the DNA sequence:

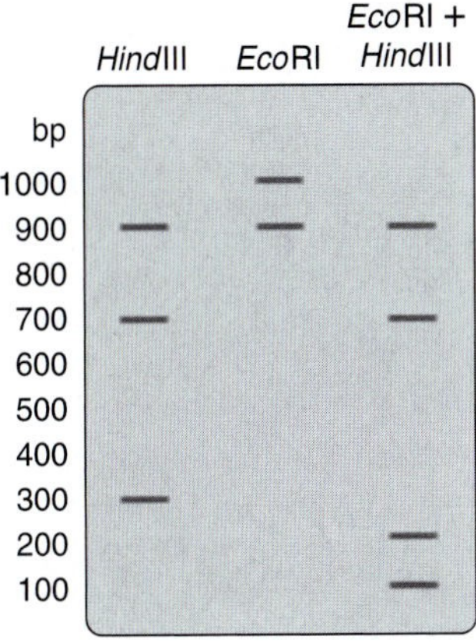

10. The separation pattern of a fragment of DNA digested separately with *Hin*dIII and *Sal*I and with both enzymes simultaneously follows. Construct a map of the fragment of DNA. Fragment size is given in kilobase pairs.

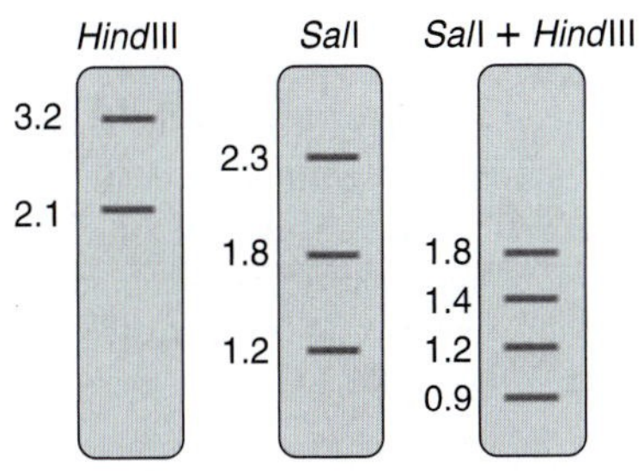

Cloned DNA Fragments Can Be Sequenced.

One of the primary uses of cloning vectors is for the eventual sequencing of the DNA insert. DNA sequencing is dependent upon obtaining large quantities of a DNA fragment, thus gene cloning and the PCR are extremely helpful to DNA sequencing. An example of a plasmid cloning vector that possesses many useful characteristics for cloning of DNA fragments for sequencing is M13mp19 (Figure 12.17).

The most useful method for sequencing DNA is the **Sanger dideoxy sequence technique** (Figure 12.18). This method, developed in the 1970s by Frederick Sanger (Medical Research Council, Cambridge, England), depends upon using a chain-terminating nucleotide that has both the 2′ and 3′ hydroxyl groups removed from the deoxyribose to produce a 2′,3′-dideoxyribonucleotide (ddNTP, where *N* refers to any of the four nucleotides). A low concentration of a different ddNTP is added to each of four separate reaction mixtures, which also contain the normal deoxyribonucleotides (dNTP). This mixture also contains DNA polymerase, the single-stranded template of the sequence to be determined, and a primer to start the reaction. Because the 3′-OH is needed for phosphodiester bond formation with the next nucleotide, the absence of a 3′-OH on ddNTP terminates chain growth when ddNTP is incorporated into the growing chain in place of the corresponding dNTP. This generates a series of DNA molecules that differ in length, depending upon where the ddNTP is incorporated. Random chain termination occurs when the DNA polymerase adds the ddNTP. For example, if ddATP were added to a mixture of primer, template (the sequence to be determined), and DNA polymerase, polymerization would continue until a ddATP is incorporated, and then it would stop at an A. Because the ddATP concentration is about 1/100th of dATP, ddATP it is not always incorporated at all locations; dATP usually is. Thus, a series of different-length polymers is formed, each ending with ddATP. After the reactions are complete, the newly formed DNA strands are separated by polyacrylamide gel electrophoresis, one lane for each nucleotide. Each band in a lane represents the varying lengths of the growing strand where synthesis stopped because of the addition of a ddNTP. As a consequence, every nucleotide in the template strand has a corresponding band in the gel, depending upon where chain termination occurred. Thus, the sequence of the nucleotides in the DNA fragment can be determined directly from the gel after autoradiography (when using a radioactive label to locate the band) because each fragment differs in length by one nucleotide. For a detailed explanation, see Figure 12.18. The elongating strand is

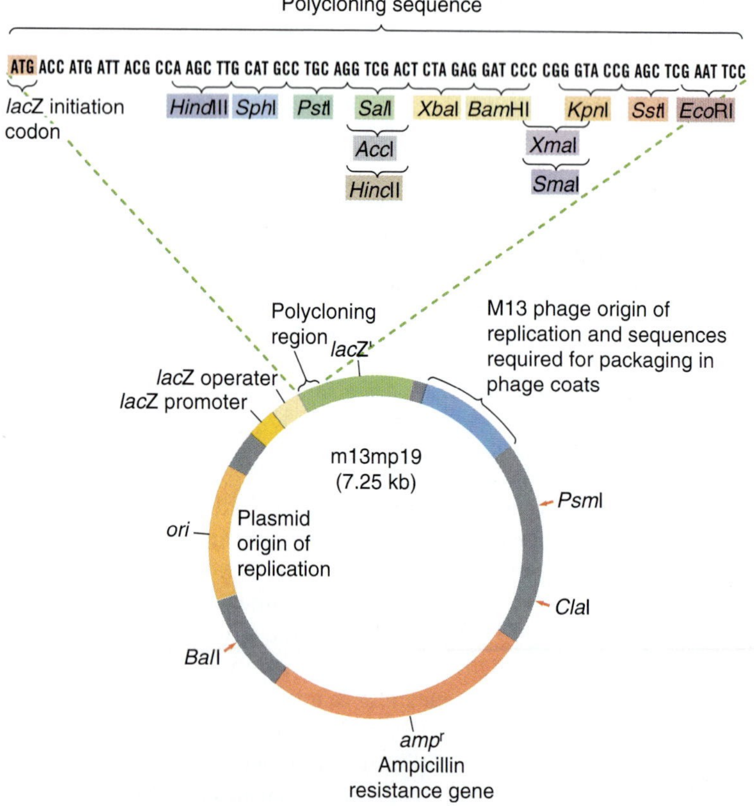

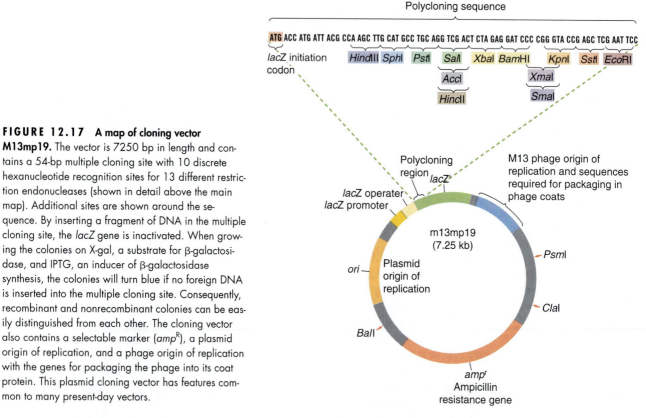

FIGURE 12.17 A map of cloning vector M13mp19. The vector is 7250 bp in length and contains a 54-bp multiple cloning site with 10 discrete hexanucleotide recognition sites for 13 different restriction endonucleases (shown in detail above the main map). Additional sites are shown around the sequence. By inserting a fragment of DNA in the multiple cloning site, the *lacZ* gene is inactivated. When growing the colonies on X-gal, a substrate for β-galactosidase, and IPTG, an inducer of β-galactosidase synthesis, the colonies will turn blue if no foreign DNA is inserted into the multiple cloning site. Consequently, recombinant and nonrecombinant colonies can be easily distinguished from each other. The cloning vector also contains a selectable marker (amp^R), a plasmid origin of replication, and a phage origin of replication with the genes for packaging the phage into its coat protein. This plasmid cloning vector has features common to many present-day vectors.

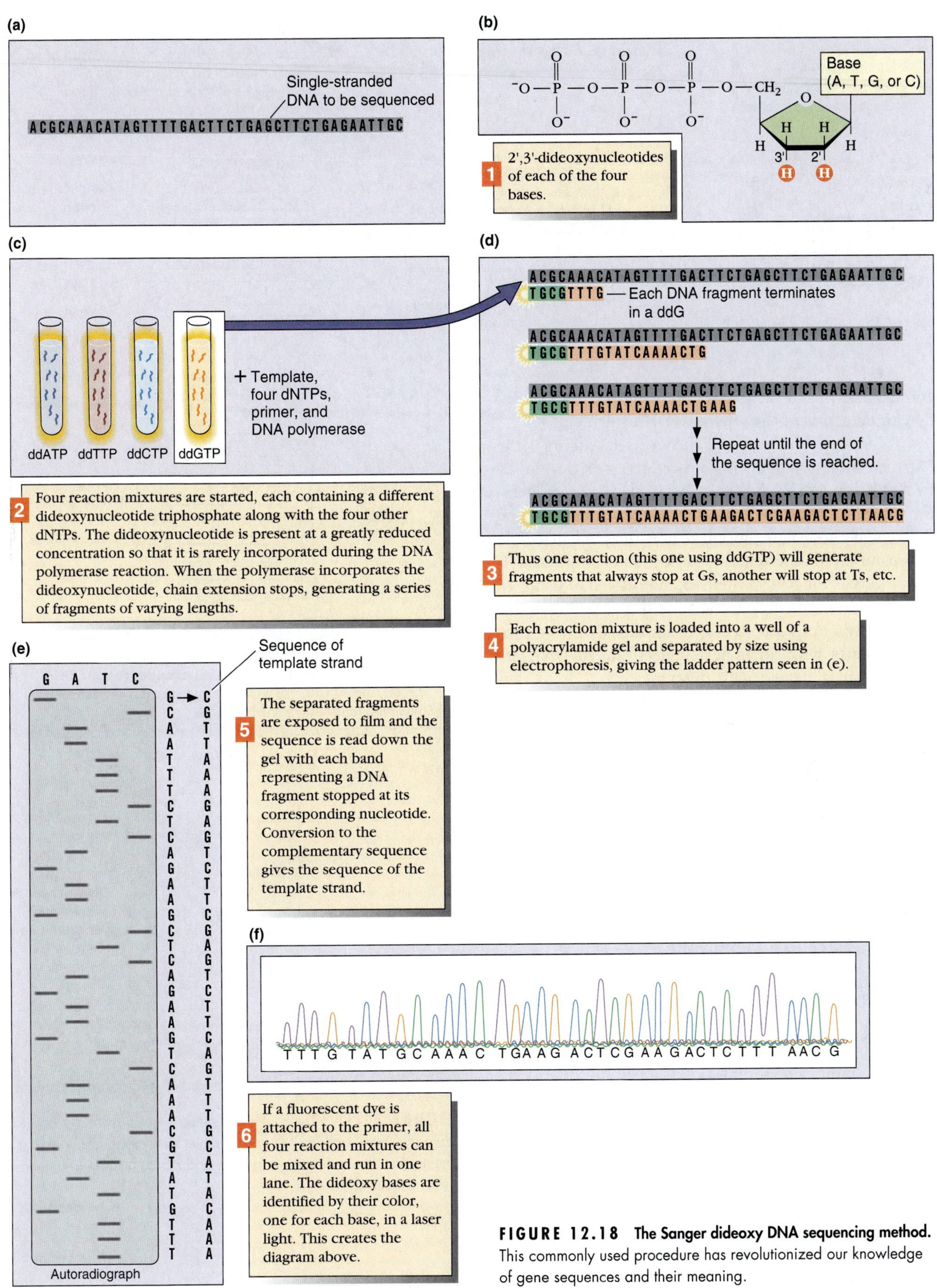

(a)
Single-stranded DNA to be sequenced
ACGCAAACATAGTTTTGACTTCTGAGCTTCTGAGAATTGC
(b)
Base (A, T, G, or C)
$^{-}O-P(=O)(O^{-})-O-P(=O)(O^{-})-O-P(=O)(O^{-})-O-CH_2$
O
H H
H 3' 2' H
H H
1 2',3'-dideoxynucleotides of each of the four bases.
(c)
ddATP ddTTP ddCTP ddGTP
+ Template, four dNTPs, primer, and DNA polymerase
2 Four reaction mixtures are started, each containing a different dideoxynucleotide triphosphate along with the four other dNTPs. The dideoxynucleotide is present at a greatly reduced concentration so that it is rarely incorporated during the DNA polymerase reaction. When the polymerase incorporates the dideoxynucleotide, chain extension stops, generating a series of fragments of varying lengths.
(d)
ACGCAAACATAGTTTTGACTTCTGAGCTTCTGAGAATTGC
TGCGTTTG — Each DNA fragment terminates in a ddG
ACGCAAACATAGTTTTGACTTCTGAGCTTCTGAGAATTGC
TGCGTTTGTATCAAAACTG
ACGCAAACATAGTTTTGACTTCTGAGCTTCTGAGAATTGC
TGCGTTTGTATCAAAACTGAAG
Repeat until the end of the sequence is reached.
ACGCAAACATAGTTTTGACTTCTGAGCTTCTGAGAATTGC
TGCGTTTGTATCAAAACTGAAGACTCGAAGACTCTTAACG
3 Thus one reaction (this one using ddGTP) will generate fragments that always stop at Gs, another will stop at Ts, etc.
4 Each reaction mixture is loaded into a well of a polyacrylamide gel and separated by size using electrophoresis, giving the ladder pattern seen in (e).
(e)
G A T C
Sequence of template strand
G → C
GCAATTTCTCAGAAGCTCAGAAGTCAAACGTATGTTT
CGTTAAAGAGTCTTCGAGTCTTCAGTTTGCATACAAA
Autoradiograph
5 The separated fragments are exposed to film and the sequence is read down the gel with each band representing a DNA fragment stopped at its corresponding nucleotide. Conversion to the complementary sequence gives the sequence of the template strand.
(f)
T T T G T A T G C A A A C T G A A G A C T C G A A G A C T C T T T A A C G
6 If a fluorescent dye is attached to the primer, all four reaction mixtures can be mixed and run in one lane. The dideoxy bases are identified by their color, one for each base, in a laser light. This creates the diagram above.

FIGURE 12.18 The Sanger dideoxy DNA sequencing method. This commonly used procedure has revolutionized our knowledge of gene sequences and their meaning.

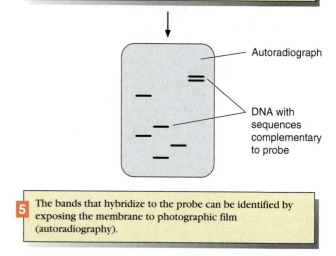

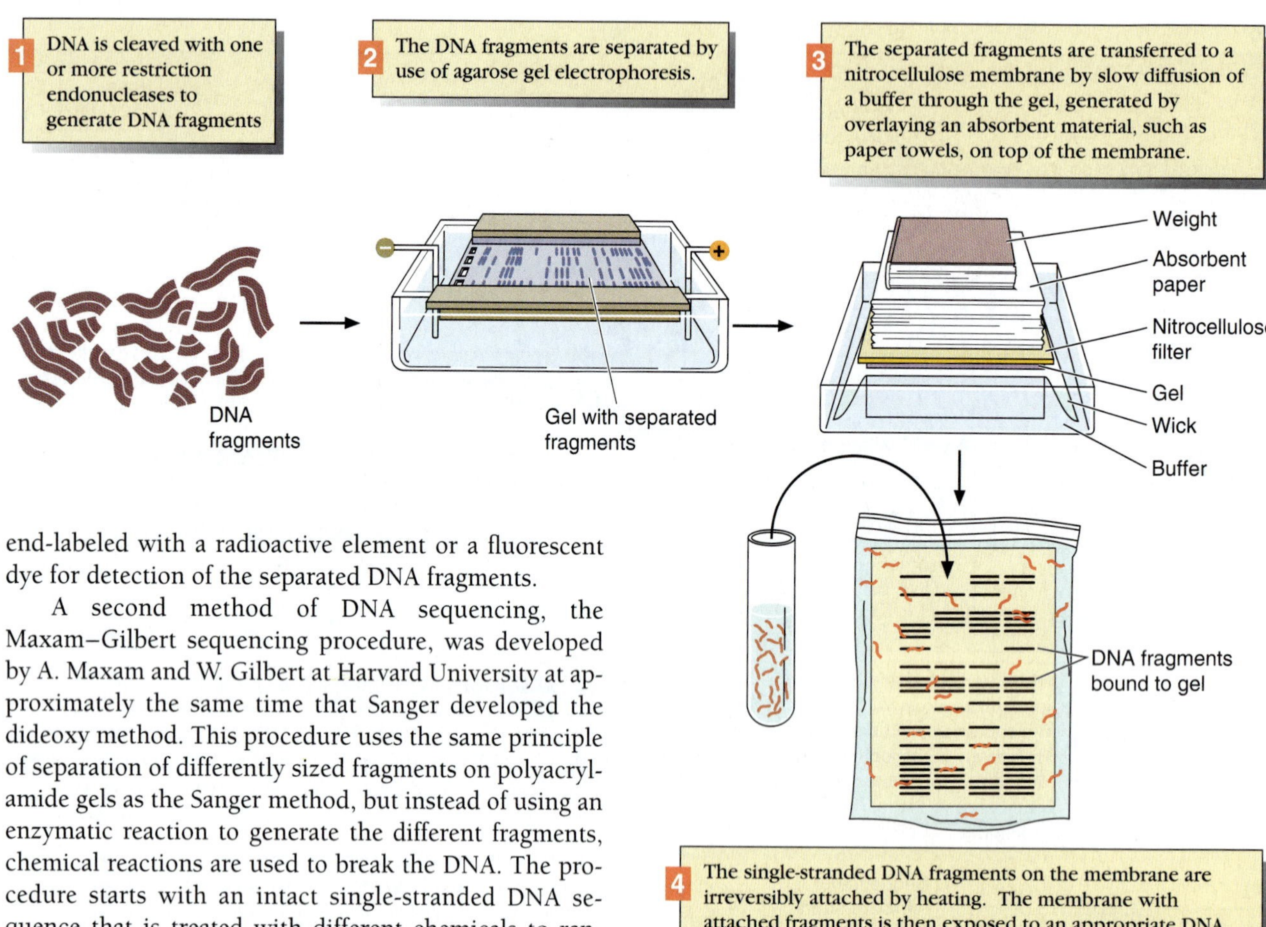

1 DNA is cleaved with one or more restriction endonucleases to generate DNA fragments
2 The DNA fragments are separated by use of agarose gel electrophoresis.
3 The separated fragments are transferred to a nitrocellulose membrane by slow diffusion of a buffer through the gel, generated by overlaying an absorbent material, such as paper towels, on top of the membrane.
DNA fragments
Gel with separated fragments
Weight
Absorbent paper
Nitrocellulose filter
Gel
Wick
Buffer
DNA fragments bound to gel
4 The single-stranded DNA fragments on the membrane are irreversibly attached by heating. The membrane with attached fragments is then exposed to an appropriate DNA probe labeled with ^{32}P where complementary sequences will bind to the specific DNA bands.
Autoradiograph
DNA with sequences complementary to probe
5 The bands that hybridize to the probe can be identified by exposing the membrane to photographic film (autoradiography).

FIGURE 12.19 The Southern blotting procedure. This procedure effectively and easily transfers DNA from an agarose gel to a membrane where a probe can be hybridized to a complementary DNA sequence.

end-labeled with a radioactive element or a fluorescent dye for detection of the separated DNA fragments.

A second method of DNA sequencing, the Maxam–Gilbert sequencing procedure, was developed by A. Maxam and W. Gilbert at Harvard University at approximately the same time that Sanger developed the dideoxy method. This procedure uses the same principle of separation of differently sized fragments on polyacrylamide gels as the Sanger method, but instead of using an enzymatic reaction to generate the different fragments, chemical reactions are used to break the DNA. The procedure starts with an intact single-stranded DNA sequence that is treated with different chemicals to randomly break the DNA strand at A, G, C, and T ends. These fragments are then separated just as in the Sanger procedure. The more commonly used procedure is the Sanger method because it is easier to use and it is adaptable to automation. The Sanger procedure has been used to sequence genomes from a number of organisms (see *CURRENT INVESTIGATIONS: The Human Genome Will Be Sequenced Soon*).

DNA Fragments Can Be Identified After Transfer to a Membrane.

DNA cloned into vectors can be used to characterize the identity of a gene, to locate the coding regions within the DNA fragment, and to study the regulatory regions flanking the coding region of the gene. With the development of gene cloning, a method was developed for identifying specific DNA fragments after digestion with a restriction endonuclease and separation on agarose gels. This procedure, called **Southern blotting,** was developed by E. M. Southern (Figure 12.19). In this procedure, DNA is first cleaved with one or more restriction endonucleases to create populations of restriction fragments of differing sizes. These DNA fragment populations are separated according to size using electrophoresis in an agarose gel. The number of bands depends upon the complexity of the DNA and the number of restriction enzymes used to cleave the DNA. After separation of the DNA, it is denatured (in the gel using low pH) and transferred to a nitrocellulose membrane by the capillary flow of a buffer through the gel toward the nitrocellulose membrane. This carries the DNA fragments from the gel onto the nitrocellulose membrane. The loosely attached fragments on the membrane are baked in a vacuum oven to tightly bind the single-stranded fragments to the membrane. These fragments can be hybridized to radioactive DNA probes that possess complementary sequences to one or more of the separated fragments. After hybridization, exposing the membrane to an X-ray film (followed by development of the film) will reveal a precise and reproducible pattern of bands. For example, the probe could be for the gene specifying the β-globin protein of hemoglobin, and the DNA fragments could be a complete digest of the entire genome of an organism. The probe for the β-globin gene will hybridize only to fragments that contain complementary sequences. Exposing the radioactivity to photographic film will reveal the exact DNA band in the original gel that corresponds to complementary sequences of the probe. Thus, the probe has identified the size and location of the band in the gel for the gene for the β-globin protein of hemoglobin, which is useful information for cloning and characterizing the gene. For example, heterologous probes could be used to find a specific band, and the copy number of the gene would be revealed from the number of bands.

A similar technique for analyzing RNA has been developed and is known as **northern blotting.** With this technique, RNA (for example, the total mRNA from an organism) is separated using an agarose gel. As with Southern blotting, the RNA is transferred to a membrane and then hybridized to an appropriately labeled probe. Autoradiography will produce bands indicating the number and size of the RNA sequences complementary to the probe or whether the gene is actually expressed. Northern blots aid in characterizing specific genes and the level of gene expression. For example, an investigator may not know whether a gene under investigation is expressed in a specific tissue or at what level the gene is expressed. RNA could be isolated from the tissue, separated by gel electrophoresis, and transferred to a membrane for analysis. A probe for the gene is labeled with a radioactive nucleotide and hybridized to the RNA on the membrane. Only those sequences that have homology will bind to the radioactive probe. In this way, the size of the RNA can be determined (by its migration rate in the gel when compared to known standards), and the level of expression can be estimated by the amount of the label attached to the RNA. It is also possible that two differently sized fragments are generated as a result of differential processing of the RNA. Northern blotting is a very valuable technique in molecular biology.

Probe DNA must have a high level of radioactivity to detect extremely small quantities of DNA or RNA on the nitrocellulose membrane using autoradiography. The probe is labeled in several different ways. A commonly used method uses a radioactive (usually ^{32}P) nucleotide triphosphate to label the DNA. The DNA fragment is first nicked with DNase, and then DNA polymerase I from *E. coli* is added, which results in $5' \longrightarrow 3'$ exonuclease degradation from the nick while simultaneously synthesizing a new strand (commonly known as **nick translation**). Another method for labeling DNA probes uses **random primer extension.** In this procedure, a mixture of chemically synthesized random oligonucleotides 6 bases in length ($4^6 = 4096$ different sequences) is used as a primer on a single-stranded segment of DNA. Sequence similarity to the 6-bp sequences will, by chance, be present, and what is called the Klenow fragment of DNA polymerase (which synthesizes DNA but lacks the $5' \longrightarrow 3'$ exonuclease portion of the DNA polymerase) will be used to add radioactive nucleotides to the DNA probe. The result is a very high level of radioactivity that can be easily detected on photographic film after Southern or northern hybridization. Finally, DNA can be labeled using nonradioactive nucleotides. Instead of radioactivity, one of the nucleotides has a side group that binds an enzyme (such as alkaline phosphatase), which reacts with specific substrates to emit a luminescent signal. The luminescent signal is used just like a radioactive signal and is detected by exposure to film.

SOLVED PROBLEM

Problem

Using the following data, determine the nucleotide sequence of a DNA insert cloned into an M13 cloning vector.

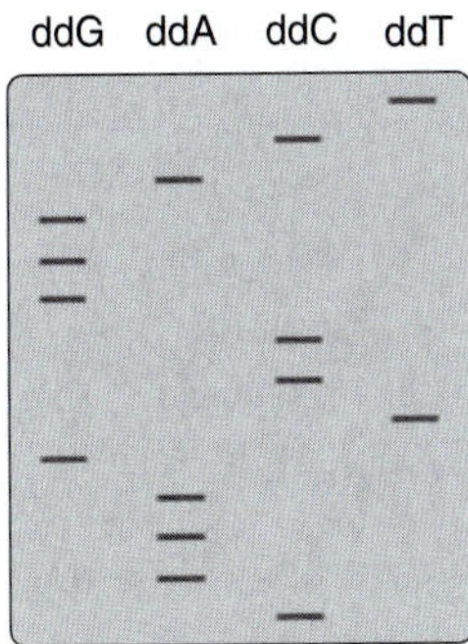

Solution

Examining the sizes of the separated fragment in each lane of the gel reveals the DNA sequence. For example, in the ddG lane, four bands appear that represent DNA fragments where synthesis stopped because of the incorporation of a ddG instead of a dG. Thus, whenever a G appears in the strand of DNA, a band will appear and reflects the location of the G in the parent strand. Similarly, bands appear in each of the other lanes because ddA, ddC, or ddT were incorporated in place of their corresponding deoxynucleotides. The sequence is read by starting at the top and identifying the largest fragment that represents the first base (T in this case). The next largest fragment represents a fragment stopped at the next base, and so forth. Reading from the top the sequence is

3′-TCAGGGCCTGAAAC-5′

SECTION REVIEW PROBLEMS

11. If you had a genomic library of an organism, let us say the horse, what would be the most straightforward method of finding the gene that codes for proline tRNA?

12. The following separation gel was obtained for the sequencing of a fragment of DNA using the Sanger method. What is the sequence of the DNA obtained from the data from the gels?

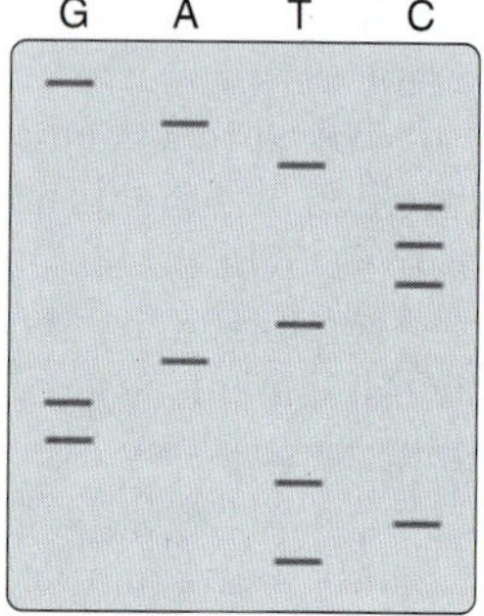

13. What is the difference between Southern and northern blotting?

CURRENT INVESTIGATIONS

The Human Genome Will Be Sequenced Soon

In the mid-1980s, instrumentation was developed for sequencing DNA at the rate of 5000–10,000 bp per day under ideal conditions. The error rate is in the ballpark of 1 in every 1000 bases, but this error is reduced by sequencing the fragments several times. This technology contrasts to that of the 1970s, when sequencing even pBR322 took several years—all 4363 bp! What is the next step? The yeast and *Escherichia coli* genomes have recently been sequenced, and *Drosophila melanogaster* and *Arabidopsis thaliana* (a small plant studied by plant geneticists) seem to be the next likely candidates because of the tremendous amount of accumulated genetic and molecular knowledge obtained for each.

Needless to say, the Human Genome Project, as it is known, has been very controversial because sequencing 3 billion bases will consume billions of dollars and the rewards have not been clearly defined. Some scientists say that it is a waste of hard-to-come-by dollars and subtracts from money that could be spent for other more worthwhile projects. Other scientists say that it would give us unparalleled information that would be invaluable in understanding and pinpointing the location of the 3500 known inherited human diseases. Already over 1000 genes of the human genome have been sequenced and stored in the computers of the National Gene Bank (along with another 12 million bases from other sequenced genes). It seems clear that the human genome will be sequenced, but the project is so immense that it will be well into the 21st century before it is completed.

The project has been broken down into several steps. First, a very extensive and complete map was generated while simultaneously developing more sophisticated and rapid techniques for sequencing DNA. Second, sets of overlapping DNA fragments that span the entire genome are being generated. The maps will be organized into 24 sets—one for each chromosome, including the X and Y chromosomes. These overlapping fragments are then sequenced, leading to a gene bank in a computer that is accessible to all scientists. Anyone can log into the database and ask any question, such as, Where does a particular cloned DNA sequence belong? The computer may answer: On chromosome number 8; 1,145,397 bases to the left of the centromere. Many practical applications for the complete genome will probably emerge. For example, simultaneous to the sequencing of the human genome, the genomes of other animals and plants are being completed or will be partially known. We can thus compare our DNA sequence directly with other animals and plants to derive evolutionary relationships. Direct comparison of DNA sequences should provide considerably more information than comparison of morphological traits. Finally, we should be able to learn much more about the subtle changes in DNA sequences between members of the human population that give rise to such differences in phenotype as the ability to be a great musician or composer.

Specialized Cloning Vectors Are Widespread.

After plasmid cloning vectors were designed and constructed, many other specialized cloning vectors were constructed, usually for one specific purpose. For example, cloning vectors have been developed to express gene products at high levels in bacteria for protein production, and others were designed to study regulatory regions of genes. In the following sections, we will discuss some of these specialized cloning vectors.

Prokaryotic Vectors Permit the Expression of Eukaryotic Genes.

An important goal of applied gene technology is the production of useful proteins from recombinant DNA molecules. However, genomic eukaryotic genes cannot be expressed in bacteria because most eukaryotic genes possess different regulatory signals and introns that bacteria cannot splice (some yeast genes can be expressed in prokaryotes). A large family of specialized vectors have been created for expressing eukaryotic gene products in large quantities using strong bacterial promoters. These vectors have an efficient prokaryotic promoter adjacent to the cloning site for the eukaryotic gene (which would have its normal promoter removed or, in the case of a cDNA clone, the promoter was never present). As a result, when a eukaryotic gene is inserted into the vector and transformed into *E. coli,* the gene will be expressed and the protein may be produced in large amounts (5% of all the protein made by the cell). This type of vector is called an **expression vector** and contains, in addition to an efficient promoter, both a ribosome binding site and a 5′-ATG-3′ start codon adjacent to the cloning site. A transcription termination signal is also present on the other side of the gene insertion site. An important feature of an expression vector is that the reading frame of the inserted gene must correctly match the reading frame of the initiation codon, thus expression vectors are frequently available in three versions, each differing by 1 bp, which shifts the reading frame.

The tryptophan (*trpE*) promoter is a very efficient promoter and is frequently used for high-level expression of foreign genes in *E. coli*. The *trpE* promoter becomes very active when β-indoyl acetic acid is added or tryptophan is removed from the growth medium. The pWT111 vector possesses a tryptophan promoter/operator/leader region (492 bp) oriented in the same direction as the tetracycline resistance gene (*tet*^{R}) (Figure 12.20). The unique *Hin*dIII site between the *tet*^{R} gene and the *trpE* regulatory region allows insertion of a gene in the same reading frame as the *trpE* gene. The protein product will have seven amino acids from the *trpE* gene, but this is unlikely to affect the function of the protein. This plasmid thus permits cloning of a DNA fragment with HindIII ends in a way that preserves the reading frame at the transition from *trpE* to the recombinant region. Two other plasmids related to pWT111 (pWT121 and pWT131) function the same except that the reading frame of the *Hin*dIII cloning site is shifted 1 bp downstream in each vector, thus accommodating any reading frame of the inserted gene.

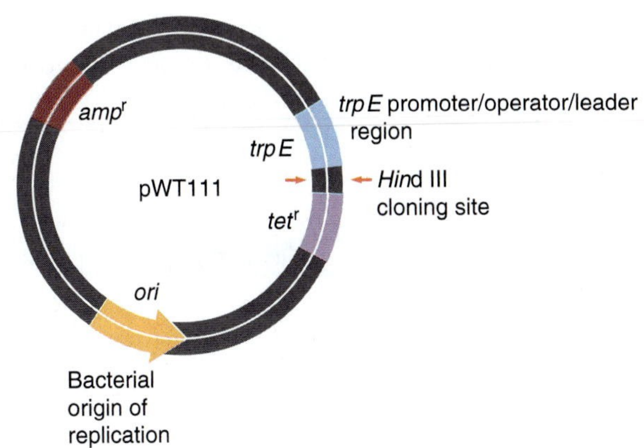

Hind III
pWT111 ATG CAA ACA CAA AAA CCG ACT CCA AGC TTT
Met Gln Thr Gln Lys Pro Thr Pro Ser Phe

Hind III
pWT121 ATG CAA ACA CAA AAA CCG ACT CCA AGC TCC AAG CTT
Met Gln Thr Gln Lys Pro Thr Pro Ser Ser Lys Leu

Hind III
pWT131 ATG CAA ACA CAA AAA CCG ACT CCA AGC TCC AAG CTC CAA GCT TGG
Met Gln Thr Gln Lys Pro Thr Pro Ser Ser Lys Leu Gln Ala Trp

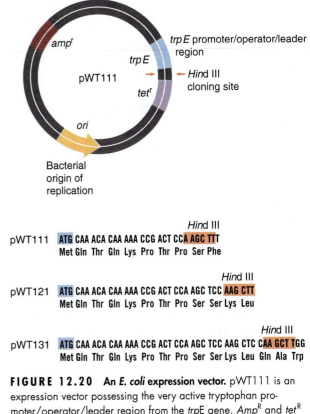

FIGURE 12.20 An *E. coli* expression vector. pWT111 is an expression vector possessing the very active tryptophan promoter/operator/leader region from the *trp*E gene. *Amp*^{R} and *tet*^{R} genes are present for selecting transformed *E. coli* host cells. Cleavage at the *Hin*dIII site, just after the start codon of the gene, leaves seven amino acids of the *trp*E gene fused to the gene of choice. The ribosome binding site and a termination signal are also present. Two other plasmids with unique *Hin*dIII sites each shift the reading frame 1 bp, respectively (pWT121 and pWT131). This enables a gene of choice to be inserted into any of the three reading frames and provides a very high level of expression after induction with 3-β-indolyl acetic acid.

The first practical application of genes fused to a highly active promoter produced human insulin, rat insulin, rat growth hormone, and human growth hormone. Animal protein hormones are especially attractive candidates for bacterial production because they are produced in such small quantities in animals. This technology allows large quantities of the hormone to be produced at a cost greatly reduced compared to isolating it from the original animal source. An example is human growth hormone, which is now commercially available at an affordable price. A strain of *E. coli* has been constructed that produces 3% of its total protein as human growth hormone, which is a considerable amount of protein product from one gene. (See *CURRENT INVESTIGATIONS: Many Different Proteins Are Currently Manufactured Using Cloned Genes.*)

SOLVED PROBLEM

Problem

You have the coding sequence of a gene cloned into a plasmid vector, and it includes seven bases before the ATG start codon and seven bases downstream from the stop codon. The restriction endonuclease *Bam*HI recognizes a region in the 7-bp sequence just before the start codon, and there is a *Sal*I recognition site just after the translation termination triplet. You would like to transfer this gene into the multiple cloning site of an expression vector pT7.7, which has a strong promoter just adjacent to its multiple cloning site. The sequence of bases and the enzymes that recognize specific sequences in and around the multiple cloning site of pT7.7 are

```
                    EcoRI        SmaI   BamHI
TAC CTA ATG GCT AGA ATT CGA  GCC CGG GGA TCC
        Met Ala Arg Ile Arg  Ala Arg Gly Ser

XbaI    SalI    PstI
TCT AGA GTC GAC CTG CAG
Ser Arg Val Asp Leu Gln
```

The start codon is shown in bold and the respective amino acids are shown for the codons. How would you put your coding sequence into this vector in the multiple cloning site? Would you expect to produce protein from your coding sequence in *E. coli*?

Solution

You can remove the gene's coding sequence from its present vector by digesting it with *Bam*HI and *Sal*I, which will retain its directional orientation, and then place it into pT7.7 in the same sites. You digest pT7.7 with the same two enzymes, which will remove a short piece of DNA from the multiple cloning site while linearizing the plasmid and exposing a *Bam*HI site on one end and a *Sal*I site on the other end. Your coding sequence can be ligated into this site using DNA ligase. The reading frame is retained by this procedure, but you now have two start codons: yours and the one present in the vector. However, a ribosome binding site is just upstream of the vector's first ATG start codon and the proper distance from it, thus translation will usually start at this site. The protein product will posses additional amino acids, but these may not affect its function.

SECTION REVIEW PROBLEMS

14. What is an expression vector, and how does it differ from a regular plasmid vector?
15. What characteristics would you look for when deciding which bacterial promoters to choose for constructing an expression vector? Identify some specific bacterial promoters that would be useful in an expression vector.
16. When expressing a eukaryotic gene in an *E. coli* expression vector, what precautions must be taken to ensure expression?

Cloning Vectors Permit the Study of Gene Regulatory Regions.

It is frequently desirable to study the regulatory regions of a gene (e.g., promoters, enhancers; these will be discussed in more detail in Chapter 14) after the gene has been cloned. For example, it is not always clear just what sequences within the upstream regulatory region are responsible for regulating gene expression. Regulatory regions can be assayed by altering the DNA sequence of the regulatory region (deletions or base-pair substitutions), inserting the altered clone into a bacterial cell (or yeast or eukaryotic cell), and determining whether expression of the gene is affected. If the changed base pair is essential, the level of expression of the gene will change. Unfortunately, many gene products are difficult to assay. For example, the gene product may be a protein such as β-globin whose presence cannot be measured enzymatically. In other cases, the enzyme assays may be difficult and tedious. This problem has given rise to another family of cloning vectors that fuse the regulatory region of one gene to the coding region of another gene that has an easily assayed product. This type of gene is referred to as a **reporter gene.** For example, the assay for β-galactosidase is a simple, rapid, inexpensive, and very sensitive colorimetric assay. As a consequence, the action of a gene's regulatory region can be studied by assaying for β-galactosidase activity instead of the normal gene product. Whenever the controlling region under study is active, the fused reporter gene protein will be made. In a practical sense, the controlling region of a gene is first identified in the cloned segment, mutations are created in the controlling region and fused to the reporter gene. This combination of mutated controlling region and reporter gene is then transformed into the cell (usually the same cell from which the original segment was cloned so that all the proper transcription factors will be present to interact with the controlling region), and the level of reporter gene product is assayed.

An example of reporter gene usage is the study of the regulatory regions of a gene family. In gene families, all the genes in the family produce a product with identical or similar functions (for example, the α- and β-globin genes that make the protein components of hemoglobin). Numerous copies of globin-producing genes are present in most animals, and they constitute a family of genes. Another example is the genes that code for rRNA. Most animals, plants, and bacteria have multiple copies of these essential genes. Using a transcriptional fusion vector with a reporter gene, each gene can be separated and studied independently from the other copies of the gene, and each gene's contribution can be clearly characterized with respect to the others (Figure 12.21). With the availability of transformation technology for many different species of plants, animals, and microorganisms, the regulatory region fused to a reporter gene can be reintroduced into the organism's genome and its expression studied during development.

Chloramphenicol acetyltransferase (CAT) is another excellent reporter gene. This enzyme cleaves a bond in the antibiotic chloramphenicol, inactivating the drug and producing two easily distinguished products. The CAT reporter gene has been used to study the light-regulated expression of the small subunit of ribulose-1,5-bisphosphate carboxylase, or rubisco. This protein is encoded by a nuclear gene and transported into the chloroplast of plants; in combination with the large subunit of the rubisco gene, it catalyzes the first step in CO_2 fixation in photosynthesis and, consequently, is extremely important to plants.

Considerable research has been devoted to understanding how light regulates rubisco gene expression. However, the rubisco gene product is difficult to assay quantitatively. Consequently, the regulatory region adjacent to the gene for rubisco's small subunit protein was cloned and fused to a CAT reporter gene (Figure 12.22). The regulatory region is made up of about 1000 bp upstream of the 5′ end of the rubisco small subunit mRNA. When the regulatory region is fused to a CAT reporter gene and the sequences are transformed back into tobacco cells, photoregulation of CAT activity takes place. It is essential to transform this gene/regulatory region construct back into tobacco cells (from whence it came) because specific transcription factors are present that interact with the promoter and enhancer regions. To determine exactly what sequences within the 1000 bp interact with these transcription factors and are responsible for photoregulation, a series of deletion mutants were prepared within the 1000-bp upstream region and each deletion was fused to a CAT reporter gene (Figure 12.22). These were then separately transformed into tobacco

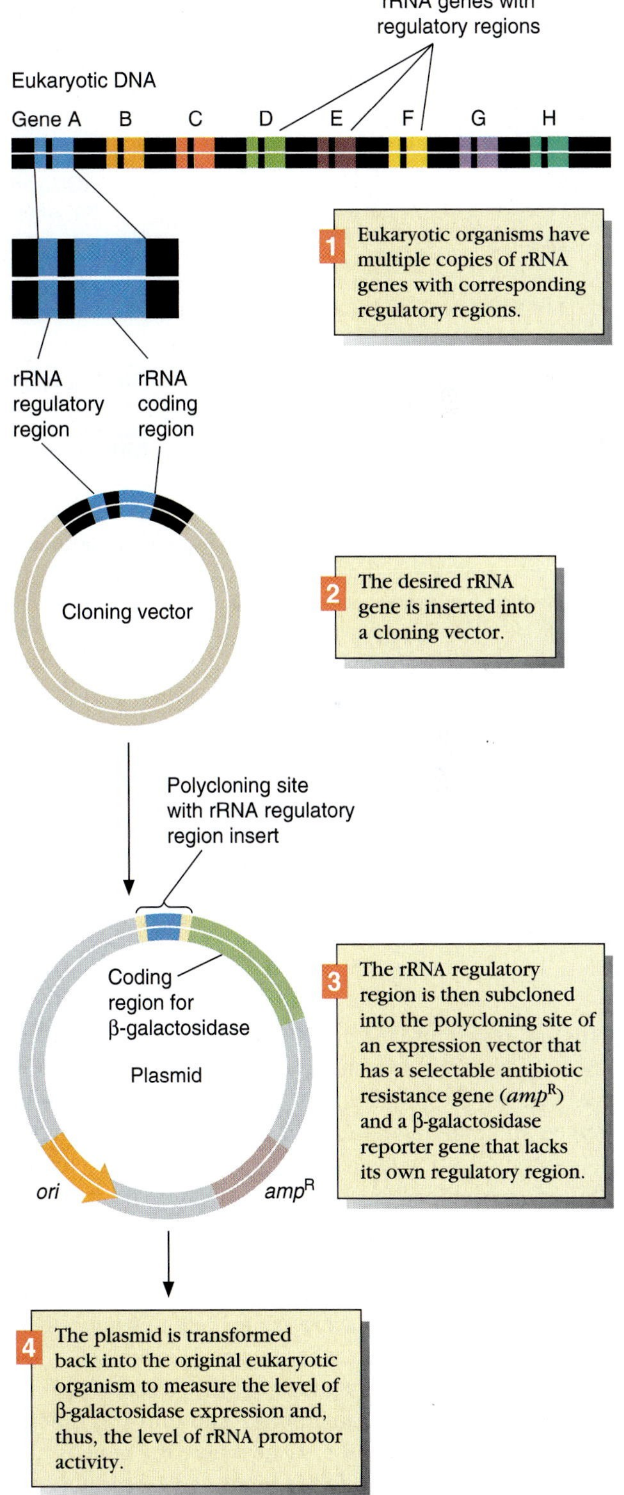

FIGURE 12.21 Vectors for studying regulatory regions of genes. In this example, the regulatory regions of several rRNA genes are tested. Within the genome, rRNA genes are present in many copies, but each gene is expressed differently, and this vector system allows separate analysis of each regulatory region. The rRNA genes are cloned into a plasmid vector, and the regulatory regions are identified by gene sequencing. These regulatory regions are then subcloned into the site next to the β-galactosidase structural gene. The regulatory region's expression can then be studied by measuring β-galactosidase activity rather than rRNA expression (which is more difficult to assay).

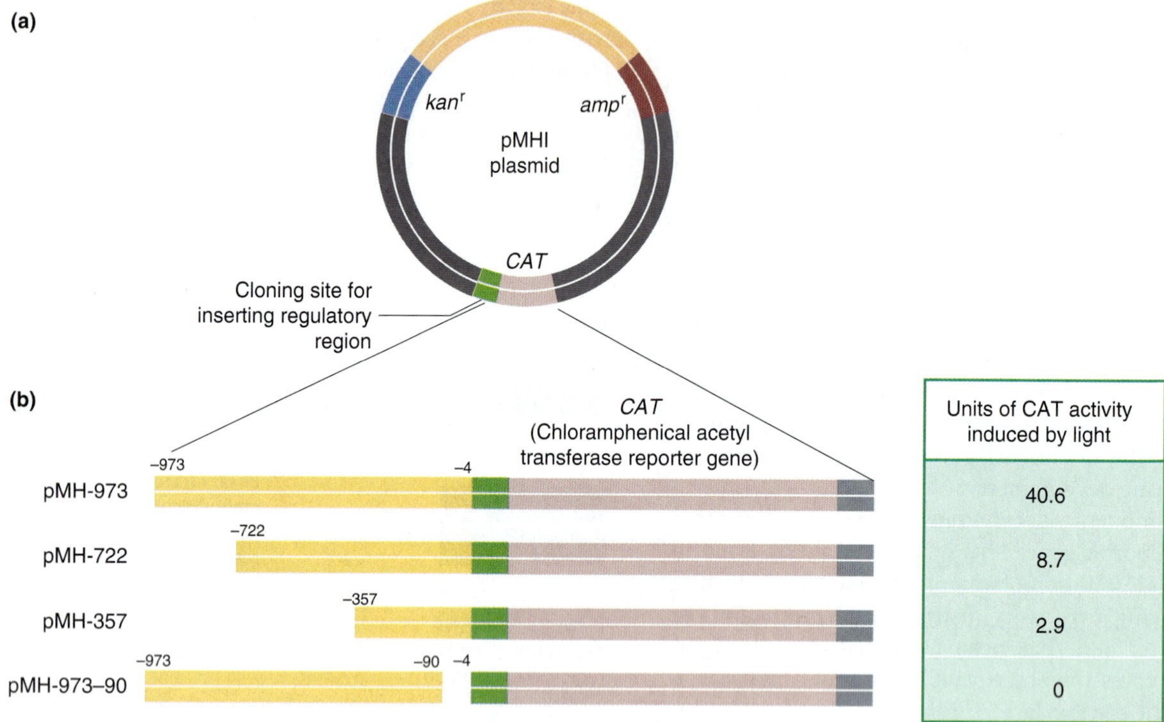

FIGURE 12.22 Analysis of a regulatory region of a light-inducible gene. (a) An expression vector, pMH-1, was used to measure the light induction capacity of sequences upstream of the *rbc*S gene (ribulose-1,5-bisphosphate carboxylase, small subunit). The vector contains two selectable markers, kanamycin resistance (kan^R) and ampicillin resistance (amp^R), in addition to the CAT (chloramphenicol acetyl transferase) reporter gene with an adjacent cloning site for insertion of regulatory regions. (b) A series of deletions were made in the 5′ upstream sequence of the *rbc*S gene to determine its capacity to promote light induction of the CAT gene (which is easily assayed). Each of the deletion sequences has been fused to the CAT gene and transformed into tobacco plants to measure activity of the partial promoters. The numbers adjacent to the sequences represent the number of the base pairs from the mRNA start site. The amount of light induction is shown at the right in CAT gene activity units. Removal of bases from −973 to −357 greatly reduced light-induced activity but not did not totally eliminate it. However, removal of the final 90 bp from this region abolishes light induction. *(Taken from Timko, M. P., A. P. Kausch, C. Castresana, J. Fassler, L. Herrera-Estrello, G. Van der Broeck, M. Van Montagu, J. Schell, and A. R. Cashmore. 1985.* Nature, *318: 579.)*

cells and tested for CAT activity in the absence and presence of light. The sequences between base pairs −4 and −973 (using the transcription start site as +1) were involved in light regulation by interaction with trans-acting transcription factors. The investigators also found that orientation of these upstream sequences was unimportant, which is frequently true for enhancer sequences.

Reporter genes have many uses in addition to studying the sequences adjacent to a gene to determine which base pairs control gene expression. Reporter genes can also be used to determine where a gene is expressed in an organism. For example, a gene may only be expressed in the leaf (or skin, if it is an animal), and thus the geneticist can easily establish the tissue where the gene is expressed, especially when the reporter gene's product catalyzes a reaction that results in a colored product. For example, the controlling region of a gene thought to be involved in cell wall function could be fused to a color-producing coding sequence, transformed into the plant under study, and examined for color production specific to the cell walls of the plant. In addition, some genes are turned on only at certain stages of development (for example, during gastrulation), and reporter genes can be used to determine the timing of gene expression during the development of the organism.

We have summarized several specialized cloning vectors, but many more exist. For example, cloning vectors have been devised for specifically mutagenizing a gene at a particular base pair. Others have been devised for introducing genes into plants, and still others are used exclusively for introducing genes into higher eukaryotic cells. As you continue your studies in genetics

CURRENT INVESTIGATIONS

Many Different Proteins Are Currently Manufactured Using Cloned Genes

Great hope has been placed on genetic engineering to increase industrial development in the United States and to keep foreign competitors from getting the upper hand—as has happened with the automobile and microelectronics industries. Development of genetically engineered products has been greatly facilitated by creative financing. Since its inception in the early 1980s, over $3 billion in public equity has been poured into the biotechnology industry, giving rise to a large number of products in a relatively short period of time (and also to a number of scientist millionaires—literally hundreds by the 1990s). Some of these products are listed in the following table. Altogether, it is estimated that sales of these protein products could amount to $20–$50 billion per year in the 1990s and beyond.

Protein	Uses or Applications
Blood factor IX	Hemophilia clotting factor
Blood factor VIII	Hemophilia clotting factor
Epidermal growth factor	Stimulates antibody production
Growth hormone (human)	Corrects pituitary deficiency, short stature
Growth hormone (pig)	Increases weight gain
Growth hormone (cow)	Increases milk production
Tissue plasminogen activator	Treats heart attacks and stroke erythropoietin, anemia
Interleukin-2	Used in cancer treatment
Interleukin-5	Stimulates antibody production
Interferons	Treats viral infections and cancer
Insulin	Treats diabetes
Monoclonal antibodies	Used in diagnosis and treatment of diseases
Tumor necrosis factor	Used in cancer treatment

and molecular biology, you will encounter many of these specialized cloning vectors.

In this chapter, we have dealt with the basic knowledge of what recombinant DNA technology is all about. However, we have not discussed many of the applications of the biotechnology revolution. In Chapter 13, some of the applications of gene cloning toward the understanding of gene function and organization of a genome are discussed, including locating a gene within a genome by mapping, cloning, and sequencing the entire genomes of humans, plants, animals, and bacteria.

SECTION REVIEW PROBLEMS

17. What is a gene fusion? Why would you want to use one?
18. What is the primary advantage of using gene controlling-region fusions with reporter genes?

Summary

In this chapter, we have examined the basic techniques used for cloning and manipulating DNA fragments. For cloning of DNA fragments, it is first necessary to have a means of cleaving DNA into discrete fragment sizes. This is accomplished with restriction endonucleases that cleave DNA at specific sequences. Restriction endonucleases have been isolated from many different bacteria, each cleaving DNA by a staggered or blunt-ended cut at a different palindromic sequence. These DNA fragments can be purified (using agarose gel electrophoresis) and rejoined, irrespective of the source of DNA (human DNA to bacterial DNA), by DNA ligase. Purified DNA fragments are joined with bacterial plasmid DNA and transformed into bacterial cells to produce large amounts of a specific DNA fragment for chemical sequencing for studying gene structure and function, as well as for mapping the cut sites of other restriction endonuclease. DNA fragments are mapped by cleaving DNA into smaller fragments, determining the molecular size of the DNA fragments by agarose gel electrophoresis, and deducing the order of the generated fragments. Such a map permits determination of the order of genes present on subfragments of a larger fragment.

DNA is also directly cloned from genomic DNA or from mRNA using reverse transcriptase, an enzyme that generates a cDNA (c = complementary) copy of the mRNA. Double-stranded DNA is synthesized from the mRNA using reverse transcriptase and DNA polymerase. This cDNA is cloned into a plasmid, and the plasmid introduced into *E. coli* by transformation. In either case, when a collection of cDNA clones is prepared, it represents the expressed mRNAs. This type of library differs from a genomic library, which represents the entire genome. A gene can be found in a library using a probe, which is a sequence of RNA or DNA that has sequence similarity to the gene that is sought. In some cases, genes are found using a probe from a nearby marker. In this method, the gene is found using chromosome walking, which uses a subclone of the original clone as a probe to identify an adjacent fragment of DNA and eventually locate the gene. After a gene or DNA fragment is found, it can be sequenced using the Sanger dideoxy or Maxim–Gilbert sequencing procedure.

The polymerase chain reaction (PCR) technique allows rapid amplification of DNA fragments without cloning but requires knowledge about the sequence of the DNA fragment amplified. The PCR reaction requires two primers complementary to the two DNA strands; the DNA between the primers is amplified by DNA polymerase in successive rounds of denaturation, cooling to anneal the primer and synthesis.

Other types of useful cloning vectors are expression vectors, which have strong promoters that can be fused to the coding region of a gene to obtain high quantities of the gene product, and those vectors that have an easily assayed reporter gene that can be attached to regulatory region of a gene. These vectors are used to study the structure, regulation, and expression of genes as well as to investigate genes that make products that are difficult to assay or are normally made in very small quantities.

Selected Key Terms

cDNA library
cloning vector
chromosome walking
expression vectors
gel electrophoresis
genomic DNA library
northern blotting
polymerase chain reaction (PCR)
positional cloning
probe
reporter gene
restriction endonuclease
reverse transcriptase
Sanger dideoxy sequence technique
shuttle vector
Southern blotting

Chapter Review Problems

1. You have isolated a rare protein from the fungus *Aspergillus* and have sequenced the *N*-terminal 35 amino acids. Now you wish to clone the gene encoding this protein. What strategy would you use to obtain a genomic clone of this gene?
2. You have isolated another rare protein, coding for a hormone protein. You have sequenced the entire protein and found that it is only 60 amino acids long. You would like to clone this gene from maize, but you know that the genomic library that you have in the refrigerator has over 100,000 clones in it, and you do not want to spend the time searching through all these clones for the one lone gene that codes for your hormone. Is there a better way? How would you obtain this gene clone without searching every clone of the library?
3. You wish to label a 23-base oligonucleotide for use as a hybridization probe. How would you do this?
4. You have cloned a *Bam*HI fragment into the *Bam*HI site of a plasmid vector pBR322. Restriction endonuclease digestion with *Bam*HI, *Eco*RI, and *Sal*I of this cloned fragment (not including the plasmid) was done and the fragments were analyzed by agarose gel electrophoresis. The following pattern of bands was revealed (after staining with ethidium bromide; the numbers refer to the molecular weight of the DNA fragments).

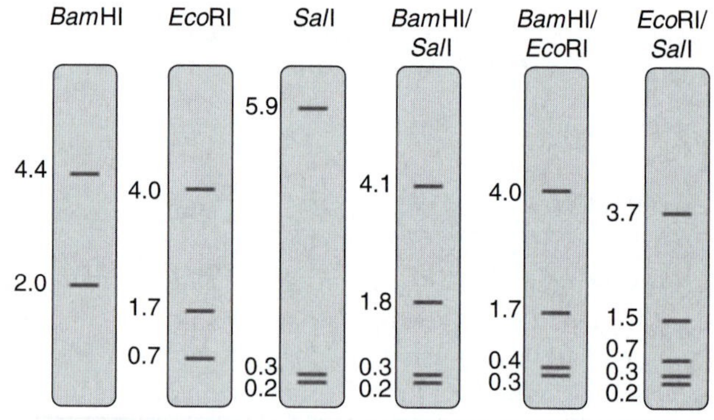

Draw a restriction map of the recombinant plasmid indicating distances between the sites.

5. Draw the product dideoxy sequencing ladder for the following template strand to be analyzed:

   3′-CCGTATAGCATGGCTGGCTGAATTC-5′

6. Which of the following restriction endonucleases would be better for preparing a genomic library of soybean: *Hin*dIII 5′-A-AGCT-T; *Not*I 5′-GC-GGCC-GC; or *Taq*I 5′-T-CG-A. Why? (Hint: Soybean has a 3×10^9 bp in its genome.)

7. You find that the DNA in your soybean plant has a very high G + C content. Does this change your strategy? (Hint: Think in terms of recognition sequences.)

8. After cloning a gene from yeast into pBR322 in *E. coli,* you find that the gene product is expressed at a very low level in *E. coli.* How can you increase the level of expression of the gene product of this gene in *E. coli?* Assume that you know the DNA sequence of the gene and the surrounding regulatory regions. (Hint: Think in terms of regulatory regions.)

9. You are involved in a research project that requires you to clone the centromere from maize. The maize chromosomes are well mapped. After examination of the map you find that on chromosome 10, two markers, *y9* (yellow endosperm) and *zn1* (zebra necrotic), are located adjacent to the centromere 2 m.u. apart and spanning the centromere. After some research, you find that 1 m.u. equals 14 kb. How would you go about cloning and sequencing the centromere from maize? Is this a realistic project that will not require too much time?

10. Suppose that you wished to synthesize an oligonucleotide to use as a probe to search for the gene coding for the protein with the following *N*-terminal amino acid sequence: Trp–Ser–Lys–Tyr. How many different sequences of RNA could be translated into that amino acid sequence? How many different DNA sequences could you make to probe for the gene? Give the sequence of one such oligonucleotide. (Hint: Think in terms of the degenerate code.)

11. In chromosome walking using a plasmid vector, how do you obtain the "next" fragment in your chromosome walk? (Hint: The fragments overlap in a walk.)

12. How do you know which direction you are going in a chromosome walk? That is, how do you know that you are headed toward the gene you seek or away from the gene?

13. You have a genomic DNA clone of a gene that has been sequenced along with 1 kb of sequence extending from each end. You find from the sequence that it does not possess any convenient cloning sites at either end, and you need one before the start and after the stop signals to subclone the coding sequence into another important vector. The 5′ end of the mRNA coding region for this gene has the following sequence:

    5′-GAGGGGUAGGAG**AUG**GGGUCUCACACA...3′

    whereas the 3′ end of the gene has the sequence

    5′-GAUAAGCUUUUGCUCCUCTGC**UAA**GGGAAUACAU...3′

    The start and stop codons are indicated in bold. With the information on restriction endonuclease recognition sequences shown here, design two primers that contain these sequences as well as information to allow PCR amplification of the entire coding region of this gene for cloning into your vector. This vector possesses a multiple cloning site with *Bam*HI and *Sal*I cloning sites adjacent to each other. (*Bam*HI recognizes 5′-GGATCC-3′ and *Sal*I recognizes 5′-CAGCTG-3′.)

14. In many bacteria, genes with related functions are regulated in a similar manner, usually by a common metabolite that acts on a protein that, in turn, acts on a promoter sequence upstream of the gene. You are interested in bacterial genes that respond to low concentrations of nitrogen. This family of promoters turns on a wide variety of genes under limiting nitrogen conditions. You speculate that you can identify new genes by how their promoters respond to low nitrogen concentrations. How would you search for promoters in *E. coli* that are regulated by a limiting concentration of nitrogen? (Hint: Think about specialized cloning vectors.)

15. The following gene configuration exists in *Rhizobium* and is involved in the process of nodule formation on the plant host:

    The *nodD* gene product is a regulatory gene and activates the other three genes (among others) to produce their protein products. The NodD protein is normally produced in very small quantities except when activated by a chemical substance produced by the plant root system. You want to study the regulation of *nodD*, but you cannot fuse this gene to a reporter gene because the fusion would destroy the protein that interacts with the chemical produced by the plant. How would you set up your experiment to study the regulation of *nodD* expression?

16. After cloning a gene for β-globin into a vector, it was placed in *E. coli* as the host. Attempts to express the gene were unsuccessful, but when the vector was transferred into a yeast host cell, the β-globin gene was expressed. Why?

17. All cloning vectors have some common features, namely
    a. they are small in size (less than 10 kb).
    b. they have a selectable marker, unique restriction sites, and the ability to replicate in the host organism.
    c. they are all circular DNA molecules.
    d. they are easily transformed into the host organism.
    e. b and d.

18. cDNA libraries differ from genomic libraries because
    a. cDNA libraries have greater numbers than genomic libraries.
    b. on average, cDNA clones are larger in size than genomic clones.
    c. cDNA library clones lack introns and controlling sequences.
    d. genomic clones lack introns and controlling sequences.
    e. cDNA libraries do not have probes.

19. A useful probe for searching a genomic library could include
    a. deduced nucleotide sequences from a known amino acid sequence.

b. rRNA sequences.
c. purified mRNA sequences.
d. a heterologous sequence from another organism.
e. all the above.

20. A 10-kb DNA clone was digested with *Eco*RI, and two fragments were found after agarose gel electrophoresis. One was 4 kb and the other was 6 kb. Digestion of the same 10-kb fragment with *Hin*dIII resulted in two fragments as well, but they were 3 and 7 kb. A double digestion with both *Hin*dIII and *Eco*RI gave three fragments: 1, 3, and 6 kb. The order of these three fragments and cut sites is
    a. 3 kb, *Hin*dIII; 1 kb, *Eco*RI; 6 kb.
    b. 1 kb, *Eco*RI; 3 kb *Hin*dIII; 6 kb.
    c. 1 kb, *Hin*dIII; 3 kb, *Eco*RI; 6 kb.
    d. 1 kb, *Eco*RI; 6 kb, *Eco*RI; 3 kb, *Hin*dIII.
    e. Either a or c.

21. Assume that you wish to clone the gene for insulin. In what order would you perform the following steps to accomplish this goal?
    1. Select a clone that hybridizes.
    2. Grow transformed cells on selective media.
    3. Ligate genomic DNA and vector DNA.
    4. Cut genomic DNA and vector DNA with restriction endonuclease.
    5. Hybridize clones with labeled probe.
    6. Transform bacterial cells with recombinant plasmids.

    a. 2, 3, 6, 5, 4, 1 b. 1, 2, 5, 6, 4, 3 c. 4, 3, 6, 2, 5, 1
    d. 4, 3, 5, 2, 6, 1 e. 2, 6, 5, 3, 1, 4

22. In using a cloning vector with a multiple cloning site spliced into the *lacZ* gene, production of β-galactosidase (a blue color on X-gal) indicates that
    a. an insert has been placed in the multiple cloning site.
    b. the cells are antibiotic resistant.
    c. the cell lacks a plasmid.
    d. the multiple cloning site is missing.
    e. there is no insert in the multiple cloning site.

23. In constructing a cDNA library, reverse transcriptase is used to
    a. clone retroviral genes.
    b. obtain RNA copies of DNA genes.
    c. obtain cDNA copies of DNA fragments.
    d. obtain cDNA copies of mRNA.
    e. generate a probe.

24. All the necessary ingredients to obtain a PCR product are
    a. a template.
    b. a template and two flanking primers.
    c. a template, two flanking primers, and NTPs.
    d. a template, two flanking primers, NTPs, and DNA polymerase.
    e. a template, two flanking primers, NTPs, DNA polymerase, and reverse transcriptase.

25. The following sequence of DNA was used for amplification using PCR.

    5′-GTCGATGGGCCCATTGCTCGAATCATTCTAGCCG-3′
    3′-CAGCTACCCGGGTAACGAGCTTAGTAAGATCGGC-5′

    The two primers had the sequence: 5′- ATGGGCC-3′ and 5′-CTAGAAT-3′. What is the sequence of the final PCR product after 20 cycles?
    a. 5′-CATTGCTCGAATC-3′
    3′-GTAACGAGCTTAG-5′
    b.
    5′-GTCGATGGGCCCATTGCTCGAATCATTCTAGCCG-3′
    3′-CAGCTACCCGGGTAACGAGCTTAGTAAGATCGGC-5′
    c. 5′-ATGGGCCCATTGCTCGAATCATTCTAG-3′
    3′-TACCCGGGTAACGAGCTTAGTAAGATC-5′
    d. 5′-CATTGCTCGAATCATTCTAGCCG-3′
    3′-GTAACGAGCTTAGTAAGATCGGC-5′
    e. 5′-GTCGATGGGCCCATTGCTCGAATC-3′
    3′-CAGCTACCCGGGTAACGAGCTTAG-5′

Challenge Problems

1. Restriction endonucleases cleave DNA at palindromic sequences of the double-helical molecule. How is it possible for the enzyme to recognize the right sequence in the DNA considering that the base pairs are internal to the DNA molecule?
2. How can you clone a gene from a eukaryotic source, which may or may not contain introns, and expect the gene to be expressed in a bacterial host?
3. You want to study genes that participate in the early developmental events of an organism, and you know that the organism has about 75,000 unique genes in the total genome. What type of gene library would you prepare that would be enriched for genes expressed in early development? Explain how this would be done.

Suggested Readings

Innis, M. A., D. H. Gelfand, J. J. Sninsky, and T. J. White. 1990. *PCR Protocols, a Guide to Methods and Applications*. Academic Press, New York.

Lewin, B. 1997. *Genes VI*. Cell Press, Cambridge MA/Oxford University Press, New York.

Maxam, A. M., and W. Gilbert. 1977. "A new method for sequencing DNA." *Proceedings of the National Academy of Sciences USA*, 74: 560–564.

Mullis, K. B. 1990. "The unusual origin of the polymerase chain reaction." *Scientific American*, 262 (April): 56–65.

Sambrook, J., E. F. Fritch, and T. Maniatis. 1989. *Molecular Cloning, a Laboratory Manual,* 2nd ed. Cold Spring Harbor Laboratory Press, Cold Spring Harbor, NY.

Sanger, R., A. R. Coulson, G. G. Hong, D. F. Hill, and G. B. Petersen. 1982. "Nucleotide sequence of bacteriophage lambda DNA." *Journal of Molecular Biology,* 162: 729–753.

Singer, M., and P. Berg. 1991. *Genes and Genomes.* University Science Books, Mill Valley, CA.

Smith, C. L., J. G. Econome, A. Schutt, S. Klco, and C. R. Cantor. 1987. "A physical map of the *Escherichia coli* K12 Genome." *Science,* 236: 1448–1453.

Southern, E. M. 1975. "Detection of specific sequences among DNA fragments separated by gel electrophoresis." *Journal of Molecular Biology,* 98: 503–517.

On the Web

Visit our Web site at **http://www.saunderscollege.com/lifesci/titles.html** and click on A/G/M Genetics for links to the following chapter-related resources on the World Wide Web:

1. **Gene cloning.** This site has a brief description of gene cloning.
2. **References.** This site has a list of citations that address issues in gene cloning.
3. **Searching for sequences.** A DNA sequence can be entered and a search made of the existing computer gene bank that has similarity or identity.
4. **Specific DNA sequence searches.** This site allows searching of the yeast and *Arabidopsis thaliana* DNA sequence databases for sequence similarity.
5. **Specific organisms.** This site will lead you to other sites dealing with the genetics and molecular biology of organisms from dogs to mosquitoes.
6. **Restriction enzyme database.** This is a link to other sites that have lists of restriction enzymes as well as suppliers.

CHAPTER

13

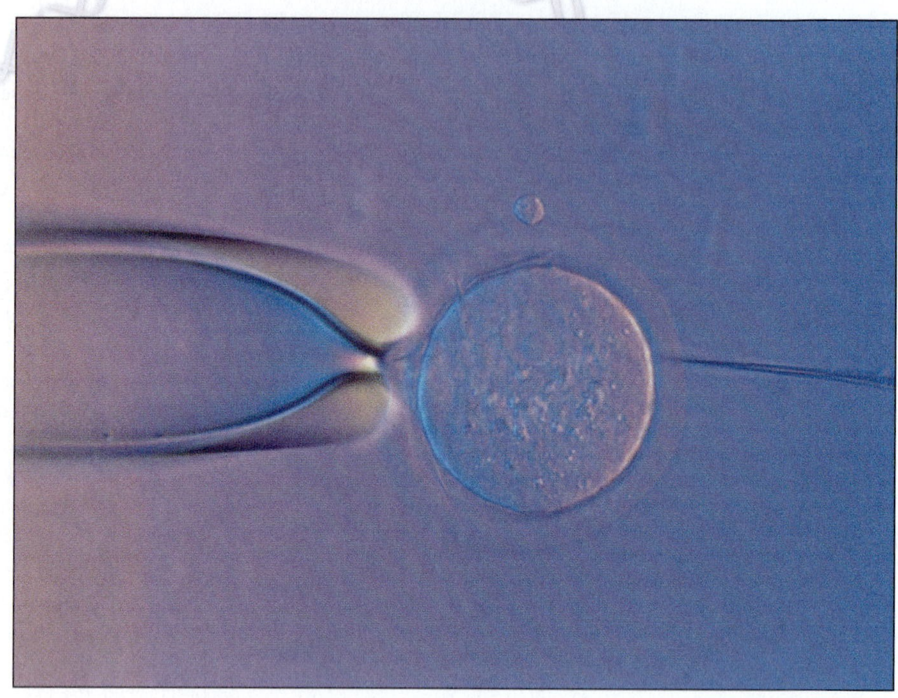

DNA from any source can be microinjected into animal cells using a very fine needle. This may result in a cell with new genes and can develop into an animal containing the new genes. *(M. Baret/Rapho/Photo Researchers, Inc.)*

Genomic Analysis and Modification

One of the ultimate uses of recombinant DNA technology is to learn how genomes are organized, where genes are located in the genome, and how genomes differ among organisms. This goal has given rise to a variety of projects to analyze genomes. Genome projects are usually a three-step process that begins with first preparing genetic maps and then physical maps and finally determining the complete base sequence of the genome. In the final analysis, genome projects are intended to elucidate the function of all the genetic material in an organism, including the noncoding repetitive sequences found in great abundance in most eukaryotes. Genome projects are now underway for many organisms, including humans, and include about two dozen prokaryotes, some of whose DNA has been completely sequenced, such as *Haemophilus influenzae* (1.8 megabases, Mb), *Mycoplasma genitalium* (0.6 Mb), *Methanococcus jannaschii* (1.7 Mb), *Helicobacter pylori* (1.7 Mb), *Methanobacterium thermoatuotrophicum* (1.7 Mb), *Mycoplasmia pneumoniae* (0.8 Mb), *Synechocystis* sp. (3.6 Mb), and *Escherichia coli* (4.7 Mb). Only one eukaryotic genome, *Saccharomyces cerevisiae* (yeast; 12,068 Mb), has been sequenced through an international effort involving some 600 scientists in Europe, Japan, and the United States. Other important eukaryotic organisms whose genomes are under intense study include *Drosophila melanogaster, Caenorhabditis elegans* (a nematode), *Arabidopsis thaliana* (a small model plant), and *Mus musculus* (the mouse). These genomes will take longer to analyze because they are considerably larger than the yeast genome. The goal of the Human Genome Project is to have a complete sequence by the year 2005, and it appears likely that this goal will be met.

Eukaryotic genomes are large and complex, and many genes in higher organisms span enormous tracts of DNA. For example, the *Bithorax* complex in *Drosophila,* which participates in the regulation of the development of the fly's segmentation pattern, spans approximately

320 kb of DNA. Recent estimates of the size of the Duchenne muscular dystrophy gene in humans suggest that it covers more than 2 million bp, whereas its exons cover only 15 kb of DNA. Efforts to clone and sequence the human genome and the genomes of intensively studied organisms require special techniques and approaches. The first step, constructing a genetic map, has evolved considerably from the traditional genetic map, which contains only morphological markers. In the second step, large segments of the genome are cloned using special cloning vectors and DNA separation techniques, which we will discuss in this chapter. The third step is the construction of an ordered set of clones of an entire genome or chromosome, and in the fourth and final step the DNA of each clone is sequenced.

In addition to understanding the genomes of many organisms, scientists have found ways to manipulate genomes to produce new gene products (plastics from plants, for example), to create precise mutations in mice and *Drosophila* (some that mimic human diseases and conditions), and to modify the genome of human cells to reverse genetic diseases. These procedures not only have profound effects on our accumulated knowledge of genomes but also affect our daily lives. As a result of these procedures and techniques, many new products that enhance our enjoyment of life and affect many aspects of our society as well as other societies around the world are on the market.

By the end of this chapter, we will have answered the following three questions to gain an understanding of how genomes of organisms are analyzed:

1. How can we improve existing genetic maps of plants and animals using recombinant DNA technology?
2. What are some of the techniques used for cloning and analyzing large DNA fragments?
3. How do you introduce genes into the genomes of plants and animals so that they are inherited by the next generation?

DNA Sequence Polymorphisms Make Useful Genetic Markers.

Detailed genetic linkage maps are fundamental tools for studying plant and animal genome organization and for eventually cloning specific genes. Genetic maps, with many genetic markers, are also extremely helpful in breeding programs for both plants and animals and have helped to select for increased food production and disease resistance.

In the past, the scarcity of genetic markers (for example, coat color, eye color, or leaf shape) has hindered the construction of useful genetic maps in plants, animals, and humans. Furthermore, constructing genetic maps for plants and animals requires extensive interbreeding of genetic stocks to ascertain the relative position of genetic markers. In humans, this approach is impractical, and thus it is impractical to construct an extensive genetic map of human chromosomes using classic genetic markers. Another disadvantage of the classic genetic map construction is that it frequently uses mutants that have a detrimental effect on the parents involved, making crosses difficult and inefficient.

An extension of classic means of producing genetic maps has been developed, and it incorporates spontaneously generated mutations in the DNA as well as the natural variations that exist at specific loci. Spontaneously generated mutations in DNA give rise to changes in restriction enzyme cleavage sites that can be readily detected. The variations of restriction endonuclease cleavage sites between individuals in a population are called **restriction fragment length polymorphisms** (RFLPs). Other forms of variation that exist between genomes is known as hypervariable loci. **Hypervariable loci** include the size and number of tandem repeats of short DNA segments (10–60 bp), termed **variable number of tandem repeats** (VNTR) **loci,** and repeats of very short DNA sequences (usually 2 bp or so, such as CACACACACACA), termed **microsatellite DNA.** Both RFLPs and hypervariable loci enable researchers to bypass some of the problems associated with morphological genetic markers. One distinct advantage of these variations is that they can be used to generate maps containing a large number of markers. Maps that saturate the genome with molecular landmarks can be created, and then these maps can be used to clone genes that are adjacent to them. It is estimated that any two humans have 1 million bp that are different between their genomes; thus, differences in restriction endonuclease recognition sites, VNTR loci, and microsatellite loci are readily available.

Methods Are Available To Detect Genomic DNA Differences.

Restriction endonucleases cleave DNA at specific nucleotide recognition sequences, and if the sequence changes by just 1 bp, the enzyme will no longer cut at that site (Figure 13.1). Variations in DNA sequence occur among individuals within a population at a frequency of about 1 in every 1000 bp, but they almost never represent detectable differences that are functionally significant. These small differences in DNA sequence can be used to construct a map of any genome and are useful in identifying an individual. Recall from Chapter 12 that the length of DNA fragments can be easily determined using agarose gel electrophoresis, where shorter DNA fragments migrate faster than longer DNA fragments. Thus, if a DNA fragment loses (or gains) a

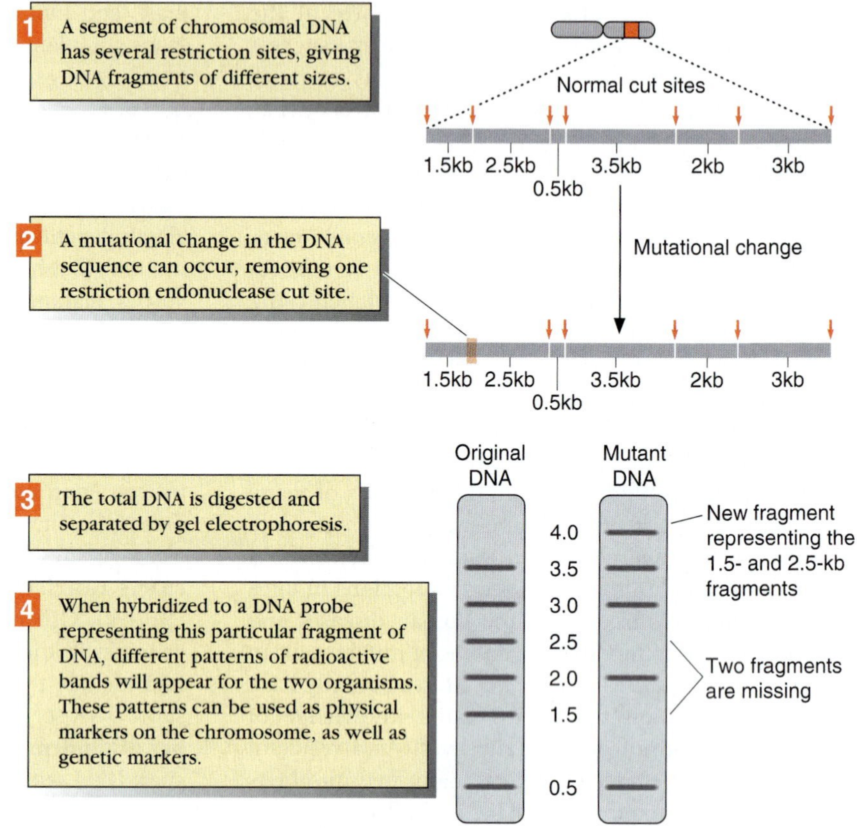

1 A segment of chromosomal DNA has several restriction sites, giving DNA fragments of different sizes.
Normal cut sites
1.5kb 2.5kb 0.5kb 3.5kb 2kb 3kb
Mutational change
2 A mutational change in the DNA sequence can occur, removing one restriction endonuclease cut site.
1.5kb 2.5kb 0.5kb 3.5kb 2kb 3kb
Original DNA
Mutant DNA
3 The total DNA is digested and separated by gel electrophoresis.
4.0
3.5
3.0
2.5
2.0
1.5
0.5
New fragment representing the 1.5- and 2.5-kb fragments
Two fragments are missing
4 When hybridized to a DNA probe representing this particular fragment of DNA, different patterns of radioactive bands will appear for the two organisms. These patterns can be used as physical markers on the chromosome, as well as genetic markers.

FIGURE 13.1 RFLP identification. RFLP variations in a DNA sequence are detected between two closely related organisms, one of which has undergone a mutational change in a restriction endonuclease recognition site. This mutation represents a true genetic change and can be used as a genetic marker.

restriction endonuclease cleavage site, the sizes of the fragments produced after digestion will change accordingly. In addition to changes in nucleotide sequences at a restriction endonuclease cut site, changes in distances between cut sites may reflect deletions, additions, or rearrangements of DNA between cut sites and result in different restriction fragment patterns after separation by agarose gel electrophoresis. RFLPs refer to the differences in the distance between adjacent restriction endonuclease cut sites of individuals in a population. Note that these RFLP changes in DNA sequence are inherited in a codominant fashion, unlike most morphological markers. Thus, if one copy of DNA from a diploid organism has an RFLP and the other does not, you will see both fragment patterns on an electrophoretic gel.

Like RFLPs, VNTRs and microsatellite DNA can be identified in a population and used in mapping experiments and for identification purposes. When DNA is digested with a specific restriction endonuclease, fragments of different sizes result, each possessing different lengths of VNTRs. These VNTRs are at different locations in the genome and thus represent genetic landmarks. After digestion, a Southern blot of the separated DNA fragments containing the VNTRs is hybridized to a probe containing sequences complementary to the VNTR sequence. In this way, the fragment pattern (usually between 10 and 50 bands) is identified. These patterns are distinctive for each of us, except identical twins. They also show some similarity between close relatives. These patterns act as "fingerprints" of our DNA and can be obtained more readily than traditional fingerprints because DNA is often inadvertently left at crime scenes.

Microsatellite DNA is extremely useful in genetic mapping of the human genome because these sequences can be easily identified using the PCR. For example, the cytosine–adenine (CA) repeat occurs thousands of times in eukaryotes and is present in many different locations. Like VNTRs and RFLPs, there is tremendous variation within the human population in the size of the repeat unit. To map these loci, first the sequences surrounding the repeated unit are determined, and then the PCR is used to amplify the sequences. The PCR is more amenable to DNA analysis because such small quantities of DNA can be used, and the separated fragments do not require Southern blotting and visualization of the bands with a radioactive probe; the pattern can be observed directly on an agarose gel after staining with ethidium bromide. Moreover, these PCR primers can be easily synthesized by an investigator or traded among investigators, thus allowing many scientists access to the same PCR primers for mapping. (See CONCEPT CLOSE-UP: *Taking the Witness Stand with DNA Fingerprinting.*) Individual variation within a population occurs because the microsatellite sequences are different sizes, depending upon how

CONCEPT CLOSE-UP

Taking the Witness Stand with DNA Fingerprinting

For many years forensic laboratory examination of biological evidence (semen, blood, tissue, and hair) has permitted identification of a suspect with about 90–95% rate of certainty. In 1986, Alec Jefferys of England devised a procedure called DNA fingerprinting, which gives a much higher degree of certainty, a value greater than 99.99%. Jefferys found that the human genome contains many microsatellite regions that consist of tandemly repeated nucleotide sequences, all of which have a short common core sequence (2–10 bp in length). The number of repeated units in the microsatellites, and thus their length, varies from one individual to the next and can be easily detected. When DNA is digested with a restriction endonuclease, separated by agarose gel electrophoresis, blotted on nitrocellulose paper, and hybridized to the microsatellite core sequence, the resulting pattern is unique to an individual and is called the variable number of tandem repeats pattern (VNTR). This unique pattern is used to identify an individual, and there is little chance that two individuals will have the same pattern (at least, none have ever been found except in identical twins).

The courtroom validity of VNTR fingerprinting was established in the case of Randall Jones of Florida. After Jones' car became stuck in the mud, he noticed a pickup truck nearby at a fishing ramp. The occupants, a man and woman in their early twenties, were sleeping. Jones shot them both in the head at close range to obtain their pickup truck to pull his car out of the mud. He dumped the dead man and woman in the nearby woods and raped the dead woman before he left. In doing so, he left behind a sample of his sperm and thereby a DNA sample. Upon hearing the evidence based on Jefferys's DNA fingerprinting pattern, it took the jury 12 minutes to convict Jones, who was condemned to death.

However, misleading conclusions can arise using VNTR fingerprinting if the DNA is partially degraded, is present in very small amounts, or is carried out by a poorly trained investigator. For example, DNA band separation is sometimes confusing and can be misleading to untrained technicians. Contamination by other DNA-containing specimens (for example, blood from another individual mixed with the victim's blood) can also give misleading results. A wrong conclusion could also be obtained if two individuals had the same DNA fingerprint. It was originally calculated that no two individuals (a chance of 1 in 9 billion, more people than are presently on the earth) could possess the same DNA pattern. This conclusion has been questioned by many leading scientists, although two individuals, other than very closely related individuals, have yet to be identified with identical "fingerprints."

Another type of fingerprinting, developed by Cetus Corporation of California, works with extremely small quantities of DNA, and the DNA sample can be partially broken down. This procedure is known as the polymerase chain reaction (PCR), and it multiplies the number of copies of a sequence to produce sufficient material for analysis. A single hair follicle is enough. An interesting example of the use of the PCR was found in the case of a very old man who had died of starvation in a rest home in Pennsylvania. The operators of the rest home were charged with negligent homicide, and the credentials of the autopsy physician were questioned. As a result, it was necessary to exhume the body and redo the autopsy. However, because the body had been buried for over a year, the DNA was degraded. When the body was exhumed, suspicions arose as to whether or not the intestines had been switched with another individual. The prosecution had to prove whether the body belonged to the intestines. The Cetus workers were able to amplify and compare an 82-bp fragment from both the internal organs and the body tissue and found them to be the same. The defendants were convicted of negligent homicide but were acquitted of tampering with a dead body.

long the repeat unit is at a specific location in the genome. After the PCR products have been separated on an agarose gel, a very distinctive DNA banding pattern for each individual in a population is seen. This banding pattern is, consequently, a fingerprint of the genome of an individual and is useful both in mapping and in forensic analysis.

It is important to note that genetic changes resulting in RFLP and hypervariable loci are synonymous with genetic changes resulting from classic genetic mutations and can be used in the same manner. An RFLP marker or hypervariable locus is not externally visible as a trait (such as albinism) but can be treated as such even though the genetic change results only in the laboratory production of DNA fragments of different size. As a result, the presence of differently sized restriction endonuclease fragments between two individuals can be treated just like any other genetic marker, and appropriate crosses can be made between these individuals to determine, using recombination, the genetic distance between these sites. After preparing an RFLP and hypervariable loci genetic map of an organism, it is possible to combine it with a classic morphological genetic map to give a composite map. In this way, genetic maps that contain a very large number of markers can be constructed. Many of these markers will have no detrimental effect on the organism and can be used for gene cloning and breeding programs.

In RFLP mapping, the differences in the sizes of restriction endonuclease-generated DNA fragments are commonly analyzed by electrophoresis in agarose gels (Figure 13.1). In reality, the detection of specific RFLP sites must be done in the presence of many other DNA fragments. With a large genome such as that for a human (about 3 billion bp), a large number of fragments are generated after restriction endonuclease digestion, and the gel separation pattern appears as a smear of DNA from the top to the bottom of the agarose gel when stained with ethidium bromide. For example, if human DNA were digested with a restriction endonuclease that recognizes a 6-bp sequence, about 180,000 different fragments would be produced. To detect the differences in size of specific individual fragments among thousands of similarly sized fragments, the digested DNA is transferred from the agarose gel to a nylon or nitrocellulose membrane by the Southern transfer technique. A small subset of specific DNA fragments can then be detected by nucleic acid hybridization with radioactively labeled DNA probes representative of a DNA sequence from the genome. The pattern of hybridization to DNA fragments on the nylon membrane is then determined by overlaying a sheet of photographic film, and the radioactive DNA strands bound to fragments on the sheet expose the film. The developed film reveals a pattern of fragments that form double-stranded DNA molecules with the DNA probe. This pattern, then, represents the RFLP pattern for only a small segment of the genome complementary to the probe used.

The specific DNA probes for identifying RFLPs are found by first cloning the entire genome (or most of it) into a cloning vector. Clones are then randomly picked that give different hybridization patterns between two individuals in the population where a genetic cross has been made or can be made. These randomly picked probes are then used to generate a map by performing genetic crosses and determining the crossover frequency as determined by the two patterns given by the probes. When microsatellite DNA polymorphisms are identified between two mating individuals, crosses are made, and the map distances are similarly generated.

RFLP and Microsatellite Maps of Chromosomes Can Be Generated for Any Organism.

Plant and animal breeders and geneticists have at their disposal maps of the genomes of many species, some with thousands of RFLP and microsatellite markers. During the late 1980s and 1990s, extensive and complete maps were developed for all the important agricultural crops, some trees, agricultural animal model organisms, as well as the human genome. Markers are still being added to generate very complete genetic maps of the genomes. Because understanding how a genetic map is generated is very important, we will consider the individual steps involved in locating a microsatellite DNA marker to a specific chromosome and mapping it in relation to a morphological marker.

The genetic maps for humans are always the most difficult to generate because it is not possible to make controlled crosses and analyze large numbers of progeny from one specific cross. Thus, to determine the genetic relationship between a specific morphological trait (such as the dominantly inherited trait of polycystic kidneys) and a microsatellite marker, we must first identify a large multigenerational family that has the disease and correlate the phenotype of these family members with the inheritance of the banding pattern of chromosome-specific PCR markers. Polycystic kidney disease is especially interesting because it accounts for 10% of all end-stage renal diseases worldwide. This disease causes renal cysts that result in renal failure between the ages of 40 and 60. In the first step of mapping a trait to a microsatellite

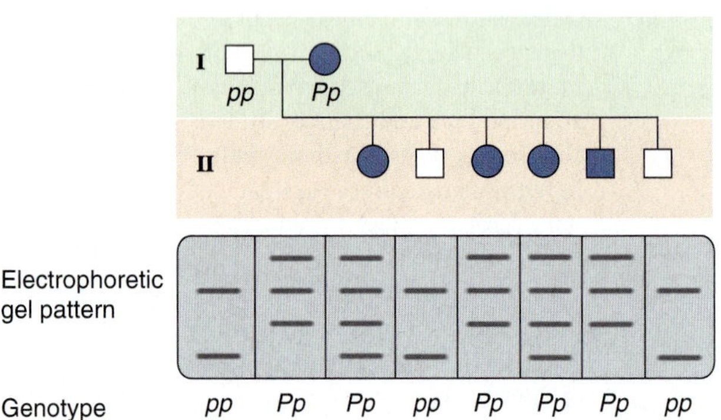

FIGURE 13.2 A family pedigree for polycystic kidneys. This figure shows the distribution of microsatellite PCR-generated bands among the members of a family with this dominant autosomal trait (many bands are usually seen, but for simplicity, only a few are shown here). Affected individuals are indicated by the blue color. The affected mother and four of her progeny are heterozygotes and show the disease. Each shows a banding pattern of separated DNA fragments that correlate with the trait, meaning that this band is linked to the trait. If enough individuals are examined, crossovers, where the disease trait is separated from the banding pattern, may be found. The map distance between the gene for polycystic kidneys and the site on the DNA that is responsible for generating the PCR pattern can then be calculated.

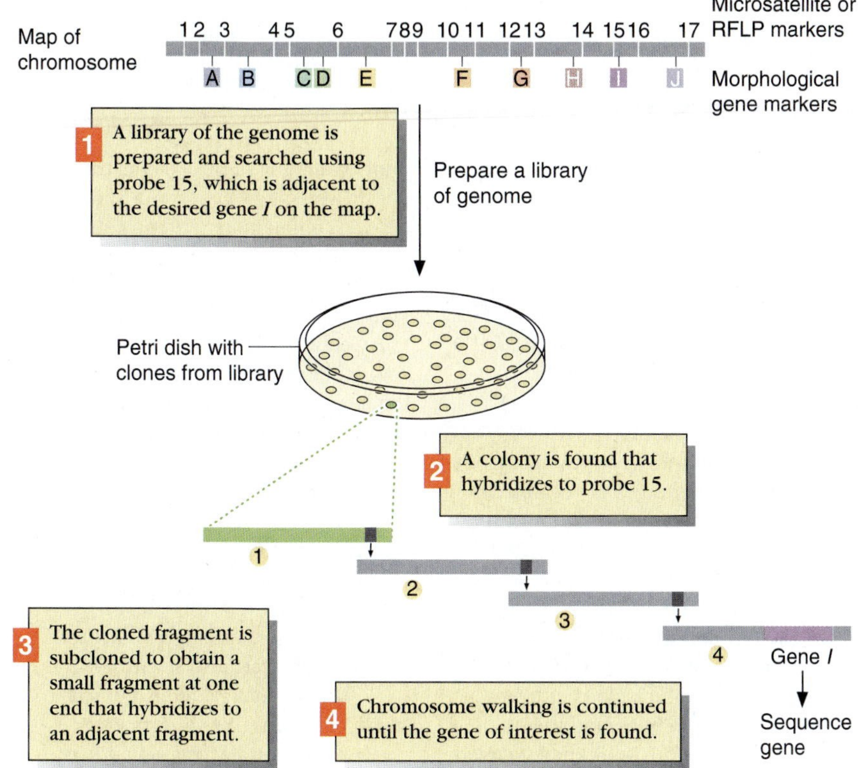
Microsatellite or RFLP markers
Map of chromosome
1 2 3 4 5 6 7 8 9 10 11 12 13 14 15 16 17
A B C D E F G H I J
Morphological gene markers
1 A library of the genome is prepared and searched using probe 15, which is adjacent to the desired gene I on the map.
Prepare a library of genome
Petri dish with clones from library
2 A colony is found that hybridizes to probe 15.
1
2
3
4
Gene I
Sequence gene
3 The cloned fragment is subcloned to obtain a small fragment at one end that hybridizes to an adjacent fragment.
4 Chromosome walking is continued until the gene of interest is found.

FIGURE 13.3 Chromosome walking. A design is outlined here for cloning a gene when the map position of the gene is known in relation to a closely linked RFLP or microsatellite marker. (1) In this example, the morphological gene *I* is desired, which, for example, could be a disease-causing gene in either a plant or an animal. (2) The walk begins with clone-containing sequences complementary to probe 15. (3) Adjacent clones are then found by using the end fragment of the first clone as a probe to find the second, and so forth. (4) Finally, a clone is found that contains gene *I*.

banding pattern, a pedigree analysis is performed to determine the pattern of inheritance for the polycystic kidney disease (Figure 13.2), and each family member's DNA sample is tested to determine the pattern of inheritance revealed by PCR gel patterns. In the second step, if the pattern of inheritance for the polycystic kidney disease matches the pattern of inheritance for a specific microsatellite marker, the disease can be assigned to the chromosome represented by the microsatellite marker (Figure 13.2). If the extended family is large enough, crossover will occur at a specific frequency, and the genetic distance in map units can be established between the microsatellite marker and the disease trait. In this way, many microsatellite markers have been linked to disease-determining genes. The same logic used to map polycystic kidney disease to a specific microsatellite marker is used to link two RFLP patterns given by two different probes. Two microsatellite markers are similarly mapped.

RFLP and microsatellite markers are very useful in cloning genes where the gene product is unknown (for example, the genes that cause muscular dystrophy or polycystic kidneys). The first step in cloning is to identify the gene to a particular chromosome and then locate a closely linked flanking microsatellite or RFLP markers. The gene is cloned by "walking" down the chromosome from one microsatellite or RFLP marker to another marker flanking the gene of interest. The gene must be in the area between the two flanking markers and be identifiable. Chromosome walking is described in Figures 12.12 and 13.3. In chromosome walking, first a clone from the genomic library that is closely linked to the morphological marker is identified. This cloned marker is then subcloned to generate an end-fragment clone and used as a probe to identify the adjacent sequence from the same genomic library. This sequence of events is then repeated with each subsequent clone. In this manner, it is possible to "walk" from the location of the first clone toward the flanking genetic marker. Using a variety of means, it is possible to identify coding sequences within these cloned fragments (after sequencing the DNA) and to deduce the amino acid sequence. These methods are time-consuming and laborious, but they have been successfully used to identify a number of human disease-causing genes, including muscular dystrophy. When the protein product of the gene is known, it is possible to study its mode of action and devise therapies to treat the disease. In addition, markers that identify the gene can be used to detect the presence of the disease in affected fetuses for treatment of symptoms or possible abortion when treatment is not possible.

As a result of studies such as these, three genetic sites that cause polycystic kidneys have been found on the human genome. Two of these sites have been cloned: PKD1, located on chromosome 16 at site 16p13.3 (about 80% of the cases and cloned in 1994), and PKD2, located on chromosome 4 at site 4q21–23 (about 15% of the cases and cloned in 1996). Analysis of the DNA and

Concept Close-Up

Eugenics: Are We Seeing a Rebirth?

Eugenics, now widely discredited, was a movement that started in earnest in the early 1900s to improve the human genetic stock. Eugenics is defined as "preventing the degeneration of the race and preserving its admirable qualities by breeding." Such things as feeblemindedness, lack of ability to speak the English language, chronic diseases, cancer, criminality, poverty, moronic behavior, and masturbation, among many others, were thought to be genetically inherited and could be eliminated with proper breeding of the most-fit families. The eugenics movement sponsored "fitter family" competitions to promote increased numbers of what they considered the best human traits. The prevailing thought at the time was that better human traits should be selected, and most of the individuals involved in the eugenics movement truly believed that the human race could be improved through these methods.

As a result of the eugenics movement, many laws and enactments were passed by state and federal governments to prevent the unfit from mating with the fit. Many of these laws were simply racist. For example, 34 states passed anti-miscegenation laws, making marriage between blacks and whites illegal, and thus preventing the infiltration of "undesirable genes" into the white population. Two factors eventually discredited the supporters of the eugenics movement: they based their logic and actions on unsound genetic principles, and the Nazis enthusiastically took up the eugenics cause in the 1930s, with disastrous consequences.

A serious problem with the eugenics movement was the lack of understanding of the exact nature of the gene, how genes were distributed in populations, and how genes could be changed by mutation and selection. This ignorance led to many claims by eugenicists that they could not live up to, and as a consequence the movement had tremendous negative social implications. After the 1940s, most scientists who had an understanding of the true nature of how genes behave in populations disassociated themselves from eugenics, and thus the eugenics movement became largely populated with crackpots and extremists, although unfortunately some were also scientists.

Are we seeing a rebirth of some form of the eugenics movement with modern molecular genetics as the tool? As a result of rapid advances in molecular genetics, RFLP and microsatellite mapping, and DNA sequencing, it is now possible

amino acid sequences of these two genes suggests that they are involved in cell–cell signaling. Eventually, a treatment may be possible or preventive measures may be developed that will alleviate the symptoms of the disease. Other genes that have been mapped and cloned include Duchenne muscular dystrophy (short arm of the X chromosome), Huntington disease (short arm of chromosome 4), cystic fibrosis (long arm of chromosome 7), and type I neurofibromatosis (chromosome 17), which causes the so-called elephant-man disease. On the downside, RFLP and microsatellite markers can be used by employers to identify individuals that possess defective genes, and these people can be discriminated against (see *Concept Close-Up: Eugenics: Are We Seeing a Rebirth?*).

RFLP and Microsatellite Maps Are Useful in Studying Genome Organization and in Breeding Programs.

The family Solanaceae contains many well-known plant species, including tomato, potato, eggplant, tobacco, and pepper. The haploid chromosome number of each of these species is 12, yet the DNA content per cell varies from 0.74 pg (pg = picogram, 1×10^{-12} g) in tomato to 2.76 pg in pepper. Why is there such a large disparity in the amount of DNA per cell in these closely related species, and is the gene order of these species conserved? A partial answer to this question can be obtained by comparing the RFLP linkage maps of the pepper, potato, and tomato. Examination of extensive RFLP maps of pepper, potato, and tomato shows that gene order is highly conserved. If gene order is conserved, then it is likely that all these related plants share a common ancestral genome. In fact, the RFLP map of the tomato, potato, and pepper are nearly identical except that the linear order of small groups of genes on the chromosomes has been modified. The species share the same number of centromeres ($x = 12$), but the chromosomal regions around the centromeres have undergone extensive rearrangements. Thus, by comparing RFLP map data, we can conclude that differences in the amount of DNA per haploid genome between pepper and tomato are probably not caused by changes in the coding regions of the genome. A more likely explanation is that a small percentage of the total DNA of most plants is composed of coding regions, and the added DNA is most easily explained by the presence of repeated sequences.

RFLP markers can also be used to locate useful genetic traits on the chromosomes of important crop

to develop prenatal tests for many genetic diseases and to ascertain whether a fetus will develop a disease. In the near future, it will be possible to screen a fetus for 50 (or more) individual traits (for example, heart disease, hypertension, cancer, alcoholism, manic depression, Alzheimer's disease, Huntington disease, and perhaps musical ability or athletic ability). These data could then be fed into a computer that would tell individuals how to alter their lifestyles to avoid the manifestations of the disease, how they could prevent the disease, or if the data were available early enough, even to influence whether the mothers wanted to carry the fetuses to term.

But, what happens when a disease is diagnosed where no known treatment or preventive steps are available? How does an individual handle the fact that they will acquire a debilitating or expensive-to-treat disease at age 45 or 50? For example, polycystic kidneys is a dominantly inherited disease that manifests itself by loss of kidney function at about age 50. How would an insurance company or employer handle the information that a 20-year-old applicant had a disease-causing allele? Would they demand that all applicants be tested before issuing a policy or make certain tests a condition of employment? In some instances, this has already happened. The US Army now tests career candidates for polycystic kidneys (when suspected from family histories) and other diseases and denies them career opportunities if they are found positive. Hypertension causes heart disease and early death, so companies don't want to invest large sums of money into potential executives before knowing these facts, and they insist on such tests. Does this information then result in discrimination because of your genetics? How do we avoid such problems? Will the genetically disadvantaged become the socially disadvantaged? As a society, we will need to learn how to deal with such questions intelligently so as not to discredit the technology and abilities of the "new biology" as the eugenics movement was discredited 50 years ago.

species (for example, the oil or protein content of the seed as well as fungal, bacterial, and viral disease-resistant genes). After an RFLP marker or family of markers has been identified and is inherited with the desired trait, the breeder does not have to conduct tedious genetic tests of each plant for the desired trait but merely must isolate some DNA from leaf segment tissue and determine whether the closely linked RFLP marker is present. The breeder can test for the RFLP marker before the plant has matured because all that is needed is a small sample of DNA, which can be obtained from the leaf tissue. Thus, the breeder can dispose of undesirable plants early in the breeding program and save space, money, and cultivating efforts.

RFLP and Microsatellite Markers Have Many Advantages over Morphological Markers.

RFLP and microsatellite markers have many advantages over morphological markers, including the ability to determine genotypes at the whole plant, tissue, or cellular level. Other advantages include the ability to distinguish multiple alleles at any locus, the absence of deleterious effects of the alleles, the codominance of most alleles, and vastly fewer epistatic or pleiotropic effects, all of which enable a virtually limitless number of markers to be evaluated in a single segregating population. Additionally, RFLP and microsatellite markers exist in vastly greater numbers than morphological markers and thus allow saturation of genetic maps with closely linked sites.

SOLVED PROBLEMS

Problem

From the following pedigree and RFLP pattern of the parents and the progeny, explain the inheritance pattern of the trait and how this correlates with the RFLP pattern seen for each individual. Affected individuals are indicated by shading.

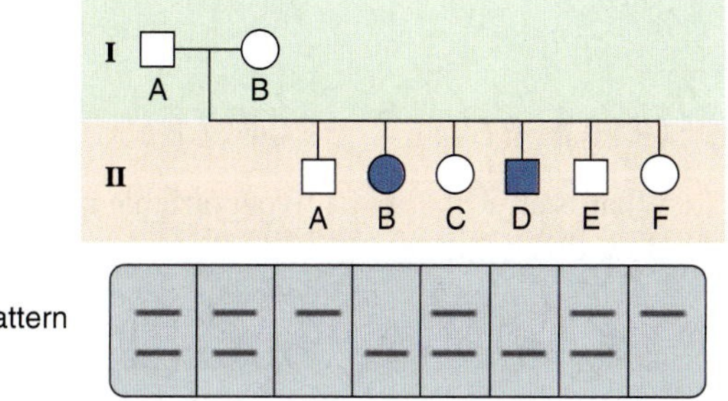

Solution

Because the trait is not found in the parents but is found in two of the progeny, the genetic trait is very likely recessive. However, the RFLP probe provides an opportunity to examine the trait as if it were codominant. The two bands in the parents represent the normal gene and the affected gene, one on each chromosome (thus, the codominant determination). The progeny II-A and II-F have only the normal allele, but II-B and II-D have only the affected allele, whereas II-C and II-E are heterozygotes for the normal and mutant allele.

Problem

You have a pair of PCR primers that have been specifically designed to amplify microsatellite markers, and you want to correlate specific agarose gel bands (using these primers) with the incidence of the dominant disease polycystic kidneys. Using DNA samples of members of an extended family, you find that one PCR product, which migrates at 0.83 kb, correlates with the presence of the disease. To further your studies, you now examine other pedigrees that show polycystic kidney disease, and you find that 867 individuals out of a total of 1933 who have the disease simultaneously have the 0.83-kb marker band. In addition, you find two diseased individuals who do not possess the band, and three individuals who have the band but do not have the disease. How do you explain these data and use them to establish the genetic distance between the polycystic kidney gene and the microsatellite marker?

Solution

It appears that the 0.83-kb marker is closely linked to the polycystic kidney disease gene. You can use these data to map the distance between these markers. The genetic distance is estimated from the frequency of crossovers between the marker and the gene within the total population. In this example, you have a total of 1933 individuals that have been examined, and about half (867) are diseased. This is to be expected if the disease-causing allele is dominant. The other two classes (with the gene and without the marker, and without the disease and with the marker) represent the crossover classes, or 5 out of 1933 individuals. The map distance is determined by dividing 5 by 1933 and multiplying by 100, or 0.26 m.u.

SECTION REVIEW PROBLEMS

1. What is an RFLP and a hypervariable marker?
2. How do you detect RFLPs? If you were searching for RFLPs in a genome, what types of DNA sequences would you choose or avoid?
3. Explain several different ways that RFLPs can be formed.
4. The following pattern is seen by hybridizing a radioactive probe to restriction endonuclease-digested DNA from *Drosophila*. What can you conclude about this probe from the hybridization pattern?

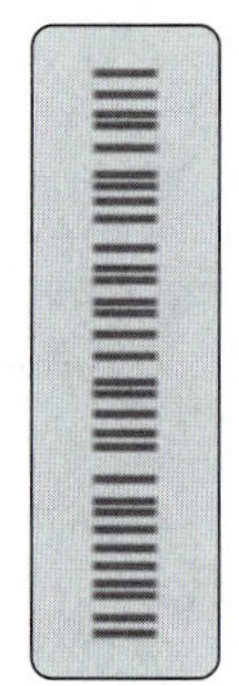

5. A microsatellite marker has been identified in humans and gives the following PCR pattern with the following three individuals (the numbers refer to the size of the fragment in kilobase pairs).

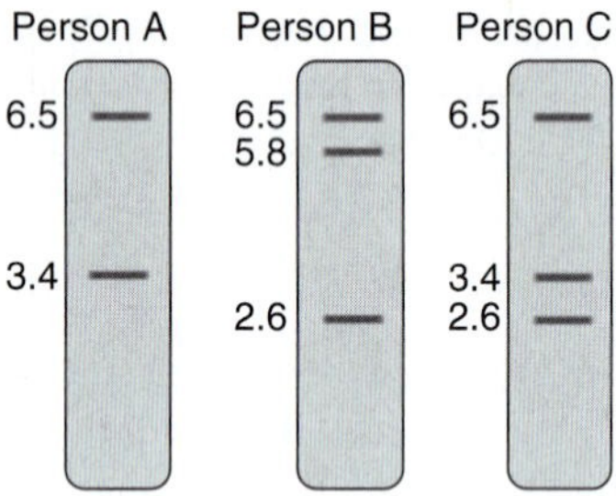

Person A is accusing Person B of fathering her child. Is this possible? Explain. Is this absolute proof that Person B is or is not the parent of the child? Why or why not?

LARGE DNA FRAGMENTS CAN BE ISOLATED AND CLONED.

In Chapter 12, we discussed the cloning of DNA into plasmids and their subsequent transformation into *E. coli* for replication. Plasmids are easy to work with in the laboratory and can be introduced into *E. coli* with relative ease [see *HISTORICAL PROFILE: Herbert Boyer (1936–) and Stanley Cohen (1935–)*]. Most plasmids used for cloning are less than 10 kb in size; thus after a DNA fragment is inserted into the plasmid, the total size rarely exceeds 25 kb. However, using traditional transformation procedures, *E. coli* transformation efficiencies decrease as the size of the plasmid increases. Because of the small size of plasmid clones, organisms with large genomes may contain several million different clones in their genomic libraries. Working with such a large number of clones is very tedious and time-consuming. In addition, many eukaryotic genes are quite large, exceeding 15 kb, and cannot be cloned as a single DNA molecule using plasmids as vectors. These problems have led to new methods of cloning DNA, as well as to ways to improve the transformation efficiency of *E. coli*. In the following sections we discuss several methods for cloning large DNA fragments.

Phage Lambda and Cosmids Are Useful for Cloning Medium-Sized DNA Fragments.

The bacteriophage lambda is one of the most studied and best characterized of bacterial viruses. In Chapter 8, we discussed the life cycle and genetics of this important bacterial virus. Lambda has served as a model for the study of bacteriophage morphology, DNA replication, and regulation of gene expression. It is not surprising, therefore, that it was one of the first molecular cloning vectors.

Lambda (λ) vectors are created by deleting portions of the central region of the lambda genome (see Figure 14.15 for a description of the lambda genetic map), which contains genes not needed for lytic replication of the phage. This middle region of the lambda phage can be replaced with a foreign DNA fragment. The size of the foreign fragment is limited by the amount of DNA deleted, but is about 15–25 kb. To prepare a genomic library using lambda as a cloning vector, lambda DNA is first digested with an appropriate restriction endonuclease to remove the central part of the lambda genome. Many lambda vectors have convenient restriction sites for this purpose (Figure 13.4). The DNA to be cloned is simultaneously digested with the same enzyme, and the DNA fragments are mixed with the digested lambda arms. These DNA fragments are joined to the lambda arms by DNA ligase. Recombinant DNA fragments that possess two *cos* sites (the cohesive ends of linear lambda DNA) at each end are then packaged into phage heads and tails to form infectious phage particles. DNA can be

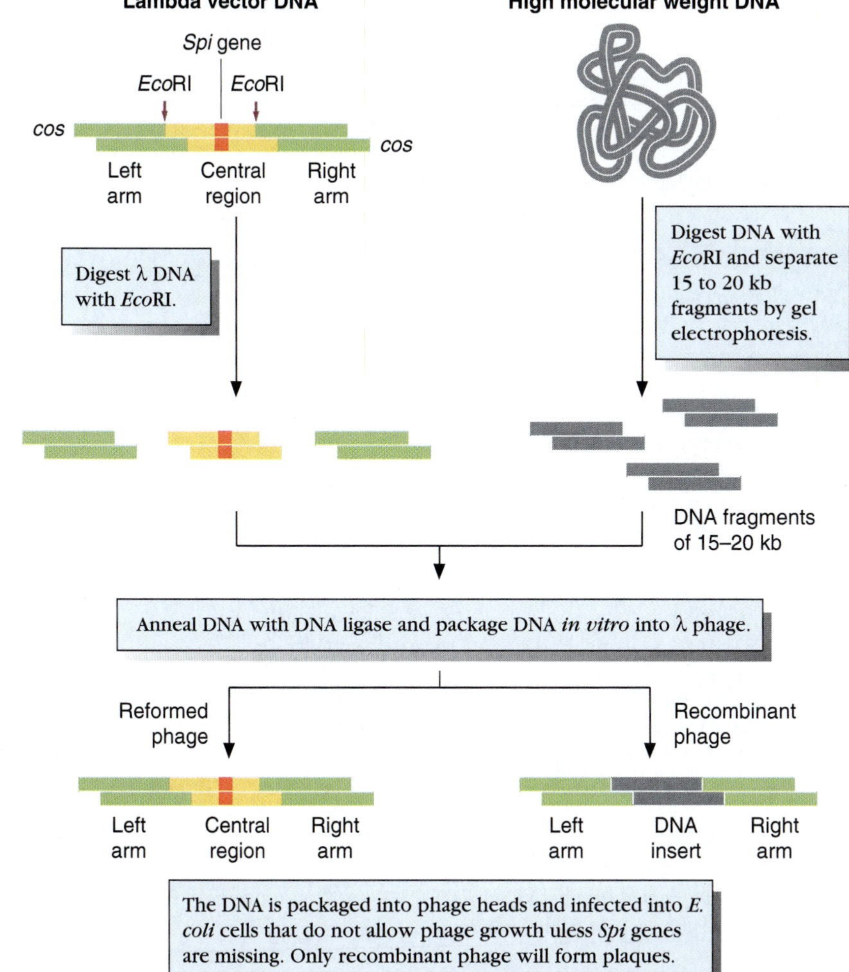

FIGURE 13.4 Cloning with lambda phage. When using lambda phage as a cloning vector, DNA fragments are inserted into a nonessential region of the phage genome. The "central region" of about 15 kb of lambda DNA is removed by *Eco*RI digestion and replaced by a DNA fragment insert. In the last step of this procedure, the newly constructed phage with insert DNA is packaged into phage heads and tails to form infectious phage particles. This occurs only if the *cos* sites are separated by at least 38.5 kb but not more than 51 kb; otherwise, they will not fit properly into the phage head and will not produce a functional phage particle.

packaged into heads only if the *cos* sites are separated by at least 38.5 kb but not more than 51 kb, or they will not fit properly into the phage head. Convenient schemes have been devised to select phage containing foreign DNA inserts and eliminate phage in which the original fragment has been reinserted. One scheme uses a gene (*spi*) normally located on the central region. When *spi* is present, the phage will not grow and form plaques on certain strains of *E. coli* (Figure 13.4). Thus, if the central region fragment is reinserted, these phage do not grow.

One advantage of using lambda-derived cloning vectors is that the chimeric (altered) DNA can be packaged to form infective phage *in vitro* when mixed with pre-prepared phage head and tail proteins. Lambda phage requires a "head-full" of DNA to package the DNA into the head proteins. This head-full of DNA must be between 38.5 and 51 kb to be packaged; thus, only DNA fragments that fill the head are selected. The size of the foreign DNA insert is determined by the size of the removed central-region fragment. From 15 to 25 kb of central-

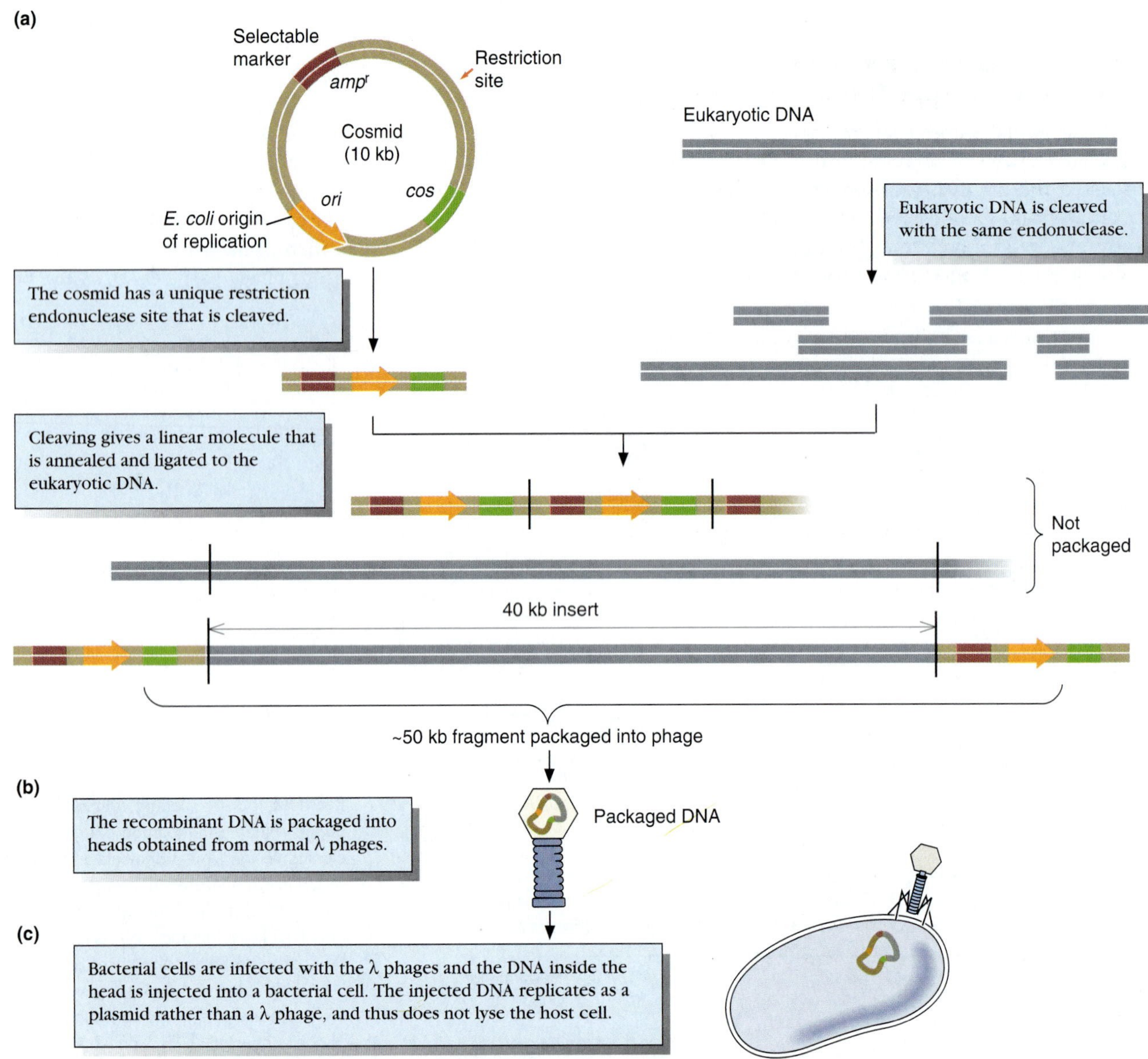

FIGURE 13.5 Cosmid cloning. (a) The cosmid DNA is digested with a restriction endonuclease to cut at a unique site, and the DNA to be inserted is cut with the same enzyme. The two populations of linear molecules are then mixed, annealed, and ligated together to form recombinant molecules. These recombinant molecules can then be (b) packaged into lambda phage heads and (c) transformed into a bacterial host.

region DNA can be replaced by foreign DNA; thus, if a gene were about 15 kb in size, you could reasonably expect that it would be present in one clone of your lambda library.

Cosmids are lambda-derived plasmid vectors that were developed to allow cloning of DNA fragments larger than 25 kb. Cosmid vectors contain a selectable marker (for example, tetracycline resistance), a plasmid origin of replication, a site into which DNA can be inserted, and the *cos* site from phage lambda. The name cosmid comes from a combination of the words for the *cos* site and plas*mid*. As you may recall from Chapter 9, lambda replicates by the rolling circle method and thus generates long strings of lambda DNA, called concatemers. For lambda to be packaged into a phage head, the DNA must be about 50 kb in length, or the size of the lambda genome. An enzyme, terminase, cuts the DNA concatemer at the *cos* site (with a staggered cut) into properly sized fragments for packaging into heads (Figure 13.5). Cosmids therefore can act as plasmids and replicate in a bacterial host; but, like lambda, they can also be cut into 50-kb fragments and packaged into phage heads. The cosmid vector is usually about 10 kb; thus the insert is approximately 40 kb. However, some cosmid vectors are as small as 5 kb, and the insert fragment is sometimes as big as 45 kb (Figure 13.5). The packaged cosmid plus its insert (which must be between 38.5 and 51 kb) can then be injected into the bacterial cell by infection rather than transformation.

Because of their larger cloning capacity, cosmids have proven to be valuable cloning vehicles for large genes or gene clusters. For example, in Figure 13.3, an RFLP probe is used to identify a clone from a genomic library and then this clone is used to chromosome-walk to an interesting nearby gene. The gene is eventually encountered (although identifying a gene is sometimes a difficult task; see Chapter 12 for a discussion) during the walk by hybridizing subclones of one fragment to overlapping adjacent clones to form the walk. Cosmid chromosome walking can be accomplished using larger steps (30–40 kb), thus making it less labor-intensive than with plasmids.

SECTION REVIEW PROBLEMS

6. Explain the difference between a plasmid and a cosmid.
7. Why are cosmids and lambda better cloning vectors than plasmids when preparing libraries of organisms that have large genomes?
8. How do antibiotic-resistant genes in cosmid cloning vectors detect when DNA has been inserted?

Artificial Chromosomes Are Useful for Cloning Large DNA Fragments.

Efforts to clone and sequence the human genome and the genomes of intensively studied organisms such as maize, *Arabidopsis thaliana, Caenorhabditis elegans, Drosophila melanogaster,* and *Saccharomyces cerevisiae* require methods of cloning that give large-sized DNA fragments. Even though cosmids are excellent cloning vectors, a cosmid library of the human genome would require over 500,000 clones! Efforts to develop cloning vectors for large DNA fragments started with the extensively studied eukaryote, *Saccharomyces cerevisiae* (yeast). This organism has many of the useful attributes of *E. coli* (such as a rapid growth rate, single cells, and a relatively small genome). Yeast has a doubling time about twice that of *E. coli,* exists as a haploid and diploid (see Chapter 3 for a description of the yeast life cycle), and has a genome only two and one half times (12 Mb) that of *E. coli.* Yeast undergoes meiosis and mitosis, has a nucleus, and possesses the enzymes for processing of RNA after transcription. This last feature is extremely important because *E. coli* is unable to process RNA. In addition, yeast can be transformed easily after treatment with a mixture of snail gut enzymes that degrade the thick, complex yeast cell wall composed of proteins and polysaccharides. Digestion with snail gut enzymes yields spheroplasts (cells minus their cell wall) that can be transformed. Spheroplasts can then be grown to regenerate the cell wall material and form normal yeast cells. Yeast cells, therefore, constitute a remarkably flexible system for the molecular biological analysis of eukaryotic genes.

Yeast Artificial Chromosomes Allow Extremely Large Fragments To Be Cloned. Methods have been developed that allow extremely large fragments of DNA to be cloned, and one of these methods uses components of the chromosomes of yeast. The vectors used in cloning large DNA fragments in yeast are called **yeast artificial chromosomes** (YACs) and can replicate fragment sizes of 1000 kb or more, 20 to 30 times the size of DNA fragments in cosmids. YACs are nothing more than a large plasmid replicating as a linear molecule in a manner similar to normal chromosomes in yeast. Thus, YAC vectors replicate in the cell in the same number of copies as all the other chromosomes.

For cloning, YAC vectors carry many of the features of other vectors, with a couple of additional characteristics. The YAC vector must have two telomeres to maintain the ends of the chromosome and a centromere for attaching to the spindle fibers. In addition, it must have yeast and bacterial selectable markers, as well as yeast and *E. coli* origins of replication (Figure 13.6) because it

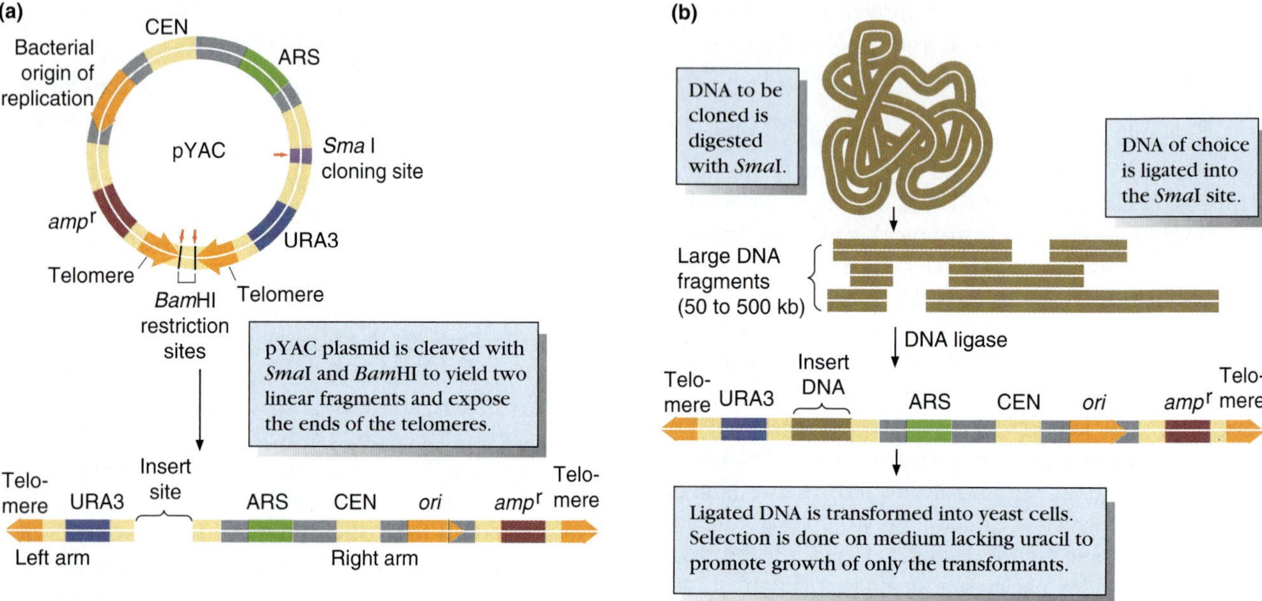

FIGURE 13.6 Yeast artificial chromosome cloning system. YACs are used for cloning large segments of DNA. (a) Diagrammed is a YAC possessing all the necessary characteristics for replication in yeast (a centromere, *CEN;* telomeres, *TEL;* and an autonomous replication sequence, *ARS*) as well as sequences for replication and selection in bacteria. To yield linear fragments, the YAC DNA is digested with *Sma*I and *Bam*HI. (b) In this YAC, the insert site is a unique *Sma*I site; thus, the DNA to be inserted is also digested with *Sma*I. The two populations of digested DNA are mixed, annealed, ligated together, and transformed into yeast cells for growth.

is replicated in both hosts. Finally, it must have unique cloning sites for inserting DNA. For convenience, YAC vectors are maintained in *E. coli* as a circular plasmid. For cloning, the vector DNA is purified from *E. coli* and digested with two restriction endonucleases to release two fragments, each containing a telomeric end. The DNA to be cloned is ligated into the cloning site *in vitro*, and the linear fragment is used to transform yeast cells. The *URA3* (a requirement for uracil in the medium) gene product is selected for in the transformed yeast cells (or some other appropriate yeast marker), and these cells survive and grow.

Large DNA Fragments Can Be Cloned into Special Plasmid Vectors. Because *E. coli* has been the traditional organism of study in both bacterial and molecular genetics, it is preferable to clone DNA fragments into this very well-understood organism. Traditional plasmids have not been useful in the cloning of large DNA fragments because large DNA fragments are not easily transformed into *E. coli*. This has limited the insert size of plasmid vectors to 10–15 kb. This limitation can be bypassed by **electroporation,** a procedure that applies brief, high-current pulses to cells. This results in a transient and reversible increase in plasma membrane permeability, allowing large DNA fragments (up to 800 kb) to cross the plasma membrane, although the efficiency of transformation decreases with the size of the DNA. Because of this advance in technology, bacterial plasmids that allow large DNA fragments to be cloned and maintained in *E. coli* have been developed.

The bacterial cloning systems that have been developed for cloning large DNA inserts are referred to as **bacterial artificial chromosomes** (BACs). The BAC system is based on the well-studied *E. coli* F-factor (discussed in Chapter 8). The replication of this plasmid is strictly controlled to maintain a low copy number per cell (usually only one or two copies per cell). This low copy number per cell reduces the potential for recombination between DNA fragments, preventing rearrangements that might alter the sequence of DNA in the clone. F-factors are capable of maintaining DNA fragments as large as 1000 kb, but usually the cloned fragments are 100–300 kb in length. An example of a BAC vector is given in Figure 13.7.

A YAC or BAC library requires only one tenth as many clones as cosmid libraries to represent the entire genome. YAC and BAC vectors have a distinctive advantage over cosmids as cloning vectors when large genomes are involved. One very distinctive disadvantage of YAC and BAC libraries is the possibility that the inserted DNA sequence may rearrange by recombination. All cloning

HISTORICAL PROFILE

Herbert Boyer (1936–) and Stanley Cohen (1935–)

Most molecular geneticists informally date the birth of recombinant DNA to a set of experiments published by Stanley Cohen and Herbert Boyer (with their students) in the November 1973 issue of the *Proceedings of the National Academy of Sciences USA*. In this article, they describe the first experiment, using *Eco*RI (discovered and characterized by Herbert Boyer), to construct a recombinant plasmid and subsequently introduce it into *E. coli* using transformation, which Stanley Cohen had been working on for some time. This publication was a milestone because it was the first time that anyone had used a restriction enzyme and a plasmid to construct a recombinant DNA molecule and introduced it into a bacterial cell, where it subsequently replicated. In a few brief sentences at the end of the article, they modestly point out the potential of this finding as a vector for cloning DNA sequences, regardless of their origin.

FIGURE 13.A Herbert Boyer (left) and Stanley Cohen. *(Courtesy of Herbert Boyer and Stanley Cohen)*

Herbert Boyer was born in a lower middle class neighborhood of Pittsburgh, where his father worked in the coal mines and on the railroads. He went to St. Vincent College in Latrobe, PA, a small Benedictine liberal arts college where he intended to prepare for medical school. However, after giving a paper on DNA in a cell physiology course, he became excited about this strange molecule, decided "to hell with medicine," and entered graduate school at the University of Pittsburgh to study bacterial genetics. After 3 years of postdoctoral work at Yale, he took his first real job as an assistant professor at the University of California—San Francisco in 1966 at the age of 30 where he began exercising his talents as a bacterial geneticist.

It was at this time that Boyer began his studies on restriction endonucleases, identifying *Eco*RI from *E. coli*. At a presentation of these findings at a scientific meeting in Hawaii in 1972, Stanley Cohen (who was, and still is, a professor at Stanford University) recognized the significance of this discovery in relation to his own efforts on the study of plasmids and their reintroduction into bacteria by transformation, and they began to work together. The culmination of this joint effort was their landmark paper.

Stanley Cohen was born in 1935 in Perth Amboy, NJ, where he attended Rutgers University, receiving his B.S. degree in 1956. He studied medicine at the University of Pennsylvania, and in 1960 he moved to Stanford University where his main area of research was the clinical application of new knowledge, particularly the molecular biology of bacterial plasmids, plasmid evolution and inheritance, and transposable elements associated with plasmids. Transposable elements possess antibiotic-resistance genes, and at that time many physicians overprescribed antibiotics, which has caused the proliferation of antibiotic-resistance genes among bacteria.

In addition to becoming millionaires, both Herbert Boyer and Stanley Cohen (Figure 13.A) have won many prestigious awards for their scientific accomplishments. Their discovery has been the subject of many articles in the popular press (winning them a spot on the cover of *Time* magazine).

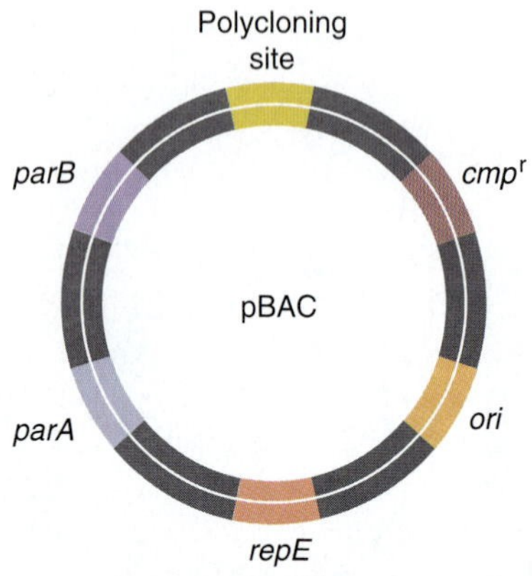

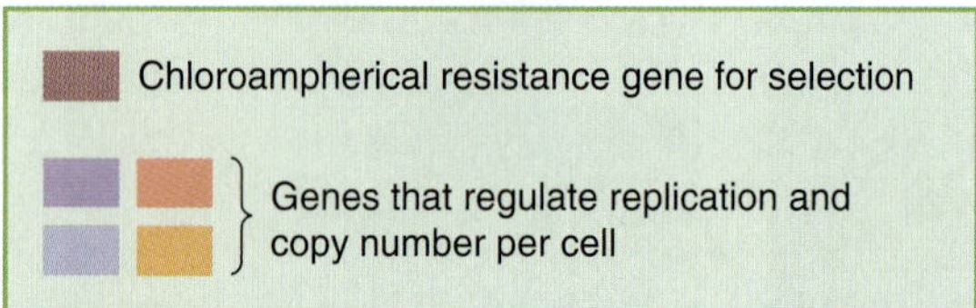

FIGURE 13.7 A BAC vector. The general features of a bacterial artificial chromosome prepared from the *E. coli* F-factor. The *parA*, *parB*, and *repE* genes are required for partition and replication of the vector in *E. coli*. BAC vectors not only replicate in *E. coli* and thus are easy to handle, but they also purify the DNA and propagate.

vectors possess the risk that the inserted fragment will rearrange, but the chances of a rearrangement of the sequences within the clone increases with the size of the inserted fragment. Thus, YAC and BAC cloned sequences are at special risk of rearranging the sequence of DNA.

SECTION REVIEW PROBLEMS

9. What is one advantage that YAC vectors possess for cloning eukaryotic genes?
10. What are the main advantages of using YAC or BAC vectors over cosmid vectors when preparing a library?
11. What specialized sequences are required in a YAC vector?

PFGE Permits Large DNA Fragments To Be Separated.

Gel electrophoresis of DNA molecules, as described in Chapter 12, is carried out by placing DNA in a well of a solid matrix, usually agarose or polyacrylamide gel, and inducing the molecules to migrate in response to an electric field. The molecular weight range that can be separated using conventional agarose gel electrophoresis is limited to sizes smaller than 50 kb of DNA. In recent years a number of instruments have been developed that allow separation of DNA molecules up to 3000 kb (or 3 Mb). These instruments use two or more alternating electric fields, called **pulsed field gel electrophoresis** (PFGE), to alter the migration properties of the DNA molecules significantly.

The separation of differently sized molecules in agarose and polyacrylamide depends on the sieving properties of the gel matrix. In conventional gel electrophoresis, molecules greater than 50 kb show essentially the same mobility; hence, separation is not possible. By using an alternating electric field, the DNA molecule migrates for a short time in the direction of the current while elongating in the same direction. When the electric field is changed, the DNA molecule must change conformation and reorient before migrating in the direction of the altered field. The time required for this reorientation has been found to be related to molecular length; larger DNA molecules take more time to realign after an electric field is switched than smaller DNA molecules. Thus, larger DNA molecules spend more time switching directions and migrate more slowly.

The chromosomes of *Saccharomyces cerevisiae*, which vary from 250 to 1500 kb, can be separated using PFGE. Shown in Figure 13.8 are 16 bands representing the 16 chromosomes of yeast. After separating the individual chromosomes, DNA–DNA hybridization with a cloned gene can be used to locate the gene to its chromosome. In addition, purifying single chromosomes in this manner allows construction of single-chromosome libraries, decreasing the number of clones per library. PFGE can also be used to determine the size of DNA inserts carried by the YACs in the transformed yeast cells. The DNA is first separated using PFGE, and the separated chromosomes are stained with ethidium bromide so that they can be visualized under ultraviolet light. The size can be determined by comparing them to standards of known size. Also, the DNA from the gel can be transferred to a membrane using the Southern transfer method. This DNA can be hybridized to a probe to identify a sequence on the YACs (Figure 13.9). For example, the separated YACs could be hybridized to individual

FIGURE 13.8 Pulsed-field gel electrophoresis of yeast chromosomal DNA. This procedure permits separation of very large DNA fragments or whole chromosomes. Yeast has 16 chromosomes, and 16 bands are resolved. The yeast genome is 12 Mb, and the chromosomes vary in size from 0.25 to 1.5 Mb. *(Bio-Rad Laboratories)*

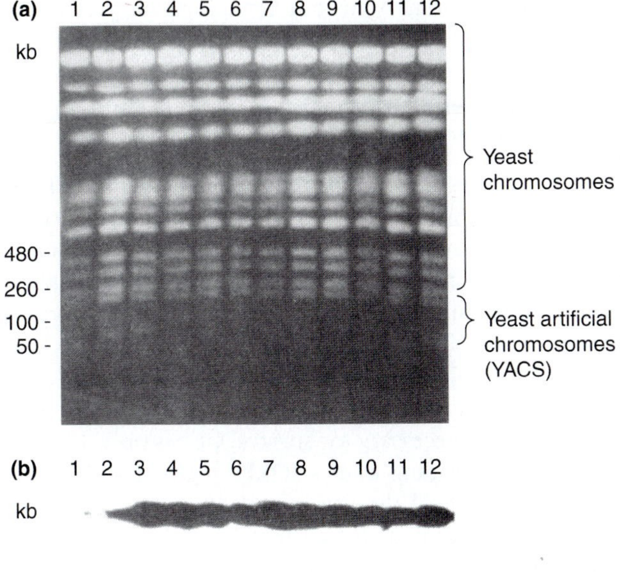

FIGURE 13.9 Identifying YAC clones by PFGE. Yeast chromosomes and yeast artificial chromosomes can be separated. The YACs were prepared from DNA of the model plant *Arabidopsis thaliana* and transformed into yeast. (a) Twelve different transformed cell lines were lysed, and their DNA were separated using PFGE. The molecular weight of the artificial chromosomes can be estimated to range from 75–260 kb. (b) The DNA from the separated chromosomes was transferred to a nylon membrane and hybridized with labeled pBR322 DNA, present in all the YACs but not the yeast chromosomes. *(P. Guzman and J. R. Ecker. 1988.* Nucleic Acid Research *16: 11091.)*

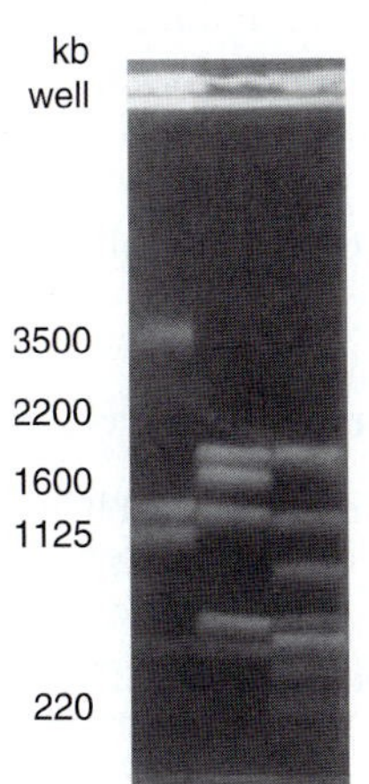

FIGURE 13.10 Estimating the size of a genome. DNA from *Rhizobium meliloti*, a nitrogen-fixing soil bacterium, was digested with *PacI*, *SwaI* and both enzymes and stained with ethidium bromide after being separated by PFGE. The size of the fragments in kilobase pairs is shown at the left of the gel and is determined by the presence of a known standard. By adding the molecular weight of each fragment to obtain a total, we can estimate the total molecular weight of the chromosome. *(B. W. Sobral, R. J. Honeycutt, and A. G. Atherly. 1991.* Journal of Bacteriology, *173: 704)*

RFLP probes to identify the YAC containing the complementary sequence. If an RFLP probe is known to be linked to genes of interest, then it is possible to locate a number of genes to a chromosome very quickly.

Another useful application of PFGE is in determining the total size of the genome. For example, the molecular size of the chromosome of *Rhizobium meliloti*, the nitrogen-fixing soil bacterium that forms a symbiosis with alfalfa plants, was found to be 6200 kb using PFGE. Figure 13.10 shows the PFGE electrophoretic migration pattern of DNA fragments from *Rhizobium meliloti* after digestion with *SwaI*, a restriction endonuclease that recognizes an 8-bp sequence within the genome. By adding the molecular sizes of the individual DNA fragments, it is possible to determine the size of the total chromosomal DNA accurately, including any plasmids that may be present.

Because DNA breaks easily, precautions must be taken to prevent the DNA sample from breaking while being prepared for PFGE. For example, before cell lysis, the bacteria containing the DNA of interest are embedded in an agarose "plug" to protect the DNA from shearing as it is handled. The embedded cells are then lysed by enzyme digestion of the cell walls, and the released DNA is digested (still in the gel) with the selected endonuclease. This plug of agarose with the digested DNA is then placed in the well for PFGE separation.

SOLVED PROBLEM

Problem

You have identified a new bacteria species and found it to have a G + C content of 80%, which is unusually high. You now want to determine the genome size of this bacterial species using PFGE. You have at your disposal a restriction endonuclease that recognizes the base sequence 3′-TATATA-5′, and you think that this enzyme may be ideal for your purpose, which is to produce a relatively low number of fragments so that you can add their molecular weights without confusing one band with another (i.e., they do not comigrate to the same spot). With this information, calculate the average DNA fragment size after the DNA is digested with your restriction endonuclease.

Solution

If the G + C content were 50% and the enzyme recognized a sequence that was exactly 50% G + C, the size of the fragments produced after digestion would be 4^6 because there are four different bases in DNA that are equally represented and the recognition sequence is 6 bp long. In this case, a recognition sequence would occur every 4096 bp. On the other hand, when the G + C content is 80%, or the A + T content is 20%, then the probability of a cut is dependent upon the number of A + T in the genome, which is 1/10 for each (10% A and 10% T) because all the bases in the endonuclease recognition sequence are either A or T. Thus, the frequency of cut sites is $(1/10)^6$, and the size of the fragment is 10^6 instead of 4^6, which equals 1 million bp. Thus, an average-sized bacterial genome, which is about 4 million bp in length, would be cut into four pieces. After separating the fragments, you could add them (using known internal molecular weight standards, which are run simultaneously with the samples) to obtain the true size of the chromosome of this new bacterial strain.

SECTION REVIEW PROBLEMS

12. When preparing DNA for PFGE gel analysis, what special precautions must be taken to ensure that the results are not misinterpreted?
13. Explain the following difference between PFGE DNA separation and gel electrophoresis using unidirectional field separation: How do the molecular weights of the fragments differ? What are the upper and lower limits of DNA fragment sizes that can be separated using PFGE?

GENOME SEQUENCES HAVE REVEALED SOME INTERESTING FINDINGS.

The availability of a large number of molecular markers in genomes has allowed the identification and sequencing of large numbers of genes. In humans, about 1000 genes have been identified, but another 16,000 sequences derived from clones prepared from mRNAs (cDNA clones) have been mapped, creating an extensive map of the human genome. Large amounts of data on the sequences of many organisms are pouring into genome data banks, and it is now possible to obtain some interesting insights into genome sequences, including an estimated minimum number of genes required for life for both prokaryotes and eukaryotes. At this writing, scientists have complete genomic sequences for 141 viruses, 51 organelles (mitochondria and chloroplasts), 11 prokaryotes, and 1 eukaryote (yeast). These numbers will swell with time, creating an immense amount of data for analysis. Eventually, databases will contain the sequences of numerous genes from many individuals within a species as well as from different species. These data will allow scientists to trace inheritance patterns to locate chromosomal regions harboring genes for common diseases. It should become possible to track human diversity at the DNA sequence level, which may lead to an understanding of disease susceptibilities. By comparing the DNA sequences of genomes from diverse organisms, we may be able to unlock the record of 3.5 billion years of evolutionary history, revealing branches in the tree of life and the timing of major evolutionary events.

Bacterial Gene Numbers Vary from 500 to 8000.

The sequences of a number of bacterial genomes (Table 13.1) reveal the approximate gene number and gene density in each of these organisms. *Mycoplasma genitalium* is particularly interesting because it has the smallest known genome of any free-living organism. This organism is a parasite and is a member of a class of bacteria that lacks a cell membrane and uses humans as one of its hosts. Consequently, this organism may represent a minimal gene set necessary for survival [it has only 473 predicted open reading frames (ORFs)]. By examining the DNA sequence of the genome of *M. genitalium*, genes were found that code for a variety of functions, such as DNA replication, transcription and translation, DNA repair, cellular transport, energy metabolism, the cell envelope, and cellular processes. Because *Mycoplasma* is a parasite, it obtains many of its biosynthetic components from its environment. *Mycoplasma* also lacks the genes for coping with stress (heat shock) as well as many regulatory genes. The lack of genes for many biosynthetic pathways might be expected in an organism that is a parasite because it can get these products from its host cell. *Haemophilus influenzae*, which has the next largest genome of organisms currently sequenced, has over three times as many genes as *M. genitalium*, but about one third of these genes are identical duplications. Of the extra one-third genes in *Haemophilus*, many are probably involved in biosynthetic pathways and coping with stress.

A comparison of the size of the 473 predicted ORFs in *Mycoplasma genitalium* to the size of the 1760 predicted ORFs in *Haemophilus influenzae* shows that both have about the same gene density. The average size of a gene in *Mycoplasma genitalium* is 1040 bp (one gene every 1235 bp), whereas *Haemophilus influenzae* possesses an average size of 900 bp (one gene every 1042 bp). These findings indicate that the reduction in genome size that has occurred in *Mycoplasma* has not resulted in an increase in gene density or a decrease in gene size. These numbers probably will be reflected in other prokaryotes as their sequences become available for analysis.

Eukaryotes Have Very Similar Gene Numbers.

The only eukaryote whose complete genomic sequence is known is yeast (*Saccharomyces cerevisiae*), and its gene number is not representative of higher eukaryotic cells that must live and develop in the complex environments of a whole organism. Examination of the yeast genome, however, reveals 5800 ORFs and a gene density similar to prokaryotes. Yeast has about 5% highly repetitive DNA.

TABLE 13.1 *Size and Probable Number of Genes for Some Organisms*

	Organism	Number of Genes	Genome Size (Mb)	Number of Base Pairs per Gene*
Prokaryotes	*Mycoplasma genitalium*	473	0.58	1235
	Haemophilus influenzae	1,760	1.83	1042
	Escherichia coli	4,405	4.7	(1146)
	Myxococcus xanthus	8,000	9.45	(1181)
Eukaryotes	*Saccharomyces cerevisiae*	5,800	12.07	2300
	Drosophila melanogaster	12,000	165	(8500)
	Oxytricha similis	12,000	100	
	Caenorhabditis elegans	14,000	100	
	Mus musculus	70,000	3300	
	Homo sapiens	70,000	3300	
	Nicotiana tabacum	43,000	4500	

* This value is an estimation.

Calculation of the size of each gene in the remaining part of the genome reveals an average gene size of about 2300 bp, only about twice that found in prokaryotes. Because yeast genes possess introns (although fewer than in higher eukaryotes), the size of an average gene in yeast is close to the average length, based on current data, of a eukaryotic mRNA (2100 nucleotides). An estimate of the number of yeast genes that are essential for viability can be made from the number of genes whose inactivation is lethal. These data indicate that only about 2000 of the 5800 genes are required for viability in yeast. This prediction is consistent with sequence analysis of the yeast genome, which indicates that nearly 70% of the genome possesses duplications.

With the limited amount of sequence data available, it is more difficult to estimate the number of genes required for more complex organisms, such as *Homo sapiens*. The estimated number of genes in *Homo sapiens* varies from 50,000–100,000, with an average estimate of 70,000 genes. However, humans may possess a quadrupled gene number inherited from predecessors; thus, a minimal gene set for a viable eukaryote may be around 14,000–15,000 genes. For example, a reasonably accurate estimate of the gene number in *Drosophila melanogaster* can be determined from a variety of data other than complete genome sequences. The length of an average gene transcript can be determined from the hundreds already analyzed and then aligned with the genomic DNA to determine the actual size of the gene. By adding an additional amount to the ends of the gene for the controlling sites, the size of an average gene can be estimated to be 8500 bp. Because *Drosophila* contains heterochromatic regions (Chapter 10; DNA that is rich in repetitive sequences) that are gene-poor, the amount of heterochromatic DNA is subtracted to obtain the coding regions within the genome. In *Drosophila*, the 115 Mb of euchromatic DNA is estimated to contain 13,200 transcriptional units, but this estimate is probably high because it contains at least 15 Mb of mobile DNA (Chapter 19). Using this calculation and others, the average number of genes in *Drosophila* is estimated to be about 12,000. Comparing this number with other organisms (a protozoan and a nematode; Table 13.1), which also have 12,000–14,000 genes, suggests that morphological complexity is not a reflection of the number of genes. The larger number of genes in *Homo sapiens* and other higher eukaryotes probably reflects gene duplications that have occurred in our evolutionary past, and the minimal set is in fact is closer to 15,000 genes. However, at present, insufficient data are available to predict the exact percentage of gene duplications in humans.

Cloned Genes Have Many Applications.

During the last decade, methods have been developed for introducing foreign genes into a large number of organisms, including humans. Most notable among the organisms that can be transformed are dicotyledonous plants (plants that have two seed leaves—soybeans, peanuts, and trees but not grasses), yeast, mammalian cells, and the favorite of many geneticists, *Drosophila*. To ensure that the introduced foreign gene is passed on to the next generation, the germline cells must be transformed. In humans, foreign genes can be added to the genome, but only to somatic cells, such as blood or liver cells. Ethical and moral reasons prevent scientists from introducing genes into germ cells. The following discussion presents several procedures for introducing foreign genes into eukaryotic cells, producing transgenic organisms, or in the case of humans, transgenic cells or organs.

Foreign Genes Can Be Introduced into Plants.

The traditional method of plant improvement depends upon plant breeding, which is a very sophisticated branch of Mendelian genetics and has had a tremendous effect on our society. Plant breeding has steadily increased yields of crops, such as wheat and corn over the last 50 years. In the 1930s, corn yields were in the range of 30–50 bushels per acre (3700 liters/hectare), but today's yields are between 100 and 140 bushels per acre (7400-10,360 liters/hectare). However, to introduce a new gene into a plant by breeding, the gene (along with many other unwanted genes) must be introduced into plants by crossbreeding with another strain or cultivar. The resulting hybrid is then repeatedly backcrossed to the original parent, selecting for the introduced gene in each cross. Backcrossing eventually eliminates most of the unwanted genes, but it usually takes six to eight generations, which with corn can take 3–4 years. Another limitation of breeding is that the breeder is restricted to interbreeding populations. That is, the gene(s) of interest must be in another corn line or in a species that will interbreed with corn. In contrast, transformation of plants with genetically engineered DNA bypasses both of these restrictions.

How do you transform a plant with a new gene? Plant tissue taken from young seedlings is capable of growing in cell cultures if the medium contains the proper nutrients and plant hormones. Tissue culture cells from some plant species (such as tobacco, tomato, carrot, potato, petunia, soybean, alfalfa, and some lines of corn, to name a few) are capable of forming roots, shoots, or regenerating whole plants. Plant regeneration from a single cell is important for transformation because regeneration from a transformed cell will give germline-transformed plants. Mere transformation of a cell within a plant will not result in a transformed plant in the next generation unless the germline cells are specifically targeted.

Several methods for introducing foreign genes into plants, to produce transgenic plants, are available, but one method offers the most promise among certain types of plants. Agrobacteria are soil bacteria that can infect plants and cause tumors. These tumors are commonly known as crown gall, and infected plant cells grow in an undifferentiated manner much like a cancer. Using Southern blotting and hybridization, the tumor-cell DNA of a crown gall is shown to contain DNA from the infective bacterial strain, and this DNA is integrated into the plant genome. In *Agrobacterium tumefaciens*, the bacterial DNA transferred into the plant DNA comes from a plasmid called the **tumor-inducing plasmid** or the **Ti-plasmid.** The Ti-plasmid is a circular DNA molecule of 200 kb, which represents about 2% of the total *Agrobacterium* DNA. Within the Ti-plasmid is a 23-kb segment of DNA called the **T-DNA,** which is the bacterial DNA transferred into the plant DNA upon infection (Figure 13.11). The two ends of the T-DNA region in the Ti-plasmid have almost identical 25-bp sequences called the T-left and the T-right sequences. These sequences are required to transfer the T-DNA to a plant genome. No other DNA sequences in the T-DNA are required to transfer the T-DNA into plant DNA. However, the genes present in the T-DNA region, when transferred to the plant, cause tumor formation by producing excessive amounts of the plant hormones cytokinins and auxins. Also present on the T-DNA are genes for synthesizing specialized chemicals called opines, which provide an energy supply for the invading agrobacteria (Figure 13.11).

If the genes between the 25-bp repeats of the T-DNA are removed (this amounts to about 22 kb of DNA) and replaced with a desired DNA sequence (for example, a sequence of DNA coding for a disease-resistance gene), this sequence is then transferred into the plant genome. The transformed plant cell can be regenerated into a whole plant, and the introduced DNA sequence will be passed on to succeeding generations in a Mendelian manner. No crown gall tumors will result because the tumor-causing genes have been removed. Figure 13.11 shows a tobacco plant constructed using this method. The plant was transformed with a gene from the firefly (*luciferase*) to demonstrate how easy it is to introduce a gene into a tobacco plant. As a result, the transformed tobacco plant expresses firefly luciferase and emits green light.

Using agrobacteria, some very useful genes have been introduced into plants. These include pathogen-resistant genes, a group of genes that synthesize poly D-3-hydroxy-butyrate (a plastic), and genes that confer resistance to the presence of herbicides. The gene for glyphosate resistance (a glycophosate is a broad-spectrum herbicide used by farmers and homeowners) was inserted into the soybean and cotton plant genomes, and glyphosate-resistant soybeans and cotton were successfully grown. Fields containing these plants can be sprayed with Roundup and only the susceptible plants (weeds) in the field are killed.

Transformed *Drosophila* Is Useful for Studying Genes and Genomes.

Drosophila melanogaster has proven to be an excellent model system for the general study of animals at the molecular and cellular level. Over the decades, geneticists

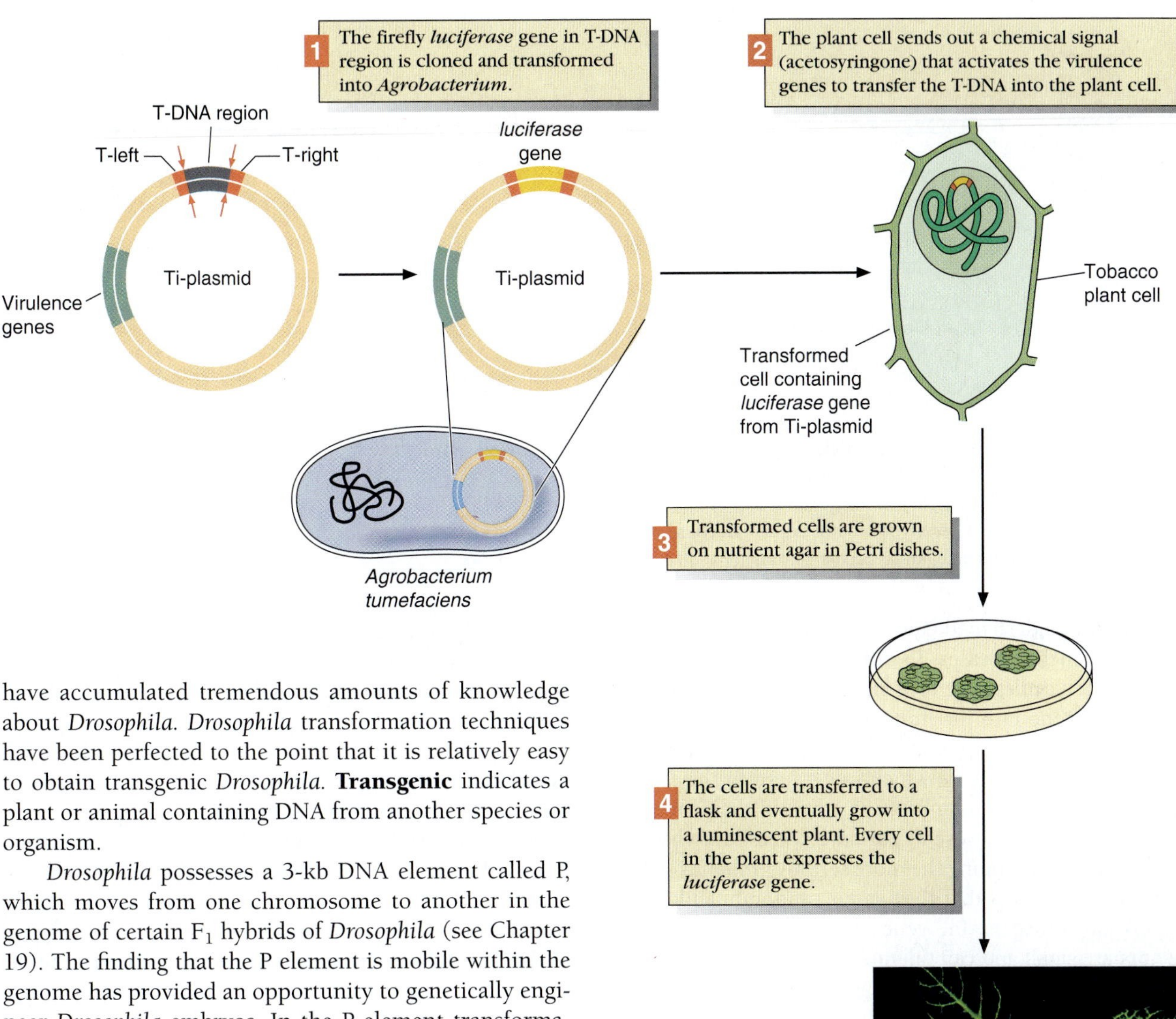

FIGURE 13.11 Preparing transgenic plants. To transform a plant cell with *Agrobacterium tumefaciens,* the Ti-plasmid is used to insert the firefly *luciferase* gene, whose product can emit green light when the plant is grown on the proper medium. The bacterial cell contains the Ti-plasmid, which can transfer the *luciferase* gene (or any desired DNA sequence) into the plant cell. The firefly *luciferase* gene is located between the flanking left and right border region. *(Keith Wood, University of California, Santa Barbara)*

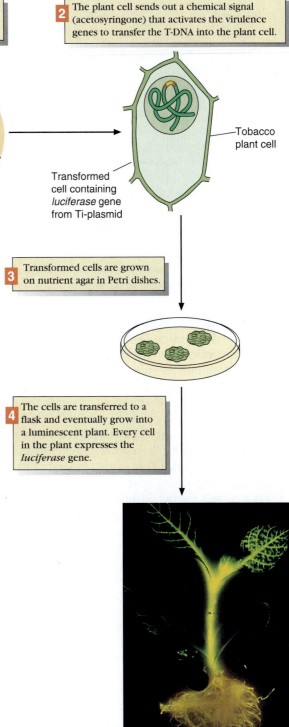

have accumulated tremendous amounts of knowledge about *Drosophila. Drosophila* transformation techniques have been perfected to the point that it is relatively easy to obtain transgenic *Drosophila.* **Transgenic** indicates a plant or animal containing DNA from another species or organism.

Drosophila possesses a 3-kb DNA element called P, which moves from one chromosome to another in the genome of certain F_1 hybrids of *Drosophila* (see Chapter 19). The finding that the P element is mobile within the genome has provided an opportunity to genetically engineer *Drosophila* embryos. In the P-element transformation procedure (Figure 13.12), the desired gene is genetically engineered into the P element, replacing the genes in the P element that produce the transposase enzyme that allows the element to move about within the genome. The genetically engineered P-element DNA is microinjected into a *Drosophila* embryo simultaneously with normal P-element DNA. The normal P-element DNA possesses a normal transposase gene necessary for genomic movement of the engineered P element. As a result, the disarmed P element carrying the new gene is inserted into the genome. Any genetically engineered P element that inserts into the germline-cell DNA is then passed on to succeeding generations.

Genes Can Be Introduced into Human Cells To Reverse the Effects of Inherited Diseases.

As soon as scientists were able to clone genes from humans, they considered the possibility of introducing normal genes to correct inherited disorders. This process is known as **gene therapy** and is being increasingly studied

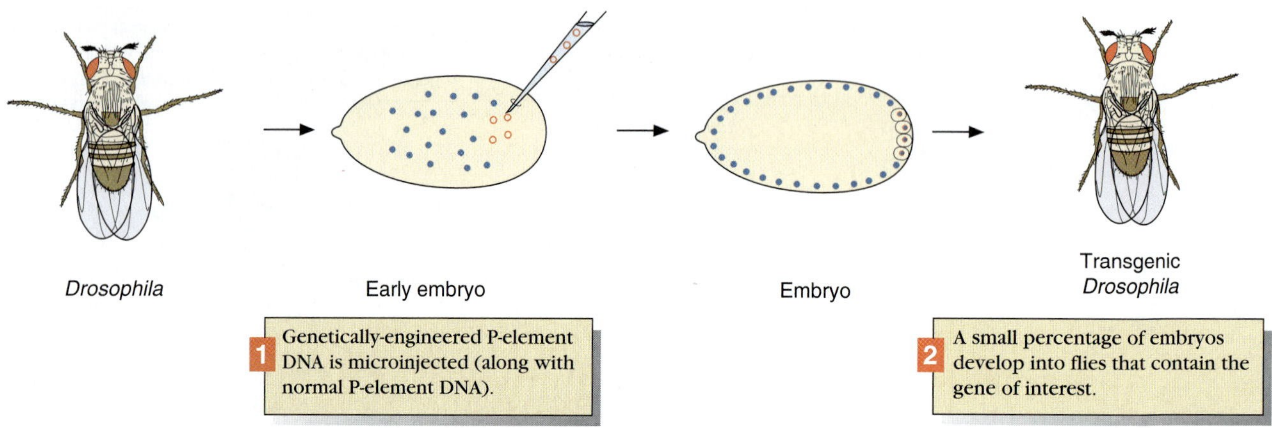
Drosophila
Early embryo
Embryo
Transgenic Drosophila
1 Genetically-engineered P-element DNA is microinjected (along with normal P-element DNA).
2 A small percentage of embryos develop into flies that contain the gene of interest.

FIGURE 13.12 Transforming *Drosophila*. The small *Drosophila* embryo is microinjected with DNA prepared from a genetically modified P element (see Chapter 19). The P element and its added DNA will integrate into the genome and be transmitted as part of the *Drosophila* genome. Not all embryos are transformed because it is difficult to get the DNA into the nucleus where recombination occurs; in addition, some of the embryos will die.

and used in modern medicine as a means of reversing the effects of genetic diseases. To introduce a gene into a human cell, a harmless vector is needed to stably integrate the cloned gene into the human genome. The first vectors that were developed for gene therapy used genetically altered retroviruses, which integrate into the DNA of the human cell and replicate as a part of the genome. Retrovirus vectors are useful but have several drawbacks for gene therapy. Among the more serious problems with retrovirus vectors is that they insert randomly and may disrupt an indispensable gene. Additionally, retrovirus vectors are small and can tolerate inserts of only about 8 kb, smaller than many human genes even with the introns removed.

Before a gene therapy trial can be conducted, many precautions are taken to ensure that the patient is protected from undue risk. At the federal level, proposals for gene therapy are reviewed by panels of scientists, lawyers, and ethicists; they make their recommendation to the director of the National Institutes of Health, who has final approval. At the local level, a similar review panel must approve the trial and then monitor the progress of the gene therapy trial to protect the interests of the patients. Two important criteria for a gene therapy trial are that (1) no other form of therapy is available and (2) the trial itself does not harm the patient. In practice, the gene of interest must be cloned and the tissue where the gene is to be introduced must be accessible (usually blood or liver cells) to the vector.

The first disorder to be treated with gene therapy was for severe combined immunodeficiency (SCID). SCID is a rare autosomal recessive disorder that leaves affected individuals with no functioning immune system. The body needs an immune system to protect itself from invading organisms, so exposure of SCID individuals to invading organisms usually results in death at an early age. One form of SCID is caused by a defective gene in adenine deaminase (see Chapter 1). A team of investigators, led by Steven Rosenberg of the National Institutes of Health, treated a young girl with this disease in 1990. They isolated some specialized white cells (T cells) from her blood, mixed these cells with a genetically modified retrovirus that contained a normal adenine deaminase gene, and allowed them to grow for a short period of time in culture. The virus infected the T cells and introduced the normal copy of the gene (the old copy remained but this did not harm the patient) into the genome (Figure 13.13). These transformed cells were increased in number, because billions are required to treat the patient, and then infused back into the patient. Patients who have been treated in this manner lead near-normal lives, but they must be reinjected with transformed T cells periodically. A more practical approach to such transformation would be to transform the patient's stem cells, which constantly produce new T cells as well as other blood cells. This is a more difficult procedure because these cells reside in the bone marrow and are difficult to isolate, but it would be a permanent correction of the defective gene. Studies are presently being conducted to accomplish this end.

Mutations in Mice That Mimic Human Diseases Can Be Made.

The traditional method of finding mutants is to mutagenize randomly a population of cells or individuals with DNA-damaging chemicals, allow the mutants to grow for a short time, and then select for the mutant phenotype in the population. The selected mutant cell type (as in yeast or bacteria) or individual (as in *Drosophila* or maize) is then carefully examined to determine changes that may reveal their exact mechanism of action for the mutant

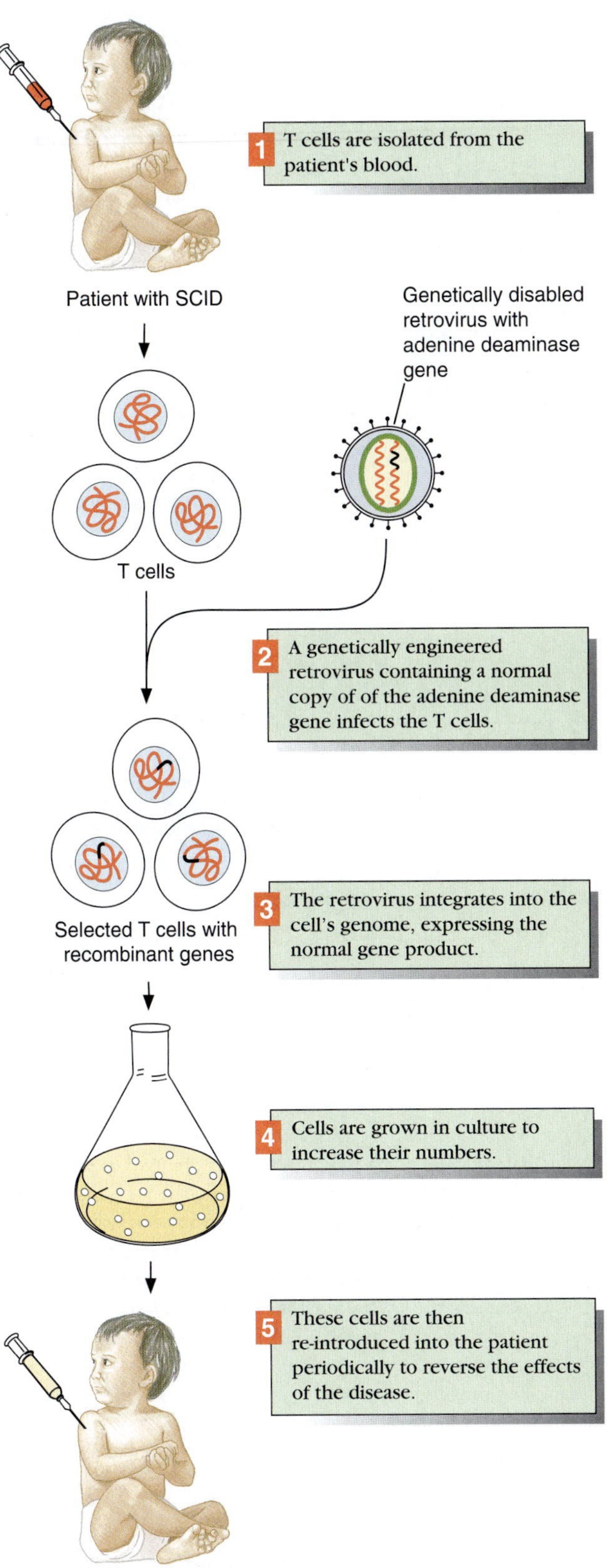

1 T cells are isolated from the patient's blood.
Patient with SCID
Genetically disabled retrovirus with adenine deaminase gene
T cells
2 A genetically engineered retrovirus containing a normal copy of of the adenine deaminase gene infects the T cells.
Selected T cells with recombinant genes
3 The retrovirus integrates into the cell's genome, expressing the normal gene product.
4 Cells are grown in culture to increase their numbers.
5 These cells are then re-introduced into the patient periodically to reverse the effects of the disease.

FIGURE 13.13 Gene therapy. This patient has a deficiency in adenine deaminase, which causes a form of SCID disease. A normal adenine deaminase gene is inserted into extracted T cells using a retrovirus vector, and the cells are put back into the individual. Expression of the normal gene reverses the phenotype of the disease as long as the cells remain in the blood.

gene. This general approach works extremely well with single-celled organisms or very small organisms that have a relatively short generation time and can be produced in relatively high numbers, such as *Drosophila melanogaster.* One hundred thousand mutagenized fruit flies constitute a fairly large experiment, but it is practical because the flies can be contained in a relatively small space. However, this approach to finding mutants works very poorly with larger animals such as mice, where the mutagenized population is limited to about 1000 animals because of the cost of housing them.

Nevertheless, if scientists are ever to decipher the many sophisticated processes present in complex higher mammals, such as the underlying mechanisms of genetic diseases, cancers, development, and neural functions, they must use a more sophisticated organism than the fruit fly. From the geneticist's viewpoint, the mouse is an ideal animal in which to study mutant gene processes because it is small and prolific and serves as a remarkably good analogue of most human biological processes. As a consequence, in the last 10 years, methods that allow the inactivation of specific genes, and even a gene in a specific tissue, in the mouse have been developed. These methods not only enable very specific and accurate research of specific genes, but they also have improved the cost-effectiveness of finding mutations in mice.

The process of inactivating a specific gene in a mouse is called **gene targeting,** and the resultant animal is called a **knockout mouse.** The first criterion for creating a knockout mouse is to have a cloned copy of the gene in question. Recombinant DNA techniques can be used to alter the cloned copy to create a mutant version, even changing different parts of the gene to create different mutants. Thus, strains of mice whose mutations mimic specific human genetic defects can be developed. The process for creating knockout mice that have cancer is now relatively straightforward. For example, a mimic of retinoblastoma (childhood cancer of the eye) has been created in mice. In another example, the tumor suppressor gene *p53* has been inactivated in a knockout mouse line. Tumor suppressor genes (Chapter 20) prevent the development of cancers, and it is known that over 60% of all human cancers have defects in the *p53* gene. After creating the *p53* knockout mouse line, the more precise determination of the responsibility of the *p53* gene product was established; it acts as a watchdog to block the division of cells that have had their DNA damaged from chemical or environmental insults. When *p53* gene function is eliminated, the damaged DNA is passed to daughter cells, increasing the possibility of cancer formation. A deeper understanding of the molecular pathology of diseases using knockout mice should permit the development of more effective gene therapy procedures.

How is a specific gene inactivated in a mouse to create a knockout mouse line? In the first step, the cloned gene is inactivated by one of several means, but a

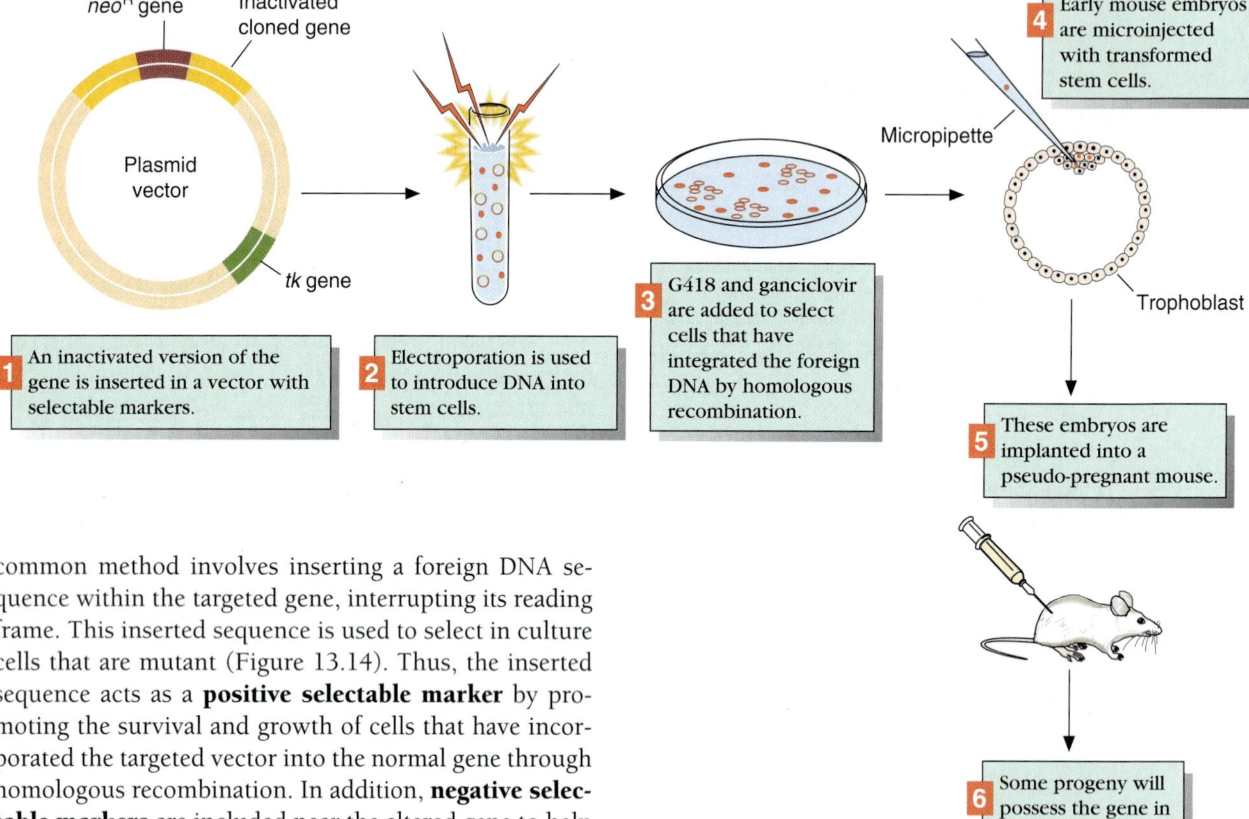

neo^R gene
Inactivated cloned gene
Plasmid vector
tk gene
1 An inactivated version of the gene is inserted in a vector with selectable markers.
2 Electroporation is used to introduce DNA into stem cells.
3 G418 and ganciclovir are added to select cells that have integrated the foreign DNA by homologous recombination.
Micropipette
4 Early mouse embryos are microinjected with transformed stem cells.
Trophoblast
5 These embryos are implanted into a pseudo-pregnant mouse.
6 Some progeny will possess the gene in their germ cells.

FIGURE 13.14 Creating knockout mice. In the first step, the normal version of the gene is altered by inserting a neo^R gene within it and a *tk* gene nearby. These genes are used as positive and negative selectable markers after introducing the gene into the genome. Some of the electroporated DNA fragments will integrate randomly, but very few (about 1 in 1000) will integrate into the homologous copy of the gene by recombination. These transformed cells can be selected by the presence of G418, an analogue of neomycin that does not harm the cell if the neo^R gene is present, and ganciclovir, a drug that does not harm the cell if the *tk* gene from a herpes virus is absent. Homologous recombination must occur to integrate the neo^R gene and exclude the *tk* gene, an unlikely event in random insertion, but likely if homologous recombination occurs.

common method involves inserting a foreign DNA sequence within the targeted gene, interrupting its reading frame. This inserted sequence is used to select in culture cells that are mutant (Figure 13.14). Thus, the inserted sequence acts as a **positive selectable marker** by promoting the survival and growth of cells that have incorporated the targeted vector into the normal gene through homologous recombination. In addition, **negative selectable markers** are included near the altered gene to help eliminate most of the cells that have incorporated the targeted vector into a random location and not directly into the exact copy of the gene by homologous recombination. The positive marker is frequently a neomycin-resistance gene (neo^R) and the negative marker is typically the thymidine kinase (*tk*) gene from a herpesvirus. When the *tk* gene is present in a cell, and the cells are grown in the presence of the drug ganciclovir, they are rapidly killed. If the neo^R gene is present, and the cells are grown in the presence of an analogue of neomycin called G418, cells that lack the neo^R gene are killed, and those that possess it survive.

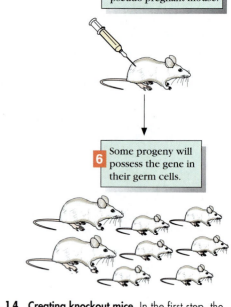

The cultured mouse cells containing the inserted gene are then introduced into the mouse. The cells of choice for transformation are an embryo-derived stem-cell line, obtained from early mouse embryos. These cells can be cultured in Petri dishes indefinitely and are pluripotent (capable of giving rise to all cell types). The vector with the gene of choice is introduced into stem cells by electroporation and integrates into the genomic DNA, either randomly or by homologous recombination (Figure 13.14). In the final step, the mutant neo^R and ganciclovir-resistant stem cells are introduced into early mouse embryos by microinjection with a very fine needle. The injected embryos are then implanted into pseudopregnant mice and allowed to develop to term. Some of the resulting mice, when mature, will produce sperm or eggs derived from the stem cells. By mating such mice to normal mice, offspring that are heterozygous for the targeted mutant gene can be generated. Homozygous offspring can be derived by matings between heterozygous siblings, and one quarter of their offspring will be homozygous for the inactivated allele. Such animals will display abnormalities that will reveal the normal functions of the target gene in all their tissue. The study of knockout mice is expanding at a very rapid rate, and many new methods for introducing genes into mice are being developed. For example, in 1994, techniques were developed to specifically knock out a gene in just the cells that a researcher wants to study, for example, in only the T cells.

SECTION REVIEW PROBLEMS

14. Explain the difference between the Ti-plasmid and the T-DNA of *Agrobacterium tumefaciens*. How is it possible to remove the genes in the T-DNA region and insert new genes of choice without disrupting the ability of *Agrobacterium* to transfer the T-DNA segment into plant cells?
15. After cloning a large fragment of DNA from *Drosophila*, inserting it into a P element, and microinjecting it into a *white-eye* mutant, you find that it complements the *white-eye* mutation. You do not know where the gene is located within your cloned large fragment of DNA, but you know from published findings that it is about 20% the size of your fragment. How do you locate the smallest possible fragment that contains the *white-eye* gene within the larger fragment?

Summary

Genetic maps are fundamental tools for geneticists. Restriction endonucleases allow construction of DNA (or chromosome) maps that take advantage of the differences in restriction endonuclease recognition sequences as well as hypervariable regions that exist between individuals within a population. These differences are caused by mutations, rearrangements, deletions, and unequal crossovers. These small differences are called restriction fragment length polymorphisms (RFLPs), microsatellite markers, and variable number of tandem repeats (VNTR) and can be detected by separating restriction endonuclease digested DNA by gel electrophoresis or with the PCR. RFLP and microsatellite differences offer a unique method of studying inheritance because they are codominant and require only a small amount of DNA to detect their presence.

Cloning very large DNA fragments offers several advantages to the geneticist: the number of clones in the prepared library is decreased (increasing the chance of finding a gene sequence), some eukaryotic genes are extremely large (up to 500 kb) and consequently do not fit into plasmid vectors, and it is very tedious and time-consuming to study large genes with small cloned fragments (5–10 kb). Also, larger DNA fragments allow easier construction of restriction maps of whole genomes.

Methods for cloning larger DNA fragments (15–40 kb) have been developed and include lambda and cosmid vectors. Nonessential genes are removed from lambda, allowing insertion of up to one third of the genome as foreign DNA. Lambda offers the advantage that the cloned DNA is packaged into phage heads for injection into the bacterial cell and thus is not dependent upon transformation. Cosmids are lambda-derived plasmid vectors containing a selectable marker, a plasmid origin of DNA replication, and a *cos* sequence from lambda for packaging into phage heads.

Yeast-cloning vectors have also been developed for cloning large DNA fragments of 100–700 kb. These vectors, known as YACs (yeast artificial chromosomes), contain telomeres, a centromere, a selectable marker, and a cloning site. Bacterial cloning vectors have also been developed that allow replication of very large cloned fragments. These are called BACs, (bacterial artificial chromosomes).

DNA fragments up to 10,000 kb or even whole chromosomes can be separated from each other by pulse field gel electrophoresis (PFGE). This technology allows much larger DNA fragments to be prepared for cloning, mapping, and sequencing.

The availability of a large number of molecular markers in genomes has allowed the identification and sequencing of a large number of genes, as well as a few complete genomes. Of the dozen prokaryotes whose genomes have been sequenced, scientists have found that the gene density is about the same in each (1 every 1200 bp), and the minimal set of genes is about 1700 for a free-living viable organism. Eukaryotes have a larger number of genes, from 5800 for yeast to about 14,000 for more complex organisms. Humans may have inherited a quadrupling of this number from their ancestors, giving the current number of about 70,000 genes.

One goal of DNA cloning is to study the function of the DNA segment in an organism. This is routinely done with yeast but is more difficult with plants and animals. Genes can be inserted into plants using the soil bacterium *Agrobacterium tumefaciens*, which infects plants and transfers a segment of its DNA into the plant host. Normally *A. tumefaciens* causes a cancerous-like growth on a plant, but when the T-DNA segment is replaced with the gene of choice, no tumor is formed and the gene of choice can be introduced into a plant, and transgenic plants are obtained. Transgenic *Drosophila* are obtained by microinjecting embryos with cloned DNA that is flanked by sequences from a P element. Normal P element DNA is also injected to supply functional enzymes for insertion of the DNA into the *Drosophila* genome. The injected DNA is randomly integrated in the chromosome and is transmitted to subsequent cell generations.

Genes can also be inserted into human cells to correct some genetic diseases, and altered genes can be inserted into mice to create knockout mouse lines. These mouse lines are useful in studying the function and processes governed by the gene.

Selected Key Terms

bacterial artificial chromosomes
cosmids
hypervariable loci
knockout mice
microsatellite markers
pulsed field gel electrophoresis
restriction fragment length polymorphisms
T-DNA
Ti-plasmid
transgenic
variable number of tandem repeats
yeast artificial chromosomes

Chapter Review Problems

1. In preparing DNA for PFGE analysis, you need 1 μg of DNA per well (or lane) to separate the sample. Too much DNA will overload the well, and poor resolution will result. With too little DNA, you will not be able to detect its presence by ethidium bromide staining. In preparing PFGE gels to separate DNA into individual fragments, whole bacteria are embedded into agarose plugs, the bacteria in the plugs are treated with enzymes, and the plug is placed in a small slot at the top of the agarose gel. The genome size of the bacteria you are studying is 3×10^9 Da. How many bacteria do you need to have in a plug of 0.1 ml to get 1 μg of DNA? (Hint: Avogadro's number = 6.022×10^{23}, and assume that each bacterium contains one chromosome.)
2. The longest recognition sequence found for a restriction endonuclease is 8 bp. The recognition sequence can be increased to 10 bp by using an adenine methylase, abbreviated M.*Cla*I, that methylates ATCGAT to give ATCGmAT. When methylation occurs at the proper sequence, cleavage of the DNA is accomplished with *Dpn*I, a methyladenine-dependent endonuclease. The reaction follows:

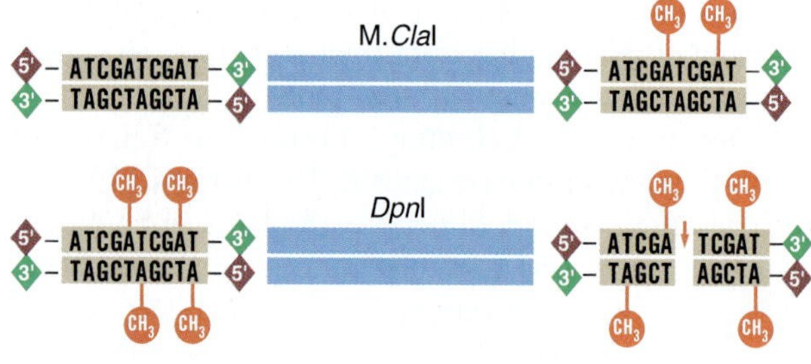

Calculate the size of DNA fragments that would arise from these reactions. In this calculation, assume that all four bases occur randomly at equal concentrations (each is one quarter of the total) and that the G + C content of the organism in question is 60%.

3. What key chromosome segments are essential to give a relatively high degree of stability to a YAC vector in the yeast host?
4. Assume that you have a hybridization probe that gives the following hybridization pattern for a mother and child where the paternity is in question. The numbers are expressed as kilobase pairs.

Child	Mother
4.5	4.5
4.1	
3.7	
3.5	
	3.1

What hybridization patterns would you predict for the genetic father?

5. Both PCR and Southern blotting have been used to analyze DNA samples in forensics (DNA fingerprinting). What are some of the advantages of PCR over Southern blotting for this application?
6. You are the defense attorney in a civil case in which your client has been sued for child support. The prosecution presents the following information concerning DNA fingerprinting of the child, mother, and your client.

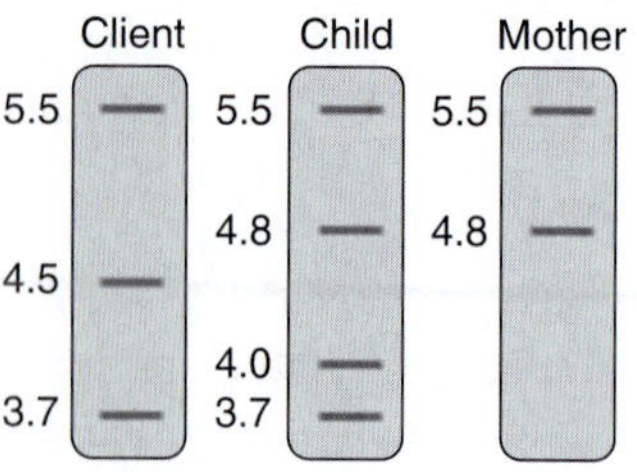

How would you argue your case?

7. You are a tomato breeder who is trying to identify and clone a gene in tomatoes (*Lycopersicon esculentum*, $2n = 24$) that causes fruit to be yellow instead of red. The yellow phenotype (y) is recessive to red (Y). Previous analysis has shown that the *y* gene is on chromosome 7. Because an RFLP map exists for tomatoes, you choose to use RFLP mapping to locate the *y* gene. Your first step is to identify DNA probes from chromosome 7 that will detect differences between strain 114, which is yellow (genotype *yy*) and strain 22, which is red (genotype *YY*). You find that two of the chromosome 7 probes detect DNA strain differences as follows.

	Probe 3	**Probe 5**
Strain 114 (*yy*)	*Pst*I: 8 kb	*Bam*HI: 5 kb
Strain 22 (*YY*)	*Pst*I: 6 kb, 2 kb	*Bam*HI: 7 kb

You make a cross between strain 114 and strain 22 to produce F_1 individuals. You then cross the F_1 plants back to strain 114 (*yy*) and examine 100 progeny for fruit color and DNA. The backcross progeny follow.

Number of Plants	**Fruit Color**	**Size of DNA (kb) Hybridized to Probe 3 (*Pst*I digest)**	**Size of DNA (kb) Hybridized to Probe 5 (*Bam*HI digest)**
41	Red	8, 6, 2	7, 5
39	Yellow	8	5
4	Red	8, 6, 2	5
3	Yellow	8	7, 5
6	Red	8	7, 5
6	Yellow	8, 6, 2	5
0	Red	8	5
1	Yellow	8, 6, 2	7, 5

a. Give the genotype of the F_1 plant including both the classic (yellow) and RFLP alleles, indicating which alleles are on the same chromosome.
b. Determine the gene order and recombination frequencies between genes and draw a genetic map including all the markers. (Hints: Follow the same steps in analysis as for any three-point mapping question. Because DNA markers reveal the genotype, they act as codominant alleles, and you can easily distinguish between homozygotes and heterozygotes.)

8. In Problem 7, what is the most likely cause of the RFLP detected by Probe 3? Support your answer.

9. How many base pairs are present in the human genome? How many kilobase pairs are present?

10. A cloned gene from *Brassica napus* is used as a radioactive probe against a DNA sample from *Brassica campestris*. Both DNAs were digested with *Pst*I and separated on an agarose gel, followed by hybridization with the radioactive probe. There were two radioactive bands on the autoradiogram of *Brassica campestris*, but one from *Brassica napus*. How can these results be explained?

11. Additional data were obtained on the cloned gene described in Problem 10. Two primers were designed to represent the two ends of the cloned gene, and PCR was done on *Brassica napus* and *Brassica campestri* DNA. Only one band was detected for both plant DNAs after PCR, and the intensity of the bands were identical after 20 cycles. Digestion of the PCR product with *Pst*I gave two bands from *Brassica campestri* but one from *Brassica napus* DNA. Do these data shed light on your answer to Problem 10?

12. You have found a polymorphism in DNA between two populations of true-breeding mice; one population is very overweight, and the other is normal weight. This polymorphism was identified from randomly selected sequences present in the mouse genome, and one was selected that showed this DNA polymorphism between the two populations. You think that this polymorphism may lead you to the cloning and identification of a gene in mice responsible for obesity, and if so, you may make a huge amount of money selling the gene product or inhibitors of the gene product. How would you determine that this polymorphism is linked to the obesity trait, and if it is closely linked, how would you clone the gene for obesity?

13. In planning an experiment, you discover that you must prepare a genome library of your organism. You are working with an animal that has a genome size of 500 Mb. How many cosmid clones, with an average insert size of 30 kb, do you require to obtain a 99% certainty that you have your clone? (Hint: See Chapter 12 for the equation used to calculate the answer.)

14. With the same data provided in Problem 13, calculate the number of clones you must have in your library with a 95% certainty that you have your clone. Is it worth going after the additional 4%?

15. If an organism has a high number of repetitive DNA sequences, what problem does this pose in chromosome walking from one YAC clone to another YAC clone, or from one BAC clone to another BAC clone? What other problems might arise as a result of repetitive sequences in a clone?

16. You would like to prepare a chromosome map of restriction cut sites of a *Pseudomonas* strain of bacteria. Because the genome is 4.6 Mb, you will need infrequent cuts by a restriction endonuclease to obtain independently migrating large fragments on your PFGE gel. The G + C content of the organism is 64%. From the following restriction enzymes, which is the most likely to give you infrequent cuts and thus large fragments?

Enzyme	**Recognition Sequence and Cut Site**
*Not*I	5′-GC↓GGCCGC-3′
*Sfi*I	5′-GGCCNNNN↓NGGCC-3′
*Sma*I	5′-CCC↓GGG-3′
*Dra*I	5′-TTT↓AAA-3′

17. If the organism in Problem 16 were *Rhizobium meliloti*, which has a G + C content of 54%, how would this change your decision as to enzyme selection?

Challenge Problems

1. Huntington disease is caused by a dominant allele that manifests its phenotype between the ages of 30 and 50. In the pedigree of the family shown here, DNA samples were taken from each of the family members, and a pair of microsatellite primers were used in the polymerase chain reaction to detect the presence of linked sites in their DNA.
   a. Which, if any, of these microsatellite markers has the potential of being linked to the Huntington disease locus?
   b. Is it possible from these data to establish whether this gene is sex-linked?
   c. How would you establish for certain whether your putative marker is actually linked to the Huntington disease locus?
   d. How would you establish the map distance between the marker and the Huntington disease locus?

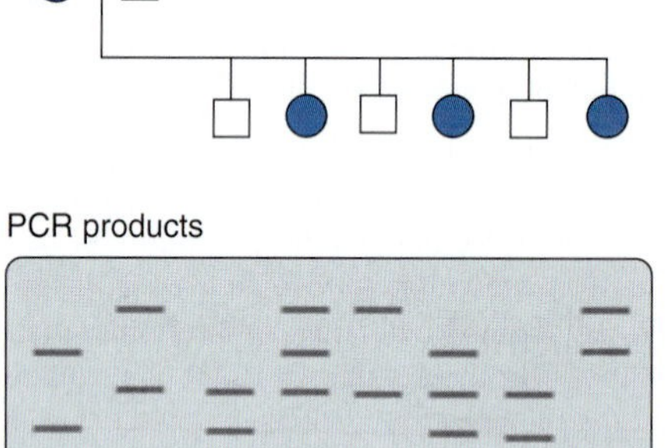

2. In your search for a 1.25-kb RFLP probe that maps near a gene of interest, you find one probe that correlates with the presence of the gene 99% of the time. As a result, you decide to characterize the sequences surrounding your probe more carefully. To do this, you search your genomic library and find a clone that your probe hybridizes to and determine the *Eco*RI map of this 8.5-kb clone. The map of this clone follows, and the area that your clone hybridizes to is identified by the line over the map.

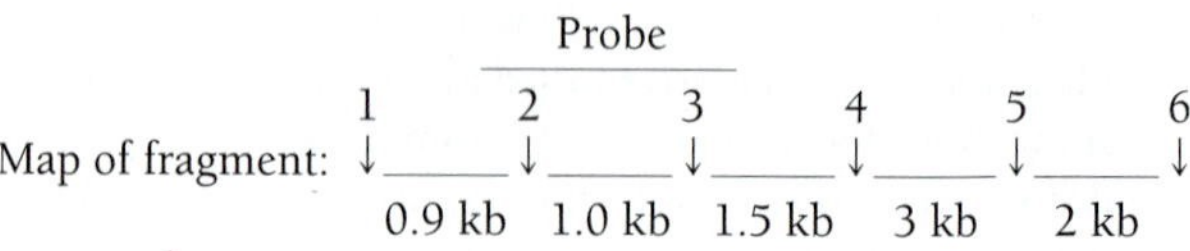

You cut the DNA from three different individuals with *Eco*RI, electrophorese the fragments in an agarose gel, transfer them to a nylon membrane, and hybridize to your probe. The following data were found showing the DNA bands from the Southern blot after hybridization to the probe.

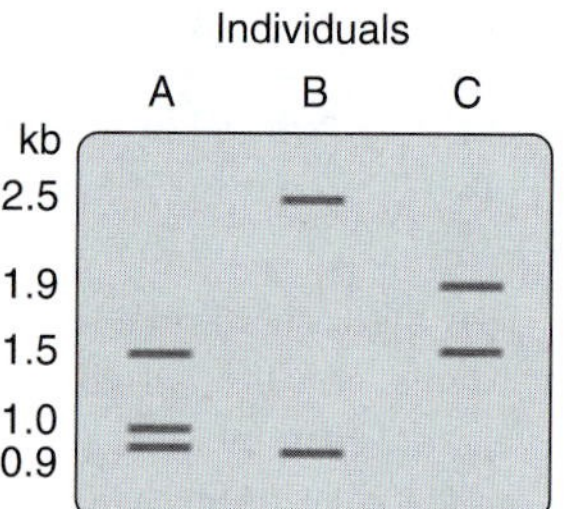

Explain these data.

3. You have cloned a gene from tobacco that is regulated by the presence or absence of light. In the presence of light, the gene is turned on and in the absence of light the gene is off. You want to determine what upstream sequences in your clone are involved in light regulation. You have an *Agrobacterium* Ti-plasmid with a β-glucuronidase gene (whose expression is easily detected in plant tissue) without a promoter. This plasmid also contains a multiple cloning site near the start site of the β-glucuronidase gene. How would you go about determining what upstream sequences of your cloned gene are important for light-regulated gene expression?

Suggested Readings

Burke, D. T., G. F. Carle, and M. V. Olson. 1987. "Cloning of large segments of exogenous DNA into yeast by means of artificial chromosome vectors." *Science,* 236: 806–812.

Capecchi, M. R. 1994. "Targeted gene replacement." *Scientific American,* 270 (March): 52–59.

Fraser, C. M., *et al.* 1995. "The minimal gene complement of *Mycoplasma genitalium.*" *Science,* 270: 297–403.

Goffeau, A., *et al.* 1996. "Life with 6000 genes." *Science,* 274: 546–567.

John, M. E., and G. Keller. 1996. "Metabolic pathway engineering in cotton: Biosynthesis of polyhydroxybutyrate in fiber cells." *Proceedings of the National Academy of Sciences USA,* 93: 12768–12773.

Lai, E., B. W. Birren, S. M. Clark, M. I. Simon, and L. Hood. 1989. "Pulsed field gel electrophoresis." *BioTechnique,* 7: 34–42.

Lander, E. S. 1996. "The new genomics: Global views of biology." *Science,* 274: 536–539.

Miklos, G. L., and G. M. Rubin. 1996. "The role of the genome project in determining gene function: Insights from model organisms." *Cell,* 86: 521–529.

Old, R. W., and S. B. Primrose. 1994. *Principles of Gene Manip-*

ulation: An Introduction to Genetic Engineering, 5th ed. Blackwell Scientific Publications, Boston.

Schlessinger, D. 1990. "Yeast artificial chromosomes: Tools for mapping and analysis of complex genomes." *Trends in Genetics,* 6: 248–258.

Schuler, G. D., *et al.* 1996. "A gene map of the human genome." *Science,* 274: 540–546.

Singer M., and P. Berg. 1991. *Genes and Genomes.* University Science Books, Mill Valley, CA.

Thomas, T. L., and T. C. Hall. 1985. "Gene transfer and expression in plants: Implications and potential." *BioEssays,* 3: 149–153.

Tinland, B. 1996. "The integration of T-DNA into plant genomes." *Trends in Plant Science,* 1: 178–183.

Velander, W. H., H. Lubon, and W. N. Drohan. 1997. "Transgenic livestock as drug factories." *Scientific American,* 276 (January): 70–75.

White, R. 1985. "DNA sequence polymorphisms revitalize linkage approaches in human genetics." *Trends in Genetics,* 1: 177–180.

White, R., and J-M. Lalouel. 1988. "Chromosome mapping with DNA markers." *Scientific American,* 258 (February): 40–48.

On the Web

Visit our Web site at **http://www.saunderscollege.com/lifesci/titles.html** and click on A/G/M Genetics for links to the following chapter-related resources on the World Wide Web:

1. **Yeast genome database.** This site allows you to search for specific DNA sequences, genes, and many other aspects of the yeast genome.
2. **Integrated human gene map.** This site allows you to find specific genes on the human genetic map.
3. **Stanford human genome center mapping site (SHGC).** This site provides data on the construction of high-resolution radiation hybrid maps of the human genome and the sequencing of large, contiguous genomic regions. The SHGC also maintains an active education program to make research results available to students, teachers, and the community.
4. **Bacterial database.** This site provides a collection of databases containing DNA and protein sequence, gene expression, cellular role, protein family, and taxonomic data for microbes, plants, and humans.
5. ***Arabidopsis* database.** This site has all the sequence and genetic data as well as mutant stocks available for this small experimental plant.
6. ***E. coli* genome project.** This site has all the sequence and genetic data from the *E. coli* genome project.
7. **Human genome projects.** This site provides links to human genome projects and databases.
8. **Primer for genome projects.** This site provides considerable background information (DNA structure, replication, cloning, etc.) to understand human genome projects and other genome projects.

CHAPTER

14

This electron micrograph of a chromosome shows an area of gene expression, or a chromosome puff. The RNA (the product of the gene) is shown in brown/violet, and the DNA is shown in blue. *(Science Source/Photo Researchers, Inc.)*

Regulation of Gene Expression

In Chapter 11, we learned that the DNA in the cells of an organism encodes all the information necessary to construct the whole organism. In this chapter, we will discuss the mechanisms cells use to manage the information in their DNA. Management systems are necessary because of the immense volume of information in the genome and because different cell types require different information at different times in their development. The DNA in the human genome contains about 100,000 genes, and if all genes were expressed equally, then all cells would be the same. It is estimated that a typical higher eukaryotic cell expresses between 10,000 and 20,000 genes, which is only about 10–20% of the total gene complement of the cell. Each particular cell type (for example, a brain cell) has a unique function, structure, and enzymes and, thus, needs to express only certain genes and not others. Consequently, complex organisms have evolved mechanisms to regulate the expression of genes so that a unique, specific set of genes is expressed in each cell type. The general phenomenon of specific gene expression in a certain cell type or during a specific stage of development is known as **differential gene expression.** The mechanisms controlling differential gene expression are a critical part of each species' genome.

Examining the gene products of two very different cell types (for example, a pollen tube and a root cell) reveals that over 95% of the proteins are identical. Thus, the functional differences between cell types are often determined by a small subset of genes. The remainder are referred to as housekeeping genes. **Housekeeping genes** are responsible for synthesizing common proteins such as histones, ribosomal proteins, metabolic enzymes, structural components, and myriad other proteins required to maintain cell life and reproduction. How the small subset of genes in each specific cell type is regulated is the subject of this chapter and of Chapters 15 and 16. Regulation of gene expression is a fundamental area of study and is critical to our understanding of basic cellular functions as well as genetic diseases.

By the end of this chapter, we will have addressed the following five questions:

1. How does a single regulatory gene or site control the expression of several structural genes?
2. How do regulatory proteins turn genes on and off in both prokaryotes and eukaryotes?
3. What structural features are common to all regulatory proteins?

4. How do posttranscriptional gene regulatory mechanisms work?
5. How can the localized rearrangement of DNA control gene expression?

The Regulation of Gene Expression Occurs at Many Levels.

It is important to distinguish between eukaryotes and prokaryotes when examining the control of gene expression. Many of the regulatory mechanisms these two organisms use are similar, but some are very different, in part because of the very different environments in which they live. Most eukaryotic cells have a specialized role in conjunction with a larger, multicellular body, whereas each prokaryotic (bacterial) cell is a self-contained organism. To fulfill their role, eukaryotic cells must produce certain specialized proteins and structures at precise times and in precise locations within the cell and the body. They do this in response to signals from the environment or neighboring cells that activate certain genes and suppress others. These precise, coordinated, and sometimes swift changes in gene activity are necessary for the organism to survive.

In contrast, prokaryotic bacterial cells live in a diverse, rapidly changing, and highly competitive environment. To survive, prokaryotes must be able to turn genes on and off in rapid response to environmental changes. Consider an *E. coli* cell that is capable of using different carbon compounds as energy sources (for example, sucrose, fructose, lactose, and galactose). This bacterial cell typically encounters only one or two of these compounds at a time. To conserve energy, the cell turns on the genes that produce the metabolic enzymes necessary to use the available carbon source(s), and it turns off all others.

There are multiple points in the steps between gene expression and protein synthesis at which gene expression can be controlled in both prokaryotic and eukaryotic cells (Figure 14.1). These points of control can be separated into two general areas: transcriptional control and posttranscriptional control. **Transcriptional control,** the primary control point, is the regulation of RNA synthesis from a DNA template. All cells possess a large set of sequence-specific **DNA binding proteins** whose main function is to turn genes on or off at the transcriptional level. DNA binding proteins possess precise structures that recognize and bind to specific DNA sequences.

Posttranscriptional controls are secondary mechanisms for controlling gene expression after transcription. Posttranscriptional controls include (1) RNA processing control, (2) translational control, (3) mRNA degradation control, and (4) protein activity control. **RNA processing control,** which is restricted to eukaryotes, determines how and when the primary transcript is spliced or otherwise processed to form a usable mRNA. For example, some RNA transcripts produce different mRNAs from the same gene in different cell types by selectively using introns. **Translational control** determines which mRNAs are translated into proteins and when they are translated. For example, translation may be inhibited by mRNA binding to specific proteins or promoted by special nucleotide sequences at the end of the mRNA that facilitate ribosome binding. **Messenger-RNA degradation control** affects the stability of certain mRNA species, selectively destabilizing some and stabilizing others. A protein that is needed in large quantities over an extended time period may have a very stable mRNA, for example β-globin mRNA, which has a half-life of over 10 hours (half-life is the time required for 50% of the molecules to

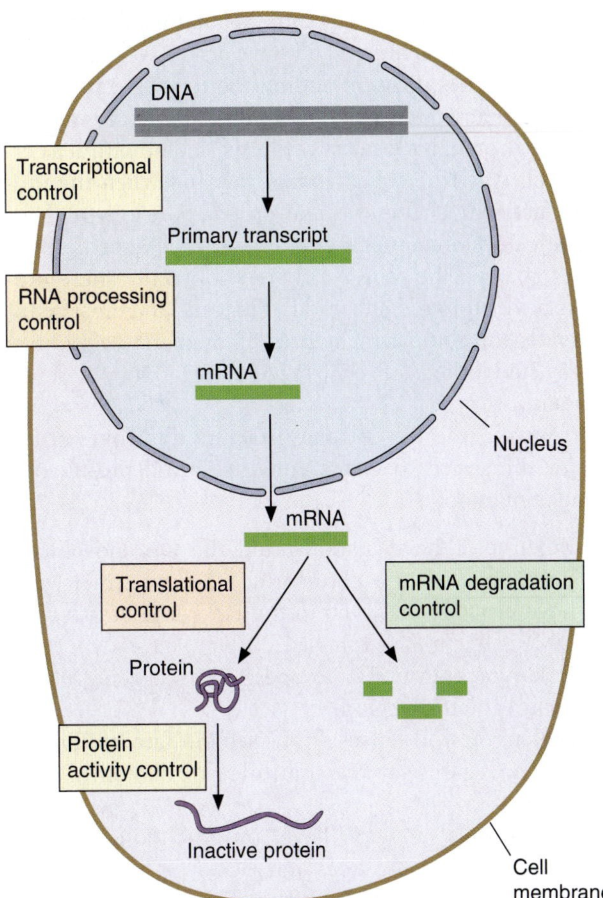

FIGURE 14.1 Control points in gene expression. Control of gene expression occurs most often during transcription, but it can also occur during processing of the primary transcript RNA into mRNA, during the translation of the mRNA into a protein, to adjust the rate of the mRNA degradation, and to regulate the activity of the protein.

be degraded). In contrast, some regulatory proteins are only transiently required and have an mRNA half-life of several minutes. **Protein activity control** selectively activates, inactivates, modifies, or compartmentalizes specific protein molecules within the cell or within a certain cell type, thereby affecting how and when the proteins act. For example, some proteins are required for specific developmental events and act only in certain cell types at a specific time during development. These proteins have profound effects on other cells and must be inactivated or compartmentalized immediately after they have acted; otherwise, they could cause abnormal development.

As you read this chapter, bear in mind some of the major differences between eukaryotic and prokaryotic gene regulation:

1. Eukaryotes have histones and the chromosome is tightly folded into chromatin, whereas prokaryotes have proteins that bind to the DNA and most of the DNA is free to be expressed.
2. Eukaryotes do not have operons; each gene has its own regulatory region.
3. Prokaryotes do not use transcription factors and enhancer regions for the controlled regulation of each gene.
4. Prokaryotes do not have introns; consequently, mRNA processing does not occur in prokaryotes. However, mRNA processing is used extensively in eukaryotic gene-expression regulation.

TRANSCRIPTIONAL CONTROL I: THE *lac* OPERON IS AN EXCELLENT MODEL OF PROKARYOTIC TRANSCRIPTIONAL CONTROL.

A curious thing happens when *E. coli* cells are grown in a mixture of glucose and lactose (a disaccharide sugar). The growing bacteria consume all the glucose before they begin to metabolize any of the lactose. The switch to lactose is accompanied by a slight pause in growth, during which time the enzyme β-galactosidase is produced in large quantities (Figure 14.2). β-galactosidase is the first enzyme involved in the metabolism of lactose, and it hydrolyses lactose (at the $1 \rightarrow 4$ β-linkage between glucose and galactose) to its constituent subunits, glucose and galactose. These monosaccharides are then further metabolized to provide carbon and energy to the cell. However, if glucose is added back to the growth medium, β-galactosidase synthesis ceases, and any enzyme remaining in the cells is slowly diluted out as the bacterial population grows. The cell uses the glucose and ceases to metabolize lactose (see CONCEPT CLOSE-UP: *Why Can't Some People Digest Milk?*)

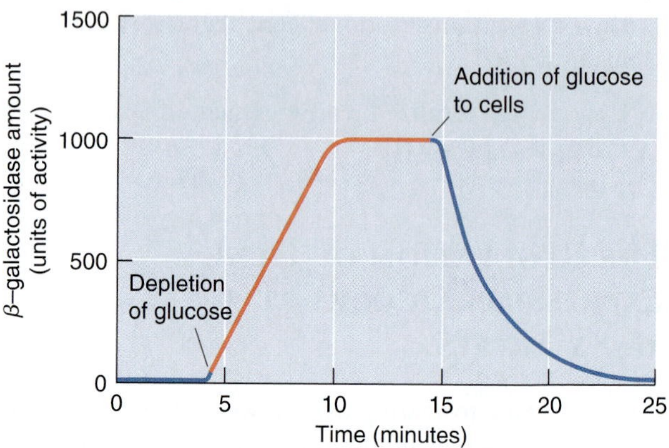

FIGURE 14.2 ***E. coli*** **synthesis of β-galactosidase.** When lactose is present and glucose is depleted from the growth medium, β-galactosidase is produced in large quantities. When glucose is added back to the growth medium, β-galactosidase synthesis ceases because the cell prefers glucose over lactose as a carbon source.

The remarkable thing about this observation is the *E. coli*'s ability to turn a specific gene on or off in response to an environmental change (in this case, the presence or absence of glucose and lactose). This observation led to understanding the regulation of β-galactosidase gene expression, and it was the first major breakthrough in understanding gene control mechanisms. This pivotal discovery was made in the late 1950s by Francois Jacob and Jacques Monod of the Pasteur Institute (Paris) after years of studying lactose metabolism and the many different mutations affecting lactose metabolism. Their work is central to our present-day understanding of gene control. They were awarded the Nobel Prize in 1965 for their work, which led to an understanding of the regulation of gene expression.

For an *E. coli* cell to metabolize lactose, it needs two enzymes, β-galactosidase and β-galactoside permease, both of which are synthesized when lactose is the sole carbon source (a third enzyme, β-galactoside transacetylase is also synthesized, but its function is not known). β-galactosidase is responsible for cleaving the $1 \rightarrow 4$ β-linkage between the subunits of lactose. The permease is necessary for active transport of the lactose into the cell (Figure 14.3). Neither of these enzymes has any function in the cell in the absence of lactose so the bacterial cell conserves energy by not synthesizing them until needed. Similar regulation of gene expression is also found for a large variety of other carbon sources (for example, xylose, raffinose, and arabinose).

▶ **FIGURE 14.3** **The action of β-galactoside permease and β-galactosidase.** β-galactoside permease is responsible for active transport of lactose into the cell, and β-galactosidase hydrolyzes the lactose $1 \longrightarrow 4$ β-linkage to produce glucose and galactose.

Concept Close-Up

Why Can't Some People Digest Milk?

The vast majority of the people living in Asia and Africa are not able to digest milk, in contrast to most Americans and individuals of European descent. Those individuals who cannot digest milk get diarrhea and can dehydrate, sometimes to the point of death, when they drink milk. More commonly, the symptoms are bloating and intestinal gas. Why can't some people digest milk? When you are born, you produce large quantities of the enzyme β-galactosidase, sometimes called **lactase,** which is responsible for converting the disaccharide lactose to the monosaccharides glucose and galactose. We all possess lactase as juveniles because milk contains about 7% lactose. But, at about age 3 or 4 years; this gene is switched off, or nearly off. This is wherein the problem lies. In some people, an adult form of the gene for lactase is switched on (lac^+), but in others it is not (lac^-). In those individuals who do not have an active adult lactase gene, undigested milk passes into the lower intestine and is digested by enteric bacteria, which creates CO_2 and causes bloating. As might be expected, this trait is inherited and passed on from generation to generation.

Fewer black people than white people can drink milk without side effects. What is the reason for this heterogeneity in the population? The answer to this question is not known for certain, but we can speculate that in our distant past a mutation that allowed individuals to digest milk may have occurred. It is likely that milk was not generally available to the adult population until some form of agriculture developed, probably about 10,000 to 15,000 years ago. When milk became available to adults, a lac^+ phenotype could easily confer an advantage to those individuals. The lac^+ trait would increase fitness, and these individuals would contribute more progeny to the next generation, giving rise to populations that possessed lactase. An alternative speculation is that all humans may have had adult lactase in the distant past. In Africa, where the tsetse fly transmission of trypanosomes has decimated the cattle populations, the speculation is that Africans have lost their adult lactase because there has been no selective pressure to retain expression of this gene.

Interestingly, yogurt can sometimes be consumed by both lac^- and lac^+ individuals because, during the fermentation process that produces yogurt from milk, bacteria convert the lactose to glucose and galactose and eventually to lactic acid. In addition, some commercially available forms of milk now include large numbers of *Lactobacillus acidophilus* bacteria, which can help make the milk more digestible for lac^- individuals. When this milk is consumed, the *L. acidophilus* bacteria end up in the gut and do the digesting for the lac^- individuals.

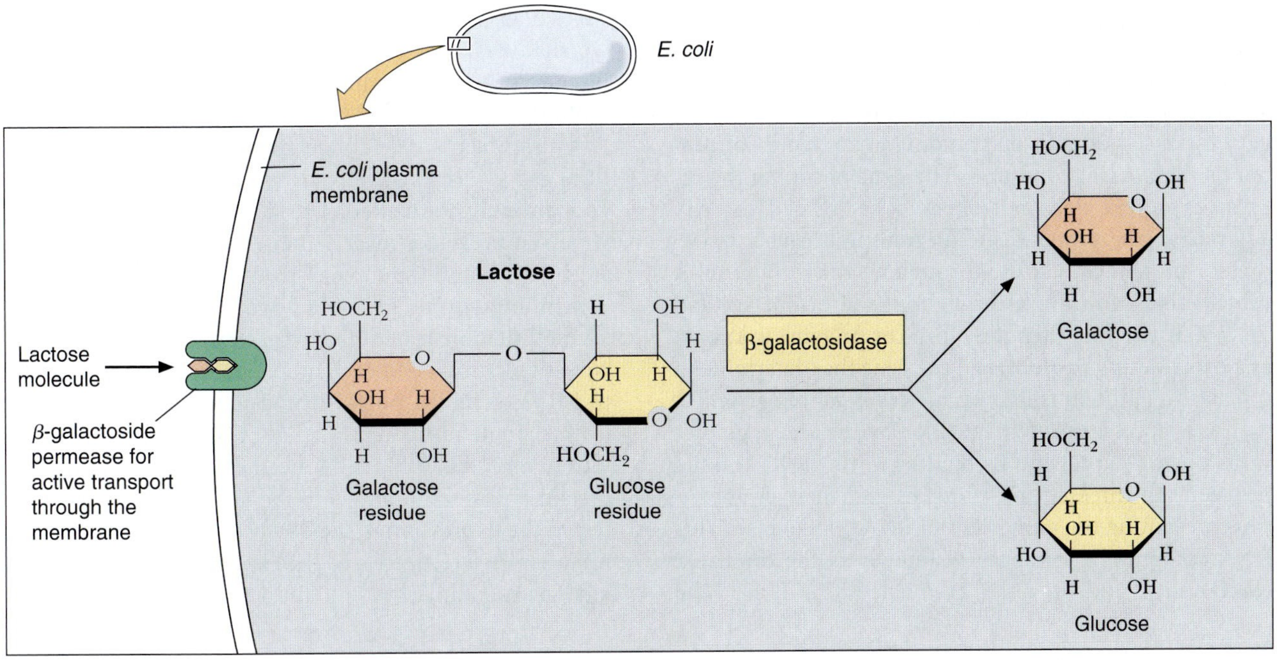

E. coli
E. coli plasma membrane
Lactose
Lactose molecule
β-galactoside permease for active transport through the membrane
Galactose residue
Glucose residue
β-galactosidase
Galactose
Glucose

Enzyme Synthesis Is Induced by Environmental Signals.

The expression of a gene in response to a substance in the environment is called **induction,** and the genes capable of environmental stimulation are referred to as **inducible genes.** Other genes that are always expressed (such as tRNA, rRNA, or RNA polymerase-determining genes) are called **constitutive genes.** Still other sets of genes are turned off when a particular substance is in the cellular environment, and these are called **repressible genes.**

In *E. coli,* the presence of lactose (in the absence of glucose) induces the synthesis of β-galactosidase and β-galactoside permease; thus, the genes for these enzymes are inducible. Four genes are involved in the regulation of β-galactosidase and β-galactoside permease synthesis, and three sites are set aside for binding regulatory proteins. Collectively, this group of genes and binding sites is referred to as an **operon** (Figure 14.4). Operons exist for many metabolic pathways and are usually named after the main metabolite in the enzymatic pathway; for example, the operon responsible for transport and degradation of lactose is called the lactose, or *lac* operon. The mRNA from the three structural genes in the lactose operon are transcribed as a single product using one promoter. An mRNA transcribed from more than one cistron is referred to as a **polycistronic mRNA.** In the case of the *lac* operon, three structural genes are transcribed as a single polycistronic mRNA, the *lacZ* gene (β-galactosidase), the *lacY* gene (β-galactoside permease), and the *lacA* gene (transacetylase).

An important gene in the *lac* operon is the *lacI,* or regulator gene, which is responsible for the synthesis of a small number (a few per cell) of regulatory proteins referred to as **repressor proteins** (Figure 14.4). The repressor protein binds to a specific DNA sequence and inhibits β-galactosidase synthesis. Normally β-galactosidase is not synthesized in an *E. coli* cell; thus, inhibition of β-galactosidase synthesis and binding of the repressor protein is the normal condition. The gene encoding the repressor protein does not need to be located directly adjacent to the protein's site of binding because the repressor protein can diffuse throughout the cytoplasm. A protein that can diffuse throughout the cytoplasm and act at a distant site (like the repressor protein) is known as a ***trans*-acting protein,** and the gene that determines a *trans*-acting protein is known as a ***trans*-acting gene.**

The repressor protein encoded by the *lacI* gene prevents operon expression by binding to the DNA at a site between the three structural genes and their promoter, thus preventing the movement of RNA polymerase. This DNA recognition sequence is known as the **operator (*lacO*) site** and is only 21 nucleotides long. The tetrameric repressor protein binds tightly (Figure 14.4) to the *lacO* DNA sequence. However, in the presence of lactose (actually **allolactose,** which has a $1 \rightarrow 6\,\beta$-linkage instead of a $1 \rightarrow 4\,\beta$-linkage and is made from lactose by β-galactosidase in small quantities in the cell), the repressor protein does not bind to the *lacO* site, and transcription of the three adjacent structural genes proceeds (Figure 14.4). Note that the *lacO* site is a protein-binding site on the DNA, and it does not make a product. For this reason, it is not a cistron as defined by the *cis–trans* test (Chapters 6 and 8). This type of sequence is known as a ***cis*-acting site** and affects only the expression of genes immediately adjacent to it.

Allolactose binding to the repressor protein changes its three-dimensional structure so that it cannot bind to the *lacO* DNA sequence. As a consequence, the lactose operon is induced, and transcription can proceed. The small molecule (allolactose) that binds to the repressor protein is referred to as the **inducer.** The end result is that the *E. coli* cell is able to make the enzymes it needs for the breakdown of lactose, but only when lactose is present in the cell. When the inducer (allolactose) is removed or consumed, the repressor binds to the operator site (*lacO*) and prevents the further transcription of the three adjacent cistrons.

Mutations Were the Key to Understanding the *lac* Operon.

How was the *lac* operon deciphered? Jacob and Monod isolated and characterized a number of *E. coli* mutants in the lactose metabolism genes and formulated the operon hypothesis to explain the behavior of these mutants [see HISTORICAL PROFILE: *Jacques Monod (1910–1976)*]. Theirs was a masterful tour de force of deductive reasoning, and we shall examine a number of these mutants and their phenotypes to determine how each mutant helped to build the operon model.

First, Jacob and Monod noted that the three structural genes (cistrons) were always controlled together, or **coordinately controlled.** The lactose metabolic enzymes were synthesized at nearly the same time and rate. This suggested that these three cistrons had a single common controlling site or promoter. Second, genetic mapping studies demonstrated that the three structural genes were tightly linked to each other, lending credence to the hypothesis that a polycistronic message was initiated from a single promoter. Further evidence to support the idea of a polycistronic mRNA came from the finding that certain types of mutations in the *lacZ* and *lacY* genes affected the expression of the downstream genes. That is, a mutation in *lacZ* decreased the expression of the *lacY* and *lacA* genes. Similarly, a mutation in *lacY* affects *lacY* and

In the Absence of Lactose

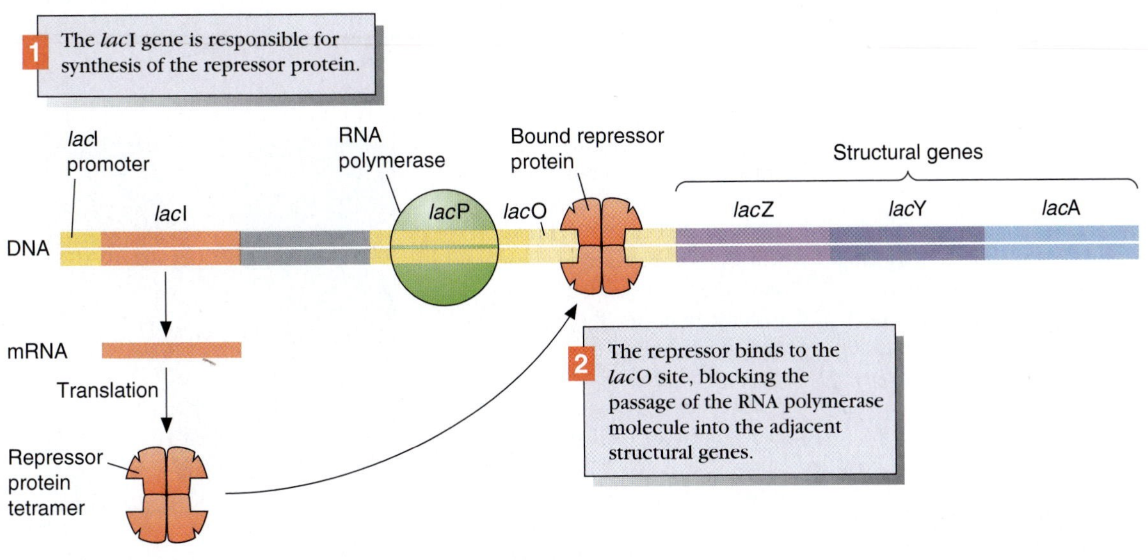

In the Presence of Lactose

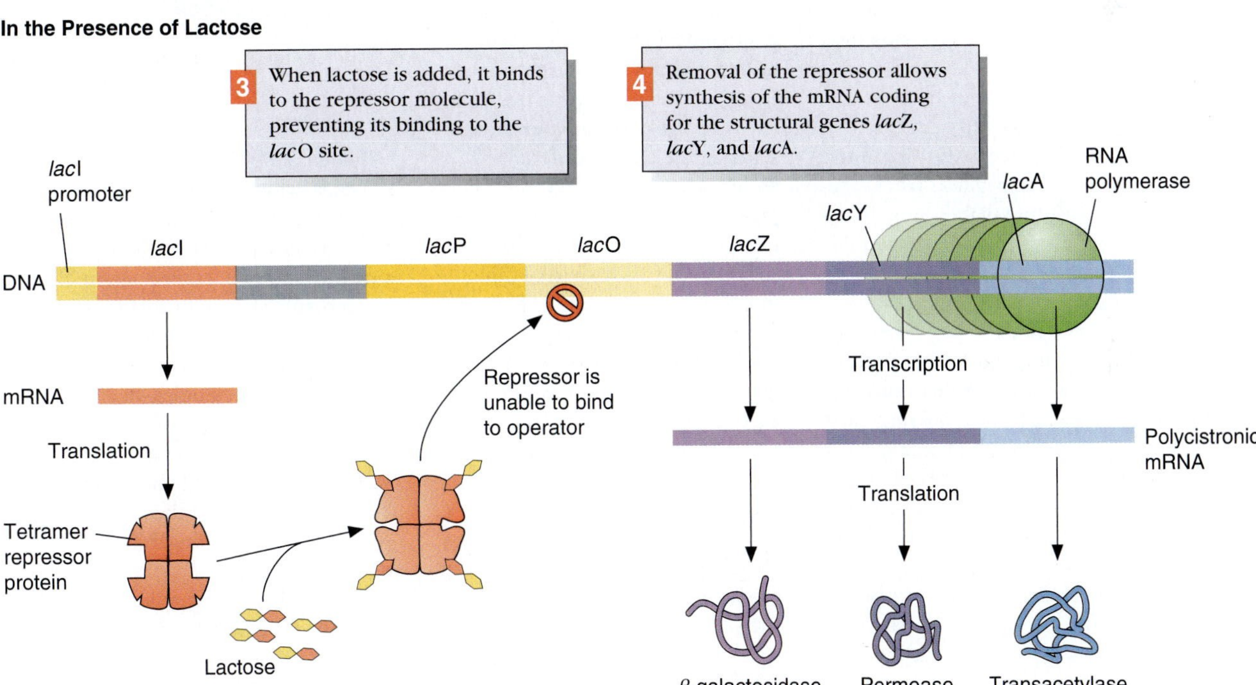

FIGURE 14.4 The lactose operon. In the lactose operon, the regulatory gene (*lacI*) and the DNA recognition sequence (*lacO*) regulate the transcription of the adjacent structural genes (*lacZ*, *lacY*, and *lacA*). In the absence of lactose, the repressor protein (lacI protein) binds to the operator site near the structural genes and blocks transcription. In the presence of lactose, the repressor protein is prevented from binding to the operator site and allows transcription to proceed.

HISTORICAL PROFILE

Jacques Monod (1910–1976)

It is the dream of most scientists to have their careers marked by at least one notable discovery—something for which they will be remembered long after their death. Few scientists realize this dream. An exceptionally rare individual will make a series of illuminating discoveries that give rise to new concepts and open new vistas. Jacques Monod was one of these rare individuals.

Jacques Monod (Figure 14.A) was born in Paris in 1910 to Sharlie Monod (formerly, Sharlie MacGregor from Milwaukee) and Lucien Monod, a painter, engraver, and art historian. Jacques Monod lived in Paris most of his life, receiving his advanced degrees from the Sorbonne and the Pasteur Institute (founded by Louis Pasteur), both very distinguished institutions for teaching and research. It was at the Pasteur Institute that Monod developed an intense interest in microbiology and the applications of genetic and biochemical principles to the study of microbes. However, after concluding his formal education, Monod was unable to carry out scientific research because Paris was occupied by the invading Nazis. Monod joined the underground resistance movement and was arrested by the Gestapo, but he managed to escape. Although at this stage of his life little time was available for research, Monod formulated some ideas on the induction of the enzyme synthesis or enzyme adaptation, as it was then called.

After the war, he began his long career at the Pasteur Institute in Paris where he later became director. This career was marked by collaborations with many great scientists, including Francis Crick and François Jacob. During the late 1940s and early 1950s, Monod's work was directed toward understanding the induction of β-galactosidase in *E. coli*. By this time he had shown that its synthesis was dependent upon an active gene (*Z*) and a product of a closely linked gene known as the *I* gene. Mutations in the *I* gene (I^-) were constitutive for β-galactosidase production (they produced the enzyme at a high level in the absence of the inducer). At this time, nothing was known about the repressor protein nor was the concept of an operator site understood. In fact, mRNA had not been discovered. At that time, most scientists believed rRNA or the ribosome somehow assembled the correct order of the amino acids in a protein. Also, conjugation in bacteria had just come into general use allowing the construction of various partial diploids in *E. coli*. The technique of conjugation in bacteria provided data for the landmark discovery by Arthur Pardee, François Jacob, and Jacques Monod that led to the hypothesis of a diffusible "repressor" that blocks the expression of the *lac*Z^+ gene. These experiments were also the key to understanding the existence of an mRNA and eventually the concept of an operon and an operator site controlling a group of adjacent structural genes. The operon hypothesis was developed as a result of the isolation and characterization of hundreds of mutants affected in β-galactosidase synthesis.

The second great contribution of Monod to scientific understanding was the concept of allostery,

FIGURE 14.A **Jacques Monod.** *(Courtesy of the Pasteur Institute, Paris)*

which he proposed in 1961. This concept, which was not accepted by the enzymologists of the day, said that the activity of an enzyme is dependent upon the conformation of the protein. The conformation was controlled at the attachment–detachment site by an effector molecule. In his own studies, he envisioned that the inducer (lactose) would be bound to the repressor which, in turn, would change the binding properties of the repressor so that it could no longer bind to the operator site.

Many honors were bestowed on Monod including the Nobel Prize for Physiology and Medicine, which he shared with François Jacob and Andre Lwoff in 1965. He died at the early age of 66 from infectious hepatitis at the southern resort town of Cannes in France.

lacA, but not *lacZ* (Figure 14.5). Because of this effect on downstream cistron expression, these mutations were called **polar mutations.** It was later found that polar mutations are restricted to nonsense mutations (creation of a chain-termination signal, see Chapters 11 and 17). In contrast, a missense mutation affects only the expression of the cistron where it is located.

Jacob and Monod also discovered a class of unusual mutants that affected all the cistrons equally. These mutants synthesized β-galactosidase, the permease, and the transacetylase at full induction levels, in the absence of the inducer. Genetic mapping data indicated that the mutations causing this phenotype were located near the lactose structural genes, *lacZ, lacY,* and *lacA*. They designated this new gene *lacI* and the mutants as $lacI^-$ mutants. From the operon model presented in Figure 14.4, you can see that if no repressor is made (or if a defective repressor is made; the $lacI^-$ mutant), the repressor does not bind to the operator, and the structural genes are constitutively expressed.

From Chapter 8 you may recall that partial diploids in bacteria can be made by the presence of an F′-plasmid that contains an extra copy of the gene of interest. With the use of partial diploids and complementation tests, Jacob and Monod established that the $lacI^-$ mutations were in a separate cistron. For example, they constructed the following genotype in an *E. coli* strain:

$$\frac{lacI,\ lacO,\ lacZ^-,\ lacY^-}{lacI^-,\ lacO,\ lacZ,\ lacY} \quad \begin{matrix}\text{F}'\text{-plasmid}\\ \text{Chromosome}\end{matrix}$$

This is a partial diploid for the lactose cistrons with the wild-type *lacI* gene present in *trans* with respect to the wild-type structural genes (*lacZ* and *lacY*). The structural genes adjacent to the wild-type *lacI* gene are all mutant. If the *lacI* cistron makes a diffusible product, this genetic configuration is still inducible, not constitutive, because the F′-plasmid *lacI* cistron repressor protein can bind to the chromosome *lacO* site to block expression of the operon on the chromosome. This was exactly what

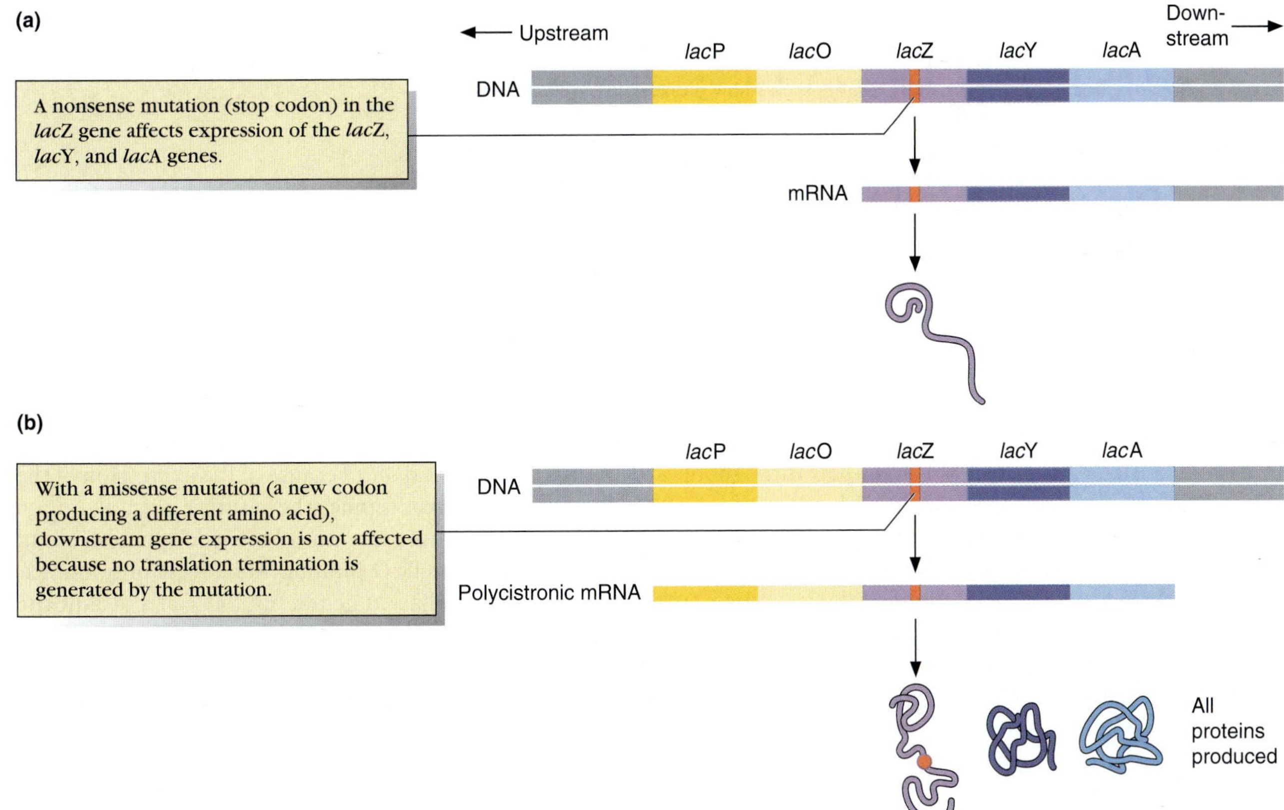

FIGURE 14.5 Polar mutations. This model shows how a nonsense mutation (a termination codon) affects expression of downstream genes, whereas missense mutations do not. (a) The presence of a nonsense mutation in *lacZ* also causes decreased expression of *lacY* and *lacA*. (b) The presence of a missense mutation affects only the expression of *lacZ*.

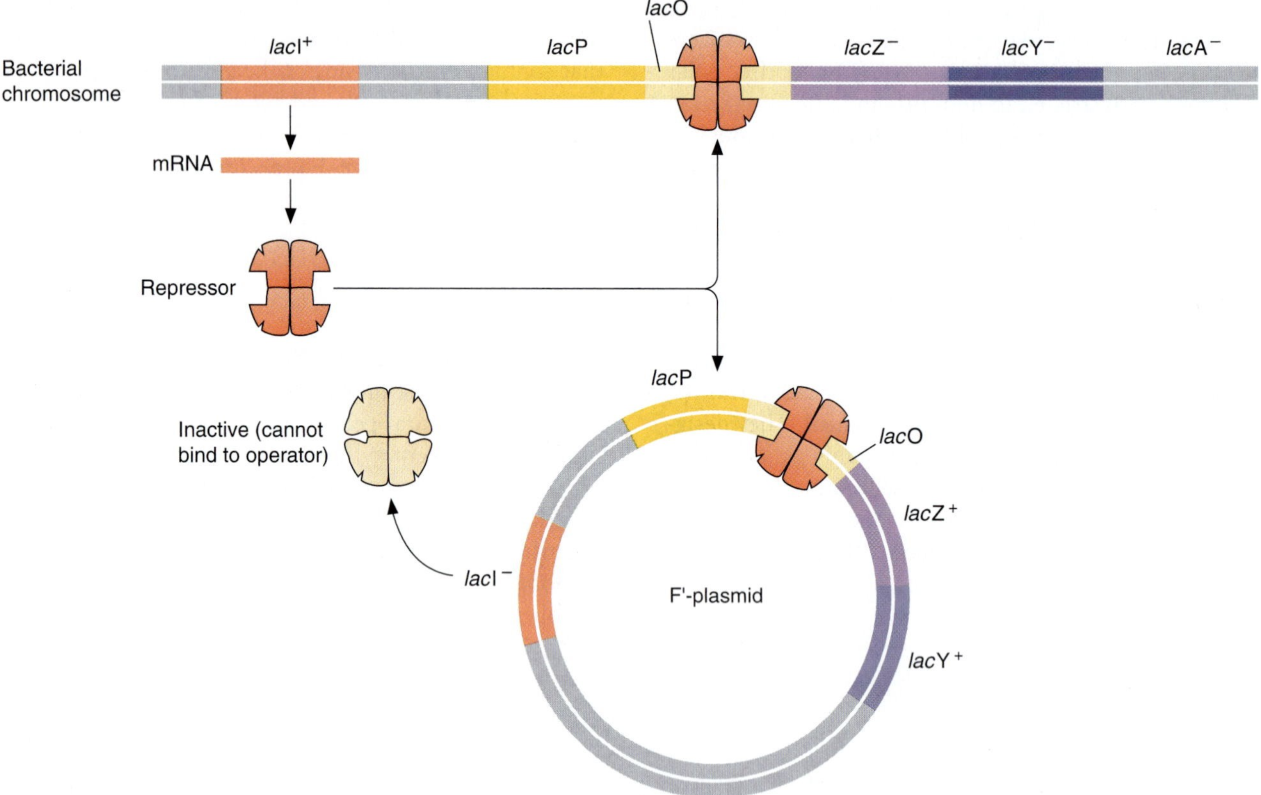

FIGURE 14.6 ***Trans*-acting genes.** This diagram shows how a diffusible product (the repressor) can affect expression of genes in *trans*. The *lacI* gene on the chromosome makes a repressor that can bind to both operators (chromosome and plasmid), but in this example, repressor binding affects only the expression of the plasmid structural genes because the chromosomal *lacZ*$^-$ and *lacY*$^-$ genes are mutant. Addition of lactose permits induction of β-galactosidase and permease from the plasmid genes.

was observed (Figure 14.6). This experiment also showed that the *lacI* gene is a separate cistron from the structural cistrons and makes its own product. The protein product of the *lacI* gene is able to diffuse throughout the cell and attach to any operator (*lacO*) binding site. As mentioned earlier, a gene that makes a diffusible product is known as a *trans*-acting gene. This type of gene can affect distantly located genes. In contrast, the *lacO* site is a *cis*-acting site and can control only the adjacent cistrons because the *lacO* site does not make a product. For example, when Jacob and Monod constructed the following partial diploid:

$$\frac{lacI,\ lacO,\ lacZ,\ lacY \quad \text{F}'\text{-plasmid}}{lacI,\ lacO^{-},\ lacZ,\ lacY \quad \text{Chromosome}}$$

they found that β-galactosidase or permease enzymes were continuously synthesized, even in the absence of lactose. In this example, the lacI protein (the repressor) is made and binds to the nonmutant (on the F′) *lacO* site, but cannot bind to the *lacO* site on the chromosome. Thus, the expression of the structural genes on the chromosome (but not on the F′-plasmid) is continuous and is said to be ***cis*-dominant.** The structural genes on the F′-plasmid are induced, but because of the expression of the genes on the chromosome, the result is seen as only a slight increase in the amount of enzyme synthesized. In this example, the *lacO* binding site affects only the expression of the adjacent genes, not those at a distant site (in this case, on the plasmid). This type of mutation in the *lacO* gene is not usually referred to as *lacO*$^-$ as indicated in the example, but as *lacO*^{c}, where the *c* refers to constitutive synthesis. We will revisit *lacO*^{c} mutations later in this section.

The repressor protein for the lactose operon has two binding sites: one that binds to the operator site on the DNA and another that binds to the allolactose inducer. In the *lacI*$^-$ mutation just described, the defective repressor protein cannot bind to the operator site. The portion of the repressor protein that binds to the *lacO* site is not the same as the binding site for the inducer; thus, another

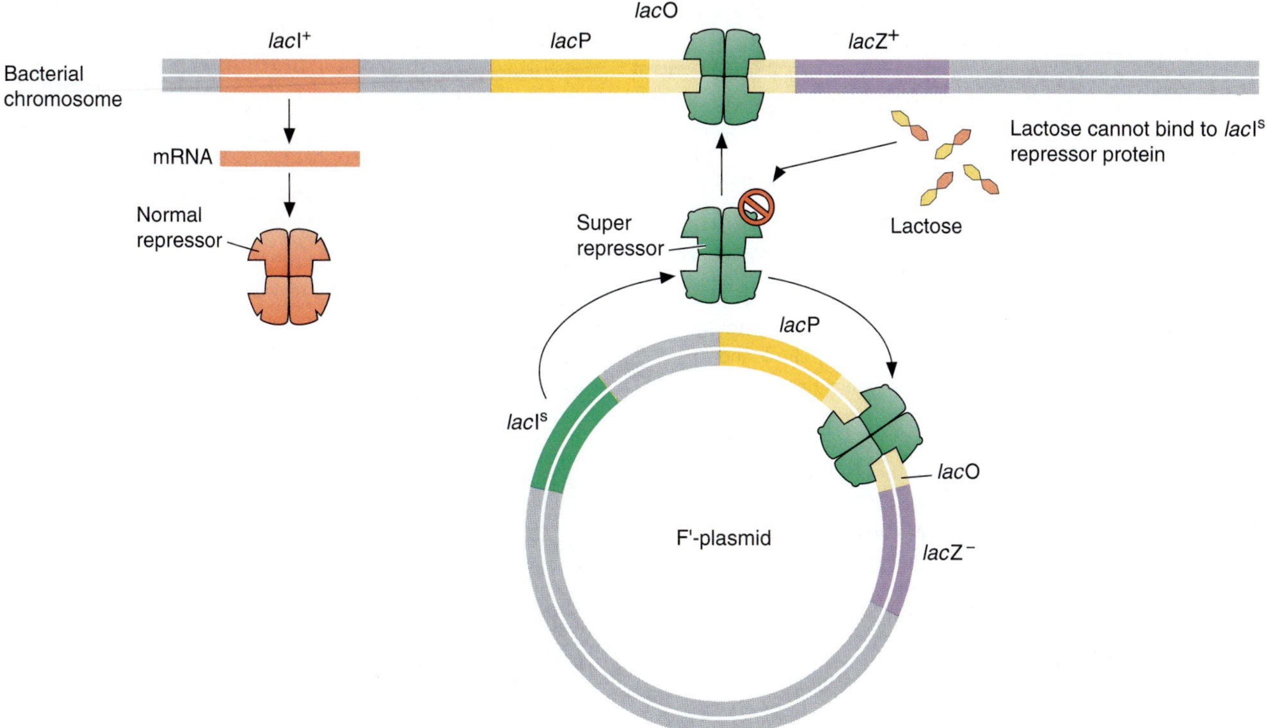

FIGURE 14.7 ***Trans*-dominant genes.** The illustration shows how the diffusible product of the plasmid *lacI* gene when mutated to *lacI*^{s} dominantly affects gene expression in a *trans* configuration. The repressor protein is altered by the mutation so the inducer (lactose) can no longer bind. As a consequence, the mutant repressor protein binds tightly to both operator sites (*cis* and *trans*), and does not permit the normal repressor to bind and unbind. It is therefore said to be superrepressed and *trans*-dominant. Although not shown in this figure, subunits of the normal repressor and the superrepressor protein can combine to form a heterodimer. The heterodimer usually does not interfere with superrepressor binding to the operator.

portion of the repressor protein can also be affected by mutations. Allolactose binding site mutations in the *lacI* gene were also found by Jacob and Monod, and they were called *lacI*^{s} (superrepressor) mutants (Figure 14.7). *lacI*^{s} mutations make a very low level of β-galactosidase and do not respond to the presence of an inducer. Because *lacI*^{s} mutations affect the ability of the repressor to bind allolactose, the repressor remains tightly bound to the *lacO* site. As a consequence, the RNA polymerase cannot bind to the promoter, blocking transcription. The repressor in a partial diploid of the F′ *lacI*^{s} allele binds tightly to both the *cis*- and *trans*-located operator binding sites, preventing induction after the addition of lactose.

$$\frac{lacI^{s},\ lacO,\ lacZ^{-},\ lacY \quad \text{F}'\text{-plasmid}}{lacI,\ lacO,\ lacZ,\ lacY \quad \text{Chromosome}}$$

From these results, we can conclude that a *lacI*^{s} allele is dominant over the wild-type *lacI*$^{-}$ allele.

In recent years, the DNA sequence of the *lacI* gene has been determined. The *lacI* repressor protein is 1040 amino acids long, and mutations in the amino terminal half of the protein cause defective *lacO* binding ability. Mutations in the carboxyl end determine its ability to bind to the inducer.

Because the repressor protein binds to the *lacO* or operator site on the DNA, another type of mutation is possible; this mutation is in the operator DNA sequence and affects the binding of the repressor to the operator site. Figure 14.8 shows the nucleotide sequence of the operator site (*lacO*) and nucleotide changes that affect repressor binding. Note that the sequence of nucleotides at the *lacO* site, which is the repressor protein recognition sequence, has a twofold symmetry. This is characteristic of many DNA sequences that bind proteins. Mutations in the operator site are referred to as O^{c} (operator constitutive) mutations, as mentioned earlier. Constitutive expression of the operon is a logical outcome if the repressor protein cannot bind to the operator site. When the operator site is not blocked by the repressor protein, RNA polymerase constitutively transcribes the adjacent cistrons. Furthermore, because the *lacO* site does not

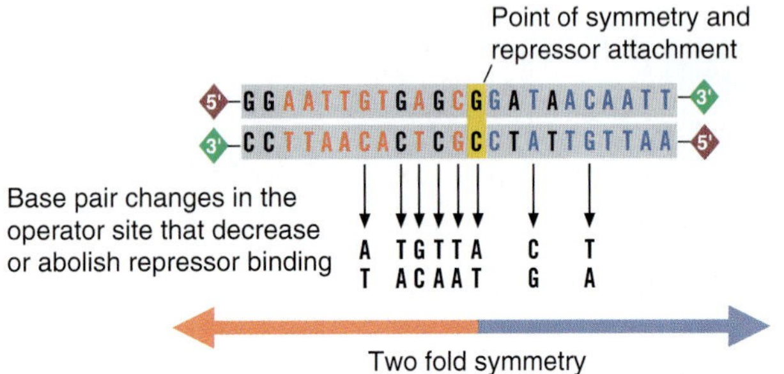

FIGURE 14.8 Operator site mutations. This is the nucleotide sequence of the lactose operon operator site, and it shows the mutations that affect repressor binding. The sequence has a twofold symmetry (red sequence on left is the mirror image of the blue sequence on the right) around the yellow GC pair in the center of the colored domains. This is the site of repressor attachment. *(Adapted from W. Gilbert, A. Maxam, and A. Mirzabekov. 1976. In N. O. Kjeldgaard and O. Malloe, eds.,* Control of Ribosome Synthesis. *Academic Press, New York.)*

make a product (unlike the *lacI* cistron), *lacO* is *cis* dominant and *trans* recessive. Thus, the genotype

$$\frac{lacO^{c},\ lacZ^{-}\quad \text{F}'\text{-plasmid}}{lacO,\ lacZ\quad \text{Chromosomal}}$$

will be inducible because the O^c mutation (on the plasmid) affects only the adjacent *lacZ*$^-$ cistron and therefore is *cis* with respect to the location of the *lacO* site and the structural genes that it controls. However, the *lacO*^{c} mutation does not affect the chromosomal *lacZ* cistron because it is in a *trans* configuration. Thus, β-galactosidase is inducible because the *lacO* site adjacent to the *lacZ* structural gene is normal. The *lacO* site controls only the *cis*-located *lacZ* gene.

Additional mutations were found that mapped between *lacI* and *lacO* and affected the expression of the lactose operon. (These and other mutations in the lactose operon are summarized in Table 14.1.) We now know that the region between *lacI* and *lacO* contains the promoter for the operon, or the *lacP* site. Promoter mutations (P^-) are *cis* dominant because the promoter does not make a diffusible product and acts only on the adjacent genes. The promoter nucleotide sequence for the *lac* operon contains promoter-characteristic –10- and –35-bp sequences (with the first base of the mRNA as +1) that affect operon gene expression when altered (Figure 14.9). Changes in promoter sequence affect RNA polymerase binding and, depending upon the change, the rate of transcription of mRNA is decreased or increased. Some promoter mutations drastically affect RNA polymerase binding; others have less effect.

From these findings, Jacob and Monod were able to propose a model, which has held up well over the years, for the control of expression of lactose metabolism genes. Molecular biologists and geneticists have further

TABLE 14.1 *Effects of Lactose Operon Mutations*

Mutation	Effect
lacI$^-$	The repressor protein cannot bind to the operator site and constitutive synthesis of the enzymes results.
lacI^s	The repressor protein is tightly bound to the operator site and cannot bind the inducer. No transcription and no induction results.
lacO^c	The repressor protein cannot bind to the operator site. Constitutive synthesis of the enzymes results.
lacP$^-$	RNA polymerase cannot bind; thus, no transcription occurs.
lacZ$^-$	No β-galactosidase synthesis occurs.
lacY$^-$	No permease synthesis occurs.
lacZ$^-$	No transacetylase synthesis occurs.
Polar mutations gene	Translational termination mutations (UGA, UAA, and UAG) in a structural affect expression of the gene and downstream genes.
crp$^-$	The catabolite activator protein cannot bind to the *crp*-binding site adjacent to the promoter, preventing RNA polymerase from binding. No induction results.

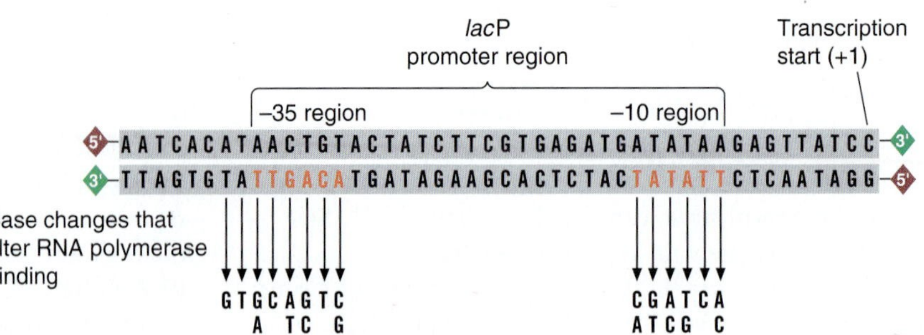

FIGURE 14.9 The promoter region of the lactose operon. The –10- and –35-bp regions are shown with the various base changes that affect RNA polymerase binding. Some of these alterations produce only slight changes in the polymerase binding, whereas others produce substantial changes.

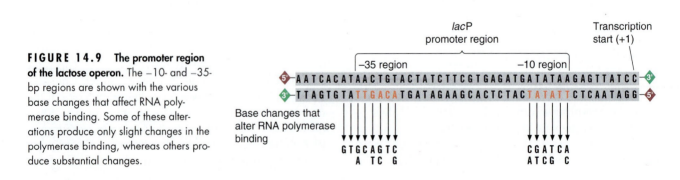

characterized this operon, and it is remarkable that, with only genetic evidence, Jacob and Monod were so precise in their predictions.

Catabolite Repression Affects *lac* Operon Expression.

At the beginning of this discussion on induction of the lactose operon, we mentioned the curious phenomenon wherein *E. coli* consumes glucose completely before it metabolizes any lactose that might be present (Figure 14.1). Why is this so? The answer is that an additional control system is superimposed upon the repressor–operator system regulating the expression of the lactose operon. This system is known as **catabolite repression** and allows a cell to use glucose preferentially. Catabolite repression is not restricted to the lactose operon. It is a general mechanism that controls operons that catabolize carbon sources, such as galactose and arabinose, and allows the cell to exercise its preference for glucose as a carbon source.

RNA polymerase acts by simply binding to the promoter and transcribing the mRNA of the structural genes. However, the mechanism is more complex for a catabolite repressible operon. For the RNA polymerase to initiate RNA synthesis at *lacP*, the cooperative effort of another protein is required. This protein, the **cyclic AMP receptor protein** (CRP), first binds to **cyclic AMP** (cAMP). Cyclic AMP is synthesized from ATP by the enzyme **adenylcyclase** (Figure 14.10). The CRP protein is a dimer composed of two identical subunits of 22,500 Da each, and it is activated by a single molecule of cAMP. The RNA polymerase will not initiate RNA synthesis unless the CRP protein–cAMP complex is bound to the CRP site on the lactose operon (Figure 14.11). Like the *lacO* site, the CRP–cAMP DNA recognition sequence has twofold sequence symmetry:

GTGAGTTAG**CTCAC**
*
CACTCAATCGAGTG

How does glucose affect β-galactosidase synthesis? Under all conditions in the *E. coli* cell, cAMP is converted into AMP by the enzyme **phosphodiesterase.** Thus, this enzyme will degrade all the cAMP if synthesis does not occur simultaneously. In the absence of glucose, cAMP is synthesized by adenylcyclase (Figure 14.10), and the CRP–cAMP complex forms, activating transcription of the operon (Figure 14.11). The presence of glucose indirectly inactivates adenyl cyclase, which results in a rapid reduction of the amount of cAMP in the cell. If cAMP is absent, the cAMP–CRP–protein complex does not form and does not activate the lactose operon or

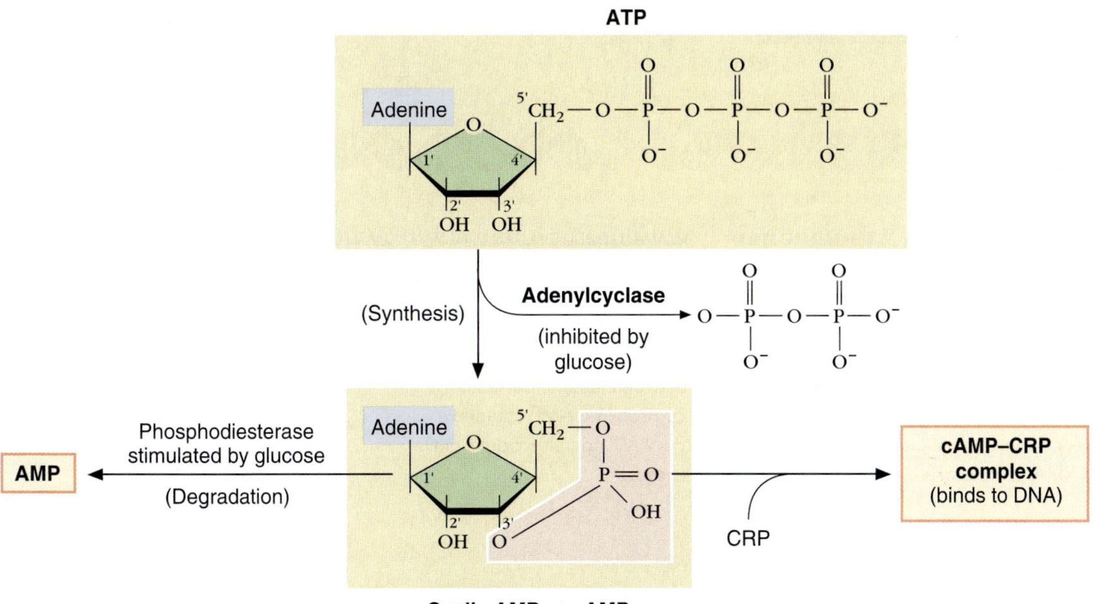

FIGURE 14.10 **Synthesis of cyclic AMP.** The enzyme adenylcyclase is responsible for the synthesis of cAMP from ATP. When glucose is present, no cAMP is synthesized because the enzyme is indirectly inhibited by glucose. In addition, the existing cAMP is hydrolyzed by phosphodiesterase. Cyclic AMP binds to the CRP, which changes its conformation, and this complex binds to the CRP-binding site adjacent to the promoter, allowing transcription to proceed.

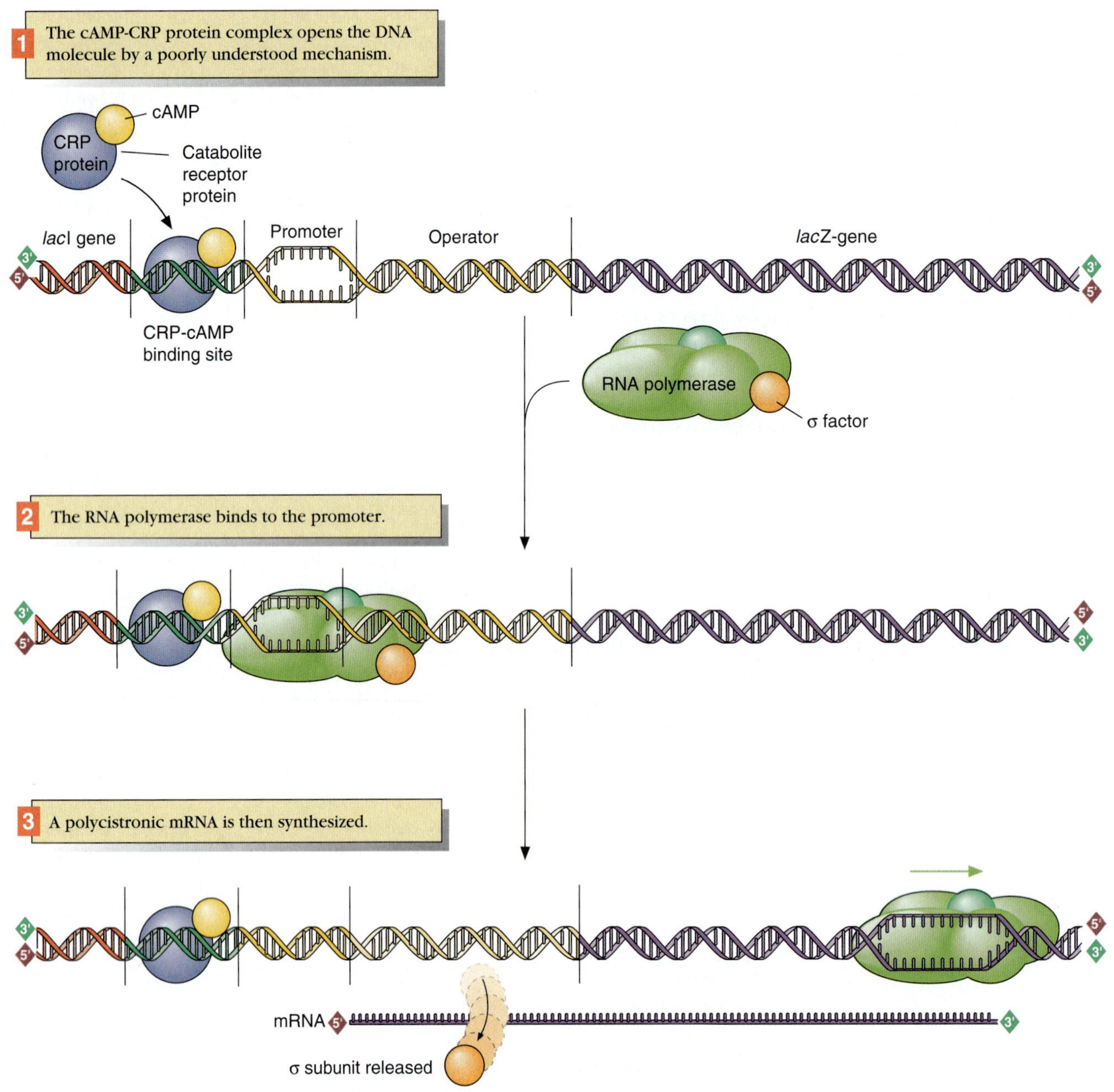

1 The cAMP-CRP protein complex opens the DNA molecule by a poorly understood mechanism.
cAMP
CRP protein
Catabolite receptor protein
lacI gene
Promoter
Operator
lacZ-gene
CRP-cAMP binding site
RNA polymerase
σ factor
2 The RNA polymerase binds to the promoter.
3 A polycistronic mRNA is then synthesized.
mRNA
σ subunit released

FIGURE 14.11 Catabolite receptor protein–cAMP complex binding. The organization of the promoter–operator region of the lactose operon is shown. Between the *lacI* gene and the promoter is the CRP–cAMP binding site. The catabolite receptor protein–cAMP complex must bind at this site for expression of the lactose operon. The CRP–cAMP complex opens the promoter to facilitate RNA polymerase binding and transcription ensues.

other catabolite repressible operons. As might be expected, a mutation in the CRP protein gene (*crp*$^{-}$) will inhibit the induction of the *lac* operon as well as other catabolite repressible operons. If no active CRP protein (bound to cAMP) is available, then it cannot bind to the CRP-protein DNA recognition sequence and activate the RNA polymerase binding site for attaching the polymerase to the promoter.

The Three-Dimensional Structure of Some Proteins Is Changed by Binding Small Molecules.

When the inducer molecule (lactose in the previous example) binds to the repressor protein, the repressor protein can no longer bind to the operator site. The repressor protein possesses two distinct binding sites on the surface of the protein, one for the inducer and one for the operator. The binding of the inducer to the repressor prompts a change in its operator-site binding properties. This is a conformational change of the three-dimensional structure of the protein, called an **allosteric transition.** Proteins that undergo allosteric transitions are called **allosteric proteins.** Allosteric transitions, or the change from one protein conformation to another, occur in many different proteins when a small molecule binds to them.

The *lac* repressor, for example, is capable of adopting two distinct conformations, depending upon whether or not lactose is bound to it. Similarly, cAMP converts the CRP protein from one conformation to another, allowing it to bind to the CRP site on the lactose operon. In addition, phosphodiesterase is stimulated by an allosteric transition to hydrolyze cAMP to AMP by the binding of glucose.

SOLVED PROBLEM

Problem

From the genotype (for the lactose operon) of the *E. coli* strain presented, predict whether β-galactosidase and permease will be induced, constitutive, or not produced in the presence and absence of lactose.

$$\text{Genotype:}\quad \frac{\text{F}'\text{-plasmid}\quad I^+O^cZ^-Y^+}{\text{Chromosome}\quad I^sO^+Z^+Y^+}$$

Solution

The first thing to examine in this problem is the condition of the *lacI* genes on both the chromosome and the plasmid and their relation to the *lacZ* cistrons. *lacI* makes the repressor protein that binds to the operator site, controlling expression of the *lacZ* cistron. For the chromosomal allele, *lacI* is I^s and its protein product can no longer bind the inducer (allolactose); thus it remains bound to the chromosome operator, regardless of whether or not an inducer is present. The plasmid copy of the *lacI* allele is active and will bind the inducer. However, because the repressor protein is a diffusible product, the mutant (*lac*I^s) repressor remains bound to the chromosome operator and cannot be pried off. This leads us to the conclusion that this genotype will be noninducible for β-galactosidase. Note that the operator site is mutant for constitutive expression of the plasmid-bearing cistrons. This has no effect on β-galactosidase expression because the operator site is *cis* dominant and the *lacZ* gene is defective. The *lac*O^c on the plasmid does not affect *lacZ* gene expression on the chromosome. Also, the *lac*O^c site on the plasmid cannot bind the repressor; thus, the permease gene will be constitutively expressed.

SECTION REVIEW PROBLEMS

1. What is an operon? What features of an operon ensure that all the genes will be coordinately controlled?
2. How does an inducible operon differ from a constitutive operon?
3. Explain how the repressor in the *lac* operon prevents RNA polymerase action.
4. A mutation in gene B in the operon shown here affects the expression of genes B and C by abolishing protein formation, but it has no effect on gene A protein production. Explain.

Transcriptional Control II: Attenuation, Antitermination, Methylation, and DNA-Binding Proteins Control Gene Expression.

The early studies on the lactose operon in *E. coli* laid the foundation for studies on other regulatory circuits in bacteria and bacterial viruses. These studies also set the stage for deciphering regulation of gene expression in eukaryotes and their viruses. Shortly after the regulation of the lactose operon was delineated, a number of other metabolic regulatory pathways in bacteria were defined, most notably the synthesis of tryptophan, a necessary amino acid in the synthesis of many proteins.

Attenuation and Positive Control Regulate Tryptophan Synthesis.

The operon that is responsible for the synthesis of tryptophan (the *trp* operon) in *E. coli* differs from the *lac* operon quite significantly. The *trp* operon was found to possess a secondary mechanism of gene-expression regulation known as **attenuation.** In the lactose operon, the level of gene expression (that is, the concentration of the products of the operon) varies from nearly zero without induction to several hundred times that amount with lactose induction. The operator and repressor control the level of gene expression. In the tryptophan operon, however, the attenuation mechanism adds another level of regulation of gene expression (over another tenfold range). Combined, induction and attenuation can vary the level of expression over 1000-fold range.

The tryptophan (*trp*) operon of *E. coli* (Figure 14.12) was characterized by Charles Yanofsky (Stanford University) and his colleagues. This rather large operon

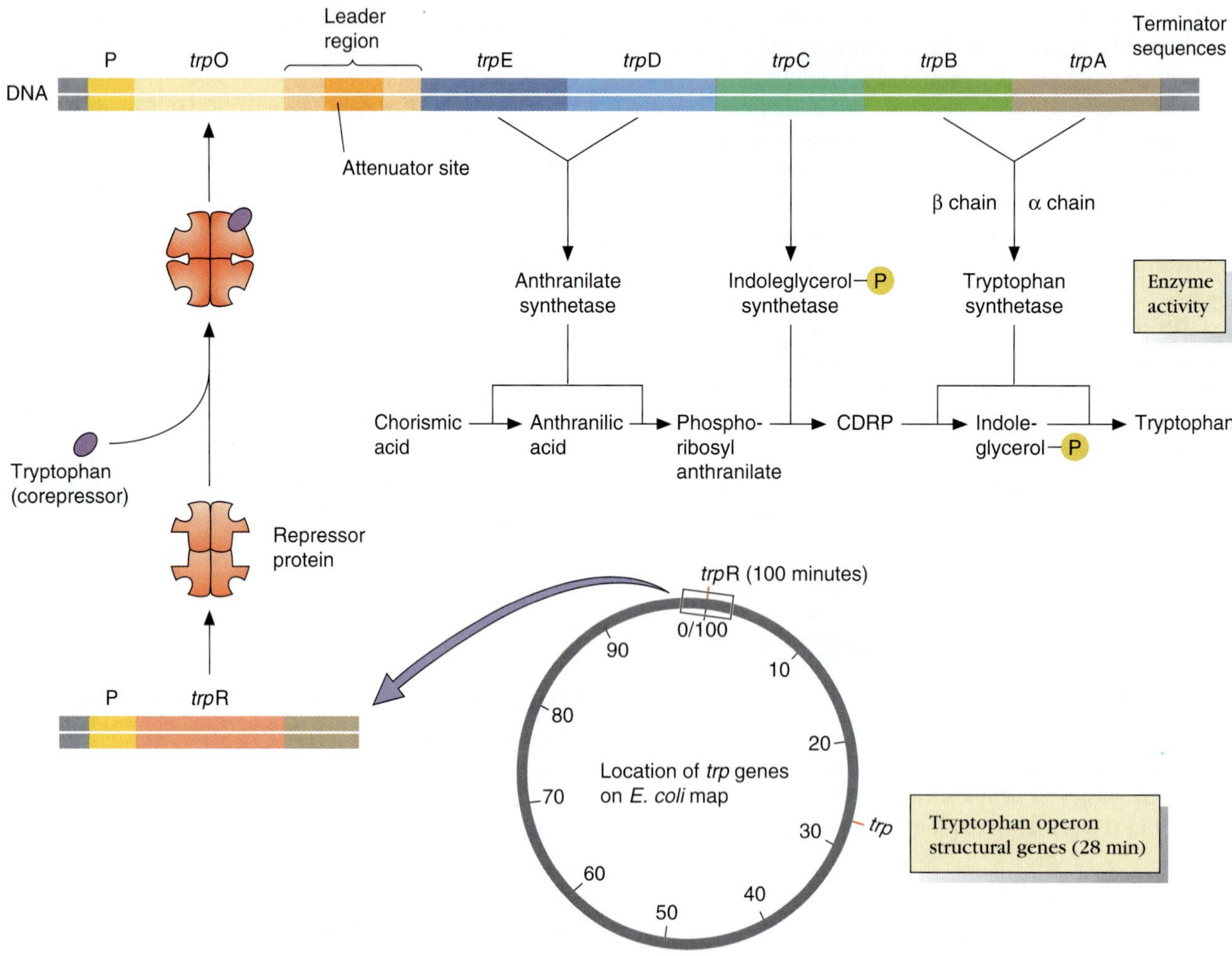

FIGURE 14.12 The tryptophan operon. The tryptophan operon codes for five structural genes, and it is controlled by an operator site (*trpO*) and an attenuator site. The *trpR* gene encodes the repressor protein. The *trpR* gene is located about one quarter of the genome away from the *trp* gene site (shown at the bottom of the figure). When the repressor protein complexes with tryptophan, it binds to the operator site and blocks the initiation of transcription. This happens when tryptophan is available in the environment and the cell does not have to synthesize it.

(7-kb transcript) has five structural gene cistrons (*trpE, trpD, trpC, trpB,* and *trpA*), a repressor gene (*trpR*), and an operator site (*trpO*). It is responsible for synthesizing the amino acid tryptophan from chorismate. *E. coli* produce a polycistronic mRNA that has a lifetime of about 3 minutes, which enables the bacteria to respond quickly to environmentally induced changes in the need for tryptophan.

The first level of control in the tryptophan operon is achieved by interaction between a 58-kDa regulatory protein, encoded by the *trpR* cistron and the *trpO* operator site. The interaction between the *trpR* protein and the *trpO* site regulates the production of tryptophan (Figure 14.12). When tryptophan is in excess in the environment, the operon is turned off by the repressor protein–tryptophan complex binding to the operator site. In this complex, tryptophan is said to be a **corepressor.** If the cell has a plentiful supply of tryptophan, there is no need to synthesize more tryptophan, thus saving energy for the cell. The binding of the repressor–tryptophan complex to the operator site prevents transcription of the operon because operator–promoter binding sites overlap. Thus, RNA polymerase cannot bind to the promoter. When there are low levels of tryptophan in the cell or the surrounding medium, the repressor–tryptophan complex does not form, and tryptophan operon transcription proceeds. In this way, the cell makes tryptophan only when it is needed.

Note that operons concerned with the synthesis of essential compounds (for example, the amino acids) are

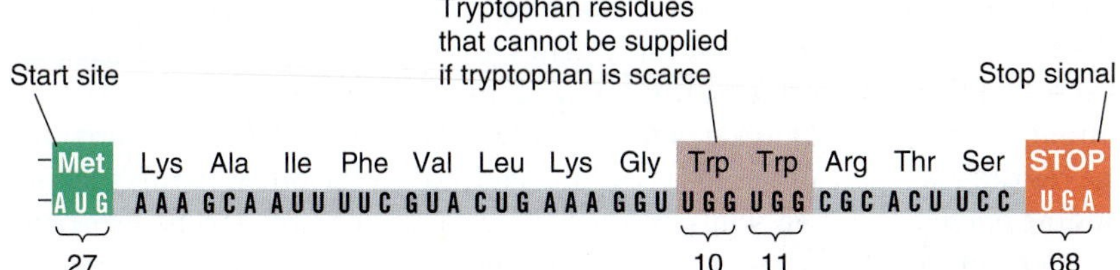

FIGURE 14.13 The attenuator site containing the *trp* leader region polypeptide coding area. This figure shows the nucleotide sequence of the *trp* mRNA leader region and the amino acid sequence of the polypeptide coded within it. The polypeptide contains two tryptophans; consequently, the availability of tryptophan in the growth medium affects the expression of downstream genes negatively or positively.

normally turned on, whereas operons controlling nonessential metabolic pathways are normally turned off (for example, lactose). Thus, the normal condition for the tryptophan operon is to be turned on (because of low levels of tryptophan in the environment).

The second level of control in the tryptophan operon, mentioned earlier, is attenuation. This control mechanism is superimposed upon the operator–repressor regulation to fine-tune the level of gene expression. Regulation of gene expression by attenuation further varies the level of gene expression but is present only in normally turned-on (or repressible) operons such as tryptophan. It is not found in inducible operons. Attenuation works because of the presence of sequences (the attenuator site, Figure 14.12) within the leader region of the polycistronic mRNA of the operon that permit special three-dimensional folding of the mRNA. This three-dimensional folding, in turn, affects transcription of downstream RNA sequences. The DNA sequence of the tryptophan operon leader region, containing the attenuator site, is shown in Figure 14.13 and possesses an AUG start codon at nucleotide 27 of the leader region, which is distinct from the *trpE* AUG start codon at nucleotide 162. Fourteen codons downstream from the attenuator AUG codon is a UGA stop codon at nucleotide 68. This small cistron within the 5′ end sequence of the mRNA codes for a 14-amino acid protein. It should be noted that this small protein contains tryptophan residues at positions 10 and 11. These two tryptophan codons are the key to how the mRNA folds, and they establish whether downstream structural genes will be transcribed.

To understand how attenuation functions, recall that translation and transcription occur simultaneously in prokaryotes. When tryptophan is scarce during translation, the ribosome stalls on the two tryptophan UGG codons (region 1) because of the limited amount of tryptophanyl-tRNA. The stalled ribosome allows regions 2 and 3 of the RNA to form a base-paired stem-and-loop structure. This preferentially formed complex in turn permits RNA polymerase to continue transcribing the structural genes beyond the attenuator site (Figure 14.14). The preferential folding between regions 2 and 3 prevents folding between regions 3 and 4. On the other hand, if tryptophan is abundant in the medium, the ribosome does not stall in region 1 and proceeds down the mRNA. This allows formation of a stem-and-loop structure between regions 3 and 4 and signals termination of transcription.

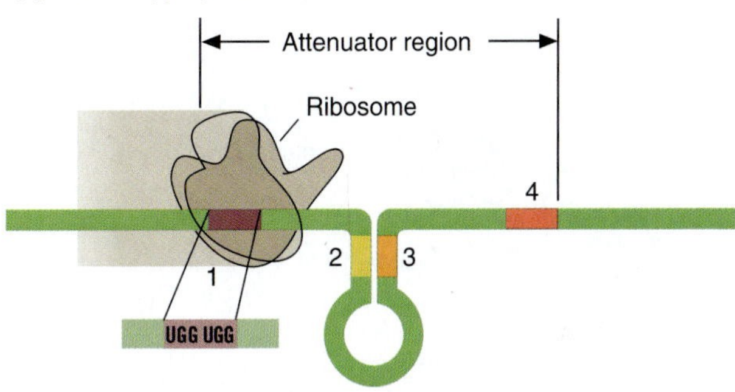

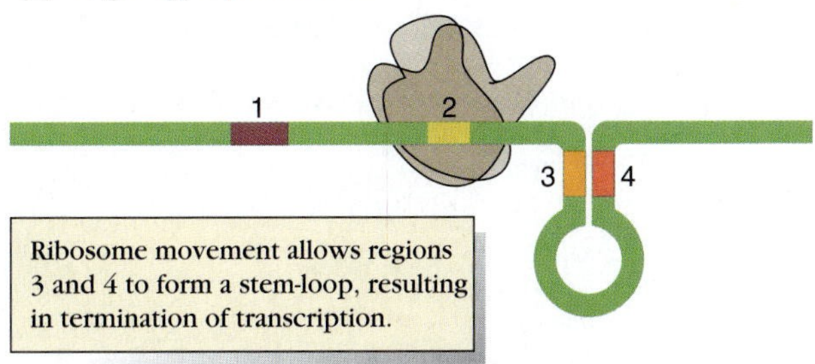

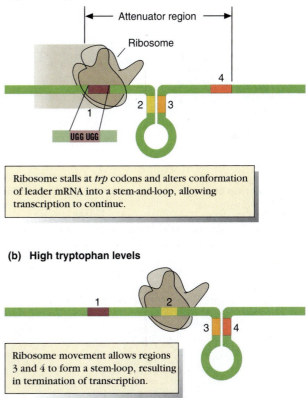

FIGURE 14.14 Attenuator-site regulation of tryptophan synthesis. Attenuation in the *E. coli trp* operon depends upon the configuration of the leader region mRNA and varies depending upon the concentration of tryptophan in the medium. (a) Low tryptophan concentrations allow transcriptional read-through, and (b) high tryptophan levels prematurely terminate some mRNA transcripts.

The attenuator control system is also present in other operons that control the synthesis of amino acids. In each instance, the leader polypeptide is short and has a number of codons for the amino acid through which the operon controls; for example, 7 histidine residues are present in the leader polypeptide of the histidine operon, and 7 phenylalanine residues (of 15) are present in the phenylalanine operon leader polypeptide. In fact, in these amino acid operons, attenuation is the primary mode of regulation.

Antiterminators and Repressors Control Gene Expression in Lambda Phage.

You may recall from our earlier discussions of DNA phage lambda (Chapter 8) that lambda is a temperate phage (it does not necessarily kill its host, but it can integrate into the host genome). Lambda has a genome of 48 kb and is surrounded by a protein coat. Its host is *E. coli*, and there are two pathways of infection: the **lytic pathway** with full expression of the phage genes, which results in the lysis of the host, or the **lysogenic pathway,** in which most genes are repressed and the phage integrates into the genome of the host (Figure 14.15). Lambda genes must have precise regulation to allow the phage to follow one pathway or the other. In this section, we will consider some of the control mechanisms the phage uses to follow the lytic or lysogenic pathways.

Many of the essential components of phage reproduction are supplied by the host cell. When a phage enters a bacterial cell, it uses host components for its own reproductive purposes. A phage can reproduce very rapidly, sometimes in as little as 20 minutes, or less time than it takes the bacterial host to double. To use the host's components for their own purposes, some phage initially express genes that affect the host's ability to synthesize its own components, effectively monopolizing the cell's reproductive machinery for their own purposes. Phage also express genes that control the phage life cycle, and these genes are usually expressed sequentially and early in the life cycle of the phage.

Gene expression in lambda phage follows a pattern common to many phage: the genes are transcribed as they are needed. In the lytic pathway, there are three stages of gene expression: **immediate-early** (DNA synthesis and regulation of the immediate- and delayed-early genes), **delayed-early** (prophage insertion, gene recombination, late-gene expression), and **late genes** (phage head and tail protein synthesis, cellular lysis) (Figure 14.15). Figure 14.15b shows all the genes in phage lambda, their functions, and the stage at which

(a) Lytic infection

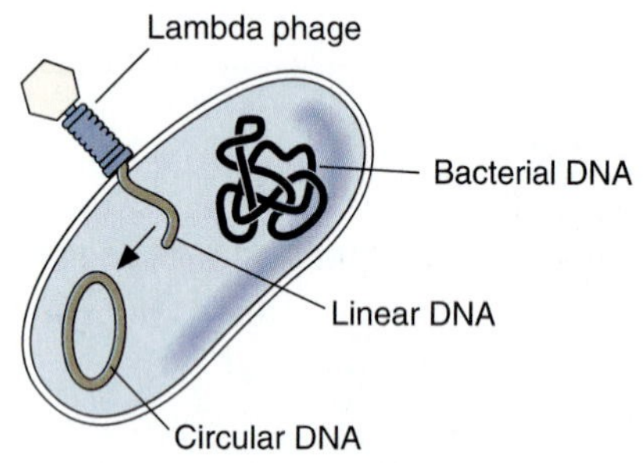

(b) Linear map of phage λ and the pattern of gene expression

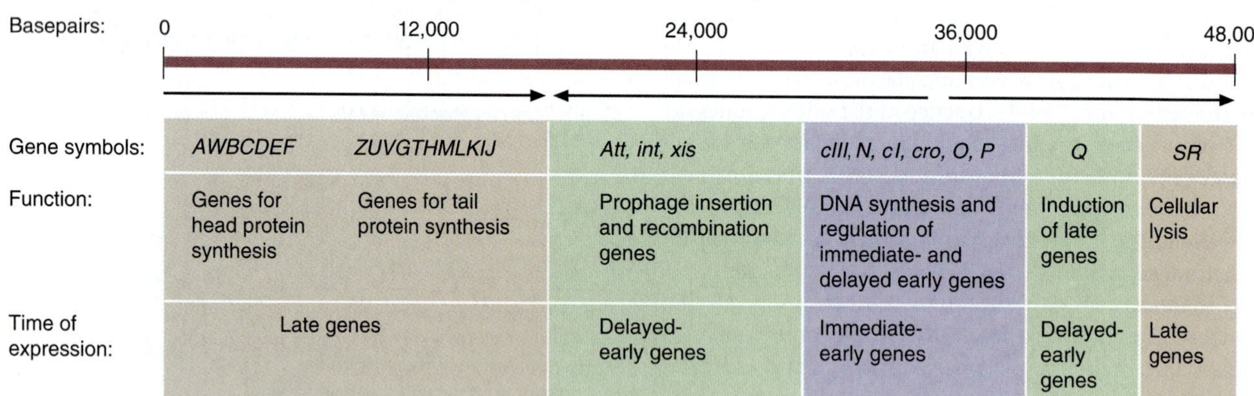

Basepairs:
0
12,000
24,000
36,000
48,000
Gene symbols:
AWBCDEF
ZUVGTHMLKIJ
Att, int, xis
cIII, N, cI, cro, O, P
Q
SR
Function:
Genes for head protein synthesis
Genes for tail protein synthesis
Prophage insertion and recombination genes
DNA synthesis and regulation of immediate- and delayed early genes
Induction of late genes
Cellular lysis
Time of expression:
Late genes
Delayed-early genes
Immediate-early genes
Delayed-early genes
Late genes

FIGURE 14.15 Lambda phage infection and genetic map. (a) In the lytic lambda-infection pathway in the *E. coli* cell, the phage DNA converts from a linear to a circular form, and it does not integrate into the bacterial genome. The phage genes are expressed, and phage multiplication occurs. (b) The organization and function of the lambda phage genes are shown in this linear map. Genes with similar functions are clustered, and gene expression occurs in groups, as the need for them arises.

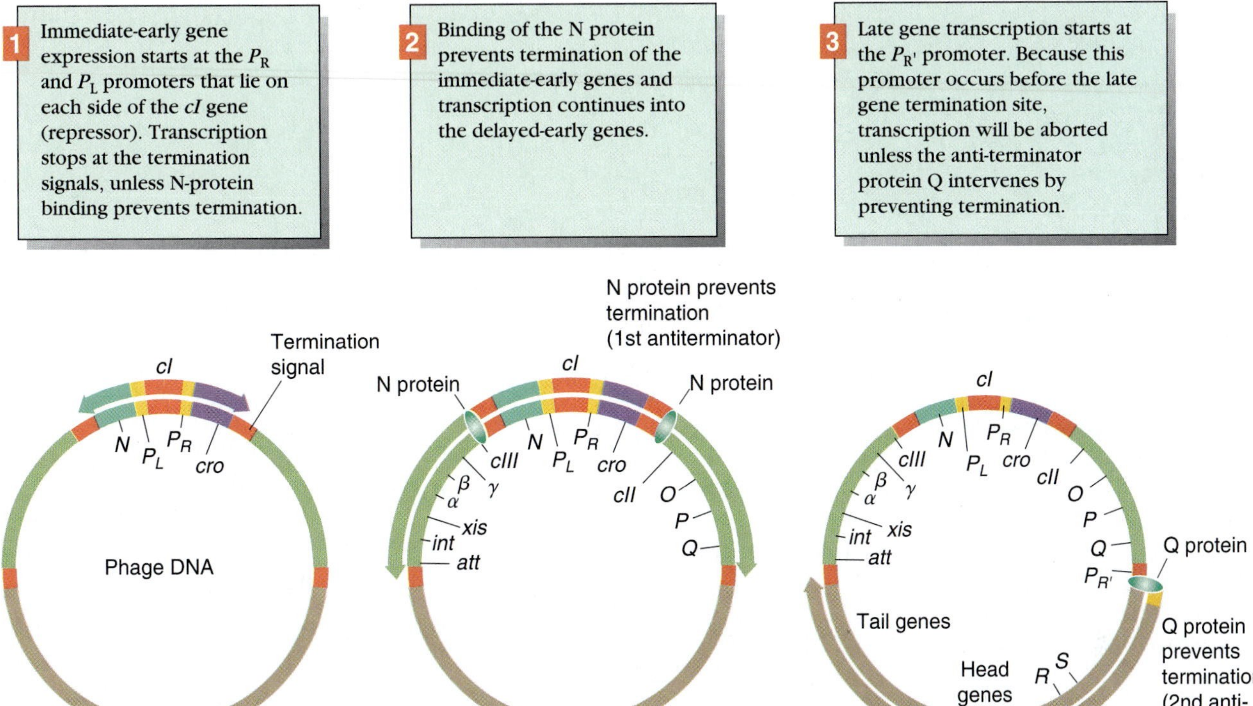

FIGURE 14.16 Lytic control of gene expression in lambda phage. This figure shows the control of transcription in phage lambda during the lytic pathway and how the genes are turned on as infection progresses.

they are expressed. Figure 14.16 shows how the three stages of gene expression interact. For example, the phage head and tail are not needed until later in the lytic pathway and are therefore transcribed late (Figure 14.16). We will discuss the expression of each of these groups of genes and how they regulate the lytic cycles of the lambda phage.

When a lambda phage infects a bacterial cell, it must decide whether to enter the lytic cycle or the lysogenic cycle. This decision depends upon a genetic switch and involves competition between the Cro protein (a repressor, the product of the immediate-early *cro* gene) and a repressor that is the product of the *cI* gene. The Cro protein repressor (Cro stands for control of repressor and other genes) turns off *cI* gene transcription, preventing synthesis of the cI repressor protein. When the Cro protein blocks *cI* gene expression, the cI repressor protein is not synthesized. In the absence of the cI repressor protein, two immediate-early genes, *cro* and *N*, are expressed, and the lytic cycle ensues (Figure 14.16). When the cI repressor protein dominates, the lysogenic pathway will be taken. About 90% of lambda infections follow the lytic pathway, and about 10% follow the lysogenic pathway.

The lytic development of phage lambda is sequentially regulated by two positive regulatory proteins, N and Q, which are called **antiterminators.** The immediate-early genes are regulated by the *N* and *cro* genes, which are read from two promoters (P_L for left promoter; P_R for right promoter) that are on each side of the *cI* gene (Figure 14.17). RNA polymerase automatically binds to these promoters. The P_L promoter transcribes the N protein leftward on one strand of the DNA, and the P_R promoter transcribes the Cro protein rightward on the other strand. Both Cro and N proteins are critical regulatory proteins. Without regulation, transcription is stopped by DNA termination signals after the *N* and *cro* cistrons are transcribed (Figure 14.16). If transcription stops after the immediate-early transcripts, the delayed-early genes (and, consequently, the late genes) will not be expressed. However, termination at these sites can be prevented by the N protein attaching at a DNA recognition sequence between the immediate-early and delayed-early genes. This attachment is called the **antitermination site.** Antitermination is required for expression of downstream genes. Thus, the N protein must be produced in order for the delayed-early genes to be expressed. The N protein is highly unstable, with a half-life of 5 minutes. For this reason, the *N* cistron must be continually expressed to maintain transcription of delayed-early genes.

A second antitermination signal, regulated by the delayed-early *Q* gene, is needed to express the late genes.

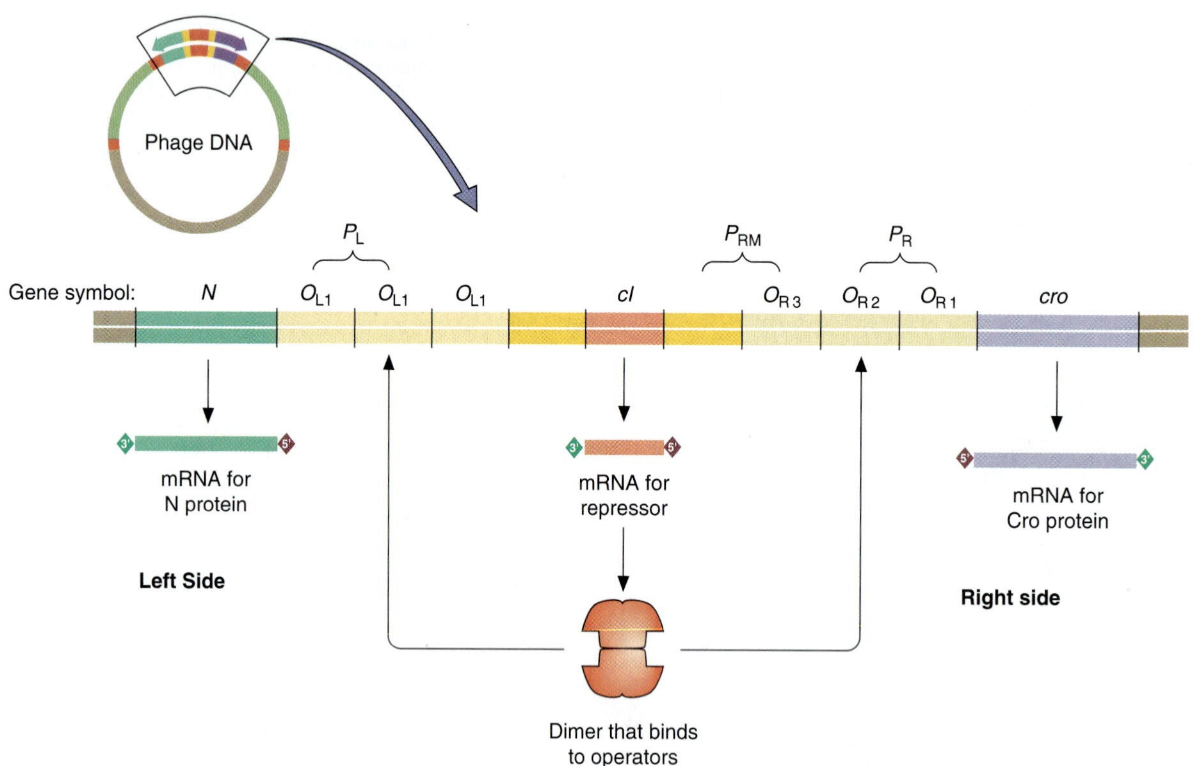

FIGURE 14.17 The central lysogenic controlling region of phage lambda. The *cI* gene is responsible for synthesizing the cI protein repressor, which in turn binds to the adjacent operators that control the synthesis of the immediate-early genes (*N* and *cro*). P_{RM} is the promoter for the repressor mRNA, and the P_R is the promoter for the *cro* gene mRNA. P_L is the promoter for the *N* gene.

The Q protein is an antiterminator protein for the delayed-early genes, and when it binds to the second antitermination site, late genes can be expressed (Figure 14.16). Late-gene transcription starts at the $P_{R'}$ promoter (located just after the Q gene but before the delayed-early gene's termination signal) but is terminated unless the Q protein is present. When late genes are expressed, all the components of the phage are present in the cell, and assembly ensues. An important late gene expresses a protein that destroys the bacterial cell wall, releasing the phage into the surrounding media.

If the phage decides to enter the lysogenic cycle and incorporate into the host's genome, the first step is to inactivate the genes responsible for lytic functions. To accomplish this, the phage produces the cI repressor protein, which blocks transcription of the *cro* gene and prevents the synthesis of the Cro protein. The cI repressor protein binds to the operators O_R and O_L to prevent transcription initiation. This repressor protein is the only protein synthesized in the lysogenic state (also called the prophage state). The cI repressor protein is synthesized because the product of the *cII* gene (just to the right of the *cro* gene, and stabilized by the presence of the cIII protein) activates the transcription of the *cI* gene from a promoter called P_{RE} (promoter for repressor establishment) located to the right of the *cI*–*cro* region. The cI repressor protein blocks the expression of the immediate-early genes, which in turn blocks the delayed-early and late gene expression necessary for chromosome replication, progeny phage assembly, and cell lysis. As a consequence, the lambda phage enters the lysogenic state and remains that way until outside factors induce a lytic cycle. The lysogenic or prophage state is useful to the phage because it essentially gets a free ride in the host bacterial cell. The phage can remain integrated into the host genome for extended periods without any expenditure of its own energy. On the other hand, if the host cell is threatened (for example, by exposure to ultraviolet light), the lytic cycle is induced and the phage escapes the cell.

As Figure 14.17 shows, the promoters for the cI repressor protein and the *cro* gene are located *within* the right-side operators (O_{R3}, O_{R2}, and O_{R1}). It would seem that the cI repressor protein would block its own synthesis by binding to the operator sites and blocking its own promoter. This does not occur, however, because the O_{R1}

operator, and to a lesser extent the O_{R2} operator, have a much higher affinity for the cI repressor than O_{R3}. When the repressor is bound to O_{R1}, it completely blocks the action of RNA polymerase, and thus blocks *cro* transcript synthesis. But when the repressor binds to O_{R2}, it actually promotes the binding of RNA polymerase to the P_{RM} promoter (P_{RM} stands for promoter for repressor maintenance), guaranteeing cI repressor protein synthesis. cI repressor protein synthesis is essential to ensuring that the phage remains in a prophage state in the host genome. Thus, the cI repressor protein has two critical functions: it represses the *cro* gene and activates its own transcription.

What factors might induce a prophage in the lysogenic cycle to switch to a lytic cycle? The lytic cycle is essentially an escape mechanism for the phage to enable it to leave a host environment that is no longer hospitable. Thus, a prophage may switch to the lytic cycle if the Cro protein concentration becomes too high. In this case, the switch occurs because the Cro protein can also bind to the to O_R, which will block transcription from P_{RM} and reduce the concentration of the cI repressor protein. When this happens, the cI repressor level cannot rise high enough to block transcription from P_L and P_R, and the N and Q proteins are made. These two antiterminators then enable production of late-gene products and eventual lysis. Damage to the DNA by ultraviolet light will also induce lysis through action of the RecA protein (Chapters 8 and 19), which possesses a protease function. For some not-yet-understood reason, damaged DNA activates the protease function of the RecA protein (perhaps by binding an oligonucleotide released during damage), which in turn inactivates the cI repressor protein by cleaving it into segments. When this happens, the Cro protein preferentially binds to O_{R3} operator, preventing further cI repressor synthesis and ensuring that the lytic cycle follows.

SECTION REVIEW PROBLEMS

5. The tryptophan operon codes for five structural genes (excluding the repressor, which is distantly located on the genome). However, only three enzymes are required to synthesize tryptophan from chorismate. Propose an explanation that would account for this fact. (Hint: See Figure 14.12.)
6. Explain the difference between an antiterminator and an attenuator.

The Yeast *GAL* Regulatory Pathway Is a Good Model of Eukaryotic Gene Regulation.

An intensively studied eukaryotic regulatory pathway is the *GAL* pathway of yeast. The eight genes of this pathway are responsible for producing the enzymes that metabolize galactose as well as genes that regulate its expression (Figure 14.18). Five of these genes (*GAL*2, *MEL*1, *GAL*4, *GAL*80, and *GAL*5) are located on different

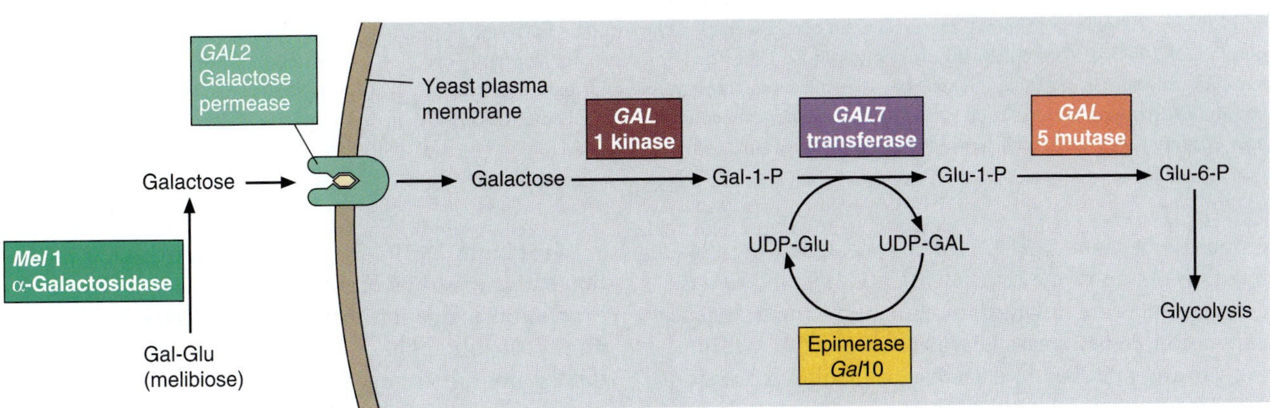

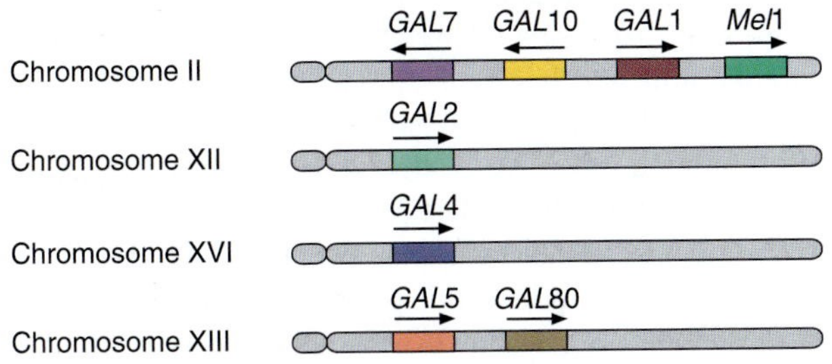

FIGURE 14.18 Galactose utilization pathway in yeast. This figure shows (a) how galactose is metabolized and (b) the location of the *GAL* genes on individual chromosomes. The enzymes and their respective genes are galactose permease (*GAL2*), galactokinase (*GAL1*), galactose-1-phosphate uridylyltransferase (*GAL7*), phosphoglucomutase (*GAL5*), α-galactosidase (*MEL1*), and uridine diphosphoglucose 4-epimerase (*GAL10*). The *GAL4* and *GAL80* genes are involved in regulating the expression of yeast structural genes.

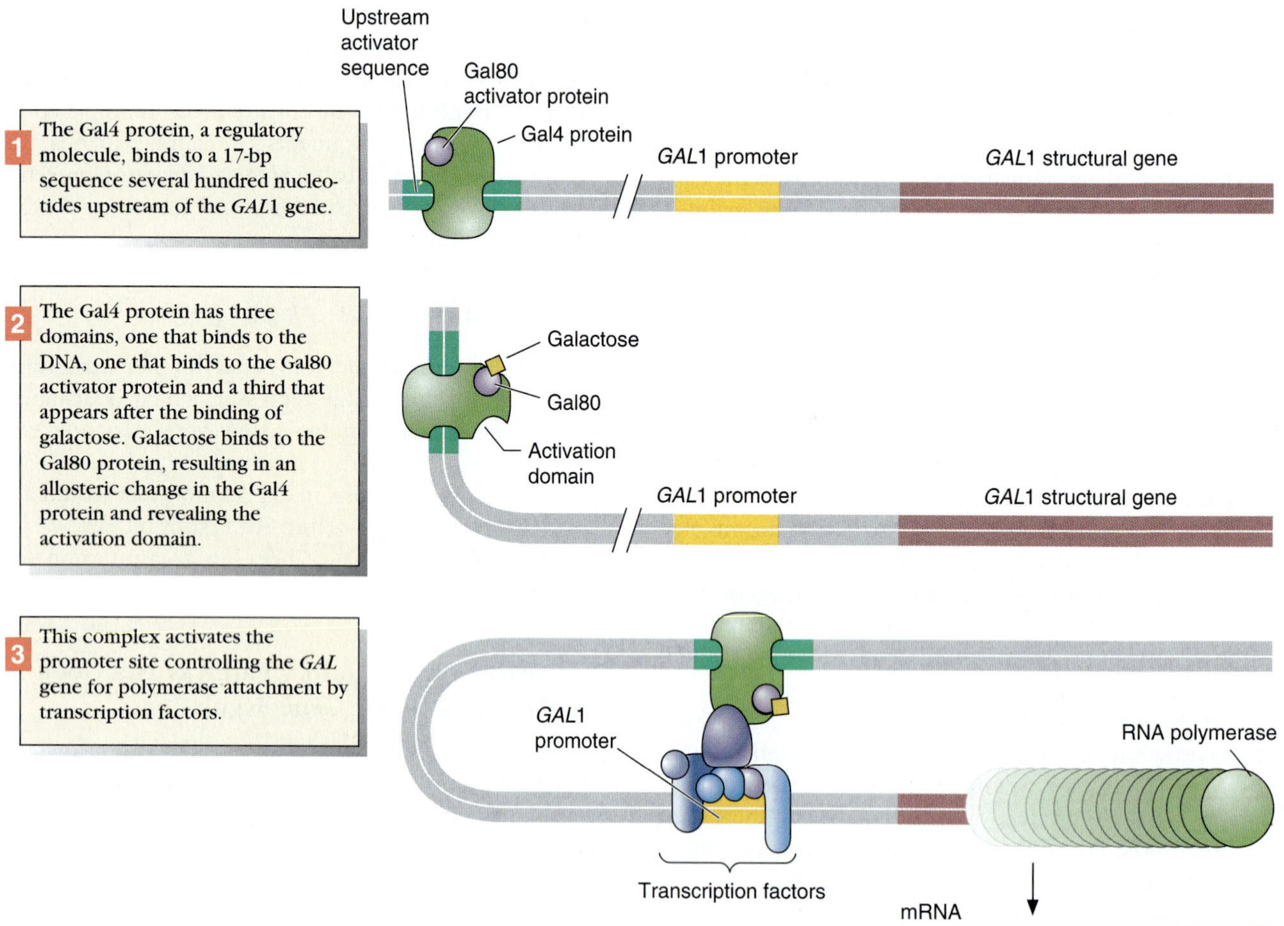

1
The Gal4 protein, a regulatory molecule, binds to a 17-bp sequence several hundred nucleotides upstream of the *GAL*1 gene.
Upstream activator sequence
Gal80 activator protein
Gal4 protein
*GAL*1 promoter
*GAL*1 structural gene
2
The Gal4 protein has three domains, one that binds to the DNA, one that binds to the Gal80 activator protein and a third that appears after the binding of galactose. Galactose binds to the Gal80 protein, resulting in an allosteric change in the Gal4 protein and revealing the activation domain.
Galactose
Gal80
Activation domain
*GAL*1 promoter
*GAL*1 structural gene
3
This complex activates the promoter site controlling the *GAL* gene for polymerase attachment by transcription factors.
*GAL*1 promoter
RNA polymerase
Transcription factors
mRNA

FIGURE 14.19 ***GAL1* gene-expression regulation in yeast.** The *GAL1* expression regulation pathway is a general model that applies to all the *GAL* gene-expression pathways. This figure assumes that regulation of gene expression occurs when the chromosome folds and puts the *GAL4–GAL80* complex into proximity with the *GAL1* gene promoter for "loading" of the RNA polymerase complex. It is still speculation that the chromosome folds during activation of *GAL* gene promoters, but it is a very likely scenario because it is difficult to envision how upstream sites could otherwise control gene expression.

chromosomes, and *GAL*7, *GAL*10, and *GAL*1 are clustered on chromosome II. The *GAL*3, *GAL*4, and *GAL*80 gene products are involved in regulating structural gene expression. When yeast is grown on glucose medium, even in the presence of galactose, catabolite repression prevents expression of the *GAL* genes, similar to *E. coli* and the lactose operon. Note that, unlike the lactose operon in *E. coli*, the *GAL* pathway for the metabolism of galactose is *not* an operon. However, as you will see, eukaryotes achieve the same result as an operon (regulation of all the enzymes in the pathway at the same time and to the same level), even though each of the structural genes has its own promoter. Coordinate eukaryotic enzyme regulation is carried out via similar regulatory sequences upstream from the promoter of each of the genes.

In Figure 14.19, the *GAL*1 gene-expression regulation pathway is shown. *GAL*1 is used as a model for gene-expression regulation in the *GAL* pathway because the same principle applies to all the *GAL* genes. All the structural genes are arranged similarly, that is, they have an **upstream activator sequence** (UAS) adjacent to the structural gene. The UAS is the binding site for the GAL4 protein. In this model, the GAL4 protein has two binding sites, one that binds to the DNA upstream activator sequence and another that binds the GAL80 regulatory protein. Without a bound galactose molecule, the GAL4 protein cannot activate any adjacent *GAL* structural gene, even when the GAL4 protein is bound to both the UAS and the GAL80 activator protein. Galactose, however, is able to induce the production of the GAL enzymes by binding to the GAL80–GAL4 protein bound to the UAS (Figure 14.19). This final binding step exposes an activation domain on the GAL4 protein, allowing the GAL4 protein to load RNA polymerase to the promoter site of each of the genes in the galactose pathway. The GAL4 protein is also phosphorylated (that is, a phosphate group is added to an amino acid of the protein) in this process, and the kinase responsible for this phosphorylation is believed to be determined by the *GAL*3 gene. Phosphorylation is required to maintain structural gene expression. In summary, galactose binds to the GAL80–GAL4 protein complex, which in turn functions as an upstream activator by promoting transcription of the *GAL*1 structural gene. This induction mechanism is the same for all the structural genes in the metabolic pathway; thus, all the genes are induced at the same time by the presence of galactose in the medium.

The GAL4 protein is quite large, containing 881 amino acid residues. The 73–amino acid residues of the GAL4 amino-terminal end form what is termed a **zinc finger** (zinc fingers and other DNA sequence-recognition motifs in proteins are discussed in the next section) that specifically binds to a 17-bp DNA recognition sequence. The DNA-binding motif, however, cannot activate transcription alone, nor can the rest of the protein function in the absence of a DNA-binding motif. Transcription is activated by the 114-amino acid carboxyl-terminal region of the protein in combination with the DNA-binding motif.

How does the GAL4 protein activate downstream promoters? Although the exact mechanism is not understood, DNA looping is one possible mechanism of enhancer or UAS action (see Figure 14.19). This proposed mechanism is consistent with the observation that UASs and enhancers are most effective when relatively close to the promoter to be activated. Effectiveness is progressively reduced as the enhancer or UAS is moved farther away from the promoter.

Another general mechanism exemplified by the GAL4–GAL80 system is the presence of **transcription factors** (TF), which are required for the synthesis of eukaryotic genes. Specifically, transcription factor IID (abbreviated TFIID) is a site-specific, DNA-binding complex required for selective initiation of transcription by eukaryotic RNA polymerase II. Promoters for RNA polymerase II often contain a TATA box (Chapter 11), recognized by TFIID, and they are located a short distance (30 bp) upstream of the transcription initiation site. A TFIID-promoter complex directs the ordered entry of RNA polymerase II and general transcription factors (TFIIA, TFIIB, TFIIE, TFIIF, TFIIH, and TFIIJ) onto the promoter to create a multiprotein complex capable of transcriptional initiation. Loading polymerase II and its transcription factors onto the promoter is the ultimate objective of this regulatory process, and it leads to transcription of specific genes. (Figure 14.19 shows the multiprotein complex loaded on the *GAL*1 gene's promoter.) These are not the only transcription factors, however. Over the last decade many additional transcription factors have been identified, and we will encounter many of them as we discuss regulation of gene expression and development. The presence or absence of a specific transcription factor can regulate the expression of a specific gene(s) during development or in the presence of external chemical signals. Many regulatory pathways in eukaryotes have not been clearly defined, and currently this is a very active area of research. It is clear that a large fraction of eukaryotic genes is regulated, and transcription factors are likely to be involved in this regulation. (See *Current Investigations: How Do Bacteria, Animals, and Plants Adapt to Stress?*)

Steroid Hormones Can Activate Gene Expression in Mammals.

A large number of genes in mammals respond to the presence of steroid hormones, and these genes are activated in a manner similar to the *GAL4–GAL80* activated genes. Steroid hormones have a major influence on growth, tissue development, and body homeostasis. They are a diverse class of low-molecular-weight organic compounds that share similarities in structure. All steroid hormones act by binding to a specific **receptor protein** that in turn activates gene transcription. Receptor proteins reside in the cytoplasm and have specific binding sites for the steroid and for a specific enhancer sequence(s). The estrogen estradiol-17-β provides a good model of steroid hormone action (Figure 14.20). Initially, a stimulus promotes the synthesis of the steroid hormone from an endocrine cell. The steroid hormone is then transported in the blood to the target cell, passes through the plasma membrane, and binds to a receptor protein in the cytoplasm. The receptor protein–steroid hormone complex then passes through the nuclear membrane pores and stimulates the transcription of one or more genes by binding to enhancer sequence(s). The products of these genes in the target cell are thus the direct result of steroid production in the endocrine cell.

Receptor proteins and the genes that respond to a large number of steroid hormones have been isolated and characterized. All steroid receptor proteins have similarities in structure, making them members of a superfamily. **Superfamilies** are groups of related proteins that have similar sequences and structures. For example, in all steroid receptor proteins the carboxyl-terminal regions, which bind the hormone, share a 30–60% similarity in amino acid sequence. The amino-terminal region of the steroid receptor proteins always possesses the DNA-binding domain. The DNA-binding domains recognize a specific DNA sequence (the enhancer) located upstream of the gene to be activated.

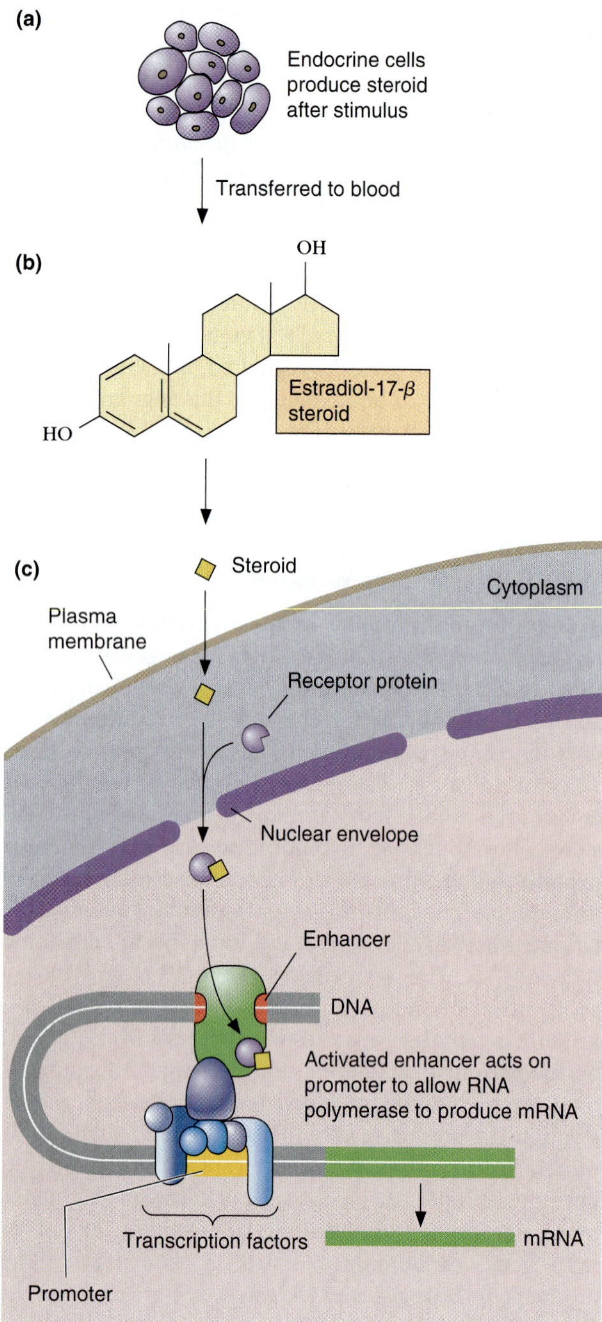

FIGURE 14.20 Steroid hormone action. This is a proposed model for the action of steroid hormones in mammalian cells. The producing cells in the adrenal gland or thyroid gland (a) synthesize a steroid hormone (b) that travels via the blood and is transported across the plasma membrane of the target cell, for example an ovary (c). The steroid hormone binds to a receptor protein (several thousand will be in the cell), forming a complex. This complex now binds to the specific enhancer sequence and activates gene expression. The receptor protein–steroid hormone complex is specific for genes necessary for ovarian function and will not interact with other enhancer sequences.

DNA Methylation May Control Some Aspects of Cell Development.

In the preceding sections, we discussed how some individual genes are controlled in response to environmental stimuli. In this section, we will address a larger question, How are groups of genes turned on or off? Individual cells in higher eukaryotic organisms can differ dramatically in both morphology and function (for example, a neuron and a liver cell). All cells in higher organisms must have mechanisms by which large groups of genes are turned on at some stage of development and remain turned on throughout the life of the cell. Exactly which genes are turned on (and sometimes off) determines whether a cell will be a neuron, lymphocyte, liver, or brain cell. What are the control mechanisms responsible for this differentiation of a fertilized egg into specialized cell types? The full answer to this question is not known, but it very likely involves a variety of mechanisms. A full discussion of differentiation is given in Chapters 15 and 16. At this point, we will discuss only one possible mechanism, DNA methylation.

As we discussed in Chapter 9, the bases in DNA can be covalently modified to add various side groups. One such modification is the addition of a methyl group at the 5 position of cytosine, producing **5-methylcytosine** (Figure 14.21). Between 2 and 7% of the cytosines of vertebrate DNA are methylated this way, and as much as 30% of plant DNA is methylated in this manner. Most of the methyl groups are concentrated in the heterochromatin, and the remainder are scattered throughout the genome as methylated CpG (^{m}CpG) doublets (both cytosines on the double helix are usually methylated; Fig-

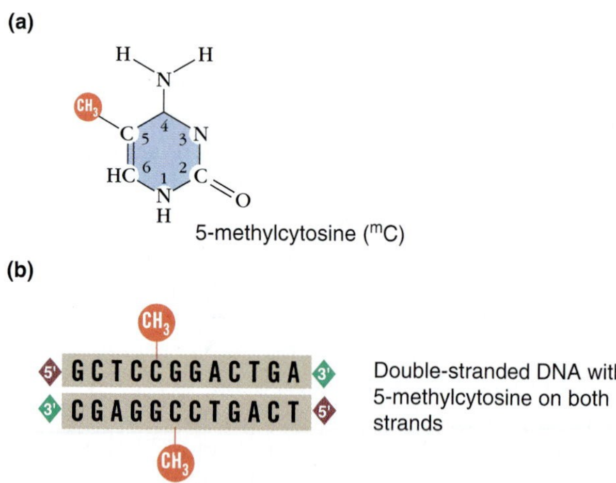

FIGURE 14.21 Methylation of DNA. (a) The structure of 5-methylcytosine, and (b) the placement of two 5-methylcytosines on opposite stands of DNA in a 5′-CpG-3′ sequence. The presence or absence of these methyl groups can affect gene expression.

CURRENT INVESTIGATIONS

How Do Bacteria, Animals, and Plants Adapt to Stress?

Both prokaryotic and eukaryotic cells quickly respond to stress or damaging stimuli. In response to a range of different stresses, including heat shock, nutrient deprivation, and metabolic disruption, cells synthesize about two dozen different proteins (called stress proteins). The most thoroughly studied stress is heat shock, in which a sudden increase in temperature induces the synthesis of a group of proteins collectively referred to as heat shock proteins. Genes that are responsible for the synthesis of heat shock proteins are among the most evolutionarily conserved genetic systems known. They are very similar in amino acid sequence in both prokaryotes and eukaryotes. Heat shock proteins accumulate to very high levels in stressed cells, accounting for as much as 15% of the total protein in *E. coli*.

The function of several of the heat shock proteins is beginning to be understood. Most of these proteins are synthesized at low levels under normal growth conditions and play a vital role in protecting the cell from the damaging effects of heat and other stresses. The heat shock proteins hsp70 and hsp60 (the number refers to the molecular weight of the protein times 1000) are involved in assembly or disassembly of protein complexes. Hsp70 is involved in translocating certain proteins through intracellular membranes and binds to DNA replication complexes. Hsp60 has been found to interact with steroid hormone receptors. In plants, hsp60 interacts with what may be the most abundant protein in the biosphere, ribulose-1,5-bisphosphate carboxylase-oxygenase, an enzyme that fixes CO_2 in chloroplasts.

FIGURE 14.B ***Cataglyphis bombycina.*** *(Courtesy of Donat Agosti, American Museum of Natural History, New York.)*

An amazing example of the use of heat shock proteins for protection is found in the desert-dwelling ant *Cataglyphis bombycina* (see Figure 14.B), which is perfectly suited for life in the Sahara desert. *C. bombycina* anticipates the high temperatures of up to 140°F found in the desert sand by producing heat shock proteins in abundance, even when in underground nests. These proteins seem to protect them from the heat. Investigators discovered this phenomena by exposing them to their natural conditions, grinding them up, and finding copious quantities of the heat shock proteins: a preemptive strike against the searing temperatures.

Invading organisms also can produce stress proteins in the host. This can be detrimental because some of these proteins are very similar to the host's stress proteins and can create an autoimmune disease in the host organism. For example, researchers have shown that the hsp60 proteins from tuberculosis bacterial and human cells are sufficiently similar that they are both recognized by antibodies from a tuberculosis patient. Thus, a bacterial invasion may induce the body's immune system to make antibodies to the stress proteins of the bacteria, and these antibodies may also attack some of the body's own vital proteins. It is believed that this may be the development process of some forms of chronic rheumatoid arthritis, which is an inflammation of the synovial membranes of joints.

ure 14.21). Methylated residues are maintained throughout DNA replications by a **maintenance methylase** that acts only on those CpG sequences base-paired with a ^{m}CpG sequence. Thus, after every replication of DNA, the maintenance methylase monitors the DNA, adding methyl groups onto CpG residues on the newly synthesized strand opposite a ^{m}CpG. As seen in Figure 14.22, the preexisting pattern of DNA methylation is maintained.

Direct evidence that methylation affects gene expression is found from experiments using the nucleoside analogue 5-azacytidine (5-azaC). 5-azaC is an analogue of cytidine and contains a nitrogen atom instead of a carbon at position 5 in the pyrimidine ring (Figure 14.23). This analogue cannot be methylated and inhibits maintenance methylation, reducing the general level of methylation. When cells are treated with 5-azaC, previously inactive genes become active. For example, when maintenance methylase was inactivated by mutation in the small model plant *Arabidopsis thaliana,* the mutant plants made more leaves, took longer to start flowering, and produced abnormal flowers. These results suggest that methylation is important for expression of some genes.

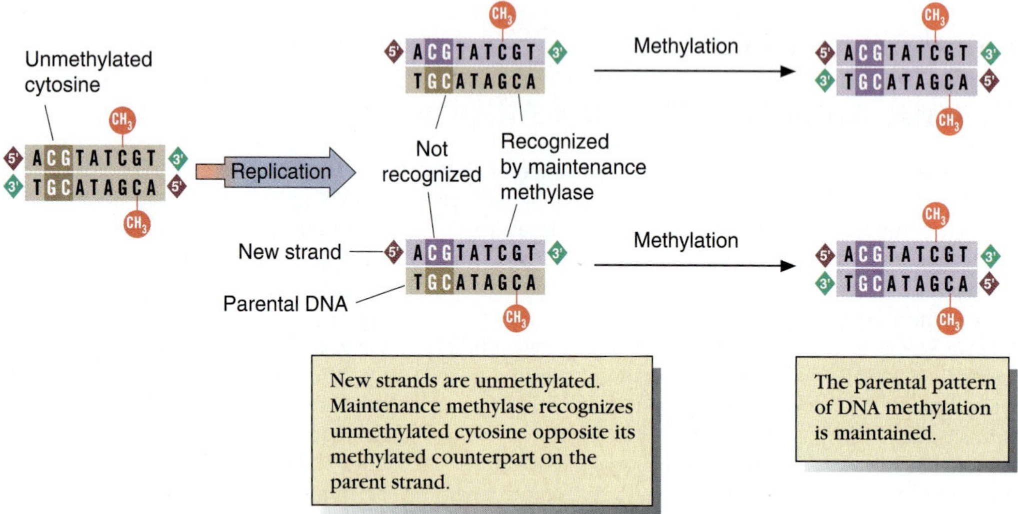

FIGURE 14.22 Inheritance of methylated DNA. DNA methylation patterns are faithfully maintained from generation to generation. After replication, the CG sequence from the template strand retains the methyl cytosine. The newly synthesized DNA strand is recognized by the maintenance methylase, and a methyl group is added to the cytosine (opposite a parental strand that retained the methyl cytosine), restoring the pattern of the parent molecule. Thus, only the site that was methylated remains methylated.

Additional evidence that methylation plays an important role in regulating groups of genes comes from the observation that there are islands of low-DNA methylation in both vertebrates and plants. These are known as **GC islands.** Within these islands are many of the housekeeping genes—those genes that encode the many proteins that are essential for cell viability and are therefore expressed in most cells.

When methylated and unmethylated genes are prepared by recombinant DNA methods and introduced into cultured mammalian cells, only the unmethylated genes are expressed. Moreover, when a methylated gene is introduced into a cell that normally expresses the gene, it is demethylated and activated. For example, the methylated actin gene from myoblast cells (cells that form muscle) is activated when inserted back into a myoblast cell. Demethylation takes place in the absence of DNA replication, suggesting that demethylation does not require replication and is able to activate genes selectively to determine specific cell types and functions. Direct evidence for a specific gene activation process has been obtained by exposing incompletely differentiated

(a)

Methylation site is blocked

5-azacytosine (5-azaC)

Cytosine

Methylation

5-methylcytosine

(b)

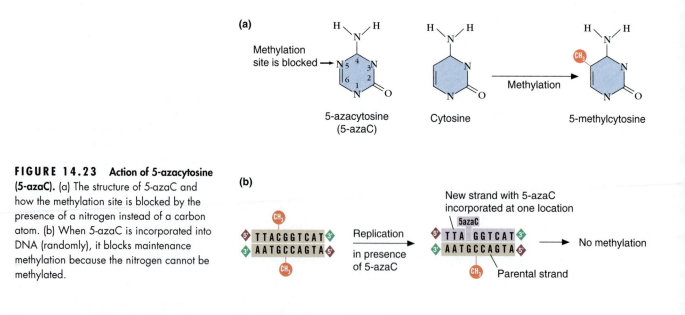

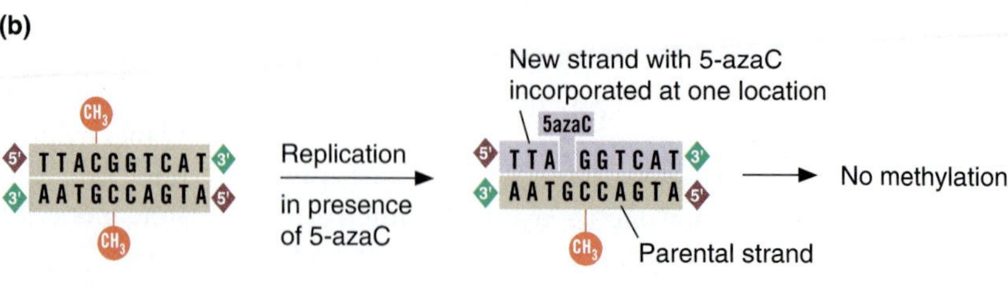

FIGURE 14.23 Action of 5-azacytosine (5-azaC). (a) The structure of 5-azaC and how the methylation site is blocked by the presence of a nitrogen instead of a carbon atom. (b) When 5-azaC is incorporated into DNA (randomly), it blocks maintenance methylation because the nitrogen cannot be methylated.

fibroblasts (which can be regarded as connective-tissue cells) to 5-azaC. Because the demethylation pattern of each cell would not be expected to be identical, one would predict that several different cell types may arise from this experiment. In fact, some cells became myoblasts (muscle-cell precursors) and then mature muscle cells that actually twitched on the cell culture plate. Other cells differentiated into fat-storing adipocytes. These findings suggest that the methylated state of genes can indeed influence the developmental pathway of cells. However, we do not know whether this mechanism is actually used by cells during development. We do know that methylation is not the only development mechanism because *Drosophila melanogaster* and *Caenorhabditis elegans,* two model invertebrates, do not possess any methylated DNA, yet both express their genes differentially according to developmental programs and environmental stimuli.

DNA-Binding Proteins Have Unique Structural Regions That Enable Transcription.

In both prokaryotes and eukaryotes, we have seen that *trans*-acting regulatory proteins (or transcription factors) bind to DNA with a high degree of specificity to turn genes on or off. This specificity of DNA recognition is the result of a general class of proteins with defined DNA-binding regions. Two examples of proteins with DNA-binding motifs are zinc finger proteins (as mentioned previously) and **helix-turn-helix** proteins. Other DNA-binding proteins have been characterized and include homeodomains, POU domains, leucine zippers, ETS domains, and Rel domains. However, for simplicity, we will discuss only zinc finger proteins and helix-turn-helix proteins. All DNA-binding proteins possess protrusions designed to interact specifically with the major and minor grooves of DNA, and their amino acid side groups interact with the individual bases.

The zinc finger proteins have regions (referred to as motifs) that vary in size from 60 to 90 amino acids and are composed of repeating groups of short amino acid sequences (Figure 14.24a). Each motif is composed of a folded structure held together by strategically located cysteine and histidine residues coordinated by a zinc molecule, from which the name zinc finger is derived. As a consequence, a 12-residue loop forms a potential DNA binding section. Some proteins that possess zinc finger binding domains are the transcription factor IID (TFIID), the GAL4 protein, and the steroid receptors.

The helix-turn-helix proteins have regions with two alpha helices connected by a short polypeptide β-turn (Figure 14.24b). This structure fits snugly into the major groove of the target DNA site (Figure 14.24c) and recognizes stretches of nucleotides. Specific amino acid residues in the helix form patterns of hydrogen bonds that uniquely recognize particular bases, conferring specificity. A specific DNA-binding protein (for example, the lambda cI repressor) may have several alpha helices, but one is the "recognition helix" and when bound, it lies deeply within the major groove of the DNA molecule. A second helix makes contacts with the sugar-phosphate backbone and bases adjoining the recognition sequence and stabilizes the protein–DNA interaction.

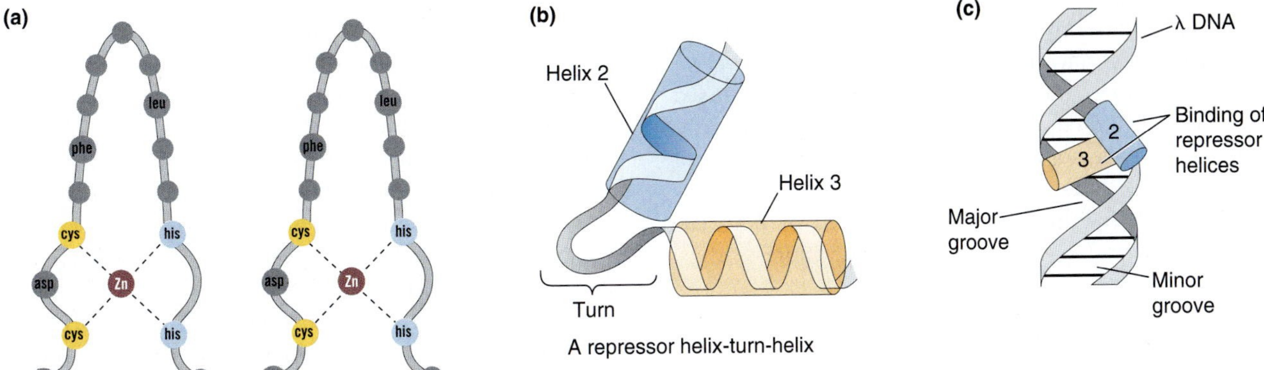

FIGURE 14.24 DNA binding proteins. Models illustrating how DNA-binding proteins fold and interact with DNA. (a) The folding is shown for a linear arrangement of repeated domains, each centered upon a zinc ligand. The amino acid residues (his = histidine, cys = cysteine, leu = leucine, phe = phenylalanine, asp = aspartic acid, tyr = tyrosine) surrounding the zinc (Zn) are highly conserved among different species. (b and c) Two alpha helices of the lambda repressor that are involved in binding to the major groove of the DNA molecule. These structures represent a few of the many different protein motifs that recognize DNA sequences.

SECTION REVIEW PROBLEMS

7. With the following mutants, predict whether the *GAL* operon in yeast is inducible, constitutive, or has no gene product expression when galactose is added:
   a. *GAL80*$^-$
   b. *GAL4*$^-$
   c. UAS deletion
8. Explain how the GAL4 protein activates a downstream promoter.
9. What are the similarities between the GAL4 protein and steroid receptor proteins?
10. Which of the following proteins would you expect to possess zinc finger or helix-turn-helix motifs? Why?
    a. GAL4 protein
    b. galactose epimerase (an enzyme in galactose metabolism)
    c. GAL80 protein
    d. steroid receptor protein

GENE REGULATION CAN OCCUR AT THE POSTTRANSCRIPTIONAL LEVEL.

In this section, we describe how expression of genes can be regulated after transcription has been initiated. This posttranscriptional control includes changes in RNA processing to obtain more than one message from a single gene, changing the coding sequence after mRNA synthesis, and variations in the stability of the mRNA. In many cases, these mechanisms have only recently been discovered and are not completely understood. It should be noted that prokaryotes do not possess introns (with a few exceptions) and thus do not process mRNA before translation. Consequently, mRNA processing is not used in prokaryotes to regulate gene expression. Additionally, prokaryotes have not been found to alter the sequence of completed mRNAs before translation. On the other hand, mRNA stability differences are used by both eukaryotes and prokaryotes to regulate gene expression.

One Gene May Code for More Than One Protein.

With the exception of overlapping genes (two genes using the same nucleotides for a short region, Chapter 8), we have assumed that one coding region specifies the information for a single polypeptide or protein. This assumption has been challenged with the discovery of alternative splicing. **Alternative splicing** uses different exons and introns from a gene's RNA product to create different mRNAs. You may recall (Chapter 11) that RNA processing or splicing removes introns from pre-mRNA to create mRNA. In alternative splicing, one group of exons may be used in one tissue and another group in a second tissue (Figure 14.25). In other cases, some introns are not spliced out and instead are used as exons, which yield quite different mRNAs and protein products. A large proportion of higher eukaryotic genes produce multiple proteins through alternative splicing. For example, several splicing possibilities may exist in a transcript, and the gene may produce a number of different proteins.

A gene can be spliced via one pathway in one tissue (for example, liver) and via another pathway in another tissue (for example, salivary gland). An example of this is the gene for amylase in the mouse (Figure 14.25). The mouse amylase gene has two promoters, separated by about 2850 bp, and four exons, two of which are adjacent to the promoters (exons S and L). In the salivary gland, the first promoter is used and the second exon (L) is spliced out of the final transcript, leaving exons S, 2, and 3 to be joined to form the mRNA. However, in liver tissue the second promoter is used leaving out exon S, and the final transcript contains exons L, 2, and 3. The two proteins differ in length by 37 amino acids as well as an amino acid sequence at the *N*-terminal end, but they are not radically different in structure. They both have the same general function in their respective tissues; they are both amylases. Closely related proteins produced by alternative splicing are called **protein isoforms,** and these isoforms usually leave the catalytic sites unchanged. However, exceptions are always found. For example, alternative gene splicing produces two very different polypeptide hormones, calcitonin in the mouse thyroid gland and calcitonin-gene-related peptide (CGRP) in neural tissue.

Alternative splicing can also generate two different proteins using the same promoter. An example is the formation of the α and β forms of troponin T of rat muscle (an accessory protein in Ca^{++} regulation of skeletal muscle). As shown in Figure 14.26, the 3′ half of the gene contains five exons, but only four are used to construct an individual mRNA. In one pattern, the α-exon is spliced into the final mRNA, and in the other pattern, the β-exon is spliced into the final mRNA.

How is RNA splicing regulated? The different choices of RNA splice sites are mediated by the binding of tissue- and gene-specific proteins to the growing RNA transcript, which selects specific splice-site junctions for the new mRNA transcript. These proteins (known as SR proteins) are expressed in a very tissue-specific manner, and they can be totally absent from one tissue and present in large quantities in another tissue. The absolute and relative amounts of SR proteins expressed by a given cell type determine the preferred splice site usage and therefore contribute to the regulation of gene expression.

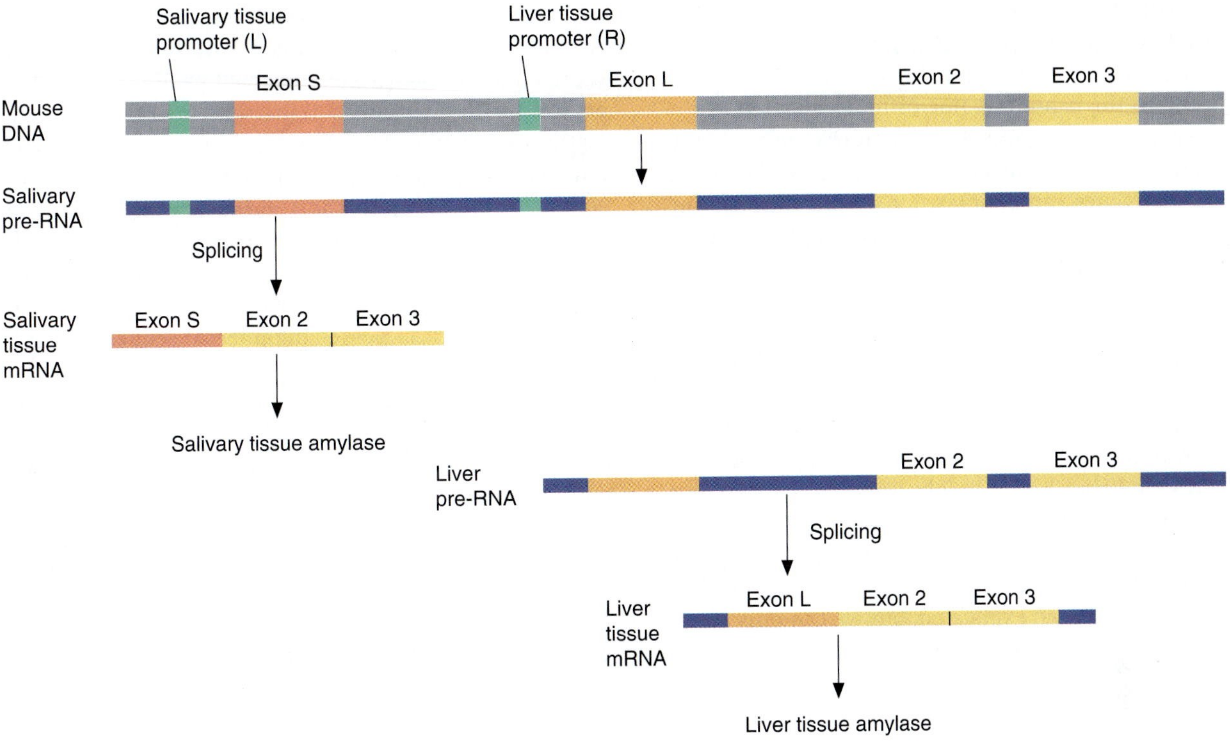

FIGURE 14.25 Alternative splicing of amylase mRNA. The gene for amylase in the mouse has alternate splicing pathways. In the salivary gland, the left promoter and exon S are used. In liver tissue, the right promoter and exon L are used. Both give mRNAs that produce amylase proteins.

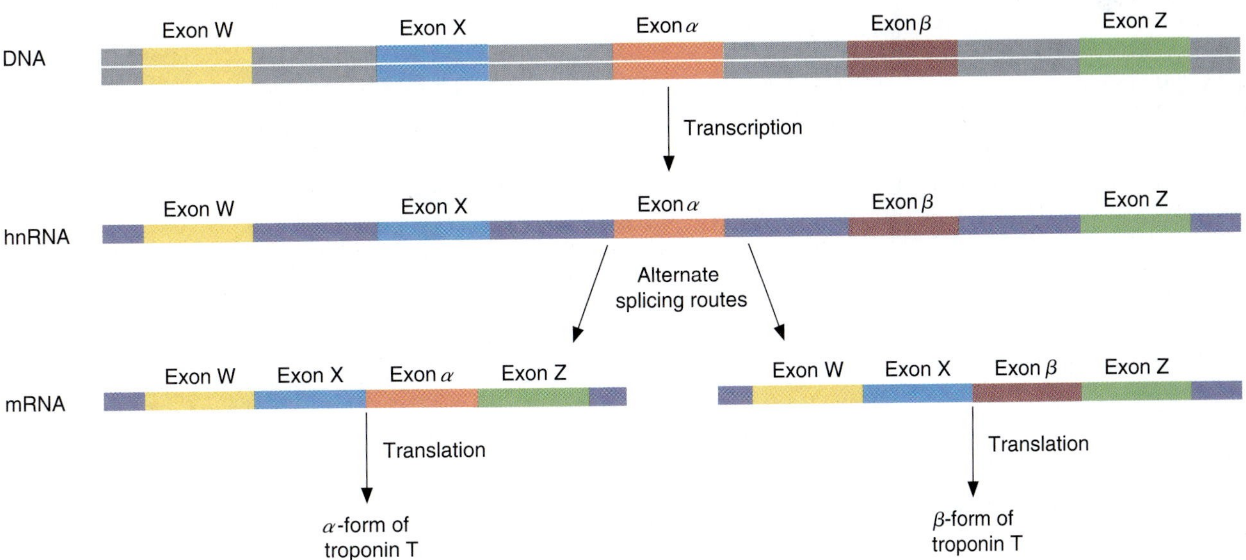

FIGURE 14.26 Alternative splicing of troponin T gene product. The 3′ end of the troponin T (a muscle protein) gene product in the rat has two alternative splicing pathways for the α and β forms. One splicing pathway gives an mRNA that includes exon α, whereas the other splicing pathway gives an mRNA that includes exon β.

The Coding Sequences of mRNAs Can Be Changed After Synthesis.

According to the central dogma of molecular biology, a coding sequence of an mRNA can represent only what is present in the DNA that codes for the gene. We have learned that genes can be interrupted with introns, but introns do not change the sequence of coding events within the gene. In the previous section, we learned that alternative RNA splicing can give a variety of gene products from a single gene. In 1989, biologists uncovered what seemed to be the ultimate surprise: a novel molecular mechanism was discovered in the mitochondria of trypanosomes (a parasite of humans and insects) that alters RNA transcripts coding for proteins. This process is called **RNA editing,** and it involves adding or removing one or more uridine nucleotides within selected regions of the transcript. This change causes shifts in the original reading frame and new codons. The RNA editing mechanism has recently been explicated, but we can only speculate as to why such a bizarre phenomenon exists.

The RNA to be edited is first transcribed from what is called a **cryptogene,** which can be considerably shorter than a comparable unedited gene from another organism. The pre-edited transcripts are corrected primarily by uridylyl (U) additions, but also by U-residue removal using the sequence information separately encoded in one or more short **guide RNAs** (gRNA). In some transcripts, U residues are inserted at over 100 different places to constitute over half the protein-coding nucleotides. The gRNA is a small RNA molecule (about the size of a tRNA) possessing internal sequences complementary to parts of the transcript, although base pairing is frequently imprecise; Us are specified by both As and Gs (Figure 14.27). The gRNAs also possess, on their 5′ end, stretches of uridylyl residues. The gRNA molecule forms hydrogen bonds to the unedited cryptogene mRNA and begins to move down the transcript, inserting or removing U nucleotides (Figure 14.27) by directed endonuclease cleavage, addition of Us from free UTP, and religation of the mRNA. The sequences of the short gRNAs correspond to patches (groups of nucleotides

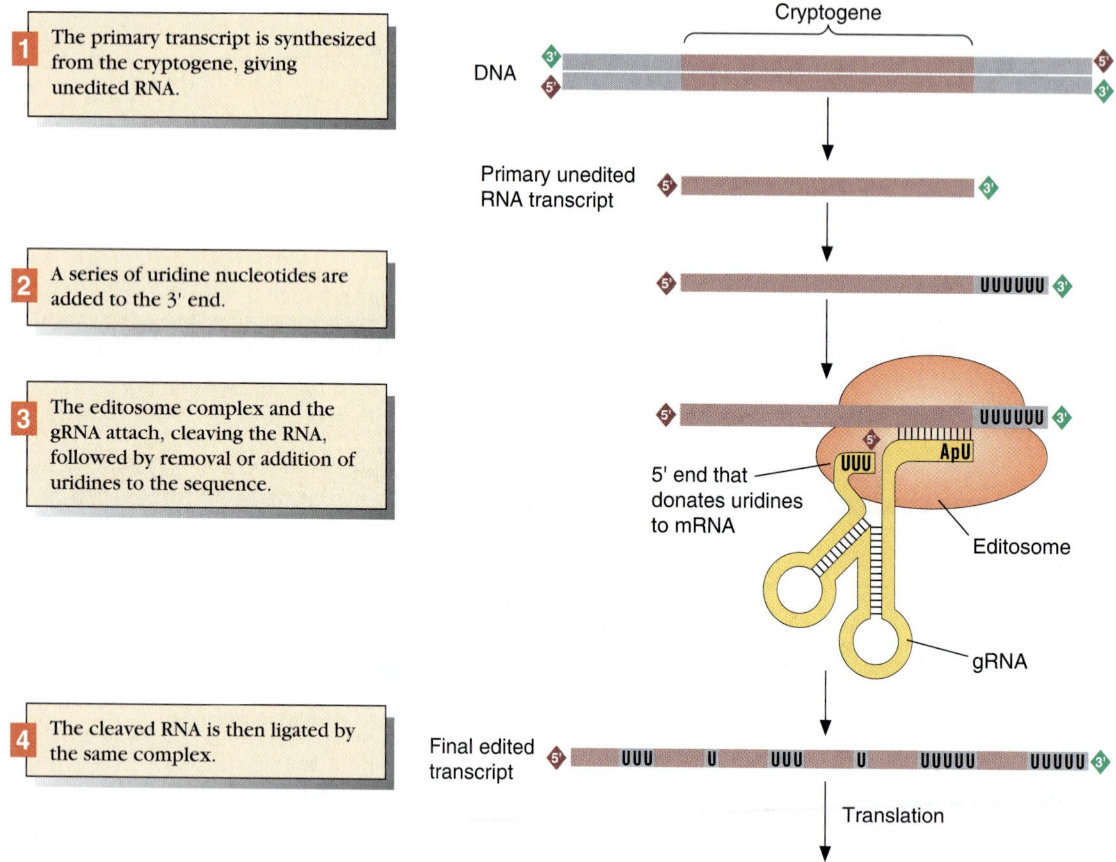

1 The primary transcript is synthesized from the cryptogene, giving unedited RNA.
2 A series of uridine nucleotides are added to the 3' end.
3 The editosome complex and the gRNA attach, cleaving the RNA, followed by removal or addition of uridines to the sequence.
4 The cleaved RNA is then ligated by the same complex.
Cryptogene
DNA
3'
5'
5'
3'
Primary unedited RNA transcript
UUUUUU
UUU
ApU
5' end that donates uridines to mRNA
Editosome
gRNA
Final edited transcript
UUU
U
UUU
U
UUUUU
UUUUU
Translation

FIGURE 14.27 RNA editing. Imperfect duplexes with the 3′-gRNA (guide RNA) occur during editing and the reaction progresses in the 5′ direction. This editing produces changes in the mRNA sequence after it is synthesized.

within the sequence) of the correctly edited mRNA sequence. The reactions are catalyzed by a complex of enzymes (necessary for RNA cleavage, addition or deletion of U residues, and RNA ligation) called the **editosome** that add or remove U residues as specified by the gRNA. An important component of this model is that the sequence of the edited RNA is specified by base pairing with the gRNA.

Across species, cryptogenes have a number of similar characteristics, which argues for a common ancestry of these unusual genes. Furthermore, RNA editing has been found in a wide variety of organisms, from the chloroplasts of flowering plants to trypanosomes. The discovery of the RNA editing phenomenon challenges the concept of the faithful transmission of genetic information from DNA through RNA to protein. From this finding, we can conclude that the gene sequence is not always a perfect predictor of the protein sequence.

The Stability of mRNA Can Control Gene Expression.

The stability of mRNA is an obvious mechanism for controlling gene expression levels, and eukaryotes, and sometimes prokaryotes, use this mechanism. The rate of mRNA synthesis and destruction (the turnover rate) influences cellular levels of mRNA in a manner analogous to filling a water bucket that has holes in it. As water flows into the bucket (mRNA synthesis), some fraction also escapes through the holes (mRNA destruction). The size of the holes (the RNA turnover rate) will affect the total accumulation of water (mRNA). Small holes (turnover is low) will allow accumulation, whereas large holes (turnover is high) will deplete the level rapidly. Bacteria have a very high mRNA turnover rate; the lifetime of an average mRNA is only about 3 minutes, although some prokaryotic mRNAs have longer lifetimes. This rapid turnover allows the bacterial cell to adapt to its changing environment rapidly.

Conversely, because eukaryotic cells have a relatively stable environment, they have many very stable mRNAs that have lifetimes of hours rather than minutes. Some of these mRNAs code for ongoing housekeeping functions, for example, the cell structural proteins and metabolic functions. In certain cells, a long-lived mRNA may make abundant amounts of only one protein, for example, a β-globin protein in the red blood cell. Nevertheless, some eukaryotic mRNAs have relatively short lifetimes of 30 minutes or less. Many of these mRNAs code for proteins involved in regulating cell growth and development, activities for which the protein is needed only transiently to perform a specific function, and after which they must be removed quickly.

Messenger RNAs are destroyed by a class of enzymes called ribonucleases (RNases). Most cells contain ten or more different ribonucleases, but it is not clear which of them degrade mRNAs and which degrade other types of RNA molecules (e.g., rRNA, tRNA). Several different regions of the mRNA can act as a signal to the ribonuclease to degrade or not to degrade an individual mRNA, including the 5′ terminal sequence, the internal secondary structure, the 3′ terminal sequence, and a poly(A) tail. Usually the RNA folds to a specific three-dimensional structure (for example, a stem-and-loop configuration), which can either stimulate or inhibit RNase digestion. In addition, a group of poorly understood proteins can bind to mRNAs to protect them from nuclease digestion.

An example of an unstable mRNA is the product of the *myc* gene, first identified in chickens but also present in all animals. Myc is a protein used to turn on the expression of other genes during differentiation of specific cell types, and thus a cell needs only the Myc protein for a very brief time and at a specific concentration. An excess of Myc protein can be dangerous because it will result in cancerous growth. As a consequence, the *myc* gene mRNA is relatively unstable and has a half-life of about 30 minutes. The 3′ untranslated leader region of this gene contains long sequences rich in A and U nucleotides, which are probably responsible for *myc* mRNA instability. Deletions within the 3′ leader region of the mRNA increases stability and causes cancerous growth, suggesting that this portion of the mRNA is responsible for the rapid degradation of the *myc* gene mRNA.

SECTION REVIEW PROBLEMS

11. By what means can a single gene produce as many as five different gene products?
12. Two proteins, A and B, have very different functions in the cell. One is a regulatory protein and the other is a housekeeping protein. What would you predict concerning the stability of the mRNAs that determine each of these proteins? Why?

DNA Sequence Rearrangements Can Permanently Alter Gene Expression.

In the examples of gene-expression regulation we have examined thus far, in no case has the change been a stable alteration in the DNA coding sequence. However, stable changes in DNA gene sequences do occur in many organisms and are an important part of the repertoire of gene regulatory mechanisms. In contrast to other forms of gene regulation, DNA rearrangements are faithfully copied during subsequent DNA replications and are inherited by all the mitotic progeny of the cell. Some of these rearrangements are reversible, and over long enough periods of time, they can produce alternating patterns of gene expression. In the following examples, we see that *Salmonella* and *Trypanosomes* can undergo DNA rearrangements to evade the immune system of their hosts (reversible rearrangements) and that the hosts undergo genome rearrangements to acquire antigenic specificity (irreversible rearrangements).

Salmonella Strains Undergo DNA Rearrangements to Evade the Immune Response.

A well-studied example of DNA sequence rearrangement is in the bacterium *Salmonella,* which has an alternating pattern of flagellin protein expression. Many bacteria, including *Salmonella,* move by waving a long whip-like structure known as the flagellum, which is composed of the protein flagellin. If *Salmonella* infects a higher eukaryote, flagellin is recognized as foreign by the host's immune system, which eventually destroys the invading bacteria. As a consequence, *Salmonella* has evolved an ingenious mechanism for evading destruction by the host's immune system by changing the amino acid sequence of the flagellin protein about every 1000 divisions. This is called **phase variation,** and it protects a bacterial population from the host's immune responses. About every 1000 divisions, a DNA sequence flip-flops, rearranging a DNA sequence that controls the synthesis of flagellin

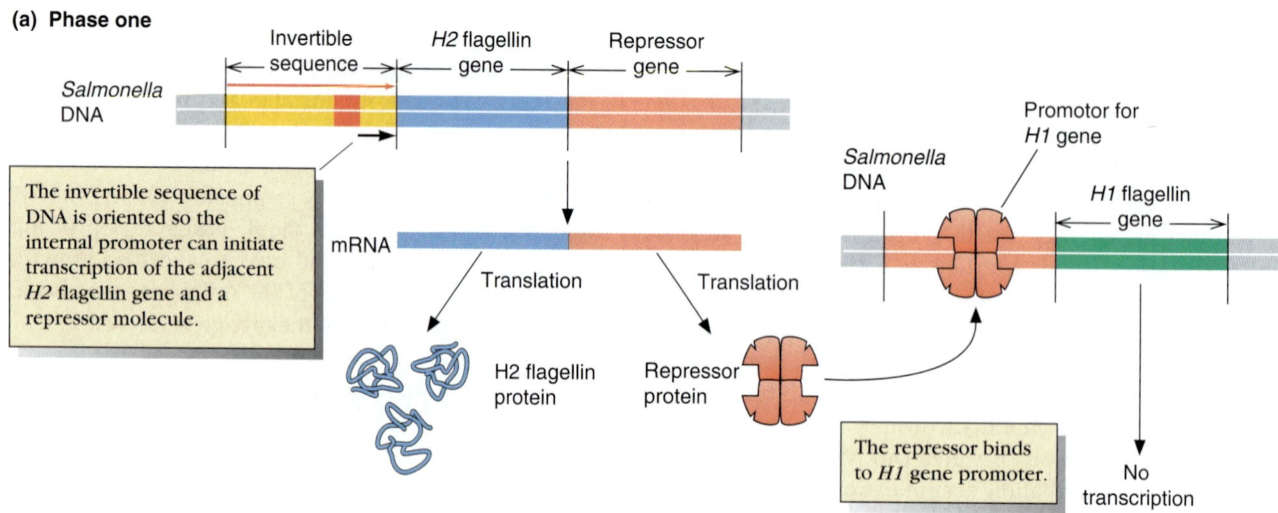

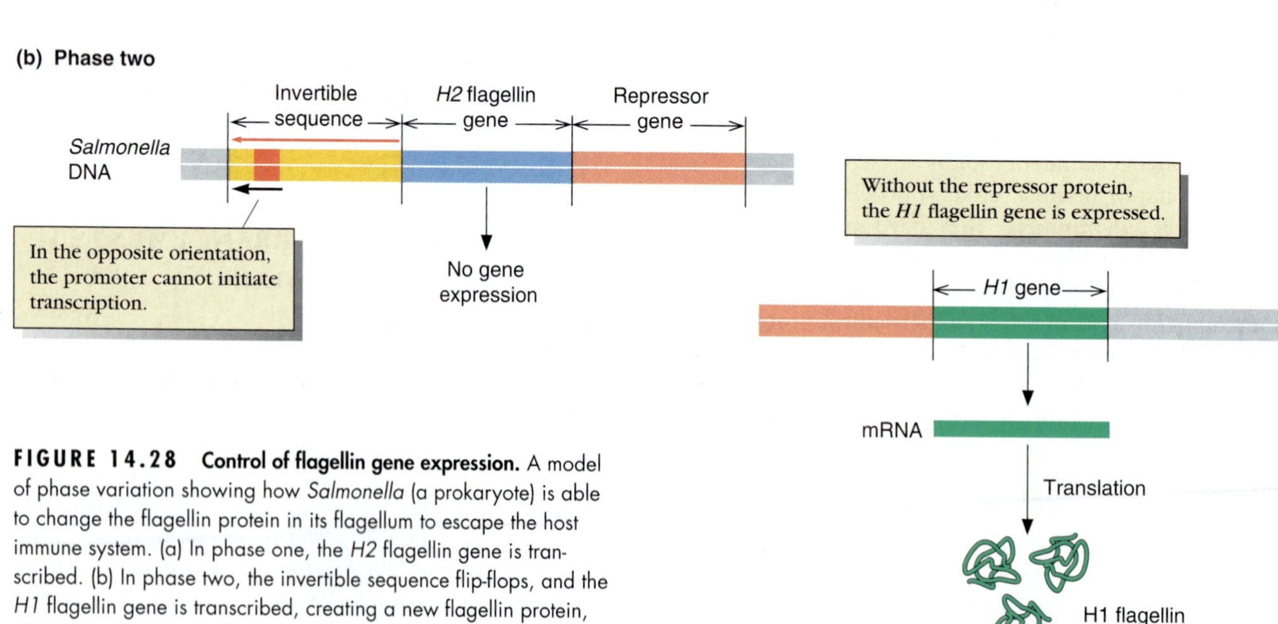

FIGURE 14.28 Control of flagellin gene expression. A model of phase variation showing how *Salmonella* (a prokaryote) is able to change the flagellin protein in its flagellum to escape the host immune system. (a) In phase one, the *H2* flagellin gene is transcribed. (b) In phase two, the invertible sequence flip-flops, and the *H1* flagellin gene is transcribed, creating a new flagellin protein, which in turn forces the host to create new antibodies.

mRNA (Figure 14.28). The flip-flop shuts off one flagellin gene while turning on the other flagellin gene. As a result, clones of this new bacterial population have a different flagellin protein in their flagellum. The host antibodies cannot immediately recognize the new protein, thus killing of the bacteria by the immune system is delayed allowing the bacteria to multiply for an additional period of time.

The control circuit for flagellin synthesis is depicted in Figure 14.28. The two genes that synthesize the two different types of flagellin proteins are found at two chromosomal locations: one flagellin gene (*H2*) is adjacent to the invertible segment (995 bp, bounded by a 14-bp sequence found at both ends; see Chapter 19 on transposable elements), and the other gene (*H1*) is distantly located. When the invertible segment is oriented in one direction (Phase 1 in Figure 14.28), the promoter within it is properly oriented to initiate the transcription of the adjacent *H2* flagellin protein gene and another gene that encodes a repressor molecule. The repressor molecule binds to the promoter of the second flagellin gene *H1* and prevents its transcription. Thus, when the invertible sequence is in one orientation, only the H2 flagellin protein is made. However, about once in every 1000 generations, the invertible sequence flip-flops, orienting the promoter in the opposite direction (Phase 2 in Figure 14.28). In this orientation, neither H2 flagellin nor the repressor protein is synthesized. Without the repressor protein, the promoter of the flagellin gene *H1* is released for transcription of the *H1* mRNA.

Trypanosomes Use DNA Rearrangement To Alter Their Surface Coats To Evade Immune Responses.

The trypanosome is a small unicellular parasite that can live in the bloodstream of mammals or the gut and salivary gland of the tsetse fly. The best investigated species, *Trypanosoma brucei,* does not grow in man, but other species of trypanosomes have caused tremendous losses in human productivity and animal life in central Africa. The trypanosome is transmitted from the tsetse fly to a mammal via a bite, whereupon it enters the bloodstream and multiplies. The fly is infected (or reinfected) by biting an infected animal and ingesting the blood, spreading the parasite from one mammal to another (Figure 14.29). In humans, the parasite causes a disease known as sleeping sickness. The infected individual becomes more and more lethargic and eventually becomes comatose. The cattle population of central Africa has been greatly reduced by trypanosome infections, restricting the availability of milk and other beef products.

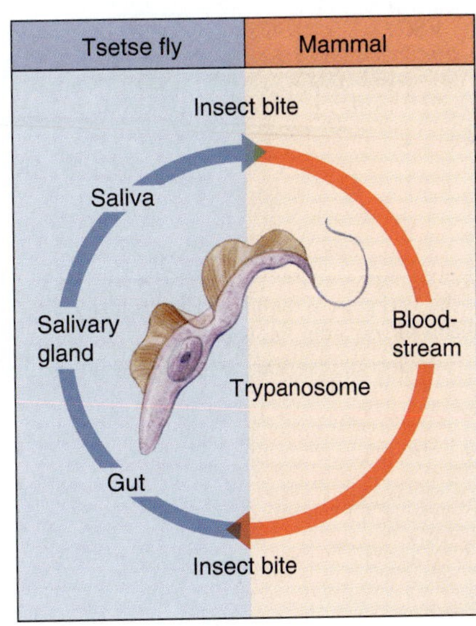

FIGURE 14.29 The transmission cycle of the trypanosome. *Trypanosoma bruci* passes through several morphological forms when it alternates between the tsetse fly and the mammalian host.

The trypanosome surface coat consists of a **variable surface glycoprotein** (VSG) that completely covers the outer surface of this unicellular organism. This protein is the only antigenic structure exposed to the host. The remarkable aspect of the trypanosome is that it is able to change its variable surface glycoprotein to evade destruction by the host's immune system. Upon infecting a host, a trypanosome will have one VSG exposed, but about once in 10,000 to 100,000 trypanosomes (about once every 2 weeks) the gene encoding the VSG coat changes. As a result, the host must synthesize a completely new antibody against the newly emerged trypanosome. About the time the host has made the new antibody, the trypanosome will again have within its population an individual that has switched to a new VSG coat. This happens before the host immune system is able to kill all members of the population. These cycles of VSG change continue until the host succumbs. A single trypanosome has the genetic capacity to make about 1000 VSGs, each sufficiently different in sequence so that antibodies against any one will not react against the others. This remarkable antigenic variation of trypanosomes saves them from destruction by the host immune system.

How is this variation possible? Each trypanosome carries the entire VSG repertoire of its strain, but only one copy is expressed at any one time. A single VSG gene is selected for expression and transcribed by virtue of its presence at an **expression site.** Many expression sites exist, but all are located near telomeres, and only one is active at a time. Unexpressed VSG genes are

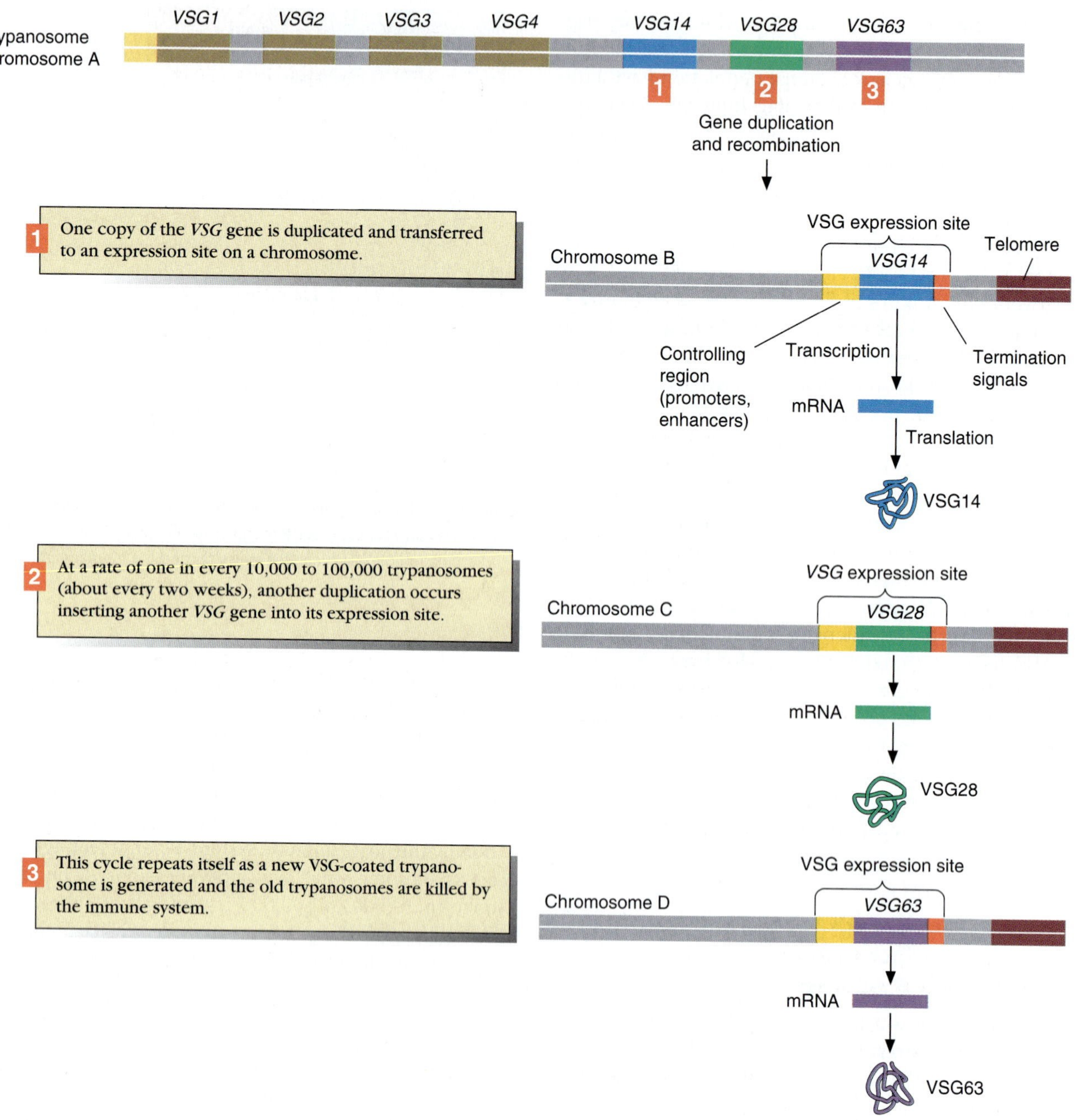

FIGURE 14.30 Trypanosome antigenic changes. The trypanosome changes its surface coat protein to avoid the immune response. The variable surface glycoprotein that covers the trypanosome is the only antigenic material exposed. Several hundred different genes encode VSG; however, only one of these genes is expressed at any one time. Periodically, a trypanosome in the population changes its expression from one VSG to another and is thus able to escape the immune system of the host by multiplying with the new VSG coat. This change allows the trypanosome to multiply and infect the host over an extended time period.

stored internally in the chromosome (Figure 14.30). The gene to be expressed is duplicated and transferred to the expression site by recombination. The expression site may change, but the previously expressed copy is always inactivated when the new copy is activated. Unexpressed VSG genes in the chromosome can be copied into nonexpressed sites as well as into expression sites. Thus, an internally located gene may undergo a two-stage activation: it is first transferred by recombination from an unexpressed site to an expression site, and then it is activated when needed. The sequence in which the different VSG genes are expressed seems to be random.

It is assumed that some change occurs at the expression site to activate the VSG mRNA expression. The exact change that occurs is not known, but it could be an alteration in chromatin structure, or it might involve a change in methylation or a rearrangement of a sequence controlling gene expression.

Antibody Diversity Is Required for Hosts to Survive.

We cannot live very long if our immune system is inactive. Any vertebrate that is immunologically deficient runs the risk of early death from infectious agents—bacteria, fungi, parasites, or viruses. Rarely, humans are born lacking a functional immune system, but they usually die soon after birth. One such child, known as "David," was born without an active immune system in the 1970s. He was sealed in an airtight chamber to protect him from invading organisms, and everything he touched or ate was sterilized. After about 10 years, he was removed from the chamber to give him an immune system through a bone marrow transplant from his sister. Unfortunately, he died soon after as a result of massive infections, clearly emphasizing our need for an immune system and the difficulty of reestablishing a defective immune system.

When your body is infected by a foreign molecule or organism, the immune system responds by initiating a series of events leading to the destruction and elimination of invading organisms. A remarkable feature of the immune system is that it distinguishes between **foreign** invaders and **self.** The immune system does not destroy what it recognizes as self molecules. Occasionally the immune system fails to make this distinction and reacts destructively against molecules of itself. This results in an autoimmune reaction or diseases that are sometimes fatal. Examples of autoimmune diseases include type I diabetes, some forms of arthritis, multiple sclerosis, and lupus (systemic lupus erythrematosis).

The cells responsible for immune specificity are a class of white blood cells known as **lymphocytes.** Lymphocytes develop from pluripotent hemopoietic stem cells, or **stem cells** for short (located in the bone marrow), and they give rise to all blood cells including red blood cells, white blood cells, and platelets. Lymphocytes are also present in the lymph (a colorless fluid in the lymphatic vessels that connect the lymph nodes in the body) and in lymphoid organs (thymus, lymph nodes, spleen). During the 1960s it was discovered that the immune response was mediated by two different classes of lymphocytes: **T cells,** which develop in the thymus and are responsible for cell-mediated immunity (for example, tissue transplants), and **B cells,** which develop in fetal liver and adult bone marrow in mammals and are responsible for antibody production. In the following sections, for simplicity, we will restrict our discussion to how B cell diversity produces the millions of different mammalian antibodies that defend against invasion by destructive organisms.

The Clonal Selection Hypothesis Explains Antibody Diversity. How is the B cell immune system able to respond to millions of different foreign antigens, each in a highly specific way? This question puzzled scientists for many years, and only recently has the answer become known. Any substance capable of eliciting an immune response is referred to as an **antigen** (for antibody generator). The body synthesizes a protein known as an **antibody** in response to the presence of an antigen; a different antibody is produced for each antigen. All antibodies are proteins, and all antigens are foreign molecules. Antigens come in many different forms: carbohydrates, proteins, lipids, nucleic acids, and a vast array of organic molecules. We now know that the remarkable ability of the immune system to respond in a specific way to such a wide variety of antigens is the result of **clonal selection.** The clonal selection hypothesis, developed in the 1950s, stated that during development of stem cells into B cells, each cell becomes committed to react with a particular antigen before ever being exposed to it (Figure 14.31). The cell expresses this commitment in the form of cell-surface-receptor proteins that bind to the specific antigens. When an antigen binds to the cell-surface receptors on one specific cell, the cell is activated to proliferate and form a clone of cells. This process is known as **clonal expansion.** This clone of cells now produces antibody molecules which are secreted into the blood and react with the antigen.

A human body possesses about 10^{11} B cells (those blood cells responsible for antibody synthesis), which is almost equal to the mass of the liver. Not every one of the B cells recognizes a different antigen because many are duplicates resulting from clonal expansion, but it is not unreasonable to assume that 50 to 100 million (10^7 to 10^8) different antigen-bearing B cells are present in each individual.

An invading organism (such as a bacterial cell) will have on its surface many antigens that elicit the production of many different antibodies. Each of these antibodies can react with a different antigenic site on the bacterial cell surface. These mixtures of different antibodies are referred to as **polyclonal antibodies.** If, by chance, a single antigen elicits a single antibody, then it is called a **monoclonal antibody.** Collectively, all the antibodies produced by the millions of B cell lymphocytes are called **immunoglobulins** (Ig).

The Molecular Structure of Antibodies Is Simple. What does an antibody molecule look like, and how is it able to recognize so many different molecules? Each

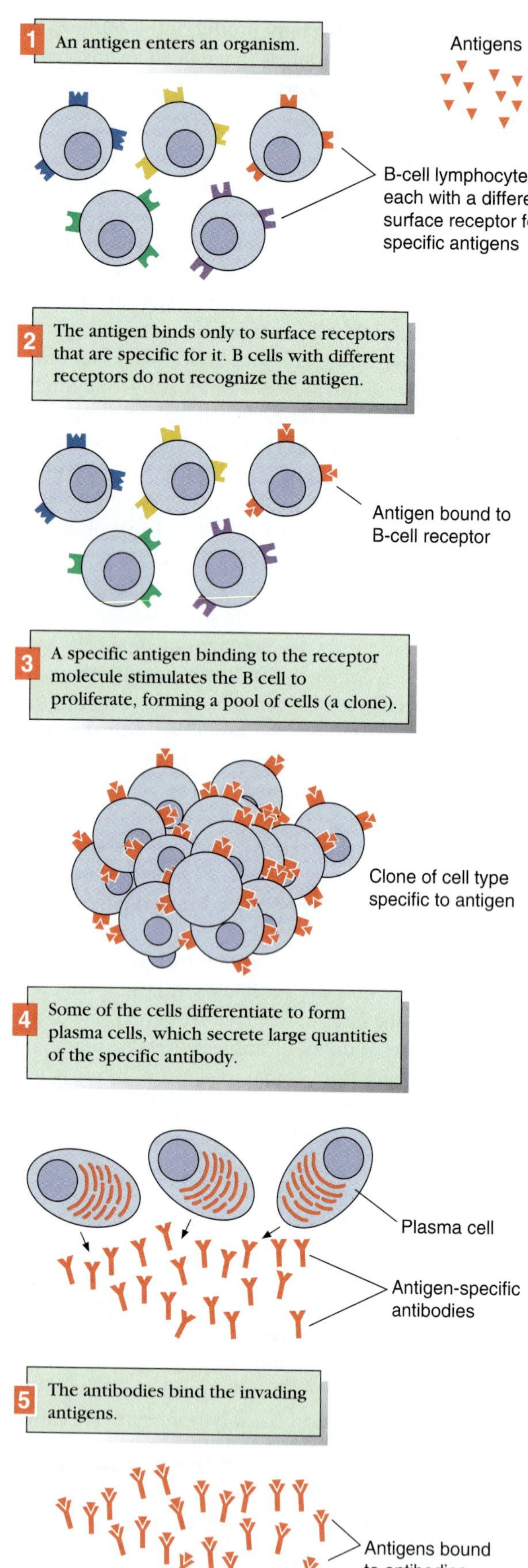

1 An antigen enters an organism.
Antigens
B-cell lymphocytes, each with a different surface receptor for specific antigens
2 The antigen binds only to surface receptors that are specific for it. B cells with different receptors do not recognize the antigen.
Antigen bound to B-cell receptor
3 A specific antigen binding to the receptor molecule stimulates the B cell to proliferate, forming a pool of cells (a clone).
Clone of cell type specific to antigen
4 Some of the cells differentiate to form plasma cells, which secrete large quantities of the specific antibody.
Plasma cell
Antigen-specific antibodies
5 The antibodies bind the invading antigens.
Antigens bound to antibodies

◀**FIGURE 14.31 The clonal selection hypothesis.** When an invading antigen binds to a receptor on a B cell, an individual antibody is produced. This cell is already committed to producing the antibody to the invading antigen, and it now proliferates, producing the antibody in large quantities. The antibody then binds the antigen to inactivate it.

antibody is a tetramer (Figure 14.32) consisting of two identical **light (L) chains** and two identical **heavy (H) chains.** All light chains and heavy chains are very similar in molecular structure; they vary only in amino acid sequence in certain regions of the molecules. Also, any light chain can associate with any heavy chain. Therefore, to produce 10^8 different antibodies, it is necessary to have 10^4 different light chains and 10^4 different heavy chains. Using this reasoning, the genome needs to code for 20,000 or 2×10^4 different antibody protein molecules. However, it is estimated that most vertebrates have only about 100,000 or 10^5 different genes in the total genome. Thus, it follows that the antibody-coding capacity of the genome would need to be made up of about 20% of all genes (20,000 of the 100,000 proposed genes). This unlikely situation was a stumbling block for the clonal selection hypothesis when it was first proposed. As we will discuss shortly, DNA rearrangements in the stem cells permit many different antibody proteins to be

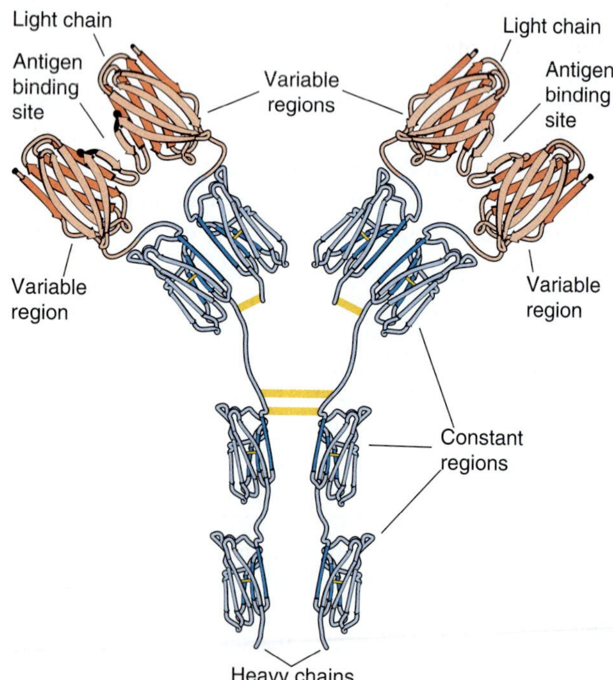

Light chain
Antigen binding site
Variable regions
Light chain
Antigen binding site
Variable region
Variable region
Constant regions
Heavy chains

FIGURE 14.32 The antibody or immunoglobin protein. Each molecule possesses two copies of a light (L) chain and two copies of a heavy (H) chain held together with disulfide bonds (the yellow lines). The variable domains of the light and heavy chains (V_L and V_H) make up the antigen-binding sites.

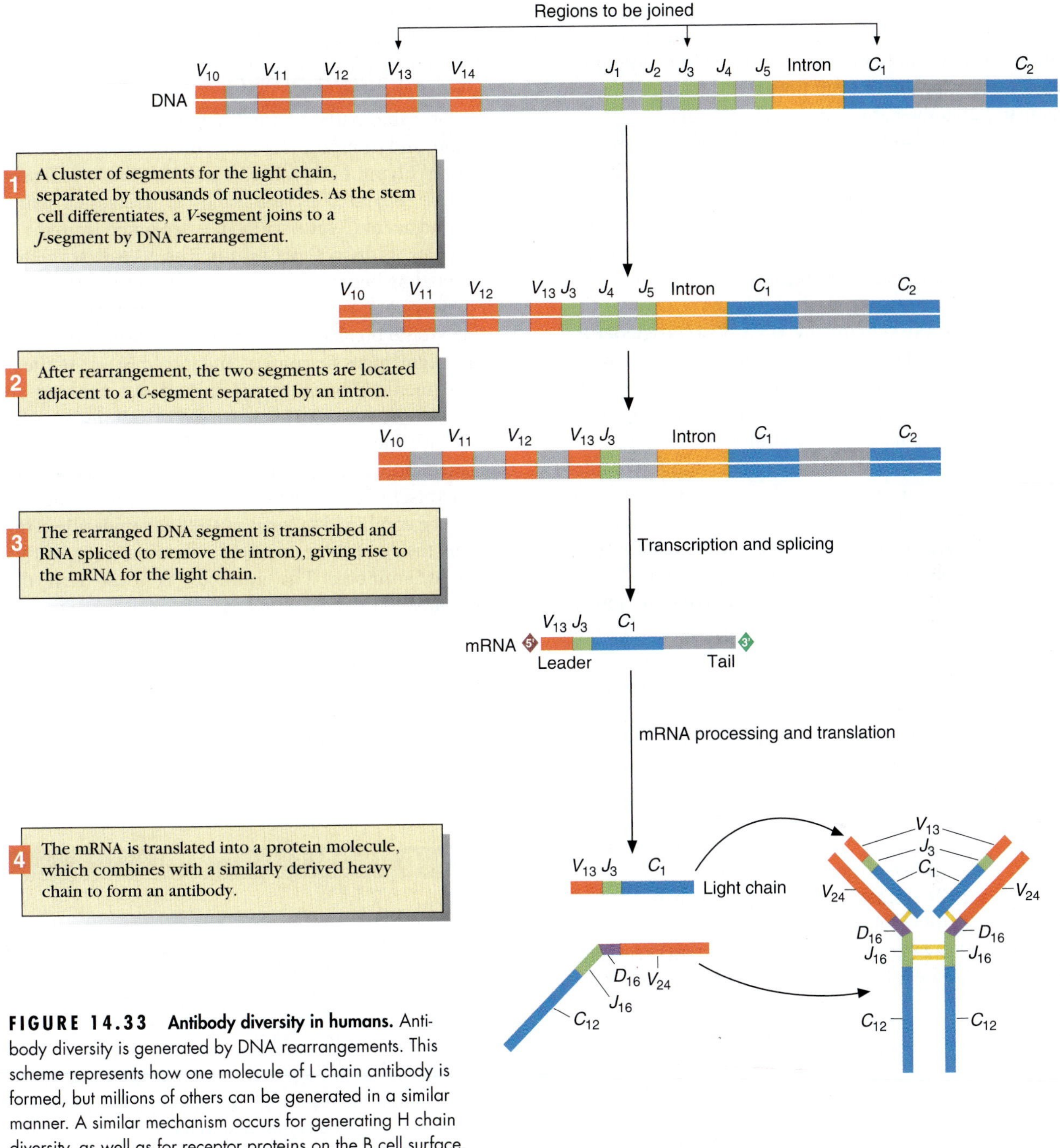

FIGURE 14.33 Antibody diversity in humans. Antibody diversity is generated by DNA rearrangements. This scheme represents how one molecule of L chain antibody is formed, but millions of others can be generated in a similar manner. A similar mechanism occurs for generating H chain diversity, as well as for receptor proteins on the B cell surface.

produced from a small number of genes. Before we discuss the mechanism by which these DNA rearrangements occur, we first need to examine some of the details of the antibody molecule itself.

Comparing the amino acid sequences of many different antibodies reveals that both the heavy and light chains have a **variable sequence** at their amino-terminal ends and a **constant sequence** at their carboxyl-terminal ends. The constant region represents about half of the light chain molecule (110 of 220 amino acids) but only one quarter of the heavy-chain sequence (110 of 440 amino acids). The constant regions between different molecules show only a very small variation in sequence, whereas the variable regions have a high degree of variation. This finding was important in explaining how so many different antibodies can be formed from a few genes because the variable regions (labeled V_L and V_H) are also the antigen-binding sites (Figure 14.33). The amino-terminal ends of the light and heavy chains come together to form the antigen-binding site, and the variability of their amino acid sequences provides the structural basis for the diversity of antigen-binding sites. Both the heavy and light chains are folded to contain regions consisting of 110 amino acids (4 on the heavy chain and 2 on the light chain), and each region contains one interchain disulfide bond (Figure 14.33). Three of the four folded segments in the heavy chain are homologous to one another and to the constant regions of the light chains. Likewise, the variable regions in both the light and heavy chains are similar to each other and, to a lesser extent, to the constant regions.

Antibody Diversity Requires DNA Rearrangements. Vertebrates have evolved a unique genetic mechanism to generate an almost unlimited number of different light and heavy chains. Instead of a continuous stretch of genetic material for many single antibody molecules, the genome codes for individual segments of antibody molecules. Thus, there is one group of gene segments for the variable region and another group for the constant region, each tandemly arranged along a stretch of chromosome. Each light- or heavy-chain gene segment is slightly different from the others, enabling the construction of many complete but different genes. During chromosome rearrangement, complete light or heavy chain genes (Figure 14.33) are constructed with randomly selected gene segments. These gene segments are rearranged in the stem cell to form new, complete genes in individual cells, each capable of synthesizing a specific and different heavy and light chain from an antibody protein. For example, humans have seven different copies of the light chain *C* region, and about 300 different copies of the *V* region. The number of potentially different light chain genes that can be formed from these segments is $7 \times 300 = 2100$. Actually, each light chain consists of a *V* gene (for variable) segment, a *C* gene (for constant) segment, and a *J* gene (for joining) segment so this number is even greater.

Each heavy chain is composed of a *V*-gene segment, a *C*-gene segment, a *J*-gene segment, and also a *D*-gene segment (for diversity). In humans, there are about 200 different *V*-segment genes, 8 different *C*-segment genes, 6 different *J*-segment genes, and 20 different *D*-segment genes. Thus, the heavy chain segments can potentially generate $200 \times 8 \times 20 \times 6 = 192{,}000$ different heavy chain genes. Over 5 million different antibody molecules can be formed by joining the various segments and combining the heavy and light chains in all possible combinations. When all the segments and gene families are taken into account, this number is about 10 to 100 times higher.

How do the DNA rearrangements occur in the lymphocyte stem cells? Currently, the exact mechanism is not known, but it involves recombination. It is known from abundant evidence that, as the clonal selection theory predicts, each individual cell derived from DNA rearrangements makes a different antibody. It is also known that the heavy and light chain gene segments to be rearranged are located on several chromosomes as families of clusters.

A probable pathway for DNA rearrangement, transcription, and translation of antibody molecules has been proposed (Figure 14.33). In this pathway, the *V–J–C* segments are separated by thousands of nucleotide pairs, and the adjacent *C*-gene segment is separated from the *J*-gene segments by a short intron. The *V–J–C* segments are flanked by heptamer-spacer sequences that are recognized by specific recombinases, introducing site-specific double-strand breaks for rejoining by ligation. One of the *V*-gene segments is rearranged and joined to one of the *J*-gene segments. The intron is spliced out, making an mRNA containing one *V*-gene segment, one *J*-gene segment, and one *C*-gene segment. This mRNA is then translated into a light chain. A similar process will simultaneously occur to generate a heavy chain, and the two will be combined to form an antibody molecule. This cell will then continue to divide, creating more of this particular antibody. The antibody is located in the plasma membrane. There, its variable region is exposed to the exterior environment to serve as a receptor site to bind to the specific antigen it alone can recognize. If an organism or molecule bearing this antigen invades the host, this specific cell will form a clone and generate more antibody molecules, which in turn will destroy the antigen-bearing invader.

SOLVED PROBLEM

Problem

The shark has a very limited immune system, but it does not become infected with bacteria or fungi, nor is any cancer known in sharks. Upon examination of the number of genes for the heavy chain, it was found that one species of shark had 12 *V*-gene segments, 3 *C*-gene segments, 5 *J*-gene segments, and 4 *D*-gene segments. The light chain had 16 *V*-gene segments, 2 *C*-gene segments, and 6 *J*-gene segments. What is the theoretical maximum number of possible antibody proteins that this shark could create by rearrangements during stem-cell development?

Solution

The solution to this problem lies in straight probability determination. That is, the probability of one event occurring simultaneously with another event is the product of each of their individual probabilities. Thus, for the heavy chain there are 12 different chances for a *V*-gene segment to combine with the other segments, or $12 \times 3 \times 5 \times 4 = 720$ different combinations of all the gene segments. Similarly, in the light chain, there are $16 \times 2 \times 6$ different combinations possible, or 192 different light chains. The final probability is the product of the chance that each of the light chains will combine with each of the heavy chains, or $192 \times 720 = 138{,}240$ different possible antibody proteins.

SECTION REVIEW PROBLEMS

13. What flagellin protein(s) would be synthesized in a *Salmonella* strain that had a deletion in the repressor gene controlling the synthesis of the H1 protein? Explain.
14. What is the primary advantage of a genome rearrangement, as compared to transcriptional regulation, when the rearrangement changes the expression of a gene?
15. Define the following terms: antigen, antibody, monoclonal antibody, and polyclonal antibody.
16. How can several hundred different genes produce over 100 million different antibody proteins? Explain.
17. Why do most of the variations in antibody protein sequence occur in the variable region?

Summary

Because cells express only a small fraction of their total gene complement and require some gene products only for specialized functions, many different mechanisms regulate the expression of genes. The most common and energy-efficient expression regulating mechanism is transcriptional regulation. Transcriptional regulation can be either positive or negative; that is, *trans*-acting proteins can turn a gene either on or off by binding to a controlling region on the DNA. An excellent example of negative regulation is the lactose operon. An operon is a cluster of genes regulated by a single controlling site (promoter and operator). The product of an operon is a single polycistronic mRNA whose synthesis is inducible or repressible. During the induction (or repression) process, the inducer (or corepressor) binds to the repressor protein, preventing (or allowing) the repressor to bind to the operator site. This, in turn, permits the RNA polymerase to bind to the promoter, and transcription begins. The key to understanding this complex system of gene regulation in the lactose operon was through the analysis of many different mutants affecting different sites in the regulatory circuit of the lactose operon. Repressible operons provide a second level of gene control through an additional control site in the leader region of the gene, known as the attenuator site. Attenuator regions regulate transcription of the gene-coding region in response to the concentration of the final product of the pathway.

Many features of prokaryotic gene regulation are also present in eukaryotes. The galactose regulatory circuit in yeast is an excellent model organism for studying eukaryotic gene expression. In this regulatory circuit, the GAL4 and the GAL80 proteins regulate the downstream structural genes by binding to an upstream activator sequence. In higher eukaryotes, the genes responsible for the synthesis of steroid hormones are regulated in a similar manner. In both prokaryotes and eukaryotes, *trans*-acting regulatory proteins binding to DNA have several common features, including zinc finger motifs and helix-turn-helix motifs. These protein motifs are geometric protrusions that recognize specific DNA sequences, allowing the regulatory proteins to bind to key sites turning gene transcription on or off.

Gene expression can also be regulated after transcription, and an important mechanism is alternative splicing. Alternative splicing allows a single gene to produce several different proteins by selectively splicing out specific exons while retaining others. For example, in one tissue (neuron), one form of the protein contains all the exons in the coding region, whereas in another tissue (skin cell), a smaller protein with a slightly different function is created by splicing out one exon.

Other mechanisms of posttranscriptional regulation include RNA editing and mRNA instability. RNA editing changes the original RNA message by adding and deleting U nucleotides from selected regions of the transcript. The stability of mRNA is also an important mechanism for posttranscriptional regulation of gene expression. Sequences located at the 3′ end, 5′ end, and internal to the mRNA, determine its stability. A more stable mRNA is translated into more proteins, and unstable mRNAs produce only a few proteins before being degraded by ribonucleases.

Some forms of gene regulation are inherited and involve actual rearrangement of the DNA in the genome. For example, *Salmonella* can alter the DNA sequence coding for its flagellin protein, allowing the bacterium to evade a host's immune response by periodically altering the flagellin protein structure. This change involves inverting a section of DNA that controls the synthesis of two different flagellin structural genes. Trypanosomes also evade hosts' immune responses by altering the molecules of their surface coat. The trypanosome has a wide variety of surface protein genes. It successfully evades the host's immune response by randomly changing its surface coat protein through DNA rearrangements. The trypanosomes that have new surface coats survive, and those with old varieties are killed by the immune response. This cycle repeats itself hundreds of times, until the host finally succumbs.

The vertebrate immune system can generate millions of different antibodies, which bind to the antigenic surfaces of invading organisms from a relatively small cluster of DNA sequence cassettes, each coding for different regions of the antibody molecule. The antibody molecule is composed of a constant region and a variable region. The variable regions of antibody proteins possess tremendous diversity to act as the antigen binding site. Antibody molecules are synthesized by lymphocytes, and different lymphocytes generate different antibodies. This variation results from differentiation and rearrangement of the lymphocyte genome. Using this mechanism, an organism can generate millions of different antibodies from a relatively small number of genes.

Selected Key Terms

allosteric transition
alternative splicing
antibody
antigen
attenuation
catabolite repression
clonal selection
corepressor
inducer
operator site
operon
polycistronic mRNA
RNA editing
transcription factors
upstream activator sequence

Chapter Review Problems

1. The salivary gland and the liver of the mouse both synthesize α-amylase and both are dictated by the same gene. However, in the salivary gland, 2% of the mRNA codes for α-amylase, whereas in the liver only 0.02% of the mRNA codes for α-amylase. The two mRNAs differ by 48 nucleotides at the 5′ end; the salivary mRNA possesses additional nucleotides. How might you explain these observations?
2. In each of the following genotypes of *E. coli,* predict whether β-galactosidase will be constitutive, inducible, or absent.
   a. $lacI^+$, $lacP^+$, $lacO^+$, $lacZ^-$
   b. $lacI^-$, $lacP^+$, $lacO^+$, $lacZ^+$
   c. $lacI^+$, $lacP^+$, $lacO^c$, $lacZ^+$
   d. $lacI^+$, $lacP^+$, $lacO^+$, $lacZ^-$ / $lacI^-$, $lacP^+$, $lacO^+$, $lacZ^+$
   e. $lacI^+$, $lacP^+$, $lacO^+$, $lacZ^-$/ $lacI^s$, $lacP^+$, $lacO^+$, $lacZ^+$
3. a. How does an attenuator differ from a terminator? What do repressors and steroid receptor proteins have in common?
   b. What do promoters, CRP binding sites, and operators all have in common?
4. Predict the effect on induction of β-galactosidase when an *E. coli* strain is grown in glucose if
   a. the synthesis of cAMP were eliminated by a deletion mutation.
   b. the breakdown of cAMP were eliminated by a deletion mutation.
5. Does it make a difference if the structural gene for the repressor protein, which controls an operon, is located adjacent to the operon, or can this gene be anywhere in the genome? Explain.
6. Would you predict that the regulation of the enzymes responsible for the degradation of glucose (the glycolytic pathway) are regulated, constitutive, or induced? Explain.
7. A mutant was found in *E. coli* that produces β-galactosidase whether lactose is present or not. Predict the genotype(s) of this mutant.
8. A mutant in yeast was found by deleting the first 70 amino acids of the GAL4 protein (the GAL4 protein controls the *gal* genes, which are responsible for the metabolism of galactose). The remainder of the protein was normal. What is the phenotype of this mutant with respect to the induction of the galactose metabolic enzymes?
9. It is estimated that mammals have about 80,000 different genes, but the immune system is capable of synthesizing considerably more than 100,000 different antibody proteins. How did this information influence thinking when the clonal selection hypothesis was formulated?
10. What is a *trans*-acting protein? Distinguish this from a *cis*-acting sequence.
11. You have been studying the system controlling the synthesis of a group of proteins important for mitosis in a newly discovered organism. You add 5-azacytidine to the cells and allow them to incorporate the 5-azacytidine into their DNA. Much to your surprise, you find mitosis no longer functions as you expect. What mechanism of gene regulation might you suspect is involved in controlling mitosis? Explain.
12. Explain the difference between alternative splicing and RNA editing.
13. You have been studying the stability of an mRNA made in the leaf cells of maize and find that the mRNA is very unstable; it has a half-life of only 15 minutes. You have cloned this gene and want to determine whether the instability of the mRNA is determined by the 5′ end of the gene or the 3′ end of the gene. How would you design an experiment to answer this question?
14. In gene regulation, a *cis*-acting element differs from a *trans*-acting factor in that
    a. the *trans* factor must be on the same DNA molecule.
    b. the *cis* element must be on the same DNA molecule.
    c. the *trans* element must not be on the same DNA molecule.
    d. the *cis* element must not be on the same DNA molecule.

e. the *cis* element may be on the same DNA molecule or on a different DNA molecule.

15. In the following partial diploid, two copies of the *lactose* operon are present. Assume that no glucose is present and indicate whether functional β-galactosidase is produced when lactose is present or absent.

genotype: $I^-P^+O^+Z^+/I^+P^+O^cZ^+$

	Lactose	No Lactose
a.	(+)	(−)
b.	(+)	(+)
c.	(−)	(−)
d.	(−)	(+)

16. In the following partial diploid, two copies of the *lac* operon are present. Assume that no glucose is present and indicate whether functional β-galactosidase is produced when lactose is present or absent.

genotype: $I^-P^+O^+Z^+/I^+P^+O^cZ^-$

	Lactose	No Lactose
a.	(+)	(−)
b.	(+)	(+)
c.	(−)	(−)
d.	(−)	(+)

17. In the following partial diploid, two copies of the *lac* operon are present. Assume that no glucose is present and indicate whether functional β-galactosidase is produced when lactose is present or absent.

genotype: $I^sP^-O^+Z^+ / I^+P^+O^+Z^+$

	Lactose	No Lactose
a.	(+)	(−)
b.	(+)	(+)
c.	(−)	(−)
d.	(−)	(+)

18. In the control of the wild-type *lactose* operon, if glucose is absent (assume that lactose is present),
   a. cAMP is high and transcription will occur.
   b. cAMP is low and transcription will occur.
   c. cAMP is high and transcription will not occur.
   d. cAMP is low and transcription will not occur.
   e. if lactose is present the operon will be off.

19. Which would not be a constitutively expressed gene(s) in *E. coli* growing in minimal medium with glucose as a carbon source?
   a. β-galactosidase
   b. rRNA
   c. tRNA
   d. amino acid synthetic enzymes
   e. nucleotide synthetic enzymes

20. What would be the effect of deleting the gene for cAMP binding protein?
   a. Glucose could not be metabolized.
   b. The genes of the *lac* operon could not be turned on.
   c. The genes of the *lac* operon would always be turned on.
   d. cAMP could not be made.
   e. There is no effect on the *lac* operon transcription.

21. Which of the following is not used in the regulation of eukaryotic gene expression?
   a. inactivating or altering DNA
   b. controlling transcription with specific transcription factors
   c. splicing precursor mRNAs differently
   d. operons with polycistronic mRNA
   e. All may be used.

22. In eukaryotic mRNA transcription, regulatory proteins
   a. recognize specific DNA sequences.
   b. bind at sites upstream from the gene.
   c. may be gene-specific.
   d. may be cell-specific.
   e. All of the above.

23. Why is it that $lacO^c$ mutations affect only adjacent cistron expression in the *lac* operon and do not affect gene expression of cistrons in a *trans* configuration (using partial diploids)?

24. Explain why a $lacI^+$ gene is dominant over a $lacI^-$ mutation when they are present as a partial diploid.

25. It was found that bacterial population A, which possessed a catabolite repression system, could successfully outcompete a second bacterial cell population, B, which did not possess a mechanism for catabolite repression. Population A completely overgrew population B when the two were mixed and grown on glucose as a carbon source. Give a molecular explanation for this observation.

26. Explain how a mutation in the CRP protein (crp^-) would affect the expression of the *lac* operon.

27. The *his*, *phe*, *thr*, and *leu* operons are primarily regulated by an attenuator mechanism quite similar to that described for the *trp* operon. In each case, based on what you know about the *trp* operon, predict the tandem codons present in the leader sequence that are important for attenuation to occur.

28. What phenotype would you expect of phage lambda or the host carrying the following mutations? Specifically, can lambda carry out a lysogenic or a lytic cycle?
   a. A deletion in *cI*
   b. A deletion in *Q*
   c. A deletion in *N*
   d. A mutation in *recA*

29. You have been working with a bacterial strain, *Rhizobium*, that forms a symbiotic relationship with legume plants and is able to convert nitrogen from N_2 to a usable form for the plant. You know that during this symbiosis, the bacterium expresses a large number of genes that are not expressed in the absence of the plant symbiotic partner. You suspect that changes in the methylation pattern of the genome may be turning on large groups of genes and turning off other genes, explaining the difference in gene expression. What simple experiment could you do to test this hypothesis?

30. A mouse has 300 different light-chain *V*-gene segments, 5 different light-chain *C*-gene segments, 1000 different heavy-chain *V*-gene segments, and 8 different heavy-chain *C*-gene segments. Ignoring the variation that might occur from the *J*-gene segment or the *D*-gene segment, how many different antibody molecules can be produced from these segments?

Challenge Problems

1. Two mutations are found in the Z gene of the lactose operon: one of these mutations (*Z1*) abolishes the production of β-galactosidase but has no effect on the production of either permease or transacetylase. Both are fully induced in the presence of lactose (and the absence of glucose). The other mutation (*Z2*) maps at nearly the same location in the Z gene and also abolishes the production of β-galactosidase. However, the *Z2* mutation also reduces the production of permease and transacetylase to about 30% of their fully induced levels in the presence of lactose. Provide a molecular explanation for this observation, including how the two mutants differ.
2. From what you know about the GAL4 protein, predict the phenotype of the four different types of mutations possible in the *GAL4* gene.
3. In an experiment, you mix a cloned gene, RNA polymerase, and transcription factor X along with ribonucleotides, and you obtain transcription. However, if histones are present in the mixture, no transcription occurs. But, if you incubate the transcription factor with the DNA, and then add the histones and polymerase, transcription occurs. What do you conclude from this experiment?

Suggested Readings

Alberts, B., D. Bray, J. Lewis, M. Raff, K. Roberts, and J. Watson. 1989. *Molecular Biology of the Cell,* 2nd. ed. Garland Publishing, London and New York.

Greenblatt, J. 1991. "Roles of TFIID in transcriptional initiation by RNA polymerase II." *Cell,* 66: 1067–1070.

Holliday, R. 1989. "A different kind of inheritance" (discusses methylation of DNA during development). *Scientific American,* 260 (June): 60–68.

Jacob, F., and J. Monod. 1961. "Genetic regulatory mechanisms in the synthesis of proteins." *Journal of Molecular Biology,* 3: 318–356.

Johnson, P. F., and S. L. McKnight. 1989. "Eukaryotic transcriptional regulatory proteins." *Annual Review of Biochemistry,* 58: 799–839.

Lewin, B. 1997. *Genes VI.* Cell Press, Cambridge, MA.

Ptashne, M. 1989. "How gene activators work." *Scientific American,* 260 (January): 41–50.

Rhodes, D., and A. Klug. 1993. "Zinc fingers." *Scientific American,* 268 (February): 56–62.

Rowen, L., B. F. Koop, and L. Hood. 1996. "The complete 685-Kilobase DNA sequence of the human β T-cell receptor locus." *Science,* 272: 1755–1762.

Singer, M., and P. Berg. *Genes and Genomes.* 1991. University Science Books, Mill Valley, CA.

Smith, W. W. J., J. G. Patton, and B. Nadal-Ginard. 1989. "Alternative splicing in the control of gene expression." *Annual Review of Genetics,* 23: 527–577.

Tjian, R. 1995. "Molecular machines that control genes." *Scientific American,* 274 (February): 54–61.

On the Web

Visit our Web site at **http://www.saunderscollege.com/lifesci/titles.html** and click on A/G/M Genetics for links to the following chapter-related resources on the World Wide Web:

1. **Transcription factors.** This site provides links to databases describing *trans*-acting transcription factors and *cis*-acting sites in eukaryotes.
2. **The lactose operon.** This site is maintained by MIT and shows the control of gene expression and mutants of the lactose operon.
3. **The tryptophan, histidine, and lambda operons.** A discription of these operons and their control is graphically presented.

CHAPTER

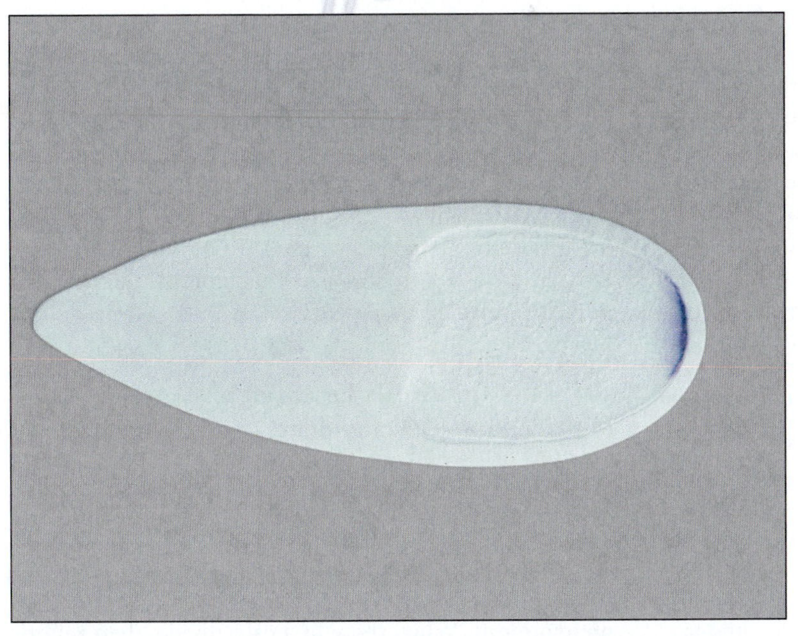

Oskar mRNA in a *Drosophila* oocyte. This *Drosophila* oocyte has been stained to show the posterior localization of *oskar* mRNA (blue). The localization of maternal mRNAs are a key step in the establishment of the embryo body pattern. *(Gwendolyn Vesenka, Iowa State University)*

Developmental Genetics: Genetic Regulation of Cell Fate

How genes are turned on and off to control the development of multicellular organisms is a very large and complex subject that includes aspects of developmental biology, embryology, genetics, and molecular biology. It is impossible to cover all this subject in two chapters of one book, and we will not attempt to make a comprehensive survey. Instead, we will concentrate on a few fundamental principles of development and how these are regulated by genes. The study of the genes that regulate developmental processes is called **developmental genetics.** Today, this is one of the most exciting, fastest-growing areas of genetics, mostly because of the application of genetic and molecular genetic techniques in a few model organisms. These studies have uncovered the genetic basis for fundamental developmental principles that underlie the development of many, if not all species, including humans.

Modern studies of development began during the 1800s and early 1900s, as biology was changing from an observational science into an experimental science (Chapter 1). Numerous elegant experiments were performed. In these experiments, manipulation of developing organisms revealed that all multicellular organisms begin life as a single cell that undergoes repeated mitotic cell divisions to generate the millions of cells in the body. As the body develops, a key process is **pattern formation,** that is, how the different types of cells are produced in the precise spatial patterns needed to form the tissues and organs characteristic of that species. Somehow, in most individuals all the necessary cell types are produced at the right time and in the right place. This intricate, marvelously complex process takes place with little or no external guidance and is remarkably accurate, successfully producing millions of normal individuals, generation after generation. The developmental process is clearly controlled by genes because parents pass the characteristics of their species to their offspring (cats always give rise to cats, and flies, to flies).

The goal of most developmental genetic investigations is to identify the genes that regulate a particular developmental process and to discover how these genes control development. During the past 20 years, many important advances have been made in our understanding of the identity of the genes that control development and of how these genes function. Many of these advances have been made in studies of model organisms, particularly *Drosophila melanogaster* and *Caenorhabditis elegans*, using the sophisticated genetic and recombinant DNA analysis techniques that were discussed in Chapters 12 and 13. The success of these investigations has encouraged additional developmental genetic investigations of other species (especially mice and zebra fish) and plants (especially *maize* and *Arabidopsis*). These studies all take advantage of certain characteristics of each species that lend themselves to particular types of genetic and/or developmental investigations. The results of these studies clearly indicate that many of the genetic mechanisms controlling development are common to all species. Thus, the results of developmental genetic investigations on model organisms are directly applicable to understanding and treating human diseases and developmental defects.

In this chapter, we will discuss some of the fundamental principles of development and how they are studied. In addition, we will consider some examples of how genes control one of the most important of these processes: how an embryonic cell becomes committed to form a particular cell type. The examples used in this chapter and Chapter 16 were chosen in part because they are common to many species and in part because the identities and functions of at least some of the genes controlling them are known. These examples illustrate the fundamental principles of how genes control development in many different species, including humans. By the end of this chapter, we will have answered four specific questions:

1. How do embryonic cells become committed to one particular cell fate?
2. Do organisms use different genetic mechanisms to generate different cell types?
3. How do cells maintain their commitment to form one cell type throughout the life of the organism?
4. How do geneticists identify and investigate genes that regulate cell fate?

Cell Determination Is an Essential Developmental Process.

By 1900, biologists understood that organisms begin as a fertilized egg and that all the cells in the individual are mitotic descendants of this first cell. As development proceeds, the mitotic progeny of the first cell form different specialized cell types and become organized into the tissues and organs of the body. Since 1900, armed with increasingly powerful microscopes and investigative techniques such as cell marking and cell transplantation, developmental biologists have undertaken numerous experiments trying to discover the details of how this process of development works. Most of their efforts have focused on trying to discover what controls the behavior of cells during development.

Cell-Fate Control Is Essential for Multicellular Development.

Perhaps the most important question about development is how cells adopt a particular cell fate (that is, how they choose which particular cell type to form). Because the first few embryonic cells can give rise to all the cell types in the body, they are referred to as **totipotent,** or **pleuripotent.** When these first cells divide, their mitotic progeny can still form all or many different cell types. However, as development proceeds step by step, cells' initial potential to form many different cell types is gradually restricted (Figure 15.1). Ultimately, cells are committed to forming to one particular cell type, called their **cell fate.** Developmental biologists refer to the steps in this process as "cell-fate decisions," as though cells consciously decide to abandon some of their potential cell types. This is simply a convenient way of describing this gradual, restrictive process. Throughout the development process, each cell-fate decision restricts the cell and its mitotic progeny to forming only one subset of its original potential choices.

Cell-fate decisions are often divided into one of two categories, depending on their stability. A cell-fate decision that gives a cell and all its progeny an irreversible, heritable developmental commitment is called a **cell determination.** A cell that has made a determination decision is called a determined cell. Determined cells hold tightly to their identity, even after severe experimental manipulations. The second class of cell-fate decisions are those that are potentially reversible. These are called **cell specifications.** For example, suppose that in early development a cell makes the cell-fate decision to be a nerve cell. If development proceeds normally, all the mitotic progeny of this cell will form only neural structures. However, if we experimentally intervene to alter development, we can test whether this cell's decision to be neural is potentially reversible. If the mitotic progeny of this cell remain neural even after they are transplanted into a nonneural tissue or are placed into tissue culture or are forced to undergo extra cell divisions, then we would conclude that the cell-fate decision to be neural was irreversible and that the decision to become neural was a determination decision. However, if experimental

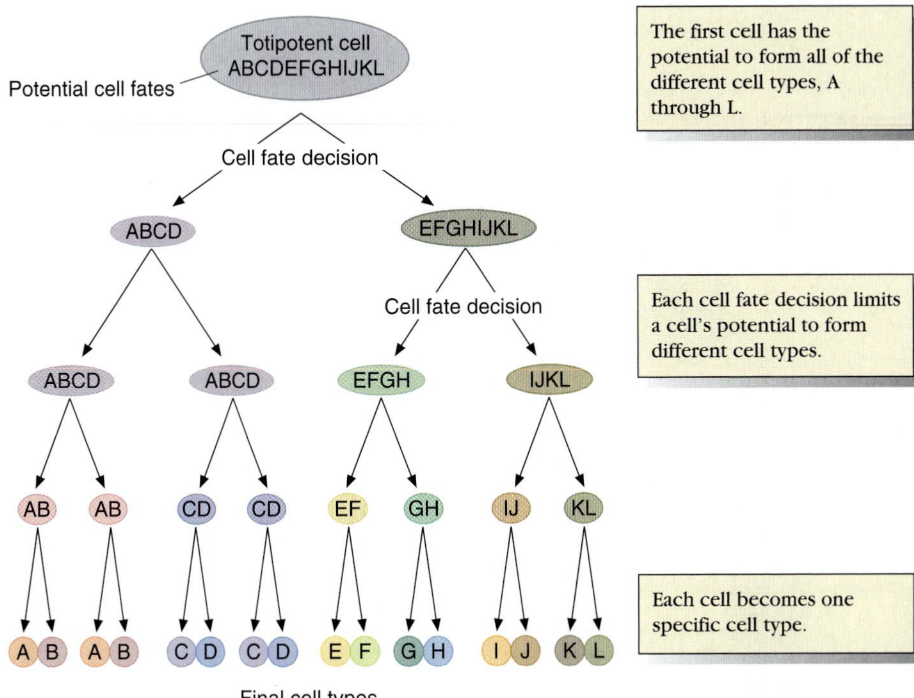

FIGURE 15.1 Cell fates in development. A totipotent cell can give rise to any of the different cell types in the mature organism. During development, cells make a series of cell-fate decisions, each of which ultimately leads cells to their final fate as specific cell types.

manipulation readily induces the cell to reverse or switch its fate and form nonneural structures, then we would conclude that the decision was a specification. Very few cell-fate decisions have been experimentally tested, and thus very few are unequivocally known to be determinations. Thus, for the purposes of this chapter, we will use the general term **cell-fate decision** for all decisions that have not specifically been shown to be determinations.

Because development is a stepwise process, a cell may be determined and still be pleuripotent. For example, a cell determined to be neural may give rise to a wide variety of different neural cell types because the commitment to form one specific nerve cell is made only after additional cell-fate decisions. As development progresses, cell-fate decisions restrict the mitotic descendants of that cell, so that they can form only some neural cell types. Later decisions further restrict the cells, until eventually each cell forms one particular cell type. Some of these cell-fate decisions may be determinations, and others may be potentially reversible.

Cells that have become committed to form one cell type eventually begin to produce the specialized structures that are characteristic of that cell type, a process called **cell differentiation.** For example, cells that become determined to be erythrocytes (red blood cells) produce large amounts of hemoglobin and do not undergo any further cell division. This type of cell is referred to as being **terminally differentiated,** and many terminally differentiated cells usually do not divide again and are eventually degraded. However, not all differentiations are terminal. Some cells differentiate into a cell type whose normal function includes the ability to divide and to give rise to other cells. For example, human blood contains stem cells that continue to divide and give rise to different types of blood cells throughout the life of the organism or in response to a specific stimulus from the environment. A more specific example is the hematopoietic stem cell, which gives rise to erythrocytes, platelet cells, and other cell types (Figure 15.2). In this chapter, we will use the term "cell differentiation" for the process whereby a cell expresses its cell fate by producing specialized products characteristic of that cell type.

In 1934, T. H. Morgan suggested that all cell-fate decisions are controlled by a sequence of differential gene actions. Morgan proposed that at each stage of development, specific sets of regulatory genes are active in each cell, and that these sets of genes define a cell's type and govern its behavior. He also proposed that the products of these genes interact with gene products from other cells, and that these interactions can feed back to the cell nucleus to turn some genes off and others on. Interactions of this sort generate a new set of active genes and give the cell a new type. Morgan's concept of sequential gene action and interaction is now widely accepted. Today, determining how cell-fate decisions are regulated is largely a question of identifying which specific set of regulatory genes controls a particular cell fate and how this set of developmental regulatory genes is activated. In this chapter, we will discuss some examples of genes and genetic mechanisms that control cell-fate decisions.

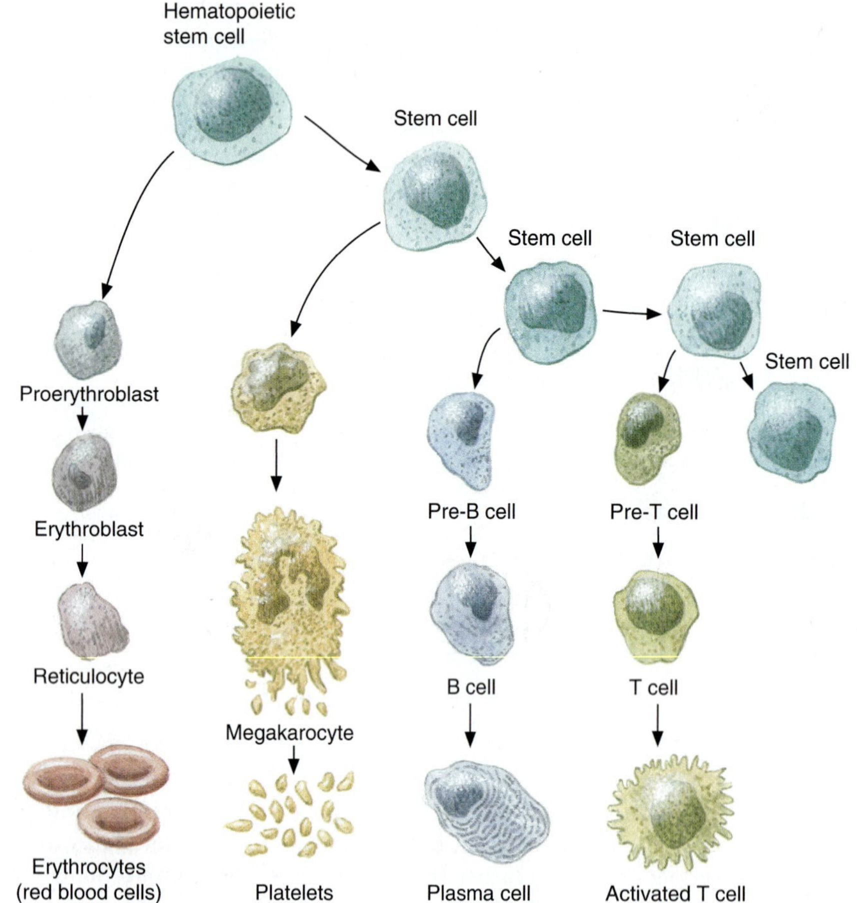

FIGURE 15.2 **Blood cell types.** A hematopoietic stem cell is a differentiated cell that undergoes additional mitotic divisions, cell-fate decisions, and cell differentiation to produce a variety of blood cell types. However, the stem cell itself remains determined, continually producing new cell types.

The potential importance of understanding the genetic mechanisms that control cell-fate decisions is enormous, especially for human medicine. Humans have a very limited ability to regenerate tissues and organs that have been damaged by accident or disease. However, the genetic program that directed cells to form these tissues and organs is still present in the cells of the individual. If these genetic programs could be reactivated, some cells could regenerate damaged tissues or organs. An enormous step forward in this area was made in 1997 in Scotland by a research team working with sheep. They were able to induce the nucleus of one somatic cell to forget all its previous cell-fate decisions, return to a totipotent state, and undergo the ultimate regeneration to form a complete individual (*CURRENT INVESTIGATIONS: A Sheep Called Dolly*). This work dramatically proved that at least some differentiated somatic cells contain all the genetic information needed for development, and that under certain experimental conditions, differentiated cells can be induced to replay the entire developmental program.

Mosaic and Regulative Development Are Two General Mechanisms for Cell Determination.

The mechanisms that species use to control cell-fate decisions can be divided into two general systems. One system uses **cytoplasmic determinants** to assign a cell its particular fate. A cytoplasmic determinant is a molecule that, when present in the cytoplasm, causes the cell to adopt a particular cell fate. For example, some cell-fate decisions are controlled by cytoplasmic determinants (mRNA or proteins) packaged in the egg cytoplasm by the genes of the mother and distributed to the blastoderm cells during the first embryonic cell divisions (Figure 15.3). The renowned German developmental biologist Wilhelm Roux called this system of cell determination **mosaic development.** Embryos that use this system of development are mosaics of cells, and each cell contains its own unique set of determinants.

The second system for controlling cell determination uses signals passed between cells to establish cell

Current Investigations

A Sheep Called Dolly

On February 27, 1997, a research team headed by Dr. Ian Wilmut announced in a paper published by the journal *Nature* that they had successfully produced a viable sheep by inserting the nucleus of a somatic cell from a mature adult sheep into an unfertilized egg. This lamb was the first mammal to develop from a cell derived from an adult tissue. The lamb was named "Dolly" in honor of the American singer Dolly Parton. This work represented only a portion of Wilmut's experiments investigating whether differentiated cells can be induced to change their fate. Their success demonstrated that at least some somatic cells do indeed retain all the genetic material needed to control the development of an individual, and also that this developmental program can be restarted under the right cytological conditions.

Wilmut's procedures were not complicated. He isolated the tissues of a 9-day-old embryo, a 26-day-old fetus, and mammary gland cells from a 6-year-old old ewe. These tissues were then dissociated into single cells and placed in cell culture. After the cells began to divide, the culture medium was changed by reducing the concentrations of the growth factors necessary for division. This reduction caused the cells to cease dividing and enter a state of arrested development called the G_0 state (see Chapter 3). Individual G_0 cells were then fused with enucleated oocytes (their nucleus removed) that had been isolated from normal sheep. The fusion produced what were called reconstructed embryos, which sometimes began to divide in culture. Embryos that reached the 8- or 16-cell stage were transferred into ewes that served as surrogate mothers. Viable lambs were obtained from reconstructed embryos begun with all three cell types, although the success rate was low. Dolly developed from an embryo whose nucleus came from an adult mammary gland cell, and she was the only success in 277 attempts. Although this process is referred to as "cloning" in the popular press, the offspring are not, strictly speaking, clones of the donor individuals. Their nuclear genome is identical to that of the nuclear donor, but their mitochondrial genotype is the same as the oocyte donor.

Wilmut suggested that he was successful because he used G_0-stage cells. Using cells in other stages of the cell cycle as nuclear donors may lead to an irreversible conflict of regulatory signals between the oocyte cytoplasm and the donor nucleus. It is easy to understand how such a conflict could arise. The nucleus of a normal dividing cell contains a set of active regulatory genes that control gene activity in the nucleus and that maintain the determined state of the cell. Contradictory signals from the oocyte cytoplasm would have to first overwhelm the existing regulatory gene activity before it could activate the genes essential for initiating embryonic development. Cells in G_0, however, are quiescent with little gene activity, and many, if not all, of the regulatory genes that normally govern the cell's determined state are repressed. For this reason, the nucleus may be more susceptible to reprogramming by the cytological signals in the oocyte. The G_0 state may, in fact, represent a nuclear state similar to the normal state of the nucleus during the first few embryonic cell divisions when few, if any, of the embryonic genes are active.

Wilmut's success raises a number of important bioethical questions. Given the similarity between development in humans and in sheep, it is likely that this same process would work with human cells. An active debate about whether this process should be used to produce human embryos has begun. Those in favor suggest that this technology may, in certain cases, be the only way to help some childless couples have children who have a biological relationship to at least one parent. Others argue that further research on this technique may lead to selective technologies that would induce differentiated cells to form stem cells that could produce replacements for diseased or damaged organs without the problems of tissue rejection that currently hinder organ transplantation between unrelated individuals. Those not in favor argue that this process eliminates essential aspects of human dignity. Society has rejected the principle of slavery, that one human being can own another and can treat that individual as property with no human rights. They argue that while the reconstructed embryo should be thought of as an individual in its own right, such an individual could be thought of as less than human, or might be created simply to serve as a reservoir of spare parts for another individual. Others reject the procedure on religious grounds. The United States has enacted a ban on using federal government funds for either research into using this process to produce humans or research to actually attempt to do it. This ban does not, however, apply to further research into using this process on animals.

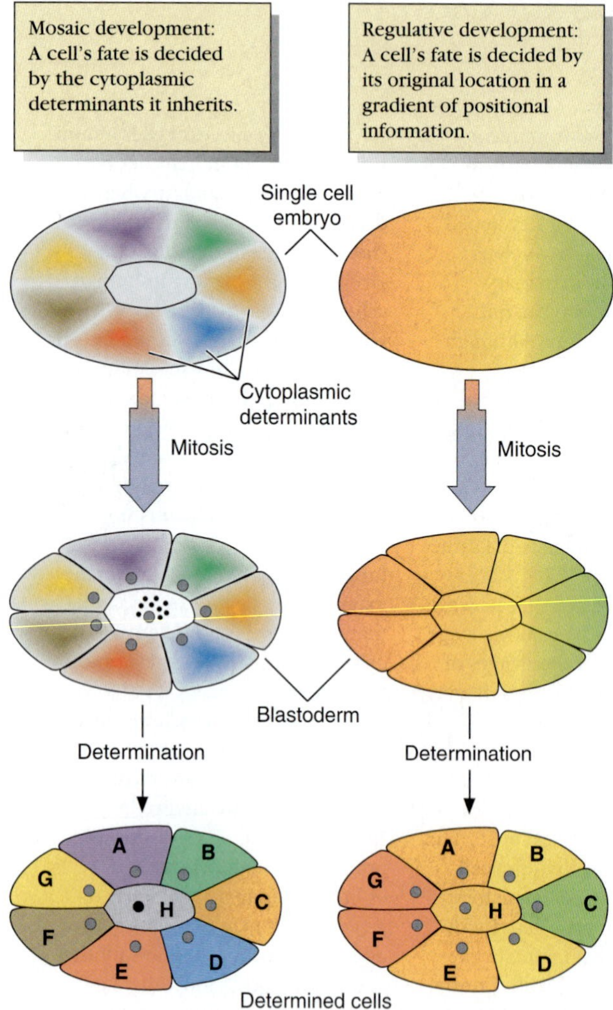
(a)
(b)
Mosaic development: A cell's fate is decided by the cytoplasmic determinants it inherits.
Regulative development: A cell's fate is decided by its original location in a gradient of positional information.
Single cell embryo
Cytoplasmic determinants
Mitosis
Mitosis
Blastoderm
Determination
Determination
A
B
C
D
E
F
G
H
Determined cells

FIGURE 15.3 Mosaic and regulative development. (a) In an embryo with mosaic development, initial cell-fate decisions are controlled by cytoplasmic determinants deposited in the cytoplasm of the egg by the mother. (b) In an embryo with regulative development, initial cell-fate decisions are controlled by a gradient of a signaling molecule. The gradient, or concentration, is a function of the cell's position in the embryo and decides its fate.

fate. This system is called **regulative development.** In 1968, Lewis Wolpert suggested that regulative development is controlled by **positional information.** A positional information system consists of a signal that informs a cell of its location or position in the developing embryo. This signal often takes the form of a "gradient" of gene products. A gradient is established by localized transcription of a gene, with a high concentration of gene product near the source and low concentrations farther away. The concentration of the gene product at any point indicates how far away a cell is from the source. Cells measure the concentration of the gene product; from this, they determine their location and then adopt cell fates appropriate for their position. For example, in an embryo with regulative development, the maternal genome may package proteins or mRNAs in the egg that generate gradients of positional information molecules in the embryo (Figure 15.3). Examples of both types of systems can often be found in the same organism.

Mosaic and regulative developmental mechanisms have fundamental differences that can be discovered by experimental testing. One important difference is the function of the molecules that control cell fate. A mosaic cytoplasmic determinant is specific for *only* one fate. In contrast, the positional information molecules of regulative development are not specific for any particular fate, and a positional information system may be used repeatedly during development to control several different cell fates. Experimental manipulations that change the concentration of positional information molecules in a cell will change its fate, but altering the concentration of a cytoplasmic determinant will not. Another difference between these two systems is that in mosaic development a cell's fate is decided without reference to its neighbors. In regulative development, however, communication between cells is an essential part of the system. In the next two sections, we will review some of the events of embryonic development in two well-studied model organisms, *Drosophila melanogaster* and *Caenorhabditis elegans*.

SECTION REVIEW PROBLEMS

1. For each of the following terms, give a brief definition:
   a. cell determination
   b. cell differentiation
   c. terminal cell differentiation
   d. positional information
   e. cytoplasmic determinant
2. The flying snouters (*Otopteryx volitans*) have mosaic embryonic development. If cytoplasm is transplanted from the head region of one egg to the abdominal region of another, what result do you predict?

THE DEVELOPMENTAL GENETICS OF *DROSOPHILA*.

Drosophila melanogaster is one of the most popular organisms for developmental genetic studies. More than 70 genes that control cell determination during the first 3–4 hours of embryonic development have been identified in *Drosophila*. All these genes were first identified by isolating mutants whose aberrant phenotypes were the result of abnormal embryonic development. The study of developmental mutations has been essential to our current understanding of the genetic basis of *Drosophila* development. Before we discuss some of these mutations, we will review the events of *Drosophila* embryonic and adult development.

The *Drosophila* Life Cycle Has Five Stages.

The *Drosophila* life cycle is divided into five stages: oocyte, embryo, larva, pupa, and adult (Chapter 3). Egg development begins in the ovaries of *Drosophila* females with an **oogonial cell** that undergoes four mitotic divisions with incomplete cytokinesis (division of the cytoplasm). These divisions produce 16 cells that are connected by cytoplasmic channels (Figure 15.4). One of the two cells that has four connecting channels becomes the **oocyte,** and the other 15 cells develop into **nurse cells.** As the 16-cell cluster matures, the nurse cells produce proteins and RNAs that are transferred into the oocyte cytoplasm. By the end of oogenesis, the nurse cells have contributed their entire cytoplasmic contents to the oocyte and they degenerate. The oocyte and nurse cells become surrounded by a single layer of up to 1000 somatic cells termed **follicle cells.** The follicle cells collect yolk proteins (and other substances) from the blood and transfer them to the oocyte, especially to the outer region of the cytoplasm, termed the **cortical cytoplasm.** This is a region where important embryonic developmental events occur. Toward the end of oocyte development, the follicle cells form a tough **vitelline membrane** and a **chorion** (egg shell) that surround the oocyte. Even before fertilization, the egg has well-defined, visible exterior and interior structures (Figure 15.4).

The mature *Drosophila* egg is about 140 μm by 500 μm in size, about 90,000 times larger than a typical *Drosophila* cell. After the egg is fertilized and laid, the oocyte nucleus completes meiosis and the oocyte and sperm nuclei fuse. The new zygotic nucleus then begins mitotic divisions. The first nuclear mitotic divisions occur rapidly (every 7–10 minutes) and are not followed by cytokinesis. These divisions produce an embryo consisting of one very large cell containing multiple nuclei. A cell with multiple, independent nuclei is called a **nuclear syncytium.** After the ninth mitotic division, most of the zygotic nuclei migrate outward, reaching the cortical cytoplasm about 2.5 hours after egg laying. The embryo is now in the **syncytial blastoderm** stage (Figure 15.5). The nuclei that reach the posterior end of the egg immediately form the first separate embryonic cells, called **pole cells.** The pole cells give rise to the germ cells. The rest of the nuclei divide three more times and then, about 3 hours after egg laying, cell membranes form simultaneously around the cortical nuclei. The embryo is now in the **cellular blastoderm** stage (Figure 15.5). Within 15 minutes after cell formation, the inward migration (invagination) of cells, termed **gastrulation,** begins. Gastrulation generates the three primary cell types (mesoderm, endoderm, and ectoderm) that give rise to all the larval and adult structures. A complex series of cell movements and cell differentiation events generate the internal and external structures of the larval body from these three initial cell types by 20 hours after egg laying (at 25°C) and the larva hatches from the egg at about 24 hours after egg laying.

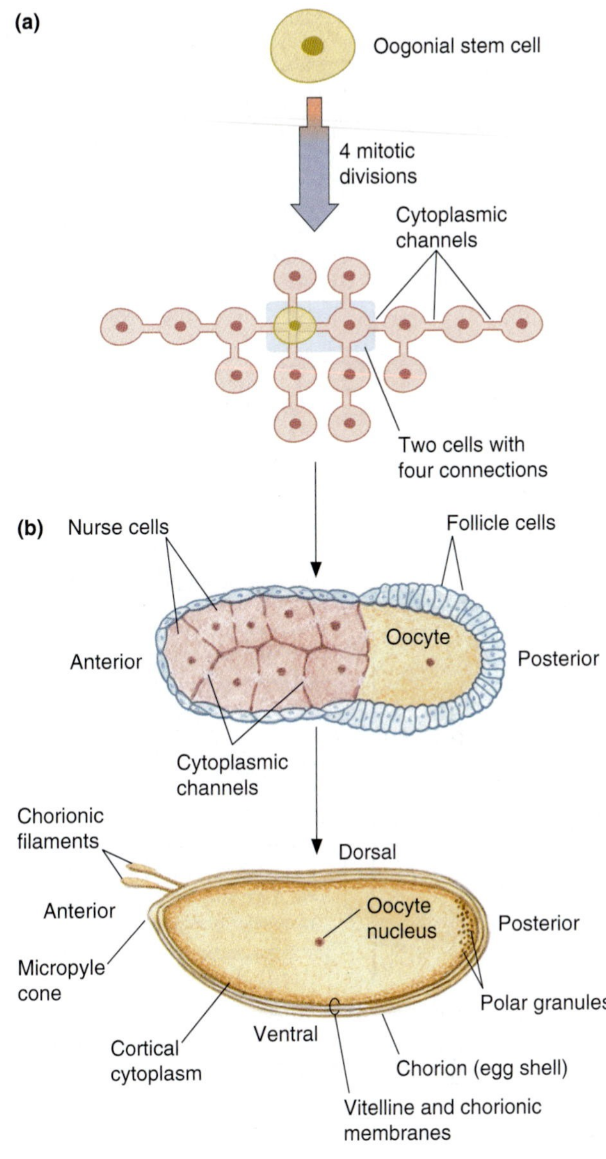

FIGURE 15.4 Oogenesis in *Drosophila.* (a) The oogonial stem cell undergoes four divisions with incomplete cytokinesis, producing 16 cells connected by cytoplasmic channels. One of the cells that has four channels becomes the oocyte, and the others become nurse cells. (b) The 16-cell cluster becomes surrounded by a single layer of somatic follicle cells. The nurse cells and follicle cells pass proteins and mRNAs to the oocyte. The mature eggs have anterior/posterior and dorsal/ventral axes, indicated by the location of the chorionic filaments (for gas exchange) and the micropylar cone (for sperm entry), which are on the dorsal surface.

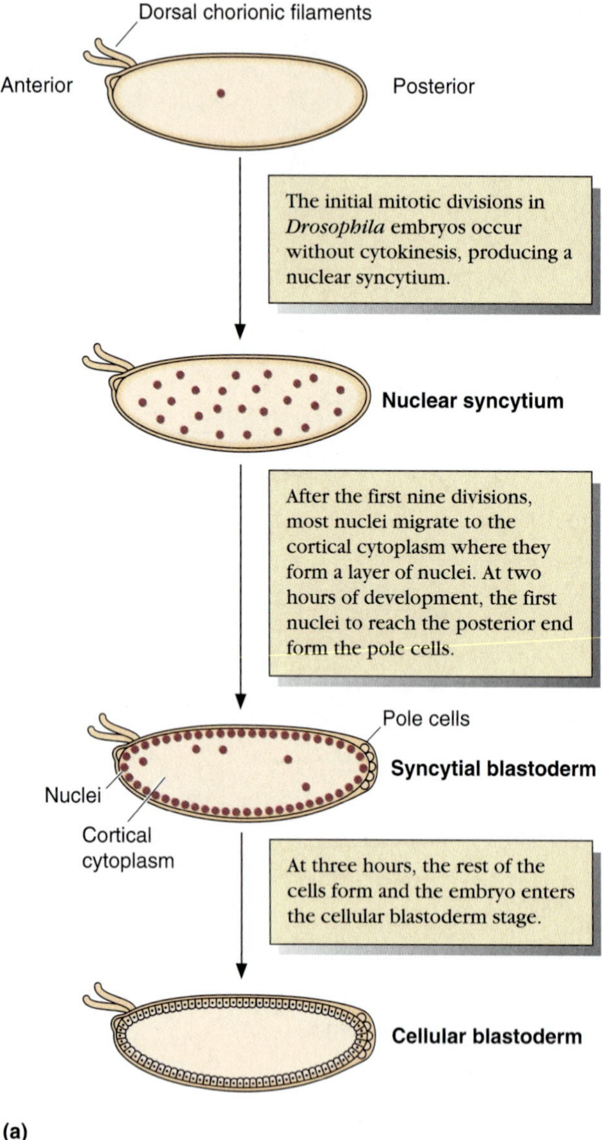
Dorsal chorionic filaments
Anterior
Posterior
The initial mitotic divisions in *Drosophila* embryos occur without cytokinesis, producing a nuclear syncytium.
Nuclear syncytium
After the first nine divisions, most nuclei migrate to the cortical cytoplasm where they form a layer of nuclei. At two hours of development, the first nuclei to reach the posterior end form the pole cells.
Pole cells
Syncytial blastoderm
Nuclei
Cortical cytoplasm
At three hours, the rest of the cells form and the embryo enters the cellular blastoderm stage.
Cellular blastoderm
(a)

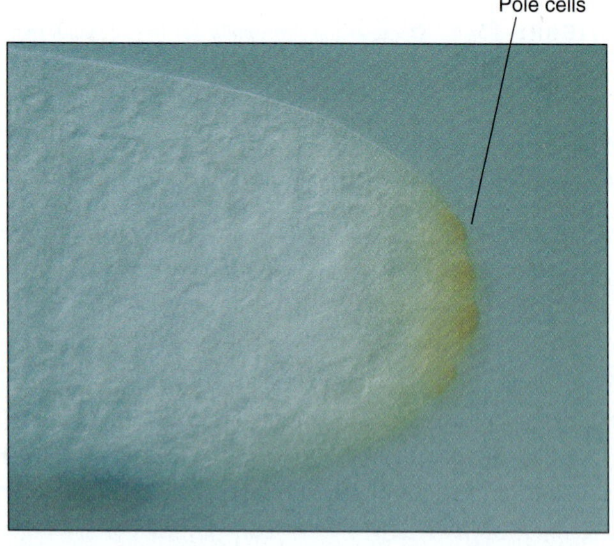
Pole cells
(b)

◄ **FIGURE 15.5** ***Drosophila* preblastoderm development.** (a) *Drosophila* embryos form a nuclear syncytium during the first 2 hours of development. During this time, the nuclei are exposed to a common cytoplasm, and regulatory molecules can pass freely from nucleus to nucleus. The nuclei become separated by cell membranes at 3 hours. (b) This photograph shows the pole cells at the posterior of an embryo at 2.5 hours. *(Gwendolyn Vesenka, Iowa State University)*

The cells in the developing *Drosophila* embryo divide into a series of segments and parasegments. Each segment is a band of cells that extends around the embryo and forms one specific set of structures. The boundaries of each segment first become visible during embryonic development (10 hours after egg laying) and are clearly visible in the larval and adult cuticle (Figure 15.6). The *Drosophila* embryo contains six head segments, three thoracic segments, and ten abdominal segments. Each segment is subdivided into an anterior compartment and a posterior compartment. Parasegments, which are discussed in more detail in Chapter 16, consist of the posterior compartment of one segment and the anterior compartment of the next. Parasegments are the action domains of the genes that regulate segment identity, and their relationship to the visible segments is an area of current investigation (Figure 15.6). The thoracic segments and the first eight abdominal segments form the visible external larval cuticle (which contains numerous cuticular structures such as hairs, sense organs, and cuticular wrinkles). The head segments and the last two abdominal segments form internal structures. Each segment is formed by a unique set of cells and has a unique arrangement of cuticular structures. These structures are often used to identify segments and to define the phenotypic effects of mutations. For example, if a mutated gene alters the cuticular structures only in the thoracic segments, the normal function of that gene must be critical for thoracic segment development.

The larva feeds and grows for about 4 days and then it **pupariates,** forming a tough outer case (called the puparium) within which it undergoes **metamorphosis.** During metamorphosis, most of the larval body degenerates, and the adult body is formed by special groups of adult precursor cells located throughout the body. The external cuticle is formed by cells in the **imaginal discs** (sometimes spelled "disks") and by groups of abdominal cells called histoblasts (Figure 15.7). The histoblast cells that form each abdominal segment are called "histoblast nests" because in some histological preparations they cluster together in a group that resembles a bird's nest. Imaginal discs are formed during embryonic development by small groups of 2 to 20 epidermal cells. Each group of disc cells forms in a characteristic location in the body and is determined to form one particular portion of the adult body (for example, a leg, wing, or eye).

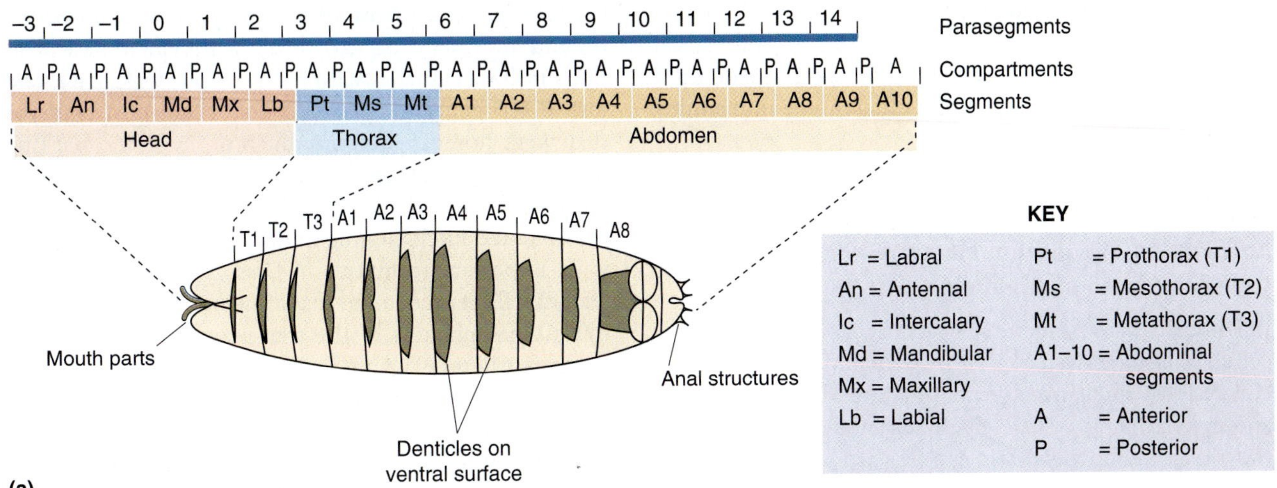

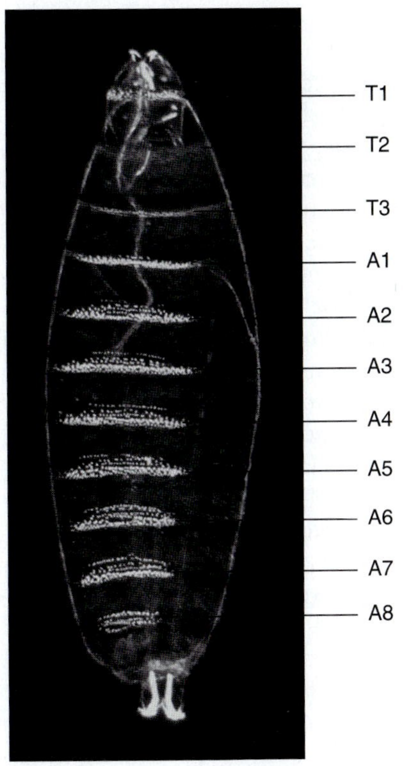

FIGURE 15.6 Segmentation of *Drosophila*. (a) Somatic cells of *Drosophila* divide into segments that become organized into compartments. In each compartment, a particular set of regulatory molecules controls cell determination. Segments may have one anterior (A) and one posterior (P) compartment, or they may subdivide into parasegments that each contain the A compartment of one segment and the P compartment of the next. (b) The external cuticle of the first instar larva is formed by the cells of three thoracic (T1, T2, and T3) and the first eight abdominal (A1, A2, and so forth) segments. The cuticle of each segment can be identified by its unique pattern of denticles and sense organs. *(Kori Radke, Iowa State University)*

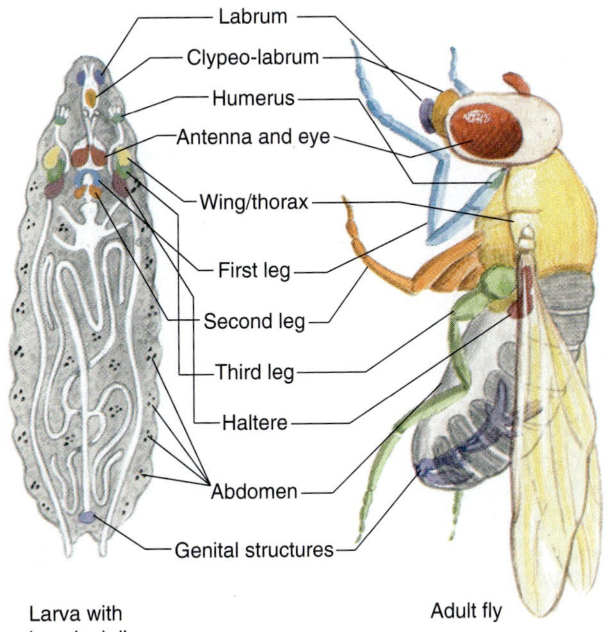

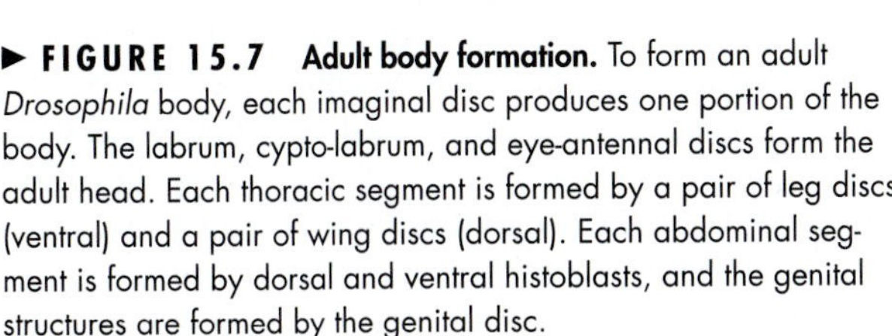

► **FIGURE 15.7 Adult body formation.** To form an adult *Drosophila* body, each imaginal disc produces one portion of the body. The labrum, cypto-labrum, and eye-antennal discs form the adult head. Each thoracic segment is formed by a pair of leg discs (ventral) and a pair of wing discs (dorsal). Each abdominal segment is formed by dorsal and ventral histoblasts, and the genital structures are formed by the genital disc.

Imaginal disc cells divide throughout the larval period, and by metamorphosis they may contain 40,000 cells. The hormones that trigger metamorphosis stimulate the disc cells to cease dividing and to differentiate into adult structures.

In *Drosophila* Embryos, Cell Determination and Pattern Formation Occur at the Cellular Blastoderm Stage.

In *Drosophila,* key somatic cell-fate decisions occur at the cellular blastoderm stage. The experiments that have discovered this are instructive not only for what they reveal about *Drosophila* cell determination but also because they illustrate how genetics can be used to study development. In 1968, Karl Illmensee transplanted syncytial blastoderm nuclei from donor embryos with an ebony (*e*) and multiple wing hair (*mwh*) genotype to new positions in host embryos with a yellow (*y*), white (*w*), and forked (*f*) genotype. He used donor and host embryos that were homozygous for different cuticular marking mutations so that he could observe the characteristic structures formed by the mitotic descendants of the transplanted nuclei. Using mutant alleles as cell markers is an important technique in *Drosophila* developmental analysis. Illmensee found that *e mwh* cells (formed by the mitotic descendants of the transplanted nuclei) always formed structures characteristic of their new location in the host and never of their original location in the donor (Figure 15.8). If these syncytial blastoderm-stage nuclei had already been determined, they would have formed structures characteristic of their original position. Thus, Illmensee's experiments indicated that embryonic nuclei are not determined at the syncytial blastoderm stage.

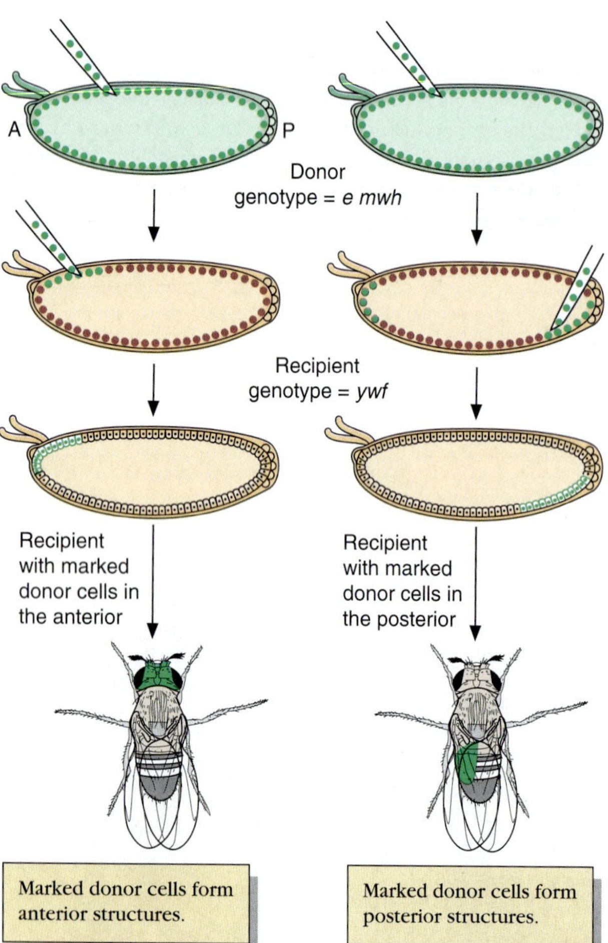

FIGURE 15.8 Estimating the stage of nuclear determination. Syncytial blastoderm-stage nuclei isolated from one region of a donor *e mwh* blastoderm are transplanted into different regions of host *y w f* blastoderms, where they eventually form cells at the cellular blastoderm stage. The resulting *e mwh* structures are characteristic of the *e mwh* nuclei's new location in the host, not their original location in the donor, indicating that embryonic nuclei are not determined in the syncytial blastoderm stage.

In 1971, Walter Gehring and his coworkers tested the determination state of cells in the cellular blastoderm stage. They isolated cells from anterior and posterior portions of cellular blastoderm-stage donor embryos, mixed them with cells from whole-host embryos, and then cultured the cell mixtures. Again, the donor and host strains had different cuticular-marking mutations. However, the mitotic descendants of cells from the anterior portions of the donor formed only anterior structures (head and anterior thorax), and the descendants of cells from posterior regions of the donor formed only posterior structures (posterior thorax and abdomen) (Figure 15.9). The commitment of these cells to form structures from their original location indicates that cellular blastoderm-stage cells are determined to form structures in particular regions of the body.

In 1976, Eric Wieshaus and Walter Gehring used induced somatic clones to investigate the timing of cell determination. If *Drosophila* heterozygous for recessive mutations are irradiated during the embryonic or larval stage, induced mitotic recombination will produce clones of homozygous somatic cells (this technique was discussed in Chapter 6). Wieshaus and Gehring irradiated embryos heterozygous for recessive alleles that gave easily observed mutant phenotypes. The treated individuals were allowed to complete adult development, and then the size, shape, and locations of the induced somatic clones were recorded. Some clones induced by irradiation at 3 hours after egg laying formed parts of both the adult leg and the wing. This meant that at 3 hours (cellular blastoderm stage), some cells were still capable of giving rise to two different imaginal discs (wing and leg). However, embryos irradiated at 7 hours after egg laying had clones that formed parts of the wing or leg, but never both. This indicates that cells become determined to form particular imaginal discs at, or a few hours after, cell blastoderm. From the size of the induced somatic cell clones, Gehring and others estimated the num-

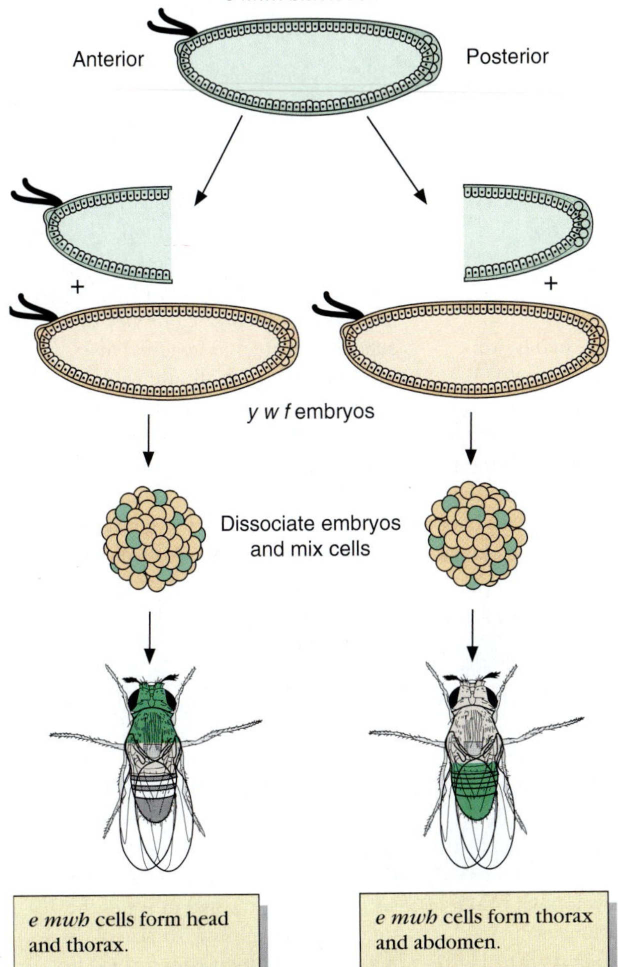

FIGURE 15.9 Cell determination at the cellular blastoderm stage. Anterior and posterior halves of *e mwh* embryos were dissociated into single cells, mixed with cells from dissociated whole *y w f* embryos, and transplanted into *y w f* larval hosts. During metamorphosis, the *e mwh* cells from the anterior half embryos formed only anterior structures (head and thorax), and the *e mwh* cells from posterior half embryos formed only posterior structures (thorax and abdomen).

ber of cells that originally formed each imaginal disc (see CURRENT INVESTIGATIONS: *Using Somatic Mosaics in Developmental Genetic Analysis* and Table 15.1).

Fate Mapping Reveals That Cell Determination Is Spatially Organized in the Blastoderm.

In *Drosophila,* blastoderm cells become determined in a detailed spatial pattern. This was first shown in an analysis of **gynandromorphs,** individuals that are part male and part female. If a female preblastoderm nucleus (XX) loses one of its X chromosomes, all its mitotic progeny cells will have only a single X chromosome (XO). These cells develop as male cells, so the individual is part male and part female (sex determination of *Drosophila* is discussed in Chapters 3 and 16). If the surviving X chromosome contains recessive cuticular marking mutations, then the male and female tissue can be distinguished (Figure 15.10). A. H. Sturtevant in 1929 noticed that gynandromorphs are about half male and half female, and that the male and female cells form discrete patches of tissue located in different positions in different individuals. Sturtevant concluded that the X chromosome was often lost during the first mitotic division in the embryo, and that the first division does not have a fixed orientation. This means that the two nuclei formed by the first mitotic division give rise to different regions of the blastoderm in each individual (Figure 15.10).

TABLE 15.1 *Estimates of the Numbers of Cells That Form* Drosophila *Imaginal Discs*

Disc Cell-Type	Cell Number
Wing	10–12
Haltere	6
Eye-antennal	11–14
Leg	10–20
Genital	2

Sturtevant noticed that structures far apart in the body, such as head and abdomen, are of opposite sex about half the time in gynandromorphs, but that structures close together, such as the wing and leg, are usually of the same sex. He suggested that the frequency with which two structures are of opposite sex in gynandromorphs reflects the physical distance between the precursor cells for those structures in the blastoderm. Sturtevant proposed that cells become determined at fixed positions in the cell blastoderm. This means that in gynandromorphs those positions that are far apart have a greater chance of being populated by migrating nuclei of a different sex than positions that are close together. Sturtevant used the frequency with which two structures are of the opposite sex in gynandromorphs as a measure of the distance between precursor cells. For example, in a sample of gynandromorphs the head and wing might be of different sexes in 49 out of 160 individuals, or 30.6% of the cases, whereas wing and the first abdominal segment might be of opposite sexes in only 29 of the 160 individuals, a frequency of 18.1%. Sturtevant suggested that the smaller fraction of difference between the wing and abdomen than between the head and wing indicates that the precursor cells that form the wing are physically closer to those that form the abdomen than to those that form the head. By measuring the sex of a large number of structures, Sturtevant made a blastoderm fate map he

CURRENT INVESTIGATIONS

Using Somatic Mosaics in Developmental Genetic Analysis

Somatic mosaic analysis is one of the most powerful of the genetic tools available for developmental genetic analysis. In *Drosophila,* induced mitotic recombination in a heterozygous cell produces two homozygous daughter cells that give rise to clones of homozygous cells. Cell marking is an old technique in developmental biology. It has long been used to discover the developmental potential of a cell by observing the range of structures formed by its mitotic descendants. Somatic genetic mosaics have several advantages for this type of study in *Drosophila,* especially for studies of imaginal disc development. A number of recessive mutations give cells a detectable phenotype without disrupting the developmental process. These mutations permanently mark cells and mark all descendants of the original cell. Induced mitotic recombination is a noninvasive technique for marking cells that does not require surgical disruption of the organism. Somatic clones can be induced at any desired stage of development, and large numbers of individuals can be treated, giving many clones for analysis.

Two types of developmental genetic analyses that use somatic cell clones are cell-lineage analysis and analysis of cell autonomy. In cell-lineage analyses, the range of cell types marked by clones and the location of clones in a tissue are used to define the developmental potential of cells at a given stage of development. Consider the hypothetical example of a tissue that contains four different cell types—A, B, C, and D. Cells with the capacity to form all four cell types (ABCD cells) become determined to form either A or B types (AB cells) or to form either C or D types (CD cells), and these then become determined to form A or B or C or D types. The pathway is thus:

$$\text{ABCD} \longrightarrow \text{AB vs. CD} \longrightarrow \text{A vs. B vs. C vs. D}$$

Clones induced early in development contain all four types (A, B, C, and D). This means that the initial marked cells of the clones were ABCD cells. Clones induced slightly later in development contain either A and B cells or C and D cells. This means that by the time these clones were induced, some of the ABCD cells have become AB cells, and others have become CD cells. Clones induced at a still later stage each contain only one cell type—A or B or C or D. This means that by the time these clones were induced the cells had become determined to form only one cell type—A, B, C, or D. The timing of the irradiation that induces the clones allows us to deduce when in development each of these determinations occur.

Cell-lineage analysis can also be used to estimate the number of cells present in a given tissue. The number of cells present in the population at the time when the clone is induced can be estimated by calculating the **average clone size.** At the time when the clone is first induced, it consists of a single marked cell, and the total population consists of n cells. The one marked cell is thus a certain fraction ($1/n$) of the total cell population. Assuming that the marked cell grows at the same rate as the other cells in the population, this fraction will not change during development, and the original cell number (n) can be estimated from the fraction of the total tissue occupied by the clone (Figure 15.A). For example, if there are 260 bristles in an average adult *Drosophila* leg, and the average number of bristles marked by clones induced during late embryonic development is 13 bristles, then the average clone size is $13/260 = 1/20$ of the total. There must, therefore, have been 20 cells in the leg disc late in embryonic development. Average clone size estimates indicate that imaginal discs are founded by small numbers of cells, and that the discs grow exponentially throughout the larval period (Table 15.1).

A second technique for calculating cell number is the **minimum patch-size** estimate. This technique uses the size of marked patches in gynandromorphs to estimate the number of cells that initially form the disc. When a disc is first formed, one, two, three, or more of the initial n founder cells might be marked male cells, and in these cases, the disc will be mosaic and have a patch of male cells. Those discs that have only one initial male cell will have the smallest-sized patch of male tissue in the adult, equal to $1/n$ of the total. By determining the size of the male patch in the tissues formed by a single disc in gynandromorphs, the minimum patch size can be determined. The fraction of the total marked by the minimum patch is assumed to be equal to $1/n$. The minimum patch size analysis may be used only when the clone of marked cells is initiated before the founder cell population arises. Minimum patch size estimates for the imaginal discs also indicate that discs are founded by small populations of cells (Table 15.1).

Cell-autonomy tests are another use of somatic mosaics. A mutation is considered to be autonomous if a single mutant cell has a mutant phenotype even when surrounded by wild-type cells. Information about autonomy has two uses. First, it can be used to make deductions about the nature of the action of the mutant gene product. For example, if a mutation blocks a signaling pathway, it might do so by

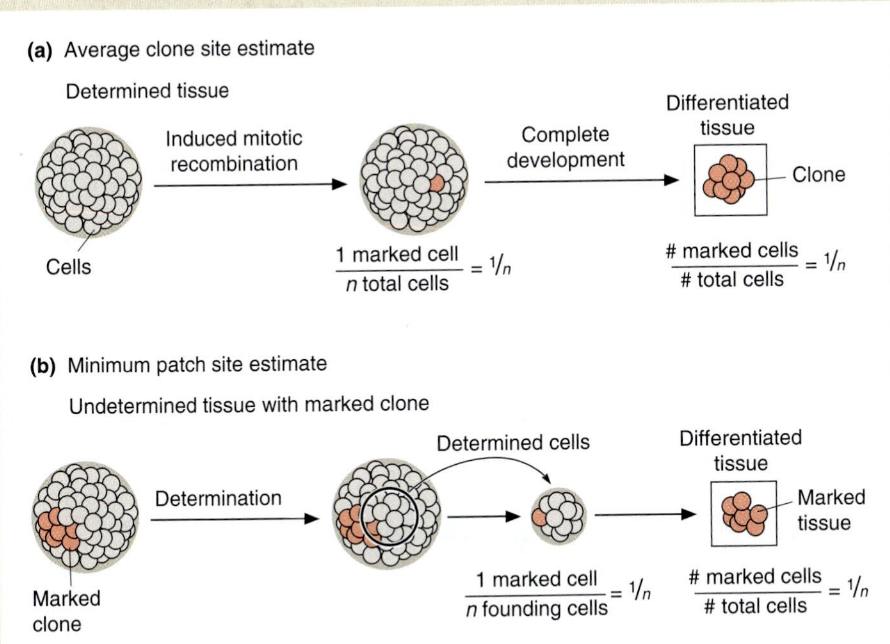

FIGURE 15.A Estimating cell number. Two techniques for estimating cell number both depend on measuring the size of the patch of marked tissue in somatic mosaics. (a) The average clone size estimate uses the average fraction of the tissue formed by marked clones that are induced after the tissue is determined. The inverse of this fraction is an estimate of the average cell number in the tissue at the time of clone formation. (b) The minimum patch size estimate uses the smallest fraction of the tissue marked by a clone that was induced before the tissue is determined. The inverse of this fraction is an estimate of the number of cells that originally form the tissue.

eliminating the signal or the signal receptor. If the mutation eliminates the receptor, the mutation will be autonomous because a cell without receptors cannot respond to normal levels of signal produced by neighboring wild-type cells. However, a mutation that blocks signal production will be nonautonomous because the mutant cell can respond normally to the signal produced by its wild-type neighbors.

A second use of cell autonomy tests is to determine when and where in development a gene is active. Genes that regulate development are active at particular tissues at particular developmental stages. If a clone of cells homozygous for a mutation is induced in a tissue in which the gene is active, then the clone will have a mutant phenotype because the cells of this tissue require the presence of the gene product to have a normal phenotype. If a clone of mutant cells is produced in a tissue in which the gene is normally not active, or in a tissue in which that particular gene has already carried out its function, then the cells of the clone will not have a mutant phenotype because these cells do not need an active copy of that gene to have a normal phenotype. By inducing clones of homozygous mutant cells at different times and in different tissues, the time and tissue in which the gene is active can be deduced. The tissue in which a gene functions is called the focus of the gene. Consider a recessive lethal mutation that is located on the X chromosome. Gynandromorphs are produced in which the surviving X chromosome carries the lethal mutation and also the marker mutation *yellow.* The focus of this mutation can be determined by examining a large number of surviving gynandromorphs and determining which tissues in these individuals are never formed by male (*yellow*) cells. For example, if the focus of action of this gene is in the head, then all gynandromorphs that have male heads will die, and all surviving gynandromorphs will have female heads. The absence of gynandromorphs with a male head indicates that the focus of action of this gene is in the head.

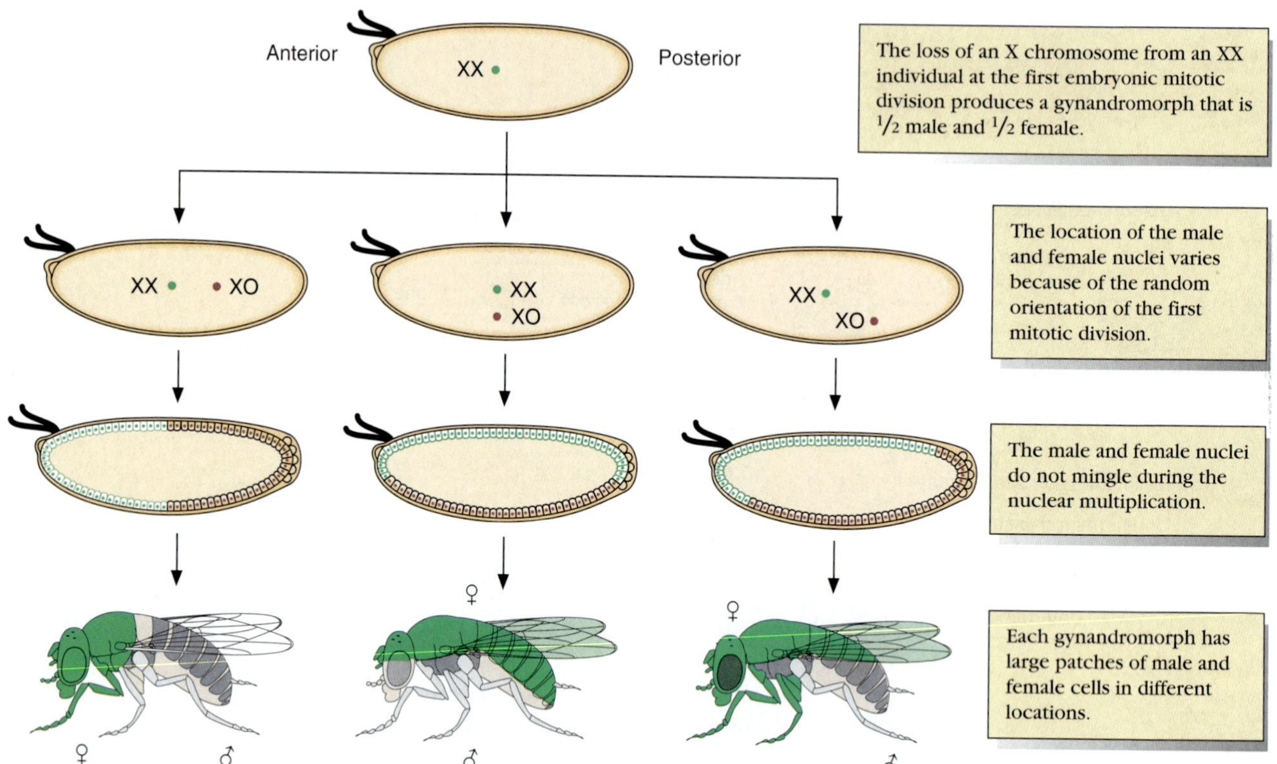

Anterior
XX
Posterior
The loss of an X chromosome from an XX individual at the first embryonic mitotic division produces a gynandromorph that is 1/2 male and 1/2 female.
XX
XO
XX
XO
XX
XO
The location of the male and female nuclei varies because of the random orientation of the first mitotic division.
The male and female nuclei do not mingle during the nuclear multiplication.
Each gynandromorph has large patches of male and female cells in different locations.
♀
♂
♀
♂
♀
♂

FIGURE 15.10 Adult cell determination in *Drosophila*. A particular structure in an adult gynandromorph is either male or female, depending upon the location of male (XO) or female (XX) nuclei in the blastoderm. The fates of the cells that form the adult are initially decided by their location in the blastoderm.

believed showed the spatial arrangement of the precursor cells in the blastoderm (Figure 15.11).

Today, gynandromorph analysis is often used to determine the phenotypic effects of mutations in mosaic individuals. In honor of Sturtevant's initial work, the unit of gynandromorph map distance is sometimes called the Sturt. One m. u. = 1 Sturt = 1% of gynandromorphs having these two tissues of different sex.

In 1984, Lois Schardin used ultraviolet laser microbeams and fine-glass needles to kill or remove single cells or small numbers of cells from embryos in the cell blastoderm stage. These embryos developed into individuals with larval and/or adult defects, usually missing structures. Schardin carefully recorded the location of the defects she induced, using a coordinate system based on the length and width of the egg. For example, a defect might be induced at 30% egg length, meaning it was 30% of the distance between the anterior and posterior poles of the egg. When the treated individual matured, she observed the location of the abnormality her treatment had induced. Treatments in the same location produced the same defects, and by plotting the missing structures on her egg coordinate scale, she was able to generate a map showing what structures were affected by treatment in specific locations of the blastoderm. The fact that treatments that killed blastoderm cells at one location also eliminated a set of larval or adult structures convinced Schardin that the treated cells were the precursors for those structures, and that her defect map in fact represented a map of the location of determined cells in the blastoderm. Her map shows that the external cuticular structures of the larva and adult are formed by the cells from two rectangular regions, each containing about 1100 cells, one on each side of the embryo (Figure 15.12). The cells within each region form half of the body segments and each half-segment is founded by about 100 blastoderm cells arranged in a stripe about 4 cells wide and 25 cells long. The blastoderm cells that give rise to the head structures are in the anterior, the thoracic precursors are in the middle, and the abdominal precursors are in the posterior. The cells outside of these regions form internal structures, such as the mesoderm and the nervous system. Comparison of her map with Sturtevant's earlier gynandromorph map showed a striking agreement. Both maps had the same structures in the same arrangements. The agreement of these two maps, produced using different techniques, convinced most researchers that embryonic cell determination does occur

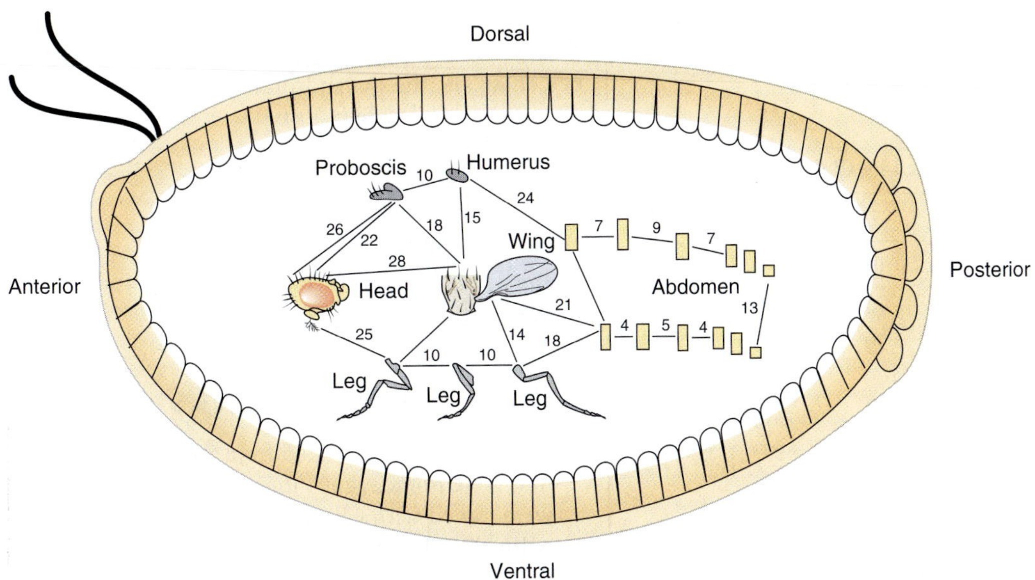

FIGURE 15.11 Gynandromorph fate map. The fate map shows the relative locations of the cells that give rise to particular adult structures. The percentage of gynandromorphs in which the two structures are of different sex is used as a map distance between the two structures. The numbers in the figure indicate the map distances between pairs of structures. This map distance is proportional to the physical distance between the precursor cells in the blastoderm.

at the blastoderm stage in *Drosophila,* and that it does occur in a precise spatial pattern, with cells determined to form particular structures always located in specific regions of the blastoderm.

These experiments give us an understanding of the process of cell-fate decisions in *Drosophila* embryos. Prior to cell formation, the syncytial blastoderm nuclei are undetermined. At or shortly after the time of cell formation, cells become committed to form either larval structures or imaginal discs that give rise to adult structures. There are two key features to remember about blastoderm determination: first, determination occurs in a precise spatial pattern. Cells in one region of the blastoderm always become committed to form structures characteristic of one region of the body. This spatial pattern of blastoderm fates is similar to the organization of the body, indicating that the blastoderm determination decisions organize the cells into the final body pattern. Second, in the blastoderm, each cell receives a segmental identity, but not its final fate. At the blastoderm stage, each cell is a member of a small group of cells that will give rise to all of the structures in one particular body segment or compartment. To generate these larval and adult structures the blastoderm cells must divide, and the progeny cells must undergo additional cell-fate decisions.

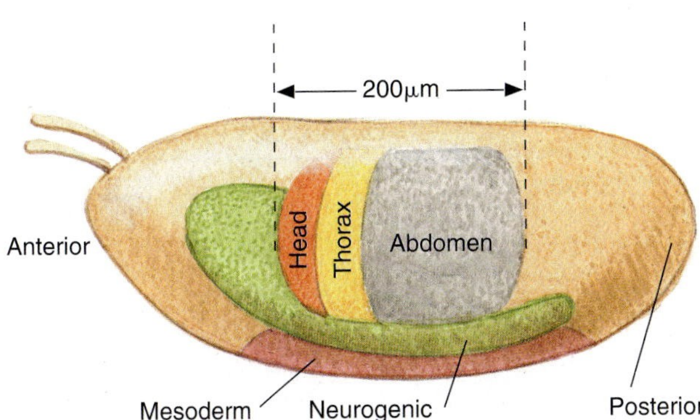

FIGURE 15.12 A defect fate map. Defect fate mapping indicates that the precursor cells that give rise to the external cuticle on one side of the *Drosophila* larva are located in a 200 μm × 200 μm region in the cell blastoderm. This region contains about 1100 cells. The locations of the precursor cells that give rise to internal structures such as the nervous system (neurogenic region) and the mesoderm are also shown.

SOLVED PROBLEMS

Problem

An hypothetical organism contains three types of cells in its epidermis—X, Y, and Z. In an effort to discover whether these types arise from determined precursor cells, a series of marked clones were induced at different stages of development. The time of induction and the cell types formed by the clones are listed. Were these cell types formed by determined cells? If so, when do the cells become determined?

Development (hours)	Cell Types Marked by Clones						
	XYZ	XY	XZ	YZ	X	Y	Z
50	10	10	10	10	15	15	15
100	20	20	5	20	30	30	30
150	0	50	0	50	75	75	75
200	0	0	0	100	150	150	150
250	0	0	0	0	300	300	300

Solution

The fact that clones induced late in development contain only one type of cell suggests that these cells do become determined to form a particular cell type (X, Y, or Z). The pattern of clones formed by intermediate induction suggests that they do so in a particular way. Clones induced at 50 or 100 hours can include all three types of cells, indicating that the cells founding these clones can give rise to all types and are not determined. Clones induced at 150 hours can include XY or YZ but not XZ or XYZ cells. This suggests that at between 100 and 150 hours the cells have become determined, and that all cells are of two types—those that will give rise to X or Y cells and those that will give rise to Y or Z cells. There are no cells left that can give rise to all three cell types. Clones induced at 200 hours can include YZ but not XY cell types, suggesting the XY cells have become determined to form either X or Y but not both, while the YZ cells can still give rise to both types. Clones induced at 250 hours include only X, Y, or Z cell types, indicating that at this time each cell is determined to form only one cell type. The determination pathway is, thus, XYZ ⟶ (150 hours) ⟶ XY, YZ ⟶ (200 hours) ⟶ YZ, X, Y ⟶ (250 hours) ⟶ X, Y, Z.

Problem

A sample of 160 gynandromorphs was examined to discover the sex of a set of tissues. (The results appear in the following table.) What are the gynandromorph fate map distances between these structures?

Structures	Number of Cases of Different Sex	Number of Cases of Same Sex
head–wing	49	111
proboscis–Pt leg	51	109
humerus–head	46	114
wing–Abd1	29	131
Mt leg–Abd1	38	122

Solution

The gynandromorph map distance is calculated as the percentage of the gynandromorphs in which the two structures are of different sex:

head–wing = 49/160 = 30.6%

proboscis–Pt leg = 51/160 = 31.9%

humerus–head = 46/160 = 28.7%

wing–Ab 1 = 29/160 = 18.1%

Mt leg–Ab1 = 38/160 = 23.7%

SECTION REVIEW PROBLEMS

3. Single cells were isolated from *Drosophila* embryos that were homozygous for *y, sn,* and *mal*. The cells were mixed with cells from wild-type embryos and cultured. After culture, the tissues formed by *y sn mal* cells were noted. From the following results, which of these cell types were determined?

Origin of *y sn mal* Cells	*y sn mal* Structures
Blastoderm neurogenic region	Neuroblast, dermoblast
Imaginal leg disc	Leg
Imaginal wing disc	Wing
Blastoderm abdominal region	Thorax and abdomen
Blastoderm head	Eye, antenna, and proboscis

4. Explain why blastoderm-cell culturing is a direct test of determination, whereas somatic mosaic analysis is an indirect test.
5. In a gynandromorph analysis, the frequencies with which pairs of structures were of opposite sex were recorded. From the following results, determine the fate-map distances between the precursor cells of each pair of structures.

Structures	Number of Gynandromorphs	
	Opposite Sex	Same Sex
A–B	4	71
B–C	23	52
A–C	30	45
C–D	19	56
A–D	15	60
B–D	34	41

6. If each segment in a *Drosophila* embryo is founded by a stripe of cells 4 cells wide, and if the gynandromorph fate-map distances between two structures 1 segment apart is 5 units, how many cells are there between the following pairs of structures?

   X–Y = 15 units, L–M = 10 units, Z–P = 8 units

A Cell's Lineage Controls Its Fate in *C. elegans*.

The nematode *Caenorhabditis elegans* has become an important organism for the genetic analysis of animal development, as noted in Chapter 1. Much of the current interest in nematode development is the result of the efforts of Sidney Brenner at the MRC Molecular Biology laboratory in Cambridge, England. Brenner chose the nematode in the early 1970s as a model organism for the study of animal developmental genetics and perfected a variety of genetic techniques specific for the nematode. One of the most interesting features of nematode development is that cell lineage plays a major role in cell determination, and this makes the nematode an outstanding organism for studying the establishment and inheritance of cell determination. After a short embryogenesis (11 hours) nematodes go through four larval stages (L1, L2, L3, and L4) of 9–12 hours each. The individual becomes an adult only 55 hours after egg laying (Figure 15.13).

Cell Determination Is Controlled by Cell Lineage in *C. elegans*.

In *C. elegans* the pattern of cell division is largely invariant. For most cells, the time, the location, and the orientation of each cell division is the same in each individual nematode, and each cell has the same fate in every individual. This means that **cell lineage,** the cell division pathway that produced a particular cell, is a major, if not the only, factor determining that cell's fate. The first four cleavage divisions in the *C. elegans* zygote occur in the oviduct before the egg is laid, producing a blastoderm with 16 blastomeres. The first three divisions occur along what will be the future body axes. The first division occurs along the anterior/posterior body axis, the second along the dorsal/ventral, and the third along the medial/lateral. These divide the embryo into a series of distinct cell lineages called the AB, MS, E, C, D, and Q lineages (Figure 15.14). Each lineage gives rise to a set of cells that always forms the same structures in the body. Because *C. elegans* is transparent to light, each cell division in each lineage can be observed directly in living individuals by placing the individual on a microscope slide. When this is done, it is seen that embryonic cell divisions continue until the embryo has 540 cells. At this point, it undergoes extensive morphogenesis; all the major organs form, and the embryo assumes the larval shape. Each individual passes through four larval stages before becoming an adult (Figure 15.13). The adult structures develop within the larval body. Most adults are self-fertilizing hermaphrodites, but a few are males. The constant pattern of cell division means that each adult has the same number of cells. Because some cells (such as muscle) are multinucleate, there are more nuclei than there are cells. There are 959 somatic nuclei in each adult hermaphrodite and 1031 in each male. The ultimate fate of any nematode cell can be accurately predicted in advance from a knowledge of that cell's lineage.

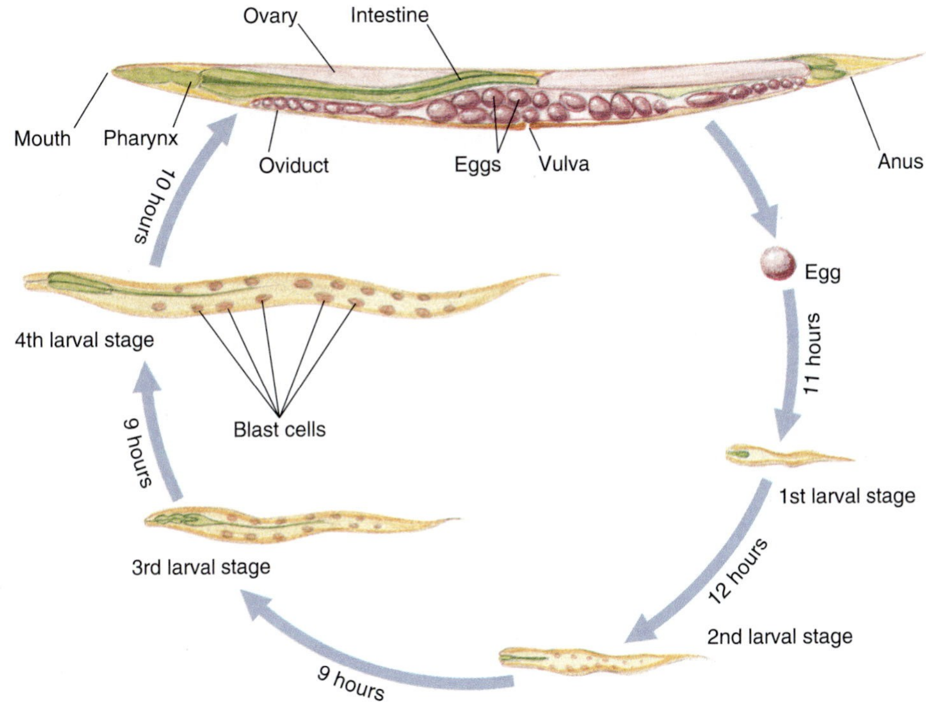

FIGURE 15.13 The life cycle of *Caenorhabditis elegans*. *C. elegans* has an embryonic (egg) stage, followed by four larval stages. Unlike *Drosophila*, the transition from larva to adult is gradual, and adult structures develop from groups of blast cells during the larval stages.

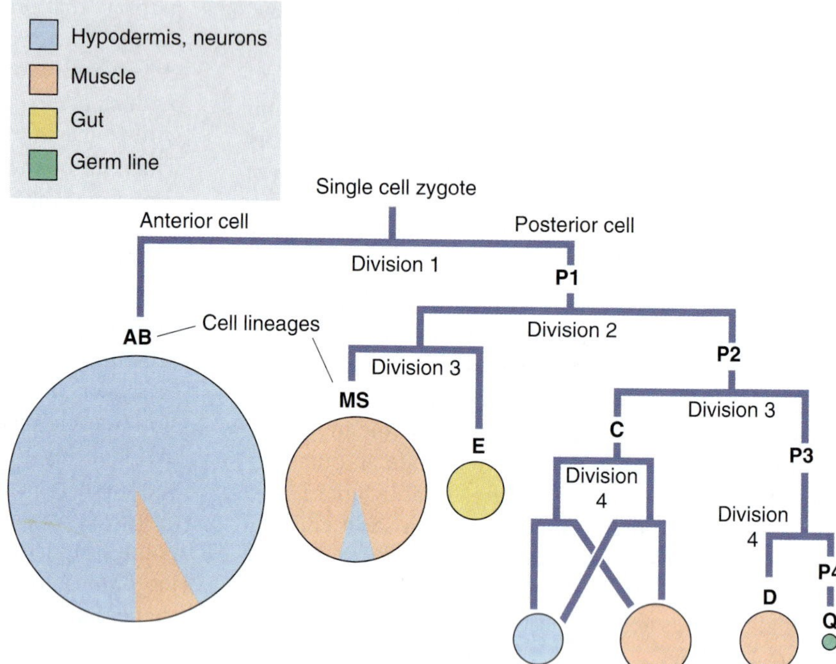

FIGURE 15.14 The first four cell divisions in *C. elegans*. The first four cell divisions of *C. elegans* establish separate cell lineages that give rise to particular cell types. The first division produces an anterior cell (AB) and a posterior cell (P1). The P1 cell division produces a P2 cell and a cell that divides to produce the first cell of the MS and the E lineages. The P2 cell divides to produce a P3 cell and the first cell of the C lineage. The P3 cell divides to produce the first cell of the D lineage and a P4 cell, which will give rise to the Q lineage. Each lineage gives rise to a specific set of larval and adult structures.

A system of nomenclature that identifies the position and cell lineage of each *C. elegans* cell has been developed to allow investigators to follow the exact fate of each cell. For example, many of the adult structures arise from a series of 12 larval **blast cells.** When a blast cell divides, the progeny cells are labeled according to their position and the orientation of the division. If the division has an anterior/posterior orientation, the anterior cell is labeled "a" and the posterior "p"; if the division has a dorsal/ventral orientation the dorsal cell is labeled "d" and the ventral "v." A cell labeled "P8apd" would be the dorsal descendent of the posterior descendent of the anterior descendent of the eighth blast cell. Although this system may seem elaborate, it gives each cell a label unique to its lineage.

Adult Development in *C. elegans* Builds on the Pattern Established During Embryonic Development.

The adult body of *C. elegans* builds upon the existing larval structures rather than replacing structures. Some adult features, such as the components of the adult reproductive system, are formed by specific groups of cells that are the progeny of the blast cells. Each major adult structure is formed by groups of interacting blast-cell progeny called **cell equivalence groups,** which have some similarities to the imaginal discs in *Drosophila.* One well-studied cell equivalence group, called the vulval precursor cells (VPC), is composed of some of the descendants of the P3, P4, P5, P6, P7, and P8 blast cells that form the vulva in hermaphrodites. Vulval development is strongly influenced by inductive signals sent from the

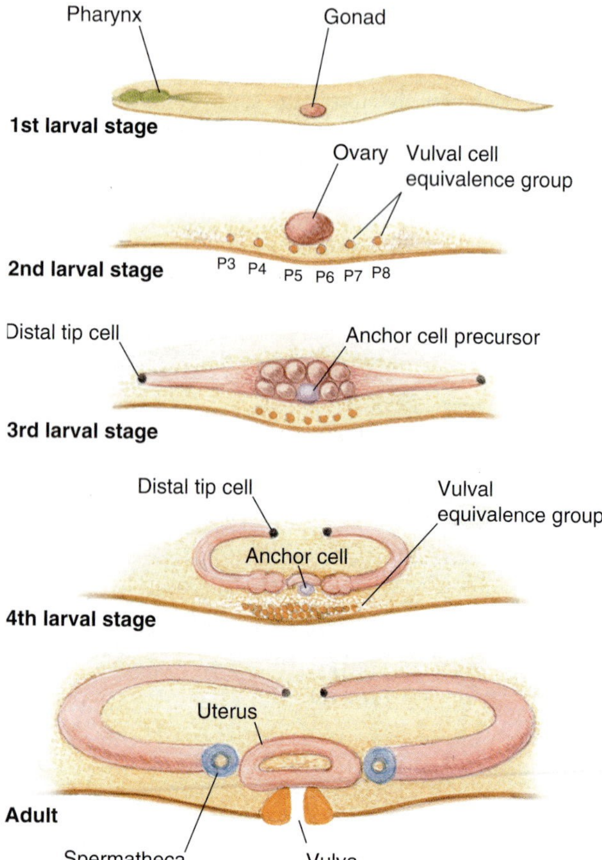

FIGURE 15.15 Vulval development in *C. elegans*. The vulva develops from some of the progeny of the vulval cell equivalence group blast cells (P3–P8). The anchor cell in the ovary sends inductive signals to the vulval precursor cells.

developing **anchor cell** in the ovary to the VPC group. The anchor cell is located between the arms of the developing ovary and sends inductive signals that are critical for organizing the ovary and vulva. In normal development, the six cells in the VPC are arranged in a longitudinal row along the ventral epidermis next to the developing ovary (Figure 15.15). During larval development, these cells adopt a spatially graded pattern of cell fates: 3°, 3°, 2°, 1°, 2°, 3°. Cells adopting a 3° fate form nonvulval epidermis, and cells adopting a 2° or 1° fate form specific sets of vulval structures. The cell closest to the ovary (P6) normally adopts the 1° cell fate. If this cell is removed, one of the 2° cells assumes a 1° fate and a 3° cell replaces the 2° cell. If all the cells are removed, no vulval structures are formed, indicating that (1) no cells outside of this group are capable of adopting a vulval cell fate, and (2) if faced with a disruption of development, these blast cells have some capacity to change their fates.

SOLVED PROBLEMS

Problem

Explain how the regulation of blast cells indicates that the vulval cell equivalence group (VPC) is not a mosaic system.

Solution

In a mosaic system, cells have a fixed fate and cannot change. Elimination of a cell removes a structure because no other cell can replace that cell. However, if a member of the VPC is removed, another cell in the group takes its place and forms the structures it would form. There are limits on the replacement, of course, and a hierarchy of ability to replace a missing neighbor. The 2° or 3° cells will replace a 1° cell, showing that these cells retain the potential to form a 1° cell. However, no cell from outside the VPC is able to replace a missing VPC cell.

Problem

The spatial arrangement of the 1°, 2°, and 3° cells in the VPC suggests that these blast cells are influenced by a positional information gradient that has its source in the ovary (cells closest to the ovary become 1° cells; those furthest from it become 3° cells). How would you test this hypothesis, and what results would you expect in your experiments if it were true?

Solution

A genetic test would be to isolate mutations that cause the ovary to not develop and then to generate somatic mosaic individuals in which the ovary precursor cells are mutant but the VPC precursor cells are not. In these individuals, the VPC should be altered and should not show the normal spatial organization of 1°, 2°, and 3° cells.

SECTION REVIEW PROBLEMS

7. Describe one similarity and one difference between the embryonic development of *C. elegans* and *Drosophila*.
8. Describe one similarity and one difference between an imaginal disc of *Drosophila* and a cell equivalence group in *C. elegans*.
9. Give a brief definition of each of the following:
   a. syncytial blastoderm
   b. cell blastoderm
   c. imaginal disc
   d. blast cell
   e. cell equivalence group

GENETIC MECHANISMS CONTROL CELL-FATE DECISIONS.

In this section, we will consider some examples of individual genes and groups of genes in *Drosophila* and *C. elegans* that control specific cell-fate decisions. These examples include both embryonic and adult cell-fate decisions, and they have been chosen to illustrate some of the different types of gene action that control cell-fate decisions. These examples also illustrate the similarities and the differences between some of the genes that control cell fate in these two organisms.

Maternal Gene Products Concentrated in the Egg Control Developmental Regulatory Gene Action.

The process of establishing the fate of the *Drosophila* blastoderm cells begins with the action of genes in the maternal genome. Several groups of maternal genes regulate cell fate in the embryo. Each group produces a localized concentration of one gene product in a particular region of the embryo, often by generating a localized concentration of mRNA molecules in the egg cytoplasm. For example, the anterior group of genes establishes a high concentration of the *bicoid* (*bcd*) mRNA in the anterior end of the embryo. The *bcd* gene is transcribed in the nurse cells, but the mRNA is not translated into a protein product before fertilization. When the nurse cell cytoplasm is passed into the oocyte *bcd*, mRNA is sequestered in the anterior cytoplasm of the oocyte (Figure 15.16). After the egg is fertilized and zygotic nuclear divisions begin, the *bcd* mRNA begins translation as the zygotic nuclei are migrating toward the cortical cytoplasm. As a result of this migration, the *bcd* gene product is highly concentrated in the anterior end of the embryo. By the syncytial blastoderm stage, the *bcd* product is diffused in a gradient in the embryo.

The *bcd* gene product is a transcriptional regulatory molecule that binds to specific sites in the promoter

(a)

Immature oocyte

Nurse cells

Oocyte

Anterior

Posterior

bcd mRNA

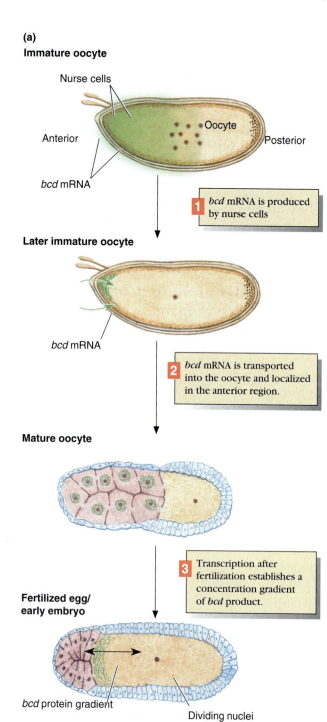

1 *bcd* mRNA is produced by nurse cells

Later immature oocyte

bcd mRNA

2 *bcd* mRNA is transported into the oocyte and localized in the anterior region.

Mature oocyte

3 Transcription after fertilization establishes a concentration gradient of *bcd* product.

Fertilized egg/ early embryo

bcd protein gradient

Dividing nuclei

(b)

bicoid mRNA

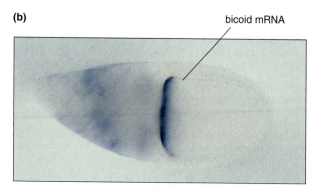

◀ **FIGURE 15.16 Localization of maternal *bcd* mRNAs.** (a) The *bcd* mRNA is produced by nurse cells and localized in the anterior cytoplasm. Translation of this mRNA in the preblastoderm oocyte produces the anterior positional gradient in the *Drosophila* embryo. (b) This shows the distribution of *bcd* mRNA in a developing oocyte. *(Jack R. Girton, Iowa State University)*

regions of a number of zygotic genes. For example, the *bcd* gene product is a transcriptional activator of the *hunchback* (*hb*) gene, a zygotic developmental regulatory gene that is involved in establishing blastoderm cell fate. There are multiple regulatory binding sites in the *hb* promoter region at which the *bcd* gene product can bind, and if enough sites are bound, the *hb* gene is transcribed (Figure 15.17). This ensures that *hb* is transcribed in those regions where the concentration of *bcd* gene product is high, but not in regions where *bcd* concentration is low. By this means, maternal gene-product concentrations control gene expression differentially.

This mechanism of concentrating mRNA in the oocyte cytoplasm for later translation is not unique to *bcd*, and more than one maternal gene product may act on the same zygotic gene. For example, the product of the *torso* (*tor*) gene can bind to the *hb* promoter to inhibit

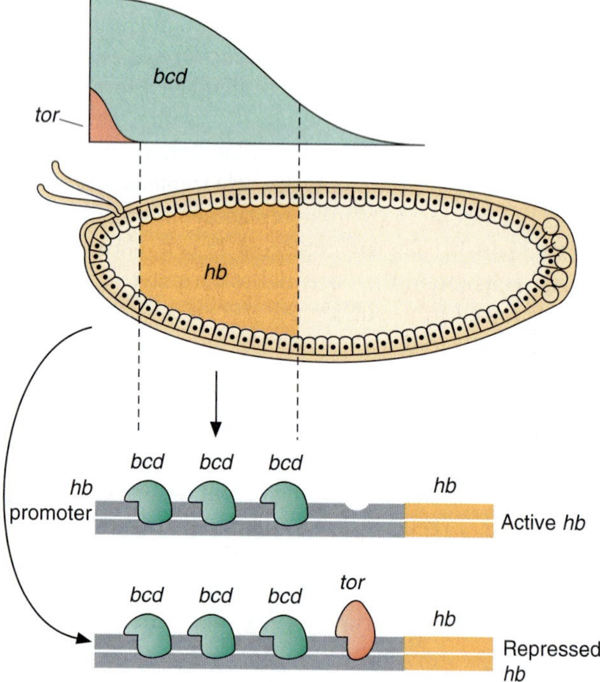

FIGURE 15.17 Transcription of hunchback. The hunchback (*hb*) gene is transcribed in a band of cells in the anterior of the blastoderm. Transcription of *hb* is activated by high concentrations of the transcription activator *bcd* gene product (short arrow) but is repressed by high concentrations of the transcriptional repressor *tor* gene product (long arrow). The spatial distributions of the products of *bcd* and *tor* thus define the location of *hb* activity.

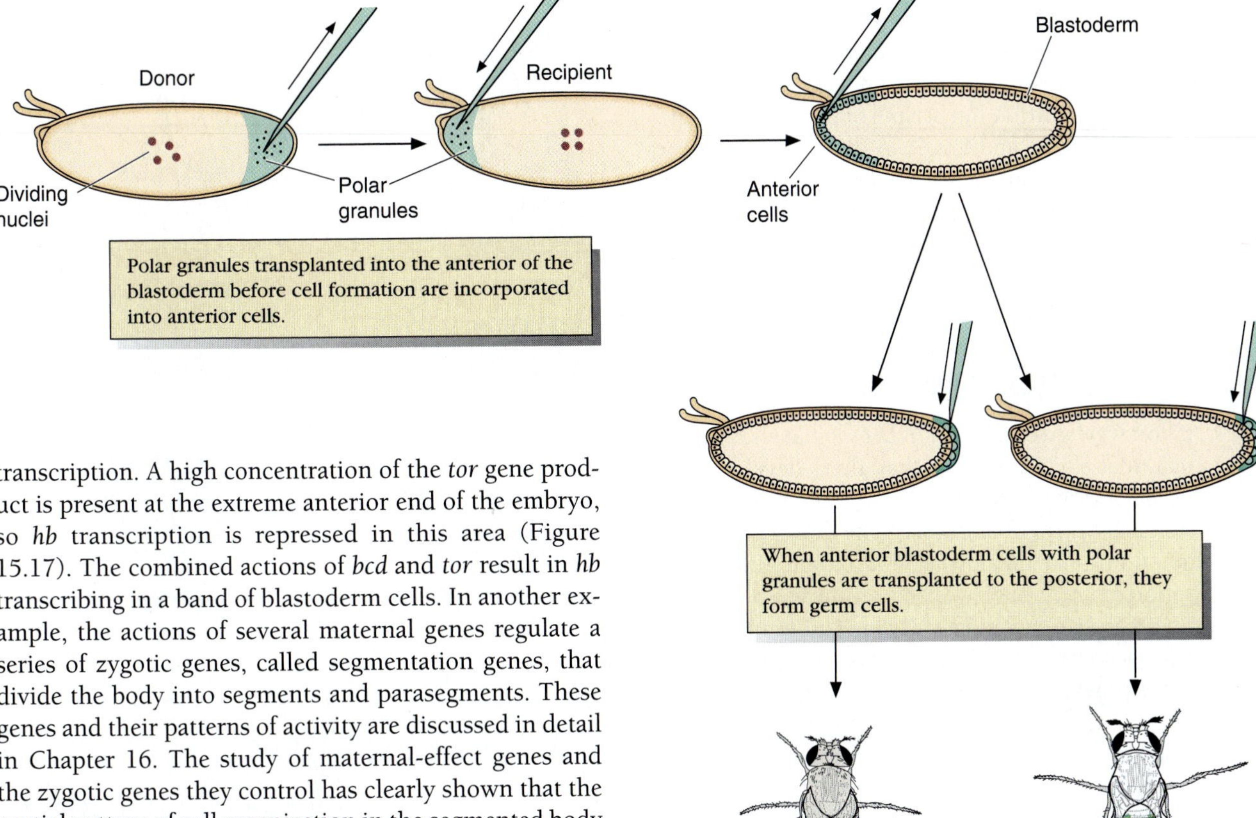

FIGURE 15.18 Polar granule transplantation. The polar granules contain cytoplasmic determinants that induce cells to form germ cells. If polar granules are transplanted to other regions of a blastoderm before cell formation, the cells that incorporate these polar granules become germ cells.

transcription. A high concentration of the *tor* gene product is present at the extreme anterior end of the embryo, so *hb* transcription is repressed in this area (Figure 15.17). The combined actions of *bcd* and *tor* result in *hb* transcribing in a band of blastoderm cells. In another example, the actions of several maternal genes regulate a series of zygotic genes, called segmentation genes, that divide the body into segments and parasegments. These genes and their patterns of activity are discussed in detail in Chapter 16. The study of maternal-effect genes and the zygotic genes they control has clearly shown that the spatial pattern of cell organization in the segmented body results from localized concentrations of mRNA deposited in the egg by the maternal genes. This mRNA activates a localized sequence of zygotic gene expression, which creates gradients of gene products that control cell-fate decisions of the blastoderm cells.

Germ-Line Cell Determination Is Controlled by Cytoplasmic Determinants Packaged in the Polar Granules.

One of the first and most fundamental cell-fate decisions is whether to form germ-line cells or somatic cells. This decision is controlled by cytoplasmic determinants packaged in the posterior of the egg. During early embryonic development, the first nuclei to reach the posterior end encounter densely staining cytoplasmic bodies, known as **polar granules,** that contain RNA and protein. Nuclei that encounter polar granules form the first embryonic cells (the pole cells) and become determined to form germ cells.

In a direct test of whether polar granules actually contain germ-cell determinants, Karl Illmensee and Anthony Mahowald in 1974 used fine-glass needles to transplant polar granules from the posterior of early cleavage-stage embryos (donor embryos) into the anterior of different early cleavage-stage embryos (recipient embryos). In the recipient embryos, donor nuclei that reached the anterior sometimes incorporated the polar granules when they formed cells. If these anterior cells were then transplanted into the posterior region of an embryo whose polar granules had been destroyed, they formed functioning germ cells (Figure 15.18). Normal anterior cells transplanted into this same region never formed germ cells, and transplanting cytoplasm from any other region into the anterior did not cause the anterior cells to form germ cells. These results indicate that the polar granules contain a cytoplasmic determinant that causes blastoderm cells to become germ cells. Thus, this experiment confirms that cytoplasmic determinants can control the cell fate of embryonic cells.

The polar granules contain a number of proteins and RNAs. Three prominent polar granule components are the products of the genes *oskar* (*osk*), *vasa* (*vas*), and *tudor* (*tud*). Mutations of these genes block pole granule formation, and an experimentally induced high concentration of *osk* mRNA in the anterior of an embryo is sufficient to induce anterior cells to form functional pole cells. It is thus likely that the *osk* gene codes for the pole cell determinant, and *vas* and *tud* code for proteins necessary for *osk* product transportation or function. Polar granules also contain other proteins and RNAs that are necessary for proper pole granule function, and the roles of these products are less well known. For example, polar granules contain RNA transcripts of the *polar granule component* (*Pgc*) gene, and this RNA does not code for a protein product. Eliminating this RNA does not prevent the formation of pole cells, but it does prevent the resulting pole cells from migrating to the gonads and forming functioning germ cells. Much is not yet understood about the function of polar granules in germ-cell formation.

SECTION REVIEW PROBLEMS

10. If you were given a strain of *Drosophila* with a new maternal-effect mutation that alters development in the embryonic head, and if you were also given a cloned fragment of the *bcd* gene, how would you determine whether the new mutation affects the action of *bcd?*
11. The recessive mutations *swallow* and *bcd* are both maternal-effect mutations. Homozygous mutant females produce embryos that are missing head structures. If you were given a strain of flies of unknown genotype that had this same phenotype, how would you determine which mutation was present?

Homeotic Genes Control Segmental Identity in *Drosophila.*

By the fourth hour of embryonic development, the somatic cells of the *Drosophila* blastoderm have been organized into segments and compartments by the action of a series of maternal gene products that activate a series of zygotic segmentation genes. This process will be discussed in Chapter 16. Here we will discuss how the cells in each segment acquire their segment-specific cell fate. The cells in each segment adopt a particular segmental identity (e.g., head, thorax, abdomen). This decision is a true determination decision, and it is controlled by a group of genes known as the **homeotic genes.** The initial blastoderm cell population that forms each segment contains only about 200 cells (100 on each side of the embryo). These cells must undergo many more cell divisions and additional cell-fate decisions to produce the final body structures. For example, the cells in the mesothoracic segment give rise to the adult wing (containing approximately 40,000 cells). The job of the homeotic genes is to direct the process of post-blastoderm development, which is different for each segment.

The homeotic genes are divided into three main groups: the ***Bithorax* Complex** (BX-C), the ***Antennapedia* Complex** (ANT-C), and the ***Polycomb* Group** (PcG). The *BX-C* genes control cell determination in the posterior thoracic and abdominal segments, and the *ANT-C* genes regulate the determination of the cells in the head and anterior thoracic segments. The *PcG* genes regulate the expression of other homeotic genes. The structure and function of these genes is discussed in Chapter 16. In this section, we will discuss some of the fundamental principles of homeotic gene function, focusing on the *BX-C* gene.

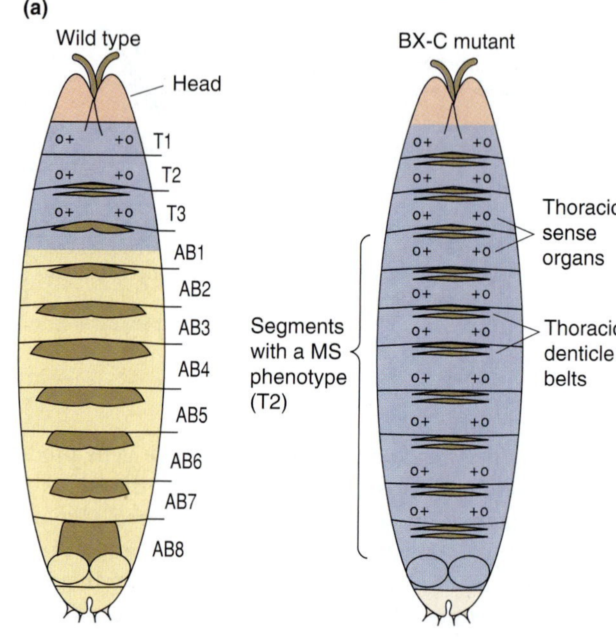

(b)

FIGURE 15.19 Mutations of *Bithorax* Complex (*BX-C*) genes. Mutations of *BX-C* genes transform embryonic and adult segments or parts of segments into other segments. (a) Embryos homozygous for a deficiency that eliminates all the genes in the *BX-C* have the embryonic metathoracic (T3) segment and the first eight abdominal segments (AB1–AB8) transformed into mesothoracic (T2) segments (MS). Each transformed segment has external cuticular structures (denticles and sense organs) normally found on the second thoracic segment. (b) Mutant alleles of *BX-C* genes also transform adult segments. This adult has mutant alleles of *abx, pbx,* and *bx* and has the metathoracic segment transformed into a second mesothoracic segment, complete with wings. *(b, E. B. Lewis, California Institute of Technology)*

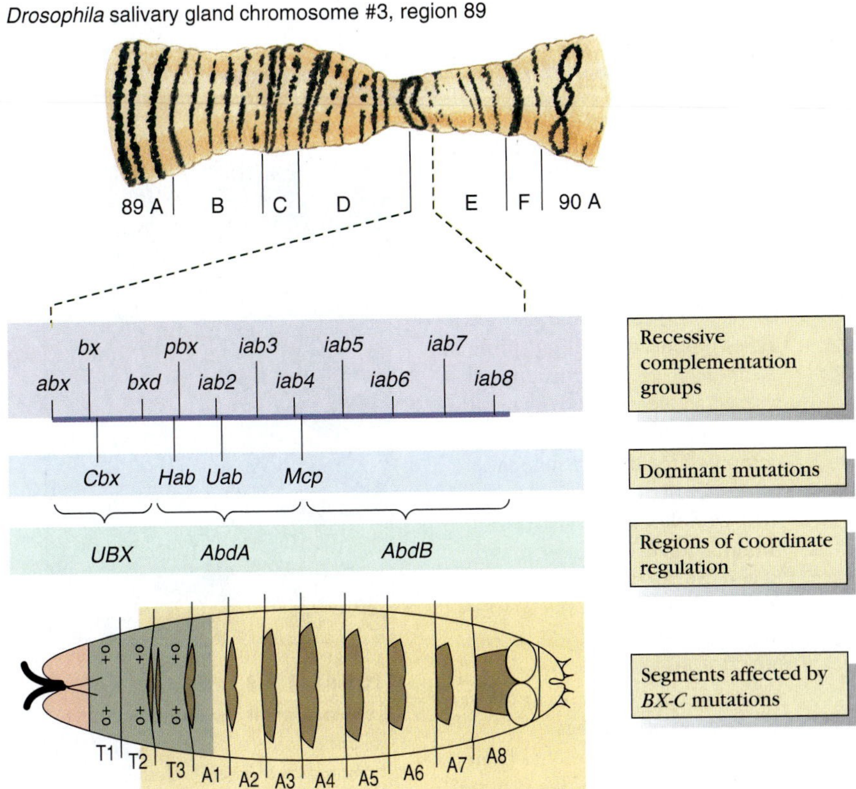

FIGURE 15.20 The *BX-C* genes. The *BX-C* genes are located in the third chromosome at the salivary gland chromosome region 89E1–2. The recessive mutations define a series of genes whose name reflects the region affected by mutant alleles: *abx* = abruptex, *bx* = bithorax, *bxd* = bithoraxoid, *pbx* = postbithorax, *iab2* = infraabdominal2, *iab3* = infraabdominal3, *iab4* = infraabdominal4, *iab5* = infraabdominal5, *iab6* = infraabdominal6, *iab7* = infraabdominal7, *iab8* = infraabdominal8. Several dominant *BX-C* mutations are known: *Cbx* = Contrabithorax, *Hab* = hyperabdominal, *Uab* = ultraabdominal, *Mcp* = Miscadestralpigmentation. The genes defined by recessive mutations are organized into three coordinately regulated groups: *UBX* = ultrabithorax, *Abd-A* = abdominal A, and *Abd-B* = abdominal B.

The *BX-C* contains three groups of genes organized into three regions of coordinate regulation, called UBX, Abd-A, and Abd-B (Figure 15.19). The *BX-C* genes are located on the third chromosome and are in salivary gland chromosome region 89E1–2 (Figure 15.20). Amazingly, genetic mapping analyses indicate that the order of the *BX-C* genes along the chromosome is the same as the order of the segments they affect (Figure 15.20). This was established by Edward B. Lewis, who also isolated several dominant mutations resulting from chromosomal breaks in the BX-C chromosomal region. Each dominant mutation fails to complement recessive alleles of several different *BX-C* genes. The different dominant mutations can be grouped into three different classes, depending on which sets of recessive alleles they do not complement. For example, several dominant *Ubx* mutations all fail to complement recessive alleles of *abx*, *bx*, *pbx*, and *bxd*. These results suggested to Lewis that the recessive complementation groups in *BX-C* are organized into three regions of coordinate regulation.

Our current understanding of the *BX-C* is largely the result of the work done by Lewis, one of the outstanding researchers in the investigation of *Drosophila* homeotic mutations [see HISTORICAL PROFILE: *Edward B. Lewis (1918–)*]. Lewis and his coworkers have studied the *BX-C* since 1948, and they have isolated and analyzed many recessive and dominant mutations of *BX-C* genes. They discovered that an individual homozygous for a deficiency that removes the entire BX-C has the metathoracic and all abdominal segments transformed into additional copies of the mesothoracic segment (Figure 15.19). Individuals homozygous for point mutations that eliminate one of the *BX-C* genes have one segment or compartment transformed into a more anterior segment or compartment. For example, mutations of the *Iab-5* gene cause the fifth abdominal segment to be trans-

HISTORICAL PROFILE

Edward B. Lewis (1918–)

Edward B. Lewis (Figure 15.B) is recognized as one of the truly outstanding figures in *Drosophila* developmental genetics. He was born on May 20, 1918, in Wilkes-Barre PA, where he grew up. From 1935 to 1939 he attended the University of Minnesota, receiving a B.A. in 1939. He had developed interests in music and in genetics, and it was only after an emotional struggle that he decided to pursue genetics. Lewis enrolled in the graduate program at Cal Tech and immersed himself in the outstanding *Drosophila* genetics group as a graduate student working with the well-known *Drosophila* geneticist C. P. Oliver. Lewis received his Ph.D. in 1942 and immediately joined the Army Air Corps (where he rose to the rank of captain), serving from 1942 to 1946. After leaving the service, Lewis returned to Cal Tech as a faculty member. Beginning as an instructor of biology in 1946, he rose through the academic ranks. In 1966, he was appointed Thomas Hunt Morgan professor. He has received numerous awards and honors, including election to the National Academy of Science and the 1995 Nobel Prize.

Lewis has made numerous important contributions to *Drosophila* genetics. When he began his career in the 1940s, the concept of the gene was different from that of today. Genes were thought to be capable of being altered by mutagenic agents, but indivisible by recombination. In 1940, Oliver had first demonstrated that recombination could occur between two alleles of the *lozenge* gene in *Drosophila,* but his finding was not universally accepted. In a classic study, Lewis analyzed two mutations, *star* and *asteroid,* that he had mapped to the same location on the second chromosome. Individuals with these mutations in *trans* (*star* + / + *ast*) had a mutant phenotype, similar to that of homozygous *ast*. Because they failed to complement and mapped to the same location, they were considered alleles. However, Lewis observed that *star/asteroid* females produced one + offspring in 3235 offspring. Intrigued, he performed an elegant analysis involving 14 different crosses that demonstrated that recombination can occur between *star* and *asteroid.* This work forced a revision in thinking about alleles. At first a new term, "pseudoalleles," was invented for mutations that appeared to be alleles but that could recombine. Eventually it was realized that recombination could occur between most alleles of all genes.

Lewis' major work has been his genetic analysis of the genes of the *Bithorax* complex (BX-C). For 40 years, he has continued to induce and analyze *BX-C* mutations, making this one of the best-studied genes or gene complexes in *Drosophila.* The details of the genetic functions present in this complex, and his model of *BX-C* function, are described in the text. It is impossible to overestimate the impact these studies have had on developmental genetics. For many years, *BX-C* has been the preeminent system for the study of the genetic regulation of cell determination. Lewis' concept of regulatory gene function revolutionized thinking about how genes control development. The cloning of BX-C has allowed the molecular structure of these genes to be examined directly and the molecular mechanisms of *BX-C* gene action to be defined. The results of these studies suggest that the BX-C action mechanism may be more complex than Lewis anticipated, but that many of his fundamental ideas were correct. One definition of a successful scientific model is that it stimulates further experiments. By this standard, Lewis' model of BX-C function has been an outstanding success. The current molecular analyses of BX-C are providing us with extraordinary insight into how genes regulate complex developmental processes and how eukaryotic genes function. It is important to understand that many of these studies depended on the wealth of alleles and chromosomal aberrations carefully collected and painstakingly analyzed by Lewis over the years. Lewis' genetic analysis of BX-C mutations have truly been the foundation of a revolution in developmental genetics.

FIGURE 15.B Edward B. Lewis. A photograph of Edward B. Lewis, the California Institute of Technology Thomas Hunt Morgan Professor of Genetics (emeritus), in his laboratory. *(E. B. Lewis, California Institute of Technology)*

formed into a second copy of the fourth abdominal segment. The *bithorax* (*bx*) mutations transform the anterior compartment of the metathoracic segment into a second copy of the anterior compartment of the mesothoracic segment. The *post-bithorax* (*pbx*) mutations transform the posterior compartment of the metathoracic segment into a second copy of the posterior compartment of the mesothoracic segment (Figure 15.20). These mutations indicate that the products of the *BX-C* genes control the segment identity of the cells in the metathoracic and abdominal segments and that each gene controls determination in one segment or compartment (Table 15.2).

Because mutations that cause a loss of all of the *BX-C* genes transform all metathoracic and abdominal segments into mesothorax segments, Lewis proposed that during development each segment is initially given a mesothoracic determination by other determination genes and that the *BX-C* genes later override this. To explain how point mutations in each *BX-C* gene can cause one segment to be transformed into the next most anterior segment, Lewis proposed that each gene is active in the segment it controls and in all more-posterior segments. He also proposed that there is a strict hierarchy of control, with each *BX-C* gene being epistatic to the genes controlling more anterior segments (Figure 15.21). This complex, elegant model was a conceptual breakthrough in our understanding of how determination genes function. It also provided an answer to one of our key questions, how determined cells maintain their state. As long as the correct *BX-C* genes are expressed in determined

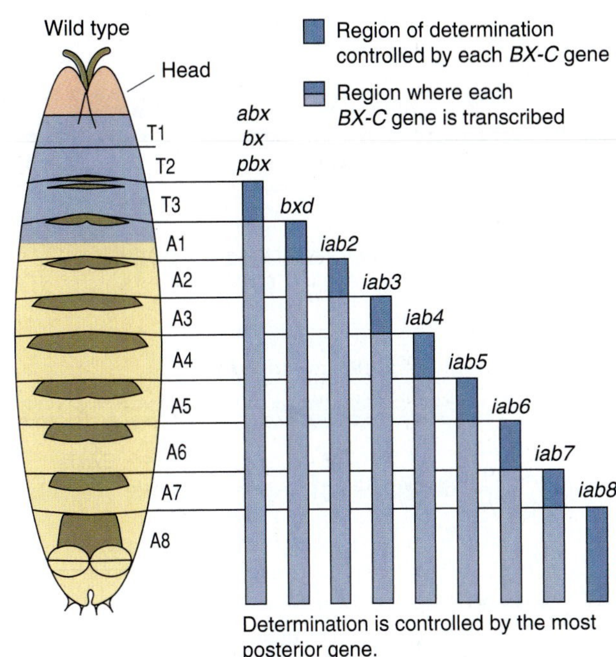

FIGURE 15.21 **Action of *BX-C* genes.** E. B. Lewis proposed that the products of the genes of the *BX-C* are determinants that control cell fate in each thoracic and abdominal segment. He proposed that there is a hierarchy of epistatic interaction among these genes, such that each gene is epistatic to the genes controlling determination in more anterior segments. Thus, the fate of cells in each segment is controlled by the action of the most posterior, active *BX-C* gene.

TABLE 15.2 *The Genes of the* BX-C *of* Drosophila *Identified by E. B. Lewis and Coworkers.*

Group		Region Affected
***UBX* group**		
Abruptex	*abx*	Posterior compartment of the mesothorax
Bithorax	*bx*	Anterior compartment of the metathorax
Bithoraxoid	*bxd*	Posterior compartment of the metathorax
Postbithorax	*pbx*	First abdominal segment
***Abd-A* group**		
Infraabdominal2	*iab2*	Second abdominal segment
Infraabdominal3	*iab3*	Third abdominal segment
Infraabdominal4	*iab4*	Fourth abdominal segment
***Abd-B* group**		
Infraabdominal5	*iab5*	Fifth abdominal segment
Infraabdominal6	*iab6*	Sixth abdominal segment
Infraabdominal7	*iab7*	Seventh abdominal segment
Infraabdominal8	*iab8*	Eighth abdominal segment

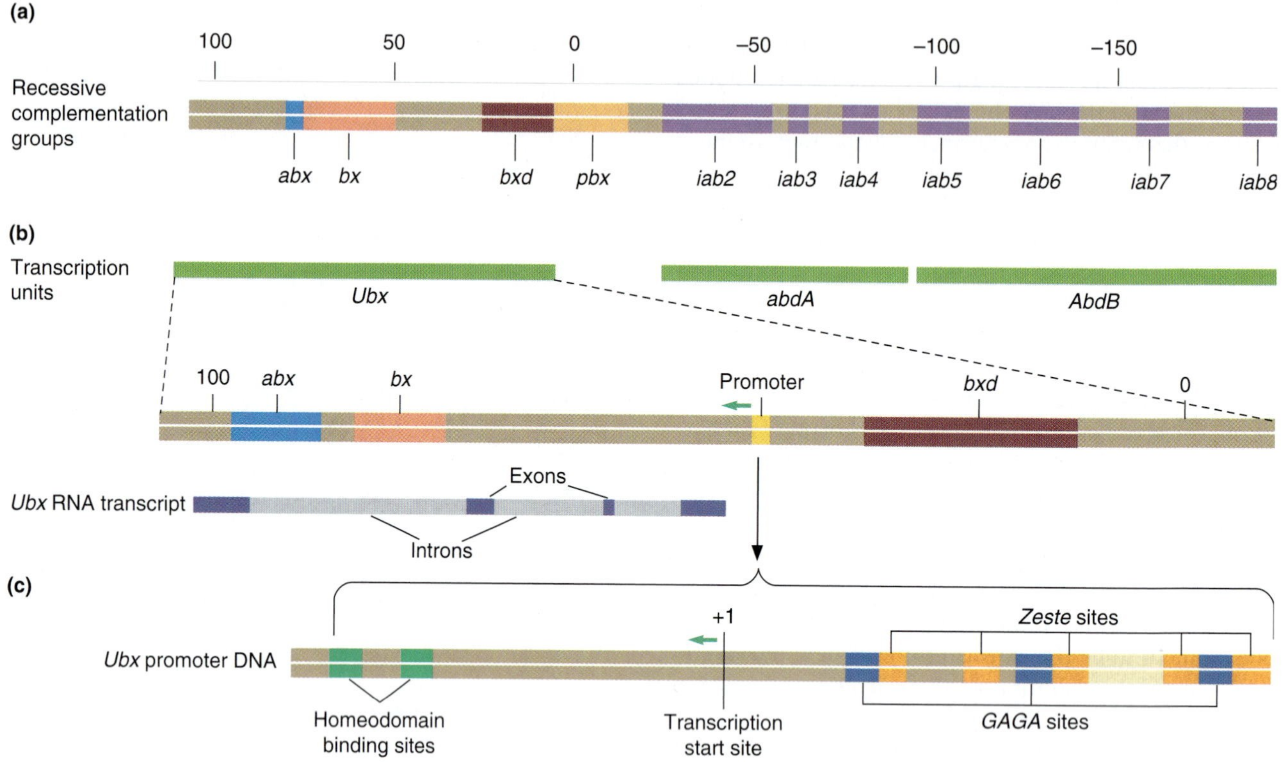

FIGURE 15.22 Molecular structure of BX-C. (a) The BX-C region has been mapped to a 300-kb region of genomic DNA. (b) This region contains three large transcription units containing several large exons. The *Ubx* transcription unit contains over 80 kb of DNA although the primary transcript is processed into one of three mRNAs of only 3.2, 4.3, or 4.7 kb. The promoter region of the *Ubx* transcription unit contains a number of regulatory protein binding sites. These include binding sites for the products of homeobox genes, *zeste*, *NTF-1*, and *GAGA*, and sites whose binding proteins are not yet known. Some of the proteins that bind to these sites enhance transcription, and others repress transcription.

cells, they will have a continuous supply of the correct determinant. This internal production of determinants by homeotic genes is the reason why determined cells maintain their state in culture or after being transplanted.

In 1983, the BX-C region was cloned and analyzed by Welcome Bender and coworkers. They discovered that the complex occupies about 300 kb of DNA in the 89E region of the third chromosome. A detailed genetic map of this region has been made, and the mutations studied by Lewis were mapped to discrete regions of the DNA (Figure 15.22). For example, *bx* mutations are located in a 20-kb region between −75 kb and −55 kb on the map. The 300-kb interval contains three large transcription units: the *Ubx* unit, the *Abd-A* unit, and the *Abd-B* unit, which correspond to the three regulative regions identified by Lewis. The *Ubx* unit has a primary transcript of more than 80 kb that is processed into either a 3.2-, 4.3-, or 4.7-kb mRNA. All three of these contain the same sequences from the 5′ and 3′ ends of the primary transcript, but they contain different internal sequences. For example, the 3.2-kb RNA contains two internal, 51-bp "microexons" surrounded by three introns totaling 73 kb. Most of the recessive mutations studied by Lewis are not located in the protein-coding regions of the transcription unit but instead are located in *cis*-acting regulatory sites that regulate *Ubx* transcription in specific thoracic and abdominal segments. The *Ubx* promoter contains a combination of transcriptional control elements and TATA box consensus sequences located immediately proximal to the RNA transcription start site and enhancer elements located farther away (Figure 15.22). The transcription factors that act on the *Ubx* promoter include general factors essential for transcribing all promoters, as well as some promoter-specific factors that may control the spatial and temporal distribution of *Ubx* products. The molecular basis for the action of the *BX-C* is not yet clear.

In the BX-C region there are three copies of a highly conserved, 180-bp DNA sequence called the **homeobox.** The homeobox codes for a 60-amino acid peptide called the **homeodomain.** The amino acid sequence coded by the homeobox has a helix-turn-helix motif similar to DNA binding proteins. In *in vitro* binding studies, homeodomain proteins bind to specific DNA target sequences (see CURRENT INVESTIGATIONS: *The Homeobox*). This

CURRENT INVESTIGATIONS

The Homeobox

The homeobox is a 180-bp DNA sequence that codes for a protein domain of 60 amino acids called the homeodomain. This sequence was discovered as a cross-hybridizing region among the DNA of the *Drosophila Antp, ftz,* and *Ubx* genes. Finding a small region of similar sequence in a few genes is not an uncommon occurrence. However, later studies using this sequence as a probe revealed that at least 20 *Drosophila* genes had a homeobox, all of which were either homeotic genes or genes that control the spatial organization of the embryo. Genes containing homeoboxes also have been found in fungi, in all higher animals including humans, and in all plant species that have been examined. They may, in fact, be present in all eukaryotes. Screenings of the *C. elegans* genome indicate that some 60 genes have homeoboxes. Different homeobox genes in *Drosophila* have different degrees of DNA sequence similarity, ranging from 48 to 81%. Homeobox genes from *Drosophila* and other species also have high degrees of DNA sequence similarity, and high degrees of similarity in the amino acid sequence in their protein products. For example, in the amino acid sequences of the *Drosophila Antp* homeodomain and the human *Oct2* homeodomain, 58 of the 60 amino acids are identical. This suggests that the homeobox sequence has been highly conserved throughout periods of evolution. In *Drosophila* and *C. elegans,* the genes that contain homeoboxes are genes that regulate cell determination and pattern formation, suggesting that the homeodomain may identify genes common to developmental regulation.

The homeodomain has been shown to be a DNA-binding domain that recognizes specific sequences as its binding site. The first indication of this function came from the observation that the homeodomain has a significant similarity to the DNA binding region of the yeast *MAT-a2* gene, a gene whose product is known to bind to DNA. Analysis of the *Drosophila Antp* homeodomain's three-dimensional structure indicates this domain contains a helix-turn-helix structure that can bind to DNA. Individual homeodomain protein-binding sites contain a variety of different DNA sequences. The crystalline structure of the *Drosophila engrailed* homeodomain, when bound to DNA, was recently determined, and specific amino acids were identified as critical for its binding. Most binding sites consist of about 12 bases and contain a TAAT "core" sequence (Figure 15.C). Evidence that the sequence specificity of binding resides within the homeodomain itself has been obtained in several studies. The DNA-binding specificities of the homeodomains of the *Drosophila Antp, deformed, Ubx,* and *Sex combs reduced* genes were found to reside in amino acids 56, 59, and 60. These amino acids are the only ones that differ among the homeodomains of these genes. If these three amino acids are changed, the specificity of DNA binding is also changed.

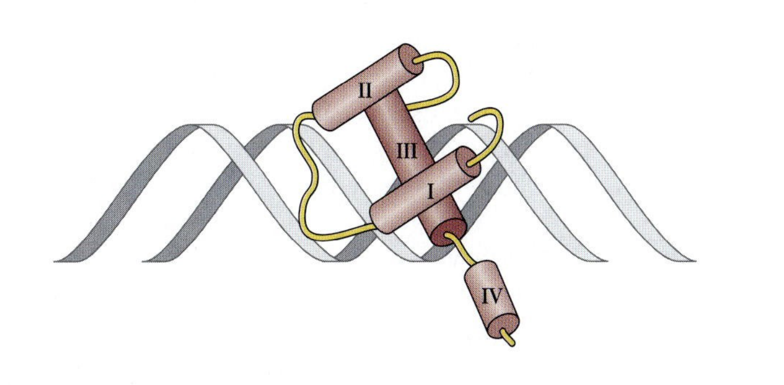

FIGURE 15.C The homeobox protein. The amino acid sequence coded for by the homeobox in the engrailed gene has three alpha helices (I, II, III) and an extended N-terminal arm (IV). The helices are arranged in a helix-turn-helix, three-dimensional structure that can bind to DNA. Specific amino acids in this sequence make contact with specific nucleotides in the homeobox target sequence.

The finding that many homeotic gene products contain a specific sequence of amino acids that recognizes and binds to a specific DNA sequence should not be surprising. These genes are, after all, regulatory genes whose function involves the activation and/or inactivation of determination or differentiation genes. This function requires the gene's products to bind *cis*-acting regulatory sites in or near the target gene promoter. It is interesting from a functional and an evolutionary standpoint that so many developmental regulatory genes in widely different species should have the same DNA-binding motif. This raises the interesting possibility that the homeodomain represents an ancient regulatory system that has been conserved.

suggests that homeotic gene products bind to specific DNA sequences in the promoter regions of other genes. Homeodomain binding sites have been found in the promoters of *Ubx* and in several other genes, suggesting that direct binding of homeotic gene products to promoter regions may be an important action mechanism in these genes. For example, to initiate transcription, the homeodomain-containing product of the *ftz* gene binds to the *Ubx* promoter. To suppress transcription, the homeodomain-containing products of *Abd-B* and *en* bind to the *Ubx* promoter. Using the homeobox as a probe, investigators have discovered homeotic genes in nearly every species of animal. Mammals have a set of homeotic genes, called the *Hox* genes, whose structures are very similar to the *Drosophila* homeotic genes. A more detailed discussion of the *Hox* genes in provided in Chapter 16.

SECTION REVIEW PROBLEMS

12. How would you determine whether a new mutation is part of the *Ubx* group of coordinately controlled *BX-C* genes?
13. Certain DNA sequences in the promoter regions of BX-C act as *cis*-acting regulators, controlling transcription in particular regions of the body or in all cells. If you had a clone containing a DNA sequence from the *Ubx* promoter region of BX-C, how could you determine which type of regulatory sequence was present?

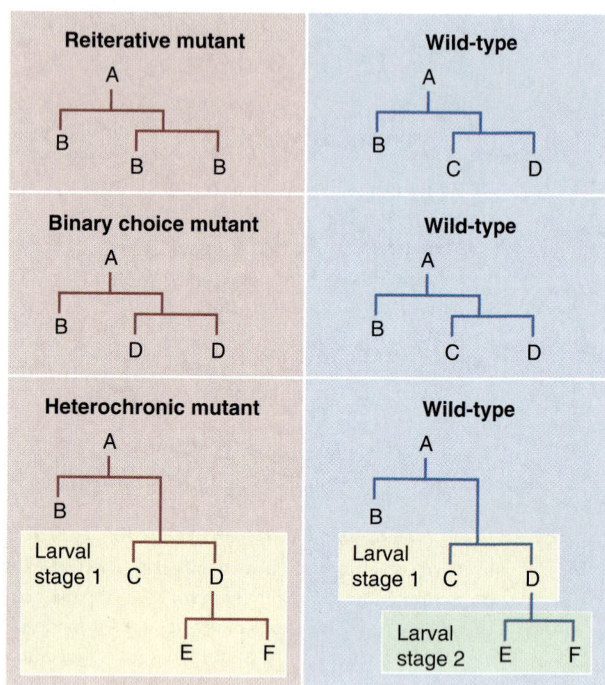

FIGURE 15.23 Homeotic mutations in *C. elegans*. There are three types of homeotic mutations. Reiterative mutations cause a cell to repeat a previous division; binary choice mutations cause a cell that normally gives rise to two different types of cells to give rise to only one; and heterochronic mutations cause cells to become determined at the wrong developmental stage, often earlier in development than is normal.

Cell Determination in *C. elegans* Is Controlled by Maternal Genes and Homeotic Genes.

The importance of cell lineage in the nematode developmental program suggests that the action of cell determination genes and cell division-control genes may be closely related. Genetic analyses of the initial steps in *C. elegans* development have revealed that a large number of genes have maternal effects, suggesting that during the early stages of development many events are controlled by maternal gene action. Consider the example of the maternal effect gene *emb-5*. Mutant *emb-5* alleles alter one particular cell lineage, the E lineage. In wild-type individuals, the first cell of the E lineage arises at the third cleavage division and gives rise to the intestine (Figure 15.14). The first E cell divides to produce two E cells, which migrate to interior positions in the body. After they reach these positions they divide, producing four E cells that each give rise to one fourth of the intestine. In embryos produced by an *emb-5* mutant parent, the two E cells divide prematurely, before they have completed migration, and the four E cells undergo a prolonged period of migration to abnormal positions. Further development of the intestine is abnormal in several ways. Because abnormal timing of the division of the first two E cells is the first effect of *emb-5* mutations, the normal function of this locus must be to delay this division until the cells reach a specific location. Because *emb-5* is a maternal-effect mutation, the maternal genome must control the timing and location of individual cell divisions in early embryonic development.

The homeotic genes in the nematode are divided into three classes of mutations according to how they affect cell determination and cell division. These classes of homeotic mutations are called reiterative, heterochronic, and binary choice (Figure 15.23). **Reiterative mutations** cause particular cell divisions to be repeated; that is, instead of following its own lineage pattern, a progeny cell repeats the parental division pattern. **Binary choice mutations** affect particular cell determination events. **Heterochronic mutations** cause particular developmental events to occur at unusual developmental times.

Cell–Cell Communication Regulates Individual Cell-Fate Decisions.

After segment determination is complete, separate fates must be established for individual cells in each segment.

Local cell–cell interactions play an important part in these fine-scale, cell-fate decisions, especially those decisions that involve the specialization of individual cells in a previously identical group. Cell–cell signaling mechanisms can be divided into two general classes. One class, **lateral specification,** involves cell–cell signaling between initially equivalent cells that leads to their adopting different fates. Lateral-specification signal pathways can cause one cell or a group of cells to adopt a different fate from surrounding cells, or they can prevent the surrounding cells from also adopting that fate. The second class, **lateral induction,** involves cell–cell signaling in which one cell influences the cell-fate choice of a second, nonidentical cell. In this section, we will consider examples of genes that control both types of cell–cell signaling systems and the cell-fate decisions they regulate.

One gene that is involved in several lateral specification cell-fate decisions in *Drosophila* development is the *Notch* (*N*) locus. *Notch* is a large gene, occupying 40 kb of genomic DNA and coding for a 300-kd protein with 2703 amino acids. This protein acts as a transmembrane receptor and has a complex structure containing a variety of repeated amino acid sequences. For example, the Notch protein extracellular domain contains 36 repeated copies of an epidermal growth factor-like sequence. The Notch protein serves as a receptor that binds to the product of the *Delta* (*Dl*) gene during neural cell–cell signaling. Delta is an extracellular protein that contains several conserved amino acid sequences that interact with the epidermal growth factor-like repeats of the Notch protein. As a consequence, loss-of-function *Delta* mutations produce the same neural cell-fate decision abnormalities as *Notch* mutations. In this case, cell–cell signaling appears to involve a Delta protein on the surface of one cell binding to a Notch protein on a neighboring cell. This binding activates the Notch signal pathway and, in a manner that is not yet understood, sends to the nucleus a signal that alters the expression of a particular set of target genes. This alteration is an essential part of the cell-fate decision.

One of best-studied examples of cell-fate regulation involving Notch signaling is the generation of the precursor cells of the central nervous system. A group of about 1800 cells in the neural region of the blastoderm gives rise to two types of embryonic cells; one type is the precursor of the nervous system (called neuroblasts) and the second is a class of epidermal cells (called dermoblasts). About 25% of the cells in the neural region become neuroblasts that are interspersed among the dermoblasts throughout the region. Initially, all the cells in the neural region express genes of the *achaete-scute* complex. The products of these genes are transcriptional control factors that must be present for cells to adopt a neuroblast fate. In cells that adopt a dermoblast fate, lateral cell–cell interactions mediated by *Notch* repress *achaete-scute* expression. Cells that adopt a neuroblast fate continue to express *achaete-scute*. In individuals homozygous for loss-of-function alleles of *Notch*, all cells in the neural region continue to express *achaete-scute*, and all adopt a neuroblast fate. This leads to a lethal overproduction of neuroblasts. If a presumptive neuroblast is removed or killed, one of the adjacent cells that normally would have formed a dermoblast forms a neuroblast instead. These results indicate that *Notch*-mediated signals sent from developing neuroblasts inhibit their neighbors from adopting a neuroblast fate, and that this inhibition involves repressing the expression of genes in the *achaete-scute* complex.

The *Notch* gene is widely expressed in proliferating cells during the embryonic, larval, and adult development. Mutational analyses indicate that *Notch* is required for proper cell-fate decisions in the mesoderm, germ-line cells, ovarian follicle cells, adult wing formation, adult peripheral nervous system formation, and adult eye formation. In each of these developing tissues, the loss of the Notch protein causes cells that have several developmental options to chose an inappropriate fate. Lateral specification involving *Notch* signaling may play an important part in a variety of cell-fate decisions.

A great deal of evidence suggests that *Notch*-like cell–cell signaling plays an important role in cell-fate decisions in other species. Genes with structures similar to *Notch* have been discovered in several species, including *C. elegans*, zebra fish, *Xenopus*, chickens, mice, rats, and humans. The *C. elegans* gene *lin-12* has a structure similar to *Notch*, and it plays a role in cell–cell signaling during ovary formation. The nematode ovary is bilaterally symmetrical, with two arms that meet in the center (Figure 15.15). The specialized anchor cell, or AC, is located between the two arms. There are normally two cells that have the potential to form either an AC cell or a ventral uterine precursor cell (VU), and the fate of both cells depends on cell–cell interactions. Both cells initially produce a certain amount of the AC determinative signal and its cell surface receptor (*lin-12*), but later one begins to produce more signal and less receptor (and develops into the AC) and the other begins to produce more receptor and less signal (and develops into VU). Mutational analysis confirms that *lin-12* signaling is essential for this cell-fate decision. In individuals with *lin-12* loss-of-function mutations, both cells produce signal but no receptor, and both become AC cells. In individuals with *lin-12* overexpression mutations, both cells have an excess of receptors and become VU cells (Figure 15.24). The role of *lin-12* is to receive a signal that causes the cell to adopt a VU fate. There is clearly a competition between the signal and the receptor to determine cell fate in the two cells.

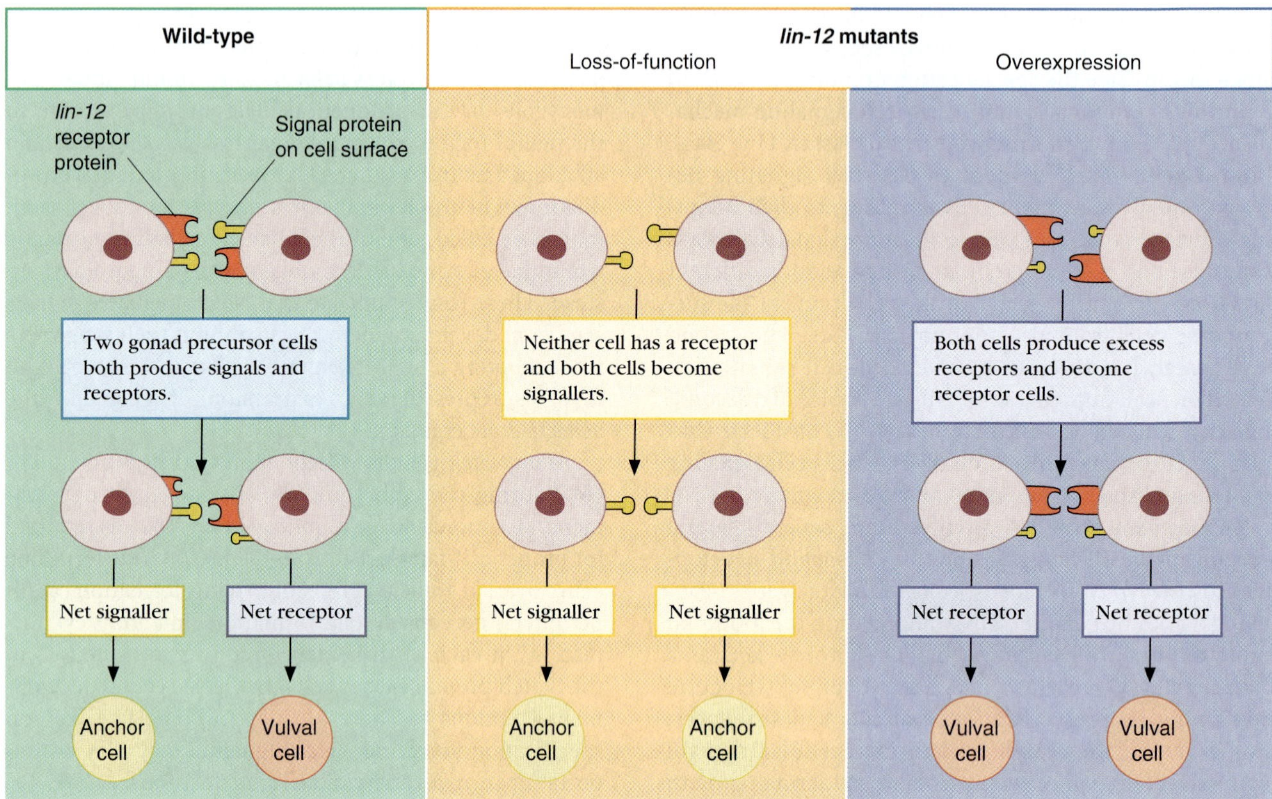

FIGURE 15.24 Cell–cell signaling. The product of the *lin-12* gene is a cell surface receptor for the anchor cell (AC) determination signal. Two gonadal precursor cells that have the potential to form an AC initially produce both signal proteins and signal protein receptors. One cell eventually produces more signal and develops into the AC, and the other produces more receptor and develops into a vulval cell. Mutations that change the function of the *lin-12* gene can change the fate of both cells.

Lateral induction is the second type of cell–cell signaling that can influence cell-fate decisions through an interaction between two nonidentical cells in which one cell causes the other to adopt a specific cell fate. An example of lateral induction in *C. elegans* is the influence of the AC on the group of cells that form the vulva (vulva precursor cells, or VPC). All the cells in the VPC are capable of adopting any of these fates, and their choice is determined by a signal sent from the AC. The particular fate each cell adopts depends on its position with respect to the AC. The closest cell to the AC normally adopts a 1° cell fate, and the two next closest adopt a 2° fate. If the AC is removed, these cells do not form a vulva, indicating an inductive signal from the AC is necessary for these cells to adopt a vulval cell fate. The vulval induction signal is encoded by the *lin-3* gene. This gene codes for a transmembrane protein that has an amino acid domain like the epidermal growth factor (EGF), similar to that of the *Notch* gene product. The receptor for this signal is encoded by the *let-23* gene. The *let-23* gene product is a transmembrane protein with a structure similar to other EGF receptors. The VPCs exposed to the highest concentrations of *lin-3* protein adopt the 1° cell fate, cells exposed to an intermediate concentration adopt the 2° fate, and cells exposed to a low concentration adopt the 3° fate.

SOLVED PROBLEM

Problem

The recessive mutation *hairy* (*h*) in *Drosophila* causes an increase in the number of bristles formed on the thorax, abdomen, and wing. The extra bristles are usually found in clumps adjacent to normal bristles. If you generate a somatic mosaic in which the marked clone of cells is homozygous for the hairy mutation (*h/h*) but the surrounding cells are not (*h*/+), then cells in the clone will form bristles, but the surrounding cells will not. If such a clone includes some of the cells immediately adjacent to a large thoracic bristle, the cells in the clone that are within three to four cells of the normal bristle site form extra bristles, but cells not in the clone that are three to four cells from the bristle site do not. What do these results tell you about how bristles are formed, and what do you suggest may be the role of the hairy gene in this process?

Solution

The bristle formation process appears to have several steps. First, a group of cells in and near the site of a bristle become competent to form a bristle. Next, one cell becomes determined to form the bristle and sends lateral inhibition signals to the surrounding cells that prevent the other cells in the group from forming a bristle. Because the cells in the group that are homozygous for a mutation of *hairy* form bristles anyway, these cells must not be inhibited. This suggests the normal function of the hairy gene is essential for inhibition.

SECTION REVIEW PROBLEMS

14. Explain the difference between lateral specification and lateral induction.
15. If you were to use a laser microbeam to eliminate the P6.p cell from a first larval-stage *C. elegans*, would that individual develop a vulva? Explain why or why not.
16. You have isolated mutations in two genes (*b* and *c*) in *C. elegans* that cause the failure of a binary choice cell-fate decision (A vs. B), and you suspect this decision is controlled by a mechanism similar to that controlling the AC–VU decision. How might you test your hypothesis?
17. *Drosophila* homozygous for the mutation *scute* do not form scutellar bristles. Curt Stern induced somatic clones of homozygous *scute* mutant cells and observed that if such a clone occupies the site where a scutellar bristle normally forms, no bristle is formed. However, a *scute*$^+$ cell adjacent to the clone may form a bristle. What does this indicate about the interaction between the scutellar bristle-forming cell and its neighbors in a normal individual?
18. If you used a laser microbeam to eliminate the precursor cell of a scutellar bristle, would a bristle form?

Summary

In this chapter, we have discussed some of the fundamental processes of eukaryotic development and some of what we now understand about the genetic mechanisms that control these processes. A common theme in eukaryotic development is that embryonic cells with the capacity to form a variety of specialized cell types become committed to form one specific cell type by making a series of cell-fate decisions. Each of these decisions is controlled by regulatory genes and involves the activation and/or inactivation of specific sets of genes. In our discussions, we have used examples from two model organisms, *Drosophila* and *C. elegans*, because these organisms lend themselves to experimental genetic manipulation. Similar cell-fate decisions are made during the development of all higher organisms, including humans. In fact, there are human counterparts of many of the genes we have discussed, and they are currently being studied. One of the most important conclusions we have made from modern developmental genetic investigations is that there is a great deal of similarity between the genes that control development in different species.

Our discussions included examples of regulation of cell fate at three different levels of development. The initial cell determination decisions that occur during the early stages of embryonic development are controlled by the maternal genome in many species. Maternal genes produce cytoplasmic determinants or positional information molecules that are packaged in the oocyte cytoplasm to control the initial cell-fate decisions of embryonic cells. Later in development, at about the blastoderm stage, homeotic genes are activated in groups of cells to commit these cells to form particular regions of the body. Homeotic gene activation occurs in a spatially defined manner to produce the basic organization of the body. The products of the homeotic genes are determinants that continuously reinforce the cell's determined state. Continuous expression of the homeotic genes throughout development ensures that determined cells never "forget" their state, even if subjected to severe experimental manipulation. Similar behaviors in other cell-fate-controlling genes may account for the stability of other cell-fate decisions. Finally, during the development of individual tissues, cell–cell signaling mechanisms regulate individual cell-fate decisions. This signaling can be lateral specification signals between two originally identical cells that cause them to adopt different fates. It can also be inductive signaling, with signals sent from one cell directing the determination of another, nonidentical cell.

Selected Key Terms

Antennapedia complex
binary choice mutation
Bithorax complex
cellular blastoderm
cell determination
cell differentiation
cell equivalence group
cell fate
cell lineage
cytoplasmic determinant
gynandromorph
heterochronic mutation
homeobox
homeotic gene
imaginal disc
metamorphosis
mosaic development
pattern formation
polar granules
pole cells
Polycomb group
positional information
regulative development
reiterative mutation
syncytial blastoderm
terminal differentiation

Chapter Review Problems

1. If all the cells of an organism have the same genes, how can different cells have different gene activities?
2. How would you test the hypothesis that the germ line in *C. elegans* is established by a cytoplasmic determinant?
3. One possible explanation for the cell-lineage specific determination of the initial *C. elegans* cell divisions is the presence of cytoplasmic determinants in the egg. Explain how you might test this hypothesis.
4. During embryonic development of *Drosophila,* a small group of brain neurons makes the cell-fate decision to become the precursor cells of the optic lobe of the adult brain. These cells become visibly distinct by the end of embryonic development. How could you discover whether this cell-fate decision is a determination or a specification?
5. In many animal species, the egg has two poles, called the animal pole and the vegetal pole. The blastoderm cells that receive animal pole cytoplasm develop into the embryonic body. How could you discover whether there are cytoplasmic determinants in the cytoplasm of the animal pole that cause animal pole cells to develop this way?
6. Give a brief definition of the following terms:
   a. cell determination
   b. cell differentiation
   c. cell-fate decision
7. Give a brief description of one difference between the following pairs of items/terms:
   a. mosaic development vs. regulative development
   b. positional information vs. cytoplasmic determinant
8. Would you classify the *bcd* gene product gradient in the embryo as an example of a mosaic or a regulative mechanism? Explain your answer.
9. Explain how you could use Illmensee's transplantation technique to discover whether the RNA or the protein component of pole cells is the cytoplasmic determinant.
10. If a UV laser microbeam is used to destroy a small group of nuclei in a *Drosophila* embryo at the syncytial blastoderm stage, defects are rarely or ever produced in the resulting larva or adult. However, if a group of cells is destroyed at cell blastoderm, defects are usually produced. What does this indicate about the timing of cell determination in the *Drosophila* embryo?
11. Would induced somatic mosaic analysis be a good technique for investigating the time of cell determination in *C. elegans?* Explain your answer.
12. *Drosophila* embryos expressing the *nanos* mutant phenotype have no abdominal segments. Embryos expressing the *bicoid* mutant phenotype have no head segments. What can you conclude about the normal functions of these genes?
13. How would you test the hypothesis that the dorsal/ventral body axis in *Drosophila* embryos is established by a gradient of positional information with the product of the dorsal gene being the positional information molecule?
14. How would you test the hypothesis that a cell-fate decision occurs at 5 hours of embryonic development that separates one precursor cell population into two, one that produces the first leg and another that produces the second leg in *Drosophila?*
15. We have assumed in our discussions of *Drosophila* development that before the cell blastoderm stage all nuclei have the same set of genes. Describe how you might test this assumption.
16. Explain why the minimum patch size analysis discussed in CURRENT INVESTIGATIONS: *Using Somatic Mosaics in Developmental Genetic Analysis* cannot be used to estimate the number of cells present in a growing imaginal disc at an intermediate stage of development.
17. The average number of marked hairs in clones induced in imaginal leg discs at different stages of development follow. Assuming that there are 30,000 hairs on an adult leg, how many cells were in the leg disc when these clones were induced? a. 1200 b. 1000 c. 20 d. 2
18. If each *Drosophila* segment is formed by a 4-cell-wide stripe of blastoderm cells, and structures 1 segment apart are 5 Sturts apart in a gynandromorph fate-mapping experiment, how many Sturts apart are the blastoderm precursor cells for the following structures? (Note: For ease of calculation, assume that the structures are formed by cells in the middle of each segment.)
    a. wing and second abdominal segment (3 segments apart)
    b. prothoracic leg and metathoracic leg (2 segments apart)
    c. haltere and eighth abdominal segment (8 segments apart)
19. How would you test the hypothesis that the Notch-like gene *lin-12* is part of a lateral signaling system between the

vulval precursor cells that is used by the 1° and 2° cells to prevent 3° cells from forming vulval structures?

20. If the primary cell in the VPC is removed, a secondary cell replaces it by becoming a primary cell. Explain how the result indicates whether the VPC is controlled by a mosaic or a regulative system.
21. Based on his findings that the *abx, bx, pbx,* and *bxd* genes have coordinate regulation, Lewis suggested the *BX-C* might resemble a bacterial operon. How would you test the hypothesis that these four genes are part of an operon that resembles the *lac* operon?
22. How could you determine whether the *bx* gene of the *BX-C* is actually being transcribed in the abdomen?
23. *Antp* is expressed in the abdominal segments of *Drosophila* individuals with *Ubx* deficiency mutations, but not in wild-type individuals. Explain how this might occur.
24. The following list gives the number of cells in imaginal leg discs at different stages of development. Leg discs contain 30,000 cells at the end of development. If a somatic cell clone is induced in a leg disc at each of these stages, how many cells will be in each clone at the end of development?
    a. 20-cell stage
    b. 100-cell stage
    c. 500-cell stage
    d. 1000-cell stage
    e. 5000-cell stage
25. How would you test the hypothesis that the Notch protein is expressed on the surface of cells that become dermoblasts in the neural region of the blastoderm?
26. The human equivalent of the *Notch* gene is expressed in early embryonic development. How would you test the hypothesis that this gene has the same function as the *Drosophila Notch* gene?
27. A new experimental treatment of *C. elegans* with antisense RNA is able to block the action of individual genes selectively. What effect on development would you expect to result from treatment of embryos with antisense RNA for the *lin-12* gene?
28. Two unique body structures arise from adjacent cells (Cell A and Cell B). Cell A is proposed to influence cell B by a process of induction. Explain how this might work.
29. Explain how you might test the induction hypothesis proposed in Problem 28.
30. A recessive, cell autonomous mutation that eliminates both cell A and cell B has been discovered. Explain how you could use this mutation in a somatic mosaic analysis to test the hypothesis that cell A induces its neighbor to become cell B, as proposed in Problem 28.
31. Describe one similarity between the following items:
    a. imaginal disc and cell equivalence group
    b. binary choice mutation and heterochronic mutation
    c. syncytial blastoderm and cellular blastoderm
32. In the *Drosophila* embryo, embryonic neuroblasts use the *Notch* signal pathway to influence the cells adjacent to them, causing these neighboring cells to become epidermal cells and not neuroblasts. Describe how you could discover whether this influence is an induction or an inhibition.
33. Give a brief definition of the following:
    a. reiteration mutation
    b. heterochronic mutation
    c. binary choice mutation
34. Explain how you could use Illmensee's transplantation technique to discover whether the imaginal discs that form the adult wing are founded by cells that receive a cytoplasmic determinant.

Challenge Problems

1. If you were to transform *C. elegans* with a fusion gene containing the regulatory region of the *hb* transcription unit and the structural region of β-galactosidase, would you expect to see β-galactosidase produced?
2. In a test of cell determination, you dissociated wing imaginal discs from an individual homozygous for the *yellow* (*y*) mutation and leg imaginal discs from an individual homozygous for the *ebony* (*e*) mutation into single cells. You then mixed these cells together and cultured the mixture in a wild-type larval host. After the host completed metamorphosis, you recovered the implant and examined it for leg and wing structures. What results would you expect?
3. You are interested in discovering the time in development of the nematode when the progeny cells of the P3 blast cell begin to form their cell equivalence group. The scientist in the next lab who works with *Drosophila* suggests that you consider doing an induced somatic mosaic analysis. Do you believe that this would be a good procedure? Explain the results you would expect from such a test, and explain why this would or would not be a good procedure.

Suggested Readings

Artavanis-Tsakonas, S., K. Matsuno, and M. E. Fortini. 1995. "Notch signaling." *Science,* 268: 225–232.

Campos-Ortega, J. A. 1990. "Mechanism of a cellular decision during embryonic development of *Drosophila melanogaster:* Epidermogenesis or neurogenesis." *Advances in Genetics,* 27: 403–453.

Cohen, S., and G. Jurgens. 1991. "*Drosophila* headlines." *Trends in Genetics,* 7: 267–272.

Gehring, W. J., and Y. Hiromi. 1986. "Homeotic genes and the homeobox." *Annual Review of Genetics,* 20: 147–174.

Lewis, E. B. 1978. "A gene complex controlling segmentation in *Drosophila.*" *Nature,* 276: 565–570.

Lewis, E. B. 1994. "Homeosis, the first 100 years." *Trends in Genetics,* 10: 341–343.

Nakamura, A., R. Amikura, M. Muki, S. Kobayashi, and P. F. Lasko. 1996. "Requirement for a noncoding RNA in Drosophila polar granules for germ cell establishment." *Science* 274: 2075–2079.

Peifer, M., and A. Bejsovec. 1992. "Knowing your neighbors: Cell interactions determine intrasegmental patterning in *Drosophila.*" *Trends in Genetics,* 8: 243–249.

Sternberg, P. W. 1990. "Genetic control of cell type and pattern formation in *Caenorhabditis elegans.*" *Advances in Genetics,* 27: 63–116.

Wilmut I., A. E. Schnieke, J. McWhir, A. J. Kind, and K. H. S. Cambell. 1997. "Viable offspring derived from fetal and adult mammalian cells." *Nature,* 385: 810–813.

On the Web

Visit our Web site at **http://www.saunderscollege.com/lifesci/titles.html** and click on A/G/M Genetics for links to the following chapter-related resources on the World Wide Web:

1. **The interactive fly.** A cyberspace guide to *Drosophila* genes and their roles in development. This site contains information on the structure, function, and history of the investigation of the developmental regulatory genes discussed in this chapter.

The following are two *C. elegans* Web sites for information on *C. elegans* genetics and development:

2. **The worm breeder's gazette.** *Caenorhabditis elegans* WWW Server, University of Texas Southwestern Medical Center at Dallas.
3. **A genetic database for *C. elegans*.** *Caenorhabditis* Genetics Center, 250 Biological Sciences Center, 1445 Gortner Avenue, University of Minnesota, St. Paul, MN 55108–1095.

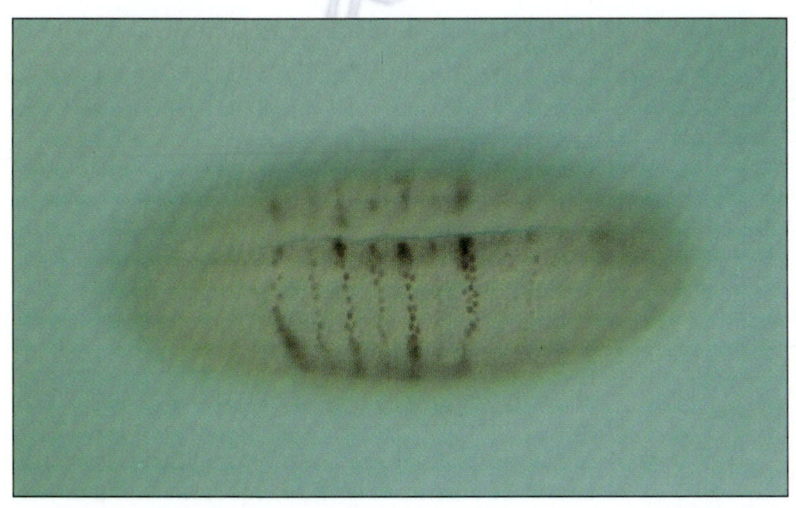

The expression of *engrailed*. The *engrailed* gene controls cell fate in each body segment of *Drosophila melanogaster*. The *engrailed* gene is first expressed in a series of heavy and light stripes during the early stages of embryonic development. *(Jack Girton/Iowa State University)*

CHAPTER 16

Developmental Genetics: Hierarchies of Genetic Regulation

Genes control a cell's decision to adopt a particular cell fate. The decision of each cell to become a particular cell type is controlled by the products of regulatory genes acting as determinants or in cell–cell signaling systems. Genes that regulate cell-fate decisions must be activated and/or inactivated in specific locations and at precisely the right times to generate new cell types when and where they are needed in the developing individual. We do not know a great deal about how this is done, but some fundamental principles are becoming clear. We do know that during development, cell fate and the organization of cells into tissues and organs are controlled by groups of interacting genes. In this respect, specific groups of genes that control particular aspects of the spatial and temporal organization of cells during development have been identified. These genes function in a carefully controlled sequence. Another rather amazing finding is that the gene groups that control development are highly conserved across species. Many genes have been found to be essentially identical in DNA sequence and/or in function in a wide range of different species. This suggests that the genetic mechanisms governing fundamental developmental processes are common to many species.

In this chapter, we will discuss several examples of the genetic systems that organisms use to generate spatial patterns of cell types during development. We will also discuss the gene systems that organisms use to make fundamental decisions that affect all the cells in the individual. The systems we will discuss include the genes that organize cells into segments in *Drosophila*, the homeotic genes of animals, and the genes that control sex determination in *Drosophila* and in the nematode. By the end of this chapter, we will have answered four basic questions about development:

1. What role do maternal gene products play in organizing the embryonic body?
2. How do genes organize embryonic cells into body segments?
3. What are the functions of the homeotic genes?
4. How do the sex chromosomes determine an individual's sex?

Gene Hierarchies Control Embryonic Pattern Formation.

One of the oldest questions in developmental biology is how cells in the developing individual become organized into tissues and organs. This organization process is called **pattern formation,** and although we do not know everything about how pattern formation is controlled, some fundamentals of the process are becoming clear. The first step of embryonic pattern formation is controlled by maternal genome genes that produce gene products, either cytoplasmic determinants or positional information systems, that are incorporated into the egg cytoplasm in precise spatial patterns. These maternal gene products activate the first developmental regulatory genes in the zygote genome, and they activate these genes in a precise, spatially correct pattern. The action of these first zygotic genes begins a sequence of gene actions that eventually results in the formation of precise spatial patterns of cell determination in the embryo, accomplished by interacting groups of zygotic genes called gene hierarchies. The sequence of actions taken by gene hierarchies is called a **developmental pathway.** The maternal gene products activate the first genes in the hierarchies, and these activate others, until eventually different groups of cells in each region of the embryo have unique sets of active determination genes.

Modern investigations of developmental genetic pathways seek to identify the gene members of particular gene hierarchies, to determine the role of each gene, and to discover how the genes are controlled. The main tools used in these analyses are mutagenesis, developmental analysis of mutant phenotypes, and molecular genetic analysis of gene structure and function. One important mutagenesis strategy is to identify all the genes in the genome that are involved in a particular pathway by **saturation mutagenesis.** In any mutagenesis experiment, mutant organisms are isolated. The aberrant phenotype of these organisms suggests that the normal function of the gene is important for one particular developmental pathway. For example, *pbx* mutations in *Drosophila* cause transformations of the first abdominal segment into a thoracic segment, suggesting that the normal function of this gene is required for normal cell determination in the first abdominal segment. In a saturation mutagenesis, the experiment is continued until researchers have recovered the mutant alleles of all the genes in the genome that can mutate to give that phenotype. Next, the mutants are analyzed using standard genetic techniques (for example, complementation tests and mapping) to determine the number and location of the genes. The mutant genes are then cloned, and their molecular structures and functions are defined.

Saturation mutagenesis of gene hierarchies is easy to describe but frequently requires an enormous amount of work. Analyses of this extent can be carried out only in certain model organisms such as *Drosophila* and *C. elegans*. In the next two sections of this chapter, we will consider some examples of *Drosophila* gene hierarchies that control early embryonic pattern formation events.

The *Drosophila* Body Axis Is Established by Gradients of Maternal Gene Products.

The embryonic body of *Drosophila* begins to be organized with the action of maternal genes that establish the body polarity (the anterior–posterior and dorsal–ventral axes). *Drosophila* embryos always develop with a certain polarity, that is, the head always forms at the anterior end of the egg; the germ cells form at the posterior; the dorsal surface (the back) is always on the upper side of the egg; and the ventral (the front) is always on the lower side of the egg. This correlation between the egg and the final body indicates that the maternal genome, which controls the polarity of the egg, also establishes the polarity of the embryo. The genes controlling the establishment of the anterior–posterior and dorsal–ventral body (called **axis-polarity genes**) are maternal-effect genes. These genes are transcribed in the mother, but their products (mRNA and/or protein) are packaged in the egg, where they control the embryonic body polarity. For example, large amounts of the *dorsal* gene product are packaged in a specific region of the oocyte, and this placement determines orientation of the dorsal–ventral

TABLE 16.1 *Genes Controlling Development of the Anterior/Posterior Axis in* Drosophila.

Gene Class	Gene Symbol	Mutant Phenotype
Anterior class		
bicoid	*bcd*	Mutations cause maternal effect transformations that produce embryos with two posteriors and no anterior.
Bicaudal-D (2)	*Bic-2*	
Bicaudal-D (3)	*Bic-3*	
swallow	*swa*	
exuperantia	*exu*	
Posterior class		
nanos	*nos*	Mutations cause maternal effect transformations. Embryos have reduced or missing segments.
oskar	*osk*	
vasa	*vas*	
staufen	*sta*	
Terminal class		
torso	*tor*	Mutations cause maternal effect transformations. Embryos have no terminal regions.
torso-like	*tsl*	
trunk	*trk*	
fs(1)Nasrat	*fs(1)N*	
pole hole	*phl*	

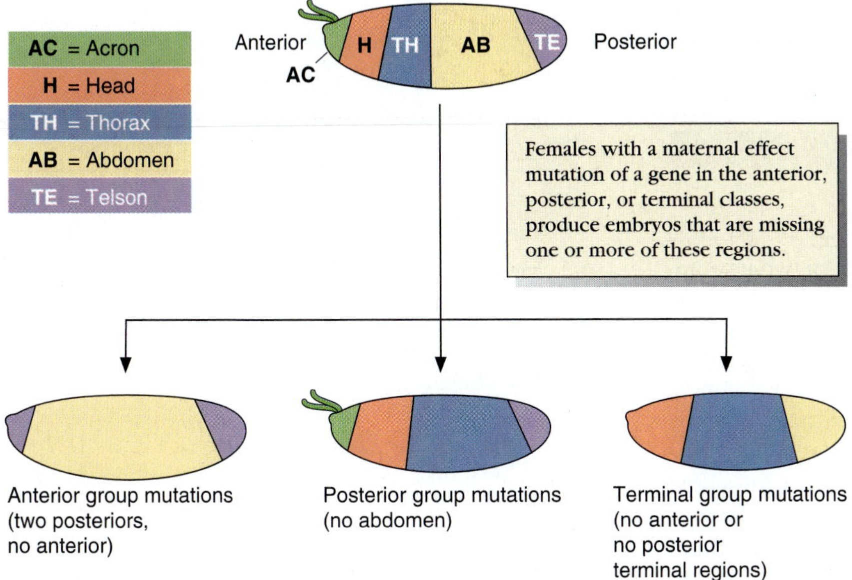

FIGURE 16.1 Maternal axis-polarity genes. The maternal axis-polarity genes of *Drosophila* divide the embryo into five regions: acron (AC), head (H), thorax (TH), abdomen (AB), and telson (TE). The cells in the head, thorax, and abdomen regions give rise to exterior body structures. Females with a maternal-effect mutation of a gene in the anterior, posterior, or terminal classes produce embryos that are missing one or more of these regions.

body axis. Mutations of these genes in the mother's genome cause alterations in the phenotype of the offspring, regardless of the genotype of the offspring. In the rest of this section, we will discuss the genes that control the anterior–posterior axis.

The genes controlling development along the anterior–posterior body axis fall into three gene classes (Table 16.1). The anterior, or **bicoid,** class contains five genes. Females homozygous for mutant alleles of any one of these genes produce embryos that have the anterior end of the embryo replaced by a mirror-image duplication of the posterior. The posterior, or **oskar,** class contains four genes, and females mutant for one of these genes produce embryos that have no abdominal segments. The third, or **terminal,** class contains at least five genes, and females mutant for any of these genes produce embryos that are missing the extreme terminal portions at both the anterior and posterior ends of the embryo (Figure 16.1). The embryo can be divided into five different zones that reflect the regions affected by mutations of these three classes of genes: the acron (anterior pole), head, thorax, abdomen, and telson (posterior pole).

Each group of maternal axis-forming genes produces a localized concentration of one gene product in a particular region of the embryo. The anterior group genes establish a concentration gradient of the *bicoid* (*bcd*) gene product in the embryo, with the highest concentration in the anterior end of the embryo and low concentrations in the middle and posterior (described in Chapter 15). The posterior group establishes a concentration gradient of the *nanos* (*nos*) gene product, with a high concentration in the posterior of the embryo. The terminal group establishes high concentrations of the *torso* (*tor*) gene product at both ends of the embryo (Figure 16.2). The products of these genes are transcriptional regulatory molecules (activators or repressors) that bind to promoter regions of zygotic genes that control segmentation of the body, either inducing transcription or repressing transcription of genes. The concentration differences of

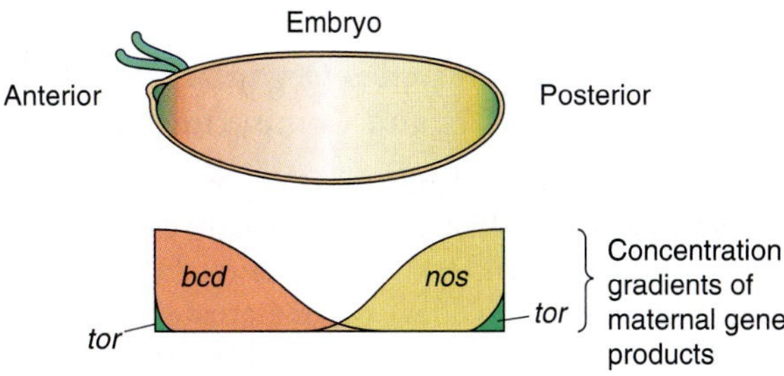

FIGURE 16.2 Gradients of maternal products. The maternal axis-polarity genes establish concentration gradients of the products of the *bcd* (anterior), *nos* (posterior), and *tor* (terminal) gene products in the embryo. These gradients control the spatial activity of zygotic patterning genes.

the maternal gene products in different regions of the egg cause different zygotic genes to become active in different regions of the embryo.

SOLVED PROBLEM

Problem

How would you discover whether a gene is an axis-forming gene in *Drosophila,* and if it is, in what group it belongs?

Solution

The axis-forming genes in *Drosophila* have two important characteristics that can be tested. First, mutations of these genes give maternal-effect phenotypes that include the alteration or elimination of one region of the embryo. Thus one test would be to isolate mutations of this gene, generate females homozygous for the mutation, and examine the embryos produced by these females for abnormalities of axis formation. The second characteristic is that axis-forming genes are active in the mother, and not in the embryo. Thus a second test would be to use a probe that will hybridize to the mRNA produced by this gene to determine whether the gene is transcribed during oogenesis by maternal cells. If the gene is an axis-forming gene, its class can be determined by examining the details of the maternal-effect phenotype it produces and comparing this with that of known axis-forming genes.

SECTION REVIEW PROBLEMS

1. If you were given a strain of *Drosophila* with a new maternal-effect mutation that is thought to be an allele of one of the anterior group genes, how would you determine whether this hypothesis is correct?
2. The dorsal–ventral body axis is thought to be established by a gene hierarchy that produces a gradient of the protein product of the *dorsal* gene. How could you test this hypothesis?

TABLE 16.2 *Segmentation Genes in* Drosophila

Gene Class	Gene Symbol	Mutant Phenotype
Gap		
tailless	*til*	Mutations cause groups of adjacent segments to be missing. Each gene affects a different group of segments.
huckebein	*hkb*	
orthodenticle	*otd*	
empty spiracles	*ems*	
buttonhead	*btd*	
hunchback	*hb*	
Krüppel	*Kr*	
Knirps	*Kni*	
giant	*gt*	
Pair-rule		
barrel	*brr*	Mutations cause defects in alternate segments.
fuzi tarazu	*ftz*	
hairy	*h*	
paired	*prd*	
runt	*run*	
evenskipped	*eve*	
oddskipped	*odd*	
Segment-polarity		
engrailed	*en*	Mutations cause portions of every segment to be missing. Mutations of different genes may affect similar or different portions of each segment.
hedgehog	*hh*	
patched	*pat*	
costal-2	*cos-2*	
cubitus interruptus	*ci*	
naked	*nkd*	
shaggy	*sgg*	
wingless	*wg*	
disheveled	*dsh*	
porcupine	*porc*	
fused	*fu*	
gooseberry	*gsb*	

Zygotic Segmentation Genes Organize the Embryo into Segments and Compartments.

Segmentation is a fundamental characteristic of the *Drosophila* body (the different segments are described in Chapter 15). The segmental pattern is established by a gene hierarchy that includes both maternal and zygotic genes. Before the cellular blastoderm stage, zygotic **segmentation genes** are activated in different regions of the embryo. Segmentation genes are organized into three groups based on the phenotype of their mutant alleles: gap genes, pair-rule genes, and segment-polarity genes (Table 16.2). Embryos homozygous for mutations of the **gap genes** are missing a group of adjacent larval segments, producing a "gap" in the segmentation pattern. For example, *Krüppel* is a gap gene, and embryos homozygous or heterozygous for the mutant allele *Krüppel* (*Kr*) are missing the thoracic and anterior abdominal segments (Figure 16.3). Embryos homozygous for mutations of one of the **pair-rule genes** are missing every other segment. For example, larvae homozygous for the pair-rule mutation *even skipped* (*eve*) are missing all the even-numbered segments, and larvae homozygous for

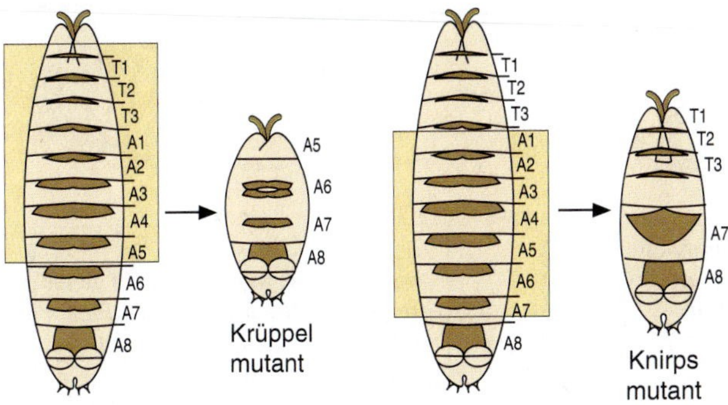

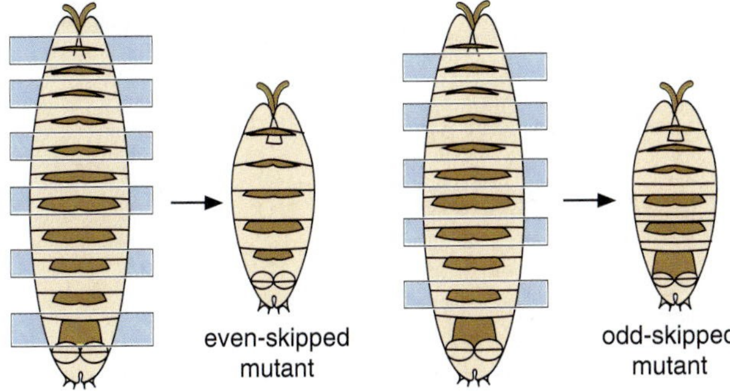

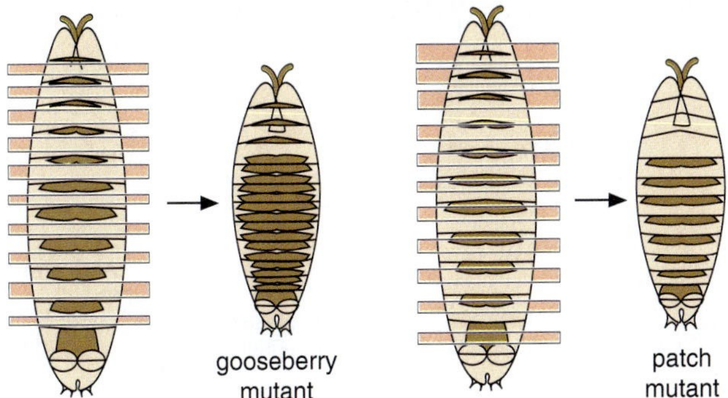

◀**FIGURE 16.3 Gap, pair-rule, and segment-polarity gene mutations.** Embryos homozygous for mutations of (a) gap genes, (b) pair-rule genes, and (c) segment-polarity genes are missing particular embryonic segments or parts of segments. Highlighted areas indicate the missing segments.

the mutation *odd skipped* (*odd*) are missing all the odd-numbered segments (Figure 16.3). Individuals homozygous for mutant alleles of the **segment-polarity genes** are missing a particular portion of every segment. For example, embryos homozygous for the mutations *gooseberry* (*gsb*) or *patch* (*pat*) are missing a specific portion of each segment.

The gap genes become active in the syncytial blastoderm stage, two nuclear divisions prior to cell formation. The blastoderm location in which each gap gene is transcribed is controlled by the products of the maternal-axis genes. The maternal gene products regulate gap-gene transcriptional activity by binding to regulatory sites in the gap-gene promoter regions. The gap-gene products in turn activate or repress pair-rule genes, which then activate or repress segment-polarity genes. For example, the *bcd* gene (a maternal gene) product is a transcriptional activator of the *hunchback* (*hb*) gene (a gap gene). The high concentrations of *bcd* product in the anterior of the embryo activates *hb* transcription (described in Chapter 15). The maternal gene products act on the gap genes to produce a series of syncytial blastoderm nuclei stripes that express (transcribe) each gap gene. These stripes are initially overlapping, but they become discrete by mutual inhibitory interactions between the gap genes. The products of the gap genes contain DNA binding domains, and each gap-gene product binds to the promoter region of other gap genes. For example, the *Knirps* (*Kni*) gene product is a negative regulator of *Krüppel* (*Kr*), meaning that if both *Kni* and *Kr* are initially activated in a nucleus, the *Kni* product blocks the transcription of *Kr*. This type of inhibition sharpens the borders of the gap genes' activity domains. The regulatory interactions limit the activity of each gap gene to a broad stripe of nuclei in the blastoderm (Figure 16.4).

The pair-rule genes are the next genes in the pathway to become active (Table 16.2). These genes have a mutant phenotype with defects in alternate segments (Figure 16.3). The pair-rule genes are transcribed in a series of seven stripes of nuclei from the anterior to the posterior end of the syncytial blastoderm. Each pair-rule stripe is a two- to four-nuclei-wide band that can be seen at the last nuclear division before cell formation. The location of the pair-rule nuclear stripes is controlled by the location of gap-gene transcription and the presence of maternal gene products. Transcription of the pair-rule genes begins in the nuclei of the cells in each stripe at about the time of the last nuclear division before cell formation. Evidence that gap genes affect pair-rule gene expression comes from mutations and experimental treatments that alter the locations of the broad stripes of gap-gene transcription products; alteration of the gap-gene products also alters the locations of the stripes of pair-rule gene transcription. Each of the pair-rule stripes of nuclei eventually becomes a two- to four-cells-wide stripe in the cell blastoderm. Some of these stripes correspond to segments, but others do not. For example, the most-anterior stripe of nuclei transcribing *hairy* (*h*) corresponds to the mandibular segment, and the most anterior stripe of nuclei transcribing *runt* (*r*) corresponds to

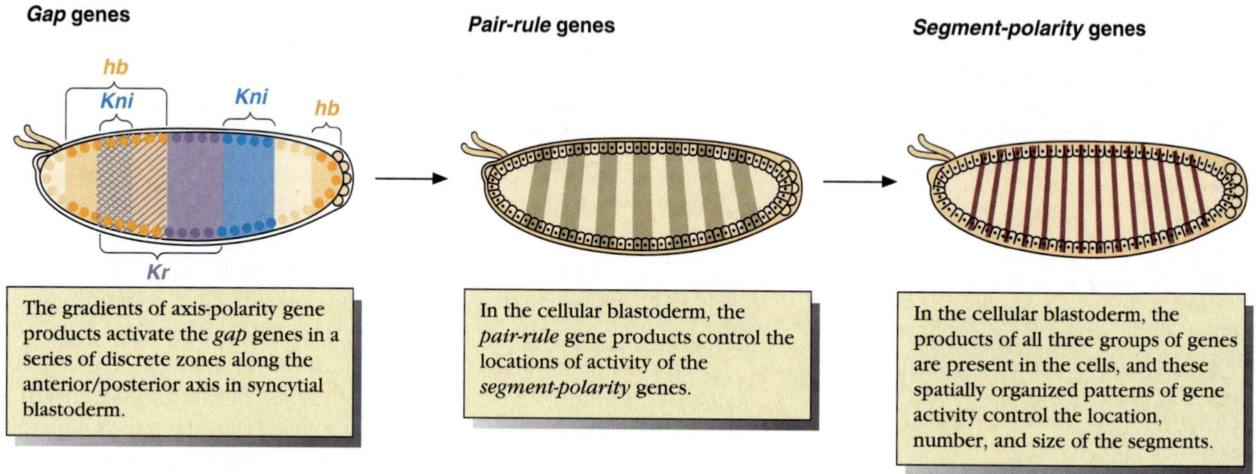

Gap genes
Pair-rule genes
Segment-polarity genes
hb
Kni
Kni
hb
Kr
The gradients of axis-polarity gene products activate the *gap* genes in a series of discrete zones along the anterior/posterior axis in syncytial blastoderm.
In the cellular blastoderm, the *pair-rule* gene products control the locations of activity of the *segment-polarity* genes.
In the cellular blastoderm, the products of all three groups of genes are present in the cells, and these spatially organized patterns of gene activity control the location, number, and size of the segments.

FIGURE 16.4 Expression of segmentation genes. Each segmentation gene is expressed in a particular spatial region, consisting of a stripe, or stripes, of blastoderm nuclei or cells. The gap genes are expressed in broad stripes, the pair-rule genes in 7 medium stripes, and the segment-polarity genes in 14 narrow stripes.

the maxillary segment (Chapter 15), but each stripe of nuclei transcribing *even skipped* (*eve*) or *fushi tarazu* (*ftz*) forms part of two different segments.

The transcriptional regulation of the pair-rule genes is accomplished when the maternal-effect gene protein products are bound with the gap genes to regulatory sites in the pair-rule genes' promoter regions. For example, the *eve* gene transcription is enhanced by the products of the *bcd* and *hb* genes and repressed by the products of the *Kr* and *Gt* genes. Certain combinations of maternal and gap-gene products activate *eve*, whereas other combinations repress *eve*. The maternal-gene products and the gap-gene products are organized in overlapping stripes of nuclei. These combinations of products activate *eve* in seven stripes in the embryo (Figure 16.5). These examples demonstrate how complex, spatially organized patterns of gene action and cell determination can be generated in an organism by the combination of spatially localized maternal-gene products that interact with regulatory binding sites in the promoter regions of the zygotic genes.

The segment-polarity genes are the last set of genes to be activated. Mutations of these genes cause a pattern of defects that include the loss of particular portions of each segment (Figure 16.3). The best studied of these genes are *wingless* (*wg*) and *engrailed* (*en*), which control the polarity of segments and divide each segment into anterior and posterior compartments. Both *wg* and *en* are initially transcribed in 14 one-cell-wide stripes in the cell blastoderm (Figure 16.6). Keeping with our hierarchical pattern of gene expression, the transcription of the segment-polarity genes is controlled by the pair-rule genes. For example, in wild-type individuals, the transcription of *en* in the cells of the odd-numbered stripes requires the expression of both of the pair-rule genes *prd* and *eve*. The transcription of *en* in the cells of the even-numbered

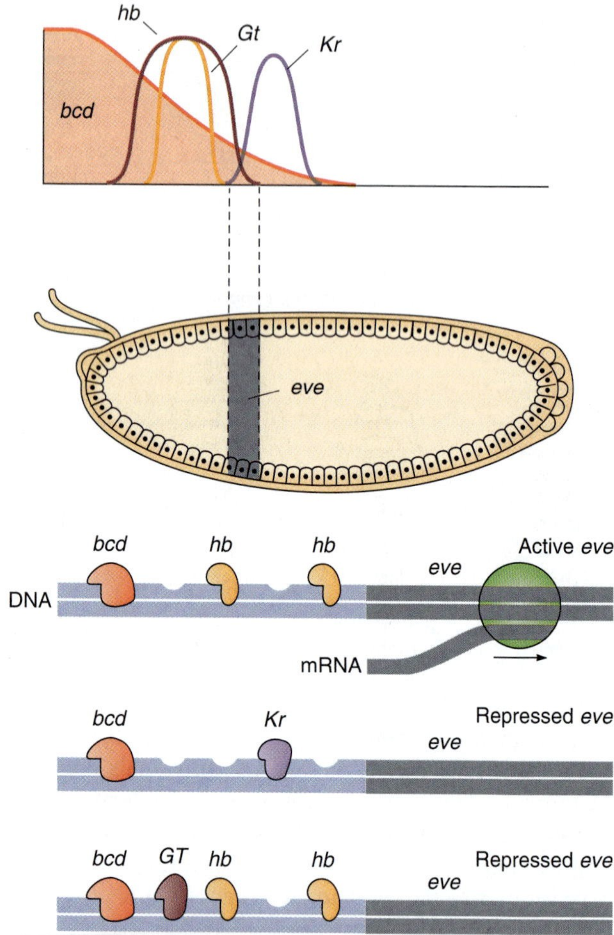

FIGURE 16.5 Transcription of *eve*. The pair-rule gene *eve* is transcribed in blastoderm cells where the products of the *bcd* and *hb* genes (activators) are present and the concentrations of the products of the gap genes *Gt* and *Kr* (repressors) are low. Other combinations of activators and repressors in different regions produce the seven stripes of *eve* transcription in the embryo.

FIGURE 16.6 Expression of *engrailed*. This photograph shows the expression of the *engrailed* gene in a *Drosophila* embryo. The *engrailed* gene is expressed in a series of 14 one-cell-wide stripes in the blastoderm. The cells expressing *engrailed* give rise to the posterior compartments of the larval and adult body segments. *(J. Girton, Iowa State University)*

stripes requires the expression of the pair-rule genes *ftz* and *odd*. The *wg* gene is transcribed in cells that are expressing either *eve* or *ftz* and that are adjacent to a cell expressing *en* (Figure 16.7). Each four-cell stripe gives rise to one segment, the cells expressing *en* give rise to the posterior compartment, and the other three cells produce the anterior compartment. This pattern is so specific that the locations of the *en* and *wg* transcripts are often used as molecular positional markers to define the locations of the segments in embryos. Transcription of the maternal axis-formation genes, the gap genes, and the pair-rule genes ceases shortly after the blastoderm stage, but the segment-polarity genes continue to be transcribed in the mitotic progeny of the blastoderm cells. The activity of segment-polarity genes is, in fact, required for the blastoderm cells to maintain their anterior and posterior identity throughout larval and adult development.

The organized segment generation in the *Drosophila* embryo illustrates how complex biological patterns can be generated. The maternal gene products packaged in the egg activate the first genes in a hierarchy of zygotic regulatory genes. These zygotic genes are activated in the nuclei in particular regions of the embryo where the correct combinations and/or concentrations of maternal gene products exist. These products interact with regulatory sites in the zygotic genes' promoters to initiate transcription. The products of the first genes of the hierarchy activate other genes, and these then activate others. Each gene's products have a characteristic pattern of spatial distribution and a specific set of genes with which they interact. These interactions generate a complex, spatially ordered pattern of gene expression. This pattern of gene expression divides the embryonic cells into discrete groups, each containing a small number of cells with a unique set of active regulatory genes. Each of these groups of cells is the founding cell population for a particular segment or compartment. Thus, for example, the segmentation genes control the organization of the cells into segmental and compartment groups. However, they do *not* assign each segment an identity; that is, they do not control which particular segment the cells of a group will form. That decision is controlled by another group of genes, the homeotic genes (discussed in the next section). When each segment has an established identity, additional cell-fate decisions occur. They divide the cells into the subgroups that serve as the founding cell populations for segment-specific structures (such as larval

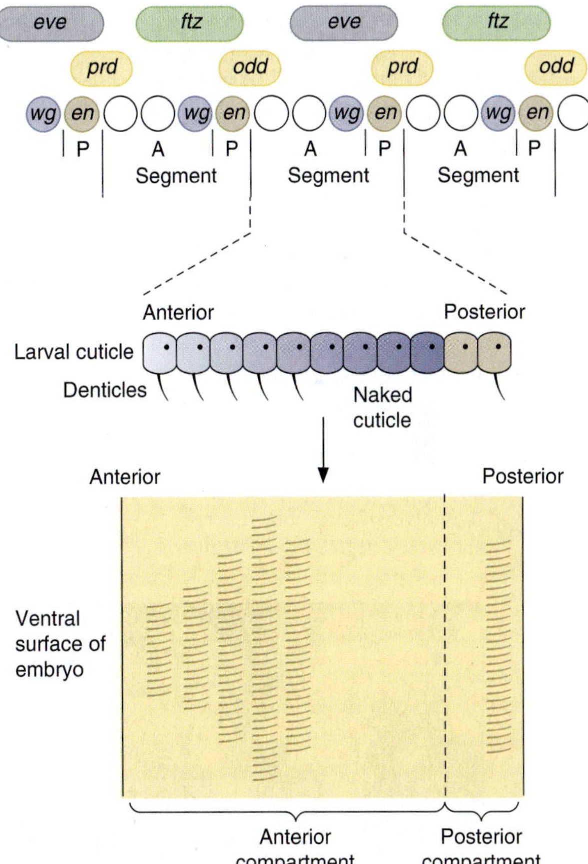

FIGURE 16.7 Activity of segment-polarity genes. The segment-polarity genes *wg* and *en* are initially expressed in a series of one-cell-wide stripes, whose locations are defined by the actions of the pair-rule genes. The *en* gene is expressed in cells in which the pair-rule genes *odd* and *ftz* are expressed or where the genes *prd* and *eve* are expressed. The *wg* gene is initially expressed in cells that are expressing either *eve* or *ftz*, and this expression is enhanced in cells that are adjacent to a cell expressing *en*. This generates a concentration gradient of *wg* product throughout the cells that develop into the anterior compartment of the segment.

HISTORICAL PROFILE

Christiane Nüsslein–Volhard and Eric Wieschaus

In 1980, Christiane Nüsslein–Volhard and Eric Wieschaus (Figure 16.A) reported the results of their successful saturation mutagenesis experiment to identify the genes that regulate *Drosophila* embryonic segmentation. Their paper had a profound impact not only because of the details of the system but also because of the revelation that saturation mutagenesis could unlock the genetic basis of complex developmental processes. They represented a new generation of developmental geneticists, individuals who understood the power of molecular biology to analyze developmental questions and the power of *Drosophila* genetics to identify genes by mutagenesis. Another key factor in their success was the successful combination of their personalities and talents.

Christiane Nüsslein–Volhard was born on October 22, 1942, in Germany. She came from a large, supportive family who encouraged her early interest in biology. After completing high school, she entered the university in Frankfurt with a desire to become a research scientist. She eventually moved to the university at Tübigen, where she took a degree in biochemistry and began her graduate career in molecular biology. In 1973, she completed her Ph.D. with a thesis involving the study of polymerase binding to phage promoters. Not satisfied to continue with this more biochemical approach to biology, she accepted a fellowship to work in the laboratory of Walter Gehring at Basel. There she learned the techniques of *Drosophila* genetics and became fascinated with the question of how genes regulate embryonic development. There she also met and became friends with Eric Wieschaus.

Eric Wieschaus was born in South Bend, IN, on June 8, 1947. His boyhood was spent in Birmingham, AL, in a southern environment that included exploring woods and studying local wildlife. In high school, he attended a summer National Science Foundation program working in a zoology laboratory at the University of Kansas. This sparked his interest in embryology, and he entered the University of Notre Dame with a desire to major in biology. At Notre Dame, he became acquainted with *Drosophila* through a job in the laboratory of Harvey Bender. Genetics did not interest him then because there was very little experimental analysis of embryonic development in *Drosophila* at that time. He decided to attend graduate school at Yale University, majoring in biology. There, he began working with Donald Poulson, the great *Drosophila* embryologist whose analysis of embryonic development in *Drosophila* was the standard reference for decades. Wieschaus's interest was sparked in embryonic development of *Drosophila*. In his second year, he switched to the laboratory of Walter Gehring, where he used somatic mosaic analysis to discover that cell determination occurs in the first few hours after the cell blastoderm stage. Wieschaus's experiments were completed in Basel, where Gehring had moved to establish a permanent laboratory. There Wieschaus met Nüsslein–Volhard, and their collaboration began.

In 1978, Nüsslein–Volhard and Wieschaus both received positions as group leaders at the European Biology Laboratory at Heidelberg. With no teaching or grant-writing obligations, they concentrated all their time on isolating mutations that affected embryonic development. Nüsslein–Volhard's determination and Wieschaus's extensive experience with *Drosophila* embryonic development quickly led to the development of novel techniques for screening large numbers of mutant lines for embryonic defects. This intense effort led to the isolation of 15 new mutations that defined the segmentation genes. In 1981, Wieschaus accepted a faculty position at Princeton, where he remains today, concentrating on the genetic regulation of gastrulation and the establishment of the cytoskeleton. Nüsslein–Volhard moved in 1981 to the Max Plank Institute in Tübigen, where she has continued to study embryonic development mutations in *Drosophila* and in zebrafish.

FIGURE 16.A Photograph of Christiane Nüsslein–Volhard and Eric Wieschaus. *(Reuters/Jan Colls/Archive Photos)*

structures or imaginal discs). We have not yet identified the genes controlling many of these later cell-fate decisions.

The segmentation genes were discovered by Christiane Nüsslein–Volhard and Eric Weischaus (see *HISTORICAL PROFILE: Christiane Nüsslein–Volhard and Eric Wieschaus*). They performed large-scale saturation mutagenesis experiments with *Drosophila,* seeking to identify the genes that control the earliest steps in the organization of the embryonic body. To do this, they selected mutations that had an embryonic lethal phenotype and then examined the lethal individuals for evidence of abnormal segmentation. They screened tens of thousands of mutant *Drosophila* and discovered mutations in 15 genes. From the phenotypes of the lethal individuals, they were able to group the mutations into the three groups discussed earlier (gap genes, pair-rule genes, and segment-polarity genes). They announced their discoveries in a paper published in 1980, and they made their mutant strains widely available for others to study. These mutants have led to some of the greatest advances in our understanding of development.

SOLVED PROBLEM

Problem

A newly isolated zygotic lethal mutation in *Drosophila* results in death during embryonic development, and the lethal embryos have abnormalities of body segmentation. How would you discover in which class of segmentation genes this mutation takes place?

Solution

There are several ways to classify a segmentation gene in *Drosophila.* First, the details of the mutant phenotype should be determined. Gap genes affect contiguous groups of segments, pair-rule genes affect every other segment, and segment-polarity genes affect a particular region in each segment. Examine the lethal embryos produced by this mutation and see if they fit one of these patterns. Second, examine the pattern of transcription of the gene by hybridizing a probe that binds to the gene's mRNA to normal embryos. Segmentation genes are transcribed in discrete patterns in the regions that are affected by their mutant alleles. Is this unknown gene transcribed in a single large stripe beginning in the syncytial blastoderm (gap gene), is it transcribed in 7 stripes early in cell blastoderm (pair-rule gene), or is it transcribed in 14 narrow stripes later in cell blastoderm (segment-polarity gene)? Third, the unknown mutation can be tested in a complementation test with alleles of known members of the three groups. Failure to complement any of these indicates its group and locus.

SECTION REVIEW PROBLEMS

3. Give a brief definition of the following:
   a. maternal-effect mutation
   b. pair-rule gene
   c. segment-polarity gene
4. For the following pairs of terms/items state one difference:
   a. maternal-effect mutation vs. zygotic mutation
   b. segmentation genes vs. determination genes
   c. gap genes vs. segment-polarity genes
5. If you were given a labeled DNA probe that is complementary to the mRNA produced by each of the following *Drosophila* genes, briefly describe the pattern of hybridization you would expect to see in cellular blastoderm stage embryos.
   a. gap gene
   b. pair-rule gene
   c. segment-polarity gene
   d. *bcd*

HOMEOTIC GENES CONTROL CELL IDENTITY IN ANIMALS.

In Chapter 15, we briefly described the **homeotic genes** in *Drosophila.* These remarkable genes give each cell in each segment an heritable segmental identity, and each has characteristic, distinguishing features. You may recall that the homeotic genes each contain a region of highly conserved DNA sequence called the **homeobox.** Genes with homeobox sequences similar to those of *Drosophila* homeotic genes have been discovered in all animal species (except sponges) and in all plant species that have been investigated. This suggests that homeobox-containing genes code for very important, highly conserved functions. In this section, we will briefly review the identity and organization of the homeotic genes in *Drosophila* and mammalian species and our present understanding of their role in development.

The Homeotic Gene Complex in *Drosophila* Contains *ANT-C* and *BX-C* Genes.

In Chapter 15, we discussed the homeotic genes of *Drosophila* and the role of one set of these genes, the *BX-C,* in controlling the segment identity of cells in the thoracic and abdominal segments. The identities of the segments in the head and anterior thorax are controlled by a second set of homeobox-containing genes in the *Antennapedia* complex (ANT-C). The ANT-C contains five homeotic genes, and evidence from mutational analyses indicates that these genes control segment identity. Mutations that cause *ANT-C* genes to be expressed in a segment where they are not normally expressed (gain-of-

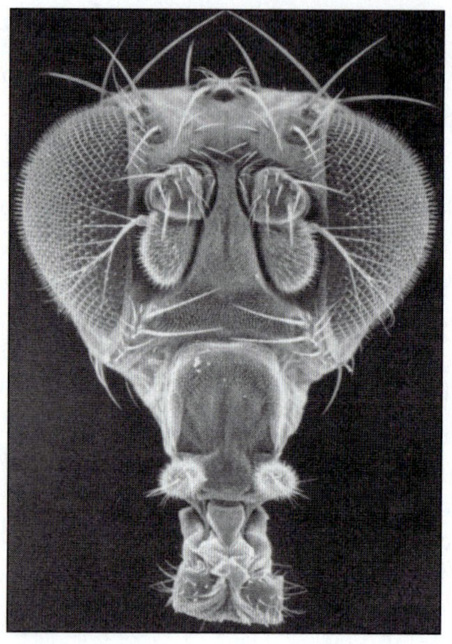

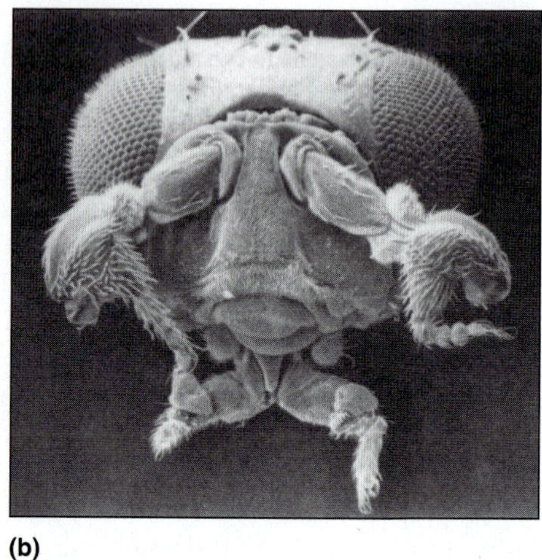

FIGURE 16.8 ***ANT-C* adult mutant phenotypes.** Gain-of-function mutations that cause the *Antennapedia* gene to be expressed in the head transform antennal and labial structures into thorax. (a) A scanning electron micrograph of a wild-type adult head showing the normal antennae. (b) A scanning electron micrograph of the head of an *Antp pb* double-mutant fly. The antennae and palps have been transformed into mesothoracic legs. *(Thomas Kaufman, Indiana University)*

function mutations) cause the transformation of that segment into another segment type. For example, a gain-of-function mutation of the *Antennapedia* (*Antp*) gene that causes *Antp* to be transcribed in the antennae will transform the antennae into legs (Figure 16.8). Loss-of-function mutations in *ANT-C* genes also cause specific segments to change their identity. A loss-of-function mutation of the *proboscipedia* (*pb*) gene causes the proboscis to develop as a leg.

The *ANT-C* genes are located in a cluster on the third chromosome at salivary gland chromosome band position 84A-B. This chromosomal region contains three developmentally important, nonhomeotic loci (*ftz, zen,* and *bcd*) interspersed with five homeotic loci: *lab, pb, Dfd, Scr,* and *Ant* (Figure 16.9). The *ANT-C* genes are not regulated coordinately; each gene is regulated individually. With the exception of one gene (*Dfd*), a distinguishing feature of the *ANT-C* genes is that the order of the *ANT-C*

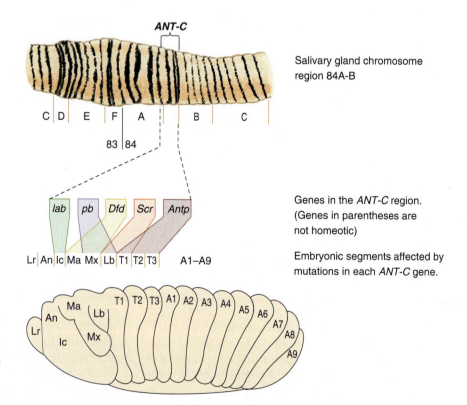

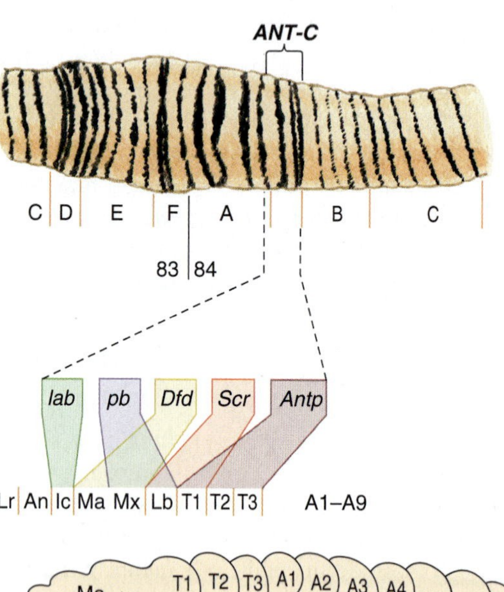

FIGURE 16.9 **The *Antennapedia* complex.** The *Antennapedia* complex (ANT-C) is located in the salivary gland chromosome region 84A-B of the third chromosome. ANT-C contains homeotic genes *labial* (*lab*), *Deformed* (*Dfd*), *Sex combs reduced* (*Scr*), *proboscipedia* (*pb*), and *Antennapedia* (*Antp*) interspersed with the nonhomeotic genes *zeu, bcd,* and *ftz.* With the exception of the *Dfd* gene, the genes are arranged in the same order along the chromosome as the segments they affect are arranged in the individual.

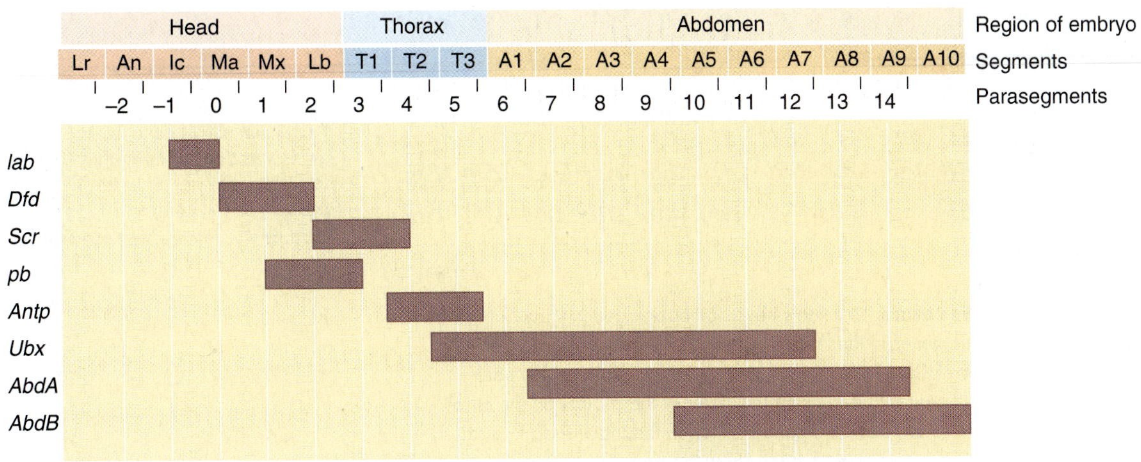

FIGURE 16.10 Homeotic gene expression. The *ANT-C* and *BX-C* genes are expressed in spatial regions that include those body regions affected by their mutations. The spatial boundaries of the expression regions correspond to the boundaries of parasegments, not segments.

genes on the chromosome is identical to the order of the segments in the individual that they control. Like the *BX-C* genes, the *ANT-C* genes are large and have complex molecular structures. For example, the *Antennapedia* gene occupies over 100 kb of genomic DNA and produces multiple RNA products. It contains two separate promoters, a proximal promoter (P2) and a distal promoter (P1). Transcripts produced by the two promoters contain some of the same exons and some different exons. Regulation of *Antennapedia* expression includes differential promoter activation.

The homeotic genes of the ANT-C and BX-C are all transcribed in specific regions of the embryo (Figure 16.10). Curiously, the regions that the *ANT-C* and *BX-C* genes regulate do not exactly correspond to the segment areas in which they are expressed. Rather, the boundaries of transcription of the homeotic genes correspond to parasegment boundaries, suggesting that parasegments are regions of coordinate gene regulation. The functional relationship between parasegments and segments is not understood. Each parasegment consists of the posterior compartment of one segment and the anterior compartment of the next most posterior segment. For example, the anterior boundary of expression of *Ubx* is the fifth parasegment, which corresponds to the posterior compartment of the mesothorax and the anterior compartment of the metathorax (Figure 16.10). We know that the genes in ANT-C and BX-C are similar, and the fact that in other species the homeotic genes are in one group suggests that they actually are parts of a single ancestral homeotic gene complex. The most likely explanation is that the homeotic genes were once a single cluster of genes that was divided by a chromosomal rearrangement during the evolution of *Drosophila*. Because of this, the ANT-C and BX-C are collectively referred to as the ***Hom-C* group.**

The *Hox* Genes Control Determination in Mammals.

The extensive analysis of *Drosophila* homeotic genes has identified certain key homeotic gene features; these include the transformation of segments by mutations, the presence of homeobox sequences, and the arrangement of genes on the chromosome in a cluster, with individual genes in the same sequence on the chromosome as the segments they control are arranged in the body. Molecular genetic analyses using the *Drosophila* homeobox sequences as probes have identified genes with sequences similar to the *Drosophila* homeoboxes in all plant and animal species studied to date (except sponges). The universality of the homeobox sequences raised the important question of whether the genes with homeobox sequences in other species shared other features of the *Drosophila* homeotic genes. Investigations showed that in some species they did.

In mammals, for example, there are 38 homeobox-containing genes arranged in four clusters. These homeobox-containing genes are all referred to as ***Hox* genes.** The sequence of each *Hox* gene is most similar to one specific *Drosophila* homeotic gene, and, within each cluster, the *Hox* genes are arranged in the same order on the chromosome as the *Drosophila* *ANT-C* and *BX-C* genes (Figure 16.11). The four sets of *Hox* genes are designated by a letter (a, b, c, d), and the members of each set are numbered (1–13) starting from the gene at the

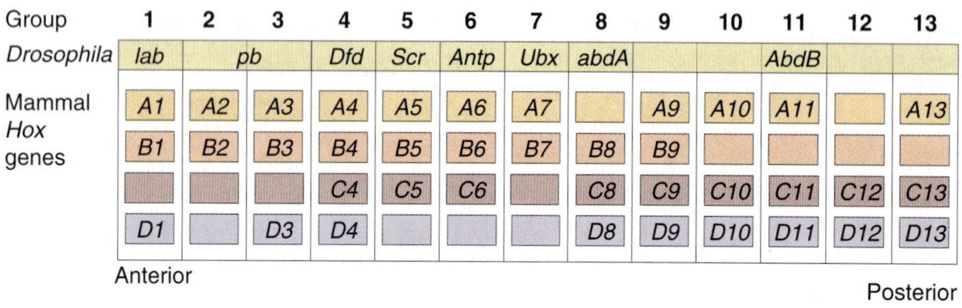

Group	1	2	3	4	5	6	7	8	9	10	11	12	13
Drosophila	*lab*	*pb*		*Dfd*	*Scr*	*Antp*	*Ubx*	*abdA*			*AbdB*		
Mammal *Hox* genes	*A1*	*A2*	*A3*	*A4*	*A5*	*A6*	*A7*		*A9*	*A10*	*A11*		*A13*
	B1	*B2*	*B3*	*B4*	*B5*	*B6*	*B7*	*B8*	*B9*				
				C4	*C5*	*C6*		*C8*	*C9*	*C10*	*C11*	*C12*	*C13*
	D1		*D3*	*D4*				*D8*	*D9*	*D10*	*D11*	*D12*	*D13*

Anterior — Posterior

Each mammalian gene has the greatest sequence homology to the *Drosophila* gene in the same position in the complex.

FIGURE 16.11 Mammalian *Hox* genes. Mammals have homeobox genes (*Hox* genes) that are homologous to the homeotic genes in the *ANT-C* and *BX-C* of *Drosophila*. The mammalian genes are organized in four clusters (named *Hoxa*, *Hoxb*, *Hoxc*, and *Hoxd*), each with up to 13 genes. A key feature of each cluster is that the order of the groups on the chromosome is the same as the spatial distribution of the gene's action in the individual. In addition, the homeobox and surrounding regions of each gene show the greatest sequence homology to the homeobox and surrounding region of genes in the same position in the other clusters. For example, *lab*, *Hoxa-1*, *Hoxb-1*, and *Hoxd-1* have greater sequence similarity to each other than they do to other *Hox* genes. The mammalian gene complexes may represent duplications of an ancestral *ANT-C/BX-C* complex.

anterior end of the cluster. The *Hoxa-1* gene, for example, is the gene in the set that is most similar to the *Drosophila lab* gene (Figure 16.11). Each mammalian *Hox* gene is expressed in a discrete region of the mouse (the mouse is often used as a mammalian model organism), and the anterior/posterior order of the expression regions is the same as the order of the genes on the chromosome. The anterior *Hox* genes are most similar to the *Drosophila lab* gene (*Hoxa-1*, *Hoxb-1*, and *Hoxd-1*) and are expressed in the most anterior regions of the mouse. The similarities between mouse and *Drosophila* homeotic gene structures and expression patterns suggest that the homeotic gene cluster is evolutionarily ancient, and that there may be functional reasons for the arrangement of these genes on the chromosome.

The structural similarity of the mouse *Hox* genes and the *Drosophila* homeotic genes suggested that the *Hox* genes might function in the mouse in a manner similar to the *Drosophila* homeotic genes (that is, as regulators of regional determination during development). Recent evidence from mutational analyses supports this hypothesis. Loss-of-function mutations have been generated in several mouse *Hox* genes, including *Hoxa-1*, *Hoxa-2*, *Hoxa-3*, *Hoxa-4*, *Hoxa-5*, and *Hoxa-6*, and all showed aberrant phenotypes with a common theme. Mice homozygous for a loss-of-function allele of one of these genes have homeotic transformations of structures in the region where that *Hox* gene is normally expressed. In all cases, the transformation is restricted to the anterior portion of the expression region and does not include regions where a more posterior *Hox* gene is expressed. This suggests that in the mouse, as in *Drosophila*, a particular *Hox* gene does not control determination throughout its entire region of expression. Determination in the posterior regions is controlled by the next-most posterior gene in the *Hox* cluster.

Gain-of-function mutations have also been generated by attaching the structural region of a *Hox* gene to the promoter region of a different *Hox* gene. Inserting such a construct into the genome can cause the expression of the *Hox* gene in a specific, but abnormal, region. This is called ectopic expression. Ectopic expression of a *Hox* gene can cause homeotic transformations, with one important provision. Ectopic expression in a region anterior to the normal region of expression of that *Hox* gene causes that region to adopt a more posterior determination. However, ectopic expression in a region posterior to the normal region of expression does not produce homeotic transformations.

These studies suggest that the homeobox-containing *Hox* genes do control regional determination in mice, in a manner analogous to the segmental determination of the *ANT-C* and *BX-C* genes in *Drosophila*. Like the *Drosophila* genes, the *Hox* genes also show a hierarchy of control, with the more posterior genes being epistatic to the more anterior genes. In mice, this relationship is called **posterior preference.** This suggests that the hierarchical control relationship may also be evolutionarily ancient. Timing-of-expression studies indicate that the more-anterior *Hox* genes are expressed earlier in development. The mouse embryo develops from anterior to posterior, with the more-anterior body regions developing first and more posterior regions later. Thus, this hierarchy of control may have evolved as a mechanism for dividing the developing individual into zones, or stripes of cells with similar determination, without the need for transcriptional suppression of the earlier homeotic genes. This model can explain the patterns of epistasis and the results of the ectopic expression experiments. There is still a great deal we do not understand about the *Hox* genes, and this is currently an area of very active investigation.

SOLVED PROBLEM

Problem

How is the function of a homeotic gene different from that of a segmentation gene? How would you discover which type a new mutation is?

Solution

A homeotic gene functions to give cells in a particular region (a segment or parasegment) an identity. They do this by producing a product that acts as a determinant. The presence of this product in a cell gives the cell its identity, that is, in which particular segment or parasegment it is. A segmentation gene acts to organize cells into groups. The action of segmentation genes in a cell does not give that cell an identity, but it does organize the cell into a group with other cells that also have the same set of segmentation genes active.

There are several ways to discover which type a particular gene is. One is to make a mutation of the gene and then to examine the mutant phenotype. If the gene is a segmentation gene, the mutant individuals will have an abnormal number, size, or structure of segments. If the gene is a homeotic gene, there will be no change in the segmentation pattern, but the identity of one or more segments or parasegments will be changed into that of a different segment or parasegment (for example, abdomen into thorax).

SECTION REVIEW PROBLEMS

6. A series of fusion constructs have been made in which cloned regulatory regions from the *Drosophila* homeotic genes listed here have been attached to the structural gene for β-galactosidase and transformed back into a wild-type strain. The activity of these constructs can be determined by a simple staining test. For each of the following genes, indicate the regions where you would expect to see activity in embryos.
   a. *Antp*
   b. *Scr*
   c. *UBX*
   d. *pb*
7. If you were given a cloned copy of the mouse *Hoxa-3* gene, how would you test the hypothesis that this gene has the same function as the *Drosophila pb* gene?

SEX DETERMINATION IS CONTROLLED BY A GENE HIERARCHY.

The homeotic genes and the genes that control cell determination are examples of gene hierarchies that are evolutionarily conserved and are found in many different species. In this section, we will consider another example of a gene hierarchy whose genetic basis is currently being actively investigated: sex determination. In most animals, the initial step in sex determination is the chromosome composition of the individual (as discussed in Chapter 3). This is called the **primary signal.** This chromosomal composition signal must be converted into a pattern of stable, sex-determination gene activity that activates sex-specific genes in different tissues. These genes must function in many different cell types to produce the physical differences between males and females. Many of the genes in the sex-determination hierarchies of *Drosophila* and *C. elegans* have been identified by isolating mutations that alter sexual development or that are lethal in one particular sex. We will discuss these two hierarchies and their very different molecular modes of action in this section.

Sex-Specific RNA Processing Controls Sex Determination in *Drosophila*.

In 1921 Calvin Bridges discovered that the primary signal for sex determination in *Drosophila* is the ratio of X chromosomes to autosomes (see CONCEPT CLOSE-UP: *The X : A Primary Signal in* Drosophila). Individuals with an XX : AA ratio of 1 develop as females, individuals with a ratio of 1/2 develop as males, and individuals with a ratio between 1 and 1/2 develop as intersexes. Cellular sex determination occurs at the cellular blastoderm stage, so syncytial blastoderm XX nuclei that lose one X chromosome become XO nuclei that form male cells, but XX embryonic cells that lose one X chromosome after cell blastoderm still develop as female cells. Sex determination in *Drosophila* is controlled by a small group of genes acting in a hierarchy of gene action. The first genes respond to a primary signal and initiate three separate pathways: X chromosome dosage compensation, sex determination in somatic cells, and sex determination in germ-line cells. Of these three genetic developmental pathways, the best understood is the pathway controlling sex determination in somatic cells.

The genes controlling sex determination have been identified by isolating sex-specific lethal mutations or mutations that cause individuals of XX or XY chromosome type to develop as intersexes or to switch sexes. The action sequence of these genes has been defined by generating double-mutant individuals and observing the patterns of epistasis (mutations of genes farther down a pathway are epistatic to mutations of genes higher up in the pathway). Four of the *Drosophila* genes involved in the sex-determination pathway are: *sex lethal* (*sxl*), *transformer* (*tra*), *transformer2* (*tra2*), and *doublesex* (*dsx*).

The sex of an individual is controlled by how these genes are expressed and spliced during the development pathway (for a review of splicing, see Chapter 11). *Sex*

CONCEPT CLOSE-UP

The X : A Primary Signal in *Drosophila*

Calvin B. Bridges discovered many years ago that the X : A ratio in *Drosophila* is the primary signal responsible for sex determination, but the genetic mechanism that measures the number of X chromosomes and converts this ratio into an action pattern for the sex-determination genes has only been discovered recently. The mechanism hinges on the *Sxl* gene, which controls the pattern of sex-specific splicing in the genes in the sex-determination pathway. The X : A signal controls *Sxl* action by acting on one of two *Sxl* promoters: Sxl_{pe} and Sxl_{pm}. In females, the X : A signal activates transcription at Sxl_{pe} during nuclear division stages 12 to 14 of the syncytial blastoderm. At the cellular blastoderm stage, initiation of transcription at Sxl_{pe} ceases and initiation of transcription at Sxl_{pm} begins in both sexes. In females, the transcriptional activity of Sxl_{pe} produces a pool of female-specific *Sxl* gene product that ensures that the Sxl_{pm} primary transcripts are spliced in the female-specific pattern. In males, this *Sxl* product is not present, and the Sxl_{pm} transcript is spliced in the male-specific pattern.

The genes that make up the primary signal thus have one specific function—to initiate or suppress transcription of Sxl_{pe}, depending on the X : A ratio. Borrowing terminology from mathematics, these genes are divided into two groups, numerator genes (genes on the X that indicate the number of X chromosomes) and denominator genes (genes on the autosomes that indicate the number of autosomes). Both types of genes have been discovered in sophisticated genetic analyses that have made use of the distinct mutational phenotypes of numerator and denominator genes. Decreasing the amount of an X-linked numerator gene's product (by inducing loss-of-function mutations or by decreasing the number of copies) kills females, whereas increasing the amount of product (by inducing gain-of-function mutations or increasing the number of copies) kills males. The autosomal denominator genes have the opposite properties. Decreasing the amount of a denominator gene's product (by inducing loss-of-function mutations or by decreasing the number of copies) kills males, whereas increasing the amount (by inducing gain-of-function mutations or increasing the number of copies) kills females. In addition, the mutant effects of numerator and denominator genes depends on *Sxl*. A gain-of-function mutation in a numerator gene kills males because it causes an inappropriate activation of Sxl_{pe} in XY : AA individuals. This male-specific lethality is suppressed by a loss-of-function mutation of *Sxl*. Likewise, a loss-of-function mutation in a numerator gene kills females because it prevents the activation of Sxl_{pe} in XX : AA individuals. This female-specific lethality is suppressed by gain-of-function mutations of *Sxl*.

Using these phenotypes and interactions as criteria, several X-linked numerator genes have been discovered: *sisterlessA* (*sisA*), *sisterlessB* (*sisB*), *sisterlessC* (*sisC*), and *runt* (*run*). All act as activators of

lethal is the first gene in the pathway, and it is transcribed in the first cells of both XX and XY individuals. In XY and XO cells (male cells), the primary *sxl* transcript is spliced in a male-specific splicing pattern (Figure 16.12). This splicing pattern produces an mRNA with a translation stop codon early in the transcribed region. This stop codon causes early termination of translation, producing a nonfunctioning, truncated *sxl* gene product. In XX (female) cells, the *sxl* primary transcript has a female-specific splicing pattern. This splicing pattern gives an mRNA that produces a full-length, functional product containing an RNA binding domain. The *sxl* product binds to a specific site on the *sxl* primary transcript and, when bound, prevents the transcript from being spliced in the male-specific pattern. This means that when the *sxl* product is present in a cell, all new *sxl* primary transcripts are processed with the female-specific splicing pattern. This establishes a stable system of continued *sxl* product production. This system eliminates the need for continued presence of the primary signal. However, if the *sxl* product is lost or inactivated in a cell after the cellular blastoderm stage, the cell will revert from the female pattern and develop as a male cell. The function of the primary signal is to establish either a male or a female pattern of *sxl* splicing, and this maintains itself throughout the rest of development. This is an example of a mechanism for establishing and maintaining particular patterns of gene expression in determined cells, and this may be a widespread process.

The ability of the *sxl* product to bind to RNA is also essential for expression regulation of other genes. The next gene in the pathway, the *tra* gene, also has female- and male-specific splicing patterns that are controlled by the *sxl* gene product. The *sxl* product binds to the primary transcript of the *tra* gene, blocking the 3′ splice site that is used in the male-specific splicing pattern. This en-

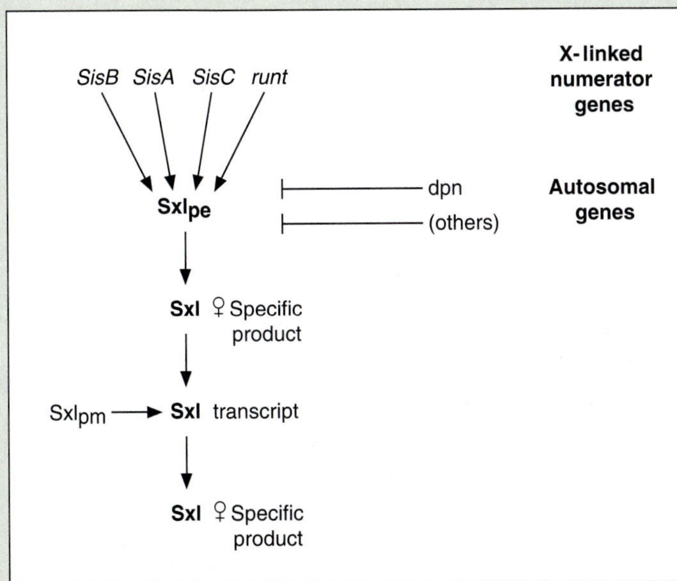

FIGURE 16.B The primary signal genes in *Drosophila*. The primary signal for sex determination in *Drosophila* is controlled by X-linked numerator genes and autosomal denominator genes. Numerator genes activate Sxl_{pe}, producing a female-specific product that directs the splicing of the Sxl_{pm} transcript in the female pattern. Denominator genes repress Sxl_{pe}, causing the Sxl_{pm} transcript to splice in the male-specific pattern. In XX individuals enough numerator gene products are present to activate Sxl_{pe}, but in individuals with only one X chromosome (XY or XO), there are not enough.

Sxl_{pe}. Only one denominator gene has been discovered so far, *deadpan* (*dpn*). Systematic screens of most of the genome have not revealed any other single gene with a significant phenotype. This suggests that there are many denominator genes spread throughout the genome, each with a small, additive effect. The action current model for these genes is shown in Figure 16.B. In XY : AA or XO : AA individuals, the balance in the amount of numerator gene products (activators) and denominator gene products (repressors) suppresses transcription of Sxl_{pe}, and the individual develops as a male. In XX : AA individuals, the greater numbers of numerator gene copies result in a higher concentration of numerator gene products. This leads to activation of Sxl_{pe}, and the individual develops as a female.

sures that the transcript will be processed in the female-specific splicing pattern (Figure 16.12). Keeping with the hierarchical control, the *tra* gene product also has an RNA-binding domain. The product of the *tra* gene complexes with the product of the *tra*-2 gene (another gene involved in sex determination) to form a dimer that binds to the primary transcript of the *dsx* gene. This binding ensures that the *dsx* gene transcript is processed in a female-specific splicing pattern. This produces a female-specific *dsx* product that activates female sex differentiation genes. If the *dsx* primary transcript does not bind a *tra/tra-2* dimer, then by default it is spliced in a male-specific pattern, producing a male-specific product that activates male-specific sex-differentiation genes. This sex-determination pathway uses a different method of gene regulation, controlling gene activity through RNA processing regulation, rather than by regulating the initiation of RNA production. Processing as a means of control is not unique to *Drosophila* nor to the sex-determination gene hierarchy.

Sex-Specific Transcriptional Regulation Controls Sex Determination in *C. elegans*.

Sex determination in the somatic cells of *C. elegans* and *Drosophila* have a number of similarities. The primary signal in *C. elegans* is the ratio of X chromosomes to autosomes. Individuals having an X : A ratio of 1 develop as hermaphrodites, and individuals having a ratio of 1/2 develop as males. Sexual dimorphism in *C. elegans* is extensive, and at least 30% of the hermaphrodite cells and 40% of the male cells show sexual specialization. Genetic analyses of sex determination began by isolating mutants that produced the opposite sexual phenotypes; XX individuals were transformed into males, and XO individuals, into hermaphrodites. Through the analysis of

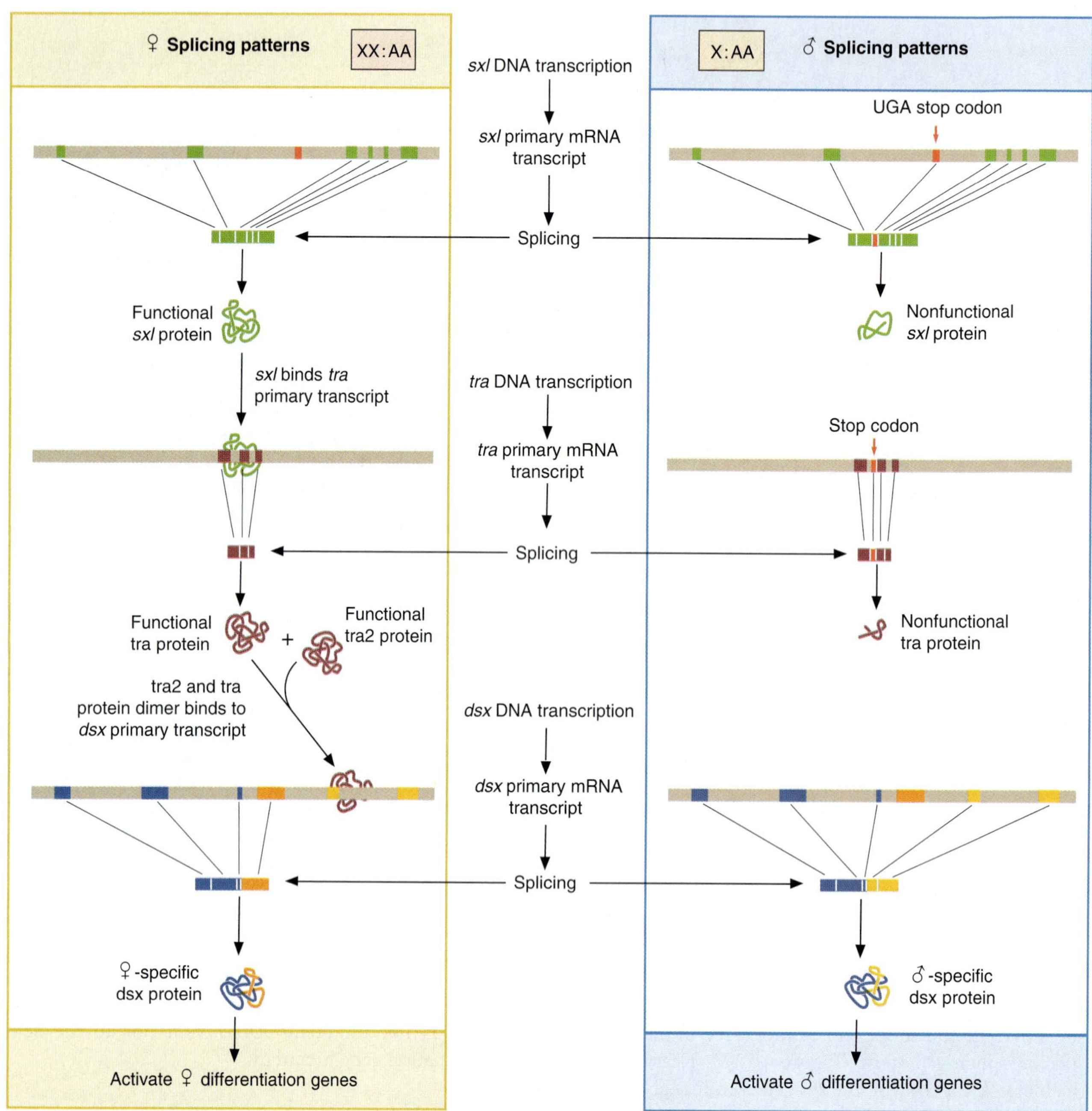

FIGURE 16.12 ***Drosophila* sex-determination genes.** The genes that control sex determination in *Drosophila* control the splicing patterns of primary transcripts of the *sex lethal* (*sxl*), *transformer* (*tra*), and *double sex* (*dsx*) genes. Each gene has a female-specific and a male-specific splicing pattern that generates an mRNA containing different exons. These mRNAs make different products. The female *sxl* product binds to the *tra* primary transcript and causes it to be spliced in the female-specific pattern. The male *sxl* product is nonfunctional. The *tra* gene product binds to the *dsx* primary transcript and causes it to be spliced in the female-specific pattern. The male *tra* protein is nonfunctional. The female- and male-specific *dsx* proteins activate different sets of sex differentiation genes.

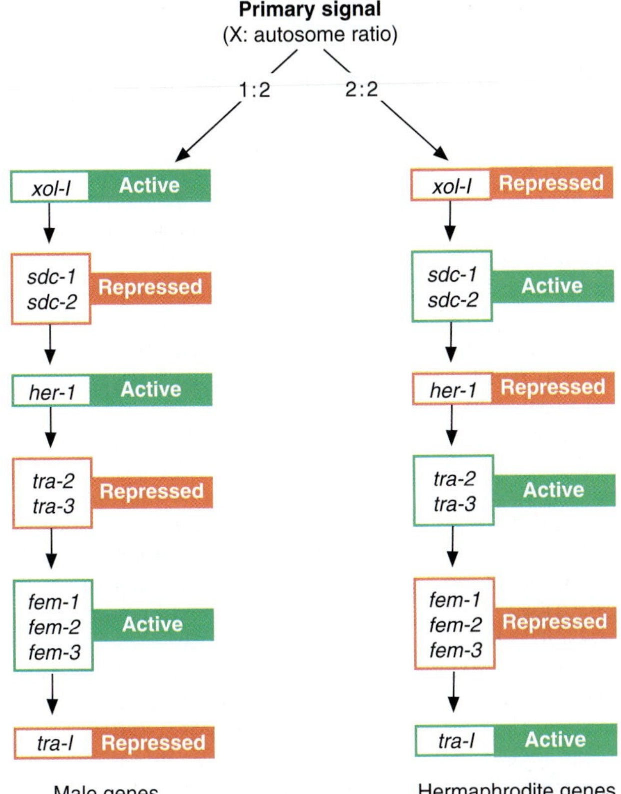

FIGURE 16.13 ***C. elegans* sex-determination genes.** A gene hierarchy controls sex determination in *C. elegans*. The primary signal (X : A ratio) activates the first gene in the hierarchy (*xol-1*) in males and inactivates it in hermaphrodites. If *xol-1* is active, it represses the next group of genes; if *xol-1* is inactive, the next group becomes active by default. Each gene, if active, represses the next gene, or group of genes, in the pathway. If the last gene (*tra-1*) is active, it activates hermaphrodite-specific differentiation genes. If *tra-l* is repressed, male-specific genes are activated by default.

epistatic interactions in double-mutant individuals, the genes involved in *C. elegans* sex determination have been found to function in a developmental hierarchy (Figure 16.13). The genes involved in sex determination can be divided into two groups: one group of genes (*tra-1*, *tra-2*, and *tra-3*) function exclusively in XX animals to generate hermaphrodite development, and a second group of genes (*her-1*, *fem-1*, *fem-2*, and *fem-3*) function exclusively in XO individuals to initiate male development. Mutations in the first group cause XX individuals to develop as males but have no effect in XO individuals, and mutations in the second group cause XO individuals to develop as hermaphrodites but have no effect in XX individuals. This pattern of mutually exclusive action is the result of the genes normally being active only in one cell type. Sex determination in *C. elegans* is a classic case of a developmental genetic switch that activates one of two sets of determination genes. In XO individuals, the primary signal represses the activity of the *tra-1* gene, whereas in XX individuals the primary signal activates the *tra-1* gene. In *C. elegans,* male development is the default state, so the normal function of *tra-1* is to suppress male somatic development and to activate genes that initiate female somatic development. The exact action mechanisms of these regulatory genes has not yet been established but likely includes transcription, RNA processing, translation, and/or modification of gene products.

Sex determination in higher organisms, including humans, is controlled by groups of interacting genes whose actions are often similar to those of *Drosophila* or *C. elegans*. There are also, of course, important differences (see CURRENT INVESTIGATIONS: *Human Sex Determination*). The goal of current studies of human sex determination are the same as studies in model organisms: to identify the genes controlling the process and to discover the nature of their actions. The difficulty of doing genetic studies in humans, and especially the inability to do mutagenesis screens, makes studying human sex determination more difficult.

SOLVED PROBLEM

Problem

How would you discover whether a newly identified species of primitive insect has a sex determination mechanism similar to that of *Drosophila* or *C. elegans?*

Solution

This can be done by examining the action of the genes that control sex determination. Identify these genes by isolating mutations that alter sexual development specifically in one sex or mutations that kill one sex preferentially. Next, examine the pattern of expression of these genes. If the sex determination is like that in *Drosophila,* these genes will be transcribed in both sexes, but each gene's primary transcript will be spliced differently in males and females. If the sex determination mechanism is like that of *C. elegans,* these genes will be transcribed only in one sex.

SECTION REVIEW PROBLEMS

8. The *sxl* gene is believed to have different patterns of RNA processing in males and females. How would you test this hypothesis?
9. What phenotype would you expect to see in *sxl tra* double-mutant individuals?
10. Describe one similarity and one difference in the genetic control of sex determination in *Drosophila* and *C. elegans.*

CURRENT INVESTIGATIONS

Human Sex Determination

Humans, like other mammals, have a sex-determination mechanism that depends on the Y chromosome. Individuals with a Y chromosome (XY and XXY) develop as males, and individuals without a Y chromosome (XO, XX, and XXX) develop as females. The X : autosome (A) ratio does not influence sex determination. The sex determining factor on the Y has been discovered to be a regulatory gene called *SRY* (for Sex Region on the Y). The *SRY* gene directs development of the embryonic gonadal primordium (the embryonic genital ridge). The gonadal primordium has the potential to develop into either a testis or an ovary. Deciding the direction of its development is the key point of mammalian sex determination. The presence of an active *SRY* gene causes the primordium to develop as a testis. After the testis is developed, it produces male-specific endocrine hormones that direct the somatic cells of the individual into a male sexual differentiation pathway. Without a functioning *SRY*, the primordium develops as an ovary and produces female-specific hormones that direct the individual into a female sexual differentiation pathway. The hormonal balance causes the development of the secondary sexual characteristics. Thus, the key question of human sex determination is how the *SRY* directs the embryonic bipotential gonadial primordium to adopt a testis instead of an ovary developmental fate.

There have been several models of how the Y directs sex determination in mammals. A popular current model is that the *SRY* interacts with another gene: *Dosage Sensitive Sex Reversal* (*DSS*). *DSS* has been mapped to the X chromosome, at position Xp21. This gene has been proposed to act as a repressor of a series of male pathway genes (*MPGs*) that direct the embryonic primordium to form a testis. The proposed actions of *DSS* and *SRY* are shown in Figure 16.C. In XY individuals, the *SRY* product is proposed to inactivate *DSS*, which results in the *MPGs* becoming active. These genes cause the embryonic bipotential gonad to develop as a testis. As part of testis development, other genes that repress female pathway genes (*FPGs*) (such as *MIS*, also called the "antimüllerian hormone gene") are activated. In XX individuals, *SRY* is not present, and the *DSS* gene on the X chromosome that is not inactivated and converted into a Barr body represses the *MPGs*. Slightly later in development, the *FPGs* are activated, leading to the development of the embryonic bipotential gonad as an ovary. As part of ovary development, other genes that permanently repress the *MPGs* are activated.

This model is based, in part, on the sexual development of individuals with mutant alleles of these genes. Unlike *Drosophila* or *C. elegans*, however, it is not possible to undertake mutagenesis screens in humans, and investigators are limited to examining the effects of rare spontaneous mutations and/or chromosome abnormalities. For example, XY individuals with a chromosomal duplication of the Xp21 region have two active copies of *DSS*. These individuals develop as females, indicating that two doses of *DSS* can overcome the action of a single *SRY* and suppress the *MPGs*. Because XXY individuals develop

Summary

In this chapter, we have discussed how groups of regulatory genes acting in pathways or hierarchies regulate some of the most important aspects of development. Our discussion has centered on three examples—the control of embryonic pattern formation by maternal and zygotic segmentation genes, the control of cell fate by homeotic genes, and the control of sex determination. The fact that these developmental processes are regulated by gene hierarchies is not exceptional. One of the fundamental discoveries of developmental genetics is that important developmental processes in all species are controlled by hierarchies of regulatory genes. Not only has this general principle been highly conserved during evolution, but certain individual genes and gene hierarchies have been conserved as well. Thus detailed analyses of the genetic regulatory hierarchies of selected developmental processes in model organisms, such as those we have discussed in this chapter, give information not only about that particular process but also about similar processes in other organisms. Much of this information is directly applicable to higher organisms, including humans.

Maternal-gene products play a direct role in generating the organization of the embryonic body. During oocyte development, the products of maternal regulatory genes (RNA and protein) are packaged into the oocyte cytoplasm in a precise spatial pattern. The maternal-gene products act in one of two ways. Some are cytoplasmic determinants that function in mosaic systems to give a

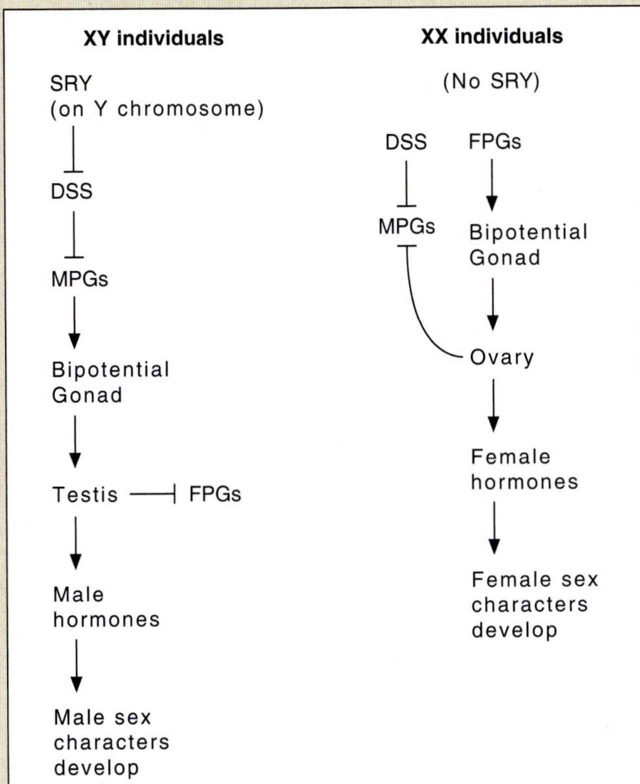

FIGURE 16.C The sex determination pathway in humans. The key to sex determination in humans is the *SRY* gene, located on the Y chromosome. In individuals with a functional *SRY*, the dosage-sensitive sex reversal (*DSS*) gene is repressed, which results in the activation of the male-specific genes (*MPGs*). These direct the bipotential gonad to develop into a testis, which produces male-specific hormones that repress female-specific genes (*FPGs*) and stimulate the development of male sex characteristics. In individuals without a functional *SRY*, *DSS* represses the *MPGs*, and the *FPGs* direct the bipotential gonad to develop into an ovary. The ovary produces female-specific hormones that repress the *MPGs* and stimulate the development of female-specific sex characteristics.

male characteristics, it is assumed that X chromosome inactivation normally occurs before *DSS* action. The inactivation of the extra X leaves these individuals with only one functional *DSS* gene. The fact that *DSS*-negative individuals develop male sexual characteristics also indicates that *DSS* is not itself a part of the male pathway but instead regulates the activation or inactivation of the pathway.

specific identity to the embryonic cells that receive them. Others establish positional information gradients as part of regulatory systems that control cell fate. In *Drosophila*, the precursors to the germ cells (the pole cells) are determined by cytoplasmic determinants (the polar granules). Embryonic segmentation is controlled by gradients of maternal-gene products activating segmentation genes. The zygotic segmentation genes act in a gene hierarchy (gap genes, then pair-rule genes, then segment-polarity genes) to divide the embryonic cells into segments and compartments. Both maternal and zygotic genes then activate the correct determination gene (homeotic) in each segment or compartment. Embryonic cell determination in *C. elegans* is regulated by cell lineage to a greater extent than in *Drosophila*. The early cell divisions establish a number of cell lines that give rise to particular cell types and structures. One feature that *Drosophila* and *C. elegans* share is that adult structures in both species develop from special sets of cells set aside during embryonic development. In *Drosophila*, these are the imaginal discs and histoblasts, and in *C. elegans* these are the blast cells. Both of these begin as small groups of embryonic cells that divide throughout the larval period and then differentiate into adult structures, and each group produces one particular structure or region of the body.

The homeotic genes are a cluster of regulatory genes whose products serve as determinants for embryonic segments and/or compartments in *Drosophila*. These genes are characterized by a particular sequential organization along the chromosome and by the presence of a highly conserved, 183-bp region called the homeobox. The homeobox codes for a DNA-binding region, which is specific for regulatory sequences found in the promoter region of many developmental regulatory genes. Genes containing homeoboxes have been found in nearly all eukaryotic species (plant and animal). Where they have been analyzed, homeobox genes often have a cell determination or other developmental regulatory function. The highly conserved structure of homeobox genes and their (apparently) equally conserved function suggests that these genes represent a nearly universal system of cell determination genes whose products control cell determination by binding to promoter sequences of a great many other genes and controlling their expression.

It has been known for many years that the sex of individuals in many species is controlled by the presence or absence of specific sex-determining chromosomes. The chromosome composition is called the primary signal for sex determination. This primary signal regulates the activity of one or a few specific genes that activate a sex-determination gene hierarchy, which activates the correct sex-specific genes in each individual. In *Drosophila,* the sex chromosome-to-autosome ratio determines the activities of several dosage-sensitive genes that regulate the splicing of the *sxl* gene transcript. In females, the transcript is spliced to give a functional product, and in males it is spliced to give a nonfunctional product. The functional *sxl* product regulates the splicing of the *tra* gene product, which in turn regulates the splicing of the *dsx* gene product. The sex-specific patterns of splicing cause different gene products to be made from identical primary transcripts in males and females. Thus sex-specific RNA splicing is the key element in regulating sex determination genes in *Drosophila.*

Sex determination also occurs in other species by a sex-specific gene hierarchy that is activated by one or a few genes. In each species there are important differences in how this hierarchy functions. In *C. elegans,* the X chromosome primary signal activates the first gene in the hierarchy by sex-specific transcription initiation. In mammals, including humans, the *SRY* on the Y chromosome activates male hormone-producing genes in the gonad. This directs the gonad to develop into a testis and to begin production of the male hormones that direct the somatic cells to develop as male. In females, the lack of an *SRY* activates female-specific genes in the gonad that direct it to form an ovary. The ovary secretes hormones that direct the somatic cells to develop as female. In mammalian species, the development of the visible somatic sexual characteristics is thus an indirect effect of the hormone production by the gonads.

Selected Key Terms

axis-polarity genes
developmental pathway
gap genes
Hox genes
pair-rule genes
pattern formation
primary signal
saturation mutagenesis
segmentation genes
segment-polarity genes

Chapter Review Problems

1. How might you determine whether there are some additional, undiscovered genes in the *Drosophila* posterior group?
2. The spatial distribution pattern of *eve* expression (seven stripes) has been proposed to be produced by the action of gap genes. How would you discover which gap genes are responsible?
3. How would you discover whether the spatial pattern of *eve* expression in the *Drosophila* embryo depends on cell–cell communication via the pathway?
4. There are nine genes whose mutant phenotypes all include the failure to form abdominal segments: *nanos, pumilio, vasa, staufer, cappuccino, spire, valois, oskar,* and *tudor.* What can you conclude about the normal functions of these genes? Describe how you might determine whether these genes act in a genetic hierarchy.
5. The second major body axis in *Drosophila* is the dorsal–ventral axis. How would you test the hypothesis that establishing this axis requires the early action of the *corkscrew* gene?
6. How could you determine whether the homeotic gene *Hoxa-7* from the mouse has the same function as one of the *BX-C* genes of *Drosophila?*
7. The continued expression of the *wg* and the *en* genes is proposed to be necessary to maintain the anterior–posterior compartment determination throughout adult development. How might you test this hypothesis?
8. The expression of the segment polarity gene *engrailed* is believed to be controlled by the expression of the pair-rule genes. How would you determine which pair-rule genes control *engrailed* expression?
9. For each of the following *Drosophila* genes, indicate in what group or class the gene is and what the distribution of the gene's product would be in the embryo.

a. *Scr*
b. *Antp*
c. *Kr*
d. *prd*
e. *nos*

10. You have isolated a homeobox-containing gene from the horse. How would you determine in which *Hox* group this gene belongs?

11. The *Hoxa-7* gene has been attached to a promoter that allows it to be expressed in every cell in the *Drosophila* embryo. Assuming that *Hoxa-7* can control cell determination in *Drosophila*, what would be the phenotype of embryos carrying this gene?

12. The mouse *Hox* group 11, 12, and 13 genes are all most similar to the *Drosophila AbdB*, the most posterior *Drosophila* homeotic gene. How would you test the hypothesis that these three genes code for determinants for segments in the mouse that are more posterior than any in *Drosophila?*

13. What phenotype would you expect of individuals carrying gain-of-function mutations that cause continuous expression of the *her-1* gene in *C. elegans?*

14. Three new genes in the springtail (a primitive insect) have been identified as genes that control sexual development. All three have been cloned. How would you use these clones to discover whether sex determination in springtails is controlled by a mechanism like that of *Drosophila* or like that of *C. elegans?*

15. In *Drosophila*, gynandromorphs male (XO) and female (XX) cells exist side by side. Why do the XX cells not influence their XO neighbors to develop as female cells?

16. A newly isolated allele of the *Drosophila tra* gene causes XY individuals to develop as females. The mutant lesion is in the transcribed region of the gene. What might be the molecular lesion that caused this mutation?

17. Would you expect the mutation in Problem 16 to be a dominant or a recessive allele? Explain your answer.

18. If you were given a strain of *Drosophila* with a new maternal-effect mutation, how would you determine whether this mutation is an allele of one of the terminal group genes?

19. What is a totipotent cell? Give an example of such a cell in an organism.

20. The terminal group of genes are thought to produce localized concentrations of the *tor* gene product in the posterior and in the anterior terminal regions of the egg. Explain how you could discover whether this is done by localizing maternal produced mRNA molecules or protein in the egg.

21. The *trunk* (*trk*) gene is a member of the terminal group. Explain how you might discover whether this gene's product functions as a repressor of the *tor* gene.

22. Give a brief definition of the following:
a. terminal class gene
b. gap gene
c. segment polarity gene

23. Describe one difference between each of the following pairs of genes:
a. *bicoid* vs. *runt*
b. *hairy* vs. *wingless*
c. *engrailed* vs. *bithorax*

24. You have discovered a tube containing DNA left behind by Ed Lewis that is labeled "cloned *engrailed* gene." Describe how you would discover whether this actually is *engrailed* DNA.

25. If *bcd* regulates the action of gap genes by binding to their promoter region (see Chapter 15), describe the effect on embryonic development you would expect a gain-of-function mutation of *bcd* to have that produces twice as much *bcd* product.

26. Describe how you would use the concept of saturation mutagenesis in *Drosophila* to identify the genes responsible for the formation of the embryonic nervous system.

27. If you were given a labeled DNA probe that is complementary to the mRNA produced by each of the following *Drosophila* genes, briefly describe the pattern of hybridization you would expect to see in cellular blastoderm stage embryos.
a. *hb*
b. *ftz*
c. *en*
d. *wg*

28. Give a brief description of each of the following:
a. homeobox
b. *Hox* gene
c. the *Hom-C* group
d. *ANT-C*

29. If you cloned a gene from the mouse that contained a homeobox sequence, describe how you would determine which *Hox* gene this was.

30. Give a brief definition of each of the following:
a. primary signal
b. dosage compensation
c. male-specific splicing
d. *sxl*

31. You have cloned the *Tribolium* counterpart of the *Drosophila dsx* gene. Describe how you would discover whether this gene functions in *Tribolium* like *dsx* functions in *Drosophila*.

32. Describe one difference and one similarity in sex determination in *Drosophila* and *C. elegans*.

33. You have isolated a new mutation that you believe is a gain-of-function (constitutively active) allele of a gene required for male development in *C. elegans*. What would you expect to be the phenotype of individuals with this mutation?

34. How would you test the hypothesis that, in *C. elegans*, sex determination occurs by the activation of different sets of genes, rather than sex-specific splicing of one set of genes?

35. Describe one difference and one similarity between the following pairs:
a. sex determination in humans and sex determination in *Drosophila*
b. sex determination in humans and sex determination in *C. elegans*

Challenge Problems

1. In *Drosophila,* a gynandromorph is a mosaic of male and female cells. In gynandromorphs, male structures form from patches of male cells and female structures form from patches of female cells. Would a human who is a mosaic of male and female cells show the same kind of phenotype? Explain why or why not.
2. Describe how you could use the saturation mutation technique to discover all the genes in *Drosophila* that respond to the male-specific spliced *dsx* gene product. Indicate in your answer what phenotype you would expect mutations of these genes to have.
3. It has been suggested that if the *Hoxa-1* gene from the mouse replaces its corresponding homeobox gene in *Drosophila,* it would function normally, giving an individual with a normal phenotype. Describe how you could test this hypothesis, and explain the results you would expect if the hypothesis were correct.

Suggested Readings

Bender, W., M. Akam, E. Karch, P. A. Beachy, M. Peifer, P. Spierer, E. B. Lewis, and D. S. Hogness. 1983. "Molecular genetics of the bithorax complex in *Drosophila melanogaster.*" *Science,* 221: 23–29.

Boncinelli, E., and A. Mallamaci. 1995. "Homeobox genes in vertebrate gastrulation." *Current Opinion in Genetics and Development,* 5: 619–627.

Cline, T., and B. J. Meyer. 1996. "Vive la différence: Males vs. females in flies vs. worms." *Annual Review of Genetics,* 30: 637–702.

Duboule, D., and G. Morata. 1994. "Colinearity and functional hierarchy among genes of the homeotic complexes." *Trends in Genetics,* 10: 358–364.

Jimenez, R., A. Sanchez, M. Burgos, and R. D. De La Guardia. 1996. "Puzzling out the genetics of mammalian sex determination." *Trends in Genetics,* 12: 164–166.

Kaufman, T. C., M. A. Seeger, and G. Olsen. 1990. "Molecular and genetic organization of the Antennapedia gene complex of *Drosophila melanogaster.*" *Advances in Genetics,* 27: 309–362.

Simon, J., M. Peifer, W. Bender, and M. O'Conner. 1990. "Regulatory elements of the bithorax complex that control expression along the anterior-posterior axis." *European Molecular Biology Organization,* 9: 3945–3956.

Steinmann-Zyicky, M., H. Amrein, and R. Nöthiger. 1990. "Genetic control of sex determination in *Drosophila.*" *Advances in Genetics,* 27: 189–237.

Sternberg, P. W. 1990. "Genetic control of cell type and pattern formation in *Caenorhabditis elegans.*" *Advances in Genetics,* 27: 63–116.

Wilkins, A. S. 1986. *Genetic Analysis of Animal Development.* Wiley, New York.

Wilson, E. B. 1927. *The Cell in Development and Heredity,* 3rd ed. Macmillan, New York.

On the Web

Visit our Web site at **http://www.saunderscollege.com/lifesci/titles.html** and click on A/G/M Genetics for links to the following chapter-related resources on the World Wide Web:

1. **The human sex determination LSRY gene.** This is a reference page at the Mendelian Inheritance of Man Web site, which contains information on the structure, function, and history of this gene.
2. **The human sex determination DSS gene.** This is a reference page at the Mendelian Inheritance of Man Web site, which contains information on the structure, function, and history of this gene.
3. **The *Drosophila* sex determination genes.** This site contains information and links to other sites about the *Drosophila* sex determination genes.
4. **The *C. elegans* sex determination genes.** This site provides information and links to other sites about the *C. elegans* sex determination genes.
5. **The *C. elegans* development site.** This site contains information and links to other sites about genes regulating *C. elegans* development.

PART

3

Considerable genetic diversity exists in all species. This photo shows the tremendous genetic diversity in the tomato. *(D. Cavagnaro/Visuals Unlimited)*

How Genes and Genomes Change and Evolve

CHAPTER 17
Gene Mutation and Repair

CHAPTER 18
Changes in Chromosome Structure and Number

CHAPTER 19
Mobile Genetic Elements and Recombination

CHAPTER 20
The Genetics of Cancer

CHAPTER 21
Population and Evolutionary Genetics

CHAPTER 22
Molecular Evolution

CHAPTER

17

A mutation in a pigment-producing gene has resulted in the lack of color, or an albino. The albino alligator and its sibling shown here were found in a Louisiana swamp. *(Gregory G. Dimijian, M.D./Photo Researchers, Inc.)*

Gene Mutation and Repair

The genetic material of an organism, either DNA or RNA, can be changed by both internal and external mechanisms or agents. The fidelity of DNA replication is extremely important in preventing unwanted mutations. However, when the genetic material is changed, and the change is inherited, a **mutation** has occurred. Overall, mutations affect organisms at one of two levels: some affect only a few base pairs (discussed in this chapter), and some affect entire chromosomes (discussed in Chapter 18). It should be emphasized that not all mutations have harmful effects on an organism. In fact, all organisms endure a normal background level of mutations as a result of normal cellular functions and random interactions with the environment. Mutations are the seeds for evolutionary change, and some mutations have beneficial effects on the organism.

What does a mutation do to an organism or cell to cause an inherited change? We know from Chapter 11 that the sequence of bases in the DNA of most genes codes for the sequence of amino acids in the protein product of the gene. The gene is first transcribed into RNA, which is then usually translated into a protein or a protein subunit. Proteins are important structural components of the cell. Proteins, for instance, make up the enzymes that catalyze nearly all the metabolic reactions within the cell. The sequence of amino acids in a protein is very important because any change in the amino acid sequence may inactivate or alter the protein's function. In turn, this change in function will probably result in a mutant phenotype. The mutant phenotype is restricted to one cell or cells derived from the original mutant cell. If a sufficient number of these mutant cells accumulate, they can have serious consequences for the organism (for example, cancer) (Chapter 20).

However, the key difference between a true mutation and other changes in an organism is its heritability. In this respect, mutations that occur in somatic cells are passed on to only a few progeny cells but not the next generation. For mutations to be inherited in the next generation, they must occur in the germ cells or in the cells that give rise to germ cells. Environmental factors such as **teratogens** (chemical agents causing malformations during development) can cause alterations in the apparent phenotype of an organism during development. For example, a frog exposed to a teratogen might develop five legs instead of four, but this change is neither inherited nor will it be passed on to the frog's offspring. Chemicals that cause these developmental changes are not mutagens and do not cause any change in DNA.

The different types of mutations and how they arise will be explained in this chapter. By the end of this chapter, we will have answered three questions:

1. How are the many different types of mutations categorized?
2. What are the molecular mechanisms that cause different types of mutations?
3. How does a cell repair damaged DNA?

There Are Many Different Categories of Mutations.

We have established that to qualify as a mutation, a change in an organism must be capable of being passed on to progeny. There are two forms of inherited mutations. Clearly, any mutation in a germline cell, either plant or animal, will pass the mutation on to the offspring of the organism. These kinds of mutations are called **germline mutations.** The second form of mutations, called **somatic mutations,** occur in cells that do not usually give rise to the next generation of sexually reproducing organisms (the exception being some somatic cells in plants). In this case, the progeny of a mutant animal somatic cell (for example, a skin cell) will give rise to a patch of mutant skin cells. The mutation has clearly been passed on to progeny at a cellular level, but it will not be passed on to the next generation of the organism's progeny.

In higher plants, some somatic cells can eventually end up in the germline. For example, a mutant somatic stem cell in a plant may give rise to a side branch. From this branch, a flower may form and all the cells of this flower will possess the mutation. Some of these flower cells undergo meiosis to form eggs and sperm to give rise to the next generation, which will possess the mutation. The Red Delicious apple is a notable result of a somatic mutation.

Mutations can be divided into two broad categories according to how they occur. Mutations can be spontaneous, or they can be induced, although the distinction between these categories is not always clear-cut. **Spontaneous mutations,** which account for a very large number of mutations, result from internal factors such as replication errors, mistakes in recombination, misrepair of DNA damage, depurination, deamination of bases, and the movement of transposons. Spontaneous, or background, mutations do not just "happen"; in every case, they have clear biochemical causes. **Induced mutations** result when organisms are exposed, accidentally or purposely, to any of the vast number of chemical agents and/or radiation that cause DNA changes that can be passed to offspring. As a result of exposure to chemical **mutagens** (agents that induce mutations), DNA is chemically changed or DNA replication is affected, which increases the rate of mutation above the background level. Radiation causes mutations by directly interacting with the genetic material or by creating chemicals that react with DNA, causing changes in the sequence of bases in a gene. In addition to these environmental mutagens, genetic engineering can now produce mutations in specific genes in transgenic plants and animals (see Chapter 13 and *Concept Close-Up: Antisense Genes Can Cause a Mutant Phenotype*). These mutations are very useful in understanding the function of a specific gene as well in commercial production of useful plants and animals.

Either spontaneous or induced mutations can be categorized according to the actual genetic change that takes place in the DNA. A large percentage of DNA mutations results from **base-pair substitutions** (for example, a GC pair replacing an AT pair). Base-pair substitutions, in fact, are the main topic of this chapter. There are two distinctly different base-pair substitution mutations. In the first type, **transition mutations,** one purine is replaced by the other purine on the same strand of DNA, or one pyrimidine is replaced by the other pyrimidine on the same strand of the DNA. In the second type, **transversion mutations,** the purine is replaced by a pyrimidine on the same strand of DNA, or a pyrimidine is replaced by a purine on the same strand of the DNA. Thus, transitions include GC ⟶ AT and AT ⟶ GC, and transversions include GC ⟶ CG or TA and AT ⟶ TA or CG. We will briefly describe the important base-pair substitution mutations next.

Not all mutations cause phenotypic changes in a cell or organism. Two types of base-pair mutations that change the DNA (and thus are inherited) but do not alter the visible phenotype of an organism are silent and neutral mutations (Figure 17.1). These two types of mutation can be either transitions or transversions. A **silent mutation** is a base-pair change in an allele's codon that produces no change in the amino acid sequence of the protein but that does change the base sequence in the DNA and, consequently, the mRNA. Because the genetic code is degenerate and because most amino acids have multiple codons, silent mutations are produced for most amino acids. We can illustrate this type of mutation with leucine, which has six different codons: a CUU codon can be changed to a CUA codon and still code for leucine in the protein.

Neutral mutations are base-pair substitutions in an allele's codon that produce a different amino acid in the protein, but this new amino acid does not change the function of the protein. These codon changes include

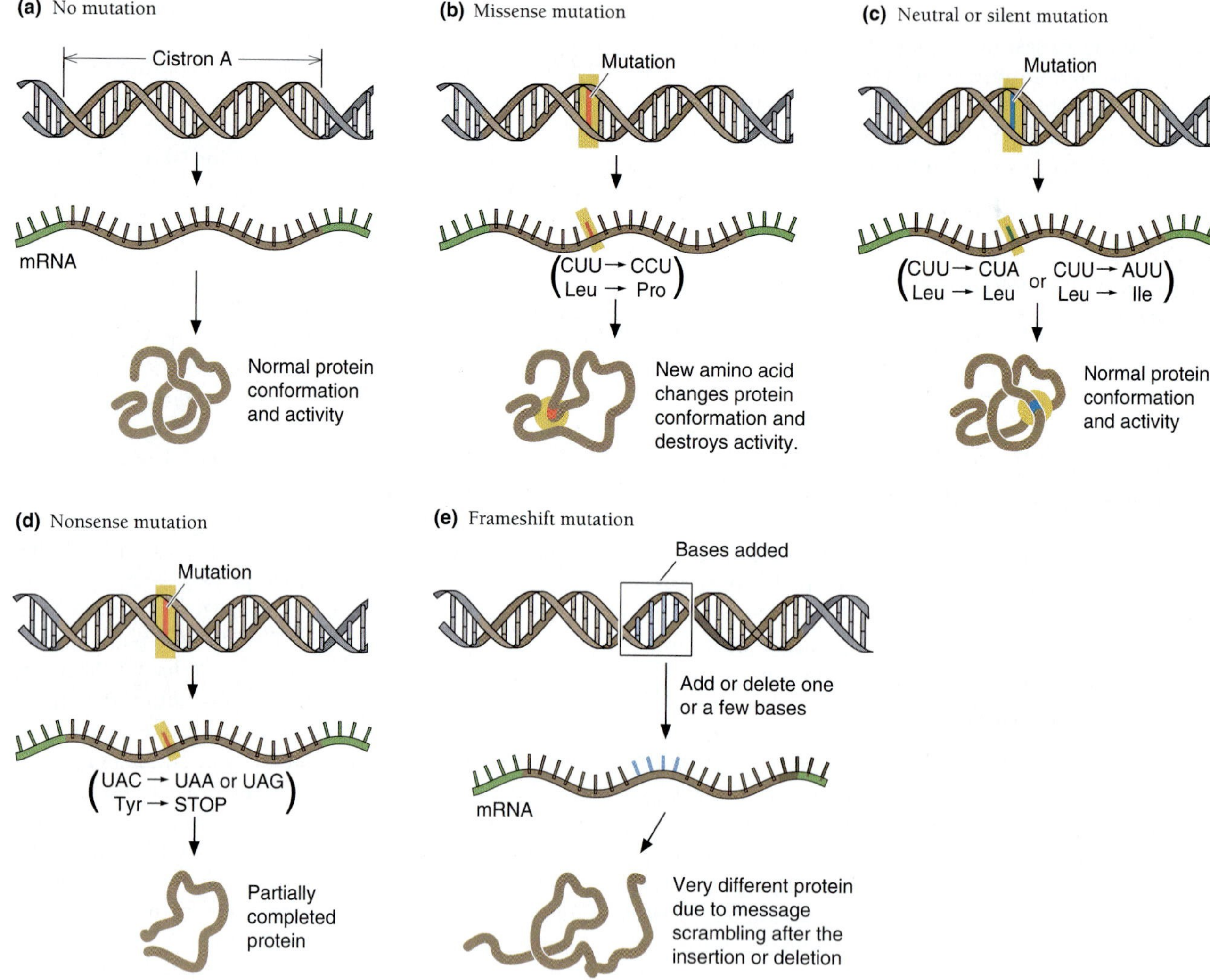

FIGURE 17.1 Different types of mutations produce different effects in the protein product. (a) The wild-type or functional protein is shown. (b) A missense mutation can cause a change in the function of a protein, or it may not if the mutation occurs in a nonessential region of the protein. (c) A neutral or silent mutation does not change the function of the protein, but a neutral mutation may also be a missense mutation. (d) Nonsense mutations cause premature termination of translation and thus yield partially completed proteins that are nearly always nonfunctional. (e) Like nonsense mutations, frameshift mutations nearly always yield nonfunctional products.

amino acid substitutions in a protein in which the new amino acid is chemically similar to the original amino acid. For example, the leucine codon CUU might be changed to the isoleucine codon AUU. The two amino acids, leucine and isoleucine, are chemically very similar, and thus the change usually would not alter the function of the protein. Neutral mutations can also occur when a codon changes from one amino acid to another, but the amino acid is not critical for the function of the protein and thus causes no phenotypic change.

Transitions and transversions of DNA base pairs can give rise to a missense mutation. In a **missense mutation,** an altered amino acid codon gives a different amino acid in the protein (for example, a change from leucine to proline) (Figure 17.1). Missense mutations are occasionally "leaky" because they produce protein products that possess partial activity and thus have a limited phenotypic change. Missense mutations are usually the cause of **temperature-sensitive mutations** in which a single base-pair change alters the function of the protein product, but only at elevated temperatures. At normal or lower temperatures, the mutant protein behaves as its wild-type counterpart. Temperature-sensitive mutations are members of a class of **conditional mutations** in

CONCEPT CLOSE-UP

Antisense Genes Can Cause a Mutant Phenotype

Anyone who has tasted an out-of-season tomato from a supermarket knows that they do not come close to the taste of garden-grown tomatoes. There are two reasons for this unpalatable situation. First, breeders have selected for commercial use varieties that can be mechanically harvested without bruising. These tomatoes have decreased moisture content and increased proteins and carbohydrates; they are ideal for making tomato sauce and ketchup, but not salads. Second, tomatoes are picked while they are still completely green in order to give shippers more time to get them to the market. Such tomatoes turn a pale red color during shipment and never develop the full flavor and texture of tomatoes picked after they have begun to ripen.

These pink monstrosities masquerading as tomatoes have been improved by researchers using antisense RNA to silence specific genes involved in the ripening process. Antisense RNA is produced by inverting the coding segment of the gene with respect to the promoter and inserting it back into the genome. When this process is followed, the other strand of DNA is read, producing an mRNA that is complementary to the normal RNA product. It is hypothesized that protein synthesis is inhibited because the two RNAs bind to each other like Velcro and prevent translation.

What gene should be silenced to produce a commercially useful tomato while retaining taste for the supermarket customers? During ripening, a tomato undergoes a series of biochemical changes that affect color and texture as well as accumulation of the red pigment lycopene. A very important enzyme that accumulates during ripening is polygalacturonase (PG). This enzyme is partially responsible for softening the fruit by helping to solubilize the pectin fraction of the cell walls. If polygalacturonase could be inhibited during ripening, then the tomato would not get soft, and the grower could allow the tomato to vine-ripen. Such tomatoes can be mechanically harvested and shipped to market without fear of the tomatoes turning to mush.

Several mutants affecting polygalacturonase levels have been found in tomatoes. One mutant, *rin*, produces little PG and does not significantly soften during ripening. Unfortunately, the *rin* mutation is pleiotropic and affects other desirable characteristics of the tomato.

An antisense copy of the PG gene has been prepared and inserted into tomatoes. These transgenic tomatoes had as little as 1% of PG activity and have a shelf-life of months. Unfortunately, these tomatoes are not commercially available because the developers had difficulty mass-producing them cost-effectively. Other enzymes have also been silenced using the antisense approach, including invertase, pectinesterase (which are also involved in cell-wall metabolism), and a gene involved in the synthesis of ethylene. Ethylene is important in controlling plant developmental processes and fruit ripening. An advantage of using antisense copies of a gene to silence the original gene is that it has no biological effect on the consumers of the product.

which the organism is wild-type or nearly so under one environmental condition (usually a low temperature) and mutant under another environmental condition (usually a higher temperature). However, the environmental conditions are not restricted to just temperature differences.

When an amino acid codon is changed to give a stop codon, UGA, UAA, or UAG, the resulting mutation is called a **nonsense mutation** (Figure 17.1). Nonsense mutations prematurely stop protein synthesis, and only a portion of the original protein is produced. Thus, nonsense mutations are usually extreme in their phenotypic effect. Unlike missense mutations, nonsense mutations rarely display partial activity because the protein product of the allele is changed so radically.

After base-pair substitutions, **insertion mutations** and **deletion mutations** are the next most-common causes of DNA mutations. Insertions and deletions of a base pair(s) in DNA cause **frameshift mutations** (discussed extensively in Chapter 11 and in Figure 17.1). The insertion or deletion of one or two (but not multiples of three) base pairs in an allele causes an altered reading frame in the mRNA; for example, if the DNA coding strand CAT CAT CAT CAT CAT has a single base-pair deletion at base pair 6, the mRNA will read CAU CAC AUC AUC AUC, and so forth. A frameshift mutation usually has a radical effect on the protein product. Almost all frameshift mutations yield protein products with a complete loss of activity. When the reading frame is shifted from the normal reading frame, the normal translational termination signal is ignored (because it is out of frame), and the mutant protein is terminated earlier or later.

Frameshift mutations and **point mutations** (the substitution of one base pair for another, or the addition or deletion of a single base pair) can be classified as either **forward mutations** or **reverse mutations.** A forward mutation changes a wild-type to a mutant phenotype, whereas a reverse mutation changes a mutant to a wild-type or pseudo-wild-type phenotype. Using the example from the previous paragraph, the reverse mutation could be the addition of another base pair to the sequence (at position 8, for example, giving CAT CAC ATA CAT CAT CAT). The two codon changes that are remaining in the middle of this sequence may be silent mutations, thus they would have no effect on reestablishing the function of the mutant protein. Because the reverse mutation did not occur at precisely the same site as the original mutation, this second mutation is said to be a suppressor mutation, in addition to being a reverse mutation.

A **suppressor mutation** is any event that occurs at a second site and restores the phenotype toward wild-type. Suppressor mutations can partially or completely restore the activity of the gene product. In this example, the suppressor mutation is an **intragenic** or **intracistronic suppressor** because the second mutation occurred in the same gene as the original mutation. Suppressor mutations outside the cistron of the original mutation can also mask or restore the primary mutation and are referred to as **extragenic** or **extracistronic suppressor** mutations. For example, if a mutation occurs in a gene and a second mutation (the suppressor mutation) occurs in another gene, these proteins must interact to form a functional protein. For example, assume that a particular protein is a dimer consisting of two protein subunits, α and β each coded by a different gene. The primary mutation could be in the α-subunit gene, and the suppressor mutation could be in the β-subunit gene. The suppressor mutation in the β-subunit protein compensates for the primary mutation in the α-subunit protein, giving a partially or completely functional dimer (Figure 17.2).

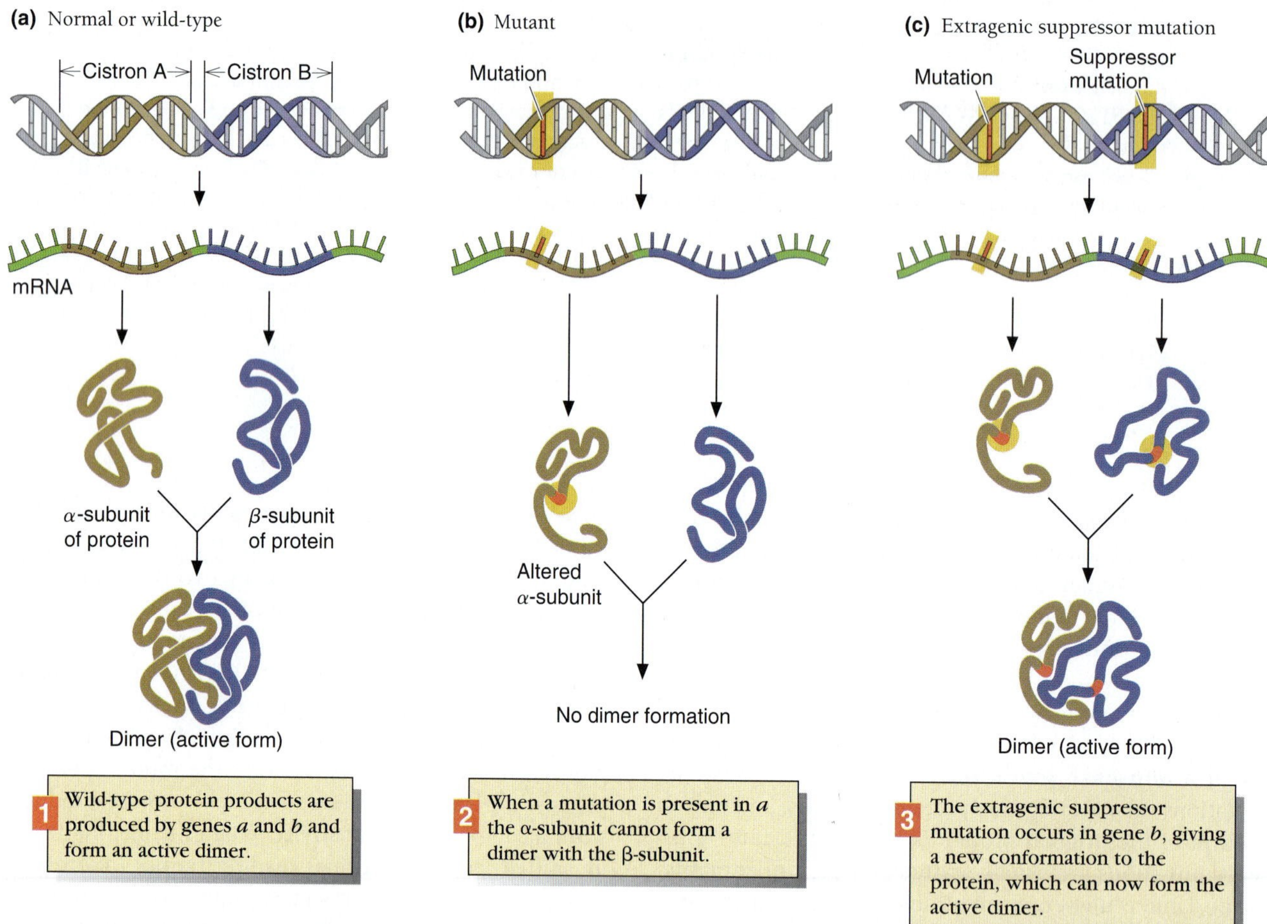

FIGURE 17.2 Extragenic suppressor mutation. (a) Some peptides must form dimers or tetramers with other peptides to be functional proteins. (b) A mutation in one peptide may result in a nonfunctional enzyme, but (c) a second mutation in another peptide can compensate for the first mutation. The second compensating mutation is a suppressor mutation.

Beyond deleting just one base pair, as in the case of frameshift mutations, deletion mutations may involve large numbers of base pairs or segments of chromosomes (Chapter 18). A deletion of one base pair typically has a radical effect on the gene product, often completely inactivating it. In addition, when a deletion shifts the reading frame of the mRNA, it is likely that a nonsense codon will be encountered. This nonsense codon, which was out of frame in the wild-type, is now recognized and will shorten the protein product. If deletions involve large numbers of base pairs, reversions rarely occur because the cell has no mechanism for restoring large segments of lost DNA.

Even though a mutation occurs in the DNA, it may not have an effect on the phenotype of the organism. Recall from Chapter 2 that we discussed recessive, codominant, and dominant mutations. In diploid organisms, a mutation will not be phenotypically visible if it is recessive. Recessive mutations are phenotypically visible in subsequent generations only after crossing with another similarly mutant organism (for example, a parent or sibling to obtain a homozygous recessive genotype). On the other hand, dominant or codominant mutations will be phenotypically visible in the first generation and may have an effect on the organism. You would not expect a homozygous recessive condition to result from two different mutations. Mutations are relatively rare, and thus two simultaneous mutations, one in each allele of a gene, would be extremely rare. For example, if the mutation rate for one allele is 10^{-7} per allele per generation, then the rate for both alleles (mutating simultaneously) is $10^{-7} \times 10^{-7}$ per allele per generation. There is an even lower chance of a mutation being expressed for some genes. For example, the rRNA genes are present in most organisms in multiple copies (sometimes 100s); thus, the chance of finding an individual defective in rRNA production (or a great reduction in production) is extremely rare unless a single mutation deleted some or all the alleles simultaneously.

SECTION REVIEW PROBLEMS

1. Distinguish between a mutagen and a teratogen.
2. What cell type in the body must be mutant if the next generation is to be affected by the mutation?
3. Some mutations do not cause phenotypic changes in the organism, even though it can be shown by DNA sequencing that an actual base change has occurred in a gene. Give several reasons why this is so (consider both prokaryotes and eukaryotes).
4. Leaky (mutations showing partial activity) and temperature-sensitive (ts) mutations are almost always missense mutations, never nonsense mutations. Explain.
5. Frameshift mutations frequently cause the protein product of the allele to be prematurely terminated at a point after the frameshift. Why?
6. A suppressor mutation can be intragenic or extragenic. Give a molecular example of an intragenic and extragenic suppressor mutation.
7. Why are mutations usually not visible in the F_1 generation but are visible in the F_2 if the mutant F_1 individuals are selfed?

Spontaneous Mutations Are Caused by Natural Phenomena.

In this section, we will discuss the causes of most spontaneous mutations—internal factors such as replication errors, depurination, deamination of bases, transposons, and mistakes in recombination. In the next section, we will discuss induced mutations, which are produced by radiation and exposure to foreign chemicals in the environment.

DNA Replication Errors May Cause Mutations.

The DNA replication apparatus of all organisms is very accurate. Recall from Chapter 9 that prokaryotic DNA polymerases have $3' \longrightarrow 5'$ exonuclease activity that allows proofreading or excision of incorrectly paired 3′-terminal nucleotides. This function, along with other repair mechanisms and base-pair selection, results in an extremely low error rate of about 1 in every 10^{10} nucleotides incorporated. In practical terms, a genome of 3 billion bp (3×10^9), as in humans, will have one mistake about every third division. However, only a small fraction of these mutations will be visible because many are in nongene areas of the genome (such as repeated DNA or stretches between genes) or are silent mutations. As you may have deduced, spontaneous mutations are relatively rare. However, the key term is "relatively"; spontaneous mutations may be rare, but they can add up over time when you consider that a human, for example, produces about 1×10^{16} cells in a lifetime.

All the bases can exist in one of two **tautomeric forms,** either the **keto** or the **enol** form if it has a hydroxyl group, or the **imino** and **amino** forms if it has an amino group. The common forms of the base are the keto or imino forms, whereas the rare forms are the enol and amino forms. Changes from a keto to an enol form, or imino to amino form are called tautomeric shifts. For example, hydrogen atoms can move from one position in a purine or pyrimidine to another nitrogen (Figures 17.3 and 17.4). Tautomeric shifts can cause mutations because the uncommon forms of the bases do not always pair properly during DNA replication. If the base is in the rare form at the moment of pairing, then an incorrect base may be incorporated, causing a mutation. The rate

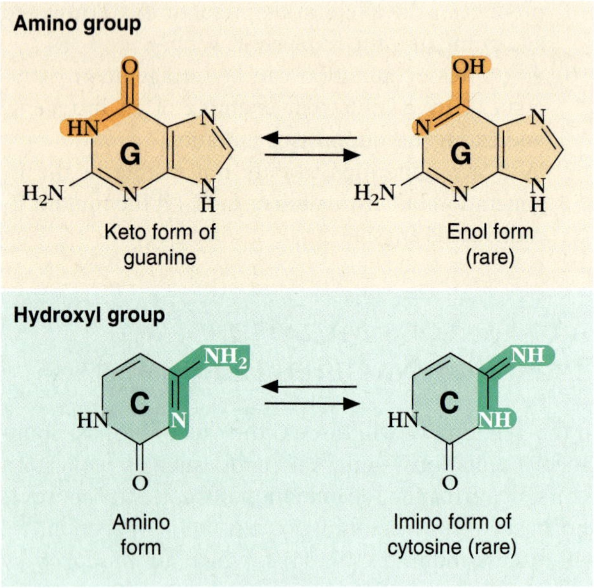

FIGURE 17.3 The rare tautomeric structures of the bases. The rare movement of a hydrogen atom results in an alternative form of the bases. This alternative structure does not pair properly with its complementary base.

of tautomeric shifts (enol to keto or imino to amino) predicts that a mutation will occur approximately 1 in every 10,000 bases incorporated. In reality, the mutation rate is one change in 10^{10} bases incorporated. This inconsistency between predicted mutations and actual mutations is caused by the error correction mechanism of the DNA polymerase (Chapter 9). DNA polymerase is extremely active in removing improperly incorporated nucleotides using its $3' \longrightarrow 5'$ exonuclease activity. If a mispairing does occur and is maintained in the replicating DNA, the result is a transition mutation in the next generation (Figure 17.5).

Another form of spontaneous replication error is DNA slippage or slipped mispairing, which can lead to frameshift mutations, small deletions, and added repeat units of nucleotides. In the mid-1960s, George Streisinger and his colleagues at the University of Oregon proposed a mechanism for the formation of frameshift mutations that involved DNA slippage. The proposal came from their understanding of the nucleotide sequence surrounding a series of mutants in the lysozyme gene of phage T4 (Figure 17.6). They found that most frameshift mutations occurred near sequences of DNA where runs of a single base pair occurred (for example a

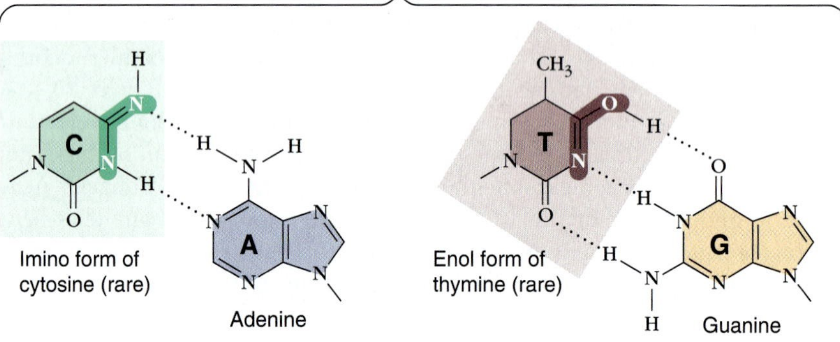

FIGURE 17.4 Mismatched bases. The alternate tautomeric chemical structures of the bases showing the changed positions of the hydrogens and mismatched pairing. This mismatching results in a mutation if it occurs during DNA replication.

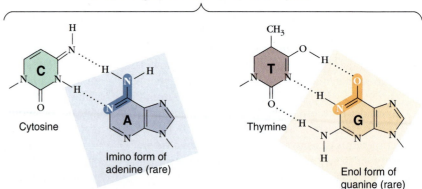

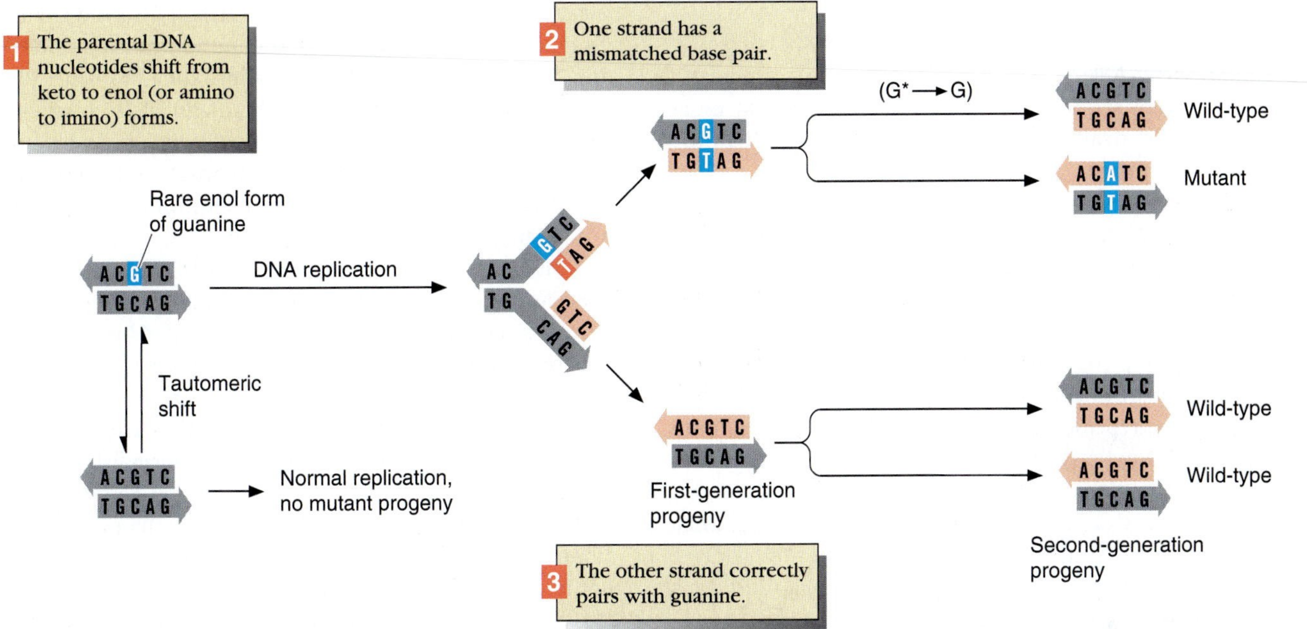

FIGURE 17.5 Mismatched bases produce mutations during DNA replication. Mismatched base pairs produce two different DNA molecules after replication, one containing the mutation or a G:C changed to an A:T, and the other identical to the parent molecule. Although this figure shows the rare form of the base being produced in template strand, the rare tautomer can also be present in the incoming base that pairs with the template strand base. The result will be the same.

series of A:T pairs). From these observations, they proposed that DNA looped out during replication at stretches of specific base pairs. Seemingly, mispairing occurred as a result of the long stretches of the same base. Subsequent studies on the *lacI* gene of *E. coli* and the *rII* locus of phage T4 have confirmed the original hypothesis proposed by Streisinger. Studies on the *rII* locus of phage T4 have shown the presence of "hot spots" that represent an increased rate of mutations found at a specific locus. At positions 131 in the *rIIA* and 117 of the *rIIB* allele, an extraordinary overrepresentation of frameshift mutations is found. Reexamination showed these sites, including several other hot spots, to be runs of six A:T pairs, which very likely are causing these

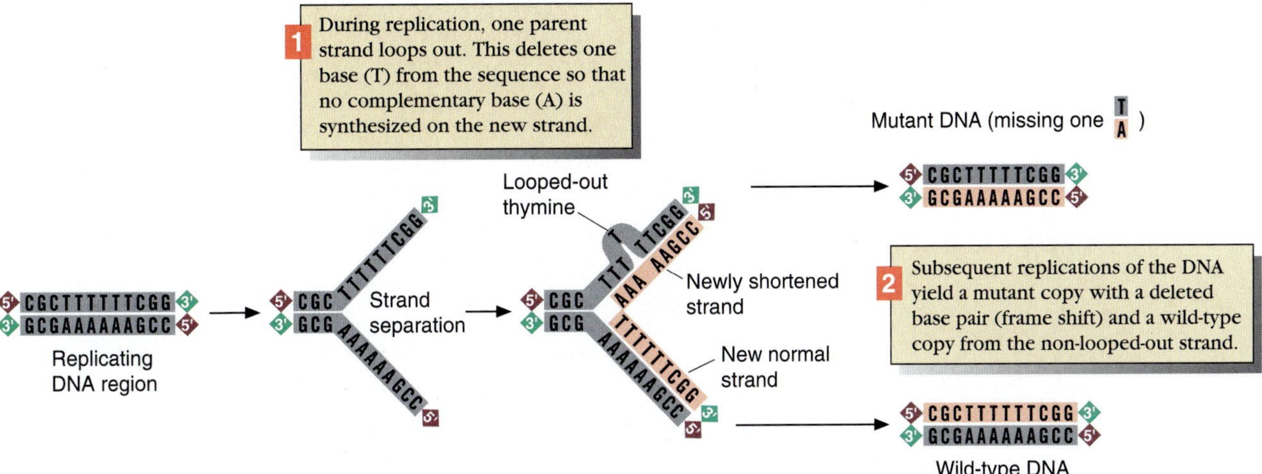

FIGURE 17.6 DNA slippage of a single base. This simplified version of the Streisinger model for DNA slippage results in a deleted base pair after replication. This occurs only when runs of the same base occur in the DNA.

mutations by DNA slippage. However, 12 other runs of 5 consecutive A:T base pairs that are not hot spots are present in the *rIIA* gene. Thus, DNA slippage may be greatly influenced by sequences surrounding the A:T runs, or, at least six identical base pairs in a sequence are required for DNA slippage.

In a separate study using the *lacI* gene from *E. coli*, it was demonstrated that hot spots are generated specifically at a 4-bp sequence repeated four times in tandem (Figure 17.7). These runs of 5′-CT *CTGG CTGG CTGG CTGG* CT-3′ sequences resulted in the loss (or gain) of groups of four base pairs at a time, suggesting a DNA slippage mechanism. In some mutants, this sequence was represented three times, but in others, only two times. Single-strand looping out during replication, which occurs during DNA slippage, causes deletions and represents a sizable fraction of spontaneous mutations. In the *lacI* gene studies with *E coli*, it was found that large deletions represented 19 of 140 spontaneous mutations and occurred at short repeated sequences.

DNA slippage mutations also occur in humans, although the exact mechanism of slippage is still not fully understood. These mutations result in a number of diseases in humans, one of which is "fragile-X syndrome." Fragile-X syndrome is the most common cause of inherited mental retardation, seen in approximately 1 in 1200 males and 1 in 2500 females. The diagnosis of fragile-X syndrome originally used cytogenetic analysis of metaphase spreads and showed sensitivity to breakage at site Xq27.3 under specific growth conditions in cell culture, hence the name of the syndrome.

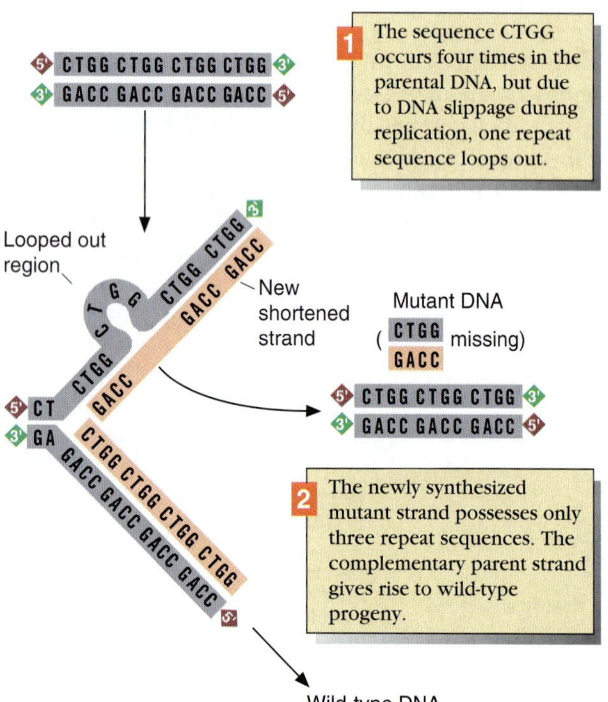
CTGG CTGG CTGG CTGG
GACC GACC GACC GACC
1 The sequence CTGG occurs four times in the parental DNA, but due to DNA slippage during replication, one repeat sequence loops out.
Looped out region
New shortened strand
Mutant DNA
(CTGG GACC missing)
CTGG CTGG CTGG
GACC GACC GACC
CT
GA
2 The newly synthesized mutant strand possesses only three repeat sequences. The complementary parent strand gives rise to wild-type progeny.
Wild-type DNA

FIGURE 17.7 DNA slippage of multiple bases. DNA slippage can cause deletions of several base pairs when repeated sequences occur in clusters. This same mechanism can also causes the loss of two repeat units by looping-out two repeat units rather than one. This type of deletion occurs only in runs of repeat units.

In 1991, the fragile-X gene (*FMR1*) was characterized and found to contain a tandemly repeated trinucleotide sequence (CGG) near its 5′ end. The mutation responsible for fragile-X syndrome involves increasing the numbers of the CGG sequence within the gene (called "expansion"), which is probably caused by a DNA polymerase slippage process. In normal individuals, the number of CGG repeats in the *FMR1* gene varies from 6 to approximately 50. Repeat numbers between about 50 and 200 cause the disease to vary in intensity; individuals with 200 repeats manifest a very severe phenotype. For poorly understood reasons, some alleles are unstable and expand from generation to generation, whereas others are stably inherited. Most, but not all, males with a full fragile-X mutation are mentally retarded and show typical physical and behavioral features. Of females with a full mutation (between about 75 and 200 repeats), approximately one third are of normal intelligence, one third are of borderline intelligence, and one third are mentally retarded.

Because of its novel nature, inheritance of the fragile-X mutation is less straightforward than classic Mendelian traits. Although passage through a female meiosis is necessary for significant expansion of the trinucleotide repeat, the expansion is most likely to occur during early embryonic development. Because expansion occurs in a multicellular embryo and the extent of expansion may vary from cell to cell, individuals often display somatic heterogeneity in allele size. Expansion of the trinucleotide repeat to more than 200 repeats is almost always associated with methylation of the promoter region of the gene and correlated gene inactivation. Although it is clear that methylation status plays a role in phenotype, its effect on clinical severity is somewhat unpredictable, especially in females.

Recently, DNA studies have improved the accuracy of testing for fragile-X syndrome. The size of the trinucleotide repeat segment can be examined using the polymerase chain reaction (PCR). PCR analysis uses flanking primers to amplify a fragment of DNA spanning the repeat region. Thus, the sizes of the PCR products are indicative of the approximate number of repeats present in each allele of the individual being tested. In a small number of fragile-X patients, mechanisms other than trinucleotide expansion, such as deletion or point mutation, are responsible for the syndrome.

Spontaneous Chemical Changes Can Cause Mutations.

Another type of spontaneous mutation is caused by either depurinations or deaminations. **Depurination** results when a purine, either adenine or guanine, is removed from DNA when the covalent bond breaks between the purine and the deoxyribose. This seemingly radical event actually occurs very frequently. For example, a mammalian cell will lose approximately 10,000 purines from its DNA during one cell division of about 20 hours at 37°C. If these purine losses (lesions) are not repaired (as most of them are), no complementary base will be present during the next round of replication to specify the incoming base for the new strand. Instead, a randomly chosen base will occupy the position opposite the depurinated site, potentially causing a mutation (Figure 17.8).

Deamination (removal of the amino group) of cytosine and 5-methylcytosine yield uracil and thymine, respectively (Figure 17.9). The other bases are not readily deaminated. Cytosine deamination is one of the most frequent reactions and causes serious damage to cellular DNA. It can occur at 37°C at a significant rate through spontaneous hydrolysis of the amino group. Uracil, produced by cytosine deamination, will mispair with adenine instead of guanine, ultimately causing a G:C pair to be converted to an A:T pair. Deamination of 5-methylcytosine (which normally pairs with guanine) produces thymine, which pairs with adenine. For the *lacI* gene of *E. coli*, it was shown that the four sites in this gene occupied by 5-methylcytosine were the source of 44 of 55 independently isolated GC-site spontaneous mutants. The remaining 11 mutations were at unmethylated cytosines, and all were G:C to A:T transitions. 5-Methylcytosine generates many more mutations than cytosine because DNA repair mechanisms do not always recognize its deaminated product, thymine, as an odd base, and consequently the mismatch is not repaired as frequently. As discussed later in the chapter, most of the depurinations and deaminations are repaired before they cause a permanent mutation.

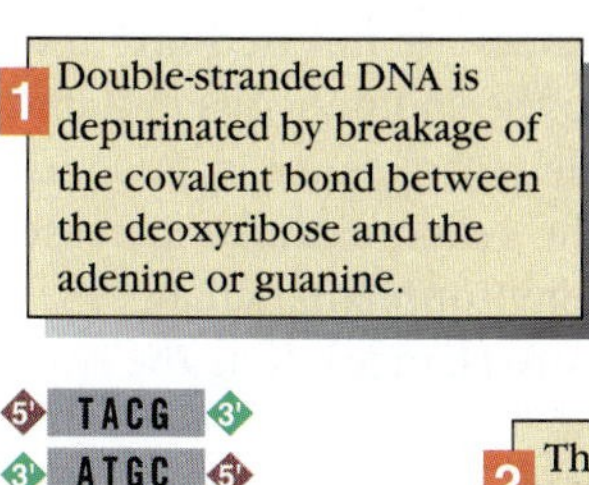

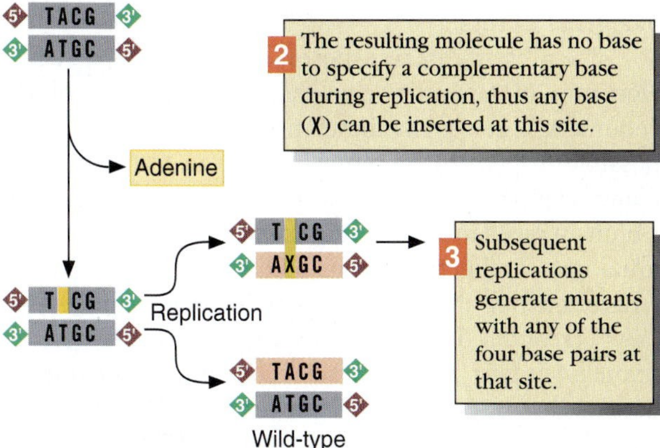

FIGURE 17.8 Depurination may result in mutations after replication. Depurination removes a purine but leaves the phosphodiester linkage intact. As a consequence, one strand has no specific base to pair with during subsequent replications and may produce mutations.

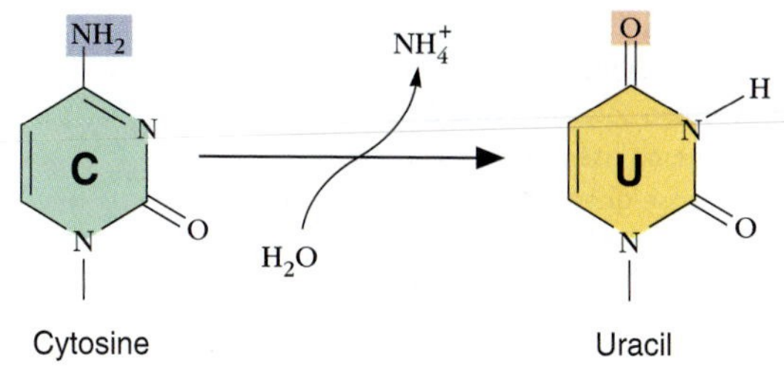

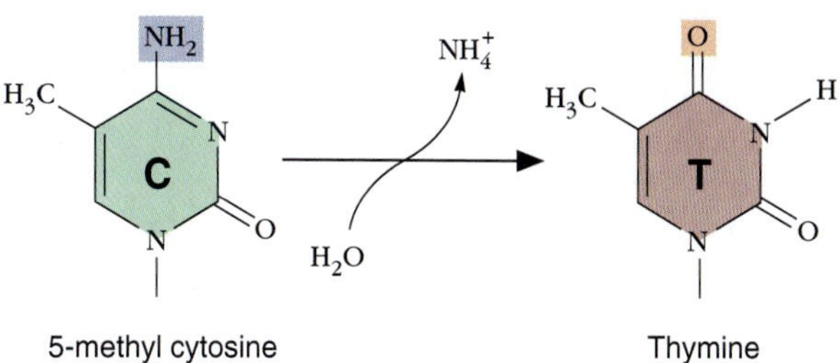

FIGURE 17.9 Deamination of cytosine. Both cytosine and 5-methylcytosine can lose their amino groups, yielding bases that have new base-pairing properties.

Transposons and Insertion Sequences Act as Mutagens.

Geneticists are accustomed to thinking of the genome as static, changing only very slowly with accumulated mutations. However, we now know that this concept is not completely true. There are mobile elements present in the genome known as **transposons,** or transposable elements, and **insertion sequences.** Transposons and insertion sequences add a degree of plasticity to the genome that was completely unexpected. Briefly (because Chapter 19 is entirely devoted to this subject), transposons and insertion sequences are 1- to 10-kb DNA elements that are capable of movement within the genome. When

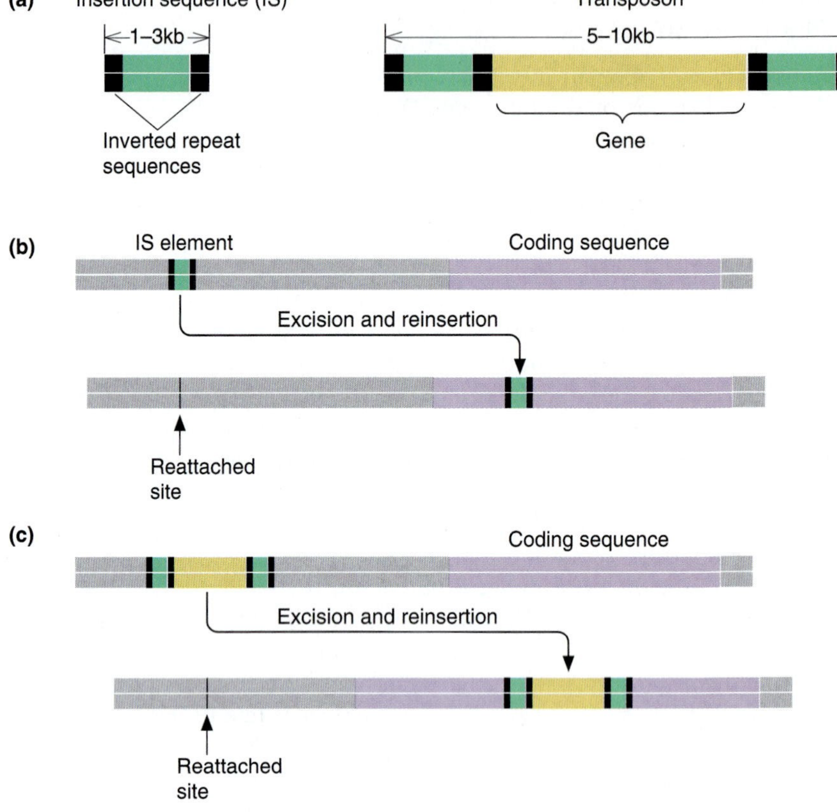

FIGURE 17.10 Insertion sequences and transposons. (a) The general structure of an insertion sequence and a transposon is shown here. Insertion sequences possess a transposase gene whose product catalyzes the actual transposition event. Some transposons have genes flanked by two insertion sequences, and others have more complex structures. Both insertion sequences and transposons possess short inverted repeat sequences. These repeat sequences are necessary for the elements to move in the genome. (b, c) A model of how a transposon or insertion sequence moves within the genome is shown here. A transposon or insertion sequence can insert in a gene at a point that will disrupt the coding sequence and cause a mutation.

they move about the genome, they are either excised or copied and inserted in another location. The possibility exists that when it is inserted, the transposon or insertion sequence may end up within an allele, disrupting the coded information and thus causing a mutation.

Transposons are found in all organisms and were first characterized at the molecular level from bacteria. The general structure of a transposon and an insertion sequence and a model of how they might cause a mutation is shown in Figure 17.10. Transposition of an insertion sequence or transposon occurs with the assistance of an enzyme, **transposase,** which is coded for by the element. Prokaryotic transposons are actually a composite of two insertion sequences (Figure 17.10) with other coding information spliced between the two elements. Both insertion sequences and transposons have terminal sequences that are required for transposition. More details on the structure of transposons and insertion sequences and how they move within the genome can be found in Chapter 19.

Examinations of known mutations indicate that a large percentage of them are produced by insertion of transposons or insertion sequences. Current estimates suggest that at some loci, between 50 and 80% of the mutations are caused by the disruption of genes by transposons or insertion sequences. The rate of transposon and insertion sequence movement from one location to another varies, but the overall rate of transposition is quite high, between 10^{-3} and 10^{-4} per copy per generation. The rate at which sequences are inserted into individual alleles is comparable to the spontaneous mutation rate of 10^{-6} to 10^{-7} per allele per generation. Also, the rate of transposition within the genome can be increased by exposing an organism to environmental changes (for example, an abrupt temperature change) or as the result of a genetic cross.

The presence of transposon and insertion sequence elements in genomes has changed our view of the genome as a device that inherently provides every means possible to prevent mutations. Much to the surprise of many scientists, evolution has provided the genome with a built-in mechanism for each species to generate its own mutations, and perhaps at its own rate, depending upon environmental changes and genetic backgrounds. This newly recognized genetic variability has, of course, had a dramatic impact on natural selection and thus on our understanding of evolutionary processes (see Chapters 21 and 22 for more discussion of this subject).

Unequal Crossing-Over Produces Mutations.

Spontaneous mutations can also be generated by mistakes made during meiotic recombination. Normally, meiotic recombination between paired homologous

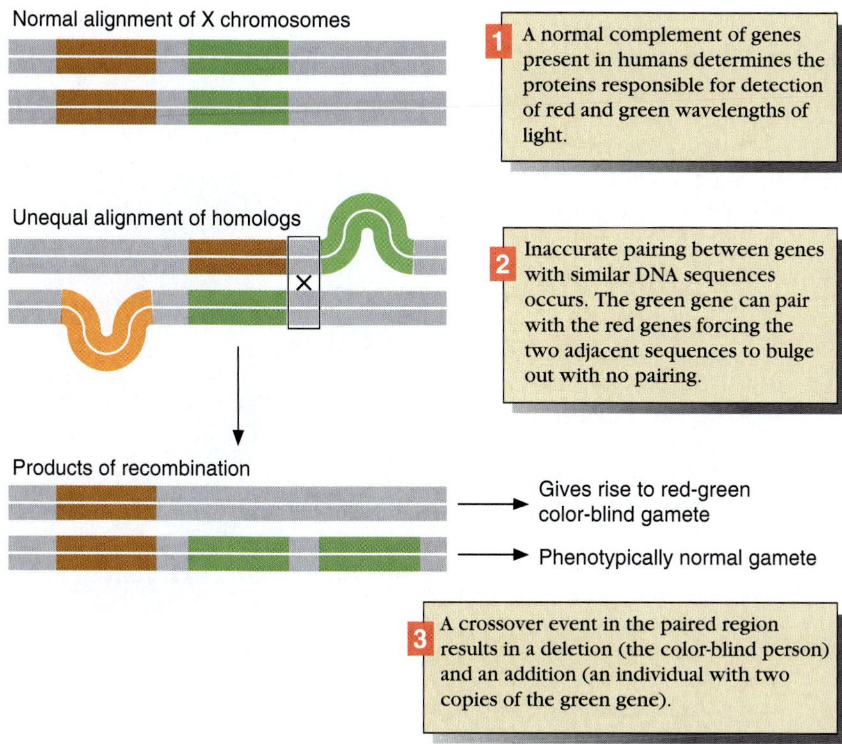

Normal alignment of X chromosomes
1 A normal complement of genes present in humans determines the proteins responsible for detection of red and green wavelengths of light.
Unequal alignment of homologs
2 Inaccurate pairing between genes with similar DNA sequences occurs. The green gene can pair with the red genes forcing the two adjacent sequences to bulge out with no pairing.
Products of recombination
Gives rise to red-green color-blind gamete
Phenotypically normal gamete
3 A crossover event in the paired region results in a deletion (the color-blind person) and an addition (an individual with two copies of the green gene).

FIGURE 17.11 Deletions generated by unequal crossing over. An unequal crossover can generate deletions and addition of genes to chromosomes during meiosis. In this case, the green gene and the red gene are nearly identical, which allows them to pair. If a crossover event occurs between them, an addition and deletion are generated.

chromosomes is very precise. However, when chromosomes contain tandem duplications (head-to-tail arrays of a single or very similar DNA sequences), mispairing between repeated sequences can occur during alignment of the two chromosomes. Recombination in the mispaired region is called **unequal crossing over,** and it produces duplication and deletion mutations.

A lack of ability to discriminate between colors in the red-to-green region of the spectrum (red–green discrimination) is found in roughly 8% of the Caucasian male population and 1% of the Caucasian female population. The genes for blue color detection lie on another chromosome, and defects in this ability are rare. Why does this group of humans have such a high frequency of red–green color vision loss? We cannot be absolutely sure of the answer, but it may lie in unequal crossing-over during meiosis between tandemly arrayed color-producing genes. The number of copies of the allele responsible for production of the red–green color detection pigment (the "green" gene) varies; some individuals have one copy, others have two copies, and still others have three copies. The red-pigment allele is nearly always present in one copy. However, only one red allele and one green allele are necessary for normal color vision.

The genes that code for the color pigments responsible for detecting red and green colors lie in a head-to-tail array on the X chromosome of humans (Figure 17.11). The "red" gene is 97% identical to the green gene; thus, they can pair during meiotic pairing. If a crossover event occurs during this mispairing, one copy of the green gene will be deleted from one gamete, and an extra copy of the green gene will be added to the other gamete. Unequal crossing over between these two tandemly arrayed genes is a possible explanation for the higher rate of color blindness in the male population. Because these genes are on the X chromosome, the frequency of male color blindness is higher than in females.

Spontaneous Mutations Occur at a Low Rate.

A variety of methods are available to detect the presence of mutations, and several of these methods are discussed later in this chapter. To detect the presence of mutations in an organism, it is usually necessary to examine a very large number of offspring to find these relatively rare events. Consequently, finding mutations in bacteria is relatively straightforward, whereas finding mutations in maize is much more laborious. Lewis Stadler [see *HISTORICAL PROFILE: Lewis J. Stadler (1896–1954)*] looked for mutations in a number of genes in maize and found values that varied from 1 to about 100 mutations per 1 million gametes. These values are expressed mathematically as 1×10^{-4} to 1×10^{-6} per generation and are known as the mutation rate. **Mutation rate** is expressed as the number of mutations occurring in some unit of time, usually per generation. Another term that is used is the mutation frequency. **Mutation frequency** is the frequency at which a specific kind of mutation (or mutant)

HISTORICAL PROFILE

Lewis J. Stadler (1896–1954)

L. J. Stadler (Figure 17.A) was a plant geneticist who made many contributions to the understanding of mutagenesis. He was born and educated in St. Louis, MO, and he attended the University of Missouri (U. of M.)(1913–1915). He later transferred to the University of Florida, receiving his B.S. degree in agriculture in 1917. He went back to the U. of M. and obtained his Ph.D. in Crop Science in 1922, and he spent most of his career at U. of M., with a brief hiatus at Cornell and Harvard. In May of 1954, he died from Hodgkin's disease, which was most likely precipitated by his exposure to X-rays during his study of X-ray induction of mutants in maize. Stadler was a brilliant, easy-going scientist and was very influential in attracting other outstanding geneticists to U. of M., including several members of the prestigious National Academy of Sciences and Barbara McClintock, who won the Nobel Prize for her work on transposons in maize.

In 1928, Stadler published a landmark paper on gene mutation in maize induced by X-rays. Up until this time, it was not known that a mutation could be induced in plants by X-rays, or for that matter by any other form of radiation or chemicals. Unfortunately for Stadler, his publication was delayed because he chose to use maize, which has a long generation time. As a result, H. J. Muller, using the much faster-generating *Drosophila*, arrived at the same conclusions as Stadler but published one year earlier a paper for which he received the Nobel Prize. Many of Stadler's contemporaries thought he should have shared the Nobel Prize for this landmark discovery.

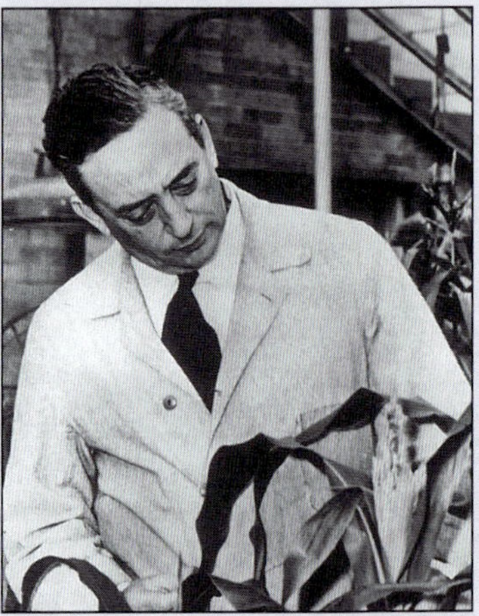

FIGURE 17.A Lewis J. Stadler.

Following Stadler's studies with X-rays, he turned in the 1930s to ultraviolet light as a possible mutagen. *Drosophila* could not be used to study the effects of UV because the overlying tissue of the testes is resistant to penetration of wavelengths between 200 and 300 nm. On the other hand, corn pollen does not have this drawback. Stadler and his students set up a series of experiments to treat pollen grains with individual wavelengths of UV light to examine the effects on gene mutation. They obtained an experimental curve showing the effectiveness of various wavelengths in inducing mutations. They found the mutation–induction curve fit well with the UV absorption curve for DNA, suggesting strongly that DNA was the genetic material. These data were presented in 1942, long before DNA was proven to be the actual genetic material.

occurs in a population. For example, Huntington disease occurs in the human population at a frequency of 1 in 10,000 individuals, but the rate of mutation is much lower (about 0.5 in 1 million). A change in the mutation rate for a specific gene or all genes can occur as a result of the presence of chemical or radiation in the environment, which is discussed in the next section.

SOLVED PROBLEM

Problem

Suppose that a series of spontaneous mutations occurred in a particular codon. These mutations resulted in the following amino acid substitutions: Normal (tryptophan) ⟶ mutant 1 (serine) ⟶ mutant 2 (leucine) ⟶ mutant 3 (valine). What is the codon for each of the mutants and the wild-type? Use the table of codons to determine the answer (Figure 11.20).

Solution

Tryptophan has only one codon, UGG. By changing the second base of the tryptophan codon to C, the codon becomes UCG, the codon for serine, and gives the first mutation. The second mutation is UCG to UUG, also by changing the second base, and now codes for leucine. The third mutation is derived by changing UUG to GUG, which codes for valine.

SECTION REVIEW PROBLEMS

8. If *E. coli* has 4.8 million bp and mutations occur at a rate of 1 in every 10^{10} bp incorporated, how many mutations would occur per generation?
9. Transposons and insertion sequences usually cause mutations by inserting into a gene (either the introns or exons), but sometimes an insertion actually increases the activity of the gene or acti-

vates a previously inactive gene. Propose an explanation for this observation.

10. In some organisms (for example, *D. melanogaster*), many spontaneous mutations have been shown to result from transposon activity. How can this be explained?
11. Suggest a possible mechanism for the high frequency of red–green color blindness in the human population.

Induced Mutations Are the Result of Our Environment.

In 1928, H. J. Muller and L. J. Stadler independently discovered that X-ray radiation could induce mutations in higher organisms. Not long after, it was found that a vast variety of chemicals also produce mutations in eukaryotes and prokaryotes. Both radiation and chemical mutagenesis have been important tools for genetic studies ever since.

It is important to understand the molecular mechanisms of how radiation and chemicals cause mutagenesis, not only for genetic studies but also for understanding how to prevent mutations from affecting the human population. For example, we now understand that most cancers (the uncontrolled proliferation of cells, see Chapter 20) are caused by multiple mutations. Clearly, agents such as chemicals and radiation that produce an increase in mutations can dramatically affect our everyday lives. In the following section, we will discuss the more common forms of radiation and chemicals and the mutations they induce. However, new chemicals that cause mutations are being added to the list every day.

Radiation Can Cause Mutations.

Radiation can cause mutations by breaking DNA, chemically changing the structure of the DNA, or by forming unstable reactive compounds that physically alter DNA. Different wavelengths of radiation produce different types of mutations. The electromagnetic spectrum varies in wavelength from 0.001 nm to 100 m (Figure 17.12). As the wavelength of electromagnetic radiation becomes shorter, the energy of the radiation increases. Ultraviolet (UV) light has a wavelength from about 10 to 300 nm (Figure 17.12) and is the most commonly encountered damaging radiation. Ultraviolet light is harmful only to cells on the surface because it has a low penetrating power. High-energy radiation (gamma and X-rays) has a wavelength between about 0.001 and about 10 nm and can cause mutations deep within the body as a result of its penetrating power.

Ultraviolet light emits most of its energy between 240 and 280 nm. Because DNA absorbs light intensely at 250–260 nm, UV light is very damaging to the DNA of all the cells it can penetrate. Fortunately, UV light has very limited penetrating power, and in multicellular organisms such as humans, only the surface layers of cells are damaged by UV radiation. However, prolonged exposure to UV light will kill most microorganisms, and it has been very successfully used by scientists and medical personnel to sterilize instruments.

The sun is an excellent source of UV light, and it can be extremely damaging to exterior cells of the body after repeated or prolonged exposure. The amount of solar UV light received at the earth's surface varies with season, altitude, latitude, and ozone levels. Levels are also affected by the distance sunlight must pass through

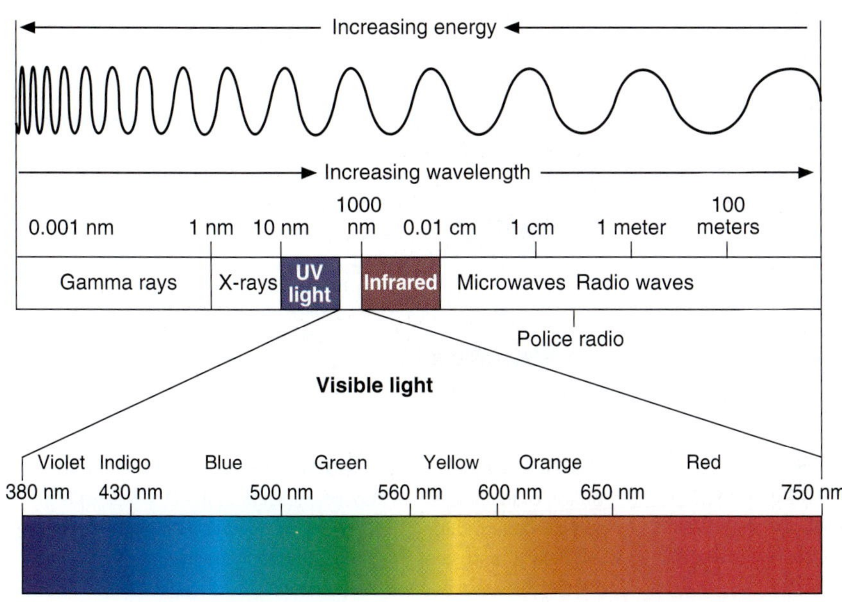

FIGURE 17.12 The electromagnetic spectrum. Energy is greatest at the shorter wavelengths. Ionizing radiation, including gamma and X-rays, is of the highest energy and can penetrate deep into tissue. The most damaging radiation is ultraviolet light in the range of 250–300 nm, which has poor penetrating power but can damage surface cells.

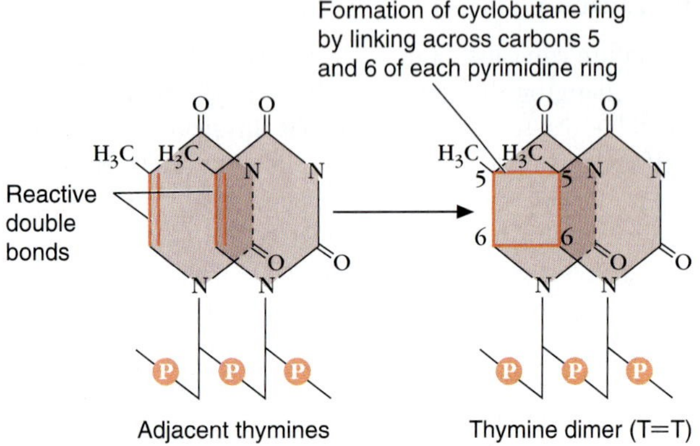

FIGURE 17.13 **Thymine dimer formation.** After exposure to ultraviolet light, the double bonds from the two adjacent pyrimidines, which can be thymine–thymine, cytosine–cytosine, or thymine–cytosine, can form a cyclobutane ring.

the atmosphere, which absorbs UV light. Most glass or clothing absorbs UV rays completely.

Exposure to UV light produces mutations in the following manner. Ultraviolet radiation can cause two pyrimidines adjacent to each other on the same strand of DNA (Figure 17.13) to form a dimer. Thymine dimers are the most common form of pyrimidine dimers, although C:T and C:C dimers also form. Dimers rarely form across two strands, only between adjacent pyrimidines on the same strand. This reaction occurs frequently in skin cells exposed to normal sunlight, and thousands of **pyrimidine dimers** can form in just a few hours. The presence of a pyrimidine dimer in the DNA affects base-pairing during complementary strand formation. The unpaired dimer disrupts the normal structure of DNA and bulges slightly from the double-stranded molecule (Figure 17.14). The dimer cannot act as a template to direct the complementary strand synthesis because the replication apparatus (DNA polymerase) cannot traverse the dimer. DNA polymerase stops before and restarts after the dimer, leaving a gap in the new DNA sequence opposite the dimer. The gap is repaired, in most circumstances, by randomly placing bases opposite the dimer. Most dimers are rapidly repaired (discussed in the last section of this chapter), but if they are not, they can cause mutations during replication.

Gamma and X-rays also cause DNA damage and mutations. The mechanism by which they cause mutations is less well-understood but is believed to involve the formation of free radicals. Free radicals are chemicals that have an unshared electron and consequently are extremely reactive to molecules in the cell. They react with DNA to cause breaks, cause deletions, and alter base-pairing characteristics. Unlike UV radiation, gamma and X-rays can easily penetrate deep into tissue. As a consequence, all cells in the body are subject to damage from high-energy radiation, including the germ cells.

Chemicals Can Induce Mutations.

Different classes of chemical mutagens have different modes of action. Some chemicals imitate bases (called base analogues) and replace them during replication, others alter base structure, and still others directly induce insertions and deletions during DNA replication. Most chemical mutagens generate a premutational lesion that can be repaired prior to replication (which will be discussed in the next section). However, if DNA replication occurs before the lesion is repaired, a mutant base sequence is produced, and an heritable mutation is formed.

Base Analogues Replace Normal Bases. Base analogues have chemical structures that are similar to the four nucleotides, and when the base analogue is phosphorylated, it can be incorporated into the DNA. The base analogue **5-bromodeoxyuridine** (BrdU) has a bromine at the 5 position of the pyrimidine ring instead of a methyl group and, thus, resembles thymine. As a result, BrdU is usually incorporated opposite adenine (Figure 17.15). However, BrdU can exist in either the enol or keto form, and when it is in the enol form, it pairs with

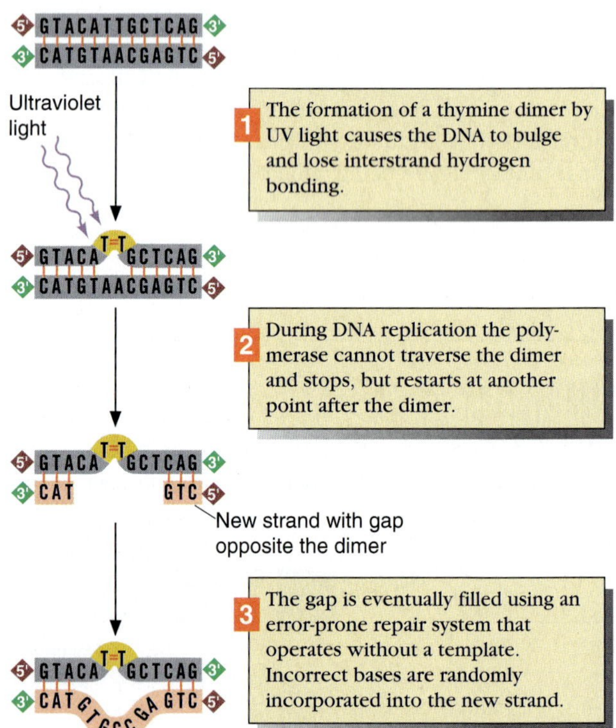

5' GTACATTGCTCAG 3'
3' CATGTAACGAGTC 5'
Ultraviolet light
1 The formation of a thymine dimer by UV light causes the DNA to bulge and lose interstrand hydrogen bonding.
GTACA T=T GCTCAG
CATGTAACGAGTC
2 During DNA replication the polymerase cannot traverse the dimer and stops, but restarts at another point after the dimer.
GTACA T=T GCTCAG
CAT GTC
New strand with gap opposite the dimer
3 The gap is eventually filled using an error-prone repair system that operates without a template. Incorrect bases are randomly incorporated into the new strand.
GTACA T=T GCTCAG
CAT GTGGCGA GTC

FIGURE 17.14 **Thymine dimer-induced mutation.** The formation of thymine dimers causes a bulge in the DNA, stopping polymerization during replication. The gap is filled by a poorly understood and sometimes inaccurate mechanism, thus creating a new DNA sequence and a mutation.

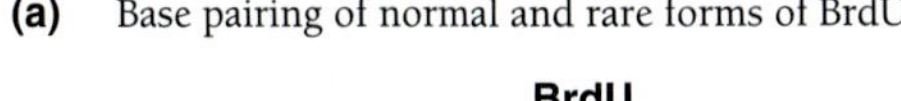

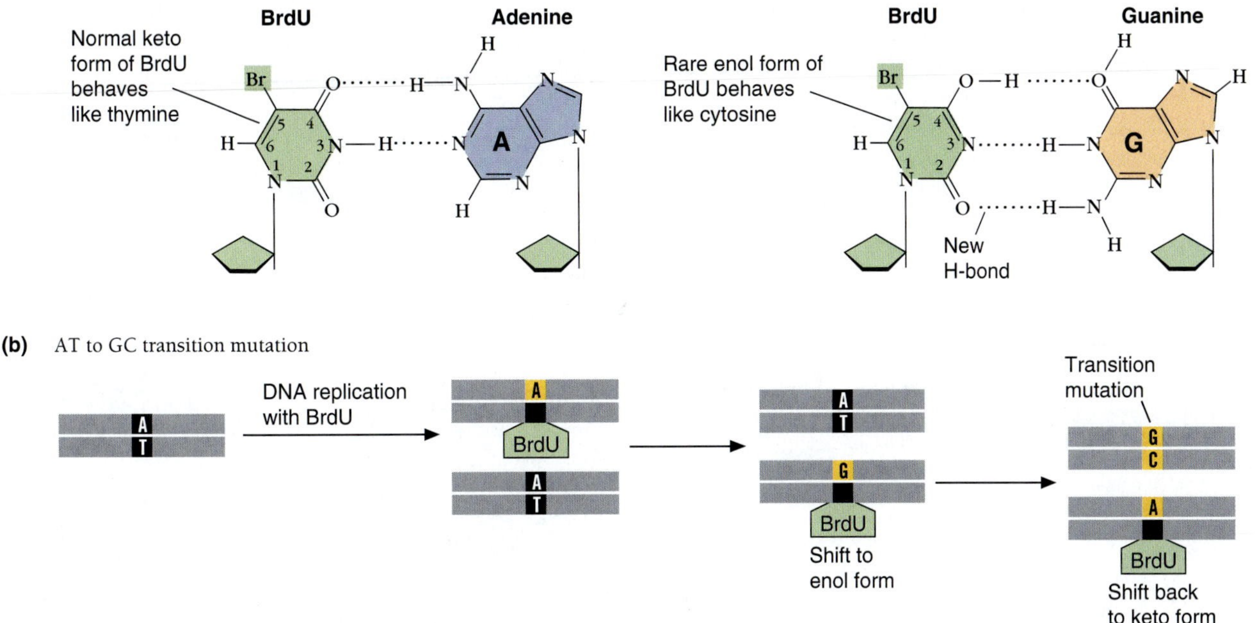

FIGURE 17.15 Mutations induced by 5-bromodeoxyuridine. (a) The normal and rare states of 5-bromodeoxyuracil (BrdU) show pairing with either adenine or guanine. (b) If BrdU is incorporated into DNA during replication, it can pair with either guanine or adenine, depending upon its configuration. Thus, BrdU can produce either a transition from an A:T pair to a G:C pair, or a G:C pair to an A:T pair.

guanine. The difference between BrdU and a normal nucleotide is that it is in the enol form for a longer time. Consequently, during the following rounds of replication, BrdU can produce frequent transitions of GC pairs to AT pairs and of AT pairs to GC pairs. DNA replication is required for base analogue mutations; nonreplicating cells exposed to base analogues are not affected (except during some repair processes).

Another common base analogue is **2-aminopurine** (2-AP), which resembles adenine but mispairs with cytosine. This base analogue causes many mutations because it is easily incorporated into DNA. When 2-AP is incorporated into DNA as adenine opposite thymine, it can generate A:T to G:C transitions when it shifts to the amino form and pairs with cytosine rather than thymine. If 2-AP is misincorporated as adenine opposite cytosine, it creates G:C to A:T transitions (Figure 17.16).

There is a large number of base analogues, some that are harmful and others that are not. For example, base analogues are sometimes used in the treatment of cancer. Cancer cells divide rapidly, whereas normal cells either do not divide or divide only occasionally. The incorporation of the base analogue into cancer cell DNA causes a large number of mutations in these cells, eventually killing them. The side effect, however, is that some normal cells (those that are dividing) are also killed, which explains the side effects of chemotherapy. Another base analogue that many of us consume on a daily basis is caffeine (a purine analogue), but it has no mutational consequences because it is not incorporated into DNA in the presence of physiological concentrations of guanine and adenine.

Chemicals Can Modify Base-Pairing Capabilities. A very large number of chemicals react with the four nucleotides and modify their base-pairing capabilities, causing transition mutations. One such chemical, **hydroxylamine** (NH_2OH), is commonly used in industry and chemistry laboratories, and it should be handled with care. Hydroxylamine specifically hydroxylates the amino nitrogen of cytosine, allowing the modified cytosine to hydrogen-bond like thymine (Figure 17.17). Hydroxylamine does not require DNA replication at the time of mutagenesis to modify the base, but it does require DNA replication to cause mispairing and the subsequent G:C to A:T transitions. Hydroxylamine is commonly used in chemistry laboratories, but it is not usually encountered in a normal day-to-day life except at extremely low concentrations. If hydroxylamine comes into contact with the skin, for example, the skin cells are mutagenzied and will eventually die. Prolonged exposure can cause cancer.

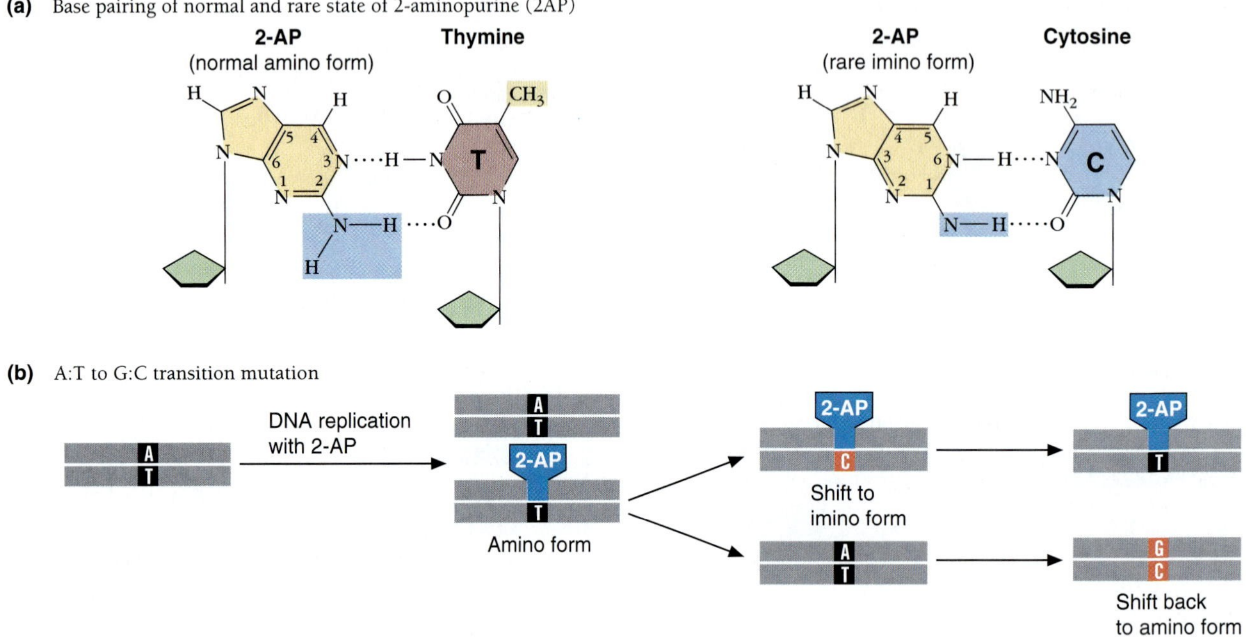

FIGURE 17.16 Mutations induced by 2-aminopurine. (a) The normal and rare states of 2-AP show pairing with thymine or cytosine here. The A:T to G:C transition shown in (b) results from the incorporation of a 2-AP opposite a thymine. The transition may also occur in the opposite direction, from a G:C to an A:T.

Nitrous acid (HNO_2) is another common chemical that modifies nucleotides so that ensuing replications cause mutations. Nitrous acid is a potent mutagen and causes deamination of adenine, guanine, and cytosine, all of which possess amino groups. As might be expected, removing the amino group changes the electric charge distribution within the molecule and consequently the base-pairing properties. For example, when adenine reacts with nitrous acid, it is converted to hypoxanthine and then base-pairs with cytosine instead of thymine; reaction of cytosine with nitrous acid converts it to uracil (Figure 17.18), which then base-pairs with adenine instead of guanine. The reaction of guanine with nitrous acid converts it to xanthine, but it retains its base-pairing with cytosine. Deamination of adenine gives A:T to G:C transitions, and deamination of cytosine results in G:C to A:T transitions.

Sodium nitrite is the salt of nitrous acid, and low concentrations of sodium nitrite are commonly used in the preservation of some foods (for example, hot dogs and cold cuts). Some controversy exists as to the use of sodium nitrite as a food preservative, but no direct link

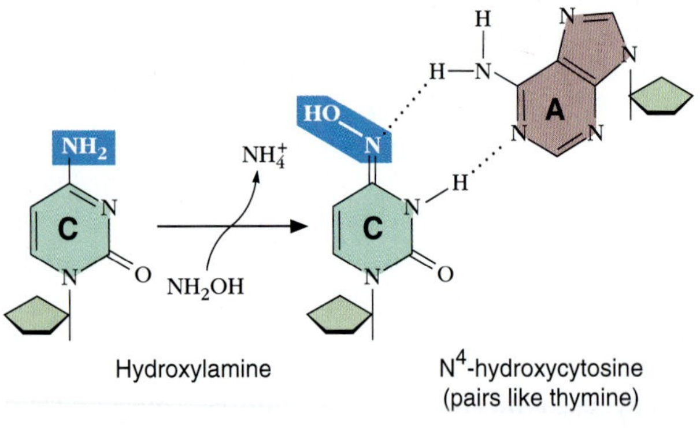

Hydroxylamine

N^4-hydroxycytosine (pairs like thymine)

FIGURE 17.17 Mutagenesis with hydroxylamine. Treating DNA with hydroxylamine converts cytosine to N^4-hydroxycytosine. This modified cytosine only base-pairs with adenine and thus creates G:C to A:T transitions.

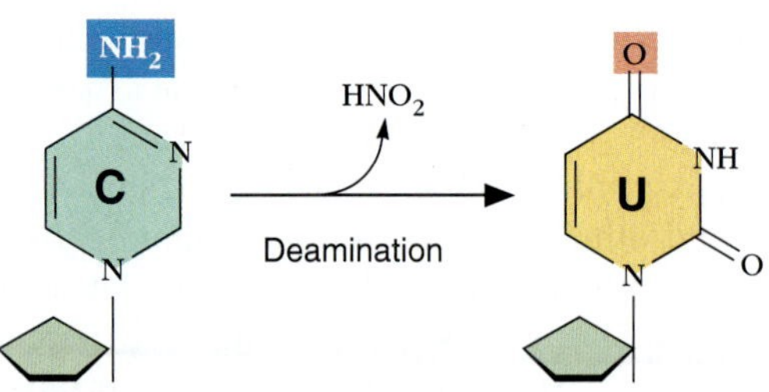

FIGURE 17.18 Deamination with nitrous acid. Nitrous acid deaminates cytosine, creating uracil. Uracil can then base-pair like thymine and gives a C:G to T:A transition upon DNA replication.

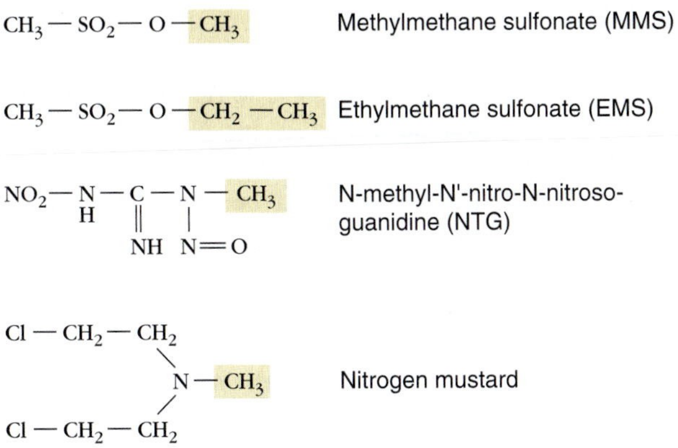

$CH_3-SO_2-O-CH_3$ Methylmethane sulfonate (MMS)

$CH_3-SO_2-O-CH_2-CH_3$ Ethylmethane sulfonate (EMS)

$NO_2-N(H)-C(=NH)-N(N=O)-CH_3$ N-methyl-N'-nitro-N-nitroso-guanidine (NTG)

$(Cl-CH_2-CH_2)_2N-CH_3$ Nitrogen mustard

FIGURE 17.19 Commonly used alkylating agents. The reactive alkyl group is highlighted in yellow. All these compounds are highly reactive with DNA, yielding many mutations with relatively low concentrations of the agent.

has been shown between the use of sodium nitrite as a food preservative and cancer formation in humans.

An extremely large group of very potent mutagens are the **alkylating agents.** All these compounds are highly reactive with DNA and yield many mutations at relatively low concentrations. These chemicals are used extensively in industry in the synthesis of many products that we use daily (for example, some plastics). They have also been used in warfare as the poisonous gases known as mustard gases, and they cause an extremely painful death caused by skin and mucosal membrane lesions and loss of lung function. The structures of some common alkylating agents are shown in Figure 17.19, and the reaction sites for these agents are shown in Figure 17.20.

Some alkylating agents can crosslink the two DNA strands, which may cause the strands to break during replication. Alkylating agents as a class cause every type of alteration: transitions, transversions, deletions, and frameshifts. DNA reactions with methylating agents (a subclass of alkylating agents that adds a methyl group to the reactant) yield primarily 7-methylguanine and methylation of the oxygen at carbon 6 of guanine to form O^6-methylguanine. Because both of these alterations are rapidly repaired, these products are not major mutagens. The formation of 3-methyladenine (a common methylation product of adenine) causes most mutations because the majority of bacterial species cannot remove 3-methyladenine using repair mechanisms. If a methylated or ethylated base is not immediately repaired, its base-pairing properties are altered, and a "wrong" base can pair opposite the modified base during DNA replication.

Two laboratory alkylating agents used to mutagenize organisms such as bacteria and *Drosophila melanogaster* are **nitrosoguanidine** (NTG) and **ethylmethane sulfonate** (EMS) (Figure 17.19). They are extremely potent mutagens that induce a high number of mutations into growing cells. NTG acts preferentially on single-stranded DNA (for example, at the DNA replication site when single-stranded DNA is transiently produced). As a result, clusters of mutations are generated on the chromosome, and thus NTG is less useful for laboratory analysis. It is difficult to attribute a mutation to the altered phenotype because more than one mutation has been created. EMS is commonly used in genetics laboratories to mutagenize bacteria, *Drosophila,* and seeds.

Intercalating Agents Cause Mutations by Inserting into DNA. Another group of potent mutagens are the **acridines** and acridine-like compounds (proflavin and acridine orange), which are planar heterocyclic compounds (Figure 17.21). The positively charged acridines intercalate, or sandwich, themselves between the stacked base pairs in double-helical DNA. In so doing, they change the rigidity and alter the conformation of the helix. The presence of acridines during replication of DNA causes frameshift mutations by adding or deleting one or two base pairs. The model of how this happens is similar to DNA slippage (Figure 17.7). Recall that Francis Crick used acridines to determine that a DNA code word is three nucleotides long (Chapter 11).

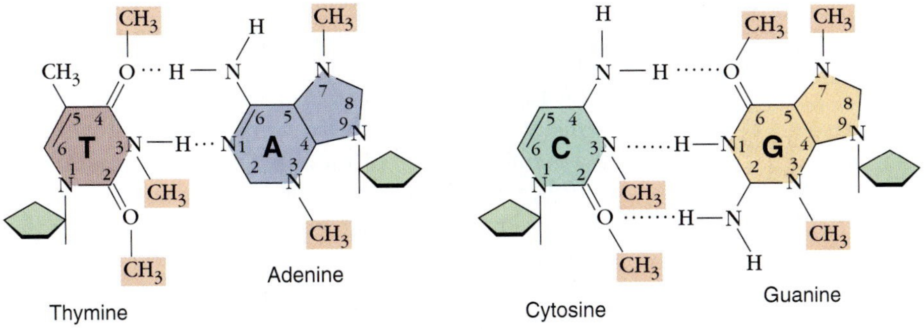

Thymine
Adenine
Cytosine
Guanine

FIGURE 17.20 Reactive methylation sites on DNA bases. The normal hydrogen bonding is shown, but this bonding may be changed when the molecules are methylated. Methylation sites are shown in pink.

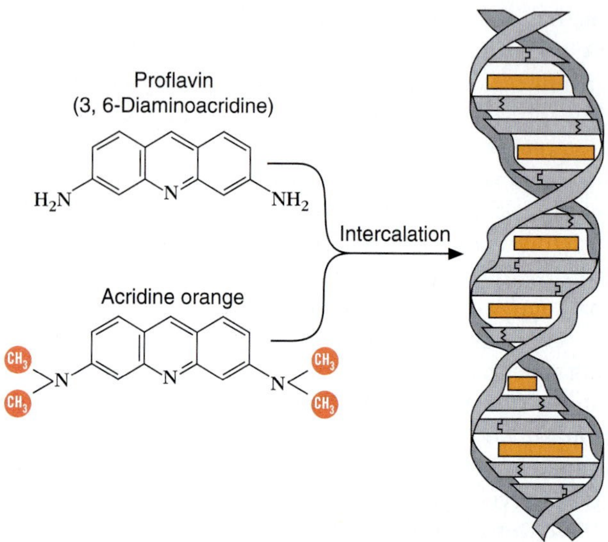

FIGURE 17.21 Intercalation of acridine-like compounds into DNA. Both proflavin and acridine orange are very strong frameshift mutagens. During replication, either can insert between the stacked bases (they have a planar structure with a basic charge on the amino group), altering the reading frame by adding or subtracting a base.

SOLVED PROBLEM

Problem

A series of four different mutations were induced in *E. coli* using four different chemical mutagens—BrdU, NH_2OH, EMS, and proflavin. The four resulting mutants were then treated with each of three mutagens to induce reversions, with the following results.

	Mutagen Used to Induce Reversion		
Mutant	**BrdU**	**NH_2OH**	**Proflavin**
A	+	+	−
B	+	−	−
C	−	−	+
D	−	−	−

(+) Indicates reversions occurred back to wild-type phenotype; (−) indicates no reversions occurred.

Indicate the type of change that occurred in the DNA for each of the four original mutants, and suggest which mutagen was originally used to induce each of the four mutants.

Solution

Mutant A was most likely induced by BrdU to give an A:T to G:C transition because it can be reverted using BrdU and NH_2OH. NH_2OH specifically hydroxylates the amino nitrogen of cytosine, allowing it to pair with adenine, thus causing the reverse G:C to A:T transition.

Mutant B is probably a G:C to A:T transition because it cannot be reverted by NH_2OH nor by proflavin, which causes frameshift mutations only. A G:C to A:T transition could be caused by BrdU, NH_2OH, or EMS.

Mutant C is either an addition or deletion of a single base pair because only proflavin can revert mutant C. Similarly, mutant C was induced by proflavin because it is the only mutagen in the group that can cause insertions or deletions.

Mutant D is very likely a transversion because both BrdU and NH_2OH cause only transitions, and proflavin inserts and deletes only.

SECTION REVIEW PROBLEMS

12. Of the following types of mutations, which can be classified as more likely spontaneous, or more likely induced, and which could be either?
    a. transition b. transversion c. frameshift d. deletion e. nonsense
13. How can ultraviolet light cause a transversion?
14. Can 5-bromodeoxyuridine and 2-aminopurine induce same-site reversions of mutations induced by themselves? Why or why not?
15. Is DNA replication required to induce a chemical change in the DNA (and thus a subsequent mutation) using (a) BrdU, (b) HNO_2, (c) NH_2OH?

Mutagenicity Can Be Measured.

Our society is extremely dependent upon a vast array of chemicals to produce the many conveniences we are accustomed to having in our daily lives, and many chemicals are quite toxic. About 70,000 chemicals are routinely used in medicine, manufacturing, food preservatives, cosmetics, and farming practices. These do not include the thousands of chemicals with which we come into contact daily that are natural products, everything from the ingredients in broccoli to charcoal-cooked hamburgers. Several recent studies have shown that a large percentage of known mutagens are also carcinogens, increasing our awareness of the correlation between mutagenicity and carcinogenicity. Mutagens not only threaten the germline cells, but they also pose a risk as carcinogens of somatic cells. Unfortunately, somatic cells, which compose most of the body, are very likely to be exposed to hazardous chemicals.

Many test systems have been developed to screen for carcinogenicity. These tests are usually very expensive and time-consuming and involve the use of animal cells in culture or whole animals—usually rats or mice. In these tests, very large doses of the test chemical are given to shorten the time span of the experiment, which does not mimic real-life situations. A less expensive and more rapid test was developed by Bruce Ames at the University of California at Berkeley in the early 1970s. This test, now known as the **Ames test,** uses *Salmonella typhimurium* auxotrophs for the amino acid histidine. The test measures the reverse mutation rate from *his*$^-$ to *his*$^+$, comparing the normal rate of reversion to the rate of reversion in the presence of a putative mutagen. To enhance the detection of mutations, the bacterial strains used in the Ames test have had their excision-repair mechanisms (see the next section for a discussion of excision repair) inactivated by a mutation. They also carry a second mutation that eliminates the protective lipopolysaccharide coating of wild-type *S. typhimurium.* Without the polysaccharide coat, the permeability of the cells is increased, and without the excision-repair mechanism, the reversion rate is increased. To make the Ames test mimic the metabolic reactions within the animal body, liver enzymes are added to the medium. Ingested compounds are normally detoxified or broken down by enzymes in the liver. However, some of these detoxification mechanisms create toxic or mutagenic compounds from compounds that are not normally toxic. Thus, by adding liver enzymes to the Petri dish, the detoxification mechanisms of mammals can be simulated.

To perform the Ames test, the cells of the *his*$^-$ strain are spread on a growth medium in a Petri dish that has liver enzymes but lacks histidine (Figure 17.22). The liver enzymes are obtained from rats by injecting them with a chemical that induces the synthesis of the detoxification enzymes. Following induction, the rats are killed, and their livers are removed. The homogenized liver tissue is treated to remove high- and low-molecular-weight materials [for example, plasma membrane material (they are not needed) and endogenous histidine molecules (so as to not promote the growth of *his*$^-$ cells)]. The putative mutagen is also added to the medium. Any mutation reverting the *his*$^-$ mutation to *his*$^+$ will result in a colony. The number of revertant colonies are counted and compared to a control dish that does not have the mutagen added. The increase in the number of revertant colonies, compared to the control, is a measure of the mutagenicity of the chemical. For example, using this method, the ratio of revertant colonies on the test Petri dish to the control dish gives 105 for methylmethane sulfonate (MMS), whereas aflatoxin B_1 gives a ratio of 1.2 million. Thus, aflatoxin B_1 is 11,500 times more mutagenic than MMS (divide 1.2 million by 105).

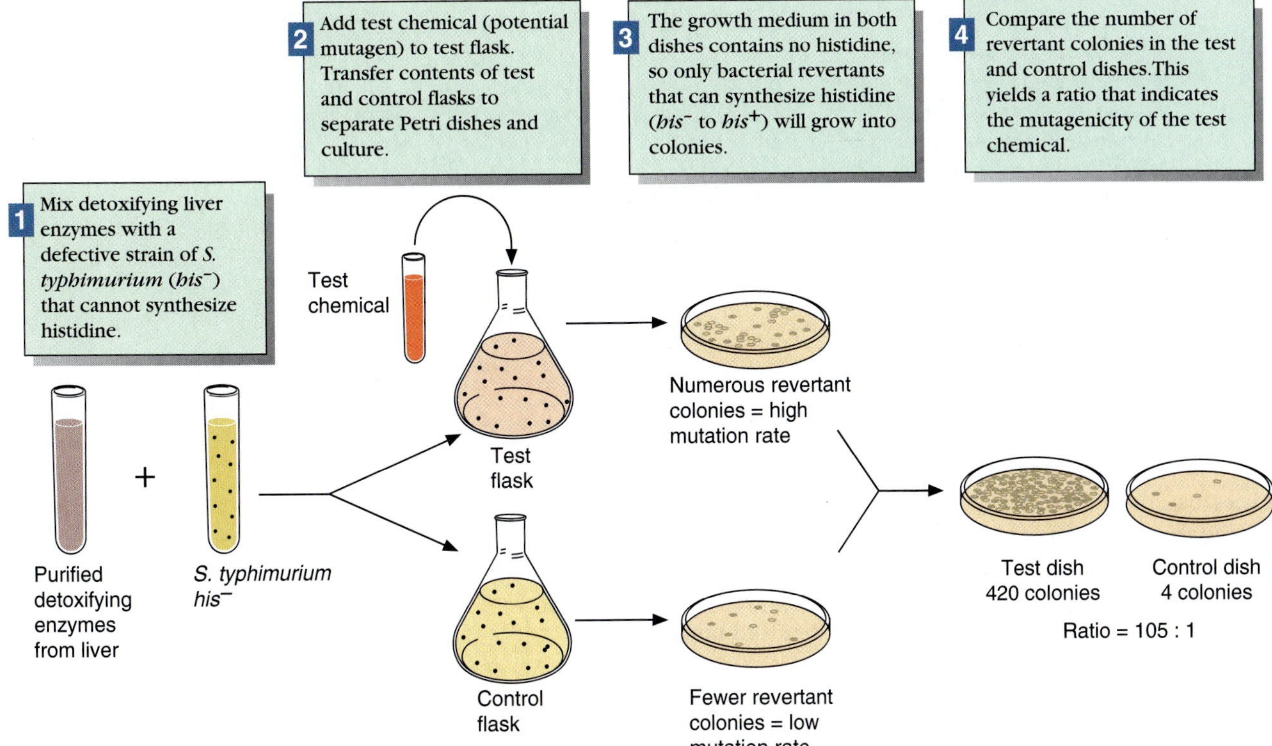

FIGURE 17.22 The Ames test. The Ames test determines whether a chemical is mutagenic and potentially carcinogenic. The relatively low cost and high level of sensitivity of the test makes it very useful.

CURRENT INVESTIGATIONS

Many Natural Products Are Mutagenic

In recent years, we have heard more and more about how our diets affect our health, especially in relation to the production of cancers. Research has shown that our diets contain an enormous variety of natural mutagens and carcinogens. It is also apparent that we are ingesting vastly greater quantities of these substances than was previously suspected. Perhaps these natural chemical products should be of primary concern rather than the mutagenicity of industrial chemicals, food additives, and pollutants in our environment. For example, in 1989 the United States had a big publicity-generated scare concerning the plant-growth regulator Alar, which is used to delay ripening of apples so that they do not drop prematurely. Alar was said to be carcinogenic, but when put in perspective with the chemicals in the apple and other chemicals in our daily diets, it does not seem to be so bad. For instance, the hydrazines in a helping of mushrooms are 60 times more carcinogenic than the Alar consumed in a glass of apple juice or 20 times greater than a daily peanut butter sandwich, which frequently contains aflatoxin B_1. Our diets contain literally millions of natural chemicals; in fact, it is not practical to test them all for carcinogenicity.

Animal tests and the Ames test have been used to evaluate cooked foods for their potential for inducing cancers; and it has been found that burnt and browned sugars or breads contain a variety of mutagens. In addition, caffeine and its close relative theobromine found in coffee, tea, cocoa, and some soft drinks may increase the risk of tumors by inhibiting DNA repair enzymes. Plants synthesize many carcinogenic or teratogenic chemicals as defense mechanisms to ward off the animals that want to consume them. Examples of plant carcinogens include psoralen and its derivatives, which are widespread in plants and have been used as sunscreens in France; solanine and chaconine are teratogens and are found in greened potatoes. Other foods that contain natural carcinogens include bananas, basil, broccoli, cabbage, cauliflower, celery, horseradish, mustard, turnips, and black pepper. In addition, red wines are believed to be responsible for the high incidence of stomach cancers among the French people, although red wine also seems to decrease the incidence of coronary heart disease. It seems that nothing can be consumed that does not contain a mutagen!

Another big problem with American diets is the consumption of excess quantities of fats. The average American consumes 40% of her/his calories in the form of fat. Comparisons of cancer deaths rates in different national populations have provided important clues to the nutritional causes of cancer. Very different types of cancers appear in the United States than appear in Japan. In the United States, colon, breast, and prostate cancers are the most prevalent, whereas stomach cancers are in excess in Japan. When the amount of dietary fat intake is plotted against the number of deaths by breast cancer, the results are striking (see Figure 17.B); the more fat in the diet, the higher the rate of breast cancer. How might fat intake cause cancer? It may be caused by rancid (oxidized) fat because it represents a sizable percentage of the fat intake. Unsaturated fatty acids and cholesterol in fat are very prone to oxidation, which produces a variety of carcinogenic compounds. Another likely explanation is that many carcinogens are soluble in fats and accumulate in the fat of the animals we eat.

Many industrial, university, and government laboratories use the Ames test to detect potential mutagens among the wide variety of chemicals used. For example, the data presented in Figure 17.23 represent the results from an Ames test to detect mutagenic and potentially carcinogenic chemicals in the urine of smokers and non-

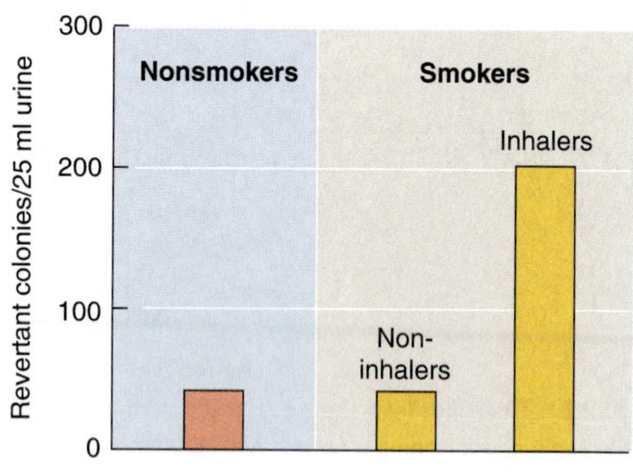

▶ **FIGURE 17.23 The Ames test and smokers.** A comparison of the urine of smokers and nonsmokers shows that inhaling smokers generate many more revertant colonies in the Ames test. This may explain why smokers suffer from a much higher level of bladder cancer (among other cancers) than nonsmokers.

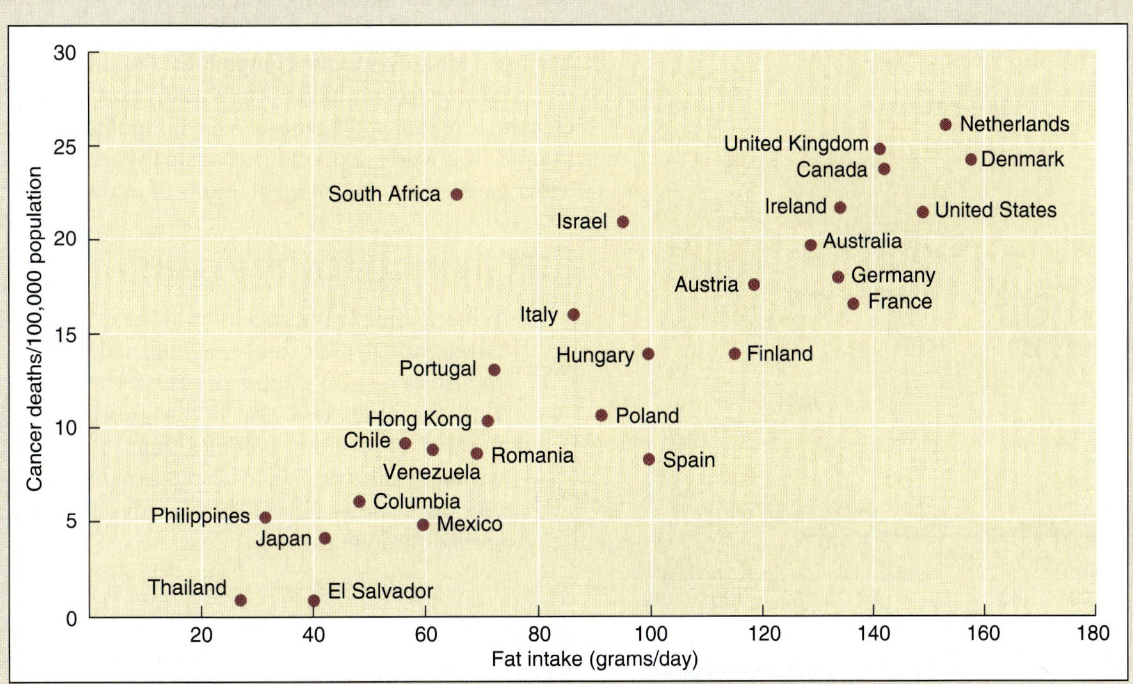

FIGURE 17.B Relationship between dietary fat intake and cancer in different countries. As fat intake increases, the number of certain types of cancer deaths correspondingly increase, most notably colon, prostate, and breast cancers. *[Adapted from L. A. Cohen. 1987. "Diet and cancer." Scientific American, 257 (November): 42.]*

How do we survive this onslaught of chemicals in our foods and in our environment? Like the plants that have evolved defense mechanisms from animal predators, we have evolved similar mechanisms to diminish the toxic impact of the chemical onslaught. Mechanisms that protect animals against carcinogenic chemicals include the facts that the surface layers of skin, stomach, intestines, and colon are shed every day; that the body produces dozens of detoxifying enzymes; and that there are a variety of small molecules in our diets that are anticarcinogenic (vitamin E, β-carotene, glutathione, vitamin C, and uric acid).

smokers. Cigarette smoke contains over 6800 different chemicals, many that are highly toxic. These include hydrogen cyanide, vinyl chloride, and nicotine. These test results very likely explain why smokers have a higher rate of lung and bladder cancers compared to nonsmokers.

A federal mandate specifies that a fire retardant must be present in children's pajamas. From 1973 to 1977 the chemical tris-2-dibromopropyl phosphate (tris, for short) was routinely added to children's pajamas as a fire retardant. Using the Ames test, it was discovered that tris is a potent mutagen, and thus by inference, a carcinogen. Tris is readily absorbed through the skin and was found in the urine of children wearing tris-treated clothing, even after the clothing had been repeatedly washed. Thus, the good intentions of the government were, in fact, potentially carcinogenic to the children who were benefiting from the fire retardant.

It can be estimated—and accurate measurements are difficult—that over 80% of all human cancers are caused by substances in the environment (see CURRENT INVESTIGATIONS: *Many Natural Products Are Mutagenic*). Aflatoxin B_1, for example, is an extremely potent mutagen and potential carcinogen. This chemical is produced by the fungus *Aspergillus flavus*, which grows on moist

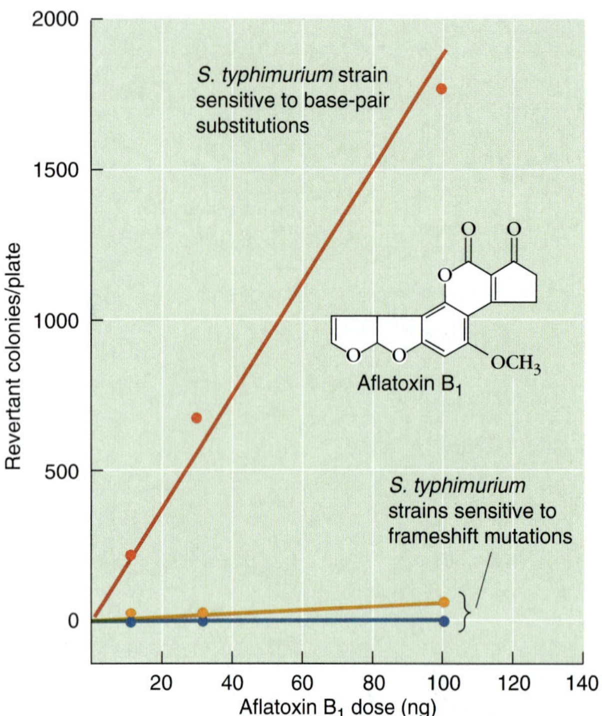

FIGURE 17.24 Aflatoxin B_1 mutagenicity using the Ames test. Aflatoxin is a potent mutagen for base-pair substitutions. This compound is a natural product produced by a fungus that grows on foods. Peanuts are sometimes contaminated with aflatoxin. *(From J. McCann and B. N. Ames. Advances in Modern Toxicology, Vol. 5, 1978. Edited by W.G. Flamm and M.A. Mehlman. Copyright by Hemisphere Publishing Co.)*

food such as nuts, corn, and beans during storage. Aflatoxin B_1 can be a contaminant in corn products and especially peanut butter. Figure 17.24 shows the mutagenic effects of the presence of aflatoxin B_1 at extremely low doses (nanogram amounts) during the Ames test. As a result, the federal government has strict rules governing the testing of stored grains that might be subject to the growth of *Aspergillus flavus.*

It is believed that air pollution may be responsible for 5–10% of all cancers of the respiratory system, and the incidence of lung cancer is higher in polluted urban than in rural areas. Carcinogenic air pollutants include alkenes, which are oxidized by the atmosphere to epoxides and are known mutagens; aldehydes, including formaldehyde; halogenated hydrocarbons; benzene, which is linked to leukemia in humans; and motor vehicle emissions, which are linked to lung cancer. Dinitropyrenes are potent mutagens and are present in diesel exhaust fumes, coal-fired furnace fumes, and certain types of photocopying toners. Water is also a source of mutagenic compounds, either as a result of groundwater contamination (from farming practices and industrial byproducts) or by the consumption of fish or shellfish who live in it and accumulate mutagenic agents. It is estimated that over 100 million persons throughout the world are presently consuming water that contains a significant concentration of mutagens. Drinking water can be tested for its mutagenic potential using the Ames test without having to identify specific compounds. Polluted river water, when tested with the Ames test, gave greatly increased levels of mutagenesis over nonpolluted control samples. Similarly, extracts from fish living in polluted water gave greatly increased amounts of mutagenesis.

SECTION REVIEW PROBLEMS

16. What is the significance of a high correlation between mutagenicity and carcinogenicity?
17. In testing a small sample of your favorite hair dye with the Ames test, you find a greatly increased number of revertant colonies than in the control without hair dye. What simple experiment could you do to determine if the hair dye is being absorbed through the skin?

BIOLOGICAL REPAIR REVERSES MANY MUTATIONS.

As we have seen, DNA is constantly subjected to damage to its fidelity during replication. It also can suffer alterations through normal exposure to the environment. As a consequence, living cells have evolved a series of mechanisms to correct damaged DNA and maintain its fidelity during and after replication. Some of these correction processes were discussed in Chapter 9 (for example, the $3' \longrightarrow 5'$ exonuclease activity of DNA polymerase). However, attacks on DNA are so numerous that myriad systems for the repair of damaged DNA have evolved, some of which are highly specific for certain types of damage. Without these repair mechanisms, the rate of mutation would be much greater. As will be discussed, the loss of a DNA repair mechanism usually has a very drastic effect on the organism, often resulting in early death from cancer and rapid aging. It seems likely that a component of aging is an accumulation of mutations (see CURRENT INVESTIGATIONS: *Accumulated Mutations May Cause Death, Aging, and Cancer*). The majority of investigations into repair have been done in microorganisms, as reflected in the following discussion. However, similar repair mechanisms are likely to be present in all organisms. Mitochondria lack most repair processes and thus accumulate mutations in their DNA at a faster rate than nuclear DNA.

Mutations Can Be Directly Repaired.

Many repair mechanisms involve "major surgery" on the DNA, but several are simple reversals of the damage. For

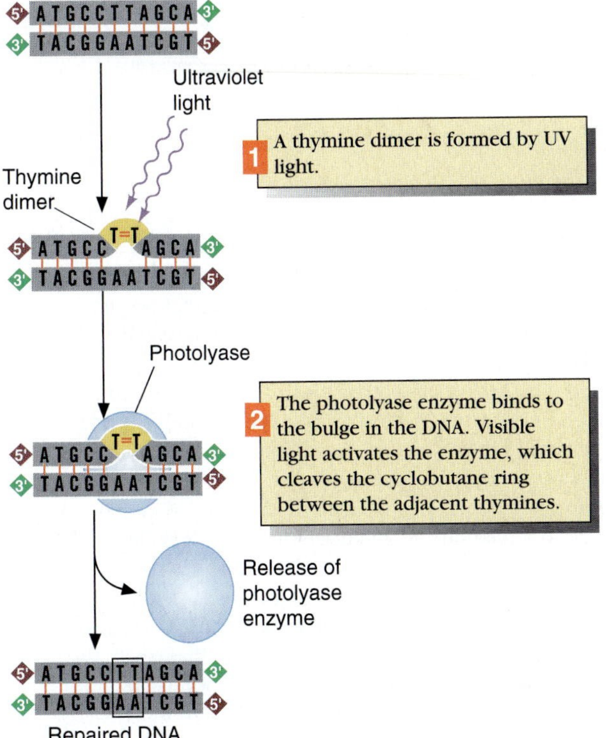

FIGURE 17.25 **Photoreactivation of DNA.** Thymine dimers are cleaved by the enzyme photolyase in the presence of visible light, correcting DNA damage without replication.

example, UV light forms pyrimidine dimers in DNA at a very high rate (Figure 17.13). An enzyme present in nearly all organisms is photolyase, coded for by the gene *phr* in *E. coli*. Synthesis of this enzyme is stimulated by visible light and it then binds to the dimer. The photolyase enzyme catalyzes a photochemical reaction that converts the cyclobutane ring back into individual pyrimidines. This process is called **photoreactivation** (Figure 17.25).

A second common type of damage to DNA occurs when the oxygen at carbon 6 of guanine is methylated to form O^6-methylguanine (Figure 17.20). This methyl group can be directly removed (corrected) by an enzyme called O^6-methylguanine methyltransferase, coded for by the *ada* gene in *E. coli*. The methyl group removed from the guanine is transferred to a cysteine SH group on the protein, which permanently inactivates it. Because each enzyme molecule can remove only one methyl group, a high level of methylation may overload the cell's ability to correct DNA methylation damage, leading to mutations. This last point emphasizes how it is easy it is to obtain misleading experimental results by feeding rats or mice high doses of a mutagen. For example, in experiments to determine whether commercially used chemicals cause cancer, the dosage levels that result in cancers are often unrealistically high and, therefore, do not reflect the actual risk.

The Complementary Strand of DNA Is Used in Repair.

A unique feature of double-stranded DNA is that the genetic information is present in both strands. Thus, if one strand is damaged, the other strand possesses the necessary information to synthesize a new molecule. Single-stranded DNA or RNA do not have this luxury.

One very efficient repair system that uses the information in the complementary strand is **excision repair.** This system of repair senses any serious damage-induced distortion of the DNA helix. In *E. coli,* one excision-repair pathway is coded by three genes, *uvrA, uvrB,* and *uvrC*. The products of these genes remove and replace short patches of DNA (12 bp). The three genes actually code for three separate subunits of one enzyme, an endonuclease. The UvrABC enzyme detects serious damage-induced distortion of the DNA helix and makes an incision in the sugar–phosphate backbone of one strand on both sides of the lesion, releasing a 12-bp oligonucleotide containing the mutated bases (Figure 17.26). The gap is filled in by

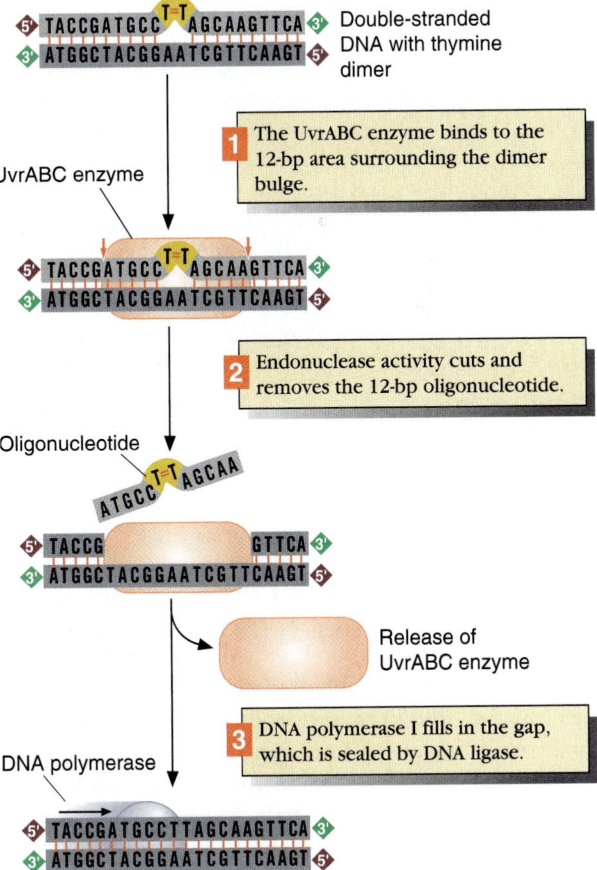

FIGURE 17.26 **Excision repair of DNA.** Damaged nucleotides in DNA can also be corrected with the UvrABC enzyme system. In this method, a fragment of DNA is excised and replaced using the complementary strand to provide the correct sequence.

CURRENT INVESTIGATIONS

Accumulated Mutations May Cause Death, Aging, and Cancer

Like bacteria, animal cells can be grown in a Petri dish culture. Unlike bacterial cells, animal cells become attached to the growth surface and form a monolayer of cells, whereupon they stop dividing. The cessation of growth is a characteristic of normal animal cells and does not indicate their forthcoming death. To continue growth in culture, such cells must be diluted and transferred to a fresh nutrient medium. The repeated dilution and transfer of animal cells to a fresh medium permits them to grow until their proliferation ceases, whereupon they undergo a process called cellular senescence followed by cell death. The number of cell divisions that occur before senescence of a human embryo cell line is approximately 50; interestingly, this number decreases if the tissue is taken from individuals of increasing age. The approximate number of cells produced in a human life span is 1×10^{16}, which is also equivalent to the number of cells from 50 cell divisions. Also, a tortoise with a life span of about 175 years will have an average number of cell doublings in tissue culture of 100, whereas a mouse with a life span of 3 years will have only 18 or so cell doublings. In contrast, many tumor cells can be grown indefinitely in culture, escaping senescence. These cells are called immortal.

Why do some cell lines grow forever, others have limited life spans, and still others very short life spans? Does the limited life span of a cell line have any relationship to aging and eventual death of whole animals? Several theories of cellular senescence have been proposed, and one, termed the error catastrophe model, proposes that accumulation of random mutations in DNA of somatic tissue results in loss of proliferative capacity.

A second hypothesis is that senescence is a genetically programmed process. Evidence for the programmed death model comes from the cell-culture-death experiments described previously. In this model, a gene or group of genes is responsible for the eventual death of an organism. If this hypothesis is true, then scientists should be able to identify genes or at least chromosomes that contain programmed death genes. Human chromosome number 1 was identified as the location of these genes by first fusing two cells lines, one that was mortal (a fetal lung fibroblast cell line that died at 50 or so doublings) and another that was immortal (a tumor cell line). Experimenters found that the fused cells, which contained chromosomes from both cell lines, were mortal and died after about 60 cell doublings. They concluded that gene(s) controlling cell mortality must be dominant.

Some clones of the cell line, however, escaped senescence. Examination of these clones showed that they had all lost both copies of chromosome 1. To further demonstrate that chromosome 1 possesses genes that control cell death, chromosome 1 was purified and microinjected into immortal cells, and the cells were again mortal. These data lend considerable credence to the programmed death model of cellular senescence and suggest that we have a gene or a group of genes located on chromosome 1 that may influence longevity.

It follows that inactivation of longevity genes in an organism

DNA polymerase I (using the information in the complementary strand) and sealed by DNA ligase. Loss of the UvrABC repair pathway in *E. coli* greatly increases the mutation rate, as would be expected.

In humans, a rare inherited disorder, *Xeroderma pigmentosum,* results from a loss of one of the enzymes in this excision-repair pathway. This disease is an autosomal recessive disorder with a frequency of 1 in 250,000 individuals. Affected individuals are extremely sensitive to light, which causes dry, flaking skin and heavy pigmentation and eventually results in disfiguring malignant tumors. Early death from cancer is common. This defective repair pathway also increases the probability that an individual carrying the defective gene will develop colon cancer. It is estimated that at least nine different genes in human are involved in excision repair, suggesting that the pathway is more complex than that of *E. coli*. A defect in any one of these genes results in an increased risk of developing some form of cancer. Perhaps not surprisingly, the most frequent form is skin cancer, which results from exposure to sunlight (UV rays).

A second method for removing damaged DNA or a modified base (for example, after alkylation) involves a family of enzymes called **DNA glycosylases.** Instead of making an incision in the phosphodiester backbone of the DNA strand, DNA glycosylases cut the bond between the sugar in the backbone and the modified base (Figure 17.27). This results in the loss of the purine or pyrimidine base, and the site is designated as an **apurinic** or **apyrimidinic site** (AP site). AP sites also result from depurination, as discussed earlier (Figure 17.8). The "hole" created by the AP site is recognized by an **AP**

might confer immortality to a cell, which is characteristic of cancer cells. This, in turn, may cause the death of the organism. It is a fact that as we age the cumulative cancer risk increases dramatically and varies with the metabolic rate of the organism. Short-lived animals with high metabolic rates (for example, mice and rats) have a 30% risk of cancer at age 3, but in humans with a relatively low metabolic rate, the 30% risk of cancer is not reached until age 85. Do we accumulate cellular damage to our DNA in relation to our metabolic rate? Four important endogenous processes lead to significant DNA damage—oxidation, methylation, deamination, and depurination. Faster metabolic rates yield faster production of oxidants in each cell. This, in turn, could lead to faster accumulation of somatic damage, carcinogenic events, and aging. To test this hypothesis, oxidative damage to DNA was measured. This was done by testing for the appearance of the products of DNA repair, which appear in the urine. One of these products is thymine glycol. Figure 17.C shows a direct correlation between excretion of thymine glycol and metabolic rate, suggesting a relationship between metabolic rate and oxidative damage to DNA. Perhaps oxidative damage may also speed up the aging process.

The real question is whether any of these findings have any bearing on aging of the whole animal. The answer is a big "maybe." It is likely that many processes in the body have an effect on aging, and no one phenomenon is completely responsible for aging and death. For example, in 1996, a group of scientists cloned the gene responsible for Werner's syndrome, a human genetically inherited disease that causes the classic characteristics of aging in individuals in their twenties: wrinkles, cataracts, and fragile bones. The gene responsible for this disease is on the short arm of chromosome 8 and seemingly affects an enzyme that has helicase properties, which infers that aging may be related to DNA repair and metabolism. This gave credence to the idea that DNA damage is related to aging, but it is not the whole story because aging is very complex and likely governed by many genes and environmental phenomena. It is certainly clear that in evolutionary terms, aging and death are important for the diversity and survival of species on earth.

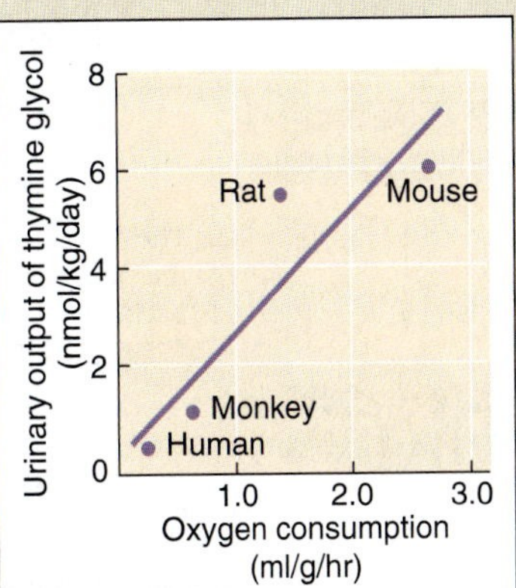

FIGURE 17.C Average urinary output of thymine glycol in relation to specific metabolic rate.

endonuclease that cuts the backbone, leaving a 3′-OH primer end from which DNA polymerase I initiates synthesis to replace the missing nucleotides. Many different glycosylases have been characterized, including those responsible for removal of alkylated bases, UV photodimers, and other damaged bases.

Ataxia telangiectasia (AT), Fanconi's anemia, and Bloom's syndrome are other human diseases that are associated with a defect in DNA repair, although the exact defects are not known. AT is an autosomal recessive condition and has a frequency of about 1 in 50,000 individuals. All these diseases are associated with chromosome breakage, and afflicted individuals are hypersensitive to X-rays or chemicals that cause DNA damage. These patients are all cancer-prone, again supporting the relationship between mutation and cancer.

In Chapter 9, we learned that DNA polymerase possesses 3′ $\longrightarrow$ 5′ exonuclease activity associated with DNA proofreading. If a base-pairing mistake is made, the DNA polymerase can back up, remove the mismatched base, and install the correctly matched base. However, even this process is not error-proof. In *E. coli*, the final responsibility for mismatch repair after DNA replication is a **mismatch correction enzyme** coded for by the gene *mut*HLSU. This repair pathway has been most studied in *E. coli*, but biochemical studies have provided evidence that eukaryotes have similar mismatch repair systems. One of the enzymes from this pathway monitors the structure of newly replicated DNA for distortions, which may include mismatched base pairs (coded by *mut*S). This enzyme binds to the DNA at the mismatch to initiate the repair process. This enzyme must distinguish

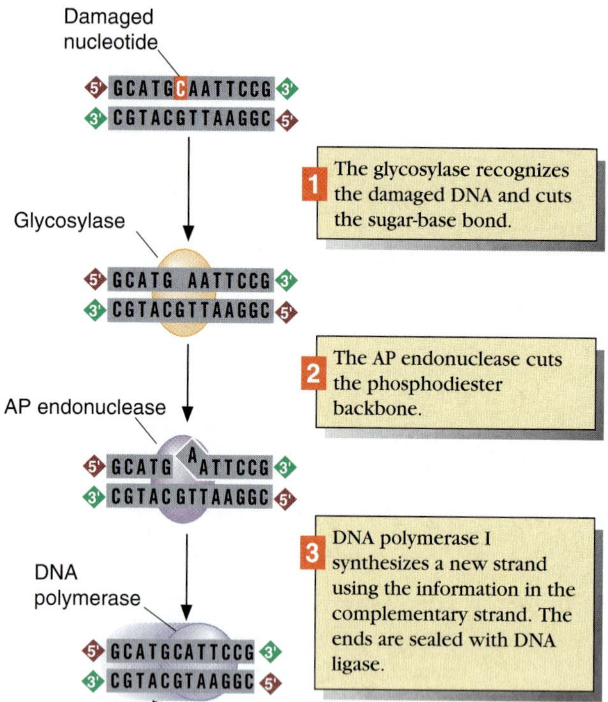

Damaged nucleotide
GCATGCAATTCCG
CGTACGTTAAGGC
1 The glycosylase recognizes the damaged DNA and cuts the sugar-base bond.
Glycosylase
GCATG AATTCCG
CGTACGTTAAGGC
2 The AP endonuclease cuts the phosphodiester backbone.
AP endonuclease
GCATG ATTCCG
CGTACGTTAAGGC
3 DNA polymerase I synthesizes a new strand using the information in the complementary strand. The ends are sealed with DNA ligase.
DNA polymerase
GCATGCATTCCG
CGTACGTAAGGC

FIGURE 17.27 Glycosylase repair of DNA. Glycosylases recognize many different kinds of modified bases that base-pair improperly. This system adds to the repertoire of DNA correction methods available to the cell, ensuring that mutations do not arise.

between the old strand and the newly synthesized strand so as to remove the mismatched base from the newly synthesized strand. In *E. coli,* the enzyme differentiates between the old and new strands by scanning for GATC sequences in the newly synthesized strand. All GATC sequences in the old strand have had the adenine modified at the N^6-position by a methylase giving N^6-methyladenine. As a consequence, the mismatch enzyme does not remove nucleotides from this tagged strand. In *E. coli,* the methylase is coded by the gene *dam;* thus, strains defective in the *dam* gene cannot discriminate between the template and the newly synthesized strands and have a high spontaneous mutation rate. After recognition of the mismatch, a GATC-specific endonuclease (*mut*H) cleaves the unmethylated strand. The subsequent excision process requires MutS, MutL, and MutU proteins (DNA helicase) and an exonuclease. DNA polymerase fills in the gap, and DNA ligase seals the two ends.

In humans, a high spontaneous mutation rate was found in a hereditary cancer known as **nonpolyposis colon cancer,** which affects 1 person in 200. This genetic defect accounts for about 15% of all human colon cancers. This fact led investigators to examine the mismatch repair pathway in humans, and they found that about half the individuals affected with this disease had a defective gene similar to the *E. coli mut*S gene, and the other half had a defective gene similar to *E. coli mut*L. These genes have now been cloned. The cloned sequence can be used for early detection of the genetic defect, so afflicted individuals can take preventive measures to reduce their risk.

DNA Recombination Can Repair Mutations.

Some lesions to DNA are extremely damaging and require completely new DNA to repair the information. For example, during synthesis of DNA with DNA polymerase, pyrimidine dimers block the progress of the enzyme. To pass a pyrimidine dimer, the polymerase dissociates and reattaches at a point downstream of the dimer. The result is a section of DNA without a complementary sequence. Also, both strands of DNA can be damaged in the same area, and during excision repair the double helix will be broken. In addition, X-rays can break DNA strands, leaving blunt ends. These drastic DNA damages must be repaired in the absence of complementary strand information; this is accomplished by joining bits and

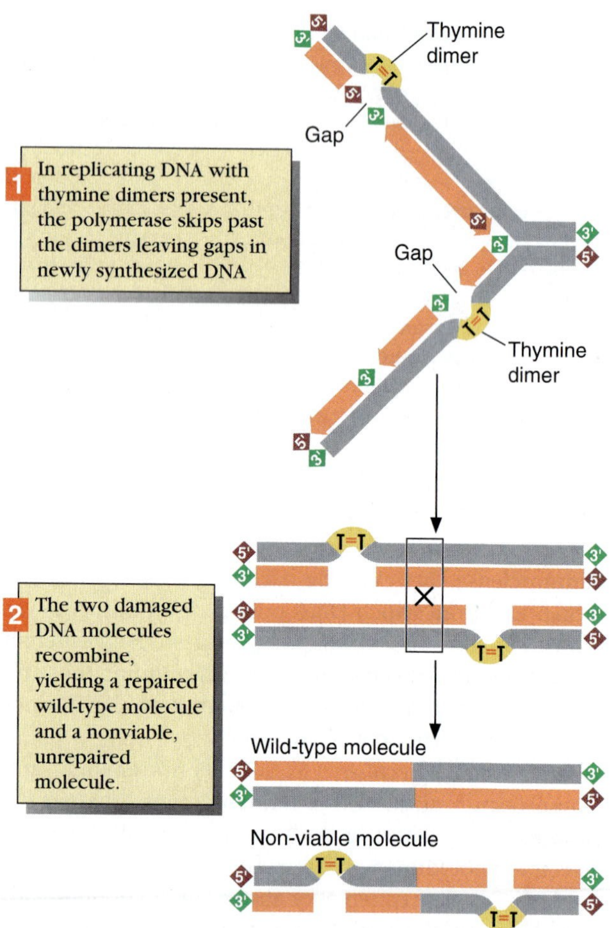

Thymine dimer
Gap
1 In replicating DNA with thymine dimers present, the polymerase skips past the dimers leaving gaps in newly synthesized DNA
Gap
Thymine dimer
T=T
2 The two damaged DNA molecules recombine, yielding a repaired wild-type molecule and a nonviable, unrepaired molecule.
Wild-type molecule
Non-viable molecule

FIGURE 17.28 Recombinational repair of DNA. One good DNA molecule is salvaged by recombination from two mutagenized molecules.

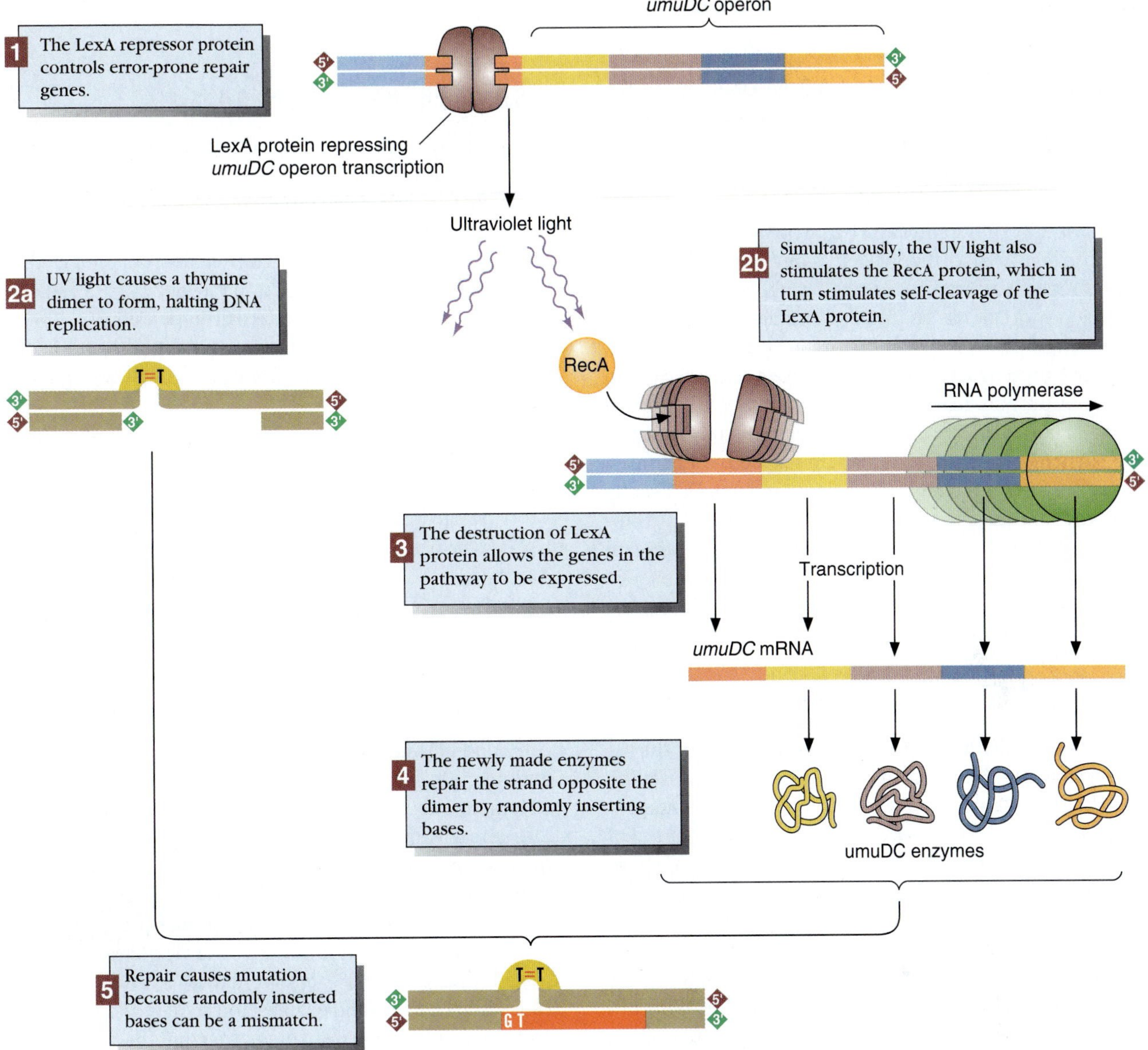

1
The LexA repressor protein controls error-prone repair genes.
umuDC operon
LexA protein repressing umuDC operon transcription
Ultraviolet light
2a
UV light causes a thymine dimer to form, halting DNA replication.
2b
Simultaneously, the UV light also stimulates the RecA protein, which in turn stimulates self-cleavage of the LexA protein.
RecA
RNA polymerase
3
The destruction of LexA protein allows the genes in the pathway to be expressed.
Transcription
umuDC mRNA
4
The newly made enzymes repair the strand opposite the dimer by randomly inserting bases.
umuDC enzymes
5
Repair causes mutation because randomly inserted bases can be a mismatch.
T=T
G T

FIGURE 17.29 **Error-prone repair of DNA.** This system is invoked when all others fail. It creates errors during the correction of the damaged DNA sequence. Thus, the error-prone repair process itself is mutagenic, but sometimes it creates less damaging mutations than the original mutation and saves the cell from death.

pieces of homologous sets of separate DNA molecules. The mechanism that accomplishes this is referred to as **recombinational repair.** In *E. coli,* recombinational repair requires the *rec*A gene. Diploid organisms have homologous chromosomes, but bacteria carry the same sequence in newly replicated DNA (two copies will be in the same cell). For recombinational repair to work, the damaged areas must be in different locations on the two separate molecules that undergo recombination. The recombination creates one good DNA molecule and one irreparable DNA molecule. For example, if replication encounters two thymine dimers, one in each strand and close to each other (Figure 17.28), the two strands together have sufficient information for one good, double-stranded molecule. Recombination generates one DNA molecule that is undamaged and one molecule with two damaged areas (Figure 17.28).

Some DNA Repair Mechanisms Produce Errors.

The final repair process we will discuss, known as the **error-prone repair** pathway, is used only as a last resort to ensure the survival of the cell. Damaged DNA is repaired without respect to complementary sequence fidelity and occurs only after the cell has undergone DNA replication. Error-prone repair inserts nucleotides opposite a pyrimidine dimer with random base selection (Figure 17.29) using a poorly understood mechanism. This insertion connects the two ends of the DNA strand but does not maintain the fidelity of information in the newly synthesized strand. When the strand is replicated, the error will be perpetuated.

In *E. coli*, error-prone repair is dependent upon a group of 20 genes controlled by a repressor protein, the LexA protein. Also involved in controlling the error-prone repair pathway is the *recA* gene, whose product, the RecA protein, has its primary function in recombination. When UV light produces thymine dimers in DNA, the RecA protein activates a self-cleavage reaction within the LexA protein, inactivating it. When the LexA protein is inactivated, the genes involved in error-prone repair are turned on, synthesizing their respective enzymes. Two of these genes in *E. coli* are *umuC* and *umuD*. Exactly how the enzyme products of these genes act is not understood, but it is known that mutations in either of these genes prevent error-prone repair.

SECTION REVIEW PROBLEMS

18. Outline three different repair pathways for a thymine dimer.
19. If you had a bacterial strain defective (by mutation) in AP endonuclease, but it still had an active DNA glycosylase system, would you predict that the spontaneous mutation rate of this strain would be higher or lower than wild-type? Why?
20. Would the mutations in the strain of bacteria described in Problem 19 cause transversion? Explain.
21. Explain why *dam*$^-$ strains of *E. coli* have a high spontaneous mutation rate.
22. If a cell has an excision repair pathway, why does it need recombinational repair and error-prone repair pathways?

Summary

A mutation is defined as an inherited alteration in the genetic material. Mutations in somatic tissue are passed on to the next generation of cells, but mutations must occur in the germ line to be passed on to the next generation of organisms. Two general classes of mutations exist: spontaneous mutations, which are restricted to mistakes made internally in the cell; and induced mutations, which are caused by external forces or mutagens (for example, chemicals or radiation in the environment).

Some mutations cause base changes in the DNA but may not alter the function of the protein product of the gene. Neutral mutations can cause minor changes in the protein structure or substitute a similar amino acid, or silent mutations may not cause any amino acid change because there are multiple codons for some amino acids. A single base change in DNA gives a point mutation, and it may be either a transition or a transversion. A transition retains the pyrimidine:purine pair, and a transversion substitutes a pyrimidine:purine pair for a purine:pyrimidine pair. Either a transversion or a transition can be a missense or a nonsense mutation, depending on whether they generate a sense codon or a nonsense codon (UGA, UAA, and UAG). Frameshift mutations are caused by insertion or deletion of one or two bases, and they alter the reading frame of the mRNA.

Spontaneous mutations may be caused by mistakes in base-pairing during replication. These mistakes are rare because of proofreading by DNA polymerase and the mismatch correction enzyme. DNA slippage during replication also causes mutations (usually frameshifts) by mispairing at a series of A:T or G:C pairs. Deaminations and depurinations also cause spontaneous mutations and are very common events but are repaired by AP endonucleases, which cut the backbone and allow DNA polymerase to fill in the correct bases using complementary strand information. Transposons and insertion sequences can also cause spontaneous mutations in all species. These small (1–10 kb) DNA elements in the genome are capable of moving from one location to another, which causes mutations. In some genes, it is estimated that over 80% of spontaneous mutations are caused by transposons or insertion sequences. Additionally, inaccurate pairing during recombination causes large deletions or additions in DNA and is usually associated with gene sequences that are tandemly repeated.

Radiation also causes mutations. The most commonly encountered damaging radiation is from UV light, specifically from 260–280 nm. DNA absorbs radiation most intensively at this wavelength, which causes chemical changes in the DNA. The most common chemical change is the formation of dimers of adjacent pyrimidines. These dimers cannot be copied accurately by DNA polymerase during replication because the dimer forms a bulge in the strand containing the dimer. Incorrect bases are added to the new strand opposite the dimer, causing a mutation. High-energy radiation also causes damage to DNA by the formation of free radicals, which react with DNA, causing breakage of the strands and chemical changes in the bases. These changes result in mutations.

2-Aminopurine and 5-bromodeoxyuridine are examples of base analogues that cause transitions by mispairing with cytosine or guanine, respectively. Many chemicals, such as hydroxylamine, nitrous acid, and alkylating agents, create modifications in the four bases that can change their base-pairing properties and cause mutations in ensuing generations. Damage can be repaired by the excision repair system or glycosylases, which make incisions in the sugar–nucleotide bond. The AP endonuclease then removes a group of bases by breaking the phosphodiester bond, and the sequence is replaced by DNA polymerase.

Tests have been developed to detect and measure mutagenic potential easily. The Ames test uses mutations in the histidine biosynthetic pathway in *Salmonella typhimurium* strains to measure the reversion rates of mutations for mutagenic potential. The Ames test has revealed that many chemicals used in everyday products are mutagenic, including flame retardants and photocopy toners.

Ultraviolet light produces pyrimidine:pyrimidine dimers of adjacent nucleotides, but dimers can be repaired using a variety of methods, including photolyase, an enzyme that cleaves the dimer structure forming individual pyrimidines. Alternatively, error-prone repair inserts nucleotides randomly opposite the dimer and may cause mutations in ensuing generations. Excision repair removes the dimer and some surrounding bases, allowing DNA polymerase to insert the correct bases using complementary strand information.

Selected Key Terms

acridines
alkylating agents
Ames test
conditional mutations
deamination
depurination
frameshift mutation
missense mutation
mutagen
neutral mutation
nonsense mutation
point mutation
pyrimidine dimer
silent mutation
spontaneous mutation
suppressor mutation
tautomeric forms
transition mutation
transposon
transversion mutation

Chapter Review Problems

1. 5-Bromodeoxyuridine (BrdU), a base analogue, is used in chemotherapy. Why can BrdU be used as in cancer therapy, whereas EMS (ethylmethane sulfonate), an alkylating agent, cannot?
2. A gardener finds in her collection a white-flowered petunia that has a small patch of deep blue on one petal of a flower. She has been looking for a mutant of this type for some time. Should she get excited or not? Explain.
3. For what types of gene products are conditional–lethal mutations (for example, one that renders an enzyme temperature-sensitive) especially useful in the study of how cells work?
4. Siamese cats are dark colored on their extremities and white on other parts of the body. What type of mutation in a pigment-forming gene could give this pattern of pigmentation?
5. A gene was sequenced and a section of its mRNA was found to have the following sequence: CGCAAGAGGAUGCGCAUG. This gives rise to the amino acid sequence Arg–Lys–Arg–Met–Arg–Met. Treatment with an acridine inserts a G at position 7, inactivating the protein product. What treatment would you suggest to create a revertant back to wild-type?
6. In all organisms is a built-in system to create mutations to aid in natural selection. Explain this system.
7. Describe two mechanisms for the formation of deletions: one for small deletions and one for large deletions.
8. What repair systems require replication, and what repair systems do not require replication of DNA?
9. It is known that a large percentage of spontaneously generated mutants are caused by insertion sequences or transposable elements inserting into genes. If you have a mutant that you suspect is caused by one of these elements, how would you go about proving or disproving your suspicion?
10. Classic hemophilia is caused by a recessive sex-linked mutation in clotting factor VIII. Neither Queen Victoria of England nor her ancestors had hemophilia, but many of her descendants did. Offer an explanation as to the origin of this long line of hemophiliacs in the European monarchy.
11. Some somatic mutations in plants have been used in agriculture, including the Red Delicious apple and the Navel orange. Why is this not true for animals as well?
12. Distinguish between a premutation lesion and a mutation.
13. Why are the mutations produced by intercalating agents likely to be more deleterious on average than are those caused by alkylating agents?
14. In the following experiment, the wild-type protein is found to have the following sequence:

    Phe–Trp–Met–Ser–Lys–Arg

    A mutant is isolated with the following protein sequence:

    Phe–Trp–Met–Trp–Lys–Arg

a. What is the most likely nature of the mutation responsible for this change?
b. Could this mutation have been the result of BrdU mutagenesis? Why or why not?

15. In organisms that have been extensively studied by geneticists over the years, many mutant strains have been isolated. Molecular analysis of these mutants, including DNA sequencing of mutant genes, is often undertaken. However, in many cases, it is found that there is more than one sequence difference between the mutant allele and a wild-type allele. How can scientists be sure which base difference is the one responsible for the mutant phenotype?

16. A portion of a wild-type protein and a mutant protein follows.

wild-type ... Phe–Arg–Lys–Leu–Ala–Trp ...
mutant ... Phe–Arg–Lys

If the mutant protein resulted from only one single mutation, the most likely specific mutation is a. insertion. b. nonsense. c. missense. d. deletion. e. not possible with a single mutation.

17. A mutation can result during DNA replication when the rare form of adenine pairs with a. adenine. b. guanine. c. thymine. d. cytosine. e. none of the above.

18. How does suppression of a mutation relate to reversion?
a. Reversion restores the wild-type phenotype; suppression does not.
b. Suppression restores the wild-type phenotype; reversion does not.
c. Reversion changes the mutant codon; suppression alters a different codon.
d. Suppression changes the mutant codon; reversion alters a different codon.
e. Both reversion and suppression alter the mutant codon.

19. Assume that an Ames test is performed on a new compound. Tube 1 contains distilled water as a control, tube 2 contains the compound, and tube 3 contains the compound and rat liver enzymes. The contents of each tube are mixed with *his*$^-$ *S. typhimurium* cells and spread on Petri dishes that contain no histidine. Tube 1 produces 5 revertant colonies, tube 2 produces 35 revertant colonies, and tube 3 produces 300 revertant colonies. What is the best conclusion from these data?
a. The compound is not mutagenic.
b. The compound and its metabolites are not mutagenic.
c. The compound and its metabolites are equally mutagenic.
d. The compound is mutagenic, but its metabolites are not.
e. The compound is mutagenic, and its metabolites are more mutagenic.

20. Ultraviolet light produces chemical changes in the DNA that include one of the following:
a. addition and deletion of nucleotides
b. purine:purine dimers
c. pyrimidine:pyrimidine dimers
d. alkylation of purine
e. breaks in the double strands

21. The alterations in DNA produced by UV light can be corrected by a. photolyase. b. excision repair. c. glycosylases. d. deamination. e. both a and b.

22. A portion of a wild-type protein and a mutant protein follows.

Wild-type ... Phe–Arg–Lys–Leu–Ala–Trp ...
Mutant ... Phe–Arg–Lys–Ile–Ala–Trp ...

If the mutant protein resulted from only one single mutation, the most likely specific mutation is a. missense. b. nonsense. c. deletion. d. addition. e. suppressor.

Challenge Problems

1. *E. coli* colonies that are *lac*$^-$ can be differentiated from colonies that are *lac*$^+$ by incorporating a dye (eosin-methylene blue) into the agar growth medium. Colonies that are *lac*$^-$ yield a pink colony, and colonies that are *lac*$^+$ yield purple colonies. In an experiment, the investigator mutagenizes some *E. coli* cells that are *lac*$^+$ by exposure to UV light. She then allows them to grow one generation and finds that in every million colonies plated on the selective medium, 678 are all pink. All others are purple. In a separate experiment, instead of allowing the cells to grow after mutagenesis, she immediately spreads the mutagenized cells on the indicator agar. After examination, she finds that in every million colonies 643 are half pink and half purple.
a. What is the rate of mutation of *lac*$^+$ to *lac*$^-$ induced by UV light in this experiment?
b. How do you explain the development of colonies that were half pink and half purple in the second part of this experiment?

2. In a series of experiments, an *E. coli* gene involved in tryptophan synthesis was repeatedly mutagenized by an agent that causes missense and nonsense mutations. Thousands of mutants were found and examined to determine exactly which amino acid had been changed or whether a termination codon had been formed. Of the 268 amino acids in this protein (tryptophan synthetase A), only 30 positions were changed by missense mutations. On the other hand, nearly every position was changed to a nonsense mutation in which termination triplets could be formed from the existing codon. Explain these findings at the molecular level.

Suggested Readings

Adelman, R. L., R. L. Saul, and B. N. Ames. 1988. "Oxidative damage to DNA: Relation to species metabolic rate and life span." *Proceedings of the National Academy of Sciences USA*, 85: 2706.

Bohr, V. A., and K. Wasserman. 1988. "DNA repair at the level of the gene." *Trends in Biochemical Science*, 13: 429–433.

Castellani, A. 1987. *DNA Damage and Repair*. Plenum Press, New York.

Cavenee, W. K., and R. L. White. 1995. "The genetic basis of cancer." *Scientific American*, 272 (March): 72–79.

Cohen, L. A. 1987. "Diet and cancer." *Scientific American*, 257 (November): 42–48.

Friedberg, E. C. 1985. *DNA Repair*. Freeman, New York.

Leffell, D. J., and D. E. Brash. 1996. "Sunlight and skin cancer." *Scientific American*, 275 (July): 52–59.

Ott, W. R., and J. W. Roberts. 1998. "Everyday exposure to toxic pollutants." *Scientific American*, 278 (February): 36–42.

Sancar, A., and G. Sancar. 1988. "DNA repair enzymes." *Annual Review of Biochemistry* 57: 29.

Singer, B. 1996. "DNA damage: Chemistry, repair and mutagenic potential." *Regulation in Toxicology and Pharmacology*, 23: 2–13.

Sorsa, M., and H. Vainio. 1982. "Mutagens in our environment." In *Progress in Clinical and Biological Research*, vol. 109, Alan R. Liss, New York.

On the Web

Visit our Web site at **http://www.saunderscollege.com/lifesci/titles.html** and click on A/G/M Genetics for links to the following chapter-related resources on the World Wide Web:

1. **Human cancers.** This site describes testing for the risk of breast and ovarian cancer mutations in humans.
2. **Human mutations.** This site is a database for human mutations.
3. **Related links.** This site has links to other DNA repair sites.
4. **List of chemicals.** This list of chemicals is consulted by laboratory workers to determine risk.

CHAPTER

18

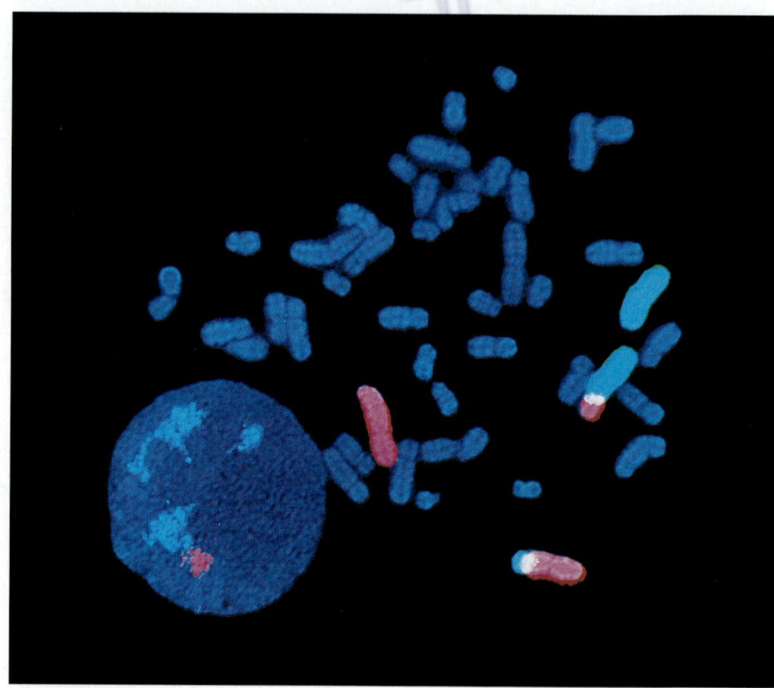

Translocation of a human chromosome. A fluorescence in situ hybridization micrograph of human chromosomes from a cancer patient, showing a translocation between chromosomes 2 and 3. *(© 1995 SPL/James King-Holmes/Science Photo Library/Photo Researchers, Inc.)*

Changes in Chromosome Number and Structure

Up to this point, our discussions of mutational changes in DNA have centered around small changes (for example, substitution of one base pair for another or deletion of a few base pairs). Usually, these small changes affect the expression of only one gene. In this chapter, we will discuss mutational changes that may involve long stretches of DNA, or even entire chromosomes. As you might expect, large alterations in chromosome structure can affect one gene or many genes. These large changes in chromosome structure are referred to as **chromosomal aberrations,** and there are two basic types of changes: the addition or loss of entire chromosomes or sets of chromosomes and the addition, deletion, or rearrangement of large portions of individual chromosomes. For many years, cytogenetic tests were used to classify mutations, and only mutations that visibly altered the structure or number of chromosomes were considered to be chromosomal aberrations. All the rest were considered to be point mutations (discussed in Chapter 17). Today we recognize that this definition is too restrictive. Many species, for which cytological techniques for chromosome examination have not been developed, have chromosomal aberrations that are not always visible with a microscope. Thus, chromosomal aberrations are distinguished from point mutations using genetic tests, and we will discuss some of these tests later in this chapter.

Chromosomal aberrations represent some of the most important mutations in genetics. Their importance in human medicine, agriculture, and genetics research is widely recognized. In humans, chromosomal aberrations may affect one gene, hundreds of genes, or thousands of genes, and they may have profound effects on an individual's phenotype. For example, the phenotype in humans known as Down syndrome results from a chromosomal abnormality in which an extra copy of chromosome 21 is present. Down syndrome individuals have a wide range of developmental abnormalities and health problems, even though they may not have any mutant alleles on chromosome 21. On the other hand, chromosomal aberrations are not always deleterious but, in fact, can produce highly beneficial phenotypes. Many of the plants we use every day for food or fiber (wheat, cotton, potatoes, bananas) and many of the flowers we enjoy in our gardens have chromosomal aberrations. For example, modern wheat, which accounts for about 20% of the food calories consumed by the entire human race, has six sets of chromosomes. These result in larger grains, greater harvests, and the ability to adapt to a wider range of environmental conditions. The study of chromosomal abnormalities has also given us considerable information about how chromosomal structure affects gene function. Some of the unique properties of chromosomal aberrations make them especially useful as tools in genetic experiments.

In this chapter, we will discuss many of the unique aspects of changes in the number and structure of chromosomes. By the end of this chapter, we will have addressed six specific questions about chromosomal aberrations:

1. What are the different types of changes that result in alterations in chromosome number?
2. How do changes in chromosome number affect an organism's phenotype?
3. What types of changes in chromosome structure alter the number and order of genes in a chromosome?
4. How do changes in the number and order of genes affect an organism's phenotype?
5. How do chromosomal aberrations affect synapsis, recombination, and segregation during meiosis?
6. How are changes in chromosome structure used in genetic analysis?

Polyploidy Is an Increase in the Number of Chromosome Sets.

All species have a normal, or standard, set of chromosomes that contains one copy of each gene normally found in that species. This set is called the **monoploid** set, and it is designated by the letter *X*. As we discussed in Chapter 3, most animal species and many plant species are diploids, which means they and have two sets (2X) of chromosomes in their somatic cells and premeiotic germ cells. Humans, for example, have a monoploid chromosome number of 23 (X = 23), and normal human cells contain 2X = 46 chromosomes. Individuals that contain one or more complete sets of chromosomes are called **euploid** (*eu-* means true and *-ploid* is a Greek word meaning fold as in twofold). Chromosomal aberrations that change the number of chromosomes in an individual are divided into two categories. Aberrations that result in individuals having different numbers of complete sets of chromosomes are called changes in ploidy or euploid changes. If individuals have three, four, or more complete sets, they are called **polyploid** individuals. Individuals with three sets (3X) are called **triploids,** individuals with four sets (4X) are **tetraploids,** individuals with five sets (5X) are **pentaploids,** and so on (Table 18.1). Aberrations that result in individuals having one of more extra chromosomes or losing one or more chromosomes are called **aneuploid** changes (*aneu-* is Greek for uneven). In this section, we will discuss changes in ploidy, and we will discuss aneuploid changes later in the chapter.

TABLE 18.1 *Chromosome Composition of Polyploids*

Name	Sets of Chromosomes	Haploid Set
Monoploid	1X	
Diploid	2X	N = X
Triploid	3X	
Tetraploid	4X	N = 2X
Pentaploid	5X	
Hexaploid	6X	N = 3X
Octaploid	8X	N = 4X

In all sexually reproducing individuals, the first meiotic division (the reduction division) reduces the number of sets of chromosomes in the cell by one half. The chromosome set present in a germ cell, after a normal first meiotic division, is called the **haploid** set and is designated by the letter *N*. In diploid individuals, the monoploid and the haploid sets of chromosomes are the same (X = N). For example, the tomato is 2N = 2X = 24, and tomato pollen cells are N = X = 12. In polyploids after the first meiotic division, the germ cells will contain more than one monoploid set of chromosomes. Thus, the haploid and monoploid chromosome sets will not be the same. For example, in a tetraploid tomato, 4X = 48, cells that have completed the reduction division of meiosis will contain two monoploid chromosome sets (N = 2X = 24) (Figure 18.1). This use of the X symbol may seem unnecessary at first glance, especially when we are accustomed to referring to the basic chromosome set as the haploid set. However, this convention is useful in dealing with polyploid individuals. Throughout this chapter, you will see that the concept of the monoploid set is very

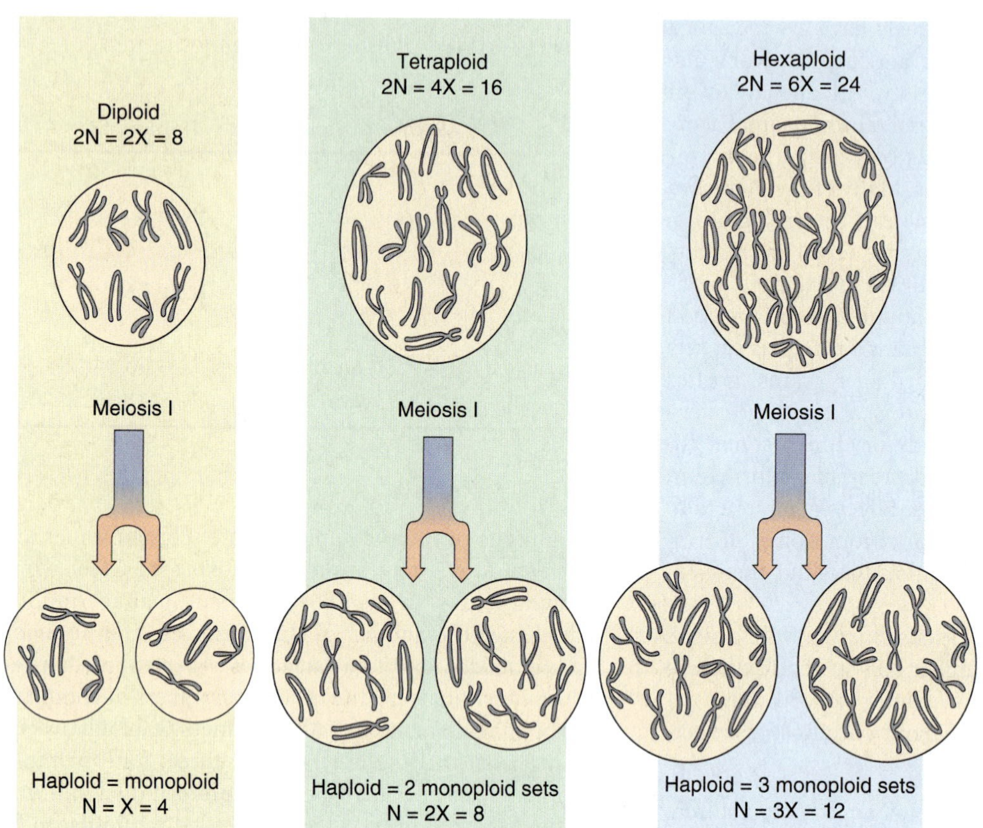

FIGURE 18.1 Monoploid and haploid chromosome numbers. The haploid chromosome number (N) is the number of chromosomes in a nucleus after the reduction division of meiosis (meiosis I). The monoploid number (X) is the minimum number of chromosomes that contain a complete set of genes. In a diploid species, the haploid chromosome number is the same as the monoploid number (N = X), but in polyploid species, N may be 2X, 3X, or more.

useful when distinguishing the number of chromosomes in a gamete (N) from the number of chromosomes in the minimum set (X).

Plants and Animals Differ in Their Ability to Tolerate Ploidy Changes.

Animals that are polyploid usually do not survive, or if they do, they have severe developmental or physical abnormalities. In contrast, polyploid plants are often larger and more vigorous than their diploid cousins, possessing more leaves, larger flowers, and bigger fruits. Why should a polyploid plant with extra sets of normal chromosomes obtain a beneficial phenotype effect, when this same phenomenon has a deleterious effect on animals? Unfortunately, we do not know the answer to this question, although the effect of polyploidy is real and increases with increasing numbers of monoploid sets. For example, in *Drosophila,* triploid females (XXX;AAA, where *X* refers to the X chromosome and *A* refers to an autosomal set of chromosomes) are not very vigorous, but will produce offspring when mated to normal diploid males (XY;AA). However, no viable tetraploid (XXXX;AAAA) *Drosophila* have ever been recovered. This result is very likely caused by the inviability of the tetraploid. The most widely accepted hypothesis about why animals are negatively affected by polyploidy is that it simultaneously alters many important aspects of gene activity, in two ways: first, by changing the actual copy number of the genes and, second, by changing the amount of regulatory gene products in the cell. Increasing the number of copies of a regulatory gene will increase the concentration of the gene's product in the cell, and a concentration that is too high will give an abnormal increase or decrease in the amount of activity of the target genes (as discussed in Chapters 14 and 15). Even if the effect on any one gene is small, the combined effect on all the genes will be considerable. Why plants should not show these same types of problems with increased gene copy number is simply not understood.

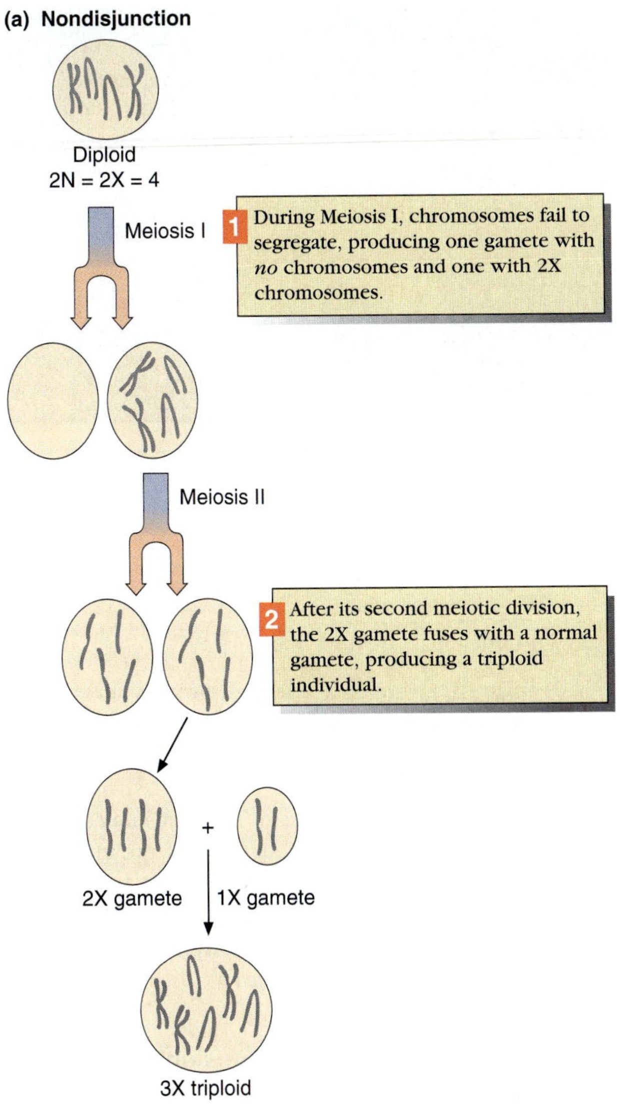

(a) Nondisjunction
Diploid
2N = 2X = 4
Meiosis I
1 During Meiosis I, chromosomes fail to segregate, producing one gamete with *no* chromosomes and one with 2X chromosomes.
Meiosis II
2 After its second meiotic division, the 2X gamete fuses with a normal gamete, producing a triploid individual.
2X gamete
1X gamete
3X triploid

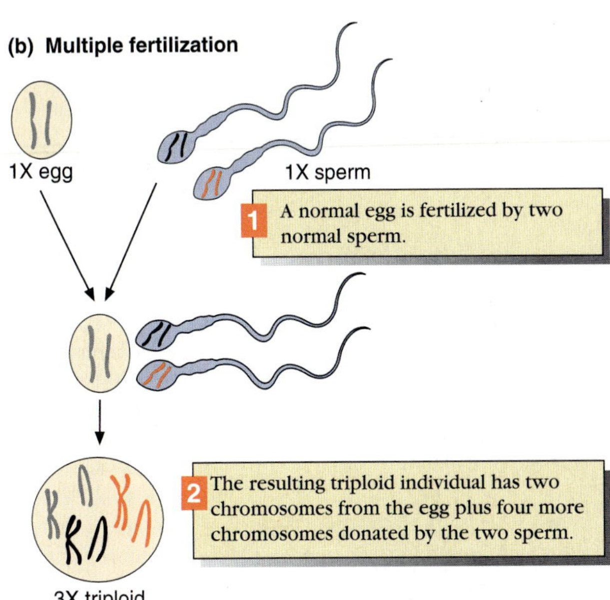

(b) Multiple fertilization
1X egg
1X sperm
1 A normal egg is fertilized by two normal sperm.
2 The resulting triploid individual has two chromosomes from the egg plus four more chromosomes donated by the two sperm.
3X triploid

FIGURE 18.2 Producing an autotriploid. (a) A triploid (3X) individual can be produced by the fusion of a gamete with 2X chromosomes with a normal (X) gamete. (b) A triploid individual may also be produced by the fertilization of a normal (X) egg by two normal (X) sperm.

SECTION REVIEW PROBLEMS

1. Several plants with different ploidy levels all have a monoploid set with five chromosomes (X = 5 chromosomes). For each of the following, give the total number of chromosomes in a somatic cell.
   a. triploid
   b. tetraploid
   c. diploid
   d. hexaploid
2. Give the haploid number for each type listed in Problem 1.

Autopolyploids Contain Extra Copies of the Same Set of Chromosomes.

Polyploids are divided into two classes, depending on the origin of their chromosome sets. Individuals with extra copies of the normal monoploid chromosome set of that same species are called **autopolyploids.** Individuals whose chromosomes come from two or more different species are **allopolyploids.** Autopolyploids are produced by several different mechanisms, including the failure of chromosomes to segregate properly during meiosis (nondisjunction) as well as the fertilization of an egg by more than one sperm (Figure 18.2). Allopolyploids can also be artificially generated by treatment of cells with the chemical **colchicine,** an alkaloid derivative of the autumn crocus *Colchicum autumale.* Colchicine blocks spindle fiber formation, which results in a mitotic cell cycle that includes chromosome replication without cell division. After one such cycle, the cell will have twice as many chromosomes. For example, a diploid cell that normally has 2X = 10 chromosomes will have 4X = 20 chromosomes. Such a cell would be an **autotetraploid** (Figure 18.3).

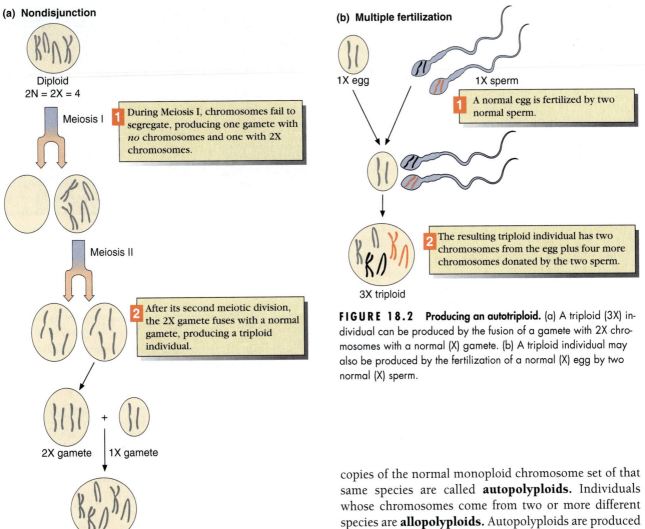

Many commercial plants are autopolyploids, which results in larger plants with bigger leaves, larger flowers, and bigger fruit. These are characteristics that farmers, plant breeders, and consumers all prize. For example, many polyploid ornamental flowers are popular because of their larger, showier blossoms (Figure 18.4). An excellent example of a polyploid crop is potatoes (*Solanum*

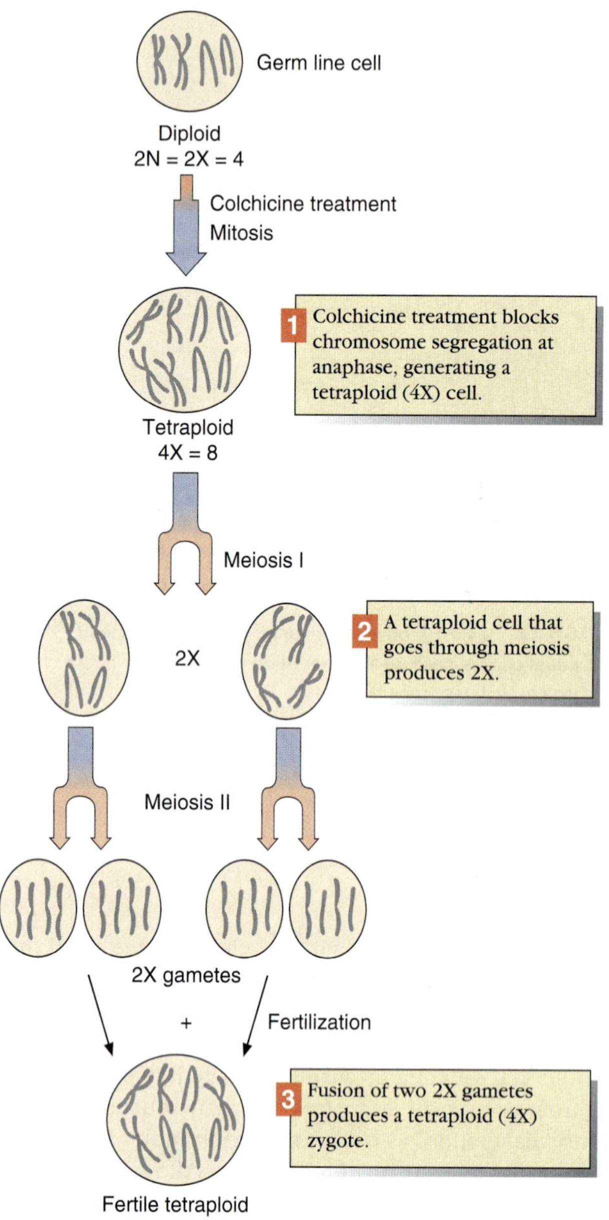

FIGURE 18.3 Colchicine-induced autopolyploidy. Treating a cell with colchicine blocks spindle fiber formation, causing failure of the chromosomes to segregate during meiotic or mitotic cell division. Failure to segregate causes a doubling of the chromosome number. In a germline cell, this produces diploid (2X) gametes that fuse to produce an autotetraploid (4X).

tuberosum), where X = 12 chromosomes. Commercial potato lines are tetraploid (2N = 4X = 48), but other lines have 24, 36, 48, 60, 72, 96, 108, 120, or 144 chromosomes, all multiples of 12. Another example is cultivated bananas, which are 3X, or triploid. Polyploid banana plants must be propagated vegetatively (for example, by cuttings) because they are sterile as a result of aberrant meiotic products.

FIGURE 18.4 Polyploid flowers. Flowers of diploid and tetraploid plants of the daylily strain Stella de Oro. The polyploid strain has larger leaves and flowers. *(Roy Klehm, Klehm Nursery)*

SECTION REVIEW PROBLEM

3. A primitive, small-headed, diploid wheat strain (*Triticum monococcum*) has 14 chromosomes (2N = 2X = 14). Strains of emmer wheat (*Triticum dicoccum*) are tetraploid (2N = 4X = 28) and have harder, larger kernels than *Triticum monococcum*. Tetraploid wheat strains are widely grown in the United States and Europe for making pasta flour. How would you experimentally test the hypothesis that *Triticum dicoccum* is actually a tetraploid derived from *Triticum monococcum*?

Allopolyploids Contain Chromosome Sets from More Than One Species.

Allopolyploids are polyploids that contain sets of chromosomes from two or more different species. To illustrate an allopolyploid, consider a cross between two plants of different species, species S with 2X = 6 and species Z with 2X = 4 (Figure 18.5). The fusion of a gamete from species S with a gamete from species Z will produce a single-celled zygote (called an **amphidiploid;** *amphi-* is Greek for "of two kinds") with five chromosomes (three from species S and two from species Z). These five chromosomes are thus composed of two different chromosome sets. Let us assume that this zygote is viable and grows into a full-sized F_1 plant. When the germ cells enter meiosis, each chromosome normally pairs with a homologous partner, and later these two chromosomes segregate in anaphase I. However, in our example F_1 plant, only one of each chromosome is present, so the five chromosomes have no homologous partners and cannot pair in meiosis. Consequently, in metaphase of meiosis I, each chromosome will line up alone on the metaphase plate, and in anaphase each will segregate independently. This random

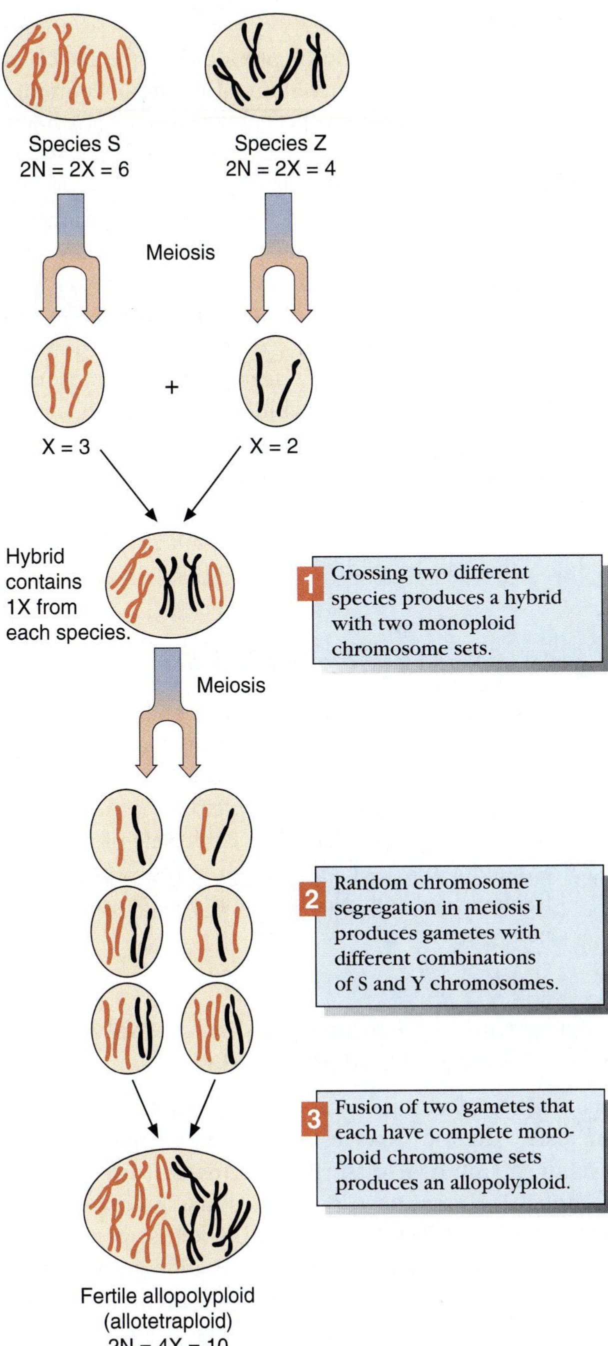

FIGURE 18.5 **Generation of allopolyploids.** An allopolyploid is produced when two different species are crossed and chromosome doubling occurs in the hybrid. Chromosome doubling can occur if, in meiosis, random segregation of the unpaired chromosomes produces gametes with two complete monoploid sets of the chromosomes. Fusion of two such gametes produces an allopolyploid.

segregation will produce meiotic products with 0, 1, 2, 3, 4, or 5 chromosomes (Figure 18.5). These meiotic products may not form functional gametes; but if they do, most such gametes will not have a complete monoploid set of either S or Z chromosomes. Instead, the gametes will possess one or more chromosomes from each set. Fusion of two such gametes will usually produce zygotes with partial sets of S and Z chromosomes, which will be inviable. The failure to produce gametes with complete monoploid sets of chromosomes is a major reason why hybrids between two different species are usually sterile. An example of a sterile animal resulting from a cross between two different species is the mule, which is the offspring of a female horse and a male donkey. Mules have one monoploid set of chromosomes from each parent.

Some of the gametes formed by the F_1 plant in our example will, by chance segregation, have complete S and Z chromosomes sets (N = 2X = 5). The fusion of two such gametes produces a zygote that has two sets of S and two sets of Z chromosomes, a total of 4X = 10 chromosomes. If this zygote is viable and develops into an individual, the individual will be polyploid. Because the two sets were derived from two different species, this plant will be an allopolyploid. More specifically, because the individual contains four chromosome sets, it is called an **allotetraploid.** Because an allotetraploid individual has two copies of each chromosome, each chromosome will have a homologous pairing partner in meiosis. Thus, during meiosis, each S chromosome will synapse with its homologous partner, and each Z chromosome will synapse with its homologous partner. Chromosome segregation will occur normally in meiosis I, and each meiotic product will receive complete S and Z monoploid sets of chromosomes. Thus, the gametes will have a haploid chromosome complement of one S monoploid set and one Z monoploid set (N = 2X = 5 chromosomes). Fusion of two such gametes will produce a fertile allotetraploid individual (2N = 4X = 10 chromosomes). Not all allotetraploids are viable and fertile because of developmental problems that arise from having two different sets of chromosomes, but many that are have phenotypes combining elements of both parental strains. For commercially important species, generating allopolyploids is an opportunity to combine valuable traits from separate species into a single strain.

One example of an important allopolyploid is wheat. Modern bread wheat (*Triticum aestivum*) has 42 chromosomes and is believed to be an allopolyploid with two copies of three distinct sets of chromosomes. These are often represented as AA, BB, and DD. The origin of these three sets is not certain, but two (AA and BB) are believed to come from primitive wheat, and one (DD) is believed to be from a species of goat grass (*Aegilops squarrosa*). This hypothesis was experimentally tested by crossing a tetraploid wheat (2N = 4X = 28) with *Aegilops squarrosa* (2N = 2X = 14). The offspring of the cross had 21 chromosomes and resembled the primitive bread wheat strain *Triticum spelta*. These plants were sterile, but treating them with colchicine produced a fertile line with 42 chromosomes that had a strong resemblance to modern

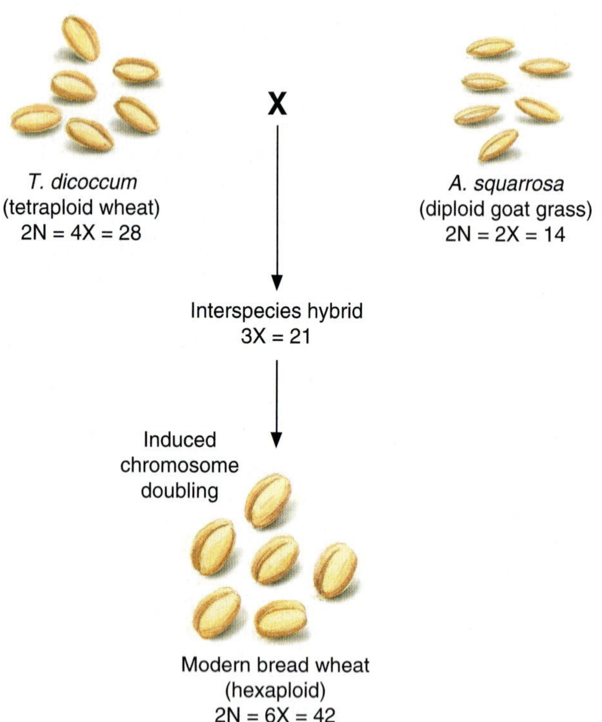

FIGURE 18.6 Hexaploid wheat. Modern bread wheat is a hexaploid (6X = 42). One hypothesis about its origin is that wheat originated from a cross between a tetraploid wheat plant (4X = 28) and a plant of *Aegilops squarrosa* (goat grass, 2X = 14). The interspecies hybrid would have had 3X = 21 chromosomes, and chromosome doubling produced a fertile hexaploid (6X = 42) plant.

strains of bread wheat (Figure 18.6). This series of events is likely rare in nature. However, the favorable characteristics of the resulting plants (higher yield, larger kernels, easier milling) meant that these rare individuals were probably highly prized by farmers. As a consequence, they probably saved seeds from these plants for the next spring. Not all allopolyploids have a favorable combination of traits; unfortunately, it is not possible to predict which traits from each parental species will be expressed in the new polyploid strain (see CONCEPT CLOSE-UP: *The Rabbage*).

SOLVED PROBLEM

Problem

Mules are the sterile progeny of a male donkey (2N = 2X = 62) mated with a female horse (2N = 2X = 64). Assume that mules are sterile because of a failure of chromosome pairing and segregation in meiosis, and ignore the possibility that polyploid animals are not viable. What would be the probability that a mating of two mules would produce a fertile polyploid offspring, and how many chromosomes would that offspring have?

Solution

A polyploid mule could be produced by the fusion of two gametes that each contain all the mule's chromosomes. The mule contains 63 chromosomes (1X horse = 32 + 1X donkey = 31). If there is no pairing between horse and donkey chromosomes, the probability that a mule gamete will receive all 63 chromosomes is $(1/2)^{63}$. The probability that two such gametes would fuse to produce a polyploid individual would be $(1/2)^{63} \times (1/2)^{63} = (1/2)^{126}$. If any of the horse and donkey chromosomes do pair in meiosis, then the number of independently segregating chromosomes will be fewer, and the probability of producing gametes with all 63 chromosomes by segregation alone will be lower. The polyploid mule would have 126 chromosomes (2N = 4X). Because animals do not tolerate such increases in chromosome number, it is very unlikely that this individual would be viable.

SECTION REVIEW PROBLEMS

4. The chromosomes in cells of a stable polyploid species were examined in meiosis I and found to have 26 chromosomes or 13 bivalents. Is this an autopolyploid or an allopolyploid?
5. In an attempt to establish a stable line of plants, a triploid with X = 6 chromosomes was selfed. What frequency of tetraploids would you expect in the next generation?

Segregation of Chromosomes in Autopolyploids Gives Different Mendelian Ratios.

As a result of chromosome pairing and segregation in meiosis, diploid species generate a characteristic Mendelian ratio of offspring in crosses (discussed in Chapter 2). In contrast, polyploids may produce different patterns of chromosome pairing and segregation in meiosis and thus may produce quite different offspring ratios. The precise ratios produced depend on the type of polyploid and on the number of sets of chromosomes present. Allopolyploids, for example, contain chromosome sets from different species, and often two copies of each set are present. Thus, each chromosome will have one regular pairing partner, just as it did in the original diploid species. If this individual is heterozygous for an allele, segregation will give a characteristic Mendelian ratio in the offspring. However, in polyploids that have more than two copies of a chromosome set, each chromosome has multiple possible pairing partners. Depending upon the number of chromosome sets, this situation will result in different offspring ratios. In the remainder of this section, we will explore some of the effects that multiple chromosome sets have on segregation patterns.

One very important factor in chromosome segregation in polyploids is whether the polyploid has even (for example, 4X, 6X, 8X) or odd (for example, 3X, 5X, 7X) numbers of chromosome sets. In a polyploid individual with an even number of chromosome sets, each chromosome has at least one homologous chromosome to pair with in meiosis. If each chromosome pairs with only one other chromosome, forming a normal bivalent, they will segregate in anaphase of meiosis I. In a polyploid individual with odd numbers of chromosome sets, at least one copy of each chromosome will not have a pairing partner. This odd chromosome will segregate randomly in meiosis, producing meiotic products that usually do not have complete sets of chromosomes.

Let us first consider some examples of chromosome pairing and segregation in polyploids with even numbers of chromosome sets. Consider the example of a hypothetical autotetraploid with four homologous chromosomes (C1, C2, C3, C4). Each chromosome has three possible pairing partners in prophase of meiosis I. The C1 chromosome can pair with C2, or with C3, or with C4 (Figure 18.7). Assuming that all chromosomes pair only in normal bivalents, any one cell will have one of three possible pairing patterns of the four chromosomes: C1 with C2 and C3 with C4, or C1 with C3 and C2 with C4, or C1 with C4 and C2 with C3. During metaphase of meiosis I, each bivalent aligns on the metaphase plate independent of the other bivalents, and each pair of chromosomes segregates independently, giving meiotic products with six genotypes: C1C2, C1C3, C1C4, C2C3, C2C4, and C3C4 (Figure 18.7). If all three pairing patterns are equally likely, then each will occur in one third of the germ cells undergoing meiosis, and all six of these genotypes will be equally frequent.

The increased number of chromosome pairing possibilities is why a heterozygous polyploid species produces gamete and offspring genotypic ratios that are different from those that result from a heterozygous diploid. If two of the chromosomes in our hypothetical tetraploid had dominant alleles for a specific gene (*A1A2*) and two had recessive alleles (*a3a4*), they would give a total genotype of *A1A2a3a4*. These heterozygotes will produce gametes

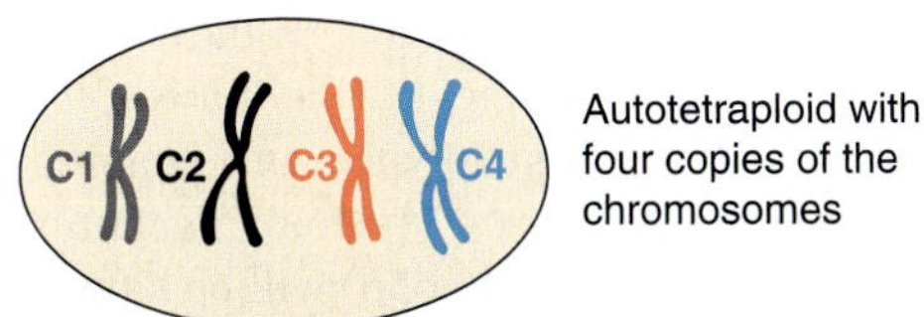

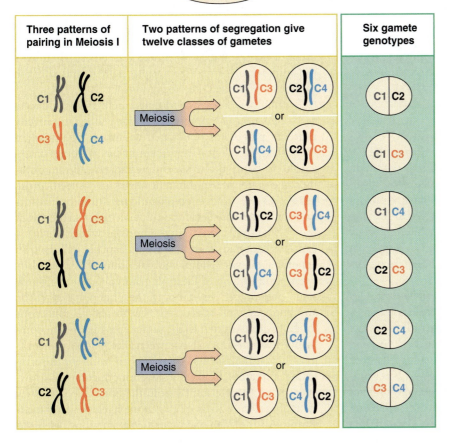

FIGURE 18.7 Tetraploid chromosome segregation. In an autotetraploid, each chromosome has three possible pairing partners in meiosis I, so in each cell the chromosomes will pair in one of three different pairing patterns. In each pairing pattern, the four chromosomes will form two bivalents. Because the two bivalents in each cell segregate independently, a total of 12 gamete classes contain six different, equally frequent, gamete genotypes.

CONCEPT CLOSE-UP

The Rabbage

The generation of fertile allopolyploids is an important part of modern commercial plant breeding, whose goal is to generate a new species that combines valuable features from each original species. Such work offers no guarantee of success because it is not possible to predict in advance which characteristics the new hybrid will express. This is illustrated by the work of the Russian geneticist Karpenchenko in the Soviet Union. In 1928, he attempted to generate a new fertile species by crossing the radish (*Raphanus sativus*) and the cabbage (*Brassica oleracea*). His goal was to create a fertile polyploid with leafy heads like a cabbage and large, edible roots like a radish; this cross would have been a great benefit to Soviet agriculture.

Both radish and cabbage are diploids with 2N = 2X = 18 chromosomes. Crossing them generated viable F_1 plants that had 18 chromosomes (X = 9 from cabbage and X = 9 from radish). Karpenchenko continued to collect large numbers of seeds from the F_1 plants, and eventually a few viable seeds were found. These produced fertile plants with 36 chromosomes (2N = 4X = 36, with 2X = 18 cabbage chromosomes and 2X = 18 radish chromosomes). These viable seeds were the result of the chance fusion of two gametes that had received all 18 chromosomes from the original hybrid genome. The new hybrid was called *Raphanobrassica*, or the "rabbage." The rabbage, unfortunately, was an agronomic failure because it had small leaves like a radish and a small root like a cabbage.

with six different genotypic classes: (*A1A2*) (*A1a3*) (*A1a4*) (*A2a3*) (*A2a4*) (*a3a4*). Let us assume for a moment that the two dominant alleles give the same dominant phenotype (*A*) and that the two recessive alleles give the same recessive phenotype (*a*). This will result in the gametes produced by each heterozygote being 1/6 *AA*, 4/6 *Aa*, and 1/6 *aa*. In a cross between two heterozygotes, the gametes will combine randomly, giving 6 × 6 = 36 genotypic progeny classes with five different genotypes. This would give the following ratio of genotypes:

AAAA	*AAAa*	*AAaa*	*Aaaa*	*aaaa*
1/36	8/36	18/36	8/36	1/36

If the *A1* and *A2* alleles are completely dominant, the offspring will have a phenotypic ratio of 35/36 dominant (*A*) to 1/36 recessive (*a*). If the *A1* and *A2* alleles are incompletely dominant, the offspring will have a phenotypic ratio of 1/36 *AAAA* to 8/36 *AAAa* to 18/36 *AAaa* to 8/36 *Aaaa* to 1/36 *aaaa*. These two ratios are strikingly different from the 3/4 to 1/4 and 1/4 to 2/4 to 1/4 ratios in the offspring of a cross between two heterozygous diploids.

Mendelian genetic analysis in tetraploids becomes more complex when segregation occurs at several loci. A tetraploid that is heterozygous at two unlinked loci (*AAaa BBbb*) will produce gametes with 6 × 6 = 36 genotypic classes. If two heterozygous individuals are crossed, the offspring will have 36 × 36 = 1296 genotypic classes. Only one (1/1296) will be homozygous recessive at both loci (*aaaa bbbb*). Likewise, only one (1/1296) will be homozygous dominant at both loci (*AAAA BBBB*). An individual heterozygous at three unlinked loci (*AAaa BBbb CCcc*) will produce 6^3 = 216 gamete genotypes, and crossing two such individuals will produce 6^6 = 46,656 genotypic classes in the offspring. One (1/46,656) will be homozygous recessive at all three loci (*aaaa bbbb cccc*), and one (1/46,656) will be homozygous dominant (*AAAA BBBB CCCC*). Heterozygous tetraploids give gametic ratios following the formula of 6^n = number of gamete genotypes, where 6 = the number of gamete genotypic classes, and *n* is the number of unlinked, heterozygous loci.

A number of other conditions can further complicate the calculation of phenotypic and/or genotypic progeny ratios in polyploids. One such condition occurs when more than two alleles are present in a polyploid. For example, in our hypothetical autotetraploid, all four chromosomes might have different alleles of a specific gene (*A1 A2 A3 A4*). The genotypic ratio produced by crossing two such individuals can be calculated. No matter what the allelic arrangement, each tetraploid will produce six gamete genotypic classes, and the offspring will have 36 genotypic classes. The phenotypic ratio is calculated by determining which of these 36 genotypic classes give the same phenotype. Determining the phenotype given by any one genotypic class is subject to all the conditions found in diploids; that is, the same types of different dominance relationships exist between alleles in polyploids as in diploids. In cases where more than one gene is segregating, the problem becomes even more complex. However, it can be determined from the phenotype given by each of the 36 × 36 = 1296 genotypic classes in the offspring. All the allelic and genetic interactions discussed earlier in the book (for example, epistasis and modifiers) that give modified Mendelian ratios in diploids can also give modified polyploid phenotypic ratios.

Analysis of allele segregation in polyploids is even more complex because during prophase of meiosis I, the homologous chromosomes in autopolyploids do not always pair neatly in bivalents. Three homologous chromosomes can pair together, forming a structure known as a trivalent, or four can pair forming a tetravalent. When such complex pairing patterns occur, it is more difficult to predict how individual chromosomes will segregate. Unusual pairing and altered segregation are important aspects of polyploidy.

Polyploids that have uneven numbers of chromosome sets (such as 3X, 5X, or 7X) have special problems with chromosomal pairing and segregation in meiosis. Consider the case of meiosis in a triploid (3X) species. If two chromosomes form a bivalent, the third will have no pairing partner. This unpaired chromosome will segregate at random in anaphase, migrating to one or the other pole of the cell. Half of the gametes will receive two copies of this chromosome, and half of the gametes will receive one copy. In a triploid individual with two chromosomes (CCC and BBB), the random segregation of the two unpaired chromosomes produces four different chromosome compositions in the meiotic products: CCBB, CBB, CCB, or CB. Assuming all meiotic products form functional gametes, selfing this individual will produce offspring with a variety of chromosome compositions:

	Female Gametes			
Male Gametes	**CCBB**	**CBB**	**CCB**	**CB**
CCBB	**CCCCBBBB**	CCCBBBB	CCCCBBB	**CCCBBB**
CBB	CCCBBB	CCBBBB	**CCCBBB**	CCBBB
CCB	CCCCBBB	**CCCBBB**	CCCCBB	CCCBB
CB	**CCCBBB**	CCBBB	CCCBB	**CCBB**

The classes written with bold letters represent euploid classes. One of these classes is diploid (**CCBB**), one is tetraploid (**CCCCBBBB**), and four are triploids (**CCCBBB**). The rest, written in regular font, are aneuploids. Aneuploids, as we will discuss in the next section, are often inviable. An individual with more than two chromosomes will have more classes that are unbalanced. For example, a triploid individual with three chromosomes would have $2^3 = 8$ different gamete chromosome compositions and $8 \times 8 = 64$ classes of offspring genotypes; 1 would be tetraploid, 1 diploid, 8 triploid, and 54 aneuploid. This illustrates why triploids with uneven numbers of chromosome sets are sterile or nearly so. A polyploid with uneven numbers of chromosome sets that reproduces sexually will have greatly reduced fertility, which places it at an evolutionary disadvantage compared with diploid or tetraploid individuals. Polyploid animals with uneven numbers of chromosome sets are very rare, but polyploid plants that have uneven numbers of chromosome sets can survive and even flourish if they can reproduce asexually by vegetative growth. As mentioned earlier, the banana is an example of a triploid plant that is sterile but is propagated vegetatively. Fruit growers consider the extra effort worthwhile because triploid bananas produce no seeds in their fruit, whereas diploid bananas have large, hard, inedible seeds.

SOLVED PROBLEM

Problem

An allotetraploid plant of genotype *Aaaa BBbb Cccc* (*A* = green leaf and *a* yellow leaf, *B* = red flower and *b* white flower, and *C* = curved stem and *c* = straight stem) was crossed with another allotetraploid plant of genotype *Aaaa Bbbb cccc*. Assuming that all the dominant alleles are completely dominant, that all chromosomes pair only in bivalents, and that meiotic products with all chromosome compositions form functional gametes, from this cross what is the probability of obtaining progeny with the following phenotypes: yellow, white, straight; green, white, straight; green, white, curved; and green, red, curved?

Solution

The probability can be determined by first obtaining the probability of the different individual phenotypes. The probability of a progeny plant being yellow, white, straight is equal to the product of the probability of being *aaaa*, of being *bbbb*, and of being *cccc*. The probability of being *aaaa* is equal to the probability of two aa gametes fusing = (1/2) (1/2) = 1/4. The probability of being *bbbb* is equal to the probability of two bb gametes fusing = (1/6) (1/2) = 1/12. The probability of being *cccc* is equal to the probability of two cc gametes fusing = (1/2) × 1 = 1/2. The probability of being *aaaa*, *bbbb*, and *cccc* is the multiple of these three = (1/4) (1/12) (1/2) = 1/96. The probability of obtaining progeny with the other phenotypes can be calculated in the same way: green, white, straight = (3/4) (1/12) (1/2) = 3/96; green, white, curved = (3/4) (1/12) (1/2) = 3/96; green, red, curved = (3/4) (11/12) (1/2) = 33/96.

SECTION REVIEW PROBLEMS

6. A tetraploid that is heterozygous for one gene (*RRrr*) has a pink flower. This plant is crossed with a homozygous recessive tetraploid (*rrrr*) that has a white flower. Assume that *R* is an incompletely dominant allele for red flower color.

a. What phenotypes would you expect in the next generation?
b. What phenotypic ratios would you expect in the next generation?

7. A tetraploid plant from the cross in Problem 6 with the genotype *Rrrr* was selfed.
a. What genotypic and phenotypic ratios would you expect in the offspring?
b. If *R* were completely dominant, what phenotypic ratio would you expect in the offspring?

Polyploid Animal Species Have Special Problems with Sex Determination.

One important reason why there are few polyploid animal species is sex determination. In many animal species, the sex of an individual is determined by the ratio between the X chromosomes and autosomes (discussed in Chapters 3 and 16) and not by merely the presence or absence of a Y chromosome. In polyploids, this ratio may be altered. Consider *Drosophila*, in which individuals with a ratio of X chromosomes to autosomal sets of 1:1 are females and individuals with a ratio of X chromosomes to autosomal sets of 1:2 are males. Triploid *Drosophila* with three X chromosomes and three sets of autosomes (XXX;AAA) have an X:autosome ratio of 1:1 and are viable females. However, neither X:AAA (1:3) nor XX:AAA (2:3) individuals have a 1:2 ratio, so there are no triploid *Drosophila* males. Triploid strains of *Drosophila* do exist and are maintained in laboratories by mating triploid females to diploid males and selecting the triploid female progeny every generation. Maintaining these strains is possible only because *Drosophila* has a small chromosome number (X = 4) and the frequency with which triploid females produce eggs with two complete monoploid chromosome sets is reasonably high $[(1/2)^4 = 1/16]$. In theory, male and female tetraploid *Drosophila* that have proper X:autosome ratios (XX:AAAA = 1:2 for males and XXXX:AAAA = 1:1 for females) could be produced, but no such individuals have ever been recovered, suggesting that such an increase in chromosome numbers cannot be tolerated.

Plants have a much greater tolerance for changes in ploidy than animals, and allopolyploid plants are produced today by two procedures. In one procedure, plants from different species are crossed. Then either chromosome doubling is induced in the hybrid, or they are screened for spontaneous chromosome doubling. This approach has been used for many years. In 1925, two species of tobacco, *Nicotiana glutinosa* (2N = 2X = 24) and *Nicotiana tabacum* (2N = 4X = 48) were crossed, and the offspring plants (36 chromosomes) were mostly sterile, but one produced a few progeny. These were fertile and had 72 chromosomes. Cytological investigations confirmed these plants were allohexaploids that had two *glutinosa* and four *tabacum* chromosome sets (2N = 6X = 72). Cytogenetic investigations of natural populations of plants suggests that this type of allopolyploidy occurs regularly in nature, especially in the grasses. The second experimental approach that is currently in use is to fuse single cells from plants of two different species and then regenerate complete plants from single, fused cells. This approach has the advantage that cells can be fused from species that do not produce fertile hybrids. The disadvantage is that it requires sophisticated cell culture techniques for regenerating the plants from fused cells (see CURRENT INVESTIGATIONS: *Generation of Polyploids by Cell Fusion*).

SECTION REVIEW PROBLEM

8. You can buy seeds of some common plants (grapes and watermelon) that will produce plants that have seedless fruit. Explain how you would produce seeds that will grow into a plant that produces a seedless watermelon.

ANEUPLOIDY IS A GAIN OR LOSS OF INDIVIDUAL CHROMOSOMES.

Individuals that have gained or lost one or more chromosomes are called aneuploids. Aneuploid is a general term that refers to any loss or gain of a chromosome, thus special terminology has been devised to convey more specific information. A normal diploid individual can be said to be **disomic** (to have two copies) for each chromosome. Individuals who are diploid but have an extra chromosome are said to be **trisomic** for that chromosome. The chromosome number is then written 2X + 1. For example, a trisomic tomato would have two normal sets (2X = 2N = 24) + 1 = 25 chromosomes. Diploid individuals that lose a single chromosome are said to be **monosomic** for that chromosome and are consequently 2X − 1. A monosomic tomato would have two normal sets (2X = 2N = 24) − 1 = 23 chromosomes. Individuals that have two different extra chromosomes are **double trisomic** (2X + 2). A doubly trisomic tomato would have 24 + 2 = 26 chromosomes. Individuals that have lost one copy of two different chromosomes are **double monosomic** (2X − 2). A doubly monosomic tomato would have 24 − 2 = 22 chromosomes. Individuals that have two extra copies of a single chromosome are said to be **tetrasomic.** A tetrasomic tomato would have 24 + 2 = 26 chromosomes. Individuals that have lost both copies of one chromosome are said to be **nullisomic.** A nullisomic tomato would have 24 − 2 = 22 chromosomes. All these individuals (trisomic, double trisomic, monosomic, double monosomic, tetrasomic, and nullisomic) are aneuploids. The chromosome content of an

Current Investigations

Generation of Polyploids by Cell Fusion

Our previous discussions of techniques for the generation of plant allopolyploids have mentioned only techniques involving sexual methods. However, allopolyploids and autopolyploids can also be generated by a technique known as somatic cell fusion (see Figure 18.A). In the first step of somatic cell fusion, tissue is isolated from two different plants, and single-cell suspensions are prepared. The cells are then treated with enzymes that remove the cell walls, after which they are referred to as **protoplasts**. The protoplasts from the two plants are mixed in a solution of polyethylene glycol, which induces cell fusion. After washing, the mixture is spread on a growth medium, where colonies will eventually appear. A key step in the process is the selection of the hybrid cells. As with animal cell fusion used for mapping human chromosomes (described in Chapter 6), the growth medium is usually selective, allowing only the fused cells to grow into colonies (referred to as callus). The callus material can then be transferred to other growth medium containing plant hormones in the proper proportions, and some will regenerate whole plants. If the parent plants are from different species, these plants will be allopolyploids.

There are several advantages to producing polyploids by cell fusion. First, each somatic cell contains a complete set of genes, so the new polyploid contains the entire chromosome complements of both original parental strains. Because no meiotic step was involved, there has been no chance for segregation of the chromosomes. This means that particular combinations of alleles on different chromosomes may be passed intact to the polyploid. Also, many potential polyploids may be quickly generated in culture without the necessity of growing several generations of whole plants. Finally, cell fusion can be used to generate polyploids from plants whose reproductive systems are so different that they cannot be induced to cross.

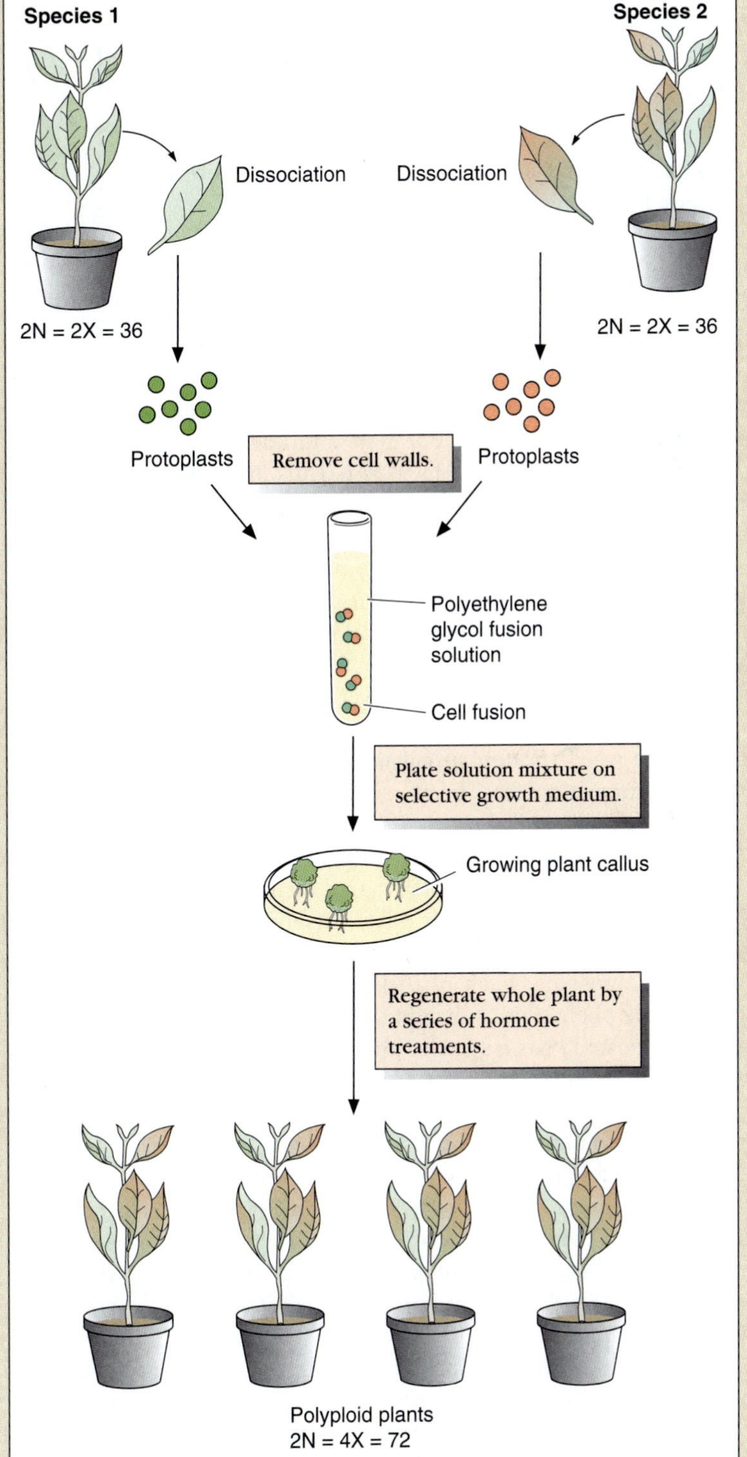

▶ **FIGURE 18.A Formation of polyploids by cell fusion.** Allopolyploid plants can be generated by the fusion of somatic plant cells, followed by regeneration of whole plants from the fused cells.

aneuploid human is designated by using a special notation that gives the total number of chromosomes present and a description of the abnormality. For example, an individual who has one X chromosome and no Y is described as 45,X. This indicates the individual has a total of 45 chromosomes (one less than the normal 46) and has a single X. A trisomic individual with three X chromosomes and a normal set of autosomes has a chromosome composition of 47,XXX, and a Down syndrome individual, who has three copies of chromosome 21, has a chromosome composition of 47,21 21 21.

Meiotic Chromosome Segregation in Aneuploids Produces Unbalanced Gametes.

Aneuploids have many of the same problems with abnormal chromosome segregation in meiosis as polyploids with uneven numbers of chromosome sets. In a trisomic individual (2X + 1), the three copies of the single chromosome may pair and segregate in several different ways. In almost every case, two of the chromosome copies will segregate to one pole and one to the other pole during anaphase of meiosis I. Thus, one half of the meiotic products will have an extra chromosome (X + 1), and one half will have the normal monoploid number (X). If these meiotic products then form functional gametes, fusion of these gametes with normal gametes (X) will produce another generation of aneuploid (2X + 1) individuals.

Trisomic strains of plants can be very useful in certain kinds of genetic analyses. In this respect, an example is the jimson weed (*Datura stramonicum*), whose chromosome composition has been extensively studied. *Datura* is normally a diploid (2N = 2X = 24). Each *Datura* chromosome has unique cytological characteristics (such as size and shape), which makes their identification possible. As a consequence, all 12 trisomic strains have been isolated, and each contains an extra copy of a different chromosome. Each of these 12 strains has a unique visible phenotype, indicating that an extra copy of one chromosome has a specific phenotypic effect. As expected, two types of female gametes are produced by a trisomic plant—one half with one extra chromosome and one half with the normal haploid complement. However, male gametes are different. In meiosis, chromosome segregation is the same as in female cells, but only the meiotic products with the normal haploid chromosome number forms functioning pollen. The extra chromosome renders the pollen either nonfunctional or unable to compete with normal pollen. In crosses between trisomic plants, the six different female gamete genetic types and three different male gamete genetic types combined to produce 18 zygotic genotypes (Figure 18.8), half of which are trisomic (2X + 1) and the other half of which are diploid (2X).

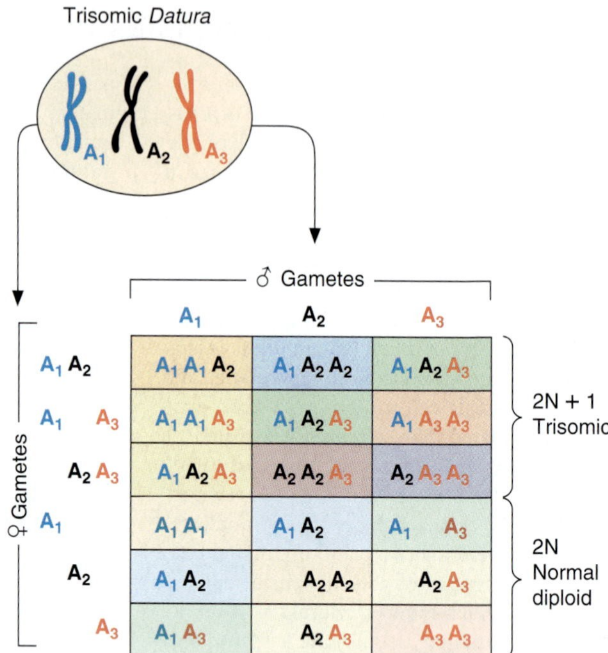

FIGURE 18.8 Trisomic chromosome segregation. In a trisomic *Datura*, there are six different female gamete genotypes for the trisomic chromosome. Because pollen with an extra chromosome is not functional, only three functional male gamete genotypes are produced. This gives 18 genotypic classes in the progeny. Of these, half are trisomic, and half are normal diploids.

Trisomic *Datura* strains can be used to map a recessive mutation to a particular chromosome. For example, if we had a diploid strain with a new recessive *white flower* (*w*) mutation, we could cross homozygous mutant (*ww*) individuals as males with plants from each of the 12 trisomic strains discussed earlier. The trisomic F_1 individuals from each cross would then be selfed and the phenotypic ratios in the resulting progeny determined. In 11 of the crosses, the *white* mutation is not on the chromosome that is trisomic. In the F_1 (+ *w*) of these crosses, the chromosome containing the *white* mutation will segregate normally from its homologue containing the dominant **wild-type allele.** This will result in gametes with a genotypic ratio of 1/2 colored to 1/2 white flowered. When these F_1 are selfed, the progeny will have the normal Mendelian phenotypic ratio of 3/4 colored and 1/4 white flowered. However, in one of the crosses, the female parent plant will be trisomic for the chromosome that contains the *white* mutation. The F_1 plant (+ + *w*) will be trisomic and will have two copies of the chromosome that contain + alleles and one that contains the mutant allele. Only 1/6 of the F_1 meiotic products will have the *white* allele alone. These meiotic products will give rise to eggs of which only 1/6 will have a *w* genotype. Because aneuploid meiotic products will not

Diploid white strain		Trisomic lines	F_1	Self-cross	F_2 phenotypes
ww	×	2N + A	→ F_1	→	3 colored : 1white
ww	×	2N + B	→ F_1	→	3 colored : 1white
ww	×	2N + C	→ F_1	→	3 colored : 1white
ww	×	2N + D	→ F_1	→	3 colored : 1white
ww	×	2N + E	→ F_1	→	3 colored : 1white
ww	×	2N + F	→ F_1	→	3 colored : 1white
ww	×	2N + G	→ F_1	→	3 colored : 1white
ww	×	2N + H	→ F_1	→	17 colored : 1white ← Cross involving the strain that is trisomic
ww	×	2N + I	→ F_1	→	3 colored : 1white
ww	×	2N + J	→ F_1	→	3 colored : 1white
ww	×	2N + K	→ F_1	→	3 colored : 1white
ww	×	2N + L	→ F_1	→	3 colored : 1white

FIGURE 18.9 Trisomic mapping. A *Datura* mutation that gives white flowers can be mapped to a specific chromosome by crossing it to a series of trisomic strains and then selfing the progeny. Most of the testcrosses give a 3 + : 1 white offspring phenotypic ratio (+ = wild-type), but the cross involving the strain that is trisomic for the chromosome on which the *white* mutation is located gives a 17 + : 1 white ratio. The 17 : 1 ratio is the result of the fact that of the 18 genotypic classes in the progeny, and only one has a homozygous *white* genotype. In this example, the ratios indicate that the *white* mutation is on chromosome H.

give rise to functional pollen, 1/3 of the functioning pollen will have a *w* genotype. Selfing these F_1 individuals will thus give progeny with a phenotypic ratio of (1/6) (1/3) = 1/18 white flowered and 17/18 colored (Figure 18.9). This phenotypic ratio results from the different segregation ratio of the trisomic chromosomes and the failure of aneuploid meiotic products to form functional pollen. Knowing that a 1/18 to 17/18 ratio will result from the one cross in which the *white* mutation is located on the chromosome that was trisomic and that a 1/4 to 3/4 ratio will result from all other crosses, we simply analyze the results of the 12 crosses to determine which gives the 1/18 to 17/18 ratio.

In our example, the results of the crosses indicate that the white mutation is on chromosome 7. Having trisomic lines whose extra chromosome has been identified allows a plant geneticist to determine the chromosomal location of an unknown recessive mutation by simply crossing, inbreeding, and then observing the phenotypic ratios in the F_2.

SOLVED PROBLEM

Problem

A new mutation in soybeans (yellow leaf vs. the normal green) is to be mapped using a series of trisomic lines. This was done using plants homozygous for yellow leaf. The yellow leaf plant was crossed to a series of plants, each of which was trisomic for a different chromosome. The offspring of these crosses were inbred, and the phenotypes of the F_2 (yellow vs. green) were determined. From the following results, on what chromosome is yellow located?

Trisomic Chromosome	F_2 Phenotype Yellow	Green
1	10	38
2	11	35
3	15	41
4	10	36
5	10	35
6	13	40
7	16	48
8	12	44
9	2	43
10	14	48
11	18	49
12	15	49
13	11	36
14	19	49
15	18	46
15	15	40
16	15	44
17	16	47
18	13	38
19	16	49
20	14	42

Solution

In a normal diploid cross, the progeny of a selfed heterozygote will be 1/4 yellow and 3/4 green. In a cross with the trisomic strain, the ratio will shift to a 1/18 yellow and 17/18 green. These results indicate that the cross with the trisomic for chromosome 9 gives this ratio, so the yellow gene must be located on chromosome 9.

SECTION REVIEW PROBLEMS

9. A plant that is heterozygous at the *P* locus (*Pp*) has a high frequency of nondisjunction.
   a. What would be the genotypes of the meiotic products produced by a cell that underwent nondisjunction in meiosis I for the chromosome that contains the *P* locus?
   b. What would be the genotypes of the meiotic products produced by a cell that underwent nondisjunction in meiosis II for the chromosome that contains the *P* locus?
10. A trisomic mouse was discovered with an *AAa* genotype for an allele on the trisomic chromosome. If the father were *Aa* and the mother were *aa*, in which parent did the nondisjunction occur?
11. In *Drosophila*, ey^D and ci^D are dominant visible and recessive lethal mutations. A male *Drosophila* has been isolated that is trisomic for chromosome 4 and has the following fourth chromosome genotype: $+/ey^D/ci^D$. If this trisomic-4 male were crossed with an ey^D/ci^D diploid female fly, what genotypic and phenotypic ratios would you expect in the progeny? Assume that recombination does not occur between fourth chromosomes in *Drosophila* and that *Drosophila* sperm with an extra fourth chromosome function as well as normal sperm.
12. A series of crosses were made between *Datura* homozygous for a small flower mutation (*s*) and several trisomic lines. The trisomic F_1 progeny were backcrossed, and the phenotypes of the F_2 follow. On which chromosome is the flower mutation?

	F_2 Phenotype	
Trisomic Line	**Small**	**Normal**
2N + A	15	14
2N + B	12	13
2N + C	4	24
2N + H	8	9
2N + J	10	13
2N + K	14	13
2N + L	11	16

Aneuploidy for Sex Chromosomes in Animals Produces Abnormal Sexual Development.

Animals are generally intolerant of aneuploidy, and individuals trisomic or monosomic for the larger autosomes either do not survive or show major developmental abnormalities. The effects of aneuploidy are generally correlated with the size of the chromosome, with the smaller chromosomes generally giving a less severe aneuploid phenotype. Exceptions to this general rule are the sex chromosomes. Aneuploidy for sex chromosomes results in abnormal sexual development, but the overall effect is less severe than aneuploidy for autosomes of similar size.

The effects of sex-chromosome aneuploidy were first investigated by Calvin Bridges [see HISTORICAL PROFILE: *Calvin B. Bridges (1889–1938)*] in his studies on the chromosomal basis of inheritance. As discussed in Chapter 3, Bridges demonstrated that aneuploid *Drosophila* with different sex chromosome compositions survived but showed a variety of different phenotypic effects. For example, XXY individuals were fertile females that had no phenotypic abnormalities associated with having a Y chromosome. The extra Y chromosome was detected only because of abnormal gametic ratios. XXX individuals were barely viable and showed abnormalities of sexual development, and they were usually sterile. Individuals that were monosomic for the X were males and indistinguishable from normal males, except that they were sterile as a result of a failure of sperm maturation. There are two reasons why sex chromosome aneuploidy is more tolerated than autosomal aneuploidy in *Drosophila*. First, the Y chromosome carries only a few genes, mostly concerned with male fertility, so extra copies of the Y do not cause much genic imbalance. Second, in *Drosophila*, as in other heterogametic species, dosage compensation mechanisms have evolved because one sex normally has twice as many copies of the genes on the X chromosome (see Chapters 3 and 16).

Aneuploidy in humans for the X chromosome produces a variety of abnormalities of sexual development (Table 18.2). The characteristics peculiar to individuals monosomic for the X chromosome (45,X) were first described in 1938 as **Turner syndrome.** Turner syndrome occurs in about 1 in every 2500 live births, although the frequency of X chromosome nondisjunction giving monosomy for the X chromosome is about ten times this value. There are fewer live births with Turner syndrome than the actual frequency with which X chromosome nondisjunction occurs because about 90% of fetuses with Turner syndrome are spontaneously aborted. Turner syndrome individuals are females (because they lack a Y chromosome) with a range of abnormalities that often includes short stature, very limited sexual development, sterility, a wide chest, a webbed neck, and reduced mental capacity. Individuals who have lost only the short arm of the X chromosome can have all the symptoms of Turner syndrome, suggesting that the genes responsible for genic imbalance are on the short arm. Individuals missing the long arm show few symptoms.

Individuals who have extra copies of the X chromosome (47,XXX or 48,XXXX) survive but have a series of

TABLE 18.2 *Human Sex Chromosome Aneuploids*

Name of Syndrome	Sex Chromosomes	Frequency
Turner	X	1/2500 female births
Normal female	XX	
Triplo-X	XXX	1/700
Triplo-X	XXXX	
Triplo-X	XXXXX	
Nullo-X (lethal)	Y	
Normal male	XY	
Male	XYY	1/1000 male births
Klinefelter	XXY	1/500 male births
Klinefelter	XXXY	
Klinefelter	XXXXY	

abnormalities known as the triplo-X syndrome. Some triplo-X individuals have no visible abnormalities, whereas others have abnormalities including mental retardation and failure to develop sexual characteristics during puberty. Individuals with four X chromosomes show more severe symptoms than trisomic X individuals. The key to identifying these individuals is cytogenetic analysis because dosage compensation in humans results in inactivation of all but one of the X chromosomes. The inactivated chromosomes become heterochromatic and can be detected in somatic cells as Barr bodies (see Chapters 3 and 16). The number of X chromosomes can be determined by counting the number of Barr bodies in somatic cells (X chromosomes = Barr bodies + 1). For example, individuals with three X chromosomes have two Barr bodies. The fact that individuals with two or more Barr bodies show an abnormal phenotype indicates that the X inactivation process is not perfect; either inactivation occurs after some genes on the extra X chromosomes have begun to function or some of the genes on the inactivated X are still active.

Humans who have a (47,XXY) karyotype have a phenotype known as **Klinefelter syndrome,** which was first described in 1942. These individuals are phenotypically male but develop a number of female sexual characteristics. In addition, they often show degrees of mental retardation. Individuals with a single X chromosome and extra copies of the Y chromosome (47,XYY) on average are taller and more gangly than (46,XY) individuals. An initial study of XYY individuals in a Scottish prison in 1965 showed a much higher than expected frequency of XYY individuals (9 in 315, or 1 in 35 compared to a normal frequency of 1 in 500), suggesting that this chromosome makeup may produce increased aggressive and antisocial behavior. This finding was highly controversial, and a number of other studies have tried to substantiate these findings. Investigators studying the effects of XYY have encountered tremendous social pressure not to conduct this type of study. At present, there is conclusive evidence that XYY individuals show certain physical traits (tall and gangly), but recent studies suggest that they do not have a genetic predisposition to abnormal behavior.

SECTION REVIEW PROBLEMS

13. A child with Klinefelter syndrome who is also red/green color-blind was born to a normal man and a normal woman.
    a. In which parent did the nondisjunction occur? Explain your answer.
    b. Did the nondisjunction occur in meiosis I or meiosis II?

Aneuploidy for Autosomes Is Usually Lethal in Animal Species.

In *Drosophila,* aneuploidy for either chromosome 2 or 3, the largest autosomes, is invariably lethal. However, aneuploidy for the small fourth chromosome is not. In fact, individuals with an extra copy of the fourth (triplo-4) chromosome have no visible abnormalities, although individuals monosomic for the fourth (mono-4) chromosome show several abnormal symptoms, including slow growth and small bristles. The mild effect of aneuploidy for the fourth chromosome is attributed to the small number of genes on this chromosome. Genetic and molecular studies on the fourth chromosome have shown that it contains only about 50 genes. On the other hand, the second and third chromosomes contain about 2000 genes each.

In humans, aneuploidy for any of the larger autosomes is usually fatal during embryonic development, leading to a spontaneous abortion. Analyses of spontaneous abortions have indicated that well over half are the result of chromosomal abnormalities in the fetus. However, humans who are aneuploid for some of the smallest autosomes sometimes survive. The best known of these are individuals trisomic for chromosome 21, or **trisomy 21.** These individuals suffer from a variety of phenotypic abnormalities called **Down syndrome,** named after Langston Down, who first described the phenotypes associated with the syndrome in 1866. (Down syndrome was previously referred to as Down's syndrome, but the possessive is no longer used.) The chromosomal cause of Down syndrome was uncovered in 1958 and was the first recognized human condition associated with aneuploidy. Most trisomic 21 fetuses do not survive until birth. Those who do survive have characteristic phenotypes that usually include broad skulls, short stature, epicauthal folds (which caused this syndrome to be called

HISTORICAL PROFILE

Calvin B. Bridges (1889–1938)

Calvin B. Bridges (Figure 18.B) was born on January 11, 1889, in Schuyler's Falls, NY. Tragedy struck Bridges early in his life as both his father, a farmer near Plattsburg, NY, and his mother died before he was four. From this early age he was raised by his grandmother, and because they moved from place to place, he had a scattered early education and did not finish high school until he was 20 years old. He was an excellent student, however, and won a scholarship that allowed him to attend Columbia University, although the scholarship did not provide much for living expenses. Bridges was determined to have a career as a research scientist. In 1909, as a freshman, he enrolled in introductory biology, which fortunately was taught that year (and only that year) by T. H. Morgan. Bridges and another student in the class (A. H. Sturtevant) were excited by Morgan's experimental approach to biological science, and both requested the opportunity to work in Morgan's lab. Morgan knew of Bridges' need for a part-time job but was not convinced that such an inexperienced scientific novice was what he needed. However, Bridges was persistent, returning often to visit Morgan's lab, and one day he spotted a new mutation with light red eyes in a bottle set aside for discarding. Impressed with his keen observational skills, Morgan took him on as a part-time laboratory assistant and paid his salary from his own private funds. Bridges quickly proved his worth and became one of the key members of the lab. Bridges received his bachelor's degree in 1912 and his Ph.D. in 1916 under Morgan's tutelage.

Bridges was a superb experimentalist who had remarkably keen eyesight. He was without a doubt the best in the Morgan lab at spotting new mutations. He also had outstanding technical skills and was an innovator who introduced a number of new techniques into the analysis of *Drosophila*. One such technique was the use of ether to anesthetize flies because it is necessary to immobilize them when searching for alterations in phenotype. He had a precise mind, enormous powers of concentration, and delighted in the rigorous execution of long, detailed experiments. Bridges was the first to discover deficiencies, inversions, and transloca-

FIGURE 18.B Calvin B. Bridges. Calvin B. Bridges was one of the pioneers of modern genetics. *(UPI/Corbis-Bettmann)*

mongolism originally), and stubby hands and feet with characteristic skin creases. Down syndrome individuals have lower intelligence but are capable of learning many skills. They have been shown to be careful workers when properly trained. In spite of their handicaps, they are often happy, affectionate people who have the ability to enjoy life.

Most cases of trisomy 21 are the result of nondisjunction during meiosis. The frequency of trisomy 21 is about 1 in 700 births in populations of European ancestry for women aged 25 years or under, but the frequency increases dramatically with age (Figure 18.10) to about 1 in 100 for women of age 40 and 1 in 10 for women of age 45. Nondisjunction occurs in both males and females,

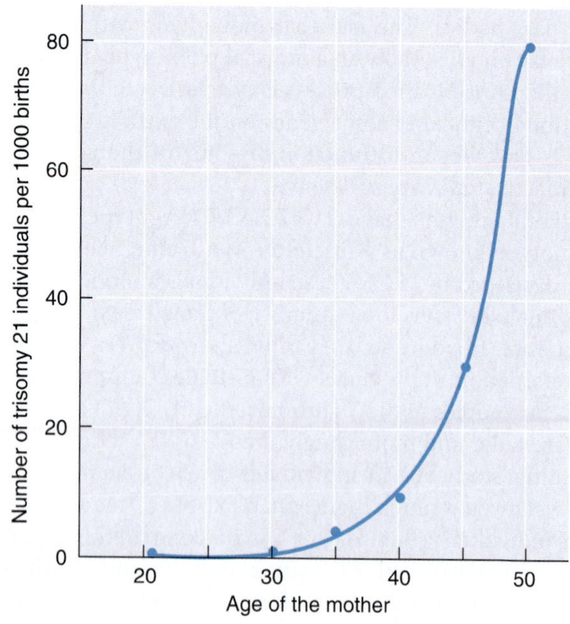

▶ FIGURE 18.10 Frequency of trisomy 21 in humans. The frequency of trisomy 21 in humans increases with increasing age of the mother.

tions in *Drosophila*. He performed numerous experiments to determine the nature of these aberrations, including cytological examination of meiotic chromosomes and genetic crosses. Bridges' remarkable work on nondisjunction was considered a key proof of the chromosome theory, and his careful experiments rapidly gained the respect of senior geneticists. In 1919, William Bateson, one of the founders of the modern discipline of genetics and a very important person in the scientific establishment, visited the Morgan lab for several days. Bateson was not convinced that the chromosome theory was correct, and his letters reveal that he did not have a high opinion of Morgan. During his visit he spent long hours with Sturtevant and Bridges discussing their actual experiments, data, and conclusions. This eventually resulted in Bateson changing his mind and accepting the chromosome theory as fact. "Bridges inspires me with complete confidence," he reported to a friend in a letter. In 1929, when Morgan moved to California to establish the Division of Biology at California Institute of Technology, he took Sturtevant and Bridges with him, and they continued their collaboration.

In 1933, the giant polytene chromosomes in the *Drosophila* salivary glands were discovered. Bridges realized their immense value in the analysis of his large collection of chromosomal aberrations and began working with them. There was a great need for a standard map of the bands seen in the polytene chromosome, and Bridges undertook the monumental task of producing an accurate drawing of all the more than 5000 bands in the chromosome set. His first effort was published in 1935, and it remains the standard used in every *Drosophila* lab today. A revised version was begun in 1938, but Bridges never completed it. At this time, he was also carrying out an extensive series of experiments, making numerous detailed crosses and maintaining over 900 strains in the lab. His plans were never completed because he contracted a bacterial infection of the heart valves and died of heart failure on December 27, 1938, at the age of 49.

Much has been said about Bridges' personality. He was lighthearted and sociable, always the center of the social scene, and held a number of nonconformist views. He had striking good looks and a passion for women, and he continually amazed the more conservative Morgan and Sturtevant with his numerous romantic escapades. Bridges was actively interested in politics and held extreme left-wing views. He supported both the 1917 Communist Revolution in Russia and the Communists in the 1936 Spanish Civil War. It is one of the more remarkable achievements of the Morgan group that Bridges and Sturtevant, with such opposite views on life, worked together in harmony for many years. When Morgan received the Nobel Prize in 1933, he divided the prize money among his own, Sturtevant's, and Bridges's children in recognition of the contributions these two outstanding scientists had made.

but studies have shown that the extra chromosome in the trisomic individuals almost always comes from the female parent. This is thought to be a result of the fact that human females are born with all of their oocytes already formed in their ovaries. These thousands of oocytes remain in a state of suspended development until sexual maturity, when one oocyte matures per month. The nuclei of the oocytes are arrested partway through meiosis (see Chapter 16), and it is believed that as females age, the maintenance process may fail, leading to higher frequencies of nondisjunction. In contrast, male spermatogenesis occurs continuously, so that male meiotic products do not have long periods of arrested development.

In addition to nondisjunction, portions of chromosome 21 may sometimes become translocated to another chromosome (translocations are discussed later in this chapter), and the segregation of such chromosomes may produce individuals with extra copies of portions of chromosome 21, which this can lead to Down syndrome phenotypes. In the United States, over 100,000 individuals have Down syndrome, and approximately $1 billion per year are spent on their training and care.

SECTION REVIEW PROBLEMS

14. Occasionally, humans trisomic for chromosomes 13, 18, or 21 survive to birth, but no individuals trisomic for chromosomes 20 or 22 have been recorded. Chromosome 20 and 22 are no larger than 13 or 18. Why do you think there are viable trisomies for 13 and 18, but not for 20 and 22? Suggest a way to test your hypothesis.
15. Thanks to modern medical care, individuals with Down syndrome are living longer. In theory, if two such individuals were to marry, what would be the genotypes of their offspring, and what would be the frequency of each?

CHROMOSOME REARRANGEMENTS ARE CHANGES IN CHROMOSOME STRUCTURE.

Chromosome rearrangements, mutations that change the structure of chromosomes, are the second major category of chromosomal abnormalities. These rearrangements usually result from chromosomes that break into fragments that subsequently rejoin improperly. We will consider four types of chromosomal rearrangements in this section: lost portions of chromosomes (deficiencies), added chromosomal material (duplications), changes in the order of genes on a chromosome (inversions), and exchanges between chromosomes that are not homologous (translocations). In addition to their value in studies of chromosome and gene behavior, chromosomal abnormalities have immense importance as tools in genetic research, in medicine, and in agriculture.

Chromosomal Breakage Can Produce Deficiencies and Duplications.

Mutations that remove a portion of a chromosome are called **deficiencies,** or deletions. We discussed some of the genetic uses of deficiencies in complementation testing and gene mapping in Chapter 6. Deficiencies may be as small as a single nucleotide pair, or they may be almost as large as an entire chromosome. Deficiencies large enough to be detected as chromosomal aberrations will usually include more than one gene. Deficiencies are produced by mutagens, especially ionizing radiation such as X-rays, that break chromosomes. When a chromosome is broken, the ends are "sticky"; that is, they tend to rejoin but they do not have a special affinity for their original counterpart and will rejoin with any other broken chromosome end.

Very rarely, deficiencies occur through loss of the telomere at the tip of the chromosome. These **terminal deficiencies** are rare because the telomere is needed for proper chromosome replication (described in Chapter 10) and possesses bound proteins and looped nucleic acids that prevent binding to other chromosomes (Figure 18.11). Terminal deficiencies were first observed by Barbara McClintock in a number of studies of chromosome breakage in corn. She observed that when broken chromosomes replicate, the ends of newly replicated sister chromatids often join. This generates a chromosome with two centromeres, called a dicentric chromosome (Figure 18.12). When the two centromeres attempt to segregate during the next cell division, the chromosomes break, resulting in two chromosomes that each have a broken end. During the next chromosome replication, the broken ends of these chromosomes join, and the cycle of chromosome breakage and fusion repeats itself. With each round of replication, the broken end is partially degraded, and eventually the chromosome is completely lost. This process is called the breakage–fusion–bridge cycle.

One common way that deficiencies are produced occurs when a chromosome is broken into three segments and the two terminal fragments join. Such a deficiency is called an **interstitial,** or intercalary, deficiency. The first interstitial deficiency was discovered by Calvin Bridges in *Drosophila* in 1917. Bridges was studying a mutation on the X chromosome that had a recessive lethal phenotype, and he noticed that this mutation had some unusual properties in heterozygous females. It failed to complement more than one recessive mutation in the *forked* to *Bar* region of the X chromosome; furthermore, heterozygous females did not produce any progeny with chromosomes recombinant in this interval. Bridges recognized this mutation as a deficiency, and it is now standard practice to identify deficiencies by failure to complement recessive mutations along with the failure to produce progeny with chromosomes recombinant in the

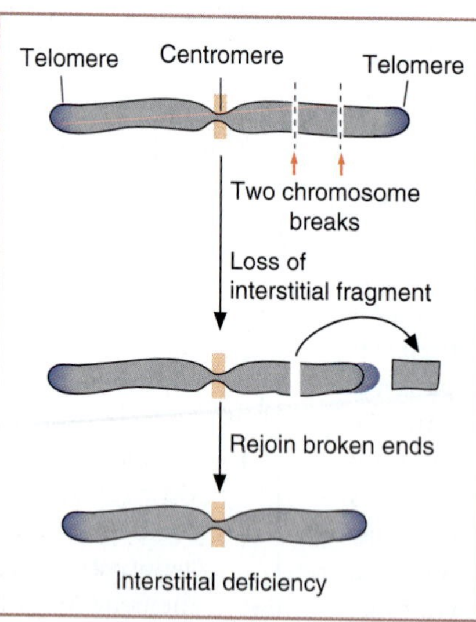

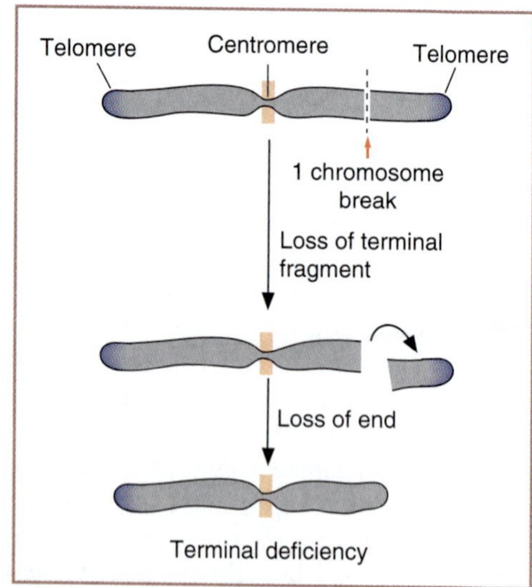

FIGURE 18.11 Interstitial and terminal deficiencies. If a chromosome is broken in two places, the two ends may rejoin, causing the loss of the intervening fragment. This generates an interstitial deficiency. If a chromosome is broken in one place, the terminal portion may be lost, generating a terminal deficiency. Terminal deficiencies are rare because a chromosome without a telomere is not stable.

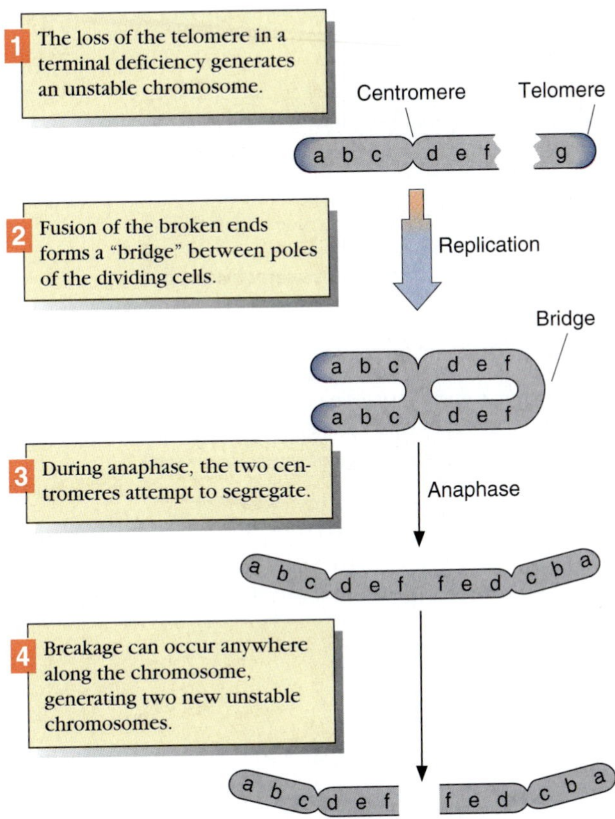

FIGURE 18.12 The breakage–fusion–bridge cycle. The breakage–fusion–bridge cycle starts with a terminal deficiency that eliminates the telomere. After replication, fusion of the broken ends of the chromatids generates a dicentric chromosome that forms a "bridge" between the poles of the dividing cell. The dicentric chromosome breaks and the cycle continues. Breakage can occur anywhere along the chromosome. Eventually the broken ends are degraded, and the chromosome is lost.

region of the deficiency. This method is especially useful in species in which cytogenetic analysis of chromosomes is not possible. When a deficiency is large enough to be recognized as a chromosomal aberration (using cytogenetic techniques), it is often lethal when homozygous because the organism is not able to tolerate the complete loss of so many genes. When a deficiency is heterozygous, it may show a dominant visible mutant phenotype because a gene is lost whose normal function requires two functional copies of the gene. Such a gene is called a **haplo-insufficient** gene. Well-known examples of haplo-insufficient genes in *Drosophila* are at the *Minute* loci; there are over 30 *Minute* loci. Mutations or deficiencies of a *Minute* locus are homozygous, lethal, and produce a dominant phenotype manifesting a slow development with small bristles. An example of a human phenotypic abnormality caused by a haplo-insufficient gene deficiency is the "cri du chat," or cat's cry, syndrome, caused by the deficiency of a portion of the fifth chromosome. Individuals with this syndrome are severely retarded and have multiple developmental abnormalities. The syndrome is named for the sound the infants make, which resembles the crying of a cat.

Deficiencies may also produce a mutant phenotype because of **breakpoint effects.** For example, a chromosome break may occur within a gene, eliminating part of the gene and producing an altered gene product. Also, when a chromosome loses a segment, subsequent joining of the two ends may bring close together two regions of the chromosome and alter gene expression. Such a change is called a **position effect.** Position effects are known to occur when chromosome aberrations shift a gene to a new location, particularly when genes are moved adjacent to a region of heterochromatin. Position effects may give a mutant phenotype without altering the base sequence of the gene itself. A third type of breakpoint effect occurs when a deficiency is generated with breakpoints in two different genes. When the two ends are joined, a hybrid gene is generated with the 5′ region of one gene attached to the 3′ region of another. The hybrid gene may have novel properties that produce a mutant phenotype.

Mutations that produce an extra copy of a portion of a chromosome are called **duplications.** Like deficiencies, duplications may be as small as a single base pair or as large as most of a chromosome. Duplications are classified according to the location and arrangement of the extra material in the chromosomes (Figure 18.13). When the duplicated segment of the chromosome is in the same sequence and adjacent to the original copy, it is referred to as a **tandem duplication.** However, if the extra chromosomal material is reversed in orientation, it is called a **reverse tandem duplication.** A **displaced duplication** is an extra copy of genetic material in another location on the chromosome. Finally, a duplication may also exist as a small, free chromosome, but it must possess a centromere to be retained during cell division.

As with deficiencies, Calvin Bridges first reported duplications in *Drosophila* (1919). Bridges discovered a duplication of a portion of the X chromosome that contained a wild-type copy of the *vermilion* gene. During crosses, this duplication behaved as though it were a suppressor of the recessive *vermilion* mutation. We now

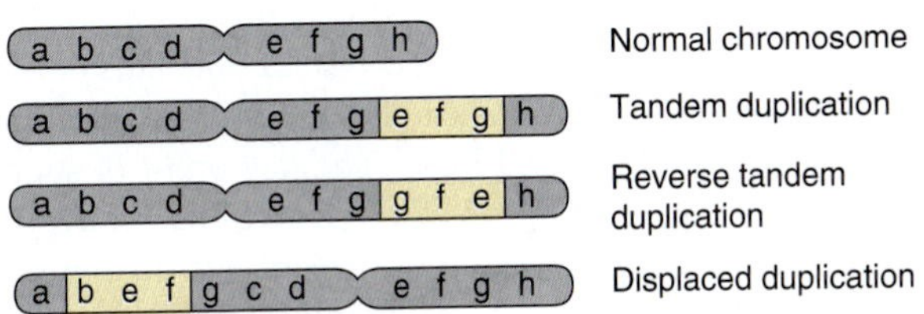

FIGURE 18.13 Tandem, reversed, and displaced duplications. The extra material in a duplication can be inserted in different orders and locations in a chromosome.

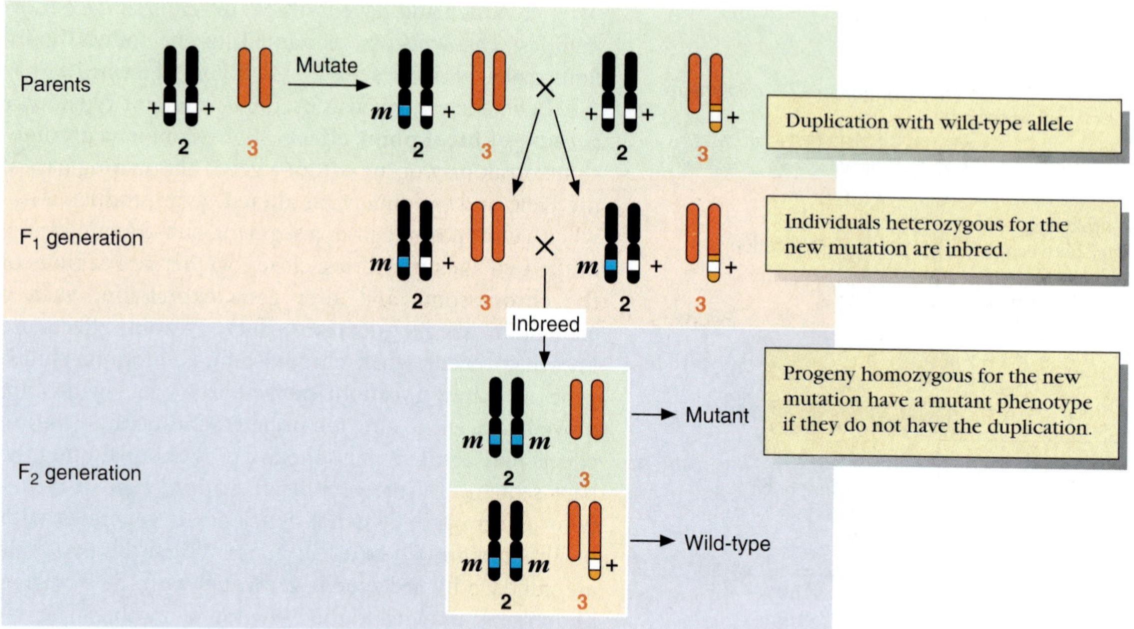

FIGURE 18.14 Mutation selection with duplications. A duplication may be used to select mutations in a particular chromosomal region. Recessive mutations that are complemented by wild-type alleles in the duplication are located in the duplicated region. This type of duplication complementation is especially useful when dealing with lethal mutations because individuals that have the mutation and the duplication survive and can be bred.

understand that this ability to suppress or complement recessive mutations is one of the key genetic characteristics of duplications. Unlike a suppressor of a single gene, a duplication contains extra copies of all genes in the duplicated region of the chromosome and will complement recessive mutant alleles at any locus in the region. This ability to suppress the phenotype of a mutation can be used in the analysis of new mutations. For example, it can establish whether a new mutation is an allele of a known locus. If a new recessive mutation is complemented by a known duplication, the mutation must be an allele of a gene included in the duplicated chromosomal region. The procedures used for such complementation tests are similar to those described previously (Chapter 6) for deficiency complementation tests. Unlike deficiency complementation tests, individuals with a duplication that complements a recessive lethal or visible mutation will survive and be fertile. Thus they can be bred to establish strains of the new mutation. This characteristic of duplications makes them useful in mutation isolation screens (Figure 18.14). The key point of such screens is to isolate mutations that give a mutant phenotype when the duplication is not present but that give a wild-type phenotype when the duplication is present. This indicates that the mutation is in a gene located in the duplicated region.

SECTION REVIEW PROBLEMS

16. Several male *Drosophila* heterozygous for multiple recessive mutations were crossed to females heterozygous (*Df*/+) for six different deficiencies. The phenotypes of the progeny from each cross follow. Assume that these mutations are located in the chromosome in the following order: *th st cp in ri* p^P *ss bxd k* e^S. Determine the location of each deficiency.

Deficiency	Offspring	Phenotypes
1	+	in ri p^P
2	+	th st cp
3	+	cp in ri p^P
4	+	st cp in ri p^P
5	+	cp in ri p^P ss
6	+	p^P ss bxd

Inversions Change the Order of Genes Within Chromosomes.

Inversions are changes in the order of genes in a chromosome. They result from a chromosome being broken at two locations, generating three fragments, and the in-

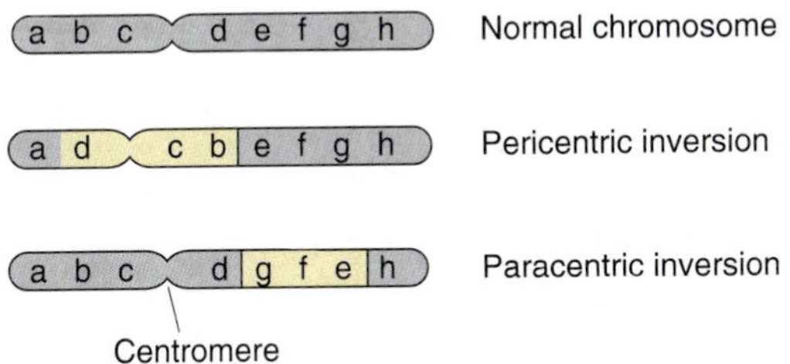

FIGURE 18.15 **Pericentric and paracentric inversions.** A pericentric inversion includes the centromere in the inverted region; a paracentic inversion does not include the centromere.

ternal fragment being reinserted in reverse order. Inversions are divided into two general classes, depending upon whether the centromere is included in the inverted region. If the centromere is included in the inverted region, an inversion is called a **pericentric inversion.** If the centromere is not included in the inverted region, an inversion is called a **paracentric inversion** (Figure 18.15). Both types of inversions change the order of genes in a chromosome. Because the genes in the inversion region are not lost, inversions usually affect only those genes at or very near the breakpoints. Inversion breakpoints may cause position effects.

Inversions can produce two types of position effects by changing the location of a gene in the chromosome: variegated position effects and stable position effects. When an inversion produces a **variegated position effect,** some genes on the inverted segment are inactivated in some somatic cells but not in others, producing a individual with a mosaic or variegated phenotype (Figure 18.16). Variegated position effects often result when a gene is brought into close proximity to a heterochromatic region of the chromosome. Inversions with one breakpoint in heterochromatin and another in euchromatin, for example, often produce variegated position effects. The molecular mechanism of variegated position

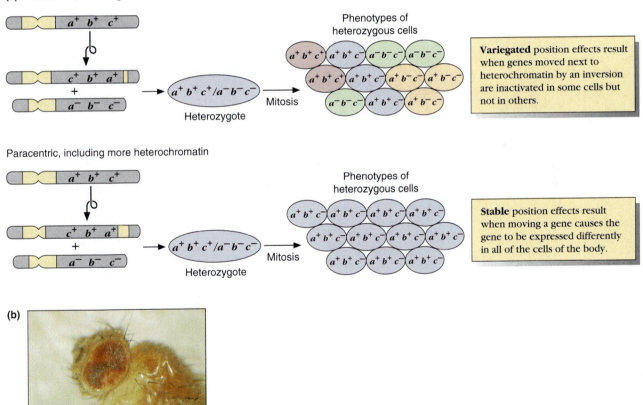

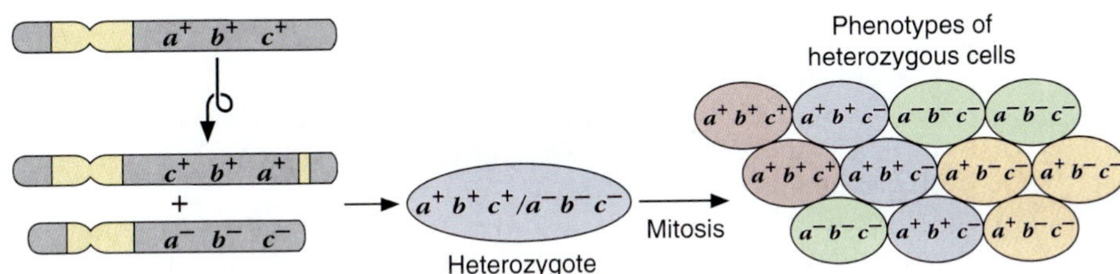

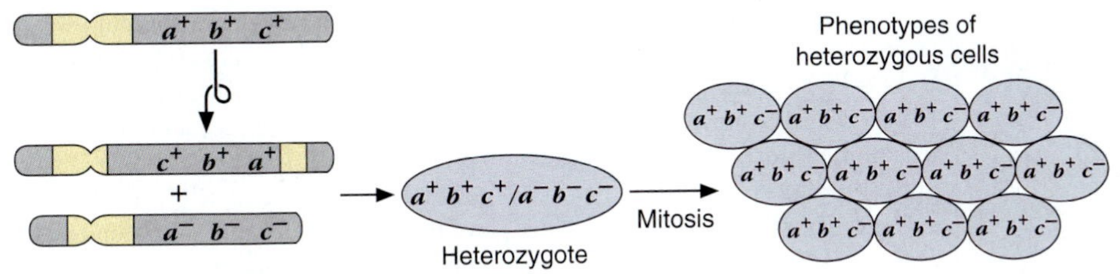

(b)

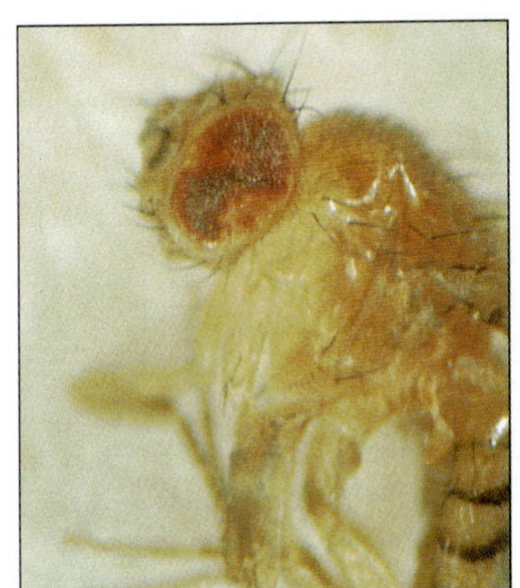

FIGURE 18.16 **Position effects.** (a) Variegated position effects or stable position effects can result when genes are moved next to heterochromatin by an inversion. (b) In a *Drosophila* that is variegated for an allele of the *white* gene, some facets are pigmented and others are not. *(b, Jack R. Girton, Iowa State University)*

CONCEPT CLOSE-UP

Bar Eye, a Stable Position Effect

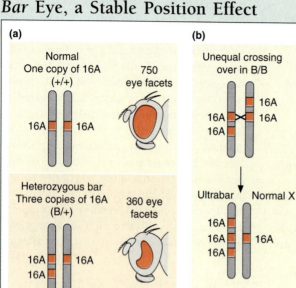

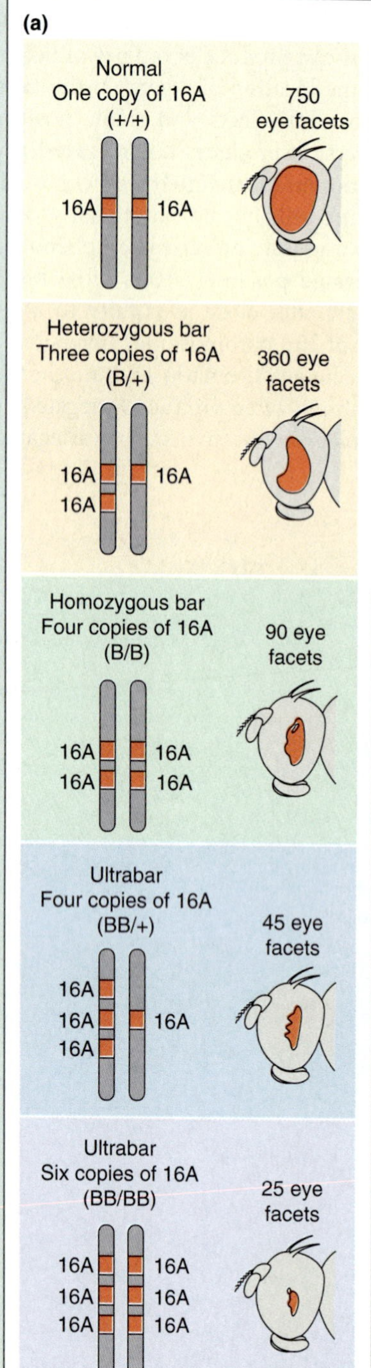

FIGURE 18.C ***Bar* eye: A stable position effect.** Changing a gene's location can affect its action. Such a change is called a position effect. (a) A duplication of the 16A region of the X chromosome in *Drosophila* causes the distinctive *Bar* eye mutation. Increasing the number of copies of the 16A region causes a more extreme *Bar* eye. However, the severity of the phenotype is affected by the arrangement of the duplicated region. An individual with three copies on one chromosome and one on the other (3 + 1 = 4) has a more extreme phenotype than an individual with two copies on each chromosome (2 + 2 = 4), even though they both have the same total number of copies. (b) Unequal crossing over between chromosomes with two copies of the 16A region can generate a chromosome with three copies and a chromosome with one.

Position effects have been intensively studied as a key to understanding the relationship between chromosome structure and gene action. One of the first examples of a stable position effect was discovered in studies of the *Bar* mutation in *Drosophila*. *Bar* (*B*) is a partially dominant X-linked mutation that reduces the number of facets present in the compound eye. Wild-type individuals have about 750 facets in each eye, but individuals homozygous for *Bar* (*B/B*) have only about 90 facets, and individuals heterozygous for *Bar* (*B/+*) have about 360 facets. Interestingly, wild-type reversions arise in homozygous *Bar* individuals with a frequency of about 1 in 1600 individuals. This value is much is too high to be a reverse mutation, which occurs at a frequency of less than 1 in 10,000. A new, more severe allele (called *Ultrabar* or *Doublebar*) also arises in homozygous *Bar* individuals with a frequency of about 1 in 1600 individuals (which is too high to be a new forward mutation). Individuals heterozygous for *Ultrabar* have 45 facets, and individuals homozygous for *Ultrabar* have only 25 facets. Both the reversions and the new mutations always arise in association with a recombination event in the *Bar* region of the X chromosome.

Cytological analysis indicates that the original *Bar* mutation was a small duplication of about seven salivary gland chromosome bands in the 16A region of the X chromosome. *Bar* chromosomes contain two copies of this region in a tandem duplication, *Ultrabar* chromosomes contain three copies of the region, and wild-type revertants contain only one copy. Reversions of *Bar* to wild-type and mutation to *Ultrabar* result from unequal crossover (see Figure 18.C) between two *Bar* chromosomes that produces one normal chromosome (the reversion) and one chromosome with three copies of the region (the *Ultrabar*). An important point about *Bar* is that individuals heterozygous for *Ultrabar* and a normal chromosome, with a total of four copies of the 16A region, have a smaller eye (45 facets) than individuals homozygous for *Bar*, who also have a total of four copies of the 16A region. The arrangement, or position, of the copies in the chromosome affects the individual's phenotype.

effects is currently not known. Many geneticists believe it results from gene inactivation caused by the presence of special types of chromatin structure in heterochromatin. Support for this viewpoint comes from the fact that genes in the mammalian X chromosome that becomes the heterochromatic Barr body are inactive. **Stable position effects** are new, constant patterns of gene expression associated with a gene's new location, and they behave much like a permanent mutation of the gene. They can be reversed by reinverting the chromosome segment and moving the gene back to its original position (see CONCEPT CLOSE-UP: *Bar Eye, a Stable Position Effect*, and Figure 18.16). One important conclusion can be made about position effects in general: the location of a gene in the chromosome sometimes has an important effect on its expression.

Inversions have important effects on meiotic recombination. Calvin Bridges isolated a series of mutations that appeared to be dominant suppressors of meiotic recombination. When individuals heterozygous for such a "crossover suppressor" mutation were testcrossed to individuals with several recessive mutations, the frequency of recombinant offspring was greatly reduced, giving decreased linkage map distances (Figure 18.17). This phenomenon was called map shrinkage. Each crossover suppressor mutation affected only the genes within a particular chromosomal region. Unlike deficiencies, crossover suppressors complement recessive mutations in the region and often are not recessive lethal in homozygotes. Using cytogenetic analyses, crossover suppressor mutations were shown to be inversions, and the mechanism of their action was related to the nature of chromosome pairing in the first meiotic division.

To explain how inversions cause crossover suppression, we must reexamine chromosome pairing during meiosis. In prophase of the first meiotic division, homologous chromosomes pair along their entire length. In a cell in which one chromosome has an inversion and the other is normal, the inverted and normal chromosomes must form an **inversion loop** to pair properly (Figure 18.18). Such loops can be seen in cytological preparations of paired chromosomes in meiosis, and they are especially visible in the salivary gland chromosomes of *Drosophila*. How does the inversion loop cause crossover suppression? When a crossover occurs between the two chromosomes in an inversion loop, it may produce chromosomes with new abnormalities. And, the type of abnormality produced depends upon the type of inversion. If a single crossover occurs between the chromosomes within the inverted region of a paracentric inversion,

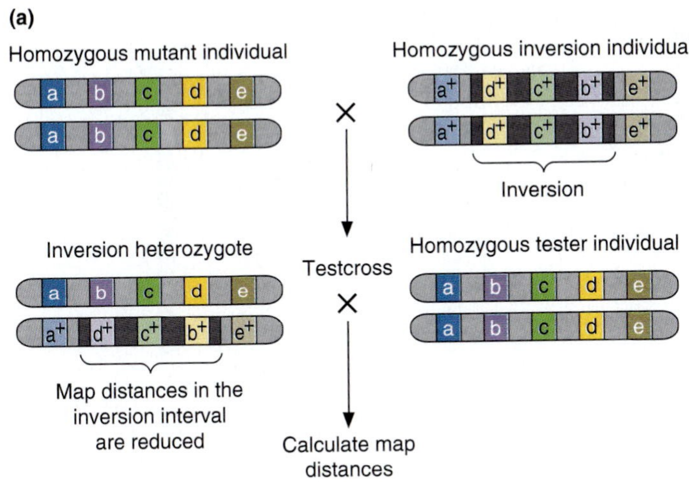

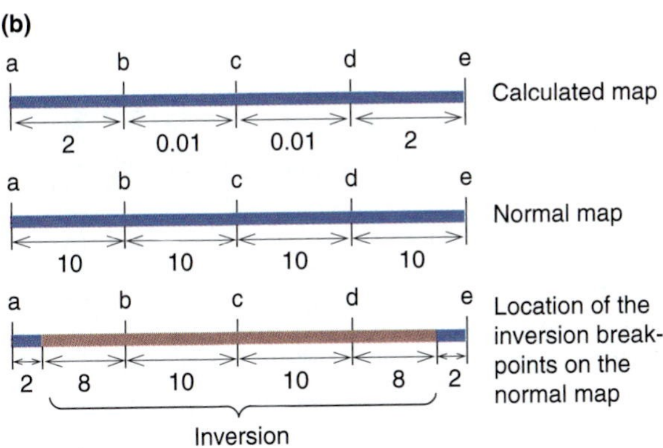

FIGURE 18.17 Crossover suppression and map shrinkage. (a) Crossing over in an inversion produces inviable gametes, which reduces the number of recombinant progeny from a cross. This effect was originally called crossover suppression. The loss of recombinant progeny reduces map distances in the inversion region, an effect called map shrinkage. This can be used to detect the presence of inversions. (b) Map shrinkage begins and ends at the breakpoints of the inversion.

The position of the inversion breakpoints can be determined by comparing the map distances calculated from the progeny of the inversion heterozygote with normal map distances.

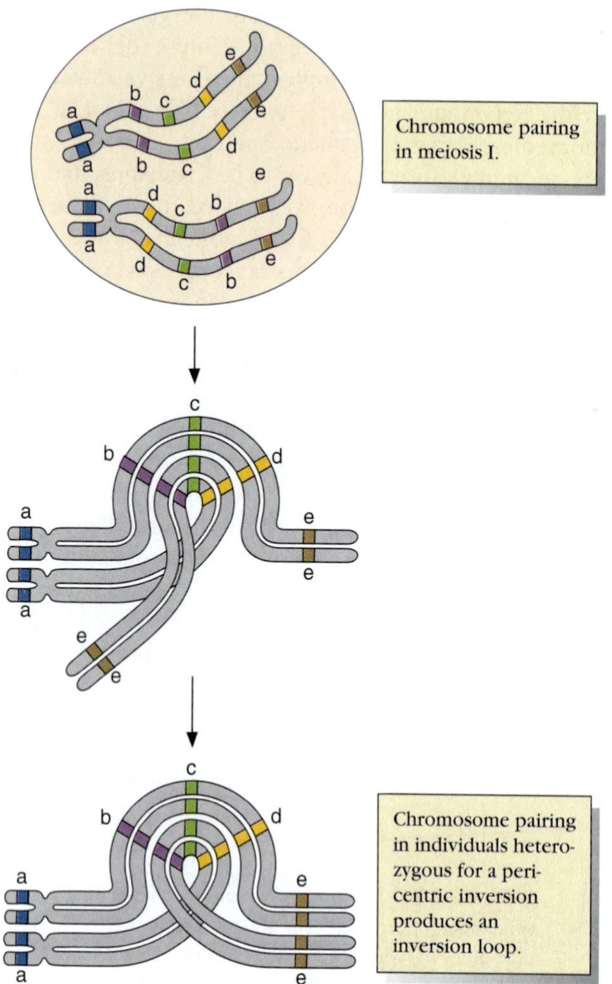

◄**FIGURE 18.18 Formation of inversion loops.** Chromosome pairing in individuals heterozygous for a paracentric inversion produces an inversion loop during meiosis.

segregation will produce four distinct types of chromosomes (Figure 18.19). The two chromatids not involved in the crossover will produce two parental-type chromosomes, one with normal gene order and the other with the inverted order. The two recombinant chromatids will form one **dicentric** chromosome with two centromeres, and one **acentric** chromosome, with no centromere. Because acentric chromosomes do not have a centromere, they do not attach to spindle fibers and do not segregate during anaphase. Acentric chromosomes are usually not incorporated into one of the progeny nuclei and are degraded in the cytoplasm. Dicentric chromosomes will

►**FIGURE 18.19 Crossing over in a paracentric inversion loop.** A crossover in a paracentric inversion loop produces two unstable recombinant chromosomes: a dicentric chromosome and an acentric chromosome. The dicentric and the acentric chromosomes cannot segregate normally during anaphase of meiosis I and are lost or degraded. The only chromosomes that are packaged into functional gametes are the nonrecombinant normal and inversion chromosomes. This is the reason why individuals heterozygous for a paracentric inversion produce few recombinant progeny and have reduced fertility.

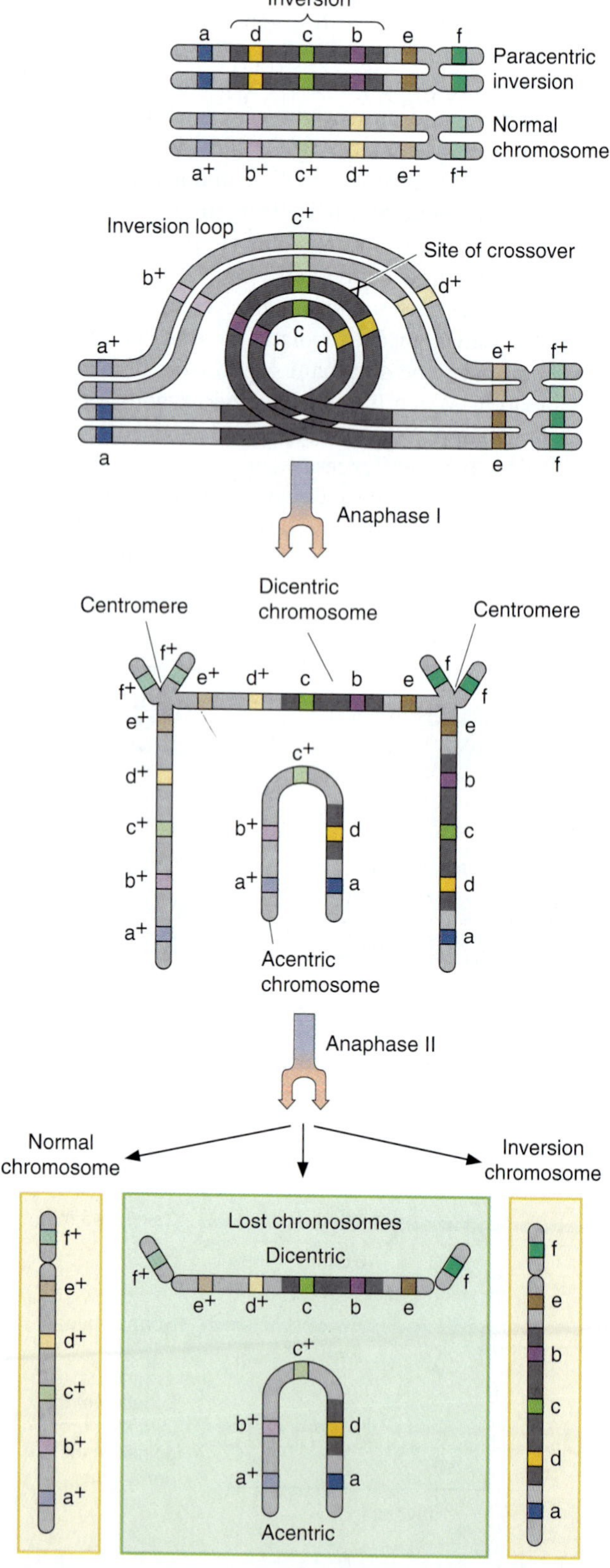

break when the centromeres attempt to segregate. If a broken fragment is incorporated into a progeny nucleus, it will have a broken end that does not have a telomere. This end is unstable, and eventually the chromosome degenerates. Because neither the dicentric nor the acentric chromosomes will segregate normally, no gametes will be formed containing recombinant chromosomes. Because linkage map distances are calculated using the number of recombinant type progeny (see Chapter 6), a reduction in the number of gametes with recombinant chromosomes reduces the map distance within the inverted region.

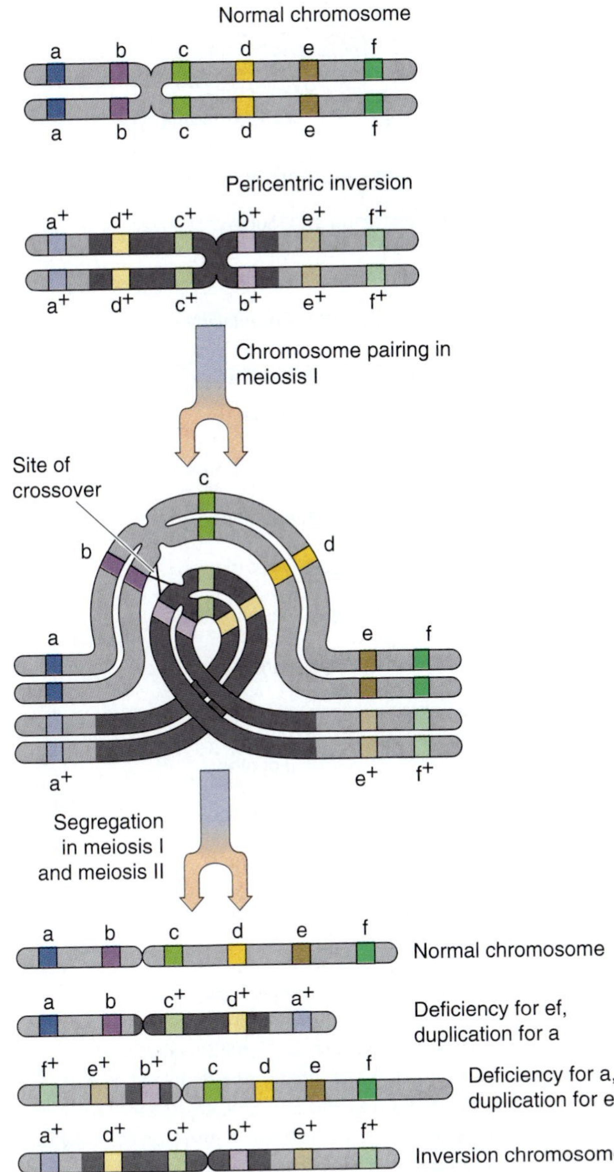

FIGURE 18.20 Crossing over in a pericentric inversion loop. A crossover in a pericentric inversion loop produces chromosomes with duplications and deficiencies for the chromosome regions distal to the inversion breakpoints, but it does not produce dicentric or acentric chromosomes.

In contrast to paracentric inversions, crossover within the inversion loop of individuals heterozygous for pericentric inversions produces chromosomes with a single centromere that segregate normally. However, these chromosomes contain duplications and deficiencies for the portions of the chromosomes distal to the inversion breakpoint (Figure 18.20). If the duplicated and deficient regions are large, the zygotes that receive these chromosomes will not be viable. Because of the reduced number of gametes with viable genotypes, inversion heterozygotes are usually semisterile. The only viable offspring have parental-type chromosomes. Thus, inversion heterozygotes have reduced genetic map distances because all offspring that receive recombinant-type chromosomes die. For both paracentric and pericentric inversions, map shrinkage results from the lack of viable recombinant-type progeny.

An interesting feature of map shrinkage is that a small number of viable recombinant-type progeny are produced by individuals heterozygous for inversions. These progeny are produced by double crossovers. For example, consider the paracentric inversion heterozygote shown in Figure 18.21. A second crossover within the inversion loop between the same two chromatids that were involved in the first crossover will give two structurally normal recombinant chromosomes. The recombinant chromosomes will segregate normally and will give

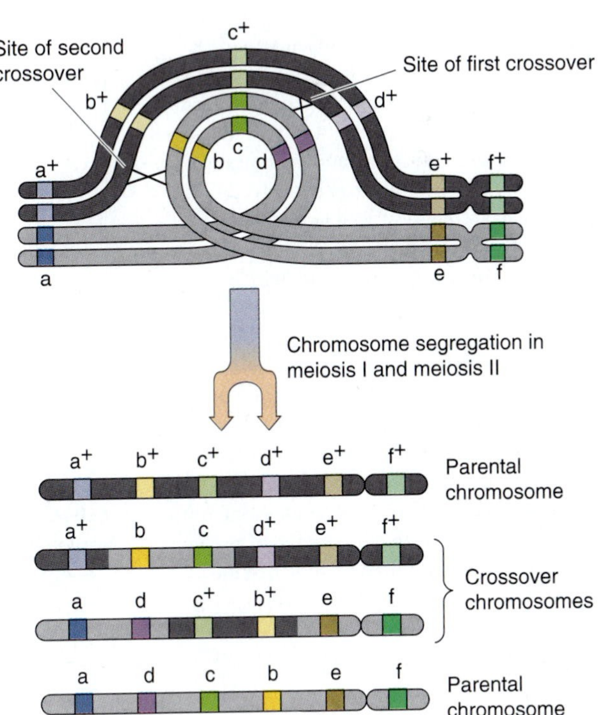

FIGURE 18.21 A double crossover in a paracentric inversion loop. If two crossovers occur between the same two chromatids within an inversion loop, the double-crossover chromosomes have a normal structure and a recombinant genotype.

CONCEPT CLOSE-UP

Double Crossovers in Inversion Loops

Double crossovers that occur between the four chromatids in a bivalent during meiosis I are classified according to which of the four chromatids are involved in the two crossovers. A **two-strand double** occurs when the two chromatids involved in the first crossover are also involved in the second crossover. A **four-strand double** occurs when two chromatids are involved in one crossover, and the other two are involved in the second crossover. A **three-strand double** occurs when one of the two chromatids involved in the second crossover is also involved in the first crossover. There are two different three-strand doubles that can occur, and if double crossovers occur randomly, the three types should occur in a frequency of 1 two-strand double : 2 three-strand doubles : 1 four-strand double. When both crossovers of a double crossover occur within an inversion loop, some viable double-crossover chromosomes may be generated. Two-strand doubles give two viable double-crossover chromosomes, three-strand doubles each give one, and four-strand doubles give none (Figure 18.D). These are the only viable recombinant chromosomes formed by crossovers in an inversion loop, which explains why inversions greatly reduce map distances (see also Chapter 6). Inversions do not prevent crossover events from occurring, but they do cause all single crossover chromosomes and many double-crossover chromosomes to be inviable. This reduces the number of viable recombinant progeny used to calculate map distance.

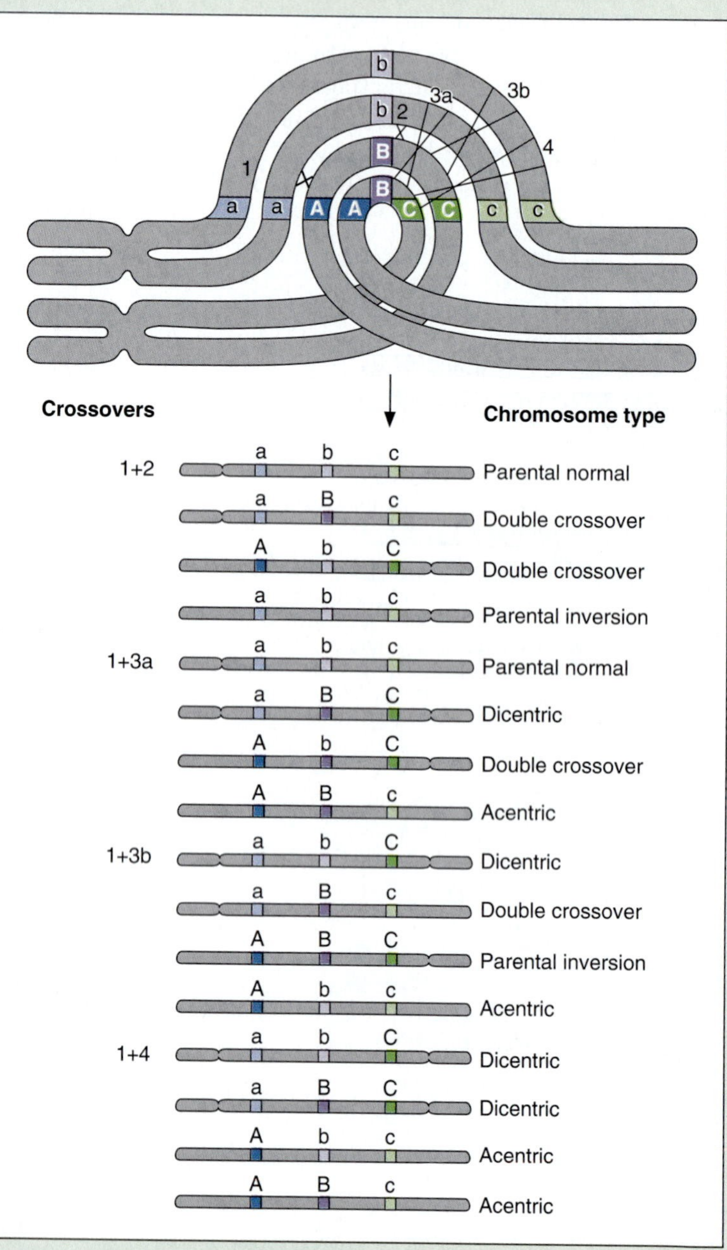

FIGURE 18.D Four types of double crossovers. A single crossover in an inversion loop (1) can be combined with four different second crossovers (2, 3A, 3B, 4) to make a double crossover. A two-strand double (1 and 2) occurs when both crossovers involve the same two chromatids. A three-strand double (1 and 3A or 1 and 3B) occurs when one chromatid is involved in both crossovers. A four-strand double (1 and 4) occurs when neither of the chromatids involved in the first crossover are involved in the second crossover. Two-strand and three-strand double crossovers in an inversion loop produce viable double-crossover chromosomes that can give recombinant progeny.

viable progeny with double-crossover genotypes in a testcross (see CONCEPT CLOSE-UP: *Double Crossovers in Inversion Loops*).

Geneticists often use inversions as tools to preserve particular combinations of alleles. For example, in *Drosophila*, several special chromosomes called **balancer chromosomes** have been constructed that contain multiple inversions covering almost the entire chromosome. Most balancer chromosomes also contain mutations with dominant visible and recessive lethal phenotypes. Individuals heterozygous for these chromosomes produce no viable recombinant progeny. A single crossover in any of the inverted regions will produce inviable chromosomes, and because the chromosome contains multiple inversions, there are very few or no double crossovers within any one inversion. Balancer chromosomes can be used to maintain a chromosome containing mutant alleles of several different genes for many generations, without recombination breaking up the desired allelic combination. This is done by continually crossing individuals who are heterozygous for the multiple mutant chromosome and a balancer chromosome. The dominant visible mutations on the balancer allow the geneticist to identify quickly and easily the heterozygous individuals in each generation. If the multiple mutant chromosome being maintained also contains a recessive lethal allele, it is called a balanced-lethal combination (Figure 18.22). The offspring that are homozygous for either chromosome die, and only the heterozygous offspring survive. A balanced lethal strain will produce only heterozygous progeny generation after generation.

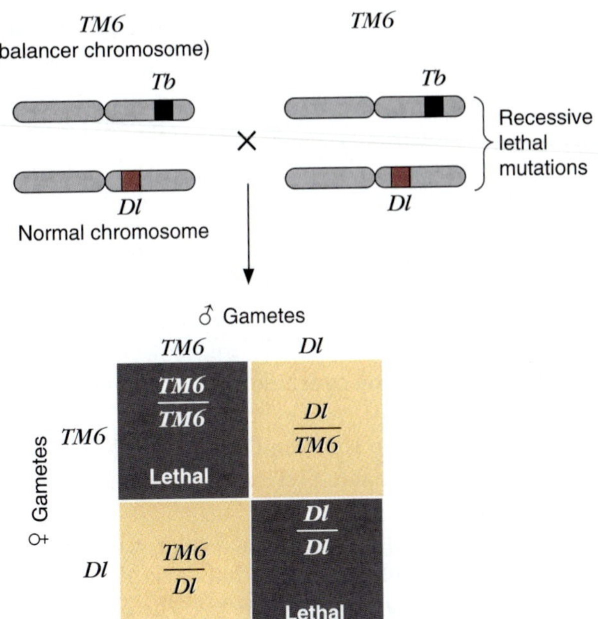

FIGURE 18.22 Balanced lethal genotypes. Individuals heterozygous for the balancer third chromosome *TM6*, which contains the recessive lethal mutation *Tb*, and a third chromosome containing the recessive lethal mutation *Dl* have a balanced lethal genotype. When two such individuals are mated, progeny that are homozygous for either of these chromosomes die and only heterozygous progeny survive. The *TM6* chromosome contains several inversions involving most of the chromosome, and no recombinant progeny are produced.

SOLVED PROBLEM

Problem

Drosophila heterozygous for the recessive, X-linked mutations *cv* (map position 13.7), *sn* (21.0), and *lz* (27.0) were also heterozygous for an X chromosome inversion. These flies were testcrossed, and the phenotypes of their progeny follow. Where are the inversion breakpoints?

Progeny Phenotype	Number
cv sn lz	710
+ + +	712
cv sn +	18
+ + lz	12
cv + +	25
+ sn lz	20
cv + lz	1
+ sn +	2
Total	1500

Solution

Calculating the map distances between these loci from these data gives the following: $R_{cv\text{-}sn} = 3.2$ cM, $R_{sn\text{-}lz} = 2.2$ cM, $R_{cv\text{-}lz} = 5.4$ cM. These numbers are smaller than the normal map distances: $R_{cv\text{-}sn} = 7.3$ cM, $R_{sn\text{-}lz} = 6.0$ cM, $R_{cv\text{-}lz} = 13.3$ cM. The fact that the number of double-crossover individuals is not significantly reduced supports the idea that the map shrinkage is caused by an inversion. The location of the breakpoints of the inversion can be determined by subtracting the observed map distances from the normal map distances.

inversion
cv sn lz
13.7 16.9 21.0 24.8 27.0

SECTION REVIEW PROBLEMS

17. A *Drosophila* homozygous for multiple recessive alleles (*cur st* p^P *red e gro*) was crossed to an unknown individual with a wild-type phenotype. The offspring were testcrossed, and the recombination distances between each pair of alleles were

calculated. These values are compared with the normal values.

	Normal	Observed
cur-st	14.0	14.0
st-p^P	4.0	2.0
p^P*-red*	5.6	0.01
red-e	17.1	0.02
e-gro	19.3	11.3

a. What can you conclude about the unknown individual?

b. Draw a map showing the type and location of any abnormality present in the unknown individual.

18. The following recombination map was made by crossing a multiply mutant line (a b c d e) with an unknown strain (+ + + + +). The offspring all had a normal (+) phenotype. The normal map distance between each locus is 10 m. u. (cM).

a — 8 — b — 0.01 — c — 3 — d — 10 — e

a. What type of chromosome aberration is present in the unknown strain?

b. Where is this aberration located on the chromosome?

c. Draw a picture of the abnormal chromosome pairing with the normal chromosome in meiosis I.

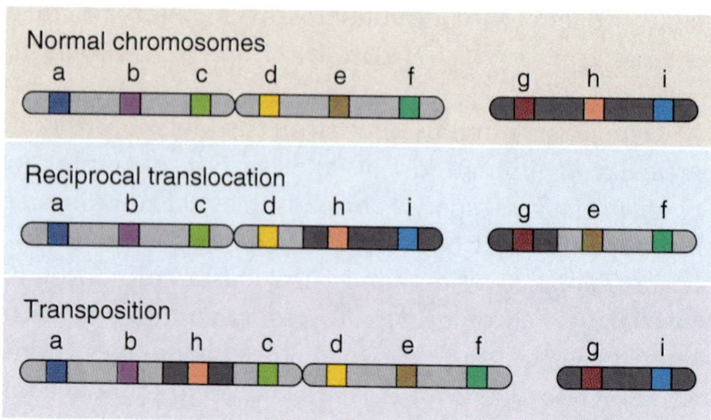

FIGURE 18.23 Reciprocal translocations and transpositions. Reciprocal translocations occur when parts of two nonhomologous chromosomes are exchanged. Transpositions occur when a part of one chromosome is inserted into another, nonhomologous chromosome.

Translocations Are Produced by Exchanges Between Nonhomologous Chromosomes.

When chromosome breaks occur in different chromosomes at the same time, the fragments may rejoin in the wrong combinations. Such exchanges of material between two nonhomologous chromosomes are called **translocations.** Two common types of translocations are **reciprocal translocations,** an exchange of the distal portions of two chromosomes, and **transpositions** (or interstitial transpositions), which are the insertion of a portion of one chromosome into another (Figure 18.23). Translocations can produce mutant phenotypes by breakpoint effects and position effects similar to duplications or inversions. In addition, translocations alter the genetic linkage groups. For example, if two genes on one chromosome are normally linked and one gene is translocated to a new chromosome, the two genes will no longer be linked but will now show independent assortment.

The breakpoint/rejoin positions in translocations have been found to be the cause of a number of important human diseases, especially cancers. For example, chronic myeloid leukemia is fundamentally a genetic disorder that is caused by a specific type of translocation in blood stem cells. This translocation is called the **Philadelphia chromosome** because it was first discovered in cases studied in Philadelphia. It involves a translocation between chromosomes 11 (breakpoint q25) and 22 (breakpoint q13). This translocation is believed to be caused by exposure to mutagenic agents such as irradiation or benzene. The fact that only translocations between chromosomes 11 and 22 at these specific breakpoints cause chronic myeloid leukemia indicates that the cancer results from a breakpoint effect that changes the activity of a gene or group of genes located at these positions.

In an individual homozygous for a reciprocal translocation, the meiotic pairing of the chromosomes generates bivalents that segregate normally. However, in an individual heterozygous for a reciprocal translocation, all four chromosomes pair during prophase of meiosis I in a pairing configuration called a **tetravalent** that has a characteristic "cross" shape (Figure 18.24). When chromosomes pair as bivalents, the alignment of the bivalent on the metaphase plate orients the chromosomes, which leads to one chromosome segregating to each of the progeny nuclei. The chromosomes in a tetravalent, on the other hand, can align on the metaphase plate in one of several different orientations, and the chromosomes have three possible segregation patterns during anaphase I: alternate, adjacent 1, and adjacent 2. In **alternate segregation,** homologous centromeres segregate from each other, and every other centromere (that is, alternate centromeres) migrates to the same progeny nucleus. Alternate segregation produces two progeny nuclei that each contain one complete parental set of chromosomes. One nucleus has both normal chromosomes, and the other has both translocation chromosomes, but both nuclei contain complete sets of genes.

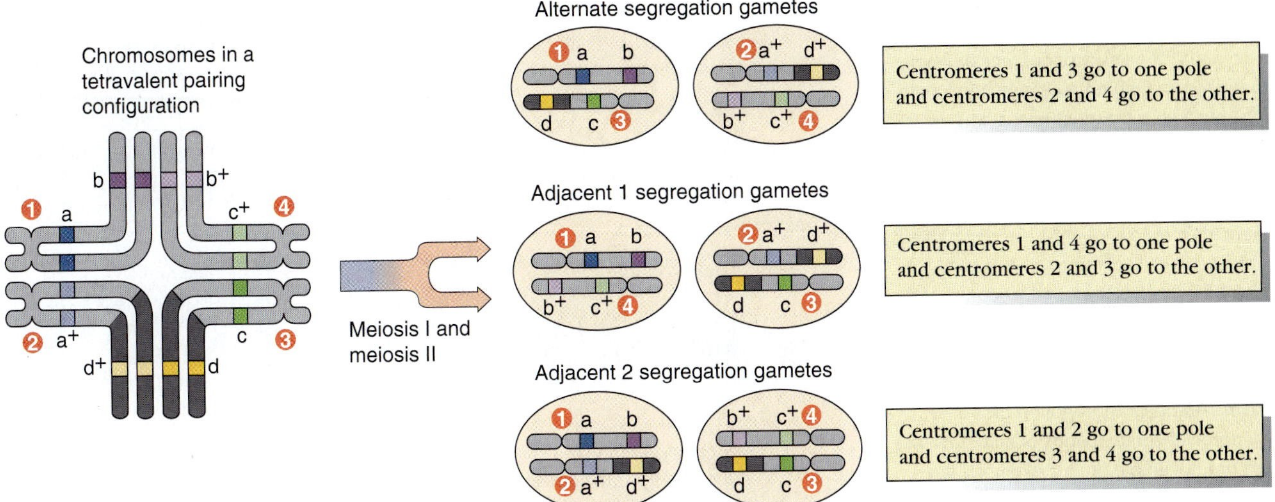

FIGURE 18.24 Chromosome segregation in a reciprocal translocation heterozygote. The chromosomes of an individual heterozygous for a reciprocal translocation will form a tetravalent pairing configuration in prophase of meiosis I. The four chromosomes in a tetravalent segregate in one of three different patterns in anaphase I. Alternate segregation produces meiotic products with balanced genomes that contain complete sets of genes. Adjacent 1 and adjacent 2 segregations produce meiotic products with unbalanced genomes that are missing some genes and have extra copies of others.

The other two patterns of segregation produce meiotic products with unbalanced genomes. In **adjacent 1 segregation,** the homologous centromeres segregate, and adjacent centromeres migrate to the same nucleus. This results in both progeny nuclei containing duplications for some portions of the genome and deficiencies for other portions. The duplicated and deficient regions are the portions of the translocation chromosomes distal to the translocation breakpoints. In **adjacent 2 segregation,** the homologous centromeres do not segregate from each other, but instead they migrate together into one progeny nucleus. This results in both progeny nuclei being duplicated and deficient for regions of the genome. In this case, the duplications and deficiencies are for the portions of the chromosomes proximal to the translocation breakpoints. Because none of these chromosomes is dicentric or acentric, the abnormal chromosome sets can be incorporated into gametes. If the duplications and deficiencies are large, the gametes with the unbalanced chromosome sets will produce inviable zygotes. In theory, all three patterns of segregation are equally frequent and should occur one third of the time. However, in some species and with some translocations, the frequency of adjacent 1 and adjacent 2 segregation is much lower. The mechanism for this is not understood.

Individuals heterozygous for translocations show two major effects related to the pairing of the translocation chromosomes in meiosis I. One effect is partial sterility, which is caused by a high frequency of inviable gametes resulting from adjacent 1 and adjacent 2 segregation. The second effect is altered linkage patterns of the chromosomes involved in the translocation. This latter effect is called **pseudo-linkage** and has been used to detect new translocations. For example, consider the results of the crosses illustrated in Figure 18.25. A *Drosophila* homozygous for recessive mutations on each chromosome (*y/y; b/b; bx/bx; ey/ey*) is crossed with an individual that has wild-type alleles for all these genes but that is homozygous for a translocation. The F_1 are testcrossed by mating them to *y/Y; b/b; bx/bx; ey/ey* individuals. The translocated chromosomes in the F_1 that undergo adjacent 1 or adjacent 2 segregation will produce unbalanced gametes and inviable progeny. The gametes resulting from alternate segregation will have parental-type chromosomes (both normal or both translocated) and will give viable progeny. Thus, no progeny will survive that possess one chromosome from each of the two original parents (one normal and one translocated). This means that there are only eight phenotypic classes in the progeny of the F_1, instead of the 16 that we expect if there were no translocations. Because the translocation in this example involves chromosomes 2 and 3, all the progeny are parental type for the genes on chromosomes 2 and 3 (*b bx* or + +). There are no progeny with *b* + or with + *bx* genotypes because all such individuals have died. This is pseudo-linkage. If we were screening unknown strains for translocations, the presence of pseudo-linkage between marker alleles on chromosomes 2 and 3 would have indicated that the original unknown was homozygous for a translocation involving chromosomes 2 and 3.

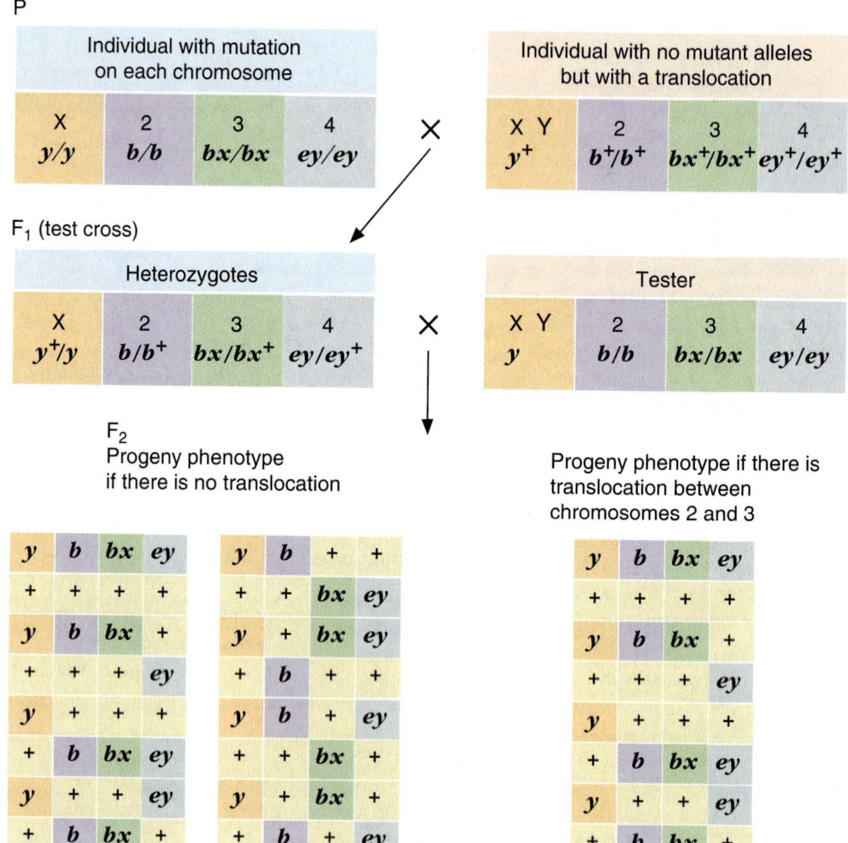

FIGURE 18.25 Pseudo-linkage. Pseudo-linkage occurs in individuals heterozygous for translocations. For example, an individual with no recessive point mutations but with a translocation is crossed to an individual with recessive visible mutations on each chromosome. The heterozygous F_1 are testcrossed, and the phenotypic ratios in the F_2 are determined. In the germ cells of the F_1, the chromosomes involved in the translocation do not assort independently. Alternate segregation produces viable gametes containing the parental combinations of chromosomes. Adjacent segregation produces nonparental combinations of chromosomes that are not viable. In this example, the translocation is between the second and third chromosomes, so there are no progeny with the nonparental *b* + or + *bx* phenotypes.

SOLVED PROBLEM

Problem

Two strains of corn were crossed that were homozygous for different alleles at several loci: straight (*Cr*) vs. crinkly (*cr*) leaves, tall (*D*) vs. dwarf (*d*), both on chromosome 3; and waxy (*W*) vs. starchy (*w*) endosperm, purple (*C*) vs. white (*c*) aleurone, both on chromosome 9. The progeny of this cross were semisterile. How would you determine if this sterility was caused by a reciprocal translocation between chromosomes 3 and 9 in the strain containing the dominant alleles?

Solution

If the progeny of this cross were heterozygous for a reciprocal translocation, then the semisterility is the result of the generation of unbalanced gametes produced by the patterns of segregation of the translocated chromosomes. These unbalanced gametes will contain recombinant chromosomes; so to test this hypothesis, the heterozygous plants can be testcrossed to plants homozygous for recessive alleles on normal chromosomes. If there is a translocation, then the progeny of the testcross will contain only parental type phenotypes (*Cr D W C* and *cr d w c*). A second test would be to examine the chromosomes in germ cells during the prophase of meiosis I. A characteristic cross-shaped tetravalent chromosome figure will be seen.

SECTION REVIEW PROBLEMS

19. A strain of corn homozygous for four recessive mutations (*a g r c*), each on a separate chromosome, was crossed with a strain of teosinte (a primitive corn). The progeny were backcrossed to the corn strain, and the phenotypes of the offspring were determined. The phenotypic classes in the offspring follow. Produce an hypothesis that can explain these results.

a	g	r	c
a	g	r	+
a	g	+	c
a	g	+	+
+	+	r	c
+	+	+	c
+	+	r	+
+	+	+	+

20. For each of the following definitions or descriptions, give the appropriate term that describes the chromosome rearrangement.

a. An exchange between two nonhomologous chromosomes in which a fragment of one chromosome is inserted into the other.
b. A chromosome aberration that produces dicentric chromosomes in heterozygous individuals.
c. A chromosome aberration that causes genes in different linkage groups to appear linked in heterozygotes.
d. A pattern of segregation in meiosis I in a translocation heterozygote in which homologous centromeres segregate and unbalanced gametes are produced.

21. The following diagram shows the pattern of pairing of meiotic chromosomes in an individual heterozygous for a reciprocal translocation.

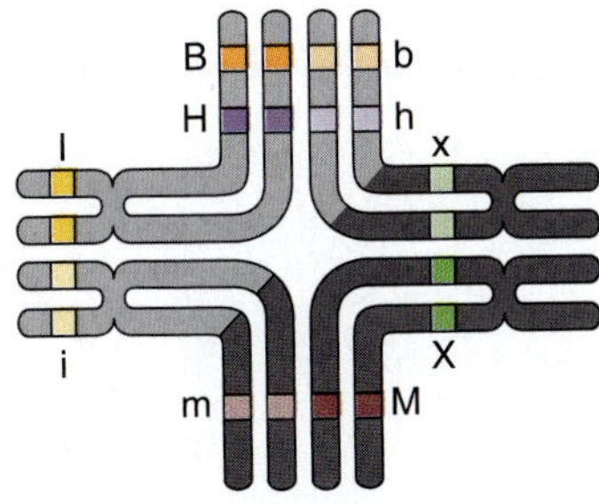

a. What gamete genotypes would be produced by alternate segregation?
b. What gamete genotypes would be produced by adjacent 1 segregation?
c. What gamete genotypes would be produced by adjacent 2 segregation?

22. What would be the gamete genotypes produced by all three patterns of segregation in the individual from Problem 21 if a single crossover occurred between the *X* locus and the translocation breakpoint?

23. What would be the gamete genotypes produced by all three patterns of segregation in the individual from Problem 21 if a single crossover occurred between the *B* and *H* loci?

Robertsonian Translocations Can Produce Compound Chromosomes and Ring Chromosomes.

Translocations that involve the exchange of an entire chromosome arm are sometimes called **Robertsonian translocations.** These translocations are important because they may lead to a change in the number of chromosomes in the monoploid set. Robertsonian translocations can occur by the fusion of two acrocentric or telocentric chromosomes into one metacentric chromosome, or by the fission, or splitting, of one metacentric chromosome into two acrocentric or telocentric chromosomes (Figure 18.26). Consider the X chromosome in

(a) Direct Robertsonian translocations

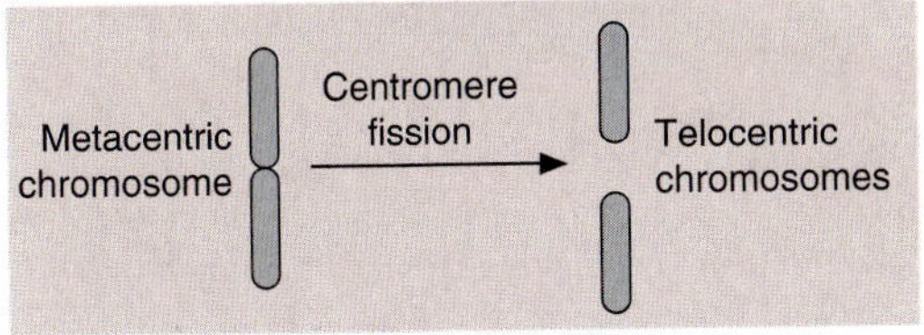

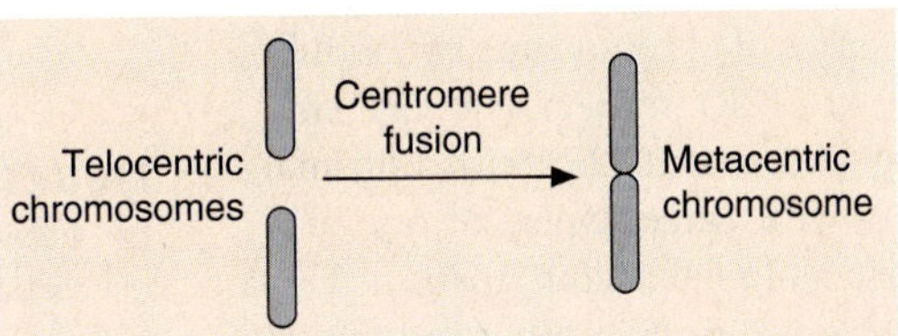

(b) Robertsonian translocation with arm exchange

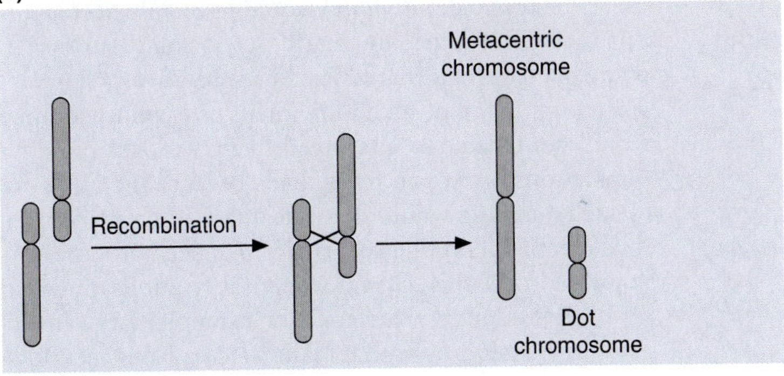

(c) Generation of an attached X chromosome

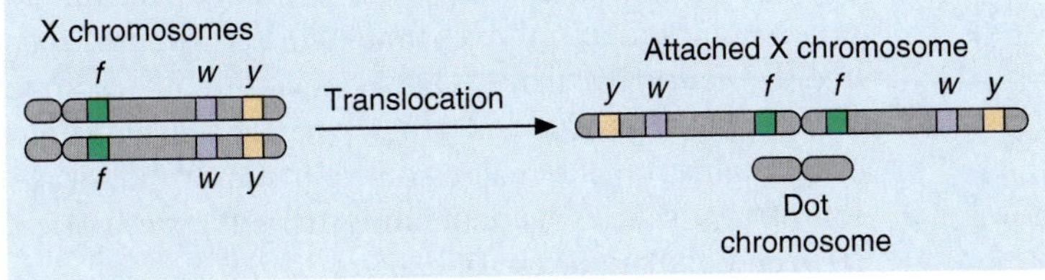

FIGURE 18.26 Robertsonian translocations. Robertsonian translocations are translocations involving whole chromosome arms, and they may alter the number of chromosomes in the monoploid set. This can be done directly by centromere fusion or fission (a) or by translocations that exchange very small arms for large arms (b). This generates one large metacentric chromosome and one small "dot" chromosome. A well-known example of this is the generation of attached X chromosomes in *Drosophila* by Robertsonian translocations that exchange the long arm of one X for the short arm of the other (c).

Drosophila. The X is normally an acrocentric chromosome with only a very small amount of material, containing no vital genes in the short arm. A Robertsonian translocation that exchanges the entire long arm of one X for the short arm of another generates two different chromosomes. One is a metacentric chromosome that has two complete copies of all of the X-linked genes (called an **attached-X chromosome**). The other is a small dot chromosome that contains no vital genes.

Several attached-X chromosomes have been generated in *Drosophila*. The stability of the attached-X chromosome depends on the orientation of the material in the two arms. The two copies of the X may be in two different orientations with respect to each other (Figure 18.27). They may be in tandem, that is, end to end, or they may be reversed. A compound tandem chromosome is unstable in organisms that undergo meiotic recombination because a single crossover between the two arms will generate a ring chromosome and an acentric fragment (Figure 18.28). Several ring X chromosomes have been generated in this manner, and some have been useful in genetic studies.

The X is not the only chromosome in *Drosophila* that can be involved in Robertsonian translocations. Translocations involving autosomes can generate chromosomes that contain two copies of one arm of a major autosome. These chromosomes are called compound chromosomes and are given the symbols *C*( ), with symbols in the parentheses indicating the identity of the attached chromosomes. For example, chromosomes consisting of two copies of the left or right arms of chromosome two would have the symbols, *C*(2*L*) or *C*(2*R*), respectively. Crossing an individual containing *C*(2*L*) and *C*(2*R*) with a normal individual gives progeny with three copies of one arm and one copy of the other, an inviable combination (Figure 18.29). Only when crossed with another individual containing compound second chromosomes will fertile diploid progeny be produced. In nature, this type of translocation would usually be lost in animal species because the original individual would have only normal individuals with which to mate. But it could survive in many plant species that reproduce vegetatively. Should a small population of such individuals be established, they would not be able to interbreed freely with surrounding populations of normal individuals because the offspring of such a mating would die. The attached chromosomes thus create a reproductive barrier that separates the two populations. Other chromosomal aberrations can also act as reproductive barriers. For example, large inversions or translocations that make heterozygous individuals semisterile serve as reproductive barriers, although they are not complete barriers. The gradual accumulation of point mutations over time in the two populations would eventually make each population a separate species. Robertsonian and other types of translocations are thought to make major contributions to the formation of new species by generating reproductive barriers between populations (see Chapters 21 and 22).

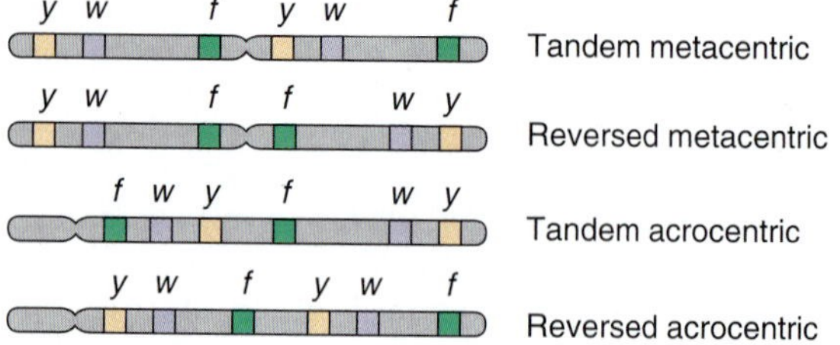

FIGURE 18.27 Attached-X chromosomes. Attached-X chromosomes are given different names depending on the arrangement of the two chromosomes: tandem (both aligned in the same direction) or reversed (aligned in different directions); metacentric (with their common centromere in the center) or acrocentric (with their common centromere at one end). The arrangement of the chromosomes has a major influence on how the attached-X chromosomes behave in meiosis.

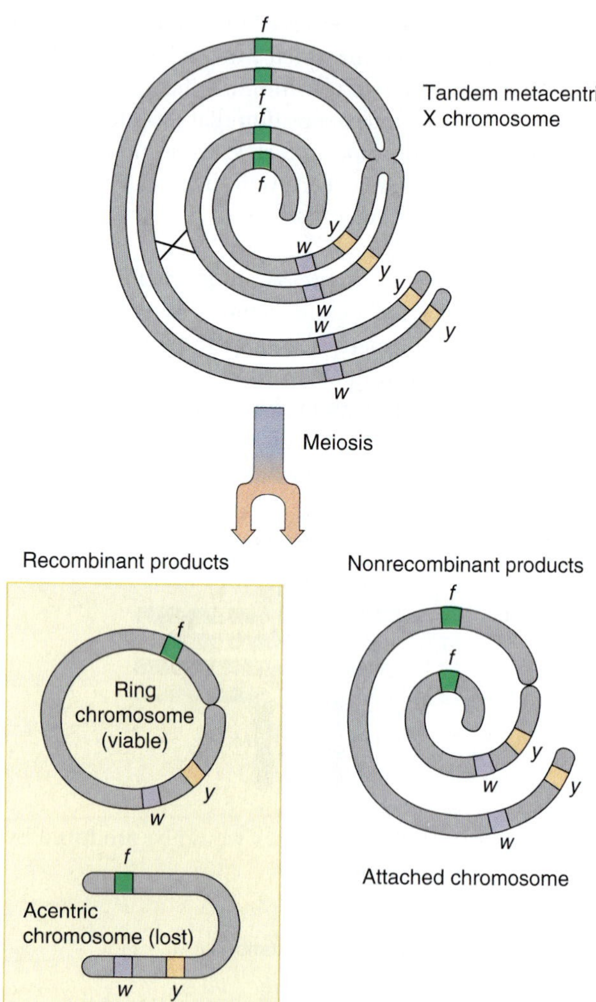

FIGURE 18.28 Generation of ring chromosomes. Recombination between two homologous regions in a compound tandem attached-X chromosome may generate a ring chromosome.

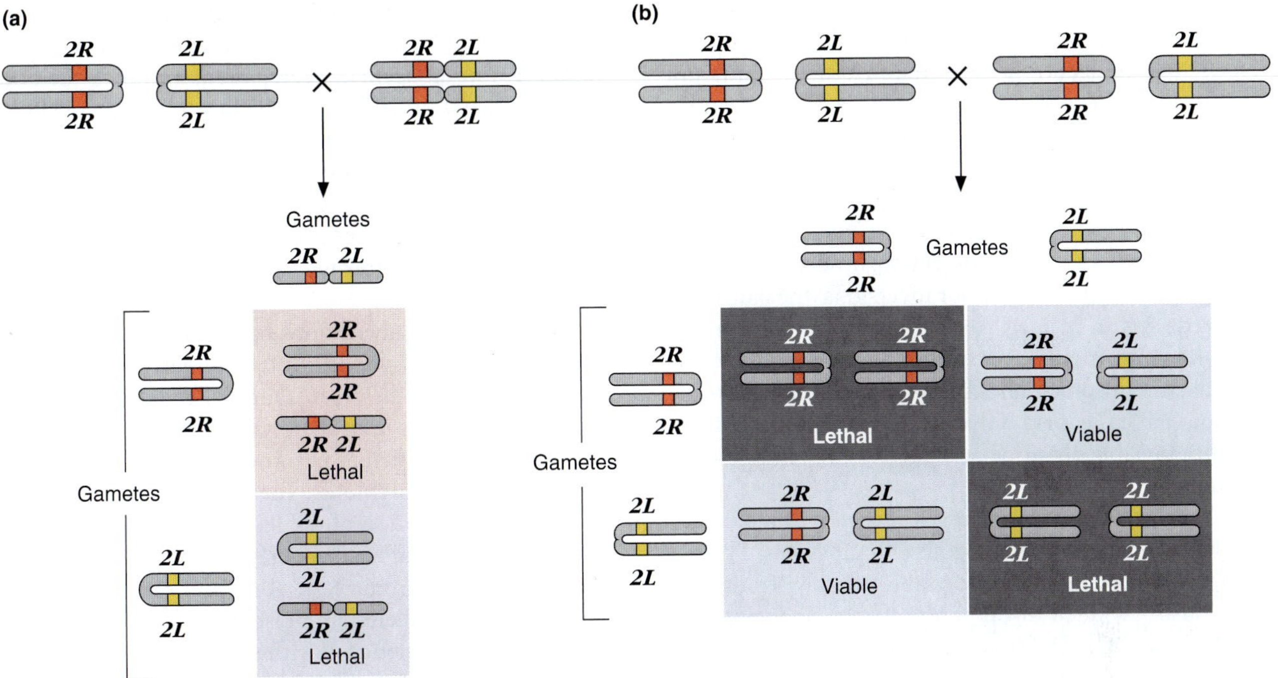

FIGURE 18.29 Compound autosomes. (a) When an individual containing *C(2L)* and *C(2R)* is crossed with a normal individual, the progeny are all aneuploid, containing three copies of one chromosome arm and one copy of the other. (b) When a *C(2R) C(2L)* individual is crossed with another *C(2R) C(2L)* individual, fertile progeny are produced.

SECTION REVIEW PROBLEM

24. A number of female *Drosophila* with one ring X chromosome and one rod X chromosome were produced.
    a. Draw the X chromosomes in anaphase I if a single crossover occurs.
    b. Draw the X chromosomes in anaphase if a two-strand double crossover occurs.
    c. Draw the X chromosomes in anaphase if a four-strand double crossover occurs.

Summary

Chromosomal aberrations include changes in both chromosome number and structure. Changes in chromosome number include increases in the monoploid sets (polyploidy) as well as changes in individual chromosomes (aneuploidy). Polyploids with even-numbered chromosome sets (4X = tetraploid, 6X = hexaploid) are often viable and fertile in plants and may have attractive characteristics such a larger leaves, flowers, or fruits. A number of important commercial crop strains are polyploids, and the experimental induction of polyploidy through crossing or cell fusion is an important area of investigation today. Polyploids with odd numbers of monoploid chromosome sets (3X = triploid, 5X = pentaploid) may be viable but are usually sterile because segregation of the unpaired chromosomes in meiosis produces mostly aneuploid meiotic products. Animals are far less tolerant of polyploidy and aneuploidy than plants. Trisomic animals usually die, and those that survive have a number of severe developmental abnormalities.

Chromosome aberrations that involve changes in chromosome structure include deficiencies, duplications, inversions, and translocations. Deficiencies and duplications are losses and gains of portions of chromosomes. Individuals homozygous for a deficiency or a duplication may have mild or severe phenotypic abnormalities, depending on the number and nature of the genes involved. Individuals homozygous for large deficiencies are not viable. Deficiencies can be of great value to geneticists in mapping the chromosomal location of particular genes in complementation tests. Inversions are reversals of the order of genes caused by breaking a chromosome in two locations and replacing the segment in the reverse orientation. Inversions are classified as either paracentric or pericentric. Paracentric inversions do not include the centromere in the inverted region, and pericentric inversions do include the centromere within the inverted region. Individuals homozygous for an inversion may have an abnormal phenotype if the breakpoints inactivate a gene, either by dividing it or by moving it into a different region where its expression is affected by adjacent genes or heterochromatic material. In individuals heterozygous for inversions, the pairing of chromosomes in meiosis can occur only by forming an inversion loop. A crossover within this loop produces abnormal chromosomes, and the progeny that result are often not viable. Because only recombinant chromosomes result in inviable offspring, a reduction in the number of recombinant progeny is seen.

Translocations are exchanges of chromosomal material between two nonhomologous chromosomes. The exchange may be either reciprocal or one way. Translocations also produce breakpoint effects and position effects. In translocation heterozygotes, reciprocal translocation chromosomes pair with their normal homologues in a tetravalent with a characteristic cross shape. These chromosomes segregate in three different patterns called alternate, adjacent 1, and adjacent 2. Adjacent 1 and adjacent 2 segregations produce meiotic products with unbalanced genomes and give inviable progeny. Because of this, translocation heterozygotes are often semisterile. The progeny that is produced usually shows pseudo-linkage. In pseudo-linkage, the progeny possess only the parental chromosome arrangements (both translocated chromosomes or both nontranslocated chromosomes).

Chromosomal aberrations have a special place in genetics because they provide unique opportunities to gain information about the organization of genes on the chromosomes as well as information about how genes are expressed when in different parts of the genome. Studies of aneuploids and polyploids, for example, indicate that proper gene regulation often depends on a proper genic balance or the number of copies of genes, and altering this balance can produce an abnormal phenotype even when no mutant alleles are present. Chromosomal aberrations provide geneticists with a collection of uniquely useful tools (such as balancer chromosomes) for genetic investigations.

Selected Key Terms

allopolyploid
amphidiploid
aneuploid
autopolyploid
chromosomal aberration
deficiency
Down syndrome
duplication
euploid
inversion
Klinefelter syndrome
paracentric inversion
pericentric inversion
polyploid
position effect
reciprocal translocation
Robertsonian translocation
translocation
transposition
Turner syndrome

Chapter Review Problems

1. Two plants from different species of roses were crossed, and the offspring were viable but sterile. For many years these sterile offspring were propagated by cuttings, when suddenly a fertile seedling appeared.
   a. How might this fertile plant have arisen?
   b. If both parental species had X = 9, how many chromosomes would you expect in the infertile offspring?
   c. How many chromosomes would you expect in the fertile plant?
2. The order of genes on two homologous chromosomes are shown here. Draw a diagram of these chromosomes paired in meiosis I.

   a (centromere) *b c d e f g h i j k l m n*

   a (centromere) *b g f e d c h i j m l k n*
3. The following are the gene orders on a normal acrocentric chromosome and on an inversion.

   Normal (centromere) 1 2 3 4 5 6 7 8 9 10

   Inversion (centromere) 1 2 8 7 6 5 4 3 9 10

   a. Draw a diagram showing these chromosomes paired with a 4-strand double crossover within the inversion.
   b. Draw the gametes that would be formed following the crossovers.
4. Modern cultivated oats are hexaploid with 42 chromosomes. How many chromosomes would a triploid oat strain have? A tetraploid?
5. A fully fertile grass species has 14 chromosomes. What is N for this species? After two cell cycles in a colchicine solution, what is the maximum number of chromosomes a cell from this species might have?
6. Explain how two genes on nonhomologous chromosomes might always segregate together in a particular line when this line is inbred but not when the line is outcrossed.

7. The following diagram represents the pairing pattern of chromosomes in a translocation heterozygote.

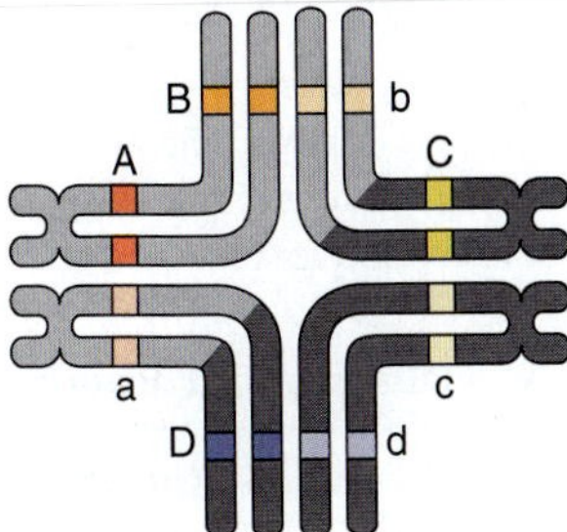

a. What will be the genotypes of gametes produced by adjacent 1 segregation?
b. What will be the gametes produced by alternate segregation?

8. The genotypes of gametes produced by an individual who was heterozygous for recessive alleles at six loci (*x y z r q s*) and heterozygous for a translocation follow. From the genotypes of the gametes, draw a diagram showing the paired chromosomes in prophase of meiosis I of the individual. Assume that the chromosomes containing the recessive alleles are normal for this species, and that no crossing over occurs.

Alternate	Adjacent 1	Adjacent 2
x y z q r s	*x y z R Q Y Z*	*x y z X S*
X S Z Y Q R	*X S r q s*	*R Q Y Z r q s*

9. A cross was made between two plants from a species that is normally diploid. One had white flowers (*w*), and the other had bushy leaves (*b*). The F_1 had a normal phenotype, and one of these F_1 plants was testcrossed as both a male and a female to a stable white bushy strain. The results follow.

Offspring female × *w b* male	Offspring male × *w b* female
+ + 5/12	+ + 2/6
+ *b* 5/12	+ *b* 2/6
w + 1/12	*w* + 1/6
w b 1/12	*w b* 1/6

Explain these results.

10. The chromosome number in diploid *Nicotiana glutinosa* is 24. How many different trisomics are possible?

11. A stable polyploid plant species was discovered that has 26 chromosomes, which form 13 bivalents in meiosis. Is this an allopolyploid or an autopolyploid?

12. Give a brief description of each of the following:
a. monoploid chromosome set
b. aneuploid
c. euploid
d. diploid
e. polyploid

13. The following represents the number of chromosomes in the somatic cells and gametes of several different plants. All these plants are descended from one original species with X = 5. Give the haploid number (N) and the ploidy of each.

Somatic Cell	Gamete
a. 20	10
b. 40	20
c. 10	5
d. 15	Variable

14. Give one difference and one similarity between an allopolyploid and an autopolyploid.

15. An animal breeder has proposed crossing a golden hamster (*Mesocricetus auratus* 2N = 2X = 44) with a guinea pig (*Cavia cobaya* 2N = 2X = 64) and then doubling the chromosome number to create a polyploid species. This polyploid species will be valuable if it retains the body shape of the guinea pig and the golden coat color of the hamster.
a. How many chromosomes will the hybrid animal have?
b. How many chromosomes will the polyploid have?
c. Is the breeder certain that the polyploid will be valuable?

16. A new grain has been discovered that is thought to be the result of a cross between two grass species (Species A with X = 6 and Species B with X = 9), followed by chromosome doubling. Explain how you might test this hypothesis.

17. A plant breeder has proposed crossing the kidney bean (*Phaseolus vulgaris* 2N = 2X = 22) with a potato (*Solanum tuberosum* 2N = 4X = 48) and doubling the chromosome number of the hybrid with colchicine. The breeder hopes to obtain a plant with large, edible roots and also large, edible seeds.
a. How many chromosomes will the hybrid have?
b. How many chromosome will the polyploid have?
c. Is the breeder assured of obtaining a polyploid plant with the desired characteristics?

18. A strange rodent has recently been found. It has the body shape of a mouse (*Mus musculus* 2N = 2X = 40) but the size and tail of a rat (*Rattus norvegicus* 2N = 2X = 42). Explain how you might test the hypothesis that this new rodent is a hybrid between these two species.

19. A child was born with Down syndrome. When another gene on chromosome 21 was investigated, the child was found to have an *AAa* genotype. The father was *aa*, and the mother *Aa*. In what parent did the nondisjunction occur?

20. In an attempt to establish a new line of stable, fertile polyploid flowers, a triploid snapdragon (*Antirrhinum majus*) with X = 8 was selfed. Assuming normal chromosome pairing, what frequency of tetraploids would you expect in the next generation?

21. A hypothetical tetraploid plant that is heterozygous for one gene (*Mmmm*) is selfed.
a. What will be the genotypes of the gametes produced by this plant?
b. What will be the ratios of these gametes?

c. What will be the genotypes of the offspring from this self cross?
d. What will be the ratios of these genotypes?
e. Assuming that the *M* allele is completely dominant, what will be the phenotypic ratio of these progeny?

22. On an expedition to South America, you discover a new plant species. These plants resemble a common lily but have much larger leaves and flowers. You have access to several strains of common lily and standard cytological equipment. How could you determine whether this new plant is a tetraploid lily?

23. Two strains of tetraploid daylily have different alleles of four genes that give valuable traits. A plant breeder wishes to cross these two strains (*aaaa bbbb CCCC DDDD* with *AAAA BBBB cccc dddd*) and self the F_1 to generate new combinations in the F_2. The goal is to generate a plant that is homozygous for recessive alleles at all four loci (*aaaa bbbb cccc dddd*). What will be the ratio of plants homozygous recesive for all four loci in the F_2?

24. Several triploid *Drosophila* females were mated with normal diploid males.
    a. What ratio of diploid, triploid, and aneuploid zygotes will be formed?
    b. What will be the ratios of the genotypes in the adult progeny?

25. Give a brief definition of the following terms:
    a. trisomic
    b. monosomic
    c. tetrasomic
    d. nullisomic
    e. double trisomic

26. A new diploid strain of *Datura* has been discovered with large, fragrant flowers. This trait is controlled by a single, recessive Mendelian allele (*f*). Explain how you could map this mutation to one chromosome.

27. Give a brief definition of the following:
    a. Turner syndrome
    b. Klinefelter syndrome
    c. triple X syndrome

28. Give a brief definition for the following:
    a. interstitial deficiency
    b. haplo-insufficient locus
    c. breakpoint effect
    d. position effect

29. Several maize plants heterozygous for multiple recessive mutations were crossed to plants heterozygous (*Df*/+) for five different deficiencies. These mutations are located on one maize chromosome in the following order: *yg* (abnormal chlorophyll), *c* (colorless kernel), *sh* (shrunken kernel), *bz* (bronze), *wx* (waxy). The phenotypes of the progeny from each cross follow. From these results, determine the location of each deficiency.

Deficiency	Offspring	Phenotypes
1	+	Colorless, shrunken
2	+	Shrunken bronze
3	+	Abnormal chlorophyll, colorless, shrunken
4	+	Shrunken
5	+	Colorless, shrunken, bronze, waxy

30. Give a brief definition for each of the following:
    a. pericentric inversion
    b. paracentric inversion
    c. reciprocal translocation
    d. varigated position effect
    e. inversion loop

31. Give a brief definition for the following:
    a. dicentric
    b. acentric
    c. alternate segregation
    d. adjacent 1 segregation
    e. adjacent 2 segregation

32. A tetraploid Stella de Oro from a true-breeding strain with yellow ruffled flowers was crossed with a plant from a true-breeding tetraploid strain with purple straight flowers. The F_1 all had purple ruffled flowers. These F_1 were crossed with plants from a true-breeding tetraploid strain with yellow straight flowers.
    a. Using allele symbols of your own choosing, explain what genotypes and phenotypes you would expect in the offspring of this cross, assuming that these genes are not linked.
    b. In what ratios would you expect these genotypes and phenotypes?

Challenge Problems

1. Plants from a triploid strain of tomato (*Lycopersicon esculentum* 3X = 36) were selfed. What ratio of diploid, triploid, tetraploid, and aneuploid progeny will be produced?

2. A multiply mutant *Drosophila* with *multiple wing hairs* (*mwh*, 0.0), *javelin* (*jv* 19.2), *sepia* (*se* 26.0), *grooved* (*gv* 36.2), *tilt* (*tt* 40.0), and *frizzled* (*fz* 41.7) was crossed with flies from an unknown strain that had wild-type alleles for these genes (+ + + + + +). The F_1 all had a normal phenotype. The F_1 females were backcrossed to males from the multiple mutant strain, and their progeny were scored for the phenotypes of each gene. From these progeny, the following linkage map was produced.

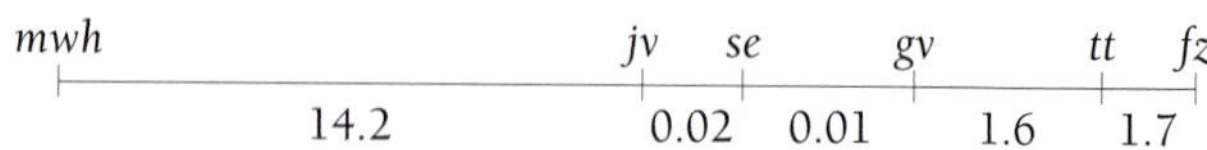

a. What type of chromosome aberration is present in the unknown strain?
b. Where is this aberration located on the chromosome?
c. Draw a picture of the abnormal chromosome pairing with the normal chromosome in meiosis I.

3. The following diagram represents the meiotic chromosome pairing configuration in a corn plant heterozygous for a reciprocal translocation. This individual was crossed with an individual with normal chromosomes who had recessive alleles at all loci. The progeny were collected as seeds, planted, and grown, and their phenotypes were scored.

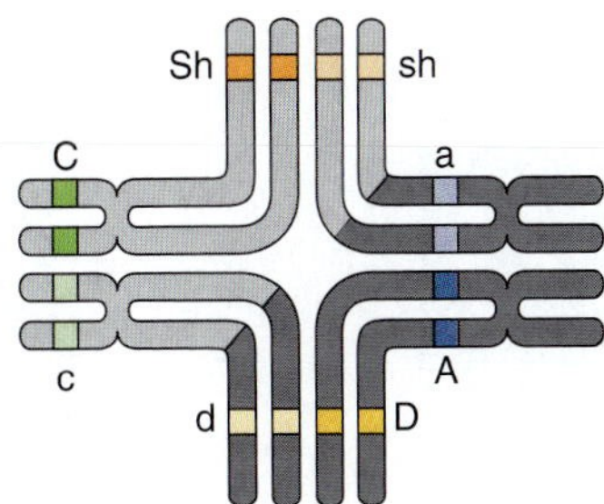

a. What would be the genotypes of the gametes produced by the three patterns of segregation in this translocation heterozygote, assuming that there is no recombination?
b. What genotypes would you expect in the adult progeny?

Suggested Readings

Abbott, C. 1995. *Somatic Cell Hybrids: The Basics.* Oxford University Press, New York.

Baker, W. K. 1968. "Position effect variation." *Advances in Genetics*, 16: 133–169.

Blakeslee, A. F. 1934. "New jimson weeds from old chromosomes." *Journal of Heredity*, 25: 80–108.

Blakeslee, A. F. 1941. "Effect of induced polyploidy in plants." *American Naturalist*, 75: 117–135.

Cummings, M. R. 1991. *Human Heredity: Principles and Issues.* 2nd ed. West Publishers, St. Paul, MN.

Epstein, C. J. 1986. *The Consequences of Chromosome Imbalance: Principles, Mechanisms, and Models.* Cambridge University Press, New York.

Gardener, R. J. M., and G. R. Sutherland. 1996. *Chromosome Abnormalities and Genetic Counseling.* Oxford University Press, Oxford.

Heim, S. 1995. *Cancer Cytogenetics.* Wiley-Liss, New York.

Judd, B. H., M. W. Shen, and T. C. Kaufman. 1972. "The anatomy and function of a segment of the X chromosome of *Drosophila melanogaster*." *Genetics*, 71: 139–156.

Lindahl, T. 1996. *Genetic Instability in Cancer.* Cold Spring Harbor Laboratory Press, New York.

Mattell, S. H., J. A. Mathew, and R. A. McKee. 1985. *Principles of Plant Biotechnology and Introduction to Genetic Engineering in Plants.* Blackwell Press, Oxford, UK.

McClintock, B. 1941. "The stability of broken ends of chromosomes in *Zea maize*." *Genetics*, 26: 234–282.

McKusick, V. A. 1994. *Mendelian Inheritance in Man: A Catalog of Human Genes and Genetic Disorders.* Johns Hopkins University Press, Baltimore.

Obe, G. and A. T. Natarajan. 1990 *Chromosomal Aberrations: Basic and Applied Aspects.* Springer-Verlag. New York.

On the Web

Visit our Web site at **http://www.saunderscollege.com/lifesci/titles.html** and click on A/G/M Genetics for links to the following chapter-related resources on the World Wide Web:

1. **The Agricultural Genome Information System Cytogenetic Terms Index.** A site with definitions for terms used in studies of chromosome abnormalities in agriculture.
2. **Down Syndrome: Understanding the Gift of Life.** This site was created by and for the families of individuals with Down syndrome.
3. **The Oregon Health Sciences University.** This is a clinical Web page on chromosome abnormalities.
4. **Oregon Health Sciences University.** This is a clinical Web page on sex chromosome abnormalities.
5. **Online Mendelian Inheritance in Man.** This is the National Institutes of Health Web site for human genetics, and it includes information on human mutations and chromosome abnormalitites.

CHAPTER

19

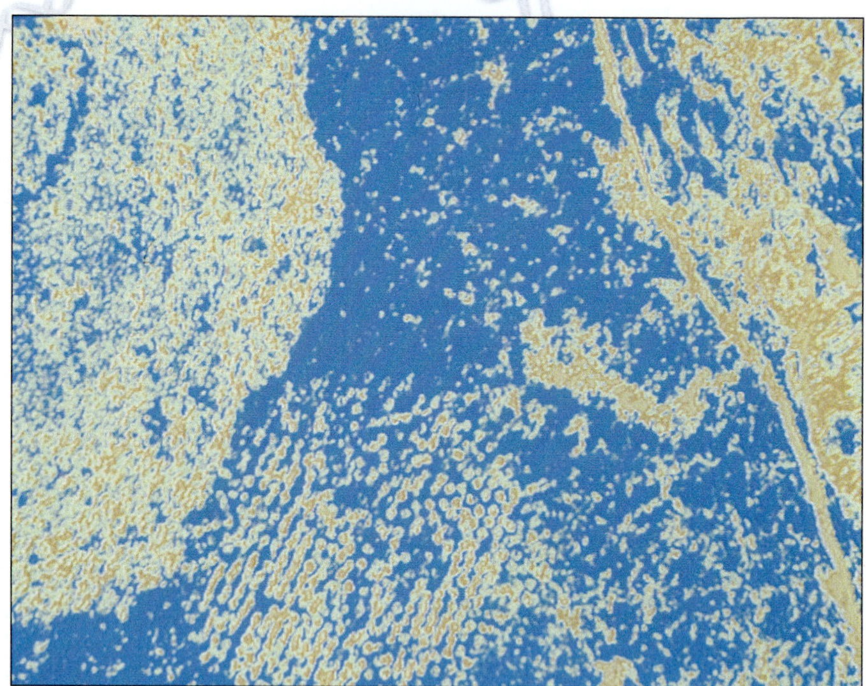

Color-enhanced electron micrograph showing *copia* retrotransposon particles within the nucleus of a *Drosophila* fat body cell. The *copia* particles, which are lined up in an array adjacent to the nucleolus, contain two molecules of *copia* RNA as well as the enzyme reverse transcriptase. The formation of such particles is an intermediate stage in the retrotransposon life cycle. *(J. McDonald, University of Georgia)*

Recombination and Transposable Elements

From the molecular perspective, recombination can be divided into two main categories: homologous and nonhomologous recombination. The term **homologous recombination** refers to genetic exchanges between two DNA molecules with identical or nearly identical nucleotide sequences. Homologous recombination is further subdivided based upon the extent of the region of homology existing between the recombining DNAs. The first category, **generalized recombination,** refers to exchanges between DNA molecules sharing extended regions of homology, such as homologous eukaryotic chromosomes that pair and recombine during meiosis. The second category of homologous recombination is **site-specific recombination,** which refers to exchanges between DNA molecules sharing limited regions of sequence homology. An example is lambda phage (Chapter 8), which inserts into *E. coli* DNA via an exchange event at a very specific position on the chromosome. The term **nonhomologous recombination** refers to the process by which a DNA molecule can integrate at a chromosomal position where there is no nucleotide sequence homology. The integration of transposable elements is an example of nonhomologous recombination.

Recall from Chapter 17 that random mutations occur in the genome, and one mechanism of random mutation is the insertion into a gene of a DNA element that inactivates the gene. These DNA elements are called insertion sequences, or transposable elements, and they can insert copies of themselves elsewhere in the genome. In addition, these elements can cause rearrangements of host DNA as well as activate other genes by inserting promoters nearby. In this chapter, we will discuss the structure and behavior of these elements as well as the experiments that led up to their discovery. Finally, we will explore the evolutionary significance of these mobile elements.

By the end of this chapter, we will have considered the molecular mechanisms that underlie homologous and nonhomologous recombination as well as the broader scope of how transposable elements move about in the genome. We will also have answered five major questions:

1. What are the molecular mechanisms responsible for generalized recombination?
2. What are the molecular mechanisms underlying site-specific recombination, and how do these mechanisms differ from those involved in generalized recombination?
3. What are transposable elements, and how were they first discovered?
4. What are the mechanisms by which transposable elements are able to integrate into new chromosomal locations?
5. What is the overall biological significance of transposable elements?

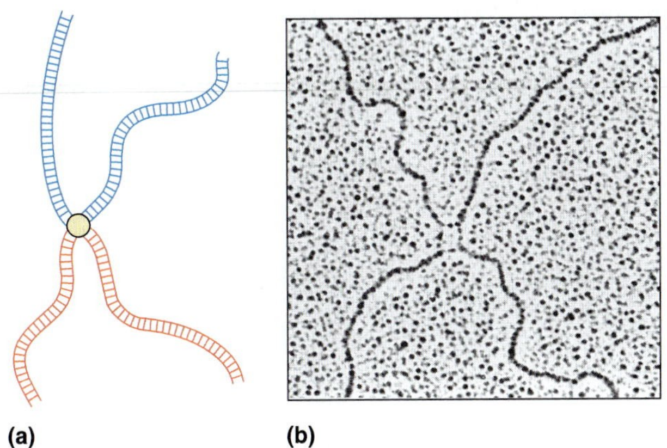

FIGURE 19.1 A Holliday structure. Holliday structures are generated at an intermediate step in the recombination process wherein the two duplexes are connected at the recombination joint. (a) This figure shows an interpretive drawing of the electron micrograph in (b), which is representative of a Holliday structure. *(Dr. Ross Inman, University of Wisconsin, Madison)*

Homologous Recombination Utilizes Sequence Complementarity.

Both general and site-specific recombination are forms of homologous recombination because they both require the pairing of homologous stretches of DNA (sequences that are complementary) in order for the genetic material to be exchanged. However, the two recombination processes differ in the extent of genetic homology that is necessary between the DNA molecules involved in the exchange.

Generalized Recombination Involves an Exchange Between Homologous DNA Molecules.

Generalized recombination is the process that enables reciprocal exchanges of genetic material between homologous chromosomes during meiosis (Chapter 3). This process is also responsible for those recombinant progeny that result from conjugation between compatible bacterial strains (Chapter 8). Several molecular models of recombination have been proposed over the past several years to account for generalized recombination. All the models account equally well for certain facts known to be associated with most recombination events. For example, it has been demonstrated that recombination involves the breaking and rejoining of homologous DNA molecules. We also know that at intermediate stages in the recombination process, the homologous duplex molecules involved in recombination covalently link to one another to produce what is termed a **Holliday structure** (Figure 19.1) in honor of Robin Holliday, the geneticist who first proposed its existence.

Beyond explaining the general features of recombination, however, the various models of recombination differ with regard to the specific molecular mechanisms involved. At the present time, two similar but distinct models are considered to be most consistent with the available facts: the single-strand break model and the double-strand break model.

The Single-Strand Break Model. According to the single-strand break model, recombination begins with single-strand nicks or breaks of homologous strands at corresponding points in the two paired DNA duplexes (Figure 19.2). These breaks, which may occur spontaneously across the chromosome or may be induced by environmental stresses such as radiation, allow each free end to pair with its complement in the other duplex. In *E. coli*, this pairing requires the protein product of the *recA* gene. The reciprocal exchange event produces an initial connection between the two participating DNA duplexes that is initially stabilized by hydrogen bonding and later by covalent bonds as a result of enzyme-mediated DNA repair of the nicks. At this intermediate point in the process, the two duplexes are connected at the so-called recombination joint to form a connected or joint molecule (Figure 19.2). This joining of the two duplexes is the physical basis of the Holliday structure referred to in Figure 19.1.

The point of contact between the two duplexes consists of hybrid or heteroduplex DNA (a **heteroduplex** is two strands of DNA from different sources that are not exactly complementary). The double-stranded DNA molecule in the heteroduplex region of contact consists of one strand from each of the parental DNA molecules.

The region of heteroduplex DNA may be extended by a process called branch migration, in which an unpaired region of one of the single strands displaces a paired region of the other single strand (Figure 19.3). There are two types of branch migrations. Spontaneous branch migration may proceed equally in both directions for up to several hundred base pairs. Protein-directed branch migration, on the other hand, is another RecA protein–controlled process and induces directional migration, which may result in a heteroduplex DNA region of up to a thousand or more base pairs.

The process by which the connected DNA duplexes are ultimately separated from one another is called **resolvement.** Resolvement requires that two additional

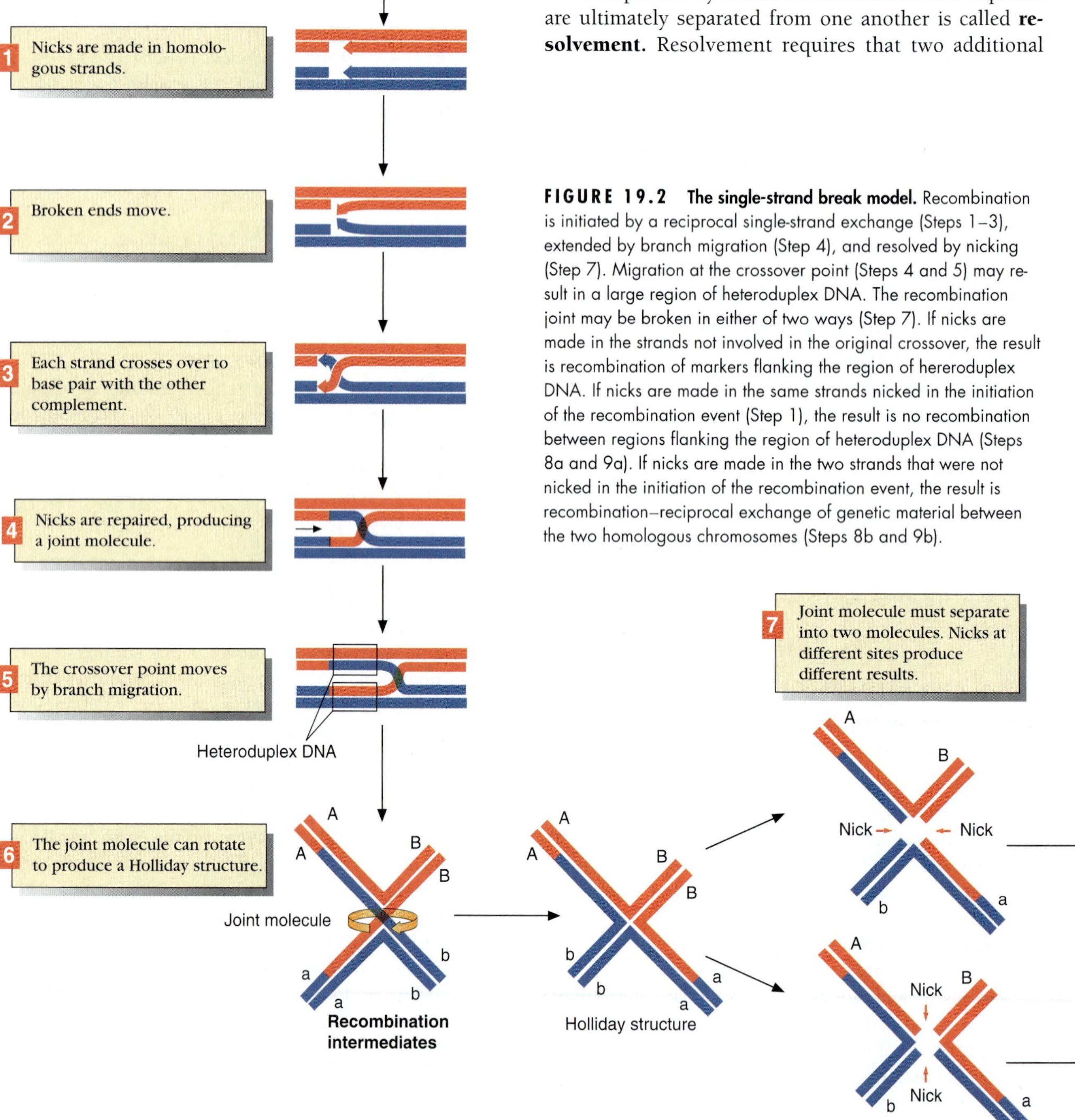
Homologous chromosomes pair
A
B
a
b
5'
3'
1 Nicks are made in homologous strands.
2 Broken ends move.
3 Each strand crosses over to base pair with the other complement.
4 Nicks are repaired, producing a joint molecule.
5 The crossover point moves by branch migration.
Heteroduplex DNA
6 The joint molecule can rotate to produce a Holliday structure.
Joint molecule
Recombination intermediates
Holliday structure
7 Joint molecule must separate into two molecules. Nicks at different sites produce different results.
Nick
Nick
Nick
Nick

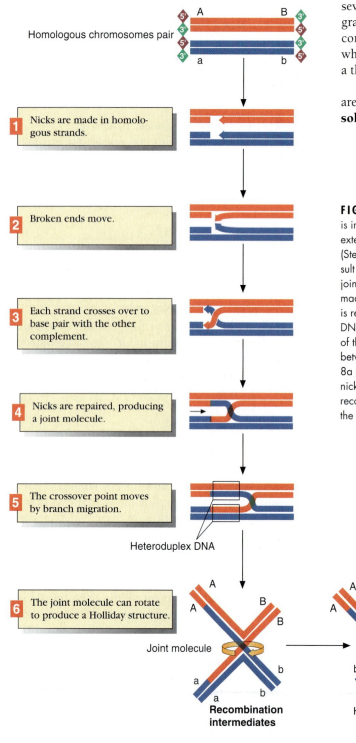

FIGURE 19.2 The single-strand break model. Recombination is initiated by a reciprocal single-strand exchange (Steps 1–3), extended by branch migration (Step 4), and resolved by nicking (Step 7). Migration at the crossover point (Steps 4 and 5) may result in a large region of heteroduplex DNA. The recombination joint may be broken in either of two ways (Step 7). If nicks are made in the strands not involved in the original crossover, the result is recombination of markers flanking the region of hereroduplex DNA. If nicks are made in the same strands nicked in the initiation of the recombination event (Step 1), the result is no recombination between regions flanking the region of heteroduplex DNA (Steps 8a and 9a). If nicks are made in the two strands that were not nicked in the initiation of the recombination event, the result is recombination–reciprocal exchange of genetic material between the two homologous chromosomes (Steps 8b and 9b).

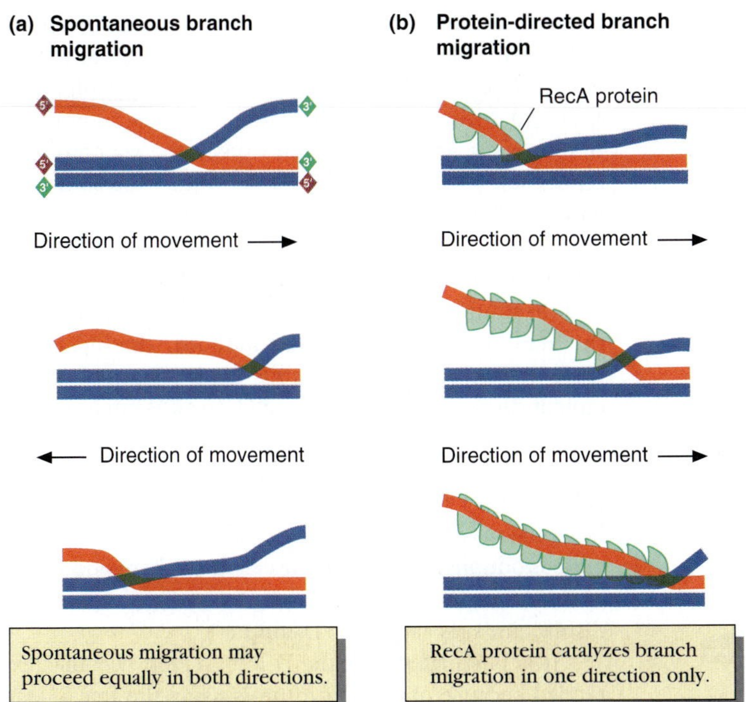
(a) Spontaneous branch migration
5'
3'
Direction of movement →
← Direction of movement
Spontaneous migration may proceed equally in both directions.
(b) Protein-directed branch migration
RecA protein
Direction of movement →
Direction of movement →
RecA protein catalyzes branch migration in one direction only.

FIGURE 19.3 The two types of branch migration. (a) Spontaneous branch migration is a random back-and-forth process typically extending less than 100 bp in either direction. (b) The RecA protein-directed branch migration proceeds at a uniform rate in only one direction, which can result in the formation of 1000 bp or more of heteroduplex DNA. Both (a) and (b) illustrate the migration of a single strand in a duplex.

nicks be made in the now-joined DNA molecules. However, these additional nicks may arise in two alternative positions. The genetic consequences of the two options are quite different (Figure 19.2). If the nicks are made in the two strands that were not involved in the initial reciprocal exchange event (that is, the two strands that were *not* nicked in the initiation of the exchange event), then the consequent separation of the joint molecule will result in a conventional recombination event (that is, reciprocal exchange of genetic material between the two homologous chromosomes). If, on the other hand, resolvement is effected by nicking the two strands that participated in the initial exchange event (that is, the two molecules that were nicked at the start of the exchange process), then the consequent separation of the joint molecule will result in the release of the original parental molecules, and no conventional recombination event will have taken place. However, it is important to note that even under this second scenario, there will be a region of heteroduplex DNA contained in both of the

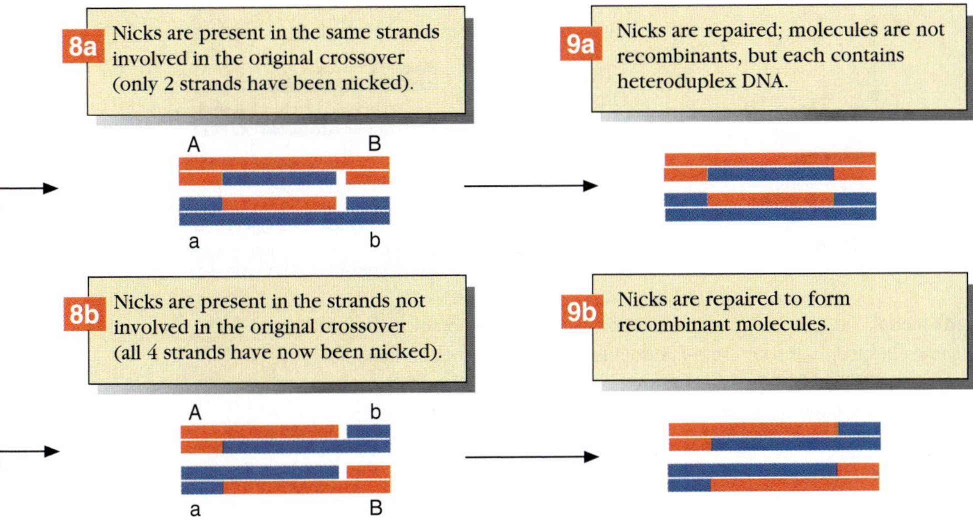
8a Nicks are present in the same strands involved in the original crossover (only 2 strands have been nicked).
A
B
a
b
9a Nicks are repaired; molecules are not recombinants, but each contains heteroduplex DNA.
8b Nicks are present in the strands not involved in the original crossover (all 4 strands have now been nicked).
A
b
a
B
9b Nicks are repaired to form recombinant molecules.

released parental strands. This region marks the point where the molecules were once joined, and it will vary in length depending upon how much of branch migration occurred during the intermediate stage of the recombination process. As mentioned previously, the region of heteroduplex DNA can extend to several thousand base pairs in length.

The Double-Strand Break Model. According to the double-strand break model, the DNA exchange event is initiated by an endonuclease-induced, double-strand break in one of the homologous duplexes called the recipient duplex. The double-strand break is enlarged to a gap by an exonuclease that digests one strand on either side of the initial break, resulting in 3′ single-strand ends (Figure 19.4). One of the free 3′ ends finds its way to the homologous duplex called the donor duplex. The region of heteroduplex DNA generates an area where one strand of the donor duplex is displaced. This region of localized displacement is called the **D-loop.**

It is postulated that the D-loop region becomes extended because of repair synthesis using the free 3′ end of the recipient strand as a primer. The D-loop region eventually encompasses the entire length of the original gap. At one end of the gap, the 3′ end of the recipient strand pairs with the displaced strand of the donor duplex, and repair synthesis is initiated using the 3′ recipi-

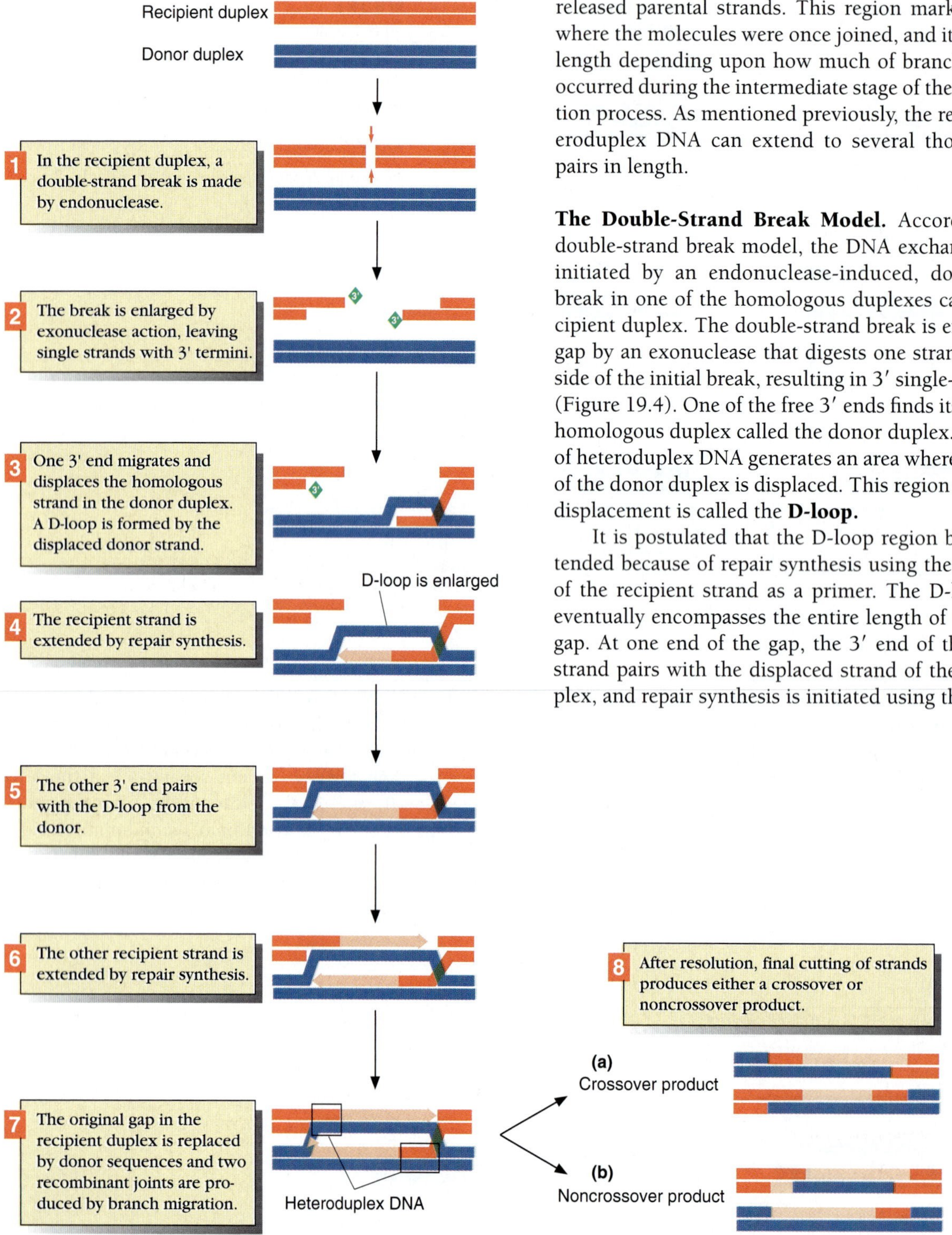
Recipient duplex
Donor duplex
1 In the recipient duplex, a double-strand break is made by endonuclease.
2 The break is enlarged by exonuclease action, leaving single strands with 3' termini.
3'
3 One 3' end migrates and displaces the homologous strand in the donor duplex. A D-loop is formed by the displaced donor strand.
D-loop is enlarged
4 The recipient strand is extended by repair synthesis.
5 The other 3' end pairs with the D-loop from the donor.
6 The other recipient strand is extended by repair synthesis.
7 The original gap in the recipient duplex is replaced by donor sequences and two recombinant joints are produced by branch migration.
Heteroduplex DNA
8 After resolution, final cutting of strands produces either a crossover or noncrossover product.
(a)
Crossover product
(b)
Noncrossover product

FIGURE 19.4 The double-strand break model. The process of recombination between two homologous chromosomes (recipient shown in red, donor shown in blue) is shown using the double-strand break model. Recombination is initiated by a double-strand break in one of the homologous duplexes (Step 1), and the break is enlarged by an exonuclease, resulting in 3′ single-strand ends (Step 2). One of the 3′ ends migrates, displacing the donor duplex in the D-loop (Step 3). Repair synthesis extends the D-loop (Step 4) and the other 3′ end pairs with the displaced strand of donor duplex, initiating repair synthesis (Steps 5 and 6). Subsequent branch migration results in two recombinant joints (Step 7) that may be broken in one of two ways. If one joint is cut on the inner strand and one on the outer strand, a recombination event results (Step 8a). If both joints are cut on the inner or outer strands, no recombination takes place (Step 8b).

ent strand as a primer (Figure 19.4). This synthesis continues until the newly synthesized strand reaches the other end of the gap and base-pairs with the displaced donor strand. At this stage in the process, there is heteroduplex DNA on both sides of the gap. Subsequent branch migration results in a molecule with two recombinant joints (Figure 19.4).

In the double-strand break model, two alternate pathways of resolvement are possible, and each has a different genetic consequence (Figure 19.4). If joints are resolved in different ways, that is, if one joint is cut on the inner strand and the other on the outer strand, the result is a conventional recombination event with the release of two crossover products (Figure 19.4). If, on the other hand, both joints are resolved in the same way, that is, both joints are cut on the inner strand or both on the outer strand, the two original parental duplexes are released, and no conventional recombination event will have taken place.

An interesting feature of the double-strand break model is that in the chromosome products of either mode of resolvement, the gap region defined by the D-loop intermediate will, after resolvement, contain sequences identical to those carried by the original donor duplex. Thus in the region of the parental duplexes originally involved in the exchange event, there will always be a replacement or "conversion" of recipient sequences by donor sequences (hence, the origin of the terms "donor" and "recipient" duplexes). This genetic phenomenon whereby one member of a heteroallelic pair is converted to the other during meiosis is well documented and is called **gene conversion.**

SOLVED PROBLEM

Problem

*Rec*A mutants are defective in generalized recombination. What may be the molecular basis of this mutant phenotype?

Solution

In the bacteria *E. coli,* in which the molecular processes underlying recombination have been intensively studied, the *rec*A gene has been shown to be involved both in the heteroduplex pairing of single strands from paired complementary duplexes and in the process of branch migration. A disruption in either one or both of these processes could be due to a defect in the *rec*A gene in generalized recombination.

SECTION REVIEW PROBLEM

1. Briefly compare and contrast the key features of the single-strand break and the double-strand break models of recombination.

Site-Specific Recombination Is Exemplified by Lambda Phage Integration and Excision.

Site-specific recombination is the process through which phage DNA is integrated into predetermined sites on the bacterial chromosome (Chapter 8, especially Figure 8.31). As stated, site-specific recombination occurs between two specific DNA sequences having only limited regions of homology. The sequences involved in this reaction provide specificity by presenting target sites for precise enzyme-mediated integration and excision events. The integration (and excision) of the lambda phage into a specific location in the bacterial chromosome is perhaps the best-studied example of site-specific recombination.

Integration and excision of lambda phage is brought about at specific attachment sites on the bacterial and phage chromosomes called *att*B and *att*P, respectively (Figure 19.5a). The genetic fine structure of *att*B consists of three contiguous regions called B, O, and B′. Similarly, the phage attachment site, *att*P, consists of the three regions, P, O, and P′. The respective flanking regions of the two *att* sites (B, B′ and P, P′) are called arms and are all sequentially distinct from one another. The arms are important as recognition sequences for the enzymes involved in the integration and excision events (Figure 19.5a). On the other hand, the O region is composed of a 15-bp core sequence that is identical in both *att*B and *att*P. Within the O region, staggered cuts in the common core sequences allow the reciprocal exchange to actually take place (Figure 19.5b). As a consequence of this exchange, two new *att* sites are generated on either side of the integrated phage. These new sites are called *att*L (L = left) and *att*R (R = right), and they have the genetic fine structure of BOP′ and POB′, respectively (Figure 19.5). The generation of these new sites ensures that the process of integration and excision are distinct, thus any given integration event will not be immediately reversed by an excision event and vice versa.

Integration requires the products of the lambda phage gene *int* and the bacterial gene product IHF (integration host factor). Excision also uses the *int* gene and IHF, but it also requires the product of a second lambda phage gene, *xis*. Lambda integration and excision are controlled by two distinct series of enzyme reactions because different DNA target sequences are involved in the two processes.

Thus, although generalized and site-specific recombination both require sequence homology between the DNAs involved in the exchange event, the enzymatics of the two processes are quite different. The sequences involved in site-specific recombination are specifically recognized by the enzymes involved in the exchange process. In generalized recombination, enzymes also help mediate the exchange, but there are no specific recognition sequences. Instead, the enzymes involved in generalized recombination are drawn to regions of paired chromosomes where breakage has occurred.

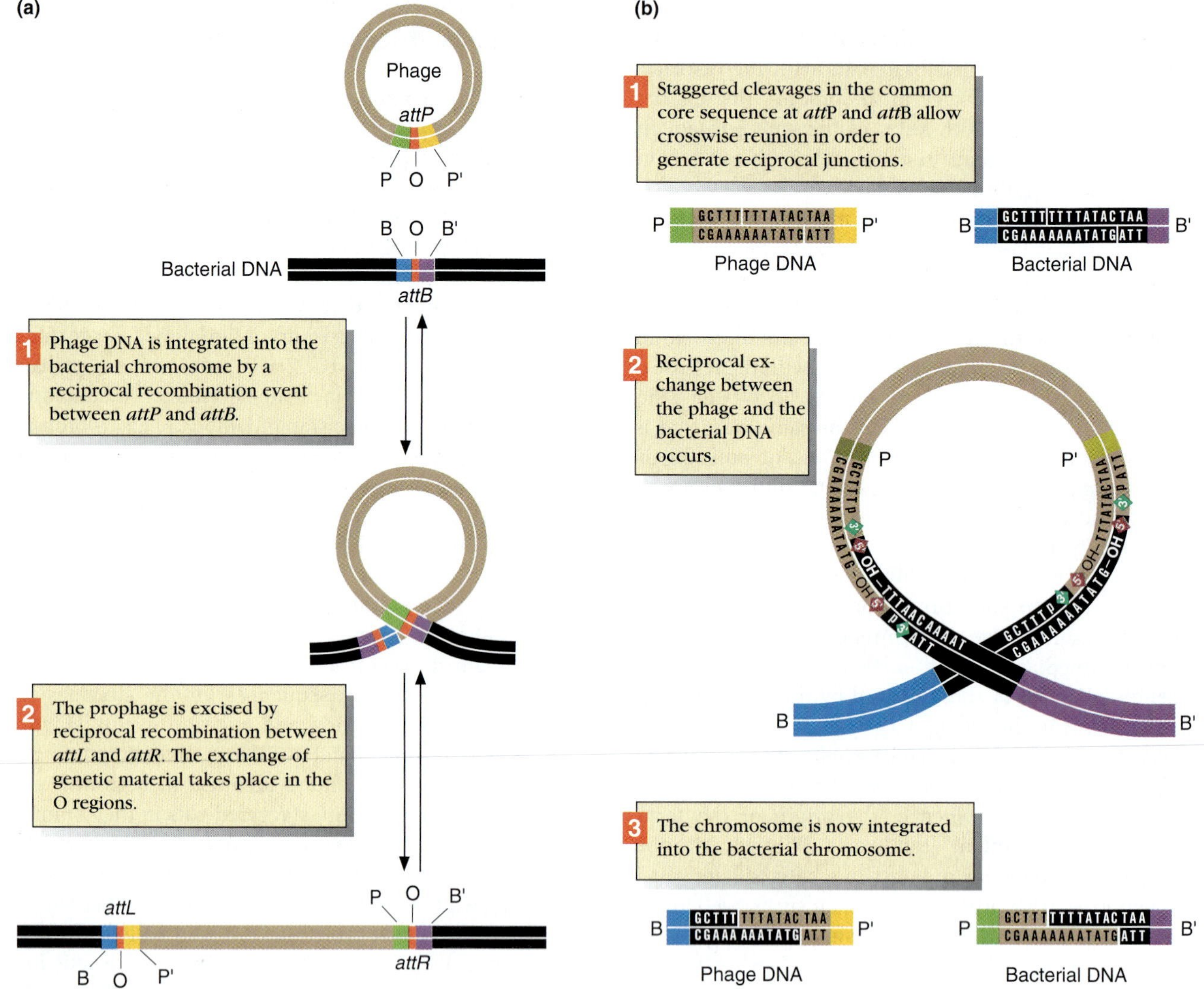

FIGURE 19.5 Lambda phage integration. (a) This figure shows how phage DNA (circular) is integrated into the bacterial chromosome by a reciprocal recombination event between the *att*P and *att*B sites. It also shows how the prophage is excised by reciprocal recombination between *att*L and *att*R. (b) Here is a closer look at the exchange of genetic material that takes place in the O regions.

SOLVED PROBLEM

Problem

What may have been the evolutionary advantage for *E. coli* to evolve a specific DNA site for lambda phage to integrate?

Solution

If lambda phage inserted randomly along the *E. coli* chromosome, it would be a dangerous potential source of insertional (frameshift) mutations. By limiting lambda insertions to a specific site, *E. coli* avoids this potential mutational load.

SECTION REVIEW PROBLEM

2. Briefly outline the process by which lambda phage is able to integrate into the *E. coli* chromosome at a specific location.

TRANSPOSABLE ELEMENTS ARE UBIQUITOUS.

A **transposable element** (TE) is a DNA sequence that is able to move and integrate throughout an organism's genome. This subject was briefly introduced in Chapter

17 when we discussed spontaneous mutations. Because the movement of transposable elements involves the integration of genetic material, the process of transposition is placed under the general heading of recombination. In contrast to homologous recombination processes that require at least some degree of sequence homology between the DNAs involved in the exchange, TE integration does not require sequence homology. For this reason, the mechanisms by which transposable elements integrate into chromosomes are classified as nonhomologous recombination. In the remainder of this chapter, we will focus on transposable elements and the molecular mechanisms that underlie their movement and integration.

FIGURE 19.6 The Mexican black strain of maize carrying the dominant *Dt* allele. The pattern results from the mutational instability associated with the *Dt* locus. *(Courtesy of M.G. Neuffer, University of Missouri.)*

Early Genetic Evidence Indicated That Genomic Instability May Be Caused by Transposable Elements.

The first indication that genomes may contain unstable and possibly mobile components was the appearance of unexpected phenotypes among the progeny of certain strains of maize. As appropriate molecular procedures have become available, first in bacteria and later in higher organisms, geneticists have been able to definitively show that these and other unusual genetic results are the consequence of the insertion of mobile DNA pieces, which today we call transposable elements.

It can be argued that the path leading to the discovery of transposable elements began in the 1940s with the experimental work of two of the founders of maize genetics, Marcus Rhoades and Barbara McClintock [see HISTORICAL PROFILE: *Barbara McClintock: A Scientific Hero (1902–1992)*]. We will, therefore, begin our overview of transposable elements with a brief historical review of the relevant work of these two scientists.

Marcus Rhoades and the *Dt* Mutant in Maize. Marcus Rhoades was engaged in analyzing the genetics of kernel color in maize when he uncovered a strain of Mexican black corn that produced unusual F_1 phenotypes when selfed. From previous studies, Rhoades knew that the black Mexican variety of corn was homozygous for a recessive allele *a* determining dark kernel pigment. In most instances, the selfing of such plants resulted in progeny showing the same dark kernel phenotype. However, when a particular strain of maize was selfed, the F_1 progeny carried a number of speckled or spotted kernels and an occasional clear kernel in addition to the expected all-dark kernels (Figure 19.6). Rhoades concluded that the appearance of clear areas on the kernels was the result of a somatic reversion mutation from the recessive dark-determining allele *a* to the dominant clear determining allele *A* during the development of the kernels. According to this model, the sooner in the developmental process the reversion occurred, the greater the clear area of the kernel.

Although Rhoades' explanation seemed consistent with his data, he noted two oddities about his results which, at the time, were unique and which he could not explain. Some kernels of the variant strain displayed multiple spots. In addition, there was an unusually large number of mutant kernels in the F_1 progeny plants, which indicated that the reversion rate was very high. Indeed, Rhoades estimated that in order to account for his observations, the rate of reversion would have to be several orders of magnitude higher than what was typically observed in maize and other experimental organisms. In addition, Rhoades noted that the exceptional genetic instability was detectable only when the *A* locus was in a particular genetic background. Rhoades attributed this to presence of a mutant dominant allele at another locus that he referred to as *Dt* (Dotted). The exceptional genetic instability observed at the *A* locus coupled with the fact that this instability was apparently under the control of an allele mapping to another locus were two unique observations that could not be explained at the time.

SECTION REVIEW PROBLEM

3. What features of the *Dt* mutant led Rhoades to believe he was dealing with a new kind of mutation?

Barbara McClintock and Controlling Elements in Maize. In the early 1940s, Barbara McClintock identified a strain of maize that was characterized by the occurrence of frequent chromosomal breaks on the short arm of chromosome 9. Chromosome breakage at this site resulted in an acentric fragment (that portion of chromosome 9

HISTORICAL PROFILE

Barbara McClintock: A Scientific Hero (1902–1992)

In mythology, the term "hero" is generally reserved for great revolutionaries whose vision and perseverance contributes to the establishment of a new world order. Barbara McClintock's pioneering and revolutionary work on transposable elements, which she persistently carried out in spite of general skepticism among many of her peers, establishes her as one of those rare scientists who truly deserves the status "scientific hero" (Figure 19.A).

McClintock's scientific greatness was apparent very early in her career. She was a pioneer in the development of maize as a prime experimental genetic system. Indeed, within only a decade of receiving her Ph.D., McClintock was recognized by many of her contemporaries as the premier cytogeneticist of the century. McClintock's early accomplishments led to her election to the prestigious National Academy of Science in 1944, only the fourth woman in history to be so elected. In 1945, she was elected president of the Genetics Society of America. There is no question that even if McClintock's scientific career had ended in the early 1940s, she would have been ensured a respected place in the history of genetics. However, McClintock's greatest triumphs were yet to come.

In 1942, McClintock moved to the Cold Spring Harbor Laboratory and began work on dicentric chromosome formation and chromosome breakage in maize. This work led, rather unexpectedly, to her discovery of an interesting class of genetic factors that appeared to move freely throughout the genome, causing unstable mutations in the process. McClintock's subsequent proposal for the existence of what we today call transposable elements came at a time in the history of genetics when the paradigm of a stable genome with genes mapping to fixed chromosomal positions was firmly entrenched. As a consequence, her data and ideas concerning movable genes were incomprehensible to many of her peers. The barriers McClintock found herself up against were as much psychological as they were scientific, a fact she well recognized. In recalling these early years, McClintock once remarked that she soon realized that no amount of published evidence would be effective in getting her ideas about movable genes widely accepted at the time. These must have been personally difficult times for McClintock, and a lesser person may well have chosen to soften her claims to prevent possible erosion of her reputation among many of her peers. However, no such face-saving option was ever considered by McClintock. She continued to persist for the next 15 years, until the discovery of transposable elements in the then popular *E. coli* bacterial system led ultimately to a vindication of her views. Today the once-revolutionary concept is widely accepted that a major component of the genome is capable of movement and thus responsible for many, if not most, spontaneous mutations. The full biological ramifications of transposable elements are still being investigated, but they portend to be monumental.

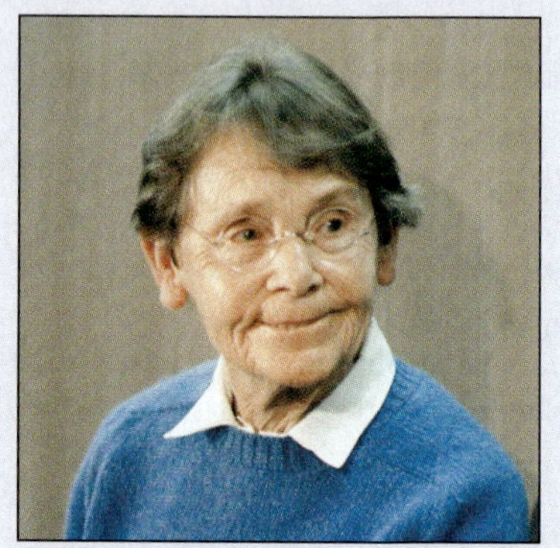

FIGURE 19.A Barbara McClintock. *(UPI/Corbis-Bettmann)*

McClintock's early work and insights on transposable elements were formally recognized 40 years later, when she was awarded a Nobel Prize in 1983. It was McClintock's ability to see in her data what others could not and her perseverance in single-handedly defending her ideas in the face of adversity that transformed her from "scientific great" to "scientific hero." She continued working at Cold Spring Harbor Laboratory on the genetics of maize until her death at the age of 90 in 1992.

distal to the breakpoint) and the subsequent loss of genes contained on the distal fragment in descendent cells (Figure 19.7). Because several genes encoding kernel phenotypes (*Colorless—Cl*, *Shrunken—Sh*, *Bronze—Bz*, and *Waxy—Wx*) were located on the short arm of chromosome 9, recessive mutations at these loci (*cl*, *sh*, *bz*, and *wx*) occurred at a high rate in this chromosomally unstable strain. McClintock was able to identify the specific site of chromosome breakage, which she designated *Ds*, for dissociation (Figure 19.7). McClintock went on to show that breaks would occur only at the *Ds* site when the allele was present in genetic backgrounds carrying another gene that she designated *Ac* for activator. McClintock realized that she was dealing with a phenomenon that on the surface appeared to be similar to that reported by Rhoades; she had identified a strain in which the appearance of high somatic mutation rates at particular loci were under the control of another locus.

McClintock's critical breakthrough came when she was able to show that the *Ds* "site" was itself subject to

(a) Microscopic detection of breakage

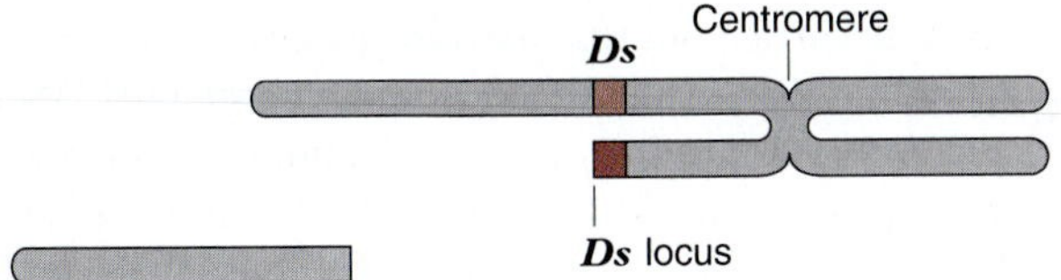

(b) Genetic detection of breakage

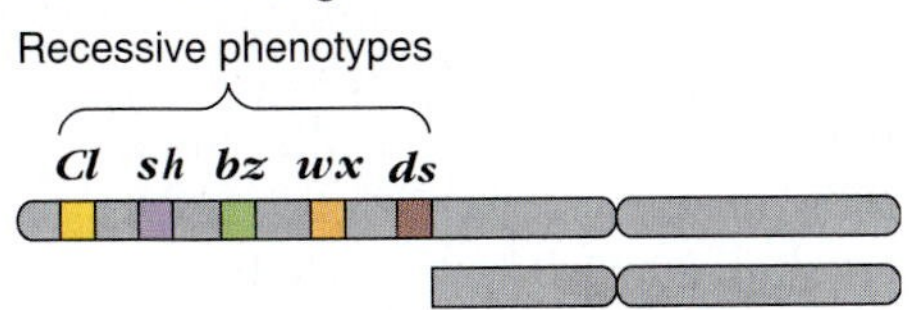

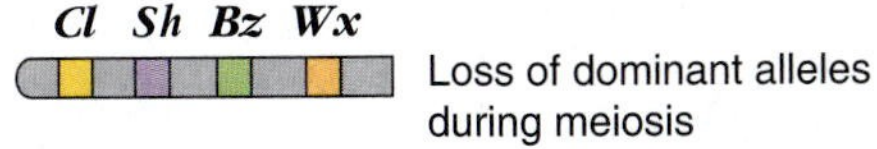

FIGURE 19.7 Genetic detection of the *Ds* element in maize. Chromosomal instability (breakage) caused by the presence of the *Ds* element in maize can be detected (a) cytologically by visual detection of the break and (b) genetically by the consequent loss of genetic markers (dominant alleles: *Cl* = *Colorless, Sh* = *Shrunken, Bz* = *Bronze, Wx* = *waxy;* recessive alleles: *cl* = *colorless, sh* = *shrunken, bz* = *bronze, wx* = *waxy*).

movement, resulting in similar types of chromosomal and genetic instabilities wherever it inserted. McClintock found that the movement of what she now referred to as the *Ds* "element" would occur only in genomes containing *Ac*. Moreover, McClintock showed that *Ac* was also capable of moving or transposing to new chromosomal locations. Thus McClintock demonstrated the existence of two distinct types of mobile or transposable elements in maize, one that acted autonomously in its ability to move about the genome (*Ac*), and another (*Ds*) whose movement was somehow dependent upon the presence of an autonomous element within the same genome. Later in this chapter, we will study the molecular processes that underlie these properties of maize transposable elements.

McClintock's findings were the first definitive evidence that genomes contain mobile or transposable genetic elements and that these elements could result in pronounced genetic instabilities. McClintock's findings were difficult for her contemporaries to understand, let alone appreciate (her papers were very complex and not clearly written). Up until this time, the position of genes in chromosomes was considered fixed and not subject to the kinds of seemingly random movements described by McClintock. Nevertheless, her elegant early work on what we today call transposable elements was suitably, if belatedly, acknowledged when she was awarded the Nobel Prize in 1983.

SOLVED PROBLEMS

Problem

What is it about the *Ds* gene that led McClintock to conclude that she was dealing with a new type of genetic entity?

Solution

McClintock identified a specific region of frequent breakage toward the end of chromosome 9 in a genetically unstable strain of maize. She named this site *Ds*, for dissociation, because this was the site where the end of chromosome 9 frequently dissociated from the rest of the chromosome. In subsequent experiments with the same unstable strain, McClintock noted that the *Ds* region in fact mapped to different chromosomal regions, causing similar breaks wherever it was located. Based on this observation, she postulated that *Ds* was in fact some sort of mobile genetic element.

Problem

Why do you think that many geneticists were initially skeptical of McClintock's findings?

Solution

The consensus view of the genome during McClintock's day was that it was highly stable. Mutations were recognized to occur at low constant frequencies at fixed genetic positions called genetic loci. These genetic loci were located at a fixed position along the chromosomes. McClintock proposed that some loci could in fact move about the genome. This proposition was so radically different from the prevailing view at that time that most geneticists thought that it was an isolated phenomenon and had no widespread significance.

SECTION REVIEW PROBLEMS

4. McClintock originally named *Ac* and *Ds* controlling elements. Why do you think she initially chose this term?
5. What is the difference between the *Ac* and *Ds* elements as you understand them at this point?

Transposable Elements Were First Isolated from Bacteria.

Although McClintock's work was the first clear indication that movable DNA sequences existed in any genome, the molecular techniques necessary to isolate and characterize these elements in eukaryotes were not sufficiently developed until the late 1970s. However, the techniques of prokaryotic molecular biology were highly developed by the 1960s, when evidence first emerged that mobile elements were also present in the genome of *E. coli*. It was for this reason that the first transposable element was isolated and characterized on the molecular level in bacteria.

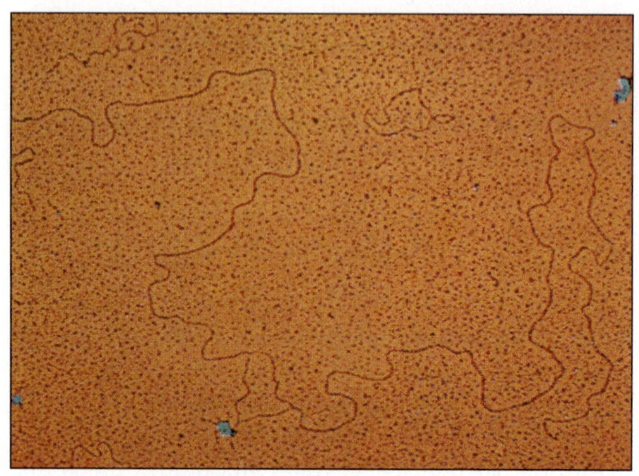

FIGURE 19.8 Electron micrograph of a *gal*⁺/*gal*⁻ DNA heteroduplex. The single-stranded loop is caused by the presence of an insertion sequence in the *gal*⁻ strain. *(© Dr. Thomas Broker/Phototake NYC)*

The galactose (*gal*) operon in *E. coli* consists of three genes under the regulatory control of a single operator. In the early 1960s, a series of polar mutants that mapped to each of the three structural genes contained within the *gal* operon was isolated. Remember that a polar mutant is one that affects the expression of one or more nonmutant genes located downstream from the actual mutation site. Polar mutants are often frameshift mutations caused by the insertion or deletion of single base pairs. Chemical mutagens such as acridine dyes, which cause a high frequency of insertion and deletion mutations, are often used to revert frameshift mutants to their wild-type. Such mutagens induce second-site lesions that restore the correct reading frame in defective genes (see Chapter 17). Curiously, treating the *gal* polar mutants with these mutagenic agents resulted in no wild-type revertants. The tentative conclusion was that the *gal* polar mutants were the result of some other class of insertion or deletion event.

In order to investigate the molecular character of the *gal* polar mutants in more detail, transduction experiments were carried out that used the capacity of phage to incorporate occasionally pieces of host (bacterial) DNA into its own genome during lytic infections (Chapter 8). For example, in one experiment, phages that had incorporated a wild-type galactose gene (gal^+) were isolated, as were some that had incorporated the mutant gal^- gene. DNA was isolated from these two classes of phage, and the samples were subjected to density gradient centrifugation (Chapter 9). It was found that the phage DNA that had incorporated the mutant gal^- gene was significantly larger than phage DNA carrying the wild-type gal^+ gene. It was concluded that the gal^- polar mutant was defective because of the insertion of a relatively large and presumably mobile piece of foreign DNA. This hypothesis was subsequently verified using electron microscopy (Figure 19.8).

SOLVED PROBLEM

Problem

What were the characteristics of the gal^- mutant that led researchers to conclude that they were not dealing with a typical frameshift mutation?

Solution

The gal^- mutant behaved as a frameshift mutation insofar as sequences located downstream of the presumed insertion site were nonfunctional. Previous studies with single-base insertion/deletion mutations frequently reverted to wild-type after treatment with acridine dyes (do you remember why?). Curiously, acridine dyes did not result in reversions of the gal^- mutant, even after repeated tries. This unusual property of the gal^- mutant led researchers to conclude that they were dealing with an atypical frameshift mutant.

SECTION REVIEW PROBLEM

6. Describe how the first bacterial transposable element was discovered and isolated.

TABLE 19.1 *Physical Characterization of Some IS Elements*

Element Name	Length (bp)	Size of Inverted Repeat (bp)	Size of Direct Repeats at Target Site (bp)	Number in Typical *E. coli* Strain
IS1	768	23	9	5–8
IS2	1327	41	5	5
IS4	1428	18	11 or 12	1–2
IS5	1195	16	4	?
IS10R	1329	22	9	?
IS903	1057	18	9	?

Bacterial Transposable Elements Share Common Features.

Bacteria contain two classes of transposable elements: small elements of about 1 kb in length called insertion sequences and larger composite elements or transposons (Tn), which range in size from about 2–9 kb in length. In the following paragraphs, we will discuss the structure and significance of these two transposable elements.

Insertion Sequences Are the Simplest Class of Bacterial Transposable Elements.

The transposable element first isolated from the *gal*$^-$ mutant in the 1960s turned out to be a representative of what is now recognized as the simplest class of bacterial TEs, called **insertion sequences** (ISs). A number of IS elements have been isolated from a variety of bacterial strains (Table 19.1). A typical *E. coli* cell may carry anywhere from one to ten IS elements.

Although the various types of IS elements are structurally distinct from one another, they share many common features. For example, all IS elements are bordered on their 5′ and 3′ ends by short **inverted terminal repeats** (IRs) typically between 15 and 25 bp in length (Figure 19.9). The two IRs carried by any particular element are identical or nearly identical in sequence but inverted in direction. Between the two IRs is an open reading frame (recall that an open reading frame, or ORF, is a coding sequence that is capable of coding for a protein product) that encodes functions necessary for IS element transposition or movement from one chromosomal location to another.

Insertion sequence elements are relatively simple, encoding only a **transposase.** This enzyme functions to excise the IS element from its existing chromosomal site and splice it into a new chromosomal position within the host chromosome. Transposases recognize specific sequences within each of the inverted repeat sequences at the end of their encoding element. They also make staggered cuts in the host DNA sequences adjoining the element. Finally, transposases make cuts at an integration site somewhere within the genome, and thus the recombination event is nonhomologous. When an IS element is integrated into a new chromosomal site, the staggered cuts are subsequently filled in, resulting in the generation of a short, direct repeat of host DNA sequences on either side of the inserted element (Figure 19.10). The size of these short, direct repeats may differ

IS element
Inverted repeat | One or more open reading frames | Inverted repeat
15–25 bp | ≈1000 bp | 15–25 bp

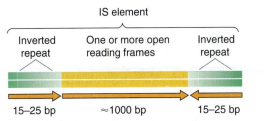

FIGURE 19.9 Structure of a bacterial insertion sequence. In this representative insertion sequence, two inverted repeats flank an internal region, which encompasses one or more open reading frames encoding transposase. The transposase recognizes sequences within the inverted repeats in order to cut the element in and out of target chromosomes.

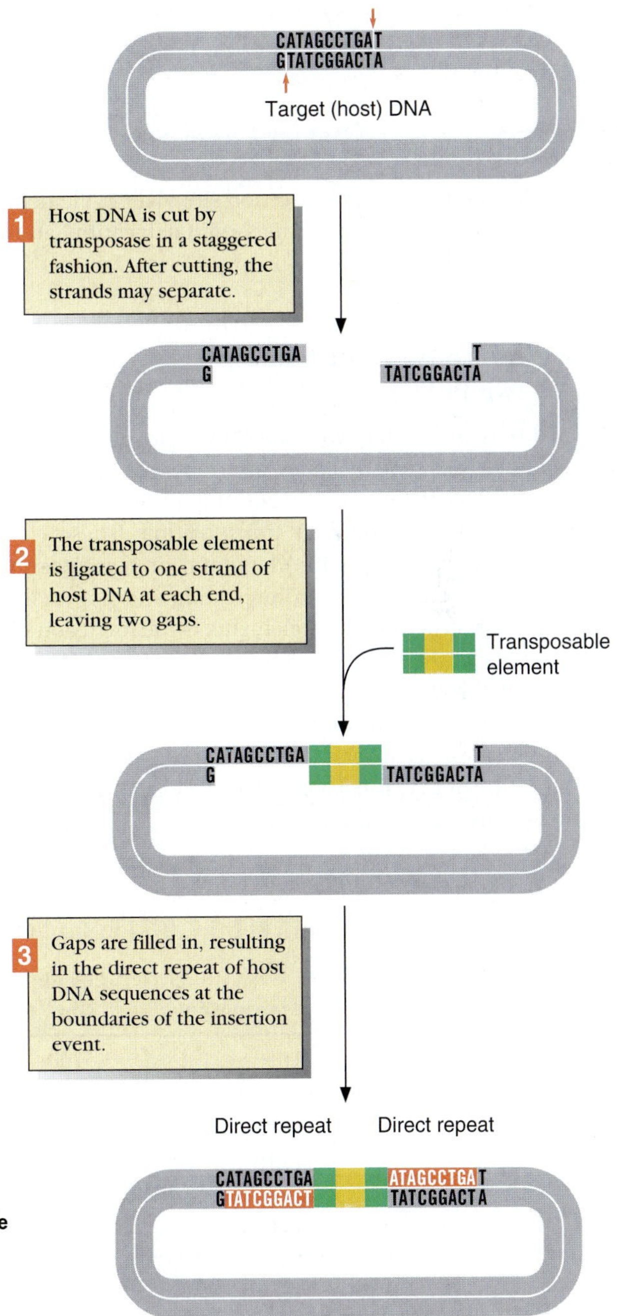

FIGURE 19.10 Transposase-mediated integration of transposable elements. Integration of transposable elements mediated by transposase results in the generation of direct repeats in the host DNA flanking the element.

between particular IS elements but are typically between 4 and 12 bp in length (Table 19.1).

SOLVED PROBLEM

Problem

Suppose that you isolate a mutant IS element that is unable to transpose. Propose hypotheses as to the possible genetic reasons for this inability.

Solution

An IS element can be nonfunctional for two reasons: (1) there may be a mutation in the element's transposase encoding sequence that renders the enzyme nonfunctional, or (2) there may be a mutation in the transposase recognition sequence located in one or both of the IRs.

SECTION REVIEW PROBLEM

7. What determines the insertion site of an IS element into the genome from its original location after transposition?

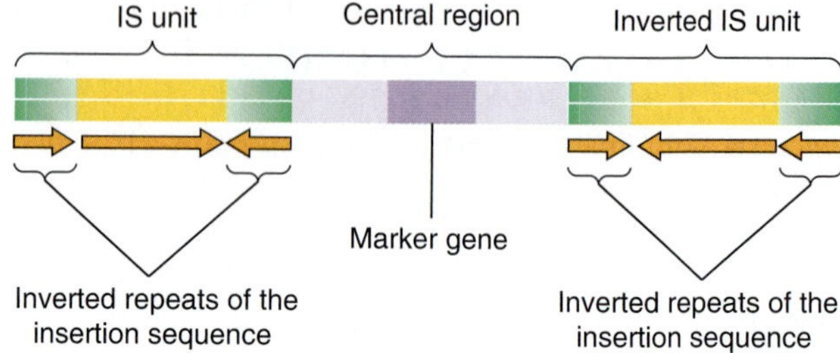

FIGURE 19.11 Tn element structure. Tn element transposons have a central region carrying markers not involved with the transposition process (for example, drug-resistance markers) flanked by IS modules. The IS modules may be in direct or indirect orientation with respect to one another. If the IS modules are in an inverted orientation (as shown), the short inverted terminal repeats at the end of the composite transposon will be identical.

Some Transposons Are Made Up of Two IS Elements.

In addition to IS elements, another class of bacterial transposable elements encodes information for drug resistance and other functions in addition to sequences encoding information for their transposition. These elements are called **composite elements,** or **transposon elements (Tn elements).** An interesting and important aspect of the structure of Tn elements is the inclusion of IS elements as the right and left "arms." These IS modules may be in either direct or indirect orientation (both going in the same direction or going in opposite directions) with respect to one another. Figure 19.11 shows the structure of a typical Tn element. Tn elements, like IS elements, are named with numbers (Tn1, Tn2, and so forth) with respect to their order of discovery. In some Tn elements, the IS components are structurally identical, whereas in others they may differ significantly. For example, both the right and left arms of Tn9 are identical IS1 elements, whereas the right and left arms of Tn10 (IS10R and IS10L) differ in sequence by 2.5% (Table 19.2).

The ability of transposons to move about the genome rests upon *trans*-acting functions (remember that a *trans*-acting function can be located anywhere in the genome; the gene encoding the function does not have to be adjacent because it encodes a product that is diffusible throughout the cell) encoded in one or the other of the IS modules and *cis*-acting elements (which are physically adjacent) present in both. If one of the IS modules of a transposon encodes a defective transposase, the element may still maintain its ability to move provided that the transposase encoded by the other IS module is functional and that the recognition sequences for the transposase in both terminal IS modules are present. The fact that only one of the two IS modules in a transposon must maintain transposase functions may explain why the sequences of the pair of IS modules in some

TABLE 19.2 *Physical Characteristics of Some Transposons*

Transposon	Length (bp)	Selectable Genes	Terminal Modules	Sequence Similarity Between Modules
Tn5	5700	kan^R	IS50R IS50L	1 bp difference
Tn9	2500	cam^R	IS1	Identical
Tn10	9300	tet^R	IS10R IS10L	2.5% divergence
Tn903	3100	kan^R	IS903	Identical

kan = kanamycin, *cam* = chloramphenicol, *tet* = tetracycline

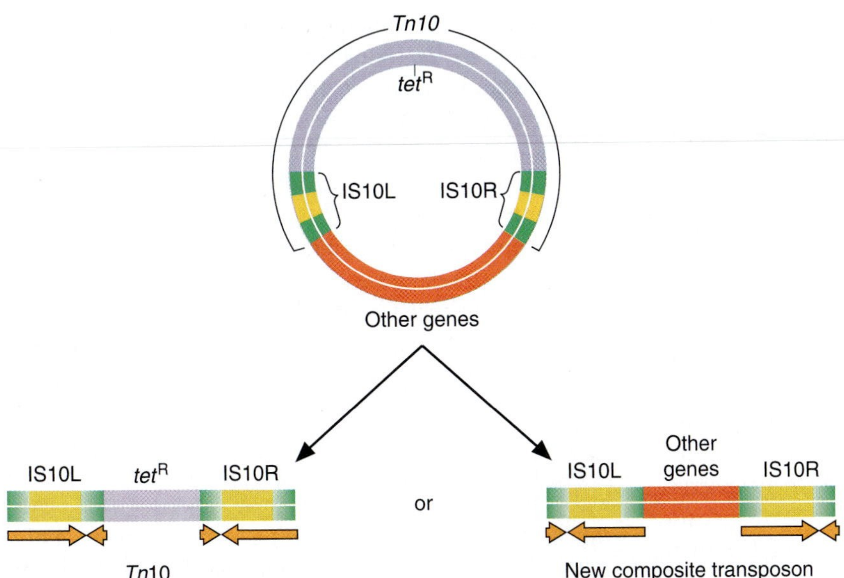

FIGURE 19.12 The structure of Tn10. The Tn element Tn10, which consists of the tetracycline resistance gene (gray) and two IS elements, IS10L and IS10R, has inserted into a small circular plasmid (blue) that contains other genes. When the two IS elements move, they may move either the tetracycline-resistance gene (maintaining the integrity of the Tn element) or the other plasmid genes (producing a new composite transposon).

transposons have diverged significantly from one another over evolutionary time.

A Tn element differs from an IS element in that it has two IS modules flanking a selectable gene (for example, a gene encoding antibiotic resistance or heavy-metal tolerance). Why do functional IS modules remain associated with particular antibiotic resistance genes and thereby maintain the integrity of Tn elements over time? At the outset, there is no molecular reason why active IS elements cannot excise from Tn elements. Indeed, such excisions are known to occur. Interestingly, it is also known that IS elements are fully capable of mobilizing any DNA sequence that lies between them regardless of the information this mobilized sequence encodes (Figure 19.12). Thus, the initial reason why IS elements become part of Tn elements encoding antibiotic resistance and similar functions is apparently because it is to their adaptive advantage to do so. In the rather uncertain environments faced by bacteria, IS elements that are associated with antibiotic resistance genes on a permanent or even semipermanent basis may have a significant selective advantage over those elements that are not. In addition, over time (on an evolutionary scale), mutations that disrupt the transposase target sequences on the inner side of the flanking IS modules may arise. Such mutations might be favored by natural selection because they would prevent excision of the individual IS modules, thus maintaining the structural integrity of the transposon over time.

SOLVED PROBLEM

Problem

Why do you think that most Tn elements contain selectable genes, such as antibiotic resistance or heavy-metal tolerance genes?

Solution

IS elements are capable of mobilizing any DNA sequence that lies between them. If by chance two IS elements mobilized an antibiotic resistance or heavy-metal tolerance gene, it may have been to their adaptive advantage because bacteria often find themselves in hostile environments containing antibiotics or heavy metals. If subsequent mutations arose over the evolutionary history of this transposon, disrupting the transposase recognition sequences on the inner side of the flanking IS modules, the structural integrity of the transposon would be stabilized.

SECTION REVIEW PROBLEMS

8. Which do you believe evolved first, IS elements or Tn elements? Why?
9. How is it possible that some functional Tn elements carry an IS module with a defective transposase encoding gene?

Bacterial Transposable Element Transposition Occurs by a Conservative, Replicative Mechanism.

Two insertion sequences, IS10 and IS50 (and the transposons of which they are a part), transpose by cutting themselves out of one chromosomal location and inserting themselves into another. This mode is called **conservative transposition** because both strands of the parental TE are conserved during the process (Figure 19.13a). Other TEs, such as IS1, IS903, and the bacterial phage mu, may transpose by conservative transposition or by an alternate pathway called **replicative transposition** (Figure 19.13b). In the replicative mode of transposition, the parental element remains at the original chromosomal site while a newly synthesized element inserts at a second chromosomal location. Replicative transposition results in a net increase in the number of TEs in a genome; conservative transposition does not.

SECTION REVIEW PROBLEM

10. What are the essential differences between conservative and replicative transposition?

Another class of transposons that encode their own transposition functions and genes for selectable traits without relying on IS elements is the TnA-type transposons (Tn1, Tn3, Tn500, and Tn1000). These are typically large (approximately 5 kb) and considerably more complex than the composite transposons considered earlier. The TnA transposons carry two genes necessary for transposition in addition to a gene for a selectable trait such as drug resistance. Perhaps the most intensively studied TnA element is Tn3 (Figure 19.14).

Tn3 carries two 38-bp inverted terminal repeats that, like the IRs carried by IS elements, contain recognition sequences (approximately 25 bp) for the action of a transposase. The transposase is encoded by the *tnp*A gene. As with IS and composite Tn elements, a direct repeat of host DNA is generated at the site insertion, which, in the case of TnA elements, is 5 bp in length.

A second gene encoded by Tn3, which is involved with transposition, is tnp^R. The tnp^R gene product has a several functions, one as a transcriptional repressor of itself and the transposase encoding *tnp*A gene, and another as a resolvase that fosters recombination. During transposition, the transposase catalyzes the formation of a **cointegrate** (when two DNA sequences simultaneously integrate) between the donor and recipient replicons, for example two plasmids (Figure 19.14b). During the formation of the cointegrate, two copies of Tn3 are generated, one at each junction. The resolvase produced by the tnp^R gene resolves the cointegrate by mediating recombination between the two Tn3 elements. The two replicons then separate, each with a copy of Tn3. TnA transposons are unique in that they undergo only the replicative mode of transposition. The resolvase-mediated recombination between the two Tn3 elements contained within the cointergrate occurs at a specific site within the element called *res* (Figure 19.14).

EUKARYOTIC TRANSPOSABLE ELEMENTS ARE HIGHLY VARIED.

TEs have thus far been identified in the genomes of yeast, fungi, nematodes, insects, mollusks, fish, reptiles, plants, and mammals (including humans). There is every reason to believe that they are constituents of all eukaryotic genomes. There are two types of eukaryotic TEs: those whose movement depends upon a transposase-mediated process, and those whose movement depends upon reverse transcription of an RNA intermediate.

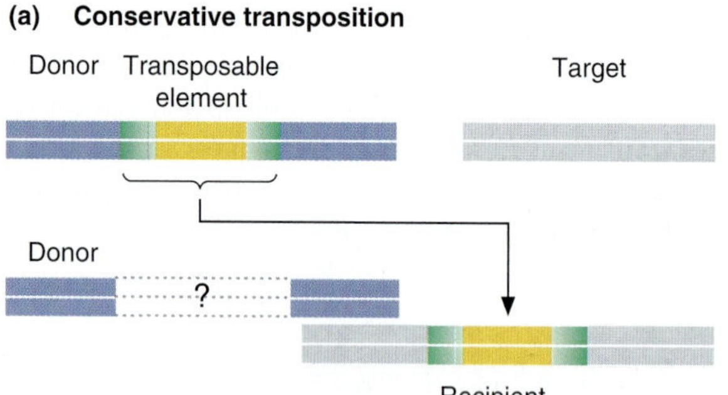

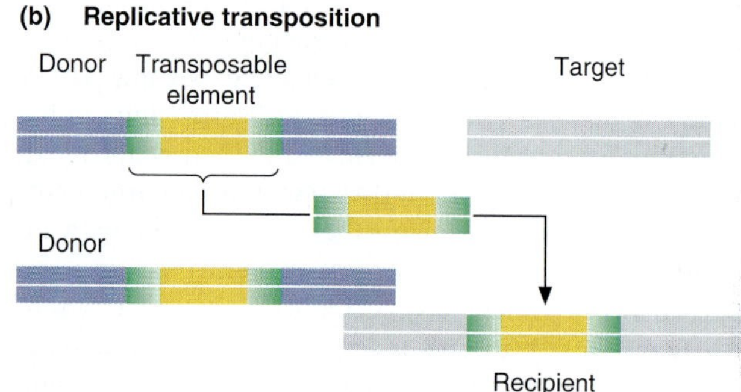

FIGURE 19.13 Conservative and replicative transposition. (a) Conservative transposition allows the transposable element to move from a donor site to a recipient site. What happens to the donor site is not presently clear. (b) Replicative transposition creates a copy of the transposable element, which is then inserted at the recipient site. The donor site remains unchanged.

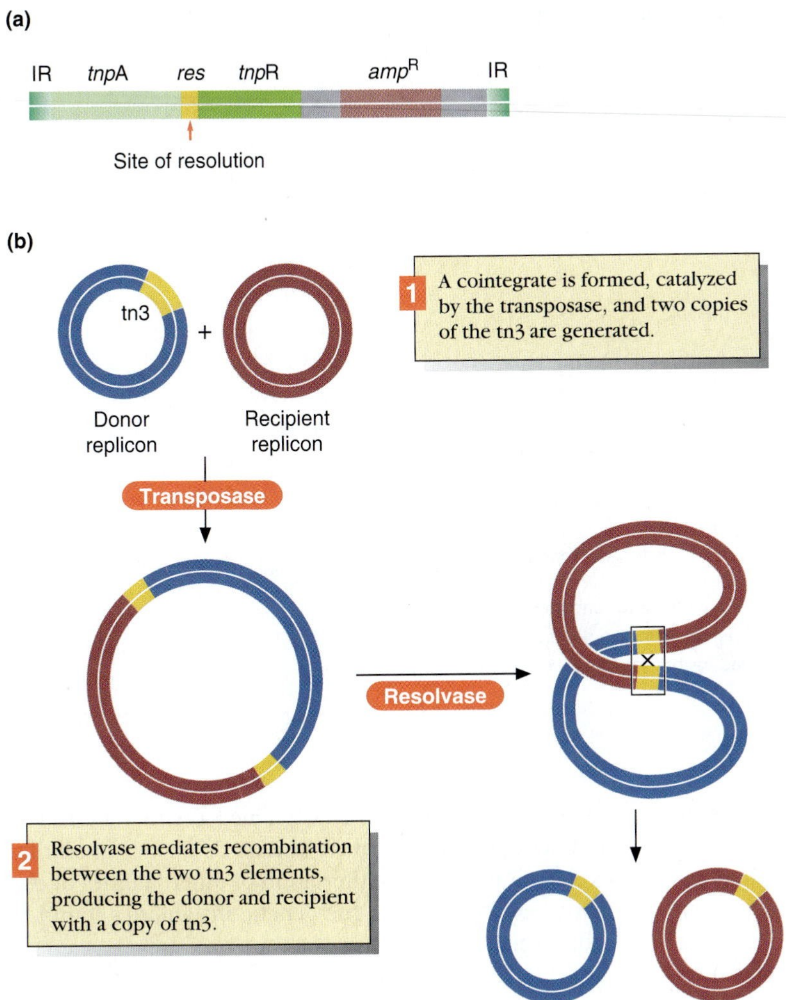
(a)
IR
tnpA
res
tnpR
ampR
IR
Site of resolution
(b)
tn3
+
Donor replicon
Recipient replicon
Transposase
1 A cointegrate is formed, catalyzed by the transposase, and two copies of the tn3 are generated.
Resolvase
2 Resolvase mediates recombination between the two tn3 elements, producing the donor and recipient with a copy of tn3.

FIGURE 19.14 The structure of Tn3. (a) Tn3 has inverted repeats (IR), an internal *res* site (the site of resolution), and two genes believed to be involved in the transposition process: *tnp*A, a gene-encoding transposase, and *tnp*R, which encodes a repressor of itself and the *tnp*A gene. Tn3 also carries an ampicillin-resistant gene (amp^R). (b) Transposition involves the cointegration of the donor and recipient replicons by the transposase, with the formation of two copies of the element. The protein product of *tnp*R functions as a resolvase.

Some Eukaryotic Transposable Elements Require a Transposase for Transposition.

Among the best-characterized eukaryotic transposable elements are those whose movement is mediated by an element-encoded transposase, similar to bacterial transposable elements. Included in this group are the maize-controlling elements that were genetically characterized by McClintock.

We now recognize that the maize genome contains many families of transposable elements whose movement is transposase-dependent. Each family contains a transposase-encoding **autonomous element** and a number of generally smaller **nonautonomous elements** that either lack a transposase-encoding gene or carry a defective transposase gene. All potentially active elements contain target sequences upon which the transposase acts. Nonautonomous elements can be mobilized only when present in a genome that also contains a transposase-encoding autonomous element. Autonomous elements, as the name implies, are self-regulating and thus may transpose within any genome in which they are present. One well-studied family of eukaryotic transposable elements that depends upon transposase to be mobilized is the maize *Ac* and *Ds* elements. This family provided the basis of McClintock's initial observations.

The *Ac–Ds* Family of Maize Transposable Elements. The maize *Ac* transposable element is 4563 bp in length and encodes a single transposase-encoding gene (Figure 19.15). The transposase gene contains four introns and is regulated in a developmental and tissue-specific manner. The *Ac* element is bordered on its 5′ and 3′ ends by two 11-bp inverted repeats that contain recognition sequences for the action of the element-encoded transposase. As with the bacterial transposons, the transposase produces staggered cuts at the target site, resulting in the generation of 8-bp, direct repeats of host DNA sequence at the site of insertion.

Different *Ds* elements vary considerably in size, but all *Ds* elements have a defective transposase-encoding

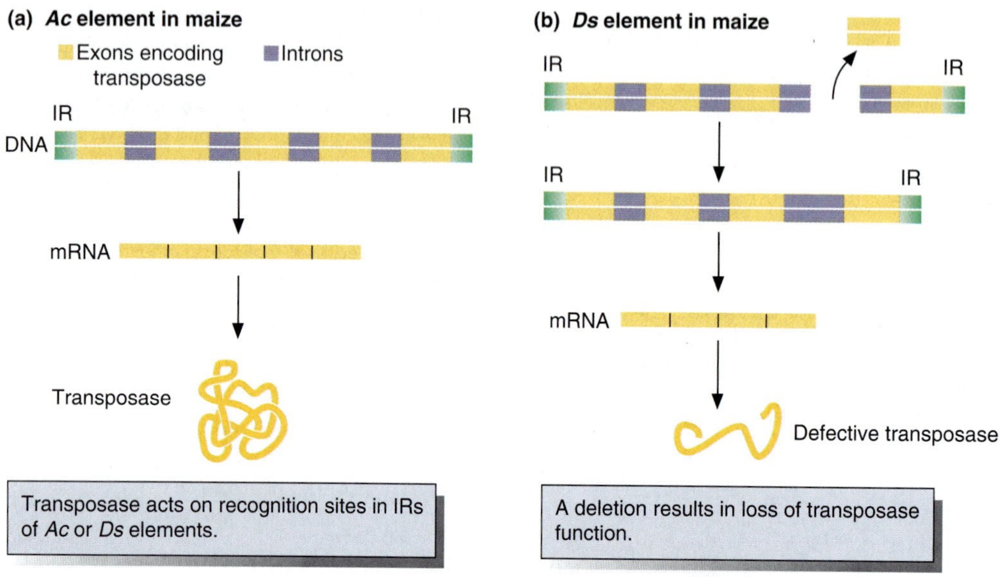

(a) *Ac* element in maize

Exons encoding transposase · Introns

IR · IR

DNA

mRNA

Transposase

Transposase acts on recognition sites in IRs of *Ac* or *Ds* elements.

(b) *Ds* element in maize

IR · IR

IR · IR

mRNA

Defective transposase

A deletion results in loss of transposase function.

FIGURE 19.15 Structure of the maize *Ac* element. The *Ac* element (a) encodes the transposase functions, whereas the *Ds* elements (b) have internal deletions that make them deficient in transposase functions. *Ds* elements can acquire the necessary transposase functions in *trans* from functional *Ac* elements.

sequence. In most instances, the inability of *Ds* elements to produce functional transposase is attributable to the deletion of all or part of the transposase-encoding sequence. Sequence analysis has demonstrated that most *Ds* elements are deletion variants of full-length *Ac* elements. The only possible exception to this generalization are the small *Ds*1 elements that are sequentially homologous to *Ac* elements only in the fact that they carry the same 11-bp inverted repeats. Hence the evolutionary origin of *Ds*1 elements is currently uncertain.

To remain potentially mobile, all *Ac* and *Ds* elements must maintain the integrity of those sequences recognized by the transposase. As mentioned previously, these essential target sequences are contained within the 11-bp inverted repeats of these elements. However, other sequences contained within the body of the element may play an auxiliary role as well.

Excision of a transposable element from a particular chromosomal position may result in a break, which can lead to the loss of alleles carried on the resulting acentric chromosomal fragment (Figure 19.16). If the transposon-induced break occurs prior to chromosome replication, a subsequent sister chromatid fusion can lead to what McClintock described as the fusion–bridge–breakage cycle. In this condition, the two centromere-containing sister chromatids fuse, resulting in a dicentric chromosome. During mitosis, these dicentric chromosomes have a high probability of breaking unevenly, resulting in genetic imbalance in the progeny cells (Figure 19.16). Thus, all the genetic abnormalities originally recognized by McClintock can now be explained as either a direct or indirect consequence of *Ac* and/or *Ds* elements. Similar explanations invoking the action of other families of elements can account for the genetic instabilities found in other maize strains.

SECTION REVIEW PROBLEMS

11. In molecular terms, what is the difference between autonomous and nonautonomous elements?
12. Which do you believe evolved first, *Ds* elements or *Ac* elements? Why?

The P-Element Family of *Drosophila* Transposable Elements. Another example of eukaryotic TEs that are dependent upon transposase for movement is the *Drosophila* P-element family. Although P elements were first isolated from *D. melanogaster,* it soon became apparent that not all *D. melanogaster* strains carried these transposable elements. Strains that do not contain P elements are called M strains, and those that do carry the element are called P strains. (The letter P stands for paternal, and the letter M stands for maternal. These letters refer to a particular set of genetic crosses that led to the discovery of P elements. A full description of these crosses is presented later in this section.) It has recently been found that none of the closely related sibling species of *D. melanogaster* carry P elements nor do strains of *D. melanogaster* established from flies collected in the

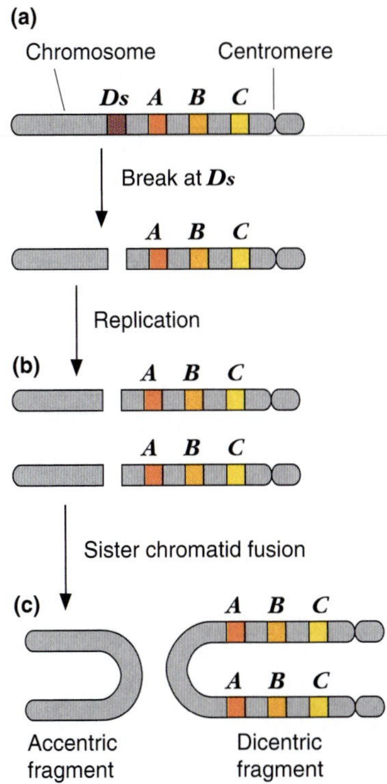

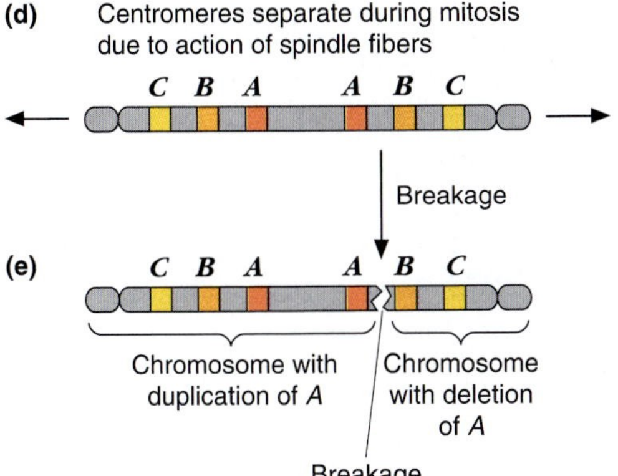

FIGURE 19.16 The chromatid fusion–bridge–breakage cycle. (a) A break at the *Ac*-activated *Ds* site in a chromosome provides an initiation site for the cycle. Subsequent replication (b) and sister chromatid fusion (c) results in the generation of acentric and dicentric fragments. Because the acentric fragment lacks centromeres, it is lost during meiosis. (d) Spindle fibers attach to the centromeres of the dicentric fragment, pulling each end of the chromosome to opposite poles, ultimately causing a break. (e) Breakage can occur at any position along the dicentric chromosome resulting in meiotic products with a chromosomal imbalance.

wild prior to about 1960. P elements began to show up in significant frequencies in wild-collected *D. melanogaster* flies beginning in the mid-1960s, and today are present in all *D. melanogaster* collected in the wild. It is currently believed that P elements were introduced into *D. melanogaster* from a distantly related *Drosophila* species by a virus- or parasite-mediated, cross-species transfer.

Fully functional P elements are approximately 2.9 kb in length and contain four open reading frames separated by three introns (Figure 19.17). Because transposition-defective P-element mutations have been mapped to each of the four ORFs, we may conclude that each ORF encodes information that is critical to transposition. As is the case with the maize-controlling elements, truncated versions of the full-length P element abound in the genomes of P strains. These truncated elements are typically defective for transposase functions but may be transpositionally activated in *trans* by a fully functional element. The typical P strain will contain 10–20 full-length P elements and 20–30 truncated elements.

Although P elements are transcriptionally active in both germline and somatic tissues, P-element transpositions normally occur only in the germline. The molecular basis of this difference in transpositional activity is caused by differential splicing patterns occurring in somatic and germline tissue (Figure 19.17). RNAs present in somatic tissue are the result of the splicing of ORF0 to ORF1 and ORF1 to ORF2. The resulting somatic transcripts give rise to a 66,000-Da protein, which acts as a

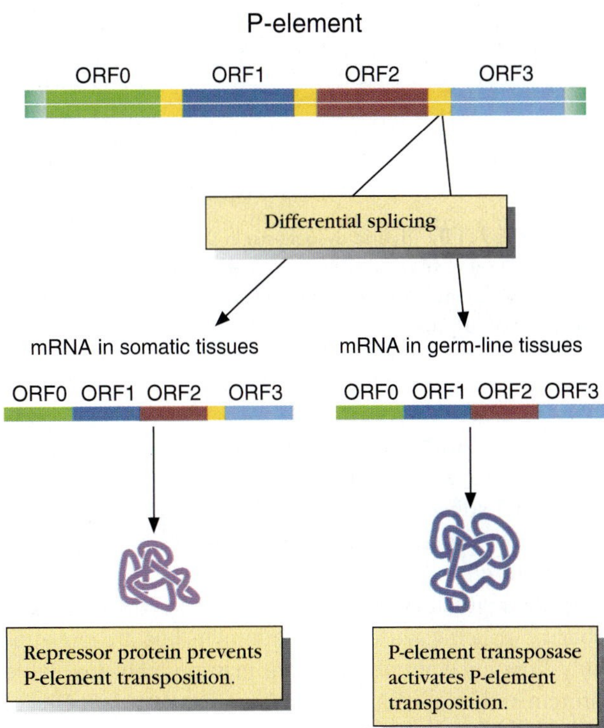

FIGURE 19.17 The structure of the *Drosophila* P element. The *Drosophila* P element has four open reading frames (ORFs). The first three ORFs are spliced together in somatic tissues encoding a 66,000-Da protein that represses transposition. All four ORFs are spliced together in germline tissue, resulting in production of an active transposase.

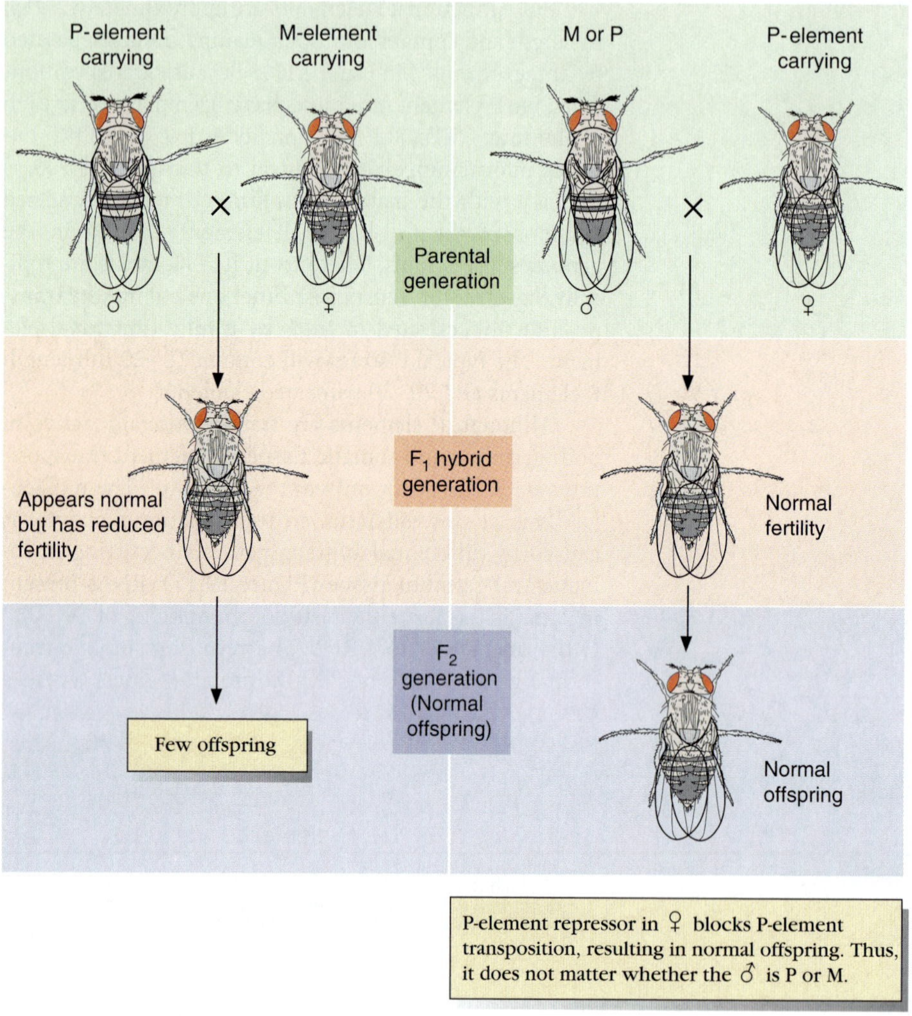

P-element carrying
M-element carrying
M or P
P-element carrying
×
Parental generation
♂ ♀ ♂ ♀
F_1 hybrid generation
Appears normal but has reduced fertility
Normal fertility
F_2 generation (Normal offspring)
Few offspring
Normal offspring

P-element repressor in ♀ blocks P-element transposition, resulting in normal offspring. Thus, it does not matter whether the ♂ is P or M.

FIGURE 19.18 Hybrid dysgenesis. In *Drosophila*, when a P-male is crossed to a M-female, P-element transpositions occur in the eggs, resulting in a high frequency of P-element insertion mutations. This syndrome, which is called hybrid dysgenesis, results in F_1 hybrids with reduced fertility.

repressor of transposition. In germline tissue, an additional splice, which joins ORF2 to ORF3, is made. This additional splice results in an RNA that gives rise to an 87,000-Da protein, which functions as the P-element transposase.

The regulatory factors responsible for the germline-specific splicing pattern are unknown but are presumably host encoded. The factors may possibly involve a protein and/or a snRNA (snRNA are small nuclear RNAs, see Chapter 11) molecule, which complexes with the newly transcribed P-element RNA to facilitate the correct splicing sequence.

Another interesting instance of regulatory control involving P elements occurs when P strains are crossed to M strains to produce P–M hybrids. If the parental P-male (paternal) is crossed to an M-female (maternal), the resulting hybrids will experience high rates of P-element transposition in their germ cells, leading to significantly reduced fertility or sterility. This condition is referred to as **hybrid dysgenesis** (Figure 19.18). However, when the reciprocal cross is carried out, that is, M-males × P-females, the F_1 progeny are normal. A P-element encoded repressor is manufactured in the somatic "nurse cells" and pumped into the egg as part of the cytoplasm (see Chapter 15). This effectively blocks P-element transposition. Accordingly, only crosses in which P-element-carrying sperm are introduced into eggs of the M-cytotype will result in dysgenesis. These facts not only explain what happens when M-males mate with P-females, but they also account for the fact that the progeny of crosses between two P-type flies (or two M-type flies) are completely fertile.

SOLVED PROBLEM

Problem

Could it be of adaptive advantage for *Drosophila* not to have active P-element transposase produced in somatic tissue?

Solution

Yes. The lack of transposase in somatic cells renders P elements transpositionally inactive in these cells, thus reducing the frequency of somatic cell mutations.

SECTION REVIEW PROBLEM

13. What is hybrid dysgenesis?

Some Eukaryotic Transposable Elements Require Reverse Transcriptase for Transposition.

The eukaryotic TEs considered thus far are similar to the bacterial IS and Tn elements in that they are dependent upon an element-encoded transposase for mobilization. In this section, we will examine a class of eukaryotic elements called **retrotransposons**, which are structurally and functionally more closely related to infectious retroviruses than to any other group of eukaryotic or prokaryotic transposable elements. The most basic and significant similarity between retrotransposons and retroviruses is the fact that reverse transcription of an RNA intermediate is an essential component of their replicative cycles. In each case, the process is mediated by an element- or virus-encoded enzyme called **reverse transcriptase** (RNA-dependent DNA polymerase).

Retrotransposons are the most widely distributed and most abundant of all eukaryotic transposable elements. In the genomes that have been most extensively studied thus far (that is, yeast, maize, *Drosophila*, and mice), retrotransposons have been estimated to make up to between 10% and 50% of the total genomic DNA. The abundance and extensiveness of retrotransposons in eukaryotic genomes make them a significant potential source of mutation. Indeed, in *Drosophila*, retrotransposons have been found to be responsible for a majority of all spontaneous mutations that have significant phenotypic effects (Table 19.3).

The insertion of retrotransposons in or near genes is frequently associated with regulatory and developmental mutations, including those responsible for the onset of cancer. For these and other reasons, extensive research has been, and continues to be, devoted to understanding the molecular biology of this important and interesting class of eukaryotic transposable elements.

SECTION REVIEW PROBLEM

14. In what ways do retrotransposons differ from the eukaryotic TEs considered previously in this chapter?

The Retroviral and Retrotransposon Life Cycles Share Many Common Features.

It is useful to view the replicative cycle of retrotransposons as an abbreviated version of the replicative

TABLE 19.3 *Proportion of* Drosophila melanogaster *Spontaneous Mutations Caused by the Insertion of Retrotransposons*

Locus	Spontaneous Mutations Caused by Retrotransposon Insertion (%)	Number Mutants Screened
v (vermilion)	80	5
ct (cut)	100	28
ry (rosy)	60	5
f (forked)	75	4
su(s) (suppressor of sable)	71	7
Bx (Beadex)	100	4
bx (biothorax)	89	9
sc (scute)	100	2
Antp (Antennapedia)	40	5
Average	79	69 loci screened

process of true infectious **retroviruses.** Because of the many structural and functional similarities between retrotransposons and retroviruses, it is generally recognized that both types of elements evolved from a common ancestor. We will consider the more involved replicative process of infectious retroviruses first and then contrast it with the process by which retrotransposons replicate and move about the genome.

The Retroviral Life Cycle. A retrovirus carries its genetic information in the form of RNA. A retroviral particle consists of two genome-length strands of RNA and the enzyme reverse transcriptase packaged in a protein capsid, which in turn is enclosed within a protein-polysaccharide envelope (Figure 19.19).

Retroviral infection begins with glycoprotein knobs located on the surface of the viral particle attaching to receptor proteins located on the surface of the target cell (Figure 19.20). For the attachment to occur, there must be compatibility between the viral glycoprotein knobs and the receptor proteins on the surface of the target cell. This explains why retroviruses show specificity with regard to the host cells they infect. After the virus has attached to the cell surface, the contents of the viral particle are introduced into the cytoplasm of the target cell.

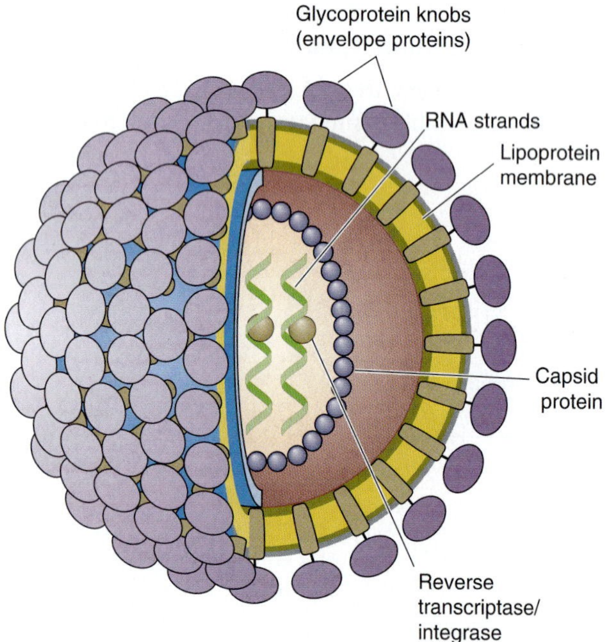

FIGURE 19.19 Schematic cross-section of a retrovirus. The two RNA strands (green) and reverse transcriptase/integrase proteins are packaged within the protein capsid. The capsid is surrounded by the viral envelope consisting of a lipoprotein membrane that contains glycoprotein knobs.

Retroviral RNA has direct repeats at each end called the R regions (Figure 19.21a). R regions vary in length from 10–80 nucleotides among different retroviruses. Adjacent to the 5′ R region is the 80- to 100-nucleotide U5 region (U = unique, 5 = to the 5′ end). The R and U5 regions contain regulatory sequences that, as we shall soon see, are important to retroviral expression. Following the U5 region in the retroviral genome are the *gag*, *pol*, and *env* sequences that encode retroviral structure and replicative functions. The *gag* region encodes proteins of the capsid, the *pol* region encodes the reverse transcriptase and enzyme functions (integrase) involved with integration into the host genome, and the *env* region encodes the protein component of the viral envelope. The 3′ end of the retroviral RNA consists of a unique 170- to 1250-nucleotide-long U3 region (U = unique, 3 = to the 3′ end), which directly precedes the 3′ (R region) direct repeat. Like the R and U5 regions, the U3 region contains regulatory sequences important to retroviral expression.

After the contents of the viral envelope are introduced into the target cell's cytoplasm, reverse transcription takes place, and the genetic information encoded in viral RNA is transcribed into a double-stranded DNA molecule. Reverse transcription of retroviral RNA is rather unique because the enzyme jumps to a new position on the RNA template part of the way through the process (a description of this jumping process is beyond the scope of this book). This switch in position results in a DNA product with the U3, R, and U5 sequences of the RNA template being repeated on either side of the DNA product (Figure 19.21b). These repeated regions are referred to as **long terminal repeats** (LTRs).

The DNA copy of the retroviral genome finds its way to the nucleus of the target cell by a currently unknown mechanism, and it is then integrated at a random position into the host genome. The integrated copy of the viral genome is referred to as the retroviral provirus (Figure 19.21c). The retrovirus may remain dormant in the proviral form for an indefinite period of time, called the latency period (see *CURRENT INVESTIGATIONS: AIDS, a Disease Associated with the Human Inmmunodeficiency Virus*). Mean latency periods can vary significantly between different retroviruses or for the same retrovirus in different hosts.

Retroviral proviruses emerge from the latency period when they become transcriptionally activated by RNA polymerase II (Figure 19.22) and other host-encoded transcription factors. Retroviral LTRs carry characteristic features of eukaryotic promoters as well as enhancer sequences that provide different retroviruses with the ability to be expressed preferentially in a tissue-specific manner. The LTR also carries termination signals and poly(A) addition sites. Although both the 5′ and 3′ LTRs

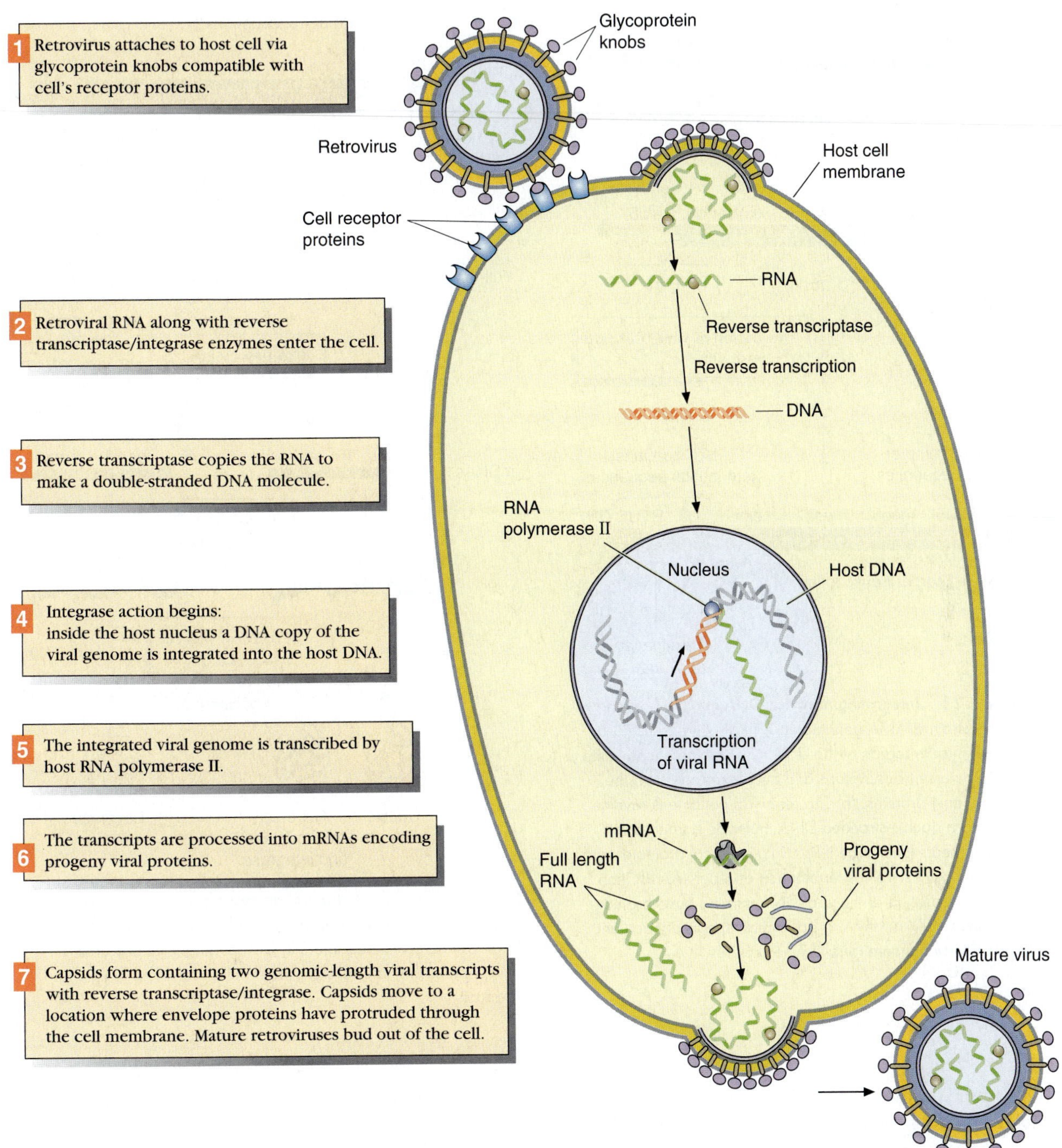

1 Retrovirus attaches to host cell via glycoprotein knobs compatible with cell's receptor proteins.
2 Retroviral RNA along with reverse transcriptase/integrase enzymes enter the cell.
3 Reverse transcriptase copies the RNA to make a double-stranded DNA molecule.
4 Integrase action begins: inside the host nucleus a DNA copy of the viral genome is integrated into the host DNA.
5 The integrated viral genome is transcribed by host RNA polymerase II.
6 The transcripts are processed into mRNAs encoding progeny viral proteins.
7 Capsids form containing two genomic-length viral transcripts with reverse transcriptase/integrase. Capsids move to a location where envelope proteins have protruded through the cell membrane. Mature retroviruses bud out of the cell.
Glycoprotein knobs
Retrovirus
Host cell membrane
Cell receptor proteins
RNA
Reverse transcriptase
Reverse transcription
DNA
RNA polymerase II
Nucleus
Host DNA
Transcription of viral RNA
mRNA
Full length RNA
Progeny viral proteins
Mature virus

FIGURE 19.20 Overview of the retroviral life cycle. The retrovirus attaches to the surface of the target cell, and the contents enters the cell. A DNA copy of the viral genome is generated via reverse transcription and is integrated into the host chromosome. The integrated viral genome is transcribed, producing full-length RNA. Some of the full-length RNA will serve as the genomes of progeny viruses and some is processed into mRNA-encoding viral proteins. Viral capsids are assembled at the periphery of the cell. Envelope proteins are embedded in cell membranes. The complete viral particle buds out of the cell, completing the retroviral life cycle.

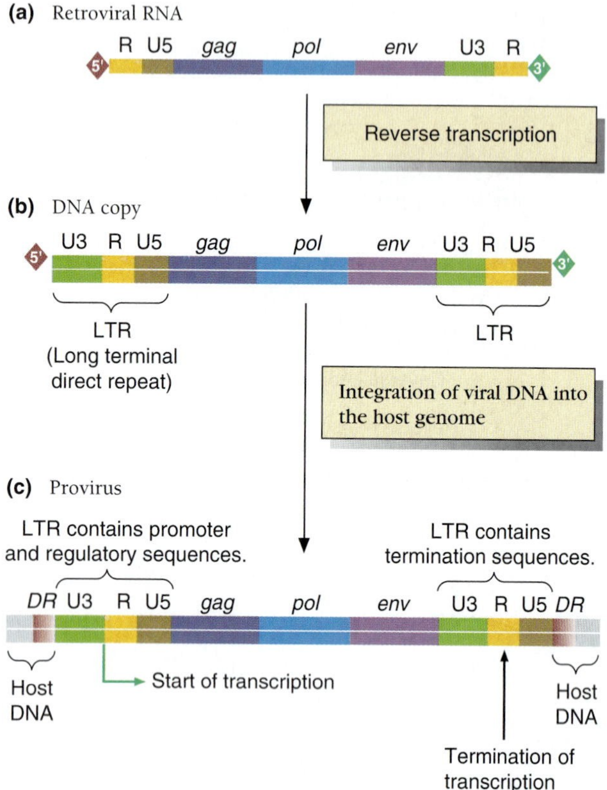

FIGURE 19.21 Integrating retroviral DNA into the host chromosome. (a) Retroviral RNA genome ends in direct repeats (R) at each end. The middle region of the retroviral genome encodes the capsid (*gag*), reverse transcriptase and integrase (*pol*), and the viral envelope (*env*) proteins. (b) The retroviral genome is reverse transcribed into a double-stranded DNA molecule bordered by long terminal repeats (LTRs). The LTRs contain critical promoter and enhancer sequences as well as termination and poly(A) addition sequences. (c) Integration of the viral DNA into the host chromosome generates small (usually < 6 bp) direct repeats at each end. The integrated viral genome is called the provirus.

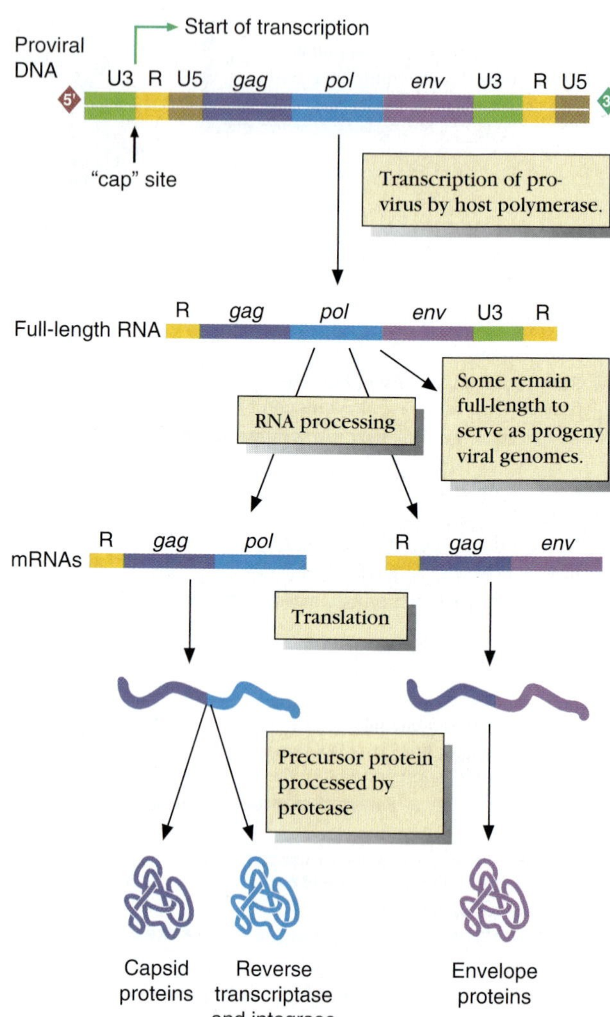

FIGURE 19.22 Retroviral processing reactions. Some full-length RNA will serve as progeny viral genomes. Some full-length RNA will be processed into mRNA-encoding viral precursor polyproteins that later are processed into individual functional products.

carry the same promoter and termination sequences, the promoter functions are typically used only in the 5′ LTR, whereas termination functions are usually active only in the 3′ LTR. The molecular basis of this specificity is not completely understood. However, it has been shown that if the 5′ LTR is rendered nonfunctional through some mutation, the 3′ LTR may serve as a promoter and transcribe adjacent 3′ host sequences. If the adjacent sequence is a gene, it may acquire a new pattern of expression, and this may have negative effects for the host. For example, proviral activation of cellular oncogenes (Chapter 20) has been found to induce the transformation of normal host cells into cancer cells.

The transcriptional start, or cap, site in the retroviral LTR is at the beginning of the 5′ R region. Some of the nascent RNAs generated from the proviral template remain full-length and are destined to be packaged into progeny viruses; the rest of the nascent RNAs become processed into mRNAs encoding the *gag*, *pol*, and *env* functions described earlier (Figure 19.22). The retroviral mRNAs are translated in host polysomes in the cytoplasm. The *gag-pol* region is initially translated as a polyprotein and later processed into independently functional proteins by the action of a protease. Two genomic-length RNAs are packaged into capsid particles along

(Text continued on page 604.)

CURRENT INVESTIGATIONS

AIDS, a Disease Associated with the Human Immunodeficiency Virus

Acquired immunodeficiency syndrome (AIDS) is a devastating disease that has been associated with a human retrovirus called HIV I (human immunodeficiency virus, type I, Figure 19.B). A glycoprotein on the surface of the HIV particle (gp120) is compatible with receptor proteins on the surface of T4 lymphocytes (CD4 proteins). T4 lymphocytes play a key regulatory role in the immune system's ability to respond to invading pathogens. When HIV has successfully gained entrance to an individual's bloodstream, the virus seeks out and infects T4 cells and other cells displaying CD4-like receptor proteins (for example, other immune cells, certain cells of the nervous system and intestine, and some bone-marrow cells).

The structure of the HIV provirus is somewhat different than the LTR-*gag*-*pol*-*env*-LTR configuration of the prototype mammalian and murine retroviruses. HIV and members of its related retroviral family (lentiviruses) contain additional sequences that autoregulate the viral expression (Figure 19.C).

Because reverse transcription is an extremely error-prone process, HIV proteins are constantly acquiring new sequences. Those substitutions that adversely affect HIV function are presumably weeded out by natural selection, whereas those that are selectively neutral or nearly neutral may contribute to a immunologically highly variable pool of HIV particles. This flexibility has hindered efforts to produce a single effective HIV vaccine. Although efforts are underway to target vaccines against functionally important and thus structurally less variable HIV proteins, most current therapies can only slow and not

(continued)

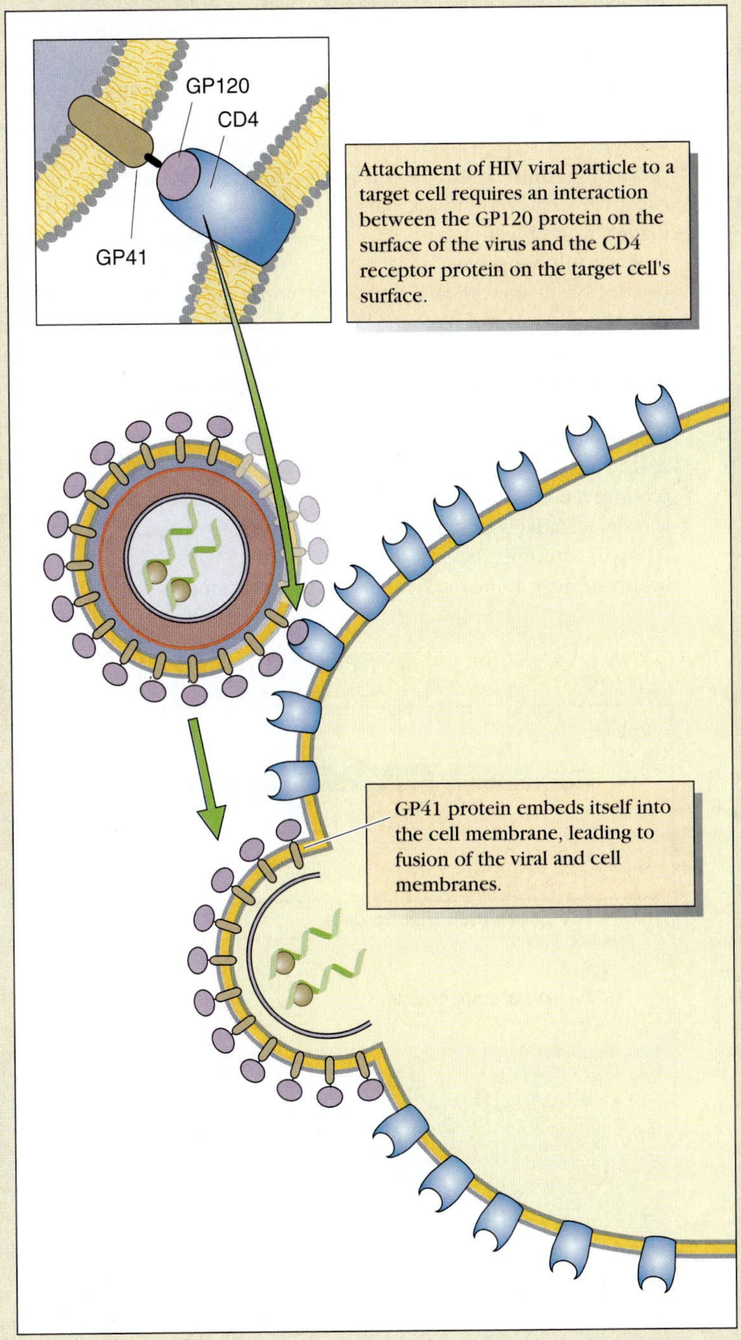

GP120
CD4
GP41
Attachment of HIV viral particle to a target cell requires an interaction between the GP120 protein on the surface of the virus and the CD4 receptor protein on the target cell's surface.
GP41 protein embeds itself into the cell membrane, leading to fusion of the viral and cell membranes.

FIGURE 19.B HIV infection. HIV preferentially infects cells with CD4-like receptors such as are found on the immune system T4 cells.

(continued)

stop progression of AIDS. As described later, base analogues such as AZT (azidothymidine) terminate HIV replication but have the disadvantage of also inhibiting DNA synthesis involved with critical body functions such as red blood cell production. A new class of drugs called protease inhibitors also interfere with key steps (such as protein processing) in the retroviral life cycle. Moderate levels of base analogues in combination with protease inhibitors have proven to be effective in slowing HIV replication in many infected patients. Because the retroviruses are especially prone to rapid evolution (see Chapter 22), HIV is likely to evolve resistance to most current therapies over time. Thus, even though efforts to produce a permanent cure for, or at least an effective vaccine against, HIV will continue, the best public health strategy at the moment is mass education programs to help prevent the spread of the disease. High-risk groups such as homosexuals and intravenous drug users are being especially targeted in these educational programs.

For those individuals already infected by the virus, the currently prescribed therapy involves drugs that interfere with viral replication and the processing of the *gag* and *pol* gene products by a protease (Figure 19.22). Base analogues such as AZT (Figure 19.D) are commonly prescribed at the present time. AZT will incorporate into the growing nucleotide chain of reverse transcribed DNA copies of the HIV genome but provides no free 3′-OH end for attachment of subsequent bases in the growing chain. Thus, incorporation of AZT results in chain termination. Unfortunately, chain terminators such as AZT interfere with DNA polymerase-based replication as well, and so critical host functions such as the production of new blood cells are common negative side effects of AZT therapy. The ideal strategy in AZT therapy is to administer a level of the drug that maximally interferes with HIV replication while minimally interfering with critical host-cell functions such as the production of new blood cells in bone marrow. Unfortunately, such optimal levels are difficult to attain.

Another class of HIV inhibitors function by preventing the protease from cleaving the *gag-pol* gene product into its component parts. This class of potential anti-HIV drugs is perhaps the most important group of experimental treatments at the present time, although the current candidates still face major problems—including the development of resistant strains of HIV, poor oral absorption, and expense and difficulty of manufacture. Protease inhibitors are drugs that re-

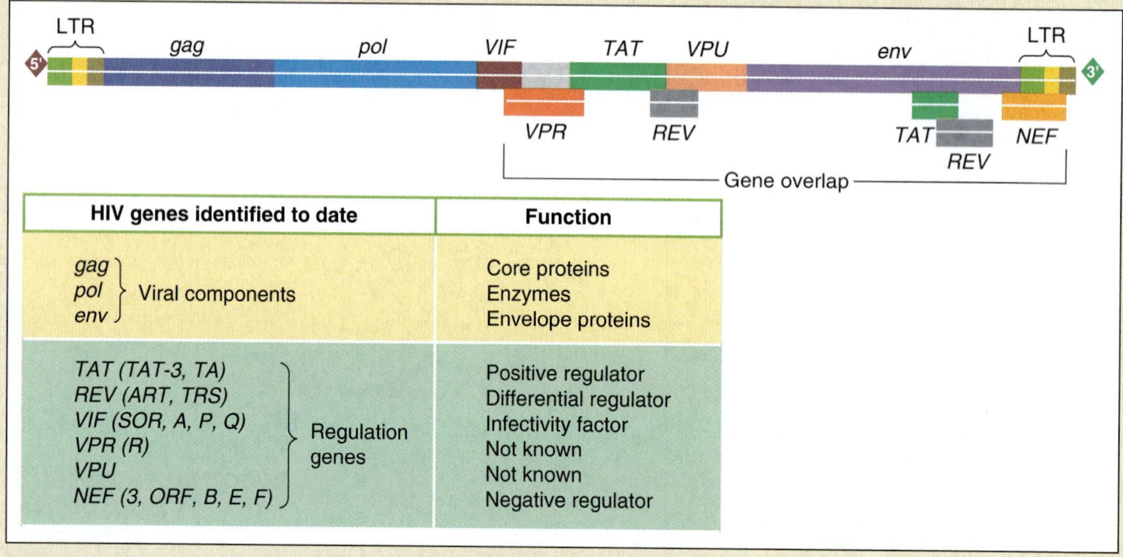

HIV genes identified to date		Function
gag	Viral components	Core proteins
pol		Enzymes
env		Envelope proteins
TAT (TAT-3, TA)	Regulation genes	Positive regulator
REV (ART, TRS)		Differential regulator
VIF (SOR, A, P, Q)		Infectivity factor
VPR (R)		Not known
VPU		Not known
NEF (3, ORF, B, E, F)		Negative regulator

FIGURE 19.C Genetic structure of HIV. More complex than the typical mammalian retrovirus, the HIV genome includes at least nine genes that are arranged along the viral DNA (top) and flanked by LTRs. Like other retroviruses, only three genes, *gag*, *pol*, and *env*, encode components of virus particles. The other genes serve to regulate the expression of the viral genes. Several of these regulatory genes are divided in noncontiguous pieces, which require splicing to produce the functional transcript.

semble pieces of the protein chain that protease normally cuts. By gumming up the protease "scissors," HIV protease inhibitors prevent protease from cutting long chains of proteins and enzymes into the shorter pieces that HIV needs to make new copies of itself. New copies of HIV are still made and still push through the wall of the infected cell even if the long chains are not cut up into the correctly sized smaller pieces. But these new copies of HIV are not completely formed, so they cannot go on to infect other cells. Thus, treatment with AZT and protease inhibitors lowers the chance that the virus will be able to bypass therapeutic intervention through constant mutational change. However, these drugs are not a cure for AIDS, only a stopgap measure intended to slow progression of the disease until more lasting and effective treatments are devised.

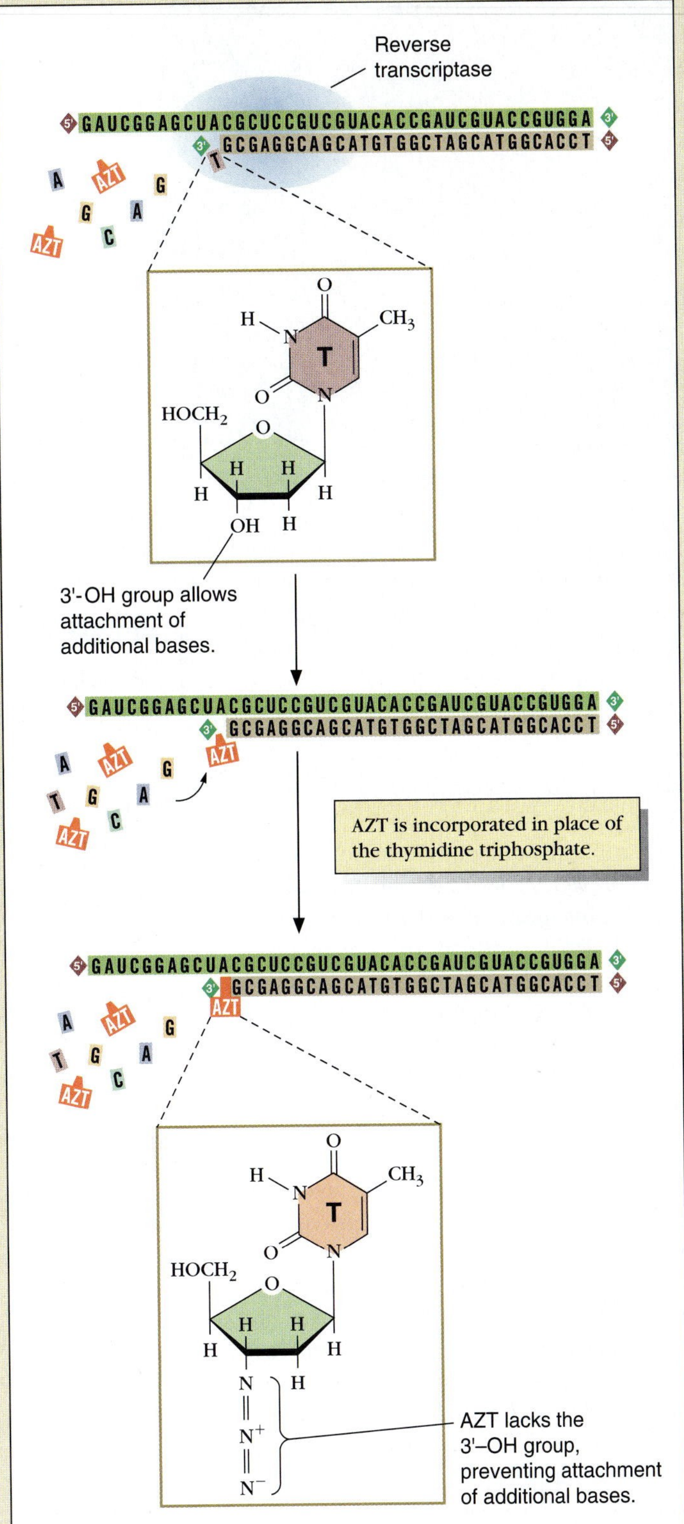
Reverse transcriptase
GAUCGGAGCUACGCUCCGUCGUACACCGAUCGUACCGUGGA
GCGAGGCAGCATGTGGCTAGCATGGCACCT
3'-OH group allows attachment of additional bases.
AZT is incorporated in place of the thymidine triphosphate.
AZT lacks the 3'–OH group, preventing attachment of additional bases.

▶ FIGURE 19.D Azidothymidine (AZT) is an analogue of thymidine triphosphate. If AZT is incorporated into a growing chain of viral DNA in place of the normal thymidine triphosphate, replication (reverse transcription) is terminated because AZT lacks the 3′-OH group necessary for the addition of the next base in the growing DNA chain.

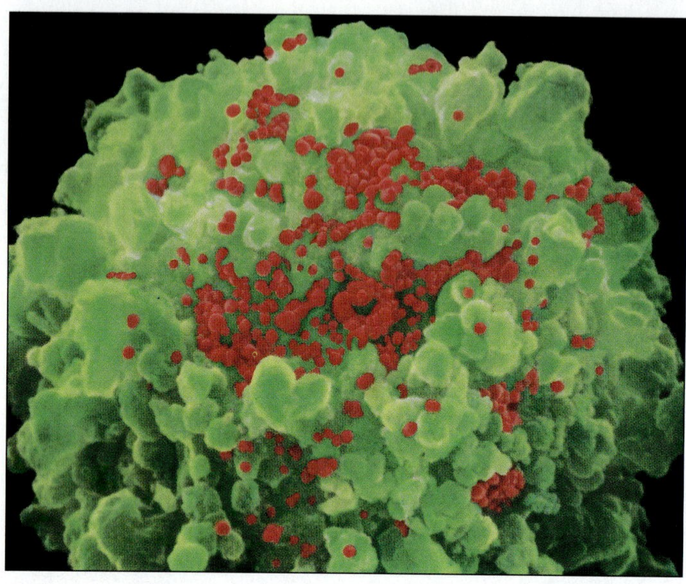

FIGURE 19.23 Human immunodeficiency virus. This electron micrograph shows the HIV particles budding from the membrane of an infected T4 cell. Portions of the plasma membrane from the infected cell along with glycoproteins encoded by the viral *env* gene form the envelope of the mature HIV. *(NIBSC/Science Photo Library/Photo Researchers, Inc.)*

with *pol*-encoded proteins. The capsid particles migrate toward regions of the inner cell surface, where the viral-encoded envelope proteins have already partially penetrated and complexed with the cell membrane. The capsid particle adjoins this region of the cell membrane, and the entire structure buds out into the intercellular region as a mature infectious particle (Figure 19.23).

SECTION REVIEW PROBLEM

15. Reverse transcriptase, unlike the DNA polymerase discussed in Chapter 9, does not have the ability to "edit" itself. This means that the mutation rate associated with DNA products of reverse transcription is much higher than the mutation rates of products of DNA polymerase. What do you think may be the significance of this fact with regard to the development of an adequate vaccine against retroviral infection?

The Retrotransposon Life Cycle. The life cycle of retrotransposons is very similar to that of retroviruses with the exception that no infectious particles are produced (Figure 19.24). The structure of the integrated retrotransposon is similar to the retroviral provirus, although sequences homologous to the retroviral *env* region may be lacking (Figure 19.25). Retrotransposon LTRs contain features of eukaryotic promoters and are transcriptionally responsive to RNA polymerase II and other host-encoded transcription factors. Some of the nascent retrotransposon RNAs remain full-length, whereas others are processed into mRNAs encoding *gag* and *pol* functions. As in the retroviral life cycle, full-length RNA and reverse transcriptase are packaged into capsid particles where reverse transcription is believed to occur. The DNA copy of the retrotransposon is then reintegrated at random positions in the host genome.

SECTION REVIEW PROBLEM

16. What is the essential difference between the life cycles of retrotransposons and retroviruses?

As mentioned previously, retrotransposons are a significant source of mutation in eukaryotes. Retrotransposon insertion mutations are especially interesting because they are often associated with altered patterns of gene regulation. Detailed studies carried out in yeast and *Drosophila* have shed considerable light on the molecular mechanisms underlying this important class of regulatory mutants.

Retrotransposon insertions located 5′ to chromosomal genes often result in novel mutant regulatory phenotypes (see CURRENT INVESTIGATIONS: *Retrotransposon Insertions Can Result in Novel Regulatory Mutations*). Retrotransposon insertion mutants often display a significant reduction in or loss of the adjacent chromosomal gene's expression. One model offered to explain this phenomenon is called transcriptional interference (Figure 19.26). According to this model, regulatory molecules involved in the transcription of the retrotransposon located just 5′ to the chromosomal gene are postulated to interfere with the transcriptional activation of the adjacent chromosomal gene. When the inserted retrotransposon is transcriptionally down-regulated, the transcriptional activity of the chromosomal gene is increased and sometimes approaches wild-type levels. In these situations, the retrotransposon insertion does not unconditionally alter the chromosomal gene's activity but rather may alter its timing or tissue-specific pattern of expression.

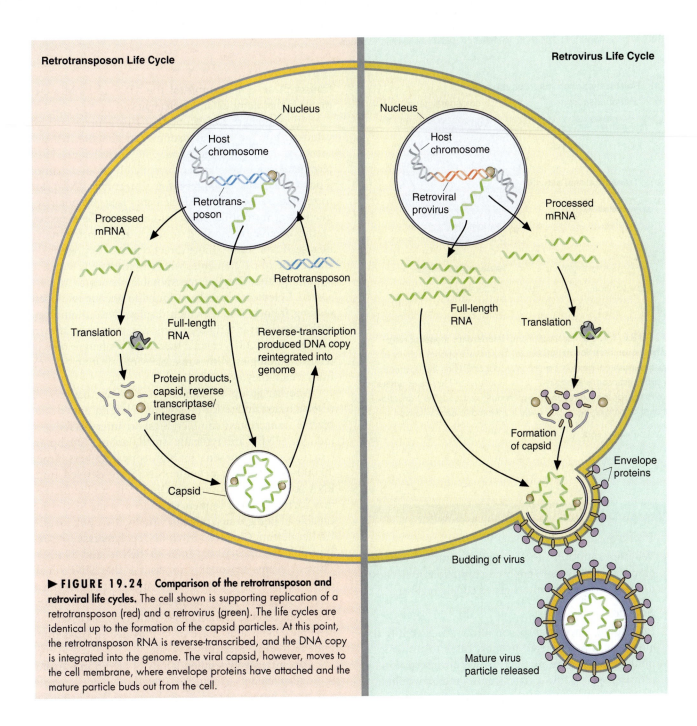

Retrotransposon Life Cycle

Retrovirus Life Cycle

Nucleus

Host chromosome

Retrotransposon

Processed mRNA

Retrotransposon

Full-length RNA

Translation

Reverse-transcription produced DNA copy reintegrated into genome

Protein products, capsid, reverse transcriptase/ integrase

Capsid

Nucleus

Host chromosome

Retroviral provirus

Processed mRNA

Full-length RNA

Translation

Formation of capsid

Envelope proteins

Budding of virus

Mature virus particle released

► **FIGURE 19.24 Comparison of the retrotransposon and retroviral life cycles.** The cell shown is supporting replication of a retrotransposon (red) and a retrovirus (green). The life cycles are identical up to the formation of the capsid particles. At this point, the retrotransposon RNA is reverse-transcribed, and the DNA copy is integrated into the genome. The viral capsid, however, moves to the cell membrane, where envelope proteins have attached and the mature particle buds out from the cell.

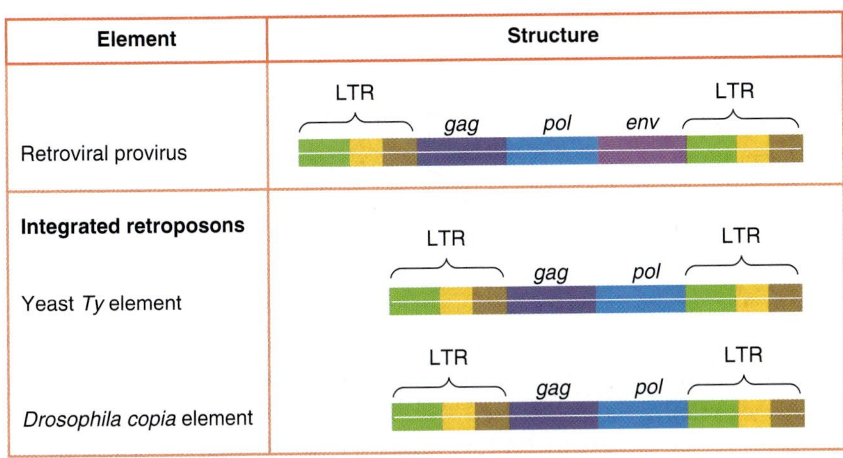

Element	Structure
Retroviral provirus	LTR *gag* *pol* *env* LTR
Integrated retroposons	
Yeast *Ty* element	LTR *gag* *pol* LTR
Drosophila copia element	LTR *gag* *pol* LTR

FIGURE 19.25 Retrotransposons are similar in structure to the integrated form of retroviruses. This figure compares the DNA structure of an integrated retrovirus with various yeast *Ty* and *Drosophila copia* retrotransposons.

(a) Retrotransposon insertion mutant

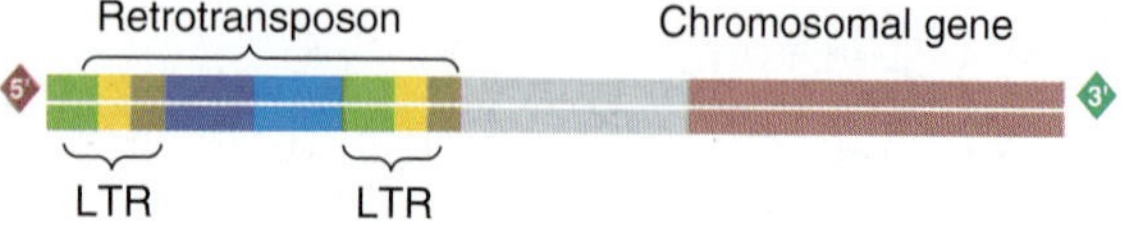

(b) Transcriptional activity

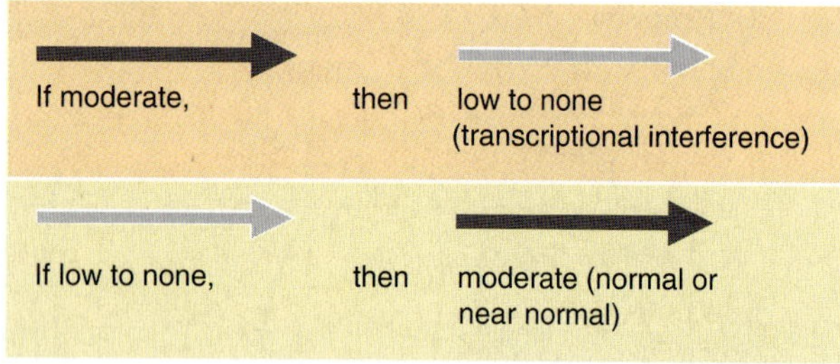

FIGURE 19.26 Transcriptional interference model of retrotransposon insertion mutations. (a) The transcriptional activity of a chromosomal gene is significantly reduced when the upstream retrotransposon is expressed. (b) In tissues or at life stages where retrotransposon expression is significantly reduced, the transcriptional activity of the mutant allele is restored to near wild-type levels.

SOLVED PROBLEM

Problem

Why are retrotransposon insertions into control regions potentially more important mutations than those caused by insertions into a gene's coding region?

Solution

Insertion of a retrotransposon into a gene's coding region would be expected in most cases to result in a nonfunctional gene product. However, insertion of a retrotransposon into a gene's control region would not destroy the function of the gene's product but rather may alter its temporal or spatial pattern of expression. Altering the temporal/spatial patterns of a gene's expression can have dramatic effects on the phenotype. For example, the normal development of an embryo is dependent upon the precise temporal/spatial expression of genes in morphological development. Disruption of this regulatory pattern of expression can result in aberrant morphologies (see Chapter 16).

Some Eukaryotic Transposable Elements Are Repetitive Sequences

In addition to retroviruses and the LTR-containing retrotransposons discussed earlier, there is another class of eukaryotic elements called LINES (long interspersed elements) whose movement is dependent upon reverse transcriptase. An example of this class of mobile elements is the mammalian L1 elements.

An estimated 20,000–50,000 L1 elements are located within the typical mammalian genome. L1 elements are relatively large, about 6500 bp in length. They terminate in an A-rich tract, presumably the remnant of a poly(A) tail that was present in a precursor RNA prior to reverse transcription. L1 elements lack promoter sequences or intron-like sequences, as would be expected of the reverse-transcribed product of a mature polymerase II transcript. Short direct repeats are present on either side of L1 elements, which is consistent with the hypothesis that they are transposable elements. At least some L1 elements have been found to encode two open reading frames, one of which displays weak homology with reverse transcriptase. It is not currently known whether these ORFs encode functional protein or merely retain homologies with some class of retrotransposons from which they evolved.

Another group of mammalian elements that rely on reverse transcriptase for mobilization but do not encode reverse transcriptase are SINES (short interspersed elements). SINES are typically small, about 300 bp in length, but are present in high-copy numbers, from 50,000–300,000 per genome. SINES are derived from RNA polymerase III transcripts and thus, unlike LINES, may carry promoter sequences. Indeed, some SINES are capable of being transcribed *in vitro* by RNA polymerase III. Recall from Chapter 11 that RNA polymerase III is responsible for transcribing snRNAs, tRNAs, and 5S RNA. Thus, it seems reasonable to assume that SINES may have originated by reverse transcription of one of these classes of cellular RNAs sometime in their evolutionary past. The reverse transcriptase functions necessary to generate SINES must have been provided in *trans* from some retrovirus or retrotransposon.

One intensively studied family of SINES in humans is the *Alu* elements. About 300,000 *Alu* elements are present in the human genome, which is more than half of all

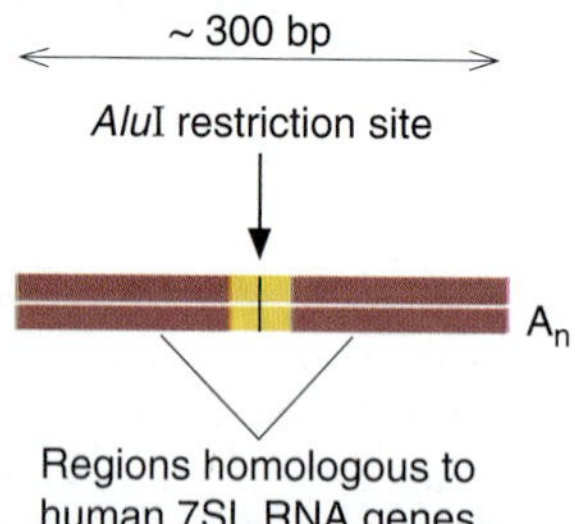

FIGURE 19.27 *Alu* is a repeated human transposable element. This figure shows genetic structure of the human *Alu* element.

CURRENT INVESTIGATIONS

Retrotransposon Insertions Can Result in Novel Regulatory Mutations

Gypsy is a *Drosophila* LTR retrotransposon that has been found to cause novel regulatory effects when inserted within a gene's regulatory region. One *gypsy* insertion mutant that has been studied intensively is the *yellow*^{2} (y^2) allele in *Drosophila melanogaster*. The X-linked *yellow* gene encodes a 1.9-kb RNA expressed in the late embryo to early larval stage of development and later during the final stages of adult body structure formation in pupae. Pupal expression of the *yellow* gene is responsible for the normal coloration of adult structures such as mouth parts, bristles, wings, body cuticle, and tarsal claws (Figure 19.E). The temporal and spatial expression of the *yellow* gene is under the control of a series of tissue-specific transcriptional enhancers that independently regulate *yellow* gene expression in different tissues and stages of development (Figure 19.F).

Flies homozygous for the y^2 allele have mutant wings and body cuticle color but are normal color for all other body parts. Interestingly, the position of the *gypsy* element in the y^2 mutant gene separates the wing and body cuticle enhancers from the gene's promoter but not the enhancers controlling expression of the gene in bristles, mouth parts, and tarsal claws. This suggests that the *gypsy* insert can interfere with enhancer function but only if positioned between the enhancer and promoter. The molecular basis of this intriguing observation has been recently worked out by Dr. Victor Corces and his collaborators at Johns Hopkins University. It turns out that the *gypsy* element contains a binding site for a host-encoded, DNA-binding protein called *suppressor-of-Hairy-wing* [*su(Hw)*]. The su(Hw) protein contains a zinc-finger DNA binding domain as well as a leucine-zipper region that allow interactions with other regulatory proteins. Corces and his collaborators proposed that when su(Hw) protein is bound to the inserted *gypsy* element in the mutant y^2 gene, it prevents interaction of proteins bound to the wing and body cuticle enhancers with the gene's promoter (apparently by altering the local chromatin structure). The enhancers located proximal to the *gypsy* insertion site are unaffected. Consistent with this model, flies homozygous for the y^2 allele and homozygous for a defective *su(Hw)* allele display normal coloration in all body parts. The mutant y^2 allele exemplifies the kinds of novel regulatory mutations that are often associated with retrotransposon insertions.

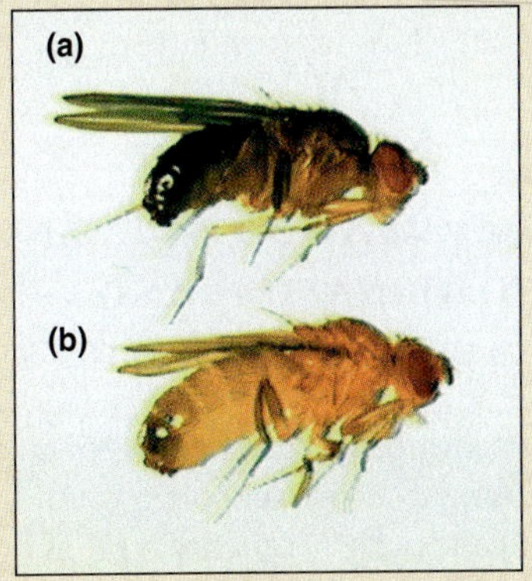

FIGURE 19.E Phenotype of y^2 mutant of *Drosophila*. A wild-type (a) and a y^2 mutant (b) of *Drosophila melanogaster*. The wild-type and mutant flies differ in the color of the wing and body cuticle. *(Dr. Wolfgang J. Miller/University of Georgia)*

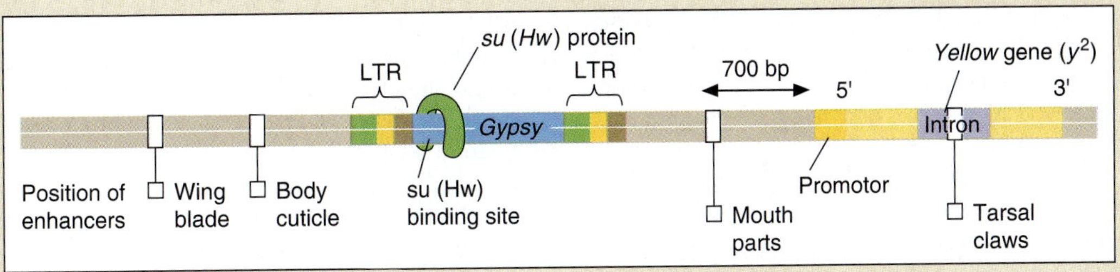

FIGURE 19.F The structure of the y^2 gene. The *gypsy* element has inserted 700 bp upstream of the transcription start site of the y^2 gene. The two exons represented by boxes are separated by an intron. The various enhancers involved in the expression of the *yellow* gene in different tissues are represented by boxes adjacent to the names of the corresponding tissues.

SINES present in humans. The name "*Alu* elements" derives from the fact that these elements contain a unique *Alu*I restriction site 170 bp 3′ to their 5′ end (Figure 19.27). *Alu* elements can be transcribed *in vitro* by RNA polymerase III into an snRNA having homology to the 7SL RNA component of the signal recognition particle. This particle plays a key role in the intracellular localization of proteins. Thus it seems likely that *Alu* elements were derived from 7SL RNA-like molecules some time in the evolutionary past.

SECTION REVIEW PROBLEM

17. What evidence is there that the replication of LINES and SINES involves reverse transcription of an RNA intermediate?

BIOLOGICAL SIGNIFICANCE OF RECOMBINATION AND TRANSPOSABLE ELEMENTS.

Although most researchers recognize the importance of recombination in mapping studies, the evolutionary significance of the process is often overlooked. If it were not for the fact that chromosomes are able to exchange genetic material by recombination, natural selection would not be able to favor (or disfavor) variant genes individually but only as a composite of a larger chromosomal unit. Under such conditions, individual genes that might increase the fitness of individuals in which they are present might well be lost from a population or species by virtue of being carried on the same chromosome with a lethal or semilethal gene. Thus, without recombination, natural selection would be a less-efficient process, and many of the well-adapted species we see on earth today (including humans!) may not have had the opportunity to evolve at all.

All classes of prokaryotic and eukaryotic TEs are a potential source of mutation. The insertion of a TE into a

CURRENT INVESTIGATIONS

The Use of Transposable Elements in Genetic Engineering

In addition to their importance as potential sources of mutation, transposable elements have recently begun to be used by geneticists to vector genes and other DNA sequences into eukaryotic genomes for research and commercial purposes. The first successful use of a TE as a genetic vector was the germline integration of a modified P element into *Drosophila* embryos by Gerry Rubin and Allan Spradling in 1982. Since that time, numerous other plant and animal TEs and viruses have been successfully engineered to serve as vectors in eukaryotes.

The P-element vector originally devised by Rubin and Spradling had most of the internal transposase-encoding sequences removed and replaced by a cloning site into which a donor gene could be conveniently inserted. Also included in the body of the modified P-element vector was the *Drosophila rosy* (ry^+) gene that could be used as a visible selectable marker when the vector was introduced into flies homozygous for the defective ry^- allele and displaying the distinctive "rosy" eye color phenotype. The left and right inverted repeats of the P element were retained in the vector, thus providing the *cis*-sequences that serve as targets for the P-element transposase (Figure 19.G).

Rubin's and Spradling's original procedure involved microinjecting the vector DNA into the pole cell region of *Drosophila* (nine nuclear division) embryos of an M-type strain homozygous for the mutant ry^- allele. Along with the P-element vector DNA, full-length P-element DNA was coinjected in order to provide the P-element transposase functions necessary for transposition. Flies developing from injected embryos were phenotypically still *rosy* (only their germ cells had been transformed), but their offspring were found to contain a large proportion of individuals with wild-type, red-colored eyes. These putatively transformed flies were shown to pass on their newly acquired phenotype in a Mendelian fashion, which indicated that the donor DNA had integrated into a chromosome. Subsequent *in situ* hybridizations verified that the P-element vector had integrated into at least one chromosome of the recipient flies.

Transposable elements can also be used to "tag" genes for eventual cloning. This procedure is especially useful because it is not necessary to know anything about the exact function of the gene that you seek. The first step in tagging a gene with a TE is to find a mutation of the gene's function that is caused by insertion of the TE. The next step is to prepare a library from the mutant organism's genome and then search the library using the TE as a probe. The clone found to hybridize to the TE probe may possess the gene that you seek, but then again it may not. To make sure, the flanking regions around the probe are subcloned (because they possess sequences of the gene in question), and the subclone is used as a probe to search the genomic library of the wild-type parent. This clone, which should be the wild-type version of the mutant gene, is then used to complement the mutant strain to determine if the wild-type phenotype can be reestablished. Using this procedure, many interesting genes have been cloned, especially from maize, in which the TEs have been very well characterized.

gene's coding region typically results in a loss of gene function. More significant, however, are those TE insertions into a gene's regulatory region. Such insertions do not usually destroy the gene's function but often alter the gene's temporal and/or tissue-specific pattern of expression. Transposable element insertion mutations displaying altered regulatory patterns of expression have been associated with the onset of cancer and other developmental abnormalities.

Recent evidence suggests that many epigenetic phenomena such as methylation, heterochromatinization, and other gene-silencing mechanisms (see Chapters 10 and 14) may have originally evolved as a defense against transposable elements. Once evolved, these silencing mechanisms were exploited by the host genome for other regulatory functions. Thus, it may be that transposable elements have played a very basic role in genome evolution.

Many bacterial transposons have been shown to play a significant role in the transmission of antibiotic-resistant genes among bacterial strains. Such transposons are a significant factor in the rapid emergence of resistant strains of pathogenic bacteria. Finally, TEs have emerged as a valuable tool in genetic engineering because they can be used to vector foreign genes into genomes. The first eukaryotic TE to be exploited as a vector was the *Drosophila* P element (see CURRENT INVESTIGATIONS: *Use of Transposable Elements in Genetic Engineering*). Today many other TEs and viruses are used by geneticists to vector foreign DNA into genomes.

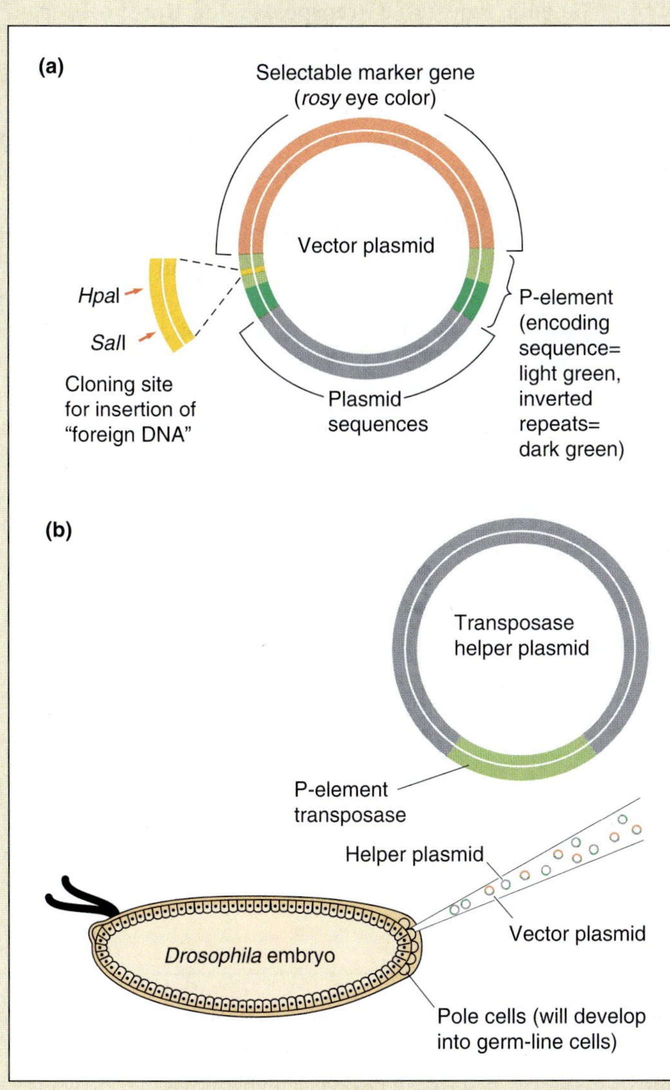

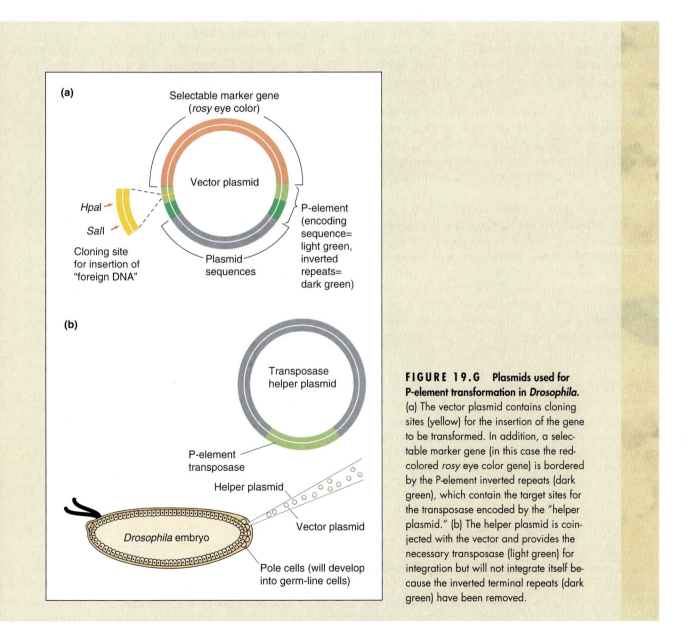

FIGURE 19.G Plasmids used for P-element transformation in *Drosophila.* (a) The vector plasmid contains cloning sites (yellow) for the insertion of the gene to be transformed. In addition, a selectable marker gene (in this case the red-colored *rosy* eye color gene) is bordered by the P-element inverted repeats (dark green), which contain the target sites for the transposase encoded by the "helper plasmid." (b) The helper plasmid is coinjected with the vector and provides the necessary transposase (light green) for integration but will not integrate itself because the inverted terminal repeats (dark green) have been removed.

Summary

Recombination can be subdivided into two general categories: homologous and nonhomologous recombination. Homologous recombination refers to exchanges that occur between two homologous DNA molecules. Exchanges of genetic material between two molecules having extensive homology as occurs between chromosomes during meiosis is called generalized recombination. Genetic exchanges between DNA molecules having only limited sequence homology is called site-specific recombination. An example of site-specific recombination is the integration and excision of lambda phage into and out of the *E. coli* chromosome.

Two molecular models of generalized recombination are consistent with the available data. The single-strand break model postulates that recombination begins with single-strand nicks of homologous strands at corresponding points in the two paired duplexes. In the double-strand break model, the exchange is initiated by a double-strand break in only one of the homologous duplexes.

The molecules involved in site-specific recombination are homologous over very short stretches of their DNA sequences. The sequences involved in site-specific recombination provide specificity by presenting target binding sites for precise enzyme-mediated integration and excision events.

Nonhomologous recombination refers to the process by which a DNA molecule can integrate at a chromosomal position where there is no sequence homology (for example, the chromosomal integration of transposable elements).

A transposable element (TE) is a DNA sequence that is able to move and integrate throughout a host genome without having sequence homology with the chromosome at the site of integration. The first evidence that transposable elements existed came from genetic research in maize conducted by Barbara McClintock. McClintock found that genes responsible for chromosomal breakage in maize were capable of moving to new chromosomal locations. The first transposable elements to be isolated and characterized at the molecular level were the bacterial insertion sequence or IS elements. IS elements encode only that genetic information necessary to bring about their own transposition. Transposons are bacterial transposable elements made out of two IS elements flanking a sequence of encoding functions. These functions are unrelated to transposition but usually encoding some selectable trait such as antibiotic resistance or heavy-metal tolerance. Transposons may move throughout the bacterial genome using a conservative or replicative mode of transposition.

The movement of some eukaryotic transposable elements requires a transposase for transposition. This group includes the maize *Ac* and *Ds* family of elements and the *Drosophila* P element. The majority of eukaryotic transposable elements require reverse transcriptase for transposition. These elements, called retrotransposons, are structurally and functionally similar to infectious retroviruses.

Homologous recombination is of major biological significance because it provides natural selection with the opportunity to favor (or disfavor) variant genes individually rather than only as a part of a larger chromosomal unit. Transposable elements are biologically significant because they are a major source of mutation. Bacterial transposable elements facilitate the spread of antibiotic-resistance genes providing a serious challenge to our ability to control bacterial diseases. Transposable elements may have played a significant role in genome evolution.

Selected Key Terms

autonomous element
composite element
conservative transposition
generalized recombination
Holliday structure
homologous recombination
inverted terminal repeats
long terminal repeat
nonautonomous elements
nonhomologous recombination
replicative transposition
retrotransposon
retrovirus
site-specific recombination
transposable element

Chapter Review Problems

1. List similarities and differences between the integration/excision processes of phage and IS elements.
2. Knowing what you now know about transposable elements, propose an hypothesis to explain the observations of Rhoades concerning the behavior of the Mexican black strain of corn. Propose experiments to test your hypothesis.
3. The first genetic evidence for transposable elements came from McClintock's work in maize, and yet the first transpos-

able element isolated and characterized on the molecular level came from bacteria. Why?

4. Suppose that you discover that mutant allele at the *Adh* locus in *D. melanogaster* contains a large segment of DNA inserted into the gene's first exon. You hypothesize that this DNA insert is a transposable element. What genetic experiment could you do to test your hypothesis? What molecular analysis could you carry out to test your hypothesis?
5. What role might Tn elements play in the rapid evolution of bacterial strains that are resistant to antibiotics?
6. The repressor protein encoded by *Drosophila* P elements may help limit the number of copies of P elements present within a genome. Why might it be advantageous to transposable elements to autoregulate their numbers within host genomes?
7. In the early days of research on retroviruses, it was thought that these viruses were destroyed upon entering the cell because no viral particle could be observed within the cell after infection. Unexpectedly, however, researchers found that retroviral particles could later be observed to bud out of these very same cells. Explain the basis of these unexpected results.
8. Propose a possible experimental strategy to determine whether the single-strand break or double-strand break model is the actual molecular mechanism by which recombination occurs.

Challenge Problems

1. We saw in Chapter 17 that organisms have evolved very efficient repair systems to minimize the effect of mutations. Why do you think that organisms have not evolved mechanisms to reduce or eliminate recombination events during meiosis?
2. Some biologists view transposable elements as examples of purely parasitic DNA; in other words, they believe the elements have no positive effect on the hosts in which they reside. Do you agree or disagree with this point of view? Why?
3. There is at least one instance of an infectious virus having been isolated and found to contain a transposable element inserted within its genome. What do you see as the significance of this finding?
4. Some maize transposable elements have been found to contain consensus splice sequences in or near their inverted repeats. What may be the adaptive advantage of this for the elements? For the host?
5. What are the negative side effects of the AIDS drug AZT? What kind of negative side effects would you expect if AZT were administered to a pregnant woman or infant? Why?
6. Given what you have learned about the retrovirus life cycle, what is the best strategy to prevent the onset of a retroviral-mediated disease such as AIDS?

Suggested Readings

Berg, D., and M. Howe (eds.). 1989. *Mobile DNA*. American Society of Microbiology Publication, Washington, DC.

Eggleston, A. K., and S. C. West. 1996. "Exchanging partners: Recombination in *E. coli*." *Trends in Genetics*, 12: 20.

Engles, W. R. 1997. "Invasions of P elements." *Genetics*, 145: 11.

Federoff, N. V. 1994. "Barbara McClintock." *Genetics*, 136: 1–10.

Keller, E. F. 1983. *A Feeling for the Organism*. Freeman, New York.

McDonald, J. 1998. "Transposable elements, gene silencing, and macroevolution." *Trends in Ecology and Evolution*, 13: 94–95.

McDonald, J. 1995. "Transposable elements: Possible catalysts of organismic evolution." *Trends in Ecology and Evolution*, 10: 123–126.

McDonald, J. 1993. "Evolution and consequences of transposable elements." *Current Opinion in Genetics and Development*, 3: 855–864.

Shapiro, J. (ed.). 1983. *Mobile Genetic Elements*. Academic Press, New York.

Stahl, F. 1987. "Genetic recombination." *Scientific American*, 256 (February): 90–101.

Szostak, J., T. Orr-Waever, and R. Rothstein, 1983. "The double-stranded-break repair model for recombination." *Cell*, 33: 25–35.

Whitehouse, H. 1982. *Genetic Recombination: Understanding the Mechanism*, Wiley, New York.

On the Web

Visit our Web site at **http://www.saunderscollege.com/lifesci/titles.html** and click on A/G/M Genetics for links to the following chapter-related resources on the World Wide Web:

1. **The dynamic genome.** This site describes Barbara McClintock's ideas in the Century of Genetics; it is edited by Dr. Nina Fedoroff.
2. **Barbara McClintock.** This site has Barbara McClintock vitae and a description of her life and works.
3. **Retrotransposons.** This site gives an overview of how retrotransposons move about in the genome.

CHAPTER

20

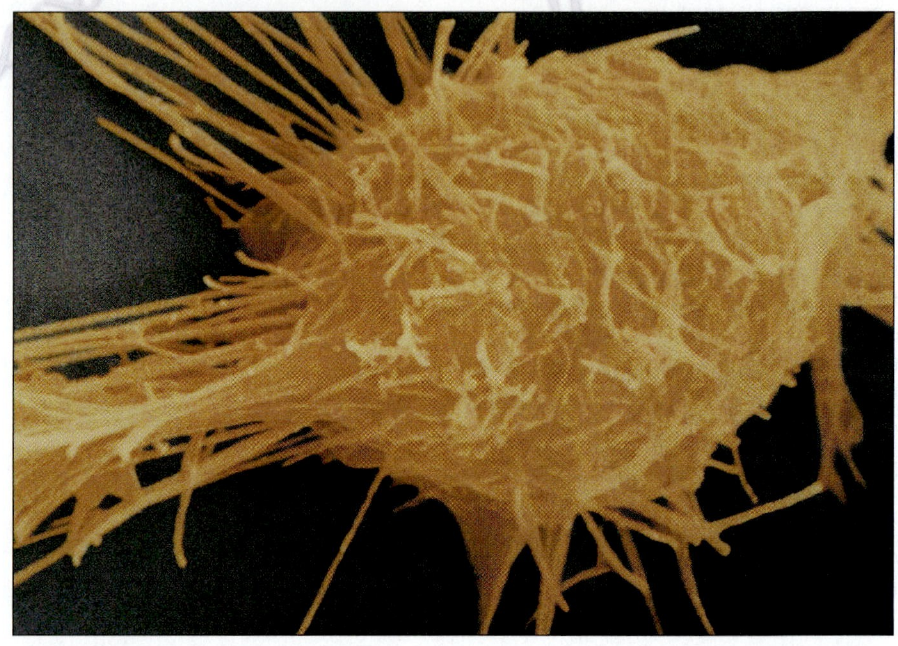

Electron micrograph of a cancer cell. *(Biological Media/Science Source/Photo Researchers, Inc.)*

The Genetics of Cancer

In Part 2, we have focused on the experimental findings that have increased our basic understanding of genetic structure and function. Knowledge gained from basic genetic research not only provides scientists with new information on gene function but also provides a better understanding of the molecular basis of genetic diseases. Such an understanding can, in turn, lead to better methods of early diagnosis and treatment. The medical impact of modern genetics is nowhere more apparent than in the diagnosis and treatment of cancer.

In this chapter, we will review the evidence that cancer has a genetic basis. A major breakthrough in cancer genetics was the discovery that many tumor viruses contain cancer-causing genes that are homologous to genes present in normal cells. First, we will focus on the function of these genes in normal cells and the types of mutations that can convert them into cancer-causing genes. Second, we will discuss one class of the genes that are responsible for maintaining the integrity and fidelity of DNA. Because cancer is a genetic disease, an increased mutation rate caused by a defect in a repair pathway will result in an increased chance of cancer. Third, we will consider a recently discovered class of genes that is capable of suppressing the tumor phenotype of cancer cells. We will analyze the function of proteins encoded by these so-called *tumor-suppressor genes* in normal cells and the consequence of mutations that cause them to stop functioning. The significance of current genetic knowledge to early diagnosis and treatment of this deadly disease will be discussed.

By the end of this chapter, we will have addressed five specific questions:

1. What is cancer?
2. What is the evidence that cancer has a genetic basis?
3. How were the genes that cause cancer (oncogenes) first discovered?
4. What are the different classes of oncogenes, and what are their cellular functions?
5. How has the study of genetics improved the prospect of curing cancer in the near future?

Cancer Is a Genetic Disease.

The term **cancer** refers to a family of more than a hundred different diseases, usually malignant (deadly), that are characterized by uncontrolled cell proliferation. The primary cause of cancer of any type is a malfunction in cell regulation that, as we will see, is caused by aberrations in one or more genes. Because the genetic changes leading to most cancers originate in a single cell, cancers are said to be clonal in origin. The **clonal theory of cancer development** states that all the cells constituting an abnormal growth or **tumor** are descendent from a single genetically altered cell (Figure 20.1). Tumors can be either malignant or benign. **Malignant tumors** are cancerous and invade and destroy adjacent normal tissue. Malignant tumors frequently **metastasize,** or slough off cells that are then spread via lymphatic channels or blood vessels to lymph nodes and other tissues in the body, where they establish secondary malignant growths (Figure 20.2). **Benign tumors,** in contrast, are encapsulated growths that do not invade and destroy surrounding tissue. Benign tumors are thus not considered cancers.

It is estimated that one out of every three Americans will develop cancer during his or her lifetime and that about one out of four Americans will eventually die of the disease (Table 20.1).

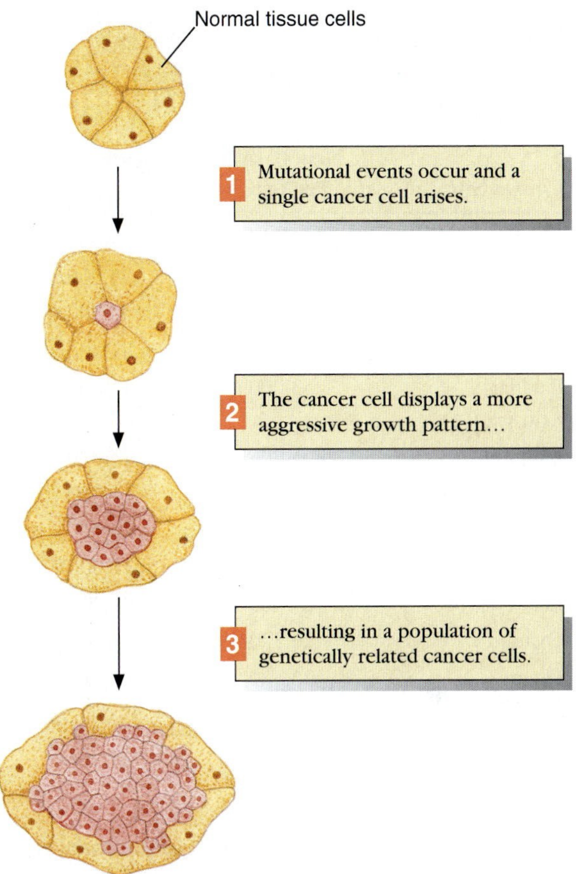

Normal tissue cells
1 Mutational events occur and a single cancer cell arises.
2 The cancer cell displays a more aggressive growth pattern...
3 ...resulting in a population of genetically related cancer cells.

FIGURE 20.1 The clonal theory of cancer development. This theory states that cancer tumors consist of the descendants of a single transformed cell.

Cancers Can Be Classified According to Tissue Type.

Cancers are usually classified according to the type of cell and tissue in which they arise. **Carcinomas** are cancers of the epithelial cells that cover the body surface and line the intestine and other internal organs (for example, the colon, liver, lung, and skin). Carcinomas are by far the most common type of cancer, accounting for about 90% of all malignant tumors. The next most frequent

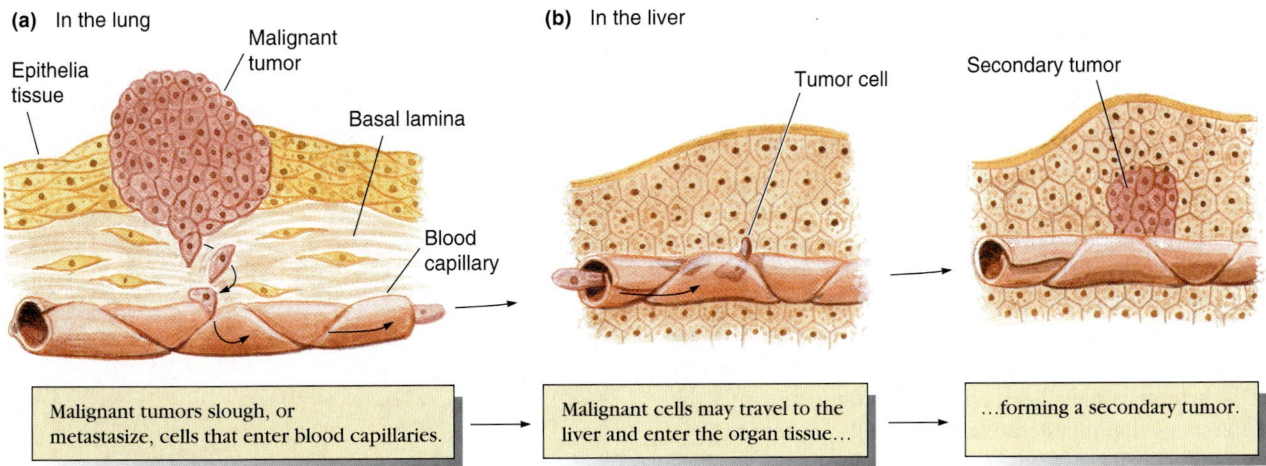

FIGURE 20.2 Metastasis. Malignant tumors may slough off cells that can be spread through the lymphatic channels or blood vessels to other tissues in the body. Depicted is the spread of tumorous cells from the lung (a) to the liver (b) by way of blood vessels. After the malignant cells enter the liver, they may adhere to capillary walls and enter the liver tissue by a process called extravasation. Once in the liver tissue, the cancer cells proliferate to form a secondary tumor.

TABLE 20.1 *Frequency of Various Cancers and the Frequency of Cancer Deaths Per Year in the United States**

Cancer Site	New Cases Per Year	Deaths Per Year
Lung	177,000	159,000
Colon/rectum	133,000	55,000
Breast	185,700	44,000
Prostate	317,100	41,400
Bladder	53,000	12,000
Non-Hodgkin's lymphoma	53,000	23,000
Uterus	49,000	11,000
Skin	38,000	7,300
Kidney	30,600	12,000
Leukemia	27,600	21,000
Ovarian	26,700	14,800
Pancreas	26,300	27,000

*American Cancer Society, Atlanta, Georgia, "Cancer Facts and Figures," 1997.

classes of cancers are **leukemias** and **lymphomas,** which arise in the blood and lymph systems, respectively. Leukemias and lymphomas account for about 8% of all cancers worldwide. The least frequent class of cancers is the **sarcomas,** which are solid tumors of the connective tissues like muscle and bone; these constitute about 2% of all cancers.

The Major Distinguishing Characteristic of Cancer Cells Is Uncontrolled Growth.

One of the technological breakthroughs that facilitated the molecular genetic analysis of cancer was the ability to grow mammalian cells in culture. A number of different types of mammalian cells can now be maintained in culture by growing them at an appropriate temperature (usually the body temperature of the organism from which the cells were derived) and in a medium supplemented with the correct combination of nutrients and other factors necessary to stimulate cell proliferation. By maintaining and analyzing cultures of normal and cancer cells, scientists have been able to catalogue precisely the differences that characterize cancer cells. In comparison to normal cells, cancer cells have unlimited life spans and undifferentiated morphologies and functions, do not adhere to growth surfaces, have no cell–cell contact growth inhibition, and are not inhibited from growing at high cell densities.

Of all these distinguishing characteristics of cancer cells, the most notable is uncontrolled growth. Normal cells grown in culture will divide until they reach a certain maximum density, at which point they will stop dividing. This property of **density-dependent growth inhibition** is typically lost in cancer cells, which continue to divide even under conditions of high cell density until the essential nutrients contained in the medium are depleted and/or the level of cell-generated toxic wastes becomes incompatible with continued cell growth. The unregulated growth of cancer cells in culture is consistent with the characteristic loss of growth controls typically associated with tumorous cells in whole organisms (Figure 20.3).

Another property of many transformed or cancerous cells grown in culture is a reduced requirement for growth factors. **Growth factors** are serum proteins that bind to receptor proteins on the outer surface of cells to stimulate cell division. An organism's cells normally depend upon growth factors synthesized by other specialized cells. For example, the proliferation of fibroblast

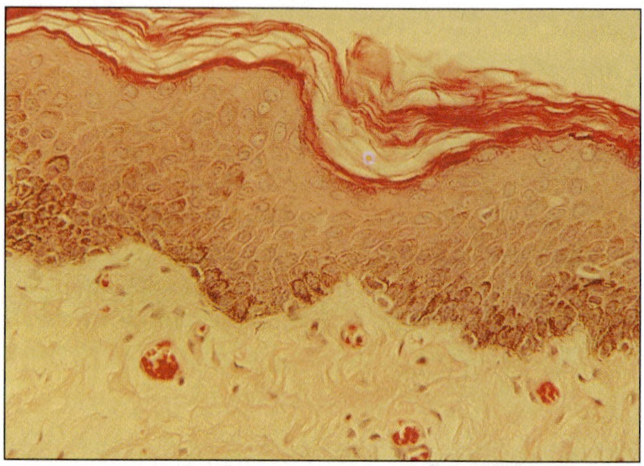

(a)

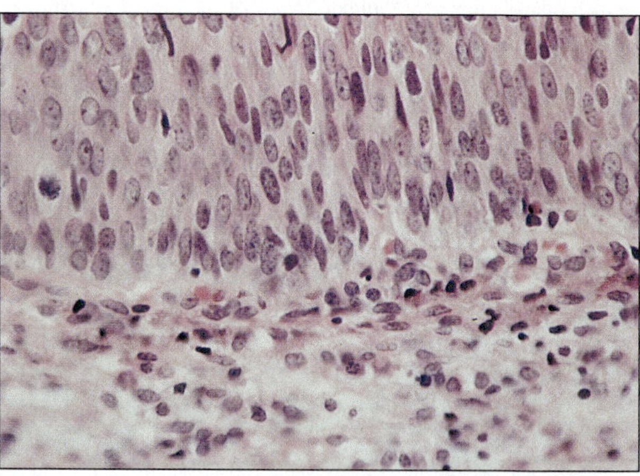

(b)

FIGURE 20.3 Normal and cancerous mouse epithelial tissue. Morphological differences between (a) normal and (b) cancerous epithelial tissue. Cancerous tissue is characteristically made up of a high density of undifferentiated cells. *(Dr. Lou Laimins, Northwestern University Medical School, Chicago, Illinois)*

cells in humans and other animals depends upon the presence of **platelet derived growth factor** (PDGF) produced by platelets. The synthesis of PDGF by platelets is usually induced as part of the wound-healing process. However, some cancer cells have the ability to produce growth factors autonomously and thereby stimulate their own proliferation. Other cancer cells simply have a significantly reduced need for growth factors and are thus able to maintain a high rate of proliferation. In these cancer cells, one or more of the intracellular signals regulating cell growth and division, which would normally be activated by growth factors binding to receptor proteins on the cell surface (Figure 20.4), are presumably being activated even in the absence of growth factors. Once again, the consequence is unregulated cell proliferation.

In addition to these aberrations in cell regulation, many cancer cells also display abnormalities in cell adhesion and other cell-surface properties characteristic of normal cells. For example, the growth and functioning of normal cells are often sensitive to physical contact with and/or adhesion to adjacent cells. The ability of cells to coordinate with one another is critical to the normal development and functioning of multicellular organisms. Normal cells adhere to one another, and division is typically inhibited by physical contact with other cells (**contact inhibition**) (Figure 20.5). Cancer cells are usually unable to interact with other cells in this way. In cultures, it is easy to observe loss of contact inhibition in cancer cells. This loss contributes to the metastasis of these aberrant cells (that is, to the ability of these cells to invade normal tissue and form a tumor).

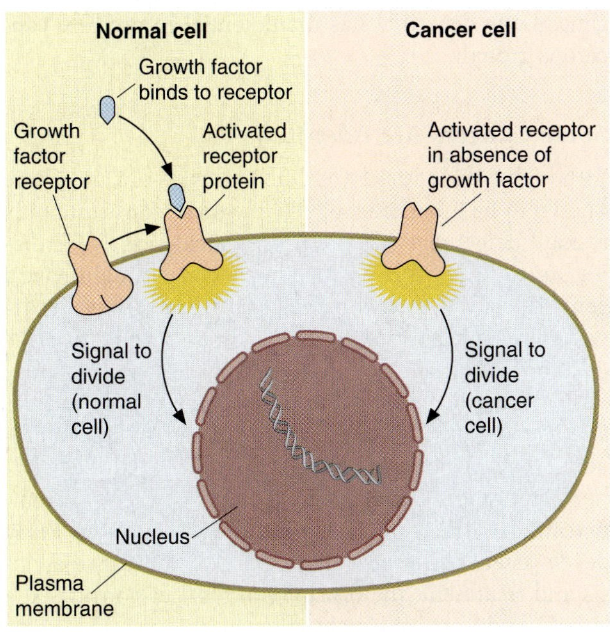

FIGURE 20.4 Aberrant growth factor receptors. Mutant growth factor receptors can induce cell proliferation even in the absence of growth factor stimulation, leading to cancer.

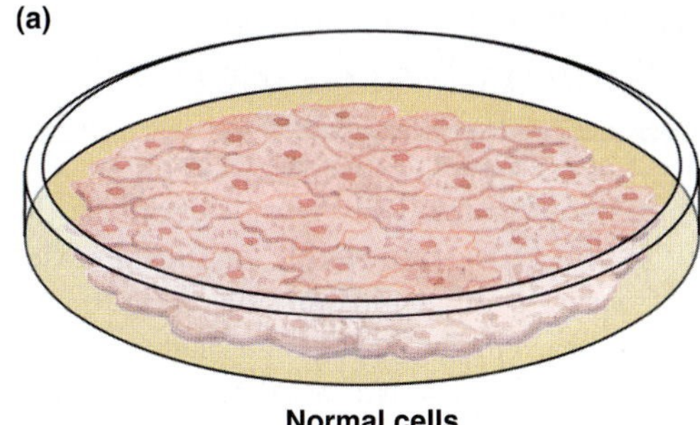

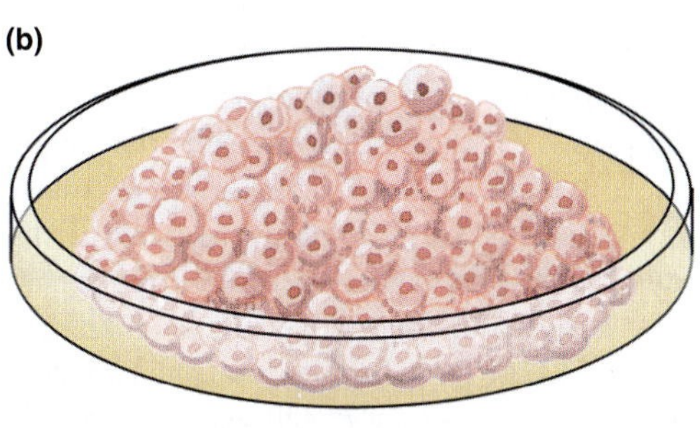

FIGURE 20.5 Normal cells and cancer cells in culture. Unlike (a) normal cells, (b) cancer cells are not inhibited by contact with other cells but continue to grow and pile up on each other.

SOLVED PROBLEM

Problem

Why was the ability to grow mammalian cells in culture a significant breakthrough in the study of cancer?

Solution

The ability to grow and maintain mammalian cells in culture allowed scientists to catalogue precisely the biochemical and molecular differences that distinguish cancer cells from normal cells. In addition, the ability of many chemicals and other treatments to retard or stop cancer cell growth are most efficiently tested on cells grown in culture before being tried on experimental animals.

SECTION REVIEW PROBLEMS

1. Distinguish among the various types of cancer.
2. What is the most common type of cancer, and why do you think it is more common than other types of cancer?
3. Compared to normal cells, what are the most distinguishing characteristics of cancer cells grown in culture?

The Evidence That Cancer Has a Genetic Basis Is Multifaceted.

The most direct and unambiguous evidence that cancer has a genetic basis is the relatively recent identification and isolation of the genes responsible for the disease. However, many years prior to this discovery, there was strong circumstantial evidence that cancer had a genetic basis (Table 20.2). First, it was known that a predisposition to many forms of cancer is inherited. This suggested that the cancer or at least a propensity for developing the disease is encoded in genes. Second, the clonal theory of tumor development (see earlier discussion) implied that the cancerous properties of the original transformed cell are genetically passed on to the progeny cells that make up the tumor. Third, it was known that many **mutagens**—compounds and radiation treatments known to induce genetic damage—are also **carcinogens,** factors capable of inducing tumors in experimental animals. This correlation strongly implied that the transformation of a normal cell to a cancer cell was caused by a mutation in one or more genes. Fourth, it was known by the middle of this century that some cancers, especially in birds and rodents, are caused by infection with certain viruses, termed **oncogenic viruses.** Further investigation of many of these oncogenic or cancer-causing viruses led to the eventual direct identification and isolation of the **viral oncogenes,** or v-*oncs*, responsible for the onset of the disease. More recently, it was found that genes homologous to v-*oncs*, called **proto-oncogenes,** are normal components of all vertebrate (and at least some invertebrate) genomes. These cellular proto-oncogenes function normally in regulating cell growth and development, but when they are mutated, they may become oncogenic themselves, at which point they are referred to as **cellular oncogenes,** or c-*oncs*. An oncogene is any gene that has the potential to produce cancerous growth.

TABLE 20.2 *Circumstantial Evidence That Cancer Is a Genetic Disease*

Phenomenon	Example	Proposed Explanation
Predisposition to many cancers	Familial retinoblastoma	Predisposed individuals are heterozygous for cancer-causing recessive allele
Tumors develop from single transformed cells	All tumors	Inheritance of abnormality by all descendent cells
Carcinogens are also mutagenic	*N*-methyl-*N*′-nitrosourea	DNA changes are responsible for transformation of normal cells
Some viruses cause cancer	Rous sarcoma virus	Deletion studies show the cancer-causing properties of many oncogenic viruses are associated with a viral gene

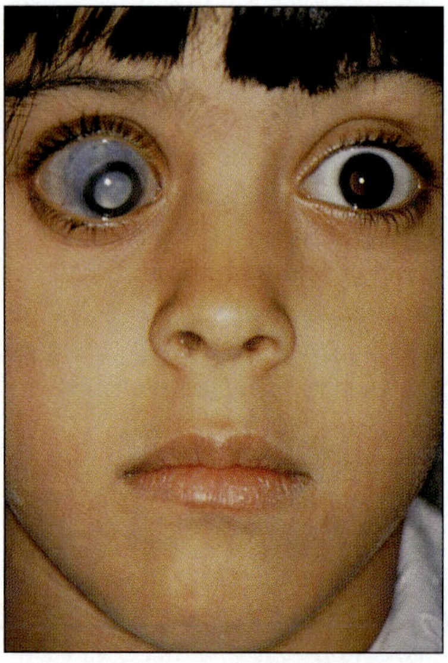

FIGURE 20.6 Retinoblastoma eye tumor. This view looks directly into the eye of a retinoblastoma patient. The tumor, which appears as an opaque mass in the center of the eye, is clearly visible and is usually manifested in childhood. *(©1991 Custom Medical Stock Photo)*

Some Cancers Are Inherited.

Although a wide variety of different types of cancers are known to be inherited, collectively these constitute only a small percentage of total cancer incidence. Nevertheless, one of the first indications that cancer might have a genetic basis was recognizing that predispositions for developing some types of cancer are passed on from one generation to the next.

The predisposition for many cancers is inherited, and usually these cancers have their onset in early childhood. One of the most intensively studied of these childhood tumors is **retinoblastoma,** which is caused by embryonic retinal cells transforming during the early development of the eye (Figure 20.6). With early diagnosis and treatment, the disease is no longer considered a lethal condition, and most afflicted children survive to have children of their own. This has allowed detailed studies to be made of the inheritance patterns associated with congenital retinoblastomas. The results of these

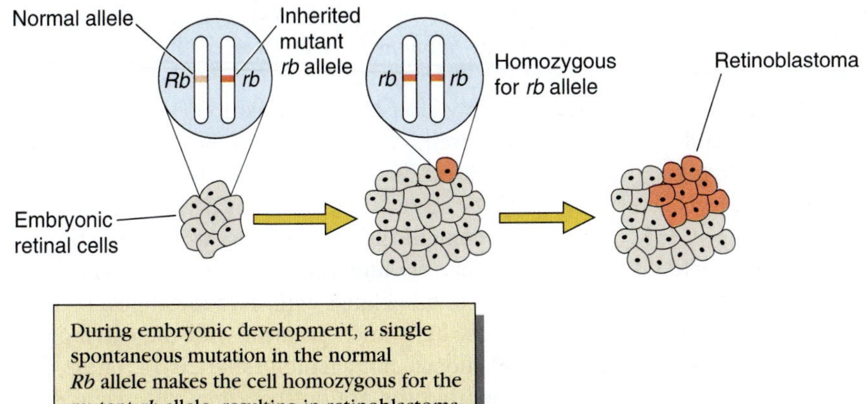

FIGURE 20.7 Inheritance of retinoblastoma. Individuals genetically predisposed to retinoblastoma are heterozygous for this cancer-causing allele and thus are only a single mutational step away from the disease.

studies indicate that the disease is the result of a recessive mutation at the *rb* locus on chromosome 13. Individuals heterozygous for the cancer-causing allele at this locus are predisposed to develop the disease because their cells are only one mutational step away from being homozygous for the cancer-causing allele. About half of the offspring of congenital retinoblastoma patients will also be predisposed to develop the disease (Figure 20.7).

In addition to the inherited form of retinoblastoma, the disease can arise spontaneously in individuals with no history of retinoblastoma in their family background. These spontaneous cases are caused by mutations in somatic (embryonic retinal) cells and thus are not heritable. Patients with the inherited form of retinoblastoma usually develop tumors in both eyes (because both eyes have one defective copy of the allele), whereas only one of the eyes becomes tumorous in the spontaneous form of the disease. The incidence of retinoblastoma is about 1 out of 20,000 children. The *rb* gene has now been cloned and characterized. It produces a protein pRB, which is 928 amino acids long and can be classified as a tumor-suppressor gene. We will consider the function of the protein encoded by the *rb* gene later in this chapter.

The frequency of other types of predisposition for inherited cancers is somewhat lower than has been found for retinoblastoma. For example, a childhood cancer of the kidney called **Wilms's disease** (a tumor of the kidney and urinary tract) affects about 1 out of every 100,000 children.

Many Carcinogens Are Also Mutagens.

Carcinogens are usually identified by one of two approaches: epidemiological surveys or experiments conducted with laboratory animals. These two approaches are often used cooperatively to identify cancer-causing agents. **Epidemiology** is the study of disease incidence in different populations or groups of individuals. For example, epidemiologists have clearly established that cigarette smokers are at least ten times more likely to develop lung cancer than are nonsmokers (Figure 20.8). Based on this result, it can be reasonably hypothesized that cigarette smoke is carcinogenic. However, there are alternative explanations, such as that lung cancer may be caused by high anxiety or stress. It may be that individuals prone to extreme stress are more likely to smoke cigarettes but that smoking per se is not the direct cause of lung cancer. Defining what constitutes stress for different individuals might make it difficult to eliminate this alternative hypothesis by epidemiological studies alone. However, the potential carcinogenic properties of the chemicals contained in cigarette smoke can be and have been unambiguously tested in experimental animals. The results indicate that tobacco smoke contains a variety of compounds that are either themselves carcinogenic or become carcinogenic after being metabolized. For example, one of the chemicals present in tobacco smoke is benzo(a)pyrene, which is metabolized in the body to the soluble product benzo(a)pyrene 7,8-diol-9,10-epoxide,

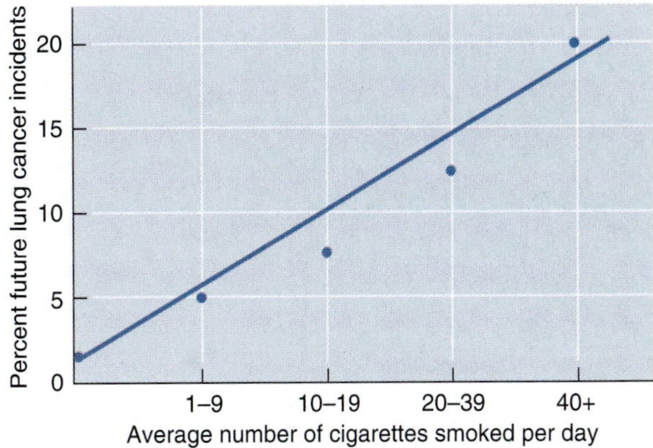

FIGURE 20.8 Lung cancer frequency in smokers. This plot shows the relationship between the frequency of lung cancer and the number of cigarettes smoked per day. *(From The Health Consequences of Smoking: Cancer—A Report of the Surgeon General, 1982. U. S. Dept. of Health and Human Services, Washington, DC)*

which can be excreted. This diol epoxide attacks the electron-rich areas of the bases contained in DNA, leading to mutations (Figure 20.9). This example illustrates the complimentary nature of epidemiological and experimental approaches to identifying carcinogens. Epidemiological studies are usually quite helpful in identifying potential carcinogens, which can then be definitively tested by laboratory experiments.

A large number of substances have now been identified as carcinogens. As this list was first being compiled, it became obvious that many compounds that were previously found to be mutagenic were also carcinogenic. This connection was made crystal clear in 1973 when a University of California biochemist, Bruce Ames, tested the ability of many known carcinogens to cause mutations in *Salmonella* bacteria (see the Ames test in Chapter 17). Because many carcinogens become actively mutagenic only after being metabolized, Ames devised a protocol whereby compounds to be tested are first treated with a rat liver extract before being subjected to his mutability screen. Ames reasoned that the liver extract would contain the enzymes that would be acting on carcinogens *in vivo* and thus mimic the normal enzymatic processes associated with liver metabolism. Using this approach, he found that nearly all the carcinogens tested were indeed mutagenic (Table 20.3). These data further supported the notion that cancer is a genetically based disease.

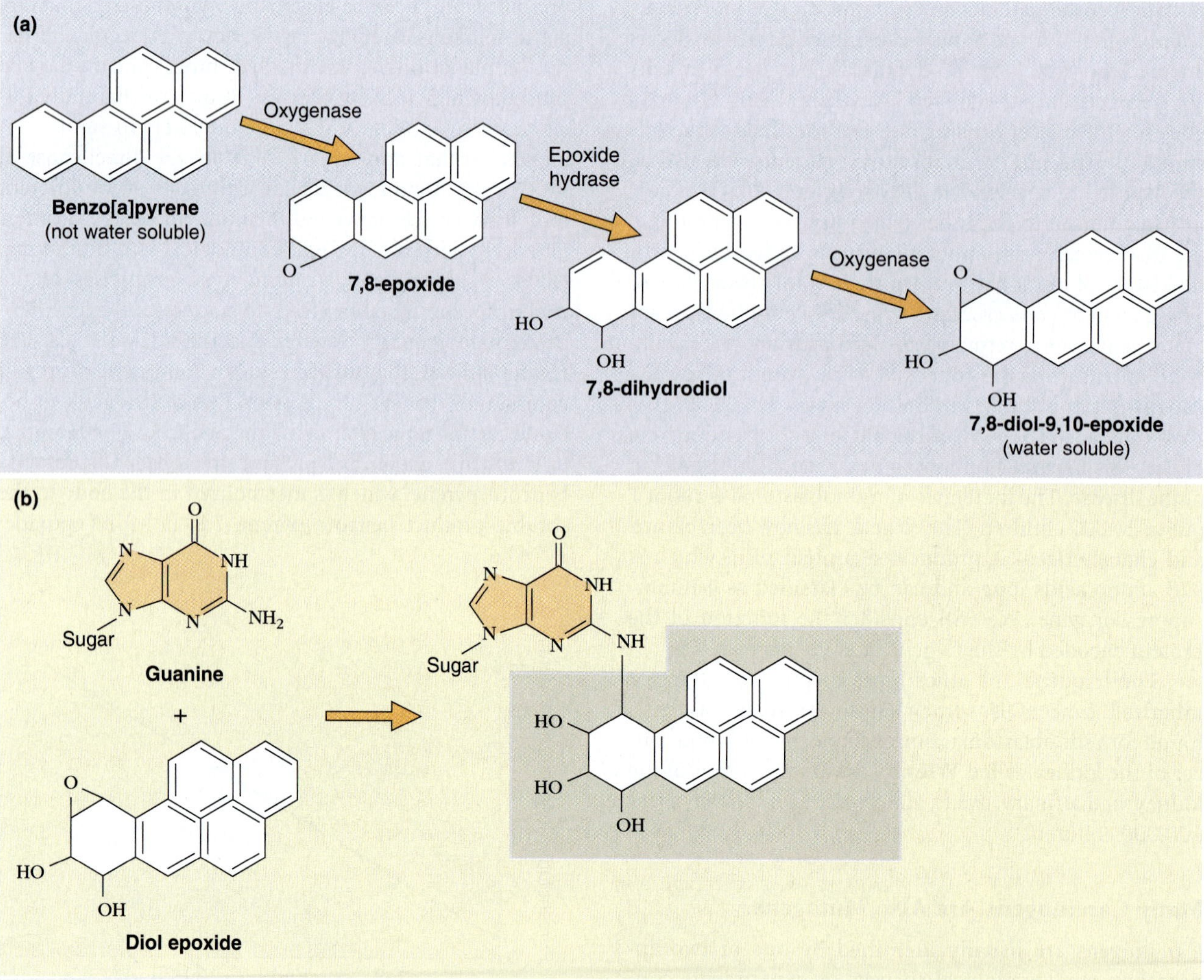

FIGURE 20.9 Carcinogens in tobacco smoke. (a) Hydrocarbons such as benzo(a)pyrene are insoluble in water; thus they are chemically modified in the body to become soluble so that they may be excreted. (b) The soluble product of the reactions in (a) is benzo(a)pyrene 7,8-diol-9,10-epoxide, which can attack the electron-rich area on a DNA base, altering the DNA structure, which leads to a mutation.

TABLE 20.3 *Ability of Known Carcinogens To Cause Mutations in* Salmonella typhimurium *in Presence (+) or Absence (−) of Rat Liver Homogenate*

	Carcinogen	Presence (+) or Absence (−) of Carcinogen in Assay	Presence (+) or Absence (−) of Liver Homogenate in Assay	No. of Mutations at *his* Locus Per Plate
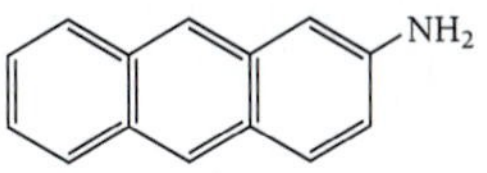	2-Aminoanthracene	+ + −	+ − +	11,200* 27 46
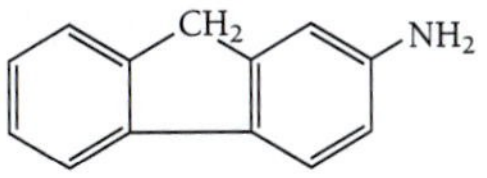	2-Aminofluorene	+ + −	+ − +	11,300* 39 29
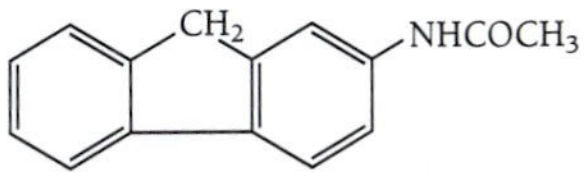	2-Acetylaminofluorene	+ + −	+ − +	13,600* 21 46
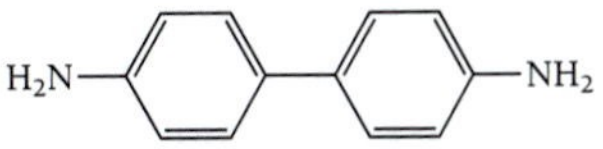	Benzidine	+ + −	+ − +	265* 16 36
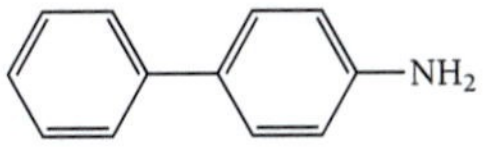	4-Aminobiphenyl	+ + −	+ − +	980* 28 46
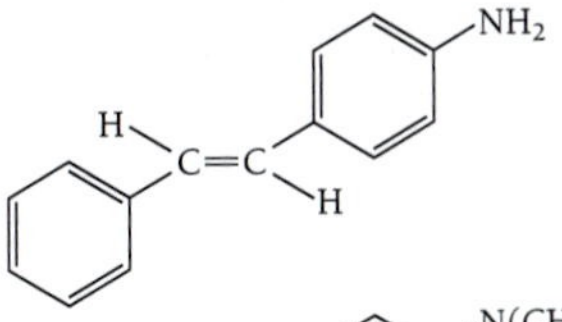	4-Amino-*trans*-stilbene	+ + −	+ − +	842* 17 42
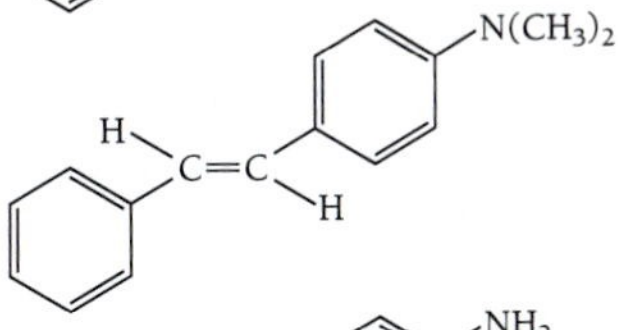	4-Dimethylamino-*trans*-stilbene	+ + −	+ − +	896* 28 53
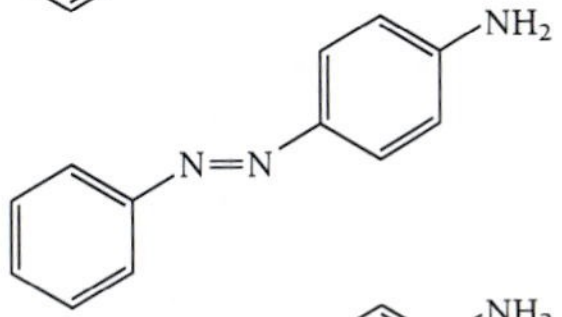	*p*-(phenylazo)-aniline	+ + −	+ − +	94* 13 31
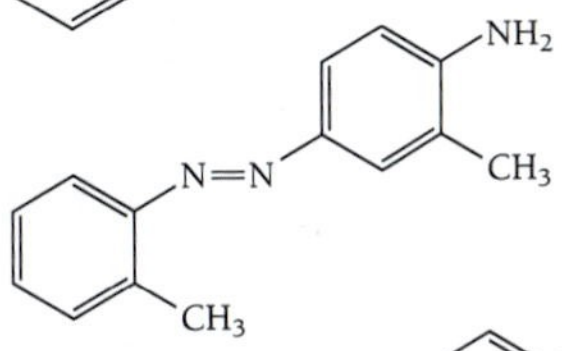	4-(*o*-tolylazo)-*o*-toluidine	+ + −	+ − +	305* 17 29
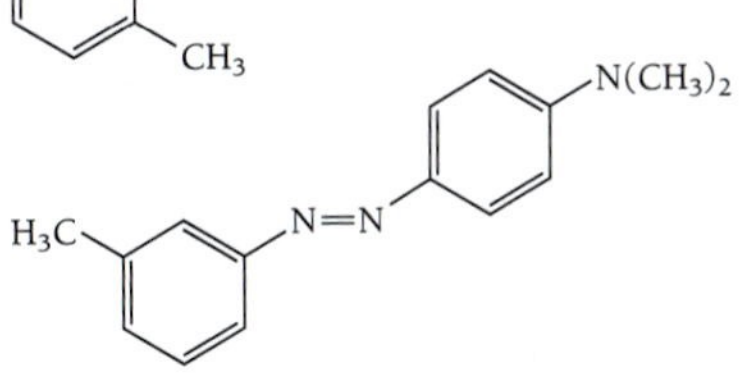	*N*,*N*-Dimethyl-*p*-(*m*-tolylazo)-aniline	+ + −	+ − +	147* 13 31
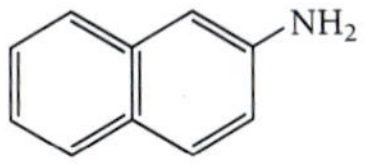	2-Naphthylamine	+ + −	+ − +	85* 15 21

continued

TABLE 20.3 *continued*

Carcinogen	Presence (+) or Absence (−) of Carcinogen in Assay	Presence (+) or Absence (−) of Liver Homogenate in Assay	No. of Mutations at *his* Locus Per Plate
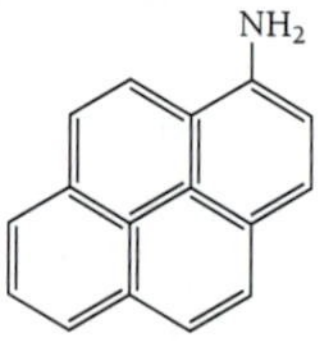 1-Aminopyrene	+	+	398*
	+	−	59
	−	+	29
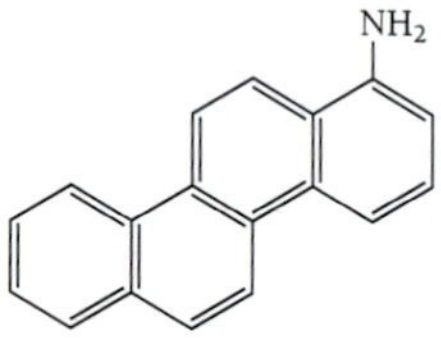 6-Aminochrysene	+	+	638*
	+	−	30
	−	+	46
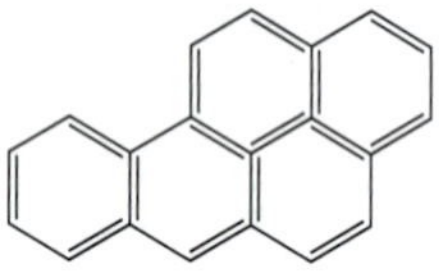 Benzo(a)pyrene	+	+	505*
	+	−	34
	−	+	44
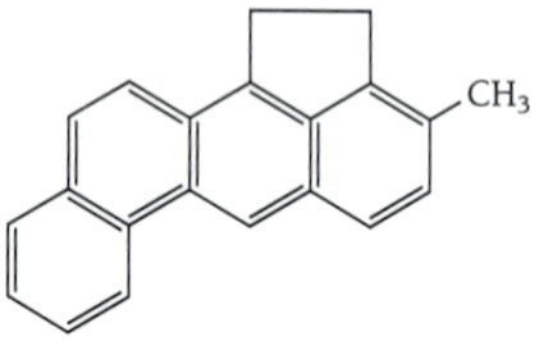 3-Methylcholanthrene	+	+	510*
	+	−	21
	−	+	27
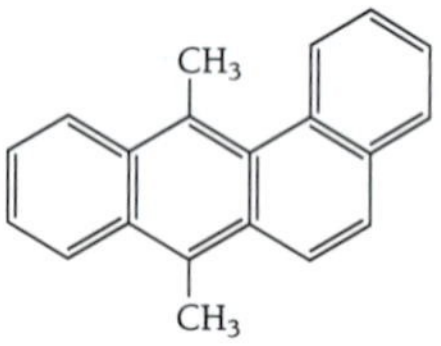 7,12-Dimethylbenz(a) anthracene	+	+	88*
	+	−	20
	−	+	36
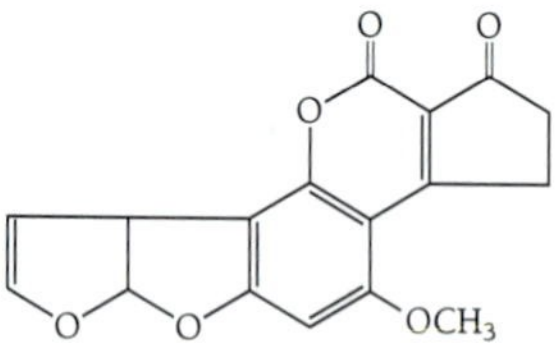 Aflatoxin B_1	+	+	266*
	+	−	26
	−	+	26
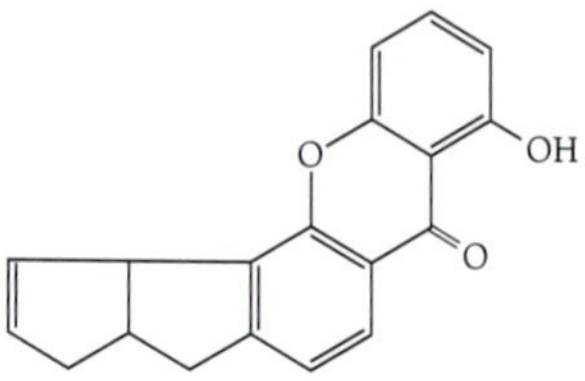 Sterigmatocystin	+	+	121*
	+	−	8
	−	+	32

*Numbers of mutations significantly higher than controls. The results indicate that many carcinogens are highly mutagenic after they are metabolized. (*From Ames, B., W. Durston, E. Yamasaki, and F. Lee. 1973.* Proceedings of the National Academy of Sciences USA, *70: 2281.*)

Inherited Defects in DNA Repair Processes Cause Cancer.

Cancer is normally a disease of old age; it takes decades to accumulate sufficient mutations to form a malignant growth. However, some individuals acquire cancers at an early age. Some of these individuals appear to possess a built-in high mutation rate, as if they were exposed to high levels of mutagens all their lives. Several genes involved in this form of early cancer formation have been identified and have been found to encode DNA repair or maintenance enzymes (Chapter 17). Among the best understood of these defective genes are those responsible for a disease characterized by extreme sensitivity to sunlight, called **xeroderma pigmentosum.** Even mild exposure to sunlight leads to many skin cancers in these patients. The disease is caused by a defect in a gene (*xp*) involved in nucleotide excision repair.

A second group of repair genes associated with cancer are those responsible for mismatch repair. Defects in these genes cause hereditary nonpolyposis colorectal cancer. Finally, two additional repair genes have recently been shown to be involved in breast cancer formation in humans: *BRCA1* and *BRCA2*.

Viruses Can Carry Viral Oncogenes That Are Responsible for Cancer.

During the early part of this century, it was demonstrated that cancerous tumors are transmissible from one organism to another. When a segment of a tumor growing in one animal was transplanted into a normal tissue of another animal, a new tumor of the same type would arise in the recipient. In 1908, a Rockefeller University biologist, Peyton Rous, refined these studies by homogenizing a sarcoma removed from the breast of an afflicted chicken and, after spinning the suspension in a centrifuge, injected the supernatant suspension into the muscle of a normal chicken. Within a matter of days, the normal chicken developed a sarcoma at the site of the injection. Subsequent experiments, in which the supernatant was passed through a filter with a pore size small enough to exclude all cells but large enough to admit virus particles, eventually led to the association of the **tumorigenic** or cancer-causing effect with a virus. The virus was named **Rous sarcoma virus** (RSV) in honor of Rous's pioneering work (Figure 20.10).

A large number of oncogenic viruses, which rapidly induce tumors (within days) after infection, have now been isolated from a variety of vertebrate species. Many of these acutely transforming viruses have been found to be retroviruses (see Chapter 19). One interesting and significant difference between oncogenic retroviruses associated with the acute onset of tumors and other retroviruses is the presence in the oncogenic retroviruses of an additional gene that, when deleted, removes the viruses' acute oncogenic properties. These cancer-causing genes are called viral oncogenes, or v-*oncs* (Figure 20.11).

Over 40 oncogenic retroviruses have thus far been isolated from various vertebrate species. Each of these retroviruses contains one or (on rare occasions) two v-*oncs* that are essential for the tumor-causing properties of the virus but not important for viral replication. In a few cases, the same v-*onc* is found in more than one virus, but as a general rule each oncogenic retrovirus contains a unique v-*onc*. About 50 different v-*oncs* have thus far been identified. These v-*oncs* have been catalogued into

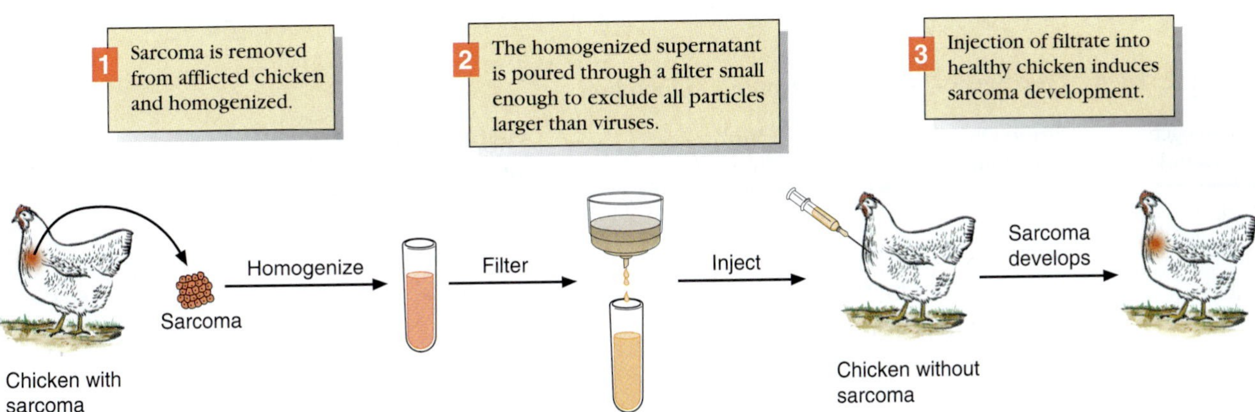

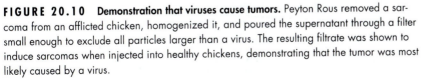

FIGURE 20.10 **Demonstration that viruses cause tumors.** Peyton Rous removed a sarcoma from an afflicted chicken, homogenized it, and poured the supernatant through a filter small enough to exclude all particles larger than a virus. The resulting filtrate was shown to induce sarcomas when injected into healthy chickens, demonstrating that the tumor was most likely caused by a virus.

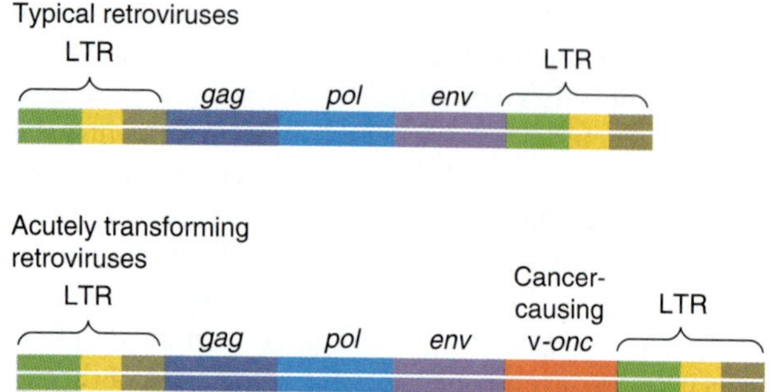

FIGURE 20.11 The structures of typical nontransforming retroviruses and acutely transforming retroviruses. The typical retrovirus contains the *gag, pol,* and *env* regions coding for the viral capsid, reverse transcriptase, and envelope proteins, respectively (see Chapter 19). In acutely transforming retroviruses, one or more additional genes are present that are not present in typical (nonacutely transforming) retroviruses. When these additional genes are deleted, acutely transforming retroviruses lose their cancer-causing abilities. The cancer-causing genes present in acutely transforming viruses are called viral oncogenes (v-*oncs*).

subgroups based upon the cellular function of the proteins they encode (Table 20.4). For example, the v-*onc* contained in RSV, called v-*src,* encodes a protein kinase called pp60^{src} (phosphoprotein, 60,000 MW) that transfers phosphate groups from ATP to protein. Protein kinases typically phosphorylate (add phosphate groups) to serine and threonine residues, but pp60^{src} preferentially phosphorylates tyrosine. As we shall see later in this chapter, protein phosphorylation is one of the key processes by which regulatory signals are transmitted through the cell. The protein kinase pp60^{src} is believed to trigger a series of reactions that signal the cell to enter into rapid cell division and hence cellular transformation.

Other v-*oncs*, such as v-*sis*, encode proteins homologous to cellular growth factors that, when expressed in infected cells, stimulate cell proliferation. Nuclear transcription factors (see Chapter 14) similar to those encoded by the viral oncogenes v-*myc* and v-*myb* are also believed to induce cell division when expressed in infected cells. Indeed, essentially every aspect of the cellular regulatory network seems to be represented by v-*oncs*. This underscores the fact that a normal cell can be transformed into one that displays the cancer phenotype in many different ways. Note that the introduction of viral oncogenes is a dominant event; that is, it is a gain-of-function above and beyond the function of the normal gene (proto-oncogene) already present in the cell. This explains why Rous obtained cancers in chickens immediately after infection with the virus.

Although oncogenic retroviruses constitute the most abundant class of cancer-causing viruses, a few other kinds of viruses are also capable of inducing tumors. These other viruses also carry oncogenes, but, unlike the retrovirus cancer-causing genes, the oncogenes carried by these other viruses are critical to viral replication, as well as being responsible for tumor induction (Table 20.5). An example is the *E6* gene of the bovine and human papilloma viruses.

When a cell is infected by an oncogenic retrovirus, the v-*onc* becomes stably integrated into a chromosome of the infected cell. This stable integration of the viral oncogene into a host chromosome not only imparts tumorigenicity to the infected cell but also ensures that the cancer phenotype is passed on to all descendent cells, thus leading to tumor formation.

TABLE 20.4 *Oncogenes Classified According to Function*

Oncogene Symbol	Proto-Oncogene Function	Viral Source	Viral-Induced Tumor
sis	Growth factor	Monkey	Sarcoma
erbB	Growth factor receptor	Chicken	Erythroleukemia
fms	Growth factor receptor	Cat	Sarcoma
ras	GTP-binding protein	Rat	Sarcoma
src	Tyrosine kinase	Chicken	Sarcoma
abl	Tyrosine kinase	Mouse, cat	Pre-B-cell leukemia
raf	Serine/threonine kinase	Chicken, mouse	Sarcoma
myc	Transcription factor	Chicken	Sarcoma
fos	Transcription factor	Mouse	Osteosarcoma
jun	Transcription factor	Chicken	Fibrosarcoma

TABLE 20.5 *Cancer-Causing Viruses That Are Not Retroviruses*

Virus	Oncogene	Affected Species
Epstein–Barr	*BNLF-1*	Human
Human papilloma	*E6*	Human
Bovine papilloma	*E6*	Cow
Adenovirus	*1A, E1B*	Rodent

SOLVED PROBLEM

Problem

Offer an hypothesis to explain why patients with the inherited form of retinoblastoma usually develop tumors in both eyes, whereas only one of the eyes becomes tumorous in the spontaneous form of the disease.

Solution

Because the recessive *rb* allele that causes retinoblastomas is relatively rare in natural populations, it is unlikely that two individuals heterozygous for the allele will mate and produce *rb/rb* homozygotes. Thus, most individuals prone to the inherited form of the disease are *rb*/+ heterozygotes. In these individuals, the *rb* allele is present in the zygote and is passed on to all somatic cells by mitosis during development. Cells heterozygous for the defective *rb* allele are only one mutational step removed from the cancerous *rb/rb* genotype. If *rb*/+ heterozygotes experience a single somatic mutation converting early embryonic cells to the *rb/rb* genotype prior to formation of the precursor cells of each individual eye, both eyes will develop retinoblastomas. Because individuals who do not inherit a defective *rb* allele are two mutational steps away from the cancer-causing genotype, it is less likely that both mutations will occur early enough in development to be present in an embryonic cell that is a precursor to cells that will develop in both eyes.

SECTION REVIEW PROBLEMS

4. What is a carcinogen? How are carcinogens usually identified?
5. What were the initial pieces of evidence that convinced early cancer researchers that cancer had a genetic basis?
6. What is meant by the term "predisposition to inherit cancer"? Name three examples of inherited cancers.
7. How does cigarette smoking cause cancer?
8. How were oncogenic viruses discovered?

Viral Oncogenes Are Homologous to Genes Contained in Normal Cells.

After scientists discovered that acutely transforming retroviruses such as RSV carried cancer-causing genes, the next question was, What is the origin of these retroviral oncogenes? Previous research by viral geneticists had demonstrated that viral genomes very efficiently store information in their nucleic acid. This efficient use of space is believed to have evolved because viral genomes are so small, and there would have been strong selective pressure to use the available space maximally. This hypothesis was based on the observation that viral genes often encoded multiple functions by using such space-saving devices as overlapping reading frames and multiple-splicing patterns (see Chapter 14). Given this premium on space-saving efficiency, it was considered rather surprising that retroviral v-*oncs* served no function critical to retroviral replication but only encoded information to transform the host cell in which they were introduced. This paradox led some geneticists to speculate that v-*oncs* may be recent evolutionary acquisitions, perhaps sequestered from the genome of infected hosts by some recombination-like mechanism (Figure 20.12).

Nonacutely Transforming Retroviruses Produce Acutely Transforming Oncogenic Retroviruses.

The first direct evidence that retroviral v-*oncs* may have been derived from the genomes of infected hosts came from studies carried out with nonacutely transforming retroviruses such as the **avian leukosis viruses** (ALV), which causes lymphomas in chickens. **Nonacutely transforming retroviruses** such as ALV are distinguished from acutely transforming viruses such as RSV by the fact that the time period between viral infection and the onset of the tumor is considerably longer for the nonacutely transforming viruses. For example, mice inoculated with ALV will, after several weeks, occasionally develop tumors. On those occasions when a tumor does form, new acutely transforming retroviruses can be isolated from the cancerous tissue. These new, acutely transforming oncogenic retroviruses isolated from the tumorous mice were derived by a recombination event between ALV and the host genome. This fact implied that the infected host cells carried the precursors of v-*oncs* as normal components of their genomes (Figure 20.13). These derived, acutely transforming retroviruses are similar to ALV except that the *env* gene has been replaced by an oncogene caused by a recombination event between the ALV *env* gene and a host (mouse) gene. This suggests that normal mice genomes contain genes with the potential to cause cancer.

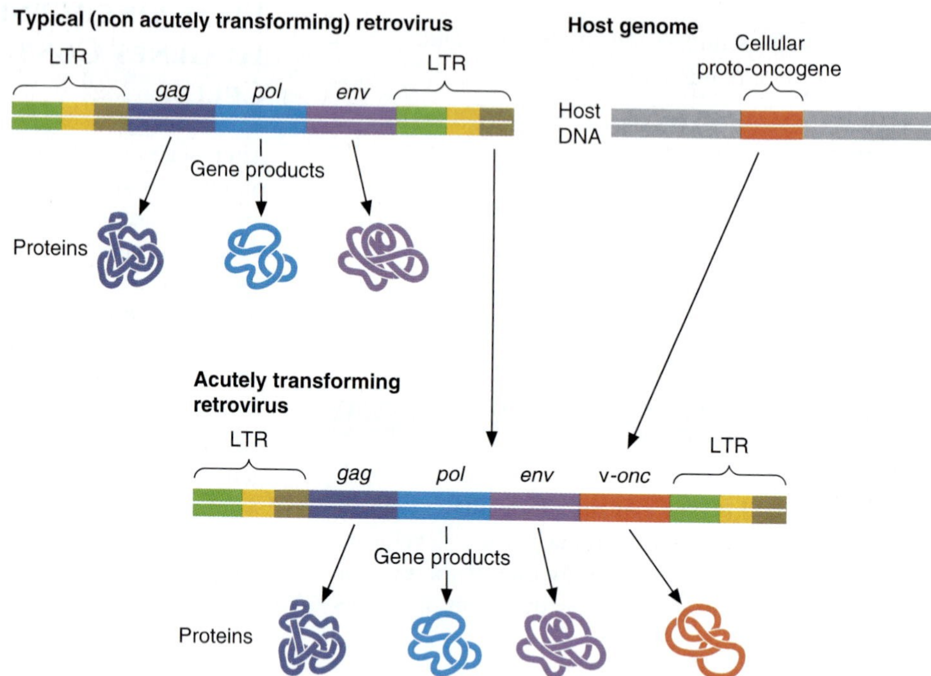

FIGURE 20.12 Acquisition of an oncogene by a virus. Acutely transforming retroviruses are believed to have evolved by acquiring copies of proto-oncogenes from host genomes that confer on these viruses their ability to transform cells.

Genes Homologous to Viral Oncogenes Are Present in the Genomes of Normal Cells.

The hypothesis that viral oncogenes are derived from genes contained in normal cells was directly tested in 1976 by two University of California biologists, Harold Varmus and J. Michael Bishop (see HISTORICAL PROFILE: *Harold Varmus and J. Michael Bishop*). Using a cDNA probe of the v-*src* oncogene isolated from the **avian sarcoma virus** (ASV), these workers showed by DNA–DNA hybridization studies that a gene homologous to v-*src* is present in normal avian DNA. Upon publication of this landmark study, similar experiments using other v-*onc* probes hybridized against DNA isolated from other

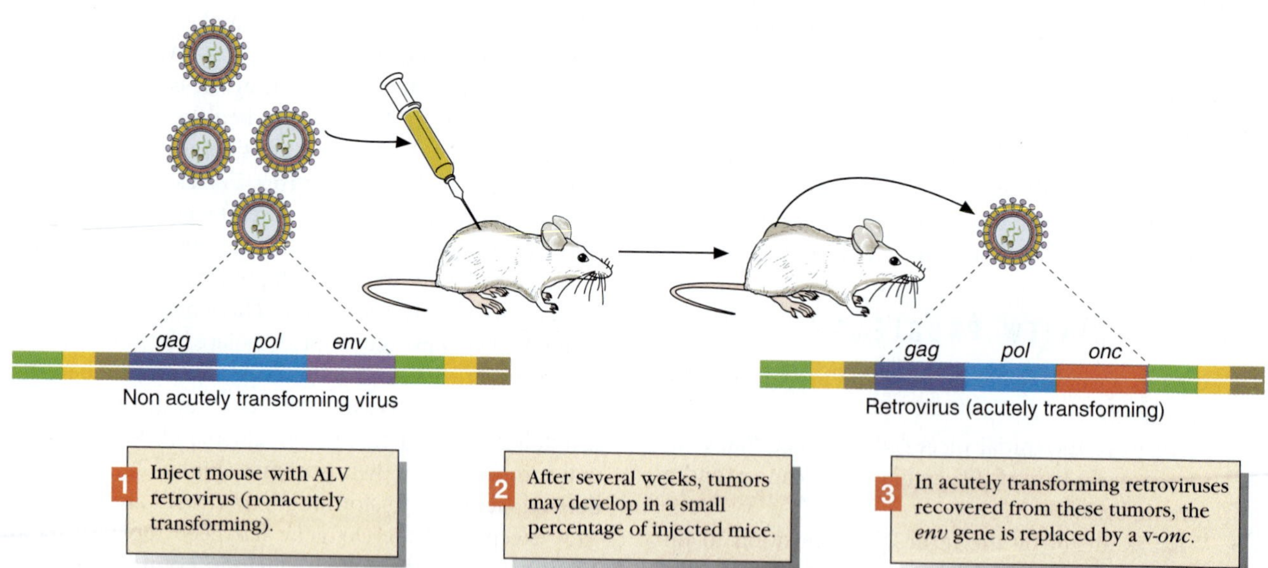

FIGURE 20.13 Tumor development from a retrovirus. Mice inoculated with the non-acutely transforming retrovirus ALV will occasionally develop tumors from which acutely transforming retroviruses can be recovered.

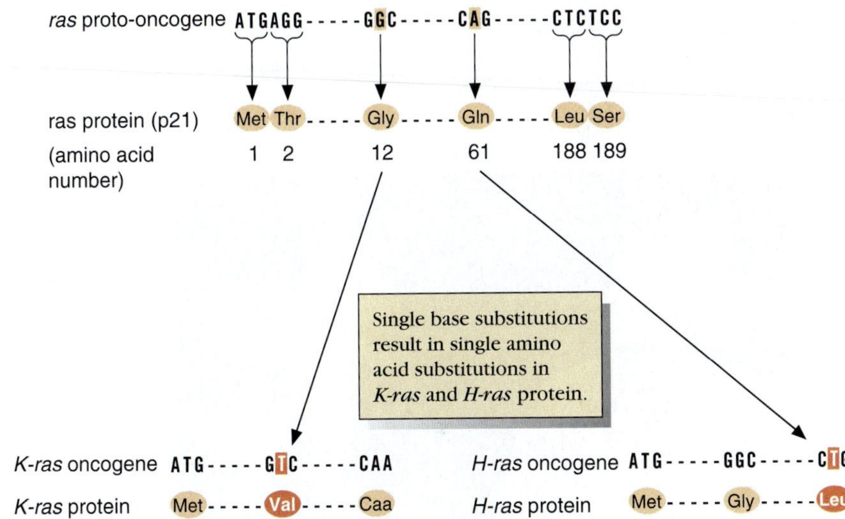

FIGURE 20.14 Transformation of a proto-oncogene to an oncogene. The K-*ras* and H-*ras* oncogenes differ from their proto-oncogene homologues by single amino-acid substitutions. The p21 protein encoded by the *ras* proto-oncogene consists of 209 amino acids; glycine is present at position 12 and glutamine, at position 61. Analysis of $p21^{ras}$ from several tumors shows a single amino acid difference at one or the other of these positions. For example, the protein encoded by the K-*ras* oncogene has arginine at position 12 rather than glycine. The protein encoded by the H-*ras* oncogene has leucine at position 61 rather than glutamine.

species demonstrated the universality of the initial finding. Collectively, these studies not only supported the contention that v-*oncs* are derived from normal cellular genes (the proto-oncogenes) but also implied that spontaneous and carcinogen-induced tumors probably involved mutations in one or more proto-oncogenes.

Viral Oncogenes and Proto-Oncogenes Are Different.

Although v-*oncs* are clearly derived from proto-oncogenes, it is important to remember that significant differences exist between these two classes of genes. V-*oncs* induce tumors, whereas proto-oncogenes carry out normal regulatory functions in the cell. One important difference between v-*oncs* and proto-oncogenes is the manner in which their expression is regulated. Expression of a v-*onc* is under the control of a retroviral promoter, which may show significant deviations in levels and cellular patterns of expression from that of its homologous proto-oncogene. In many cases, this alteration in the level or pattern of expression is sufficient to induce cellular transformation. For example, members of the *myc* proto-oncogene family are normally not expressed or are expressed at very low levels in nondividing, differentiated cells. Virus-mediated overexpression of v-*myc* in differentiated cells typically contributes to the loss of the differentiated phenotype and onset of cell proliferation.

In addition to sometimes being expressed at higher levels and/or in different tissues than their proto-oncogene counterparts, v-*oncs* frequently encode proteins that are structurally distinct from their proto-oncogene homologues. In some cases, these structural differences are sufficient to induce cellular transformation. For example, a number of proto-oncogenes encode proteins that function within intracellular signaling pathways. These pathways normally serve as mediators of information transfer between the growth factors that bind to cell surface receptors and the internal functions that lead to cell proliferation. The *ras* proto-oncogene encodes one of these signaling proteins and is homologous to the *ras* oncogene carried by the **Harvey rat sarcoma virus** (H-*ras*) (originally isolated from a thyroid carcinoma) and the **Kirsten rat sarcoma virus** (K-*ras*) (isolated from several cancers, including colon, lung, and pancreatic tumors). A single nucleotide change between the *ras* proto-oncogene and its retroviral counterpart results in a single amino acid change in the protein (a glycine to valine in K-*ras* and a glutamine to leucine in H-*ras*). This protein change alters the properties of the v-*ras* encoded protein p21 (21,000 MW) sufficiently to cause it to become oncogenic (Figure 20.14).

Another distinctive difference between v-*oncs* and their proto-oncogene homologues is that v-*oncs* lack introns, whereas their proto-oncogene homologues display the typical interrupted structure of most eukaryotic genes.

SOLVED PROBLEM

Problem

What is the difference between an acutely transforming and nonacutely transforming oncogenic retrovirus? Give an example of each.

Solution

An acutely transforming retrovirus is one that carries a v-*onc* sequence as part of its viral genome. The v-*onc* sequence is under the regulatory control of the viral promoter. When expressed, it leads to the rapid transformation of the infected cell. The Rous sarcoma virus is an example of an acutely transforming retrovirus. A

HISTORICAL PROFILE

Harold Varmus and J. Michael Bishop

The discovery of oncogenes was to the field of cancer biology what the discovery of the structure of DNA was to molecular biology: it opened up completely new and productive areas of scientific investigation. Although Harold Varmus and J. Michael Bishop (Figure 20.A), the two scientists who were awarded the Nobel Prize in 1989 for the discovery of oncogenes, came from quite different personal and educational backgrounds, they were brought together by a mutual desire to understand the molecular basis of cancer.

FIGURE 20.A J. Michael Bishop and Harold Varmus. *(Reuters/Corbis Bettmann)*

Harold Varmus was born in Oceanside, NY, in 1939. He attended Amherst College, a private college in western Massachusetts, where he majored in English Literature. Although Varmus also studied biology during his undergraduate days at Amherst, he remained primarily interested in literature. After his graduation in 1961, Varmus decided to attend graduate school at Harvard University, where he studied Anglo-Saxon and metaphysical poetry. During his time at Harvard, Varmus became fascinated with the emerging field of molecular biology and the potential it held for medical science. He completed a masters degree in literature in one year; then he applied and was accepted to Columbia University Medical School in New York.

Varmus received his medical degree in 1966 and went on to do postdoctoral research on the role played by cAMP in the regulation of the bacterial *lacZ* gene (see Chapter 14). In 1970, Varmus traveled to the University of California in San Francisco to join J. Michael Bishop's lab in studying how specific classes of retroviruses caused cancer.

J. Michael Bishop was born in York, PA, in 1936. He attended grammar school in a two-room rural school. Bishop's interest in

nonacutely transforming retrovirus, such as the avian sarcoma virus, does not contain an oncogene as part of its viral genome. However, if the virus happens to insert near a proto-oncogene, it may alter the normal pattern of expression of this host gene leading to cellular transformation. Because it may take an extended period of time for a nonacutely transforming retrovirus to insert fortuitously near a proto-oncogene, the time from infection to the onset of cancer is usually considerably longer than for acutely transforming retroviruses.

SECTION REVIEW PROBLEM

9. List some structural and functional differences that distinguish v-*onc*s from proto-oncogenes.

PROTO-ONCOGENES CAN BE CONVERTED INTO CELLULAR ONCOGENES.

The high degree of structural homology between proto-oncogenes and v-*onc*s implies that many nonviral mediated spontaneous and carcinogen-induced tumors may involve mutations in one or more proto-oncogenes. This, in fact, appears to be the case. The types of mutations capable of converting a proto-oncogene to a cellular oncogene (c-*onc*) appear to vary from situation to situation, but overall they represent the typical spectrum of mutations previously associated with other eukaryotic genes (that is, point mutations, translocations, insertions, and gene amplifications).

medicine was stimulated early in his life by his family doctor, who allowed Bishop to assist him during summer vacations. Bishop attended Gettysburg College, where he majored in chemistry. After graduating from Gettysburg in 1957, Bishop attended Harvard Medical School, where, in addition to the usual medical curriculum, he became particularly interested in microbiology and immunology. After medical school, Bishop carried out postdoctoral research on viruses at the National Institutes of Health (NIH) in Bethesda, MD. He subsequently left NIH to take a position at the University of California Medical School in San Francisco, where he was the leader of a research team studying oncogenic retroviruses. In 1970, Varmus arrived.

By the early 1970s, researchers had established that the cancer-causing potential of the Rous sarcoma virus was attributable to a viral oncogene called v-*src*. In 1976, Bishop and Varmus published a paper reporting that DNA complementary to v-*src* could be found not only in avian cells that had been infected with RSV but in noninfected cells as well. At first this rather startling observation was explained by many skeptical observers as most likely to be caused by the presence of cryptic viral genes in avian cells, which were the remnants of previous infections. The alternative hypothesis, favored by Varmus and Bishop, was that normal cells contain genes homologous to viral oncogenes as normal constituents of their genome. In order to convince their critics, Bishop and Varmus showed that genes homologous to v-*src* were also present in vertebrate cells that had never been exposed to RSV. This finding demonstrated convincingly that v-*src* homologous genes are indeed normal components of the vertebrate genome. Subsequent works by Varmus, Bishop, and others demonstrated that numerous other viral oncogenes also had homologues, not only in the genomes of vertebrates but within the genomes of many invertebrates as well. Bishop and Varmus postulated that viral oncogenes were originally derived from their cellular homologues by a process similar to viral transduction in bacteria (see Chapter 8). Today, the cellular homologues of viral oncogenes are called proto-oncogenes, and they are known to play essential roles in cellular regulation of all organisms. Thus, the original discovery by Varmus and Bishop has not only led to a more complete understanding of the molecular basis of cancer, it has also provided tremendous insight into the mechanisms that control cellular regulation in normal cells.

Some Proto-Oncogenes Become Cellular Oncogenes Through Point Mutations.

The *ras* proto-oncogene is a common target for mutations associated with carcinogen-induced tumors in experimental animals. For example, when the carcinogen/mutagen *N*-methyl-*N*-nitrosourea (MNU) is administered to mice, breast tumors are found to arise at a high frequency. In these tumors, the *ras* proto-oncogene often undergoes a G-to-A transition, converting the normally functioning gene into a c-*onc*. Interestingly, G-to-A transitions are precisely the type of mutation known to be induced by MNU. A variety of other carcinogens have been shown to induce characteristic DNA lesions in the *ras* proto-oncogene, suggesting that this gene is a direct target for mutagens.

Point mutations in *ras* proto-oncogenes have been associated with a number of human cancers, such as colon and rectal cancer, lung and thyroid tumors, and several types of leukemia and lymphomas (Figure 20.14). In many instances, *ras* mutations are detectable in precancerous growths such as colon **adenomas,** which suggests that *ras* point mutations may represent early steps in the tumor-forming process.

Some Proto-Oncogenes Become Cellular Oncogenes Through Translocations.

One of the best-studied examples of a tumor being induced by a translocation is **Burkitt's lymphoma,** which is characterized by uncontrolled proliferation of lymphocytes. Classic cytological examination of Burkitt's

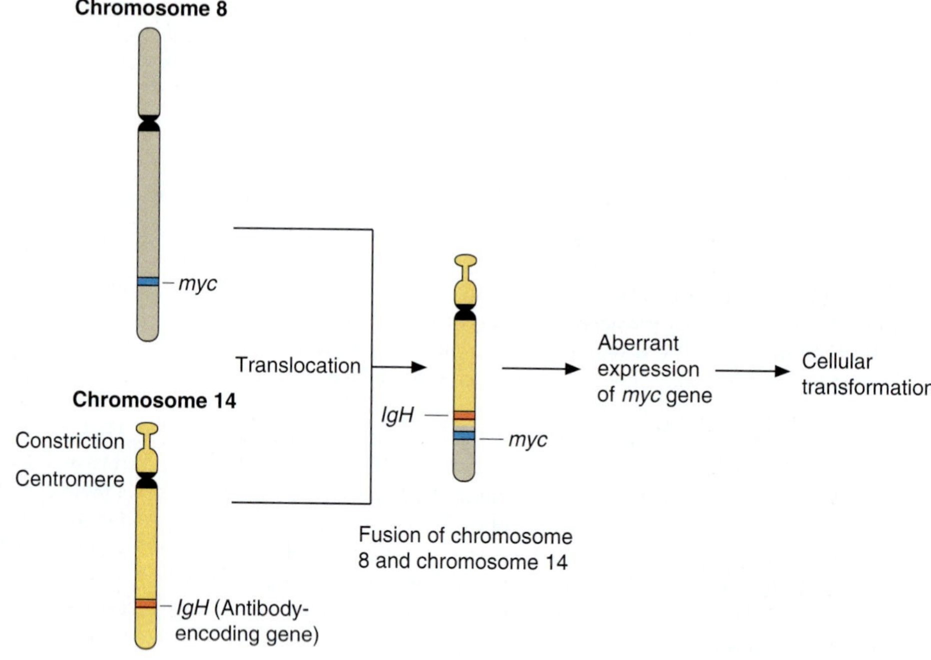

FIGURE 20.15 Cancer caused by a translocation. A translocation between chromosome 8 and chromosome 14 brings the *myc* gene under regulatory control of an antibody-encoding gene, which results in aberrant *myc* expression and consequent cellular transformation.

lymphoma cells established early on that this tumor is almost always associated with a translocation of a piece of chromosome 8 to either chromosome 2, 14, or 22 (Figure 20.15). Translocation to chromosome 14 occurs most frequently. It was later established that the translocation sites on chromosomes 2, 14, and 22 are immunoglobin gene loci. Because these antibody-encoding genes normally undergo translocations (see Chapter 14), the mutation involving chromosome 8 is believed to be an aberration of this process. The segment of chromosome 8 involved in the translocation associated with Burkitt's lymphoma contains the *myc* proto-oncogene. The translocation process is believed to place the *myc* gene under abnormal regulatory controls, which results in atypically high levels of *myc* expression. The high *myc* levels in turn lead to cellular transformation and tumor formation.

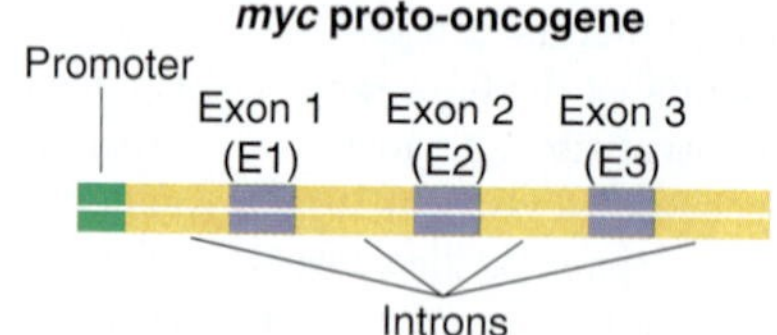

(a) ALV insertion between Exon 1 and Exon 2

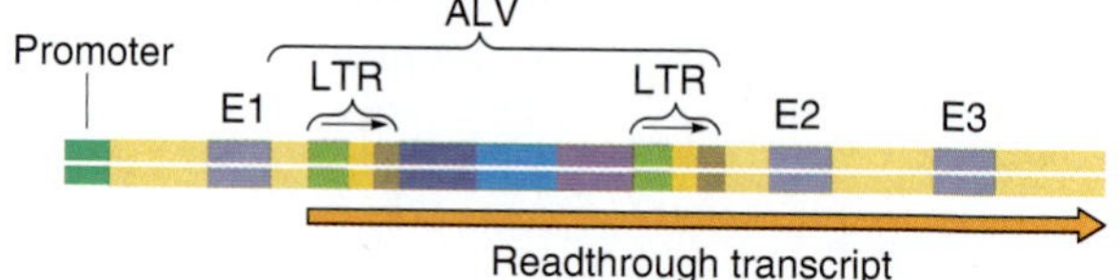

(b) ALV insertion in reverse orientation

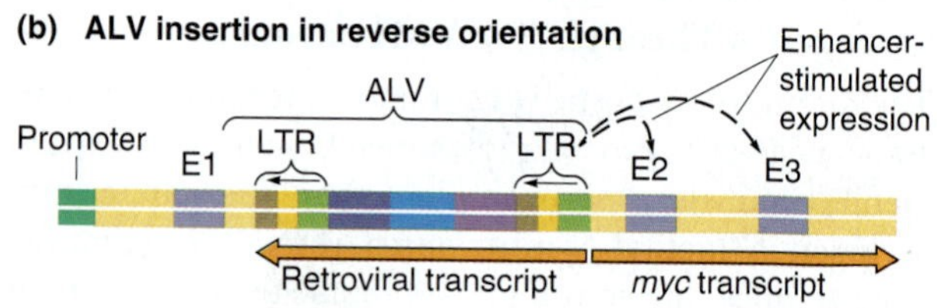

(c) ALV insertion downstream of *myc*

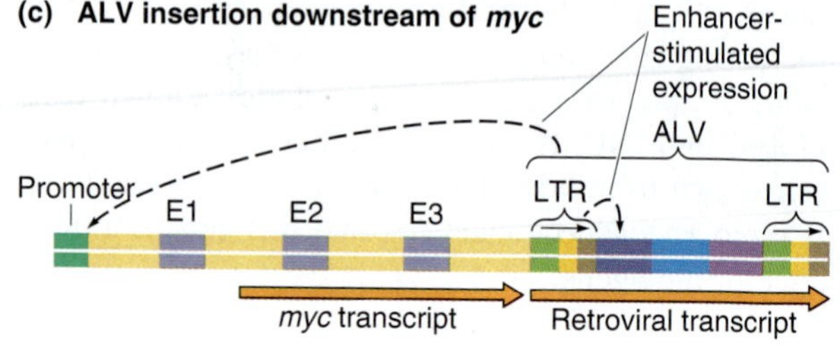

► **FIGURE 20.16 Cellular transformation caused by a retroviral element insertion.** When ALV is inserted adjacent to the *myc* gene-coding sequence, the *myc* gene is overexpressed in bursal lymphomas. Different ALV insertion events associated with three distinct lymphomas are shown. The *myc* proto-oncogene consists of three exons: the first encodes a long nontranslated leader sequence and the second two encode the MYC protein. (a) The ALV provirus has inserted between the first and second exons of the *myc* gene. The two coding exons of the gene are expressed as part of a long transcript initiated within the viral LTR (long terminal repeat). (b) The ALV provirus is inserted within the first intron in reverse orientation to the *myc* gene. Expression of the coding exons of *myc* is stimulated by enhancer sequences within the viral LTR. (c) The retroviral element is inserted upstream of *myc*. Expression of the *myc* gene is induced by enhancer sequences contained within the LTR.

A Retroviral Element Can Induce a Proto-Oncogene To Become a Cellular Oncogene.

In Chapter 19, we considered regulatory mutations that are induced by the insertion of retroviral proviruses and retrotransposons adjacent to chromosomal genes. Analogous insertional events adjacent to proto-oncogenes have been associated with overexpression of the c-*onc* (that is, a dramatic increase in the level of c-*onc* gene product) and the subsequent transformation of normal cells to cancer cells. For example, the avian leukosis virus is a retrovirus associated with bursal lymphomas in chickens. Unlike the acutely transforming retroviruses, ALV does not contain a v-*onc*. In addition, the time interval between ALV infection and tumor onset is considerably longer than for acutely transforming retroviruses such as RSV. A clue about how ALV induces tumors was that in several independently derived tumors, the virus had inserted in the same chromosomal position. This position, as it turned out, is adjacent to the *myc* proto-oncogene (Figure 20.16). This insertion had caused the gene to overexpress through the change in the promoter. Other oncogenes that are known to be activated in tumors by retroviral element insertion are *erbB*, *myb*, *mos*, H-*ras*, and *raf*. These findings are consistent with the general conclusion established in Chapter 19 that retroviral-like elements are an important source of regulatory mutation in eukaryotic genomes.

Some Proto-Oncogenes Become Cellular Oncogenes Through Gene Amplification.

A fourth mechanism by which proto-oncogenes may become oncogenic is **gene amplification** (Figure 20.17). Because amplification is a process by which the number of copies of a gene is increased, the result is usually overexpression; that is, the amount of gene product being produced increases significantly. A number of tumors have been found to contain cells that have experienced gene amplification. In a number of cases, it has been shown that the amplified regions contain a known proto-oncogene. For example, amplification of the *myc* gene located on chromosome 8 has been associated with transformation of a normal cell to a cancer cell. Such amplification may be located in the tandem arrays at the

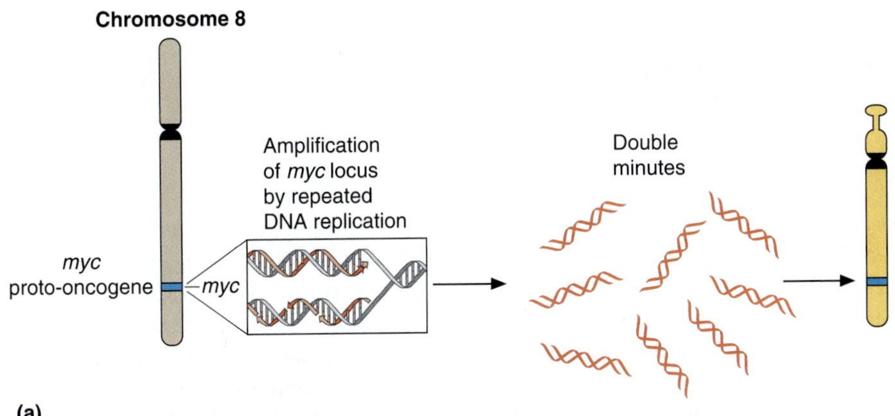

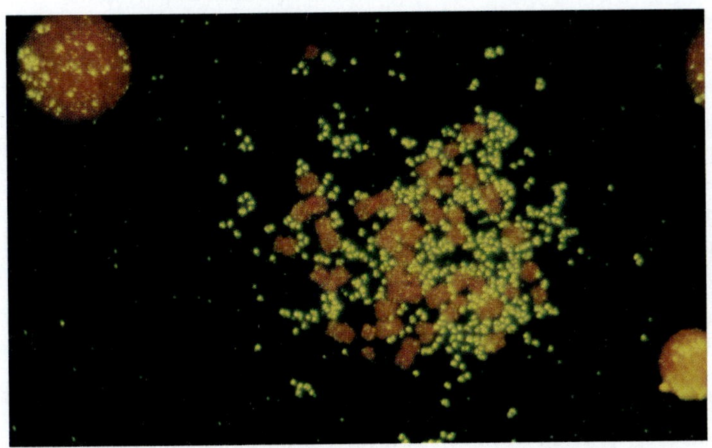

FIGURE 20.17 Gene amplification can cause cancer. Increased expression of a proto-oncogene can result from the formation of multiple copies of the gene caused by gene amplification. (a) The *myc* gene is amplified by repeated rounds of DNA replication. The amplified copies may be located in tandem at the *myc* locus or may excise from the chromosome and exist as double minutes. (b) Fluorescence *in situ* hybridization with c-*myc* DNA probe on a metaphase cell showing hybridization to multiple extra-chromosomal copies ("double minutes") of the *myc* gene (green signals). The chromosomes are counterstained with propidium iodide (orange). *(Dr. A. Mohamed, Pathology Department, Wayne State University Medical Center, Detroit, Michigan).*

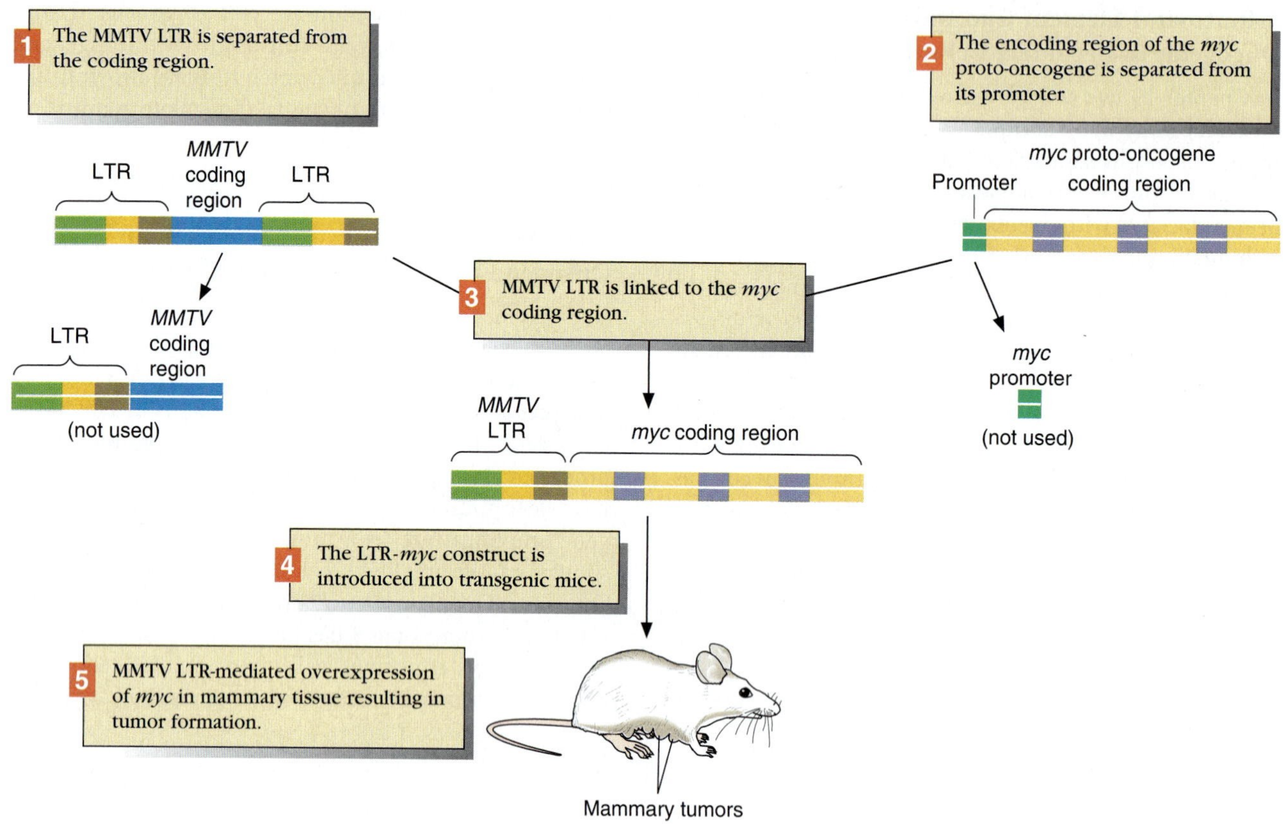

1 The MMTV LTR is separated from the coding region.
2 The encoding region of the myc proto-oncogene is separated from its promoter
MMTV
LTR
coding
region
LTR
myc proto-oncogene
Promoter
coding region
3 MMTV LTR is linked to the myc coding region.
LTR
MMTV
coding
region
(not used)
myc
promoter
(not used)
MMTV
LTR
myc coding region
4 The LTR-myc construct is introduced into transgenic mice.
5 MMTV LTR-mediated overexpression of myc in mammary tissue resulting in tumor formation.
Mammary tumors

FIGURE 20.18 Overexpression of a gene can cause cancer. When the *myc* proto-oncogene is fused to the MMTV LTR (containing the promoter) and introduced into the mouse genome, overexpression of *myc* results, but only in mammary tissue where the MMTV LTR promoter is very active.

proto-oncogene locus or may excise from the parental chromosome and exist as small extrachromosomal copies called *double minutes* (Figure 20.17).

Gene amplification usually occurs during later stages of tumor cell development. For example, a type of brain tumor called a **neuroblastoma** is classified into four stages based on tumor size and how extensively the tumor has spread from its site of origin. A member of the *myc* proto-oncogene family, N-*myc*, is rarely expressed in early-stage neuroblastomas but is amplified in over 50% of late-stage tumors. Moreover, these late-stage tumors associated with N-*myc* amplifications are much more likely to progress and metastasize than those in which N-*myc* has not been amplified. Gene amplification is also associated with oncogenes and tumors such as other *myc* family genes (c-*myc*, N-*myc*, and L-*myc*) in small-cell lung carcinomas and the *erbB-2* gene in breast and ovarian cancers.

The fact that overexpression of *myc* family genes occurs either by retroviral-element insertion, translocation, or amplification suggests that overexpression of the MYC protein in cells is tumorigenic. This hypothesis was directly tested by placing the *myc* gene under heterologous promoter control and introducing the construct to produce transgenic mice (Figure 20.18). The **mouse mammary tumor virus** (MMTV) LTR is known to be extremely active in mammary tissue. Transgenic mice carrying the *myc* gene under regulatory control of MMTV LTR experienced overexpression of *myc* in mammary tissue and the consequent development of mammary gland carcinomas. Experiments with transgenic mice carrying the *myc* gene linked to a lymphocyte-specific enhancer also resulted in overexpression of *myc* in lymphocytes and the consequent development of lymphomas at high frequency. Collectively, these studies support the view that overexpression of *myc* is oncogenic.

SOLVED PROBLEM

Problem

The mutations associated with the *ras* proto-oncogene that convert it to an oncogene (c-*ras*) are not distributed randomly over the length of the *ras* gene but are clustered within specific regions of the coding sequence. Offer an hypothesis to explain this observation.

Solution

From what we know of the mutation process, there is no reason why certain regions of a gene should be more prone to point mutations than any other. The fact that only mutations arising in certain regions of the *ras* proto-oncogene convert it to a cancer-causing cellular oncogene (c-*ras*) seems best explained by hypothesizing that these regions of the gene encode amino acids critical to normal functioning of the *ras* proto-oncogene. This hypothesis is in fact consistent with the recently established three-dimensional structure of the protein product of the *ras* proto-oncogene, which shows that the regions mutated in c-*ras* are precisely those regions of the protein that are associated with its normal function.

SECTION REVIEW PROBLEMS

10. Each of the cancers associated with *myc* family of oncogenes involves overexpression. For each of the cancers associated with *myc* discussed in this section, describe the genetic aberration involved and how the aberration may be expected to result in *myc* overexpression.
11. How was the hypothesis that *myc* overexpression is carcinogenic directly tested?

THE PROTEINS ENCODED BY PROTO-ONCOGENES FUNCTION AS CELL GROWTH REGULATORS.

As noted at the beginning of this chapter, it was recognized very early in the history of cancer biology that a major difference between normal cells and cancer cells is the loss of regulated control of cell proliferation. During the early days, many geneticists working on the cancer problem reasonably assumed that a complete understanding of the molecular workings of the cancer cell would be possible only after our knowledge of the molecular processes regulating cell division in normal cells was complete. Ironically, the situation has turned out somewhat the reverse. In fact, the discovery of oncogenes and how their (mis)functioning in cancer cells leads to uncontrolled cell proliferation has contributed greatly to our understanding of how genes (proto-oncogenes) regulate growth and proliferation in normal cells.

Growth and differentiation of normal cells are now known to be controlled by hormones, growth factors, and other external signals that bind to receptor proteins on the surface of target cells. These activated receptors, in turn, stimulate a series of reactions within the cell, ultimately leading to the transmission of the signal to the nucleus, resulting in a consequent cell behavior change. We also know that cells are able to control their division cycle internally to prevent uncontrolled division. Many of the genes involved in cell division control have been cloned and their function characterized, and they are collectively known as **cell division control**, or *cdc*, genes (discussed in Chapter 3). Different oncogenes have been found to encode proteins such as growth factors, membrane proteins, and protein kinases that act at each stage in the process of cellular information flow, allowing scientists to contrast and elucidate the function of these proteins in transformed and normal cells.

Some Proto-Oncogenes Encode Growth Factors.

Some proto-oncogenes have been shown to encode growth factors that are normally made and secreted by specialized cells for the purpose of signaling a response in other cell types. For example, it was noted earlier in this chapter that blood platelet cells normally synthesize and excrete PDGF (platelet-derived growth factor), which is encoded by the *sis* proto-oncogene. PDGF binds to receptors on fibroblast cells, stimulating them to proliferate as part of the wound-healing process. Under normal conditions, cells that produce growth factors do not also contain receptors for growth factors. Similarly, those cells that contain receptors and thus respond to a particular growth factor do not normally produce the factor. A major problem arises when this distinction is disrupted. For example, if fibroblasts that contain PDGF receptors are infected with a retrovirus carrying the *sis* oncogene or experience a regulatory mutation in their endogenous *sis* proto-oncogene resulting in gene expression, they will begin to synthesize PDGF, which in turn stimulates their own proliferation (Figure 20.19). Abnormal situations in which a cell responds to a growth factor that it also produces is called **autocrine growth stimulation,** and it is a common cause of cancer.

Other growth-factor-encoding genes that have been found to be associated with cancer when abnormally expressed are *hst* and *int*-2. These proto-oncogenes are frequently found to be overexpressed in human breast tumors caused by gene amplification. The *int-3* gene, which encodes another growth factor called interleukin-3, has frequently been found to be overexpressed in some human leukemias because of a translocation event.

Some Proto-Oncogenes Encode Growth-Factor Receptor Proteins.

As discussed in the previous section, receptor proteins are the external sensors associated with the plasma membranes of some cells to which growth factors bind. Bound receptor-proteins transmit external signals into the cell, which initiates a series of reactions ultimately leading to cell division. A defective receptor protein that sends such

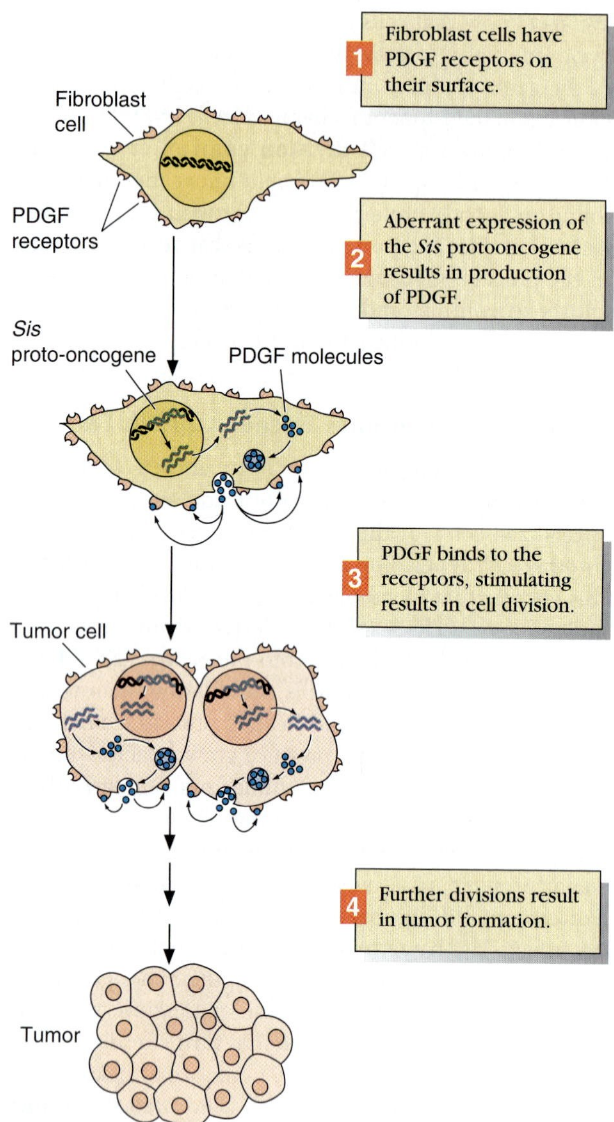

FIGURE 20.19 Overexpression of a proto-oncogene. Tumor formation is the consequence of aberrant expression of *sis* proto-oncogene in fibroblast cells.

a signal into a cell without having been stimulated by the binding of a growth factor will result in abnormal cell proliferation and consequent tumor formation.

Growth-factor receptor proteins consist of three domains.

1. The external domain is the part of the protein that actually binds to the growth factor.
2. The transmembrane domain penetrates the plasma membrane and connects to the internal domain.
3. The internal domain is exposed to the interior of the cell (Figure 20.20).

When a growth factor binds to the external domain, the entire receptor protein undergoes a conformational change that activates the internal domain. The activated

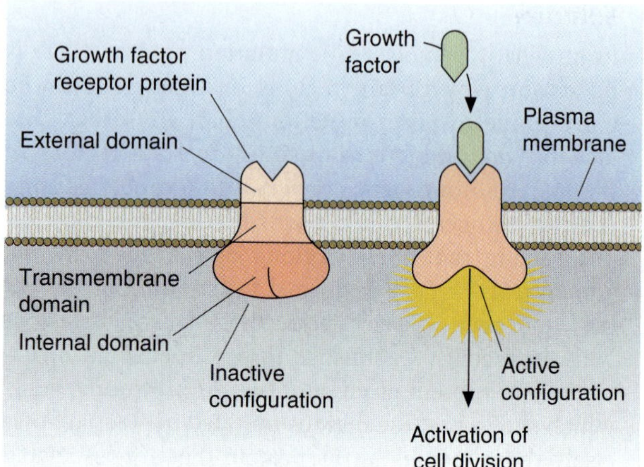

FIGURE 20.20 Growth-factor receptor proteins. These proteins consist of external, transmembrane, and internal domains. When growth factor binds to the external domain, a conformational change results in activation of the internal domain.

internal domain of the receptor protein in turn activates one or more molecules contained within the cell by a process called **intracellular signal transduction.** The mechanisms that underlie intracellular signal transduction are most often enzymatic (Figure 20.21). For example, the internal domain of the PDGF receptor is a protein-tyrosine kinase that, when activated, catalyzes the phosphorylation (that is, the attachment of phosphate to the amino acid tyrosine) of at least four distinct intracellular targets. These targets in turn pass the growth-factor-initiated signal on to the nucleus.

Proto-oncogenes that encode the domains of growth-factor receptor proteins can become oncogenes in two ways (Figure 20.22). Regulatory mutations that dramatically increase the expression level of these genes result in an overabundance of protein product, which can result in the malfunctioning of receptor proteins. For example, it was noted previously that breast and ovarian tumors have been frequently associated with amplification of the *erbB* gene. The *erbB* gene encodes the internal domain of the **epidermal growth factor** (EGF) receptor, which is associated with protein-tyrosine kinase activity. The best evidence indicates that the excess numbers of the internal domain of the EGF receptor become attached to the internal surface of the plasma membrane in a permanently activated configuration. This results in the continuous activation of intracellular targets, which leads to continuous cell proliferation. Growth-factor receptor proteins may also become oncogenic by mutations that alter their structure in such a way that the intracellular domain is active even in the absence of growth-factor binding. For example, mutations causing the loss of the external domains of the receptor proteins encoded by the *ret* and *trk* oncogenes result in a conformational change in the receptor so that the internal domain is continually activated. This, in

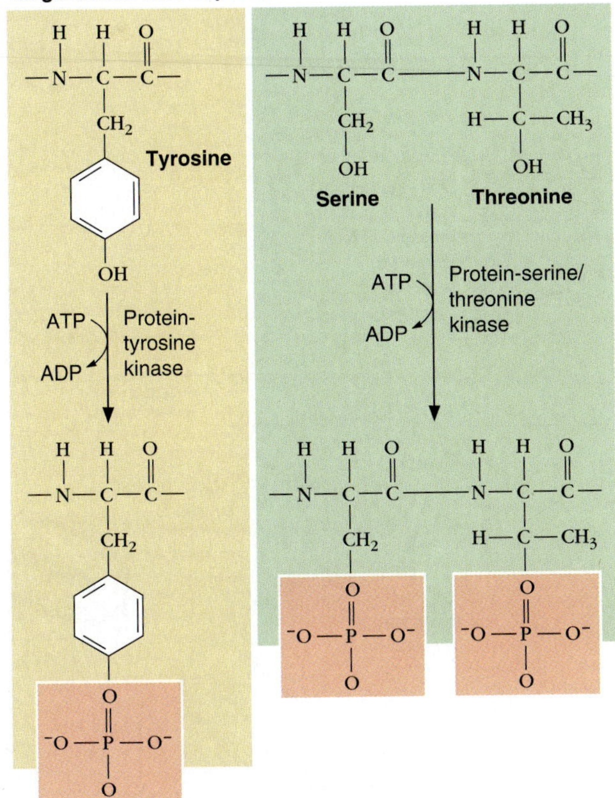

FIGURE 20.21 Protein kinases. Kinases catalyze the transfer of phosphate groups from ATP to specified amino acids (tyrosine or serine/threonine).

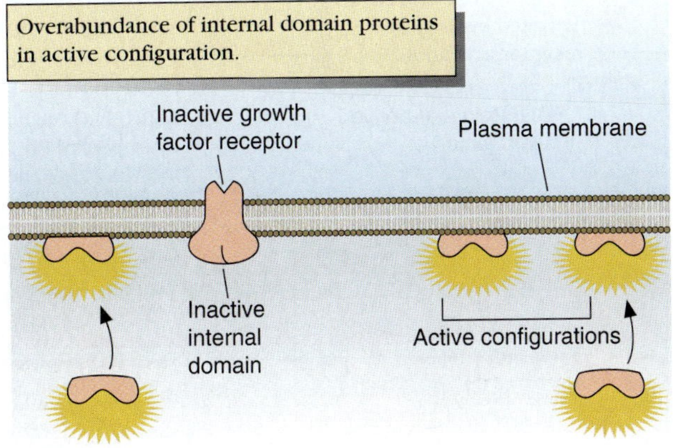

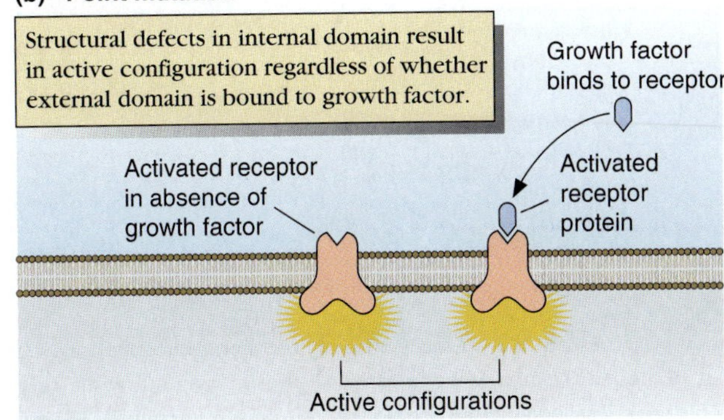

FIGURE 20.22 Proto-oncogenes encoding domains of growth-factor receptor proteins. These domains can become oncogenic in either of two ways. (a) Regulatory mutations that dramatically increase the level of expression of internal domain encoding genes can result in an overabundance of product, which binds to the internal surface of cells in an active configuration. (b) A point mutation in a gene encoding an internal domain may result in a protein that is in a permanently active configuration.

turn, results in continuous cell proliferation and consequent tumor formation.

Some Proto-Oncogenes Encode Intracellular Signaling Proteins.

In the previous section, we learned that growth-factor receptor proteins frequently transfer information to intracellular target molecules by phosphorylation. Phosphorylation is a common mechanism by which information is transmitted within a cell. As it turns out, many of the target proteins phosphorylated by activated growth-factor receptors are themselves protein kinases. Thus, we can envision the flow, or cascade, of information from receptor proteins through the cell to the nucleus as a cascade of enzymatic reactions, many of which involve protein kinases. For example, one of the target proteins phosphorylated by the PDGF receptor is an enzyme called phospholipase C (Figure 20.23). When activated, phospholipase C catalyzes the formation of diacylglycerol (DAG), which is an activator of protein kinase C. Protein kinase C is a protein serine/threonine kinase that initiates a series of reactions, often referred to as the kinase cascade, leading to cell proliferation. The gene encoding protein kinase C is a proto-oncogene because certain mutations that alter the protein's structure have been shown to result in uncontrolled cell proliferation and consequent tumor formation.

Another example of proto-oncogenes that encode a special class of intracellular signaling proteins (membrane-associated G proteins that bind GTP and GDP) is the *ras* family of genes. As discussed previously, *ras* oncogenes have been implicated in a number of cancers, including colon and lung cancers. The *ras* proto-oncogene encodes a protein, $p21^{ras}$, that binds guanine triphosphate (GTP) in its active form and guanine diphosphate (GDP) in its inactive form (Figure 20.24). Activation of $p21^{ras}$ occurs when bound GDP is replaced with GTP. This activation reaction has been shown to be catalyzed in mammalian cells by a family of **guanine-nucleotide-releasing factors** (GRFs) (Figure 20.25). Deactivation occurs by the hydrolysis of GTP to GDP. The deactivation reaction is stimulated by GTPase-activating

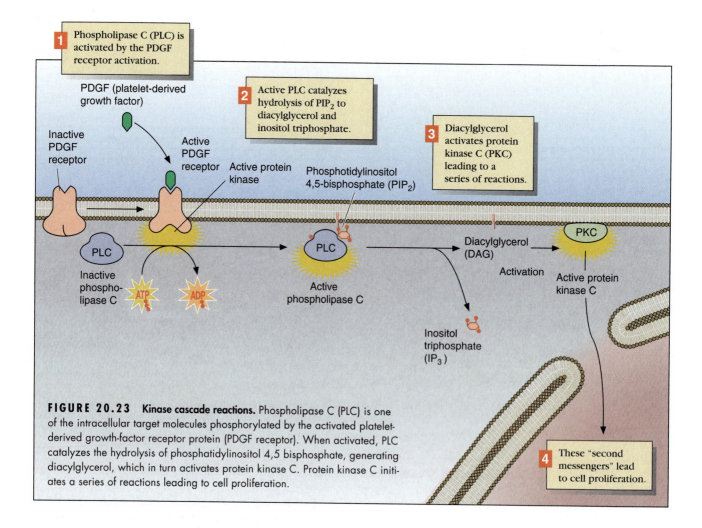

1 Phospholipase C (PLC) is activated by the PDGF receptor activation.

PDGF (platelet-derived growth factor)

Inactive PDGF receptor

Active PDGF receptor

Active protein kinase

2 Active PLC catalyzes hydrolysis of PIP_2 to diacylglycerol and inositol triphosphate.

Phosphotidylinositol 4,5-bisphosphate (PIP_2)

3 Diacylglycerol activates protein kinase C (PKC) leading to a series of reactions.

PLC

Inactive phospho-lipase C

ATP

ADP

PLC

Active phospholipase C

Diacylglycerol (DAG)

Activation

PKC

Active protein kinase C

Inositol triphosphate (IP_3)

4 These "second messengers" lead to cell proliferation.

FIGURE 20.23 Kinase cascade reactions. Phospholipase C (PLC) is one of the intracellular target molecules phosphorylated by the activated platelet-derived growth-factor receptor protein (PDGF receptor). When activated, PLC catalyzes the hydrolysis of phosphatidylinositol 4,5 bisphosphate, generating diacylglycerol, which in turn activates protein kinase C. Protein kinase C initiates a series of reactions leading to cell proliferation.

proteins (GAPs). GAP is one of the proteins phosphorylated by the PDGF receptor protein. Phosphorylation of GAP inactivated by the PDGF receptor helps maintain $p21^{ras}$ in an active state, thereby leading to cell proliferation. In general, the relative activities of GRFs and GAPs operating on $p21^{ras}$ determine its state of activation in the cell at any given moment.

Mutations in the *ras* oncogenes that interfere with the action of GAPs or GRFs and thus affect the rate of GTP hydrolysis on $p21^{ras}$ have been associated with about 30% of human malignancies.

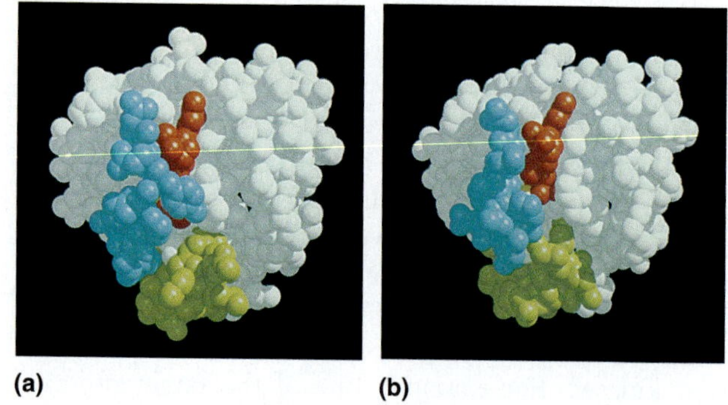

FIGURE 20.24 A three-dimensional computer image of $p21^{ras}$ protein. The protein is "active" (a) when bound to GTP (shown in blue) and "inactive" (b) when GTP is hydrolyzed to GDP. Switching from the active to the inactive state is brought about by changes in the protein's confirmation. Mutations in the *ras* proto-oncogene result in the protein being in a permanently active state and thus oncogenic. *(Dr. Sung-Hou Kim, University of California, Berkeley)*

Some Proto-Oncogenes Encode Transcription Factors.

The signals initiated by activated growth-factor receptor proteins are transmitted to the nucleus, where they induce changes in gene expression (Figure 20.26). Many of these regulatory changes are mediated by modulations in the activity of transcription factors (see Chapter 14), which represent the last step in the intracellular signaling process. For example, it was mentioned earlier that

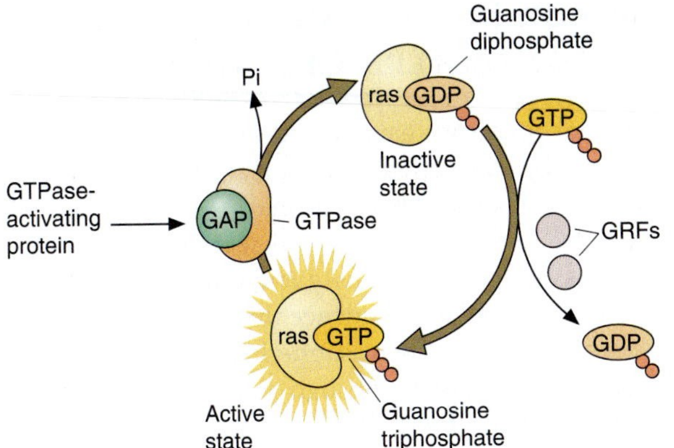

◀ **FIGURE 20.25 Regulation of p21^{ras} by guanine-nucleotide binding.** In its active form, p21^{ras} is normally bound to GTP. It binds to GDP in its inactive form. The exchange of bound GDP to GTP is catalyzed by a family of GRFs. The p21^{ras} activity is terminated by hydrolysis of bound GTP to GDP, which is stimulated by GAPs.

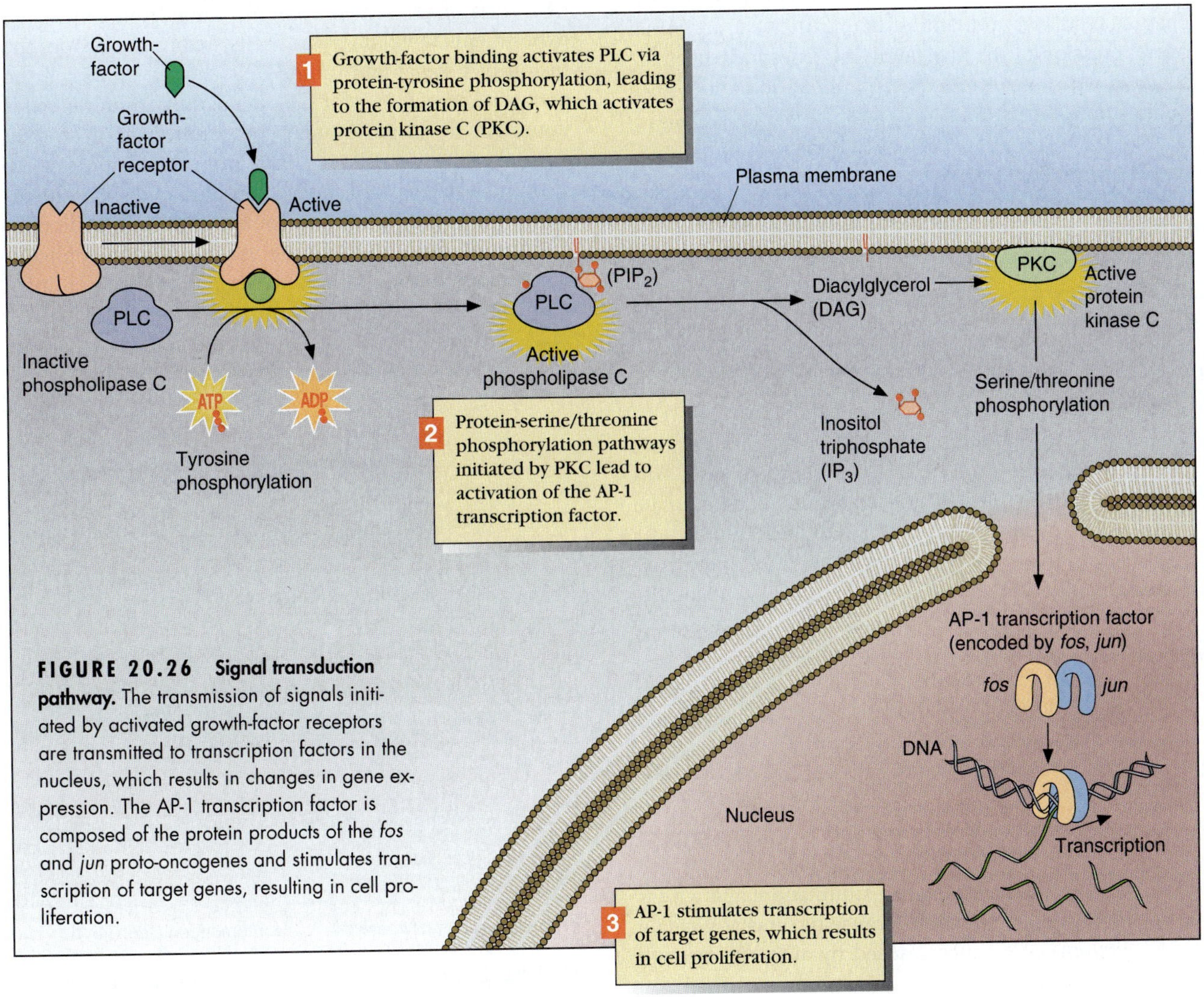

1 Growth-factor binding activates PLC via protein-tyrosine phosphorylation, leading to the formation of DAG, which activates protein kinase C (PKC).

2 Protein-serine/threonine phosphorylation pathways initiated by PKC lead to activation of the AP-1 transcription factor.

3 AP-1 stimulates transcription of target genes, which results in cell proliferation.

FIGURE 20.26 Signal transduction pathway. The transmission of signals initiated by activated growth-factor receptors are transmitted to transcription factors in the nucleus, which results in changes in gene expression. The AP-1 transcription factor is composed of the protein products of the *fos* and *jun* proto-oncogenes and stimulates transcription of target genes, resulting in cell proliferation.

the PDGF receptor results in activation of protein kinase C. Protein kinase C goes on to stimulate the AP-1 transcription factor to initiate the expression of a series of target genes, which consequently leads to cell division. The AP-1 transcription factor consists of two different proteins encoded by the *fos* and *jun* proto-oncogenes. Mutations in either of these genes, which result in transcriptional activation of the target genes normally regulated by AP-1, result in abnormal cell division and subsequent tumor formation.

Another group of proto-oncogenes that regulate gene expression is the *myc* family of genes discussed previously. For example, c-*myc* expression in fibroblasts is normally activated in response to PDGF stimulation. The c-*myc*-encoded protein then turns on a series of target genes required for cell division. As noted earlier in this chapter, regulatory mutations that result in the abnormal overexpression of the *myc* family of genes have been associated with uncontrolled cell proliferation and consequent tumor formation.

SOLVED PROBLEM

Problem

List three examples of proto-oncogenes that encode regulatory proteins. Would you expect mutations that result in overexpression of these genes to be classified as dominant or recessive mutations? Explain.

Solution

The *fos* and *jun* proto-oncogenes encode proteins that combine to form the AP-1 transcription factor. Another group of proto-oncogenes known to regulate gene expression is the *myc* family of genes. It is difficult to make any sweeping statements concerning whether or not a mutation that results in overexpression of a regulatory protein will act as a dominant or recessive mutation. However, in most situations, it is likely that overproduction of the regulatory protein will lead to dominant overexpression of target genes (if the regulatory protein is an inducer) or underexpression (if the regulatory protein is a repressor).

SECTION REVIEW PROBLEMS

12. What is autocrine growth stimulation, and why is it considered a common cause of cancer? Give an example of a cancer caused by autocrine growth stimulation.
13. In what ways can a receptor protein misfunction to stimulate uncontrolled cell proliferation? Give examples to support your answer.
14. What are intracellular signaling proteins, and how can they malfunction to cause cancer? Give examples to support your answer.

TUMOR-SUPPRESSOR GENES PREVENT UNCONTROLLED CELL GROWTH.

The oncogenes we have considered thus far in this chapter may be considered dominant alleles in the traditional Mendelian sense. That is, it takes only one copy of the oncogene to induce changes that result in uncontrolled cell proliferation and consequent tumor formation. However, there was early evidence that at least some forms of inherited cancers, such as inherited retinoblastoma, were being passed down through family lines as recessive traits. Initially, it was thought that retinoblastoma was a dominant trait (which is why the gene symbol is sometimes capitalized), but it has since been found that it has incomplete penetrance. An indication that there might be a class of recessive oncogenes came from cell fusion experiments in which hybrid cells are produced from the fusion of normal and tumor cells (Figure 20.27). If the oncogene responsible for the phenotype of a tumor cell is dominant to its proto-oncogene counterpart in normal cells, then hybrid cells resulting from the fusion of a tumor and a normal cell should continue to display the tumor phenotype. However, contrary to this prediction, tumor/normal hybrid cells are usually normal in phenotype, at least to the extent that they are unable to initiate tumors when inoculated into experimental animals. These results suggested to early researchers that normal

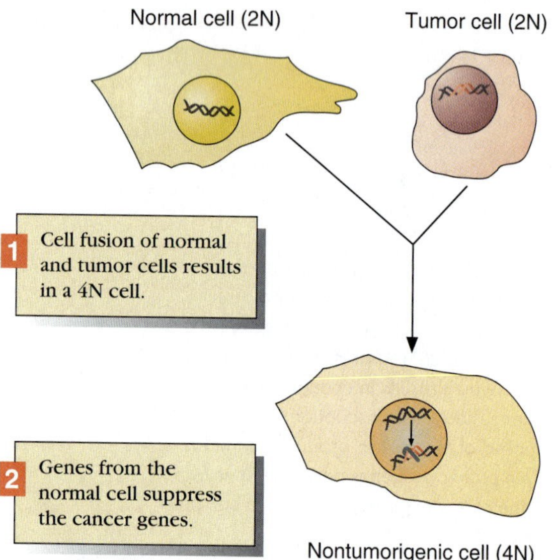

FIGURE 20.27 Testing for tumor-suppressor genes. When normal cells and tumor cells are fused, the hybrid cell is usually nontumorigenic in phenotype. This indicates that the genes responsible for the cancer-like properties of the cell are being suppressed by genes contained in the normal cell. Thus, the tumor-causing gene is dominant over the normal gene. This phenomenon was one of the early pieces of evidence for the existence of tumor-suppressor genes.

TABLE 20.6 *Some Tumor-Suppressor Genes and Their Functions*

Gene	Cellular Function	Tumors Produced
DCC	Cell adhesion	Colon/rectal carcinomas
DPC4	Signal transduction	Pancreas
NF1	Intracellular signal transduction	Neurofibrosarcomas
MST1	Cell division break	Wide range of cancers
p53	Transcriptional regulation, induces abnormal cells to kill themselves	Brain, breast, colon/rectal, lung, cervix osteosarcomas, leukemia, and lymphomas
RB	Master break in cell cycle	Retinoblastomas, osteosarcomas, bladder, breast, lung, and liver
WT1	Transcriptional regulation	Wilms's tumor of kidney

cells contain one or more genes which are able to "suppress" the tumor-forming potential of transformed cells. Such genes are called **tumor-suppressor genes.** They may encode a variety of gene products including regulatory proteins, intracellular signaling proteins, and cell adhesion proteins (Table 20.6).

Some Tumor-Suppressor Genes Encode Regulatory Proteins.

As stated earlier, mutations in the *rb* tumor-suppressor gene have been shown to be responsible for retinoblastoma. Extensive research has been carried out on the *rb* gene in recent years. For example, it has been found that the protein product of *rb* is activated by the extracellular growth factor TGF (transforming growth factor), which is a strong inhibitor of the proliferation of skin epithelial cells. Present evidence suggests that *rb* may mediate inhibition of epithelial cell growth by turning off *myc* expression. This interaction illustrates that at least some tumor-suppressor genes may function by regulating the expression of dominant-acting oncogenes such as *myc.*

Other tumor-suppressor genes that seem to be able to regulate the expression of dominant-acting oncogenes are *MST1, WT1,* and *p53. WT1* is the gene responsible for the inherited form of Wilms's disease. Mutations in *p53* have been implicated in several leukemias, bladder cancer, and breast tumors. In fact, the *p53* gene is probably the most important cancer-suppressor gene so far identified. It is associated with over 60% of all known cancers. A healthy cell normally keeps a small number of *p53* (the name comes from protein of 53,000 Da) molecules, degrading them and replenishing them on a regular basis. If a chemical damages the cell's DNA in a way that threatens to set it on its way to uncontrolled growth, the supply of *p53* will build up and stop cell division until the DNA is repaired. It halts cell division by acting as a transcription factor to activate the expression of other genes that control cell division, some of which are the *cdc* genes mentioned earlier. If the cell is damaged beyond repair, *p53* will signal the cell to kill itself (known as **apoptosis**), thus eliminating it from forming a cancerous clone of cells. A mutation in the *p53* gene is recessive because only one copy of the gene is necessary to prevent uncontrolled cell growth. However, if one copy of the *p53* gene is defective, the chances of totally losing *p53* function are greatly increased. Most defects in *p53* are not inherited but are acquired by exposure to mutagens or because of a defective DNA repair pathway. The metabolites of benzopyrene (Figure 20.9) in cigarette smoke frequently result in a change of only one G to a T in the *p53* gene. If both *p53* alleles are mutated in this way, it will lead to lung cancer.

Some Tumor-Suppressor Genes Encode Intracellular Signaling Proteins.

As discussed previously, the GAP protein plays a critical role in regulating the activity of $p21^{ras}$. Inactivation of GAP results in an increase in intracellular levels of activated (that is, GTP bound) $p21^{ras}$, which leads to cell proliferation. The tumor-suppressor gene *NF1* encodes a product that is functionally analogous to GAP in that it appears to function normally as an inactivator of $p21^{ras}$. Mutations in the *NF1* gene that result in the loss of its encoded protein's ability to hydrolyze $p21^{ras}$-bound GTP to GDP behave as recessive oncogenes. Neural cells that are homozygous for the mutant *NF1* allele are associated with the formation of neurofibromas, which are tumors in the connective tissue of nerves.

Some Tumor-Suppressor Genes Encode Cell Adhesion Proteins.

As mentioned at the beginning of this chapter, one of the distinguishing characteristics of cancer cells is the ability of these transformed cells to invade surrounding tissues and metastasize to other body regions. Recent studies indicate that the *DCC* tumor-suppressor gene may be directly involved in these processes. *DCC* is a member of a gene family that encodes cell-surface proteins involved in cell–cell interactions. Mutations in *DCC* reduce cell adhesion and thus contribute to tumor spread. For example, it has been shown that *DCC* inactivation occurs during late stages of colorectal cancer, when the tumor begins to acquire the ability to invade surrounding normal tissue. Future research should more precisely define the role of *DCC* mutations in tumorigenesis.

SECTION REVIEW PROBLEMS

15. How were tumor-suppressor genes first discovered?
16. Describe the functions encoded by tumor-suppressor genes and give an example of each.

CANCER DEVELOPMENT IS A MULTISTEP PROCESS.

In this chapter, we have been considering the spectrum of genetic defects that are associated with cancer development as if they were independent occurrences. In fact, this is an oversimplification, at least with regard to nonviral mediated cancers that occur spontaneously or as a result of a carcinogen-induced mutation. As a general rule, the development of non–viral-induced tumors appears to be a multistep process in which a number of genetic defects accumulate progressively as the tumor develops.

Perhaps the best-documented example of the progressive nature of tumor development is colon cancer in humans. Two very early events associated with benign colon growths (called adenomas) are the inactivation of the tumor-suppressor gene *APC* and the occurrence of one or more point mutations in the *ras* proto-oncogene, converting it to an oncogene. The progression of these colon adenomas to malignant tumors has been associated with additional inactivating mutations in K-*ras* and the *p53* and *DCC* tumor-suppressor genes (Figure 20.28). This pattern of a progressive series of mutations at proto-oncogene and tumor suppressor loci has also been associated with the development of human breast tumors, neuroblastomas, and neurosarcomas, and it supports the hypothesis that nonviral mediated tumor development is a stepwise process.

THE PROSPECTS FOR THE FUTURE ARE PROMISING.

It should be clear from the information presented in this chapter that tremendous strides have been made over the past 20 years in elucidating the molecular genetic mechanisms that underlie the cancer process. The question remains, however, as to how and when this information will translate into a reduced frequency and possibly even an eventual cure for this deadly disease.

One benefit of the discovery of proto-oncogenes and the type of genetic lesions that cause them to become oncogenic is the development of genetic screens for the early diagnosis of cancers. For example, the level of amplification of the N-*myc* gene in neuroblastomas can be detected by Southern blots (see Chapter 12). This methodology is being used to determine the development stage of these tumors. Such information is useful to clinicians because early-stage and advanced-stage neuroblastomas are optimally treated with quite different therapies. Genetic screens have also been developed that can detect the *abl* oncogene translocation associated with some forms of leukemia. The screen is used as a sensitive assay to monitor the response of patients to chemotherapy. Other similar screens continue to be developed using other oncogene loci, and this trend will undoubtedly continue for the foreseeable future.

The application of our current molecular understanding of cancer to the development of new and effective cancer therapies is proving to be more problematic than the development of genetic screens. For example, we now know many of the oncogene products that stimulate cells to undergo uncontrolled cell proliferation. Thus, in principle, it may be possible to design drugs that

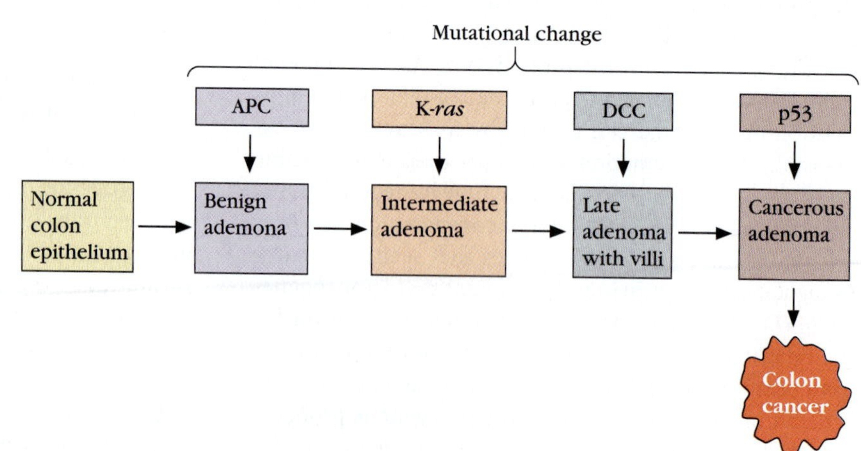

FIGURE 20.28 Colon cancer production. A proposed series of events that cause colon cancer. This model demonstrates why multiple mutational events must occur before cancer is acquired.

Current Investigations

Suppression of a Malignant Tumor by Gene Transfer

As in other areas of genetics research, the use of experimental organisms has proven to be extremely useful in studying certain aspects of the genetic basis of cancer. For example, rodents and chickens have historically played important roles in the discovery and study of oncogenic retroviruses. In more recent years, yeast has become an important experimental system for studying the molecular mechanisms by which oncogenes regulate cell proliferation. *Drosophila melanogaster* is also used in addressing basic questions in cancer genetics.

There are more than 50 genes thus far identified in *Drosophila* that, when mutated, produce tissue-specific tumors. These genes, displaying a recessive tumor phenotype, have been designated as tumor-suppressor genes. Twenty-three *Drosophila* tumor-suppressor genes have thus far been cloned and characterized on the molecular level. The first *Drosophila* tumor-suppressor gene to be studied was the *lethal(2)giant larvae* (l(2)gl) gene. In the early 1970s, Elizabeth Gateff of the University of Mainz in Germany, one of the pioneers in the area of *Drosophila* cancer genetics, demonstrated that *Drosophila* homozygous for mutant *l(2)gl* alleles develop malignant tumors of the neuroblasts of the presumptive adult optic centers of the larval brain and in imaginal disk cells (Figure 20.B). These tumors result in a complex syndrome characterized by atrophy of several tissues, bloating, and death by the late larval stage of development.

In the mid-1980s, the *l(2)gl* gene was isolated and characterized by Bernard Mechler and his colleagues, also working at the University of Mainz in Germany. Using a P-element transformation vector (see Chapter 19), Mechler and his colleagues were able to introduce a functional *l(2)gl* gene into the genome of embryos homozygous for the mutant allele and to thereby suppress the cancer phenotype in the F_1 transformants and their descendants. This was the first example of a cancer being cured by genetic intervention in any animal system.

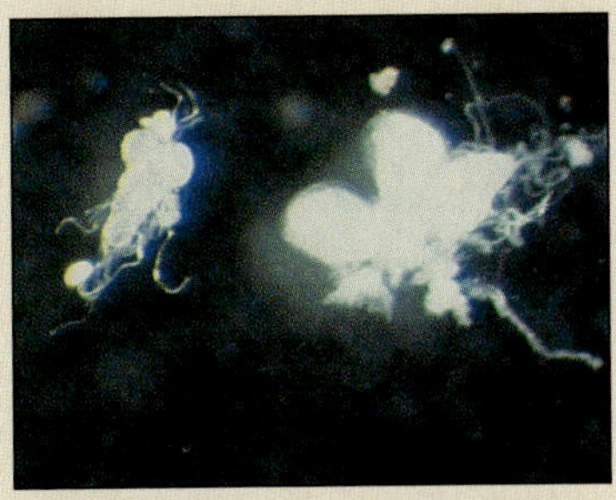

FIGURE 20.B Comparison of whole-mount preparations of wild-type and mutant *l(2)gl* brain. Whole-mount preparations of larval wild-type brain (*left*) and enlarged tumorous brain of the *l(2)gl* mutant (*right*). (*J. McDonald, University of Georgia*)

The *l(2)gl* gene displays many of the characteristics associated with tumor-suppressor genes such as the *rb-1* gene, which when mutated has been associated with childhood retinoblastoma. In the future, it is hoped that inherited human cancers such as retinoblastoma may be cured by gene therapy.

specifically interfere with these gene products. The drawback, however, is that many oncogene and tumor-suppressor gene products are also important in the control of cell proliferation in normal cells. Developing oncogene-specific drugs may be possible in those cases in which the oncogene product is structurally distinct from its proto-oncogene homologue. For example, the point mutations that transform the *ras* proto-oncogene into an oncogene provide a structural difference at the protein level that may be a potential target for a *ras* oncogene-specific drug.

The clinical treatment of cancer by the use of gene therapy is a great hope for the future. For example, the idea of being able to introduce one or more tumor-suppressor genes into cancer cells to revert them back to a noncancerous state has great appeal. However, one major hurdle involved in this process is how to deliver these genes into the cancer cells that make up the tumor. There is growing enthusiasm among many cancer geneticists that the development of viruses as vectors for such "anti-oncogenes" may prove useful. For example, recent experiments carried out in mice, in which retroviral vectors have been used to transfer genes into brain tumor cells in order to stop the cancerous growth, have been remarkably successful and have stimulated the initiation of similar experimental trials in humans. Another possibility is the use of gene therapy to "cure" certain inherited forms of cancer, such as heritable neuroblastoma. The basis of this optimism is the fact that such "cures" have been accomplished in experimental organisms such as *Drosophila* (see Current Investigations: *Suppression of a Malignant Tumor by Gene Transfer*). For example, in the future it may be possible to introduce a normal copy of the *rb* gene into gametes of individuals identified by diagnostic genetic screens as being genetically predisposed to developing neuroblastomas (that is, identified as being heterozygous for a defective *rb* allele). Again, the development of efficient gene delivery systems, such as may be provided by viral vectors, will be a prerequisite to the realization of such therapies.

Other approaches to preventing cancerous cell growth are activating the immune system to attack the cancer or using a very specific antibody to deliver a toxin to the growing tumor. Monclonal antibodies (see Chapter 14) are useful in delivering a specific toxin to tumor cells, but their drawback is that normal cells usually possess the same array of antigens as do tumor cells.

Growing tumors require a considerable amount of food; thus, depriving the tumor of its blood supply is another method of curtailing tumor growth. There is a group of chemicals that are able to prevent the vascularization of tumors, and many are in clinical trial at the moment. These include interleukin-12, which increases the production of an inhibitor called inducible protein 10.

Summary

Cancer is a family of more than 100 different diseases, all of which are characterized by a malfunction in cellular regulation leading to uncontrolled cell growth and proliferation. Cancers are classified according to the type of cell and tissue in which they arise.

Evidence that a predisposition to cancer has a genetic basis is both circumstantial and direct. Cancer cells pass the information encoding their aberrant phenotype to their descendent cells. Many mutagens known to cause genetic damage to cells are also known to be carcinogenic, that is, to cause cancer. The cancer-causing properties of oncogenic retroviruses are attributable to particular viral genes called viral oncogenes, or v-*oncs*.

V-*oncs* are homologous to normal cellular genes called proto-oncogenes, which can mutate to become cancer-causing cellular oncogenes, or c-*oncs*. V-*oncs* and proto-oncogenes may differ both structurally and in terms of their regulation. V-*onc* expression is under the control of a retroviral promoter, which may show significant deviations in both levels and cellular patterns of expression from those of its homologous proto-oncogene. These regulatory differences may be sufficient to induce cellular transformation. V-*oncs* may also be sequentially different from their proto-oncogene homologues. V-*oncs* lack introns that are typical of their proto-oncogene counterparts.

Proto-oncogenes can become c-*oncs* by point mutations, which alter the structure of their gene products, or by regulatory mutations, which alter the quantity and/or distribution of their gene products. Point substitutions in the *ras* gene altering the structure of its $p21^{ras}$ protein product have been associated with colon and breast cancers in humans. Regulatory changes in *myc* gene expression have been associated with several human cancers, including lymphomas and neuroblastomas. Regulatory changes in *myc* expression leading to cellular transformation have been associated with translocations (Burkitt's lymphoma), retroviral element insertion (ALV), and gene amplification (neuroblastomas).

Analysis of the change in cellular functions associated with cancer-causing mutations in proto-oncogenes has helped define the function of these genes in normal cells. Proto-oncogenes encode proteins involved in a variety of cellular functions controlling the flow of information from the cell surface through the cytoplasm into the nucleus. Some proto-oncogenes, such as *sis*, encode growth factors. Other proto-oncogenes, such as *erbB*, *ret*, and *trk*, encode growth-factor receptor proteins, many of which display tyrosine-kinase activity. Some proto-oncogenes encode proteins involved in intracellular signaling processes. Some of these are plasma-membrane-associated G proteins, such as the product of the *ras* gene ($p21^{ras}$), whereas others are cytoplasmic protein-serine kinases, such as the products of the *raf* and *mos* genes. Another important class of proteins encoded by proto-oncogenes is nuclear transcription factors and other DNA-binding proteins involved in transcriptional regulation. Examples are those encoded by the *jun*, *fos*, and *myc* oncogenes.

Cell fusion experiments between normal cells and cancer cells have led to the recent discovery of another class of genes called tumor-suppressor genes. As their name implies, these genes encode products that are capable of suppressing the cancer phenotype of transformed cells. Loss-of-function mutations at these loci are generally recessive but in the homozygous condition can lead to cancer. Like c-*oncs*, tumor-suppressor genes have been shown to code for a variety of cellular functions. Some, such as *rb*, encode nuclear regulatory proteins that may interact with oncogene products such as Myc to control cell proliferation. Some suppressor genes such as *NF1* encode proteins involved in intracellular signaling processes, whereas others, such as *DCC*, encode cell adhesion proteins.

Non–viral-mediated transformation of normal cells to cancer cells appears to be a stepwise process involving a series of mutations at proto-oncogene and tumor-suppressor loci. For example, the onset and development of human colon cancer involves a series of mutational steps.

Our current understanding of the genetic basis of cancer has already led to the development of a number of genetic screens that aid in the early diagnosis of the disease and help monitor the success of chemical and radiation treatments. The implementation of gene-based cancer therapies is an exciting prospect for the future, but it will require further development of site-specific drug technologies as well as the design of efficient delivery systems whereby genes can be introduced into cancer cells.

Selected Key Terms

adenoma
apoptosis
autocrine growth stimulation
cancer
carcinogen
carcinoma
cellular oncogene
epidemiology
gene amplification
intracellular signal transduction
leukemia
lymphoma
metastasize
neuroblastoma
oncogenic virus
proto-oncogene
retinoblastoma
sarcoma
tumor-suppressor gene
viral oncogene

Chapter Review Problems

1. Distinguish between benign and malignant tumors.
2. Which do you think came first, c-*oncs* or v-*oncs*? Why?
3. Are v-*oncs* dominant or recessive to their proto-oncogene counterparts? How do you know?
4. The frequency of lung cancer has been decreasing in men over the past decade but increasing in women. Propose an hypothesis to explain this trend and describe how you might test your hypothesis.
5. What contribution has molecular genetics made thus far to the clinical diagnosis and treatment of cancer? What contributions do you think are likely in the future?
6. What is the distinction between viral oncogenes (v-*oncs*), cellular oncogenes (c-*oncs*), and proto-oncogenes?
7. Fair-skinned people are more prone to develop skin cancer than are dark-skinned people because fair-skinned people have less pigment in their skin to absorb mutagenic ultraviolet radiation. Offer an hypothesis to explain the original geographical distribution of skin color in ancestral human populations.
8. What are the *cdc* genes, and how are they related to cancer?
9. What is apoptosis, and how is it related to cancer?
10. How have studies on the molecular basis of cancer contributed to our understanding of the molecular processes underlying cell division?

Challenge Problems

1. Analysis of brain tumors often reveals that the tumor derived from a transformed epithelial cell. How is this possible?
2. Suppose that you discover a new retrovirus that can cause tumors when injected into the muscles of experimental mice. How would you determine whether or not your newly discovered retrovirus carries an oncogene (v-*onc*)?
3. Many proto-oncogenes have been found to be most heavily expressed during early stages of embryonic development and either not expressed at all or expressed at significantly reduced levels when the developmental process is complete. Given what you know about the cellular function of oncogenes, propose an hypothesis to explain this observation.
4. Why do you think that most inherited cancers are caused by recessive mutations?

Suggested Readings

Aaronson, S. A. 1991. "Growth factors and cancer." *Science*, 254: 1146–1153.

Cooper, G. M. 1995. *Oncogenes*. Jones and Bartlett, Boston.

Culver, K. W., and R. M. Blaese. 1994. "Gene therapy for cancer." *Trends in Genetics*, 10: 174–178.

Kinzler, K. W., and B. Vogelstein. 1997. "Gatekeepers and caretakers." *Science*, 386: 761–762.

Knudson, A. C. 1993. "Antioncogenes and human cancer." *Proceedings of the National Academy of Sciences, USA*, 90: 10914–10921.

Old, L. J. 1996. "Immunotherapy for cancer." *Scientific American*, 275 (September): 136–148.

Trichopoulos, D., F. P. Li, and D. J. Hunter. 1996. "What causes cancer?" *Scientific American*, 275 (September): 88–97.

Vogelstein, B., and K. W. Kinzler. 1993. "The multistep nature of cancer." *Trends in Genetics*, 9: 138–141.

Weinberg, R. A. 1996. "How cancer arises." *Scientific American*, 275 (September): 62–70.

Weiss, R., N. Teich, H. Varmus, and J. Coffin (eds.). 1985. *Molecular Biology of Tumor Viruses* (2nd ed.). Cold Spring Harbor Press, New York.

On the Web

Visit our Web site at **http://www.saunderscollege.com/lifesci/titles.html** and click on A/G/M Genetics for links to the following chapter-related resources on the World Wide Web:

1. **Diet and cancer.** This site provides information on how you can reduce your risk for cancer by improving your diet.
2. **Cancer site links.** This site provides links to other sites, specifically for ovarian cancer.
3. **More cancer site links.** This site provides additional links to sites about all kinds of cancers, their prevention, and diagnosis.

CHAPTER

21

The diversity of stature and facial features existing in human populations. *(Jeff Greenberg/Visuals Unlimited)*

Population and Evolutionary Genetics

At this stage in your study of genetics, you have learned how genes function at the molecular and cellular levels, how genetic changes occur at the nucleotide and chromosomal levels to bring about new alleles, and how different alleles are precisely transmitted from parent to offspring. In this chapter, we will begin to inquire into the behavior of genes on the level of populations. Geneticists define populations as communities of individuals that are united by bonds of mating and/or parenthood. Whereas populations made up of asexual organisms are united only by common parenthood, populations of sexually reproducing organisms are united by bonds of mating as well. Populations of sexually reproducing organisms are often referred to by geneticists as **Mendelian populations** to reflect the fact that genes are passed from one generation to the next according to the Mendelian principles. Geographic and thus genetic discontinuity may exist between local Mendelian populations of the same species, but these separations are not rigid and fixed because they are frequently compromised by the migration of individuals from one population to another. The most inclusive Mendelian population is the **species** because, as a general rule, the genetic discontinuity that exists between species is absolute. In other words, mutations that arise within local Mendelian populations can, in principle, spread through an entire species but cannot ordinarily be transferred across species boundaries (Figure 21.1). This fact is the basis of the so-called **biological species concept** in which species are defined as a group of Mendelian populations that are reproductively isolated from other such groups. The biological species concept is distinct from the **typological species concept,** commonly used by taxonomists, which defines species on the basis of their phenotypic (for example, morphology or anatomy) similarity.

Evolutionary geneticists who work primarily on the population or "microevolutionary" level frequently define **evolution** as changes in gene frequencies within and between populations over time. Although this definition is in many respects overly simplistic, it nevertheless conveys the very basic notion that all the major morphological, physiological, and behavioral changes by

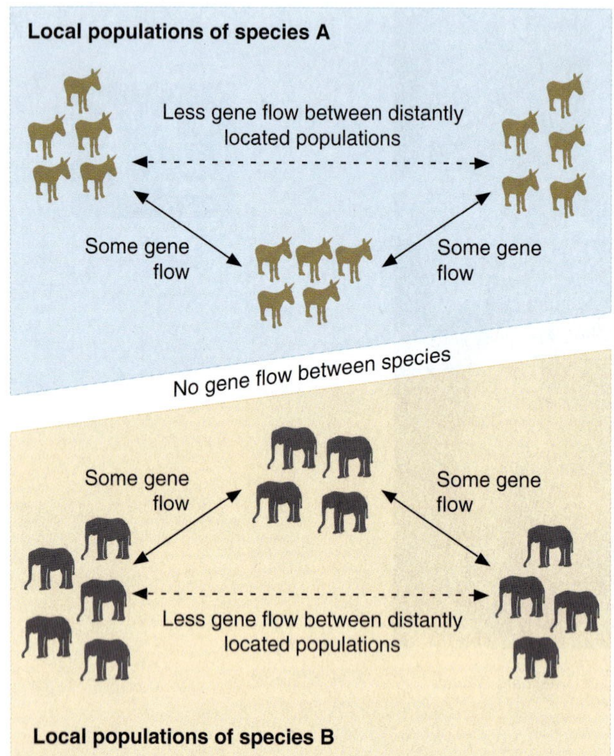

FIGURE 21.1 Different species do not exchange genes. Gene flow may exist among populations of the same species, but no gene flow exists between species. Matings and, consequently, the flow of genes usually occur within local populations. Some gene flow (migrations) occurs between neighboring populations, but the rate of gene flow decreases as populations become more distant or isolated from one another.

which we characterize different species begin with simple changes in the number and kind of genes present in populations. Populations are thus considered the smallest evolutionary unit because, unlike individuals, populations do not have fixed relatively short life spans but may genetically change or evolve over many generations.

Although individual organisms can live for only a limited period of time, populations and species may persist for many thousands of generations. The genetic changes that occur in populations and species over time are the basis of evolution. By the end of this chapter, we will have considered the mechanisms that shape the genetic character of populations and species over time by addressing three basic questions:

1. How much genetic variation is present in populations, and what is its significance?
2. What are the factors that change the genetic character of populations over time?
3. What are the mechanisms that lead to the formation of new species?

GENETIC VARIATION IS THE RAW MATERIAL FOR EVOLUTIONARY CHANGE.

The raw material for evolutionary change is genetic variation. The term "variation" denotes differences. Individuals making up natural populations may differ among themselves and members of other populations in a variety of morphological, physiological, and biochemical traits. Much of the observable variability that exists among individuals making up natural populations reflects heritable differences, or **genetic variation,** that exists among members of the population. The sum total of the genetic variation present in a population or species is often referred to as the **gene pool.**

Although the ultimate source of all genetic variation present in a gene pool is mutation, evolutionary change may not depend upon the amount of new mutants introduced into a population. This is true, as you will see later in this chapter, because most populations already contain substantial levels of genetic variation. Indeed, the typical population may contain many allelic forms of a gene, and thus there may be a large number of different homozygote and heterozygote combinations carried by the members of the population.

The factors that determine the frequency of different alleles and genotypes in a population can be complex and may involve such factors as the number of individuals making up the population, the particular mating preference of the individuals in the population, and whether or not an adaptive advantage is imparted to individuals carrying one form of a given gene. **Population genetics** is the branch of genetics that is concerned with understanding the factors determining how the genetic character of populations is maintained and can change over time. **Evolutionary genetics** is a more inclusive area of inquiry and focuses on the factors that may lead to new species and the genetic changes that are associated with speciation events.

Different Techniques Are Used To Detect Genetic Variation in Populations.

In principle, any method that enables genetic differences to be detected between individuals can be applied to population level surveys of genetic variation (Table 21.1). However, because each technique has strengths and weaknesses, the method of choice may vary from one situation to another. For example, simple visual observation can be used to quantify the population frequency of easily discernible phenotypic traits (for example, the color and banding pattern in snails) (Figure 21.2). Before using this approach, however, it is necessary to establish that the phenotypic variation being observed is in fact the product

TABLE 21.1 *Some Methods Used to Detect and Quantify Genetic Variation in Natural Populations*

Method	Advantages	Disadvantages
Visual observation	Inexpensive and easily applicable to large sample sizes	Limited to readily observable traits Accuracy may vary among observers Often difficult to distinguish between genetically vs. environmentally determined variation
Protein gel electrophoresis	Relatively inexpensive Detects most changes in protein shape, size, and charge	Unable to detect all variation, including third codon base changes, introns, and promoter regions
Restriction fragment length polymorphisms	Moderately inexpensive Detects genetic variation throughout genome	Detects only nucleotide variation at restriction enzyme cut sites or size variation in regions adjacent to cut sites
DNA sequencing	Allows precise measurement of all genetic variation	Moderately expensive Presently not practical for large surveys

of underlying genetic differences and not merely the result of environmental effects (see Chapter 17).

Biochemical techniques such as protein gel electrophoresis (Figure 21.3) have been used extensively by population geneticists to detect polymorphisms that are not readily apparent to the naked eye. Protein gel electrophoresis separates molecules on the basis of size and charge. It is a relatively accurate yet inexpensive technique for detecting many amino acid differences between proteins encoded by alternate alleles of the same gene. However, a distinct limitation of protein electrophoresis is that not all nucleotide substitutions result in changes that are detectable by the technique. For example, because of the degeneracy of the genetic code, nucleotide changes at most third-base positions cause "silent substitutions" that do not result in changes on the amino acid level.

DNA restriction enzymes have been used in population surveys in recent years to generate restriction fragment length polymorphisms (RFLPs). As you learned in Chapter 13, variation in the ability of a series of restriction enzymes to cut homologous pieces of DNA indicates nucleotide differences at the recognition sites for the enzymes. Nucleotide differences between homologous

FIGURE 21.2 Banding pattern variation in *Cepaea nemoralis.* Many of the banding patterns that exist among snails of the species *Cepaea nemoralis* are caused by genetic differences or polymorphisms segregating in natural populations. *(Chip Clark)*

(a) Contact wick Gel Sample slot

Electrode Electrode

Buffer tanks

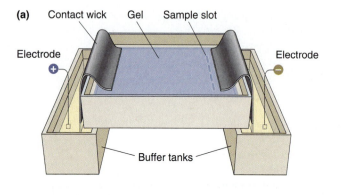

(b)

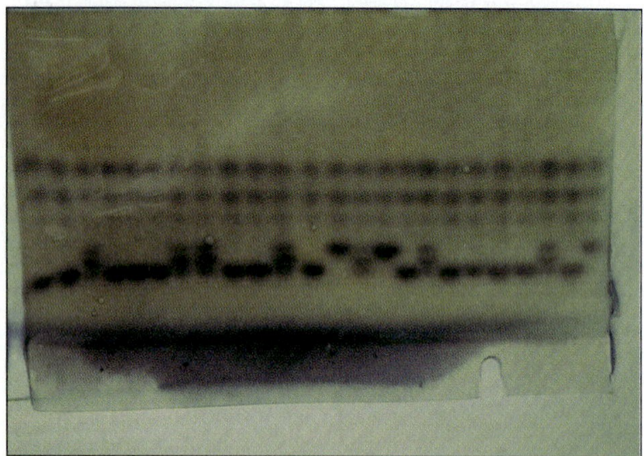

FIGURE 21.3 Total protein extracted from the "swamp pink" plant (*Helonias bullata*). (a) The proteins were separated by electrophoresis in a starch gel and stained for triose phosphate isomerase activity (*Tpi*). (b) The variable banding pattern is reflective of genetic polymorphism at the *Tpi* locus in this species. *(Mary Jo Godt)*

genes at these recognition sites will result in variation in the length of restriction fragments, which can be easily resolved by gel electrophoresis. Although RFLP studies do provide a relatively easy and inexpensive way to monitor variation at the DNA level, the drawback is that only those nucleotide substitutions mapping to restriction enzyme recognition sites can be detected. Any insertion or deletion (INDEL) variation within the region bordered by a restriction site will also be detected by RFLP analysis.

In recent years, DNA sequencing techniques have been applied by population geneticists to detect genetic differences between individuals. Although DNA sequencing is certainly the most direct and unambiguous method to detect genetic variation, it remains even today a relatively expensive and time-consuming procedure. It is not yet widely used in large-scale population surveys, in which hundreds of samples may be routinely examined. These limitations, however, are rapidly being overcome as the techniques of DNA amplification by the polymerase chain reaction and direct genomic sequencing methods (see Chapters 12 and 13) become more generally applicable to population genetic surveys. These new methodologies considerably reduce the labor and costs involved in sequencing large numbers of samples.

In summary, population geneticists have a variety of methodologies available by which they can detect genetic variation in natural populations. The particular technique employed in any given instance will depend upon the survey being conducted. For example, if you were interested in determining the frequency of the gene responsible for sickle-cell anemia in human populations around the world, you might choose protein gel electrophoresis. Protein electrophoresis is a relatively inexpensive technique, which makes it compatible with large-scale surveys, yet the technique is fully capable of distinguishing the disease-causing hemoglobin allele (*S*) from the one encoding normal hemoglobin. If, on the other hand, you were interested in the specific issue of whether variation might be present in a particular human population within the nonencoding promoter regions of the hemoglobin gene, you might choose to use DNA sequencing or RFLP analysis. Your choice would be based on the fact that you are dealing with a relatively small sample size and a single population and that you are interested in monitoring DNA sequence variation in a region of gene that is not translated into protein and thus is undetectable by protein gel electrophoresis.

SOLVED PROBLEM

Problem

For each of the three hypothetical situations that follow, identify the most appropriate methodology for measuring genetic polymorphism and justify your decision.

a. Find the frequency of genes determining two alternate color variants in a natural population of moths.
b. Find the genetic data to support or refute the plaintiff's allegation in a paternity suit.
c. Find the frequency of variation mapping to the heat shock enhancer sequence in the promoter region of the *Drosophila hsp70* gene (see Chapter 13).

Solution

a. Visual observation may be a sufficient method to monitor color morph polymorphism in a natural population of moths, provided prior experiments have established the genetic basis of the phenotype being monitored (for example, the polymorphism may be caused by a single allelic difference at a single locus).
b. Protein-level polymorphism (such as blood proteins) analysis may be sufficient if the protein being examined differs between the alleged parents. If the protein does not differ, it may be necessary to ana-

lyze the DNA polymorphism using either RFLP or DNA sequence analysis.

c. Because upstream regulatory sequences such as enhancers are nonencoding, protein level studies will not be useful. RFLP analysis may be able to detect restriction site and/or size variation within the regulatory region of interest, but the most definitive analysis would require DNA sequence analysis.

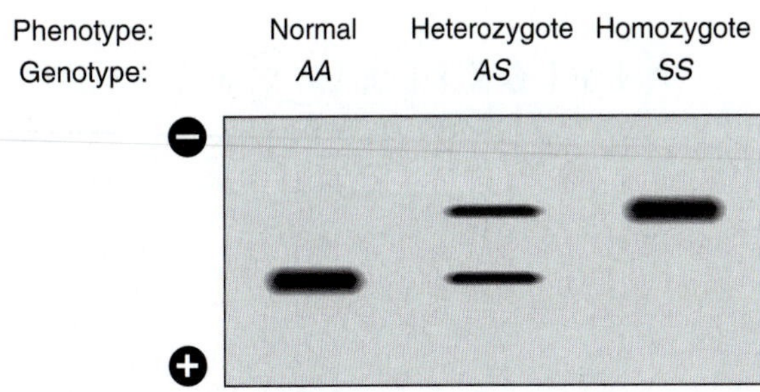

FIGURE 21.4 Electrophoretic differences among hemoglobin variants. When blood samples are separated by electrophoresis (see Figure 21.3), normal hemoglobin (*AA*) migrates toward the cathode and sickle-cell anemia hemoglobin (*SS*) migrates toward the anode. Individuals with the sickle-cell trait show both kinds of hemoglobin. A single base substitution is responsible for the difference between the beta chain of a normal hemoglobin molecule and that of hemoglobin *S*, causing valine to substitute for glutamic acid in the sixth position of the peptide.

Geneticists Are Interested in Understanding Genetic Variation in Natural Populations.

Geneticists study the genetic variation present in natural populations for a number of reasons. In the case of disease-causing genes such as the hemoglobin *S* allele, which causes sickle-cell anemia in individuals homozygous for the allele (Figures 21.4 and 21.5), analyzing the gene's distribution pattern between populations may provide insight into the origins and maintenance of the disease. For example, surveys of the frequency of the hemoglobin *S* allele around the world reveal a trend that was first noted in 1949 by one of the founders of population genetics, J. B. S. Haldane [see HISTORICAL PROFILE: *J. B. S. Haldane (1892–1967)*]. Haldane observed that the frequency of the *S* allele is highest in areas of the world with a high incidence of the malarial parasite ***Plasmodium falciparum*** (Figure 21.5). This correlation led Haldane to hypothesize that the *S* form of the hemoglobin protein might, in addition to causing anemia, also be associated with resistance to malarial infection. Medical studies have subsequently provided direct evidence that supports this hypothesis. Thus, it was a population genetics survey that contributed to our current understanding of why the disease-causing sickle-cell allele has been maintained for so long in human populations. The survey also helped to explain why the frequency of the *S* allele is exceptionally high in contemporary ethnic groups whose ancestors can be traced to tropical regions of the world plagued by the malarial parasite.

Another example of an area where gene frequency data is of growing importance is forensic science. As discussed in Chapter 12, very small amounts of DNA (even

(a)

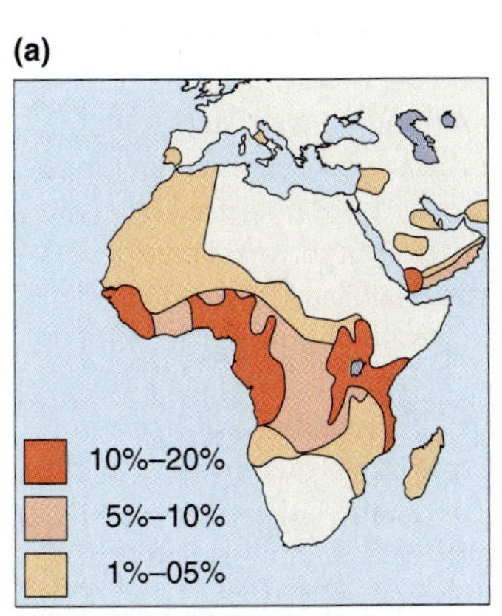

Frequency of sickle-cell gene

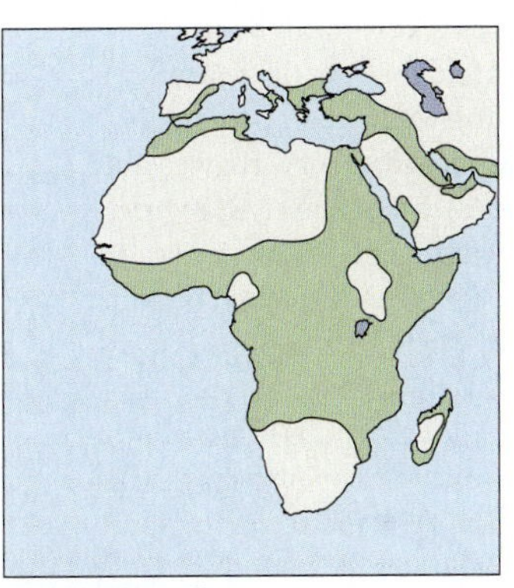

Distribution of *Plasmodium falciparum* malaria

(b)

Normal and sickled red blood cells

FIGURE 21.5 Sickle-cell anemia. (a) The frequency of the sickle-cell allele (left) and distribution of *Plasmodium falciparum* malaria (right) in Africa and European countries bordering the Mediterranean are shown. The distribution of the sickle-cell allele is closely correlated with that of this malarial disease. (b) The sickle-cell allele causes red blood cells to have jagged shapes, which in turn causes them to pile up and block small blood vessels, thus starving regions of the body for oxygen and nutrients. The jagged shape also makes the sickle cells highly resistant to the *falciparum* parasite that causes malaria. *(b, Visuals Unlimited)*

HISTORICAL PROFILE

J. B. S. Haldane (1892–1967)

J. B. S. Haldane (Figure 21.A) was born in England in 1892. He was the son of a famous physiologist, and his early childhood was extremely stimulating. Haldane's father often encouraged him to accompany him to scientific lectures. In 1901, when Haldane was only 9 years old, his father took him to a lecture presented by the English biologist A. Darbishire on the then recently discovered Mendelian principles. The young Haldane is reported to have been very impressed. As a result, while in his teens, Haldane and his sister shared in the hobby of raising guinea pigs at home in an effort to discover evidence of Mendelian inheritance in this species.

Although Haldane's formal training at Eton and Oxford was in classics and mathematics (he received degrees in both subjects simultaneously), his interest in genetics never waned. He was especially interested in the research being conducted on genetic linkage and mapping by T. H. Morgan and his collaborators at Columbia in the early 1900s because this aspect of genetics required direct application of mathematical principles. While serving in World War I, Haldane kept up with the work of Morgan's group by reading scientific journals in New Delhi, where he was stationed. In 1921, after returning from the war, Haldane published his first two papers on theoretical genetics, in which he introduced a formula that could be used to deduce reliable estimates of the linkage relationships among genes from recombination frequency data.

In another early publication, Haldane introduced a principle that is now called Haldane's Rule. He noted that in those rare instances when members of closely related but different species attempted to interbreed, either the male or female progeny of the interspecific cross were completely absent or, if present, were usually rare and sterile. Haldane noted that the sex that was absent, sterile, or rare was the heterogametic sex, that is, the sex that had the pair of dissimilar sex chromosomes (see Chapter 3). His genetic explanation of the rule was that in the heterogametic sex, alleles of all genes located on the X chromosome are present in the hemizygous condition. Hence, deleterious recessives that produce disharmony as a result of epistatic interaction with genes located on autosomes will typically have a deleterious effect. In the homogametic sex, these disruptive genes are more likely to be "masked" by dominant alleles present on the X chromosome of the opposite parent.

Between 1924 and 1934, Haldane published a series of ten landmark papers in which he applied mathematical principles to the study of natural selection. By the 1920s, the fact of evolution as postulated by Charles Darwin was generally accepted by the scientific community, as were the basic principles of Mendelian genetics. However, there remained considerable confusion as to how these two basic bodies of theory could be integrated. Haldane's work focused on integrating Mendelian principles into Darwinian theory. He was able to demonstrate precisely how the frequency of genes in a population would be expected to change under different types of selection. For example, based on the frequency of changes that were observed in populations of the peppered moth (*Biston betularia*) in industrial regions of England during the Industrial Revolution, Haldane estimated that the dark morph must have been enjoying an approximately 50% selective advantage over the white morph. This was considered a wholly unrealistic estimate at the time, but when the phenomenon was studied experimentally by Kettlewell some 30 years later (see discussion of Kettlewell's experiments in this chapter), Haldane's early estimates were found to be essentially correct. The work published by Haldane in these ten early papers served as the core of his most famous work, *The Causes of Evolution*, which stands as one of the cornerstones of modern population genetics theory.

FIGURE 21.A J. B. S. Haldane. *(UPI/Corbis-Bettmann)*

Haldane was an eccentric character who always maintained a variety of outside interests. As a young man, Haldane was a vocal supporter of the teachings of Karl Marx, although this support declined in his later years and was displaced by a strong interest in the teachings of Hinduism. Haldane's fascination with the Hindu philosophy led him to move to India in 1957, where he continued his teaching and research in genetics. In his later years, Haldane adopted the Indian form of dress and a vegetarian diet and became an Indian citizen.

Haldane's work on the integration of evolutionary biology with mathematics, along with similar efforts by the English statistician R. A. Fisher and the American geneticist S. Wright, serves as the mathematical foundation of modern population genetics.

DNA from a single cell) can be amplified by the polymerase chain reaction technique. PCR yields sufficient DNA for DNA sequencing or RFLP analysis of the amplified sequence. These new technologies have significance for forensic science because the minute quantities of DNA contained within traces of blood or other body tissues found at the scene of a crime or on a victim's body may help narrow the list of potential suspects in a criminal investigation. The potential utility of this DNA technology to forensic science, however, rests heavily upon the availability of reliable data on gene frequencies in natural populations. For example, imagine that DNA isolated from a hair follicle attached to the murder weapon is amplified by PCR and subjected to RFLP analysis. Suppose further that an RFLP pattern associated with this DNA does not match that of the murder victim's DNA but is identical to the RFLP pattern carried by the victim's estranged spouse. Whether or not this data could be presented by the prosecution to implicate the spouse in the murder would rest heavily upon the frequency of this particular RFLP pattern in the population of which the spouse is a member (for example, racial or ethnic group). If the RFLP pattern in question were found to be extremely rare in the population (1 in 1000 or 1 in 1 million), then it would support an association between the murder weapon and the estranged spouse. On the other hand, if the RFLP pattern were found to be even moderately frequent in the population (1 in 100 or 1 in 10), the RFLP data would certainly not constitute reliable incriminating evidence.

Perhaps the most common reason why geneticists are interested in quantifying the amount of genetic variation in natural populations is because this information can be used to test hypotheses regarding the genetic basis of evolution. The neo-Darwinian theory of evolution is a modern-day extension of the ideas first presented by Charles Darwin in 1859. According to the Darwinian view, adaptive evolution proceeds by natural selection favoring the most "fit" of the genetic variants present in a population. In principle, the most fit genotype in a population is the one that proportionally contributes the most genes to the next generation. In practice, the best adapted genotype is usually the one that leaves the greatest number of reproducing offspring. Darwinists believe that the amount of genetic variation in a population may be a measure of the population's adaptive evolutionary potential. For this and other reasons that will be discussed later in this chapter, a precise knowledge of the amount of genetic variation present in populations is necessary for testing many evolutionary hypotheses.

TABLE 21.2 *Genotypes at the Hemoglobin A Locus in a Sample of Individuals from West Africa*

	Genotypes			
	AA	*AS*	*SS*	Total
Number of individuals	168	80	2	250
Observed genotypic frequencies	0.67	0.32	0.01	1.00

	Alleles		
	A	*S*	*Total*
Allelic frequencies	0.83	0.17	1.00

The Amount of Genetic Variation in a Natural Population Can Be Quantified.

There are several numerical ways to describe the amount of genetic variation that is detected in natural populations. One way is to determine the relative numbers of individuals falling into each of the genotypic classes represented in the population. For example, consider the results of an electrophoretic survey of the types of adult hemoglobins carried by individuals in a West African population. Blood samples were taken from 250 randomly selected adults (Table 21.2). Electrophoretic techniques were used to determine the type of hemoglobins carried by each individual. Because each hemoglobin variant is known to be encoded by a different allele, these data can be used to determine each individual's genotype at the hemoglobin locus. Of the 250 individuals sampled, 168 carried the normal hemoglobin protein; that is, they were homozygous for the normal *A* allele (*AA*). A total of 80 individuals carried both the normal hemoglobin protein and the variant *S* form of the protein. These individuals were genetically heterozygous (*AS*) at the hemoglobin locus. Only two individuals were homozygous for the *S* allele (*SS*), and they were severely anemic.

The relative frequency of each hemoglobin genotype in this population is calculated by dividing the number of individuals having each genotype by the total number of individuals sampled. Thus, the frequency of the *AA* homozygotes is 0.67 (168/250), the *AS* heterozygote frequency is 0.32 (80/250), and the *SS* homozygote frequency is 0.01 (2/250). Note that the sum of the frequencies of each genotype in a population must equal 1.00 ($0.67 + 0.32 + 0.01 = 1.00$).

If we envision the West African population not as a group of individuals but rather as a "pool" of genes (the gene-pool concept), then we might express the variability present at the hemoglobin locus in terms of gene rather than genotypic frequencies. To calculate the gene or allelic frequencies at the hemoglobin locus in this sample, we would divide the number of each allele by the total number of alleles in the sample. Being a diploid organism, every human carries two alleles at each locus.

Thus, in a sample of 250 individuals, 500 alleles are represented at the hemoglobin locus. Each *AA* individual carries two *A* alleles, and every heterozygote carries only one *A* allele. Thus, the frequency of *A* alleles in our sample, $f(A)$, is $[2(168) + 80]/2(250) = 0.83$. The frequency of the *S* allele, $f(S)$, is 0.17. The sum of the frequencies of each of the alleles in a population will equal 1.00 ($0.83 + 0.17 = 1.00$). Thus, when only two alleles are in the population and we have computed the frequency of one of the alleles, we need only to subtract the computed frequency of one of the alleles from 1.00 to arrive at the frequency of the second allele ($1.00 - 0.83 = 0.17$).

We can generalize these computations by letting $p(A_1)$ and $q(A_2)$ represent the frequencies of two alleles A_1 and A_2, respectively. Let A_1A_1 represent the number of A_1 homozygotes, A_1A_2 the number of heterozygotes, and A_2A_2 the number of A_2 homozygotes. *N* will represent the total number of individuals in the population. Thus, the frequency of the A_1 allele in a population is represented by the following formula:

$$p(A_1) = \frac{2(A_1A_1) + A_1A_2}{2N} = \frac{A_1A_1 + (1/2)\,(A_1A_2)}{N}$$

After the frequencies of genes in a sample have been determined, how reliable is the assumption that the computed values are a true reflection of the frequencies in the population as a whole? The answer rests primarily upon two factors. First, the sample analyzed must have been chosen randomly. In the previous example, 250 West Africans were sampled to determine the frequencies of the hemoglobin *A* and *S* alleles. If these 250 individuals were all patients hospitalized for anemia, the computed gene frequencies could certainly not be expected to be representative of the population as a whole. Thus, care must be taken in gene frequency surveys to ensure, insofar as it is possible, that the sample is chosen randomly and fairly represents of the genetic makeup of the population as a whole.

A second factor that can significantly influence the reliability of gene frequency estimates is the size of the sample. In general, small samples are less reliable than large samples. To accommodate this discrepancy, population geneticists usually calculate the standard deviation (*s*) of each frequency estimate. Recall from Chapter 5 that the standard deviation is the square root of the variance. For normally distributed quantities (for example, gene frequency estimates), the true value of the estimate (for example, the computed frequency of an allele) will lie within two standard deviations of the estimated value 95% of the time. For estimates of the frequencies of two alleles, *p* and *q*, the standard deviation is given by the formula $\sqrt{pq/2N}$, where *N* is the size of the sample used to estimate the frequencies.

SOLVED PROBLEMS

Problem

In a recent survey of a US human population, 75 individuals were found to be homozygous for the hemoglobin *A* allele, and 25 were found to be heterozygous for the *A* and *S* alleles. No *SS* homozygotes were detected in the 100 individuals sampled. Compute the genotypic frequencies as well as the frequencies of the *S* and *A* alleles present in this population.

Solution

We are told that 75 individuals are *AA*, 25 are *AS* and 0 are *SS*. The total number of individuals surveyed is 100. Therefore, the frequencies of the three genotypes (*AA*, *AS*, *SS*) are

AA	*AS*	*SS*
75/100 = 0.75	25/100 = 0.25	0/100 = 0

To compute the frequency of the *A* allele, we need to divide the number of *A* alleles by the total number of alleles sampled in our survey. Remember that *AA* individuals carry two *A* alleles ($2 \times 75 = 150$), whereas *AS* individuals carry only one *A* allele ($1 \times 25 = 25$). The total number of alleles surveyed is $2N$, where *N* is the number of (diploid) individuals surveyed. Thus,

$$f(A) = \frac{2(\text{Number of } AA \text{ individuals}) + \text{Number of } AS \text{ individuals}}{2(\text{Total number of individuals sampled})}$$

$$= \frac{2(75) + 25}{2(100)}$$

$$= 0.875$$

Similarly,

$$f(S) = \frac{2(\text{Number of } SS \text{ individuals}) + \text{Number of } AS \text{ individuals}}{2(\text{Total number of individuals sampled})}$$

$$= \frac{0 + 25}{2(100)}$$

$$= 0.125$$

Check: The frequency of both alleles together should total 1.00, which they do ($0.875 + 0.125 = 1.00$).

Problem

In the West African population discussed earlier, the frequency of the hemoglobin *A* and *S* alleles were 0.83 and 0.17, respectively, and *N* was 250. What are the 95% confidence limits of these gene frequency estimates ?

Solution

The true gene frequency values will lie within two standard deviations of the estimated values 95% of time. Thus, in this example, $s = \sqrt{(0.83)(0.17)/500}$, or 0.02. This means that there is a 95% probability that the true value of the frequency of the hemoglobin *A* allele lies within two standard deviations of 0.83 [0.83 ± 2(0.02)], between the values 0.87 and 0.79. Similarly, there is a 95% probability that the true value of the frequency of the *S* allele lies within two standard deviations of 0.17 [0.17 ± 2 (0.02)], between the values 0.15 and 0.21.

SECTION REVIEW PROBLEMS

1. Derive the general formula for determining the frequency of the A_2 allele, $q(A_2)$.
2. How reliable are gene frequency estimates?
3. To convince yourself of the importance of sample sizes in estimating gene frequencies, calculate the reliability of the frequency estimate of the hemoglobin *A* allele in West Africa in the preceding Solved Problem if it had been based upon a sample size of 25 instead of 250.

THE LEVEL OF GENETIC VARIATION IN NATURAL POPULATIONS CAN BE ESTIMATED.

Up until now, we have been dealing with the frequency of alleles and genotypes at only one locus. Organisms, however, carry thousands of loci. When considering evolutionary questions, population geneticists may want to know how genetically variable a population is, on average, across many loci. Experimentally, this means that surveys need to be conducted on the same population at many loci. Many such surveys have been carried out on a variety of species using differences in protein migration during gel electrophoresis.

Table 21.3 presents the results of one survey conducted on the marine worm *Phoronopsis viridis*. In this study, 120 worms were assayed for genetic variation at 39 enzyme-encoding loci. The allelic frequencies at each of the 27 variable loci are presented in Table 21.3. It is relevant to note at this point that although individual (diploid) organisms carry only two alleles per locus, any number of alleles may be segregating at a locus in the population as a whole. In the population characterized in Table 21.3, more than two alleles were found to be segregating at 12 of the 39 loci surveyed; only two alleles were segregating at 15 of the surveyed loci; and 12 loci were found to be **monomorphic,** that is, all individuals sampled at these 12 loci were found to be homozygous for the same allele.

The Levels of Genetic Variation in Natural Populations May Be Expressed in Two Ways.

The overall genetic variation in a population can be expressed in two common ways. One way to express overall variation is to designate a locus as polymorphic (having two or more alleles segregating in the population in significant frequencies) or monomorphic (having only one allele in significant frequency) and to then compute the percentage of polymorphic loci in the population. The higher the percentage of polymorphic loci in a population or species, the more genetic variability in that population or species.

In order to use this measure of variability, some arbitrary criteria must be set to establish whether a locus is polymorphic or not. Most population geneticists define a locus as polymorphic if the frequency of the most common allele is 0.99 or less. Using this criterion, the population of *Phoronopsis,* characterized in Table 21.3, is polymorphic at 20 of the 39 loci, or 51.2% (percentage of polymorphic loci) of the loci surveyed. However, some population geneticists consider a locus polymorphic only when the frequency of the most common allele is 0.95 or less. Under this more stringent criterion, only 11 out of 39, or 28.2%, of the loci in the *Phoronopsis* population would be considered polymorphic. Thus, under the first criterion this population might be characterized as variable, whereas under the second criterion the population might be considered relatively devoid of genetic variation. This discrepancy underscores the shortcoming of the percentage-of-polymorphism statistic for describing the genetic character of a population: it is based upon an arbitrary definition of what constitutes a polymorphic locus. A less arbitrary and thus more desirable method to measure the genetic variation in a population is to calculate the average heterozygosity over all loci.

The **heterozygosity** at a locus is defined as the frequency of heterozygotes at that locus in the population. Thus, in the West African population with the *S* allele, considered earlier, the frequency of heterozygotes, or heterozygosity, was 0.32. In this case, only two alleles were segregating in the population (*A* and *S*), and thus only one class of heterozygotes (*AS*) was present. In cases in which more than two alleles are segregating at a locus, the heterozygosity value would be the sum of the frequencies of all the heterozygotes in the population. For example, three alleles were found segregating at the *Est-5* locus in the *Phoronopsis* population. This means that there could be as many as three heterozygote classes. The heterozygosity at the *Est-5* locus would be the sum of the individual frequencies of the three heterozygote classes (not shown in Table 21.3). The overall heterozygosity value at the *Est-5* locus is 0.443, as shown in Table 21.3.

If we average heterozygosities over all loci analyzed in a population survey, we come up with a statistic called

TABLE 21.3 *Allele Frequencies of 27 Variable Loci in 120 Individuals of the Marine Worm* Phoronopsis viridis.

Gene Locus	Frequency of Allele 1	2	3	4	5	6	Heterozygosity
Acph-1	0.995	0.005					0.010
Acph-2	0.009	0.066	0.882	0.014	0.024		0.160
Adk-1	0.472	0.528					0.224
Est-2	0.008	0.992					0.017
Est-3	0.076	0.924					0.151
Est-5	0.483	0.396	0.122				0.443
Est-6	0.010	0.979	0.012				0.025
Est-7	0.010	0.990					0.021
Fum	0.986	0.014					0.028
αGpd	0.005	0.995					0.010
G3pd-1	0.040	0.915	0.017	0.011	0.011	0.006	0.159
G6pd	0.043	0.900	0.057				0.130
Hk-1	0.996	0.004					0.008
Hk-2	0.005	0.978	0.016				0.043
Idh	0.992	0.008					0.017
Lap-3	0.038	0.962					0.077
Lap-4	0.014	0.986					0.028
Lap-5	0.004	0.551	0.326				0.542
Mdh	0.008	0.987	0.004				0.025
Me-2	0.979	0.021					0.042
Me-3	0.017	0.824	0.159				0.125
Odh-1	0.992	0.008					0.017
Pgi	0.995	0.005					0.010
Pgm-1	0.159	0.827	0.013				0.221
Pgm-3	0.038	0.874	0.071	0.017			0.185
Tpi-1	0.929	0.071					0.000
Tpi-2	0.008	0.004	0.962	0.013	0.013		0.076
Average heterozygosity (*H*)							0.072

From Ayala, F.J., et al. 1974. *Biochemical Genetics*, 18: 413.

the **average heterozygosity,** or *H*, which is used by geneticists as a relative index of the level of genetic variation present in the population as a whole. The average heterozygosity of the *Phoronopsis* population characterized in Table 21.3 is 0.072.

When the results of protein electrophoretic surveys from many different species are compared, a number of interesting trends become apparent (Table 21.4). For example, vertebrates appear to be significantly less genetically variable on average than invertebrates. Similarly, inbreeding plant species seem to be significantly less variable than outcrossing species. We will consider a likely explanation of this last observation later in this chapter.

DNA Sequence Variation Provides Information on Natural Populations.

The bulk of the survey data currently available on genetic variation levels in natural populations derives from studies using gel electrophoretic separation of proteins. However, survey data collected using molecular techniques such as DNA sequencing and restriction analysis

TABLE 21.4 *Genetic Variation in Natural Populations of Some Major Groups of Animals and Plants*

Organism	Number of Species	Average Number of Loci Sampled Per Species	Average Heterozygosity
Invertebrates			
Drosophila	28	24	0.150
Wasps	6	15	0.062
Other insects	4	18	0.151
Marine invertebrates	14	23	0.124
Land snails	5	18	0.150
Vertebrates			
Fishes	14	21	0.078
Amphibians	11	22	0.082
Reptiles	9	21	0.047
Birds	4	19	0.042
Mammals	30	28	0.051
Plants			
Self-pollinating	33	14	0.058
Cross-pollinating	36	11	0.185
Overall averages			
Invertebrates	57	22	0.134
Vertebrates	68	25	0.060
Plants	69	13	0.121

have become more numerous in recent years and will no doubt become commonplace over the next decade.

DNA sequence surveys can potentially uncover much more genetic variation than surveys that use gel electrophoretic separation of proteins. Protein gel electrophoresis can, in principle, detect only nucleotide substitutions that affect amino acid replacements. Such replacements that have an effect on a protein's net charge or shape will in turn modify the protein's migration rate in an electrophoretic gel. DNA sequence variability in introns, third-base positions in many codons, and other regions of a gene that do not result in amino acid replacements will go undetected in protein gel electrophoretic surveys. For example, Figure 21.6 summarizes the results of a survey

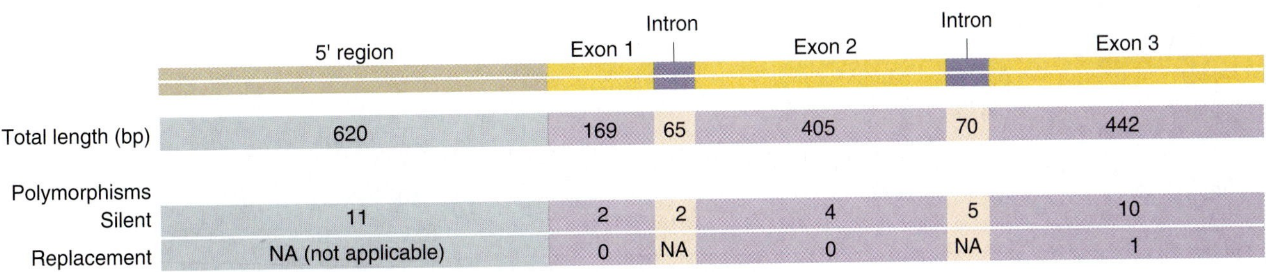

FIGURE 21.6 Nucleotide variability in alcohol dehydrogenase. The distribution of nucleotide variability among 11 encoding *Adh* genes isolated from a population of *Drosophila melanogaster* is shown.

of DNA sequence variation present among 11 alcohol dehydrogenase (*Adh*) genes isolated from a natural population of *Drosophila melanogaster*. Although more than 67 variable nucleotide positions were among the 11 genes surveyed, only one of these nucleotide differences resulted in an amino acid replacement that could be detected by protein gel electrophoresis. The presence of much higher levels of naturally occurring genetic variation at silent sites rather than at replacement sites is a trend that is apparent in all the DNA surveys thus far carried out.

SOLVED PROBLEM

Problem

The results of a protein electrophoretic survey of 36 enzyme-encoding loci in the Antarctic krill are presented in the following table. What is the average heterozygosity (*H*) for this population?

Locus	Frequency of Allele a	b	c	d	e	f	Observed Frequency of Heterozygotes
Acph-1			0.996			0.004	0.008
Ao-1		0.012	0.960	0.028			0.081
Ald-1		0.012	0.988				0.024
Ald-2		0.169	0.831				0.274
Aph		0.004	0.996				0.008
Est-1	0.138		0.850	0.012			0.291
Est-4		0.012	0.988				0.024
Est-5		0.028	0.972				0.065
G6pdh-1		0.008	0.992				0.016
Got		0.402	0.592	0.004			0.449
Hk-1	0.028		0.969	0.004			0.063
Hk-2		0.004	0.996				0.008
Idh			0.996	0.004			0.009
Lap		0.004	0.996				0.008
Mdh-2	0.020		0.980				0.039
Mdh-3	0.004	0.123	0.874				0.236
Me-2		0.007	0.993				0.014
Odh		0.039	0.957	0.004			0.087
Pgi	0.020		0.787	0.178	0.016		0.323
To-2			0.988	0.012			0.024
Xdh	0.004	0.996					0.008

Fifteen loci were monomorphic.
(After Ayala F., J. Valentine, and G. Zumwalt. 1975. *Limnology and Oceanography*, 20: 635.)

Solution

To compute the average heterozygosity (*H*) for this population, we must add the observed heterozygosity values at each locus and divide by the total number of loci sampled.

$$H = \frac{\text{Sum observed heterozygosity values at each locus}}{\text{Total number of loci sampled}}$$

$$= \frac{2.059}{36}$$

$$= 0.057$$

SECTION REVIEW PROBLEM

4. For the population in the preceding Solved Problem, what proportion of loci are polymorphic by the 0.95 criterion of polymorphism? By the 0.99 criterion?

Different Factors Shape the Genetic Variation in Natural Populations.

If a population's reproductively active individuals mate with one another without regard to their respective genotypes, the population is said to be **randomly mating.** If we adopt the gene-pool concept and envision the population as a collection of individual genes, and if this population undergoes random mating, then the probability that any two alleles will be combined in a (diploid) zygote at fertilization will be the product of the respective frequencies of those alleles in the population. Thus, if you know the frequencies of the alleles segregating at any given locus in a randomly mating population, you can easily compute the expected genotypic frequencies in that population. This fact was first demonstrated independently by the English mathematician Godfrey Hardy and the German physician Wilhelm Weinberg in separate papers published in 1908. Today it is referred to as the **Hardy–Weinberg Principle.**

To see how the Hardy–Weinberg Principle was derived, consider a population segregating for two alleles *A* and *S*. Assume further that the frequency of the *A* allele is p and that the frequency of the *S* allele is q. If males and females are mating with each other randomly, we can assume that the male and female gametes meet at random in the formation of zygotes. As a consequence, the expected frequency of a given genotype will simply be the product of the frequencies of the two alleles p and q. Thus, the probability that one individual will have the *AA* genotype is the probability of receiving the *A* allele from the mother (p) multiplied by the probability of receiving the *A* allele from the father (p), or $p \times p = p^2$. By the same reasoning, the probability that an individual will have the *SS* genotype is $q \times q = q^2$.

The *AS* genotype can arise in two possible ways. The mother could provide the *A* allele, and the father could provide the *S* allele. The probability of this event would be $p \times q = pq$. Alternatively, the mother could provide the *S* allele and the father could provide the *A* allele. The probability of this event would also be $p \times q = pq$. Thus, the expected frequency of *AS* heterozygotes in the population is the sum of the probabilities of these two alternative events or $pq + pq = 2pq$ (Figure 21.7). Thus, the expected genotypic equilibrium frequencies at a locus where there are two alleles segregating in a randomly mating population is, in fact, a simple binomial expansion: $(p + q)^2 = p^2 + 2pq + q^2$. As mentioned earlier, it is quite possible that a population may have more than two alleles segregating at a locus. The expected genotypic equilibrium frequencies at a locus having any number of alleles (n) segregating in a randomly mating population can be generated by applying the appropriate squared expansion, $(a_1 + a_2 + a_3 + \ldots a_n)^2$, where $a_1, a_2, \ldots a_n$ are the relative frequencies of the alleles segregating at the locus in question. Thus, if there are three alleles segregating at a locus, we could represent these respective frequencies as p, q, and r and derive the expected genotypic frequencies via the trinomial expansion: $(p + q + r)^2 = p^2 + q^2 + r^2 + 2pq + 2pr + 2qr$.

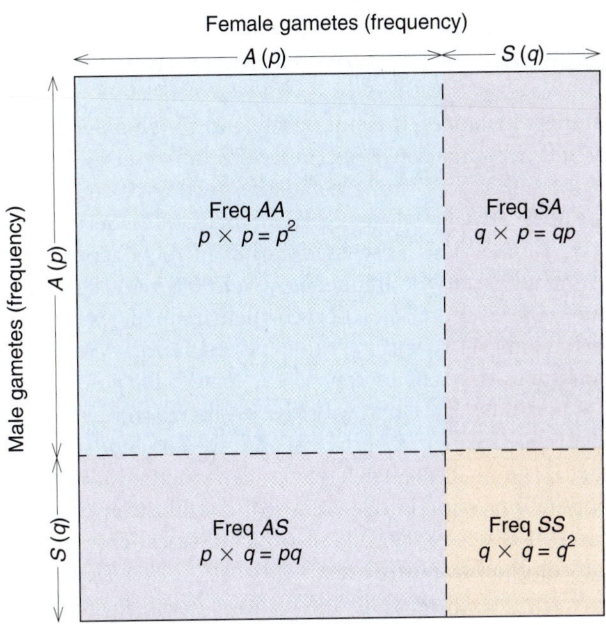

FIGURE 21.7 Hardy–Weinberg proportions. This unit square demonstrates the derivation of the Hardy–Weinberg proportions as generated from the random union of gametes.

When the observed genotypic frequencies in a population are not significantly different from the expected Hardy–Weinberg frequencies, the population is said to be in **Hardy–Weinberg equilibrium.** The Hardy–Weinberg Equilibrium Principle calls our attention to the inherent genetic stability of large, randomly mating populations. Gene and genotypic frequencies are expected to remain constant from generation to generation unless disturbed by additional factors, such as inbreeding or natural selection. In some cases, it may be obvious whether or not a population is in Hardy–Weinberg equilibrium. For example, in populations of plants that are only able to self-fertilize, there will be no heterozygotes even at loci where two alleles are present in the population at substantial frequencies. This lack of heterozygotes in an inbred population is obviously different from the expected $2pq$ heterozygote frequency and indicates that the population is not in Hardy–Weinberg equilibrium.

The Chi-Square Test Will Determine Whether a Population Is in Hardy–Weinberg Equilibrium.

In most instances, it is not possible to conclude whether or not a population is in Hardy–Weinberg equilibrium without first performing a statistical test. For example, in the West African survey of allelic frequencies at the hemoglobin-*A* locus, discussed earlier in this chapter, the observed genotypic frequencies were 0.67 *AA*, 0.32 *AS*, and 0.01 *SS*. From the observed allele frequencies of $p(A) = 0.83$ and $p(S) = 0.17$, the expected Hardy–Weinberg genotypic frequencies can be calculated: 0.69 *AA*, 0.28 *AS*, and 0.03 *SS*. Obviously, the observed and expected genotypic frequencies are close, but they are not so close that it can be immediately concluded whether or not the population is in Hardy–Weinberg equilibrium. In such cases, it is necessary to perform a statistical **chi-square** (χ^2) or **goodness-of-fit test.**

The chi-square value is a measure of the deviation of the observed numbers from those expected under a given model. In Chapter 2, the chi-square test was used to test Mendelian principles; in this chapter, we will use it to determine whether the observed genotypic frequencies in a population are significantly different from those predicted under the assumptions of the Hardy–Weinberg Equilibrium Principle. Recall that the formula used to compute the chi-square value is

$$\chi^2 = \Sigma[(O - E)^2/E]$$

where O and E are the observed and expected numbers of a particular genotype and Σ means that the computed values are summed over all genotypic classes represented in the population. Thus, the chi-square value (χ^2) is the sum of the squared difference between the observed and expected genotypic values divided by the expected genotypic values. After the chi-square value has been calculated and the degrees of freedom are determined, the probability that the observed values will deviate from the expected values by chance can be obtained from the chi-square table (Table 2.3).

Recall from Chapter 2 that the chi-square test requires a comparison of expected and observed numbers rather than frequencies. The expected number of each genotype can be computed by multiplying the expected frequencies by the sample size, N. The degree of freedom for the chi-square test is $n - 1$, provided that nothing from the observed data other than the sample size (N) is used in calculating the expected frequencies. When computing the expected Hardy–Weinberg genotypic frequencies, however, we typically employ the observed gene frequency values in addition to the sample size. Thus, when applying the chi-square test to determine whether observed genotypic frequencies are significantly different from Hardy–Weinberg expected genotypic frequencies, the degree of freedom is $n - 2$ rather than $n - 1$.

SOLVED PROBLEM

Problem

Determine whether the genotypic frequencies at the hemoglobin-*A* locus in the West African population previously considered in this chapter (Table 21.2) are in Hardy–Weinberg equilibrium.

Solution

The first step is to compute the Hardy–Weinberg expected genotypic frequencies. For a locus with two alleles segregating in a population (such as the hemoglobin-*A* locus in the West African population represented in Table 21.2), the expected frequencies of *AA*, *AS*, and *SS* genotypes are p^2, $2pq$, and q^2, respectively. Thus, the expected frequency of *AA* genotypes is $(0.83)^2$ or 0.69; of *AS* genotypes is 2(0.83)(0.17) or 0.28; and of *SS* genotypes $(0.17)^2$ or 0.03. Before we can apply the chi-square test, we must convert these expected frequencies to expected numbers by multiplying each expected frequency by 250, the number of individuals sampled. Thus, the expected number of *AA* genotypes is (0.69)(250) or 172.5; of *AS* genotypes is (0.28)(250) or 70, and of *SS* genotypes is (0.03)(250) or 7.5. We are now ready to apply the chi-square formula:

$$\chi^2 = \Sigma\left[\frac{(\text{Observed} - \text{Expected})^2}{\text{Expected}}\right] \quad \text{or} \quad \Sigma\left[\frac{(O - E)^2}{E}\right]$$

	AA	*AS*	*SS*
Observed	168	80	2
Expected	172.5	70	7.5
$\frac{(O - E)^2}{E}$	20.25	100	30.25
$\frac{(O - E)^2}{E}$	0.117	1.429	4.03
$\Sigma\frac{(O - E)^2}{E}$ =	5.578		
$P \leq 0.05$			

Looking at the chi-square table (Table 2.3), we see that a value of 5.58 with one degree of freedom ($n - 2$) is significant at the 5% level. That is, the deviation that exists between the observed and expected values is greater than would be expected by chance. We can therefore reject the hypothesis that the expected and observed numbers are the same and conclude that this West African population is not in Hardy–Weinberg equilibrium with respect to the frequency of its hemoglobin genotypes. The primary reason for this is the lower-than-expected number of *SS* homozygotes, which are in low frequency because of the high mortality rate among these severely anemic individuals.

Equilibrium Values of X-Linked Genes Are Different in Males and Females.

Thus far, our discussion of the expected equilibrium frequencies in randomly mating populations has been limited to autosomal genes in which the number and frequencies of the different genotypes can be assumed to be the same in males and females. This assumption does not hold for X-linked genes in which the expected genotypic frequencies between males and females can be quite different. For example, consider the gene responsible for hemophilia. This human blood disease is caused by a recessive X-linked gene and causes uncontrolled bleeding in afflicted individuals when they are wounded. Because males carry only one X chromosome, the frequency of hemophilic males in a population will be equal to the frequency of the hemophilia-causing allele q. Similarly, the frequency of unaffected males in the population will be the frequency of the normal allele p. Females, on the other hand, carry two X chromosomes and will thus display the familiar p^2, $2pq$, q^2 genotypic frequencies, where q is the frequency of the hemophilia-causing allele and p is the frequency of the normal allele.

SOLVED PROBLEM

Problem

You learned in Chapter 2 that color blindness is caused by a recessive, sex-linked allele. Suppose that the frequency of the color-blind allele is 0.01. Assuming random mating, compute the expected genotypic frequencies of color-blind males in this population. What is the expected genotypic frequency of females?

Solution

The Hardy–Weinberg equation allows us to compute the expected genotypic frequencies in a randomly mating population under the assumption that each individual carries two copies of each gene. This is a correct assumption for all autosomal loci in diploid organisms. With regard to genes carried on the X chromosome, however, there is an important difference between the sexes. Females carry two copies of X-linked genes (such as the gene responsible for color blindness), and males carry only one copy. Because females are diploid for X-linked genes, the equation used to compute the expected genotypic frequencies will be the same as that used for genes at autosomal loci. We are told that the frequency of the color-blind allele in a population is 0.01. This means that the frequency of the normal vision allele must be 1.00 − 0.01, or 0.99. Thus, the expected frequency of female genotypes at this locus is p^2, $2pq$, q^2 where $q = 0.01$ and $p = 0.99$:

p^2	$2pq$	q^2
$(0.99)^2$	$2(0.99)(0.01)$	$(0.01)^2$
0.98	0.02	0.0001 (color blind)

Because males carry only one X chromosome, the Hardy–Weinberg equation does not apply. Rather, the expected frequency of males carrying any X-linked gene will be the same as the frequency of the gene in the population. Thus, the expected frequency of color-blind males will be 0.01. This means that color-blind males will be 100 times more frequent in this population than color-blind females.

SECTION REVIEW PROBLEM

5. Three genotypes were observed at the *Adh* (alcohol dehydrogenase) locus in a *Drosophila* population. In a sample of 1110 flies, the three genotypes occurred in the following numbers. (Note that *Adh* alleles are named according to the relative electrophoretic mobility of their protein products; hence, we have *Adh* "fast" (*F*) and "slow" (*S*) alleles.)

Genotypes	*FF*	*FS*	*SS*
Number	634	391	85

a. Calculate the observed genotypic and allelic frequencies.
b. Assuming random mating, calculate the expected genotypic and allelic frequencies.
c. Use the chi-square goodness-of-fit test to determine whether or not this population is in Hardy–Weinberg equilibrium.

Nonrandom Mating May Alter the Genotypic Frequencies in a Population.

The Hardy–Weinberg principle rests upon the assumption that the matings that occur among members of a population are random with respect to genotype. In reality, this assumption is often violated. For example, stature is often an important consideration for humans in choosing mates. More often than not, tall individuals will marry other tall individuals, and short individuals will marry other short individuals. Such mating preferences among individuals of similar phenotype is called **positive assortative mating.** Other documented examples of positive assortative mating include the blue-and-white snow geese, which prefer to mate with geese of the same color as their parents, and some species of song birds, such as sparrows, which prefer to mate with individuals

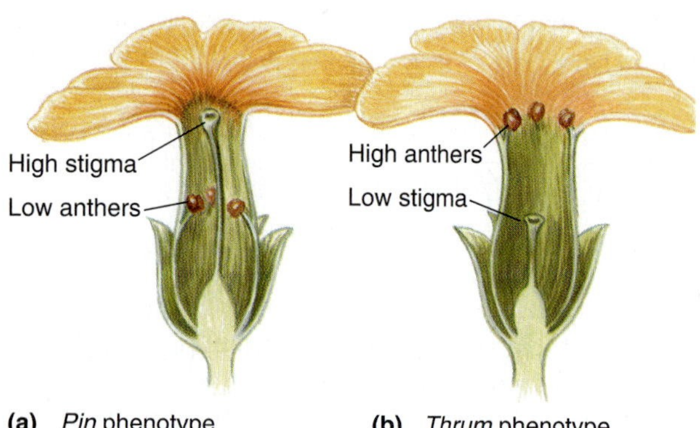

FIGURE 21.8 Primrose species phenotypes. (a) In the pin phenotype of *Primula officinalis,* the stigma is placed high in the flower and the anthers placed low. (b) The thrum phenotype is just the opposite, having the anthers high and the stigmas low. These opposite designs facilitate cross-fertilization between the two types.

who sing the same song as themselves. **Negative assortative mating** refers to preferential mating among individuals of different phenotypes. For example, the primrose, *Primula officinalis*, is a plant species that displays two main flower types, *pin* and *thrum*. Because of the structural design of these two main flower types, cross-pollination between *pin* and *thrum* types is greatly favored over pollination between flowers of the same type (Figure 21.8).

Because phenotypic similarity is generally reflective of genotypic similarity, assortative mating usually results in nonrandom mating between the various genotypes present in the population. As a consequence, a deviation from expected Hardy–Weinberg genotypic proportions will occur at those loci that encode the phenotypes upon which the mating preferences are based.

Perhaps the most common form of nonrandom mating observed in natural populations is inbreeding. **Inbreeding** occurs when matings between relatives is more frequent than would be expected if couples were selected from the population randomly. Whereas assortative mating results in a deviation from the expected Hardy–Weinberg genotypic proportions at the locus or loci encoding the phenotype upon which the mating preference is based, inbreeding results in a deviation from the Hardy–Weinberg proportions at all loci.

The most extreme form of inbreeding is **selfing,** which is a characteristic of some species of plants such as the garden pea (*Pisum sativum*) or soybean (*Glycine max*). In selfing plants, pollen from a given flower pollinates its own ovules. Selfing plants, such as the pea, can generally be recognized because they have flowers that never really open and in which the anthers grow into physical contact with the stigma (Figure 21.9).

Populations of selfing organisms are, in effect, a collection of inbred lines that are genetically homozygous. To see why homozygosity exists, consider an hypothetical starting population consisting entirely of *Aa* heterozygotes. Selfing among the *Aa* heterozygotes will result in three progeny genotypes: *AA, Aa,* and *aa,* in proportions 1/4, 1/2, and 1/4, respectively. Thus, the proportion of heterozygotes is halved in one generation of selfing. The proportion of heterozygotes in the F_2 generation will again be reduced by 1/2, resulting in 1/4 the proportion of heterozygotes that were present in the starting population [$(1/2)^n$, where n = number of generations of selfing]. The dramatic effect selfing can have on a population's genotypic frequencies is displayed in Table 21.5. A population that begins with 100% heterozygotes will be reduced to 12.5% (1/8) heterozygotes in only three generations of selfing.

After one generation of selfing, the two classes of homozygotes (*AA* and *aa*) carry alleles that are identical by descent because both alleles in the homozygous classes are copies of the only allele of that type (*A* or *a*) present

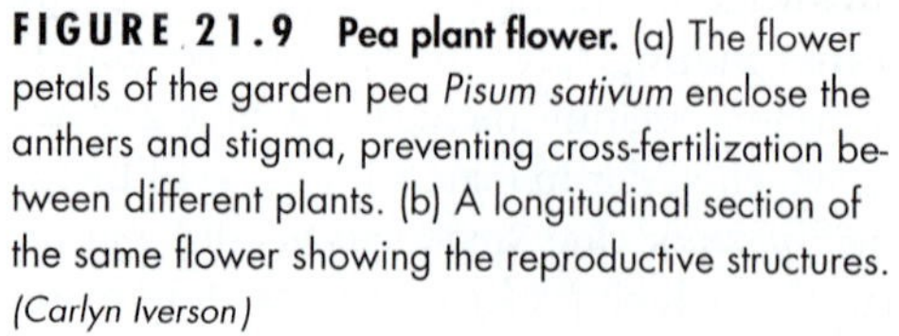
FIGURE 21.9 Pea plant flower. (a) The flower petals of the garden pea *Pisum sativum* enclose the anthers and stigma, preventing cross-fertilization between different plants. (b) A longitudinal section of the same flower showing the reproductive structures. *(Carlyn Iverson)*

(a) (b)

TABLE 21.5 *Results of Selfing in a Population Started Entirely from Heterozygotes*

Generation	Genotypic Frequencies *AA*	*Aa*	*aa*	Proportion of Heterozygotes	*F*	Frequency of *a* Allele
0	0	1	0	1	0	0.5
1	1/4	1/2	1/4	1/2	1/2	0.5
2	3/8	1/4	3/8	1/4	3/4	0.5
3	7/16	1/8	7/16	1/8	7/8	0.5
n	$\frac{1-(1/2)^n}{2}$	$(1/2)^n$	$\frac{1-(1/2)^n}{2}$	$(1/2)^n$	$1-(1/2)^n$	0.5

in the selfed heterozygous parent. The effect of inbreeding on genotypic frequencies is typically measured by a variable called the **inbreeding coefficient,** or *F*, which is formally defined as the probability that an individual receives two alleles, at any given locus, that are identical by descent. At this point, it is important to distinguish between two genes that may be structurally identical and those that are identical by descent. Two alleles that are identical by descent were originally copied from the same single allele carried by an ancestor and thus (barring mutation) will be structurally identical. However, not all genes that are structurally identical necessarily have been inherited from the same immediate ancestor. The value of *F* can vary from 0 to 1. *F* is 0 when there are no individuals homozygous by descent in the population, that is, when there is no inbreeding; *F* is 1 when all individuals in the population are homozygous by descent or when the population is made up of all inbreeding individuals.

To see how *F*-values are computed in a selfing population, consider the population depicted in Table 21.5. After one generation of selfing, the proportion of individuals carrying two alleles identical by descent is the same as the total frequency of homozygotes, which is 1/2. Because selfed homozygotes will produce only homozygous offspring, the 1/2 inbreeding acquired in the first generation of selfing will remain. In the second generation of selfing, one half of the progeny of the remaining heterozygotes will again be homozygous and carry two alleles identical by descent. Thus, the *F* value in the progeny of the heterozygotes will again be 1/2. However, the remaining heterozygotes in this second generation of inbreeding represent only one half of the total population. For this reason, the *F* value, 1/2, needs to be multiplied by 1/2 [that is, (1/2) (*F* value of interbreeding heterozygotes) × (1/2) (Proportion of heterozygotes in second generation of selfing)], which equals 1/4. Adding the second-generation *F* value (1/4) to the preexisting *F* value from the first generation of selfing (1/2), we come up with a total *F* value of 3/4; in other words, after two generations of selfing, three fourths of the individuals will be homozygous for alleles that are identical by descent. In each succeeding generation, the value of *F* will increase by 1/2 × the frequency of the heterozygotes in the previous generation (Table 21.6). Eventually, nearly all individuals in a selfing population will carry two alleles identical by descent, and the *F* value will approach 1.

As long as all genotypes in the population mate, on average, the same number of times, the allele frequencies will not change from generation to generation. Thus, although inbreeding will significantly affect changes in genotypic frequencies over succeeding generations, it will not, by itself, directly influence allelic frequencies.

F values can also be computed for less extreme forms of inbreeding. For example, consider the pedigree of a mating between a brother (C) and sister (D) shown in Figure 21.10, where each arrow represents the transmission of one gamete. A and B represent the two unrelated parents of C and D. The progeny of this brother–sister or **full-sib mating,** represented by E, receives one gamete each from the full-sibs C and D. Because E's grandparents, A and B, were unrelated, we can assume that their alleles at any given locus are not identical by descent. If we represent the two alleles at a locus in

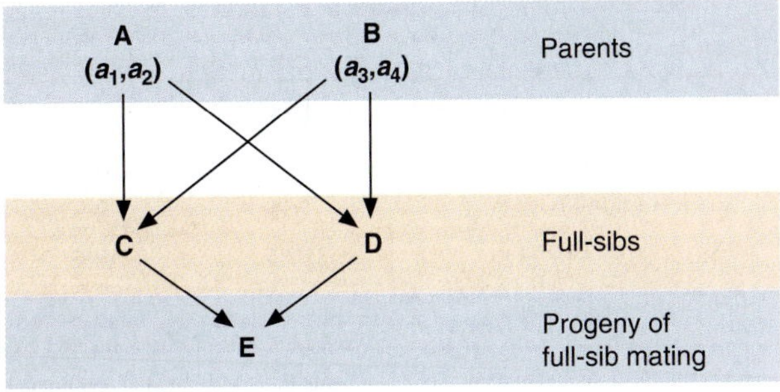

FIGURE 21.10 **Pedigree of an offspring (E) of a brother–sister (full-sib) mating.**

grandparent A as a_1 and a_2 and in grandparent B as a_3 and a_4, the probabilities of the four types of progeny from mating A × B are $(1/4)(a_1a_3)$, $(1/4)(a_1a_4)$, $(1/4)(a_2a_3)$, and $(1/4)(a_2a_4)$. Thus, individuals C and D each have a probability of 1/2 of carrying the a_1 allele (that is, the probability of being either (a_1a_3) or $(a_1a_4) = (1/4) + (1/4) = 1/2$).

What is the probability that E will be homozygous for grandparent A's a_1 allele? To determine this, we multiply the probability that C carries the a_1 allele, which is 1/2, by the probability that C will pass this allele on to E, which is also 1/2. Thus, the probability that E will inherit grandparent A's a_1 allele from C is $(1/2)(1/2) = 1/4$. Likewise, the probability that E will inherit the a_1 allele from D is also 1/4. Thus, the probability that E will be homozygous for a_1 is $(1/4)(1/4) = 1/16$.

In some human cultures, marriage among relatives (for example, first cousins) is not uncommon. This practice can result in a significant increase in the frequency of individuals homozygous for disease-causing alleles (Table 21.6). In small populations, the effective level of inbreeding can be quite high. For example, it is traditional among the Yanomami Indians of South America for the village chief to have several wives. As a consequence, many of the sexually mature members of the village are likely to be the grandchildren of a previous village chief. In most large populations, however, extensive inbreeding is not common. It is more likely that only a small fraction of the population is mating with relatives, usually because of spatial proximity as might occur in small isolated villages. Thus, in most large populations F values are expected to be small.

It should be obvious by now that all forms of nonrandom mating, whether assortative mating or inbreeding, are likely to result in a population having genotypic frequencies that are significantly different from those expected under conditions of Hardy–Weinberg equilibrium. In cases of inbreeding and positive assortative mating, there may be an increase in homozygosity and a corresponding decrease in heterozygosity. These changes require adjustment of the expected Hardy–Weinberg genotypic frequencies (homozygotes: $p^2 + q^2$; heterozygotes: $2pq$), which are computed on the assumption of random mating between unrelated individuals.

The expected frequencies of genotypes in the noninbreeding (randomly mating) component of the population are the expected Hardy–Weinberg proportions (p^2, $2pq$, q^2) multiplied by the probability that these individuals are indeed part of the randomly mating part of the population; $1 - F$ (if the probability of being part of the inbreeding component is F, then the probability of being part of the random mating component must be $1 - F$) (Figure 21.11). Thus, the expected genotypic frequencies in a composite population (including inbreeding and randomly mating individuals) with two alleles A and a in frequencies p and q, respectively, are

$$AA: p^2(1 - F) + p(F)$$

$$Aa: 2pq(1 - F)$$

$$aa: q^2(1 - F) + q(F)$$

The actual F value for any natural population can be estimated relatively easily by comparing the observed frequency of heterozygotes in a population to the Hardy–Weinberg expected frequency. As shown previously, inbreeding is expected to reduce the frequency of heterozygotes relative to that predicted from random mating. This fact is reflected in the following mathematical formula:

$$F = 1 - \frac{\text{Observed frequency of heterozygotes}}{\text{Hardy–Weinberg expected frequency of heterozygotes}}$$

If the observed and expected frequencies of heterozygotes are the same, then $F = (1 - 2pq)/2pq = 0$, and there is no evidence of inbreeding. In contrast, if all indi-

TABLE 21.6 *Inbreeding Depression in Human Populations*

Population	Unrelated Parents		First Cousins	
	Sample Size	Frequency (%)	Sample Size	Frequency (%)
United States (1920–1956)	163	9.8	192	16.2
France (1919–1925)	833	3.5	144	12.8
Sweden (1947)	165	4	218	16
Japan (1948–1954)	3570	8.5	1817	11.7
Average		6.5		14.2

The values represent frequencies of various diseases and physical and mental defects among children of unrelated parents and children whose parents were first cousins. The chances that the children will be homozygous for recessive defective alleles are greater when there is inbreeding.

(From Stern, C. 1973. *Principles of Human Genetics*, 3rd ed. Freeman, San Francisco.)

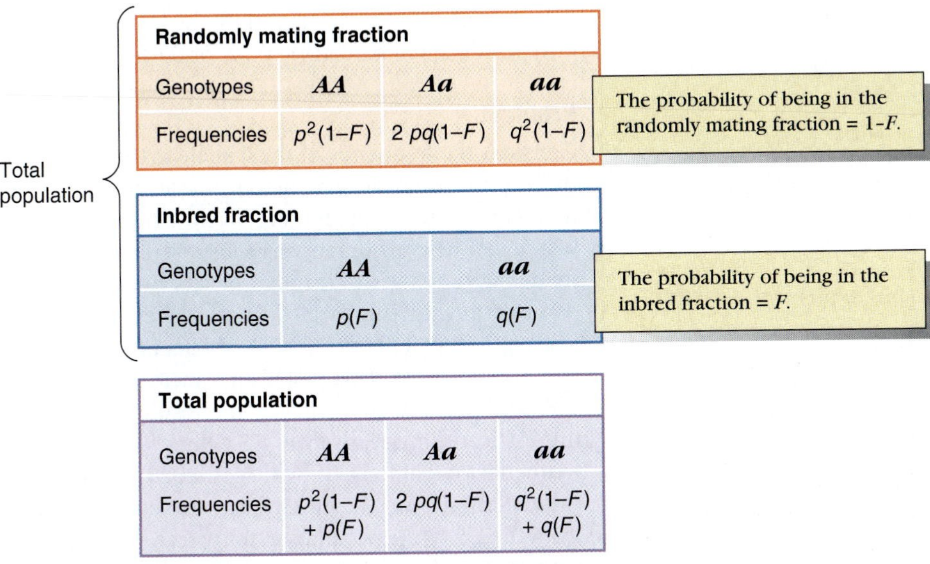

Total population
Randomly mating fraction
Genotypes AA Aa aa
Frequencies p2(1–F) 2 pq(1–F) q2(1–F)
The probability of being in the randomly mating fraction = 1–F.
Inbred fraction
Genotypes AA aa
Frequencies p(F) q(F)
The probability of being in the inbred fraction = F.
Total population
Genotypes AA Aa aa
Frequencies p2(1–F) + p(F) 2 pq(1–F) q2(1–F) + q(F)

FIGURE 21.11 Large populations typically may consist of an inbreeding fraction and a randomly mating fraction. In this population, two alleles (*A* and *a*) are segregating at a locus in a fraction (*F*) of inbreeding individuals and another fraction ($1-F$) of randomly mating individuals. Only the two homozygote genotypes *AA* and *aa* will be present within the inbreeding fraction. The expected frequency of *AA* genotypes in this inbreeding fraction of the population is $p(F)$, whereas the expected frequency of *aa* genotypes will be $q(F)$. Three genotypes (*AA, Aa, aa*) are represented in the randomly mating fraction of the population in the expected frequencies $p^2(1-F)$, $2pq(1-F)$, and $q^2(1-F)$, respectively.

viduals in a population are homozygotes, as would be expected in a population of self-fertilizing plants, then $F = (1 - 0)/2pq = 1$.

SOLVED PROBLEM

Problem

Imagine that the frequency of a recessive allele in a very large population of plants is 1%. What would be the expected frequency of individuals homozygous for this allele if

a. all members of the population were inbreeding?
b. half of the individuals in the population were inbreeding and the other half were randomly mating?

Solution

a. If all members of the population are inbreeding, then $F = 1$. The frequency of the recessive allele is 0.01. The expected frequency of individuals homozygous for the recessive allele in this population is given by the following equation: $q^2\ (1 - F) + q(F)$. Thus,

$$(0.01)^2\ (1 - 1) + 0.01(1) = 0.01$$

(i.e., the frequency of the recessive allele)

b. If half the individuals in the population are inbreeding, $F = 0.5$. Using the formula in part a, we can compute the expected frequency of individuals homozygous for the recessive allele:

$$(0.01)^2(1 - 0.5) + 0.01(0.5) =$$
$$(0.001)(0.5) + 0.005 = 0.0055$$

SECTION REVIEW PROBLEM

6. What would be the frequency of individuals homozygous for the recessive allele described in the Solved Problem if only 10% the members of the population were inbreeding?

Migration Affects Gene Frequencies in Populations.

Gene flow, or **gene migration,** occurs when individuals move from one population to another and interbreed. The effect of migration on the gene frequencies in the recipient population is dependent upon two variables: (1) the degree of genetic difference between the donor and recipient populations and (2) the rate of migration from the donor to the recipient population. If we assume that the frequency of a particular allele in the donor population is $p(d)$, and the frequency of this allele in the original recipient population is $p(r)$, then the expected frequency of the allele in the recipient population after migration p' is given by

$$p' = (1 - m)\, p(r) + mp(d)$$
$$= p(r) - m[p(r) - p(d)]$$

The change in gene frequency in the recipient population (Δp) is $p'(r) - p(r)$. Substituting the value of $p'(r)$ obtained previously, we obtain

$$\Delta p = p(r) - m[p(r) - p(d)] - p(r)$$
$$= -m\,[p(r) - p(d)]$$

If the two populations are identical with regard to gene frequencies [that is, if $p(d) = p(r)$], then the movement of individuals from one population to the other will have no effect [$p(r) = 0$]. However, if the donor population is genetically quite different from the recipient population, the genetic character of the recipient population may undergo significant genetic change even if the migration rate m is low.

If migration between two populations continues each generation, the allelic frequencies in the two populations will become similar with time. This relationship is represented by the following equation, in which t represents the number of generations of migration:

$$p^t(r) = [1 - m]^t\,[p(r) - p(d)] + p(d)$$

SOLVED PROBLEMS

Problem

Assume that the frequency of the allele encoding the O blood type is 0.60 in an isolated island population of humans and that the frequency in a nearby mainland population is 0.40. Although the populations have been isolated for years, a weekly ferry service has been recently introduced, resulting in migration from the island to the mainland. It is estimated that 10% of the parents giving rise to the next generation on the mainland population will be immigrants from the island ($m = 0.1$). What will be the expected change in frequency of the O-type encoding allele in the mainland population in the next generation?

Solution

The expected change in frequency of the O-type encoding allele in the recipient population is given by

$$\Delta p(r) = -m\,[p(r) - p(d)]$$

where $p(r)$ is the frequency in the recipient population before migration, $p(d)$ is the frequency of the O-type allele in the donor population, and m is the migration rate from the donor to recipient population. Thus,

$$p(r) = -0.1(0.40 - 0.60) = 0.02$$

After one generation of migration, the frequency of the O-type allele will be 0.40 + 0.02, or 0.42.

Problem

Consider the frequency of the *R* allele at the locus controlling the Rh (Rhesus) blood type in humans. The frequency of this gene among black Africans is 0.630, which can be considered very close or identical to the frequency among the original blacks brought to this country during the early days of slavery about 300 years ago (ten generations). The frequency of the *R* allele among present-day African Americans is 0.446. The difference in frequency is the result of gene flow into African Americans from US whites. Given that the frequency of *R* among US whites is 0.028, compute the estimated rate (m) at which genes have been flowing from the US white population to African Americans by rearranging the equation immediately preceding this set of Solved Problems.

Solution

Given the formula

$$p^t(r) = [1 - m]^t\,[p(r) - q(d)] + p(d)$$

we can use simple algebra to rearrange this equation so that we can solve for the migration rate m:

$$(1 - m)^t = \frac{p^t(r) - p(d)}{p(r) - p(d)}$$

where $p(r)$ = allele frequency in the recipient population before migration, $p^t(r)$ = allele frequency in the recipient population after t generations of migration, and $p(d)$ = allele frequency in the donor population. Continuing on,

$$(1 - m)^{10} = \frac{0.446 - 0.028}{0.630 - 0.028}$$
$$(1 - m)^{10} = 0.694$$
$$1 - m = 0.964$$
$$m = 0.036$$

Thus, the average rate of gene flow from US whites into the US African-American population has been 3.6% per generation over the past 300 years.

SECTION REVIEW PROBLEMS

7. What would be your estimate of the frequency of the O-type allele in the mainland population described in the preceding Solved Problem after one generation of migration if the migration rate was only 1%?

8. Although the average frequency of the *R* allele in African Americans is 0.446, there is considerable variation in this value in different parts of the United States. For example, in Claxton, GA, the frequency of the *R* allele in African Americans is 0.533, whereas the frequency in the Oakland, CA, population is 0.486. Compute the rate of gene flow (m) from US whites into each of these two African American populations over the past 300 years. Propose one or more hypotheses to explain the differences in rate.

Mutation Is the Ultimate Source of All Genetic Variation.

Evolution requires genetic variation as its raw material, and the process of mutation is the ultimate source of all genetic variation. In Chapters 17, 18, and 19, the nature of mutations was discussed. Because different types of mutations can have quite different phenotypic consequences, it is reasonable to conclude that different types of mutations may have quite different evolutionary consequences, as well. We will consider this issue in more detail in Chapter 22. For now, however, we are interested only in the population dynamics of mutations after they arise. We will not, therefore, distinguish between the various types of mutations, for most of them can be described by the same population genetic equations.

Consider a situation in which there are two alleles, *A* and *a*, at a locus and the rate of mutation from *A* to *a* is some frequency expressed per gamete per generation and represented by the Greek letter μ. Assume that at the time we begin to observe this population, the frequency of the *A* allele is p_0. In the next generation, a proportion (μ) of the *A* alleles will mutate to *a* alleles. Thus, the frequency of *A* alleles after one generation (p_1) will be the initial frequency (p_0) less the frequency of the newly arisen *a* alleles (μp_0):

$$p_1 = (p_0 - \mu p_0) = p_0\,(1 - \mu)$$

In the following generation, a proportion μ of the remaining *A* alleles (p_1) will again mutate to *a* alleles and so the frequency of the *A* alleles in the second generation will be

$$p_2 = p_1 - \mu p_1 = p_1\,(1 - \mu)$$

If we replace the value p_1 in this equation with its equivalent $p_0(1 - \mu)$, we get

$$p_2 = p_1(1 - \mu) = p_0(1 - \mu)(1 - \mu) = p_0(1 - \mu)^2$$

Thus, after t generations, the frequency of the *A* alleles will be

$$p_t = p_0(1 - \mu)^t$$

SOLVED PROBLEMS

Problem

Calculate the frequency of *A* in 100 generations assuming a starting frequency of 1.00 (i.e., no *a* alleles present) and a mutation rate of $\mu = 10^{-9}$ per gamete per generation (a typical value for nucleotide substitutions in eukaryotes).

Solution

The expected frequency of any allele after t generations of mutation is given by

$$p_t = p_0(1 - \mu)^t$$

where p_0 is the initial frequency of the allele, μ is the mutation rate, and t is the number of generations of mutations.

We are given a starting frequency of 1.00 and a mutation rate of 10^{-9}, and we are told to compute the expected frequency after 100 generations. Thus,

$$p_t = 1.00\,(1 - 10^{-9})100$$
$$= 0.990$$

Problem

Calculate the frequency of the same allele in the same population described in the previous Solved Problem after 10,000 and 100,000 generations.

Solution

You will find that although the frequency of the *A* allele will eventually go to zero, it will take a very long time to do so if mutation is the only factor involved.

Although the rate of single nucleotide mutations is quite low, recent data suggest that the rates at which transposable elements insert into new genomic positions may under some circumstances be $10^{-2} - 10^{-3}$ or higher. Some geneticists have even suggested that rates of transposition may significantly increase in response to stress or other environmental stimuli. Thus, the evolutionary significance of simple nucleotide mutations and transposon-mediated mutations may be quite different. This question will be examined in Chapter 22.

The Frequency of Genes in Populations May Fluctuate by Chance.

Have you ever been in a situation where you decided to resolve some issue by flipping a coin? If you have, you probably did so because you believed that the outcome would have a 50% chance of coming out in your favor. In other words, you believed the coin had an equal probability of coming up heads or tails. Suppose, however, that for some reason you suspected the coin was biased. You might ask to have the coin flipped several times to see how often heads and tails actually appeared. The expectation is 50–50, but you probably would not be too upset if in ten trials, heads came up seven or eight times instead of the expected five. After all, such a deviation from expectation could occur by chance and not necessarily be indicative of a biased coin. On the other hand, you might become suspicious if in 100 flips, heads came up 70 to 80 times or more.

You do not need to be a statistician to realize that some degree of deviation from random expectations can occur and indeed should be expected, especially if the number of trials is small. However, statistical theory can be very useful in determining what degree of deviation from an expectation is to be reasonably expected in any particular situation. For example, the expected deviation from the anticipated 50–50 ratio in a series of coin flips is given by the formula for the standard deviation, $s = \sqrt{ht/n}$ where h and t are the expected frequencies of heads and tails (expected to be 0.50 each for an unbiased coin) observed in n flips of the coin; specifically, $s = \sqrt{(0.5)(0.5)/n}$. The observed frequency of heads (or tails) is expected to fall within the range of one standard deviation from the expected value (0.50) 68% of the time and within two standard deviations of the expected value 95% of the time.

Reconsider the example in which heads came up eight times in ten flips of the coin. The standard deviation expected for this experiment would be $s = \sqrt{(0.25)/10}$, or 0.16 ($2s = 0.32$). Thus, for ten flips of the coin, the frequency of heads (or tails) would be expected to lie between 0.18 and 0.82 (i.e., 0.50 ± 0.32) 95% of the time. The observed frequency of heads, 0.80, being less than 0.82, is then no cause for alarm. However, in the second example, in which the coin was flipped 100 times, the expected standard deviation would be 0.05 $s = \sqrt{(0.25)/100}$. We would have expected the frequency of heads to fall between 0.4 and 0.6 95% of the time ($0.5 \pm 2s$ or 0.10). The fact that heads appeared in a frequency of 0.80 is indeed reason to be suspicious of the fairness of the coin.

Sampling error is the formal name used by statisticians to describe the fact that the actual frequencies can deviate from expectations by chance. Sampling error is not limited to events such as a flip of the coin and, in fact, can be equally well applied to situations such as genes in a population. Suppose that a certain locus has two alleles segregating in a large population of 1000 individuals, and that each allele is in 50% frequency. The expected frequency of each allele in the next generation should also be 0.50, but it may be expected to deviate by chance. As with the example of the coin, the expected frequency of the allele in the next generation will have a 95% probability of being 0.50 ± 2 standard deviations. The formula for standard deviation in this context is $\sqrt{pq/2N}$, where p and q represent the expected frequencies of the two alleles (equal to their observed frequencies in the previous generation), and N is the number of individuals in the population ($2N$ = the number of genes, assuming we are dealing with a population of diploid individuals). Thus for the preceding example, $s = \sqrt{(0.25)/2000}$, or 0.01 (thus, $2s = 0.02$), meaning that it is likely that the observed frequency of our hypothetical allele after one generation of random mating will lie somewhere between 0.48 and 0.52. Gene frequency changes in a population that are caused by sampling error are called **genetic drift.** The effect of genetic drift is generally very slight and is often unnoticeable in very large populations (analogous to very many flips of the coin). However, chance deviations may not be so insignificant in small populations (analogous to what happens with only a few flips of the coin) and may result in substantial changes in gene frequencies and even in the complete loss of an allele from the population over time (Figure 21.12).

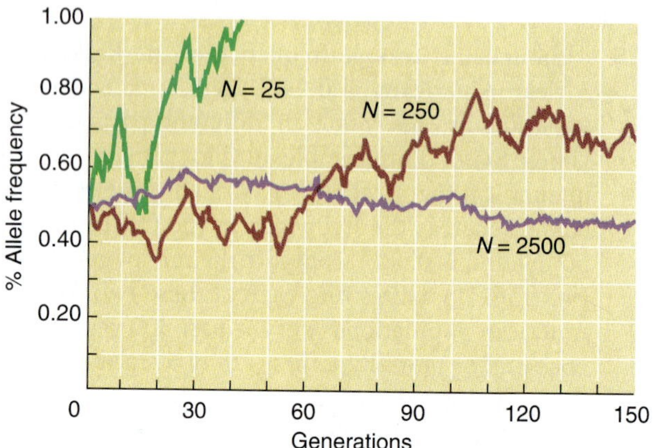

FIGURE 21.12 **Genetic drift.** Chance fluctuations in gene frequency are more pronounced in small populations than in large populations. The graph displays the results of computer simulation experiments that show the changes in gene frequency at a particular locus as a result of genetic drift in three populations differing only in population size. The frequency of the allele being monitored was 0.5 in all populations at the start of the experiment. Note that the random changes in frequency were most dramatic in the population consisting of only 25 individuals, and the least amount of random change in frequency occurred in the population with 2500 members.

If we assume that the new allele arises by mutation in a population of N individuals, the initial frequency of this new allele will be $(1/2)N$. If this new allele imparts no selective advantage or disadvantage to its carriers, most likely it will be eliminated from the population by genetic drift. However, there is a small chance that it will not be lost but rather will increase in frequency and perhaps drift to fixation, that is, 100% frequency in the population. The probability that a new mutant allele will

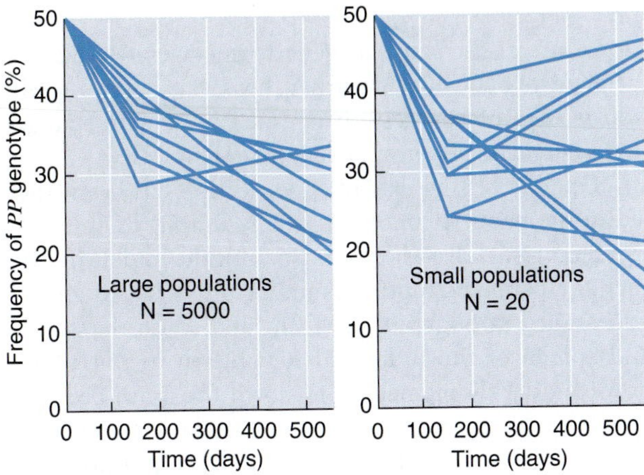

FIGURE 21.13 Founder effect in laboratory populations of *Drosophila pseudoobscura*. Populations were begun with the variant *PP* at 0.5 frequency. Ten replicate, large populations were started with 5000 individuals each, and ten replicate, small populations were started with only 20 individuals each. After 500 days (about 18 *Drosophila* generations), the range of frequencies (variance) between different populations was much greater among the small populations than among the large populations. The high variance in frequencies among the populations started with only 20 individuals is attributable to founder effects.

drift to 100% frequency and become fixed in the population is equal to its initial frequency (that is [1/2]*N*). Thus, the larger the population size, the less likely it is that a newly arisen allele will eventually become fixed in a population by genetic drift. It can be shown that the average number of generations needed for a newly arisen neutral allele to become fixed in a population is 4 times the number of reproducing individuals in the population (4*N*). Thus, for example, if the number of breeding individuals in a population is 10,000, the process of fixing a newly arisen neutral allele requires about 40,000 generations.

Perhaps the most dramatic change in gene frequencies resulting from genetic drift occurs when a large population is suddenly reduced in size to relatively few individuals. Such reductions may occur when a few individuals leave the main population and inhabit an unoccupied (by members of the same species) and isolated area. If these few pioneers interbreed and establish a new population, the new population will, of necessity, reflect the genetic character of the founders and may differ substantially from the original ancestral population. Changes in gene frequencies caused by such colonizations are said to be the result of **founder effects.**

Another way in which gene frequencies may suddenly change as a result of sampling error is if a large randomly mating population is catastrophically reduced in size because of, for example, a flood, a fire, or a volcanic eruption. If the few surviving individuals happen to be genetically atypical of their parental population, their progeny population will also be genetically dissimilar to the ancestral population. Changes in gene frequency that occur in this way are said to be the result of a **genetic bottleneck.**

A laboratory demonstration of the effect of sampling error on gene frequencies was provided by Theodosius Dobzhansky in 1957. Dobzhansky started 20 laboratory populations of *Drosophila pseudoobscura* all with the chromosomal genotype *PP* in 0.5 frequency. Ten of the populations were started with 5000 individuals, and the other ten were started with only 20 individuals each. The frequency of *PP* was periodically monitored over the period of 1.5 years (about 18 generations). The results, represented in Figure 21.13, demonstrate that the variation in frequency of *PP* among the small populations was considerably larger than that observed among the large populations. This difference is due to the fact that the effect of sampling error is more pronounced among populations started with relatively few individuals.

Outside the laboratory, rapid changes in the genetic character of populations also occur. For example, the Dunkers are a religious sect who immigrated to the United States from Germany in the mid-18th century. Twenty-seven families of Dunkers established communities in central Pennsylvania, and their descendants have continued to live in small, relatively isolated communities. The effects of sampling errors can be noted at several loci. For example, the frequency of blood type A is 40–45% in German populations and in US populations of German descent, whereas among the Dunkers, the frequency is 60%.

Sampling errors may even contribute to the emergence of new species. For example, founder effects as well as genetic bottlenecks brought about by frequent volcanic activity are believed to have contributed to rapid genetic changes that have led to the tremendous diversity of species endemic to the Hawaiian Islands (Figure 21.14).

FIGURE 21.14 Diversity of Hawaiian *Drosophila* species. *Drosophila heteroneura* (right) and *Drosophila silveshis* (left) are two closely related Hawaiian *Drosophila*, which show tremendous morphological diversity. *(Ken Kaneshiro, University of Hawaii at Manoa)*

SOLVED PROBLEM

Problem

Suppose that, as a class project, you establish a small laboratory population of *Drosophila* with the alleles *B* and *b* in 0.20 and 0.80 frequencies, respectively. Assume that the size of your experimental population is 1000 individuals. If the *B* and *b* alleles are selectively neutral with respect to one another (neither allele is favored by natural selection), what is the probability that the *B* alleles will eventually become fixed in this population?

Solution

The probability that a neutral allele will become fixed because of chance alone is simply its frequency in the population. Thus, the probability that the *B* allele will eventually go to 100% frequency in the population is currently 0.20.

SECTION REVIEW PROBLEM

9. Suppose that, as a class project, you establish a small laboratory population of 1000 *Drosophila* with most of the individuals homozygous for the *B* allele, and with one individual heterozygous (*Bb*). Assume that the *B* and *b* alleles are selectively neutral with respect to one another (neither allele is favored by natural selection).
   a. What is the probability that the *b* allele will eventually become fixed in this population?
   b. How long would it take for the *b* allele to become fixed in this population by genetic drift? (Note: The generation time for *Drosophila* is about 2 weeks.)

Natural Selection Directs Changes That Increase a Population's Adaptiveness to Its Environment.

It is often the case in populations that one genotype is associated with structural or functional characteristics that allow it to cope relatively better with its environment and thus leave proportionally more offspring than other genotypes in the population. The differential reproductive ability of alternative genotypes in a population is called **natural selection.**

Differential reproduction may be subdivided into a number of component parts, such as differential mating success, differential fertility, and differential developmental time. But because all these components will eventually affect relative reproductive ability, it is convenient when considering overall population dynamics to treat everything under a single heading. Like genetic drift, mutation, and migration, the process of natural selection can result in a change in gene frequencies. Unlike the mechanisms considered earlier, however, natural selection is the only process that results in genetic changes that systematically promote adaptation.

One interesting example of selection has been studied in the peppered moth (*Biston betularia*) by the English biologist J. D. Kettlewell. Up until the mid-19th century, the peppered moth was uniformly speckled gray in appearance throughout all of England. With the advent and spread of the Industrial Revolution in the 1760s, however, darkly pigmented forms of the moth began to appear in the regions where the local vegetation had been blackened by industrial soot and other pollution. In heavily contaminated regions, the dark variety nearly replaced the lightly pigmented forms, which continued to be most common in unindustrialized areas. The sudden prevalence of the dark variety was caused by a rapid increase in frequency of a single dominant allele controlling pigment formation. At first this increase in frequency was mistakenly attributed to a pollution-induced increase in the mutation rate. Further study by Kettlewell and his colleagues, however, made it clear that the actual cause was natural selection.

Kettlewell showed that the peppered moth spends most of its time resting and feeding on the trunks of trees. In unpolluted regions, these tree trunks are often covered with a pale-colored lichen that provides the lightly pigmented moths with camouflage protection from birds. In those industrialized regions where pollution had killed the lichen cover, the grayish moths became easy prey for hungry birds, whereas the dark forms were relatively protected (Figure 21.15). The result was an increase in the dark pigment coding gene.

FIGURE 21.15 Peppered moth species. Members of *Biston betularia* display two color morphs, dark and light, which are the result of a simple allelic difference at the locus encoding body color. The dark allele (*B*) is dominant to the light allele (*b*). Dark moths are more readily visible to bird predators on light lichen-covered trees, whereas light moths are more easily seen on the darker bark of trees not covered by lichen. *(M. Tweedie/Photo Researchers, Inc.)*

TABLE 21.7 *Allelic Frequency Changes After One Generation of Selection Against Recessive Homozygotes (bb)*

	Genotypes			
	BB	*Bb*	*bb*	Total
A.				
Initial zygote frequency	p^2	$2pq$	q^2	1.00
Fitness	1.0	1.0	$1.0 - S$	
Contribution of each genotype to next generation	p^2	$2pq$	$q^2(1 - S) = q^2 - q^2S$	$1 - q^2S$*
Normalized frequency after one generation of selection	$\frac{p^2}{1 - q^2S}$	$\frac{2pq}{1 - q^2S}$	$\frac{q^2 - q^2S}{1 - q^2S}$	$\frac{1 - q^2S}{1 - q^2S} = 1.00$

*Because $(p^2 + 2pq + q^2) = 1$, it follows that $(p^2 + 2pq + q^2) - q^2S = 1 - q^2S$.

B. The computation of the frequency of the peppered moth's genotypes after one generation of selection against the *bb* genotype.

	Genotypes			
	BB	*Bb*	*bb*	Total
Initial zygote frequency	0.01	0.18	0.81	1.00
Fitness (W)	1.0	1.0	1.0 − 0.33 = 0.67	
Contribution of each genotype to next generation	1(0.01) = 0.01	1(0.018) = 0.018	0.67(0.81) = 0.543	1 − (0.81)(0.33) = 0.73
Normalized frequency after one generation of selection	$\frac{0.01}{0.73} = 0.01$	$\frac{0.18}{0.73} = 0.25$	$\frac{0.54}{0.73} = 0.74$	$\frac{0.73}{0.73} = 1.00$

Note that the frequency of *bb* homozygotes has decreased 7%. The frequency of the *b* allele will continue to decrease each generation.

As is the case for the other processes that effect changes in gene frequencies in populations, the action of natural selection has been mathematically formalized by population geneticists. **Fitness** is a term used to describe the reproductive efficiency of a genotype relative to other genotypes in the population. Fitness is usually symbolized by the letter *W* and represents a genotype's relative probability of contributing genes to the next generation. Relative fitness values range from 0 to 1.00. A related parameter is the **selection coefficient,** symbolized by the letter *S*. The selection coefficient is an expression of the extent to which a genotype is selected against relative to the most fit genotype in the population and is mathematically defined as $S = 1 - W$ (thus $W = 1 - S$). If a particular genotype results in lethality or sterility, its *W* and *S* values would be 0 and 1, respectively. By contrast, a genotype that has the highest reproductive efficiency in the population would have $W = 1$ and $S = 0$.

Table 21.7a shows how the effect of selection against the homozygous recessive genotype (*bb*) can be mathematically represented over one generation. We begin by assuming that the starting population is in Hardy–Weinberg equilibrium and that the starting frequencies of the three genotypes *BB*, *Bb*, and *bb* are p^2, $2pq$, and q^2, respectively. The sum of these frequencies is 1.00. To compute the expected frequencies after one generation of selection, we multiply the frequency of each genotype by its fitness value (that is, its expected probability of contributing genetically to the next generation). For genotypes *BB* and *Bb*, the fitness values are 1.00; for the *bb* genotype, the fitness value is less than 1.00 because the genotype is being selected against. The degree to which the *bb* moths are selected against relative to the *BB* and *Bb* genotypes is $1 - S$. Thus, to determine the expected frequency of the *bb* genotype after one generation of selection, we must multiply $q^2 \times (1 - S)$, which is $q^2 -$

q^2S. Thus, the sum of the frequencies after one generation of selection ($p^2 + 2pq + q^2 - q^2S$) has been reduced from what it was prior to selection (i.e., $p^2 + 2pq + q^2$) by the quantity q^2S. Because the sum of the genotypic frequencies prior to selection was 1.00, the sum after selection is $1 - q^2S$. In order to normalize the equation so that the sum of the genotypic frequencies again equals 1.00, we need to divide both sides of the equation by $1 - q^2S$. Thus, the normalized expected genotypic frequencies after one generation of selection are

$$\frac{p^2}{1 - q^2S} + \frac{2pq}{1 - q^2S} + \frac{q^2}{1 - q^2S} = 1.00$$

In order to see how these algebraic computations can be applied to a real situation, let's reconsider the example of the peppered moth originally studied by Kettlewell (Table 21.7b).

To complement his field studies, Kettlewell carried out a laboratory experiment to determine the percentage of each type of moth that actually survived on light and dark backgrounds. In one experiment it was found that on a dark background, 57 out of 70 (or 83%) of the darkly pigmented forms survived, but only 37 out of 70 (or 56%) of the lightly pigmented forms avoided predation by birds. Based on these data, we can estimate the relative fitness of the dark genotype to be 1.00 (0.83/0.83) and the light genotype to be 0.67 (0.56/0.83). Because the dark-pigment coding allele (*B*) is dominant to the light allele (*b*), the fitness coefficients for the three genotypes *BB*, *Bb*, and *bb* will be 1.00, 1.00, and 0.67, respectively. As in the algebraic derivations described earlier, we assume that the population is initially in Hardy–Weinberg equilibrium. The starting frequency of the *B* allele is 0.10, and that of the *b* allele is 0.90. Thus, the starting frequency of the three genotypes *BB*, *Bb*, and *bb* will be 0.01, 0.18, and 0.81, respectively (that is, p^2, $2pq$, q^2).

We begin by multiplying the frequency of each genotype by its respective fitness value. For genotypes *BB* and *Bb*, the fitness value is 1.00; for genotype *bb*, the fitness value is 0.67. After multiplication we come up with the following values:

BB		*Bb*		*bb*
(0.01)(1.0)	+	(0.18)(1.0)	+	(0.81)(0.67)
		or		
0.01	+	0.18	+	0.54 = 0.73

To normalize the results so that the total frequency of all three genotypes in the population is again 1.00, we must divide both sides of the equation by 0.73:

BB		*Bb*		*bb*		Total
$\frac{0.01}{0.73}$	+	$\frac{0.18}{0.73}$	+	$\frac{0.54}{0.73}$	=	$\frac{0.73}{0.73}$
			or			
0.01	+	0.25	+	0.74	=	1.0

Thus, the frequencies of the three genotypes *BB*, *Bb*, and *bb* after one generation of selection are 0.01, 0.25, and 0.74, respectively. From this information, we can compute the frequency of the *b* allele in the usual way.

$$f(b) = 0.74 + (1/2)(0.25) = 0.86$$

Although we have calculated the change in gene frequency for only one generation of selection in the preceding example, it is possible to calculate what the expected changes in gene frequency would be over much longer periods of time. Such reiterated computations are tediously repetitive and so are usually carried out by computer. Population I in Figure 21.16 depicts what happens to gene frequencies of the peppered moth's light allele (*b*) over 100 generations. Eventually, the light allele is selected out of the population. Figure 21.16 also presents the change in gene frequency of the *b* allele assuming different dominant–recessive relationships and selective coefficients existing between the three genotypes

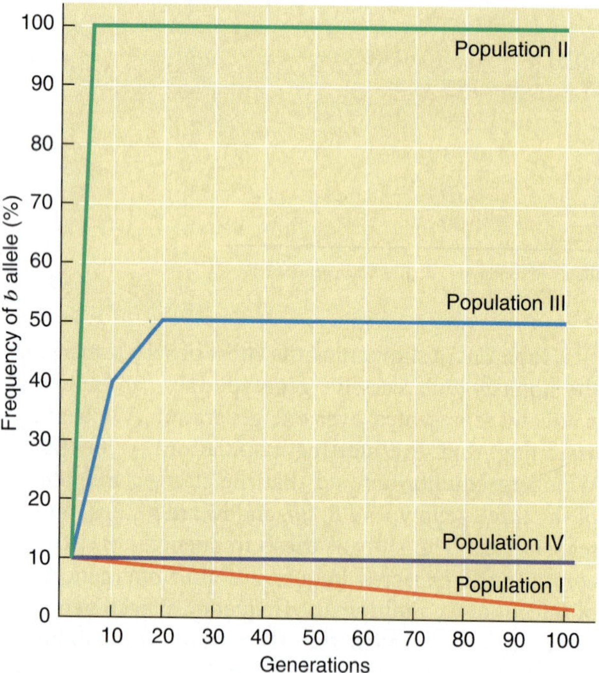

FIGURE 21.16 Gene frequency changes in the peppered moth's *b* (light) allele. Computer simulations show changes in the frequency of the *b* allele in different dominance–recessive relationships between the two alleles *B* and *b* and different fitness relationships between the three genotypes *BB*, *Bb*, and *bb*. (See text for details.) *(Simulations courtesy of Dr. A. J. Cuticchia.)*

TABLE 21.8 *Fitness of Three Genotypes at the Hemoglobin A Locus in a Sampling of 12,387 Adults from Nigeria*

	Genotypes				
	AA	*AS*	*SS*	**Total**	**Frequency of S**
Observed numbers	9365	2993	29	12,387	
Observed frequencies	0.76	0.24	0.002	1.00	0.12
H–W expected frequencies	0.77	0.22	0.01	1.00	0.12
Survival efficiency	0.76/0.77 = 0.98	0.24/0.22 = 1.12	0.002/0.01 = 0.15		
Relative fitness	0.98/1.12 = 0.88	1.12/1.12 = 1.00	0.15/1.12 = 0.13		

represented at this locus. In population II, for example, *BB* and *Bb* individuals are assumed to be lethal or sterile (that is, having selective coefficients of 0.0). In such a case, the *b* allele becomes fixed (that is, it goes to a frequency of 1.00) by **directional selection** in only one generation. Populations III and IV are particularly interesting because they reach an equilibrium in which both *B* and *b* alleles are actually being maintained in the population by virtue of the fact that the heterozygotes are the most-fit genotype. These latter two cases are examples of what is termed **balancing selection.** An actual instance of balancing selection operating in a natural population is the polymorphism that exists in many African populations for the hemoglobin *S* and *A* alleles, which were considered earlier in this chapter.

The relative fitnesses (*W*) of the hemoglobin *AA*, *AS*, and *SS* genotypes can be computed by comparing the genotypic frequencies in newborns with frequencies present in late middle-aged adults because selection will have operated by that age. If we assume random mating, the genotypic frequencies in newborns should be identical to the expected Hardy–Weinberg genotypic frequencies. Thus, the ratios of the observed adult genotypic frequencies to the expected Hardy–Weinberg frequencies will be an estimate of the relative survival efficiency of each hemoglobin genotype. These estimates can then be converted to relative fitnesses by dividing them by the largest value of the survival efficiency (Table 21.8).

The fact that individuals heterozygous for the *A* and *S* alleles are more resistant to death caused by malarial infection than *AA* homozygotes is balanced by the fact that *SS* homozygous individuals are likely to die at an early age because of severe anemia. Because *AS* heterozygotes have a net selective advantage under these conditions, balancing selection maintains the *A* and *S* alleles in populations in which malaria is a life-threatening disease (as in Population IV in Figure 21.16).

Gene Frequency Changes in Natural Populations Are the Result of a Combination of Processes.

We have considered a number of processes that are capable of changing the frequency of genes and/or genotypes over evolutionary time. For convenience, we have considered each one of these processes in isolation. However, in reality, they all may be operating simultaneously to bring about evolutionary change over time. In any given population, the influence of one or a few processes may predominate. For example, the effect of random genetic drift is always pronounced in small populations, whereas in very large populations, the effect of drift may be relatively slight. Often, the effect of natural selection is pronounced in populations subjected to significant environmental change. For example, insect populations being treated with insecticides are likely to exhibit selective increases in insecticide-resistant alleles. Thus, one of the major challenges to experimental population geneticists is to discern the relative contributions of factors such as selection, drift, and migration to changes in the genetic character of natural populations over evolutionary time. (See CONCEPT CLOSE-UP: *The Neutralist–Selectionist Debate.*)

SOLVED PROBLEM

Problem

Consider a situation in which an insecticide-resistant gene in the house fly *Mustica domestica*, (*R*), is dominant to the nonresistance coding allele (*r*). Suppose that a new insecticide spraying program is instituted in a community to eradicate this pest. Although the insecticide chosen has a very short half-life, it is found to be extremely effective for a period of 2 weeks. From

CONCEPT CLOSE-UP

The Neutralist–Selectionist Debate

We have considered the various processes capable of bringing about changes in the genetic composition of populations. In reality, of course, all these processes may, to a greater or lesser extent, be acting simultaneously to shape the genetic character of populations and species. A determination of the actual contribution of each of these various processes in shaping the genetic character of natural populations requires the analysis of real data, and even then there can exist differences of opinion as to which model or models best fit the data. An interesting case in point is the so-called neutralist–selectionist debate that consumed population geneticists during the latter part of the 1970s and into the 1980s. Indeed, even to this day the controversy has not been completely resolved.

The roots of the neutralist–selectionist debate can be traced back to the 1950s and two opposing world views on the way selection operates in natural populations. On the one hand were those predominately laboratory geneticists who observed that when a new mutant allele arose in laboratory stocks of *Drosophila* or other experimental species, they were almost always less fit than the wild-type allele and were quickly lost from the population. Advocates of this so-called classic point of view believed that relatively little genetic variation should be present in populations because over evolutionary time natural selection would have purged natural populations of all but the most-fit genes. On the other hand were those primarily field biologists who frequently observed that a great deal of phenotypic variation is present in most natural populations and believed that this phenotypic variation was a reflection of an underlying abundance of naturally occurring genetic variation. Advocates of this point of view believed that the most likely explanation was that heterozygotes were the most-fit genotype at many loci and that, as a consequence, genetic variation would be preserved at these loci in natural populations by balancing selection.

When the results of the protein gel electrophoresis surveys conducted in the late 1960s and early 1970s made it apparent that most natural populations contained substantial levels of genetic variation (Table 21.4), the advocates of the "balancing selection" point of view felt vindicated. This euphoria, however, was short-lived. A counterargument was very quickly proposed. It hypothesized that the variation being uncovered by the electrophoretic surveys was not being eliminated by natural selection because it was selectively neutral and thus imparted no adaptive advan-

laboratory studies, you determine that 80% of flies carrying the *R* allele survive when exposed to the insecticide, but only 10% of the *rr* homozygotes survive the same treatment. If the initial frequency of the *r* allele in the population is 0.95 before the insecticide is administered, what will be its frequency one generation (2 weeks) after insecticide treatment? (Assume that the relative sensitivities to the insecticide as determined in the laboratory are an accurate prediction as to the relative fitnesses experienced in the wild populations.)

Solution

The first step is to compute the expected genotypic frequencies prior to selection. We are told that the initial frequency of the *r* allele is 0.95 before the insecticide was administered. Because there are only two alleles segregating at this locus in the population (*R* and *r*), the frequency of the *R* allele will be 1 minus the frequency of the *r* allele, or 0.05 [$f(R) = 1 - 0.95 = 0.05$]. From the gene frequencies, we can compute the expected genotypic frequencies using the Hardy–Weinberg equilibrium equation p^2, $2pq$, q^2. Thus,

RR		*Rr*		*rr*		
$(0.05)^2$	+	$2(0.95)(0.05)$	+	$(0.95)^2$	=	1.00
0.0025	+	0.0950	+	0.9025	=	1.00

The next step is to assign relative fitness values to each genotype. We are told that in laboratory studies the *RR* and *Rr* genotypes survived eight times better than *rr* genotypes. If we assign *RR* and *Rr* genotypes a relative fitness (*W*) of 1.00 (80% out of 80%), the relative fitness of the *rr* genotypes will be 0.125 (10% out of 80%). To compute the expected frequencies of the three genotypes after one generation of selection, we need to multiply the relative frequencies by their respective fitness values. Thus, we have the following values:

RR		*Rr*		*rr*		
(0.003)(1.00)	+	(0.095)(1.00)	+	(0.903)(0.125)	=	
0.003	+	0.095	+	0.113	=	0.211

We need to normalize these values by dividing both sides of the equation by 0.211 (the sum of all the genotypic

tage to individuals carrying the variant alleles. Thus, the original debate suddenly shifted gears and pitted the intellectual descendants of the balanced school (now referred to as selectionists), who believed that natural selection was maintaining most of the genetic variation present in populations, against the descendants of the classic school (now called neutralists), who believed that most of the genetic variation segregating in populations was nothing more than "genetic noise" and of no adaptive significance.

According to the neutralists, most, if not all, of the naturally occurring genetic variability uncovered by the electrophoretic surveys was nothing more than selectively neutral allelic variants responding to the vagaries of genetic drift. The neutralists believed that these alleles were on their way to eventual fixation or extinction but, in any case, were of no adaptive evolutionary significance.

The neutral theory has represented and continues to represent a serious conceptual challenge to many of the long-held notions of population and evolutionary geneticists. If most of the genetic variation present in natural populations is indeed adaptively neutral, then where is the raw material upon which natural selection is believed to act in order to bring about adaptive change? If the neutralists are correct, the process of adaptation may depend upon a much smaller reservoir of genetic variation than previously suspected. Under such a scenario, mutation may be a relatively more important rate-limiting step in adaptive evolution than believed by most neo-Darwinians. Unfortunately, it has proven extremely difficult to acquire data that can unambiguously resolve the selectionist–neutralist debate. It is the general consensus among most contemporary population geneticists that at least some of the genetic variation that exists in natural populations is most certainly selectively neutral or nearly so, and some of it is most likely being maintained by natural selection. There remains, however, a considerable difference of opinion among population geneticists as to what proportion of naturally occurring genetic variation falls into each of these categories. We will briefly consider the neutralist–selectionist debate again in Chapter 22, when we examine the rates at which various DNA sequences have changed over long spans of evolutionary time.

frequencies must equal 1.00). Thus, the expected genotypic frequencies after one generation of selection follow.

RR		***Rr***		***rr***		
$\frac{(0.003)}{(0.211)}$	+	$\frac{(0.095)}{(0.211)}$	+	$\frac{(0.113)}{(0.211)}$	=	$\frac{(0.211)}{(0.211)}$
0.01	+	0.45	+	0.54	=	1.00

We can now calculate the $f(r)$ one generation after selection:

$$f(r) = 0.54 + (1/2)(0.45) = 0.76$$

SECTION REVIEW PROBLEM

10. Suppose that, in an insecticide-free environment, the relative survivorships of the genotypes discussed in the preceding Solved Problem are such that *rr* homozygotes have a fitness value of 1.0 and *Rr* and *RR* genotypes have a fitness value of only 0.8. If the housefly's generation time is about 2 weeks, and if the insecticide already described is administered only one time, what would you expect the frequency of the *r* allele to be 4 or 5 weeks after insecticide treatment (that is, one generation after the insecticide had degraded)?

SPECIATION HAS A GENETIC BASIS.

At the beginning of this chapter, we described biological species as the most inclusive Mendelian population. In sexually reproducing species, reproductive isolation is the essential defining characteristic of biological species. The process by which groups of organisms acquire reproductive isolation from other groups and thereby attain species status is an important area of inquiry for evolutionary geneticists.

There are two basic models of speciation, the allopatric model and sympatric model. The term **allopatric speciation** refers to populations that inhabit separate geographic locations. Thus, the allopatric model

TABLE 21.9 *Description of Reproductive Isolating Barriers*

A. Prezygotic RIBs

1. *Ecological isolation.* Organisms live in the same geographic region but occupy different habitats and thus do not come into contact.
2. *Temporal isolation.* Organisms mate (or, if plants, flower) at different times (either different seasons or at different times of the day) and thus do not come into contact.
3. *Behavioral isolation* (ethological isolation). The sexual attraction between males and females of different groups is weak or absent.
4. *Mechanical isolation.* Copulation or pollen transfer between two groups of organisms prevented by incompatibly shaped or sized genitalia or different structures of flowers.
5. *Gametic isolation.* Female and male gametes do not attract each other, or spermatozoa (or pollen) are inviable in sexual ducts of animals (or stigma of flowers).

B. Postzygotic RIBs

1. *Hybrid inviability.* Hybrid zygotes fail to develop or otherwise fail to reach sexual maturity.
2. *Hybrid sterility.* Hybrids fail to produce viable gametes.
3. *Hybrid breakdown.* The F_2 (or backcross) progeny of hybrids display reduced viability or fertility.

of speciation describes how two or more groups of organisms belonging to the same species that are geographically isolated from one another may acquire reproductive isolation. The term **sympatric speciation** refers to populations that occupy the same geographic region. The sympatric model of speciation describes how reproductive isolation may evolve among groups of organisms living in the same area.

Speciation Is the Acquisition of Reproductive Isolating Mechanisms.

The process of speciation under either model is the acquisition of **reproductive isolating barriers,** or RIBs. There are two general classifications of RIBs, prezygotic and postzygotic (Table 21.9). **Prezygotic RIBs** are mechanisms that prevent the formation of hybrid zygotes. An example of a prezygotic RIB is **behavioral isolation,** which is the failure of two or more groups of organisms to interbreed because of a lack of sexual attraction between the opposite sexes of the different groups. Another example of a prezygotic RIB is **temporal isolation,** which exists when organisms mate or fertilize at different times of the day (e.g., different flowering times in plants) or during different seasons of the year.

Postzygotic RIBs are mechanisms that reduce the viability or fertility of hybrid zygotes after they are formed. There are three different types of postzygotic RIBs (Table 21.10). An example is hybrid inviability, which exists when hybrid zygotes are inviable and fail to develop or otherwise fail to reach sexual maturity.

Speciation May Occur When Populations Are Geographically Isolated.

According to the allopatric model of speciation, the establishment of reproductive isolating barriers is a two-step process (Figure 21.17). The first step requires that gene flow between two populations of the same species be interrupted. This cessation of gene flow is usually assumed to result from some physical or geographic barrier, such as a mountain range or significant stretch of water. The proviso that gene flow be interrupted is important because it ensures that any genetic differences that arise within the separated populations will not spread and thus homogenize throughout the species range. It is further envisioned that, during the period these populations are genetically isolated, genetic differences will accumulate between them. These differences may be the gradual consequence of natural selection genetically adapting the isolated populations to their particular environments, or they may arise perhaps more rapidly because of the chance fixation of alleles, especially if one or the other of the newly isolated populations were established by relatively few individuals. It is postulated that the genetic differences that arise between the two isolated populations during this stage of the process are sufficient to result in the establishment of postzygotic RIBs that become manifest should members of the two populations ever again come into contact and attempt to mate.

Step two in the isolating process is predicted to occur if and when the physical barriers that initially isolated the populations are removed and the two popula-

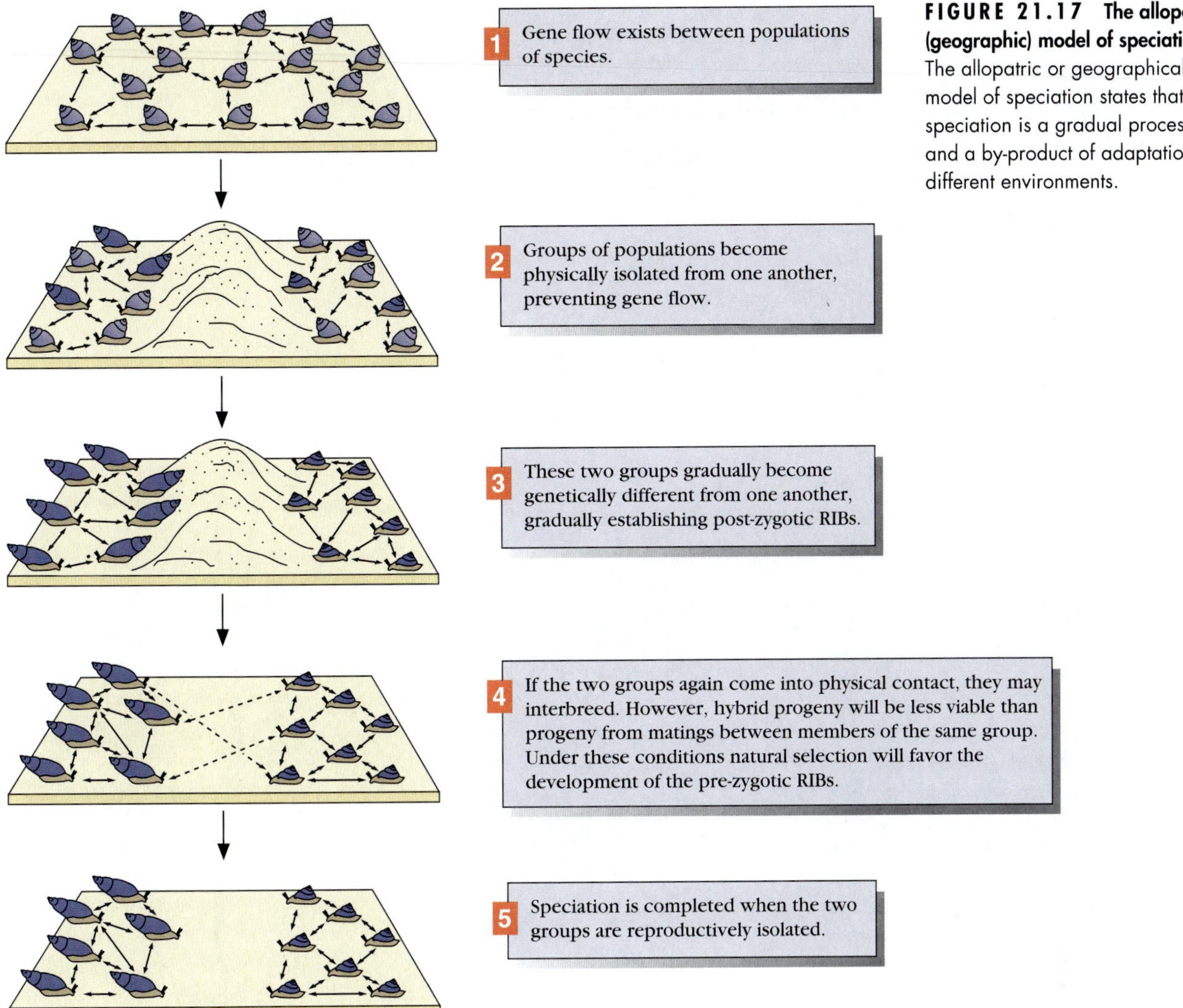

1 Gene flow exists between populations of species.
2 Groups of populations become physically isolated from one another, preventing gene flow.
3 These two groups gradually become genetically different from one another, gradually establishing post-zygotic RIBs.
4 If the two groups again come into physical contact, they may interbreed. However, hybrid progeny will be less viable than progeny from matings between members of the same group. Under these conditions natural selection will favor the development of the pre-zygotic RIBs.
5 Speciation is completed when the two groups are reproductively isolated.

FIGURE 21.17 The allopatric (geographic) model of speciation. The allopatric or geographical model of speciation states that speciation is a gradual process and a by-product of adaptation to different environments.

tions are once again brought into physical contact. As stated earlier, the genetic differences between the populations are now believed to be sufficient to result in a significant reduction in the number of progeny resulting from hybrid matings. This situation is believed to favor those genotypes not inclined to engage in cross-population matings, which in turn should favor the establishment of postzygotic RIBs.

Speciation May Occur by Sudden Genetic Changes.

According to the sympatric model, reproductive isolation can sometimes occur among groups of organisms without the groups being physically isolated from one another. The cessation of gene flow among the speciating groups is envisioned to occur relatively rapidly and while the two groups occupy the same geographic region. Perhaps the most common form of sympatric speciation is polyploidy, that is, the sudden amplification of entire chromosome complements (see Chapter 18). Polyploidy, which has occurred relatively frequently over the evolutionary history of many plant species, can arise in one or a few generations. The cytological differences between the polyploid individuals and the progenitor species result in an immediate postzygotic RIB and thus the establishment of the new species. Several examples of polyploid species were described in Chapter 18.

Other chromosomal aberrations, such as translocations, may suddenly arise within members of a species and reduce the fertility of progeny produced from all matings except those occurring between individuals carrying the same chromosomal aberration. Such a situation would favor the establishment of prezygotic RIBs between individuals carrying the new chromosomal aberration and other members of the population (see *Current*

CURRENT INVESTIGATIONS

Clarkia biloba and *Clarkia lingulata:* Two Plant Species That Arose by Sympatric Speciation

Two plant species that appear to have evolved through sympatric speciation are *Clarkia biloba* and *Clarkia lingulata* (Figure 21.B). *C. biloba* is relatively widespread throughout northern California, and *C. lingulata* has only been found in two sites in the central Sierra Nevada at the periphery of *C. biloba's* southern range. Morphologically, the two species are quite similar in appearance, differing only in petal shape. Chromosomally, however, the two species differ by a translocation, several pericentric inversions, and the presence of an additional chromosome in *C. lingulata* that was derived by the fusion of parts of two chromosomes present in *biloba*. Although both *C. biloba* and *C. lingulata* are normally outcrossers, they are also capable of self-fertilization.

It is believed that *lingulata* arose from a series of chromosomal aberrations in *biloba* that were initially increased in frequency in the population by means of self-fertilization. After this founding population was established, outcrossing to the normal *C. biloba* plants would have been less productive than crosses between plants of the same aberrant genotype. This is believed to have led to the eventual establishment of prezygotic RIBs and the establishment of the two distinct species that we observe today.

FIGURE 21.B The annual plant species *Clarkia lingulata* is believed to have evolved from its progenitor species *Clarkia biloba* sympatrically. *C. biloba* has bilobed petals and *C. lingulata* does not.

INVESTIGATIONS: Clarkia biloba *and* Clarkia lingulata: *Two Plant Species That Arose by Sympatric Speciation*).

Many Questions Remain Concerning the Genetic Basis of Speciation.

The allopatric and sympatric models of speciation are not mutually exclusive; it seems clear that even though some species arise in allopatry, other species (especially certain plant species) arise in sympatry. What remains poorly understood at the present time, however, is what genetic changes are necessary and sufficient for the establishment of pre- and postzygotic RIBs and what conditions, if any, may increase the likelihood that these critical genetic changes will take place.

Some evolutionary geneticists believe that the initial genetic changes necessary to establish postzygotic RIBs arise as a by-product of the genetic changes brought about by natural selection to better adapt the population to its environment. Secondary contact is believed to result in natural selection favoring the establishment of prezygotic RIBs. Advocates of this point of view see natural selection as playing a major role at each stage in the speciation process. Other evolutionary geneticists believe that speciation may have little to do with natural selection. They believe that the formation of RIBs is triggered by chance events that may have a higher probability of occurring in small, isolated populations in which rapid genetic changes caused by genetic drift and inbreeding are expected to be high. Continuing research into these issues may help resolve some of the remaining questions over the next decade.

Summary

Genetic variation is the raw material for evolutionary change. The sum of the genetic variation present in a population or species is referred to as the gene pool. A variety of techniques, ranging from simple visual inspection to protein gel electrophoresis, RFLP analysis, and DNA sequencing, may be used to estimate the level of variability in a population or species. The amount of genetic variation in a population may be expressed as the percentage of polymorphic loci or average heterozygosity (*H*). Extensive surveys conducted in a great number of species over the past several decades indicate that most populations harbor substantial levels of genetic variability.

Several factors may combine to shape the quantity and character of genetic variability present in populations. If mating is random among members of a population, the genotypic frequencies will maintain a stable equilibrium frequency that will be determined by the frequency of the genes in the population. This fact was demonstrated independently by the English mathematician Godfrey Hardy and the German physician Wilhelm Weinberg and is thus referred to as the Hardy–Weinberg principle. If mating within a population is not random, genotypic frequencies may deviate significantly from Hardy–Weinberg expectations, but allele frequencies will be unaffected. Assortative mating refers to the preferential mating among individuals having the same (positive) or different (negative) phenotypes. Inbreeding is the most common form of nonrandom mating and occurs when matings between relatives are more frequent than is expected if couples were selected from the population randomly. Selfing is the most extreme form of inbreeding and can lead to a population that is effectively devoid of heterozygotes within very few generations.

Migration or gene flow occurs when individuals move from one population to another and interbreed. The effect of migration on the gene frequency of a population depends upon how genetically different the donor population is from the recipient population. Mutation is the ultimate source of all genetic variation and thus evolutionary change. Newly arising alleles that impart no net advantage or disadvantage to their carriers are called neutral alleles. Random fluctuations in the frequency of neutral alleles in a population is called genetic drift. Genetic drift is most pronounced in small populations. Natural selection is the differential reproductive output of alternative genotypes in a population. Alleles that impart an adaptive disadvantage to their carriers will be eliminated from the population by natural selection. Alleles that impart an adaptive advantage to their carriers will increase in frequency within the population. At some loci, heterozygotes are the most fit genotype in the population. In such cases, genetic variability is maintained at the locus by balancing selection. Some population geneticists, called selectionists, believe that most of the genetic variability present in natural populations is being maintained by balancing selection. Other population geneticists, called neutralists, believe that most of the genetic variability in populations is nothing more than the random drifting of neutral alleles over time.

Biological species are defined as groups of interbreeding populations that are reproductively isolated from other such groups. There are two basic models that describe how new species arise. According to the allopatric model, speciation begins with geographic isolation between groups of organisms belonging to the same species. Over time, these isolated groups become genetically different to such an extent that, when they resume physical contact, they produce inferior hybrids. In time, such a situation is believed to favor the establishment of complete reproductive isolation by natural selection and thus the emergence of new biological species. According to the sympatric model, reproductive isolation can arise very quickly among members of a species living in the same geographic area. Examples are the numerous polyploid species of plants. The allopatric and sympatric models of speciation are not mutually exclusive. Many aspects of the genetic mechanisms that underlie the speciation process remain to be determined.

Selected Key Terms

allopatric speciation
average heterozygosity
biological species concept
evolutionary genetics
fitness
founder effect
gene pool
genetic bottleneck
genetic drift
Hardy–Weinberg principle
inbreeding
inbreeding coefficient (F)
natural selection
negative assortative mating
population genetics
positive assortative mating
reproductive isolating barriers
selection coefficient
sympatric speciation
typological species concept

Chapter Review Problems

1. The following numbers of electrophoretic genotypes (1 and 2 represent two different alleles) were found at the following three enzyme-encoding loci in a human population. Calculate the genotypic and allelic frequencies at each locus.

	Genotypes (numbers)		
Locus	**1/1**	**1/2**	**2/2**
Mdh	620	380	0
Adh	120	250	230
Pgm	250	500	250

2. Assuming random mating, calculate the expected number of heterozygotes at the *Mdh* locus from the data presented in Problem 1. Use the chi-square test to determine whether the observed and expected numbers of individuals are significantly different.

3. The most common form of hemophilia is caused by a sex-linked recessive allele. If the hemophilia allele is in a frequency of 0.001 in a population, what are the expected genotypic frequencies at this locus for males and females in the population?

4. Tay–Sachs disease is caused by an autosomal recessive allele. About 1 per 100,000 newborns are homozygous for this recessive allele and usually die by the age of 4 years. Assuming that this locus is in Hardy–Weinberg equilibrium, compute the frequency of the Tay–Sachs allele in humans and the frequency of heterozygotes.

5. Why are estimates of the levels of genetic variation in populations based on electrophoretic surveys likely to be underestimates of the actual levels of genetic variation present?

6. DNA sequence surveys of the genetic variation present in *Drosophila* populations indicate that introns are more genetically variable than exons. Hypothesize why this is the case.

7. Companies that sell seed to farmers are always trying to breed for superior strains (for example, strains that give higher yield or more disease resistance). Suppose you are hired to produce homozygous strains of corn from a plant heterozygous at the *A* locus (A_1A_2). What will be the genotypic frequencies after three, five, and seven generations of repeated self-fertilization (inbreeding)?

8. Assume that, in a population of birds, two alleles exist for tail color: red, which is recessive, and blue, which is dominant. Out of a total population of 1000 birds, 640 have red tails, and 360 have blue tails. If the population is in equilibrium, what is the frequency of the red alleles? a. 0.8 b. 0.64 c. 0.36 d. 0.2 e. 0.4

9. What is the frequency of the blue allele in Problem 8? a. 0.8 b. 0.64 c. 0.36 d. 0.2 e. 0.4

10. In Problem 8, what percentage of the birds are heterozygous for the tail color gene? a. 64% b 32% c. 48% d. 4% e. 46%

11. In Problem 8, what percentage of the birds is homozygous for the dominant allele? a. 64% b. 32% c. 48% d. 4% e. 46%

12. Which of the following, when acting alone, does not have any significant effect on the allele frequency in a population? a. selection and mutation b. mutation c. inbreeding d. migration e. all the above

13. Mutation as a force in genetic diversity serves to
    a. increase genetic diversity
    b. rapidly change gene frequencies
    c. maintain a genetic equilibrium
    d. work against selection
    e. none of the above

14. The trend toward increased mobility of the world's population will probably lead to
    a. increased gene frequency differences between populations
    b. decreased gene frequency differences between populations
    c. a decrease in mutation rate
    d. no change in gene frequency between populations because of selection
    e. all the above

15. In a population of 1000 golfers, a genetic difference exists with respect to playing the game. The allele that governs golfing ability is incompletely dominant. In this population, 60 were excellent golfers (*gg*), 440 were average golfers (*Gg*), and the remaining 500 were true hackers (*GG*) and were destined to remain that way. What is the frequency of the g allele in the population?

Challenge Problems

1. Based on electrophoretic surveys, mammalian species have been found to have about half as much genetic variation as invertebrates. Come up with an hypothesis to explain this finding. (Hint: Invertebrate population sizes are usually quite large.)
2. Propose an experimental program to help resolve the issue of whether or not the majority of the genetic variation present in natural populations is selectively neutral.
3. Two related picture-wing species of Hawaiian *Drosophila, D. silvestris* and *D. heteroneura*, are morphologically quite distinct but essentially identical with regard to the frequencies of alleles encoding electrophoretically detectable proteins. Is this fact most consistent with a gradual or rapid mode of speciation? Why?

Suggested Readings

Ayala, F., and J. Valentine. 1979. *Evolving: The Theory and Processes of Organic Evolution.* Benjamin/Cummings, Menlo Park, CA.

Dobzhansky, T., F. Ayala, L. Stebbins, and J. Valentine. 1977. *Evolution.* Freeman, San Francisco, CA.

Futuyma, D. 1997. *Evolutionary Biology.* Sinauer, Sunderland, Massachusetts.

Kimura, M., and J. Crow. 1964. "The number of alleles that can be maintained in a finite population." *Genetics*, 49: 725–738.

Kreitman, M. 1983. "Nucleotide polymorphism at the alcohol dehydrogenase locus of *Drosophila melanogaster.*" *Nature*, 304: 412–417.

Lewontin, R. 1974. *The Genetic Basis of Evolutionary Change.* Columbia University Press, New York.

On the Web

Visit our Web site at **http://www.saunderscollege.com/lifesci/titles.html** and click on A/G/M Genetics for links to the following chapter-related resources on the World Wide Web:

1. **Links to evolution.** This site lists sites related to evolution and should lead you to almost any subject of interest.
2. **Evolution.** This site links you to other Web resources in molecular evolution and systematics.

CHAPTER

22

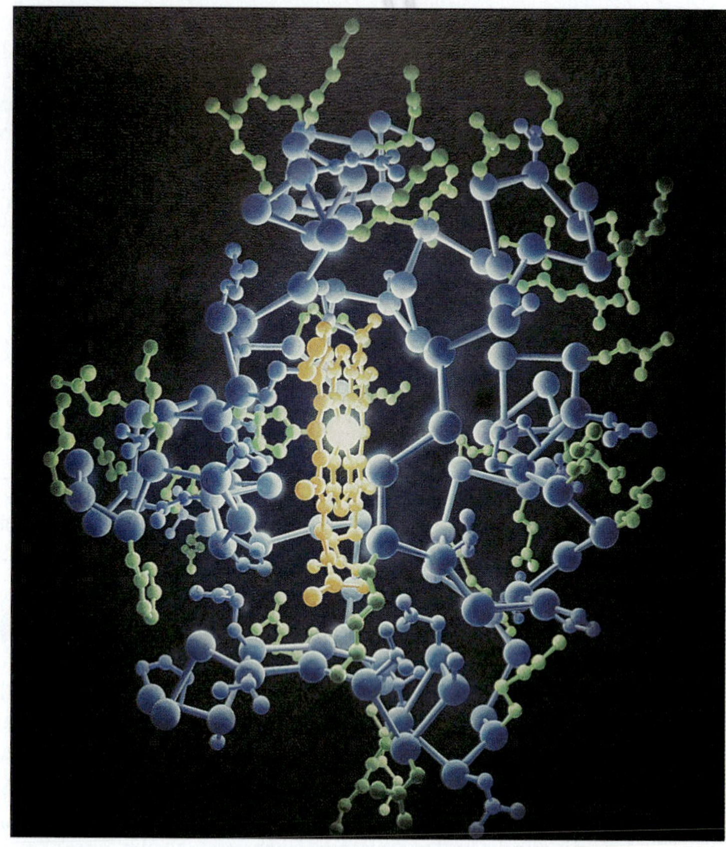

Cytochrome-c is a slowly evolving molecule that is used by molecular evolutionists to estimate rates of evolutionary change among a variety of invertebrate and vertebrate species. *(Irving Geis, NYC)*

Molecular Evolution

We began our study of the evolutionary process with a survey of the mechanisms that shape the genetic character of populations and species over time. We will continue our examination of the evolutionary process by focusing our attention primarily on those changes that have occurred in genomes over the last several hundred million years. Comparative studies of DNA structure and function in various genomes have provided considerable insight into the kinds of molecular changes that have occurred. These changes have accompanied the emergence of the vast array of phenotypic diversity that we associate with eukaryotic species. We will see that some of the evolutionary changes that have occurred on the molecular level seem to have been adaptively neutral, whereas others have clearly played important roles in the acquisition of new and significant molecular structures and functions.

Molecular evolution is a very broad area of investigation that includes the study of changes in genomic sequence, structure, and function over evolutionary time. This study area also examines the consequences of these changes on morphology, physiology, behavior, and other aspects of organismic evolution. The field of molecular evolution encompasses such diverse topics as the origin of self-replicating molecules, the subsequent emergence of primitive organisms, and the evolutionary history of particular eukaryotic genes. Clearly, such an array of subjects cannot be covered in a single introductory chapter. Thus, in this chapter, we limit our discussion of molecular evolution specifically to those aspects of genome evolution directly related to topics covered elsewhere in this text. We concentrate on three areas of current research interest: the evolution of multiple gene families, the evolution of introns, and DNA sequence evolution. By the end of this chapter, we will have addressed the following three questions:

1. What are gene families, and how did they evolve?
2. How did introns evolve, and what is their evolutionary significance?
3. What are the relative roles of genetic drift and natural selection in the DNA sequence changes that have occurred over evolutionary time?

MULTIGENE FAMILIES HAVE ARISEN OVER EVOLUTIONARY TIME.

Eukaryotic genomes contain related groups of genes. These related gene groups, consisting of two or more genes with similar or identical DNA sequences, are called **gene families.** Gene families, such as the genes encoding rRNA or the histone proteins, have descended by duplication and divergence from common ancestral genes. The DNA sequence similarity within a family can range from identical, or nearly identical, to quite different. In fact, within a family, some sequences may have as little as 50% DNA sequence identity yet still be similar enough to have clearly evolved from a common ancestral gene. Most members of a gene family are clustered in close chromosomal proximity to one another; however, some are located on different chromosomes. These dispersed gene family members are presumed to have been translocated to their different locations subsequent to, or possibly during, the process of gene duplication.

Generally speaking, members of a gene family have the same or related functions. For example, all the members of the mammalian hemoglobin gene family encode proteins whose job is carrying oxygen. However, even when members of a gene family have the same basic function, they are not always expressed at the same time during development. Different members may be expressed at different life stages and/or in different tissues, reflecting the fact that evolutionary divergence has occurred at the level of gene regulation. For example, some of the members of the mammalian hemoglobin gene family are expressed in adults, whereas others are expressed only at the fetal stage of development.

Given sufficient time to evolve, the DNA sequences of some members of a gene family may diverge to the point that the encoded protein acquires a new function. For example, the lactalbumin gene, which encodes a subunit of the enzyme that catalyzes the synthesis of lactose, is in the same family as the gene-encoding lysozyme, an enzyme that degrades the polysaccharide component of certain bacterial cell walls. These two enzymes do have a functional commonality, however; they both act on carbohydrates.

Finally, some members of gene families are not transcribed or produce transcripts that are not properly processed and translated into functional proteins. In a number of instances, these nonfunctional gene family members have been found to lack the necessary promoter sequences and one or more of the introns that are characteristic of active members of the gene family. These nonfunctional gene family members are called **pseudogenes.**

Gene Families Arise by Gene Duplication.

In its most general sense, DNA duplication refers to an increase in the number of copies of a segment of DNA (see Chapters 17 and 18). Depending upon the extent of the genomic region involved, DNA segment increases may result in duplication of an entire gene (gene duplication), part of a gene (partial gene duplication), an entire chromosome (aneuploidy), or an entire genome (polyploidy). Gene families are believed to arise from a succession of gene duplications.

As discussed in Chapters 17 and 18, **unequal crossing over** is one of the primary mechanisms of gene duplication. Unequal crossing over can occur when two adjacent regions of a chromosome contain similar DNA sequences. During meiosis, mispairing between the homologous chromosomal regions may occasionally occur. If genetic exchange or recombination occurs within these regions of mispairing, nonreciprocal chromosomal products will result (Figure 22.1). Depending upon where the regions of sequence identity lie along the chromosome, unequal crossing over may result in duplication of all or part of a gene (Figure 17.11 also shows unequal crossing over and its resultant products).

Unequal crossing over can generate deletions as well as duplications. For example, **thalassemia** is a disease resulting from an inability to make functional α- or β-globins. Many of the mutations that cause thalassemia are believed to be caused by unequal crossing over among regions of the α- or β-globin gene family. For example, the mutant allele resulting in one of the α-thalassemias (*α-thal-2L*) is missing 4.2 kb of DNA, including the entire *α2* gene. The endpoints of this deleted stretch of DNA lie within regions of the α-globin cluster, which contains homologous sequences (Figure 22.2). It is therefore most likely that the *α-thal-2L* allele is one of the products of an unequal crossover event that occurred within these regions of sequence homology. The fact that a large number of thalassemia alleles are products of

FIGURE 22.1 Unequal crossing over results in gene duplication. If two homologous chromosomes mispair, recombination will result in one chromosome having two copies of the gene (duplication) and the other having none (deletion). Unequal crossing over usually occurs when two adjacent DNA sequences are identical or nearly identical.

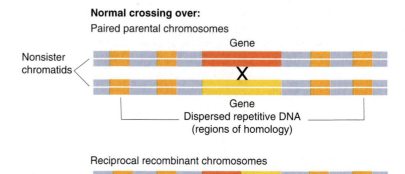

Normal crossing over:

Paired parental chromosomes

Gene

Nonsister chromatids

X

Gene

Dispersed repetitive DNA (regions of homology)

Reciprocal recombinant chromosomes

+

Unequal crossing over:

Mispaired parental chromosomes

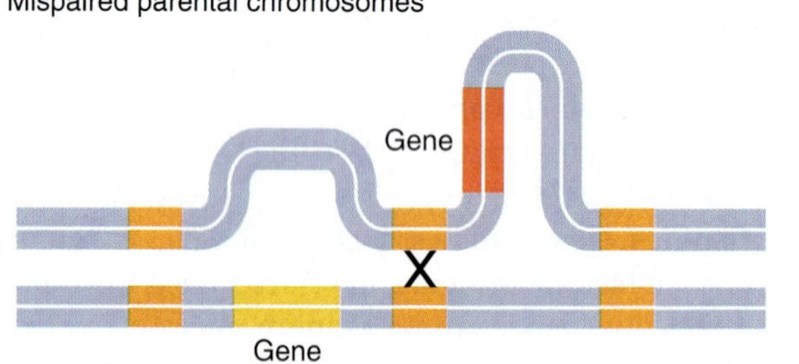

Nonreciprocal recombinant chromosomes

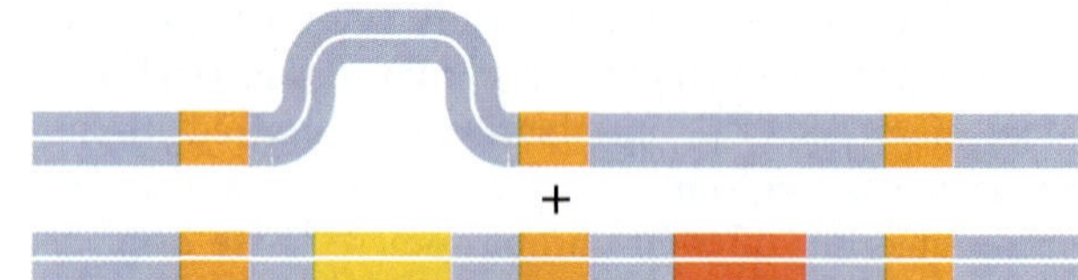

unequal crossover implies that unequal crossing over is not an uncommon phenomenon. This lends credence to the hypothesis that unequal crossing over plays a significant role in gene duplication and thus in the evolution of **multigene families.**

The Globin Gene Family Evolved by Gene Duplications.

The mammalian hemoglobin gene family consists of eight active globin genes encoding hemoglobin proteins that, on the basis of relative degree of sequence homology, can be subdivided into two clusters, the α-globin cluster and the β-globin cluster. The α-globin cluster consists of the *ζ-globin* gene, which is expressed during early embryogenesis, and two nearly identical α-globin genes, $\alpha 1$ and $\alpha 2$, which are expressed in the fetus and adult life stages, respectively. The θ gene has only recently been discovered, and its protein product has yet to be identified. Thus, at this point, it is not known whether or not the θ gene is functionally active. The β-globin cluster consists of a single embryonic *ϵ-globin* gene, two nearly identical fetal globin genes ($G\gamma$ and $A\gamma$), and a minor adult *δ-globin* in addition to the major adult *β-globin* gene.

α-globin gene cluster:

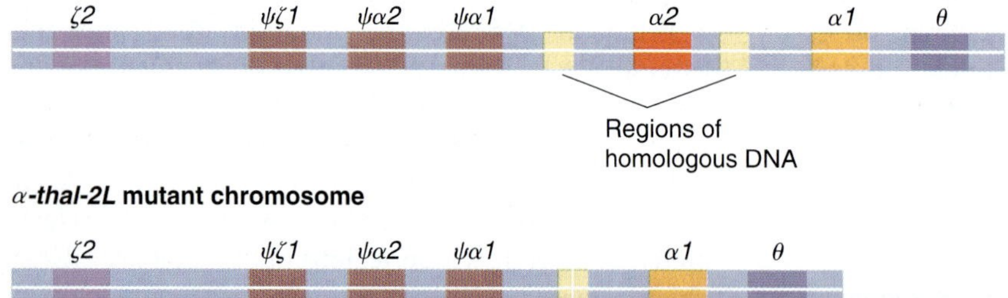

FIGURE 22.2 Deletion of the α2 gene in humans. The *α-thal-2L* deleted chromosome is most likely the product of unequal crossing over because the ends of the deletion contain sequences homologous to sequences flanking the *α2* gene. Pairing among these sequences could have resulted in an unequal crossover event, as depicted in Figure 22.1. This resulted in the *α-thal-2L* chromosome, which lacks 4.2 kb of DNA, including the entire *α2* gene.

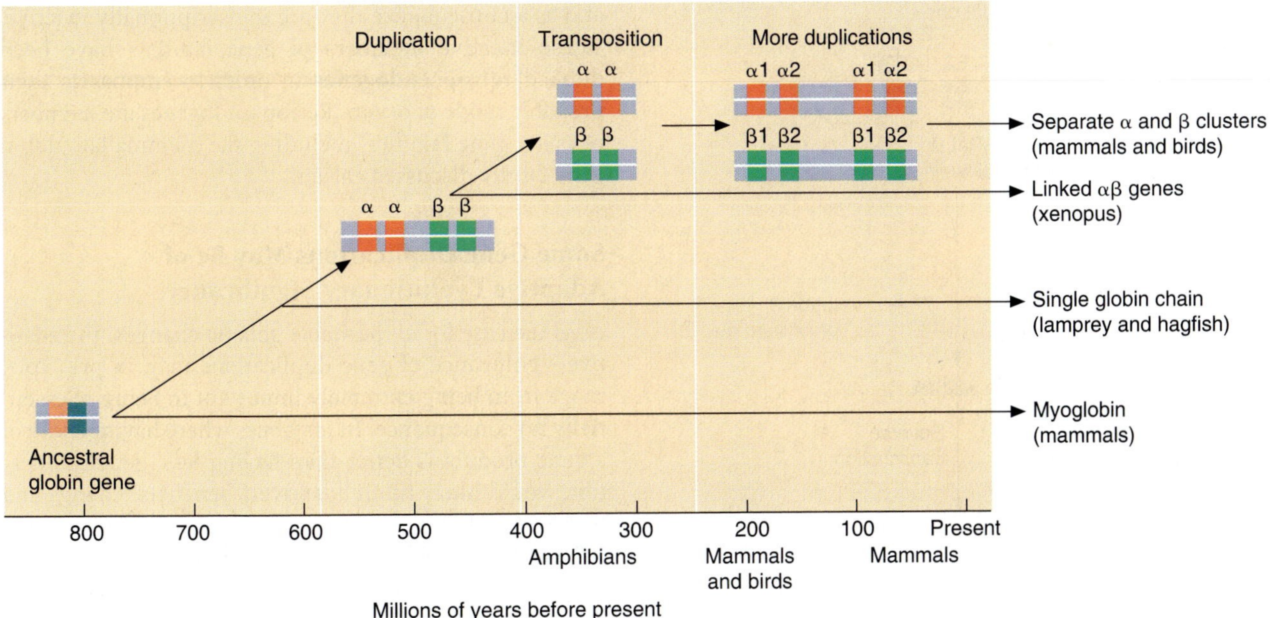

FIGURE 22.3 Proposed evolution of the mammalian globin gene family. The mammalian globin gene family appears to have evolved by a series of duplications, transpositions, and point mutations from a single ancestral gene over the past 400–500 million years.

Although all the members of the mammalian globin gene family display a significant degree of sequence homology, some of the members are nearly identical, whereas others are more divergent. These variable degrees of sequence identity have been used by molecular evolutionists to establish the most likely series of duplication events that have transpired over evolutionary time to result in the current family of genes. From these analyses, it is estimated that an initial duplication of an ancestral globin gene occurred about 600 million years ago, establishing the precursors of the α and β clusters. This ancestral gene may have been related to the myoglobin gene (mammals) or the leghemoglobin gene found in plants. The leghemoglobin gene has one more intron than the myoglobin gene; otherwise, it is nearly identical to the myoglobin gene. Our current complement of mammalian hemoglobin genes was attained by a series of duplications of this ancestral gene, followed by transpositions and mutations within each cluster over the last 400 to 500 million years (Figure 22.3).

Some Duplicated Genes Are the Product of Reverse Transcription.

The hormone insulin is derived by a series of processing reactions from a precursor protein called preproinsulin. Although most vertebrates have a single preproinsulin gene, rats, mice, and a few fish species (toadfish, bonito, and tuna) have been shown to carry two copies of the gene (designated I and II). The human, rat II, and chicken preproinsulin genes contain two introns, whereas the rat I gene contains only one intron. Because the majority of preproinsulin genes contains two introns, it seems reasonable to conclude that two introns represent the ancestral condition among this group of genes. It is thus most likely that during or shortly after the duplication of the rat gene, one of the gene copies lost one of its introns. What is particularly remarkable about this duplication event, however, is that the intron was precisely excised, leaving the coding region of the gene intact. Such a precise excision of the intron argues against the hypothesis that the intron was lost as a result of unequal crossing over or some other imprecise recombination-like event. The most likely scenario is that the intron was precisely cut out by some enzymatic mechanism analogous to, if not identical to, that involved in intron processing (see Chapter 11). According to this hypothesis, the intron was processed out of the rat II gene's nascent transcript followed by reverse transcription of this partially processed transcript and reinsertion of the DNA product (the rat I gene) back into the genome (Figure 22.4). Replicate genes that are believed to have been generated via reverse transcription of partially or completely processed progenitor gene transcripts are called **retrogenes.**

Sequence analysis has demonstrated that the rat II gene transcript that served as the template for the rat I retrogene was initiated upstream of the rat II gene's

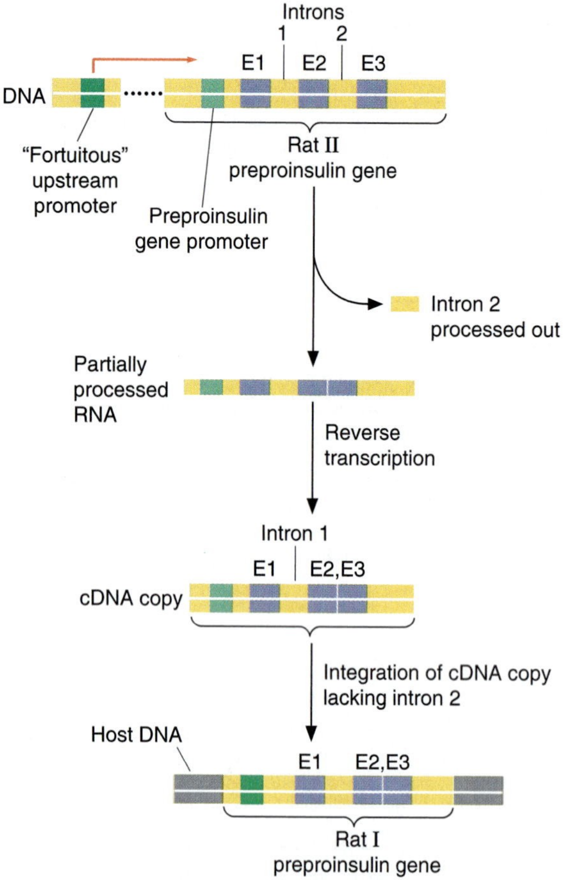

FIGURE 22.4 Hypothetical origin of the rat I preproinsulin gene. It is very likely that the rat I preproinsulin gene is the product of reverse transcription of a partially processed rat II preproinsulin transcript that was initiated from a "fortuitous" promoter (dark green) located upstream of the normal preproinsulin promoter (light green). The fortuitous promoter is necessary; without it, the new gene would not possess its own promoter because promoters are not present in transcribed RNA. The bottom part of this figure illustrates the structure of the new gene after it has reintegrated into the genome.

normal transcriptional initiation site. The apparent explanation of this unusual occurrence was the presence of an additional transcriptional initiation sequence (a "fortuitous" promoter) located upstream of the rat II gene's normal promoter (Figure 22.4). Thanks to this fortuitous circumstance, the template RNA that was reverse-transcribed to produce the rat I retrogene contained the promoter sequences necessary for it to be transcriptionally active. In most instances, however, retrogenes would not be expected to contain promoter sequences and thus would not be transcribed unless inserted by chance downstream of an active promoter. Consistent with this expectation, most retrogenes do in fact lack promoters, and as a consequence they are transcriptionally inactive. These inactive members of gene families have been dubbed **retropseudogenes** in order to emphasize their probable mode of origin. Retropseudogenes are members of many gene families, including the mammalian globin gene family discussed earlier.

Some Gene Duplications May Be of Adaptive Evolutionary Significance.

As is the case for all heritable genetic changes, the adaptive significance of gene duplications may, in principle, range from being extremely important to being of essentially no consequence. In instances where having more of a gene product is better than having less, gene duplication would likely be an adaptively beneficial change and thus be favored by natural selection. For example, when eukaryotic chromosomes replicate, large quantities of the DNA-binding histone proteins need to be rapidly produced. It is thus not surprising that multiple copies of histone-encoding genes are characteristic of all eukaryotic genomes. Other examples of multicopy genes that produce large volumes of essential gene products are the ribosomal RNA (rRNA) and transfer RNA (tRNA) encoding genes.

In many cases, however, having more of a particular gene product is likely to be of little or no adaptive advantage. For example, we have seen that some species have two copies of the preproinsulin gene, whereas others have only one. There appears, however, to be no significant advantage or disadvantage associated with having excess levels of insulin. In such instances, gene duplications are most likely adaptively neutral or nearly neutral events, and the ultimate fate of the duplicate gene within the species will be subject to random genetic drift (see Chapter 21).

Interestingly, even gene duplications that are adaptively neutral at their outset may acquire adaptive significance over evolutionary time. Because only one of the duplicate copies of a gene is required to maintain the gene's established function, any additional copies are free to accumulate new and possibly adaptive mutations. For this reason, many molecular evolutionists believe that gene duplication is a critically important process in the evolution of new gene functions. It is impossible to imagine, for example, that the functionally and developmentally diverse array of vertebrate globins could have evolved without the process of gene duplication.

SOLVED PROBLEMS

Problem

What are the possible adaptive advantages of multigene families?

Solution

A gene duplication could be of selective advantage if more of the gene's product would be of functional benefit to the organism. For example, the number of ribosomes that can be produced in a cell is limited by the levels of rRNA present. One way to ensure rapid and abundant synthesis of rRNA is to have multiple copies of rRNA genes. In fact, rRNA genes are repeated in the genome presumably for this very reason. Other examples of multicopy gene families providing the cell with a needed abundance of gene product are the histone gene family and the tRNA gene family.

Problem

Describe a situation in which more than one copy of a gene might be of a selective disadvantage.

Solution

Situations in which significant changes in the levels of regulatory proteins may have dramatic effects on patterns of gene expression. For example, if a repressor protein were suddenly made twice as abundant in a cell because of a duplication event, it might lead to a reduced expression level of genes critical to normal cellular function.

SECTION REVIEW PROBLEMS

1. What are the two major classes of pseudogenes, and how might they have arisen over evolutionary time?
2. What are gene families and how do they evolve?

THE EVOLUTIONARY ORIGIN OF INTRONS IS CONTROVERSIAL.

Introns are another feature of gene and genome structure of active interest to molecular evolutionists. As you learned in Chapter 11, eukaryotic genes are typically discontinuous in structure, consisting of alternate stretches of coding regions, called exons, and noncoding regions, called introns. Because nearly all prokaryotic genes that have been studied lack introns, there are two possible hypotheses for intron origin: (1) the ancestral protein-coding genes arose initially as interrupted structures, or (2) the ancestral protein-coding genes were initially uninterrupted sequences of DNA into which introns were subsequently inserted. In other words, "Which came first, the intron or the gene?" As is the case with many evolutionary questions, the answer to this one does not appear to be straightforward and may, in fact, be a combination of these two hypotheses.

Some Introns May Be Older Than the Genes in Which They Are Contained.

In 1977, the Nobel Prize–winning molecular biologist Walter Gilbert postulated that genes in eukaryotic cells arose as collections of exons brought together by recombination within intron sequences. According to Gilbert's hypothesis, introns provide large stretches of DNA in which recombination events may occur while preserving a gene's reading frame and thus its encoded function. In this way, sequences encoding previously evolved protein functions could be juxtaposed to sequences encoding other functions in novel combinations that would presumably accelerate the rate at which complex proteins evolve (Figure 22.5).

Since it was proposed in the 1970s, Gilbert's hypothesis has been supported by at least two important findings.

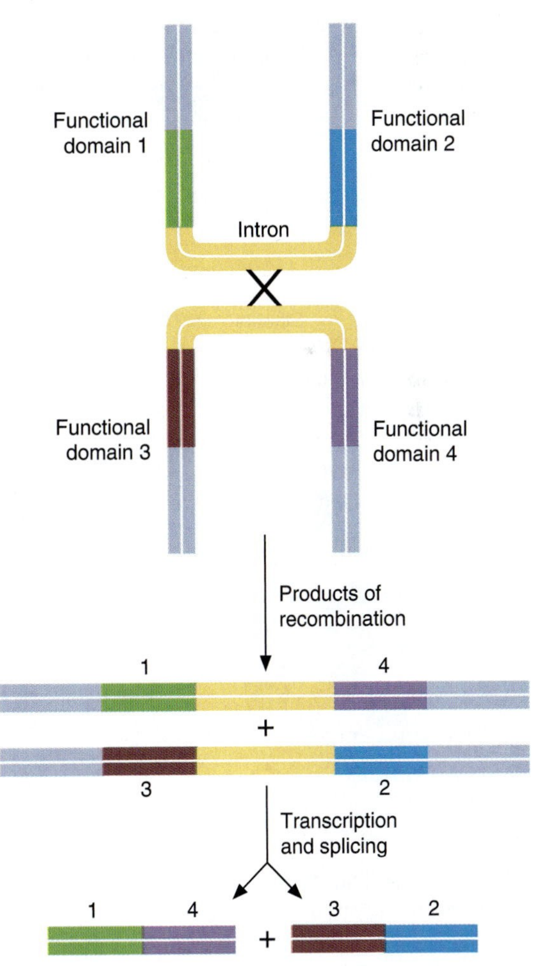

FIGURE 22.5 The exon shuffling hypothesis. According to the exon shuffling hypothesis, introns allow recombination between previously evolved functional domains, accelerating the rate of protein evolution. This can result in new combinations of protein domains yielding new enzyme functions.

First, in a number of genes, introns have been found to separate regions of DNA sequences that encode distinct functional and/or structural domains. A **domain** is a region of a complex protein's sequence that can be associated with a particular structure or function. For example, the enzyme alcohol dehydrogenase catalyzes the removal of a hydrogen atom from an alcohol's –OH group and facilitates its passage to the cofactor NAD^+. The active site of alcohol dehydrogenase consists of an alcohol-binding domain as well as an NAD^+-binding domain (Figure 22.6). The observation that, for at least some complex proteins, the encoding domains of the DNA sequence are separated from one another by introns is precisely the structure one would predict if introns were significant factors in allowing pre-evolved domains to become juxtaposed by recombination events over evolutionary time.

The second body of observations that has been offered in support of Gilbert's hypothesis is that in many cases in which two closely related genes contain variable numbers of introns, the common ancestral genes have been found to contain at least as many introns (and in the same positions) as the descendent gene. Collectively, these observations support the contention that, for at least some gene families, the presence of introns is a very old state of affairs and intron loss may be a common evolutionary phenomenon.

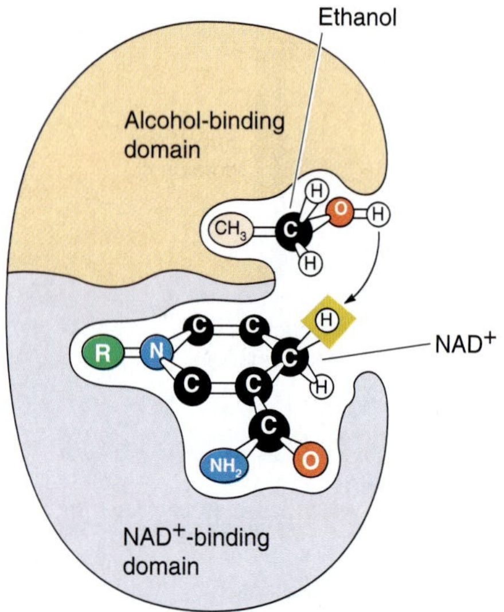

FIGURE 22.6 The three-dimensional structure of alcohol dehydrogenase. The enzyme alcohol dehydrogenase consists of an alcohol-binding domain and an NAD^+-binding domain. The NAD^+-binding domain in alcohol dehydrogenase is very similar in sequence and structure to the NAD^+-binding domains in other dehydrogenases (for example, malate dehydrogenase). It is likely that the various dehydrogenases evolved by recombination between a gene encoding a primitive NAD^+-binding protein with different genes encoding different substrate binding proteins (for example, an alcohol-binding protein).

Some Introns May Be Younger Than the Genes in Which They Are Contained.

Despite the fact that introns are apparently ancient components of some eukaryotic genes, there is emerging evidence that, in a number of gene families, introns have been acquired relatively recently in evolutionary history. For example, the serine protease gene family consists of several related genes that can be grouped according to their DNA sequence similarity (Figure 22.7). All the serine protease genes share two introns in common within the protease-like encoding region. As one follows the branches of the family tree, however, new and unique introns appear in different genes. The *tPA* and *uPA* genes, for example, share an intron that is not present in other family members and that is missing from more distantly related precursors to the trypsin gene family, such as thrombin. In addition, chymotrypsin has two and elastase three unique introns. These introns are not shared by other family members nor are they present in genes ancestral to the trypsin gene family. Such extreme variability in intron patterns is most likely to have arisen from the acquisition of introns over the evolutionary history of the serine protease gene family. Thus, although some of the introns present in eukaryotic genes are apparently quite ancient, others may have been acquired relatively recently in evolution.

Molecular Mechanisms May Explain How Introns Were Acquired over Evolutionary Time.

Although we may never have a complete answer, it seems likely that some introns are ancient and others have been acquired more recently (see *CURRENT INVESTIGATIONS: What Is the Function of an Intron?* in Chapter 11). For those introns that likely have been acquired relatively recently in evolutionary history, there are plausible molecular models of explanation. It is now known, for example, that some classes of introns have the ability to cut themselves out of RNA transcripts either by possessing unique autocatalytic activities or by encoding specific endonucleases that, in turn, perform the excision function (see Chapter 11). This ability of some classes of introns to control their own excision from nascent transcripts has prompted the hypothesis that at least some introns may have evolved from semiautonomous entities similar to what today we refer to as transposable elements (see Chapter 19).

Further evidence in support of the hypothesis that at least some introns evolved from transposable elements is the fact that some transposable elements in higher eu-

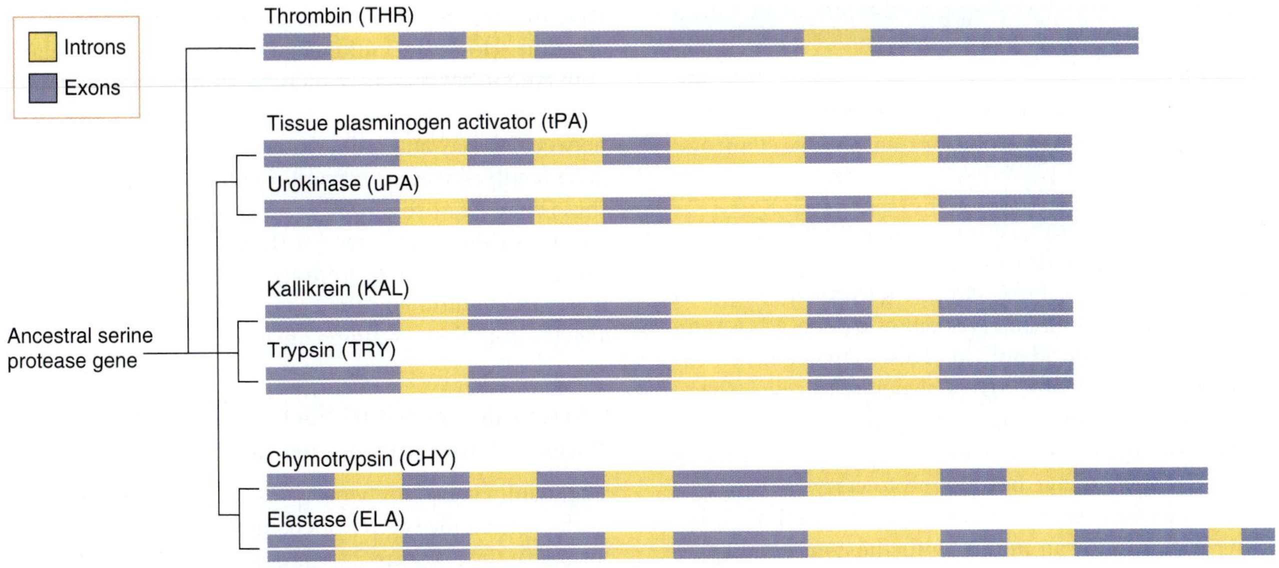

FIGURE 22.7 Coding regions of the protease-like portion of the serine protease gene family. The serine protease gene family tree was deduced from DNA sequences. Note that many of the members of the gene family have introns in the same locations. THR = human thrombin, tPA = human tissue plasminogen activator, uPA = pig urokinase, KAL = mouse kallikrein, TRY = rat trypsin, CHY = rat rhymotrypsin, ELA = rat elastase.

karyotes can function as introns. For example, a number of mutant genes in maize that produce normally sized transcripts, despite the fact that transposable elements are inserted within their exons (Figure 22.8), have been identified. It has been found that these maize transposable elements carry consensus-splice sequences at their borders that allow transposable element sequences to be spliced out at the RNA processing stage. Thus, mature mRNAs of these insertional alleles are, in effect, indistinguishable from those of the wild-type genes. Other examples of transposable elements behaving as introns have been reported in *Drosophila* as well. These findings lend credence to the hypothesis that at least some recently acquired introns may have evolved from transposable elements.

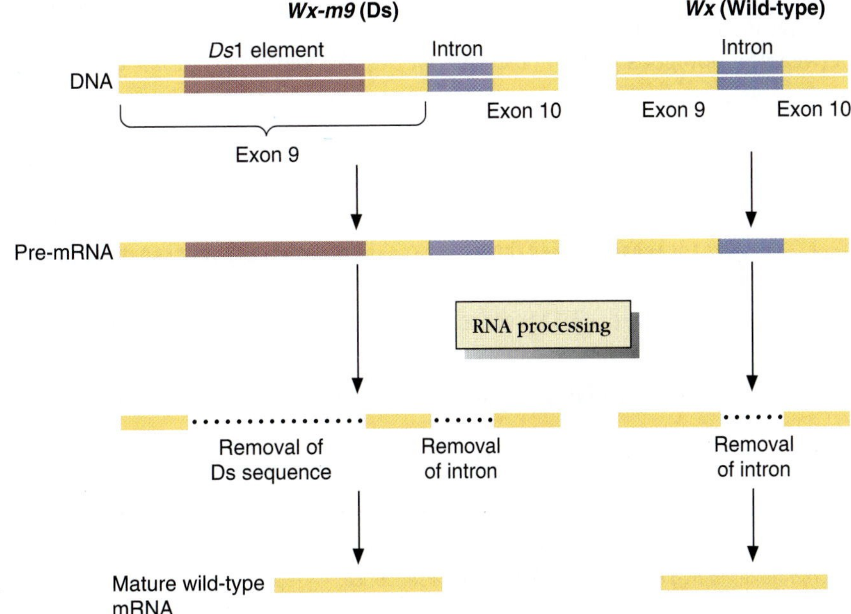

FIGURE 22.8 The splicing of *Ds* elements from pre-mRNA. The insertion of maize transposable elements into genes does not always produce nonfunctional alleles. Certain transposable element insertion alleles in maize, like the waxy gene (*Wx*) shown here, display a wild-type mRNA after splicing. Shown in this example is the *Wx-m9* insertion allele in which a *Ds*1 element has inserted into exon 9 of the *wx* gene in maize and is properly spliced out like a normal intron.

SOLVED PROBLEM

Problem

What is the evidence that at least some introns may have evolved from viral-like or transposable elements?

Solution

The introns of some primitive eukaryotes display a number of properties that are characteristic of transposable elements. For example, introns in fungi that are able to splice themselves out of nascent RNA transcripts have been identified. This is analogous to the ability of some transposable elements to splice themselves out of DNA. In addition, certain transposable elements found in some higher eukaryotes such as maize and *Drosophila* have been found to carry at their borders consensus splice sequences that result in their being mistaken for an intron by the host-encoded splicing machinery. These transposable elements are spliced out of nascent RNA transcripts just as if they were introns. These similarities between some transposable elements and introns suggest that the two may have evolved from a common ancestor.

SECTION REVIEW PROBLEMS

3. What is the evidence that at least some introns are ancient?
4. What is the evidence that some introns have been acquired relatively recently?

SEQUENCE CHANGES ARE USED TO PREDICT EVOLUTIONARY TIME.

Molecular evolutionists are also interested in the origin and significance of the nucleotide differences that exist between homologous genes present in species. All genes have undergone DNA sequence changes over evolutionary time. However, multigene families are a particularly convenient context in which to study DNA sequence evolution. By comparing the sequence changes that exist between closely and more distantly related members of a gene family, scientists can approximate the rate at which these changes arise as well as an appreciation of the consequences they may have for the evolution of new gene functions. Using these findings, scientists can also estimate the evolutionary time that separates contemporary species from their common ancestors.

Nucleotide Substitutions Change at Different Rates in Different Regions of the Gene.

The insulin gene family provides a useful demonstration of the nucleotide sequence changes that occur over evolutionary time. Figure 22.9 presents an alignment of the nucleotide sequences of a region of the rat I, rat II, human, and chicken preproinsulin genes. As you might expect, for some positions, the nucleotides are identical among all four preproinsulin genes, but the majority of the sites vary among the different genes. The nucleotide differences between the preproinsulin genes represent substitutions that have become fixed since these species diverged from a common ancestor about 300 million years ago.

For a group of genes encoding proteins that have the same function, such as the vertebrate preproinsulin genes, it seems reasonable to predict that rates of substitution will generally be higher at nucleotide positions where substitutions have little or no functional consequence than at positions where substitutions are apt to have a detrimental effect. The sites that have a detrimental effect could be acted on by selection, eliminating them from the population; consequently, we would not see these changes. Table 22.1 summarizes the percentage of nucleotide substitutions that have occurred among the human, rat, and

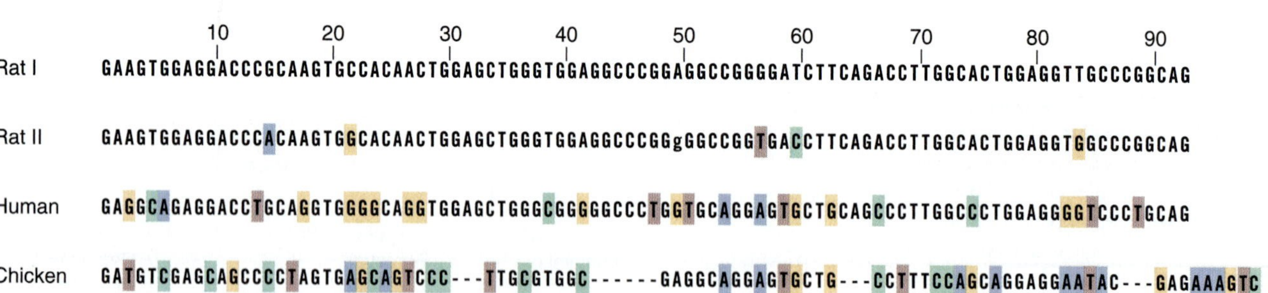

FIGURE 22.9 DNA sequence comparisons of the insulin gene. Alignment of the C-peptide encoding region of the rat I, rat II, human, and chicken preproinsulin genes is shown. The colored boxes indicate a mismatch with the rat I sequence; — represents a deletion.

TABLE 22.1 *Percent of Nucleotide Substitutions Among Genes in the Insulin Family*

Pairwise Comparison of Genes	Replacement Sites: A- and B-peptides	Replacement Sites: C-peptide	Silent Sites: A- and B-peptides	Silent Sites: C-peptide
Rat I/rat II	1.8	3.2	32	18
Human/rat	5.2	21	76	110
Human/chicken	8	63	122	140
Rat/chicken	10.7	49.4	64	150

chicken preproinsulin genes over evolutionary time. The percentage of substitutions has been computed separately for different regions of the gene so that differences that may be related to function can be more easily discerned. Consistent with the prediction that changes are more likely to occur at functionally unimportant sites, we see that percentage substitutions have been substantially higher at nucleotide sites where changes do not affect amino acid sequence (for example, third-base positions or silent mutations) than at those sites where nucleotide changes result in new amino acids. Furthermore, overall rates of nucleotide substitution have been higher within nonencoding introns than within exons. Another interesting and informative comparison is between regions of the preproinsulin gene that encode different aspects of the function of the protein.

The proinsulin protein is produced by proteolytic cleavage of the first 24 amino acids of the preproinsulin polypeptide, called the *preregion*. Proinsulin consists of three regions: A, B, and C. The A and B regions, which will become the two chains of mature insulin, are connected by a middle peptide called the C-peptide (Figure 22.10). A major function of the C-peptide is to bring together properly the A and B regions of the polypeptide so that the appropriate disulfide bonds can be formed. After this is accomplished, the C-peptide is enzymatically removed, and the mature insulin protein is formed. Based on this information, we might not expect the precise amino acid sequence of the C-peptide to be as critical as those of the A- and B-peptides. Consistent with this hypothesis, we find that the percentage of nucleotide substitutions within the C encoding region have been greater than within the A and B encoding regions of the preproinsulin gene (Table 22.1).

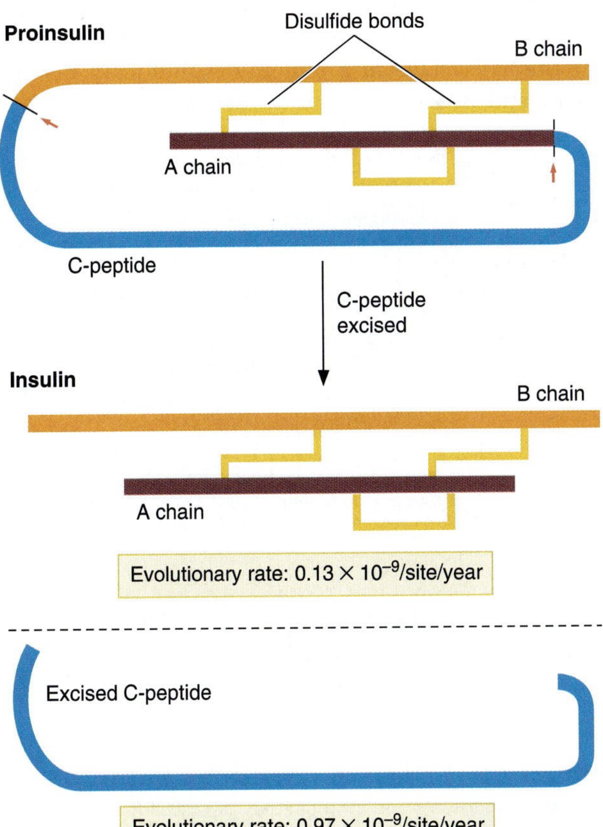

▶FIGURE 22.10 **The structure of insulin.** A mature insulin molecule consists of one A- and one B-peptide linked by disulfide bonds. The C-peptide serves as a spacer to help orient the A- and B-peptides for proper sulfide-bond formation and is removed from the protein when it is no longer needed.

Comparative analysis of the gene sequences among other gene families has revealed trends similar to those observed among the preproinsulin genes; that is, substitutions are typically more frequent at nucleotide positions where changes have little or no function consequence (Figure 22.11).

High Substitution Rates Characterize Genes That Are Evolving a New Function.

α-Lactalbumin is one component of the lactose synthesis system. The gene responsible for the α-lactalbumin sequence is generally believed to have evolved from the precursors of the lysozyme genes. The fact that lactalbumin has not yet been found in nonmammalian vertebrates has led to the hypothesis that the lactalbumin protein evolved from the lysozyme gene through a gene duplication event that took place at the onset of mammalian evolution (when the mammary gland originated), or approximately 175 million years ago. If this evolutionary scenario is correct, we can compute the average rate of lactalbumin gene evolution and compare it with the average rate of lysozyme gene evolution in vertebrates prior to the presumed duplication event. The results suggest that the lactalbumin gene evolved extremely fast during the period when α-lactalbumin was diverging in function from the precursors of modern-day lysozymes. Once the α-lactalbumin function was evolved, the rate of substitution within this family of proteins was considerably reduced and proceeded, thereafter, at a constant average rate. These findings suggest that the dynamics of protein evolution may change at different stages in a protein's evolutionary history (Figure 22.12). During the period when a protein is acquiring a new function, rates of amino acid substitution (and thus substitution rates of the encoding DNA sequences) may be high and primarily driven by positive selection (see Chapter 21). In contrast, once a protein has acquired its basic functional configuration, natural selection may act predominately to maintain the protein's functional integrity, primarily by eliminating deleterious mutations. At this later stage in a protein's evolutionary history, many and perhaps most of the sequence substitutions that occur are those that do not significantly disrupt the protein's established function; they may be changes that are selectively neutral or nearly so.

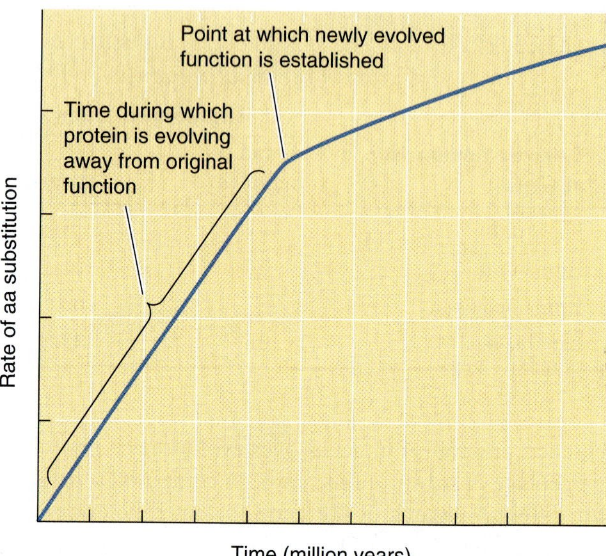

FIGURE 22.12 Rate of evolution of a hypothetical gene. The rates of nucleotide substitution in a gene may be high when the protein product is evolving a new function and may slow down after the new function is acquired. Directional selection may increase the rate of amino acid substitutions for a protein evolving a new function. After the evolving protein acquires the function, rates of substitution may decrease, reflecting the fact that natural selection is conserving the newly established function.

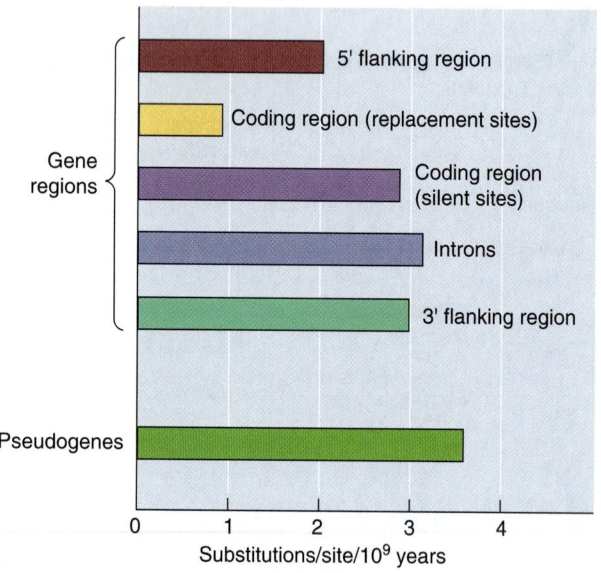

FIGURE 22.11 Substitution rates in different segments of genes. Comparison of the average rates of nucleotide substitution in different regions of genes and in pseudogenes is shown.

SOLVED PROBLEM

Problem

Explain why different regions of a protein-coding gene may evolve at significantly different rates.

Solution

Not all regions of a protein are of equal functional significance. In general, natural selection does not tolerate nucleotide substitutions at positions that encode amino acids critical to the function of a protein. On the other hand, substitutions at positions encoding amino acids nonessential to protein function are more readily tolerated by natural selection; consequently, these positions may evolve relatively rapidly over evolutionary time. There is an abundance of data consistent with this view of protein evolution. For example, preproinsulin consists of three regions: A, B, and C. After post-translational processing, the A and B regions become the two chains of the functional insulin protein, and the C region serves a transient structural role in allowing the proper disulfide bonds to be formed between the A- and B-peptides. The C-peptide is enzymatically removed from the mature insulin protein. Because the C-peptide is playing a transient structural role, many of the amino acid positions may equally well be occupied by a variety of amino acids. As a consequence, it has been observed that rates of amino acid substitution have been much higher for the C-peptide than for the A- and B-peptides over evolutionary time.

SECTION REVIEW PROBLEMS

5. From what you know of the function of the preproinsulin C-peptide, would you expect rates of amino acid substitution to be approximately equal over the entire length of the peptide, or would you predict that some regions will be more constrained?
6. Is it likely or unlikely that a gene will evolve at the same rate at different stages of its evolutionary history? Why?

Amino Acid Substitution Rates Are Constant Among Functionally Homologous Groups of Proteins.

We can plot the number of replacement substitutions (those substitutions resulting in amino acid changes) among the human, rat, and chicken preproinsulin genes against the time since these species have diverged. When this is done, we find that an approximately linear relationship exists between these two variables (Figure 22.13). This same linear relationship exists among other groups of functionally related genes as well. Thus, as stated previously, it appears to be a general rule that replacement substitutions among groups of functionally related proteins occur at approximately constant average rates over evolutionary time.

Although rates of amino acid substitution appear, on average, to be constant among groups of functionally related proteins, rates can differ substantially between functional groups. For example, the cytochrome c protein (a protein essential for energy production, and thus very widespread) has evolved at an extremely slow rate over the past 100 million years, whereas other proteins such as hemoglobin and fibrinopeptide have evolved much faster (Figure 22.14). Molecular evolutionists explain these differences in evolutionary rates by using essentially the same logic used to account for the rates of substitutions within the preproinsulin gene regions: replacement substitutions (among related proteins) having a well-defined function are more likely to occur at positions where the change will have relatively little effect on

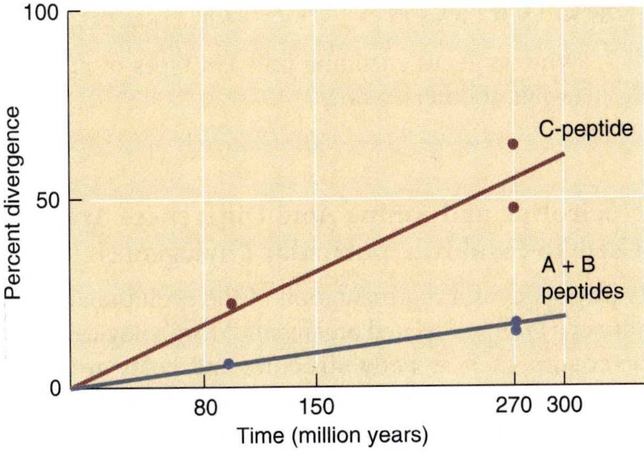

FIGURE 22.13 Sequence divergence of replacement (nonsilent) substitutions. The sequence divergence substitutions in the C-peptide encoding region vs. the A- and B-peptide encoding regions of the preproinsulin gene are plotted against time. Each data point represents the percentage of sequence divergence between the C or A and B preproinsulin peptides isolated from pairs of species vs. the time that has transpired since the two species were separated from a common ancestor.

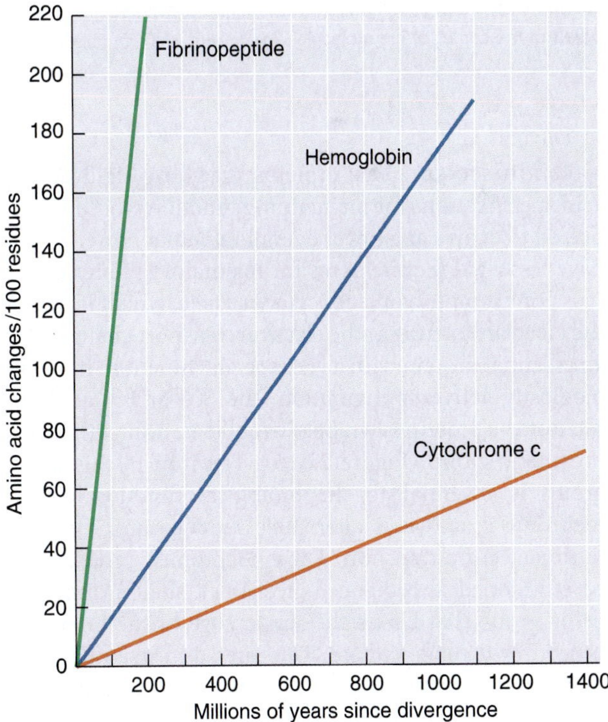

FIGURE 22.14 The rate of amino acid substitution over time. A comparison of the average rates of amino acid substitution during the evolution of three different proteins is shown. Cytochrome c, being the most functionally constrained molecule, evolves at the slowest rate; fibrinopeptide, being the most functionally unconstrained molecule, evolves at the fastest rate.

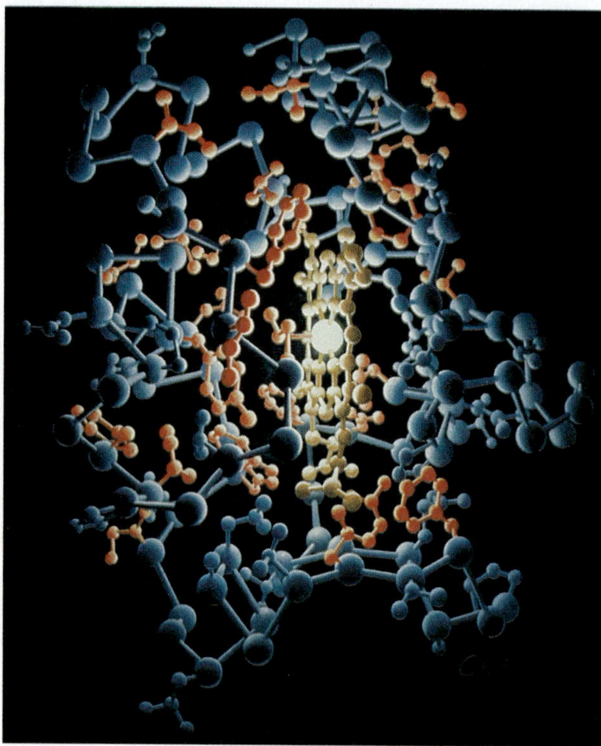

FIGURE 22.15 The three-dimensional structure of cytochrome c. The heme group (red) is an iron-containing prosthetic group located at the center of the molecule. *(Irving Geis, NYC)*

protein function. Thus, proteins that have rigid amino acid specifications for maintaining function would be expected to evolve at a lower overall rate than proteins that have less rigid requirements for maintaining function.

For example, consider cytochrome c, which is one of the electron carriers in the electron transport chain and is an essential protein with a very specific function. In its biologically active configuration, the 104-amino-acid cytochrome c protein is wrapped around an iron-containing prosthetic group (Figure 22.15). The iron in this heme group is essential to the biological function of cytochrome c, and it is alternately reduced and oxidized during electron transport. The cytochrome c protein may be envisioned as a one-cell-layer-thick shell around the heme group that leaves the single edge of the heme exposed in the heme-crevice. The three-dimensional structure of the cytochrome c molecule and its configurational relationship to the heme group are critical to its overall function. We would expect amino acid substitutions in cytochrome c that disrupt the internal integrity of the molecule to be disfavored by natural selection. Consistent with this expectation, it turns out that the internal hydrophobic amino acid groups have been highly conserved in cytochrome c proteins from all species.

We might expect somewhat less selective pressure to be exerted against amino acid substitutions occurring on the surface of the cytochrome c molecule. This, in fact, appears to be the case, not only for surface amino acid positions in cytochrome c but also for surface positions in most other proteins. The extent to which surface positions are tolerant of substitutions varies among proteins and depends upon whether the configuration of amino acids on the surface are critical to the protein's function. For example, proteins that are required to bind or interact with other proteins or molecules on their outside surface may be more constrained in their tolerance for amino acid substitutions than proteins that are not involved in such interactions.

The cytochrome c molecule is required to interact with cytochrome oxidase and cytochrome reductase, both of which are large macromolecular complexes (Figure 22.16). To accommodate this binding potential, the surface of the cytochrome c molecule has relatively strict structural requirements, especially with regard to the surface distribution of charged groups and the location of aromatic residues. The surface requirements of proteins that do not bind to other molecules (fibrinopeptide) or proteins that bind relatively small molecules such as oxygen (hemoglobin) (Figure 22.16) would be expected to have proportionally less stringent requirements for the conservation of surface structures. Consistent with this expectation, overall rates of substitution are proportionally greater for hemoglobin and fibrinopeptide than for cytochrome c (Figure 22.14).

SECTION REVIEW PROBLEM

7. Why do genes encoding different types of proteins evolve at different rates?

Nucleotide and Amino Acid Differences Are Used To Establish Molecular Phylogenies.

A phylogeny is a representation of the evolutionary history of a group of related organisms. Morphological characteristics such as body structure and color are often used to construct phylogenies of related animals. Nucleotide or amino acid differences among homologous DNA or protein sequences can be used to reconstruct **molecular phylogenies.**

Several computational methods have been devised to construct molecular phylogenies. Perhaps the simplest method is the **UPGMA method,** which stands for unweighted pair group method with arithmetic mean. The correct application of the UPGMA method requires that rates of substitution are approximately constant among

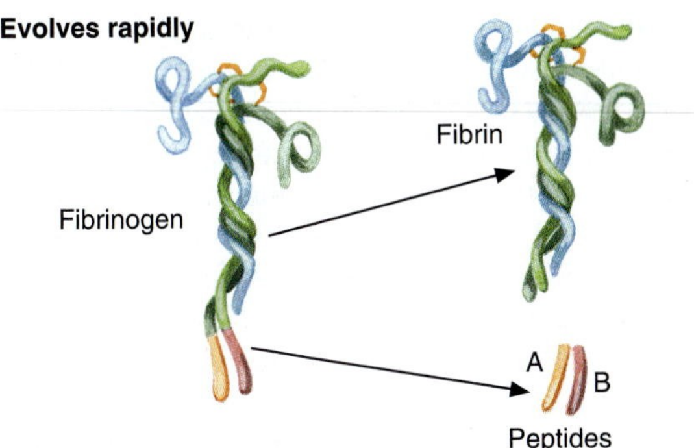

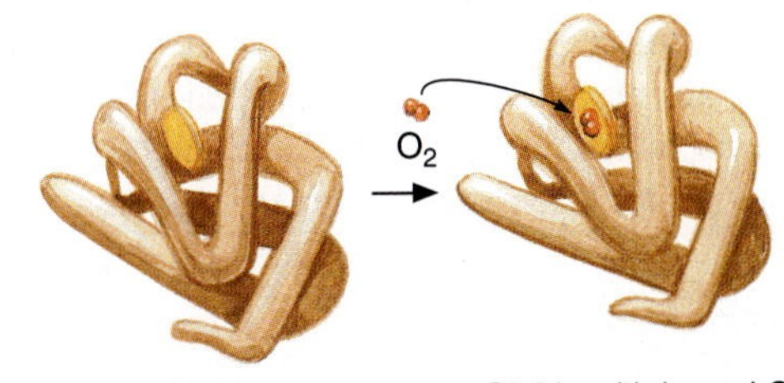

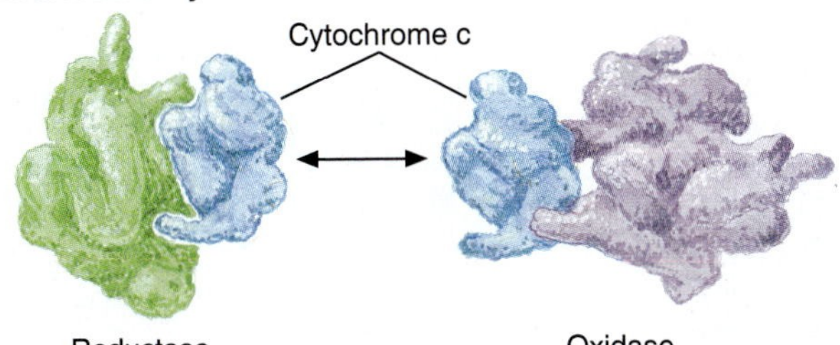

◄ **FIGURE 22.16 Protein function and rates of evolution.** The more complex the interactions of a protein with other molecules or macromolecules, the slower its rate of evolution, because natural selection is intolerant of amino acid substitutions on the protein surface. Fibrinopeptides have no interaction with other molecules and thus evolve very rapidly; hemoglobin binds oxygen and carbon dioxide, which are relatively small molecules, and thus evolves at a moderate rate; and cytochrome c interacts with large macromolecular complexes and evolves very slowly.

the different lineages (see Figure 22.11) so that an approximately linear relationship exists between substitution differences and divergence times. If this requirement is fulfilled, then sequences with fewer differences between them can be assumed to be more closely related to one another than sequences that have more differences between them.

To see how a molecular phylogeny might actually be constructed, consider the data presented in Table 22.2. Shown are the average number of nucleotide differences (per 100 sites) that exist between humans, chimpanzees, gorillas, orangutans, and rhesus monkeys over a 5.2-kb stretch of noncoding DNA. Because this stretch of noncoding DNA is considered to be selectively neutral, we can assume that the rates of substitution in this region have been, on average, constant over the time span the species have been evolving away from each other. Thus our data can reasonably be assumed to fulfill the UPGMA requirement that substitutional differences be linear with respect to divergence times.

To construct a phylogenetic tree based on these data, we first identify the two species separated by the fewest substitutional differences—in this case, humans and chimpanzees. Humans and chimps are joined at a branch point, called a node, positioned halfway between them. This node represents their most immediate common ancestor. Because we assume that rates of substitution were approximately equal in the lineages leading from the common ancestor to present-day humans and chimps, the nucleotide differences accumulated in each lineage are considered to be half the number of substitutions currently separating the two species, or 0.73 (1.45/2) (Figure 22.17). We next compute the differences

TABLE 22.2 *Nucleotide Substitutions in a Noncoding Region of DNA*

	Species				
Species	**Human**	**Chimpanzee**	**Gorilla**	**Orangutan**	**Rhesus monkey**
Human		1.45	1.51	2.98	7.51
Chimpanzee			1.57	2.94	7.55
Gorilla				3.04	7.39
Orangutan					7.10

The values are mean numbers of nucleotide substitutions per 100 sites over a 5.3-kb stretch of DNA from each of the organisms.

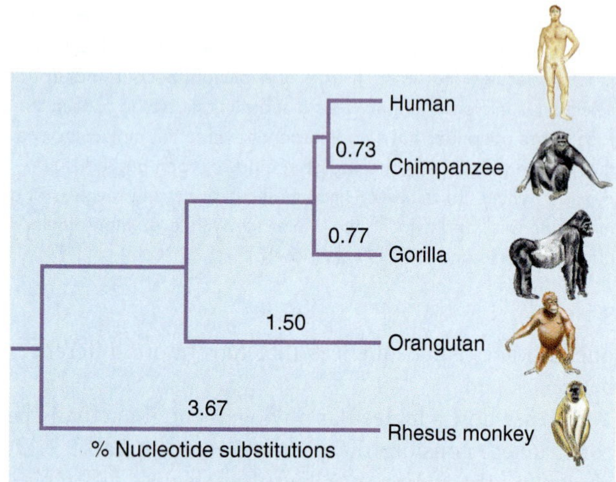

FIGURE 22.17 Primate evolutionary tree. This phylogenetic tree for humans, chimps, gorillas, orangutans, and rhesus monkeys was computed by the UPGMA method using the data presented in Table 22.2.

between the remaining species and the composite of humans and chimps. This is done by averaging the differences between each of the remaining species and chimps and humans, respectively. For example, the nucleotide difference between gorilla and human is 1.51, and between gorilla and chimp is 1.57. Thus, the average difference between gorilla and the human–chimp composite is 1.54 [(1.51 + 1.57)/2].

After establishing a matrix of these newly computed values (Table 22.3), we determine that the human–chimp composite and gorilla are separated by the shortest distance and are, therefore, the next group of species to be joined together. The number of nucleotide differences separating gorillas from their common ancestor with humans and chimps is 0.77, or half the average differences separating gorilla from the human–chimp composite (1.54/2). Continuing with this process, we eventually come up with the phylogenetic tree presented in Figure 22.17.

Species	Sequence (positions marked 10, 20, 30, 40, 50)
Human, chimpanzee	----------- G D V E K G K K I F I M K C S Q C H T V E K G G K H K T G P N L H G L F G R K T G Q A
Rhesus monkey	----------- G D V E K G K K I F I M K C S Q C H T V E K G G K H K T G P N L H G L F G R K T G Q A
Horse	----------- G D V E K G K K I F V Q K C A Q C H T V E K G G K H K T G P N L H G L F G R K T G Q A
Bovine, sheep	----------- G D V E K G K K I F V Q K C A Q C H T V E K G G K H K T G P N L H G L F G R K T G Q A
Dog	----------- G D V E K G K K I F V Q K C A Q C H T V E K G G K H K T G P N L H G L F G R K T G Q A
Rabbit	----------- G D V E K G K K I F V Q K C A Q C H T V E K G G K H K T G P N L H G L F G R K T G Q A
Gray whale	----------- G D V E K G K K I F V Q K C A Q C H T V E K G G K H K T G P N L H G I F G R K T G Q A
Kangaroo	----------- G D V E K G K K I F V Q K C A Q C H T V E K G G K H K T G P N L N G I F G R K T G Q A
Penguin	----------- G D I E K G K K I F V Q K C A Q C H T V E K G G K H K T G P N L N G I F G R K T G Q A
Chicken, turkey	----------- G D I E K G K K I F V Q K C S Q C H T V E K G G K H K T G P N L H G L F G R K T G Q A
Pigeon	----------- G D I E K G K K I F V Q K C S Q C H T V E K G G K H K T G P N L H G L F G R K T G Q A
Pekin duck	----------- G D V E K G K K I F V Q K C S Q C H T V E K G G K H K T G P N L H G L F G R K T G Q A
Snapping turtle	----------- G D V E K G K K I F V Q K C A Q C H T V E K G G K H K T G P N L N G L I G R K T G Q A
Bullfrog	----------- G D V E K G K K I F V Q K C A Q C H T C E K G G K H K V G P N L Y G L I G R K T G Q A
Tuna	----------- G D V A K G K K T F V Q K C A Q C H T V E N G G K H K V G P N L W G L F G R K T G Q A
Samia cynthia (moth)	------G V P A G N A E N G K K I F V Q R C A Q C H T V E A G G K H K V G P N L H G F Y G R K T G Q A
Screwworm fly	------G V P A G D V E K G K K I F V Q R C A Q C H T V E A G G K H K V G P N L H G L F G R K T G Q A
Wheat	A S F S E A P P G N P D A G A K I F K T K C A Q C H T V D A G A G H K Q G P N L H G L F G R Q S G S T
Baker's yeast	---- T E F K A G S A K K G A T L F K T R C E L C H T V E K G G P H K V G P N L H G I F G R H S G Q A
Candida krusei (yeast)	-- P A P F E Q G S A K K G A T L F K T R C A E C H T I E A G G P H K V G P N L H G I F S R H S G Q A
Neurospora crassa (mold)	----- G F S A G D S K K G A N L F K T R C A E C H G E G G N L T Q K I G P A L H G L F G R K T G S V

FIGURE 22.18 Comparison of the amino acid sequences of cytochrome c proteins. In the 21 different species represented, the amino acid positions have been numbered according to the wheat sequence, which has 112 amino acids. At 18 positions (black) the same amino acid is found in all the sequences. Closely related species share in common more amino acids than more distantly related species. For example, human and rhesus monkey differ at only one site. (A = alanine, C = cysteine, D = aspartic acid, E = glutamic acid, F = phenylalanine, G = glycine, H = histidine, I = isoleucine, K = lysine, L = leucine, M = methionine, N = asparagine, P = proline, Q = glutamine, R = arginine, S = serine, T = threonine, V = valine, W = tryptophan, and Y = tyrosine.)

TABLE 22.3 *Distance Matrix Between Species*

Species	Species: Human–chimp (composite)	Gorilla	Orangutan
Human–chimp (composite)			
Gorilla	1.54		
Orangutan	2.96	3.04	
Rhesus monkey	7.53	7.39	7.10

Distances are nucleotide substitutions computed between the human–chimp composite and then each of the other species.

Because the rates of amino acid substitutions among homologous proteins appear, on average, to accumulate at nearly constant rates, molecular evolutionists can also use these substitutional differences to construct molecular phylogenies. For example, consider the cytochrome c sequences from 22 different species that are presented in Figure 22.18. At 22 out of 104 positions, the same amino acid sequence is found in all the proteins. However, at the other positions, the proportion of sites having the same amino acid varies significantly among the different proteins. As a rule, however, closely related species share more similarities than distantly related species. For example, cytochrome c of human and monkey differ at only one site (position 66), whereas the cytochromes of horse and human differ at 11 sites. The amino acid differences that exist among the cytochromes presented in Figure 22.18 can be used to reconstruct a molecular phylogeny such as that depicted in Figure 22.19.

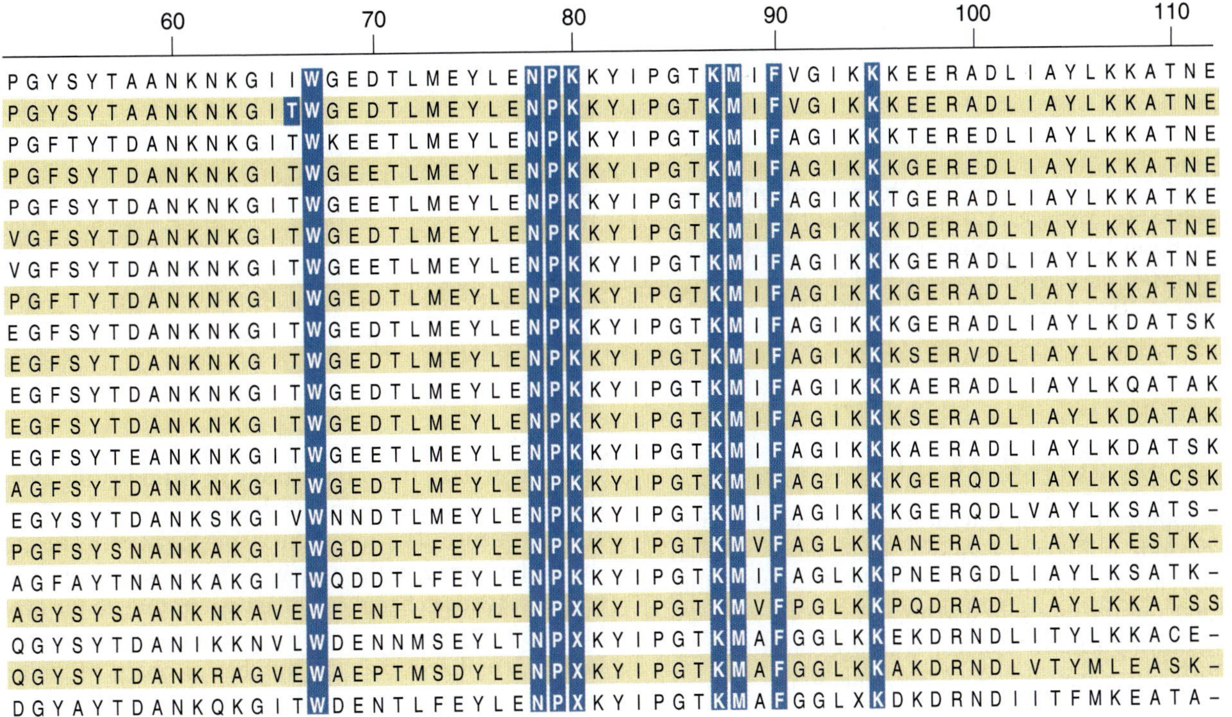

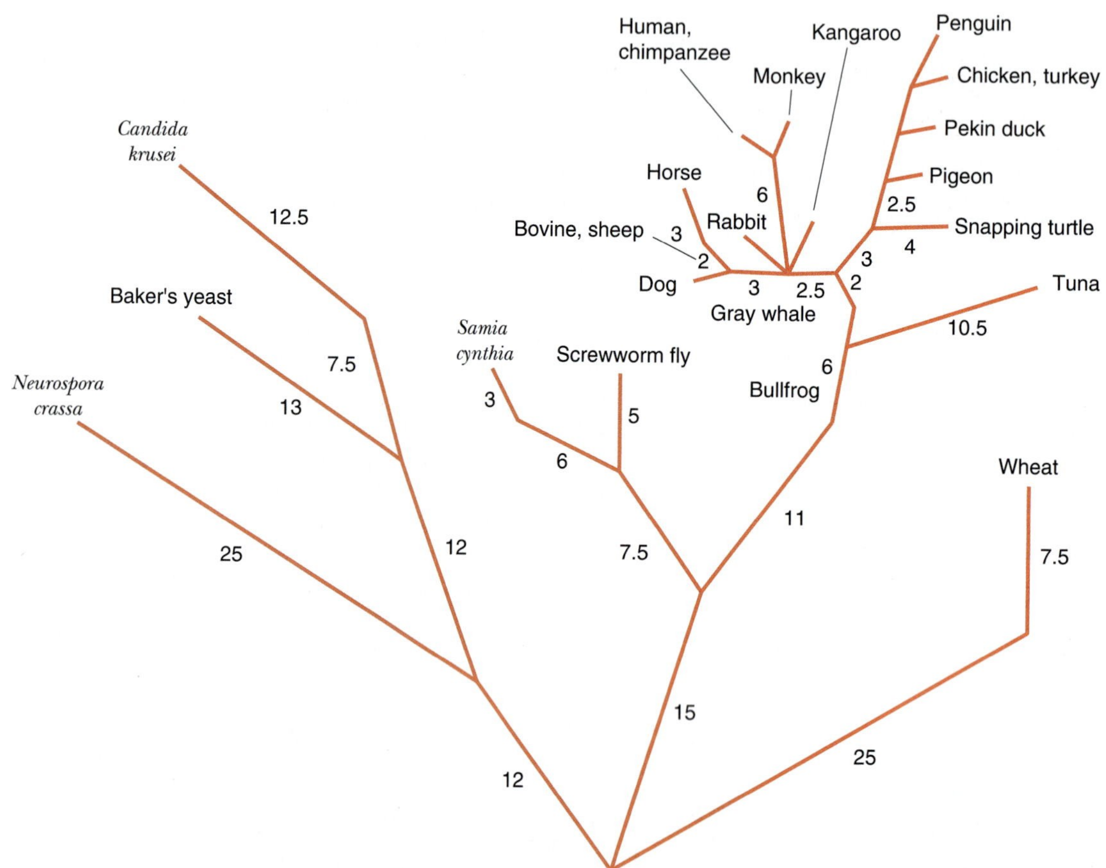

FIGURE 22.19 Phylogenetic tree based on cytochrome c sequence comparison. This phylogenetic tree, consisting of 21 species, is based on the cytochrome c alignments presented in Figure 22.18. The numbers along the branches give the amino acid changes between a species and a hypothetical progenitor. A slowly evolving molecule like cytochrome c is appropriate for establishing phylogenetic relationships among distantly related species.

SECTION REVIEW PROBLEM

8. Suppose that you determine the number of nucleotide differences among five closely related species of sunfish over a stretch of nonencoding and presumably selectively neutral DNA. Use the following data to establish a phylogenetic tree for these species using the UPGMA method.

Mean Number of Base Substitutions per 100 Sites

	Species				
	A	**B**	**C**	**D**	**E**
A		1.00	1.31	1.95	2.27
B			1.37	1.85	2.57
C				2.00	2.36
D					2.18

Molecular Data Has Limitations in Establishing Phylogenies. One limitation involved in using protein or DNA sequence data to establish phylogenies relates to the evolution rate of the protein or gene under study. The rate of sequence change is not always relative to the evolutionary time span separating the species in the phylogeny. It is likely that the number of observed amino acid or nucleotide differences between distantly related proteins or genes may be an underestimate. This is a result of the fact that the actual number of substitutional changes are not truly represented because, over evolutionary time, some of the substitutions have reverted, or "back"-mutated, to their original form. For example, imagine that you determined that two distantly related genes carry a guanine at the same nucleotide position. You might reasonably conclude that no evolutionary change has occurred at this site over evolutionary time. But, in fact, the guanine may have been replaced by cytosine and then changed back to guanine again.

Because the probability of a back mutation substitution increases with time, the tendency for observed substitutional differences to underestimate the actual numbers of changes increases between distantly related species over evolutionary time. This is especially true if the gene or protein under consideration is known to be evolving at a high rate. For example, it is very unlikely that the number of substitutional differences that exist between the fibrinopeptide molecules carried by snake and humans is an accurate account of the actual number of substitutions. Because fibrinopeptide genes evolve at such a high rate, it is likely that at least some back mutations have become fixed between the snakes and human fibrinopeptide genes over the 300 million years that they have been evolving away from each other. Although statistical correction procedures exist to help alleviate the potential errors introduced by this problem, the best solution is to not use a rapidly evolving protein or DNA sequence to estimate phylogenetic relationships among distantly related species.

Similarly, slowly evolving proteins, such as cytochrome c, are not useful for establishing evolutionary relationships between closely related species such as the primates. In this case, the cytochrome c molecules are structurally identical and do not yield useful information. A more rapidly evolving family of proteins, such as the mammalian carbonic anhydrases, is more appropriate for establishing phylogenetic relationships between the closely related primate species (Figure 22.20).

In recent years, evolutionists have successfully established phylogenetic relationships between populations of organisms that are members of the same or closely related species by using very rapidly evolving molecules such as vertebrate mitochondrial DNA. Vertebrate mitochondrial DNA (mtDNA) is a circular, double-stranded molecule 15–17 kb in length. Vertebrate mtDNA evolves at a rate of about 6×10^{-8} substitutions per year, which is approximately 10 times faster than the average rate of nuclear gene evolution in vertebrates. The reason for this unusually high rate of mtDNA evolution is believed to be caused by vertebrate mitochondria having a highly error-prone replication process and a poor or nonexistent repair system. Regardless of the cause, the exceptionally high evolutionary rates associated with vertebrate mtDNA makes the molecule especially useful for determining evolutionary relationships among closely related species and other even more closely related groups such as geographical races and subspecies. Another attractive feature of mtDNA for phylogenetic analyses is the fact that it is maternally inherited. Thus, allelic segregation need not complicate establishing evolutionary relationships. In some instances, the high level of resolution attributed to mtDNA has allowed molecular evolutionists to address important questions in evolutionary genetics and conservation biology (see *Current Investigations: The Utility of mtDNA Surveys in Conservation Biology*, and *Current Investigations: Are Neanderthals Our Ancestors?*).

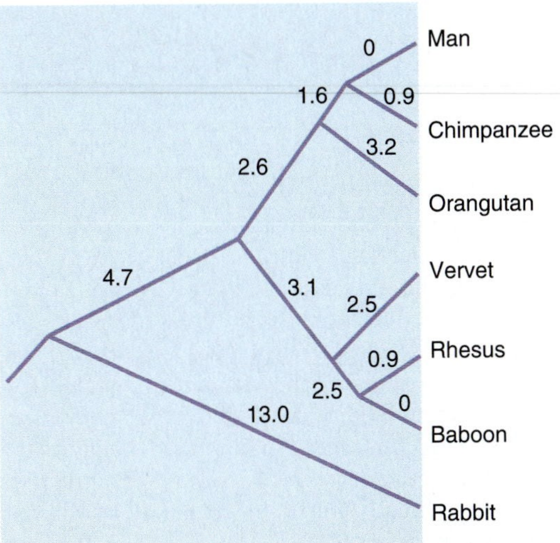

FIGURE 22.20 Primate phylogenetic tree based on carbonic anhydrase I. A phylogenetic tree of various primates based on the sequence of 115 amino acids in carbonic anhydrase I is shown. A rapidly evolving protein like carbonic anhydrase is more appropriate for establishing phylogenetic relationships between closely related species than a slowly evolving protein such as cytochrome c. The numbers on the branches are estimated numbers of nucleotide substitutions that have occurred over the evolution of the different carbonic anhydrase I molecules. *(Adapted from Tashian, et. al., 1976, Molecular Anthropology, M. Goodman and R. Tashian, Plenum Press, NY)*

In conclusion, scientists need to select judiciously those molecules upon which they plan to base molecular phylogenies. Rapidly evolving molecules are appropriate for establishing phylogenetic relationships between closely related species or races that can be presumed to have diverged from a common ancestor within the past few million years or less. Slowly evolving molecules are appropriate for establishing phylogenetic relationships between distantly related species that likely diverged from a common ancestor hundreds of millions of years ago. An inappropriate choice of molecule in either situation could result in molecular trees that greatly distort true phylogenetic relationships.

Sequence Differences Can Be Used to Estimate Times of Evolutionary Divergence.

If rates of substitutional change were indeed constant over evolutionary time, protein and DNA molecules would, in effect, be **molecular clocks,** and substitutional differences could be used to determine the times at which different molecules (and thus species) diverged from common ancestors. Times of evolutionary

CURRENT INVESTIGATIONS

The Utility of mtDNA Surveys in Conservation Biology

One area of contemporary biology in which mtDNA analysis is having a particularly significant impact is conservation biology. The results of a number of recent mtDNA studies have revealed that, in at least some instances, traditional taxonomic criteria used to identify species may be inadequate. These inadequacies may significantly undermine legislative as well as biological efforts intended to protect endangered species.

The red wolf (*Canis rufus*), which on the basis of traditional taxonomic criteria was considered by many experts to be a separate wolf species, has recently been shown by mtDNA analysis to actually be a hybrid between the gray wolf and the coyote. Recent mtDNA studies carried out on other animal groups suggest that such naturally occurring, interspecific hybridization events may not be as infrequent as previously believed. For example, recent mtDNA studies have revealed that natural hybridization events have frequently occurred between two separate pairs of critically endangered species: northern spotted owls with barred owls and blue whales with fin whales. These and subsequent findings are not only forcing evolutionists to reconsider the reliability of traditional taxonomic criteria, but they are making it difficult to implement government laws originally designed to protect endangered species.

In 1973, Congress passed the Endangered Species Act with the well-intentioned aim of protecting plant and animal species whose number and habitat were so depleted as to threaten their very survival. To discourage the artificial production of hybrids and thus help preserve species' integrity, it was decided that hybrid species would be excluded from protection under the Endangered Species Act.

This exclusion provision was instituted before the recent finding that at least some endangered species (and perhaps nonendangered species as well) consist of significant numbers of naturally occurring hybrid individuals. Thus, the law as it is currently interpreted could result in many threatened species losing their protected status. Obviously, the Endangered Species Act will need to be reinterpreted to recognize the fact that naturally occurring hybridization events among closely related species may not be legitimate grounds for removing a species from the endangered list.

In addition to providing data that should lead to more informed conservation legislation, future application of mtDNA studies in conservation biology should help to improve the ability of biologists to make correct decisions on how best to preserve species that are on the verge of extinction. For example, consider the case of the dusky seaside sparrow (*Ammodramus maritimus nigrescens*), which until recently inhabited a small region on

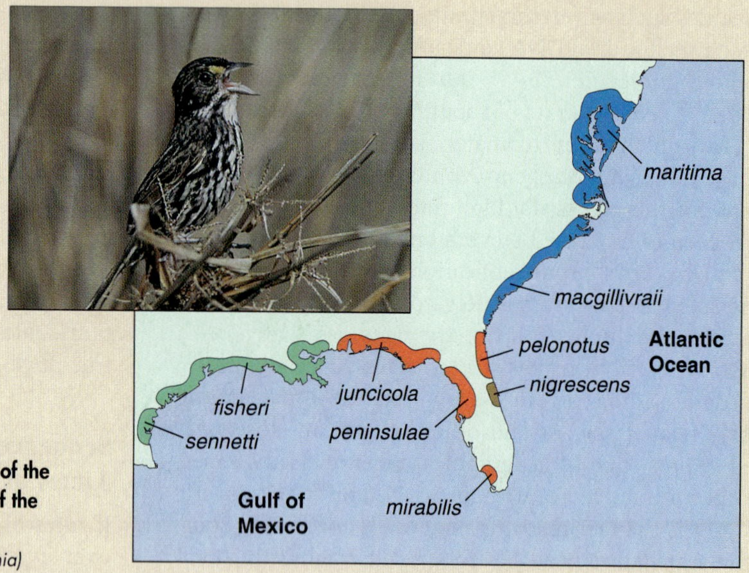

FIGURE 22.A Geographic distribution of the nine taxonomically recognized subspecies of the seaside sparrow *(Ammodramus maritimus).*
(VIREO, Academy of Natural Sciences, Philadelphia)

the eastern cost of Florida (Figure 22.A).

In 1980, it was determined that only six dusky seaside sparrows, all males, were left in nature. In a last-ditch effort to preserve the gene pool of this otherwise doomed species, the decision was made to capture the remaining males and to cross them with females of the most closely related subspecies available. The plan was that female offspring of this hybrid cross would be backcrossed to the parental males and that this backcrossing procedure would be continued until the original dusky seaside sparrow males became too old to reproduce. In this way, it was hoped that a "reconstructed" dusty seaside sparrow breeding stock would eventually be established and reintroduced to its original natural habitat.

A critical aspect of this plan was deciding which subspecies of seaside sparrows was phylogenetically most closely related to the endangered one. In 1980, mtDNA data were not available, so the decision about which subspecies to use was based on morphological and behavioral criteria alone. The best taxonomic information available at the time indicated that the closest subspecies to the dusky seaside sparrow was Scott's seaside sparrow (*A. m. peninsulae*), which inhabits Florida's Gulf shore (Figure 22.A).

In 1989, John Avise and Bill Nelson published the results of a restriction enzyme survey (the DNA from different species was digested with different restriction enzymes and the banding patterns compared) of mtDNA from birds belonging to the subspecies of *A. maritimus*, including mtDNA from the last surviving male dusky seaside sparrow. Using the UPGMA phylogenetic tree-building method, the authors were able to reconstruct the evolutionary relationships among the seaside sparrows (Figure 22.B). The results indicate that the now-extinct dusky seaside sparrow (*A. m. nigrescens*) is molecularly indistinguishable from the other two Atlantic subspecies (*A. m. maritima* and *A. m. macgillivraii*) but quite distinct (differing at about 1% of their mtDNA sites) from the Gulf subspecies, including Scott's seaside sparrow (*A. m. peninsulae*), whose females had been chosen for the breeding program. Thus, in this case, the use of traditional taxonomic criteria led to an incorrect choice of subspecies for crossing to the endangered subspecies. The consequence was the creation of a new artificial subspecies rather than the eventual preservation of a naturally occurring, endangered one. The use of mtDNA analyses in the design of conservation strategies should help to avoid such errors in the future.

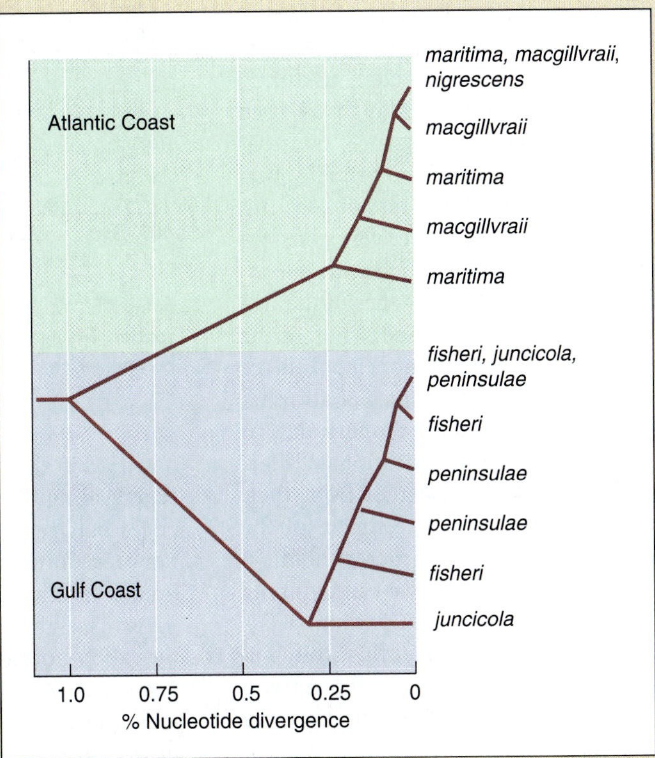

FIGURE 22.B Evolutionary relationship among seaside sparrows. This phylogenetic tree was computed using the UPGMA method and shows the distinction between mitochondrial DNA genotypes of the Atlantic coast vs. Gulf coast populations of the seaside sparrows. *(Adapted from Kimura and Ohta, 1972, in* Proceedings of 6th Berkeley Symposium on Math, Stat., and Prob., *p. 43)*

divergence between species are usually first determined by examining fossil remains. Fossil remains, buried in rock formations, can be accurately dated by radiological techniques. However, there are many instances in which these techniques, because of a poor fossil record or other problems, cannot accurately estimate species divergence. In such cases, the ability to estimate times of species divergence by molecular techniques is especially useful.

To see how substitutional differences might be used to estimate times of evolutionary divergence, consider the distance values used to construct the primate phylogeny presented in Figure 22.17. If the rate of substitutional changes used to compute this tree were constant over evolutionary time, the number of nucleotide substitutions that occur in each one of the branches of the tree should be proportional to the time elapsed. Thus, if the actual geological time of any one of the branch points are known from some outside source, the times of all other divergences can be computed by simple proportion. For example, the average number of substitutional differences separating humans and chimpanzees from their common ancestor is 0.73. The current estimate of the time of the human–chimpanzee divergence is about 7 million years. Given only these values, we can compute that it must have taken about 9.6 million years to accumulate 1.00 substitution per 100 sites over this stretch of DNA (7 million years/0.73 substitutions = 9.6 million years/substitution). Given this rate of substitution, we can now estimate that the lineage leading to African monkeys split from the lineage leading to humans, chimpanzees, gorillas, and orangutans about 35 million years ago ($9.6 \times 3.69 = 34.6$). In fact, this value is very close to the 30-million-year estimate based on traditional dating techniques.

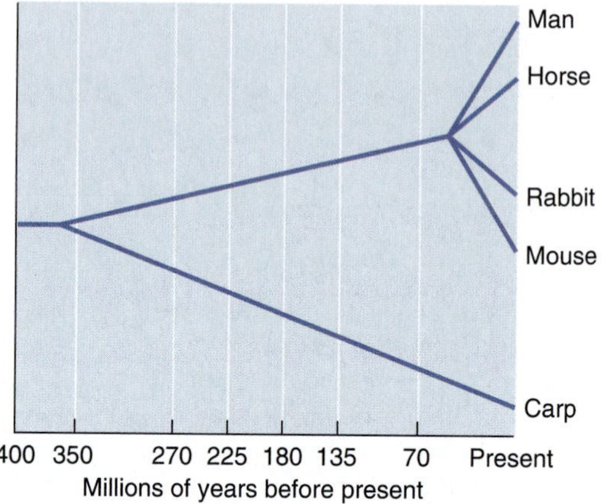

FIGURE 22.21 Phylogeny of carp and four mammalian species. The tree displays the number of amino acid differences between each of the species in the alpha chain of hemoglobin.

Molecular Data for Estimating Times of Evolutionary Divergence Has Limitations. The idea of using nucleotide and amino acid substitutions as molecular clocks is sound in principle, but its utility rests heavily upon the assumption that sequence substitution rates have been constant within the lineages being compared. Although rates of substitution usually appear to be constant when differences are examined among a broad spectrum of species over long spans of evolutionary time (for example, Figure 22.14), some notable inconsistencies in rates have been observed.

Consider, for example, the α-hemoglobin phylogeny of carp and four mammalian species presented in Figure 22.21. The number of amino acid differences among the four mammalian species range from 17 between human and mice to 28 between rabbits and mice. Thus, the difference between mice and rabbits is 65% greater than the difference between mice and humans. Because the branch of the tree leading to mice is the same in both the human-to-mouse and rabbit-to-mouse comparisons, we can conclude that 11 more amino acid substitutions have occurred in the lineage leading to rabbits than in the lineage leading to humans. Similarly, the number of differences between humans and rabbits is 25, whereas that between humans and mice is only 17. Because the branch leading to humans is the same in both these comparisons, eight more replacements must have occurred in the rabbit lineage than in the mouse lineage. Thus, it appears that rates of amino acid substitution at the α-globin locus have not been constant among these five mammalian species.

These kinds of inconsistencies in substitution rates are not uncommon. For example, it has been observed that generally lower rates of substitution have occurred in humans than in monkeys and higher rates in rodents than in primates. Several explanations have been offered to account for these observed differences in sequence substitution rates. Natural selection is perhaps the most obvious explanation. As we have already seen, natural selection may significantly increase the rate of substitution in genes that are in the process of acquiring new functions. On the other hand, it can be argued that selection is probably not the full explanation for overall higher rates of sequence substitution in a species. This is because it seems unlikely that all or most of the genes carried by a particular species are in the process of acquiring new or improved functions simultaneously.

One alternative to the possibility that natural selection is responsible for inconsistencies in substitution rates is the so-called **generation-time effect hypothesis.** According to this hypothesis, rates of mutation and hence substitution may be a function of generation time rather than total elapsed time. For example, because rodents experience more rounds of DNA replication per year than primates, it follows that their rates of mutation and thus substitution rates per year may be higher as well. The same argument has been applied to account for the overall rate differences observed between monkeys

CURRENT INVESTIGATIONS

Are Neanderthals Our Ancestors?

There are currently two general hypotheses on the evolutionary origin of modern humans. The out-of-Africa hypothesis claims that modern humans arose in Africa and subsequently replaced existing human populations around the world. The opposing regional continuity hypothesis maintains that modern humans evolved continuously in many parts of the world. According to advocates of the regional continuity hypothesis, the African ancestors of modern humans repeatedly interbred with other ancestral human lineages present in Europe and Asia before branching into the lineage leading to modern humans around 120,000 years ago. The two hypotheses were put to a dramatic test in 1996, when a group of German and American geneticists led by Svante Pääbo at the University of Munich were able to isolate and sequence a mitochondrial DNA (mtDNA) sample from a Neanderthal bone.

It is extremely difficult to isolate and analyze DNA from 30,000-year-old fossilized bone, not only because of the problems associated with DNA degradation but also because of the ease with which contaminating DNA can corrupt fossil samples. Pääbo's group was successful in its efforts because it had an exceptionally well-preserved bone sample to work with and because it took extensive measures to reduce the possibility of contamination. For example, the bone sample Pääbo's group used was repeatedly washed with bleach to destroy contaminating DNA on its surface. In addition, sterile techniques were used during the entire DNA extraction procedure, which was carried out in a room where no other DNA work had ever been done. They chose to investigate a region of mtDNA because it is known to evolve at an especially rapid rate, thus providing researchers with the possibility of ultimately computing a reliable evolutionary distance estimate among closely related human lineages. To amplify the extremely small samples of DNA from the bone, the group used the polymerase chain reaction (see Chapter 12) and specifically designed primers that flanked the region to be cloned. The PCR products were individually cloned and sequenced, which allowed the researchers to distinguish any contaminating sequences from the Neanderthal sequences.

Their results conflict with the continuity hypothesis and agree with the out-of-Africa hypothesis. The sequences that the group found were compared with 968 different sequences from living humans and found to be very distinct. Pääbo and his collaborators estimated that Neanderthals and modern humans last shared a common ancestor some 550,000–600,000 years ago, which is long before the estimated time of divergence (120,000 years) of the lineage leading to modern humans (Figure 22.C). Although these remarkable new findings clearly favor the out-of-Africa hypothesis, it remains a formal possibility that genetic evidence for Neanderthal mixing has been lost from the modern human gene pool as a result of founder effects and genetic drift or of natural selection. However, by employing the careful techniques pioneered by Pääbo's group, it seems highly likely that future analysis of fossil DNA from other primitive peoples will soon lead to a final resolution of the evolutionary origins of modern humans.

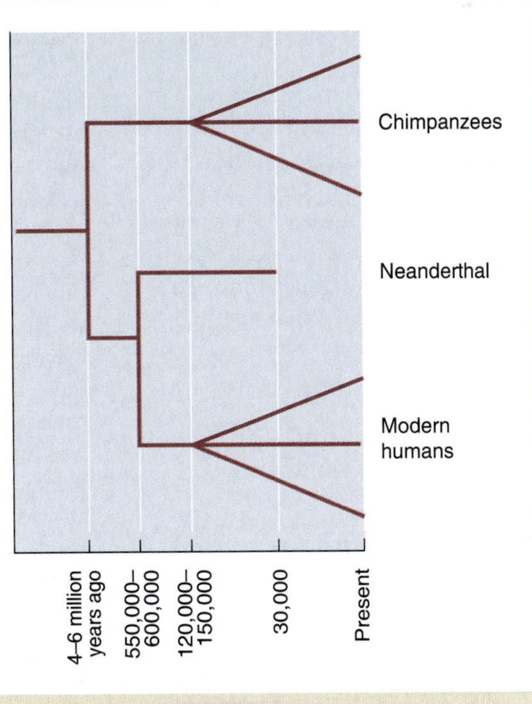

FIGURE 22.C **Estimated time of divergence of humans from their ancestors.**

and humans. Although the generation-time effect hypothesis may account for some of the variation in rates between species, it cannot be the whole story. For example, we noted above that rates of substitution at the α-globin locus seem to be higher in rabbits than in mice. This difference, at least, cannot be explained by a generation-time effect because mice have a shorter generation time than rabbits.

Another possible explanation that has been offered to account for differences in substitution rates between species is the possibility that the efficiency of DNA repair systems may vary between species. There is, for example, preliminary evidence that suggests that rodents may be less efficient at repairing DNA replication errors than humans. If substantiated, this difference in repair efficiency might imply that rodents will accumulate relatively more mutants per replication cycle than humans and therefore have a relatively higher rate of substitution.

In general, then, amino acid and nucleotide substitutions can be employed as internal molecular clocks and used to determine the approximate time when species have diverged from a common ancestor. However, care must be taken in using this dating method because it assumes that the molecular clock has been ticking at the same constant rate in and between all lineages being compared. In fact, some data indicate that the molecular clock does not always tick at a constant rate even within the same lineage, let alone between lineages. Thus, estimates of divergence times based on this approach may not always be accurate.

FIGURE 22.22 Rates of morphological evolution are highly variable. Although the morphologies of frogs representing different suborders (*Xenopus* and *Rana*) have remained very similar for 150 million years, those of chimps and humans have changed dramatically in only 10 million years or less.

SECTION REVIEW PROBLEMS

9. If, by paleontological evidence, species A and B in Section Review Problem 8 have been estimated to have diverged 500,000 years ago, estimate the time of divergence among species A and D.
10. Suppose that your estimate of the divergence time between species A and D in Problem 9 is a considerable underestimate of the divergence time determined by traditional methods. What might account for the discrepancy?

Sequence Evolution Rates Are Not Always Correlated with Rates of Morphological Evolution.

In the mid-1970s, Allan Wilson [see *HISTORICAL PROFILE: Allan Wilson—A Pioneer in the Application of Molecular Data to Evolutionary Questions (1937–1991)*] and his colleagues examined the amino acid differences that exist among a variety of related proteins within two major vertebrate species groups: frogs and placental mammals. The reason for studying these two groups was the fact that their rates of morphological evolution are known to be dramatically different (Figure 22.22). As a group, frog species are morphologically and anatomically similar despite the fact that they have been distinct species for about 150 million years. In contrast, placental mammals are among the most morphologically diverse species groups, even though they separated from each other only about 75 million years ago. Wilson and his collaborators reasoned that if amino acid substitutions that accumulate over evolutionary time are causally related to evolution on the organismic level, then rates of morphological and molecular evolution should be correlated. If this is true, then morphologically diverse placental mammals should have higher average rates of amino acid substitutions than morphologically similar frog species.

Rather surprisingly, no such correlation was found. Wilson and others found that rates of protein evolution are essentially the same in both placental mammals and frogs. Because the frog species have been separated from each other considerably longer than placental mammals, the frog proteins are, on average, more divergent from each other than are mammalian proteins. This is true despite the fact that, as a group, frogs are more morphologically similar to one another than are placental mammals.

As you might imagine, these findings created quite a stir among evolutionary geneticists when they were first reported in the late 1970s. If, as the results of Wilson's

Historical Profile

Allan Wilson—A Pioneer in the Application of Molecular Data to Evolutionary Questions (1937–1991)

Allan Wilson (Figure 22.D) was born in Ngarawahia, New Zealand, in 1937. He came to the United States in 1955 to study biochemistry at the University of California at Berkeley, where he received his Ph.D. in 1961. After postdoctoral study at Brandeis University in Waltham, MA, Dr. Wilson was invited to return to U. C. Berkeley in 1964 to join the Biochemistry Department faculty. He remained on the faculty at Berkeley for the rest of his scientific career.

From very early in his graduate student career, Wilson was convinced that large biological molecules such as proteins and nucleic acids encoded a wealth of information on the evolution of the organisms in which they were contained. Although this idea was considered rash and ungrounded by many of the more classically oriented biologists who dominated the field of evolutionary biology in the middle to late 1960s, Wilson persisted in his views and went on to substantiate his claims with hard data. Today, Allan Wilson is acknowledged as one of the founders of the field of molecular evolution.

Although Wilson's work has focused predominately on questions of primate evolution, his major findings have always been of significance to all areas of evolutionary biology. During the late 1960s and early 1970s, Wilson, in collaboration with his colleague Vince Sarich, pioneered the concept that most amino acid changes that accumulate in proteins over evolutionary time do so at a nearly constant rate and thus constitute a type of molecular evolutionary clock. Using the molecular clock, Wilson and Sarich caused quite a controversy in 1967 when they published a paper concluding that humans and the African apes probably diverged from one another as recently as 5 million years ago. This estimate was in sharp contrast to the prevailing view among paleontologists of the day, who maintained that the human–ape divergence occurred at least 15 million years ago.

In 1977, Wilson became involved in another controversy when he and his collaborators, including Mary-Claire King, Linda Maxson, and Vince Sarich, presented evidence that rates of organismic (for example, morphological or behavioral) and protein sequence evolution were not correlated with one another and thus were probably causally unrelated as well. Wilson and his colleagues postulated that organismic evolution was predominately the consequence of regulatory gene changes, which affected the timing and spatial patterning of gene expression over development. Wilson's views have done much to stimulate current research interests in regulatory evolution.

In more recent years, Wilson once again became involved in controversy. Based on a survey of mtDNA restriction polymorphism from various races and ethnic groups, Wilson and his colleagues Rebecca Cann and Mark Stoneking hypothesized that modern humans may have descended from a single female who lived in sub-Saharan Africa about 220,000 years ago. If substantiated, the data of Wilson and his collaborators will negate the previously popular hypothesis that the ancestors of modern human have been present in both Asia and Africa for at least 1 million years. Wilson and his collaborators have claimed that their data suggest that the primitive, human-like

FIGURE 22.D **Allan Wilson.** *(University of California, Berkeley)*

Asian lineages that are represented by so-called Java human and Peking human did not contribute to the gene pool of modern humans. At the present time, these claims remain controversial and it may be several years before the controversy is finally resolved.

Regrettably, Allan Wilson died in 1991 from cancer. His scientific career stands as a vindication of his early belief that proteins and nucleic acids are a storehouse of information on the evolutionary history of the organisms in which they reside. Indeed, this very basic notion now serves as the intellectual cornerstone of the field of molecular evolution.

group and others suggest, widespread changes in protein sequence are not the cause of evolutionary change on the morphological level, what then is the molecular basis of morphological evolution? Wilson and his colleagues hypothesized that morphological and other aspects of organismic evolution may be the result not so much of changes in the nucleotide sequences of genes but in the regulation of gene expression during development.

Whether or not this hypothesis is correct remains an open question. However, there is no doubt that these early studies of Wilson and others represent a significant landmark in the study of molecular evolution. A good deal of research effort has been, and continues to be, devoted to understanding the mechanisms and significance of regulatory changes in evolution. For example, recent studies on the evolution of homeobox genes (see Chapter 16) in both *Drosophila* and vertebrates indicate that *cis*-regulatory changes have played an important role in the evolution of this important class of developmental genes. This relatively new area of study is known as **developmental evolutionary biology,** and it is devoted to understanding how the genes involved in early development are responsible for large morphological differences between species, both vertebrates and invertebrates.

SOLVED PROBLEM

Problem

What is the evidence that regulatory changes may be the basis of evolutionary change at the organismic level (morphological, behavioral, and so forth)?

Solution

The indirect evidence is that rates of morphological evolution are often not correlated with rates of amino acid and nucleotide substitution. This implies that rates of amino acid/nucleotide substitutions are not the major causal determinants of morphological evolution. In the late 1970s, A. C. Wilson and his collaborators proposed that changes in the regulation of gene expression rather than changes in the gene sequence are the primary determinant of morphological evolution. Recent findings that homeobox genes are intimately involved in morphological development have prompted studies into how these genes have evolved. Preliminary results indicate that changes in the regulated patterns of expression of homeobox genes have been of primary importance in morphological evolution.

Summary

Molecular evolution is a broadly encompassing area of investigation that includes the study of changes in genomic sequence, structure, and function over evolutionary time and the consequence of these changes on morphology, physiology, behavior, and other aspects of organismic evolution. Three areas of molecular evolution research directly related to genetics are the evolution of gene families, the evolution of introns, and DNA sequence evolution.

Gene families are related groups of genes that have evolved by duplication and divergence from a common ancestral gene. A typical example of a gene family is the mammalian globin gene family, which evolved by a series of gene duplications over the past 600 million years. Members of gene families are usually in close chromosomal proximity to one another but may contain members that map to diverse chromosomal locations. The primary mechanism by which members of a gene family duplicate is unequal crossing over. Some gene families contain nonfunctional members called pseudogenes. Some pseudogenes have arisen by reverse transcription of an mRNA product transcribed by an active member of the gene family. These retropseudogenes typically lack a promoter and introns and can be found at diverse chromosomal locations. The duplication of some genes may be favored by natural selection when, for example, more of the gene product is an adaptive advantage (for example, histone genes, rRNA, and tRNA genes). Gene duplication may also be an evolutionary advantage because it provides a mechanism by which genes may evolve new functions. Because only one of the duplicated copies of a gene is required to maintain a gene's established function, any additional copies are free to accumulate new and possibly adaptive mutations.

There are currently two views on the evolutionary origin of introns. One point of view maintains that introns provide large stretches of DNA where recombination may occur without disrupting a gene's reading frame and thus its encoded function. In this way, sequences encoding a previously evolved function can be juxtaposed next to a sequence encoding another function, which might accelerate the rate at which complex proteins evolve. According to this model, introns are as old as or older than the genes in which they reside. The alternative view is that many and perhaps most introns have been acquired by genes over evolutionary time by insertion of sequences of DNA perhaps analogous to transposable elements. These two views are not mutually exclusive, and data exist that are consistent with both views.

All genes have experienced DNA sequence changes over evolutionary time. In most cases, nonencoding regions of genes (for example, introns) evolve at a faster rate than encoding regions. Substitutions that accumulate in nonencoding regions are the result of the random fixation of neutral alleles, and thus their rate of fixation should be constant and a function of mutation rate and population size (see Chapter 21). During that period of a gene's evolutionary history when it is evolving a new function (for example, lactalbumin gene evolving from the lysozyme gene), natural selection may favor change and thus the rate of substitution in encoding regions may be greater than rates of substitution in nonencoding regions. However, after a gene has acquired a basic function, natural selection will disfavor changes in functionally important regions of the gene, resulting in a slower rate of substitution in encoding than in nonencoding regions.

Rates of substitution are, on average, constant among functionally homologous proteins, but they can vary significantly between groups. This is a result of the fact that the proportion of functionally important and thus selectively constrained sites varies among genes. For example, the genes encoding cytochrome c in various species have evolved away from one another at a constant average rate. This is also true for globins. However, the rate of substitution among globin genes is significantly greater than among cytochrome c encoding genes because of the proportionally greater number of functionally important sites in cytochrome c proteins than in globins.

Nucleotide or amino acid differences between molecules carried by different species can be used to establish molecular phylogenies. Phylogenetic trees that are established on the basis of sequence differences between molecules are usually quite accurate and often agree with trees established by more traditional criteria. Insofar as molecules evolve at constant rates, they can be viewed as molecular clocks and used to estimate times of divergence between species.

Sequence evolution rates are not always correlated with rates of morphological, behavioral, and other aspects of organismic evolution. It has been proposed that the major differences in morphology, anatomy, physiology, and behavior that we associate with some species and most higher taxonomic groups are caused by changes in patterns of gene regulation over development.

Selected Key Terms

gene family
generation-time effect hypothesis
molecular clock
molecular evolution
molecular phylogeny
multigene family
pseudogene
retrogene
retropseudogene
thalassemia
UPGMA method

Chapter Review Problems

1. Proteins evolve at different rates at different stages of their evolutionary development. Do you agree or disagree with this statement? Why?
2. What are some of the advantages and possible dangers of using sequence differences to establish phylogenies?
3. What are some of the variables that may influence the reliability of estimated divergence among species based upon molecular data?
4. What is the evidence that regulatory changes may be especially important in the evolution of new morphologies, behaviors, and other aspects of oganismic evolution?
5. Some molecular evolutionists maintain that gene duplication is a prerequisite for the evolution of new proteins. Do you agree or disagree with this statement? Why?

Challenge Problems

1. Suppose that you determine the number of nucleotide differences existing among five related *Drosophila* species over a 100-nucleotide region of a known pseudogene. The results of your study are presented in the following table. Use the data to establish a phylogenetic tree for these species using the UPGMA method.

	Species				
	A	**B**	**C**	**D**	**E**
A		1.00	1.50	1.70	2.00
B			1.40	1.65	2.50
C				1.90	2.90
D					3.00

2. Some molecular evolutionists have hypothesized that introns evolved from transposable elements. Can you hypothesize why, from the perspective of the transposable element, it might be of selective advantage to evolve consensus splice sequences within its terminal repeats?

3. Why might it be of selective advantage to the host to evolve splicing enzymes that are able to recognize cryptic splice consensus sequences within the terminal repeats of transposable elements?

Suggested Readings

Akam, M., P. Holland, P. Inghram, and G. Wray (eds). 1994. *The Evolution of Developmental Mechanisms.* The Company of Biologists Limited, Cambridge, U.K.

Carrol, S. B. 1995. "Homeotic genes and the evolution of arthropods and chordates." *Nature,* 376: 479.

Erwin, D., J. Valentine, and D. Jablonski. 1997. "The origin of animal body plans." *American Scientist,* 85: 126.

Gilbert, W. 1978. "Why genes in pieces?" *Nature,* 271: 501.

Kahn, P., and A. Gibbons. 1997. "DNA from an extinct human." *Science,* 277: 176–178.

Krings, M., A. Stone, R. W. Schmitz, H. Krainitzki, M. Stoneking, and S. Pääbo. 1997. "Neandertal DNA sequences and the origin of modern humans." *Cell,* 90: 19–30.

Li, W., and D. Graur. 1991. *Fundamentals of Molecular Evolution.* Sinauer Press, Sunderland, MA.

McDonald, J. (ed). 1993. *Transposable Elements and Evolution.* Kluwer Academic Press, Dordrecht, The Netherlands.

Pennisi, E., and W. Roush. 1997. "Developing a new view of evolution." *Science,* 277: 34–37.

Raff, R. A. 1996. *The Shape of Life.* University of Chicago Press, Chicago.

On the Web

Visit our Web site at **http://www.saunderscollege.com/lifesci/titles.html** and click on A/G/M Genetics for links to the following chapter-related resources on the World Wide Web:

1. **Organellar genome sequences.** This site will link you to other sites that are sequence databases for organelle DNAs.
2. **Evolution and systematics.** This site links you to other Web resources in molecular evolution and systematics.

Appendix
Answers to Section Review Problems

*Visit our Web site at **http://www.saunderscollege.com/lifesci/titles.html** and click on A/G/M Genetics to find short answers to all Section Review Problems, Chapter Review Problems, and Challenge Problems.*

Chapter 1

1. The term "heredity" refers to the fact that "like begets like," that is, offspring generally tend to resemble their parents. The term "variation" refers to the fact that despite the similarity between parents and offspring, heritable differences do exist.

2. Transmission genetics deals with the inheritance pattern of a specific trait(s) during a cross. Molecular genetics is devoted to studying the biochemical and molecular mechanisms by which hereditary information is stored in DNA and subsequently transmitted to the proteins. Population and evolutionary genetics elucidate the factors that determine the genetic composition of groups of interbreeding or related organisms, called populations, and how genetic changes in populations can lead to the formation of new reproductively isolated groups of populations, or species.

3. According to the epigenetic theory, organisms develop from the material contained in eggs, which is stimulated to develop by the contribution of the male. According to the preformationist theory, sex cells contained a fully developed (albeit miniaturized) adult called the homunculus, which had only to increase in size within the mother's womb until birth. There were two versions of this theory, depending upon one's perspective. The "spermists" believed that the homunculus resided in sperm, and the "ovists" believed that it was to be found within the egg. Each homunculus was believed to contain another homunculus within its germ cells and so on.

4. Often a son or daughter will look more like one or the other parent rather than appearing to be an equal mixture of both.

5. Remove seedlings from the top and bottom of the hill and grow them in a constant "greenhouse" environment. If the difference between the two sets of plants is genetic, yields should continue to be different even though the plants are grown under the same conditions. To specifically test the hypothesis that the differences in yield are the result of differences in the amount of light available, seedlings collected from the top and bottom of the hill can be grown in a constant shade or constant light to see if the differential in yields disappears.

6. Perhaps the amount of available water is different at the top and bottom of the hill.

7. The widely held belief that the earth must be "the center of the universe" presented a barrier to the general acceptance of the heleocentric theory of our solar system.

Chapter 2

1. a. Two genotypes and two phenotypes: Cc = color, cc = white. **b.** 500 Cc and 500 cc.

2. a. The red-orange fish is heterozygous for alleles of two pairs of genes, one for color (red-orange is dominant to green) and the other for the appearance of the crescent spot (spot is dominant to no spot). **b.** R = red-orange, r = olive green, S = spot, s = no spot. The red-orange, spotted parent is $Rr\ Ss$. The olive green parent is $rr\ ss$.

c.

Number	Color	Spot	Genotype
15	red-orange	crescent	$Rr\ Ss$
12	red-orange	none	$Rr\ ss$
18	olive green	crescent	$rr\ Ss$
14	olive green	none	$rr\ ss$

3. The curved-wing phenotype is given by a dominant allele, and the straight-winged phenotype, by a recessive allele of the same gene.

4. a. The F_1 should have long flowers. **b.** There should be 25 stunted plants.

5. The shape (oblong or round) and color (yellow or red) are controlled by alleles of two independently assorting genes. The allele giving oblong shape is dominant to the allele giving round shape, and the allele giving yellow color is dominant to the allele giving red color.

6. a. 4/16. **b.** 2/16. **c.** 2/16.

7. a. $(4/16)^5 = 0.00098$. **b.** $(1/16)^5$. **c.** 7.6×10^{-6}. **d.** 6.1×10^{-5}.

8. a. yellow round = 1/4, green round = 1/4, probability = 1/2. **b.** The probability that the first offspring is $rr\ yy$ or $rr\ Yy$ is 1/2. The probability that the second offspring is $rr\ yy$ or $rr\ Yy$ is 1/2. The probability they both are $rr\ yy$ or $rr\ Yy$ is 1/4. **c.** 2/64. **d.** 2/1024.

9. $\chi^2 = 1.17$, df = 2, $0.5 < p < 0.7$. The results do fit a 1 : 2 : 1 ratio.

10. $\chi^2 = 2.55$, df = 3, $0.05 < p$. Yes, they do show independent assortment.

11. a. The normal function is required for wing formation. **b.** The normal function is required for bristle formation. **c.** The normal function is required for the production or accumulation of storage protein.

12. a. *wingless wg.* **b.** *bristless brl.* **c.** *empty kernel emk.*

13. A recessive allele. 1/24.

14. a. 1/4. **b.** 1/2. **c.** 2/3. **d.** 0. **e.** 1/4.

Chapter 3

1. Similarity: Both G_1 and G_2 are "gap" phases in interphase, and cells can arrest in G_1 or G_2. Difference: G_1 occurs before DNA replication, and the chromosomes have only a single chromatid. G_2 occurs after DNA replication, and chromosomes have two chromatids.

2. Extract the chromosomal DNA and measure the amount per cell. Compare the amount in these cells with the amount in cells known to be in G_1. Cells in G_1 have half as much DNA as cells in G_2.

3.

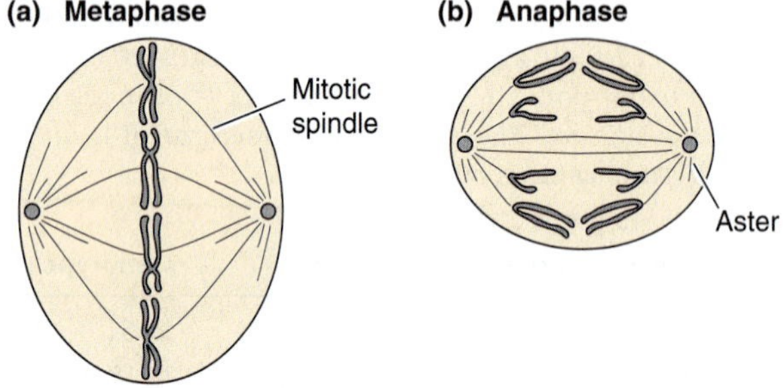

4. a. A chromosome arm is the part of a chromosome on one side of the centromere. A chromatid is one copy of an entire chromosome. Difference: a chromatid may contain more than one chromosome arm. **b.** A submetacentric chromosome has two arms, one slightly larger than the other. A telocentric chromosome has only one visible arm. Difference: the centromere is nearly in the center in the submetacentric, but it is at one end in the telocentric. **c.** The centromere and the kinetochore have very different functions. The centromere holds the two chromatids together until anaphase of mitosis, whereas the kinetochore is the site where the spindle fibers attach to the chromosome.

5.

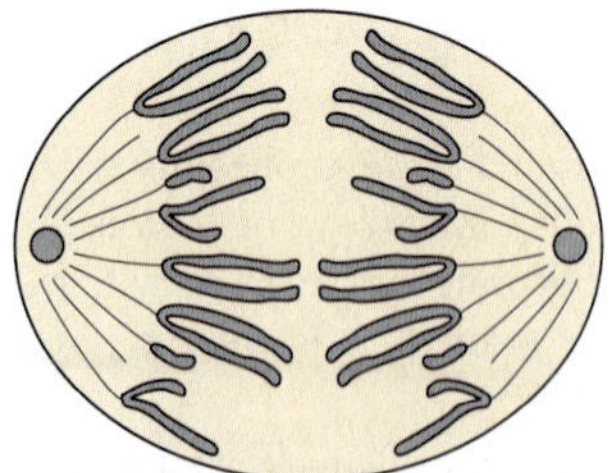

6. Late prophase or early metaphase. The chromosomes are condensed and have a characteristic size and shape.

7. Euchromatin is not condensed during interphase and contains active genes. Heterochromatin is highly condensed during interphase and prophase, and it contains few active genes.

8. Facultative chromatin is variable. A particular region may be facultative heterochromatin in one cell and euchromatin in another. Constitutive heterochromatin is a permanent feature of chromosomes, and it is always heterochromatic.

9. a. 46. **b.** 69. **c.** 12. **d.** 4. **e.** 30.

10. *Drosophila* are diploids ($2N = 8$). This individual has ten chromosomes; there are two additional large metacentric chromosomes present.

11. a. Both are part of prophase I; Leptotene is the first stage of prophase I, and Pachytene is the third stage of prophase I. **b.** Both the synaptonemal couplex and chiasma form during prophase I and involve physical association between homologous chromosomes. **c.** Both of these terms refer to paired homologous chromosomes. The four chromatids of two synapsed homologous chromosomes are a tetrad. Two synapsed homologous chromosomes are called a bivalent.

12. Similarity: In both, chromosomes segregate to opposite poles of the dividing cell. Difference: The chromosomes do not divide in anaphase I, but in anaphase of mitosis each chromosome divides into two progeny chromosomes.

13. a. Difference: Oogonia are diploid, whereas polar bodies are haploid. **b.** Difference: Primary spermatocytes are diploid cells in the first meiotic division, whereas spermatids are haploid meiotic products. **c.** Difference: The gametophyte stage is the haploid, and sporophyte is the diploid stage of plants. **d.** Difference: Synergids and antipodals are found at opposite ends of the embryo sac. **e.** Difference: The endosperm is the triploid tissue of the seed. The embryo sac is a special maternal structure within which the embryo forms. **f.** Difference: A bud is the beginning of asexual reproduction in yeast. An ascospore is a meiotic product in yeast.

14. a. In meiosis II and mitosis, the centromere of each chromosome divides during anaphase, and one of the two new progeny chromosomes segregates to each pole of the cell. **b.** (1) Meiosis includes two nuclear divisions and mitosis includes only one; (2) in meiosis the chromosome number is reduced by half, whereas in mitosis the chromosome number is not reduced; and (3) only germ cells undergo meiosis, but both germ cells and somatic cells undergo mitosis.

15. a. In males, both meiotic divisions are followed by equal cytoplasmic divisions producing four spermatids. In females, unequal cytoplasmic division produces one ootid and three polar bodies. **b.** Maturation of the pollen grain includes two mitotic nuclear divisions producing three haploid nuclei. Maturation of the egg includes three mitotic divisions producing eight haploid nuclei.

16. In sexual reproduction, diploid yeast cells undergo meiosis to produce haploid meiotic products. The meiotic products form ascospores, which germinate to produce a new generation of haploid cells. In asexual (vegetative) reproduction, diploid or haploid cells undergo mitosis and budding to produce progeny cells that are identical to the parent cell. During periods of favorable conditions, asexual reproduction allows genetically superior strains to produce large numbers of individuals. Under unfavorable conditions, sexual reproduction generates individuals with different, possibly superior genotypes. It also produces ascospores that can remain dormant for long periods until conditions have improved.

17. In moths, males are homogametic (*ZZ*) and females are heterogametic (*WZ*). Hypothesis: The dark male is *DD*, and the light female is *dW*. The F_1 are dark females *DW* and dark males *Dd*. The F_2 are *dW* and *DW* females, and *Dd* and *DD* males. To test this hypothesis, mate the dark-bodied F_1 males to light-bodied females. The progeny should be: females 1/2 dark (*DW*) and 1/2 light (*dW*); males 1/2 dark (*DD*) and 1/2 light (*dd*).

18. The F_1 male moths are heterozygous (*Dd*), whereas the parental male moths are homozygous (*DD*). Mate each dark moth separately with several light female moths. An F_1 moth will give both light and dark progeny, whereas a parental moth will give all dark progeny.

19. a. The *yellow* gene is sex-linked. y = yellow, y^+ = gray. The *rosy* gene is autosomal. ry = rosy, ry^+ = red. The mutant alleles of both genes are recessive. **b** and **c.**

Parental	(female) $yy\ ryry$	X	(male) $y^+Y\ ry^+ry^+$
F_1	(female) yy^+ryry^+	X	(male) $yY\ ryry^+$
	gray-brown red		yellow red
F_2	(female)		(male)
	$yy\ ry^+ry^+$		$yY\ ry^+ry^+$
	$yy\ ryry^+$		$yY\ ryry^+$
	$yy\ ryry$		$yY\ ryry$
	$y^+y\ ry^+ry^+$		$y^+Y\ ry^+ry^+$
	$y^+y\ ry^+ry$		$y^+Y\ ryry^+$
	$y^+y\ ryry$		$y^+Y\ ryry$

20. Nondisjunction in the males could produce sperm with no Z chromosome. Fusion of such a sperm with an egg containing a Z chromosome with a *d* allele would give a female (*dO*) with light color.

21. It should be passed from father to son. Females could never have this trait, and they could never be carriers of this trait.

Chapter 4

1. a. Complete dominance: homozygotes and heterozygotes have the same phenotype. Overdominance: heterozygotes have a more extreme phenotype than either homozygote. **b.** Incomplete dominance: heterozygotes have an intermediate phenotype different from either homozygote. Codominant: heterozygotes have the phenotypes of both alleles.

2. The new mutation is an allele of *sparkling* that shows overdominance.

3. a. There are two genes, one controlling flower color and the other controlling stem color. **b.** Flower color: *PP* = purple, *Pp* = blue, *pp* = white. Stem color: *DD* = *Dd* = dark, *dd* = light. Thus, each gene has two alleles. **c.** Blue flower, dark stem.

4. A codominant allele generates its full phenotype regardless of the other allele present. An incompletely dominant allele generates an intermediate phenotype when heterozygous with certain alleles. Incomplete dominance is a relative relationship between alleles.

5. a. L^ML^N **b.** L^ML^M, L^ML^N **c.** L^ML^M, L^ML^N, L^NL^N **d.** L^ML^N, L^NL^N

6. a. $A^2\ B^2\ C^1\ D^1\ E^1\ F^2\ G^2\ H^2$ **b.** No. This test indicates only that this man might be the father. It is possible that another man with these same alleles is the real father.

7. a. Same. **b.** Not the same. **c.** Same. **d.** Same. **e.** Same.

8. a. An amorph is an allele with no function; an antimorph is an allele with an new function antagonistic to the wild-type function of that gene. **b.** A null allele produces no gene product. An allele that produces a nonfunctional product would be an amorph, but not a null. **c.** A hypermorph is an allele that produces more than the normal amount of gene function. A hypomorph is an allele that produces less than the normal amount of gene function.

9. A segregation test should be used. Individuals heterozygous for different alleles should be test-crossed and the progeny analyzed to determine whether they always have only one of the parental alleles.

10. a. *R* = dominant round, *r* = recessive long, *B* = incomplete dominant blue, *b* = recessive white. Parent blue round = *BB RR*, parent white long = *bb rr*, F_1 = *Bb Rr*.

F_2 white, long flowers = *bb rr*

pale blue, long flowers = *Bb rr*

blue, long flowers = *BB rr*

white, round flowers = *bb RR* or *bb Rr*

pale blue, round flowers = *Bb RR* or *Bb Rr*

blue, round flowers = *BB RR* or *Bb Rr*

b. The color allele (*B*) is incompletely dominant.

11. a. Two. **b.** There is an interaction between the recessive alleles of these two genes such that individuals homozygous recessive for both genes have a long shape. **c.** *dd disdis* × *DD DisDis*

12. a. Epistasis and complete dominance both involve the complete masking of alleles by other alleles. In dominance, one allele masks another allele of the same gene. In epistasis, one allele or pair of alleles masks alleles of another gene. **b.** Epistasis and codominance are different, opposite relationships. Codominant alleles express their phenotype regardless of which other alleles are present. Epistasis involves one set of alleles masking another set.

13. a. Parents: white = *ww BB*, black = *WW bb*

F_1: agouti = *Ww Bb*
F_2: agouti = *W– B–*
black = *W– bb*
white = *ww ––*

b. White and black are produced by recessive alleles of two independently segregating genes. The *white* allele is epistatic to the *black* allele.

14. a. Two genes. **b.** Gene one: *TT* = *Tt* = triangle, *tt* = oval. Gene two: *OO* = *Oo* = round, *oo* = oval. Triangle is epistatic to round and oval; round is epistatic to oval. The F_2 are 9/16 *T– O–* = Triangle; 3/16 *T– oo* = Triangle; 3/16 = *tt O–* = round; and 1/16 *oo tt* = oval.

15. a. Partial penetrance and incomplete dominance are both instances of incomplete expression of a mutant phenotype by a mutant allele. The difference is whether all individuals with the allele have an intermediate phenotype (incomplete dominance) or whether some of the individuals express the phenotype and others do not (partial penetrance). **b.** Variable expression: Individuals with the same genotype have different degrees of severity of mutant phenotype. Codominance: individuals who are heterozygous show the phenotype of both alleles. **c.** Suppression and epistasis: One allele or pair of alleles prevents the expression of alleles of a different gene. Suppression involves the restoration of a normal phenotype. Epistasis involves masking with a different mutant phenotype. **d.** Enhancer mutations increase the severity of the mutant phenotype of another mutation. Overdominant alleles show a more extreme phenotype when heterozygous than either allele does when homozygous.

16. The penetrance of some of the phenotypes of Down syndrome is less than 100%.

17. The original curved black males were heterozygous for a dominant *Curved wings* (*C*) mutation that is recessive lethal and homozygous for a recessive *black body* (*b*) mutation. The F_1 were *Cc Bb* and *cc Bb*. The F_2 phenotype ratio is altered by the death of all *CC* individuals.

18. There are two genes segregating in these strains, both with recessive, temperature-sensitive alleles. Individuals homozygous for recessive alleles of both genes have an embryonic lethal phenotype. The F_2 contain 1/16 lethals (62.5/1000 = 1/16).

Chapter 5

1. a, b, and e are quantitative, and c and d are qualitative. a, b, and c show continuously variable phenotypes and do not give defined ratios in a cross.

2. The samples could be crossed and the F_1 inbred. The distribution of the phenotypes in the F_1 and F_2 will show whether it is a qualitative or a quantitative trait.

3. 9 classes.

4. 41 classes.

5. Mean = 157.1, $V_p = 300.32$, $s = 17.33$.

6. Compare the means of the two strains using a t test: $t = (11.87 - 11.81) / (3.58/15 + 5.77/16) = 0.10$. df = 29, probability > 90%. The means are not significantly different.

7. A t test indicates the two means are significantly different. $t = 2.72$, df = 28, probability < 2%.

8. $r = 0.889$.

9. $b = 0.68$.

10. The 75-in. parent gives 71.1-in. offspring. The 55-in. parent gives 57.5-in. offspring.

11. $V_e = 1.30$, $V_p = V_{F2} = 5.98$, $V_g = V_p - V_e = 4.68$.

12. $V_a = 2[5.98 - (4.54 + 4.03)/2] = 3.39$, $V_d = 4.65 - 3.39 = 1.26$.

13. $h^2 = 3.39 / 5.98 = 0.57$.

14. Tibia $h^2 = 0.15$, tarsus $h^2 = 0.58$. The tarsus will respond faster.

15. $R = 33.7$, $S = 72.7$, $h^2 = 0.46$.

16. cov (xy) = 72.47, $V_p = 93.71$, $H^2 = 72.47 / 93.71 = 0.77$.

Chapter 6

1. a. $\dfrac{y\ \ w\ \ +}{+\ \ +\ \ os}$ $\dfrac{p^p\ \ +\ \ +}{+\ \ bx\ \ pbx}$

b. $\dfrac{A\ \ D\ \ rg}{a\ \ d\ \ Rg}$ $\dfrac{y\ \ sh\ \ C}{Y\ \ Sh\ \ c}$

c. $\dfrac{wnt7\ \ +}{+\ \ cyc}$ $\dfrac{br^3}{+}$

d. $\dfrac{Cy\ \ tri\ \ +}{+\ \ +\ \ wt}$ $\dfrac{ttr\ \ +}{+\ \ wz}$ $\dfrac{pho\ \ +}{+\ \ ey}$

2. a. Yes. **b.** *A X / A X* and *a x / a x*.

3. Yes, they are linked, because the parental types (black cinnabar and gray red) are present more often than the recombinant types.

4. *A* and *C* are linked, *A* and *B* are not linked, and *B* and *C* are not linked.

5. 50%.

6. a. *m P aw / m P aw* and *M p Aw / M p Aw*. **b.** *m p aw*.
c. *m* — 8.0 — *p* — 17.0 — *aw*

7. CC = 0.49.

8. *a* — 10 — *e* — 30 — *c* — 18 — *b* — 5 — *d*

9. 3.5 cM.

10.

a.	**b.**	**c.**
s aw p / s aw p	4918	4717
S Aw P / s aw p	4918	4717
s aw P / s aw p	1007	1208
S Aw p / s aw p	1007	1208
s Aw P / s aw p	1307	1508
S aw p / s aw p	1307	1508
s Aw p / s aw p	268	67
S aw P / s aw p	268	67

11. *ala* — 10.1 — *mt* — 4.3 — *o* — 3.3 — *p*

12.

ala	+		*p*
ala	+	*o*	*p*
+	*mt*	**X**	+
+	*mt*	*o*	+

13. *o* — 45 — *e* — 10 — *b* — 15 — *c* — 20 — *d* — 10 — *a*

14. Chromosome 4.

15.

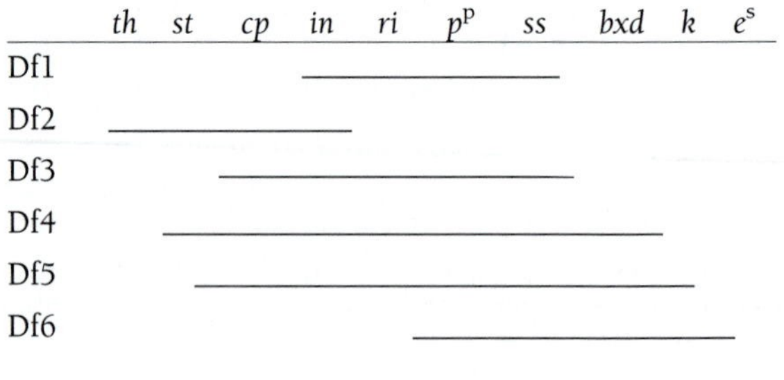

	th	*st*	*cp*	*in*	*ri*	p^p	*ss*	*bxd*	*k*	e^s
Df1				—	—	—	—			
Df2	—	—	—	—						
Df3			—	—	—	—	—			
Df4		—	—	—	—	—	—	—		
Df5			—	—	—	—	—	—	—	
Df6						—	—	—	—	—

16.

```
|  b  |  c  |  e  |
       ---- a ----
  -- d --
```

17.

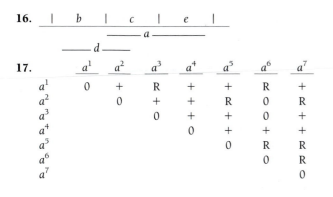

	a^1	a^2	a^3	a^4	a^5	a^6	a^7
a^1	0	+	R	+	+	R	+
a^2		0	+	+	R	0	R
a^3			0	+	+	0	+
a^4				0	+	+	+
a^5					0	R	R
a^6						0	R
a^7							0

Chapter 7

1. Backcross the F_1 individuals to plants with straight leaves. If twisted leaves are given by a dominant allele, then the progeny will be 1/2 twisted leaves and 1/2 straight leaves. To test for maternal effects, do reciprocal crosses. When twisted-leaf plants are used as females, all the progeny will have twisted leaves. When straight-leafed plants are used as females, all the progeny will have straight leaves.

2. 1/2 short leaf and 1/2 long leaf.

3. Do reciprocal crosses. Cross individuals homozygous for the suspected maternal-effect mutation (generated by crossing two heterozygotes) to a known wild-type strain. If this is a maternal effect, females from the strain will give lethal offspring and males from the strain will give normal offspring.

4. a. Cytoplasmic inheritance: the F_2 have the same genotype and phenotype as the parental females. Dominant mutation: the F_2 are 1/2 heterozygous (show the trait) and 1/2 homozygous for the normal allele (do not show the trait). **b.** Cytoplasmic inheritance: the F_2 have the same genotype and phenotype as the parental females. Dominant epistasis: in a dihybrid cross, the F_2 phenotypic ratio is 12 : 3 : 1.

5. a. All green leaf. **b.** All variegated (some white, some variegated, some green).

6. Male sterile but female fertile.

7. Male fertile.

8. F_1 = male fertile. F_2 = 9 fertile : 7 sterile ratio.

9. Determine whether the infective agent has the structure of a virus or a viroid. Isolate cytoplasm from infected plants and purify the infective agent. A virus has a genome with several genes. A viroid does not contain genes that code for any proteins.

10. Determine whether the trait is infectious by injecting a small amount of cytoplasm.

11. a. Backcross female sensitive flies to nonsensitive flies for two generations. If this is a maternal effect, there should be nonsensitive offspring. **b.** Do reciprocal crosses, and then backcross the F_1 females to males of the opposite phenotype. The sensitive phenotype will be inherited only from sensitive females, not sensitive males, and the offspring of sensitive females will always be sensitive, even after many generations of backcrossing to nonsensitive males. **c.** Backcross the sensitive F_1 females to nonsensitive males. The offspring should show a 1 sensitive : 1 nonsensitive ratio. **d.** Isolate cytoplasm from sensitive flies and inject it into nonsensitive flies. Some of them should become sensitive or give sensitive offspring.

Chapter 8

1. $2^{24} = 1.67 \times 10^7$.

2. a. *tet*^{s} **b.** *phe*$^-$ **c.** *lac*$^-$ **d.** *kan*^{r} **e.** *rrn*^{ts}
a, c, d, and e are prototrophs, and b is an auxotroph.

3. Conjugation requires contact between two mating bacteria, transferring the entire chromosome if not interrupted. Transformation is the uptake of a small fragment of DNA present in the surrounding environment and does not require physical contact between mating cells. Transduction requires the participation of a third party (phage) to transfer the DNA from one cell to another.

4. A conjugative plasmid has a set of genes that allows it to be mobilized into another bacterial cell.

5. Acquisition of an F-plasmid converts an F^- into an F^+. Integration of the F-plasmid into the bacterial chromosome gives an *Hfr*.

6. There are two reasons: the conjugation tube is usually broken, and even if the entire chromosome were transferred, multiple recombination events would preclude its total integration into the recipient strain.

7. They differ by their points of integration and orientation.

8. By mapping many different genetic markers with *Hfr* strains having different points of integration and orientation.

9. The *str*^{r} marker is present in the F^- strain to allow selection against the donor strain using streptomycin.

10. Otherwise the conjugation would continue after the cells are spread on the Petri plate, bringing along other markers.

11. These two mutations are in separate genes; otherwise, the *leu* gene on the F′ would have allowed both to grow on media lacking leucine.

12. You are seeing recombination between the two markers generating a *leu*$^+$ genotype.

13. A plasmid is not required for growth of the cell, has its own origin of replication, has a variable copy number, and can contain almost any gene(s) that can be cloned into the plasmid.

14. *pheA* and *leuC* are distant from each other and did not co-transform. The predicted frequency would be $0.00016 \times 0.00009 = 0.0000000144$.

15. A virulent phage does not integrate into the genome of the host, whereas a temperate phage can integrate or go into a lytic cycle.

16. A prophage is integrated into the bacterial chromosome, and the phage is not.

17. 8.4×10^9.

18. 160,461.

19. The overlapping deletions divide the loci into a series of regions. Only those mutations that are in the same region need to be mapped by crossing them with each other.

20. The steps are attachment, entrance of phage DNA into the cell, phage DNA replication, transcription and translation of phage genes, and packaging of DNA into the phage particle.

The accident occurs during packaging of the DNA into the phage heads when bacterial DNA is packaged instead of phage DNA.

21. The phage contains only bacterial DNA, not phage DNA that codes for the lytic cycle.

22. Minute 17 is the integration site of lambda phage, and it transduces only those markers surrounding the integration site.

23. This new phage is likely a specialized transducing phage, and the integration site is near *art*. This could be further tested by attempting to transduce other markers near *art* and also by physically looking for integration of the phage at this site during a prophage stage.

Chapter 9

1. Transformation is the transfer of exogenous DNA into a cell through the cell membrane followed by incorporation into the genome. Avery, MacLeod, and McCarty showed that the transforming factor of Griffith's experiments was susceptible to DNase and resistant to RNase and protease, indicating that DNA was the genetic material.

2. Chromosomes, the site of genetic information, contain both protein and DNA. Protein was characterized earlier than DNA and was thought to be more complex in composition and sequence than was DNA, and thus a more likely candidate as the genetic material.

3. Avery, MacLeod, and McCarty's DNA was contaminated with trace amounts of protein, and other scientists thought that DNA might be mutating the host genetic material. Conclusive evidence was provided by Hershey and Chase (1952). See text for details.

4. Deoxyribose is a pentose sugar lacking the 2′-OH group, whereas ribose possesses the 2′-OH.

5. 5′ to 3′.

6. A nucleoside is a base + sugar. A nucleotide is a base + sugar + phosphate. The phosphate is bonded to a C at the 5′ position.

7. C = 22%, A = 28%, and T = 28%.

8. The DNA with the higher G + C will be more stable because it has more H bonds.

9. DNA is a long, thin molecule with 3.4 nm/10 bp and a diameter of 2.6 nm, and it has a large groove and a small groove as it forms a double helix.

10. DNA longer than 3×10^7 tends to shear into two pieces of about 10^7 Da when purified.

11. Semiconservative replication means that the progeny DNA contains one parental (conserved) and one newly synthesized strand of DNA. The proof was provided by Meselson and Stahl (1958) and is described in the text.

12. Bidirectional replication proceeds in both directions simultaneously from one site of initiation, which is necessary to complete DNA replication in a reasonable length of time. Because of the slow speed with which DNA polymerase acts, it is also necessary to have multiple replication points to ensure that replication takes minutes, not weeks.

13. There will be 1,388 origins of replication spaced 2,161,383 bp apart.

14. A primer provides a free 3′ -OH group for extension of the growing chain, and a template provides the information to specify the sequence of bases in the DNA.

15. One strand is synthesized continuously (the leading strand), whereas the other strand is synthesized discontinuously (the lagging strand) in the opposite direction from the growing point.

16. The high degree of fidelity in DNA replication is maintained by the template, which specifies base selection; mismatches are corrected by 3′ → 5′ exonuclease activity of DNA polymerase III or polymerase I.

17. In prokaryotes, DNA polymerase III is the main enzyme of synthesis; DNA polymerase I is responsible for primer removal and fill-in during joining of Okazaki fragments. In eukaryotes, at least four different enzymes exist with specialized functions. The general mechanism of polymerization is otherwise identical.

18. Although both DNAs are circular at the time of replication, phage lambda replicates by a rolling circle method, whereas mitochondria replicate by D-loop expansion.

Chapter 10

1. It has two nucleoid structures just before fission.

2. DNase or RNase action result in uncoiling; protease digestion does not. This implies that the proteins are buried within the structure where the enzyme is unable to digest them.

3. A plasmid is 1–5% the size of the chromosome, is frequently present in multiple copies, and is not necessary for the survival of the cell.

4. A linking number is the number of twists in supercoiled DNA.

5. The two classes are type I, which makes a transient single-strand break, and type II, which makes a transient double-strand break.

6. It unwinds. Nothing remains to constrain the double-helical molecule in the supercoiled state.

7. Supercoiling makes the molecule much more compact.

8. In 1974, Ruth Kavenoff and Bruno Zimm measured the molecular weight of the longest *Drosophila melanogaster* chromosome and found a value of 0.41×10^{11} Da. The haploid content of *D. melanogaster* is 1.2×10^{11} Da, and it has four chromosomes. Thus, dividing by 4 yields 0.3×10^{11}, so these two values are sufficiently close (0.41×10^{11} and 0.30×10^{11}) to conclude that the largest chromosome from *Drosophila* is composed of one double-helical molecule of DNA.

9. The average chromosome is 20.6 mm (Table 10.1); thus, 20.1/0.01 = 2060 ratio of condensed to extended.

10. No one knows for sure, but highly repetitive DNA may be important in attachment of spindle fibers during meiosis and mitosis. Also, it may have some evolutionary importance.

11. 875,000 nucleosomes/genome.

12. Histone proteins comprise the protein portion of the nucleosome around which DNA wraps. They also neutralize the negative charge of the DNA and enable DNA to coil into a more compact structure.

13. The scaffold is the central protein core of the chromosome that holds the DNA loops to a fixed point.-

14. Euchromatin is generally active and contains primarily genetic material that does not stain darkly with Feulgen stain. It contains unique sequences. Heterochromatin is darkly staining inactive (i.e., it does nothing) genetic material containing repetitive DNA sequences and methylated CG.

15. Facultative heterochromatin is a transient form of genetically inactive material methylated at CG doublets. Constitutive heterochromatin is a stable form and is composed of repetitive sequences as well as methylated CG doublets.

16. 5-Methylcytosine is believed to represent inactive areas of the genetic material, but this point is still debatable.

Chapter 11

1. A gene is responsible for the amino acid sequence of a protein, which may or may not be an enzyme or part of an enzyme. The gene also controls the amount of the protein synthesized at any one time.

2. The bacteria has a mutation prior to the last step of the synthetic pathway for the synthesis of cysteine.

3. The sigma subunit is needed to recognize the promoter of the genes to avoid random initiations.

4. In bacteria, each promoter is composed of a −10 TATAAT consensus sequence and a −35 TGTTGACA consensus sequence. The −10 and −35 refer to the distance from the start site of RNA initiation.

5. A polycistronic mRNA is made from several adjacent cistrons, all controlled by one promoter. A cistron is a genetic term used interchangeably for a gene.

6. The upstream sequences that control gene expression in eukaryotes are the so-called TATA box located at −35 bp and the CAAT box located at about −75 bp from the start site of RNA synthesis.

7. An enhancer is an additional controlling element for a gene and can be located either upstream or downstream of the promoter and in either sequence orientation. The enhancer is responsible for control of the level of expression of the gene, whereas the promoter is primarily responsible for RNA polymerase binding.

8. After synthesis, the product of RNA polymerase II, messenger RNA, is capped with 7-methylguanosine, a poly(A) tail is added, and introns are removed.

9. An intron is a noncoding sequence within a gene that is spliced out after RNA synthesis, and an exon is a coding sequence.

10. Autocatalytic splicing reactions require the participation of a free guanosine to supply the attacking –OH group, whereas enzyme-catalyzed splicing reactions use an adenosine within the intron.

11. Transcription is the synthesis of RNA using DNA as a template, and translation is the synthesis of proteins using the coded information in mRNA.

12. mRNA contains the coding information for the gene and thus acts as a template for the generation of proteins. It also varies considerably in size, reflecting the size of the gene and gene product. tRNA is a small molecule (about 70 bp) and 63 different types exist, each with a different anticodon specifying a different amino acid. tRNA participates in protein synthesis by bringing the amino acid to the ribosome and placing it properly as specified by the mRNA. rRNA is a structural component of the ribosome, and three types exist, varying in molecular weight and function. In prokaryotes, they include 23S, 16S, and 5S RNAs. The exact function of each is not totally clear, but the 23S rRNA possesses sequences that position the mRNA properly on the ribosome.

13. met-$tRNA^{init}$ is used only to initiate protein synthesis and differs from $tRNA^{met}$.

14. The fidelity of translation needs to be high so that it does not synthesize a mixture of protein products from a single gene. This would seriously affect gene function and the function of the organism.

15. Met–Pro–Gly–Asn–Thr–Ala–Gly–Asn–Ser
Met–Pro–Gly–Lys–His–Ser–Arg–Glu

16. $4^4 = 256$.

17. Six, each from a different reading frame and from both strands.

18. Overlapping codes affect two or more protein products during a mutational event, which may be more severe. Overlapping coding sequences allow compaction of information. More information can be stored in a shorter sequence.

Chapter 12

1. Restriction endonucleases can cleave DNA at specific sites and generate single-stranded tails that can anneal to another piece of DNA cut with the same enzyme.

2. The universality of the genetic code is essential.

3. The *Eco*RI and *Hind*III sites are probably blocked by the presence of methyl groups, whereas the *Sal*I sites are not.

4. Plasmid cloning vectors are small, self-replicating DNA molecules that contain unique cleavage sites and an antibiotic resistance gene(s).

5. Blunt-ended fragments can be joined by increasing the amount of DNA ligase in the reaction mixture and increasing the time of incubation.

6. The cDNA library represents cloned sequences from active genes and does not contain introns, which allows expression in bacteria when the proper promoters are present. cDNA cloned fragments will be smaller because of the lack of the introns. Regulatory sequences are not present in cDNA cloned fragments, which may be a disadvantage.

7. The starting material, mRNA, is not representative of the entire genome.

8. 808.

9.

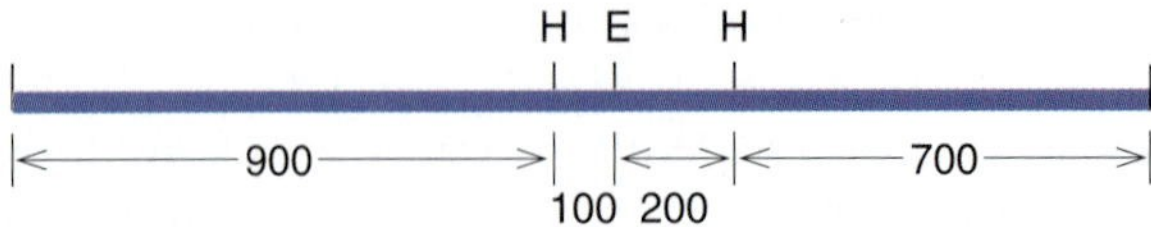

10.

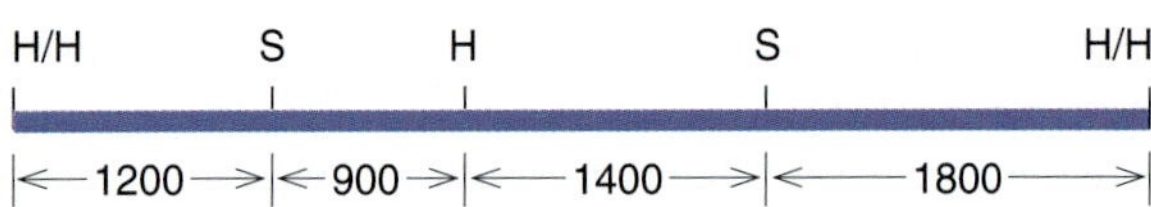

11. Use labeled proline tRNA as a probe to search the library.

12. 5′-TCTGGATCCCTAG-3′.

13. Southern blotting uses DNA, and northern blotting uses RNA as the target.

14. An expression vector differs from a regular plasmid vector in that it contains a strong promoter with an adjacent cloning site.

15. An ideal expression vector might contain a strong promoter that is inducible. Inducibility allows researchers to control gene expression so that the host cell is not harmed or slowed during normal growth. Some good promoters include *lac*, *trp*, and *taq*.

16. The gene must have the bacterial controlling sequences (promoter, terminator, and so on), and the eukaryotic gene cannot have any introns or other processing events for activity.

17. A gene fusion is the fusion of a reporter gene to the controlling sequences of a gene. They are useful for the study of gene expression.

18. Gene controlling-region fusions enable gene product assays with greater ease and sensitivity, enabling a study of the regulatory region.

Chapter 13

1. An RFLP is a DNA-fragment size difference generated by a given restriction enzyme. A hypervariable marker is representative of differences in the size of repeated DNA regions between individuals.

2. RFLPs are detected by Southern blotting using a labeled probe, either RNA or cloned DNA, or by PCR.

3. RFLPs can be formed by translocations, inversions, deletions, insertions, and point mutations.

4. The probe has repetitive DNA sequences within it and is detecting these sequences in the total genome digest.

5. The child may be the product of a mating between Person A and Person B. These data do not absolutely prove paternity, but they cannot exclude it.

6. Both plasmids and cosmids have bacterial origins of replication, selectable markers, and cloning sites. Cosmids also have a *cos* sequence that allows packaging into lambda phage heads.

7. They give bigger insert sizes.

8. An antibiotic marker can have the insert site located within it. Thus, when the insert is present, the marker is not expressed.

9. YAC vectors allow cloning of extremely large DNA fragments, and eukaryotic genes are frequently very large as a result of the presence of introns.

10. The size of the insert is much larger in BACs and YACs than in cosmids.

11. A YAC vector requires telomeres and a centromere.

12. The DNA must be handled with care to prevent breaking. If shearing occurs, the gel separation pattern will be smeared.

13. PFGE will easily separate fragments up to 3000 kb, whereas gel electrophoresis will separate DNA fragments up to 50 kb. Both methods can separate fragments of just a few nucleotides in length or differing in length by one nucleotide.

14. T-DNA is a component of the Ti-plasmid of *Agrobacterium tumefaciens*. The T-DNA actually represents about a quarter of the total size of the Ti-plasmid and possesses nonessential genes for transferring the T-DNA region into plant cells. The only region in the T-DNA that is essential are the two 25-bp sequences at each end of the T-DNA region.

15. This large DNA fragment can be cut into smaller fragments using restriction endonucleases, and each of these fragments can be inserted into the P element. The smallest-sized fragment that complements it is then used for study.

Chapter 14

1. An operon is a group of adjacent structural genes controlled by a contiguous operator gene. The unique feature of an operon is the synthesis of a polycistronic mRNA and a single promoter.

2. An inducible operon can be turned on and off, depending upon environmental factors, and a constitutive operon is always expressed.

3. The repressor prevents RNA polymerase from binding because their binding sites overlap on the DNA.

4. The mutation must be a polar or nonsense mutation affecting downstream translation of the mRNA.

5. Two of the enzymes are heterodimers.

6. An antiterminator protein allows gene expression beyond a termination site by binding to the termination site. An attenuator can turn up or turn down the level of gene expression in response to the presence of environmental factors.

7. **a.** constitutive. **b.** absent. **c.** absent.

8. The GAL4 promoter most likely activates a downstream promoter by looping the DNA and at the same time loading the polymerase and its transcription factors onto the promoter.

9. They both have DNA-binding sites that are similar in structure.

10. a and d, because they bind to DNA.

11. By alternative splicing of the various exons and introns.

12. The housekeeping genes will probably produce very stable mRNAs, whereas the regulatory protein mRNAs would probably be less stable. In this way, the regulatory proteins can rapidly change their concentrations in the cell. Housekeeping genes do not need to change their concentrations; in fact, they

need to maintain them at a relatively constant level because the products of the genes are in constant demand.

13. If the repressor is absent, both the H2 flagellin and the H1 flagellin proteins could be made, but only if the invertible sequence were in the right orientation. Normally, the repressor inhibits the synthesis of the H1 flagellin protein, but without the repressor, the expression of the H1 flagellin gene would become constitutive. On the other hand, if the invertible sequence were in the other orientation (and did not act as a promoter for H2), then only the H1 protein would be made, and it would be made constitutively.

14. A genome rearrangement is inherited from generation to generation, whereas transcriptional control is not.

15. An antigen is a foreign molecule that elicits the production of an antibody in higher eukaryotes. An antibody is a protein made by the immune system that specifically binds to the foreign antigen. A monoclonal antibody is a single antibody species that binds to one site on the foreign antigen. A polyclonal antibody is a mixture of monoclonal antibodies, each of which bind to different sites on the foreign antigen.

16. Genes can produce so many antibody proteins because they have many segment variations. These segments are rearranged to produce gene products or antibodies.

17. Variation occurs in the variable region because this area of the protein interacts with the antigen. Because many different antigens exist, the *V*-gene segment must have great variation.

Chapter 15

1. a. Cell determination: the irreversible, heritable adoption of a specific cell fate. **b.** Cell differentiation: the generation of specialized products by a cell as part of the expression of its cell fate. **c.** Terminal cell differentiation: the differentiation of a cell into a cell type that does not divide any further. **d.** Positional information: a signal that gives a cell information about its location in the organism, information the cell uses in making a cell fate decision. **e.** Cytoplasmic determinant: a molecule that directs the cell to adopt a particular cell fate.

2. A second copy of head structures will form in the abdomen where the head cytoplasm is inserted.

3. Imaginal leg disc cells, imaginal wing disc cells.

4. Culturing cells tests whether cells change type when in a new environment. In somatic mosaic experiments, we observe the structures formed by the progeny of a cell in their normal environment, without challenging them to change.

5.

Structures	Fate Map Distance
A–B	5.3
B–C	30.7
A–C	40
C–D	25.3
A–D	20
B–D	45.3

6. X–Y = 12 cells, L–M = 8 cells, Z–P = 6.4 cells.

7. A similarity: in both systems cell determination is controlled by genes, with each gene controlling determination within a particular region. A difference: in *Drosophila* blastoderm cells are determined at blastoderm in a spatially organized pattern. In *C. elegans* cells are determined by cell lineage starting from the first division.

8. A similarity: both imaginal discs and cell equivalence groups contain all the cells determined to form one particular structure. A difference: imaginal discs grow throughout the larval period but do not begin to form adult structures until metamorphosis. The cells of a cell equivalence group begin to form the adult structure during the larval stages.

9. a. Syncytial blastoderm: the stage of *Drosophila* embryonic development where the embryo consists of a nuclear syncytium, with most of the nuclei in the cortical cytoplasm. **b.** Cell blastoderm: the stage of embryonic development where the embryo consists of a ball of cells; in *Drosophila,* it is the stage at which cells first become determined. **c.** Imaginal disc: a tightly associated collection of cells that are all determined to form one portion of the adult. **d.** Blast cell: a cell in *C. elegans* that divides to produce adult-specific structures. **e.** Cell equivalence group: a group of blast cells in *C. elegans* whose descendants are the only cells capable of forming a particular adult structure.

10. Use the cloned fragment as a probe in an *in situ* hybridization experiment to determine directly whether *bcd* mRNA distribution is altered in mutant embryos.

11. Do complementation tests, cross individuals from this strain with individuals carrying known *swallow* or *bcd* mutant alleles. If the strain of flies has a *bcd* mutation, it will fail to complement the known *bcd* mutant allele; if it has a *swallow* mutation, it will fail to complement the known *swallow* allele.

12. Test for complementation with known mutations of the *Ubx* group. If the new gene is part of this group, it will fail to complement with *Ubx* mutations, yet will complement recessive mutations of genes in other groups.

13. Attach the cloned *BX-C* sequence to an indicator gene, such as β-galactosidase, transform the construct back into the organism, and observe the location of β-galactosidase transcription. If the sequence is a tissue-specific regulator, then β-galactosidase will appear in the cells in particular regions of the body.

14. Lateral specification: a signal sent between two previously identical cells that causes them to adopt separate fates. Lateral induction: a signal sent between two nonidentical cells that is used by one cell to influence the fate of the other.

15. Yes, it would. Other cells in the vulval cell equivalence group would form the structures normally formed by the descendants of P6.p.

16. Isolate loss of function and gain of function (overproducing) mutations of both genes. Gain of function mutations of the receptor gene should cause both cells to adopt the net receptor fate, and loss of function mutations of the receptor gene should cause both cells to adopt the net signaler fate.

17. In a normal individual, the scutellar bristle-forming cell inhibits the neighboring cells from forming bristles.

18. Yes, one of the neighboring cells would adopt the bristle-forming fate and would form the bristle.

Chapter 16

1. Do complementation tests with this allele and recessive alleles of all other known anterior class genes. If a female with this allele in trans with a known allele of an anterior class gene produces embryos with a mutant, anterior class phenotype, the unknown mutation must be an allele of that gene.

2. Inject blastoderm stage embryos with *Dorsal* gene product to establish an abnormal gradient of product. This should change the location of the body axis.

3. a. Maternal-effect mutation: a mutation that affects the offspring of a mutant female, no matter what the genotype of that offspring. **b.** Pair-rule gene: a segmentation gene that is expressed in seven stripes in the blastoderm and whose mutation causes the loss of alternate segments. **c.** Segment polarity gene: a segmentation gene that is expressed in 14 stripes of cells and whose mutation causes the loss of some structures in each segment.

4. a. Maternal-effect mutations affect the offspring of a mutant female, whereas zygotic mutations affect the mutant individual. **b.** Segmentation genes organize cells into spatially discrete groups, each of which gives rise to a segment or compartment. Determination genes give each segment or compartment its unique identity. **c.** Gap genes are the first zygotic genes in the segmentation gene hierarchy. They are activated by maternal gene products and are expressed in broad stripes in the syncytial blastoderm. Segment polarity genes are activated late in the segmentation gene hierarchy. They are expressed in 14 narrow stripes in the cellular blastoderm.

5. a. A broad stripe of cells. **b.** Seven intermediate-sized stripes of cells. **c.** Fourteen narrow (one-cell-wide) stripes of cells. **d.** A cluster at the anterior end of the embryo.

6. a. Parasegments 4 and 5. **b.** Parasegments 2, 3, and 4. **c.** Parasegments 5–12. **d.** Parasegments 1, 2, and 3.

7. Clone the *Hoxa-3* gene into a *Drosophila* transformation vector and transform it into a strain with a *pb* mutation. If the clone can restore the normal phenotype, then it is functionally equivalent.

8. Hybridize a labeled *sxl* probe to mRNA isolated from individuals of different sexes. The *sxl* probe should hybridize to a different-sized RNA in males and in females.

9. All should develop as males.

10. Similarity: in both organisms, sex determination is controlled by a gene hierarchy whose regulatory activities are sex-specific and are triggered by a primary signal set by the ratio of the X chromosome to the autosomes. Difference: *Drosophila* genes in the pathway are regulated by sex-specific patterns of primary transcript splicing, whereas in the nematode gene transcription is regulated.

Chapter 17

1. A mutagen causes alterations in DNA that are inherited, and a teratogen causes abnormalities in developmental patterns that are not inherited.

2. Germ cells.

3. Silent mutations, including missense mutations, cause genetic changes but not phenotypic changes. Recessive mutations in a heterozygous state may be present in eukaryotes and would not be manifested until selfing when they become homozygous. Several types of silent mutations occur: changes in the third base of a codon resulting in the same amino acid; changes to codons that code for chemically similar amino acids (leucine to isoleucine); and alterations in the protein that are nonessential to function.

4. Nonsense mutations terminate protein synthesis prematurely, but missense mutations substitute one amino acid for another. Thus, the drastic structural changes associated with nonsense mutations rarely produce temperature-sensitive or leaky mutations.

5. Out-of-frame nonsense codons are encountered.

6. Intragenic suppressors could be downstream frameshifts from the original frameshift or structurally compensating mutations. Extragenic suppressors can be mutations in anticodons in tRNAs suppressing nonsense mutations or in another protein that forms a dimer with the first protein, restoring its activity.

7. Most mutations are recessive and therefore not expressed unless they are homozygous. Selfing the F_1 will yield 1/4 homozygous recessives in the F_2.

8. 4.8×10^{-4} mutations/generation.

9. Insertion sequences and transposon elements sometimes possess promoter activity in their termini, and by inserting at the correct position, the gene will use the promoter from the insertion or transposon element instead of its own promoter.

10. The rate of transposition is much higher (one event for every 1000 to 10,000 divisions) than the rate of base-pair mutation.

11. The mutation rate of color-blindness is high because of unequal crossing over of tandemly arranged color pigment genes during meiotic recombination. Occasionally, some genes are deleted. Mutant alleles are not eliminated by selection.

12. a. both; **b.** induced; **c.** both; **d.** spontaneous; **e.** both.

13. Repair of UV damage allows random bases to be inserted opposite the original bases.

14. Yes. Both of these base analogues can pair with their respective bases; thus, they can mutate in either direction.

15. a. yes; **b.** no; **c.** no.

16. Cancer has a mutational origin.

17. Use the Ames test with some of your urine before and after dying your hair.

18. Three ways of repairing a thymine dimer: photoreactivation cleaves the pyrimidine–pyrimidine bond; mismatch repair excises a misincorporated base with endonucleases and replaces them using the information in the other strand; and excision repair also excises mutated bases and replaces them using the information in the other strand.

19. Higher. If DNA glycosylases remove a purine or pyrimidine, without the AP endonuclease there is no way to correct the lost bases. Thus, on subsequent replications, random bases will be added opposite the gaps left by the glycosylases, creating mutations.

20. Yes, any base can be added opposite the gap.

21. Without the methylation reaction, the mismatch repair system cannot discriminate between the new and old strands after replication and may correct the wrong strand.

22. As backup systems in case the other repair mechanisms do not repair the mistakes.

Chapter 18

1. **a.** 15 chromosomes. **b.** 20 chromosomes. **c.** 10 chromosomes. **d.** 30 chromosomes.

2. **a.** There is no set haploid number. The random segregation of unpaired chromosomes in meiosis I produces meiotic products with different numbers of chromosomes. **b.** 10 chromosomes. **c.** 5 chromosomes. **d.** 15 chromosomes.

3. Determine whether the chromosomes in *Triticum dioccum* are multiple copies of *Triticum monoccum* chromosomes. A cytogenetic analysis could be done in which the chromosomes are stained and compared with *Triticum monoccum* chromosomes.

4. Allopolyploid.

5. 1/4096.

6. **a.** Three phenotypes: pink (RRrr), light pink (Rrrr), white (rrrr). **b.** 1/6 pink (RRrr), 4/6 light pink (Rrrr), 1/6 white (rrrr) progeny.

7. **a.** Genotypic ratio = 1/4 RRrr, 1/2 Rrrr, 1/4 rrrr. Phenotypic ratio = 1/4 pink, 1/2 light pink, 1/4 white. **b.** 3/4 red, 1/4 white.

8. Cross a tetraploid plant (e.g., watermelon, 2N = 4X = 48) with a diploid (2N = 2X = 24) plant, and the progeny will be sterile triploids (3X = 36 chromosomes). The triploid is sterile because all offspring are inviable aneuploids and do not form seeds.

9. **a.** *Pp* and _. **b.** *P P*, _, *pp*.

10. Nondisjunction occurred in the father. This produced an *AA* sperm that fused with an *a* egg to produce an *AAa* individual.

11. The genotypes of the progeny are:

Gametes of the Diploid	Gametes of the Trisomic					
	$+\ ey^D$	$+\ ci^D$	$ey^D\ ci^D$	$+$	ey^D	ci^D
ey^D	$+/ey^D/ey^D$	$+/ci^D/ey^D$	$ey^D/ey^D/ci^D$	$+/ey^D$	ey^D/ey^D	ci^D/ey^D
ci^D	$+/ey^D/ci^D$	$+/ci^D/ci^D$	$ey^D/ci^D/ci^D$	$+/ci^D$	ey^D/ci^D	ci^D/ci^D

These 12 classes are of equal frequency. Two of the classes are lethal (ey^D/ey^D and ci^D/ci^D), so there are ten genotypic classes in the adult progeny. These give three phenotypic classes: 6/10 *ey ci*, 2/10 *ey*, and 2/10 *ci*.

12. Chromosome C.

13. **a.** The nondisjunction must have occurred in the woman. The XXY child is color-blind (an X-linked trait), so both X chromosomes must have the color-blind allele. The man has normal vision, so his X cannot contain the color-blind allele; therefore, the child must have received both X chromosomes from the woman, who must be heterozygous for the color-blind allele. **b.** The nondisjunction must have occurred in meiosis II because the child received two copies of one of her mother's X chromosomes.

14. Chromosomes 13, 18, and 21 contain fewer genes that are sensitive to changes in chromosome dosage than chromosomes 22 or 20. Thus extra copies of chromosomes 20 or 22 cause more disruption of embryonic development. This could be tested by examining fetuses that are spontaneously aborted. Karyotyping these fetuses would indicate which were trisomic for chromosomes 20 or 22. Analysis of the physiology of these fetuses would indicate whether they had more severe developmental abnormalities than individuals trisomic for chromosomes 13, 18, or 21.

15. Genotypic ratio at conception 1/4 2N + 2 : 1/2 2N + 1 : 1/4 2N. There should be very few (if any) live births of 2N + 2 individuals, some live births of 2N + 1 individuals, and more live births of 2N individuals.

16.

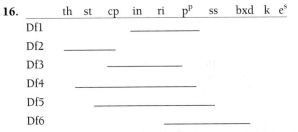

	th	st	cp	in	ri	p^P	ss	bxd	k	e^s
Df1				—	—	—				
Df2	—	—	—							
Df3			—	—	—					
Df4		—	—	—	—	—				
Df5			—	—	—	—	—			
Df6						—	—	—		

17. **a.** The unknown individual had an inversion with breakpoints between *st* and p^P, and between *e* and *gro*.

b.

cur	*st*	|	p^P	*red*	*e*	|	*gro*
14	2	2	5.6	17.1	8		11.3

inversion (from the first | to the second |)

18. **a.** Inversion. **b.** One breakpoint is 8 m.u. to the right of a, and the second is 7 m.u. to the right of c.

c.

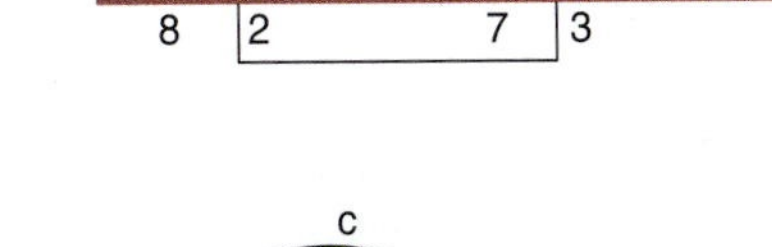

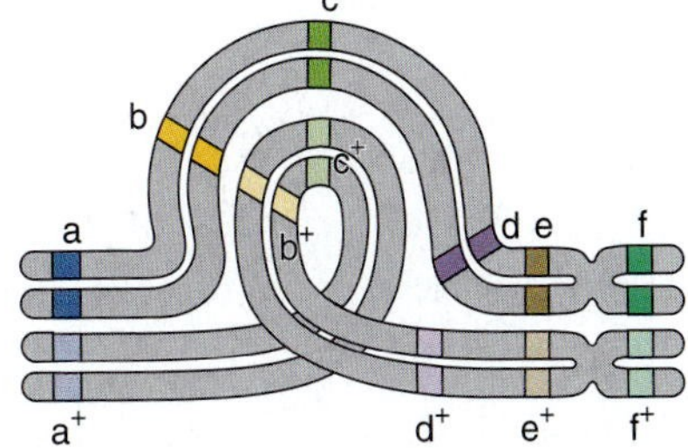

19. The *a* and *g* alleles are not segregating independently. This could be pseudolinkage caused by a translocation between the chromosome and *a* and the chromosome with *g*.

20. **a.** Transposition or insertional translocation. **b.** Pericentric inversion. **c.** Translocation. **d.** Adjacent 1 segregation.

21. a. alternate = IHB, XM xhb, im. **b.** adjacent 1 = IHB, xhb im, XM. **c.** adjacent 2 = IHB, im xhb, XM

22. alternate = IHB, Xhb im, xM IHB, XM xhb, im.
adjacent 1 = IHB, xM im, Xhb IHB, xhb im, XM.
adjacent 2 = IHB, im xM, Xhb IHB, im xhb, XM.

23. alternate = IHb, XM im, xhB IHB, XM xhb, im.
adjacent 1 = IHb, xhB XM, im IHB, xhb im, XM.
adjacent 2 = IHb, im XM, xhB IHB, im xhb, XM.

24.

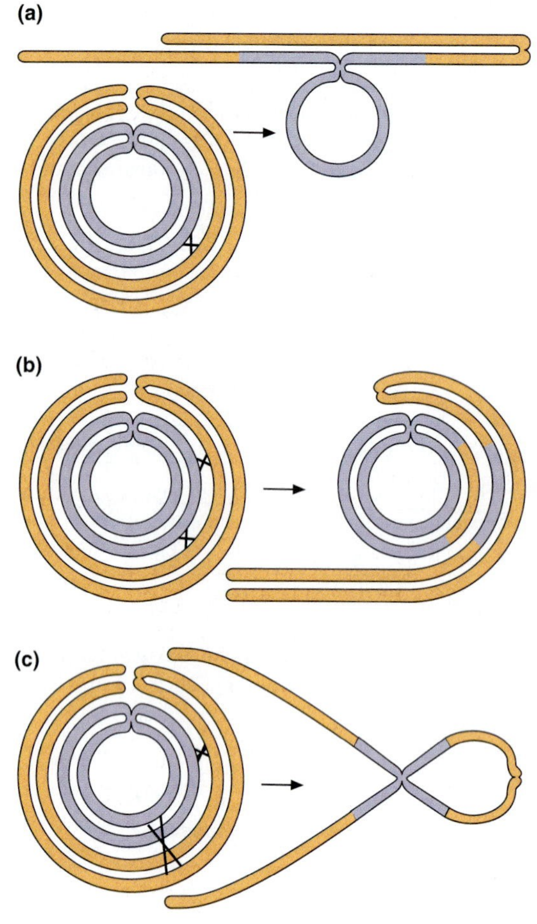

Chapter 19

1. Both models require endonuclease action (one is a double-strand, and the other is a single-strand cut), both have branch migration after rejoining the donor and recipient duplexes, and both result in the nearly identical products. Experimentally, it is difficult to differentiate between the two models.

2. Both lambda and the *E. coli* chromosome have a specific site of sequence homology, the O-region, and recombination occurs only at this site. The two sequences align, a staggered cut in the O-region allows exchange of the duplex strands, which are then resealed. Integration requires the products of the lambda phage gene *int* and the bacterial gene product IHF (integration host factor). Excision also uses the *int* gene and the IHF but additionally requires the product of a second lambda phage gene, *xis*. The fact that lambda integration and excision are controlled by two distinct series of enzyme reactions is the result of the fact that different DNA target sequences are involved in the two processes.

3. Rhodes estimated that in order to account for his observations on *Dt* reversion, the rate of reversion would have to be several orders of magnitude higher than what was typically observed in maize and other experimental organisms. In addition, Rhodes noted that the exceptional genetic instability was detectable only when the *A* locus was in a particular genetic background.

4. McClintock identified a maize strain in which the appearance of high somatic mutation rates at particular loci were under the control of another locus.

5. The *Ac* element is complete, that is, it possesses the transposase gene, whereas the *Ds* element does not and thus depends upon the presence of *Ac* in *trans* to effect transposition.

6. The first bacterial transposable element was found because of a *gal* mutation that was polar, and this polar mutation was not revertible by acridines mutagens. Further investigation, using heteroduplex mapping, showed the presence of a large piece of DNA in the *gal* gene.

7. Transposition of IS elements is random and does not depend upon sequence homology.

8. IS elements likely evolved first because transposons are composed of two copies of an IS element flanking a gene.

9. Transposons possess two IS elements. Only one copy of the functional transposase is required for transposition. Therefore, if one is defective, it should not harm the transposon's ability to transpose.

10. In conservative transposition, the copy of the IS element moves from one location to another, but it does not generate any new copies. In replicative transposition, a new copy of the element is always generated.

11. Autonomous elements possess their own transposase, whereas nonautonomous elements do not and transpose only in the presence of an autonomous element that can supply the transposase in *trans*.

12. *Ac* elements probably evolved first and then suffered from mutations or deletions within the transposase encoding region, resulting in *Ds* elements.

13. If the parental P-male is crossed to an M-female, the resulting hybrids will experience high rates of P-element transposition in their germ cells, leading to significantly reduced fertility or sterility. This condition is referred to as hybrid dysgenesis.

14. To transpose from one site to another, a retrotransposon requires the participation of RNA polymerase II for expression and a reverse transcriptase to create a DNA copy for integration. TEs require only a transposase that recognizes the terminal sequences of the TE to effect transposition.

15. The lack of editing by reverse transcriptase means that errors will occur during replication at a greatly increased rate compared to DNA polymerases. This means that the sequence of the retrovirus will change rapidly and by chance produce new versions that may escape the immune system and new vaccines.

16. Retrotransposons do not possess the *env* gene and therefore are not packaged into a virus particle to be passed to another cell. Consequently, they are re-inserted into the genome.

17. They terminate in A-rich tracts, presumably remnants of a poly(A) trail that was present in a precursor RNA prior to reverse transcription.

Chapter 20

1. Carcinomas are cancers of the epithelial cells; leukemia and lymphoma are cancers of blood and lymph; and sarcomas are cancers of the connective tissue (bone and muscle).

2. Carcinomas are the most common type of cancer. This is a result of the fact that these tissues are most exposed to the environment: UV light and chemical mutagens that are present in our food, drink, and air. Epithelial cells are our first line of defense against the outside world.

3. Cancer cells are immortal, do not possess contact inhibition, and have a reduced need for growth factors.

4. A carcinogen is a chemical or ionizing radiation that results in the formation of cancerous growth of cells. Carcinogens are nearly always mutagens, and the same tests can be used. For example, the Ames test (Chapter 17) uses mutants of *Salmonella* to test for reversion to wild-type, but mice, rabbits, or any animal can also be used to determine whether a particular chemical causes cancer.

5. Evidence for the genetic basis of cancer is (1) mutagens are frequently carcinogens, (2) some cancers are inherited, (3) clonal theory suggested that tumor development was passed to future cells, and (4) some viruses cause cancer and these viruses contain oncogenes.

6. A predisposition to inherit cancer means that an individual has a higher than average chance of developing a particular cancer (e.g., breast cancer, colon cancer, or retinoblastoma).

7. The chemical benzopyrene, present in cigarette smoke, is metabolized in the liver to a derivative, which is very carcinogenic.

8. Peyton Rous discovered that viruses were oncogenic by taking an extract of a chicken tumor, grinding it up, and filtering it through a membrane that permitted only organisms as small as viruses to pass through the membrane. He then injected the filtrate into another chicken, and it immediately developed cancer.

9. Viral oncogenes and proto-oncogenes are derived from the same gene, but they can differ by the level of expression of the gene product or by actually producing a gene product that is slightly different. They may differ only by one nucleotide, however. C-oncs usually contain introns; v-oncs do not.

10. Translocation of the *myc* locus to chromosome 14 places *myc* under abnormal regulatory control, resulting in overexpression. Insertion of the ALV protovirus into chromosome 8 adjacent to the *myc* proto-oncogene results in overexpression of the *myc* proto-oncogene. Amplification of the *myc* proto-oncogene results in overexpression. The amplified copies may exist in tandem at the *myc* locus or may exist extrachromosomally as "double minutes."

11. The *myc* proto-oncogene was placed under promotional control of the MMTV LTR. Because the MMTV promoter, which is contained in the LTR, is strongly expressed in many tissues, transformation of a mouse with the genetically engineered construct resulted in overexpression of the *myc* in the mouse mammary tissue and induced tumor formation.

12. When a cell responds to a growth factor that it also produces, it is called autocrine growth stimulation. Normally, cells that produce growth factors do not also contain receptors for growth factors; likewise, cells that contain receptors to a particular growth factor do not normally produce the factor. When this state of affairs is disrupted, cellular transformation may occur because cells become capable of stimulating their own growth. For example, when a retrovirus causing the *v-sis* oncogene (encoding PDGF) infects a fibroblast, the cell will transform, resulting in a sarcoma.

13. Regulatory mutations that dramatically increase the expression level of genes encoding receptor proteins can result in cancer formation. For example, overexpression of the *erbB* gene, which encodes the internal domain of the EGF receptor has been associated with the onset of ovarian and breast tumors. Point mutations that alter the structure of growth-factor receptor proteins may also lead to cancer if the alteration is such that the intracellular domain is active even in the absence of growth-factor binding. Examples of this are point mutations in the *vet* and *trk* oncogenes, which lead to cellular transformation.

14. Intracellular signaling proteins transfer information by phosphorylation and thus fall into the general category of kinases. Mutations in proto-oncogenes encoding intracellular signaling proteins can disrupt the normal informational flow in the cell, leading to cancer. An example is phospholipase C, which when activated catalyzes the formation of diacylglycerol, which is an activator of protein kinase C, which initiates a series of reactions leading to cell proliferation.

15. Initial evidence that tumor-suppressor genes might exist came from cell fusion experiments in which hybrid cells were produced from the fusion of normal and tumor cells. The fact that these hybrid cells displayed a normal phenotype suggested that normal cells contain one or more genes that are able to "suppress" the cancer phenotype.

16. One form of tumor-suppressor gene encodes gene products that regulate growth (e.g., *rb* and p53). Another form encodes intracellular signaling proteins (e.g., *NF1*), whose product inactivates p21ras, a growth stimulator. Other tumor-suppressor gene products are cell adhesion proteins. For example, the *DCC* tumor-suppressor gene is a cell surface protein involved in cell–cell interactions, and mutations in this gene reduce cell adhesion and contribute to tumor spread.

Chapter 21

1.
$$q(A_2) = \frac{2(A_2A_2) + A_1A_2}{2N} = \frac{A_2A_2 + (1/2)A_1A_2}{N}$$

2. The reliability of a gene frequency estimate depends upon two factors. First, the individuals surveyed should be a random sample of the population at large. Second, the larger the sample, the better. The larger the sample size (N), the smaller the standard deviation (s), and thus the greater the likelihood that the true gene frequency lies close to the estimated value.

3. The formula to compute standard deviation is

$$s = \sqrt{pq/2N}$$

In this case, if $N = 25$, then $s = 0.05$.

The 95% confidence interval is 2X s, or in our case 2 (0.05) = 0.1. Thus, there is a 95% chance that the true gene frequency of the S allele lies between 0.73 and 0.93.

4. To be defined as a polymorphic locus under the 95% criteria, the most frequent allele must be less than 0.95 in frequency. Of the loci, 5 out of the 36 (or 13.9%) were polymorphic by this criteria. To be defined as a polymorphic locus under the less stringent 99% criteria, the most frequent allele must be less than 0.99 in frequency. Of the loci, 13 out of the 36 (or 36.1%) were polymorphic under this criteria.

5. a. Observed genotypic frequencies:

FF	*FS*	*SS*
0.57	0.35	0.08

Observed allelic (gene) frequencies:

$f(F) = \quad = 0.75$
$f(S) = \quad = 0.25$

b. Expected genotypic frequencies:

FF	*FS*	*SS*
0.56	0.38	0.06

c. 7.58. Looking at the chi-square table (Table 2.3), we see that a value of 7.58 with one degree of freedom ($n - 2$) is significant at the 5% level (i.e., the deviation that exists between the observed and expected values is greater than what would be expected by chance). Therefore, we can reject the hypothesis that the expected and observed numbers are the same and conclude that this population is not in Hardy–Weinberg equilibrium.

6. 0.0019.

7. 0.002. Therefore, after one generation of migration, the frequency of the *D* allele will be 0.40 + 0.002, or 0.402.

8. a. Claxton, GA: $m = 0.017$. **b.** Oakland, CA: $m = 0.027$. Thus, migration rates of *R* are higher in the Oakland area than in Claxton. These differences may reflect differences in socio/economic barriers to interracial relationships in these two areas of the country.

9. a. In this population, $1/2N = 1/(2 \times 1000) = 0.0005$. **b.** The average time, in generations, required to fix a new allele is $4N$. In this population, $4N = 4000$. It would take 4000 generations, and because each generation is about 2 weeks, that means 8000 weeks, or about 153 years and 10 months.

10. The starting genotypic frequencies (as computed in the solved problem above) are

$$\frac{RR}{0.009} + \frac{Rr}{0.396} + \frac{rr}{0.595} = 1.00$$

We can now calculate the frequency of the *R* allele after one generation in the insecticide-free environment to be 0.207.

Chapter 22

1. Pseudogenes and retropseudogenes. Pseudogenes are defined as nonfunctional members of a gene family. The basis of the loss of gene function may be point mutations, small insertions or deletions, or other mutational events typically associated with gene inactivation. In these cases, the pseudogene is a remnant of a once-active gene. Another possibility is that the pseudogene is the product of reverse transcription of a partially or completely processed progenitor gene transcript. The reverse transcriptase gene product, which may be inserted at a genomic position distant from its progenitor gene, typically lacks a promoter and introns and is thus usually incapable of producing a gene product. This class of nonfunctional genes is called retropseudogenes to reflect their mode of origin.

2. Gene families are groups of two or more genes having similar or identical DNA sequences and encoding the same or related functions. Gene families are believed to arise by gene duplication resulting from unequal crossovers.

3. In many cases where two closely related genes contain variable numbers of introns, the common ancestral genes contain at least as many introns in the same positions as the more "intron rich" decedent gene. This finding indicates that the introns present in many host genes represent an ancestral state.

4. Recent studies have shown that in some cases (e.g., the serine protease gene family) the position and number of introns in genes are not represented in ancestral genes. This suggests that introns can be acquired over evolutionary time. Consistent with this view is the fact that some classes of introns behave as mobile DNA elements, suggesting that at least some introns may have evolved from transposable-like elements.

5. The primary function of the C-peptide is to position the A and B regions of the proinsulin polypeptide properly so that the appropriate disulfide bonds can be formed. Regions of the C-peptide that are involved in forming the proper secondary structures in the polypeptide would be functionally critical; other regions of the C-peptide would not. These critical regions of the C-peptide would be expected to evolve at a slower rate than other regions of the C-peptide.

6. It is unlikely that a gene will evolve at the same rate at different stages of its evolutionary history. At early stages of a gene's evolutionary history, when the gene product is evolving a new function, high evolutionary rates would be expected. After a gene product has evolved a critical function, natural selection may permit only those substitutions that do not interfere with function. Thus, evolution rates might be expected to slow down later in a gene's evolutionary history.

7. Because proteins with different functions can be expected to vary in the number of specific amino acids that must be maintained at specific sites in order to preserve function, natural selection will exert different levels of constraints on proteins. For example, nearly every amino acid in the cytochrome c molecule is essential to the proper functioning of the protein. Thus, natural selection has not allowed many substitutions over evolutionary time. In contrast, there is little functional constraint on the amino acids that are present throughout most of the fibrinopeptide molecule. Fibrinopeptides are among the most rapidly evolving class of proteins.

8. We begin by identifying the two species that are separated by the least number of substitutional differences, which in this case is species A and B. Because the rates of substitution are assumed to be equal in the lineages leading from the common ancestor to species A and B, the differences accumulated in each branch are 1.00/2, or 0.50. We next compute the differences between the remaining species and species A and B taken as a composite pair. This is done by averaging the distances between each of the remaining species and species A and B, respectively. For example, the difference between species A and C is 1.31 and between species B and C is 1.37. Thus, the average difference between species C and the A–B composite is 1.34 (i.e., 1.31 + 1.37/2). Using this same approach, the distance between species D and composite species A–B–C (1.95) and between species E and composite species A–B–C (2.33) can be computed. When all the distance values are computed in this way, the following distance tree can be constructed to show the phylogenetic relationship among all of the species.

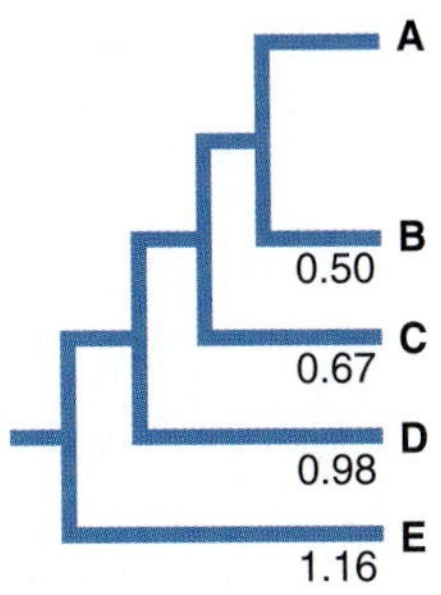

9. 1.17 million years.

10. The average rate of substitution between species A and D may have been significantly lower than between species A and B.

Glossary

acridines A group of planar heterocyclic compounds that can intercalate between stacked bases in DNA and cause frameshift mutations.

acrocentric A chromosome with the centromere near one end.

adaptation The process by which organisms are modified to function in a given environment.

additive alleles Alleles that make an independent, additive contribution to an individual's phenotype for a quantitative trait.

additive rule A rule used to calculate the probability that either of two independent events will occur. The probability that event A or event B will occur is the sum of the probabilities that each event will occur separately. This is represented symbolically as $P(\text{A or B}) = p(\text{A}) + p(\text{B})$.

adenoma A benign growth in the colon often associated with one or more mutations in the *ras* proto-oncogene. Progression of adenomas to malignant tumors have been associated with inactivation of the *p53* tumor-suppressor gene.

adjacent 1 segregation The segregation pattern in translocation heterozygotes in which adjacent homologous centromeres segregate. Adjacent 1 segregation gives unbalanced gametes.

adjacent 2 segregation The segregation pattern in translocation heterozygotes in which adjacent nonhomologous centromeres segregate. Adjacent 2 segregation gives unbalanced gametes.

albinism The inability to produce melanin. Albinism is a recessive mutation, but mutations in several different gene can result in this phenotype.

alkaptonuria A human genetic defect caused by a single recessive mutation that causes the accumulation of homogentisic acid, a metabolic by-product of the amino acid tyrosine.

alkylating agent A chemical that can add alkyl groups (for example, methyl or ethyl groups) to another molecule. Ethyl methane sulfonate is a mutagen that adds methyl groups to the bases.

allele frequency The percentage of all alleles in a population that are represented by one particular allele. Same as gene frequency.

alleles Different forms of the same gene. According to the Mendelian concept of a gene, a gene is an inherited factor controlling the phenotype of a trait, and alleles are copies of genes with some modification that alters this phenotype in some way. Alleles are detected only when the differences in phenotype can be detected.

allele series The total set of alleles for a given gene.

allopatric speciation Speciation occurring primarily as a result of geographic isolation.

allopolyploid A polyploid individual that has sets of chromosomes from two or more different species.

allosteric transition The binding of a small molecule, such as an inducer to a protein, that affects the conformation of the protein and its binding at other sites.

allotetraploid A polyploid individual with four sets of chromosomes from two different species.

allozyme Distinct allelic forms of an enzyme that can be physically separated by electrophoresis.

alternate segregation The segregation pattern in translocation heterozygotes in which alternate homologous chromosomes segregate together. Alternate segregation gives balanced gametes.

alternative splicing RNA splicing of the primary transcript using different exons to produce different mRNAs, which in turn give different proteins during translation.

amber codon UAG, a translational termination codon.

Ames test A test for potential mutagens, developed by Bruce Ames. A mutant strain of *Salmonella* is exposed to a potential mutagen, and the rate of reversions to wild-type is measured and represents the chemical's mutagenic capacity.

amino acid The monomeric unit of proteins. Amino acids are composed of a carboxyl group and an amino group attached to a central carbon atom, which is also attached to 1 of 20 different side chains or R groups.

amino acyl-tRNA synthetases A group of 20 different enzymes, each specific for the attachment of a specific amino acid to a tRNA possessing the correct anticodon for that amino acid.

amniocentesis A technique for genetic testing of a fetus by isolating fetal cells from amniotic fluid. Amniocentesis is often used for determining the fetal karyotype.

amorphic One of the functional classifications of alleles devised by H. J. Muller. An amorphic allele has no gene function.

amphidiploid Another term for an allotetraploid, an individual with four sets of chromosomes from two different species.

anaphase The phase of mitosis or meiosis when the chromosomes leave the metaphase plate and migrate toward the two poles of the cell.

anchor cell A special cell in *C. elegans* that controls cell determination during the development of the vulva.

aneuploid An unbalanced chromosome set that contains extra chromosomes or is missing individual chromosomes but not complete monoploid sets of chromosomes.

annealing of DNA Situation that occurs when two complementary strands of DNA reform a double-stranded molecule.

anthers Structures in a plant flower that contain pollen.

antibody A protein that recognizes specific molecules and binds to them. Each antibody protein is composed of four subunits, two heavy chains and two light chains.

anticodon The codon complementary nucleotide sequence present on tRNA.

antigen A chemical or substance bound by an antibody.

antimorphic One of the functional classifications of alleles devised by H. J. Muller. An antimorphic allele has an antagonistic function to the normal, wild-type allele. An individual heterozygous for an antimorphic allele and a normal allele has less of the normal gene function than an individual heterozygous for a deletion and a normal allele.

antiparallel The relative polarity of the two strands in a DNA double helix; one strand goes in the 5′ to 3′ direction and the other strand goes in the 3′ to 5′ direction. The numbers refer to the carbon atoms in the deoxyribose.

antipodal cells Three cells containing haploid nuclei in the embryo sac that are located at the opposite end of the sac from the gamete nucleus.

antisense strand The complement of the sense strand.

antiterminator A protein that antagonizes the termination of transcription at the termination site.

AP endonuclease An enzyme that breaks the phosphodiester bond near a purinic or a pyrimidinic (AP) site, allowing DNA polymerase to repair the damaged segment.

apoptosis Programmed cell death; a process whereby cells kill themselves for the good of the organism.

AP sites Sites in a DNA strand that lack purines or pyrimidines attached to the sugar phosphate backbone; apurinic or apyrimidinic sites.

apurinic site A deoxyribose in a DNA strand that has lost its purine base.

Aristotle A Greek philosopher who lived between 384 and 322 B.C. He is recognized as one of the greatest of the Greek philosophers, and his writings were accepted as the final authority on many subjects for many years. He was noted for his attempts to explain the workings of the natural world by observation and deduction.

ascospores Fungal spores resulting from meiosis within asci.

ascus (plural, asci) The sac within which meiosis and spore formation take place in *Neurospora*, yeast, and other fungi.

aster The centriole with attached spindle fibers.

***att* site** A site on lambda phage genomes and bacterial genomes where recombination occurs, allowing integration of the phage genome into the host genome.

attenuator A region or sequence of nucleotides in the leader region of several operons that act in the presence of the product of the operon (for example, tryptophan) to reduce the rate of transcription from the structural genes.

autoimmune disease A condition that occurs when the immune system does not recognize a structure (for example, a protein) within the organism as self and makes antibodies against this protein as if it were a foreign invader.

Autonomous element Transposable element that encodes all the genetic information necessary for its own transposition.

autonomously replicating sequence, or ARS element An 11-bp sequence of DNA present in yeast and some other organisms that may be the site of initiation of DNA replication.

autopolyploid A polyploid that contains extra sets of chromosomes, with all chromosomes originating in the same species.

autoradiography Exposure of radioactively labeled material to a film or radiation-sensitive emulsion to determine the extent and location of label.

autosome All chromosomes except the sex chromosomes.

autotetraploid A polyploid with four sets of chromosomes, with all chromosomes originating in the same species.

auxotroph A mutant bacterial strain that requires a specific supplement in order to grow (for example, the amino acid alanine).

auxotrophic mutation A mutation that causes an organism to be unable to synthesize a molecule required for growth and/or development. If this molecule is supplied in the growth medium, the individual will grow and may have a normal appearance.

average clone size A method of calculating the number of cells in a population (n). The average size of induced somatic cell clones, measured as a fraction of the total population, is used as an estimate of $1/n$.

avian leukosis virus (ALV) A retrovirus that has frequently been found to cause bursal lymphomas in chickens by inserting adjacent to the proto-oncogene.

axis formation The establishment of overall body polarity in the embryo along the anterior/posterior, dorsal/ventral, or medial/lateral axes.

axis formation genes Genes whose normal function is required for the normal establishment of the embryonic body axis in *Drosophila*. These genes include the gap genes, the pair-rule genes, and the segment polarity genes.

axis polarity The orientation of the body in space. Each three-dimensional body has three axes, the anterior/posterior body axis, the dorsal/ventral axis, and the medial/lateral axis. The body of each organism normally has a set polarity, or sequence in which structures are arranged, along each body axis. In *Drosophila*, the head is more proximal than the thorax, which is more proximal than the abdomen.

axis polarity genes The genes controlling the establishment of anterior/posterior and dorsal/ventral body orientation. Many of these genes are axis-formation genes.

B-chromosome A small chromosome found in maize that has no known function and carries no known genes.

B-DNA The naturally occurring form of DNA that is the same as the model proposed by Watson and Crick.

β-galactosidase An enzyme that catalyzes the conversion of lactose to galactose and glucose.

BAC A bacterial artificial chromosome; a plasmid that is capable of replicating in a bacterial host (usually *E. coli*) with a very large insert of foreign DNA, perhaps as large as 300 kb of nonplasmid DNA.

backcross A cross in which an offspring is mated to one of his or her parental types.

bacteriophage or **phage** A bacterial virus composed of nucleic acid and a protein outer coat. It is not capable of replication except when in a bacterial cell.
balanced lethal combination A heterozygous genotype in which homologous chromosomes contain different recessive lethal mutations, so that individuals homozygous for either chromosome die and only heterozygotes are viable.
balancer chromosome A chromosome containing inversions or other chromosome aberrations that prevent recombination. Viable offspring produced by an individual heterozygous for a balancer and a normal chromosome that contains mutant alleles of several genes contain only parental-type chromosomes.
Barr body An X chromosome that has become heterochromatic.
base Either a purine or a pyrimidine, usually linked to a ribose or deoxyribose and a component of DNA and RNA.
base analogues Compounds that are structurally similar to the four natural bases and can be incorporated into DNA. Base analogues can cause mutations by imprecise base-pairing during replication.
base pair A purine opposite a pyrimidine in DNA or RNA.
benign tumors Encapsulated tumors that do not invade and destroy adjacent tissues and do not metastasize.
bidirectional replication Replication of DNA in two directions simultaneously from one growing point (but always in the 5′ to 3′ direction).
binary choice mutation A homeotic mutation in nematodes that affects cell fate decisions.
binomial theorem A mathematical theorem dealing with two mutually exclusive events $(a + b) = 1$. In genetics, the binomial is often used to calculate probabilities of multiple events, such as groups of offspring with particular genotypes.
biological species A group of organisms that are able to interbreed and produce fertile offspring.
bithorax complex A large locus in *Drosophila* containing homeotic genes that control segmentation in the thorax and abdomen.
bivalent The structure formed when two homologous chromosomes pair during meiosis I.
blast cells Cells in the larval nematode that produce adult structures.
blastomere An early embryonic cell formed by cleavage of the egg.
blastula The hollow ball of cells formed during animal embryogenesis.
blending inheritance Theory of heredity popular in the 19th century that postulated that inherited traits were determined by fluid-like determinants that became mixed or "blended" at conception, resulting in progeny that displayed characteristics intermediate between those of the parents.
bottleneck A drastic decrease in the size of a population, usually resulting in a reduction in genetic variability.
branch migration A process by which a DNA strand, which is partially paired with its complementary strand in a duplex, is able to extend its pairing by displacing the resident homologous strand.
breakage and reunion A model of genetic recombination in which two DNA molecules are broken at analogous positions and rejoined in a crosswise fashion; breakpoint effect; a mutant effect caused by the breakage of the chromosome that occurs during the formation of one of several different types of chromosomal aberration.
breakpoint effect A mutant effect caused by the breakage of the chromosome that occurs during the formation of one of several different types of chromosomal aberration.
bromo-deoxyuridine An analogue of thymidine that can be incorporated into DNA, causing mutations.
Burkitt's lymphoma A cancer characterized by uncontrolled proliferation of lymphocytes caused by a translocation of a piece of chromosome 8 containing the *myc* oncogene to either chromosome 2, 14 (most frequently), or 22. This translocation event is believed to place *myc* under abnormal regulatory controls.

C_0t plot A graphical representation of the kinetics of renaturation of a sample of single-stranded DNA molecules. The graph is a plot of the fraction of remaining single-stranded DNA molecules versus the product of DNA concentration at time zero and time *t*. The percentage of unique, moderately repeated, and highly repeated sequences can be determined from these data.
CAAT box A DNA sequence located about 75 bp upstream of the translational start site in eukaryotic genes. It affects the level of gene expression.
cAMP Cyclic adenosine monophosphate plays a key role in the regulation of many metabolic processes but is necessary for operon expression in the presence of glucose.
cancer A family of more than a hundred different often deadly diseases that are characterized by uncontrolled cell proliferation.
cap The altered 5′-end of eukaryotic mRNA that includes a 7-methylguanosine.
capsid The protein coat immediately surrounding nucleic acids and internal proteins of a virus particle.
carcinogen Compounds capable of inducing cancer.
carcinoma Cancers of epithelial cells that cover the body surface and line the intestine and other internal organs. Carcinomas account for about 90% of malignant tumors.
catabolite activator protein (CRP or CAP) A protein, which, together with cAMP, activates operons in bacteria. These operons are subject to catabolite repression.
catabolite repression The repression of an operon by glucose. Catabolite repression is released by the presence of a positive control compound, cyclic AMP.
cDNA A DNA copy of mRNA made by reverse transcriptase.
cell determination The adoption of a specific cell fate. A determined cell always passed its determination to its mitotic progeny and will not change its fate if it is transplanted to a new environment. During development, a cell undergoes a series of determinations.
cell differentiation The generation of specialized products by a cell as part of the expression of its cell fate; cell equivalence group; a group of interacting cells in the nematode that collectively form a particular structure.
cell equivalence group A group of interacting cells in the nematode that collectively form a particular structure.
cell fate The cell type that a progenitor cell is committed to form in an organism.

cell furrow A contractile ring of microfilaments that separates dividing animal cells.

cell lineage The cell division pathway that produces a particular cell type.

cell lineage analysis Analyzing the fate of individual cells by using induced somatic recombination to mark the cells and all their progeny.

cell lineage restriction A developmental restriction that limits the progeny of a cell to forming certain structures.

cell plate A cell membrane that forms in the equatorial region of a dividing plant cell.

cellular blastoderm The stage in *Drosophila* embryogenesis at which the zygotic nuclei become surrounded by cell membranes. This occurs at about 3 hours of development.

cellular oncogenes (c-*onc*) Proto-oncogene that has undergone a genetic change contributing to transformation of the cell in which it is contained.

centimorgan (cM) The unit of distance in linkage mapping, named in honor of T. H. Morgan.

centriole The cellular organelle that serves as the focus for the mitotic and meiotic spindle fibers.

centromere A region of the chromosome where the spindle fiber attaches to allow segregation of chromosomes during meiotic and mitotic cell divisions and does not divide until the beginning of anaphase. This region is referred to as the *CEN* region; in yeast it is about 250 bp long and is composed of three regions of conserved sequences.

Chargaff's rules The concentration of adenine always equals the concentration of thymine, and the concentration of cytosine always equals the concentration of guanine.

chiasma (plural, chiasmata) A cross-shaped structure of chromatids. Chiasmata are believed to indicate the sites of recombination between chromosomes.

chi-square (χ^2) A mathematical test used to determine the goodness of fit of observed and expected results of genetic crosses. The value of the chi-square statistic indicates whether the amount of difference between the observed and the expected is too large to be the result of chance.

chloroplast An organelle in plants and some lower eukaryotes with the ability to photosynthesize.

chorion The egg shell. The chorion surrounding *Drosophila* eggs is formed by the follicle cells.

chromatid One complete copy of a chromosome produced during S phase of mitosis. A replicated chromosome has two chromatids attached at the centromere.

chromatin A complex of RNA, DNA, and proteins that make up the chromosomes.

chromomeres Tightly coiled DNA found in the bands of polytene chromosomes of diptera.

chromosomal aberration A mutation that visibly changes chromosome structure. Chromosome aberrations include deficiencies, duplications, translocations, and inversions.

chromosome A densely staining cellular entity, composed of DNA complexed with proteins. Chromosomes contain, or consist of, a linear sequence of genes.

chromosome puffs Regions of chromosomes that expand greatly during particular developmental stages when genes in that region are active. Puffs are specially noticeable in the *Drosophila* polytene chromosomes.

chromosome satellite A small region of chromatin that is connected to the rest of the chromosome only by a thin stalk.

chromosome scaffold A central core structure of a condensed chromosome. It is composed of protein.

chromosome theory of inheritance The theory that suggests that genes are physically located on chromosomes and that Mendelian inheritance can be explained in terms of chromosome behavior during cell division.

chromosome walking After a fragment of DNA is cloned, one end is subcloned and used as a probe in the gene library to find other cloned DNA fragments that overlap the original clone. This procedure is then repeated over and over again, generating DNA fragments that overlap each other and "walk" down the chromosome.

***cis*-acting site** A genetic site, represented by a binding site on the DNA, that influences genes immediately adjacent to the site, such as promoters, enhancers, and operators.

***cis–trans* test** A test to determine if two mutant sites are in the same functional gene (or cistron) or in separate genes. Same as complementation test.

cistron A genetic unit defined by the *cis–trans* test. It is used interchangeably with the term *gene*.

clonal selection Antibody receptors for all possible antigens are present on B cells, and when stimulated by binding of the antigen to the receptor stimulates cell proliferation. The product is a clone of cells that produces antibodies specific for the stimulatory antigen.

clone A population of individuals or DNA fragments that are identical.

cloning vector A plasmid or other DNA sequences capable of self-replication that can be used for insertion of foreign DNA sequences.

coding strand The strand of DNA that has the same polarity and sequence (in RNA bases) as the transcriptional product.

codominance A form of dominance relationship between two alleles of one gene in which the heterozygote shows the phenotypes of both alleles.

codon Unit of the genetic code consisting of three nucleotides that specify one amino acid in the genetic code.

codon degeneracy The specification of an amino acid by more than one codon.

coefficient of regression The slope of the line that most closely relates two correlated variables.

cointegrate structure Structure produced by the fusion of a replicon containing a transposon with another not containing a transposon.

colinearity The sequence of codons in a gene corresponds to the sequence of amino acids in the protein product of that gene, without any breaks or punctuation.

compartment A region of the adult delimited by cell lineage restrictions, which are also called compartment boundaries.

compartment boundary A cell lineage restriction that divides the cells of imaginal discs into two populations. Compartment boundaries are located by using somatic cell clones.

competence A state or condition where bacteria are susceptible to uptake of DNA for transformation.

complementary base-pairing The hydrogen bonding between a purine and a pyrimidine in double-stranded DNA.

Adenine pairs with thymine, and guanine pairs with cytosine.

complementary DNA or cDNA clone A recombinant DNA clone complementary to an mRNA sequence.

complementary strand When the sequence of the two strands of DNA have base-pairing of G : C and A : T.

complementation group A group of mutations that all fail to complement (that is, that give mutant phenotypes in *trans* heterozygotes); also called a cistron.

complementation test See *cis–trans* test.

composite element Bacterial transposon that has a central region encoding functions unrelated to transposition flanked by two IS elements. At least one of the IS elements encodes an active transposase. (See also *transposon.*)

concordance Identity of matched pairs or groups for a given trait.

conditional lethal A mutation in an organism that confers lethality under one environmental condition but is nearly normal under another condition. Temperature differences are a common form of conditional lethals.

conditional mutation A mutation that gives different phenotypes in different environments or in response to one particular environmental factor (such as temperature).

conjugation The process of transferring genetic material from an *Hfr* bacterial strain to an F′ via a conjugation tube; consanguineous marriage; a marriage between two related people.

consanguinity Related by common descent from an ancestor.

consensus sequence The average of several similar sequences.

conservative replication During DNA replication, both strands of the parent DNA are transferred to one daughter molecule, whereas the other molecule has two newly synthesized strands.

conservative transposition The movement of a transposable element from one location in the genome to another.

constitutive genes A continually expressed gene (for example, a housekeeping gene).

constitutive heterochromatin Regions of heterochromatin that are heterochromatic in all cells.

contig A group of cloned DNA sequences that have sequence overlap.

continuous variation Variation not represented by distinct classes.

coordinate repression The situation that occurs when a group of genes are regulated together, usually in an operon.

corepressor A product in a metabolic pathway that interacts with the repressor protein forming a complex, preventing the repressor from binding to the operator.

correlation The tendency of two variables to vary together.

correlation coefficient A statistic that measures the strength of association of two variables.

cortical cytoplasm The region of cytoplasm at the outer edge of the egg.

cosmids A plasmid that has a *cos* site in it. A *cos* site is a 12-nucleotide sequence that acts as a sticky end for packaging DNA into lambda phage heads. The plasmid can then act as lambda but can replicate in the host cell and also as a DNA segment for packaging into lambda heads.

***cos* site** A base sequence that is needed at flanking ends of the DNA segment for insertion into the lambda head proteins. The enzyme terminase recognizes this sequence and creates a staggered cut.

cotransduction Transduction of two or more genetic markers in the same phage.

cotransformation When two or more genetic markers are transformed into a cell simultaneously.

covariance of x and y A statistic used to measure the correlation coefficient of the two variables (x and y).

CpG islands Clusters of this sequence that often occur upstream of eukaryotic genes.

crossing (cross-breeding) A controlled mating between selected parents. This is a fundamental tool of transmission genetics.

crossover The reciprocal exchange of genetic material between chromosomes during meiosis. It is responsible for genetic recombination.

crown gall A tumorous growth on plants that results from infection of the plant with *Agrobacterium.*

CRP Cyclic AMP receptor protein, necessary for activation of catabolite repressible operon. Also known as the CAP protein.

C-value The total amount of DNA per haploid genome for an organism.

C-value paradox Organisms have more DNA than is required to code for the necessary functions of the cell. In the C-paradox the amount of DNA does not correlate with the complexity of the organism. For example, amphibians have more DNA than do humans.

cytidine A nucleoside containing the cytosine base.

cytogenetics An area of genetics concerned with the study of chromosomes and their behavior.

cytokinesis The division of the cytoplasm during a cell division.

cytoplasmic determinant A physical substance located in the cytoplasm that directs the cell to adopt a particular cell fate. Cytoplasmic determinants are assumed to act by interacting with the nucleus to activate a particular set of determination genes.

cytoplasmic inheritance Heredity determined by DNA in the organelles of the cell; chloroplasts or mitochondria.

cytosine The pyrimidine base in DNA.

D-loop The structure resulting from the displacement of a region of one strand of duplex DNA by a single-stranded invading strand in a *recA*-mediated recombination process.

dalton The mass of a hydrogen atom.

daughter chromosomes The two chromosomes formed by the division of a two-chromatid chromosome during mitosis.

deamination The removal of an amino group from a chemical compound. Deamination of adenine or cytosine causes base-pairing changes.

defect fate maps Maps showing the location of the precursor cells for particular structures. These maps are generated by killing or removing particular blastoderm cells, allowing the individual to complete development and then noting which structures are missing or abnormal.

deficiency The loss of a segment of a chromosome. Same as deletion.

degenerate code A code in which one amino acid is specified by more than one codon.

degree of genetic determination Another name for heritability in the broad sense (H^2).

degrees of freedom The number of independent variables in an experiment. Knowing the number of degrees of freedom is essential for certain types of statistical tests (for example, chi-square).

deletion The removal of a segment of DNA from a chromosome.

denaturation The alteration of the physical and three-dimensional structure of proteins and nucleic acids by mild treatment that does not break covalent bonds. Activity is usually destroyed.

deoxyribonucleic acid (DNA) The genetic material of most organisms. DNA is a long double-helical molecule composed of the deoxyribonucleotides—deoxyadenylic acid, deoxythymidylic acid, deoxyguanylic acid, and deoxycytidylic acid. The two strands are held together by hydrogen bonds between A : T and G : C pairs.

depurination Removal of the purine ring by breaking the base-sugar bond.

determinants Substances that direct cells to adopt or maintain a particular cell fate. Determinants are the products of determination genes.

determination genes Genes whose products cause cell determination; the products of the *BXC* gene are determinants for posterior thorax and abdomen.

developmental genetics The branch of genetics concerned with the genes that regulate organismal development and how these genes function.

developmental pathway A series of gene actions that leads to a particular determined state.

diakinesis The final stage of the first meiotic prophase. During diakinesis, the chiasmata terminalize, the nuclear membrane breaks down, and the spindle fibers attach to the kinetochore.

dicentric chromosomes The product of fusing two chromosome fragments that carry centromeres. The structure is unstable and is usually broken when the two chromosomes are pulled to opposite poles during cell division.

dideoxynucleotide A nucleotide missing the hydroxyl group at both the 2′ and 3′ position of its deoxyribose.

dihybrid An individual heterozygous for two allele pairs.

dihybrid cross A cross between individuals who have different alleles of two genes.

diploid An individual or cell with two sets of chromosomes.

diplotene The fourth stage of the first meiotic prophase. During diplotene, the synaptonemal complex disappears, and the chromosomes can be seen to be held together by chiasma.

discontinuous variation Phenotypic variation with distinct classes.

disjunction Separation of chromosomes during anaphase of mitosis or meiosis.

displaced duplication A duplication in which the extra portion of the chromosome is not located adjacent to the normal portion.

dizygotic twins Fraternal twins derived from the fertilization of two eggs.

DNA fingerprint The use of highly variable regions of the DNA sequence to identify an individual.

DNA gyrase A topoisomerase II that creates negative rotations of DNA.

DNA ligase An enzyme that will join the 3′-OH end of a DNA strand to a 5′-PO_4 end when in double-stranded form.

DNA polymerase(s) A group of enzymes responsible for various steps in the synthesis of a complementary strand of DNA from a template.

DNA primase A DNA-dependent RNA polymerase capable of synthesizing short stretches of RNA for initiating DNA synthesis during replication.

DNA sequencing Determining the order of nucleotides in a piece of DNA.

dominance An interaction between alleles in heterozygotes in which one allele controls the phenotype. The allele controlling the phenotype is the dominant allele, and the phenotype produced is the dominant phenotype.

dosage compensation The activity of a gene is increased or decreased depending upon the number of active copies of the gene.

double fertilization The fusion of two pollen (male) nuclei with two nuclei from the embryo sac (female). One nucleus fuses with the female gamete nucleus to form the zygotic nucleus, and the other fuses with two endosperm nuclei to form the triploid endosperm nucleus.

double helix In DNA, two polynucleotide strands running in opposite directions (one is 5′ to 3′ and the other 3′ to 5′) to form a double-stranded molecule that takes a helix formation. A helix is a spiral turning in a left-handed manner or a right-handed manner.

Down syndrome The collection of human phenotypic effects caused by being trisomy for chromosome number 21.

Ds A defective transposable element found in maize. It cannot transpose without an Ac element, which produces a transposase in *trans*.

duplication The occurrence of a chromosomal segment more than once in a chromosome; the multiplication of cells.

eclosion The emergence of an adult *Drosophila* from the puparium.

editing The process of checking a nucleotide or amino acid to determine whether it is correct during the synthesis of DNA and proteins.

electrophoresis The movement and separation of charged molecules in solution in an electrical field. Agarose and polyacrylamide are porous matrices that retard the migration of the charged molecules of DNA in an electrical field. Larger fragments move more slowly than smaller fragments.

electroporation The application of an intense electrical current to a cell to allow uptake of DNA through induced pores.

elongation factor A protein involved in the elongation process during translation.

embryogenesis Embryonic development.

embryo sac The gametophyte stage in the female flower of plants.

encode A unit that contains the information to specify an RNA molecule or a polypeptide.

endonucleases An enzyme that breaks the internal phosphodiester bond of DNA.

endosperm The triploid tissue in seeds that surrounds the embryo. Endosperm is formed from the fusion of two female and one male nuclei.

endosymbiotic hypothesis A theory on the origin of chloroplasts and mitochondria in eukaryotic cells. The theory states that primitive eukaryotic cells formed a symbiosis with prokaryotic cells, which supplied them with energy through the conversion of sunlight to energy or carbon sources to energy.

enhancer element A sequence located upstream or downstream of a promoter that increases the level of gene expression. They are found only in eukaryotes.

environmental variance The variability in the phenotypes of individuals in a population that is caused by differences in the environment.

enzyme A protein (sometimes RNA) capable of catalyzing the conversion of one compound to another without itself being altered.

epidemiology A branch of medicine that investigates the cause and control of epidemics.

epidermoblasts Dermoblasts arising in the ectodermal neurogenic region of the *Drosophila* embryo.

epigenetic A change in the expression of a gene that changes the phenotype without permanently changing the gene itself.

epistasis Situation that occurs when an allele of one gene affects the expression of alleles at another location on the genome.

epitope A specific structural feature of an antigen that stimulates the production of an antibody.

equational division The second meiotic division. The chromosomes divide during this division.

equilibrium A dynamic system where no net change exists.

equilibrium density gradient centrifugation Centrifugation of molecules in a high-density solution (usually CsCI) until a density gradient forms. Sedimenting molecules will form a band in the gradient where their density equals that of the solution.

error-prone repair An inducible repair system that randomly inserts bases opposite damaged DNA strands. Also called SOS repair.

ethidium bromide A complex organic molecule that binds tightly to double-stranded DNA. The DNA will fluoresce under ultraviolet light after exposure of the DNA to ethidium bromide.

euchromatin The chromosome regions that condense later in prophase, that stain lighter, and that contain active genes. Chromosome regions are either heterochromatic or euchromatic.

eugenics The application of genetics to improve mankind.

eukaryote An organism whose cells contain a nucleus.

euploid A balanced genome, one that contains only complete sets of chromosomes.

evolutionary genetics Field of genetics devoted to the study of the genetic basis of adaptation and speciation. See *population genetics*.

excision repair The removal of a 12-nucleotide damaged segment of DNA by the UvrABC enzyme, followed by resynthesis with DNA polymerase using complementary strand information.

exconjugant An F^- *E. coli* strain that has received a fragment of DNA from a male strain.

exons The coding sequence within a gene transcript.

exonuclease An enzyme capable of degrading DNA or RNA from the end inward.

expressed sequence tags (EST) Short cDNA sequences used to link physical maps of genomes.

expression vector A plasmid that possesses a strong promoter, a ribosome binding site adjacent to a start codon, but only a small fragment of the original gene. The coding sequence for a gene can be inserted in the correct reading frame to allow production of the gene product in large quantities.

expressivity The degree of phenotype expressed for a particular genotype.

extrachromosomal inheritance DNA units in the mitochondria and chloroplast that determine specific phenotypes.

extranuclear inheritance The inheritance of genes or other factors that control a phenotype through the cytoplasm.

F_1 First filial generation; the first generation of descent from a given mating.

F_2 Second filial generation; the second generation of descent from a given mating.

facultative heterochromatin Regions of heterochromatin that are euchromatic in some cells.

fertilization The fusion of a male gamete with a female gamete to form a zygote.

F-factor A conjugative plasmid present in *E. coli* capable of self transfer to other bacterial cells. Its presence confers a F^+ phenotype to the cell.

fine-structure map A map that shows the location of mutant lesions within a gene. These lesions may be at very close intervals, sometimes only nucleotides apart.

first filial (F_1) generation The first generation of offspring in a cross. If inbred or selfed, the offspring of F_1 individuals will be the F_2 generation.

fitness The number of offspring left by an individual. This value is often compared to the average of the population or of another genotype as a standard.

fitness value Relative reproductive efficiency of a genotype relative to other genotypes in a population.

fixation An event that occurs when all the alleles at a locus, except one, are eliminated from a population. The remaining allele is then said to be fixed.

flagellum (plural) A whiplike structure used for locomotion in unicellular organisms.

follicle cell One of the somatic cells that surrounds and assists in the development of an oocyte.

founder effect Reduction in genetic variability and/or dramatic change in gene frequency resulting from genetic drift when a new population is established by a small group of individuals.

founder principle The possibility of forming a new population from a larger population, derived from a small random sample of individuals. This new population is probably not genetically identical to the larger population with respect to gene frequencies.

F′-plasmid A plasmid that has acquired a segment of the host bacterial chromosome resulting from imprecise excision from the *Hfr* state.

frameshift mutation The addition or deletion of bases that cause a change in the reading frame of a gene.

frequency distribution A summary of the value of the phenotypes for a quantitative trait of a population of individuals. Each individual is placed in a class or category with others of similar phenotype, and the numbers of individuals in each class are displayed.

fuelgen reaction A technique for staining DNA with acid-fuchsin after hydrolysis with hydrochloric acid.

fusion protein A protein made by recombinant DNA technology where one gene is fused to another with respect to reading frame and forms one protein from two genes.

gag The gene in retroviruses that encodes the viral coat proteins.

GAL4 A transcription factor in yeast that activates the genes responsible for galactose utilization.

gametophyte The stage in a plant life cycle during which gametes are produced. In most plants, this is the haploid stage.

gap genes *Drosophila* genes active in early embryogenesis that control an early step in segmentation. Mutation of a gap gene causes the loss of a contiguous group of segments.

gastrulation The stage of embryonic development at which invagination of the precursor cells of the endoderm and mesoderm occurs. This marks the end of the blastoderm stage of embryonic development.

G-bands Chromosome bands of heavy staining that appear after staining chromosomes with Giemsa stain.

GC box A hexamer having the sequence GGGCGG (on one strand) occurring in a number of structural genes. The transcription factor Sp1 binds to this site.

gel electrophoresis The separation of molecules by charge and molecular weight in a gel matrix subjected to an electrical field. See *electrophoresis*.

gene A hereditary determinant of a specific biological function located at a fixed position on a chromosome. A gene can be defined either genetically or physically. The genetic test for a gene is the *cis–trans* test or complementation test. In physical terms, the gene is defined as the coding region of DNA that determines a protein product.

gene amplification A process by which a number of copies of a gene is increased, resulting in overexpression of the gene's product. Amplification of some *c-oncs* has frequently been associated with cancer.

gene conversion A process by which the DNA sequence of one allele is converted to that of the other allele.

gene family A set of genes having sequence homology (usually greater than 50%), which have evolved by duplication of an ancestral gene. Members of the same gene family are usually in close chromosomal proximity to one another and possess similar functions.

gene flow Movement of genes by migration from one population to another.

gene frequency Same as allele frequency.

gene hierarchy A group of genes that act in a particular sequence in a developmental pathway. The activity of one gene is regulated by the genes higher in the hierarchy and serves to regulate the activity of the genes farther down the pathway.

gene imprinting The differential expression of a gene depending on whether it was maternally or paternally inherited.

gene pool The total genetic variability carried in a population or species.

generalized recombination See homologous recombination.

generation-time effect An hypothesis that states that the rate of mutation and evolution is a function of the number of total generations elapsed, not years. Thus, species with short generation times should evolve faster.

genetic bottleneck See *bottleneck*.

genetic code The 64 different sequences, 3 nucleotides long, that code for specific amino acids during translation. One code word (codon) specifies a start signal (AUG) while three specify a termination signal (UAA, UAG, and UGA). The 3 termination codons do not specify an amino acid.

genetic drift Changes in gene frequencies caused by random fluctuations from one generation to the next.

genetic marker Genes, usually mutant, used to experimentally track an individual during a genetic cross or selection procedure.

genetic mosaics Individuals containing cells with different genotypes.

genetic variance The differences between individuals in a population that are the result of differences in genotype.

genic balance The concept that the number of copies of genes influences their regulation and that a balance exists between regulatory genes and their targets that can be disrupted by changing the numbers of copies of genes.

genome A complete set of chromosomes inherited as a unit from a parent.

genomic library A collection of DNA fragments in a cloning vector that is representative of the total genome of an organism.

genotype The total set of genes in an organism. For genetic crosses concerned with one or a few particular genes, all other genes are considered to be normal, and the genotype is written as the combination of alleles of only those genes being considered.

germ cell A reproductive cell that, when mature, is capable of fertilization and development into an organism.

glycosylases A group of enzymes that can remove damaged bases by breaking the sugar-base bond, leaving an AP site.

G_0 phase A specialized form of the G_1 phase. Somatic cells that have ceased to divide remain permanently in G_0.

G_1 phase The gap or growth phase of the cell cycle between the end of mitosis and the beginning of the S phase.

grana Stacked complexes of thylakoids in chloroplasts that contain the photosynthetic enzyme complexes.

guanine A purine base found in DNA and RNA.

gynandromorph An individual that is a mosaic of male and female cells.

gynandromorph fate map A map showing the relative location of adult and larval precursor cells on the blastoderm surface. These maps are made by calculating the relative distance between the precursor populations using the frequency that the two populations are of different sex as a measure of distance.

hairpin structure An antiparallel helix structure formed from single-stranded nucleic acids with a stem and a loop. The stem is formed by complementary base-pairing within the molecule.

half-life The time required for one half of a population of molecules to be converted to another form (for example, conversion of ^{32}P to ^{31}P).

haplo-insufficient A locus that must be present in more than one copy to give a normal phenotype.

haploid One complete set of chromosomes (*n*) from a diploid organism.

Hardy–Weinberg Principle A rule stating that in an infinitely large, randomly mating population, when there is no selection, migration, or mutation, genotypic frequencies can be computed from gene frequencies, and these will remain constant from one generation to the next.

Harvey, William English physician who lived from 1578 to 1657. He is noted for being the first to make an accurate description of the circulation of the blood.

helicases A group of enzymes that are capable of unwinding double-stranded DNA.

helix-turn-helix A structural motif present on a DNA binding protein. It recognizes a specific nucleotide sequence.

hemizygous A condition when only one allele of a pair is present; opposite a deletion or on the X chromosome of males.

hemophilia An individual that lacks a clotting factor and consequently bleeds easily. Some of the genes for clotting factors are sex-linked.

heredity Transmission of traits from one generation to the next.

heritability The degree to which a given trait is controlled by inheritance.

hermaphrodite An animal with both male and female reproductive organs.

heterochromatin Highly condensed inactive DNA that stains darkly with Giemsa stain. Heterochromatin is either facultative or constitutive. Constitutive heterochromatin is composed of repeated sequences of DNA. Facultative heterochromatin is a transient form of inactive DNA (for example, an inactive X chromosome) that is usually the result of methylation of cytosine.

heterochronic mutations Homeotic mutations in nematodes that cause cell divisions or determinative events to occur at inappropriate times.

heteroduplex The formation of double-stranded DNA from two single-stranded sequences from different sources (for example, one strand from *E. coli* and the other from a phage). If the two sequences are not identical, they may form bulges where no pairing occurs.

heteroduplex mapping Hybridization between two molecules of nucleic acids that are not entirely complementary, identifying the location of the noncomplementary sites using an electron microscope.

heterogeneous nuclear RNA (hnRNA) The same as pre-mRNA. The immediate product of transcription in an eukaryote, containing both introns and exons.

heterokaryon A cell containing two or more different nuclei.

heteroplasmic A cell or organism that contains mitochondria or chloroplasts that have more than one genotype.

heterosis When two inbred lines (homozygous for many alleles) are crossed, giving rise to a heterozygote that has superior traits for many alleles.

heterozygote An individual with two different alleles of one gene.

Hfr A bacterial cell that contains a conjugative plasmid integrated into its genome; in *E. coli* this plasmid is usually the F plasmid. As a consequence, during transfer of the F plasmid, the chromosome is transferred as well.

histone Very basic positively charged low-molecular-weight proteins associated with the first level of DNA coiling in eukaryotes to form nucleosomes.

Hogness box A sequence of TATAAA that defines where the RNA polymerase II will bind to DNA in eukaryotes and is located about 30 bp upstream of the start codon. Also known as the TATA box.

Holliday structure An intermediate structure in crossover that involves strand exchange between homologous double-stranded NA molecules. It was first hypothesized to exist by Robin Holliday.

holoenzyme The entire molecular structure required to carry on an enzymatic function (for example, polymerase III with all of its seven subcomponents).

homeobox A sequence of 183 nucleotides that form a conserved 61 amino acid DNA binding protein domain. The homeobox was discovered in *Drosophila* homeotic genes and has been found to be highly conserved in many species. It is predominantly found in genes that regulate cell determination and pattern formation.

homeodomain The conserved 61 amino acid sequence coded for by a homeobox.

homeosis The transformation of one tissue or organ into another.

homeotic genes Genes that control the determination of cells. Many, but not all, homeotic genes contain homeoboxes.

homeotic mutation A mutation that causes homeosis, the transformation of one cell type into another. The *Antenneopedia* mutation, for example, transforms antenna into mesothorax.

homologous chromosomes A pair of chromosomes that resemble each other and have nearly the same DNA sequence. One comes from the male parent and the other, from the female parent.

homologous recombination Recombination between two chromosomes that have extensive sequence identities.

homoplasmic A cell or individual whose chloroplasts or mitochondria all have identical genotypes.

homozygote An individual with identical alleles of a gene.

host range The spectrum of host bacteria that can be infected by a virus.

housekeeping genes Genes that code for proteins that are used by nearly all cells.

***Hox* genes** All homeobox-containing genes in mammals that correspond to the BX-C and ANT-C in *Drosophila.*

Human Genome Project An international effort to determine the sequence of nucleotides in the human genome.

Huntington disease An hereditary degenerative disorder of the nervous system caused by a dominant mutation. Symptoms usually become visible in middle age. Formally known as Huntington chorea.

hybrid The offspring of a cross between two unrelated individuals.

hybrid dysgenesis Situation that occurs in *Drosophila* and results in mutations, chromosome breakage, and sterility. This results from activation of a transposable element.

hybridization Reforming double-stranded DNA from complementary sequences.

hybrid vigor See heterosis.

hydrogen bond A weak noncovalent bond between two atoms formed when a hydrogen atom is shared between two atoms, one of which is usually oxygen. Hydrogen bonds hold the two complementary strands of DNA together.

hydrolysis Degradation of a covalent bond by the addition of water.

hypermorphic One of the functional allele classifications of H. J. Muller. A hypermorphic allele has more gene function than the standard, wild-type allele.

hypervariable regions The sequence region within an antibody that determines the specificity of the antibody.

hyphae (singular, hypha) A threadlike fungal vegetative structure composed of cells aligned end to end. Hyphae form the main vegetative mass in fungi such as *Neurospora.*

hypomorphic One of the functional allele classifications of H. J. Muller. A hypomorphic allele has less gene function than the standard, wild-type allele.

hypothesis Possible explanation of a given trend or regularity apparent in a body of scientific data.

illegitimate recombination Process by which transposable elements are able to insert into host chromosomes without having homology with chromosomal sequences at the site of insertion.

imaginal disc The population of cells set aside during *Drosophila* embryonic development to form adult structures. Each disc forms one particular set of structures—leg, wing, genital structures, etc.

imaginal disc fate map A fate map showing the location of the imaginal disc cells that will give rise to specific structures during metamorphosis.

immunoglobulins A collective term referring to all the antibodies produced by an organism.

inbreeding The mating of related individuals.

inbreeding coefficient (f) The probability that the two homologous alleles in an individual are identical by descent.

incomplete dominance Progeny that possesses a phenotype that is approximately intermediate between the homozygous parents.

independent assortment The independent segregation of alleles of different genes.

inducer An environmental agent that triggers transcription from an operon.

induction The alteration of a cell's fate by an outside influence, usually a signal sent by another cell. Alternatively, this term is used to indicate the turning on of the expression of a gene by the presence of an inducer.

infectious inheritance A form of extranuclear inheritance. This involves the inheritance of an agent that can be inherited with cytoplasm but that can also be an infectious agent.

initiation codon A codon that directs the reading of the transcript in the correct reading frame. In most cases, it is AUG, but occasionally it is GUG.

insertion sequence (IS) Smallest bacterial transposable element that carries only sequences encoding functions necessary for its own transposition.

***in situ* hybridization** The hybridization of a DNA probe to homologous sequences in a chromosome while still in the cell. The location of hybridization in the chromosome indicates the location of DNA sequences complementary to the probe.

interaction variance The difference in phenotype of individuals in a population that is caused by an unequal interaction between select genotypes and environments.

interbands Regions of light staining between chromosomes, usually seen in polytene chromosomes.

interference Value that is a measure of the ability of two crossover events to occur independently. It is calculated by determining the coefficient of coincidence and subtracting this value from 1.

interphase The G_1, S, and G_2 phases of the cell cycle.

interrupted mating Breakage of the conjugation tube during mating of an *Hfr* with an F^- to limit the amount of donor chromosome transferred and thereby facilitate mapping.

interstitial deletion A deletion that removes material between the centromere and the telomere.

introns The noncoding sequences within a RNA transcript.

inversion loop The loop formed by synapsis between an inverted and a normal chromosome.

inverted terminal repeats Short related or identical DNA sequences in reverse orientation at the boundaries of many transposable elements.

iso-accepting tRNAs Situation that occurs when more than one tRNA exists for a single amino acid.

karyokinesis The division of the nucleus during a cell division.

karyotype The complete complement of metaphase chromosomes of a cell or an individual.

kinetochore The site of attachment of the spindle to the chromosome. Usually the kinetochore is located at the centromere.

Klinefelter syndrome The collection of human phenotypes produced by a genome with more than one X chromosome and a Y chromosome.

knockout mice Mice that have had a normal gene replaced by a nonfunctional allele using recombinant DNA methods.

lagging and leading strands The lagging DNA strand is synthesized discontinuously, and the leading strand is synthesized continuously during replication.

Lamarck, J. P. A French (1744–1829) naturalist that proposed that acquired characteristics can be inherited.

lampbrush chromosomes A chromosome that is highly expanded and uncoiled with respect to its DNA and found in the oocytes of amphibians.

leader region A sequence of bases on the transcript between the promoter and the start codon.

leaky mutation A mutation that leaves partial enzyme activity and biological function.

legitimate recombination Process by which two homologous DNA molecules exchange genetic material.

leptotene The first stage of the first meiotic prophase.

Lesch–Nyhan An hereditary degenerative disorder.

lethal mutation Usually refers to recessive mutations that, when homozygous, result in the death of the individual; can also refer to dominant mutations that, when heterozygous, result in the death of the individual.

leukemia Cancers of blood cells. Together with lymphomas, they account for about 8% of all malignant tumors.

ligase An enzyme that joins the ends of two strands of DNA.

ligation The joining of two DNA molecules with a covalent bond.

LINE (Long interspersed nuclear elements) Families of long (about 6000 bp) moderately repetitive transposable elements in eukaryotes.

linkage groups Groups of genes that are on the same chromosome.

linkage map A map based on recombination frequency that shows of the relationship of genes on a chromosome.

locus A fixed position of a chromosome occupied by a gene or one of its alleles.

long terminal repeat (LTR) A sequence of 500–880 bp directly repeated at the ends of a retrovirus or retrotransposon. The LTR contains promoter and termination sequences typical of eukaryotic genes.

lymphomas Cancers of the lymph system. Together with leukemia, they account for about 8% of all malignant tumors.

lysis The destruction of a cell membrane and the release of its contents.

lysogen A bacterial cell harboring a prophage.

lysogenic bacteria A bacteria harboring a temperate phage.

lysogenic cycle The phage life cycle in which the phage is incorporated into the bacterial genome and is propagated with the bacterial genome.

lytic cycle The phage life cycle in which the phage infects the bacterial cell and produces more phage particles that are released by lysis of the bacterial cell.

macromolecule A very large molecule (for example, a protein or nucleic acid).

malignant tumors Tumors that invade and destroy adjacent tissues and frequently metastasize.

Marfan syndrome A disorder of the connective tissues caused by a dominant mutation.

maternal effect Traits controlled by the mother and expressed in the progeny. Maternal effects are usually caused by a transcription factor or its mRNA that is produced by the mother and included in the egg.

maternal effect mutation A mutation that affects the phenotype of the offspring of mutant females, but not mutant males. Usually maternal effect mutations act by altering proteins or RNA molecules packaged in the egg that are essential for embryonic development.

maternal inheritance Inheritance controlled by extrachromosomal DNA and transmitted through the egg cytoplasm.

Maxam–Gilbert sequencing A method of sequencing of DNA that uses chemical agents to break DNA into fragments that are then separated by size. Each size represents a fragment ending at a specific base; thus the sequence can be read directly from the separated fragments.

mean The arithmetic average; the sum of the values in a sample divided by the sample size.

megaspore Premeiotic germ cells in the female flowers of plants.

meiosis Two cell divisions that reduce the number of chromosomes by half.

Mendel, Gregor The father of genetics; an Augustinian monk who discovered the fundamental rules governing the inheritance of traits.

Mendelian population A naturally interbreeding group of individuals that share a common gene pool.

Mendelism When the progeny phenotypes of a cross reflect the operation of Mendel's laws.

merozygote A partial diploid strain. A bacterial strain that contains a chromosome and an F′ -plasmid containing an extra segment of the chromosome.

messenger RNA (mRNA) The transcript of a gene or genes that acts as template for protein synthesis.

metacentric A chromosome with the centromere in the center.

metamorphosis The replacement of larval organs and tissues with adult organs and tissues that occurs in insects during the pupal stage.

metaphase The second portion of mitosis or meiosis during which the chromosomes align on the metaphase plate.

metaphase plate The single, equatorial plane in the center of the spindle where the chromosomes are aligned during metaphase.

metastasis The spread of cancer cells to other parts of the body.

microsatellite markers A type of repetitive DNA that contains short repeat sequences interspersed throughout the genome and is used in DNA fignerprinting.

microspore Premeiotic cells in the male flowers of plants.

minimum patch size estimate A method for estimating the number of somatic cells (n) found in a developmental cell population. This estimate uses the smallest of a population of clones induced before the cell population is founded as a measure of $1/n$.

mismatch correction enzyme An enzyme that scans the DNA immediately after the duplication region for base mismatches.

mismatch repair The repair of DNA sequences that are not properly hydrogen bonded.

missense mutation A mutation changing a codon to another codon that is not a termination codon.

mitochondria An organelle in the cytoplasm of eukaryotes where energy production takes place.

mitosis The process of nuclear division in eukaryotes that produces two identical daughter nuclei.

modified dihybrid ratios Phenotypic ratios in the F_1 or F_2 of a cross that differ from the standard ratios obtained by Mendel in his pea crosses. Modified ratios result when genes are linked or when alleles show different dominance relationships.

molecular clock The rate at which nucleotide (amino acid) substitutions occur in a given DNA (protein) sequence.

molecular clock hypothesis Hypothesis that the rate of nucleotide (amino acid) substitution in a DNA (protein) sequence is constant in all evolutionary lineages provided the DNA (protein) sequence retains its original function.

molecular evolution The study of evolution using DNA and protein sequence data to analyze sequence changes that have occurred over time.

molecular genetics Field of genetics devoted to the study of the biochemical mechanisms by which heredity information is stored in nucleic acid and transmitted to proteins.

molecular phylogeny The determination of phylogenetic relationships between organisms by the use of DNA sequence data.

monoclonal antibody A pure B cell line that produce a single antibody. Polyclonal antibodies are produced by a mixture of cell lines and thus produce a mixture of antibodies.

monohybrid A cross between parents that differ by one trait.

monohybrid cross A cross between two individuals that are heterozygous for the same allele pair.

monomer A single molecule that can form polymers with other molecules.

monoploid The smallest set of chromosomes that contains one copy of each gene.

monosomic An individual or cell that is diploid for all but one chromosome. The chromosome complement of a monosomic is 2N-1.

monozygotic twins Identical twins derived from a single fertilization.

mosaic An organism that has cells of different genotypes.

mosaic development Development in which the initial determination decisions in the blastoderm are controlled by cytoplasmic determinants. Such a blastoderm will have little or no ability to regenerate after the loss of cells.

multigene family See gene family.

multiple alleles Situation that occurs when more than one allele can exist for a particular locus in a population.

multiplication rule A probability rule used to calculate the probability that two independent events will both occur. The probability that event A and event B will both occur is the product of the probabilities of each event occurring separately. This is written symbolically as $p(A + B) = p(A) \times p(B)$.

multiplicity of infection or **moi** The ratio of infecting phage to bacteria in an experiment.

mutagen An agent that causes mutations (chemicals, radiation, etc).

mutant An organism or cell that possesses a mutation in a gene.

mutant allele An allele that gives a mutant phenotype (clearly different and usually inferior). Not all alleles with a DNA sequence different from the normal or wild-type allele give mutant phenotypes.

mutational analysis The deduction of the function of a gene by the analysis of the phenotypes produced by a series of mutant alleles.

mutations Alterations in the DNA sequence of an allele; an inherited genetic change.

mycelium The mass of hyphae that form the main vegetative tissue in *Neurospora* or other fungi.

narrow sense heritability (h^2) The proportion of the phenotypic variance that is caused by additive effects of alleles.

natural selection The process of differential reproduction of alternative genotypes in a population.

negative assortive mating A system of mating in which phenotypically dissimilar individuals tend to mate more frequently than would occur under random mating.

negative control A gene expression system that requires a protein(s) to turn off gene expression.

neomorphic One of the functional allele classifications devised by H. J. Muller. Neomorphic alleles are gain-of-function alleles that have new or modified functions different from the function of the normal or wild-type allele.

neuroblastoma Cancerous growth of neuroblast cells.

neuroblasts Neural precursor cells formed in the ectodermal neurogenic region of the *Drosophila* embryo.

neurogenic genes A group of six *Drosophila* genes whose function is required for normal cell-fate decisions in the neurogenic region of the embryo.

neurogenic region The region of the cell blastoderm in which neuroblasts and epidermoblasts arise.

neutral mutation A mutation that has no effect on phenotype and is therefore neutral with respect to natural selection.

nick-translation A procedure for labeling DNA with radioactive nucleotides. The DNA is nicked with an endonuclease, and DNA polymerase is added to use the nicks to degrade and incorporate radioactive nucleotides.

nonautonomous element Transposable element that contains the target sequences for the action of a transposase but does not encode an active transposase itself and thus requires that this function be provided in *trans* in order for transposition to take place.

nondisjunction The failure of segregation of sister chromatids during mitosis or homologous chromosomes during meiosis.

nonhistone proteins Proteins associated with chromosomes that are not histones. For example, they may be enzymes involved in DNA replication or structural components of the chromatin.

nonhomologous recombination Recombination between two DNA sequences that do not possess sequence homology.

nonsense mutation A mutation that alters a gene to produce a nonsense codon, either UAG, UGA, or UAA.

northern blotting The transfer of separated RNA from an agarose gel to a membrane for eventual hybridization using a labeled probe.

nucleic acid DNA and RNA, a polymer composed of pentoses, bases, and phosphates.

nucleoid A condensed bacterial chromosome composed of protein, RNA, and DNA.

nucleosome About 145 bp of DNA coiled around an octomer of histone proteins.

nucleotides and **nucleosides** A nucleotide is composed of a base (A, T, C, G, or U) covalently bonded to a ribose or deoxyribose, which is covalently bonded to a phosphate (for nucleotides). No phosphate is present in nucleosides.

nullisomic A genotype that contains no copies of a particular chromosome.

null mutation A mutation that completely abolishes the expression of a gene.

nurse cell One of the support cells of the oocyte. Mitotic division of one oogonial precursor cell produces 15 nurse cells and one oocyte in *Drosophila*.

Okazaki fragments Short segments of DNA synthesized on the lagging strand of DNA. They are usually about 1000–2000 bp in length.

oncogene A gene that participates in the production of cancer.

oncogenic viruses Viruses that induce cellular transformation in infected cells; oocyte; the germline cell that becomes the female gamete.

oocyte The germline cell that becomes the female gamete.

oogonia A population of dividing cells that produce primary oocytes.

oogonial cells Stem cells that produce the oocyte and its mitotic sister nurse cells.

ootid One of the four meiotic products in females that produce the gametic nucleus.

open reading frame (ORF) A putative gene that is uninterrupted by any stop codons.

Operator site A sequence of nucleotides in an operon where the repressor protein binids.

operon A group of adjacent structural genes whose mRNA is synthesized as one piece. The operon is controlled by a group of regulatory sites (promoter and operator) and is present only in prokaryotes.

origin of replication A site consisting of a base sequence where DNA synthesis always starts. *E. coli* has one origin of replication, whereas eukaryotes have many.

overdominance A dominance relationship in which the heterozygote has a greater or more extreme phenotype than individuals homozygous for either allele.

p-arm The short arm of a human chromosome.

pachytene The third stage of the first meiotic prophase. During pachytene, chromosome pairing is completed.

pair rule genes *Drosophila* genes that regulate one step in the segmentation process. Mutation of one of these genes causes the loss of alternating segments.

palindrome A sequence of DNA that is symmetrical and reads the same in both directions.

paracentric inversion An inversion with the centromere included in the inverted region.

parasegment A four-cell-wide stripe of blastoderm cells. Each parasegment is the domain of activity of specific combinations of homeotic genes and consists of the posterior compartment of one segment and the anterior compartment of the following (more posterior) segment.

parental generation (P generation) The first generation in a multigenerational controlled breeding scheme.

partial digestion Digestion of DNA with restriction endonucleases to yield fragments that are not cleaved at all existing sites.

particulate inheritance The theory that inheritance is controlled by discrete heritable units (genes).

pattern formation The coordination of cell division, cell determination, and cell differentiation in space and time to generate a particular three-dimensional body or structure.

pedigree A diagram showing the phenotypes and relationship between members of a family.

pedigree analysis The deduction of the genetic basis for a trait from a study of the recorded phenotypes of a family organized in a pedigree.

penetrance The proportion of individuals with a given genotype in a population that expresses the phenotype.

peptide bond A covalent bond that joins two amino acids together through the carboxyl group of one amino acid and the amino group of a second amino acid.

peptidyl transferase An enzymatic reaction function of the large subunit of the ribosome that catalyzes the formation of a peptide bond during translation.

pericentric inversion An inversion in which the centromere is not included in the inverted region.

perithecia Fruiting bodies in fungi such as *Neurospora* that contain the asci.

PFGE See *pulse field gel electrophoresis*.

phenocopy Situation that occurs when an environmental agent induces a phenotype that resembles a particular mutant phenotype.

phenotype The physical characteristics of an organism. Individuals carrying different sets of alleles will have different physical characteristics, or phenotypes.

phenotypic selection Survival or reproductive success that is determined by an individual's phenotype.

phenotypic variance Differences between the phenotype of individuals in a population.

phenylkaptonuria (PKU) A human inherited disorder caused by a mutant allele of the gene responsible for the enzyme phenylalanine hydroxylase. In individuals homozygous for the allele that causes PKU, this enzyme is faulty and the accumulation of phenylalanine produces defects in central nervous system development, mental retardation, and epileptic seizures.

phosphodiester linkage A covalent bond between the 3′-OH and the 5′-PO_4 of two nucleotides.

photoreactivation An enzyme (photolyase) that cleaves the cyclobutane ring of ultraviolet-induced pyrimidine dimers. Its activity is induced by visible light.

pilus (plural, pili) A structure on the surface of bacteria coded for by the presence of a F-plasmid. Similar structures are present on bacteria lacking F-plasmids referred to as fimbriae. One pilus forms a conjugation tube to the F^- cell that acts as a bridge for transfer of the DNA from the donor *E. coli* strain to the recipient.

pistil The part of the plant flower that receives the pollen.

plaque A clear area on a lawn of bacteria that represents lysed bacterial cells from a phage infection. A single plaque will represent a clone of a specific phage.

plasmid A small autonomously replicating circular double-stranded DNA present in bacteria and some eukaryotes. Many different types of plasmids exist, some conferring antibiotic resistance, and others, the ability to metabolize unusual organic compounds.

pleiotropy The production of numerous phenotypic effects by a single mutation.

ploidy The number of complete sets of chromosomes present in a cell or individual.

point mutation A mutation that does not visibly alter the chromosome. Point mutations are usually very small alterations in chromosomal DNA, often single base changes.

polar bodies The three meiotic products in females that do not become the ootid nucleus.

polar granules Cytoplasmic bodies containing protein and RNA that are found in the posterior cytoplasm of the *Drosophila* egg. Polar granules are cytoplasmic determinants for the germline cells.

polar mutations A mutation in a structural gene affecting the expression of downstream structural genes. The mutations are usually frameshift mutations.

pole cells The embryonic cells that give rise to the germline cells. In *Drosophila,* pole cells are the first cells to form, and they appear at the posterior pole of the egg.

poly(A) tail A sequence of adenosines added to a eukaryotic mRNA after transcription.

polycistronic message One mRNA molecule made from one promoter, but for several genes.

polycistronic mRNA A messenger RNA synthesized in prokaryotes that is composed of the genetic information of more than one gene or cistron.

polyclonal antibodies A mixture of different antibodies, each capable of binding to a single organism but specific for a different site on the surface.

polycloning sequence See polylinker sequence.

polygenic inheritance A phenotype that is controlled by multiple genes.

polylinker sequence A short DNA sequence containing multiple cleavage sites for different restriction endonucleases.

polymer A long chain of a repeating unit. Examples of polymers include protein, nucleic acids, and carbohydrates such as starch and cellulose.

polymerase chain reaction (PCR) A method that allows amplification of DNA using DNA polymerase and a primer for both strands. This results in an exponential increase in the DNA copy number with cycles of synthesis.

polymorphic loci See *polymorphism.*

polymorphism The presence of two or more alleles at a locus in an interbreeding population.

polynucleotide kinase An enzyme that phosphorylates the 5′-OH terminal of DNA.

polyploid An individual or cell with more than two sets of chromosomes.

polytene chromosomes Large chromosomes resulting from several rounds of DNA replication without mitosis. The polytene chromosomes from cells of the salivary glands of *Drosophila* are used to locate the position of genes in chromosomes.

population A group of individuals that are potentially capable of interbreeding.

population genetics Field of genetics that investigates the factors that determine the genetic composition of interbreeding or related organisms called populations.

position effect The alteration of a gene's function caused by changing its location in the chromosome.

positional cloning The isolation or cloning of a DNA sequence based on its map position.

positional information According to the hypothesis of L. Wolpert, information passed between cells that informs a cell of its relative location in the body. Cells then use this information to activate particular sets of determination genes.

positive assortive mating A system of mating where phenotypically similar individuals mate more frequently than they would under random mating.

posttranscriptional control Control of gene expression that occurs after the pre-mRNA is synthesized (for example, mRNA stability, splicing, and modification).

preformationism Religiously motivated 19-century hereditary theory that proposed that a small, fully formed individual or "homunculus" is present in one of the parental gametes.

pre-mRNA The immediate product of transcription in eukaryotes, including introns and exons. It is also called hnRNA or heteronuclear RNA.

Pribnow box An *E. coli* promoter sequence located about −10 bp upstream of the start of transcription and the site of RNA polymerase binding.

primary oocyte An oocyte in the first meiotic division.

primary signal In sex determination, the initial or starting signal for the establishment of one sexual type. In *Drosophila,* the ratio of X chromosomes to autosomes.

primary spermatocytes Male germ cells in the first meiotic division.

primase or **primosome** The primase is a component of the primosome that is responsible for synthesis of short sequences of RNA for priming of DNA synthesis.

primer A sequence of nucleotides that possess a free 3′-OH group for extension of the new DNA polymer.

prion An infectious agent that causes degenerative disorders of the central nervous system (scrapie in sheep, Creutzfeldt–Jacob disease in humans); thought to be a protein.

probability The likelihood or chance that a certain event will occur.

probe A fragment of DNA or RNA labeled with radioactivity, a dye, or an antigen for detecting the presence of complementary sequences of DNA or RNA.

product rule A rule stating that the probability of both of two independent events occurring is the product of their individual probabilities.

prokaryotes Microorganisms that lack a true nucleus.

prometaphase A transition period between prophase and metaphase during which the chromosomes become aligned on the metaphase plate.

promoter A sequence of bases that defines where the RNA polymerase will bind and initiate transcription.

proofreading The ability of DNA polymerases to remove mismatched nucleotides with 3′ to 5′ exonuclease activity during DNA synthesis.

prophage The genome of a temperate phage that is integrated into a bacterial chromosome and replicated as part of the bacterial host chromosome.

prophase The first portion of mitosis or meiosis during which the chromosomes become visible and the centriole divides.

proto-oncogene A normal gene that when mutant participates in the formation of cancer.

prototroph A bacterial strain with the ability to grow on minimal media (compare to auxotroph).

pseudogene An inactive DNA sequence derived from an active gene by the accumulation of inactivating mutations (see *retropseudogene*).

pseudo-linkage The tendency of alleles to segregate together in translocation heterozygotes that results from some segregation patterns giving unbalanced, inviable gametes.

pulse field gel electrophoresis (PFGE) A method for separating very large fragments of DNA or whole chromosomes up to 10,000 kb in length. The procedure uses an alternating electric field during separation of the DNA, and it has been found that the rate of migration varies with the size of DNA.

pulse labeling Experimental incorporation of a radioisotope into a molecule (perhaps DNA) for a short time, followed by dilution with excess unlabeled precursor to stop the incorporation.

Punnet square A diagram in which all the possible genotypic combinations are arranged in an array. Punnet squares are used to predict the progeny ratios that will occur from a particular mating.

pupation The formation of puparium by an insect larva. This is the first step in metamorphosis.

purine A two-ringed nitrogenous base. In DNA and RNA, they are adenine and guanine.

pyrimidine A single-ringed nitrogenous base. In DNA and RNA they are cytosine, thymine, and uracil. Pyrimidines always pair with purines in DNA.

pyrimidine dimers The cyclobutane ring formed between adjacent pyrimidines in DNA and induced by ultraviolet light.

q-arm The long arm of a human chromosome.

Q bands Chromosome bands of ultraviolet light fluorescence that appear after staining with quinacrine mustard.

qualitative trait A discontinuous trait that is controlled by Mendelian alleles.

quantitative trait A trait determined by many genes. It has a continuous distribution of phenotypes.

quaternary structure When two or more subunits of a protein interact to form a complete protein or enzyme.

random mating Every organism in a population has an equal chance of mating without regard to genotype.

random primer extension Using a mixture of chemically synthesized DNA fragments (6 bases long), with each possible sequence represented, to prime DNA synthesis for radioactive labeling.

random sample A sample that is obtained by a procedure that gives every member of the population an equal chance of being included.

reading frame The genetic message read in triplets from a defined point, usually from the AUG start codon to a stop codon.

recA A gene that encodes the RecA protein, which participates in homologous recombination.

recessiveness and **recessive** An interaction between alleles in a heterozygote. A recessive allele does not control the phenotype in heterozygotes, and a recessive phenotype is not expressed in heterozygotes.

reciprocal crosses Two crosses done with identical genotypes in which each genotype is present in the male parent in one cross and the female parent in the other. Reciprocal crosses are done to determine whether the sex of the parent influences the outcome of the cross.

reciprocal translocation A translocation involving an exchange of material by nonhomologous chromosomes. The amount of chromosomal material exchanged during the translocation need not be equivalent.

Recombinant DNA technology Techniques for separating, uniting, and amplifying heterologous DNA molecules.

recombination The reassortment of genes in new combinations following exchange of DNA sequences.

recombinational repair A repair system that uses a separate but identical source of DNA to correct mistakes by recombination.

reductional division The first meiotic cell division. Chromosome number is reduced during this division.

reductionist A philosophy of science that advocates breaking questions or systems into their smallest parts for study.

regeneration The formation of a whole individual or organ from a single cell or a few cells. Plants have a much greater ability to regenerate than animals.

regulative development A pattern of development in which cell fate is controlled by cell position. A regulative embryo can regenerate or replace cells or nuclei that are lost or killed.

regulatory gene A gene that encodes for an RNA or protein product whose function is to control the expression of other genes.

reiterative mutations Homeotic mutations in the nematode that cause particular cell divisions to be repeated.

renaturation The reformation of a double-stranded DNA or RNA structure by the formation of hydrogen bonds between complementary sequences.

repetitive DNA Sequences of DNA repeated many times in the genome of an individual. Sometimes there are over a million repeats.

replicative transposition Transposition of a transposable element where the parental element remains at the original chromosomal site, whereas a newly synthesized element is inserted at a new site.

replicon A unit of DNA capable of replication (for example, a plasmid or a chromosome).

reporter gene A gene that codes for a protein that is easily assayed. This gene is usually fused to a foreign promoter/regulatory region whose gene is very difficult to assay. The reporter gene facilitates the study of the regulatory region.

repressor protein A molecule that binds to the operator and prevents transcription of an operon.

reproductive isolating barriers (RIBs) Barriers that give rise to new species. RIBs can be prezygotic or postzygotic (for example, behavioral isolation is prezygotic and hybrid infertility is postzygotic).

restriction endonucleases A family of enzymes that cleave double-stranded DNA at specific sequences, leaving blunt ends or overhanging single-stranded sequences.

restriction fragment A short fragment of DNA generated after digestion with a restriction endonuclease.

restriction fragment length polymorphism (RFLP) A variation of the restriction endonuclease cut sites between individuals that results in variously sized fragments between different individuals.

restriction map A map of the location of restriction endonuclease cut sites in a piece of DNA.

reticulocytes An immature erythrocyte actively synthesizing hemoglobin.

retinoblastoma A childhood cancer of the retina.

retrogene Gene that has been generated via reverse transcription of a partially or completely processed ancestral gene transcript.

retroposon or **retrotransposon** A transposable element whose replicative cycle requires passage through an RNA intermediate. The information encoded in the RNA is copied to DNA via reverse transcriptase.

retropseudogene A pseudogene derived from the reverse transcription of an RNA molecule to form a cDNA copy, which is then incorporated into the genome. Retropseudogenes typically lack promoters and introns and are usually not physically linked to other members of their gene family.

retrovirus An RNA virus that spends part of its life cycle as an integrated DNA in the host chromosome. The genomic information in the RNA is copied to DNA via reverse transcriptase.

reverse mutation A mutation producing wild-type or near-wild-type phenotype from a mutant phenotype. Reverse mutations can be a same-site or second-site mutation.

reverse tandem duplication A duplication in which the duplicate region is adjacent to the original copy but in inverse order.

reverse transcriptase An RNA-dependent DNA polymerase that can synthesize a complementary copy of an RNA molecule.

rho factor A protein factor in bacteria that is involved in the termination of transcription at specific DNA sites.

ribonucleic acid (RNA) A single-stranded polymer composed of the ribonucleotides, adenosine, guanosine, cytosine, and uridine joined by a linking phosphate group. RNA comes in many different lengths and has a variety of functions within the cell.

ribosomal proteins Proteins that are involved in determining the structure of a ribosome. Each ribosome is about half protein and half RNA.

ribosomal RNA (rRNA) A large RNA molecule, encoded by the genome, that is an integral part of the ribosome.

ribosome A particle composed of rRNA and protein where protein synthesis occurs.

RIBs See *reproductive isolating barriers.*

RNA editing A process involving addition, removal, or substitution of one or more nucleotides from selected regions of the transcript.

RNA polymerase A family of large multimeric enzymes responsible for synthesis of RNA from the DNA template.

Robertsonian translocation A translocation involving a complete chromosome arm.

rRNA Ribosomal RNA. A component of the ribosome and in bacteria is represented by 5S, 16S, and 23S molecules.

Sanger dideoxy sequence analysis A method of DNA sequencing that depend upon DNA polymerase to incorporate bases that block further chain growth. This results in DNA fragments of varying lengths that are separated according to size by electrophoresis. The sequence can then be read from the sized fragments.

sarcoma Solid tumor of the connective tissue (muscle and bone). Sarcomas are the least frequent type of tumor, constituting about 2% of all malignant tumors.

satellite DNA A class of DNA that appears on a CsCl-density gradient centrifugation pattern representing a small fraction of the total DNA with a different density.

saturation mutagenesis The mutagenesis of an organism's genome that, on the average, will produce a mutation in all loci. These mutations produce at least one mutation in all genes that can mutate to produce the selected phenotype.

scaffold protein The protein core attached to the DNA in condensed eukaryotic chromosome.

scatter plot A graph showing the relationship between two variables.

scientific method Logical method of discovery used by scientists to establish new theories. The process begins by careful observation, followed by the formulation of hypotheses to explain observed trends. Experiments are performed to distinguish between alternative hypotheses.

secondary spermatocytes Male germ cells in the second meiotic division.

second filial (F_2) The second progeny generation in a Mendelian cross. The F_2 are produced by inbreeding the F_1.

second-site mutation Same as suppressor mutation.

segmentation The division of the body into a series of regions, each formed by a unique set of cells.

segmentation genes Genes controlling the division of the body into segments. Segmentation genes in *Drosophila* do not control segment determination (that is, the decision as to what each segment will form).

segment determination genes Genes that establish a particular identity, or determined state, in each segment of a developing insect. They do not control the establishment of the segments.

segment polarity genes *Drosophila* genes that regulate one step in the segmentation process. Mutation of one of the segment polarity genes causes all segments to lose a particular region of each segment.

segregation Mendel's first rule, that the alleles of one gene separate so that only one goes into each gamete.

selection response The amount of phenotypic change in one generation when selection is applied to a group of individuals.

selective coefficient The difference in fitness between a particular genotype and another genotype chosen as a standard of reference.

selector gene A developmental control gene that is active in the cells of a particular compartment.

selfing A self cross. In plants, fertilization of eggs with pollen from the same plant.

self-splicing The excision of introns from RNA by a mechanism that does not require the participation of a protein.

semiconservative replication A form of DNA replication where one strand of the parent molecule ends up in each of the progeny molecules. In conservative replication, both parent strands would end up in one progeny molecule.

sex chromosome A chromosome pair in eukaryotes that usually differ in size and shape and control sex determination. The presence of identical pairs is usually present in the female and nonidentical pairs in males, although the reverse also occurs.

sexduction The transfer of an F′-plasmid (containing bacterial genetic information) from one bacterial strain to another.

sex-influenced trait A trait that appears in both sexes, but there is a difference in phenotype (for example, milk production).

sex linkage Genes that are located on the X, or sex chromosomes.

sex-linked trait A trait that is located on the sex chromosome and is phenotypically present in the heterogametic sex when recessive.

shotgun cloning A random collection of cloned DNA fragments representative of the total genome of the organism.

shuttle vectors A cloning vector that has two or more origins of replication for replication in different hosts (for example, *E. coli* and yeast).

sigma factor or **subunit** In prokaryotes, the subunit of RNA polymerase that directs the enzyme to start synthesis at the promoter sequence of each gene.

silencer element A sequence of bases upstream of the transcriptional start site of a gene. When activated, it turns down the rate of transcription.

silent mutation A base change causing a substitution of one codon for another but still coding for the same amino acid.

SINE (short interspersed repeated sequence) A class of short (< 500 bp) interspersed, highly repeated sequences present in eukaryotes.

single stranded binding (SSB) protein A small protein that does not have enzymatic activity and binds to single-stranded DNA to stabilize it for subsequent replication.

sister chromatids The two chromatids of one chromosome after synthesis has been completed.

site-specific recombination Recombination between two specific sequences as in lambda phage integration and excision in *E. coli*.

small nuclear RNAs (snRNA) RNAs 100–300 nucleotides in length that are involved in the splicing reaction in eukaryotes and present in spliceosomes.

somatic cell A body cell whose mitotic descendants do not form gametes.

somatic mutation A mutation in a non-germ cell that can give rise to a spot of cells that are mutant.

Southern blotting The transfer of DNA fragments from an agarose gel to a membrane for hybridization with a radioactive probe.

specialized transduction A type of transduction in which only specific genes are transferred.

species A group of animals, fungi, or plants that are capable of interbreeding.

specifications Cell-fate decisions that, unlike determination decisions, can be altered if the cell is moved or transplanted. These decisions are stable during normal development. An example is the decision of a *Drosophila* imaginal leg disc cell to form a particular adult structure, such as a bristle. If that cell is cultured, its descendants may form different parts of the leg.

spermatids The meiotic products of male germ cells that form sperm.

spermatogonia A population of dividing cells in males that gives rise to the spermatocytes.

S-phase The portion of the cell cycle during which DNA is synthesized.

spindle A collection of microtubules radiating from the centriole that attach to and direct the movements of the chromosomes.

spliceosome A large complex of nucleic acids and proteins that catalyzes the splicing reaction in eukaryotes.

splicing The cutting and rejoining of RNA molecules that occurs during the processing of eukaryotic gene transcripts.

splicing reaction The removal of introns from an RNA transcript in eukaryotes.

spontaneous mutation A mutation that occurs without the intervention of chemicals or radiation.

sporophyte The stage in a plant life cycle during which meiosis occurs. In most plants, this is the diploid stage.

stable position effect A position effect that gives a uniform, consistent phenotype.

standard deviation The square root of the variance. For normally distributed quantities, approximately 95% of all values will lie within two standard deviations of the mean.

stop codon A codon that ends translation (UGA, UAA, and UAG) of mRNA.

strain A population of individuals that all share the same trait. Strains are maintained by interbreeding individuals within the same strain.

structural gene A gene that codes for an mRNA molecule and consequently the protein.

stuffer fragments A fragment of DNA in lambda phage that can be replaced with a foreign DNA fragment, thus using the lambda phage as a cloning vehicle.

submetacentric A chromosome with the centromere slightly off center.

sum rule The probability of either of two mutually exclusive events occurring is the sum of their individual probabilities.

supernumary element A chromosome that is present in only some individuals, and is thus not vital for the growth and development of the individual.

suppressor mutations A second-site mutation that suppresses the mutant phenotype back to wild-type. Suppressor mutations can be intergenic or extragenic.

sympatric model of speciation A description of how reproductive isolation may evolve among groups of organisms living in the same area.

synapsis The pairing of homologous chromosomes during prophase I of meiosis.

synaptonemal complex A complex structure that forms during the pairing of homologous chromosomes in meiosis. This complex is believed to assist in precisely pairing the chromosomes.

syncitial blastoderm The stage in *Drosophila* embryogenesis when the zygotic nuclei have reached the cortical cytoplasm but cell membranes have not formed.

syndrome A collection of phenotypes in humans that are associated with, and are usually produced by, the same cause.

synergids Haploid nuclei in the embryo sac of maize that are located adjacent to the gamete nucleus.

syngamy The fusion of two gametes to form a zygote.

synkaryon A fusion nucleus following the fusion of cells with genetically different nuclei.

tandem duplication A duplication in which the duplicate region is adjacent to the original and is in the same order.

TATA element Element present in eukaryotes at about −30 bp from the transcription start site; the likely promoter site. The consensus sequence is TATAAAAA.

tautomeric shift The spontaneous isomerization of a base to an alternate hydrogen-bonding state, either enol or keto and imino or amino, which may lead to a mutation.

T-DNA A segment of the Ti-plasmid of *Agrobacterium tumefaciens* that determines a group of genes that code for plant hormone production or tumor induction in plants. This group of genes is flanked by 25-bp sequences that are recognized by proteins and transfers this segment of DNA into a plant cell upon infection.

telocentric A chromosome with the centromere at one end.

telomerase The enzyme responsible for synthesis of telomeres. It is a ribonucleoprotein of about 220 kDa and is characteristic of each species in specifying the sequence of the telomere repeat unit.

telomere The terminal region of a chromosome that prevents random fusions to other chromosomes. The telomere is composed of multiple repeats of a basic consensus sequence of $C_{18}(T/A)_{14}$.

telophase The terminal phase of nuclear division. During telophase, the chromosomes unwind and reform an interphase nucleus.

temperate phage A phage that can enter a lysogenic phase, becoming a prophage.

temperature-sensitive mutation A mutant event changing the protein so that function is present at a high or low temperature but absent at the inverse temperature. It is a form of conditional lethal mutation.

template strand A strand of DNA that dictates the sequence of bases to be polymerized into a complementary strand.

teratogen A chemical that causes developmental abnormalities in an organism.

terminal deletion A deletion that includes the tip of the chromosome. Terminal deletions usually produce unstable chromosomes because they eliminate the telomere.

terminal differentiation A differentiated state with no further cell division.

terminal transferase An enzyme that will add homopolymer tails to DNA.

termination codons A three-nucleotide sequence in the mRNA used for stopping translation at the end of a gene (UAA, UGA, and UAG).

terminator sequence A sequence of bases in DNA that stops transcription and releases the RNA product.

testcross A cross in which one of the parents is homozygous for recessive alleles. Testcrosses are often used to determine whether an individual with a dominant phenotype is homzygous or hetrozygours.

test statistic A numerical measure in a statistical test whose value depends on the amount of difference between two sets of results. The numerical value of the test statistic is often used in probability calculations.

tetrad The four meiotic products of a single cell. Also used to describe the four chromatids in a synapsed pair of homologous chromosomes during the first meiotic division.

tetrad analysis Genetic analysis in fungal species that relies on the counting of genotypes and phenotypes of ascospores.

tetrasomic A genome that is mostly diploid but that contains four copies of one chromosome.

thalassemia An anemia-causing disease resulting from a mutation in one of the hemoglobin genes.

third filial (F_3) The third generation in a Mendelian crossing scheme. The F_3 is produced by inbreeding or selfing F_2 individuals.

thylakoid A structure in the chloroplast that contains the photosynthetic complexes.

thylakoid membrane The membrane that surrounds the thylakoid.

thymine A pyrimidine base that is found in DNA but not in RNA and that pairs with adenine.

Ti plasmid A plasmid in *Agrobacterium tumefaciens* that contains the T-DNA and the virulence genes.

Tn element See *transposable element.*

topoisomerases A class of enzymes that can confer twists into closed circular DNA to give or relieve supercoiling. Topoisomerase I makes one incision in double-stranded DNA and allows rotation of the other strand and reseals the end. Topoisomerase II makes double-stranded cuts, allows rotation, and then reseals.

totipotent cell A cell that has the capability to give rise to all structure or cell types.

***trans*-dominant** When a protein product of one gene controls other genes that are not physically adjacent to the controlling gene.

transcription The synthesis of RNA using a DNA template using the enzyme RNA polymerase.

transcription factor DNA sequence-specific proteins that bind to enhancers or promoters regions of genes and affect transcriptional initiation. Hundreds of different transcription factors exist and possess known structural motifs that bind to DNA (for example, zinc fingers, helix-turn-helix, and homeobox domains).

transdetermination A spontaneous change of cell fate in a cultured imaginal disc cell. Transdetermination usually occurs only after a long period (weeks to months) of culture.

transducing phage or virus A phage that can carry host DNA from one cell to another.

transduction The transfer of genetic information from a donor cell to a recipient bacterial cell by a phage. The phage accidentally packages donor cell DNA in place of phage DNA, and upon infection of a recipient cell, the donor DNA is injected and subsequently undergoes recombination.

transfer RNA (tRNA) Any of a series of small RNA molecules that act as an adapter molecule between an anticodon and a codon on mRNA during protein synthesis.

transformation A process by which extracellular DNA is transported into a cell for eventual integration into the genome. Bacteria are the usual recipients, but transformation can be done with animal, fungal, and plant cells. Transformation also means genetic and consequent cellular changes that occur in a normal cell leading to its change into a cancer cell.

transgenic plant or animal A plant or animal containing DNA from another species of organism.

transition mutation A substitution of any pyrimidine : purine or purine : pyrimidine pair for another (for example, a GC to AT transition).

translation The conversion of an mRNA sequence into a protein sequence on the ribosome.

translational frameshifting Two proteins produced from the same mRNA start site where one protein stops at a termination site, but a second protein is made by shifting the reading frame upstream from the stop codon, allowing the synthesis of a fusion protein.

translocation The movement of DNA from its normal location in a chromosome to a different chromosome.

transmissible spongiform encephalopathy A disease that affects the central nervous system of animals, causing a "spongy" phenotype of the brain of the infected animal. The agent that causes this disease is a prion.

transmission genetics The oldest area of genetic investigation that (1) infers the function of genes by observing the effect of heritable changes in the phenotype, (2) establishes the precise rules by which heritable traits are passed on from parents to their offspring, and (3) maps the location of genes on chromosomes.

transposable elements (transposon) A segment of DNA that can exit from one site and reinsert at an alternate site, inactivating a gene if it inserts within an exon, intron, or promoter sequence.

transposase The enzyme required for the insertion of (non-retroposon-like) transposable elements into the host chromosome.

transposition A unilateral translocation. DNA is removed from one chromosome and inserted into another.

transposon A short sequence of DNA (3–15 kb) that is capable of movement within the genome from one location to another.

transversion A substitution of a pyrimidine : purine pair for a purine : pyrimidine pair, or vise versa, for example an AT-to-CG transversion.

trichogynes Specialized mating filaments extending from the fruiting bodies of fungi-like *Neurospora.*

trihybrid cross A cross between individuals differing for three characteristics.

trisomic An individual or cell that has three copies of one chromosome and two of all others. The chromosome complement of a trisomic is $2N + 1$. Trisomy 21. A human genome that contains three copies of chromosome 21. Individuals with trisomy 21 have Down syndrome.

trisomy 21 A human genome that contains three copies of chromosome 21. Individuals with trisomy 21 have Down syndrome.

true breeding A strain that repeatedly produces only offspring like itself. Mendel's strains of homozygous peas were true breeding because they produced homozygous offspring generation after generation.

tumor Abnormal growth made up of cells descendent from a single genetically altered cell.

tumor suppressor gene A gene present in the genome of normal cells that is able to suppress the tumor-forming potential of transformed cells. Mutant alleles of tumor suppressor genes act as recessive determinants of cancer and are associated with inherited human cancers.

Turner syndrome A human syndrome caused by abnormal numbers of sex chromosomes. Individuals with a single X chromosome and no Y chromosome have Turner syndrome.

unequal crossover A recombination event in which the two recombining sites lie at nonidentical but locally homologous chromosomal locations.

unique sequence A class of DNA sequences that is present in the genome once.

universal code The genetic code is exactly the same in all species.

unweighted pair group method with arithmetic mean (UPGMA) One of several computational methods that have been devised to construct molecular phylogenies.

UPGMA (unweighted pair group method) A relatively simple computational method to construct molecular phylogenies from comparative sequence data.

upstream activator sequence (UAS) In yeast, a regulatory sequence several hundred or several thousand base pairs removed from a gene that enhances the expression of the gene. UASs are identical in function to enhancers found exclusively in yeast.

uracil A pyrimidine base found in RNA but not in DNA.

variable numbers of tandem repeats (VNTR) Each individual has a distinct pattern of these repeat sequences in their genome, and with the use of PCR, the sequences can be amplified and separated by gel electrophoresis. The resulting pattern is a fingerprint of the DNA from that individual.

variance A statistical measure of the deviation of individual measurements about their mean, calculated as the sum of the square of the deviation of each measurement from the mean, divided by the total number of measurements.

variant An allele or phenotype that differs from the standard or wild-type but that is not deleterious or abnormal.

variation Differences between individuals in a population.

variegated position effect A position effect that varies from cell to cell in mutant individuals.

variegation Patches or clones of somatic cells within one tissue or organism that show differing phenotypes.

vegetative reproduction Propagation by cuttings from the parent; asexual reproduction.

viral oncogenes (v-oncs) Constituents of the genomes of oncogenic viruses that are responsible for their cancer-causing properties.

viroid A small virus-like agent with a circular DNA or RNA genome tt does not code for any proteins.

virulent phage A phage that lyses its host (see *lytic cycle*).

Vitelline membrane The membrane surrounding the cytoplasm of the egg in *Drosophila*.

wild-type The standard or normal genotype or phenotype for that species.

wobble hypothesis The ability of a base in the third position of an anticodon to hydrogen-bond to different bases, so as to read more than one codon.

Wolf, Casper An embryologist (1733–1794) who disproved the doctrine of preformationism.

X chromosome A chromosome present in two copies in the homogametic sex and one copy in the heterogametic sex.

X-linked Genes that are located on the X chromosome.

YACs (yeast artificial chromosomes) A linear DNA molecule that contains a centromere, origins of replication for bacteria and yeast, telomeres, and a selectable marker. YACs replicates as a chromosome in the yeast cell.

Y chromosome A sex chromosome that is present in one copy in the heterogametic sex and not present in the homogametic sex.

Z-DNA, B-DNA, and **A-DNA** Various forms of DNA, depending upon ionic conditions and degrees of hydration. The natural form of DNA is the B-form.

zinc finger A structural motif on a DNA binding protein that recognizes specific DNA sequences.

zygote The initial, single cell of an organism formed by the fusion of two gametes.

zygotene The second stage of the first meiotic prophase.

Index

A
aberrations, chromosomal, 540–577
  and speciation, 673–674
abortive transduction, 239
acentric chromosomes, 564
acquired immune deficiency syndrome (AIDS), 601–603
  and DNA sequencing, 364
  and HIV, 201–202, 329, 601–604
  and Kaposi's sarcoma, 125
acridines, 525–526, 588
acrocentric chromosomes, 54
activator proteins, 499
adaptation
  and bacteria, 213–214, 214*t*
  and gene duplication, 682–683
adapter molecules. *See* transfer RNA.
addition rules, 28–29, 29*i*
additive alleles, 118
adenine (A), 257, 258*t*
adenine monophosphate (AMP), 421
adenomas, 627, 638
adenosine deaminase deficiency, 4
adenosine diphosphate, 192
adenosine triphosphate, 192, 276
adenylcyclase, 421
adjacent 1 and 2 segregations, 569
Aegilops squarrosa (goat grass), 545
aflatoxin B_1, 528–530
agar, 211
agarose. *See* gel electrophoresis.
agglutination, 91–92, 91*i*
aging, 530, 532–533
agriculture, 2. *See also* breeding.
  and antibiotic-resistant bacteria, 227
  application of transmission genetics to, 36, 41
  and autopolyploidy, 543
  and chromosomal aberrations, 540
  and genetic diversity, 135
  and genetic engineering, 383, 386, 388–389, 400
  and infectious inheritance, 188, 203, 205
  and mitochondrial mutations, 197–199
  and structure of DNA, 282
Agrobacterium tumefaciens (bacteria), 400–401
AIDS. *See* acquired immune deficiency syndrome.
Alar, 528
albinism, 40, 311–312, 313*i*
alcohol dehydrogenase, 684
algae. *See Chlamydomonas; Chlorella vulgaris.*
alkaptonuria syndrome, 39, 311, 313*i*
alkylating agents, 525
alleles, 21. *See also* chromosomes; deoxyribonucleic acid; genes; *specific types of alleles.*
  dominance relationships among, 86–99
  and gene linkage, 143–147
  multiple, 85–113
  recombination of, 474
  segregation of, 19–26
allelic complementation analysis, 176
allelic series, 86
alligators, 508*i*
allolactase, 414
allopatric speciation, 671–673
allopolyploids, 543–546, 550
allosteric proteins, 423
allosteric transitions, 423
allostery, 416
allotetraploids, 545
alternate segregations, 568–570
alternative splicing, 436
Altman, Sidney, 329
ALV. *See* avian leukosis viruses.
Ames, Bruce, 527, 618
Ames tests, 527–530, 618
amino acids
  as building blocks of proteins, 251, 310–311
  and mRNA translation, 328–338
  substitution rates in, 689–690, 700
aminoacyl tRNA synthetase, 334
amino forms, 513–514
amino group, 310
2-aminopurine, 523–524
Ammodramus maritimus (sparrows), 696–697
amniocentesis, 57–58
amorphic alleles, 97–98
AMP. *See* adenine monophosphate.
amphibians. *See Notophthalmus viridescens; Xenopus laevis.*
amphidiploids, 544
anaphase
  of meiosis, 61–62, 63*i*
  of mitosis, 49, 51*i*
anchor cells, 469
Anderson, French, 4, 4*i*
Anderson, Russell, 153
aneuploids, 541, 549–558, 679
Antennapedia complex (ANT-C), 472, 493–495, 494*i*
anthers, 17
antibiotics, 213
  resistance to, 227, 591, 609
antibodies, 91, 433
  diversity of, 443–447
  tests for, 4–5
anticarcinogens, 530
anticodon loops, 333
antigen–antibody reactions, 91–94
antigens, 91, 443
antimorphic alleles, 98–99
antiparallel DNA strands, 260
antisense mutations, 511
antitermination sites, 427
antiterminators, 426–429
ants. *See Cataglyphis bombycina.*
apoptosis, 637
apples, 509
apurinic or apyrimidinic endonuclease, 532–533
apurinic or apyrimidinic sites, 532. *See also* depurination.
Arabidopsis thaliana (plants), 11*i*
  and chromosome walking and jumping, 87
  databases for, 12
  and developmental genetics, 452
  genome of, 372, 382, 393
  and maintenance methylase, 433
  overdominance in, 87
  as research organisms, 10–12
  telomeres in, 301
  and yeast artificial chromosomes, 397
Arber, Werner, 350, 354
Aristotle, 7
arthritis, rheumatoid, 433, 443
artificial selection, 133
Ascaris (nematode worms), 65
asci, 71, 159
ascospores, 71
Aspergillus flavus (fungi)
  gene mapping in, 169–170
  mutagens in, 529–530
asters, 49, 50*i*
Astrachan, Lazarus, 341
ASV. *See* avian sarcoma viruses.
Ataxia telangiectasia (disease), 533
attached-X chromosomes, 153, 572
attenuation, 423–426
autocatalytic splicing, 324, 326–327, 328*i*
autocrine growth stimulation, 631
autoimmune diseases, 443
autonomous elements, 593
autonomous replication sequences, 276
autopolyploids, 543–544
  segregation of chromosomes in, 546–550
autoradiography, 265, 299
autosomes, 72
autotetraploids, 543–544
auxotrophic mutations, 106–107, 107*i*, 213
average clone size, 462
average heterozygosity (*H*), 652
Avery, Oswald T., 251, 253
avian leukosis viruses (ALV), 623, 628–629
avian sarcoma viruses, 624
Avise, John, 697
axis-polarity genes, 486–488
azidothymidine (AZT), 602–603

B

Bacillus subtilis (bacteria), 228
backcrossing, 25–26, 697
back mutations, 694–695
bacteria. *See Escherichia coli; other specific bacteria.*
bacterial artificial chromosomes, 394–396
bacteriophage, 211, 230*i*, 231*i*
  and DNA, 252–254
  gene mapping in, 230–239, 233*i*, 238*i*, 343*i*
  incorporation of host DNA by, 588
  infection by, 230–231, 232*i*
  lambda (λ), 230–231, 231*i*, 242–243
  life cycle of, 254*i*
  mutant, 233–234
  and restriction endonuclease, 354
  and site-specific recombination, 578
  supercoiling in, 285*i*
  T1, 213–214
  T2, 253
  T4, 338
  T-even, 230, 231*i*, 233
  and transduction, 239–244
balanced-lethal combinations, 567
balancer chromosomes, 567
balancing selection, 669–671
bananas, 541, 544, 549
Barr, Murray, 80
Barr bodies, 80–81, 299–300, 502, 555, 563
base analogues, 522–523
base-pair substitutions, 509. *See also* mutations.
bases, 256–257, 258*t*
  mismatching of, 514–515
  sequences and pairs of, 260
  sequencing, 382
  tautomeric forms of, 513–514
Bateson, William, 99, 116, 142, 557
B cells, 443
Beadle, George, 215, 312–314
beetles. *See Tribolium castaneum.*
behavioral isolation, 672
Bender, Harvey, 492
Bender, Welcome, 476
benign tumors, 613
Benzer, Seymour, 177, 235–238
Bessman, M. J., 270
bicoid genes, 487
bidirectional replication, 265–268, 272*i*
binary choice mutations, 478
binding proteins, DNA, 411, 435
binomial theorem, 29–31, 655
bioethics, 455
biological species concept, 643
biometricians, 116
bipotential gonads, 502–503
Bishop, J. Michael, 624, 626*i*
  life history of, 626–627
Biston betularia (peppered moths), 648, 666, 668
Bithorax complex (BX-C), 472–476, 475*t*, 493–495
bivalents, 61
blast cells, 468
blastoderms, 457, 460–465
blending inheritance, 7
  vs. transmission genetics, 21
blights, 198
blood, 443
blood types
  ABO, 94–96, 94*t*, 95*t*
  MN, 92–93
Bloom's syndrome, 533
Bouchard, T. J., 137
Boveri, Theodore, 47, 63, 65*i*, 72, 142
  life history of, 65
bovine spongiform encephalopathy, 205
Boyer, Herbert, 395*i*
  life history of, 395
Brachydanio rerio (zebra fish), 452, 492
Bradyrhizobium japonicum (bacteria), 211*i*
Bragg, Sir Lawrence, 340
branch migrations, 580–581
branching method, 25, 25*i*
Brassica oleracea (cabbages), 548
breakpoint effects, 559
breeding
  and genetic engineering, 383, 386, 388–389
  selective, 41, 129, 511, 548
Brenner, Sydney, 12, 338, 340–341, 467
Bridges, Calvin, 144, 554, 556*i*
  and chromosome mapping, 175
  and chromosome theory of heredity, 77–79
  and crossover suppressor mutations, 563
  and interstitial deficiencies, 558–559
  life history of, 556–557
  and sex determination, 497–498
5-bromodeoxyuridine, 522–523
budding, 71
Burkitt's lymphoma, 627–628
BX-C. *See Bithorax* complex.

C

C. elegans. See Caenorhabditis elegans.
CAAT box, 320
C-values, 288, 292
cabbages. *See Brassica oleracea.*
Caenorhabditis elegans (nematode worms), 13*i*
  databases for, 12
  developmental genetics in, 435, 467–469, 478–480
  genome of, 382, 393
  homeoboxes in, 477
  life cycle of, 467–469, 467*i*
  and maternal-effect genes, 190
  as research organisms, 9–10, 12–13, 452
  and saturation mutagenesis, 486
  sex determination in, 499, 501
  transposable elements in, 592
caffeine, 523, 528
Cairns, John, 265–267, 270
callus, 551
cancer, 612–642
  classification of, 613–614
  development of, 638
  as genetic disease, 612–615, 616*t*
  in humans, 529–530, 614*t*
  and long terminal repeats in retroviruses, 529–530
  in mice as human analogues, 403
  and mitotic trigger genes, 48
  and mutations, 508, 521, 532–533
  and *p53* genes, 403, 637
  and telomerase activity, 303
  and translocations, 568
  and transposable element mutations, 609
  treatment of, 523, 638–640
Canis rufus (red wolves), 696
Cann, Rebecca, 701
capsids, 598
carbon fixation, 193, 375, 433
carboxyl group, 310
carcinogens, 526, 528–530, 616
  as mutagens, 616–620
  natural, 528
carcinomas, 613
carriers, 75
catabolite repression, 421–422
Cataglyphis bombycina (ants), 433
catalytic RNA, 329
cat's cry syndrome, 559
Causes of Evolution, The [Haldane], 648
cdc genes. *See* cell division control genes.
cDNA. *See* complementary DNA.
Cech, Thomas, 263, 329
cell adhesion proteins, 637–638
cell autonomy tests, 462–463
cell–cell communication, 478–481
cell cycle, 47–60, 48*i*
  and chromosome structure, 298
cell determination. *See* cell fate.
cell differentiation, 453
cell division control (cdc) genes, 48, 631
cell equivalence groups, 468
cell fate, 452–454
  as developmental process, 452–456, 495–497
  genetic regulation of, 451–484, 493–497
  mapping of, 461–465
cell furrows, 50, 51*i*
cell fusion. *See* somatic cell hybridization.
cell growth regulation, 631–636
cell lineage, 467–469
cell markers, 460, 462
cell numbers, 461*t*, 463*i*
cell plates, 50, 51*i*
cells
  energy use in, 192–193
  origins of, 263
cell specifications, 452
cellular blastoderms, 457
cellular oncogenes, 616
  conversion from proto-oncogenes, 626–631
centimorgan (cM), 154
centrifugations, equilibrium density gradient, 265
centrioles, 49, 50*i*
centromeres, 48, 301, 304
Cepaea nemoralis (snails), 645*i*
chains, 444
Chargaff, Erwin, 257–259
Chargaff's rule, 258
Chase, Martha, 253, 255
chemicals, as mutagens, 521–526, 588
chemotherapy, 523
chiasmata, 61, 62*i*
chickens
  and cancer research, 621, 639
  evolution of insulin in, 681, 686–687, 689
  sex-linked inheritance in, 74, 75*i*
chimpanzees. *See* primates.
chi-square (χ^2) tests, 32–34, 33*t*. *See also* statistics.
  and population genetics, 656–657
Chlamydomonas (algae), 194
chloramphenicol, 213
chloramphenicol acetyl transferase, 375–377
Chlorella vulgaris (algae), 200
chlorophyll, 192–193
chloroplasts, 187
  DNA replication in, 276–277
  function of, 192–193, 276, 375
  genomes of, 193–194
  and introns, 326
  maternal inheritance of, 194–197
chorionic villi sampling, 57–58
chorions, 457
chromatids, 53

chromatins, 294*i*, 296*i*, 297
chromomeres, 305
chromosomal aberrations, 540–577
  and speciation, 673–674
chromosome arms, 53
chromosome jumping, 361–362
chromosome mapping
  through gene mapping, 153–159, 219–223
  through tetrad analysis, 159–166
  through trisomic chromosome segregation analysis, 552–553
chromosome morphology, 52–54. *See also* karyotypes.
chromosome puffs, 305–306, 305*i*, 306*i*, 410*i*
chromosomes, 47. *See also* alleles; deoxyribonucleic acid; genes.
  abnormalities in, 57–60
  artificial, 393–396
  changes in number of, 540–558
  circular, 193–194, 572
  and DNA, 261, 287–293
  DNA packing in, 282–286, 293–299, 294*i*
  lampbrush, 304*i*, 305
  during mitosis, 53*i*
  molecular structure of, 282–308
  morphology of, 52–54
  relationship to genes of, 142–143
  sex, 72–82
  during somatic cell cycle, 47–51, 298
  staining of, 55–57
  structural changes in, 558–573
  structure of, 55–57
  and transmission genetics, 46–84
chromosome theory of heredity
  history of, 47, 65
  proof of, 72–81
chromosome walking, 361–362, 387
ciliates. *See Paramecium; Tetrahymena thermophila.*
cis-acting sites, 414, 418
cis configurations, 144, 149*i*
cis-dominant genes, 418
cis–trans tests. *See* complementation tests.
cistrons, 177–181, 236, 319, 414. *See also* genes.
Clarkia biloba and *lingulata* (annual plants), 674
clonal expansion, 443
clonal selection, 443–444
clonal theory of cancer development, 613, 616
clones, 1*i*, 211–212, 349
  applications for, 399–405
  average size of, 462
  of DNA sequences, 356*i*, 390–398
  of genes, 348–381
  information from, 365–372
  techniques for creating, 348–360, 608
cloning vectors, 353, 355–358
  specialized, 373–377
CMS. *See* cytoplasmic male sterility.
coding strands, 317
codominance, 87
  detection of, 91–94
  and mutations, 513
codons, 328, 331
  frequency of, 340–342, 342*t*
coefficients. *See* coincidence coefficient; correlation coefficient; heritability coefficient; inbreeding coefficient; regression coefficient; sedimentation coefficient; selection coefficient.
Cohen, Stanley, 395*i*
  life history of, 395
coincidence coefficient, 156
cointegrates, 592
colchicine, 543–545
Colchicum autumale (flowering plants), 543
cold-sensitive mutations, 106
colon cancer, 638
color-blindness, 519
competent cells, 228
complementary base-pairing, 260
complementary DNA (cDNA), 359–360
complementation group. *See* cistrons.
complementation mapping, 176–181
complementation matrix, 178–179, 179*i*
complementation tests, 96–97, 96*i*, 178*i*, 560
  in bacteria, 223–225, 225*i*
  in bacteriophage, 235–236
complementing genes, 224
complete dominance, 86
complete penetrance, 103–104
complete recessiveness, 86
complete transduction, 239
composite elements. *See* transposons.
conditional mutations, 106–108, 510–511
  in bacteria, 213
conidia, 313
conjugation, 210–211
  in bacteria, 214–227, 216*i*
conjugation tubes, 217
conjugative plasmids, 217
consanguineous marriages, 36
consensus sequences, 316
conservation biology, 696–697
conservative replication, 264
conservative transpositions, 592
conserved sequences, 293
constant expressivity, 103–104
constant sequences, 445
constitutive genes, 414
constitutive heterochromatin, 55, 299
contact growth inhibition, 615
contig sets, 362
continuous DNA synthesis, 271
contractile rings, 50, 51*i*
coordinately controlled genes, 414
core enzymes, 316
corepressors, 424
corn. *See Zea mays.*
correlation, 124–125. *See also* statistics.
correlation coefficient (r_{xy}), 125. *See also* statistics.
Correns, Carl, 47, 87, 196
cortical cytoplasm, 457
cosmids, 391–393
co-transformations, 229
cotton, 400, 541
covariance of x and y [cov(xy)], 125, 134. *See also* statistics.
cows, 205
coyotes, 696
creativity in science, 8
Creighton, Harriet, 150
Crick, Francis, 341*i*
  and genetic code, 338–339, 525
  life history of, 340–341, 416
  and structure of DNA, 259, 264, 269–270
cri du chat syndrome, 559
crimes, solving of. *See* forensic analysis.
cristae, 192
crossing, 18, 20*i*. *See also* backcrossing; dihybrid crossing; monohybrid crossing; reciprocal crossing; testcrossing; trihybrid crossing.
  and dominance relationships, 88
  and linkage detection, 148–149, 151*i*
crossing over. *See* double crossing over; meiotic crossing over; mitotic crossing over; unequal crossing over
crossover suppression, 563
cross-pollination. *See* crossing.
Creutzfeld–Jacob disease, 204
cryptogenes, 438–439
cyclic AMP, 421, 421*i*
cyclic AMP receptor protein, 421
cystic fibrosis, 361–362, 388
cytochrome c, 690–695, 692*t*–693*t*
cytogenetics, 52
cytokinesis, 49–50, 51*i*, 52*i*
  incomplete, 457
cytology, 46
cytoplasmic determinants, 454, 471–472
cytoplasmic inheritance. *See* maternal inheritance; mitochondrial DNA.
cytoplasmic male sterility, 197–199
cytoplasmic organelles. *See* chloroplasts; mitochondria; organelles.
cytosine (C), 257, 258*t*
cytoskeletons, 49

D

D-loops, 582–583
D. melanogaster. See Drosophila melanogaster.
Danna, Kathleen, 354
Darbishire, A., 648
dark reactions, 193
Darwin, Charles, 648–649
databases, genetic, 12
Datura stramonicum (jimson weeds), 552
Davis, Bernard, 216
deamination, 517
death and accumulated mutations, 532–533
defect mapping, 464–465
deficiency complementation, 174–176
deficiency mutations. *See* deletion mutations.
degenerate codes, 339
degree of genetic determination (H^2), 129–130, 134, 136
degrees of freedom, 32, 656. *See also* statistics.
delayed-early gene expression stage, 426–428
Delbrück, Max, 213–214, 233
deletion mapping, 237–239
deletion mutations, 511, 558–559
  vs. point mutations, 179–180
denaturation, 260, 290
denominator genes, 498–499
density-dependent growth inhibition, 614
deoxyribonuclease, 251, 283
deoxyribonucleic acid (DNA), 3, 250*i*. *See also* alleles; chromosomes; genes; *specific types of DNA.*
  amplification of (*See* polymerase chain reaction.)
  and antibody diversity, 443–447
  and cancer, 621
  chemical nature of, 256–264
  and chromosomes, 287–288
  cloning of, 356*i*
  content per nucleus, 252, 253*t*
  content per organism, 287–288, 287*t*
  databases for, 12
  digestion of, 349–353, 366
  errors in replication of, 513–516
  excess, 282, 288
  and flow of genetic information, 309
  and genes, 86, 143
  as genetic material, 250–256, 520

methylation of and gene imprinting, 191
mitochondrial, 93, 695–697, 699, 701
origins of, 263
packing of, 283–287, 293–299
physical nature of, 250–281
repair of, 270, 514, 700
repetitive, 287–293, 289*t*
replication of, 264–278
and restriction endonuclease, 349–353, 645
sequencing of cloned, 368–372
single-stranded, 261
sizes of, 262*t*
deoxyribose, 257
depurination, 517
developmental biology. *See also* embryonic development.
of *C. elegans*, 13
and cell fate, 452–456
developmental evolutionary biology, 702
developmental genetics, 451
and genetic regulation of cell fate, 451–484
and hierarchies of genetic regulation, 485–506
developmental mutations, 456–457
developmental pathways, 486
De Vries, Hugo, 47, 144
diabetes, 443
diakinesis phase, 61, 62*i*
dicentric chromosomes, 564
diets, 528
differential gene expression, 410
dihybrid crossing, 22–24
and dominance relationships, 88
and gene interactions, 99–108
and gene linkage, 148–149, 151*i*
dinosaurs, DNA from, 6
Diplodomys ordii (kangaroo rats), 300*i*
diploids, 59, 541
and aneuploidy, 550
diplotene phase, 61, 62*i*
directional selection, 669
discontinuous DNA synthesis, 271
diseases. *See also* prions; viroids; viruses.
diagnosis of, 3
prevention of, 227
diseases, inherited human. *See also specific diseases.*
and allele dominance, 89–91
and aneuploidy, 554–557
and antibiotic-resistant bacteria, 227
and cloning, 361–362
and defects in DNA repair, 532–534
and DNA sequencing, 364–365, 398
and DNA slippage, 516
and genes, 39–41
and genetic engineering, 401–405
and genetic mapping, 386–388
and haplo-insufficient genes, 559
and Human Genome Project, 372
and infectious inheritance, 201–202, 203–205
and metabolic defects, 314
and mitochondrial mutations, 195–196
and mutations, 99, 106, 534
and pedigree analysis, 36–39
and recessive mutations, 311
sex-linked, 75–76, 657
DiSilva, Ashanti, 4*i*
disomic diploids, 550
displaced duplications, 559
distal arms, 53–54, 157
divergence, timing of evolutionary, 690–700
divergent selection, 132
DNA. *See* deoxyribonucleic acid; *specific types of DNA.*
DNA binding proteins, 411
and transcription, 435
DNA fingerprinting, 385
DNA glycosylase, 532
DNA gyrase, 274, 286
DNA helicase, 273–274
DNA ligase, 270, 272, 274, 295
DNA loops, 261, 582–583
DNA methylation. *See* methylated DNA.
DNA polymerase, 269–271, 273, 295, 533
and polymerase chain reaction, 362–365
DNase. *See* deoxyribonuclease.
DNA sequencing, 3, 348*i*, 364
of cloned DNA, 368–372
evolutionary time and changes in, 686–702
and genome projects, 382, 398–399
and population genetics, 646, 652–653
DNA slippage, 514–516
Dobzhansky, Theodosius, 665
Dolly ("cloned" sheep), 1*i*, 455
domains, 284–285, 297, 326, 684
domestication of plants and animals, 5
dominance relationships
among multiple alleles, 89–91
types of, 86–88, 86*t*, 87*i*
dominant lethal mutations, 106
dominant traits, 20
complete, 86
and mutations, 513
donkeys, 545
donor duplexes, 582–583
Doppler, Christian, 18
dosage compensation, 79–81
double crossing over, 565–567
double fertilizations, 69
double monosomic diploids, 550
double-strand break model, 582–583
double trisomic diploids, 550
Down, Langston, 555
Down syndrome, 80, 540–541, 552, 555–557
Drosophila melanogaster (fruit flies), 13*i*
and alkylating agents, 525
aneuploidy in, 555
and *Antennapedia* gene, 99
attached-X chromosomes in, 572
axis-polarity genes in, 486–488
balancer chromosomes in, 567
and chromosome theory of heredity, 47
and chromosome walking and jumping, 362
and complementation tests, 97
and crossing over, 153
databases for, 12
and deficiency complementation, 175
developmental genetics of, 456–466, 469–480, 486–493
development in, 435
distinguishing recessive mutant alleles in, 100*i*
and DNA replication, 268, 276
dominance relationships among alleles in, 86, 89
duplications in, 559–560
expressivity in, 103–104, 103*i*
and fine-structure gene mapping, 176–177
and foreign genes, 399
and gene linkage, 148–149
and gene mapping, 153–159, 158*i*, 166–169
gene numbers in, 399
genes controlling development in, 486*t*
gene symbols for, 23, 35
and genetic engineering, 400–402, 608–609
genome of, 287, 293, 372, 382, 393
haplo-insufficient genes in, 599
homeobox genes in, 477, 702
interstitial deficiencies in, 558
inversion loops in, 563
life cycle of, 68*i*, 457–460
and maternal-effect genes, 190
and methylated DNA, 301
modifier mutations in, 102, 102*i*
mutagenesis in, 402–403, 520
and neutralist–selectionist debate, 670–671
penetrance in, 103–104, 103*i*
phenotypic variance in, 128–129
and ploidy change, 542
polytene chromosomes in, 305–306, 305*i*, 306*i*
and population genetics, 654
position effects in, 561–562
pseudo-linkage in, 569–570
as research organisms, 10, 13, 142, 145, 348, 452, 556–557
retrotransposons in, 597, 597*t*, 604–605, 607
Robertsonian translocations in, 571–572
satellite measurement in, 289*i*
and saturation mutagenesis, 486
segmentation development in, 487–493, 488*t*
and sex-chromosome aneuploidy, 554
and sex-chromosome nondisjunction, 77–78
sex determination and sex-linked inheritance in, 26, 72–75, 73*i*
temperature-sensitive mutations in, 106
transposable elements in, 592, 594–597, 685
and tRNA, 334
tumor-suppressor genes in, 639
Drosophila pseudoobscura (fruit flies)
founder effects in, 665
Duchenne muscular dystrophy genes, 383, 387–388
Dunkers, 665
duplications, 559
of genes, 679–681
Dussoix, Daisy, 354
dyes for chromosomes, 55

E

E. coli. See Escherichia coli.
ectopic expression, 496
editing, RNA, 342, 438–439
editosomes, 439
eggs, maternal effects in, 187–188
electromagnetic spectrum, 521*i*
electroporation, 228, 394
electrostatic bonds. *See* hydrogen bonds.
elephants, 364
embryonic development. *See also* pattern formation.
and gene hierarchies, 485–493
and maternal-effect genes, 189–191
Endangered Species Act, 696–697
endonuclease, 349, 684
endosperm, 69
endosymbiosis, 199–200, 326
energy, cellular, 192–193
Enfield, F. D., 132
enhancer mutations, 102
enhancer sequences, 321, 607
enol forms, 513–514

environmental mutagens, 508–509
environmental variance (V_e), 126–127
  estimates of, 127–129
  twin studies of, 134, 136–137
enzymes. *See also* regulation of gene expression.
  and capping of mRNA, 322
  core, 316
  mismatch correction, 533
  and protein synthesis, 334
  synthesis of, 414
epidemiology, 617
epidermal growth factor, 632
epigenesis, 7
epistasis, 100, 497
equational division. *See* meiosis II.
equilibrium density gradient centrifugations, 265
error catastrophe model, 532
error-prone repair, 535–536
erythrocytes, 453
Escherichia coli (bacteria), 217*i*, 283*i*
  and adaptations, 213–214
  attenuation of gene expression in, 424
  auxotrophic mutations in, 106
  and bacterial artificial chromosomes, 394–395
  bacteriophage and, 230–239, 278, 426, 578
  and catabolite repression, 421
  and cDNA libraries, 359
  chromosome packing in, 283–285
  cloning vectors for, 353
  codon frequency in, 341–342
  conjugation in, 215–219
  databases for, 12
  deamination in, 517
  and DNA replication, 265–268, 270, 272–274, 276
  and DNA slippage, 515–516
  and expression vectors, 373
  gene mapping in, 219–224, 222*i*, 228–229
  and genetic code, 338–339
  genome of, 261, 262*i*, 372, 382
  and heat shock, 433
  and host restriction, 350
  and human growth hormone, 373–374
  and *lac* operon, 412–423
  mistakes during protein synthesis in, 337
  mutations of, 212–213
  nucleoid of, 284*i*
  and Okazaki fragments, 271–272
  overlapping genetic sequences in, 344
  and plasmid vectors, 390–392, 394
  and polymerase chain reactions, 364
  primase in, 272
  protein synthesis in, 331–335
  and recombinant DNA, 395
  repair of mutations in, 531–536
  as research organisms, 10–11, 211, 348–349, 393
  restriction map for, 367
  RNA polymerase in, 315–316
  and shuttle vectors, 358
  and single-strand break model, 579
  transcription and translation in, 330
  transduction in, 230, 239–244
  transformation in, 227–228
  transposable elements in, 586–588
  and yeast artificial chromosomes, 393–394
estradiol-17β, 431–432
estrogen, 431–432
ethidium bromide, 366, 383
ethylmethane sulfonate, 525
euchromatin, 55
  composition of, 299–301
eugenics, 388–389
eukaryotic organisms, 11, 315
  DNA packing in, 287–293
  gene numbers in, 398–399
  gene regulation in, 429–431
  transcription in, 320–321
  vs. prokaryotic organisms, 411
euploids, 541
evolution, 643, 648. *See also* natural selection.
  and gene duplication, 682–683
  genetic basis of, 649
  and genetic variation, 644–651, 663
  and genome sequences, 398
  and human genome project, 372
  molecular, 678–704
  molecular *vs.* morphological, 700, 702
  and mutations, 508, 518, 663
  and neutralist–selectionist debate, 670–671
  rates of, 686–702, 691*t*–693*t*
  and recombination and transposable elements, 608–609
  tools for study of, 188
  and triploids, 549
evolutionary genetics. *See* population and evolutionary genetics.
excision
  of introns, 684
  of lambda phage, 583–584
excision repair, 531–532
exconjugates, 218
exons, 323, 683, 687
exonuclease, 5′→3′, 272, 533
expansion mutations, 516
experimental techniques, 3–5. *See also* Ames tests; branching method; chorionic villi sampling; complementation tests; DNA sequencing; gel electrophoresis; Maxam–Gilbert sequencing procedure; polymerase chain reaction; recombinant DNA techniques; research; research organisms; Sanger dideoxy sequence technique; scientific method; segregation tests; statistics; UPGMA method.
expression sites, 441–442
expression vectors, 360, 373–374
expressivity, 103–104
extracistronic suppressors, 512
extragenic suppressors, 512
extranuclear inheritance, 187–209
  and infectious inheritance, 199–206
  and maternal inheritance, 188–192

F
F-factors. *See* F-plasmids.
F_1 generation. *See* first filial generation.
F_2 generation. *See* second filial generation.
F_3 generation. *See* third filial generation.
F-pili, 217
F-plasmids
  and bacterial artificial chromosomes, 394–396
  and bacterial conjugation, 216–219
F′-plasmids, 223–225
factorial, 31. *See also* statistics.
facultative heterochromatin, 55, 299
Fanconi's anemia, 533
Farber, Susan, 136
fate mapping, 461–465. *See also* cell fate.
fats in the diet, 528–529
Feulgen reaction, 55
fibrinopeptides, 690–691, 695
fibroblasts, 631, 636
fine-structure gene mapping, 176–181
  in bacteria, 228–230
fingerprinting, DNA, 385
first-division segregations, 164, 164*i*
first filial (F_1) generation, 20
first gap (G_1) phase, 47–48
fish. *See also Brachydanio rerio.*
  evolution of insulin in, 681
  phylogeny of, 698
  transposable elements in, 592
Fisher, Sir Ronald, 33–34, 116, 119, 120*i*, 648
  life history of, 120
fitness (*W*), 667, 669*t*
fixation, 664–665
flour beetles. *See Tribolium castaneum.*
flowers. *See Colchicum autumale; Primula officinalis.*
fluorescent *in situ* hybridization, 299
follicle cells, 457
foreign invaders, 443
forensic analysis, 3, 93, 385, 647–649
forward mutations, 512
fossils. *See* paleontology.
founder effects, 665, 699
four-strand stage crossing over, 152, 152*i*
  double, 566
Fraenkel-Conrat, Heinz, 255
fragile-X syndrome, 516
frameshift mutations, 338, 510–511, 514, 588
Franklin, Rosalind, 258–259
frequency distributions. *See also* statistics.
  and population genetics, 649–650, 663–671, 667*t*, 669*t*
frogs. *See Xenopus laevis.*
fruit flies. *See Drosophila melanogaster; Drosophila pseudoobscura.*
full-sib mating, 659
fungi. *See also Aspergillus flavus; Neurospora crassa.*
  transposable elements in, 592
fusion. *See* somatic cell hybridization
fusion–bridge–breakage cycles, 594, 595*i*

G
G_0 arrest
  and "cloning," 455
  in mitotic cell cycle, 48
G bands, 55–57, 56*i*
G_1 phase. *See* first gap phase.
G_2 phase. *See* second gap phase.
gain-of-function mutations, 98–99, 493–494, 496, 498
galactose
  in bacteria, 588
in yeast, 429–431
β-galactosidase, 412–414, 416
Galton, Sir Francis, 115–116
gametogenesis, 66–72
gametophytes, 69
gamma-ray radiation as a mutagen, 521–522
gap genes, 488–489
Garrod, Sir Archibald, 311–312
gastrulation, 457
Gateff, Elizabeth, 639
GC boxes, 320
GC islands, 361, 434
Gehring, Walter, 460–461, 492

gel electrophoresis, 365, 383–384, 645, 647. *See also* pulsed-field gel electrophoresis.
and population genetics, 645, 647, 652, 670–671
gene amplification, 629–631. *See also* polymerase chain reaction.
gene conversion, 583
gene duplications, 679–683
gene expression. *See* regulation of gene expression.
gene families, 679
gene flow, 661–663
gene focus, 463
gene imprinting, 191–192
gene libraries. *See* genomic DNA libraries.
gene mapping, 142–186
in bacteria, 219–223, 239–244, 608
in bacteriophage, 235–239
and chromosome mapping, 153–159
fine-structure, 176–177
and genome projects, 382
in humans, 170–174, 173*i*
limitations of, 244
gene migration, 661–663, 669
gene numbers, 398–399
gene pools, 644, 649, 655, 699
gene products in germ cells, 188–189
generalized recombination, 578–579, 583
generation-time effect hypothesis, 698
generative nuclei, 69
genes, 2, 21, 86. *See also* alleles; chromosomes; cistrons; deoxyribonucleic acid.
cloning of, 348–381
evolution of, 327
interactions among, 85–113
locating through G-banding, 56
mutations and function of, 97–99
regulatory regions of, 374–377
relationship to chromosomes of, 142–143
switching in response to environmental stresses, 412
symbols for, 21, 23, 35, 146–147, 212–213
and transmission genetics, 46–84
as units of inheritance, 16
gene targeting, 403
gene therapy, 4, 401–403, 639–640
genetic bottlenecks, 665
genetic code, 309–347, 329*i*
deciphering of, 338–342, 525
exceptions to, 342–344
genetic diversity, 135
genetic drift, 664, 669, 682, 699
genetic engineering
and DNA synthesis, 270
and mutations, 509
and plasmids, 226–227
and restriction endonuclease, 354
and structure of DNA, 270
and transposable elements, 608–609
uses for, 382–383, 399–405
genetic fine-structure analysis, 176
genetic markers, 35, 383–390, 404
genetic regulation. *See* regulation of gene expression.
genetics, 1–14
defined, 1–2
as experimental science, 7–9
history of, 5, 7, 16, 46–47, 349
and racism, 388–389
genetic stability, 585–587
genetic variance (V_g), 126
genetic variation
detection of, 644–647, 645*t*
estimation of, 651–654, 653*t*
and evolution, 644–651
factors affecting, 655–671
and mutations, 644, 663
genetic vectors, 608–609
genome projects, 2, 372, 382–383
genomes, 2, 382–409
detecting differences among, 383–386
sizes of, 397
genomic DNA libraries
constructing, 358–360, 372
using, 360–365
genotypes, 2, 21
germ cells. *See also* somatic cells.
cell-fate decisions for, 471–472
and gametogenesis, 69, 71
meiotic division of, 60–72
mutations in, 508
transposable elements in, 595–596
germline mutations, 509
germplasm collections, 135
Gerstmann–Straussle syndrome, 204
Gilbert, Walter, 370, 683–684
globin. *See also* hemoglobin; myoglobin.
molecular evolution of, 679–681
Glycine max (soybeans), 399–400, 658
glycosylase, DNA, 532
goat grass. *See Aegilops squarrosa.*
Goldschmitt, Richard, 74
gonadal primordia, 502–503
goodness-of-fit tests, 656. *See also* statistics.
gorillas. *See* primates.
grana, 192
grasses, 550. *See also Aegilops squarrosa.*
Green, Melvin, 176
Griffith, Frederick, 251–252
gRNA. *See* guide RNA.
growth, cell
regulation of, 631–636
and tumor-suppressor genes, 636–638
uncontrolled, 614–615
growth-factor receptors, 615*i*, 631–633
growth factors, 614, 631
growth media, 171, 211–213, 614
GU–AG rule, 325
guanine (G), 257, 258*t*
guanine-nucleotide-releasing factors, 633, 635
guide RNA (gRNA), 438
gynandromorphs, 461, 463–465
gyrase, DNA, 274, 286

H
H proteins. *See* histone proteins.
Haemophilus influenza (bacteria), 210*i*, 228, 382, 398
hairpins, 261
Haldane, J. B. S., 120, 647, 648*i*
life history of, 648
Haldane's Rule, 648
Hallauer, Arnel, 132
handedness in chromosome coiling, 285
haploid chromosome set, 58, 541–542
haploid number, 58
haplo-insufficient genes, 559
Hardy, Godfrey, 655
Hardy–Weinberg equilibrium, 655–657, 660, 667–669
Hardy–Weinberg Principle, 655, 657–658
Harvey, William, 7
Harvey rat sarcoma viruses, 625
Hayes, William, 218
heat-sensitive mutations, 106
heat shock, 433
heavy (H) chains, 444
helicase, DNA, 273–274
Helicobacter pylori (bacteria), 382
α-helix, 310, 311*i*
helix-turn-helix proteins, 435, 476–477
Helonias bullata (swamp pink plants), 646*i*
hematopoietic stem cells. *See* stem cells.
hemizygous sex, 74
hemoglobin, 309*i*, 312*i*, 313–314, 327, 375
function of, 690–691
and mutation for sickle-cell anemia, 89–91, 647
and phylogenies, 690–691, 698
and population genetics, 649*t*, 669*t*
hemophilia, 75, 76*i*, 377*t*, 657
herbicides, 400
heredity, 2, 5, 7. *See also* chromosome theory of heredity; diseases, inherited human; transmission genetics.
heritability, 129–137
heritability coefficient (h^2), 129–132, 130*t*
heritability in the broad sense. *See* degree of genetic determination.
heritability in the narrow sense. *See* heritability coefficient.
Hershey, Alfred D., 235, 253, 255
heterduplexes, 228
heterochromatin, 55, 399
composition of, 299–301
heterochronic mutations, 478
heteroduplexes, 579–580
mapping of, 323
heterogametic sex, 72
heterogeneous nuclear RNA (or pre-mRNA), 320, 323–326
heterokaryons, 169–170
heterologous probes, 360
heteroplasmic genes, 195
heterozygosity, 651, 652*t*
heterozygotes, 21
Hfr cells. *See* high frequency of recombination cells.
hierarchies of genetic regulation, 485–506
high frequency of recombination (*Hfr*) cells, 216–218
highly repetitive DNA sequences, 289, 292–293
histoblasts, 458
histone proteins, 293–295, 294*i*, 295*t*
history of genetics, 5, 7, 16, 46–47, 349. *See also specific researchers.*
HIV. *See* human immunodeficiency virus.
Hoagland, Mahlon, 340
Hodgkin's disease, 520
holandric inheritance, 74, 79
Holliday, Robin, 579
Holliday structures, 579
holoenzymes, 273, 316
Hom-C group, 495
homeoboxes
in *Drosophila melanogaster,* 476–478, 493
in mammals, 495–497
homeodomains, 476–478
homeotic gene complexes, 493–495
homeotic genes, 472–478, 491
and control of cell identity, 493–497
homogametic sex, 72
homogenistic acid, 311
homologous chromosomes, 60–61
homologous recombination, 578–584
homologues. *See* homologous chromosomes.
homoplasmic genes, 195

Homo sapiens (humans). *See also* diseases, inherited human; Human Genome Project.
- aneuploidy in, 554–557
- DNA slippage in, 516
- evolution of, 691–695, 698–700
- evolution of insulin in, 681, 686–687, 689
- gene mapping in, 170–174, 173i
- gene numbers in, 399
- genetic databases for, 12
- genome of, 261, 287, 289, 292
- lethal mutations in, 106
- lifespan of, 532
- number of cell divisions in, 513
- phenotypic variances in, 134, 136–137
- repair of mutations in, 532–534
- sex determination in, 501–503
- transposable elements in, 592

homozygotes, 21
homunculus, 7
horses, 545, 698
Hosta montana (variegated plants), 196–197
host-range mutant bacteriophage, 234
hot spots, 238–239, 515–516
housekeeping genes, 410, 434
Hox genes, 495–497
Human Genome Project, 372, 382–383
- contig set for, 362
- mapping for, 384

human growth hormone, 373–374, 377t
human immunodeficiency virus (HIV), 255
- and AIDS, 201–202, 329, 601–604
- and DNA sequencing, 364

humans. *See Homo sapiens.*
hunchback gene, 470, 489
Huntington disease, 40, 106, 388, 520
HU proteins, 284, 293
hybrid dysgenesis, 596
hybrids, 7
Hydra virilis (hydra), 200
hydrogen bonds, 260
hydroxylamine, 523–524
hypermorphic alleles, 98–99
hypervariable loci, 383–386
hypomorphic alleles, 97–98
hypotheses, 8. *See also* generation-time effect hypothesis; models; null hypothesis; out-of-Africa hypothesis; regional continuity hypothesis; scientific method; wobble hypothesis.

I

Illmensee, Karl, 460, 471
imaginal discs, 458, 461t, 462, 639
imino forms, 513–514
immediate-early gene expression stage, 426–427
immune system
- function of, 443, 601

immune system disorders
- ADD, 4

immunoglobulins, 443–444
Inborn Errors of Metabolism [Garrod], 311
inbreeding, 36, 38, 658–661, 660t
inbreeding coefficient *(F)*, 659
inchworm model of telomeric action, 303
incomplete dominance, 86–87, 87i
independent assortment, rule of. *See* Mendel's second rule.
induced mutations, 509
inducer molecules, 414, 422
inducible genes, 414
induction, 414
infectious inheritance, 188, 199–206
information
- genetic storage of, 2, 309
- and the Internet, 12

inheritance. *See also* heredity; transmission genetics.
- blending, 7
- extranuclear, 187–209
- holandric, 74, 79
- infectious, 188, 199–206
- maternal, 188–192, 189i, 190i, 194–197
- paternal, 191, 195

initiation codons, 329, 334
initiation factors, 335
insects. *See Drosophila melanogaster; Drosophila pseudoobscura; Tribolium castaneum.*
insertion mutations, 511
insertion sequences, 218, 517–518, 578, 588t, 589–591. *See also* transposable elements.
in situ hybridization, 299, 300i
insulin, 342, 373, 377t
- molecular evolution of, 681–683, 686–687

integration of lambda phage, 583–584
intelligence, 134, 136–137
interaction variance (V_i), 126–127
intercalary deficiencies. *See* interstitial deficiencies.
intercalating agents, 525–526
intercellular communication, 478–481
interference (I), 156
interferon, 377t
Internet, 12
interphase, 47–49
interrupted mating, 221
intersexes, 497
interspecific hybridization, 170, 696
interstitial deficiencies, 558
interstitial transpositions, 568
intracellular signaling proteins
- and proto-oncogenes, 633–634
- and tumor-suppressor genes, 637

intracellular signal transduction, 632
intracistronic suppressors, 512
intragenic suppressors, 512
introns, 323
- function of, 326–327, 683
- molecular evolution of, 681, 683–687
- removal of, 326–327

inversion loops, 563, 564i, 565i, 566i
inversions, 560–568
inverted terminal repeats, 589
IQ tests, 134, 136–137
IS elements. *See* insertion sequences.
isoaccepting tRNAs, 339

J

Jacob, François, 221, 341
- and lactose metabolism, 412, 414, 416–421

Janssens, Franz A., 150
Java human, 701
Jeffreys, Alec, 385
Jeon, Sang-Hak, 190
jimson weed. *See Datura stramonicum.*
Juel-Nielsen, Niels, 136
Jurassic Park [movie], 6

K

kangaroo rat. *See Diplodomys ordii.*
Kaposi's sarcoma, 125
kappa particles, 199–202
karyotype analysis, 52
- and chromosomal abnormalities, 57–60

karyotypes, 52–54
Kavenoff, Ruth, 287
keto forms, 513–514
Kettlewell, J. D., 648, 666, 668
Khorana, Gobind, 339
kinase cascade reactions, 634i
kinetochores, 49, 51i
King, Marie-Claire, 57, 93, 93i, 701
Kirsten rat sarcoma viruses, 625
Klenow fragments, 371
Klinefelter syndrome, 79, 81, 555
knockout mice, 403–404
Kölreuter, Josef, 115–116
Kornberg, Arthur, 269, 270i
- life history of, 270

Kuempel, Peter, 266–268
kuru disease, 204

L

lac operon, 412–423
- and mutations, 414, 418–421

lactalbumin, 679, 687–689
lactase. *See* β-galactosidase
Lactobacillus acidophilus (bacteria), 413
lagging strands, 271
Lamarck, Jean Baptiste, 5–6
lambda (λ) bacteriophage, 230–231, 231i, 242–243
- cloning with, 391–393
- control of gene expression in, 426–429
- and rolling circle DNA replication, 278
- site-specific recombination in, 583–584

lambda (λ) vectors, 391–393
lampbrush chromosomes, 304i, 305
Landsteiner, Karl, 92, 94
late gene expression stage, 426–428
lateral induction, 479–480
lateral specification, 479
Lathyrus odoratus (sweet peas), 99–100, 143
leader regions, 335
leading strands, 271
Lederberg, Joshua, 215i
- and genetic transfer in bacteria, 210, 214–216
- life history of, 215

leghemoglobin, 327
Lehman, I. R., 270
lentiviruses, 601
leptotene phase, 60, 62i
Lesch-Nyhan syndrome, 40
lethal mutations, 104–106
lethal periods, 105
leukemias, 530, 568, 614
Lewis, Edward B., 473–476, 474i
- life history of, 474

libraries. *See* genomic DNA libraries.
life, origins of, 263
lifespans, 533
ligase, DNA, 270, 272, 274, 295
light-dependent reactions, 193
light (L) chains, 444
lilies, 287
Limnaea peregra (snails), 188–189
Lincoln, Abraham, 40–41
linear regression, 126
LINES. *See* long interspersed elements.
linkage groups, 149
linkage maps, 153
- construction of, 157–163
- and crossing over frequency, 154

mitotic vs. meiotic, 169, 169*i*
and prediction of crosses, 160
linkages, 142–186
detection of through F_1 crosses, 147–149, 151*i*
and segregation of alleles, 143–147
linking numbers, 285
Linn, Stewart, 350
loci, 153
crossing over within, 176–177
long interspersed elements (LINES), 606–608
long terminal repeats, 598
loops. *See* anticodon loops; DNA loops; inversion loops.
loss-of-function mutations, 97–98, 494, 496, 498
luciferase, 400–401
luminescence, 371
lung cancer, 617–618
Luria, Salvadore, 213–214
Lwoff, André, 416
lymphocytes, 443, 601, 627
lymphomas, 614
Lyon, Mary, 80
lysogenic infection pathways, 231, 232*i*, 426, 428–429
lytic infection pathways, 231, 232*i*, 426–428

M

M phase. *See* mitosis.
McCarty, Maclyn, 251, 253
McClintock, Barbara, 150, 520, 586*i*
life history of, 586
and maize, 585–587, 593–594
MacLeod, Colin M., 251, 253
mad cow disease, 205
Mahowald, Anthony, 471
maintenance methylase, 433
maize. *See Zea mays*
major grooves, 260
malaria, 90, 647
malignant tumors, 613
mammoth, woolly, 364
mapping functions, 157, 159*i*
map shrinkage, 563, 565
map units, 154
Marfan syndrome, 40–41
markers, genetic, 35, 404
maternal-effect mutations, 188
maternal effects, 188–192
maternal gene products, 469–472, 478, 486–488, 490
maternal inheritance, 188–192, 189*i*, 190*i*
of organelle mutations, 194–197
and rates of evolution, 695
Mathaei, Heinrich, 339
mating, nonrandom
and population genetics, 657–661
Maxam, A., 370
Maxam–Gilbert sequencing procedure, 370
Maxson, Lisa, 701
mean ($\bar{x}$), 121. *See also* statistics.
Mechanism of Mendelian Heredity, The [Morgan], 47, 145
Mechler, Bernard, 639
medicine, 2
application of transmission genetics to, 36–41
and chromosomal aberrations, 540
diagnostic tests in, 3
and gene therapy, 4, 401–403, 639–640
and genetic engineering, 400–405
and infectious inheritance, 188
and structure of DNA, 282
and treatment of AIDS, 601–603
megaspore nuclei, 69
meiosis
and gene linkage, 143–147
and Mendel's rules, 65–66
mispairing during, 679
timing of, 66–72
meiosis I, 60, 62*i*–63*i*
in polyploids, 541, 544–545, 549, 552–554
and segregation of homologous chromosomes, 60–61, 579
meiosis II, 60, 62*i*–63*i*
and segregation of sister chromatids, 61–63
meiotic crossing over, 146, 146*i*
and chromosome exchange, 150–153
linkages and frequency of, 154
mutations during unequal, 518–519
meiotic products, 63
Mendel, Gregor, 18*i*
and beginnings of genetics, 5, 7, 16
life history of, 18–19
and rules of inheritance, 17–27, 65–66
statistical fit of data collected by, 33–34
Mendelian inheritance. *See* transmission genetics.
Mendelian populations, 643
Mendel's first rule of segregation, 22
and chromosome theory of heredity, 65–66
effects of multiple chromosome sets on, 546–550
predictions using, 26–27
violations of, 187
Mendel's second rule of independent assortment, 24
and chromosome theory of heredity, 66, 67*i*
predictions using, 26–27
mental retardation, 516
merozygotes, 223
MERRF. *See* myoclonic epilepsy and ragged-red fibers disease.
Meselson, Matthew, 264–265
messenger-RNA degradation controls, 411
messenger RNA (mRNA), 315, 317, 340–341
and genomic DNA libraries, 359–360
modification of after transcription, 321–328, 438–439
purification of, 323*i*
translation of, 328–338
metabolic rates and lifespan, 533
metacentric chromosomes, 54
metamorphosis, 458
metaphase
of meiosis, 61, 62*i*
of mitosis, 49, 50*i*, 51*i*
metaphase plates, 49, 50*i*
metastasis, 613, 615, 637
Methanobacterium thermoautotrophicum (bacteria), 382
Methanococcus jannaschii (bacteria), 382
methods. *See* experimental techniques; scientific method.
methylated DNA, 299–301
and cell development, 432–435
and gene imprinting, 191
and host restriction, 350–351, 354
and mutations, 525, 531
5-methylcytosine, 432
mice. *See Mus musculus.*
microsatellite DNA, 383–390
and breeding programs, 388–389
and chromosome mapping, 386–388
microspores, 69
migration, gene, 661–663, 669
milk digestion, 413
minimum patch size, 462
minor grooves, 260
Mirabilis jalapa (variegated plants)
incomplete dominance in, 87, 87*i*
variegated leaves in, 196–197
mismatch correction enzymes, 533
missense mutations, 510
mitochondria, 187, 192*i*
DNA replication in, 276–277
function of, 192, 276
genetic codes, exceptions in, 342
genomes of, 193–194
and introns, 326
and mutations, 530
paternal inheritance of, 195
mitochondrial DNA, 93
and rates of evolution, 695–697, 699, 701
mitosis, 49–51, 50*i*–51*i*. *See also* cell cycle; meiosis.
mitotic crossing over, 167*i*
and gene mapping, 166–170
mitotic spindles, 49, 50*i*
blockage of formation of, 543
mitotic trigger, 48
models, 8. *See also* double-strand break model; error catastrophe model; inchworm model; programmed death model; single-strand break model.
moderately repetitive DNA sequences, 288–289
modified dihybrid ratios, 88
and gene interactions, 99–104
modifier mutations, 102
molecular clocks, 695, 698, 700
molecular evolution, 678–704
molecular genetics, 3, 244
and developmental genetics, 486
molecular phylogenies, 690–695
monkeys. *See* primates.
monoclonal antibodies, 443, 640
Monod, Jacques, 341, 416*i*
and lactase metabolism, 412, 414, 416–421
life history of, 416
monohybrid crossing, 19–22, 22*t*
and dominance relationships, 86–87
monomorphic loci, 651
monoploids, 541–542, 571
monosomic diploids, 550
monospecific hybridization, 170
monozygotic twins, 134, 136–137
Morgan, Thomas Hunt, 65, 145*i*, 556–557
and chromosome theory of heredity, 72–74, 86
and crossing over, 150–151
and developmental biology, 453
and gene linkage, 142–143, 148, 648
and gene mapping, 154
life history of, 144–145
and study of *Drosophila melanogaster,* 13, 47
mosaic development, 454–456
moths. *See Biston betularia.*
mouse mammary tumor virus, 630
mRNA. *See* messenger RNA.
mtDNA. *See* mitochondrial DNA.
mules, 545
Muller, Hermann Joseph, 98*i*, 144, 520–521
and interference, 156
life history of, 98
and mutagens, 97
Mullis, K. B., 362

multigene families, 679–683
multiple cloning sites, 353
multiple origins of replication, 268, 269*i*, 276
multiple sclerosis, 443
multiplication rule, 28–29, 28*i*, 29*i*
muscular dystrophy, 383, 387–388
Mus musculus (mice)
  and alternative splicing, 436
  and cancer research, 623–624, 630, 639
  chromosomes of, 261, 262*i*, 289, 292
  and developmental genetics, 452
  evolution of insulin in, 681
  genome of, 382
  homeotic genes in, 496
  and inheritance of paternal mitochondria, 195
  lethal mutations in, 105–106, 105*i*
  lifespan of, 532
  mutagenesis in, 402–405
  mutation rates in, 700
  phylogeny of, 698
  recessive epistasis in, 100–101, 101*i*
  retrotransposons in, 597
  and tests for carcinogens, 526
  and transformation experiments, 251, 252*i*
mussels. *See Mytilus.*
mutagenesis, 508–539
  and developmental genetics, 486, 492
  measurement of, 526–530
mutagens, 509
  as carcinogens, 616–620
  natural, 528
mutant alleles, 35
  as cell markers, 460
  classification of, 97–99, 97*t*
  and gene function, 97–99
  and gene interaction, 85
mutant lesions, 177
mutational analysis, 35
mutation rates (μ), 519–520, 663
  and molecular evolution, 686–702
  and natural selection, 698
mutations, 2, 508–539. *See also specific types of mutations.*
  and cancer, 626–627
  categories of, 509–513
  and genetic variation, 644, 663
  induced, 521–530
  modifier, 102
  repair of, 530–536
  and retrotransposons, 604, 607
  spontaneous, 513–521
  and syndromes, 39
  and transposable elements, 608–609
Mycoplasma capricolum (microorganisms), 342
Mycoplasma genitalium (microorganisms), 382, 398
Mycoplasma pneumoniae (microorganisms), 382
myoclonic epilepsy and ragged-red fibers (MERRF) disease, 195–196
myoglobin, 327, 681
Mytilus (mussels), 195

N

Nathans, Daniel, 354
natural selection, 133, 666–669. *See also* evolution.
  and gene duplication, 682–683
  and mutations, 518
  and population genetics, 648–649, 669, 674–675, 699
  and rates of gene substitution, 698
  and recombination and transposable elements, 608–609
Nature *vs.* Nurture controversy, 137
Neanderthals, 699
negative assortative mating, 658
negative interference, 156
negative selectable markers, 404
Nelson, Bill, 697
nematodes. *See Ascaris; Caenorhabditis elegans.*
neomorphic alleles, 98–99
neuroblastomas, 630, 639
neurofibromatosis, 388
Neurospora crassa (fungi)
  and chromosome mapping, 159, 161, 163–166, 215
  and genes and enzymes, 312–314
neutralist–selectionist debate, 670–671
neutral mutations, 509–510
neutral selection, 670–671
Newman, Horatio H., 136
nick translation, 371
Nicotiana glutinosa and *tabacum* (tobacco), 550
Nilsson-Ehle, Herman, 117
Nirenberg, Marshal, 339
nitrites, 524–525
nitrogen-fixing bacteria. *See Rhizobium meliloti.*
nitrosoguanidine, 525
nitrous acid, 524
nonacutely transforming retroviruses, 623
nonautonomous elements, 593
nondisjunction, sex-chromosome, 77–78, 543, 556–557
nonhistone proteins, 293–295
nonhomologous recombination, 578
nonoverlapping genetic code, 341
nonparental ditype, 160–161
nonpolyposis colon cancer, 534
nonrandom mating, 657–661
nonsense mutations, 510–511
nonsilent mutations, 689
nontranscribed spacers, 327
northern blotting, 371
Notch loci, 479–480
Notophthalmus viridescens (amphibians), 304*i*
nuclear syncytia, 457
nuclei, 46–47
  in plants, 69
nucleic acids. *See* deoxyribonucleic acid; ribonucleic acid.
nucleoids, 283
nucleosides, 256
nucleosome core particles, 295, 296*i*
nucleosomes, 294*i*, 295, 296*i*
nucleotides
  as building blocks of DNA, 250, 256–258
  substitution rates of, 686–687, 691*t*
null alleles, 97–98
null hypothesis (H_0), 123
nullisomic diploids, 550
numerator genes, 498–499
nurse cells, 457, 596
Nüsslein–Volhard, Christiane, 492*i*, 493
  life history of, 492
Nutrasweet, 314–315

O

Ochoa, Severo, 339
Okazaki, Reiji, 271–272
Okazaki fragments, 271–272
Oliver, C. P., 176, 474
oncogenic viruses, 616, 621–623, 622*t*. *See also* cellular oncogenes; proto-oncogenes; viral oncogenes.
oocytes, primary, 69, 457
oogonial cells, 457
ootids, 69
open reading frames (ORF), 335, 398, 589, 606
operator sites, 414
  mutations in, 420
operons, 414, 416, 424
  mutations of, 420*t*
orangutans. *See* primates.
ordered tetrads, 163–166, 164*i*
ORF. *See* open reading frames.
organelles, 192–199, 202, 276. *See also* chloroplasts; mitochondria.
  genetic code exceptions in, 342
  and introns, 326
organs, replacing diseased or damaged, 455
Orgel, Leslie, 340–341
oskar genes, 487
out-of-Africa hypothesis, 699
ovalbumin, 323–324
ovaries, 502–503
overdominance, 87
overexpression of genes, 629–630, 632, 636
overlapping genetic code, 341, 343–344
ovists, 7
owls, 696
oxidative phosphorylation, 192

P

Pääbo, Svante, 699
P generation. *See* parental generation.
p53 genes, 403, 637
pachytene phase, 61, 62*i*
pair-rule genes, 488–490
paleontology
  and DNA from fossils, 3, 6, 699
  and polymerase chain reactions, 6
  and rates of evolution, 698
palindromes, 351
paracentric inversions, 561, 563–564
Paramecium (ciliate protozoa), 199–202
Pardee, Arthur, 341, 416
parentage testing, 92–93
parental ditypes, 160
parental (P) generation, 20
parental types, 144–145
p-arms, 56
partial diploids, 223
partial linkages, 146
partial penetrance, 103–104
Pasteur, Louis, 416
paternal inheritance, 191, 195
paternity testing. *See* parentage testing.
pattern formation, 451
  control of by gene hierarchies, 485–493
  and developmental genetics, 460–461
Pauling, Linus, 90
PCR. *See* polymerase chain reaction.
peanuts, 399, 528, 530
peas. *See Lathyrus odoratus; Pisum sativum.*
pedigree analysis, 36–41, 36*i*, 37*i*, 38*i*
Peking human, 701
penetrance, 103–104, 636
penicillin, 213, 227
pentaploids, 541
peppered moths. *See Biston betularia.*
peppers, 388
peptide bond, 310, 311*i*

pericentric inversions, 561, 565
permissive temperature, 106, 107*i*
Perutz, Max, 340
pests, agricultural, 135
PFGE. *See* pulsed-field gel electrophoresis.
phage. *See* bacteriophage.
phase variations, 440
phenocopies, 107
phenotypes, 2, 21
  determination of, 309
  distributions of, 117–119
  and proteins, 310–315
phenotypic selection, 129
phenotypic variance (V_p), 121, 122*i*
  genetic and environmental components of, 126–129, 134–137
phenylketonuria (PKU) syndrome, 36–37, 39–40, 313*i*, 314
Philadelphia chromosomes, 568
Phlox pilosa (flowering plants), 87
Phoronopsis viridis (marine worms), 651, 652*t*
phosphodiesterase, 421
phosphodiester linkage, 258
phosphorylation, 431, 622, 632–634
photoreactivation, 531
photoregulation, 375–376
photosynthesis, 192–193, 375
phylogenies, 690–695
Physarum (slime molds), 327
Pisum sativum (peas), 16*i*, 17*i*, 21*i*, 658
  and rules of inheritance, 17–27
PKU. *See* phenylketonuria syndrome.
plants. *See also* agriculture; *specific plants.*
  life cycle of, 66, 69, 70*i*
  polyploid, 550
plaque, 232
plaque assay, 231
plaque-forming units, 232
plaque-morphology mutant bacteriophage, 234
plasmids, 217, 277, 353, 390–391
  and antibiotic-resistant bacteria, 227
  and bacterial conjugation, 214–225
  as cloning vectors, 353, 355–357, 368, 394
  DNA coiling in, 283
  DNA replication in, 277–278
  types of, 226, 226*t*
Plasmodium ciparum (malarial parasites), 647
platelet derived growth factor, 615
plating, 212, 215, 357
pleiotropic alleles, 39
pleiotropic mutations, 102–103
pleuripotent cells, 404, 452
ploidy, 59
Pneumococcus (bacteria), 290*i*
point mutations, 179, 511–512. *See also* chromosomal aberrations.
  and oncogenes, 627
polar bodies, 69
polar granules, 471–472
polar mutations, 417, 588
pole cells, 457
pollen grains, 69
pollen tube nuclei, 69
pollution and cancer, 530
polyacrylamide. *See* gel electrophoresis.
polyadenylation, 322
polycistronic mRNAs, 319, 319*i*, 414
polyclonal antibodies, 443
polycloning site, 353
polycystic kidney disease, 386–387, 389
polygalacturonase, 511
polygenetic quantitative traits, 114–119
polylinker, 353
polymerase. *See* DNA polymerase; RNA polymerase.
polymerase chain reaction (PCR), 3–6, 362
  DNA sequence amplification using, 362–365, 363*i*, 385, 516, 699
  and population genetics, 648–649
polymorphic loci, 651
polypeptides. *See* proteins.
polyploids, 59, 541
  segregation in, 546–550
  sex determination in, 550
polyploidy, 541–550, 679
  and speciation, 673–674
polytene chromosomes, 305–306, 305*i*
population and evolutionary genetics, 3, 643–677
  and natural selection, 666–669
  and neutralist–selectionist debate, 670–671
  and speciation, 671–675
  and statistics, 649–651
populations, 3, 643
positional cloning, 361
positional information, 456
position effects, 559
positive assortative mating, 657
positive interference, 156
positive selectable markers, 404
posterior preferences, 496
posttranscriptional controls, 411
  and gene regulation, 436–439
postzygotic RIBs, 672, 674
potatoes. *See Solanum tuberosum.*
potato spindle tuber viroids, 203
Poulson, Donald, 492
Prakash, O., 57
preformationism, 7
pre-mRNA. *See* heterogeneous nuclear RNA.
preproinsulin, 681–682, 686–687, 689
Prescott, David, 266–268
prezygotic reproductive isolating barriers, 672, 674–675
primary nondisjunction, 77–78
primary oocytes, 69, 457
primary signals, 497–499
primary structures, 310, 311*i*
primase, 272
primates. *See also Homo sapiens.*
  comparison of chromosomes among, 56–57, 56*i*
  evolution of, 691–694, 698–700
primers, 269
  synthesizing, 264
primosomes, 272, 274
Primula officinalis (primroses), 658
prions, 203–205, 204*i*
probability, 27–34. *See also* statistics.
  and binomial theorem, 31, 655
  and chi-square test, 32–34
  and pedigree analysis, 38
  rules of, 28–29
  and t tests, 122–124
probes, 360
  labeling of, 371
proflavin, 338
programmed death model, 532
proinsulin, 687
prokaryotic organisms, 315
  DNA packing in, 283–286
  transcription in, 315–320
  vs. eukaryotic organisms, 411
promoters, 316, 317*i*, 373, 420, 607
proofreading, 275–276, 513, 533
prophage, 231
prophase
  of meiosis, 60–61, 62*i*, 563
  of mitosis, 49, 50*i*
proplastids, 192
protease, 251, 284
protease inhibitors, 602–603
protein activity controls, 411
protein isoforms, 436
protein kinase, 631–633
proteins. *See also* antibodies.
  in chromosomes, 293, 295
  databases for, 12
  and genetic information flow, 309
  and introns, 326
  mistakes during synthesis of, 335, 337–338
  and phenotype, 310–315
  structure and function of, 310–311, 422–423
  synthesis of, 331–335, 343–344, 377*t*
  vs. DNA as genetic material, 250–251
proto-oncogenes, 616, 622, 627
  conversion into cellular oncogenes, 626–631
  vs. viral oncogenes, 625–626
protoplasts, 551
prototrophs, 213
protozoa. *See Paramecium; Tetrahymena thermophila.*
proximal arms, 53–54, 157
Prusiner, Stanley, 204
pseudogenes, 679
pseudo-linkage, 569–570
Pseudomonas aeruginosa (bacteria), 228
  culturing of, 211–214
  genetics of, 210–248
  mutations in, 212–214
  as research organisms, 10, 211
  and resistance to antibiotics and drugs, 226–227
pulsed-field gel electrophoresis (PFGE), 396–398
pulse labeling, 266
Punnett, R. C., 24, 99, 143
Punnett squares, 23*i*, 24–25, 24*i*
pupariation, 458
purine bases, 256–257, 257*i*
pyrimidine bases, 256–257, 257*i*
pyrimidine dimers, 522

Q

Q bands, 55
q-arms, 56
qualitative traits, 114
quantitative traits
  and polygenic inheritance, 114–119
  properties of, 114–115, 115*t*
quaternary structures, 310, 311*i*

R

rII locus, 235–237, 237*i*
  and genetic code, 338–339
"rabbages," 548
rabbits, 698, 700
racism and eugenics, 388–389
radiation as a mutagen, 521–522, 568, 579
radioactive probe labeling, 371, 383
radishes. *See Raphanus sativus.*
random mating, 655
random primer extension, 371
random samples, 121. *See also* statistics.

Raphanobrassica ("rabbages"), 548
Raphanus sativus (radishes), 548
Rattus norvegicus (rats)
  evolution of insulin in, 681–682, 686–687, 689
  lifespan of, 533
  and tests for carcinogens, 526–527, 639
realized heritability, 131, 131*i*
reassociation. *See* renaturation.
receptor proteins, 431
recessive traits, 20
  complete, 86
  and diseases, 311
  and mutations, 513
recipient duplexes, 582–583
reciprocal crossing, 73–74
reciprocal translocations, 568–569
recombinant DNA techniques
  history of, 348–349, 377, 395
  and plasmids, 226
  and restriction endonuclease, 353*i*
  uses for, 382–383
recombinant types, 145
recombination
  in bacteria, 217–218
  in bacteriophage, 234–235
  of chromosomes, 142–143
  and mutation repair, 534–535
  and natural selection, 608–609
recombination repair, 534–535
red wolves. *See Canis rufus.*
Redi, G. P., 87
reduction division. *See* meiosis I.
regional continuity hypothesis, 699
regression analysis, 125–126. *See also* statistics.
regression coefficient *(b)*, 126. *See also* statistics.
regression line, 124*i*. *See also* statistics.
regulation of gene expression, 410–450
  and cancer, 622, 625, 627, 631–637
  and developmental biology, 451–484
  and evolution, 701–702
regulative development, 454–456
reiterative mutations, 478
relaxation, 285
release factors, 335
renaturation, 260, 290–292
repetitive DNA, 287–293, 289*t*
repetitive sequences
  and transposable elements, 606–608
replica plating, 357
replicative transpositions, 592
reporter genes, 374–377
repressible genes, 414
repressor proteins, 414, 416, 426–429, 499, 596
reproductive isolating barriers (RIB), 572, 672, 672*t*
research, genetic, 3–5, 3*t*
  and social pressures, 388–389, 555
research organisms. *See also* experimental techniques.
  choosing of, 9–10, 9*t*
  examples of, 11–13, 142, 145, 202, 211, 348, 586
resolvement, 580–581
response to selection (R), 130–134
restriction endonuclease, 349–353, 352*t*, 645
  discovery of, 354
restriction fragment length polymorphisms (RFLP), 383–390
  and breeding programs, 388–389
  and chromosome mapping, 386–388
  and population genetics, 645, 652–653
restriction maps, 365–367
restrictive temperatures, 106, 107*i*
retinoblastomas, 616–617, 636–637, 639
retrogenes, 681–682
retropseudogenes, 682
retrotransposons, 578*i*, 597, 597*t*, 606–608
  life cycle of, 597, 604–606
retroviruses, 402, 597–598
  and AIDS, 601–603
  life cycle of, 597–605
  and oncogenes, 621–626, 628–629
reverse mutations, 512
reverse tandem duplications, 559
reverse transcriptase, 359
  and gene duplication, 681–683
  and transposition, 592, 597, 601, 606
RFLP. *See* restriction fragment length polymorphisms.
rhesus monkeys. *See* primates.
Rhizobium meliloti (bacteria), 11*i*, 397
  as research organisms, 10–11
Rhodes, Marcus, 197, 585–586
rho (ρ)-factors, 317
RIB. *See* reproductive isolating barriers.
ribonuclease, 251, 283, 439
ribonucleic acid (RNA). *See also specific types of RNA.*
  chemical nature of, 262–264
  and gene expression, 436–439
  and genetic information flow, 309
  as genetic material, 254–256
  origins of, 263
  splicing of, 436–437
ribose, 257
ribosomal RNA (rRNA), 315, 327–328, 333*i*
  transcription of, 332–333
ribosomes, 330–335
ribozymes, 329
ribulose bisphosphate carboxylase, 193
ring chromosomes, 193–194, 572
RNA. *See* ribonucleic acid; *specific types of RNA.*
RNA editing, 342, 438–439
RNA polymerase, 295, 316*i*, 318*i*, 320*t*
  and genetic information transfer, 315–321, 606
RNA processing controls, 323, 411
RNase. *See* ribonuclease.
RNA splicing, 323
Robertsonian translocations, 571–573
rodents. *See Mus musculus; Rattus norvegicus.*
rolling circle replication, 277–278
Rosenberg, Steven, 402
"Roundup" (herbicide), 400
roundworms. *See Ascaris; Caenorhabditis elegans.*
Rous, Peyton, 621–622
Rous sarcoma virus, 621, 627
Roux, Wilhelm, 454
rRNA. *See* ribosomal RNA.
RSV. *See* Rous sarcoma virus.
Rubin, Gerry, 608
rubisco, 375–376
rules of transmission genetics. *See* Mendel's first rule; Mendel's second rule.
Russell, Lianne, 80
Ryan, Francis, 215

S

S. cerevisiae. See Saccharomyces cerevisiae.
S phase. *See* synthesis phase.
Saccharomyces cerevisiae (yeasts), 11*i*
  artificial chromosomes in, 393–394, 397
  centromeres in, 301, 304*i*
  and DNA replication, 276
  foreign genes in, 399
  galactose regulation in, 429–431
  genome of, 372, 382, 393, 398–399
  life cycle of, 71*i*
  and pulsed-field gel electrophoresis, 396
  reproduction of, 69, 71
  as research organisms, 11, 393
Salmonella typhimurium (bacteria), 227
  and Ames tests, 527, 618
  carcinogens for, 619*t*–620*t*
  immune response to, 440–441
sampling errors, 122, 664–665
Sampson, Lilian, 144
Sanger, Fred, 343, 368, 370
Sanger dideoxy sequence technique, 368–370
sarcomas, 614
Sarich, Vince, 701
satellite DNA, 54, 289
saturation mutagenesis, 486, 492–493
scaffolds, 294*i*, 297, 297*i*
scatter plot, 124–125, 124*i*
Schardin, Lois, 464
scientific method, 8, 8*t*. *See also* experimental techniques.
  and statistical tests, 32
  and violations of accepted theories, 187
scrapie, 204–205
secondary nondisjunction, 78
secondary structure, 310, 311*i*
second-division segregations, 164, 165*i*
second filial (F_2) generation, 20
second gap (G_2) phase, 48–49
sedimentation coefficient, 331
seeds
  banks of, 135
  maternal effects in, 187–188
  mutagenesis in, 525
segmentation genes, 488–493, 488*t*
segment-polarity genes, 489–491
segregation, rule of. *See* Mendel's first rule.
segregation tests, 94–96, 95*i*, 96*i*
selection. *See* artificial selection; balancing selection; breeding; clonal selection; directional selection; divergent selection; natural selection; neutral selection; phenotypic selection.
selection coefficient *(S)*, 667
selection differential (S), 130–132
selection pressures, 133
self, 443
selfing, 17, 658–659
selfish DNA, 327, 340–341
self-pollination. *See* selfing.
semiconservative replication, 264, 266*i*
senescence, 532
sequence complementarity, 579–584
sequencing. *See* DNA sequencing.
serine protease, 684–685
severe combined immunodeficiency, 402
sex chromosomes, 72
  and aneuploidy, 554–555
  and facultative heterochromatin, 299–300
sex determination, 78–81
  control of by gene hierarchy, 497–502
  in polyploids, 550
sex-linked inheritance, 72–76
sex-region-of-the-Y (*SRY*) genes, 502–503
sexual development, abnormal, 554–555
sheep
  "cloning" of, 1*i*, 454
  and scrapie, 204–205

Shields, James, 136
Shine–Dalgarno sequences, 335
short interspersed elements (SINES), 606–608
shuttle vectors, 357–358, 639
sickle-cell anemia, 89–91, 89*i*, 646–647
sigma factors, 316, 316*t*
silent mutations, 509–510
  and population genetics, 645, 654
Simms, E. S., 270
SINES. *See* short interspersed elements.
Singer, B., 255
single-copy DNA sequences, 288
single-strand break model, 579–582
single-stranded binding proteins (SSB), 274, 295
sister chromatids, 48
site-specific recombination, 578, 583–584
skin
  and cancer, 621
  and solar radiation, 521–522, 532
sleeping sickness, 441
slime molds. *See Physarum.*
slippage, DNA, 514–516
small nuclear RNA (snRNA), 315, 325–326
Smith, Hamilton, 350, 354
smoking, 528–529, 617–618, 637
snails. *See Cepaea nemoralis; Limnaea peregra.*
snakes, 695
snRNA. *See* small nuclear RNA.
sodium nitrite, 524–525
Solanum tuberosum (potatoes)
  and chromosomal aberrations, 541
  and polyploidy, 543–544
  and RFLP linkage maps, 388
solar radiation, 521–522
solenoids, 294*i*, 295, 296*i*
somatic cell clones, 167
somatic cell hybridization
  generation of polyploids through, 551
  and human gene mapping, 170–174
somatic cells, 46. *See also* germ cells.
  life cycle of, 47–60, 48*i*
somatic mosaic analysis, 462–463
somatic mutations, 509
Sonneborn, Tracy M., 199–200, 202*i*
  life history of, 202
Southern, E. M., 370
Southern blotting, 370–371, 638
soybeans. *See Glycine max.*
sparrows. *See Ammodramus maritimus.*
specialized transducing phage, 243
speciation, 671–675
  and reproductive isolating barriers, 572, 672, 672*t*
  and sampling errors and founder effects, 665
species, 3, 643–644
  evolutionary relationships among, 57
spermatocytes, primary, 69
spermists, 7
spliceosomes, 324–326
splicing reactions, 323–324, 497–498
spontaneous mutations, 509, 586
sporophytes, 69
spotted owls, 696
Spradling, Allan, 608
SRY genes. *See* sex-region-of-the-Y genes.
SSB. *See* single-stranded binding proteins.
stable position effects, 562–563
stacking forces, 260
Stadler, Lewis J., 519–521, 520*i*
  life history of, 520
Stahl, Franklin, 264–265
staining of chromosomes, 55–57
standard deviation (s), 122, 650, 664. *See also* statistics.
Statistical Methods for Research Workers [Fisher], 120
statistics, 9. *See also* chi-square tests; correlation; probability; standard deviation; test statistics; t tests.
  and quantitative traits, 116, 119–129
  reliability factors for, 650
  in transmission genetics, 27–34
stem cells, 261, 425, 443, 453–455, 454*i*
sterility, 545, 549
Stern, Kurt, 166–167
steroid hormones, 431–432
sticky ends, 352
stigmata, 17
Stoneking, Mark, 701
stop codons, 329, 335
strains, 18
Streisenger, George, 339, 514–515
Streptococcus marcasens (bacteria), 290*i*
Streptococcus pneumoniae (bacteria), 228
  and DNA as genetic material, 251, 253
streptomycin, 213
stress
  adaptation to, 433
  and gene expression, 412
stroma, 192, 193*i*
Student's t tests, 122–124, 123*t*
Sturtevant, Alfred H., 144, 153–154, 556–557
  and fate mapping, 461, 464
submetacentric chromosomes, 54
substitution rates. *See* mutation rates.
substitutions. *See* mutations.
sucrose gradient centrifugation, 331
Summerfelt, Robert, 130
supercoiling, 285
superfamilies, 431
suppressor mutations, 102, 512
Sutton, Walter, 47, 65, 72, 142
swamp pink plants. *See Helonias bullata.*
sweet peas. *See Lathyrus odoratus.*
symbionts, 199–200, 326
sympatric speciation, 671–672, 674
synapsis phase, 60, 62*i*
synaptonemal complex, 60, 62*i*, 64*i*
syncytial blastoderms, 457
syndromes, 39. *See also specific syndrome.*
  caused by recessive alleles, 39–40
syngamy, 66
synkaryons, 169–170
synthesis phase, 48
synthetase. *See* aminoacyl tRNA synthetase.

T

t tests, 122–124, 123*t*
tandem duplications, 559
TATA box, 320, 431, 476
Tatum, Edward
  and genetic transfer in bacteria, 210, 214–216, 312–314
tautomeric forms, 513–514
Tay–Sachs disease, 364
T cells, 443
T-DNA, 400
techniques, experimental. *See* experimental techniques
telocentric chromosomes, 54
telomerase, 302
telomeres, 53, 301, 301*t*
  shortening and synthesis of, 302–303
telophase
  of meiosis I, 61–62, 63*i*
  of mitosis, 49, 51*i*
temperate phage, 231
temperature effects, 106
temperature-sensitive mutant bacteriophage, 234
temperature-sensitive mutations, 106, 107*i*, 510
  in bacteria, 213
template strands, 269, 317
temporal isolation, 672
teratogens, 508, 528
terminal deficiencies, 558
terminal differentiation, 453
terminal genes, 487
termination codons, 329, 335
terminator sequences, 317, 319*i*
tertiary structures, 310, 311*i*
testcrossing, 25
  and gene mapping, 154–156, 162–163
  linkage detection through, 147–148
  and maternal-effect mutations, 188–189
testes, 502–503
tests. *See* experimental techniques.
test statistics, 32. *See also* statistics.
tetracycline, 213, 227
tetrad analysis, 159
  and chromosome mapping, 159–166
tetrads, 61, 63, 159
Tetrahymena thermophila (ciliate protozoa), 327, 329
tetraploids, 541, 547
tetrasomic diploids, 550
tetratypes, 161
tetravalent pairing, 568–569
thalassemia, 679
thalidomide, 107
theobromine, 528
theories. *See* chromosome theory of heredity; hypotheses; models; scientific method.
Thermus aquaticus (bacteria), 364
third filial (F_3) generation, 20
three-strand double crossing over, 566
threshold levels, 97
thylakoid membranes, 192
thylakoids, 192
thymine dimers, 522
thymine glycol, 533
thymine (T), 257, 258*t*
Ti-plasmids. *See* tumor-inducing plasmids.
Tn elements. *See* transposons.
toads. *See Xenopus laevis.*
tobacco. *See also Nicotiana glutinosa* and *tabacum.*
  genetically engineered, 400
  and lung cancer, 617–618
  and reporter genes, 375–376
tobacco mosaic virus, 254–255
tomatoes, 511
  and RFLP linkage maps, 388
topoisomerase, 276–277, 286. *See also* DNA gyrase.
tortoises, 532
totipotent cells, 452
T-phage, 233
traits, 114–119
trans-acting genes, 414, 418*i*, 590
trans-acting proteins, 414
trans configurations, 144, 149*i*
transcribed spacers, 327
transcription, 309, 315
  and information transfer, 315–321, 318*i*

transcriptional controls, 411
by *lac* operon, 412–423
transcriptional interference, 604, 606
transcription factors, 320, 431, 622
and proto-oncogenes, 634–636
trans-dominant genes, 419*i*
transduction, 211, 215. *See also* bacteriophage.
in bacteria, 230, 239–244
signal, 632, 635*i*
transfer RNA (tRNA), 315, 329, 333*i*, 339–341
and protein synthesis, 331–334
transformation, 211
in bacteria, 227–230
and DNA, 251, 252*i*
transgenic organisms, 401
transition mutations, 509
translational controls, 411
translations, 309, 328–338, 336*i*
translocations, 568–573
and oncogenes, 627–628
and speciation, 673–674
transmissible spongiform encephalopathies, 203–205
transmission genetics, 3, 16–45
applications of, 36–41
basis for, 46–84
exceptions to (*See* extranuclear inheritance.)
polygenetic, 114–141
transposable elements, 289, 578–611. *See also* insertion sequences; transposons.
in bacteria, 589–592
in eukaryotes, 592–608
examples of, 584–588
and natural selection, 608–609, 663
transposase
and transposition, 518, 589, 592–597
transpositions, 568, 589, 592–597
transposons, 517–518, 520, 590–591, 590*t*
transversion mutations, 509
Tribolium castaneum (flour beetles), 132–133, 133*i*
trihybrid crossing, 24–25
triploid nuclei, 69
triploids, 541
trisomic diploids, 550, 552–553
trisomy 21, 555–556. *See also* Down syndrome.
Triticum aestivum (wheat)
and allopolyploidy, 545–546
and chromosomal aberrations, 541
polygenetic traits of, 117–119, 117*i*, 118*i*
Triticum spelta (spelt), 545
tRNA. *See* transfer RNA.
true-breeding strains, 18
Trypanosoma brucei (tsetse-fly parasites), 441–443
tryptophan
promoters for, 373
synthesis of, 423–426
tsetse flies, 441
tuberculosis, 433
tumorigenic effects, 621
tumor-inducing plasmids, 400
tumors, 613
tumor-suppressor genes, 636–639, 637*t*
Turner syndrome, 79, 81, 554
twin studies, 134, 136–137
two-strand stage crossing over, 152, 152*i*
double, 566
typological species concept, 643
tyrosine, 311, 313

U
ultraviolet light as a mutagen, 520–522, 531–532, 536
unequal crossing over, 518–519, 562, 679–680
unidirectional replication, 266
unique DNA sequences, 288
UPGMA method, 690–691
upstream activator sequences, 430
uracil (U), 257

V
vaccines, 601
van der Waals forces. *See* stacking forces.
variable expressivity, 103–104
variable number of tandem repeats loci, 383–386
variable sequence, 445
variable surface glycoproteins, 441
variant alleles, 35
variation, 2, 5, 7, 35
among alleles, 86–99
and chromosome segregation during meiosis, 66
variegated position effects, 561
variegation, 561
and maternal inheritance, 196–197
Varmus, Harold, 624, 626*i*
life history of, 626–627
Victoria, Queen (of Great Britain), 75, 76*i*
viral oncogenes
and cancer, 616, 621–623
and normal cells, 623–627
vs. proto-oncogenes, 625–626
viroids, 202–203, 203*i*
virulent phage, 231
viruses, 211
and bacteria, 210–248 (*See also* bacteriophage.)
cancer-causing, 623*t*
DNA replication in, 277–278
and infectious inheritance, 201–202
vitelline membranes, 457
Volkin, Elliot, 341
v-*oncs*. *See* viral oncogenes.
von Tschermak, Erich, 47

W
Watson, James, 259, 264, 269–270, 340
Weinberg, Wilhelm, 655
Werner's syndrome, 533
whales, 696
wheat. *See Triticum aestivum; Triticum spelta.*
Wieschaus, Eric, 460, 492*i*, 493
life history of, 492
wild-type alleles, 35, 552
Wilkins, M. H. F., 259
Wilms's disease, 617
Wilmut, Ian, 455
Wilson, Allan C., 57, 700–702, 701*i*
life history of, 701
wobble hypothesis, 339
Wolff, Caspar, 7
Wollman, Elie, 221
Wolpert, Lewis, 456
wolves, 696
worms. *See Ascaris; Caenorhabditis elegans; Phoronopsis viridis.*
Wright, S., 648

X
X chromosomes, 72
inactivation of, 79–81
Xenopus laevis (frogs)
DNA replication in, 267–269
rates of evolution in, 700
tRNA in, 334
xeroderma pigmentosum, 532, 621
X-linked genes, 74, 562, 657
X-ray radiation as a mutagen, 520–522, 534, 558

Y
Y chromosomes, 72
role of, 79
Yanofsky, Charles, 423–424
Yanomami Indians, 660
yeast artificial chromosomes, 393–394, 397
yeasts. *See also Saccharomyces cerevisiae.*
and cancer research, 639
and codon frequency, 341
conjugation in, 218
genome of, 293
retrotransposons in, 597, 604–605
and shuttle vectors, 358
transposable elements in, 592
yogurt digestion, 413
Yunis, J. J., 57

Z
Zea mays (corn or maize)
allele designations in, 147
cloning of, 608
crossing over in, 151–152
cytoplasmic male sterility in, 197–199
databases for, 12
and developmental genetics, 452
genetic diversity of, 135
genome of, 393
life cycle of, 70*i*
mutation rates in, 519–520
as research organisms, 9–10, 586
response to short-term selection of, 132–133, 133*i*
retrotransposons in, 597
and southern corn leaf blight, 198
transposable elements in, 585–587, 593–594, 685
zebra fish. *See Brachydanio rerio.*
Zimm, Bruno, 287
zinc fingers, 431, 435, 607
Zinder, Norton, 215
zygotene phase, 60, 62*i*

L-Amino acids	Abbreviation	Properties of side chains
Alanine	Ala (A)	
Valine	Val (V)	
Isoleucine	Ile (I)	
Leucine	Leu (L)	Nonpolar
Phenylalanine	Phe (F)	
Proline	Pro (P)	
Methionine	Met (M)	Contains sulfur
Glycine	Gly (G)	
Serine	Ser (S)	
Threonine	Thr (T)	
Tyrosine	Tyr (Y)	Polar
Tryptophan	Trp (W)	
Asparagine	Asn (N)	
Glutamine	Gln (Q)	
Cysteine	Cys (C)	Contains sulfur
Aspartic acid	Asp (D)	Acidic
Glutamic acid	Glu (E)	
Lysine	Lys (K)	
Arginine	Arg (R)	Basic
Histidine	His (H)	